PRONGHORN
Ecology and Management

PRONGHORN
Ecology and Management

Bart W. O'Gara and Jim D. Yoakum

technical editor Richard E. McCabe
illustrated by Edson Fichter *and* Daniel P. Metz

produced in cooperation with the
Cooperative Fish and Wildlife Research Unit Program of the U.S. Fish and Wildlife Service
and the North American Pronghorn Foundation

UNIVERSITY PRESS *of* COLORADO
A WILDLIFE MANAGEMENT INSTITUTE BOOK

The Wildlife Management Institute (WMI) is a private, nonprofit, scientific and educational organization, based in Washington, DC. WMI's sole objective, since its founding in 1911, has been to help advance restoration and sound management of North America's wildlife resources. As part of WMI's program, scientific information generated through research and management experiences is consolidated, published and used to strengthen resource decision making, management opportunities and methodologies, and general understanding of and appreciation for wildlife and their habitats. *Pronghorn: Ecology and Management* is one of 30 books produced by WMI since 1942, including the award-winning *Ducks, Geese and Swans of North America, Big Game of North America, Mule and Black-tailed Deer of North America, Elk of North America, White-tailed Deer: Ecology and Management, Ecology and Management of the Mourning Dove, Ecology and Management of the Wood Duck, Ecology and Management of the North American Moose,* and *North American Elk: Ecology and Management.* For additional information about WMI, its publications, programs and membership, write to Wildlife Management Institute, 1146 19th Street, NW, Suite 700, Washington, DC 20036. Or, see the website *www.wildlifemanagementinstitute.org.*

Published by the University Press of Colorado
5589 Arapahoe Avenue, Suite 206C
Boulder, Colorado 80303

Printed in the United States of America

The University Press of Colorado is a proud member of the Association of American University Presses.

The University Press of Colorado is a cooperative publishing enterprise supported, in part, by Adams State College, Colorado State University, Fort Lewis College, Mesa State College, Metropolitan State College of Denver, University of Colorado, University of Northern Colorado, and Western State College of Colorado.

The paper used in this publication meets the minimum requirements of the American National Standard for Information Sciences—Permanence of Paper for Printed Library Materials. ANSI Z39.48-1992

Library of Congress Cataloging-in-Publication Data

O'Gara, Bart W.
Pronghorn : ecology and management / by Bart W. O'Gara and Jim D. Yoakum ; technical editor Richard E. McCabe ; illustrated by Edson Fichter and Daniel P. Metz.
p. cm.
Includes bibliographical references and index.
ISBN 0-87081-757-4 (hardcover : alk. paper)
1. Pronghorn antelope. 2. Wildlife management. I. Yoakum, James D. II. Title.
QL737.U52O39 2004
639.97'9639—dc22

2004001640

Design and layout by Daniel Pratt

Dedication

This book is dedicated to Dr. Edson Fichter, who died in 1994. He was a professor, philosopher, poet, researcher, author, mentor, curator and artist who made our lives richer. Never once did he hesitate to share information or art that might be used to enhance the welfare of pronghorn.

Edson was born in 1910 and raised on a small farm in Iowa. After receiving a Ph.D. at the University of Nebraska, he traveled west with his wife (Ardith) and daughters (Nancy and Sarah) and became a professor at Idaho State College, Pocatello. There he taught biology, zoology and wildlife management for 26 years. He also was curator of mammals for the Idaho Museum of Natural History.

Edson made studying the life history of pronghorn his career. Year after year he trekked to his beloved Pahsimeroi Valley to observe and record pronghorn behavior. He was mentor to dozens of students and published more than 20 scientific papers on pronghorn. His volumes of notes, meticulously recorded, will be a source of knowledge for biologists to explore for years to come.

Edson was totally dedicated to pronghorn and other wildlife. His scientific writings will remain a cornerstone of biological knowledge. In addition, his many drawings will demonstrate biological accuracy and beauty for eons. Edson's drawings graced the covers of the 1976 and 1978 Pronghorn Antelope Workshop proceedings and the first edition of the Workshop's *Management Guidelines* published in 1978. His poetic writings of wilderness and wildlife are eloquent, evoking a profound awareness of nature—to which we all belong.

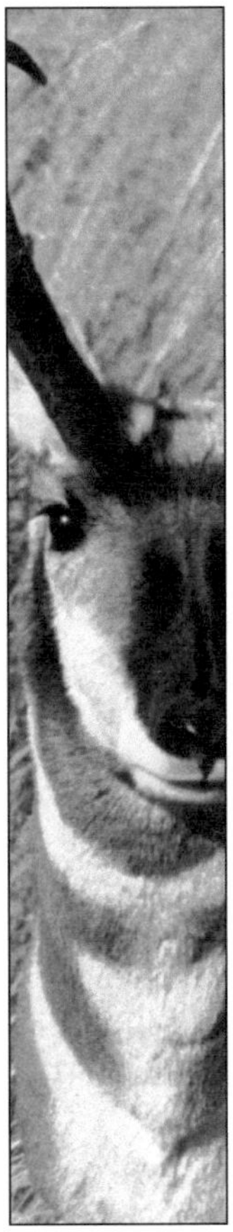

Contents

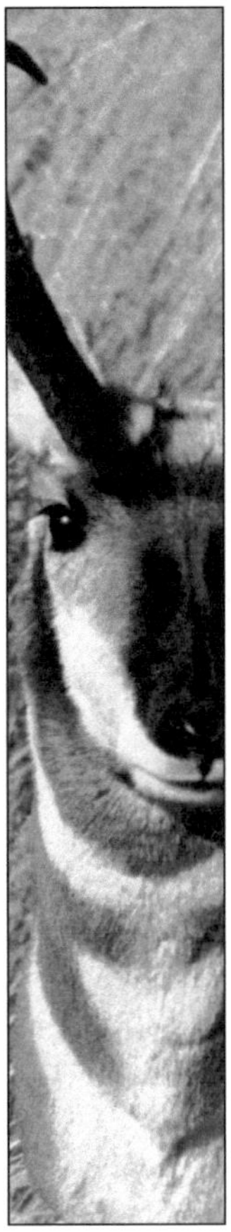

Figures

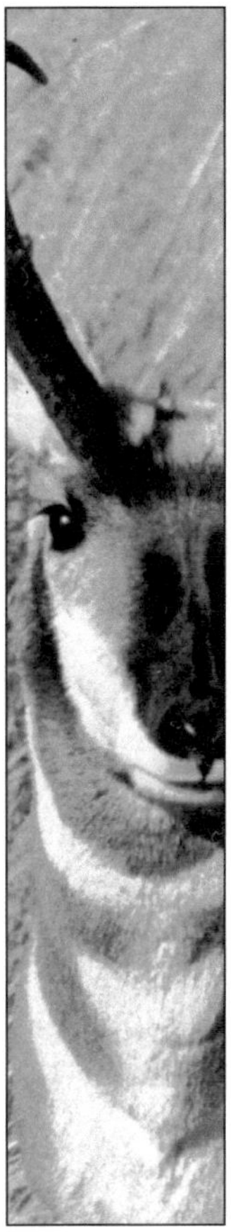

Tables

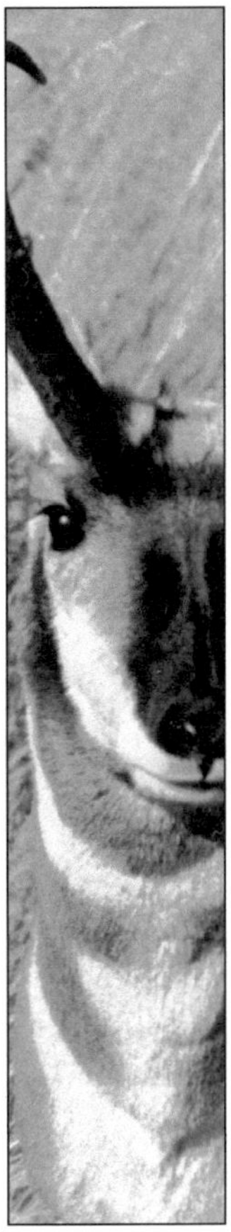

In Memoriam

Oscar "Ock" V. Deming was the first wildlife management biologist at the Hart Mountain/Sheldon National Wildlife Refuge Complex in Lakeview, Oregon. His encouragement to "stand by your convictions" was quite instrumental in Jim D. Yoakum's recognition of the need to document pronghorn habitat requirements, assess food habit investigations and report manipulation practices contributing to the enhancement of pronghorn habitats.

Philip L. Wright, Bart W. O'Gara's major professor and mentor, introduced Bart to the scientific method, pronghorn biology and the mysteries of reproductive physiology. Their relationship ripened into a lifelong friendship, involving numerous interactions dealing with science and conservation—liberally interspersed with hunting trips.

Lucille O'Gara, Bart's wife of 37 years, died too young. She encouraged him to retire from the U.S. Navy and, despite the lack of a high school diploma, go to college. She assisted in field studies and homework—even in frigid winter encampments in Yellowstone National Park.

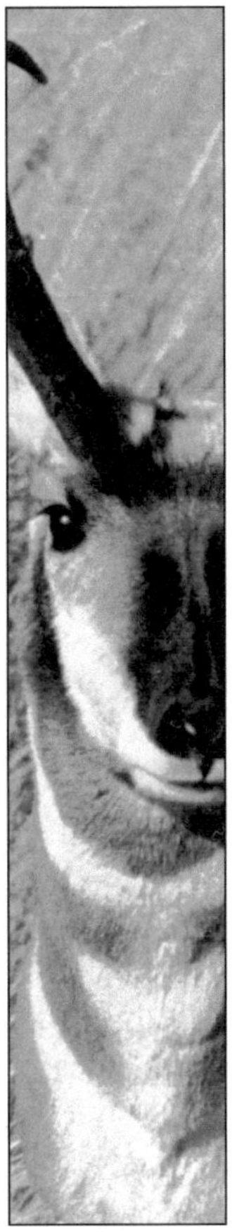

Foreword

The history of pronghorn shows that humans can increase numbers of a game species through scientific management and sustained-yield use of that species. Pronghorn numbers increased rapidly—from near extirpation in 1910 to more than 1 million in the 1980s. That is a truly remarkable success story. Even more remarkable is the fact that, during that span, about 4 million pronghorn were legally harvested. However, because of fences, energy extraction, loss of habitat and competition with livestock, pronghorn probably never again will be witnessed in the millions that roamed the frontier of western North America.

Since the days of the early explorers, pronghorn have been called antelope, and both terms have been used interchangeably. But, pronghorn is the taxonomically correct name in scientific literature, analogous to bison instead of buffalo and Native American rather than Indian. The name pronghorn differentiates this uniquely North American species from African antelope, some of which have been translocated into pronghorn habitat.

During a trip to southern Africa to study, photograph and hunt many species of antelope approximately 30 years ago, we decided to write a comprehensive book on pronghorn. Fortunately, Daniel A. Poole, then-president of the Wildlife Management Institute (WMI), agreed that WMI would produce it, as it did Arthur Einarsen's classic, *The Pronghorn Antelope and Its Management*, in 1948. Our objective was to devote approximately half of the book to the scientific aspects of the species and its management and the rest to habitat and rangeland conservation necessary for pronghorn. Both are important to managers responsible for decisions that affect pronghorn, and to researchers studying North America's fastest land animal. We, having worked with pronghorn during our entire professional careers, bring more than 90 years, combined, of field and laboratory experience. Jim worked primarily with the habitat; Bart worked primarily with the animal. Our other authors have added considerably to the measure of pronghorn biology and ecology expertise that this book has to offer.

A goal of this book was to synthesize the accumulated biological and management data about the species. Although a wealth of scientific information exists for some North American ruminants, comparatively little is available for pronghorn. Also, differences in behavior and needed habitat exist between other North American ruminants and pronghorn. This might have contributed to a number of prevalent myths and misconceptions concerning the latter. Another goal was to reduce such misunderstanding by making as much scientific and empirical information available as possible, in a useful fashion. Finally, we intended to identify and highlight subjects and areas for which important information is limited.

This book was planned as a 200- to 250-page, semitechnical text. Subsequently, Laurence R. Jahn, WMI's vice president at the time, suggested that we work with Richard McCabe, WMI's then-director of publications, to expand and include it as part of WMI's ecology and management monograph series. A number of decades and generations of pronghorn, as well as numerous drafts of this work, have passed since then, and we hope that this product of that time and effort will serve the wildlife management profession to benefit pronghorn as well as Einarsen's book did for many years.

During the early 1930s, Murray (1932:45) wrote: "The pronghorn should certainly be preserved for its unique scientific significance. It is the only living species of its genus. That genus is the only living genus of a family. Fossil remains indicate that it is one of the oldest of American families. It occurs nowhere else on the globe. When the passenger pigeon and the heath hen were exterminated through man's cupidity and ignorance, they left near relatives somewhat similar in appearance. But when the last antelope is gone, there will be no similar living creature on all the face of the earth." We agree.

Fortunately, wildlife scientists, managers, hunters and other conservationists—persons of foresight and forebearance—not only have saved the pronghorn from the fate of passenger pigeons and heath hens, but have given them the opportunity to thrive. We hope such people—of this and future generations—will care equally about the well-being of pronghorn and the landscape and human values that sustain them. If this book helps that endeavor, then the decades of investment will have been entirely and gratefully worthwhile.

—Bart W. O'Gara
Jim D. Yoakum

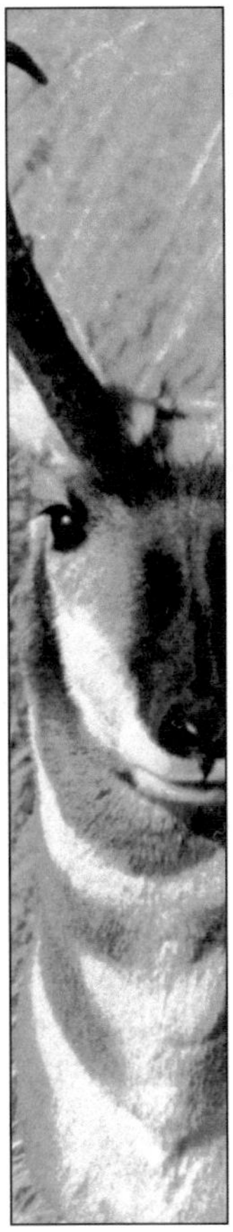

Preface

This comprehensive book, decades in construction, comes at an important time for the future of North America's unique pronghorn. The last work of such substance on the species—Arthur Einarsen's *The Pronghorn Antelope and Its Management*—dates back nearly six decades, a period in which pronghorn made a remarkable, perhaps unparalleled, recovery. This revival, mainly by virtue of science-based wildlife management, was paid for largely by sportsmen and sportswomen. A classic example of North America's wildlife management success, pronghorn emerged from near extinction in the first quarter of the past century to relative abundance entering the new millennium.

How this happened and what it has taken to do the job are well described here. But, this documentation of successes might delude one into thinking the job is done. The strongest contribution of this work may be the foundation it provides to assist in the next round of pronghorn research, management and conservation.

And there will be another round. The landscape that pronghorn inhabit is as unique and wonderful as the big-eyed speedsters themselves. Where pronghorn habitats have been coveted by people for agriculture, residences, recreation and other uses, pronghorn usually merely are tolerated and thrive only through human benevolence. Their strongholds, in the new millennium, are the high plains and sage deserts of Wyoming, northern Rockies, Great Basin and arid portions of the Southwest. In Wyoming, where as much as 90% of pronghorn live, their formerly wild environs—a core of their existence as a species—now are changing, perhaps too much and too rapidly for science and management to keep pace.

Arguably, no place feels more like pronghorn country than Wyoming's Red Desert or South Pass, where travelers on the Oregon, California and Mormon trails crossed the continental divide and encountered and used pronghorn for their very survival. It was truly wild pronghorn country, where free-ranging herds have migrated north and south more than 100 miles each way for thousands of years. Pronghorn herds traversed these high plains, survived droughts and blizzards, and adapted to much change. Grazing, ranching, fencing and developing for towns, highways and homes have affected the pronghorn, but not overcome their resilience. It still is pronghorn country.

However, a human value—a 100-year-old one—that threatens pronghorn country is becoming more pervasive. The invasion of pronghorn habitat for natural energy resources (coal, oil, natural gas and coal-bed methane) is progressing at a rate that dwarfs the severity of other intrusions. In addition to the human influences and obstacles that pronghorn have navigated successfully, now new roads, well pads, processing complexes, pipelines and much higher levels of human activity have added ecological insult to this formerly wild, forever historic country. It and a thousand other locations that contributed to the recovery of pronghorn now are being exploited without a clear plan for the needs of pronghorn.

This story also is a classic example of the dilemma of North American wildlife, proving that no success story is guaranteed forever. Suppose that the new human imposition on pronghorn habitat shifts available wintering areas to ones less accessible or less nutritious. Suppose that it modifies spring movements so that fawns encounter more predators. Uncertainties such as these are the *yin* and *yang* of wildlife science. For those for whom the sacrifice of pronghorn and pronghorn country is not an option, successful resolution needs a scholarly foundation. That is what this book has to offer—to the scientists, the land managers, the wildlife managers, the researchers, the exploiters, the sportsmen and sportswomen, and to other interested North Americans, all of whom have a vested interest in their pronghorn and their pronghorn country.

The wisdom of experience with pronghorn in this entertaining and enlightening book can help chart a course to sustain the most indigenous of North American wildlife and the myriad values that the pronghorn represents.

—ROLLIN D. SPARROWE
PRESIDENT
WILDLIFE MANAGEMENT INSTITUTE

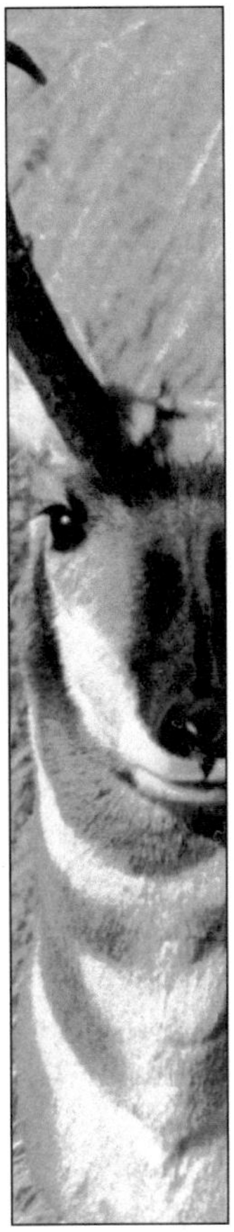

Acknowledgments

No one works in a vacuum. Hundreds of people contributed to this book in many different ways. Bart O'Gara and Jim Yoakum thank the dozens of fellow wildlife biologists whose publications and research constitute the lion's share of the contents herein. We especially appreciate the input of pronghorn biologists, managers and students who frequently attended the pronghorn workshops and shared unpublished data, experiences, insights and photos with us.

We sincerely appreciate the support, collaboration, numerous technical editings and endurance of Richard E. McCabe, Executive Vice President of the Wildlife Management Institute. Over the years, Dick and this endeavor were very fortunate to have the organizational, research and word-processing skill of three talented assistants—Lyla Baumann, Kelly Stockwell and Jennifer Rahm.

The U.S. Bureau of Land Management (BLM) provided Jim with support to document habitat characteristics and enhancement. John Crawford, Chief of Wildlife from 1974 to 1985, recognized the need for technical contributions to support field biologists through publications. Edward Spang, Chief of Resources for the Nevada State Office, provided logistical and moral support to "make contributions to wildlife science, and public land management." This book commenced before word processors and computers were available, and early drafts were laboriously typed. Chapters were reviewed and retyped as many as 15 times.

Jim is grateful for editorial assistance and critical evaluation of chapter drafts by Jack Richardson, Richard E. Genelly, Marshall White, Michael C. Hansen, Donald E. Spalinger, David E. Brown, Jerry L. Holecheck, Fred H.

Wagner, Raul Valdez and Bruce Morrison. Jim also is indebted to BLM employees Dwan Berryman and Mary Clark for typing and proofing services.

The U.S. Fish and Wildlife Service, through the Cooperative Wildlife Research Unit office in Washington, DC, encouraged Bart to pursue basic and applied research. He was given the flexibility to work part-time with pronghorn throughout his career. Raymond Murray, Vice President for Research and Sponsored Programs at the University of Montana, Missoula, facilitated research funding and was always helpful. Numerous individuals with the Montana Department of Fish, Wildlife and Parks physically or financially aided Bart's research.

Personnel of the Montana Cooperative Wildlife Research Unit helped Bart in many ways. Former Assistant Unit Leader (current Leader) Joe Ball carried a good deal of the teaching and administrative load. Secretaries/office managers Ginger Schwarz and Virginia Johnston helped in more ways than they knew; besides typing and proofing, they bore much of the office personnel burdens. Graduate students (many working with pronghorn) and work-study students assisted in numerous ways, besides keeping life interesting. The entire Unit crew were friends who enable Bart to spend a great deal of time on pronghorn research and writing. Incidentally, Virginia Johnston typed Bart's pronghorn thesis and continued to process his manuscripts until this book went to press.

Wilma O'Gara, Bart's childhood sweetheart and the love of his life, worked many hours with him during the final phase of the book. Checking references and proofreading wasn't what she had planned for retirement but she took it in stride.

The principal authors and editor of this book gratefully acknowledge the contributions and assistance of the following individuals and many others who helped us. It should be noted that many of the persons below contributed also or mainly to this book's companion volume, *Prairie Ghost: Pronghorn and Human Interaction in Early America.*

Terry Abraham, Washington State University, Pullman
Sonja Aerni, Nebraska Game and Parks Commission, Lincoln
Jean Ake, James Jerome Hill Reference Library, St. Paul, Minnesota
Steve Alastuey, Utah Division of Wildlife Resources, Cedar City
William Allman, The White House, Washington, DC
Durward Allen, Purdue University, West Lafayette, Indiana
Ronald Anglin, Oregon Department of Fish and Wildlife, Eugene
Patricia Ansley, Glenbow Museum, Calgary, Alberta
Robert Anthony, Oregon State University, Corvallis, Oregon
Brooke Arkush, Weber State University, Ogden, Utah
Connie Armstrong, Rawlins, Wyoming
L. L. Arnevick, Minneapolis, Minnesota
Alan Arsenault, Saskatchewan Fish and Wildlife Branch, Saskatoon
Bob Autenrieth, Idaho Department of Fish and Game, Jerome
Susan Kraft Ball, Missoula, Montana
Joe Ball, Montana Cooperative Wildlife Research Unit, Missoula
Carol Barber, Wyoming State Archives, Cheyenne
Morley Barrett, Alberta Fish and Wildlife, Edmonton
Jeannie Baumann, The Walters Art Gallery Baltimore, Maryland
Donald M. Beal, Utah Division of Wildlife Resources, Cedar City
Donald Bedunah, University of Montana, Missoula
Ted Benzon, South Dakota Game, Fish and Parks Department, Pierre
Charles Berger, Smithsonian Institution, Washington, DC
Margaret (Mrs. Junius) Bird, Bronx, New York
Craig Birhle, North Dakota Game and Fish Department, Bismarck
Floyd M. Blunt, Wyoming Department of Game and Fish, Cheyenne
Walt Bodie, Idaho Department of Fish and Game, Boise
Natasha Bonilla, Museum of the American Indian, New York, New York
Clay Brewer, Texas Parks and Wildlife Department, Alpine
Dave Brewster, Saskatchewan Fish and Wildlife Branch, Saskatoon
Kim Brinkley, Los Angeles Zoo, Los Angeles, California
Russell Bromby, Colorado Division of Wildlife, Denver
Peter T. Bromley, Virginia Polytechnic Institute and State University, Blacksburg
David W. Brown, Arizona State University, Phoenix
Ed Bry, North Dakota Game and Fish Department, Bismarck
Jim Burant, Public Archives of Canada, Ottawa, Ontario
Carol Burke, Museum of Northern Arizona, Flagstaff
Vanetta Burton, University of Montana, Missoula
Charles Cadieux, Albuquerque, New Mexico
Jorge Cancino, Centro de Investigaceons Biologicas del Noroeste, La Paz, Baja California Sur
Rodney Cannutt, U.S. Forest Service, Portland, Oregon

Ken Canon, Texas Tech University, Lubbock
Kay Carlson, Dull Knife Memorial College, Lame Deer, Montana
John Carter, Nebraska State Historical Society, Lincoln
Herbert Cartright, The Petroleum Museum, Midland, Texas
Stephen Catlett, American Philosophical Society, Philadelphia, Pennsylvania
Liz Clancy, Denver Museum of Natural History, Denver, Colorado
Michael Coan, Caldwell Veterinary Teaching Center, Caldwell, Idaho
E. Lendell Cockrum, University of Arizona, Tucson
Jack Coffrin, Coffrin's Old West Gallery, Miles City, Montana
Kevin Conway, Utah Division of Wildlife Resources, Salt Lake City
Judd Cooney, Pagosa Springs, Colorado
Allen Cooperrider, U.S. Bureau of Land Management, Denver, Colorado
Mike Cox, Nevada Division of Wildlife, Carson City
Deanna Cross, The Metropolitan Museum of Art, New York, New York
Wallace Dailey, Harvard University, Cambridge, Massachusetts
Rick E. Danvir, Deseret Land and Livestock Ranch, Woodruff, Utah
Ray Dasmann, Humboldt State University, Arcata, California
Robert Deblinger, Colorado State University, Fort Collins
Kenneth Diem, University of Wyoming, Laramie
James Dinsmore, Iowa State University, Ames
Diana Doan-Crider, Texas A&I University, Kingsville
Dwight Dobson, Saskatchewan Fish and Wildlife Branch, Maple Creek
Gary Doster, Southeastern Wildlife Disease Study, Athens, Georgia
Mike Dunbar, U.S. Fish and Wildlife Service, Lakeview, Oregon
William C. Edwards, Colorado State University, Fort Collins
Arthur S. Einarsen, Oregon State University, Corvallis
Ralph Elder, Eugene C. Barker Texas History Center, Austin
Jean Elkington, Oregon Historical Society, Portland
Jim E. Ellis, Colorado State University, Fort Collins
Glenn Eriksen, Montana Department of Fish, Wildlife and Parks, Helena
Catherine Engel, Colorado Historical Society, Denver
Peg Zane Evers, U.S. Bureau of Land Management, Reno, Nevada
Erin Ferri, Buffalo Bill Historical Center, Cody, Wyoming
Edson Fichter, Idaho State University, Pocatello
Terry Fingerhut, Nebraska Game and Parks Commission, Lincoln
John Fischer, California Department of Fish and Game, Sacramento
Lloyd Fox, Kansas Department of Wildlife and Parks, Emporia
Heidi Frautsch, Kunstmuseum, Bern, Switzerland
George Frison, University of Wyoming, Laramie
Laura Furney, University Press of Colorado, Boulder
Bonnie Gardner, South Dakota Historical Resource Center, Pierre
Eleanor Gehres, Denver Public Library, Denver, Colorado
Adrian Gibbons, Joslyn Art Museum, Omaha, Nebraska
K. D. Goodman, Annapolis, Maryland
Campbell Grant, Carpinteria, California
Ken Greer, Montana Department of Fish, Wildlife and Parks, Helena
Richard Guenzel, Wyoming Department of Game and Fish, Laramie
Pamela Haas, American Museum of Natural History, New York, New York
Barry Hale, New Mexico Department of Game and Fish, Santa Fe
Katherine Hanna, The Taft Museum, Cincinnati, Ohio
Edward L. Hansen, Cooperative Wildlife Research Unit, Corvallis
H. J. Harju, Wyoming Department of Game and Fish, Cheyenne
Lois Harrington, The Petroleum Museum, Midland, Texas
Frank Hayes, University of Georgia, Athens
Jim Heffelfinger, Arizona Game and Fish Department, Tucson
William G. Hepworth, Wyoming Department of Game and Fish, Laramie
Jeff Herbert, Montana Department of Fish, Wildlife and Parks, Helena
Ellen Herring, Agoura, California
Mike Hess, Nevada Division of Wildlife, Reno
Ruth Beebe Hill, Friday Harbor, Washington
Vic Hill, Lakeview, Oregon
Robb Hitchcock, North American Pronghorn Foundation, Casper, Wyoming
Pat Hollinshead, Arizona State Museum, Tucson
Doug Humphreys, Texas Parks and Wildlife Department, Austin
David Hunt, Joslyn Art Museum, Omaha, Nebraska
Elizabeth Jacox, Idaho State Historical Society, Boise
Laurence Jahn, Wildlife Management Institute, Washington, DC
Jim Jeffress, Nevada Department of Wildlife, Winnemucca

Bill Jensen, North Dakota Department of Game and Fish, Bismarck
David Johnson, The Taft Museum, Cincinnati, Ohio
Rolf Johnson, Washington Department of Fish and Wildlife, Olympia
David Johnston, Arlington, Virginia
Michael Kaplan, The Newberry Library, Chicago, Illinois
William Keller, U.S. National Park Service, Yellowstone Park, Wyoming
Amy Kelly, Amon Carter Museum, Fort Worth, Texas
Robert Kercher, Great Falls, Montana
Rebecca Kohl, Montana Historical Society, Helena
Robert Kindschy, U.S. Bureau of Land Management, Vale, Oregon
Jeanne King, The Thomas Gilcrease Museum, Tulsa, Oklahoma
David W. Kitchen, Humboldt State University, Arcata, California
Alice Koch, California Department of Fish and Game, Templeton
Michael Korn, Montana Department of Fish, Wildlife and Parks, Helena
Lonn Kuck, Idaho Department of Fish and Game, Boise
Lawrence Landis, Eugene C. Barker Texas History Center, Austin
Parry Larsen, New Mexico Department of Game and Fish, Roswell
James Lavender-Telia, Colorado Historical Society, Denver
Raymond Lee, Arizona Department of Game and Fish, Phoenix
Tom Lesson, Vancouver, Washington
Martin Levitt, American Philosophical Society, Philadelphia, Pennsylvania
Heather Linville, University of Montana, Missoula
Katherine Lipsett, Glenbow Museum, Calgary, Alberta
Samantha Loos, Wildlife Management Institute, Washington, DC
Don Luce, James Ford Bell Museum of Natural History, Minneapolis, Minnesota
Chris Madson, Wyoming Game and Fish Department, Cheyenne
Michael Mahoney, Trinity College, Hartford, Connecticut
Neil Mangum, Custer Battlefield Historical Museum, Crow Agency, Montana
Brent Markham, Alberta Fisheries and Wildlife Management Division, Edmonton
Randy Matchett, U.S. Fish and Wildlife Service, Lewistown, Montana
Michael K. McCabe, Omaha, Nebraska
Robert McCabe, University of Wisconsin, Madison
Clint McCarthy, U.S. Forest Service, Lee Vining, California
Bob McCarty, U.S. Bureau of Land Management, Pinedale, Wyoming
Nancy McClure, Wyoming State Parks and Historic Sites, South Pass City
Doug McChristian, Hardin, Montana
James McKenzie, North Dakota Game and Fish Department, Bismarck
Bette McKown, Wildlife Management Institute, Washington, DC
Mark McNay, Alaska Department of Fish and Game, Fairbanks
Wyman Meizner, Texas Parks and Wildlife Department, Austin
Larry Mensching, Joslyn Art Museum, Omaha, Nebraska
Karl Menzel, Nebraska Game and Parks Commission, Bassett
Joan Metzger, Arizona Historical Society, Tucson
Eugene Miller, Mill Valley, California
George Mitchell, University of Regina, Regina, Saskatchewan
Melissa Moore, The Minneapolis Institute of Arts, Minneapolis, Minnesota
Anne Morand, The Thomas Gilcrease Museum, Tulsa, Oklahoma
Leotine (Tina) Nappe, Foresta Institute, Carson City, Nevada
Cathy NcNassor, Natural History Museum of Los Angeles County, Los Angeles, California
Susan Near, Montana Historical Society, Helena
Lou Nelson, Idaho Department of Fish and Game, Boise
Richard Ockenfels, Arizona Department of Game and Fish, Phoenix
Wilma O'Gara, Lolo, Montana
Lloyd E. Oldenburg, Idaho Department of Fish and Game, Boise
Kerry Olson, Wyoming Game and Fish Department, Cheyenne
LuRay Parker, Wyoming Game and Fish Department, Cheyenne
Carol Jean Peddicord, Wildlife Management Institute, Washington, DC
Jeff Pederson, New Mexico Department of Game and Fish, Santa Fe
Matt Peek, Kansas Department of Wildlife and Parks, Topeka
Mercedes Penarowski, Colorado Historical Society, Denver
John S. Phelps, Arizona Department of Game and Fish, Phoenix
Daniel Pond, University of Montana, Missoula
Danielle Ponsolle, Orlando, Florida

Daniel Poole, Wildlife Management Institute, Washington, DC
Joseph Porter, Joslyn Art Museum, Omaha, Nebraska
Chuck Post, South Dakota Game, Fish and Parks Department, Pierre
George Post, Colorado State University, Fort Collins
Peter Powell, Foundation for the Preservation of American Art and Culture, Chicago, Illinois
Daniel Pratt, University Press of Colorado, Boulder
Darrin Pratt, University Press of Colorado, Boulder
William H. Pyle, U.S. Fish and Wildlife Service, Lakeview, Oregon
Duane Pyrah, Montana Department of Fish, Wildlife and Parks, Lewistown
L. "Bud" Pyshora, California Department of Fish and Game, Redding
Susan Rapps, Museum of the American Indian, New York, New York
Michael Ratliff, U.S. Bureau of Land Management, Washington, DC
Barry Reiswig, Lakeview, Oregon
Jose Maria Reyes, Mexico City, Mexico
Phil Richmond, Nebraska Game and Parks Commission, Lincoln
Richard Ridenhour, Humboldt State University, Arcata, California
Archy Roach, FTE Enterprises, New York, New York
Mimi Roberts, Joslyn Art Museum, Omaha, Nebraska
Tom Rogers, Wildlife Ltd., Edgefield, South Carolina
Charles H. Rouse, U.S. Fish and Wildlife Service, Lakeview, Oregon
Bill Rudd, Wyoming Department of Game and Fish, Green River
Leonard Lee Rue, III, Blairstown, New Jersey
Kenneth Sabol, Wildlife Management Institute, Washington, DC
Judy Schinz, Penticton, British Columbia
Loretta Schmidt, U.S. National Park Service, Oyster Bay, New York
Chuck Schluter, South Dakota Game, Fish and Parks Department, Pierre
Charles C. Schwartz, Colorado State University, Fort Collins
Keith Sexson, Kansas Department of Wildlife and Parks, Emporia
Bob Shaw, Shelburne Museum, Shelburne, Vermont
Michael Shaw, Oklahoma Department of Wildlife Conservation, Oklahoma City
Dona Sieden, Minnesota Historical Society, St. Paul
David C. Simpson, Texas Tech University, Lubbock
Shari Small, Buffalo Bill Historical Center, Cody, Wyoming
Murphy Smith, American Philosophical Society, Philadelphia, Pennsylvania
Maeve Spain, Glenbow Museum, Calgary
Rollin Sparrowe, Wildlife Management Institute, Washington, DC
Juan J. Spillett, Utah State University, Logan
Christie Stanley, Kansas State Historical Society, Topeka
Bruce Stillings, North Dakota Game and Fish Department, Dickinson
Todd Strand, State Historical Society of North Dakota, Bismarck
Bob Strom, National Wildlife Federation, Vienna, Virginia
Robert Streeter, U.S. Fish and Wildlife Service, Washington, DC
Charles Sundstorm, U.S. Forest Service, Deer Lodge, Montana
Leonard Sutton, Denver, Colorado
Danny Swepston, Texas Parks and Wildlife Department, Austin
Doug Thayer, California Department of Fish and Game, Alturas
Melissa Thompson, Amon Carter Museum, Fort Worth, Texas
E. Tom Thorne, Wyoming Game and Fish Department, Wheatland
Leonard Thorson, State Historical Society of North Dakota, Bismarck
Cindy L. Ticer, Arizona Department of Game and Fish, Phoenix
Henry Tobias, University of Oklahoma, Norman
Marshall Townsend, University of Arizona Press, Tucson
Charles E. Trainer, Oregon Department of Fish and Wildlife, Corvallis
Brent Treichel, Alberta Fisheries and Wildlife Management Division, Edmonton
Jose "Pepe" C. Trevino, Chihuahua, Chihuahua
Jan Trumm, Joslyn Art Museum, Omaha, Nebraska
George K. Tsukamoto, Nevada Department of Wildlife, Reno
Dale Looks Twice, Pine Ridge, South Dakota
Connie Valdex, Pagosa Springs, Colorado
Joe Van Wormer, Bend, Oregon
Jeff Ver Steeg, Colorado Division of Wildlife, Denver
Amy Verone, Sagamore Hill National Historic Site, Oyster Bay, New York
Denise Wagorn, Public Archives of Canada, Ottawa, Ontario
Brian Wakeling, Arizona Game and Fish Department, Phoenix
Murray Walton, Wildlife Management Institute, Dripping Springs, Texas
Danny Watson, Oklahoma Department of Wildlife Conservation, Forgan

S. David Webb, University of Florida, Gainesville
Mike Welch, Utah Division of Wildlife Resources, Salt Lake City
Chuck Weschler, Sporting Classics, Camden, South Carolina
Chuck Darrel Weybright, New Mexico Department of Game and Fish, Santa Fe
Susan Whetstone, Utah State Historical Society, Salt Lake City
Donald Whittaker, Oregon Department of Fish and Wildlife, Portland
Larry Wing, University of Wyoming, Laramie
John Witherbee, Union Pacific Railroad Company, Omaha, Nebraska
Terry L. Yates, University of New Mexico, Albuquerque
David Zezulak, California Department of Fish and Game, Sacramento
Jason E. Zimmer, University of Wyoming, Laramie
Stephen Zimmer, Philmont Scout Ranch, Cimarron, New Mexico
Rex S. Zobell, U.S. Bureau of Land Management, Cheyenne, Wyoming

PRONGHORN
Ecology and Management

SECTION ONE

ECOLOGY

C H A P T E R O N E

Scientific Classification

Bart W. O'Gara and Christine M. Janis

The prong-horned antelope is found only in North America, and it possesses so many anatomical peculiarities, found in no other animal, that zoologists have created for it a separate Family, which it occupies in solitary state. It is like an island in a vast sea, unrelated . . . an animal more wonderful than the rarest orchid that ever bloomed.

Hornaday (1904:116)

The placing of the genus Antilocapra *as the sole representative of a family, Antilocapridae, on an equality with the Cervidae and the Bovidae is unnatural, as was pointed out by Cope long ago. The visceral anatomy, the osteology, and the dentition, as shown by Dr. James Murie, associate* Antilocapra *with Bovidae. The only essential difference between* Antilocapra *and the other Bovidae is the casting off of the annual increment of its horn.*

Lyon (1908:398)

It would be unusual in science if the great dinosaur debate were resolved quickly. Science isn't like that. While individual scientists may advocate their interpretations with great passion, the true progress of science usually follows a long period of evidence gathering on both sides that eventually yield a preponderance fitting one theory. Even then, of course, there are always diehards who find themselves outside the mainstream. . . . Debate, however, is the engine of scientific advancement.

Rensberger (1986:78)

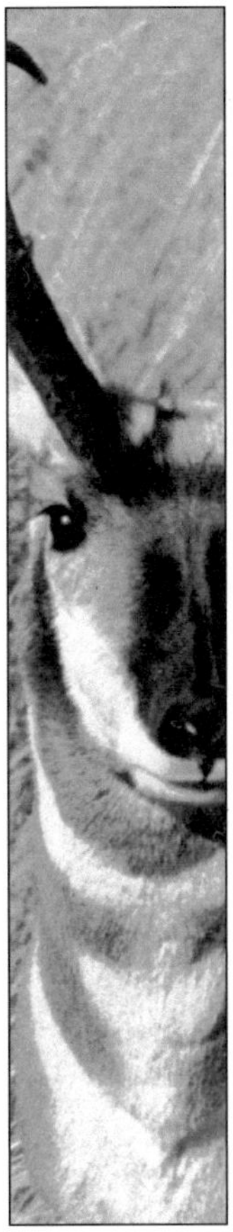

TAXONOMIC RELATIONS

Pronghorn belong in the order Artiodactyla—hoofed mammals that have a unique double-pulley type of astragalus, the bone in the ankle joint that articulates with the upper part of the leg (Schaeffer 1947). Artiodactyls walk on two toes—digits three and four. Digit one is lost in all living forms; digits two and five may be complete, as in the hippopotamus, reduced, as in deer, or absent entirely, as in camels and some true antelope (Bovidae). Members of the other living order of hoofed mammals, the Perissodactyla, center their weight on the third toe and usually have lost digits one and five (although digit five is retained in the front foot in tapirs and some fossil forms). They either may walk on the third toe alone, as does the horse (where toes two and four are reduced to splint bones), or retain all three digits and incorporate a footpad under the toes, as in the rhinoceros (Figure 1).

The earliest known artiodactyls lived about 55 million years ago in North America and Europe. These were primitive, rabbit-sized omnivores commonly known as dichobunids, and no representatives have survived to the present. By the late Eocene (about 45 million years ago), the first members of the three living suborders were recognizable in the fossil record. These suborders are Suina (piglike forms), Tylopoda (including the camelids, as well as a number of extinct forms) and Ruminantia (which includes pronghorn and other living horned ruminants, as well as more primitive hornless forms, such as the mouse deer or chevrotains).

Living members of both Tylopoda and Ruminantia possess a three- or four-chambered stomach for the digestion

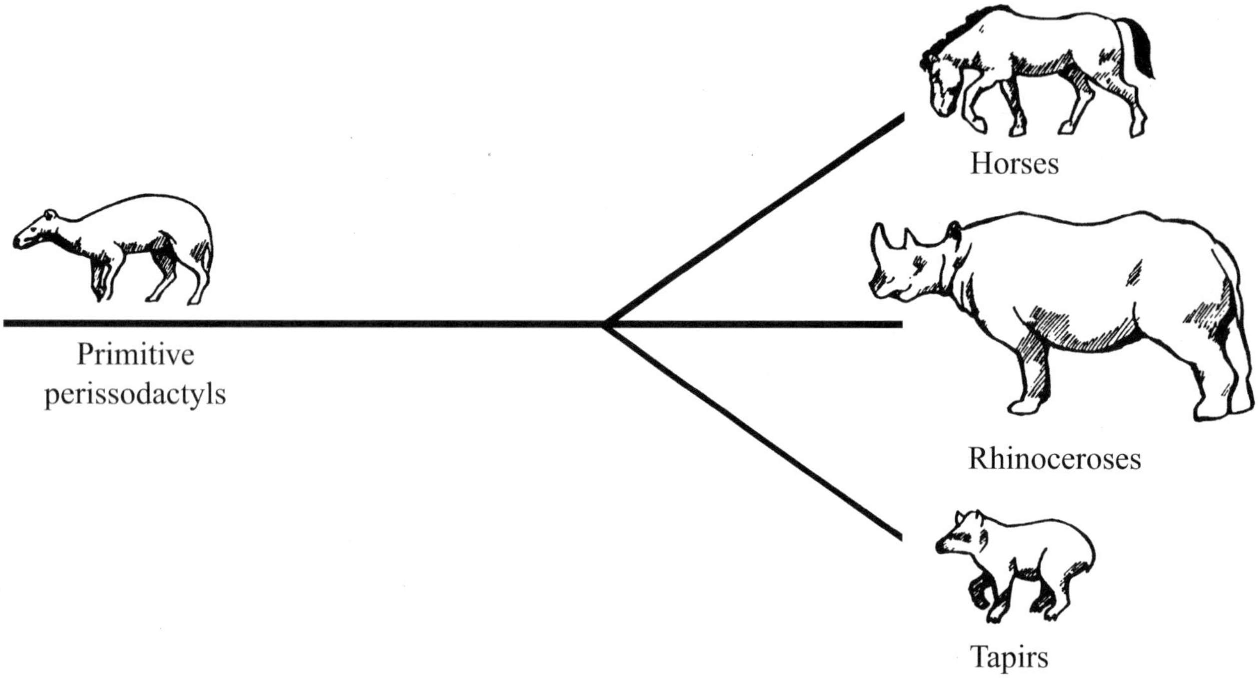

Figure 1. *A simplified family tree of the odd-toed ungulates, the perissodactyls (adapted from Romer 1962). This does not represent phylogenetic structural evolution.*

of fibrous vegetation, although the elaboration of these stomachs from a simple, enlarged condition probably evolved in parallel in the two suborders (Langer 1974). Both suborders have selenodont cheek teeth (teeth with high-crested lophs, designed to cut herbage). These teeth contrast with the bunodont teeth of most suines (which are like human teeth, with low, rounded cusps, and indicative of a more omnivorous diet) and the lophodont cheek teeth of perissodactyl herbivores (which are analogous to selenodont teeth in function, but have a different pattern to the lophs on the crown). Selenodont means moon tooth and refers to the characteristic crescent shape of the lophs.

Tylopods and ruminants show a number of other convergences in their evolution, such as the evolution of long-legged running forms with high-crowned cheek teeth (camels in Tylopoda, pronghorn, and Old World antelope in Ruminantia), the evolution of bony horns (in the extinct North American tylopod family Protoceratidae); and the evolution of a giraffelike camel (*Alticamelus*) from the North American Miocene epoch (see Janis 1982 for further discussion of ruminant and tylopod parallelisms).

Whereas living and fossil members of Ruminantia are characterized by a unique anatomical feature—fusion of the cuboid and navicular bones in the ankle joint—Tylopoda cannot be characterized so and perhaps should be regarded merely as an ill-defined assemblage of ruminant-like animals that possess selenodont cheek teeth. However, camels and protoceratids are related more closely to true ruminants than are other tylopods (Webb and Taylor 1980). Tylopods and ruminants have different evolutionary histories. Ruminantia probably originated in Asia in the late Eocene, although some primitive families also were found in the early Tertiary of North America. Camelid and protoceratid tylopods appeared at the same time in North America, and other tylopods were found at this time in Europe. The European families met their demise in the early Oligocene (around 35 million years ago), when true ruminants invaded Europe from Asia. However, ruminants coexisted with camels and protoceratids throughout the Tertiary in North America.

The suborder Ruminantia is divided into the infraorder Tragulina (small, hornless forms) and the infraorder Pecora (larger, later-appearing forms that usually possess horns of some description). However, horns are lacking in many pecoran fossil taxa and in the living genera *Moschus* (musk deer) and *Hydropotes* (Chinese water deer). The most primitive living ruminant family is the Tragulidae (infraorder Tragulina), with two genera (one in the African tropics and one in the Asian tropics) of so-called mouse deer, or chevrotains. Although tragulids are not known as fossils from before the Miocene (possibly because they occupied tropical forests where fossilization was rare), they are very similar to other traguloid families that were found in the early Tertiary of Eurasia and North America. None of the other traguloids survived past mid-Tertiary times, and living tragulids are confined to tropical forest environments, which were more widespread in the early Tertiary. Tragulids have a three-chambered stomach because they lack the omasum of higher ruminants. And, they retain complete, if reduced, side toes. They lack any form of horns or antlers. They also lack completely fused cannon bones, which occurred in some of the extinct traguloid families. The other living ruminant families in the infraorder Pecora have four-chambered stomachs, reduced or absent side toes, antlers or horns (usually), and completely fused cannon bones that have better-developed articulatory surfaces for the foot bones on their lower ends. Fossil pecorans might not possess all of these characteristics; the infraorder is defined by some advanced features of the base of the skull and a more sophisticated ankle joint, the latter suggesting greater capacity for running (Webb and Taylor 1980).

The living pecoran families are as follows: Cervidae (deer), Moschidae (musk deer, which may be considered with the Cervidae by some taxonomists), Bovidae (cattle, sheep and Old World antelope), Giraffidae (giraffe and okapi) and Antilocapridae (pronghorn). With the exception of the Moschidae, the living pecoran families usually possess horns of some sort, although these appear to have evolved independently in each case (Janis 1982, Janis and Scott 1987). Traditionally, families that possess low-crowned (brachydont) cheek teeth (Cervidae and Giraffidae) or retain the side toes (Cervidae) have been considered primitive because their morphology more closely resembles the condition seen in the tragulids. In contrast, families with high-crowned (hypsodont) cheek teeth (Bovidae and Antilocapridae) or long legs and the loss of the side toes (Giraffidae, Bovidae and Antilocapridae) have been considered advanced (see Romer 1966). However, hypsodont cheek teeth and long legs with the loss of the side toes are morphological characteristics that are functionally associated with a fibrous diet (such as grass) or an open habitat (such as the plains), respectively. Both features have convergently evolved with the bovid and antilocaprid condition in camels and horses. Thus, it is difficult to say which of the pecoran families is the most primitive. Some evidence from tooth structure and stomach anatomy suggests that the Giraffidae might be the most primitive living family, with the Cervidae, Bovidae and Antilocapridae all being equally derived, albeit with different types of specializations reflecting the different environments that they have inhabited (Janis and Scott 1987). Specialization does not equate necessarily with evolutionary advancement. Egg-laying mammals, for example, are highly specialized in their own way, despite being revolutionarily primitive.

Cervidae possess deciduous antlers, whereas Giraffidae possess bony, skin-covered, nondeciduous horns, termed ossicones. However, the most specialized pecoran families—Bovidae and Antilocapridae—are characterized by hypsodont cheek teeth and permanent bony horn cores covered by keratinous sheaths. Much controversy exists concerning the taxonomic position of the pronghorn within the Pecora. The traditional classification is in the family Antilocapridae, allied with the Bovidae in the suprafamily Bovidoidae (Simpson 1945, Romer 1966). O'Gara and Matson (1975) place pronghorn as the subfamily Antilocaprinae within the family Bovidae. In contrast, other investigators have allied the family Antilocapridae with the Cervidae in the suprafamily Cervidoidae (Leinders and Heintz 1980).

Many characteristics of bovids and cervids are similar and overlapping. It is not surprising that pronghorn have been argued to belong within Cervidae, Bovidae or a subfamily (Antilocaprinae) in the Bovidae. Lack of fossils of both early bovids and antilocaprids makes classification

The relationship of pronghorn to other ruminants has long been a source of disagreements among taxonomists. Many saw pronghorn as the possible link between bovids and cervids. Murie (1870:338) summed the attitude of his time: "If I were asked by a single term to denote what the animal is, I should be obliged to Germanize the English phraseology and name it a Giraffe-hoofed, Sheep-haired, Deer-headed, Goat-glanded Antelope—an expression however rugged, yet explicit enough to baffle those who are skeptical of gradational forms." Photo courtesy of the U.S. Bureau of Biological Survey.

difficult. In effect, taxonomists are looking analogously at the twigs of a tree and trying to reconstruct the trunk and branches (Figure 2).

The presence of annually deciduous horns was the principal reason that pronghorn were placed in a family separate from other bovids. A common impression concerning the horns of bovids was expressed by Romer (1966), who reported that all bovids have true horns (usually in both sexes), consisting of an unbranched horn core of bone covered with a sheath of horn, and no part of this structure is ever shed. However, exfoliation of outer horny material has been described for some members of six subfamilies of Bovidae (Buffon 1764, Ogilby 1840, Hall 1935, Biddulph 1936, D'Abreau 1936, Khan 1936, Chandrabhanusinhji 1937, Krammer 1964, Mohre 1965, O'Gara and Matson 1975). Buffon's (1764) observations are noteworthy because they indicate that more than 200 years of selective breeding have changed the growth characteristics of domestic cattle horns. Buffon wrote that oxen cast their horn sheaths at 3 years of age, and new sheaths, like permanent teeth, were never shed again. He then described subsequent growth in which annual nodes were formed so that taking the first node as 3 years and adding 1 year for each of the intervals between succeeding nodes revealed the age of an ox. However, if Antilocapridae evolved from Merycodontinae, this discussion is moot—antilocaprid and bovid horns could not be homologous.

Horn cores of female pronghorn or castrated bucks sometimes are cone shaped, and the sheaths usually are

The presence of annually deciduous horn sheaths was the main reason pronghorn were not considered bovids when first classified. This buck, shot in Montana during early November, was nearing the time to cast horn sheaths. New horn growth at the tips of the horn cores was beginning to loosen the sheaths. By twisting, the left sheath was removed, but the right was too solid to move. Photo by David Kitchens.

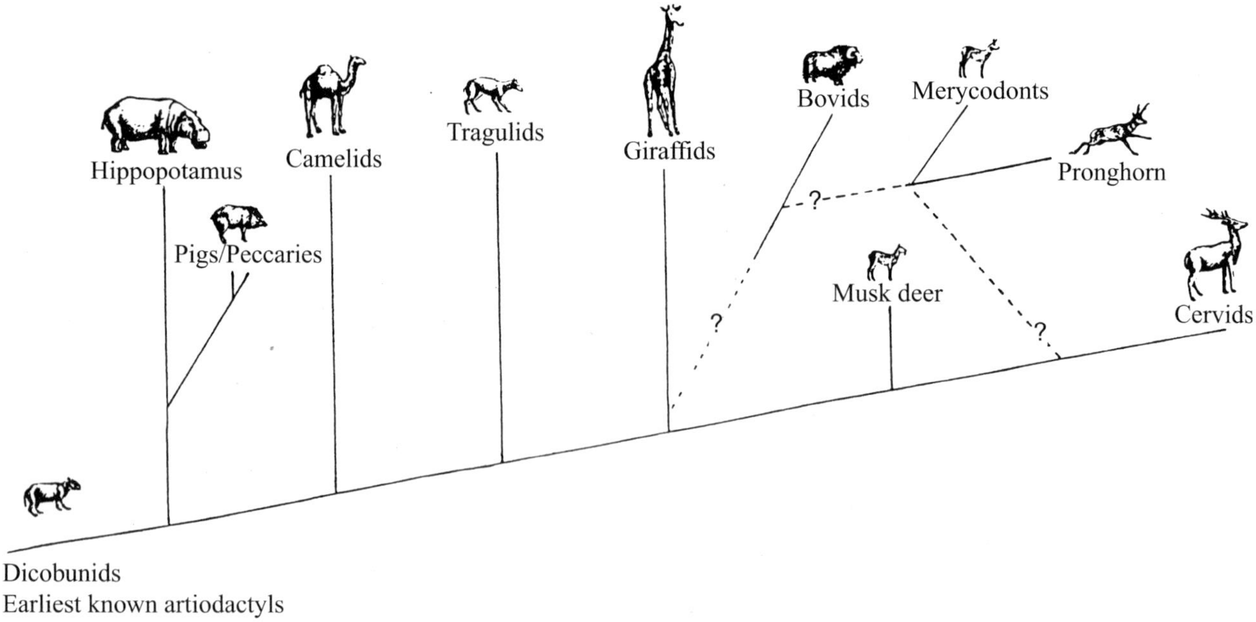

Figure 2. *A structural view of artiodactyl relationships—only those discussed in this chapter—that reflects the current debate concerning the relationship of pronghorn to other artiodactyls (adapted from Romer 1962 and information presented in this chapter). The sequence of taxa shown here does not represent phylogenetic structural evolution.*

without prongs. Individual pronghorn of those two categories sometimes produce nested annual increments of horny sheaths, superficially similar to those of sheep (Pocock 1905, O'Gara 1969b). Some bovids, with slim horn cores nearly reaching the tips of their sheaths (subfamilies Hippotraginae, Antilopinae and Cephalophinae), exfoliate large strips or complete sheaths from the smooth tips (which may be three-fourths of the horn), but only flakes from the annulared portions of their horns close to the head (O'Gara and Matson 1975). Pronghorn remain the only recent mammals with annually deciduous horns.

Cope (1888) expressed the opinion that the shedding of horns by antilocaprids, even if normal (which was still in doubt at that time), did not furnish a character of sufficient value to justify its separation as a distinct family from Bovidae. Lyon (1908) also maintained that pronghorn should be considered a subfamily—Antilocaprinae of the Bovidae.

Horn shedding by pronghorn bucks is controlled by the annual testicular cycle (O'Gara et al. 1971). As pointed out by Leinders and Heintz (1980), this parallels the growth and casting of antlers in cervids and probably has the same function in relation to the rut. The horns of female or castrated pronghorn are not shed on a regular basis (Pocock 1905, O'Gara 1969b)—further evidence of control by male hormones.

Bovid horns generally have annual starts and stops in growth; such growth apparently is related to annual periods of stress brought on by severe winters, as well as by male hormones. Growth rings are more evident in mountain and northern species than in those of lowlands or tropical areas and occur in females as well as males. However, growth in the horns of sexually mature mountain sheep rams comes to an abrupt halt just before the rut, causing a distinct break in horn fibers (V. Geist personal communication:1996). This break is not seen in the horns of ewes and juvenile rams. Rather, a thickening occurs without a break, suggesting that a sudden hormonal change terminates annual horn growth in mature rams.

The timing of exfoliation in bovids has not been studied extensively, but it seems to occur primarily during early growth (Ogilby 1840, Krammer 1964) or at the time of maturity (Buffon 1764). According to Bubenik (1982), some ancestors of pronghorn may have had completely deciduous horns, i.e., the horn cores were shed.

Horn sheaths of fused hairs are another characteristic of pronghorn often cited as a reason for putting them in a separate family. However, histological studies have shown that the horn-forming process in pronghorn involves formation of keratin similar to that of bovid horns (Morrison 1961, O'Gara and Matson 1975). Pronghorn horn sheaths grow faster than those of bovids, and considerable hair is trapped in the growing sheaths. Roots of hairs are embedded in the skin under the basal portions of sheaths, and hair shafts course obliquely toward their tips. Hairs do not occur in the hooks and prongs. The base of a horn sheath from a young Rocky Mountain goat that O'Gara examined histologically had about the same amount of hair incorporated as was in the bases of pronghorn sheaths.

Horn sheaths often are retained for 2 to 5 years by pronghorn does because the horn cores generally are oval, instead of bladelike as in bucks. Therefore, prongs are not grown and the annual growth of sheaths can rest, as in bovids. Photo by Bart O'Gara.

The horn cores of bovids arise embryologically as dermal ossifications (*os cornu*) under the skin covering the frontal bones, according to Bubenik (1982) and Solounias (1988). According to the literature (see Janis and Scott 1987), only domestic bovids heterozygous for the polled gene and giraffes grow dermal ossifications. The antlers of cervids and horn cores of pronghorn and bovids arise as outgrowths of the frontal bones (Goss 1983, Janis and Scott 1987).

After considering the O'Gara and Matson (1975) statements that the horn-forming process in *Antilocapra* involves formation of keratin similar to that of nails and claws, Leinders and Heintz (1980:158) concluded that, "Combining these statements with the fact that horn formation is common in mammals, one might wonder if horn formation is a significant character for classification." Scott and Janis (1987) contended that a number of lines of evidence suggest that cranial appendages have evolved in parallel at least twice, and they may have evolved in parallel within each pecoran family.

The horn sheaths of pronghorn differ from those of bovids by their forked shape, as well as the annual shedding of sheaths. Forked horn sheaths might not seem an important characteristic for taxonomy, especially when compared with the knobs, spirals and other modifications seen on bovid horns. Such modifications serve as aids for fighting and for visual identification among male and female conspecifics. The prong of pronghorn serves the same purpose. However, forked horns have never appeared among the wide diversity of living and fossil bovids. Developmental studies in the early part of the 20th century suggested that it is not possible, embryologically, for a bovid to grow a forked horn, which indicated that horns evolved in parallel between pronghorn and bovids (Janis and Scott 1987). Photo by E. P. Haldon; courtesy of the U.S. Fish and Wildlife Service.

Pronghorn have spaces between their nasal and lacrimal bones below and in front of their eyes, and this condition has been likened to that in Cervidae. Koopman (1967) indicated that bovids did not have these lacrimal vacuities, apparently because most North American bovids have very solid skulls with the nasal bones fused to the lacrimals. However, some bovids of the subfamilies Antilopinae and Hippotraginae also have large lacrimal vacuities (Brooke 1878, O'Gara and Matson 1975), as do the okapi (Giraffidae) and a number of fossil pecorans; this characteristic is not valid for taxonomic classification (Janis and Scott 1987).

Twelve characteristics that distinguish Cervidae from Bovidae were listed by Brooke (1878:58), who noted that "none of these characters can be taken singly as distinctive of a Cervine from a Bovine animal." One of the features Brooke listed to distinguish between the two families was the number and position of the orifices to the lacrimal duct. Flower (1875) already had noted that the number of orifices was different between the two families. Cervids usually have two orifices on the rim of the orbit; bovids ordinarily have only one inside the orbit. O'Gara and Matson (1975) mistakenly reported that pronghorn had only one orifice, as in the Bovidae. Leinder and Heintz (1980) pointed out that pronghorn usually have two orifices, differing from those in cervids only by the position of the upper one inside the rim of the orbit. The lower orifice is placed, as in cervids, near the margin of the rim. Two of 36 skulls at the University of Montana Museum have only one orifice; the other 34 have two. Among the 34 skulls with two orifices, the distance between them varies from about 0.28 inch (7 mm) to a tiny spicule of bone less than 0.04 inch (1 mm) wide. About 20% of the pronghorn skulls in the American Museum of Natural History collections have a single orifice (Scott and Janis 1987).

The layers of tissue that make up the placenta are considered significant in taxonomy and morphology. Of the 12 characteristics Garrod (1877) used to separate Cervidae from Bovidae, Brooke (1878) noted that only placental morphology (taxonomically significant soft tissue) was consistent in all the species he considered. Members of Cervidae have oligocotyledonary placentas, with 7 to 10 large placentomes for one or two fetuses; members of Bovidae have polycotyledonary placentas, generally with 60 to 140 placentomes for one or two fetuses, although the four-horned antelope has only 20 to 30 (Mossman 1937). Pronghorn have an average of 92 placentomes, with a range of 46 to 180, for (usually) two fetuses (O'Gara 1969a). However, Moschus *resembles bovids and pronghorn in its placental anatomy, again suggesting that, in this case, cervids show a derived characteristic, and the other families merely retain the original primitive pecoran condition. Above are two near-term fetuses that are attached by their umbilical cords to approximately 80 placentomes.* Photo by Bart W. O'Gara.

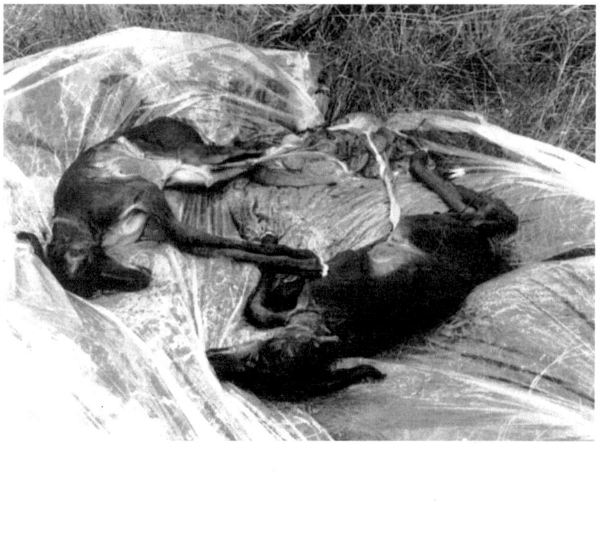

Leinders and Heintz (1980) concluded that the presence of two orifices in pronghorn supported the idea of a cervoid ancestry, but they reported that several species in two tribes of Bovidae (Tragelaphini and Bovini) also have two orifices, as do occasional individuals in the tribes Alcelaphini and Rupicaprini. These exceptions to the rule somewhat lessen the importance of this characteristic and serve as a reminder of Brooke's (1878) warning about taking any characteristic singly.

Grooves on anterior sides of the metatarsals (cannon bones) were emphasized by Heintz (1963) and Leinders (1979) as a distinctive characteristic between bovids and cervids. In bovids, the grooves are open for the full length of the bones; in cervids, they close above the distal ends of the shafts. Metatarsals of pronghorn resemble those of cervids. Leinders (1979) indicated that the grooves of *Eotragus*, the earliest known bovid, were similar to those of present-day bovids. The earliest Merycodontinae had metatarsals with closed gulleys, although some later species secondarily reverted to an open gulley. All known fossil antilocaprines have closed gulleys (Janis and Scott 1987).

The interpretation of the metatarsal gulley is problematical on at least two levels. Possession of a closed gulley is assumed to be a derived state within the Pecora, an open gulley being the primitive condition. But, closed gulleys have evolved separately within the traguloids (in the living genus *Tragulus* in the family Tragulidae), although the closed gulley develops in a different fashion in the tragulids (Janis and Scott 1987). The most primitive pecoran, the extinct *Gelocus*, did not have fused metatarsals. Thus, the open and closed gulleys of living pecorans may, in fact, represent separate evolutionary events, neither more derived than the other, although the open gulley is the more usual condition (Janis and Scott 1987). A problem also exists in assigning the limbs to fossil species that have been identified from skulls and teeth. Most fossil postcranial material is referred rather than associated (Scott and Janis 1987), although it is possible to reduce the degree of uncertainty in assignation by study of fossil taxa from localities where they are the only ruminant present or the only ruminant of that body size.

The upper and lower first molars are brachydont in most species of deer. The constricted necks of these short-crowned teeth are on a level with or a little above the borders of the sockets in dry skulls, and the upper ends of the roots are visible. In bovids (with the exception of *Cephalophus*, *Litocranius* and some *Tragelaphus*), the same molars are hypsodont. A large portion of these teeth (at least in the unworn condition) are encased in the sockets without constrictions between the long crowns and the short roots (Brooke 1878). The hypsodont dentition of antilocaprines was considered by Frechkop (1955) and O'Gara and Matson (1975) as an indication of close relationship to the bovids. However, hypsodont cheek teeth have evolved independently within a large number of mammalian families (where the less-specialized members of these families have brachydont cheek teeth); the families with hypsodont members include other hoofed mammals (horses, rhinos, camels, pigs, hippos, elephants and hyraxes) and a variety of nonungulate mammals (such as wombats, sloths and certain rodents). Hypsodonty clearly is an adaptive characteristic for an abrasive diet, and, whereas hypsodonty in bovids and antilocaprids certainly is suggestive of ecological similarities, it is not necessarily indicative of a close taxonomic relationship.

Dental and skeletal characteristics are studied more extensively than are those of soft viscera to determine the taxonomic status of mammals because many researchers have only museum fossils with which to work. Also, skeletal elements fossilize, allowing the evolutionary forms of many animals to be traced. However, soft tissues also have value in ascertaining relationships among groups. Cervids (except *Moschus*) lack gallbladders (Simpson 1984); pronghorn and bovids (to our knowledge) have gallbladders. The presence of a gallbladder in *Moschus* led Leinders and Heintz (1980) to speculate that it still was present in primitive cervoids, and its presence in pronghorn could be explained as the retention of an original characteristic.

The penis of the pronghorn resembles that of cervids by the absence of a processus urethras and Cowper's gland, according to Frechkop (1955). However, Cowper's glands are present in some cervids and absent in others, according to Garrod (1877). Murie (1870) also noted that Cowper's glands appeared to be absent in pronghorn, but he thought they were replaced or represented by glands within the urethral cavity.

Little attention has been paid to soft tissues to determine relationships among families and less has been paid to physiology and behavior. Similarities between pronghorn and Old World bovids, especially those of the subfamily Antilopinae, were noted by Gilbert (1973). The structural affinities he noted were hypsodont teeth, complete loss of side toes, well-developed digital pads and erectile rump patches with associated glands. Similar physiological adaptations included diurnal habits under intense solar radiation. Behavioral similarities included territoriality, formation of dung piles, stereotyped linked sniffing, pawing, urination and defecation, facial glands used in marking vegetation, male courtship rituals and the mannerism of males lying down in females' bedding sites or marking the sites with urine.

Bovids and antilocaprids are the only ungulates that scrape the ground with a forefoot before urinating and defecating in sequence and in striking postures. At least 11 species of bovids do this, but no cervids do, according to

Walther (1984:345), who added that, "the details of this urination–defecation sequence in pronghorn agree so strikingly with the corresponding behavior of gazelles and certain dwarf antelope . . . that I find it hard to believe a mere analogy." However, no living species of cervid dwells in truly open habitat, and such behaviors are not evident in the supposedly primitive forest-dwelling Cephalophinae bovids (the dwarf antelope). Possibly, antilocaprids and open-habitat bovids evolved their similar behavioral repertoires in parallel.

By 1965, about 50 species of Bovidae had been studied cytogenetically. Although the diploid number varied from 30 to 60 among these species, the fundamental number varied only from 58 to 62, with three exceptions (Wurster and Benirschke 1968). The fundamental number in pronghorn is 58 or 60. Seven species of deer showed 68 to 70 chromosomes, but the muntjac had only 46 (Wurster and Benirschke 1967).

The chromosome numbers and configurations of 44 artiodactyl species were investigated by Todd (1975), who found both very similar in *Antilocapra* to those in several bovids. In another study of 10 species (Baccus et al. 1983:117), the data on chromosome numbers supported, "the closer relationship between the Antilocapridae and Bovidae than between the Cervidae and either of the bovid families." However, based on genetic distances, Baccus et al. (1983:117) maintained that, "the pronghorn is genetically more similar to the species in the Cervidae (I = 0.305) than to those in the Bovidae (I = 0.215)." Bovids that may be related closely to pronghorn, subfamilies Antilopinae and Hippotraginae, were not included in their study and the only noncervids they had were five pronghorn, seven bison and six cattle.

Protein electrophoresis and sequence divergence of mitochondrial deoxyribonucleic acid (mtDNA) are being used to elucidate relationships among taxa, but like more conventional taxonomic criteria, no study or data set can be considered the final word. Studies using protein electrophoresis often survey 30 or more genes, a minute fraction of the genome (Cronin 1993). Also, those genes represent only a specific part of the genome.

The main drawback of using sequence divergence of mtDNA for quantification of genetic differentiation is that it reflects a limited part of the gene pool of populations. Nineteenth century taxonomists used only a few individuals to represent an entire taxon. If this practice is applied with molecular genetic data, important relationships will be missed. N. Cronin (personal communication:1996) indicated that 20 specimens from each population or subpopulation from each taxon should be sampled when using mtDNA to estimate taxonomic relationships. Thus, in most cases, hundreds of individuals should be sampled during biochemical taxonomic studies.

Most studies have used only a few individuals from a few species of any given family. Results can be misleading. When amino acid sequence studies on fibrinopeptides of only pig, reindeer, sheep, goats and cattle had been conducted, results indicated that sheep and goats were more closely related to reindeer than to cattle (Mross and Doolittle 1967). Comparisons of antilocaprids with bovids and cervids that do not include numerous individuals from a number of species of Old World antelope will always be suspect.

Good assessment of overall genetic differentiation apparently cannot be provided by mtDNA sequencing alone. Methods for study of other parts of the genome are rapidly being developed with nuclear DNA technologies, but application to wildlife are few (see Cronin 1993). This does not minimize the importance of genetic studies. As more studies are reported concerning the genome of any given family, species or other taxon, a more complete understanding of the relationship of that taxon to others will result.

Studies of 19 amino acid sequences on fibrinopeptide A of 18 artiodactyles indicated that the amino-terminal half of the pronghorn fibrinopeptide is different from that of either reindeer or sheep by only a single amino acid residue (Mross and Doolittle 1967). The carboxy-terminal half, on the other hand, has undergone remarkable changes. It is unique from other artiodactyl fibrinopeptides A that had been studied, having no phenylalanine at position A9, no glutamic acid at A6 and no valine at A2. Some of these traits had been considered indispensable for the attachment of thrombin.

Some early biochemical research indicated that pronghorn were closely related to bovids. The cross-reactivity of six blood antigens was examined in 21 species from members of what were generally considered five families of artiodactyls (Suidae, Camelidae, Cervidae, Antilocapridae and Bovidae) by Curtain and Fudenberg (1973). The results showed distinct differences between all the other families, but not between Antilocapridae and Bovidae. Although the different antigens clearly evolved at different rates, the relative ranking of cross-reactivity correlated well with the accepted taxonomic order of the ruminants. Curtain and Fudenberg noted that *Antilocapra* was closer in the reaction of all six antigens to Bovidae than to Cervidae, including the North American subfamily, Odocoileinae. They maintained that, although it may be possible for anatomical parallelism to be deceptive, the serological relationships, which should be dependent on completely different factors, suggest that *Antilocapra* is related more closely to bovids than to cervids. Only one of the gazelles, the springbok, was included in Curtain and Fudenberg's experiments, but it showed the closest relationship with pronghorn.

In another study, pancreatic ribonuclease from pronghorn was compared with that of 11 bovid species, five cervids and a giraffe by Beintema et al. (1979), who wrote that pronghorn and giraffes were closely related and placed with the bovids, leaving deer as a separate taxon. However, because the condition of the original pecoran will never be ascertained, it is difficult to determine whether these investigative conclusions mean that the antilocaprids and bovids are specialized in a similar way or if, again, they represent the primitive condition and the cervids are specialized. Comparison with data from the mouse deer and musk deer might help in this instance.

The phylogenetic relationships among the generally accepted families of higher ruminants remain contradictory when judged by results of modern molecular studies, genetic studies and traditional taxonomic criteria. For instance, sequences of the mitochondrial cytochrome b gene indicate a clade (the set of species descended from a particular ancestral species) containing antilocaprids, cervids and giraffids separate from a bovid clade (Irwin et al. 1991). Within the higher ruminants of this study (cow, deer, giraffe, goat, pronghorn and sheep), most species appeared to radiate from one another 20 to 25 million years ago. By the same criteria, goats, sheep, fallow deer and pronghorn, radiated from each other only about 10 million years ago.

Sequences of mitochondrial 12s and 16s genes indicate a clade containing antilocaprids and bovids, with cervids and giraffids in separate clades (Gatesy et al. 1992). This study included 17 species of bovids from nine tribes—a greater variety than seen in other studies. Of special interest, pronghorn clustered with the nilgai of India. Other mitochondrial ribosomal results indicated the families of higher ruminants descended from a somewhat different group (Allard et al. 1992).

Analysis of kappa-casein DNA sequences, including 434 nucleotides, suggests that antilocaprids form a sister group to a bovid/cervid/giraffid clade (Cronin et al. 1996). This supports the traditional position of antilocaprids in their own family.

Based on divergent times estimated from the fossil record, Lou (1998) calculated a mutation rate of 0.135% per million years. Phylogenetic analysis of six species representing the families Antilocapridae, Cervidae and Bovidae indicated that pronghorn is a sister taxon relative to a clade comprised of Cervidae and Bovidae, also supporting retention of the monophyletic family Antilocapridae.

Some of the puzzling aspects of bovid ancestry were discussed by Simpson (1945). Although pronghorn are very similar structurally to bovids, suggesting a common ancestry, all of the known fossils are from North America, whereas all Bovidae were exclusively Old World until the Pleistocene. This poses the question of geographic location of the common ancestor and how antilocaprids became separated from bovids, if that is the case. However, the first bovid predates the first antilocaprid by several million years.

The earliest known bovid—*Eotragus*—lived about 20 million years ago in Europe and East Africa. The teeth, skeleton and size of this animal were quite similar to those of *Merycodus*—generally considered one of the first fossil antilocaprids in North America, although the first known antilocaprid fossil dates from about 18 million years ago (see Chapter 2). *Eotragus* possessed small, simple, backwardly slanting horns situated behind the orbits. *Merycodus* had small, forked, upright horns situated directly over the orbits. According to Pilgrim (1946), *Eotragus* was the oldest member of the bovid tribe Boselaphini, which today contains the Indian nilgai and the four-horned antelope.

Antilocaprids could have diverged early from the main bovid line, as indicated by Matthew (1934). If so, the ancestors of pronghorn apparently reached the New World long before bovids, which were not apparent in North America until the Pleistocene. Many mammals migrated to North America from Eurasia in the early Miocene (approximately 20 million years ago), including an extinct cervid-related subfamily (Dromomerycinae), an extinct moschid subfamily (Blastomerycinae), and certain bear (*Amphictis, Cepalogale, Hemicyon*) and rhino (*Aphelops, Teleoceras*) taxa. At the same time, browsing three-toed horses (anchitheres) migrated in the other direction, from North America to the Old World. The fact that antilocaprids do not show up in the fossil record until slightly after this date probably is an artifact of preservation. It seems most likely that antilocaprid ancestors evolved in the Old World and migrated to North America at this time with the other mammals. A diversity of hornless pecoran taxa is found in the early Miocene of Eurasia, but as it is difficult to know what would determine a primitive, hornless antilocaprid or bovid (with or without hypsodont cheek teeth), identification of any ancestors among this assemblage currently is not possible. In contrast, both cervoids and giraffoids have a number of uniquely identifying features, and some of these hornless forms have been identified as early cervoids or giraffoids (Hamilton 1978, Janis and Scott 1987). Certainly, the traguloids that inhabited North America in the Oligocene were far specialized in their own right to be possible ancestors of any pecoran families (Webb and Taylor 1980). The first antilocaprids and bovids (as identified by their horns) appear to have been highly specialized when they first appeared in the fossil record, which further complicates the possibility of identifying a plausible Old World ancestor.

Along with early biochemical and chromosomal studies, the many similarities between pronghorn and present-day bovids led O'Gara to regard antilocaprids as bovids.

However, recent DNA investigations indicate Antilocapridae is a legitimate family (Cronin et al. 1996). O'Gara still considers pronghorn more closely related to Bovidae than to other families, although many of its bovidlike characteristics are influenced by habitat. On the other hand, minor osteological differences cited by Heintz (1963), Leinders (1979) and Leinders and Heintz (1980) as evidence of a cervoid ancestry also are problematical. Further DNA sequencing or fossil finds may prove that bovids and antilocaprids are not related. If true, pronghorn and gazelles can be cited as one of the best examples of convergent evolution among mammals.

Janis considers the present evidence supportive of a cervoid relationship for antilocaprids, although she agrees that the issue is problematical and that, if antilocaprids are not closely related to bovids, the behavioral convergences are remarkable. According to present-day phylogenetic systematics theory (cladistics), gross similarities or dissimilarities have little bearing on evolutionary interrelationships. More important is whether or not two taxonomic lineages share uniquely derived features. As previously mentioned, the physiological and biochemical characters are hard to evaluate; on the basis of current evidence, it is difficult to know which are primitive and which are derived, especially as they cannot be compared with the condition in fossil forms. Despite the overall similarity of bovids and antilocaprids, there is no unique feature that unites all the members of these two families.

The features of the closed metatarsal gulley and the double lacrimal orifice may be designated as unique features that unite antilocaprids and cervids (Leinders and Heintz 1980), but, as previously noted, these characteristics are not without problems. The fossil record shows that a closed metatarsal gulley can give rise to an open one, but not vice versa (Janis and Scott 1987). A closed metatarsal gulley possibly was present in the common ancestor of bovids, antilocaprids and cervids, with bovids secondarily reverting to an open type of gully, also seen in some merycodontinae. The same also may be true for the double lacrimal orifice because it is present in a variety of forms in several bovid tribes. In contrast, no living or fossil giraffoid possesses a closed metatarsal gulley or a double lacrimal orifice. Certainly, the area of antilocaprid affinities is wide open for future research.

Individual pronghorn in a population show wide variation in osteological characteristics. Size and number of nerve foramina are especially variable. Because herds are somewhat exclusive breeding units, in which a few dominant bucks do most of the breeding, and because doe bands are essentially sister groups (Pyrah 1987), skeletal characteristics could shift quickly when populations are reduced to low numbers. This is a plastic (capable of undergoing metamorphic changes) species, in which numbers of chromosomes, teats and ribs can vary. Mutations (white color, four horns, drooping horns and four prongs on two horns) may rapidly (over a short evolutionary time period) become commonplace in a herd. The unusual animals generally are considered prime trophies by hunters who selectively remove them from the population, so evolutionary potential is not known.

Bubenik (1982:182) cogently observed that, "much more study must be done before the classification of antilocaprins is clear." We cannot improve that evaluation. Debate on the classification of pronghorn will continue for some time.

SPECIES

The family Antilocapridae has one surviving species. Its first published binominal nomenclature *(Antilope americanus)* was by Ord (1815:292). Ord (1818:149) redesignated pronghorn *Antilocapra americana*, the generally recognized species name since that time. However, Smith (1827) maintained that Ord published after M. de Blainville noticed a pair of horns in the Royal College of Surgeons, in London, under the erroneous name *Cervus hamatus*.

Smith (1827:162) retained Antilope, the usually recognized genus for most Old World antelope at that time. Smith (1827:167–168) wrote: "the compound word *Antilocapra* offers a vicious structure, being constructed of two different languages, and the first part not even classical Greek. . . . Next, if the names of genera and species are intended to convey some notion of the form and characters of the subject they intend to designate, the compound *Antilocapra* would express an intermediate between Antilope and Goat . . . whereas the resemblance is between the Antelope and Deer."

Smith (1827) also objected to *americana* on the basis that a whole continent was not restrictive enough for the name of a middle-sized mammal, especially as he believed another species of pronghorn existed in the far north. Thus, he used the specific name *Antilope furcifer*.

Gray (1855:10) named an *Antilocapra anteflexa* with the type specimen, "most probably from America." This classification was based on an unrecorded set of horns in the collection of the Derby Museum, Liverpool, England. They resembled the tipped-ahead horns occasionally seen on pronghorn (see Chapter 5).

Smith (1827) described an *Antilope palmata*, which he did not know whether to designate a new species or subspecies of pronghorn. The horns of the specimen, for which he gave measurements, were 11 inches (28 cm) long and 2.50 by 1.25 inches wide (6.4 by 3.2 cm) at the base. Smith indicated that the sheaths of several specimens he inspected were compressed anteriorly and posteriorly so the hollow bases formed elongated diamonds, not the oval of

*Shortgrass prairie grasslands (*left*) support the majority of pronghorn and are the most productive area for the species. However, shrubs often are required for pronghorn to survive harsh, prolonged winters and droughts. Shrubsteppe habitat (*right*) often serves as a year-round pronghorn range, but seasonal herd movements may be necessary if winter snow accumulations are great.* Left photo by Tom Pojar. Right photo by Steve Boyle.

pronghorn. The prongs supposedly widened as they ascended, becoming larger and more turned in than in pronghorn. Smith claimed to have seen a head and mutilated skin, from near Baffin Bay. Compared with those of pronghorn, the horns were more voluminous in relation to the animal's size; the head was wider and shorter with little or no white hair, the fur was soft and wooly, the limbs were shorter and more robust, and more white covered the underparts. Smith maintained that palmated antelope inhabited mountainous areas near frozen oceans and brushed snow with their horns from the mosses and lichens upon which they fed. He did not state if or where specimens of this presumed pronghorn—or relative of the pronghorn—were archived.

This confusing situation may have been placed in proper perspective by Richardson (1829:264) when he wrote: "Major Smith has fallen into unaccountable mistake in supposing that the palmated antelope inhabits the bleak regions near the frozen ocean; and that specimens had been brought from Bafins Bay. No specimens whatever of this antelope were obtained on any of the expeditions to Bafins Bay, nor is there any mention made of the animal either in the narrative or zoological appendices of Capt. Ross' or of any of Capt. Parry's voyages."

SUBSPECIES

Antilocapra americana is a plastic species occupying a large and diverse land area (see Chapter 4). Anatomical characteristics used to separate subspecies are slight and intergrades occur. Thus, the concept of a species with many clines and gradients seems more logical than does the concept of subspecies. However, descriptions will be given for the generally accepted five pronghorn subspecies, as identified by the investigators who named them. Minor differences between subspecies are not clearly evident, but photographs of pronghorn in the habitats from which supposed subspecies were named will be presented.

Antilocapra americana americana

Lewis and Clark collected the pronghorn-type specimen on the "Plains and highlands of the Missouri River" (Miller and Kellogg 1955:816), and Ord (1818) named it *Antilocapra americana. A. a. americana* (*americana*) is the subspecies described under physical characteristics in this book (see Chapter 5).

A. a. americana were found as far south as southern Texas, central New Mexico, central Arizona and the Colorado Desert of southern California (Miller and Kellogg 1955). Although pronghorn apparently were absent from Washington State during the 20th century (except for a few that were released there and did not survive), Osborne (1953) maintained they were present in prehistoric times, possibly as late as the early 1800s, and Graham and Lundelius (1994) indicate pronghorn fossils in central, northeastern and southeastern Washington during the Late Holocene (500–4,000 years before present). From diaries of early French and Spanish expeditions and from mission accounts, Simpson and Leftwich (1978)

More than 90% of all pronghorn belong to the presumed A. a. americana *subspecies. It formerly was found on the prairie steppes of Canada, from the Red River and other areas of southwestern Manitoba, westward to the Rocky Mountains, as far north as the present city of Edmonton, Alberta. To the east, in what is now Saskatchewan, pronghorn were seasonally found as far north as 50 degrees north latitude (Mitchell 1980). Pronghorn fossils estimated at 835, 930 and 1,045 years old were found at the Stendal site, west of Virdan in southwestern Manitoba (Morlan 1999).* Photo by Peter T. Bromley.

determined that pronghorn occurred along the Gulf Coast of Texas in the vicinity of Galveston and Lavaca bays. Pronghorn ranged over at least two-thirds of Texas during presettlement times (Leftwich 1977), but they did not occur in the state's eastern coniferous forest and only marginally occupied the hill country of the central Edwards Plateau.

Concerning the pronghorn in Mexico, Goldman (1945) noted that three somewhat isolated, fingerlike southern extensions carry the general range of pronghorn into Mexico, and these extensions involve geographic races differing from one another only in slight details of size, color and structure.

Antilocapra americana mexicana

Merriam (1901:31) named *A. a. mexicana* (*mexicana*) and collected the type specimen in "Sierra en Media, Chihuahua, Mexico." He described it as similar to *americana,* but paler, with the mane absent or reduced to a narrow line of dark hairs. A median dark streak usually was present on the necks, sometimes reaching from shoulder to ears. Head markings were sharply defined, and the backs of the heads were distinctly white or whitish. Although the skulls were similar to those of *americana*, the orbits protruded less and the muzzles were thinner. The measurements Merriam gave for the male type specimen fall within the range for males of *americana* collected in Alberta by Mitchell (1971).

All pronghorn in Mexico have been listed as Appendix I (which is roughly equivalent to endangered) by the Convention on International Trade in Endangered Species of Wild Fauna and Flora (CITES), since 1975 (U.S. Fish and Wildlife Service 1997a).

Using mtDNA and allozyme analyses, Lee (1992) considered his findings consistent with recognition of *mexicana* as a subspecies. He also thought that pronghorn from Marathon Basin, Texas, should be considered *mexicana,* but the rest of western Texas and southern New Mexico was a natural intergrade zone between *americana* and *mexicana*. Lee et al. (1989) recommended that pronghorn not be translocated into *mexicana* herds of Marathon Basin, Texas. Lee (1992) extended this recommendation to the Sears Ranch, Jeff Davis County, Texas, and De Baca County, New Mexico, if *mexicana* is to be preserved genetically pure.

In the United States, American pronghorn ranged from western Iowa, Kansas and Oklahoma westward to western Montana, southern Idaho, Nevada and California (Miller and Kellogg 1955). The eastern range distribution of pre-1850 pronghorn is poorly documented and little physical evidence is available. According to Grinnell (1929), records of the U.S. Biological Survey indicate that pronghorn occurred in southwestern Minnesota and western Iowa; he reported a sighting of pronghorn near Tracy, Minnesota, in 1885. Hall (1981) listed records of pronghorn sightings in Kittson and Lyon counties, Minnesota, and in Harrison County, Iowa. Fossil pronghorn dating to the Late Holocene (500–4,000 years ago) (Graham and Lundelius 1994) have been reported from eastern and central Missouri, but McKinley (1960) concluded that pronghorn were not present in that state during historic times. A pronghorn skull, dating from the late 1800s, recently was found in Grand Forks County, North Dakota (Jensen and Seabloom 1993). The skull supports the assertion that pronghorn ranged as far east as the rivers that form the eastern border of North Dakota. Photo by LuRay Parker; courtesy of Wyoming Department of Game and Fish.

Antilocapra a. mexicana *formerly ranged from southeastern Arizona and southwestern New Mexico eastward to the Jornada and Tularosa desert valleys and regions west of the Organ Mountains, and southward through western Texas, Chihuahua and Coahuila to northeastern Durango, Mexico (Miller and Kellogg 1955). The pronghorn that were common west of the Pecos River, in Texas, were thought to be* A. a. americana. (Left) *Mexican pronghorn generally seek habitats comprised of grasses, forbs and shrubs—as in this Texas valley—similar to habitat occupied by* A. a. americana *farther north. Because of the habitat they occupy, Mexican pronghorn are as apt to make long movements in response to droughts as they are to avoid deep snow. Such movements frequently are blocked by livestock fences in this subspecies' range, resulting in extensive mortality and reduced productivity.* (Right) *During years of average to high precipitation,* A. a. mexicana *may have a great variety of forbs at their disposal. However, they often are fenced with domestic stock, and the succulent forage they need is overused during droughts. Fawn survival is reduced and adults may be poisoned by having to subsist on tarbush or other toxic plants.* Photos courtesy of the Texas Department of Parks and Wildlife.

Mexican pronghorn bucks use shrubs (top) *during a dry spell in Texas when forbs are scarce. Like all pronghorn, Mexican pronghorn prefer flat terrain with short vegetation* (below) *so that danger can easily be detected and outdistanced.* Photos courtesy of the Texas Department of Parks and Wildlife.

Antilocapra americana peninsularis

This subspecies' numbers fell to a dangerously low level, with only two remnant groups totaling about 50 head surviving in the Viscaino Desert by 1984 (A. Castellanos and D. Biggins personal communications:1984). That number apparently doubled by 1986 because of coyote control and favorable weather (A. Castellanos personal communication:1986). Favorable weather continued, and in 1993, 175 pronghorn were counted (Cancino et al. 1994). The total probably was higher. However, a census during June 1997 indicated another decrease in numbers and dispersal of the herd over a large area (Sanchez-Sotomayor et al. 1999).

The peninsular pronghorn was named from 12 specimens, all from the type locality and neighboring district (Nelson 1912). The type specimen came from 45 miles (72 km) south of Calmalli, Baja California, Mexico. The facial markings and ears are darker than those of other subspecies; the black border is wider. The terminal third of the back of the ears are blackish, with the terminal inch nearly pure black. The strip of rufus between the tail and back is broader on *A. a. peninsularis* (*peninsularis*) than on *mexicana*. Horns of the peninsular pronghorn are short, and surfaces up to the prongs are rough with many small protuberances. These horns are thicker laterally at the bases and less divergent than are those of the other subspecies. The molars are heavier than in other subspecies (Nelson 1912). That this southern race of desert-dwelling prong-

Peninsular pronghorn are being raised in captivity to restock areas of historic, but unoccupied, habitat in Baja California and southern California. Fawns of captive does are hand-fed to provide easily handled embryo donors for transplants into American pronghorn surrogate dams. Photo by Ramón Castellanos.

Antilocapra a. peninsularis *formerly occurred from central, lower California, south to the head of Ballenas Bay at approximately 27 degrees north latitude, and north on the west coast to approximately 29 degrees 30 minutes north latitude. On the Gulf side, these pronghorn ranged to beyond 32 degrees north latitude, up to the southern end of the Colorado Desert (Nelson 1912). Peninsular pronghorn were restricted to the Viscaino Desert of lower California at the beginning of the 21st century. They were extirpated from most habitat with year-round, free-standing water by competition with livestock and by poaching. They occupy coastal desert where ocean fog occurs almost nightly. Few succulent plants, even cacti, are available. The octillo plants are covered with a gray lichen that soaks up moisture from the fog. The pronghorn appear to feed extensively on such lichens during the morning hours—securing both high protein food and moisture from the lichens.* Photo by Aradit Castellanos.

During dry weather in the Viscaino Desert, peninsular pronghorn have to frequent arroyos and high, dense vegetation where they can find palatable forbs and brows. This puts both adults and fawns at risk from predators. Bobcats are most apt to take the former and coyotes the latter. Surprisingly for desert-adapted animal, the peninsular pronghorn are comparatively large and dark colored. Photo by Aradit Castellanos.

horn should be dark colored whereas the other two southern subspecies are light colored, as would be expected of desert animals, is somewhat surprising.

The peninsular pronghorn was officially listed by the U.S. Department of the Interior as an endangered subspecies in 1975 (U.S. Fish and Wildlife Service 1999b). Reasons for its reduced range and numbers are related to drought, habitat deterioration, predation and poaching (Centro de Investigaciones Biologicas de Noroeste 1994).

Antilocapra americana oregona

The type specimen came from Hart Mountain (Warner Mountains), Lake County, Oregon. Bailey (1932) described the adult male type specimen as about the size of *americana* or slightly larger, with relatively larger feet and longer horns, slightly paler with less black about the face and on the mane, and less white on the crown and shoulder stripes. Measurements Bailey gave for the type specimen fall within the range for adult males collected in Alberta by Mitchell (1971), and the horns were not unusually long—slightly less than 15 inches (38 cm). However, Bailey's measurements for an adult female slightly exceeded the range for Alberta pronghorn.

Mitochondrial DNA and allozyme analyses led Lee (1992) to conclude that populations within the presumed historic range of *oregona* (*oregona*) are not genetically distinct from *americana*. He contended that, because the name *americana* (Ord 1818) is older than *oregona* (Bailey 1932), *americana* has priority. Thus, all populations of pronghorn in Oregon, California, Idaho, Nevada and those translocated into Washington from Oregon should be considered *americana*.

Although Lee (1992) did not consider *oregona* a valid subspecies, he recommended that, when considering translocations, managers recognize the importance of preserving populations that are best adapted to local ecological conditions.

When naming A. a. oregona, *Bailey (1932:45) noted that "The animals show only slight and gradual variations over their entire range, and no sharp lines of difference between described forms can be found." The Oregon pronghorn supposedly inhabits the open sagebrush country of eastern Oregon, but the exact limits of its range were not determined by Bailey or anyone else. Despite their close proximity to the location that* A. a. oregona *was collected, Hall (1946) considered the pronghorn of Nevada to be* A. a. americana. *The shape of the nose of this buck, photographed in Oregon during 1940, seems to be a more common characteristic of pronghorn in that state than of those farther east.* Photo by William L. Finley; courtesy of Oregon State Game Commission.

Antilocapra a. oregona *occupy high, cold, semiarid rangelands. Some herds must be free to change elevations to avoid deep snow and obtain preferred forages in accordance with plant phenology. Oregon pronghorn often are in shrubsteppe habitat* (top)*, where the vegetative cover averages 50% or less—allowing for unobstructed running. Shrubs, such as those in the background, can be taller than pronghorn typically like to enter. An adult doe* (bottom) *measured by Bailey (1932), when naming* A. a. oregona, *slightly exceeded the range of size recorded by Mitchell (1980) for pronghorn in Alberta. This was the only published record found for a presumed subspecies falling outside the range of measurements Mitchell published for* A. a. americana *in Canada.* Photos by Jim D. Yoakum.

Antilocapra americana sonoriensis

Goldman (1945) named the *A. a. sonoriensis* (*sonoriensis*); the type specimen was collected 40 miles (64 km) north of Costa Rica, from a ranch on the northern side of the Rio de Sonora, southwest of Hermosillo, Sonora, Mexico. He portrayed the Sonoran pronghorn as the smallest subspecies, generally similar to the Mexican subspecies, with colors about the same. The skull resembles that of *mexicana,* but is smaller. The frontal depression is shallower; premaxillae are less extended posteriorly along the median line, and auditory bullae are flatter and project less below the basioccipital. Measurements of the female type specimen fall within the range of those for an adult female *americana* collected in Alberta by Mitchell (1971).

The Sonoran subspecies was listed by the U.S. Fish and Wildlife Service as endangered in 1967 (U.S. Fish and Wildlife Service 1999b), and a recovery plan was completed (U.S. Fish and Wildlife Service 1998a).

Confusions and Diffusions

In his discussion of the validity of *sonoriensis*, Cockrum (1981:5–9) made a good case for viewing subspecies of pronghorn with suspicion: "Living pronghorn are the result of the combined interactions of the survival of the fittest or, especially in small populations, the random fixation of genes as the result of the Sewell Wright effect. In brief, the Sewell Wright effect is the result of the chance survival of a given combination of genes in a small population, although the combination may include genes with no apparent survival benefits.

"To statistically evaluate the similarities of 2 or more populations, measurements of some 15 individuals of the same sex and a given age from a given locality should be compared with individuals of identical ages and sex from a second locality. If we define a given locality as the geographic range occupied at various times of the year by a given population, the probability of acquiring a statistically significant sample size is almost zero. Normal population size and age structure do not permit such large sample sizes—even with total population extermination.

"With a clear understanding of the above concepts as background, let us examine the available facts concerning the so-called Sonoran pronghorn, *Antilocapra americana sonoriensis*.

"This subspecies was first named in 1945 by Major E. A. Goldman on the basis of only 2 specimens:

1. Adult female, skin and skull (age not given) No. 256938, U.S. National Museum (Biological Survey Collection) from 40 miles N of Costa Rica, a ranch on the northern side of the Rio de Sonora, southwest of Hermosilla, Sonora (type).
2. Skull only of female ('collected many years ago'), No. 3691 in the U.S. National Museum, at Fort Buchanan, now Crittenden, Arizona ('share some of the characters of the type') (Camp Crittenden is now an abandoned site just west of Sonoita, Santa Cruz County, Arizona).

"We now know that age, sexual, and individual variations in a given population may be large. Statistically significant variations may occur between populations separated by small areas or by short periods of time.

The Sonoran pronghorn ranged the desert plains of west-central Sonora and north to southern Arizona (Miller and Kellogg 1955). Thompson-Olais (1944) estimated that there were fewer than 300 Sonoran pronghorn in Arizona. A 2002 survey showed just 21 to 33 of these animals in Arizona. Castillo et al. (2000) estimated 313 for Sonora, Mexico. Historically, the Sonoran pronghorn decline was due primarily to habitat deterioration and competition with livestock. During the late 1800s, farming and irrigation practices and town developments along the Gila River and Rio sonoyta created intermittent, often dry, riverbeds (Wright and deVos 1986). Drying of the area, coupled with grazing by several thousand livestock, degraded habitat available to pronghorn. Livestock were removed in Arizona by the early 1980s, but the area probably will not completely recover because of the lowered water table. Also, once shrubs become dominant, 100 to 200 years may be required to regain productivity of forbs and grasses. Most recently in Arizona, prolonged drought has been the limiting factor. (Top) *In desert habitat of southern Arizona,* A. a. sonoriensis *often have to enter areas of dense vegetation taller than 15 inches (38 cm) to find succulent forage.* (Bottom) *Sonoran pronghorn have to enter areas of dense vegetation taller than 15 inches (38 cm) to find succulent forage. In such habitat, they are relatively vulnerable to predators, such as bobcats and coyotes.* Photos courtesy of the Arizona Department of Game and Fish.

Mammalogists have demonstrated that statistical morphological differences can be demonstrated between the geographically (and socially) separated populations of house mice living on the ground floor and in the loft of the same barn. Thus, the presence of differences between the pronghorn of Sonora, Baja and Arizona should not be a surprise to anyone. There may be cause to reexamine the whole concept of listing a subspecies on a list of 'international trade in endangered species of wild fauna and flora'—as shown on page 10473 of Federal Register, Tuesday, February 22, 1977, Part IV.

"Pronghorn are now extinct in the region of the type locality. The type is the only example known to be preserved from the population that formerly occurred at the type locality. I examined this specimen in the U.S. National Museum. It is a young female with only slightly worn teeth. The size is somewhat smaller than females of similar age from Chihuahua (equals *A. a. mexicana*) and northern Arizona (equals *americana*).

"Examination of a series of skulls in the U.S. National Museum reveals that there appears to be some correlation between age and height of prong, with older animals having higher prongs. However, there appears to be some evidence of a geographic cline in the trait, with the higher prongs occurring in northern populations. Exceptions were found in all areas, with low prongs being observed even on specimens from Wyoming and Montana. Variability was also found in the amount of hair on the base of the horn. Either trait is too variable for use as a subspecies indicator.

"It is probable that the antelope now found in southwestern Arizona and northwestern Sonora are somewhat different, morphologically, from those that formerly existed at the type locality of *A. a. sonoriensis*. Further differences exist between skulls from both areas and the skull of the female from Fort Buchanan. Certainly the habitats at the three localities are different. Just as microgeographic morphological variations have been demonstrated between populations of house mice living in a single barn, we could probably identify differences between each of these pronghorn populations. My opinion, as well as that of many other taxonomists, is that such differences do not warrant the application of separate subspecific names. The specimens from near Caborca, Sonora, differ more from the type specimen of *A. a. sonoriensis* than that specimen differs from animals taken in northern Arizona.

"Another factor that should be considered in evaluating minor morphological differences associated with different geographic locations is that, with time, shifts in gene frequencies, microhabitats, and even nutritional resources can result in major shifts in morphological features. For example, Dingman (1966) demonstrated that morphological differences of a magnitude equal to that used by some taxonomists occurred in pocket gopher populations in Yuma County, Arizona, in less than 50 years.

"The female pronghorn utilized by E. A. Goldman as the basis of named *A. a. sonoriensis* was taken on 11 December 1932, 48 years ago. As far as is known, this is the only specimen to be preserved from the population. Even if the population had survived to the present, the microhabitat, as a result of different land use, would have resulted in differences in abundance and species of food plants. Nutritional studies have clearly shown that diets influence morphology. Perhaps modern descendants would be different.

"The outstanding trait emphasized by Goldman as characterizing *A. a. sonoriensis* was small size. He wrote: 'Size smallest of the subspecies of *Antilocapra americana*. Similar in general to *Antilocapra americana mexicana* of northwestern Chihuahua, but smaller, color about the same . . . [d]iffers from *Antilocapra americana americana* of South Dakota and *Antilocapra americana peninsularis* of central Baja California in smaller size and paler coloration.'

"I compared measurements of the type of *A. a. sonoriensis* with measurements of 4 skulls from near Caborca deposited in the U.S. National Museum. The Caborca specimens are all young with the third upper molar not fully erupted in three and unworn in the fourth. All are younger animals but in cranial measurements, all are much larger than the type of *sonoriensis*. In fact, they can be easily matched in size by specimens of similar age from northern Arizona or from Chihuahua. I have not seen any other skull of a similar aged animal that is comparable to the type of *sonoriensis*.

"In contrast to the type specimen, the specimen from Fort Buchanan is not small. Goldman, in his original description, stated that this specimen 'shares some of the characteristics of the type of *sonoriensis*, and may be referenced to the same form. On geographic grounds, however, specimens from southeastern Arizona may be expected to exhibit gradation toward *mexicana*.'

"In the absence of any other specimens from the type locality, it is unlikely that we will ever be able to determine whether the small female designated by Goldman was characteristic of a now extinct local population or whether she was simply an aberrant individual of small size.

"Paradiso and Nowak (1971:855–8) examined some of these same specimens and published selected measurements. Their conclusions differ markedly from those expressed above. Their conclusions concerning the identity and characteristics of *A. a. sonoriensis* were based primarily on the skulls of the 4 males taken near Caborca. They reported that one (347452) was an adult male 'with full development of horns, and with all teeth erupted and showing considerable wear, and 3 juvenile males with little development of horns and with third upper molars

Lack of herbaceous vegetation, which often is caused by droughts or heavy livestock use, sometimes forces the presumed Mexican subspecies into shrubs, which proliferate during dry weather and grazing pressure, in search of succulent forage. This predisposes them to predation as well as poor nutrition. Malnutrition early in life can lead to cranial characteristics approximating those characterizing Mexican pronghorn. Photo by Rich Ockenfels; courtesy of the Arizona Department of Game and Fish.

partly erupted.' I concur that numbers 347453–5 are all young, with the third upper molars not erupted. However, in my judgment, number 347452 (the only one for which Paradiso and Nowak published measurements) is a young adult with only slightly worn teeth. Many of the males listed as being measured for the calculation of average and extreme values for the other 'subspecies' have teeth that are much more worn and have other signs of greater age. Thus, it is not surprising that the single male '*sonoriensis*' from near Caborca is smaller than the mean values given for the other subspecies. However, in almost all cases, it is larger than the smallest value given in the range of values reported for the other subspecies. The values reported for Goldman's type of *A. a. sonoriensis* (a young adult female, with slightly worn teeth) are small but those reported for the female from Camp Crittenden fall within the ranges reported for other subspecies. As already discussed, the single known preserved example from the type locality of *A. a. sonoriensis* presents several problems, none of which can be answered with currently available data. Critical questions include: (a) Was the type of *A. a. sonoriensis* representative of others of the same age and sex that existed in the region of the type locality? (A sample of one has highly limited predictive values. Genetic aberrations, hormonal malfunctions, disease, dwarfs, nutritional runts—all are known to cause unusually small individuals in a population.) and (b) Did the range of variability in the population at the type locality of *A. a. sonoriensis* include the 'Fort Buchanan' specimen?

"A close examination of the table of measurements given by Paradiso and Nowak (op. cit.) reveals that extreme values given for *A. a. americana* include most values for specimens of other subspecies.

"In any case, it appears that the recognition of *Antilocapra americana sonoriensis* as a separate subspecies is unwarranted at this time. The few available specimens from Sonora and adjacent southwestern Arizona indicate that pronghorns there were made up of a series of morphologically differing local populations. Further, most of these differences can be found in some individuals in populations in northern Arizona, New Mexico, and Chihuahua.

"The subspecies map published by Hall and Kelson (1959, volume II, p. 1023) ascribing southern California to the (former) range of *A. a. americana* (type locality, Great Plains), central and southern Arizona to *A. a. mexicana* (type locality, Chihuahua), and extreme southwest Arizona and adjacent Sonora to *sonoriensis* is not compatible with any known biographic affinity patterns in plants or other animals.

"If *A. a. sonoriensis* Goldman is not a recognizable biological entity, the question becomes what to call these populations. *A. a. mexicana* Merriam would appear to be the logical answer. However, it appears that, at best, this is a very weakly marked subspecies. C. H. Merriam (1901:31) originally named this on the basis of one specimen from Sierra en Media, Chihuahua, and characterized it as being 'similar to *Antilocapra americana americana* but paler.' In

1907, Mearns (Mammals of the Mexican Boundary) assigned all specimens from southern California and Arizona south of the Colorado River, as well as specimens from southern New Mexico and Chihuahua, to this subspecies.

"In other words, Mearns labeled the paler animals from the more arid portions of the range of the pronghorn as *A. a. mexicana* and, by implication, everything else was *americana*. This arrangement has more meaning, biogeographically, than the patterns presented by Hall and Kelson (op. cit.).

"However, the recognition of a subspecies simply on the basis of color appears questionable. Other morphological features present a mosaic of geographic distributions over which a north–south cline appears to be superimposed. These findings force me to the conclusion that the use of subspecific names for the various populations of pronghorn is not justified. Mosaic patterns of variability in color as well as clinal patterns in other features make separation into subspecies meaningless."

The following facts strengthen Cockrum's assumptions. The horn characteristic he noted for presumed *sonoriensis* bucks probably was the result of time of collection and the breeding phenology of pronghorn in that area. A very small adult pronghorn occasionally occurs in populations of normal-sized animals.

Male Sonoran pronghorn collected have displayed poor horn sheath development and an overall hairy appearance that, according to Phelps (1981), were thought to be a characteristic of *sonoriensis*. However, Phelps indicated that most collections have been made during autumn and winter, and those specimens probably possess maturing horn sheaths. He noted that birth normally occurs during the first or second week of March. Providing that the gestation period is similar to that of northern pronghorn, the rut should occur during late June and early July, with horn casting during August. If that is the situation, the bucks should be in hard horn from about April until early August. Animals collected during October through December should have hooks and prongs close to the head with hair on the proximal 30% to 60% of their horns.

The small size of the type specimen used by Goldman (1945) when naming *sonoriensis* could easily have been an aberrant specimen of small size. In Yellowstone National Park during 1963, O'Gara (1968) collected a 7-year-old doe that was very fat and weighed 68 pounds (31 kg). Other adult does in the area averaged 110 pounds (50 kg). The small doe was delicately built and fine boned. Regrettably, the skull was not kept, but the animal certainly was smaller than the type specimen for *sonoriensis*.

Whether all five subspecies are valid is a moot question, complicated by translocations of *americana* into ranges of the other supposed subspecies. More than 4,000 pronghorn were translocated in New Mexico between 1936 and 1957 (Russell 1964), so subspecies are mixed there. All of the pronghorn herds in southern Arizona that occur in grasslands have received translocated animals from northern Arizona. The desert populations on the Cabeza Prieta Game Range (*sonoriensis*) and at Lake Havasu City (*mexicana?*) are the only ones in southern Arizona that have not been supplemented by translocations (K. Jackson personal communication:1976).

Considerable variation is common among pronghorn collected in the same area. Differences in configuration of skulls, number of ribs, coloration and color patterns are obvious upon close examination.

Paradiso and Nowak (1971) acknowledged that the number of specimens available was insufficient to allow complete appraisal of the taxonomic status of *sonoriensis*. However, they concluded that, from investigation of the six skulls available to them, *sonoriensis* was more distinct from the other presumed subspecies they checked (*americana*, *mexicana* and *peninsularis*) than any of the others were from each other. Characteristics Paradiso and Nowak found to be different, in addition to those Goldman (1945) had noted when he named *sonoriensis,* were: molars of *sonoriensis* are shorter and narrower, the rostrum is more slender (reflecting the narrow skull), and there is practically no ridge extending from the interior edge of the alveolus of the first premolar to the anterior tip of the maxilla, as it does in the other subspecies.

Hoffmeister (1986) observed that the type specimen might be smaller than average for the presumed subspecies, and the distinctiveness of *sonoriensis* remained to be ascertained if more specimens became available. Ron Nowak, U.S. Fish and Wildlife Service Office of Scientific Authority, examined 13 skulls from the generally accepted range of the Sonoran pronghorn (U.S. Fish and Wildlife Service 1998a). One was too fragmented to assess. Six appeared *sonoriensis*-like and six appeared *mexicana*-like. Nowak indicated that the only available specimens that could definitely be assigned to the Sonoran subspecies were from Sonora. He speculated that the part of Arizona within the accepted range of *sonoriensis* is a zone of integration between that subspecies and others.

Nowak maintained that *sonoriensis* should be retained as a subspecies. He stated that the entire species should be studied to determine how much variation occurs within and between geographically separated samples. Then, a judgment could be made whether or not the specimens from southwestern Sonora fall beyond the range of variation shown by other populations.

Electrophoretic studies were pursued at Cornell University in 1988 to aid in taxonomic verification of *sonoriensis* (U.S. Fish and Wildlife Service 1998a). No differences were found between serum proteins of Sonoran and Mexican pronghorn.

Sonoran pronghorn seldom are able to congregate in herds because their habitat will not support large numbers of these selective feeders in one area. This presumed subspecies are similar to A. a. mexicana, *and characteristics of the two presumed subspecies grade into one another.* Photo courtesy of the Arizona Department of Game and Fish.

A sequence comparison of the mtDNA control region of presumed Sonoran and Mexican pronghorn was conducted at the National Fish and Wildlife Forensics Laboratory during 1996. No variation was found among 10 individuals of each presumed subspecies. Comparing the two presumed subspecies indicated less than 1% sequence divergence (1 substitution per 185 bases compared). At the same laboratory, a portion of the nuclear SRY gene sequence was compared between a single Sonoran and a single American pronghorn. The gene sequences were identical (U.S. Fish and Wildlife Service 1998a).

Lee (1992) provided evidence supporting, but not proving, subspecific differences between *americana* and *mexicana*. However, the apparently natural, broad zone of interbreeding between these presumptive subspecies, as noted by Buechner (1950a) and Lee et al. (1989), suggests a cline with no discreet break between the two forms. Analyses of skulls also led Nowak (U.S. Fish and Wildlife Service 1998a) to postulate a zone of gradation between *mexicana* and *sonoriensis* in Arizona. Whitaker (1970) maintained that the extremes of such clines should not be considered subspecies.

Lee's (1992) data indicated, but again did not prove, that *americana* and *oregona* were the same subspecies. Protein, mtDNA and nuclear DNA studies—all involving small samples—indicated little genetic differences among *americana, mexicana* and *sonoriensis* (see U.S. Fish and Wildlife Service 1998a). Amor (2000) conducted the first mtDNA study that included all five presumed subspecies of pronghorn. She compared 93 individuals from Mexico and the United States, using the mitochondrial d-loop control region as a molecular marker. Her results did not support subspecies classification in pronghorn.

Lee et al. (1989) recommended that a modern, systematic study be designed to address the phylogenetic history of pronghorn, based on a number of independent data sets. These would include techniques, such as protein electrophoresis and restriction enzyme mapping of nuclear and mtDNA. Nowak (U.S. Fish and Wildlife Service 1998a) recommended essentially the same method, using cranial characteristics. Such studies would be complicated by the low numbers of *sonoriensis* and *peninsularis*, the extinction of intermediate populations that may have formed clines and the many translocations that have occurred. Nevertheless, these scientific techniques seem to be the only way to elucidate the true taxonomic relationships among pronghorn populations throughout North America.

Local gene pools with differences in color, size and form should be expected. Pronghorn social distribution results from discreet, autonomous herds developing seasonal movements and habitat-use traditions in distinct geographic areas; low dispersal of female offspring and polygamy of territorial males cause doe bands to be, essentially, sister groups (Pyrah 1987). Line breeding and inbreeding of domesticated animals are used to speed the selection of desirable physical qualities—not very different from what happens in the wild with pronghorn. Thus, differences

To survive harsh, northern winters, such as above in Montana, pronghorn must be able to reach areas where sagebrush protrudes above the snow. Historically, such climate-induced movements caused extensive mixing of herds and contributed to gene flow. Photo courtesy of Montana Department of Fish, Wildlife and Parks.

among herds often are noticeable and constant. However, before migration routes were blocked by artificial structures and cultivated lands and before populations were drastically reduced, pronghorn made extensive movements in response to droughts, severe winters, population pressures and other natural phenomena. Such movements should have caused some mixing of herds and maintained gene flow among most populations. Also, Pyrah (1987) pointed out that dispersal (pioneering) is uncommon, but it does occur, sometimes without severe weather to promote it. Whether minor differences in color, size and form should qualify for classification as subspecies or clines within a species depends on one's definition of a subspecies.

CONCLUSIONS

Controversy exists concerning the taxonomic position of pronghorn. They usually are placed in the family Antilocapridae within the suprafamily Bovidoidae. Lack of bovid and antilocaprid fossils complicate classification. Annually deciduous horns and (supposed) hair horns were the main reasons that early taxonomists did not place pronghorn in the family Bovidae. Present-day researchers generally discount horns as important taxonomic characters. Two (in most cases) lacrimal orifices and closed grooves above the distal ends of the cannon bones have led some taxonomists to consider pronghorn to be related closely to cervids. Extensive DNA studies may reveal the true relationship of pronghorn to other ruminants in time.

Minor differences in color, size and form have led to the naming of five subspecies. Mitochrondrial DNA and protein electrophoresis testing have indicated practically no difference between *americana* and *oregona*, an intergrade zone between *americana* and *mexicana*, and a similar zone between *mexicana* and *sonoriensis*. Thus, the concept of clines within a wide-ranging species seems more legitimate than five subspecies.

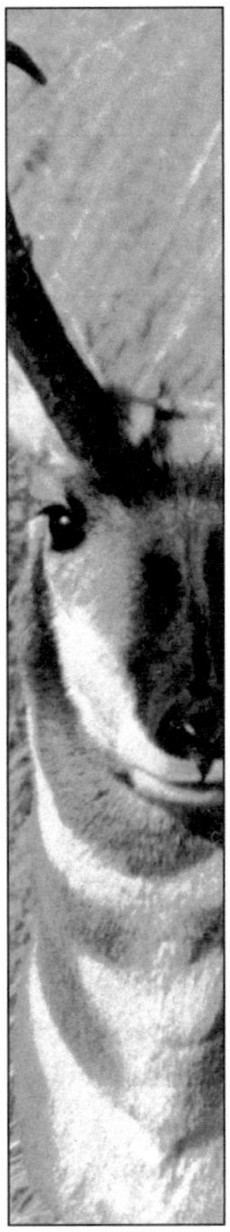

C H A P T E R T W O

The Fossil Record

Bart W. O'Gara and Christine M. Janis

Recent pronghorns (Antilocapridae) are represented by only one species, Antilocapra americana *Ord, restricted to western North America. During the late Cenozoic, however, a dozen genera of Antilocapridae inhabited North America, some ranging east to Florida. Thus, the flowering of this fascinating family must be comprehended by way of paleontological study. Hopefully such studies will shed light not only on the origins and radiation of the family, but also on the variety of its adaptations.*

Webb (1973:203)

To understand the evolutionary radiations of pronghorn and their position as North American parallels to the gazelles of the Old World, it is necessary to review the Cenozoic Era of the past 65 million years, commonly known as the age of mammals (Table 1). Small, insectivorous mammals mainly were present from the end of the Triassic Period of the Mesozoic era, and they did not become large, dominant animals until extinction of dinosaurs at the end of the Cretaceous period. The world of the Paleocene and Eocene epochs of the Tertiary period of the Cenozoic era was quite different from that of today. Through continental drift, the continents were approaching their present positions, diverging from the single landmass (Pangea) of the early Mesozoic. Vegetation throughout the world was tropical, even within the Arctic Circle (McKenna 1980).

Some interchange of mammals occurred between Old and New Worlds at this time, although North America already was separated from South America, and, in the Eocene (about 50 million years ago), the same types of mammals roamed the Northern Hemisphere (McKenna 1975). For example, the earliest horse, *Hyracotherium*, was from North America and England; it disappeared from the Old World at the end of the Eocene. The interchange probably was through a continental connection across the Greenland area. Artiodactyls that were present at that time, commonly known as dichobunids, were broadly ancestral to the present-day families. They were small—2 to 5 pounds (1–2 kg)—apparently omnivorous, forest-dwelling, and would have been similar in appearance and behavior to some of the modern South American rodents, such as the agouti, or

TABLE 1. Geologic time chart from the Mesozoic Era to present

Era	*Period*	*Epoch*	*Time span*
		Holocene	10,000 years to present
	Quaternary	Pleistocene	1.8 MYA[a] to 10,000 years
		Pliocene	5.2 to 1.8 MYA
		Miocene	23 to 5.2 MYA
		Oligocene	33.4 to 23 MYA
		Eocene	55.5 to 33.4 MYA
Cenozoic	Tertiary	Paleocene	65 to 55.5 MYA
	Cretaceous		144 to 65 MYA
	Jurassic		208 to 144 MYA
Mesozoic	Triassic		245 to 208 MYA

[a] MYA = millions of years ago.

Source: From Woodburne 1987, D. Prothero personal communication: 1994, Janis et al. 1998.

to the African elephant shrews, than to any modern artiodactyl.

In the late Eocene, the climate of northern latitudes changed from nonseasonal to seasonal, probably with cold winters. Woodland and heathland replaced tropical forests across North America and northern Eurasia (Wolfe 1985). At that time, the primitive dichobunids mostly disappeared, and most of the artiodactyls of the late Eocene can be identified as members of the living suborders Suina (pigs and their relatives), Tylopoda (camels and their relatives) and Ruminantia (giraffes, true antelope, deer, pronghorn, cows, goats, etc.). A connection probably existed between North America and Asia, a land bridge; suines, tylopods and ruminants were in both areas, although rarely represented by the same genera or even by the same families. The early North American ruminants—members of the traguloid (mouse deer-related) families, Leptomerycidae and Hypertragulidae—also had their counterparts in Asia. However, intercontinental migration routes apparently were restricted during the Oligocene; the North American traguloid families continued their evolutionary radiation on this continent but were too specialized to have given rise to any of the later North American pecorans. The larger selenodont (having crescent-ridged molars for chewing coarse or dry vegetation) artiodactyls of the North American Oligocene were the tylopods—camelids, protoceratids (deerlike, horned animals) and oreodonts (animals that ranged from rabbit to tapir size, resembling present-day hyraxes in body form).

In contrast, during the Oligocene of Eurasia, the first pecoran ruminants (members of the hornless family Gelocidae) were present, and the first members of the higher pecoran families, such as the Bovidae, Cervidae and Giraffidae, appeared early during the Miocene of Eurasia and Africa. The first antilocaprids likely originated there at the same time, although their fossils have not yet been identified. However, it is probable that they had a Eurasian origin, because the North American traguloids are not their ancestors, and the other pecoran families had an Old World origin. Possible Eurasian antilocaprid ancestors might exist in a variety of small, hypsodont taxa known from the late Oligocene and early Miocene of Mongolia. These include the taxa *Palaeohypsodontus asiaticus* (Trofimov 1958) and *Gobiocerus mongolicus* (Sokolov 1952), which often are assumed to be early bovids, but might just as easily represent early antilocaprids or another precociously hypsodont pecoran lineage, such as moschids. Unfortunately, it is impossible to identify either bovids or antilocaprids from dental remains alone, because many diagnostic tooth characters of ruminants are lost in hypsodont (high-crowned) cheek teeth, which have a relatively simple occlusal morphology (Janis and Scott 1987).

During the early Miocene, 18 to 20 million years ago, a route of interchange again occurred between North America and Eurasia, likely through the Bering Straits. At that time, ancestral stocks for the North American mid-Tertiary endemic pecoran ruminants—the hornless moschid-related blastomerycids, the horned deer-related dromomerycids and the antilocaprids—entered North America and began to diversify.

By the middle Miocene, about 15 million years ago, Earth was becoming drier; rainfall was more seasonal worldwide, causing the spread of grassland savannas in both North America and the Old World. However, the treeless prairie or steppe typical of much of the Northern Hemisphere today did not appear until the Pleistocene (Wolfe 1985). These emerging North American savannas were the focus for radiation of the Antilocapridae.

For about 18 million years, apparent ancestors of modern pronghorn roamed western North America, playing much the same role as gazelles did in the Old World. However, the diversity of antilocaprids declined in the Plio-Pleistocene era (5 million years ago) when the diverse savanna-type vegetation was replaced by treeless prairie.

Pronghorn remain animals of paleontological mystery. Although possible ancestors date approximately 19 million years ago, these animals already were highly developed. No older fossils have been found to tell where and from what stock early antilocaprids evolved. Photo by Irene Vandermolen; courtesy of Leonard Rue Enterprises.

We know about this earlier diversity because extensive fossil beds from the late Tertiary period are found in western North America, and a tremendous record of diverse antilocaprid taxa has been compiled from them. Fortunately, some taxa are represented by large samples and others by whole skulls or even entire skeletons. Without such a rich fossil record, there would be no hint that *Antilocapra* was the last of a great radiation of antilocaprids, or that sheath-horned prongbucks developed after velvet-horned pronglets—names coined by Frick (1937).

Most vertebrate fossil deposits in western North America are found in channel and bank sediments at stream sites. Ancient quicksands, such as those along Nebraska's Niobrara River, trapped and preserved many specimens, enabling a keyhole view of what the fauna were like.

In the latter part of the Cenozoic era, roughly 3 to 20 million years ago, ungulates, rodents, carnivores and many other groups of animals were approaching their modern forms. Camels, horses, mastodons and a diversity of other large mammals roamed North America. The vast prehistoric savannas teemed with such mammals—a scene that now seems more African than American. Yet, something was missing—the tremendous numbers and variety of African antelopes. In the place of gazelles were antilocaprids that were ancestors of modern pronghorn.

A land connection was established between North and South America across the Isthmus of Panama by 2.5 million years ago, opening the way for giant ground sloths, bony-shelled glyptodonts and armadillos to move north and mingle with the native North American fauna. In reverse, North American camels, tapirs and cats (such as jaguars) spread into the habitats of South America's endemic large mammal fauna. Horses (*Hippidion* and *Onohippidium*) and elephant-related gomphotheres

(*Stegomastodon*) also spread south at this time, but became extinct in South America by the end of the Pleistocene epoch. At the same time, some reconnection was established between North America and the Old World. Three-toed grazing horses (hipparionids) appeared in the Old World during the late Miocene epoch; camelids appeared in the Old World for the first time in the Pliocene epoch; and true deer (subfamily Odocoilinae) appeared in North America during the Pliocene epoch and subsequently spread into South America. By that time, the North American endemic deerlike dromomerycids were extinct.

About 2 million years ago, the Ice Age began. Ice accumulated in polar regions, drawing so much water that ocean levels dropped 200 to 300 feet (61–91 m). The Beringian land bridge, between Siberia and Alaska, once again was passable, allowing extensive movements of animals. True (one-toed) horses spread into Asia, whereas bison (some with horns spanning 10 feet [3 m]), mammoths, moose, caribou, musk-oxen, mountain sheep, mountain goats and Eurasian bears reached the Americas. Although some species, such as the saiga antelope and yak, only reached Alaska, others dispersed over much of the continent. Communities of animals that had evolved on separate continents intermingled, turning North America into an ecological battlefield. Native and immigrant animals not only competed with and preyed on each other but often were forced into unfavorable situations by rapid environmental changes. Grizzly and black bears from Asia replaced the native short-faced and spectacled bears, whereas gray wolves took over from bone-eating dogs and bear-dogs of earlier epochs. If the saiga antelope had spread southward into the Great Plains, it might have competed with pronghorn for an environmental niche. Humans also came to the New World through the Bering land bridge and gradually spread over the Americas (Martin 1984).

At least four major ice sheets, up to 2 miles (3 km) thick at their centers, expanded southward as far as the present cities of New York, Cincinnati and St. Louis. In contrast, the climate in the higher latitudes during interglacial periods, which lasted for thousands of years, was milder than that in continental conditions today. Each advance of ice altered the landscape and drove animal life southward or, in some instances, toward unglaciated areas of the north. The last retreat of glaciers, which started approximately 14,000 years ago, is still in progress. Coincident with climatic warming and drying, a great wave of extinction occurred, and the largest mammals of North and South America ceased to exist (Guilday 1971).

Paleo-Indians may have played a role in the extinction of many large but clumsy beasts.

The classification of genera and species within Antilocapridae is confusing because all but one species of this family are extinct. The two principal classifications were by Frick (1937) and Simpson (1945). Frick, in his monograph on horned ruminant fossils, recognized many genera and subgenera. Simpson's classification recognized only 12 extinct genera. Since then, an additional genus, *Ottoceros*, has been described (Miller and Downs 1974), and *Hexameryx* has been separated from *Hexobelomeryx* (Webb 1973). Eighteen genera are described in this chapter. Both Frick's and Simpson's taxonomic classifications recognized two distinct subfamilies (Merycodontinae and Antilocaprinae). Frick (1937) indicated that the Antilocaprinae evolved from the Merycodontinae, and subsequent researchers have accepted that interpretation. It is used here, but data supporting or refuting Frick's contention (see also Janis and Scott 1987) have not been found.

MERYCODONTINAE (PRONGLETS)

Fossils of the subfamily Merycodontinae have been found throughout much of the historic range of pronghorn in western North American deposits, dating from about 8.8 million to 18.8 million years ago (Figure 3). Their highly specialized teeth and skeletons are very different from those of contemporary North American fossils of other ruminant families. Pronglets are well documented, with more than 4,000 specimens‐some of which are nearly complete and well-preserved skeletons—in various collections. All pronglets are uniform in skeletal and dental structure. They are separated mainly by the form of the horns, which, in contrast to the Antilocaprinae, were carried only by males. The teeth, vertebrae and limbs of pronglets resemble those of modern pronghorn, but differ in the form of the horns, the retention of side toes on the front legs and in their overall size.

The merycodontines were approximately 20 inches (51 cm) high at the shoulder, barely half the height of modern pronghorn (Figure 4), and weighed between 26 and 44 pounds (12–20 kg)—of comparable size to such present-day neotragine antelope as the steenbok or the oribi. Dewclaws, present in the front legs of some genera, were smaller than those of deer. The legs were long, slender and laterally compact, like those of modern pronghorn, suggesting that they also were fleet-footed (Furlong 1927).

Merycodontine skulls are pronghornlike, with prominent tubular eye sockets and long, slender nasal bones and muzzles; the lacrimal depressions (indentions of the lacrimal bone that accommodate skin glands in front of the eyes) of deer are absent. The cheek teeth are highly hypsodont; the upper incisors and the upper and lower first premolars have been lost, as is typical of most pecorans. Small upper canines are retained in the early genus, *Paracosoryx*, but lost in later genera (Janis and Scott 1987).

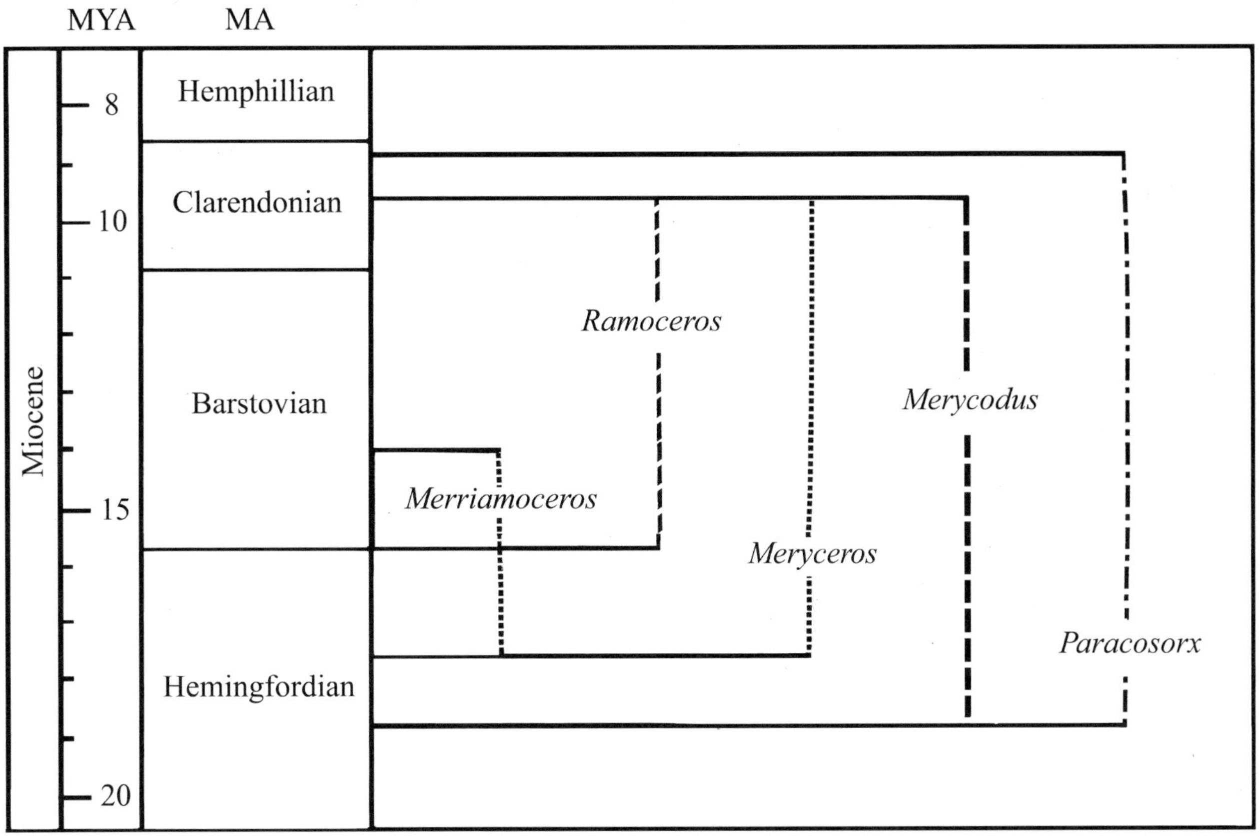

FIGURE 3. *Approximate times, in millions of years ago (MYA) and in relation to North American land mammal ages (LMA), when merycodontines appear in the fossil record, according to Janis and Manning (1998).*

The bony outgrowths from the skull are long and forked, sometimes with several tines, making pronglets appear deerlike. However, like the horns of pronghorn, the outgrowths are located over the eyes, whereas deer antlers arise posterior to the eyes. These outgrowths are called horns simply to be consistent with nomenclature in the prongbucks. Goss (1983) presented evidence that they should be called nondeciduous antlers. The horns of merycodonts were deciduous according to Matthew (1904), who believed they were repeatedly shed because some pronglets had more than one burr (flange, coronet, pseudoburr or pseudocoronet) on their "antlers." Matthew later decided, however, that they were nondeciduous horn cores, and he regarded the burrs as terminal points of the skin that covered the head. Furlong (1932) maintained that no specimens or records showed horns shed above the burrs. However, Bubenik and Voorhies (unpublished data in Bubenik 1982) reportedly found pedicels of merycodonts after casting.

Cope (1874, 1877) believed that the burrs were exostosed bone around beams that had broken off and been reunited. He indicated that the fractured beams were not lost because some kind of covering retained them. Cope further ventured that the covering was skin, like

FIGURE 4. *Modern pronghorn (*Antilocapra*) are about twice as tall as pronglets (*Merycodus*) were.* Illustration by Susan Kraft Ball.

that of the pronghorn, when the horn sheath was immature, or giraffe and concluded that *Merycodus* represented an ancestral type of cervid. He postulated that antler shedding originated because, during combat, beams frequently broke off above the bases, and this being an annual event, shedding became periodical. Such unlikely explanations no doubt influenced the thinking of early taxonomists trying to place pronghorn in a classification system.

The question of the type of covering the bony outgrowths had remains. If horn sheaths were present, they could not have been shed as they are by pronghorn, nor could they have grown from the bases as do bovid horns. To shed in one piece or grow from the bases, horn cores must be larger at their bases than anywhere else along their lengths. Matthew (1924) suggested that they could have been covered with horny sheaths that split or peeled periodically after new protective sheaths formed under them. However, the surfaces of merycodont horns above the burrs are covered by shallow anastomosing grooves (sulci) resembling those on deer antlers. Voorhies (1969) interpreted these sulci as channels for blood vessels under furred skin, much like deer antlers in velvet.

The presence of more than one burr around the basal portion of merycodont horns is a condition not encountered in modern deer antlers. Burrs of deer antlers are cast along with the antler. If merycodont horns were lost annually, each set would have had to separate from the pedicel above the burr, not below it, and a new burr would be expected to have formed each year distal to the previous ones. Although such a sequence of events would not be impossible, little or no reason exists to conclude that the burrs were associated with the casting of those outgrowths (Goss 1983).

Also, the fossil record does not yield a series of progressively more complex structures within any single species, as might be expected to have been produced during the first few years of life while the animal was maturing, as seen in modern deer. The apparent absence of unbranched horns is particularly significant. Only in the most immature specimens are short, unbranched horns found; these structures were clearly in the juvenile stages of development. Thus, although pedicels increased in length, they evidently did not become increasingly complex with age, as do present-day antlers, suggesting that merycodonts may have attained their full stature relatively early in life. Considerable individual variation exists in the shapes and sizes of these horns from one specimen to another (Goss 1983).

The burrs tend not to be so much outgrowths from the underlying bone as depositions of bone on the surface. In this respect, they differ from the burrs of antlers. In some specimens, they are relatively unattached and easily broken off (Furlong 1927). Each of the pseudoburrs is associated with a separate outer layer of bone (Voorhies 1969). Possibly, the pedicel increased in diameter each year by the addition of bony material outside the existing bone (Goss 1983). Such an interpretation supports the contentions of Furlong (1927) and Frick (1937) that skin of the horn regressed proximally each year—the burrs representing successive levels to which the viable skin of the pedicel receded. One specimen of *Merycodus* has obvious wear on its anterior prongs, meaning that, at least for a time, the horns were uncovered (Frick 1937), unless the wear occurred after death. In pronghorn, although the sheath is deciduous, the core is always covered by skin.

The original horns apparently grew to their full dimensions as skin-covered appendages. Shedding of the skin could have been induced hormonally, as in deer antlers, and dead bony protuberances probably resulted. The animals whose fossils lack burrs presumably died while their horns were still viable. Those with a single burr are thought to have survived until the horns matured and their skin covering was lost as far proximally as indicated by the burr. Those with two or more burrs probably survived for several more years, losing a distal ring of pedicel skin each winter and laying down successively more bony annuli superficially resembling the burrs of deer antlers (Goss 1983). Voorhies (1969) indicated that the pedicels lacked burrs until the second year of growth. He also noted that the bone of merycodont appendages was histologically similar above and below the burrs. Deer antlers differ greatly from the structure of the pedicle bone below the burr (Janis and Manning 1998).

Pronglet horns may have been similar in some respects to the skin-covered horns of modern giraffes, although the mode of development of giraffe horns is different from that seen in other horned pecorans and, within the Giraffoidea, horns evidently evolved independently in at least two families (Janis and Scott 1987). *Merycodus* appears to have had bony, skin-covered horns, although the pattern of branching was reminiscent of deer antlers. The skin, or even portions of the horn, may have shed at times, although these horns were not truly deciduous in the manner of deer antlers. The forked horns of *Meryceros* (Figure 5), with burrs at the bases of the points, could represent specimens having skin (similar to that on their heads) covering the trunks of their horns. The points, on the other hand, probably were encased by velvetlike skin or deciduous horn sheaths like those of the later prongbucks. The diversity of pronglet horns is indicated in Figure 5.

New finds may broaden times of existence for some genera. Merycodontines are known from the Hemingfordian land mammal age (LMA)—about 18.8 million years ago—through the Clarendonian LMA—about 8.8 million years ago. The merycodontines were at their peak

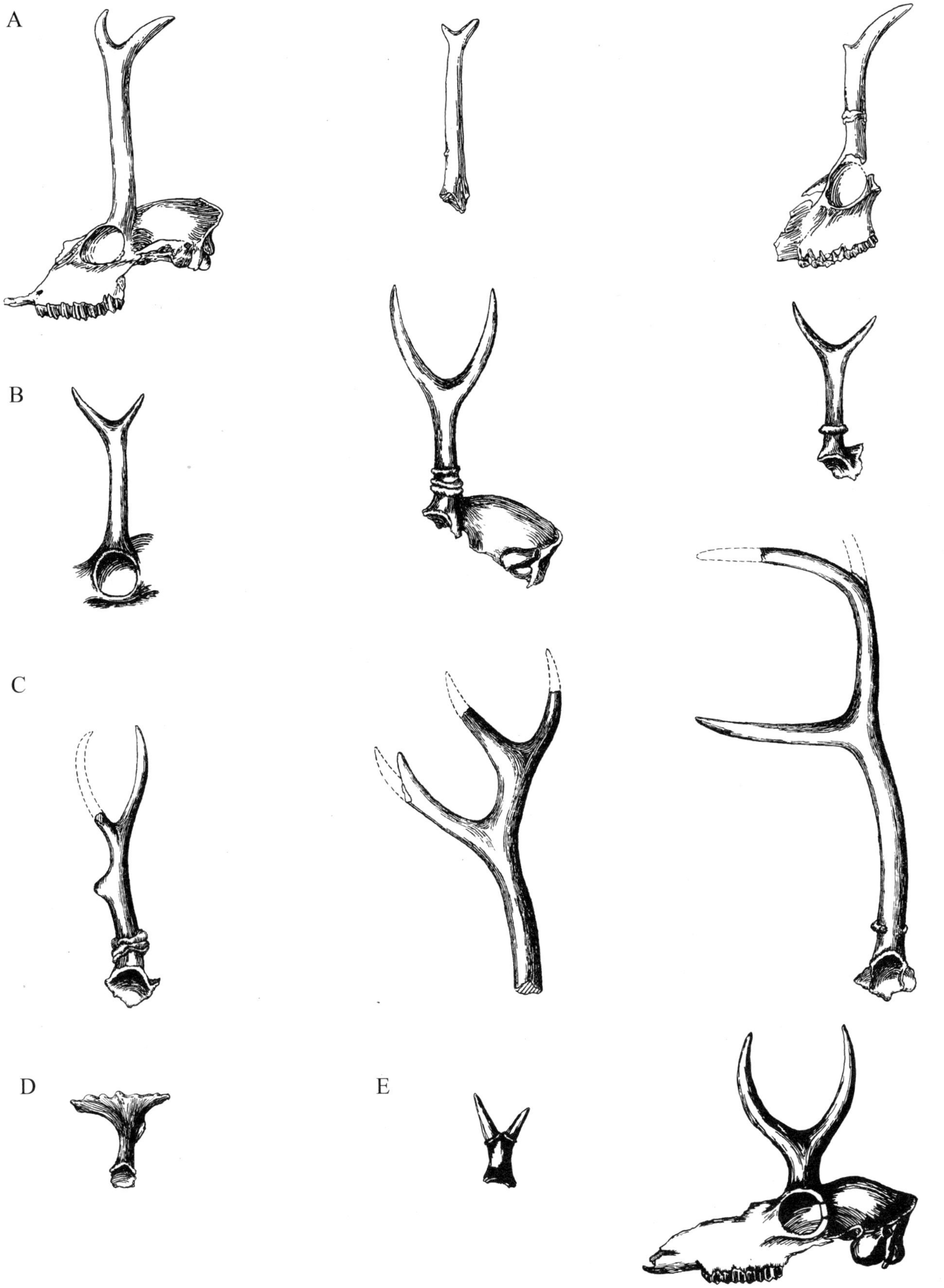

FIGURE 5. *Pronglet (Merycodontinae) skulls and horns: (A)* Paracosoryx, *(B)* Merycodus, *(C)* Ramoceros, *(D)* Merriamoceros, *(E)* Meryceros. Illustrations by John Eiler; all from Frick (1937).

of diversity in the Barstovian LMA—approximately 13 million years ago—and the Clarendonian forms overlap with the earliest antilocaprines (see Janis and Manning 1998).

Paracosoryx (Frick 1937) is the most primitive merycodontine genus; it is known from the entire time range of the Merycodontinae (early Hemingfordian to late Clarendonian LMAs). The cheek teeth of this genus were less hypsodont than those of other merycodontines; they retained a number of the styles and pillars seen in more primitive forms of brachydont ruminants, such as deer, and they also retained a small upper canine. These features were lost in later merycodontines (Janis and Scott 1987).

Merycodus (Leidy 1854) (= *Cosoryx*) fossils are among the most ancient ones, commonly considered to have been antilocaprids, first appearing in the Hemingfordian LMA. This genus had a similar temporal range to *Paracosoryx*, lasting until the Clarendonian LMA. The horns forked, and the angle of the fork on many were 90 degrees or more. The horns ranged from 3 to 6 inches (7.6–15.2 cm) long and had from zero to three burrs near the bases.

Ramoceros (Frick 1937) bore deerlike horns generally with three or four points. Some were dichotomously branched, resembling the antlers of mule deer. These were the longest of the pronglet horns, with a few specimens exceeding 10 inches (25 cm) in length. *Ramoceros* is known from the Barstovian and Clarendonian LMAs.

Merriamoceros (Frick 1937) had the most atypical horns of all the merycodontines. They were short and palmate, with multiple points, and looked somewhat like miniature moose antlers. The shafts or pedicels, typical of other genera, were nearly missing. Young specimens had elementary anterior and double posterior prongs somewhat similar to horn cores of the prongbucks *Hexobelomeryx* and *Hexameryx*. Older animals had 10 or more small points. This genus is known only from complete horns, partial crania and complete and partial lower jaws from a restricted number of localities in the late early and middle Miocene.

Meryceros (Frick 1937) had many species with horns quite similar to those of *Merycodus*. Six inches (15.2 cm) was about the maximum length for *Meryceros* horns. *Meryceros* is known from the late Hemingfordian and the early Clarendonian LMAs.

ANTILOCAPRINAE (PRONGBUCKS)

Fossils of Antilocaprinae date from about 14 million years ago to the early late Barstovian LMA, but they were not common until the Hemphillian LMA (Figure 6). With the loss of the front lateral toes (dewclaws), prongbucks had truly pronghornlike skeletons, but most were smaller and stockier than *Antilocapra*. The Antilocaprinae had proportionally longer limbs than the Merycodontinae, suggesting that they were adapted to a more arid habitat than were the earlier forms. Antilocaprinae teeth also were more hypsodont than those of Merycodontinae. The first antilocaprines, *Sphenophalos* (medial Barstovian LMA) and *Proantilocapra* (early Clarendonian LMA), were contemporaneous with the latest merycodontines, and antilocaprines were at their most diverse in the Hemphillian LMA, about 7 million years ago. Although the La Brea tar pits of Los Angeles preserved some excellent specimens, Antilocaprinae fossils are fewer and less complete than those of Merycodontinae. Roosevelt and Burden (1934) found pottery and a bone cracked by fire along with fossils of *Tetrameryx* in the rubble of an Arizona cave.

Prongbucks possessed varied and unusual horns, as indicated in Figure 7. Some had multiple or forked cores; others had twisted horn cores that flared widely or twisted tightly and grew almost straight up. Burrs, typical of pronglets, were absent. Merriam (1911) pointed out that the horn cores in some specimens terminated in spongy bone. Possibly, solid horn was extruded upward from such tips. In *Antilocapra*, the tips of horn cores are spongy, with the hook at the top of the horn jutting from this spongy area. The sheaths cannot be seen because they did not fossilize. Today, gnawing animals, insects and carnivores consume horn sheaths, or the sheaths rapidly disintegrate from exposure to the elements. It was no different millions of years ago.

Horn cores of Antilocaprinae differ from those of Merycodontinae in the smooth texture of their surfaces. The smooth core surfaces and absence of burrs imply the presence of heavy horn sheaths. For sheathed horns to be shed in one piece, each horn core must bear a separate sheath, be nearly straight and, in cross-sectional dimensions, must decrease toward the tip (Webb 1973). The assumption can be made, then, that the bases of forked horn cores of prongbucks must have been covered with skin and the prongs or forks covered by sheaths. If so, the horns would have looked somewhat like those of a pronghorn buck during winter. At that time, the horn cores of pronghorn are covered with skin and hair, but the terminal hooks and prongs are hard horn.

Given the mode of horn sheath development in *Antilocapra*, Janis and Manning (1998) deemed it plausible that merycodont-type horns transformed into antilocaprine-type horns. Retention of a juvenile skin layer (somewhat like the velvet on deer antlers) and its later keratinization, could have gradually resulted in a thin sheath. Eventually, this could have led to a thicker sheath with tines extending beyond the bony core. With a sheath and permanent layer of skin over the horn core, the merycodont burr would be lost. By thickening and shortening the horn shaft, as well as by adding a sheath,

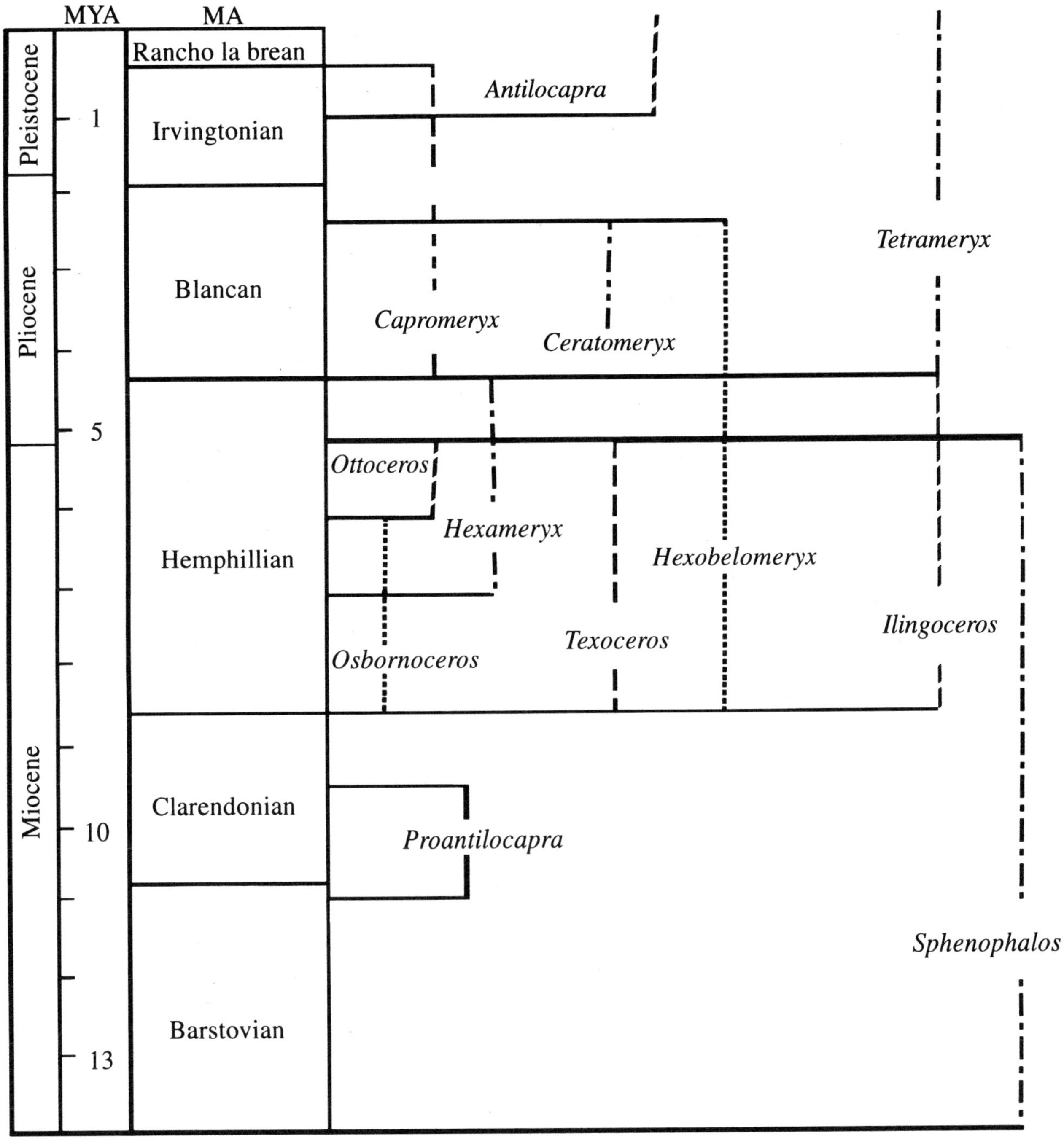

FIGURE 6. *Approximate times, in millions of years ago (MYA) and in relation to North American mammal ages (MA), when antilocaprines appear in the fossil record, according to Janis and Manning (1998).*

antilocaprines would have greatly increased the strength of the horn. This may be reflected in the locked-horn type of combat characteristic of pronghorn bucks. Such combat involves a great deal of stress on the shaft.

Sphenophalos (Merriam 1909), or *Plioceros* (Frick 1937), is known from the middle to the latest Miocene. The horn cores are heavily proportioned, laterally compressed and bifurcated near the tip. These horn cores are shorter and broader than those of present-day pronghorn. They are situated above and slightly behind the eyes, and they slope backward and slightly outward. The posterior is the longer of the two branches, and the anterior parts of the horn cores are slightly narrower than the posterior parts. The largest cross section is near the base; this and the smooth surface indicate heavy horn sheaths, according to Webb (1973). Considerable variation has been demonstrated

FIGURE 7. *Prongbuck (Antilocaprinae) skulls and horns: (A)* Sphenophalos (Pliceros), *(B)* Proantilocapra, *(C)* Texoceros, *(D)* Osbornoceros, *(E)* Ilingoceros, *(F)* Hexobelomeryx, *(G)* Ottoceros, *(H)* Hexameryx, *(I)* Ceratomeryx, *(J)* Capromeryx, *(K)* Antilocapra, *(L)* Subantilocapra, *(M)* Tetrameryx. Illustrations by John Eiler: A, C, D, E, J, K-b and M from Frick (1937); B from Barbour and Shultz (1934); F from Furlong (1941); G from Miller and Downs (1974); H and L from Webb (1973); I from Gazin (1935); and K-a from Richards and McCrossin (1991).

between horn cores of various species and animals of different ages. A female skull attributed to this genus lacks true horns, but has what Frick (1937) termed supra-orbital excrescences. *Sphenophalos* was a small animal, not much larger than the contemporaneous merycodontines, as also was true of *Proantilocapra*.

Proantilocapra (Barbour and Schultz 1934) is known from the Clarendonian LMA, was about the height of, but heavier than, pronglets (as indicated by more robust limb bones), and was described from fossils found in Nebraska. A hornless (presumably female) skull of this genus also is known. The horn cores have a bladelike cross section and truncated tips resembling the growing core tips of male *Antilocapra* fawns 6 to 10 months of age. The prongs of *Antilocapra* grow over the narrow front edges of somewhat similarly shaped cores.

Texoceros (Frick 1937) had widely forked horn cores with elongated forks that are almost round in cross section. The anterior fork is flexed outwardly and the posterior fork inwardly. The horns are positioned slightly more posterior to the eyes than are the horns of most antilocaprines. Webb (1969) suggested that *Texoceros* is synonymous with *Sphenophalos*. *Texoceros* is known from the Hemphillian LMA.

Osbornoceros (Frick 1937), from fossil beds in New Mexico, was on an evolutionary side branch, not the main ancestral line leading to modern pronghorn. The horn cores are relatively long and slightly twisted, with a backward and outward curvature that must have given the animal an appearance similar to that of some African antelope. *Osbornoceros* is known from the Hemphillian LMA.

Hexobelomeryx (Furlong 1941) fossils have been found in deposits of northern Mexico and the southern United States. The horn cores do not extend as far forward with reference to the eyes as they do in other genera of *Antilocaprinae*. The base is comparatively tall, and two prongs (the rear ones in males and the forward ones in females) lie close together, have a long common base and are parallel. They were sheathed by a common horny covering (Webb 1973). *Hexobelomeryx* is known from the Hemphillian, latest Hemphillian and early Blancan LMAs.

Ilingoceros (Merriam 1909), another evolutionary side branch, had long, slender, twisted horns that grew upward and slightly outward. The tips widen slightly and are notched so each core ends in two distinct prongs (Merriam 1911). Early paleontologists looked for relationships between *Ilingoceros* and Old World tragelaphine antelope such as the kudu (Merriam 1909). Some species were nearly as large as *Antilocapra*. *Ilingoceros* is known from the early to late Hemphillian and the latest Hemphillian LMA.

Ottoceros (Miller and Downs 1974) fossils have been found in California deposits, and indicate a small—22 inches (55.9 cm) at the shoulder—slender animal with less protruding eyes than *Antilocapra*. The horn cores have long bases with two short prongs at the distal ends; the anterior horn cores are slightly shorter than the posterior ones. The slightly twisted cores have lateral flanges beginning just above the orbits and terminating well below the prongs. Horn cores of females were smaller than those of males (Miller and Downs 1974). *Ottoceros* is known from the Hemphillian LMA.

Hexameryx (White 1941) fossils have been found in Florida—the only area outside the historic range of the present species where fossil antilocaprids have been found. Three horn cores diverge widely from a short base, and each had a separate sheath, making this a six-horned animal. The same sexual dimorphism in horn pattern occurs as in *Hexobelomeryx* (Webb 1973). *Hexameryx* is known from the Hemphillian LMA.

Ceratomeryx (Gazin 1935), found in Idaho, has four flattened horn cores and a skull appreciably smaller than that of *Antilocapra*. The anterior horn cores, located directly over the eyes and directed slightly backward, are much larger than the posterior pairs; the posterior ones are quite small. The eyes apparently were very large for the size of the animal and set farther forward in the head than in *Antilocapra* (Gazin 1935). *Ceratomeryx* is known from the early Blancan LMA, about 4 million years ago.

Capromeryx (Matthew 1902) fossils have been found from Nebraska to California and Mexico. This was a small animal, with the largest animals standing no more than 2 feet (61 cm) at the shoulder; the legs are long and slender. This early prongbuck was built much like the pronglets. A pair of nearly parallel horn cores stand erect from a common base over each eye. *Capromeryx* is known from the early Blancan LMA, approximately 4 million years ago, and Rancholabrean LMA, about 0.5 million years ago.

Tetrameryx (Lull 1921), another four-horned genus, was first seen in the early Blancan LMA. It persisted to within a few thousand years ago (Roosevelt and Burden 1934). It was smaller and somewhat heavier than *Antilocapra* and apparently not as swift a runner (Colbert and Chaffey 1939). Many researchers consider the tetra-horned animals, with anterior and posterior horn cores of approximately equal size, as a distinct genus—*Stockoceros* (Dalquest 1974). Frick (1937) also named an early Pleistocene epoch subgenus of *Tetrameryx* as *Hayoceros*.

At least six species of *Tetrameryx* are known; one species with all horn cores nearly equal in size (Figure 7K-b) was contemporary in Mexico with one that had short, flat, anterior cores and long, slender, posterior ones (Mooser and Dalquest 1975; Figure 7K-a). Fossils of some of the larger, slimmer species of *Tetrameryx* could be mistaken for *Antilocapra* if the horn cores were missing.

Subantilocapra (Webb 1973) was found in Florida deposits of the Hemphillian. Webb (1973) proposed *Subantilocapra* as a subgenus of *Antilocapra* and designated the species name as *A.* (*S.*) *garciae*. He noted that the horn cores seemed intermediate to those of *Sphenophalos* and *A. americana*. Richards and McCrossin (1991) considered *A.* (*S.*) *garciae* to belong within the genus *Sphenophalos*, and Janis and Manning (1998) agreed. However, they placed this taxon within their temporal sequence of antilocaprid genera as *Proantilocapra*, as is done here. The horn cores of *Subantilocapra* are compressed in a transverse plane, tall-based and forked with an anterior prong about one-seventh the length of the posterior one. There is little divergence between prongs. The shaft of the horn resembles that of *Antilocapra*, but with narrower anterior edges, more convex medial sides, flatter lateral sides and the whole shaft curving slightly outward (see Janis and Manning 1998).

Antilocapra (Ord 1818) includes the living species *A. americana* (Ord 1818) and an extinct species, *A. pacificus* (Richards and McCrossin 1991). *A. americana* fossils are found only in the species' historic range and only in deposits less than 1 million years old. *A. pacificus* fossils, recovered from the San Joaquin River delta near San Francisco, California, date from the Pleistocene epoch. Horn core, orbit and temporal fossa morphology differs from that of *A. americana*. Furthermore, the new specimens diverge from a demonstrated allometric relationship between cranial and horn core size in *A. americana*. The new specimens have horn cores that are well outside the modern range but have a cranium in the median size range for *A. americana*.

By about 10,000 B.C., Native Americans were carrying pronghorn carcasses into rock shelters and other locations where bones were preserved (Graham and Lundelius 1994a, 1994b, Morlan 1999, Arroyo-Cables et al. 2001). Locations of such sites have been found to coincide closely with the species' range when Europeans reached North America. McCabe et al. (2004) mapped 278 archaeological sites with pronghorn remains, shown in relation to other, cumulative evidence of the late prehistoric pronghorn range. They also described the 19th century fate of pronghorn at the hands of western Indians and Euro-American pioneers.

CONCLUSIONS

Antilocaprinae might have evolved from the Merycodontinae, about 12 million years ago, but the two subfamilies coexisted during the late Miocene epoch. *Sphenophalos*, a widely distributed genus, probably is on the direct ancestral line to *Antilocapra* (Furlong 1932, Webb 1973). *Subantilocapra* seems an intermediate form in such a sequence.

Perhaps too much importance has been attached to shape and number of horn cores. They seem to be the most variable of all the skeletal elements, as the following indicates. J. Gay (personal communication:1975) photographed the skull of a four-horned pronghorn and wrote "We receive several four-horned heads every year, but in almost all cases the second pair is small, vestigial, and does not have a bony core as the regular horns do." Figure 8 is a sketch of the mounted head of the four-horned animal. It probably would be considered a different species or even genus if it had been found in the La Brea tar pits. There are several reports of similar skulls; one found in Montana featured posterior horn cores about a third as long as those in Figure 8, and another shot in Montana had posterior sheaths and horn cores about three-fourths as long.

The reasons for extinction of all but one genus of antilocaprid are speculative. Romer (1966) believed that the appearance of bison in the Pleistocene epoch was related to extinction of some genera of prongbucks. However, they had been declining since the Pliocene epoch, as were many North American mammals, and it is doubted that direct competition for food was important. Bison represented a large food resource and undoubtedly broadened the food base for carnivores, thus increasing the number of predators, including humans. New predators, such as wolves, increased with the bison, but antilocaprids had long been exposed to a diversity of North American carnivores.

FIGURE 8. *The mounted head of the four-horned pronghorn (perhaps it should be called a forked-horn) pronghorn shot in Wyoming. The only fossil genus known with a small horn behind the main one is* Ceratomeryx *(see Figure 7I), which generally is not considered to be on the direct line of evolution to* Antilocapra. *However, members of* Tetrameryx *(considered* Stockoceros *by some investigators) (see Figure 7M) with anterior and posterior horn cores of approximately equal size (see Figure 7Kb) would resemble the condition pictured in this skull, if the posterior horns were reduced in size (most posterior horn sheaths grow backward instead of curling upward as occurred in this specimen). The consistency in location (and wide distribution) of the extra horn cores suggests a genetic basis for the four-horned condition in modern pronghorn.* Sketched by Susan Kraft Ball from a photo by Jim Gay.

The skull of a four-horned pronghorn (see Figure 8). Such animals are reported from somewhere in the species' range every year, and smaller (barely noticeable) bumps are common on the posterior bases of buck horn cores. Photo by Jim Gay; courtesy of Taxidermists, Inc., Laramie, Wyoming.

As they dispersed across North America and into South America, primitive people may have eliminated some or many large mammalian species (Martin 1984). However, Bromley (1977) believed it unlikely that hunting influenced the extinction of most of the pronghorn line because only three genera of the Antilocaprinae possibly survived long enough to be affected by humans. He suspected that *Antilocapra americana* survived primarily because of preadaptations that prevented overheating and desiccation when the species was subjected to climatic extremes.

The ancestry of pronghorn remains a mystery. This group cannot be uniquely North American, as often stated in popular articles and the older scientific literature. They obviously had Eurasian ancestors that reached North America during the early Miocene epoch, 18 million years ago. From sequencing of of mitochondrial deoxyribonucleic acid (mtDNA), Kraus and Miyamoto (1991) reinforced conclusions long held by morphologists that antilocaprids, bovids, cervids and giraffids were derived from a common ancestor. Radiation apparently took place over a period of 5 million years, some 23 to 28 million years ago. Such evidence is inconsistent with a uniquely North American origin for pronghorn.

C H A P T E R T H R E E

From Exploitation to Conservation

Bart W. O'Gara and Richard E. McCabe

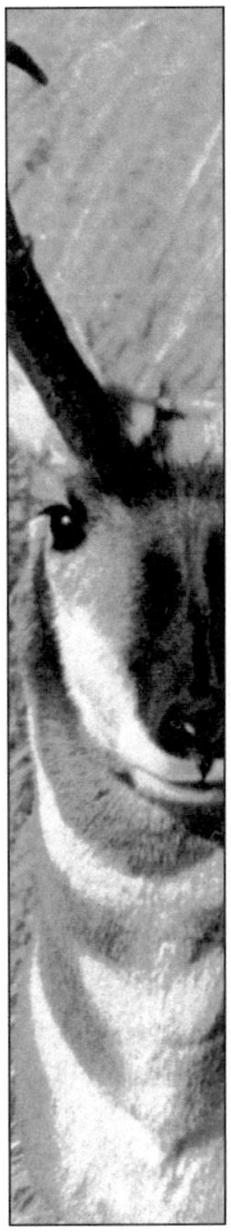

1500 TO EARLY 1800s

The history of antelope in Colorado typifies the tragic story of so many of our game animals. It's a story that began with unlimited abundance in the early part of the last [19th] century, only to reach near-annihilation of the species by the end of that era. Relentless slaughter, lack of management and encroachment of civilization played their ruthless role throughout the 1800s.

Thomas L. Kimball *in* Hover et al. (1959:2)

Kimball penned the above in specific regard to Colorado, but he accurately described the fate of pronghorn throughout Canada and the United States. However, data from archaeological sites in Mexico containing pronghorn remains (Arroyo-Cabrales et al. 2001) and early Spanish accounts indicate that pronghorn numbers were declining in the southern extremity of the species range well before the 19th century (see McCabe et al. 2004). Whether the early decline was related to climate change, competition with introduced livestock, efficient harvests by mounted hunters or some other factor is unknown.

Native American hunters pursued pronghorn for 10,000 years or more before Europeans invaded the continent (McCabe et al. 2004). The arrival of horses increased Indians' ability to harvest big game (Jablow 1951), and many thousands of pronghorn were killed with little effect on total numbers. The speedy pronghorn reproduce rapidly and populations in healthy habitat can withstand high levels of predation. Indians barely altered—except by fire—the western deserts, prairies and steppes, and fires improved vegetation for pronghorn consumption.

The first wildlife law in what is now the United States, adopted in 1623, provided that hunting would be free to all members of Plymouth Colony, but landowners retained the right to control animals on their property. During the remainder of the 17th century and most of the 18th, at least a dozen wildlife laws were passed by colonies or states—mostly concerned with the hunting of white-tailed deer. In 1789, the U.S. Constitution was adopted, establishing federal authority to make treaties, set federal land policy and regulate interstate commerce. This established the basis for federal involvement in wildlife conservation and management (Mackie 2000).

The earliest written records about pronghorn are in Spanish and French, with the former predating English accounts by approximately 150 years. Linne (1948) and Townsend (2000) published maps depicting pronghorn hunts from the time of New Spain, beginning in 1519.

As the Corps of Discovery ascended the Missouri River, William Clark wrote a letter, dated April 2, 1805, to William Henry Harrison, in which he advised that the region was abundant in "antelope or goat" (Thwaites [VII] 1969:316). Specimens collected by the corps introduced pronghorn to science. Reports of and from the Lewis and Clark expedition and later government surveys facilitated settling of the western United States, eventually leading to near extirpation of many prairie- and steppe-adapted species, including pronghorn (see Chapter 4).

McCabe (1982, 2002) ranked by species the amounts of animal protein eaten by Native Americans in various regions of North America. Pronghorn occurred in eight regions, and they failed to rank as most important in any. In the Southwest, deer and pronghorn tied for second. The latter were third on the southern plains and the Great Basin and fourth in the central plains and central California. Bison were the most important food animal on the plains, as were jackrabbits in the Great Basin and Southwest. However, pronghorn were hunted extensively for skins, dietary variety, etc. (McCabe et al. 2004). As bison were nearly exterminated in the late 1800s, pronghorn became increasingly important to plains Indians.

MID- AND LATE 1800s

The closing decades of the nineteenth century were the darkest in the history of wildlife in America. The few regulations with regard to hunting that were in existence at that time were seldom enforced, and many animals were pushed to the brink of extinction. Selling wild game was carried on openly, even in those states that had prohibited this commerce; and it was only through the concerted efforts of a few farsighted naturalists and sportsmen that such regulations were at last enforced, shorter hunting seasons initiated, and lower bag limits established, thereby ensuring the future existence of most of the game animals that survive to the present day.

Rossi and Hunt (1971:190)

Before 1870, Native Americans and Euro-Americans generally harvested pronghorn lightly because bison, elk and deer provided more meat and larger hides and were relatively easy to hunt (Huntington 1904). The breech-loading rifle and metallic cartridges greatly improved hunting efficiency in the West (Dodge 1959). They hastened the demise of bison, which caused settlers as well as hide and market gunners, to turn their sights on pronghorn.

In 1866, Colonel R. B. Marcy opined: "The wild animals that abound on the Great Plains today will soon be as unknown as the Indian hunters who have for centuries pursued them. The world is fast filling up" (Marcy 1963:xiii). In the next decade, John Mortimer Murphy (1879:325) penned, "This interesting animal, like some others, is destined to disappear in a short time from the list of American fauna, for it cannot live in the thickly inhabited country, and its favorite haunts are now being occupied so rapidly by stock-raisers that their herds and flocks are pushing it further into wild and inhospitable regions." Thus, to those who cared about pronghorn, the proverbial writing was on the wall soon after the mid-19th century.

In 1849, the U.S. Department of Interior was established to administer the General Land Office and other federal agencies. In 1862, Congress established the U.S. Department of Agriculture, and its Homestead Act authorized largely unrestricted settlement on 160 acres (65 ha) of public land (Mackie 2000).

"Cattlemen and sheepmen gave little thought to owning the land—only the grass served their purpose. Now, another wave of land hungry colonists poured over the horizon. Under the new Homestead Act sodbusters began moving in with their women, children and plows. These newcomers intended to put down roots and stay awhile—they immediately set about the business of plowing up the prairie and fencing the waterholes" (Blair 1987:v).

In 1885, Congress passed the Unlawful Enclosures Act. At the time, the act was intended to prevent ranchers from fencing public land to prevent use of that land by livestock other than their own (Donahue 1999). During the 20th century, the act became important to prevent ranchers from restricting pronghorn movements (see Chapter 17).

In 1877, John Dean Caton published *The Antelope and Deer of America* (Caton 1877). About 48 pages were devoted to the pronghorn, and a fairly complete description of its anatomy and habits was presented. This work, especially the section on anatomy, would be referred to frequently by later authors.

By the mid-1800s, the "cabri," "prairie goat," "American antelope" was in trouble. "The taking over of the semiarid West for cattle raising and agriculture resulted in

When pioneers lived in soddies, or sod houses on the prairie, there were few structures from which a game animal could be hung for cooling, skinning or photographing. These fellows used a wagon tongue to serve the purpose. Similar ingenuity nearly led to extirpation of pronghorn where homesteaders, numbering four families to the square mile (1/0.4 km²), occupied much of the prairie. Photo courtesy of the State Historical Society, North Dakota Heritage Center, Bismarck.

the elimination of the pronghorn . . . from much of its primitive range" (Allen 1962:181). The plight of pronghorn in Arizona, for example, was described as follows: ". . . declines in pronghorn populations did not occur until sometime after the coming of the railroads and large herds of cattle in the 1880s. More detrimental, according to most observers, was the arrival of increasingly large herds of domestic sheep after 1890. Feeding on forbs as well as grass, the 'woolies' (and later, goats) competed directly with the pronghorn for food" (Brown 1994a:1). Domestic sheep totaled about 40 million on rangelands of the 11 western states during 1890 to 1900 (Wagner 1978).

"Even more importantly, the end of the open range and the fencing off of pastures restricted the movements of the antelope and made them susceptible to human and animal predators. Woven sheep fences were especially detrimental, as the antelope would follow these impassable barriers for miles before succumbing to lack of forage, deep snow or drought.

"The most serious inroads on pronghorn numbers, however, were more direct. Antelope, whenever and wherever they were encountered, were shot for food, as potential competitors with domestic animals or just for the fun of it" (Brown 1994a:1).

Each province or state naturally had different problems or sequences of events, but Arizona and Wyoming provide good examples of approximately what was happening throughout western Canada and the United States with regard to wildlife laws. During 1893, the Arizona game code was amended to close the pronghorn season from January 1 through September 1, and made it illegal to ship game out of season. Game could still be sold and there was no restriction on sex or age of animals killed. As pronghorn numbers continued to dwindle, the legislature

By the 1870s, the nation's industrial revolution was well underway, and improvements in transplantation and equipment made game more valuable to market gunners and sportsmen alike. Some of the latter looked to the British Isles for proper sporting methods and socially acceptable methods. Some 19th century sporthunters traveled by train and had railway cars sidetracked to serve as hunting camps for a variety of game. This group was living in a Northern Pacific "hunting car" in Montana during autumn 1878. Photo by F. Haynes; courtesy of Haynes Foundation Collection, Montana Historical Society, Helena.

Categorizing hunting by purpose—such as food, market and recreation—is difficult because categories often overlap. Hunting to supply food, or to sell to others frequently provides recreation as well. Likewise, hunting for sport often provides food, according to Fleharty (1995). Almost every human activity that had anything to do with pronghorn during the 19th century must share some blame for their population plunge toward extinction. It is problematic to quantify blame, and it makes little difference now. However, of all the user groups, enlightened sportsmen clearly did the most to save the few, remaining pronghorn to start their recovery. Even in the 1800s, successful pronghorn hunters drew the attention of onlookers, if this photo is any indication. The photo is labeled "Tom Arnold—Bernard Newman with Badlands Antelope Circa: 1896." Courtesy of the State Historical Society, North Dakota Heritage Center, Bismarck.

prescribed a year-round closed season on does and fawns in 1897. More importantly, game could no longer be sold. And, to prevent Indians from killing game, it became illegal for them to hunt off of the reservations to which they were assigned (Brown 1994a).

When Wyoming attained statehood in 1869, an act to prohibit the sale of pronghorn hides or meat between February 1 and August 1 each year was passed by the first legislature. However, no provision was made for enforcement, and the act was unpopular with Wyoming residents (Thomas 1966).

"American sportsmen, those who hunted and fished for pleasure rather than commerce or necessity, were the real spearhead of conservation. . . . With establishment in the early 1870s of national newspapers like *American Sportsmen*, *Forest and Stream*, and *Field and Stream*, outdoorsmen acquired a means of communicating with each other, and a rapid growth of group identity was the result. Increasingly, gunners and anglers looked upon themselves as members of a fraternity with a well-defined code of conduct and thinking. In order to obtain membership in this order of 'true sportsmen' one had to practice proper etiquette in the field, give game a sporting chance, and possess an aesthetic appreciation of the whole context of sport that included a commitment to its perpetuation. . . . The most obvious manifestation of the new self-awareness was the rapid growth of sportsmen's clubs and associations" (Reiger 1975:21–22).

In 1832, more than half a century before the Boone and Crockett Club was founded, artist George Catlin proposed that the federal government establish a park to protect wildlife and wilderness—to be viewed by citizens and the world. In 1872, Yellowstone National Park was established,

In 1875, a hunting season was established in Wyoming for pronghorn from August through January, and sales of hides and meat were outlawed. But, no provision was made for enforcement, and the law was ineffective. A new law in 1888 set the season for pronghorn hunting as September 1 to January 1. Nonresidents were prohibited from hunting in the state—an attempt to discourage market hunters. However, residents could hunt anytime to feed themselves and their families. Still, no provision was made to enforce the law (Greenquist 1983). Wyoming had more pronghorn in 1902, when these photos were taken, than did any other state—as it did 100 years later. Game laws were on the books but enforcement was negligible. Photo courtesy of Wyoming State Archives, Museums and Historical Department, Cheyenne.

but more than two decades passed before it became a sanctuary for wildlife. In 1891, members of the Boone and Crockett Club fought to prevent railroads from entering the park. The railroad lobby was powerful, and only strong public sentiment—fanned by George Bird Grinnell's editorials in *Forest and Stream*—kept the rails out (Reiger 1975). The Act to Protect Birds and Animals in Yellowstone National Park, which probably saved the pronghorn herd there, was signed by President Cleveland in 1894.

Howard Douglas was appointed superintendent of Rocky Mountains Park in 1897 and was among the first in government to comprehend the role and importance of wildlife in a dominion park. As superintendent of Canada's first national park, he was in a good position to influence government involvement with wildlife preservation (Foster 1978).

Attempts to exploit Yellowstone National Park for private gain and the defeat of those attempts, contributed to the emergence of the U.S. Forest Reserve System, now the National Forest System. In 1891, President Harrison issued a proclamation, setting aside 1,239,040 acres (501,439 ha) in Wyoming as The Yellowstone National Park Timberland Reserve. This was the beginning of the system. A Division of Forestry, now the U.S. Forest Service, within the U.S. Department of Agriculture, was established in 1881. The understaffed and underfinanced division accomplished little during the 19th century (Meyer 1966, Reiger 1975).

Before efficient game laws could be passed, the public had to accept the necessity for legislation and be motivated to abide by it once enacted. An early problem was the status resentment of nonsportsmen. Some farmers condemned game laws as encroachments on the liberties of the people for the welfare of the aristocracy. Grinnell's editorials were instrumental in convincing the majority of the public that laws prohibiting the destruction of game represented the greatest good to the greatest number of people. Getting local people, including those who favored the new game laws, to report infractions was difficult. Some informers were visited in the dark of night by barn

(Left) *The Shields party hunted near Bennett Creek, Montana, in 1882. G. O. Shields founded the League of American Sportsmen and, with the backing of a group of New York sportsmen, the Campfire Club of America—both in 1897. The latter wielded considerable influence on the national conservation scene by the first decade of the 20th century, as did Shields himself as the editor of* Recreation. (Right) *The one-day bag of the Shields party hung to cool and be photographed. During the late 1800s, even sportsmen of Shields' caliber hunted during summer and shot fawns, judging by the size of the one in this photo.* Photo by L. A. Huffman; courtesy of Coffin's Old West Gallery, Miles City, Montana.

burners (Reiger 1975). The idea that traditionally free and unstructured hunting should be financially supported by sportsmen and regulated by states—which became the cornerstone of game management—was new in the late 19th century.

Two states outlawed pronghorn hunting during the 19th century—California in 1883 (Popowski 1959) and North Dakota in 1899 (Cadieux 1986). Some states enacted other regulations concerning pronghorn during the late 1800s, generally with little effect. Among them, the California Legislature closed the season on pronghorn for 6 months each year in 12 counties during 1852 (Popowski 1959). Idaho provided seasonal protection for pronghorn in 1864 (Mackie 2000). Other instances of early attempts to protect pronghorn were reported by Cadieux (1986) as follows: Colorado—coursing outlawed in 1877 and a limit of five pronghorn per hunter set in 1891, this was reduced to one in 1897; Montana—season closed from February 15 to August 1, in 1873, with a limit of eight pronghorn per year.

By 1895, pronghorn were the chief targets of Wyoming's hide and meat hunters. The state game warden reported that pronghorn were still numerous but were being killed by the wagonload, and Indians—deprived of their staple food source, bison—were killing large numbers. Game laws were revised in 1899, and for the first time, deputy game wardens were paid—$500.00 per year. Also for the first time, state residents were required to purchase hunting licenses for $1.00 if they were going to hunt in a county other than that of their residence. Out-of-state hunters were required to buy $40.00 licenses. Also, for the first time, the number of pronghorn a hunter could kill was limited to three per year. Penalties for game law violations were set at $25.00 to $100.00 or imprisonment for 10 to 90 days in the county jail or both. Fines collected went to the state game fund. Wyoming's game laws finally had teeth (Greenquist 1983).

Intrinsic to the 19th century concept of game laws was the belief that wildlife was to be preserved as a food supply. Aesthetics and ethics were of little concern to most shooters. However, by the end of the century, the mood of the public was changing. "The idea of 'ripeness' is useful: Americans were *ready* to be concerned about their environment," according to Nash (1968:37), who described the emergence of the conservation movement as the "ending of the Frontier." "Conservation would be, in a sense, the new frontier, keeping the nation young, vigorous, prosperous, democratic, replete with opportunity for the individual and, because of its relation to nature, wholesome and moral" (Nash 1968:38).

Federal wildlife regulations before 1900 were limited in scope, but the doctrine of public trust (decided in 1842

During the late 19th century, Edgar A. Mearns explored the area on both sides of the Arizona–Mexico border and took this picture there in 1887. Mearns was a naturalist of some note and described much desert flora and fauna for science. He regarded pronghorn as an uncommon and incidental species, due to the region's limited plant food and water, and because the animals were readily killed, wounded or simply harassed by settlers and travelers. Photo by Edgar A. Mearns; courtesy of the U.S. National Archives, Washington, D.C.

that the public's right to fish superseded a landowner's claim of personal property) and of a state's right to regulate wildlife within its borders (decided in 1896) had been established (Bean 1978).

1900 TO 1930

For years and years, the antelope millions of Montana and Wyoming grass-lands fed the scout and Indian-fighter, freighter, cowboy and surveyor, ranchman and sheep-herder; but thus far I have yet to hear of one western state that has ever spent one penny directly for preservation of the antelope!
Hornaday (1913:2)

The 20th century started off on a positive note with passage of the Lacey Act on May 25, 1900. Congressman John F. Lacey, who had carried the Boone and Crockett Club banner in the fight to protect Yellowstone National Park, wrote the bill that made interstate shipment of game killed in violation of state laws a federal offense. It also made the importation of foreign wildlife illegal, except under permit of the Secretary of Agriculture, and vested the administration and enforcement of the act in the Biological Survey (Matthiessen 1959). This act severely curtailed market hunting as a big business and, thereafter, game animals were increasingly regarded as a recreational resource rather than a commodity (Trefethen 1961). Sport hunting was to create an independent agency with a single mission, wildlife management, and thereby move wildlife off a permanent position on the back burner of public concern (Lund 1995). After more than 100 years, the Lacey Act still is the most efficient law on the books for prosecuting poachers who transport wildlife between states or countries.

When Theodore Roosevelt took the presidential oath of office on September 14, 1901, a new era in American land and wildlife management began. He used his powers to set aside numerous natural areas. During his presidency, 100 million acres (40 million ha) were added to the forest reserves, called national forests after 1907; five national parks and 17 national monuments were established; and an extensive system of wildlife refuges, containing more than 50 units, was created (Reiger 1975).

In 1903, Wind Cave National Park was established near the Black Hills of South Dakota. During October 1914, six pronghorn fawns and seven adults were shipped from Alberta to Wind Cave, where 5,000 acres (2,024 ha) had been fenced to receive them (Trefethen 1961). Fencing was used in other big game refuges of limited size.

Also in 1903, Roosevelt appointed a public lands commission to investigate range conditions and develop proposals for stabilizing rangelands. The commission sent a questionnaire to stockmen. Of the 1,400 replies received, 78% favored some governmental control of grazing on public lands (Donahue 1999).

Gifford Pinchot had visited Roosevelt soon after he became governor of New York in early 1899 and offered suggestions for reforming the administration of the state's natural resources (Reiger 1975). From their close association as leaders of the Boone and Crockett Club, Roosevelt had received a thorough exposure to Grinnell's ideas for handling the nation's natural resources. There is little question that he was the original source of many of the concepts Roosevelt established as national policy (Clepper 1966, Nash 1967).

Pinchot lacked interest in any resource except forests and originated few ideas. His great contribution to conservation lay in his talents as a manager who implemented Roosevelt's concepts fostered by Grinnell.

During the early 20th century, the U.S. Forest Service assumed responsibility for millions of acres of wildlife habitat that had been altered by logging, livestock grazing and irrigation. The first national forests were created in a climate of alarm. Harvesting of natural resources had reached such staggering proportions that historians characterized the era as the "Great Barbecue" (Catton and Mighetto 1998). Such unregulated use and its devastation promoted the rising conservation movement, which had focused first on market hunting. Although much of the national forest land was not pronghorn habitat, some of the national grasslands administered by that agency were. Also, multiple-use and other laws later promulgated for the U.S. Forest Service were applicable to the U.S. Bureau of Land Management (BLM) and impacted a vast portion of pronghorn rangeland.

By 1905, the wildlife conservation movement had begun to take form. Most of its elements—laws and enforcement, refuges, protection of breeding stocks, state and federal administration—were recognizable, but the whole was uncoordinated and loosely organized. The Conference of Governors, called by President Theodore Roosevelt in 1908, was the catalyst that pulled the various parts into perspective and molded them into a unified structure (Matthiessen 1959). State officials, who had gone their separate ways, began to compare data and borrow progressive ideas from each other. This strengthened a liaison that had begun in 1902 with establishment of the International Association of Game, Fish and Conservation Commissioners (Trefethen 1966), now the International Association of Fish and Wildlife Agencies.

In 1905, the first national game preserve—Wichita National Game Preserve, Oklahoma—was dedicated (Mackie 2000). Early attempts to establish pronghorn there on historic range failed despite numerous tries with hand-raised fawns (Cadieux 1986).

In 1907, the American Bison Society, backed by the Boone and Crockett Club, proposed establishment of the National Bison Range. The bill to do so was quickly passed by Congress and signed by President Roosevelt in 1908 (Trefethen 1961). Land was acquired on the Flathead Indian Reservation in Montana for the range, and this was the first federal refuge specifically planned to preserve a

An idea of the scarcity of pronghorn on western rangelands during the late 1800s and early 1900s is indicated by the fact that the first pronghorn brought to Wichita National Forest, Oklahoma, were fawns that were captured and hand-raised in Alberta. Translocations met with little success until techniques were developed for capturing and moving comparatively large numbers of adults. Photo courtesy of the U.S. Forest Service, Washington, D.C.

big game species. The fenced, 18,542-acre (7,504 ha) range was primarily designed to preserve bison at a time when they were threatened with extermination. However, it supported a small (variable population of 100–200) but thrifty introduced herd of pronghorn that proved valuable for translocations and research on the species. However, the herd, on nonhistoric rangelands, requires occasional coyote control to remain viable (O'Gara and Malcolm 1988).

Pronghorn received little protection until well into the 20th century. Lantz (1910:11) summarized the situation: "The present distribution and numbers of pronghorn are a sad commentary upon earlier game protection in the West. A few fugitive bands are now to be found in cattle country from longitude 101° westward." He cited the 1906 annual report of Wyoming state game warden D. C. Nowlin, which indicated that pronghorn had decreased to an alarming extent throughout the state. In 3 years, for example, the Green River herd diminished from about 6,000 to less than 2,000. Hundreds starved during storms, and others were slaughtered by Ute Indians.

(Top) *Hand-raised fawns quickly became pets and seemed to enjoy riding in or racing vehicles and playing with children or dogs. This tameness facilitated raising fawns, but became a detriment when they were released.* (Bottom) *Numbers of captive fawns bedded and played together freely. However, the more fawns kept together the greater the chance of pathogens spreading among them. They were especially vulnerable to respiratory diseases.* Top photo courtesy of the U.S. National Archives, Washington, D.C. Photo taken at the Slim Buttes Antelope Preserve, South Dakota, by W. D. Parker; bottom photo courtesy of the U.S. National Archives, Washington, D.C.

During the first decade of the 20th century, D. C. Nowlin was the conservation conscience of Wyoming. He forcefully pressed the governor and public to increase funding for law enforcement and closure of pronghorn seasons. By 1906, he reported more hunting in the state than ever, and residents had turned from meat to trophy hunting. Despite or perhaps because of this, public attitudes became more favorably disposed toward game protection. Nowlin, as well as much of the Wyoming populace, was influenced by President Roosevelt's stance on conservation. This apparently caused a decrease in poaching. The pronghorn season was closed in 1908, setting the stage for pronghorn recovery. Thereafter, wardens' reports indicated increasing pronghorn populations.

Pronghorn, never numerous in Minnesota, apparently were extirpated there by 1900 (Swanson 1940). Pronghorn received protection in Kansas from 1903 to 1973, except for the period between 1921 and 1924, according to Funk (1988).

The Canadian provinces fared no better than the states in conserving pronghorn. In Manitoba, subsistence hunting eliminated them by 1900 (Cadieux 1986). Concerning pronghorn in Alberta and Saskatchewan during the second decade of the 20th century, Hewitt (1921:72) wrote: "It is the same story of extermination following the advent of man armed with rifles and the extension of agriculture. The settlement of the country and the construction of railroads have also introduced a new factor, namely the wire fence, that has had a very marked effect in confining the remaining herds to restricted areas and thus preventing their normal migration."

Ernest Thompson Seton, a Canadian naturalist, artist and author of more than a dozen books on wildlife between 1896 and 1925, produced the first extensive scientific paper on pronghorn—*Life Histories of Northern Animals* (Seton 1909). The 37-page chapter, plus five plates, on pronghorn was surprisingly accurate. Seton's estimates of "pristine" pronghorn numbers generally were used throughout the 20th century. A later four-volume set of books (Seton 1925) were widely used as textbooks in wildlife and zoology courses for some 30 to 40 years. Volume Three of that set contained a 56-page chapter on pronghorn.

One of the major problems confronting the Federal Parks Branch after 1912 was the declining number of pronghorn on the Canadian prairies. A disastrously hard winter in 1906–07 had reduced their numbers greatly and pronghorn extirpation in Canada seemed almost inevitable. Their numbers declined further with the loss of rangeland and indiscriminate shooting by Indians and whites alike (Foster 1978). Until seasons were finally closed, Hornaday (1913) chastised Alberta and Saskatchewan for allowing the shooting of two pronghorn per year. Only about 1,000 pronghorn remained in Alberta after the severe winter, and conditions worsened as more and more rangelands were taken over by homesteaders (Cadieux 1986).

Nemiskam National Park, 5,440 acres (2,202 ha) under fence in southeastern Alberta, was formed in 1922 by the simple expedient of building a fence around a herd (Gabrielson 1943). The 42 pronghorn in the park, when fenced, increased to 235 by 1935. The park was closed in 1947 because pronghorn populations had reached satisfactory levels outside of it (Banfield 1974).

Mexico also was losing pronghorn. According to E. W. Nelson, Director of the U.S. Bureau of Biological Survey, pronghorn formerly ranged over nearly the entire length of Baja California. By 1912, they were gone from a large part of their ancient rangelands, and their steadily decreasing numbers indicated their imminent extinction (Hornaday 1913).

The U.S. government took heed of farmers' and ranchers' requests for assistance to reduce predation on livestock and assigned responsibility for predator control to the Bureau of Biological Survey in 1915. This somewhat clouded the assertion of authority for management of resident wildlife, including pronghorn, at that time (Almand and Zagata 1979).

During the early 20th century, William Temple Hornaday, a well-to-do and well-connected outdoorsman, wrote numerous books on taxidermy, natural history, the desert, Canadian Rockies, the founding of Montana, war on wildlife, etc. He was an outspoken campaigner for reducing bag limits, closing seasons, and so on, at a time when such measures were needed. His voice was influential in the closing of pronghorn seasons in some states.

An antihunting movement existed even in early 20th century North America. By the 1920s, Hornaday was director of the New York Zoological Society and had become the standard bearer of all the antihunting, preservationist and sentimentalist forces in North America. Through the *Peoples Home Journal*—one of the more widely circulated family magazines, but one scarcely qualified to comment impartially on problems of wildlife management. Hornaday started an intensive and rather successful campaign for closing hunting seasons on common game. He offered prizes to farm children who were responsible for posting land against hunting. He had support for his views in the editorial pages of some of the leading newspapers of the time, whose editors and publishers knew little and cared less about hunting. In his younger years, Hornaday himself had been an ardent hunter (Trefethen 1961). He particularly abhorred pronghorn hunting because he was sure the species was on the brink of extinction (Hornaday 1913).

Beginning with California in 1883, all Canadian and United States jurisdictions to which pronghorn were in-

(Top) *Hunters, thought to be near Gleichen, Alberta, bring in the game.* (Bottom) *Seemingly, gas-powered buggies quickly replaced horse-drawn ones for pronghorn hunting. These photos were not dated, but this one, taken near Brooks, Alberta, shows a 1912 Franklin air-cooled automobile. The season had been closed to all pronghorn hunting in 1906. Either some Canadian hunters of that era were about as law-abiding as their counterparts south of the border, or the air-cooled vehicle was still in use when the season reopened in 1935.* Top photo by of Hugh Forester. Bottom photo by F. R. Pike. Both photos courtesy of the Glenbrow Archives, Calgary, Alberta.

Table 2. Years when provincial and state hunting seasons for pronghorn were closed and reopened, as reported in various publications

Province or state	*Season closed*	*Season reopened*	*References*
Alberta	1906	1935	Popowski (1959)
Arizona	1905	1941	Brown (1994a, 1994b)
California	1883	1942	Popowski (1959)
Colorado	1899	1945	Cadieux (1986)
Idaho	?	1934	
Kansas	1903	1920	Funk (1988)
Mexico	1922	Continues	Cancino et al. (2000)
Montana	1903	1934	Mussehl and Howell (1971)
Nebraska	1907	1953	Popowski (1959)
Nevada	1909	1921	Lantz (1910), Popowski (1959)
New Mexico	?	1932	
North Dakota	1899	1951	Cadieux (1986), Popowski (1959)
Oklahoma	1914	1966	Cadieux (1986)
Oregon	1913	1938	Popowski (1959)
Saskatchewan	1914	1940	Popowski (1959)
South Dakota	1911	1936	Popowski (1959)
Texas	1903	1944	Popowski (1959)
Utah	1908	1945	Popowski (1959)
Wyoming	1908	1927	Greenquist (1983)

When two references are given, the first reports closing date, and the second indicates reopening.

digenous outlawed pronghorn hunting by 1914 (Table 2); seasons were not closed in Mexico until 1922 (Cancino et al. 2000). The measures seemed too late because remnant bands continued to decline (Popowski 1959).

None of the proclamations related to pronghorn in Arizona during the late 1800s had the desired effect. Thus, in 1905, the season was closed. However, as in other states, game laws were ignored and pronghorn hunters often were game hogs—"flock shooting" running animals, killing or wounding more does and fawns than bucks. Despite game laws, pronghorn shooting was a popular and unregulated pastime. During the second decade of the 20th century, "sports" were riding motorcycles and automobiles to pursue pronghorn.

World War I provided a market for every cow, sheep and goat raised in America. Arizona rangelands, already suffering from overuse, were stocked at ever higher levels. When the United States entered the war in 1917, what few grazing regulations were in effect on public lands were cancelled. Killing pronghorn and other big game that competed with livestock was seen as almost a patriotic duty. Beef and mutton were bringing high prices, and those commodities were touted as just as important for the war effort as ammunition (Brown 1994a).

Wildlife problems at the beginning of the 20th century were primarily those of scarcity. With protection, segments of some species expanded so rapidly that they threatened their own environments. Yellowstone elk and Kaibab mule deer are well-known examples (Trefethen 1961). That factors other than hunting influenced pronghorn numbers was well illustrated with the herd in Yellowstone National Park. In 1908, approximately 2,000 head roamed the park. By 1912—after 25 years of protection—the herd numbered only 500 (Hornaday 1913).

By the 1920s, approximately 15 million cattle joined the hoards of sheep already devouring forage in the 11 western states. Declines in condition of the semiarid western rangelands indicated they were overstocked, and forage was used beyond its sustainable capabilities (Wagner 1978, Donahue 1999).

Hornaday (1913) reported that lumpy jaw, or actinomycosis (see Chapter 10) was common among pronghorn. He predicted that the disease was destined to play an important role in the extinction of the species. He also predicted that pronghorn would be one of the first species of North American big game to become extinct. The high incidence of lumpy jaw probably reflected unhealthy rangelands because of overuse by livestock.

After World War I, returning soldiers were encouraged to homestead the few remaining public lands with water and forage. In much pronghorn habitat, the peak period of homesteading was during the 1920s. To make matters worse, a postwar depression struck the West in 1919, exacerbating already depleted rangelands and encouraging the poaching of pronghorn (Lewis 1996).

Around 1920, homesteading and sodbusting reached a peak, and pronghorn reached their lowest ebb. A series of wet years had stimulated homesteading to such an extent that little space and tolerance were left for the animals, and it seemed as if the species might be on an irrevocable

As early as 1915, one Wyoming warden's report indicated that pronghorn were overpopulating some local spots and recommended an open season in those areas. This photo of pronghorn in a Wyoming alfalfa field during 1915 illustrates the problem. Irrigated alfalfa provided the most succulent nutritious forbs available in arid landscapes. Also, the diurnal, piebald pronghorn were more visible than deer that also were attracted to such fields. Pronghorn hunting was permitted in 1927, closed in 1928 and reopened in 1929. More than 6,000 (of approximately 26,000 thought to roam the state) were killed that year, and the public reaction was markedly unfavorable. The season was closed again, but crop depredations became a problem once more and hunting was permitted in again 1933 and for the rest of the 20th century (Greenquist 1983). Photo by V. Bailey; courtesy of the U.S. Biological Survey, Washington, D.C.

decline (Gabrielson 1941). By 1921, many conservationists despaired of saving the pronghorn. The American Bison Society estimated that only 11,749 pronghorn remained in Canada and the United States. Even the most dedicated sportsmen believed that, if pronghorn could be saved from extinction, they would always be confined to a few remote desert regions and special refuges. Protective legislation failed and efforts to secure suitable pronghorn refuges were generally unsuccessful. Also, the few fledgling attempts to translocate pronghorn to national parks and other potential sanctuaries failed (Brown 1994a).

When pronghorn numbers reached a low about 1920, a national conference was called to develop a plan for recovery, then postponed at the suggestion of E. W. Nelson, chief of the U.S. Biological Survey. A survey from 1921 through 1923 indicated that, conservatively, about 30,000 pronghorn remained—more than had been feared—and most were in Wyoming, Montana and northern Nevada. Only a handful were reported from other former strongholds, such as Kansas and Nebraska. The long-awaited conference was held during December 1923. Most of the major conservationists of the day plus many state game wardens attended. Pronghorn already were protected throughout North America; the problem was enforcement. The participants decided that creation of pronghorn refuges and a program to restock former habitat was needed. Most opposition to refuges was removed by a decision that any proposed pronghorn refuge needed only to be closed to hunting and not to livestock grazing. Arizona alone established 76 (Brown 1994b).

What the 1923 conference participants did not know was that pronghorn already were increasing. Livestock grazing declined after World War I, and homesteaders began moving off their small ranches. Coyote control by federal and state agents may have had a positive effect, and pronghorn may have been learning to cope with fences and other human-created impositions (Brown 1994b).

The Arizona Game Protective Association was founded in 1923. It adopted a rather stylized pronghorn buck as its logo (a pronghorn in various guises remained the association's herald until the 1950s, when it was replaced by a desert bighorn) (Brown 1994b).

During the early days of pronghorn restoration, it appeared that coyotes might be preventing the buildup of herds where pronghorn had been drastically reduced (Allen 1962). Such may have been the case, and extensive predator control—mostly for the benefit of wool growers—may have speeded pronghorn recovery.

*Restocking historic pronghorn rangelands proved difficult. The skittish animals did poorly in confinement, and attempts to trap and hold adults invariably maimed or killed them. Therefore, early restocking attempts concentrated on capturing newborn fawns and raising them to adulthood. However, such animals proved vulnerable to predators and accidents, and they also died in large numbers during inclement weather (Brown 1994b). (*Left*) Fawns were captured on the Washoe County State Game Refuge, Nevada, in 1924. They were translocated to a private ranch. (*Right*) Fawns were raised at Sully's Hill Game Preserve, North Dakota, in 1927.* Left photo by S. Riley. Right photo by Weseley D. Parker. Both photos courtesy of the U.S. National Archives, Washington, D.C.

Gabrielson (1941) described a long-term predator control program in the area where California, Nevada and Oregon join. In 1920, a small population of pronghorn persisted there. It was protected by state game laws, and little evidence of poaching was detected. However, they had experienced poor fawn survival and population numbers had not increased for several years. Between January 1, 1921 and June 30, 1934, 7,595 coyotes and bobcats were removed from the area by the Biological Survey. The pronghorn population increased from about 500 specimens to more than 7,000. Gabrielson concluded that pronghorn were numerous enough for the same or greater predator pressure to be of less importance, and other factors might become more serious. Apparently, no attempt was made to compare these increases with those during the same years in similar habitat without predator control.

Gabrielson's (1941) account caused Allen (1962:261) to conclude: "As in this example, predator control is practical as a temporary, pinch-hit measure for the protection of a breeding stock on an adequate range from which a game animal has been nearly or wholly eliminated. This would be justifiable even at high cost if the final result was the establishment of a productive population" (see Chapter 11).

In 1928, the death of Charles Sheldon—adventurer and hunter—came as a shock to fellow members of the Boone and Crockett Club. E. R. Sans, a predator control worker in the Bureau of Biological Survey, was interested in the future of a hard-pressed little band of pronghorn eking out a sparse existence around a few springs in a great expanse of semidesert in northwestern Nevada. The idea of a refuge there, commemorating Sheldon, was presented to the club, which agreed to raise $10,000—half the price of the private land—for use as a sanctuary. The other half of the needed land was to be donated by the Audubon Society. Learning of this, President Hoover made additional allowance for 30,000 acres (12,140 ha) of adjacent public land and signed an executive order creating the 34,325-acre (13,890 ha) Charles Sheldon National Antelope Range (Trefethen 1961).

Pronghorn may have influenced their own survival in ways seldom considered. Early in the 20th century, Theodore Roosevelt (1969:136) maintained that: "In some places the [pronghorn] now seem to have learned wisdom, for the slaughter among them has been so prodigious that the survivors have radically changed their character. Their senses are as keen as ever, and their wits much keener. They no longer give way to bursts of panic curiosity; they cannot be attracted by any amount of flagging, or by the appearance of unknown objects, as formerly." Photo by Peter T. Bromley.

Leopold (1918:406) wrote: "Witness game refuges, game farming, and countless new departures in game laws. The time has come for *science* to take the floor, *prepared to cope with the situation*." He expounded a theory that became the foundation of 20th century scientific wildlife management. In essence, he believed that quality and quantity of habitat governed the number of wild animals that a given area could support, and each wild population produced a surplus beyond that needed to maintain breeding stock. The surplus can be removed by hunters without endangering the species. If it is not removed by hunters, it will die of other causes, some of which may endanger the breeding stock more seriously than regulated hunting (Trefethen 1966). Leopold also proposed formation of a new discipline—game management. It received impetus from *Game Survey of the North Central States* (Leopold 1931).

Provincial and state wildlife biologists in Canada and the United States heeded Leopold's teaching and, after pronghorn populations began to build, seasons were reopened (see Chapter 20). This started a flow of money to pay for law enforcement, research, translocations and other management practices of benefit to pronghorn and other species.

Sportsmen aided pronghorn recovery efforts by actively opposing poaching, because scofflaws stole from the pool of animals that others could hunt legally and, without the local goodwill and revenue generated by hunting, pronghorn populations surely would not have increased as they did.

When pronghorn populations were at their lowest numbers during the 1920s, those in Mexico were comparatively numerous (Nelson 1925). All pronghorn hunting was outlawed in Mexico during 1922, and the season remained closed throughout the rest of the 20th century (Cancino et al. 2000). Without the prospect of a legal opportunity to hunt pronghorn, poaching has been a problem. Without income from hunting licenses, few wardens were available to curb that poaching; Mexico had other problems. Pronghorn mostly occupied arid and easily overgrazed rangelands and rural poverty prevailed.

As big game populations increased, hunters sometimes objected to reopening seasons and, especially, to harvesting females (Trefethen 1966). Such objections were shared by some state wildlife agencies (Catton and Mighetto 1998). In response to Wyoming's 1929 pronghorn season, *The New York Times* of October 26, 1929, published the following: "Accounts have just reached New York of what is described in Wyoming newspapers as the 'most senseless and atrocious' slaughter of wild animals that has ever occurred in that state or perhaps in the nation. The only thing approaching it is the wholesale destruction of buffalos a half century ago. For a period of six days, early this month, the protection of antelope which had been in force

When pronghorn seasons reopened during the late 1920s, automobiles had become the principal means of transportation for hunters, as this October 1929 photo indicates. It was taken on the Phelps Ranch in Wyoming. Photo courtesy of the Wildlife Management Institute, Washington, D.C.

for years was removed, with the result that the animals were butchered by the hundreds. Medicine Bow alone reported more than a thousand carcasses carried through that town in the automobiles of the hunters. Cody reported upward of 200 killed in that neighborhood on a single day. If the reports in the *Cody Enterprise* and the *Wyoming State Tribune* and *Cheyenne State Leader* are unbiased, the affair was a disgrace alike to the State of Wyoming and to the self-styled hunters who took advantage of the blunder of the Wyoming State Game Commission in opening the season on antelope" (Quinn 1930:2–3). Wrote Clark (1928:221): "Wyoming is engaged in a brutal and ridiculous campaign of extermination . . . the State is selling licenses for pronghorned antelope . . . we may wake up before long to a realization of the fact that the pronghorn is extinct."

"The attitudes of the American public toward wildlife and wilderness shifted dramatically during the early 1900s with changing immigration patterns and an exodus of rural dwellers to the city. For the first time in North American history, a majority were liberated from largely utilitarian dependence on wildlife and came to appreciate nature from a more remote and distant setting" (Kellert and Smith 2000:49).

1931 TO 1960

America cannot support 200 million people in prosperity and still have 200 million buffalo. But we can have millions of deer, wild turkey, antelope, . . . and other species that can thrive in an environment molded by human beings. . . . America's wildlife biologists and administrators have the knowledge that is needed to accomplish these results and much of the necessary legal authority. All they need are the understanding and support of the general public.

Trefethen (1966:36–37)

In 1931, Congress passed the Animal Damage Control (ADC) Act to destroy animals injurious to agriculture,

Lands acquired by the Boone and Crockett Club for a pronghorn refuge in Nevada were turned over to the U.S. Bureau of Biological Survey and, in January 1931, President Herbert Hoover signed the executive order creating the Charles Sheldon National Antelope Range, totaling 34,325 acres (13,891 ha) (Trefethen 1961). In August 1932, the Audubon Society transferred its Nevada holdings to the federal government, but its administration by the Bureau of Biological Survey did not begin until 1933. Water developments, primarily for cattle, which continued to graze on the Sheldon for most of the 20th century, were used extensively by pronghorn. Photo courtesy of the U.S. Bureau of Biological Survey, Washington, D.C.

forestry, game animals, livestock, etc. The ADC trappers worked on both public and private land, establishing a close working relationship with stockmen that lasted throughout the 20th century (Donahue 1999).

By the 1930s, sport hunting associations put pressure on the U.S. Forest Service to reevaluate livestock grazing in many areas (Catton and Mighetto 1998). This was the beginning of public involvement in the use of public lands—involvement that would accelerate during the rest of the 20th century.

In the first wildlife textbook, *Game Management*, Leopold (1933:20–21) wrote, "The hope of the future lies not in curbing the influence of human occupancy—it is already too late for that—but in creating a better understanding of the extent of that influence and a new ethic for its governance." This book was to have a tremendous impact on perceptions of ecology as a science and land ethics as an imperative (Meine 1988). Also in 1933, Leopold inaugurated game (wildlife) management in the University of Wisconsin's Department of Agricultural Economics (McCabe 1987, Meine 1988). He soon had his own department at the university.

The worst drought in American history struck the prairies during the 1930s. The dirty '30s saw summers filled with searing winds all over the Great Plains. At times, it was unsafe to drive at high noon. Dust drifted in around windows and doors of Plains homes, and drifting soil buried buildings. A farmer producing a little grain made practically nothing from it, for this also was the Great Depression.

As dust clouds darkened much of the continent, many wild populations were threatened. Because of the national economic depression, the U.S. Congress was reluctant to provide wildlife mitigation. Income tax revenues dropped, and general funds could not be spared for what many regarded as frills. However, Congress came to understand that hunters could pay for wildlife restoration and most were willing to do so (Trefethen 1966).

Conversion of rangeland to cropland halted when homesteaders by the thousands, from northern Texas to Canada, were forced to abandon marginal lands. This allowed large areas to revert to native vegetation (Cadieux 1986). Droughts generally reduce pronghorn numbers and impair their habitat. This one, in the long run, benefited them by clearing settlers from the land, allowing the animals to reclaim some of their former rangelands.

"The public domain, administered by the Department of Interior, was not subject to effective grazing control until 1934 when the Taylor Grazing Act was passed and the Grazing Service was established. The act was intended to protect public lands from overgrazing and soil erosion by regulating their use and occupancy. The Grazing Ser-

During the 1970s, remains of homesteads abandoned 30 to 50 years earlier were evident in many areas, such as this one (top) *in the Missouri River breaks of eastern Montana. Ten or more homestead sites were to be found on many western ranches at the beginning of the 21st century, indicating the greater rural population—much of it feeding on pronghorn—a century earlier. Drought, windstorms and year-after-year crop failures during the dirty thirties cost many families their livelihoods and homes* (bottom)*. A change in weather and farming practices eventually prompted a recovery. Droughts ordinarily depress wildlife populations, but drought during the 1930s ultimately benefited pronghorn by driving homesteaders from the land.* Top photo courtesy of the Montana Department of Fish, Wildlife and Parks, Helena. Bottom photo (eastern Colorado) courtesy of the U.S. Soil Conservation Service, Washington, D.C.

vice was merged with the General Land Office in 1946 to form the Bureau of Land Management" (Wilson 1966:140). The BLM lands would support more pronghorn than that of any other land management agency. Only private lands harbor more of the animals. Thus, management policies on BLM lands (public domain) are extremely important to the species.

The Taylor Act states the following purposes: to stop injury to public lands by preventing overgrazing and soil deterioration; to provide for their orderly use, improvement and development; and to stabilize the livestock industry dependent on public lands (Donahue 1999).

The chief of the Bureau of Biological Survey, appointed under the administration of Franklin D. Roosevelt, was a sportsman, newspaper reporter and Pulitzer Prize winning editorial cartoonist from Iowa, Jay N. "Ding" Darling (Lendt 1989). In 1935, Darling launched the Cooperative Wildlife Research Unit program—a system of training and research stations for advanced students in wildlife management at various colleges. The units were financed and administered jointly by the Bureau of Biological Survey, the respective state wildlife agency and college, plus the American Wildlife Institute, forerunner of the Wildlife Management Institute. The units conducted wildlife research while training graduate students to enter professional service (Trefethen 1966). Much of the research findings reported in this book were gathered through those units.

In December 1936, President Roosevelt issued an executive order creating a 549,000-acre (222,180 ha) game range adjacent to the Sheldon refuge, and E. R. Sans was appointed the first refuge manager. At the same time, Roosevelt established the Hart Mountain National Antelope Range (Hart Mountain) as a range and breeding ground for pronghorn and other wildlife. The range was established through the interest of the Oregon State Game Commission and Ira N. Gabrielson, Darling's successor as chief of the Bureau of Biological Survey. The Sheldon and Hart Mountain ranges (now refuges) complemented one another as homes to the tri-state (California, Nevada and Oregon) pronghorn herd. Hart Mountain ordinarily provides summer rangelands and Sheldon serves as the principal wintering ground (Trefethen 1961).

During the 1920s and 1930s, public sentiment favored establishment of wildlife refuges, and Darling, among other dynamic conservationists, pushed hard for funding acquisition of at least one refuge in the natural habitat of each big game species. "A satisfactory big-game refuge should be large enough to provide adequate year-around food and cover and to allow scope for the normal movements of the species to be protected. It must provide sufficient forage and water for a herd large enough to insure perpetuation of the species" (Gabrielson 1943:35). The Sheldon/Hart Mountain complex met Gabrielson's criteria for a pronghorn refuge.

Darling was responsible for President Roosevelt's calling of the First North American Wildlife Conference, a broadened version of the old American Game Conference and precursor of the continuing North American Wildlife and Natural Resources Conference. The inaugural North American Wildlife Conference, held in 1936, convened administrators and sportsmen to discuss new approaches to wildlife conservation. Darling, other leading conservationists and the American Wildlife Institute also established the National Wildlife Federation at that time, with the intention of unifying and coordinating the activities of local and state conservation groups (Trefethen 1966).

A biological survey of the area now known as the Charles M. Russell National Wildlife Refuge (CMR) was conducted in 1935 by famed naturalist Olaus Murie, reporting on the proposed Fort Peck Migratory Bird Refuge, identifying existing wildlife populations and making recommendations for management in the area if it was designated a refuge. Nearly 18 months later, President Franklin D. Roosevelt signed Executive Order 7509, establishing the Fort Peck Game Range. That order established the primary purpose for the range: the natural forage resources therein would be utilized first for the purpose of sustaining a maximum of 400,000 sharptail grouse and 1,500 pronghorn, the primary species, and such nonpredatory secondary species as might be necessary to maintain a balanced wildlife population. Furthermore, it provided that all forage resources within the range would be available, except as otherwise provided with respect to wildlife, for domestic livestock (U.S. Fish and Wildlife Service 1985).

One of the most far-reaching pieces of wildlife legislation was the Federal Aid in Wildlife Restoration Act of 1937, better known as the Pittman-Robertson, or P-R, Act, after its legislative sponsors (Kallman 1987). It allocated to the states all revenues collected under an 11% excise tax on sporting arms and ammunition. Each state received its annual apportionment under a formula based on the number of licensed hunters and the area of the state. These monies could be used for wildlife research, land acquisition and development. The Pittman-Robertson funds paid 75% of the total cost of such projects. An important clause in this law required the states to apply all of their hunting license revenues to the operation of the wildlife agency and its programs. Before 1937, state legislatures commonly diverted those revenues to highway and school programs (Trefethen 1966).

"The taking over of the semiarid West for cattle raising and agriculture resulted in the elimination of the pronghorn. . . . When land use became more stabilized, it appeared that there were large areas that could be producing pronghorns without compromising other interests. Accordingly, the animals were systematically trapped and moved to new territory. . . . Research indicates that the pronghorn feeds much more on weeds and sage [sagebrush] than upon grass and only occasionally competes to any significant extent with range livestock" (Allen 1962:181).

The new science of wildlife biology focused on the enhanced welfare of selected species (usually ones that were hunted or trapped). The Wildlife Society, an international organization oriented toward professionals in wildlife management and research, was founded in 1937. One of its primary missions was to disseminate results of scientific studies and management experiences on wildlife populations and their habitats. To that end, the Society started publishing the quarterly *Journal of Wildlife Management* (*Journal*) almost immediately (Thomas and Pletscher 2000). The *Journal* only published short papers, so wildlife monographs were authorized, and the first of these was published in 1958. The monographs were not financed by The Wildlife Society's members' dues, but by sponsors—usually the organizations that conducted the research. By the 1960s, some wildlife managers perceived the *Journal* as too research-oriented. In 1972, The Wildlife Society began publishing *The Wildlife Society Bulletin* to address better the needs of managers.

The Cabeza Prieta Game Range (about 860,040 acres or 348,058 ha) was established by executive order in 1939, to be jointly administered by the Grazing Service and the U.S. Fish and Wildlife Service under the conditions of the Taylor Grazing Act (Gabrielson 1943). Initially, protection of a remnant herd of bighorn was the principle function of the range, but it later became the main hope for preventing extirpation of the Sonoran pronghorn in the United States. In 1978, the classification of game range was changed to wildlife refuge.

Canada developed an almost similar number and total acreage of national parks and refuges, as did the United

Pronghorn recovery was fortuitous in 1937. During that year, the Federal Aid in Wildlife Restoration (Pittman-Robertson) Act was passed, and the New Mexico Department of Game and Fish developed a corral system that was capable of capturing numerous adult pronghorn with acceptable levels of injuries and mortality (Russell 1964). This system facilitated the translocation of more than 30,000 pronghorn during the 20th century. Thus, during one year, funding and the means for moving large numbers of pronghorn were realized. These circumstances greatly accelerated the return of pronghorn to many rangelands where they had been extirpated and likely could not have reinhabited otherwise. Photo by Jeff Pederson; courtesy of the New Mexico Department of Game and Fish, Santa Fe.

States. All Canadian parks protect wildlife at all times, although few Canadian parks or refuges have pronghorn (see Gabrielson 1943).

During his four terms as president, from 1933 to 1945, Franklin D. Roosevelt provided a strong executive leadership in conservation not seen since the administration of his older cousin, Theodore Roosevelt. At the same time, Congress demonstrated unprecedented support for conservation initiatives. Outstanding measures included the Wildlife Coordination Act of 1934—which directed federal land management agencies to take remedial action for fish and wildlife wherever water impoundments or pollution affected habitat—and the Pittman-Robertson Act of 1937 (Catton and Mighetto 1998). In 1940, the U.S. Fish and Wildlife Service was established in the Department of Interior by consolidating the former Bureau of Biological Survey from the Department of Agriculture and the Bureau of Fisheries from the Department of Commerce (Trefethen 1966). The Fish and Wildlife Service would become second only to the BLM in importance to pronghorn conservation.

By midcentury, provinces in Canada and states in the United States had well-staffed administrative agencies entrusted with protecting and managing game animals and regulating hunting within their borders. Wildlife management had attained professional status with training through graduate programs at many leading universities. For the greater part, sensible, flexible and well-enforced laws protected game (Trefethen 1961). Arthur S. Einarsen, leader of the Oregon Cooperative Wildlife Research Unit, published *The Pronghorn Antelope* in 1948. This was the first scientific monograph concerning the species. The restoration of pronghorn was one of the more impressive accomplishments of the wildlife conservation movement and pronghorn themselves (see Chapter 4).

In the United States, state agencies were supported primarily by hunter's fees, and most of the authority for wildlife decisions was delegated entirely or partially to nonsalaried, politically appointed commissions or, in a few cases, to single commissioners serving in the executive branch of the state government. Resource management decisions often must be made contrary to public opinion. When popularity and wise resource use are in conflict, the resource must come first. The role of representing the public and, at the same time, making wise policy decisions for a resource management agency is uniquely a commissioner's task. No better system than the commission form of wildlife resource government has been devised (Toweill 1979). The USDA Forest Service and Soil Conservation Service, through their own staffs of wildlife specialists, maintained close liaison with state wildlife officials. Nearly all projects that benefited game animals also benefited nongame species (Trefethen 1966).

Canadian approaches to wildlife management, research and education differed significantly from those in the United States during the 20th century for four major reasons.

1. The ratio of people to resources was remarkably low in Canada, which influenced professional organization and management flexibility.
2. Large expanses of Canada's land were publicly owned; government policies had widespread impacts.
3. Most responsibility for resource management was assigned by the British North American Act of 1867 to provincial governments. Therefore, government policy and impacts across the country are highly variable.
4. Canadian universities had no clear mandate for extension to resource users and managers; wildlife research in Canada tended to be more curiosity-driven than applied.

"Combined, these features have produced an approach to shepherding wildlife resources that is distinctly Canadian" (Bunnell et al. 1995:138). The Canadian approaches to wildlife management, research and education facilited transition to the new discipline of conservation biology.

"In 1960 the Multiple Use-Sustained Yield Act directed that the national forests should be managed for the production and use of all their goods and services—wood, water, forage, wildlife, and recreation" (Meyer 1966:53). The BLM administered the public domain lands under policies generally similar to those applied by the U.S. Forest Service to the national forests (Meyer 1966). Thus, pronghorn would be affected by such legislation. The act supplemented the purposes given in the Organic Act of 1897—protection of watersheds and the national timber supply. The new act seemed to signal greater emphasis on recreation, wildlife and fish in national forest management (Catton and Mighetto 1998).

By 1960, about 56 million acres (22.7 million ha) were owned by or dedicated to use by Native Americans. Military reservations in the United States totaled some 32 million acres (13 million ha). The Department of Defense adopted conservation practices consistent with military use and sought the aid of other federal agencies to develop management plans. It has added wildlife managers to its civilian staff (Meyer 1966). Indian reservations and military posts in the western states supported substantial numbers of pronghorn.

Paul Errington (1945, 1946, 1956, 1967), through studies of game birds, furbearers and rodents, reinforced Leopold's (1925, 1933) contentions that habitat governed the amount and kind of game an area could support, and that surpluses could be harvested without endangering the breeding stock. Errington's provocative ideas concerning carrying capacity, predator/prey relationships and inversity formed a broad foundation of ecological theory that still influences management practices, such as hunting seasons and bag limits, in the early 21st century.

By the mid-20th century, the Wyoming Game and Fish Commission's policies had shifted from protecting pronghorn to controlling the ever-increasing numbers. Agency personnel spent years learning pronghorn management methods. Census techniques were improved and altered, hunter harvests were monitored, and management of hunter harvests became the primary focus of pronghorn management in Wyoming, as in other states with substantial pronghorn populations (Greenquist 1983).

Wildlife management became a mixture of public and private resource management. Wildlife is a public resource, owned by all citizens, yet the habitat upon which wildlife depends is not always publicly owned, and wildlife often moves freely between public and privately owned lands. Likewise, the public has the right to use wildlife for recreational purposes, but they may not be able to exercise that right because the wildlife may be on privately owned lands (Mangun 1992). As Lindzey and Wingard (1979:17) expressed: "Wildlife is unique among natural resources. The public owns it, government administers it, private landowners help produce it, business benefits from it, recreation-minded people use it, and many misunderstand its place, uses, and values."

1961 TO 2000

As regards resource conservation, the record of the twentieth century contrasts happily with that of the nineteenth. When we look at the debit side of the record, we see the careless, wasteful blundering exploitation of our resources as the penalty paid by a new nation to settle a raw continent. On the credit side is the gradual development of an American conservation policy, not yet perfectly realized to be sure, but continually promoted by the strenuous and idealistic action of an aroused public conscience.

Clepper (1966:14)

By 1961, hunting became management for most game species in Canada and the United States. Science had become involved with wildlife studies and wildlife came to be managed by provincial, state and federal governments. Populations were regulated by sport hunting, and hunters supported protection, research and habitat acquisition or improvement through license fees and taxes on sporting goods. Sport hunting, as used in this book, refers to legal, recreational harvesting of animals on a sustained-yield basis—whether mainly for meat, recreation, trophies or other values—as opposed to commercial or subsistence hunting.

During the 1960s, public sentiment began to shift because of *Silent Spring* (Carson 1962), the National Environmental Policy Act, the Endangered Species Act and changing public opinion concerning environmental health and aesthetic values associated with outdoor experiences. The latter included strong sentiments concerning the morality of hunting and trapping (Bolen 1989).

A lesson learned early in the evolution of wildlife management in North America was that any conservation plan should take the aspirations of local people into account; the survival of wildlife and their habitats ultimately depends on those people. Local interests bear the burden of game costs, such as support of law enforcement and expenses of wildlife damage (Lund 1995). Protecting wildlife is difficult if local people do not have a stake in the welfare of that wildlife. In the case of pronghorn, sport hunting and its benefits to local peoples provided that stake.

"For a long time leaders in resource enterprises, public and private, have been searching for a reconciliation of two approaches: resource exploitation and resource conservation. Many have advocated *wise use* of resources as the goal to be sought. The problem is to figure out what this means in specific situations" (Fisher 1966:170). In the case of game animals, the foregoing description of sustained-yield sport hunting appears to qualify as wise use.

Congress took action to fulfill the land classification provisions of the Taylor Grazing Act with passage of the Classification and Multiple Use Act of 1964. This act directed the Secretary of the Interior to develop and promulgate criteria to determine which of the public lands exclusively administered by the BLM should be disposed of for certain purposes or retained and managed for certain purposes. During the interim, BLM lands were to be managed under the principle of multiple use as defined in the Multiple Use–Sustained Yield Act of 1960 (Wilson 1966). Fences on BLM lands generally were modified to allow pronghorn passage (see Chapter 17), and water developments provided a necessity of life that was scarce on many parts of western plains. By the late 1970s, about 45% of all pronghorn were on the 474 million acres (192 million ha) managed by the BLM. Those lands constituted about 50% of the land area in 11 western states. The BLM concentrated on maintaining existing productive habitats and restoring to satisfactory conditions the 83% of the 150 million acres (61 million ha) of grazing lands considered to be in poor condition. In fiscal year 1976, slightly more than a $0.01 per acre ($0.004/ha) was available for wildlife habitat management and research. Management of wildlife habitats and populations in cooperation with state agencies and users of BLM lands began in the 1960s (Jahn and Trefethen 1978).

Congress enacted the Federal Land Policy and Management Act (FLPMA), generally known as the Organic

Act, in October 1976. The act amended certain provisions of the Taylor Grazing Act and repealed the homestead laws, with an exception for Alaska. The Taylor Act remained mostly intact, and grazing on BLM lands today is subject to both FLPMA and Taylor Grazing Act regulations. Grazing permits on BLM lands and national forests are privileges, not compensable property interests. The FLPMA begins with Congress's unequivocal policy declaration that the public lands be retained in federal ownership, unless as a result of the land-use planning procedure provided in the act, it is determined that disposal of a particular parcel will serve the national interest (Donahue 1999).

The Public Land Law Review Commission (PLLRC), assigned the specific task of determining how public lands could best be managed to achieve the maximum benefit for the general public, was critical of the Taylor Grazing Act. Its 1970 report, *One Third of the Nation's Lands*, covered all federal lands, but the report's greatest significance was for the BLM. The PLLRC recommended that forage be allocated by statute to game and nongame wildlife. It also implied that, when livestock grazing needs and wildlife management concerns collided, the wildlife objectives should take precedence (Donahue 1999). Such a recommendation has great significance for the welfare of pronghorn.

The Public Rangelands Improvement Act (PRIA) of 1978 showed Congress's awareness of the ongoing degradation of western rangelands. PRIA urged that livestock grazing on certain BLM lands should be temporarily or permanently discontinued to improve rangeland conditions. The PRIA thereby strengthened the case for a biodiversity conservation plan on arid lands unsuited to livestock grazing (Donahue 1999). If followed, this should favor pronghorn survival in arid habitats.

In 1963, the Fort Peck Game Range was redesignated the Charles M. Russell National Wildlife Range (CMR) in honor of Montana's great cowboy artist. The BLM, however, continued to administer the livestock grazing program, subject to the policy requirements of the U.S. Fish and Wildlife Service. In 1976, the administrative status of CMR and all other game ranges in the nation was changed by Public Law 94-223. This ended the joint management between the BLM and the U.S. Fish and Wildlife Service (Service). Public Land Order 5635 of 1978 changed the name of the game range to Charles M. Russell National Wildlife Refuge (U.S. Fish and Wildlife Service 1985). Forty years of joint management by the Service and the BLM had not worked to the advantage of wildlife. Of the total animal unit months (AUMs) available, 61% were allotted to livestock. The livestock allotment on the range as a whole worked out to 10 acres (4 ha) per AUM, twice the rate recommended by Murie in 1935. This is the stocking rate that CMR was managed under during most of the 20th century. Wildlife interests clearly were secondary to livestock, and wildlife and their habitats continued to deteriorate.

In 1966, the Endangered Species Preservation Act authorized the Secretary of the Interior to maintain a list of native endangered species. This was forerunner to the Endangered Species Conservation Act of 1969 that authorized the Secretary to generate a list of species threatened with extinction. The Endangered Species Act of 1973, signed into law by President Nixon, extended the acts of 1966 and 1969 to prohibit the taking of species listed as endangered, protected look-alikes or species threatened over only part of their range. Also in 1973, the United States ratified the Convention on International Trade in Endangered Species (CITES), which became effective in 1975 (Mackie 2000).

The Endangered Species Act imposed on the BLM, as on all federal agencies, an affirmative duty to conserve threatened and endangered species and their habitats. The act also directed agencies to ensure that any action authorized, funded or carried out by such agencies was not apt to jeopardize the existence of any threatened or endangered species. Furthermore, agencies were directed by Congress to use all methods and procedures deemed necessary to bring threatened and endangered species to a point at which the measures provided by the Endangered Species Act were no longer necessary (Donahue 1999). In the case of pronghorn, only two presumed subspecies—the Sonoran pronghorn in Arizona and Sonora plus the peninsular pronghorn in Baja California Sur—have been classified as endangered (see Chapter 4). All pronghorn in Mexico are listed as Appendix I (essentially the same as endangered) by CITES. At the end of the 20th century, it could not be said that classification under the Endangered Species Act had speeded recovery of endangered pronghorn.

Desimone (1995:85) wrote: "In the past endangered or threatened species were considered individually. The broad approach is more ecologically sound and has become the necessary approach. . . ."

The 1960s and 1970s saw numerous environmental laws passed. Those that had the potential to benefit pronghorn are listed in Table 3. Taken collectively, they reduced the grip that commodity users, such as the timber, livestock and mining industries, had on public lands. However, it takes active enforcement to make those acts serve their purpose.

Sometimes, especially if the state and federal administrations strongly support industry over environment, the people whose jobs are to enforce laws look the other way. For instance, during a Pronghorn Antelope Workshop field trip in Wyoming in 1992, author O'Gara visited

Congress was not satisfied with the history of dual administration of wildlife ranges and, through Public Law 94-223, intended the U.S. Fish and Wildlife Service (Service) to institute necessary changes to ensure that wildlife received first priority. The Service began to manage the Charles M. Russell National Wildlife Refuge under the provisions of the National Wildlife Refuge System Administration Act of 1966. Grazing on the refuge continued, but only under provisions of that act. Many areas of the refuge were being overgrazed by Bureau of Land Management-licensed livestock seeking the cool water and influence of Fort Peck Lake. It was apparent that the only long-term solution to that problem was construction of a fence on the refuge boundary, but such fencing was stopped by political pressure. Photo courtesy of the Charles M. Russell National Wildlife Refuge, Lewiston, Montana.

pronghorn winter rangelands where natural gas wells were being drilled. Drill pads and a network of roads appeared to usurp about 20% of the pronghorn winter range on both public and private lands. Personnel from the gas drilling company and the local BLM office were there to answer questions. When asked about drilling at an angle so that fewer pads and roads might be required, the company representative said mitigating damage to the habitat was cheaper and preferred by the company. In response to a question concerning mitigation for damages to pronghorn habitat on public lands, the company representative said they complied with the BLM recommendations in revegetating roadsides and drill pads. The BLM representative said they required planting three species of grasses in such areas. Despite federal laws (National Environmental Policy Act of 1969, Federal Land Policy and Management Act of 1976, and Surface Mining Control and Reclamation Act of 1977) that require restoration of disturbed public lands to natural conditions with reseeding mixtures that include native forb and shrub species. The recommendations of Plummer et al. (1968) generally are considered ideal for restoring rangelands used by pronghorn and call for 10 to 30 species of grasses, forbs and shrubs. When asked about discrepancies between local BLM requirements and the law, the BLM representative said local sheep ranchers—who ran sheep on public and private land on pronghorn winter range—were content with the grass plantings. Also, restoring rangelands to natural conditions was expensive and a great inconvenience.

Perhaps this situation will improve. Anonymous (1995) reported that standards for postmining reclamation plans had finally been passed by the Wyoming legislature after a threat by the National Wildlife Federation and others to sue Wyoming to comply with federal regulations. These new shrub standards included oversight provisions by the Wyoming Game and Fish Department and directed that postmining density, composition and distribution of shrubs would be based on site-specific evaluations of premining vegetation and wildlife use. Except where a lesser density would be justified by premining conditions, at least 20% of the eligible land would have to be restored to shrub patches supporting an average density of one shrub per 10.8 square feet ($1/m^2$), and these patches could be no less than 0.05 acre (0.02 ha) each and arranged in a mosaic that would optimize habitat interspersion and edge effect. By the late 20th century, the National Wildlife Federation and its state and regional affiliates increasingly forced compliance with federal and state laws through lawsuits or threats to sue.

Answers to questions concerning mitigation for natural gas development on public pronghorn winter range became more obscure the longer the subject was pursued. According to a Wyoming Game and Fish Depart-

TABLE 3. Federal laws and court decisions during the 1960s and 1970s, with potential to influence pronghorn and pronghorn management

Year	*Laws and court decisions*
1964	The Wilderness Act established a National Wilderness Preservation System.
1964	Congress established the Public Land Law Review Commission to suggest improvements in policies, laws and regulations affecting federal lands, their resources and their uses.
1966	Endangered Species Preservation Act authorized the Secretary of the Interior to maintain a list of native endangered species (forerunner to Endangered Species Conservation Act of 1969 that authorized the Secretary of the Interior to generate a list of species threatened with extinction).
1969	National Environmental Policy Act (NEPA) required an environmental impact statement for any federal action affecting the quality of the human environment (assumed wildlife was part of human environment) and formed the Environmental Protection Agency.
1969	Classification and Multiple Use Act mandated multiple-use management on lands administered by the BLM.
1970	The U.S. Dept. of the Interior issued a regulatory statement similar to U.S. Dept. of Agriculture W-2 of 1940 that recognized state authority over resident wildlife and encouraged cordial state-federal relations.
1971	Wild Free-Roaming Horses and Burros Act directed the secretaries of agriculture and interior to protect those animals on public lands.
1972	President Nixon signed Executive Order 11643 prohibiting use of compound 1080 (sodium monofluoroacetate) and other toxicants in federal predator control programs or on federal lands and banning interstate shipment of toxicants for use in predator control.
1973	Endangered Species Act (ESA) of 1973, extending the ESAs of 1966 and 1969, further prohibited the taking of species listed as "endangered," protected "look-alikes," or species threatened over only part of their range, and designated "critical" habitats.
1973	The United States ratified the Convention on International Trade in Endangered Species (became effective in 1975).
1974	Federal Court ruled that public hunting was an authorized use of a national wildlife refuge and that hunting was a legitimate tool for management of wildlife under state control.
1974	Forest and Rangeland Renewable Resources Planning Act called for preparation of land management plans for protection and development of national forests.
1974	Sikes Act provided for wildlife, fish and game conservation and rehabilitation programs on military and other public lands in cooperation with the states.
1976	Supreme Court ruled that, under the property clause of the Constitution, Congress exercises complete control over public lands including the power to regulate and protect wildlife on those lands.
1976	Federal Land Policy and Management Act (FLPMA) recognized state authority over fish and resident wildlife on federal lands, reserving specified limited controls to federal administrators. Mandated land use plans by the BLM.
1976	National Forest Management Act (NFMA) of 1976 stipulated that national forest management plans must comply with NEPA and that management would maintain viable populations of native vertebrates on national forests.
1978	Supreme Court affirmed a state's right to charge unequal hunting license fees for residents and nonresidents.

Source: From Mackie 2000.

ment biologist, the environmental impact statement (EIS) for drilling stipulated that mitigation would be provided, but did not specify the price or what would be done. The biologist had contacted gas drilling company personnel, but was unable to negotiate mitigations. He was told by a drilling company employee that the company had agreed to mitigation to get the EIS approved, but did not expect to do or pay for any mitigation.

When workshop participants discussed mitigation, it seemed little could be done to benefit wintering pronghorn, even if the company honored its commitment. New winter rangelands simply could not be created, the BLM would not enforce restoration of disturbed habitat and the network of roads would be almost impossible to close to the public on the flat, low vegetation area. One option seemed feasible—use mitigation money to reduce competition on the winter range by buying grazing permits on BLM land or buying private land that was being used for grazing sheep. The BLM representative said the agency could not take AUMs from livestock and transfer them to wildlife, even if the ranchers were willing to relinquish their leases. State personnel said Wyoming agriculturists would no doubt block transfer of private land to a state or federal agency even under a willing seller situation. In short, one could only conclude that pronghorn and their habitat could only be reduced and degraded by this type of resource extraction.

Another example of well-meaning laws failing to achieve the desired results came to light during the late 1980s. Congress attempted, through the 1985 Food Security Act, to control excessive erosion rates, discourage the conversion of highly erodible native prairie for agricultural commodity production and reduce crop surpluses. The Conservation Reserve Program (CRP) and Sodbuster provisions were included in the act to help meet these objectives. The CRP involves planting vegetative cover on highly erodible croplands for a 10-year period. This had multiple advantages including erosion control, crop reduction, and provision of cover, food and nesting sites for wildlife. Sodbuster provisions attempt to prevent the plowing of native vegetation, especially on highly erodible land, for crops.

During 1987 to 1989, at least 56,000 acres (22,663 ha) of native prairie were converted to grain fields in five Montana counties (Montana Farm Bill Coordinator 1989). Documenting the magnitude of sodbusting is difficult. The above data were gathered by U.S. Fish and Wildlife Service personnel who tediously searched through approximately 4,000 individual conservation plans in county offices.

Although the Soil Conservation Service, now the Natural Resource Conservation Service, authorized plowing of these highly erodible native prairies through conservation plans, apparently no agency was charged with recording statistics on sodbusting. In response to an inquiry, the state conservationist responded that the Soil Conservation Service did not maintain accurate records concerning conversion of native prairie to croplands. He also stated that, at that time, gathering such data was not a high priority (Montana Farm Bill Coordinator 1989). The Sodbuster Provision of the 1985 Food Security Act had failed to protect Montana's highly erodible native prairies from conversion to croplands at a time when landowners in Montana were being paid more than $200 million by the federal government to plant cover crops on erodible croplands under the CRP. Sodbusting negates the beneficial aspects of the 1985 Food Security Act by increasing erosion and food production. It also is very destructive for species, such as pronghorn and sage hens, which are dependent on native vegetation.

Montana already had more than 18 million acres (7.3 million ha) of highly erodible cropland (Montana Farm Bill Coordinator 1989). Adding to this total by sodbusting seems shortsighted, but may be profitable for some landowners. Approximately one-third of the landowners who broke native sod in the five Montana counties had enrolled cropland in the CRP. It is easy to imagine a wheat farmer with time on his or her hands and idle equipment because the fields were in the CRP, seeing profit to be made by sodbusting. It is more difficult to rationalize Soil Conservation Service personnel authorizing such sodbusting of highly erodible soils. However, at the time, the vice president of the United States was head of the Council on Competitiveness, and environmental protection generally was ignored if it interfered with profits.

"The self-image of the hunter with respect to conservation was put in capsule form in President Richard M. Nixon's message of May 2, 1972, proclaiming the first National Hunting and Fishing Day: 'Through a deep personal interest in our wildlife resources, the American hunter and fisherman have paved the way for the growth of modern wildlife management programs. In addition, his purchase of licenses and permits, his payment of excise taxes on hunting and fishing equipment, and his voluntary contributions to a great variety of conservation projects are examples of his concern for wildlife populations and habitat preservation.

" 'His devotion has promoted recreational outlets of tremendous value for our citizens, sportsmen and nonsportsmen alike. Indeed, he has always been in the forefront of today's environmental movement with his insistence on sound conservation programs' " (Scheffer 1974: 31–32).

From 1937 through 1983, ADC trappers killed approximately 3.7 million coyotes, 500,000 bobcats, 50,000 red wolves, 26,000 bears, 8,000 mountain lions and 6,600 gray wolves (Donahue 1999). The ADC programs became highly controversial as environmental awareness grew. This led to President Nixon's 1972 ban of the predicide 1080, which resulted in vigorous complaints by wool growers and praise by groups such as Defenders of Wildlife. Collectively, such controversies contributed to a decision by President Reagan's political appointees to remove animal damage control from the Department of the Interior to the Department of Agriculture, where the environmental movement had a less responsive audience (Mangun 1992).

Again, using the example of pronghorn in Arizona, a problem the animals faced after 1961 was changing land-use patterns. More than half of the animals resided on, or at least made seasonal movements across, private land. Due to the increase in retirement capital, developers acquired a number of Arizona ranches and carved them into 40-acre (16 ha) ranchettes (Brown 1994b)—probably the most efficient way to eliminate wildlife with the fewest number of people. Buyers are attracted to ranchette developments partly because they like to watch wildlife. However, as roads, fences, dogs, horses, and so on increase, pronghorn and other large animals are squeezed out of their former habitat. Fortunately, other states have had less subdivision problems in pronghorn habitat than has Arizona. But, with increasing human numbers and wealth, it can happen anywhere and is happening in many parts of pronghorn range.

As pronghorn continued to increase in numbers after 1960, hunter harvests in states like Montana and Wyoming seemed to have little influence on population trends; severe winters—often exacerbated by droughty summers and fences that hindered movements—did. Pronghorn populations in Canada and northern states of the United States were knocked back by severe winters in 1964–65, 1977–78 (Cadieux 1986), 1983–84 (Crowe and Strickland 1984) and 1996–97 (Yoakum et al. 1999). This vacillating history reflected the facts that pronghorn there are on the northern limits of their range, and their movements are impeded by fences and other human developments (see Chapters 12 and 17). "The gravest risk to the survival of pronghorns in Canada is the occurrence of severe winters, since they are prevented from migrating southward by woven wire fences across their routes" (Banfield 1974:404).

Severe winter storms also can cause extensive pronghorn mortality much farther south. A storm in east-central Arizona during the winter of 1967–68 killed hundreds of pronghorn and put several populations in danger of extirpation (Brown 1994b).

By the 1980s, some biologists and some segments of the public were frustrated with the applicability of most research reported in the *Journal of Wildlife Management* to management and conservation of wildlife. Of particular concern was lack of attention to complete ecosystems. Gavin (1989:345) expressed such dissatisfaction. "I contend that biological questions asked by wildlife biologists remain narrow in scope and potential applicability, and that our decisions to do site-specific, problem-oriented research in the short-term may have sold short our ability to solve the very problems we thought we were attacking directly."

Six decades after founding of the *Journal of Wildlife Management*, the first issue of *Conservation Biology* was published in 1987. In the first issue, Soulé (1987:4) wrote that "The Society is a response . . . to the biological diversity crisis that will reach a crescendo in the first half of the twenty-first century."

"For various reasons, emergence of the Society for Conservation Biology has piqued the interest of wildlife biologists. Some enthusiastically greeted the new society because of what seemed to be a fresh approach to the conservation of wildlife resources, whereas others wondered about more mundane matters such as overlap and purpose, perhaps not excluding a tinge of professional jealousy stemming from the 'glitz' and instant acceptance apparently afforded to the new society" (Bolen 1989:351).

"Some traditional wildlifers look on the conservation biologists as too theoretical, too ivory-towerish, too unwilling to soil their hands in the day-to-day political, economic, and land-use dirty work of managing wildlife on the land, and belatedly discovering a cause which the wildlifers have been committed to for decades. Some conservation biologists look on wildlifers as too oriented to the consumptive uses, and not sophisticated enough scientifically" (Wagner 1989:358).

Like most ruminants—if not hunted—pronghorn will lose their fear of humans. If people feed such animals, a dangerous situation is created. A food-conditioned buck, such as this one in Wind Cave National Park, South Dakota, has the potential, especially during the rut, to gore anyone. Photo by Steve Gniadek.

"Now long after the sheepmen, cattlemen and farmers have learned to live with each other and have homogenized, another wave of speculators and profiteers have appeared upon Wyoming's landscape. The rush to extract our energy resources from the earth is upon us. Uranium, oil, gas and coal must be pumped and dug from mother earth as quickly as possible. To heck with the future—we need all of these now" (Blair 1987:v).

Blair (1987:245) wrote of the late 1980s: "Changes in natural resource management especially in wildlife management, were now escalating at such a rate that problems seemed to be stacking atop one another. Wildlife managers would almost get on top of one problem to become mired in two or three more.

"Environmental degradation was occurring at such a rate and in forms never dreamed of 20 years before. Wyoming was well on the way to paying the supreme price for 'progress.' Resource management agencies, especially the Wyoming Game and Fish Department, were taxed to the limit trying to maintain a balance and keep precious resources from being swallowed up by the new wave of exploitation. At the same time, professional wildlifers feared that they were only seeing the tip of the iceberg—this thing was going to keep building and snowballing."

Because Wyoming has more pronghorn than any other province or state, one would expect populations and habitat there to be more secure than any place else. Examples of how pronghorn welfare is affected by innocuous-seeming activities were revealed during a field trip sponsored by the Wyoming Department of Game and Fish in 1992. We were dismayed to see how little consideration pronghorn were given in the state. The one-day field trip was for participants in the Pronghorn Antelope Workshop held in Rock Springs that year.

The first stop was on a state secondary road (Wyoming Highway 28) that transected a pronghorn movement route between winter and summer rangelands. Mining activities had increased traffic on the road, and Game and Fish personnel were fighting a Highway Department proposal to fence the road with livestock and game-proof fences. Highway Department officials were sympathetic to the pronghorn problem, but felt they needed fencing because of liability, although Wyoming is an open range state. Attorneys advised state highway officials that they are more liable if they fence and livestock get on the highway than if the roads are left unfenced (P. Riddle personal communication:1994). Thus, the fences used generally are substantial barriers to pronghorn as well as livestock. A speeding car had recently hit a cow on the road, and it appeared another pronghorn movement route would be severed. In many cases, livestock interests push for road fencing to help their livestock operations.

At another stop, we were shown pronghorn winter rangeland north of Rock Springs, which the city proposed fencing to deflect pronghorn around the city and away from traditional winter rangeland, so housing and city buildings could expand into the area. This was on checkerboard BLM and Union Pacific (private) land. The Wyoming Department of Game and Fish had been trying fruitlessly to get a corridor declared for pronghorn movements and have BLM trade for private land to facilitate such a corridor.

Much of the rangelands in southern Wyoming are managed by BLM. However, a strip paralleling Highway 80 and the Union Pacific Railway consists of a checkerboard of private and public lands. This occurred when the federal government gave every other section, in a strip 40 miles (64 km) wide along the railroad, to the railroad company as an incentive to forge a rail link between the eastern and western United States. Railroaders marked off sections of track about every 100 miles (160 km) and put maintenance people at settlements (sections) where supervisors were in charge of maintaining the tracks. According to Game and Fish personnel conducting the field trip, railroad surveyors were instructed to locate sections on pronghorn winter rangelands when possible. Such rangelands generally were in areas of shallower snow, milder temperatures and flatter land than surrounding lands. Originally, sections were surrounded only by water and fuel facilities for the trains and a few dwellings for railroad workers. Towns naturally grew around the sections. Thus, expansion of towns encroached on critical winter rangelands for pronghorn.

The result of habitat losses and blocked movement routes often are not seen until natural vagaries of weather turn against pronghorn. The cumulative effects of drought, severe winters, overgrazing and loss of habitat left many Wyoming pronghorn populations severely reduced from all-time highs of just a few years before, according to Anonymous (1994). Not long before, population estimates exceeded 500,000 in Wyoming. This figure dwindled to 384,000 in 1994. Effects of the particularly severe winter of 1992–93 were still being felt in 1994.

Endangered pronghorn subspecies in Mexico received added protection in 1988 when the habitat of the peninsular pronghorn was included in the 6,293,118-acre (2,546,790 ha) Vizcaino Biosphere Reserve (J. Cancino personal communication:2001). Also, the Sonoran pronghorn's rangelands in Sonora were included in the 1,765,668-acre (714,556 ha) El Pinacate y Gran Desierto de Alar Biosphere Reserve. The principle objective of the reserves is to protect the indigenous flora and fauna.

By the end of the 20th century, the BLM was entrusted with stewardship responsibilities for multiple-use management on more than 264 million acres (107 million ha) of public land. Most of that land was in Alaska and the 11 western states. "It is the mission of the Bureau of Land Management to sustain the health, diversity and productivity of the public lands for the use and enjoyment of present and future generations" (U.S. Bureau of Land Management 2000:ii).

By the late 20th century, the U.S. Forest Service was responsible for wildlife habitat on 155 national forests and 19 national grasslands, totaling more than 191 million acres (77 million ha). "Although early conservationists intended that the first national forests would protect game and fish as well as other resources, the agency devoted much of its attention . . . to timber harvesting and reforestation after World War II. From the outset, wildlife and fisheries biologists in the Forest Service had to compete with livestock and lumber interests. In recent years, public criticism of forestry, grazing, fire management, and mineral extraction practices and their effects on wildlife have further complicated relations within the agency as well as with local communities. Even so, the Forest Service's interest in protecting wildlife, fish, and rare plants has increased throughout the 20th century, gaining considerable momentum since the 1970s" (Catton and Mighetto 1998:i).

Pronghorn adapt well to human-related activities as long as they can still roam and secure their year-round needs without undue harassment. However, residential and energy development have displaced many bands by reducing the amount of habitat available to them and blocking seasonal movements. Photo by LuRay Parker; courtesy of Wyoming Department of Game and Fish, Cheyenne.

Regarding pronghorn, such momentum included removal of cattle from Organ Pipe Cactus National Monument by 1978, from Cabeza Prieta National Wildlife Refuge in 1983 and from Barry M. Goldwater Air Force Range by 1986 for the benefit of Sonoran pronghorn. Livestock had contributed to the changing vegetative composition of the desert region (U.S. Fish and Wildlife Service 1998). The number of AUMs allotted to livestock on the CMR was reduced about 33% in 1985 (U.S. Fish and Wildlife Service 1985). A 15-year program of no domestic livestock grazing was initiated on Hart Mountain and Sheldon national wildlife refuges (U.S. Fish and Wildlife Service 1994).

During the late 1990s, the volume of timber harvest fell to about 25% of what it had been during the 1980 to 1988 (Catton and Mighetto 1998) period. The last decade of the 20th century saw a shift to ecosystem management. It also saw the emergence of two foundations devoted to the welfare of the species (Figures 9 and 10).

FIGURE 9. *The North American Pronghorn Foundation was established in 1991 with headquarters in Casper, Wyoming. "Objectives of the foundation are to ensure the future of free-roaming, wild pronghorn by providing financial and other assistance to continue research, public education, interagency cooperation, habitat development, sound management practices and issue advocacy through protective legislative policy" (Foundation Statements 2001:6).*

FIGURE 10. *In 1992, a group of concerned sportsmen and wildlife conservationists founded the Arizona Antelope Foundation, "dedicated to increasing the antelope population within Arizona through active participation in advocacy of antelope management and habitat enhancement programs" (Lewis 1996:4).*

Advances in computer technology, global positioning system collars and other high-tech advances during the last several decades of the 20th century accelerated a trend toward more complex wildlife management systems based on analyses of wildlife and habitat relationships. For example, Ockenfels and Wennerlund (1994) found a powerful tool for analysis of pronghorn habitat with a geographic information system. High-tech electronic devices appear to be the only way to discover levels of poaching or crippling losses—problems that have plagued pronghorn managers for years.

Antihunting sentiment seemed to wax and wane during the 20th century. In a poll of a large number of college students, 77% of the women and 55% of the men believed that hunting endangered species—despite the fact that no species had ever become extinct in the United States as a result of sport hunting (Shaw 1973).

By the end of the 20th century, Kellert and Smith (2000) reported that the number of people opposed to hunting was considerable, but the right to hunt, especially for meat, continued to be supported by the public. However, nonconsumptive use of wildlife declined 17% from 1991 to 1996.

Hutton et al. (1995) noted that the "green" movement was an eclectic coalition of interest groups that made a significant impact in developed countries by raising public awareness concerning environmental problems generated since the industrial revolution. By well-organized campaigning, the movement gained significant political influence, much of it sensibly directed and deserving of support. However, many of the groups in the Green coalition have mutually exclusive interests. Those favoring animal protection philosophies have had negative effects on wildlife conservation.

Against what is known of the dynamics of wildlife populations, many people believe that the killing of an animal means a permanent reduction in the number of that species. To those people, the concept that "the species is everything and the individual is nothing" (Reiger 1975:14) seems patently wrong.

Despite what many nonhunters believe, the destiny of nearly all wild creatures is death by starvation, disease or predation. This is the reason that the science of wildlife management has proven so successful in the 20th century, and it is the reason pronghorn could be restored to the capacity of their remaining habitat while millions were being legally harvested. "[E]ven if we all became vegetarians, we would be killing animals right and left. The intensive production of any crop means clearing land and removing it as habitat for a myriad wildlife species" (Reiger 1975:15).

The first meeting of the Antelope States Workshop was held in New Mexico during 1965. Biologists attending from nine western states were interested in continuing technical sessions concerning pronghorn management, so a second meeting was held in Colorado during 1966. Since then, meetings have been conducted during even-number years, with the 21st held in Bismarck, North Dakota, in 2004.

During the mid-1970s, at least half of the United States residents opposed hunting, and women and urban residents had the most negative attitudes toward the sport, according to Kellert (1978). Bissell and Duda (1993) found that the values surrounding hunting were centered on the family and social circles of hunters. With increases in urbanization and the percentage of female-headed households, the recruitment potential for hunters was declining. Ruh (1994) also reported widespread opposition to hunting, but the impact of antihunting sentiment on hunters' numbers was slight; only 3 of 35 state agencies surveyed indicted a reduction in hunters in the field because of such pressure. In addition, Ruh found a prevalent view that antihunting sentiment possibly made hunters more conscientious. Photo by Judd Cooney, Pagosa Springs, Colorado.

During the fifth meeting, in Montana during 1972, affiliation with the Western Association of Game and Fish Commissioners was suggested, and a committee was appointed to draft workshop objectives. On July 13, 1973 the workshop was sanctioned by the commissioners. The goal of the workshops, as adopted during 1974 in Utah, is to provide information relative to and encourage perpetuation of wild stocks of pronghorn as an ecological, aesthetic and recreational resource on western rangelands, public and private, at the species' most productive levels consistent with other proper land use.

A slight name change was adopted—from Antelope States Workshop to Pronghorn Antelope Workshop—during the business meeting in 1974. This title better reflected representations; Canadians had become active in the organization.

Workshop proceedings contain a vast amount of information to the conservation of pronghorn. As of 2004, a complete set of these proceedings were available at the University of Montana's library in Missoula.

During the 1961 to 2000 period, consumptive users provided funding and important population management tools. Wildlife management provided the harvestable surplus for recreational use (Peyton 2000). However, by the 1990s, numerous biologists, especially conservation biologists, came to believe that cultural mores in Canada and the United States had changed to such an extent that this arrangement was out of step with mainstream public opinion (Holsman 2000, Organ and Fritzell 2000). In the face of social change, professionals and stakeholders can be pawns or planners in what happens, reacting separately to each new crisis or investing in strategic planning as a community of professionals (Peyton 2000).

Heberlein (1991:528) wrote: "Our current and highly successful system of scientific, rational and government-centered wildlife management . . . [is like] a heavily loaded Great Lakes freighter, constructed in the latter half of the nineteenth century with its course set in the first decades of the twentieth." Heberlein predicted that the half-century relationships between sport hunting and wildlife management was coming to an end and advised wildlife managers to accept that fact rather than deny the existence of major social change. He recommended accepting a reduction in the number of hunters, which would prove easier to manage and be a more affluent and probably more supportive constituency.

Not everyone agreed that a reduction in number of hunters was desirable because money will be necessary for continuous research and management if wildlife is to flourish. For instance, the International Association of Fish and Wildlife Agencies developed the Hunting As a Choice project, to help its member state agencies address trends that eschew consumptive uses of wildlife. Di Camillo and Schaefer (2000) assessed the effectiveness of the program (website and computer games) to determine whether it could increase the interest of 10- to 12-year-old suburban children in hunting or hunter education classes. In Jacksonville, Florida, nearly half of 91 suburban school children exposed to the website and computer game requested more information about taking a hunter education course from the Florida Fish and Wildlife Conservation Commission. Participants also became more tolerant of other people who hunted.

Organ and Fritzell (2000:780) "recommend that the wildlife management profession engage in dialogue and investigations to determine what the implications are to the profession and to wildlife conservation of a declining hunting subculture." Actually, "Big game hunting in America is more popular today than it has been at any other time in our history. Surprising news for sure, given the undercurrent of anti-hunting attitudes fueled by animal-rights propaganda" (Duda 2000:38).

Peyton (2000:777) noted out a situation that was becoming common at the end of the 20th century: "In the enthusiasm to produce a wildlife surplus and then to harvest it, a critical ingredient often missing is the relationship of the hunter with the hunted and the ecosystem involved." Clarke (1979:7) also maintained that: "In many highly touted hunts, the so-called hunter is merely an executioner whose function, so far as quality is concerned, has been to subsidize the skills of others." Indeed, by the end of the 20th century, guided hunters did not need to know much about the game they shot. Often, the guide was the hunter and the client was simply the shooter. The situation was even more pronounced when canned hunts, on grounds surrounded by game-proof fences, were involved. Such shooting may do more to tarnish the image of sport hunting than all the antihunting literature has accomplished. Fortunately, the vast majority of pronghorn hunting involves open landscapes and wild animals.

"Many Americans are surprised to learn that hunters have been the mainstay of wildlife conservation during the past 60 years. Long before such terms as 'ecology' and 'environment' entered popular vocabulary, taxes on firearms and ammunition—willingly supported by hunters and the hunting industry—were rescuing wildlife species and their habitats that would otherwise have disappeared" (Duda 2000:41).

"Federal taxes imposed on the sale of sporting goods added $293 million to the coffers of state fish and wildlife agencies in the first three quarters of fiscal year 2000. The Pittman-Robertson Act of 1937 and the Dingell-Johnson Act of 1950 established the taxes on gear and ammunition to fund maintenance and restoration of fish and wildlife. The U.S. Fish and Wildlife Service . . . reports that hunters and other outdoor recreationists have over the years contributed more than $5.2 billion through those two tax programs alone" (Cromrich and Burgess 2001:81).

Greenquist (1983:iv) summarized reasons for the 20th century increase of pronghorn. "The increase can be attributed to the timely protection given the pronghorn by the Wyoming Game and Fish Commission, the pronghorn's rapid reproductive rate, a reduction in livestock numbers and numerous livestock watering developments."

Overharvest was the principal problem for game animals during the 19th century. Loss and degradation of habitat became the main problem during the 20th century. By the early 1980s, more than 1 million acres (400,000 ha) of prime farmland, together with their food and wildlife production potential, were being lost annually to undirected urban sprawl (Cutler 1982).

CONCLUSIONS

The principal thrust has been to tame the wilderness, provide for economic growth, and raise the level of material welfare for all Americans.

Fisher (1966:168)

Although Fisher (1966) admitted that tremendous progress was made in restoring wildlife populations during the 20th century, he evaluated the overall situation as above.

Pronghorn were nearly extirpated by 1920, but increased to more than a million animals by 1983 (see Chapter 4). This is a remarkable story of a species expanding its numbers while its habitat often was being lost or degraded. Expansion of agricultural and energy-extractive operations, barriers to seasonal movements and other human-related activities took a toll. However, regulated hunting and a minimum of consideration in land-management planning facilitated the proliferation of a desirable animal.

A big problem facing game species during the 19th and early 20th centuries was unregulated hunting and overharvest, much of it for the market. Hunters, guided and helped by conservation greats, such as George Bird Grinnell, Theodore Roosevelt, Aldo Leopold, Ding Darling, responded with restrictions and money for species protection and habitat management. By the late 20th century, the biggest problem was loss and modification of habitat. Because hunters financed acquisition of habitat and influenced management, especially on public lands, hunted species often fared better than nonhunted ones.

Although sport hunters primarily financed protection and management, others also became involved in decisions related to pronghorn management during the 20th century. Entrepreneurs in nearby communities, landowners, photographers, wildlife watchers and many others developed aesthetic and economic interests in the animals. Managers attempted to satisfy, as nearly as possible, the dynamic and often contrary desires of the hunting and nonhunting public. They enabled the recovery of pronghorn to substantial numbers after the 1930s, allowing for a legal harvest of about 4 million animals (see Chapter 20). That harvest garnered support for maintaining healthy pronghorn populations from entrepreneurs, hunters and landowners who benefited from pronghorn management as practiced.

The recovery of pronghorn to thriving, huntable numbers was accomplished and sustained, in large part, by hunter enthusiasm and dollars. At least in the early years of its reemergence as big game, any pronghorn legally taken was both novelty and trophy. Above, noted author and sportsman, Ernest "Papa" Hemingway (left), poses with his kill. It is not known if this was Hemingway's 1941 pronghorn hunt in the Pahsimeroi Valley of eastern Idaho, which provided the incident that was the basis for his short piece, The Shot. Photo courtesy of the Trophy Room Books, Agoura, California.

During the 20th century, people in Canada and the United States realized the benefits of rational, government-centered wildlife management systems (Heberlein 1991). Those systems had evolved during the latter half of the 19th century and early 20th century as leisure and wealth, resulting from the industrial revolution, spawned urban sportsmen and sportswomen. They believed the appropriate use of wild animals was to provide recreation and meat through regulated hunting, and they were willing to pay for the privilege. Also, during the last half of the 20th century, public land management agencies received mandates for multiple-use management through a variety of statutes. Collectively, they reduced the grip of commodity users on public lands and put more equitable emphasis on other uses. Also, greater emphasis was placed by managers on economics, aesthetics, user attitudes and preferences, as well as interests of citizens other than direct users of resources. However, for various reasons—

The recovery of pronghorn numbers during the 20th century demonstrated the efforts of early conservationists, such as George Bird Grinnell, Theodore Roosevelt and Aldo Leopold. Voluntary funding by hunters, dedication of wildlife conservationists, and cooperation and tolerance of landowners also contributed to the animals' restoration. The return of pronghorn especially pleased hunters, but many can appreciate the life, interest and beauty the animals add to many arid landscapes. Photo courtesy of the Wyoming Department of Game and Fish, Cheyenne.

the profit margin, politics, ignorance or apathy, among others—laws are not always effective on the ground where pronghorn live.

Despite increased urbanization, industrialized agriculture, urban sprawl, habitat loss, pollution and human population growth, game and nongame species generally have flourished. Only species that depend on pristine habitat or were especially vulnerable to pollution—problems difficult and expensive to deal with in consumption-oriented societies—have fared poorly. Leakey and Lewin (1995:232) summarized the general situation: "Blessed with reason and insight, we move toward the twenty-first century in a world of our own creation, an essentially artificial world in which (for some, at least) technology brings material comfort and leisure brings unprecedented artistic creation. So far, unfortunately, our reason and insight have not prevented us from collectively exploiting Earth's resources—biological and physical—in unprecedented ways." Fortunately, human reason and insight helped pronghorn recover from near extinction, and pronghorn themselves have been adaptable enough to survive and even thrive in an essentially artificial world.

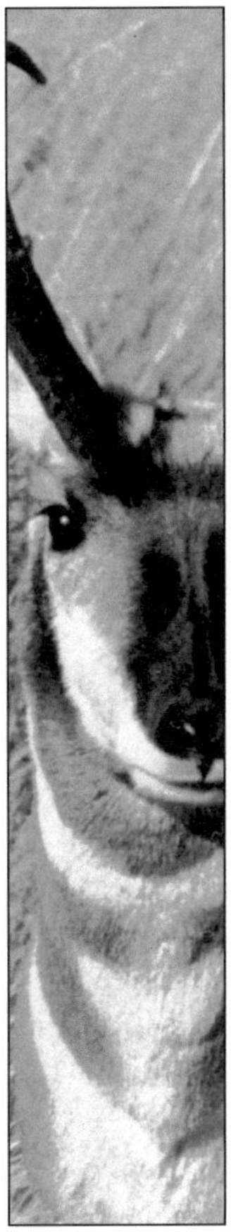

C H A P T E R F O U R

Distribution and Abundance

Jim D. Yoakum

Those of us who in the early days traveled the prairie from north to south, and east to west, used to wonder constantly at the great numbers of the Antelope which day after day were seen on all sides of the command, feeding or resting, or running, in troops of 10 or 20, or 30; and later, when winter approached, gathering into great droves, which always numbered hundreds, and sometimes thousands.

George Bird Grinnell *in* Seton (1927:423)

When compared with other large herbivores, such as deer or elk, pronghorn are a more visible species. They mainly inhabit open, flat rangelands, rather than forests or mountains. They do not hide in tall dense vegetation; they often travel in herds; they sometimes can be seen from miles away. During recent years, pronghorn have been surveyed comparatively easily from fixed-wing aircraft and helicopters, and their numbers have been documented more accurately than those of many other large terrestrial mammals because they are so visible.

DISTRIBUTION

The pronghorn and its progenitors lived for eons on the prairies, shrubsteppes and semiarid lands that now comprise portions of Canada, Mexico and the United States. At least within late prehistory (see Chapter 2), pronghorn coexisted with bison, but the two species' geographic ranges did not overlap perfectly. Both species were most abundant on the shortgrass prairies of the Great Plains.

The range of the bison once extended eastward across the tall grasslands and deciduous forests to Pennsylvania, whereas pronghorn seldom were reported east of the Mississippi River. Evidently, forests with dense shrub understories restricted pronghorn occupation. In pre-Columbian time, bison ranged as far west as Oregon and California, but by the time of that region's exploration by fur trappers, traders, missionaries and other white civilization vanguards in the early 1800s, bison were living only in small isolated areas west of the Rocky Mountains (Seton 1927). Whether bison there were extirpated

Most modern surveys to estimate pronghorn numbers are conducted with fixed-wing aircraft or helicopters. Pronghorn are easier than most wild ungulates to count from the air because they are diurnal, gather into winter concentration herds and occupy rangelands with low-growing vegetation (such as the sagebrush steppe above)*—all characteristics that favor counting them. Winter surveys help to provide counts when the ground is completely covered with or completely free of snow; small patches of snow can make it difficult to spot piebald animals. Agencies increase aerial survey accuracy by taking photographs of bands and later counting the animals when more time is available.* Photo courtesy of Edson Fichter.

because of semiarid areas drying out, the Indian horse culture emerging or other cause is unknown.

Except in the Central Valley of California, pronghorn were less numerous west of the continental divide than on the Great Plains; they occupied most of the grasslands and were scattered in the steppes and deserts. In Canada, bison roamed into arboreal forests of the Northwest Territories. The northern limit of pronghorn was the northern terminus of open prairie in Alberta, Assinibonia (now within Saskatchewan), Manitoba and Saskatchewan. To the south, pronghorn lived on the peninsula of Baja California and the central high plains, from the United States border to the states of Hildago and Mexico—approximately 100 miles (161 km) north of present Mexico City (see Chapter 1).

Grinnell and Hornaday Compilations

At least three early attempts to map the distribution of pronghorn involved requesting readers of popular magazines to report localities where the animals were observed during a given year (Grinnell 1897b, Hornaday 1904, 1914) (Figure 11). As editor of *Forest and Stream*, George Bird Grinnell reached outdoors men and women throughout pronghorn habitat. William T. Hornaday, director of the New York Zoological Society, was a regular contributor to the *Peoples Home Journal*—a widely circulated family magazine popular in rural areas.

1900 Roosevelt and Seton Maps

The pronghorn distribution map published by Theodore Roosevelt et al. (1902) was included in a book discussing pronghorn life history and Roosevelt's hunting experiences in North America (Figure 12). The pronghorn range was delineated for 1900, as reported to Roosevelt by biologist C. Hart Merriam. It included central Alberta and Saskatchewan in Canada, and the states of Durango, Zacatecas and San Luis Potosi in Mexico. Nelson (1925) found no pronghorn in these areas 24 years

FIGURE 11. (Top) *Provisional depiction of pronghorn herds reported by readers of* Forest and Stream *to its editor, George Bird Grinnell, during 1896. The results appear to be incomplete, but may indicate the most densely populated areas shortly before the end of the 19th century (from Grinnell 1897).* (Center) *Locations of pronghorn herds in 1903, as reported to William T. Hornaday (from Hornaday 1904).* (Bottom) *Isolated herds of steadily diminishing pronghorn, as depicted by Hornaday in 1914 (from Hornaday 1914). Note the disappearance of animals from the Dakotas, Nebraska, Kansas, Oklahoma, Texas and Mexico in less than 20 years, indicated by these three depictions.*

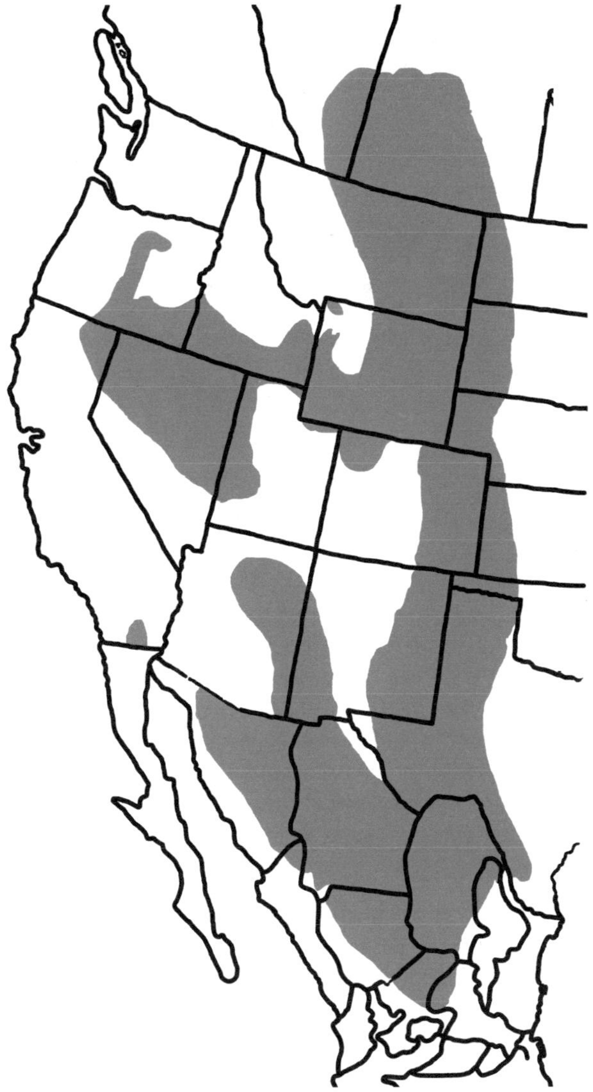

FIGURE 12. *Rangelands (tinted area) occupied by pronghorn in 1900, according to C. Hart Merriam (from Roosevelt et al. 1902).*

later. Merriam was well traveled in Mexico and knowledgeable of the country's biological conditions, but he failed to report pronghorn in Baja California. Ernest Thompson Seton (1909) also published a map depicting pronghorn distribution for 1900 (Figure 13). He drew on information from Merriam and Hornaday as well as material

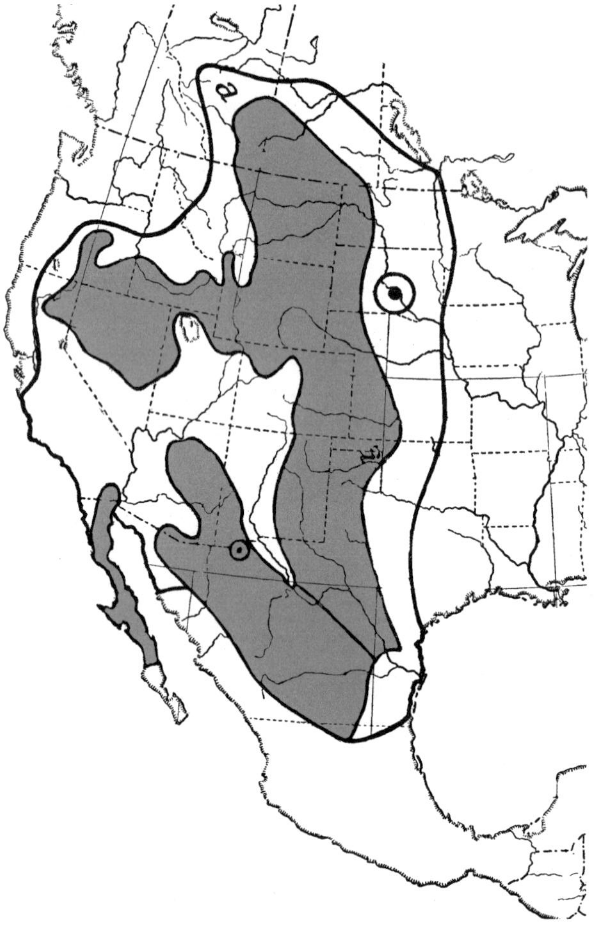

FIGURE 13. *Based on reports by many early travelers, personal observations, C. Hart Merriam, William T. Hornaday, the Bureau of Biological Survey and U.S. Department of Agriculture, Ernest Thompson Seton (1909) produced a map representing pronghorn distribution in 1900 (tinted area). He also delineated, by bold line, what he called the "primitive" (macro) range, which agrees somewhat with present knowledge concerning the species distribution when Euro-Americans reached North America.*

possessed by the Bureau of Biological Survey and other agencies of the U.S. Department of Agriculture. The 1900 maps were similar, but Seton showed pronghorn in Baja California and separated the two subspecies named at that time. He also provided an outer-limits line for the species' primitive, or macro, range, which agrees reasonably well with present knowledge (see McCabe et al. 2004).

1925 Nelson Map

Nelson's (1925) survey provided information regarding the habitats occupied by pronghorn when the species was near its population nadir. He compiled the information from reports received from wildlife workers of each province and state within the species' range. Herd locations were plotted on 19 separate maps that were consolidated as a single composite map in Figure 14. Occupied habitat had been reduced to isolated and relatively small areas, except for the region where California, Nevada and Oregon join and for two regions in Mexico.

Mexican Map, 1949 and 1984

Gonzalez and Lafon (1993) provided a map (Figure 15) that illustrates the shrinkage of pronghorn range in Mexico. Note the general extirpation in the southern distribution and the small, isolated populations by 1984.

Translocations to Nonhistoric Rangelands

Since the early 1900s, a number of attempts were made to introduce pronghorn to habitat outside the species' ancestral range. The first took place in 1911, when young pronghorn were released on the National Bison Range in the Flathead Valley of Montana (Nelson 1925). They soon were killed by coyotes and domestic dogs. A subsequent translocation there also failed (Byers 1997a). Finally, during 1951 and 1952, after dogs had been controlled and

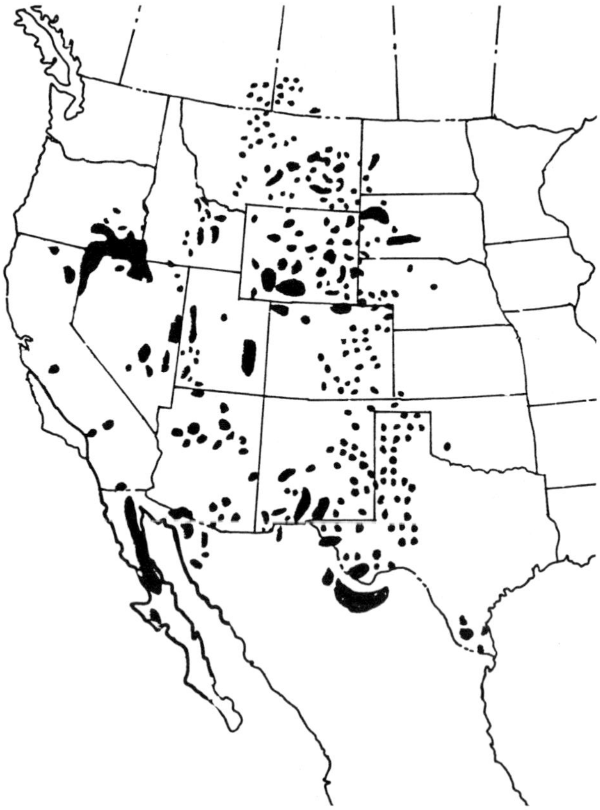

FIGURE 14. *Composite map of pronghorn herds reported by provincial and state conservation agencies during 1922 to 1924 (redrawn from Nelson 1925).*

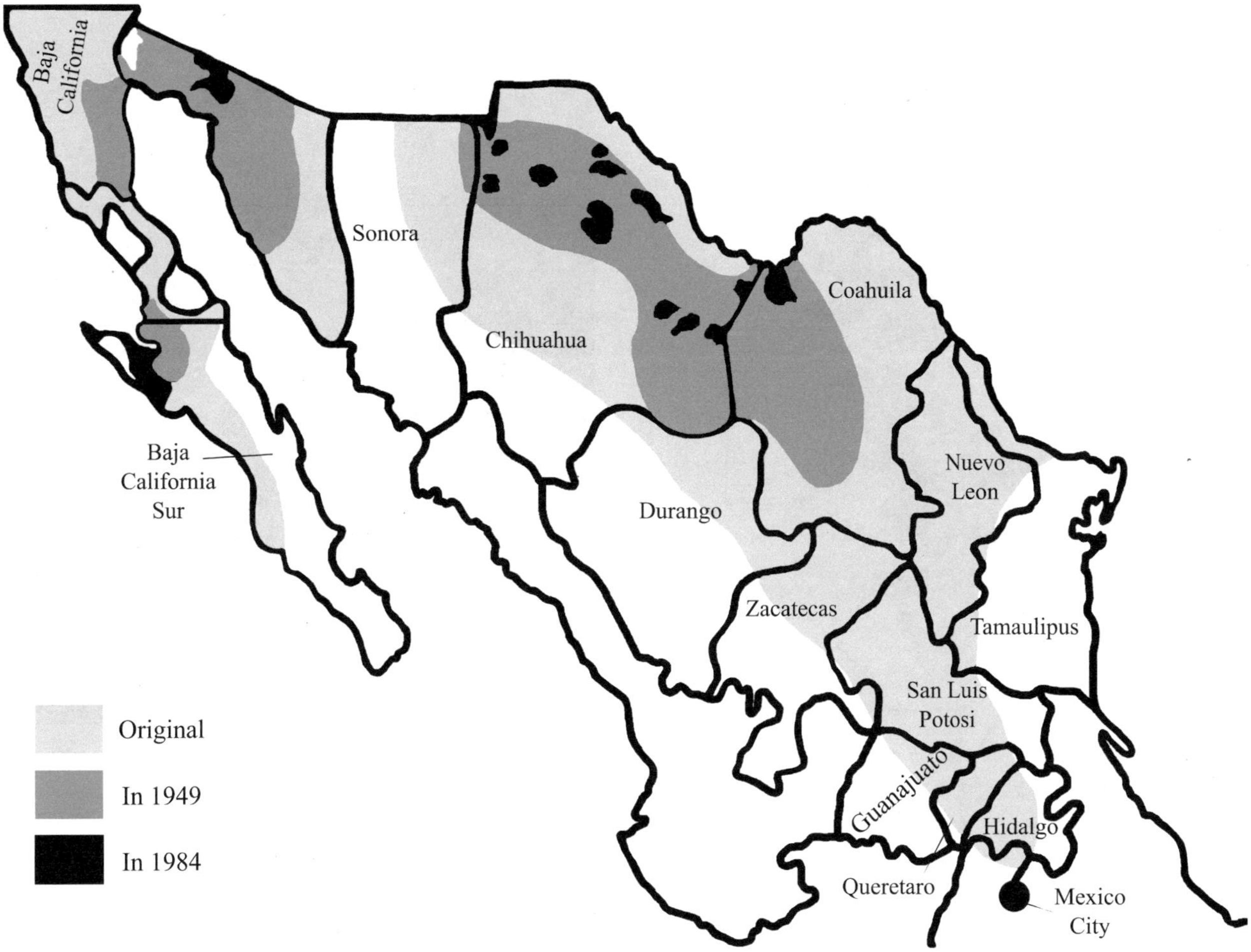

FIGURE 15. *Original, 1949 and 1984 distributions of pronghorn in Mexico (Gonzalez and Lafon 1993).*

coyotes reduced by poison, pronghorn were translocated to the range (Dow and Wright 1962, Byers 1997a).

During the 1940s, pronghorn from Oregon and Nevada were transported to the shrubsteppes of east-central Washington (G. Johnson personal communication:1999). It appeared the animals would thrive. After predator control, the population quickly increased to more than 200. However, it slowly decreased and were gone by the late 1990s.

In 1959, 18 pronghorn were moved from the National Bison Range to the Hawaiian island of Lanai (Nichols 1960). The herd increased to more than 150 animals by 1967 and used most of the available 35 square miles (91 km^2) of habitat. Recruitment appeared to cease (N. Ueoka personal communication:1972). By 1990, the herd was extirpated.

Thirty-eight pronghorn from Colorado were released in Okeechobee County, Florida, in 1966. They disappeared within 18 months after release, although the area—a prairie-type habitat—had been selected for its presumed suitability (Elliott 1966).

1999 Map

With the exception of translocations to the National Bison Range, pronghorn occupy most of the rangelands where the species was located when Europeans first came to America. For several decades before the 1980s, pronghorn increased numerically more than 3,500%—from about 30,600 to more than 1 million—since Nelson's (1925) survey; all growth was within the historic range (Figure 16). The 1999 map is based on reports received from provincial and state wildlife agencies and was prepared by the North American Elk Foundation, in Missoula, Montana. Many populations were fragmented at the end of the 20th century.

ABUNDANCE

Since Columbus

Several wildlife historians estimated pronghorn numbers, at one time, equal to or exceeding those of the American bison. Nelson (1925) suggested 30 to 40 million as a

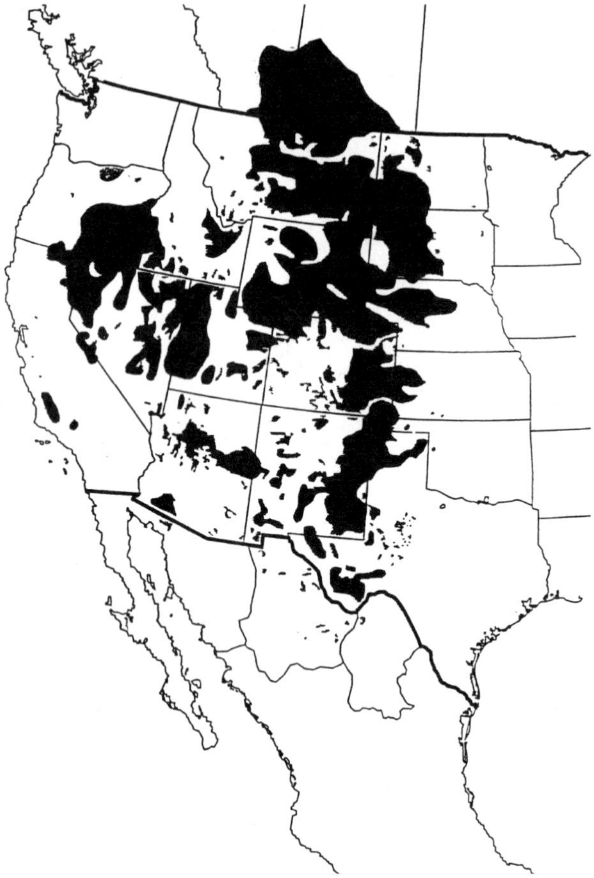

Figure 16. *Not all habitats within the shaded distribution area were occupied by pronghorn. Distribution of pronghorn in Canada, Mexico and the United States in 1999.*

The translocation of pronghorn from Montana to Lanai Island, in the Hawaiian Islands, was an unsuccessful venture. Recruitment occurred on the island as long as native vegetation was short. After cattle—which had been raised on the island during World War II and for a few years thereafter—were removed, pronghorn gradually declined until none remained approximately 15 years after the initial release. Normal plant succession apparently choked the pronghorn habitat and made them vulnerable to predation by dogs. Guinea grass, molasses grass and haole koa, a shrubby tree, had been heavily used by cattle and responded quickly to their removal. Attempts to establish pronghorn outside the species historic range seldom are successful. Photo courtesy of the Hawaii Division of Fish and Game, Honalulu.

conservative estimate, which has been repeated so often it is now considered fact by many people. Grinnell (1929:135–136) wrote: "In old times in summer one traveled day after day over the prairie with antelope constantly about him, not in great herds but in the little groups. . . . It was in winter, when the antelope collected in immense herds, that their numbers became impressive to the casual observer, and of these great herds, seen for the most part only by the winter meat hunter, not much has been written. . . .

"I have more than once spoken of the very great numbers of antelope found on the plains in early days, and have expressed the belief that they were more abundant than the buffalo. With this statement most old-time prairie men agree. Yet, except in winter when they gathered in herds of thousands, the antelope were not conspicuous, since usually they were found in small groups from three to fifty or sixty, and when alarmed ran away in these small groups. The antelope were small, protectively colored, and scarcely noticeable a mile or two away except by their white rump patches, while the buffalo, large and black, were visible at a great distance and when alarmed fled in companies, gathering to themselves as they ran into other herds, which at last formed a great black mass unaccountable and sometimes astounding. The antelope when they fled did not gather up in herds and soon disappeared, but the buffalo were seen for a long distance. . . ."

Early explorers, such as Lewis and Clark (Burroughs 1961, Moulton 1987b) and Fremont (Egan 1977) observed the numbers of animals sighted and of big game killed for food. Their journals indicate pronghorn were most numerous on the prairies (see McCabe et al. 2004). The Lewis and Clark journals portrayed the abundance of pronghorn when Euro-Americans crossed the continent's western plains and mountains. They first saw pronghorn in the vicinity of Bon Homme Island, below the mouth of the Niobrara River in Nebraska. From that time, their

Twelve hand-raised pronghorn from northwestern Nevada were translocated to Hermit Basin Camp in Grand Canyon National Park, Arizona, during 1924 (Nelson 1925). The area was not historic pronghorn habitat. A herd failed to become established because the topography, vegetation and other ecological factors proved not to be conducive to pronghorn survival. Photo by E. A. Goldman; courtesy of the U.S. National Archives, Washington, D.C.

journals made frequent reference to the abundance of pronghorn and bison across the Missouri River highlands and westward to the Rocky Mountains. The following was written by Lewis, on April 25, 1805, while the expedition was near the mouth of the Yellowstone River: "the whole face of the country was covered with herds of Buffalo, Elk, & Antelopes; deer are also abundant, but keep themselves more concealed in the woodland. the buffalo, Elk, and Antelope are so gentle that we pass near them while feeding, without appearing to excite any alarm among them, and when we attract their attention, they frequently approach us more nearly to discover what we are, and in some instances pursue us a considerable distance apparently with that view" (Moulton 1987b:67).

Pronghorn were reported almost daily from Nebraska to Three Forks, Montana. Although the expedition primarily obtained other big game for food (bison, elk, deer), they shot 62 pronghorn for consumption (Burroughs 1961). Of these, 90% were obtained east of the Rocky Mountains. A few were taken in the Beaverhead Valley, Montana, and three were taken in the Lemhi Valley, Idaho, west of the continental divide, but none was shot west of Lolo Pass, Montana.

Lewis remarked in his journal, of February 22, 1806, at Fort Clatsop on the Pacific Coast (Moulton 1990:336): "The Antelope is found in the great plains of Columbia and are the same of those on the Missouri found in every part of the untimbered country. They are by no means as plenty on this side of the Rocky Mountains as on the other."

The Lewis and Clark expedition helped to open the West to trappers, pioneers, ranchers and settlers. Some of these early travelers recorded numbers of animals in their observations of wildlife (see McCabe et al. 2004). One was trapper Osborne Russell, who kept a journal from 1834 to 1843 as he trapped beaver in the Rocky Mountains. His entry for several days travel stated: "On my way down the bear river I met thousands of antelope traveling towards their winter quarters, which is generally in the Green River valley" (Haines 1955:112).

Captain John C. Fremont made sketchy notes relative to wildlife abundance during his trips across the West. He documented pronghorn abundance on the Great Plains with such remarks as "Antelope were seen frequently," but few references were made to them west of the Rockies, except in California (Egan 1977:16).

Pronghorn apparently were most abundant on the short grasslands of North America and the Central Valley of California. J. S. Newberry (1855:71) in his *Explorations and Surveys to Ascertain the Most Practical Route for a Railroad from the Mississippi River to the Pacific Ocean*, reported to the U.S. Senate with reference to the pronghorn: "Though found in nearly all parts of the territory of the United States west of the Mississippi, it is probably most numerous in the valley of the San Joaquin, California. There it is found in herds literally of thousands."

California's abundant pronghorn populations were referred to by a Dr. Canfield in his writings of 1828 (Caton 1877). He remarked on their abundance not only in the Central Valley, but also in Salinas Valley and other valleys extending to the Pacific Ocean.

The most widely accepted estimate of pristine pronghorn numbers is the analysis by Nelson, Chief of the U.S. Bureau of Biological Survey (now the U.S. Fish and Wildlife Service). Nelson (1925:3–4) stated: ". . . it may be considered a conservative estimate to place its probable original numbers at not less than thirty to forty millions, and possibly more."

From 1850 to 1900, great numbers of trappers, miners, loggers, ranchers, pioneers, settlers and homesteaders moved westward. With them came an industrialized civilization with many hungry humans to feed, which resulted in market hunting and conversion of native grasslands to cultivated crops. All contributed to the pronghorn's rapid decline, brought on primarily by: (1) plowing and seeding of native grasslands to crops, eliminating the preferred native forage, water and space needed by pronghorn for reproduction and survival; (2) blocking of seasonal routes, causing large die-offs during severe winters and droughts; and (3) unregulated killing of all sexes and ages on a year-round basis by market and sport hunters, settlers, and others.

Decimation was rapid. An early settler remarked that pronghorn were abundant near Whitewater, Manitoba, during the 1870s, but hunters killed the last one by 1881. Seton (1927) traveled the area in 1882 and neither saw nor heard of pronghorn there at that time. Einarsen (1948) cited early records of pronghorn being numerous in south-

According to Newberry (1855), pronghorn probably were more numerous in the San Joaquin grasslands of southern California than anywhere else. However, native herds were extirpated by the late 1930s, and the grasslands were converted for agriculture and other human developments. A small herd was translocated to the adjacent Carrizo Plain during the 1980s and became established. Reintroduction of these native mammals to the state's grasslands reflects their adaptability to some altered, historic landscapes if other limiting factors—such as unregulated hunting in this case—are removed. Photo by Alice Koch.

eastern Oregon before 1858, but many of those herds had vanished by 1890. Hoover et al. (1959) estimated the pronghorn's precipitous decline in Colorado as: 1860—2 million; 1898—25,000; and 1918—less than 1,000.

Sundstrom (1967) provided chronological records for nearly a 100-year period (1869 to 1967) for Wyoming (Table 4). He noted the pronghorn's decline, beginning in 1869—when Wyoming was a part of the Idaho territory—as Europeans colonized the new state, and the animals subsequent increase from 2,000 in 1906 to 150,000 in 1963.

Above are a few examples of records kept of the rapid decline of pronghorn and early increases. All of the records of the late 1800s, from throughout pronghorn range, indicate a tremendous decrease from vast, free-roaming herds to scattered bunches of dozens or a few hundred.

20TH CENTURY SURVEYS

The 20th Century

Surveys of pronghorn populations generally are estimates of numbers for given areas at specific times. They commonly are accomplished in modern times by experienced wildlife biologists skilled in sophisticated survey procedures and statistical analysis (Pojar et al. 1995). They

TABLE 4. A history of pronghorn relationships with Euro-Americans in Wyoming from 1869 to 1963

Year	*Estimated statewide population*	*Notation*
1869		The territory adopted its first law making it illegal to offer pronghorn for sale between February 1 and August 15.
1873		An epizootic devastated 75% to 90% of pronghorn between Yellowstone and Missouri rivers.
1875		The first hunting season established from August 15 to January 15. Animals could be killed for food only.
1881		An estimated 53,000 pronghorn were shipped down the Yellowstone River by market hunters.
1882		The hunting season was decreased 40% (now August 1 to November 1). Any citizen could kill pronghorn for food if hunted within 10 miles (16.1 km) of the citizen's residence.
1885		Pronghorn reported still in large numbers.
1886–87		The severe winter caused tremendous mortality among cattle (estimated 50%) and wildlife such as pronghorn on the plains.
1887		It was reported that pronghorn were extensively hunted within the Yellowstone National Park boundaries.
1888		Nonresidents were barred from hunting within the state. This was enacted as a means of controlling heavy market hunting.
1890		Wyoming becomes a state. Enacted a law this year governing hunting and fishing.
1895		Enacted laws to levee fines for fish and game violations. Bag limit for pronghorn was three. Nonresidents allowed to hunt again.
1900		By now, pronghorn had been greatly reduced to a few isolated and diminishing herds. Poaching was a serious problem with much public sentiment favoring such activities in this wide open, sparsely settled country.
1901		The bag limit reduced from three to two pronghorn for nonresidents. Residents could still harvest three.
1902		The state game warden warned that pronghorn were being harvested so rapidly that they were on the point of extirpation.
1904	5,000	It was estimated the total population in Wyoming was 5,000. The limit for residents was reduced to two.
1905		The state game warden recommended the season be closed for 6 years; the state legislature enacted a law to close the pronghorn season for 6 years. All wildlife declared property of the state. A license was required to hunt or to photograph wild game.
1906	2,000	Lowest population estimate for the state.
1912		Pronghorn herds reported increasing; however, a hard winter caused high mortalities. The state game warden reported pronghorn were suffering due to heavy use of vegetation by livestock.
1915		Due to the closed season, herds began to increase. A growing public sentiment favoring protection defeated a proposal to reopen the season.
1916–17		Populations decreased due to a severe winter. The state game warden stated, "Hard winters and short range have something to do with the disappearance of these animals and I am of the opinion that it will be a long time, if ever, when an open season should be allowed on them."

continued on next page

TABLE 4—*continued*

Year	Estimated statewide population	Notation
1917		Wyoming sets aside its first State Antelope Preserve (actually a refuge). The 85-square mile (219 km^2) Carter Mountain Preserve in Park County.
1919	8,387	The state game warden estimated the statewide pronghorn population at 8,387 animals.
1923	13,895	Statewide herds had increased to an estimated 13,895. This was the largest number estimated in the state since the turn of the century.
1927	20,610	The first pronghorn hunting season in 18 years was held. The harvest was estimated at 307.
1929	26,625	A total of 6,353 pronghorn harvested in 12 counties.
1930–32		Complete closed season for pronghorn during these years.
1933–38		An estimated 1,800 harvested each year.
1934	33,825	Landowners granted $2.00 of each license for pronghorn harvested on the landowner's property. This practice continues to partially remunerate the landowners for loss of pasture grazed by pronghorn.
1939	34,765	The first statewide aerial census commenced. Before this, all population census data were obtained from ground counts and estimates.
1941		The first pronghorn were trapped and translocated to augment other herds.
1942	55,000	For the first time in almost 40 years, the total number of pronghorn to be harvested was not restricted. Special "two-antelope" permits were issued to residents of the state.
1948–49		Severe winter resulted in emergency feeding for pronghorn and other big game. Many pronghorn died as a result of exposure, exhaustion and starvation.
1950	76,856	Populations increasing.
1952		Annual harvest of 41,000.
1953	108,027	Herd numbers greatly increased.
1954–58	140,000	Herds increasing despite harvests.
1963	150,000	A tremendous comeback since the low of 2,000 in 1906.

Source: Modified from Sundstrom 1967.

provide trend information that, when computed over a period of years, indicate whether populations are increasing, static or decreasing.

Provincial and state wildlife agencies use an array of means to survey and estimate pronghorn populations, including ground counts using horses or vehicles, aerial flights with fixed-wing, rotary or glider aircraft, hunter harvest data computations and, at time, multiple means (see Firchow 1990, Pojar et al. 1995, Gunzel 1997, Lee et al. 1998). Because surveys are accomplished through various means, the resulting information varies in precision; all should be recognized as estimates, not as precise counts of all animals on the rangelands surveyed.

Methods for counting pronghorn were investigated as early as the late 1940s in the Intermountain West, to coordinate field procedures and decrease duplication among state agency censuses (Springer 1950). These procedures—issued by Griffith (1962) and later updated by Salwasser (1980)—were improved and documented with recommended guides for implementation. The Interstate Antelope Conference region traditionally used two annual surveys: winter counts to report survival and total population numbers, and summer surveys to record spring recruitment.

Different techniques conducted for estimating pronghorn populations are reported in Chapter 18. The 1998

Before the 1950s, pronghorn surveys usually relied on the use of horses and vehicles. Obtaining estimates of numbers was difficult because of pronghorn wariness, the open terrain and some inaccessible rangelands. Nevertheless, ground surveys were and often still are important to provide sex and age ratios for estimating population dynamics. In some states, ground surveys are the only survey techniques used. In other states, ground surveys are employed mainly where pronghorn densities are low. Photo by Jim D. Yoakum.

Pronghorn Management Guides also has a review of current pronghorn survey procedures (Lee et al. 1998). These different survey practices are a result of the magnitude and

Ground surveys to estimate pronghorn numbers often lacked accuracy, as evidenced by the above winter scene on Drakes Flat, Oregon, with the Hart Mountain National Antelope Refuge escarpment in the background. The herd is bunched and totals more than 200 milling animals. An accurate count was obtained only after the herd started to run in a narrow line over a nearby tableland. Photo by Jim D. Yoakum.

economics of counting pronghorn in differing regions. Agencies of provinces or states with large numbers of pronghorn (50,000 or more animals), compared with agencies with smaller numbers (less than 25,000 animals), generally expend more time, funds and effort surveying pronghorn numbers to help set harvest regulations. Based on expenses and necessity, agencies with low numbers (California, Nevada, Oregon and others) have conducted aerial trend transects. Some agencies complete only a portion of their area in 1 year and finish the rest of the state in succeeding years. Other states spend large quantities of time and monies for intensive aerial line transects to obtain as precise an estimate as possible to determine numbers that are available for harvests (Colorado, Wyoming). Still other states derive their total estimated numbers from statistical analyses of previous-year harvest. With so many agencies accomplishing surveys with varying techniques, it often is difficult to compare results of surveys.

One restriction can apply to pronghorn surveys. At times estimated numbers are calculated for pre- or posthunting seasons. When the latter reports are provided, it is necessary to compensate when comparing population numbers with those of states reporting preseason estimates. For example, some wildlife agencies harvest up to 30% of pronghorn populations. This percentage, coupled with additional data regarding crippling losses and illegal kills, can result in up to 40% fewer animals following the harvest season. Adjusting data for these different procedures is imperative when compiling and comparing estimated populations for different areas.

Eleven surveys of pronghorn numbers have been conducted since 1920. All provide information for the United States, and some list data for Canada and Mexico. Results of those surveys follow.

Pronghorn were at their lowest level by 1915, at an estimated total of 13,000 (Trippensee 1948, Hoover et al. 1959). Most provinces and states had recognized by then the imminent danger of their pronghorn population's extirpation and passed laws forbidding hunting.

1920 and 1921 Surveys. In 1920 and 1921, the National Bison Society conducted a pronghorn survey (Nelson 1925). M. S. Garretson, secretary of the society at the time, compiled the survey results and estimated a total of 11,750 pronghorn in Canada and the United States. No estimate was provided for Mexico.

1922 to 1924 Survey. During the 1920s, the U.S. Bureau of Biological Survey initiated another pronghorn survey. It was conducted from 1922 to 1924, with the cooperation of personnel from provincial and state wildlife agencies, conservation organizations and interested private individuals. Efforts were made to locate and record each surviving herd in Canada, Mexico and the United States. The details and results of those surveys were reported by Nelson (1925). Estimated populations for each province, state and country are shown in Table 5. The 1924 pronghorn population for Canada, Mexico and the United States was estimated at 30,600, indicating an increase in numbers from 1915 to 1924.

Nelson's (1925) report was a remarkable accomplishment for that period; similar surveys of other large terrestrial mammal populations were few. Nelson's report has become the base survey of pronghorn for the early 20th century. In addition, he reported management strategies to enhance pronghorn welfare (e.g., methods of capturing and translocating animals; establishment of refuges; conferences for state, national and international cooperation; etc.).

1937 to 1955 Surveys. Public concern for the conservation of wildlife in the United States was high from 1900 to 1925. At the insistence of many concerned groups and individuals, the chief of the Bureau of Biological Survey issued a call for a conference on pronghorn. It was conducted at the National Museum in Washington, D.C., on December 15, 1923, and was important to the future of pronghorn. Ten conservation organizations, seven state wildlife agencies, four U.S. federal agencies, the Canadian government and several private ranchers were represented. Discussions between ranchers and government agencies led to the establishment of the Hart Mountain National Antelope Refuge, in south-central Oregon, the first large national refuge dedicated to the preservation and perpetuation of pronghorn in the United States. Furthermore, the Bureau of Biological Survey agreed to serve as management coordinator, gathering and integrating numerical information continually. In compliance, the U.S.

Table 5. Population surveys for pronghorn in Canada, Mexico and the United States from 1922 to 1999 (rounded to closest 100)

Political unit	*1922–24*[a]	*1964*[b]	*1970*[c]	*1976*[d]	*1983–84*[e]	*1989*[f]	*1995*[f]	*1997*[g]	*1999*[h]
Canada									
Alberta	1,000	16,000	10,000	13,300	21,500	22,300	17,300	13,200	20,100
Saskatchewan	300	4,300	8,000	9,000	10,000	27,300	22,000	10,000	12,800
Subtotal	1,300	20,300	18,000	22,300	31,500	49,600	39,300	23,200	32,900
Mexico	2,400	1,200	1,000	1,000	600	500	800	600[i]	1,200[j]
United States									
Arizona	700	10,000	6,000	7,300	9,000	9,000	11,000	10,200	7,800
California	1,100	2,700	3,000	5,000	6,800	7,000	6,500	5,600	5,200
Colorado	1,200	15,200	20,000	31,000	57,500	58,000	68,800	72,600	64,300
Idaho	1,500	4,700	6,000	13,300	21,500	21,800	12,500	12,000	8,500
Kansas	100	100	400	1,100	1,200	2,000	1,800	2,500	2,200
Montana	3,000	95,500	95,000	71,200	161,500	300,000	225,000	77,500	77,800
Nebraska	200	9,000	6,900	9,800	9,000	2,500	4,100	6,500	6,000
Nevada	4,300	4,500	4,500	6,500	9,800	15,000	15,600	14,500	16,000
New Mexico	1,700	22,500	15,000	26,900	30,000	30,000	36,000	35,000	30,000
North Dakota	200	14,200	6,900	8,100	5,700	7,800	10,000	5,200	7,500
Oklahoma	100	200	300	200	400	600	1,500	1,500	2,000
Oregon	2,000	8,900	10,000	11,300	14,000	22,600	17,100	15,000	19,900
South Dakota	700	27,400	25,000	35,500	67,000	35,000	47,900	20,500	33,300
Texas	2,400	9,400	15,900	10,500	12,000	14,500	12,000	9,600	10,900
Utah	700	1,000	1,200	2,600	6,000	7,000	11,000	10,500	14,000
Wyoming	7,000	140,000	200,000	168,000	608,000	450,800	441,500	370,300	467,000
Subtotal	26,900	364,800	416,100	408,300	1,019,400	968,600	922,300	669,000	765,200
Total	30,600	386,300	435,100	431,600	1,051,500	1,018,700	962,400	692,800	799,300

[a] Nelson (1925).
[b] Yoakum (1968).
[c] Sundstrom et al. (1973).
[d] Yoakum (1978).
[e] Yoakum (1986).
[f] O'Gara (2000).
[g] Yoakum et al. (1999).
[h] Wildlife Management Institute (2001).
[i] Ticer and Munig (1999).
[j] Instituto Nacional de Ecologia (2000).

Soon after World War II, wildlife agencies began using fixed-wing aircraft to survey pronghorn. This greatly improved data collection, by covering more land area systematically. Wildlife biologists were trained in aerial survey methodology, and population estimates were checked by repetitive procedures. Although aerial surveys only provided estimates of populations, they were more accurate than previous ground surveys. Fixed-wing airplanes now are used more than any other aircraft for surveying herds of pronghorn. Above is an airplane and pilot from Arizona assisting a pronghorn biologist to plan a survey of peninsular pronghorn on the Vizcaino Desert located in Baja California Sur, Mexico. Before the use of airplanes during the mid-1980s, pronghorn were counted by ground surveys in Baja California Sur. Use of aircraft has increased the ability to obtain a relatively accurate and timely survey of these pronghorn that are few in number and widely scattered. Use of ultralight aircraft is much less expensive than use of fixed-wing airplanes or helicopters. Ultralights also can provide the pilot with opportunities for close aerial photographs of animals, but their use in covering large landscapes is limited. Photo by Patrick Robles.

Fish and Wildlife Service (formerly the Bureau of Biological Survey) assembled information and estimated pronghorn populations for 30 years (Table 6). Most of the data were obtained from reports by state wildlife agencies, although some findings were obtained from private sources and federal agencies. At times, the validity of estimates were questioned, but the population trend of pronghorn in the United States increased tremendously. Based on available information, estimated pronghorn numbers increased from 26,600 in 1924 to 284,600 in 1955 (Table 6)—more than a 10-fold increase in 30 years. By 1955, the future of pronghorn had brightened considerably and the species was no longer considered on a course toward extinction.

1957 Survey. Since the mid-1950s, pronghorn trends have been analyzed by independent surveys. The first such endeavor was conducted in 1957 by Hoover et al. (1959), who tabulated a total of 259,300 in the United States. No data were reported for Canada or Mexico.

1964 Survey. Not until 1964 was a comprehensive survey of pronghorn populations conducted, similar to that of Nelson (1925), for Canada, Mexico and the United States. This was accomplished by Yoakum (1968), who contacted the wildlife agencies of provinces and states in Canada and the United States that had pronghorn populations. Data for Mexico were obtained from Starker A. Leopold at the University of California, Berkeley, who had compiled status reports for game birds and mammals of Mexico (Leopold 1959a). Results of the survey are listed in Table 5, with totals of 20,300 for Canada, 1,200 for Mexico and 364,800 for the United States. Populations increased in Canada and the United States, but a downward trend prevailed in Mexico.

1970s Surveys. Sundstrom et al. (1973) conducted a survey for each province or state with pronghorn in Canada, Mexico and the United States. Results were 18,000 in Canada, 1,000 in Mexico and 416,100 in the United States (Table 5). Pronghorn in Mexico were limited to the states of Baja California, Chihuahua and Sonora. For the United States, 100 animals were reported for Washington and 130 for Hawaii—both populations the result of translocations. The Sundstrom et al. survey also indicated that 46% of the pronghorn population in the United States was located in Wyoming.

During spring 1977, a questionnaire was sent to provinces in Canada and western state wildlife agencies in the United States to obtain the 1976 pronghorn population estimates for the *Big Game of North America* (Yoakum 1978a). Canada's pronghorn population continued an upward trend to 22,300 animals; United States' herds totaled 408,300; Mexico's population was estimated at 1,000 (Table 5).

1983 and 1984 Survey. A comprehensive survey of pronghorn was compiled for Canada, Mexico and the United States in 1983 and 1984 (Yoakum 1986). Most of the population figures were obtained from the *Proceedings of the 11th Pronghorn Antelope Workshop,* with additional information obtained by correspondence for areas not provided in the proceedings. The survey indicated an all-time high pronghorn population for the 20th century—1,084,500—with subtotals of 31,500 for Canada, 600 for Mexico and 1,019,400 for the United States (Table 5). It appears that, in 1984, pronghorn were second in numbers only to deer (mule and white-tailed) for big game in the United States and probably second or third to caribou and deer continentally.

In the United States, the Great Plains states (Colorado, Montana, South Dakota and Wyoming) contained 89% of all pronghorn. Wyoming and Montana alone had 76%. The Intermountain West and Pacific states (California, Idaho, Nevada, Oregon and Utah) had 4%; the remainder were in Arizona, Kansas, Nebraska, New Mexico, North

TABLE 6. Estimated pronghorn numbers from U.S. Fish and Wildlife Service big game inventories, 1922 to 1955

Year	*Estimated population*	*Notes*	*References*[a]
1922–24	26,000[b]	First major survey	Bulletin 1346
1937	131,600[b]	Inventory for 15 states	Wildl. Research and Manage. Leaflet BS-172
1938	186,100[b]	Inventory for 15 states	Wildl. Leaflets BS-142 and 175
1940	176,600[b]	Inventory for 16 states (Washington added)	Wildl. Leaflet 207
1941	199,400	Inventory for 16 states	Research Report 8
1942		No inventory conducted	
1943	246,000	Inventory for 16 states	Wildl. Leaflet 283
1944		No inventory conducted during these years of World War II	
1946	233,900	Inventory for 16 states	Wildl. Leaflet 303
1947	181,200	Inventory for 15 states; no figures given for one state	Wildl. Leaflet 321
1948	187,900	Inventory for 15 states; no figures given for one state	Wildl. Leaflet 321
1949	208,000	Inventory for 16 states	Wildl. Leaflet 321
1950	225,800	Inventory for 15 states; no figures given for one state	Wildl. Leaflet 342
1951	236,000[b]	Inventory for 14 states; no figures given for two states	Wildl. Leaflet 342
1952	248,100	Inventory for 16 states; no figures given for one state (Kansas added for first year)	Wildl. Leaflet 345
1953	273,200	Inventory for 17 states	Wildl. Leaflet 364
1954	262,300	Inventory for 17 states	Wildl. Leaflet 376
1955	284,600	Inventory for 17 states	Wildl. Leaflet 387

[a] All published by the U.S. Fish and Wildlife Service.
[b] Figures rounded to closest 100.

Dakota, Oklahoma and Texas. Hawaii and Washington were surveyed, but each had less than 100 animals, so were not tabulated. The entire population for Canada was located in the Great Plains region of Alberta and Saskatchewan.

From 1924 to 1984, states in the United States with grassland environments experienced pronghorn increases of 2,000% to 11,000%. During the same time, states with shrubsteppe habitat had increases of 130% to 5,200%. Canada's grasslands increased 2,300%. These long-term trends appear to indicate that population densities were related to differing carrying capacities of grasslands and shrubsteppes.

The increase in pronghorn from 1924 to 1984 was—a virtual irruption. Table 7 and Figures 17 and 18 provide data regarding this 60-year trend, with additional information for the period 1984 to 1999. With 3% of the continental pronghorn population, Canada had a 2,400% increase, whereas the United States, with 97% of the population, experienced a 3,900% increase. Mexico, conversely, with less than 1% of North America's pronghorn, showed a 75% decline. From 1924 to 1984, the total population of the species increased 3,500%.

Pronghorn estimates for Mexico were obtained from a survey conducted in 1984 by Gonzalez and Lafon (1993). It apparently was the first comprehensive survey of pronghorn in Mexico. Herd locations and numbers for the states of Baja California Sur, Chihuahua, Coahuila and Sonora are provided in Figure 15 and Tables 5 and 8. Gonzalez and Lafon demonstrated that the population had been shrinking for 60 years at a rate of about 30 animals per year, and these investigators expressed concern for the population's future.

TABLE 7. Differences in pronghorn numbers in North America from 1924 to 1984

			Difference between 1924 and 1984		
Country	*1924 estimated population*	*1984 estimated population*	*Number of animals*	*Percentage change*	*Percentage of total population in 1984*
Canada	1,300	31,500	+30,200	+2,400	3
Mexico	2,400	600	–1,800	–75	T
United States	26,900	1,052,400	+982,300	+3,900	97
Total	30,600	1,084,500	1,010,700	3,500	100

T = trace.
Source: Yoakum 1986.

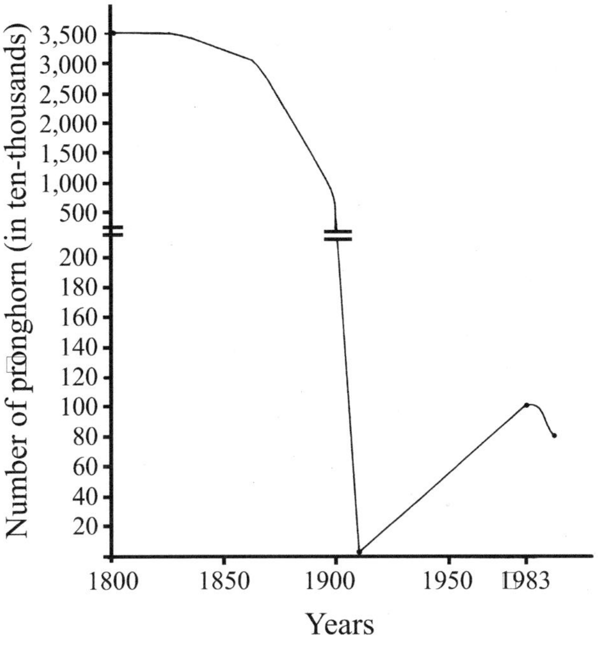

FIGURE 17. *Speculated pronghorn population trends in North America, 1800 to 2000. Numbers are rounded to closest thousand.*

Reasons for the total population increase during that 60-year span include: (1) protection from overhunting by laws favoring conservative harvests; (2) translocation of herds from well-populated habitats to unoccupied historic rangelands; (3) major decreases in domestic sheep partially eliminated significant competition for forage classes preferred by pronghorn;

(4) development of thousands of water improvements, primarily for livestock, has made seasonal habitats available that previously had limited use due to lack of drinking water; (5) a depressed livestock economy for several decades has decreased the number of cattle using western rangelands, resulting in more forage for pronghorn; and (6) the emergence of wildlife science and management practices based on science more than trial and error.

1989 and 1995 Surveys. During the business meetings for the 15th and 16th Pronghorn Antelope Workshops, the need was expressed for periodic pronghorn population updates. O'Gara (2000) conducted such surveys in 1989 and 1995 for Canada and the United States (Table 5). Estimated numbers were requested for prehunt figures.

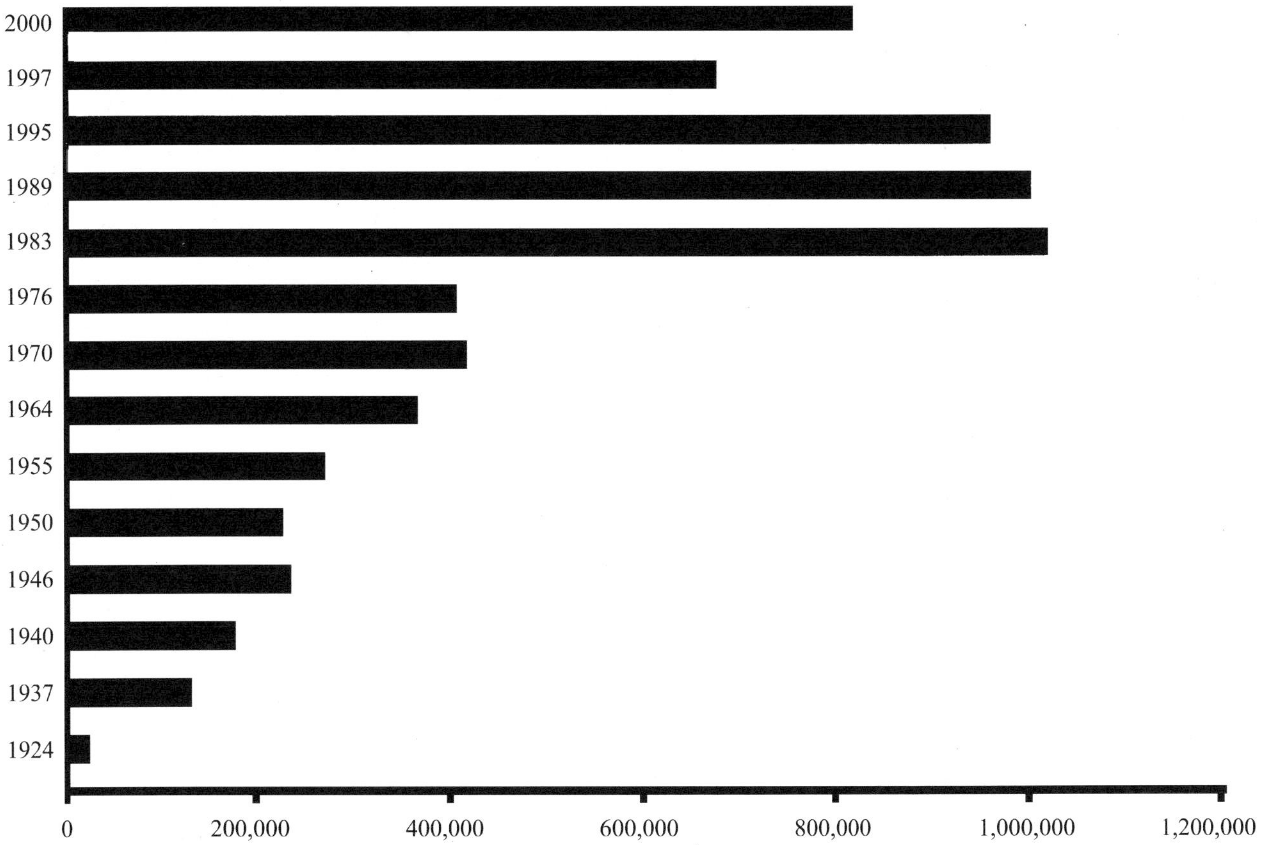

FIGURE 18. *Population estimates of pronghorn in North America (Canada, Mexico and the United States) from 1924 through 2000. Numbers are rounded to closest thousand.*

After the Great Plains grasslands, the shrubsteppe biome is where pronghorn are most abundant. Fawn recruitment and herd trends vary greatly in relation to the condition of preferred forage, which is related directly to weather patterns, mainly precipitation. Highest pronghorn numbers are associated with a mixture of grasses, forbs and shrubs. Forbs high in nutritional values during spring and summer help produce healthy fawns. Shrubs during winter are staple survival forage. Photo by Jim D. Yoakum.

Table 8. Trend in pronghorn populations from 1924 to 1984 in Mexico

Time period	*Estimated population*	*Change in years* — *Number of animals*	*Change in years* — *Percentage*	*Average number of animals lost per year*
1924 to 1973	2,400 to 800	−1,600	67	33
1973 to 1978	800 to 650	−150	19	30
1978 to 1984	650 to 446	−204	31	34

Source: Gonzalez and Lafon 1993.

Therefore, data from states submitting posthunt data were adjusted by adding the year's harvest plus 10% of harvest for crippling losses and illegal kills. California and Oregon submitted incomplete estimates, but the majority of population numbers were submitted. Data for states in Mexico were included by representatives of Mexico at the 16th Pronghorn Antelope Workshop. Total numbers in 1989 and 1995 were, respectively, 17,800 and 89,100 less than the population of 1,051,500 for 1983, reported by Yoakum (1986).

The total pronghorn population remained relatively stable from 1983 to 1995, whereas certain provincial and state populations fluctuated extensively. For example, pronghorn numbered 67,000 in South Dakota in 1983, but dropped to 35,000 in 1989 and increased to 47,900 in 1995. Reasons for these population oscillations are not well understood, but indications are that they were related to varying weather patterns, which directly affected forage production and nutritional values. However, there is continued permanent loss of habitat that affects populations as documented with recent geographic information system habitat surveys in Arizona (Ockenfels et al. 1996). Long-term research is needed to ascertain the relationships among weather patterns, forage production and population fluctuations for pronghorn herds in grasslands, shrubsteppes and deserts. Such information would provide some ability to predict trends in population numbers.

1997 Survey. An extensive survey was accomplished by mailing questionnaires to wildlife agencies in provinces and states with pronghorn herds in Canada, Mexico and the United States for 1997 (Yoakum et al. 1999). Estimated numbers were received from all contacted government agencies in Canada and the United States (Table 5). No response to the questionnaire was received from Mexico, but comparable data were obtained from the 18th Pronghorn Antelope Workshop Proceedings (Ticer and Munig 1999).

When population numbers for 1997 were compared with data for 1995, major changes were evident. Colorado, Kansas, Nebraska and New Mexico indicated increases, whereas Alberta, Saskatchewan, Montana, North Dakota and South Dakota reported major decreases (Table 9). Total numbers indicated a 25% to 30% decrease from 1995 to 1997 and a 34% decrease from 1984.

2000 Survey. The last survey conducted in the 20th century was obtained for this book (Wildlife Management Institute 2001). Provincial and state agencies responsible for pronghorn management in Canada, Mexico and the United States were contacted for estimates in 1998 and 1999. Responses were received from all agencies in Canada and the United States. The questionnaire requested data for prehunt estimated numbers. Survey results are shown in Table 5.

Montana was unable to provide a population number, so an estimate was made by applying the same proportion (11.2%) of the 1997 population to the 2000 survey. The substituted number denotes a minimum number (G. Erickson personal communication:2000). However, Montana's trend information seems consistent with that for surrounding similar areas.

Estimated pronghorn numbers in Mexico were obtained from a management report prepared by the federal government of Mexico (Secretaria de Medio Ambiente, Recursos Naturales y Pesca 2000).

During the last half of the 1990s, the North American pronghorn population numbered between 650,000 and 800,000 animals, with a slight increasing trend for the 4 years at the end of the 20th century.

Twentieth century reports indicate that pronghorn numbers increased dramatically from 1915 to 1983, remained relatively static until 1995, then plummeted approximately 30% in 2 years (Figure 15). Severe winters, droughty summers, lack of quality forage and coyote predation on the northern plains, where pronghorn suffered the most dramatic declines, were reported as major decimating factors.

Piecemeal attrition of pronghorn habitat, particularly in critical wintering and fawning areas, as well as seasonal travel corridors, was accelerating. During the economic surge of the 1980s and 1990s, intensified agriculture, subdivision, energy developments, upgraded roads and other activities prevailed. Such habitat modification and loss may influence negatively the recovery of pronghorn from inevitable weather-induced die-offs. Clearly, loss of pronghorn habitat translates to loss of pronghorn. Unfortunately, habitat losses tend to be insidious, go unreported and are not analyzed before the loss is substantial.

Wyoming prevails as the geographic center of pronghorn abundance. Sundstrom et al. (1973) reported Wyoming sustained 46% of all pronghorn 30 years ago, whereas the state in year 2000 maintained closer to 60%. Apparently, rangeland with an abundance of shrubsteppe/grasslands, low human densities and relatively low farming acreage retain the combination of ecological factors that produce large numbers of pronghorn.

DENSITY

The density of pronghorn is determined by dividing the area of occupied habitat by the number of animals and, usually, is reported as the number of animals/square mile. Density data for landscapes may be compiled for a local site, management unit, region, state, province or nation. Some examples follow. Density also has been estimated for pronghorn occupying different vegetative communities, and such data can aid in determining the values of vegetative characteristics in relationship to animal carrying capacities (see Chapter 13).

TABLE 9. Trend (percentage change) in pronghorn numbers in Canada and the United States from 1995 to 1997

Increased	*Stable*	*Decreased less than 25%*	*Decreased 25% or more*
Nebraska (37)	Oklahoma	Texas (20)	Montana (66)
Kansas (28)		Wyoming (16)	South Dakota (56)
New Mexico (14)		California (14)	Saskatchewan (55)
Colorado (6)		Oregon (12)	North Dakota (48)
		Arizona (7)	Alberta (25)
		Nevada (7)	
		Utah (5)	
		Idaho (4)	

Source: Yoakum et al. 1999.

An area encompassing the western area of the Dakotas and eastern Montana and Wyoming is reported to produce consistently the highest recruitment level for pronghorn (Vriend and Barrett 1978). This is the ecotone for mixed and short grasslands of the Great Plains, and sustains the highest density of pronghorn for any biome in North America. These prairies are dominated by grasses and forbs, with low-growing shrubs primarily in coulees. Pronghorn populations fluctuate in relation to winter severity and tend to recover quickly after winter decimation. Photo by Harold Umber; courtesy of the North Dakota Department of Game and Fish, Bismarck.

1800s Reports

Seton (1927) estimated two densities for the late 1800s. He reported that a 1,000-square mile (2,600 km^2) area in Yellowstone National Park had an estimated 1,000 pronghorn in 1896— an average of one animal per square mile (0.4/km^2) of habitat. He noted that hunters considered pronghorn scarce at that time for the area compared with other rangelands of real "Antelope Country." Seton cited another record of a concentration of about 2 million pronghorn seen from a train traveling from Denver, Colorado, to Cheyenne, Wyoming, during the winter of 1868–69. Based on these and other records, Seton estimated that some regions of pronghorn habitat during the 1800s maintained densities of more than 20 pronghorn per square mile (7.7/km^2) for grasslands on the Great Plains. However, the possibility that anyone could tally 2 million animals while passing them on a train seems remote. Also, calculating without the benefit of marked animals and an extensive study or the area over which a wintering herd would disperse, would produce mere speculation.

1964 Report

During 1964, Yoakum (1968) contacted state wildlife agencies in the western United States regarding rangelands occupied by pronghorn. At that time, the total population was estimated at 364,800. Wyoming had the highest density, with more than 2 pronghorn per square mile (0.8/km^2). Montana, Nebraska, New Mexico, North Dakota and South Dakota averaged 1 to 2 per square mile (0.4–0.8/km^2). All other states had less than 1 animal per square mile (0.4 km^2) of habitat (Table 10). Thus, states with and adjoining grasslands had the highest pronghorn

Table 10. Estimated density of pronghorn on rangelands in the United States during 1964

More than 2 per square mile (0.8/km^2)	*1–2 per square mile (0.4–0.8/km^2)*	*Less than 1 per square mile (0.4/km^2)*
Wyoming	Colorado	Arizona
	Montana	California
	Nebraska	Idaho
	New Mexico	Kansas
	North Dakota	Nevada
	South Dakota	Oklahoma
		Oregon
		Texas
		Utah
		Washington
(140,000 pronghorn)	(183,400 pronghorn)	(41,700 pronghorn)

Source: Yoakum 1968.

densities, whereas states in the Intermountain West and the deserts supported lower population levels.

The pronghorn population densities in Table 10 were derived from statewide, yearlong data. They do not reflect seasonal variations, which are most pronounced during winter concentrations for northern populations. Likewise, states maintaining less than 0.1 pronghorn per square mile (0.04/km²) can be classified as peripheral habitat, although they may have sustained herds historically.

The biome with the lowest density of pronghorn is the desert. "Most of the time, their food is much less digestible than the food enjoyed by pronghorn farther north. Consequently, they will likely have fewer and smaller fawns, less milk or run out of milk sooner. The fawns will likely grow slowly into adulthood with low performance for life" (Geist and Francis 2001:46). This density rating existed before the arrival of Euro-Americans and remains the same at the beginning of the 21st century. Photo courtesy of the U.S. Fish and Wildlife Service, Washington, D.C.

1982 Report

In 1982, a report of pronghorn densities in the United States was released by the Southeastern Cooperative Wildlife Disease Study Team, at the University of Georgia, Athens. This analysis was performed around the time pronghorn were at their population zenith in the 20th century (Figure 19). Population data were obtained from state wildlife agencies. Four densities were designated—more than 10 pronghorn per square mile (3.9/km²), 5 to 10 per square mile (1.9–3.9/km²), 1 to 5 per square mile (0.4–1.9/km²) and areas where pronghorn were rare or absent (Table 11). The areas were computed with a planimeter to determine area of occupied habitats, and these findings also are listed in Table 11. Such densities may fluctuate for some areas because of seasonal movements. Areas in Montana depict distribution only, because density figures were not available, except for a 30-square mile (78 km²) area in Yellowstone National Park.

Only Wyoming reported densities of 10 or more per square mile (3.9/km²) for 1,000 acres (405 ha) or more, and this accounted for 2% of pronghorn habitat in the state. Four states reported areas with 5 to 10 pronghorn per square mile (1.9–3.9/km²); 12 states reported 1 to 5 per square mile (0.4–1.9/km²), and 15 states reported herds for areas with less than 1 pronghorn per square mile (0.4/km²) (Table 11). The total pronghorn population was estimated at the following densities in the United States (excluding Montana because data were unavailable): 0.5% for more than 10 pronghorn per square mile (3.9/km²); 5.2% for 5 to 10 per square mile (1.9–3.9/km²); 29.6% for 1 to 5 per square mile (0.4–1.9/km²); and 51.0% for fewer than 1 pronghorn per square mile (0.4/km²) for the remaining herds. Colorado, Montana and Wyoming had 49.4% of pronghorn habitat and reported estimated populations totaling close to 82% for the early 1980s in the United States.

Wyoming Report

A study of the 3,530-square mile (9,143 km²) Red Desert region in south-central Wyoming during 1968 and 1969 provided information regarding density differences for summer and winter rangelands (Sundstrom 1970). Summer habitat averaged 3.1 pronghorn per square mile (1.2/km²), ranging from 0.2 to 8.6 pronghorn per square mile (0.08–3.31/km²) in 20 different counting blocks used by 11,000 animals (Table 12). The same study area contained 12 winter units within 729 square miles (1,888 km²) and 8,000 pronghorn at densities of 2 to 26 per square mile (0.8–10.0/km²). Sundstrom considered forage quality and water distribution as conditions most affecting summer numbers. Factors influencing winter densities were average winter snow depths, availability of shrubs for forage and low, rolling hills providing cover from chilling winds.

Montana Report

Working on a shrubsteppe in the Yellow Water Triangle area of central Montana during December 1967 through 1978, Pyrah (1987) reported average densities of 3.8 pronghorn per square mile (1.5/km²) for 10 herds. Average densities for a 386-square mile (1,000 km²) area were computed based on adult and fawn numbers and years (Table 13). Pyrah (1987:44) stressed density as an important population parameter, but hastened to qualify that, "Because *density* is the result of the effects of a num-

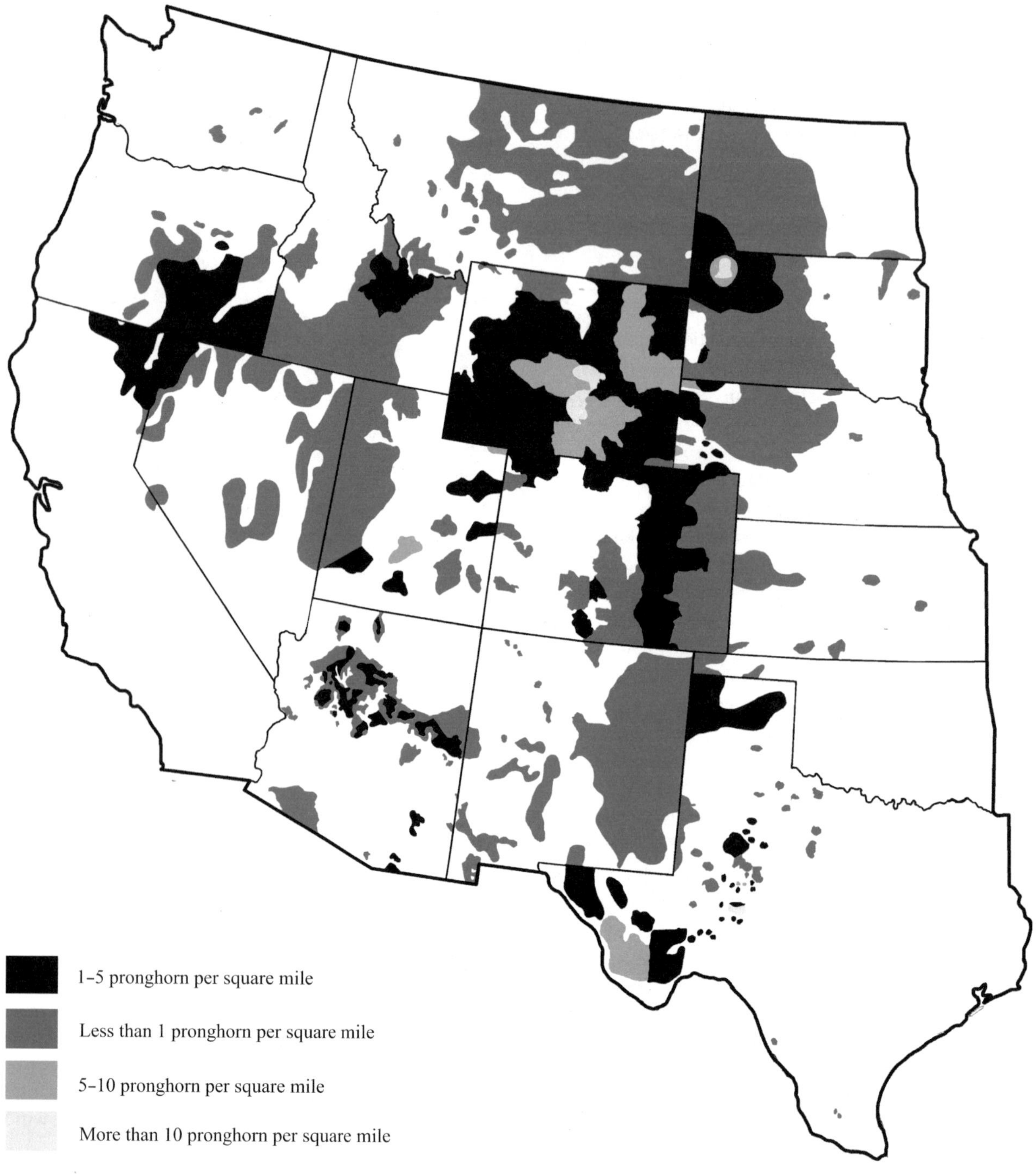

Figure 19. *Distribution of pronghorn by population density where suitable habitat existed in the United States in 1982 (Southeastern Cooperative Wildlife Disease Study 1982).*

ber of interacting factors, density *per se* may not be as important as other factors: food abundance, competition, security, juxtaposition and interspersion of habitat needs, and a multitude of others." Montana experienced a severe winter die-off during 1977–78. However, after the die-off, herd numbers rapidly returned to 1968 densities. It also was noted that densities were more similar among herds at low population levels than among herds at high levels.

The Yellow Water Triangle report is unique because it provides separate data regarding doe and fawn densities. Table 14 provides information for adult females. Adult

TABLE 11. Suitable pronghorn habitat by pronghorn density in the United States in 1982, as reported by the Southeastern Cooperative Wildlife Disease Study, University of Georgia, Athens[a]

	Suitable habitat by pronghorn density[b]						
State	*More than 10*	*5–10*	*1–5*	*Less than 1*	*Suitable habitat*[b]	*% of total area*	*% of total pronghorn habitat in United States*
Arizona	— —	— —	5,920 (15,332)	12,910 (33,436)	18,830 (48,769)	16.6	3.7
California	— —	— —	7,230 (18,725)	690 (1,787)	7,920 (20,512)	5.0	1.6
Colorado	— —	— —	24,630 (63,791)	29,610 (76,689)	54,240 (140,481)	52.2	10.6
Idaho	— —	— —	5,030 (13,027)	27,590 (71,458)	32,620 (84,485)	39.4	6.4
Kansas	— —	— —	— —	5,360 (13,882)	5,360 (13,882)	6.5	1.1
Montana	— —	— —	30 (78)		68,780 (178,140)	47.2[c]	13.5
Nebraska	— —	640 (1,657)	840 (2,175)	16,680 (43,201)	18,160 (47,034)	23.6	3.6
Nevada	— —	— —	3,200 (8,288)	27,740 (71,846)	30,940 (80,134)	28.0	6.1
New Mexico	— —	— —	— —	44,020 (114,011)	44,020 (114,011)	36.2	8.6
North Dakota	— —	— —	2,500 (6,475)	29,490 (76,379)	31,990 (82,854)	46.1	6.3
Oklahoma	— —	— —	— —	1,550 (4,014)	1,550 (4,014)	2.2	0.3
Oregon	— —	60 (155)	16,490 (42,709)	7,610 (19,709)	24,160 (62,574)	25.1	4.8
South Dakota	480 (1,243)	850 (2,201)	9,980 (25,848)	26,310 (68,142)	37,620 (97,435)	49.1	7.4
Texas	20 (51.8)	4,100 (10,619)	17,130 (44,366)	3,560 (9,220)	24,810 (64,257)	9.5	4.9
Utah	— —	1,360 (3,522)	5,040 (13,053)	22,890 (59,285)	29,290 (75,861)	35.6	5.8
Wyoming	2,020 (5,231)	19,420 (50,297)	53,120 (137,580)	3,230 (8,365)	77,790 (201,476)	79.9	15.3
Total	2,520[d] (6,526)	26,430[d] (68,453)	151,140[c] (391,452)	259,240[d] (671,431)	508,080 (1,315,927)		
Percentage of total habitat	0.5	5.2	29.6	51.0			

[a] Hawaii and Washington State populations were less than 100 animals, so were excluded.

[b] Square miles (km^2).

[c] Montana density data not provided, except for 30 square miles (78 km^2), as noted.

[d] Lacks density data for Montana.

female density varied over years among herds and was believed related to habitat quality. Average doe density for the 10-year study was 1.9 per square mile (0.7/km^2). Fawn density averaged 1.3 per square mile (0.5/km^2) (Table 15); yearly averages varied from a low of 0.9 per square mile (0.3/km^2) in 1974 to a high of 1.8 per square mile (0.7/km^2) in 1972. The highest fawn density in any one year was 4.6 per square mile (1.8/km^2). However, fawn density on a nearby rangeland reached 8.8 per square mile (3.4/km^2) during 1971 and 1972.

Pyrah's (1987) findings relative to fawn/doe densities and population dynamics may be of importance to managers of pronghorn for all biomes. This is especially true for areas where managers reported an increasing herd population, although fawn ratios were reported as static or declining. Pyrah (1987:50–51) stated: "Herd production more closely reflected the number of productive does in the herd than the number of fawns each doe raised. Fawn density was positively correlated with doe density within herds as well as overall . . . especially during years of population increase (1968 to 1973), suggesting that successful fawn production causes high density. Conversely, fawn/doe ratios generally were negatively correlated with densities of does within and over all herds. Correlations between fawn density and fawn/doe ratios were varied, but generally negative during 1968 to 1973 and positive during 1974 to 1978. Because of this, fawn densities for most herds and the total number of fawns produced in the Yellow Water population continued to increase or remained high into the mid-1970s, while doe/fawn ratios declined."

This phenomenon exists for herds in other areas, especially northern Great Basin environments. It may be appropriate for wildlife biologists to evaluate Pyrah's findings further and determine if they have biological significance in other areas.

1984 Assessment

Using data from Yoakum's (1986) comprehensive survey on population size for 1984, and factoring in square

TABLE 12. Pronghorn densities for summer and winter rangelands on the Red Desert, Wyoming, for 1968–69

Counting block	*Summer rangelands (1969)* Square miles (km^2)	Number of pronghorn	Density per square mile (km^2)	*Winter rangelands (1968–69)* Square miles (km^2)	Number of pronghorn	Density per square mile (km^2)
1	153 (396)	1,321	8.6 (3.3)	79 (205)	2,017	26.0 (9.8)
2	300 (777)	1,281	4.3 (1.6)	109 (282)	685	6.0 (2.4)
3	133 (344)	858	6.5 (2.5)	8 (21)	144	14.0 (6.9)
4	396 (1,026)	772	1.9 (0.8)	49 (127)	174	4.0 (1.4)
5	275 (712)	1,343	4.9 (1.9)	16 (41)	92	6.0 (2.2)
6	206 (534)	1,547	7.5 (2.9)	13 (34)	55	4.0 (1.6)
7	145 (376)	323	2.2 (0.9)	186 (482)	3,968	21.0 (8.2)
8	92 (238)	113	1.2 (0.5)	42 (109)	325	8.0 (3.0)
9	44 (114)	96	2.2 (0.8)	16 (41)	115	7.0 (2.8)
10	118 (306)	118	1.0 (0.4)	21 (54)	295	14.0 (5.5)
11	299 (774)	1,866	6.2 (2.4)	10 (26)	70	77.0 (2.7)
12	135 (350)	143	1.1 (0.4)	180 (466)	344	2.0 (0.7)
13	75 (194)	134	1.8 (0.7)			
14	165 (427)	342	2.1 (0.8)			
15	144 (373)	169	1.2 (0.5)			
16	126 (326)	20	0.2 (0.1)			
17	104 (264)	122	1.2 (0.5)			
18	202 (523)	46	0.2 (0.1)			
19	117 (303)	102	0.9 (0.3)			
20	304 (787)	62	0.2 (0.1)			
Total	3,533 (9,150)	10,778	3.1 (1.2)	729 (1,888)	8,284	11.0 (4.4)

Source: Sundstrom 1970.

TABLE 13. Late July to early September densities of adult and fawn pronghorn per square mile (km^2) by herd in the Yellow Water Triangle, Montana, 1968 through 1978

Year	*Herd* I	II	IV	V	VI	VII	*Average*
1968	1.89 (0.73)	2.55 (0.98)	4.38 (1.69)	3.22 (1.24)	2.14 (0.83)	3.65 (1.41)	3.38 (1.30)
1969	2.48 (0.96)	3.66 (1.41)	6.04 (2.33)	3.58 (1.38)	3.01 (1.16)	2.99 (1.15)	3.38 (1.30)
1970	2.23 (0.86)	6.14 (2.37)	7.69 (2.97)	3.40 (1.31)	3.71 (1.43)	5.48 (2.12)	4.05 (1.56)
1971	4.34 (1.68)	6.81 (2.63)	5.56 (2.15)	3.76 (1.45)	3.51 (1.35)	5.04 (1.95)	4.31 (1.66)
1972	3.78 (1.46)	10.55 (4.07)	11.36 (4.38)	5.27 (2.03)	3.33 (1.29)	4.10 (1.58)	5.10 (1.96)
1973	5.18 (2.00)	7.18 (2.77)	11.48 (4.43)	5.54 (2.14)	3.71 (1.43)	6.75 (2.60)	5.32 (2.05)
1974	2.70 (1.04)	7.11 (2.74)	9.23 (3.56)	4.57 (1.76)	2.99 (1.15)	5.87 (2.27)	4.24 (1.64)
1975	1.64 (0.63)	6.66 (2.57)	6.27 (2.42)	5.60 (2.16)	3.38 (1.30)	7.69 (2.97)	4.59 (1.77)
1976	2.98 (1.15)	5.59 (2.16)	6.39 (2.47)	5.91 (2.28)	3.13 (1.21)	6.53 (2.52)	4.44 (1.71)
1977	2.79 (1.08)	5.63 (2.17)	4.38 (1.69)	4.12 (1.59)	3.27 (1.26)	5.15 (1.98)	3.87 (1.49)
1978	2.67 (1.03)	5.92 (2.29)	4.38 (1.69)	3.38 (1.27)	2.63 (1.02)	4.59 (1.77)	3.44 (1.33)
Average	2.97 (1.15)	6.16 (2.38)	7.01 (2.71)	4.40 (1.69)	3.16 (1.22)	5.26 (2.03)	3.86 (1.49)

Source: Pyrah 1987.

TABLE 14. Late July to early September densities of adult female pronghorn per square mile (km²) by herd in the Yellow Water Triangle, Montana, 1968 through 1978

	Herd						
Year	*I*	*II*	*IV*	*V*	*VI*	*VII*	*Average*
1968	0.71	1.15	1.30	1.19	0.75	1.22	0.95
	(0.27)	(0.44)	(0.50)	(0.46)	(0.28)	(0.47)	(0.36)
1969	1.12	1.41	2.72	1.67	1.42	1.55	1.48
	(0.43)	(0.54)	(1.05)	(0.64)	(0.55)	(0.59)	(0.57)
1970	1.08	2.66	4.26	1.69	1.78	2.55	1.93
	(0.42)	(1.03)	(1.64)	(0.65)	(0.69)	(0.98)	(0.74)
1971	2.20	2.96	2.72	1.37	1.56	2.10	1.89
	(0.84)	(1.14)	(1.05)	(0.53)	(0.60)	(0.81)	(0.73)
1972	1.92	4.59	4.73	1.89	1.65	1.94	2.26
	(0.74)	(1.77)	(1.83)	(0.73)	(0.63)	(0.75)	(0.87)
1973	2.91	3.29	4.73	2.41	1.89	2.82	2.52
	(1.12)	(1.27)	(1.83)	(0.93)	(0.73)	(1.09)	(0.97)
1974	1.61	3.37	4.85	2.12	1.71	3.04	2.23
	(0.62)	(1.30)	(1.87)	(0.82)	(0.66)	(1.17)	(0.86)
1975	1.21	3.89	3.08	2.93	1.83	3.71	2.43
	(0.47)	(1.50)	(1.19)	(1.13)	(0.71)	(1.43)	(0.94)
1976	1.77	3.18	3.08	2.99	1.61	2.88	2.29
	(0.68)	(1.23)	(1.19)	(1.15)	(0.62)	(1.11)	(0.88)
1977	1.89	2.55	2.01	1.80	1.57	2.32	1.87
	(0.73)	(0.98)	(0.78)	(0.69)	(0.61)	(0.90)	(0.72)
1978	1.36	2.63	1.66	1.37	1.40	2.10	1.62
	(0.52)	(1.02)	(0.64)	(0.53)	(0.54)	(0.81)	(0.63)
Average	1.62	2.88	3.19	1.95	1.56	2.38	1.95
	(0.63)	(1.11)	(1.23)	(0.75)	(0.60)	(0.92)	(0.75)

Source: Pyrah 1987.

TABLE 15. Late July to early September pronghorn fawn density per square mile (km²) by herd in the Yellow Water Triangle, Montana, 1968 through 1978

	Herd						
Year	*I*	*II*	*IV*	*V*	*VI*	*VII*	*Average*
1968	0.96	1.07	1.78	1.17	0.92	1.66	1.09
	(0.37)	(0.41)	(0.69)	(0.45)	(0.36)	(0.64)	(0.42)
1969	0.87	1.30	2.60	1.42	1.02	1.16	1.18
	(0.34)	(0.50)	(1.00)	(0.55)	(0.39)	(0.45)	(0.46)
1970	0.77	2.18	2.01	1.10	1.37	1.99	1.40
	(0.30)	(0.84)	(0.78)	(0.42)	(0.53)	(0.77)	(0.54)
1971	1.36	2.33	1.78	1.26	1.19	2.05	1.47
	(0.52)	(0.90)	(0.69)	(0.49)	(0.46)	(0.79)	(0.57)
1972	1.21	3.48	4.62	2.14	1.03	1.27	1.76
	(0.47)	(1.34)	(1.78)	(0.83)	(0.40)	(0.49)	(0.68)
1973	1.46	1.92	3.55	2.14	1.16	2.82	1.74
	(0.56)	(0.74)	(1.37)	(0.83)	(0.45)	(1.09)	(0.67)
1974	0.53	1.33	1.42	1.17	0.64	1.60	0.93
	(0.20)	(0.51)	(0.55)	(0.45)	(0.25)	(0.62)	(0.36)
1975	0.28	2.00	1.89	1.71	0.94	2.21	1.28
	(0.11)	(0.77)	(0.73)	(0.66)	(0.36)	(0.85)	(0.49)
1976	0.77	1.48	2.37	1.51	0.99	2.21	1.28
	(0.30)	(0.57)	(0.91)	(0.58)	(0.38)	(0.85)	(0.49)
1977	0.65	1.48	2.01	1.55	0.92	1.88	1.21
	(0.25)	(0.57)	(0.78)	(0.60)	(0.36)	(0.73)	(0.47)
1978	0.93	1.92	2.13	0.83	0.83	1.77	1.12
	(0.36)	(0.74)	(0.83)	(0.32)	(0.32)	(0.68)	(0.43)
Average	0.89	1.86	2.38	1.45	1.00	1.93	1.32
	(0.34)	(0.72)	(0.92)	(0.56)	(0.39)	(0.74)	(0.51)

Source: Pyrah 1987.

Table 16. Pronghorn density in the United States, 1984

State	Pronghorn population	Habitat in Square miles	Habitat in (km²)	Density in Square miles	Density in (km²)
Wyoming	608,000	77,790	(201,460)	7.8	(3.0)
Montana	161,500	68,780	(178,126)	2.3	(0.9)
South Dakota	67,000	37,620	(97,428)	1.8	(0.6)
Colorado	57,500	54,250	(140,496)	1.1	(0.4)
California	6,800	7,920	(20,511)	0.9	(0.3)
New Mexico	30,000	44,020	(114,002)	0.7	(0.3)
Idaho	21,500	32,600	(84,479)	0.7	(0.3)
Oregon	14,000	24,160	(62,569)	0.6	(0.2)
Arizona	9,000	18,830	(48,765)	0.5	(0.2)
Nebraska	9,000	18,160	(47,030)	0.5	(0.2)
Texas	12,000	24,810	(64,252)	0.5	(0.2)
Oklahoma	400	1,550	(4,014)	0.3	(0.1)
Nevada	9,800	30,940	(80,128)	0.3	(0.1)
Kansas	1,200	3,360	(8,701)	0.2	(0.1)
North Dakota	5,700	31,990	(82,847)	0.2	(0.1)
Utah	6,000	29,200	(75,622)	0.2	(0.1)
Total	1,019,400	505,980	(1,310,430)		

miles of utilized habitat, pronghorn densities for U.S. states were calculated (Table 16).

ENDANGERED SPECIES STATUS 2000

Unregulated killing of pronghorn coupled with major alterations of pronghorn habitats, decreased herds from millions to a barely sustainable remnant. Fortunately, wildlife management programs stabilized populations, and pronghorn gradually increased until common again, but at numbers far below their presumed levels before European occupation of the New World. The species certainly is not threatened today, but three presumed subspecies and certain isolated herds are officially listed as threatened or endangered or are otherwise considered as "at risk."

Among the various listing agencies and organizations, those citing pronghorn are:

1. the federal government of Mexico, listing all endemic populations of the Mexican, peninsular and Sonoran subspecies as endangered (Secretaria de Desarrollo Social 1994);
2. the federal government of the United States, listing two subspecies—the peninsular in Baja California Sur, Mexico, and the Sonoran in Arizona and Sonora, Mexico (U.S. Fish and Wildlife Service 1999);
3. the Convention on International Trade in Endangered Species of Fauna and Flora (CITES), listing all pronghorn in Mexico (U.S. Fish and Wildlife Service 1997a);
4. the International Union of Conservation of Nature and Natural Resources (1996) listing three pronghorn subspecies in Mexico—peninsular pronghorn as critically threatened, Sonoran pronghorn in Mexico and the United States as endangered, and the Mexican pronghorn in Mexico and the United States at lower risk and not threatened, but survival is dependent on conservation.

Canada

In December 2002, The Canadian parliament passed its first federal law establishing an endangered species program. It is referred to as the Species at Risk Act (SARA) and went into effect in late 2003. According to Taubman (2003:5): "Unlike the U.S. Endangered Species Act, SARA protects only 'federal species' such as fish, migratory birds, and plants and animals living on federal land or in the oceans. While the new law will protect 233 species and their federal habitat, provincial governments will not have authority over all other species." It is not known at this time how the extirpated pronghorn population in Manitoba reported by Seton (1929) will be affected by this new program.

Mexico

Gonzalez and Lafon (1993) reported the history of pronghorn numbers and their continued decline (Table 8). The federal government recognized this dilemma early on and decreed laws in 1922 for the protection of all pronghorn (Secretaria de Agricultura y Fomento 1922). The future of pronghorn in Mexico appeared so bleak that Villa (1955) asserted that "la ultima manado" (the last herd) was doomed. Fortunately, this did not prove correct. By 1985, populations remained at less than 1,000 and the federal government declared all endemic pronghorn subspecies as endangered (Secretaria de Desarrollo Social 1994). Following are ecological and management data pertaining to each subspecies.

Mexican Pronghorn. The Mexican pronghorn is the most abundant and widely distributed subspecies in Mexico. It once ranged over seven or more states (Figure 15). Pronghorn were observed by Euro-Americans during the 1540s—more than 250 years before Lewis and Clark on their 1804 to 1806 expedition described the species for science.

During the early 1900s, Grinnell (Sheldon 1925) observed abundant herds in small bands, diminishing in numbers toward the south in Chihuahua. Populations were most numerous on the eastern rangelands of Chihuahua, where estimated numbers were several thousand. Lumholtz (1902) confirmed their abundance in the state.

Nelson (1925) listed 700 pronghorn in Chihuahua and 600 in Coahuila from 1922 to 1924. He noted that others inhabited Durango, but gave no estimation of number. Since the early 1920s, pronghorn were extirpated from Coahuila and Durango, but were reintroduced and appear to have reestablished in Coahuila (Miranda 2000).

Information concerning pronghorn distribution and abundance for the high plateau region has been limited. No terrestrial or aerial surveys were conducted until the 1970s (Anderson 1972).

During 1977 and 1978, pronghorn in Chihuahua were investigated to determine numbers, composition and distribution by Trevino (1978). He conducted aerial and ground surveys and reported the total minimum population in 1978 was 595 on 10 areas in northern and eastern regions (Table 17). Average sex ratios were one adult male to two adult females, which may be average for populations with little harvesting (Trevino and Schemnitz 1982). Average fawn/doe ratios were 17 per 100, which is low compared with that of northern herds in the United States that range from 30 to 100 fawns (Vriend and Barrett 1978). Illegal killings and barbed-wire fences were noted as factors affecting populations.

The National Ecology Institute, Chihuahua State University and the Mexican Air Force cooperatively conducted a survey during the winter of 1995–96 (Ramirez et al. 1999). They observed 172 pronghorn.

Sonoran Pronghorn. The first extensive survey of pronghorn in Sonora was by Nelson (1925), who reported four separate populations in the northwestern region. A total of 595 animals were counted in November 1924. Nelson noted that animals traveled back and forth across the border between Arizona and Sonora. Villa (1951) estimated there were more than 1,000 animals in Sonora during the early 1950s.

The *Sonoran Pronghorn Recovery Plan* for the United States reported between 200 and 300 animals in Sonora, during the early 1980s (U.S. Fish and Wildlife Service 1982) (Figure 20). A subsequent estimate of Sonoran pronghorn numbers was 313 in 1993 (Snow 1994). Ticer and Munig (1999) reported 88 in 1999.

In 1998, the *Final Revised Sonoran Pronghorn Recovery Plan* in the United States contained limited data on populations in Mexico (U.S. Fish and Wildlife Service 1998a). The plan estimated 200 to 500 individuals in Sonora. Two major reasons were provided for decreasing populations: competition for forage with livestock and illegal hunting.

Peninsular Pronghorn. More information is published on the peninsular pronghorn than any other presumed

TABLE 17. Study areas and population dynamics for Mexican pronghorn in 1977 for Chihuahua, Mexico

Area	Total area in acres (ha)	Number of pronghorn				Percentage of total in Mexico	Adult sex ratio[a]	Survival rate fawns/doe	Immature sex ratio[a]
		Males	Females	Fawns	Total				
El Berrendo	153,202 (62,000)	9	23	2	34	6.1	1:2.5	0.09	1:1.0
El Cuervo[b]	74,130 (30,000)	10	21	4	35	6.2	1:2.1	0.19	1:1.0
Villa Ahumada	17,297 (7,000)	2	6	0	8	1.4	1:3.0	0.00	0
Moctezuma	201,139 (81,400)	27	39	4	70	12.7	1:1.5	0.10	1:1.0
El Sueco	222,390 (90,000)	55	131	18	204	36.4	1:2.4	0.14	1:2.0
Tres Castillos	79,813 (32,300)	13	25	3	41	7.3	1:1.9	0.12	1:2.0
Coyame	134,175 (54,300)	33	44	17	94	16.8	1:1.3	0.39	1:1.5
Julimes	104,276 (42,200)	5	10	1	16	2.8	1:2.0	0.10	1:1.0
Benavides	64,246 (26,000)	4	14	3	21	3.7	1:3.5	0.20	1:0.5
Camargo	174,205 (70,500)	11	22	4	37	6.6	1:2.0	0.18	1:1.0
Total	1,224,875 (495,700)	169	355	56	560	100.0	Mean = 1:1.99	Mean = 0.17	Mean = 1:1.4

[a] Male to female.

[b] Does not include two additional herds of pronghorn—30 at El Oso and Iglesias and 32 at Doce de Octubre—observed in autumn 1978.

Source: Trevino and Schemnitz 1982.

The Mexican pronghorn is native to the high plateau grasslands of Mexico (above)*. The southern high plateau plains of the state of Chihuahua are short grasslands that sustain the greatest current densities of the Mexican pronghorn subspecies.* (Bottom) *Two wildlife biologists observe a distant herd of pronghorn in a grass/yucca community in Chihuahua, Mexico. An abundance of grasses, forbs and shrubs indicates a healthy habitat. However, the presumed Mexican subspecies has experienced major declines during the 20th century throughout Mexico, and now it is classified as endangered there.* Top photo by Patrick Robles. Bottom photo by Jim D. Yoakum.

subspecies in Mexico. Consequently, a detailed accounting of ecological conditions and population dynamics is available.

The peninsular pronghorn is the most endangered megafauna in Mexico (J. Cancino personal communication:2001). The subspecies' historic distribution encompassed approximately half of the Baja California peninsula, but its present range is a fraction of the former distribution, with all herds now on the Vizcaino Biosphere Reserve (Figure 21). The climate is dry and warm, with a mean annual precipitation of 3.3 inches (8.4 cm) and a mean annual temperature of 64 to 72 degrees Fahrenheit

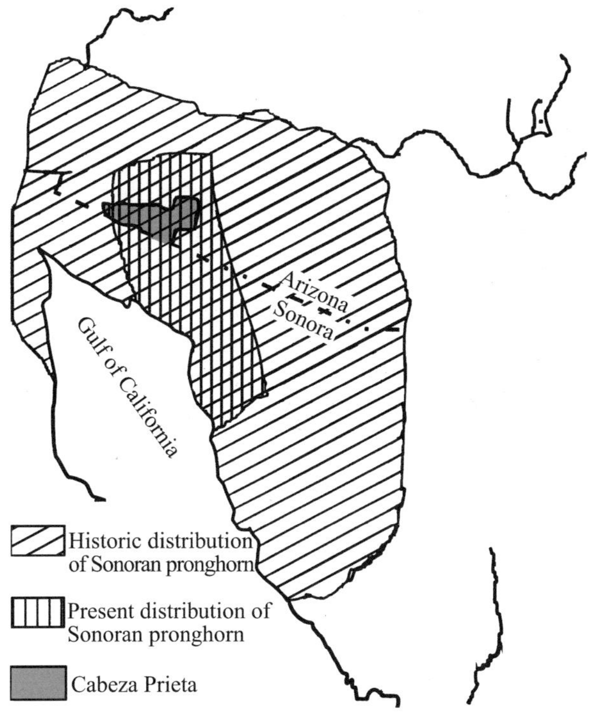

FIGURE 20. *Historic and 1998 distribution of Sonoran pronghorn in Arizona, United States, and Sonora, Mexico (U.S. Fish and Wildlife Service 1998a).*

FIGURE 21. *Historic and current range of the peninsular pronghorn and location of the Vizcaino Biosphere Reserve in Baja California Sur, Mexico (Cancino et al. 1998).*

(18°–20°C). The region experiences heavy coastal fog, which often blankets the area in spring and contributes moisture to the growth of epiphytic plants. Topography consists of plains, low hills, sand dunes and dry water courses, with elevations from sea level to 820 feet (250 m). Vegetation is mainly a forb and shrub community, often succulent and tolerant of salt; few grasses exist (Cancino et al. 1998).

There are no known permanent fresh, free water sources for pronghorn (Cancino 1988). It appears that the peninsular pronghorn adapted to meeting requirements for water primarily through preformed water (water obtained from forage) and, to a lesser degree, metabolic water (obtained through physiological processes). Some drinking water may be obtained from ephemeral pools following rains. Observations indicate that animals have developed foraging habits of feeding during early mornings when dew often is abundant on plants.

Peninsular pronghorn maintain low densities compared with herds on grasslands of the high plateau in north-central Mexico (Gonzalez and Lafon 1993). In 1924, 500 animals were estimated to compose the peninsular population (Nelson 1925). Commencing in 1977, periodic surveys were conducted for the remaining herd on the Vizcaino Desert (Cancino et al. 1998). Numbers fluctuated from lows in the early 1980s to highs in the mid-1990s and lows again in the late 1990s (Table 18).

National Pronghorn Recovery Plan. In 1996, representatives of the federal government of Mexico announced they were developing a recovery plan for all endemic pronghorn subspecies (Cancino et al. 2000) (see Chapter 16). The completed plan estimated the following number for states with pronghorn at the end of the 20th century: Baja California Sur 200, Chihuahua 550, Coahuila 200 and Sonora 250—a total of 1,200 (Secretaria de Medio Ambiente, Recursos Naturales y Pesca 2000). These estimates have been added to the year 2000 in Table 5.

United States

The Endangered Species Act of 1973 (16 U.S.C. 1931) provides authority and procedures for listing, protecting and restoring endangered wild animals and plants. It requires the best available scientific information to determine the status of species or subspecies, and provides programs to conserve the ecosystem on which endangered species subsist. These procedures have resulted in the listing of one pronghorn subspecies as endangered in the United States—the Sonoran pronghorn.

Peninsular pronghorn have occupied the Vizcaino Desert of Baja California Sur for centuries. However, numbers of this presumed subspecies were greatly reduced by the end of the 20th century, and the subspecies now is classified as endangered by the federal government of Mexico. The entire peninsular population (estimated at 100 to 200 at the end of the 20th century) occurs in desert habitat within sight of the Pacific Ocean. Gray whales were sighted at both ends of aerial transects in 1984. These are the only pronghorn found close to sea level in modern times. Such surveys also disclosed herds of cattle, burros and horses cohabiting that desert. Photo by David E. Brown.

The Sonoran pronghorn—the only presumed subspecies of pronghorn classified as endangered in the United States—inhabits the Sonoran Desert in scattered, small groups. The adult male at left was photographed on the Cabeza Prieta National Wildlife Refuge in southwestern Arizona. Although most of the population is protected on federal lands, numbers have not increased after more than 60 years of protection. The subspecies may be endangered because of vegetative alterations, resulting from disturbances by Euro-Americans during the past 200 years. Vegetative communities in unhealthy condition contribute to endangered pronghorn status. Sonoran pronghorn inhabit broad alluvial valleys and foothills surrounded by block-faulted mountains in the Sonoran Desert. Vegetation is predominantly creosote and bursage-covered flats bordered by washes supporting paloverde, mesquite and ironwood. Herbaceous plants generally are sparse, but forbs can be abundant during years of above average precipitation. Cacti are common and are consumed readily by pronghorn for nutrients and water. The climate generally is arid, with average temperatures ranging from 63° to 90°F (17°–32°C) annually. Average annual precipitation is around 5 inches (13 cm) in a bimodal pattern from December to February and during monsoons, which can occur between July and August. Surface drinking water is scarce. Pools after rains provide moisture until they evaporate. Permanent free water sources are limited to constructed reservoirs, wells and water catchments. Pronghorn use of these artificial structures is limited at best (Cutler 1996, U.S. Fish and Wildlife Service 1998). Nowak and Paradiso (1983) commented that pronghorn may drink freely when water is available, but can derive sufficient moisture from forage. Photo courtesy of the U.S. Fish and Wildlife Service, Cabeza Prieta National Wildlife Refuge, Arizona.

Sonoran Pronghorn. The Sonoran pronghorn population has been listed as endangered, since March 11, 1967, because of low numbers and loss of habitat. The subspecies exists in southwestern Arizona and on adjacent rangelands in Sonora, Mexico.

A recovery plan for the subspecies was completed in 1982 (U.S. Fish and Wildlife Service 1982). It was revised in 1998 and contains new information on population numbers and habitat conditions (U.S. Fish and Wildlife Service 1998a).

The Sonoran pronghorn was described for science in 1945 (Goldman 1945). Herds were observed in the late 1800s, ranging on open valleys on both sides of the Ari-

TABLE 18. Surveys of peninsular pronghorn in the Vizcaino Biosphere Reserve, Baja California Sur, Mexico

Year	*Month*	*Survey method*	*Total counted*	*Male/female ratio*
1977	February	Ground	83	96:100
1978	June	Ground	80	60:100
1979	October	Ground	54	28:100
1980	April	Ground	26	
1981	June	Ground	47	83:100
1982	May	Ground	39	
1983	December	Ground	46	50:100
1984	April	Aerial	53	62:100
1985	March	Ground	97	38:100
1986	March	Ground	97	
1988	March	Ground	48	65:100
1991	March	Ground	60	88:100
1993	April	Ground	51	94:100
1993	April	Aerial	121	80:100
1993	April	Aerial and ground	135	89:100
1995	November	Aerial	149	
1996	May	Aerial	39	
1996	November	Aerial	78	
1997	April	Aerial	13	
1997	June	Aerial	3	
1998	February	Aerial	52	
1999	November	Aerial	88	
2000	November	Aerial	161	
2001	No survey conducted			
2002	November	Aerial	151	

Source: Cancino et al. 1998, J. Cancino personal communications: 2003.

zona/Sonora border, but were rare by the early 1900s (Mearns 1907). The subspecies' historic and current distributions are illustrated in Figure 20.

A review of the literature provides the following numbers for Sonoran pronghorn in the United States (U.S. Fish and Wildlife Service 1998a): 1925—105 estimated; 1941—60 in southwestern Arizona excluding the Organ Pipe National Monument; 1957—less than 100; 1968—50; 1968 to 1974—50 to 150 observed; 1981—100 to 150 by Arizona Game and Fish Department; 1992—121 transect survey; 1994—109 transect survey; 1996—216 transect survey; and 1998—fewer than 300.

A November 30 to December 4, 2002 aerial survey by the Arizona Game and Fish Department (2002), in cooperation with the U.S. Fish and Wildlife Service, produced an estimate of 21 or 23 Sonoran pronghorn remaining in Arizona, according to a December 9, 2002, news release by the state agency. Almost no rain during the preceding 13 months resulted in zero fawn survival in 2002 and, from June to mid-August, 80% of the radio-collared adults succumbed to drought-related causes.

Yellowstone National Park Population. To clarify terminology in the Endangered Species Act referring to the phrase "distinct population segment of any species," policy procedures were later adopted. One of these procedures is recognition of a subspecies that may be in peril and "loosely bounded and regionally distributed." The criteria may be applicable for the pronghorn population in the northern range of Yellowstone National Park.

A population of pronghorn occupies part of the northern range in Yellowstone National Park. These animals comprise a native herd that has occupied the ecosystem since Pre-Columbian times. Its numbers have oscillated during the past century: 550 to 700 in the 1930s; 600 to 800 in the 1940s; averaged 140 in the 1970s; increased to a peak of 594 in 1991; and dropped to 229 in 1996 (Yellowstone National Park 1997). Ground and aerial surveys during April 2001 indicated 206 pronghorn in the park (K. Murphy personal communication:2001).

Because of the herd's historic and biological significance and its small size, there is concern about its viability. Mortality factors contributing to low numbers are believed to include coyote predation, increasing human use of winter rangelands, invasion of exotic vegetation and competition with more abundant elk, bison and mule deer for preferred forage (Clark 1999).

Prospects for Yellowstone National Park pronghorn were evaluated through population viability analysis by Goodman (1996). He concluded that the population was

The Cabeza Prieta National Wildlife Refuge in southwestern Arizona was established, in part, to perpetuate native wildlife. The endangered Sonoran pronghorn lives there in precariously low numbers (Hervert et al. 2000). Investigations are needed to determine whether the endemic vegetation has changed significantly since Euro-Americans arrived and whether the ecosystem has sufficient nutritious forage to sustain a viable pronghorn population. Photo by John Hervert.

extremely vulnerable to wide swings in numbers and the risk of extinction was high. Stochastic factors, such as disease or severe weather, could eliminate the herd. Goodman indicated that the probability of extinction within 100 years was 18%. He noted that it is conventional to consider a population to be severely endangered when its probability of extinction within a century is above 5%.

Recognition of the Yellowstone National Park population's threat of extinction may qualify it for protection under the Isolated Vertebrate Population Policy Act (U.S. Fish and Wildlife Service and National Oceanic and Atmospheric Administration 1996) because it is an isolated distinct vertebrate population and possesses unique physical features described by Goodman (1996).

At a February 2002 meeting of researchers, scientists and managers, a several-decade trend of declining pronghorn numbers in Yellowstone National Park was identified. Concern was expressed that, should the trend continue for another decade, the park's pronghorn population likely would be extirpated. A significant contributing factor was said to be the park's burgeoning elk population (Boccadori 2002). The elk consume large quantities of limited, preferred, succulent, nutritious forage needed to sustain thrifty pronghorn herds. In addition, intensive elk and bison foraging may be contributing to inadequate security cover for pronghorn neonates, because the vegetation structure, composition and volume are annually decreased, leaving the fawns unusually vulnerable to aerial and ground predators.

SUMMARY

Pronghorn originally ranged from the south-central grasslands of Canada, south to the high plains north of Mexico City, and from the Mississippi River and Gulf Coast of Texas westward to the Pacific Ocean. Today, herds occupy much of the same area, although they often are found in fragmented habitats. Several attempts have been made to establish herds outside their historic range, but all have failed to produce thrifty herds except one that is periodically augmented.

It has been estimated there were 35 million or more pronghorn during pristine times. Those herds were re-

Pronghorn in Yellowstone National Park are unique because they occupy historic habitat free of translocated herds. Consequently, this pronghorn population should be genetically pure. For more than a century, visitors to the park have observed pronghorn at close range, foraging and bedding near the stone arch at the north entrance near Gardiner, Montana. The opportunity to observe pronghorn at this site can be experienced now, just as it was in 1915 when this picture was taken. Photo by Elliot W. Hunter; courtesy of the Haynes Foundation Collection, Montana Historical Society, Helena.

duced by 99% during the latter half of the 19th century by hunting and loss of habitat to agriculture and other human uses. The population was reduced to 13,000 by 1915—the lowest in history. Conservation practices ended this decline in the early 20th century. Within 10 years (1915 to 1925) populations increased more than 100% to 30,500.

From 1924 to 1984, pronghorn increased more than 3,000% to more than 1 million—the highest number in the 20th century. Populations remained at a plateau from 1984 to 1995 and decreased approximately one-third by 1997. Although populations skyrocketed for Canada and the United States during the 20th century, a continual decline of more than 80% occurred in Mexico. For 2000, pronghorn estimates were 32,900 for Canada, 1,200 for Mexico and 765,200 for the United States, for a total of 799,200.

The Mexican, peninsular and Sonoran subspecies of pronghorn are on national and international lists as endangered. All are located in Mexico, although the Sonoran also is found in southwestern Arizona. Endangered subspecies comprise less than 1% of the total pronghorn population in North America.

Pronghorn densities and biomass were computed for the United States. Such information was not complete for Canada and Mexico. Wyoming has the highest densities, followed by Montana, South Dakota and Colorado. Apparently, states with grasslands had the highest densities, shrubsteppes were next, and deserts had the lowest densities and biomass.

An assessment is provided relative to habitat characteristics and conditions of populations listed as threatened or endangered. Based on this evaluation, recommendations are provided for research and management programs. Emphasis is given first to restoring the habitat—then restoring the subspecies.

S E C T I O N T W O

BIOLOGY

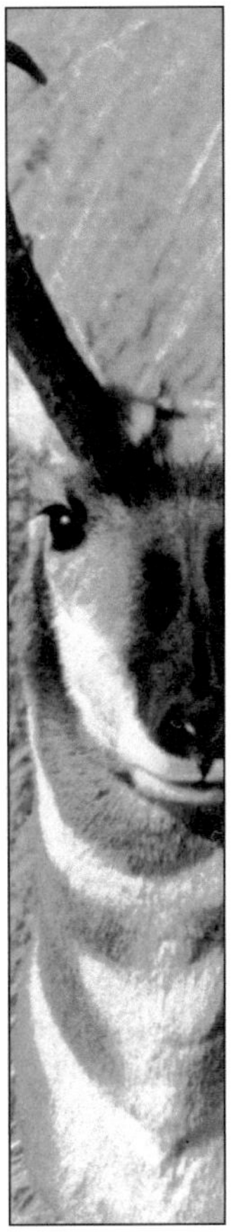

C H A P T E R F I V E

Physical Characteristics

Bart W. O'Gara

Nothing can surpass the delicate elegant finish of their limbs, in which lightness, elasticity, and strength are wonderfully combined. All the attitudes and movements of this beautiful animal are graceful and picturesque; and it is altogether as fit a subject for the fanciful uses of the poet as the oft-sung gazelle of the East.

Irving (1868:276)

The dry interior of North America west of the Mississippi was a land of tallgrass and shortgrass prairies and shrub-steppes, grading into aspen parklands to the north and deserts to the south. When Europeans first gazed upon this vast expanse, great herds of pronghorn roamed at will to satisfy their daily and seasonal needs. The decimation of those herds, as well as the restoration of North America's pronghorn populations, are chronicled in *Prairie Ghost: Pronghorn and Human Interaction in Early America* (McCabe et al. 2004). As amazing as the roller-coaster history of pronghorn numbers is the animal itself. The dry, open country molded the pronghorn into an animal quite different from other North American ruminants. It is designed to move easily and speedily over the wide, open spaces of the West and survive without benefit of dense or tall cover.

"This is peculiarly a plains animal, loving the high, dry prairie, and being frequently found so far from water as to raise doubt in the minds of many plains men whether he may not be able to live without drinking. Absurd as this is, the hypothesis has more of plausibility than would first appear. There is scarcely a desert so barren and arid that the antelope cannot find a means of existence, and apparently a very comfortable existence, as, wherever found, he is invariably at the proper season in splendid condition" (Dodge 1959b:193).

A surprisingly complete characterization of the pronghorn was contained in the journals of the Lewis and Clark expedition (Moulton 1987a). Captain Clark killed a buck on September 14, 1804, and Lewis recorded measurements of the animal. Description of anatomy, coloration, eyes, feet, movements, speed, numbers, vulnerability to predation, and so on, are scattered through the journals. Caton (1877) provided further information on physical characteristics as well as natural history of this unique species.

Full-grown pronghorn are smaller than most North American deer, as noted by Caton (1877), and their robust and somewhat chunky bodies are predominantly white and rusty brown to tan, with black and dark brown markings about the head and neck. The large head is carried upright, and the ears, frequently edged by a thin blackish line, stand erect when a pronghorn is at ease, giving the animal an alert appearance. When a pronghorn is curious, its ears are projected forward to catch the slightest sound, whereas a retreating pronghorn keeps one ear pointed at the source of its displeasure. An irritated pronghorn lays its ears back, as does a horse. The ears, 5 to 6 inches (13–15 cm) long, terminate in curved points and are covered inside and out with hair.

Antilocapra americana seems to be a plastic species, with considerable variation in color, horn shape and behavior from area to area. Other less well-known physical inconsistencies include a variable number of chromosomes (Hsu and Benirschke 1969), occasional specimens with extra horns or teats, females with or without horns, some does with a different type of uterus than most (O'Gara 1968), and individuals with different numbers of ribs (M. Barrett personal communication:1978).

Unlike deer, but similar to bovids and most other warm-blooded animals, pronghorn have gallbladders. Other visceral differences between pronghorn and deer include larger tracheae, hearts and lungs for their size—adaptations for rapid, sustained running. Another adaptation for speed is the somewhat smaller size of the stomach.

The white, brown and tan body of the pronghorn is somewhat chunky, and the head is large, with black markings. Does have less black on the head, and markings are variable enough in both sexes that individuals can be recognized by a keen observer. Photo by Joe Van Wormer; courtesy of the Denver Museum of Natural History.

The kidneys are large—probably to facilitate elimination of toxins in many plants that are poisonous to domestic livestock but seem relished by pronghorn.

EYES

> *Many have compared the eyesight of the pronghorn to that of a man aided by eight-power binoculars. . . . Even the location of the eyes, close to the crown of the head, provides added protection enabling them to peer over terrain features with a minimum of exposure to would-be enemies. Perhaps some of the acuteness of the pronghorn's sight may be attributed to their large size, which greatly exceeds that of other animals of similar body proportions.*
>
> Hoover et al. (1959:17)

Indeed, pronghorn eyesight has often been compared with that of a human looking through binoculars. For instance: "Lacking the bison's strength, the pronghorn has other assets. Endowed with vision as acute as that of a human equipped with eight-power binoculars, it keeps a sharp watch for enemies" (*Time-Life Books* 1985:61). Proving or disapproving the theory would be difficult, but most who have hunted or studied the species do not find it hard to believe.

Captain Meriwether Lewis wrote of the first pronghorn shot by Captain William Clark in 1804: "The eye is large and prominent, the sight is circular, deep sea green, and occupyes one third of the width of the eye" (Moulton 1987a:72). Captain Lewis apparently examined the new specimen by a campfire that was reflected by the tapetum at the backs of the eyeballs. This reflective layer appears green in most mammals and is what makes eyes glow in headlights.

Pronghorn eyes are purplish black and expressive. In relation to its size, the pronghorn has the largest eyes of any North American ungulate, reflecting its use of open habitat and heavy dependence on visual acuity. The eyeballs of adults of both sexes are about 1.4 inches (36 mm) in diameter, and several investigators have stated that pronghorn eyes are as large as those of horses (Einarsen 1948, Van Wormer 1969). I have no data on horses, but the eyes of large mule deer bucks I measured were the same size as those of pronghorn, and the eyes of elk were only 0.1 to 0.2 inch (3–5 mm) larger. The pronghorn's eye color and relative size are very similar to those same characteristics in gazelles on grasslands, steppes and deserts of Africa and Asia.

Townsend (1978) maintained that a pronghorn doe he shot in 1834 shed great tears that rolled down her face as he approached. Having shot and collected many pronghorn, I have never observed such a phenomenon and doubt that it occurs.

Besides being large, the eyes of pronghorn protrude somewhat, giving the animals a wide field of vision. As a

For its size, the pronghorn has the largest eyes of any North American ungulate. Like the eyes of ungulates in general, those of pronghorn move little within their sockets. However, the sockets protrude from the skull, and the eyes bulge somewhat from them. Although the eyes appear to be on the sides of the head, they are oriented forward enough to allow limited binocular vision forward, leaving uniocular vision to the sides. Because of its protruding eyes, pronghorn have nearly a 300-degree circle of vision without moving their heads or eyes (Van Wormer 1969). A slight turn of the head allows them to see anything in the posterior "blind spot." This animal had shed horn sheaths 2 days earlier. Photo by Joe Van Wormer; courtesy of the Denver Museum of Natural History.

general rule, whenever a person can see a buck's cheek patch, that pronghorn is capable of detecting the human's movements. Not only do pronghorn eyes protrude, but the tubular, bony rims of the eye sockets also protrude—far enough to protect the eyes from damage when an animal feeds on shrubs or short forbs growing beneath dry vegetation. For further protection, pronghorn often close their eyes while feeding. In addition, the eyes are shielded from intense sunshine by thick, black eyelashes about 0.5-inch (1.3 cm) long, giving the animals the same sort of innocent, appealing appearance that some people seek to effect with artificial lashes. The large, lustrous eyes are set far back in the head, and the nose is very long.

Pronghorn detect rapid movements at great distances, an asset vital to survival. In open land situations typical of pronghorn habitat, the ability to detect predators at a distance is important. Kitchen (1974) presented evidence that pronghorn have excellent eyesight—as one would expect of a prairie animal—but lack visual acuity. He noted that motionless humans in plain sight often were ignored, and that pronghorn seemed unable to detect movements slower than 1 minute of arc per second.

Like most mammals, pronghorn have long been considered color-blind. However, investigation has shown that pronghorn eyes have cones (color receptors) as well as rods (black-and-white receptors), and that they have a "visual streak" in which the receptors are aligned in a band for especially acute vision along the surface of the flat prairie (A. Rusoff personal communication:1978).

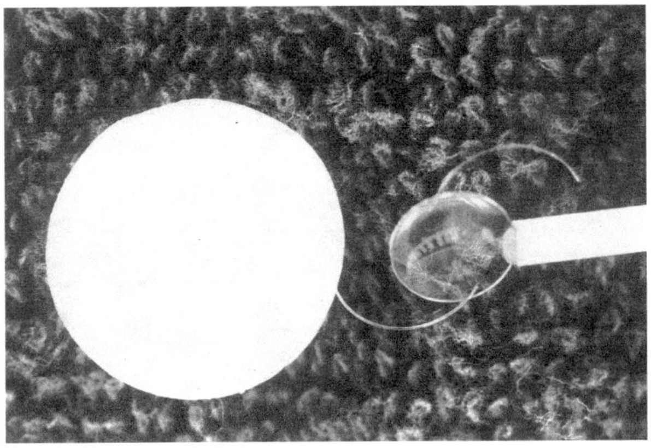

The eye lens (left) of a 2-year-old, 110-pound (50 kg) pronghorn buck and a plastic lens (right) that replaced the lens of a 175-pound (79 kg) man. The pronghorn lens was clear when removed from the eyeball, but became opaque in the fluid used to preserve it. Photo by Bart W. O'Gara.

HOOVES

Pronghorn feet are long and slim, lack dewclaws common to most other ruminants and offer little resistance to the grass and shrubs of western prairies and deserts. Unlike hair on the rest of a pronghorn's body, that on the feet is short, fine and flexible; it resists wear admirably. The muscles are concentrated in the legs, with only tendons in the feet, reducing weight and leaving little to bruise when an animal is running. This arrangement reduces the required circulation, which usually is sufficient to prevent freezing but not extensive enough to lose much body heat during cold weather.

The pointed hooves are double and black, with cartilaginous padding to cushion the shock of running over hard ground and rocks. Interdigital glands open to the outside through holes about 0.12 inch (3 mm) in diameter on the fronts of all four feet, between the toes and about 1 inch (25 mm) above the hooves. The front hooves, larger than the rear ones, seem to carry most of a pronghorn's weight when the animal is running.

Average lengths of hooves on Oregon bucks reported by Einarsen (1948) were 2.9 inches (7.4 cm) front and 2.5 inches (6.4 cm) rear. Less difference was found between front and rear hooves of Colorado pronghorn (Bear et al. 1973); the hoof length of 31 mature does averaged 2.8 inches (7.1 cm) front and 2.7 inches (6.9 cm) rear, and those of 15 adult bucks averaged 2.6 inches (6.6 cm) front and 2.4 inches (6.1 cm) rear.

The legs and feet of pronghorn are long and slim, with large muscles close to the body—adaptations for speed and endurance. The similarity in build to gazelles and other small antelope of Africa and Asia is evident. Conversely, caribou and wildebeest have chunky bodies, large heads and long legs, and are noted long-distance runners. Pronghorn legs and feet appear fragile but are quite tough. Photo by Jim D. Yoakum.

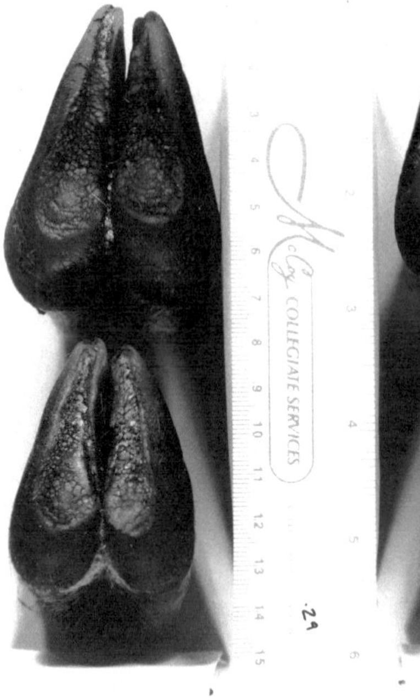

Front hooves (top) and rear hooves (bottom) of a 6-year-old doe (left) and a 4-year-old buck (right) from Montana. The front hooves are noticeably larger than the rear—a characteristic shared to a lesser degree with other North American ruminants (Halfpenny and Biesiot 1986). Pronghorn gallop more than do other North American ruminants, and their center of gravity is near the forequarters (Bullock 1974). Most of the power for running is generated by the hindquarters, but when a pronghorn is running, its weight is carried principally on the front hooves. The rear hooves do not strike the ground as forcefully, so the larger front hooves are an advantage. The pronghorn is the only North American ruminant that does not have dewclaws (side toes). Photo by Gerald Kling.

Pronghorn tracks show narrow, pointed tips and broad bases. The hind footprint usually covers that of the front, leaving a tidy, alternating track. The straddle (width apart) is about 3.5 to 9 inches (8.9–23.0 cm). Pronghorn tend to drag their feet in snow. The toe tips spread wide when the animals are galloping. The stride when walking generally is 8 to 19 inches (20–48 cm), but the tracks of galloping pronghorn may be 14 feet (4.3 m) or more apart. Deer tracks have narrower bases, shorter strides between gallop groups and often show dewclaws.

Bucks use their front hooves to scratch depressions where they deposit urine and feces, and both sexes paw to uncover snow-covered forbs or spring grasses. Some researchers have stated that pronghorn do not dig through snow, but—although they avoid deep snow—craters made in snow by pawing pronghorn were described and photographed by Brockley (1949), and I often have watched them pawing snow away as they fed. When snow is dry and light, pronghorn pawing can cause it to fly up higher than the animals' backs. Bucks generally use their horns during fights or to threaten other species. Does will threaten with their heads or horns, but usually strike with their front hooves when actually contacting predators, fawns or other does in an agonistic encounter.

COLORATION AND PELAGE

The conspicuous color pattern of pronghorn has drawn attention and speculation for some time. Cockrell (1900) maintained that if the animals inhabited Africa, their coloration would be cited as evidence supporting the theory of recognition marks—to assist members of a species in distinguishing their fellows from other species. He maintained, however, that the markings of pronghorn were guide marks that assisted the animals in keeping together during flight. More recent observers generally have maintained that pronghorn coloration made individual animals difficult to focus on as a band turned or shifted as they ran—thus, presenting a confusing picture to pursuing predators (Kitchen 1974).

Coloration differs between individual pronghorn and according to season, sex and age. Shufeldt (1920) reported a large buck in Wyoming with a jet black head and shoulders, but the rest of its body was normally colored. A Wyoming rancher told me of two bucks with completely black heads on his ranch in 1950, and W. Hepworth (personal communication:1983) told me that melanistic individuals had been reported near Laramie, Wyoming. He saw one that was essentially black on body surfaces where pronghorn usually are tan, and the normally white areas were rufous buff.

A melanistic female fawn was born on May 20 or 21, 1978, at the Desert Experimental Range in Utah (Clary et al. 1985). It had a normally colored twin. The melanistic fawn was radio-marked when 3 to 4 days of age and observed many times during the summer. The color remained a charcoal black throughout its juvenile pelage, but the adult coat contained much more brown. The most striking difference in the adult pelage was an absence of white bands on the neck. The adult hair contained some charcoal color but not nearly as dark as the juvenile hair. This fawn's mother had been collared during a previous study. On June 4, 1980, a second melanistic fawn was observed with a marked doe in the same general area. The mother was from the same group of pronghorn as the one observed in 1978, but the collar mark was thought to be different. The collar had been on nearly 8 years when the 1980 observation was made. Deterioration of collars over time would make it difficult to be 100% certain that the does were not, in fact, the same, but the colors were described differently by the observer. A third dark fawn was observed on May 21, 1982. When first seen soon after birth, all of the melanistic fawns were within one area of valley floor less than 1.5 miles (2.4 km) across. The mothers of the first and third dark fawns definitely were different does. No difference was evident in size or behavior among melanistic and other fawns.

Melanistic pronghorn fawn born at the Desert Experimental Range in Utah. Three such fawns were observed in the same general area from 1978 through 1982. They became lighter colored as adults. Melanism is an abnormal development of dark pigment in the skin and hair. It is uncommon in ruminants, but has been fixed in springbuck and fallow deer by selective breeding, indicating a genetic basis. Photo by Don Beale.

Spotted or piebald pronghorn occasionally have been reported in Wyoming. W. Hepworth (personal communication:1982) described variations for pronghorn west of Casper, Wyoming: a fawn with white spots; a female fawn with a nearly white head; and a female fawn with one white side, the other side with a white dorsoventral band across the body and the front leg all white. About 25 miles (40.2 km) northwest of Rawlins, Hepworth witnessed an old buck with white spots. The lower right side of the head was white back to the cheek patch. White coloration was on the left front leg all the way to the body, on the left hind leg to the stifle joint, on the right foreleg to the knee and on the right rear leg to about halfway between the hock and stifle joint.

Near Dillon, Montana, in 1972, I observed five light-colored fawns and yearlings that had dark horns, eyes, eyelashes, lips and noses but were otherwise yellowish white in areas that normally are brown or tan. A large, nearly white buck had been shot in the area several years before, presenting the possibility that the sires and dams of the fawns and yearlings were all offspring of that large buck. If so, a simple recessive mutation was indicated.

A white pronghorn fawn with dark nose, eyes and horns. This buck was seen several times in southeastern Wyoming. White animals that are not albinos seem to be more common in pronghorn than in other ruminants. However, as with melanism, the trait has been fixed in springbuck and fallow deer. White fawns undoubtedly are easier for predators to see and kill than are cryptically colored "normal" fawns, and white adults may even be easy prey. Human hunters, the principal predator of adult pronghorn today, tend to select for odd-colored animals. However, these white animals do not completely lack coloring matter, as do albinos, so they probably do not suffer from the weak eyes that plague albinos. Photo by Bud Smith; courtesy of the Colorado Division of Wildlife, Denver.

Several all-white pronghorn were reported from west of Casper, Wyoming, during the 1970s (W. Hepworth personal communication:1982). J. Johnston (personal communication:1982) observed another Wyoming fawn that was white with a black nose, and a white doe was shot in eastern Montana in 1982 (R. Schoening personal communication:1982). All of these animals had dark eyes, eyelashes, hooves and horns. However, a fawn picked up in central Wyoming, and reported in *Nature Magazine* (Richard 1940), had light blue eyes, white eyelashes and a black nose, and the inside of its mouth was red. Supposedly, this animal could not erect its hair as can normally colored pronghorn.

Two albino bucks were reported between Glenrock and Douglas, Wyoming, by B. Trebelcock (personal communication:1982). One was shot and its head mounted (R. Gordon personal communication:1994). That animal had pink eyes and nose. Its horns—which tipped ahead over the nose—and hooves were yellowish amber and slightly translucent. This is the only reliable description I have been able to find of an albino pronghorn.

An albino pronghorn was seen by many people between Verde Valley and Whipple Barracks, Arizona, in 1885 according to Mearns (1907). It was not known to have been killed up to the time of Mearn's departure in 1888. Because no one handled the animal, and most reported albinos turn out to be nearly white pronghorn with some normal pigmentation, this cannot be considered a positive sighting of an albino. It makes sense that a diurnal animal without pigment protecting its eyes would be handicapped in the intense sunlight of the western Plains.

Color patterns usually conform to the following description in the northern subspecies. Southern subspecies, except *A. a. peninsularis*, generally are lighter, as is typical of desert animals.

The pronghorn head, which varies from creamy white on the sides to various shades of brown on top, is marked in the male with brownish black patches that start just below the ears and extend downward 3 to 4 inches (7.6–10.2 cm). These patches cover subauricular scent glands and vary in size seasonally. Before and during the rut, the patches are at least twice as large as during late winter. During the rut, they appear darker than at other times because of greasy glandular material on the hairs. Seton (1927) reported bucks with missing cheek patches—a condition not noted since.

Blackish markings sometimes occupy the entire face below a line connecting the horns. Often, this "mask" is T-shaped with a horizontal line between the horns and a vertical line down the nose. Some individuals have black only near the ends of their noses. The philtrum, a narrow strip of black skin about 0.1 inch (0.3 cm) wide and devoid of hair, extends upward from the anterior edge of the upper lip between the nostrils, which also are black.

Distributed along the boundaries of the mouth opening and on the ventral surface of the chin are stiff, fine,

A dominant male (right), his ears back and head slightly lowered, gives a moderate intensity cheek patch display while in a typical threat posture. Dominant bucks ordinarily have large, very black cheek patches by the beginning of the rut, apparently because hormones that promote aggressive dominance behavior also promote proliferation of cheek patches and their glandular secretions, which, in turn, darken the patches. The buck on the left is 2.4 years old; the dominant buck is 3.4 years old. Photo by Peter T. Bromley.

white hairs 0.4 to 1.2 inches (1–3 cm) long and interspersed among the more typical soft, fine hairs. They probably are tactile in function. Shorter, more typical hairs grow profusely on the epidermis that extends into the mouth cavity where the dorsal and ventral lips meet when the mouth is closed. The dorsal lips are more muscular and adapted to the grasping of forage than are the ventral lips, which they overhang laterally and anteriorly because of differences in size of the upper and lower jaws. The lips are very flexible, but the anterior part of the lower lip is not as pliable as other portions. Angles of union between the upper and lower lips are about an inch (2.5 cm) anterior to the premolars (Wenzel 1955).

The body skin of the two sexes is of similar thickness, but the head and neck skin of a male pronghorn is nearly twice as thick as that of the female (Kitchen and Bromley 1974). Apparently, the male's thick skin acts as a shield in body areas where horn penetration during fighting is most probable.

The pronghorn throat is elegantly marked with a white crescent above a white shield; the point of the shield usually joins the white underparts at the base of the neck. To the casual observer, these bands and shields may appear to be similar on all pronghorn. However, several behaviorists have observed that the widths of bands, shapes of shields and comparative amounts of white and tan vary enough that a keen observer accordingly can recognize individual animals (Bromley 1967, Kitchen 1974, Byers 1997a). The lips, chin, areas around the bases of the horns, and on the cheeks and inner surfaces of the ears are whitish. Erectile 3- to 4-inch (7.6–10.2 cm) hairs of the mane on the back of the neck are russet, usually tipped with varying amounts of black.

Females 5 to 10 months old sometimes are difficult to differentiate from adults, but because fawns erect their manes more often than do adults (with the exception of courting bucks), the mane can be used to tell the difference. Bucks of the same age are easy to distinguish because the black cheek patches show they are males, and the 2-inch (5.1 cm) horns indicate that they are younger than yearlings.

The pronghorn spends most of its life on the open plains where its conspicuous markings are important for visual communication with other pronghorn. The bold head and neck markings are important in intraspecific encounters according to Kitchen (1974). An aggressive buck advances with depressed ears. The blackish horns in combination with dark facial markings exaggerate the size of the horns. A courting buck advertises his secondary sexual characteristics by presenting his cheek patch to does; the horns and face mask form a right angle and the ears are in a neutral position, accentuating the horns. The mane is partly erected, increasing the apparent size of the neck, which is set off by white bands. Territorial bucks usually face the direction from which encroaching bucks are most likely to approach. The striking neck and head markings doubtlessly warn approaching bucks at some distance that they will have to face a territorial male if they proceed.

A narrow, tan to rusty strip connects or nearly connects the upper surface of the short tail—about 4 inches (10.2 cm) long—colored back. On bucks, a dorsal or median gland is located about 1 foot (30 cm) anterior to the base of the tail. It is covered by thin, scattered hairs that are whitish to tan and smaller than the dark adjacent hairs. Rusty brown to tan hairs—the longest measuring 1.5 to 2.0 inches (3.8–5.1 cm)—cover the back, most of the neck and the outer sides of the limbs. Outer surfaces of the lower limbs are covered with short, tan hair. White hair clothes the undersurface and extends down the inner sides of limbs and well up the sides of the body, forming a rectangular area between shoulder and hip. Bromley (1977) postulated that the white undersurface could be an adaptation to reflect heat radiated from the ground during hot weather. Under cold conditions, the same pelage could function to prevent heat loss.

White hairs of the twin rump patches are about 3 inches (7.6 cm) long and almost gleam when erected, fanning beyond the normal contours of the body. In the bright sunshine of the Plains, one can see the flashing rump patches several miles away. From observations of wild pronghorn early in the 20th century, Seton (1927:432) aptly described the erected rump patches as spread radially into "great white blooming twin chrysanthemums, more or less flattened at the center. . . ." A rump gland is on each side of the diminutive tail in those flattened centers. These glands can be detected from the outside only by spreading the hairs to find the patches of tan greasy material on the bases of hairs that cover the glands. On the inside of the skin, the glands show as raised, thickened circles of skin.

Kitchen (1974) contended that rump patches and white side markings confuse pursuing predators. Cursorial predators generally single out and concentrate on one animal in a herd. Keeping track of an individual pronghorn in a running group is difficult; the flashing rump patches and white sides present a confusing display, especially when a herd makes a sudden turn. According to Hirth (2000), large rump patches evolved as a signal to keep ungulate groups together in open habitat.

During autumn, female pronghorn fawns often are difficult to distinguish from adults. The nose is shorter than that of an adult doe, but the difference is slight. Fawns erect their rump patches and, especially, their manes more frequently than do adults. After several minutes of observation, a person may be able to recognize a fawn by the latter characteristics. The frequent erection of rump patches and manes reflect excitement, fear, playfulness and skittishness, just as screaming and laughing are more characteristic of human children than of human adults. Piloerection seems involuntary—the automatic result of such emotions as excitement and fear. The oft-seen cartoon of hair standing on end depicts the condition in humans that actually happens in pronghorn. When erected, a pronghorn's rump patches fan beyond the normal contours of the animal's body. Photo by Len Clifford; courtesy of Leonard Rue Enterprises, Blairstown, New Jersey.

During late winter or early spring, long hair around the pronghorn face and where legs chafe the body begins shedding, revealing new, short hair. By late summer, the old hair of healthy animals is completely shed and the new coat is prime and sleek, displaying the darkest, brightest colors of the year. A few hairs are shed continually throughout the year. Annual shedding occurs primarily during March to May in northerly areas, and pelage is in prime condition by mid-July.

In Oregon, Dinges (1958) dyed a captive doe entirely black with Nyanzol D when it became apparent that shedding was almost complete. This made observation of shedding easy because the dyed hairs were replaced by new,

Conspicuous markings are important for visual communication (aggression, identification, contact within bands, warnings, etc.) among pronghorn. Against some backgrounds, such coloration makes pronghorn easy to see, even at great distances. Against other backgrounds, especially with snow on the ground, pronghorn colors are quite cryptic. Predators are the greatest threat to adult pronghorn during periods of snow. Photo by Cathy and Don Jenni.

naturally colored hair. Annual shedding by the doe began during late February with long hair areas of the back. Short hair areas of the belly, lower sides and legs began to shed when the long hair portions became prime. The short hair was completely replaced by the end of July but had not attained full growth.

During winter, outer hair bleaches and once rusty brown areas become light tan. Thus, although pronghorn appear to have two coats each year because of color changes, only one new coat is grown annually. Yearling does are particularly light colored in spring before the molt and usually can be distinguished from older individuals by this characteristic, plus their slightly smaller heads and slimmer body contours than those of adults. (Also, adult does generally have enlarged udders that are noticeable before and for at least a month after parturition.)

New hair is short, smooth and flexible, but it lengthens as winter approaches. The stalks of the long hairs are pithy and thicker than the tips. Once bent, a hair does not straighten again, and hairs are easily rubbed off. The microscopic structure of pronghorn hairs—a tough, reticulated surface of cells with serrated edges and large central air cells—was illustrated by Murie (1870). Coarse and lifeless as the long hair may look by midwinter, it provides excellent protection from wind and cold. The large cells provide dead-air spaces, just as a down-filled jacket, and a pronghorn can regulate the insulating properties of its pelage by erecting or flattening hair. Consequently, pronghorn seldom need to seek shade or the shelter of timber even during hot desert summers or severe winters on the northern Plains. A scattered undercoat of fine fur lies next to the skin, but apparently provides little thermal protection. Also, pronghorn hair has an oily film that readily sheds water, and the animals can shake their skins violently and disperse water from their coats without appearing to move their bodies. Being good swimmers, pronghorn readily ford rivers and, buoyed up by the air cells in their hair, they ride very high in the water.

Biting flies are a source of irritation to all wild and domestic ruminants. The flies are most active on sunny, calm days during summer. Forbs often are desiccated by midsummer, and pronghorn feed in and around patches of snowberry, wild rose and other woody species that shade forbs from desiccation by sun and wind or provide succu-

Pronghorn chase away flies and other inset pests by twitching small patches of hair erect. Those patches, as seen of the left hip of this animal, can be moved about the body, creating a somewhat comic rippling effect, but it appears to be effective for warding off biting insects. Photo by Peter T. Bromley.

lent leaders. Flies often are numerous in such areas, and feeding pronghorn shake their heads, kick at their sides and frequently erect and lower their hair, apparently displacing biting insects or making it difficult for them to reach the skin. As soon as feeding is completed, the animals seek open breezy knolls or ridges to bed and ruminate.

Resting or feeding pronghorn occasionally erect small patches of hair, sometimes no more than 1 inch (2.5 cm) in diameter, and sometimes twice as large as a man's hand, apparently in response to an insect or other source of irritation. Such erected patches can be moved vertically or horizontally across the body, creating an odd effect.

HORNS

The prongs that give pronghorn their name are unique, as is annual shedding of horn sheaths. Spots where horns will develop can be detected on fetuses about 4 months before birth. Normally, all adult bucks and about 70% of adult does have horns. The horns of does usually are small and begin growing during the second year of life; those of males start in the second month.

Pronghorn fight horn-to-horn, using a series of thrusts and counterthrusts as they try to gore their opponents. When horns lock, each buck attempts to push the other off balance and gain the uphill advantage. Bucks seldom engage in serious fights except during the rut. Although fights usually are of short duration, injuries are fairly common. The style, frequency and severity of fights match the predictions of Geist (1966), based on horn morphology. He maintained that, if horns were fairly short and daggerlike, combat would be rare but serious.

McCullough (in Kitchen 1974) postulated that shedding was related to the prong. He pointed out that, if pronghorn added length to their horn sheaths by adding increments at the bases, growth centers at the bases of the prongs eventually would shear off and render further growth impossible. Prongs cease growing when skin proximal to them becomes hard horn, so casting seems the only way to enlarge the prong in successive years. Byers (1997a) believes that deciduous horn sheaths have advantages over deciduous antlers and permanent horn sheaths (see Chapter 8).

Males

Horns with anterior prongs project from ridges just above the eyes of adult pronghorn bucks. Horn shape, inclination and curvature vary greatly among individuals, but horns of a representative buck protrude upward, outward and slightly forward to form a V. They then curve into hooks, often bearing whitish tips, that usually curve inward or backward. The prongs, extending out front, serve to engage an opponent's prongs, lessening the danger from the sharp tips as two males test each other's strength.

Seton (1927:454) watched pronghorn bucks fighting in the Washington Zoo and "was made to realize how exactly each detail of the apparently harmless horn had a purpose, offensive or defensive, for which it was highly specialized . . . as they thrust and parried, the purpose of the prong was clear. It served the Antelope exactly as the guard on the bowie knife or the sword serves a man; for countless thrusts that would have slipped up the horn and reached the head, were caught with admirable adroitness in their fork. And the inturned harmless-looking points? I had to watch long before I saw how dangerous they might be when skillfully used. After several minutes of fencing, one of the bucks got under his rival's guard; making a sudden lunge, which the other failed to catch in the fork,

Prongs are the basis for the common name of the pronghorn. They generally are absent on doe horns and are quite variable on those of bucks. The principal variations involve the size of the prong and its position between the base and tip of the horn. The prong may be as low as one-fourth of the way from base to tip or as high as two-thirds of the way. Photo by Bart W. O'Gara.

he brought his inturned left point to bear on the unprotected throat of his opponent, who saved himself from injury by rearing quickly, and throwing himself backward." Seton noted further that, in their fights, wild pronghorn usually were struck and sometimes ripped in the throat or neck.

During 1987 to 1994, Byers (1997a) watched 82 fights among pronghorn bucks on the National Bison Range. Although fights only lasted, on average, about a minute and a half, 10 (12%) proved fatal to one of the contestants. They died as a direct result of wounds (pierced trachea, internal injuries, dislocated limbs and sprains) or lost mobility or condition, becoming vulnerable to coyotes.

Prongs normally are single, compressed triangles, but double points or the presence of several small auxiliary prongs are not unusual. Below the prongs, sheaths are compressed laterally and contain some hair; above the prongs, the hooks are cylindrical, tapering and curved. The hooks normally incline inward or backward, some appear lyrate, and occasionally the tips are directed forward or nearly straight up. Pronghorn horn sheaths usually are described as black, but close inspection often reveals streaks of brown, and the bases of some sheaths are yellowish brown.

Horn prongs serve as guards, to catch and parry horn thrusts of other pronghorn bucks, exactly as the hilt on a sword serves a fencer. The adult buck above has fairly typical horns of average size. Photo by Peter T. Bromley.

The horn sheaths rest on bony cores (like sheaths on knife blades), the two being separated by a living layer of skin. The cores are somewhat triangular in cross section, and the prongs on the sheaths develop over the sharp front edges. Horn cores of male fawns and most yearlings are nearly round in cross section, but when yearlings grow prongs, the cores are triangular.

Horn lengths vary with subspecies and area, and nearly maximum horn development generally is reached by 3 years of age, before all teeth are fully grown and before bucks acquire and hold prime territories. During field and aerial surveys, biologists generally classify as mature (2 years old or older) those bucks with horns approximately 1.5 times longer than their ears. Those with horns nearly the same length as their ears usually are yearlings. However, a yearling with 14-inch (36 cm) horns was shot on the Fort Belknap Indian Reservation in Montana during 1994 (M. Fox and C. Mitchell:personal files).

The lengths and sequence quoted in this paragraph represent usual measurements during autumn for bucks in the northern part of their range. Buck fawns grow about

Prongs normally are single, compressed triangles, but double points or the presence of several small auxiliary prongs are not unusual. Below the prongs, sheaths are compressed laterally and contain some hair; above the prongs, the hooks are cylindrical, tapering and curved. The hooks normally incline inward or backward, some appear lyrate and occasionally the tips are directed forward or nearly straight up. Pronghorn horn sheaths usually are described as black, but close inspection often reveals streaks of brown, and the bases of some sheaths are yellowish-brown. This unusual horn sheath was picked up in central Montana. Photo by Bart W. O'Gara.

Horn and bony horn core of male pronghorn fawn are growing rapidly during December (as above left*). Only the very tip of the sheath, above the white ring, is hard horn. Yearlings usually do not grow prongs, but when they do* (right)*, the prongs generally are relatively small.* Photos by Bart W. O'Gara.

2-inch (5.1 cm), nearly round, cone-shaped horns by their first autumn. The few times I observed shedding of the sheaths from such horns, they were cast during late November. Horn cores of male fawns grow rapidly after the sheaths are shed. Horns of most yearlings are 7 to 9 inches (18–23 cm) long and may or may not be pronged, and 2-year-old males have pronged horns roughly 11 to 13 inches (28–33 cm) long. Bucks 3 years old or older generally have horns 12 to 14 inches (30–36 cm) in length. As of 1996, the world record horns—about 17.5 inches (44.5 cm) in length with basal circumferences of about 7 inches (17.8 cm) (see Reneau and Reneau 1996)—were carried by a 3-year-old (P. Wright personal communication:1996). Most mature male pronghorn in Alberta have 13- to 15-inch (33–38 cm) horns that are 5.5 to 6.3 inches (14–16 cm) in basal circumference during autumn (Mitchell 1980).

Lengths of pronghorn horns in the record books are about as representative of their species as are the heights of professional basketball players of theirs. Data on horn sizes, by age classes, were gathered from 2,013 hunter-killed bucks by field checks in North Dakota, at checking stations in Nebraska, on Texas ranches and by Native American guides in Montana (Table 19). Horn development in this fairly large sample was nearly at its maximum by 3 years of age. Sample sizes of animals older than 5 years are too small to judge at what age horns regress in size. However, testes of 7-year-old bucks are smaller than those of 2- to 6-year-olds (O'Gara 1968, Mitchell 1980). This implies that testicular hormones, which promote horn growth, would be somewhat lacking in 7-year-old and older bucks. Although the sample size was small, Firchow's (1986) data indicated poor horn growth by yearlings in Colorado. Overall, yearling horn lengths were the most variable of those reported for any age class during the five studies in Table 19.

With many species of big game, such as Rocky Mountain sheep and goats, the largest males often are past breeding age, and trophy hunting has little effect on the species. However, pronghorn may become trophies before they enter the breeding pool.

Although the Boone and Crockett Club asks for a tooth from each trophy animal entered in their records program, it had only six pronghorn teeth to submit for sectioning and study by the most refined techniques (Wright

One horn sheath often is cast a day or two before the other. Because the sheath is light, this unilateral casting does not seem to unbalance a pronghorn buck as does such dropping of an antler for large mule deer or elk. Photo by Joe Van Wormer; courtesy of the Denver Museum of Natural History.

1994). Three of these were determined to be 3 years old; one each was 4, 5 and 7 years old. One of the 3-year-olds turned out to be the current world's record taken by Mike O'Haco in Arizona in 1985. It may seem incredible that the O'Haco trophy could be that young, but these results are not surprising because pronghorn grow rapidly, and yearling bucks may weigh more than 80% of their adult weights. By comparison, a yearling bull elk weighs only about 50% of its potential adult weight.

Apparently none of the states with large pronghorn populations have done very complete or thorough sampling of the ages of their trophy animals. The Kansas Game Department, a state with relatively few pronghorn, sectioned incisor teeth from more than 50 bucks taken during 1992 and 1993. Three-year-old bucks there have grown horns 14 or 15 inches (35.6 or 36.1 cm) long, and the num-

TABLE 19. Average horn length and basal circumference for hunter-killed pronghorn bucks in Colorado, Montana, Nebraska, North Dakota and Texas

		Length		*Basal circumference*		
Area	*Age*[a]	*Number*	*Inches (cm)*	*Number*	*Inches (cm)*	*Source*
North Dakota	1	18	7.8 (19.8)			McKean (1951)
	2	9	10.0 (25.4)			
	3	3	13.0 (33.0)			
	4 and older	12	10.8 (27.4)			
Nebraska	1	309	7.4 (18.8)	323	4.7 (11.9)	Menzel (1980)
	2	241	11.9 (30.2)	264	5.5 (14.0)	
	3	119	13.4 (34.0)	122	5.9 (15.0)	
	4 and older	118	13.5 (34.3)	126	6.0 (15.2)	
Texas	1	37	9.1 (23.1)	37	4.9 (12.4)	Bone (1984)
	2–3	507	12.4 (31.6)	507	5.5 (14.0)	
	4	338	13.7 (34.8)	338	5.7 (14.6)	
	5 and older	193	13.9 (35.3)	193	5.8 (14.7)	
Colorado	1	9	5.3 (13.5)			Firchow (1986)
	2 and older	30	13.0 (33.1)			
Montana	1[b]	2	9.3 (23.6)	2	5.4 (13.7)	Mitchell and Maher (2001)
	2	28	13.2 (33.5)	25	5.6 (14.2)	
	3	53	13.6 (34.5)	50	5.9 (15.0)	
	4	15	13.6 (34.5)	15	5.9 (15.0)	
	5	6	13.2 (33.5)	6	6.0 (15.2)	
	6	4	13.5 (34.3)	4	5.8 (14.7)	
	8	1	12.0 (30.5)	1	5.0 (12.7)	

[a] Ages during hunting seasons were 3 to 5 months greater than whole-year numbers given.
[b] A yearling with 14-inch (35.6 cm) horns was not included.

ber of horns of that size do not appear to be better represented in 4-year-old or older bucks.

During a study in southern New Mexico, Brown and Verhelst (2002:10) found, "no significant annual difference between mean sheath sizes in animals aged 2 to 6. Although pronghorns from 2 and 3 years of age had the same average sheath mass and lengths as 5 and 6 year old bucks, the sheath sizes of the younger animals were more varied. Pronghorn sheaths in bucks age 7 and older were significantly smaller than those of animals younger than 7 ($P = 0.007$). Pronghorn sheath mass, but not length, was significantly influenced by the amount of precipitation falling during the winter (October–March) season prior to the hunt ($P = 0.003$)."

Variability in horn size is great; therefore considerable overlap is obvious among age classes. Year-to-year variations have not been studied extensively, but horn size seems to correlate positively with food quality and quantity, as well as negatively with severe winter weather. Pronghorn bucks born during 1994 in central Montana had smaller horns as adults than those of bucks born during 1992 or 1993 (Mitchell and Maher 2001). Their study area received about 50% less precipitation in 1994 than during the previous 2 years.

Body weight generally is equated with good nutrition. Danvir (2000) found that, of the 68 three-year-old or older bucks he examined, average weights of bucks having Boone and Crockett scores greater than 75 were 101.1 pounds (49.4 kg); those scoring less than 75 weighed an average of 95.8 pounds (46.8 kg). Most horn (especially sheath) growth takes place during December through March, when animals may be stressed by cold, deep snow and inferior or hard-to-reach forage. Thus, growth of horn sheaths should be greatest during mild winters after growing seasons with optimum conditions for plant production, as should horn core growth of fawns and yearlings.

Variability in horn shape also is great. Horn spread does not closely correlate with horn length. The 1,298 record-class pronghorn reported by Reneau and Reneau (1996) averaged 7.3 inches (18.5 cm) tip-to-tip, with the narrowest being one-eighth inch (0.3 cm) apart and the widest 20 inches (50.8 cm). The inside measurement for those trophies averaged 10.8 inches (27.4 cm) and ranged from 5.3 to 20.6 inches (13.0–52.4 cm).

The timing of sheath casting varies somewhat between northern and southern animals and is variable over comparatively short distances. Sheath casting commenced in late October during an Alberta study, and casting was slightly earlier in the southern part of the study area than in the northern part (Mitchell 1980). On November 10, 1962, 42% of the bucks checked in the southern area had cast their sheaths, in contrast to 30% of those from the northern area. Shedding in Texas began in mid-October and was essentially completed by early November, according to Buechner (1950a). The latest record he had of a buck carrying old sheaths was November 26—very similar to the latest date I observed in Montana. T. Yates and P. Fry (personal files), of New Mexico State University in Las Cruces, found an apparent 1-month difference in times of sheath casting for bucks in Hidalgo and Otero counties, New Mexico. The animals were within about 150 miles (241 km) of each other. Such variations apparently occur because local climate and vegetational phenology causes variations in times of breeding. Condition, especially hormonal balance, of a pronghorn can affect the timing of growth or shedding. Thus, animals occasionally growing or shedding sheaths at abnormal times should be expected. See Chapter 8 concerning horn growth and casting.

Cast antlers of deer and elk bleach and are easily seen from long distances. On the other hand, pronghorn sheaths are inconspicuous and not often found unless especially sought—a possible reason for early doubt that sheaths were cast. Pronghorn sheaths, particularly the bases, are soft and soon disintegrate, even if not chewed on by carnivores and rodents. Cahalane (1947:66) reported that sheaths "softened by rain or snow" appeal to badgers and coyotes. He added that they also were gnawed by porcupines, ground squirrels, mice, gophers, prairie dogs and "even the antelope have been seen chewing their own horns after shedding them." In Montana, the side next to the ground is reddish and crumbles easily by the spring after a sheath is shed, and the side exposed to the sun is cracked and splintered. The hooks are tougher and last one to several years exposed to sun and rain. Shed sheaths, which animals find easily because of the

Cast sheaths of pronghorn horns seldom are found because they are eaten or disintegrate rapidly. The side next to the ground crumbles easily by the spring after it was cast, and the side exposed to the sun is cracked and splintered. The pronghorn is the only species worldwide that annually sheds its coverings over permanent horn cores. Photo by Mike Francis.

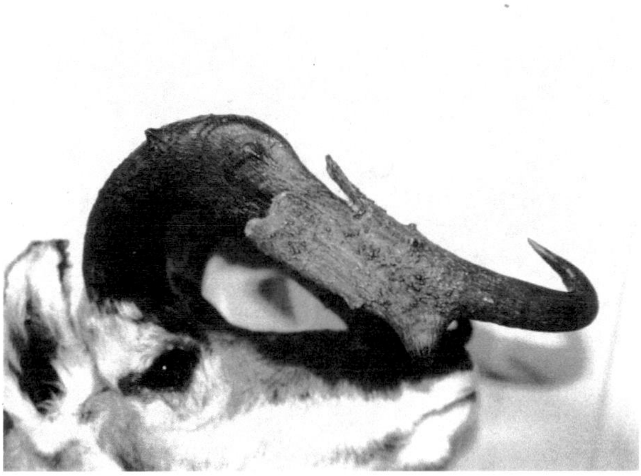

The curled-ahead shape of the basal horns of a buck taken in Montana (left) *are typical of castrated pronghorn bucks. However, they are unusual because they seem to have grown that way after a normal set of sheaths had formed.* (Right) *The distal sheath obviously was pushed off by a developing sheath under it, but the two never separated.* Photos courtesy of Willie Womack.

The skull and horns of a 2-year-old, four-horned pronghorn buck shot in Montana during autumn 1988, with the horn sheaths removed from the left horn cores to show their sizes. The four-horned circumstance seems a genetic anomaly (somewhat similar to white coat color) that occurs infrequently but over a wide geographic area. It is tempting to think of four-horned animals as evolutionary throwbacks, but no fossils are known with smaller posterior horns. Photo by Gerald Kling.

Crumpled horns occasionally are seen. Apparently, when a speeding fawn with soft, growing horn cores strikes a fence wire or other obstacle, the core can be so mangled that growth of a normal sheath becomes impossible. Photo by Bart W. O'Gara.

strong odor, immediately become food for coyotes, badgers, porcupines and mice.

Atypical or "freak" horns are fairly common. Horns projecting forward, almost parallel to the nose, occasionally are seen, and bucks with extra horns often have been reported. A buck with a third horn at a 90-degree angle behind its left horn and about one-third as long as the normal two was pictured in *Wyoming Wildlife* (Anonymous 1964). Lewis (1995) reported a four-horned pronghorn

The head of a pronghorn buck shot in Oregon (left) *features a small horn on the bridge of its nose. A much larger nose horn was reported from South Dakota. Such horn anomalies are not common, but neither are they rare. In addition to a nose horn, a pronghorn taken in Montana in 1993* (right) *also had small horns below and behind the bases of the normal horns. The small horns came off during handling. Several other cases of nose horns were reported from the same proximity. None had a bony core.* Left photo by Darrel Gretz. Right photo by Frank Martin.

"Swoop," a territorial buck on the National Bison Range, had horns that tipped forward more than normal. This may have been a genetic characteristic inasmuch as several does that shared his territory had similarly tipped horns. Horns tipped ahead so far that the prongs touch the nose occasionally are seen. Often, such horns are large and their owners tend to be especially successful in fights. Above, Swoop has partially erected his rump patches and the hair around his median gland (several inches forward of the rump patch on the midline of his back). Photo by Bart W. O'Gara.

A pronghorn buck (show-mounted) shot on the Fort Huachuca military reservation in Arizona had abnormal prongs projecting backward. Each measured 3.75 inches (9.5 cm). The normal prongs on this excellent head measured about 6 inches (15.2 cm), and the horn length was about 16 inches (40 cm). Whether the additional prongs were an advantage or disadvantage when sparring or fighting is unknown. However, for agonistic or courting displays, the additional mass would seem to have been an advantage. Other bucks with backward projecting prongs have been seen at Fort Huachuca. Photo by Jim Tomlin.

The skull of Swoop demonstrates the characteristic shape for skulls of bucks with tipped-ahead horns. Strangely, horn cores intermediate to this and normal cores are rare. Photo by Bart W. O'Gara.

In the same area of eastern Oregon there were a number of reports of pronghorn with unusual horn waves above very small prongs. Animals with similar horns were reported from the same area. Photo by Joe Van Wormer; courtesy of the Denver Museum of Natural History.

buck from Wyoming. In the Big Hole Valley of Montana, R. Nichols picked up a subfossil pronghorn skull with a small horn core behind each of the normal ones. Hoover et al. (1959) described a four-horned buck from Colorado with similar horn cores. I have seen or seen photos of at least 20 four-horned bucks from Montana and Wyoming. Bucks have been reported in Arizona with four prongs (J. Tomlin personal communication:1985). W. Hepworth (personal communication:1979) of the Wyoming Game and Fish Department has seen small horns on the bridge of the nose and as many as four small horns attached to the skin behind the normal horns. D. Gretz shot a buck in Oregon that had a small horn on the bridge of its nose. Another buck with a nose horn, which appeared two or three times as large as the one shot by Gretz, was shot in South Dakota (Anonymous 1983). Nose horns are simply sheath material. None were reported to have cores.

Females

Of 95 adult does collected on the National Bison Range and in Yellowstone National Park, 29 were hornless (O'Gara 1968). The hornless does had tiny nipples of bone about 0.1 inch (0.25 mm) high under whorls of hair. Horned

A yearling pronghorn doe in autumn, with horns just beginning to develop. Although buck fawns develop horns, does ordinarily do not do so until their second year of life. Female horns do not seem to have a specific function; hornless does sometimes are dominant to horned does. Photo by Bart W. O'Gara.

A female pronghorn, with unusually large horns and prongs (top), *stretches after standing. About 30% of does in Montana remain hornless. For those with horns, as above (with unusually large horns and prongs), the average horn length is only 1.6 inches (4.1 cm).* (Bottom) *An adult doe with slightly smaller than average horns. The horn cores of almost all does are nearly round in cross section. The cores of bucks are triangular in cross section, and the prongs develop over the sharp leading edges.* Photos by Peter T. Bromley.

does, on the other hand, had bony cores 0.5 to 3.0 inches (1.3–7.6 cm) long. Einarsen (1948) wrote that does in Oregon occasionally had horns comparable to those of mature bucks, and doe horns 12 to 13 inches (30.5–33.0 cm) long occasionally are reported. However, horn sheaths of does in my collection averaged 1.6 inches (4.1 cm) long and ranged from 0.3 to 5 inches (0.76–12.7 cm) (O'Gara 1968). "Does" with horns longer than 9 inches (23 cm) that have been investigated proved to be pseudohermaphroditic bucks (see Chapter 9).

Only 5 of the 66 horned does in my collection had rudimentary prongs. Left and right horns of several does differed in size and shape, some being two to three times longer than their counterparts; in contrast, buck horns normally differ only slightly. Yearling does had comparatively small horns, 0.3 to 1.5 inches (0.8–3.8 cm) long. I have never seen a doe fawn with horns. The horn cores of almost all does are round to oval in cross section.

Four of 95 does I collected had double horn sheaths and one had a triple sheath. Buechner (1950a) reported a doe with double sheaths collected on March 11 in Texas and said that Fisher (1942a) trapped a doe during January that carried four concentrically arranged horns. Multiple sheaths probably are retained because they are not large

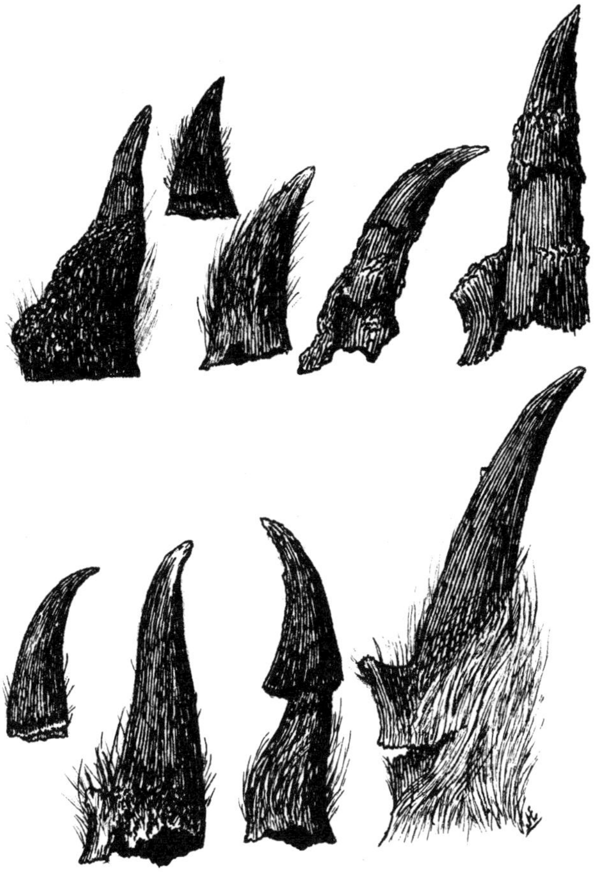

FIGURE 22. *Representative horns from nine pronghorn does. Note that three horns have double sheaths and the largest horn is separating from a horn under it.* Illustration by John Eiler.

enough or heavy enough to fall off as a result of jarring or by their own weight. Also, they usually have no prongs to interfere with stacking, and does ordinarily do not use their horns as weapons.

Variability in horn size prevents casual observers from determining whether or not a doe has recently cast her sheaths (Figure 22). Although admitting that critical observations were lacking, a number of biologists presumed shedding was similar in does and bucks (Lyon 1908, Skinner 1922, Rush 1944, Einarsen 1948). Horn sheaths of does collected in Montana and Yellowstone National Park, apparently ready to be cast, came off during handling in October, November, December, February and March (O'Gara 1969b). A tame doe on the National Bison Range carried double sheaths throughout the spring of 1966 and shed both sets on July 15, 1966. The sample size is small, but these observations, plus those made by F. Blunt (personal communication:1974) of captive does in Wyoming, indicate that sheath casting by does is not as seasonal as that of bucks. However, Byers (1997a) reported that closely observed does on the National Bison Range usually shed old sheaths and began to grow new ones during midsummer. He also noted that the size and orientation of doe horns (straight up, leaning forward or back) varied from year to year.

SIZES AND WEIGHTS

Members of the southern pronghorn subspecies (*A. a. mexicana*, *A. a. sonoriensis* and *A. a. peninsularis*) generally are considered smaller than those of the northern subspecies, and *A. a. oregona* supposedly is slightly larger than *A. a. americana* (Bailey 1932). Few weights are available for comparison among subspecies, but some measurements are recorded in Table 20. These measurements do not indicate any great size differential between presumed subspecies. Collections and measurements of any subspecies by different researchers probably would show as much variation.

Measurements not commonly taken, but of importance to biologists who install identification collars on pronghorn, are neck sizes (Table 21). Necks of bucks swell during the rut. I measured bucks in mid-September with necks as large as 27 inches (68.6 cm) at their bases and 21 inches (53.3 cm) behind the ears.

TABLE 20. Measurements of adult pronghorn[a]

Subspecies	*Area*[b]	*Sex*	*Total length*	*Length of tail*	*Length of hind foot*	*Length of ear*	*Shoulder height*
Antilocapra americana americana	Montana(1)	M	52.7 (133.8) [36]	5.3 (13.5) [36]	14.8 (37.6) [36]	5.9 (15.2) [36]	34.4 (87.4) [36]
		F	53.5 (135.9) [12]	5.0 (12.7) [12]	14.4 (36.6) [12]	5.9 (15.0) [12]	32.7 (83.1) [12]
	Montana and Yellowstone Nat'l Park(2)	M	56.9 (144.5) [27]	4.3 (10.9) [27]	16.8 (42.7) [27]	5.7 (14.5) [27]	37.0 (94.0) [27]
		F	57.0 (144.8) [93]	4.3 (10.9) [93]	16.5 (41.9) [93]	5.7 (14.5) [93]	35.6 (90.4) [93]
	Alberta(3)	M	55.7 (144.5) [16]	4.1 (10.4) [15]	15.9 (40.4) [15]	5.6 (14.2) [15]	34.4 (87.4) [11]
		F	55.4 (140.7) [9]	3.8 (9.7) [8]	15.6 (39.6) [9]	5.6 (14.2) [9]	33.9 (86.1) [9]

continued on next page

Table 20. Measurements of adult pronghorn[a]

Subspecies	Area[b]	Sex	Total length	Length of tail	Length of hind foot	Length of ear	Shoulder height
	Colorado(4)	M	54.8 (139.2) [15]	5.1 (13.0) [15]	16.5 (41.9) [15]	5.7 (14.5) [15]	35.8 (90.9) [15]
		F	55.0 (139.7) [31]	4.7 (11.9) [32]	16.5 (41.9) [31]	5.9 (15.0) [32]	35.0 (88.9) [32]
	(5)	M	55.2 (140.1) [53]	4.3 (10.9) [30]	16.6 (42.2) [31]	5.6 (14.3) [30]	35.0 (89.0) [43]
		F	57.9 (146.5) [47]	4.2 (10.7) [4]	15.5 (39.3) [4]	5.3 (13.5) [3]	35.4 (90.0) [21]
	Idaho(6)	M	55.2 (140.2) [18]	4.4 (11.1) [5]	16.6 (42.1) [18]	5.9 (15.0) [18]	35.2 (89.5) [18]
		F	53.4 (135.6) [13]	5.0 (12.7) [5]	16.5 (41.8) [13]	5.7 (14.5) [13]	34.4 (87.3) [13]
	Arizona(7)	M	57.0 (144.8) [4]	4.8 (12.2) [4]	17.5 (44.5) [4]		
A. a. oregona[c]	Oregon(8)	M	58.8 (149.4) [1]	4.2 (10.7) [1]	17.0 (43.2) [1]	5.5 (14.0) [1]	
		F	58.5 (148.6) [1]	4.0 (10.2) [1]	17.4 (44.2) [1]	5.3 (13.5) [1]	
	California(9)	M	57.4 (145.9) [275]	4.4 (11.2) [251]	17.8 (45.2) [251]	6.4 (16.3) [359]	
	Oregon and Nevada(10)	M	52.9 (134.4) [21]	4.1 (10.4) [19]	16.9 (42.9) [19]	6.1 (15.5) [21]	37.3 (94.7) [21]
		F	50.5 (128.3) [1]	4.0 (10.2) [1]	16.5 (41.9) [1]	6.0 (15.2) [1]	36.0 (91.4) [1]
	Oregon and Nevada(11)	M	57.3 (145.5) [67]	4.3 (10.9) [70]	17.2 (43.7) [49]	6.3 (16.0) [88]	35.9 (91.2) [65]
A. a. mexicana	Chihuahua(12)	M	55.9 (142.0) [1]	5.7 (14.5) [1]	16.1 (40.9) [1]	32.7 (83.1) [1]	
	Texas(13)	M	53.0 (134.6) [1]	2.5 (6.4) [1]	15.5 (39.4) [1]	6.0 (15.2) [1]	36.3 (92.2) [1]
		F	53.0 (134.6) [1]	2.5 (6.4) [1]	15.5 (39.4) [1]	5.9 (15.0) [1]	34.5 (87.6) [1]
	Southern Arizona(7, 14)	M	54.2 (137.7) [10]	5.1 (13.0) [10]	16.4 (41.7) [6]	6.2 (15.7) [4]	32.9 (83.6) [10]
		F	56.4 (143.3) [6]	5.1 (13.0) [6]	16.6 (42.2) [5]	7.0 (17.8) [1]	33.7 (85.6) [5]
A. a. sonoriensis	Sonora(15)	M	55.9 (142.0) [1]	4.7 (11.9) [1]	16.1 (40.9) [1]		
	Sonora(7)	F	52.8 (134.1) [1]	4.7 (11.9) [1]	16.1 (40.9) [1]	5.7 (14.5) [1]	
A. a. peninsularis	Lower California(7)	M	54.0 (137.2) [14]	3.3 (8.4) [7]	15.8 (40.1) [2]	37.6 (95.5) [14]	
		F	52.1 (132.3) [9]	3.4 (8.6) [2]	16.2 (41.1) [4]	36.5 (92.7) [7]	

[a] Measurements in inches (cm); [sample size].

[b] *Source:* (1) Buck 1947; (2) O'Gara 1968, personal files; (3) Mitchell 1971; (4) Bear et al. 1973; (5) Firchow 1986; (6) Bodie 1979; (7) U.S. National Museum; (8) Bailey 1932; (9) McLean 1944; (10) Mason 1952; (11) J. O. Meeker, personal files; (12) Merriam 1901; (13) Buechner 1950; (14) New Mexico Museum; (15) Goldman 1945.

[c] No longer recognized as a valid subspecies; presently, included as *A. a. americana.*

Table 21. Neck circumference measurements of pronghorn

				Neck circumference		
				Greatest	Least	
Age	Area	Sex	Sample size[a]	Inches (cm)	Inches (cm)	Source
Adult	Colorado	M	12–13	22.2 (56.4)	15.2 (38.6)	Bear et al. (1973)
	Montana	M	10–10	21.5 (54.6)	16.0 (40.6)	B. W. O'Gara (personal files)
	Oregon	M	29–28	23.2 (58.9)	19.2 (48.8)	J. O. Meeker (personal files)
	Nevada	M	50–50	22.4 (56.9)	18.7 (47.5)	J. O. Meeker (personal files)
	Colorado	F	24–26	19.3 (49.0)	13.4 (34.0)	Bear et al. (1973)
	Montana	F	12–12	18.5 (47.0)	13.0 (33.0)	B. W. O'Gara (personal files)
Yearling	Colorado	M	15–15	20.7 (52.6)	15.9 (40.4)	Bear et al. (1973)
	Colorado	F	7–7	18.6 (47.2)	13.5 (34.3)	Bear et al. (1973)
3 months old	Colorado	M and F	8–8	16.3 (41.4)	11.8 (30.0)	Bear et al. (1973)
1 day old	Nevada	M and F	0–18		6.8 (17.3)	McNay (1980)
3 to 4 hours old	Idaho	M	0–5		6.5 (16.5)	Autenrieth (1984)
	Idaho	F	0–11		6.1 (15.5)	Autenrieth (1984)

[a] Only a single measurement was available for some animals.

Northern pronghorn are heavier than often indicated in the popular literature. Weights vary tremendously because of the influence of time of year, range conditions and weather, especially droughts and severe winters (Table 22). Generally, during autumn hunting seasons in Montana, does weigh slightly more than 100 pounds (45 kg), and bucks weigh 10 to 15 pounds (4.5–6.8 kg) more.

FAWNS

Everything about a newborn fawn seems exaggerated. Its eyes seem too big, its legs too long, its face too narrow and, except when the fawn is hiding, its pointed ears stand up as if on constant alert. The color pattern of pronghorn less than a month old is somewhat obscure—white areas (on adults) are stained with dull buff, and blackish parts

TABLE 22. Body weights of adult pronghorn 2 years of age or older

		Sample size		*Weight in pounds (kg)*			
Subspecies	*Area*	*Male*	*Female*	*Male*	*Female*	*Season*	*Source*
Antilocapra americana americana	Wyoming	—	60	— (—)	111 (50)	Autumn/ spring	Edwards (1958)
	Wyoming	52	76	107 (49)	105 (48)	All year	W. G. Hepworth (personal files)
	Colorado	6	8	92 (42)	90 (41)	Winter	Hoover et al. (1959)
	Montana, Yellowstone National Park	30	90	129 (59)	111 (50)	All year	O'Gara (1968, personal files)
	Alberta	35	40	119 (54)	105 (48)	All year	Mitchell (1971)
	Colorado	15	32	111 (50)	102 (46)	All year	Bear et al. (1973)
	Idaho	11	12	106 (52)	102 (50)	Autumn	Bodie (1979)
A. a. oregona	California	67	—	113 (51)	— (—)	Spring/ summer	McLean (1944)
	Oregon	Unknown	Unknown	114 (52)	92 (42)	Autumn	Einarsen (1948)
	Oregon Nevada	19	1	124 (56)	105 (48)	All year	Mason (1952)
A. a. mexicana	Texas	114	128	94 (43)	87 (40)	Autumn	Buechner (1950)
	New Mexico	1	1	112 (51)	109 (49)	Spring	U.S. National Museum

With remarkably well-camouflaged coloration, pronghorn fawns lie hidden in the vegetation during most of their first 3 weeks of life, primarily to avoid detection by predators. As seen above, this often is among wildflowers and forbs that provide prime nutrition for their dams. Photo by Jim D. Yoakum.

Just as its eyes seem too big and its face seems too narrow, a pronghorn fawn has exceptionally long legs for its overall size. Such legs facilitate rapid growth by the fawn and enable it to run quickly at an early age. Photo by William and Irene Finley.

Anogenital grooming, practiced by pronghorn does until fawns are about 3 weeks of age, apparently reduces fawn odors and the possibility that the neonates will be detected by mammalian predators. After nursing, fawns elicit such grooming by walking in front of their dams and assuming a rump-up posture. Grooming induces evacuation, and the mother ingests urine and feces. Photo by Cathy and Don Jenni.

are only faintly indicated. The backs and sides are grayish brown, almost perfect camouflage with prairie soil, rocks and patches of dry, matted vegetation—especially bunchgrasses. Bromley (1977) noted that a fawn with its neck, head and ears lowered looks so much like a pile of dried bison dung as to suggest mimicry. Fawns only a few hours old can erect their rump patches, and little whorls of dusky hair mark locations where horns will grow.

Fawns, wearing camouflaged coats, lie hidden in vegetation during most of the first 3 weeks of their lives. The hiding strategy serves to deny predators information concerning the presence and location of a fawn. This requires coordinated behavior by mother and young (see Chapter 6). Nearly all aspects of the behavior that support this strategy show evidence of refinement by natural selection. Hiding has been reported for all pronghorn populations in which neonatal behavior has been studied (see Byers 1997a).

About the time fawns join doe/fawn bands, the neonatal pelage begins molting and acquiring the conspicuous colors displayed by adults, according to Bromley (1977). Thus, the smallest members of a herd do not draw the attention of predators by their colors. Dinges (1958) found that captive fawns he worked with in Oregon did not completely shed their juvenile pelage until mid-August, at about 10 weeks of age.

Many people contend that pronghorn fawns, and fawns of other big game animals, lack odor. Body odor may be slight but it apparently exists. A miniature poodle I once owned, which was well trained and had an excellent nose, found fawns every time he was taken to a general area where fawns were bedded. The dog appeared to find fawns by scent, just as he did game birds. W. Bodie (personal communication:1977) watched a coyote turn at a right angle from its course on a ridge top directly into the wind blowing from a bedded fawn. Sometimes, bedded

Except for short periods with their dams, fawns generally remain solitary until about a week of age. At that time, they begin to form sibling bonds. During a study in Alberta, fawns consistently bedded together by the time they were 2 weeks old. Photo by Jim D. Yoakum.

At about 1 month of age, pronghorn fawns begin to acquire adult coloration and frequently flare their rump patches and erect their manes. The color change is important because fawn colors would tend to draw the particular attention of predators to the smallest members of a band. Deer fawns do not lose their spots and appear like small adults, but they also are not in the open where predators can readily see them. Photo by Bart W. O'Gara.

Pronghorn fawns emit odor, but it is a very slight amount. In these scenes, a dog was being used to find fawns, but was unaware of the bedded fawn that researchers already had found. It actually lay down next to the fawn without detecting it and was startled when researchers picked up the fawn. Photos by Joe Van Wormer.

Elk and deer fawns generally have spotted, dark coats that camouflage them well in dappled light under shrubs and trees. In contrast, pronghorn fawns have uniform, light-colored coats that are more suitable camouflage in open bed sites exposed to the sun. Because of its relatively open habitat, the pronghorn fawn attains adult size, coloration and speed relatively quickly. Photo circa 1925 by Sam Woodring; courtesy of Yellowstone National Park, Wyoming.

fawns are surprisingly easy to locate from a distance of 20 to 30 feet (6.1–9.1 m) by the sound of their breathing.

Pronghorn fawns are precocious. Hepworth and Blunt (1966) maintained that fawns could run at speeds up to 25 miles per hour (40 km/hr) for short distances when only a few days old. Two fawns I collected at about 3 weeks of age had rumens full of green vegetation. Two other fawns orphaned on July 19, at approximately 7 weeks of age, not only survived but maintained their position in the hierarchy of the herd (Bromley and O'Gara 1967).

When captured young, pronghorn fawns become doting, playful and intelligent pets.

GROWTH AND DEVELOPMENT

The pronghorn gestation period is long (about 8.4 months), but nearly half of the fetal weight is gained during the last month of pregnancy. The reproductive tract of a doe with twin fetuses and amniotic fluid at parturition time in Montana weighs approximately one-fifth the weight of a doe without its reproductive tract.

Pronghorn fawns are large at birth, considering the size of the does and the fact that they usually have twins. Many weights of neonatal fawns have been published, but exact ages of the fawns often are not known or given. Weights in Table 23 are for fawns that were observed being born, were still wet or had bloody umbilical cords and white, fleshy "bumpers" on their hooves.

Barrett (1982) reported the mean weight of 465 fawns captured during his study in Alberta as 9.4 pounds (4.2 kg), and no difference was noted between the sexes. Those fawns ranged from 4 hours to 10 days in age (mean = 2.8 days).

D. Kitchen (personal communication:1976) found that if one of a set of twins was especially large, its sibling also was large. He also noted that fawns of does that had low

TABLE 23. Weights of pronghorn fawns on the days they were born

Area	*Sex*	*Sample size*	*Pounds (kg)*	*Source*
Montana	M	56	7.4 (3.4)	D. W. Kitchen (personal files)
	F	47	7.2 (3.3)	
	M and F	15	8.2 (3.7)	Von Gunten (1978)
	M and F	21	7.7 (3.5)	Showers-Corneli (1980)
Oregon	M	1	8.0 (3.6)	Hansen (1955)
	F	2	8.9 (4.0)	
Colorado	M	14	7.1 (3.2)	Hoover et al. (1959)
	F	17	6.5 (2.9)	
	M and F	28	6.7 (3.0)	Tileston and Yeager (1962)
	M and F	10	7.6 (3.4)	Firchow (1986)
Idaho	M and F	42	8.0 (3.6)	Bodie (1979)
	M	15	8.7 (3.9)	Autenrieth (1984)
	F	25	8.8 (4.0)	
Nevada	M and F	18	8.2 (3.7)	McNay (1980)

social status averaged almost 1 pound (0.45 kg) less than fawns of does that were high in the hierarchy. Fawns on the National Bison Range in Montana usually weigh slightly less than 8 pounds (3.6 kg), about average for neonate pronghorn on healthy rangelands elsewhere. However, J. Reichel (personal communication:1977) recorded one newborn fawn on the range that weighed 11 pounds 4 ounces (5.1 kg); its twin was 1 pound (0.45 kg) lighter.

Mean measurements of 3- to 4-hour-old fawns in Idaho (15 female and 25 male) (Autenrieth 1984) were as follows: total length—females 23 inches (58.4 cm), males 21.7 inches (55.1 cm); hind foot length—both sexes 10.2 inches (25.9 cm); and ear length—females 3.57 inches (9.1 cm), males 3.55 inches (9.0 cm). When measuring newborn fawns, I have noticed that females often are slightly longer than their male siblings, but the males have slightly larger chests and necks and are slightly taller at the shoulder.

Nutritional levels of does during gestation, which are influenced by weather and health of vegetation, undoubtedly influence birth weights. Weights of neonates in Idaho were significantly higher ($P < 0.05$) during spring 1977 than during 1976 and 1978 (Bodie 1979). The difference of 1 pound (0.45 kg) probably was related to the extremely mild winter of 1976–77. A lack of snow that winter allowed pronghorn to use all seasonal rangelands throughout the winter.

Nutritional intake strongly influences weight gain in pronghorn neonates. During spring 1993, body weights of 1- to 2-day-old female fawns at the Sybille Wildlife Research and Conservation Education Center in Wyoming averaged 7.40 ± 0.5 pounds (3.36 ± 0.21 kg). Five fawns bottle-fed a mean peak intake of 1.36 ± 0.07 quarts (1,190 ± 63 mL) of milk daily gained 5.4 ± 0.2 ounces (152 ± 5 g) per day. Five others fed a mean peak intake of 1.10 ± 0.05 quarts (1,044 ± 50 mL) of milk daily gained only 3.9 ± 0.2 ounces (110 ± 5 g) per day (Martin and Parker 1997). Thus, the fawns ingesting 12% less milk were gaining 28% less weight than were their better-fed counterparts. Beale and Smith (1970) reported that fawns nibbled vegetation 2 weeks after birth, and little difference existed between food habits of adults and those of 5- to 19-week-old fawns.

During Tileston's and Yeager's (1962) study in Colorado, fawns weighed approximately twice as much at 2 weeks of age as at birth. Firchow (1986) recorded weights of Colorado fawns during their first 5 to 6 days of life (Table 24).

Ages of fawns on the National Bison Range, Montana, could be determined fairly accurately for the first week of life. Data from two studies there indicated the rate of growth for neonates (Table 25). Early growth of fawns in Colorado seemed quite similar to that in Montana.

Growth of fawns was recorded during a study in Texas by Tucker (1979). Ages were estimated in that study by growth of the hooves (Table 26). Comparison of total lengths in Tables 25 and 26 indicates that, on the National Bison Range in Montana, fawns are larger and grow faster than those in Texas. If it can be assumed that the techniques for estimating ages were equally accurate among the studies, newborn fawns on the National Bison Range were as long as 2-week-old fawns in Texas, and 1-week-old National Bison Range fawns were longer than 3-week-old Texas fawns. Growth of fawns in Colorado (Table 24) was similar during the first week to that of fawns in Montana (Table 25).

Pronghorn fawns not only are large at birth, they also develop rapidly—faster than the young of most American artiodactyls. In this respect, North American pronghorn resemble the antelope of Africa that also inhabit open grasslands. These adaptations reduce the amount of time the young are most vulnerable to predation and

TABLE 24. Mean weights and measurements for pronghorn fawns in southeastern Colorado during the fawns' first 1 to 6 days of life, 1947 to 1954, 1983 and 1984

Year	Days old[a]	Sample size	Weight in pounds (kg)	Length in inches (cm)	Girth in inches (cm)	Hindfoot length in inches (cm)	Ear length in inches (cm)
1947–1954	1 or less	31	6.8 (3.1)	23.0 (58.4)	14.8 (37.6)	9.5 (24.1)	3.4 (8.6)
1983	1 or less	4	7.6 (3.4)	25.3 (64.3)	15.3 (38.9)	10.0 (25.5)	
	1–2	5	7.7 (3.5)	24.8 (63.0)	15.2 (38.7)	10.2 (25.9)	
	3–4	3	9.8 (4.5)	26.1 (66.3)	17.0 (43.3)	10.4 (26.3)	
	5–6	1	11.4 (5.2)	27.2 (69.0)	16.9 (43.0)	11.4 (29.0)	
1984	1 or less	6	7.6 (3.5)	25.5 (64.7)		9.8 (24.8)	3.3 (8.4)
	1–2	8	8.4 (3.8)	26.1 (66.4)		10.2 (26.0)	3.4 (8.6)
	3–4	3	9.5 (4.3)	27.1 (68.8)		10.4 (26.5)	3.7 (9.3)
	5–6	2	11.2 (5.1)	28.9 (73.5)		11.3 (28.8)	3.7 (9.4)

[a] Firchow (1986) estimated age using a combination of behavioral and physical characteristics described by Bromley (1977).

Source: 1947 to 1954: from Hoover et al. 1959; 1983 to 1984: from Firchow 1986.

TABLE 25. Mean weights and measurements for pronghorn fawns on the National Bison Range, Montana, during the fawns' first week of life, 1977 and 1978

Year	Age	Sample size	Weight in pounds (kg)	Length in inches (cm)	Girth in inches (cm)
1977	1 or less	15	8.2 (3.7)	25.5 (64.8)	14.9 (37.8)
	2–3	11	8.4 (3.8)	26.2 (66.5)	15.2 (38.6)
	4–5	3	10.4 (4.7)	28.2 (71.6)	16.8 (42.7)
	6–7	1	12.8 (5.8)	29.1 (73.9)	16.9 (42.9)
1978	1 or less	21	7.7 (3.5)	26.5 (67.3)	14.9 (37.8)
	2–3	7	8.4 (3.8)	25.8 (65.5)	15.1 (38.4)
	4–5	1	9.9 (4.5)	29.5 (74.9)	16.0 (40.6)
	6	5	11.9 (5.4)	27.0 (68.6)[a]	16.8 (42.7)

[a] 3 in sample.

Source: From Von Gunten 1978, Showers-Corneli 1980.

TABLE 26. Measurements of Texas fawns 1 to 24 days of age[a]

Age in days	Sample size	Sex	Total length	Length of tail	Length of hind foot	Length of ear	Neck circumference	
							Least	Greatest
1–2	1	F	22.05 (56.0)	1.58 (4.0)	10.20 (25.9)	3.35 (8.5)	6.50 (16.5)	6.69 (17.0)
3	1	F	21.85 (55.5)	1.58 (4.0)	9.41 (23.9)	3.66 (9.3)	6.69 (17.0)	6.89 (17.5)
6	2	F	23.23 (59.0)	1.69 (4.3)	9.53 (24.2)	3.70 (9.4)	6.14 (15.6)	6.46 (16.4)
8	3	M	20.55 (52.2)	1.69 (4.3)	10.40 (26.4)	3.58 (9.1)	6.34 (16.1)	6.89 (17.5)
8	2	F	22.44 (57.0)	1.85 (4.7)	10.43 (26.5)	3.80 (9.6)	6.18 (15.7)	6.81 (17.3)
10	3	M	23.15 (58.8)	1.69 (4.3)	10.24 (26.0)	3.82 (9.7)	6.57 (16.7)	7.20 (18.3)
10	3	F	25.16 (63.9)	2.05 (5.2)	10.31 (26.2)	3.86 (9.8)	6.26 (15.9)	6.93 (17.6)
13	1	M	22.05 (56.0)	1.77 (4.5)	9.53 (24.2)	3.86 (9.8)	6.10 (15.5)	7.48 (19.0)
13	1	F	25.87 (65.7)	2.17 (5.5)	11.14 (28.3)	4.06 (10.3)	6.69 (17.0)	7.24 (18.4)
15	1	M	24.02 (61.0)	2.36 (6.0)	10.63 (27.0)	3.94 (10.0)	6.69 (17.0)	7.48 (19.0)
17	1	F	24.02 (61.0)	1.58 (4.0)	9.76 (24.8)	3.86 (9.8)	6.30 (16.0)	7.09 (18.0)
22	1	M	24.80 (63.0)	2.17 (5.5)	10.94 (27.8)	4.09 (10.4)	6.50 (16.5)	7.09 (18.0)
22	1	F	28.35 (72.0)	2.36 (6.0)	10.63 (27.0)	4.13 (10.5)	6.69 (17.0)	7.87 (20.0)
24	1	F	26.38 (67.0)	2.17 (5.5)	10.63 (27.0)	4.13 (10.5)	7.09 (18.0)	8.27 (21.0)

[a] In inches (cm).

Source: From Tucker 1979.

allow them to travel with their bands or herds to the best feeding areas.

Pronghorn are smaller than northern white-tailed deer, but reach full weight at a proportionately earlier age. Working with a sample of more than 1,000 deer, Roseberry and Klimstra (1975) found that both sexes reached near-maximum weights at 4.5 years of age, but that females reached full weight earlier than did males. In contrast, Mitchell (1980), who collected pronghorn throughout the year in Alberta, reported that newborn male and female fawns showed similar weights during May and June and similar weight gains of about 60 pounds (27.2 kg) during their first summer and autumn. By their second winter, yearlings weighed 90 to 117 pounds (40.8–53.1 kg) and corresponded in weight to mature animals. In Wyoming, female yearlings weighed 77 to 109 pounds (34.9–49.4 kg) and males weighed 68 to 113 pounds (30.8–51.3 kg) (W. Hepworth:personal files). The ranges in weights were more than twice what I observed with yearlings in Montana. W. Hepworth (personal communication:1982) indicated that drought and bad winters occurred during the collections. Apparently, that accounted for at least part of the variation.

Skeletal tissues of mammals generally achieve maximum growth sooner than muscle or adipose tissue. The Roseberry and Klimstra (1975) data indicated that the hind feet of white-tailed deer reach maximum length by 3.5 years of age in males and 2.5 years in females. Comparable data on pronghorn indicate maximum hind foot growth by 1.5 years of age (Figure 23). Hansen (1955) found that 1-month-old pronghorn fawns in Oregon were nearly one-half the height of their dams. By 2.5 months of age, they appeared to be a little more than three-quarters the height of their mothers; at 3.5 months (September), the fawns were nearly as tall as the does.

At a mean age of 38 days, the brisket height of five captive fawns was 15.5 ± 0.2 inches (39.3 ± 0.5 cm)—80.2% the length of adult legs. By 7 months of age, the average for two fawns was 17.9 ± 0.2 inches (45.5 ± 0.4 cm)—92.9% that of adults (Martin and Parker 1997). Lengths of brisket, tarsus and carpus of these bottle-fed fawns increased linearly within the narrow range of weights monitored.

During their first summer, pronghorn fawns spend considerable time playing. Miller and Byers (1991) estimated that fawns on the National Bison Range, Montana, would weigh 55.8 pounds (25.3 kg) at 12 weeks of age instead of 51.8 pounds (23.5 kg) if less time was spent in play. Because mortality during severe winters disproportionately affects fawns, such a difference in September weights may not be trivial. However, fawns that do not play may be less fit to avoid predation.

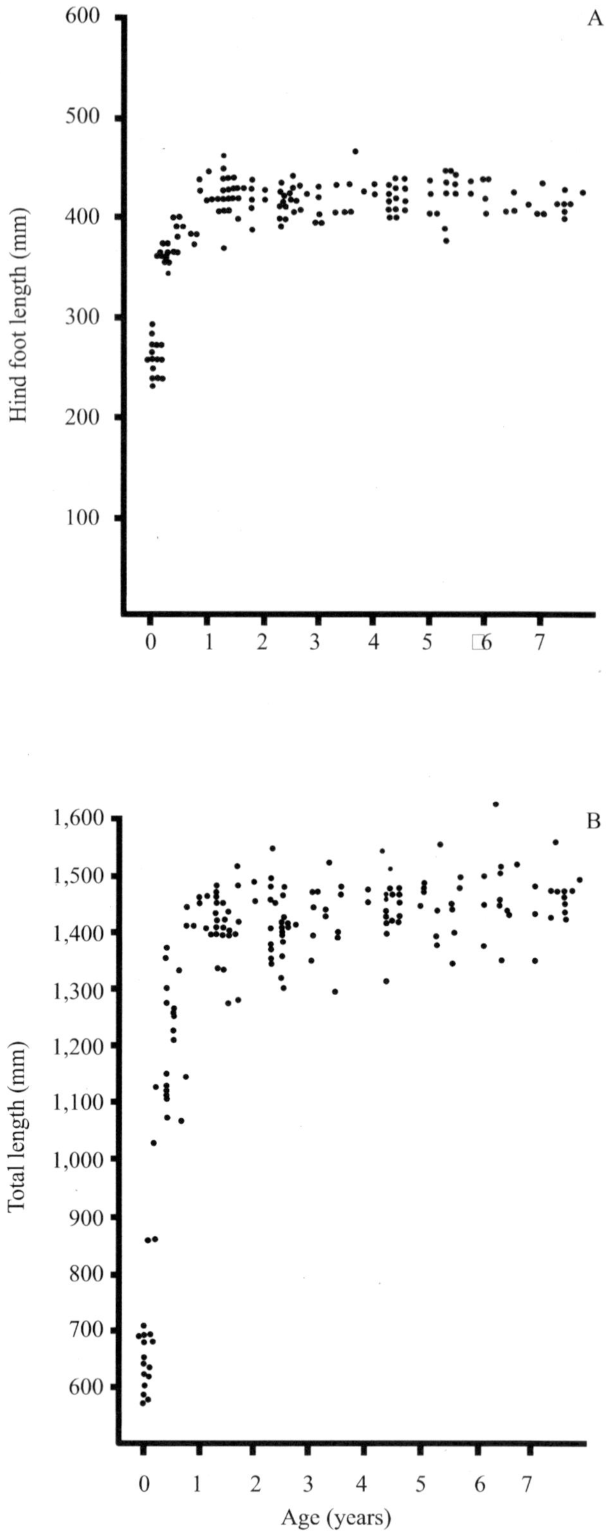

Figure 23. *Hind foot (top) and total (bottom) lengths versus age in years for 166 pronghorn of both sexes from Yellowstone National Park and the National Bison Range (O'Gara 1968, P. Wright:personal files).*

After pronghorn fawns join doe/fawn groups, they associate mostly with other fawns and spend much of their time playing. Although they commonly are called fawns, some workers believe "kids" would be a better name, inasmuch as the animals' mischievous behavior and curiosity are suggestive of typically rambunctious, human youngsters. Note the horns starting to grow on the 6- to 7-week-old buck on the left. Photo by Peter T. Bromley.

LONGEVITY

Pronghorn generally are not considered long-lived animals. I collected nearly 200 jaws in Yellowstone National Park and on the National Bison Range—where old animals were common—but only two jaws had missing incisiform teeth and excessive wear on molariform teeth, indicating extreme old age. Using the tooth eruption and wear method, 15 jaws were classified in the 9-year or older category, one of which was from a known-age 9-year-old. All but two of the jaws had good solid teeth with about the same amount of wear on both the incisiform and molariform teeth as did the known-age 9-year-old. Several jaws I picked up were from animals that had died "naturally" and had teeth in similar condition. Yoakum (1957a) found 4 of 89 jaws picked up on a refuge in Oregon with missing or "pegged" incisors and extensively worn molars, indicating ages beyond 9 years.

These observations indicated that most pronghorn (which, until then, had survived predation, diseases, starvation, and so on) died of natural causes in their ninth year, and Hepworth (1965) also maintained that most pronghorn did not live beyond 9 years of age. The oldest jaw in the Wyoming Fish and Game Department collection was from a 10.5-year-old doe, according to Hepworth and Blunt (1966).

However, by the cementum annulus technique, W. Hepworth (personal communication:1979) later determined the ages of 23 pronghorn that died during the severe winter of 1971–72. Seven exceeded 9 years of age. All were animals that succumbed to malnutrition, exposure or injury. Three were 11 years old, and one each was 10, 13, 14 and 18. Only one of those old animals—the 14-year-old—was a buck. Also, animals assigned to the 9-year class by Kerwin and Mitchell (1971) on the basis of tooth wear ranged from 8.5 to 15.5 years of age by the cementum annulus method.

These examples of unusual longevity depend on proper techniques for determining ages. However, at least two marked animals in Montana lived beyond 9 years of age. I collected a doe that had been tagged as an "adult" 9 years before. Assuming that she may have been a yearling at the time of tagging, she was at least 10 years old. She was very thin, had only three incisiform teeth, and had lost the right, upper first molar—resulting in abnormal growth of the corresponding lower molar. She probably would not have survived the winter, but when collected on October 15, she was carrying twin fetuses of normal size. J. McLucas (personal communication:1982) captured an adult doe near Shawmut, Montana, on December 1, 1948. He ear tagged and released her in the vicinity of Boulder, Montana. She was killed near Boulder during the hunting season of 1962, still wearing the tag that was placed in her ear 14 years earlier.

From 1981 through 1994, Byers (1997a) knew the birth dates and year of death for 7 doe and 11 buck pronghorn on the National Bison Range that had survived as fawns and yearlings (Table 27).

Even if the animals in Table 27 were a year younger at death than indicated by the method used, the data indicate that deaths from old age do not fall within one annual age class as earlier workers had believed. Individual differences probably occur, just as in humans, but vegetative

TABLE 27. Approximate ages at death for unhunted doe and buck pronghorn on the National Bison Range, Montana, 1981 to 1984

Sex	*Age at death*	*Number*
Does	5	1
	10	1
	11	2
	14	2
	15	1
Bucks	4	3
	8	3
	9	3
	10	1
	12	1

Source: From data gathered by Byers 1997.

conditions and weather seem to strongly influence how far into their dotage pronghorn will survive. Winters were comparatively mild during 1981–1994 on the National Bison Range. As Byers (1997a:203) noted: "Sexual selection is likely to create male characters that enhance mating success at the expense of longevity." It is tempting to speculate that 4 years, the age at which most bucks acquire territories, is a dangerous time. However, the sample size probably does not warrant such speculation.

SPEED AND GAITS

When Captain Meriwether Lewis watched pronghorn running along a ridge in 1804, he wrote: "So soon has these antelopes gained the distance at which they had again appeared to my view I doubted at ferst that they were the same that I had just surprised, but my doubt soon vanished when I beheld the rapidity of the flight along the ridge before me it appeared reather the rappid flight of birds than the motion of quadrupeds. I think I can safely venture the assertion that the speed of the animals is equal if not superior to that of the finest blooded courser" (Moulton 1987a:82).

Pronghorn often race alongside and cross in front of vehicles passing through their range—as if to prove they can. Sex, age, physical condition, characteristics of the terrain and other factors undoubtedly influence the speed of individual pronghorn. I have raced many pronghorn and never had one keep up with me at 50 miles per hour (80.5 km/hr). The top speed for most pronghorn probably is near 45 miles per hour (72.4 km/hr), although there is no way of knowing whether the animals I "raced" were running at top speed. At any rate, pronghorn have no serious contenders for the title of North America's fastest land mammal. They often run from one place to another, apparently because it is as easy for them to run as walk. Bullock (1974) concluded that pronghorn can maintain speeds of 30 to 40 miles per hour (48.3–64.4 km/hr) for several miles, and Carr (1927) reported pronghorn racing his car at 30 miles per hour (48.3 km/hr) for 7 miles (11.3 km). Einarsen (1948:45) indicated that about 30 miles per hour (48.3 km/hr) "probably is their usual cruising rate."

Nichol (1942:13) wrote: ". . . an old buck essayed to race the car, and for a few rods little attention was paid to him. Then the writer noticed the characteristic sinking of the animal's back which indicated that he was about to extend himself. The car was speeded up, and for one and two-tenths miles the animal ran at the rate of 45 miles an hour. The constancy of his speed was as remarkable as the pace. At no time could an appreciable variation either way from 45 miles be detected on the speedometer, until a final spurt when the buck crossed the road in front of the car. The speed of the car was not changed, and it is presumed the animal had to add something to his speed in order to cross. For fully two-thirds of this distance the race was on an upgrade steep enough to cause a car to roll back freely if not braked. The surface over which the antelope ran was densely covered with rocks to the size of a milk-bucket or larger. At no time during the race was the antelope more than 30 yards from the car and the play of its shoulder and flank muscles was remindful of smoothly and swiftly moving shuttles. A few days later the speedometer of the car was checked and found to be accurate."

Pronghorn often race alongside and cross in front of vehicles, as if to prove they can. Such behavior also is characteristic of open-country bovids of Africa and Asia. Although pronghorn ordinarily are sure-footed, their hooves do not grip icy surfaces well; if this road was ice-covered, the animals could take nasty falls or be hit by an automobile. Animals should not be chased with vehicles, but most "races" with pronghorn are instigated by the animals and can hardly be avoided. Photo by LuRay Parker; courtesy of the Wyoming Department of Game and Fish, Cheyenne.

Although the length of their strides may reach 18 feet (5.5 m) at a fast gallop, pronghorn gaits normally are smooth, without excessive vertical body movements. In comparison, deer waste more energy in leaps and bounds, perhaps because they are adapted to running through thicker and taller vegetation. Pronghorn do run over and around rocks, burrows and low brush, invariably without missing a step. Their keen eyesight and maneuverability make this possible. Bighorn sheep, deer and elk seem unable to reverse directions at a tangle net when stampeded by an aircraft. Pronghorn groups, 50 to 100 strong, will do an about-face when the leaders are within a few feet of a net, often sending a shower of stones and dirt through the net. All members of the group seem to know when to turn or reverse direction in unison, like a school of fish, without bumping into one another. Photo by Mike Francis.

Hewitt (1973:44) recounted this memoir of one Jesse A. Applegate who was traveling with his parents on the Oregon Trail during 1843.

"Antelope and long-eared rabbits were everywhere. Father had two black dogs called greyhounds; they were very fast runners and could soon pick up a rabbit, but when they chased an antelope it was quite different. One day an antelope had in some way been separated from the herd and ran through the train. One of the dogs, Fleet by name, pursued the antelope, and the chase led across a level plain. The black dog as he sped on with all his might, looked like a crane flying along the plain. We were all excited, for the dog was gaining on the antelope at every bound and would no doubt soon overtake him. The dog thought so too, for when he was within a few yards of the antelope and expected in another bound or two to seize its prey, he gave a yelp, but that yelp seems to have been a fatal mistake, for that antelope, in a few seconds after that bark, was fifty yards away from the dog and flying over the plain as if he had been shot out of a gun. He actually passed over many yards before we could see the dust rise behind him. The dog was so astounded that he stopped short, and after gazing at the antelope for a moment, no doubt amazed beyond expression, turned about and trotted back to the train. It was said that dog would never chase an antelope afterwards."

Morphological and physiological adaptations have permitted the pronghorn to develop as an agile, running species. A light body weight for its height and a streamlined body shape are enhanced by the relatively small food volume capacity of the stomach and cecum (Sundstrom et al. 1973). Pronghorn begin breathing through their mouths almost as soon as they start to run. Their hearts, lungs and tracheae are large for their body size—apparently adaptations for extended running. The increased vital capacity enables the pronghorn to cover long distances rapidly on a daily basis to obtain food and water and to outrun mammalian predators with a minimum of effort (see Chapter 8).

Specializations that contribute to the running ability of pronghorn include long, slender limbs with light distal parts. According to Byers (1997a), past predators exerted selection to enhance quick acceleration, as well as running speed and endurance. Seemingly, predators did influence the pronghorn's tremendous running abilities, but the animal apparently was preadapted for speed before predation became a driving force. Janis and Wilhelm (1993) present evidence from limb morphology of fossil species that pursuit predation did not exist until the latest Tertiary. In contrast, ungulates evolved long legs, similar to those of present-day cursorial taxa, by the middle Tertiary—some 20 million years earlier. Janis and Wilhelm concluded that the long legs of ungulates served to decrease transport costs in association with climatic changes. They also concluded that predation by long-distance pursuit is a Plio-Pleistocene phenomenon related to development of cold and arid climates.

Figure 24. *Pronghorn occasionally bound, but their gaits normally are smooth without excessive vertical body movements. Bounding is seen primarily when fawns are playing or when bucks are entering the rut. When pronghorn move about slowly, they do not seem graceful. In fact, they often appear to have sore feet or be slightly arthritic. Even when trotting leisurely to keep up with a band, a pronghorn is not especially graceful. Only when one is scared, excited or curious does it show its grace and athleticism. The "pompous, alarmed walk," as described by Einarsen (1948:47), is a high-stepping gait accompanied by flared rump patches, while the head is held high and ears are directed forward. What Einarsen called the "elegant trot" seems to be an extension of the pompous walk, with both the forefeet and hind feet on first one side of the body and then the other striking the ground nearly simultaneously.* Illustration by Shirley Briggs; from Einarsen 1948.

While turning at a gallop, pronghorn flex their inside front legs to shorten them and allow proper placement of the outside feet. When chased or frightened into running and then confronted with a strand fence, the animals may virtually dive beneath the bottom wire at nearly full speed (gallop). The small pronghorn in this band ducked safely under a fence, but larger, trailing animals veered away, perhaps because the vision of the fence line was obscured by dust thrown up by the diving animals or because they were intimidated by the small space below the fence wire. Top photo courtesy of the U.S. Fish and Wildlife Service, Charles M. Russell National Wildlife Refuge, Lewistown, Montana. Bottom photo by Robb Hitchcock

As a pronghorn accelerates, it passes through a canter or lope to a full gallop (Figure 24). In most pronghorn gaits, acceleration is accomplished by increasing both the length and rate of the stride. However, at a fast gallop, speed is increased by increasing the stride length to as much as 18 feet (5.5 m) but slightly decreasing the rate of stride (Bullock 1974).

Although capable of jumping vertical barriers as high as 8 feet (2.4 m) (Spillet et al. 1967) and broad jumping as far as 25 feet (7.6 m) (Bullock 1974), few pronghorn learn to jump fences. They usually crawl under fences if the bottom wire is at least 16 inches (40.6 cm) above the ground. Crawling, however, hardly describes the way running pronghorn duck under fences or dive between wires at full speed, often losing considerable hair in the process. See Chapter 17 for further information concerning pronghorn and fences.

VOCALIZATIONS AND MECHANICAL SOUNDS

Pronghorn produce a wide variety of vocalizations and mechanical sounds, each of which play a role in communication with others of the species. For example, Gregg (1955) noted that sounds made by running pronghorn alarmed others.

Mature pronghorn of both sexes snort or blow through their noses when they note something strange in their habitat. Einarsen (1948:45) quite accurately described the alarm snort: "It is done through the nostrils, sounding much like *Cha-oo,* with accent on the first syllable, and a long pause before sounding the *oo* (as in boot)." Other pronghorn within hearing distance are immediately alerted by the "Cha-oo." Waring (1969), who analyzed the sounds with electronic equipment, found they were single, unvoiced blows given at 9-second intervals. Each sound was brief (200 msec) and produced by the turbulence of a sudden expulsion of air. Maximum amplitude occurred early in the blow. He observed that dogs, prairie dogs and even men revealed themselves to investigating pronghorn by leaving concealment to look for the source of the Cha-oo sound.

Fawns produce soft, mewlike bleats when interacting with their dams (Bromley 1967). Fawns utter this mew as their mothers approach, and it probably helps the does to locate their concealed offspring (Kitchen 1974). When caught and handled, young fawns sometimes emit loud bleats. The bleat is a continuous, high-pitched wail that can be heard more than 0.5 mile (0.8 km) away and brings other pronghorn rushing to the scene. Occasionally, adults utter similar bleats when they are grabbed during trapping operations.

Does emit low grunts that Kitchen (1974) described as clicking sounds. He observed that females grunted most often when approaching their hidden fawns but also when pursued by bucks. In response to the grunt, fawns jumped up and approached their mothers.

The buck roar sounds much like a loud stomach growl and probably is caused by exhaling air. It is used principally during chases of either bucks or does (Kitchen and Bromley 1974). The snort–wheeze is another strictly male vocalization. It begins with a typical Cha-oo and is followed by a series of shorter bursts of sound descending in pitch and volume until termination. Gregg (1955:105) called this the "laugh" call. As they call, bucks pump their sides and probably their diaphragms; the latter bursts are wheezelike. This is the call hunters imitate to attract bucks or stop them when running (see Chapter 22).

Territorial pronghorn exhibited "dear enemy" (a neighbor is less feared than a stranger) reactions after being subjected to playbacks of snort–wheeze vocalizations from other bucks. Sonograms of snort–wheeze recordings from 11 bucks appeared to show distinctive interindividual frequency patterns (Chao 1985). Territorial pronghorn showed more urgency in defensive activities, as well as greater increases in the duration of vigilance and defensive activities, after territorial non-neighbor than territorial neighbor playbacks.

Both sexes grind their teeth during agonistic encounters. A human can hear such grinding at a 9- to 15-foot (2.7–4.8 m) distance (Kitchen and Bromley 1974).

Courting males make three distinct sounds. One is vocal; the other two are created by movements of the mouth. The pitch of the whine starts high and then smoothly decreases. If a doe starts to move away, the whine grades into the roar. During courtship approaches, bucks either lip smack or tongue flick, both of which create a low, staccato, sucking sound, much like that of a nursing fawn (Bromley and Kitchen 1974).

SENSES

Although the senses of hearing and smell are greatly overshadowed by the highly developed vision of this species, these senses also are acute.

Hoover et al. (1959:17)

Little is known concerning pronghorn acuity of hearing. However, fawns responded to soft grunts uttered by does 150 to 180 feet (45.7–54.9 m) away, whereas Kitchen (1974) could hear those sounds only when he was within 75 feet (22.9 m) of their source. Nine pronghorn reportedly reacted to a quiet camera motor 700 to 800 feet (213.4–243.8 m) away, but which people could not hear at a 50-foot (15.2 m) distance (Viliani 1984).

The pronghorn's sense of smell is well developed. By defecating, urinating and marking with scent glands, pronghorn maintain their territories and identify one another. They use odors (chemical messengers) to give alarm, to select mates and to feed. Odors that induce physiologic or behavioral changes in the same species (such as fright and sexual arousal) are called pheromones. Pronghorn have a large amount of olfactory epithelium. Olfactory epithelium differs from the receiving ends of other senses in that its nerve cells extend from the outside environment directly into the brain (Gibbons 1986).

Pronghorn occasionally take flight upon scenting an unseen person, but more often they mill about or even approach the source of scent, apparently trying to get visual confirmation of danger. Müller-Schwarze et al. (1974) found that males can distinguish their own subauricular gland scent from that of others. Einarsen (1948:53) observed: "Although asleep, these animals will respond to a strange scent. Often when a sleeping buck is observed at close range, he will leap into action when his nose tells him of the observer's presence."

CONCLUSIONS

Full-grown pronghorn seldom exceed 125 pounds (57 kg) in weight, and southern, lowland animals are smaller. The eyes are large, reflecting heavy dependence on eyesight. The somewhat chunky bodies are contrasting white and rusty brown to tan, with black and dark brown markings about the neck and large head. In bright sunshine, one can see the flash of the white rump patches several miles away. The feet are long and slim and lack dewclaws. Pronghorn hairs contain large air cells, providing excellent insulation, and the animals can regulate temperature by erecting or flattening their coat. Consequently, they seldom seek shade or shelter. The prongs that give pronghorn their name and the annual shedding of horn sheaths are unique. A newborn fawn's eyes seem too big, its legs too long and its face too narrow. Pronghorn generally live to 9 years of age, but ages twice that have been recorded. Top speed for most pronghorn is near 45 miles per hour (70 km/hr), and they can maintain 30 to 40 miles per hour (45–65 km/hr) for several miles. Pronghorn produce a wide variety of vocalizations and mechanical sounds, and each plays a role in communication between animals. The pronghorn's vision has been compared to that of a man with eight-power binoculars, and they detect movement at great distances; senses of smell and hearing also are well developed.

The long limbs with reduced numbers of toes, lightweight bones and comparatively small digestive tracts are adaptations to use a minimum of energy during the rapid locomotion necessary to small ungulates in open habitat. For ruminants, pronghorn have small stomachs and correspondingly large tracheae, lungs and hearts. The large tracheae and lungs provide for rapid intake of oxygen and elimination of gasses produced during extreme exertion. The large heart provides the pumping power needed for oxygenating blood in the lungs and speeding it to all parts of the body. In combination, all of these factors provide the requisites for a swift runner with great endurance.

These characteristics indicate a beautiful, interesting and well-adapted animal, but do not dwell on its uniqueness among North American ruminants or its dependence on native vegetation and wide open space (see Chapters 13 and 14). Being a small ruminant with a small stomach, the pronghorn must have succulent vegetation high in protein and other nutrients.

To secure the proper food at the proper time, pronghorn must move extensively; however, pronghorn evolved on open plains where speed was necessary and the ability to jump was not. Thus, they cannot (or will not) cross many fences. As open spaces shrink and more impediments to free movement are erected, this unique little speedster of the western Plains will require unique management techniques to ensure its long-term survival (see Chapter 17).

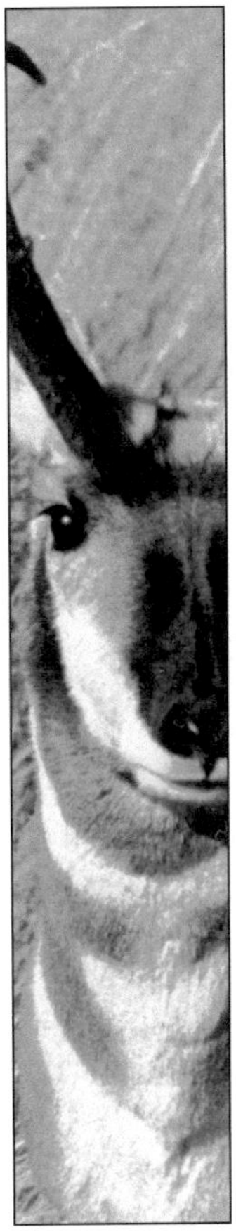

C H A P T E R   S I X

Behavior

Bart W. O'Gara

Their habits are shy and capricious; they keep on the open plains, are quick to take the alarm, and bound away with a fleetness that defies pursuit. When thus skimming across a prairie in the autumn, their light gray or dun color blends with the hue of the withered herbage, the swiftness of their motion baffles the eye, and they almost seem unsubstantial forms, driven like gossamer before the wind.

While they thus keep to the open plain and trust to their speed, they are safe; but they have a prurient curiosity that sometimes betrays them to their ruin. When they have scud for some distance and left their pursuer behind, they will suddenly stop and turn to gaze at the object of their alarm. If the pursuit is not followed up they will, after a time, yield to their inquisitive hankering, and return to the place from whence they have been frightened.

Washington Irving (1868:276)

Pronghorn can be wary, but innate curiosity sometimes overpowers what would seem "common sense." Dodge (1959b:107) wrote of the foibles of the little beauties thus: "There is a phase of curiosity peculiar to this animal, which is in itself more remarkable than that ordinary curiosity which it has in common with all plains animals, the black-tail especially; and this has added more antelope to my bag than any other of its characteristics. It is a panic produced by a combination of curiosity and terror. In a country where antelope have not been much hunted, they become, as it were, beside themselves at the appearance of any object very unusual and strange.

"Thus a waggon train crossing such a country will attract every herd within the range of vision. They will rush at it with every indication of extreme terror, and, passing within a few yards, will sometimes make a complete circuit of the train, going off at last in the direction from whence they came. I have known a few instances where the panic was so great that the herd passed between waggons, although the whole wide plain was open to them. . . .

"On another occasion, when passing in a spring waggon over a broad, barren, and apparently lifeless plain, three specks appeared on the far horizon, and, making directly for us, soon developed into antelope. The waggon was stopped. A companion and myself seized our rifles, sprang out, and made all ready to give them a warm reception. The antelope came on, passed completely round us, within fifty yards, and went off in the direction they came, unhurt, though we each fired eight or ten shots at them."

A pronghorn's curiosity is not easily satisfied. As some 70 men and 250 horses of the Wyeth expedition trekked through Nebraska and Wyoming in 1834, a pronghorn doe ran bleating toward the column as though she wished to join it. She followed along for nearly an hour, at times approaching within 30 to 40 yards (28–37 m) and standing to gaze at the slowly moving procession until she was shot (Townsend 1978).

Pronghorn appear flighty when confronted with possible danger, but are intelligent animals, adapting rapidly to disturbances that do not threaten their welfare. They seem able to distinguish danger from nondanger. During the 1960s to 1980s, I collected numerous big game animals on the National Bison Range for studies. Those animals were habituated to tourists in vehicles, and shooting from a utility vehicle allowed standing shots at reasonable distances from a solid rest—until the animals became frightened of stopping vehicles. Pronghorn became wary of my utility vehicle quicker than did deer and elk, but the latter became fearful of all vehicles. Pronghorn still allowed the close approach of tourists or me in a car or pickup of different color and shape from that used in previous collections. Suffield, a large military training base in southern Alberta, is home to many pronghorn. They merely trot out of the way of charging infantry and tanks and go back to feeding or resting. Cannon, machine gun and rifle fire are ignored (Geist and Francis 2001). On the Fort Belknap Indian Reservation in central Montana, pronghorn are hunted during autumn, and prairie dogs are hunted year-round—with most shooting by some 8,000 permit holders during April through June. Pronghorn become accustomed to such harmless (to them) rifle fire and pay little attention (Maher and Mitchell: unpublished manuscript).

Coloration, behavior and habitat preference place pronghorn among the most observable of North American big game. Consequently, the graceful pronghorn has been studied extensively by behaviorists and is enjoyed by a multitude of nature lovers. I have watched thousands during every season of the year and always feel a tingle of excitement and pleasure when witnessing the madcap antics of a courting buck, a group bedded on a grassy slope under a bright western sky or a doe nursing her playful fawn. The pronghorn's contrasting colors are reminiscent of those of Old World gazelles that also inhabit grasslands and steppes. Such coloration tends to make the animals conspicuous rather than camouflaging them.

Pronghorn, like most plains ungulates, are gregarious. Group sizes vary throughout the year and under differing environmental conditions. Names for the various groups have not been used consistently. Pyrah (1987) provided names for social groups in Montana, and those names are used here. Herds are discrete populations—all of the resident pronghorn on a particular land area that have routine social interactions and often winter as a single group. Each herd functions independently. Its population characteristics are related to habitat and population influences only in its own herd's range, and genetic exchange with pronghorn of adjacent herds is minimal. Herd ranges include all areas used seasonally by all social units of a herd, except during extremely severe winters (and possibly droughts). In some cases, combinations of streams, highways, railroads, fences and livestock use act as barriers to pronghorn movements and define boundaries of herd ranges. In some habitats, herds may not be as well defined as they were in central Montana where Pyrah studied them.

"Bands" are subdivisions of herds. Doe bands are composed of female associates that have nearly exclusive use of a traditional summer range. Summer bands also include bachelor buck bands that associate during the interval between spring dispersal and the onset of rut. Whereas herds often are separated by environmental barriers, bands within herds are isolated primarily by behavioral mechanisms. Doe bands apparently know when they are within another band's range by scent or previous experience. Two doe bands fled from other bands' ranges into their own at the appearance of a pickup or airplane (Pyrah 1987). Distribution apparently reflects the combined influences of genetics, social behavior, tradition and habitat responses.

In flight, bucks normally run with their noses pointed toward the ground, whereas does tend to hold their heads high. Does generally lead herds or bands and, if a buck or bucks are present, they usually bring up the rear. These behavioral differences help to distinguish between sexes at long range.

FEEDING HABITS

Feeding selectively on the most succulent vegetation, pronghorn drift across the prairies, shrubsteppes and deserts, selecting not only certain species of plants, but also particular parts of those plants. They use their mobile and dexterous upper lips to select plant parts, draw them into their mouths and hold them in place until cut free by the incisiform teeth, which are jerked forward and upward by movements of the head. Unlike domestic livestock, pronghorn move constantly while feeding. They make no attempt to use all of the available palatable forage in any one small area; instead they leave an abundance of food distributed over their range (Buechner 1950a). According to Einarsen (1938), pronghorn make thrifty use of what they eat, leave hardly a trace during their dainty foraging and feed on plants that seldom are used by cattle or sheep. Most studies indicate that pronghorn primarily consume forbs and shrubs instead of grasses as do cattle or bison.

(Top) *Pronghorn will feed on hay when little palatable forage is available above the snow, but if it is used to tide starving pronghorn during a period of severe weather, feeding must begin when the animals still are in good condition.* (Bottom) *During periods of deep snow, pronghorn must spend more time pawing for food on croplands than they do on rangelands.* Photos courtesy of the North Dakota Game and Fish Department, Bismarck.

Selective feeding is necessary for pronghorn because they have a smaller ratio of rumen volume to body weight than do cattle or sheep. Thus, pronghorn need a diet of higher quality, in terms of crude protein, than do domestic ruminants (Schwartz et al. 1977). The small rumen results in inefficient use of high-fiber foods (Hanley 1982). The leafy parts of forbs and shrubs are digested more rapidly, contain more crude protein and phosphorus, and have less fiber than do the same parts from grasses (Smith et al. 1972, Short et al. 1974, Huston et al. 1981, Wofford and Holechek 1982). During severe winters, pronghorn readily use haystacks.

Schwartz and Nagy (1976) discovered just how selective pronghorn were when they watched four tame animals feeding on the Pawnee National Grasslands in northeastern Colorado. In 203,725 bites, 87 plant species were consumed, including 18 grasses or sedges, 64 forbs, 4 shrubs and 1 lichen. Of the 87 species, only 26 accounted for 3% or more of the total number of bites. Percentage species composition in the diet bore little resemblance to percentage availability of plant species. Thus, the prairie on which pronghorn fed was not a homogenous resource, and the animals were very selective.

Feeding by pronghorn often is referred to as dainty because the animals usually select small plants or new leaves that are especially succulent. However, I have watched pronghorn feeding greedily on musk thistles, gulping down leaves and chunks of stems 8 to 10 inches (20–25 cm) long. The thistles were growing along a busy road, and the animals obviously were eager to get away from the disturbance and go where they could regurgitate and chew the food in peace.

Byers (1997a) described feeding pronghorn as walking slowly, occasionally pausing and holding their noses just above or brushing through vegetation. Food items appeared to be found by scent as well as sight. Individuals commonly passed their noses along the length of a forb and then removed one or two leaves from the plant. At other times, individuals temporarily specialized in single items—such as flowers—walking briskly from plant to plant, obviously guided by vision. Plant parts were gathered rapidly, but a pronghorn sometimes fed for a minute or more before accumulating a mouthful, which it chewed briefly before swallowing.

Working in the Red Desert of Wyoming, Taylor (1972a) reported that summer feeding seemed leisurely, with the animals using small portions of the available plants. Winter feeding was much more intense. Pronghorn spent more time at each plant and less time at other activities while feeding. The portions of plants selected during winter seemed to be larger, with the whole tips of shrub leaders being consumed instead of just leafy parts, as in summer. Pronghorn generally spend more time feeding than resting, and foraging activity peaks near sunrise and sunset.

For both sexes, the proportion of the day devoted to feeding varies seasonally. All pronghorn on the National Bison Range spent more time feeding during winter than at most other times of year, apparently in response to the increased costs of thermoregulation and locomotion through snow (Byers 1997a). During spring, does spent more time feeding than did bucks, apparently reflecting the energy demands of late gestation and early lactation. During the rut, bucks spent less time feeding than at other times of year.

Being selective feeders, pronghorn generally remove the preferred food items as they go, causing herd or band members behind to spread out to form a fanlike grazing pattern. In contrast, less selective feeders, such as bison, capable of eating almost all parts of a plant, only reduce the size of each food item. Therefore, animals following the leaders can stay close together in a concentrated herd. Animals with selective feeding styles often occur singly or in relatively small groups. From the standpoint of nutrition, solitary pronghorn have an advantage. However, for animals in open country, numbers mean security. When animals group, there are many pairs of eyes and ears to detect danger and many other animals to hide behind.

Selective feeding has a negative effect on scanning and vice versa. Thus, lone animals risk an energy deficit or lengthened exposure to predation. Although small ruminants can afford to be more vigilant—or more selective while maintaining vigilance—than can larger ones (Illius and Fitzgibbon 1994), herding behavior reduces the time each member must spend scanning for danger.

Height of vegetative cover (ability to see around while foraging) influences the amount of time that pronghorn spend looking about them while feeding. During a study of five recently translocated pronghorn does in California, Goldsmith (1988) found that they lifted their heads to scan more often, spent more time scanning and the length of each scan was significantly longer when they were foraging in sagebrush habitat than when they were foraging in open meadow. The height of vegetation in the sagebrush averaged 30 inches (67 cm), and that in the meadow averaged less than 12 inches (30 cm). The does were vigilant an average of 11.6% of the time in the shrubs and 6.3% of the time in the meadow. Individual pronghorn increased their vigilance in different ways. Most increased frequency and duration of scanning, one increased the length of each scan, and one did not change her vigilance between habitats. Vigilance apparently was related to detection of predators; the does did not seem to be watching other pronghorn.

On the prairie of southeastern Alberta and northern Montana, Bruns (1977) observed feeding habits of pronghorn during winter. They were adept at pawing away snow to reach food, and such pawing was observed during any feeding session when snow covered the ground. Pawing was last observed in spring (March 18) when snowmelt revealed sufficient bare spots to allow continuous feeding without pawing. Interactions at feeding craters in the snow indicated a social hierarchy, with adult bucks at the top, followed by adult does and then fawns. The mean depth of snow on rangeland feeding areas was 10.8 inches (27.4 cm). One herd fed primarily on croplands that had an average snow depth of 5.9 inches (15 cm). Pronghorn spent 70% of their winter feeding sessions ingesting food.

Although less snow accumulated on the croplands than the rangelands, the sprouted oats and wheat the pronghorn sought were covered by snow. The animals spent almost twice as much time pawing for their food as did those feeding on rangelands. Although sagebrush protruded above the snow, pronghorn pawed for it below the snow level. Bruns (1977) found that the sagebrush above snow level was almost odorless in contrast to the strong aromatic smell of that below the snow. Analysis of samples taken in March suggested the forage above the snow line had lower levels of moisture, protein, calcium and potassium than did forage under the snow.

(Top) *The leaves and leaders of shrubs contain more nutrients than do grasses during winter, and forbs often are nonexistent or unavailable under snow. Consequently, northern pronghorn often are seen feeding on shrubs.* (Bottom) *Pronghorn will paw away snow much deeper than that shown here to reach palatable vegetation. However, if possible, the animals will move long distances to reach more shallow snow whenever depths exceed about 10 inches (25 cm).* Photos by Cathy and Don Jenni.

Bruns (1977) concluded that activities of pronghorn were "conservative" during winter, although feeding bouts were longer at that time of year. However, Taylor's (1972a) study of feeding patterns indicated that pronghorn rested less during winter than at other times of year. Behaviors vary from area to area and year to year, but extended feeding bouts probably allow concentration of the mandatory feeding time during the most environmentally favorable parts of the day and still allow low levels of overall activity (Amstrup 1978).

Pronghorn are adapted to move long distances quickly and with a minimum of effort to find and use the best available forage. Gaits and speeds are discussed in Chapter 5.

MOVEMENTS

The sun has barely risen over the Wind River Range, and the hunters have positioned themselves to intercept their quarry. A small group of antelope meanders through the sagebrush, slowly advancing closer and closer. The hunters find it hard to remain calm as the antelope continue their approach. Finally, the antelope are within range. It sounds like a modern archery hunt for antelope, but this scenario played itself out on the sagebrush hills of western Wyoming over 6,000 years ago.

These prehistoric hunters were taking advantage of a natural migration corridor, a low ridge bordered on one side by the Green River and the New Fork River on the other. The antelope had begun their annual northerly migration to summer ranges when they were intercepted by the hunting party. The development of bones of unborn antelope found at the site indicate the kills occurred in late March or early April, the same time of year pronghorns still pass Trappers Point on their way north.

Sawyer and McWhirter (2000:36)

According to Seton (1927), pronghorn of Jackson Hole traveled 150 miles (240 km) southward to the Red Desert. Those on the prairies of Saskatchewan moved 100 miles (160 km) south and west. Those on the plains near the Rockies went toward the foothills, and those on open prairies about the Black Hills flocked there from all directions.

Seton's (1927) observations of movements between Jackson Hole and the Red Desert were later confirmed by radio-tracking members of what Wyoming Game and Fish Department biologists call the Sublette herd. In fact, movements were even longer during some years than those Seton reported (Riddle 1990, Sawyer and McWhirter 2000).

Most migration routes have areas where topography, vegetation or (today) development create bottlenecks. Apparently, the most important bottleneck for the Sublette pronghorn herd is near Trappers Point, 7 miles west of

Pinedale at the junction of U.S. Highway 191 and Wyoming Highway 352. This naturally occurring bottleneck was approximately 1 mile (1.6 km) wide, restricted to the southwest by the Green River riparian complex and to the northeast by the New River riparian area. Development has narrowed this bottleneck to approximately half a mile (0.8 km) of sagebrush rangeland administered by the Bureau of Land Management (Sawyer and McWhirter 2000). The elk migration through this area has ceased to exist.

Radiocarbon dates from the lowest layer of a pronghorn kill site at Trappers Point indicate that Indians were using the bottleneck between 7,880 and 6,180 years ago to trap pronghorn. Thus, the Sublette herd may have migrated past Trappers Point for 8,000 years (Sanders 2000) or more.

Theodore Roosevelt et al. (1902) observed that pronghorn from most of North and South Dakota gathered in the Black Hills during winter. Such movements became impossible when the country was settled. Simpson and Leftwich (1978) noted that many pronghorn in Texas, trapped by drift fences, froze to death during blizzards of the 1880s and 1890s. Such drift fences played an important role in the decline of pronghorn in that state. According to Haley (1949), settlers killed 1,500 pronghorn trapped by a drift fence near the Panhandle town of Clarendon in 1882.

Environmental conditions and tradition both are involved in the seasonal movements of pronghorn. Because environmental factors can influence or trigger moves, the timing of movements is variable. A seasonal movement, although somewhat traditional, might be modified by the availability and succulence of forage and influenced or dictated by topography. Some pronghorn move very little.

The length of historic seasonal movements of pronghorn apparently depended on forage conditions and weather. Mills (1932) believed that pronghorn he watched in Colorado lived on the same ranges year-round. However, one snowy winter they were gone. He found them 15 miles (24 km) away, where less snow had fallen or the wind had partly swept it away.

Grinnell (1929) maintained that pronghorn shifted between winter and summer ranges annually, using well-established migration routes. He pointed out two crossings—one was on the Missouri River below Fort Berthold, North Dakota, and the other was a pass not far from old Fort Steel, near the head of the North Platte where the animals crossed to winter farther south in Colorado. If pronghorn migrated as predictably as Grinnell believed, using certain river crossings and passes annually, most such mass migrations ended as humans reduced numbers of animals and erected barriers to movements. Today, opportunistic seasonal movements dictated by habitat conditions, weather and barriers have mostly replaced true migratory patterns. Movements of the Sublette herd of Wyoming seem an exception.

Yoakum (1978a) reported that three pronghorn herds living within sight of the Hart Mountain National Antelope Refuge in south-central Oregon had three separate movement patterns. First, a small nonmigratory herd occupied an area about 5 miles (8 km) in diameter in Warner Valley. The site had sufficient forage and year-round water, and snow depth seldom exceeded 6 inches (15 cm). Second, a herd with altitudinal seasonal movements occupied the nearby Drakes Flat tableland, which is about 15 miles (24 km) in length. Pronghorn used the higher elevations throughout some years; however, deep snows occasionally forced the animals to lower elevations. As soon as snow depths receded, the herd moved back to higher elevations to feed on preferred plants. Third, pronghorn moved longer distances on Hart Mountain. The deeper the snow, the farther the herd traveled to lower elevations with less snow. These movements were related to annual snow depths. In the Red Desert of Wyoming, pronghorn chose two distinct habitat-use strategies (Deblinger 1988). Most moved seasonally, but some remained sedentary.

The timing and length of seasonal movements vary with altitude, latitude, weather and rangeland conditions (Yoakum 1978a). Generally, on rangelands that receive substantial amounts of snow, animals from large wintering herds disperse during spring. Today, deep snow sometimes forces pronghorn to move 200 miles (320 km) or more from summering areas (Riddle 1990). Before settlers constructed numerous barriers to pronghorn travel, the animals apparently made even more extensive movements (Roosevelt et al. 1902). Sheep-proof fences became especially troublesome because pronghorn refused to jump over and could not crawl under them.

Observations of pronghorn, which had been tagged in Saskatchewan, on Montana winter ranges, indicated movements of 140 miles (225 km) from summer ranges (Hnatiuk 1972). Amstrup (1978) found that radio-collared pronghorn along the Montana/Wyoming border did not use the same winter ranges in successive years, and most shifted their summer activity areas significantly. Pronghorn are opportunistic; if possible, they will move about and take advantage of the best vegetation available each year. Conceivably, Fichter (1976) wrote, places (i.e., winter and summer ranges and the routes between) are the traditional entities, not the act of migration. I agree, but would add that the places used may vary from year to year because of weather conditions, available forage or land-use practices. Before the near extermination of bison in the late 1800s, those great shaggy beasts probably altered vegetation enough on a somewhat cyclic basis to cause pronghorn to shift seasonal ranges.

In recent time, traveling pronghorn invariably encounter numerous obstacles—especially fences. Despite their leaping ability, pronghorn are reluctant to jump fences and few learn to do so. However, if the bottom of a fence is high enough, most pronghorn will crawl underneath. When making an unalarmed approach to a fence to be negotiated (top left)*, a pronghorn will lower its head and drop to its knees (wrists, actually), then, with its chest on the ground and forelegs bent back under the brisket* (top right)*, it will arch its back (to avoid the bottom strand) and push forward with the hind legs until the animal's head and shoulders are on the far side of the fence. The pronghorn then scooches forward by purchase gained from extension of the hind legs* (bottom left) *until recovery is made on the other side (bottom right). Pronghorn often are found with streaks of bare skin and long cuts on their backs and shoulders from overanxious or overcasual maneuvering under fences, or from reckoning with fences too low. When snow is deep or fences are overbuilt, they can block pronghorn movements.* Photos by Robb Hitchcock.

A study conducted in Idaho illustrated how variable seasonal movements can be (Table 28). The distances moved during spring 1977 were much shorter than those shown in the Table because the animals wintered higher in the valleys during the mild winter of 1976–77 than they did during the severe winter of 1975–76. Summer ranges generally were near the heads of valleys, and some pronghorn spilled over into the heads of adjacent valleys. As a result, two of the Crooked Creek does summered in Montana (Hoskinson and Tester 1980). All of the 22 radioed pronghorn followed for 2 years returned to the same general areas during summers 1976 and 1977.

As snow melts, northern animals generally move back to summer rangeland. Data gathered by Hoskinson and Tester (1980) in Idaho suggest that snow cover affects the location of winter rangelands, the start of spring movements and the rate at which pronghorn move to summer rangelands. However, during their study, fall movements

Table 28. Straight-line distances moved between winter centers of activity[a] to those of summer by radio-collared pronghorn in three Idaho valleys during spring 1976

		Average		Range	
Valley	*n*	*Miles*	*Kilometers*	*Miles*	*Kilometers*
Little Lost River	12	29.3	47.1	3.0–40.3	4.8–64.8
Birch Creek	12	33.3	53.6	7.1–60.7	11.5–97.7
Crooked Creek	10	20.0	32.9	4.2–38.6	6.8–62.0

[a] Determined by method of Hayne (1949).

Source: From Hoskinson and Tester 1980.

occurred without snow or storms, and coincided with a shift down-valley in location of the most succulent vegetation. Throughout summer, autumn and early winter, radio-collared pronghorn were located on or moving toward that area of their entire range that had the most succulent vegetation. During summer in Idaho, Bodie (1979) also found that the movements of females were related to succulence of vegetation. As forage dried, does moved their fawns into the sagebrush/foothill habitat type, where the vegetation was tall and the moisture content of new-growth vegetation was consistently highest. The animals reversed their movements into sagebrush/flat habitat type, where the vegetation was short, after a soaking rain and generally stayed there if forage moisture content did not drop much below 60%. In the Red Desert of Wyoming, summer range was used traditionally, but winter ranges were chosen opportunistically (Deblinger 1988).

During a study in Colorado, does made longer seasonal movements between range centers during all seasons than did bucks (Firchow 1986), but none of the movements was extensive. The greatest mean distance moved between seasonal activity centers was 4.8 miles (7.8 km) by does between winter and spring ranges. Fidelity to summer range was high; the mean distances between activity centers for 1983 and 1984 were 0.9 mile (1.4 km) and 1.5 miles (2.4 km) for bucks and does, respectively. Twenty-four percent of the movements between summer and winter activity centers were less than 1.2 miles (2 km); 48% were 1.9 to 6.2 miles (3–10 km); 28% were greater than 6.2 miles (10 km). Long-distance movements did not appear to be spurred by weather. Two yearling does made long-distance moves off the study area and did not return. Sixty-five percent of all weekly movements within home ranges were less than 1.9 miles (3 km). The longest movement within a 1-week period was 25.5 miles (41 km) by a pregnant female during early May. Most long-distance moves by bucks occurred during February to April.

Natural barriers, such as steep, timbered mountains and large bodies of water, impede pronghorn movements and determine, to some extent, their distribution. Artificial barriers, such as fences, interstate highways, reservoirs and railways, further complicate movements and reduce, often drastically, the carrying capacity in areas

Pronghorn occupy high, cold grasslands in the northwestern United States. These two bucks, with horn development suggesting January or February, are wintering in Yellowstone National Park where adult bucks disperse during spring from large winter herds to become territorial. Photo by William S. Keller; courtesy of the U.S. National Park Service, Washington, D.C.

where pronghorn must move long distances to procure the year-round necessities of life (see Chapter 13).

Harsh winters cause pronghorn to attempt to reach winter rangeland where shrubs–especially sagebrush—protrude above the snow, often with tragic results. The winter of 1977–78 was one of the most severe of the century within pronghorn habitat in Fergus and Petroleum counties, Montana. Continuous snow cover from mid-November through mid-March, with depths as great as 56 inches (142 cm), caused about 6,000 pronghorn normally found there to leave what had been traditional winter range. About half moved north or east into the timbered breaks of the Missouri or Musselshell rivers. They made considerable use of the major drainages, feeding on silver sagebrush sites with associated greasewood and green rabbitbrush. The steep slopes, often more than 60 degrees, were extensively used, as were intermittent ridges where the animals fed on Rocky Mountain juniper. The other half moved south, stopping at islands of closed-canopy ponderosa pine within sagebrush/grassland habitat. These islands were north of highways that paralleled streams with riparian vegetation. Whether the pronghorn stopped because of the trees, roadside fences, roads or streams is unknown. They stayed in the islands and fed on ponderosa pine for more than a month. The level of mortality was not quantified but may have approached 50% in both of those unusual wintering areas. Some small herds experienced complete mortality (Olson et al. 1978).

During the same winter (1977–78), pronghorn in Alberta suffered extensive mortality. Snow covered the sagebrush flats and snowberry stands, and 77 inches (196 cm) remained on the pronghorn range from November through the end of March. Pronghorn moved to any place that vegetation protruded above the snow; about 50% of the animals left traditional winter ranges and many moved into willow stands (Group Discussion 1978). Apparently, nontraditional movements often occur during severe weather. Attempts were made to feed the starving animals, and they readily fed on second-cutting alfalfa, pea vine and timothy. However, the survival rate was about the same between fed and unfed herds.

Tracks indicated that many pronghorn crossed the ice of Fort Peck Reservoir, in eastern Montana, from north to south during the winter of 1977–78; some may have been Canadian animals (Group Discussion 1978). They probably were trapped south of the reservoir when the ice broke up. On May 22, 1978, three biologists boating along the south shore of the reservoir saw three pronghorn swimming toward that shore (J. Wambaugh personal communication:1980). The reservoir was 1.00 to 1.25 miles (1.6–2.0 km) wide, and the animals were first seen about a quarter of a mile (0.4 km) from shore. They swam high in the water and moved slowly, taking 4 to 5 minutes to reach land and did not appear tired from the swim. More than 100 pronghorn were in timber along the shoreline. Apparently, these were animals that had crossed on the ice and were trying to return north. It could not be determined whether the swimming pronghorn had been frightened into turning back by the sound of the boat motor or had turned back because the reservoir was too wide.

Pronghorn swim high in the water buoyed by their hollow hair. Supposedly, on annual migrations in historic times, pronghorn swam large rivers, such as the Missouri. Photo by Jim Wambaugh; courtesy of the Montana Department of Fish, Wildlife and Parks, Helena.

Pyrah (1987) found that pronghorn on good rangeland in central Montana ordinarily had winter ranges near the centers of their herd ranges. However, movements to nontraditional winter rangelands during the severe winters of 1977–78 and 1978–79 were documented during his 13-year study of marked pronghorn. The winter of 1951–52 was the most recent one approaching those two in severity. During two other relatively severe winters, 1964–65 and 1968–69, the animals remained on their traditional herd winter ranges. Thus, the movement of herds outside their traditional ranges to emergency winter ranges could not have been the result of previous experience or knowledge of those ranges. An instinctive urge to move under extremely adverse conditions and recognition of emergency winter ranges could have survival value for pronghorn in areas where severe winter conditions occur infrequently (Pyrah 1987). This situation also exists in Nevada; pronghorn use the Railroad Point area every 30 to 50 years when extremely harsh winter weather forces them to this wintering ground. Such areas should be considered crucial and classified as winter ranges in management plans (J. Yoakum personal communication:1989).

As pronghorn try to reach emergency winter ranges, they often lose their fear of people and move into towns or around ranch buildings. At such times, they are vulnerable to harassment and mortality by dogs. Photo courtesy of Montana Department of Fish, Wildlife and Parks, Helena.

An extensive study of pronghorn movements was conducted in south-central Wyoming during 1980 through 1984 (Deblinger et al. 1984). Pronghorn were fitted with individually recognizable neckbands (776) and radio collars (57). More than 6,600 aerial and ground observations of marked animals were recorded. Movement patterns were relatively constant on summer rangelands but showed great variation during winter. Radio-collared bucks showed strong affinity to spring–summer–autumn home ranges, whereas does tended to return only to general summer areas. Winter distributions appeared to be dependent on weather, forage and individual behavior. Some individuals were sedentary residents in small areas, whereas others emigrated far from their trapsites without returning.

The majority of marked animals moved between summer and winter rangelands. Variations in winter pronghorn distributions were reflective of winter weather and snowfall patterns. Marked pronghorn traveled an average of 11 miles (18 km) from their respective trapsites throughout the 1980–81 winter. Their movements were in a southerly direction from their summer rangelands and farther south than where they had wintered the previous year. During the 1982–83 winter, a season of normal snowfall, pronghorn traveled approximately 19 miles (30 km) in a south and southwesterly direction. During the severe winter of 1983–84, pronghorn traveled approximately 50 miles (80 km). Their movements were variable but mostly southerly and southwesterly until reaching an interstate highway fence, the most formidable barrier in the study area. Pronghorn traveled east or west along the fence until mid-January and then stayed in place until May. The majority of marked pronghorn wintered in different areas during the 1983–84 winter, as compared with the previous three winters.

Movements of southern pronghorn have not been studied as extensively as those of northern populations. However, winter weather apparently necessitates movements by pronghorn as far south as Texas. Buechner (1950a) reported that 150 pronghorn froze to death during a blizzard in the southern part of the Panhandle. Fences prevented the animals from moving to protected areas as they had when the range was unfenced. He also noted that hundreds starved during a severe drought in the 1930s.

In central Arizona, individual pronghorn varied greatly in their movements; 47 were captured during 3 years and typically located 1 to 2 days apart. However, 3 to 4 days sometimes elapsed between locations (Ockenfels et al. 1994). Does made the longest movements, and 24.9 miles (4 km) was the longest move between successive locations. Differences in movement patterns were related to barriers—fences, paved highways and dense, tall vegetation. Fifteen of 47 pronghorn (all in one game management unit) had maximum consecutive location (i.e., single) movements of 6.2 miles (10 km) or more; eight had single movements of 12.4 miles (20 km) or more.

Sonoran pronghorn appeared to move to seasonal rangelands in response to forage—not climatic conditions (Wright and deVos 1986). However, movements between seasonal rangelands occurred throughout fall and just before summer. Two does moved slightly more than 26 miles (42 km) when traveling to and from their summer rangelands. One buck traveled 18.6 miles (30 km) or more to and from his summer range, and another traveled about the same distance to and from Mexico.

Before artificial barriers were created, pronghorn in arid and semiarid areas probably moved considerable distances in search of water and succulent vegetation. Now, some southern pronghorn are restricted in their movements and may die of thirst or from eating poisonous plants where fences prevent them from reaching water or succulent vegetation during droughts (Hailey et al. 1966).

GROUP SIZES AND COMPOSITION

The spatial distribution of pronghorn in any given area is partly a function of subgrouping—departure, return, exchange between bands and coalescence, according to Fichter (1972). He noted fluidity of group cohesion by observing recognizable individuals during spring and summer in the upper Pahsimeroi Valley of Idaho, and added that "geographic differences, or perhaps more appropriately, *ecologic differences*" occur in behavior. Such differences must be expected among pronghorn at different population densities inhabiting rangeland of varying quality and plant composition from sea level to 11,000 feet (3,350 m).

Pronghorn seem to herd primarily for protection from predators and for seasonal movements. However, few present-day predators—with the possible exception of golden eagles and, in some areas, mountain lions—are a serious threat to a healthy, adult pronghorn. An individual reduces its own risk of death by remaining in a band. The survival value of being close to conspecifics apparently outweighs potential disadvantages, such as increased competition for food and exposure to parasites (Byers 1997a).

"[S]pecies that typically occur in open prairie or tundra habitat, such as bison, pronghorn, or caribou, do not have physical cover with which to conceal themselves from predators, and therefore they use groups of conspecifics as cover. This is the 'selfish herd' concept. . . . Although the group as a whole is conspicuous to predators, the probability that any one individual will be singled out for predation is reduced by a factor roughly equal to the number of animals in the group. Each member of the group benefits from the presence of other group members, hence the term 'selfish' " (Hirth 2000:180).

The relationship between pronghorn group size and vigilance was examined in Grand Teton National Park, Wyoming (Lipetz and Bekoff 1982). Sociality seemed to provide increased protection through greater awareness and faster detection of predators. As group size increased, the proportion of vigilant pronghorn decreased. Individuals on the periphery of a group appeared to be more wary than those in the center of a group. Apparently, the larger the pronghorn group, the greater the energy efficiency, because as group size increases, the individuals can feed more steadily and spend less time watching for predators. This should be especially important during winter when pronghorn gather into large herds. Byers (1997a) maintained that the grouping tendency of pronghorn is an example of a behavioral trait that was adapted when more efficient predators were present.

Data are scarce concerning group sizes of southern pronghorn, but bands are smaller and vary less seasonally than in the North (Wright and deVos 1986, Hughes 1991). In a discussion of the endangered Sonoran pronghorn in Arizona, Phelps (1978:77) described what probably is one extreme for band sizes on marginal range: "It probably approaches three animals. . . . Down there the behaviour is not oriented towards bachelor groups or territorial bucks . . . it's oriented toward survival through the summer . . . they scatter out, they break into unit herds, of one, two, three animals." In such habitat, no climatic or habitat conditions seem to exist to regularly cause an entire herd to come together.

Wright and deVos (1986) reported an average annual mean group size of 5.1 for radio-collared Sonoran pronghorn located from the air during 1983 to 1985. Average monthly group size varied from 2 to 8.6. Congregations of 6 to 15 animals were observed during late autumn and winter. The largest group observed was 21 pronghorn in December 1984. During March through August, the average monthly group size was 3.9. Hughes (1991) found smaller group sizes (2.5 in 1988 and 2.4 in 1989; range 1–7) for the same months on Organ Pipe Cactus National Monument and Cabieza Prieta National Wildlife Refuge. Hughes located his radio-collared animals from the ground and may not have been able to see all of the members of some bands in tall vegetation.

During a study in the Trans-Pecos region of Texas, Buechner (1950a) provided further insight into composition and sizes of southern bands. He noted that groups varied markedly with the seasons. Females were scattered widely while caring for their young. During late spring, when the fawns were 4 to 6 weeks old, the does came together again, bringing their fawns to form bands of about 6 to 20. The size of summer bands was extremely variable, usually consisting of one buck, 1 to 8 does and 2 to 14 fawns. Most of the males were in bachelor bands at that season, and as many as 15 in a band was common. Buechner observed one band of 30 in April. Yearling does were found with mature does and fawns. During the rut, females were mostly in harems, young males tarried around the harems and fawns formed large bands. After the rut, herds of 20 to 60 animals commonly were seen during the late autumn and winter months. Buck bands began forming the last week of March, and many bucks and some does were seen singly. This description of Texas social groups by season is very similar to what I observed with territorial pronghorn in Montana. Buechner thought the animals he studied were harem breeders at the time; however, after reviewing Bromley's (1967) thesis, H. Buechner (personal communication:1973) realized that the Texas animals he studied were territorial.

The largest group Trevino (1978) saw in Chihuahua, Mexico, was composed of 52 pronghorn of both sexes during mid-February; 10 to 24 animals in a group were common during spring. Nursery bands of 3 to 15 females were seen in midsummer, and these does with their kids often were joined by lone females and young males during late summer and early autumn. Bachelor bands of two to seven bucks were seen during summer, and groups of both sexes and all ages congregated in autumn. Trevino did not speculate on the breeding system.

On semiarid rangeland in southwestern Utah, pronghorn aggregated into herds containing 20 to 35 head of mixed sex and age classes (Smith and Beale 1980). Seldom were more than 40 animals observed together. By April, pronghorn began segregating into smaller groups, and group sizes and composition were fluid, changing from day to day. Homogeneous sex and age classes were common for short periods. Usually, one or two yearling does would be with a group of mature does, and a yearling male would accompany a group of mature males. Mature males were seen singly or in pairs. During May, expectant does separated from other animals and appeared to seek seclusion. After parturition, doe/fawn groups formed, but sizes of groups varied from year to year and seemed to be influenced by abundance of forbs and forage production on shrubs. When forbs were abundant and relatively large amounts of palatable forage were available, groups of five or more does and their fawns fed together. During years of drought, doe/fawn groups were smaller. Frequently, a doe and her young would feed in washes where plants were still succulent, and predation was severe under such circumstances. By late June, doe/fawn groups often included a mature buck that tried to hold the does and fawns together and exclude other bucks. Small groups of young bucks banded together during the summer. The rut began in August and peaked in early September. Bucks herded does into harems, and the largest groups usually were controlled by old bucks. Younger bucks had one or two does or none at all.

Bayless (1969) noted that group sizes in central Montana increased as snow covered the ground. Working in the same area, Freeman (1971) also observed a steady increase in average group size from 6 during early autumn to 55 during late winter, when herds of 90 to 98 pronghorn were common. Freeman also reported that group sizes were larger on some habitat types than on others. For approximately 3 months, from about the first of November, pronghorn gather in large herds for mutual protection, according to Skinner (1922). Skinner further noted that this period coincides with the growing season of the bucks' horns, and the herds are led by does during those 3 months. Actually, groups of pronghorn usually are led by does during all seasons.

In the Pahsimeroi Valley of Idaho, mean group size from January 15 to March 15 was 48 animals, with an average density of 10.7 pronghorn per square mile (4.1/km^2) of occupied winter range (Bernt 1976). Spring dispersal began during the last half of March, and group size dropped to 16. By early May, mean group size was 3.8 animals, with a density of 1.7 pronghorn per square mile (0.66/km^2).

During three winters of observations in central Wyoming, Easterly et al. (no date) found large differences in mean group sizes between December and February (inclusive) periods (Table 29). The disparity did not seem related to winter severity. Weather conditions were considered more severe than average during 1988, less severe than average during 1989 and average during 1990—the year of small group sizes. Densities of pronghorn on the study area (13–16/square mile [5–6/km^2]) were nearly similar among years. Vegetation on the study area was primarily sagebrush grasslands.

During a 19-month period, 571 pronghorn groups—involving 6,713 individuals—were observed and recorded in the Powder River Basin of Wyoming (Gamo 1997). No differences were noted in location patterns between years, so they were combined in Figure 25. Average group sizes were slightly larger on native grass/shrublands than on reclaimed mining land.

When northern, territorial pronghorn arrive on summer ranges, does generally collect into bands of a dozen or less, young bucks form slightly larger bachelor bands and old bucks claim territories. Pyrah (1970) described this annual regrouping as the "spring shuffle." On ref-

Table 29. Mean number of pronghorn observed and average group size during aerial surveys over the Rattlesnake Hills winter range, Wyoming, during 1988 to 1990

	1988		*1989*		*1990*	
Month	*Mean number*	*Average group size*	*Mean number*	*Average group size*	*Mean number*	*Average group size*
October	—	—	2,411	17	—	—
November	—	—	2,879	52	—	—
December	2,523	59	1,329	53	2,415	29
January	2,298	42	2,256	53	1,148	29
February	2,196	56	1,958	52	2,004	29
March	1,802	17	2,427	24	2,287	17
April	1,307	10	1,612	13	1,638	8
November–February	2,302	50	2,216	52	1,834	20

Average of two survey flights per month.
Source: From Easterly et al. no date.

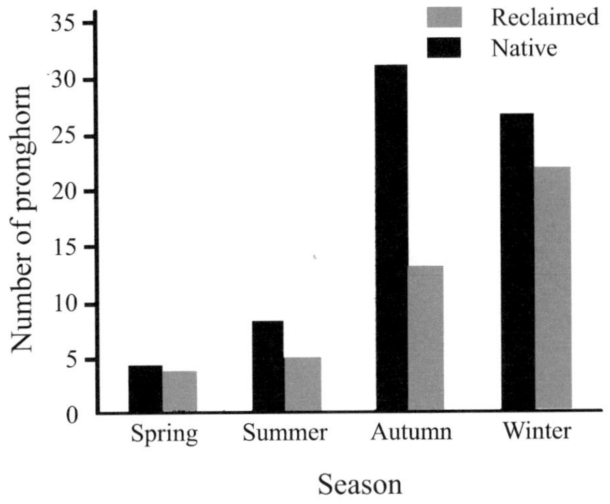

FIGURE 25. *Average seasonal sizes of pronghorn groups on native and reclaimed rangelands in the Powder River Basin of Wyoming (from Gamo 1997).*

uges where pronghorn are not hunted, bucks seldom acquire territories before 4 years of age.

Fluidity of pronghorn bands has been reported by many researchers. Gregg (1955) wrote that, although pronghorn were strongly social, the associations were constantly changing, and such changes were common because permanent attachments among adults were lacking. Gregg also noted that changes in band sizes were most likely to happen when a band was in some attractive habitat, especially one that allowed bands to see other bands easily. Folker (1956) found that groups varied from day to day and even in one day. Ingold (1969) maintained that pronghorn groups appeared to lack stability, and large bands usually were temporary aggregations of small bands. He thought the apparent lack of stability was not usually a result of random movement by individuals, but it resulted from the movement of bands. Because pronghorn can maintain visual contact over considerable distances and come together quickly when threatened, pronghorn herds are well organized even when widely scattered. Thus, although doe bands are flexible in size and composition, each should be considered a unit of a larger group (Fichter 1972). Kitchen (1974) found that pronghorn bands, particularly does, established specific home ranges, but association between individuals generally was loose. Bands of about the same size were seen in the same areas for several days at a time, but individuals in those bands were not always the same. Composition changed frequently.

Swanger (1977) noted that the size and composition of doe bands changed daily or even hourly during summer. Does formed moderately strong bonds with a few other females, but did not stay with them exclusively. Does moved about as individuals rather than as groups. When chased, or at the appearance of danger, smaller pronghorn bands came together to form a large group. When the danger disappeared, these larger groups broke up into smaller bands immediately or stayed together for awhile before disbanding.

Doe bands accompanied by a territorial buck were larger than unaccompanied bands, indicating that herding attempts by bucks were successful. Wentland (1968) reported that does in central Montana moved 59% farther between successive observations when not accompanied by a territorial buck during summer.

After 20 years of observations on free-ranging pronghorn in east-central Idaho, Fichter (1987) concluded that their spacing within groups is much more variable than brief or casual observations might suggest. He reported that the instability of doe/fawn groups was most obviously and frequently expressed by moves of one or two members of a group away from the others. At times, these moves seemed aimless. On other occasions, they appeared to be quite purposeful. Fichter suggested that an increase in population density might cause an increase in the fluidity of pronghorn groups and further speculated that adaptations to open habitat were relevant to such behavior. He concluded that (1) widely spaced or dispersed individuals and subgroups can be an integrated group, (2) loose organization may be more apparent than real and (3) pronghorn can communicate visually over distances of 1.2 miles (2 km) or more; therefore, coordination is possible among widely scattered individuals and subgroups. E. Fichter (personal communication:1987) provided flow charts depicting the varying combinations (subgroups) of pronghorn bucks during a short time period in summer (Figures 26 and 27) and another chart illustrating changes in composition of groups involving does, fawns and bucks during summer (Figure 28).

In central Montana, Pyrah (1970) found more stability in pronghorn doe bands than has been reported in other studies, high coefficients of association among members of a band and low coefficients of association between does from different bands. Also, when summer bands mixed, they almost invariably separated into original bands. Pyrah (1987) considered doe bands as discrete social units of herds that occupied separate ranges within herd ranges during the spring–summer–autumn period. Some marked females, observed for periods up to 7.5 years, used essentially the same areas and associated mostly with the same marked animals year after year. Among 17 doe bands that each included at least one marked animal, fidelity to bands averaged 86% and fidelity to band ranges averaged 77%. Pyrah's (1987) study area was in high-quality pronghorn habitat, and winter and summer ranges were not far apart. Also, a highway, railroad and agricultural

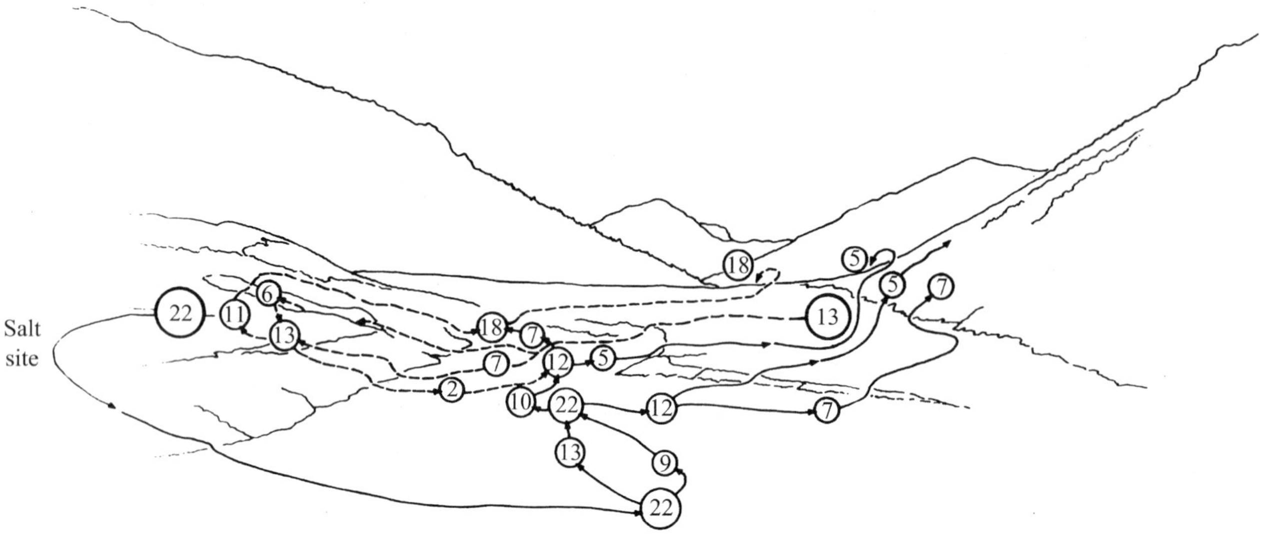

Figure 26. *A flowchart showing the varying combinations (subgroups) of 35 pronghorn bucks from 12:20 to 6:41 P.M., July 2, 1965, over an area of about 250 acres (1 km²), in the Pahsimeroi Valley of Idaho. The numbers refer to the number of pronghorn at a given location according to the movement pattern (arrowed lines). (from Fichter 1987).*

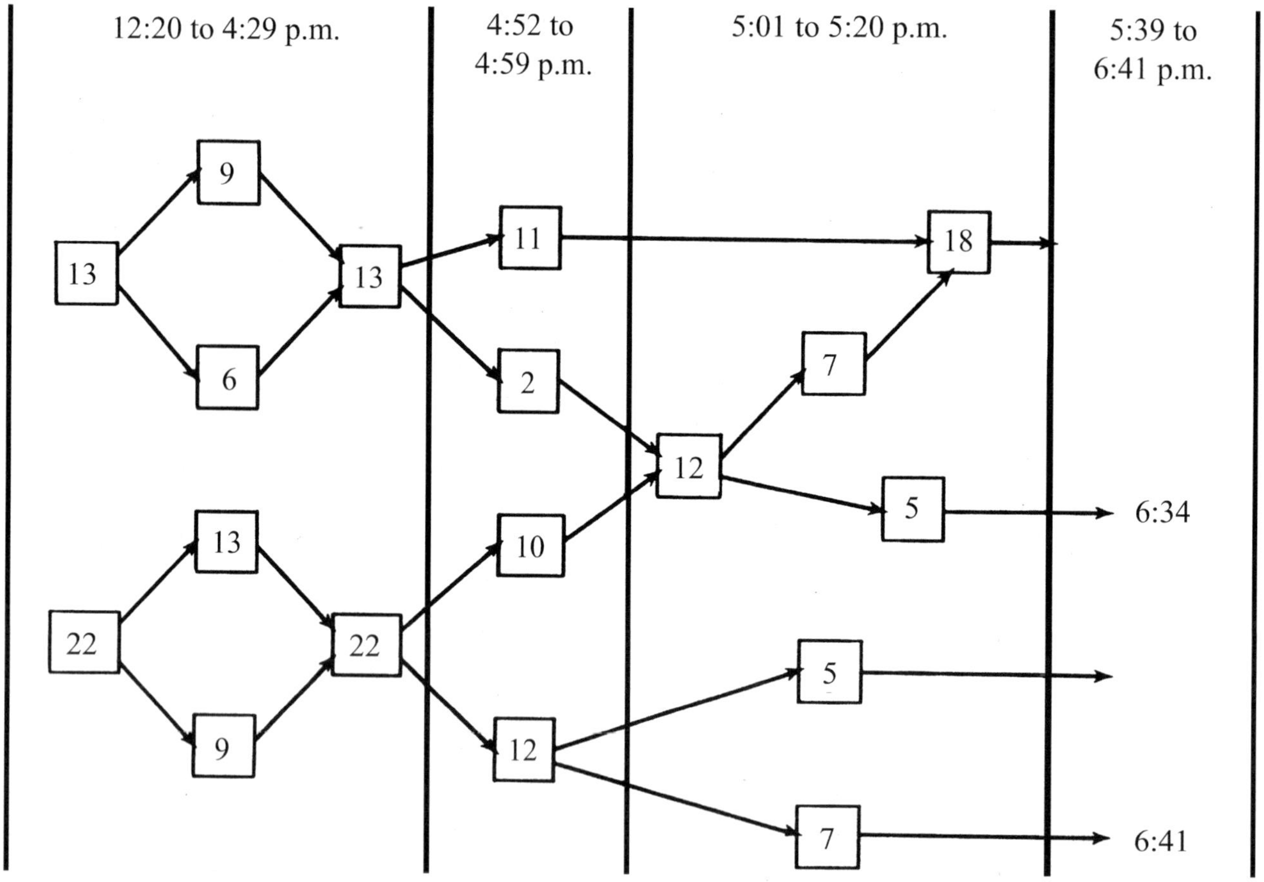

Figure 27. *A formalized flowchart diagramming the flexible social structure among 35 pronghorn bucks over a period of 6 hours and 21 minutes on July 2, 1965, in the Pahsimeroi Valley of Idaho (from Fichter 1987). Boxed numbers represent subgroups and the arrows lead to regroupings that often characterize groups of pronghorn and reveal their flexibility.*

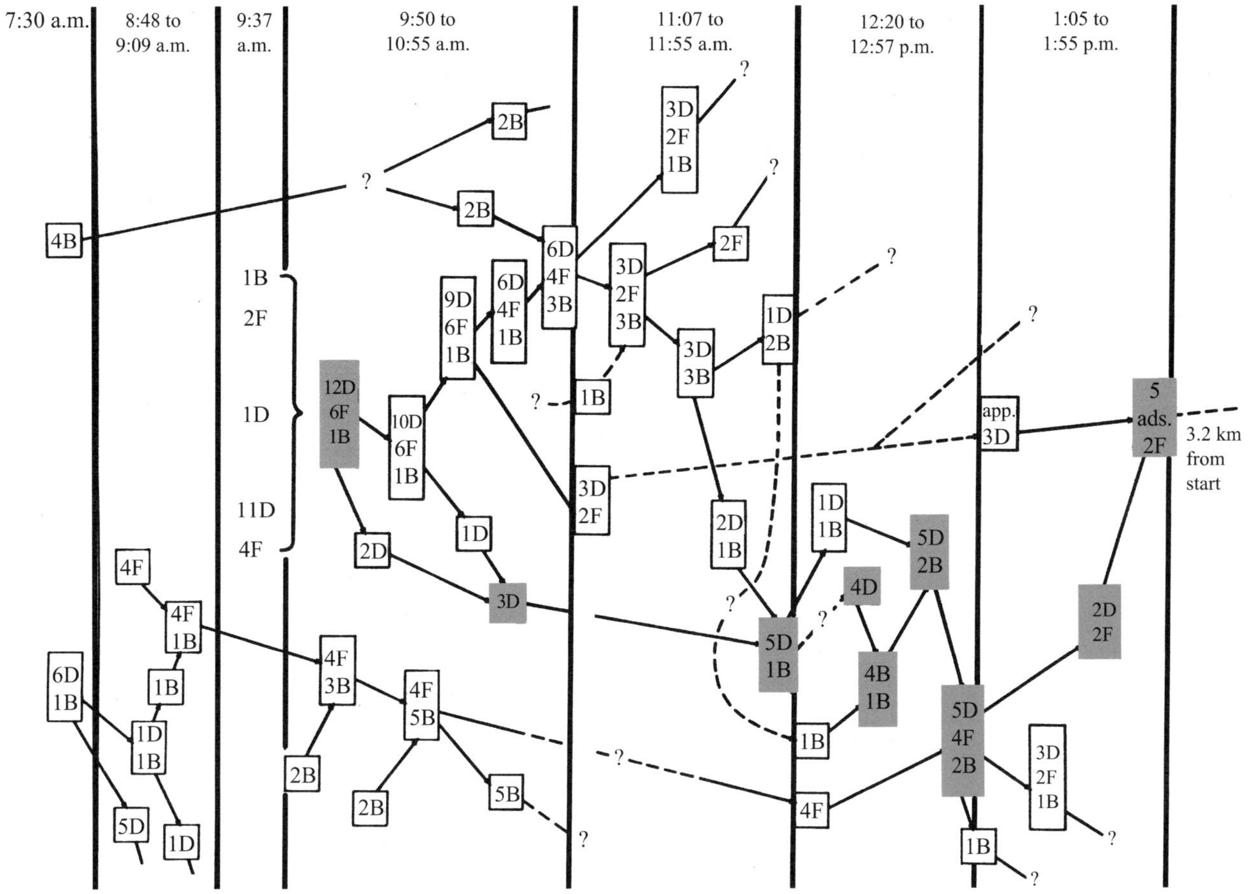

Figure 28. A flowchart illustrating the changes in composition of groups or subgroups of pronghorn, involving 18 does, 10 fawns and 11 bucks, during a period of 6 hours and 15 minutes on July 13, 1965, in the Pahsimeroi Valley of Idaho (from Fichter 1987). The shaded boxes identify the "route" of a marked, mature doe. Abbreviations: ads. = adults; B = buck; D = doe; F = fawn.

valley to the south were barriers to movement in that direction, except during the occasional very severe winter. All of this may have contributed to the stable social structure and a close relationship of pronghorn in that area.

Stability of social structure may originate from close kinships within each doe band, according to Pyrah (1987). He maintained that low dispersal rates of female offspring and polygamy of territorial bucks cause doe bands to be essentially sister groups. Because little genetic exchange occurs between herds, each herd theoretically represents a different gene pool. Genetic closeness may strengthen group social bonds and decrease agonistic behavior. Near-identical genotypes may produce similar glandular odors, external appearances, behavioral traits, food preferences and habitat-use patterns, including migration. Unrelated females may not be fully accepted into resident female social orders. Two of four marked immigrant females in central Montana were observed to have unstable home ranges, residing part time in two different doe band ranges. However, Byers (1997a)—working with a population in which he could recognize all individuals and mother/daughter pairs—found that daughters were not in their mothers' groups more frequently than would be expected by chance.

On the National Bison Range, doe band sizes vary greatly during any time period, but a definite seasonal trend exists—peaking in late spring, dropping drastically during June and early July, increasing through August, and peaking again in early September. These trends coincide with the major events in the life cycle. During parturition in late May and early June, does seek semi-isolation and move away from bands. After fawns are born, does return to groups, leaving periodically to nurse their hidden fawns. During June, large numbers of does are seen singly or in groups of two or three. Several weeks after birth, fawns join their dams in a doe band. Does with fawns apparently form larger bands than do those without fawns, for protection from predators. The size of doe bands increases throughout July and August, peaking in early September as pockets of succulent vegetation attract and concentrate the animals. As the rut starts in mid-September and progresses into October, doe band size drops sharply, reflecting increased harassment by territorial and bachelor bucks (Swanger 1977). Decreases in doe band size during the rut also were noted by Einarsen (1948), Prenzlow (1965a) and Fichter (1972).

Byers (1997a) also found that group sizes on the National Bison Range varied predictably with seasons. During winter, pronghorn formed large, mixed-sex aggregations, and the entire population sometimes was found in a single group. Because pronghorn show limited sexual dimorphism in feeding habits and susceptibility to predation, the sexes need not segregate during winter. Strangely, social interactions are at an annual minimum in the large, crowded wintering herds.

Byers (1997a) noted that does seldom were seen alone except during the birthing season. He observed does leaving bands a few days before giving birth. They occasionally associated with other does passing through their chosen birthing sites. More often, they aggressively drove such visitors away.

Formation of summering bands of does and fawns takes place gradually over a period of a few weeks. Soon after fawns are born, lactating and dry does band together. When they join those bands, fawns spend much of their time with other fawns. By this association, they develop balance and speed while playing during summer. Autenrieth and Fichter (1975) described playing fawns as performing exaggerated and excessive movements, and had the impression that the amount of play reflected the condition of the animals or their environment. Bromley (1967) concluded that fawns, by 6 weeks of age, are functional members of the summer bands and have their own hierarchical order in which males are dominant to females. Cavorting fawns sometimes chase other species. I have seen 2- to 3-month-old fawns chase white-tailed deer to the point of apparent exhaustion, forcing the deer to seek thick shrubs. Tame fawns sometimes play with dogs and even learn to chase cars. Play generally continues until autumn but seldom is seen on winter ranges.

Although doe/fawn bands, in herds with territorial bucks, frequently move from one territory to another while feeding, they spend much of their time on bucks' territories; consequently, such bands often appear to include one adult male. Mature females usually lead groups in flight and, if present, adult males often bring up the rear. The usual shape of a fleeing group is an ellipse, sometimes very elongated (Gregg 1955).

Female fawns may be separated from their dams during the rut but usually join them again after the latter have bred. These mother/daughter bonds lead to closely related doe/fawn bands. Photo by Len Clifford; courtesy of Leonard Rue Enterprises, Blairstown, New Jersey.

The play of fawns, recognized primarily by exaggerated and excessive movements, shows four patterns—leaping, running, sparring and butting (Autenrieth and Fichter 1975). Practice (for running, both flight and intraspecific chasing) and socialization seem most relevant to the pronghorn's adaptation to open habitat. Leaping is the neonate's first attempt at play and may occur within the first few hours of life. The amount of play during summer reflects the condition of the animals and their environment. Photo by Joe Van Wormer; courtesy of the Denver Museum of Natural History.

(Left) *Fawns join doe/fawn bands at about 3 weeks of age. After that, they spend much of their time playing with other fawns, and mock battles, as well as exaggerated and excessive movements, are common components of such play.* (Right) *Part of a typical doe/fawn band on a buck's territory. Note horns starting to grow on the buck fawn, farthest right in the picture.* Left photo by Mike Francis. Right photo by Peter T. Bromley.

Rutting activities in unhunted populations usually separated fawns from their dams long enough to stop milk production; thus, the fawns were weaned (O'Gara 1968). Small bands of fawns were common during the rut, but the weaned fawns rejoined the does as soon as does left the harems. Only about 10% of the does I handled in October were still lactating, but most of them had heavy udders, indicating lactation within the preceding month. Becker (1972) observed fawns staying with does in harems during the rut in central Montana. I interpreted this to suggest that the bucks were younger in the hunted population and did not control the harems as completely

The usual shape of a fleeing group of pronghorn is an ellipse, which may become very elongated if the animals run a long distance. Photo courtesy of the Montana Department of Fish, Wildlife and Parks, Helena.

A fawn nursing in September. On good range, 3- to 4-month-old fawns, such as above, are 75% to 90% as large as their mothers and are close to being weaned. Photo by Peter T. Bromley.

as did the older bucks. However, J. Meeker (personal communication:1983) found males on the lightly hunted Sheldon National Wildlife Refuge in Nevada herding fawns into harems along with adult females.

Before the rut in Yellowstone National Park, Gilbert (1973) spent 7 weeks observing a bachelor band with a stable dominance hierarchy. The band was socially cohesive but intense, prolonged displays and sparring were common. Subordinate males were courted and mounted by dominant males, and attempts to follow and mount a departing male, often after a sparring match, resulted in strings of dominant males following subordinate ones. Thus, a "following" behavior pattern was observed that correlated with the dominance hierarchy. The bachelor band occupied relatively less productive habitat adjacent to three territories held by large bucks, but commonly entered those areas during the rut to court estrous females. This encroachment naturally led to confrontations, because nonterritorial bachelors were not tolerated in territories established by large bucks, although those territories often were shared by does and doe/fawn bands. Byers (1997a) suggested that bachelors cooperated in sparring, and the activity was a form of motor training. Although young bucks would attempt to court estrous females, Byers described them as cautious and not attempting to breed seriously.

In heavily harvested populations where trophy bucks are selected by hunters, bachelor bands may be smaller and have younger members, and 3-year olds sometimes acquire territories. In the North, size of bachelor bands decreases during August; by the rut in September, lone bachelors are common.

After the rut, territorial bucks leave their territories, males cast their horn sheaths and large herds form with little social interaction. Sizes of wintering herds are determined by geography, range conditions and winter severity. When snow gets deep, pronghorn usually are forced to reach browse or perish (Martinka 1967); therefore, large herds gather on northern sagebrush rangelands during

In bachelor bands, subordinate bucks sometimes are courted and mounted by dominant bucks. Thus, while traveling, dominant males follow subordinate ones as seen here. The following 2-year-old ordinarily would not allow the yearling to follow him. Photo by LuRay Parker; courtesy of the Wyoming Department of Game and Fish, Cheyenne.

Before the rut, buck behavior, especially that of lone bachelors, becomes erratic. An apparently calm animal, feeding or traveling, sometimes jumps suddenly into the air and begins "fighting" with a bush. Photo by James R. Good.

When sparring with vegetation, bachelor bucks can acquire interesting temporary headwear. Photo by Peter T. Bromley.

some winters. In 1964, I saw a cohesive herd of approximately 1,300 pronghorn in the Centennial Valley of Montana.

Prenzlow et al. (1968) described the annual cycle for pronghorn with a harem system of mating in northeastern Colorado. During October, after the rut, the pronghorn formed large aggregations containing both sexes and remained in those herds throughout winter. Pregnant does left doe bands before parturition in early May. Separate doe/fawn and bachelor bands then formed and were maintained through early summer. During mid- to late August, adult males began herding females and fawns into breeding harems that they held until the rut terminated in October. Thus, the grouping and movements described by Prenzlow et al. for harem-breeding pronghorn in Colorado were quite similar to those described for territorial animals.

ACTIVITIES

Pronghorn sleep often, but without continuity or regularity and usually for short periods, according to Einarsen (1948). They sleep both day and night and may be as alert and active at midnight as at midday. However, activity is greatest just after sunrise and before sunset. These peaks are greatest during hot weather (Kitchen 1974). Pronghorn generally follow a daily cycle of feeding, bedding, ruminating and movements to and from water (when available). Actual schedules of these activities vary from day to day as well as season to season and show regional variations as well (Kitchen and O'Gara 1982).

When well fed, pronghorn are able to withstand extreme temperatures, but malnutrition, resulting from unhealthy vegetative conditions, deep snow or droughts makes them vulnerable to excessive cold. Because pronghorn usually stay on the prairie, a casual observer might think that they do not seek protection from extreme temperatures, particularly cold. However, they do have behavioral adaptations that enhance their survival during severe northern winters.

(Top) A bachelor band in summer. These bands generally are composed of yearlings and 2- and 3-year-olds, plus a few older bucks that have not acquired territories. (Bottom) Territorial bucks influence the size of doe bands during summer patrolling their territory by herding stray does onto territories and actively trying to stop them from leaving. Photos by Peter T. Bromley.

Bruns (1969) found that an established hierarchy at feeding craters pawed in the snow reduced expenditure of energy. Pronghorn selected microhabitats with softer and less dense snow, lower wind velocities and less snow than the average for the area studied in Alberta. They conserved energy by reducing daily travel, traveling single file in snow and lying down during postdawn periods of hard snow and low temperatures. Social interactions are rare in winter. During 4,555 hours of pronghorn observation in the Pahsimeroi Valley of Idaho, only 2.1 incidents and 2.6 minutes of social activities per 100 hours were seen (Bernt 1976). During April, after spring dispersal had begun, 14.6 incidents and 5.8 minutes of social behavior per 100 hours were observed.

Olson et al. (1978) observed that pronghorn moved into conifer cover during the severe winter of 1977–78 in eastern Montana. Leftwich and Simpson (1977) found considerable reference to pronghorn in the Texas Panhandle using shallow depressions on the lee side of low hills during blizzards in the late 1800s. With fencing of open range, animals, drifting with blizzards, died alongside fences (Simpson and Leftwich 1978). Amstrup (1978) and Buechner (1950a) noted that pronghorn along the Montana/Wyoming border and in Texas, respectively, sometimes rested in the shade of trees during hot weather.

Byers (1997a) provided year-round activity budgets for bucks and does on the National Bison Range. Pronghorn spent the majority of their time feeding and the next greatest proportion of time reclining and ruminating. Time up—standing, walking, and so on—almost equaled time reclining and greatly exceeded both feeding and reclining time for bucks during the rut. For does during the rut, time up also exceeded that for reclining and almost equaled feeding.

Feeding also was the most common behavior during all seasons on native and reclaimed rangelands in the Pow-

Pronghorn often bed in compact groups during winter, especially during high winds, to reduce heat loss. They also lie with their heads curled back along their bodies. Photo by Ed Bry; courtesy of the North Dakota Game and Fish Department, Bismarck.

Although shade seldom is available in much of the pronghorn's habitat, the animals will use it when available—even if it is a thin streak across the prairie. Photo by Richard Guenzel; courtesy of the Wyoming Department of Game and Fish, Cheyenne.

der River Basin of Wyoming (Gamo 1997). Bedding was the next most common behavior (Table 30). This Table suggests slight differences in behaviors on native and reclaimed habitats. However, the greatest differences were noted during autumn and winter, when sample sizes were small.

Instrumented pronghorn along the Montana/Wyoming border were found by Amstrup (1978) to be inactive 29%, 22%, 20% and 24% of the time during winter, spring, summer and autumn, respectively. Most of these observations involved females. Kitchen (1974) observed that activity levels of territorial bucks increased dramatically during the rut, therefore Amstrup's (1978) data probably would have been different if more bucks were included. During eight, 24-hour tracking periods during summer and autumn, Amstrup found pronghorn to be largely diurnal and crepuscular. The tendency for the animals to be active primarily during daylight hours did not vary between July and November. His observations indicated that, during the rut, bucks were more active at night than females, but they still were less active than during the day. During summer, pronghorn ceased activity abruptly at the onset of darkness and usually resumed activity at first light. Daily peaks of activity occurred shortly after sunrise and shortly before sunset. In the Red Desert of Wyoming, Taylor (1972a) also observed

TABLE 30. Seasonal percentages of pronghorn behaviors visually observed by Gamo (1997) on reclaimed and native areas in the Powder River Basin of Wyoming, 1995 and 1996

Habitat	*Number*	*Percentage of observed behaviors*				
Season	*of groups*	*Feeding*	*Bedded*	*Walking*	*Running*	*Standing*
Reclaimed	160	69	16	6	1	8
Spring	45	60	24	11	0	5
Summer	94	68	14	4	0	14
Autumn	16	94	0	0	6	0
Winter	5	80	20	0	0	0
Native	242	57	26	7	0	10
Spring	88	57	28	8	0	7
Summer	116	54	22	9	1	14
Autumn	30	70	23	3	0	4
Winter	8	50	50	0	0	0

a midday peak. During Amstrup's (1978) study, daylight periods of inactivity were highly variable, but usually longest in midafternoon. Occasionally, pronghorn would have an activity bout before dawn and then remain inactive for much of the early morning. During winter, they seemed to remain inactive longer after sunrise than they did during summer. In Texas, D. Simpson (personal communication:1978) observed pronghorn moving around during moonlit nights. One herd of 32 animals moved 0.75 mile (1.2 km) during the hours of darkness.

Daily movements and longest distance moved per day vary with seasons and generally are shortest during spring, when forage is succulent and abundant. Daily movements are longest during autumn, when forage is dry and scarce. Such movements are in response to phenology of certain plants and to changes in weather (Kitchen and O'Gara 1982). During winter, shifts in wind direction or new snowfall caused Montana pronghorn to move 0.6 to1.2 miles (1–2 km) to areas protected from wind and with taller vegetation (Kitchen 1974). Lactating does shifted foraging areas in response to plant succulence and plant phenology. Daily movements of 10 Texas fawns averaged 1,570 yards (1,436 m) at 1 week of age and increased to 4,785 yards (4,375 m) at 12 weeks of age (Tucker and Garner 1984).

In Nevada, movements from birthsites to first bedsites were measured for five pronghorn fawns (McNay 1980). The distances ranged from 109 yards (0.1 km) to 875 yards (0.8 km) and averaged 438 yards (0.4 km). The average distance, up to 9 days of age, between consecutive beds was 0.44 mile (0.7 km); the greatest and least distances, by two 3-day-old fawns, were 1.2 miles (2.0 km) and 0.12 mile (0.2 km). The relative position of consecutive locations varied from nearly linear to nearly circular. The average distance between 12 fawn capture sites and final locations was 2.2 miles (3.5 km); one fawn was last located only 0.19 mile (0.3 km) from its capture site, another was 5.9 miles (9.5 km) from its capture site. In Alberta, Barrett (1984) documented progressively longer movements of 62 radio-collared fawns with increasing age (Table 31).

A 3-week-old Texas fawn shifted its home range approximately 2.8 miles (4.5 km), possibly because another fawn, that had been bedded about 82 yards (75 m) away, was killed by a coyote. Another fawn, captured at about 15 days of age, moved 1.6 miles (2.6 km) when released and did not return to the area where it was captured. Diel movements of 10 fawns averaged 1,569 yards (1,435 m) at 1 week of age and 4,785 yards (4,375 m) at 12 weeks. By 4 weeks of age, fawn movements closely resembled those of their mothers (Tucker and Garner 1984).

SIZES OF RANGES

[A]ll home-range estimators should be used with caution. Use of these estimators without some means of verifying results obtained may produce inaccurate home range estimates. . . . Researchers should verify results obtained in home-range studies with a combination of practical knowledge of animals on the study area, plotting locations on maps when questionable results are produced, and common sense.

Canon (1993:120)

Canon estimated home ranges of Texas pronghorn using five methods (Table 32). The results indicated the differences that can be expected when using different estimators with the same set of data.

The home range of most mammals consists of a few areas of intensive use interconnected by pathways, according to Sanchez (1993). Burt (1943) defined home range as the area traversed by an individual in its normal activities of feeding, mating and caring for young.

Early 20th century reports indicated that pronghorn had small home ranges. A level stretch of open prairie 2 miles (3.2 km) across seemed ample range for a band of 20 throughout a whole season in western Wyoming, according to Seton (1927). Seton also cited other observations that pronghorn's home localities were 3 to 4 or 6 to

TABLE 31. Mean minimum daily movements of 62 pronghorn fawns on a mixed-grass study area of Alberta

Age in days	*Number of locations*	*Mean minimal daily movement, in yards (m)*
1–3	49	587 (537)
4–7	182	542 (496)
8–15	347	719 (658)
16–24	306	861 (787)
25–40	434	1,094 (1,000)
40+	327	1,961 (1,793)

Source: From Barrett 1984.

TABLE 32. Average size of home range for 28 adult pronghorn does located 100 or more times each during 1990 to 1993 in Hudspeth County, Texas, as estimated by five methods

Estimator	*Average home range size, in square miles (km^2)*
Grid count[a]	15.5 (40.2)
Minimum convex polygon[b]	17.3 (44.7)
85% harmonic mean[c]	12.9 (33.4)
90% harmonic mean[c]	16.4 (42.4)
95% harmonic mean[c]	23.9 (61.9)

[a] Based on plotting each animal's location on topographic maps and counting 1 km^2 grids. A minimum area polygon then was constructed, based on perimeter locations, fence-line locations and preferred travel lanes. This method approximated use of a polar planimeter to estimate ranges (see Canon 1993).

[b] After Mohr (1947).

[c] After Stuwe and Blohowick (1988).

Source: From Canon 1993.

8 miles (4.8–6.4 or 9.7–12.9 km) in diameter. However, all of those reports were of unmarked animals. Often when animals are marked, researchers are surprised to find that bands that appear to be the same day to day or week to week have different animals cycling through them. Apparently, the social tolerance, general behavior and feeding habits of a species dictate that a certain number (and sometimes age and sex ratio) of individuals can use any habitat at a given time. Evidence of this was found on winter rangelands in Wyoming by Ryder (1983), who indicated that laissez-faire selection of habitat could no longer occur once preferred habitat reached an animal-perceived carrying capacity. Pronghorn apparently can assess forage supplies and densities of conspecifics unless snow restricts habitat choices. Studies focusing on the responses of individually marked animals tend to overlook the importance of animal density.

In central Montana, as indicated by mapped locations during a 13-year study, pronghorn herds occupied discrete ranges to which marked individuals exhibited high fidelity; 88% were consistently observed within their own herd ranges (Pyrah 1987). Little contact or mixing was observed among herds, and that which did occur largely involved dispersal. Five herd ranges varied in size from 18 to 81 square miles (46–208 km^2) and averaged 41 square miles (105 km^2). Those ranges consisted of upland plateaus or benches between drainages and had long axes parallel to drainages. Tall and dense vegetation, steep-sided stream channels and contact with predators or predator sign probably reduced pronghorn use of those areas and, thus, mixing of herds. Herds and herd ranges apparently are not as well defined in many areas as they are in central Montana. Doe band ranges varied in size from 4 to 21 square miles (10–54 km^2) and averaged 10 square miles (26 km^2) (Pyrah 1987).

Territorial bucks generally have the smallest ranges of any sex or age class during summer (Table 33). Doe/fawn bands usually use the summer ranges of several territorial bucks plus areas between and around them, and bachelor bands tend to have the largest summer ranges. Using sightings of known does, Swanger (1977) determined that doe home ranges on the National Bison Range covered an average of four buck territories. Home ranges in central Arizona appeared to be determined partially by fences (Ockenfels et al. 1992).

Marked and otherwise individually recognizable pronghorn were studied in central and southeastern Montana during the late 1960s and early 1970s (Table 34). All five of those studies were short-term research projects for Master of Science theses. Most animals were not radio-collared, and some seasonal ranges were estimated from only five locations. Thus, actual ranges probably were larger than reported. Eight of the animals located by Bayless (1969) shifted to winter ranges 1.6 to 5.0 miles (2.6–8.0 km) from those used during early winter and did not return.

Summer and winter ranges of radio-collared pronghorn on the shrubsteppes of Idaho varied greatly between years and among valleys (Hoskinson and Tester 1980) (Table 35). Also, during early summer, yearlings wandered back through familiar areas rather than remain on typical summer range. Their ranges were two to five times as large as adult ranges, and shifts of 6 to12 miles (10–20 km) were common. Such movements by young animals might lead to dispersal from the herd and, ultimately, to pioneering of unoccupied habitat, as well as maintaining gene flow. Pyrah (1987) postulated that the more-or-less random wandering of small groups of immature and socially unattached pronghorn that he observed, especially during spring when does and territorial bucks were returning to traditional ranges, might reflect some innate pioneering behavior. Some of those groups consisted of yearlings of both sexes following a 2- or 3-year-old male.

Ockenfels et al. (1994:1–2) maintained that: "Pronghorn move within their home ranges in response to influences such as water and food availability, irritating disturbances,

Table 33. Sizes of territories defended by bucks, as indicated by intensive, multiyear studies

Area	*Season*	*Number of territorial bucks*	*Average*	*Territory range, in acres (ha)*	*Source*
National Bison Range, Montana[a]	3 summers 2 ruts	12[b]		62–1,072 (23–434)	Kitchen (1974)[c]
Wind Cave National Park, South Dakota[a]	Summers and ruts				Bromley (1977)[c]
	Year 1	9	148 (60)	49–257 (20–104)	
	Year 2	10	168 (68)	59–309 (24–125)	
	Year 3	10	178 (72)	42–289 (17–117)	
Middle Park, Colorado[d]	4 summers	10	570 (234)	20–1,964 (8–795)	Min (1997a)[f]
	4 ruts	5–8[e]	358 (145)	89–835 (36–338)	

[a] Grassland habitat.

[b] Twelve territories.

[c] Territories mapped by observing defense of boundaries by individually recognizable bucks.

[d] Shrubsteppe habitat.

[e] Eight territorial bucks during the rut in 1990 and 1991, five in 1992 and 1993.

[f] Size of territories estimated by the minimum convex polygons method (Southwood 1966). Focal animals were radio-collared, wore numbered collars, had unique horns or morphology of ears.

and weather conditions." Hoskinson and Tester's (1980) study confirmed this and showed how much weather, cultivated crops and fences can influence sizes and locations of pronghorn ranges (Table 35).

The first winter of Hoskinson and Tester's (1980) study—1975–76—was severe, and the second winter—1976–77—was mild with less snow. During the first winter, pronghorn occupied the lower ends of three valleys at the edge of the Snake River plain. During the second winter, 1976–77, 13 radio-collared pronghorn stayed an average of 7.3 miles (11.7 km) up-valley from their previous winter's ranges; the maximum distance was 17.5 miles (28.1 km).

In the Little Lost River valley, ranges were much larger during the mild than during the severe winter. In the Crooked Creek valley, ranges were about the same size

Table 34. Sizes of seasonal ranges, as reported for individually recognizable pronghorn in Montana

Area	*Age-sex*	*Season*	*Number of pronghorn*	*Territory size in acres (ha)*		*Source*
				Average	*Range*	
Central Montana[a]	Ad-F	Summer	4	2,854 (1,155)	— —	Wentland (1968)[b]
	BB	Summer	8	4,115 (1,665)	— —	
	TB	Summer	7	986 (399)	— —	
Central Montana[a]	Ad-F	Winter	6	2,841 (1,150)	1,250–3,840 (506–1,554)	Bayless (1969)[c]
	Yrlg-F	Winter	1	5,574 (2,256)	— —	
	Yrlg-M	Winter	1	4,160 (1,684)	— —	
Central Montana[a]	Ad-F	Summer	14	672 (272)	77–2,406 (31–974)	Roberts (1970)[d]
	BB	Summer	11	1,741 (704)	230–4,851 (93–1,963)	
	TB	Summer	15	448 (181)	154–774 (62–313)	
Southeastern Montana[e]	Ad-F	Summer	3	2,970 (1,202)	531–4,928 (214–1,994)	Campbell (1970)[d]
	Yrlg-F	Summer	1	544 (220)	— —	
	BB	Summer	2	2,058 (829)	1,203–2,912 (487–1,178)	
	TB	Summer	2	576 (233)	403–800 (163–324)	
	Yrlg-M	Summer	2	1,949 (789)	1,888–2,010 (764–813)	
Southeastern Montana[e]	Ad-F	Winter	5	1,523 (616)	768–2,240 (311–907)	Freeman (1971)[d]
	Ad-M	Winter	4	672 (272)	192–896 (78–363)	

Ad = adult; Yrlg = yearling; F = female; M = male; BB = bachelor buck; TB = territorial buck.

[a] Primarily sagebrush/grasslands.

[b] Ranges were estimated by drawing a line around outside observations on a map and calculating the enclosed area with the aid of a polar planimeter.

[c] Ranges were estimated in the same manner as by Wentland (1968), but three does were radiocollared.

[d] Ranges were estimated using a method similar to that of Hayne (1949).

[e] Primarily sagebrush/grasslands and agricultural fields.

TABLE 35. Average sizes of home ranges, in acres (ha), of radio-monitored pronghorn, by season and area in southeastern Idaho[a]

	Little Lost River Drainage		*Birch Creek Valley*		*Crooked Creek Valley*	
	Number of pronghorn	*Range*	*Number of pronghorn*	*Range*	*Number of pronghorn*	*Range*
Winter 1975–76						
Females	5	3,598 (1,456)	7	13,529 (5,475)	4	6,729 (2,723)
Males	4	2,590 (1,048)	4	13,675 (5,534)	4	3,875 (1,568)
Summer 1976						
Females	5	3,598 (1,456)	6	2,424 (981)	3	4,873 (1,972)
Males	4	3,993 (1,616)	2	2,360 (955)	5	3,264 (1,321)
Yearlings	3	7,193 (2,911)	4	13,358 (5,406)	2	5,827 (2,358)
Winter 1976–77						
Females	4	12,160 (4,921)	6	7,401 (2,995)	5	7,139 (2,889)
Males	3	3,714 (1,503)	3	2,728 (1,104)	2	4,767 (1,929)
Yearlings	1	5,733 (2,320)	1	9,531 (3,857)	1	2,044 (827)

[a] Ranges estimated by the minimum area method (Mohr 1947).
Source: From Hoskinson and Tester 1980.

during both winters. In the Birch Creek valley, ranges were smaller during the mild than during the severe winter (Hoskinson and Tester 1980).

During the severe winter, ranges in the Little Lost River valley were about 18% as large as those in Birch Creek valley and about 62% as large as those in Crooked Creek valley (Hoskinson and Tester 1980). During the mild winter, sizes of ranges were nearly equal among the valleys (Table 35). Irrigated farms occupied much of the mouth of the Little Lost River valley, leaving only narrow strips of undisturbed land along the foothills. During the severe winter, the small ranges were in these strips, and only limited movements were possible between mountains and partially fenced farmlands. The Crooked Creek valley had only minimal agricultural development, and Birch Creek valley had none.

Little Lost River pronghorn were in better physiological condition when captured than were those from the

Pyrah (1987) noted some movement by individuals or small groups of pronghorn between herds or off of his study area in north-central Montana. Most cases of such dispersal involved more or less random wandering by small groups of immature animals and apparently socially unattached adults. Pyrah believed that dispersal could be an important factor in pioneering unoccupied habitat and population regulation. Photo by Cindy Ticer; courtesy of Arizona Department of Game and Fish, Phoenix.

other drainages (Seal and Hoskinson 1978). High-protein agricultural crops probably contributed to the welfare of the Little Lost River pronghorn during the severe winter, although fences reduced the sizes of winter ranges (see Chapters 8 and 14). Logically, farms were located in these areas because of good soil and growing conditions. Thus, native vegetation may have been more nutritious at the mouth of Little Lost River valley than at the mouth of the other valleys.

Up valleys 6 to 9 miles (10–15 km), where pronghorn ranged during the mild winter, most of the rangeland was not fenced or cultivated. Thus, during the mild winter, pronghorn in the three valleys occupied somewhat similar habitat, and ranges were nearly equal in size.

Some radio-collared pronghorn studied by Hoskinson and Tester (1980) were not faithful to one winter range. An adult doe spent the first winter in Little Lost River valley and the next winter 12 miles (19 km) southeast in Birch Creek valley. Another adult doe wintered first in Birch Creek valley and moved 22 miles (35 km) northeast to spend the next winter in Crooked Creek valley. A Crooked Creek valley buck spent the second winter 12 miles (20 km) southwest in Birch Creek valley. An adult doe that wintered in Crooked Creek valley during 1975–76 remained 11 miles (18 km) southeast on her 1976 summer range during winter 1976–77. Also, two Crooked Creek pronghorn traveled to Birch Creek valley and returned during the severe winter. An adult doe made the trip once, and an adult buck made it three times.

In Colorado, Firchow (1986) found the seasonal home ranges of radio-collared, adult pronghorn highly variable. Convex polygon home ranges were twofold and fivefold larger during winter than during summer for males and females, respectively. For both sexes, spring and fall ranges were slightly smaller than summer ranges. Ninety-five percent harmonic mean transformation estimates were, on average, 45% greater than the 95% convex polygon estimates. Also on average, 50% of core areas were contained within 13% of the 95% harmonic mean transformation home ranges. Mean fawn summer range size was approximately 20% smaller than adult doe summer range size. Sample sizes were comparatively small during this study, but because the animals were monitored year-round by radiotelemetry—and ranges were calculated by two estimates—it is a good example of the variability of pronghorn habits as well as methods for judging sizes of home ranges (Table 36).

Easterly et al. (no date) investigated sizes of pronghorn winter ranges in the Rattlesnake Hills (Table 37). The sagebrush/grassland study area was considered crucial winter range for pronghorn in central Wyoming by personnel of the Wyoming Game and Fish Department. During the study, pronghorn densities were comparatively high—13 to16 per square mile (5–6/km^2).

The sample size was small, but a study of the Sonoran pronghorn indicated larger home ranges than those of other populations that have been studied (Wright and deVos 1986)—as might be expected for desert animals (Table 38). In the Trans-Pecos area of Texas, Canon (1993) found the mean, annual home range consistently larger for does than bucks. He considered the 90% harmonic mean the best estimator of normal activity and used it to compile home ranges. In central Arizona, few differences existed between home ranges of bucks and does (Ockenfels et al. 1994). Sizes, using minimum convex polygons, varied greatly and were bimodally distributed, reflecting the presence of migratory and nonmigratory individuals. In southern New Mexico, yearlings had larger home ranges than adults, but ranges did not differ between sexes (Clemente et al. 1995) (Table 38).

TABLE 36. Average seasonal home range sizes and ranges of radio-marked adult pronghorn in grasslands, yucca grasslands and cholla grasslands of southeastern Colorado during January 1983 to December 1984, as estimated by two methods[a]

Sex/season	*Number of pronghorn*	*95% convex, in square miles (km^2)*		*95% harmonic, in square miles (km^2)*	
		Average	*Range*	*Average*	*Range*
Does					
Winter	10	19.6 (50.9)	7.8–43.1 (20.2–111.6)	40.5 (104.9)	14.6–134.3 (37.7–347.8)
Autumn	15	7.5 (19.5)	2.1–34.8 (5.5–90.0)	8.9 (23.1)	1.9–33.2 (4.9–86.1)
Spring	8	9.5 (24.7)	1.1–40.4 (2.8–104.7)	14.0 (36.2)	3.9–32.8 (10.1–85.1)
Summer	20	9.9 (25.6)	3.1–36.4 (8.0–94.2)	19.4 (50.3)	4.9–47.5 (12.6–123.0)
Bucks					
Winter	6	25.9 (67.1)	6.6–50.6 (17.0–131.1)	34.0 (88.0)	9.6–69.6 (24.8–180.4)
Autumn	9	3.8 (9.8)	1.8–5.6 (4.7–14.5)	7.2 (18.6)	3.2–15.2 (8.2–39.4)
Spring	3	4.5 (11.6)	1.9–6.6 (4.8–17.1)	6.6 (17.2)	3.2–8.9 (8.4–23.0)
Summer	15	5.0 (13.0)	1.3–18.7 (3.4–48.4)	9.3 (24.2)	1.9–31.8 (4.8–82.5)

[a] A minimum of 22 radio locations per sex–season–year combinations was used for each home range calculation.

Source: From Firchow 1986.

TABLE 37. Size of pronghorn winter ranges in the Rattlesnake Hills of central Wyoming, 1988 through 1990

Year	*Number of pronghorn*	*Winter range size, in square miles* (km^2)
1988	17	14.6 (37.9)
1989	20	13.2 (34.2)
1990	13	11.2 (29.0)

Radio-collared pronghorn were located two to three times a month; universal transverse mercator (UTM) coordinates were entered into a computer database and home ranges were calculated using the TELDAY microcomputer program (T. H. Lonner and D. E. Burkharter unpublished).

Source: From Easterly et al. no date.

Ockenfels et al. (1997) radio-tracked 37 adult pronghorn in northern Arizona (Table 38). They found that fenced, paved two- and four-lane roads and highways, as well as fenced railroad rights-of-way were barriers to pronghorn movements and influenced the shapes of home ranges. Ward et al. (1976) did not map home ranges per se. However, they noted that Interstate 80 in southern Wyoming was a barrier to pronghorn, and wintering herds trying to move south remained close to the highway on its north side. Conversely, summering herds were stopped on their movements northward. A young doe, trapped and radio-collared on April 5, 1975, near the highway, was located 34 different days before she was shot by a hunter on September 16, 1975. Her range covered an area 15 miles (24 km) long and 3 miles (4.8 km) wide on the south side of Interstate 80 and parallel to it. Obviously, the highway dictated the shape of her range.

Six does and one buck, radio-collared and located at least 19 times on the Dugway Proving Ground in Utah, had average home ranges of 48.7 and 23.3 square miles (126.2 and 60.4 km^2), respectively (Table 38). The largest and smallest doe ranges were 105.6 and 16.3 square miles (273.6 and 42.3 km^2), respectively. The two largest ranges

TABLE 38. Average size of annual home ranges as reported in the literature for both sexes and various age classes of radio-collared pronghorn

State	*Age-sex*	*Number of pronghorn*	*Average home range size, in square miles* (km^2)	*Source*
Arizona[a]	Ad-F	6	212.6 (550.6)	Wright and deVos (1986)[b]
	Ad-M	3	308.8 (799.8)	
Texas[c]	Ad-F	28	16.4 (42.4)	Canon (1993)[d]
	Ad-M	8	9.7 (25.1)	
Arizona[e]	Ad-B[f]	9	104.5 (270.6)	Ockenfels et al. (1994)[g]
	Ad-B[h]	20	20.2 (52.4)	
New Mexico[i]	Ad-F	2	4.6 (11.9)	Clemente et al. (1995)[j]
	Ad-M	2	6.4 (16.6)	
	Yrlg-F	2	13.0 (33.7)	
	Yrlg-M	1	13.1 (34)	
Arizona	Ad-F	15[k]	47.9 (124)	Ockenfels et al. (1997)[l]
	Ad-M	5[k]	31.7 (82)	
	Ad-F	13[m]	50.2 (130)	
	Ad-M	4[m]	22.8 (59)	
Utah	Ad-F	6[n]	48.7 (126.2)	Bates (2000)[o]
	Ad-M	1[n]	23.3 (60.4)	

Ad = adult; Yrlg = yearling; B = both sexes; F = female; M = male.

[a] Presumed Sonoran subspecies in desert habitat.

[b] Range size calculated from UTM locations on USGS maps and a computer program adapted for Arizona Department of Game and Fish, based on 30 or more locations of each animal during an 18 to 24 month period.

[c] Yucca grasslands and desert shrub habitat.

[d] 90 percent harmonic mean (Stuwe and Blohowick 1990).

[e] Shrubsteppe habitat.

[f] Migratory pronghorn.

[g] Minimum convex polygon (Ackerman et al. 1990).

[h] Nonmigratory pronghorn.

[i] Grasslands, grass/mesquite and mesquite in Chihuahua Desert.

[j] Minimum convex polygon (Jennrich and Turner 1969).

[k] Great Basin grassland and juniper woodland.

[l] Home range sizes estimated by the 100 percent minimum-convex polygon method (Ackerman et al. 1990), based on 3,407 aerial and ground locations by radio-telemetry.

[m] Shortgrass prairie.

[n] Pickleweed, salt flats, annual weeds, etc.

[o] Outer convex polygon.

were those of does that moved to a traditional wintering area (Bates 2000).

During spring and early summer in Texas, home ranges of doe pronghorn were relatively small while fawns were hiding (Canon 1993). During a wet year, with succulent vegetation readily available, the does' ranges increased about 20% with fawns at heel. During a dry year, when succulent vegetation was not readily available, the does' ranges increased 89% (Table 39).

Eleven pronghorn fawns were captured in Arizona during 1991 and collared with mortality-sensing telemetry units (Ockenfels et al. 1992). Six fawns survived to 6 months. Signals were monitored on nearly a daily basis for 1 month, then three times per week until the fawns were 6 months of age. Use areas were very small for the first 21 days (Table 40), but increased significantly between four consecutive 21-day periods. Only two fence crossings were documented, and home ranges appeared to be partially determined by fences.

Individual and mean home range sizes of fawns in the tallgrass prairie of eastern Kansas increased from 49 acres (20 ha) at 7 days of age to 971 acres (393 ha) at 60 days (Rothchild 1993), coinciding with the findings of other researchers. Mean range sizes of Kansas fawns were smaller than those reported from Nevada (McNay 1980) (Table 40). Home ranges of fawns seem to be smaller in areas where fences of any type are numerous. During all studies, home ranges of 5- to 6-week-old fawns began to resemble parental doe home ranges.

Home range sizes of 12 radio-collared fawns, less than 3 weeks old, in Nevada indicated larger ranges for young does than for bucks (McNay 1980)—apparently a function of sample size. Large home ranges found during this study probably reflected a scarcity of preferred forage in some areas that caused adult does to move considerable distances while foraging. McNay believed that, because large home ranges implied extensive movement by fawns, large home ranges contributed to mortality. Long movements would make fawns highly visible to predators and increase the fawns' energy demand—a factor that could be significant under adverse weather conditions.

Table 39. Average home ranges of pronghorn mothers during fawning (April 15 to June 17) and postfawning (June 18 to August 20) periods in Hudspeth County, Texas, as estimated by Canon (1993) using the 90% harmonic mean estimate and 15 or more locations per season

Year	*Season*	*Average home ranges, in square miles (km²)*
1990[a]	Fawning	6.6 (17.1)
	Postfawning	12.5 (32.5)
1991[b]	Fawning	6.6 (17.2)
	Postfawning	7.9 (20.4)

[a] A dry year.

[b] A wet year.

Source: Stuwe and Blohowick 1988.

The average size of home ranges for 10 instrumented fawns in Texas increased rapidly with increasing age (Tucker 1979). Fawns' ranges began to resemble their dams' home ranges when fawn/doe bands were formed beginning at 5 to 6 weeks postpartum. The average home range of fawns at 16 weeks of age was used to compare home ranges of males with those of females (Tucker and Garner 1984), and males were shown to have significantly larger average home ranges than did females—again, apparently a function of sample size (Table 40). Home ranges of these Texas fawns were comparatively small, possibly because the animals were restricted to a 15,006-acre (6,073 ha) pasture.

Fawns fitted with radio collars in Alberta were monitored until they were 58 days of age (Barrett 1984). A nearly linear relation existed between increases in areas occupied and ages from 3 to 52 days. Range sizes increased dramatically during days 53 through 58, probably reflecting seasonal range shifts of doe/fawn bands. Similar behavior was indicated during Canon's (1993) study (Table 39). Fawns older than 7 months often had larger home ranges than those of adults—compare Tables 34, 34 and 40.

The large variation in sizes of pronghorn home ranges results from differing habitat quality, population and group sizes, density, past history of land use, and season, according to Kitchen and O'Gara (1982). Thus, sizes of pronghorn home and seasonal ranges and distances moved daily or seasonally vary so much with these factors and weather conditions that results from studies in one area seldom have application to another area, or even to another year. Some of the smaller home and seasonal ranges reported may be the result of low numbers of relocations. Also, most instrumented pronghorn were tracked for only 1 to 2 years. Thus, the sizes of lifetime home ranges have not been estimated. They apparently are huge for some pronghorn.

BREEDING SYSTEMS

Pronghorn males are not simply "territorial" or "nonterritorial." Males exhibit different degrees of defense of exclusive areas, and this defense appears related to characteristics of their food source, population density, and sex ratio, perhaps through the mediating influence of testosterone.

Maher (2000:336)

Mating systems often are classified with respect to the focus of defense by males. They can defend females directly or indirectly by defending a resource (Emlen and Oring 1977). The former generally has been labeled a harem-

TABLE 40. Home range size as reported in the literature for pronghorn fawns of various ages

State	Age	Number, by sex[a]	Average home range size, in acres (ha)	Source
Arizona[b]	1–21 days	6	262 (106[c])	Ockenfels et al. (1992)[d]
	6.0 months	6	3,721 (1,510)	
Kansas[e]	7–60 day mean			Rothchild (1993)[f]
1991		5-F	321 (130)	
1991		4-M	432 (175)	
1992		12-F	539 (218)	
1992		9-M	319 (129)	
Nevada[g]	<3 weeks	4-F	571 (231)	McNay (1980)[h]
		8-M	1,337 (541)	
Texas[i]	3 weeks	10	143 (58)	Tucker and Garner (1984)[j]
	12 weeks	10	2,550 (1,032)	
Montana[k]	0–4 months	5-F	704 (285)	Campbell (1970)[l]
		2-M	1,216 (492)	
Colorado[m]	1–4 months	11[n]	805 (1,990)	Firchow (1986)[p]
		11[o]	1,356 (3,350)	
Montana[k]	7–10 months	3-F	2,417 (978)	Bayless (1969)[q]
	7–10 months	5-M	1,580 (639)	
Montana[k]	7–10 months	1-F	2,688 (1,088)	Freeman (1971)[l]
	7–10 months	1-M	960 (389)	
Idaho[g]	7–12 months			Hoskinson and Tester (1980)[r]
Little Lost River valley		4	3,709 (1,501)	
Birch Creek valley		5	26,553(10,746)	
Crooked Creek valley		3	5,636 (2,281)	

[a] F = female, M = male; if not designated, both sexes are included.
[b] Grassland habitat.
[c] Use area tended to double with each additional 21-day period.
[d] Ranges estimated by the minimum convex polygon method imported into software program Home Range, based on a minimum of 36 locations by radiotelemetry.
[e] Tallgrass prairie habitat.
[f] Ranges estimated by the minimum convex polygon method (Mohr 1947) using McPAAL Software (Stuwe and Blohowick 1988), based on a minimum of six locations by radiotelemetry. Visual locations were attempted daily or every other day.
[g] Shrubsteppe habitat.
[h] Ranges estimated by minimum convex polygon method based on a minimum of four sightings of each radio-collared fawn.
[i] Desert grassland habitat.
[j] Ranges estimated by the minimum area method of Mohr (1947) using a compensating polar planimeter, based on daily locations by radiotelemetry.
[k] Sagebrush grassland and agricultural fields.
[l] Ranges estimated by a method similar to that described by Hayne (1949), based on a minimum of five locations.
[m] Grasslands, yucca grasslands and cholla grasslands.
[n] 95% convex polygon.
[o] 95% harmonic mean.
[p] Locations were plotted on maps using the UTM grid system, and 93% of all telemetry locations were confirmed visually.
[q] Home ranges were estimated by drawing a line around the outside points of observations for each animal, and the enclosed area was calculated with the aid of a polar planimeter, based on a minimum of six locations per animal.
[r] Seasonal home ranges were estimated by the minimum area method of Mohr (1947), based on 14 or more locations for each animal.

type breeding system and the latter territoriality. Pronghorn use both systems, but territorial bucks must defend harems on their territories, as well as the territories.

Early and some subsequent literature concerning the breeding behavior of pronghorn described the animals as having a harem-type mating system in which dominant bucks controlled and defended does without regard to specific locations (McLean 1944, Einarsen 1948, Buechner 1950a, 1950b, Mace 1954, Gregg 1955, Folker 1956, Prenzlow et al.1968, Deblinger and Ellis 1976, Deblinger and Alldredge 1989, Maher 1991, 1994). Other studies indicated that pronghorn were territorial—defending specific areas rather than bands of females (Cole 1956, Cole and Wilkins 1958, Bromley 1969, 1977, 1991, Pyrah 1970, Gilbert 1973, Kitchen 1974, Kitchen and Bromley 1974, Copeland 1980, McNay 1980, Yates 1986).

Maher (1994) observed that pronghorn populations were qualitatively categorized as nonterritorial or territorial. This could lead to the assumption that all nonterritorial populations are the same. However, Maher (1991, 1994) used operational definitions and quantitative descriptions to detect differences between two nonterritorial populations that she may not have noticed using qualitative descriptions of their social organization. Unfortunately, quantified data were not available for comparison with other populations. Maher (1994) observed that such data are required if more is to be learned concerning the effects of ecology and demography on social organization.

At the 1971 Conference on Ungulate Behavior and Its Relation to Management (Geist and Walther 1974), in Calgary, Alberta, Buechner pointed out that the pronghorn he observed during the late 1940s were, in fact, territorial, and not harem breeders as he had earlier reported (Buechner 1950a, 1950b). Einarsen (1948) studied pronghorn in Oregon and reported that they had a harem-type breeding system. D. Kitchen (personal communication:1978) observed pronghorn in the same areas and found that they were territorial at that time.

Ingold (1969) described a social organization involving pronghorn harems and territories in Wyoming, and he suggested that only the most dominant bucks held harems on territories, whereas less dominant males held harems without regard to area. Kitchen and Griep (1978) reported that another Wyoming population displayed a system in which breeding was restricted to specific areas, or "dominions," that were not defended by large males. Females were herded within dominions, but harems were not formed. The does frequently moved between breeding areas during the rut. Large males even moved between dominions with each other, but dominance between bucks shifted as they moved from one area to another.

P. Griep (personal communication:1978) studied two hunted populations of Wyoming pronghorn, one on heavily grazed rangelands. Bucks displayed little aggression in either population during summer. When the rut began, more energy was spent in herding does than in expressing dominance or territorial behavior. Bucks on the heavily grazed rangeland apparently did not defend a specific land area. Griep believed that water availability, poor vegetation, hunting pressure and, perhaps, fences played some part in causing modifications of behavior. It stands to reason that a buck on depleted rangeland cannot defend a territory large enough to support himself and a doe/fawn band throughout the summer.

Deblinger and Ellis (1976) described the behavior of a small group of captive pronghorn in a natural shortgrass

These pronghorn, photographed near Ejida Juan Alvarez, Sonora, Mexico, occupy hot desert country where a buck cannot defend a territory large enough to support himself and a harem. Photo courtesy of the Arizona Game and Fish Department, Phoenix.

prairie environment in Colorado. Their observations of spatial organization closely paralleled those made about 35 miles (56 km) away by Prenzlow et al. (1968) of pronghorn that displayed a harem-breeding system. Deblinger and Ellis (unpublished data) conducted an intensive, quantitative study of specific behaviors of the formerly tame animals. These behaviors were consistent with those reported for territorial bucks in South Dakota and Montana by Bromley (1969, 1977) and Kitchen (1974), respectively. Deblinger and Ellis described one display (the stiff-legged walk) that Bromley and Kitchen did not describe. Possibly, this was a question of interpretation rather than of behavioral variation, because the gait is used frequently and might or might not be part of a display. Kitchen described the parallel walk—a mutual display of territorial bucks that Deblinger and Ellis did not observe. Deblinger and Ellis described an approach-circle-chase sequence of male/male interactions that may be an analogue of the parallel walk.

Kitchen (1974) reported spatially oriented scent marking, but Deblinger and Ellis (unpublished data) observed random orientation in scent marking. Male/male aggression was initiated during early spring among the territorial animals studied by Kitchen, whereas Deblinger and Ellis observed those interactions to commence much later, in September, among the pronghorn that displayed a harem-breeding system in Colorado. Male/male interactions were frequent and intense among the territorial animals, but male/female interactions were most frequent among the animals studied by Deblinger and Ellis (unpublished data).

These comparisons suggest that differences in pronghorn social organization are not caused by major differences in specific behaviors or displays; these seem to be constant in pronghorn social systems. Rather, social organization is influenced by differences in the density and spatial dispersion of individuals and the quality of their habitat. In other words, the same behavioral repertoire might result, under different environmental conditions, in different social systems. When these results are interpreted in terms of possible genetic and environmental contributions to social organization, they suggest two influences: a strong genetic contribution to specific behaviors and displays, and a major environmental contribution to spatial dispersion, frequency, intensity and timing of individual interactions.

Intraspecific social variation is an important sociobiological concept, and use of different types of mating systems under different ecological circumstances has been reported for three species of African antelope (Deblinger and Alldredge 1989). Such variation indicates social flexibility and can provide insights into relationships between social systems and ecological barriers.

Two nonterritorial populations studied by Maher (1991, 1994) occupied habitats considered to be of low productivity—one on the Sheldon National Wildlife Refuge, Nevada, the other in central California (Carrizo Plains). When food resources are scarce, widely distributed or of low quality, the cost of maintaining a territory should prevent territoriality (see Maher 1994). Comparison of bucks in the two populations indicated similar social patterns, but behavior of Sheldon bucks was closer to the territorial end of a continuum between territoriality and undefended home ranges than was that of Carrizo bucks. Differences could be based on population densities. The Carrizo Plains population was small, widely dispersed and recently reintroduced.

Variability of pronghorn breeding systems obviously has an ecological basis. According to Kitchen and O'Gara (1982), when forage quality, especially succulence, varies between areas and the best resources are clumped, pronghorn tend to be territorial, and males on the best territories do most of the breeding. As resource distribution becomes more uniform and of relatively poor quality, the system shifts toward dominions. The breeding system shifts to harem formation when unusual resource conditions occur, when population levels are low or when sex ratios are unusually skewed (ratios of 1 male to 10 or more females).

McNay (1980) observed territorial bucks on the Sheldon National Wildlife Refuge of northwestern Nevada during the summer of 1978. Shortly before the rut, shallow lakes (playas) dried up, and the does congregated on succulent vegetation growing there. Breeding seemed to take place on the playas amid considerable confusion, as bucks tried to defend and hold harems. However, J. Meeker (personal communication:1983) studied the behavior in more detail and discovered that little breeding actually took place on the playas, and the bucks did not attempt to hold harems. Instead, the territorial bucks herded groups of does back to territories, sometimes breeding them along the way. Maher (1991, 1994) found bucks on untended home ranges on the same refuge.

During a drought in southwestern Montana, I saw large pronghorn groups, containing both sexes, in irrigated alfalfa fields at the time of the rut. The large bucks showed little aggression toward each other and often fed or bedded together, although some bucks were associated with doe bands and, by means of low-intensity displays, seemed to exclude other bucks from those bands. These were casual observations, so I could not be sure whether the numerous large males were territorial bucks from the surrounding dry country or large bachelors. However, I could not find territorial bucks in the dry country where they were evident during years of normal rainfall. I believe the territorial bucks moved into the succulent fields because doe bands could not be held on dry territories, and the dominant, formerly territorial bucks, were holding harems without regard to area.

Deblinger and Ellis (unpublished data) conducted systematic observations in Shirley Basin, about 25 miles (40 km) northeast of Medicine Bow, Wyoming, and reported the simultaneous occurrence of two distinct mating systems within a single population of pronghorn. No evidence of territoriality was noted. Groups with a single male and multiple females displayed typical harem organization. These groups moved freely across the prairie, often to be replaced shortly by a similar group. Groups with multiple males and females (not previously reported in the literature) showed signs of male hierarchical organization. Subdued aggressive interactions were observed among the bucks. Dominance appeared to be related to size and age.

Deblinger and Ellis (unpublished data) were impressed by the low frequency and intensity of aggressive male/male interactions in this population, especially in the bands with multiple males and females. Up to six adult males frequently associated with and simultaneously courted females (although not the same one) in close proximity, within the same cohesive band.

Why should two types of social structure occur in the same population? To answer this question, Deblinger and Ellis (unpublished data) examined the density, distribution and social organization of the pronghorn related to the habitats. Overall density for the 76-square-mile (250 km^2) study area was 15.6 per square mile (6/km^2). Within 96 census blocks, density varied from 0 to 80 pronghorn per square mile (0–33/km^2). Seven upland prairie blocks without water did not contain pronghorn; 17 blocks transected by flowing streams averaged 31 individuals per square mile (12/km^2). In addition to providing water, the flowing streams influenced vegetation. During autumn, upland prairie vegetation was senescent and generally low in nutritional quality. However, the flowing streams and some receding ponds were bordered by green vegetation. Hence, the surface water bodies provided two attractive resources not generally available within the habitat—water and growing, succulent, nutritious forage. The pronghorn bands with multiple males and females occurred exclusively in nine census blocks containing surface water and high pronghorn densities (average = 38 per square mile [14.5/km^2]). Harems with a single male occurred in 62 of the 96 census blocks, with average pronghorn densities of 19.8 per square mile (7.6/km^2). Deblinger and Ellis hypothesized that pronghorn distribution and density were determined by forage and water distribution and that the social systems were directly influenced by animal density. Bands with multiple males and females occurred only in areas of high density, whereas harems generally occurred in areas of more moderate densities.

Pronghorn on the National Bison Range in Montana maintained a territorial mating system from 1965 through 1978 (Bromley 1969, Kitchen 1974). Preliminary observations in 1981 suggested that the mating system had changed, and data from the ruts of 1982 through 1984 revealed a progressive decay in territoriality. Byers and Kitchen (1988) compared data from the "territorial years," 1969 to 1978, with data from the "decay years," 1982 to 1984. They found that fewer territories were defended during 1982 to 1984 than during 1969 to 1978 because of a small proportion of bucks defending territories, not a smaller number of bucks available. Most territorial bucks abandoned their territories early in ruts during 1982 to 1984 or lost control of females on them after frequent and persistent intrusions by nonterritorial bucks. Many does left territories and mated with nonterritorial bucks. Territorial bucks that maintained control of females did so by limiting the size of their territories.

The mating system change followed mortality during the winter of 1978–79 that eliminated 75% of the males, including all males older than 5 years, and all male fawns. The age structure of the population was skewed toward old animals before the winter of 1978–79 by coyote predation on fawns beginning about 1970. During 1982–1984, the number of males present was not different from the number of males during 1969–1978, but the frequency distribution of male ages was strongly shifted toward younger ages.

The small number of older males during 1982 to 1984 probably resulted in smaller proportions of males initially defending territories and in less effective territorial defenses. When females clustered on the few territories where defenses were at first successful, they attracted large numbers of nonterritorial males. The resulting high rates of raids on these territories, coupled with reduced defense radii by territorial males, allowed females only 12% reclining time during the rut. This increased energy cost, plus an apparently greater risk of injury on weakly defended territories, appeared to prompt many females to seek calmer matings elsewhere. Also, if female pronghorn practiced mate selection based on horn or body size, they might have reduced their efforts to remain on territories during 1982 to 1984. Males from 1969 to 1978 were larger than males from 1982 to 1984, and showed greater variance in horn size (Byers and Kitchen 1988).

Byers and Kitchen (1988) postulated that at least two conditions influence the tendency of males to be territorial. First, males must be at least 3 years old before they attempt to defend a territory. Second, the declining proportion of males defending territories during 1982 to 1984—which coincided with an increasing number of males

3 years old and older— suggested that, based on the frequency of territorial defense in the population, males decide whether or not to attempt such defense.

The foregoing studies and observations indicate that pronghorn display different breeding systems in different areas and at different times in one area. A species must maintain flexibility to occupy diverse habitats. Such flexibility aids in adapting to specific areas, but the species also must be flexible enough to alter systems in a very short time when environmental conditions change rapidly.

Variation in pronghorn spatial organization is well documented, but underlying ecological or physiological reasons for differences are poorly understood. Maher (2000) collected behavioral data from populations in South Dakota (Wind Cave National Park in the southern Black Hills) and Montana (Bar Diamond Ranch north of Billings) to determine whether bucks differed in use of space, response to intruders and behaviors indicative of territorial defense. Maher measured sex ratios and population densities. She examined characteristics of food resources, including forb species diversity, richness, coverage, biomass and nitrogen content, plus how they changed during the growing season. She also collected and analyzed fecal samples to determine whether bucks differed in testosterone concentrations.

Pronghorn bucks at Wind Cave were more territorial than those at Bar Diamond, although bucks at Bar Diamond became more territorial during the second year (Maher 2000). The forb community was more diverse at Wind Cave, contained greater amounts of forbs later in summer and had higher nitrogen content later in summer than the forb community at Bar Diamond. Population density was lower at Wind Cave, although density dropped at Bar Diamond during the second year, and sex ratios were skewed toward bucks at Bar Diamond. Finally, males at Wind Cave tended to have higher testosterone concentrations than did Bar Diamond bucks. With lower population density and higher forb abundance and quality, food resources were more economically defensible at Wind Cave, and bucks were more territorial there than at Bar Diamond. Analyses using these and other pronghorn populations indicated that population densities and sex ratios correlated weakly with spatial organization. Precipitation correlated strongly, suggesting that plant productivity has a powerful role in determining pronghorn territoriality.

According to Maher (2000:335), "Spacing systems are affected as much by the animals' internal state, i.e., physiology, as by external conditions, e.g., population density and food abundance. Physiological condition is reflected in metabolic processes and hormone levels, and hormones have important effects on reproduction and aggression."

TERRITORIAL BEHAVIOR AND COURTSHIP

General wisdom has it that the dominant buck rounds up does more or less against their will, herds them until they come into season, then badgers them into breeding. . . . It's hard to believe that any buck, no matter what his testosterone level, could keep a doe in a harem if she decided she wanted out. There is good reason to believe that she sizes up a buck while the buck sizes her up. The choice of mates is probably mutual.

Madson (2000:31)

National parks and refuges are ideal locations for observing animal behavior, and intensive studies of pronghorn territoriality and courtship have been conducted in such areas. Because many adult bucks are present and rangeland ecological conditions generally are good, those behaviors often are more pronounced than they are among hunted populations or in areas with less available forage.

Territoriality is a behavioral strategy by which a limited number of animals in a population each sequester some limited, defendable resources within an exclusive space—the territory. Although advantageous, territoriality is a risky, time-consuming and energetically expensive undertaking. An animal must restrict some or all of its activities to the territory, patrolling and defending it, as well as repeatedly advertising its presence. Such behavior diverts time and energy from other activities, such as feeding (see Chao 1985).

From their earliest interaction with peers, pronghorn bucks compete for social status. Dominance rank was established by 6 weeks of age and persisted until at least 3 months of age on the National Bison Range, Montana (Bromley 1991). Working on the same range, Byers (1997a) found fawns' social ranks closely correlated with their relative birth dates in a social group. Age differences of only a few days created size asymmetries that resulted in this correlation. Bromley (1991) observed that competition appeared playful among fawns, but deadly serious among mature bucks. A dominance hierarchy existed among territorial bucks in Wind Cave National Park, South Dakota. Bromley believed bucks claimed territories with the greatest abundance of preferred food, as Kitchen (1974) had reported. However, Byers (1997a) maintained that resource-based territoriality did not hold up when data were inspected critically.

Working in parks and refuges, Bromley and Kitchen (1974) found that territorial bucks did almost all of the breeding. Establishment of territories was related to sexual behavior, but bucks were territorial for a much longer period than was required to breed females.

Pronghorn territories in Montana were established in relation to topography, being small in areas with intervening ridges and large in flatter areas (Kitchen 1974).

"The only North American ungulate that exhibits resource-defense polygyny is the pronghorn, although this type of breeding behavior occurs in many African antelopes. . . . In fact only certain populations of pronghorns are territorial, and others are harem breeders" (Hirth 2000:184). Photo by Len Clifford, courtesy of Leonard Rue Enterprises, Blairstown, New Jersey.

Males with small territories interacted more with their neighbors than did bucks with large territories. None of the studied territories overlapped and all were widely spaced, generally with a no-man's land between them. All territories used by does during the rut had open water, and bucks were most successful on territories where the terrain aided in preventing the escape of does. Sizes of rutting territories in Colorado were correlated negatively to shrub cover and positively to grass cover (Min 1997a). Reasons for the difference were not speculated on.

Does were sequestered by territorial bucks deep within their territories in South Dakota, using swales or other areas hidden from neighboring territorial bucks and bachelors (Bromley 1977). The most clearly defined boundaries were in open grasslands and often in contested areas, suggesting that the bucks adjusted their use of space so as to best defend that space against intrusion by other males. Bromley postulated that bucks expended more energy marking and defending open habitat than they did along forest edges and other areas. Bucks that used fences and forests as partial boundaries defended larger territories than did those in grasslands. Territorial bucks sometimes ventured to locate and breed does that did not come within their territories.

Walther (1984) noted that territorial bucks often separate themselves somewhat from their harems during the rut, and stand or rest above them on the slope of a hill in strategic position to keep the females, as well as the surrounding area, under watch. In those positions, territorial bucks also are readily recognizable to potential rivals.

On the National Bison Range, Montana, during 1975 and 1976, does showed differential use of buck territories in Alexander Basin, whereas does on the north side used all territories about equally (Swanger 1977). The territorial buck in Alexander Basin with whom does associated most during the summer did almost all of the observed breeding in that area. On the north side, however, one buck did most of the observed breeding, although does had used north-side territories equally during the summer.

On the National Bison Range during early summers 1988 to 1990, does with surviving fawns tended to remain in temporarily stable bands of other mothers, young and one mature male at a spatially discrete location (Byers et al. 1994). Does that had lost their fawns to predation moved between groups, and does with fawns occasionally switched locations.

Bucks on the National Bison Range, Montana, in Wind Cave National Park, South Dakota, and in Yellowstone

Normally acutely aware of humans and other conspecifics visible as is the doe above, courting bucks tend to reduce their alertness in proximity of a prospective mate. Note the horn punctures on the doe's left hip where she apparently had been hooked by an amorous buck. Photo by Bart W. O'Gara.

National Park are territorial from March to October (Gilbert 1973, Kitchen 1974, Bromley 1977). Bucks on the Red Desert of Wyoming establish territories during July near water sources after most surface water has dried up, and the same bucks are seen on some territories year-to-year (E. Taylor personal communication:1986).

Territories provide territorial bucks and their does with sufficient forage, and the former mark vegetation with subauricular gland secretions. Danvir (2000) noted that doe/fawn groups favored forb-rich areas in Utah. However, bachelor herds did not use those areas, apparently because they avoided dominant bucks' territories.

"Several hypotheses for the function of scent marking by territorial [African] antelope were tested with observational data. Of these, the hypotheses that scent marking is driven by intrasexual competition between neighbouring males, and that marks serve as an honest advertisement of a male's ability to defend his territory from rivals, were supported best. . . . [M]arking rate was unrelated to territory size or the number of females on adjacent territories, but males with many male neighbours marked at higher rates than those with fewer male neighbours" (Brashares and Arcese 1999a:1).

Captain Meriwether Lewis reported concerning pronghorn and silver sagebrush in Montana on April 14, 1805: "They [pronghorn] feed on it, and perfume the hair of their foreheads and necks with it by rubbing against it" (Moulton 1987:35). Captain Lewis apparently saw bucks marking sagebrush with their cheek patches.

Marking (elimination of urine and feces) by territorial bucks is accomplished during a linked sniff–paw–urinate–defecate (SPUD) sequence. As performed by these territorial bucks, the postural changes are extreme and each element of the sequence is performed in exaggerated form—in contrast to the often incomplete performances by bachelor bucks. "Faeces may be ideal substances for scent marking because they have a minimal energetic cost to the signaller. However, marking with faeces is also constrained by the animal's ability to produce faeces" (Brashares and Arcese 1999b:11).

Territorial bucks frequently perform the SPUD sequence on the spot where a female has urinated. Moodie and Byers (1989) maintain that such behavior is part of a concealment strategy. Territorial and harem-holding bucks make obvious attempts to hide females from sight by herding them into small ravines or swales. Moodie and Byers noted instances in which such hidden harems were found by other bucks only after a doe in the group was in

A pronghorn buck rubbing the secretions of his cheek patch on vegetation so other bucks will recognize his defended territory. The same plants sometimes are marked many times. Photo by Peter T. Bromley.

(Top) *A territorial buck with long horns holds does in a swale during the rut. Concealment is part of the male pronghorn's mating strategy.* (Bottom) *A buck, uttering a high-pitched whine, springs to meet and court a doe. His body becomes rigid with legs straight, neck vertical and nose horizontal as he approaches her. Note the erected hair of the mane and around the median gland.* Photos by Peter T. Bromley.

estrus, and suggested that intruding bucks found hidden does by scent.

Urination by pronghorn typically is followed by defecation. Females squat to urinate, then defecate after rising and usually while walking away from the urine spot (Autenrieth and Fichter 1975). Flehmen, sometimes called lip curl, involves retraction of the upper lips, which opens an orifice to the Jacob's organ in the upper palate where urine is tested for estrous hormones. Flehmen was used by all bucks on the National Bison Range in response to odors from doe urine, according to Kitchen (1974).

A buck performing the SPUD sequence precedes urination by pawing and, sometimes, by flehmen. He then moves each front foot ahead to take an extended stance, urinates on the pawed spot, then moves his hind feet ahead, squats and defecates. The feces fall on or close beside the urine spot. Before defecation, each hind foot may be moved forward one step separately or simultaneously in a hopping motion. While feeding, bucks and does sometimes will defecate without urinating.

Defenses of territories by bucks vary, but most encounters consist of some or all of five phases: (1) stare at intruder, (2) vocally advertise presence, (3) approach intruder, (4) interact (displays) with intruder and (5) chase intruder (unless there is withdrawal). Occasionally, fights occur.

All medium-sized animals, regardless of species, are watched closely as they enter a territory; territorial bucks stare for 10 seconds to 25 minutes (or even longer) before reacting to the presence of other animals. If the intruder is a pronghorn, a territorial buck may utter a snort–wheeze and erect his mane, the upper portion of his rump patches and hairs around his median gland. Intruding bucks sometimes retreat after only a simple snort–wheeze; yearlings tend to retreat quicker than do older bucks; fewer bucks

(Left) *Females squat to urinate and then defecate after rising, usually while walking away. In contrast, bucks tend to urinate and defecate on the same spot.* (Right) *Even during the nonbreeding season, a buck will immediately sniff the urine spot a doe has left. When odors of estrogen or related compounds are present, the buck will attempt to mask those odors with his own urine and feces by performing the SPUD on the doe's urine spot.* Photos by Cathy and Don Jenni.

flee during the rut than before or after it. Dominant bachelors sometimes use the snort–wheeze, and territorial bucks use it during courtship and while herding does (Kitchen and Bromley 1974).

Encroaching males that do not leave a territory in response to a snort–wheeze are confronted by territorial bucks, whose speed of approach for such confrontations depends on the depth of intrusion into the territory and age of the intruder. Encroaching bucks well inside territories, and yearlings, elicit rapid approaches by territorial bucks and usually result in running chases as intruders are driven from the area. On the other hand, adult bucks that remain near the edge of a territory are approached cautiously. Territorial bucks often lope or trot to within 40 yards (37 m) of an intruder, then lower their heads, depress their ears and walk to within 15 yards (14 m). At this point, the walk becomes a slow, stiff, deliberate gait, ending in a broadside threat similar to the threat display noted by Prenzlow et al. (1968). If both bucks display, they usually are parallel or nearly so, and parallel displays may lead to parallel walking in a head-low posture or sometimes to a closer approach. By this time, intruding bucks usually have given ground. During some slow approaches, territorial bucks thrash vegetation with their horns in redirected aggression, mark with their cheek patches or perform the SPUD sequence. They often grind their teeth during a slow approach and occasionally stop to feed or snort–wheeze. While this is going on, territorial neighbors and sometimes other intruders mark, thrash or feed as a territorial buck approaches. When presenting a horn threat to an opponent, pronghorn bucks do not tuck in their chins; their prongs and entire horns point more forward than do the horns of most bovids.

When a territorial buck and an invader are within 15 yards (14 m) of each other, both usually perform the same behavioral acts. Consequently, many acts are used and reused as the two bucks feed–thrash–mark–walk–threaten. To intensify an encounter, a territorial buck moves closer to the intruder. In most interactions, the intruder no longer responds by this time to the territorial buck's initiative, and simply watches. Such inaction usually leads to a strong threat by the territorial buck, which generally results in the intruder being herded away at a walk or chased away at a run.

Kitchen and Bromley (1974) observed 15 fights, of which 14 occurred during the rut. All were prefaced by a slow approach with heads low, ears back and manes and rump patches compressed. Between 0.5 and 1.5 yards (0.5–1.5 m), bucks stood facing and staring at each other for as long as 6 seconds. In nine fights, the bucks clashed suddenly, whereas in five others, the bucks slowly engaged their horns. When one yearling presented his cheek patch, his yearling opponent lunged and delivered a serious shoulder injury. Fights consisted of a series of thrusts and counterthrusts resembling the actions of humans fencing with swords. Because prongs were not always large enough to engage an opponent's prongs, some fights were horn-to-horn and some head-to-head. Bucks tried to push their adversaries off balance by twisting their necks, and each tried to gain an uphill advantage during

Not only does a territorial buck execute the SPUD (sniff, paw, urinate, defecate) behavioral routine in response to does and intruding bucks, but he sometimes performs it when a human enters his territory. (Top left) *This buck has just sniffed the spot where a doe urinated and is doing the flehman—a behavior that involves drawing back the lips, raising the head slowly and swinging it laterally while licking the nose.* (Top right) *The same buck paws vigorously with each front foot before completing the SPUD sequence behavior. Urination* (bottom left) *and defecation* (bottom right) *on the pawed spot follow in sequence to complete SPUD sequence behavior.* Photos by Peter T. Bromley.

battle. Although the average length of the 15 fights was only 2 minutes, five animals were seriously injured. Kitchen and Bromley (1974) saw 11 injured bucks, probably victims of fights, and found two large dead bachelor bucks with deep puncture wounds in their lungs and hearts.

Byers (1997a) observed 82 fights between pronghorn bucks; 10 (12%) resulted in the death of one of the contestants. Fights lasted a mean of 95 seconds with a range of 3 to 720 seconds. With rare exceptions, fights occurred in the presence of a doe that was ready to mate, indicating that bucks do not engage in this potentially lethal activity without an obvious, immediate reward. Bucks may die as a direct, nearly immediate, result of wounds, or injuries may result in poor condition or reduced mobility—making the animal susceptible to coyotes.

While conducting a 3-year study in South Dakota, Bromley (1977) observed traditional use of territories by bucks. Six bucks held the same general territories in each

The SPUD mark of a territorial buck at the side of a road that was one boundary of his territory. Although such markings are not always on specific boundaries, the same scrapes sometimes are marked many times. Photo by Bart W. O'Gara.

of the 3 years. One territorial buck on Bromley's study area was killed by a car and, about 1 week later, the dominant bachelor buck from a nearby band was observed performing space-claiming behavior in the western half of the vacated territory. Dominant bachelors did not necessarily wait for territorial bucks to die before attempting to claim a territory. Bromley observed what he thought probably was a more typical territorial acquisition when a 3.5-year-old buck left a bachelor herd and localized his activities near the edge of a prairie dog town. He performed space-claiming behavior, but was clearly subordinate to his two neighbors, both firmly established territorial bucks. The 3.5-year-old persisted in using part of one territorial buck's territory whenever he could do so with impunity. The next year, the territorial aspirant was more

(Top) *When approaching an intruder, territorial bucks depress their ears, proceed in a slow, stiff, deliberate gait, often grind their teeth, and occasionally stop to feed or snort–wheeze. When a territorial buck and an invader are within 15 yards (14 m) of each other, the intruder simply watches. Such inaction usually leads to a strong threat and the intruder being chased away.* (Bottom) *A territorial buck may chase another buck for a long distance after a territorial dispute. If a harem is on the territory, bachelors will drive away as many does as possible from the territory before the territorial buck returns.* Photos by Joe Van Wormer; photo courtesy of the Denver Museum of Natural History.

(Top) *A territorial buck, on left with ears back, forcefully evicts bachelor males from his territory during rutting season. Territorial bucks occasionally utter low guttural roars before or during a chase, especially preceding attempts to gore fleeing males. Chases range from a few yards to 3 miles (5 km), and when the animal being chased also is a territorial buck, dominance usually is reversed when the intruder's territory is entered (Kitchen and Bromley 1974).* (Bottom) *Fighting by bucks during the rut occurs when threats fail to scare an intruder from a territory. These fights are serious and sometimes fatal. Note the horn tip (arrow) hooked through the near buck's neck.* Photos by Peter T. Bromley.

confident and interacted aggressively with both of his neighbors until a new territory was established. Bromley summarized that bucks become territorial by acquiring vacated territories, gradually laying claim to the edge of existing territories or pioneering new ones in undefended areas.

Herding of does is a prominent aspect of a territorial buck's behavior. Bucks herded does throughout the year on the National Bison Range, and such herding peaked during the rut with a secondary peak before fawning in April (Kitchen 1974). Although territorial bucks herded a great deal during the rut, does frequently moved from territory to territory within their home range.

As the rut approaches, a territorial buck redoubles his effort to hold does on his territory. Harassment of does by territorial and bachelor bucks increases as bachelor bucks intensify their attempts to associate with females. Territorial bucks try to keep bachelors away from females but often leave doe groups on their territories unattended to pursue single females. The unattended doe groups then are broken up by bachelor bucks that invade the territories (Cole and Wilkins 1958).

A sexually aroused buck announces his intentions to the doe by emitting a high-pitched whine. If the doe is receptive, she raises her tail and the buck approaches her rump with short prancing steps. At about 5 yards (4.6 m), the buck starts waving his head laterally, while flicking his tongue and making a low sucking sound and chewing motion. No other species shows such striking and pronounced head waving during courtship, according to Walther (1984). If the doe is in estrus, she stands with her hind legs spread slightly, occasionally looks at the approaching male and then allows him to mount. Copulation is brief—one forceful pelvic thrust with little if any weight placed on the doe.

According to Bromley and Kitchen (1974), does usually lose interest in bucks after being bred, and Pojar and Miller (1984) found no evidence that copulation occurred more than once during an estrous period. In both studies, the animals were tame or nearly so. Observations were from short distances and presumably accurate. The occasional reports of wild pronghorn copulating repeatedly may be from distant observations of bucks attempting to mount near-estrous does. However, during four intensively

(Top) Bachelor bucks are quick to steal a doe if the territorial master leaves his harem to chase another bachelor or a straying doe. (Bottom) If a bachelor successfully cuts a doe from a harem, he will chase it for a considerable distance and try to hold it in a draw or other location not easily seen by the territorial buck. Photos by Joe Van Wormer; photo courtesy of the Denver Museum of Natural History.

watched ruts on the National Bison Range, Byers et al. (1994) observed one instance of multiple copulations by a doe. In this case, a 2-year-old buck copulated by taking advantage of a tactical blunder by an older buck. After a harem's defense was breached, the older buck seemed to be successfully courting an estrous doe, but left her temporarily to herd a nonestrous doe. The 2-year-old male slipped behind the estrous doe and managed to copulate. The doe subsequently refused courtship from the older buck, but 24 hours later, accepted three copulations during 1 hour from a 3-year-old buck. Byers (1997a) indicated that does ordinarily mate once and rapidly turn off the olfactory signals that attract bucks.

Folker (1956) noted prerutting activity during August in Owyhee County, Idaho. Bachelor bucks engaged in pushing and butting duels and frequent chases. Yearlings were especially active, engaging in sham battles, pawing the ground and sparring with sagebrush. One buck gave vent to his feelings by jumping into the air, twisting in a half circle and kicking his heels into the air.

On the National Bison Range, mate choice by pronghorn does is clearly observable (Byers et al. 1994). Males cannot force copulation, only temporarily block female movement, and do not monopolize resources critical to does. Does apparently act independently, and do not copy the mating choice of other does. Females there practice three distinct strategies of mate choice.

Sampling does visit several harem-holding bucks, remain with each a short time (mean 4.5 days) and switch between bucks at an increased rate as estrus approaches. Switches often appear to be energetically expensive, as the buck attempts to hold does in his harem. The actual mating visit is brief, about 1.5 days (Byers et al. 1994). Sampling does always leave bucks that fail to defend a zone of tranquility around their harems; however, 71% of exits from males are for no apparent reason. Most sampling does return to and mate with a buck they visited within the week before estrus.

Inciting does behave as do samplers until estrus, then they run away from the harem male (Byers et al. 1994). By so doing, these does incite vigorous chasing, aggressive displays and fighting. Inciting females watch competition closely and then copulate with the male that emerges from the competitive swarm as the individual able to defend access to her; this may or may not be the harem buck she had just left. Choice by the inciting doe is clear; she vigorously refuses courtship by other bucks until the controlling buck returns from a prolonged chase or fight.

Finally, a small proportion of does practice a "quiet" strategy each year (Byers et al. 1994). More than a month before estrus, they move to an isolated, peripheral location occupied by a single buck and remain with him throughout estrus. This strategy may be used when does have low energy reserves. The proportion of does that used the quiet strategy was greater in 1989 than in 1988 or 1990. Apparently, this was a consequence of the very dry

(Top) A pronghorn buck courting a doe approaches with short, prancing steps while making low sucking sounds and waving his head laterally, displaying his cheek patches and wafting scent from glands that are located under the patches. Hairs of the buck's median gland, in the middle of his back above the hips, are flared, adding another scent. (Center) *A doe in estrus raises her tail and occasionally looks at a buck, approaching with short prancing steps. This is practically the only time a doe looks directly at a buck. A stare is an act of aggression in pronghorn society.* (Bottom) *Copulation is essentially one forceful pelvic thrust, with little if any weight placed on the doe (J. Meeker, personal communication:1982). In rare instances, a mistimed or misaimed penetration thrust can cause the male to flip onto his back.* Photos by Peter T. Bromley.

summer of 1988, followed by the unusually snowy winter of 1988–89. Fawns were lighter and born later during the spring of 1989 than during 1988 or 1990.

Doe strategy was not associated with dominance rank, summer lactation or age. Min (1997a) postulated that, if pronghorn does choose their mates, variations in morphology or behaviors of the bucks—or their territories—must be evident to the does. However, mating success of bucks 3 to 8 years of age on the National Bison Range was not correlated with body size, horn size, age, rate of display or with activity budgets of harems they controlled (Byers et al. 1994). Buck mating success was strongly correlated with each day's harem size during the rut. Because harem defense demands sustained, often strenuous activity by bucks, the behavior of does is a direct assay of buck vigor and, therefore, a seemingly appropriate means for selecting a mate with good genes.

In Colorado, Min (1997b) found that does increased their home ranges nearly fourfold from summer to rut and visited 88% of the territories in the study area. Several buck traits were related to doe congregation size, but the importance of the traits varied across years. However, buck courtship, cheek patch size and scent marking behaviors were consistently related to doe congregation patterns across years. But no correlation was found between former territory owners' and newcomers' attraction of does, indicating that does were primarily assessing bucks and not their territories.

Dominance rank and aggression frequency are not synonymous or necessarily related, and they may independently influence female reproductive success (Fairbanks 1992). Dominance rank represents an individual's performance, in terms of the likelihood of winning encounters with any other female, in relation to the performance of all other individuals. Aggression frequency is the rate at which an individual is involved in aggressive interactions. Unlike other female ungulates, dominance rank in pronghorn is not related to age. Because pronghorn generally give birth to twins, a greater potential exists for variability among does in number of fawns successfully reared annually than in species in which females bear a single offspring each year. Thus, differences in reproductive success among individual pronghorn may be detectable even in short-term studies.

The offspring of 11 pronghorn does were caught and monitored during 1989 and 1990 in Colorado; thus, the data set for analysis consisted of 22 fawning events with each doe contributing equally to the data set (Fairbanks 1992). Ten does were 3 years old or older; the other was a 2-year-old at the time of parturition, but she bred as early as older does. Of the 11 does, 4 were dominant (in the upper half of their respective dominance hierarchies) during both years, 4 were subordinate each year, 2 were subordinate in 1989 but dominant in 1990 and 1 was dominant in 1989 but subordinate in 1990. Thus, 11 fawning events involved dominant does and 11 involved subordinate does.

The median parturition date for dominant does was 10 days past the birth date of the first fawn; for subordinate does, it was 16 days past the first fawn's birth date (Fairbanks 1992). The difference is not statistically significant, but the 6-day difference may be biologically significant because the fawning season lasted only 17 days in 1989 and 21 days in 1990.

All dominant does were observed nursing and grooming twin fawns within 4 days of parturition; only 6 of the 11 (54%) subordinate does were observed with twins (Fairbanks 1992). Dominant does successfully weaned twins in 4 of 11 (36%) fawning events; subordinates weaned twins in only 1 (9%) of 11 fawning events. Dominance and aggression affected reproductive parameters differently. Dominance affected prenatal investment in offspring; dominant does tended to give birth earlier and raised more fawns than did subordinate does. Aggression affected postnatal survival of fawns; does that received more aggression weaned fewer fawns.

Does not in estrus move away from courting bucks. A retreating female lowers her head and twists her neck, often flicking her tail and slightly lowering her whole body. These acts are in sharp contrast to the raised head and stationary stance of a receptive female in estrus. Bucks sometimes chase unreceptive does through a series of loops and curves while frequently uttering deep guttural roars and hooking at them. Bucks occasionally knock does down during such chases (Bromley and Kitchen 1974).

Even does in estrus run from groups of courting bucks, which usually are bachelors. Chases by bachelors often are long and attract other bachelors. When a bachelor chases a doe into a territorial buck's territory, the latter invariably challenges the former, giving the doe sanctuary (Bromley and Kitchen 1974). This behavior undoubtedly contributes to the high breeding success of territorial bucks.

Bachelors molest females throughout the summer and become very aggressive as the rut approaches. During September in Yellowstone National Park, I saw an estrous doe harassed by seven bachelors. The hair was ripped from at least a third of her shoulders and sides, and blood trickled from several of the bare patches. As a buck mounted her, others tried to dislodge him and, as often as not, the doe received the horn thrusts. When the doe attempted to flee, the bachelors pursued her, hooking at her with their horns. Twice, I saw the doe knocked down; once she did not get up immediately, and a buck tried to mount her on the ground while others horned her and struck her with their front hooves.

Does not in estrus move away from courting bucks. A retreating doe often lays her ears back and turns her head just enough to see the buck. Photo by Bart W. O'Gara.

The pronghorn's territorial system appears to be beneficial to the species in at least three ways. First, it ensures that the largest, most aggressive bucks do most of the breeding (Bromley and Kitchen 1974). Second, the territory provides a haven where does can escape the overzealous courtship of bachelor bucks during parturition, lactation and the rut (Cole and Wilkins 1958). And finally, territories defended by dominant bucks keep bachelors from competing with pregnant or lactating does on the best rangelands (Gilbert 1973). These benefits seem real, but the harem-breeding system may be as effective in restricting breeding to the most dominant bucks. Harem breeders also should protect does from harassment by bachelors during the rut. Thus, territoriality probably is most beneficial during spring and summer.

DOE/FAWN INTERACTIONS

Autenrieth (1976) reported that pronghorn on his study area in Idaho did not fawn in the same place each year and suggested that variations in snow conditions influenced fawning site selection. See Chapter 13 for characteristics of fawning areas.

From several hours to a full day before giving birth, a pronghorn doe selects a birth site. Those in Owyhee County, Idaho, sought rocky, sagebrush-covered areas (Folker 1956). Topography there often took the form of shallow, bowl-like depressions that provided some concealment but allowed does good visibility. Usually the band moves on, leaving the doe alone, but sometimes pronghorn of all ages and both sexes remain in the area.

Autenrieth and Fichter (1975) made a number of significant observations regarding birthings and doe/fawn interactions. Siblings were born from 9 to 31 minutes apart; the does usually began to lick each fawn immediately after delivery, generally eating the amnions as they cleaned up the newborn. The neonates righted themselves and oriented on their mothers within a few minutes and nursed for the first time from 24 to 107 minutes after birth. Both mothers and young frequently got up and laid down, probably establishing mother/infant bonds by touch, smell and sight.

A doe I watched on the National Bison Range ate the amnions as she cleaned her newborn fawns, but the placenta (afterbirth) was expelled after she left her bedded fawns and moved away from the area. Apparently, such behavior is not always the rule, as Autenrieth and Fichter (1975) reported a doe eating afterbirth while still with her newborn fawns. Although all the observed Nevada does that gave birth to twins ate the afterbirth, McNay (1980) witnessed one doe (possibly a yearling) give birth to a single fawn and not eat any of the fetal membranes.

After giving birth, a doe licks the entire fawn, but grooming of the anogenital region soon takes precedence over generalized licking and continues after each suckling bout until the fawn or fawns are 2 to 3 weeks old. During such grooming bouts, fawns assume a distinctive rump-up posture to urinate and defecate, while the mother ingests the urine and feces (Autenrieth and Fichter 1975). Tucker (1979) reported that Texas fawns elicited grooming by walking in front of the does and assuming the rump-up posture. Prenzlow (1964) and Bromley (1967) postulated that anogenital grooming reduced the fawns' odors and decreased the possibility of their being found in their beds by mammalian predators.

The fawn's behavior during the hiding strategy represents a specialized ability. To recline motionless for hours on a cold, wet spring day demands a huge energy expenditure in heat production. Also, to freeze into complete immobility at the approach of a predator demands that a fawn respond in a way opposite of how it will react to the same situation 3 weeks later and for the rest of its life (Byers 1997a).

One or two yearlings, especially females, are the most common companions of adult females during parturition in many areas, and yearlings, marked as fawns, often are seen with their dams. Pyrah (1987) postulated that does and their neonate daughters forge strong social bonds. However, yearling females accompanied parental does in only 3% of observations at 1,359 fawn locations in Alberta (Barrett 1984).

Yearlings in Idaho butted fawns, sometimes in an aggressive manner and sometimes playfully; yearlings also

(Top) *Several hours before giving birth, a doe walks with her tail upright and stands with her hind legs stretched backward more than normal. During this time, she moves nervously about the chosen birth site, lying down and getting up.* (Bottom) *An amnion extruding from the vulva of a pronghorn doe about to give birth. Note the front hoof just appearing at the top of the vulva. The white tip is a soft rubbery bumper that keeps the sharp hooves from puncturing the fetal membranes. These bumpers dry and wear away by the time fawns are 2 to 4 hours old.* Photos by Peter T. Bromley.

(Top) *Does stand up and lay down many times during labor, but fawns usually are born while the doe is recumbent. The first fawn of a pair of twins has just slid free of the vulva in the top picture. Immediately after giving birth, does begin to lick each fawn and eat the amniotic membrane that surrounds the fawn at birth* (center and bottom). Photos by Peter T. Bromley.

Postpartum interaction of a pronghorn doe and her newborn fawns. This is an important period during which they establish bonds by touch, smell and sight. Licking by the doe rapidly dries and fluffs the coat of a newborn fawn. Photo by Joe Van Wormer; courtesy of the Denver Museum of Natural History.

Generalized licking of the entire fawn soon gives way to anogenital grooming after each nursing, a behavior that lasts until the fawns are 2 to 3 weeks old. Such grooming reduces the fawns' odors, reduces the chances they will foul their bed, decreases the possibility of their being scented by mammalian predators and helps the fawn/dam imprint. Photo by Mike Francis.

were aggressive toward and frequently dominant to the older does (Autenrieth and Fichter 1975). Autenrieth and Fichter noted that fawns sometimes exhibited a fear response toward nonmaternal does, and fawns appeared more inclined to withdraw from adult pronghorn that only threatened to butt than from yearling does that made physical contact.

Eight instances of aggressive behavior by yearling pronghorn toward newborn fawns were recorded in Idaho (Bodie 1978, 1979). Sometimes, yearlings butted fawns hard enough to knock them down. Twice, a yearling placed itself between a newborn and its mother, apparently trying to prevent the fawn from reaching its dam. Does either butted adult pronghorn showing interest in their offspring or crowded between the fawns and adults but did not interfere with the abusive yearlings. Those yearlings may have been the previous fawns of the does concerned and were exhibiting sibling rivalry. Bodie believed that the yearlings interfered with nursing and caused imprint confusion. One radio-equipped fawn apparently died from injuries received from butting. Necropsy showed extensive bruises and subcutaneous hemorrhages on most of the back and sides, the liver had abrasions and extravasated blood was present in the abdominal cavity. No external wounds were present to indicate predator involvement.

During another Idaho study, a fawn that had recently nursed apparently was stepped on by an adult pronghorn, rupturing the abomasum and causing shock (Autenrieth 1984).

With their mothers watching, fawns select their own bedding sites from the first day. A doe usually leaves her fawns for the first time 3 to 6 hours after parturition. During the next 3 weeks or so, she returns at intervals varying from about 1 to 6 hours to the area where her young are bedded. Byers and Byers (1983) observed an average intervisit time of 225.5 minutes. Success of such hiding strategy depends, in part, on the mother's ability to minimize information she transmits about her young's hiding place while remaining close enough to distract or drive away predators.

Byers and Byers (1983) predicted that does would (1) maintain minimum distances from their hidden fawns sufficient to cause the expected energy gain for a ground predator systematically searching around the doe to fall below that expected when searching for some other prey, (2) orient the axis of either head or trunk toward the hidden fawn no more frequently than would be expected by chance and (3) schedule behavior so that no activity is more likely than another to occur when a visit to a hidden fawn is imminent. Pronghorn does on the National Bison Range, where coyotes frequently took pronghorn fawns, fulfilled the first and third predictions. However, the does pointed both head and trunk toward their hidden fawns more frequently than in other directions. Byers and Byers concluded that pronghorn does distribute their activities in ways as close to optimal as can be expected.

Wild ungulates are vulnerable to predation early in life. This selective pressure has led to two strategies, following and hiding, for neonatal defense or predator avoidance (Lent 1974). To determine relative costs and benefits of following and hiding, Carl and Robbins (1988) studied neonatal mountain goats and mule deer and their dams. For the first 4 weeks postpartum, the deer fawns (hiders) had a higher growth rate than did the goat kids (followers). However, the lactating deer lost weight during early lactation, whereas lactating goats maintained their weight. Pronghorn fawns are extreme hiders for the first 2 weeks of life, undoubtedly contributing to their extremely rapid

(Left) *Before a fawn seeks a bed, its dam seems especially alert.* (Right) *The doe often still is grooming a fawn as it starts for the hiding place it will choose.* Photos by Joe Van Wormer; courtesy of the Denver Museum of Natural History.

While its mother stands watch, a young fawn finds its bedding site after nursing. Does appear to look for predators and fix on where their fawns bed. Photo by Peter T. Bromley.

Fawns do not move very far on their own during the first 2 weeks of life. However, does may move fawns for 2 miles (3 km) or more when only a couple of days old. Such movements generally occur after some type of disturbance, especially by a predator. While traveling, fawns seem mildly excited as evidenced by the erected rump patch of this one. Photo by Joe Van Wormer; courtesy of the Denver Museum of Natural History.

For about a week, except for short periods with their dams, fawns remain solitary. By 2 weeks of age, pronghorn siblings bed together and interact with each other. Siblings older than 3 weeks seem inseparable. Photo by Jo Meeker.

growth. However, pronghorn does studied in Montana and Yellowstone National Park gained weight during the time their fawns were in hiding (O'Gara 1968).

According to Hirth (2000:179–180), "Species with hider-type young have neonates that are altricial and remain bedded and concealed in dense cover for 2–4 weeks following birth. . . . Species with follower-type young have neonates that are precocial and able to follow their mother soon after birth, such as caribou and bison. Species with this strategy typically are plains dwellers (e.g., savannah, prairie, tundra) and do not have access to cover for their

(Left) *A young fawn may rise from its bed and go to its dam while she still is several yards away, apparently in response to a low call.* (Right) *Fawns apparently recognize their mothers primarily by sight and sounds, but verification may depend on smell as the two make contact. Recognition by the doe depends on sniffing the fawn's hindquarters.* Photos by Cathy and Don Jenni.

young." Pronghorn fawns do not fit into either of these categories. They are precocial and able to follow their dams soon after birth, but are hiders. They are plains animals without access to dense cover, but hide anyway.

During a study in Montana, doe fawns were more active than buck fawns (Byers and Moodie 1990). Across all ages, females stood, moved and interacted more than males, but the latter spent more time eating than did the females. Although fawns usually do not move about a great deal during the first 2 weeks, an instrumented fawn in Idaho was moved by its dam about 2 miles (3 km) a day during the first 3 days of its life (W. Bodie personal communication:1979). This was in an area with considerable disturbance by predators.

In Alberta, siblings exhibited, with age, progressively greater interactions; their feeding and activity bouts usually occurred at the same time, and they bedded closer to each other (Barrett 1984). They first bedded together at 7 days of age and by 14 days such behavior was common. Siblings older than 24 days seemed inseparable. Minimum daily movements of Alberta fawns increased progressively after 1 week of age.

Fawns often rise from their beds and go to the doe when she is still several yards away, the distance increasing as the fawns get older. Six behavioral features characterize this mother-to-fawn reunion period: recognition, nursing, anogenital grooming, play, the relocation move and the move to lie secluded (Autenrieth and Fichter 1975).

Young fawns at ease lie with their heads up and appear alert. When they see movement, they flatten out with their heads on the ground, remain perfectly still, their ears lowered and eyes open. While lying in this position, fawns less than about 3 days old, if approached carefully, can be touched or picked up. A newborn fawn I kept in my backyard flattened out when a dog or person walked through a pasture 50 to 200 yards (46–183 m) away, but when a dog or person came around the corner of the house, within 5 to 25 yards (5–23 m) of the fawn, it remained motionless in its alert pose. On one occasion, Van Wormer (1969) heard a doe snort at a watching coyote, and her fawn dropped where it was nursing. Similar behavior, with or without the threat of a predator, was noted frequently by Autenrieth and Fichter (1975). I have seen

(Left) *By the time they are a month old, fawns begin to look like miniature adults.* (Right) *A doe's role in nursing is largely passive—standing still with back arched, hindquarters lowered and knees slightly spread.* Photos by Joe Van Wormer; courtesy of the Denver Museum of Natural History.

Twin fawns bedded in the alert (foreground fawn) and hiding (background fawn) positions. Usually, siblings are in similar positions when found. Photo by Jim D. Yoakum.

fawns in full flight, either alone or with their dams, suddenly drop out of sight. Fawns that take this evasive action are very difficult to find and, when found, are always in the hiding position.

Fawns' flight distances and tendencies for flight from approaching people increased with age during an Alberta study (Barrett 1984). The transition from bedding to fleeing at the approach of danger occurred at approximately 24 days of age. Average flight distances increased from 16 yards (15 m), at 4 to 7 days of age, to 234 yards (214 m), when fawns were more than 40 days of age.

Fawns apparently recognize their mothers largely by sight, although visual contact may be preceded by vocal communication and verified by smell during the nose-to-nose greeting. For does, identification of their offspring begins visually, but confirmation often, if not always, depends on sniffing the fawns' hindquarters. Her role in nursing being largely passive, the doe stands still with her back arched in a way that lowers her hindquarters and spreads her knees slightly. Nursing fawns do not wag their tails or root the udder. The length of nursing bouts decreases significantly during the 4 months before weaning (Autenrieth and Fichter 1975). Females frequently leave their fawns unattended, leading to the idea that nearby does are "babysitting" (Einarsen 1948).

Does frequently are seen with several fawns, leading to the assumption the does are "babysitting." However, does associated with fawn groups take an interest only in their own young in an alert-alarm, flight situation. Babysitting does sometimes leave fawn groups completely unattended for more than 2 hours (Kitchen 1974). Photo by Mike Francis.

CONCLUSIONS

Pronghorn are dainty feeders, selecting small plants or parts of plants that are especially succulent and nutritious. Timing and length of seasonal movements vary with environmental conditions. Deep snow forces some pronghorn to move up to 200 miles (323 km) from summering areas. Movements probably were much greater before white men erected barriers to movements. Pronghorn are gregarious, and group sizes depend on season and environmental conditions. When northern, territorial pronghorn arrive on summer ranges, the does collect into bands of a dozen or less, young bucks form larger bachelor bands and old bucks claim territories. Interchanges between bands are common. When the best forage is clumped, pronghorn tend to be territorial. As resources become more uniform in distribution and of poorer quality, the system shifts toward dominions. The breeding system shifts to harem formation when population levels are low or sex ratios are unusually skewed. Recent research indicates that habitat quality, which affects the physical condition of pronghorn, ultimately influences testosterone levels of bucks. The higher the testosterone level, the more apt a buck is to be territorial. Territorial bucks defend territories that are used by does and fawns, but other bucks are excluded. Many behaviors, short of fighting, are used in territorial defense, but when fights do occur, deaths are common. Fawns hide for the first 3 weeks of life, and their dams feed them about three times a day. This is a predator avoidance behavior involving both does and fawns.

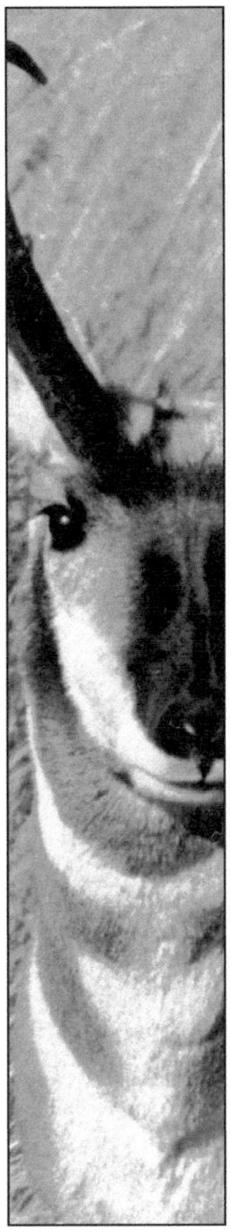

C H A P T E R   S E V E N

Skeletal and Visceral Anatomy

Bart W. O'Gara and Susan Kraft Ball

Considering the body as the complex and wonderful machine that it is, ANATOMY is the study of how that machine is put together. It deals with the structure of the parts, ranging from the molecular components of the tiniest cells to the whole individual, and their relationship to one another and their environment.

Landau (1923:3)

For this chapter, firsthand observations are used to describe bones and viscera. In addition, there is substantial reliance on the descriptive work of others, especially Murie (1870) and Gilbert (1980) for skeletal characteristics, Wenzel (1955) for the digestive apparatus, and Dow and Wright (1962) for teeth.

Terminology is primarily from Evans and Christensen (1979), but terms describing the four-chambered stomach are from Sisson and Grossman (1953). Anatomical terms often are followed by common names or older, frequently used, anatomical names in parentheses. Anatomy is primarily the study of form; function is discussed under physiology (Chapter 8).

Unless otherwise stated, measurements given are those of *Antilocapra americana americana* from the northern portions of its range. Average measurements were taken from the largest sample size available in the literature or from O'Gara's unpublished data. Sizes are variable from area to area, even over short distances, and among individuals from a given area. Numbers of foramina, shapes of bones and organs and even numbers of bones of pronghorn are more variable than those of most other mammalian species.

SKELETON

Pronghorn have long legs and thin, lightweight bones (Figure 29). The skeleton serves as support and protection, while providing levers for muscular action. It functions as a storehouse for minerals and as a site for fat storage and blood cell formation. Bones may be grouped according to shape, structure, function, origin or position.

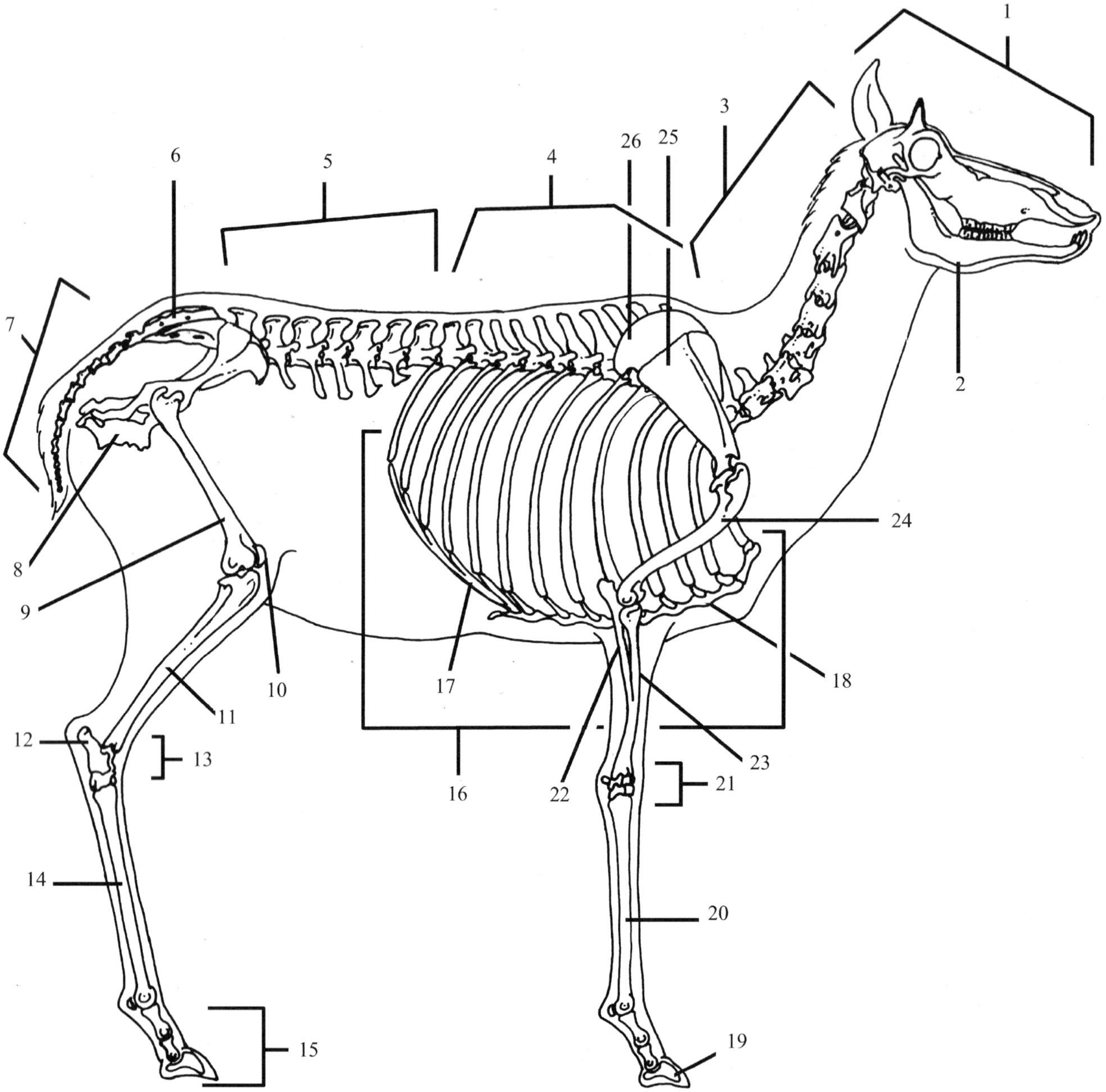

FIGURE 29. *The skeletal structure of a pronghorn doe featuring: 1 = skull; 2 = mandible; 3 = cervical vertebrae; 4 = thoracic vertebrae; 5 = lumbar vertebrae; 6 = sacrum; 7 = caudal vertebrae; 8 = pelvic girdle; 9 = femur; 10 = patella; 11 = tibia; 12 = calcaneus; 13 = tarsus; 14 = cannon; 15 = phalanges; 16 = ribs; 17 = costal cartilages; 18 = sternum; 19 = hoof; 20 = cannon; 21 = carpus; 22 = radius; 23 = ulna; 24 = humerus; 25 = scapula; 26 = scapular cartilage.* Illustration by Susan Kraft Ball.

Anatomists have long grouped bones according to shape, although borderline forms exist. Five general bone divisions are recognized on this basis—flat, irregular, long, short and sesamoid. The flat and irregular bones are characteristic of the skull and vertebral column, whereas long, short and sesamoid bones are found in the limbs.

Flat bones form the limb girdles, where they provide surfaces for muscle attachment. In the head, they surround and protect the sense organs and brain, and provide muscle attachment. Bones of the face are flat, streamlining the head and providing maximum shielding without undue weight. The large, lightweight heads of pronghorn overhang the animals' centers of gravity; a heavy head would be a handicap in locomotion. The flat bones of the cranium consist of outer and inner plates of compact bone, with spongy bone uniting them.

Irregular bones are principally those of the vertebral column, but the term also includes the three parts of the hip bone and bones of the skull not of the flat type. Jutting processes characterize irregular bones. Most of these processes are for muscle and ligamentous attachments, but some are for articulation.

Long bones are characteristic of the limbs. Bones of the thigh (femur) and arm (humerus) are good examples. Typically, a long bone, during its growth, possesses a shaft (diaphysis) and two ends (epiphyses). Each end is separated from the shaft by a plate of growing cartilage during development. At maturity, the epiphyseal cartilage ceases to grow, the epiphysis fuses with the shaft and no distinguishable division exists between epiphysis and diaphysis.

Long bones form levers and possess great tensile strength. They are capable of resisting many times the stress to which they normally are subjected. The ends of most long bones are enlarged and smooth, and enter into the formation of freely movable (synovial) joints. These joints permit the greatest degree of movement and are the ones most commonly involved in dislocations. All synovial joints are characterized by a joint cavity, a joint capsule—including an outer fibrous and an inner synovial membrane—synovial fluid and articular cartilage. A layer of hyaline cartilage provides smooth articulating surfaces on the ends of long bones. The enlargement of each extremity of a long bone serves a dual purpose. It diminishes the risk of dislocation and provides a large bearing surface for articulation. The distal end of the terminal phalanx of each digit is an exception to the stated rule. It is covered by a hoof and is not articular, therefore is neither enlarged nor smooth.

Short bones are confined to the carpal (wrist) and tarsal (ankle) regions. They vary in shape from cuboidal with six surfaces to irregularly compressed rods with only one flat, articular surface. In those bones having many surfaces, at least one surface is nonarticular. This surface provides an area where ligaments may attach and blood vessels may enter and leave the bone.

The term sesamoid is derived from the Greek word for a seed that is small, flat and obovate. Sesamoid bones vary from tiny spheres to the slightly bent, ovoid patella (kneecap). Sesamoid bones are present near freely moving joints. They usually are formed in tendons, but they may be developed in the ligamentous tissue over which tendons pass. They usually possess only one articular surface, which glides on a flat or convex surface of one or more of the long bones of the extremities. Their chief function is to protect tendons at the places where greatest friction exists.

Many bones are united by cartilaginous joints, which sometimes are referred to as synchondroses. Unions of this type may be formed by hyaline cartilage, fibrocartilage or a combination of the two; the unions are subject to change with increasing age. Hyaline cartilage joints, or primary joints, usually are temporary and represent persistent parts of the fetal skeleton or secondary cartilage of growing bones. Fibrocartilaginous joints, or secondary joints, sometimes are referred to as amphiarthroses. The best examples of such joints are those of the pelvic symphysis, mandibular symphysis, sternum and vertebral bodies. The fibrocartilage uniting these bones may have an intervening plate of hyaline cartilage at each end. Occasionally, these joints may ossify, as do hyaline cartilage joints.

Although bones appear to be rigid and not easily influenced by the soft tissues that surround them, soft tissues contour the bones. Smooth indentations in the form of grooves are produced by blood vessels, nerves, tendons and ligaments that lie adjacent to them. Roughened elevations (eminences) or depressions are produced by the attachments of tendons and ligaments.

Much can be learned about the role in life of a specific bone by studying its eminences and depressions. There is a functional, embryological or pathological reason for the existence of almost every irregularity. Most eminences serve for muscular and ligamentous attachments. Grooves and fossae (shallow cavities in bones with large openings) in some instances serve a similar function. Facets are small, articular surfaces that may be flat, concave or convex. Trochleas and condyles usually are large, articular features of bone. The roughened, enlarged parts that lie proximal to the condyles on the humerus and femur are known as epicondyles.

Bone, unlike cartilage, has a nerve and blood supply. Long bones and many flat and irregular bones have a conspicuous nutrient artery and vein passing through the compact substance to serve the marrow within. Such arteries pass through a nutrient foramen. Not all of the blood supplied by the nutrient artery is returned by the nutrient vein or veins; much of it, after traversing the capillary bed, returns through veins that perforate the compact bone adjacent to the articular surfaces at the extremities of these bones.

Skull

The skull (cranium) is the most complex and specialized part of the skeleton. It houses the brain and the sensory organs for hearing, equilibrium, sight, smell and taste, while providing attachment for the teeth, tongue, larynx and a host of muscles. It basically is divided into facial and neural (braincase) portions.

The facial region is specialized to provide a large surface area serving respiratory and olfactory functions and a long mandibular surface for the implantation of teeth. This elongation results in a pointed rostral end (apex), and a wide, deep base that imperceptibly blends with the braincase.

The braincase or cranial cavity is separated from the cavity of the nose by a perforated plate of bone (cribriform

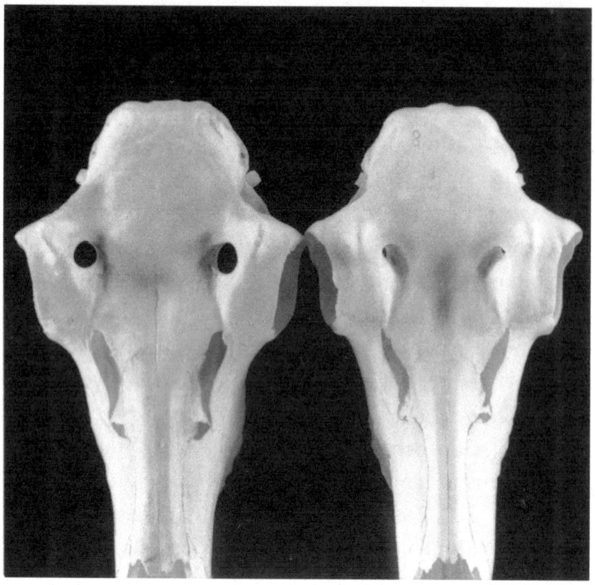

Skulls of adult does show variation in size and shape of foramina. Photo by Bart W. O'Gara.

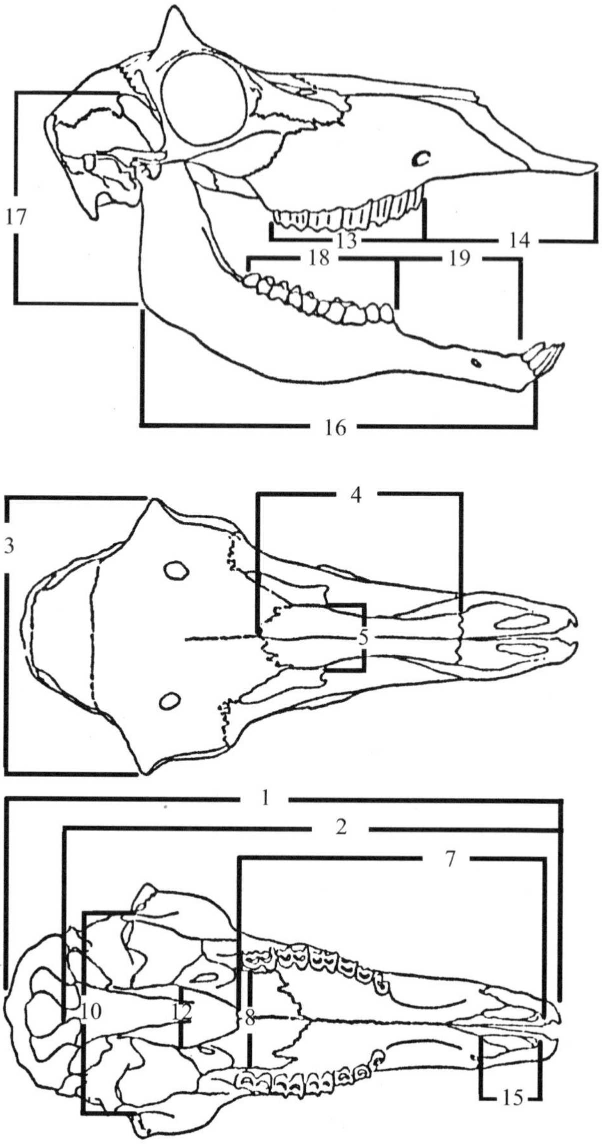

FIGURE 30. *Adult pronghorn cranial and mandibular aspects. Diagram numbers correspond to measurements in Table 41: 1 = total length; 2 = basilar length; 3 = orbital width; 4 = nasal length; 5 = nasal width; 7 = palatal length; 8 = palatal breadth at m3; 10 = zygomatic width; 12 = postnares; 13 = alveolar length; 14 = prealveolar length; 15 = incisive foramen; 16 = mandibular length; 17 = mandibular height; 18 = mandibular tooth row; 19 = mandibular diastema. Not shown are the premaxilla width, rostral height, mastoid width, horn core length, horn core base circumference and horn core spread.* Illustration by Susan Kraft Ball, adapted from Mitchell (1980).

plate). Caudally, the large opening through the occipital region, the foramen magnum, allows for passage of the spinal cord and its associated vessels.

Pronghorn skulls, especially of females, are very light and fragile. Those of eight adult does we collected in Montana weighed an average of only 7.5 ounces (213 g) without the jaws after they were cleaned and dried. The skulls of 14 bucks averaged about twice as heavy, 15.7 ounces (446 g)—still very light for the size of the animals. The skulls of two female and four male yearlings shot during early autumn averaged 6.3 ounces (180 g) and 12.3 ounces (350 g), respectively; those of two females and a male fawn, also taken during early autumn, averaged 3.9 ounces (110 g) and 5.1 ounces (145 g), respectively. Measurements of adult skulls are illustrated in Figure 30 and listed in Table 41. Comparable measurements by Paradiso and Nowak (1971) were similar for *A. a. americana* from Montana and North and South Dakota. Their measurements of *A. a. mexicana*, *A. a. peninsularis* and *A. a. sonoriensis* indicated slightly smaller and narrower skulls.

In profile, the skull is elongated, and with the top and the base forming almost parallel lines, it appears shallow or flat (Figure 31). The anterior edge of the upper jaw is long, with the prealveolar length always exceeding the length of the upper tooth row. If the skull is divided into five equal perpendicular segments, the occipitotemporal area occupies one, the orbitojugal another, and the anterior three consist of lacrimal, nasomaxillary and incisive bones. Therefore, three-fifths of the length of the skull is nose. William Clark, while inspecting the first pronghorn collected (September 14, 1804) on the Lewis and Clark Expedition, described this phenomenon as "his brains on the back of his head" (Moulton 1987:70). The large eye orbits are directed outward but slightly upward and forward, and the horn cores arise from the front bones above the posterior half of the orbits.

TABLE 41. Mean skull measurements in inches (cm) of mature pronghorn from Alberta

	Males		*Females*	
Skull aspect[a]	*Sample size*	*Mean*	*Sample size*	*Mean*
1 Total length	22	11.2 (28.4)	7	10.9 (27.7)
2 Basilar length	22	9.8 (24.8)	7	9.5 (24.2)
3 Orbital width	27	5.5 (14.0)	7	5.0 (12.7)
4 Nasal length	27	4.0 (10.2)	6	3.8 (9.7)
5 Nasal width	27	1.3 (3.2)	7	1.1 (2.9)
6 Premaxilla width	19	1.2 (2.9)	7	1.0 (2.5)
7 Palatal length	22	6.4 (16.3)	7	6.3 (16.1)
8 Palatal breadth at m3	28	2.0 (5.0)	7	1.9 (4.8)
9 Rostral height	26	2.7 (6.8)	7	2.5 (6.3)
10 Zygomatic width	28	4.2 (10.6)	7	3.8 (9.8)
11 Mastoid width	28	2.7 (7.0)	7	2.5 (6.4)
12 Postnares	21	1.1 (2.9)	4	1.1 (2.7)
13 Upper tooth row	28	2.8 (7.2)	5	2.9 (7.3)
14 Prealveolar length	22	3.6 (9.3)	5	3.5 (9.0)
15 Incisive foramen	21	0.9 (2.4)	5	1.0 (2.5)
16 Mandibular length	16	8.3 (21.1)	3	8.3 (21.1)
17 Mandibular height	16	1.8 (4.6)	3	1.9 (4.8)
18 Mandibular tooth row	16	3.1 (7.9)	4	3.0 (7.7)
19 Mandibular diastema	16	3.2 (8.1)	4	3.1 (7.9)

[a] Preceding numbers correspond to numbered parts in Figure 30.

Source: Mitchell 1980.

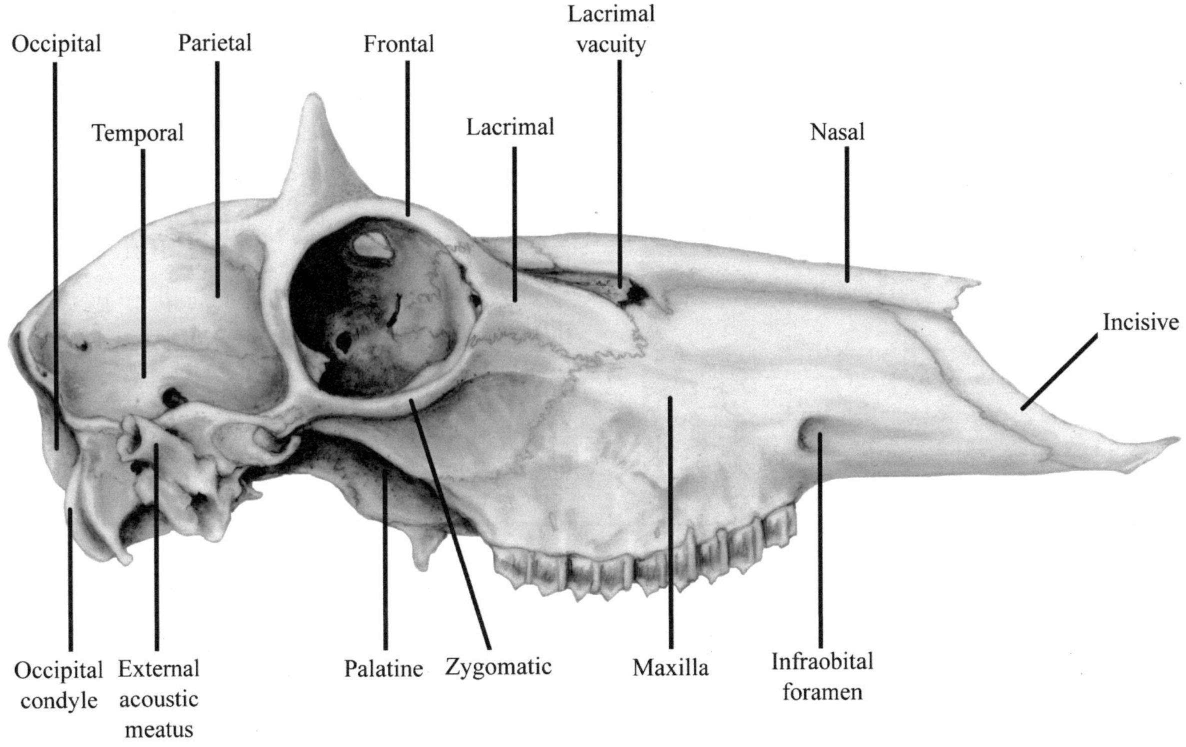

FIGURE 31. *Profile of an adult pronghorn doe skull.* Illustration by Susan Kraft Ball.

As in all mammals, frontal bones form the upper borders of the orbits. Anteriorly, the frontals articulate with the paired nasal bones, which are concave on their internal surfaces. Lateral to the nasals and articulating with the frontals, maxillae, zygomatics (also known as jugal, malar or cheek bones) and ethmoids comprise the lacrimal bones that house the lacrimal (tear) ducts. Each lacrimal duct usually opens through two orifices—an upper one inside

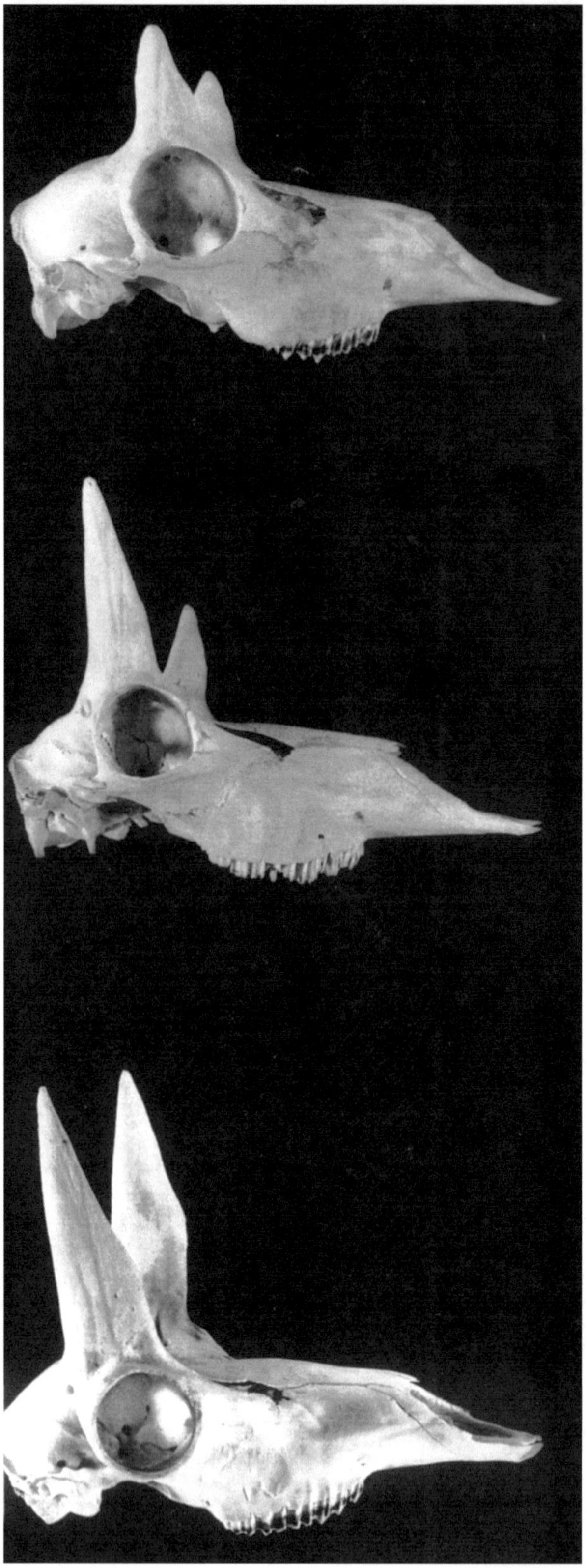

Skulls (not to scale) of a 5-month-old fawn (top)*, a 17-month-old yearling* (center) *and a 29-month-old adult* (bottom)*, indicating the general growth pattern for pronghorn bucks in Montana. Ages are approximate, based on tooth replacement and time of year of collection.* Photo by Bart W O'Gara.

the rim of the orbit and a lower one near or on the margin of the rim. Two of 36 skulls examined had only one orifice, and six more had only a threadlike spicule of bone separating the two orifices. In others, the orifices were separated by as much as 0.3 inches (7 mm). About 20% of the pronghorn skulls in the American Museum of Natural History have a single orifice (Scott and Janis 1987). The segment of the lacrimal bone contributing to the orbital ring forms about one-eighth of the circle; it tapers forward almost to a point where it reaches the maxilla. The surface of the bone is smooth and not impressed with a crumen (tearpit) as it is in most cervids. Its superior border is prevented from touching the nasal by the intervention of a large lacrimal vacuity. This vacuity is floored by the thin, smooth, dorsal nasal concha posteriorly and the fenestrated ventral nasal concha anteriorly. The vacuities are continuous with the frontal sinuses. The vacuity has an elongated sinuous shape, about 1.8 inches (46 mm) in length and 0.5 inches (12 mm) in breadth at the widest point. Anteriorly, it terminates in a truncated (squared-off) manner; the posterior end diverges from its fellow of the opposite side, and the frontolacrimal suture passes outward near its posterior, pointed end.

Inferior to the nasals are the paired maxillae that form the bulk of the face. They contain the upper teeth and exhibit large facial or infraorbital foramina (holes), through which the facial nerves pass. Incisive bones form the anterior portion of the upper jaw. The ascending limb of each articulates with and between the nasal and maxilla in a narrow wedge. The unusual forward extension of the maxilla presents a level cheek surface, giving a rather cylindrical contour to the muzzle.

The lateral margin of the orbit is formed by the zygomatic bone, which lies between and articulates with the maxilla anteriorly and the temporal posteriorly. The posteriorly projecting temporal process of the zygomatic articulates with the anteriorly projecting zygomatic process of the temporal. Together, they form the zygomatic arch, constituting a cheek buttress. A wide foramen, situated at the root and upper surface of the zygomatic arch, is continuous with the mastoidal cells.

The width at the posterior orbits, seen when looking down at the skull, is about half the length, giving the skull a somewhat triangular shape (Figure 32). However, skulls of pronghorn in the southern portions of their range, especially those of the presumptive subspecies *A. a. sonoriensis*, are narrower (Paradiso and Nowak 1971). The width between bases of the horn cores is 2.7 to 3.0 inches (70–76 mm), and the cores of adult males usually slant out far beyond the orbital periphery. At their inner bases, well forward, are elliptical or round supraorbital foramina that vary in size and enter through the roofs of the orbits. The paired parietal bones appear to be a single, bilaterally

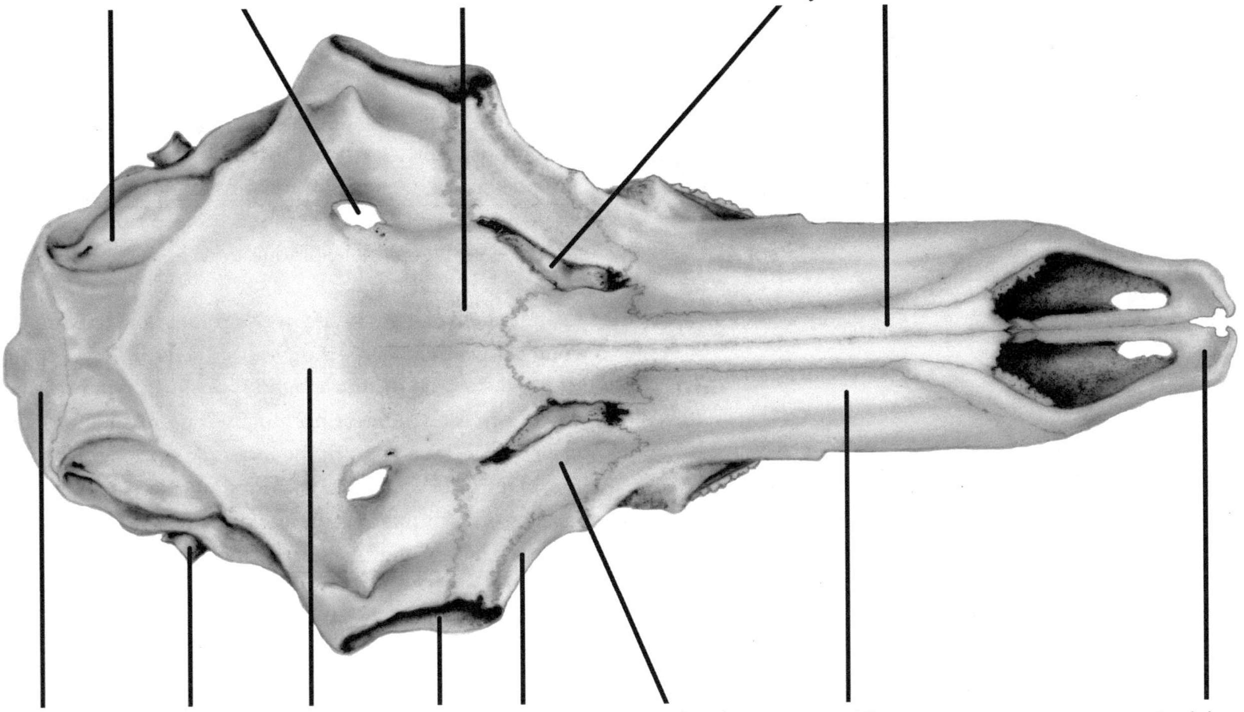

FIGURE 32. *Dorsal view of an adult pronghorn doe skull.* Illustration by Susan Kraft Ball.

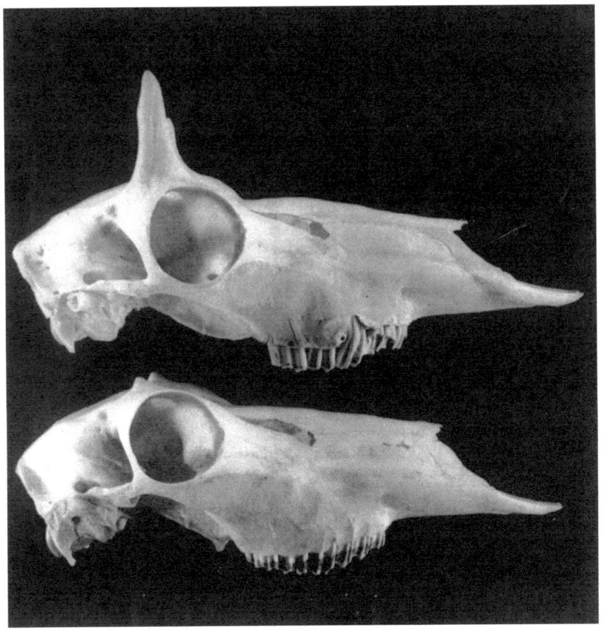

Skulls of adult Montana pronghorn does. The upper skull has larger than average horn cores. The lower has the most common size (approximately) of horn cores. Such cores may or may not support horn sheaths. Photo by Bart W. O'Gara.

symmetrical bone constituting a low arch. The front of the braincase is broad and flat, rising very little toward the coronal suture, and the anterior nasal openings are large.

The long nasal bones meet each other along the midsagittal plane at the straight, internasal suture. They insert into the frontals by a crescent-shaped frontonasal suture, and their other margins are bounded by the large lacrimal vacuities, maxillae and a small portion of the incisive bones. The nasal bones are widest posteriorly, with some lateral constriction before their tips gently broaden. In front, each is half-moon shaped. Together, they present a biconcave edge.

Viewing the bottom of the skull, the mandibular (glenoid) fossae that accommodate the condyloid (articular) processes of the lower jaw, forming the joints of the jaw, is inferior to the zygomatic processes of the temporal bones (Figure 33). Posterior to this are the auditory (tympanic) bullae and the external acoustic canals leading to the ear holes. The canals are large and stick prominently upward and backward.

The basisphenoid is about 1.3 inches (33 mm) broad posteriorly, and the orbital wings of the sphenoid bones are less depressed than in most similarly sized ruminants.

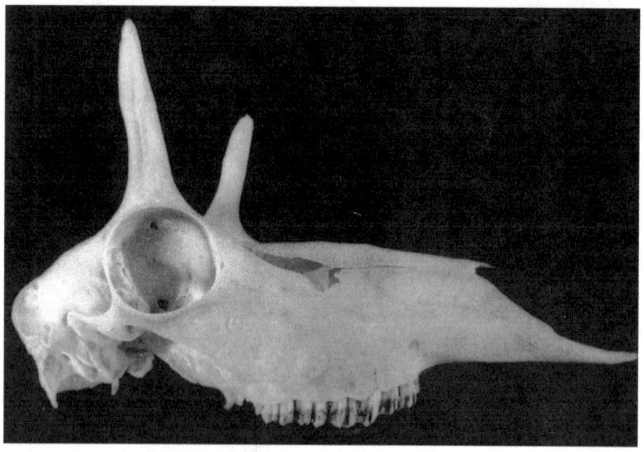

The skull of an adult Montana doe, with the largest horn cores seen among about 200 pronghorn doe skulls examined from Montana and Yellowstone National Park. Photo by Bart W O'Gara.

It continues forward with a steady reduction in width to the junction of the presphenoid, which also narrows moderately forward in direct proportion to the breadth of its occipital end. The narrow, forward end disappears under the vomer, which is about 5 inches (13 cm) long and reaches the incisive bones.

The pterygoid plates and processes are thin and widely spaced. The temporal wings are narrow anteriorly and posteriorly. The horizontal plates of the palatines, together, form the posterior fourth of the hard palate and meet the maxillae at a level between the first and second molars. The suture between them presents a wide arch, interrupted posteriorly by holes through which nerves pass. The posterior edges of the palatine bones end the hard palate at the level of the last molar. A deep notch or interspace lies between the last molar and the wall of the posterior nares.

The maxillae meet in the midsagittal plane to form the front part of the roof of the mouth; the anterior portions are characterized by an elevated sharp ridge running forward from each anterior premolar. The maxillae fork anteriorly and dovetail with the incisive bones, which form the lower border of the nasal opening and contain the incisors of the upper jaw in most mammals. Pronghorn, however, lack upper incisors, so their incisive bones support a toothless pad.

A posterior view of the cranium (Figure 34), minus the mandible, illustrates the horn cores of bucks spreading outward from their bases and those of does standing

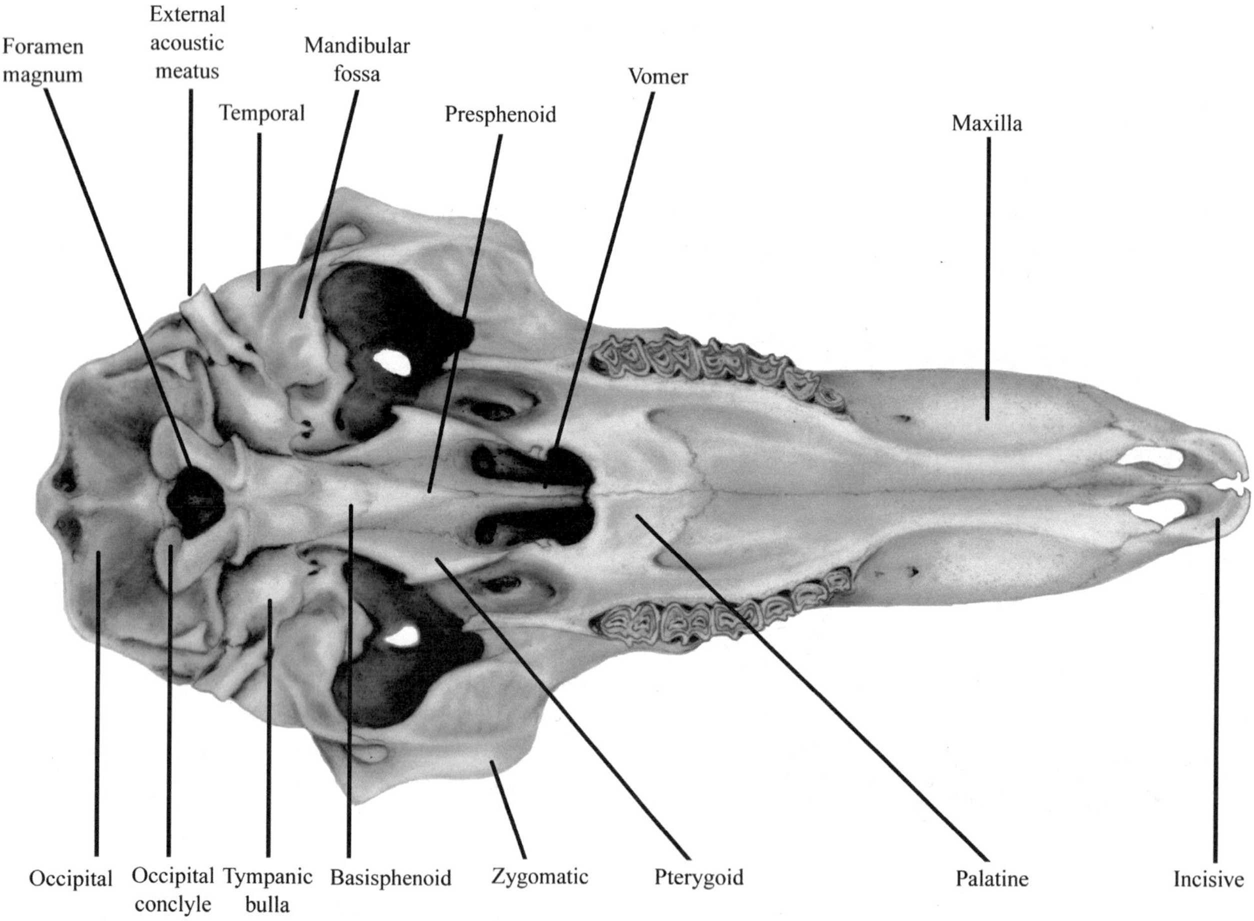

Figure 33. *Ventral view of an adult pronghorn doe skull.* Illustration by Susan Kraft Ball.

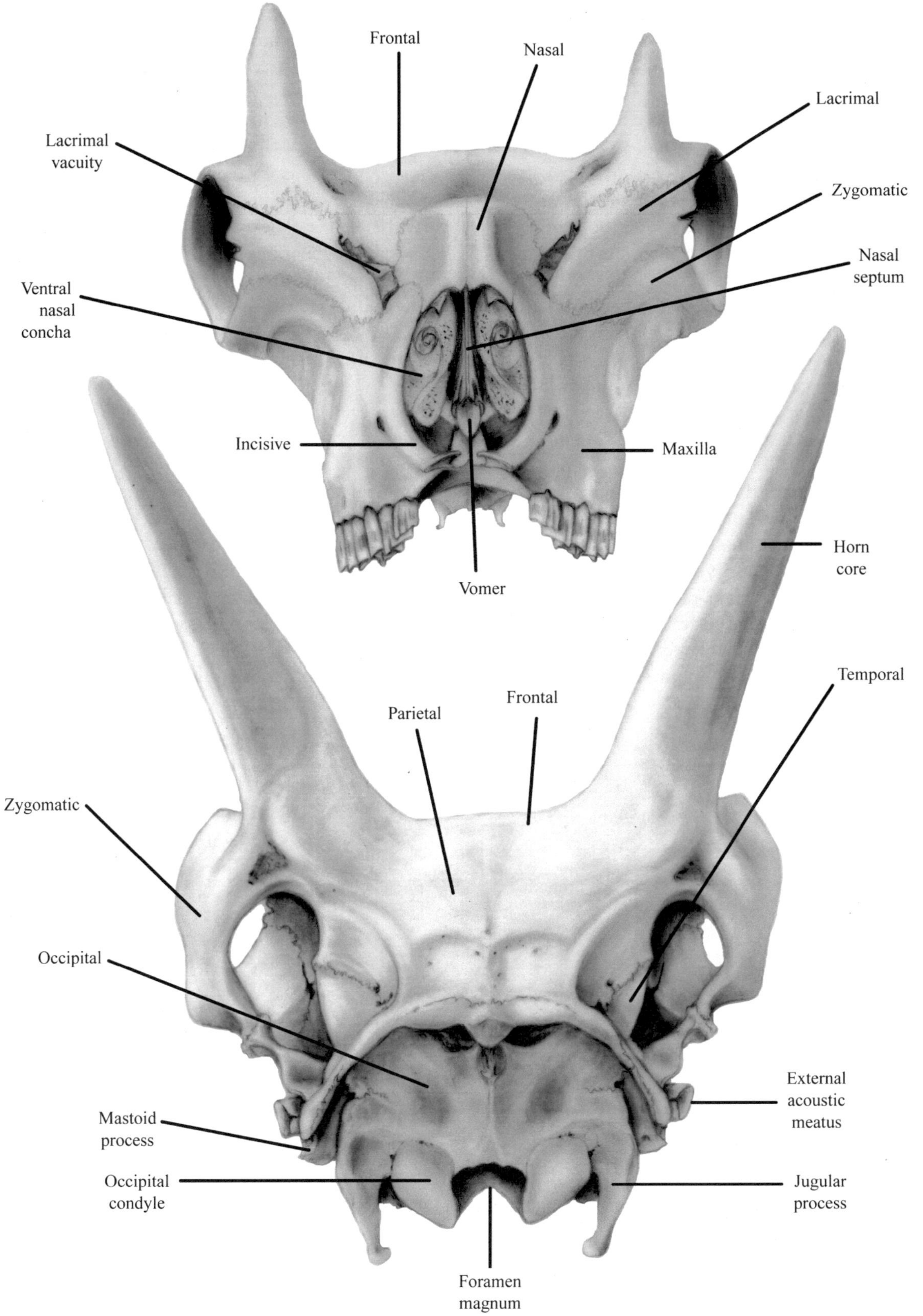

FIGURE 34. *Anterior view* (top) *of an adult pronghorn doe skull and posterior view* (bottom) *of an adult pronghorn buck skull. Different sizes of left and right horn cores on does, as illustrated here, are quite common. Disparities in sizes of horn cores are uncommon on bucks.* Illustration by Susan Kraft Ball.

nearly vertical, prominent post-frontal flattening and the square outline of the orbits. On mature bucks, the tips of the horn cores usually are 8 or 9 inches (20 or 23 cm) apart, and the cores, thickest behind and narrow-edged in front, are about 5 to 6 inches (13–15 cm) long. However, the horn cores of two large "trophy" bucks from New Mexico measured 7 inches (18 cm) in length. We have measured skulls of bucks in which the tips of the horn cores were as wide as 12 inches (30 cm) apart and as narrow as 4 inches (10 cm). In the latter, the tips of the sheaths touched. At the broadest part of the horn cores, where the prongs diverge, they usually measure 1.5 to 2.0 inches (38–51 mm) from front to back. However, configuration and size of horn cores vary greatly. The mean length, circumference and spread at the tips of the horn cores of nine adult bucks in Alberta were 5.2 inches (133 mm), 4.0 inches (102 mm) and 8.8 inches (224 mm), respectively (Mitchell 1980).

The occipital is a single bone lying in the midsagittal plane at the back of the skull, pierced by the foramen magnum through which the spinal cord passes. The occipital condyles possess two distinctive articular planes, which meet in a median, raised, acute line. Both facets are relatively flat—the posterior one nearly vertical, the anterior one directed obliquely forward and inward. The triangular jugular (paramastoid) process underhangs the condyles. The mastoid process is comparatively small.

The anterior view shows the turbinates or nasal conchae—so named because of their snaillike shape—internal to the nasals. The ventral conchae are uncommonly long, inflated, extremely thin, fragile and highly fenestrated bones. The vomer (ventral partition between the two sides) and nasal septum are unusually stout. The septum is cartilaginous for only the anterior 1.5 inches (38 mm), and the remaining 8- by 1.5-inch (200 by 33 mm) portion is ossified. The ossified septum bridges the space between the vomer and nasal bones and undoubtedly adds strength to the long but lightweight rostrum.

The sagittal view shows the external auditory meatus (ear canal) leading internally to the tympanic bulla—a bulbous structure inferior to the meatus. Internal to the meatus and superior to the bulla is the petrosal portion of the temporal (Figure 35). Lateral and superior to these are the flat, fan-shaped, squamous portions of the temporals that overlap the parietals from below. Articulating with the frontals anteriorly and the occipital posteriorly, the parietals form the top of the cranial vault and are scalloped internally to accommodate convolutions of the brain.

Endocranially, the sphenoid is somewhat saddle-shaped to accommodate the pituitary gland (hypophysis). Anterior to the sella turcica (Turkish saddle), the bone is pierced by the optic canals (foramen). Anteriorly, the sphenoid articulates with the cribriform plate of the ethmoid. The olfactory nerves pass through the many small holes of the large and fragile cribriform area.

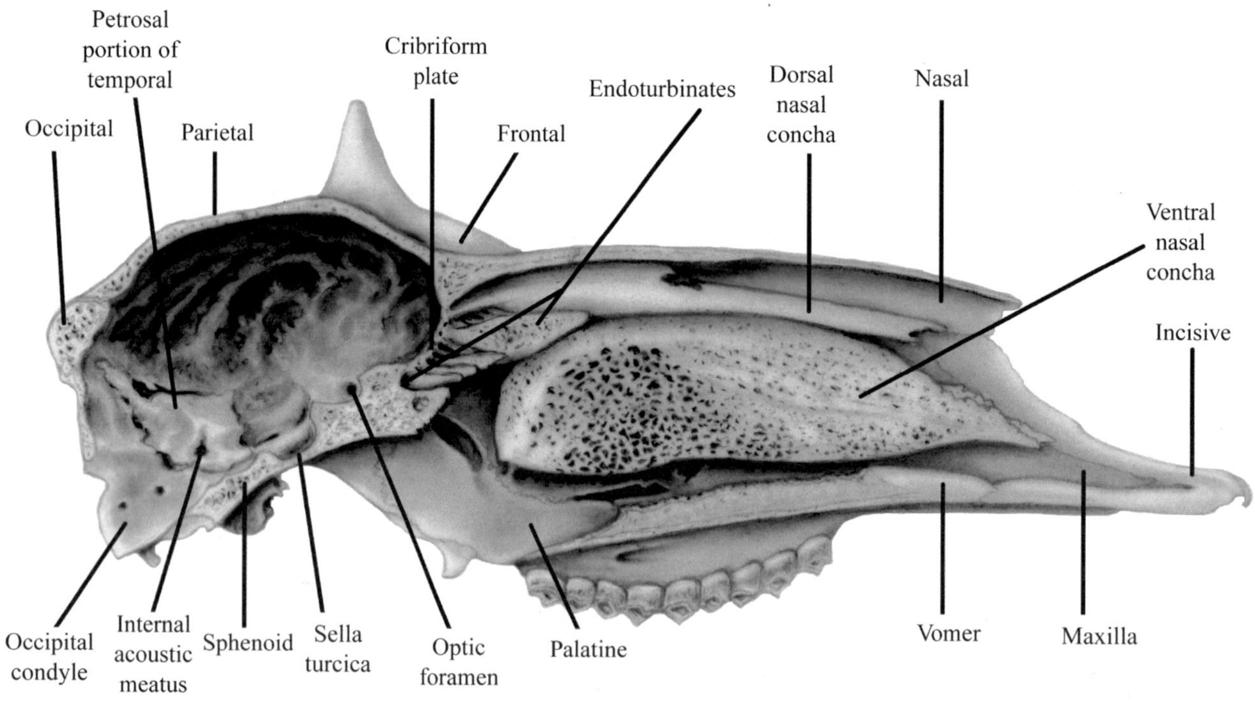

Figure 35. *Sagittal section of the skull of an adult female pronghorn with the vomer and nasal septum removed.* Illustration by Susan Kraft Ball.

Lower Jaw

The lower jaw consists of paired symmetrical bones, the mandibles, fused at their anterior ends by the mandibular symphysis (Figure 36). The anterior or horizontal part that supports the teeth generally is called the body. The ramus rises and recedes from the body in a more backwardly oblique manner than in most ruminants; this is rendered more apparent by the rounding off of the angle. A large foramen for passage of a branch of the mandibular nerve lies on the lingual (tongue) side of each ascending ramus. One and sometimes several holes for the passage of nerves lie on the lateral side of the body near the anterior mandibular symphysis.

The coronoid process is long, narrow and straight. The mandibular notch is shallow, and the flattened, slightly concave facet of the small condylar head (process), which articulates with the cranium, is almost without a neck. Its transverse length is about three times that of its anteroposterior width. Ventrally in most specimens, the ramal angle possesses a wide but shallow emargination toward the body end. In young animals, up to 3 years of age, roots of the molars, especially molars 2 and 3, often pen-

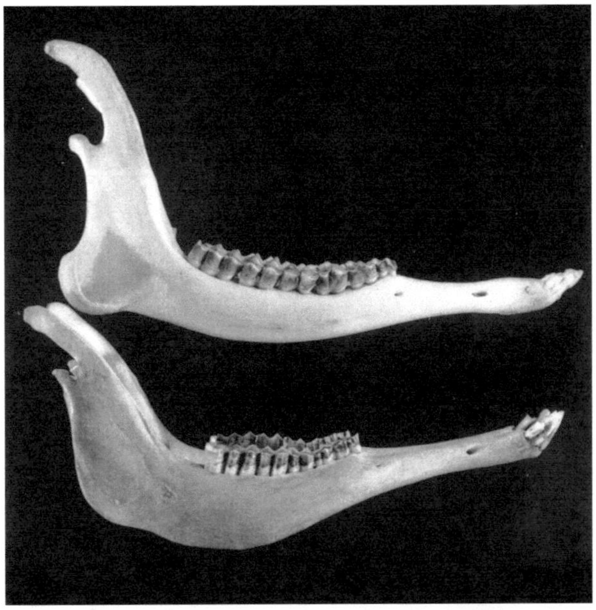

Comparison of typical lower jaws of an adult female white-tailed deer (top) and an adult female pronghorn (bottom). Note the rounded-off angle between the body and the ramus of the pronghorn compared with that of the white-tailed deer. Photo by Dewayne Williams.

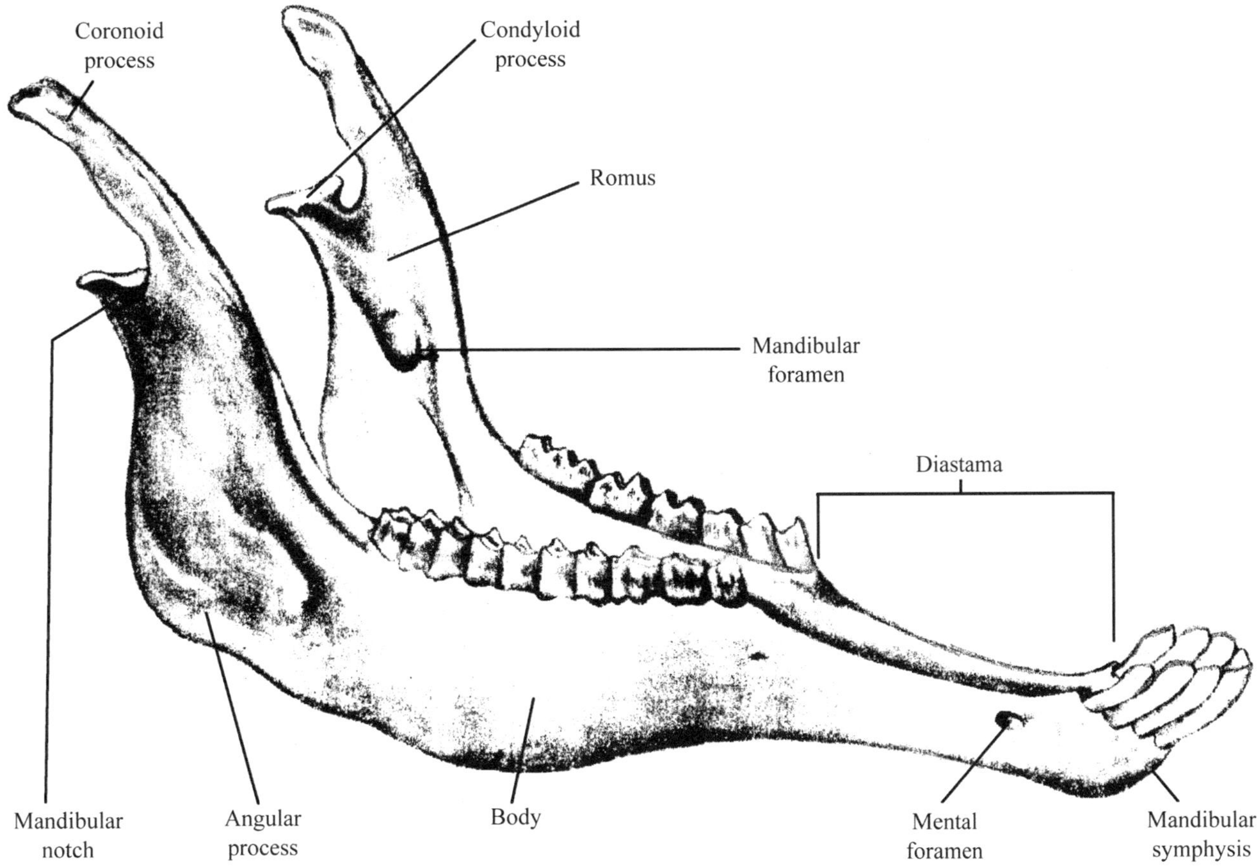

FIGURE 36. *The lower jaw of an adult female pronghorn.* Illustration by Susan Kraft Ball.

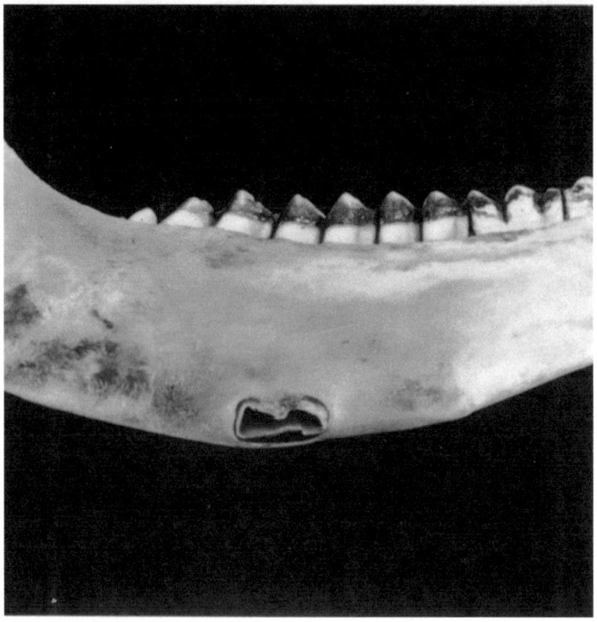

The lower jaw of a 2-year-old pronghorn shows penetration of the ventral surface by the roots of m2. Photo by Dewayne Williams.

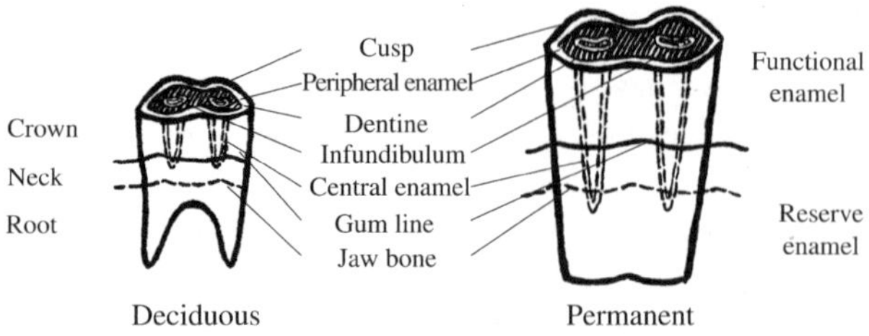

FIGURE 37. *Anatomy of a typical deciduous premolar and molar of a pronghorn. The grinding surface is called the occlusal surface, the buccal side is next to the cheek and the lingual side is next to the tongue. An infundibulum is a funnel-shaped cavity surrounded by central enamel extending downward from the occlusal surface.* Illustration by John Eiler, from Dow and Wright (1962).

etrate the ventral surface. The dental portion of the body is slightly concave below. The diastama is long and laterally compressed, with a sharp upper border and a terminal incisive expansion.

Teeth

Pronghorn teeth are hypsodont (Figure 37), having long crowns and either no roots or roots that develop months or years after the teeth erupt. Thus, they do not have a belt or ridge at the gum line. Such teeth continue to grow throughout the animals' lives and are especially adapted for grinding rough vegetation and compensating for wear from grit and dirt. Abbreviations for teeth generally are as follows: D = deciduous; P = permanent; i = incisor; c = canine; p = premolar; and m = molar. Thus, the first permanent incisor is Pi1 and deciduous premolar 2 is Dp2. Molars are designated merely as m because all are permanent.

The dental formula for pronghorn is incisors 0-0/3-3, canines 0-0/1-1, premolars 3-3/3-3 and molars 3-3/3-3. Thus, 20 of the pronghorn's 32 teeth are deciduous and 12 are permanent. Pronghorn retain 32 of the 44 teeth that compose the primitive dentition of mammals. The remaining teeth are as follows: upper—premolars 2, 3 and 4; molars 1, 2 and 3; lower—incisors 1, 2 and 3; canine; premolars 2, 3 and 4; and molars 1, 2 and 3.

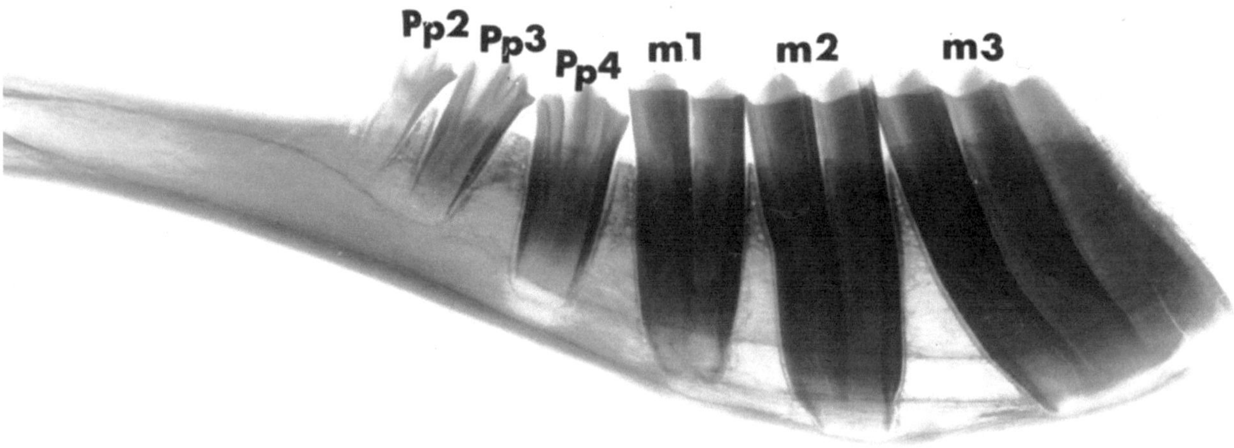

Radiogram of the lower jaw of 3.3-year-old pronghorn, including permanent premolars (Pp) 2–4 and molars (m) 1–3. All teeth are almost fully formed, yet distinct roots are visible only on Pp3 and, in only Pp4 and m1 have the bases of the teeth begun to move away from the lower edge of the jaw; m1 and m2 have two cusps on the lingual side and two cusps on the buccal side with two infundibula between; m3 is the largest tooth in the jaw, with three lingual cusps, three buccal cusps and three infundibula. The cusps of p2 and p3, in both the deciduous and permanent forms, are represented by an irregular ridge. Dp4 possesses three lingual and three buccal cusps in addition to three infundibula, although Pp4 has three infundibula, the cusps are not distinct. Radiogram by Phil Wright.

The six incisors and two canines (together called incisiform teeth) are simple, spadelike teeth with rounded tips that become flat with wear. Permanent incisiform teeth of pronghorn are larger and longer than those of white-tailed or mule deer, and much larger than their deciduous predecessors (milk teeth).

Hyoid Apparatus

The hyoid apparatus consists of four paired bones that support the base of the tongue and articulate with the larynx. It is located between the rami of the mandibles and along the floor of the pharynx (Figure 38) and acts as a suspensory mechanism for the larynx. The stylohyoid is the first and largest of the hyoid bones and extends from the base of the petrosal part of the temporal bones to the root of the tongue. It is slender, about 0.2 inch (5 mm) wide, 0.08 inch (2 mm) thick and 3.1 to 3.5 inches (8–9 cm) long. An expansion of thin bone about 0.7 inch (17 mm) deep by 0.6 inch (14 mm) wide with a hook pointed anteriorly extends from the ventral side of the stylohyoid near its connection with the skull. The whole expansion, especially the hook, apparently is a point of attachment for muscles that move the apparatus. The other three bones are small and connected together to form a half-circle, with the last one, the thyrohyoid, adhering to the upper border of the thyroid cartilage of the larynx. The short (0.5 in [12 mm]) epihyoid is connected to the ventral anterior end of the stylohyoid. The slightly curvilinear ceratohyoid is next. It is about 0.8 inches (2 cm) long, broadens from about 0.16 to 0.24 inches (4–6 mm) in width and is followed by the thyrohyoid. The thyrohyoid is about 0.7 inches (18 mm) long and tapers from 0.31 to 0.16 inches (8–4 mm) in width. The basihyoid is not ossified, and is represented by a cartilaginous rod connecting the two sides of the hyoid apparatus at the junction of the ceratohyoids and thyrohyoids.

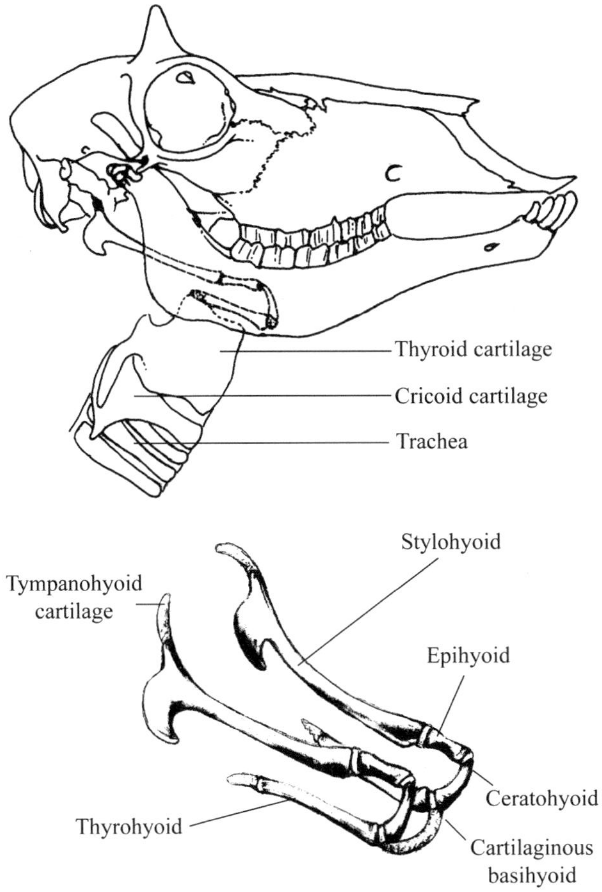

FIGURE 38. *Diagrammatic representation of hyoid bones connecting the skull and larynx of a pronghorn.* Illustration by Susan Kraft Ball.

Vertebrae

The vertebrae protect the spinal cord and furnish a relatively incompressible bony column through which the propelling force generated by the legs is transmitted to the rest of the body. The axial and abdominal muscles use the vertebral column in locomotion. The vertebrae also partly support and protect the abdominal and thoracic viscera and give rigidity and shape to the body in general. In support of the viscera of the trunk, the vertebral column has been compared to a bow, and the abdominal muscles and linea alba to a string. As the string, the abdominal muscles do not attach to the ends of the bow, but at some distance from them. Cranially, the attachment is to the rib cage; caudally, it is to the pelvis. The amount of movement between any two adjacent vertebrae is small, but the combined movement permitted in all the intervertebral articulations is sufficient to allow considerable mobility of the whole body in any direction.

The vertebral elements in the spinal column of pronghorn are 7 cervical, 13 thoracic, 6 lumbar, 4 sacral, and 10 or 11 caudal. The cervical vertebrae are heavy in comparison with those in other spinal regions, although the latter have longer spinous processes (Figure 29). The flat, broad atlas supports the skull (Figure 39). Its condylar articular surface is low, and the transverse process is a thin, wide plate of bone, ending posteriorly in a rounded, flat process. The lateral vertebral foramen pierces it vertically and then passes horizontally through the vertebral foramen. The transverse foramen pierces the transverse process vertically and has a common dorsal opening with the lateral vertebral foramen. A rudiment of a spinous process exists at the middle of the bone, flanked forward by two grooves, and the body has a moderate-sized ventral tubercle. The vertebrae generally diminish in size from the atlas posteriorly.

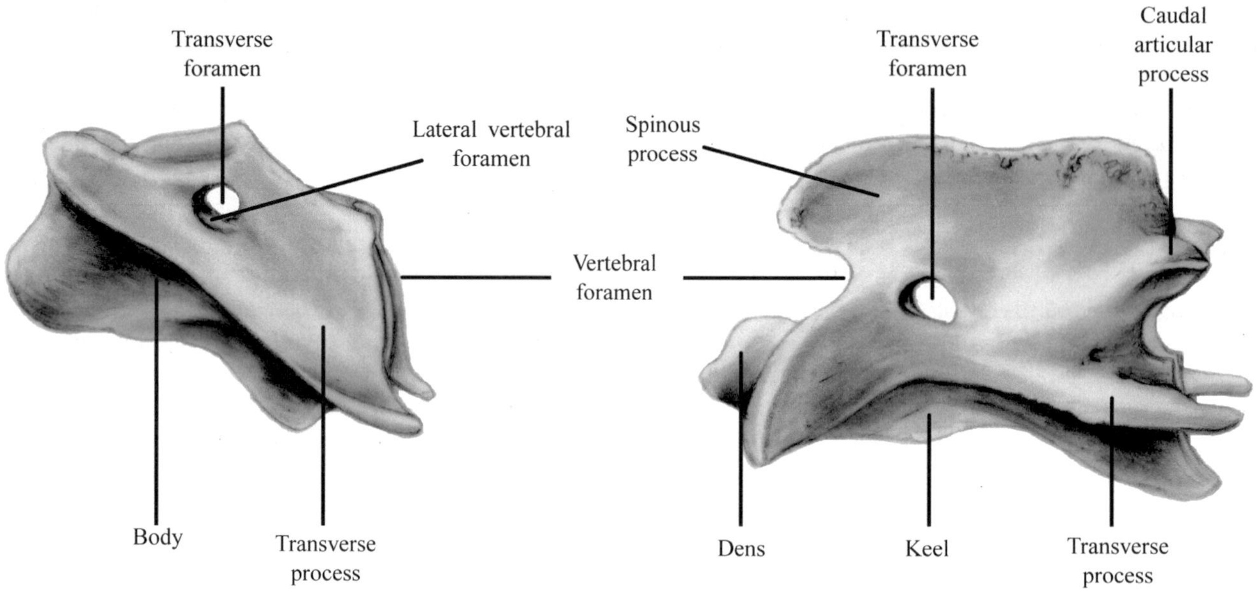

Figure 39. *The atlas* (left) *and axis* (right) *of an adult pronghorn doe.* Illustration by Susan Kraft Ball.

The axis (Figure 39) has a spinous process (neurapophysis or neural spine), about 0.6 inches (1.5 cm) high in females and 0.9 inches (2.2 cm) high in males, that runs the entire length of the vertebra. Anteriorly, the spine projects as a process forward; posteriorly, it is truncated. A sharp-pointed, barblike, transverse (accessory) process is directed backward and overlaps the third vertebra. A well-marked keel is fully developed in the second (axis), third, fourth and fifth cervicals. This keel is rudimentary in the first (atlas) and sixth cervicals, but absent from the seventh. Transverse foramina perforate the vertebral arches of vertebra three through six, from front to back (Figure 40).

Spinous processes are absent in the third and fourth cervical vertebrae, but are developed in the fifth and lengthened in the sixth and seventh. The vertebral arches of the third and fourth vertebrae are marked by a posteromedial depression, and the third through seventh have lateral ones. The second cervical is the longest, and lengths reduce progressively to the seventh, which is only one-third as long as the axis. The pleurapophyseal elements of the transverse processes (corresponding to rib articulating processes in the thoracic region) of the sixth cervical vertebra are broader than on other neck vertebrae. The five posterior cervical vertebrae interlock with each other by well-developed opisthocellian (posteriorly hollow) articulations.

The thoracic vertebrae are relatively uniform in size. The center of movement of the spine hinges on the tenth thoracic vertebra, where the spinous processes change direction from backward to forward. The first thoracic spinous process is of considerable length, about 4 inches (10 cm), and from it to the third, they increase about 0.2 inches (0.5 cm) in length. Between the third and the sixth, they remain nearly uniform in length and then slope posteriorly (Figure 40). The next four gradually shorten to 1.6 inches (4 cm). The spinous processes on the last three thoracic vertebrae are broader, also 1.6 inches (4 cm) long and are directed anteriorly. Mammillary processes (metapophyses) in the form of tubercles are developed on the first thoracic vertebra, enlarge on the second, and then remain nearly uniform as far as the eleventh. The tubercles on the twelfth and thirteenth increase in magnitude, articulate with adjoining vertebrae and approach the prominent platelike form found in the lumbar region. The thoracic vertebral region measures about 14 inches (36 cm).

The bodies of the six lumbar vertebrae are nearly uniform in size. Their transverse processes are of moderate breadth, but very long, almost riblike (Figure 40). The first is about 2.2 inches (55 mm) long, and the second to the fifth are about 2.6 inches (65 mm) in length, and the sixth is about 2.2 inches (55 mm) long. All slant anteriorly and curve forward. The spinous processes are broad, of nearly uniform height, about 1.2 inches (3 cm) high, and the first five have dorsal hooks pointed forward. The mammillary processes are prominent and appear rounded. Between the last lumbar and the first sacral vertebrae is a large lozenge-shaped interspace.

The funnel-shaped sacrum is about 4 to 5 inches (10–13 cm) in length, and its anterior transverse diameter is about three-fourths as long as its length (Figure 41). The sacroiliac fusion is formed by a single articulation of the fore-

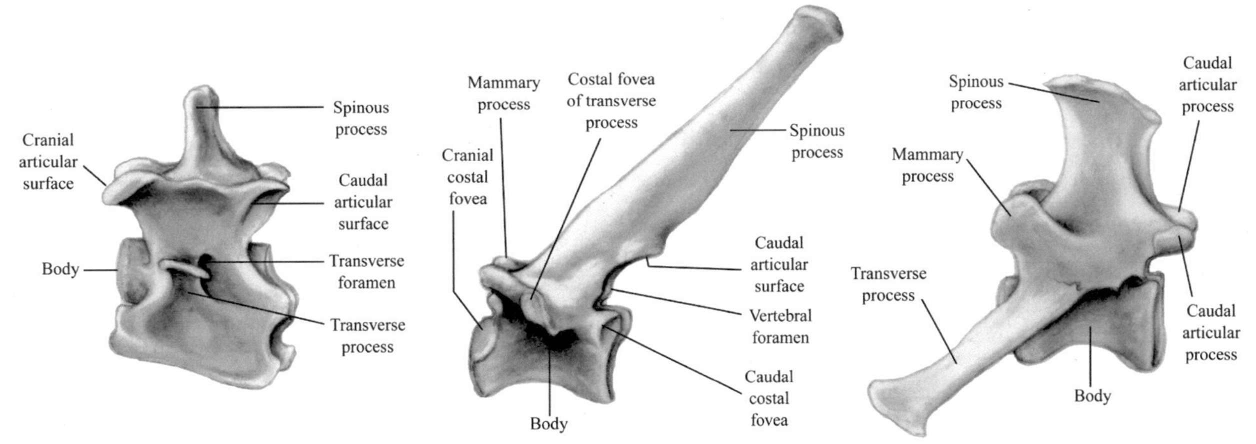

FIGURE 40. *The sixth cervical* (top)*, fourth thoracic* (center)*, and third lumbar* (bottom) *vertebrae of an adult pronghorn doe.* Illustration by Susan Kraft Ball.

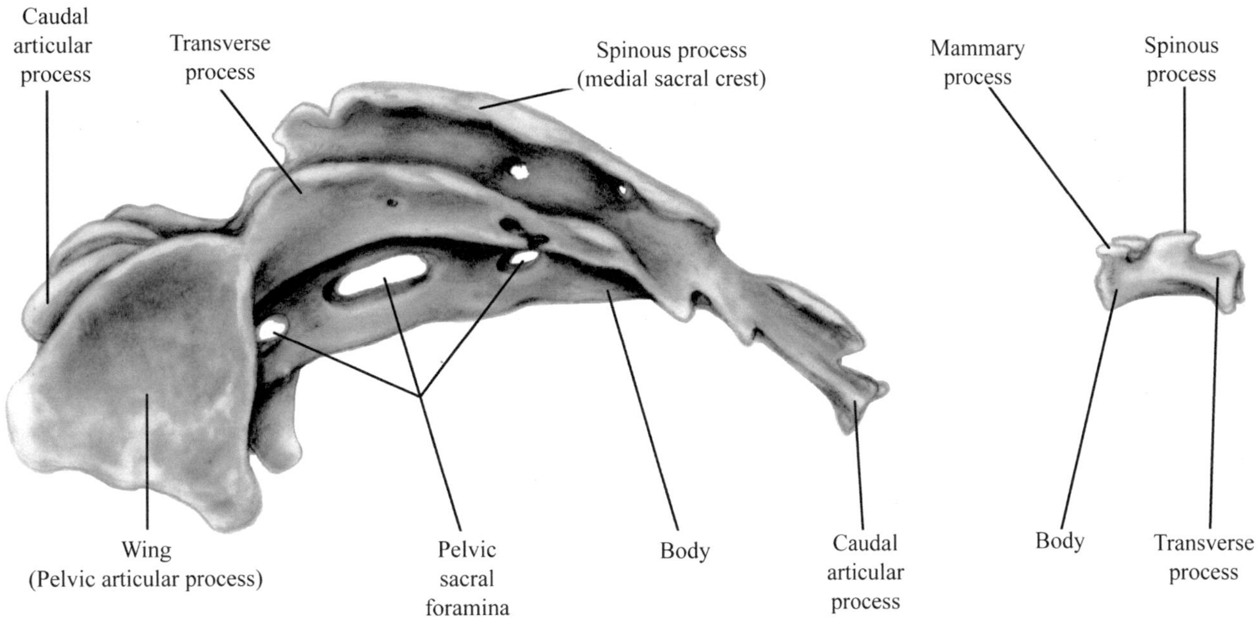

FIGURE 41. *The sacral* (left) *and third caudal vertebrae* (right) *of an adult pronghorn doe.* Illustration by Susan Kraft Ball.

most sacral vertebral element that possesses a small spinous process. The three posterior vertebrae belonging to the sacrum are fused, and their fused spinous processes decrease in height from front to back. Their transverse processes also are fused, forming a tapering vertical plate of bone parallel with and slightly lower than the neural spine.

The first or anterior caudal vertebra possesses broad transverse processes of bony plates, but give off no backward-extending zygapophysis. The remaining vertebral bodies diminish in size, becoming little more than rods or spicules (Figure 41).

Ribs

The ribs form the thoracic skeleton, except for narrow mid-dorsal and midventral strips formed by the vertebral column and the sternum. A typical rib has a head, neck, tubercle, body and costal cartilage (Figure 42). A wedge-shaped articular surface on the head articulates with two contiguous thoracic vertebrae and the intervening fibrocartilage. The articular areas, corresponding to those of the vertebrae with which they articulate, are of about equal size. The anterior one is convex, the posterior one is concave and a groove separates the two. At about the twelfth thoracic vertebra, the posterior articular area disappears,

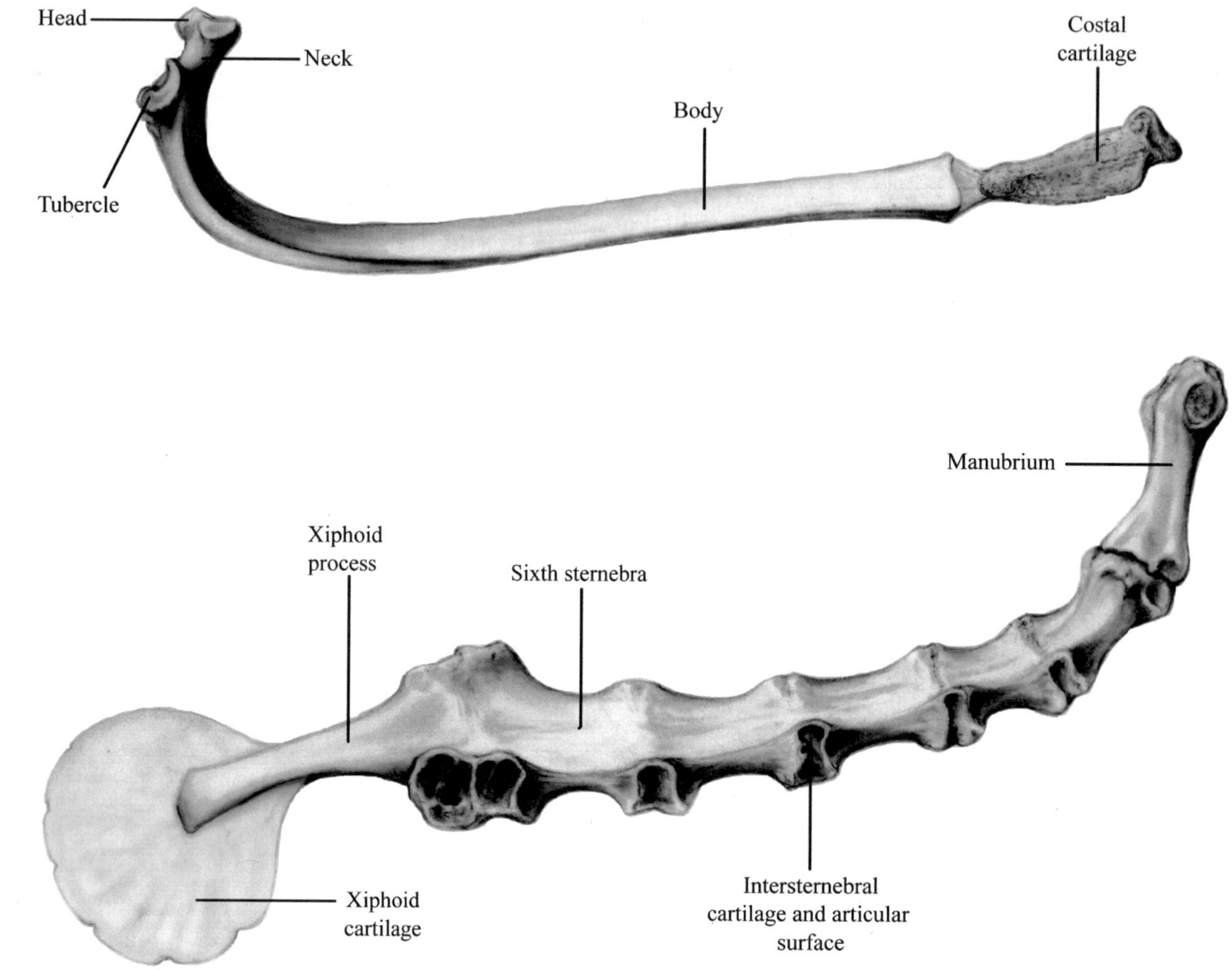

FIGURE 42. *The rib* (top) *and sternum* (bottom) *of an adult pronghorn doe.* Illustration by Susan Kraft Ball.

and the last one or two ribs articulate only with their corresponding vertebrae. The heads of these ribs are modified accordingly and lack the groove separating the two articular areas. The neck, between the head and the tubercle, is somewhat triangular. The tubercle bears an articular facet for articulation with the corresponding vertebrae. The tubercle and its articular facet are most pronounced in the forward ribs. The neck progressively shortens in the more posterior ribs, and the articular areas of the heads and tubercles become confluent on the last two ribs. However, the tubercles remain for muscular attachment.

Each rib is divided into a bony dorsal part and a ventral cartilaginous part—the costal cartilage. Costal cartilages are somewhat cylindrical continuations of the bony ribs. They are smaller in diameter than the bony ribs and may become calcified in mature animals. The bony ribs have a smooth surface, compared with the rough, pitted surface of the costal cartilages.

The chests of many mammals are flattened dorsoventrally, so the ribs are sharply curved throughout their length. In pronghorn, as in most artiodactyls, the thorax is elongated dorsoventrally, and the ribs are only sharply curved at their dorsal (vertebral) ends. The first two ribs are nearly straight, and the three or four posterior ones are more slender than the anterior ones, which are comparatively broad. The bodies of the central ribs are nearly square for about the dorsal one-fourth of their lengths, then become flattened. The ninth rib is longest.

Pronghorn ribs usually number 13 pairs, of which the anterior eight (sternal ribs) articulate with the sternum; the last five (asternal ribs) do not reach the sternum directly. The costal cartilages of the ninth through the twelfth ribs unite with the cartilages of the ribs anterior to them to form the costal arch. The costal cartilages of the thirteenth ribs end freely in the musculature of the flank, hence the name floating ribs.

Some pronghorn have more than 13 sets of ribs. M. Barrett (personal communication:1980) necropsied pronghorn fawns that had 14 sets of ribs instead of 13. Since then, O'Gara has had the chance to look at 63 pronghorn carcasses: 45 had 13 sets of ribs; 12 had 14 sets; and 6 had 14 on one side and 13 on the other. In animals with 14 ribs, the fourteenth articulated with the first lumbar vertebrae and was floating.

Sternum

The eight stout bones (sternebrae) of the unpaired sternum are partly fused so that the sternum appears to be composed of only five pieces in most mammals (Figure 42). In the pronghorn, seven bones can be distinguished in fetal sternums, and the sutures between sternebrae are easily seen in adult animals, although the fifth, sixth and seventh are firmly fused. Of these, the anterior one is cylindrical and the middle and hindmost are wider and thinner. The sternum is sharply turned up anteriorly and slightly down posteriorly. The consecutive sternebrae are joined by short blocks of intersternebral cartilage. Costal cartilages of the ribs articulate with the intersternebral cartilages, with the exception of the first pair, which articulate with the first sternebra.

The foremost sternebra, the manubrium, is a short, stout bone, expanded anteriorly. The thick, medially flattened and narrowed second element and the somewhat-wider third element are ossified together. The fourth and fifth expand in breadth but diminish in thickness; the fifth, sixth and seventh are firmly united. The sixth is the broadest of the whole series. The posterior segment (xiphisternum or xiphoid process) is a long, flat and strong process of bone, narrowing considerably toward its distal, truncated extremity. The extremity is surrounded by a discoid cartilage measuring about 2.3 inches (5.8 cm) craniocaudally and 3 inches (7.6 cm) across.

Forelimb

Each forelimb is composed of the scapula (shoulder blade), humerus, radius and ulna (forearm), carpus (wrist), cannon bone (fused metacarpals or hand bones) and two toes with three phalanges each and associated sesamoid bones (Figure 43). Although the limbs appear delicate, Einarson (1948) found that bones in the foreleg of a pronghorn resisted greater weight load per square inch than did a comparable section of cow bone.

The scapula is the most dorsal bone of the forelimb. It is not articulated with the skeleton, supporting the animal's body by muscular attachments only. It is triangular, with a flat and smooth blade, short neck and well-developed spine or ridge about an inch (2.5 cm) high on the lateral (outside) surface. The bone is about 8 inches (20 cm) long, 4.5 inches (11 cm) broad at the vertebral border and only about 0.3 inches (0.8 cm) thick; a semiossified cartilage extends about 3 inches (7.6 cm) beyond the border. A cup (glenoid cavity or fossa) at the distal end of the scapula articulates with the humerus. The spine runs from the vertebral border of the scapula to within an inch (2.5 cm) of the level of the glenoid cavity as a pronounced, elevated ridge. The edge of the infraspinous fossa forms a shallow ridge that is more pronounced than those of deer and increases in height with age. The ridge is anterior to the mesial line, so the supraspinous fossa is less than half as wide as the infraspinous fossa. The acromion process, present in many mammals, is inconspicuous or absent. However, the outer surface of the spine is flattened into a narrow shelf to varying degrees and somewhat overhangs the infraspinous fossa for most of its length. A tuberous but compressed coracoid process projects beyond the deeply scooped glenoid cavity.

Humeri are the largest bones of the forelimbs, and their shape reflects something of an animal's build and mode of travel. Artiodactyl humeri have very large proximal ends, indicating the need for support of deep-chested running animals. The pronghorn humerus is shorter than the scapula by about 0.3 inches (0.8 cm) and has a moderately stout, smooth body (shaft). A depressed, articular, semilunar-shaped head diverges backward at almost right angles to the axis of the body. The lesser tubercle is flattish; the intertubercular groove is broad, with a wide excavation to its inner side for the insertion of the subscapularis muscle. The greater tubercle is a massive three-sided eminence raised about 0.5 inches (1.3 cm) higher than the head; it partly arches over the bicipital groove. A smooth broad boss marks the place of insertion for the supraspinatus muscle; an oblique deltoid tuberosity is amply represented. A minute nutrient foramen enters about midway along the outside of the body. The medial-lateral width of the condyles of the humerus is slightly greater than the anterior-posterior width.

The radius and ulna are partly fused, reducing rotational movement—an advantage in running. Their proximal ends form the elbow joint with the humerus, and their distal ends form the carpal joint (often called knee or elbow by hunters, although it really is the wrist) with the carpal bones. The radius measures about 8 inches (20 cm) long and is slender in comparison with that of deer. The shaft of the ulna is a thin, bony splint, attached to the radius by strong connective tissue that terminates in a well-developed trihedral styloid process. The olecranon (the part of the ulna projecting beyond the elbow joint and forming the point of the elbow) is large.

The radial (scaphoid), lunare, ulnar (cuniform) and accessory (pisiform) are present in the first (proximal) row

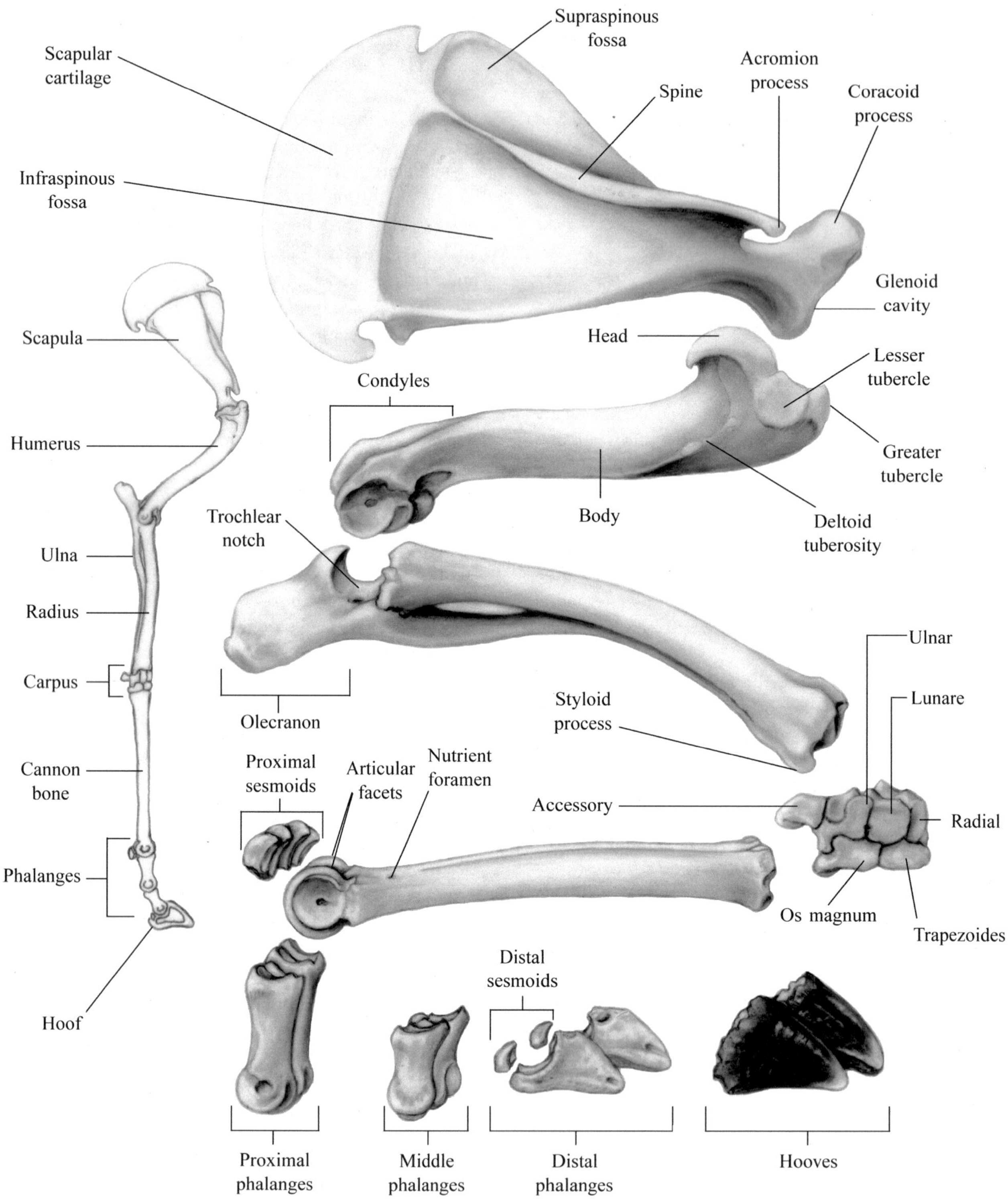

FIGURE 43. *The forelimb of an adult pronghorn.* Illustration by Susan Kraft Ball.

of carpal bones. The trapezoides and os magnum are found in the distal row.

In humans and many other animals, the five separate bones of each hand and foot are short, compared with the long bones of the limbs such as the radius. In ungulates, the functional metacarpals and metatarsals in each foot have become fused into a single, long cannon bone, about the same length as the radius. The cannon bone of the pronghorn is a subcircular rod. The anterior aspect is slightly grooved and the posterior is broadly grooved for nearly the entire length. The grooves terminate in foramina at their distal ends, and the posterior groove terminates proximally in a foramen. The proximal end of each cannon bone is divided into two large articular facets. The distal end has two pulleylike condyles separated by a cleft and which articulate with the phalanges.

All North American artiodactyls, except pronghorn, have well-formed, splintlike metacarpal and metatarsal bones for each of their side toes (dewclaws). The lack of such splints in pronghorn corresponds with the outer aspect of the feet and the lack of dewclaws.

The proximal (first) phalanx is about 1.8 to 2.0 inches (4.6–5.1 cm) long; the middle (second) is 1.0 to 1.1 inch (2.5–2.8 cm); and the distal (third) is 1.2 to 1.3 inch (3.0–3.3 cm). The distal phalanges are triangular, nearly the shape of the hooves, and the front (lateral) ones are slightly larger than the rear. Small sesamoid bones are located behind the articulating surfaces between the cannon bones and proximal phalanx, and between the middle and distal phalanges.

Hind Limb

Each hind limb consists of its half of the pelvic girdle fused as the os coxae (hip bone), the femur (thigh), the stifle (knee joint) with its large sesamoids (knee cap), tibia and fibula (leg), and the hind foot (pes). The pes includes the ankle (tarsus) with its digits, consisting of a cannon bone, two toes with three phalanges each, and the sesamoid bones associated with the phalanges (Figure 44).

The pelvic girdle is composed of three distinct parts on each side that unite to form a single bone in the adult animal. The ilium is a broad fan or blade above and anterior to the acetabulum (hip socket). Below the acetabulum and on either side of the large, oval obturator foramen lie the ischium and pubis. They articulate with their opposites at the pelvic symphysis on the midline. The elongated neck of the ilium has a blade with moderately broad wings. The acetabular fossa is shallow, with a deep synovial notch. The ischium is moderately expanded; the ischiatic tuberosity is thick and rounded posteriorly and has a prominent triangular eminence laterally.

The pronghorn pelvis is long (about 8.5 inches [21.5 cm]), narrow (about 2.3 inches [6.5 cm]) and lightly built. The brim presents a blunt oval figure, the conjugate diameters being nearly equal. The symphysial portion of the ischial and pubic bones is flat, wide and broad dorsally. It is about 3 inches (7.6 cm) long.

The relative size of the pelvic orifice of pronghorn does, through which the large and well-developed young must pass, is naturally greater than that of bucks. This results in the pelvis of the doe, when viewed from above, appearing broader than that of the buck. The conjugate measurement (diameter) has been described differently in various publications. It is described here as the distance from the sacral promontory to the cranial border of the pubic symphysis. The average conjugate diameter of 36 adult does collected in Montana was 4.4 inches (11.1 cm), with a range of 3.9 to 4.7 inches (10–12 cm). That of nine adult bucks collected in the same area was 3.7 inches (9.3 cm), with a range of 3.5 to 3.8 inches (9.0–9.6 cm).

Although measurements of the conjugate diameter did not overlap between the sexes in two comparatively small samples of animals collected in the same area, such probably would be unlikely throughout the pronghorn's range and should not be used to differentiate positively between the sexes. Denny (1958) found the conjugate diameters of bull elk to average about 1 inch (2.5 cm) smaller than those of cow elk. Because of within-sex variation, however, he recommended that conjugate diameter be used in combination with other measurements to identify posively the sexes of elk carcasses.

Relative proportions, expressed as ratios, can be used to compensate for differences in sizes of animals. Taber (1956) used the ratio of length to depth of the pubic symphysis to differentiate between the sexes of black-tailed and white-tailed deer. However, for pronghorn, the depth of the pubic symphysis is difficult to measure accurately. The length of the pubic symphysis averaged 2.8 inches (7.0 cm), with a range of 2.7 to 3.0 inches (6.8–7.6 cm) for 36 adult does; that of nine adult bucks averaged 3.7 inches (9.3 cm), with a range of 3.5 to 3.8 inches (9.0–9.6 cm). Dividing the length of the pubic symphysis into the conjugate diameter indicated that the conjugate diameter of does averaged 1.56 (range, 1.43–1.70) times the length of the pubic symphysis. The same value for bucks was 1.26 (range, 1.21–1.35). Thus, although the differences between the sexes are quite large on average, individuals nearly overlap and no doubt would in a large sample. Conjugate diameters relative to the length of the pubic symphysis seem to be indicators of sex in pronghorn, but overlaps would preclude use in court cases or any other situation that required positive differentiation.

The shape of the ischiatic arch, when viewed from the ventral aspect, indicates the sex of pronghorn (Figure 45). The arch of the buck is V-shaped, with a keyholelike form evident where tuberosities provide attachment points for suspensory ligaments of the penis. The arch of the doe is smooth and U-shaped. Thus, the general shape of the ischiatic arch and the presence of tuberosities provide positive evidence of sex in pronghorn.

The head of the femur (thigh bone) is small and nearly round. The greater trochanter is laterally flattened and rises about 0.5 inch (1.3 cm) higher than the head. The lesser trochanter is a pyramid-shaped eminence projecting from the caudomedial surface of the proximal extremity

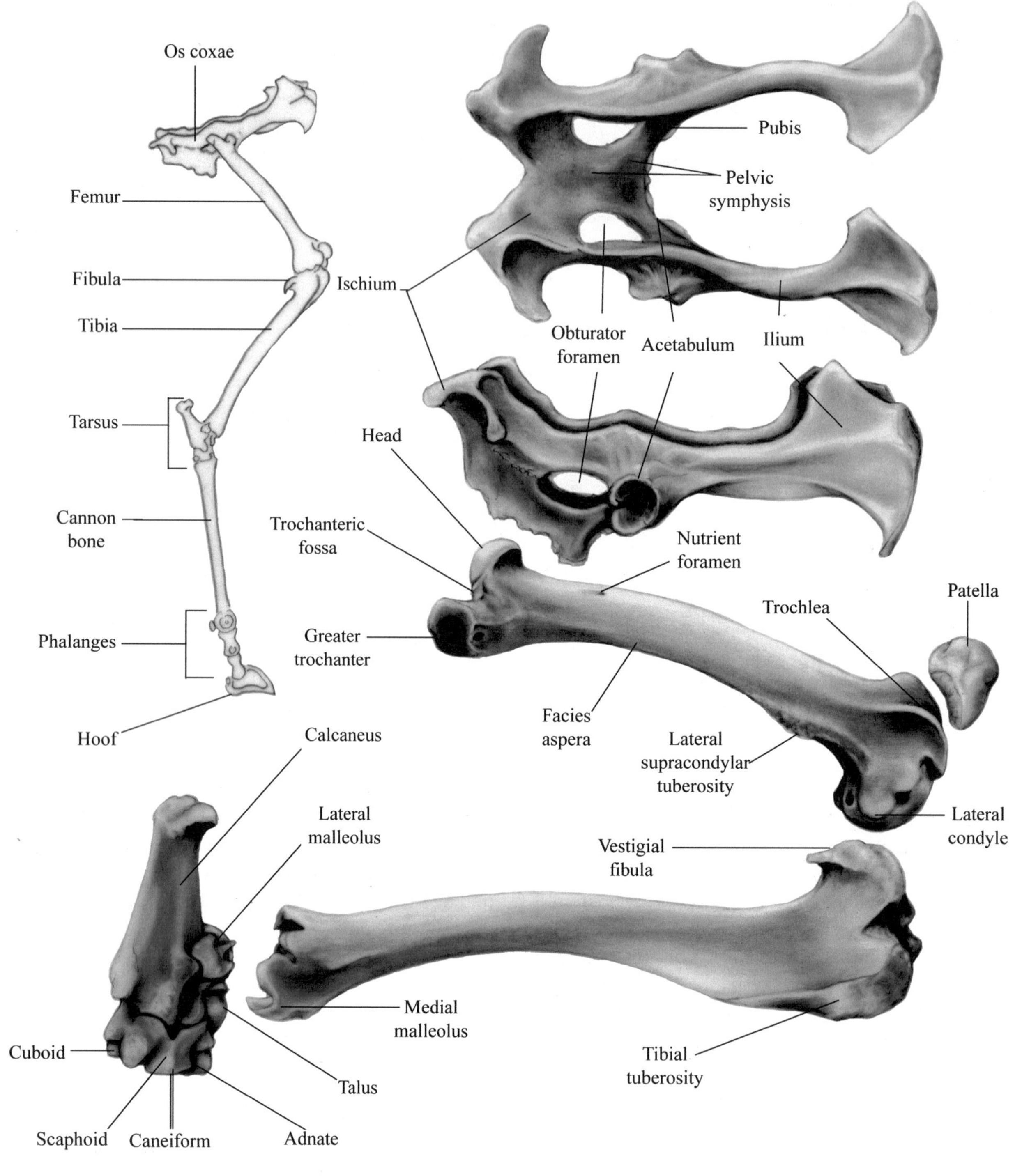

FIGURE 44. *The hind limb of a pronghorn doe minus digits that are similar to forelimb.* Illustration by Susan Kraft Ball.

near its junction with the body. The third trochanter, present in most mammals, is absent in the pronghorn, as in most other artiodactyls. The trochanteric fossa is wide, but only moderately deep. The stoutish body has a slight forward bend, and the nutrient foramen penetrates its front obliquely at its upper fourth. The facies aspera (rough surface) is broad and well defined. The lateral and medial supracondylar tuberosities are prominent. The condyles of the distal extremity of the femur are large and extend backward; the lateral one is largest and most tuberous.

The patella (knee cap) protects the stifle joint and provides a greater bearing surface for the tendon of the

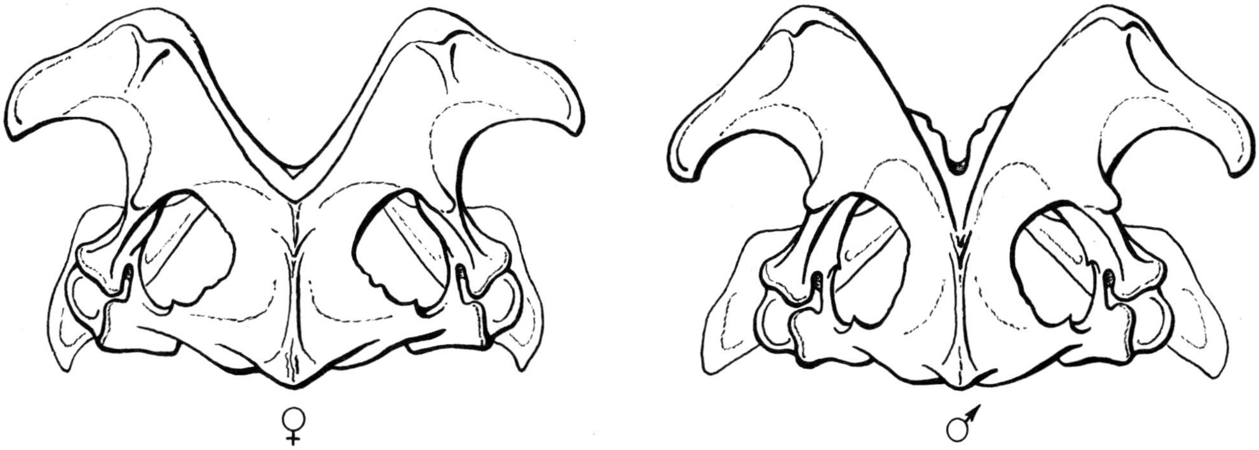

FIGURE 45. *The V-shaped ischiatic arch* (right) *of a pronghorn buck compared with the U-shaped arch* (left) *of a doe.* Illustration by Susan Kraft Ball.

greater extensor muscle to play on the trochlea of the femur. It is the largest sesamoid bone in the body, being about 1.3 inches (3.3 cm) long by 1 inch (2.5 cm) wide by 0.8 inch (2.0 cm) thick in an adult buck. The base of the bluntly trihedral patella is located proximally, and the distally located apex ends in a truncated point. The curved, smooth inner surface articulates with the patellar surface of the femur, and the irregular outer surface has many attachment surfaces for the tendon.

The tibia, which forms the stifle joint with the femur and the talocrural (hock) joint with the tarsal and metatarsal bones, is about 10.3 inches (26 cm) long. The head is large, and the tibial tuberosity is sharp-edged in front. The body is stout and convex anteriorly, but doubly ridged posteriorly for the attachment of the long flexor muscles. At the distal end of the tibia, the articular facets for contact with the talus are deep. The vestigial fibula consists of a depending spicule that represents its upper extremity, or head, and the inferior crown-shaped lateral malleolus.

The talocrural joint consists of calcaneus, talus, adnate cuneiforms, scaphoids and cuboid. The calcaneus is elongated anteroposteriorly and has a distinct waist between the heel at the posterior end and the larger anterior end, which is the site of all articular facets. The most distal articular facet faces down slightly and is lateral. The talus has a convex surface for articulation with the calcaneus and two prominent trochlear ridges at its proximal end for articulation with the tibia. The lateral side of the bone is somewhat concave; its medial side is flat.

The cannon bones of the forelimb and hind limb are strikingly similar in size and shape. Those of the hind limb are about 0.2 inch (0.5 cm) longer and slightly flattened laterally compared with those of the forelimb. The anterior groove of the hind limb is more distinct, running for almost the entire length of the body and terminating in a larger foramen at its distal end. The proximal end is divided into three articular facets, which articulate with the distal surfaces of the tarsals, rather than two as are present on the front cannon bones, which articulate with the carpals.

The phalanges of the hind feet are indistinguishable from those of the front.

VISCERA

Heart

The cone-shaped heart of the pronghorn is slightly elongated (Table 42). It is the muscular pump that circulates blood throughout the body and is divided internally by a curved, longitudinal septum into right (cranioventral) and left (caudodorsal) parts. These, in turn, are partly divided into the blood-receiving chambers (atria) and the pumping chambers (ventricles). The internal partitions are indicated on the surface of the organ by grooves. The left atrium receives oxygenated (arterial) blood from the lungs, and the left ventricle pumps it to all parts of the body. The right atrium receives deoxygenated (venous) blood from all parts of the body, and the right ventricle pumps it to the lungs. The atria are small compared with the ventricles (Figure 46).

A firm cartilaginous body (the fibrous base or cardiac skeleton) lies at the anterior base of the aorta, beneath the right atrioventricular valve. The heart is enclosed in a membranous sac, the pericardium, which often is nearly covered with fat bodies. The pericardium, in turn, lies within the mediastinum—a space between the right and left pleural sacs. The apex of the pronghorn heart is directed slightly posteriorly toward the anterior wall of the diaphragm, and the pericardium is attached to the sternum between the fourth and fifth ribs. This allows flexibility but holds the apex in place. The caudal (inferior) vena

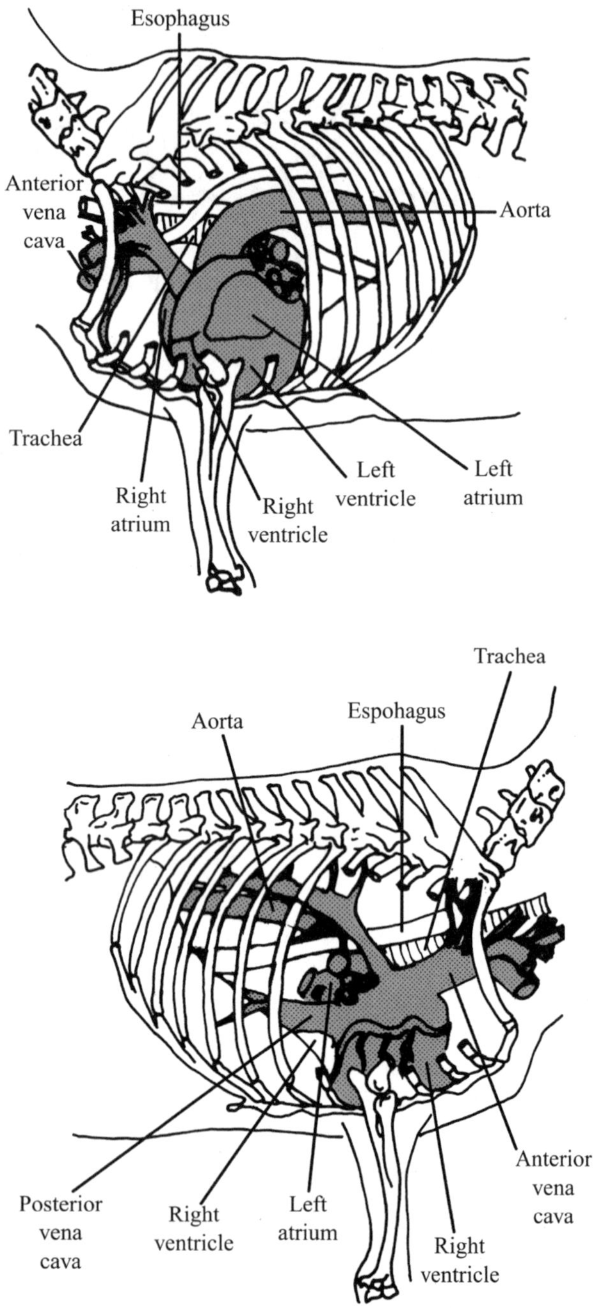

Figure 46. *Left side (*top*) and right side (*bottom*) of a pronghorn's cardiovascular system.* Illustration by Victoria Kurnat.

cava enters from behind and to the left, the fossa ovalis is closed, and the eustachian valve is large. The cranial (superior) vena cava has a thick circular layer of muscle where it enters the atrium.

Pulmonary System

The external nares or nostrils of the pronghorn form a shallow V, and each measures about 1.2 inches by 0.5 inch (3.0 by 1.3 cm). The nasal passages are long and convoluted, probably adaptations for temperature regulation and water conservation. The upper airways of the mouth, nose and trachea help regulate body temperature. The membranes are kept moist by diffusion of water from a dense network of small blood vessels just beneath the surface, which also warms the in-rushing air. The trachea and the larger bronchial tubes are lined with cells that have hairlike appendages, cilia, which constantly sweep a film of mucus and inhaled debris upward to the mouth to be swallowed. By the time outside air has moved only a short distance, it is saturated with water, warmed and generally cleansed of all but the smallest particles. The nasopharyngeal cavity is about 1.2 inches (3 cm) wide and 1.6 inches (4 cm) long. The posterior medial portion of the nasal septum projects downward from the roof of the pharynx into this cavity for about 0.2 inch (0.5 cm). Six openings occur in the region of the pharynx. Two internal nares above the soft palate communicate with the nasal cavity. Slitlike orifices of the eustachian tubes open under thin, semicircular, flaplike membranes on each side near the posterior, dorsal border above the soft palate. The opening into the larynx occurs on the posterior floor of the pharynx, and the orifice of the esophagus forms the most posterior opening (Figure 47).

The larynx guards the entrance to the trachea, serves as an air passageway, aids in vocalization and prevents inspiration of foreign material. The valvular function of the larynx, by means of the epiglottis, is vital because all substances swallowed must pass across its inlet in their course from the oral pharynx through the laryngeal pharynx to the esophagus.

The epiglottis, located on the anterior part of the pharynx, is elevated and bilaterally symmetrical. Its tip lies immediately behind the root of the tongue. When food is swallowed, the epiglottis curves posteriorly and covers the glottal opening of the trachea. The larynx is located directly caudal to the root of the tongue and the soft palate, ventral to the level of the atlas.

Laryngeal cartilages in the floor of the pharynx include the thyroid cartilage, about 1.8 inches (4.5 cm) wide, which is the largest and most anterior (Figure 38). It is formed by two quadrilateral plates that meet ventrally at an angle and then fuse. The thyroid cartilage is blunt in front, and the superior (anterior) horn is short, barely projecting above the shallow concave upper border. The posterior border is deeply scooped out above, but reversely arched below. The inferior horn is a cartilaginous rod. Except for a narrow notch close to the inferior horn, the lower border of the thyroid cartilage is straight, and the front is firmer than the flat, lateral plates.

The cricoid, also about 1.8 inches (4.5 cm) wide, is the second cartilage of the larynx. Its posterior shield is con-

TABLE 42. Measurements in inches (cm) of organs from adult Colorado pronghorn

	Sample size		*Mean*	
Organ measurement	*Male*	*Female*	*Male*	*Female*
Heart	14	29		
Longitudinal			5.5 (14.0)	5.2 (13.3)
Transverse			4.3 (10.9)	4.0 (10.1)
Liver	14	25		
Longitudinal			12.6 (32.0)	11.5 (29.3)
Transverse			7.3 (18.5)	6.7 (17.0)
Vertical			1.9 (4.8)	1.8 (4.6)
Spleen				
Longitudinal	15	24	6.0 (15.3)	5.7 (14.6)
Transverse	15	24	4.2 (10.6)	3.9 (10.0)
Vertical	11	16	0.5 (1.3)	0.4 (1.1)
Kidney	15	30		
Longitudinal			3.3 (8.5)	3.3 (8.5)
Transverse			1.9 (4.8)	1.9 (4.7)
Vertical			2.1 (5.3)	2.0 (5.0)
Brain[a]				
Longitudinal	9	16	3.0 (7.7)	3.0 (7.5)
Transverse	9	16	2.7 (6.8)	2.6 (6.5)
Vertical	8	15	1.5 (3.9)	1.5 (3.9)

[a] Beddard (1909:173) noted that "The outline of the brain as seen from above is remarkable for the squareness of the anterior portion."
Source: From Bear et al. 1973.

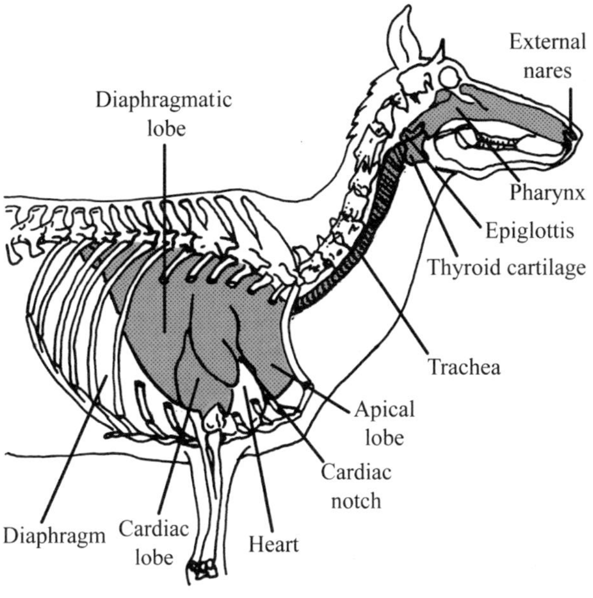

FIGURE 47. *Diagram of a pronghorn's pulmonary system.* Illustration by Victoria Kurnat.

vex; the upper and lower margins are rounded. The former is medially concave, but the latter is convex. The anterior ring completing the cricoid is bent slightly toward the trachea. The cricoid catilage is thicker than the thyroid cartilage.

The paired arytenoid cartilages rest on the anterior border of the cricoid and consist of two small elastic cartilaginous structures on each side of the glottal opening. Each arytenoid is a solid cartilaginous body of trihedral shape, with its apex directed posteriorly and curved medially. The faces and borders are slightly concave. The arytenoids are not fused along their medial borders but are covered by a mucous membrane that also extends over the cricoid cartilage.

The corniculate process of the arytenoid cartilage (cartilage of Santorini), composed of soft, yellow, elastic cartilage that is narrow and curved, reaches behind the arytenoid cartilage. The same yellow, elastic substance continues as two thin bands, downward and forward, from the anterior apices of the arytenoids to the inner thyroid fossae. These bands are the vocal ligaments that approximate in front, but leave behind them a wide wedge-shaped inferior aperture of the larynx.

The tracheae of pronghorn have variable numbers of tracheal cartilages (C-shaped cartilaginous bands). In 10 pronghorn from Montana and Wyoming, the average number to the bifurcation was 55 (range, 50–59). At the high division of the bronchus (a small branch leaving the main trachea about 2 inches [5 cm] before it divides), the rings averaged 44 (range, 40–47) (O'Gara: unpublished data).

Oxygen from the atmosphere and carbon dioxide from the blood are exchanged in the lungs. The two lungs possess many features in common. Each has a slightly concave base adjacent to the diaphragm and an apex that lies in the narrow anterior part of the thoracic cavity. The lungs are passive during respiration. Air is drawn into them because of negative pressure produced by the intercostal muscles drawing the ribs cranially and the diaphragm contracting and moving caudally.

The left lung is divided into three lobes (Figure 47). From anterior to posterior, they are the apical, cardiac and diaphragmatic. The right lung is divided into five lobes. From the anterior end, they are the apical (partly divided into two parts), anterior and posterior cardiacs, diaphragmatic and intermediate. The uppermost of these receives a separate bronchus at about 2 inches (5 cm) above the ordinary bifurcation. The intermediate lobe is located in a medial position posterior to the heart. It is separated from the other lobes by a mesentery (caval fold). The postcaval vein runs along the dorsal part of this mesentery under the intermediate lobe. The diaphragmatic lobes are the largest in both lungs.

The diaphragm essentially is a sheet of muscle with a ligamentous center about 4 inches (10 cm) in diameter around the esophagus. The muscle is attached ventrally about 1 inch (2.5 cm) anterior to the bony posterior tip of the sternum. Attachment along the costal cartilages and tips of the ribs lends flexibility to the diaphragm to become flat or dome-shaped during breathing.

Spleen

The spleen stores and concentrates red blood cells and releases them during times of need. It filters the blood and removes old, red blood cells from circulation. It also produces many white blood cells and functions in the production of antibodies. However, the spleen is not essential to health because, if it is removed, its functions are taken over by other tissues.

The pronghorn spleen lies on the dorsal anterior surface of the rumen to the left of the entrance of the esophagus and is firmly attached to the rumen by its medial border. It is nearly purple in color, with tracings of gray. The medial third of the organ, where it adheres to the rumen, is somewhat dome-shaped on its dorsal surface and may be as thick as 1 inch (2.5 cm). The rest of the spleen is only 20% to 50% as thick, and the front and left edges are very thin and somewhat scalloped.

Digestive Apparatus

The digestive apparatus consists of the oral cavity, pharynx, alimentary canal and accessory organs. The accessory organs include the teeth, salivary glands, liver, gallbladder, pancreas and anal sac. The wall of the digestive tube is richly supplied with secretory epithelium and intrinsic glands. It is lined throughout by a mucous membrane that is continuous with the surface integument at the mouth and anus.

The following description proceeds from the pronghorn's mouth to its anus and closely follows that presented by Wenzel (1955). Measurements in the following text also are from Wenzel unless otherwise stated.

The lips are about 0.4 inch (1 cm) thick and consist largely of white connective tissue infiltrated with muscular fibers. The lower lip fits around the incisors with a shallow space about 0.6 inch (1.5 cm) deep, lined with smooth tissue, between it and the teeth. The upper lip is almost twice as thick as the lower lip and overhangs it by about 0.2 inch (0.5 cm). A lobe of tissue inside the lower lip and immediately behind each canine tooth precedes papillae that line both lips and become progressively longer and thinner back to the angles of union between the lips. Conical papillae continue inside the cheeks to about 0.4 inch (1 cm) behind the last molar. The soft lining of the mouth is pink and gray, and the hard gums consist of tough, fibrous, connective tissue.

The spatular tongue, slightly narrow about the middle and thin at the broadly rounded tip, fills most of the closed mouth. In a fresh specimen, the tip of the tongue can be pulled about 4 inches (10 cm) beyond the front teeth. Anteriorly, the tongue has a grayish hue, and is dark at the sides of the root where the whitish papillae are shortest and sparsest. About 12 craterlike papillae are distributed laterally at its base. Slender, whitish (filiform) papillae with thick bases form a V-shaped pattern on the base of the tongue, with the point directed anteriorly. Mushroom-shaped (fungiform) papillae are sparsely but evenly scattered among filiform papillae that cover the dorsal surface. Papillae are absent on the undersides of the tongue. The ventral surface is fastened to the floor of the oral cavity by a medial fold of mucous membrane marking the anterior margin of attachment. The root of the tongue curves sharply downward anterior to the space between the soft palate and the epiglottis.

A pair of parallel sublingual folds on each side of the tongue near its anterior end indicates the relative position of the sublingual salivary glands that lie beneath them and, like all salivary glands, pour their secretions into the oral cavity. The major portions of these glands lie under the posterior part of the sublingual folds at the level of the last molars. They decrease in size anteriorly and terminate about 0.8 inch (2 cm) anterior to the first premolars. They are about 3 inches (8 cm) long, 0.6 inch (1.5 cm) deep at the level of the last molar and are never more than 0.2 to 0.3 inch (0.5–0.8 cm) thick. Numerous small ducts lead to the area between the papillae of the sublingual folds along the base of the tongue. Mandibular lymph nodes, located below the angles of the mandibles, are easily confused with the ventral extremities of the submaxillary glands, which are medial to them.

Lateral sublingual folds, composed of a single row of large blunt papillae, extend posteriorly along the underside of the tongue between it and the gums of the cheek teeth. Medial folds without papillae extend back as far as the middle of the diastema.

The pronghorn palate is narrow and elongated. Anteriorly, the dental pad, of tough cornified epithelium, meets the incisiform teeth of the lower jaw when the mouth is closed. Posterior to the dental pad is a papilla between the openings of the incisive ducts that communicate with the nasal cavity. Posterior to the dental pad and continuing to the level of the last molar is the hard palate. On its surface, just posterior to the dental pad and extending posteriorly to the first premolars, are 12 to 15 regularly spaced, transverse ridges with notched edges. These ridges are bisected by a wavy, depressed medial line. The remainder of the hard palate has a smooth surface punctured by numerous ducts of the parotid salivary glands.

The dark red parotid glands are the largest and most prominent of the three pairs of salivary glands. They extend ventrally from the coronoid process of the mandible to the larynx and posteriorly from the ramus of the mandible for about 1.6 inches (4 cm). Ducts emerge along the ventral surface of these glands and open inside the cheeks between the second premolars and first molars. Numerous other openings are present; when Wenzel (1955) injected the ducts, small droplets of fluid exuded from numerous points on the posterior, smooth part of the hard palate.

The mandibular (submaxillary) salivary glands are light reddish-yellow and extend posteriorly from the rami of the mandibles lateral to the thyroid and cricoid cartilages, where they nearly join each other on the dorsal side of the esophagus. They extend dorsally under the parotid gland. Ducts of the submaxillary glands open on the floor of the mouth under the tongue.

Posterior to the molars is the soft palate, with a posterior membranous curtain that separates the cavity of the mouth from that of the pharynx. In its normal position, the soft palate slopes ventrally and posteriorly from its junction with the hard palate and is about 1.8 inches (4.5 cm) long. The uvula, a small fleshy pendulum, is at its center and continues laterally with a raised musculomembranous ring guarding the pharyngeal opening. Near the middle of the soft palate and about 0.4 inch (1 cm) on either side of the medial line are orifices with underlying palatine tonsils. Each tonsil measures about 0.2 by 0.4 inch (0.5 by 1.0 cm).

The mouth is continuous with the large anterior part of the funnel-shaped pharynx, which tapers toward the opening of the esophagus. Immediately above the soft palate and posterior to the internal nares, the pharynx is subdivided. The esophagus—the first part of the alimentary canal—is a muscular tube about 23 inches (58 cm) long that extends from the laryngeal part of the pharynx to the rumen. The opening of the esophagus, about 1.2 inches (3 cm) wide, is readily apparent at the posterior, dorsal side of the cricoid cartilage. As the esophagus passes posteriorly, it lies dorsal to the trachea for about 2 inches (5 cm) and then, at the level of the thyroid gland, gradually shifts to the left side of the trachea. It stays in that same relative position throughout the remainder of the neck and is held in close contact with the trachea by fascia. At the entrance of the thoracic cavity at the sternum, the esophagus lies dorsal to the major veins, arteries, nerves and lymph ducts. Its average diameter in the neck is about 0.6 inch (1.5 cm) in females and 1 inch (2.5 cm) in males. The esophagus continues as a straight tube slightly to the left of the medial line and dorsal to the thoracic organs in an enfolding mesentery all the way to the diaphragm. The esophagus passes through the diaphragm just ventral to the descending aorta. It immediately empties into the rumenoreticular orifice of the stomach.

The pronghorn stomach consists of four compartments—rumen, reticulum, omasum and abomasum—placed in relation to each other in the typical ruminant fashion (Figure 48).

The rumen occupies most of the space within the left side of the abdominal cavity. It is partly divided into two sacs by the right and left internal, longitudinal pillars—muscular thickenings of the rumen wall—indicated externally by grooves on the lateral surfaces of the rumen. Internally, these pillars are comparatively free of papillae. The dorsal sac is about 12 inches (30 cm) long and 6 inches (15 cm) deep, and the ventral one is about 14 by 5.5 inches (36 by 14 cm). The internal surfaces of both sacs are covered with numerous papillae about 0.1 to 0.3 inch (0.2–0.8 cm) long. The largest papillae are found in hollows and corner pouches. The texture of the rumen lining reminds one of a shag rug.

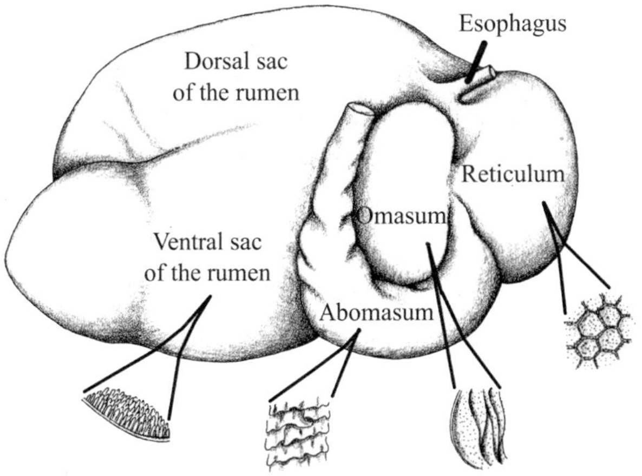

FIGURE 48. *Diagram of the pronghorn's four-chambered stomach and its internal surfaces (inserts).* Illustration by Susan Kraft Ball.

The rumen is attached dorsally to the body wall by mesenteries and to the diaphragm by thick connective tissues where the esophagus passes through. The anterior end of the dorsal rumen is rounded, and its junction with the reticulum coincides with the rumenoreticular groove. The posterior end is broadly rounded. It does not extend as far posteriorly as the ventral rumen, which it joins at the posterior groove. Although the ventral portion of the rumen is only about 2 inches (5 cm) longer than the dorsal portion, it extends about 4 inches (10 cm) farther posteriorly because its anterior end does not extend as far anteriorly. The ventral portion can be distinguished easily from the dorsal rumen because the posterior end of the former is more conical in shape than that of the latter.

The reticulum lies anterior to the rumen in a median concavity of the diaphragm. It is oval, about 6 inches (15 cm) long with a maximum width of 5 inches (13 cm), and its longest axis is directed dorsoventrally. The junction of the reticulum with the rumen is indicated externally by a groove on the dorsal surface. Internally, the point of junction is indicated by the rumenoreticular fold on the ventral margin of the large rumenoreticular orifice. The shallow, reticulate pattern of ridges on its internal surface is the most characteristic feature of the reticulum. These ridges are about 0.04 inch (1 mm) high and arranged in hexagonal or honeycomb patterns of irregular size and form. The entire internal surface, including the ridges, is covered by numerous small, pointed papillae.

The reticular groove is located dorsally on the internal surface of the reticulum. This groove, between the rumenoreticular orifice and the reticulo-omasal orifice, is about 2 inches (5 cm) long and bordered by muscular flaps about 3.5 inches (9 cm) long. The area enclosed by the flaps of the groove is smooth and about 1.0 to 1.2 inches (25–30 mm) wide.

The omasum is located on the right side of the anterior part of the rumen just posterior to the reticulum. It is about 4 inches (10 cm) long and 2 inches (5 cm) wide. The long axis is directed ventrodorsally and lies to the right of the medial plane at about the level of the seventh rib. The interior is lined with a series of leaflike projections covered by horny papillae. As food passes between these laminae, it is finely ground by the numerous papillae. The omasum is continuous with the reticulum and the abomasum by way of two orifices located close together on the left or medial surface of the omasum.

The U-shaped, thin-walled abomasum, or "true stomach" (the glandular part of the ruminant stomach) is ventral to the omasum. Beginning at the medial side of the omasum, it extends ventrally for about 3 inches (8 cm), curves to the right and continues transversely for 6 inches (15 cm). At this point, it curves dorsally for about 3.5 inches (9 cm) and joins the small intestine. The greater omentum is attached along the outside, greater curvature. The inside surface of the anterior end is characterized by a series of longitudinal rugae (folds). The rugae are much smaller in the pyloric end and run in a transverse direction.

The greater omentum, a fold of mesentery, extends from the rumen to the distal portion of the ascending colon. From the proximal points of attachment on the rumen and adhesions on the adjacent viscera, it extends posteriorly and then folds back on itself to form an apronlike sheet of mesentery lying ventral to and covering the intestines. The greater omentum often has fat bodies distributed uniformly between its layers.

The lesser omentum is much smaller than the greater omentum and extends from the right anterior borders of the greater omentum to the liver. It is attached along the anterior or parietal surface of the omasum, the pyloric portion of the abomasum, the surface of the proximal portion of the duodenum and the visceral surface of the liver.

The pylorus contains a thickened, circular sphincter muscle at the junction of the abomasum and the small intestine. It can be located by feeling for the thickening in the wall. The small intestine extends from the pylorus to the ileocolic orifice leading into the large intestine. It is the longest portion of the alimentary canal and is more or less coiled. After a pronghorn's death and the cessation of peristaltic contractions, its intestine increases in length,

The pronghorn digestive anatomy in relation to other visceral organs: A abomasum; C = colon; Cs = colic spiralis; Ce cecum; D = diaphragm; D–J = duodenum and jejunum; E = esophagus; G = greater omentum; H = heart; I = ileum; K = kidney; L = liver; Lu = lung; O = omasum; R = rumen; Re = reticulum; S = spleen; and T = trachea. (Opposite top) *The left side of a pronghorn carcass with left legs, ribs and flank removed. The rumen is so large and heavy that viscera are displaced from their normal configuration as soon as the restraining ribs and abdominal wall are removed. Also, gas builds rapidly within the gastrointestinal tract after death and contributes to displacement (note the slit in rumen [arrow] allowing gas to escape).* (Opposite bottom) *The right side of a pronghorn carcass with right legs, ribs and flank removed. This is the position in which a ruminant's carcass should be placed for necropsy. As can be seen in the top photo, the rumen obscures much of the abdominal viscera when the carcass is placed on its right side. The colon is protruding anteriorly from beneath the rumen, almost to the heart, because it is distended by gas. The greater omentum (G) shows as a thin membrane with a few streaks of fat because this is a fawn. In most adults, the greater omentum would be so engorged with fat that it would obscure the rumen and small intestines.* Photos by Bart W. O'Gara.

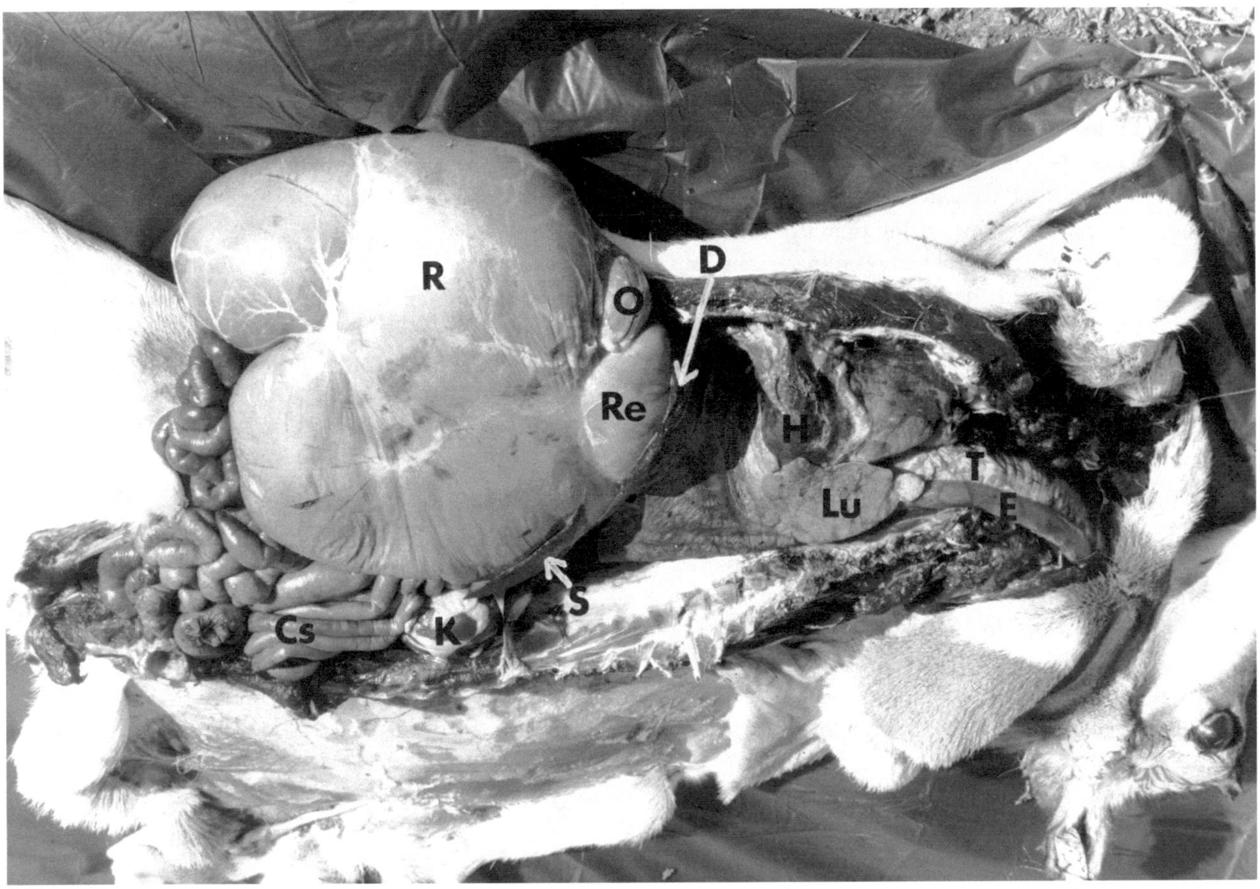
R
D
O
Re
H
T
Lu
E
S
Cs
K

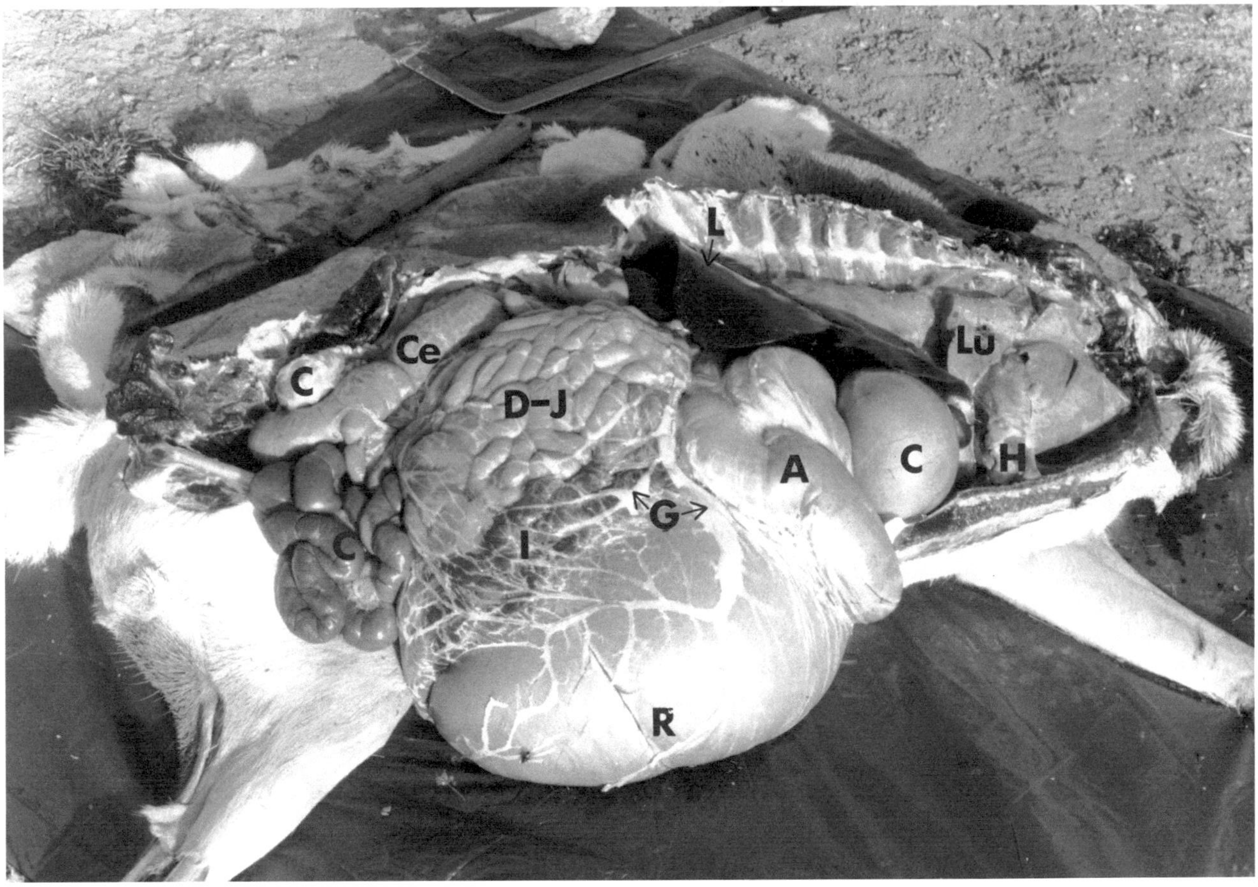
L
Lu
Ce
C
D-J
C
H
A
G
I
C
R

owing to the loss of muscular tone, and measurements can be little more than approximations.

Wenzel (1955) reported that the small intestine is about 44 feet (13.4 m) long, and its diameter varies from about 0.8 to 1.6 inches (2–4 cm) throughout most of its length. Specimens we measured were shorter—about 35 feet (10.7 m)—and much narrower, measuring from 0.5 to 1.0 inch (1.2–2.5 cm). Wenzel's material had been frozen and had lost all muscle tone and turgidity. Ours was measured fresh and soon after the animal was killed. Considerable individual variation occurs. For instance, Warwick and Williams (1975) reported that small intestines of 100 humans, measured shortly after death, ranged from 11 to 23 feet (3.4–7.2 m) for women and 16 to 26 feet (4.9–7.9 m) for men.

Mammals on high-fiber diets, such as ruminants, have longer small intestines than do omnivores or carnivores. Sisson and Grossman (1953) listed lengths of small intestines as 80 feet (24.2 m) for domestic sheep, 50 feet (15.2 m) for pigs and 13 feet (4.0 m) for dogs. These lengths are not directly comparable because these researchers do not give the sizes of the animals, except for "medium-sized" for dogs.

The proximal and most fixed portion, the duodenum, is about 2 feet (61 cm) long. It joins the convoluted area of the jejunum near the anterior flexure of the colon. The jejunum—the middle portion of the small intestine—usually appears empty or relatively so. It is about 18 feet (5.5 m) long, with an average width of approximately 0.6 inch (1.6 cm) except for the distal end, which is about 1 inch (2.5 cm) wide. It appears light colored in fresh specimens and comprises approximately the anterior half of the convoluted small intestine that lies right of the medial line and just posterior to the caudal lobe of the liver.

Distinctive differences, particularly in the mucosa, exist between a typical portion of the jejunum and that of the ileum. However, no marked gross differences in the character of the walls differentiate the jejunum from the ileum. The ileum is about 14 feet (4.3 m) long and 1 inch (2.5 cm) wide except for about the last foot (30 cm) of its length, which has thick walls and is only about 0.4 inch (1 cm) wide. It is dark colored in fresh specimens and comprises the posterior half of the convoluted small intestine posterior to the jejunum.

At the junction of the ileum with the colon is the cecum, a conspicuous blind pouch about 14 inches (35 cm) long with a maximum width of 2 inches (5 cm). The proximal 6 inches (15 cm) lie medially in the lumbar region, and the remainder lies free in the sacral region or pelvic cavity.

The colon, or large intestine, is about 18 feet (5.5 m) long. At its junction with the small intestine, it is approximately the same diameter as the cecum. Its diameter gradually decreases until it is about the same as the small intestine where the ascending colon begins. From this point, the diameter gradually increases until the descending colon is 1.6 inches (4 cm) wide at its distal end in the sacral region, where it joins the rectum.

The colon is divided into three parts. The colic spiralis, with its seven coils, lies medial to the ileum and lateral to the cecum. The ascending colon extends anteriorly and dorsally from the colic spiralis to a position immediately dorsal to the junction of the duodenum and the jejunum. Many lymph nodes, some nearly black, occur along this section of the colon. The descending colon extends posteriorly and dorsally from the ascending colon and turns to the right so that it comes to lie about midway between the kidneys. As it extends posteriorly, it is to the right of the cecum in the sacral region.

Resembling a pouch terminated by a 2.5-inch (6.4 cm) muscular tube, the rectum is about 2 inches (5 cm) in diameter and 8 inches (20 cm) long. It passes through the semicircular sacral cavity, completely surrounded by bones, from the colon to the anus. The anus is formed by the outwardly turned sphincter muscle at the end of the rectum, and indicated exteriorly by a circular area about 0.8 inch (2 cm) in diameter and devoid of hair immediately ventral to the tail.

Liver

The liver lies in a concavity of the diaphragm anterior to the reticulum. Its anterior surface follows the curvature of the diaphragm, causing it to "wrap around" the front of the rumen, and it extends as far posteriorly as the twelfth rib. It is attached dorsally to the diaphragm by a strong ligament and is divided by an anterior marginal fissure into larger right and smaller left lobes. The right lobe contains the gallbladder and, to the right of it, the triangular caudate lobe. The lobus spigelius is a very thin, small lappet near the transverse fissure, immediately to the left of the left hepatic duct. This lobe is nearly vestigial on some pronghorn livers. The vena cava passes along the inferior margin of the liver.

Weights and measurements of livers change rapidly with the condition of an animal; measurements are included in Table 42.

Gallbladder

The distal end of the gallbladder is about 1.5 inches (38 mm) from the distal edge of the right lobe of the liver. The cystic duct from the gallbladder extends about 1.0 to 1.5 inches (2.5–3.8 cm) dorsally along the visceral surface of the right lobe. It is joined there by the hepatic ducts from the right and left lobes of the liver to form the common bile duct, which empties into the duodenum about 8 inches

(20 cm) distal to the pylorus. The point where the common bile duct empties into the duodenum is about 3 inches (8 cm) anterior to the opening of the duct from the ventral lobe of the pancreas.

Pancreas

The pancreas is whitish or light yellow to pink and consists of two lobes. The ventral lobe occurs on the dorsal side of the descending portion of the small intestine; its anterior end begins at the level of the bile duct and projects posteriorly toward the right kidney. The dorsal lobe actually is a continuation of the anterior end of the ventral lobe and extends to the left as far as the most posterior edge of the spleen.

The ventral lobe is about 3 inches (8 cm) long and 1.2 inches (3 cm) wide, whereas the dorsal lobe is about 2.5 by 3.5 inches (6 by 9 cm). The duct from the dorsal lobe of the pancreas joins the common bile duct about 0.4 inch (1 cm) before the bile duct empties into the duodenum. A duct from the ventral lobe empties into the duodenum about 3 inches (8 cm) posterior to the entry of the common bile duct.

Urogenital System

The urinary organs include the kidneys, which remove metabolic wastes from the blood and excrete them in the urine; the ureters, which carry urine to the bladder; the urinary bladder, which stores urine; and the urethra, which serves to void urine from the bladder (Figure 49).

Pronghorn kidneys are reddish brown in color and situated against the dorsal wall of the abdomen on either side of the spine posterior to the last ribs. The ventral surface is covered by peritoneum. The surfaces of these bean-shaped organs are smooth. They are round in cross section and about 2.5 inches (6 cm) long. The medial border of each kidney is indented by an oval opening, the hilus, through which the ureter opens into the renal sinus. The sinus contains the renal pelvis—a funnel-shaped, saclike structure that accumulates urine from the collecting ducts of the kidney before the urine passes into the ureter. The renal pelvis is subdivided into calyces, and the renal pyramids (pyramids of Malphigi) are rounded. The cortical substance is about 0.3-inch (0.8 cm) thick. When a pronghorn is in good condition, each of its kidneys is embedded in a capsule of fat. The ureters appear as white tubes beginning at the pelvis of each kidney and penetrating the undersurface of the wall of the urinary bladder at commencement of the neck and outside the ductus deferens of males.

The urinary bladder is a hollow organ that varies in form, size and position depending on the amount of urine

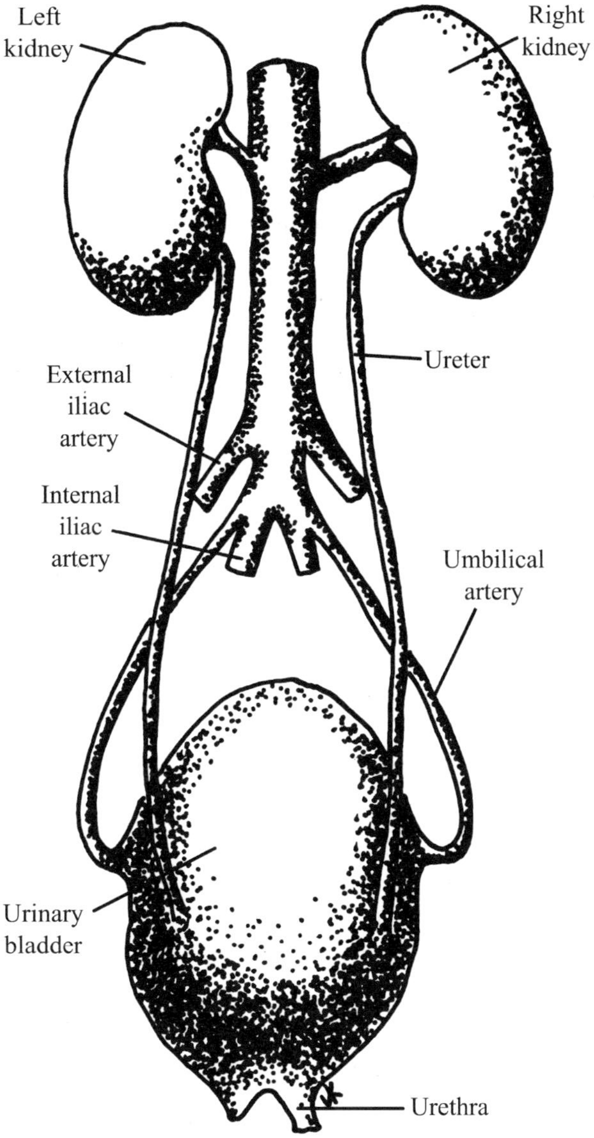

FIGURE 49. *Diagram of a pronghorn's urinary system.* Illustration by Victoria Kurnat.

it contains. It receives urine from the kidneys, through the ureters, and stores it until it is disposed of through the urethra. The bladder is situated on the pelvic floor when empty or nearly so. It does not protrude beyond the anterior margin of the pelvic girdle except when distended by urine. When empty, it is more than twice as long as wide, but as it fills, it becomes oval to nearly round. In one freshly killed specimen with a moderately distended bladder, the bladder was 3.1 inches (8 cm) wide and 3.9 inches (10 cm) long. The largest pronghorn bladder we have measured was 7 inches (17.8 cm) by 3.5 inches (8.9 cm). The pronghorn bladder is very small if examined when in a contracted condition. This may have been the condition reported by Seton (1927), who stated that the urinary bladders of these animals were so small that they could not hold 20 drops of water. He attributed the small size to the dry environmental conditions where a minimum of water was available.

In bucks, the dorsal surface of the urinary bladder is associated with the rectum, the prostate gland and the terminal portions of spermatic cords where they enter the prostate gland. The pelvic portion of the urethra is flattened and about 2.5 inches (6.4 cm) long, and the portion in the penis is about 5.6 inches (14.3 cm) long.

In females, the urinary bladder is in contact with the uterus, cervix and vagina, and the urethra is slightly flattened and fused dorsally to the wall of the vagina. The length of the urethra varies with that of the vagina from about 3.5 to 5.0 inches (8.9–12.7 cm).

Female Reproductive Tract

The reproductive tract of the female pronghorn consists of two ovaries, two uterine tubes (oviducts), a bicornuate uterus, a muscular cervix, vagina and vulva (Figure 50). Ova are produced in the ovaries and conveyed by the oviducts to the uterus. Each oviduct has a funnel-like infundibulum that receives the ova from the ovary and passes it through an ostium into the ampulla, the thin-walled portion of the oviduct with a comparatively large lumen where fertilization takes place. The isthmus is thick-walled and muscular, with a comparatively small lumen; it joins the ampulla to the uterine horn through the uterotubal junction. The ova becomes an embryo and then a fetus in the uterus, which is sealed from the outside during pregnancy by the muscular cervix. The vagina is a dilatable passage through which fetuses are expelled. The terminal segment of the tract, the vulva, also serves for the expulsion of urine. Udders are composed of skin glands, but are so closely associated functionally with the reproductive tract that they usually are described with it.

The ovaries, oviducts and uterus are attached to the dorsolateral walls of the abdominal cavity and to the lateral walls of the pelvic cavity by paired double folds of peritoneum called the right and left broad ligaments. Each broad ligament contains an ovary, uterine tube and uterine horn. It does not support or suspend the genitalia in the body cavities, but rather unites its components.

Female reproductive tracts, especially gravid ones, cannot be measured in a meaningful way after removal from the animal because they flatten and stretch. Measuring them in situ is difficult because of variability between individuals, and they change shape drastically with advancing pregnancy. Figure 50 was used as a guide for the measurements shown in Figure 51. Measurements 1 and 2 (of Figure 50) were taken on both sides, but left and right measurements were similar and, consequently, averaged.

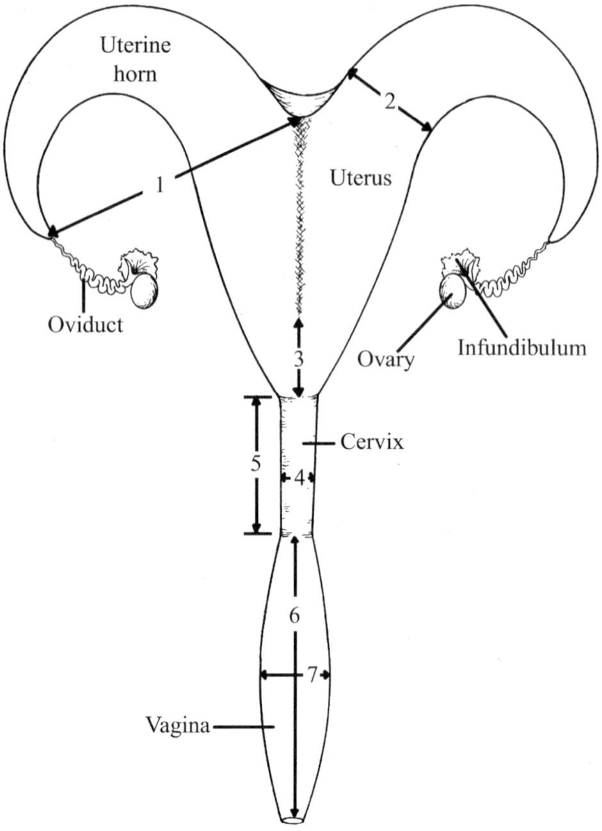

Figure 50. *Diagram of the female pronghorn reproductive tract showing measurements taken in situ for Figure 51: 1 = spread of uterine horns; 2 = diameter of uterine horns; 3 = length of uterine body (between fused uterine horns and cervix); 4 and 5 = diameter and length of cervix; and 6 and 7 = length and width of vagina.* Illustration by Susan Kraft Ball.

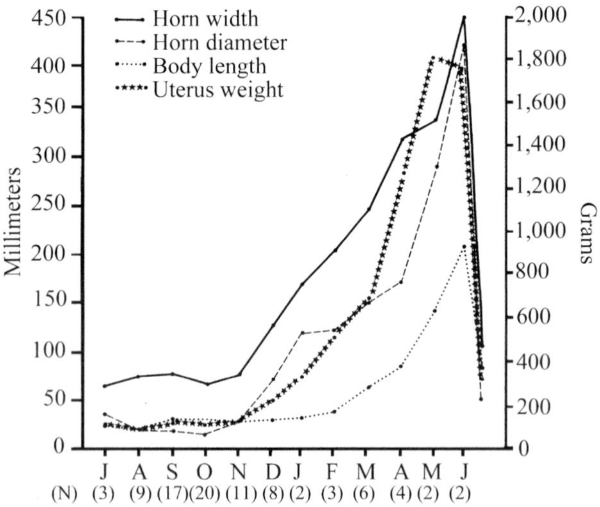

Figure 51. *In situ measurements (see Figure 50) by month of collection, of pronghorn uteri and weights of uteri without fetuses and fetal membranes. The numbers in parentheses represent the number of samples.* Illustration by Susan Kraft Ball.

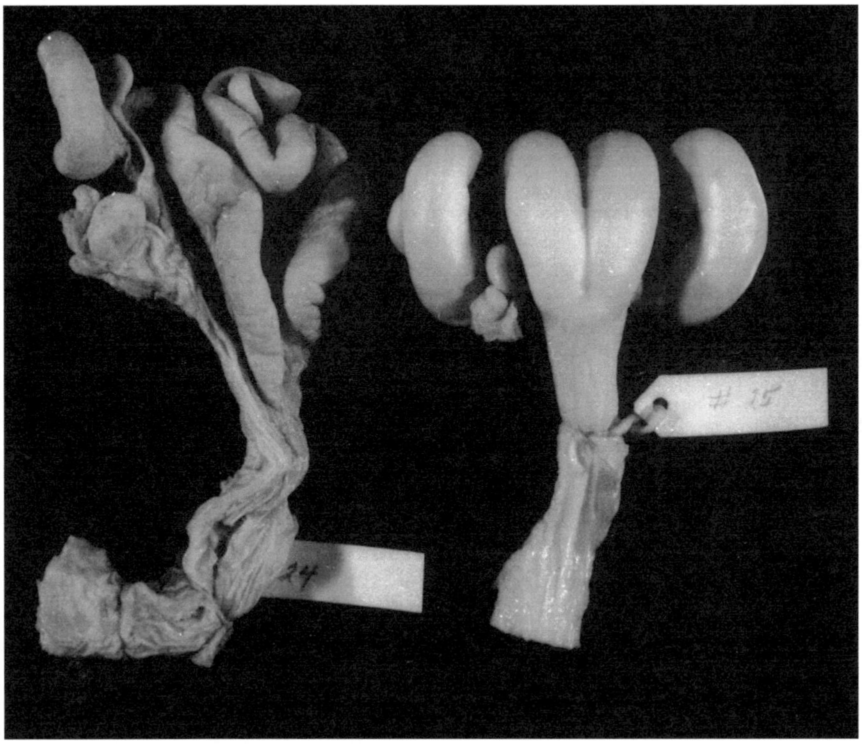

The uterine horns of young and prime-aged does (right) *curl, putting the uterotubal junction close to the ovaries; the oviducts are very convoluted. The horns of does 8-year-old and older* (left) *are flabby and less curled, and the oviducts appear stretched with no convolutions. The curve of the uterine horns and the connective tissue holding them together cause the pronghorn uterus to appear bipartite. When the connective tissue is removed, the two horns are seen to fuse along their lower portions, forming a short, single tube as the body of the uterus (corpus uteri). Thus, the pronghorn uterus is bicornuate, and each horn tapers gradually toward its oviduct.* Photo by Bart W O'Gara.

The ovaries of the pronghorn are variable in shape. They generally are spherical, but those of nongravid does are somewhat flattened. They are attached by mesovaria to the broad ligaments about one-fourth of the way from the uterine body to the uterotubal junctions. They are pinkish when fresh and somewhat translucent when they contain many vesicular (Graffian) follicles. Ovaries of pregnant pronghorn in Alberta ranged in diameter from 0.4 to 0.9 inch (11–23 mm). They generally were ovid but sometimes irregular in outline because of bulging corpora lutea. Weights of individual ovaries showed marked variations among females of similar age and between gravid and nongravid immature females. Mean ovarian weight of yearlings was statistically similar to those for 2- and 3-year-old animals, but smaller than the means for older does (Mitchell 1980). Mitchell recorded a mean weight of 145 milligrams for 47 ovaries of nongravid fawns to a maximum of 2,168 milligrams for an ovary from an 8-year-old pregnant doe.

The mean length of 16 oviducts from eight does was 2.9 inches (7.3 cm). An oviduct that was 2.6 inches (65 mm) long had an ampulla 1.1 inches (2.9 cm) and an isthmus 1.5 inches (3.6 cm) in length. The ampulla had a lumen diameter of 0.087 inch (2.2 mm), with complex branching villi and numerous secretory cells. The isthmus had a lumen diameter of only 0.035 inch (0.9 mm), with simple blunt villi and few secretory cells. The wall of the ampulla was only 0.0004 inch (0.10 mm) thick; that of the isthmus was 0.015 inch (0.38 mm) with many muscle fibers. Cilia were more abundant in the ampulla than in the isthmus (O'Gara 1968).

The lining of the uterus is characterized by four longitudinal rows of caruncles, thickenings of the uterine epithelium for implantation of fetal villi in each horn. Caruncles are not found in the body of the uterus or in the last inch (25 mm) of each uterine horn next to the oviduct. In these areas, the epithelium forms many narrow ridges running parallel to the length of the uterine horns. The caruncles lie against one another when the nongravid uterus is intact so that the lumen is an irregular slit with the tops of the caruncles touching or nearly touching each other. The mean number of caruncles in 87 uteri collected in Montana and Yellowstone National Park was 92.2, with a range of 46 to 180 (O'Gara 1968). The average numbers for left and right uterine horns were similar. The mean number of caruncles in nine does collected in Colorado was 119, with a range of 87 to169 (Firchow 1986). In fixed uteri, the surface of the tunica mucosa (endometrium) usually is white, but a few are light salmon to light brown in color. The caruncles are more pigmented than the rest of the tunica mucosa, with pigmentation being heaviest in postpartum animals. The caruncles are of the flat or intermediate type. They are extremely variable in size and number, so measurements mean little. The volume of the caruncles removed from uteri varied from 2.6 cubic centimeters for four females collected during August shortly before the breeding season to 654 cubic centimeters for a female collected during early June, immediately before giving birth, indicating a 250-fold increase during pregnancy (O'Gara 1968).

The cervix of the pronghorn is thick-walled compared with the uterus and vagina it separates. Cervices in 48 nonpregnant and early pregnancy animals collected in Montana during July through October averaged 1.3 inches (34 mm) in length and 0.7 inch (1.8 cm) in width. The cervix of a near-parturient doe had increased to 5.3 by 0.9 inch (13.5 by 2.3 cm) in preparation for passage of the fetuses in early June (O'Gara 1968). Cervices in 26 does collected throughout the year in Colorado were 2.7 inches (7.0 cm) in length and 0.7 inch (1.7 cm) in width (Bear et al. 1973). Nine cervices of does collected in Montana each had four or five valves (Figure 52).

The first (anterior) valve opens into the terminal end of the uterus during early gestation and periods of nonpregnancy. In late gestation, the cervices acquire a ventral flexure and the uterus forms a cul-de-sac that reaches about 1.2 inches (3 cm) posterior to the cervical openings. The openings of the valves into the uteri are very small and located in the centers of small mounds. The first valves are the longest, being about 0.5 inch (1.2 cm) long in nonpregnant animals. In nonpregnant animals, intervalvular spaces about 0.4 inch (1.0 cm) long are found posterior to the first valves. The three (or four) posterior valves are only about 0.12 to 0.16 inch (0.3–0.4 cm) long and close enough together that papillae extending backward from one almost touch the next one. The lumens of the first valves are lined with ridges of tissue that form papillae 0.12 to 0.16 inch (0.3–0.4 cm) long at the posterior ends. The papillae on the second valves are about 0.08 inch (0.2 cm) long; and on the third, 0.12 to 0.16 inch (0.3–0.4 cm) long; on the fourth, 0.20 to 0.24 inch (0.5–0.6 cm) long; on the fifth (posterior), 0.31 to 0.47 inch (0.8–1.2 cm) long. All cervical measurements are taken on preserved material, so some shrinkage probably occurred. With advancing gestation, lengthening of the cervices seemed to occur mostly in the posterior portions as the three valves there became farther apart. The flexure of the cervices, typical of late gestation, took place in the intervalvular spaces between valves one and two (O'Gara 1968).

Length of the vagina does not vary much with the stage of pregnancy. The mean length for 30 animals collected during October and November in Montana was 4.5 inches (11.5 cm), and for four pronghorn collected in April, it was 6.5 inches (16.4 cm) (O'Gara 1968). The average length and width of the vaginas of 26 does collected in Colorado were 6.1 inches (15.5 cm) and 0.7 inch (1.9 cm), respectively (Bear et al. 1973). Of 88 does collected in Montana throughout the year, those averages were 5.0 inches (12.7 cm) and 0.6 inch (1.6 cm), respectively (O'Gara 1968).

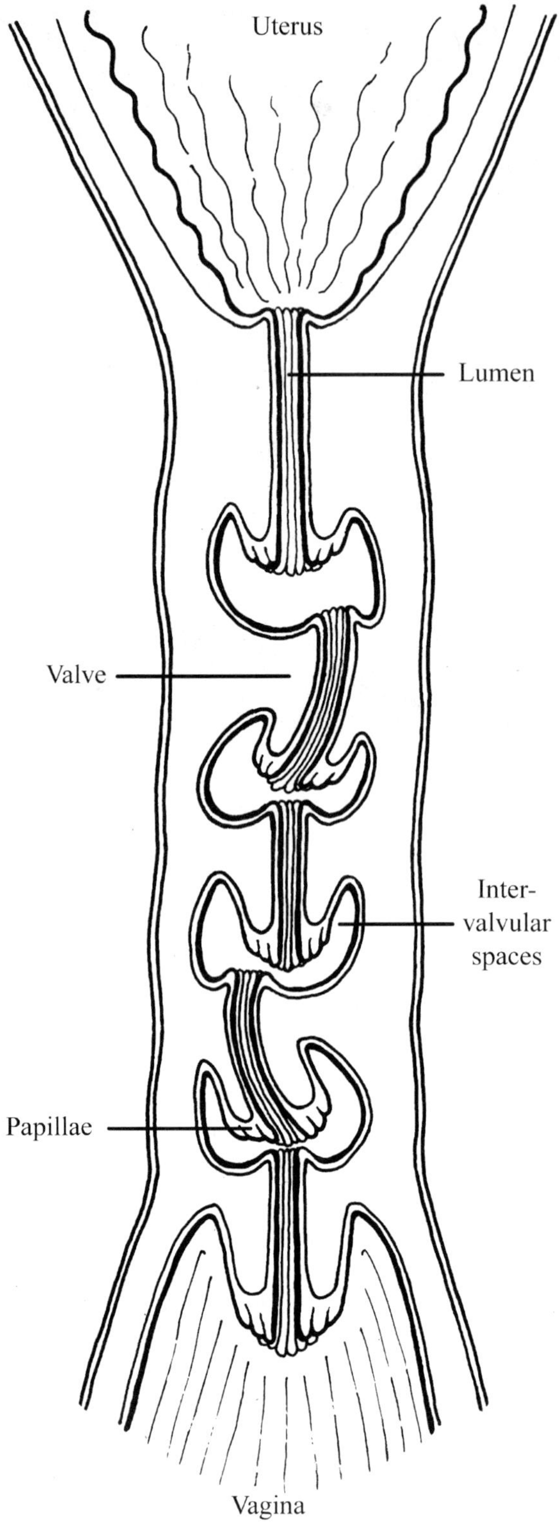

Figure 52. *Cross-section diagram of a pronghorn doe's cervical valves.* Illustration by Susan Kraft Ball.

As measured from 18 does collected in Colorado (Bear 1967a), the vulva is slightly more than an inch (2.5+ cm) long and about 1.2 inches (2.9 cm) wide. The labia are

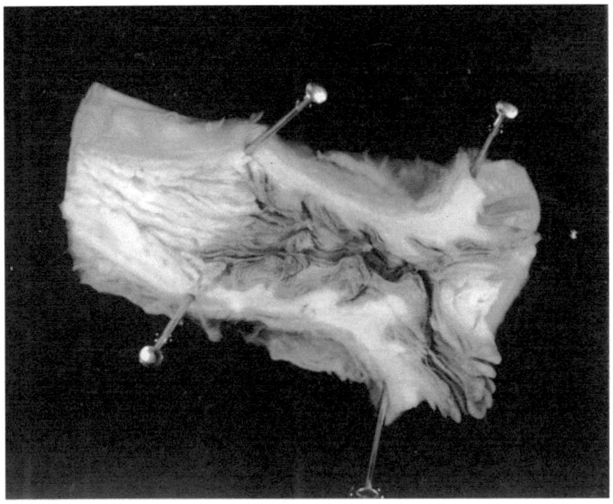

The openings through the cervical valves alternate dorsally and ventrally and the middle ones are offset horizontally so the resulting passageway resemble a corkscrew. The valves consisted of thickenings of the cervical wall, and longitudinal folds of tissue form papillae at the posterior (vaginal) ends of the valves. Photo by Bart W O'Gara.

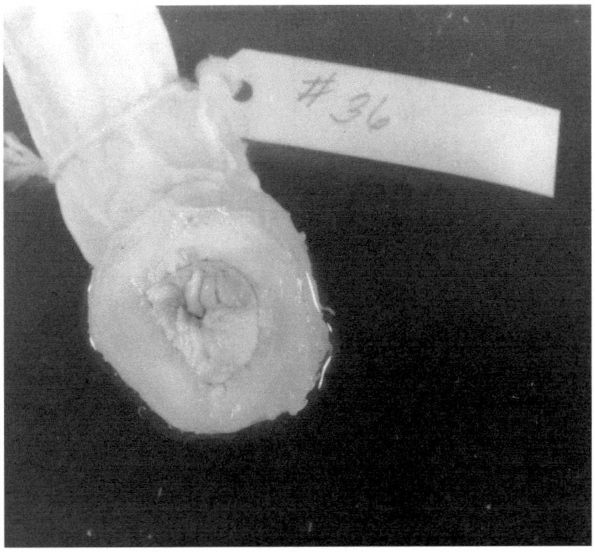

The papillae of the posterior cervical valve extend their entire lengths into the vagina. The papillae tips are always folded tightly together so openings cannot be seen until the papillae are forced apart. Photo by Bart W. O'Gara.

thick, and the ventral commissure is pointed and projected downward.

The udder, with its four teats located between the hind legs, is rather small and compact for the size of the pronghorn doe. It is difficult to delineate for measurements. Four lactating does collected in Colorado during June, July and August had udders with mean measurements of 8.3 (length) by 6.0 (width) by 3.1 (depth) inches (21.2 by 15.3 by 7.9 cm); the same measurements for eight does during September through April were 2.7 by 1.7 by 0.9 inches (6.8 by 4.3 by 2.4 cm) (Bear 1967). The udders of 71 adult does collected throughout the year in Montana had an average weight of 7.5 ounces (212 g)—the lightest mean weight was that of nine udders collected during January through March (2.9 ounces [82 g]), and the heaviest was that of four udders collected during the peak of lactation in June (30.9 ounces [875 g]). The mean length of teats of 93 does collected throughout the year was 0.9 inch (2.2 cm), with little difference between sizes of teats of lactating and nonlactating does (O'Gara 1968).

Male Reproductive Tract

The reproductive tract of a male pronghorn consists of the scrotum, two testes, the epididymides, ductus deferens, prostate gland, male urethra and penis (Figure 53). Sperm is produced in the testes, stored in the epididymides and conveyed to the urethra by the ductus deferens for ejaculation. The accessory prostate gland discharges its secretions into the urethra where they mix with fluid and sperm from the deferens ducts immediately before ejaculation.

The scrotum is a membranous pouch divided by a median septum into two cavities, each of which is occupied by a testis, an epididymis and the distal part of the spermatic cord. The median partition is made up of all the layers of the scrotum except the skin. The scrotum is covered with white hair and is located approximately two-thirds of the distance from the preputial opening to the anus. It lies between the thighs, has a spherical shape and is indented in an oblique craniocaudal direction.

The testes (male gonads), are located within the scrotum. Each oval testis is thicker dorsoventrally than from side to side. In normal position, the testis is situated obliquely, with the long axis running dorsocaudally. The surface is covered by the visceral vaginal tunic—a serous membrane continuous with the parietal peritoneum of the abdominal cavity. Internal to the vaginal tunic, the *tunica albuginea*—a dense, white, fibrous capsule—adheres tightly to the testis. The gland is suspended in the scrotum by the spermatic cord.

The size and weight of pronghorn testes during the first year of life increase through normal maturational growth. During subsequent years, changes in testicular weights are hormonally controlled (Mitchell 1980). Seasonal testicular weights for northern pronghorn are given in Table 43.

The epididymis lies along the dorsolateral border of the testis. The head (caput) of the epididymis begins on the medial surface of the testis, but immediately twists around the cranial extremity to attain the lateral side. It is slightly larger than the remainder of the epididymis. It

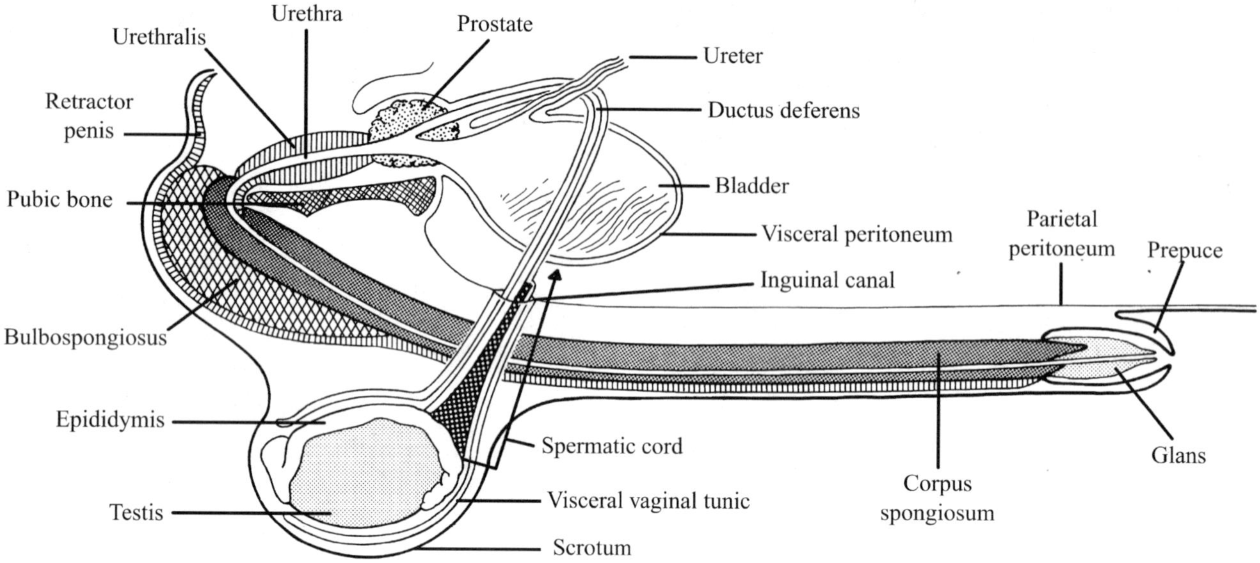

FIGURE 53. *Cross-section diagram of the male pronghorn genitalia.* Illustration by Susan Kraft Ball.

continues as the body (corpus), which runs along the dorsolateral surface of the testis, and then as the tail (cauda epididymis), which is attached to the caudal extremity of the testis. It continues craniodorsally up the spermatic cord as the ductus deferens. The curved epididymis has its concave surface in juxtaposition with the testis.

The deferent duct (ductus deferens) is the continuation of the duct of the epididymis. Beginning at the tail of the epididymis, it runs along the dorsomedial border of the testis, ascends in the spermatic cord and enters the abdominal cavity through the inguinal canal. The two deferent ducts enter the craniodorsal surface of the prostate. They lie adjacent to each other, one on either side of the median plane. They run caudoventrally through the dorsal part of the gland to open into the urethra by two slits.

Each spermatic cord is composed of the ductus deferens and the testicular vessels and nerves, along with their membranous coverings. These structures pass through the inguinal canal during the descent of the testis. The inguinal canal is a sagittal slit between the abdominal muscles. The spermatic cords are about 1 foot (30 cm) long. On reaching slightly beyond the middle of the base of the urinary bladder, they lie alongside each other on the median line. They proceed, like a single flattened tube, on the neck of the bladder, toward and between the lobes of the prostate. The bilobed prostate gland surrounds the proximal end of the membranous portion of the urethra. The gland looks much like half a walnut meat. The two lobes form a horseshoe that embraces the terminal portions of the spermatic cords.

The urethra is the canal that conveys urine from the bladder to the exterior of the body. The male urethra carries both urine and seminal secretions to the distal end of the penis. The prostatic portion passes through the prostate gland. It extends from the urinary bladder to the caudal edge of the prostate, where the membranous urethra begins. The membranous portion of the urethra tapers at either end and runs through the urethralis muscle between the prostate gland and the bulb of the penis (bulbospongiosus). The urethralis muscle constitutes a continuous arch from the neck of the urinary bladder to the bulb of the penis. Cowpers glands seem to be absent, but their function may be supplied by a bilinear, long patch of minute ducts or crypts on the floor of the anterior half of the urethral cavity. Glandular tubes open within the urethra (Murie 1870). The cavernous (penile) portion of the urethra begins at the entrance of the membranous

TABLE 43. Seasonal changes in the mean combined weights of both testes from yearling and adult pronghorn

Period	*Sample size*	*Mean weight in ounces (g)*
January/February[a]	7	0.56 (16)
January/February[b]	3	0.71 (20)
March/April[a]	13	0.85 (24)
March/April[b]	2	1.09 (31)
May/June[a]	8	1.02 (29)
June[b]	2	1.48 (42)
July/August[a]	8	1.55 (44)
July/August[b]	4	2.36 (67)
September/October[a]	10	1.30 (37)
September/October[b]	4	2.05 (58)
November/December[a]	9	0.92 (26)
November[b]	38	1.27 (36)

[a] Animals from Colorado, Montana and Yellowstone National Park (O'Gara et al. 1971).

[b] Animals from Alberta (Mitchell 1980).

urethra into the bulb of the corpus spongiosum, and extends to the external opening in the distal end of the glans (penis).

The penis is about 9 inches (23 cm) long from its tip to the flexure where it attaches to the pubic bones. The glans is relatively thick posteriorly, but flattened in front. The tip of the penis is somewhat spatular in shape, and the mouth of the urethra barely projects beyond the rounded tip. The prepuce is a complete tubular sheath of integument continuous with the skin of the ventral abdominal wall that, when the penis is not erect, contains and covers the glans. It is attached by a fold of membrane about 1 inch (2.5 cm) from the tip of the penis.

Endocrine System

The endocrine system differs from other body systems in that the component organs are not in direct continuity. They are positioned throughout the body in widely separated locations. Hormone synthesis is the one principal function shared by all components of the system. These organs and glands generally have an extensive blood vascular and lymphatic supply and an absence of secretory ducts. Endocrine glands exercise a major control over the organism. They supplement and augment the function of the nervous system in response to stimuli from both the internal and external environments. These stimuli directly or indirectly affect the specific metabolism of epithelioid cells of the endocrine tissues and organs, causing them to release relatively small quantities of hormones. The hormones may diffuse directly to other loci through the interstitial fluids or pass into blood or lymphatic vessels to be distributed to more distant receptors. Receptors, which may be cells, tissues or entire organs, are modulated in their activity by the hormone.

Adrenal glands are composed of two structurally and functionally different tissues. Each gland has an outer cortex and an inner medulla. The cortex produces steroids that regulate mineral balance by kidney function and regulate carbohydrate metabolism. The medulla modulates the immediate response of the nervous system to stress. The cortex and medulla together influence an organism's response to both acute and chronic stress.

The adrenal glands of pronghorn are small (Table 44), somewhat flattened bodies that are dark reddish in color. The right adrenal gland lies in contact with the anterior end of the right kidney and the caudate lobe of the liver. This gland is pyramidal in form, with a rounded apex and the base tightly adherent to the vena cava. The left adrenal gland is longer than the right, and both ends are rounded. It lies about 1 inch (2.5 cm) anterior to the left kidney in connective tissue and fat, and does not appear to be associated with any large blood vessels.

The thyroid glands are the largest of the ductless glands that perform only an endocrine function. The thyroids exert major control on the metabolic processes of the body and affect most body systems. The thyroid hormone is synthesized by the glandular parenchyma, stored intercellularly and released from the gland into the circulation. A second hormone produced by the thyroid, called thyrocalcitonin, or calcitonin, apparently lowers blood calcium by stimulating calcium uptake by the skeleton.

Thyroid glands of adult pronghorn are pinkish to yellowish in color; those of fawns are dark red. The glands of does are slightly larger than those of bucks, and left glands are slightly larger than right ones (Table 44). They are embedded in connective tissue against the first five or six rings of the trachea in the anterior portion of the neck. The connective tissue is thicker on the lateral side of the gland than on the medial side adjacent to the trachea. No isthmus is present between the two glands, as is found in many mammals. Structures, darker in color than the thyroids and about 0.2 inch (0.5 cm) in diameter, lie about 0.4 inch (1 cm) posterior to each thyroid in the same kind of

TABLE 44. Measurements of adrenal and thyroid glands from adult pronghorn collected in Colorado

Gland	*Sex*	*Sample size*	*Measurements*			
			Longitudinal[a]	*Transverse*[a]	*Vertical*[a]	*Weight*[b]
Left adrenal	Male	14	1.4 (3.5)	0.5 (1.3)	0.3 (0.7)	0.071 (2.0)
	Female	27	1.4 (3.7)	0.5 (1.3)	0.3 (0.7)	0.071 (2.0)
Right adrenal	Male	11	1.2 (3.0)	0.5 (1.3)	0.3 (0.8)	0.063 (1.8)
	Female	26	1.37(3.3)	0.5 (1.3)	0.3 (0.7)	0.067 (1.9)
Left thyroid	Male	13	1.4 (3.6)	0.6 (1.5)	0.2 (0.7)	0.085 (2.4)
	Female	27	1.6 (4.1)	0.6 (1.6)	0.3 (0.8)	0.092 (2.6)
Right thyroid	Male	12	1.4 (3.5)	0.6 (1.5)	0.3 (0.8)	0.078 (2.2)
	Female	21	1.4 (3.6)	0.6 (1.6)	0.3 (0.7)	0.085 (2.4)

[a] In inches (m).

[b] In ounces (g).

Source: Bear et al. 1973.

supporting fatty tissue; Wenzel (1955) thought these structures may be parathyroids.

The thymus gland lies ventral to the trachea and heart in fawns, but the exact age at which it is lost and how large it becomes before regressing is unknown. The thymus gland apparently produces white blood cells during early life. Yearlings shot during hunting season (about 16 months of age) appear to have thymus glands, but two of the suspected glands O'Gara sectioned and stained revealed only fat and connective tissue. However, size and presence or absence of thymus glands in fawns seem to depend on the health and condition of the fawns. Thus, more animals should be checked. The thymus gland of a fawn actually is two glands separated by about 0.5 inch (1.2 cm). The anterior portion (cervical) lies ventral to the trachea, and the posterior portion (cardiac) is on the ventral, anterior side of the heart. In fawns about 4 months of age, it projects about 3 inches (7.5 cm) anteriorly and posteriorly from the anterior point of the sternum. In young fawns, both portions are dark reddish. In older ones, the cervical portion is whitish, and the cardiac portion is pinkish. The cervical portion is slightly more cylindrical than the flatter cardiac portion.

Fifteen thymus glands were collected from fawn (1–15 days of age) carcasses, and mean weights for the cardiac portions increased from 0.02 to 0.03 to 0.08 ounce (0.58–0.80–2.3 g) during 3 consecutive years of study in Idaho. Apparently, range conditions were poor during the first year and improved throughout the study, resulting in better physical condition of does and their offspring (Bodie 1979). Weight data collected from the cervical portions during that study were minimal because the lobe was absent in some animals and dispersed into small sections in others.

The thymus glands of eight fawns about 4 months of age, shot in Wyoming during September 1983, had average weights and measurements as follows: cardiac—0.18 ounce (5 g; range, 3–8 g), length 2.7 inches (6.9 cm; range, 5.5–8.0 cm) and width 0.6 inch (1.6 cm; range, 1.3–2.2 cm); cervical—0.18 ounce (5 g; range, 2–10 g), length 3 inches (7.6 cm; range, 6.0–9.2 cm) and width 0.6 inch (1.6 cm; range, 1.4–2.0 cm). The sample size was small, but no sex differences were evident; the glands seemed to be slightly larger in larger fawns. Another fawn shot at the same time in the same place was less than half as large as any of the other eight. That female also had rough hair and no visceral or carcass fat. Her thymus appeared as scattered fragments of glandular tissue in a matrix of connective tissue. The scattered nature of this gland made weighing it impractical. The outline of the fragments resembled the shape of the thymuses in the larger fawns, but the cardiac portion was 30% shorter and 25% narrower than the above averages, and the cervical portion was 49% shorter and 25% narrower. A yearling buck shot at the same time appeared to have fat bodies slightly larger and of similar shape to the thymus glands of the fawns. These fat bodies were located exactly as were the thymus glands.

Other endocrine glands are described elsewhere in this chapter. The pancreas, which produces insulin that regulates the level of sugar in the blood, was described with the digestive system. The gonads, which produce hormones that promote sexual behavior and secondary sex characteristics as well as maintain pregnancies, were described with the urogenital system. Tissues known to produce hormones in the epithelial lining of the digestive tract, kidneys, pineal and pituitary glands, and placentas of other mammals have not been studied in pronghorn.

CONCLUSIONS

The long limbs with reduced numbers of toes, lightweight but strong bones and comparatively small digestive tracts of pronghorn are adaptations to use a minimum of energy during rapid locomotion. For ruminants, pronghorn have small stomachs and correspondingly large tracheae, lungs and hearts. The large tracheae and lungs provide for rapid intake of oxygen and elimination of gasses produced during violent exercise. The large heart provides the pumping power needed for oxygenating blood in the lungs and speeding it to all parts of the body. In combination, these factors provide the requisites for a swift runner with great endurance.

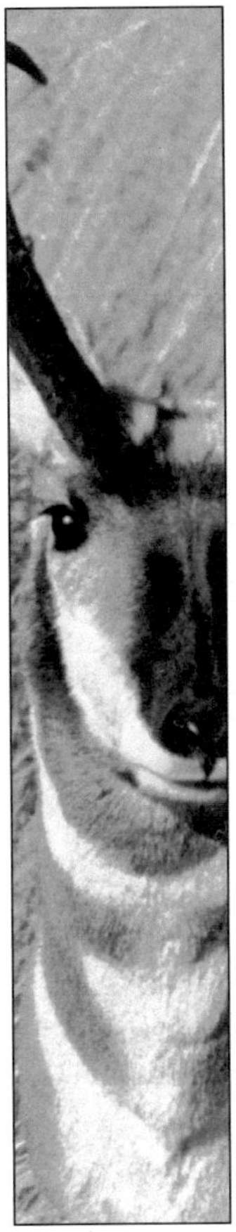

CHAPTER EIGHT

Physiology and Genetics

Bart W. O'Gara

The exceptional motor abilities and stamina of the species, however, may be reflected in part by relatively high erythrocyte, hematocrit, and hemoglobin values. Furthermore, the trachea, which is large in diameter relative to body size, and characteristic open-mouth running suggest that pronghorns have a capacious, efficient tidal exchange.

Barrett and Chalmers (1977a:453)

Curiously enough the first printed statement regarding the shedding of the horns of the American Antelope was a denial by Audubon and Bachman that such occurrence could take place.

Lyon (1908:393)

METABOLIC RATE

The energy requirements of an animal are related to its fasting metabolic rate. The average fasting metabolic rate of four young pronghorn was 92 kilocalories per kilogram per day, multiplied by $W^{0.75}$, where W is the total weight in kilograms, according to Wesley et al. (1970). This rate is above the interspecies mean of 70, but high rates also have been recorded for other wild ruminants. Wesley et al. indicated that the total energy demand of the pronghorn, relative to its body size, is greater than that of most domestic ruminants. As a result, pronghorn could undergo greater stress during food shortage. However, according to Wesley et al. (1973), the reported influence of cold temperature on the energy uptake of cattle is greater than on that of pronghorn. Therefore, pronghorn appear to be better adapted to climatic conditions on shortgrass prairie than are cattle. Wesley et al. also postulated that pronghorn are energetically more efficient than their domestic counterparts, but, because pronghorn are more selective feeders than are cattle, a given range probably would not have a higher carrying capacity for pronghorn than for cattle.

In an Idaho study, the heart and respiratory rates of 41 pronghorn fawns that were 4 hours or less old averaged 130 and 39 beats per minute, respectively; those of 34 fawns 8 to 168 hours old were 155 and 37.6 beats per minute, respectively (Autenrieth 1984). Whether such changes in rates between age groups are related to age or a greater fear response among older fawns is difficult to determine under field conditions. A mean pulse rate of 166.9 beats per minute was recorded for young fawns at capture in Alberta; Barrett and Chalmers (1979) proposed

(Left) *The pronghorn is one of the most athletic of North American mammals. Its anatomy tells the keen observer that this animal is swift and well adapted to its open, dry environment.* (Right) *Some physiological characteristics and adaptations—such as selective cooling of the brain during heat stress, large respiratory organs and airways, high buffering capacity and hemoglobin levels of the blood and high blood volumes—cannot be seen. However, they complement the pronghorn's long legs with heavy proximal muscles and lightweight feet that indicate a swift and enduring runner. Physiological adaptations to help the animals cope with semiarid environments include the long nose and low urine output, which conserves water through countercurrent exchange of respiratory air and reduces food intake during periods of dehydration, respectively. These and other physiological adaptations, many of which need further study, combine with behavior to make pronghorn unique.* Photos by Jim Jeffress; courtesy of the Nevada Department of Wildlife.

that probably was representative of normal rates when fawns were excited.

Because pronghorn have a high respiratory capacity, they may be good bioindicators, providing a sensitive measure of atmospheric pollutants in their habitats. Plutonium from atmospheric effluent at the Idaho Chemical Processing Plant in southeastern Idaho was found in the lungs and possibly other tissues of pronghorn (Markham et al. 1979). As would be expected with a highly mobile species, only some of the pronghorn collected in close proximity to the plant had detectable quantities. Concentrations were near the detection limits and apparently did not constitute a health hazard to the animals.

BODY TEMPERATURE

Temperatures of animals traditionally have been obtained using a rectal thermometer. However, temperatures obtained from wild mammals in that manner are not of value in establishing a norm for the species because of the marked temperature increases associated with the subject's stress (Thorne 1975). The mean rectal temperature of 41 drive-trapped pronghorn in Alberta was 105.4°F (40.8°C) (Barrett and Chalmers 1977a). Taking rectal temperatures of newborn fawns is practical because the neonates seem unafraid of their handlers.

Eighteen 1-day-old fawns handled in Nevada by McNay (1980) averaged 102.7°F (39.3°C), and 100 Idaho fawns averaged 100.8°F (38.2°C) (Autenrieth 1982). In the latter study, a fawn 48 to 72 hours of age had a temperature of 105.0°F (40.6°C) after running. Ashton and Olin (1988) took rectal temperatures of 1- to 14-day-old pronghorn fawns every morning in the Minnesota Zoological Garden. Normal temperatures seemed to be 101.0° to 102.0°F (38.3°–38.9°C), but they ranged from 100.0° to 104.0°F (37.8°–40.0°C). Another 76 fawns checked in Idaho by Autenrieth (1984) averaged 100.3°F (37.9°C).

The mean rectal temperature of 391 fawns captured in Alberta was 103.3°F (39.6°C) and ranged from 98.8° to 108.0°F (37.1°–42.2°C). Those fawns varied from 4 hours to 10 days in age (Barrett 1982).

Lonsdale et al. (1971) surgically implanted temperature-sensitive telemetry units in the flanks of two pregnant pronghorn does confined in a small pen. An average body temperature of 101.9°F (38.8°C) was recorded. Over a 10-minute period, the animals exhibited a temperature rise of as much as 3.2°F (1.8° C) due to handling or the excitement of nearby activity—such as the cleaning of the corral—although they were raised in captivity and accustomed to the attendants. A rise of 2.7°F (1.5°C) was noted for one instrumented doe during parturition, with a return to normal after 4 hours.

Body temperature is a basic and potentially valuable physiological parameter for identifying an infection or determining an animal's reaction to stress, excitement and other environmental influences. Variation in body temperature of pronghorn was associated with significant changes in six parameters of blood collected from drive-trapped animals (Barrett and Chalmers 1977a). Photo by Chuck Cadieux.

Using implanted transmitters, Thorne (1975) recorded temperatures of 98.4° and 100.9°F (36.9° and 38.3°C) for a mature and yearling doe pronghorn, respectively. He found that pronghorn temperatures were erratic; rarely were identical temperatures recorded an hour apart on the same animal. Changes of 2.7° to 3.6°F (1.5°–2.0°C) were not uncommon. Resting, standing, walking and eating had no marked influence on body temperature, but excitement resulting from induced stress produced notable increases. Pronghorn temperatures fluctuated more than those of mule deer.

Little difference was found between muscle and rectal temperatures in a large sample of pronghorn just after they were shot in Colorado (Bear et al. 1973). A few minutes lag always occurred after the animals were shot and before the temperatures could be taken. However, the temperatures—101.7°F (38.7°C) for muscles and 101.1°F (38.4°C) rectal—were similar to those recorded by Lonsdale et al. (1971) for live animals. Muscle temperatures recorded just after pronghorn were eviscerated showed a substantial decrease during the first 3 hours. After that, the heat loss was much slower.

Some East African antelope have the ability while running to store large amounts of heat without overheating their brains. The oxygen affinity of blood decreases as the temperature increases, promoting oxygen release to meet the demands of the muscle mitochondria for hydrogen removal. Dhindsa et al. (1974) pointed out that the retention of heat by pronghorn deserves exploration as a possible adaptation to and indicator of the species' prowess as a runner.

Heat exchange between arterial and venous blood is a well-established mechanism for temperature regulation. Netlike vascular structures, termed retia (singular = rete), function as heat exchangers by increasing the surface area and time available for transfer of heat from warmer arterial blood to cooler venous blood. This conserved and recycled heat contributes appreciably to the maintenance of constant body temperature during periods of cold temperature stress. Retia are found in the long, thin legs of wading birds, the flukes of aquatic mammals and the extremities of terrestrial mammals. Intracranial carotid retia of certain artiodactyls also function as heat exchangers to cool the brain selectively during heat stress.

Anatomical adaptations for selective cooling of the brain during heat stress were investigated in pronghorn, deer and elk. Carlton and McKean (1977) found that arteries in the carotid rete of pronghorn are of a uniformly small caliber. Not only is density of the carotid rete substantially increased by the twisting and coiling of these small vessels, but veins in the cavernous sinus of pronghorn also are small and branch profusely to intertwine among the arteries. Arterial blood traveling in any one vessel of the carotid rete of pronghorn has both a larger surface exposure and a longer contact time with venous blood than does that of either elk or deer and, thus, is better adapted to keep the brain temperature below deleterious levels during running. The orbital rete of pronghorn also is better adapted to function as a countercurrent exchange than are those of the cervids, being more highly vascularized and involving the supraorbital artery as well as the ciliary artery. The orbital rete also could conserve heat during periods of cold stress.

Although their stomachs are smaller, pronghorn use their favored forage at least as efficiently as sheep use theirs. Investigations at Wamsutter, Wyoming, showed that sagebrush-consuming pronghorn survived winters without supplemental food better than did domestic sheep in the same pastures (Sundstrom et al. 1973). The average weight loss of the clean-dressed carcasses of sheep was extreme, dropping 41% during a severe winter and 28% by March of milder winters. These were not just losses of fat, but also muscle protein due to catabolism. Pronghorn, by comparison, lost only 27% of their weight by March during the severe winter and 12% in mild winters. Photo by Steve Gallizioli; courtesy of the Arizona Game and Fish Department.

NUTRITION

Pronghorn share with other ruminants the ability to digest complex carbohydrates by a highly efficient fermentation process involving symbiotic populations of bacteria and protozoa in the rumenoreticular compartments of the stomach. These microorganisms convert fibrous forage, indigestible by most nonruminants, to volatile fatty acids, which are absorbed directly through the rumenoreticular walls, and to microbial protoplasm, which is digested in the abomasum and small intestine. The relative efficiency of these processes is largely dependent on diet and environmental conditions (see Cowan and Clark 1981).

Hoffman (1968) classified ruminants as selective, transition or roughage feeders. Selective feeders consume forages lower in fiber content and have smaller rumens with densely arranged papillae and smaller omasums and abomasums than do roughage feeders. Behaviorists generally classify pronghorn as selective feeders; the size of the animals' stomachs and other evidence confirm this classification.

The anatomy of the pronghorn digestive system was described in detail by Wenzel (1955; see also Chapter 7). Pronghorn stomachs are about half the size of stomachs found in domestic sheep, which apparently evolved primarily as grass eaters. The pronghorn cecum has less than two-thirds the capacity of a cecum of a sheep of the same weight. Forbs, when available, plus sagebrush and other browse species that remain highly nutritious during winter, provide the nourishment necessary when a relatively small volume of forage is consumed. This means that fewer pounds of forage are necessary for pronghorn to obtain their nutrient requirements than would be needed if they relied on grass (Sundstrom et al. 1973). In Saskatchewan, chemical analysis of food plants consumed by pronghorn throughout the year revealed a positive correlation between protein content and degree of utilization (Dirschl 1963). Plant species with the highest protein levels were the preferred forage during all seasons. During summer in Utah, pronghorn selected forage with protein contents of 10% to 22% and digestibility coefficients of 71% to 77% (Smith and Malechek 1974).

Digestive energy is that portion of the energy consumed that does not appear in the feces. Therefore, it is that portion absorbed from the gastrointestinal tract. Thus, digestible energy represents the total potential energy available for metabolism. Forages high in digestible energy usually are immature plants that are high in protein, minerals and other essential nutrients but relatively low in fiber (Cowan and Clark 1981).

Comparison of diets among four species of ruminants in Colorado indicated that pronghorn—followed in order by domestic sheep, cattle and bison—selected the highest quality diet (Schwartz et al. 1977). In that study, pronghorn maintained similar dietary quality on lightly and heavily grazed pastures. Rumen inocula from six species of ruminants (cattle, goats, sheep, mule deer, elk and pronghorn) were used for in vitro digestion trials on 25 forages, but no differences were detected in the digestive efficiency of the six inocula (Welch et al. 1983). Poor quality alfalfa was digested without difficulty by domestic sheep and goats, but caused impaction with inflammation in the omasal/abomasal region of mule deer. The deer had smaller omasums and abomasums than did the sheep or goats (Schoonveld 1971). Thus, the small third and fourth stomachs of pronghorn probably cannot handle large amounts of dry grasses, which constitute a small percentage of the animals' diet.

During feeding trials in Wyoming, 18 pronghorn consumed an average of 1.7 pounds (0.77 kg) of dry weight forage per day, compared with an average of 2.9 pounds (1.3 kg) for 12 slightly smaller sheep (Severson et al. 1968). The average wet weights of rumen contents during that study were 8.3 pounds (3.8 kg) for pronghorn and 15.7 pounds (7.1 kg) for sheep. Body weights of the animals averaged 98.2 pounds (44.5 kg) for pronghorn and 95.6 pounds (43.4 kg) for sheep. Severson et al. noted that rumen fills were not strictly comparable because collections were made at midmorning, after the extensive morning feeding period for sheep. Pronghorn fed for shorter time periods throughout the day, so probably had less of their daily food intake in their rumens at midmorning than did the sheep.

Feeding pronghorn moved more than did feeding sheep, covering about 1.5 times the linear distance in an equal period of time. The pronghorn also acted independently, even when in groups. This behavior probably was related to selective feeding. Sheep were more gregarious—generally, when one fed, all fed. Sheep foraged early and late in the day, especially during summer, whereas pronghorn—apparently less affected by the heat—fed off and on all day.

During another study in Wyoming, the average weight of pronghorn stomach contents was 6.7 pounds (3.0 kg) and that of sheep was 21 pounds (9.5 kg) (W. Hepworth personal communication:1979). However, those sheep weighed nearly half again as much as the pronghorn.

J. Bailey (personal communication:1981) provided data on comparative capacities of stomach compartments from 11 pronghorn collected in Colorado and Wyoming. Contents of pronghorn omasum and abomasum averaged 9.3% of the total stomach contents. This compares with 16% for four domestic sheep, 15% for four domestic goats and 13% for two mule deer, all weighed by Schoonveld (1971). However, the comparative weights of rumenoreticulum to omasal and abomasal weights, including contents, in stomachs weighed by Bear et al. (1973), were similar to those reported by Schoonveld for other ruminants (Table 45).

This suggests that pronghorn are adapted and limited to use of highly digestible forages. They are relatively small ruminants with small stomach capacity and somewhat small omasums and abomasums.

Volatile fatty acids (VFA) are metabolites produced by the microbial fermentation of forages in the rumenoreticulum of ruminant animals. These acids supply the major source of energy to the ruminant. Forages low in fiber are fermented more rapidly than are forages high in fiber, thus the former should produce higher rumenoreticular concentrations of VFA.

The concentrations and percentage distribution of VFA obtained from various age classes and sexes of prong-

TABLE 45. Stomach weights from free-ranging pronghorn in Colorado

Type weight	*Age or sex*	*Sample size*	*Pounds (kg)*	*Percentage of total*
S[a] and contents	3 to 4 months	7	5.4 (2.5)	100
R/R[b] and contents		7	4.6 (2.1)	85
S empty	Male and female	1	1.0 (0.5)	
S and contents	Yearling	7	10.2 (4.6)	100
R/R and contents	Males	7	8.7 (4.0)	85
S empty		4	1.4 (0.6)	
S and contents	Adult	14	12.5 (5.7)	100
R/R and contents	Males	14	10.8 (4.9)	86
S empty		13	1.8 (0.8)	
S and contents	Yearling	6	10.4 (4.7)	100
R/R and contents	Females	6	8.6 (3.9)	83
S empty		6	1.3 (0.6)	
S and contents	Adult	30	11.0 (5.0)	100
R/R and contents	Females	30	9.2 (4.2)	84
	S empty	25	1.9 (0.9)	

[a] Entire stomach.

[b] Rumenoreticulum.

Source: Bear et al. 1973.

horn corresponded to data obtained from domestic ruminants on somewhat comparable diets (Nagy and Williams 1969). No significant statistical differences were found in the total VFA content among various age classes and sexes of pronghorn. Percentage distribution of the different acids was the same for males and females but differed significantly with age in some cases. Yearlings showed low butyric and valeric acid levels, whereas fawns had high levels of isovaleric acid. These data should be interpreted with caution because the time lapse between killing each animal and freezing its rumen contents and the timing of kills after feeding could not be controlled.

Pronghorn fawns are functional ruminants by 4 to 5 months of age and probably have the same type of microorganisms as do the adults, according to Nagy and Williams (1969). Nagy and Williams proposed that, because VFA content and percentage distribution can reflect previous dietary regimes, analysis of rumen contents of wild animals might prove a valuable tool to determine the nutritional levels of the animals' diets.

Rumenoreticular contents from forages low in fiber apparently have lower pH values than do contents from forages high in fiber (Short et al. 1966). The pH of the rumenoreticular contents from nine pronghorn shot in Colorado and Wyoming was analyzed by Bailey (unpublished data) and averaged 5.8 with a standard error of only 0.1. This approximates the pH reported for North American deer (Browman and Sears 1955, Short et al. 1966, 1969a, 1969b), which also are selective feeders.

The livers of 131 adult pronghorn and 50 adult sheep were examined by Sundstrom et al. (1973), who found average liver weights to be 45 ounces per 100 pounds (1,276 g/45 kg) body weight for pronghorn and 31 ounces (880 g) for sheep. A large liver apparently is an important digestive and filtration adaptation, as well as a large and readily available source of energy in stored glycogen. Sundstrom et al. speculated that such a liver facilitates digestion of the fats (and perhaps essential oils) in sagebrush. Also, the large liver may be partly responsible for the ability of pronghorn to use, with little ill effect, plants higher in selenium and alkaloids than can livestock. Sundstrom et al. proposed that, although the large kidneys are primarily an adaptation to conserve body fluids—a necessity for living in a dry habitat—their large size also may allow greater excretion of toxic substances acquired from plants than do the relatively smaller kidneys of domestic sheep. Actually, large kidneys probably have little to do with their concentrating ability per se. A basic research study of the medullary area and lengths of the renal papillae would be informative.

Mean pH values for 196 pronghorn fecal samples in New Mexico were 8.66 (±0.12) during autumn and 8.01 (±0.11) during winter (Beasom et al. 1982a). The diet averaged 75% forbs during both seasons, but use of browse increased from 10% to 18% and use of grass decreased from 10% to 1% from autumn to winter. Beasom et al. proposed that the change in pH was related to the mineral content of the forages used. Forbs generally have a higher mineral content than do grasses, and browse tends to be lower than either one. The decline in pronghorn fecal pH from autumn to winter probably was due to maturation of the foods taken, as well as a doubling of the amount of browse consumed. On the other hand, Ganey (1997) did not find seasonal variation in fecal pH on the tallgrass prairie of Kansas. The pH during that study was 8.12—similar to that reported for white-tailed deer, but possibly different enough from that of mule deer to be used in differentiating between the species.

Pronghorn were observed on the National Bison Range during June through September 1977 to determine whether defecation rates were constant in different habitat types and at different levels of activity (Irby 1981). Individuals and groups that were active during 50% or more of the observation periods had defecation rates approximately 10 times greater than did groups or individuals that were active during less than 50% of the observation periods. Habitat type was less closely related to defecation frequency than was activity level. Thus, pellet counts could provide an adequate index of time spent in an area if the animals used the same habitats while resting and feeding, as they did on the Bison Range. However, if pronghorn bedded in one habitat type and fed in another, pellet groups apparently would be most numerous where the animals fed.

VITAMINS

Even though vitamins are usually found in very minute amounts, they are crucially important to animal health. Deficiencies produce signs that the wildlife nutritionist should be encouraged to recognize. While most of the deficiencies have been detected in captive animals, free ranging wildlife may also suffer.

Robbins (1983:93)

Vitamins are organic compounds that usually occur in food and are essential for health, maintenance and production. Although many vitamins cannot be synthesized by animals and, therefore, must be obtained from food, bacterial synthesis of some vitamins in the gastrointestinal tract can reduce requirements. Vitamins generally are classified as fat soluble (A, D, E, K) or water soluble (thiamin [B_1], riboflavin [B_2], niacin, pyridoxine [B_6], pantothenic acid, biotin, folic acid, cyanocobalamin [B_{12}], choline, ascorbic acid [C] and myoinositol).

Fat-soluble vitamins are absorbed from the gut with fat and often are stored in high concentrations to be used during periods of dietary inadequacy. Excesses, especially of A and D, can be highly toxic. Conversely, water-soluble vitamins (with the exception of B_{12}, in the liver, and riboflavin) generally are excreted readily in urine. Thus, they are needed in rather constant dietary supply, and toxic overdoses seldom occur, if ever. Unless otherwise stated, general information on vitamins is from Robbins (1983). Concentrations of vitamins in pronghorn tissues have been evaluated in only a few studies.

Vitamin A is important to vision as well as maintenance and growth of epithelial derivatives. A captive white-tailed deer being fed winter browse deficient in carotene became blind and paralyzed. This vitamin is not found in plants and, therefore, must be of animal origin. Fortunately, beta-carotene and other provitamin A sources are widely distributed in plants. Thus, pronghorn that have access to green forbs and grass during growing seasons and evergreen shrubs during winters and droughts would not be expected to suffer deficiencies.

The average vitamin A potency of fresh pronghorn liver was investigated in Oregon by Weswig (1956). During winter, potency was 1,024 international units per gram; during late spring and early summer, potency dropped about 20%. In September, the average of 25 livers was more than 2,200 international units per gram. Weswig concluded that the possibility of vitamin A deficiency was remote and that pronghorn browse contained considerable quantities of carotene. In Wyoming, the livers of eight pronghorn collected during spring averaged 266 micrograms of vitamin A per gram of wet liver (Edwards 1957).

The primary function of vitamin E is to maintain the functional integrity of cellular and subcellular membranes by preventing peroxidation of unsaturated fatty acids. This vitamin is widely distributed in green plants and seeds and can be stored in relatively high concentrations in the animal body. Thus, the chances of deficiencies in pronghorn seem remote. Vitamin E can compensate for selenium deficiencies—probably an important factor in pronghorn welfare on selenium-deficient rangelands. Dunbar and Velarde (1998) found 2.57 ± 0.48 micrograms of vitamin E per milliliter (µg/mL) of sera from pronghorn does during March 1997 in Oregon. During December 1998, does appeared in better condition, and Dunbar (personal files) found an average of 4.0 µg per mL (range, 3.2–4.6 µg per mL) of vitamin E in their sera. During spring 1998, sera from 22 fawns (<1–3 days old) had a mean of 1.6 µg per mL (range, 0.3–3.4 µg per mL) of vitamin E (Dunbar 1999a). Twenty-four fawns of the same age during spring 1999 had an average of 1.3 µg per mL (0.5–4.4 µg per mL) of serum vitamin E (M. Dunbar unpublished 1999a data).

Thiamin (vitamin B_1) is a necessary coenzyme in carbohydrate metabolism and possibly is important in nerve or neuromuscular transmission. A lack of B_1 can result in weakness, diarrhea, lethargy, weight loss, unsteady gait, impaired digestion, liver degeneration, seizures and death. Microorganisms in the rumen produce enough thiamin under most conditions to satisfy the health requirements of ruminants.

All cuts of meat from pronghorn shot in California had high thiamin content during May that decreased gradually to a low point in September, after which it increased (Cook et al. 1949). Thiamin intake may have increased during May because of the new growth of forage. However, pronghorn probably maintain their B vitamin economy from the products of intestinal bacteria more than from the vitamin content of their food. The low value in September—during breeding—compares with the decreased amount of B vitamin found in venison during the rut. It probably results from increased activity of the animals at that time. Average levels of thiamin during the Cook et al. (1949) study were 0.36 ± 0.03 mg/100 g of fresh tissue (hams), 0.35 ± 0.02 mg/100 g of fresh tissue (loins), 0.35 ± 0.05 mg/100 g of fresh tissue (livers) and 0.27 ± 0.03 mg/100 g of fresh tissue (shoulders). For muscles, thiamin levels were higher than those found in mule deer and about double those for beef and lamb. However, pork had two to four times the levels found in pronghorn. Thiamin levels in pronghorn liver were higher than those in beef liver but quite a bit lower than those in mule deer, lamb and pork livers.

Riboflavin (vitamin B_2) functions in two coenzymes—flavin adenine dinucleotide and flavin mononucleotide. Lack of this vitamin can result in loss of weight, weakness, retarded growth, inefficient food utilization, rough hair coat, corneal vascularization, skin atrophy, heart muscle damage, diarrhea, leg paralysis, reduced fertility and death. Riboflavin is not produced to any significant extent by most animals and must come from plants or microorganisms. Robbins (1983) did not specify whether or not rumen microorganisms produce this vitamin.

On the basis of milligrams per 100 grams of fresh tissue, riboflavin averaged 0.58 ± 0.05 in hams, 0.63 ± 0.04 in loins, 0.54 ± 0.04 in shoulders and 5.1 ± 0.80 in livers (Cook et al. 1949). These levels were slightly higher than those found in mule deer and two or three times higher than those in beef, lamb and pork. Riboflavin in pronghorn liver averaged about eight times the amount found in muscles. In contrast to thiamine, the lowest riboflavin levels were found during October, after—not during—the rut.

WATER CONSUMPTION AND KINETICS

Some pronghorn herds in Colorado lived and died without taking a drink of water, according to Hoover et al.

(1959). However, subsequent studies indicated that pronghorn need water during dry weather, and the distribution of drinking water affects the distribution of the animals, especially when succulent vegetation is scarce (see Chapter 13). Water generally was taken only once a day by pronghorn in Texas, according to Buechner (1950), who maintained that the animals may go for several days or even weeks without drinking water. He saw pronghorn drinking water on only 2 of 11 days when daylight-to-dark activity studies were made. However, Beale and Holmgren (1975) reported that decreasing the water available to pronghorn resulted in weight loss and inactivity.

On semidesert range in western Utah, Beale and Smith (1970) found that water consumption by pronghorn varied inversely with the quantity and succulence of preferred forage. When forbs were abundant and their moisture content was at least 75%, pronghorn did not drink although water was readily available. As vegetation lost succulence, water consumption began. During extremely dry periods, water consumption reached 3 quarts (2.8 L) per day per animal (see Chapter 13).

The pronghorn's long nose apparently functions as a water- and heat-conserving device. On exhalation, fully vaporized gas from the lungs cools as it flows over the extensive turbinal bones of the nose. As it cools, water content of the exhaled gas decreases by condensation on the surfaces of the turbinals. Inhaled dry air then picks up water vapor during inhalation, retaining body water. During cold weather, cold dry air is warmed and moistened as it travels through the long nose, whereas expired air gradually loses its water and heat, conserving both (Bromley 1977).

Working with two male and two female pronghorn 108 to 182 days of age that were raised in captivity from birth and trained to laboratory conditions, Wesley et al. (1970) investigated water turnover rates, percentage body water and water flux. All three were similar for animals of the same sex, but differences were noted between sexes. Females consumed more than half as much water as did males. Turnover rates and water flux were similar to those of mule deer and domestic sheep. Under the test conditions, water content was slightly higher than reported from other ruminants. Water content of the body varies inversely with fat content, and younger animals tend to have a higher percentage of body water than do older animals. Wesley el al. believed that age accounted for the high values they found in pronghorn.

Whisler and Lindstedt (1983a, 1983b) hypothesized that, because they are found in arid and semiarid areas of the western United States, pronghorn possess physiological water-conservation mechanisms. Whisler and Lindstedt measured water-loss parameters in pronghorn given all the water they would drink and in pronghorn maintained on one-fourth the amount they drank when it was available. The deprived animals lost about 15% of their body weight in 7 to 10 days. The water turnover rate was relatively low, even in the animals that had free access to water. During the dehydration period, the pronghorn reduced food intake, lowering urine output, and water turnover dropped from 35% to 50% of control values. Exhaled air temperatures were lowered—which alone would result in a 30% respiratory water saving—and evaporative water losses declined markedly. Basal respiratory rates decreased and panting ceased during rumination. Urine osmolality, plasma osmolality and hematocrit values remained constant, although urine's urea nitrogen levels elevated slightly. Fecal water loss fell to 10% of control values due to a smaller quantity of drier feces. Whisler and Lindstedt concluded that pronghorn are particularly suited to their dry, unpredictable environment, and, when water is unavailable, the animals' primary response is behavioral—conserving urinary, fecal and respiratory water.

Water is obtained from three sources—free (drinking) water, preformed water in food and metabolic water produced by oxidation of organic compounds containing hydrogen (Robbins 1983). Preformed water may be less than 10%, by weight, of air-dried forage to more than 70% of succulent vegetation (Taylor 1975). Estimates of preformed water intake can be made from detailed knowledge of food intake, but such intake can only be measured accurately under laboratory conditions (Wesley et al. 1970, Whisler 1984).

Laboratory experiments with adult pronghorn does measured contributions of free, preformed and metabolic waters (Whisler 1984). Average daily intake was 0.95 gallon per day per 100-pound (3.61 L /day/45 kg) doe (Table 46). This was slightly higher than the rate found by Wesley et al. (1970). The animals used by Wesley et al. were confined in a laboratory. Those investigated by Whisler were more active and exposed to higher solar radiation at the Sybille Experimental Unit in Wyoming.

Whisler's (1984) pronghorn who were given all the water they wanted to drink received 87.7% of their water by drinking, 9.5% as preformed water in food and 2.8% as metabolic water (Table 46). The same percentages for pronghorn given one-fourth of what they would drink were 84.0, 11.9 and 4.1, respectively. Considering the amount they drank, the food ingested by these animals (alfalfa hay and grain) must have been comparable in succulence to forage available in the wild during dry seasons. If the forage provided had been succulent enough that the experimental animals did not drink, preformed water would have provided most of the required water.

The long nose of the pronghorn, as seen on this doe watching her fawn on the National Bison Range in Montana, may help conserve water. On exhalation, air from the lungs cools as it flows over the extensive turbinal bones of the nose and condenses there. Inhaled dry air then picks up moisture, retaining water. Such an adaptation helps pronghorn survive in arid areas, especially during droughts. Photo by Peter T. Bromley.

TABLE 46. Calculated mean water intake for three hydrated and three dehydrated adult pronghorn during summer in Wyoming

	Hydrated animals		*Dehydrated animals*	
Water intake	*Gallons*	*Liters*	*Gallons*	*Liters*
Drinking	0.752	2.847	0.140	0.530
Preformed	0.082	0.310	0.019	0.075
Metabolic	0.024	0.091	0.007	0.026
Total	0.858	3.248	0.166	0.631

Source: Modified from Whisler 1984.

SALT UTILIZATION

The first published reference to the use of salt by pronghorn probably was by Audubon and Bachman (1851), who commented that pronghorn were remarkably fond of saline water and knew where all the salt licks were found. Audubon and Bachman also noted that pronghorn returned daily to licks if near their grazing grounds, and they lay down nearby after licking salty earth or drinking saltwater. A lick used by pronghorn in Sweetwater County, Wyoming, proved strongly alkaline and rich in sodium. Sulfate and probably bicarbonate were the important anions. Chloride was insignificant. Apparently, this lick was used primarily for its sodium content and alkalinity (Jones and Hanson 1985).

Several more recent publications have pointed out the pronghorn dependence on salt. Skinner (1922) observed that those in Yellowstone National Park not only were fond of "soda licks," but also used salt put out for bison. At times, pronghorn in Texas become a nuisance to ranchers by depleting salt placed on rangelands for livestock (Buechner 1950). Buechner observed pronghorn licking salt on 20 occasions and reported that one band spent from 7:20 to 10:15 A.M. at a salt trough. Pronghorn on the National Engineering Laboratory in Idaho acquire about 75% of their sodium intake from ingestion of some 1.7 ounces (48.7 g) of soil daily (Arthur and Gates 1988). Pronghorn in Colorado ingested soil around salt blocks and seemed to prefer it to direct use of blocks (Prenzlow et al. 1968). On the northern plains, pronghorn, particularly bucks, exhume shallow Indian graves (Frison 1978). Presumably, these excavations are associated with osteophagia of small bones or geophagia of soil surrounding the bones. Such soil is calcium- and phosphate-enriched by dissolution of the bones (Jones and Hanson 1985).

Ruminants require the most salt when they are on a succulent diet. Pronghorn use more succulent forage throughout the year than do most other ruminants, therefore they probably use more salt.

ADAPTATIONS FOR SPEED AND ENDURANCE

Pronghorn are among the world's most aerobic mammals, capable of maintaining high speeds over long distances (Lindstedt et al. 1988). In the process of adaptation, the reduced stomach capacity of pronghorn was accompanied by increases in size of the heart, lungs, liver and kidneys (Sundstrom et al. 1973). I weighed hearts and lungs from 10 domestic ewes, 2 to 6 years of age, for comparison with the average weights of 85 hearts and 27 sets of lungs from pronghorn does that were 2 to 9 years old. The ewes weighed about 150 pounds (68 kg) and the pronghorn does averaged 111 pounds (50 kg). The weight of sheep lungs averaged 24 ounces (680 g); those of pronghorn averaged 27 ounces (765 g). There were great differences in heart weights, considering the sizes of the animals—sheep hearts averaged 9 ounces (257 g); pronghorn hearts averaged 14 ounces (400 g). Sundstrom et al. did not find as great a difference in sizes between the hearts of sheep and pronghorn in Wyoming; the hearts and lungs of pronghorn averaged about 20% larger than those of domestic sheep. Weights of some organs are listed in Table 47. Weights of kidneys in O'Gara's (1968) collection increased more than 8% from the 2- to 3-year to the 8- to 9-year age class.

Data gathered by McKean and Walker (1974) and Walker (1974) indicated large differences between pronghorn and domestic goats of similar size in a number of parameters influencing delivery of oxygen to muscles and removal of respiratory and metabolic acids. They concluded that pronghorn are physiologically adapted as runners by virtue of: large hearts, which are almost twice the size of those of goats; large blood volume, which is 1.4 times that of goats; high hemoglobin levels, which are 50% greater than in goats; high buffering capacity of the blood; and reduced airway resistance. Lindstedt et al. (1988) found that maximum oxygen consumption by pronghorn was the highest recorded for any large mammal and was five times greater than that by goats the same size as pronghorn.

Pronghorn have large tracheae that allow rapid intake of oxygen while running. I measured the tracheae from the midneck regions of 10 domestic ewes and 10 pronghorn does. The tracheae of sheep were nearly round, with diameters averaging 0.75 inch (19 mm), whereas those of pronghorn were oval and averaged 1 by 1.5 inches (25 by 38 mm). Thus, the cross-sectional area of pronghorn tracheae was about three times larger than that of the larger ewes. However, other wild ungulates showed less difference. The trachea of a very large mule deer buck well over twice the size of pronghorn had a cross-sectional area of approximately 2 square inches (13 cm^2)—33% larger. The trachea of a cow elk, with a body size about four times that of the pronghorn, had a cross-sectional area of 2.7 square inches (18 cm^2)—less than twice as large.

Fast-running mammals have longer legs relative to their body lengths than do less-swift mammals. The lengthening of distal segments—feet—is particularly evident in slender artiodactyls such as pronghorn. The femur and hind foot of the pronghorn each constitute 31% of the total length of the hind leg; the humerus is 28% and the forefoot is 39% of the total length of the front leg. Thus, the strongly muscled, heavy proximal segment of each leg is relatively short, and the center of mass of the leg is held close to the body where it moves slowly, conserving energy. A study of running cheetahs and pronghorn has indicated that this marked proximal distribution of the mass centers of the leg, with extreme lengthening of the slender distal segments, is not essential for bursts of superlative speed, but does correlate with capacity to maintain that speed (Hildebrand and Hurley 1985). Legs of cheetahs, with heavier, shorter feet than those of pronghorn, produce significantly more energy, and the cheetah can

Table 47. Weights of organs from adult *A. a. americana*

	Sample size		*Mean weight in ounces (g)*		
Organ	*Male*	*Female*	*Male*	*Female*	*Reference*[a]
Heart	14	29	15.0 (423)	12.5 (355)	A
	14	82	16.6 (471)	14.3 (404)	B
Lungs					
Left	10	16	10.7 (303)	8.7 (248)	A
Right	10	16	14.9 (422)	12.8 (362)	A
Both	10	16	25.6 (725)	21.5 (610)	A
Both	6	27	31.4 (890)	27.0 (765)	B
Liver	14	25	40.4 (1,144)	35.7 (1,013)	A
	13	72	41.8 (1,185)	33.2 (941)	B
	11	38	48.0 (1,359)	43.5 (1,233)	C
Spleen	15	29	3.5 (98)	2.8 (80)	A
	13	85	3.9 (111)	3.3 (94)	B
	11	38	3.1 (87)	3.1 (89)	C
Kidneys	15	30	4.0 (114)	3.9 (111)	A
	12	76	4.2 (119)	3.5 (99)	B
	13	17	5.0 (142)	4.2 (118)	D
Brain	12	19	4.0 (114)	3.8 (108)	A

[a] A = Bear et al. 1973, Colorado; B = O'Gara 1968, Montana and Yellowstone National Park; C = Hepworth, unpublished, Wyoming; D = Bodie 1979, Idaho.

Pronghorn begin breathing through their mouths almost as soon as they start to run. Their hearts, lungs and trachea are relatively large for the size of the animals—adaptations for running several miles without undue stress. Other North American big game, except caribou, generally do not breathe through their mouths until they are nearly exhausted. African ruminants, such as the wildebeest, who run long distances during migrations or when pursued by cursorial predators also begin breathing through their mouths soon after they begin to run. Photo courtesy of the Charles M. Russell National Wildlife Refuge.

Another adaptation that helps pronghorn with a speedy start and energy-efficient running is the springing ligament. A large tendon or ligament located along the posterior surface of the cannon bone anchors to four sesamoid bones (much like small kneecaps). Ligaments extend from the sesamoids and insert on bones of the toes. When an animal's weight is removed from the toes, they are pulled into a plane similar to that of the cannon bone, as seen in the left front foot of the buck above. As the animal's weight is placed on the toes, the toes are forced forward (right, front foot above), stretching the ligaments and storing potential energy. Therefore, the toes are spring-loaded. When the load is diminished, the toes flip posteriorly, adding power and speed to the animal's stride. During running, springing ligaments absorb shock as the toes strike the ground and again add stored energy to the next stride. Photo by Len Clifford; courtesy of the Leonard Rue Enterprises.

accelerate and run faster than the pronghorn. However, as would be expected of carnivores that hunt using a stalk and high-speed dash, these cats have much less endurance than pronghorn.

FAT METABOLISM

Fat in muscle tissue of pronghorn in California averaged 3 g per 100 g of meat (Cook et al. 1949). Fat in the muscles increased as the general condition of the animals increased. Ether extract of fat in the liver varied in the opposite direction. This might mean that as fat disappeared in the muscles, it was mobilized in the liver, or it might merely indicate the presence of other ether-soluble substances at that time. The fat found in pronghorn muscle compares with values of 27 to 32 g per 100 g for cattle, sheep and pigs (Eaton and Konner 1985).

Bucks generally are heaviest during late summer in the northern parts of pronghorn range (Hepworth and Blunt 1966). In Alberta, adult males were heaviest during late summer, and weights decreased slightly between October and March, according to Mitchell (1980). A 154-pound (70 kg) buck shot on September 1 probably approximates the maximum whole weight attained by free-ranging pronghorn in Alberta. Bucks collected in Oregon showed a similar pattern of weight gains and losses, except for two heavy animals collected during March (Mason 1952). Two territorial bucks that I handled in Montana during July and August weighed 156 and 160 pounds (71 and 73 kg)—heavier than any weights I have seen published. Bucks taken in the same area during October and November averaged 128 pounds (58 kg), an indication of weight loss during the rut. Such a weight loss would partly reflect less forage in the digestive tract. Bucks ordinarily have about 10 pounds (4.5 kg) of forage in their rumens, but during the rut, their rumens are nearly empty. The rest of their digestive tract also contains less food, but the difference is not obvious.

In Nebraska, weights of field-dressed pronghorn bucks—2 years old and older—indicated substantial weight loss during the rut (Menzel 1980). During 1954 and 1955, when the hunting season started on the second Saturday in September, 449 field-dressed bucks weighed at checking stations averaged 89.9 pounds (40.8 kg). During 1978, when the season opened on the last Saturday in September, 47 bucks averaged 81.3 pounds (36.9 kg). The difference of 8.6 pounds (3.9 kg) in field-dressed weights during a 2- to 3-week period of the rut undoubtedly was less than differences in whole weights would have been. Much of a pronghorn's fat is removed with the viscera. Also, differences in rumen fill would not be reflected by field-dressed weights.

Pronghorn were collected throughout the year in Wyoming (W. Hepworth: unpublished data). Fat indices were not investigated during many of those collections, but whole weights indicated that both sexes were in best condition during summer and autumn, and poorest during late winter and early spring.

During year-round collections in Montana and Yellowstone National Park, several indices of physical condition were measured by fat storage or ratios of certain glands to whole weights (O'Gara 1968). The indices—live weight ratio (Taber et al. 1962), adrenal weight ratio (Hughes and Mall 1957), kidney fat index and back fat depth (Riney 1955), heart fat index (patterned after Riney's kidney fat index), colon fat and brisket fat depth (O'Gara 1968), percentage marrow fat (ether extraction) and marrow compression (Greer 1968)—all indicated the same pattern of weight loss and gain.

The kidney fat index (perirenal fat weight/kidney weight x 100) was judged the best all-around index for determining condition of pronghorn throughout the year, but femur marrow compression was best when animals were in extremely poor condition (O'Gara 1968). However, doubt was cast on the use of the kidney fat index by Dauphine (1975). He found that weights of caribou kidneys varied by as much as 80% annually, making those weights an unsuitable correction for body size. The kidney weights of 103 pronghorn, collected throughout the year, indicated considerable variation between monthly averages—from 5.3 ounces (150 g) in February to 7.6 ounces (215 g) in October. As a percentage of total body weight, kidney weights varied from 0.32 in February to 0.43 in August. Sample sizes were small during some months—only two animals in May and four in February. Thus, no conclusions can be drawn. Variations seemed similar between bucks and does, and individual variations within a month were greater than those between months. Every easily used condition index is subjective, and all are rough estimates. The wide range of the kidney fat index compensates somewhat for variations in kidney weights. Starving pronghorn have an index of about 20, and very fat pronghorn have indices as high as 300.

Ransom's (1965) comparison of kidney fat indices and fat content of marrow from 34 white-tailed deer does indicated that kidney fat was not a reliable indicator of physical condition during mobilization of femur marrow fat. Ransom suggested the complementary use of the two measurements to assess physical condition over the species' entire range.

Hall (Group Discussion 1978) indicated that—during a severe winter in Alberta—when a pronghorn's kidney fat index reached 30%, its bone marrow still was in the 70% to 80% range. Once kidney fat dropped to 20%, bone marrow fat dropped rapidly.

All of the indices used by O'Gara (1968) indicated that does in the National Bison Range and Yellowstone National Park were in best condition during December and poorest in May. The bucks also were in the poorest condition during May, but were fattest during August and early September.

A different pattern of weight losses and gains was found by Bear (1971) for pronghorn in Colorado, where males stored little fat until after the breeding season. The relatively mild winters of Colorado may allow bucks to gain more weight after the rut than they do farther north, and lush northern summers may promote fattening. Yet, why Colorado bucks do not fatten during the time when does are able to maintain themselves and suckle fawns is difficult to understand. Perhaps the fact that many Colorado pronghorn are not territorial (Prenzlow 1965a) has some bearing on bucks not gaining weight during summer. Also, high harvest levels may have eliminated many of the mature, no-longer growing bucks that normally would fatten during summer. If the fawns were not weaned during the rut in harem-breeding systems, the does would not be "released" to fatten during autumn.

Pronghorn does in Colorado maintained higher fat levels throughout most of the year than did bucks (Bear 1971). In the National Bison Range and Yellowstone National Park, does began gaining weight during vegetative green-up, but did so much slower than did bucks because of the stresses of parturition and lactation (O'Gara 1968). The does gained weight rapidly during October and November, apparently because most fawns were weaned during the rut in September. In Colorado, both sexes reached maximum fat levels during November and December (Bear 1971). Although does in the National Bison Range and Yellowstone Park also were fattest at that time, bucks were at about their annual average.

In the Southwest, weights of pronghorn are depressed by the same stresses of the rut and lactation, but for those animals, times of weight gains apparently correlate with rainy seasons.

DERIVATION OF WHOLE FROM PARTIAL WEIGHTS

Whole weights appear to be the best indicators of condition for big game animals and, indirectly, their habitat. Generally, however, only eviscerated or carcass weights are available. Eviscerated weight is derived from animals with all viscera removed—from larynx to anus; carcass weight is derived from animals with viscera, heads, feet and skins removed (O'Gara 1968). An 18% loss of weight was reported for eviscerated pronghorn in California by McLean (1944), and Einarsen (1948) stated that a dressed pronghorn with its feet removed weighed 20% less than its whole weight, apparently basing this conclusion on McLean's weights.

Studies in a number of states and provinces have indicated evisceration losses of 26% to 34% (Table 48). Adult pronghorn in Wyoming averaged about 29% loss from evisceration (Hepworth and Blunt 1966)—probably a good average for hunting seasons. Variations in Table 48 resulted primarily from times of year when animals were collected. O'Gara (1968) found consistent weight losses in year-round collections when the reproductive tracts of does were removed before whole weights were taken. The high evisceration loss reported by Mitchell (1980) for females in Alberta resulted because 9 of the 11 females were pregnant. He found a loss of 27% for females in November and 40% for a female at term pregnancy in June.

The light weights for viscera reported by McLean (1944) might have resulted from weighing adult bucks during the rut. His weights all were taken during a bucks-only hunt in California. Rutting bucks brought to a checking station in Idaho seldom had anything in their rumens, and the viscera only was 19% to 20% of the buck's whole weight (personal observations 1977). The light weights sometimes reported for pronghorn possibly resulted from using McLean's (1944) or Einarsen's (1948) 18% or 20% weight loss calculation to compute whole weights from weights of eviscerated animals.

Because percentage of weight lost by evisceration is not a convenient way to determine whole weights, the whole, eviscerated and carcass weights of 104 adult pronghorn were used to devise a multiplier that would be easy to use (O'Gara 1970). The eviscerated weight of an animal of either sex multiplied by 1.39, and the carcass weight of a female multiplied by 1.73 or that of a male multiplied by 1.75 reliably estimated whole weights with less than 2% error. Female reproductive tracts were removed before whole weights were taken because those animals were collected throughout the year. Removing reproductive tracts would make little difference during most hunting seasons.

O'Gara (1968) noticed that hunters did not always remove pronghorn heads at the atlas as he had done, but sawed the necks off just in front of the shoulders. To find a compensating factor for this difference, O'Gara weighed nine more animals and found that the appropriate adjustment could be made by adding 0.03 to either multiplier

Table 48. Loss of weight by evisceration for pronghorn from several populations

Source[a] *and locality*	*Sex and age*	*Sample size*	*Whole weight in pounds (kg)*	*Eviscerated weight in pounds (kg)*	*Percentage lost*
A	Adult male	6	116.7 (52.9)	85.7 (38.9)	27
	Adult female	15	108.7 (49.3)	75.3 (34.2)	31
B	Adult male	14	126.9 (57.6)	91.4 (41.5)	28
	Adult female[b]	90	110.7 (50.2)	79.4 (36.0)	28
C	Adult male	16	119.0 (54.0)	87.5 (39.7)	26
	Adult female	11	110.9 (50.3)	73.2 (33.2)	34
D	Adult male	52	112.2 (50.9)	81.6 (37.0)	27
	Adult female	98	105.9 (48.0)	73.3 (33.2)	31

[a] A = Bear 1967, Colorado; B = O'Gara 1968, Montana and Yellowstone National Park; C = Mitchell 1980, Alberta; D = Hepworth, unpublished, Wyoming.

[b] Female reproductive tracts were removed before whole weights were taken.

for carcass weights of animals with necks sawed off immediately in front of the shoulders. The multipliers for estimated live weight were adjusted to correspond with pronghorn having average amounts of food in their rumens. Thus, weights estimated from partial weights may involve less bias related to rumen fill than would whole weights if rumen fill was not considered.

Whole weights derived by multipliers were slightly biased because the pronghorn were bled before they were weighed. Eviscerated and carcass weights were taken soon after the death of the animals in O'Gara's collection. Thus, dehydration of an eviscerated animal or a carcass would increase the bias toward a low estimate of whole weights based on the multiplier. Hunter-killed pronghorn usually are transported considerable distances through dry prairie country. Consequently, this bias could be large if an eviscerated carcass was weighed after a hunter returned home with the carcass exposed to dehydration.

Mitchell (1980) investigated the use of multipliers to estimate whole weights for pronghorn in Alberta. His results for bucks (1.36 times eviscerated weight and 1.78 times carcass weight) were quite similar to O'Gara's (1970): 1.39 and 1.75. Mitchell's results for does (1.51 times eviscerated weight and 1.95 times carcass weight) differed considerably from O'Gara's 1.39 and 1.73. This difference resulted from O'Gara's removal of does' reproductive tracts before taking whole weights, whereas Mitchell left them intact.

Mitchell (1971) developed a method for determining whole weights from heart girths (chest girth taken immediately behind the shoulders). His regressions of whole weights on heart girths ($y = 0.3043x - 179.4817$, where y and x represent whole weight in pounds and heart girth in millimeters, respectively) were computed with combined data for pronghorn of all ages and both sexes. Mitchell also derived separate regressions for adult and yearling animals of both sexes: adults $y = 0.2545x - 126.6853$, yearlings $y = 0.2484x - 136.6747$. When I checked Mitchell's equations against individual animals in my collection, the combined regression gave more accurate results with adult animals than did the adults-only regression. However, the yearling regression, applied to yearlings, gave more accurate results than did the combined regression. All gave slightly high estimates of whole weights for pronghorn from Montana and Yellowstone National Park. The multiplier produced better results in most instances and was easier to use, but whole weights can be estimated satisfactorily from heart girth when eviscerated or carcass weight is not available.

BLOOD PARAMETERS

Background conditions for blood parameters must be quantified under controlled conditions with captive animals. For that reason, the usefulness of most blood chemistry parameters for evaluating the nutritional status of free-ranging large mammals is relatively low.

Spalinger (2000:129)

Publications reporting on the analyses of pronghorn blood have indicated considerable variation. Such conclusions must result, partly from differences in collecting techniques (trapped after long or short drives, shot, taken from tame animals, etc.), methods of laboratory analysis, processing delays and many other factors. Infections, parasites and trauma are sources of physiological changes and can influence blood parameters used for nutritional assessments (Seal et al. 1981). Most investigators noted that values cited should be regarded as preliminary information until additional studies can be conducted on pronghorn from different areas and under different conditions.

Most pronghorn blood is collected during trapping operations (Hoskinson and Tester 1980). Pronghorn are highly excitable and extremely active during any type of capture, so some blood parameters undoubtedly vary widely according to how tired, frightened or excited the animals are when blood is drawn. Tame captive animals should not be affected significantly by drawing blood, but parameters related to nutrition, normal exercise, and so on, may be quite different from those of wild animals. Young fawns do not exhibit fear of their handlers, therefore blood parameters should not be affected greatly by capture.

Chemical analysis of blood may hold the key to better understanding pronghorn physiological condition, including seasonally changing metabolic patterns and habitat quality. Because blood parameters are useful in making management decisions, much of the available data are brought together in Tables 49 and 50. Blood reference values, once established, can be used to detect subsequent physiological changes in a population, compare two or more populations and distinguish subgroups within a population—healthy versus sick fawns or pregnant versus nonpregnant does (Seal et al. 1981).

Significant differences in blood parameters were not noted between singles and twins at birth in Alberta, but considerable differences were found between the sexes. Leukocyte count, sodium, potassium and phosphorus levels had higher values in the blood of pronghorn neonates when the ambient temperature was 50.0°F (10°C) or higher than when it was 68 .0°F (20°C) or less, whereas blood urea–nitrogen (BUN) was higher at 68.0°F (20°C) or less than at 50.0°F (10°C) or higher (Barrett and Chalmers 1976).

Pronghorn who were shot had much lower volumes of both red and white blood cells than did trapped pronghorn in other studies. When a pronghorn is shot, its blood

Pronghorn become excited and stressed during trapping, so parameters of blood drawn at that time will be quite different from those of calm animals. Thus, blood parameters should be established for calm animals using tame pronghorn. However, for practical field use, parameters also should be established under normal trapping conditions. Allowing drive-trapped pronghorn bands to rest for at least 1 hour in a curtained trap and blindfolding the animals as soon as they are hand-captured reduces stress. One or two pronghorn should not be isolated; they feel more secure with their band. Photo by Jeff Pederson; courtesy of the New Mexico Department of Game and Fish.

pressure decreases and body fluids rapidly dilute the blood. Lymphocyte ratios of shot pronghorn (Bear et al. 1973) were so different from those of trapped animals that they are not included in Table 49.

Hematocrit and hemoglobin values of adult pronghorn in Alberta were significantly greater for females than for males, and significant ($P < 0.01$) differences in 9 of 16 hematological parameters were associated with year of sampling (Barrett and Chalmers 1977a). Hematocrit levels of blood from neonatal pronghorn (Barrett and Chalmers 1979) were 75% to 80% of those recorded for adults by most investigators. Neonates may have naturally low hematocrits, but they seem unafraid of their handlers and are handled for a very short time—generally without pursuit.

Excitement, exertion and stress apparently elevate hematocrit levels in older animals during capture and handling. Increased circulating red cells result from splenic contractions during handling. This probably explains some of the differences seen in Table 49. Pronghorn stressed the most during trapping in Idaho had the highest hematocrit levels (Seal and Hoskinson 1978). Pronghorn with the highest temperatures (an indicator of stress) in Alberta had the highest hematocrit levels, and those levels dropped as the animals became accustomed to being in traps (Barrett and Chalmers 1977a).

Barrett and Chalmers (1977a) maintained that the relatively high erythrocyte, hematocrit and hemoglobin values for pronghorn, coupled with the animals' large tracheae and characteristic open-mouth running, suggest that the species has a capacious, efficient, tidal exchange. However, Dhindsa et al. (1974) constructed oxygen dissociation curves for pronghorn blood, but could not find any

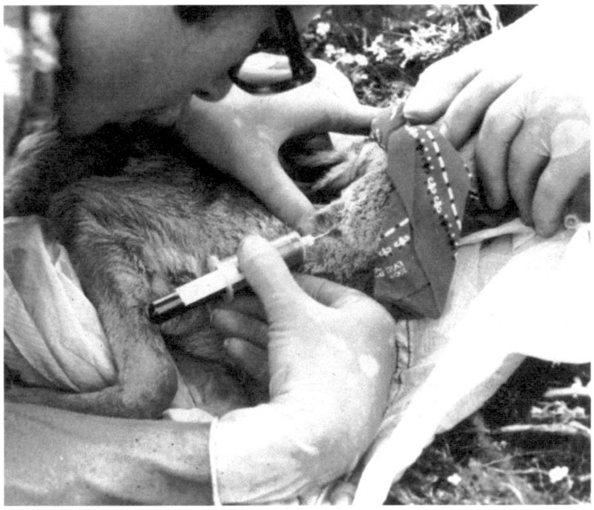

If handled gently, neonate pronghorn are not overly stressed by drawing blood. Carrying syringes and needles in 95% alcohol and soaking the skin with alcohol before inserting the needle are good precautions. Rubber gloves should be worn to reduce the amount of human odor imparted to the fawn, and blindfolding also reduces stress. Photo by Jim D. Yoakum.

Table 49. Baseline hematological values of neonatal, fawn and adult pronghorn[a]

Source	Number tested[b]	Mean	Standard deviation or range
Erythrocytes RBC ($x10^6/\mu L$)			
Id-Ad-CC	46	12.6	—
Alta-Ad	112	11.5	1.3
Alta-Ad-Male	24	11.3	1.2
Alta-Ad-Female	81	11.5	1.4
Alta-Ad-RT-low	14	11.7	1.2
Alta-Ad-RT-med	12	10.2	1.8
Alta-Ad-RT-high	15	10.8	1.4
Alta-N	122	9.7	1.0
LL	23	14.6	1.4
LL-Ad	9	13.7	0.6
LL-Fawn	6	15.6	1.0
BC	25	13.7	1.5
CC	37	13.9	1.2
CC-Ad-Male	8	13.5	0.6
CC-Ad-Female	15	13.7	1.2
CC-Fawn	9	14.5	0.6
Wyo-Ad	4	11.6	—
Col-Ad-Male	10	9.5	—
Col-Ad-Female	19	9.5	—
Ad-ISIS	24	10.5	1.8
Or-Ad-Female-Dec	18	11.7	0.9
Or-Ad-Female-Mar	19	11.3	0.7
Or-Fawn	66	9.8	1.0
Or-Ad-Female-98	7	12.1	11.2–13.1
Or-Fawn-98	28	8.7	6.6–10.7
Or-Fawn-99	25	9.2	—
Hematocrit (%)			
Id-Ad-LL	23	56.9	—
Id-Ad-BC	26	54.0	—
Id-Ad-CC	46	52.0	—
Alta-Ad	110	51.7	4.6
Alta-Ad-Male	23	50.3	4.5
Alta-Ad-Female	80	52.4	4.5
Alta-Ad-RT-low	14	52.9	3.2
Alta-Ad-RT-med	12	54.3	4.0
Alta-Ad-RT-high	15	56.1	3.1
Alta-Ad-O-Day	55	53.3	4.4
Alta-Ad-l-Day	36	51.0	4.3
Alta-Ad-2-Day	11	48.5	3.8
Alta-N	110	39.7[c]	3.9
LL	23	56.0	5.3
LL-Ad	9	54.5	3.6
LL-Fawn	6	58.5	4.7
BC	25	52.0	2.5
CC	37	51.4	2.4
Wyo-Ad	4	42.7	—
Col-Ad-Male	10	49.0	—
Col-Ad-Female	19	49.0	—
Nebr-1-24Mo	6	44.4	—
Ad-ISIS	53	40.9	5.8
Or-Ad-Female-Dec	18	54.5	4.6
Or-Ad-Female-Mar	19	53.9	3.6
Or-Fawn	67	44.0	4.5
Or-Ad-Female-98	7	57.2	54.1–64.1
Or-Fawn-98	28	40.3	29.7–50.8
Or-Fawn-99	25	—	38.0
Hemoglobin (g/dL)			
Id-Ad-LL	23	18.3	—
Id-Ad-BC	26	16.8	—
Id-Ad-CC	46	16.0	—
Alta-Ad	110	18.3	2.2
Alta-Ad-Male	23	17.6	2.0
Alta-Ad-Female	80	18.6	2.2
Alta-Ad-O-Day	55	19.0	2.6
Alta-Ad-l-Day	36	17.9	1.5
Alta-Ad-2-Day	11	17.7	1.4
Alta-N	116	14.6	1.7
LL	23	21.4	1.9
LL-Ad	9	20.8	1.2
LL-Fawn	6	22.8	1.5
BC	25	21.2	1.5
CC	37	18.6	1.2
Wyo-Ad	4	16.8	—
Col-Ad-Male	10	16.5	—
Col-Ad-Female	19	16.5	—
Ad-ISIS	42	13.4	2.0
Or-Ad-Female-Dec	18	19.7	1.3
Or-Ad-Female-Mar	19	19.1	1.3
Or-Fawn	67	14.9	1.3
Or-Ad-Female-98	7	19.8	18.7–20.6
Or-Fawn-98	28	13.0	9.7–16.3
Or-Fawn-99	25	14.3	—
P50-oxygen (mm [Hg]) 38°C, plasma, pH 7.40			
Wyo-Ad	4	28.3	0.7
Mean corpuscular hemoglobin (pg)			
Id-Ad-CC	46	12.5	—
Alta-Ad	110	16.2	2.8
Alta-Ad-Male	23	15.7	2.1
Alta-Ad-Female	80	16.5	3.0
Alta-Ad-RT-low	14	15.5	1.8
Alta-Ad-RT-med	12	19.1	4.9
Alta-Ad-RT-high	15	16.7	3.9
Alta-Ad-O-Day	55	17.0	3.0
Alta-Ad-l-Day	36	15.6	2.8
Alta-Ad-2-Day	11	15.3	1.1
Alta-N	120	15.1	1.6
Or-Ad-Female-Dec	18	16.9	0.6
Or-Ad-Female-Mar	19	16.8	0.3
Or-Fawn	66	15.2	1.1
Mean corpuscular hemoglobin concentration (g/dL = percentage)			
Id-Ad-CC	46	31.6	—

continued on next page

TABLE 49—*continued*

Source	Number tested[b]	Mean	Standard deviation or range
Alta-Ad	110	35.5	3.3
Alta-Ad-Male	23	35.2	3.6
Alta-Ad-Female	80	36.6	3.4
Alta-N	150	37.1	4.5
Alta-N-Male	91	36.3	4.1
Alta-N-Female	59	38.3	4.8
LL	23	37.6	2.4
BC	25	40.8	2.0
CC	37	36.3	2.4
Or-Ad-Female-Dec	18	36.3	1.7
Or-Ad-Female-Mar	19	35.4	0.7
Or-Fawn	66	33.9	1.8
Mean corpuscular volume (μm³) (fL = μm³)			
Id-Ad-CC	46	39.7	—
Alta-Ad	110	45.8	7.7
Alta-Ad-Male	23	45.1	6.6
Alta-Ad-Female	80	46.4	8.2
Alta-Ad-RT-low	14	45.4	4.1
Alta-Ad-RT-med	12	55.0	11.7
Alta-Ad-RT-high	15	52.8	7.7
Alta-Ad-O-Day	55	48.1	7.8
Alta-Ad-l-Day	36	44.5	8.3
Alta-Ad-2-Day	11	42.0	2.8
Alta-N	120	41.4	4.2
LL	23	38.4	2.4
BC	25	38.5	3.5
CC	37	37.1	2.4
CC-Ad-Male	8	38.2	1.7
CC-Ad-Female	15	37.2	1.5
CC-Fawn	9	35.2	1.8
Ad-ISIS	24	41.5	4.6
Or-Ad-Female-Dec	18	46.6	1.0
Or-Ad-Female-Mar	19	47.5	1.1
Or-Fawn	66	44.9	3.1
Or-Fawn-98	28	46.1	44.3–47.4
Erythrocyte sedimentation rate (1 hr), mm			
Alta-Ad	63	0.1	0.3
Alta-Ad-Male	11	0.1	0.3
Alta-Ad-Female	51	0.1	0.4
Alta-N	48	1.9	2.1
Alta-N-single	24	1.1	1.2
Alta-N-twin	24	2.7	2.6
Erythrocyte sedimentation rate (24 hr), mm			
Alta-Ad	63	12.3	9.4
Alta-Ad-Male	11	14.5	12.3
Alta-Ad-Female	51	11.9	8.9
Alta-Ad-RT-low	14	20.2	12.4
Alta-Ad-RT-med	12	14.9	7.9
Alta-Ad-RT-high	15	10.1	7.6
Alta-Ad-O-Day	42	10.4	9.2
Alta-Ad-l-Day	20	16.6	8.8

Source	Number tested[b]	Mean	Standard deviation or range
Alta-N	48	21.5	13.4
Alta-N-single	24	17.0	12.7
Alta-N-twin	24	26.0	12.9
Erythrocyte fragility (min), % saline			
Alta-N	56	0.8	0.0
Erythrocyte fragility (max), % saline			
Alta-N	56	0.4	0.0
Fibrinogen (mg/dL)			
LL	23	160.0	71.2
BC	25	247.0	85.0
CC	37	193.0	121.7
CC-Ad-Male	8	285.0	141.4
CC-Ad-Female	15	171.0	73.6
CC-Fawn	9	151.0	90.0
Leukocytic series, hematocrit (%)			
Alta-N	37	0.6	0.2
Leukocytes (x10³/μL)			
Id-Ad-LL	23	4.4	—
Id-Ad-BC	26	3.1	—
Id-Ad-CC	46	3.3	—
Alta-Ad	112	5.0	1.7
Alta-Ad-RT-low	14	4.4	1.1
Alta-Ad-RT-med	12	5.4	2.0
Alta-Ad-RT-high	15	3.7	1.6
Alta-Ad-O-Day	56	4.7	1.8
Alta-Ad-l-Day	36	5.1	1.4
Alta-Ad-2-Day	12	5.8	1.7
Alta-N	122	4.0	1.8
Alta-N-Male	75	3.6	1.7
Alta-N-Female	47	4.5	2.0
LL	23	4.8	1.9
BC	25	3.9	1.0
CC	37	3.9	1.8
Col-Ad-Male	10	1.8	—
Col -Ad-Female	19	1.8	—
Ad-ISIS	54	5.3	4.0
Or-Ad-female-Dec	18	5.2	1.7
Or-Ad-Female-Mar	19	5.4	2.2
Or-Fawn	64	0.9	3.1
Or-Ad-Female-98	7	4.2	2.3–5.5
Or-Fawn-98	28	3.8	1.1–8.3
Or-Fawn-99	25	2.8	—
Neutrophils (%)			
Alta-Ad	108	65.8	12.3
Alta-Ad-O-Day	56	64.4	12.5
Alta-Ad-l-Day	35	69.7	12.1
Alta-Ad-2-Day	10	61.4	9.9
Alta-N	120	62.9	14.5

continued on next page

TABLE 49—*continued*

Source	*Number tested*[b]	*Mean*	*Standard deviation or range*
Segmented neutrophils (%)			
Alta-Ad	63	63.6	12.2
Alta-N	120	59.9	14.6
Col-Ad-Male	9	29.0	—
Col-Ad-Female	16	40.0	—
Nebr-1-24-mo	6	48.8	—
Or-Ad-Female-Dec	18	45.4	20.5
Or-Ad-Female-Mar	19	56.2	17.1
Or-Fawn	66	66.2	17.2
Or-Ad-Female-98	7	54.0	43.0–63.0
Or-Fawn-98	28	74.0	41.0–89.0
Or-Fawn-99	25	74.0	—
Segmented neutrophils (x10³/µL)			
Or-Ad-Female-Dec	18	2.5	1.7
Or-Ad-Female-Mar	19	3.0	1.4
Or-Fawn	62	3.4	2.2
Or-Ad-Female-98	7	2.3	—
Or-Fawn-98	28	2.8	1.6–3.4
Band neutrophils (%)			
Alta-Ad	63	2.1	2.2
Alta-N	120	3.1	2.7
Ad-ISIS	20	0.1	0.2
Or-Ad-Female-Dec	18	0.3	0.5
Or-Ad-Female-Mar	19	0.8	1.4
Or-Fawn	66	1.2	1.3
Or-Ad-Female-98	7	0.4	—
Or-Fawn-98	28	—	0.0–3.0
Or-Fawn-99	25	2.0	—
Band neutophils (x10³/µL)			
Ad-ISIS	20	0.12	0.23
Or-Ad-Female-Dec	18	0.02	0.03
Or-Ad-Female-Mar	19	0.04	0.08
Or-Fawn	62	0	0.1
Or-Ad-Female-98	7	0.2	—
Or-Fawn-98	28	—	0.0–0.1
Lymphocytes (%)			
Alta-Ad	108	31.5	11.6
Alta-Ad-O-Day	56	33.5	11.8
Alta-Ad-l-Day	35	28.0	12.0
Alta-Ad-2-Day	10	32.2	8.6
Alta-N	120	33.9	14.2
Nebr-1-24-Mo	6	40.1	—
Or-Ad-Female-Dec	18	54.0	20.4
Or-Ad-Female-Mar	19	38.1	17.9
Or-Fawn	66	30.7	18.0
Or-Ad-Female-98	7	42.0	32.0–51.0
Or-Fawn-98	28	25.0	11.0–57.0
Or-Fawn-99	25	22.0	—
Lymphocytes (x10³/µL)			
Ad-ISIS	53	0.9	0.9
Or-Ad-Female-Dec	18	2.7	0.9
Or-Ad-Female-Mar	19	1.9	0.9
Or-Fawn	62	1.3	1.0
Or-Ad-Female-98	7	1.8	—
Or-Fawn-98	28	1.0	0.4–2.2
Neutrophil to lymphocyte ratio			
Id-Ad-LL	23	2.7	—
Id-Ad-BC	26	1.8	—
Alta-N	120	2.6	2.1
Or-Ad-Female-Dec	18	1.2	1.1
Or-Ad-Female-Mar	19	2.3	1.9
Or-Fawn	66	4.4	6.7
Or-Ad-Female-98	7	1.3	—
Or-Fawn-98	28	2.8	—
Monocytes (%)			
Alta-Ad	108	0.3	0.6
Alta-N	120	0.8	1.3
Nebr-1-24-Mo	6	3.9	—
Or-Ad-Female-Dec	18	2.1	3.3
Or-Ad-Female-Mar	19	1.5	2.4
Or-Fawn	66	3.0	2.3
Or-Ad-Female-98	7	2.0	1.0–4.0
Or-Fawn-98	28	—	0.0–6.0
Or-Fawn-99	25	2.0	—
Monocytes (x10³/µL)			
Ad-ISIS	36	0.2	0.4
Or-Ad-Female-Dec	18	0.1	0.2
Or-Ad-Female-Mar	19	0.1	0.1
Or-Fawn	62	0.1	0.1
Or-Ad-Female-98	7	0.8	—
Or-Fawn-98	28	—	0.0–0.2
Eosinophils (%)			
Alta-Ad	108	1.8	2.3
Alta-N	120	1.8	2.1
Alta-N-single	49	2.4	2.6
Alta-N-twin	71	1.4	1.6
Col-Ad-Male	9	6.0	—
Col-Ad-Female	17	2.0	—
Nebr-1-24-Mo	6	2.8	—
Or-Ad-Female-Dec	18	0.2	0.4
Or-Ad-Female-Mar	19	3.4	5.9
Or-Fawn	66	0.6	1.3
Or-Ad-Female-98	7	0.9	—
Or-Fawn-98	28	—	0.0–2.0
Or-Fawn-99	25	0.2	—
Eosinophils (x10³/µL)			
Ad-ISIS	39	0.2	0.3
Or-Ad-Female-Dec	18	0.1	0.3
Or-Ad-Female-Mar	19	0.2	0.4
Or-Fawn	62	0	0

continued on next page

TABLE 49—*continued*

Source	Number tested[b]	Mean	Standard deviation or range
Or-Ad-Female-98	7	0.04	—
Or-Fawn-98	28	—	0.0–0.1
Or-Fawn-99	25	0.5	—
Basophils (%)			
Alta-Ad	108	0.4	0.6
Alta-N	120	0.4	1.0
Nebr-1-24-Mo	6	1.5	—
Or-Ad-Female-Dec	18	0.7	0.3
Or-Ad-Female-Mar	19	0.4	1.5
Basophils (x10³/μL)			
Ad-ISIS	16	—	0.03–0.04
Or-Ad-Female-Dec	18	0	0.01
Or-Ad-Female-Mar	19	0.02	0.08
Platelets (x10³/μL)			
Ad-ISIS	1	678	0
Or-Fawn	44	722	274
Or-Fawn-98	28	759.3	362.0–1,300.0
Or-Fawn-99	25	482.0	—
Nucleated RBCs (%)			
Or-Fawn-99	25	1.0	—

[a] Values were rounded to one decimal place unless total value was less than 0.1. Where possible, units were converted to comparable values. The validity of values, methods and units rests with the original authors.

[b] Some animals were tested multiple times.

[c] Values were reported as hematocrits in the erythrocytic and leukocytic series. Inclusion or exclusion of leukocytes (buffy layer) in the erythrocytic hematocrit was not clear.

Id-Ad-CC = adults, Crooked Creek, Idaho (Trout 1976); Alta-Ad = adults, Alberta (Barrett and Chalmers 1977a, 1977b, 1979); Alta-Ad-Male = adult males, Alberta (Barrett and Chalmers 1977a, 1977b); Alta-Ad-Female = adult females, Alberta (Barrett and Chalmers 1977a, 1977b); Alta-Ad-RT-low = adults with rectal temperatures less than or equal to 104.7°F (40.4°C) at time of sampling, Alberta (Barrett and Chalmers 1977a); Alta-Ad-RT-med = adults with rectal temperatures between 104.7° and 106.3°F (40.4° and 41.3°C) at time of sampling, Alberta (Barrett and Chalmers 1977a); Alta-Ad-RT-high = adults with rectal temperatures greater than 106.3°F (41.3°C) at time of sampling, Alberta (Barrett and Chalmers 1977a); Alta-N = neonates (mean 2.7 days old, SD = 1.5), Alberta (Barrett and Chalmers 1979); LL = fawns (about 7 months old), yearlings and adults, Little Lost River, Idaho (Seal and Hoskinson 1978); LL-Ad = adults, Little Lost River, Idaho (Seal and Hoskinson 1978); LL-Fawn = fawns (about 7 months old), Little Lost River, Idaho (Seal and Hoskinson 1978); BC = fawns (about 7 months old), yearlings and adults, Birch Creek, Idaho (Seal and Hoskinson 1978); CC = fawns (about 7 months old), yearlings and adults, Crooked Creek, Idaho (Seal and Hoskinson 1978); CC-Ad-Male = adult males, Crooked Creek, Idaho (Seal and Hoskinson 1978); CC-Ad-Female = adult females, Crooked Creek, Idaho (Seal and Hoskinson 1978); CC-Fawn = fawns (about 7 months old), Crooked Creek, Idaho (Seal and Hoskinson 1978); Wyo-Ad = tame adults, Sybille unit of the Wyoming Game and Fish Dept. (Dhindsa et al. 1974, 1975); Col-Ad-Male = adult males, blood samples from shot animals, Colorado (Bear et al. 1973); Col-Ad-Female = adult females, blood samples from shot animals, Colorado (Bear et al. 1973); Ad-ISIS = 78 available records from 66 individuals at five institutions (International Species Information System 1995); Or-Ad-Female-Dec. = adult females net-gunned in Oregon during December (Dunbar and Velarde 1998); Or-Ad-Female-Mar. = adult females net-gunned in Oregon during March (Dunbar and Velarde 1998); Or-fawn = <1–3-day-old fawns captured in Oregon (Dunbar and Velarde 1998); Or-Ad-Female-98 = adult females net-gunned in Oregon during December 1998 (M. Dunbar personal files); Or-Fawn-98 = <1–3-day-old fawns captured in Oregon (Dunbar 1999); Or-Fawn-99 = 4–72-hour-old fawns captured in Oregon (M. Dunbar unpublished 1999 data); Id-Ad-LL = adults, Little Lost River, Idaho (Trout 1976); Id-Ad-BC = adults, Birch Creek, Idaho (Trout 1976); Alta-Ad-O-Day = adults, no delay in captivity before blood sampling, Alberta (Barrett and Chalmers 1977a, 1977b); Alta-Ad-l-Day = adults, 1-day delay in captivity before blood sampling, Alberta (Barrett and Chalmers 1977a, 1977b); Alta-Ad-2-Day = adults, 2-day delay in captivity before blood sampling, Alberta (Barrett and Chalmers 1977a, 1977b); Nebr. 1–24 mo. = six captive pronghorn in Nebraska tested 86 times by jugular punctures—biweekly during first 5 months, monthly during next 8 months and approximately every 3 months during second year of captivity (Clemens et al. 1987); Alta-N-Male = male neonates, Alberta (Barrett and Chalmers 1979); Alta-N-Female = female neonates, Alberta (Barrett and Chalmers 1979); Alta-N-single = single neonates, Alberta (Barrett and Chalmers 1979); Alta-N-twin = twin neonates, Alberta (Barrett and Chalmers 1979); RBCs = red blood cells.

characteristic of oxygen transport that related to pronghorn ability to run at exceptionally high speeds for long distances. Dhindsa et al. found that pronghorn erythrocytes do not sickle under high oxygen tension and high pH environments, as has been observed in many species of deer.

In comparison with other indigenous North American ruminants, pronghorn have exceptionally low serum cholesterol and high serum protein and albumin values (Barrett and Chalmers 1977b). Lower rectal temperatures, serum glucose, serum aspartate aminotransferase (AST; formerly serum glutamic oxaloacetic transaminase [SGOT]) and serum creatine phosphokinase (CPK) values for neonates compared with those of adults may indicate less stress from capture on neonates than adults. However, these values were much higher in wild fawns than in hand-reared pronghorn (Barrett and Chalmers 1977a). Alkaline phosphatase levels were 6 to 12 times higher in neonatal pronghorn than in adults. Alkaline phosphatase is associated with active skeletal development and would be expected to be high in very young animals (Barrett and Chalmers 1979).

Humoral stress indicators are of value only if they can be associated with the level of the stressor—something difficult to interpret from wild-trapped animals (Pojar and Spraker 1986). Pojar and Spraker reported on values of blood urea nitrogen (BUN), AST, CPK, glucose and cortisol from pronghorn subjected to four levels of handling stress. Stress levels ranged from tame animals bled within 2 minutes of restraint to wild-trapped animals in a high state of excitement for approximately 1 hour. Of the five parameters tested, cortisol and CPK showed a strong positive relation with stress; glucose and BUN showed a weak positive relation; AST showed a very weak negative

Table 50. Biochemical blood parameters of neonatal, fawn and adult pronghorn[a]

Source	Number tested[b]	Mean	Standard deviation or range
Total protein (g/dL) (considered same as serum protein)			
Id-Ad-LL	22	6.4	0.5
Id-Ad-BC	27	6.3	0.4
Id-Ad-CC	47	6.6	0.5
Alta-Ad	48	7.3	1.1
Alta-Ad-O-Day	15	6.9	0.7
Alta-Ad-l-Day	8	8.0	1.1
Alta-Ad-2-Day	7	6.2	1.8
Alta-Ad-0.5-Yr	11	7.4	1.1
Alta-Ad-1.5-Yr	8	7.6	0.8
Alta-Ad-2.5-Yr	6	6.0	1.0
Alta-Ad-3.5+-Yr	21	7.6	1.1
Alta-N	46	4.8	0.6
LL	23	6.8	0.5
BC	25	6.7	0.4
CC	37	6.5	0.4
Wyo-Ad	4	8.2	1.7
Total Protein (g/dL)			
Ad-ISIS	66	5.5	1.1
Or-Ad-Female-Dec	20	6.2	0.6
Or-Ad-Female-Mar	20	5.9	0.5
Or-Fawn	74	4.6	0.6
Or-Ad-Um-88	30	7.3	—
Or-Ad-Bv-96	34	6.5	—
Or-Ad-Female-98	9	6.5	—
Or-Fawn-98	28	3.9	2.9–5.9
Or-Fawn-99	25	4.0	—
Albumin (g/dL)			
Alta-Ad	48	4.8	0.7
Alta-Ad-O-Day	15	4.3	0.5
Alta-Ad-l-Day	8	5.1	0.4
Alta-Ad-2-Day	7	3.8	0.5
Alta-Ad-0.5-Yr	11	4.8	0.5
Alta-Ad-1.5-Yr	8	5.2	0.5
Alta-Ad-2.5-Yr	6	3.9	0.4
Alta-Ad-3.5+-Yr	21	4.8	0.7
Alta-N	46	2.4	0.3
Alta-N-Male	24	2.3	0.2
Alta-N-Female	21	2.5	0.4
LL	23	4.5	0.5
BC	25	4.4	0.7
CC	37	4.8	0.3
Wyo-Ad	4	4.5	1.0
Or-Ad-Female-Dec	20	4.3	0.4
Or-Ad-Female-Mar	20	3.9	0.3
Or-Fawn	74	1.9	0.4
Or-Ad-Um-88	30	4.1	—
Or-Ad-Bv-96	34	5.0	—
Or-Ad-Female-98	9	4.1	—
Or-Fawn-98	28	1.5	1.2–2.2
Or-Fawn-99	25	1.7	—
Globulin (g/dL)			
Or-Ad-Female-Dec	20	1.9	0.3
Or-Ad-Female-Mar	20	2.0	0.5
Or-Fawn	74	2.8	0.6
Or-Ad-Um-88	30	3.1	—
Or-Ad-Bv-96	34	1.5	—
Or-Ad-Female-98	9	2.4	—
Or-Fawn-98	28	2.4	—
Or-Fawn-99	25	2.3	—
Alpha globulin (g/dL)			
Alta-Ad	48	0.6	0.8
Alta-N	46	0.6	0.1
Alta-N-Male	24	0.6	0.1
Alta-N-Female	21	0.5	0.1
Alpha-1 globulin (g/dL)			
Wyo-Ad	4	0.2	0.0
Alpha-2 globulin (g/dL)			
Wyo-Ad	4	0.5	0.1
Beta globulin (g/dL)			
Alta-Ad	48	0.7	0.2
Alta-N	46	0.7	0.1
Wyo-Ad	4	1.3	0.1
Gamma globulin (g/dL)			
Alta-Ad	48	1.3	0.7
Alta-N	46	1.1	0.5
LL	23	0.4	0.2
BC	25	0.5	0.2
CC	37	0.5	0.2
Wyo-Ad	4	1.6	0.6
Albumin to globulin ratio			
Alta-Ad	48	2.0	0.5
Alta-N	46	1.0	0.3
Alta-N-Male	24	0.9	0.2
Alta-N-Female	21	1.1	0.3
Or-Ad-Female-Dec	20	2.3	0.3
Or-Ad-Female-Mar	20	2.1	0.5
Or-Fawn	74	0.7	0.3
Or-Ad-Um-88	30	1.2	—
Or-Ad-Bv-96	34	3.6	—
Or-Ad-Female-98	9	1.7	—
Or-Fawn-98	28	0.6	—
Or-Fawn-99	25	0.7	—
Gamma-glutamyl transferase (μL)			
Or-Ad-Female-Dec	20	20.1	11.5
Or-Ad-Female-Mar	20	16.4	8.9
Or-Fawn	73	231.2	352.4
Or-Ad-Um-88	30	16.6	—
Or-Ad-Bv-96	34	7.6	—

continued on next page

TABLE 50—*continued*

Source	Number tested[b]	Mean	Standard deviation or range
Or-Ad-Female-98	9	16.0	—
Or-Fawn-98	28	397	7.0–3,972.0
Or-Fawn-99	25	362	—
Blood (serum) urea nitrogen (mg/dL)			
Id-Ad-LL	22	39.9	6.4
Id-Ad-BC	27	11.0	2.8
Id-Ad-CC	47	17.8	3.2
Alta-Ad	106	42.8	16.6
Alta-Ad-O-Day	48	33.2	12.9
Alta-Ad-l-Day	23	51.6	12.8
Alta-Ad-2-Day	7	38.4	2.3
Alta-N	63	21.3	10.6
LL	23	44.0	8.6
BC	25	11.3	3.0
CC	37	18.6	4.3
Wyo-Ad	4	35.0	8.0
Or-Ad-Female-Dec	20	15.1	2.8
Or-Ad-Female-Mar	20	31.2	5.4
Or-Fawn	74	15.8	6.7
Or-Ad-Um 88	30	43.7	—
Or-Ad-BV-96	34	23.6	—
Or-Ad-Female-98	9	17.0	—
Or-Fawn-98	28	18.0	11.0–36.0
Or-Fawn-99	25	15.0	—
Uric acid (mg/dL)			
Wyo-Ad	4	0.4	0.1
Or-Fawn-98	28	0.5	<0.1–1.9
Cholesterol (mg/dL)			
Id-Ad-LL	22	37.0	—
Id-Ad-BC	27	35.0	—
Id-Ad-CC	47	60.5	—
Alta-Ad	100	43.0	14.4
Alta-Ad-Male	20	49.8	17.6
Alta-Ad-Female	77	40.8	12.7
Alta-Ad-O-Day	46	40.6	11.7
Alta-Ad-l-Day	21	44.2	12.4
Alta-Ad-2-Day	7	68.6	15.5
Alta-N	89	67.4	33.8
Alta-N-Male	47	60.5	29.1
Alta-N-Female	41	75.5	37.5
Wyo-Ad	4	37.0	4.0
Or-Ad-Female-Dec	20	42.0	6.3
Or-Ad-Female-Mar	20	37.4	6.8
Or-Fawn	65	55.9	23.3
Or-Ad-Bv-96	34	40.5	—
Or-Ad-Female-98	9	52.0	—
Or-Fawn-98	28	43.0	20.0–100.0
Or-Fawn-99	25	42.0	—
Glucose (mg/dL)			
Id-Ad-LL	22	216.01	62.0
Id-Ad-BC	27	185.31	39.7
Id-Ad-CC	47	188.31	34.8
Alta-Ad	99	250.0	74.0
Alta-Ad-O-Day	46	261.4	85.2
Alta-Ad-l-Day	21	251.3	48.7
Alta-Ad-2-Day	7	173.3	98.7
Alta-N	60	203.5	87.3
Alta-N-single	28	228.1	92.7
Alta-N-twin	32	182.0	77.5
Wyo-Ad	4	97.0	6.0
Or-Ad-Female-Dec	20	203.0	40.7
Or-Ad-Female-Mar	20	187.0	30.5
Or-Fawn	73	160.7	39.6
Or-Ad-Um-88	30	110.4	—
Or-Ad-Bv-96	34	262.3	—
Or-Ad-Female-98	9	178.0	—
Or-Fawn-98	28	192.0	129.0–314.0
Or-Fawn-99	25	161.0	—
Creatinine (mg/dL)			
Alta-Ad	54	11.6	48.9
Alta-N	11	2.4	2.8
Wyo-Ad	4	0.9	0.1
Or-Ad-Female-Dec	20	1.8	0.3
Or-Ad-Female-Mar	20	1.4	0.2
Or-Fawn	74	0.9	0.5
Or-Ad-Um-88	30	0.7	—
Or-Ad-Bv-96	34	2.1	—
Or-Ad-Female-98	9	1.6	—
Or-Fawn-98	28	0.9	0.3–1.9
Or-Fawn-99	25	0.9	—
Calcium (mg/dL)			
Id-Ad-LL	22	10.6	1.1
Id-Ad-BC	27	11.8	0.7
Id-Ad-CC	47	10.5	0.8
Alta-Ad	106	11.6	2.2
Alta-Ad-O-Day	48	12.0	2.4
Alta-Ad-l-Day	23	11.1	1.8
Alta-Ad-2-Day	7	9.2	2.6
Alta-N	85	12.4	1.4
Wyo-Ad	4	9.6	—
Or-Ad-Female-Dec	20	9.1	1.3
Or-Ad-Female-Mar	20	8.9	0.6
Or-Fawn	74	10.5	0.7
Or-Ad-Um-88	30	10.0	—
Or-Ad-Bv-96	34	9.5	—
Or-Ad-Female-98	9	8.9	—
Or-Fawn-98	28	9.8	8.4–10.6
Or-Fawn-99	25	10.5	—
Magnesium (mg/dL)			
Alta-Ad	106	2.5	0.9
Alta-Ad-O-Day	48	2.7	0.8

continued on next page

TABLE 50—*continued*

Source	Number tested[b]	Mean	Standard deviation or range
Alta-Ad-l-Day	23	2.1	1.1
Alta-Ad-2-Day	7	1.7	0.7
Alta-N	85	2.2	0.3
Or-Ad-Female-Dec	20	2.9	0.2
Or-Ad-Female-Mar	20	3.0	0.8
Or-Fawn	73	2.0	0.2
Magnesium (ppm)			
Or-Ad-Um-88	30	28.5	—
Or-Ad-Bv 96	34	23.1	—
Or-Ad-Female-96	20	28.9	—
Or-Ad-Female-97	21	27.2	—
Phosphorus (mg/dL)			
Alta-Ad	103	6.4	2.6
Alta-N	83	10.0	2.1
Wyo-Ad	4	6.4	—
Or-Ad-Female-Dec	20	6.6	1.6
Or-Ad-Female-Mar	20	7.7	0.9
Or-Fawn	73	10.4	1.2
Or-Ad-Um-88	30	6.4	—
Or-Ad-Bv-96	34	3.1	—
Or-Ad-Female-98	9	6.8	—
Or-Fawn-98	28	10.3	7.6–13.9
Or-Fawn-99	25	9.8	—
Phosphorus (ppm)			
Or-Ad-Um-88	30	74.7	—
Or-Ad-Bv-96	34	37.9	—
Or-Ad-Female-96	20	70.1	—
Or-Ad-Female-97	21	85.0	—
Potassium (mEq/L)			
Alta-Ad	106	7.0	1.7
Alta-N	85	6.2	0.8
Wyo-Ad	4	5.5	—
Or-Ad-Female-Dec	20	4.6	0.6
Or-Ad-Female-Mar	20	4.6	0.5
Or-Fawn	67	4.6	0.6
Or-Ad-Female-98	9	5.4	—
Or-Fawn-98	28	4.3	3.3–5.3
Or-Fawn-99	25	4.2	—
Potassium (ppm)			
Or-Ad-Um-88	30	332.5	—
Or-Ad-Female-96	20	189.2	—
Or-Ad-Female-97	21	191.0	—
Sodium (mEq/L)			
Alta-Ad	106	167.0	12.1
Alta-Ad-O-Day	48	166.5	11.5
Alta-Ad-l-Day	23	169.5	12.4
Alta-Ad-2-Day	7	145.8	13.6
Alta-N	85	145.2	14.1
Alta-N-Male	47	147.8	11.8

Source	Number tested[b]	Mean	Standard deviation or range
Alta-N-Female	36	141.5	16.4
Wyo-Ad	4	155.0	—
Or-Fawn	67	147.2	2.9
Or-Fawn-98	24	147.0	141.0–151.0
Or-Fawn-99	25	147.0	—
Sodium (ppm)			
Or-Ad-Um-88	30	3,756	—
Or-Ad-Female-96	20	3,703	—
Or-Ad-Female-97	21	3,714	—
Sodium to potassium ratio			
Or-Ad-Um-88	30	11.5	—
Or-Ad-Bv-96	34	33.8	—
Or-Ad-Female-96	20	19.8	—
Or-Ad-Female-97	21	19.6	—
Chloride (mEq/L)			
Wyo-Ad	4	111.0	—
Or-Ad-Female-Dec	20	113.2	2.2
Or-Ad-Female-Mar	20	112.2	2.7
Or-Fawn	67	109.0	2.6
Or-Ad-Um-88	30	103.5	—
Or-Ad-Bv-96	34	107.7	—
Or-Ad-Female-98	9	112.0	—
Or-Fawn-98	28	105.0	98.0–108.0
Or-Fawn-99	25	108.0	—
Carbon dioxide (mEq/L)			
Wyo-Ad	4	17.8	—
Nonesterified fatty acids (mEq/L)			
LL	23	390.0	182.2
BC	25	285.0	110.0
CC	37	456.0	206.8
Triglycerides (mg/dL)			
LL	23	182.0	76.7
BC	25	93.0	50.0
CC	37	98.0	30.4
Lactic dehydrogenase (IU = mIU/mL) (equivalence of units assumed)			
Id-Ad-LL	22	935.0	—
Id-Ad-BC	27	705.8	218.3
Id-Ad-CC	47	755.7	222.0
LL	23	846.0	383.7
BC	25	612.0	290.0
CC	37	440.0	152.1
CC-Ad-Male	8	370.0	50.9
CC-Ad-Female	15	359.0	73.6
CC-Fawn	9	459.0	90.0
Or-Ad-Female-Dec	20	491.8	161.9
Or-Ad-Female-Mar	20	441.5	80.8
Or-Fawn	71	914.4	231.0

continued on next page

TABLE 50—*continued*

Source	*Number tested*[b]	*Mean*	*Standard deviation or range*
Or-Ad-Female-98	9	460.0	—
Or-Fawn-98	28	907.0	563.0–1,820.0
Or-Fawn-99	25	1,002.0	—
Lactic dehydrogenase (Berger-Broida)			
Wyo-Ad	4	1,315.0	292.0
Serum aspartate aminotransferase (IU = mIU/mL) (equivalence of units assumed)			
Id-Ad-LL	22	384.0	—
Id-Ad-BC	27	306.2	56.2
Id-Ad-CC	47	273.0	78.5
LL	23	302.0	105.5
BC	25	198.0	75.0
CC	37	155.0	54.7
CC-Ad-Male	8	150.0	31.1
CC-Ad-Female	15	120.0	31.4
CC-Fawn	9	191.0	45.0
Serum aspartate aminotransferase (Reitman-Frankel U/mL)			
Alta-Ad	103	251.4	990.6
Alta-Ad-O-Day	45	87.4	53.4
Alta-Ad-l-Day	23	127.4	48.3
Alta-Ad-2-Day	7	115.4	29.0
Alta-N	92	106.5	94.7
Glutamic oxaleacetic transaminase (Sigma-Frankel)			
Wyo-Ad	4	112.0	19.0
Serum glutamic pyruvic transaminase (R.F.U/mL)			
Alta-Ad	106	18.2	24.8
Alta-N	36	2.0	1.0
Serum creatine phosphokinase (IU/dL)			
Alta-Ad	28	432.1	621.2
Alta-N	21	40.1	15.5
Creatine phosphokinase (mIU/mL)			
LL	23	2,996.0	1,961.5
BC	25	1,623.0	1,340.0
CC	37	982.0	699.5
Thyroxine (µg/dL)			
LL	23	5.0	1.9
BC	25	7.2	3.0
CC	37	6.4	1.8
Cortisol (µg/dL)			
LL	23	6.1	1.4
BC	25	5.8	2.0
CC	37	4.3	1.8
Total bilirubin (mg/dL)			
Id-Ad-LL	22	0.8	0.2
Id-Ad-BC	27	0.7	0.1
Id-Ad-CC	47	1.0	0.2
Wyo-Ad	4	0.4	—
Or-Ad-Female-Dec	20	0.3	0.2
Or-Ad-Female-Mar	20	0.2	0.1
Or-Fawn	74	2.5	1.2
Or-Ad-Um-88	30	0.3	—
Or-Ad-Bv-96	34	1.0	—
Or-Ad-Female-98	9	0.2	—
Or-Fawn-98	28	2.2	0.9–4.9
Or-Fawn-99	25	2.1	—
Alkaline phosphatase (IU = mIU/mL) (equivalence of units assumed)			
Id-Ad-LL	22	86.9	61.2
Id-Ad-BC	27	60.2	21.2
Id-Ad-CC	47	50.0	18.4
LL-Ad-Male	6	74.0	14.4
LL-Ad-Female	6	40.0	20.1
LL-Fawn	8	200.0	82.0
BC-Ad-Male	3	67.0	31.2
BC -Ad-Female	5	62.0	19.2
BC-Fawn	7	91.0	50.3
CC-Ad-Male	9	68.0	20.4
CC-Ad-Female	17	35.0	12.4
CC-Fawn	10	81.0	13.9
Or-Ad-Female-Dec	20	49.3	25.0
Or-Ad-Female-Mar	20	75.8	39.1
Or-Fawn	73	743.9	428.0
Or-Ad-Um-88	30	32.8	—
Or-Ad-Bv-96	34	44.6	—
Or-Ad-Female-98	9	54.0	82.2
Or-Fawn-98	28	960.0	353.0–4,840.0
Or-Fawn-99	25	847.0	—
Alkaline phosphatase (King-Armstrong U/mL)			
Alta-Ad	96	4.6	17.3
Alta-Ad-Male	20	3.6	1.7
Alta-Ad-Female	73	2.7	1.6
Alta-Ad-0.5-Yr	18	3.5	1.7
Alta-Ad-1.5-Yr	14	3.8	1.6
Alta-Ad-2.5-Yr	8	2.7	1.7
Alta-Ad-3.5+-Yr	52	2.4	1.5
Alta-N	74	29.6	20.0
Alta-N-single	39	34.6	22.1
Alta-N-twin	35	24.2	15.9
Alanine aminotransferase (U/L)			
Or-Ad-Female-Dec	20	25.2	9.6
Or-Ad-Female-Mar	20	24.0	5.2
Or-Fawn	74	15.8	8.6
Or-Ad-Bv-96	34	23.9	—
Or-Ad-Female-98	9	29	—
Or-Fawn-98	28	16.0	8.0–39.0
Or-Ad-Female-96	20	25.2	—
Or-Ad-Female-97	21	24.0	—
Or-Fawn-99	25	16.0	—

continued on next page

Table 50—*continued*

Source	Number tested[b]	Mean	Standard deviation or range
Aspartate aminotransferase (U/L)			
Or-Ad-Female-Dec	20	226.6	116.1
Or-Ad-Female-Mar	20	234.2	132.7
Or-Fawn	7	115.5	32.2
Or-Ad-Um-88	30	59.1	—
Or-Ad-Bv-96	34	139.8	—
Or-Ad-Female-98	9	232.0	—
Or-Fawn-98	28	142.0	53.0–504.0
Or-Fawn-99	25	116.0	—

[a] Values were rounded to one decimal place unless total value was less than 0.1. Where possible, units were converted to comparable values. Blood enzymes often were measured in different, nonconvertible units and are presented as originally expressed by the authors. The validity of values, methods and units rests with the original authors.

[b] Some animals were tested multiple times.

Id-Ad-LL = adults, Little Lost River, Idaho (Trout 1976); Id-Ad-BC = adults, Birch Creek, Idaho (Trout 1976); Alta-Ad-O-Day = adults, no delay in captivity before blood sampling, Alberta (Barrett and Chalmers 1977a, 1977b); Alta-Ad-1-Day = adults, 1-day delay in captivity before blood sampling, Alberta (Barrett and Chalmers 1977a, 1977b); Alta-Ad-2-Day = adults, 2-day delay in captivity before blood sampling, Alberta (Barrett and Chalmers 1977a, 1977b); Alta-Ad-0.5-yr = 0.5-year olds, Alberta (Barrett and Chalmers 1977b); Alta-Ad-1.5-yr = 1.5-year olds, Alberta (Barrett and Chalmers 1977b); Alta-Ad-2.5-yr = 2.5-year olds, Alberta (Barrett and Chalmers 1977b); Alta-Ad-3.5+ yr = 3.5-year olds and older, Alberta (Barrett and Chalmers 1977b); Ad-ISIS = 78 available records from 66 individuals at 5 institutions (International Species Information System 1995); Or-Ad-Um-88 = adults (apparently both sexes) trapped on the Umatilla Army Depot in northeastern Oregon (Whittaker et al. 1999); Or-Ad-Bv-88 = adults (apparently both sexes) captured in Baker Valley of northeastern Oregon (Whittaker et al. 1999); Or-Ad-Female-98 = adult females net-gunned in Oregon during December 1998 (M. Dunbar personal files); Alta-N-Male = male neonates, Alberta (Barrett and Chalmers 1979); Alta-N-Female = female neonates, Alberta (Barrett and Chalmers 1979); Alta-N-single = single neonates, Alberta (Barrett and Chalmers 1979); Alta-N-twin = twin neonates, Alberta (Barrett and Chalmers 1979); Or-Ad-Female-96 = adult females, Hart Mountain, Oregon, December 1996 (Whittaker et al. 1999); Or-Ad-Female-97 = adult females, Hart Mountain, Oregon, March 1997 (Whittaker et al. 1999); LL-Ad-Male = adult males, Little Lost River, Idaho (Seal and Hoskinson 1978); LL-Ad-Female = adult females, Little Lost River, Idaho (Seal and Hoskinson 1978); BC-Ad-Male = adult males, Birch Creek, Idaho (Seal and Hoskinson 1978); BC-Ad-Female = adult females, Birch Creek, Idaho (Seal and Hoskinson 1978). Other acronyms as in Table 49.

relation. Cortisol concentrations have been proposed as indicators of physiological stress and body condition, but most cortisol assays are more easily performed on urine samples than on blood (Spalinger 2000).

During a study of three Idaho pronghorn populations, capture effects were reflected in levels of lactic dehydrogenase (LDH), CPK, AST and perhaps serum cortisol. Age differences were observed for hematology, fibrinogen, LDH and AST (Seal and Hoskinson 1978). Sex and age differences were noted in alkaline phosphatase levels. Differences in blood parameters attributable to differential nutrition between the three populations were observed for BUN, nonesterified fatty acids, serum triglycerides and alkaline phosphatase (Table 50). All of these latter parameters may be useful in assessing the condition of pronghorn habitat.

Dunbar and Velarde (1998) did not attribute any differences in blood parameters of Oregon fawns to sex in a sample of 34 buck and 27 doe fawns that were 1 to 3 days of age. One-day-old fawns had higher mean corpuscular volume, gamma-glutamyl transferase and hematocrit levels than did 3-day-old fawns. The BUN and creatine levels also were negatively correlated to age, but total protein, globulin and albumen concentrations increased steadily during the first 3 days.

The striking differences in BUN levels among three pronghorn populations in Idaho (Table 50) caused Seal and Hoskinson (1978) to examine the relationship between those levels and the habitats occupied by the three populations. The animals with high BUN levels (Little Lost Creek) were feeding in agricultural fields adjacent to sagebrush-dominated wintering areas. The pronghorn with low levels of BUN (Birch Creek) wintered entirely on native rangeland, and those with intermediate levels (Crooked Creek) wintered in an area with only minimal agricultural development but high protein content in native plants. Trout (1976) noted similar differences in BUN levels for the same three populations. Whittaker et al. (1999) also noted large differences in BUN levels among Oregon populations (Table 50) and correlated those levels to perceived nutritional levels, judged by vegetative species and conditions. As in Seal and Hoskinson's (1978) study, BUN of pronghorn feeding on alfalfa (Baker Valley) was approximately twice as high as that of animals living in high-desert shrubland (Table 51).

Table 51. Comparison of blood urea nitrogen (BUN) levels among pronghorn populations considered to be on high, medium or low levels of nutrition

Nutrition level	Number of animals tested	BUN (mg/dL)	Study
High	23	44.0	Seal and Hoskinson (1978)
Medium	37	18.6	
Low	25	11.3	
High	34	43.7	Whittaker et al. (1999)
Medium	21	31.0	
Medium	30	23.6	
Low	20	15.1	

The BUN generally is considered a good indicator of protein availability in the diet. However, BUN levels also increase as an animal depletes its fat reserves and begins to catabolize muscle tissue (Spalinger 2000). This complicates the use of BUN as a nutritional indicator because both very high and very low quality diets may result in similar BUN concentrations.

The roles of strenuous muscular activity, renal failure and stress in producing high BUN values in pronghorn remains speculative. Drive-trapped adults in Alberta showed higher BUN values after a day in the trap than did those tested soon after being caught—51.6 mg/dL compared with 33.2 mg/dL. This must be kept in mind when judging range adequacy (protein availability) by using BUN from trapped animals as a criterion.

Urea is produced in the liver and excreted by the kidneys. Thus, liver impairment generally leads to decreased BUN levels, and kidney disease results in increased levels. Urea is distributed through the body's water and tends to be low in nursing animals because of their high fluid intake and urine output. Starvation, fever, burns or other processes that result in rapid tissue catabolism may increase BUN levels (Carlson 1996a).

For adult does in Oregon, mean BUN, phosphorus and alkaline phosphatase levels were higher in March than December (Table 50). In contrast, mean albumen, creatine, cholesterol, sodium and total bilirubin levels were lower in March than December (Dunbar and Velarde 1998).

Thyroxin levels directly influence metabolic rate, and several researchers have found that concentrations decline during winter. As expected, these correspond to a seasonal depression in metabolic rate and food intake. Hence, the relative concentrations of thyroxin in different populations should reflect their relative nutritional status. The use of urea/creatine ratios also is becoming popular for evaluating the nutritional stress of animals, but its measurement is performed mostly on urine instead of blood (Spalinger 2000).

TRACE ELEMENTS

Many mineral elements occur in living tissue in such small amounts that early workers were unable to measure their precise concentrations. They frequently were described as occurring in traces, and the phrase "trace elements" arose to describe them. Of the 90 naturally occurring elements, 26 are known to be essential for animal life, and 15 are accepted as trace elements. These are iron, zinc, copper, manganese, nickel, cobalt, molybdenum, selenium, chromium, iodine, fluorine, tin, silicon, vanadium and arsenic (Underwood 1977). Essential trace elements are toxic if ingested or inhaled at high levels. However, Clemens et al. (1987) maintained that pronghorn naturally selected plants high in selenium and have adapted to levels that are toxic to other mammals. Clemens et al. found blood selenium levels almost twice as high in captive pronghorn as in captive bison and white-tailed deer.

Arthur and Gates (1988) found that on the Idaho National Engineering Laboratory in southeastern Idaho, pronghorn ingested 1.7 ounces (48.7 g) of soil a day. They estimated that—in that area—the animals acquired 75% of their iron, vanadium and fluorine from the soil. Also, 10% to 50% of the pronghorn manganese, chromium, molybdenum, nickel and zinc intake was from soil, not vegetation.

Essential trace elements are needed to activate specific enzymes that act as catalysts to chemical reactions occurring in cells, where they serve a wide range of functions. Desirable trace element levels have been determined for livestock, but levels essential to the health of pronghorn generally are unknown. However, selenium deficiency was suspected in wild pronghorn from Idaho (Stoszek et al. 1978), and copper deficiency was suspected in captive pronghorn in the Los Angeles Zoo (K. Brinkley personal communication:1992). Results of studies relative to these two problems, along with other monitoring studies, are presented in Tables 52 and 53.

Selenium is needed to activate the enzyme glutathione peroxidase. In selenium deficiency, accumulation of highly toxic peroxides leads to destruction of cell membranes, resulting in low fertility, white muscle disease and generally poor health. Pronghorn rangelands in the southwestern United States and east of the Continental Divide generally have adequate to high levels of selenium in the soil. In localized areas, some plants accumulate excessive amounts. Most of the rangelands west of the Continental Divide have low to adequate levels. However, southwestern Montana, western Oregon, northern California and an area north of Mexico straddling the Arizona/New Mexico border have low concentrations (Maas et al. 1996). Prince's plume, two-grooved milk vetch and woody asters require high selenium level in the soil for growth. Thus, areas high in selenium are indicated by their presence (Galey 1996).

Liver samples from areas of poor fawn survival in Idaho and from an area of good survival in Montana were analyzed for concentrations of selenium (Table 52). Selenium levels in Idaho pronghorn livers were deficient when compared with levels found in Montana pronghorn and with normal levels of other species (Stoszek et al. 1978). Individual pronghorn had liver selenium levels as low as 0.2 ppm—the level often found in selenium-deficient farm livestock suffering from clinical signs of white muscle disease. However, comparing trace element levels in pronghorn with those of livestock or other wild ruminants has little value. As more pronghorn studies are conducted,

TABLE 52. Mineral concentrations in livers of wild and captive pronghorn, in parts per million wet weight

Location	Number of animals	Mineral content (range or standard deviation in parentheses when given)						
		Selenium	*Zinc*	*Cobalt*	*Iron*	*Cadmium*	*Copper*	*Molybdenum*
Montana[a]	16	0.32	28.7	0.07	104	—	—	—
Idaho[a]	39	0.16	24.5	0.05	103	—	—	—
Los Angeles Zoo[b]	1	—	50.9	—	201	0.18	4.9	1.1
Arizona[c]	100	0.19	—	—	—	8.3	—	—
	—	(0.01–0.81)	—	—	—	(1.1–36)	—	—
Oregon[d]								
Males	9	0.11	30.61	0.13	111.23	—	12.02	1.8
		(0.05)	(7.19)	(0.52)	(27.84)	—	(6.86)	(9.17)
Females	8	0.13	40.75	0.17	143.87	—	8.16	1.54
		(0.02)	(11.73)	(0.03)	(39.26)	—	(2.66)	(0.35)

[a] Stoszek et al. (1978); from wild pronghorn. Original concentrations were dry weight. Concentrations given here were converted to wet weight by dividing by 3.5. From report listed in Puls (1994) and reported by Dunbar and Velarde (1998).

[b] Unpublished report (4/9/92), Los Angeles Zoo by F. D. Galey; from captive pronghorn.

[c] Heffelfinger and Olding (1997) and Heffelfinger et al. (1999); from wild pronghorn.

[d] Dunbar and Velarde (1998); from adult wild pronghorn.

adequate levels for the species should become evident. Until then, the few documented levels available from productive pronghorn herds seem the only legitimate criteria for comparison.

Preliminary results of the Stoszek et al. (1978) study indicated that the Idaho rangelands from which the pronghorn were collected did not supply adequate mineral nutrients. Selenium levels—about 50% of those found in productive Montana pronghorn—could have been responsible for low survival of pronghorn in the Idaho study area. However, liver samples were taken only during one hunting season. Soil moisture and other factors can cause the plant uptake of minerals to vary from year to year, season to season and plant to plant.

Domestic sheep in northwestern Nevada require selenium in their salt blocks for normal reproduction and health. Analysis of selenium levels in pronghorn from the same area indicated normal levels of selenium—normal presumed from levels in productive herds in Montana. Apparently, pronghorn used plants low in selenium more efficiently than did sheep, or moved about and selected plants high in selenium (J. Meeker personal communication:1985). Dunbar and Velarde (1998) reported marginal selenium in pronghorn (Tables 52 and 53) from nearby rangelands in Oregon, but no clinical signs of deficiencies. They speculated that adequate levels of vitamin E might compensate for the low selenium levels.

A 1996 study, involving 100 pronghorn statewide in Arizona, indicated low (0.188 ppm) liver selenium levels in that state (Table 52). Arizona Veterinary Diagnostic Laboratory personnel considered 74% of those animals to be below normal levels for selenium (Heffelfinger and Olding 1997a).

Selenium and other trace element requirements of pronghorn warrant further research. The little research done with selenium is complicated by year-to-year changes in availability and the fact that an animal can be depleted of selenium in less than a month if it is fed deficient forage. However, Clemens et al. (1987) found that blood selenium levels of pronghorn were not affected by 24-hour fasting periods.

Copper deficiency can occur as a primary problem in milk-fed animals or in animals foraging on copper-deficient rangeland. More commonly, it occurs secondarily in association with other trace mineral imbalances, such as excess molybdenum, and it is influenced by the dietary levels of sulfur and zinc (Carlson 1996b).

Sugár et al. (1992) used neck hair from red deer in southwestern Hungary to monitor their copper levels. Concentrations in the hair of wild red deer were at or above 7 mg/kg. Levels were the same in a farmed red deer herd during 1986 and 1987, but decreased to 1 to 4 mg/kg in 1988 and remained low during 1989. Reduced keratinization of hair and hooves, low weaning weights of calves, and reduced fertility of yearlings and adults were noted in 1989. The use of hair seems a practical, noninvasive way to monitor levels of trace elements (see Franzmann et al. 1975).

In 1989, copper deficiency was diagnosed in hand-reared captive pronghorn fawns at the Los Angeles Zoo (Table 52). Clinical signs were primarily manifested in growing fawns, and included anemia, anorexia, progressive paresis/recumbency and one death—attributed to acute aortic rupture. Due to the recurrence of this syndrome and minimal hand-rearing information for this species, an investigation of trace mineral levels in captive pronghorn was undertaken. Serum copper levels in 12 fawns (0.08–0.67 ppm) raised on the original formula were below levels considered normal for ruminants (0.7–1.2

TABLE 53. Mineral concentrations in blood (whole or serum) of wild and captive pronghorn, in parts per million unless otherwise noted

Location	Number of animals	*Mineral content (with range when given)*				
		Selenium	*Zinc*	*Iron*	*Magnesium*	*Copper*
Wyoming[a]	18	—	1.26	2.21	34.6	0.94
California[b]	13	0.142	0.94	1.3	25.0	0.71
Los Angeles[c]	7	—	0.81	4.05	21.4	0.90
Arizona[d]	7	—	—	—	—	0.64 (0.39–0.72)
Arizona[d]	99	—	—	—	—	0.59 (0.22–1.8)
Oregon[e]	20	98.55[f]	1.39	2.47	—	0.55
Oregon[e]	20	49.4[g]	1.45	2.94	—	0.56
Oregon[h]	65	—	1.03	3.52	—	0.41
Oregon[h]	17	84.6 (51–120)	—	—	—	—
Oregon[h]	44	50.6 (25–104)	—	—	—	—
Oregon[i]	30	46.1[g]	0.9	2.4	28.5	0.6
Oregon[i]	34	69.1[g]	0.8	2.6	23.1	0.8
Oregon[j]	27	25[g] (16–40)	—	—	—	—
Oregon[j]	22	—	—	2.1 (0.73–3.3)	—	0.33 (0.21–0.49)
Oregon[k]	8	98[g] (72–142)				
Oregon[k]	8	—	—	2.0 (1.3–2.6)	—	0.76 (0.64–0.91)
Oregon[l]	16	—	—	2.1 (0.98–4.0)	—	0.38 (0.26–0.66)
Oregon[l]	25	46[m] (31–61)	—	—	—	—

[a] Unpublished report (06/15/92), Los Angeles Zoo by K. H. Plumlee; from sera of wild pronghorn.
[b] Unpublished report (04/15/92), Los Angeles Zoo by F. D. Galey; from sera of wild pronghorn.
[c] Unpublished report (03/19/92), Los Angeles Zoo by F. D. Galey; from sera of captive pronghorn.
[d] Heffelfinger and Olding (1997a) and Heffelfinger et al. (1999); from sera of wild pronghorn.
[e] Dunbar and Velarde (1998); from blood of wild pronghorn.
[f] Nanogram (10^{-9} gram) per milliliter (ng/mL) in whole blood.
[g] ng/ml in serum.
[h] Dunbar and Velarde (1998); from sera of wild pronghorn fawns (<1–3 days old).
[i] Whittaker et al. (1999); from sera of wild pronghorn.
[j] Dunbar (1999); from sera of wild pronghorn fawns (<1–3 days old).
[k] M. Dunbar personal files; from blood (whole or serum of wild adult does).
[l] M. Dunbar (unpublished 1999 data); from whole blood or sera of wild pronghorn fawns (<1–3 days old).
[m] ng/mL in whole blood.

ppm) by Maas and Smith (1996). Copper sulfate supplementation of the hand-rearing formula, along with changes in dietary copper levels for the adult herd, resulted in no additional clinical cases. Comparison of presupplementation and postsupplementation serum copper levels showed a significant decrease in incidence of below normal values (from >95% to 30%). However, the incidence of hypocupremia in captive pronghorn still indicates a need for further investigation into the management of this species in captivity (Miller-Edge et al. 1994). Also, although tested for, no trace of arsenic or lead was found.

Fawns in Oregon were born with low serum copper levels ($\bar{x}$ = 0.28 ppm), but reached the lowest limit of the adult range in about 3 days (Dunbar and Velarde 1998). No differences were attributed to sex.

Evidence from Australia indicated three genetic potentials in domestic sheep for copper metabolism—strong evidence that differences exist within a species (Stoszek et al. 1978). Furthermore, animals adapted for generations to an area where trace elements were deficient in the soil and vegetation would be expected to do better in that area than would animals translocated to the area. Seldom has this been considered when moving pronghorn from one habitat type to another.

Levels of copper and selenium reported in Tables 52 and 53 indicate possible deficiency problems, especially because those animals with the highest levels generally were from the most productive herds. Comparison with levels deemed adequate in other species seems unwarranted. The need for experimental manipulation of mineral levels during long-term studies seems evident. Also, results from the blood of shot animals are suspect. When an animal bleeds to death, concentrations in the remaining blood may be diluted by physiological processes.

TOOTH REPLACEMENT AND WEAR

Pronghorn teeth, like those of bovids, are hypsodont, with long crowns and either no roots or roots that develop months or years after the teeth erupt (i.e., grow through the surface of gums). The roots of hypsodont teeth continue to grow throughout the life of an animal—an adaptation to counteract wear from coarse vegetation and grit, as well as the side-to-side chewing motion pronghorn use to grind vegetation. Thus, pronghorn teeth do not shorten noticeably with wear, and age has not been estimated by height above the gum line, as with deer. However, crown height for a given tooth decreases with age once enamel growth discontinues and the tooth begins to grow away from basal border of the mandible. Thus, a measure of wear, crown or enamel height may be feasible according to Lubinski (2001). Pronghorn tooth wear and replacement resemble those of most bovids. Hypsodont dentition generally is considered an adaptation for a diet of grass, but pronghorn subsist mostly on shrubs and forbs. As in bovids and most cervids, pronghorn do not have upper incisors or canines.

At birth, fawns have six teeth pushing through the gums—two front incisors and two upper and two lower premolars. Sometimes, the incisors still are covered by a thin, translucent membrane that breaks when touched. Teeth grow very rapidly, and a 44-day-old fawn in Montana had 20 deciduous (milk) teeth and 2 molars that were about 50% grown (Dow and Wright 1962). By 1 year of age, pronghorn have their full complement of 32 teeth; the six permanent upper molars and six permanent lower molars have erupted, but the six lower incisors, two lower (incisorlike) canines and the six upper and six lower premolars still are deciduous. At about 14 months, the first deciduous teeth (incisors) are replaced by permanent teeth. In more than half of the pronghorn investigated, all teeth were permanent by 3.5 years of age, but 1 of 50 five-year-old pronghorn still had deciduous canine teeth (Table 54). No variation in tooth growth, replacement or wear has been noted between the sexes.

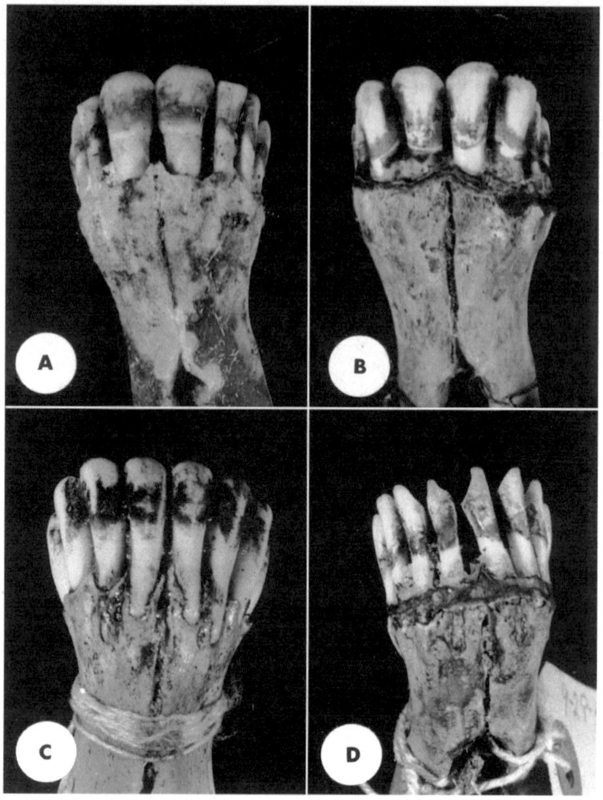

Incisiform teeth of pronghorn. During hunting seasons, pronghorn generally are 4 to 5 months older than the whole-year ages listed here. (A) A typical yearling during hunting season. About 85% of the yearlings have two large teeth (Pi1) in the center flanked by three smaller teeth on each side. If the Pi1s are not erupted, the deciduous teeth are somewhat spread by jaw growth. (B) A typical 2-year-old. Approximately 67% of 2-year olds have four large teeth (Pi1 and Pi2) flanked by two small ones on each side. Although Pi2 of the other 33% may not be visible, the milk teeth usually are pushed out or loose. (C) The "full mouth" of a prime-aged pronghorn, 3 years of age or older. About 67% of the 3-year olds have eight large teeth (Pi1–3 and Pc); 33% have six (Pi1–3) flanked by two peglike canines. About 80% of 4-year olds have eight large incisiform teeth, although the canines do not erupt until 5 years of age in occasional animals. If six or more front teeth are permanent, one must look at the cheek (molariform) teeth to determine age. (D) Incisiform teeth of a pronghorn about 9 years of age. Popowski (1959) indicated that older pronghorn (8 or more years of age) could be aged by the length of the incisors, which shorten with advancing age. Incisor length may reflect range conditions. However, incisors of pronghorn in Montana and Yellowstone National Park remained long up to 10 years of age but became narrow. Many old pronghorn existed in those populations; two jaws were from marked animals—one 9 years old and one at least 10. However, W. Hepworth (personal communication:1980) noted that the only Pi1 remaining in a doe with 18 annuli was only 0.14-inch (3.5 mm) long. The narrow incisiform teeth of fawns are close together, but those of animals 9 or more years of age have spaces between them, indicating narrowing with age. Photos by Phil Wright.

TABLE 54. Tooth eruption and replacement in the lower jaw of the pronghorn

	Incisors			*Canine*	*Premolars*			*Molars*		
Age	*1*	*2*	*3*	*1*	*2*	*3*	*4*	*1*	*2*	*3*
Birth	D	—	—	—	—	—	—	D	—	—
6 weeks	D	D	D	D	D	D	D	P	—	—
15–17 months	P	D	D	D	D	D	D	P	P	P
27–29 months	P	P	D	D	P	P	P	P	P	P
39–41 months	P	P	P?	(P)	P	P	P	P	P	P

D = milk or deciduous tooth; P = permanent tooth; parentheses indicate that tooth is in process of erupting.
Source: After Dow 1952, Dow and Wright 1962.

If high accuracy is required for age determination of individual pronghorn, especially for animals in the older age classes, sectioning of teeth to identify growth rings in the cementum seems to be the best method (McCutchen 1969, Kerwin and Mitchell 1971). Ages of pronghorn also can be estimated from tooth eruption and wear, although Dimmick and Pelton (1994) considered techniques based on wear invalid. Hoover et al. (1959) reported on criteria for determining ages up to 3 years. Dow and Wright (1962) developed a method that is suitable for most purposes, gives fair accuracy to 9 years of age and commonly is used by wildlife managers. Dow and Wright had few old, known-age animals to work with—the oldest in their study was 6.3 years old—but they estimated tooth condition for 7-, 8- and 9-year-old animals by assuming that tooth wear continued at the rate established for younger animals. A known-age 9-year-old pronghorn that I collected in 1969 and that had been tagged as a fawn in 1960 had exactly the amount of wear that Dow and Wright estimated for that age class. Much of the following information on pronghorn tooth wear and replacement is from Dow and Wright (1962).

Abbreviations for teeth are D for deciduous, P for permanent, i for incisor, c for canine; p for premolar and m for molar. Thus, the first permanent incisor is Pi1; deciduous premolar 2 is Dp2. Molars are designated merely as m because they all are permanent.

Deciduous premolars are shed at about 2.3 years, when 12 cavities, or infundibula are visible on the grinding surfaces of the fully erupted molariform teeth of each side of a jaw. Gradual disappearance of these cavities provides the main criterion used to distinguish ages for pronghorn of the older age classes. More than half of 3.5-year olds still have 12 infundibula, which then are lost by wear on the teeth at the rate of about two per year. Pronghorn with no remaining infundibula are estimated to be 9 years of age or older.

Whenever and wherever researchers are concerned with the characteristics of a species, individual variations pose a problem. Variability in pronghorn tooth eruption can be dealt with only by determining the percentages of animals that erupt teeth during given periods of time.

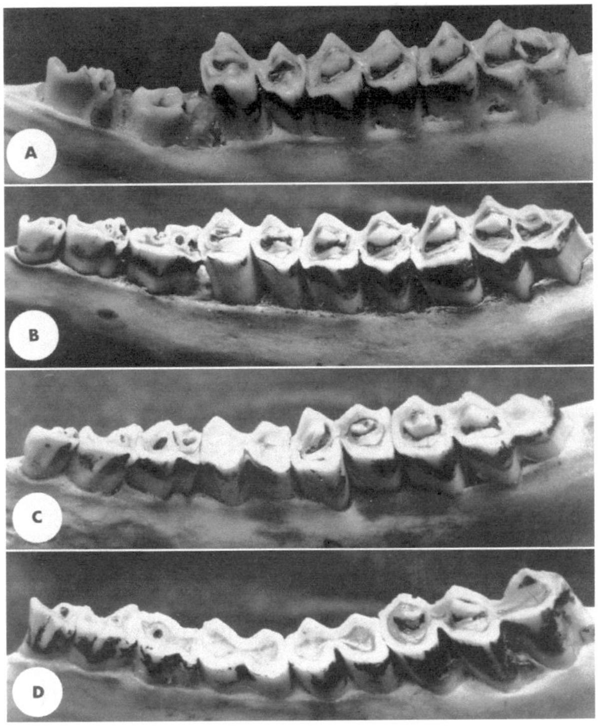

Molariform teeth of known-age pronghorn. (A) The jaw of a 2.3-year-old, with permanent premolars in the process of eruption and molars with all infundibula present; (B) the jaw of a 3.3-year-old, showing a full set of permanent teeth with 12 infundibula; (C) the jaw of a 4.3-year-old, showing m1 with wear such that both infundibula are gone, as well as the rear one from m3; (D) the jaw of a 6.3-year-old pronghorn, with only 5 of the original 12 infundibula remaining. Photos by Phil Wright.

Tooth wear is even more variable because the food preferences of individual animals differ, health or pollution may affect the durability of teeth, and range and soil conditions influence the rate of wear. In Texas, Hailey (1979) found the sequence and timing of tooth eruption similar to those observed by Dow and Wright (1962) in Montana. Although the sample size was small, photographs published by Hailey indicated that wear was greater on the teeth of the Texas pronghorn than on those of the Montana pronghorn.

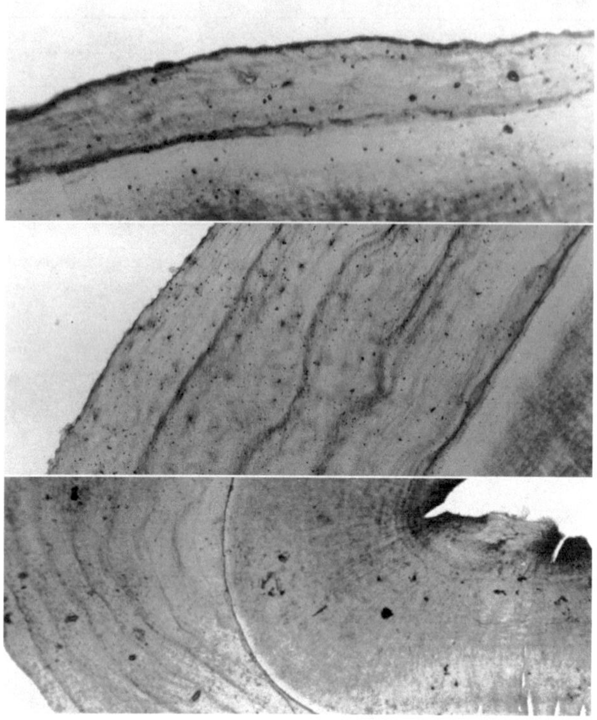

Sectional views of pronghorn incisors. (Top) *Cementum annulations in the incisor from a 2.3-year-old;* (center) *sagittal section of the incisor from a 5.3-year-old;* (bottom) *cross section of the incisor of a pronghorn estimated to be 9.3 years old.* Photomicrographs by H. McCuchins.

Bones and teeth often form annual growth rings somewhat like those of trees, especially in temperate to frigid zones where the animals undergo winter stress from cold and reduced quantity or quality of food. McCutchen (1969), who examined stained root sections from Pi1 of 10 known-age pronghorn 1.5 to 6.3 years old, showed that the numbers of cementum annuli (growth rings) were directly correlated with age. He determined that cementum begins to form just before eruption of Pi1, which occurs at 14 to 17 months of age. The initial summer growth shows in sections as a light-stained line. A narrow, dark-stained band is added during the following late August through November, and succeeding light bands form during spring. Thus, the first light–dark band of cementum outside the dentine indicates an age of 1.3 to 1.5 years, and each succeeding light–dark band represents another year.

A thin dark line on the very outside of the root of each tooth in pronghorn taken during the hunting season represents the end of summer (light staining) growth and the beginning of winter growth, because pronghorn generally are hunted during autumn about 4 months after their average birth dates. McCutchen (1969) selected Pi1 because it is (1) easy to remove, (2) the first permanent incisiform tooth to erupt and (3) erupts at a fairly predictable time.

Numbers of cementum annuli are more accurate criteria of ages than tooth wear, but interpretation sometimes is difficult, especially in teeth with discontinuous annuli. Furthermore, cementum sometimes may be reabsorbed during old age. I collected a marked female pronghorn known to be at least 10 years of age, yet its Pi1 showed only three annuli. From the teeth of other pronghorn, McCutchen (1969) also found some sections containing two dark-stained bands in each annulus. Such a condition will not confuse anyone experienced in aging by tooth sections, but it may trouble the beginner.

Kerwin and Mitchell (1971) sectioned Pi1s of 184 mature pronghorn collected in Alberta and compared ages derived by McCutchen's (1969) method with those assigned previously on the basis of tooth eruption and wear. The overall agreement for all age classes (2.5–8.5 years) was only 60.9%; 26.6% were judged to be younger and 12.5% were considered older by tooth eruption and wear techniques than by cementum annuli analysis. Agreement between the two methods generally was better in young (2.5–4.5 years) than in old (5.5–8.5 years) age classes. In comparison with age estimates based on cementum annuli, young age classes were underaged most frequently on the basis of eruption and wear, and old age classes were overaged most often.

The tooth eruption and wear technique is useful for hunters and the cheapest and easiest method for wildlife managers. Histological sections are expensive to prepare but have the advantage of greater accuracy; also, more hunters will allow the pulling of harvested pronghorn Pi1 at a checking station than the removal of jaws or splitting of cheeks. Useful information on estimating ages of pronghorn and seasons of their deaths from fossils is present in Lubinski (1997, 2001).

For further improvements in estimation of age by tooth growth and wear, a larger sample of known-age mandibles is needed.

EYE LENS GROWTH

Eye lenses grow throughout a mammal's life, and Kolenosky and Miller (1962) found that pronghorn lenses grow at a predictable rate. In lenses fixed in 10% formalin and dried 72 hours at 140.0°F (60°C), the mean difference between weights of right and left lenses from each specimen was 1.1%, but the heavier lens was not consistently on the left or right. Graphs of lens weights prepared by Kolenosky and Miller indicated three phases of growth: (1) extremely rapid growth, with a straight line relation between lens weight and age from 3 months after conception to 5 months after birth, (2) a negative acceleration phase from 5 months to 3.5 years of age and (3) a steady but very slow growth rate from 3.5 years until death.

The feasibility of aging pronghorn by lens weights versus aging by tooth eruption and wear was investigated by Hnatiuk (1970), who noted that differential tooth wear sometimes biases aging by that method. He believed that lens weights had potential for determining ages of pronghorn, but like Kolenosky and Miller (1962), Hnatiuk concluded that a series of lenses should be obtained from a natural population of known-age pronghorn to verify the growth curves they established. Mitchell (1965) noted that eye lenses increase in weight with advancing age, but he found overlaps in weights of lenses from pronghorn older than 1.5 years. Doubt was cast on the usefulness of lens weights for age determinations of animals by Friend and Severinghaus (1967), who found that such weights in white-tailed deer were affected by nutrition.

HORN GROWTH AND DEVELOPMENT

Some early investigators thought pronghorn horn sheaths were composed of consolidated hair (Bailey 1920, Seton 1927). However, Morrison (1961) showed that the horn sheath was composed principally of compressed, cornified epithelial cells, and hairs were rare or absent in the upper portions. Hair grows from the skin covering horn cores before it produces hard horn, and hair is incorporated into the horn—mostly near the bases. Roots of such hair are embedded, at least initially, in dermis under the horn sheaths, and the shafts follow an oblique course distally through the sheaths. Although hairs contribute little to the bulk or strength of horn sheaths, sebaceous glands and sweat tubules emptying into the canals of these hairs may prevent horns from becoming dry and brittle, and they contribute to the odor (O'Gara and Matson 1975).

Formation of hard horn begins about at the time of the rut, just after maximum testicular development (Figure 54). Horn sheaths are shed about 2 months later. Thus, growth and casting of horns apparently are controlled by male hormones in a manner similar to that of most cervids.

Horns grow and harden during 9 to 10 months of the year, and horn formation resembles that of skin. Although the outer layer of any skin feels soft, it consists of a layer of cells called the corneum (horny layer), which is composed of a hard protein substance (keratin) that protects underlying, delicate skin. Claws, nails, hooves and horns are specialized extensions of the corneum. Cuboidal or columnar cells in basal layers of the epidermis, next to the dermis, give rise to cells that flatten as they approach the surface. These flattened cells lose their nuclei and incorporate into the horn as layers of flat keratinized cells.

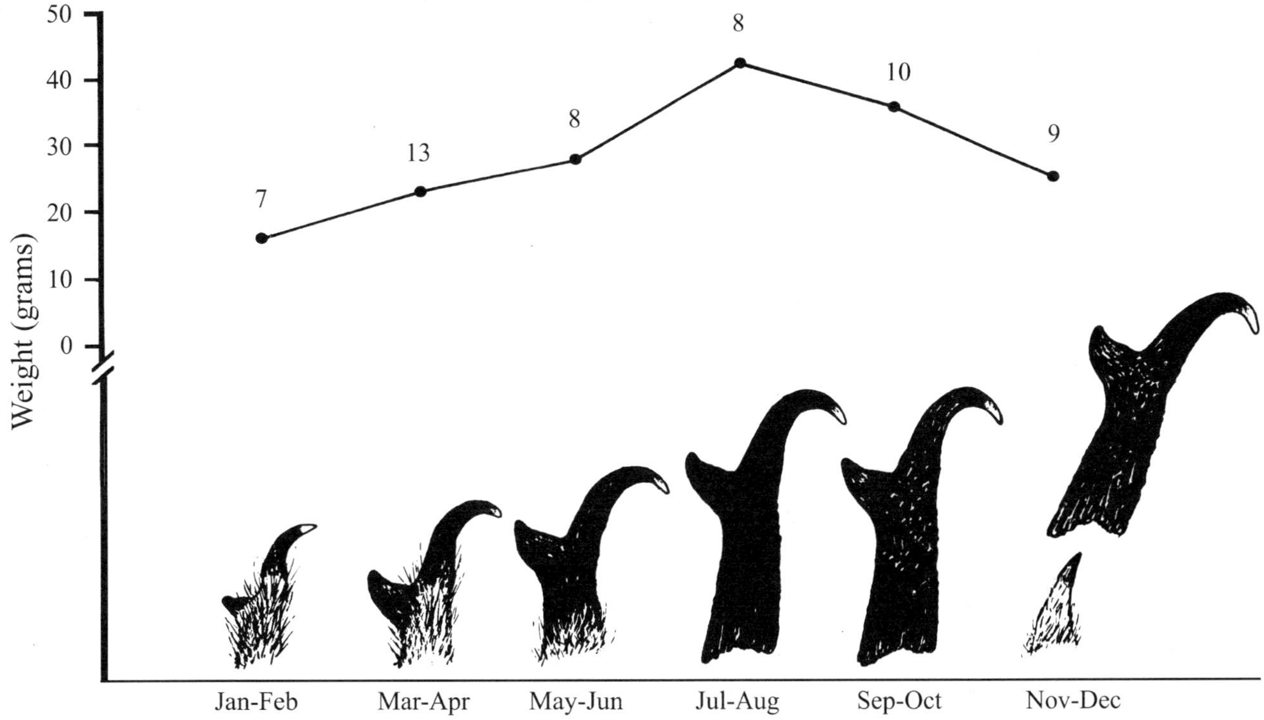

FIGURE 54 *Seasonal changes in the combined weights of both testes from yearling and adult pronghorn (from O'Gara et al. 1971). The dots represent mean weights, and numbers above them represent the sample sizes. To obtain samples large enough to yield meaningful averages, weights were grouped by 2-month increments. The drawings represent corresponding, seasonal horn development.* Illustration by Susan Kraft Ball.

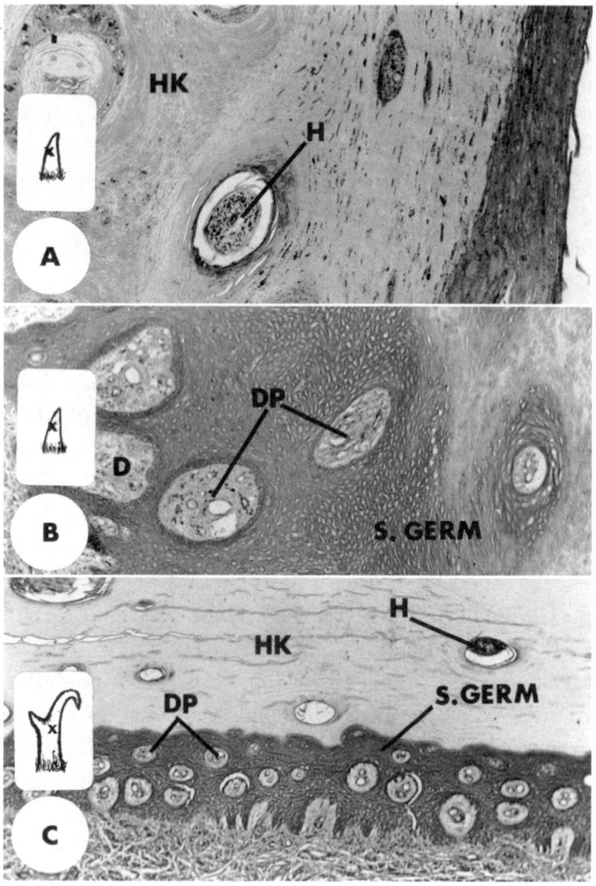

Photomicrographs of pronghorn horns (HK = hard keratin [horn], D = dermis [skin], H = hair, DP = dermal papillae and S. GERM = stratum germinativum [epidermis or outer skin]). (A) A cross section of a young growing horn, collected on November 20, is from near the horn surface. Growth and keratinization of epidermis results in the incorporation of hairs within the hard keratin of the horn sheath. (B) A cross section of a growing horn, collected on November 20, is from the interior of the horn close to the tip of the horn core. Cells next to the dermis and around the dermal papillae acquire keratin and are incorporated into the horn sheath. (C) A cross section of a mature horn collected on September 23. The hard keratin layer was thick, and little growth was taking place. The horn and underlying skin were fused. Photomicrographs by Gary Matson.

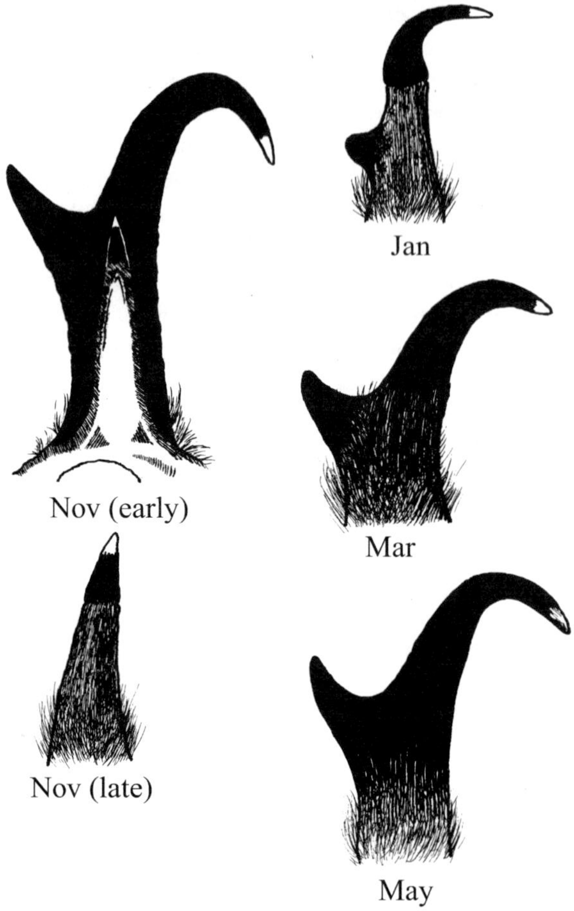

Figure 55. *Growth sequence of pronghorn horns, featuring lateral views of left horns approximately one-third natural size and the relation of hard horn (black) formation to soft skin (gray) on the bony cores. The Nov (early) horn is shown in diagrammatic longitudinal section; the white center represents the bony core. Note the wedge of new hard horn growing between the old horn and the horn core. This new material aids in horn casting and is the only hard horn remaining after the old sheath is cast, as seen on Nov (late).* Illustration by John Eiler.

Horn growth ceases in Montana pronghorn during late summer, before the rut. At that time, hard horn and underlying skin are closely associated and somewhat fused to each other. New growth begins during or after the rut—before old horns are cast—as the underlying skin begins producing new horn. This new growth contributes to the loosening of sheaths, as layers of epithelial cells slough from the old horn sheath.

New hard horn first appears as cones formed distal to the tips of bony cores (Figure 55 [Nov-early]). These hard tips slowly lift old sheaths and aid in casting, but long hairs rooted in the dermis and embedded in the sheaths hold them in place for some time. The pushing-off process is gradual and one sheath almost invariably falls before the other. Sheaths often are thrown off by sudden jumps or as bucks duck under fences. Embedded hairs usually are broken off, although some pull free from the sheath. Shed sheaths are fringed at the bottom, with ends of hairs still fast in the horny material.

Cones of hard horn are 0.7 to 1.0 inch (18–25 mm) long by the time a sheath is cast and may already display the "ivory" tips seen on many mature horns. Soft keratin, a forerunner of hard keratin (horn), covers the remainder of the future horn surface. Thus, after old sheaths are cast, new horns consist of bony cores covered by hairy gray skin topped by cones of hard black horn—often with white tips (Figure 55 [Nov-late]). Shufeldt (1920) presented good sketches, first published in 1880, of horns 1 day and 1 month after the old sheaths were cast.

At shedding, conspicuous annular swellings of skin can be seen and felt between the tips of bony cores and cones of horns. During December, shortly after horn casting, the annular swellings, which are warm to the touch and bleed profusely if cut or bruised, become even more conspicuous than they were at casting. Rapid differential growth from the swellings extrude horn tips upward and usually inward or backward, forming hooks. Frontal prongs begin approximately 2 inches (5 cm) above the head and develop rapidly as formation of hard horn proceeds (Figure 55 [Jan–Mar]).

By late March, prongs are fully developed and in the position (relative to the hooks) they maintain in mature horns. The next phase of development involves hardening of the soft keratin covering the bony cores and growth at the bases—like that of bovids. This lengthens the bases and pushes the prongs and hooks upward away from the head (Figure 55 [May]).

Growth slows in nearly mature horns and formation of hard horn ceases by August in Montana. The skin under horn sheaths remains comparatively inactive until after the rut, when new growth again results in the casting of old horn sheaths. Bucks shot during November in Montana often lose one or both horn sheaths from the shock of the bullet or when the hunter attempts to move the animal. Initiation and cessation of horn growth by male pronghorn apparently are regulated by testicular hormones (Morrison 1961, O'Gara et al. 1971).

Horn growth in pronghorn is rapid compared with that in bovids because horn formation, unrestricted by outer sheaths, takes place along the entire length of the cores. Nutrients for horn development probably are supplied primarily through skin over the cores in much the same manner that nutrients are supplied by the velvet and blood vessels under the velvet of antlers. However, the horn cores also are involved.

During autumn in Montana, when bucks are in hard horn, the surfaces of the cores are fairly smooth and dense, and the cores slope directly to sharp tips. From late November through January, when hooks and prongs are growing rapidly, the bone under the growing areas becomes spongy, porous and enlarged. Nutrients undoubtedly are supplied by blood flowing through the porous bone. The interiors of horn cores and sheaths were described by Noback (1932).

Castration has a marked effect on the horns of pronghorn bucks. Describing the horns of a castrated buck, Pocock (1905:193) wrote: "Instead of rising from the forehead as upright, laterally-compressed, bony prominences, the horn-cores bend obliquely forwards in a vertical plane, their axes inclining to the plane of the forehead at an angle of about 45°. . . . The horn-sheaths cover the core to the root, becoming gradually softer proximally, and passing into the hairy integument of the head. They project nearly horizontally forwards in the direction of the nose for a distance of about 3.5 inches, then curve downwards for about 2.5 inches, and then backwards towards the eye for about 3.5 inches, the terminal inch curving lightly inwards and downwards to a point close to the eye. Their total length along the outer or convex curve is thus about 9.5 inches. They also present a spiral twist forming about one-fourth of a complete turn. . . . Although the horn-sheaths have been described above as if each corresponded to a single fully-formed horn-sheath of an adult Prongbuck, closer examination shows that they are in reality composite—that is to say,

When a hunter attempted to roll a pronghorn buck over that was shot in Montana during early November, using the horn as a handle, the left horn sheath came off. Note the smooth, almost hairless skin covering the horn core. About 0.7 inch (18 mm) of hard horn was present distal to the horn core, including the white (ivory) tip. Hair will grow on the basal two-thirds of the skin before it starts to cornify. Such separation is not uncommon during late hunting seasons, and it does not impair a head for mounting unless a horn sheath is lost at the impact of a projectile or during a running fall. Photo by Lu O'Gara.

(Left) *A pronghorn buck has just shed its right horn; the other will go soon.* (Right) *A mature buck just after casting its old horn sheaths. Both photos were taken during November in Montana. After casting their horn sheaths, bucks join doe/fawn bands. Little aggression is seen among individuals of such bands, although adult bucks remain dominant to does and fawns. Bands coalesce into larger groups as winter progresses.* Photos by Peter T. Bromley.

An adult male pronghorn collected during November in Montana immediately after one horn sheath was cast. The white (ivory) tip already is evident. Such tips are favored by hunters because of their appearance. Some bucks of a given population will have them, others will not. The ivory is the first hard horn formed under the old sheath. Bucks older than 3 years of age seem more apt to show ivory than do younger animals, although yearlings occasionally have ivory tips on their horns. Photo by Bart W. O'Gara.

By late winter (mid-February to mid-March), pronghorn hooks and prongs are hard and nearly developed. Skin and hair cover the horn core. That skin will harden, entrapping much of the hair, and growth from the bases will extrude the prongs and hook upward away from the head. Photo by LuRay Parker; courtesy of the Wyoming Department of Game and Fish.

they consist of a series of horn-sheaths partially severed from each other. The exact number of sheaths involved in the formation of the whole is not easy to determine. There appear, however, to be six."

A castrated pronghorn buck brought to the National Bison Range from eastern Montana had horns quite similar to those described by Pocock. However, it only had three nested sheaths and the horns were hard right down to the skull. The buck caught its horns in a fence and pulled two sheaths off each horn. I intended to immobi-

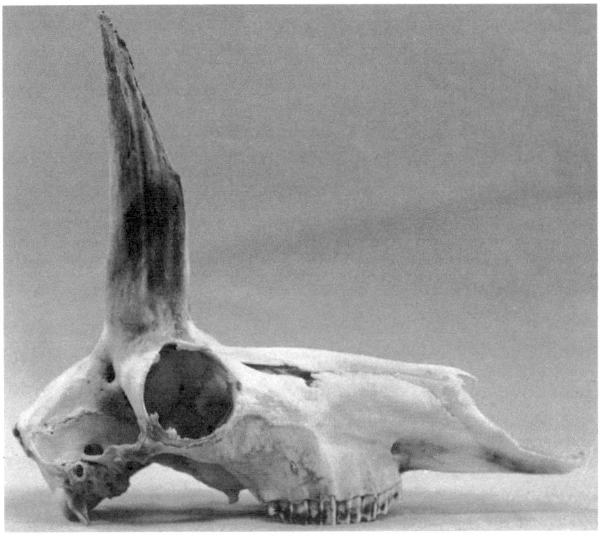

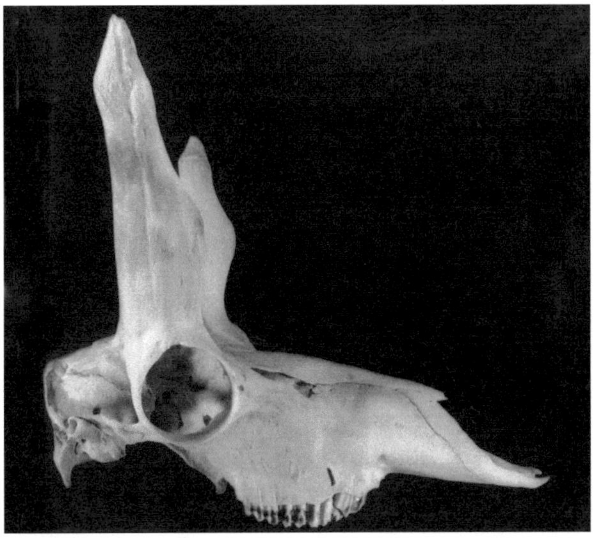

(Left) Collected in early September in Montana, this pronghorn buck had hard horn sheaths and no new horn growth was evident. The horn core, shown here, was composed of dense, smooth bone and sloped evenly to the tip. (Right) Collected in the same area during January, this buck had soft skin covering the horn cores and hard hooks and prongs that were being extruded rapidly. Note the "hips" below the tips of the horn cores and the bulges along the leading edges. Bone between the hips and tips and in the bulges was porous and full of blood when fresh. Hooks were extruding from the tips and prongs from the bulges, about as seen in the earlier photo of a buck eating hay, indicating that nutrients for horn formation were supplied through the bone as well as the skin. Photos by Bart W. O'Gara.

This skull of a winter kill picked up on the National Bison Range, Montana, illustrates the growth of hard horn while most of the horn core still was covered by skin. Judging by horn growth, the buck apparently died during early January. Photo by Bart W. O'Gara.

This 4-year-old pronghorn buck was castrated the day he was born. Only about 1.5 inches (38 mm) of the horn tips were hard horn. The animal was shot during August, when intact bucks have very large, actively secreting, subauricular glands. Note that the glandular area on this buck is very small with no sign of secretory material on the dark hair covering the gland. Photo by Bart W. O'Gara.

lize the animal to determine its age and whether the number of sheaths corresponded to its age. However, the buck disappeared before I got to it. The buck's age and when he was castrated or when given to the National Bison Range were unknown.

Another buck on the Bison Range—castrated the day it was born and shot at 4 years of age—did not grow hard horn sheaths. The horn cores leaned forward over the

The horns of this castrated (at unknown age) buck pronghorn are hard all the way to the skull, although considerable hair covers and grows through the bases. Typically, "replacement" horn sheaths still nest on top of the existing ones. Although this tame buck's subauricular glands were not developed, it manifested considerable sexual interest in does and women—unlike another buck that was castrated the day he was born. Photo by Bart W. O'Gara.

nose and had caps of hard horn at their tips, but most of the cores were covered by skin and hair. The horn sheaths resembled those of a normal buck about 2 weeks after horn casting.

The shedding of horn sheaths may have adaptive significance. Permanent horn sheaths, such as those of bovids, cannot be completely "repaired" when damaged. Antlers are shed after the rut and, thus, are repaired when regrown, and the yearlong cost of transport is lowered. However, regrowth of antlers demands a large mobilization of calcium, phosphorus and protein. Byers (1997a) indicated that casting of sheaths by pronghorn represents an ideal compromise that minimizes the cost of transport, avoids the mineralization and protein costs and allows for seasonal repair. Byers reported that one-third to one-half of mature bucks annually suffer some damage to horns. After the rut, bucks discard their used and often damaged weapons. They then are able to join and be fairly inconspicuous in mixed-sex herds while growing new sheaths.

SKIN GLANDS

Pronghorn have been thought to have 11 scent glands. A statement by Caton (1877) doubtlessly was responsible for acceptance of this exaggerated number. However, discrete glands are not present on the hocks, as reported by Caton, and the interdigital glands are not scent glands in the true sense of the word (O'Gara and Moy 1972). Apparently, pronghorn does have only two scent glands and bucks have five.

Mammalian scent glands are epidermal derivatives consisting of apocrine sudoriferous (sweat) and sebaceous (grease) glands. Many scent glands maintain constant size and activity throughout the year. However, some vary seasonally in size and activity and appear to be regulated by or in conjunction with the annual cycle of the gonads. Basically similar components—the same as those found in most mammalian skin—are present in all pronghorn skin glands, yet they serve different functions and are controlled by different processes. Sweat tubules produce odoriferous material that varies from gland to gland and species to species—the musk of some species; the sebaceous glands produce sebum, a greasy or waxy substance. When the two products are combined, sebum causes the odor to linger. Odors of materials originating in the sweat glands serve (in different species) for individual recognition, warning and as a sexual attractant. Only interdigital glands have a single large duct opening to the outside from each gland; other pronghorn glands consist of thickened skin with many minute openings on the surface.

The following descriptions apply to *Antilocapra americana americana* from Montana and Yellowstone National Park. Little has been done with other subspecies, and seasonally active glands would be regulated by the time of the breeding season in any particular area. Unless otherwise stated, descriptions of subauricular and rump glands are from Moy (1970), and those of interdigital and median glands are from Moy (1971).

Subauricular glands are located in thickened skin beneath the ears of male pronghorn. The sebaceous and apocrine regions are most highly developed at the center of the glands, with a gradual transition from the highly glandular center to the nonglandular periphery. This transition is less evident in the quiescent glands of winter specimens. Areas of the glands vary seasonally—reaching a maximum (about 13 square inches [815 cm^2]) during June and July and a minimum (about 6 square inches [373 cm^2]) during December and January.

The epidermis also varies in thickness with the seasons and is thinnest during winter months. The thickest layer of epidermis—the stratum germinativum—fluctuates significantly during the different seasons. The hair-raising—arrector pili—muscles of the subauricular glands are small and not always associated with hair follicles. Rather than to erect hair, they might serve to force secretion from the sebaceous gland by a milking action, or sebum might be secreted by pressure flow.

The peak of secretory activity is related closely to that of testicular activity, shortly before breeding takes place.

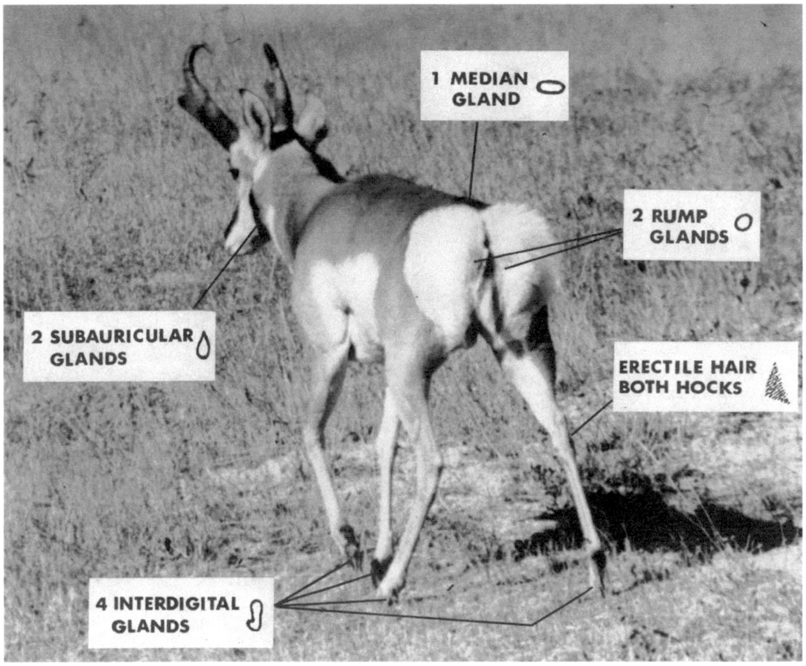

Of its skin glands, the pronghorn male has five scent glands—two subauricular, two rump and one median. The rump glands are the pronghorn doe's only scent glands. Subauricular glands produce a greasy, odoriferous material that bucks rub on vegetation within their territories; odors directly from the glands serve for sexual stimulation of does. Odors of median glands also are important in buck/doe interactions. Rump glands provide olfactory warnings in conjunction with visual warnings of rump patches. Interdigital glands secrete sebum that apparently keeps the hooves pliable, but they are not true scent glands. Areas of erectile hair on the hocks formerly were considered scent glands (see Caton 1877), but the skin is quite similar to that on other parts of the body and produces little odor detectable to humans. The buck featured is retreating with its left ear cocked toward the perceived potential threat (photographer). Its partly erect rump patches indicate mild fear or excitement. Photo by Bart W. O'Gara.

Subauricular glands of castrated bucks are similar throughout the year to the inactive glands of normal bucks during December.

Caton (1877) suspected that subauricular glands were important to the mating behavior of bucks. Histological and behavioral studies have confirmed Caton's suspicions (Moy 1970, O'Gara and Moy 1972).

Rump—or ischiadic—glands, present in both sexes, are circular areas of thickened skin about 0.7 to 1.6 inches (18–41 mm) in diameter. Glandular areas are not conspicuous and can only be distinguished externally by the presence of tan sebum that is made pungent by secretions from the sweat glands at the bases of hair stalks. However, the thickened areas can be seen easily on the inside of the skin of a freshly skinned animal. Hairs of the rump patch and the glands are white and large. They pass diagonally into the dermis at an angle of approximately 60 degrees. As well as erecting hair, the large arrector pili muscles, associated with each hair, can instantly force secretory material from the glands.

Rump glands can discharge airborne scent at any time. Once, while I was giving a tame pronghorn an immobilization shot with a hand syringe, the rump patch flared, and spray from the glands felt as if it had come from an aerosol can held 2 feet (60 cm) from my bare arm. Müller-Schwarze and Müller-Schwarze (1972) described the odor from rump glands as reminiscent of buttered popcorn.

Rump glands of castrated bucks are similar to the rump glands of intact animals. Gross measurements of these glands do not indicate seasonal variations. The centers of the glands are more highly developed than the peripheries.

Dark hair below the eyes and ears of pronghorn covers their subauricular glands. During months when the glandular area is large due to imminent or actual rut, the gland secretes large amounts of greasy, pungent material. The dark and odoriferous patches may serve as sensory cues both to estrous does and competing bucks. Photo by Peter T. Bromley.

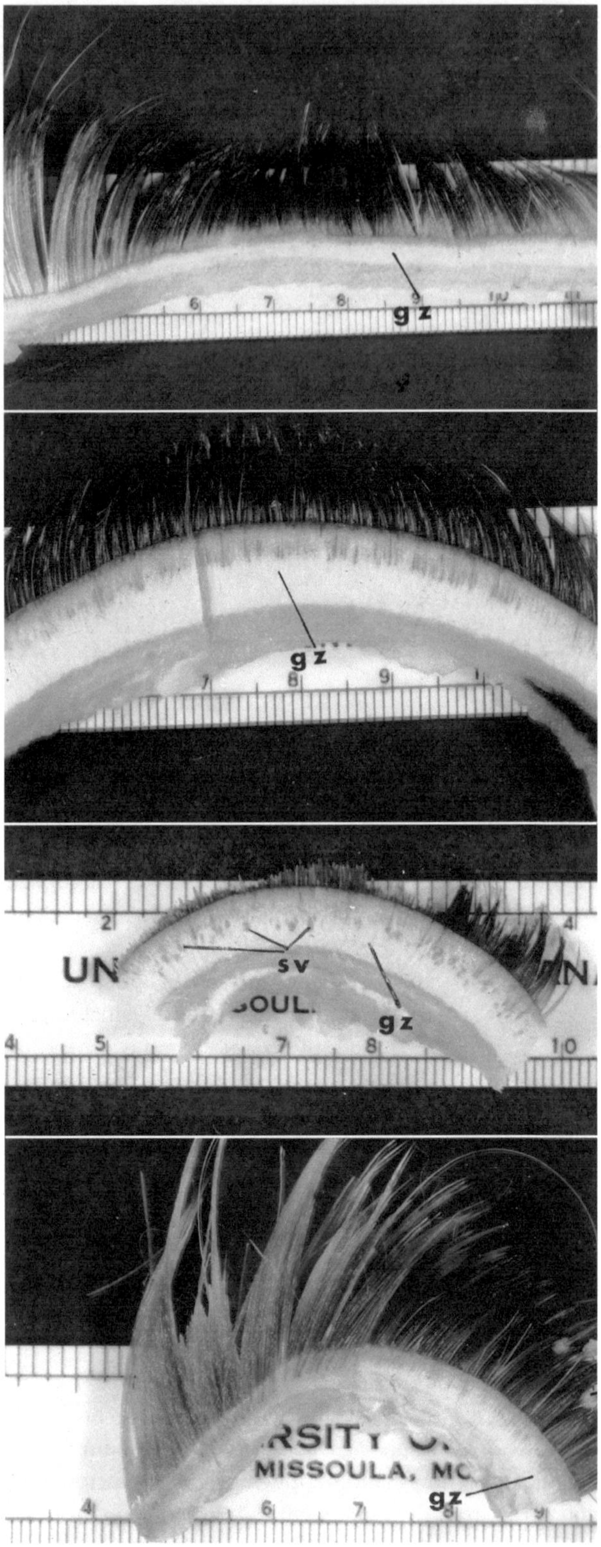

Pronghorn subauricular glands differ significantly in size, especially thickness, throughout the year in Montana. The numbers on the accompanying rulers represent inches (top) and centimeters (bottom). From top to bottom: a very thin glandular zone (gz) from a gland collected in January when secretions are minimal; a thick well-developed gz from a June collection; a thick gz and numerous large storage vesicles (sv) that are typical of an actively secreting gland, collected in August; and a regressing gz with considerable secretory material on hairs surrounding a gland collected in November. Annual proliferation and regression of subauricular glands follow similar activities of the testes. Apparently, the activity cycle of these glands is controlled by gonadal hormones. Odors from subauricular glands serve for sex recognition, stimulation of does and marking of territories. Photos by Randall Moy.

Caton (1877) suggested that the pungent odor from rump glands protects pronghorn from insects. This could be true, but these glands certainly have other functions. Seton (1927) believed that rump glands function as an olfactory alarm system for intraspecific communication. Histological data support Seton's speculation (Moy 1970).

Pronghorn does appear to use their rump patches to decoy mammalian predators from their young by alternately flaring and compressing their rump patches while dashing between the predator and bedded fawns. Young fawns have little scent, and the does' strong smell released by this action may make predators' olfactory detection of fawns more difficult (Kitchen 1974).

The forefoot and hind foot interdigital glands of pronghorn appear similar. Located between the toes, each gland resembles a sock about 2.5 inches (64 mm) long, with a short top that opens directly onto the surface of the skin. Interdigital glands of bucks and does are similar in size and sebum storage capacity, and the large central lumens usually are filled with whitish sebum, suggesting extreme activity. The thick epidermis lining the central lumens forms numerous large ducts into which many highly active sebaceous alveoli empty their contents. The glandular area consists primarily of numerous, large, lobulated and actively secreting sebaceous alveoli, with a few apocrine tubules surrounding the alveoli. The few scattered apocrine tubules are large, but little secretory material is found in them. Numerous blood vessels supply the glands and are oriented dorsoventrally in the dermis surrounding the glands. Associated with the arteries are occasional sensory corpuscles and several large nerves.

Caton (1877) believed that interdigital glands left scent on the ground with a pronghorn's every step; Seton (1927) indicated that the scent left a trail for other pronghorn to follow or avoid. Because of their locations, these glands could, if scent was present, taint the ground or low vegetation. However, sebum is the main product and little scent is produced. Bucks paw before urinating and defecating, and the resulting scrapes serve as scent markers. Sebum might be deposited during pawing and it might mix with urine to preserve the scent, but only the glands on front hooves of bucks would serve to preserve such odor. Müller-Schwarze and Müller-Schwarze (1972), who

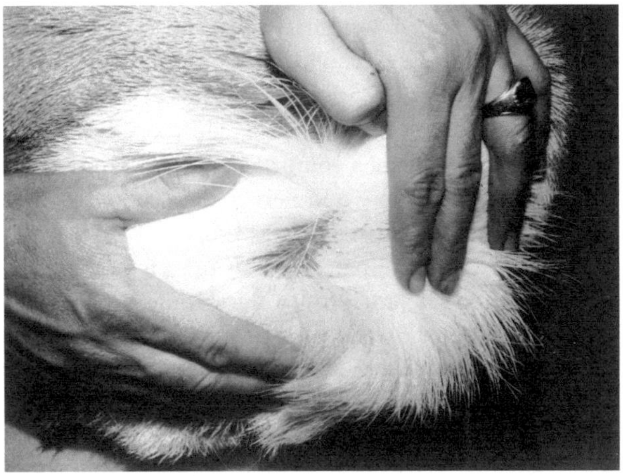

Glandular secretions indicate the locations of pronghorn rump glands in the middle of each rump patch, one on each side of the tail. Rump glands of bucks and does are similar. When rump patches are erected, a spray of odorous material is ejected and serves, in conjunction with the visual warning of the white rump patches, as an olfactory warning to other pronghorn. The tan sebum on the hairs serves to prolong the scent, which may be an important factor in recognition between individual pronghorn. Photo by Bart W. O'Gara.

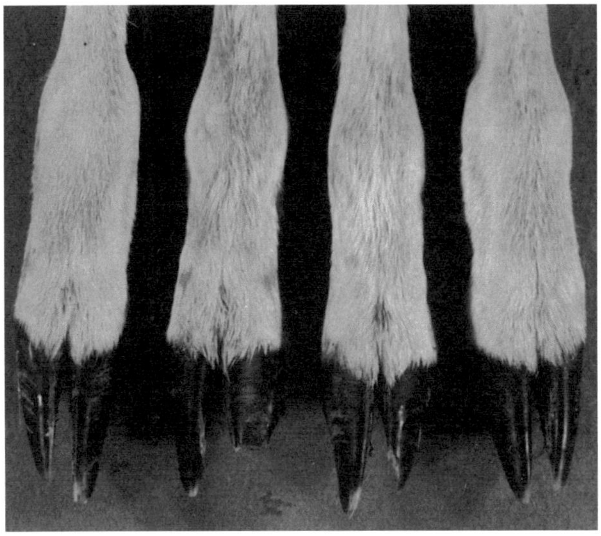

The feet of a lame pronghorn doe shot in Montana with dry, brittle, cracked and broken hooves. Careful analysis showed no evidence of arthritis or other bone problems that could have caused the abnormal conditions. However, the interdigital glands were found not to have been secreting sebum. This raised the possibility that sebum conditions the hooves of pronghorn, much as oil from the patagial glands conditions the bills of ducks. Photo by Bart W. O'Gara.

intensively observed captive pronghorn, never detected responses to the interdigital gland secretions. A lame doe, collected during April, had brittle hooves that were cracked and broken, and its interdigital glands were inactive; the collapsed lumens and an absence of sebum sug-

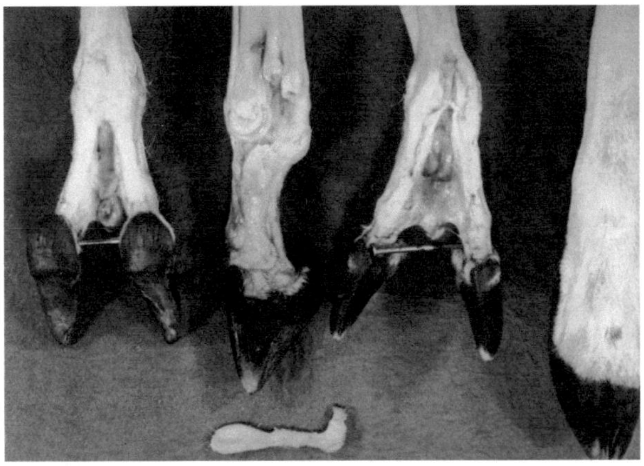

Interdigital glands of the four feet of a pronghorn doe. The foot at far left was skinned, to show the gland from the rear. The foot at middle left was skinned and the left toe was removed to reveal the gland as it lies between the toes. The foot at middle right was skinned, and the gland was excised and laid below the foot. The foot at far right shows the opening of the gland as a spot between the toes about 1 inch (25 mm) above the hooves. The interdigital glands of the pronghorn are similar between front and back feet and between sexes. Sebaceous alveoli produce much sebum, but sweat tubules are few and do not produce an odor perceptible to the human nose, leading to the assumption these are not scent glands. Photo by Bart W. O'Gara.

gested that sebum conditions hooves, keeping them flexible (O'Gara and Moy 1972).

During August 1997, five, free-ranging, adult pronghorn bucks were collected in Modoc County, California. Secretions from their interdigital glands contained tetradecanal, 1-tetradecanol, (Z)-9-hexadecen-1-ol, 1-hexadecanol and 1-hexadecyl acetate. 1-Tetradecanol had been reported as a growth inhibitor for three species of bacteria—*Propionibacterium acnes*, *Brevibacterium ammoniagenes* and *Streptococcus mutans*. (Z)-9-Hexadecen-1-ol also inhibited growth of *S. mutans* and 1-hexadecanol the growth of *P. acnes* (Wood 2001). It seems unlikely that production of antibacterial compounds is the primary function of interdigital glands. However, the deep, moist glands seem easily infected, therefore production of antibacterial compounds seems advantageous.

The unpaired median gland, sometimes called the dorsal gland, or supracaudal gland, is on the mid-dorsal line about 1.0 foot (30 cm) anterior to the base of each buck's tail. In adults, the glands are covered by thin scattered hairs, whitish to tan and smaller than the dark adjacent hairs; the stalks of these hairs usually are matted with sebum. Large, numerous arrector pili muscles are associated with the hairs.

Each gland resembles an ellipse measuring about 2.4 by 1.3 inches (60 by 32 mm) and is thickest (0.17 inch [4.2 mm]) at the center. Preliminary histological studies indicate

The median gland of pronghorn bucks is located anterior to the rump patches. Courting bucks open or flare their median glands, and does often sniff or lick them. Pheromones produced by the glands, apparently, prepare does for breeding. Bucks also flare the glands as they erect their rump patches when excited. The patches also open momentarily at death, probably because of muscle spasms that sometimes simultaneously erect the rump patches. Photo by Bart W. O'Gara.

that median glands undergo seasonal cyclic changes somewhat similar to those of subauricular glands. Median glands of adult males secrete sebum throughout the year, although more sebum and odor are discernible during late summer and early autumn than at other times (O'Gara and Moy 1972). This relation of activity to season and the fact that these glands occur only in males suggest periodicity controlled by the testicular cycle. However, the median gland of a castrated buck on the National Bison Range, unlike its subauricular glands, appeared as large as those of intact animals, and the castrate often opened (flared) his median gland.

When a pronghorn buck erects its rump patches, it sometimes simultaneously erects hairs around the median gland—so the median gland's function could be to supplement that of the rump glands. Müller-Schwarze and Müller-Schwarze (1972) believed the median gland provided females with a "priming" pheromone. They also reported that males erect the hair around their dorsal glands, enlarging the odor-carrying surface immediately before and during copulation.

Seton (1927) postulated that scent from hock glands exudes on plants or the ground when pronghorn lie down. However, skin on the hocks contains sweat tubules that are only slightly more developed than those in normal skin; no measurable thickening of skin is discernible and the hocks seem no more odoriferous than any other part of the animal (O'Gara and Moy 1972).

Pronghorn often urinate on their hind feet. The skin behind the toes of females has the strongest and most disagreeable odor of any part of a pronghorn's body. Because skin from the toes of males and females is similar and not especially glandular, odor comes from an external source, logically urine. Sebum from interdigital glands may spread to hair behind the toes, combine with urine and, thereby, assist in maintaining the strong odor. I have not noticed deeply stained or especially pungent toes on wild bucks, but Müller-Schwarze and Müller-Schwarze (1972) noted dark staining on a captive buck at 4 months of age. The darkest staining and strongest odor I have detected from wild does has been during spring, and such odor probably is important for recognition in doe/fawn relations.

GENETICS

Four races of the pronghorn have been described, but they differ in such slight characters that they may be best considered together.

Allen (1942:323)

If I were going to begin a serious study of the taxonomic status of sonoriensis, *I would want to look at the entire species* Antilocapra americana, *to see how much variation there is within and between different geographically separated samples, and to determine whether the specimens of southwestern Arizona and northwestern Sonora fall beyond the range of variation shown by other populations.*

Letter by Ron Nowak in
U.S. Fish and Wildlife Service (1998a:57)

Populations of social mammals are characterized by several demographic features that may increase the magnitude of genetic drift relative to other evolutionary forces. In particular, polygynous mating and social mediated constraints on gene flow have been proposed to foster random genetic differentiation among social groups, thereby accelerating rates of evolutionary change.

Storz (1999:553)

A survey such as that proposed by Nowak (U.S. Fish and Wildlife Service 1998a) is needed over the entire pronghorn range if the subspecific status and consequences of mixing populations are to be understood.

Karyotypes from three pronghorn shot in Montana and one shot in Texas showed a 2N chromosome number of 58, but one from the Catskill Game Farm in New York had a bimodal distribution, with some cells possessing 56 chromosomes (Hsu and Benirschke 1969). The fundamental number of chromosomes in pronghorn is 58 or 60; that of most bovids is 58 to 60 as well (Wurster and Benirschke 1968), and seven species of deer have 68 to 70 chromosomes (Wurster and Benirschke 1967). The X (female) chromosomes are acrocentric (i.e., pointed or V-shaped), and the Y (male) chromosomes of pronghorn are subtelocentric (i.e., X-shaped but with two arms of the X very short). The autosomes (54) are mostly acrocentric, but two are subtelocentric.

During the late 1960s and early 1970s, chromosome mapping was considered valuable to determine the relationship of certain species to one another. Chromosome mapping indicated a closer relationship of pronghorn with bovids than with cervids. However, advances in the various banding techniques, correlation between gene activities and chromosomes in somatic cell hybrids, and localization of special classes of DNA in chromosome regions make classic mapping nearly obsolete today.

By protein electrophoresis of allozyme markers at 24 presumptive gene loci, genetic variation was assessed in 65 pronghorn from six geographically isolated populations in west Texas (Lee et al. 1989). Pronghorn in Texas often are restricted by fences to single, large ranches. Five of the 24 presumptive genetic loci were polymorphic; heterozygosity ranged from 0.012 to 0.046 ($\bar{x}$ = 0.027). Analysis indicated close affinities among the large Upper Concho Valley herd and four of five Trans-Pecos populations. The Marathon Basin herd was significantly different from the other five herds. In summary, this study indicated that west Texas pronghorn were characterized by relatively low levels of genetic diversity within populations, but moderately high levels of genetic differentiation among populations. Lee et al. considered application of genetic information essential to conservation of genetic variation in translocations between isolated pronghorn populations and recommended no translocations into Marathon Basin.

Allozyme variation was assessed by Lee (1992) for 330 pronghorn from 29 populations across the northern and central portions of the species' range, as well as Texas. Twenty of 33 presumptive genetic loci were polymorphic. Clustering of Rogers' distance values using unweighted pair group method analysis indicated close affinities between the populations within the presumed subspecies of Oregon pronghorn (*A. a. oregona*) and the American pronghorn (*A. a. americana*). Populations from southwestern Texas and southern New Mexico clustered in several groups outside of the Oregon–American pronghorn cluster. This pattern is consistent with recognition of the Mexican pronghorn (*A. a. mexicana*) as a subspecies.

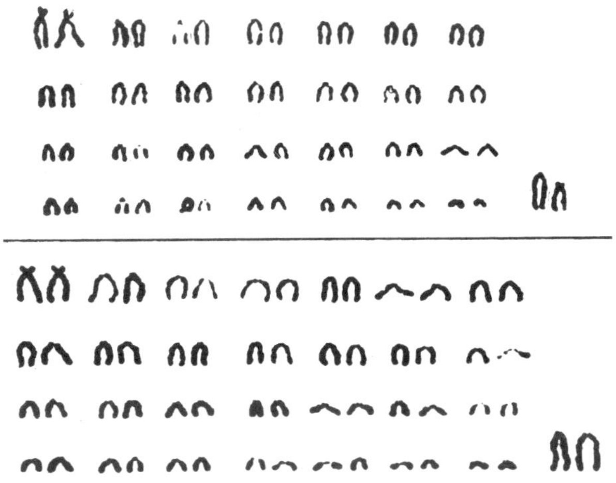

Chromosome map of a pronghorn buck (top) *and doe* (bottom), *illustrating a 2N chromosome number of 58. The fundamental chromosome number actually is 58 or 60—the same as that of most bovids.* Photo by Doris Wurster.

Restriction-fragment analysis was used to measure mitochondrial DNA variation in 110 individuals representing most of the range of pronghorn (Lee 1992, Lee et al. 1994). Nine mitochondrial haplotypes (A, B, C, D, F, G, H, I and J) were discovered, and seven of them were rare in most populations. C was most common, and A was second. Consequently, in comparison with other mammalian species, pronghorn showed a reduced amount of mitochondrial DNA variation within most populations. An exception to this was a population in Yellowstone National Park that possessed four haplotypes, two of which were considered unique to the park at that time. The Mexican pronghorn subspecies was discernible from all other subspecies in that study by one restriction site that was common in some west Texas populations. The Oregon pronghorn were not discernible from the American subspecies.

The nucleotide sequence of a clone satellite DNA (remaining DNA after most has been removed by centrifuging) from pronghorn was presented by Denome et al. (1994). The 1,477 base pair satellite is composed of 31 base pair subrepeats that are very similar in sequence to those of the major satellite DNAs from cattle, sheep and muntjac. However, the subrepeat sequence is more degenerate and variable than those of cattle and sheep.

The herd in Yellowstone National Park is one of the few remaining undisturbed herds (i.e., no translocations into or known contact with other herds in at least 100 years) of American pronghorn (Scott 1990). This herd showed a larger variation in mitochondrial DNA than did any other pronghorn herd that had been investigated, which may indicate that it contains much of the past amount of variation in DNA found in the species (Lee 1992). This genetic resource should be conserved.

Limited electrophoretic studies indicated no difference between serum proteins of Mexican and Sonoran pronghorn from Arizona. A sequence comparison of mitochondrial DNA from the control regions of 10 individuals from each of these two presumed subspecies indicated less than 1% divergence. Also, the gene sequences of the nuclear *SRY* genes from a single American and a single Sonoran pronghorn were identical (see Chapter 1, U.S. Fish and Wildlife Service 1998a).

To estimate the genetic diversity within and among pronghorn populations and to investigate the phylogenetic relationships among populations, Lou (1998) examined nucleotide sequences from a 282-base-pair segment of the mitochondial control region. Genetic variation also was analyzed among all five presumed subspecies using four microsatellite loci. DNA sequence analysis was used to search for variability in a paternally inherited genetic locus (intron 6 in the *Zfy* gene).

Mitochondrial DNA control-region sequences showed a moderately high level of genetic variability and a total of 28 haplotypes were observed. Three of four microsatellite loci (BM6444, ADCYC and BM6506) showed high levels of genetic variability with 22, 15 and 10 alleles, respectively. Only locus BM1824 was monomorphic. No variation was found in the intron of the Zfy locus. Using mitochondrial and microsatellite markers, two genetically differentiated groups of populations (northern and southern) were observed. The average Fst (0.086) estimated from microsatellite loci and Gst (0.111) from the mitochondrial DNA control-region sequences indicates moderate genetic differences among populations and suggests that gene flow across the continent is restricted (Lou 1998).

Arizona is considered home to three subspecies of pronghorn—American, Mexican and Sonoran. Translocations have compromised the genetic integrity of some herds in the state (Lee 1988), but others have remained undisturbed by introductions. Rhodes et al. (1999) surveyed genetic variation at 22 biochemical loci by starch gel electrophoresis of allozymes. Heterozygosity, number of alleles per loci and percentage of polymorphic loci were estimated for 320 pronghorn from five areas—the Arizona strip (19), northwest (19), central (166), northeast (97) and southeast (19). Using liver or muscle tissue, 37 biochemical loci were surveyed, and 22 were selected for further analyses of genetic variation. Objectives were to evaluate levels of genetic differentiation among regional herds within Arizona to assess genetic parameters of Arizona pronghorn relative to those reported by Lee et al. (1994) for herds in other areas.

Rhodes et al. (1999) found mean values for the overall Arizona population as follows: multilocus heterozygosity, 0.064; average number of alleles per locus, 2.14; and percentage of polymorphic loci, 0.23 (95% criterion). Regional herds containing unique alleles (not reported from other populations) ranged from zero in the Arizona strip and southeast to four in the northeast. Extensive polymorphism was revealed at the mannose phosphate dehydrogenase locus in Arizona pronghorn, unlike the populations surveyed by Lee et al. (1994). The only highly variable locus examined by Lee et al., which was not surveyed in Arizona, was the esterase-4 locus. This locus was resolved in only 30% to 40% of the Arizona samples and therefore, was considered unsuitable for further analysis.

In comparison with data from other pronghorn herds presented by Lee et al. (1994), the statewide Arizona population exhibited relatively high levels of genetic variation (Rhodes et al. 1999). The mean multilocus heterozygosity value of 0.064 was substantially higher than that for other populations (0.024) but decreased within the range of values observed (0.006–0.076) by Lee et al. Values for the average number of alleles per locus (2.14) and percentage of polymorphic loci (0.23) for Arizona pronghorn also were high, relative to mean values in other populations (1.15 and 0.13, respectively). The average number of alleles per locus for Arizona pronghorn was outside the range of values observed in other herds (1.00–1.50). The percentage of polymorphic loci was within the range observed in other populations (0.03–0.33).

Analysis of genetic diversity from five regional herds indicated that levels of heterozygosity and gene polymorphism varied across Arizona. The southeastern herd exhibited the highest level of heterozygosity, the lowest number of alleles per locus and the highest proportion of polymorphic loci of any examined regional herd (Rhodes et al. 1999). This herd was thought to be composed entirely of introduced pronghorn from multiple sources, including various locations in central Arizona and two in Texas. Likely, each group contributed different polymorphic loci, and the high level of heterozygosity stems from the inflated proportion of polymorphic loci in the introduced herd.

A high number (4) of low-frequency alleles were distributed over three loci in the northeastern herd (Rhodes et al. 1999). Although this region received introductions from Wyoming, Lee et al. (1994) did not indicate the presence of these rare alleles in 12 individuals sampled from Wyoming. One individual, from an area that received pronghorn from Wyoming, was responsible for two of the four rare alleles. The other two were from individuals in two different game management units with no recorded transplants. Among the five regions, highly significant differences in allele frequencies were found over all loci, and at least 10 loci contributed to this differentiation. Rhodes et al. did not consider this a surprising result, given the large area over which samples were collected and introductions from Colorado, Montana, Texas, Utah and Wyoming.

A total of 389 pronghorn were analyzed for mitochondrial DNA haplotype variation, using polymerase chain reaction protocols, to elucidate overall levels of genetic diversity and regional differentiation of Arizona pronghorn (Reat et al. 1999). Lee et al. (1994) had surveyed pronghorn from the majority of their range and found nine mitochondrial DNA haplotypes, including one (A) that appeared informative to distinguish American from Mexican pronghorn.

Four haplotypes were observed in Arizona pronghorn, including three of the nine (A, C and J) Lee et al. (1994) had described and one (K) not previously recorded (Reat et al. 1999).

Lee et al. (1994) had found haplotype C to be the most common throughout the areas they sampled. The A haplotype was fixed in pronghorn populations of Marathon Basin and occurred at low frequencies in other Texas populations, as well as in one individual each from Colorado and New Mexico. A was considered rare across most pronghorn ranges and was proposed by Lee et al. as a potential diagnostic marker for Mexican pronghorn. However, it proved to be the most common haplotype observed in Arizona pronghorn, suggesting that A may be common in the southern portion of the species' range (Reat et al. 1999).

The third haplotype (J) observed in Arizona by Reat et al. (1999) had only been recorded from Yellowstone National Park, Wyoming. The possibility of transplants from northern states bringing J to northeastern Arizona and the Arizona strip was not speculated on. Yellowstone National Park pronghorn were relocated to the National Bison Range and an area near Townsend, Montana (P. Wright personal communication:1968). Thus, genes from Yellowstone National Park could have reached Arizona in transplants from areas other than the park.

No documented pronghorn introductions were made to the central Arizona area where haplotype K was observed. Thus, it is likely that K is an ancestral marker potentially unique to Arizona. Reat et al. (1999) proposed that central Arizona populations containing this haplotype should be protected from future translocations. The occurrence of K in southeastern Arizona, an area considered devoid of pronghorn at one time, probably is a consequence of introductions from central Arizona.

The diversity of haplotypes (29) that she found during a mitochondrial DNA study of all five presumed subspecies of pronghorn led Amor (2000) to conclude that the bottleneck, which occurred early in the 20th century, did not greatly reduce genetic variability of the species. Amor further concluded that pronghorn populations originated from small reproductive reservoirs, which quickly expanded, due to reduced competition and predation, as megafauna disappeared during the late Pleistocene and prairie habitat expanded during the mid-Holocene.

Surveys of biochemical loci by electrophoresis of allozymes and mitochondrial DNA haplotype diversity indicated relatively low levels of genetic diversity within pronghorn populations and moderately high levels of genetic differentiation among populations. More extensive studies, including nuclear DNA, are needed to confirm or refute the validity of generally accepted subspecific designations and to provide guidance relative to mixing of pronghorn herds through translocations.

An enlarged genetic study designed to address the biogeographic history of pronghorn could provide a standard with which to evaluate the effects of translocation. Population genetic studies also hold promise for determining whether *A. a sonoriensis* and *A. a. peninsularis* are valid subspecies. Problems in obtaining biological material from populations of presumed subspecies that are classified as endangered largely have been resolved. The indiscriminate and sometimes unrecorded translocations of pronghorn that have taken place among many populations remain a problem.

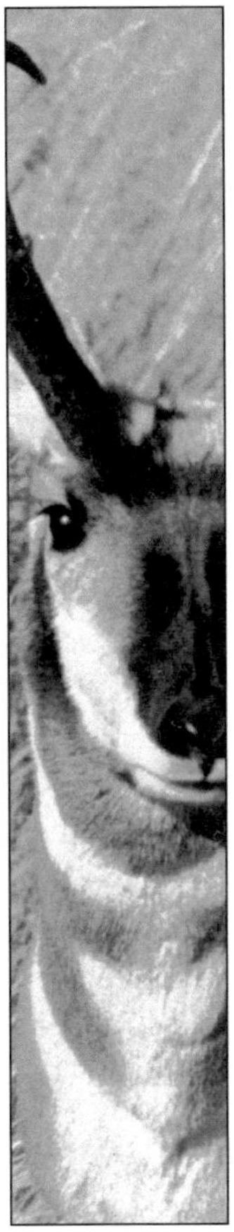

C H A P T E R N I N E

Reproduction

Bart W. O'Gara

I suggest that pronghorn have evolved a litter size of two under a long history of predation on fawns by coyotes and other ground predators. Rates of predation apparently have been high enough to create strong selection for twinning because of the low probability of a single offspring's survival. However, litter size could not increase at the expense of birth weight, or postnatal growth rate, because predators also create selection for rapidly maturing offspring. Consequently, pronghorn females annually produce two large fawns, whose combined weight at birth represents close to 18% of maternal weight. Human females investing at this rate would produce 12-pound [5.9 kg] twins. By late August, when weaning occurs, a pronghorn mother that has raised twins has supported the production of two 27-kg [55 pound] young, a mass equal to 120% of her own.

Byers (1997a:56–57)

FEMALES

Breeding Age

Pronghorn females usually breed for the first time when 16 to17 months old. Fawns occasionally conceive. One-third (6) of 18 pronghorn yearlings collected during September in North Dakota had corpora lutea of pregnancy in their ovaries, indicating they had been pregnant as fawns (McKean 1954). One of those fawns produced only one corporus luteum, but the other five produced two or more and could have given birth to twins. Also, of 17 yearling does checked in North Dakota, 6 (35%) had milk in their udders and 11 (65%) did not (McKean 1951). This was at a time when pronghorn numbers were low but increasing rapidly. Possibly, these circumstances contributed to the high percentage of North Dakota fawns conceiving.

Abundant milk in the udder of a yearling that had two fawns on the National Bison Range, Montana, led Wright and Dow (1962) to conclude that some females there bred during their first year of life. Earlier suggestions of such a possibility were made by Buechner (1950a), who reported that some females in Trans-Pecos, Texas, appeared to become pregnant when 5 to 8 months old, although females usually did not conceive until they were yearlings.

A fawn collected on January 31, 1963, in Alberta, was pregnant with two viable fetuses. Each of her relatively heavy ovaries—0.12 and 0.15 ounces (355 and 418 mg)—contained a corpus luteum of pregnancy. The forehead to rump measurements of her fetuses were 3.5 and 3.6 inches (88 and 91 mm), respectively. A 4-year-old female that was shot a week earlier contained twin fetuses with forehead to rump measurements of 6.3 and 6.7 inches (160

Initial breeding by pronghorn females usually is when they are 16 to 17 months old. Photo by Bart W. O'Gara.

and 170 mm), respectively. Comparison of the forehead to rump measurements of her fetuses with those of a growth curve suggested that the fawn conceived during the first week of November, rather than during the usual period of conception in late September and early October (Mitchell 1967). A necropsied fawn collected on the National Bison Range, Montana, carried twin fetuses, which appeared to be a month younger than those carried by yearling and adult does at that time of year. Of seven does on the National Bison Range whose ovulation incidences as fawns could be determined during a histological study, one bred at about 4 months of age; two others had ovulated and probably bred (Roseland 1977).

In Alberta, most female pronghorn give birth to fawns during late May and early June (Mitchell 1965). If the incidence of late breeding by fawns was significant, neonatal fawns should be observed as late as August. However, during 10 years of aerial surveys in July and August, less than 1% of all fawns observed was judged to have been born after June 30. Thus, the evidence suggests that, although some females may be capable of producing fawns at about 1 year of age, few actually do. This may be a function of morphological and physiological immaturity of most female fawns, the absence of suitable behavior patterns leading to coitus at that time or a reduced number of breeding males exhibiting active spermatogenesis during midwinter. Occasionally, parturition occurs as early as May 12 in Alberta. Females born 2 to 3 weeks in advance of the peak of parturition presumably would be more capable of conception in early November than would later-born fawns (Mitchell 1967).

Pronghorn in North Dakota appear to have a longer birthing period than has been reported in other northern states. Fawns there started to appear in mid- to late May and continued to be born into July (Jensen 1997). The latest date a newborn has been observed in North Dakota was early August. B. Jensen (personal communication:1999) speculated that low buck to doe ratios and intense archery hunting during September extended the breeding and, thus, the birthing seasons.

O'Gara (1968) found an increase in corpora lutea—presumed ovulation rates—from fawns to does 4 years of age. Presumed ovulation rates then remained nearly constant for 5- to 9-year-old does. Numbers of fetuses in utero did not vary with age. Byers (1997a) also found no evidence for changes in reproductive effort with age in pronghorn does. However, fawn survival did increase with age of dams, probably because of learning and improvement in the mothers' antipredator skills.

Gestation Period

Known gestation periods for captive pronghorn were 245 to 255 days (average 250 days) at the Wyoming Game and Fish Department's Sybille Wildlife Research Station (W. Hepworth personal communication:1981). Thus, the pronghorn's gestation period ranks among the longest in the ruminants of North America, exceeded only by bison and elk (Table 55). Note that gestation periods generally shorten as artiodactyl species become smaller, except for the pronghorn. Perhaps the inordinately long gestation period is necessary because the does are stressed to the limit of their capabilities to produce two large fawns.

TABLE 55. Gestation periods of North American artiodactyls

Animal	*Gestation (days)*
Bison	270
Elk	249–262
Pronghorn	245–255
Moose	240–250
Musk ox	240
Caribou	210–240
White-tailed deer	197–222
Mule deer	199–207
Rocky Mountain sheep	180
Rocky Mountain goat	147
Peccary	125–150

Source: From Asdell (1964), except for pronghorn, which is from extensive data gathered at the Wyoming Game and Fish Department's Sybille Wildlife Research Station.

Length of the gestation period on the National Bison Range, Montana, was affected by precipitation, according to Byers and Hogg (1995) and Byers (1997a). After dry summers, gestation lengths of young, prime-aged and old does were 1, 4 and 10 days longer, respectively, than those after wet summers—mean difference across all ages was 3.63 days. What differences might be in arid areas is unknown. However, M. Dunbar (personal communication:2000) speculated the opposite was true on the Hart Mountain National Antelope Refuge in Oregon.

Geist (1998) postulated that pronghorn and roe deer, another small ruminant adapted to a cold climate, evolved long gestation periods for another reason. Roe deer breed during summer, delay embryonic development and give birth at the optimum time in spring. Thus, these two small-bodied species bear the cost of mating during late summer when the cost can be paid almost directly from food ingested at the time. To mate later would involve carrying extensive fat deposits into autumn and entering winter in poor condition. Very fat pronghorn probably would be vulnerable to predators. Whether or not this is the reason for evolution of the long gestation period, such a period and the early breeding season seem to have advantages.

Pronghorn does have the highest level of maternal investment in their offspring of any North American ruminant (Byers and Moodie 1990). That investment (percentage litter birth weight and daily fawn growth rate of maternal weight) is 18 pounds (8.2 kg) and 0.07 inch (0.18 cm), respectively. In comparison, the same for bighorn is 5 pounds (2.27 kg) and 0.04 inch (0.1 cm) (Byers and Hogg 1995).

Fifty does were captured with a corral trap and helicopter on University of Texas land, in Hudspeth County, during March 1990. At that time, about 160 days postconception, Cannon et al. (1997) found transrectal ultrasonic evaluation a reliable technique for determining pregnancy in the field. Forty-three does were diagnosed as pregnant and seven were not. Four diagnosed-pregnant does died within 2 weeks, presumably from trap-related stress, and one never was seen after release, nor was a radio signal received. The remainder survived at least 1 month beyond fawning season, and all but one were alive a year later after the 1991 fawning season. Thus, a maximum 10% mortality rate was indicated for capture and handling. Of the 38 surviving, presumed pregnant does, 37 were confirmed to be pregnant—through field observations during late gestation—or had given birth. Of the seven not considered pregnant, one was determined to have been pregnant through field observation, and later was seen nursing a fawn. See Wild (1995)—under *Implantation and Intrauterine Mortality* in this chapter—concerning earlier detection of embryos through ultrasound.

During a study in Texas, serum progesterone levels in 37 pregnant pronghorn does averaged 1.6 nanograms per milliliter (ng/mL). That of six nonpregnant does averaged 0.7 ng/mL. However, the minimum observed in pregnant does was 0.8 ng/mL, and the maximum in nonpregnant does was 1.2 ng/mL (K. Cannon personal communication:1998). The does had been drive-trapped before blood was drawn, and the overlap in levels probably resulted from elevated progesterone levels related to stress in some nonpregnant does.

Breeding Seasons

Southern pronghorn have longer rutting periods that begin earlier than those of pronghorn on northern prairies. Northern pronghorn, including those in Oregon, generally breed during mid- to late September. The rut continues into October in most southwestern states, including Colorado. Breeding activity as early as July and as late as October was reported in Texas by Lehman and Davis (1942) and Buechner (1950a). Breeding behavior was observed during late September in Chihuahua (Trevino 1978), and it was observed in Sonoran pronghorn during early July along the Mexican border in Arizona (Phelps 1981). In New Mexico, local weather and altitudinal differences influence the time of breeding over fairly short distances (T. Yates and P. Fry personal communication:1980).

In semiarid Owyhee County, Idaho, the majority of fawns were born during the last 2 weeks of May, the peak coming about May 20 (Folker 1956). During a study there, the first fawn was observed on May 18, and 3- to 4-day-old twins seen on May 27 represented the latest fawn drop recorded.

Although pronghorn in some parts of Texas and Mexico have breeding seasons as long as 3 months (Lehman and Davis 1942, A. Castellanos personal communication:1984), those in most of the species' range have a comparatively short and intense rut (Einarsen 1948, Hoover et al. 1959, Prenzlow et al. 1968). Based on observed copulations in Montana, the breeding season lasted 3 weeks, and rutting behavior ended abruptly 2 days after the last observed copulation (Kitchen 1974). However, late breeding, especially of female fawns or yearlings, apparently occurs. Two yearling does were bred a month after the normal rut in Colorado and 3 weeks after a subordinate buck, previously excluded by a dominant buck, was allowed to rejoin the group (Ellis and Deblinger 1975). Also during a Colorado study, the mean date of the first estrus of two mature does (September 30) during 2 years was significantly earlier than that of four yearlings (October 22) (Pojar and Miller 1984). Young bucks that previously had been excluded from breeding territories by territorial bucks in Montana were observed courting female fawns (Kitchen

Knowing breeding and fawning periods facilitates management decisions, such as timing of hunting seasons and regulations of spring grazing by livestock on fawning grounds. (Top) *Pronghorn in most of the species' range have a comparatively short, intense rut. Where these animals were photographed in Montana, the breeding season lasts only about 3 weeks.* (Left) *In this Montana grassland, practically all fawns are born within a 3-week period and the majority are dropped during the last week of May and the first few days of June.* Photos by Bart W. O'Gara

1974). Thus, late matings may be with young, subordinate bucks.

Ovaries and Ovulation

The enclosing membrane (tunica albuginea) of the pronghorn ovary is fibrous, ranging from 0.004 to 0.014 inch (0.10–0.35 mm) thick. The surface epithelium consists of a single layer of cells—generally cuboidal; the outer zone of the cortex is distinct and fibrous, but the inner zone is less so, and the border between cortex and medulla is indistinct (Roseland 1977). Follicles and corpora lutea frequently extend into the medulla. A corpus luteum (plural: corpora lutea) is a mass of endocrine tissue filling the cavity remaining in an ovary after ovulation. Consequently, the number of corpora lutea in the ovaries of a doe should correspond to the number of fetuses in her uterus unless ova or embryos die or, as rarely happens, two ova are ovulated from one follicle.

Oogonia (germ cells) are indistinguishable from cells of the surface germinal epithelium. When oogonia attain a diameter of about 20 microns, they are at the primary oocyte stage and readily recognized under the microscope. Oocytes (ova) grow as follicles develop; in the primary follicle they average 31 microns, in the secondary 83 and in the vesicular (Graafian) follicle 92. Ova grow little, if at all, in the Graafian follicle. A zona pellucida (a noncellular, refractile layer), about 4 microns thick around the ovum, is first seen in secondary follicles. The zona pellucida is thinner and more delicate than that found in carnivores, rodents and most other mammals. An antrum (space) forms when follicles are about 0.02 inch (0.40 mm) in di-

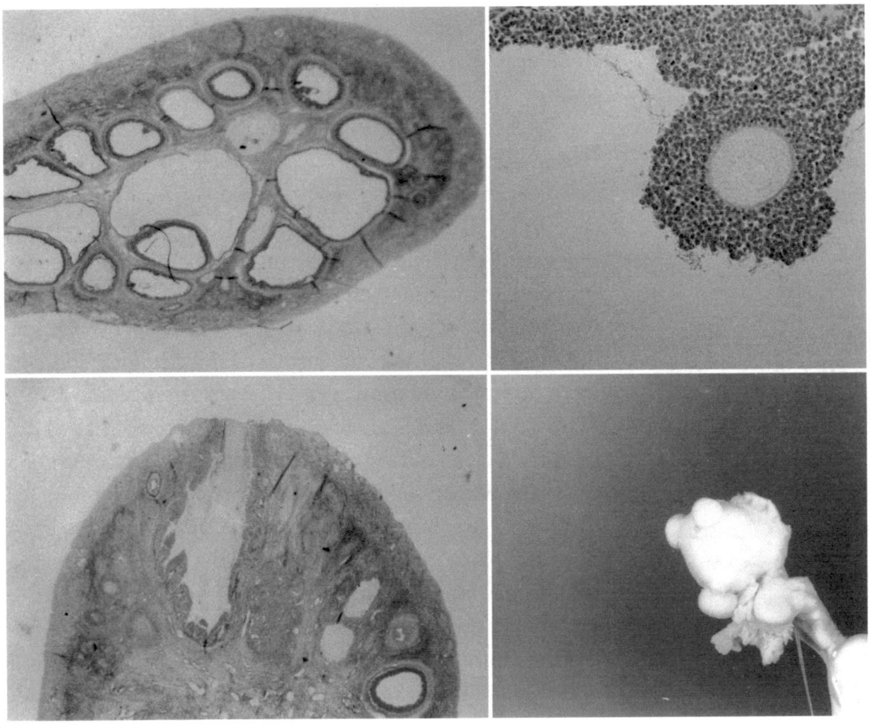

A sectioned pronghorn ovary during the graafian follicle stage before ovulation. (Top right) *An ovum in a graafian follicle.* (Bottom left) *Ovulation site on a sectioned pronghorn ovary just after ovulation. Blood will soon fill the cavity, forming a corpus hemorrhagicum.* (Bottom right) *The ovaries of a pronghorn doe late in her second month of pregnancy. The corpora lutea are partly extruded at the ovulation sites (stigma). Nearly half of a corpora may extrude from the stigma of an ovary with four or more corpora during early pregnancy, resulting in figure 8 or hourglass-shaped corpora. By the fourth month of pregnancy, the corpora lutea will be spherical and internal, and their presence will be difficult to detect from the outside.* Photos by Bart W. O'Gara.

ameter. Graafian follicles with fluid-filled spaces grow rapidly and attain a maximum diameter of about 0.18 inch (4.5 mm). The number of Graafian follicles per ovary varies with age, time of year and individual differences in hormone levels (Roseland 1977).

Ovarian events leading to ovulation were summarized for pronghorn on the National Bison Range by Roseland (1977), who prepared 26 pairs of ovaries histologically and studied them microscopically. Toward the end of August, under the influence of follicle-stimulating hormone, the number of developing follicles increases, including the follicles that produce ova for the next pregnancy. However, average diameters of those follicles do not differ markedly throughout the year. The theca interna and the theca externa differentiate from the cortex, and granulosum cells proliferate rapidly. In the preovulatory follicle of early September, small quantities of luteinizing hormone synergize with follicle-stimulating hormone to stimulate secretion of estrogen by the thecal cells. The estrogen level peaks in preparation for estrus and ovulation, and Graafian follicles move near the surfaces of the ovaries. High levels of blood estrogen produce mating behavior in the does, and positive feedback (increased production of luteinizing hormone controlled by the hypothalamus) causes ovulation, usually of two to seven ova. Dilation of lymph vessels in the thecae at this time indicates that increased fluid pressure has a mechanical influence on expulsion of the ovum. The theca externa thickens, probably because of contractile properties of smooth musclelike cells, and probably contributes to follicular rupture. During collapse of the follicle, the ovum and other debris wash out of the follicle with the escaping follicular liquid. At ovulation, the theca externa degenerates, releasing blood into the antrum for the stigmatic clot. The increased production of luteinizing hormone causes luteinization of the granulosum cells that fill the former central space and become the corpus luteum, whose function is synthesis of progesterone. Vessels of the thecae invade the forming corpus and form complex vascular systems that nourish the new structures and serve to transport the hormones produced therein. The ova are picked up by fimbrae of the oviducts and moved to the distal ends of the ampullae, where they are fertilized and begin cleavage.

Roseland (1977) observed two instances of polyovulation (more than one ovum from an ovulation site). They were in ovaries from different animals, and both follicles involved had two normal-appearing ova. Analysis of ovulation incidence indicated that silent heat (ovulation without overt behavior or odor to attract a male) occurred in one doe. If so, the resulting corpora lutea were the only accessory corpora found during Roseland's study.

More atretic (degenerating) Graafian follicles than growing ones are seen in pronghorn ovaries according to Roseland (1977). Sometimes, especially during pregnancy, all Graafian follicles are atretic. Accessory corpora lutea (luteal tissue in Graafian follicles that did not ovulate) seldom are found in pronghorn ovaries.

Estrus

Females appear to signal estrus with compounds in urine and also with a potent airborne compound. Sustained courtship by males begins about 24–48 hours before the female permits copulation. During this time, males make many courtship approaches, which progress slowly toward intromission. Males ejaculate when they gain intromission. Females mate once and rapidly turn off olfactory signaling after mating.

Byers (1997a:232)

With rare exceptions, does mate once per estrus. Einarsen (1948) suggested that pronghorn have only one estrous period each year. He reasoned that, because of the tremendous amount of energy expended during the intense rut, the pattern would not be repeated several times in a season. Most northern does are impregnated within a 2- to 3-week period annually, and most fawns are born during a similarly short period, giving the impression of only one estrus. However, Mitchell (1980) found circumstantial evidence of recurrent estrus in pronghorn does in 1962. The largest and smallest embryos collected from two does on November 11 had crown to rump measurements of 2.00 and 0.49 inches (51 and 12.4 mm), respectively. This length difference could represent an age difference of approximately 50 days (O'Gara 1968). Mitchell postulated that half that amount of time would be sufficient for a pronghorn doe to experience a second estrus if she had not been successfully bred during the first one. Based on weights of fetuses carried by 46 pregnant does, Edwards (1958b) found that the fetuses of five does were approximately 1 month behind the others in development.

A doe necropsied during September, just after breeding, had five bloody ovulation sites on her ovaries and five ova in her oviducts (O'Gara 1968). In addition, her ovaries contained four well-defined corpora lutea, indicating she had ovulated previously and that more than one estrus per year might occur if a doe did not become pregnant during the first estrus. Three does examined at a checking station in Idaho each had solid corpora lutea, hollow corpora lutea and corpora hemorrhagica, indicating that three ovulation cycles had occurred (Bodie 1979).

To settle the question of recurrent estrus, a sexually mature (2-year-old) male pronghorn was vasectomized and kept with two sexually mature (2-year-old) females during the first breeding season of a Colorado study. Each doe had given birth to twins at 2 years of age. The same two females and four yearling females were kept with the same buck during the second year of the study. All of the animals were hand-reared and tame. The buck was fitted with a leather breeding harness that held a bar of colored, wax-based marking material between his front legs so he would leave a mark on a doe as he mounted. Date of estrus was determined by markings and behavioral observations (Pojar and Miller 1984).

During the first year of the experiment with the two mature females, one of them cycled five times (including one possible short cycle), and the other cycled three times. The second year, they exhibited four and three estrous

The estrous periods of pronghorn does apparently determine the breeding seasons, because bucks seem capable of breeding throughout much of the year. This doe is standing with erected tail for the buck—a sure sign that she is in estrus. Nonestrus does clamp their tails down and run from bucks. Photo by Peter Bromley.

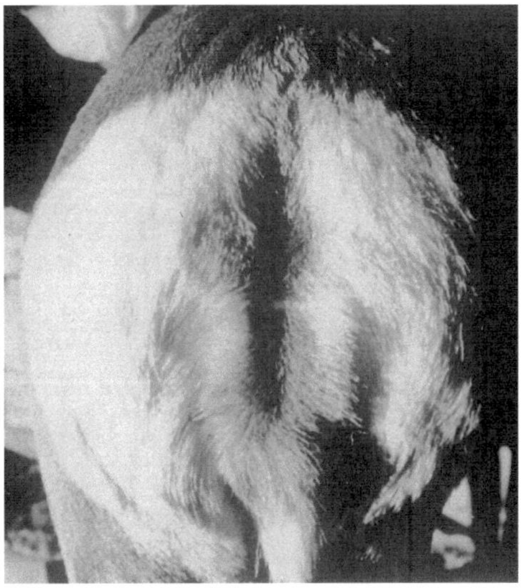

(Left) *A vasectomized buck fitted with a leather breeding harness that holds a bar of colored, wax-based material between the front legs. The bar leaves a mark on a mounted doe. This buck helped to settle the question of recurrent estrus in pronghorn.* (Right) *A tame doe with a mark on her rump, indicating she had been mounted by the vasectomized buck.* Photos by Tom Pojar.

periods, respectively. Only two of the four yearlings cycled a second time. Four irregular cycles were observed—two short (12 and 14 days) and two long (38 and 40 days). The mean length of all cycles was 27.7 days—28.4 without the irregular cycles. Pojar and Miller (1984) attributed some of the variation in cycle length to their detection methods.

Ovulation apparently takes place a short time before copulation. Three does shot during or shortly after breeding had bloody ovulation sites on their ovaries (O'Gara 1968). Another, not seen breeding, had bloody ovulation sites, uncleaved ova in her oviducts and sperm in her cervix.

Superovulation and Ovulation Rates

The presence of more corpora lutea than fetuses (superovulation), a condition found in few other mammals, was noted by Wislocki and Fawcett (1949). Mitchell (1965) found seven embryos in utero, indicating that the source of extracorpora lutea was multiple ovulation, followed by fertilization and development of excess embryos. Superovulation is the outstanding feature of pronghorn reproduction (Table 56). It may be accounted for as the remnant of a previous tendency to produce large numbers of young at one time (O'Gara 1968). It also may be a mechanism to produce extensive luteal tissue, which would in turn produce large amounts of progesterone for maintenance of pregnancy throughout the long gestation period (Roseland 1977). Another possibility—D. Mock's (personal communication:1994) insurance hypothesis—is that excess embryos may be replacements in case others died

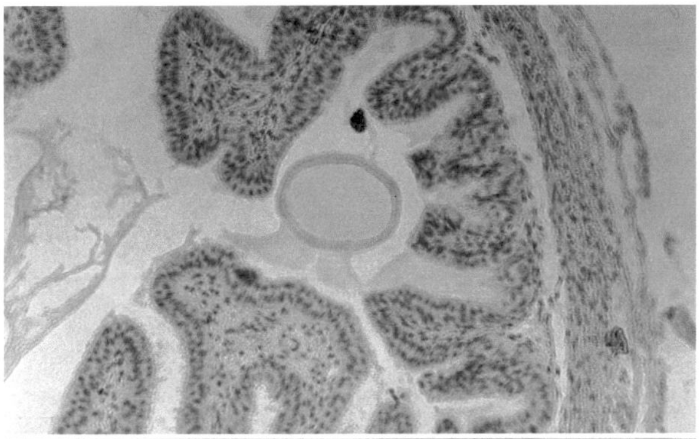
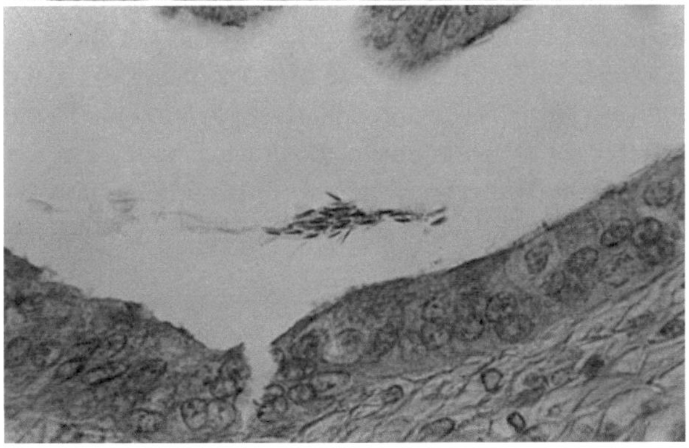

(Top) *An uncleaved egg in the oviduct of a pronghorn doe collected on September 29, 1967, in Montana. The egg was about 25% of the way down the oviduct from the ovary.* (Bottom) *Sperm in the anterior cervical valve of the same doe as in the top photo. She was being courted by a buck during the evening of September 28 and showing interest in the buck. The pair was watched until 8* P.M. *When shot the next morning at about 5* A.M., *the doe was with the same buck, but neither animal showed interest in the other. The doe apparently was bred 9 hours or less before the reproductive tract was preserved.* Photomicrographs by Bart W. O'Gara.

TABLE 56. Average number of corpora lutea (with ranges, if given) found in the ovaries of pronghorn does during studies

Area	*Age*	*n*	*Average*	*Range*	*Source*
North Dakota	Adult[a]	22	4.7	—	McKean (1954)[b]
Wyoming	Adult	54	5.6	3–9	Edwards (1958)[b]
	Yearling[b]	10	3.7	1–7	Edwards (1958)
New Mexico	Pregnant[c]	37	3.4	—	Larsen (1964)
Montana/ Yellowstone	Adult	66	5.1	2–8[d]	O'Gara (1969)[a]
	Yearling	10	3.9	2–6[d]	O'Gara (1969)[a]
Idaho	Adult	10	5.9	—	Bodie (1979)
	Yearling	6	3.9	—	Bodie (1979)
Alberta	Adult	65	5.6	3–9	Mitchell (1980)
	Yearling	38	4.2	2–7	Mitchell (1980)
Oregon	Adult	23	6.1	—	Trainer et al. (1983)
	Yearling	4	3.8	—	Trainer et al. (1983)
Colorado	Adult	11	4.6	3–7	Firchow (1986)
	Yearling	2	4.0	3–5	Firchow (1986)

[a] Adult (animals that bred at about 28 months of age or older).
[b] Yearling (animals that bred at about 16 months of age).
[c] Pregnant (29 adults and 8 yearlings).
[d] Bart W. O'Gara (personal files).

before or during implantation. These hypotheses need not be mutually exclusive.

Secretion of follicle-stimulating hormone generally is lower in young ungulates than in older ones, resulting in fewer shed ova—as indicated for yearlings in Table 56. Also, pregnant fawns seldom have more than one or two corpora lutea. Rates of ovulation apparently are lower (3.4) in the southwest than in the north (Larsen 1964). The difference could be due to poor range conditions in semiarid areas or the genetic makeup of southwestern pronghorn. Larsen did not section the ovaries during his study and could have overlooked a few that did not show on the outside. The slightly low number of corpora lutea found in North Dakota pronghorn (Table 56) could be the result of counting the scars of the previous year's corpora lutea during September. The rest of the studies in Table 56 counted active corpora lutea during pregnancy, which are easier to discern.

Wild (1995) stated that the evolutionary explanation for superovulation may be: (1) that the initial overproduction is adaptive and the dam avoids missing a reproductive opportunity (essentially Mock's insurance hypothesis); (2) the mother overproduces for quality control (i.e., as an early means of identifying genetically superior offspring); or (3) the siblicide may be counter to maternal fitness, but is a selfish action to improve an embryo's personal fitness. The first possibility (aces-in-the-hole in case of defective embryos or embryonic mortality from other causes) seems logical. However, an embryo's ability to strangle other thread-stage embryos or select implantation sites is doubtful. They apparently are manipulated by the muscular uterus under the influence of hormones from the numerous corpora lutea, and the other possibilities seem unlikely.

Corpora Lutea

The corpora lutea of mammals with advanced placentas grow rapidly, reach full size in about a week, then regress in some species after the first third or half of pregnancy. Growth of corpora lutea in Montana pronghorn is rapid during the first month of pregnancy; it slows during winter and increases again in early spring until a maximum is reached at parturition (O'Gara 1968). The mean diameters of corpora lutea tend to decrease as their number within an ovary increases. The corpora lutea of pronghorn grow to approximately 0.2 inch (5 mm) in diameter during the first week after ovulation and, if the doe becomes pregnant, slow but steady growth continues until birth.

At parturition, corpora lutea measured more than 0.3 inch (8 mm) (O'Gara 1968). The largest corpus luteum examined during an extensive study in Alberta was 0.4 inch (9.5 mm) in diameter (Mitchell 1980).

Corpora lutea begin to decline at parturition, and degenerating luteal cells result in reduced size and loss of the spheroid shape. After about a month, no luteal cells remain and the structures become corpora albicantia. In animals taken during autumn, two types of corpora albicantia can be distinguished. The scars from the last pregnancy are relatively large and often have small cavities delineating the spaces once occupied by luteal cells. Older corpora are orange, less than 0.1 inch (2 mm) in diameter and without cavities (Roseland 1977).

Cystic ovaries occasionally are found in pronghorn. The cysts often contain brownish fluid and apparently do not interfere with reproduction. I collected pregnant does with corpora lutea in cystic ovaries (O'Gara 1968).

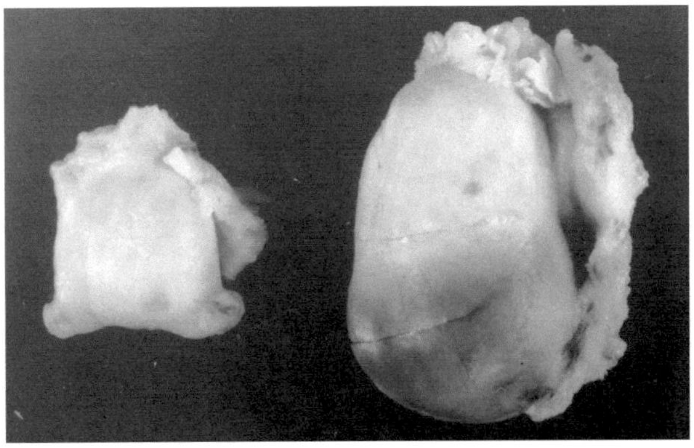

A normal (left) *and a cystic* (right) *ovary taken from a pregnant pronghorn doe. Polyovulation would result in conception for pronghorn with only one ovary. Cystic ovaries with corpora lutea have been found, indicating that cysts do not always prevent ovulation in the affected ovary.* Photo by Bart W. O'Gara.

Developing Ova and Blastocysts

Fertilized ova reach the uterus as tiny balls of cells and rapidly expand into hollow oblong vesicles (blastocysts) about 0.2 inch (5 mm) long. Blastocysts often are moved from one uterine horn to the other by muscular action of the uterus, which contracts violently under the influence of estrogen from the many follicles and developing corpora lutea in the ovaries. At this stage, the walls of the blastocysts, which eventually become fetal membranes, are one cell thick. These thin walls absorb nutrition from mucus, which is plentiful in the lumen of the uterus during early pregnancy, allowing ova to develop for nearly a month before implantation. An inner cell mass—composed of a tiny ball of cells on the wall of each blastocyst—becomes the embryo. The walls soon enter the thread stage, elongating to form tubes about 5 inches (13 cm) long and 0.02 inch (0.5 mm) in diameter (O'Gara 1968).

Intrauterine Mortality During the Thread Stage

The bicornuate uterus—Y-shaped with elongated horns—still is very active, and the fragile thread-stage blastocysts are kneaded together, often tangling. These blastocysts sometimes form overhand knots in their tubes, or two tubes form grainy or square knots. The ends of the tubes—with knots between them and the inner cell masses—bud off and become free vesicles, of which more than 20 sometimes are found in one uterus. One-fourth to one-third of the ova generally die during the thread stage because their fetal membranes are so reduced by knotting and budding off that they cannot absorb enough nutrition to survive.

Implantation and Intrauterine Mortality

Next, tubal walls thicken and acquire brush borders (small villi) to assist in absorption of food, and tubal diameters increase to those of the uterine lumens. The inner cell masses elongate, become recognizable embryos and sink through the walls of the tubes. As each sinks through, the walls encase it in a thin membrane—the amnion. Embryos in their amnions then float free within the original tubes—now called chorions. As new fetal membranes bud, embryos develop rapidly. The yolk sac, from the midgut, is a Y-shaped, ribbonlike tube several times longer than the embryo before implantation—very different in appearance from the yolk sac of most ani-

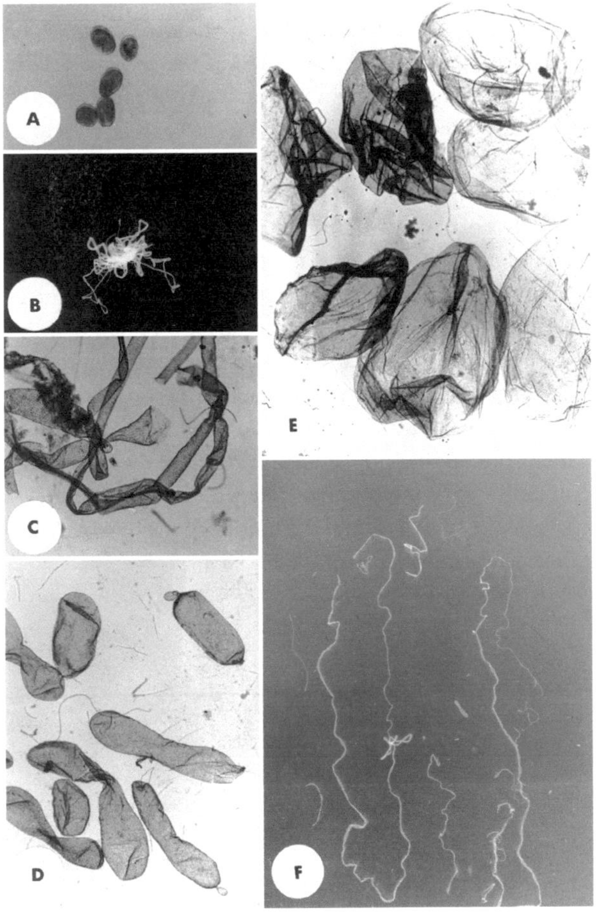

Early blastocysts of pronghorn. (A) Five blastocysts flushed from one uterus during the first week of October; (B) two thread-stage blastocysts as they were flushed from a uterus in mid-October; (C) knot in thread-stage blastocyst; (D) trophoblastic vesicles from a uterus collected during mid-October; (E) blastocysts recovered from a uterus during mid-October; and (F) the same blastocysts as in (B) teased apart. Photomicrographs by Bart W. O'Gara.

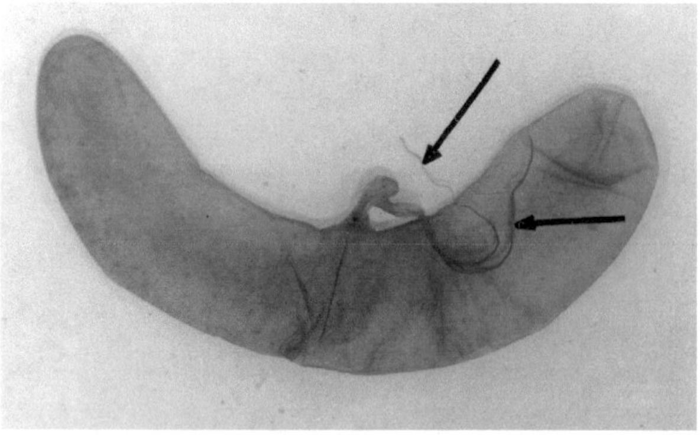

A pronghorn embryo of late October flushed out of its chorion. The large crescent-shaped vesicle is the allantois, and the Y-shaped structure is the yolk sac (arrows). This stage immediately precedes the beginning of the fusion of the allantois with the chorion. Photo by Bart W. O'Gara.

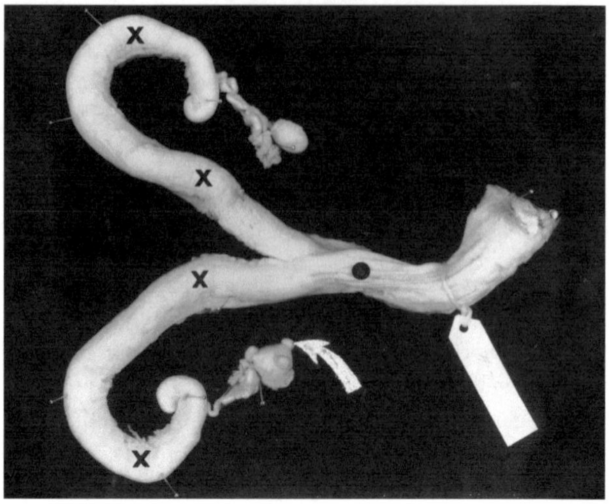

A typical pronghorn uterus with its connective tissue dissected away. The solid circle marks the anterior end of the cervix, and Xs indicate the four implantation chambers. Portions of three corpora lutea are extruded from ovulation sites on the ovary (arrow). Photo by Bart W. O'Gara.

mals. The forks of the Y lie tightly against the chorion, absorbing nutrients and transferring them to the gut. The allantois buds from the hindgut, and the embryo briefly looks like a baby whale with flukes. These flukes grow rapidly, and the vesicle of the allantois soon is as long as the chorion and starting to fuse with it. As an allantois and chorion fuse, blood vessels from the embryo extend through the allantois to the chorion. The stage now is set for implantation.

Description of implantation in the pronghorn requires definition of the terms used. The linings of some ungulate uteri have thickenings where the fetal membranes interdigitate (implant) to form placental bodies (placentomes). Strahl (1906), who described such uteri, named the uterine thickenings caruncles; the patches of villi on the fetal membranes that plug into the caruncles are called cotyledons; and the unions of the two parts are called placentomes.

The normal placement of caruncles in the pronghorn uterus is four longitudinal rows in each horn—two on each side of the mesometrial border. In some specimens, one long row lies on each side of the mesometrial border with a scattering of smaller caruncles. For example, four of 114 uteri in my collection appear bipartite. The uteri of late pregnancy and of old animals were difficult to classify. Counting only animals less than 9 years of age with nongravid uteri or uteri of early pregnancy, classification of 67 of the 114 uteri was simple. Four of these were short-horned, and the horns were elevated somewhat into a domed fundus, with the location of the dividing septum between the lower parts of the horns visible from the exterior. The other 63 uteri were typical bicornuate. No

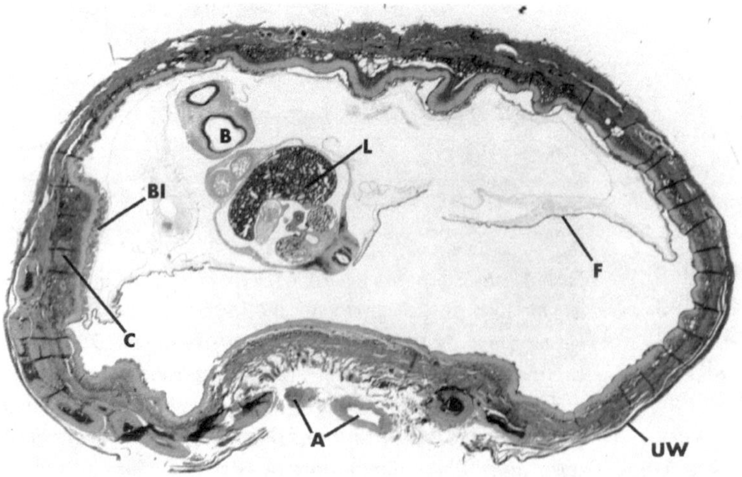

Cross section of one uterine horn of a pronghorn with the fetal membranes of the embryo just starting to implant. A = arteries; B = brain of embryo; BI = beginning of implantation; C = caruncle; F = fetal membrane (chorioallantois); L = liver; UW = uterine wall. Photo by Bart W. O'Gara.

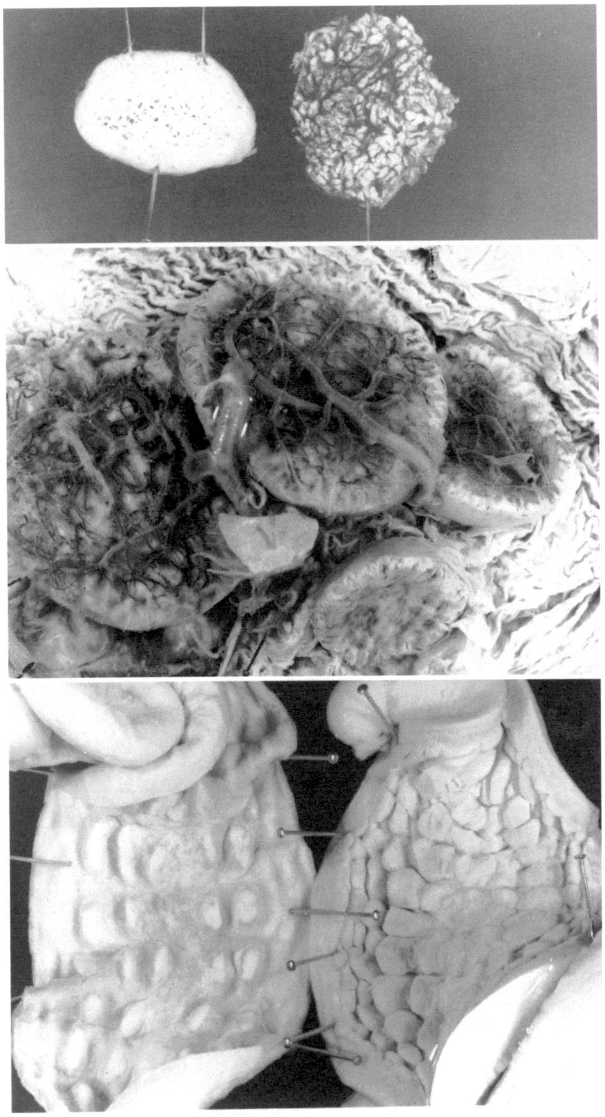

(Top) A caruncle of midgestation (left) *from which the cotyledon* (right) *has been pulled.* (Center) *Placentomes of midgestation showing vascularization from the allantois.* (Bottom) *Rows of caruncles in the opened horns of nongravid uteri. The uterine horn on the left is that of a typical bicornuate pronghorn uterus with four to five rows of caruncles. The horn on the right is from an unusual bipartite pronghorn uterus. Note the closely packed caruncles in the much shorter uterine horn.* Photos by Bart W. O'Gara.

intermediate forms were found. In the short-horned uteri, the caruncles were closer together in the rows, which seemed to be perpendicular to the longitudinal axis of the uterine horn instead of parallel to it as in the typical bicornuate uteri (O'Gara 1968). A search of the literature failed to reveal similar uterine variations in other species.

Each horn of the uterus has two expansions of the lumen, one a proximal chamber—near the body of the uterus where the two horns join—and one a distal chamber—near the oviduct. Many blood vessels serve the uterine wall at the proximal chamber, but few supply it near the distal chamber. Embryos with expanded chorionic and allantoic vesicles commonly are found in the proximal and distal chambers on both sides; thus, quadruplets often survive the thread stage. If only two embryos survive, one attaches in each of the proximal chambers.

Data from field collections indicated that implantation begins about 1 month after fertilization, when embryos are approximately 0.2 inch (5 mm) long (O'Gara 1968). Expansion of the fused chorioallantois causes it to press against the caruncles, which respond by softening and flattening until they almost disappear. The entire uterus becomes flabby as its lumen expands with the rapidly growing membranes. Surfaces of the caruncles become wavy, forming shallow pockets filled with outpocketings or folds of chorions.

Fetal membranes of embryos in proximal chambers grow extremely fast. As an embryo's membranes are pushed toward the oviduct, the end collapses into a dead, brownish cord with a knoblike necrotic tip. Before the membranes of all embryos in a horn expand enough to touch one another, those of embryos in the proximal chambers are three to four times larger than those of embryos in the distal chambers.

When a proximal embryo's necrotic tip reaches a distal embryo, the tip usually carries the distal membranes and embryo to the oviduct, where the embryo perishes from lack of nutrition. Among pronghorn in late pregnancy, necrotic wads of embryonic membranes often are found pushed into the oviducts, and sometimes distal embryos and their membranes are enfolded and mummified in the chorions of proximal embryos (O'Gara 1969a).

Wild (1995) documented losses of embryos in vivo during early gestation. Six captive pronghorn does were examined with ultrasonography at weekly intervals between an estimated 3 and 9 weeks of pregnancy. Losses of embryos were documented in five of the six does. In three cases, one of three embryos in utero appeared unhealthy during an examination and was absent on the next one. By early December, each of the six does carried only two embryos. Therefore, in vivo evidence collaborated O'Gara's (1968) findings from postmortem examinations.

Thus, the pronghorn usually bears only two of the many young that begin development. To my knowledge, it is the only animal in which the number of embryos normally is reduced during pregnancy by the two methods just described. Wild (1995) also stated that prenatal sibling competition had not been documented in any other mammalian species.

Although most displaced embryos perish quickly, some apparently survive until birth. Bromley (1967) observed a doe giving birth to twins on the National Bison Range. He caught and tagged them after the doe ate the amnions

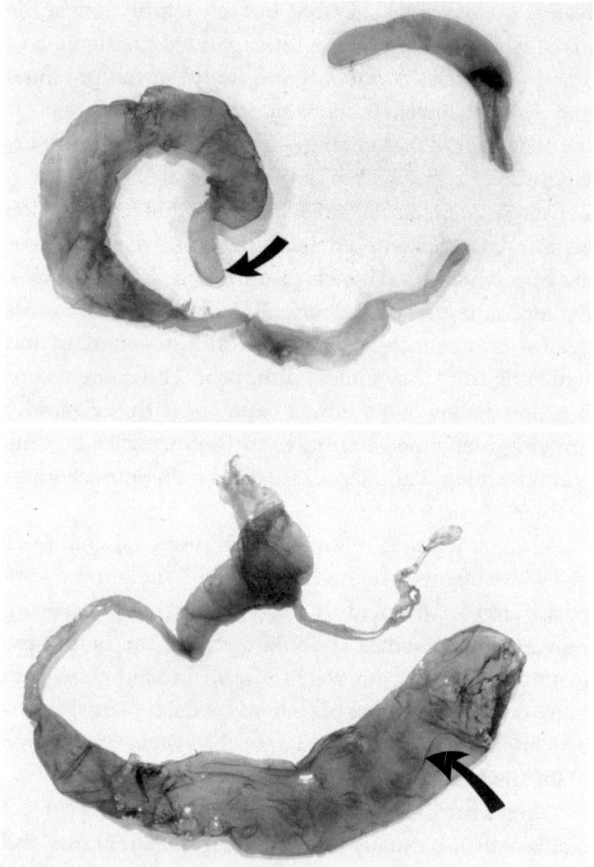

Pronghorn embryos and their fetal membranes in the approximate positions they occupied in uterine horns. (Top) *Late October embryos with an arrow pointing to the allantois protruding through the proximal end of the chorion, which was ruptured during handling; the other end of the fetal membranes was forming a necrotic tip that would soon reach and displace the smaller embryo from the distal chamber.* (Bottom) *A small (distal) embryo with its fetal membranes pierced by the necrotic tip of a larger embryo during the second week of November. The arrow points to a patch of villi (cotyledon) where implantation was beginning.* Photos by Bart W. O'Gara.

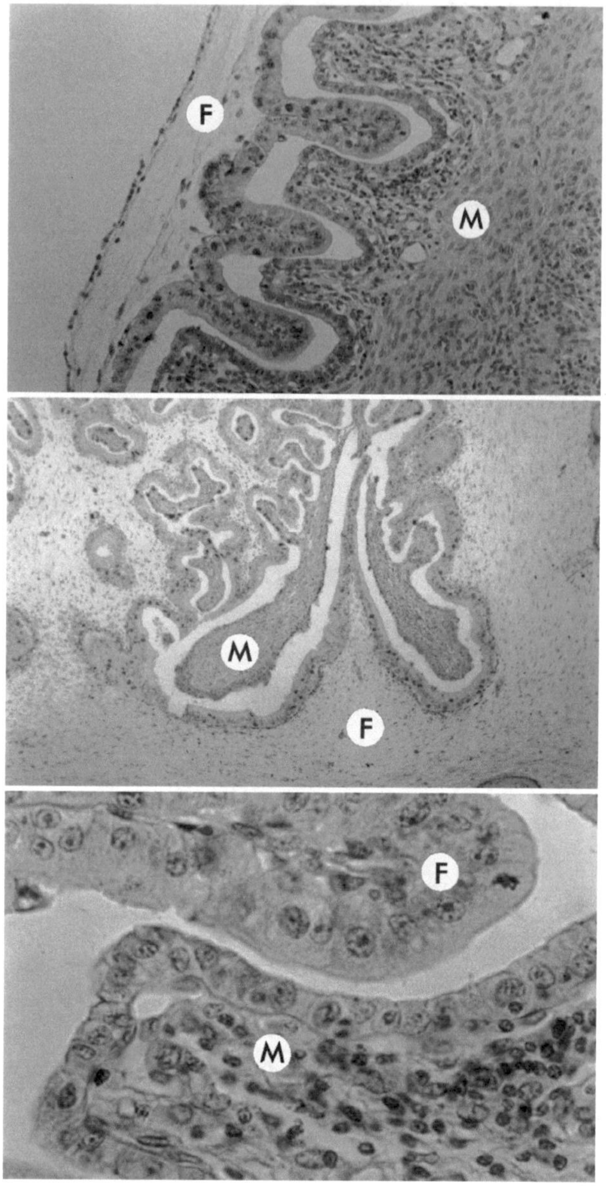

Stages of placentation in the pronghorn. (Top) *Early implantation, about 5 weeks after fertilization. Chorionic villi are simple foldings of the chorioallantois.* (Center) *Cross section of a placentome of midgestation showing branching of the chorionic villi.* (Bottom) *Fetal/maternal junction of late gestation showing the lack of local destruction of the maternal epithelium. F – fetal membrane (chorioallantois); M = maternal tissue (caruncle).* Photomicrographs by Bart W. O'Gara.

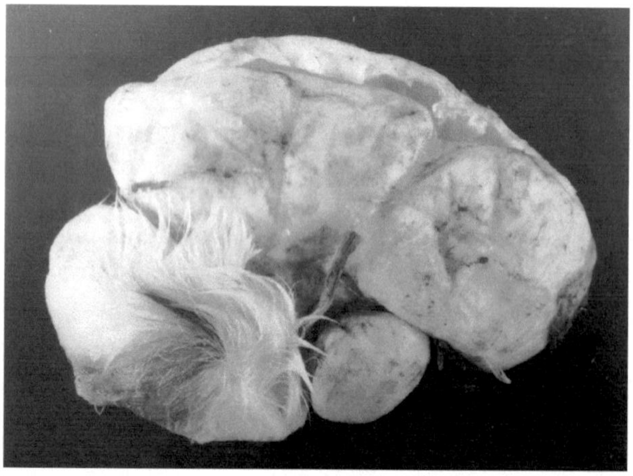

A teratoma, or acardiac monster, born with twin pronghorn fawns on the National Bison Range, Montana. Such monsters have not previously been reported from pronghorn, either because they are extremely rare or because so few births are closely observed. Photo by Bart W. O'Gara.

Near-term twin pronghorn fetuses with the uterus peeled from the fetal membranes. The visible, outer membrane is the chorion, and the dark spots are the villi of the cotyledons that were pulled from the caruncles. Photo by Bart W. O'Gara.

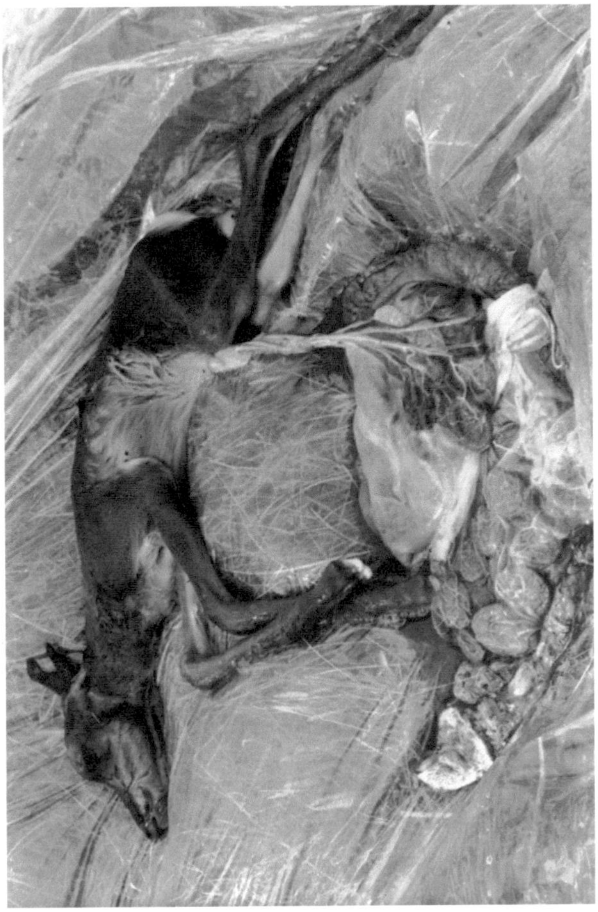

A near-term pronghorn fetus with the uterine horn and chorioallantois opened and the amnion removed to show the connection of the fetus, by its umbilical cord, to approximately 50 placentomes. Photo by Bart W. O'Gara.

and nursed the fawns. Where the ground was wet with amniotic fluid, he found an organic object about 5 inches (127 mm) long, with hair on one end. It appeared somewhat similar to acardiac monsters that are occasionally born with human babies, and consisted primarily of a membranous bag filled with bloody fluid—one end was solid, covered with hair and contained teeth. This apparently was one of the crowded or pierced embryos that had survived, although grossly deformed, as a parasite on the fetal membranes of its sibling. The smallest fawn weighed by Bromley—5 pounds (2.3 kg)—was in utero with this monster; other fawns in the same area that year averaged 7.4 pounds (3.4 kg) at birth.

As implantation proceeds, chorionic villi give off short, leaflike secondary villi at right angles to the primary ones, and the structure becomes treelike in cross section. In most ruminants, local destruction of the uterine surface occurs between the fetal villi, allowing the chorion to come into direct contact with vascular maternal connective tissue. This destruction apparently does not happen in pronghorn, therefore its placenta is of a primitive, epitheliochorial type (O'Gara 1969a). Pigs and horses have similar placentation but do not have caruncles—the villi instead are distributed all over the chorion rather than localized in cotyledons. In this respect, the gross structure of the pronghorn placenta is like that of bovids, but the fine structure is more like that of the pig and horse. However, the fine structure of placentas has only been investigated in a few species of bovids, and a more complete histological study of pronghorn placentas might reveal some local destruction of uterine epithelium.

Because maternal tissue is not destroyed during placentation, little or no bleeding occurs at birth until the umbilical cord is parted, unless membranes or the birth canal are torn. At any time during pregnancy, little traction is required to pull the fetal and maternal elements apart. The 90 or so caruncles (range, 45–180) are of the flat or intermediate type and, during pregnancy, they look like short-stemmed, flat-topped mushrooms arranged in rows. The volume of the caruncles increases about 250-

fold from the nonpregnant condition to that of late pregnancy (O'Gara 1969a). Fetal membranes of twins join and fuse in the body of the uterus when the fetuses are approximately 1.6 inches (4 cm) long and 75 days old (O'Gara 1968). The ends of chorions, in apposition to one another, degenerate and the amnions fuse.

The fetal membranes and corpora lutea of the pronghorn reflect the primitive state of its placenta. Allantoes persist through midgestation as vesicles containing nitrogenous wastes; mammals with more advanced placentas retain only the blood vessels of the allantoes.

Fetal Development

Weights are not good indicators of fetal age because fetuses gain weight rapidly if they are left in amniotic fluid after death, and nutritional stress and health of does affect the weights of fetuses. Measurements should be more dependable than weights, but even variations in measurements are tremendous. Apparently healthy fawns from the same population may differ by as much as 4 inches (102 mm) in length at birth. No one has collected a series of timed embryos to establish growth curves. Approximate rates of growth for fetuses collected in Montana and Wyoming are shown in Figure 56. All of those fawns were twins; triple or single births undoubtedly result in even greater differences. G. Post and W. C. Edwards (personal communication:1975) provided data on 36 Wyoming fetuses. They noted that one yearling doe they collected was carrying a single fetus, in contrast to twins carried by all other females. The single was much larger than twins collected at about the same time. Obviously, singles should not be used with twins in growth curves.

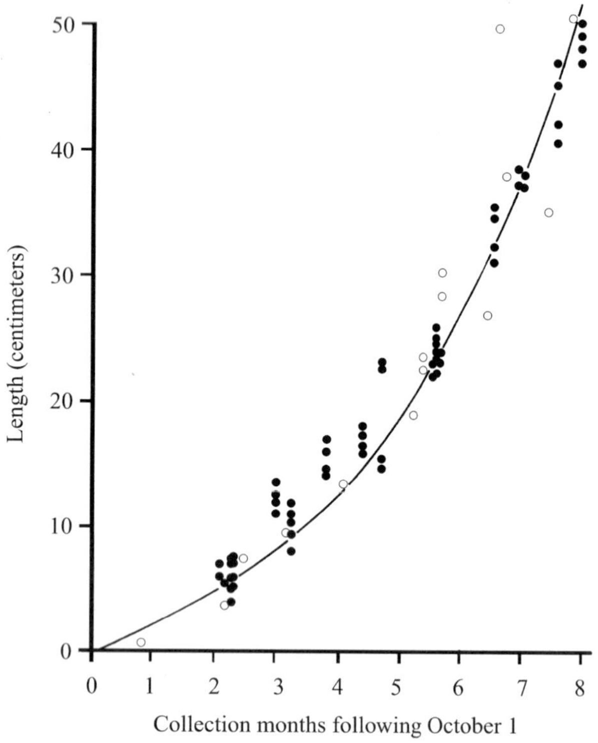

FIGURE 56. *Approximate growth curve (forehead/rump length versus time) for pronghorn fetuses. Open circles represent fetuses from Yellowstone National Park and black circles represent those from the National Bison Range in Montana.*

Morphological changes in pronghorn fetuses between approximately 54 and 229 days postconception. Fetuses shown indicate relative development, not comparative size. Photo (composite) by George Post and William Edwards.

Perhaps the acquisition of adult characteristics is the best indicator of fetal age. The sequence of such characteristics during one of my studies is shown in Table 57, but the ages are only estimated. Observed copulations and births, plus bloody ovulation sites on ovaries, indicated breeding dates between September 8 and September 20, and a mean breeding date of September 15; at best, the ages vary by 2 weeks.

Birth

The onset of reproduction in species whose birthseason lengths are adapted to seasonal fluctuations in food supply and weather should be governed by reliable predictors of these conditions, such as photoperiod and nutritional status. However, species adapted to synchronized births in order to reduce predation on newborns may also respond to lunar cycles, social signals, and other clues that do not predict environmental conditions.

Rutberg (1987:706)

Bromley (1977) noted that the birth process normally takes place within a few days, including the time from when the doe begins to seek out a place to give birth until her young are born and have left her for their first bedding sites on the prairie. Before fawns are due, the abdomens of pregnant does are noticeably distended and, just before parturition, their udders are visibly swollen. Pronghorn almost always have four teats, although I have seen two does with six; one was lactating and had milk in all six.

From several hours to a full day before giving birth, the doe selects a birth site. Usually, the band moves on, but sometimes pronghorn of all ages and both sexes remain in the area. Does with considerable fetal membrane showing, as above, still are able to stand, retract the membranes and chase a predator from the area. The birth process—from the onset of intense labor until the newborn fawns separate themselves from their mother and locate bedding sites—lasts only about 6 hours. Photo by LuRay Parker; courtesy of the Wyoming Department of Game and Fish, Cheyenne.

The onset of labor is characterized by humping of the back, elevation of the tail, licking of the flanks and a straddled stance. Tucker (1979) reported occasional dragging of the hind legs.

TABLE 57. Time of appearance of adult characteristics in pronghorn fetuses from a Montana and Yellowstone National Park population, with approximate ages in months

	Date of appearance			
Characteristic	*In first fetus*	*Age*	*In all fetuses*	*Age*
Genitalia recognizable	December 2	2.5	December 2	2.5
Hooves (white)	December 2	2.5	December 2	2.5
Gray line on hooves	January 27	4.5	January 27	4.5
Horn spots	January 27	4.5	January 27	4.5
Bumps caused by incisors under skin	January 27	4.5	January 27	4.5
Vibrissae on upper lip	February 12	5.0	March 21	6.2
Vibrissae over eye	February 12	5.0	March 21	6.2
Vibrissae on chin	February 24	5.8	March 21	6.2
Pigmented nose patch	February 24	5.8	March 24	6.3
Incisors erupted	February 24	5.8	March 21	6.2
Black hooves with white tips	March 22	6.2	March 22	6.2
Eyelashes	March 21	6.2	March 21	6.2
Black line on back of neck	March 21	6.2	March 24	6.3
Black cheek, bucks	March 21	6.2	April 21	7.2
One hair on each cheek	March 21	6.2	March 22	6.2
Hair on top of head	March 21	6.2	March 24	6.3
Entire body pigmented	March 21	6.2	March 21	6.2
Hair on outside of legs	March 24	6.3	March 24	6.3
Hair on belly	April 21	7.2	April 21	7.2
Hair on rump	April 21	7.2	April 21	7.2
Completely haired	April 30	7.5	April 30	7.5

Does lie down during most of the intense labor, sometimes on their sides, and rock back and forth. An amnion usually extrudes from the vulva for 10 minutes or more before the first fawn is born. When a fawn makes its first appearance, the front hooves, within the fluid-filled amnion, usually are seen first. Fawns often are born while does are recumbent, but does sometimes stand up after the fore parts of a fawn clear the vulva, allowing the fawn to slide free. Kitchen (1974) observed five fawns that hung by their hips from standing does for 2 to 6 minutes before sliding free. I saw a doe give birth to twins without lying down—she had been bothered by a coyote that she chased for nearly a mile (1.6 km) just before giving birth. Probably a more normal situation was described by Howard (1966). He watched a doe give birth while lying down. She got up, smelled and licked the fawn, lay back down, got up, and lay down again before giving birth to another fawn 9 minutes later. Does in Texas walked about in a wandering manner between periods of recumbency (Tucker 1979). While wandering, they sometimes assumed a squatting position and, in some instances, the hind legs appeared dislocated.

In most of the pronghorn's range, fawns are born within a 3-week period during the early growing season (Autenrieth and Fichter 1975), and the majority are dropped within a 10-day period. The availability of energy for lactation and juvenile growth probably is the ultimate factor responsible for seasonal breeding in long-lived mammals of northern latitudes (Bronson 1989). Young born too early risk hypothermia or starvation; those born too late may enter their first winter too small to survive the cold weather, deep snow, poor food and predators (Clutton-Brock et al. 1982).

Birth synchrony is a conceptual theory that proposes that an animal species that produces young in a short time period, overcomes predators' ability to kill as large a proportion of young as they would if young were born over a more prolonged period. This phenomenon often is called predator swamping. Birth synchrony was found to be an important factor influencing predation on caribou calves in Canada (Dauphiné and McClure 1974) and wildebeast calves on the grasslands of Africa (Estes 1976). Pronghorn exhibited tightly synchronized birth dates on the National Bison Range in Montana (Byers 1997a).

On the Hart Mountain National Antelope Refuge in Oregon, Gregg et al. (2001) monitored the survival of 104 radio-marked fawns from mid-May to mid-July during 1996 and 1997. Eighty-four percent (87/104) of the fawns died during the monitoring period. Average age at death was 8.4 days, and 83 (95%) of the fawns that died were less than 18 days old. Predation accounted for 86% (75/87) of the fawn deaths, and coyote predation was the greatest single cause of mortality each year. Fawns born during the peaks of the birthing periods lived longer than those born during the nonpeak periods (Figure 57). Of 17 surviving fawns, 14 (82%) were born during birthing peaks. Gregg et al.'s results indicate that birth synchrony may be an adaptation whereby losses of pronghorn fawns to predation are reduced.

Also on Hart Mountain, unpublished information suggested that pronghorn fawns were born during a very short time period in 1999 compared with 1998. More fawns

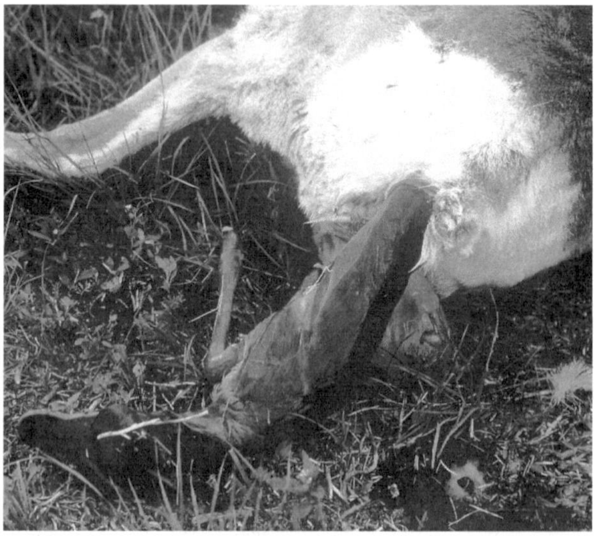

(Left) *Pronghorn does usually lie down to give birth, but they sometimes stand to allow the fetus to slide free. Fawns sometimes hang by their hips for several minutes before being expelled from the birth canal. Siblings generally are born 9 to 31 minutes apart (Autenrieth and Fichter 1975).* (Right) *A newborn pronghorn fawn will soon raises its head, try to stand and orient on its mother. Fawns usually nurse for the first time 30 to 90 minutes after birth.* Photos by LuRay Parker; courtesy of the Wyoming Department of Game and Fish, Cheyenne.

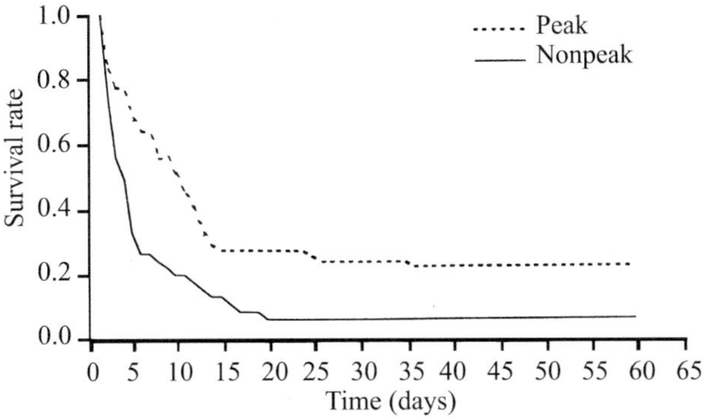

FIGURE 57. *Survival curves for radio-marked pronghorn fawns born during peak and nonpeak fawning periods at Hart Mountain National Antelope Refuge, south-central Oregon, 1996–97 (from Gregg et al. 2001).*

survived until mid-July in 1999 than in 1998 (38/100 versus 12/100). Radio-tracking data indicated that fawns born during the peak of parturition survived longer than others (Dunbar: unpublished data). The summer of 1998 was unusually wet on Hart Mountain, and Dunbar speculated that birth synchrony was related to the nutritional status of does during the breeding season. He maintained that a population of does in good health would breed at about the same time, consequently giving birth at about the same time. Conversely, if does were in poor health, the less healthy—especially the young and possibly the old—would breed later, extending the birthing time.

A short fawning period allows more fawns to avoid predation than does a long one. However, the differences among birthing times noted indicate that variations in factors affecting nutritional status of does and duration of spring green-up ultimately determine the duration of fawning periods.

Fawning occurs during late May and early June in the northern part of pronghorn range (Hepworth and Blunt 1966, Mitchell 1967, O'Gara 1968, Autenrieth and Fichter 1975). However, I have heard of April fawns in eastern Montana, and newborn fawns have been seen during July. In Idaho, 3- to 4-day-old fawns were seen on July 8 and 25 (Fichter 1958), and a newborn was observed in North Dakota during early August (Jensen 1997). Such late fawns could be the result of fawn and yearling pregnancies or a recurrence of estrus in older does that did not conceive during their first estrus.

Fawns generally are born later and over a greater span of time in the southern part of pronghorn range. Two apparent fawning peaks—one in mid-June and another in mid-July—were detected during weekly helicopter surveys in southeastern Colorado (see Pojar and Miller 1984). Texas fawns have been observed from early April to early June in the Trans-Pecos area (Buechner 1950a), and as early

Drying amniotic fluid causes the coats of newborn pronghorns to take on a wavy, marcelled appearance. Photo by LuRay Parker; courtesy of the Wyoming Department of Game and Fish, Cheyenne.

Pronghorn does usually begin licking each fawn immediately after each birthing. Drying and fluffing their coats helps to prevent hypothermia and probably contributes to bonding between mother and young. Does also eat the fetal membranes as they groom newborn fawns, which undoubtedly reduces odors at the birth site, reducing olfactory clues for mammalian predators. Protein and hormones in fetal membranes may be nutritionally beneficial to does stressed by parturition and early lactation. Photo courtesy of the Vizcaino Biosphere Reserve.

as February 23 and commonly by March 15 in south Texas. Birth of Sonoran pronghorn fawns occurs during the first or second week of March in southern Arizona (Phelps 1981), and fawns judged to be less than 1 month old were seen as late as mid-April.

In the southeastern corner of New Mexico, fawns are born during late May (Larsen 1964). Fetuses taken from does in Hidalgo County, New Mexico, seemingly were a month younger than those taken in Otero County at the same time (T. Yates and P. Fry personal communication:1981). Climatic conditions probably were making a difference of a month in breeding and birthing times in populations less than 200 miles (322 km) apart. Trevino (1978) reported seeing pronghorn breeding in Chihuahua, Mexico, during September. However, he did not observe fawns until July—probably indicating fawning in June.

Data were compiled on fawning dates in Arizona by Ticer et al. (2000). Fawning dates varied from late February to April in southwestern Arizona to May through June in the northern part of the state. Differences in fawning dates for seven sites seemed related to differences in elevation, mean number of frost-free days, mean minimum spring air temperatures and rainfall patterns that contribute to the timing of spring green-up. Comparing their data with those from other states, including Mexico, Ticer et al. concluded that the earliest fawning occurred during February on the Baja Peninsula and in southwestern Arizona and Texas. Fawning did not begin until May in northern Arizona and northern states.

Fawns are large, and pronghorn show the highest known rate of maternal investment among ungulates according to Byers and Moodie (1990). That rate (ratio), derived by the equation 2× birth weight$^{0.75}$/maternal weight$^{0.75}$, is 0.288. For comparison, similar ratios for other North American artiodactyls are as follows: bison, 0.022; Rocky Mountain sheep, 0.106; Rocky Mountain goats, 0.114; elk, 0.157; mule deer, 0.172; and white-tailed deer, 0.203.

Such high maternal investment apparently makes pronghorn does and their fawns especially sensitive to forage conditions. Hockley (1968) and Beale and Smith (1970) found recruitment related to precipitation in Wyoming and Utah, respectively. In the Powder River Basin of Wyoming, Boyse and Tate (1982) found significant year-to-year fluctuations in numbers of fawns in the harvest. Those numbers positively correlated with precipitation, hence nutrition.

Data on birth weights, suckling rates, age-specific frequency of maternal termination of suckling bouts, age at weaning and rate of rejected suckle attempts showed either no sex differences or a slight excess investment in daughters by pronghorn does (Byers and Moodie 1990). Among ungulates for which data are available, the best predictor of differential investment is not degree of sexual dimorphism (which is slight in pronghorn); it is comparative rate of maternal investment.

Litter Size and Secondary Sex Ratios

Twins were the rule during first pregnancies and occurred in 98% of the subsequent pregnancies during early Wyoming studies, but singles were three times as common as triplets (Hepworth and Blunt 1966). Numbers of fetuses per 100 does, from studies in seven states, are listed in Table 58. Yearling does are included in this Table, but only pregnant does are considered. Yearlings generally have smaller litters than do older does. For instance, Edwards (1958b) found 18 fetuses in 10 yearlings and 117 fetuses in 59 two-year-old and older does. Triplets occasionally are reported in the field, but proof of triplets demands that they are seen being born. Three embryos in utero were found by Edwards (1958b), on March 13, 1958.

TABLE 58. A comparison of reported numbers of fetuses per 100 pronghorn does

	Number			
Area	*Does*	*Fetuses*	*Fetuses per 100 does*	*Source*
California	18	35	194	Chattin and Lassen (1950)
Wyoming	43	80	186	Edwards (1958a)
Colorado	42	81	193	Hoover et al. (1959)
Idaho	56	105	188	Fichter (1962)
New Mexico	62	117	189	Larsen (1964, 1965, 1966, 1967)
Montana/Yellowstone National Park	33	65	197	O'Gara (1968)
Oregon	27	50	185	Trainer et al. (1983)
Colorado	10	19	190	Firchow 1986
Oregon	36	67	186	J. D. Yoakum (personal communication:1989)
Totals	327	619		
Average			190	

(Left) *Newborn fawns often can stand before they are completely dry, sometimes in less than half an hour after being born. Does continue to groom fawns until they are dry.* (Right) *Fawns seek their mother's udder as soon as they are able to stand. Nursing fawns do not wag their tails or root the udder. Licking of the entire fawn soon gives way to anogenital grooming.* Photos by LuRay Parker; courtesy of the Wyoming Department of Game and Fish, Cheyenne.

Byers (1997a) suggested that pronghorn evolved a litter size of two under a long history of predation on fawns by coyotes and other ground predators. Rates of predation apparently were high enough to select strongly for twinning because of the low probability of survival by a single offspring. However, litter size could not increase at the expense of individual birth weights, or postnatal growth rates, because predators also create selection for rapidly maturing offspring. Consequently, pronghorn does produce two large fawns each year, and the combined weight of those fawns at birth is about 18% of maternal weight.

Of a collection of 120 does from three populations in Wyoming, Zimmer and Lindzey (2002) reported that fetal rates, which were comparatively low, varied among populations and between the 2 years of collection. Drought conditions or other habitat problems seemed to have impaired reproduction during that study. One of the collected does proved to be a fawn and was removed from the sample. Thus, 119 does carrying 212 fetuses were examined and indicated a fetal rate of only 1.78 per doe (J. Zimmer personal communication:2002). A breakdown by weights and ages indicated that underweight and yearling does caused the unusually low fawn/doe ratio but this does not consider the relative weights of pregnant (naturally heavier) versus barren does. Ninety-seven pregnant adults averaged 114 pounds (52 kg); six carried only one fetus, but two (9.9 years and 10.9 years of age) carried triplets. Three barren does (3.9-year olds) averaged only 83 pounds (38 kg), but a nonpregnant 9.9-year-old weighted 105 pounds (48 kg). Fourteen pregnant yearlings, with a fetal rate of 1.5, averaged 103 pounds (47 kg). Four barren yearlings averaged 83 pounds (38 kg). The oldest collected does (12.9-year olds) both carried twins. Zimmer and Lindzey (2002) concluded that two fawns born per doe older than 1 year of age may lead to an overestimation of fawns produced in many populations.

"Of 300 kids taken at the Charles Sheldon and Hart Mountain refuges to be raised for restocking purposes, the sex ratio was approximately 45 males to 55 females" (Einarsen 1948:123). However, secondary sex ratios found during 13 studies—in which only fetuses in utero, observed births or both members of a set of twins were handled during the first week of life—indicated 113 males per 100 females at birth (Table 59). Studies based on older fawns are unreliable because one sex may be harder to catch than the other or the sexes may have suffered differential mortality.

Social rank affects the birth sex ratio according to Byers (1997a); subordinate does produce an excess of daughters, whereas dominant females produce an excess of sons.

Factors in Early Fawn Survival and Development

Milk intake by a pair of pronghorn twins gradually declined throughout the summer from an average of 15.6

Table 59. A comparison of reported sex ratios for pronghorn fetuses and fawns (fetuses in utero, observed births or both members of a set of twins handled during the first week of life)

Area	*Male*	*Female*	*Males per 100 females*	*Source*
California	16	19	84	Chattin and Lassen (1950)
Colorado	27	25	108	Hoover and Ogilvie (1954)
Wyoming	69	53	130	Edwards (1958b)
Montana/Yellowstone National Park	34	31	110	O'Gara (1968)
Montana and South Dakota (unpublished)	64	53	121	Kitchen and Bromley
Idaho	29	24	121	Autenrieth and Fichter (1975)
Montana	8	8	100	Von Gunten (1978a)
Idaho	23	19	121	Bodie (1979)
Montana	22	10	220	Corneli (1980)
Oregon	26	24	108	Trainer et al. (1983)
Colorado	12	6	200	Firchow (1986)
Montana	148	150	99	Byers (1997)
Wyoming	113	99	114	Zimmer and Lindzey (2002), J. Zimmer (personal communication:2002)
Totals	591	521		
Average			113	

± 0.7 fluid ounces (468 ± 22 mL) per day during the first 2 weeks of life. Estimates of milk intake were based on concurrent measurements of the maternal water turnover determined by tritium levels in the dam, transfer of tritium from dam to fawn and total water turnover in the fawn—as determined by deuterium levels (see Martin 1995).

During early lactation (4–25 days postpartum), the average percentage dry matter in pronghorn milk was 23.0 ± 0.2; fat was 8.8 ± 0.03; ash was 1.4 ± 0.1; carbohydrates were only 4.1 ± 0.2. Energy content was 5.7 ± 0.1 kilojoules (KJ) per gram (Martin 1995, Martin and Parker 1997).

During midlactation (32–53 days postpartum), milk had lower levels of dry matter, at 19.0% ± 0.5%; fat was 5.3% ± 0.12%; energy was 4.3 KJ/g. Milk composition from late lactation (60–74 days postpartum) varied little from that of midlactation. The single colostrum sample contained 3.0%, 8.8%, 5.2%, 17.9% and 1.4% fat, protein, carbohydrates, dry matter and ash, respectively; energy content was 4.2 KJ/g. Output of milk energy, based on the dam's weight, decreased from 141.7 ± 0.1 KJ/kg/day during the first 2 weeks of lactation to 22.3 KJ/kg/day at 74 days postparturition.

It is advantageous for pronghorn fawns to attain adult proportions rapidly. The efficiency of locomotion increases with body weight and leg length. Pronghorn flee from predators in compact groups. The sooner neonates can match adult running speeds and stay with a group, the better their chances of avoiding predators. Growth rates of neonates have been directly related to milk energy intake. Relative to maternal weight, peak milk volume and energy output are higher in pronghorn than for other

An imprinting period usually lasts 3 to 5 hours as the pronghorn mother and young bed together, often getting up to nurse and exchange low, soft mewing sounds as they nuzzle each other. The fawns and the doe apparently become familiar with each others' odors, appearances and voices during this time. Does often get blood on their noses, as seen here, when the umbilical cord is severed or as the afterbirth is eaten. After the imprinting period, fawns find their own beds and remain nearly motionless for hours as their dams feed far enough away to avoid attracting predators. Photo by LuRay Parker; courtesy of the Wyoming Department of Game and Fish, Cheyenne.

species of North American big game that have been studied (see Martin 1995).

Byers (1997a) found yearling bucks more likely than yearling does to die during winter, apparently because the bucks sacrificed fat accumulation in favor of skeletal growth.

MALES

During the 1952 breeding season, there were no male antelope on the Bison Range older than 1⅓ years. In the spring of 1953, several does that were present during the breeding season in 1952 produced fawns, clearly showing that yearling male antelope are capable of impregnating females.

Wright and Dow (1962:100)

The reproductive tracts of male pronghorn are quite similar to those of most ruminants. The male genital organs consist of the scrotum, two testes (located within the scrotum), epididymides, deferent ducts, prostate (an accessory gland), penis and urethra (the passageway of urine and semen).

Yearling males are capable of reproducing (Wright and Dow 1962), but fawns generally are considered to be incapable (Einarsen 1948, Hoover et al. 1959). However, O'Gara et al. (1971) found that the testes of Montana fawns were in the late prepubertal stage by November. It was thought that an occasional male fawn probably could breed after the main rutting season, a month or so later than do adults—as does an occasional female fawn. Although yearling males are capable of breeding, they seldom get a chance because older bucks hold territories where most of the breeding takes place or, in a mobile harem situation, older bucks keep yearlings away from the does. However, yearling bucks may breed does that come into estrus a month or so after the peak of the breeding season (Ellis and Deblinger 1975). Breeding behavior and behavior of bucks toward does and fawns are discussed in Chapter 6.

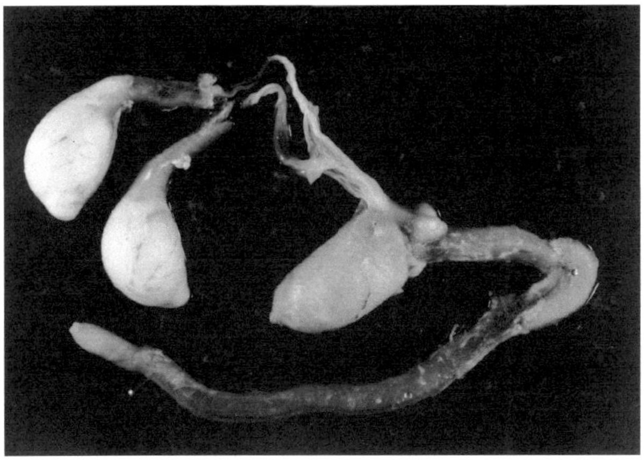

The reproductive tract of a 2-year-old pronghorn buck shot during September. By December, the testes will be less than half the size seen here. Photo by Bart W. O'Gara.

Testes

Testes of mature males vary seasonally. Those of bucks collected in Colorado, Montana and Wyoming weighed about 2.5 times as much during July and August as in December and January (O'Gara et al. 1971). This fluctuation is relatively small compared with those reported for the seasonally active testes of many mammals. However, a fivefold change was reported by Mitchell (1980) in Alberta. That extreme was between a 2-year-old (3.114 ounces [88.3 g]) collected on September 1, and a 7-year-old (0.547 ounce [15.5 g]) collected on February 3. The heaviest and lightest testicular weights also were noted in 2-year-old and 7-year-old bucks, respectively, by O'Gara et al. (1971). Although the extremes in testicular weights found by Mitchell (1980) were greater, the seasonal trends were similar during both studies. Testicular weights of 18 fawns collected during November in Alberta averaged 0.534 ounce (15.13 g) per pair, and those from 25 yearlings averaged 1.035 ounces (29.34 g) (Mitchell 1980). Mitchell also weighed 38 pairs of adult testes collected in November; they averaged 1.269 ounces (35.98 g)—about half the weight of those collected in September.

The seminiferous tubules appear as a large collection of twisted canals. Spermatozoa are formed from the epithelial lining, which contains spermatogenic cells and sustentacular (Sertoli) cells. Straight tubules are formed by the union of the numerous seminiferous tubules. These congregate in the rete testis, a network of confluent spaces and ducts. For a detailed description of spermatogenesis in pronghorn, see Mitchell (1980). Spermatogenesis by yearlings and adults continues through most of the year in Montana, but is much reduced during December through February.

Spermatozoa are stored in the epididymides before ejaculation. An epididymis is an elongated structure composed of a long, convoluted tube, the coils of which are joined by collagenous connective tissue. In addition to being a storage place for spermatozoa, the epididymis slowly transports spermatozoa to the ductus deferens. Circular smooth muscle fibers aid seminiferous tubule secretion in moving the germ cells. The length of the epididymis and the slowness of spermatozoa movement are important in allowing the spermatozoa to complete their maturation process.

Deferent ducts—continuations of the ducts of the epididymides—begin at the tails of the epididymides, extend along the dorsomedial border of each testis, ascend in the spermatic cord and enter the abdominal cavity through

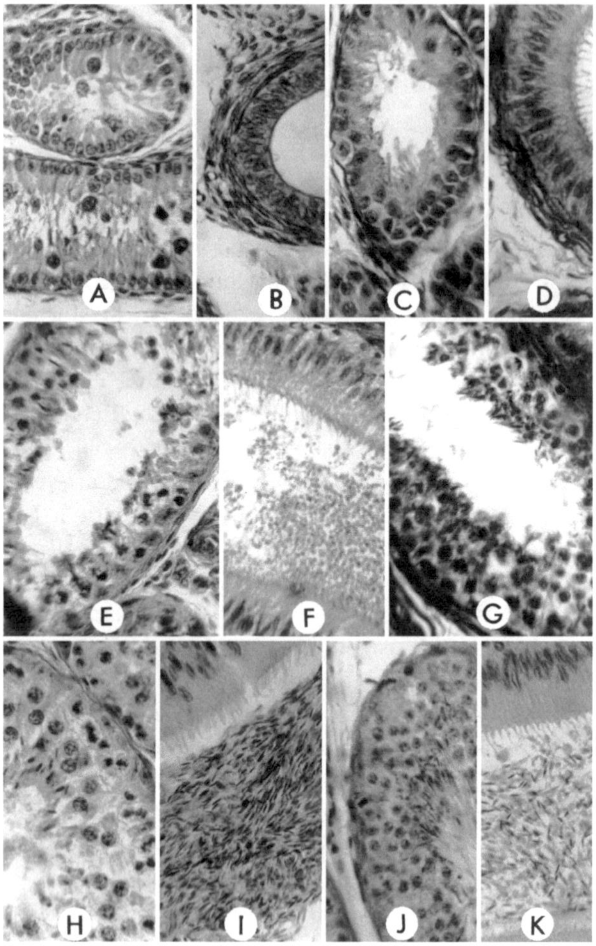

Pronghorn testes and epididymides. (A) Seminiferous tubule of a 3- to 3.5-month-old fawn; (B) epididymis of the same fawn as (A); (C) seminiferous tubule of a 5- to 6-month-old fawn; (D) epididymis of the same fawn as (C); (E) seminiferous tubule of an adult buck killed in January; (F) epididymis of the same buck as (E); (G) seminiferous tubule of an adult buck killed in March; (H) seminiferous tubule of an adult buck killed in August; (I) epididymis of the same buck as (H); (J) seminiferous tubule of an adult buck killed in December; and (K) epididymis of the same buck as (J). Photomicrographs by Gary Matson.

the inguinal canal. The deferent ducts run parallel to each other through the prostate gland and open into the urethra by two slits. By late March, tubules of the epididymides of Montana pronghorn contain scattered to numerous spermatozoa. From April through November the epididymides are packed. During December, they again become scattered and leukocytes appear, apparently to remove degenerating spermatozoa (O'Gara et al. 1971).

Accessory Sex Glands

Little research has been conducted on the accessory sex glands of pronghorn; however, the prostate glands of mature bucks appear to be in condition to facilitate breeding from March through November. Weight changes of prostate glands collected in Montana correlated closely with those of testes, and secretory activity—inferred from lengths of secretory cells in the alveoli of the prostates and vasa deferentia—followed a similar curve (Figure 58). However, sample sizes were small, only one pair of testes with associated prostate glands was collected per month, and work remains to be done on glandular material within the penis.

The prostate gland is a musculoglandular mass that completely encompasses the proximal portion of the male urethra at the neck of the bladder. The prostate is semioval in transverse section; the dorsal surface is flattened. The prostatic part of the urethra passes through the gland somewhat dorsal to its center. A prominent median septum divides the gland into right and left lobes. Each lobe is divided further into lobules that consist of numerous compound tubuloalveolar glands lined by columnar epithelium. Ducts from these glands enter the urethra throughout its circumference. The capsule of the prostate is comparatively thick, and smooth muscle fibers are found throughout the capsule. The function of the prostate is poorly understood, but prostatic secretion is believed to

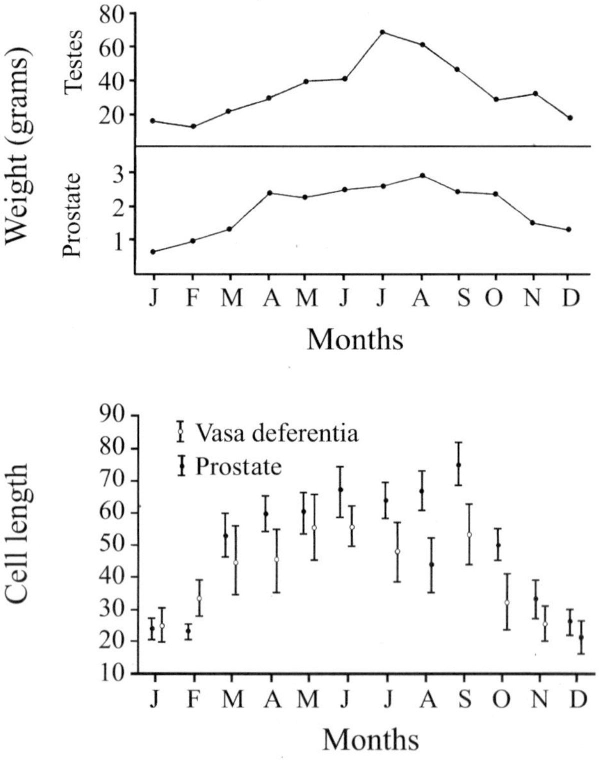

Figure 58. (Top) *Weights of pronghorn prostate glands and both testes by month of collection.* (Bottom) *Length of secretary epithelial cells in vasa deferentia and prostate glands by month of collection (from Knowles and O'Gara unpublished).*

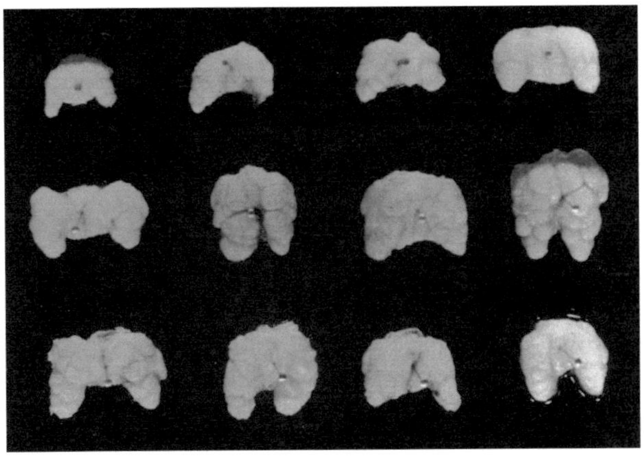

Prostate glands from adult pronghorn bucks collected in Montana. Those in the top row, from January, are at minimum size. Those in the bottom row, from August, are at maximum size, and appear to be in condition to facilitate breeding. Secretion is believed to provide an optimum environment for sperm survival and motility. Photo by Bart W. O'Gara.

provide an optimum environment for sperm survival and motility.

PSEUDOHERMAPHRODITES

Pseudohermaphroditic pronghorn occasionally are reported. I examined the reproductive tracts of two pronghorn initially described to me by the hunters who killed them and saw pictures of a third; both that I examined lacked ovaries, so they were bucks. Thus, although they externally resembled hermaphrodites, they were what is commonly termed pseudohermaphrodites. Each had a small udder, no scrotum, small (about 1 inch [25 mm]) inguinal testes and a vulva. All had poorly developed cheek patches. Two from Wyoming had 12-inch (30.5 cm) horns with average girth and prongs (W. Hepworth personal communication:1983, K. Brawley personal communication:1984). One, shot in Montana by Pyrah, had one thin upright horn about 10 inches (25 cm) long; the other horn was curled sharply forward over the eye. Two that I investigated had very small vaginas, cervixes and uteri with rudimentary oviducts and no ovaries. In each, the very short penis terminated at the vulva and resembled a greatly developed clitoris. Unfortunately, one tract was not preserved until several days after the animal was killed, and the other was preserved in shaving lotion. Neither was suitable for sectioning, so I could not positively determine whether the testes—judged by their location and shape—actually were undifferentiated ovotestes. The reproductive tracts of these male pseudohermaphroditic pronghorn closely resembled those of white-tailed deer does with polished antlers, described by Marburger et al. (1966) and Kent et al. (1983).

This apparent pseudohermaphrodite was shot near Sheridan, Wyoming, by John Pryor during October 1997. The animal was considered a horned doe and was not examined by a biologist. However, it seems typical of the animals described in the text. (Top) *Note the lack of penis and scrotum along with the bucklike horns, which scored 53 Boone and Crockett points.* (Bottom) *The cheek patch on the same animal appears normal for a post-rutting season buck.* Photos by John Pryor.

Occasional reports are received of pronghorn does with 10 to 13 inch (25–33 cm) horns. Such horns often are tipped forward over the eyes, somewhat like those of castrated bucks. I suspect that many of these animals really are pseudohermaphroditic bucks. Upon seeing a vulva and an udder instead of a scrotum and penis, most hunters would not notice the male scent glands or the small subcutaneous testes.

CONCLUSIONS

Pronghorn females usually breed for the first time when 16 to 17 months old, but fawns occasionally conceive. The gestation period (250 days) is unusually long for a small ruminant, apparently because pronghorn does have the highest level of investment in their offspring of any North American ruminant. Does are seasonally polyestrus. The uterus is bicornuate, and the placenta is polycotyledonary, the same as in bovids. Two to six embryos ordinarily are conceived, but some mortality occurs through knotting and tangling of the fetal membranes. Other excess embryos are killed when displaced by the two embryos that implant closest to the body of the uterus. About 98% of births are twins. Pronghorn fawns are highly developed at birth and grow rapidly. In favorable habitat, especially in northern areas, the birthing season is short. This apparently results in predator swapping, and a smaller proportion of young are taken by predators than they otherwise would kill. About 119 males to 100 females have been reported at birth.

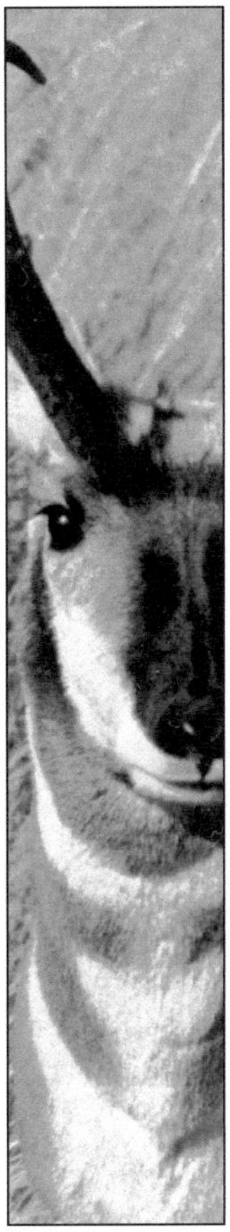

C H A P T E R T E N

Diseases and Parasites

Bart W. O'Gara

Diseases in wildlife populations have long been recognized, but their influence on populations remains mostly unknown. Other than a few obvious, devastating epidemics and occasional location of an affected individual, the impact of disease organisms on free-ranging populations has received little scientific attention. Wildlife diseases are both difficult to study and to manage. Diseases are usually viewed as occasional or incidental sources of mortality, when in fact, they may be a major influence in population regulation. Pathogens may contribute to density-dependent regulation of natural populations by magnifying effects of poor nutrition or other stresses.

Lance and Pojar (1984:1)

Many disease agents and their hosts evolved together. Of those that have associated for long periods of time, little damage ordinarily occurs to the host. Therefore, an agent and its host coevolve to a point where the host provides food and shelter for the bacterium, virus or other disease agent and usually is capable of repairing damage inflicted by that agent. An animal enjoying a high plane of nutrition can overcome or tolerate most parasites and many diseases. "Compared to the relatively feeble or less responsive immune mechanisms of domestic ungulates, the recuperative powers of cervids are a constant source of amazement" (Kistner 1982:183). The same certainly can be said of wild antilocaprids.

A distinction must be made between detection of antibodies (indication of past infection) and the carrier state. Serologic studies on wild ruminants frequently are cited to incriminate them as important reservoirs of cattle and sheep diseases, despite the fact that the disease agents have not been isolated from the healthy wild hosts in question. Such circumstantial incrimination of wildlife can be a mask for poor management. Failure to vaccinate, indiscriminate new introductions to a herd and frequent trips to the sale barn are more likely sources of diseases. Also, serological tests used to detect antibody responses of pronghorn to disease agents generally are based on methods developed for livestock, and the results are questionable. Controlled studies are needed before there is understanding and recognition of the clinical and serological responses of pronghorn to most disease agents.

Six publications (Siegmund 1967, Levine 1968, Soulsby 1968, Davis et al. 1973, Mohanty and Dutta 1981, Thorne et al. 1982) provided life histories, measurements, distri-

High mortality during the first month of life is common in many pronghorn populations. However, before the advent of miniaturized radio collars, fawn carcasses seldom were found, and those found generally were too decomposed or dehydrated for the cause of death to be determined. Controlled studies are needed to understand the clinical and serological responses of pronghorn to most disease agents. However, necropsy and collection of blood and other biological samples from hunter-killed pronghorn often are the only ways possible to gather information on the incidence of diseases and parasites in pronghorn populations. Photo by Jim D. Yoakum.

Collections of pronghorn to study the general condition of the animals, blood titers to diseases and parasite loads are less common today than they were during the mid-20th century. This is partly because such collections are less acceptable to the public and partly because sample sizes often were small and results were less than satisfactory. Photo by Lu O'Gara.

butions and general information concerning disease agents. Because at least one of the six was consulted on every disease and parasite listed, they are acknowledged here instead of in every paragraph. Thorne et al. was especially useful.

The various diseases and parasites and their effects on pronghorn are discussed unevenly because volumes have been written on a few of the most important pathogens, but other pathogens are barely known. Consequently, in certain cases, little more can be reported than the fact that a certain pathogen has been found in a pronghorn. This chapter brings together most of what has been published concerning pronghorn diseases and parasites. Thus, anecdotal reports, unsupported by laboratory tests, are included because they may spur investigation of diseases that might affect the health of pronghorn.

VIRUSES

Viruses grow only within the living cells of their hosts and cause a variety of acute or chronic disease patterns. Infected animals usually serve as reservoirs or sources of infection for other animals. Arthropods may serve either as mechanical carriers or biological vectors for many viruses. Mechanical transport occurs when virus particles adhere to the mouth parts of an arthropod feeding on an infected animal and are transferred to another host at a subsequent feeding. Biologic vector status involves reproduction of the virus in the arthropod's body, with large numbers of virus particles introduced into a host at subsequent feedings. Evolution of viral reproduction in an arthropod with accumulation in salivary glands provides an efficient mode of virus distribution. Although mechanical transfer may produce some disease outbreaks, the viruses transmitted by biologic vectors are most important in devastating epizootics. Viruses commonly transmitted by arthropods frequently are referred to as arboviruses.

Adenoviruses

Adenoviruses were so named because they originally were isolated from human adenoids. These viruses are widespread in nature, and the 77 viruses generally accepted as members of the group are distinguished by specific antigen–antibody reactions. Bovine adenovirus 3 is considered one of the most important respiratory tract pathogens of cattle in Europe. It may cause severe respiratory infections in newborn calves. Bovine adenovirus 7 was isolated from a cow in Japan, with diarrhea and respiratory distress, and from a calf in the United States, suffering "weak calf syndrome." But the role of the virus was not clear in either case (Mohanty and Dutta 1981).

Eleven (34%) of 32 adult pronghorn and 4 (29%) of 14 fawns tested in Idaho had antibodies to bovine adenovirus 3. Also in Idaho, 42 (41%) of 103 adult pronghorn and 20 (42%) of 48 fawns had antibodies to bovine adenovirus 7 (Stauber et al. 1980). Stauber et al. postulated that, although serum from pronghorn neutralized bovine adenovirus 3, the low titers in adults and fawns suggest nonspecific virus inhibition or cross-reaction with a related pronghorn adenovirus rather than specific bovine adenovirus 3 antibody. The prevalence of adenovirus 7 antibody in pronghorn was lower than in cattle.

Bluetongue and Epizootic Hemorrhagic Disease

Bluetongue (BT) and epizootic hemorrhagic disease (EHD) are caused by viruses that are so closely related that they cross-react on some blood tests. Both are infectious, arthropod-borne viruses capable of infecting domestic and wild ruminants, and both produce acute and frequently fatal hemorrhagic disease in many ruminant species. Both viruses may cause disease in some species, such as pronghorn and white-tailed deer, whereas other species appear susceptible only to BT. Because both viruses might infect some species and the infections are difficult to distinguish, the diseases caused frequently are referred to collectively as hemorrhagic disease (Thomas 1981).

Cattle are the primary reservoir of BT virus and, probably, EHD virus, with infected cattle serving as reservoirs for several years. Hemorrhagic disease can occur in cattle, but it is infrequent. Elk and other hosts that survive infection may serve as short-term reservoirs.

Seven strains of BT virus and two strains of EHD virus had been isolated in the United States by 1985. Little or no cross-immunity is known between strains of BT, so animals that recover from one strain still are susceptible to the remainder. Conceivably, outbreaks of BT could occur in consecutive years if reservoir animals for another strain were introduced into an area. However, with wild susceptible ruminants, a population usually is reduced below the level of efficient transmission during the year of an epizootic, so consecutive year mass die-offs seldom are seen (T. Kistner personal communication:1985).

Bluetongue generally occurs during late summer and early autumn, and all sex and age classes may be affected. The most important vectors in domestic and, presumably, wild animals are no-see-um gnats of the genus *Culicoides*. Die-offs can be expected to terminate shortly after subfreezing temperatures reduce the gnat population.

Bluetongue primarily is a disease of domestic sheep, with white-tailed deer, pronghorn, bighorn sheep and mule deer also being susceptible to clinical disease. Trainer and Jochim (1969) found evidence through a serologic survey in 1963 that 8 (8%) of 96 pronghorn from Wyoming and 34 (35%) of 97 from Colorado had experienced infections by the virus.

Two of four experimentally infected pronghorn died 7 and 8 days after inoculation with BT virus. They lost their appetite, became lethargic, lost coordination and had difficulty breathing. The primary lesions were those of pneumonia and hemorrhage. The two that did not die had BT antibodies before subcutaneous inoculation; both were killed at the end of the experiment, and lesions indicative of BT were not observed (Hoff and Trainer 1972). Thus, the presence of antibody titers may imply immunity. Bluetongue or EHD were diagnosed as the cause of death of two captive pronghorn in Oregon (Kistner et al. 1975).

At least 3,200 pronghorn died during a BT outbreak in eastern Wyoming during late September and early October of 1976. During August and September 1984, another outbreak occurred in northeastern Wyoming resulting in 300 known pronghorn deaths. The BT virus, serotype 17, was isolated from pronghorn in both outbreaks. Seventy-six (81%) of 94 pronghorn killed by hunters during the latter portion of the 1976 die-off and 14 (58%) of killed 1 year after the die-off had serologic evidence of exposure to BT. Mule deer also died of what was presumed to be BT during both outbreaks. The outbreak apparently followed the Cheyenne River drainage. Early pronghorn mortality was noted along drainages or near stock ponds, but subsequent aerial transects revealed dead pronghorn scattered throughout the area, with no apparent association between them and water courses. Mortality was greatest during the last 2 weeks of September and subsided with the advent of cooler weather in early October. During 1984, carcasses of 288 pronghorn, 83 mule deer and 13 white-tailed deer were found in the area of the epizootic. The total number lost was unknown, but an estimated 2% to 3% of the pronghorn population in the area died. In one pronghorn hunting area, 4% to 5% mortality was estimated. Fawn/doe ratios were obtained during August of 1977 and 1978 in the BT area and a similar area that was not affected by BT. In 1977, 1 year after the die-off, the fawn/doe ratio in the BT area was 47:100; outside the BT area, it was 101:100. Two years after the die-off, fawn/doe ratios were 93:100 in the BT area and 86:100 outside the BT area (Thorne et al. 1988).

Observable signs in free-ranging pronghorn with BT persisted from 1 to 6 days. Some animals, which had shown no previous signs of illness, died suddenly when disturbed. Others seemed to cease feeding for several days, during which time they traveled less and finally remained in one area. Many reclined and were reluctant to move or run when approached. Tracks and observations indicated that some died suddenly while walking or running, and others struggled in lateral recumbency for at least several hours before dying. All sex and age classes were affected. Hemorrhages or edema were present in and around the heart, large associated vessels and trachea. Hemorrhages also were frequent in various portions of the digestive tract, and the lungs usually were edematous. Large amounts of clear yellow fluid, which clotted on exposure to air, apparently resulted from viral damage to the endothelial lining of the vascular systems and accumulated in the thoracic cavities (Thorne et al. 1988).

Bluetongue seems the most serious disease of pronghorn. The extensive die-off during the summer of 1873, described in chapter 12 (Allen 1874), could have been BT. As stated, cattle are reservoirs for the disease because they often do not develop symptoms, but are chronic carriers. The virus is widespread throughout the United States. Soft muddy margins of ponds or slow-moving streams contaminated with cattle feces produce large numbers of no-see-ums. Thus, the probability for a pronghorn die-off is high when they congregate with carrier cattle during late summer to seek succulent vegetation and water at those ponds and streams (Thorne et al. 1982).

A pronghorn doe with bluetongue. The animal was weak, inactive and salivating copiously—common early symptoms of the disease. Bluetongue probably is the most serious disease of pronghorn. At least 3,200 died during an outbreak in eastern Wyoming during 1976. Domestic cattle are reservoirs of the virus, and bloodsucking insects spread it to pronghorn and deer, especially during droughts when the animals gather at stagnant water holes. Photo by E. T. Thorne; courtesy of the Wyoming Department of Game and Fish.

Epizootic hemorrhagic disease is difficult to distinguish from BT, but is especially virulent in white-tailed deer. It has been reported in several western states and occasionally associated with reports of pronghorn mortality. Richards (1964) provided one of the first positive records of this disease affecting pronghorn in North Dakota during 1962 and 1963. Barrett and Chalmers (1975) and Stauber et al. (1980) conducted antibody tests for EHD on the sera of 33 pronghorn from Saskatchewan and 146 from Idaho, but no reactors were found. In Yellowstone National Park, where pronghorn associate with numerous other wild ruminants, 17 tests for BT during 1987 to 1991 proved negative (D. Scott: personal files).

In Nebraska, 92 and 103 (27% and 30%, respectively) of 339 pronghorn tested positive to BT and EHD viruses, but attempts to isolate the viruses were futile. Both diseases seemed to be enzootic in the population because of the positive correlation between higher prevalences of antibodies and increasing ages of the animals. Both viruses showed the greatest prevalence of antibodies in the 3- to 6-year-old animals, indicating cycling of the viruses among the population in an enzootic situation. Johnson et al. (1986) and Trainer and Jochim (1969) noted that the prevalence of these viruses correlated with that of cattle in the area. In Colorado, 67 (52%) of 129 pronghorn tested positive for exposure to hemorrhagic disease (Firchow 1986).

Between September 20 to October 2, 1996, 288 blood samples were collected statewide in Arizona from hunter-killed pronghorn. Of these, 226 (78.5%) tested positive for exposure to BT at the Arizona Veterinary Laboratory (Heffelfinger and Olding 1997, Heffelfinger et al. 1999). This indicates that Arizona pronghorn can be exposed to BT without suffering high rates of mortality. Perhaps some strains are more virulent than others. Yearlings had lower rates of exposure than did adults. A subsample of 186 positive samples were sent to the National Veterinary Services Laboratory in Ames, Iowa, for serotyping by serum neutralization. Samples were tested for exposure to serotypes 2, 10, 11, 13 and 17. Twenty-six (13.9%) were found negative when analyzed by serum neutralization, and 21 to 23 were unsuitable for testing among various serotypes. Titer levels ranged from 1/20 to more than 1/160. Levels greater than 1/10 are considered indicative of past exposure. Higher titers indicate more recent exposure. Serotype 11 was most prevalent (107/139 = 77%), followed by serotype 17 (62/127 = 48.8%), serotype 13 (30/135 = 22.2%), serotype 10 (28/137 = 20.4%) and serotype 2 (23/126 = 16.9%).

Analysis by geographic area revealed no definite pattern of distribution, and exposure to BT appeared widespread throughout Arizona. The effects on pronghorn were unknown. Heffelfinger and Olding (1996) speculated that BT had the potential to influence reproduction because pronghorn breed during September when BT infections often occur. Thus, a viremic doe might not go into reproductive condition, or her behavior might be altered enough to disrupt breeding.

The high rate of exposure of Arizona pronghorn was unexpected. The usual vector—no-see-um gnats—thrives in moist areas with standing water, and no known, recent epizootics had been reported (Heffelfinger and Olding 1997). To what extent an alternative vector might be involved is unknown.

In Oregon, 7 (35%) and 6 (30%) of 20 adult pronghorn does tested positive to BT and EHD viruses, respectively (Dunbar and Velarde 1998, Dunbar et al. 1999) in an area without cattle. In Montana, outbreaks of hemorrhagic disease are common in white-tailed deer from areas where only about 5% of the cattle test positive for exposure to the viruses (R. Levesque personal communication:1997). The possibility of carriers other than cattle seems probable.

It should be remembered that documentation of reactors alone does not provide evidence of pathogenicity or transmission of pathogens between wildlife and livestock (Stauber et al. 1980). Kreeger (1999:18) summed up the situation well: "Although losses to hemorrhagic disease can be significant, animal populations generally recover quickly, and there is little wildlife management agencies can do to prevent their occurrence anyway."

Bovine Respiratory Syncytial Virus

Bovine respiratory syncytial virus (BRSV) also is known as RS virus or Nom's virus. Isolations of BRSV were reported from cattle with acute respiratory tract disease in Switzerland, Belgium and Japan in 1970. The Japanese outbreak affected 40,000 or more cattle and caused about 100 fatalities. Fatal cases had emphysema and lung edema. Severity of the disease in Japan probably indicates a new infection in a nonimmune population (Woods 1974). The pathogenic mechanisms at work in BRSV virus-caused diseases are not well understood in any species. The differences between human and bovine strains appear to be minor. Apart from their potential as a primary infecting agent, BRSV viruses may be important in combination infections with other viral or bacterial agents. When disease is produced, it usually is on a herd basis. The explosive nature of the disease and rapid production of antibodies suggest that the virus is carried as a persistent infection for some time before the acute disease develops (Frey 1983).

The BRSV titers in the blood of pronghorn shot in Nebraska varied from very high to low. The prevalence of BRSV antibodies in pronghorn (66 of 121 tested: 55%) was similar to that of Nebraska's cattle population. It raises questions concerning species susceptibility and transmission between cattle and pronghorn. Johnson et al. (1986) indicated that the positive findings, although not previously reported, were not a surprise because pronghorn had tested positive for other bovine respiratory viruses. Johnson et al. reported that an endemic condition apparently exists in the pronghorn population because 60% of the fawns and yearlings and 51% of the 2- to 7-year olds tested positive. On a refuge in Oregon, where pronghorn apparently had not had contact with cattle during the preceding 5 years, 18 adult does tested negative for BRSV (Dunbar and Velarde 1998, Dunbar et al. 1999).

To maintain antibodies to BRSV in adult animals, a continuous circulation of the virus in partly immune adults or reinfection of immune-deficient pronghorn, followed by multiplication and shedding of the virus, needs to exist. Bovine calves apparently become serologically positive through ingestion of infected colostrum, and titers could be increased by reinfection. This also could occur in pronghorn (Johnson et al. 1986).

Bovine Virus Diarrhea

Bovine virus diarrhea (BVD) has caused disease in all regions of the United States and in most other countries where cattle are raised. Clinical similarities to infectious bovine rhinotracheitis, foot and mouth disease and other maladies make field recognition of the disease difficult.

Serologic tests for BVD in Saskatchewan revealed 3 (4%) reactors of 84 free-ranging pronghorn (Barrett and Chalmers 1975). Tests in Idaho resulted in 2 (2%) reactors of 102 adults, but no reactors among 41 fawns (Stauber et al. 1980). Antibodies to BVD were detected in only 2 (1.4%) of 143 pronghorn tested, although cattle in the same area tested higher than 50%. In Alberta, 88 of 201 pronghorn (44%) had antibodies to BVD; about 95% of the cattle in the study area had antibodies to the same virus, and cattle could have served as a source of infection to the pronghorn (Kingscote and Bohac 1986). Nine (75%) of 12 pronghorn killed by cars during 1987 to 1991 in Yellowstone National Park tested positive to BVD (D. Scott: personal files). Two (11%) of 18 pronghorn in Arizona tested positive to BVD (Heffelfinger and Olding 1996). On a refuge in Oregon without cattle, 18 adult pronghorn does tested negative for exposure to BVD (Dunbar and Velarde 1998,

Dunbar et al. 1999). Stauber et al. (1980) speculated that BVD virus either does not replicate readily in pronghorn and exposure would not lead to an immunologic response or is highly fatal in pronghorn and very few animals survive infection. They hastened to add that proof of these possibilities awaits experimental documentation.

Contagious Ecthyma

Contagious ecthyma is a common disease of domestic sheep and goats that often infects wild Rocky Mountain bighorn sheep and goats. The lesions caused by the virus generally are only skin deep, but secondary bacterial infections sometimes occur. Sheep ranchers in the United States know this disease as sore mouth or scab mouth. The virus generally is spread by direct contact through breaks in the skin or mucous membranes. Rough, dry vegetation and salt blocks have been blamed for spreading the disease. Affected animals appear restless, lick their noses and nostrils, and rub infected parts on nearby animals or objects to relieve the irritation. The disease generally is recognized only after scabs form on the lesions, which are most common on the lips, nostrils and teats. The virus can remain viable for 10 years in scabs shed from recovered animals. Hence, once an area is contaminated, the infection frequently persists indefinitely. Reservoir animals also appear important. Pronghorn have been experimentally infected with virus isolated from Rocky Mountain bighorn sheep, but pronghorn suffered less severe infections than did sheep, indicating that contagious ecthyma would not seriously affect free-ranging animals (Lance et al. 1983).

California Encephalitis

The California encephalitis (CE) virus belongs to the genus *Bunyavirus* and has been isolated from mosquitoes and humans (Mohanty and Dutta 1981). It was first isolated from mosquitoes in California during 1952. Encephalitis, sometimes fatal, developed in rodents after infection, but a calf and monkey exhibited no signs of disease. The virus developed in squirrels and rabbits to levels that could infect mosquito vectors. A case of nonfatal encephalitis in a human apparently was caused by this virus (Bruner and Gillespie 1973). Bruner and Gillespie indicated that CE finds a natural reservoir in wild and perhaps domestic animals—the infection being propagated by mosquitoes.

The virus has not been isolated from pronghorn, but 12 (16%) of 76 pronghorn that were tested in Wyoming had antibodies to CE (Trainer and Hanson 1969), and CE reactors were found from almost every animal species and geographic site examined. The significance of CE in wild ruminants is unknown.

St. Louis Encephalitis

St. Louis encephalitis (SLE) is a warm-weather disease that occurs in central and western United States. A number of mosquito species are known to be vectors, and they apparently can convey the disease to humans. Mortality may be high, especially in people more than 45 years of age. Chickens are easily infected, but show no signs of disease. Chicken mites infected with the virus might be the means by which it is maintained from one season to the next. Whenever this disease appears in humans, antibodies to it may be found in horses, which do not suffer any recognizable symptoms as a result or play a significant role in propagation of human outbreaks (Bruner and Gillespie 1973). One (1.3%) of 76 pronghorn tested in Wyoming had antibodies to the SLE virus (Trainer and Hanson 1969).

Eastern and Western Encephalomyelitis

Eastern encephalomyelitis (EE) and western encephalomyelitis (WE) viruses are most frequent in areas with large mosquito and bird populations because they are maintained in nature by an arthropod–bird reservoir cycle from which infection is transmitted to mammalian hosts by biting insects, principally mosquitoes. Both domestic and wild birds develop severe enough infections to infect mosquitoes. The viruses multiply in the mosquitoes' bodies and persist in their salivary glands—ready for injection. Seventeen (16%) reactors were found to EE antigens and 9 (8%) to WE among 110 pronghorn in Canada (Barrett and Chalmers 1975). No sex difference was noted, but more adults than fawns were reactors. None of 76 Wyoming pronghorn reacted to EE, but 1 (1.3%) of 76 reacted to WE (Trainer and Hanson 1969).

Infectious Bovine Rhinotracheitis

Infectious bovine rhinotracheitis (IBR) is a disease, sometimes found in range cattle, that usually affects the genital organs and upper respiratory tract. It has been found in white-tailed and mule deer, and pronghorn carry antibodies to the virus. But, it seems to be of little importance to populations of these wild hosts. Venereal transmission is the usual method by which the disease is spread. However, in feedlots or other concentrations of cattle, the virus, in passing rapidly through many animals through the respiratory route, acquires increased virulence.

Of 33 pronghorn tested in 1970, 29 (88%) were IBR reactors; all were members of a single herd moving toward their winter range in Saskatchewan. The highest percentage of samples with IBR antibodies and the highest titers were in mature pronghorn (Barrett and Chalmers 1975). An additional 77 pronghorn from Alberta, tested

In Yellowstone National Park during 1989 to 1991, 3 (18%) of 17 pronghorn tested positive to infectious bovine rhinotracheitis (D. Scott: personal files). Photo courtesy of the U.S. National Park Service, Yellowstone National Park.

by Barrett and Chalmers (1975) during 1971 and 1972, yielded no reactors. Serum samples from 75 pronghorn shot in Texas were sent to commercial laboratories for IBR analyses. Fifteen duplicate samples were sent to different laboratories from a ranch where the loss of 18 head of cattle to IBR had been reported. Five samples had low to medium titers for IBR, according to one laboratory. However, the duplicates to these five samples were reported negative by the other laboratory (Hailey 1979). In Idaho, 27 (27%) of 101 adult pronghorn and 9 (22%) of 42 fawns showed antibodies to IBR antigens (Stauber et al. 1980). Although the IBR virus occurs in cattle throughout Alberta, 201 pronghorn tested there by Kingscote and Bohac (1986) did not have antibodies to the virus. Kingscote and Bohac concluded that the pronghorn had not been exposed or exposure had not evoked a measurable response. On a refuge in Oregon without cattle, 18 adult does tested negative for exposure to IBR (Dunbar and Velarde 1998, Dunbar et al. 1999).

Parainfluenza

Parainfluenza (PI) virus antibodies, indicating previous infections, are found in most adult humans. Parainfluenza viruses cause illness in young children, ranging from mild upper-respiratory infections to croup or pneumonia. The type 3 virus (PI_3) is common, and almost all children have had it by 6 years of age. It also is highly infectious for cattle and usually nonpathogenic. But for cattle under severe stress, it sometimes predisposes or works in conjunction with bacteria—especially *Pasturella haemolytica*—to cause an acute febrile upper-respiratory disease in animals herded and transported together, hence the phrase "shipping fever."

The PI_3 infections were indicated by serological tests from pronghorn in Alberta, Saskatchewan and Idaho. All 33 free-ranging pronghorn tested in Saskatchewan during 1970 were positive, but all 51 tested in Alberta during 1972 were negative (Barrett and Chalmers 1975). Also in Alberta, 102 (51%) of 201 pronghorn tested had antibodies to PI_3 (Kingscote and Bohac 1986). In Idaho, 79 (76%) of 104 adults and 22 (52%) of 42 fawns tested were positive reactors (Stauber et al. 1980). These studies indicated considerable variation in the prevalence of antibody titers among pronghorn from different areas or even from year to year in the same area. Thorsen et al. (1977) examined nasal secretions, leukocytes and preputial or vaginal swabs from 6 captive and 50 free-ranging pronghorn for viral agents. None of the study animals showed clinical signs of disease, but the nasal swabs from the free-ranging pronghorn yielded four isolations of PI_3 virus. Seven (44%) of 16 pronghorn in Yellowstone National Park tested positive to PI_3 (D. Scott: personal files). Ten (56%) of 18 serum samples from Arizona pronghorn indicated exposure to PI_3; titers ranged from very low to very high (Heffelfinger and Olding 1996). In Oregon, 12 (67%) of 18 adult does tested positive for exposure to PI_3, with low to high titers (Dunbar and Velarde 1998, Dunbar et al. 1999).

Potential sources of PI_3 virus for pronghorn include other big game and livestock with which pronghorn are sympatric. The PI_3 virus is infective for pronghorn, but its virulence is unknown.

Rabies

Rabies affects all warm-blooded animals and usually is fatal. The disease frequently results in one of two syndromes, which have been termed furious and dumb. During the furious stage, the animal is restless, may bite at inanimate objects and attack anything that moves. During the dumb stage, the animal may exhibit incoordination, tremors and convulsions, paralysis and death. Carnivores are apt to have furious rabies, and ungulates more often display the dumb rabies. Carnivores and bats are the most common reservoirs and disseminators of the disease in the United States. Striped skunks probably are responsible for most of the cases of rabies in domestic and wild ruminants in western provinces and states, such as the four cases in bison from 1960 to 2000 reported by Stoltenow et al. (2000). A tame 10-week-old pronghorn was bitten by a striped skunk—an incident observed by the owner (Wempe 1976). The fawn became lethargic 33 days after the bite and died 2 days later. Rabies was confirmed by the fluorescent antibody test, inoculation into mice and histopathological studies.

Vesicular Stomatitis

Vesicular stomatitis (VS) is a viral disease found in a wide range of hosts—domestic and wild—including humans. It is characterized by raised vesicles or blisters inside the mouth and sometimes on the skin of the feet and teats. It usually causes a mild influenzalike disease in humans, although it can be severe and require an extended convalescence. Vesicular stomatitis is caused by one of two serovars (strains or serotypes)—New Jersey and Indiana—of VS virus. The disease is endemic in parts of South America, Central America and southern North America, with periodic epizootics involving Appalachia, the Upper Mississippi Valley and the Rocky Mountain region. Mechanisms of transmission are poorly understood. Direct contact between susceptible and infected animals may be important. Epizootics occur during warm weather, when biting insects are abundant, and usually cease after killing frosts. This suggests an arthropod transmission, and VS virus, New Jersey serovar, has been isolated from *Simulium* black flies and *Hippelates pusio* eye gnats. The VS virus, Indiana serovar, has been found in *Phlebotomus* midges and mosquitoes (Thorne et al. 1983b). Two (2.7%) of 73 pronghorn tested in Wyoming for serologic evidence of arbovirus infections were positive for VS (Trainer and Hanson 1969).

One hundred thirty-nine suitable sera were obtained from hunter-killed pronghorn in Colorado; 83 (60%) were positive to VS, New Jersey serovar. Of these, 20 (14%) also were positive to VS, Indiana. Only two samples (1.4%) were positive to VS, Indiana serovar alone (Thorne et al. 1983b).

An experimentally inoculated adult doe pronghorn developed clinical signs of VS. Signs observed in the doe began on the second day after inoculation with licking of the lips and nasal discharge. By the third day postinoculation, when the doe was examined under anesthesia, mucosa of the dorsum of the tongue was blanched, edematous and sloughing with erosions along the sides of the tongue. On the fourth through the tenth days, the disease was characterized by licking of the lips, hypersalivation, anorexia, fever of 104.5°F (40.3°C) and lethargy. The doe began coughing, although she was beginning to take food and her general condition was improving. At 12 days after inoculation, she died of apparent aspiration pneumonia. The VS, New Jersey serovar, was isolated from a specimen of tongue collected 3 days postinoculation, but no virus was isolated from blood collected 3 days postinoculation or from cheek, tongue, lung or fetal tissues collected at death (Thorne et al. 1983b).

An experimentally inoculated yearling doe had no lesions when examined 72 hours after inoculation. On the fourth day, she was slightly lethargic and had a fever of 104.9°F (40.5°C), but showed no other clinical signs of VS. On day 32 postinoculation, she was euthanized because of chronic diarrhea unrelated to VS. Tongue mucosa obtained from the yearling 3 days after inoculation was negative for VS.

Although the study involved only two animals, it showed that pronghorn are susceptible to VS and that some infected animals may be seriously affected, although the infection may be mild and unapparent in others. The clinical signs and gross and microscopic lesions in the adult doe were similar to those described in domestic species and deer. However, death associated with VS is considered unusual in those species (Thorne et al. 1983b).

BACTERIA

Bacteria occur in almost every habitat capable of supporting life. Some types require high carbon dioxide levels, others need abundant oxygen and some forms (anaerobes) grow in the absence of oxygen. Many bacteria are free-living, but others inhabit animals. The presence of most of these nonpathogenic organisms on or in an otherwise healthy host does not cause disease. Pathogenic bacteria typically cause disease in an otherwise healthy individual. Many opportunistic bacteria cause disease only under certain circumstances, such as debilitation of the host. All wild animals are susceptible to infection by bacteria, and many bacterial species cause abscesses, especially in puncture wounds.

Any portion of a carcass containing an abscess should not be used as food. However, if the host is in good condition, the remaining portions are considered edible (Howe 1965).

Actinobacillosis

Actinobacillosis is a chronic infectious disease chiefly affecting ungulates and is caused by the bacterium *Actinobacillus lignieresi.* It mainly affects the soft parts of the heads of ruminants, causing abscesses and fistulas or draining tracts with thick connective tissue walls. It occasionally affects bones. The phrase "wooden tongue" is used to denote infections caused by *A. lignieresi.* Mechanical injury to the oral cavity sometimes results from ingesting rough forage, and awns of certain grasses, particularly foxtail, bearded wheat or barley, may penetrate the soft tissues of the mouth and allow bacteria to enter.

A pronghorn doe that I shot in eastern Montana had a small abscess in her cheek at the level of the first molar. A culture of the pus yielded *Actinobacillus* sp.; a fragment of vegetation, apparently a barley beard, was in the pus. *Actinobacillus* sp. was reported in South Dakota pronghorn by Bever (1950). *Actinobacillus* sp. also was cultured from abscesses of two Wyoming pronghorn killed by hunters; one had numerous splenic abscesses (Post 1960), and the other had necrotic stomatitis and pneumonia (Thorne et al. 1982). In Oregon during 1997, an *Actinobacillus* spp. and *Pasturella* spp. mixture was cultured from the tonsillar sinuses of 5 (42%) of 12 fawns tested (Dunbar and Velarde 1998). All of those fawns died of coyote predation or unknown causes by 16 days of age.

Actinomycosis

Actinomycosis is a chronic infectious disease chiefly affecting ungulates and is caused by the bacterium *Actinomyces bovis.* It generally attacks the bones in the head, causing rarefying osteitis. The typical bony tissue destruction sometimes involves the upper jaw or bones of the nasal cavity, but most commonly occurs in the lower jaw. The characteristic lump on the ventral aspect of the mandible results from bone destroyed by liquefaction necrosis being replaced by irregular excessive bone. The phrase "lumpy jaw" most often is used to denote infections caused by this species of bacteria, but infection by other bacteria, such as *A. ligneresi, Corynebacterium* spp. and *Fusobacterium necrophorum,* sometimes causes similar-appearing lesions.

One of the first records of lumpy jaw in pronghorn was provided by Skinner (1922), who stated that Yellowstone National Park elk, deer and pronghorn died from this disease, but it was particularly virulent in pronghorn. *Actinomyces* sp. was reported in South Dakota pronghorn by Bever (1950), and *A. bovis* infected four pronghorn at the Sybille Wildlife Research Unit in Wyoming (Thorne 1969).

Twelve percent of the jawbones of 238 pronghorn in Oregon had rarefying osteitis (Hansen 1955, Yoakum 1957). A high incidence of the disease was noted on rangelands that had been invaded by cheatgrass (J. Yoakum personal communication:1989).

Five percent of 140 adult pronghorn I collected in Yellowstone National Park and on the National Bison Range had jaws that showed mild cases of rarefying osteitis. These were mostly prime animals from 1.5 to 7 years

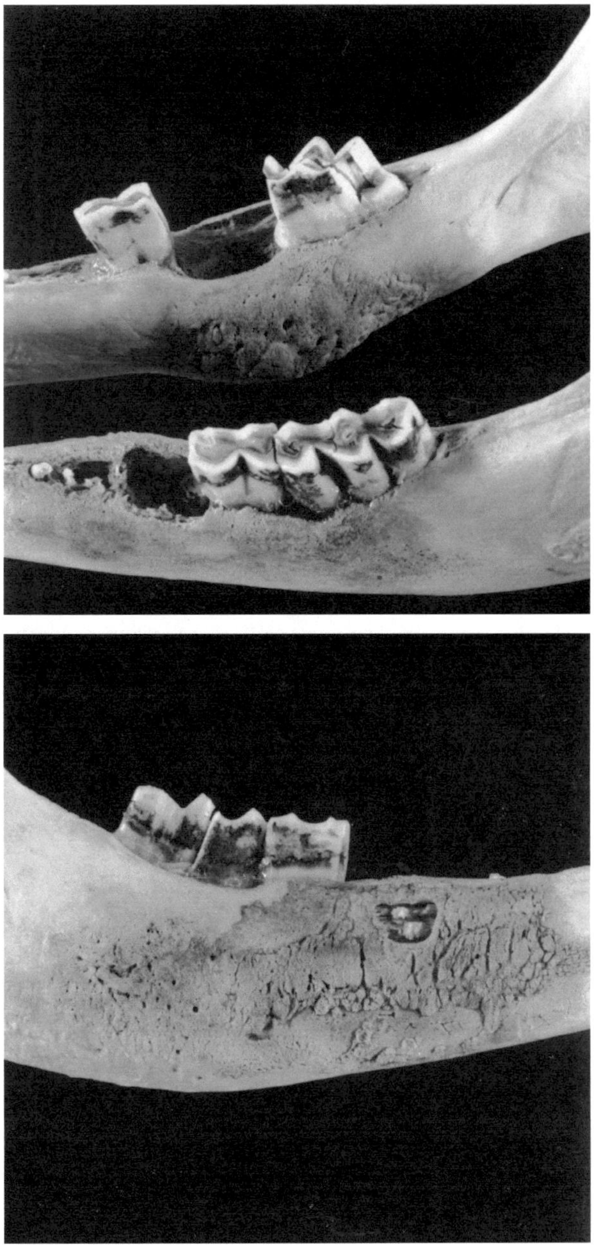

Two views of a pronghorn jaw with the typical bony tissue destruction and loss of teeth caused by Actinomyces bovis. *Actinomycosis kills or debilitates a few, especially old, pronghorn in most populations. However, it only becomes important in populations if range conditions are so deteriorated that pronghorn are forced to feed on rough vegetation that abrades gum tissue, allowing pathogens to enter.* Photos by Bart W. O'Gara.

"Lumpy jaw" can be caused by several species of bacteria. A lump on the bottom or side of the lower jaw is the first external evidence of the disease. In advanced cases, hair covering abscesses may slough and pus may drain from fistulas. Photo by Jim D. Yoakum.

of age. However, 13 jaws (48%) from 27 pronghorn that died during a severe winter on the National Bison Range showed advanced deterioration of bony tissue and loss of teeth. The 27 animals were 7 to 10 years old and had died of malnutrition complicated by bad teeth or old age, or they were killed by coyotes while in a weakened condition.

Actinobacillosis and actinomycosis apparently kill or debilitate a few, especially old, pronghorn in most populations. They only become important in a population if range conditions are so deteriorated that pronghorn are forced to feed on rough vegetation that abrades gum tissue and allows the infective agents to enter.

Alcaligenes spp.

Alcaligenes is a common, saprophytic (lives on decaying organic matter) genus of bacteria that inhabits the intestinal tracts of many animals. It is not considered pathogenic, and its presence in diseased tissue probably represents extensive postmortem contamination (Thorne et al. 1982). *Alcaligenes faecalis* was recovered from the eye of a highly parasitized pronghorn in Wyoming with multiple infections (Thorne 1968a).

Anaplasmosis

Most North American wild ruminants are susceptible to anaplasmosis, or gall sickness. Although considered noncontagious, it is of concern to western stockmen. The pathogen, *Anaplasma,* infects and destroys red blood cells, might cause anemia and presently is classified under the order Rickettsiales. The organism appears as round, marginal bodies in erythrocytes.

The causative agent of bovine anaplasmosis, *A. marginale*, has received the greatest attention. Mortality can be high among susceptible cattle, especially in nonenzootic areas. In enzootic areas, calves often become infected during their first year while they are protected from disease by maternal antibodies and have ample marrow, which produces red cells for rapid recovery from the anemia produced by the infection. Immunity is induced and maintained by constant cycling of the organism between the cattle and arthropods. Older cattle do not possess a large reservoir of active bone marrow. Hence, they frequently are affected by acute anemia resulting from many causes, including anaplasmosis. Unable to produce erythrocytes rapidly, they can die or be severely debilitated (T. Kistner personal communication: 1985).

Anaplasma marginale is spread by bloodsucking arthropods and most widely distributed throughout the tropical and subtropical regions of the world. Although mortality is high among adult cattle, it is not high among wild ungulates. Many reports have been published concerning the presence of *A. marginale* in New World deer and Old World wild bovids, but only a few have indicated that pronghorn were susceptible. When pronghorn were experimentally injected with blood from infected cattle, typical marginal bodies developed in their blood cells, but the animals did not show clinical signs of anaplasmosis (Howe et al. 1964).

Three (2%) reactors to *A. marginale* antigens were reported among 146 pronghorn tested in Idaho (Stauber et al. 1980). However, Howe et al. (1964) and Howe and Hepworth (1965) found that the test used (complement fixation [CF] test) gave false-positive and suspicious reactions when applied to serum samples of known *Anaplasma*-free pronghorn. Positive reactions occurred in 9 of 10 tests, and Howe et al. concluded that CF tests had little value for diagnosing anaplasmosis in pronghorn. The sera of 26 Montana pronghorn were examined for anaplasmosis by the CF test, resulting in 19 (73%) false-positive reactions. Inoculation of reactors' blood into calves did not produce evidence of anaplasmosis, and marginal bodies were not found in stained blood smears of the pronghorn or the injected calves (Jacobson et al. 1977). Seventeen pronghorn from Yellowstone National Park tested negative to anaplasmosis during 1987 to 1991 (D. Scott: personal files).

Although *A. marginale* can be experimentally transmitted between cattle and pronghorn, pronghorn do not seem to be important carriers. Pronghorn are not seriously affected and have been shown to remain carriers for only about 66 days (Howe et al. 1964).

Another, but less pathogenic, species is *A. ovis*. Infections have been reported from several western states, and the discovery of anaplasmosis in range sheep stimulated speculations concerning wildlife reservoirs of *A. ovis* in the western United States (Zaugg 1987). Zaugg inoculated three pronghorn fawns about 10 weeks of age with fresh, whole blood from three domestic ewes known to be *A. ovis* carriers. Whole blood was later collected from the pronghorn and inoculated into five splenectomized ewes. Clinical illness was not observed in the pronghorn, but two of the three exhibited signs of *A. ovis* infection. Four of the five splenectomized sheep inoculated with pronghorn blood developed clinical anaplasmosis within 43 days of inoculation, and one died of acute anaplasmosis 55 days after exposure. Thus, a sheep–pronghorn–sheep sequence was established. Pronghorn took longer than sheep to develop the disease, and the lack of clinical signs suggested a degree of resistance or adaptability to a familiar infection. Zaugg concluded that further work is warranted to determine (1) if infected pronghorn remain long-term carriers of *A. ovis*, (2) the reservoir status of pronghorn and (3) implications of cross-transmission of *A. ovis* between pronghorn and sheep by natural vectors.

Anthrax

Anthrax is a highly fatal disease of ruminants that can affect humans and other animals. It is caused by *Bacillus anthracis*. It may have reached the New World with European colonists and now occurs throughout the United States. *B. anthracis* produces spores in carcasses of animals that die of anthrax. These spores may infect other animals or survive in soil for long periods and again become infective in incubator areas that generally involve drying alkaline ponds or water courses with decaying vegetation and high temperatures.

Infected animals may experience bleeding or expulsion of clotted blood from the anus. Carcasses distend with gas and putrefy rapidly, and blood drains from all body orifices. Complete necropsy is not advised because of the human health hazard and because development and release of spores are enhanced. Pronghorn have not been positively implicated in die-offs from anthrax. However, a dead doe in excellent flesh was found where deer and cattle had recently died of anthrax in Texas, and Lehman and Davis (1942) assumed from the circumstances that the pronghorn had died from the same disease.

Brucellosis

Brucellosis, a contagious disease affecting many species of mammals and commonly called undulant fever when transmitted to humans, was first detected in North American wild ruminants when it was found in bison during 1917. It is caused by members of the genus *Brucella*. The signs in livestock are rather inconsistent, but abortion is the most common evidence of infection. Because brucellosis is a serious disease for western livestock, speculation exists that pronghorn might be carriers. Pronghorn were tested in the following numbers and states and provinces: 50 in Oregon during the 1950s (J. Yoakum personal communication:1980), 506 and 233 in Nebraska (Bailey 1962, Johnson et al. 1986), 33 in Saskatchewan and 77 in Alberta (Barrett and Chalmers 1975), 5,400 in Colorado (Adrian and Keiss 1977, Firchow 1986), 827 in Texas (Hailey 1979), 146 in Idaho (Stauber et al. 1980), 210 in Alberta (Kingscote and Bohac 1986), and 20 in Oregon (Dunbar and Velarde 1998, Dunbar et al. 1999). One serological reactor was found in Colorado. In Yellowstone National Park, where bison and elk infected with the disease mingle with pronghorn, no reactors were found among 17 pronghorn tested for brucellosis during 1987 to 1991 (D. Scott: personal files).

Apparently, pronghorn do not pose a threat to livestock, or vice versa, as far as brucellosis is concerned. *Brucella* has not been isolated from pronghorn, but controlled studies of the clinical response of pronghorn to *Brucella* spp. have not been conducted.

Chlamydial Infections

Chlamydial infections are caused by bacteria of the genus *Chlamydia*, and only a single species, *C. psittaci*, is recognized. Diseases caused by this pathogen include trachoma and conjuctivitis in humans, ornithosis in birds and ornitosis in mammals occasionally. Because they are small and reproduce only within host cells, members of the family Chlamydiae long were thought to be viruses.

Barrett and Chalmers (1975) found 5 reactors (7%) to chlamydial group antibodies among 77 apparently healthy pronghorn from southeastern Alberta. The strain of *Chlamydia* involved and what effect, if any, it had on the pronghorn were unknown. Two (13%) of 15 sera samples from free-ranging pronghorn from Colorado also tested positive (see Lance and Pojar 1984). In neither case was an attempt made to isolate the organism.

Clostridium

Clostridium is a genus that includes the pathogenic species responsible for botulism, tetanus, gangrene, blackleg and enterotoxemia. However, because many pathogenic and nonpathogenic species are quick to grow in fresh carcasses, the significance of *Clostridium* spp. isolated after death must be evaluated with caution. Such an isolate was made from a Wyoming pronghorn fawn with a tentative diagnosis of blackleg (Howe 1966a). Although not proven, it seems safe to assume that pronghorn are susceptible to most pathogenic clostridia.

Enterotoxemia results when animals consume highly digestible forage, grain or milk. The accumulation of partly digested, low-roughage feed in small intestines of ruminants allows the rapid proliferation of *C. perfringens* type D. This organism produces toxins that cause hemorrhages throughout the gastrointestinal tract and degeneration of the kidney and liver structure. The disease is common in domestic lambs that feed on grain, hence the common name overeating disease. It also is called pulpy kidney disease because of the chocolate pudding appearance of the kidneys at necropsy.

Enterotoxemia, resulting in the death of one fawn and debilitation of several others, was reported from tame pronghorn fawns in Colorado (Nagy et al. 1969). The fawns were about 2 months old and on a diet of milk, leaves of alfalfa hay and commercial calf starter that was being used as a transition from milk to solid food. A second confirmed, but unpublished, case of *C. perfringens* type D enterotoxemia occurred in a mature pronghorn doe brought into confinement. This animal developed bloody diarrhea 24 hours before death, and necropsy showed severe hemorrhagic enteritis (see Lance and Pojar 1984). The chance of enterotoxemia in wild pronghorn seems small, but gleaning of waste grain or other high carbohydrate, low roughage foods could result in death of a few animals. Rumen overload and polioencephalomalacia also involve consumption of high carbohydrate foods, especially grain, and might involve *C. perfringens* type D.

Corynebacterium

Corynebacterium is a genus of bacteria that may cause generalized illness or localized infections. *C. pyogenes* frequently infects wild and domestic animals, and the resulting disease often is called caseous lymphadenitis. This small bacterium is a normal inhabitant of many animals and can cause lesions that usually take the form of abscesses (often from puncture wounds) or pneumonia. In the latter, *C. pyogenes* usually is a secondary invader and often is found with bacteria of the genus *Pasteurella*. If *C. pyogenes* enters the bloodstream (bacteremia or blood poisoning), as often happens during pneumonia, abscesses can occur in widely scattered parts of the body. These abscesses usually develop slowly, have thick, fibrous tissue capsules and pus that is thick, greenish-white and odorless (Thorne et al. 1982).

Deaths of two free-ranging pronghorn fawns in Wyoming, due to *Corynebacterium* spp., were reported by Honess and Winter (1956). A 7-year-old pronghorn doe that I collected in Montana appeared listless, rough-haired and thin. Her chest cavity contained yellowish pus and fluid, and the ventral lobes of her lungs adhered tightly to the ventral ends of her ribs. Her lungs contained more abscesses than tissue. The abscesses, mostly 1 to 3 inches (25–76 mm) in diameter, were firm and greenish-white. *C. pyogenes* was the only agent that grew on cultures of material from the abscesses. *C. pyogenes* was found in an old wound and in abscesses in the upper hind legs of pronghorn in Wyoming (Howe 1965, Thorne et al. 1982).

Like many infective bacteria, *C. pyogenes* infects a few pronghorn, causing death or debilitation, but usually produces only localized abscesses. The overall effect on populations, as well as danger to humans or livestock, seems minimal.

C. ovis (pseudotuberculosis) commonly infects domestic sheep from contamination of shearing wounds. This usually results in abscessed lymph nodes with occasional extension via the lymphatics to visceral organs. The resulting condition variously has been called caseous lymphadenitis, ulcerative lymphadenitis, pseudotuberculosis or Preisz-Nocard disease. Most infections are in lymph nodes or along lymph vessels. When fatalities occur, they usually are the result of lung abscesses. The characteristic structure of a caseous core surrounded by concentric lamellations of fibrous tissue is a distinguishing feature of abscesses caused by *C. ovis*.

During May 1965, a prostrate adult buck was observed thrashing about on Drakes Flat, Oregon. When approached by humans, the buck attempted to escape but was unable to travel because of uncoordinated hindquarters. The buck was euthanized, and necropsy revealed an abscess, from which a Corynebacterium*-like organism was isolated in the thoracic cavity near the dorsal aorta. The abscess extended into the spinal canal, causing damage to the cord. The buck also suffered from pneumonia, possibly a secondary infection caused by the animal's prostrate position.* Photo by Jim D. Yoakum.

Two fat, adult pronghorn bucks in Texas, apparently normal before death, had numerous hard yellow nodules of various sizes on and in their livers (Clark et al. 1972). The nodules were diagnosed as caseous lymphadenitis. Clark et al. postulated, because of circumstances they observed, that the disease was being transmitted from domestic sheep to the pronghorn. Caseous lymphadenitis of the lymph nodes draining the mandibular region and oral cavity commonly is seen in captive pronghorn, apparently because the oral mucosa is abraded by coarse feed (see Lance and Pojar 1984).

Escherichia coli

Escherichia coli has many strains that vary in pathogenicity and it frequently is a secondary invader complicating primary infections. *E. coli* commonly is found in the lower intestinal tracts of warm-blooded animals, becoming pathogenic only when introduced into other parts of the body or when the intestinal tract is affected by other agents. Disease caused by this bacterium is called colibacillosis. Clinical disease (mild to severe diarrhea and occasionally septicemia and death) is produced by highly pathogenic forms of *E. coli* invading the intestine and becoming systemic. Clinically affected animals develop a severe, whitish diarrhea and frequently die from dehydration and acidosis.

An adult pronghorn buck found in south-central Oregon with paralyzed hind quarters had a temperature of 92°F (33°C), 8° to10° (4.4°–5.6°C) below normal. After its death, abscesses in his liver were found to contain *E. coli* (Yoakum 1957). *E. coli* has been isolated from pronghorn in Wyoming with enteritis, blood poisoning and multiple infections (Howe 1966a, Thorne 1967, 1978). A Nebraska pronghorn with pneumonia also yielded *E. coli* (Howe 1966a). Occasional infections involving this pathogen should be expected, but animals already weakened from other causes are the usual victims. Beta hemolytic strains frequently cause scours or diarrhea in newborn domestic animals and captive-reared ruminants.

Leptospirosis

Leptospirosis of cattle may be caused by any of several serovars of *Leptospira interrogans*. Leptospires are shed in urine and can survive in water or moist soil for several weeks; they may penetrate abraded skin and mucous membranes, or they might be consumed in drinking water. Exposure to contaminated water and food is the most common source of infection. Leptospirosis also may be transmitted venereally or transplacentally. Acute infections in cattle are characterized by high fever, blood-tinged urine, jaundice, loss of appetite and abortion.

Nearly 100 serovars (serum varieties—roughly similar to subspecies but referred to without *L. interrogans* preceding the name) of *L. interrogans* have been isolated worldwide, and as many as 27 known pathogenic strains are present in the United States. A wide variety of host species may become infected, and many become inapparent carriers. Wild rodents, as well as raccoon, opossum and skunk are proven reservoirs (Davis et al. 1981a). Several serovars cycle intraspecifically among prey species and their predators. Many serovars are specific for the host species, but wild ruminants also host specific serovars, such as *hardjo* and *pomona*, which are primarily leptospires of cattle. Accidental infections frequently occur in humans and other susceptible mammals. The effect on the natural host is unknown for many serovars. Disease ranges from unapparent infection through abortion in pregnant animals to death resulting from septicemia.

Study of the prevalence of leptospirosis is complicated. The results of tests made before the late 1970s are debatable because of the questionable methods on which they were based. Currently, the microagglutination test using live antigens of five major groups is recognized as most accurate in detecting leptospiral antibodies (T. Kistner personal communication:1985). Many serologic surveys have been conducted to help determine whether pronghorn carry leptospirosis because it is a serious disease of livestock on the western prairies. Fifty pronghorn tested in Oregon (J. Yoakum personal communication:1980), 809 in Texas (Hailey 1979) and 110 in Canada (Barrett and Chalmers 1975) tested negative to various serotypes.

Hoover (1971) observed an ill 1-year-old pronghorn doe in Colorado on June 4, 1970. The animal was taken to Colorado State University Veterinary Hospital. Its pulse was 98 beats per minute. Respiration was 50 per minute and labored. Rectal temperature was 105.5°F (40.8°C). The nose was dry, nostrils were coated with thick mucous deposits and sclera of the eyes were yellow. Although the animal was alert, it was listless, had trouble standing, appeared weak and frequently trembled. The urine was coffee-colored, and the animal appeared doubled up with abdominal pain. Despite antibiotics, corticosteroids, dextrose and other veterinary treatments, the pulse increased to 142 per minute and respiration to 75 per minute with shallow, intermittent cessations on June 5. The animal was very listless, not alert and exhibited considerable abdominal movement—making grunting noises during pauses in respiration. The animal died on June 6, and *Leptospira* spirochetes were found in its kidneys.

Leptospirosis was studied at the Sybille Wildlife Research Unit by Wyoming Game and Fish Department personnel who artificially infected pronghorn. Howe (1966b) inoculated five fawns with *pomona*, the serotype considered most important in cattle at that time. Four developed the disease and three died, 11, 12 and 14 days after inoculation. In a second study, three pregnant and three nonpregnant pronghorn does were inoculated (Thorne 1968b). Leptospires were detected in the circulating blood of three but none showed clinical symptoms, and leptospires were not detected in their urine. Titers to *pomona* antigens were low and of short duration, and Thorne concluded that *pomona* did not produce an important disease in adult pronghorn.

Further trials demonstrated that some pronghorn fawns were susceptible to infection by *hardjo*, the serovar considered most important to cattle by the early 1980s. The infected fawns shed leptospires in their urine for up to 167 days and were capable of infecting susceptible fawns if in close contact. No overt signs of leptospirosis were noted and pathological changes were mild or nonexistent (Thorne et al. 1983a). Sera from 155 pronghorn killed by hunters in Wyoming during 1976 and 1977 tested negative to five serotypes, including *hardjo*. However, 16 (25%) of 64 serum samples collected from trapped pronghorn during 1978 tested positive to *hardjo*. None of 81 serum samples collected in Carbon County during 1979 reacted to *hardjo* or *pomona* antigens. In 1980, sera from 116 pronghorn from various parts of Wyoming were checked for reactions to the same serovars. None reacted to *pomona*, but six (5.2%) reacted to *hardjo* (Thorne et al. 1982). Lance and Pojar (1984) noted that leptospirosis tests are based on standard bovine tests, and all serological tests of pronghorn should be viewed with caution.

Six pronghorn does in their seventh month of pregnancy also were inoculated with *hardjo*. All developed significant titers, and leptospires were recovered from their urine. One aborted her fetuses, but *hardjo* was not recovered from the fetal tissues. Recovery of *hardjo* from apparently healthy fawns 8 hours after they were born demonstrated that *hardjo* may be capable of in utero infection. During the same study, the kidneys from 33 Colorado and 12 Wyoming pronghorn killed by hunters did not test positive for *hardjo* by the culture techniques used (Thorne et al. 1984).

In Colorado, 1,405 pronghorn sera were tested against four serovars; 80 reacted to various serovars, including *icterohaemorrhagiae*, *hardjo* and *pomona* (Adrian and Keiss 1977). Subsequent serologic surveys revealed higher prevalences of antibodies against *hardjo* as follows: 0.9% of 646 pronghorn tested during 1974, 11% of 666 during 1977, 11.3% of 479 during 1978, 9.6% of 351 in 1979, and 2.8% of 576 in 1980 (Collins et al. 1981). Collins et al. (1981) also found titers in a few pronghorn to serovars *ballum*, *copenhagen*, *szwajizak* and *wolffi*. In another Colorado study, 129 pronghorn tested negative to serovars *canicola*, *icterohaemorrhagiae*, *grippotyhphosa*, *pyrogenes*, *pomona* and *hardjo* (Firchow 1986).

Seven (14%) of 506 pronghorn killed by hunters in Nebraska tested positive to *pomona*, and 9 (1.8%) of 484 tested positive to *icterohaemorrhagiae*. Baily (1962) indicated that a high proportion of the reactors coreacted to both serovars. He speculated that cross-agglutination occurred between the two, or antibodies of a third serotype were present that resulted in cross-agglutination with antigens of both *pomona* and *icterohaemorrhagiae*.

No antibodies were found to serovars *pomona*, *hardjo* or *grippotyphosa* in the sera of 210 pronghorn tested in Alberta by Kingscote and Bohac (1986). Kingscote and Bohac indicated that the absence of antibodies to *Leptospira* in all pronghorn tested was remarkable, considering the known occurrence of infections by serovars *pomona* and *hardjo* in nearby cattle. They noted that the absence of these antibodies was consistent with the findings of Barrett and Chalmers (1975) and suggestive of the existence of a transmission barrier between cattle and pronghorn in Alberta. Such a barrier might be due to aridity of the range and lack of close contact between the species. Transmission probably is more dependent on sexually related contacts.

On a refuge in Oregon without cattle, 20 adult pronghorn does tested negative to serovars *canicola*, *grippotyphosa*, *hardjo*, *icterohaemorrhagiae* and *pomona* (Dunbar and Velarde 1998, Dunbar et al. 1999).

Although leptospirosis may cause some mortality among pronghorn, the disease probably is not a major population-limiting factor (Howe 1966b, Thorne 1968b, T. Thorne personal communication:1985). Concentrations of pronghorn around shallow, warm water late in a dry summer might enhance exposure to leptospires, but such exposure probably would be too early to affect pregnancy. However, pronghorn and cattle often share habitat, and use of common water sources results in transmission of diseases, such as leptospirosis, to which both species are susceptible. Pronghorn may be relatively resistant to infection and act as carriers. Except that serovars *hardjo* and *pomona* generally are regarded as leptospires of cattle, insufficient evidence exists to implicate either pronghorn or cattle as the primary reservoir of infection. The existence of an unidentified reservoir species cannot be ruled out (Collins et al. 1981). Thorne (1973) concluded that, because only a few experimentally infected pronghorn had leptospires in their kidneys and urine, they do not play an important role in the spread of the disease. He added that pronghorn probably do not harbor leptospires in their kidneys for prolonged periods, and they are poor shedders of the organism.

Micrococcus spp.

Micrococcus spp. are spherical bacteria that develop with or without oxygen. They are widespread in nature and often occur in association with animals and animal products, such as milk, urine or skin secretions. *Micrococcus* spp. also are found in soil, decomposing organic material and water. Many species survive drying and exposure to sunlight. Members of this genus usually are not considered highly infectious, but *Micrococcus* sp. was isolated from an abscess found in a Wyoming pronghorn (Thorne 1968a).

Necrobacillosis

Necrobacillosis is an infectious disease caused by the bacterium *F.necrophorum* (formerly called *Actinomyces necrophorus*, *Bacillus necrophorus* and *Spherophorus necrophorum*). This bacterium is pathogenic to a variety of animals, including humans, and produces a wide array of diseases, including foot rot, pneumonia and abscesses in almost any organ or joint. Infections of the oral cavity, usually resulting from consumption of overbrowsed, rough vegetation causing penetrating wounds, result in diphtheria in young animals and necrotic stomatitis in older ones. Many jawbone enlargements classed as lumpy jaw actually result from localized periodontal infections by this organism, as well as *A. bovis*. *F. necrophorum* normally occurs in many animals' gastrointestinal tracts, is widespread, grows without oxygen and thrives in puncture wounds. It has been reported in North American big game since 1903, with most records for deer and elk. Necrobacillosis sometimes occurs when ungulates are crowded together in feedlots or concentrated around mud holes and seeps during hot, dry summers (Rosen 1970).

Four cases of this disease in pronghorn were reported from South Dakota by Bever (1950). Yoakum (1957) found an adult buck carcass in Oregon with visible lesions that yielded *F. necrophorum*. Whether this disorder caused the death of the animal was not known. Thorne (1971) reported necrotic stomatitis due to *Fusobacterium* sp. in a pronghorn fawn. Chalmers and Barrett (1974) diagnosed a case of infectious pododermatitis, or foot rot, in the front limb of a free-roaming female pronghorn in southeastern Alberta. The doe had marked swelling of the foot, with fistulous tracts above the hoof and pale green pus in the surrounding tissue. The causative agent grew at 98.6°F (37°C) on bovine blood agar and was presumed to be *F. necrophorus*. Two of three pronghorn necropsied after they were found dead in Saskatchewan had necrotic areas in their oral cavities, mandibular osteomyelitis, pneumonia and foot rot due to necrobacillosis (Wobeser et al. 1975).

I observed foot rot in a captive pronghorn buck kept in a wet, irrigated pasture and in two wild pronghorn that frequented wheat stubble to glean waste grain. The thin, greenish pus cultured from those animals' feet contained *Fusobacterium* sp. Two days after the tame buck was noticed limping on one hind foot, the foot was extremely swollen, and the animal acted very sluggish. A veterinarian administered antibiotics, but the buck died within 4 days of the time the limp was first noticed.

Openings of the interdigital glands probably allow bacteria to enter a favorable environment when pronghorn walk on wet ground or run over stubble-type vegetation. I have tracked running pronghorn across stubble fields and noted a few blood- or sebum-tipped wheat stalks. Apparently the short, stiff stalks left at harvest occasionally penetrate the openings of the glands between the toes, injuring tissue and inoculating bacteria. The deep glands apparently can provide ideal conditions for rapid growth of certain bacteria.

A pronghorn fawn of about 25 pounds (11 kg) found dead on the National Bison Range was sent to me for necropsy. An eagle had struck the fawn, and the claw of the hallux had penetrated the abdominal cavity. The fawn apparently died of peritonitis. The liver was swollen with purple and whitish yellow streaks alternating on the dorsal surface. The peritoneum was thickened and covered by a yellowish film. A similar film was found on the endocardium and heart valves, and several cultures of the film yielded only *F. necrophorum*.

Necrobacillosis should be expected when poor range conditions force pronghorn to feed on very rough vegetation or when droughts force them to concentrate with other ungulates around stagnant water. However, only occasional individuals, not populations, are apt to be affected.

Neisseria spp.

Neisseria spp. are not generally regarded as pathogens of animals, and the significance of an isolate of *N. catarrhalis* from the eye of a pronghorn fawn with multiple infections is unknown (Thorne et al. 1982).

Pasteurellosis

Pasteurellosis refers to certain types of pneumonia, blood poisoning and other infections caused by bacteria of the genus *Pasteurella*. Virulence might be expressed when the host is subjected to stress of almost any kind. Once clinical illness begins, large numbers of *Pasteurella* spp. are shed in saliva and feces, resulting in direct transmission from animal to animal. Thus, crowding and stressing livestock may result in many animals becoming infected—the reason that the disease often is called shipping disease.

Jessup and Boyse (1995) indicated that pronghorn appear quite susceptible to *Pasteurella* infections, and pasteurellosis has been diagnosed more frequently in pronghorn

Although pasteurellosis is well documented as a disease of wild mammals, the signs and pathology are more thoroughly described for domestic sheep and cattle than for wildlife. In domestic animals, pasteurellosis occurs as pneumonia or hemorrhagic septicemia. Pasteurella *spp. occur as obligate parasites of the respiratory tracts of apparently healthy animals.* Photo by Peter T. Bromley.

than in other wild ruminants of Wyoming, but the disease seems sporadic in nature. Two cases were described as hemorrhagic septicemia—one both pneumonic and septicemic, and the other attributed to predation, but with pasteurellosis as a possible predisposing factor. *Pasteurella* sp. was recovered from the blood of one animal examined, but was not considered the main pathogen. *P. multocida* was identified in two of the five cases from which *Pasteurella* was recovered from pronghorn (Thorne et al. 1982). One mature buck with pneumonia and septicemic pasteurellosis was described by a rancher as listless and breathing laboriously the day before it died (Howe 1965). Because pasteurellosis can kill quickly, it probably was responsible for the deaths of three fawns (9, 22 and 74 days of age) in Utah that appeared healthy a day or two before death (Beale and Smith 1973).

Twelve tonsillar sinuses of fawns (less than 1–3 days old) on Hart Mountain National Antelope Refuge, Oregon, were swabbed and cultured during 1997 (Dunbar and Velarde 1998, Dunbar et al. 1999). Two were positive for *P. multocida*, one of which also was positive for *P. haemolytica*. Five additional fawns were positive for a *Pasteurella* spp. and *Actinobacillus* mixture, and five were negative. Of the fawns that cultured positive for *Pasteurella* spp., three were killed by coyotes (two at 13 days and one at 15 days of age), and two died of unknown causes at 10 and 16 days of age. Tonsillar sinus swabs of 17 fawns tested during 1996 all were negative for *Pasteurella* spp.

Necropsies were conducted on 55 fawns that died during May through July of 1996 and 1997 in Oregon. Two died of septicemia during 1997, when 13 and 17 days old, and *P. multocida* (serotypes A:3, 4 and B:1) was isolated from both (Dunbar et al. 2000). Necropsy findings included widely scattered petechial and ecchymotic hemorrhages in the meninges of the brain, lung, epicardium and skeletal muscle. Nothing can be done to reduce the widespread presence of *Pasteurella* spp. in the respiratory tracts of mammals. However, reduction of predisposing primary invaders such as lungworms is possible. Good management should minimize stress by providing adequate high-quality forage, regulating numbers of animals on overcrowded ranges through hunting, and reducing human harassment (Thorne et al. 1982).

Plague

Plague is an acute, infectious disease caused by *Yersinia pestis*. Rodents, rabbits and associated carnivore and scavenger species are affected. The agent is transmittable to humans after the bite of an infected flea or by direct contact with an infected carcass. *Y. pestis* is a relatively fragile organism, killed by drying and most of the common disinfectants; it does not survive well outside a host.

Y. pestis most commonly is transmitted between rodents by the bites of fleas that have acquired the organism by feeding on hosts with this bacterium in their blood. Most species of fleas can transmit the organism. Other routes of plague transmission include direct contact (skinning of infected carcasses and dressing of infected animals for meat), ingestion of infected carcasses (the common route for infection of carnivores) and inhalation of infective, airborne droplets released by hosts with pneumonic plague during coughing or sneezing.

Most susceptible animals die rapidly after acquiring *Y. pestis*. Experimentally infected laboratory rodents show lethargy, loss of appetite, labored respiration, huddling and ruffled fur. Buboes (inflamed, swollen lymph nodes) and subcutaneous suppurative lesions may be found in animals surviving long enough for these to form. In humans, pustules might form at the site of infection, and buboes usually develop in the groin after flea bite infection and in the axilla after direct contact to the hands. However, affected lymph nodes have been reported from many body areas. In untreated infections, the bubo undergoes necrosis and abscess formation, and organisms are disseminated through lympathic channels or the bloodstream to other organs throughout the body. Due probably to one or more of the plague endotoxins, intravascular coagulation and petechial hemorrhages occur, leading to further tissue damage and necrosis. When these events occur in the lungs, death follows.

Widespread epidemics of human plague have occurred repeatedly, spreading from the ancestral home of the organism in the Yunnan Province of China. The "Black Death" of the 14th century—believed to have caused the death of at least one-fourth the population of Europe at that time—was caused by this pathogen.

In the United States, the first recognized human plague occurred in San Francisco during 1900. Although plague was introduced by shipborne rats and caused local outbreaks of human and rat plague in several areas of the eastern states, long-term foci in native wild rodent populations were not established in those areas. Plague, however, did spread from infected commensal rats to native rodents in the West. The first suspected epizootic in native ground squirrels occurred during 1903 in Contra Costa County, California. By the middle of the 1930s, plague had spread eastward beyond the Rocky Mountains, and it occurs periodically in 15 of the 17 contiguous western states (Thorne et al. 1982).

Although reports of plague in big game and domestic animals are limited, the disease was reported in one- and two-humped camels in Russia and one-humped camels in Libya. Also, serologic evidence of plague was found in 26 of 391 cape buffalo and 1 of 330 African elephants tested a year after a plague epidemic in Wankie National park, Zimbabwe (see Thorne et al. 1987).

An unusual case of plague in a free-ranging mule deer was reported by Thorne et al. (1987). The doe fawn was observed in a weakened condition near a cattle feedlot in eastern Wyoming during April 1980. It was taken to the Wyoming Game and Fish Laboratory, where *Y. pestis* was identified.

The incidence of human plague cases in the United States showed a marked increase during the 1970s. Of the 158 cases reported during 1925 to 1979, 67% occurred from 1970 to 1979. The increased incidence might be attributed, in part, to three factors: a substantial increase in human populations in western plague areas, heavier use of recreational areas and a trend of modern real estate developers to maintain a "natural" landscape (Thorne et al. 1982).

Between 1924 and 1987, all naturally acquired cases of human plague in the western states apparently were associated directly or indirectly with wild rodents or lagomorphs. However, during September 1987, a bowhunter was admitted to a Bozeman, Montana, hospital with symptoms of plague. He was critically ill, and blood cultures indicated *Y. pestis* in his blood. He had killed and dressed a pronghorn a short time before. *Y. pestis* also was cultured from the popliteal and prescapular lymph nodes of the pronghorn (Niccolucci 1987). Such an infection of a pronghorn appropriately was called a freak of nature. The chance of other pronghorn getting the plague is almost nonexistent, but this well-documented case serves as a reminder that improbable occurrences can happen where diseases and parasites are concerned. The chance of contracting plague from any wild animal is very low, but it can be reduced further by not handling dead animals with bare hands if cuts, abrasions or hangnails are present, by washing hands as soon as possible after handling a carcass and by cooking wild game thoroughly. The disease can be treated successfully with antibiotics.

Pseudomonas spp.

Members of the genus *Pseudomonas*, a bacteria that may cause abortion, are common in the environment, and a few are specific or opportunistic pathogens. *P. putida* was cultured from an aborted pronghorn fetus in Wyoming, but whether it caused the abortion was unknown (Hepworth 1962). *Pseudomonas* spp. were cultured from nasal and rectal swabs of pronghorn fawns, both at birth and postmortem, in Idaho (Bodie 1979). Although some of those fawns later died (or had died) of what Bodie phrased starvation disease, the study did not prove a connection between the bacterium and the deaths.

Salmonellosis

Salmonellosis is caused by members of the genus *Salmonella*. More than 81 serovars of *Salmonella* have been reported from domestic animals in the United States. The species most commonly responsible for disease in ruminants are *S. typhimurium*, *S. dublin*, *S. anatum* and *S. newport*. *Salmonella* are ubiquitous in the environment and capable of prolonged survival outside hosts. The disease most commonly is transmitted by carrier animals shedding the organism in their feces. Acute salmonellosis produces diarrhea, depression and fever. Death may occur within 48 hours or animals might linger for several days. Animals that are chronically affected often develop clinical signs after some form of stress. In general, young, old or stressed animals are most likely to develop the disease (Lance and Pojar 1984). Humans acquire salmonellosis by ingestion of contaminated food or water. Typhoid fever is caused by *S. typhi*; other species cause septicemia and gastroenteritis (Davis et al. 1973).

Two instrumented pronghorn fawns died of salmonellosis in Utah. Both appeared to be healthy a day or two before death, and body weights obtained after death indicated normal growth. A 27-day-old fawn was lethargic and appeared to have diarrhea one morning. It was found dead that afternoon, and *S. cubana* organisms were isolated from its internal organs. A 13-day-old fawn also died of salmonellosis, and *S. saka* organisms were identified (Beale and Smith 1973). *Salmonella* sp. was isolated from the small intestine of a doe fawn that died of starvation and weak fawn syndrome at 6 days of age in Oregon (Dunbar and Velarde 1998). Whether *Salmonella* contributed to the demise of the fawn was not speculated upon.

Staphylococcosis

Staphylococcosis in wildlife is caused primarily by *Staphylococcus aureus* and *S. epidermidis*. Both species may be found in healthy animals, but they can cause abscesses in all mammals, including humans. When only *Staphylococcus* sp. is present in an abscess, the pus usually is thick and creamy white or whitish green. These abscesses usually have thick, fibrous capsules, although those in livers or kidneys may be poorly encapsulated. Most abscesses are found incidentally in hunter-killed animals or during necropsies. Thus, little is known concerning clinical signs. However, a pronghorn fawn with a *Staphylococcus* sp. infection was extremely lame and reluctant to move. It had enlarged, pus-filled joints, thickened joint capsules, eroded articular surfaces and multiple abscesses in the kidneys (Thorne et al. 1982). *S. epidermidis* apparently caused eye disease in a pronghorn from Nebraska (Thorne 1970). The same species was found in nasal and rectal cultures from seemingly healthy fawns in Idaho (Bodie 1979).

Abscesses in the lungs and chest cavities of hunter-killed pronghorn and from badly infected eyes in Wyoming yielded *S. aureus*. It may have been a secondary invader in the latter (Thorne et al. 1982). *S. aureus* was cultured from three abscesses—two from the neck and one from a flank—that I found in adult bucks on the National Bison Range. The abscesses seemed to have resulted from horn punctures.

Staphylococcosis is not of great importance in pronghorn populations, but it draws considerable attention because the large abscesses are obvious in hunter-killed animals.

Streptococcus spp.

Streptococcus is a complex genus with many species, some of which are pathogenic. It usually induces pus formation and often is a secondary invader. *S. faecalis* was recovered from blood, liver and pus of an adult female pronghorn that died of extensive peritonitis after 2 days of apparent illness (Hepworth 1961). A pronghorn with chronic pneumonia was infected with *Streptococcus* sp. (alpha hemolytic) and another with chest abscesses yielded *S. zooepidemicus* (Thorne et al. 1982).

Tuberculosis

Tuberculosis is an infectious disease caused by *Mycobacterium* spp.; it affects practically all mammal species. The name comes from tubercles—small rounded nodules caused by the bacteria. These are the characteristic lesions of tuberculosis, usually gray in color. Tubercles undergo necrosis and tend to spread, especially in the routes of least resistance such as lymph or blood vessels. The lungs are especially prone to infection, and ulcers occur on the surfaces of other organs. Body cavities become filled with pus and degenerate tubercles and tissue. Almost any organ can be infected, but muscular and fibrous tissues are more or less exempt. Symptoms vary with the location of the infection and amount of destruction. When not strictly localized, symptoms in humans include fever, emaciation and night sweats. Miliary (having the appearance of millet seeds) tuberculosis is a form in which small tubercles are formed in a number of organs after the bacilli are disseminated in the bloodstream.

The death of a 2-year-old captive pronghorn buck was attributed to miliary tuberculosis (Rush 1927). The animal's lungs were a mass of hard ulcers and almost black. Pus was present around the lungs, and the lymphatic glands were greatly enlarged. The liver was covered with ulcers, but the stomach, kidneys and intestines appeared normal.

Tuberculosis now is uncommon in domestic animals in the United States and is unlikely to cause more than a very few deaths in pronghorn.

Vibriosis

Vibriosis is an infectious disease sometimes found in the genital tracts of cattle and sheep, and is especially prevalent in the Rocky Mountains. The mode of transmission in cattle is primarily by coitus; in sheep and presumably pronghorn, it is by ingestion. Vibriosis causes abortions during late gestation and increased neonatal mortality. It is caused by members of the genus *Campylobacter* (formerly *Vibrio*).

C. fetus was recovered from 5 (18%) of 28 does, or their fetuses, collected in an area of south-central Wyoming with low fawn survival during 1956 to 1958 (Trueblood and Post 1959). Trueblood and Post made isolations from blood, amniotic fluid, livers and fetal stomach contents. Domestic sheep ranged the same area, and the stomach contents and liver tissues of pronghorn fetuses (from which *C. fetus* was recovered) caused abortions when administered orally to domestic ewes. *C. fetus* was recovered from the aborted lambs. Although transmission of *C. fetus* between sheep and pronghorn was not demonstrated in the field, such transmission might occur (Thorne et al. 1982).

The pronghorn herd from which the infected does were collected had experienced ratios of 17 to 48 fawns per 100 does during 1954 to 1956, although pregnancy rates were normal. The implication was that vibriosis in pronghorn resulted in increased fetal or neonatal deaths (Howe 1970). Personnel of the Wyoming Game and Fish Laboratory conducted a study to determine the affect of *Campylobacter* from domestic sheep on pronghorn (Hepworth 1961). Suspensions of viscera from aborted *Campylobacter*-infected lambs were administered orally to eight presumably pregnant pronghorn does during April 1960. Some subsequent mortality to fawns and does occurred, but no *Campylobacter*-like organisms could be cultured from them.

Prevention of the disease in pronghorn probably could be accomplished by limiting their contact with infected sheep.

MISCELLANEOUS DISEASES

Any condition that causes a departure from a state of health is considered a disease. Many diseases debilitate or kill wild animals, but are not necessarily the result of just one species of bacterium or parasite, or the disease agents are unknown or pathogens are not involved.

Diarrhea

Diarrhea, or scours, is the too rapid passage of intestinal materials and liquids through the bowels—an effective, but occasionally debilitating, method of discharging bacterial, parasitic or other irritating substances from the intestines. It appears to be fairly common among young pronghorn fawns and occasionally afflicts adults, especially those on abnormal diets. Of 80 young, wild pronghorn fawns examined in Oregon, 12 (15%) had diarrhea. In one case, a fawn caught within a day after birth had severe symptoms of diarrhea. It was recaptured a week later and only small flakes of dry fecal material remained as evidence of the previous condition (Yoakum 1957). Yoakum cited reports from the Oregon Game Commission and California Game Department personnel stating that scours was common among young pronghorn fawns.

Escherichia coli apparently causes diarrhea in captive newborn ruminants and might do the same in free-ranging animals. *E. coli* is most apt to cause diarrhea in pronghorn fawns that have heavy loads of parasitic flatworms or roundworms. As with most diseases, good range and population management should minimize the problem.

Reoviruslike particles also have been found in feces from captive pronghorn fawns with diarrhea. Tests on the ileum of a dead fawn and on feces from two live ones indicated that the virus in pronghorn was serologically related to the calf diarrhea (neonatal calf diarrhea [NCD]) reoviruslike agent (Reed et al. 1976). Gross examination of the dead fawn revealed dehydration (indicated by sunken eyes), tail hair pasted with feces, and flaccid, dilated small intestines containing yellowish fluid. Bacteria resembling *E. coli* were found in the intestinal epithelium. Reed et al. believed the virus was transmitted to the fawns from cattle and that NCD reovirus vaccine might help prevent viral enteritis in pronghorn fawns raised in captivity.

When pronghorn are stressed by snow or drought conditions, they sometimes are forced to eat unsuitable forage that can lead to diarrhea. A number of disease organisms can contribute to the condition, which causes many problems with captive fawns. Photo courtesy of Montana Department of Fish, Wildlife and Parks.

Hemolytic Anemias

Hemolytic anemia, characterized by an accelerated rate of erythrocyte destruction, can be caused by a number of chemical agents, including chlorates, lead, phenols, benzene and naphthaline. Under natural conditions, pronghorn seem unlikely to contact such poisons. Hemolytic anemias also are caused by poisonous plants or excessive intake of certain plants. However, pronghorn generally seem resistant to plant toxins and consume many plant species, without harm, that poison domestic livestock. Bacterial hemolysins from *Clostridium perfringens* and *C. haemolyticum*, plus certain strains of *Leptospira*, *Staphylococcus* and *Streptococcus*, cause hemolytic anemias. All of these bacteria—except *C. haemolyticum*—have been reported in pronghorn.

A 4-year-old, 75-pound (34 kg) pronghorn doe—which shared a zoo exhibit with another pronghorn, two white-tailed deer and two wild turkeys—was found recumbent. Examination revealed lethargy, depression, dehydration, grade II systolic murmur, prolonged capillary refill time, red urine and icterus of the sclera and mucous membranes (Gentz et al. 1995). This combination indicated hemolytic anemia. The animal had been vaccinated with a multivalent clostridial vaccine, making an infection by *Clostridium* spp. unlikely. Red blood cell parasites, leptospirosis and other common causes of hemolytic anemia were ruled out by appropriate tests. In the absence of a definitive diagnosis, a broad-spectrum treatment regime was initiated through a catheter in the right jugular vein. The animal's condition continued to deteriorate, and (after 5 days) she was given a 22.5-ounce (750 mL) whole-blood transfusion from her unaffected sibling. Her condition improved and she was returned to the exhibit within 12 days. Hemolytic anemia had not previously been reported in pronghorn, but this incident indicates it can occur.

Hemorrhagic Pancreatitis

A pronghorn buck died of hemorrhagic pancreatitis at the Philadelphia Zoological Garden (Fox 1929). The condition was reportedly as similar to hemorrhagic septicemia, but no cultures were made for bacteria. Degeneration of gray matter of the spinal cord was noted. An association between disease of the pancreas and spinal cord had been seen several times at the Philadelphia Zoological Garden in "deer and antelope," but the species were not given. Fox considered it a toxic degenerative disease connected with food.

Hepatic Fatty Cirrhosis

Hepatic fatty cirrhosis, sometimes known as hard yellow liver disease, commonly has been found in livestock, particularly domestic sheep, in the Big Lake area of Texas (Hailey 1979). The causative agent of this noncommunicable disease has not been determined, but the lush plant growth around playa lakes of that part of Texas following periods of high rainfall seems to precede outbreaks.

Hepatic fatty cirrhosis was identified in pronghorn of the lower plains area of Texas during 1966. In 1967, 66 (67%) of 98 pronghorn inspected near Big Lake had the disease (Hailey 1979). Hailey reported that the disease alone was not fatal to pronghorn, but when combined with parasites and poor rangeland conditions, the combination may prove fatal.

Infectious Keratitis

Infectious keratitis (also known as infectious keratoconjunctivitis, vascular keratitis and pinkeye) is fairly common in domestic and wild ruminants, and several species of bacteria and viruses have been implicated. Many things about infectious keratitis are unknown: its incidence in populations, the number of animals that recover or become permanently blind, the similarity between causative agents in wild ruminants and cattle, and the possibility of transmission between wild and domestic ruminants. The etiologic agent may be a normal inhabitant of the healthy eye that becomes pathogenic after irritation. Irritants may include insects, dust and tall grass. The face fly is an important disseminator of infectious keratitis, according to T. Kistner (personal communication:1985). Nutritional deficiencies also may be important predispositing factors.

A bacterium, *Moraxella* (*Hemophilus*) *bovis*, is the most important agent of infectious keratitis in cattle and possibly in wild ruminants of North America. However, it frequently can be isolated from the normal bovine eye, and *M. bovis* administered to undamaged eyes usually fails to produce the disease (Thorne et al. 1982). The causative agent or agents in wildlife have not been clearly demonstrated, but a virus may be involved in its manifestation (Honess and Winter 1956). *Mycoplasma*, *Rickettsia* and *Chlamydia* have been cited as causative agents in sheep and goats.

The internal lining of the eyelids (conjunctiva) and cornea usually are affected. Symptoms include a watery, purulent discharge from the eye, swelling of the eye, gray or reddish and opaque cornea, and the inability to travel easily because of partial or complete blindness. In advanced cases, a dark-colored mass of numerous blood vessels develops in the middle of the eyeball, beneath the outer layer of the cornea. The mass emerges from a rupture in the cornea; it appears as a growth on the eyeball.

Whatever the agent, infectious keratitislike eyes have been noted in pronghorn throughout the West. Eleven pronghorn with pinkeye were reported in Oregon dur-

A blind yearling pronghorn found in Oregon with infectious keratitis (pinkeye). Watery discharge from the eye, as seen in this photo, is one of the first symptoms of this disease. Little is known concerning the incidence of pinkeye in wild animals, or whether infected animals often recover. Photo by Frank Grogan.

ing 1937 to 1955; nine were reported during winter and spring, and both sexes and all age classes were reported (Yoakum 1957). This disease was reported in an area of El Paso County, Colorado, where several cattle also were infected (Hoover et al. 1959). Pinkeye was reported in free-ranging pronghorn of western Texas by Buechner (1950). A published description of this disease in pronghorn involved an adult female in California during 1957. Gross examination revealed corneal ruptures and cloudiness in both eyes; in the right eye, ulcers had penetrated deep enough for aqueous humor to escape, resulting in prolapse of the iris (Mankins 1962).

According to Thorne et al. (1982) the slow, visible death of wild animals afflicted with infectious keratitis detracts from their aesthetic value and causes concern, justified or not, by stockmen. Thorne et al. recommended that affected free-ranging animals be destroyed for humane reasons and to limit possible spread of infection.

Polioencephalomalacia

Polioencephalomalacia (PEM), or cerebrocortical necrosis, is a presumably noninfectious, degenerative disease of the central nervous system of ruminants. It has been recognized in domestic cattle, sheep and goats and in fallow, white-tailed and mule deer. It generally is accepted that thiamine (a vitamin synthesized within the rumen) deficiency plays a role in the disease. *Clostridia* spp. and other gram-positive bacteria can produce thiaminases (thiamine-destroying enzymes) in the rumen, and such bacteria proliferate on high-concentrate rations, such as grain. PEM in cattle also has been associated with water deprivation and diets high in sulfate salts. Also, bracken fern and horsetail contain thiaminase and can produce a PEM-like disease in sheep. The disease causes necrosis of the cortex and swelling of the brain, often forcing a portion of the cerebellum into the foramen magnum and causing hemorrhages in the brain. Leighton and Wobeser (1999) believed the disease occurred in wild ungulates of Saskatchewan during autumn and winter because they scavenge waste grain extensively at that time of year.

Four wild pronghorn in Saskatchewan, one dead and three recumbent or staggering, were diagnosed as having a PEM-like disease (Wobeser et al. 1983). Lesions were confined to the brain and consisted of brain swelling with herniation, hemorrhages in the thalamus and brainstem and necrosis of the cortex. All of the animals had been eating grain before their deaths, raising the possibility of bacterial production of thiaminase. Many surface waters in the area where the animals were found were saline and contained large amounts of sulfate salts, another possible source of thiaminase. Plants that contain thiaminase were not known to occur in the area or to be eaten by pronghorn, but that possibility could not be ruled out. Three additional pronghorn with similar lesions were examined by Wobeser (1984). During winter 1997–98, an ill adult pronghorn buck was seen in a field near Zealandia, Saskatchewan. It had difficulty standing and moved in circles before collapsing. Lesions—including necrosis—were restricted to the brain, and hemorrhages were found in the ventral portion of the cerebellum (Leighton and Wobeser 1999).

Rumen Overload

Rumen overload (RO) is a metabolic disease that occurs when ruminants consume a large quantity of readily fermentable carbohydrate. Rapid proliferation of certain microorganisms in the rumen, with fermentation and production of lactic acid, follows the ingestion of suitable carbohydrates. The acidity kills other microflora in the rumen, causes the rumen to stagnate and increases osmolarity of the rumen fluid, so that water is drawn from the blood and other tissues into the rumen, causing dehydration and damage to the rumen wall, allowing bacteria and fungi to invade. If acid absorbed from the rumen is sufficient to overcome the body's buffering system, the animal becomes acidotic, its blood pressure decreases, kidney function ceases and the animal may die of shock. Animals that survive the acute stage may die later of fungal infection in the damaged rumen wall.

Rumen overload was found in pronghorn that had fed on wheat in Saskatchewan (Wobeser 1984). Most of the animals diagnosed as having RO were in good body condition with abundant fat, the rumens contained wheat, the rumen pH varied between 5 and 6, and the rumen mucosa was tightly adherent. The skin was tightly adherent, underlying tissues were dry and the blood was black and tarlike suggesting dehydration.

A yearling buck was observed shivering, in January, and found dead the next day. At necropsy, the animal's body was in fair condition, but its perineal region was pasted with feces, indicating diarrhea. Portions of the linings of the rumen and reticulum were hemorrhagic. The mucosa in these areas was red-black. The rumen contained a small amount of wheat and fluid, and the intestines contained only fluid. Microscopically, the affected areas were hemorrhagic and necrotic, with many fungal hyphae invading both the walls and arteries. A diagnosis of necrotizing mycotic rumenitis, probably as a sequel to earlier RO, was made.

Tumors

Tumors occur when cells undergo changes that release them from the normal mechanisms that control their growth and differentiation. Tumors may be benign or malignant (cancerous). Benign tumors are only life threatening if they occur in or crowd vital organs. They generally grow slowly and do not spread to other parts of the body. Malignant tumors generally grow rapidly and may metastasize and induce new tumors in other parts of the body. Little is known concerning the frequency of tumors in wild animals. Viruses that gain entrance through a break in the skin may cause the benign fibromas (also called papillomas or warts) occasionally seen on wildlife.

Relatively small fibromas were noted on pronghorn from three Wyoming counties by Honess and Winter (1956) and Sundberg et al. (1983). Honess and Winter reported a pronghorn having a fibroma under its cheek that was firm enough to compress the premolars inward. Seven pronghorn that I necropsied in Yellowstone National Park and on the National Bison Range had round fibromas about half as high as wide; none was larger than 1 inch (25 mm) in diameter. They were light to dark gray, had surfaces reminiscent of those on cauliflower florets and the cut inner surfaces were smooth and pinkish white. Two were just below the ear and on the throat of mature bucks. The other five were around the anuses and vulvas of does, where winter ticks, especially embedded ones, were found most often. On another doe, a deeply embedded tick was surrounded by a dark gray mass of tissue, shaped like a doughnut, about 0.2 inch (5 mm) high with a surface like those of the fibromas. Perhaps it was only an immunological reaction, or perhaps ticks carry the tumor virus or provide for its entrance.

Viruses that cause fibromas are quite species-specific, therefore pronghorn should not be carriers for livestock. The small number of fibromas reported for this species indicates that they are uncommon enough to be of little importance. Their appearance may worry a hunter, but they usually can be discarded with the skin and do not impair the quality of the meat. If a malignant tumor is suspected, the meat should not be eaten. Malignant tumors on wildlife often have a reddish color and appear meaty. Also, animals with malignant tumors usually are thin, and the stalk of the tumor often is infected.

One report exists of a malignant skin tumor (squamous cell carcinoma) in a pronghorn (Effron et al. 1977).

Weak Calf/Lamb/Fawn Syndrome

This syndrome was first recognized in the Bitterroot Valley of Montana in 1964. Calf losses reached near epizootic proportions during the 1969 calving season (Stauber 1976), not only in western Montana, but also in the nearby high mountain valleys of Idaho. Losses generally were due to late abortions and deaths of calves before 6 weeks of age. Affected young were reluctant to stand or move. They arched their backs when they stood, failed to nurse and had diarrhea. At necropsy, fibrin and hemorrhages were found in the tarsal joints, and the thymuses were small and sometimes absent. An unknown virus isolated from weak calves was incriminated by Januszewski (1972). Ozoga and Verme (1978) noted that the thymus glands of penned deer fawns dying from disease or malnutrition within a month of birth were one-sixth as large as those of fawns that died from accidents. T. Spraker stated that severe stress, induced by weather or malnutrition, causes atrophy of the thymuses of neonatal fawns within 2 to 3 days (Autenrieth 1984). High corticosteroid levels produced by the adrenal glands in response to stress are the causative agent.

I observed all of the signs of weak fawn syndrome in two fawns that died shortly after birth on the National Bison Range (see Reichel 1976), and Bodie (1979) found similar signs in some fawns in Idaho. Autenrieth (1984) reported that 28% of pronghorn fawns fitted with transmitters in Idaho during 1981 died of hypothermia or starvation. Thymus glands of those fawns were small. Dunbar and Velarde (1998) reported that 2 of 104 radio-collared pronghorn fawns in Oregon died from starvation and weak fawn syndrome, possibly because of maternal abandonment.

The small thymus of a fawn that died shortly after birth. The animal also had fibrin and hemorrhages in its tarsal joints, typical symptoms of weak calf/fawn syndrome. Such fawns are weak at birth, and does often refuse to let them nurse, seemingly sensing they will not survive. Photo by Bart W. O'Gara.

PROTOZOA

Protozoa are microscopic, one-celled animals, some of which are parasitic. Those that are parasitic in wild animals often occur in domestic animals and in humans, but the ability to cause disease in one host may not extend to another.

Multiple parasitism, whether by protozoa or any of the following parasites, is a normal condition for any wild animal. Many parasites are so well adjusted to their hosts that they seldom produce visibly harmful effects; others are harmful only when present in large numbers. A few species are damaging even in small numbers. Most parasites can predispose their hosts to other mortality factors, especially severe weather, malnutrition and predation. In well-nourished herds of pronghorn, at or below the carrying capacity of their range, parasites may debilitate or kill a few individuals. However, such debilities or deaths usually are inconsequential to the overall welfare of a herd.

Byers (1997) collected pronghorn feces on the National Bison Range, Montana, during September 1992 and 1993 for analysis of gut parasite loads. Both sexes were infected by *Nematodirus*, *Strongylus* and *Trichuris* nemotodes, as well as coccidian protozoa. The density of parasite eggs in the feces of bucks was about three times as great as in those of does. Age of bucks had no apparent influence on parasite load, and parasite load did not affect mating success.

Byers (1997) postulated that testosterone levels and reproductive behavior (stress) may have shunted physiologic resources away from the bucks' immune functions, allowing parasites to proliferate. He did not speculate on the possibility that the greatly reduced intake of forage by rutting bucks could lead to an apparent increase in parasite eggs in the reduced amount of feces.

Ciliophora

The ciliates are protozoa that move about by vibrating microscopic hairlike protrusions—cilia—on their bodies. A very large number of ciliate species exist, and the majority are free-living in water. Some are parasites of the large intestines of many mammals. Ciliates were found in filtrate from the stomachs of New Mexican pronghorn (Russell 1964). Apparently no positive identification was made of the species involved. Many species of ciliates occur in rumens and the large intestines of horses. These are not parasitic in the sense that they produce disease; rather, they are concerned in the digestive processes. However, their exact role has not been determined. The ciliates found in New Mexico pronghorn probably were a normal part of the gastrointestinal fauna.

Ciliate protozoa were not found in the rumens of pronghorn that were deprived of food for 10 to 14 days by a snowstorm and later died with full stomachs. The rumen of a pronghorn killed during the next hunting season contained 800 protozoa—apparently *Entodinium dubardi*— per milliliter of rumen fluid (Pearson 1969).

Coccidiosis

Coccidiosis is the disease caused by protozoa of the genera *Eimeria* and *Isospora*. Diarrhea—a characteristic of the disease—is caused when the parasites develop in and damage the mucosal lining of portions of the digestive tract. Eimerian species are identified by means of their oocysts, which are passed in the feces of host animals. These species generally infect only a single host species and complete their life cycle in one individual.

Eimeria antilocapra was briefly described by Huizinga (1942) from three pronghorn fecal samples collected near Laramie, Wyoming, and by Honess and Winter (1956), who reported the occurrence of the parasite in Carbon County, Wyoming. Sixteen (25%) of 63 pronghorn collected in Yellowstone National Park and on the National Bison Range were infected with *E. antilocapra*. Three oocysts of a different species of *Eimeria* were found in one pronghorn fecal sample from the National Bison Range (Todd et al. 1967). Determining whether it represents a new species would require additional information. Dunbar and Velarde (1998) reported 35 to 95 coccidial oocysts per gram of feces in 3 (15%) of 20 fecal samples from adult pronghorn in Oregon.

Details of the life cycle of *E. antilocapra* are unknown. No problems have been attributed to infection of pronghorn by this parasite, and it apparently cannot be transferred to other species.

Entamoeba antilocapra

Entamoeba antilocapra was described from a Wyoming pronghorn by Noble (1953). These are typical amoebae that move by forming pseudopods, or false feet, through which they flow directionally. They seem to cause minimal or no health problems in their hosts. Diarrhea may be associated with *Entamoeba* spp. infections, but the parasites also have been found in firm stools. Little is understood concerning the pathogenicity of this group or whether cross transmission occurs between different host species.

Entamoeba bovis

Amoebae resembling *E. bovis* were recovered from soft stools from a number of captive pronghorn fawns at the Sybille Wildlife Research Unit in Wyoming (Kingston et al. 1982). Chronic or intermittent diarrhea was observed in most individuals in a group of 21 hand-raised, 1- to 6-month-old pronghorn. At necropsy of nine fawns, ileocecal lymph nodes were enlarged and, in one animal, multifocal hepatic necrosis was present. Microscopic lesions showed necrosis and inflammation in lymph nodes, necrosis and diffuse nonsuppurative enteritis in ileum and cecum, and liquifactive necrosis in the liver. In six of nine pronghorn, trophic stages of amoebae were observed in the submucosa of the cecum and the cortex of the ileocecal lymph node; in one case, amoebae were present in the liver. This is the first report of *E. bovis*-like organisms invading the tissues of their host and being associated with pathological changes.

Giardia

Giardia is a flagellated genus that has only two species, according to current data, and is found worldwide. The Rocky Mountain area is considered a hot spot because of frequent cases in humans, usually associated with drinking water contaminated by feces. Animals, including humans, with giardiasis seldom die as a result, although diarrhea, abdominal pain and nausea are typical signs of the disease in humans.

Giardia-infected animals can be potential sources of infection for humans. Conversely, human-derived *Giardia* spp. can be transmitted to pronghorn (see Davies and Hibler 1979). However, dryland mammals, such as pronghorn, are not apt to serve as reservoirs for the parasite; aquatic species, such as beavers, often have been implicated.

Trypanosomes

Trypanosomes are flagellated parasites that inhabit the blood of vertebrates. Of 57 pronghorn trapped in Wyoming during 1980, 8 (14%) were infected with trypanosomes (see Thorne et al. 1982). Little is known concerning this parasite in pronghorn, but flagellated protozoans (*Selenomonas ruminantium*) were found in blood smears from 20 (35%) of 57 pronghorn in California (Chattin et al. 1944); the organisms were extremely variable in size and shape. Investigators have not suggested any pathogenic significance of infections by this parasite.

Sarcocystis

Sarcocystis spp. differ from most other coccidians in that the latter pass in the feces as sporocysts and require two hosts (a predator/prey complex) for completion of their life cycles. Asexual reproduction occurs in the prey host and sexual reproduction in the predator. Prey animals become infected by ingesting sporocysts passed in carnivore feces, and cysts form in their muscle tissue. The predator becomes infected by eating meat of prey containing cysts.

Muscle cysts have been reported from pronghorn in Wyoming (Thorne 1970) and Montana (Dubey 1980). The parasitic species involved has not been identified or named, and nothing is known concerning the carnivorous host, life cycle, distribution or pathology. Sporocysts or oocysts were not found in the feces or intestine of a dog fed infected pronghorn meat. Study apparently is hampered by the scarcity of *Sarcocystis* in pronghorn.

Toxoplasma gondii

Toxoplasma gondii may be found wherever felids exist, due largely to the ability of the parasites to use many different species of animals as intermediate hosts. Cats may harbor the asexual (schizonts) as well as the sexual (gametocytes) stages of *T. gondii*, depending on the stage at ingestion. Oocysts are passed in felid feces and are highly resistant in the environment, but tissue phases may be killed by freezing or cooking.

Toxoplasma spp. are the only coccidians known to cross the placenta of intermediate hosts to infect the developing fetus; this method of transmission has very important human health implications. Cats generally are unaffected by *T. gondii* infection, but some become ill and die due to pneumonia or inflammation of the liver, pancreas, skeletal muscles, brain or spinal cord. Humans may become infected by eating poorly cooked meat of infected wild or domestic animals or by ingesting oocysts from cat feces. Humans seldom suffer serious effects and usually overcome toxoplasmosis. However, women who become infected in the last trimester of pregnancy sometimes abort, experience stillbirth, or give birth to infected children with visual or aural defects, mental retardation, megacephaly (abnormal enlargement of the head), microcephaly (abnormal smallness of the head) and other abnormalities. This problem is associated more with domestic cats allowed to hunt mice and birds than with wild felids.

Two mature pronghorn—one inoculated with 10,000 oocysts and another with 1,000 oocysts—became ill and died at the Wyoming Game and Fish Department Sybille Wildlife Research Unit. The pronghorn showed clinical signs of diarrhea (sometimes bloody), loss of appetite, lethargy, necrotic enteritis, hepatitis, mesenteric lymphadenitis, and subcutaneous and pulmonary edema (Dubey et al. 1982). Dubey et al. concluded that pronghorn are among the most susceptible ruminants to *T. gondii*.

Approximately 50% of the human population in the United States has antibodies to *T. gondii*, and ingestion of infected meat is one source of human infection. *T. gondii* was isolated from muscles of a naturally infected pronghorn in Montana (Dubey 1981). Therefore, pronghorn meat should be well cooked before consumption.

FLATWORMS (PHYLUM PLATYHELMINTHES)

Flukes

Mature flukes, of the class Trematoda, exist as parasites in many organs. Typically, they have two suckers—one around the mouth and the other on the ventral midline. The digestive tract is incomplete, and most species have a complete set of male and female reproductive tracts.

Fasciola hepatica is a cosmopolitan liver fluke often found in the bile ducts of domestic and wild herbivores. It is an important parasite of domestic and wild ruminants of the northwestern, midwestern and southwestern United States in areas where the intermediate host—an amphibious snail—is found (Davis et al. 1981b). Occasionally, it is found in unusual hosts, including humans. In such hosts, the parasite may be in the lungs, under the skin or in other locations. The fluke can cause fascioliasis (liver fluke disease, liver rot), especially in sheep and cattle. *F. hepatica* is leaf-shaped, broader anteriorly than posteriorly, with an anterior cone-shaped projection followed by broad shoulderlike projections. It is grayish brown, and the ventral sucker, about as large as the oral sucker, is situated at the level of the shoulders. The cuticle is armed with sharp spines. These flukes may reach a size of 1.18 by 0.51 inches (30 by 13 mm).

The eggs of *F. hepatica* enter the duodenum with the bile, and they leave the host in the feces. Miracidia (broad anteriorly, with a small nipplelike protrusion and two eye spots) hatch from the eggs and penetrate snails of the genus *Lymnaea*. After further development, they are released from the snail as free-swimming cercariae that eventually sink to the bottom or encyst on vegetation just below water level. Hosts are infected by swallowing the cysts with plants or water. The parasites excyst in the duodenum and migrate through the abdominal cavity or are carried by the bloodstream to the liver. Migration in the liver occurs for 5 to 6 weeks before entering the bile duct, where they reach sexual maturity. Eggs then are found in the bile and feces of cattle for about 8 weeks, but they may persist for 3 years in domestic sheep.

Fasciola hepatica has been reported from pronghorn in only one location—Drakes Flat in southwestern Oregon (Yoakum 1957). Two pronghorn does collected during December 1955 and April 1956 carried the parasite. One doe had seven adult flukes in bile ducts of her liver and five eggs in her gallbladder. Five flukes were recovered from the other doe. Lesions in the livers were reported as not pathologically significant. Both does appeared in good physical condition and carried twin fetuses. Apparently, conditions were just right for pronghorn to acquire this parasite on rangeland shared with cattle, horses and sheep during the autumn and winter of 1955–56. *F. hepatica* has not been reported from pronghorn since that time.

Tapeworms

Two classes of tapeworms—Cotyloda and Eucestoda—are common in ruminants. Cotyloda has two grooves, or bothria, on its scolex (head piece), whereas Eucestoda has four suckers. Both classes usually are found in the intestines of their host, and only Eucestoda has been found in pronghorn. The flattened bodies of these worms are without digestive tracts. The body, or strobila, of an adult tapeworm usually is long and segmented into proglottids. The anterior end, or scolex, is modified as an organ of attachment. It is provided with four suckers and, in some species, with small hooks. Growth takes place just behind the scolex, in the unsegmented "neck" region. Maturity of the genital organs advances as the proglottids (both sexes in each) move farther from the scolex. The oldest proglottids, near the posterior end of the chain, are filled with eggs. The wall of a proglottid may rupture, liberating the eggs, or the proglottid may detach and pass intact with the feces.

Moniezia is a cosmopolitan genus of tapeworms in its distribution and range of hosts. It occurs in most domestic and wild ruminants throughout the range of the pronghorn. Adult tapeworms of this genus shed eggs or proglottids that pass with the host's feces. Soil-dwelling mites ingest eggs, and the larval parasite (oncosphere) penetrates the body cavity of the mite. The cysticercoid stage, infective to a vertebrate host, is reached in 3 to 4 months. If the infected mite is ingested with forage, development of the adult tapeworm takes about 5 to 7 weeks in the gut of the herbivore (Figure 59). Infection in domestic sheep is eliminated spontaneously within 5 to 6 months.

Moniezia expansa and *M. benedeni* are large tapeworms, 3 to 12 feet (1–4 m) long and 0.6 inch (15 mm) wide. The scolex has four prominent suckers, up to 0.04 inch (1 mm) in diameter, but no rostellum or hooks. The proglottids are broad, rather than long, and each has two complete sets of male and female reproductive organs.

Fawns heavily infected with *Moniezia* sometimes die or become predisposed to predation. Adults seldom are affected (Thorne et al. 1982). In North Dakota, Goldsby and Eveleth (1954) found *Moniezia* spp. only in pronghorn grazing on cattle range, and Gilmore and Allen (1960) reported *M. expansa* from pronghorn collected on ranges in New Mexico that were used primarily by cattle. Dunbar and Velarde (1998) found light infestations of *Moniezia* spp. (36–40 eggs/g of feces) in 2 (10%) of 20 adult pronghorn on a refuge in Oregon without domestic livestock.

Moniezia expansa and *M. benedeni* infections have been reported from most western states, but little can be done to protect free-ranging pronghorn. Most infections can be expected during dry weather in overgrazed pastures. However, abnormally wet years may enhance survival of tapeworm eggs and mites, and drive the mites higher than normal on the vegetation.

Thysanosoma actinioides, the fringed tapeworm, is found only in the western hemisphere. It infects domestic sheep and goats and practically all wild ruminants, but elk seem to be the most heavily parasitized of the wild animals. The worm derives its common name from the fringed posterior border of each segment, which gives it a fuzzy appearance. This segmented, almost colorless tapeworm usually is no more than 1 foot (30 cm) in length. The life cycle and mode of transmission are unknown, but larvae will develop in certain insects. Fringed tapeworms often are found in the small intestines of elk close to the pyloric valve, but no pathological changes are seen. In domestic and bighorn sheep, the common bile duct sometimes is enlarged by the worms, and the walls of the duct and biliary tree are greatly thickened and inflamed.

Three (11%) of 27 pronghorn examined in New Mexico had *T. actinioides*; two harbored only one worm each (Allen and Kyles 1953, Gilmore and Allen 1960). Those investigators concluded that *T. actinioides* undoubtedly is better adapted to domestic sheep than to pronghorn. Later investigations in Texas and New Mexico showed heavy infestations (Allen and Samson 1960, Hailey et al. 1966).

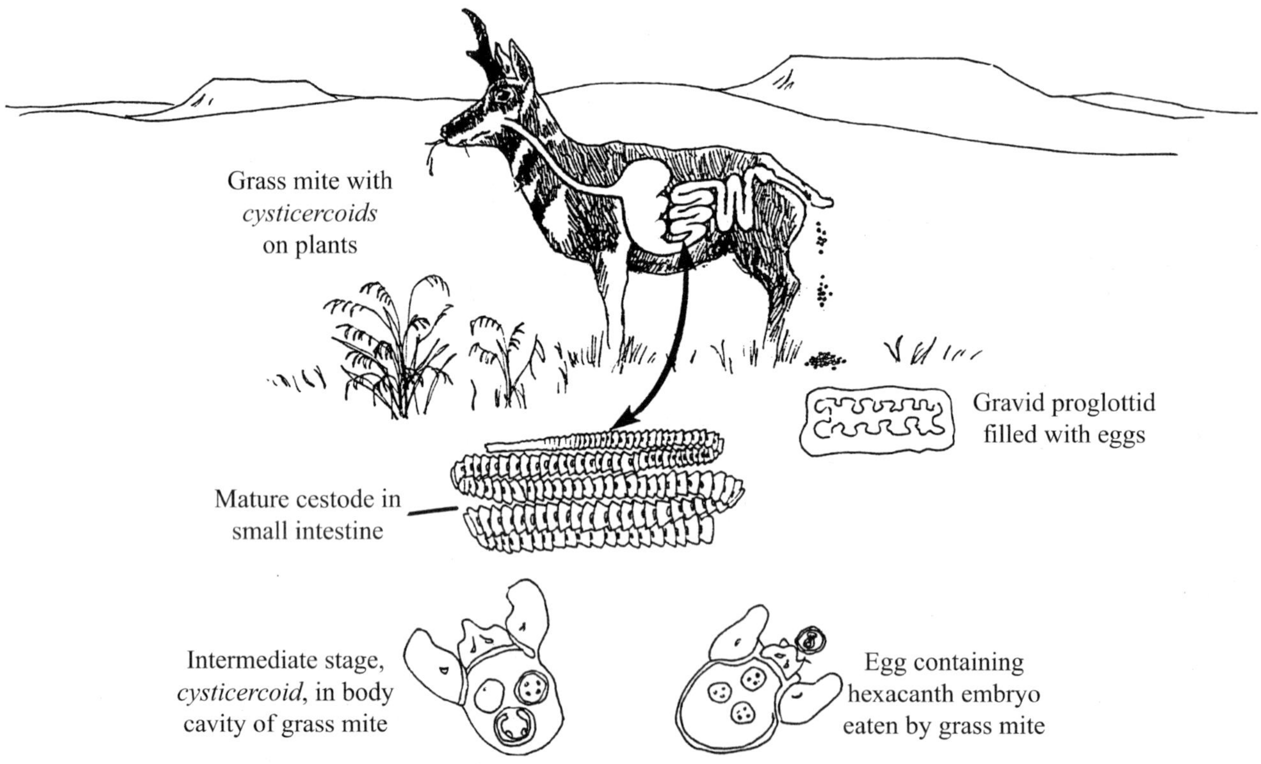

FIGURE 59. *The life cycle of the broad tapeworm,* Moniezia expansa, *sometimes found in the small intestines of pronghorn.* Illustration by Neal Wiegert from Honess and Winter (1956).

(Left) *A pronghorn fawn severely infected with* Moniezia *tapeworms (monieziasis). It was very weak and easily captured; the hind legs and rump were heavily caked with feces. The fawn was emaciated and dehydrated and probably would not have lived long. In Wyoming, such infections were found to be most common between midsummer and late autumn on heavily grazed rangelands.* (Right) *Fifty or fewer tapeworms have little apparent effect on domestic sheep, but greater numbers (which can occlude the intestine) possibly interfere with normal digestion. Illness and possible deaths of pronghorn, especially fawns, have been attributed to* Moniezia *spp. infections. In Wyoming, these infections were most prevalent between midsummer and late autumn on overgrazed ranges. Infected fawns were characterized by small body size, dehydration, chronic diarrhea and weakness. At necropsy, emaciation was a consistent feature, and the abdominal cavities in some specimens contained 1.7 to 2.5 ounces (50–75 mL) of brownish fluid. Spleens often were small, kidneys were congested and the small intestine distended as seen in the above intestine of a North Dakota pronghorn. Some tapeworm infections were complicated by heavy roundworm infestations. Fawns heavily infected with tapeworms are vulnerable to predation even if they survive the parasites.* Photos by E. T. Thorne; courtesy of the Wyoming Department of Game and Fish.

Apparently, pronghorn can be expected to harbor many *T. actinioides* where they share heavily grazed rangelands with infested livestock.

Taenia hydatigena

Cysticerci, or bladder worms, frequently are found in herbivores. They are the larval stages of tapeworms that occur in the gut of a carnivore. The thin-necked bladder worm, formerly called *Cysticercus tenuicollis*, is the larval stage of *T. hydatigena*, a tapeworm about 16 feet (5 m) long with double rows of 26 to 44 rostellar hooks. It commonly is found in the intestines of dogs, coyotes and wolves. The larval stages are fluid-filled bladders (cysts) up to 3 inches (8.5 cm) in diameter containing an inverted scolex. The gravid proglottids or eggs of *T. hydatigena* pass in canid feces, and the eggs are ingested by ruminants. Eggs hatch in the small intestine, and the emerging oncospheres penetrate the intestine and can pass through the abdominal cavity to the liver. They develop to the infective cyst stage there or, more commonly, in the body cavity. If an appropriate carnivore eats a cyst from a ruminant, the cyst wall is digested, and the worm evaginates its scolex to attach to the carnivore's gut wall by means of suckers and hooks. Development to sexual maturity and release of proglottids and eggs follow.

Significant pathological changes and secondary bacterial infections have been reported for domestic and wild ruminants, but the light cases usually seen in wild animals are of little consequence. Most infections are acquired during summer and autumn, and the cysts persist for several months. At necropsy, one may see a swollen liver, red from hemorrhages, with bore channels (10–20 days after infection), or yellow from leukocytic infiltration (25–35 days after infection). Later, the only visible pathological change may be white, irregular-shaped scars on the liver surface.

Thorne et al. (1982) reported *T. hydatigena* from bighorn sheep, mountain goats, mule and white-tailed deer, elk and moose. A sterile, unidentified cyst was found on the liver of a Colorado pronghorn (Hoover et al. 1959). A single cyst was found in 1 of 60 pronghorn that I necropsied on the National Bison Range. It was about 1.5 inches (38 mm) long and 0.5 inch (13 mm) wide, and it was identified at the University of Montana as *T. hydatigena*. Of the hundreds of deer, elk, mountain sheep and mountain goats that I necropsied on the National Bison Range, almost every one had at least a few cysts, and some apparently healthy deer had 30 to 50. The single case in a pronghorn, and that from the animal's liver, indicates pronghorn are little affected by this parasite.

Taenia ovis krabbei is found as an adult in the intestine of certain wild canids. The larval stages generally are found in the musculature of wild cervids. The cysts are small, 0.20 to 0.35 inch (5–9 mm) in length, and heavy infections in muscles cause a condition commonly referred to as measles. Cysts often are located in the heart or diaphragm, as well as in skeletal muscles. Frequently, the cysts are overlooked because of their small size. This parasite has been found in mule deer, elk and bighorn sheep in western North America (Thorne et al. 1982).

Gravid segments of this tapeworm are expelled with feces of infected canids, and the eggs are ingested by ruminants. Oncospheres released from the eggs penetrate the small intestine and enter the portal circulation of the liver. They traverse the liver and migrate to cardiac and skeletal muscles where they develop into cysts within 2 to 3 months. Although the larvae are killed by cooking, heavily infected muscles are aesthetically objectionable and probably should not be used as human food (Thorne et al. 1982).

Two cases of this parasite were found in pronghorn. The first, in a host collected near Laramie, Wyoming, was identified on the basis of a single cysticercus. The second, from a hunter-killed animal, came from east-central Wyoming and also was identified from a single parasite. The two specimens had 26 and 28 hooks, respectively. Large hooks measured 143 to 162 μm and small hooks measured 100 to 115 μm—measurements within the ranges for this species (Shults and Stanton 1986). *T. ovis krabbei* and *T. o. ovis* of domestic sheep are morphologically indistinguishable, but Sweatman and Henshall (1962) considered them different subspecies with little or no cross transmission. If that is correct, the two Wyoming specimens mentioned could be either subspecies. Pronghorn exchange other parasites with both deer and sheep, so only experimental infections, or possibly electrophoresis, could prove those to which the pronghorn is susceptible.

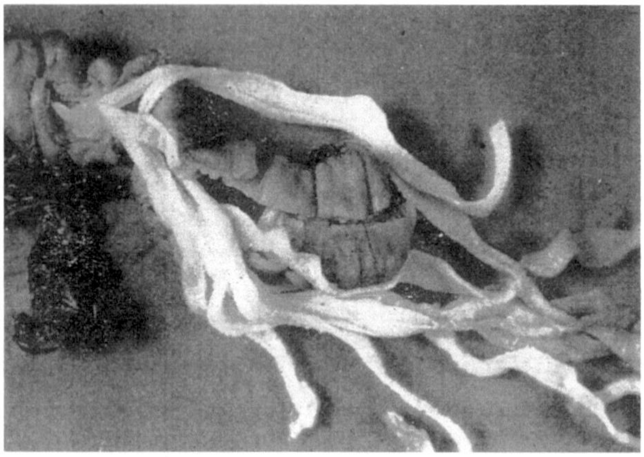

A Taenia ovis krabbei *cyst in the heart of a 2-year-old pronghorn buck shot in Wyoming. This parasite has seldom been found in pronghorn and apparently is of little importance to the species. If cysts are present in small numbers, they can be trimmed away and the rest of the carcass can be eaten, but infected meat should not be fed uncooked to dogs.* Photo by Bart W. O'Gara.

ROUNDWORMS

Roundworms, of the class Nematoda, include numerous species, both free-living and parasitic, around the world. The general morphology is an elongated, threadlike tube-within-a-tube, and sizes vary greatly. Although most nematodes are free-living, numerous species are parasites of plants or animals. Roundworms generally are less specialized physically and physiologically for parasitic existence than are flatworms. Infections in wildlife usually are detected by identifying eggs and larvae from fresh feces. Almost all wild ruminants harbor various nematodes, but only in areas of high animal concentrations are the worms numerous enough to produce visible signs of disease.

Loss of 30% to 60% of the annual increment of pronghorn fawns in South Dakota, primarily due to parasitic worms, was reported by Bever (1950). Few pronghorn, if any, escape parasitism throughout their lifetimes (Bever 1957). Bever stated that, with proper management of rangeland used by domestic livestock in South Dakota, no parasite-caused mortality was found. He presented evidence that heavily grazed sheep rangeland resulted in higher parasite loads in pronghorn than did properly grazed cattle rangeland.

The parasite load seemed to vary with the habitat of pronghorn in North Dakota (Goldsby and Eveleth 1954). Those on rangelands grazed by sheep had more abomasal parasites, whereas those on range grazed by cattle had more intestinal worms. Use of phenothiazine salt blocks on sheep range and abandonment of close sheepherding practices apparently alleviated the pronghorn parasite problem in South Dakota (Bever 1965). Proper management of livestock and pronghorn populations and rangelands should minimize health problems associated with these parasites.

To this point of the chapter, diseases and parasites have been listed alphabetically. The following lists nematodes by areas of the host in which they are found—lungworms and then those inhabiting the gastrointestinal tract. The latter are in the general order, from anterior to posterior, that they occur in pronghorn. The names for many nematodes have been changed over the years, and disagreements persist concerning the proper names for some. Used here are the most recently accepted names, with others in parentheses.

Lungworms

Lungworms infect many wild ruminants and probably are most devastating to bighorn sheep. Two subfamilies of nematodes commonly are found in the lungs of ruminants in western North America—Dictyocaulinae and Protostrongylinae. The former are yellowish white worms, and the areas infected by them are not readily apparent on the surfaces of the hosts' lungs. The latter are brown to reddish-brown worms, and areas in the lungs damaged by them usually are noticeable and sometimes pigmented.

Dictyocaulus spp. measure about 1.0 to 3.1 inches (25–80 mm) in length, and occur in the air passages of herbivores. This genus has been reported from all species of cervids in western North America, but generally is not believed to occur in pronghorn. However, during a study of game animals on heavily used rangeland in northern Yellowstone National Park, Rush (1932) reported *Dictyocaulus* spp. from the lungs of 4 (31%) of 13 pronghorn carcasses. Otherwise, hundreds of pronghorn lungs have been examined since 1932 without finding this parasite. Possibly, on the heavily used range in Yellowstone National Park shared by mule deer and elk, pronghorn did acquire *Dictyocaulus*, but the chance of infection by this parasite seems remote in pronghorn.

Protostrongylus macrotis (*Orthostrongylus macrotis*) are lungworms 1 to 1.6 inches (26–41 mm) long that infect elk, mule deer and pronghorn. This parasite was first reported from pronghorn in Yellowstone National Park by Dikmans (1932). Ninety-seven (97%) of 99 pronghorn collected in the same area during 1965 and 1966 harbored the worm (Greiner et al. 1974). *P. macrotis* was found in animals of all ages and during all months of the year, with a mean density of 22.1 worms (range 0–133 worms). Lungworms were not recovered from 26 pronghorn collected on the National Bison Range during the same study. Generally, the mature worms are found in the trachea and bronchi, and only the eggs and larvae occur deep in the lung, according to Honess and Winter (1956). However, the pronghorn collected in Yellowstone National Park had *P. macrotis* in all lobes of their lungs and in bronchioles deep in the lungs, possibly from postmortem migration (E. Greiner unpublished data:1965). Eight percent of the pronghorn investigated by Boddicker (1966) in South Dakota were infected.

Heavy infestations of lungworms can lead to unthriftiness of the lung or pneumonia, but I have never heard of a pronghorn with such a heavy infestation. The life cycle of *P. macrotis* is poorly known, but Pillmore (1958, 1959) experimentally transmitted the parasite from mule deer with terrestrial snails.

Gastrointestinal Worms

Gastrointestinal nematodes are common in all wild and domestic animals. According to Jessup and Boyse (1995), many of these parasites exist as two separate populations—those inside hosts and those on pastures.

Marshallagia (*Ostertagia*) *marshalli*, Marshall's worms, often are found in abomasums of domestic sheep and probably can be transmitted from domestic sheep to pronghorn and vice versa (Bergstrom 1975a). Males are 0.39 to 0.5 inch (10–13 mm) long, and females are 0.47 to 0.79 inch (12–20 mm) in length. The worm was reported as a parasite of pronghorn in Montana by Lucker and Dikmans (1945). During a survey of wild ruminants in Wyoming, Bergstrom (1975b) found Marshall's worms to be common in pronghorn and numerous in bighorn sheep. The parasite load usually did not exceed 600 female worms per pronghorn.

Haemonchus contortus (*H. atectus*, *H. pseudocontortus*, *H. fuhrmanni*, *H. cervinus*, *H. lunatus*, *H. santomei*, *Strongylus filicollis*, *Filaria denticulata*, *Abomesi ovis*), the large stomach worm, usually is found in small numbers in abomasums of wild ruminants, including pronghorn (Figure 60). The uterus of the female is lighter in color than the digestive tract and winds around it, hence the common names—barber pole worm and twisted stomach worm. The life cycle is direct, and blood is taken from the submucosa of the abomasum. If an infected animal is examined shortly after death, the reddish worms, 0.31 to 1.1 inches (8–27 mm) long, may be seen on the mucosa or in the lumen of the abomasum. This parasite has pathogenic potential because of its bloodsucking life style, but desiccation and cold winters on most pronghorn ranges probably prevent extensive buildup of the free-living larvae. However, *H. contortus* was listed by Bever (1957) and Boddicker and Hugghins (1969) as one of the roundworms that was numerous in pronghorn during studies in South Dakota, and it has been found in the pronghorn of most western states.

The strain found in pronghorn is not as pathogenic to domestic sheep as is the normal sheep strain (Allen et al. 1959). Allen and Samson (1961) postulated that the pronghorn strain of *Haemonchus* would be an asset to the sheep industry if lambs were inoculated with the strain or acquired it naturally. However, this possibility apparently has not been pursued.

Ostertagia spp., the medium stomach worms or abomasal worms, are common and sometimes numerous parasites that occur in abomasums, but rarely in the small intestines of ruminants. They sometimes are called brown stomach worms because they are brown when fresh. Clinical cases of disease from these slender worms are rare in pronghorn. They have a direct life cycle, as do most nematodes, with the eggs excreted in the host's feces and the infective larvae ingested by another susceptible herbivore. The ingested larvae exsheath in the omasum or abomasum of the host. They penetrate the abomasal mucosa

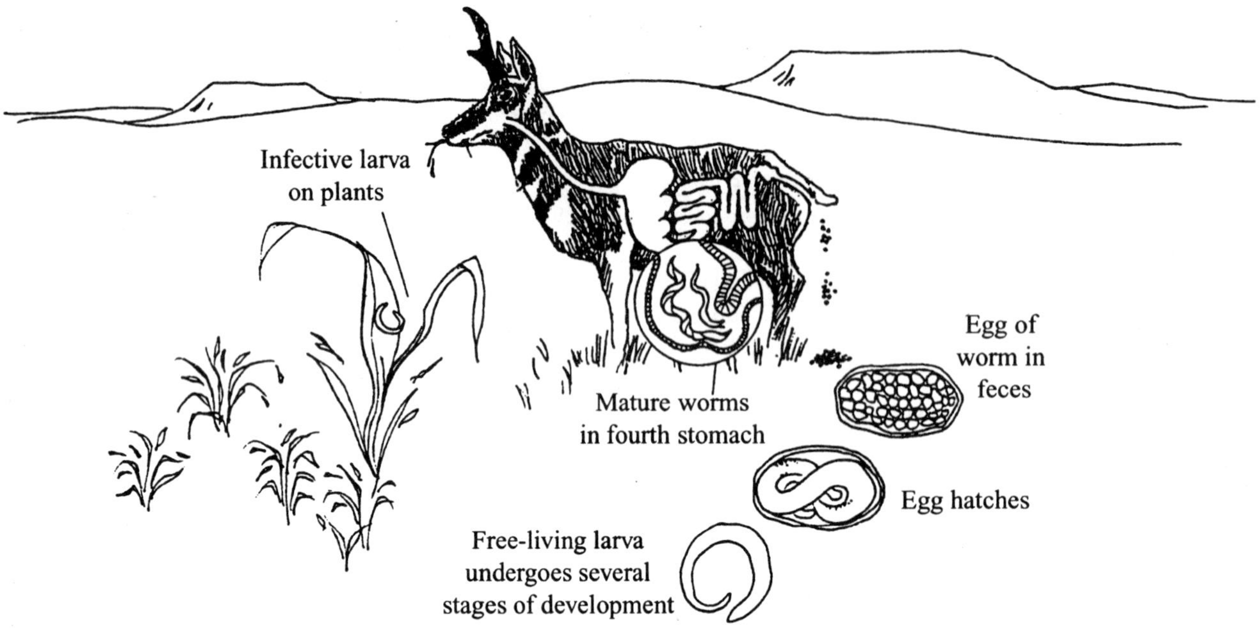

Figure 60. *The life cycle of the common stomach worm,* Haemonchus contortus, *which lives in the abomasum of pronghorn. This parasite feeds on blood by thrusting its head through the epithelium to reach the blood supply in the wall of the stomach. Eggs laid by the females leave the host in the feces. When conditions are right, a larva develops in the egg, hatches and undergoes two molts on the ground before it can infect another ruminant. After a rain or heavy dew, infective larvae crawl up vegetation to further their chance of being ingested by grazing animals.* Illustration by Neal Wiegert, from Honess and Winter (1956).

and spend one to several days developing beneath the mucosa, causing raised nodules. They return to the lumen of the abomasum, where they undergo further molts and mature. Small numbers cause no clinical symptoms, but large numbers induce diarrhea, poor digestion and loss of appetite.

Ostertagia bisonis (*Skrjabinagia bisonis*) originally was described from the abomasum of a bison in Canada (Cram 1925), but it also occurs in cattle and mule deer in Colorado, Montana and Wyoming. Males are 0.24 inch (6.2 mm) in length, and females are 0.26 inch (6.7 mm) long. This parasite was reported from the abomasum of a pronghorn in North Dakota (Goldsby and Eveleth 1954).

Osteragia bullosa (*Pseudostertagia bullosa*) primarily is a parasite of pronghorn as reported by Lucker and Dikmans (1945) and Gilmore and Allen (1960). It was the dominant parasitic species in pronghorn examined in North Dakota by Goldsby and Eveleth (1954). The males are about 0.28 inch (7 mm) long, and females are 0.35 to 0.39 inch (8.8–9.8 mm) in length. The occurrence of this parasite in domestic sheep probably is dependent on pronghorn sharing the range (Ransom and Hall 1912, Lucker and Dikmans 1945). It was found in more than 70% of the pronghorn investigated in South Dakota by Boddicker (1966). As many as 1,000 worms were found free in the chyme of the abomasum and in lesions of the mucosa with the heaviest concentrations in the pyloric area. Aside from tiny lesions in the mucosa, other pathological symptoms were not associated with these worms.

Osteragia circumcincta (*Strongylus vicarius, S. cervicornis, S. instabilis*) occurs in abomasums and occasionally in the small intestines of many ruminants throughout the world. In China, where many sheep are infected, it has been found in marmots. This is the most important species of *Ostertagia* in sheep, and it is found wherever sheep graze. The males are 0.30 to 0.33 inch (7.5–8.5 mm) in length, and females are about 0.39 to 0.48 inch (9.8–12.2 mm) long. They have been reported from pronghorn in Montana, North Dakota, South Dakota and Wyoming (Murie 1940, Bever 1950, Goldsby and Eveleth 1954, Honess and Winter 1956, Boddicker and Hugghins 1969).

Osteragia occidentalis (*O. trifida*) occurs in abomasums and rarely in the small intestines of sheep and goats in Eurasia, Australia and North Africa. It also is found in many wild ruminants, including bighorn sheep, mule deer and pronghorn in Wyoming (Thorne et al. 1982). The male is 0.47 to 0.63 inch (12–16 mm) long, but the female is undescribed.

Osteragia orloffi occurs in abomasums of sheep, cattle and wild ruminants in Eurasia and North America. Males are 0.20 to 0.35 inch (5–9 mm) long, and females are 0.22 to 0.35 inch (5.5–9.0 mm) long. It was reported from abomasums of pronghorn in South Dakota by Boddicker (1966).

A typical example of gastrointestinal worms was collected by Boddicker (1966) in South Dakota. Of the 60 pronghorn examined, only 1 was free of worms and only 1 harbored a single species. Mixed infections of three to eight species were common. These mixed infections usually included Haemonchus contortus, Ostertagia bullosa, 0. ostertagi *and* Nematodirella antilocapra *(the most numerous species in most pronghorn studies) in the abomasums (true stomachs). The latter two also were in the small intestines.* Photo by Peter T. Bromley.

Osteragia ostertagi (*Strongylus convolutus, S. cervicornis, S. harkeri*) occurs worldwide in abomasums of many species of ruminants and in the stomachs of a few other herbivores, including horses. It primarily is a parasite of cattle and is found occasionally in sheep and goats. The live worms are reddish-brown. Males are 0.26 to 0.30 inch (6.5–7.5 mm) long, and females measure 0.33 to 0.36 inch (8.3–9.2 mm). They have been reported in low numbers from pronghorn in North Dakota, South Dakota, Oregon and New Mexico (Bever 1950, Goldsby and Eveleth 1954, Yoakum 1957, Russell 1964).

Osteragia trifurcata (*O. pinnate*) occurs throughout the world in sheep, goats and wild ruminants. The male is 0.26 to 0.28 inch (6.5–7.0 mm) long, and the female—indistinguishable from that of *O. circumcincta*—is 0.39 to 0.48 inch (9.8–12.2 mm) long. This species was reported from abomasums and small intestines of pronghorn in Oregon and South Dakota by Yoakum (1957), Boddicker (1966), and Boddicker and Hugghins (1969).

Cooperia is a genus of small, reddish worms usually found in the small intestines and rarely in abomasums of ruminants. They are characterized by distinctive cuticular inflations anteriorly. Small, isolated areas of hemorrhage might be present in the small intestines, but because few worms usually are present, this parasite seldom debilitates its host.

Cooperia bisonis occurs primarily in the small intestines of bison. It rarely has been found in cattle or sheep in the western United States. The male is 0.28 inch (7 mm) long, and the female is 0.31 to 0.37 inch (8.0–9.5 mm) in length. Some investigators consider this parasite to be the same as *C. oncophora* (Boddicker 1966). It was reported from pronghorn in New Mexico by Lucker and Dikmans (1945).

Cooperia oncophora (*C. harkeri, Strongylus radiatus, S. ventricosus*) are small worms found in the small intestines of ruminants. Males are 0.20 to 0.35 inch (5–9 mm) long, and females measure 0.24 to 0.31 inch (6–8 mm). They were reported from pronghorn in Wyoming by Honess and Winter (1956), in New Mexico by Gilmore and Allen (1960) and in South Dakota by Boddicker and Hugghins (1969). Fawns apparently have less resistance to infection than previously infected adults; small, isolated areas of hemorrhages might be noted along the length of the intestine in infected animals, but clinical signs have not been noted (Thorne et al. 1982).

Cooperia pectinata (*C. nicolli*) occurs in the small intestines and rarely in abomasums of cattle, sheep, the dromedary, roe deer and various other ruminants throughout the world. In North America, it is found principally in southern states. Males are about 0.28 inch (7 mm) long, and females are 0.30 to 0.35 inch (7.5–9.0 mm) long. This worm was found in the abomasums and the small intestines of pronghorn in New Mexico by Gilmore and Allen (1960).

Cooperia punctata (*C. fieldingi, C. laterouniformis, Strongyloides bovis*) occurs in the small intestines and rarely abomasums of cattle, sheep, water buffalo and several wild ruminants. It is more common in cattle than in sheep and, although distributed throughout North America, it is especially prevalent in Hawaii. Males are 0.19 to 0.23 inch (4.7–5.9 mm) long, and females are 0.22 to 0.30 inch (5.7–7.5 mm) long. It also was reported from abomasums and the small intestines of pronghorn in New Mexico by Gilmore and Allen (1960).

Cooperia surnabada occurs primarily in the small intestines of cattle in Eurasia and North America. It also has been reported from sheep in the eastern United States and South America and from bighorn sheep. The males are 0.19 to 0.23 inch (4.9–5.8 mm) in length, and females are 0.29 to 0.48 inch (7.4–12.3 mm) long. This parasite was reported from the abomasums of two pronghorn in South Dakota by Boddicker and Hugghins (1969).

Strongyloides, called threadworms, is a genus containing several species that live part of their life cycles as parasites in the small intestines of wild and domestic mammals (mostly herbivores). The parasitic females produce eggs that pass with the hosts' feces and give rise, without fertilization, to infective larvae of another parasitic generation or to a free-living generation of minute worms. These free-living worms reproduce sexually, producing infective larvae of another parasitic generation.

Strongyloides papillosus (*S. contortus, S. longus, S. ovocinctus, S. suis, S. vituli, Trichosomum papillosum*) occurs in the mucosa of the small intestines in numerous species of herbivores. The greatest numbers are found near the pylorus. The parasitic parthenogenic females are 0.14 to 0.24 inch (3.5–6.0 mm) long. The free-living males are tiny—about 700 to 800 μm long. This species was reported from a captive pronghorn by Ransom (1911) and from pronghorn in Colorado by Hoover et al. (1959).

Trichostrongylus colubriformis (*T. instabilis, T. subtilis*) is found throughout the world in ruminants, rodents and primates, including humans. It is very common in sheep but less so in cattle. Most trichostrongylids are pink in recently killed hosts, but rapidly lose color. Both sexes are threadlike, with males measuring 0.18 to 0.30 inch (4.3–7.7 mm) in length and females 0.20 to 0.34 inch (5.0–8.6 mm). The intensity of clinical signs depends on the number of larvae ingested. Poor appetite and weight loss can result, but deaths due to the parasite are rare. Mild hyperemia and small hemorrhages can occur from the anterior duodenum through the midjejunum. This parasite was reported by Bever (1957) as one of the three species of roundworms found in a high percentage of South Dakota pronghorn. It also has been reported from pronghorn (numerous in some individuals) in many other states.

Trichostrongylus axei (*T. gracilis, T. extenuatus, T. tenuissimus*) occurs throughout the world in abomasums and occasionally in the small intestines of cattle, sheep and other ruminants. This small worm is the most common species of the genus in cattle. Occasionally, it is found in the stomachs of nonruminants. Males are 0.09 to 0.24 inch (2.3–6.0 mm) long, and females measure 0.13 to 0.31 inch (3.2–8.0 mm) in length. *T. axei* was found in abomasums and the small intestines of New Mexican pronghorn (Gilmore and Allen 1960).

Trichostrongylus capricola (*T. andreevi*) occurs in sheep, goats and roe deer of Africa, Eurasia and North America. Males measure 0.14 to 0.23 inch (3.5–5.8 mm), and females are 0.20 to 0.25 inch (5.0–6.4 mm). These rather small worms were found by Ransom (1911) in a captive pronghorn.

Nematodirus spp. are commonly called thread-necked strongyles, and are well adapted for surviving cool, dry summers and long, cold winters. Little is known concerning the effect of these parasites on their hosts, except that large numbers (e.g., 30,000 worms) of some species cause deaths of domestic lambs. Worm burdens generally are lower in wild than in domestic ruminants. Secondary bacterial infections—a result of allowing bacteria access to tissues—might be one of the most significant effects of these parasites. Diarrhea, weight loss and weakness are signs of heavy infections, and small hemorrhages in the intestines are seen at necropsy. Dunbar and Velarde (1998) reported 2 to 16 *Nemotodirus* eggs per gram of feces in 7 (35%) of 20 fecal samples from adult pronghorn on the Hart Mountain National Antelope Refuge in Oregon.

Nematodirus abnormalis occurs in the small intestines of domestic and wild ruminants in North and South America, Eurasia and New Zealand. Infective larvae within egg membranes can endure more than 4 years of desiccation and cold. Males are 0.43 to 0.67 inch (11–17 mm) long, and females are 0.71 to 0.98 inch (18–25 mm) long, resembling those of *N. filicolis*. It occasionally is found in bighorn sheep, mule and white-tailed deer, and pronghorn in Wyoming (Thorne et al. 1982). It was reported from the small intestines of pronghorn in South Dakota by Lucker and Dikmans (1945).

Nematodirus filicolis (*N. furcatus*), an important pathogen of sheep in New Zealand, occurs in the small intestines of sheep, cattle and numerous wild ruminants throughout the world. Males are 0.39 to 0.59 inch (10–15 mm) long, and females are 0.59 to 0.90 inch (15–20 mm) in length. In Wyoming, it occasionally is found in the small intestines of elk, mule and white-tailed deer, Rocky Mountain goats and sheep, and pronghorn (Honess and Winter 1956, Thorne et al. 1982).

Nematodirus oiratianus (*N. lanceolatus*) commonly is found in domestic and bighorn sheep and occasionally in pronghorn and elk. Males are 0.43 to 0.65 inch (11.0–16.5 mm) long, and females are 0.57 to 0.69 inch (14.5–17.6 mm). It was reported in pronghorn from New Mexico by Gilmore and Allen (1960).

Nematodirus spathiger occurs in the small intestines of sheep, cattle, camels and many species of wild ruminants. Third-stage larvae are infective year-round. It is a common parasite of sheep and goats, and is easily confused with *N. filicolis*. Males are 0.39 to 0.59 inch (10–15 mm) long, and females are 0.47 to 0.79 inch (12–20 mm) long. These worms were found in the small intestines of pronghorn in North Dakota, South Dakota and Wyoming (Lucker and Dikmans 1945, Bever 1950, Goldsby and Eveleth 1954, Honess and Winter 1956, Boddicker and Hugghins 1969). These parasites seldom are numerous enough in wild animals to cause serious problems.

Nematodirella spp. also are called thread-necked strongyles. They frequently are found in the small intestines of wild ruminants of high arid plains and arctic tundra, but transmission between domestic and wild ruminants in those areas is doubtful because *Nematodirella* spp. apparently are more host-specific than other trichostrongylid nematodes.

Nematodirella antilocapra (*N. longispiculata*, *N. longispiculata antilocapra*, *N. longissimespiculata*, *N. alcidis*, *N. longispicularis*, *Nematodirus antilocapra*) commonly is found in pronghorn and occasionally in mule deer and domestic sheep. Males are 0.64 to 0.96 inch (16.2–24.5 mm) long, and females are 1.1 to 1.7 inches (27.6–42.7 mm) long. Large numbers of red blood cells often highlight the worm's intestines (Boddicker 1966). *N. antilocapra* primarily is a parasite of pronghorn, and it is considered the most common parasitic worm in that species (Gilmore and Allen 1960). As many as 10,000 have been recovered from a fawn in Wyoming (Thorne et al. 1982), and the infection rate within herds may be as high as 80% (Boddicker and Hugghins 1969). In South Dakota, 1,000 to 2,000 *N. antilocaprae* commonly were found in the duodenums of pronghorn. Infestations of more than 5,000 were observed in some cases, accompanied by hemorrhaging and lesions. Large numbers of worms also were reported from pronghorn in North Dakota, New Mexico and Texas (Goldsby and Eveleth 1954, Gilmore and Allen 1960, Hailey et al. 1966).

Nematodirella antilocapra, in combination with tapeworms (*Moniezia expansa*), frequently causes diarrhea and debilitation in pronghorn fawns (Honess and Winter 1956). It was one of three roundworms considered most pathogenic to pronghorn in South Dakota by Bever (1950). The best (and probably only) methods of controlling such parasites require maintaining low population densities of pronghorn, freedom of movement for the animals and good rangeland conditions.

Capillaria spp. is a very large genus of small, slender worms with 37 known species in fish, 13 in amphibians, 13 in reptiles, 104 in birds and 88 in mammals. The species occurring in mammals have direct life histories, but those of some species occurring in birds are indirect, with earthworms serving as intermediate hosts.

Capillaria bovis (*C. longipes*, *Trichosoma bovis*) is a threadlike worm occurring within the mucosa of the small intestines of numerous ruminants in the Americas and Eurasia. The males are 0.43 to 0.51 inch (11–13 mm) long and females are 0.71 to 0.98 inch (18–25 mm). They were found in the small intestines of captive pronghorn by Ransom (1911). If it occurs in wild pronghorn, it must be very scarce.

Capillaria brevipes occurs in the small intestines of sheep throughout North America and in cattle in Texas, Virginia and Puerto Rico. It has not been reported outside the United States. The males are 0.31 to 0.35 inch (8–9 mm) long, and the females are about 0.47 inch (12 mm) long. This worm was reported from North Dakota pronghorn by Goldsby and Eveleth (1954).

Capillaria hepatica generally is found in the livers of hares and rabbits, but it can infect humans. Mature worms measure 1.6 to 2.0 inches (40–50 mm) long. Females lay eggs in bile ducts or liver tissue. When exposed to air, the eggs embryonate and become infective to a new host. If ingested, the larvae leave the egg in the host's intestine, move into the bloodstream and circulate to the liver to complete the life cycle. Adults, larvae and eggs can induce cirrhosis of the liver and other pathological changes, often resulting in severe debilitation. Livers of four pronghorn does (of 41 pronghorn carcasses examined) in Alberta contained white-to-yellow lesions. Examination of the lesions revealed numerous *C. hepatica* eggs surrounded by fibrous tissue; many were fragmented and contained calcium and iron salts, suggesting that they were not viable and that pronghorn are accidental and unsuitable hosts for *C. hepatica* (Barrett and Chalmers 1972).

Toxascaris leonia are common intestinal roundworms of dogs and cats and related carnivores. Adult male worms range in length from 2.8 to 3.5 inches (7–9 cm), and females are 4.0 to 6.7 inches (10–17 cm) long. This species does not use an intermediate host in its life cycle. Infections usually are obtained by eating the colorless eggs, which hatch and complete their first molt in the host duodenum. They then enter the wall of the intestine, where they continue their development before returning to the lumen of the gut to mature.

Although the genus *Toxascaris* is a common parasite of canines and felines, it occasionally is found in ruminants. A pronghorn collected in Wyoming—because it was weak, emaciated and probably would have died within a short time—had nematodes identified as *T. leonia* in the posterior portion of its small intestine (Bergstrom et al. 1985).

Trichuris spp., whipworms, have been found in the cecums or colons of pronghorn in Wyoming (Thorne et al. 1982), but the exact species and measurements apparently have not been determined. The anterior of the body is long and slender in this genus. The posterior is thick, and the hind end of the male is curled. They likely pose little or no threat to free-ranging animals.

Trichuris ovis has been recorded from the cecums of many species of domestic and wild ruminants. The males are 2.0 to 3.1 inches (50–80 mm) long, and the females are 1.4 to 2.5 inches (35–70 mm). *T. ovis* was reported from the large intestines of pronghorn from North Dakota and South Dakota by Bever (1950) and Goldsby and Eveleth (1954).

Trichuris discolor (*T. parvispiculum*, *T. giraffae*, *Trichocephalus discolor*, *Salamia discolor*) is found in ruminants throughout the world. The males are 1.77 to 2.32 inches (45–59 mm) long, and the females are orange-yellow and 1.69 to 2.17 inches (43–55 mm) long. The slender anterior is about three-fourths of the body length. Lucker and Dikmans (1945) found one male worm, tentatively identified as *T. discolor*, in the large intestine of a South Dakota pronghorn.

Abdominal Cavity Worms

Setaria labiato papillosa (*S. cervi*, *S. buxi*), the abdominal threadworm, occurs in the peritoneal cavity of many ruminants and is cosmopolitan in distribution. The worm also has been found in the small intestines or embedded in various organs, including eyes and udders. Males are 1.6 to 2.4 inches (40–60 mm) long and females are 2.4 to 4.7 inches (60–120 mm). The life cycle is unknown, but an arthropod vector is likely. No pathogenic conditions have been ascribed to the presence of the parasite in the peritoneal cavity, but erratic migrations may occur.

Setaria sp. was reported from the abomasum and abdominal cavity of South Dakota pronghorn by Bever (1950), and 12 *S. labiato papillosa* were reported by Boddicker and Hugghins (1969). The first nematode I found in the abdominal cavity of a pronghorn was identified as *S. cervi*. Hunters brought more *Setaria* to me at the Montana Cooperative Wildlife Research Unit than any other pronghorn parasite. The size of the worm, plus the fact that anyone gutting a pronghorn was likely to see it, seemed the reasons—not that *Setaria* was common in Montana pronghorn.

Subcutaneous Worms

Onchocerca cervipedis (*O. flexuosa*, *Wehrdikmansia cervipedis*) is a species of long, thin worm that lives in connective tissue beneath the skin of deer and domestic animals. Although called hockworms, they may be located above or below the hock. The bodies of both sexes are threadlike; the males are 1.2 to 2.4 inches (30–60 mm) long, and the females are 5.5 to 7.9 inches (140–200 mm in length). A specimen was found on the undersurface of a piece of skin from an Idaho pronghorn by Dikmans (1933). These long nematodes occasionally are found by hunters while skinning big game, but are more of a curiosity than a threat to wild animals.

ECTOPARASITES

Ectoparasites are arthropods—segmented animals such as ticks, mites, flies, lice and fleas—that commonly are found on or in the skin, hair, feathers or ears of their hosts. Most do not produce disease, but they can cause mechanical injury or irritation. Some serve as vectors of disease.

Hard Ticks

Body sizes of hard ticks, the family Ixodidae, depend on the time since their last meal. Engorged adult females of all of the species found on pronghorn may exceed 0.6 inch (15 mm) in length and 0.4 inch (10 mm) in width. All hard ticks have a hard shield (scutum) on the upper surface of the thorax. The scutum is not influenced by feeding and is, thus, the most constant indicator of size of a tick. The body of an unfed adult is about 2.5 times as long as the scutum. All measurements of hard ticks are from Brinton et al. (1965). The scutum usually has colors differing from the rest of the tick's body. Colors given in the following descriptions of hard ticks are the most common for a species but may vary from area to area or even between individuals in an area. Also, the seasonal biology may vary between northern and southern limits of a species' range or by altitude.

Pronghorn seem to harbor fewer ectoparasites than do other ungulates sharing rangelands with them. Caton (1877) speculated that the odors supplied by pronghorn rump glands were deterrents to ectoparasites. I suspect that the pronghorn's choice of habitat—bedding and feeding on and among short, sparse vegetation—offers crawling ectoparasites limited opportunity of access. Even flying insects, although they cause harassment during summer, do not ordinarily deposit eggs on pronghorn. I have never found nasal or skin bots in pronghorn—even where bighorn, bison, cattle and deer were heavily infected—nor have I heard of anyone else finding them. Photo courtesy of the U.S. Fish and Wildlife Service, Charles M. Russell Wildlife Refuge.

Dermacentor albipictus, winter or moose ticks, are widespread in the United States and Canada, especially in hilly or wooded areas. Although horses, cattle and moose are the usual hosts, the winter tick occasionally is taken from other closely related mammals. Under laboratory conditions, it will feed reluctantly on rodents. These ticks generally are more numerous on animals in poor condition than on those in robust health. The life cycle is interesting ecologically because the ticks feed in winter rather than during the spring and summer months as do nearly all other hard ticks (Howell 1939). Essentially, three peaks of adult presence on animals occur, with the first (autumn) and second (early winter) peaks representing infestations by larvae that were produced the preceding spring. Howell believed that the third (late winter) adult peak may result from progeny produced by female ticks that fed during early autumn.

Unfed females are 0.13 to 0.2 inch (3.2–5.3 mm) in length and 0.08 to 0.15 (2–3.8 mm) in width. Males measure 0.15 to 0.24 inch (3.7–6.2 mm) in length and 0.07 to 0.15 (1.7–3.8 mm) in width. The leathery integument is brown, but the back of the male is covered with a grayish scutum splotched with brown. The scutum of the female covers only the anterior fourth of the body. Engorged females are yellowish gray above and light gray below.

The winter tick is a one-host species, that is, six-legged larvae and eight-legged nymphs and adults usually develop, molt and feed on a single host. Adult females are found on hosts during early autumn through late winter, and drop to the ground after engorgement. There, when temperatures allow activity, they lay eggs that hatch under rocks or debris during early spring. The young search for hosts in autumn and early winter. Adult ticks, due to growth and engorgement, are readily seen during spring. Winter ticks not only irritate wild animals, they stress them as the tick burden increases and adult females engorge. Heavily infested animals may be weakened, emaciated, show nervous signs and lose hair. The taking of large blood meals by females begins during midwinter, when their hosts are least able to spare blood (Thorne et al. 1982).

The taxonomic status of the winter tick has been in question for many years. Differences in coloration and size of various geographic variants or forms have created confusion and led to the naming of another species—the brown winter tick, *D. nigrolineatus*. Ernst and Gladney (1975) crossed representatives of supposed *D. albipictus* and *D. nigrolineatus* in an attempt to clarify the relationship between the two forms. In reciprocal crosses, isolated pairs of the two forms mated readily, and the cross-mated females engorged, detached and produced viable egg masses. Females from a series of 10 different hybrid crosses produced viable eggs. No reproductive isolating mechanism, including hybrid sterility, was detected. Ernst and Gladney concluded that the two geographic forms of winter ticks they studied seemed to be conspecific.

Winter ticks are the ectoparasites most commonly found on pronghorn. They have been reported from pronghorn in Montana (Rush 1932), Utah (Coffey 1954, Beck 1955), Oregon (Yoakum 1957) and Texas (Hailey et al. 1966). The only references to heavy tick infestations on pronghorn are from a small area in the southwestern United States. Brown winter ticks were found in abundance crawling on pronghorn in the Trans-Pecos area of Texas (Buechner 1950). A picture of numerous ticks engorged on a dead pronghorn doe in New Mexico during

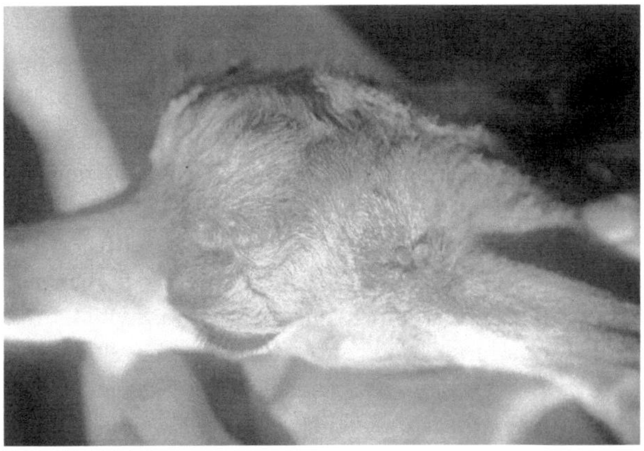

*Fourteen of 80 young fawns captured on Hart Mountain National Wildlife Refuge, Oregon, during the 1955 and 1956 fawning seasons were parasitized by ticks (Yoakum 1957). Parasite loads generally were low, but one fawn hosted 33 ticks and another fawn hosted 17. Ten adult pronghorn from the same area hosted 1 to 11 ticks each. Twelve ticks were collected alive from two adult does; 10 were identified as Rocky Mountain wood ticks (*Dermacenter andersoni*) and two were winter ticks (*D. albipictus*).* Photo by Jim D. Yoakum.

1940 was shown by Russell (1964:88). He also reported that heavy infestations of ticks, assumed to be brown winter ticks, were observed on the withers, necks and other portions of the bodies of certain pronghorn captured during trapping operations in March 1938. The ticks were crowded over sizable portions of the animals' necks and withers, "like grains of corn on a cob. The hair of the victimized individuals lacked the usual luster and smooth arrangement." Russell further stated that hundreds of pronghorn had been handled during all seasons in New Mexico, but ticks had not been noticed except in the two cases noted above. Also, all the ticks observed were in one section of the state during winter and spring. This was in shrub or borderline shrub types west of the Pecos River in sheep pastures. The heaviest tick loads I have seen on pronghorn were much lighter than those on other ungulates sharing the rangeland. Apparently, winter ticks are little more than an occasional irritant to pronghorn in most areas.

Dermacentor andersoni (*D. venustus*), the Rocky Mountain wood tick or Rocky Mountain spotted fever tick, is another of the hard ticks. Unfed females measure 0.12 to 0.19 inch (3.0–4.9 mm) in length and 0.08 to 0.11 inch (1.9–2.8 mm) in width. Males are 0.1 to 0.22 inch (2.5–5.6 mm) in length and 0.07 to 0.15 inch (1.8–3.8 mm) in width. The back of an adult male is covered by a grayish white scutum with bluish markings. The reddish brown females have only the anterior third of their backs covered by scuta. Engorged females are mottled grayish, and their scuta are indistinct because they do not expand with the abdomen.

This tick requires blood from two or more hosts during its life cycle. Larvae feed on small rodents; nymphs parasitize small rodents and some slightly larger mammals, such as rabbits. And, adults usually feed on large mammals, including humans. All three stages may be found on porcupines and jackrabbits.

Engorged females drop from their hosts during spring and early summer and lay several thousand eggs beneath the ground litter. These ticks hibernate both as nymphs and adults. The adults that come up in the spring may find hosts, feed, drop off and lay eggs. Other individuals, failing to find a host, must go back into hiding, pass the remainder of the summer and winter, and try again next spring (Cooley 1932). Wilkinson (1968) provided definitive proof of 1-year and 2-year cycles. Larvae hatch from June through August, seek a blood meal from a small mammal (necessary to continued development) and drop off to molt into nymphs. Some nymphs must overwinter in the unfed stage, but earlier emerging cohorts may locate a blood meal, drop off and find shelter where they molt to the adult stage and overwinter. During late winter and early spring, the adults and unfed nymphs climb grass or shrubs to wait for a suitable mammalian host. When they find a host, nymphs feed, drop off and produce unfed adults that aestivo-hibernate until the following year. Overwintering adults usually find a host on which they feed and mate, thereby producing the next overwintering nymph and adult siblings. While off the host, adults seek attachment until about early August, depending on the elevation and climatic conditions. Adult females usually will not be found on hosts beyond mid-August, but males may remain through October (M. Coan personal communication:1985).

Loss of blood to Rocky Mountain wood ticks usually is not great. However, tick paralysis of the host can occur when a female tick has fed for several days, usually near the base of the skull. Paralysis is caused by a neurotoxin released by the feeding tick, and the affected animal generally dies unless the tick is removed. After removal, recovery tends to be rapid. Female ticks from both 1-year and 2-year cycles are capable of paralyzing sheep (Wilkinson 1968).

Rocky Mountain wood ticks were reported from pronghorn in Oregon (Chamberlin 1937), Alberta (Holland 1940) and Colorado (Hoover et al. 1959). However, the chance of many pronghorn being seriously debilitated by this parasite seems remote.

Dermacentor parumapertus, the rabbit *Dermocenter*, is confined to arid and semiarid regions of the southwestern United States. Unfed females measure 0.12 to 0.13 inch (3.1–3.2 mm) in length and 0.08 to 0.09 inch (2.0–2.2 mm) in width. Males are 0.10 to 0.11 inch (2.6–2.9 mm) in length and 0.06 to 0.08 inch (1.5–2.0 mm) in width. Specimens may range from a pale yellowish-orange to liver brown, and from ornate (with white on the dorsal surface) to inornate. Generally, *D. parumapertus* looks like a small *D. andersoni* with more spiderlike legs (longer in relation to the body than other *Dermocenter* species) (M. Coan personal communication:1989).

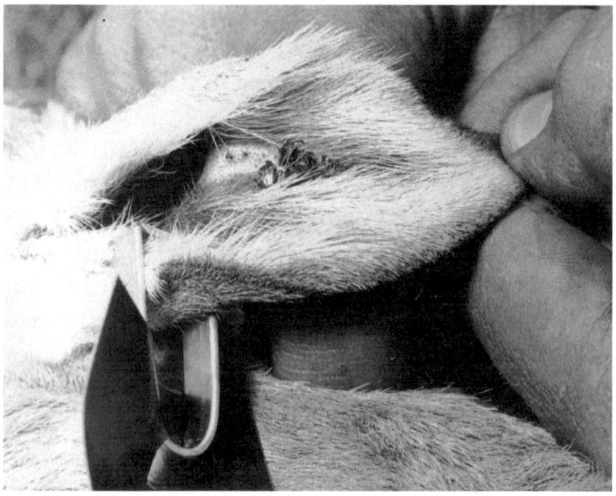

Rocky Mountain wood ticks embedded in the outer ear of a pronghorn fawn captured, tagged and released on May 28, 1964, in Alberta. Ticks seem less numerous on pronghorn than on deer and elk that share their habitat. Photo by George Mitchell.

Nymphs are most abundant during April, May and June. Larvae also occur most frequently during spring. Rodents are more heavily infested with larvae than with nymphs, and the converse is true for black-tailed jackrabbits. The adult population of ticks reaches its peak of activity on black-tailed jackrabbits in the latter part of July and is the product of overwintering adults and newly emerged females from spring-feeding nymphs. Known hosts of the immature stages of this species include antelope ground squirrels, kangaroo rats, kangaroo mice, wood rats and mice of the genera *Perognathus*, *Peromyscus* and *Reithrodontomys*. The number of eggs laid by each female varies from 30 to 6,000. Egg production is directly related to quantity of blood imbibed. The rabbit *Dermocenter* can serve as a reservoir for Colorado tick fever and tularemia (Arthur 1960).

During pronghorn fawn trapping and tagging operations in Texas, rabbit *Dermocenters* were found on a few fawns (Hailey 1979). Whether the ticks were embedded or just crawling on the fawns was not reported. When these ticks are found on cattle, they usually are on the eyelids where the thin skin allows them to engorge (Strickland and Gerrish 1965).

Soft Ticks

Otobius megnini commonly is called the spinose ear tick. Soft ticks (family Argasidae) have no scuta on their backs and are covered by leathery integument that is dark and shiny. The second, or nymphal stage, has four pairs of legs and usually is found as a parasite in the ears of animals. The body is about 0.24 inch (6 mm) long, usually constricted behind the third pair of legs and might be oval or mask-shaped. Heavy spines are conspicuous from about the level of the third legs forward, and relatively slender spines are found on the posterior of the body. Larvae have been found in caves; they find a host and attach deep in an ear, where they feed and molt to the nymphal stage in about 5 days. The nymphs feed for several months, detach, then molt to free-living, unspined adults. Without feeding again, mating occurs, and the females lay eggs in which the larvae develop.

Most animals parasitized by spinose ear ticks show annoyance. Heavily parasitized animals become heavy headed—hanging their heads and swaying them from side to side. The tick may cause paralysis resulting in death. Affected animals lose their appetites in the initial phase, then lose coordination of leg movements and collapse. Death may occur in a few days to 2 weeks. The intensity of individual infestations generally is proportional to the concentration of host animals in a given area.

This tick has been reported from numerous species in western North America, including pronghorn in Texas (Buechner 1950, Hailey et al. 1966), Oklahoma (Ellis 1955) and New Mexico (Schad 1958, Colbenson 1962, Meleney and Roberts 1970). In pronghorn, these ticks are found deep in the auditory canal. Meleney and Roberts found them in 13 (54%) of 24 pronghorn shot near Mesa, New Mexico. However, soft ticks are occasional and unimportant in pronghorn, although a few animals may be heavily infested.

Keds

Keds, family Hippoboscidae, are flies with flattened leathery bodies. Wings may be absent, reduced or lost in the adult. Adults feed on their hosts' blood. Keds frequently are called ticks by laymen.

Neolipoptena ferrisi, the louse fly or deer ked, is a brownish black insect 0.1 to 0.8 inch (3–21 mm) long. It has a distinct head, a narrow thorax bearing six legs and a rounded abdomen. Mating takes place on the host throughout the year. Larvae develop in the females and pupate in the pelage. Winged adults escape from puparia, find a host, lose their wings, feed on the host's blood, then mate. These keds are common parasites of deer in the western United States. The only record of them on pronghorn came from California (Herman 1945). A winged specimen was found on a pronghorn at a checking station where infected deer had been handled, and the occurrence was considered accidental. The next year, two keds were found firmly attached (wings already broken off)—one feeding on the upper lip and the other on the eyelid of a pronghorn. Biting and blood sucking causes distress to the host, but the few that might infest a pronghorn probably cause minimal damage.

SPECIAL PUBLIC CONCERNS

Two animal diseases received extensive media coverage during the late 19th and early 20th centuries. They are briefly addressed here because of their potential seriousness and concerns of wildlife managers, livestock raisers, sportsmen and sportswomen, and the general public—not because pronghorn have been involved.

Chronic Wasting Disease

Very little is known about chronic wasting disease (CWD). An abnormal protein called a prion interacts with normal brain proteins, causing conformation change. This change eventually leads to breakdown of brain tissue, so that it appears spongelike when viewed through a microscope. Animals that develop CWD lose weight, drink a lot and seem unafraid of humans. About 5% of deer and 1% of elk examined in southeastern Wyoming had CWD. It can be acquired by ingestion of prions. Occurrence is higher in captive herds of deer and elk than in the wild, so anything that concentrates animals—such as feeding grounds—probably increases the rate of transmission (Kreeger 1999).

The principal concerns with CWD include possible effects on wildlife populations, livestock and humans. The latter two stem from similarities to the mad cow disease in Great Britain, which is caused by a prion different from the CWD form. No evidence indicated that CWD affects cattle or humans. Researchers have been unable to transmit CWD to cattle under experimental conditions. Currently, no test is available that detects CWD in a living animal, and no cure is known (Kreeger 1999).

Because CWD is poorly understood, is of recent origin or recent recognition and is diagnosed primarily in cervids that die in captivity, it generally is thought of as a game farm disease. This, coupled with similarities to Great Britain's infamous mad cow disease, has created grave concerns among some segments of the North American public, especially hunters. A statement often appearing in newspaper and magazine articles, which increases concern, is that CWD is always fatal. With no way of diagnosing the disease before death, it is impossible to be sure of this.

Chronic wasting disease having been found in deer of parts of Colorado, South Dakota, Wyoming, Utah, Wisconsin, Illinois, New Mexico and Nebraska; pronghorn certainly have been exposed in the pronghorn states. If they were susceptible and some died in the wild, there is little chance of finding the carcasses in time to make a correct diagnosis. However, pronghorn have been kept in captivity for years within the affected area in Colorado, at Colorado State University in Fort Collins, and in Wyoming at the Wyoming Game and Fish Research Laboratory at the University of Wyoming in Laramie. In fact, mule deer at the Colorado facility have been diagnosed with CWD. Highly qualified veterinarians work with captive wildlife, including pronghorn, at both universities, but no reports of pronghorn dying from CWD have been forthcoming. Thus, pronghorn, like cattle, may be immune. Research to determine whether they are or are not is in order. Current (2004) information on CWD is available on-line at <www.cwd-info.org> and <http://www.cdc.gov/ncidod/EID/vol10no6/03-1082.htm>.

Foot-and-Mouth Disease

Foot-and-mouth disease (FMD) is caused by a highly transmissible virus that produces blisterlike sores in the mouth, between the hooves, and on the tongue and teats of cloven-hooved animals. The blisters quickly turn into very painful erosions and ulcers. Early signs of FMD may include listlessness, lameness due to foot lesions and drooling because of mouth lesions. Affected animals rapidly lose weight and often die (Thorne 2001).

During a 1924 FMD outbreak in California, wild deer were infected and more than 22,000 were killed to curtail the disease. Foot-and-mouth disease last occurred in the United States in 1929, and safeguards also have protected Canada and Mexico for many years (Thorne 2001). Recent outbreaks in Europe, and their affects on livestock, agricultural economics and tourism during the early 21st century have brought FMD to the attention of North Americans. The disease could devastate livestock and game animals if it was introduced again.

No references were found to pronghorn having been infected with FMD, but there seems no reason to doubt they would be. When the disease swept through eastern and southern Africa, carried by cattle from India in the late 19th century, practically all wild ungulates were affected and reduced to low numbers.

Anyone, particularly hunters, traveling from FMD-affected countries, including Argentina, France, Great Britain, and many African and Asian countries, should attempt to keep the disease out of North America. When hunting overseas, I carry powdered Clorox and laundry detergent to scrub my boots and hunting clothes before returning. What cannot be laundered is given away.

CONCLUSIONS

Pronghorn are relatively free of diseases and parasites, perhaps because of the dry, open habitat they usually occupy. Although occasional die-offs have been reported, they have not depressed any population for an extended period. Under some circumstances, pronghorn and livestock share diseases and parasites. The challenge is to manage pronghorn to minimize negative effects of these pathogens. Jessup and Boyse (1995:411) offer the following suggestions to meet this challenge: "In general, the healthier the livestock and wildlife populations are, the less risk there is of disease and parasite transmission. Good nutrition, reasonable stocking rates, and maintenance of animals at or below carrying capacity are the obvious basics. In general, healthy resident livestock are less likely to harbor pathogenic strains of disease agents than are animals brought from long distances or a variety of locations. Separating newly purchased stock or wildlife from a herd for a period of observation before introduction is a good practice. Preventive medical programs including routine locally appropriate vaccination, prophylactic antibiotic treatment, nutritional supplementation, and worming and parasite control programs will be a sound investment for the stockman or game rancher and may have benefits for wildlife. Serologic and pathogen isolation surveillance of free-ranging wildlife, particularly animals being relocated, should be routine. All these programs help insure healthy wildlife and livestock.

"Although there may be some circumstances where separation of wildlife and livestock is advantageous, the existence of shared diseases and parasites should not be routinely used as an argument for exclusion of either wildlife or livestock from grazing lands. The key to minimizing the potential negative impacts of diseases and parasites that may be shared by wildlife and livestock is a knowledge of their existence and a willingness on the part of wildlife and livestock interests to take appropriate management actions."

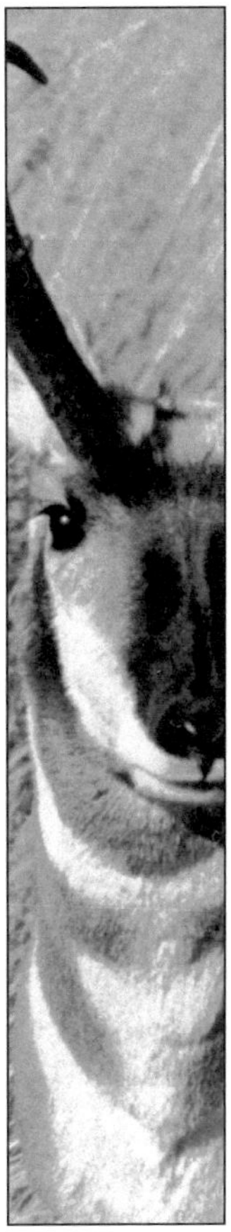

C H A P T E R E L E V E N

Predation

Bart W. O'Gara and Harley G. Shaw

Predators kill pronghorn, especially fawns, but predation is only one of many factors that influence pronghorn population trends. Predator control is not a panacea to be applied whenever pronghorn populations are lower than desired. Predator control seldom is intensive enough to actually control predators. Each herd of pronghorn is governed by controls endemic to that herd and the habitat in which it lives. The quantity and quality of habitat in which a pronghorn lives is the overriding influence on all inimical factors, including predation.

Lee et al. (1998:61)

The annual production of pronghorn fawns represents an energy bonanza available to coyotes during a short season when additional food is essential.

Byers (1997a:54)

Aside from man and his activities in hunting and livestock grazing, the coyote is the greatest menace that the pronghorn has to face.

Gabrielson (1935:576)

The above statement by famed conservationist Ira Gabrielson reflected the general sentiment of the time. Predation generally was thought to be important in the regulation of pronghorn numbers, and the roles of weather, rangeland conditions and nutrition seldom were considered. Although coyotes and other predators can be important decimating factors under certain circumstances, few biologists considered predation a major problem facing pronghorn at the beginning of the 21st century.

LEVELS OF PREDATION

Before radiotelemetry, studies concerning causes of low fawn survival did not account for most of the fawns that simply disappeared. However, studies summarized by Connolly (1978a) implicated predation as a significant factor: coyote predation on fawns in Texas prevented pronghorn from increasing despite predator control (Jones 1949); predator control with 1080 (sodium monofluoroacetate) allowed a previously static pronghorn population to increase enough to be hunted (Arrington and Edwards 1951); pronghorn herds in Oregon recovered from about 1,000 in 1920 to 10,000 by 1935 because of harvest protection and the elimination of 7,500 predators (Young and Jackson 1951); removal of coyotes increased fawn/doe ratios in small remnant herds of Utah (Udy 1953); the number of fawns per 100 does was 142% and 129% higher in successive years after coyotes were removed, compared with a nearby range in Texas without predator control (Knowlton et al. 1971); and 17 pronghorn translocated to the Umatilla, Oregon, Ordnance Depot in 1969 failed to increase until 1971, when 9 does were seen with

13 fawns, after removal of 135 coyotes and, with continued coyote control, the herd increased to about 100 by 1976 (Oregon State Game Commission 1972b).

Einarsen (1948) recorded several instances of coyotes and bobcats killing fawn and adult pronghorn in Oregon. He saw a coyote stalking a fawn during a blizzard, only to be put to flight by does. Summarizing his interpretation of coyote/pronghorn relations, Einarsen (1948:79) wrote: "It is naive to assume that this predator will graciously extend immunity to pronghorn kids when guarded only by its kind, for which the coyote has no ingrained fear. Coyotes uncontrolled, may be a serious limiting factor to the kid crop."

Eighteen telemetry studies, involving 995 pronghorn fawns whose fates were known, are identified in Table 60. The mortality for those instrumented fawns was 71%, with predation accounting for 76% of the total mortality. Mortality from predation was higher than indicated because little remained of some carcasses, predation was uncertain and the cause of death was listed as unknown. Also, most radio transmitters were designed to fall off by the time the fawns were about 2 months of age. Beale's (1978) study showed that fawns at 4 months of age were being killed by predators. Some of the deaths credited to coyotes were reported as "coyote involved" or "possible coyote" by the authors because little was left of the carcasses. In many cases, the apparently healthy fawns were seen 24 to 48 hours before death. Only a few teeth or strips of hide and a bloody radio collar, usually with tooth holes and near a coyote den, were found. Carrion feeding seemed unlikely in such instances. In experiments with carrion feeding by coyotes on domestic lambs, the coyotes seldom ate carrion and sometimes made fresh kills near 1-day-old lamb carcasses (O'Gara et al. 1983).

In the 18 studies, deaths due to causes other than predation included starvation, disease and research-induced (Table 60). Generally, instrumented animals had similar survival rates to those in the study populations. In cases where total predation was listed as greater than the combined predation of coyotes, bobcats and golden eagles, a small number of deaths were due to predation by mountain lions, small raptors or badgers, and some deaths were by unknown predators. Fawns whose radios were lost or dysfunctional were not included in Table 60.

TABLE 60. Mortality of pronghorn fawns monitored by radiotelemetry

		Mortality[b]	*Predation*[b]	*Predator kills*[b]			
Province or state	*n*[a]	*Number (%)*	*Number (%)*	*Bobcat*	*Coyote*	*Eagle*	*Study*
Montana	19	16 (84)	5 (26)	—	5	—	Reichel (1976)
Alberta	62	50 (65)	27 (44)	13[c]	14[c]	—	Barrett (1978)
Utah	200	127 (64)	97 (48)	36	39	21	Beale (1978)
Montana	23	20 (87)	20 (87)	3	16	1	Von Gunten (1978)
Idaho	42	26 (62)	15 (36)	2	4	7	Brodie (1979)
Texas	18	12 (67)	9 (50)	3	6	—	Tucker (1979)
Montana	24	16 (67)	14 (58)	3	9	2	Corneli (1980)
Nevada	34	22 (65)	13 (38)	—	12	1	McNay (1980)
Idaho	48	30 (63)	24 (50)	3	7	8	Autenrieth (1982)
Oregon	131	91 (70)	83 (63)	7	55	8	Trainer et al. (1983)
Idaho	67	26 (39)	13 (19)	7	3	3	Autenrieth (1984)
Colorado	26	17 (65)	12 (4.6)	—	12	—	Firchow (1986)
Arizona	11	5 (45)	1 (9)	—	1	—	Ockenfels et al. (1992)
Texas	78	66 (85)	63 (81)	5	49	3	Canon (1993)
Kansas	43	35 (81)	25 (58)	—	25[d]	—	Rothchild (1993)
Oregon	104	86 (83)	62 (60)	—	50	—	Gregg et al. (1999)
Oregon	26	26 (100)	16[e] (62)	1	14	1	Dunbar (1999)
Oregon	39	31 (79)	29 (94)	—	281	—	M. R. Dunbar (unpubl. 1999 data)
Totals	995	702 (71)	528 (53)	83 (16)	352 (67)	56 (11)	

[a] n = number of fawns instrumented.

[b] Percentages in parentheses are total mortality of radioed fawns due to all causes and total mortality of radioed fawns due to predation. The percentages listed as totals under three species of predators are the percentages of predation mortality that were attributed to that species of predator. They do not add up to 100% because of kills by unknown or lesser predators.

[c] Bobcats and coyotes reportedly were nearly equally involved. For this table, 13 were assigned to bobcats and 14 to coyotes for the totals.

[d] Remains of 19 fawns could not be found for 2 to 5 days. Apparently the fawns were buried by coyotes, and the solar-powered transmitters did not transmit because they were not receiving light. Thus, the transmitters were not located until coyotes returned, consumed the fawns and exposed the transmitters to light. Thus 19 of the 25 "coyote kills" were labeled coyote involved because only small portions of hair, bones and transmitters could be found at coyote caches.

[e] Two other deaths were probable predation, and five more were probable coyote predation.

Some of the studies reported in Table 60 were conducted because productivity was low in the study population. Thus, fawn mortality probably was higher than should be expected in most populations. Other studies were conducted primarily to ascertain selection of bedding habitat by fawns, and productivity was considered adequate. High levels of predation on neonates (first summer) are not unique to pronghorn. Linnell et al. (1995) reviewed 111 reports concerning neonatal mortality of 10 species of ruminants in temperate climates. They found that, in environments where predation occurred, mortality averaged 47%. Predators accounted for 67% of that mortality. In contrast, first summer mortality averaged 19% for studies in environments lacking predators.

Although predators cause substantial fawn mortality in some populations, such mortality is compensatory in many instances (see Pyrah 1987, Kohlmann 1998a, Danvir 2000, Chapter 12). Many conservationists, including some biologists, maintain that predators cull weak and sick animals to the benefit of prey population. In fact, under ordinary circumstances, young animals killed by predators are likely to have been as healthy as the ones that escaped predation. Sickly young actually may be less susceptible to predation because they tend to move less, so draw less attention (O'Gara et al. 1986).

In most areas, human activities and human-induced circumstances, such as strip mining, subdivisions, fences, livestock grazing and irrigation projects, destroy more pronghorn—through habitat destruction and restricted access to critical habitats—than predators take. But wherever pronghorn populations are static or declining, predation on fawns can accelerate a downward trend. Under most circumstances, pronghorn fawns are vulnerable to predators about 1 month out of 12. Thus, coyotes, bobcats, golden eagles and other predators generally are not tied to population fluctuations of pronghorn or any other big game populations. These predators are generalists, depending on the prey base variety of large mammals, rodents, lagomorphs, birds, reptiles, invertebrates and, in the case of coyotes, occasional vegetable matter. Thus, the importance of predation may increase when pronghorn populations are small or decreasing. Naturally, predators taking 100 fawns from a population in a valley where 1,000 fawns are born probably is biologically insignificant, but their taking 100 fawns from the same valley when only 150 fawns are born is significant.

Predators kill some pronghorn throughout bighorn's range, but significant predation is thought to occur only on marginal pronghorn ranges or other areas where the number of predators is high in relation to the number of pronghorn (Lee et al. 1998). Red foxes in Australia depressed numbers of European rabbits on one study site, but not on another with a higher density of rabbits—apparently because of better habitat (Banks 2000). Rabbit populations escaped predator regulation once they exceeded a critical density. The same sort of relationship may be true for pronghorn. Most fawns are killed before 3 weeks of age, when separated from their dams. When

Predators kill far more pronghorn fawns than adults, but remains of fawn carcasses are relatively hard to find. Since the advent of miniaturized radio transmitters, as shown on the hiding fawn above, determining the number of pronghorn fawns killed by predators has been possible and the percentages killed in some areas are significant to the population. The first published biotelemetry study of predation on pronghorn fawns was by Beale and Smith (1973) in western Utah. Of 44 instrumented fawns recovered after death, 27 had been killed by bobcats. Bobcat predation was cited as the greatest single factor reducing the pronghorn population, although only a few bobcats were involved. Photo by LuRay Parker; courtesy of the Wyoming Department of Game and Fish, Cheyenne.

does are present, they usually will protect their fawns from predators.

It is assumed sometimes that degraded or overgrazed rangelands make fawns vulnerable to predators and that big sagebrush or other cover is needed for hiding fawns (Autenrieth 1982, see Chapter 13). In Idaho, fawns bedding in the tall (30 to 54 inches [76–137 cm]) sagebrush/foothill community type suffered significantly higher predator-related mortality (1/10.6 days use) than did fawns using the short (8 to 10inches [20–25 cm]) sagebrush/grass community type (1/101.5 days use) (Bodie 1979, personal communication:1999). Aggressiveness of the does toward predators, whose approaches were easily seen in the short vegetation, may have contributed to the difference. However, another Idaho study indicated that hiding cover somewhat alleviated predation (Autenrieth 1984), and poor, limited cover was indicated as a factor in the probability of coyotes locating bedded fawns in Nevada (McNay 1980). Other factors, such as lookout points for bobcats and eagles, denning sites for coyotes and updrafts for soaring eagles, may influence levels of predation more than do the actual bedding sites.

Also poorly understood is the relationship between predation on pronghorn fawns and availability of alternate prey. Beale (1986) presented data indicating that abundant lagomorphs and rodents greatly alleviated fawn predation on 50 square miles (130 km^2) of semi-desert rangeland in western Utah. Lagomorphs were scarce during 1975, 1976 and 1977, and predators took 56%, 80% and 83%, respectively, of radio-marked fawns during summer of those years. In 1978, black-tailed jackrabbits increased 400% and remained relatively numerous through 1980. Cottontails, kangaroo rats and deer mice increased during the same period, but not as significantly as jackrabbits. In 1978, only 9% of the marked pronghorn fawns were lost to predators. Fawns were not radio-marked during 1979 or 1980, but total mortality did not exceed 35% either year (Beale 1986). After three summers of high fawn survival, pronghorn on Beale's (1986) study area increased from 85 animals during autumn 1977 to 231 by autumn 1980. Although, these data suggest an important relationship between levels of fawn predation and abundance of alternate prey, the pronghorn, rabbit and rodent populations may all have been reacting to favorable weather and improved vegetative conditions. Also, abundant lagomorphs and rodents surely would invite and support large numbers of predators. The former are subject to drastic, rapid die-offs, especially during late winter. That leaves a profusion of hungry predators to focus on pronghorn fawns the following spring. Thus, plentiful alternate prey may be a short-term blessing for pronghorn fawns, followed by catastrophe.

Dunbar and Giordano (2000) suggested that sagebrush voles were the primary winter food of coyotes on Hart Mountain National Antelope Refuge in Oregon. They proposed that prescribed burns could reduce sagebrush and sagebrush voles, thereby reducing coyotes' winter food supply and providing a nonlethal means of controlling them.

Millions of bison and pronghorn once occupied expansive prairies where Indians, wolves and golden eagles were the dominant predators. The Indian hunters and wolves are mostly gone; bobcats and coyotes have inherited much of their altered role, and sheep and cattle now are the predominant range animals. Pronghorn, although they have made an impressive comeback, often are restricted in their movements by agricultural areas, highways and fences. Thus, some herds are localized and relatively small. Under such restricted circumstances, predators may keep pronghorn populations from increasing or even eliminate them. For instance, pronghorn introduced to the National Bison Range in Montana were eliminated by predators (P. Wright personal communication:1965). With predator control, the next pronghorn translocation animals there increased rapidly (see Byers 1997a).

Coyotes occasionally kill adult pronghorn. Grinnell (1897a) watched three coyotes chase a doe in relays and kill her. They took turns; as one chased her, the others rested. If the doe had run straight away, it would have left the predators behind. Instead, she ran in large circles, as pronghorn are prone to do. The resting coyotes ascended hills or knolls to keep the chase in sight. When the doe turned and ran back toward a waiting coyote, it would sneak into her path and take up the chase. Two coyotes, working as a team in Texas, were seen stalking an adult pronghorn buck (Harley 1979). They caught the buck after a chase of about 1 mile (1.6 km). During spring 2001 in Yellowstone National Park, J. Byers (personal communication:2002) saw a pronghorn charge a coyote that was hunting close to her bedded fawn. The coyote killed the doe. During the next week or so, six more does with fawns were killed in the same general area.

During the 1950s, when many pronghorn were being translocated into unoccupied ranges in Montana, success increased with size of translocations, up to about 100 animals, especially where predators were numerous (J. McLucas personal communication:1970). In populations smaller than 100, social bonds seemed to break down, normal herding and predator defense mechanisms ceased to operate, and territoriality was reduced.

The Anderson Mesa pronghorn herd was one of the largest and most productive in Arizona. It numbered about 3,000 animals for a short period during the 1930s. Aerial surveys begun in 1944 indicated a peak of 1,800 animals in

While surveying pronghorn from a helicopter in the Owhyee Desert of Nevada, biologists and the pilot observed two coyotes pursuing a pronghorn doe. After completing that segment of the transect, the crew returned to the scene of the chase and obtained this photo of one coyote killing the pronghorn. Investigation revealed the victim was an adult doe in apparently good condition. Photo by Jim Jeffress; courtesy of the Nevada Department of Wildlife, Carson City.

1950, followed by a gradual decline, then a sharp drop during the severe winter of 1967–68. Since 1968, the herd has remained stable between 250 to 400. Fawn survival has been consistently low, except during years when coyotes were controlled by 1080. Six years of observing fawning activity indicated a normal birth rate, healthy vigorous fawns and normal maternal care (Neff and Woolsey 1980). Malnutrition, disease and parasites appeared to be insignificant factors in low fawn survival, and coyote predation appeared to be largely responsible for the lack of growth in the herd. This study is discussed in more detail later in the chapter.

Pronghorn numbered about 10,000 in Nebraska during the mid-1970s (K. Menzel personal communication:1992). Numbers were reduced to about 3,000 by the late 1980s, apparently because of low fawn/doe ratios. The low ratios appeared to be related to coyote predation.

Radiotelemetry was used to estimate survival of 58 adult pronghorn and 54 fawns in Texas during 1990 and 1991 (Canon and Bryant 1992). Adult survival estimates during the 2-year period were 0.906, 1.0 and 0.888 for all adults, bucks and does, respectively. Cause of death in adults was attributed to mountain lion predation, dietary problems, parturition complications and unknown causes. Fawn survival estimates were 0.097 in 1990 and 0.166 in 1991. Fawn losses in 1990 were attributed to predation by coyotes (50%), mountain lions (20%), golden eagles (5%) and unknown causes (15%). In 1991, predation also was the major cause of fawn mortality, with 53% attributed to coyotes, 9% to bobcats, 6% to mountain lions, 6% to golden eagles and 6% unknown.

The effects of predation usually are most pronounced when associated with other adverse conditions for a prey population. Seton (1927) reported that, on the National

The pronghorn buck above was found alive but hamstrung, near Brooks, Alberta, on February 3, 1965. Tracks in the snow indicated coyotes were the attackers. On that day, Chuck Gordon, Alberta Wildlife warden, shot this animal and four other pronghorn that were hamstrung similarly in the immediate area. He reported that 35 to 40 hamstrung pronghorn had to be destroyed in Newell County during January and February 1965. Gordon related that the coyotes chased the pronghorn, hamstrung and downed those they caught, then ran off in small groups to attack other pronghorn. Such behavior is uncommon. It probably occurs primarily when conditions are such that hunger is satiated and attacks continue for the thrill of the chase. In this case, snow conditions gave the coyotes a distinct advantage over the pronghorn during a 6-week period. Photo by Chuck Gordon.

Bison Range, a band of 64 pronghorn that normally would outrun predators was destroyed completely by them during a severe winter. Predators were able to drive the pronghorn into snowdrifts (where their movement was restricted) and kill them there. Fawn survival in Oregon decreased from 53 fawns per 100 does in 1987 to 21 fawns per 100 does in 1991. The decrease in fawn survival was related to drought and coyote predation (Van Dyke and Willis 1992). Although they had no direct evidence to support it, Oregon biologists believed the drought had a negative impact on overall productivity. At the same time, prices of coyote pelts were low, and coyote removal by trappers and hunters declined accordingly. As a result, coyote numbers increased, and pronghorn fawn survival rates decreased.

Another example of weather influencing predation was provided by Firchow (1986) in Colorado. There, 14 of 43 radio-collared adult pronghorn died during 1983 and 1984. Deaths were attributed to hunting (36%) and, coyote predation after a severe March snowstorm (29%). Under normal circumstances, coyote predation was only 14% and unknown causes were 21%. Thus, twice as many adult pronghorn were killed by coyotes in a short time after a snowstorm as were killed by the same predators during the rest of the 2-year period.

Bobcats are the second largest predation threat to pronghorn fawns. Bobcats seem to be more efficient predators of older fawns than coyotes, probably because bobcats are relatively proficient at sneaking close to a victim before attacking. Generally, bobcats live in more rocky, brushy country than do coyotes. When pronghorn entered such country in Utah during dry weather, seeking succulent vegetation, their speed and fields of vision probably were moderated by the terrain and vegetative cover, and increased predation resulted (Beale and Smith 1973). Nelson (1925) reported that a full-grown pronghorn buck was killed by a bobcat during a snowstorm in Montana.

Golden eagles are third in importance as predators of pronghorn fawns (Table 60). Sometimes, does can save their fawns from eagles. Woods (1925) watched a doe fight off an eagle that was attacking her fawns. Eagles may vie with mountain lions as the primary natural predator of adult pronghorn today. The pronghorn's speed, which foils most mammalian predators, is easily matched by the eagle. Open-field running by pronghorn also makes them easy prey for the big birds. O'Gara has re-

(Top) *While surveying wildlife from a light aircraft in eastern Montana, U.S. Fish and Wildlife Service biologists observed a golden eagle pursuing a yearling (21- to 22-month old) pronghorn buck. The eagle bound to the buck, apparently stabbing repeatedly through the rib cage into the lungs. The buck, with blood streaming down its shoulders and ribs, stopped at a fence and appeared to have given up as a photo was snapped from the passing plane.* (Left) *Biologists landed at the kill site the next morning and found the buck almost consumed.* Photos by Bob Burkholder.

ceived about 10 reports of eagles attacking or killing pronghorn for every one received concerning coyotes chasing pronghorn.

Wolves may once have been efficient predators of pronghorn. Along the Missouri River below the mouth of the Yellowstone, on April 29, 1805, Merriwether Lewis wrote: "Game is still very abundant we can scarcely cast our eyes in any direction without perceiving deer, Elk, Buffalo or Antelopes. The quantity of wolves appear to increase in the same proportion; they generally hunt in parties of six eight or ten; they kill a great number of the Antelopes at this season; the Antelopes are yet meager and the females are big with young; the wolves take them most generally in attempting to swim the river; in this manner my dog caught one drowned it and brought it on shore; they are but clumsy swimmers, tho' on land when in good order, they are extremely fleet and durable. We have frequently seen the wolves in pursuit of the Antelope in the plains; they appear to decoy a single one from a flock; and then pursue it, alternately relieving each other until they take it" (Moulton 1987b:85).

Today, the ranges of wolves and pronghorn seldom overlap, except in Yellowstone National Park. There, after reintroduction in 1995, 1,209 kills of ruminants were observed by personnel of the wolf recovery project as of April 2000 (K. Murphy personal communication:2000). Only seven kills were pronghorn—one unknown and six fawns. Since wolves became established in the park, pronghorn fawn survival has been much better where wolves are numerous than where they are not, apparently because of wolf predation on coyotes (J. Byers personal communication:2002).

Since 1970, mountain lions generally have increased in number throughout the western states and provinces. Most western states made lions game animals and regulated the sporting take. During the 1980s and 1990s, O'Gara received reports of mountain lions stalking pronghorn in the grasslands of central Montana. A few pronghorn probably were taken. Knipe (1944) recorded five pronghorn kills by lions in Arizona and indicated that lions occasionally preyed on pronghorn in certain areas. Shaw (1977) noted only one pronghorn kill among 62 prey animals during a lion study in central Arizona, where pronghorn were scarce at the time. A single lion killed more than 60 pronghorn during a 2-year period on the Edwards Plateau in Texas, according to Engstrom and Maxwell (1988).

Mountain lion predation markedly reduced survival rates of adult pronghorn does in rugged Arizona terrain (Ockenfels 1994a). During a 4-year study in central Arizona, Ockenfels (1994b) collared 47 adult pronghorn. Of 29 mortalities suffered by the collared animals, 11 were identified as lion predation. Three additional mortalities appeared to be lion kills, but too little remained for positive identification. Most lion kills were within 109 yards (100 m) of rugged terrain and usually in woody or relatively tall vegetation. Accounts of mountain lions taking pronghorn have been mostly from areas where pronghorn, seeking preferred vegetation, moved into fairly rough country or tall vegetation.

At the 15th Pronghorn Antelope Workshop in 1992, three speakers referred to mountain lion predation on fawns during recent or ongoing studies. In central Arizona, 8 of 15 radioed fawns were killed by lions (S. Schuetze personal communication:1992). Survival of 58 adults and 54 fawns was estimated with the help of radiotelemetry during 1990 and 1991 in the Trans-Pecos region of Texas (Canon and Bryant 1992). One adult doe was killed by a lion, and fawn survival was only 9.7% in 1990 and 16.6% in 1991; 20% of the 1990 deaths and 9% of the 1991 deaths were attributed to lion predation. Pronghorn fawns were killed by mountain lions during a study in Yellowstone National Park (D. Scott personal communication:1993). Also, K. Murphy (personal communication:1993) reported that a collared lion killed a pronghorn doe in a sagebrush area of Yellowstone National Park during April 1990. The carcass was disemboweled, dragged to a wooded gully and buried. Another doe and her fawns were killed by a lion on the National Bison Range during spring 2000 (B. West personal communication:2000). The doe and her fawns were on a high point near timber. Lions had been taking a substantial number of mountain goats and bighorn sheep in the area.

Black and grizzly bears have been known to kill fawns, as reported by Murie (1935) for Yellowstone National Park, but adult pronghorn (unless they were giving birth or otherwise debilitated) would not ordinarily be in much danger from bears. However, a pronghorn buck was chased and killed by a young grizzly in Yellowstone National Park during 1993 (D. Scott personal communication:1993).

Red foxes apparently took some pronghorn fawns when and where coyotes were scarce because of 1080 poisoning; the foxes could venture into the grasslands. A rancher in Wyoming told O'Gara of seeing foxes catch two fawns and finding many fawn parts at fox dens during the 1960s.

Badgers killed two fawns during a study in Oregon (Trainer et al. 1983). Badgers hunt predominantly in open country and take almost any small animal they can catch, so it would be probable that a few small pronghorn fawns were taken.

Domestic dogs may kill a few pronghorn fawns that are too young to outrun their pursuers. However, most reports of dogs killing pronghorn involve starving animals forced close to farmyards or towns by deep snow. Numbers of kills seldom are reported, but harassment of malnourished pronghorn probably leads to more deaths than do outright kills.

During the severe 1964–65 winter in Alberta, predation by dogs reached previously unrecorded levels. Snow cover forced pronghorn onto croplands and into the vicinity of human habitations, where they were subjected to harassment and predation by dogs. Within the city limits of Medicine Hat, at least 33 pronghorn were killed by dogs. Coyotes and dogs killed an additional 75 near that city (Mitchell 1980).

PREDATOR CONTROL

Before action is taken to control wildlife damage, it is important to understand the laws covering the targeted wildlife species. The management of most wild mammals, reptiles, and amphibians in the United States and Canada is the responsibility of the individual states and provinces. The capture, possession or killing of these vertebrates to achieve control of damage or nuisance situations is regulated by state or provincial laws.

Dolbeer et al. (1994:474)

Most predator control in the United States is for the benefit of livestock. Connolly (1978a) reported that about 250 times more money was spent for protection of livestock during fiscal year 1976 than was spent to benefit wildlife. However, control to protect livestock also may benefit pronghorn. Herds have increased in areas during predator control and decreased when control was curtailed. Einarsen (1948) noted that coyote control demanded by stockmen in Oregon contributed to increased pronghorn numbers.

A questionnaire was sent to provincial and state wildlife agencies in 2000 concerning predator control, or lack of it, for the benefit of pronghorn (Table 61). Neither of the two Canadian provinces with pronghorn expended funds for reduction of predators.

The only Mexican state responding to the questionnaire, Baja California Sur, indicated that predators were removed for the benefit of the endangered peninsular pronghorn. Personnel of Centro de Investigaciones Biologicas del Noroeste direct control measures themselves, but the amounts of time and money expended were not divulged.

Ten of the 16 states of the United States that harbor pronghorn contributed funds to other agencies (mostly the U.S. Department of Agriculture's Wildlife Services) in 1999 for predator control. Apparently, most of the money

Table 61. States contributing funds to predator control, amounts contributed in 1999 and percentage of possible benefit to pronghorn

State	*Dollars contributed*	*Percentage of possible benefit to pronghorn*
Arizona	22,900	100
California	22,500	<1
Idaho	100,000	<5
Montana	100,000	66
Nevada	20,000	0
New Mexico	50,000	0
North Dakota	184,000	10
Oregon	80,000	≅15
Utah	100,000	?
Wyoming	100,000	?
Total	779,400	?

Starving pronghorn sometimes are forced into towns to feed, often on ornamental shrubs or haystacks near human habitations. At such times, dogs often harass or attack the weakened animals. Photo by Fred Yates.

from wildlife agencies was used for livestock protection. Such control may or may not benefit wildlife.

No predator control was conducted by federal or state agencies in Kansas. Training in control techniques was being provided by a state extension agent for landowners and stockmen. The Kansas Department of Wildlife and Parks paid half of that agent's salary. The extent of the resulting predator removal benefits pronghorn was unknown.

Wildlife agencies of five pronghorn states—Colorado, Nebraska, Oklahoma, South Dakota and Texas—did not pay for predator control.

Predator control can make a difference in pronghorn fawn survival, but not necessarily in the stability or growth of the population. The situation on the Sheldon National Wildlife Refuge in Nevada (Sheldon) and Hart Mountain National Antelope Refuge in Oregon (Hart Mountain) is a case in point.

Between 1955 and 1967, an average of 160 coyote and 53 bobcat carcasses were recovered annually from trapping and gunning on Sheldon. The additional and no doubt significant effect of 1080 use was unknown. Officially, predator control has not been practiced since 1967. Average numbers of fawns per hundred does, from late summer censuses during periods with and without predator control, were 61 during 1955 through 1967 and 25 during 1968 through 1980 (McNay 1980). Much of the improved fawn production during the period of predator control on Sheldon apparently was due to 1080 bait stations. Presently, 1080 poison is outlawed, and aerial gunning is considered the most practical, authorized method of control (Connolly 1978a).

Without predator control, the average midwinter pronghorn population (indicated by aerial surveys) on Sheldon increased from 1,265, during 1970 to 1974, to 2,082, during 1985 to1989, indicating that, with adequate rangeland conditions and low adult mortality, pronghorn populations can tolerate high levels of predation on fawns. However, the count averaged only 966 during 1990 to 1993 (Pyle and Reiswig 1994).

The number of pronghorn seen during midsummer surveys on Hart Mountain increased from about 300 during the late 1960s to nearly 1,400 during the early 1990s (U.S. Fish and Wildlife Service 1994)—a period of minimal predator control and comparatively low fawn survival (Table 62).

TABLE 62. Number of pronghorn seen during midsummer aerial surveys of Hart Mountain National Antelope Refuge during 1955 to 1999

Period	*Total number seen*	*Fawns per 100 does*
1955–59	347	65
1960–64	348	67
1965–69	292	35
1970–74	316	30
1975–79	503	36
1980–84	712	24
1985–89	816	44
1990–94[a]	1,394	25
1995–99	1,244	20

[a] Incomplete survey in 1992.

Extensive predator control was practiced during 1955 to 1967. From the U.S. Fish and Wildlife Service (1994), with current data from U.S. Fish and Wildlife Service Headquarters, Lakeview, Oregon.

The early history of predator control on Hart Mountain is sketchy, but the practice was well underway by the mid-1950s. At first, professional trappers were hired and refuge personnel participated. Then, government trappers from the U.S. Fish and Wildlife Service's Branch of Predator and Rodent Control (later the Division of Wildlife Services) took the task. Coyote-getters, 1080 poison, leg-hold traps, denning and aerial gunning were used, usually year-round. The following Hart Mountain narratives were compiled by Mike Dunbar.

The Hart Mountain narrative reported, "After having given the method a try, it appears that a few months of intensified predator control before the young of the year of the sage grouse, antelope and deer are born is more effective in retarding predation than a year-round program in which the refuge is only a portion of the trapper's territory" (1957:5). In the summer of 1957, the highest recorded fawn/doe ratio of 108:100 was observed. In 1966, refuge personnel eliminated the use of 1080, traps and coyote-getters. The predator control program was limited to no more than 3 days of aerial gunning during the early part of the fawning season. During May 15 to18, Division of Wildlife Services personnel flew the refuge and shot 38 coyotes and 2 bobcats in 24.5 hours of flight. This compares with earlier days of year-round coyote hunting that killed 63 coyotes, for instance, in the 6-month winter/spring period of 1952–53.

The refuge's narrative noted, "After many years of coyote control of varying intensities we have gradually reduced the program to about 40 hours of aerial gunning on the antelope kidding grounds when both the coyote and antelope are there" (1967:6). In 1967, the Division of Wildlife Services killed 35 coyotes in May at a cost of $25.15 per animal. That year, concentrated efforts to conduct predator control were halted on Hart Mountain. However, anecdotal information suggests that refuge staff occasionally may have shot coyotes in the late 1960s through the early 1990s. Although this activity was within the U.S. Fish and Wildlife Service's policy (50CFR31.14), no records were kept. Since 1994, there has been no authorized coyote control on Hart Mountain.

During 1996 and 1997, 43 of 52 radio-marked (83%) pronghorn fawns died on Hart Mountain (Gregg et al. 1999). Predation accounted for 31 (72%) of those deaths, and coyotes were responsible for 81% of the predation.

Control of predators to benefit a big game population often involves wholesale reduction of predators over a large area. Even if desirable, such control seldom is economical. However, Smith et al. (1986) indicated that selective, time-specific application of aerial gunning was a reasonable and economical means of increasing numbers of pronghorn on Anderson Mesa in north-central Arizona in the presence of high coyote densities. However, this study (discussed later) did not monitor pronghorn populations beyond the last year of coyote control.

Pronghorn fawn survival was measured relative to aerial coyote removal during 1985 to 1987 on the Jackass Creek Study Area, Harney County, Oregon. August fawn/doe ratios for the study area during preremoval years, 1976 to 1984, averaged 13.6:100. The same ratio during removal years was 61.1:100 (Willis 1988). Coyote removal entailed two sessions per year—1 and 3 weeks before the normal peak of fawn drop on May 20. High-intensity, short-duration coyote removal was deemed an appropriate and cost-effective management application for pronghorn populations with poor fawn survival during periods of medium to high predator abundance.

An aerial gunning program to control coyotes was conducted in areas of the panhandle of Nebraska during May and June 1990 and during April and May 1991 due to low pronghorn fawn/doe ratios. During 1990, 112 coyotes were removed; 88 were removed in 1991. Aerial surveys in 1990 indicated 70 fawns per 100 does on the coyote kill area. Forty-four fawns per 100 does were on adjoining nonkill areas during July, and 62 per 100 versus 25 per 100 does were on those areas, respectively, during August (Menzel 1992). However, the effects of aerial coyote gunning did not carry over to the next year. In 1992, no difference was found in fawn survival between the area where coyotes were killed during 1990 and 1991 versus the area where they were not (Menzel 1994).

Census and coyote control information on University of Texas lands revealed substantial increases in fawn crops during years following coyote control (Canon 1993). The pronghorn population on university lands was estimated at 847 in 1982, when coyote control by helicopter was initiated. By 1984, after three seasons of control, the population had increased 75% to 1,467 animals. By 1986, the increase was 140% to 2,019. The average population size during 1983 through 1993 was 1,844 pronghorn, more than double precontrol levels.

Controlling one species of predator may be compensated for by increased predation by other species, as happened on the National Bison Range when coyotes were reduced and predation by bobcats and golden eagles increased (Corneli et al. 1984). In fact, coyote removal may affect the overall structure and function of a predator/prey community.

Henke and Bryant (1999) evaluated numerical changes to rodent, lagomorph, raptor and medium-sized mammalian predator populations in a shortgrass prairie ecosystem of western Texas. Two treatment and two comparison 12,355-acre (5,000 ha) sites were monitored for one year before coyote control and for 2 years during removal of 354 coyotes by aerial gunning. Faunal populations were similar on treatment and comparison sites during the year before coyote control. Coyote numbers remained constant on the comparison sites and were reduced by half on the treatment sites. Rodent and black-tailed jackrabbit densities, as well as relative abundance of badgers, bobcats and gray foxes, increased on treatment sites. Henke and Bryant's (1999) findings were consistent with the predator-mediated coexistence hypothesis, which suggests that a keystone predator can influence faunal community structure.

Today, coyotes kill more pronghorn fawns than do any other species of predator (Table 60). The fawns often are killed before they are 1 month old. During November 1970 through January 1971, 1080 poison was placed in a study area north of Valentine, Texas, for coyote control to protect pronghorn fawns. Fawn production on the area increased from 29% during summer 1970 to 79% during summer 1971 to 65% during summer 1972 (Hailey 1979).

Experimental control of coyotes during studies involving radio-collared fawns on the National Bison Range also demonstrated that control can greatly improve fawn survival (O'Gara and Malcolm 1988). In Alexander Basin, where many does gave birth, 90% of the fawns died during the first month of life in 1975 (Reichel 1976); 93% died during 1977 (Von Gunten 1978). Mortality during those years was attributed primarily to coyote predation. Removal of 19 coyotes just before fawning time during spring 1978 resulted in a fawn mortality of just 27%. During 1978, bobcats killed fawns in Alexander Basin for the first time during those studies—probably because removal of coyotes allowed the bobcats to use the grasslands (Corneli et al. 1984).

Although coyote numbers on the National Bison Range seemed to be as high during 1979 and 1980 as they were in 1977 before limited control, 22 fawns survived in 1979, despite the herd's reduction from 138 to 78 by extensive winter mortality. Coyote predation on fawns had resulted in an old-age herd, and 23 of the 32 dead animals found were 8 years old or older (Corneli 1980). Thirty fawns survived in 1980. Apparently, coyotes replacing the experienced ones that were killed during 1978 were not efficient predators of fawns. By 1982, survival was down to five fawns.

Although trappers were taking coyotes at the outside perimeter of the National Bison Range, fawn survival was one in 1984. The possibility again existed of a severe winter drastically reducing the herd with its high percentage of old animals, and fawn survival would not have been adequate for the population to rebuild. Most bucks on the range were not territorial after the die-off in 1979, and territoriality had not been reestablished by 1984, an indication of social instability resulting from a skewed age structure.

A decision was made to begin casual coyote control. Range personnel began carrying rifles in their vehicles to shoot coyotes as chances occurred. Twelve coyotes were shot during 1984, 15 were taken in 1985 and 8 were shot during 1986. Eighteen fawns survived through summer 1985, 20 through 1986 and 27 through 1987. The pronghorn population increased from 116 during winter 1984–85 to 146 during winter 1987–88, and about 65 were young animals. The level of control in no way endangered the coyote population; 26 were observed during a helicopter census in January 1988 (O'Gara and Malcolm 1988). Apparently, removal of a few coyotes caused a turnover in the population, reduced the number of predators accustomed to hunting fawns and reduced predation despite a large number of coyotes being present.

Writing of mule deer, Connolly (1981:246) advanced a concept that applies to pronghorn as well as other big game: "No one factor operates independently of the others. A single factor may be chiefly responsible for an observed trend in deer numbers, but rarely is it possible to conclude that this factor alone caused the trend. Even if, in theory, the effects of various influences could be segregated, in practice interactions are too complex to be separated by existing research methods."

The Arizona Experience

In a special report for the Arizona Antelope Foundation, Shaw (2000) assessed the need for additional research on coyote control to benefit pronghorn in Arizona. In review of that state's past pronghorn research and statewide survey data in relation to coyote control programs, he focused on two questions:

1. Does enough information exist to guide the Arizona Game and Fish to make decisions on coyote control to benefit pronghorn populations?
2. If not, what studies would provide the necessary information?

The following discussion is based on Shaw's report.

Critical Review of Arizona's Pronghorn Research History. McGregor (1935) summarized the history and status of pronghorn in Arizona. He noted that presettlement

From 1978 to 1984, the pronghorn population on the National Bison Range decreased from 138 to 116 (Range Narrative Reports). The decrease, by itself, was not cause for concern. However, the age structure of the herd was skewed to old animals by 1984. More than half of the herd had been born during 1978 to 1980, after the limited control action taken during 1978. Thus, about half of the herd was 4 to 6 years old, and almost half was older than 6 years. The coyote control conducted during 1978 allowed good fawn survival just before and after the herd reduction, and the herd soon regained its lost numbers, such as seen above. Without that control, the herd probably would have dwindled as coyotes hunted harder for the few fawns available. Another severe winter could have led to extirpation (O'Gara and Malcolm 1988). Photo by Len Clifford; courtesy of Leonard Rue Enterprises, Blairtown, New Jersey.

pronghorn densities were high, but heavy hunting nearly extirpated the species by 1907. Effects of unregulated hunting were exacerbated by the presence of many small ranches and farms scattered throughout pronghorn habitat. By the late 1920s, failure of many of the small homesteads, better game law enforcement and heavy control of predators to benefit livestock resulted in a rapid increase in pronghorn. Both trapping and poison baits were used to control predators. The Anderson Mesa herd, believed to have dropped below 100 animals by 1907, was estimated at between 3,000 and 4,000 animals by 1935. McGregor considered this density to be in excess of carrying capacity and recommended regulated hunting or allowance of an increase of predators to limit pronghorn numbers. That remarkable buildup of pronghorn occurred during the era when sheep and cattle grazing had reduced the condition of Arizona grasslands to an all-time low. Taylor (1936) also noted the buildup of pronghorn in Arizona and attributed it, at least in part, to heavy predator control. He considered failing range conditions to be a problem.

The papers of both McGregor (1935) and Taylor (1936) were based on general observations of conditions and discussions with U.S. Forest Service personnel, game wardens and ranchers. With the exception of limited counts provided by the U.S. Forest Service, the authors had no field data on pronghorn numbers. Nonetheless, they provide an early discourse on pronghorn history and perceived predator effects.

Knipe (1944) carried out the first systematic survey of pronghorn in northern Arizona. He traversed all of the grasslands by vehicle and found pronghorn numbers to be down from the high estimates of the 1930s. He did not routinely record fawn survival, but was concerned with the nearly even buck/doe ratios he observed. Knipe concluded that buck hunts were needed to spread the ratio. He recorded a fawn survival rate of 0.50 fawns per doe, for a small sample of about 150 animals. Predator control was active throughout the state, with 779 coyotes and 208 bobcats known to have been trapped on pronghorn range in 1942. Other problems Knipe noted were fragmentation through isolation of pronghorn habitat north of the Highway 66 and Santa Fe Railroad corridor and poor range conditions due to grazing. He found no pronghorn north of Highway 66, east of Flagstaff. He noted that the grasslands were largely composed of snakeweed and pinque, both poor forage, with a very low component of grasses. His recommendations included predator control, life history studies, range studies, disease studies, improved law enforcement, supplemental feeding, evaluation of fencing effects, translocations to suitable empty habitats, ungrazed pronghorn refuges, water developments and salting to control distribution. In a sense, Knipe set in motion all of the pronghorn management and research that has happened since 1944.

As a result of Knipe's (1944) report, Arrington and Edwards (1951) evaluated the effects of predator control on pronghorn fawn survival in the Anderson Mesa herd. Between 1944 and 1950, they surveyed pronghorn from aircraft with emphasis on fawn production. They found a fawn survival rate of 0.25 per doe in 1944 and concluded that poor productivity was a major cause of a decline in pronghorn numbers during 1944 and 1945. No predator control had occurred in the area during that period. They monitored the fawn survival in two populations presumed to be separated by Diablo Canyon. In winter 1945, predators on winter and summer ranges north of Diablo Canyon were heavily controlled. Predators on pronghorn range south of Diablo Canyon were not controlled. July fawn survival on the controlled area was 0.52 fawn per doe; on the uncontrolled area, it was 0.04 fawn per doe. During winter and spring 1946, both areas were controlled, with the northern zone having a fawn survival of 0.64 and the southern zone having 0.59.

Arrington and Edwards (1951) noted that strychnine baits were used on virtually all northern Arizona pronghorn ranges in 1945 and 1946. With the exception of Chino Valley and Boquillas Ranch, fawn survival was up from the preceding year, with an average of 0.64. The researchers continued to monitor fawn survival throughout the northern Arizona pronghorn ranges, with heavily controlled areas showing relatively high fawn survival. Compound 1080 became available in 1948. For winter 1947–48, all units were treated at the rate of one bait per township. Average fawn survival was 0.74. On two small areas that were not treated with 1080, survival was 0.49. These other two units, however, may have been treated with thallium chloride earlier in the winter. In 1949 and 1950, 1080 was used exclusively. Survival in treated units was 0.86 and, in untreated units, it was 0.48. This study, although weak in design, makes it difficult to ignore the short-term increase in fawn survival as a result of predator control. The researcherss, in fact, noted that effects are short-lived, with fawn crops declining the first year after control ceased. However, even in untreated areas, average fawn survival was higher than now experienced in most areas, even where aerial gunning has occurred.

After Arrington and Edwards (1951), no research on pronghorn occurred in Arizona for 24 years. This reflects, in part, the relatively good condition of the herds and lack of concern for the species. Two events resulted in renewed anxieties over pronghorn and, ultimately, new studies. First was the severe snowstorm during December 1967. More than 3 feet (91 cm) of snow fell across most of northern Arizona's pronghorn range and decimated much of the herd. Second was the 1972 presidential ban on use of compound 1080. The 1967 snowstorm was especially hard on the Anderson Mesa herd. Although other herds in northern Arizona slowly recovered during the following decade, the Anderson Mesa pronghorn population plateaued at about 400 animals. Annual surveys disclosed continued poor fawn survival, and predation was implicated as the primary problem.

In 1974, Neff and Woolsey (1979) initiated a 5-year evaluation of the factors affecting fawn survival on Anderson Mesa. They noted, in their historic review (Table 63), that the herd had experienced a downward trend since about 1950 and that fawn survival had declined from an average of 0.70 to 0.44 fawns per doe. They found that fawn survival (average = 0.76) was significantly higher when control had occurred during the preceding winter. It declined to 0.60 one year after control and to 0.35 two or more years after control. As shown in the work of Arrington and Edwards (1951), the effectiveness of coyote control in short-term elevation of pronghorn fawn survival was evident.

From 1974 to 1979, Neff and Woolsey (1979) estimated that an average of 76% of fawns born on Anderson Mesa were lost by July of each year. Almost all of the loss (95%) was attributed to coyote predation. Due to high fur prices, private trapping of coyotes continued throughout the entire study period. Some spring trapping also was carried out by Animal Damage Control personnel. Between 1976 and 1979, 228 coyotes were known to have been killed on the study area. Hence, Neff and Woolsey were not able to evaluate predation in an area with an unexploited coyote population. Insofar as possible, range condition, forest and woodland encroachment, alternative prey densities and effects of nutrition and disease were assessed, in addition to effects of coyote density and behavior. Coyote predation stood out as the main factor influencing fawn survival in the Anderson Mesa herd.

One report speculated on the effects of decreasing densities of pronghorn does on intensity of coyote predation on fawns (Neff and Woolsey unpublished). They suggested that, at high pronghorn densities, pronghorn does cooperated to harass coyotes and, thereby, reduced fawn loss. At lower densities, coyotes were able to override harassment by does, thus having a proportionately greater impact on fawns. Such a mechanism, if real, would create a situation in which a pronghorn population that had reached

TABLE 63. Summary of pronghorn population trends on Anderson Mesa, 1850 to 2000

Period	*Trend or status*	*Major influences*
1850–1876	High density, wide distribution	No human residents. No livestock except transient sheep bands. Wolf and coyote predation. Drought and blizzard losses occasional
1876–1907	Decline	Introduction of sheep and cattle, fully stocked and overstocked ranges. Early water development efforts. Construction of transcontinental railroad and Mineral Belt railroad south to Mormon Lake. Construction of Bullock and Peavine railroads south through Chino Valley into Prescott and Prescott Valley
1907–1924	Continued decline? Slow increase?	Forest Service grazing management started. Active homestead farming on and east of Anderson Mesa. Active predator control. No market hunting, no open season but law enforcement limited
1924–1933	Very rapid increase	Grazing allotments fenced; forest boundary fenced. Grazed by both cattle and sheep. Homestead farming declining. Intensive predator control. Improved game law enforcement
1933–1950	High density, variable trend	Grazing management improved but sheep still competing with antelope. Use of bulldozers for stock tank building. Intensive predator control (including eagles). Legal hunts resumed in 1941
1950–1979	Slow population decline and declining fawn survival	Grazing by cattle only. Introduction of bison to Raymond Ranch resulted in antelope-proof fencing of 26 sections of Unit 5B winter range during the late 1940s. Increased fencing for rotation grazing. Continued development of stock waters. Improved roads, increased recreational traffic. Winter die-off of 1967–68. End of coyote control by toxicants. Building of 345 KV power line
1980–2000	Continued decline in average fawn survival. Unit 5B antelope population nearing extirpation, while statewide total antelope count increased from approximately 5,000 to near 9,000	Grazing by cattle only. Improved range management on national forest and BLM lands. Sporadic aerial gunning. Ban of steel traps on public lands. Development of collaborative ranch management

Source: Adapted and updated from Neff and Woolsey 1979.

a critically low density would never be able to recover without effective coyote control. However, Byers (1997a) suggested that crowding of pronghorn does due to fencing was the cause of high coyote predation on fawns at the National Bison Range. Byers indicated that, during the fawning period, does were more inclined to be solitary. These two studies represent two extremes and suggest that protective behavior of does may function best at some intermediate density.

The results of the initial 5 years of work by Neff and Woolsey led to a more intensive study of coyote control on Anderson Mesa (Neff et al. 1985). The most important phase of this study involved evaluation of aerial gunning of coyotes from helicopters immediately before the fawning season. The study was enabled by legislation passed in Arizona in 1980 that legalized aerial gunning of coyotes for approved control projects. For 1981, 1982 and 1983, 62, 60 and 70 coyotes, respectively, were removed by aerial gunning. These removals were in addition to trapping and opportunistic shooting on the ground. A definite increase in fawn survival occurred between 1980 and 1983 (0.43–0.67) and was attributed to intense coyote control immediately before fawning. The conclusion was supported by the facts that no other pronghorn herd in northern Arizona received intensive coyote control and no other herd showed a comparable gain in fawn survival.

Using the data from 1977 to 1983, Smith et al. (1986) created a population model to project effects of coyote control on pronghorn fawn survival, then extrapolated this model to predict trends in the pronghorn buck population on Anderson Mesa. They concluded that fawn/doe ratios were negatively correlated with length of time since the last effective coyote control (for 1951 to 1979, $r = -0.84$, $P < 0.001$). Smith et al. also carried out a benefit-to-cost analysis of aerial gunning compared with other coyote control measures, based on projected increase in mature pronghorn bucks in the herd. They concluded that, in terms of production of pronghorn bucks and benefit-to-cost ratios, the most effective schedule of aerial gunning would be every other year. They also concluded that 1080 baits were less expensive and had been more effective historically than aerial gunning in terms of elevating fawn survival. Summarizing their results, Smith et al. (1986:230) noted: "The accumulated net benefits for most schedules did not exceed the accumulated net benefits for the no-control alternative until the fifth or sixth year of the simulated run, indicating that long-term planning of coyote management programs is needed. The favorable net benefit ratios at the end of the 10-year cycle appear to reflect the fact that as pronghorns increase as a result of coyote reduction, the total number of harvestable bucks increases and there is an increasing payoff for the fixed annual cost of the control operations."

Those investigators studied the descriptive, problem–identification phase and experimental phase. Their evaluation on the basis of the final product (harvestable bucks) was unique. The research, however, had problems that were beyond the biologists' control, which weakened their model (Shaw 2000). These problems included:

1. No area without coyote control was available for comparison with the area where coyote populations were reduced for the experimental phase of the project. As a result, some level of coyote reduction occurred on the study area throughout the entire study period. This affected coyote social behavior and, ultimately, the fawn survival model created by the project.

2. The study was truncated at the end of the period of experimental aerial gunning, and the fawn survival model and buck population projections were carried out without post-treatment data. Models are hypotheses that must be tested and refined by constant comparison with real-world conditions. Unless such a reiterative process is used, inaccuracies in preliminary models can cause inappropriate management decisions that, over time, can lead wildlife managers astray. Unfortunately, until now, such application of subsequent data and feedback to refine the model has not been done.

3. Although an effort was made to assess coyote populations during the experimental phase of the study, funds and manpower were not adequate to allow this to happen at a level that clearly documented population changes. As a result, conclusions regarding the effects of gunning on coyote numbers were based strictly on observed responses of the fawn survival data to gunning programs, and all effects were assumed to have been due to a reduction in coyotes. Unfortunately, because of the abrupt ending of the study, the researchers were unable to do a full assessment of the coyote population and behavior data gathered, much information went unpublished. Again, the project was truncated at the end of experimental gunning. Assumptions made regarding the extended effects of gunning for 2 or more years after gunning ceased have proven to be wrong. Had the study been continued for 5 years after gunning, with close monitoring of pronghorn fawn survival and the coyote population, this error in prediction might have been discovered and refinements of the model made.

It should be noted that a factor affecting development and testing of models during the 1970s and early 1980s was the limited availability of computers. Personal computers were rare among field biologists and did not have the power and memory to handle large data sets needed to accomplish model testing. Also, good statistical programs were not available, and access to the few computer specialists was limited. Simply getting data analyzed required a large amount of time.

Neff (1986), while acknowledging the seriousness of coyote predation effects on fawn survival on Anderson Mesa, also foresaw the problems of declining habitat area and quality statewide. He recommended that a statewide survey of habitats be undertaken. This, along with a detailed study of habitat use and mortality factors in wildlife management unit (WMU) 21, resulted from Neff's recommendations and was carried out by Ockenfels et al. (1994, 1996). The 1996 report is a tool for evaluating pronghorn range for either translocations or habitat improvement needs. By-products of the work of Ockenfels et al. include an initial assessment of fawn bedding ground characteristics (Ticer and Miller 1994) and identification of mountain lions as a major predator on adult pronghorn in areas where pronghorn are forced to use broken terrain with a fairly large shrub component (Ockenfels 1994a, 1994b).

The problem of urbanization and habitat loss also was anticipated in the early 1980s by Glenn Dickens and Steve Andrews, of the Arizona Game and Fish Department, who created a slide presentation aimed at alerting the public to the problem of habitat loss in the Prescott area. Subsequent events near Prescott show that the warnings of Dickens and Andrews were ignored.

Brown (1982) presented the first assessment of disease in pronghorn populations in Arizona. He noted the presence of titers for bluetongue, epizootic hemorrhagic disease, leptospirosis and anaplasmosis. More recently, Heffelfinger et al. (1999) found a 78.8% incidence of positive bluetongue titers in Arizona pronghorn. They also found that copper and selenium in pronghorn livers were below levels considered minimum for domestic livestock. Minimums for pronghorn are not known. Heffelfinger et al. also noted that bluetongue and deficiencies in trace elements can depress fawn survival.

Brown (1982) indicated a negative correlation between fawn survival and the previous year's drought severity index. This relationship, contrary to general belief that good precipitation aided fawn survival, has not been explored further. Heffelfinger et al. (1999), quoting Brown, observed that attempts to correlate precipitation patterns with pronghorn fawn recruitment in Arizona have been unsuccessful so far.

Overall, pronghorn research in Arizona, although spread over a long period (1944 to 1996), has stayed on track. Each effort has resulted in recommendations that have been used to design the next generation of research. The Anderson Mesa work clearly delineated the problems of coyote predation on fawns and initiated experimental implementation of predator control. However, this work was interrupted by other studies deemed more pressing, resulting in a failure to follow the pronghorn and coyote population trends long enough to evaluate management recommendations adequately.

Review of Arizona's Pronghorn Management Data, Related to Aerial Gunning of Coyotes. Shaw (2000) based the following review, in part, on an unpublished table created by John Phelps of the Arizona Game and Fish Department, which showed the fawn survival and aerial gunning history for most WMUs with pronghorn populations. Shaw's review also is based on the statewide pronghorn survey database maintained by Arizona Game and Fish Department.

With the exception of a detailed evaluation of the performance of Neff's model for Anderson Mesa (WMU 5B), Shaw focused on analysis of combined survey data from all appropriate units. This was done mainly to build adequate sample sizes to allow patterns to become apparent in the data, which might not be evident within individual units. Shaw acknowledged the hazards of using such a large data set and drawing conclusions from patterns that evolve from analysis, but considered it the only data available for developing hypotheses that might be tested under more stringent conditions.

History of Fawn Survival in Arizona. Shaw (2000) plotted Arizona's data on fawn survival for the period from 1950 to 1999 (Figure 61). Data for 1948 and 1949 were eliminated to give uniform 5-year intervals for plotting, but these years were similar to the 1950 to 1954 interval. All fawn survival data presented in this report are annual fawn/doe ratios from Arizona Game and Fish Department surveys, adjusted for the previous year's fawn crop to account for yearlings that were assumed to be nonproductive. These adjusted fawn survival figures are used for all graphs in the analysis. Adjustments assumed an equal buck/doe ratio in yearlings, and estimated yearling doe numbers were subtracted from the following year's doe population before fawn survival was calculated.

Average fawn survival for the state declined through the 1950s and 1960s (Figure 61), implying that factors other than the effects of 1080 were at work. The 1950s now are known to have included one of the most serious droughts in recorded history (Swetnam and Betancourt 1998), and this may have contributed to the decline of pronghorn populations during that period. The decline during the early 1960s is more difficult to explain. Interestingly enough, the rate of decline did not change in 1968, following the severe snows of the preceding winter. Clearly, a more detailed assessment of all factors that may have affected pronghorn survival over the past 50 years is needed.

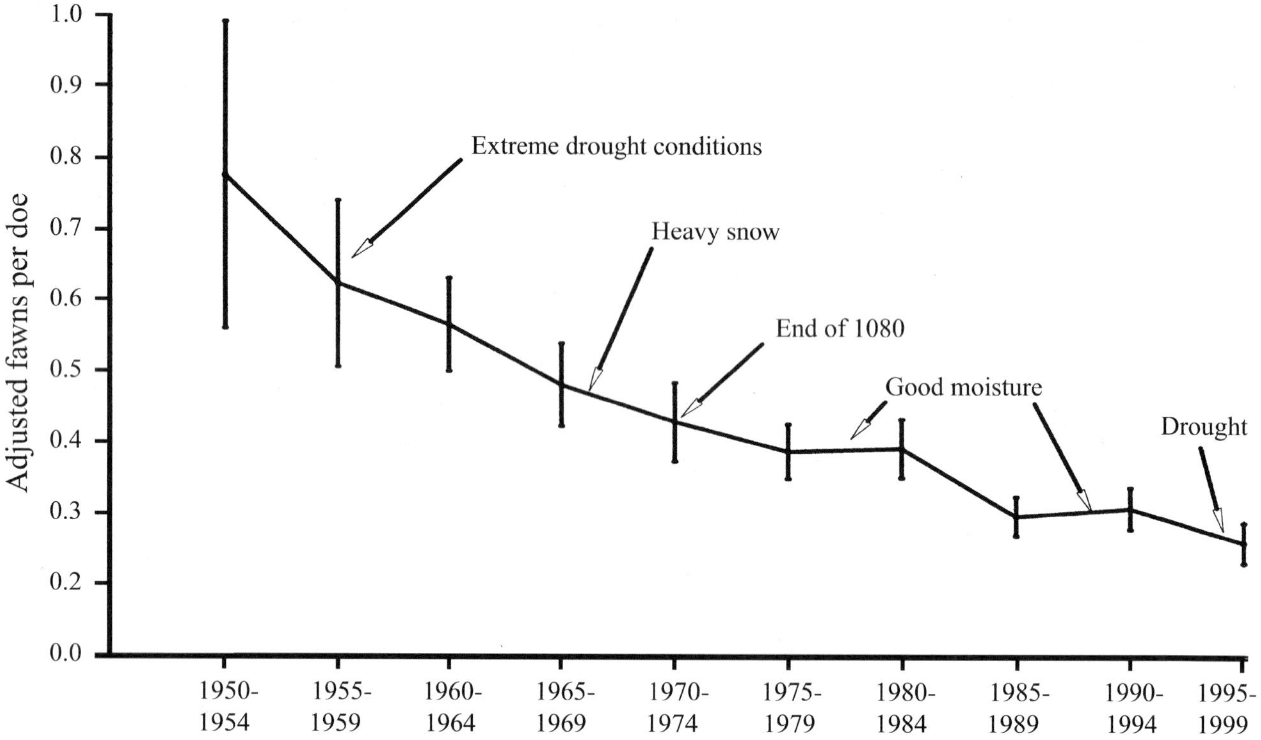

Figure 61. *Trends in Arizona pronghorn fawn/doe ratios plotted as 5-year averages, 1950 to 1999 (Shaw 2000). Adjusted ratios assume a nonproductive yearling class with a 50:50 sex ratio.*

Evaluation of Aerial Gunning of Coyotes. To assess aerial gunning of coyotes in all treated WMUs, the data were sorted into categories of untreated (all WMUs before gunning, plus all WMUs that were never gunned), year of successive treatment (first, second or third—data for 2 years having fourth year treatments were combined with third year), 1 year after treatment, 2 years after treatment and 3 or more years after treatment (Figure 62). For all WMUs combined, mean increase in fawn survival after 1 year of treatment was about 0.02 fawn per doe. For second and third consecutive years, increase was approximately 0.05 fawn per doe. None of the classes is significantly different from other classes at the 95% confidence level. This is not a large increase and, unless other mortality factors (adult mortality, moisture-associated impacts, disease, etc.) remained relatively constant or were reduced, no significant increase in pronghorn populations might be expected from such results. Of particular interest is the apparent decline in fawn survival the second and third years after gunning. Fawn survival actually dropped below the initial, ungunned levels. Beyond 3 years post-treatment, fawn survival shifts back toward the pregunning level. If this pattern is representative of the true effects, continued benefits can be accomplished only if gunning is maintained permanently at frequent intervals. The relatively small benefit of the first year suggests that, to be effective, gunning should continue for at least 2 years consecutively. The rapid postgunning decrease suggests that no more than 1 year should be allowed between gunning events. These data also suggest that the pattern of gunning sustained in the various WMUs, with 2 or more years between gunning bouts, may have contributed to the lowered fawn survival.

The next question is if any gain in fawn survival as a result of gunning is apparent. Figure 63 compares gunned with ungunned data from 1960 to present. Ungunned data include WMUs that never received gunning; gunned data include all WMUs that had any level of aerial gunning. Note that WMUs selected to be gunned may have been those with a history of low fawn survival, thus introducing a bias. Both groups started at about the same level in 1960, and the divergence between the groups began before any gunning took place. Whatever the cause, the gunned WMUs tended to average lower than ungunned WMUs, even after gunning began. An interesting aspect of this graph is that both categories declined in a parallel manner, suggesting that factors in addition to aerial gunning affected pronghorn survival. The ungunned WMUs were significantly higher, at 95% confidence interval, than were the gunned WMUs in three of the four periods after 1981, when gunning was implemented. To assess overall gunning effects within gunned WMUs, Shaw (2000) evaluated average fawn survival for all pretreatment years, all years when treatment occurred and all years

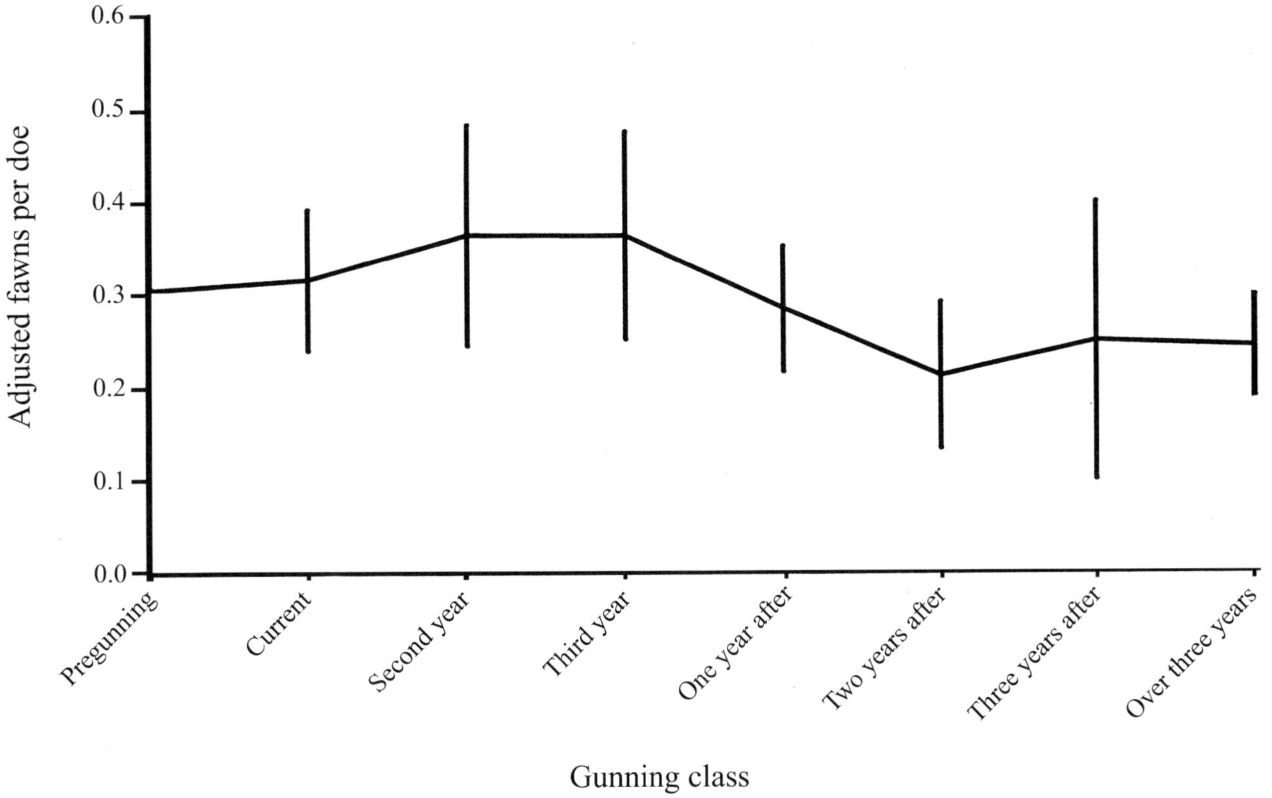

FIGURE 62. *Mean adjusted pronghorn fawn/doe ratios in Arizona, in relation to time of aerial gunning of coyotes. Vertical bars represent 95% confidence intervals.*

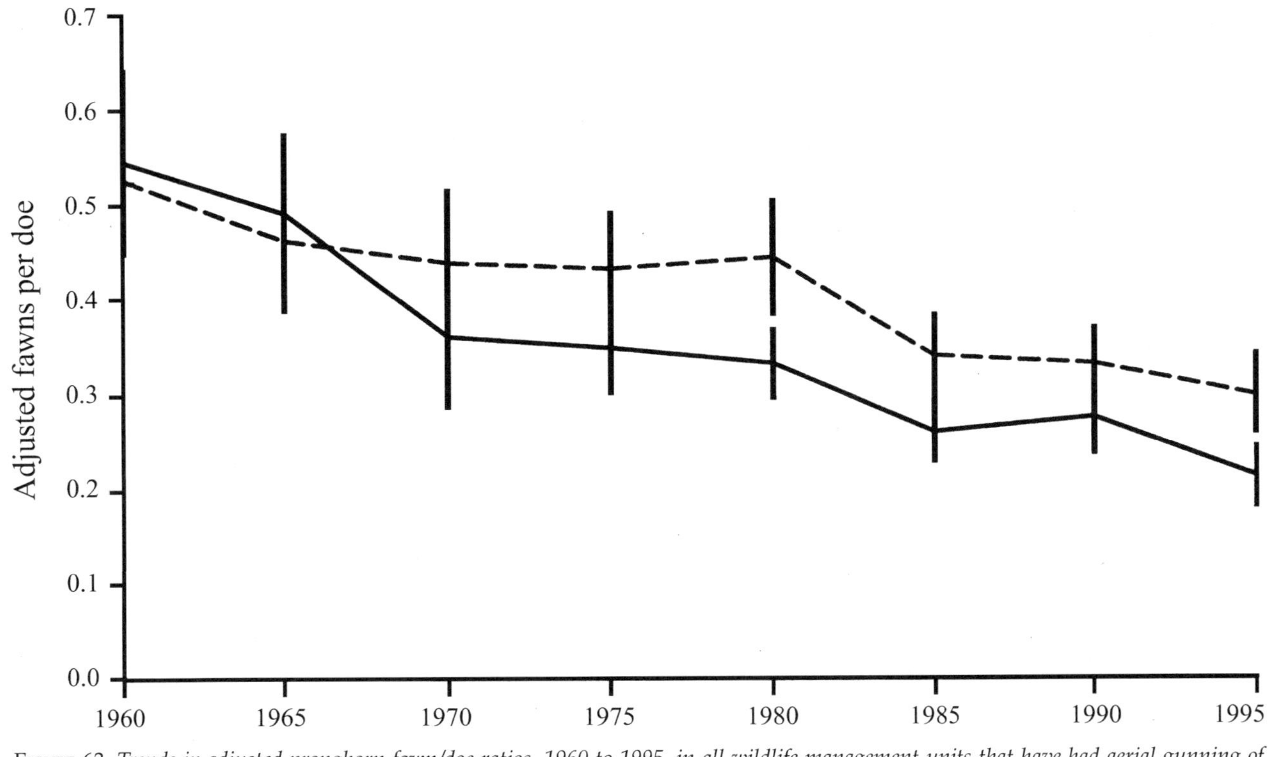

FIGURE 63. *Trends in adjusted pronghorn fawn/doe ratios, 1960 to 1995, in all wildlife management units that have had aerial gunning of coyotes versus those with no aerial gunning. The solid line represents gunned units; the broken line represents units with no aerial gunning. Vertical bars represent 95% confidence intervals.*

after treatment (Figure 64). Here, again, the short-term effect of gunning is apparent, followed by what seems to be a postgunning depression.

On the basis of the data, Shaw speculated that effective gunning must be maintained for extended periods at frequencies no lower than every other year. If periods between gunning exceed 2 years, the long-term effect may be a temporary depression of fawn survival. If this is true, then the difference between gunned and ungunned WMUs, as shown in Figure 63, may be partly a result of the schedule of intermittent gunning. How this may happen is unclear. Reviews of research on coyote population dynamics and response to control (Connolly 1978b, Knowlton et al. 1999) provide ample evidence that a coyote population can return rapidly to precontrol levels. In relatively small areas where treatment is moderate, this can occur within months. Even where control is heavy and over relatively large areas, coyote populations can return to precontrol levels within 2 to 5 years. This may explain why treatments during 3 to 5 successive years fail to show dramatic increases in fawn survival during second or third years of treatment. In most cases, the pregunning numbers of coyotes are about the same each year, regardless of previous year's gunning. Factors contributing to the rapid recovery of controlled coyote populations include: (1) increased average litter size at lowered densities, (2) increased number of yearling females breeding at lower densities, (3) increased survival of both adult and young coyotes and (4) increased colonization from surrounding areas. So far, no empirical evidence supports the idea that coyotes temporarily rebound to a post-treatment level greater than the pretreatment population.

Only one mechanism was found that might explain an increase in coyote predation on fawns during post-treatment years. In controlled populations, coyote packs tend to be smaller and more widely dispersed than in uncontrolled populations (Knowlton et al. 1999). This could result in widely dispersed predation by coyotes over fawning grounds and increase the rate of fawn loss. However, Byers (1997a) noted that coyotes learn to hunt for pronghorn fawns and will congregate in fawning areas during the fawning period.

The greatest weakness in all of Arizona's coyote gunning efforts was a failure to monitor adequately the response of coyotes, alternate prey (rodents, rabbits, etc.) and other predator populations to aerial gunning.

The above discussion centers on an overview of statewide-average fawn survival in response to aerial gunning of coyotes. Use of such a general data set, although having the advantage of providing a large sample, obscures responses of pronghorn fawn survival within specific units. In many cases, aerial gunning has occurred over only a small portion of a management unit, and few wild-

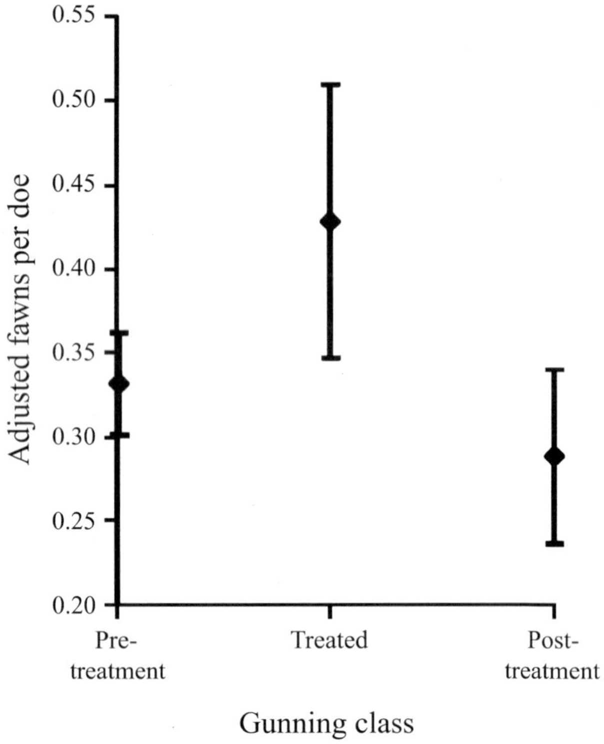

FIGURE 64. *Mean adjusted pronghorn fawn/doe ratios for all wildlife management units that have had aerial gunning for coyotes, showing averages for all years before gunning, for year of gunning and for all post-treatment years. Vertical bars represent 95% confidence intervals.*

life managers have maintained data on fawn survival in the treated and untreated portions of their units. In addition to these units, WMU 5B, because of its long history of intense surveys and research on Anderson Mesa and because of aerial gunning applied in the unit, has special value in the analysis of gunning effects. Also, the preliminary model developed by Smith et al. (1986) for that unit has not been reevaluated since it was originally published. Review and possible updating of that model may provide additional insights into gunning effects.

As a result of the Anderson Mesa study, Smith et al. (1986) developed a model describing fawn survival in relation to aerial gunning in WMU 5B ($Y = 0.57 - 0.114X$, where Y = predicted fawn survival, and X = number of years since gunning; where X is greater than 3, Y is constant at 0.23. Figure 65 presents this model plotted against the survey data for the unit. The model predicts the timing of increases resulting from gunning, but fails to track actual fawn survival rates. Care is necessary when applying such a model. Unless the mechanism of all factors affecting fawn survival can be understood, no rationale exists for modeling effects of gunning alone. Theoretically, fawn survival can vary widely between 2 years, without regard to gunning. A series of wet winters, for ex-

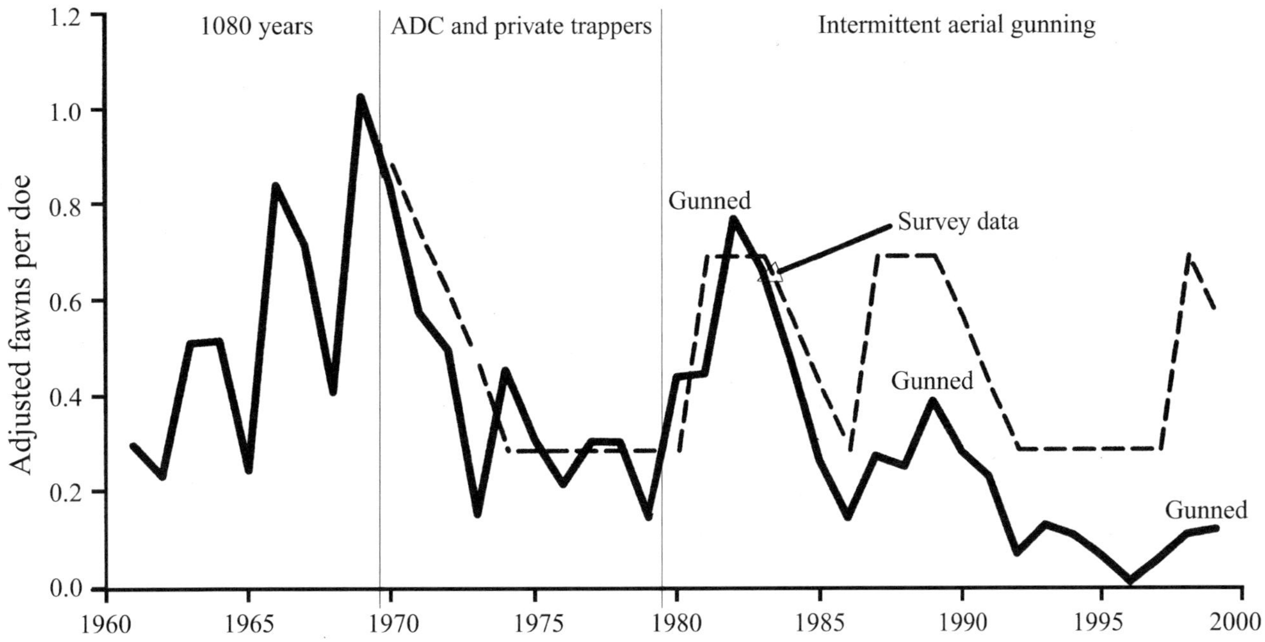

FIGURE 65. *Annual adjustment pronghorn fawn/doe ratios in Arizona's Wildlife Management Unit 5B (solid line), 1960 to 2000, compared with prediction from model developed by Smith et al. (1986) (broken line). ADC = animal damage control. Thin vertical line separates control treatment period.*

ample, might improve range conditions to the point that fawn survival will improve without gunning. As a result, the model may find its greatest utility in identifying periods when factors, other than predation, aid fawn survival. Needed modifications include:

1. The model does not provide a mechanism to track recovery of fawn survival resulting from treatment of a population that begins with a survival rate lower than 0.57 fawn per doe. That is, it does not account for a slow buildup, rather it assumes an immediate return to 0.57 during the first year of gunning. Data (Figure 65) indicate that such a rapid response does not occur. Note that 3 years of gunning from 1987 to 1989 did not result in a recovery to 0.57 fawn per doe. Such a correction should be developed by calculating percentage of initial population added each year of treatment and refined through reiteration as more data become available.
2. The constant 0.57 in the model may require refinement as more data become available. So far, it appears to be adequate, but application of gunning for several years continuously or for an extended period of alternating years might yield higher fawn survival.
3. The constant of 0.23 as the lowest possible fawn survival value has proven to be wrong. It can be eliminated from the model entirely. Experience in WMU 5B has shown that fawn survival can approach zero. Also, as already noted, a depression in fawn survival seems to occur in the second and third years after gunning, with a possible trend toward pregunning conditions after 3 years without treatment. If this pattern holds, the constant of 0.23 needs to be replaced with an equation acknowledging a short-term decline followed by gradual increase.

Perhaps the most revealing aspect of Figure 65 is the rapid decline in fawn survival to levels lower than the pregunning condition within 2 years after gunning ceased. If this proves to be consistent, then the temporal pattern of gunning probably has been a major factor in the decline in fawn survival. The mechanism at work here is obscure, but may be in the way that coyote populations respond to reduced densities after the gunning pressures are released. Research of controlled and uncontrolled coyote populations is needed.

Since 1981, 14 WMUs, representing 22 treatment periods, have had aerial gunning of coyotes to benefit pronghorn during 1 or more years (Table 64). Results have varied but generally support the conclusion that a short-term increase in fawn survival results from aerial gunning. Of the 22 treatment periods represented in Table 64, fawn survival for 18 showed some increase during the year of treatment during the average of 3 years before treatment, and fawn survival showed a decline in three periods. A test against a theoretical, even distribution of 10.5 decrease or increase each yielded a χ^2 of 10.7, df 1, where P is less than 0.001 (Table 65). If a minimum change requirement of 10% from the pretreatment average is imposed to eliminate

TABLE 64. Average fawn survival on wildlife management units (WMU) with aerial gunning to benefit pronghorn

WMU	*Years*	*Three years pregunning*	*Gunned*	*One year postgunning*	*Two years postgunning*
1	1984–1986	0.29	0.47	0.41	0.25
2A	1988–1991	0.17	0.26	0.16	0.41[a]
2B	1995–1997	0.08	0.28	0.33	0.08
2C	1984–1986	0.34	0.41	0.29	0.07
2C	1992–1994	0.13	0.26	0.31	0.05
3A	1995–1997	0.19	0.20	0.38	0.17
4A	1990	0.29	0.17	0.18	NA[b]
4A	1992–1994	0.18[c]	0.47	0.25	0.07
4A	1999	0.19	0.17	NA	NA
4B	1995–1997	0.25	0.28	0.43	0.17
5A	1990	0.17	0.06	0.10	NA[b]
5A	1992–1994	0.07[c]	0.22	0.01	0.03
5B	1981–1983	0.27	0.57	0.48	0.25
5B	1987–1989	0.27	0.29	0.23	0.22
5B	1998	0.04	0.10	0.12	NA
7	1998–1999	0.23	0.29	NA	NA
10	1984–1985	0.18	0.19	0.25	0.22
10	1990–1992	0.21	0.36	0.45	0.25
13A	1987–1988	0.13	0.28	0.22	0.13
17A	1991–1992	0.05	0.53	0.23	0.13
18B	1991	0.22	0.41	0.40	0.37
30A	1985–1988	NA	0.40	0.16	0.00

[a] Unit treated for four subsequent years. Also, second year postgunning was 1993, with exceptional winter rains.
[b] Unit was gunned again after skipping one year.
[c] Includes one gunned year.

TABLE 65. Results of χ^2 tests of direction of pronghorn fawn survival change by length of time after aerial gunning

Comparison	*Increased*	*Decreased*	*Probability*[a]	*Change*
Average of 3 years prior vs. treatment year	18	3	<0.001	Increase
Average of 3 years prior vs. 1 year after	13	6	0.11	Increase?
Average of 3 years prior vs. 2 years after	6	10	0.32	Decrease?
Average of 3 years prior vs. 3 years after	6	7	0.78	None
Treatment year vs. 1 year after	9	11	0.65	None
Treatment year vs. 2 years after	2	16	<0.001	Decrease
Treatment year vs. 3 years after	3	11	<0.04	Decrease

[a] Probability values are odds that the divergence from an equal distribution of change between increase and decrease is due to chance alone.

very small changes, 15 increase, 4 remain the same and 2 decrease, yielding a χ^2 of 14, where P is less than 0.001, tested against an even distribution of seven in each of the three categories. If the unchanged WMUs are divided equally into the other two categories, the χ^2 is 5.76, df 1, where P is less than 0.02. It appears that such consistent, positive, short-term responses to gunning are not due to chance alone.

Gunning effects may carry through the first year after gunning, but the effect is weaker. Comparing the 3 years prior with gunning with the first year after gunning, 13 WMUs increased and 6 decreased, yielding a χ^2 of 2.58, df 1, where P is 0.11. Comparing fawn survival for the gunning year with 1 year after, 9 WMUs increased and 11 decreased. This change was not significantly different than an even 10 increase and 10 decrease dispersal.

By the second year after treatment, any benefits of gunning had disappeared. Compared with the year of gunning, fawn survival for 16 treated units decreased and 2 increased, yielding a χ^2 of 10.8, df 1, where P is less than 0.001 (expected for no change = 9 increase/9 decrease). This is strong support for the conclusion that aerial gunning effects are short lived. The two treatments showing increase were during 1987 and 1993. The winter of 1992–93 had good rainfall; the winter of 1986–87 was near average, making it difficult to attribute high fawn survival during these years to precipitation and, hence, to nutritional effects.

Data from the individual WMUs do not support the conclusion that the hypothesized second year postgunning depression of fawn survival occurs, as suggested by Figure 62 and by the WMU 5B data. Comparing fawn survival the second year after treatment with the average of 3 years pretreatment, nine WMUs decreased, two remained unchanged, and five increased. Assigning each of the unchanged years to one of the other two categories yields an increase of 6 and a decrease of 10. Tested against an even distribution of eight in each category, χ^2 equals 1, df 1, where P is less than 0.32. Although not significant at normal testing levels (0.05), this probability remains strong enough to support further testing of the hypothesis. A sample of several more treatment periods would be required. Of the five WMUs showing an increase during the second year after gunning, one was during 1987, one was during 1993 and three were during 1994. Again, 1992–93 had exceptionally heavy rains, but the other 2 years were unremarkable. Both of the periods showing no change (1995 to 1997 and 1987 to 1988) were during periods of low winter precipitation. Here again, precipitation is not evidenced as a factor that strongly attenuates the effects of gunning.

For the third year after treatment, seven areas were lower and six were greater than the 3-year pretreatment average. This is not significantly different than a theoretical equal distribution. Testing the third year after treatment against the treated years, fawn survival data for 11 areas were lower and 3 were higher, giving χ^2 equals 4.5, where P is less than 0.04. Again, this strongly supports the conclusion of a short-lived effect for gunning.

One problem of comparing treatment effects at the WMUs level lies in the fact that aerial gunning often covers only a small portion of the WMUs, whereas annual surveys are done unitwide. Thus, fawn survival data from untreated portions of a WMU may dilute observable effects from treated portions. Recently, Arizona Game and Fish Department's Regions 1 and 2 have assessed fawn survival in gunned and ungunned portions of a few WMUs (Table 66). These comparisons do not assess fawn survival beyond the year of gunning. Of 14 instances in which subunit comparisons were made, 12 showed an increase in fawn survival and 2 showed a decrease. Compared with a theoretical distribution of seven increase, seven decrease, this yielded a χ^2 of 7.14, df 1, where P is less than 0.01, again supporting a conclusion that increased fawn survival during the year of gunning is not due to chance alone.

Actual benefits, however, appear to be small. To assess quantitative differences, a paired t-test was run, comparing gunned and ungunned subunits within WMUs. This yielded a probability of 0.07 that the differences within WMUs were due to chance, still supportive of a

TABLE 66. Comparison of pronghorn fawn survival on gunned and ungunned portions of wildlife management units (WMU) in Regions 1 and 2

WMU	*Year*	*Ungunned*	*Gunned*	*Difference*
2A	1990	0.06	0.14	0.08
2A	1991	0.44	0.57	0.13
2B	1995	0.24	0.30	0.06
2B	1996	0.22	0.09	−0.13
2B	1997	0.32	0.48	0.16
3A	1990	0.04	0.07	0.03
3A	1991	0.33	0.14	−0.19
3A	1992	0.12	0.14	0.02
3A	1995	0.23	0.29	0.06
3A	1996	0.00	0.10	0.10
3A	1997	0.26	0.32	0.06
5B	1998	0.05	0.10	0.05
7	1998	0.27	0.47	0.20
7	1999	0.18	0.37	0.19

Effect of aerial gunning of coyotes immediately before pronghorn fawning was extremely short-lived, with an apparent decline in survival the second year after gunning ceased to a level lower than pretreatment levels. If this is consistently the case, intermittent aerial gunning with lag times of 3 to 4 years between treatments might actually contribute to the decline of fawn survival. The mechanism involved here is unclear, but hypothetically resides in the post-treatment response of coyote populations to gunning. Photo by Don and Cathi Jenni.

positive effect of gunning. However, the mean fawn survival for ungunned subunits combined was 0.20 fawn per doe; for gunned subunits, it was 0.26, yielding a mean gain of 0.06 fawn per doe during the year of gunning. At this low average rate of benefit, any significant increase in the population would require several continuous years of gunning.

Discussion and Conclusions. From the above assessment of historic information and research and management data, Shaw (2000) concluded that effective coyote control, by whatever method, temporarily increased pronghorn fawn survival. For aerial gunning, this increase normally was small for any given year. Although historic data are largely anecdotal, seemingly high pronghorn densities in Arizona developed during the era of extremely heavy predator control that was carried out over a long period to benefit livestock during the first half of the 20th century. Between 1948 and 1972, when 1080 was in use, fawn survival averaged higher than post-1080 years. However, a long-term decline in statewide average fawn survival began during the years of 1080 application, suggesting that either it was not as effective as the all-out control efforts before World War II or some other undefined factor affected fawn survival during subsequent years. One limitation of this report is the unavailability of a history of application patterns of 1080. S. Gallizioli (personal communication:2000) recalled that only a third of the state was treated with 1080 each year. If so, alternation of treated areas might have had the same negative effect as sporadic gunning and, thereby, stimulated the decline in fawn survival due to a post-treatment response of the coyote population.

It may be possible to increase pronghorn populations in Arizona with coyote control. However, this cannot be done with small, intermittent control programs. Only continuous coyote reduction over large areas, with treatments no less frequent than every other year will be effective. This could be accomplished with a return to heavy use of toxicants, extremely heavy trapping efforts or intensive aerial gunning applied at least biennially over large areas. If such heavy treatment is not feasible ecologically, economically or politically, as Shaw suspected would be the case, the best alternative is to do no coyote control at all, hoping that the apparent gradual trend of postcontrol survival toward pretreatment levels is real. This conclusion should not preclude coyote control programs in WMUs such as 5B, where the pronghorn population is nearing extirpation and perhaps needs all forms of help.

An assessment of other factors affecting fawn survival is badly needed. At least two studies in other areas have noted density dependence functioning in fawn survival—Hess (1999) and Kohlmann et al. (1999a). Density dependence acknowledges the summation of factors affecting productivity and speculates that fawn survival decreases as pronghorn populations increase. An argument for density dependence might be made from Arizona's statewide pronghorn data, in which the total number of pronghorn counted on surveys has increased as the average fawn survival declined (Figure 66). Density dependence,

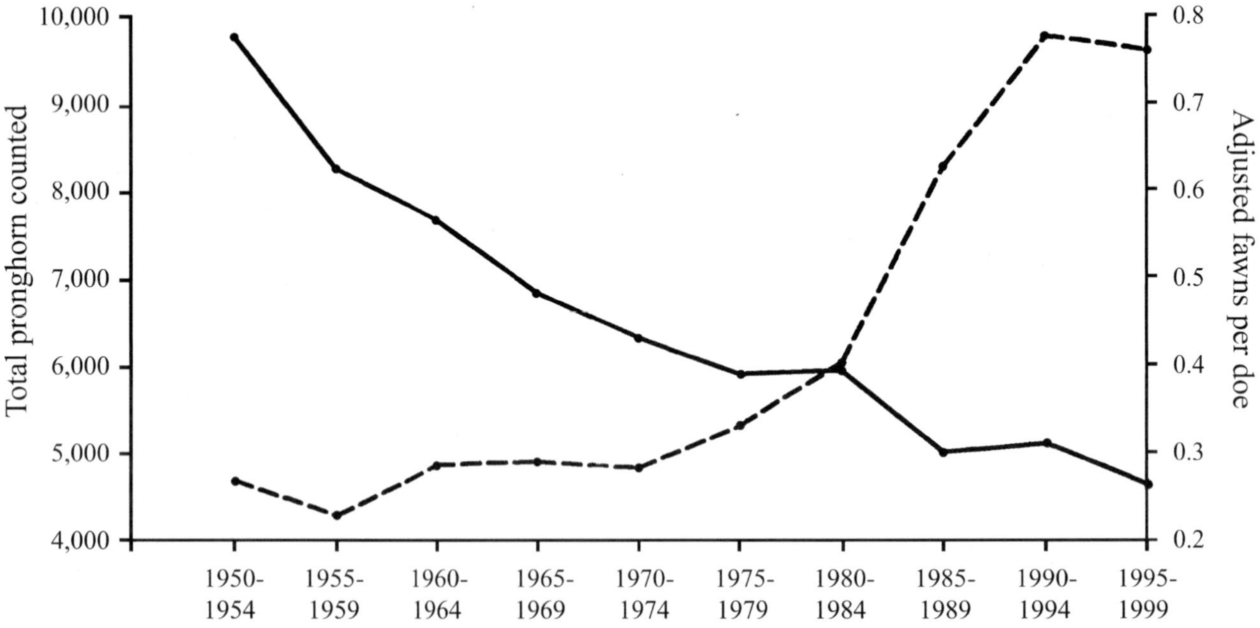

FIGURE 66. *Five-year averages of adjusted, pronghorn fawn/doe ratios (solid line), compared with 5-year averages of total pronghorn counted during annual surveys in Arizona (broken line).*

however, does not explain current failing fawn survival and population declines in WMU 5B. Also, it is a vague theoretical concept that does not identify clearly the mechanisms within the population that cause reduced fawn survival. To understand these mechanisms will require long-term population research on an established study area where adequate control exists to establish and maintain experimental predator control, pronghorn hunting and habitat manipulations. At present, no such area has been delineated in the state nor does support of all interest groups exist for such long-term study.

Based on statewide pronghorn survey data, average pronghorn fawn survival in Arizona has declined steadily for the past 50 years. This decline has continued through one of the driest periods recorded for the state (mid-1950s) and one of the wettest (1973–1993). It also has continued through a predator control regime that began with application of 1080, continued through a period of sporadic aerial gunning of selected WMUs in conjunction with widespread fur trapping, to the present condition of periodic aerial gunning in only a few WMUs and trapping only on private lands.

Aerial gunning usually elevated fawn survival in the gunned area during the years of gunning and possibly had a small residual effect during the first year after gunning ceases. A postgunning depression in fawn survival occurred the second year after gunning, with a slow return to pregunning levels, beginning in the third year after gunning. For aerial gunning to be effective, it must be maintained on at least a biennial basis. This probably is true of other forms of coyote control as well. Shaw (2000) hypothesized that intermittent coyote control may cause a downward trend in fawn survival by artificially elevating the coyote population during the second year after control. To date, no one has monitored a coyote population through an extended period of intermittent reduction to observe the actual effect of such control programs on coyote numbers.

Based on the above analysis, Shaw recommended:

1. not initiating a study focused directly on the simple effects of aerial gunning on fawn survival because it seems futile. Ample evidence demonstrates that a short-term elevation of fawn survival normally results from aerial gunning, but this benefit is limited to the treatment year or, at best, the year following treatment.
2. designing studies so that changes in coyote density and behavior, as well as changes in alternate prey and other predator species populations, can be monitored adequately in both treated and untreated environments.
3. the Arizona Game and Fish Department committing to long-term studies designed to evaluate both natural and human-caused factors affecting the populations. Assessment of the interaction of predation with precipitation and resultant nutritional conditions is especially important, as is assessment of habitat fragmentation. Any effort should try to identify all of the factors at play in the extended downward trend in fawn survival.
4. establishing a grassland research preserve, ideally with long-term agreements with landowners, to give the Arizona Game and Fish Department and cooperating research entities adequate authority to create and sustain grassland ecology, and coyote, pronghorn and alternate prey population research. This area should be centrally located in northern Arizona pronghorn range and fairly representative of all northern Arizona pronghorn habitat. It should be large enough to provide untreated and treated areas for any habitat or predator management studies that are initiated. Short-term, 3- to 5-year studies addressing simplistic, single-factor effects on pronghorn should be avoided. A logical, long-term research plan should be devised before any new field work begins, with collaboration of participating ranchers, land management agencies, concerned public and scientists. Any such plan should undergo intense peer review by appropriate specialists, both within and outside of Arizona. Techniques applied to research should be determined only after specific hypotheses to be tested have been clearly delineated. Opportunistic, or shotgun, use of intrusive methods, such as short-term predator control or infrequent aerial radio-tracking, should be avoided. Capturing and marking animals should be directed toward answering specific questions and include (a) adequate planning to provide a suitably large sample of appropriate animal classes and (b) adequate personnel to provide necessary monitoring of marked animals. Research should be accomplished by biologists and technicians residing on or near the preserve, so that they can be on the area daily and respond quickly to mortality or other significant events in the populations being studied.

Low, Summer Fawn/Doe Ratios

Little is known about the long-term compensatory or additive nature of neonatal predation, and therefore its effects on ungulate population dynamics are unclear.

Linnell et al. (1995:218)

Low fawn/doe ratios in summer were considered justification for predator control throughout much of the 20th century. This is not necessarily wise management as evidenced by four long-term sets of data—comparable to that presented in Figure 66 for Arizona.

First, 45 annual summer surveys of population structure were conducted on Hart Mountain in Oregon (Table 62). They showed higher fawn/doe ratios with than without predator control, but a large pronghorn population increase occurred after predator control was discontinued (Figure 67).

Second, 41 years of data collected in northeastern California (California Department of Fish and Game 2000) indicated a negative correlation between summer fawn/doe ratios and population size (Figure 68).

Also, data from Colorado and Utah showed a lack of correlation between fawn/doe ratios, population sizes and growth (Figures 69 and 70). Although these two studies were relatively short (Colorado = 11 years, Utah = 13 years), they involved easily studied herds isolated from other pronghorn. Both herds pioneered historic habitat

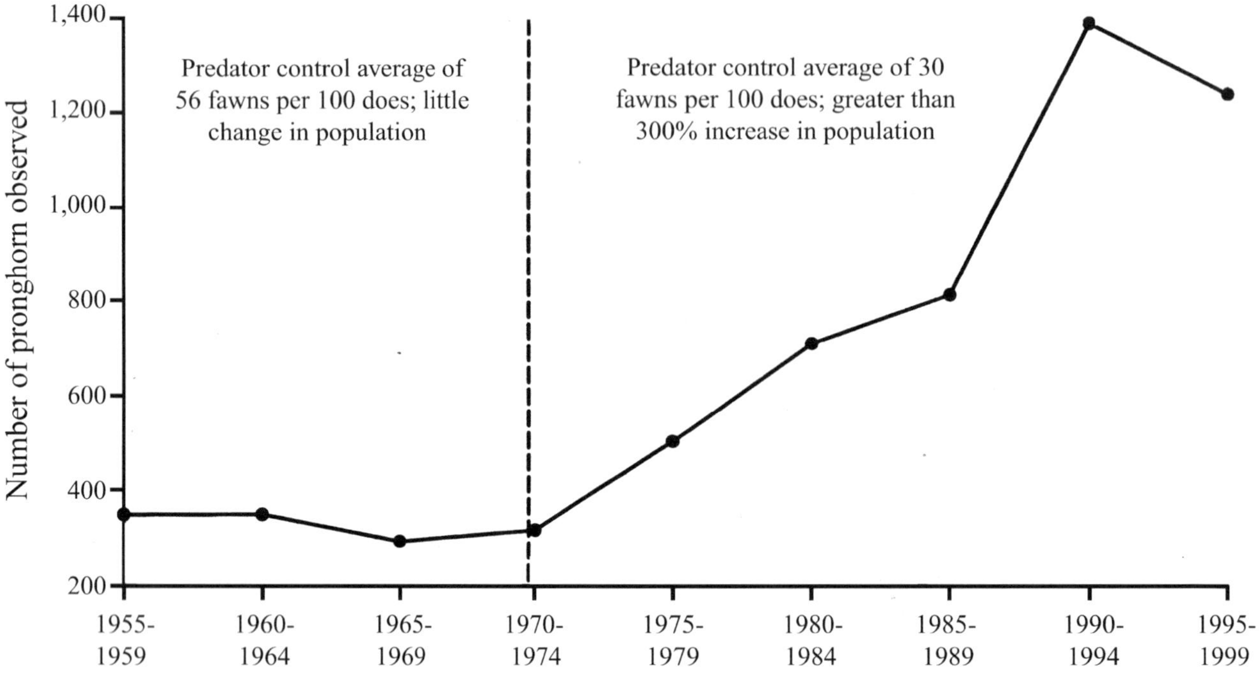

Figure 67. *Numbers of pronghorn and fawns per 100 does observed during midsummer flights (1955 to 1999) on Hart Mountain National Antelope Refuge, Oregon (from U.S. Fish and Wildlife Service 1994, unpublished files).*

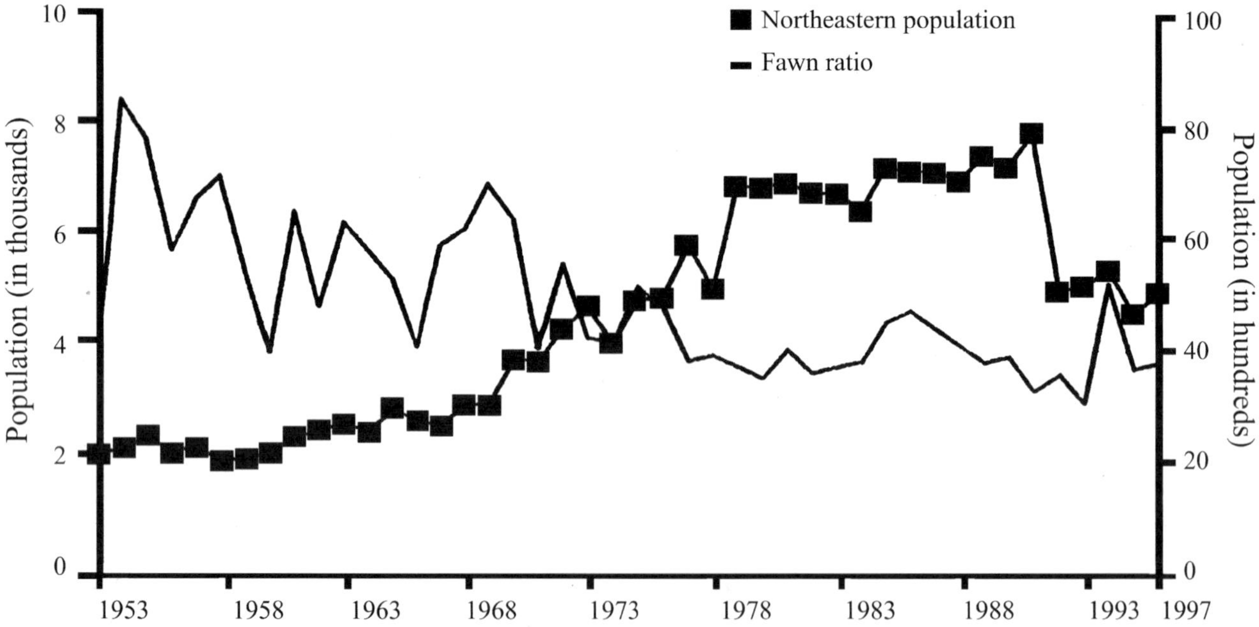

Figure 68. *Summer pronghorn fawn/doe ratios compared with herd size in northeastern California during 1953 to 1997 (from California Department of Fish and Game 2000).*

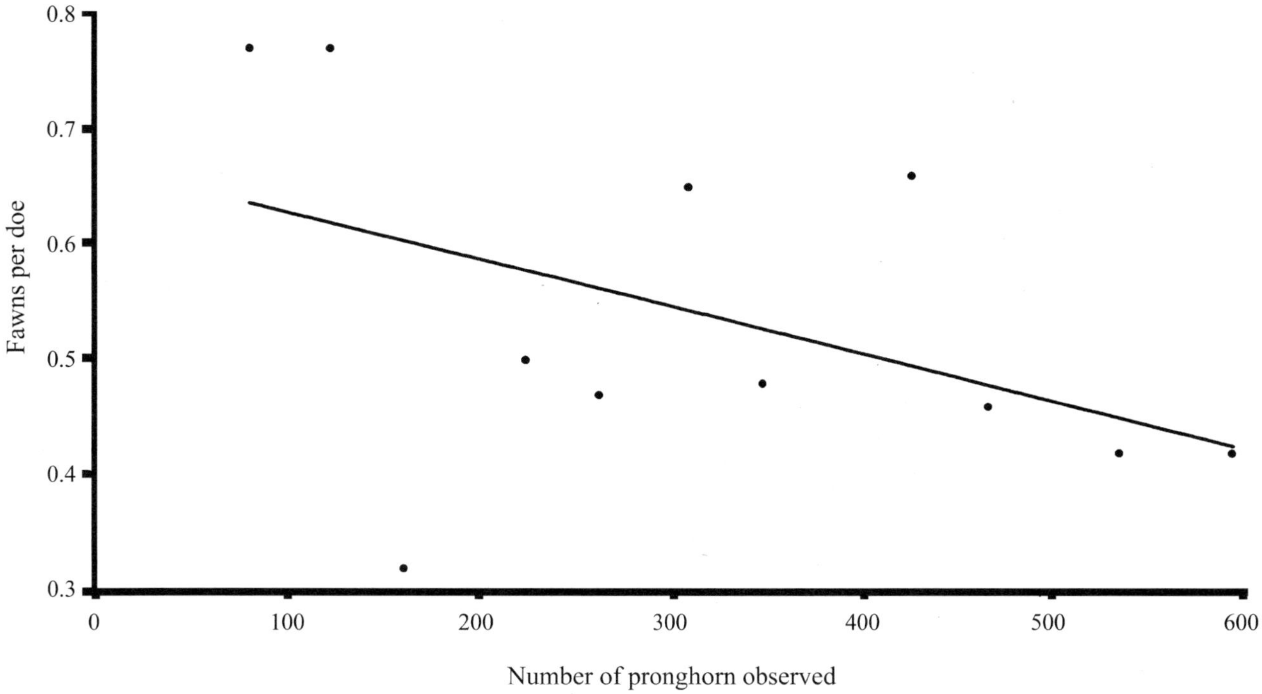

FIGURE 69. *Pronghorn population sizes and fawn/doe ratios (1986–1996) of pronghorn in Middle Park, Colorado (from Pojar 1997).*

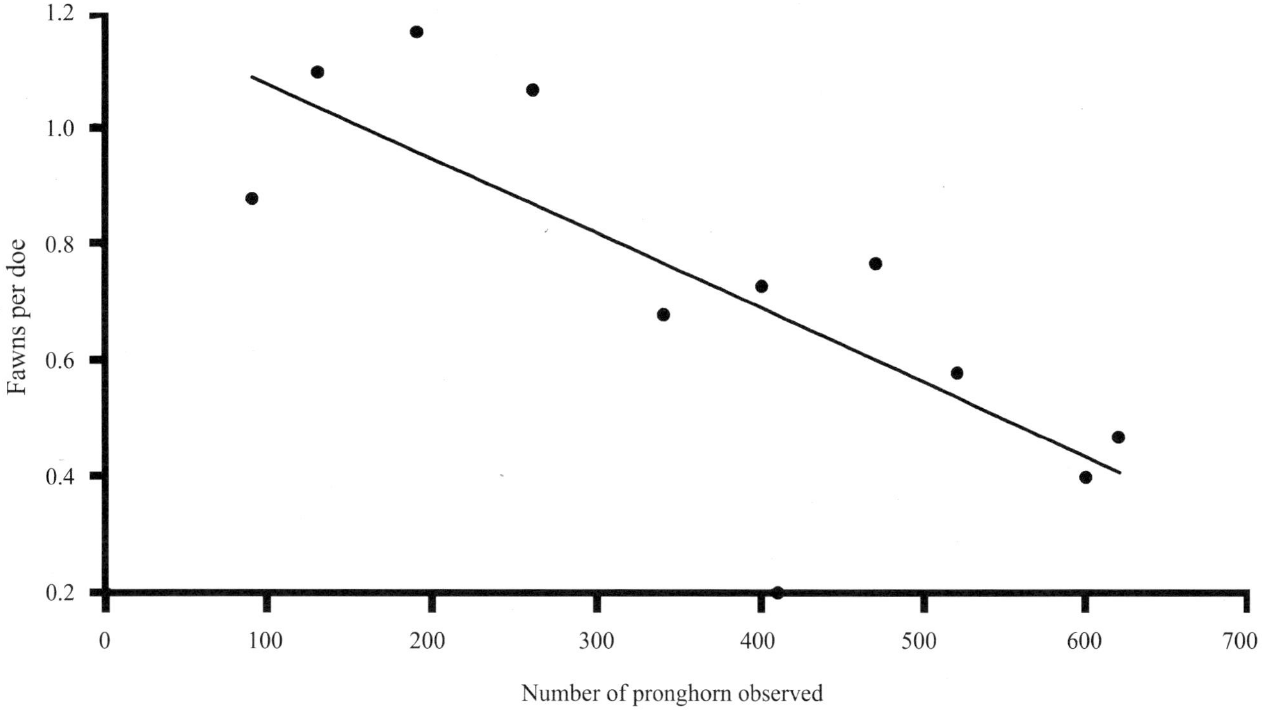

FIGURE 70. *Relationship between autumn fawn/doe ratios and prior year pronghorn population size (1984–1996) on the Deseret Land and Livestock Ranch, Utah (from Danvir 2000).*

and were monitored from the first year they entered unoccupied (by pronghorn) habitat.

Those four studies reveal that low summer fawn/doe ratios do not, of themselves, warrant predator control. Comparatively few studies document increased populations because of predator control (see O'Gara and Malcolm 1988, Canon 1993).

Summer fawn/doe ratios are poor indicators of actual

recruitment. During discussion after a presentation by Vriend and Barrett (1978), B. Helms related that, near Lusk, Wyoming, a pronghorn herd that had 100 fawns per 100 does during a July survey was down to 57 per 100 by the end of February. Spring greenup—when most mortality from malnutrition in northern habitats occurs—probably had not begun. Also, about 54% of fawns born on a central Montana study area during 1966 to 1978 survived until summer surveys; only 31% were alive a year later (Pyrah 1987).

Because neonates are easily captured and fitted with lightweight radio transmitters that fall off within 2 to 3 months, considerable information is available on early mortality (Table 60). Similar data should be gathered on fawns 3 to 12 months of age.

Much of the data here are presented in tabular form in Chapter 12, as support for the concept of density dependence in pronghorn. They illustrate that population growth and size in a given area, under most conditions, are related to habitat quality (mostly nutrition), not summer fawn/doe ratios. This is very important to remember when considering predator control as a management option.

Considerations before Instigating Predator Control

Because of conflicting public attitudes over ungulate and predator management objectives, as well as the increasing costs of predator control, careful biologic and economic justification is required for the exercise of predator control. In general, predator control is justified in big game management only when it will produce substantial ungulate increases at reasonable cost without undue damage to other environmental values and when the increased production will be used. In some cases, control may produce long-range benefits for the predator itself.

Connolly (1978a:394)

Managers should not expect to find strong public support for predator control as a broad management strategy. Our results suggest that the interested public's support of predator control may be relatively high in emergency situations when control is recommended to reverse population declines of desirable species or when it is used in conjunction with indirect management strategies, such as habitat restoration, that are likely to have greater long-term benefit.

Messmer et al. (1999:83)

Stressed populations whose numbers are already dangerously low as a result of habitat loss, overharvesting, competition with non-native species, environmental perturbations, and other causes may no longer be able to withstand "natural variations" caused by predators.

Mason (2000)

After analyzing 38 years of pronghorn winter census data and 43 years of summer composition data from California, Nevada and Oregon, Hess (1999) concluded that survival rates of pronghorn fawns in the northwestern Great Basin were density dependent. His final conclusion was that: "Fawn predation, starvation, abandonment, and weak fawn syndrome are symptoms of DD [density dependence] in the northwestern Great Basin, not causes of low fawn survival. With the pronghorn population >4x that of the 1950's, it is questionable if fawn predation is a real biological problem. Either a relocation program or a liberal hunting strategy reducing pronghorn antelope population density would produce more fawns much cheaper than predator control" (Hess 1999:107). He had reached the same conclusion 12 years earlier.

The following recommendations were proposed by Lee et al. (1998:61–62) with input of biologists and managers from all of the provinces and states with free-ranging pronghorn, as well as the U.S. Bureau of Land Management, U.S. Fish and Wildlife Service, U.S. Forest Service, university professors and others: "In treating a problem situation where pronghorn populations are reduced and predators are prevalent, the following guides should be used:

"Determine the pronghorn herd parameters that are desirable. This may be in terms of total number, rate of recruitment to the herd, age classes, etc. Determine the year-round distribution of pronghorn and the habitat types involved. Consider other population influences including, but not limited to, mortality factors.

"If a factor such as predation is determined as a significant inhibitor of a pronghorn herd, the cost of actually controlling predators in the short-term must then be balanced against the long-term return.

"If it is determined that the increase in pronghorn justifies the cost, predator control should be done on those herd units where documentation indicates predator reduction would be beneficial in meeting management objectives."

In addition to the above, evidence in this chapter and Chapter 12 concerning lack of correlation between low pronghorn fawn/doe ratios and population growth and size, as well as the previously detailed Arizona experience, should be considered. If a decision is made to instigate a control program, aerial gunning on and near the fawning grounds during spring generally is the most efficient and cost-effective method for coyotes. To be effective, such gunning must be conducted at least every other year. Timing is important; control should precede pronghorn fawning, but be late enough that coyotes have selected their whelping dens. Otherwise, other coyotes may move in after residents are killed. Gunning also should precede coyote whelping for humane reasons. Orphaned pups ordinarily will starve or be killed and eaten by other coyotes.

IDENTIFYING PREDATOR KILLS

Predation is rarely observed; therefore, accurate assessment of losses to specific predators often requires careful investigative work. The first action in determining the cause of death of an animal is to check for signs on the animal and around the kill site.

Dolbeer et al. (1994:494)

The predator that fed on or killed an animal usually can be determined, but overlaps and exceptions occur in feeding and killing behavior of various predator species. Predators sometimes learn individual feeding or killing styles and, thereafter, vary somewhat from those typical of their species. Feeding by a certain species should not be considered evidence that species made the kill. The sizes of holes made by canine teeth or talons are diagnostic, but width between punctures is not; the prey's skin often is stretched or wrinkled when an animal is seized. Extensive bleeding is one characteristic of predation. Even when external bleeding is apparent, the carcass should be skinned completely after a thorough examination of the kill site and exterior of the carcass. Subcutaneous hemorrhaging at the site of a tooth or talon puncture or bruise is proof that an animal was alive when seized. The following information on identifying kills is from O'Gara (1978) or his personal experience unless otherwise indicated.

Coyotes

Coyotes commonly feed from the viscera of large mammals after opening the abdominal cavity just behind the ribs. They seem to prefer the fat of the greater omentums, kidneys with surrounding fat, heart with pericardium and liver, in that order. When feeding on muscles, coyotes generally eat hams or the fatty meat on ribs first. They often eat the cartilage of the brisket and chew the soft lower ends of the ribs, leaving jagged portions attached to the backbone. A hungry coyote usually eats 5 to 6 pounds (2.3–2.7 kg) of meat and fat at a time, but adults often regurgitate to pups at the den and return to feed on a carcass a second or third time. Small fawns occasionally are carried intact to dens where they are almost completely consumed; a few teeth may be the only remains. Coyotes sometimes cover prey remains with dirt, according to studies in Idaho, Montana and Nevada, and the burials are neater and more complete than those of felids, which almost always cover their prey.

Sighting of a coyote feeding on a fresh pronghorn carcass should not be considered proof that the coyote killed the pronghorn. Subcutaneous hemorrhaging at the site of tooth or talon punctures, characteristic of a predator species, is good evidence of what predator killed the animal. In the case above, in Oregon, throat wounds led the researchers to believe that the pronghorn had been killed by a predator. Photo by Jim D. Yoakum.

Carcasses of radio-collared pronghorn fawns can be found, but determining the cause of death often is difficult. It requires patient, systematic skinning and dissecting plus knowledge concerning diseases, parasites and characteristics of wounds inflicted by various species of predators. Photo by Walt Bodie.

This photograph shows the remains of a week-old pronghorn fawn found at a coyote den in Montana. The radio-collared pronghorn fawn was seen at about 5:00 P.M., and the blood-stained collar was found at a coyote den the next morning with the ear tags and bits of bone and teeth. Before miniaturized radios were available, such remains were seldom found. Photo by Bart W. O'Gara.

During a study in Texas, almost all coyote kills were buried (Canon 1993). Coyotes tend to dig a hole (leaving claw marks on the sides or bottom), put the prey in and cover it.

Coyotes generally kill cleanly with bites to the throat, leaving punctures by upper canine teeth just behind or below one ear and opposing punctures on the opposite side of the trachea. The punctures are round and measure about 0.13 to 0.19 inch (3–5 mm) in diameter. The hyoid bones often are crushed, and the larynx and anterior end of the trachea sometimes can be lifted free of the neck musculature after skinning. Death apparently is caused by suffocation, not by hemorrhaging. Occasionally, coyotes bite repeatedly. Such chewing lacerates the neck muscles and paralyzes the prey, which sometimes is fed upon before death. When this happens, spurts of blood often are evident on snow, earth or vegetation. These spurts occur because arteries are torn as internal organs are pulled from the living animal. Coyotes catch small running prey, up to 15 pounds (7 kg), by biting across the back, and subcutaneous hemorrhaging occurs, ribs often are broken and internal organs sometimes are ruptured. Unless the prey is killed quickly, the coyote then takes a throathold. On occasion, a coyote will bite a fawn in the head and totally crush its skull.

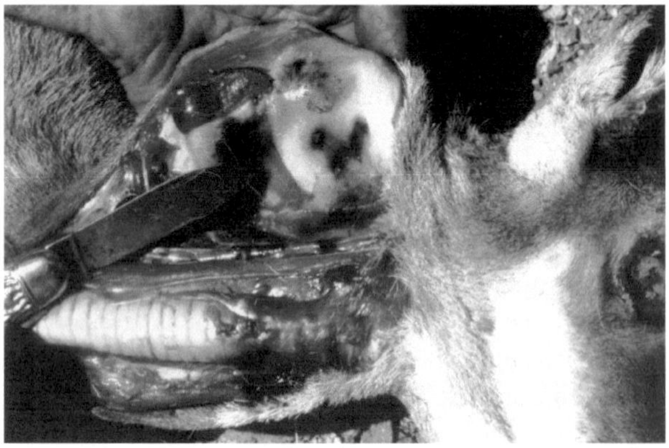

A coyote typically kills pronghorn by crushing the trachea. Its canine tooth marks show in the victim's throat skin. Photo by Don Beale.

Bobcats

On Drakes Flat, Oregon, three fawns less than 1-month-old were found covered with debris under a bitterbrush plant. A bobcat was scared from the carcasses, one of which was intact (Yoakum 1957a). Bobcats seldom eat entire fawns, but they often eat the back half and then bury the shoulders and head. Carcasses usually are cached in natural depressions and almost always covered to some extent before being left. If the prey is not too large, it may be completely buried where the soil is loose, or the remains may be pulled or carried for a considerable distance to be placed under bushes or overhanging rocks.

Bobcats usually kill their prey with a bite or bites to the back of the neck near the base of the skull, and death results from dislocated neck vertebrae. A cat often catches and holds its prey with its claws before making a killing

The carcass of a pronghorn fawn killed and partly eaten by a bobcat. Such remains usually are buried and almost impossible to find if they do not have a radio collar. Photo by Bart W. O'Gara.

bite. Unless fawns are quite small, scratches usually are evident on the shoulders, back or sides. To analyze such kills, the carcasses should be skinned to reveal the claw marks, which seldom are obvious from the outside. However, bobcats sometimes bite fawns' heads, fracturing them (but not crushing them as do coyotes), without leaving claw marks. The tooth punctures generally are smaller than those left by coyotes, measuring only 0.06 to 0.13 inch (1.6–3.0 mm). Tooth marks on the back of an animal's neck usually are teardrop shaped or slitlike. Bobcat teeth seem to slash and cut more than those of the canids.

A large pronghorn fawn killed by a bobcat at the Desert Experimental Range, Utah, with only the head covered by the cat. Photo by Don Beale.

When other material is not handy to cover its kill, bobcats (as well as mountain lions) often scratch hair from the carcass as camouflage. This breaks the outline so the corpse is less easily found, especially by carrion-eating birds. Usually, the entire body is covered. Photo by Rod Canutt.

Golden Eagles

Golden eagles generally feed first on viscera, eating everything but the stomach and intestines. They hollow out a carcass, leaving the skin and large bones mostly intact. The ribs of a young fawn or lamb generally are clipped off close to the backbone and swallowed by the birds. They do this neatly, similar to the cuts of pruning shears, and the sternum often is found nearby. A small carcass, heavily fed on by eagles, appears to be turned inside out; the backbone often is intact, although devoid of ribs, and the leg bones are pulled out of the skin with most of the meat gone. Usually, the neck and head are intact, although eagles sometimes open the skull without crushing it and eat the brain. Some birds hook their beaks into the foramen magnum and lift off the top of the skull. Others eat through the throat and tongue and open the skull from the bottom. Splashes of whitewash on the ground and small downy feathers on vegetation usually are evident where eagles have fed.

Golden eagles may strike anywhere on their prey, but often seize a fawn by its head. Examining the pattern of wounds without removing the skin is easier with eagle kills than with those of mammalian predators, but skinning often is necessary. Talon punctures generally are deeper than those made by teeth. They can be as much as 2 inches (5 cm) deep, as wide as 0.5 inch (13 mm), without tearing and more triangular than round. Crushing of bones between talon wounds is seen only on an occasional fawn skull. The only sure way to ascertain an eagle kill is to examine the pattern of wounds. Eagles grasp their prey with three toes opposing the hallux (rear toe) of each foot. The middle toe is the longest, so if the prey is large enough, the toe talons leave wounds 1 to 2 inches (25–50 mm) apart in a row or shallow v. The hallux wound may be 4 to 6 inches (10–15 cm) from the middle toe wound. However, with animals the size of a small fawn, finding fewer talon marks is common—one made by the hallux and one or two by opposing toes. With mammalian predators, four punctures or bruises from the canine teeth almost always are evident. Eagles often scissor the hemorrhages right off the skin of carcasses that have been almost completely eaten. Eagles kept at the Montana Cooperative Wildlife Research Unit did not puncture the skins of dead animals given to them as food. However, a talon puncture without a subcutaneous hemorrhage should not be used as evidence of a kill because different birds kill and feed in different ways.

The birds must repeatedly stab large prey until it bleeds out and is immobilized. Adult pronghorn usually are killed by multiple talon stabs through the tops of the ribs or into the small of the back. Eagles' talons are ideally shaped to encircle a mammal's spinal column, to puncture the aorta. The victims bleed internally after the large arteries are punctured, often in the vicinity of or just anterior to the kidneys.

Ravens and other carrion-eating birds hollow out carcasses similar to the manner eagles do, but are unable to reach the meat on the side next to the ground or to break off ribs. In contrast to eagles, which cut tendons cleanly, these birds pull meat away from the tendons leaving frilly white tufts on the bones.

The remains of twin fawns killed and partly eaten by a golden eagle. The eagle was observed to kill one fawn and begin to eat it. The doe returned and drove the eagle from the carcass. The eagle then picked up the other fawn and flew over the ridge to feed on it. Photo by Bart W. O'Gara.

Other Predators

Wolves seem to prefer fatty portions of a carcass, so their feeding patterns can appear similar to those of coyotes. However, the larger canid eats more than the smaller, and wolves often kill and feed in packs. Thus, feeding may commence in more than one area, and consumption almost always is greater than that by coyotes. Larger bones may be broken or chewed, and almost all parts of a carcass may be used.

Wolves sometimes run large animals to near exhaustion before catching them. The point of attack during such a chase usually is the rear or sides of the prey. Wolves also drive prey animals toward other wolves, and the nose or throat may be the first site grabbed. Thus, wounds with subcutaneous hemorrhages may be found almost any place on a carcass. Wounds on the flanks and hind legs often are large tears inflicted as several wolves pull in opposite directions.

Mountain lions, like coyotes, often feed behind the ribs, then the hind quarters. The big cats also place their prey in natural depressions and cover them with light debris or snow. Lions bite and scratch off tufts of hair so a cached carcass may be almost buried in hair. Lions may return to feed on a kill several times and sometimes completely devour a carcass during multiple visits, crushing and swallowing the largest bones.

Mountain lions kill in a manner similar to that described for bobcats, but the tooth punctures are larger, about 0.25 to 0.31 inch (6–8 mm) in diameter, and tend to be rounder

A 36-pound (16 kg) fawn killed by a golden eagle. Note the slash between ribs and clotted blood from the sliced lung. Wounds sometimes are so deep that the toes as well as the talons penetrate the prey's body. Photo by Bart W. O'Gara.

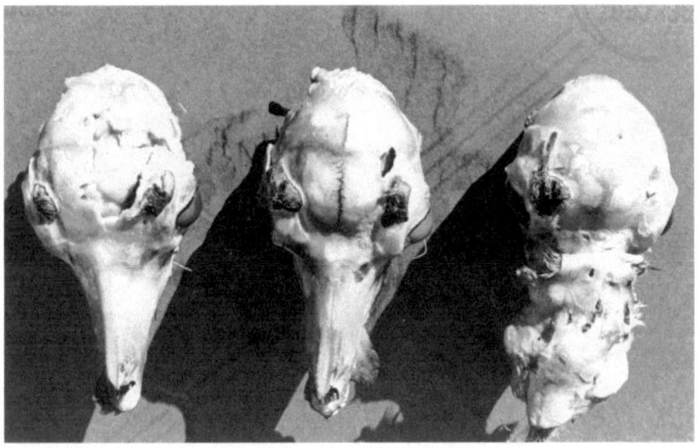

The skulls of three pronghorn fawns killed by predators. The skull on the left was lightly crushed by a bobcat. The one in the middle was punctured by two talons of an eagle, one of which caused a large hemorrhage in the brain. The skull on the right shows four small talon marks through the skin and one into the skull, but the fawn died from hemorrhaging of a bruised liver. Photo by Bart W. O'Gara.

than those of the smaller felid. The necks of the prey may be completely disjointed. Strips of hide as well as scratched out hair generally are found at the site of the attack. Lacerations caused by claws during the struggle to secure a killing bite are much larger than those caused by bobcats and usually are obvious from the outside. On larger prey, such as elk and moose, lions sometimes use a throathold to choke their victims.

Bears wallow down vegetation around a kill or carcass and often vomit or defecate close by. They use their feet to hold carcasses while feeding, so the carcasses are not slid and scattered like those fed on by coyotes. However, bears often pick up carcasses and carry them for long distances from the kill site. Bears skin carcasses neatly. They often eat the udders of lactating females first, and the genital area is a favored feeding spot on all carcasses. However, bears may begin feeding where they first taste blood—often on backs of necks, tops of shoulders or along backs where they first bit. This usually leads to extensive feeding on the loins. Bears may break bones during a kill, but they seldom chew bones while feeding as canids often do and large felids sometimes do. Black bears and grizzly bears sometimes cover carcasses, raking considerable dirt and heavy material over them, but grizzlies do so more often than black bears.

Grizzly bears sometimes break the backs of even large prey just in front of the hips as they throw their weight onto an animal. Large bruises, not accompanied by claw lacerations, often indicate that a grizzly slapped an animal while catching it. Animals the size of pronghorn may be killed by such blows, sustaining broken ribs and backs in the process. Bites often are delivered to the top of the prey's shoulders or back, and bone crushing sometimes accompanies a bite. The canine tooth punctures of black and grizzly bears are comparatively large, measuring 0.25 to 0.5 inch (6–12 mm) in diameter.

Black bears kill by neck biting more than do grizzlies, and their curved claws often leave punctures through the skin of the flanks or hams where the prey was grabbed. When a black bear takes prey with a neck bite, the kill looks very similar to predation by a mountain lion. Thus, patterns of feeding on and covering prey should be investigated carefully when an animal is killed by a large bite to the back of the neck, especially during summer, when bears are not hibernating and tracks may be hard to see.

Red foxes usually feed from the back or rear of a carcass, and they often lay down and feed from the undersides of large carcasses. Foxes can make straight, knifelike cuts, often cutting off the lips, udders or genitals. Prey, weighing almost as much as the fox, often is carried to the den. Foxes chew more, but kill in the same manner as do coyotes. The tooth punctures are small, elliptical and about 0.06 inch (1.6 mm) long and half that wide (O'Gara 1978).

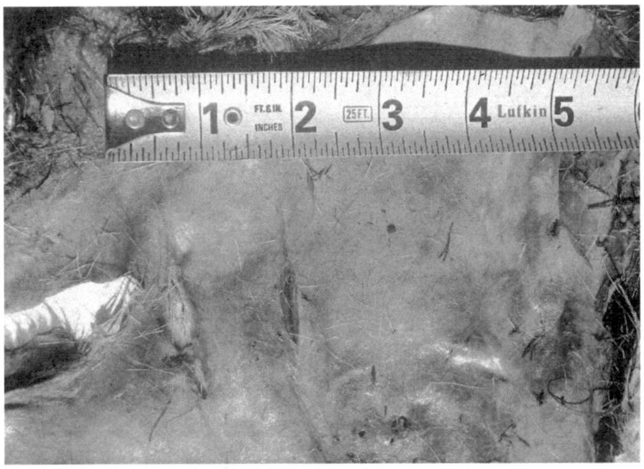

The skin of a fawn killed by a mountain lion in the Trans-Pecos region of Texas. Note the knifelike cuts, typical of cat kills, in the skin. Bobcats usually scratch out considerable hair when they grab prey as large as a pronghorn. Lions sometimes scratch out strips of skin as well as hair. Photo by S. Kemble Canon.

Badgers, like foxes, feed from the undersides of large carcasses, but they burrow completely underneath and often stay there until the meat is gone or rotten. Animals the size of young fawns usually are partly eaten and the remainder pulled 2 to 3 feet (60–90 cm) down a hole. Tooth holes in the prey's skin are rounder than those made by foxes but about the same size. Two fawn carcasses found in badger holes by Trainer et al. (1983) had numerous bite wounds, with hemorrhages, over extensive areas on the hind legs of one specimen and on the rib cage of the other.

Raptors that can take jackrabbits can take small pronghorn fawns easily. A small raptor (probably a prairie falcon judging by the closeness of a nest and foot and talon sizes indicated by wounds) killed two pronghorn fawns in Idaho. Both were bruised on one side, their livers were ruptured and talon wounds were found in the heads and necks (W. Bodie, personal communication:1979).

Most dogs do not feed, or feed minimally, on the carcasses of large animals they kill. When feral dogs do feed on big game, they usually feed on the hams because they have already lacerated them during the chase and kill. If dogs eat freshly killed big game, they usually regurgitate close by.

Dogs generally are inefficient killers of large ungulates. They usually make slashing, tearing attacks and maim as many animals as they kill. Intestines pulled out, hams skinned and chewed, and noses or faces lacerated are common results of dog attacks. More than one dog usually is involved in an attack. Large dogs occasionally gain experience and kill almost as cleanly as coyotes. However, skinning the carcasses of large animals killed by dogs generally reveals that 80% or more of the body surface was bruised by bites that did not penetrate the skin. A German shepherd killed three mule deer fawns of about 15 pounds each by biting them across their backs and apparently crushing them to death, but the dog did not attempt to eat the fawns. A similar situation would be expected with pronghorn fawns.

PRONGHORN/PREDATOR INTERACTIONS

Pronghorn have developed several reproductive and behavioral traits in response to the sustained history of predation on fawns. Affected traits include offspring birth weight, litter size, maternal behavior that supports the hiding strategy, and possible maternal cognitive capacity.

Byers (1997a:56)

Mammalian predator–prey systems are behaviorally sophisticated games of stealth and fear.

Brown et al. (1999:385)

Pronghorn tend to occupy good lookout positions and usually are watchful enough to prevent the close approach of predators. One member of a band often watches while the rest feed. A band, when climbing a low elevation, may stop before reaching the crest, permitting the leading doe to look over and examine the country ahead for some minutes before leading the band up and over. With such watchfulness and the pronghorn's speed, few predators other than eagles ordinarily catch adults. Skinner (1922) maintained that gray wolves killed a few pronghorn, although, as a rule, they confined their attacks to larger and slower animals. Also, owing to the broken character of the Yellowstone National Park winter range affording good lurking places, mountain lions managed to kill a few under favorable circumstances. Skinner doubted a bear could catch a pronghorn unless it found a very young fawn hiding. In a report to the Chief of the Corps of Topographical Engineers in the U.S. Army, Francis T. Bryan noted that, during 1846, pronghorn along the Sweetwater River in Wyoming were compelled to shelter themselves from wolves by mingling with bison (Long 1965).

In comparison with other times in its life, during and shortly after birth are the times when a pronghorn is vulnerable to predation. Also, predators have influenced the evolution of precociousness in pronghorn fawns and the behavior of the mother and young. A doe must meet the physiological needs of her fawns immediately after birth by drying and fluffing their coats to prevent hypothermia and by nourishing them. She also must protect them from predators. Bromley (1977) postulated that the optimal time at the birthsite for care of the young would be continual, whereas the optimal time spent at the birthsite to avoid predation would be zero. The birth process—from the onset of intense labor until the newborn fawns separate themselves from their mother and locate bedding sites—lasts only about 6 hours.

Being selective feeders, pronghorn logically should not feed in herds, but they do, apparently, as a defense against predators. An eagle overhead often will cause pronghorn to congregate in large groups and remain that way for as much as an hour (Bruns 1969). Coyotes and bobcats do not cause the same type of grouping, but adults are fairly secure from them. In Utah, pronghorn form sizable bands on flat desert areas when vegetation is succulent enough to support them (D. Beal personal communication:1980). However, when the country is dry, pronghorn split into small bands, often just a doe and her fawns, to feed along dry water courses where some green forbs remain. There, many fawns are killed by bobcats or coyotes.

Pronghorn and coyotes seem to have co-evolved to the point where coyotes employ a variety of hunting strategies designed to capture fawns, and pronghorn use a number of countermeasures (Byers 1997a).

Pronghorn mothers show heightened vigilance when away from hidden fawns. Compared with does whose fawns have died, those with hidden fawns spend less time feeding, but more time standing and reclining. Thus, mothers have their heads erect and stationary at least 70% of the time, compared with 50% for other does. Such heightened vigilance is adaptive; the sooner a predator is detected, the sooner appropriate defensive maneuvers can be affected. Once a doe returns to her fawns, which stand, they become conspicuous and visible from a distance. Thus, the reunion period is a time of heightened vulnerability for fawns. In the hour before reunion, mothers may show more than simple vigilance. Often, they appear to search for predators. For example, a mother may have a fawn hidden about midway up a ridge that is 0.9 mile (1.5 km) from top to bottom. During the 45 minutes before returning to a fawn, she may run to the drainage below the ridge, stand motionless with head erect for minutes at various locations there, run to the top of the ridge, scan again and then quickly return to her fawn. If a fawn is reclining in a relatively flat bowl surrounded by ridges, its mother may run to a ridge top, scan, run to the opposite ridge top, scan again and go to the fawn—traveling 1.9 to 3.1 miles (3–5 km) during the half hour before reunion. Mothers do not always appear to search for predators in this manner, but those that recently have lost a fawn to predators or have seen a predator since the fawn last bedded usually do (Byers 1997a). Byers stated that, when pronghorn mothers act in this purposeful way, it is difficult to avoid the conclusion that they intend to return to fawns, anticipate that a coyote may be in the vicinity and want to detect the coyote if it is present. Pronghorn dams usually move their fawns to a new location after reunion periods. Those that have lost a fawn or were harassed by coyotes may move fawns 0.6 mile (1 km) or more. Pronghorn mothers often move fawns back and forth across ridge tops during successive reunion periods (Byers 1997a).

Daily mortality rates are greatest for pronghorn fawns 11 to 20 days of age—the period of transition out of hiding. Fawns become increasingly visible and vulnerable to coyote attacks. Beyond 20 days, fawns are almost continuously visible, but their running ability improves their chances of surviving an attack (Byers 1997a).

After 15 years of observing pronghorn and coyotes on the National Bison Range, Montana, Byers (1997a) reported that coyotes hunted fawns in three ways—by observing, searching and cooperative group searching.

1. A coyote observed from hidden vantage points where a solitary doe could be watched. After the doe fed her fawns, the young reclined and the doe left the area; the coyote ran to the bedsite and made the kill with a bite and shake to the neck. This tactic was used infrequently.
2. A searching coyote moved close to a doe and trotted back and forth, searching a rough quadrant. Single coyotes generally did not search for extended periods because: the doe exhibited increasingly aggressive responses; the doe moved away, indicating the coyote was not close to a hidden fawn; or the doe lured the coyote away.
3. Cooperative group searching generally involved two or three coyotes. Cooperation allowed searching in an area where a doe attempted to prevent it. While the mother chased one searcher, another picked up the search where the first one was forced to break off. When the mother attacked the new searcher, the first coyote resumed the search. In this manner, two or three coyotes could search an area, and the dam could do little to prevent it.

Davis and Putnam (1951) watched from an aircraft as a small coyote chased a pronghorn doe and fawn in South Dakota. After 1 mile (1.6 km), the coyote gained on the fawn and its dam dropped back and ran into the predator, rolling it end over end. The coyote recovered and continued the pursuit. For the next 2 miles (3 km), the doe ran both ahead of and behind the coyote, repeatedly crossing between the fawn and its pursuer. Twice more she upset the predator by running into it. After the third upset, the coyote started dodging the dam and was slowly outdistanced by the fawn, which was soon joined by another doe. With the dam harassing the coyote, the fawn and second doe separated after passing the crest of a hill. The coyote followed the leading doe, and the fawn sought refuge in a nearby depression. The observed chase took 45 minutes and covered about 5 miles (8 km).

Does are not always able to protect their fawns, even from a single coyote. During May 1978 on Sheldon, a doe with fetal membranes still attached was observed in the company of a yearling doe, a buck and two newborn fawns (McNay 1980). Doe/fawn imprinting had been observed for about an hour when a large coyote trotted along a fence line 100 yards (94 m) south of the pronghorn group. The fawns were bedded, and the doe was aware of the coyote. When directly opposite the pronghorn, the coyote began a slow stalk toward the area where the fawns were bedded, breaking into a run when 50

(Left) *A pronghorn doe with young fawns nearby moves to intercept a coyote along a gravel road.* (Right) *The coyote flees and is pursued by the doe for some distance. Van Wormer saw several such interactions during spring in Yellowstone National Park. The coyotes took refuge in dense shrubs or a road culvert.* Photos by Joe Van Wormer; courtesy of the Denver Museum of Natural History.

yards (46 m) from the group. Immediately, the doe charged the coyote and a series of circular chases began in which the aggressiveness of the doe seemed to be matched by that of the coyote. Breaking away from the encounter, the coyote ran toward the fawns, gaining a hold on one as he passed over it at a run. With the doe in pursuit, the coyote continued past the fawns, turned and ran a zigzag pattern to the fence line.

Immediately after passing under the fence, the coyote stopped, turned and sat. Stopping short of the fence, the doe stared at the coyote but did not attempt to pursue it further. Returning to her fawns, the doe began moving to the northwest across the flat. The coyote again crossed under the fence to stalk the pronghorn. The doe confronted the coyote, which avoided her and "hit" one of the fawns, dragging it a few feet before being driven off by the doe. The coyote ran for the fence line where the pursuit ended as before. The second attack crippled the fawn, which then was unable to keep up with the doe and the other fawn. Within 40 minutes, the coyote made four more tries for the crippled fawn. Each encounter ended at the fence line—a refuge for the coyote and barricade to the doe. The coyote finished the kill on the sixth attempt, but was driven off by the doe and yearling buck. About 10 minutes later, a final stalk was made, and the dead fawn was not defended.

The coyote had rushed the fawns six times in 1 hour and concentrated its efforts on one fawn. It contacted that fawn on all but the third attempt. The yearling buck stood passively by during the first three encounters, joined the pursuit of the coyote to the fence line in the fourth encounter and led the chase during the fifth and sixth encounters. He periodically attempted to mount the doe, possibly in response to odors produced during parturition. The carcass was retrieved by the observers, and necropsy indicated that the fawn had been in good nutritional condition. Bruises, but no punctures, were evident beneath the skin of the back. The fatal wounds were in the throat and head.

Also on Sheldon, J. Meeker (personal communication:1981) watched two pronghorn does pursue coyotes and simply run over them. Those does did not appear to strike with their front feet, but the coyotes took considerable punishment from being stepped on. They were knocked down, but would be running again almost instantly. The does would wheel and repeat the procedure. Meeker also saw a doe butt a coyote and throw it into the air. This is the only report O'Gara has received of pronghorn butting predators. In the encounters he has seen and all others reported to him, the pronghorn used their feet to contact predators.

In Idaho, W. Bodie (personal communication:1981) noted two types of behavior by pronghorn does with fawns hidden nearby when coyotes approached. On three occasions, does or groups of adult and yearling does ran about 100 yards (91 m) past coyotes and stopped to watch the predators. The pronghorn repeated this action several times, apparently trying, without success, to entice the coyotes into chasing them away from the fawns. On one occasion, a coyote lay down and watched the procedure with evident interest. When the predator resumed hunting, its course took it toward a hiding, radio-collared fawn. A doe, followed by a yearling, ran toward the coyote. At about 20 feet (6 m), the doe dropped her head to the level of the coyote and increased her speed. The chase ended after about 150 yards (137 m) when the coyote entered dense shrubs in a creek bottom. Bodie saw four similar charges by females with hiding fawns. One doe chased a coyote more than a mile (1.6 km), and another chased one up a steep mountainside and through a pass at 11,000 feet (3,353 m) elevation.

Three times during spring, O'Gara saw does intently pursue coyotes. One doe was about to give birth; the other two had fawns bedded nearby. The predators made every effort to escape, but two were struck by the does' front hooves. Wringing their tails violently, both coyotes yelped and scrambled frantically to get away; one dragged its hind legs for approximately 30 yards (27 m) after being struck.

Tucker (1979) reported five similar observations in Texas. He also saw nine females form a ring around five fawns at the approach of three coyotes. Three does then walked toward the coyotes to watch them from 160 yards (146 m). The coyotes stopped to watch the pronghorn for about 10 minutes before wandering off.

Autenrieth (1984) observed a doe (doe_1) locating its bedded fawn, which nursed and then ran vigorously with the doe. Another doe (doe_2) chased a coyote toward doe_1 and its fawn and overran the coyote. The fawn dropped, but the coyote had seen it and ran toward it. The fawn got up and ran as both does chased and overran the coyote, which got up and chased the fawn out of sight. About 3 minutes later, Autenrieth moved to where he could see the pronghorn. Doe_2 walked away, but doe_1 circled an area of big sagebrush, rushing about and horning the brush. The coyote was not in sight. She walked about, watched the big sagebrush, and occasionally walked into it and flared her rump patches. After almost 2 hours, the doe trotted east. She had gone less than 75 yards (68 m) when the coyote walked out of the brush headed west. The doe turned, saw the coyote and ran toward it. The coyote trotted over a hill, and the doe walked into the brush, milled about and extended her muzzle as if sniffing at a spot where a dead fawn was subsequently found. She left the patch after about 30 minutes and went to her other fawn, which rebedded within 200 yards (180 m) of where its twin was killed. The doe returned to the kill site twice more. Autenrieth found the dead fawn (a female) intact with a cut near the left eye and a cut on the left front leg. Apparently, the coyote had lain near the dead fawn for nearly 2 hours without feeding on it, while waiting to get away from the doe.

Autenrieth (1984) also watched as three does chased his Labrador retriever and pinned the dog in a patch of big sagebrush on a flat with 72 does distributed across it. The most aggressive doe also behaved aggressively toward the other two does. The dog would not leave the cover until Autenrieth brought his truck within 150 yards (135 m). The does ran at the dog not more than 50 yards (45 m) from the truck. It ducked and the does nearly hit it, but did not make contact. When the dog got to the truck, the does came within 20 yards (18 m) to watch him. It walked behind the truck, and one of the does ran over the dog, making contact with at least two hooves. The dog ran back to the truck and the does stood within 10 yards (9 m), horning bushes. One of the does had a fawn bedded within 200 yards (180 m) of where the dog was first pinned down. The survival value of does fawning in the same area seemed obvious. Any ground predator entering the flat would come under attack by one or several does. Does, however, react differently to ground predators at different times or in different places. During a 3-year tagging operation in Oregon, a Labrador retriever was used to catch fawns (J. Yoakum personal communication:1988). The biologist with the trained dog caught about 75 fawns during one season, whereas biologists without dogs averaged about 25 fawns. The dog caught individual fawns as many as four times during a month, dispelling fears that tagging operations induced mortality. Does sometimes ran after the dog when it was chasing a fawn, but none of the does struck the dog, although it was holding a bleating fawn down with a paw.

On numerous occasions, O'Gara watched does either ignore or cautiously approach coyotes during summer. The pronghorn seemed more curious than belligerent, although they sometimes seemed to herd the coyote, or several does were trying to surround it. The coyotes usually moved away toward shrub patches or draws. Swanger (1977) watched coyotes pass within 10 feet (3 m) of does with large fawns during summer. Those does appeared to ignore the predators except to stay between them and the fawns.

In Yellowstone National Park, Wyoming, Gese (1999) observed that ungulate attacks on coyotes occurred mostly during summer, when calves and fawns were most vulnerable to predation. He also noted that bison and elk directed aggression primarily at alpha coyotes, whereas pronghorn directed aggressive equally toward all coyotes.

A Texas doe with a full udder, indicating fawns nearby, approaches a photographer with her mane and rump patch partially erected. She had chased a cocker spaniel, which was cowering behind the photographer, and chased the canine from the area although confronted by a human. Photo by S. Kemble Canon.

Although pronghorn does are aggressive toward predators during spring, bucks usually ignore them. Twice during spring, O'Gara watched bucks stare for a considerable time at coyotes entering their territories before running to intercept the intruders. One coyote immediately ran from the territory, herded by the territorial buck for approximately 0.5 mile (0.8 km). The other coyote ducked into a snowberry patch. The buck either could not or did not want to run in the shrubs, so it herded the coyote at a walk for about 20 minutes, until it was forced into the open and chased from the area. Although both territorial bucks evicted coyotes, the coyotes did not appear frightened, and the bucks did not press them. This kind of pursuit was quite different from that of does chasing and striking coyotes during spring, but much like the summer behavior of does.

O'Gara never saw bucks confront coyotes other than on territories. Even there, the bucks were not hostile, but seemed playful or curious. During October in Yellowstone National Park, O'Gara watched six coyotes chase two pronghorn bucks for more than 1 mile (1.6 km). The bucks ran to and mingled with a band of 13 does and fawns. The herd bunched up and two does met the six coyotes head on. The coyotes avoided the does, but circled the band twice; two of the coyotes howled several times. Every time a coyote approached within 40 yards (32–37 m) of the band, a doe advanced and the coyote retreated. After an hour, the coyotes left the band.

A pronghorn buck feeds with apparent lack of concern for a nearby coyote. Although does are aggressive toward predators during spring, bucks usually ignore them. Only when a coyote enters a buck's territory is the buck likely to chase the predator. Photo by Peter T. Bromley.

Coyotes ordinarily are reluctant to get close to pronghorn does during spring and summer. If fawns accompanied this band, or were bedded nearby, the does would attack the coyote. Photo by Tony Debold.

O'Gara found coyote-killed pronghorn against fences that impaired their running ability during winter. Tracks indicated that the pronghorn had made no attempts to face the attackers. He also saw coyote-killed deer that, when pursued, had hit fences they ordinarily would have jumped. Apparently, an animal panicked by predators is hampered by unnatural barriers that it ordinarily would circumvent.

During January, Thompson (1949) saw coyotes catch two Montana pronghorn does without a fight. However, those pronghorn already were harassed by an aircraft and may not have reacted to the predator in a normal manner.

Bruns (1969) watched three approaches, two stalks and one kill, by coyotes during winter in Alberta. The pronghorn did not attack or follow the coyotes as they generally do during other seasons. An apparently healthy female fawn that was killed by coyotes was closest to them when they attacked. A badly crippled male fawn in the same herd was not molested.

O'Gara has seen, and W. Bodie (personal communication:1978) informed him of instances in which coyotes blundered into bedded fawns. In other cases, despite many stories of fawns being odorless, coyotes turned into the wind to smell fawns and go directly to them. A coyote was observed sight-hunting on the National Bison Range by Byers and Byers (1983). They first noticed the coyote crouched on a ridge watching a pronghorn doe nursing two fawns. The coyote watched until one fawn had reclined and the doe had moved approximately 44 yards (40 m) away, then ran to the fawn and killed it within a few seconds. Does can do little to combat such a hide-and-watch strategy except lengthen the times between visits to fawns. However, the amount of time a doe can stay away is limited by the nutritional needs of the fawn.

Bobcats apparently hunt pronghorn by sight—watching fawns go to their beds after being fed. O'Gara saw a bobcat crouched on a knoll intently watching two does in a basin below and subsequently found a bobcat-killed fawn in the basin.

Although bobcats are known to kill pronghorn, they seldom are observed in the act. Nelson (1925) related an account of a bobcat killing a full-grown pronghorn buck in good condition. The buck had been trailed through open pine cover during a snowstorm, and the bobcat had leapt on its back before killing it.

Predation by golden eagles probably is observed more often than coyote attacks because the great birds hunt only by day and draw attention by swooping from the sky. Accounts of eagles killing adult pronghorn during winter indicate that pronghorn try to escape by open-field running (Lehti 1947, Thompson 1949). This allows the big birds to bind to the running prey and repeatedly penetrate the lungs and aortas with talon thrusts between and behind the ribs while riding the animals' backs. Talons are well adapted for encircling backbones and puncturing blood vessels in body cavities. O'Gara has seen these types of wounds a number of times on deer and domestic sheep. Eagles strike almost any part of their prey's body. He necropsied many domestic lambs and a few pronghorn fawns that were caught by the heads. Sometimes, the spinal cord was severed, as apparently happened when eagles killed an adult pronghorn doe in about 1.5 minutes (Lehti 1947).

Storer (1902:413) was hunting for meat in southwestern Nebraska during 1880 and watching eight pronghorn approaching into the wind: "The animals finally approached within 300 yards of me when they suddenly bunched and started on a run in the direction whence they had come. They ran some distance before I discovered the cause of the stampede. Then I saw, about 10 feet behind and about the same distance above them, an eagle. For half a mile he pursued them, maintaining his position perfectly, and I could not see that he varied an inch. The antelope were just touching the high places and were getting far away, but, watching them through my field glass, I saw the eagle swoop. One antelope left the band; the great bird followed him, and for a moment both disappeared from sight. Then they emerged from the swale and started up the opposite ridge. The eagle had gained on its prey until now it was directly over the poor beast. The race was neck and neck, with the odds all in favor of the cruel bird.

"Finally he folded his wings and dropped like a chunk of lead on the antelope's neck. The latter fell, apparently stunned by the blow. Then there was a struggle, during which the terror stricken quarry several times rose to his feet; but the bird's talons were sunk deep in his neck and the life blood was flowing rapidly. The great bird pounded the antelope with its mighty wings and apparently blinded him.

"I ran toward them, hoping to get a shot at the bird, but as I approached he arose, soared away and lit on a butte, a quarter of a mile distant. He had not yet had his well earned dinner, so I took only the hind quarter of the poor antelope and returned to the ranch without firing a shot."

If a pronghorn does not run, an eagle sometimes swoops from considerable height and hits the standing prey in the lower back, inflicting deep puncture wounds but not binding to the animal (C. Langdon personal communication:1979). If cover is available, pronghorn may use it to escape an eagle. Hailey (1979) observed a perched golden eagle near Valentine, Texas, watching an adult pronghorn doe. The eagle took flight and struck the doe in the neck or shoulders, knocking her down. She regained her feet and ran in an erratic manner. The eagle renewed its attack, and the doe took refuge under thick acacia bushes. The eagle then alighted and walked around the edge of the bushes, but did not go into them. Hailey approached and the eagle flew away. The doe had several lacerations along the top of her shoulders and neck and was reluctant to leave the cover of the bushes.

While censusing pronghorn on a winter concentration area 30 miles (45 km) west of Laramie, Wyoming, on December 31, 1974, Goodwin (1977) saw a golden eagle attack a buck fawn. The fawn was in a herd of 135 pronghorn, feeding midway up a small hill. The eagle came over the hill about 33 feet (10 m) above the ground and struck the fawn in the back, momentarily stunning it. Momentum carried the eagle to the ground, and the herd ran about 1,300 feet (400 m), trailed by the fawn, before the eagle again became airborne. After the fawn had run about 0.6 mile (1 km), the eagle again hit it in the back. Both fell to the ground, but the eagle could not hold onto the fawn, which got up and again ran slowly after the herd. The eagle struck again after the fawn had run about 656 feet (200 m), pulling the fawn to the ground with its talons embedded in the fawn's back. After about 30 seconds, the fawn—with the eagle on its back—regained its feet and ran in a circle and fell to the ground. Another eagle alighted on the fawn as Goodwin drove up to examine it. Examination revealed cuts and punctures in the lumbar region. The fawn was alive but in shock, and some blood was running from its nose and mouth. An area about 2.4 to 3.9 inches (6–10 cm) in diameter had been eaten at the base of the neck, exposing the trachea and several undamaged blood vessels.

Two additional instances of eagle predation on pronghorn were evidenced in the same general area during 1975 (Goodwin 1977). One, was a fresh trail of blood on the snow, indicating the pronghorn ran approximately 2,625 feet (800 m) before succumbing. The other a pronghorn that left a blood trail about 984 feet (300 m) long between where it was attacked and where it was found with three eagles feeding on it.

Grinnell (1897a) saw a golden eagle attempt to attack a pair of 7- or 8-month-old fawns that were racing toward a herd of perhaps 1,000 pronghorn. The eagle made three swoops at them before they reached the herd. Each time, both fawns stopped, reared and struck at the bird with their forefeet. When they joined the herd, the eagle flew away.

During spring, does with fawns attack eagles, as they do coyotes. W. Bodie (personal communication:1979) watched a doe charge and strike at a golden eagle that was attempting to take her fawn. She kept the bird from her mortally wounded offspring although she simultaneously had to chase a coyote from her second fawn. The coyote apparently had seen the eagle attack and rushed to the scene. W. Rudd (personal communication:1979), working on Bodie's study area in Idaho, watched a golden eagle hovering above the plain and a pronghorn doe racing along a ridge toward the bird. Flapping slowly, the eagle swooped toward the ground with extended talons. A fawn stood up just beneath the eagle as the doe arrived. Another fawn jumped up within 20 yards (18 m) and ran to the doe. Both fawns took shelter under the doe, and the eagle flared off and circled twice before flying away. The doe moved the fawns (both underneath her) about 600 yards (550 m), stopping frequently to look in the direction the eagle was last seen.

Byers (1997a) observed that, when golden eagles appeared overhead, does ran to their fawns to stay close to them. Does scanning the skies for eagles or watching a flying eagle tipped their heads back in a rather odd position not seen at other times.

Byres also witnessed a confrontation between a pronghorn and an eagle that indicated does anticipate the presence of predators. A doe approaching a pair of fawns reclined together close to a ridge top was about 98 feet (30 m) away when a pair of golden eagles suddenly appeared, flying low over the ridge. The doe became agitated, as pronghorn mothers do when eagles are detected overhead, and she moved to within 49 feet (15 m) of her fawns. The eagles circled for about 5 minutes. The doe tilted her head back and watched them. The eagles did not attack the fawns and eventually flew away. The doe moved away and waited about 30 minutes before again approaching her fawns. She was less than 49 feet (15 m) from them when the eagles suddenly streaked into view over the ridge, flying about 6.6 feet (2 m) above the ground. The doe was startled and, again, watched overhead as the eagles searched for about 5 minutes and flew away. The doe again moved away from her fawns and waited another 45 minutes before approaching to within 9.8 feet (3 m) of them. The eagles were not in sight, but she tilted her head back and moved it from side to side, searching the sky for approximately 25 seconds. Then she quickly stepped to her fawns. They rose and suckled briefly before she led them over the ridge.

No recent account of eagles attacking adult pronghorn indicated aggressive action by the prey. However, Roosevelt et al. (1902) reported an instance of a rancher watching two eagles worrying and pouncing on a pronghorn—seemingly a yearling. The eagles hovered over it with spread wings, now and then swooping down, their talons outthrust, to strike at the head or to settle on the loins. The animal reared and struck with hooves and horns but to no avail. Illustration by Henry Sandham, from Roosevelt (1803).

O'Gara necropsied a 20-pound (9 kg) fawn that had been struck in the hindquarters by an eagle on the National Bison Range in 1977. The fawn had escaped the eagle only to die of peritonitis because one talon had penetrated the abdominal cavity. Presumably, the doe came to the fawn's rescue because, judging from the talon patterns through the skin and into muscles, the bird had had a solid grip on the fawn with both feet. It is unlikely that so small a fawn could have escaped by itself.

Tourists watched an eagle kill two pronghorn fawns on the National Bison Range in 1978. The doe ran back and forth trying to defend both, and the eagle did not get a chance to eat either one. If the doe had not intervened, the bird probably would have killed only one and eaten it.

W. Bodie (personal communication:1979) saw 17 interactions between golden eagles and pronghorn during two springs. Adult pronghorn detected flying eagles at great distances and watched them intently. As the distance decreased, the pronghorn paced nervously before running toward the approaching birds, occasionally stopping. Lactating does frequently looked toward the location of their bedded fawns, and Bodie found several unmarked fawns by the actions of such does. On one occasion, a doe nursing two 7-day-old fawns saw two eagles approaching. She came to the alert position and ran toward the birds. The fawns immediately dropped into hiding positions. After circling the area several times, the eagles settled on a bluff about 600 yards (550 m) across a river from the bedded fawns. Before Bodie left the area, the doe paced for 30 minutes watching the eagles. Adult pronghorn, sometimes followed by yearlings, often charged eagles that were sitting on the ground. One such charge involved two eagles and nine pronghorn of both sexes.

Seven radio-instrumented fawns were classified as eagle kills by Bodie and O'Gara. One fawn, when located, was being carried by an eagle and was dropped from about 100 feet (30 m) in the air, breaking the radio. Three of the other fawns were not fed on by the eagles that killed them. Tracks at the kill sites indicated that adult pronghorn had been running and turning. Evidently, the does arrived in time to defend the dead fawns from being eaten or removed by their attackers.

In the Lamar Valley, wolves have been observed near pronghorn on numerous occasions. Ordinarily, the two species ignored each other, but tourists reported about 10 chases of pronghorn by wolves. Also, project personnel watched two unsuccessful pursuits—one of an adult doe, which was coursed about 656 yards (600 m) by five wolves, and the other of a fawn. The latter involved a very fast, intense chase by a lone wolf on July 8, 1998. The wolf was unable to close the 11-yard (10 m) gap between the two, and an adult pronghorn seemed to be pursuing the wolf (K. Murphy personal communication:1998). The fawn should have been 5 to 7 weeks old at that time, indicating an age by which fawns can outrun wolves.

On July 7, 1996, Wayne Kendall observed four wolves (two black and two gray) pass within 98 feet (30 m) of two bison. The wolves then changed course and walked toward a band of pronghorn that was aware of their presence. At one point, as the wolves closed the distance between them, all animals halted, as if each was evaluating the situation. One gray wolf suddenly loped toward the band, which ran south toward trees bordering the valley. The two black wolves joined in the chase at a full sprint. One pronghorn broke from the band and headed west, down valley. The wolves concentrated on it, but appeared incapable of closing the distance until the animal turned left and attempted to rejoin the band. This allowed the gray wolf, running to its left, to cut the angle of pursuit. In an instant, predator and prey joined. The black wolves arrived a second or two later. After a frantic 5 or 6 seconds, the ordeal was over. The chase took less than a minute and covered between 656 and 984 feet (200–300 m). The observer could not estimate the age or sex of the pronghorn because the event was over so quickly, but at the moment of contact, the gray wolf definitely was smaller than its prey (D. Smith personal communication:1997).

The death of one of six killed pronghorn fawns was observed by R. Crabtree (personal communication:1998). An alpha wolf bitch chanced past a bedded fawn, smelled it from about five yards (4.6 m), killed and consumed it. An adult doe and a second fawn remained in the vicinity.

The other five kills of pronghorn fawns were observed by park rangers or other credible observers. Five fawns were killed during June—one on the 6th, one on the 13th, one on the 19th and two on the 27th. One fawn was killed on July 14 (K. Murphy personal communication:1998). These fawns probably varied in age from 2 to 6 weeks. Most fawns killed by coyotes are killed during their first 3 weeks of life. Apparently, fawns are vulnerable to the larger, faster canids for a week or two longer—after they have joined doe/fawn bands and are conspicuous. Also, does often save fawns from coyotes. That probably is not the case with wolves. "Sometimes, however, the wolves discover and attack the young when they are too feeble to escape, and the mother then displays the most devoted courage in their defense. She rushes on them, butting and striking with her short horns, and sometimes tosses a wolf heels over head, she also uses her forefeet, with which she deals severe blows, and if the wolves are not in strong force, or desperate with hunger, puts them to flight," according to Audubon and Bachman (1851:197).

D. Scott (personal communication:1993) reported that a hiker in Yellowstone National Park watched a grizzly bear chase a pronghorn buck up a slope covered by short sagebrush. The buck ran back downhill, around the bear and into a shallow pond. The bear pursued the buck into the pond. The buck jumped over the bear and ran back up the hillside, where the bear caught and killed it. Park personnel confirmed the bear was a small grizzly, little larger than its prey, and the carcass was that of a pronghorn buck.

Does about to give birth appear easily excited and aggressive, perhaps sensing that they and their young will be vulnerable to predators. O'Gara witnessed a parturient doe nearly kill a coyote. Bromley (1977) saw a doe in early labor leap from her bed and run out of sight when a sharp-tailed grouse walked by her, and he watched other does that had commenced labor chase away birds that landed nearby. He also watched does in labor spar with other does. Bromley noted that does with fluid-filled fetal membranes and the tips of fawns' hooves protruding from their vulvas were able to halt the birth process by standing and running. He also reported that does in Wind Cave National Park not uncommonly gave birth near highways or other areas of human activity, and he believed that selection for such areas was taking place because coyotes avoided them.

Trainer et al. (1983) observed does chasing badgers from areas where fawns were bedded. Two fawns killed by badgers were found during the same study.

Pronghorn does quickly moved away and stayed an average distance of 77 ± 33 yards (70 ± 30 m) from their reclined fawns during a study on the National Bison Range, in Montana (Byers and Byers 1983). Based on a model developed during that study, it was concluded that pronghorn mothers stayed at optimal distances from their fawns to avoid predation. Byers and Byers postulated that does ideally might have stayed farther away if coyotes were cueing on the does searching for fawns and if coyotes were the only predators involved. Because eagles can search large areas quickly with little expenditure of effort, the crucial part of the hiding strategy becomes fawn immobility, not maternal distance. Inasmuch as pronghorn does can do little to prevent eagles from detecting their fawns, the ideal distance when an eagle is present is that which allows the doe to reach her fawn before the eagle does. Therefore, pronghorn mothers besieged by coyotes and eagles are expected to reduce the distance between them and their fawns below the coyote-defined optimum.

During a study in central Arizona, does apparently kept their fawns, up to 2 weeks of age, close enough to water that the does could reach it easily, but the fawns were bedded far enough from water (average 0.47 mile [0.76 km]) that predators could not find them by searching around water sources (Ockenfels et al. 1992). Coyotes killed some radio-collared fawns during this study, so coyote predation probably was the driving force for such behavior.

Berger (1979) concluded that predator harassment is beneficial to the ungulate prey because it gives naive individuals the opportunity to recognize predators in a low-risk situation, allows the prey to monitor the predator's position safely and makes the predator reluctant to attack in the future. After watching six incidences of pronghorn chasing or attacking coyotes, Reynolds (1983) proposed a spectrum of pronghorn responses to hunting coyotes. The responses most likely to occur included retreat or formation into larger groups by individual pronghorn or small groups, and attack by does with fawns. Reynolds also postulated that, in some situations, pronghorn might participate in the chase as a playlike exercise.

That pronghorn mothers attempt to protect their fawns from almost any possible danger was indicated by Rush (1944:321), who observed a doe in Montana with fawns bedded nearby: "She jumped ahead and up as if powered by steel springs, came down with all four feet bunched closely together, and we saw a bit of turf fly. Again she jumped, and again, until she had alighted on the same spot a half dozen times. When she had gone away to graze, we went down to see what had happened. A sizeable rattlesnake lay cut to pieces in the trampled sod." Illustration by Frederick Remington, from McCracken (1966).

Although territorial bucks drive coyotes from territories, and other bucks as well as does without young sometimes harass eagles and coyotes, the males and nonlactating does usually display less aggression toward predators than do females with fawns. Thus, highly aggressive behavior toward predators seems to have evolved primarily for protection of the young. This conclusion was strengthened by Lipetz and Bekoff (1980), who saw pronghorn chase coyotes in Grand Teton National Park, Wyoming, on 25 occasions. They observed that does initiated chases more often than did bucks, and they hypothesized that predator harassment protected fawns hidden or grazing near the sites of the encounters.

Pronghorn does apparently have a greater fear of humans than of coyotes, bobcats or eagles. As a human approaches a bedded fawn, the fawn's dam generally watches from a distance of 200 to 500 yards (180–460 m) and demonstrates anxiety by short runs back and forth. The distance the doe stays from the human depends on how wary she is. In parks or on refuges, a doe sometimes will approach to within 30 yards (27 m) as the human nears the fawn. If a fawn bleats as it is caught by a human, its dam will charge but veer past the human at 2 to 5 yards (1.8–4.6 m). Although a doe may stay within 50 yards (46 m) as a fawn is handled, even if the fawn continues to bleat, she seldom passes as close to the human as she did the first time.

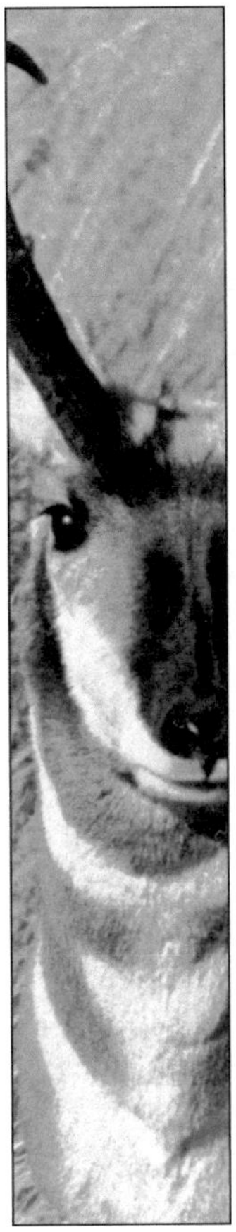

C H A P T E R   T W E L V E

Mortality Factors

Bart W. O'Gara

Periodic large die-offs during winters reduce population densities to levels at which density-dependent mortality has no consistent impact. . . . Predation on fawns may limit recruitment and retard population recovery, but it does not regulate population size. Although predation, disease, winter severity, and regulated hunting are important sources of pronghorn mortality . . . their future appears ultimately related to land-use practices that influence the availability and distribution of quality habitat within their historical range.

Barrett (1984:549)

We have no way of knowing how pronghorn coped with severe winters and droughts during prehistoric times when they were free to move long distances, or what the levels of mortality were from diseases, parasites and predators. Today, humans are directly or indirectly responsible for much pronghorn mortality. Indirect mortality is not always easy to evaluate. When a vehicle strikes an animal or an animal becomes entangled and dies in a fence, a direct human-caused death is evident. However, the significance of a highway or fence in reducing pronghorn movements, leading to poor nutrition or other problems, is considerably more difficult to identify and quantify.

Mortality studies are expensive and often inconclusive. Diseases, predators and weather are unpredictable mortality agents, and a planned study may not coincide with events that periodically reduce populations. For example, I once solicited funding for a study of a declining population and received it after that population was increasing. Also, an apparently obvious cause of death might depend on predisposing conditions not particularly evident. For example, poor nutrition resulting from reduced winter range, competition with livestock or blocked routes between seasonal or emergency rangelands may lead to winterkills or increased deaths from diseases, parasites or predators. Thus, ascribing deaths to specific factors sometimes is arbitrary.

Although the population was on an upward trend, severe weather was a major contributing factor to pronghorn losses on the Red Desert of Wyoming during some years (Sundstrom 1970). The carcasses of 434 pronghorn were collected from 1966 through 1970, and classified as victims of predation, vehicle accidents, diseases, fencing

Although pronghorn carcasses are found only occasionally, the cause of death usually goes undetermined because feeding by scavengers or deterioration of tissues destroy needed evidence. Photo by Mike Francis.

losses, birth complications and poaching. Hunter kills or crippling losses were excluded, and the cause of death was determined for 48% of the carcasses examined. The highest percentage (35%) of known seasonal losses occurred during winter. Neonatal mortality would not be detected in such a study because small fawn carcasses sometimes are consumed by predators or scavengers; even if they are not consumed, not much remains to be found after a week or two.

During Sundstrom's (1970) study, the sex and age of 330 of the 434 carcasses were determined. Because these animals apparently died of natural causes, Sundstrom considered them ecological mortality. I prefer to call them nonhunting mortality, but in some cases, causes of death were not determined and crippling by hunters could have been responsible. The data should roughly represent attritional deaths at rates normal to the populations and be an indication of the sequence of mortality to be expected in a hunted population.

Although available data often fall short of the ideal, an attempt to construct a life table can lead to useful insights, and the completed table provides estimates of values that cannot be measured directly. Because somewhat different patterns of mortality are common between sexes, life tables generally are separated according to sex.

A life table represents the age distribution of a population, usually by year class. For each age class (x), there is one value (l_x) for the number of live individuals entering the age class, and another value (d_x) for the number dying during the class interval—usually 1 year. The l_x value for any age class less the d_x value equals the l_x of the succeeding age class. Any d_x value divided by the l_x value for that age class gives the proportion dying (q_x), which is expressed as a percentage (Taber et al. 1982).

Of the 376 pronghorn for which Sundstrom (1970) obtained sex or age information, 111 fawns were found per 100 females. This ratio was quite similar to the reproductive ratio in the area—117 fawns to 100 females. Also, the buck/doe ratio of carcasses was 67:100, whereas the average prehunting season ratio during the same time was 62:100 in one part of the study area and 58:100 in another. In comparison, ratios of 114 adult and 26 fawn carcasses found on Hart Mountain National Antelope Refuge—where no hunting was allowed—and Drakes Flat, Oregon—where hunting was permitted—were 61:100 and 73:100, respectively (Hansen 1955, Yoakum 1957a). For various reasons, the sex of 46 fawns was not obtained during Sundstrom's study. Assuming that the sexes of the 46 fawn carcasses represented equal numbers of bucks and does, and excluding the 58 carcasses for which sex and age were unknown, Sundstrom constructed a dynamic life table (Table 67) and survival curves (Figure 71). Other assumptions for such a life table are that the carcass ages are a representative sample of ages at death from all causes, and the age structure has remained constant for at least 9 years, implying that birth and death schedules have remained unchanged for that long. Both of these assumptions apparently were invalid.

Hunting must have affected these life tables, especially the buck segment, because about 90% of the harvest consisted of males, and those males would not be available to die of causes other than hunting. Sundstrom believed that

TABLE 67. Dynamic life table for pronghorn in the Red Desert of Wyoming

Age class (x)	*Number dead in age class x* (d^1_x)	*Number dead per thousand* (d^1_x)	*Number survivors per thousand* (l_x)	*Number dying as a percentage of survivors in a given age class* (q_x)	*Average number of animals living between two age classes* (L_x)	*Mean expectation of life in years* (e_x)
Females						
Fawn	82	378	1,000	378	811	3.99
1	16	74	622	119	585	5.11
2	11	51	548	93	522	4.74
3	7	32	497	64	481	4.17
4	16	74	465	159	428	3.43
5	10	46	391	118	368	2.98
6	16	74	345	214	308	2.31
7	12	55	271	203	243	1.80
8	17	78	216	361	177	1.14
9	30	138	138	1,000	69	0.50
Total	217					
Males						
Fawn	68	429	1,000	429	785	2.32
1	33	208	571	364	467	2.69
2	15	94	363	259	316	2.94
3	15	94	269	349	222	2.80
4	8	50	175	286	150	3.04
5	1	6	125	48	122	3.04
6	4	25	119	210	106	2.18
7	4	25	94	266	81	1.63
8	5	31	69	449	53	1.04
9	6	38	38	1,000	19	0.05
Total	159					

Source: From Sundstrom 1970.

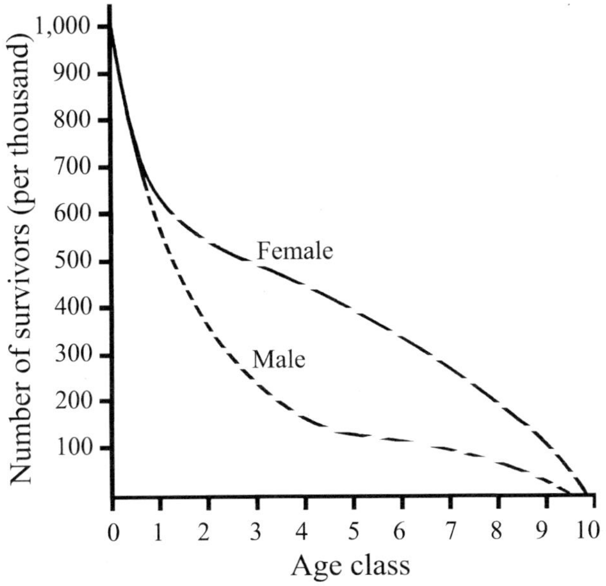

FIGURE 71. *Survival curves for male and female pronghorn in the Red Desert of Wyoming (from Sundstrom 1970).*

hunting was partly responsible for the low buck mortality after age 4 years. The following are some of the interpretations he derived from Table 67.

1. Male fawns have a higher rate of ecological mortality than do females; 42.9% and 37.8%, respectively.
2. The mean expectation of life for male fawns is 2.32 years and that of female fawns is 3.99 years. Hunting probably is one of the reasons for a lower life expectancy in bucks.
3. From 1 to 4 years of age (inclusive), male pronghorn have an average mortality of 31%, compared with 11% for females. From 5 years of age and older, both males and females experience approximately the same mortality, which averages 39% for bucks compared with 38% for does. This greater mortality in males explains why, although males and females are born in approximately the same ratio on the study area (104 males:100 females), prehunting season field classifications indicate almost twice as many females as males.
4. Only 17% of male fawns born in any given year reach their fourth year of life, compared with 46% of females.
5. Approximately 14 females of 100 reach their ninth year of life, compared with approximately 4 of 100 males.
6. The life table indicates a loss of 64% of the male segment in the fawn and yearling age classes. This loss compares with mortality estimated by herd composition on the study area, which indicated 70% mortality for males in those age classes.

A long-term (at least 10- to 15-year) study of a large number of radio-collared pronghorn is needed to develop more credible life tables. Effects of varying weather conditions on longevity could be documented during such a study. Sister studies in grassland, shrubsteppe and desert habitats would be especially informative for wildlife managers who must estimate the levels of harvests that are sustainable. Photo courtesy of the Wyoming Department of Game and Fish, Cheyenne.

Mitchell (1980) constructed life tables for breeding-age pronghorn killed by hunters in three Alberta study areas. He considered hunters there to be mostly nonselective for age classes older than fawns, although a trophy buck occasionally was taken rather than a doe or young buck. A comparison of results from those three life tables is presented in Table 68. The higher productivity at Ronalane than at Pakowski suggests that low population density, resulting from increased mortality, caused an increase in fawn survival.

After a 25-year hunting closure, the Newell study area population exhibited a relatively low turnover rate, long life expectancy and good fawn survival—but a low percentage of yearlings. Perhaps more than 1 year of data was needed to know exactly what was happening with this population. Also, the Ronalane population probably could not maintain itself with the 52% harvest. In Montana, harvests of 45% resulted in population declines (see Mitchell 1980).

DENSITY-DEPENDENT MORTALITY

High rates of fawn predation, starvation, abandonment and weak fawn syndrome are more likely symptoms of relatively high population densities than causes of low fawn survival rates. Pronghorn populations that are frequently subjected to winter related episodic mortality often have higher summer fawn ratios consistent with density-dependency.

Hess (1986:53)

Research was carried out by the Oregon Cooperative Wildlife Research Unit from 1953 through 1958 to determine the cause of low doe/fawn ratios (less than 100:50 for more than 10 years) and low fawn recruitment. This study resulted in four theses (Hansen 1955, Yoakum 1957a, Compton 1958, Dinges 1958) that dealt with pronghorn losses to diseases, predation, weather conditions, natural accidents and human-influenced accidents. No single agent could be authenticated as a limiting factor, but densities and doe-to-fawn ratios were highest on rangelands where vegetation was diverse and productive. High fawn mortality rather than depressed reproduction was indicated. Of 225 pronghorn carcasses found, 60% represented animals less than 1 year old. Not only were neonatal losses high, but almost half the adult doe mortalities were noted during fawning season. Since the advent of lightweight telemetry equipment, numerous studies of fawn mortality have been conducted in areas of low productivity (see Chapter 11). Generally, those studies have indicated high mortality within the first month of life.

Classification counts for pronghorn throughout their range were obtained from published accounts by Vriend and Barrett (1978). Mean rates of fawns per 100 does over several years ranged from 43 in Texas to more than 100 in South Dakota. High fawn survival was noted in southeastern Montana, southwestern North Dakota, western South Dakota and eastern Wyoming. In the remainder of the pronghorn range, midsummer ratios were much lower. Vriend and Barrett concluded that average fawn

Table 68. Vital statistics of three pronghorn populations in Alberta subjected to different harvest levels

Populations and dates	*Mean percentage summer population harvested*	*Mean life expectancy[a] (years)*	*Mean percentage annual adult mortality*	*Fawns per 100 does*	*Fawns per 100 mature does*
Ronalane 1960–1963	52.2	1.7	45.0	102.2	188.7
Pakowski 1960–1964	18.0	2.1	37.9	85.3	150.4
Newell 1964	7.9	2.9	29.0	105.6	179.6

[a] Mean lifetime remaining to animals that attained 1 year of age.

Source: From Mitchell 1980.

losses during the first 2 to 3 months after parturition were 25% to 65%, and low recruitment is a reality for pronghorn populations throughout most of their range.

The findings of these two studies should be kept in mind when evaluating pronghorn mortality. In Oregon, densities and doe/fawn ratios were highest on diverse and productive rangelands (Yoakum 1957a). Over a variety of rangelands occupied by pronghorn, average fawn losses of 25% to 65% during the first 2 to 3 months after parturition was a reality for pronghorn populations (Vriend and Barrett 1978). In other words, mortality—no matter what the proximate factor—was tied to habitat (especially vegetative) conditions. Thus, different levels of mortality, particularly among fawns, should be expected in different habitat types as well as under varying vegetative and weather conditions.

According to Pyrah (1987), his 13-year study yielded evidence supporting a cyclic species designation for pronghorn. He concluded that severe winters on his Montana study area did not cause extensive mortality on the upswing of a cycle, but did on a downswing. Pyrah's data indicated strong density dependence—fawn survival was highest when herd numbers were low, and winter mortality among bucks was lowest when harvest was high.

Pronghorn were absent from the Middle Park area of the Colorado Rocky Mountains for many years. During the mid-1980s, the animals returned to the area and increased rapidly. The Colorado Division of Wildlife monitored the population's performance with the aid of radiotelemetry and found that the rate of increase declined as the population size increased (Table 69), exhibiting strong density dependence (Pojar 1997). Summer distribution remained similar throughout the tracking period. Density-dependent feedback seemed linked to some factor other than fawn/doe ratios because there was no apparent influence of density on this parameter. Sixty 6-month-old fawns (30 bucks and 30 does) were radio collared during 1993 to 1995 for a mortality investigation. Mortality rates were nearly even between the sexes; 24 bucks and 25 does were still alive at the end of 1996.

During 1984 to 1996, climatic, population and habitat-use data were collected from a relatively closed population of pronghorn on the Deseret Land and Livestock Ranch, Utah. Approximately 100 pronghorn moved to the ranch during the severe winter of 1983–84. Although pronghorn were indigenous to the area, they were extirpated from the ranch during the early 1900s. Stepwise multiple linear regression was used to assess the effects of pronghorn density, coyote abundance, snow depth, summer precipitation and temperature on pronghorn population characteristics (Danvir 2000).

The pronghorn population on the Deseret Ranch increased from 90 to 620 adults during Danvir's (2000) study (Table 70). Estimates of population growth varied from 0.68 to 1.46. It correlated positively with prior-year fawn/doe ratio and negatively with winter snow depth, prior-year population size and number of pronghorn harvested the prior year. Prior-year population size and snow depth best explained variability in population growth, but growth was correlated more strongly with population size than snow depth.

Autumn fawns per 100 does ranged from 20 to 117 during Danvir's (2000) study (Table 70). Fawn numbers correlated positively with the prior year's April through September precipitation, as well as cool temperatures, and negatively with prior year population size and with coyote numbers. Prior year population size and April through

TABLE 69. Size of the Middle Park, Colorado, pronghorn herd during winters and calculated rates of increase

Winter	*Population size*	*Rate of increase*
1986–87	80	—
1987–88	122	0.52
1988–89	160	0.31
1989–90	223	0.39
1990–91	246	0.10
1991–92	292	0.19
1992–93	332	0.14
1993–94	410	0.23
1994–95	453	0.10
1995–96	520	0.15
1996–97	579	0.11
1997–98 projected	622	0.07

Beginning in 1990, 13 to 15 animals were harvested per year (i.e., the 1996–97 winter population was 594 before harvest and 579 after).

Source: From Pojar 1997.

TABLE 70. Sizes and rates of increase for the pronghorn herd on the Deseret Land and Livestock Ranch and adjacent U.S. Bureau of Land Management land in Utah, 1985 to 1996

Year	*Population size*	*Fawns per 100 does*	*Population growth*	*Pronghorn harvested*
1985	90	88	—	—
1986	130	110	1.44	—
1987	190	117	1.46	—
1988	260	107	1.37	—
1989	340	68	1.31	—
1990	470	77	1.38	78
1991	620	47	1.32	141
1992	600	40	0.96	104
1993	410	20	0.68	10
1994	400	73	0.98	14
1995	520	58	1.30	56
1996	490	—	0.94	—

Source: From Danvir 2000.

September precipitation best explained annual variability in fawn/doe ratios, but they were correlated more strongly with population size than precipitation. Fawn/doe ratios correlated with habitat types. They were highest in sparse native sagebrush/grassland and lowest in Wyoming big sagebrush/crested wheatgrass.

Because only 6 and 4 years of harvest data were available, for pronghorn bucks and does, respectively, Danvir (2000) did not formulate multiple regression models for hunting mortality. Mean annual harvest weight of mature (more than 3 years old) bucks was negatively correlated with harvest-year population size and harvest-year April through July mean temperature. Mean annual harvest weight of does was correlated most strongly with prior winter snow depth. Mean annual Boone and Crockett scores of harvested bucks correlated negatively with prior-year density and positively with prior-year harvest weights of mature bucks. Adult bucks with large horns weighed more than those with small horns.

Danvir (2000) concluded that, although coyote abundance (number observed per 100 hours) was negatively correlated with same-year fawn/doe ratios, and coyotes were seen pursuing and killing pronghorn during the study, his analyses suggested fawn survival was influenced more strongly by pronghorn density, plus summer precipitation and temperature, than by predation. His analyses also suggested that pronghorn condition, fawn survival and population growth were limited by winter severity and intraspecific competition for summer forbs and winter browse. Removal of vegetation by livestock, as practiced on the Deseret Ranch (time-controlled grazing strategy of Savory 1988), was not correlated with any of the investigated variables.

This buck fawn died of malnutrition during winter 1978–79 in North Dakota. Note the rough coat and emaciated condition of the animal. The skin appears too big for the body, robbed of fat and muscle mass by cold weather and a lack of nutritious forage that could be reached through the snow. Photo by Ed Bry; courtesy of the North Dakota Game and Fish Department, Bismarck.

Aoude and Danvir (2002) hypothesized that population size and fawn production on the Deseret Ranch were limited by a scarcity of forbs. Thus, from 1995 through 2001, they increased forb production on 6,177 acres (2,500 ha)—5% of the pronghorn habitat—by burning and chemical and mechanical manipulations. Population size and fawn production were negatively correlated with pretreatment ($r^2 = 0.89$), suggesting density dependence. Conversely, a positive correlation between population size and fawn production post-treatment ($r^2 = 0.74$) suggested an increase in carrying capacity. This supported the Aoude and Danvir hypothesis.

The Great Basin of Oregon is characterized by variable climatic conditions and a mosaic of vegetation and forage conditions. Pronghorn in this area are characterized by fluctuations in population size and herd composition. Density-dependent pronghorn fawn survival was evident in the northwestern Great Basin population (Hess 1999). California, Nevada and Oregon collected comparable data, except that Nevada dropped its winter survey in 1994. With 38 years of winter census data and 43 years of summer composition counts, the correlations among all state samples were robust. A linear regression analysis by Hess, comparing the natural logarithms of the winter census totals with the succeeding summer fawn/doe ratios, resulted in a substantial negative relationship. This strongly affirmed density dependence in this pronghorn population.

Kohlmann (1998a) analyzed standard time series data from 18 annual population surveys to evaluate the roles of density dependence and predation on population dynamics. He used a discrete time, stochastic logistic population model and a parametric bootstrap likelihood ratio to test for density dependence. Pronghorn in the area were managed primarily by buck-only harvests.

A coyote population index was derived from the number of coyotes observed per linear mile traveled during spring pronghorn counts and spring and summer game bird surveys. A predator/prey ratio was calculated as the ratio of the coyote population index to an annual jackrabbit index (Kohlmann 1998a).

Tests conducted by Kohlmann (1998a) showed a negative relationship between pronghorn population growth rates and density. Also, the population exhibited a strong return tendency and, thus, density dependence. Their analysis identified density dependence as the most influential factor affecting population growth rates in the Great Basin of eastern Oregon. Kohlmann was unable to detect a significant effect of predator population indices or predator/prey ratios on pronghorn population growth rates, although coyote predation appeared to affect fawn survival. Fawn/doe ratios appeared to be poor predictors of population growth.

The foregoing information was gained through long-term studies. Their results indicate that vegetative conditions and densities of pronghorn themselves ultimately determine trends in population numbers. Thus, most of the following causes of mortality seem to be proximate factors that actually kill pronghorn but do not ultimately determine the welfare of the species. That is not to say hunting, predation, diseases, droughts or other factors—under special circumstances—cannot drastically reduce or even eliminate populations.

Human activities, such as usurping the most fertile habitat or blocking movement routes, apparently have reduced the level at which density dependence dictates population trends. Perhaps 40 million was a density-dependent number during the 18th century and 1 million was the same for the 20th century. What might the number be for the 21st century?

SEVERE WEATHER, MALNUTRITION AND BARRIERS TO MOVEMENTS

In four years we haven't documented a single crossing of a paved road or of the boundary fence with Mexico.

John Hervert in riederici (1999)

Information such as reported by Hervert, concerning radio tracking and aerial observation of Sonoran pronghorn bodes ill for the herds during severe winters or droughts, as well as for the long-term genetic welfare of isolated herds.

Droughts, cold winters with deep crusted snow, malnutrition and human-made barriers to movements each cause many deaths among pronghorn. Yet, these factors hardly can be discussed separately. In modern-day North America, two or three factors often combine when natural disasters are reported in pronghorn populations. Severe weather and malnutrition likely have held some pronghorn populations to the limits of their habitat for eons. Now, human-made barriers to seasonal movements impose another limit that reduces, manyfold, the number of pronghorn that can survive.

Pronghorn mortality apparently was high during severe winters even before fences and other human obstructions to migrations were common. James Parotti—a game warden, hunter and guide in Choteau County, Montana—told Munson (1897b) that the winter of 1893, when the thermometer registered 61°F (51.7°C) below zero, killed off four-fifths of the pronghorn. Thousands starved to death on account of the deep snow. After that winter, he found an estimated 900 carcasses where the animals had drifted into a deep ravine and evidently did not have the strength to get out. Before that time, pronghorn were plentiful according to Parotti, but that winter killed nearly all. Although they were shot by the thousands, Parotti believed the number so destroyed was only an insignificant fraction of the total. He also believed that the animals came from the Red Deer River Country, 250 miles (400 km) to the north, where immense bands inhabited the unsettled country.

Pronghorn were numerous in Alberta at the beginning of the 20th century. The winter of 1906–07—from October until well into May—had the most severe weather recorded in western Canada (see Mitchell 1980). Much domestic livestock perished and uncounted pronghorn succumbed to exposure and entrapment in deep snow. Many animals were forced into Montana and failed to return during spring 1907. Many segments of Alberta rangeland remained virtually unoccupied during autumn 1907, and pronghorn remained scarce in Alberta until the early 1930s.

Neonatal deaths related to severe weather undoubtedly occur in all northern or high-altitude pronghorn populations. Storms during late May and June are common in those areas and often bring freezing to near freezing temperatures and snow or cold rain. High mortality of newborn fawns on the Red Desert of Wyoming, caused by freezing temperatures and precipitation during June 12 to14, 1969, was postulated by C. Sundstrom (personal communication:1989). Average doe/fawn ratios in the area during the 1960s were 100:86. The ratio dropped to 100:70 after freezing rain and snow about 6 to 7 days after the peak of fawning. Wardens in the area told Sundstrom that the ratio fell to 100:65 in 1955 after a severe mid-June storm.

Hypothermia was reported as a major mortality factor for caribou calves in some areas (Kelsall 1968). Specific conditions that would cause hypothermia in newborn pronghorn fawns are unknown, but mean daily temperatures of 34.7°F (1.5°C), accompanied by rain and wind, caused hypothermia deaths in caribou calves after 5.5 hours of exposure (Hart et al. 1961). Pronghorn fawns weigh about 64% as much as caribou calves and are not expected to maintain thermoregulatory balance under similar conditions.

Climatic conditions were more severe on the Sheldon National Wildlife Refuge, Nevada, during and for 2 weeks after the 1978 fawning period than during 1979. In 1978, snow fell on 5 days and rain on 4 days during the 30-day period. Rain fell on only two occasions during the same period in 1979, and mean estimated wind velocities were lower than in 1978. The mean rectal temperature of 12 very young fawns captured in 1978 was 102.0°F (38.9°C) compared with 103.5°F (39.7°C) for 21 captured in 1979 (McNay 1980). Whether this difference was related to ambient temperature was unknown, but seemed likely. The mean ambient temperature during May 15 to 31, 1978, was 41°F (5°C). In 1979, the mean temperature was 48.9°F (9.4°C). During 1978, 5 days had mean temperatures lower than 37.4°F (3°C). During May 23 to 25, late in the fawning season, mean daily temperatures averaged 33.8 to 36.5°F (1°–2.5°C); snow fell and accumulated up to 3 inches (7.6 cm); and winds varied from 10 to 25 miles per hour (16–40 km/hr), creating conditions that could have induced hypothermia in caribou calves.

Pronghorn normally give birth to twins, and 550 fetuses were found in 290 does during seven studies for a ratio of 190 fetuses per 100 does (Chapter 9). During 1978, on the Sheldon National Wildlife Refuge, 7 of the 12 fawns captured at 1 to 2 days of age were singletons. Severe weather apparently contributed to extensive early mortality (McNay 1980).

On a study area in New Mexico, the pronghorn population was static for many years. Severe weather was the factor thought to be limiting the population (Bradybaugh and Howard 1982). During September and October 1979, precipitation was 2.6 inches (7.6 cm) below normal, resulting in poor autumn forage production. During January 1980, heavy snows caused some pronghorn starvation. Bradybaugh and Howard concluded that decreased precipitation during several months can create problems in reproduction and survival. Heavy snows also can promote luxuriant spring vegetation that benefits pronghorn.

Research is needed to evaluate the effects of weather, especially timing and amount of precipitation as well as temperatures during the growing season for rangeland vegetation, on pronghorn fawn survival.

Malnutrition or starvation occurs to some degree almost every winter on some pronghorn ranges and is the ultimate population control in most unhunted northern populations. The same can be said about prolonged droughts on desert rangelands. Determining what percentage dies of malnutrition would require a long-term study. During years with mild winters and good precipitation, little malnutrition-related mortality would be expected. However, after a series of such seasons, populations on healthy rangeland tend to build rapidly, and 50% or more of a population can easily perish during the next severe winter or drought (see Pyrah 1987).

Adult ruminants store tremendous amounts of nutrients in their bodies. However, inadequate food intake causes many physiologic changes. During experimental starvation, the number of rumen microorganisms declines rapidly as protozoa disappear and bacteria are greatly reduced. Naturally, volatile fatty acids produced by the protozoa and bacteria, from which ruminants obtain about 70% of their energy, likewise decline. The rumen pH increases, and the strength and magnitude of rumen contractions decrease. A couple of days to 4 weeks are required for the digestive tract to readjust to food, depending on the length of the starvation period. This explains why many wild ruminants die after they receive supplemental food or after spring green-up. They enter those situations in weakened condition, with digestive tracts that are not fully functional. Also, their rumen flora and fauna are adapted to high-fiber diets, and several days are required to adjust to a diet rich in the carbohydrates available in new green growth or formulated emergency rations. In addition, new forage is high in water content and consuming enough to meet energy demands is nearly impossible before new growth reaches a length of about 2 inches (5 cm). Therefore, malnourished ruminants changing to new growth must still live in negative energy balance for up to 2 weeks. Failing to do so, some die although food is available (Thorne et al. 1982).

Seven feet (2.1 m) of snow covered the vegetation in parts of northern Arizona between December 13 and 21, 1967. Pronghorn were subjected to low temperatures and forced to move through deep snow. When the storm abated and roads were opened, the Arizona Game and Fish Department provided alfalfa hay to the herds. After snowmelt, native shrubs again became available. Soon thereafter, pronghorn were found dead on the rangelands, with their stomachs full to the point of impaction. Some animals had ingested native browse, others mainly alfalfa hay. Red gelatinous bone marrow and bloody fetuses indicated starvation. These pronghorn were without feed for 10 to 14 days before either native or artificial feed was available. Microscopic examination of rumen samples from six dead pronghorn revealed no ciliate protozoa. Bacteria observed were quite similar to those from rumen contents of a pronghorn killed by a hunter during September 1968. Pronghorn that visited a haystack during the storm survived. The hay apparently prevented rumen malfunction. Thus, if supplemental feeding is considered as a management option, it should be started before the starvation process has become irreversible (Pearson 1969).

Susceptibility of ruminants to winter malnutrition is influenced greatly by adequacy of vegetation on summer and autumn ranges, as well as winter range. Undernourished animals are susceptible to diseases, parasites and predators, but such susceptibility is difficult to quantify.

Feeding starving pronghorn during a severe winter might succeed if begun early enough. However, if the animals have been without food for a week or more, little can be done to reduce mortality. In nearly all cases, emergency feeding is expensive as well as risky. Photo courtesy of the Nebraska Game and Parks Commission, Lincoln.

Fairbanks (1992) documented that birth weights were a factor in fawn mortality. Malnutrition that extends into late gestation causes birth of small young with reduced chances for survival. Pregnant females protect developing fetuses by catabolizing much of their own fat and protein. Eventually, they abort or resorb fetuses; soft tissues are resorbed into the mother's system and bones are expelled or retained in the uterus. Retained bones might prevent subsequent pregnancies. I collected a very fat pronghorn doe during November that was not pregnant and did not show signs of having nursed a fawn the summer before. She had teeth and parts of bones from two fetuses in her uterus and many stigma of ovulation on her ovaries.

Large and small die-offs from malnutrition have been reported on numerous occasions (Martinka 1967, White 1969, Compton 1970, McKenzie 1970 and 1986, Wishart 1970, Oakley and Riddle 1974, Simpson and Leftwich 1978, Hailey 1979), and poor rangeland conditions along with cold weather, deep snow or drought usually are blamed. These factors almost always are made worse today by human-made impediments to seasonal movements. His-

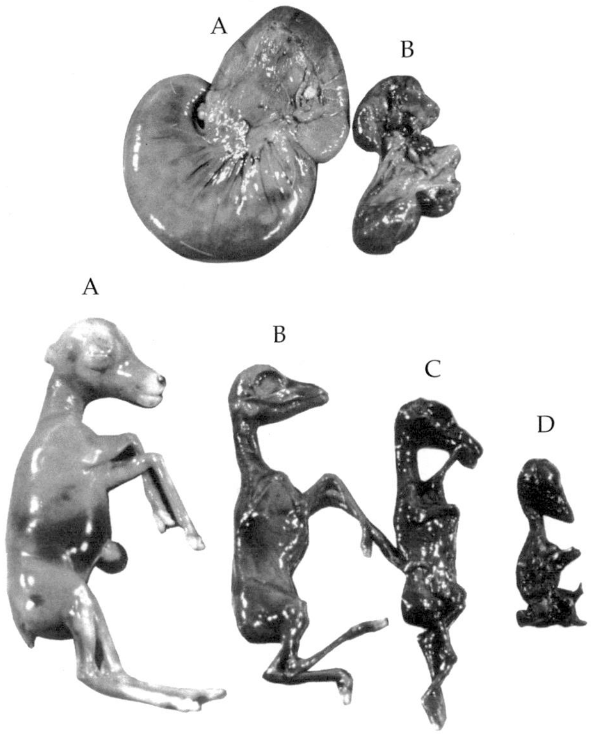

*Pronghorn does that survive a severe winter or drought may fail to produce young that year, or even in subsequent years. These reproductive tracts and fetuses were collected late in the winter of 1977–78 in Alberta. (*Top*) Reproductive tract A appears to have a still-viable fetus surrounded by amniotic fluid in one horn of the uterus; the other horn appears to contain a dead fetus. Reproductive tract B obviously contains degenerating fetuses. (*Bottom*) Ebryo A appears to have been viable before death of the doe, but fetuses B, C and D have been dead for varying lengths of time, with resorption and possible mummification taking place. If remains of dead fetuses are not expelled from the uterus, the retained parts may act as intrauterine devices that prevent subsequent pregnancies.* Photos by Morley Barrett.

toric records indicate that pronghorn moved long distances before the prairies and deserts were crisscrossed by roads, railroads, fences and farms. Roosevelt et al. (1902) reported that pronghorn were found throughout the Dakotas during summers. However, they were only found in the Black Hills of South Dakota during winter.

Pronghorn sometimes appeared in large numbers during severe winters where they had not been seen for years after settlement began. Such movements probably were normal reactions to weather. From Fort Assinaboine, Montana, Munson (1897a:7) wrote: "The snow and intensely cold weather have made antelope—never seen here in years—a drug on the market. A competent observer told me that between Havre and Glasgow, 125 miles, there were probably 40,000 antelope in the Milk

River bottom alone, a bunch every half mile. Every ranch has two or three hanging up, and in Havre they were stacked up in front of the market like cordwood. One man killed fifteen in ten minutes; another got twenty-seven in a morning; another, twenty-two in a day. The Indians are killing them by the thousands over the Fort Peck and Fort Belknap reservations to the east of here. They have evidently drifted south for a long distance, probably from the Saskatchewan Country."

Pronghorn that summer in southwestern Montana commonly move about 150 miles (240 km) to winter on the sagebrush flats of the Snake River plain in Idaho. They gather in large numbers on that plain during most winters. The Snake River, with its deep, steep canyon, generally was considered a barrier to further movements southward. However, as pronghorn numbers increased, some crossed the Snake River. During December 1983, about 200 pronghorn crossed the river west of American Falls. They only traveled a few hundred yards and encountered an interstate highway fence. The pronghorn remained in that area for several months, then recrossed the river. During mid-December 1985, 450 to 500 pronghorn crossed the Snake River to the south near King Hill, Idaho. They went to a sagebrush area near Hammett where they remained until mid-March before crossing the Snake River again and returning to spring and summer ranges (L. Oldenburg personal communication:1986). That pronghorn could cross a steep and rugged part of the Snake River Canyon—but were then apparently unwilling to brave fences, roads and other human-made barriers that confronted them south of the river—indicates the problems this species faces today.

Severe winters with deep snow tend to occur about every 5 years on the Great Plains (Pyrah 1987), resulting in reduced energy intake and increased energy demand. During a long, cold winter with deep snow, the duration of negative energy balance is more than many pronghorn can withstand. Cold winds can increase stress, but wind also can be beneficial because it clears snow from ridges, making travel easier and exposing forage.

"The pronghorn's counterpart on the steps of Eurasia is another swift, antelope-like creature, the saiga" (Time-Life Books 1985). Its history also seems similar. Estimates of pristine numbers for each species, although only guesses, are about 40 million animals. Some indication of former numbers of saiga can be gathered from Vereshchagin's (1959:355) account: "In the Ukraine saiga was common in the later Middle Ages . . . run across from the steppes into the forests in winter and from the forests to the steppes in summer in such large numbers that each peasant kills up to 2,000 of them each year."

Early in the 20th century, each species was reduced to very low numbers. With protection and management, Americans and Canadians brought pronghorn populations back to about 1 million animals by the 1980s. The Russians brought the saiga back to perhaps twice that number (Heptner et al. 1961) and harvested them for meat and skins on a commercial basis. Bekenov et al. (1998) indicated that they numbered approximately 1,200,000 in Kazakhstan alone during 1974. On the right bank of the Volga, in what is now the Autonomous Republic of Kalmykia, Russian Federation, populations began to recover during the late 1920s and increased to about 715,000 by 1978 (Teer et al. 1996), despite extensive mortality during winter 1953–54. Livestock grazing takes place on the Eurasian steppes, but fewer roads, fences and agricultural valleys impede migrations than is the case in North America.

The domain of the saiga shrunk from west to east during every century of recorded history. Sixteenth and 17th century farmers killed thousands of them in the environs of Kiev. By the late 1880s, few remained in Europe. Between 1920 and 1925, the number and distribution of saiga were at all-time lows, and many investigators predicted their speedy and total annihilation. Decrees signed by Lenin prescribing conservation measures and prohibiting harvest of rare animals were promulgated in 1919 and 1920. People found guilty of having unlawfully killed a saiga were fined or imprisoned (Adolph 1957). Also, cattle grazing was prohibited in parts of Kazakhstan to benefit saiga.

From the 1930s onward, saiga increased. During the 1940s, the All-Russian Society for Conservation of Natural Resources sponsored expeditions to estimate numbers in the European portion of the USSR. Censuses were conducted during 1944 to 1951 from automobiles and airplanes. In later years, counts were exclusively from the air, with up to 10 airplanes flying parallel. Commercial harvests began in 1954 and, thereafter, poaching—especially shooting from vehicles at night—became common (Adolph 1957). With the collapse of the Soviet Union, money for law enforcement became nearly nonexistent, and numbers of saiga in Kazakhstan were reduced to 810,000 by 1994 (Bekenov et al. 1998). Numbers in Kalmykia were down to about 150,000 by 1993 (Teer et al. 1996).

During 1950 to 1970, saiga split into small summering bands of 10 to 20 animals on mixed grasslands, very similar to what we see with some pronghorn when they are in good numbers. During normal winters, saiga traveled short distances, generally 20 to 50 miles (30–80 km), to sagebrush wintering ranges and gathered into herds of 150 to 500, again quite similar to pronghorn. However, during severe winters, the whole population shifted southward to winter in herds as large as 150,000. One marked animal moved 1,200 miles (1,900 km) south.

A wildlife agency employee has the onerous task of collecting for disposal the carcasses of pronghorn that died of malnutrition during a severe winter in Nebraska. During earlier times, pronghorn would have moved to wintering areas where shrubs protruded above the snow. Such movements often are impossible today because fences and other human-made barriers prevent them. Also, some winter ranges have been replaced by agricultural fields or towns. Photo courtesy of the Nebraska Game and Parks Commission, Lincoln.

These are soft data gathered through conversations with Russian biologists with inadequate English language skills or who communicated through interpreters with little comprehension of biology. However, information concerning group sizes is collaborated by Vereshchagin (1959) and Heptner et al. (1961).

I can only wonder whether pronghorn once made occasional movements as great as those documented for saiga. If they did make such extensive movements, it seems barriers to movement is one of the reasons North America now supports one pronghorn where 40 roamed historically. Pronghorn certainly try to move farther south during severe winters than they are able to today.

A minimum loss of 500 pronghorn associated with severe weather was reported near Glasgow, Montana, during April 1965. Those losses were on foothill grasslands, and fawn production was 39 to 55 fawns per 100 does the next summer, as compared with a norm of 90:100. On a nearby winter range where sagebrush, although heavily used, was abundant, losses were minor and fawn production normal (Martinka 1967).

During winter 1977–78, pronghorn range in Alberta received below-normal temperatures, early snowfall and frequent and severe blizzard conditions. Only 34% of the animals observed during January and February surveys were on traditional winter ranges; many herds selected sites with reduced snow accumulation but insufficient woody vegetation. Estimated mortality among the approximately 14,360 pronghorn entering the winter was 48.5%. Necropsies revealed extreme losses of body fat and mummification in 68 of 82 fetuses examined. Fawns of both sexes and adult bucks were more susceptible to winterkill than were adult females. Wishart (1970) suggested that severe winters in Alberta increased emigration, starvation, losses from predation and depressed reproduction the next year. Similarly, the severe winters of 1964–65 and 1968–69 resulted in high losses of pronghorn in Montana and the Dakotas, often because of lack of access to shrubs (Compton 1970, McKenzie 1970, West 1970).

Severe winters lead to high mortality from several causes—malnutrition, chilling and predation. Pronghorn herds on the Red Desert of Wyoming have experienced harsh winters for centuries, but human-made obstacles to free movement apparently have increased the magnitude of mortality. The winter of 1971–72 was one of deep snow, low temperatures and ground blizzards that forced pronghorn to move toward winter ranges used only during particularly harsh winters. Recently constructed fences blocked many migration routes, and the estimated loss to the pronghorn population in the Chain Lakes area was about 4,000 animals (62% of the population). The loss was

During January 1965, about 50 pronghorn were found dead on the ice of the Bow River near Bow City, Alberta. The animals attempted to cross the river on glare ice that had sagged in the middle because of the low water level. They could not negotiate the incline to either shore. When they slipped on the ice, they splayed onto their bellies. In that position, they lost body heat rapidly and rolled onto their sides as they became moribund. They were unable to right themselves and died of exposure. This probably was a rare occurrence, resulting from circumstances that produced snow-free glare ice, ice sloping from the banks because of a lessened water flow and extremely cold temperatures. Photo by Chuck Gordon; courtesy of George J. Mitchell.

attributed to the severity of weather coupled with the many fences constructed by state and federal agencies and ranchers. Most of the animals died along woven-wire fences, and mortality was almost twice as high in fenced as unfenced areas (Oakley and Riddle 1974).

Since traditional movement routes have been blocked, severe winters commonly have devastated pronghorn in the Dakotas. A winter with above-normal temperatures and below-normal snowfall and accumulation usually is a harbinger to a good year for pronghorn. During the winter of 1964–65, about 15,000 pronghorn survived in North Dakota. Between 1965 and 1980, the state experienced five major winterkills, and hunting seasons were closed during 1978 through 1981, but pronghorn numbers were reduced to 1,800. The five winters from 1980 to 1985 were comparatively mild. Winter mortality was low, and the population increased to 5,000 to 6,000 animals despite an annual harvest of about 1,000 animals during 1982 through 1985 (McKenzie 1986).

At least once every 10 years, weather conditions in South Dakota are so severe that pronghorn populations are greatly reduced. The winter of 1978–79 brought herds to the lowest numbers of the 1970s. Drought conditions during the summer of 1985 were followed by extreme cold and heavy snowfall during November and December. Pronghorn tried to reach vegetation not covered by snow. Some moved into heavily forested areas of the Black Hills, and more than 200 entered the city of Sturgis to feed on ornamental shrubs. Losses ranged from 50% to 90% on most ranges in the state, and pronghorn were reduced to all-time lows for recent times (Rice 1986).

During 13 years of study, some winter mortality occurred every year, regardless of weather conditions, in a central Montana pronghorn population. Such mortality averaged 15% (range 7%–25%) of summer populations. Winter losses in this hunted population appeared to comprise about 21%, 56% and 74% of the annual mortality of adult bucks, adult does and fawns, respectively (Pyrah 1987).

A survey of pronghorn was conducted for the years 1996 and 1997 by mailing questionnaires to provincial and state wildlife agencies in Canada and the United States. When numbers for 1997 were compared with like data from previous surveys, major differences were evident (see Chapter 4). Total numbers indicated a 25% to 30% decrease from 1995 to 1997 and a 36% decrease from the record high in 1983 (Yoakum et al. 1999).

Telephone calls to various wildlife agencies to determine reasons or postulations for the high losses produced varied responses. No agency personnel contacted reported scientific findings to explain major declines, but each expressed intuitive explanations. For northern herds,

During winters, pronghorn often are forced to make long movements or starve. Fences designed to restrict movements of cattle and sheep, with woven wire and barbed wire, tend to be fatal to weakened pronghorn. Photo by John Wood; courtesy of the Wyoming Cooperative Wildlife Research Unit, Laramie.

where the largest declines took place, the severe 1992–93 and 1996–97 winters, coupled with droughty summers, were reported as factors in Montana, North Dakota, Saskatchewan, South Dakota and Wyoming (Yoakum et al. 1999).

Pronghorn on the northern Plains usually have regained numbers within a couple of years after extensive, climate-induced die-offs (Pyrah 1987). In Wyoming, the pronghorn population increased by 95,000 head from 1990 to 1991 (Madson 1999). The possibility exists that habitat losses and barriers to movements may slow or even preclude such population expansions in the future. If this happens, total pronghorn numbers could decrease substantially.

Severe winters can cause havoc with pronghorn even in southern regions. Many pronghorn, trapped by drift fences in Texas, froze to death during the blizzards of the 1880s and 1890s (Simpson and Leftwich 1978). Such drift fences apparently played an important role in the decline of pronghorn in that state. Settlers killed 1,500 pronghorn trapped by a drift fence near the Panhandle town of Clarendon in 1882 (Haley 1949).

During 1959, winter blizzards killed 700 to 800 pronghorn in Texas, but losses of this nature in Texas were only suffered during 1 year in the last 30 years according to Hailey (1979). Ordinarily, winter mortality of pronghorn in Texas is correlated to the amount of precipitation received during late summer and early autumn. If moisture is insufficient for good forb production, pronghorn are unable to regain body condition after the rut. This problem becomes acute where net-wire fences restrict pronghorn from winter ranges with shrubs. When winter mortality occurs due to malnutrition, buck losses normally exceed doe losses by 35%, and few fawns are born or survive the following summer. The high loss of bucks was related to the animals going through the rut before winter and being unable to rebuild their fat reserves. This was caused by lack of forage during the dry winter months. Starvation and poisonous shrubs reduced pronghorn northwest of Marfa by 82% and those in an area west of Valentine by 76% from June 1967 to June 1968 (Hailey 1979).

During 1967, a December storm in northern Arizona dropped 29 to 94 inches (74–239 cm) of snow. Snowfall was followed by low temperatures, resulting in a 2- to 3-inch (51–76 mm) crust of ice on the snow. Pronghorn became immediate victims of malnutrition because most forage was buried and unavailable. This was compounded by fence entanglements and predation by coyotes and domestic dogs. Some entire bands died within 2 or 3 days and, in one area, a herd of 1,300 was reduced to 300. Starving pronghorn were forced into the town of Winslow and many were killed by dogs. An aroused public and the state wildlife agency killed more than 50 dogs. It was estimated that 4,000 of the state's 10,000 pronghorn died as a consequence of this storm (White 1969).

Die-offs due to droughts occur but are not as well documented as those caused by severe winters. Hailey (1979) presented an account of low precipitation affecting pronghorn in Texas. At the end of a drought in 1957, pronghorn in the Trans-Pecos region numbered about 7,300 animals. They increased to more than 12,000 by 1961 because of more than average moisture during the ensuing 4 years. The population dropped to 5,000 between 1961 and 1964 because of another drought. In the Marfa Flat area, 65% of the population died during 1 year from malnutrition and plant poisoning. Historically, these animals had moved to ranges with better forage conditions during years of low precipitation. However, traditional movements were restricted by woven-wire fences and preferred forage was lost to livestock; pronghorn had to eat poisonous plants or starve.

During a study in central Arizona, radio-marked pronghorn seldom crossed fenced, paved highways or fenced railroads. Ockenfels et al. (1994:43) concluded: "As populations become more isolated and thus smaller, they are more susceptible to extirpation from effects of weather, continual habitat loss, poaching or over-harvest. Small populations also are more affected by negative changes in gene frequency. . . ." Ockenfels et al. (2000) radio-collared 114 pronghorn and tested their reactions to three right-of-way types in Arizona and Mexico: (1) unfenced, paved roadways; (2) fenced, paved highways; and (3) railroad right-of-ways. Unfenced, paved roadways did not

Adult males often die before other sex and age classes because they lose weight during the rut and do not get it back before severe winter weather begins. This is less of a problem for pronghorn than for deer because of the pronghorn's early breeding season. Fawns also are vulnerable because they have higher energy requirements for their body weights and less fat reserves than do adults. Photo by Len Clifford, Leonard Rue Enterprises.

fragment pronghorn herds, as crossings were common. No crossings of fenced, paved highways were observed within or between study areas, and no fenced railroad tracks were crossed. Thus, fragmentation isolated herds, creating subpopulations. Translocating pronghorn may become necessary to maintain gene flow.

Human activities reduced pronghorn numbers about 40-fold during the last 150 years (McCabe et al. 2004; see also Chapter 3), primarily by appropriating and changing pronghorn habitat, blocking seasonal movements, and commercial and subsistence hunting. Associated with primary development is the influx of people who alter pronghorn habitat with housing developments, shopping centers, roads, etc. This is further compounded by all types of disturbances—dogs, recreational activities, poaching, etc. Few people would argue that pronghorn habitat should take precedence over human needs, but as natural and seminatural areas shrink, land-use planning becomes necessary if crucial habitat and populations of wildlife are to be maintained. Management practices aimed at improving rangeland conditions for livestock can be beneficial or detrimental to pronghorn. If such management results in a monoculture of exotic, coarse grasses, it is almost sure to be detrimental (see Chapters 15).

PREDATION

Over the years, opinions concerning the effects of predation on pronghorn populations have included the full realm of extremes. Gabrielson (1935:576) expressed a belief of many of his generation when he wrote: "Aside from man and his activities in hunting and livestock grazing, the coyote is the greatest menace that the pronghorn has to face." During the early 1900s, reintroductions into former ranges began and predator control was intensified—particularly in areas where pronghorn were transplanted. As pronghorn numbers increased, so did research into the biology of the animal itself (see Autenrieth 1978).

New information instilled new thinking; it was discovered that mortality from predators was but one of myriad pronghorn limitations. Predator control for protection of pronghorn became limited primarily to transplant areas. By the 1950s, predator control was not a standard practice for pronghorn management, although extensive predator control—including poisoning—was conducted to protect livestock on many rangelands occupied by pronghorn (Autenrieth 1978).

Predators kill manyfold more pronghorn fawns than adults, and coyotes generally are the most important predator. Bobcats, golden eagles, mountain lions and domestic dogs—except in unusual local instances—are of less importance (see Chapter 11).

Nineteen radiotelemetry studies in one province and nine states indicated that predators killed 54% of 1,097 instrumented pronghorn fawns. Of the predator kills, coyotes, bobcats and golden eagles were responsible for 65%, 14% and 12% of the predator kills, respectively. The other 9% were taken by undetermined or less important (to pronghorn) predators (Chapter 11).

Effects of predation usually are most pronounced where predators are numerous compared with their prey, or when associated with adverse conditions for the prey population. Adverse conditions usually involve droughts, severe winters or lack of nutritious forage from a variety of causes. A 25% to 30% decline in the continental pronghorn population during 1995–1997 was centered in the northern Great Plains. It was blamed partially on coyote predation in conjunction with adverse weather conditions by four of five biologists who were contacted from the areas of greatest declines (Yoakum et al. 1999).

HUNTING HARVESTS AND WOUNDING LOSSES

> *During the 19th century, market hunters provided a wide array of wild animals for purchase in the nation's butcher shops, inns, and restaurants. These included bison, elk, deer, caribou, pronghorn, bighorn sheep, mountain goat, and black bear.*
>
> Catton and Mighetto (1998:7)

During the second half of the 20th century, direct harvests through regulated recreational hunting were carefully controlled and—in my opinion—often were necessary to maintain the health of a herd and keep the herd in balance with the available habitat or prevent excessive crop damage. Although hunting accounted for a large percentage of pronghorn deaths (4%–27% of the adult populations in various provinces and states during the mid-1990s; see Chapter 20), such harvest is primarily a management tool and not ordinarily responsible for drastic reduction in population numbers. To the contrary, controlled hunting prevented drastic reductions and overpopulations (see Chapter 22). Thus, recreational hunting will occupy less space in this chapter than might be expected on the basis of the numbers of pronghorn harvested.

Pronghorn harvests are impressive reminders that managed hunting programs are compatible with increasing the number of animals. After a long period of no legal pronghorn hunting in the United States and Canada, seasons were reopened in Idaho, New Mexico and Wyoming during 1934. Less than 2,000 pronghorn were harvested that year. As pronghorn increased and more states opened seasons, the harvest increased. That increase (minimum numbers reported by provinces and states), at 10-year intervals, was 1,563 pronghorn in 1935, 19,991 in 1945, 72,418 in 1955, 64,787 in 1965, 93,511 in 1975, 133,300 in 1985 and 100,060 in 1995.

From 1955 through 1999, more pronghorn were harvested annually in Wyoming than existed in North America during the mid-1920s. The Wyoming harvest during 1983 (98,680) was nearly four times the number that existed in all of North America during 1915 (Hoover et al. 1959). That more than 2,300,000 pronghorn were harvested in one state (Wyoming) from 1934 through 1999, when less than 26,000 survived in 1925 (see Chapter 20) reflects the species' high reproductive potential and sensible management.

The total, legal Canadian and United States harvest since 1934 exceeded 4 million pronghorn by 1998. Populations and harvests may have reached a peak about 1983, after a series of mild winters on the northern Plains. Populations and harvests apparently will increase and decrease as mild and severe winters, as well as droughts, dictate. At any rate, wildlife managers and biologists can take pride in the history of pronghorn management. With public support, recreational hunting will continue to regulate, not endanger, pronghorn populations. In Mexico, where a comparatively large proportion of the remaining pronghorn resided during the 1920s (see Nelson 1925), hunting seasons have been closed since 1922 (Gonzalez y Laffon 1993). Consequently, the public has received few legal benefits from the species and generally has scant interest in it. Without legal hunting, little money has been available for protection or management. Pronghorn now are in very low numbers in Mexico, and the survival of some populations is in doubt.

Wounding losses should be included in hunter harvests, but documenting such losses is difficult. Wounding losses were investigated by aerial and ground surveys after hunting seasons in Colorado (Bear 1967a) and included not only crippled animals that escaped but also badly shot-up animals and illegal kills (yearling bucks mistaken for does by hunters with doe-only permits) that were abandoned. During 1963 and 1964, on two rangelands dominated by shrubs, wounding losses were estimated at 15% and 25%, respectively. Theoretically, in 1964, one of every four hunters killed two pronghorn. The wounding losses in the same area during 1965 and 1966 were 5% and 4%, respectively. A heavy snowstorm during the 1965 season compelled hunters to hunt on foot. Consequently, they were inclined and better able to follow up on crippling shots. Another possibility is that hunters on foot secured more standing shots than did the road hunters. Reasons for the low losses during 1966 were undetermined, but they may have been related partly to changes in regulations making yearlings with horns shorter than their ears legal on doe permits. During the same 4 years, wounding losses

During settlement of the West, meat and uncontrolled sport hunting took a tremendous toll of pronghorn after bison had been eliminated or greatly reduced in numbers. During this period, hunting was a year-round business, and hunters were unselective of age or sex in their harvests. Photo courtesy of the Wyoming State Archives, Museums and Historical Department, Cheyenne.

on the shortgrass Plains were less than 5%, apparently because the short vegetation and flat terrain allowed hunters to easily detect a crippled or dead pronghorn and retrieve it (Bear 1967a).

Oregon opened a hunting season on pronghorn in 1938 for the first time since 1913. During the first hunt, telescopic sights were prohibited because it was feared they would encourage long-range shooting. The next year, scope sights were allowed, but few flat-shooting rifles were available. Loss by crippling and delayed death was estimated at 20% (Einarsen 1948).

Crippling losses apparently were reduced in South Dakota through information and educational efforts concerning improved hunting methods and increased popularity of flat-shooting, scope-sighted rifles. Crippling seemed to average about 10% in that state by the 1960s (South Dakota Game Division no date).

Wounding losses were investigated on various ranches in Texas from 1965 through 1967 (Hailey 1979). On ranches that provided assistance to hunters, very few wounded animals were found. On those ranches with virtually no hunter control or assistance, losses from wounding averaged 25% of the legal harvest. About 20% of the ranches fell into this category.

I found a few dead pronghorn with bullet wounds during or after hunting season in Montana and took two cripples on my licenses. One hunter, who said he had shot at a running buck and hit a doe running 10 yards (10 m) behind it, was abandoning the doe as I arrived. This and the fact that almost all the carcasses were those of does and fawns indicate that more abandonment than crippling may take place. Pronghorn are in such open country and so easy to see that few cripples should get away from a competent hunter. I believe that crippling and abandonment losses are not as high in Montana as those reported for Colorado and Oregon.

Successful hunters coming through the Dad, Wyoming, checking station on the opening weekend (September 1–2) of the 1991 hunting season were interviewed. Fifty-five hunters in 29 vehicles who had harvested 62 pronghorn were asked the number of shots they fired, number of pronghorn wounded but not recovered, and wounded or dead pronghorn observed. The wounding loss was calculated by dividing number wounded by number of legal kills and multiplying by 100. Results of this initial survey indicated the overall wounding loss was 12.9% of the harvest, but the wounding loss of bucks equaled 28.5% (Jahnke 1991).

During 1992, hunters were interviewed on August 31 and on September 1 and 21 at checking stations at Dad, north of Baggs, the Bitter Creek Road, Powder Wash and along Highway 430 south of Rock Springs. Usable interviews were provided by 140 groups involving 219 hunters who had harvested 224 pronghorn. Up to 24 wounded pronghorn were reported, but some of the animals apparently were reported more than once, and 16 seemed the actual number seen. Thus, estimates of wounding loss ranged from 7% (based on what was considered an actual number) to 11% for all reports (Jahnke 1992).

Observed and reported wounding loss may depend on the density of hunters. When many hunters are in the same area, individual hunters are relatively likely to see an animal crippled and lost by someone else. Most reports of wounded animals came from the area with the highest density of hunters (Jahnke 1992). It seems that hunters would be most numerous where pronghorn are most numerous. In such a situation, an uninterrupted stalk of undisturbed animals becomes unlikely. Thus, hunters probably chance more shots at running animals—an almost sure formula for wounding pronghorn.

More dead pronghorn are found near roads and other human enterprises than in areas distant from such activities. Illegal hunting is relatively light in the Trans-Pecos region of Texas, where ranches are large and landowners are watchful. Illegal kills involve occasional target practice from public roads. In the panhandle, landowner units are smaller, with more miles of public roads, and poaching losses are somewhat higher. However, the losses are not considered high enough to have significant effects on the populations. Also, on the lower plains of Texas, most of the occupied pronghorn range is traversed by thousands of miles of roads that are necessary for oil well operations. Because of the numbers of dead animals found adjacent to roads in this area and the number of complaints concerning illegal hunting registered by landowners, it is felt that many of the herds are kept at low levels by poaching (Hailey 1979). Photo by LuRay Parker; courtesy of the Wyoming Department of Game and Fish, Cheyenne.

POACHING

With small antelope populations, as occur in some areas of Wyoming, poaching may be an important factor in preventing herd number expansion.

Creek (1967:14)

Little information is available concerning poaching of pronghorn (see Chapter 20). The very nature of poaching makes estimating its level extremely difficult.

Einarsen (1948) believed that illegal killing of pronghorn was seasonal and that few local residents poached pronghorn in Oregon during the 1930s and 1940s. However, he believed transients, tempted by the sight of pronghorn and lulled by a sense of security on isolated rangelands, sometimes took a chance. He also believed violations of one kind or another during open seasons accounted for a considerable number of kills.

Illegal kills apparently were a minor decimating factor for pronghorn in Oregon (Yoakum 1957a). Yoakum speculated that sparse human populations and the opinion that pronghorn meat was unsavory contributed to this situation. During 1954 and 1955, he authenticated 17 presumed illegal kills. Most of those occurred during open hunting seasons, as predicted by Einarsen (1948). Yoakum believed that high accidental killing of does and fawns resulted from hunters shooting at mature bucks in mixed sex and age bands.

Perhaps the percentage of instrumented animals killed—or apparently killed—by poachers during telemetry studies gives some insight. However, a poacher might avoid shooting a radio-collared animal, thinking it would increase the chance of arrest. On the other hand, the type of person who poaches might dislike conservation agency personnel and selectively shoot instrumented animals for spite.

During a study in Idaho, one radioed pronghorn doe was poached, and three other pronghorn probably were (Hoskinson and Tester 1980). Two of the three were adult bucks, and their radio collars were found unopened. The collars could not have slipped over the bucks' heads and fallen off. This was out of 54 radioed animals. Seven were legally shot during hunting seasons. If this were a typical situation, it would indicate that the poacher kill is about half that of the legal kill in that particular area.

As with poachers, whether legal hunters select for or against radio-collared animals is unknown. Knowing levels of illegal kills would help to refine harvest strategies. Implanting mortality-sensing, intrauterine transmitters into a large number of pronghorn does seems the most practical way to secure information concerning this problem.

OLD AGE

The extent of mortality from old age is poorly known for pronghorn. However, Einarsen (1948) believed that old age accounted for many of the adult carcasses found on the range. He based that belief on finding several dead pronghorn with pegged or missing incisors, emaciated bodies and rough pelage. Pronghorn usually live about 9 years, according to some researchers (Hepworth and Blunt 1966), but during periods of good rangeland conditions and mild weather, animals, especially females, twice that age occasionally are reported (see Chapter 5).

Four of 89 (4%) jaws picked up on Drakes Flat and Hart Mountain National Antelope Refuge, Oregon, by Yoakum (1957a) had pegged or missing incisors, indicating the animals had died of old age. Seven of 84 (8%) deaths of adults reported in Texas were from old age (Buechner 1950a). Fifteen of 189 adult pronghorn jaws I collected in Yellowstone National Park and on the National Bison Range were classified as 9 years old or older. Using 9 years as the average life span, my collection also indicated about 8% of the adult pronghorn in those unhunted populations probably would die of old age. However, that does not mean 8% of the total populations would die of old age. Mortality during the first month or two of life varied from 39% to 87% during 15 studies of radio-collared fawns (Table 68). The 39% may represent a fairly typical situation in which fawns die, and jaws seldom would be found. If so, old age deaths would be expected for about 5% of the populations studied, not 8%.

During a study in the Red Desert of Wyoming (Sundstrom 1970), data from 330 pronghorn carcasses found in the field were used to construct a dynamic life table (Table 67). According to that Table, 14% of the does and 4% of the bucks reached their ninth year of life. Again, using 9 years as the average life span, the same percentages (14% and 4%) would die of old age in that hunted population. This, of course, is an inflated estimate of old age deaths because an unknown number of young fawns perished before they were large enough for their carcasses to be readily found.

Bad teeth appear to be a contributing cause to most old age deaths. By 9 years of age, the molariform teeth have smooth grinding surfaces, the roots of incisiform teeth have narrowed and the teeth often are loose.

Old animals, especially those with bad teeth, naturally die during the time of year when food is least plentiful and of poorest quality. Extremes of cold or heat add to the stress. Annual mortality of 5% to 6% in animals that produce single offspring and 10% to12% in those that commonly produce twins is not unusual and probably is acceptable according to Thorne et al. (1982). Thus, mortality due to old age does not seem unusually high in either the hunted or unhunted populations cited above.

On the National Bison Range in Montana, Byers (1997a) found that the maximum life span of pronghorn bucks was less than that for does (see Chapter 5). Apparently, this is because bucks lead more dangerous lives. Seven does for which Byers knew birth dates and year of death died at approximately the following ages: one at 5 years, one at 10, two at 11, two at 14 and one at 15.

MISCELLANEOUS CAUSES

Malnutrition, severe weather, barriers to movements, predation, hunting, poaching and old age appear to be the principal mortality factors for pronghorn. These factors have been discussed roughly in their perceived order of importance. The following miscellaneous causes of mortality may be of some interest, but they seem to be insignificant as limiting factors except perhaps locally. These lesser mortality factors are not listed in order of importance because that order is unknown.

Diseases and Parasites

Symptoms of mild disease conditions in wild animals are difficult to recognize. Thus, no one really knows the levels of disease in pronghorn populations. Pronghorn have not been experimentally infected with many of the diseases or parasites they harbor. Until that is done under controlled conditions, the effect of each organism on the health of its host and the densities of infestation that are beyond normal levels and deleterious will remain speculative (see Chapter 10). Disease levels in prehistoric pronghorn herds are unknown, but those levels probably increased drastically with the advent of Euro-Americans and their livestock. "In 1858 a disease called bloody murrain killed many cloven-hoofed animals—work oxen, antelope, deer, and buffaloes—along the western trails. The stretch of the Overland Trail between Fort Laramie and Bridger was one long offense to the nostrils" (Sandoz 1978:46). Bloody murrain apparently was synonymous with rinderpest (a viral disease primarily of cattle, found mostly in Africa and Asia). Apparently, oxen were exposing wildlife to diseases for which those animals had no resistance. Also, Boller (1972:318) reported that, about 1859, "When making their trade with us the Sioux reported that all through their country toward the Platte, the elk, deer, buffaloes, bears, and wolves were dying in great numbers from sickness among them, caused by medicine of the whites."

Stratton (1981:206) wrote: "Until 1867, a quarantine against Texas longhorns had existed in Kansas due to the frequent outbreaks of Texas fever. Although the Texas cattle themselves were immune to the deadly disease, they carried the tick, which often dropped to the ground and infected other breeds of cattle." Cattle tick fever also may have had a debilitating effect on pronghorn populations. West (2001) maintained that anthrax, brucellosis, tuberculosis and a variety of parasites were acquired by wildlife from Pacific-bound livestock early in the 1800s. All except brucellosis had the potential to kill or debilitate pronghorn.

Apparently the largest pronghorn die-off reported affected herds in nearly the entire area between the Yellowstone and Missouri rivers during the summer of 1873 (Allen 1874). This epidemic apparently destroyed 75% to 90% of the pronghorn in the area, and the greatest mortality occurred during July. Allen reported that large numbers of pronghorn had ranged over this section of the country during autumn of the previous year, and bands of 200 to 300 frequently were seen by the Yellowstone Expedition of 1872 on its return eastward. That massive die-off during a short period of time indicates a viral disease, and the timing (July) makes one suspect bluetongue, transmitted by arthropods. Murphy (1879:313) also wrote of a great epidemic in 1873 or 1874 that "almost cleared" pronghorn from one section of the country.

The large number of pronghorn on the plains during pristine times undoubtedly contributed to extensive die-offs. Today, most disease-related die-offs result from animals being crowded into small pockets of habitat, and die-offs usually are restricted to those pockets because animals in the surrounding areas are not numerous or wide-ranging enough to spread the disease. Thus, management, through adequate harvests, generally precludes the kind of widespread die-offs described by Sandoz (1978), Allen (1874) and Murphy (1879).

Researchers have remarked how relatively free of diseases pronghorn have been compared with other North American ungulates (Yoakum 1957a, Hoover et al. 1959, Russell 1964, Hailey 1979). Most wild pronghorn appear to be in good health, and few wildlife managers or biologists consider diseases to pose serious threats to the species. However, a large number of pronghorn apparently died of bluetongue in eastern Wyoming during 1976 (Thorne et al. 1988). Also, pronghorn were the most heavily parasitized of seven species (white-tailed and mule deer, elk, bison, bighorns and mountain goats) sampled in South Dakota. This apparently was because the pronghorn shared overgrazed range with domestic sheep (Boddicker and Hugghins 1969). Pronghorn feed on short vegetation, increasing their chances of picking up nematode eggs and larvae, as well as mites containing larval tapeworms, under such conditions.

Although the level of pronghorn diseases and parasites is unknown, they seem to be important on a population level only in crowded situations or areas of intensive livestock grazing. Thus, good management of rangelands, proper harvest levels and maintenance of corridors for seasonal and emergency movements are the important factors affecting pronghorn production and survival (see Chapter 9).

Harassment

Disturbances, whether by humans or predators, reduce foraging efficiency of pronghorn. Such reductions result partly from time lost while running from or watching sources of disturbances. The greatest loss of efficiency probably is the result of individuals remaining together in large, alert groups for enhanced protection. This herding-together results in lower foraging efficiency according to Berger et al. (1983). Berger et al. predicted that, as the human population, recreation and construction activities increase, decreases in disturbance-free habitats are assured. They also postulated that foraging stress and escape efforts may increase to the point of energy imbalance and declining reproductive rates in most species. However, pronghorn seem to adapt readily to disturbances and should do so if given reasonable consideration during construction projects, strip-mining operations, etc.

Harassment of pronghorn by hunters and others using vehicles—particularly all-terrain vehicles—may lead to later deaths or debilitation of some pronghorn that consequently fall easy prey to predators. Pronghorn seem so well adapted to running that few people consider this possibility. However, Chalmers and Barrett (1977) documented deaths of fawns from capture myopathy after they were pursued by a vehicle and deaths of adults after extended chases by aircraft during trapping operations. Also, Copeland (1980) indicated that long hunts were disruptive to pronghorn breeding.

Another aspect of harassment that seldom is considered is the effect on behavior. Fichter and Autenrieth (1978) stated that behavior is genetically influenced and variable and, therefore, subject to selection. They pointed out that the pressure of trophy hunting during the rut may cause selection for smaller, less vigorous and nonterritorial pronghorn bucks. The function of territoriality is that of determining which males breed females. Dominance, apparently reflected in the ability to hold a choice territory, probably is based primarily on age, weight and horn size. Body size (weight) of bucks may be an expression of genetically based, individual variation, as well as age of the animal and condition of its habitat.

Autenrieth (1966) observed territorial bucks in the Upper Pahsimeroi Valley of Idaho. During 3 years of study in the same area, Bodie (1979) saw only one buck displaying territorial behavior. Bodie believed that selectivity by hunters during the rut had reduced the numbers of mature bucks and disrupted territorial behavior by harassment. Thus, deferring hunting seasons until after the rut in some areas may be necessary for the long-term welfare of populations.

Although pronghorn are easily disturbed by irregular human activities—particularly those such as hunting that may harm them—they seem to adjust rapidly to disturbances that have not harmed them. On a military training base in southern Alberta, pronghorn merely trot out of the way of infantry and tanks and seem to ignore cannon, machine gun and rifle fire (V. Geist personal communication:1997; see also Chapter 6). Also, the Phoenix Ecological Services Field Office of the U.S. Fish and Wildlife Service issued a Biological Opinion in 1997 that the use of ground surface and airspace for military training on the Barry M. Goldwater Range was not likely to jeopardize the continued existence of the Sonoran pronghorn. This opinion stemmed from observations of the animals drinking and spending considerable time where high explosive bombs were dropped (U.S. Fish and Wildlife Service 1998a).

During three winters of study in central Wyoming, densities of pronghorn within oil fields generally were lower than densities outside of oil fields (Easterly and Guenzel 1992). Apparently, activities associated with oil operations disturbed pronghorn enough to alter their use of the landscape. Easterly and Guenzel speculated that even slight disturbances may lower feeding efficiency and increase energy costs to animals, especially during winter. Annual winter mortality of radio-collared pronghorn that was attributable to natural causes (22%) was unusually high for open winters with mild temperatures. Based on condition of bone marrow, such mortality seemed attributable to exposure and malnutrition. Age at time of death averaged 7 years (range 5–11 years).

Poisonous Plants

Certain plants are toxic to animals and, under certain circumstances, can cause extensive losses. Plant toxins cause illness, birth defects, abortion and death. Plant poisons may be defense systems against herbivores, especially insects. Observations regarding the effects of poisonous plants on wildlife can be misleading because the level of poison varies with seasons, climate, soil and plant part ingested.

Ruminants have adapted to poisonous plants through a generalized diet, using a variety of plants, an ability to detect and avoid plants that have made them ill or caused discomfort and an ability to detoxify many plant toxins. Also, wild ruminants generally are not confined to a pasture and can choose from a great diversity of plant species. Native big game animals, including pronghorn, occasionally are poisoned by plants, but large losses usually occur only in overpopulated or overgrazed areas where nonpoisonous species have been depleted (Laycock 1978) (see Chapter 14). One survey of plant toxication indicated that the following plants adversely affected pronghorn: Russian thistle, bluegreen algae, chokecherry, tarbush and locoweed (Williams 1982). Selenium intoxication from inorganic material absorbed or concentrated by plants also was diagnosed for pronghorn in Wyoming (Thorne et al. 1982).

Pronghorn in Texas moved from Marfa Flats to nearby hills to seek browse when droughts reduced forbs on the flats (Hailey et al. 1966). After ranchers changed from barbed-wire to woven-wire fences, pronghorn were restricted to pastures. Three years of below average precipitation during the mid-1960s, combined with heavy livestock use, caused severe depletion of forbs and grasses in some Texas pastures. Livestock then were provided supplemental forage, but pronghorn were forced to subsist on tarbush and creosotebush. Pronghorn densities decreased from 4 per square mile ($1.5/km^2$) before the drought to 1.5 per square mile ($0.6/km^2$) after. Lesions characteristic of tarbush toxicity were found in 83% of the pronghorn necropsied, and resorbed embryos were found in three of four pregnant does. Fawn/doe ratios were reduced from 52:100 in 1964 to 17:100 in 1965 (Hailey et al. 1966, Hailey 1979).

During rapid growth, chokecherry bushes accumulate high levels of cyanide in buds and leaders. At such times, cyanide poisoning—often fatal—of livestock and deer is not uncommon. The animals killed usually are in good condition at necropsy, an odor of bitter almonds sometimes is discernible and the blood tends to be bright red. A pronghorn buck in Colorado that appeared weak and stiff was collected and necropsied. The rumen contained a quantity of chokecherry buds and leaders (Ogilvie 1955), and it was believed that chokecherry caused the buck's condition.

Locoweeds include legumes of the genera *Astragalus* and *Oxytropis*, many of which are not toxic. However, some locoweed (or crazyweed) species accumulate high levels of selenium or nitrogen and cause acute or chronic

A pronghorn buck in creosote/tarbush habitat of the Trans-Pecos, Texas, ordinarily could choose from a great diversity of plant species. However, woven-wire fences now restrict movements of some pronghorn herds. Overgrazing by livestock during droughts may force pronghorn to use an excessive amount of tarbush, resulting in death or inhibiting reproduction. Photo by S. Kremble Canon.

poisoning of animals that eat them. Clinical signs (emaciation, visual impairment, neurologic abnormalities, birth defects) appear only after 2 weeks or more of feeding on the plants. Animals may acquire a taste for and selectively ingest large quantities of locoweed.

Two cases of pronghorn poisoned by locoweed were reported by Buechner (1950a), but tests were not conducted to verify the diagnosis, and a later investigation by Hailey et al. (1966) indicated that tarbush poisoning was more likely. A doe was described as running aimlessly; its depth perception and direction apparently were uncontrollable and it was killed with a rock. Although Buechner's evaluation of locoweed pronghorn may have been correct, it is doubtful that such poisoning is a serious problem. However, a 1-month-old fawn apparently died from eating locoweed in an enclosure in Mexico, where adult pronghorn were consuming the same plant (J. Cancino personal communication:2001).

Russian thistles can accumulate potassium nitrate. When consumed by ruminants, potassium nitrate is reduced by rumen bacteria to potassium nitrite. The nitrite causes hemoglobin to be converted to methemoglobin, which is unable to transport oxygen to the tissues, and the animal dies from asphyxia. Pronghorn died around a dry reservoir in a Wyoming desert after feeding on Russian thistles that contained 1.5% potassium nitrate, apparently because that was the only succulent vegetation available (Honess and Winter 1956). Pumping fresh water into the reservoir solved the problem.

This pronghorn buck was a member of a band of eight or nine pronghorn that died after feeding in an area where elm trees had been sprayed with 2,4-D. Otherwise nonpoisonous plants concentrate nitrates while "growing themselves to death" after application of this herbicide. The nitrate is changed and becomes poisonous when acted on by a pronghorn's rumen bacteria. Photo by Jim D. Yoakum.

An adult pronghorn in a Colorado enclosure apparently was poisoned by toothed groundsel (Schwartz et al. 1976a). The animal's rumen was filled with groundsel, and diffuse hepatic degeneration was noted.

Bluegreen algae are common in natural bodies of water and, of all algae, only certain types of bluegreen algae are toxic. These algae grow rapidly during warm weather, and wind may concentrate a bloom where animals drink. Death occurs in a few to less than 24 hours. Thus, intoxicated animals usually are found near water. Diagnosis is difficult, but algal toxicity may have been the cause of pronghorn deaths in Weston County, Wyoming, during 1952 (see Williams 1982).

Pronghorn are better able to cope with native poisonous plants than are livestock or introduced wild animals. Nonetheless, even the most efficient detoxification or degradation system may be overpowered in an animal forced to consume poisonous plants because of malnutrition or other nutritional imbalances. The potential for plant poisoning should be considered along with nutrition and diseases when developing wildlife management plans according to Fowler (1983).

Selenium

Selenium poisoning usually occurs in areas where the soil selenium is greater than 10 parts per million (ppm); this often is on shale. Plants containing high levels of selenium are odorous and unpalatable, so they usually are a problem only where little else is available. Certain plants may concentrate selenium to levels as high as 15,000 ppm; however, many parts of pronghorn range are selenium deficient. Animals that ingest one massive dose of selenium may lose their appetites, become lethargic, drink and urinate excessively, go into a coma with labored breathing and die. Severe gastroenteritis is seen at necropsy. Chronic selenium poisoning may cause aimless wandering, weakness, emaciation, arthritis, hoof deformities, partial hair loss and death. Demonstrating high selenium levels (5–10 ppm is suggestive and greater than 10 ppm indicative of selenosis) in animal tissues is necessary to diagnose selenium poisoning (Thorne et al. 1982). A pronghorn buck found ill in Wyoming jumped from a moving vehicle on the way to a laboratory and was killed. Necropsy findings indicated the possibility of selenium poisoning (Post 1960). Analysis for selenium showed 4.3 ppm in the liver and 2.3 ppm in a kidney. Post concluded that the animal was suffering from selenium poisoning (see Chapter 8 concerning selenium deficiency in pronghorn).

Injuries Caused by Vegetation

Plants such as foxtail grass, cheatgrass, bearded wheat and squirreltail barley may cause mechanical injury to wild and domestic ruminants that can lead to debilitation or death. Awns occasionally penetrate the mucous membranes of the mouth, causing abscesses there or migrating to other locations to initiate abscesses or granulomas. Bacteria that cause lumpy jaw often enter with plant parts. Oral injuries by awns or impaction of vegetation around teeth were reported in Wyoming pronghorn by Thorne and Williams (unpublished data in Thorne et al. 1982).

The anatomy and habits of pronghorn make them vulnerable to other types of plant-induced injuries. I have seen evidence that stiff, stubble-type vegetation sometimes penetrates the deep glands between the toes of pronghorn. Thus, bacteria is introduced into a favorable environment.

Inhaling Debris

Pronghorn breathe through their open mouths while running. Three pronghorn I necropsied had small, soft abscesses containing grayish pus deep in their lungs. Two of these abscesses were centered on portions of wheat heads with three or four grains of wheat on 0.3- to 0.4-inch (8–10 mm) pieces of stems. The other abscess contained a stone about 0.4 inch (10 mm) in diameter. Running pronghorn apparently inhale debris kicked up by animals ahead of them. Another doe, killed during a trapping operation in which the animals were driven by helicopter, had a stone that measured 0.8 by 1.0 inch (20 by 25 mm) in its esophagus. The stone was halfway between the animal's jaw and brisket and could easily be felt from the outside.

Capture Myopathy

Thousands of pronghorn have been captured and translocated. They are delicately built and excitable, and may break their necks by hitting fences or otherwise kill themselves during trapping operations. A more subtle form of mortality, capture myopathy (CM) (see Chapter 21), resulting from capturing and handling pronghorn, was documented by Chalmers and Barrett (1977). Using the drive-trap technique, 594 adult pronghorn were captured, of which 475 were processed and the remainder were released without being handled. Acute trauma caused approximately 29 mortalities. A CM-like syndrome was associated with an estimated 20 additional deaths at the trap sites before or during processing of the 475 animals. An additional CM-related mortality of 18.8% of the processed and released pronghorn was recorded.

Seventy drive-trapped Alberta pronghorn were transported to range enclosures and four to holding pens. Three of the 70 animals were injured during transport and subsequently euthanized. Seventeen animals died of CM. Most animals died from 2 to 5 days after capture and transport; two died after 13 days (Chalmers and Barrett 1977).

Pronghorn that died after being chased during a trapping operation in Colorado had inflamed lungs and mucosa of their entire respiratory tracts; one also had inflamed esophagus. Another pronghorn, collected in an area of Colorado that had been subjected to severe sand and dust storms, had pneumonia, and its entire respiratory system was inflamed (Hoover et al. 1959). The possibility of damage from dust and debris should be considered when trapping pronghorn. Avoiding chases during dry, dusty conditions is prudent, as is spreading hay in corral traps to reduce dust during handling. Photo by Jim Jeffress.

Chalmers and Barrett also examined two 2-week-old pronghorn that died after being pursued by a vehicle. Both animals displayed clinical signs typical of CM.

I necropsied a fawn that was chased by a golden eagle and died of CM-like symptoms; its sibling was killed by the bird.

Road and Railroad Accidents

> *From the standpoint of game management, highway losses are unimportant except where peculiar and specific conditions such as application of mineral salts to highways, as a dust preventative, result in unusual concentrations of animals in heavy automobile traffic.*
>
> Dickerson (1939:107)

Little information is available on the levels of road and railroad kills among pronghorn. When water collects in barrow pits along roads and produces succulent vegetation that is attractive to pronghorn, drawing them to some roads during summer and early autumn, the result may qualify as Dickerson's (1939) specific conditions. In Texas, road kills usually occurred during winter and spring months when forbs were available along highway right-of-ways. A number of such occurrences were noted each year, but the total loss was comparatively low (Hailey 1979). I have found pronghorn carcasses of all age classes and both sexes along highways.

Eighteen pronghorn were killed, or crippled so seriously that they had to be destroyed, by a speeding vehicle near Millican, Oregon (Yoakum 1957a). Ten of the does were carrying 18 fetuses, so 36 pronghorn actually were lost in this one accident. During a 3-year study in Oregon, 24 pronghorn were killed on an area used by 10,000 to 12,000 pronghorn and crisscrossed by public roads (J. Yoakum personal communication:1987). Apparently, road kills are not a significant mortality factor.

Roads fenced to exclude livestock and game probably cause more (indirect) pronghorn mortality than do accidents on unfenced roads. Ward et al. (1976) noted that Interstate 80 in southern Wyoming formed a nearly complete barrier across what once were movement corridors between seasonal ranges for pronghorn. The animals still gathered north of the highway during winter and south of it in spring during the Ward et al. study. Some found holes through fences or crossed them on snowdrifts. Once on the fenced highway, pronghorn were especially vulnerable to traffic. At least 153 were killed along a 55-mile (88 km) section of Interstate 80 west of Laramie during a 5.5-year period. Levels of mortality from malnutrition and other factors related to separation of summer from winter ranges was unknown.

"Francis E. Warren Air Force Base, in Wyoming, contains a population of pronghorn that inhabit residential and other developed areas. This coexistence trait contributes to a high incidence of pronghorn/motor vehicle accidents. Pronghorn roadkill and related mortality and other data from the base were recorded from 1992 to 2001. Annual direct motor vehicle-attributable mortality of pronghorn average 8% of the base population. For the same period, mortality of pronghorn indirectly related to motor vehicle interactions averaged an additional 3% of the population" (Smith 2002:26).

Serious road accidents involving many pronghorn are rare. Traffic mortality is less for pronghorn than for deer using the same habitat. Pronghorn are less nocturnal than deer, so are less likely to be blinded by automobile lights and more likely to be seen by drivers in time to avoid a collision. When pronghorn/vehicle accidents do happen, it is usually with an automobile traveling at high speed and a band or herd of pronghorn on an extended seasonal movement. Above, the skidding automobile killed 18 pronghorn on Highway 20 about 13 miles (21 km) east of Millican, Oregon, during March 1956. Photo courtesy of the Oregon Historical Society, Eugene.

Railroad accidents involving pronghorn, although infrequent, draw considerable attention because many animals are often involved. In southern Idaho, for example, 132 pronghorn, including 43 does with 85 fetuses, were killed in a single train accident (R. Autenrieth personal communication:1976). During periods of deep snow, traveling pronghorn occasionally follow railroad tracks where snow has been removed, thus providing relatively easy mobility. A herd caught on tracks typically will try to outrun a train instead of breaking through the snow piled on both sides. (Left) *Pronghorn carcasses exposed by melting snow; the animals were killed by a train in southern Wyoming.* (Above) *Carcasses of pronghorn killed by a train and hauled to a Wyoming dump.* Left photo by LuRay Parker: courtesy of the Wyoming Department of Game and Fish, Cheyenne. Above photo by Rob Deblinger.

A mature buck dying on a railroad track near Medicine Hat, Alberta, during January 1965. This animal and seven others were maimed and had to be shot, or were killed outright by a train. This particular animal was hemorrhaging internally as evidenced by the frozen blood hanging from its lower jaw. Photo by George J. Mitchell.

During December 1964 in Alberta, snow depths of 20 to 30 inches (50–75 cm) forced many pronghorn to exist—day and night—on the narrow right-of-ways and edges of highways. There, snow removal equipment kept snow depths to a minimum, exposing some shrubby vegetation. Numerous traffic mortalities were suffered throughout southern Alberta. In excess of 300 pronghorn were killed by trains and vehicles within 3 weeks between Lethbridge and Medicine Hat and east to the Alberta/Saskatchewan border (Mitchell 1980).

Parturitional and Prenatal Losses

Parturitional and prenatal losses seldom are documented but do occur. Siamese twins, found in the uterus of a pronghorn doe, were reported by Shufeldt (1920). The siblings were largely united, but only the skulls were saved. They were joined just behind of the orbits.

Hansen (1955) and Yoakum (1957a) uncovered evidence in Oregon that complications at parturition (dystocia) were responsible for seven fatalities. These involved fetuses found within the pelvic girdle of a doe carcass, fetuses within doe carcasses after fawning season and doe and fawn carcasses found together. Also, a large fawn died 33 hours after being born, due to a strangulated hernia that could have been congenital or caused during birth.

In the same area, J. Meeker (personal communication:1982) watched a yearling doe trying to give birth late one evening. The next morning, she was dead. One front hoof of a very large fetus was visible where the amnion extruded from her vulva. The other leg was folded back within the uterus. From the same area in Oregon, one of 55 fawns necropsied for cause of death had died of birth complications (Dunbar and Velarde 1998). Another fawn was observed being born and was found to be dead upon expulsion from the doe. The fawn did not take a breath, but no abnormalities were detected upon necropsy (Dunbar 1999a).

Congenital deformities were noted in fetuses and neonates on the National Bison Range during 1965 to 1966, possibly due to inbreeding (O'Gara 1968). These included a fetus with an open neural pore exposing its brain, another's lower jaw and legs were about half normal length and another's head resembled a cauliflower. Another fawn that died in 1968 at about 4 days of age was found to have its small and large intestines improperly joined.

No more abnormalities were noted after translocating "new blood" into the range, until two fawns died soon after birth during spring 1984. One was almost without a nose and the lower jaw was wrapped around the front of its face. The other had legs of only about 20% normal length and a huge head. The feeding area used by their dams had been sprayed with 2, 4-D and Torodon herbicides during late summer 1983. Whether that caused the abnormalities is unknown (O'Gara 1999).

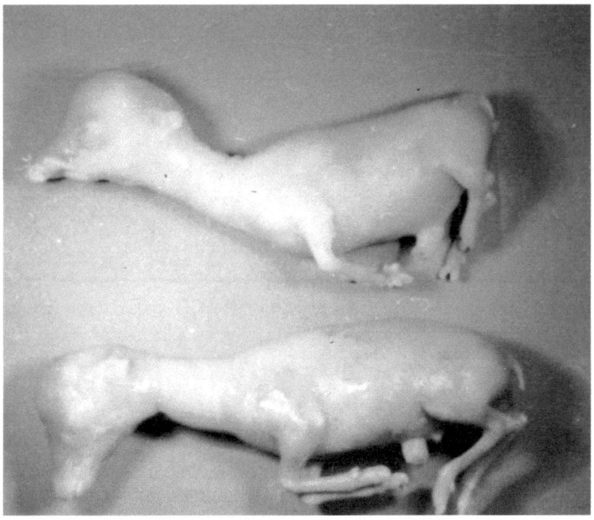

Pronghorn fetus (top) *with an underdeveloped lower jaw and slightly short legs compared with its sibling* (bottom). *Fawns of sika deer with such abnormalities occasionally are reported from England where they are called "Andy Gumps." These fetuses were carried by a pronghorn doe in a herd that has been maintained at comparatively low numbers (50 to 150) for about 35 years.* Photo by Bart W. O'Gara.

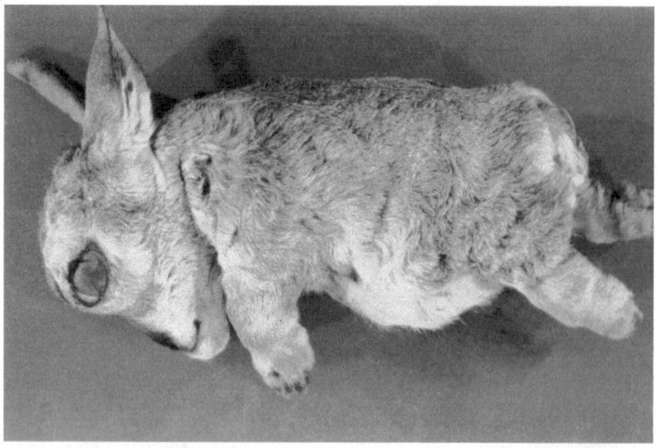

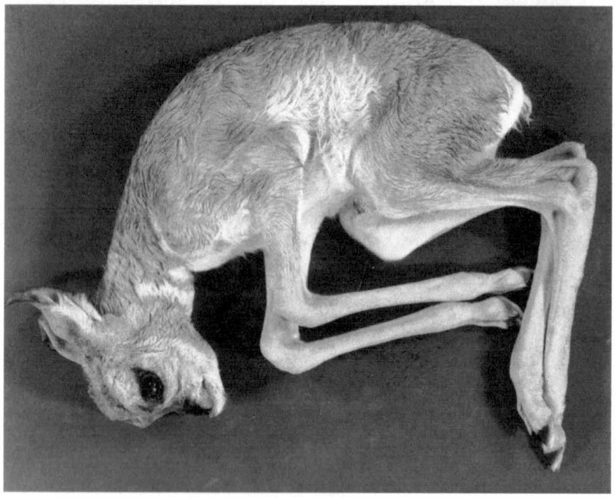

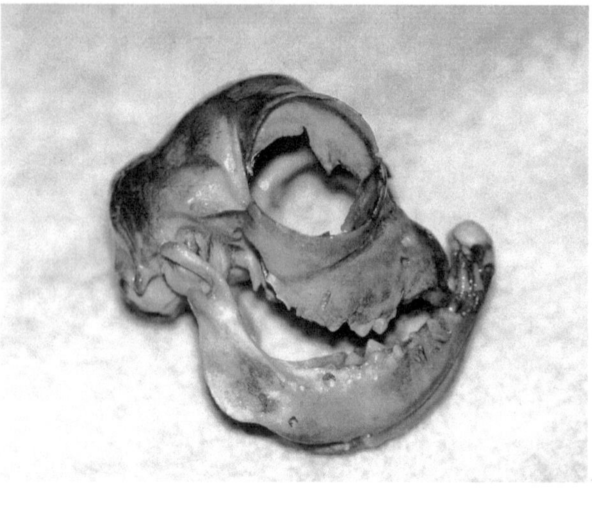

*(*Top*) An abnormal pronghorn fawn that was born with a normal appearing one that also died soon after birth. Note the huge head and very short legs. The heart was nearly round and double the size of that of a normal newborn fawn. (*Center*) Another grossly deformed fawn was found in the same area during the same spring. That one had a normal body but a short skull. (*Bottom*) The skull of a fawn that had nearly normal development of the braincase and the eyes but extreme shortening of the nasal bones. The lower jaw appeared of normal size but was bent to "fit" the malformed upper jaw. Dams of these fawns had foraged on rangelands sprayed with herbicides for weed control during late summer of the previous year. Whether spraying caused the abnormalities is unknown.* Top and bottom photos by Bart W. O'Gara. Center photo courtesy of John A. Byers.

Drowning

You can't stop an Antelope when it gets settled on going anyplace; might as well not try. I've seen them swimming the Red Deer River by thousands in the spring, when the ice was still floating down. Most of them were drowned, and those that managed to get across were killed by half-breeds. but they kept on trying to cross.

Charles Blazier, "Antelope expert of Brook, Alta," quoted in Seton (1927:451)

A pronghorn doe crossing a swollen river in Oregon was washed against a clump of willows and drowned (Einarsen 1948). In Idaho, a young fawn drowned while attempting to follow its mother across a small river (Bodie 1979). Twenty drowned pronghorn of mixed ages and both sexes were found floating in a Wyoming reservoir. Apparently, they broke through thin ice and were unable to regain the shore (W. Hepworth personal communication:1969). Pronghorn commonly cross rivers; they are good swimmers and drownings are uncommon.

Hail

Severe hailstorms periodically sweep across pronghorn habitat. Seven dead fawns and a severely crippled doe were found in a field in Texas after a hailstorm had defoliated much of the vegetation (Hailey 1979). Fourteen pronghorn killed by hailstones were found in Wyoming (B. Helms personal communication:1978). A rancher in the area contacted the Wyoming Game and Fish Department because he had seen many carcasses and suspected a bluetongue epidemic. The animals had massive hemorrhages (eagles also were found dead in the area). Arthur (2000) reported that a number of pronghorn fawns fell victim to violent hailstorms in southeastern Wyoming during early July 1999. Because they do not seek shelter, pronghorn seem especially vulnerable to hail.

Miring in Mud

Guano Lake, located east of the Hart Mountain National Antelope Refuge in Oregon, contains water only during years of above-average precipitation. Then, the lake is 1 to 2 feet (30–60 cm) deep. As it dries, the lakebed is transformed into thick muck, creating a hazard for pronghorn moving from winter to spring rangelands on Hart Mountain. During the spring of 1921, 75 pronghorn perished there, 15 were victims during spring 1938, and

The Dry Creek Flume, carrying water for irrigation in Idaho, presented a hazard to pronghorn and other wildlife (McCarty 1982). In that dry area, animals were attracted to the water. Once a deer (left) *or pronghorn entered the flume, it was trapped by fast water and steep, slippery sides. A pronghorn buck* (right)*, carried by water to a pond below, died in the 10-mile (16 km) long flume by drowning or because the front feet and legs were broken or worn off by rubbing on the bottom of the flume as the animal was carried along.* Photos by Bob McCarty.

A line of fast-moving thunderstorms crossed the Nebraska Panhandle during early July 2000, carrying marble- to baseball-sized hail. The storms killed at least 22 pronghorn in one 4- to 5-acre (1.6–2.0 ha) area of a U.S. Forest Service pasture (Grier 2000). One doe, unable to walk and blinded in one eye, was humanely destroyed. No standing vegetation remained where the pronghorn were found, but hail casualties were limited to a small area. The dead included 1 adult buck, 12 adult does and 9 fawns. The pronghorn's injuries were consistent with impacts from large hail stones, trauma associated with trampling and striking a fence when they attempted to escape the storm. Photo by Bob Grier; courtesy of the Nebraska Game and Parks Commission, Lincoln.

27 more deaths were reported in 1939 (Branson 1939). Miring probably took many pronghorn when millions of them made long movements, but extensive mortality from this cause seems a thing of the past.

Eating Trash

An emaciated 2-month-old pronghorn fawn, which had occasional tetonic contractions of the large muscles before death, was necropsied at the Wyoming Game and Fish Laboratory. The undernourished animal apparently had been eating from a trash dump. Coal ashes, sandstone brick, linoleum, plastic ribbon, nylon bristles, tinfoil and human hair were found in its stomachs, which, along with the small intestines, were greatly inflamed (W. Hepworth personal communication:1978). This animal was being kept as a pet, and a lack of proper foods probably led to the strange diet. This was an unusual situation, but I have found bits of plastic and ashes in the rumens of wild pronghorn. Such findings may indicate a deficiency of some mineral in the diet.

Fighting

Bachelor bucks spend considerable time during summer jousting and fighting, but injuries then are uncommon. As the rut approaches, fights among both bachelors and territorial bucks become more serious. Two deaths caused by bucks fighting in Texas were reported by Buechner (1950a). Two bucks in Oregon died when their horns locked during a fight, and two others found with locked horns were untangled and apparently survived (Yoakum 1957a). The right horn of one buck in Colorado pierced the underside of a second buck's jaw, and the left horn locked behind the second buck's right horn, killing both (Spencer 1942). Two bucks with locked horns also were reported from Utah.

Although the average length of 15 fights witnessed by Kitchen and Bromley (1974) during the rut in Montana and South Dakota was only 2 minutes, five bucks were seriously injured. Kitchen and Bromley saw 11 other injured bucks, apparently victims of fights, and found two large dead bachelor bucks with horn punctures in their chests. I necropsied two territorial bucks that had been

Two dead pronghorn bucks with the horns of one locked around the second buck's neck. During a fight, one buck, with heart-shaped, 15.5-inch (39.4 cm) horns and in-curving tips 1.5 inches (38 mm) apart, evidently thrust his horns upward on the underside of the second buck's throat. The horns flexed far enough apart to allow the second buck's neck to pass through. The second buck's neck was rubbed raw and scabbed, indicating that the two animals remained locked together for some time before dying. Photo by David Smith; courtesy of the Utah Division of Wildlife Resources, Salt Lake City.

killed during fights and was amazed at the extensive damage to lungs and hearts caused by their rather innocuous-appearing horns. During 82 fights observed by Byers (1997a) during the rut on the National Bison Range, 10 (12%) resulted in the death of one of the combatants.

Large Ungulates

Large ungulates kill an occasional pronghorn fawn. F. Martin (personal communication:1977) saw a bull bison attack and kill fawns, and B. Lesley (personal communication:1981) saw mules pursue and kill fawns. Aggression toward pronghorn fawns by livestock also was observed during a study in Alberta, and two fawns were found dead by trampling (Barrett 1978). Such instances seem rare and have little importance to the species. In rare instances, fawns also are killed by yearling pronghorn, as indicated by Bodie (1979).

Falls

On November 3, 1991, 150 pronghorn plunged over a 100-foot (30 m) precipice near the Green River in Wyoming. The reason they stepped into thin air is uncertain. Game warden Allen Round and biologist Tom Christiansen backtracked the herd and believe the animals were not being chased by human or beast before going over the cliff (Ottman 1992). The 100-foot (30 m) fall did not produce instantaneous death for all of the animals who tumbled from the rim. Game and Fish Department personnel had the unpleasant, yet necessary, task of humanely ending the agony of several maimed pronghorn when they arrived on the scene on November 4. They immediately noted the site as a possible learning opportunity and alerted the archeology department at Western Wyoming Community College. Professors there intend to study the site as the natural order plays its role in reclaiming the bodies. The effect of scavengers and natural decay will allow a greater understanding of both human-caused and accidental jump sites from prehistoric times.

This was not the first known case of pronghorn falling to their death in Wyoming. Einarsen (1948:85) quoted from a March 1, 1944, news release with the headline: "100

This old buck, shot in Wyoming during October 2001, had a healed hole in its nose that likely was the result of being "hooked" by another buck. Once a curved horn tip penetrates skin or deeper, struggles are inclined to increase the size of the wound before the victim is unhooked. O'Gara was hooked, barely, through the forearm skin, by a buck recovering from an immobilizing drug. The buck dragged him some 40 yards (36 m) before he could gain control of the animal and unhook the small strip of skin. Photo by Rollin D. Sparrowe.

These pronghorn (more than 30) apparently followed the white rump hair ahead of them, like a driver follows the tail lights of a car ahead. Not unlike cars going off a freeway, one after another, the pronghorn followed to their deaths. A trail down the face of the bluff, which pronghorn habitually used, was missed by only a few feet. Why the leader made this blunder, even in pea-soup fog, will never be known. Photo by LuRay Parker; courtesy of the Wyoming Department of Game and Fish, Cheyenne.

antelope killed in plunge over bluff." The animals were found in southern Campbell County at the base of a 45-foot (14 m) bluff, from which they apparently had tumbled. Officials theorized the pronghorn were blinded by a storm or stampeded by an airplane. The area where the bodies were found was virtually inaccessible to automobiles. Einarsen added that pronghorn have a tendency to stampede, and—so long as the group leader follows a course—pronghorn that have foraged together will stubbornly follow, even to their destruction.

CONCLUSIONS

Pronghorn die from myriad factors, and different factors become dominant at different times and in different places. Density-dependent population regulation has been demonstrated in several long-term studies. Thus, under most conditions, other forms of mortality seem to be proximate factors that actually kill pronghorn but do not ultimately regulate population numbers. Severe winters or droughts in combination with deteriorated rangelands or barriers to movements often are the proximate causes of die-offs today. Predators kill a high percentage of fawns, especially in areas where predators are numerous and pronghorn are scarce. The high reproductive rate of pronghorn compensates for predation in most instances.

Hunting has accounted for the death of more than 100,000 pronghorn per year during recent years, but hunting is carefully controlled to maintain the health of herds. Sport hunting is needed to keep herds in balance with available habitat and prevent excessive crop damage.

Bluetongue apparently is the most serious disease of pronghorn on adequate rangelands, especially when the animals become concentrated around stagnant water during a dry summer. Parasites become detrimental to pronghorn populations at times when the range is overgrazed by livestock, particularly domestic sheep. Land-use practices that influence the availability of quality habitat, including movement corridors within historic pronghorn range, will determine the future of the species much more than predation, regulated hunting or diseases.

C H A P T E R T H I R T E E N

Habitat Characteristics and Requirements

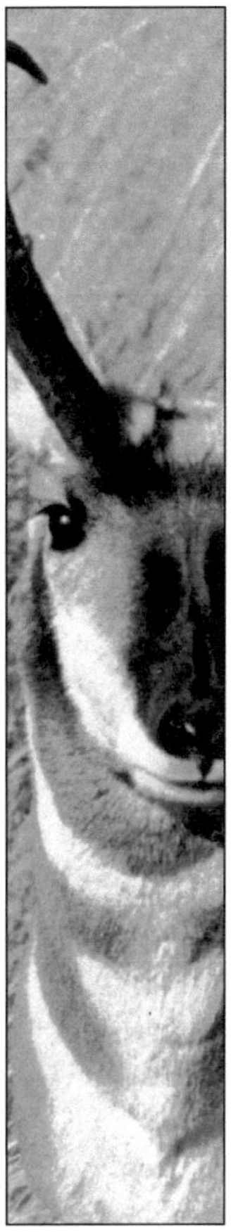

Jim D. Yoakum

When the game manager asks himself whether a given piece of land is suitable for a given species of game, he must realize that he is asking no simple question.

Leopold (1933:124)

There is enough evidence to indicate the preference of pronghorn for rolling plains. Historical accounts of the days where these animals were first seen picture them at home on the grasslands and sage flats of the plains country or plateaus. Thus do they live today. . . .

Einarsen (1948:131)

Assuming that a species can find the necessary combination of climate, substrate, and vegetation to permit it to occupy an area, the number which can then be supported in that area are determined by the amount and distribution of food, shelter, and water in relation to the mobility of the animal. . . . None of these general requirements is simple in itself. Each species requires a particular kind of combination.

Dasmann (1964:60)

Habitat requirements of ungulates are analyzed in terms of their need for food, water, cover, and reproduction as well as the way they must be arranged in space and time.

Boyd et al. (1986:520)

For centuries, pronghorn have roamed the western plains and prairies of North America. They occur in flat or rolling areas, along foothills, in mountain valleys and on plateaus; they avoid rock cliffs, steep terrain, dense woody vegetation and certain agricultural crops. They are an ungulate of semiarid landscapes—few live where annual precipitation averages less than 5 inches (12.7 cm) or more than 30 inches (76.2 cm). Pronghorn prefer ecosystems with a mixture of grasses, forbs and shrubs to provide forage requirements and fawn bedding cover.

The pronghorn copes with its enemies through sight and flight. Habitats with low-growing vegetation—less than 25 inches (63.5 cm) in height—and undulating landscapes provide optimal physiographic conditions for such behavior. These habitat characteristics differ markedly from those of other North American wild ungulates. Deer, for example, generally seek dense shrubs and trees to hide from enemies. Wild sheep and goats seek rocky cliffs and steep, precarious terrain for security cover.

The terms *habitat* and *landscape* denote a combination of abiotic (nonliving) and biotic (living) characteristics. Pronghorn occupy a broad spectrum of habitats from the Pacific coastal desert in Mexico to alpine meadows at more than 11,000 feet (3,353 m) elevation in Wyoming. At times, pronghorn have been stereotyped as desert animals (Jaeger 1961, Wagner 1980). This is a misconception because they thrive in greatest densities on grasslands and shrub-steppes, and they occur in deserts only in low numbers.

Past descriptions of pronghorn habitat often have been sketchy, anecdotal and rarely accompanied with quantified data. For better understanding and assessment of pronghorn habitat characteristics and requirements, examinations were made in areas occupied by pronghorn in 2 provinces of Canada, 4 states in Mexico, and 17 states in the United States from 1952 to 2001. The areas represent landscapes now occupied by pronghorn, and most of the areas were occupied when Anglo-Americans first saw the species. Data were collected and analyzed regarding pronghorn distribution in three biomes, which were divided into major vegetative communities. Next, the biotic and abiotic factors were identified and quantified—the focus of this chapter.

BIOMES AND VEGETATIVE COMMUNITIES

Landscapes with similar climate, soils, flora and fauna are classified into large-scale regions termed biomes (e.g., deserts, grasslands, shrubsteppes, forests and tundra) (Dasmann 1964, Bailey 1984). Biomes can be subdivided into smaller regions based on characteristics of major plants: the Great Plains into tall, mixed or short grasslands; the shrubsteppes into sagebrush and grass, juniper and grass, mesquite and grass, and salt shrublands. Certain woodlands or savannas with sparse trees and abundant herbaceous vegetation can be classified into the grassland or shrubsteppe biome, depending on the abundance and structure of woody species.

A total of 116 vegetative types are identified for the United States by Küchler (1964), of which 21 are reported as habitat occupied by pronghorn. Küchler's manual of vegetative communities is based on dominant and codominant plants identified by aspect and structural characteristics (Table 71). The manual provides descriptions and photographs of plant types, and contains a separate 40- by 60-inch (107 by 152 cm) foldout color map.

Vegetative classification systems at the landscape level inherit variation problems as noted by Küchler (1964:5): "The types of variation are therefore not uniform throughout their area, and this lack of uniformity applies to both life forms and taxa. The small scale of the map requires such a degree of generalization that a given vegetative type may vary considerably in its different sections. Frequently, a type occurs in its 'pure' form but often there are variations, inclusions and complexes. These make a type more heterogeneous than appears on the map, but they are too small to be shown at the given scale. . . . The inclusions and complexes within a vegetative type are the result of local conditions." Another liability results from mapping vegetation, which depicts the communities at differing map scales. Küchler's large map illustrated silver sagebrush communities associated with stringer watershed drainages in eastern Montana, but when this region is reduced to 8.5 by 11 inches (22 by 28 cm) or smaller pages in textbooks, the smaller communities most frequently are not included (see Allen 1967, Hylander 1966).

Then too, Küchler's (1964) system was based on the natural potential of vegetative communities and may not always conform to existing vegetative types that have developed in response to human alterations, for example, changes due to repetitive fires or long-term intensive, domestic or feral livestock grazing that have changed vegetation composition and structure.

Sundstrom et al. (1973) warned that when associating estimated pronghorn numbers with vegetative commu-

A small band of pronghorn rest and graze on a mixed grassland in Wind Cave National Park, South Dakota. These animals live yearlong in this vegetative community with bison, elk and mule deer. Photo by Jim D. Yoakum.

TABLE 71. Biomes and vegetative communities (with dominant and codominant plants) for rangelands occupied by pronghorn[a]

Biome	*Vegetative community*	*Dominant plants*	*Codominant plants*
Grasslands	Grama-buffalo grass	Blue grama, buffalo grass	Plantain groundsel, Indian rice grass, prickly pear
	Wheatgrass-needlegrass	Western wheatgrass, blue grama, needle-and-thread grass, green needlegrass	penstemon, aster, twayblade, leather root
	Wheatgrass-bluestem-needlegrass	Western wheatgrass, big bluestem, needlegrass	goldenrod, wild rose
	Bluestem-grama prairie	Little bluestem, side-oats grama, blue grama	Dalea, evening primrose, Virginia's bower, ragweed
	Sandsage-bluestem prairie	Little bluestem, sand bluestem sandsage, hairy grama	Sunflower, dropseed, Spanish bayonet
	Bluestem prairie	Big bluestem, little bluestem, switchgrass, Indian grass	Phlox, fleabane, bedstraw, sunflower,
	Nebraska Sandhills prairie	Big bluestem, sand bluestem, little bluestem, sandreed, needle-and-thread grass	Gilia, Indian rice grass, fleabane, dropseed
	Grama-galleta steppe	Blue grama, galleta	Indian rice grass, Spanish bayonet, locoweed, prickly pear
	Grama-tobosa prairie	Blue grama, tobosa	Snakeweed, prickly pear
	Grama-needlegrass-wheatgrass	Western wheatgrass, blue grama, needle-and-thread grass	Snakeweed, June grass, twayblade
Shrubsteppe	Great Basin sagebrush	Big sagebrush	Rabbitbrush, lupine, locoweed, phacelia
	Sagebrush steppe	Bluebunch wheatgrass, big sagebrush	Indian rice grass, June grass, lupine, puccoon
	Wheatgrass-needlegrass shrubsteppe	Western wheatgrass, big sagebrush, plains bluegrass, needle-and-thread grass	Winterfat, saltbush, sedge, greasewood
	Saltbush-greasewood	Shadscale, greasewood	Winter fat, red sage, salt grass, desert thorn
	Juniper steppe woodland	Bluebunch wheatgrass, big sagebrush, western juniper	Low sagebrush, arrowleaf, lupine, bitterbrush
	Juniper-pinon woodland	Utah juniper, pinon pine, western juniper	Ceanothus, bitterbrush, western wheat-grass, rabbitbrush
	Trans-Pecos shrub savanna	Tarbush, creosote bush	Catclaw, galleta, magney, Spanish bayonet
	Shinnery	Little bluestem, shin oak	Sumac, catclaw, needlegrass
	Mesquite-buffalo grass	Buffalo grass, mesquite	Juniper, needlegrass, Spanish bayonet
Desert	Creosote bush-tarbush	Tarbush, creosote bush	Prickly pear, agave, galleta, nolina
	Palo verde-cactus shrub	Palo verde, prickly pear	Catclaw desert thorn, burroweed

[a] Vegetative communities classified by Kuchler (1964).

nities, pronghorn frequently move between various vegetation types during home range and seasonal movements. Consequently, for their study, it was necessary to combine two or more vegetation types where separation by type was not practical or possible. Other studies used similar methodology.

1960s Investigation

One of the earliest attempts to quantify estimated pronghorn populations for biomes and vegetative communities was conducted during the 1960s by Yoakum (1972). Pronghorn population data were obtained from Canadian and U.S. wildlife agencies, with corresponding occupied rangelands delineated on maps (Figure 72). This information was placed on overlays depicting vegetative communities. Pronghorn occupying vegetative types and biomes were scanned and computed. The base vegetation map was adapted from Küchler's (1964) portrayal of natural vegetation. Relating pronghorn populations to all vegetative types was not possible because herds made seasonal movements and occupied different vegetative

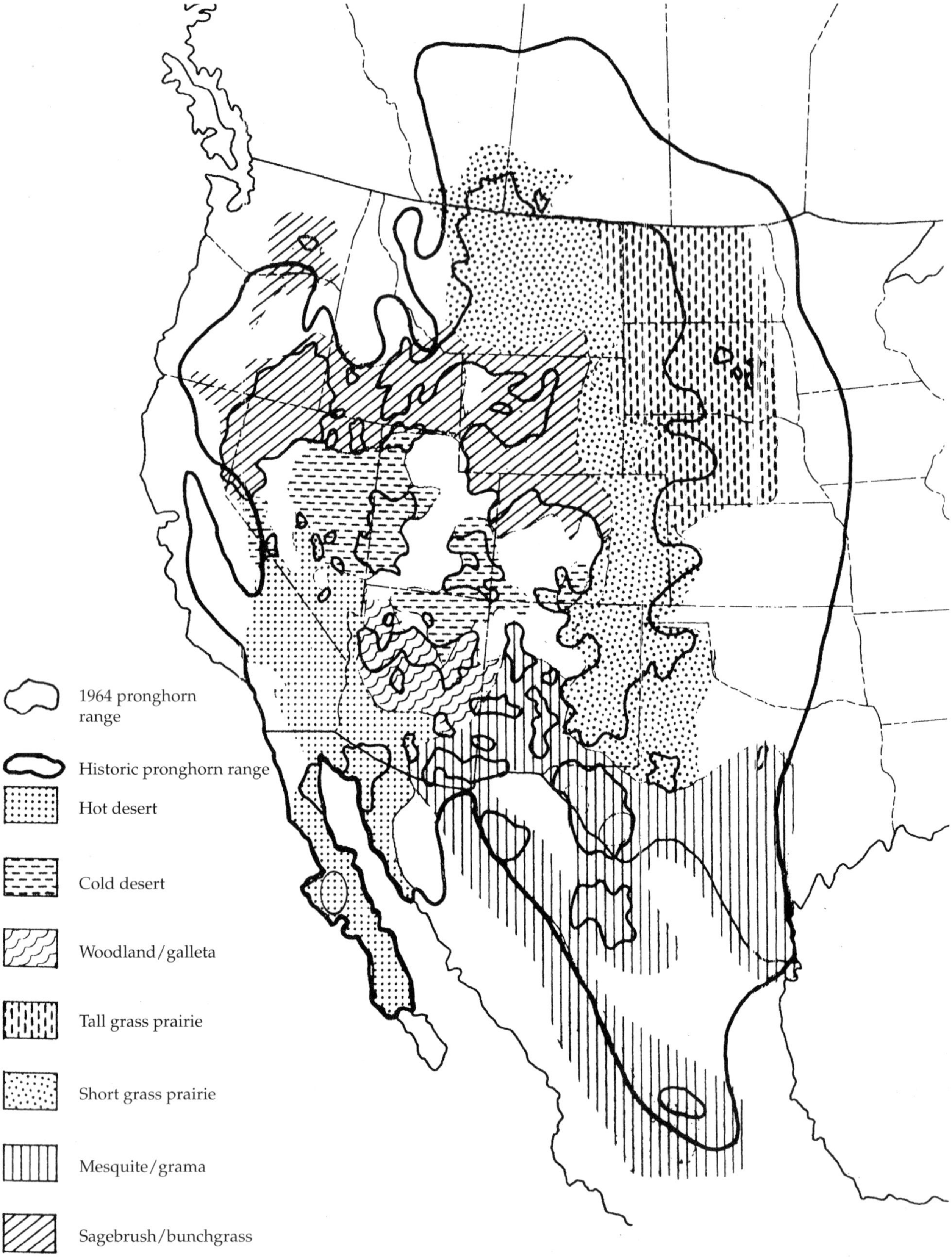

Figure 72. *Pronghorn distribution during 1964 in relation to biomes and major vegetative communities (Yoakum 1972).*

Studies during the 1960s and 1990s indicated that pronghorn were most numerous in the grassland biome. This biome sustains vegetation characteristics favorable to pronghorn: height, composition and density. Although grasses are the most productive herbaceous cover, forbs are abundant during spring and early summer—providing highly nutritious, succulent and preferred forage during the pronghorn's reproductive season. Rangelands with vegetative heights from 15 to 30 inches (38–96 cm) provide forage and fawn security cover. These habitat characteristics complement the pronghorn behavior to see and flee from its enemies. Photo by Ed Bry; courtesy of the North Dakota Department of Game and Fish, Bismarck.

communities seasonally—or in different years, depending on weather and forage conditions. Populations were computed for the 21 vegetative communities listed in Table 71, then grouped into seven categories listed in Table 72. Major plant categories listed in biomes were: (1) grassland biome, the short, mixed grasslands of the Midwest; (2) shrubsteppes, sagebrush/grasslands, mesquite, grama, woodland/grass and arid salt shrublands; and (3) cold deserts/hot deserts, including the Mojave, Sonoran, Chihuahuan and Vizcaino.

Using an estimated population of 385,000 pronghorn for 1964, Yoakum (1972) reported 68% of pronghorn occupied the grassland biome and 32% lived in the shrubsteppes (Tables 72 and 73). Less than 1% inhabited deserts. Those findings were similar to pronghorn use of grasslands and shrubsteppes made some 30 years earlier by Aldo Leopold (1933).

1970s Investigations

Less than a year after Yoakum's (1972) published report documenting estimated pronghorn numbers in relation to vegetative communities, a similar study was completed by Sundstrom et al. (1973). They sent a questionnaire to wildlife agencies in Canada, Mexico and the United States requesting pronghorn numbers in relation to vegetative types identified by Küchler (1964). The estimated total pronghorn population for 1970 was 433,000. Included were Hawaii (120 animals) and Washington (100 animals); however, data were not available for plant types.

Sundstrom et al. (1973) associated pronghorn numbers with 21 plant communities listed in Table 71. The pronghorn-associated vegetative types were listed for provinces and states, but totals for pronghorn numbers for each type were not compiled. Using the same criteria, established for biomes and vegetative communities, Yoakum later calculated the totals and indicated 65% for the grassland biome and 35% for shrubsteppes. Deserts contained less than 1%.

1990s Investigations

To update information on pronghorn to vegetation relationships for this book, the Wildlife Management Institute mailed questionnaires to all western province and state wildlife agencies during 1997, requesting pronghorn numbers for vegetative types listed in Table 71. All agencies responded, and the sum of the data is provided in Table 73: 53% for grasslands, 47% for shrubsteppes and less than 1% for deserts. Decreased pronghorn herds in the Great Plains were consistent with decreased populations reported in Chapter 4.

A review of historic pronghorn numbers in Chapter 4 substantiates pronghorn were most abundant on the

TABLE 72. Estimated pronghorn populations inhabiting vegetative communities in Canada, Mexico and individual states of the United States, 1964

		Sagebrush and bunchgrass	*Mesquite and grama*	*Short-grass prairie*	*Mixed-grass prairie*	*Shrubsteppe woodlands*	*Hot desert*	*Woodland/ galleta*
Location	*Total pronghorn population*	*Pronghorn population (precentage)*	*Pronghorn population (precentage)*	*Pronghorn population (precentage)*	*Pronghorn population (precentage)*	*Pronghorn population (precentage)*	*Pronghorn population (precentage)*	*Pronghorn population (precentage)*
Arizona	10,000	—	100 (1)[a]	—	—	—	100 (1)	9,700 (98)
California	2,700	2,700 (100)	—	—	—	—	—	—
Colorado	15,200	3,100 (20)	—	11,200 (74)	—	900 (6)	—	—
Idaho	4,700	4,700 (100)	—	—	—	—	—	—
Kansas	100	—	—	100 (100)	—	—	—	—
Montana	95,000	15,300 (15)	—	75,600 (80)	4,100 (5)	—	—	—
Nebraska	9,000	—	—	7,000 (75)	2,000 (25)	—	—	—
Nevada	4,500	3,500 (77)	—	—	—	900 (21)	100 (2)	—
New Mexico	22,500	—	2,200 (10)	18,100 (85)	—	—	—	1,100 (5)
North Dakota	14,200	—	—	—	14,200 (100)	—	—	—
Oklahoma	200	—	—	200 (100)	—	—	—	—
Oregon	8,900	8,900 (100)	—	—	—	—	—	—
South Dakota	27,400	—	—	—	27,400 (100)	—	—	—
Texas	9,400	—	1,100 (10)	8,300 (90)	—	—	—	—
Utah	1,000	—	200 (20)	—	—	800 (80)	—	—
Washington	100	100 (100)	—	—	—	—	—	—
Wyoming	140,000	65,000 (45)	—	49,400 (35)	24,000 (19)	1,400 (1)	—	—
Canada	20,300	—	—	20,300 (100)	—	—	—	—
Mexico	1,200	—	1,000 (98)	—	—	—	200 (2)	—
Total	386,400	103,500 (27)	4,300 (0.1)	190,200 (49)	71,700 (19)	4,000 (0.1)	400 (0.0)	10,800 (0.3)

[a] ( ) denotes percentage of population in province or state.
Source: Yoakum 1972.

TABLE 73. Estimated pronghorn populations occupying biomes and vegetative communities based on investigations during the 1960s, 1970s and 1990s

	Estimated pronghorn populations					
Biome and	*1960s*[a]		*1970s*[b]		*1990s*[c]	
vegetative community	*Number*	*Percentage*	*Number*	*Percentage*	*Number*	*Percentage*
Grassland						
Short-grass	191,200	49	230,200	53	255,800	38
Mixed-grass	71,400	19	53,000	12	102,700	15
Subtotal	262,600	68	283,200	65	358,500	53
Shrubsteppe						
Sagebrush/grass	107,700	29	133,000	31	290,100	44
Mesquite/grass	4,600	1	7,200	2	7,700	1
Oak/grass	4,700	<1	4,700	<1	7,900	1
Juniper/grass	5,000	1	5,000	1	9,400	1
Subtotal	122,000	32	149,900	35	308,800	47
Hot desert	400	<1	500	<1	800	<1
Total	385,000	100	433,600	100	675,000	100

[a] Yoakum (1972).
[b] Sundstrom et al. (1973).
[c] Wildlife Management Institute (1997 survey).
All estimated populations rounded to closest 100.

Great Plains before 1850 (see also McCabe et al. 2004). At that time, pronghorn numbers were greater than at present on the shrubsteppes and deserts, although densities there were far less than on the Great Plains. The large pronghorn herds adapted to the expansive Midwest plains, experienced rapid declines by the early 1900s, but increased

Although more than 98% of the California Central Valley grasslands have been converted to agricultural crops, remaining grasslands are occupied by pronghorn just as they were before the arrival of Spanish explorers more than 250 years ago. This 1999 scene shows a band of pronghorn and livestock peacefully grazing on the Carrizo Plain National Monument in San Luis Obispo County. Photo by Jim D. Yoakum.

Pronghorn inhabiting the shrubsteppe biome are second in abundance to those in the grassland biome. Shrubsteppes are dominated by various shrub species with an understory of grasses and forbs. Shrubsteppes with vegetation taller than 35 inches (88 cm) are not as frequently occupied as shrublands with shorter plants. Pronghorn habitats need adequate vegetation as security cover for fawns during their first 2 weeks of life. Newborn fawns lie prone, lacking the ability to outrun enemies. Pictured here is a week-old fawn partially concealed from horizontal view by low sagebrush, about 15 inches (38 cm) higher than the neonate. Photo by Jim D. Yoakum.

Energy developments are increasing on western rangelands. Pronghorn adapt to certain developed environments, but energy activities can increase survival problems, such as reduction in natural forage, increased human activity during fawning seasons, potential increase of illegal kills and alterations in seasonal movement corridors. Wildlands, characteristic of natural pronghorn habitats, are decreased when human development of energy sources alter landscapes. Photo by Robb Hitchcock.

to high numbers by the 1980s. However, these herds may continue to experience an insidious decline at the beginning of the 21st century—possibly due to increased native rangelands being altered to agricultural crops and human developments. Pronghorn may find a stronghold for habitat in the shrubsteppe biome in the future, for these western shrublands are less adapted for agricultural crops, and human society does not concentrate on these semiarid, cold environments.

PHYSICAL CHARACTERISTICS OF PRONGHORN HABITAT

Abiotic Characteristics

Abiotic characteristics are nonliving conditions or circumstances. For describing pronghorn habitat, physiography, soil, climate, and water are those characteristics reviewed.

Physiography. Pronghorn are typical plains or prairie herbivores associated most frequently with open, undulating or flat terrain, seldom exceeding 10% in grade. A schematic cross section of the United States, in Figure 73, illustrates the variety of physiographic regions relative to pronghorn occupation.

Historically, pronghorn inhabited differing ecosystems in central North America from the Pacific Ocean, in California, to the Missouri River, in Nebraska, from the northern prairies of Saskatchewan, Canada, to the southern highland plateau of Mexico (see Chapter 1). The southern environments supported herds from the Pacific Ocean desertlands of Baja California to the Texas grassland coastal region bordering the Gulf of Mexico. Herds ranged from below sea level in Death Valley, California (McLean 1944), to 11,000 feet (3,353 m) above sea level in Wyoming (Yoakum 1974). However, the greatest concentrations occupied landscapes from 3,000 to 8,000 feet (914–2,438 m) above sea level (Table 74).

Pronghorn can be found in foothills and lower elevations of mountains. Pronghorn readily use lowlands in Yellowstone National Park during winters, but during the summer and autumn, they move up the mountain to forage vegetation located in and around open patches of aspen and conifers. The animals have not been seen in high, steep and thickly forested adjacent areas (D. Scott personal communication:1989).

Landscape studies in Arizona disclosed that pronghorn tended to select for slopes of less than 10% and made limited use of terrain of 10% to 40% slope (Ockenfels et al. 1994). Kindschy et al. (1982) reported that herds in Oregon used topography with less than 30% slope. Rouse (1959) discussed how pronghorn use topography as shelter during winter; depressions, ridges, rims and breaks serve as shields against severe winds during blizzard conditions. During the severe winter of 1948–49, pronghorn were located in coulees or canyons where the wind was less severe.

Conversely, landscape features can adversely affect pronghorn movement and occupancy of potential habitat. Examples are large bodies of water or rivers, abrupt escarpments, mountain ranges, dense, high shrubs and trees, and deep canyons. Einarsen (1948) described two natural barriers to pronghorn—the Columbia River Gorge and a heavily forested area through which pronghorn did not travel to reoccupy nearby ancestral habitat. Large rivers are not necessarily a deterrent by themselves; Lewis and Clark witnessed herds crossing the Missouri River (Moulton 1987). Likewise, pronghorn have been known to penetrate forests with little understory and open vistas, but forests with dense, tall shrubs, or small trees appear to limit pronghorn movements (Yoakum 1978b, Lee et al. 1998). Ockenfels et al. (1994) reported that pronghorn avoided vegetation more than 24 inches (61 cm) high and selected for sites with an extended vista and few trees.

Artificial barriers, such as fences (Spillet et al. 1967, Howard et al. 1990), highways (Buechner 1950c, Ward et al. 1976, 1980, Deblinger et al. 1984, Ockenfels et al. 1994) and residential developments (Ockenfels et al. 1994, 1996) can block or restrict pronghorn movements. The barrier effects of highways and railroads are increased when they are fenced to prevent livestock encroachment (Ockenfels et al. 1994, 1996).

Soils. Soils indirectly affect animals through influences on the composition, amount, palatability and nutritive values of vegetation. Although the concept of relating soils to wildlife distribution is well established (see Leopold 1933), soil typing relative to pronghorn distribution or abundance has not been studied.

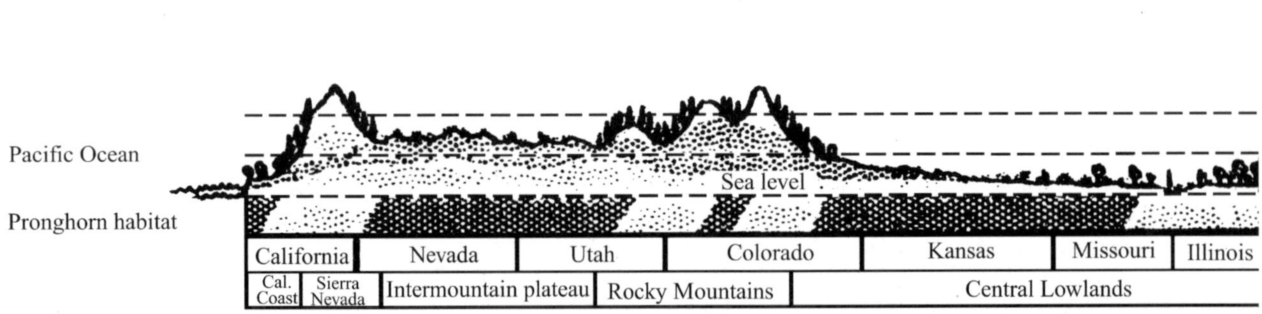

Figure 73. *Schematic cross section of the United States showing the physiographic areas occupied by pronghorn (modified from Hylander 1966).*

Table 74. Abiotic habitat characteristics for certain habitats occupied by pronghorn

Biome	*Location*	*Mean elevation in feet (m)*	*Mean temperature in degrees Fahrenheit (°C)*	*Mean precipitation in inches (cm)*	*Source*
Grassland	Southeastern Alberta	2,700 (823)	40 (4.4)	14 (35.6)	Mitchell (1980)
	Matador, Saskatchewan	3,200 (975)	— —	13 (33.0)	Dirschl (1960)
	National Bison Range, Montana	2,500 (762)	— —	14 (35.6)	Kitchen (1974)
	Pawnee Grasslands, Colorado	5,200 (1,585)	— —	12 (30.5)	Hoover (1971)
	Southeastern Colorado	5,600 (1,707)	55 (12.8)	17 (43.2)	Barrington (1975)
	Flint Hills, Kansas	— —	— —	31 (78.7)	Rothchild et al. (1994)
	Roswell, New Mexico	— —	64 (17.8)	12 (30.5)	Howard et al. (1990)
	Trans-Pecos, Texas	— —	— —	14 (35.6)	Buechner (1950c)
	Prescott, Arizona	5,000 (1,524)	— —	12 (30.5)	Ockenfels et al. (1992)
	Chihuahua, Mexico	5,600 (1,707)	58 (14.4)	14 (35.6)	Hinshaw (1990)
Shrubsteppe	Pahsimeroi, Idaho	8,000 (2,438)	— —	8 (20.3)	Autenrieth (1976)
	Yellow Water Triangle, Montana	6,250 (1,905)	45 (7.2)	13 (33.0)	Pyrah (1987)
	Hart Mountain Antelope Refuge, Oregon	5,000 (1,524)	— —	10 (25.4)	U.S. Fish and Wildlife Service (1994)
	Southeastern Oregon	6,500 (1,981)	— —	13 (33.0)	Kindschy et al. (1982)
	Wamsutter, Wyoming	6,500 (1,981)	43 (6.1)	7 (17.8)	Spillette et al. (1967)
	Rawlins, Wyoming	5,500 (1,676)	46 (7.7)	8 (20.3)	Taylor (1975)
	Desert Range Experiment Station, Utah	5,500 (1,676)	— —	6 (15.2)	Beale and Smith (1970)
	Sheldon National Wildlife Refuge, Nevada	6,000 (1,829)	54 (12.2)	13 (33.0)	McNay (1980)
	Adobe Valley, California	6,500 (1,981)	— —	11 (27.9)	Goldsmith (1988)
Desert	Cabeza Prieta National Wildlife Refuge, Arizona	5,600 (1,707)	To 120 (48.9)	<5 (12.7)	Hughes and Smith (1990)
	Vizciano, Lower California, Mexico	sea level (0)	— —	3.5 (8.9)	Cancino et al. (1994)

Differing shrubsteppe sites support differing densities of pronghorn—often in relation to the abundance of preferred herbaceous forage. Not all shrubsteppes are occupied by pronghorn, particularly those with dense, tall shrubs that would outcompete forbs and grasses and impede a pronghorn's ability to see predators as well as to run from them. Photo by Jim D. Yoakum.

Climate. The climate of North America's pronghorn rangeland is nearly as varied as is the region's topography. A great range of temperatures exists in various ecoregions used by pronghorn. Northern areas often experience lows of –20°F (28.9°C), whereas southern ranges can experience temperatures of 115°F (46.1°C) (Hylander 1966).

As with temperature, the continent can be divided into zones based on the amount of precipitation received. Rains and snows that furnish water, vital for animals and their habitats, do not fall uniformly over North America. Rangelands in the 8- to 15-inch (20.3–38.1 cm) annual precipitation

Pronghorn usually are associated with vast open plains and prairies, but they also are found on the alluvial fans and foothills of mountains. Pictured here is an aerial view of a herd in Idaho running on undulating terrain, with nearby high mountains covered with forests and cirques of snow. Photo courtesy of the Edson Fichter Collection.

zone include primary areas occupied by pronghorn today (Table 74). Ecosystems such as the Vizcaino Desert in Baja California Sur receive an average of 3.6 inches (9.1 cm) annual precipitation, and the area sustains less than one pronghorn per square mile (0.4/km^2) (Cancino et al. 1994c). Regions of high rainfall, such as the Flint Hills of Kansas, which receives annual precipitation of about 31 inches (78.7 cm), also have low pronghorn densities (Rothchild et al. 1994). This substantiates Buechner's (1950b) contention that areas with more than 15 inches (38.1 cm) of precipitation annually do not sustain high numbers of pronghorn.

The time of year that precipitation falls can be correlated to pronghorn abundance. The central prairies receive snow during winter and rain during spring and summer, favoring the growth of forbs and high pronghorn production (see Tables 71 and 72). In the Great Basin and other cold desert regions, snow and rain are received in winter and spring, but few rains occur during summer, and herbaceous forage becomes desiccated. Pronghorn densities are low in these regions of arid summers.

The right combination of temperature and precipitation is vital for the production of certain forage classes preferred by pronghorn. Such areas are west of the Mississippi River. When pronghorn were translocated to areas of higher temperature and precipitation, such as Florida (Elliott 1966) and Hawaii (Bromley 1972), they succumbed.

About 70% of pronghorn range receves snow on an annual basis. When depths exceed 12 inches (30.5 cm), animals experience difficulty obtaining forage. Where this occurs, the animals tend to make seasonal movements to areas of less snow where forage is more readily obtainable (Guenzel et al. 1982). Prolonged seasons of deep snow usually result in a shortage of available forage and obstacles to free movement. At such times, protruding shrub species become important survival food. Bailey (1986) wrote that snow depth greater than 18 inches (45.7 cm) restricted pronghorn mobility. Observing forage utilization after recent snows, Rouse (1959) concluded that, when snow was deeper than 6 inches (15.2 cm), pronghorn experienced difficulties in reaching preferred food. Deep, encrusted snow can be devastating to pronghorn because of the scarcity of available forage and the reduced ability of pronghorn to reach it or areas of better supply.

Climatic patterns affect pronghorn directly and indirectly. Adequate precipitation triggers increased forage production, which is beneficial. Severe winters with prolonged deep snow can be disastrous and are a major natural influence on pronghorn population dynamics. Left photo by LuRay Parker; courtesy of the Wyoming Department of Game and Fish, Cheyenne. Right photo courtesy of the South Dakota Game and Parks Department, Pierre.

Water. Surface water occurs sparsely in most pronghorn habitats. Springs, ponds, streams, rivers, lakes, snow and rain are sources of drinking water. In addition, within the 20th century, humans made spring improvements and constructed pipelines, windmills, troughs, reservoirs and water catchments. Most have been created for livestock or agricultural purposes, but they serve pronghorn and other wildlife as well (Yoakum et al. 1980, Kie et al. 1994).

Landscapes receiving 5 to 15 inches (12.7–38.1 cm) of precipitation annually support more than 98% of the pronghorn population (Table 74). Within these regions, some areas produce higher pronghorn densities than others.

Studies in Wyoming indicated that pronghorn maintained highest numbers when water was available during summer, and, 95% of more than 12,000 pronghorn used areas within 4 miles (6.4 km) of water (Sundstrom 1968). Occasionally, adult bucks were observed as many as 7 miles (11.3 km) from a water source. Where surface water was available at a rate of 0.2 source per square mile (0.08/km^2), the average number of pronghorn per square mile was 7.1 (2.7/km^2). As the number of water sources per square mile decreased, pronghorn density correspondingly decreased. At one water source—the only one for more than 30 square miles (77.7/km^2)—the average number of pronghorn per square mile decreased to 0.5 animals (0.19/km^2). Similar results were reported from the same area by Boyle and Alldredge (1984), and Kindschy et al. (1978, 1982) reported a similar trend in shrubsteppes of southeastern Oregon (Figure 74).

Surface waters used by pronghorn on Arizona grasslands were investigated by Ockenfels et al. (1994). They reported that pronghorn did not regularly remain within about 1,312 feet (400 m) of surface water, but more than 90% occurred within less than 1 mile (1.6 km) of water. One reason for this high rate of pronghorn proximity to water was because the study site was well "watered," and the animals simply were always within sufficient proximity to water such that its scarcity was not a problem. Observations and readings disclosed that distance to water sources was not correlated to ambient air temperatures.

According to Ockenfels et al. (1992), water on grasslands in Arizona may be an important, if not critical, factor in determining fawn bedsites for at least the 2-week postnatal period. Most neonates were located between 0.3 and 0.6 miles (0.5–1.0 km) of water. Barrett (1981) found fawn bedsites in Alberta to average 0.4 mile (0.64 km) from water in one study. However, he later found no

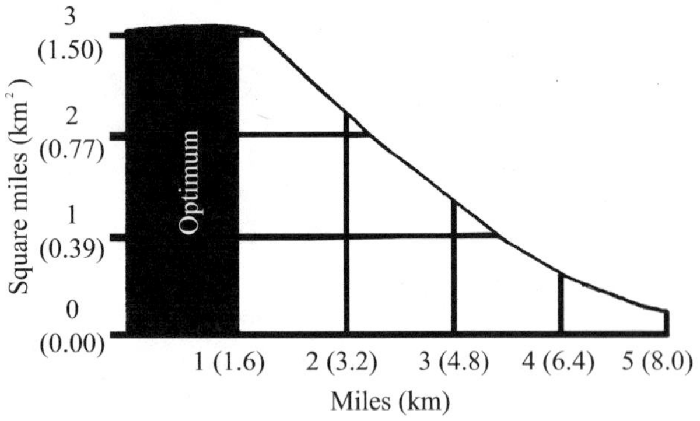

Figure 74. *Relationship of water source proximity and pronghorn density in shrubsteppe habitat (Kindschy et al. 1978).*

apparent relationship of such beds and close proximity to water at another site.

For an area of approximately 20 square miles (52 km^2) in the Sonoran desert in southwestern Arizona, Hughes and Smith (1990) found seven permanent water facilities provided drinking water for pronghorn. It was estimated the average distance of pronghorn from drinking water was more than 3.7 miles (5.95 km), and no significant difference was detected in distances to water sources between the wet and dry seasons.

For the Vizcaino Desert in Baja California Sur, Mexico (the most southern population of pronghorn in North America), herds occupy an area of approximately 30 by 90 miles (48.3 by 144.8 km). The desert receives an average of 3.6 inches (9.1 cm) annual precipitation, primarily during winter. For this region, Cancino et al. (1994c) reported no known extant permanent drinking water sources, but dew and fog from the adjacent Pacific Ocean provided frequent moisture deposits on vegetation.

Pronghorn on the Vizcaino Desert in Baja California Sur, Mexico, are the only ones now living within sight of the Pacific Ocean. Fog from the ocean moves across the desert and settles on the vegetation—thereby providing moisture to pronghorn in a habitat with no known sources of natural drinking water. Photo by Patrick Robles.

Biotic Characteristics

Biotic characteristics are factors of living matter. For describing pronghorn habitat, the characteristics of vegetation, other wildlife, livestock and humans are reviewed.

Vegetation. Possibly more than any other environmental factor, vegetation influences pronghorn distribution and density because it provides forage and water as well as cover for neonates. Habitats with specific characteristics of vegetative quality and quantity produce and maintain different pronghorn densities (see Tables 72 and 73). Among plant communities, the qualitative and quantitative factors that sustain pronghorn at various densities (plant production, diversity of plant species, herbaceous plants, etc.) are more uniform in northern than southern climes.

Vegetation sampling methodologies have evolved within several ecological disciplines (e.g., botany, forestry, rangeland science) and for a variety of management or research objectives (e.g., estimating forage for ungulates, describing habitat use by birds) (Higgins et al. 1994). Descriptions of the various methods that have been used to inventory vegetation are beyond the scope of this book, but many of the basic methods may be found in Cooperrider et al. (1986), Heady and Child (1994), Higgins et al. (1994), U.S. Bureau of Land Management (1996a), and Holechek et al. (1998).

Because pronghorn consume many forage species, classifying such plants according to nutritional and physiological characteristics and by region and season is necessary. Boyd et al. (1986) reported that many biologists and managers have found it convenient to classify vegetation as grasses, forbs and shrubs. This classification is based primarily on structure and phenology of the plants, but also correlates with seasonal nutritional qualities; shrubs have comparatively high nutrient levels in winter, whereas forbs are most nutritious during spring and summer.

Investigators have used different methods to classify vegetation, and this prerogative is justified under varying circumstances. However, the use of a standard classifica-

Rangelands with a mosaic of differing shrub/grassland communities (reflecting physiography, soils and climate) are attractive to pronghorn. Top photo by Robb Hitchcock. Bottom photo by LuRay Parker; courtesy of the Wyoming Game and Fish Department, Cheyenne.

tion system facilitates relatively reliable comparison of vegetation in different seasons and regions. Cactus is an example of a plant that has been classified as both forb and shrub. However, for most pronghorn diet studies, it has been identified as a forb (Yoakum 1990), and continuation of this classification is recommended. If the cactus or any other plant is to be listed in a forage class other than listed in Table 75, a note to that effect should be provided, making it possible for readers to recompute the data.

Vegetation characteristics, succession, condition and trend can remain stable for long periods or change quickly (Vavra et al. 1994). Some changes in vegetation trend may take decades. Habitat occupied by pronghorn on the Jornada Experiment Station in southwestern New Mexico was primarily grasslands in the 1860s. Due to invasion of mesquite, tarbush and creosotebush, the landscape gradually changed to shrublands by the 1960s. Such vegetative changes were followed by similarly gradual changes of the density and distribution of pronghorn (Ghaly 1980). On the other hand, in a matter of minutes or hours, fires can alter a shrubland to bare ground, which can be a grassland the following growing season (Pyle and Yoakum 1994). In historic times, 6-foot (1.8 m) tall stands of grass were grazed to 3 inches (7.6 cm) by bison (Costello 1969), and the same occurs today with cattle grazing (Eccles et al. 1994).

Vegetation covers, on average, 40% to 60% of pronghorn arid rangeland, with the balance of the terrain being

Table 75. Classification of plants by forage class for grasses, forbs and shrubs

Class	*Description*	*Taxonomy*	*Notes*
Grass and grasslike plants	True grass and grasslike sedges and rushes	Family Gramineae and grasslike Cyperaceae and Juncaceae	This category often is abbreviatedto grass, but still includes sedges and rushes
Forbs	All herbaceous plants other than grass and grasslike plants	Most monocots other than three families listed above; herbaceous dicots	Since classification of forbs andbrowse is not based on taxonomy, classification of some semiwoody species is somewhat arbitrary
Shrubs	All woody plants	Woody dicots	

Source: Modified from Boyd et al. 1986.

bare ground, rock or litter (Yoakum 1972, Ockenfels et al. 1994). Plant coverage for shortgrass prairies averaged 60% to 90% (Schwartz et al. 1977, Jatnieks 1983). For the shrubsteppes, the average was 40% to 60% (Yoakum 1972, Kindschy et al. 1982). And, for the Sonoran Desert, Hughes and Smith (1990) recorded that plant cover averaged 7% to 16%, and habitats occupied by pronghorn had greater vegetative cover than did adjacent unoccupied sites.

On grassland biomes, where pronghorn thrive, composition of vegetation averaged 60% to 80% grasses, 10% to 20% forbs and less than 3% shrubs (see Table 73). For shrubsteppe in Oregon, it was 15% to 35% grasses, 5% to 20% forbs and 20% to 60% shrubs. Figures 75 and 76 indicate optimal percentages of forbs and shrubs for pronghorn in shrubsteppes. Pronghorn rangeland in the desert biome was low in vegetation, showing average percentages of 2 for grasses, 2 for forbs and 7 for shrubs (Hughes and Smith 1990).

A wide variety of plant species occur in pronghorn habitats. On 1 square mile (2.6 km^2) of shortgrass prairie in eastern Colorado, Costello (1969) identified 22 grasses, 141 forbs, 10 shrubs and 4 trees (see also Schwartz et al. 1977, Jatnieks 1983). The shrubsteppes maintain a plant richness of approximately 5 to 15 grasses, 30 to 90 forbs and 5 to 10 shrubs (Yoakum 1972, Goldsmith 1988). A shrubsteppe site in natural condition with no agricultural fields in south-central Oregon produced 14 grass and grasslike plants, 73 forbs (53 perennial and 20 annual), and 12 shrubs and trees (Trainer et al. 1983).

Wet springs and summers produce an abundance and variety of succulent forbs, which are preferred forage. Movement of pronghorn from desiccated rangelands to areas of succulent forage has been observed (Cole 1956, Long 1965, Tsukamoto and Diebert 1968, Good 1977, Pyrah 1987).

Beale and Smith (1970) noted that, when vegetation had a moisture content of 75% or more, pronghorn did not drink water, although it was readily available. As the vegetation became desiccated, pronghorn commenced drinking water.

For the Sonoran Desert, Hughes and Smith (1990) determined that the moisture content of vegetation sampled in areas occupied by pronghorn averaged 36%, with 54% for trees, 42% for shrubs, 22% for forbs other than cacti, 79% for cacti and 19% for grasses.

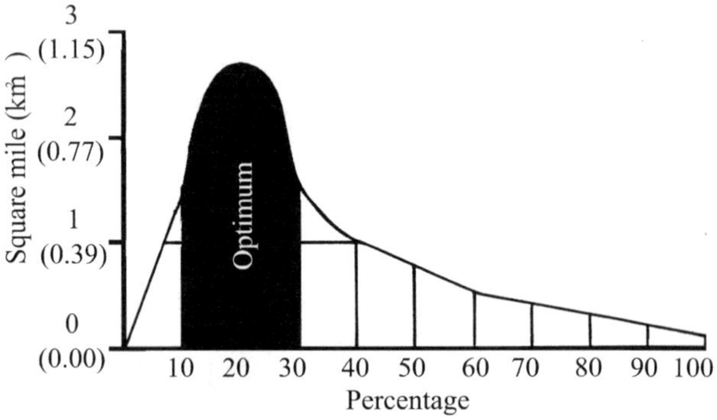

Figure 75. *Percentage of forbs in groundcover relative to pronghorn density on shrubsteppe in southeastern Oregon (Kindschy et al. 1978).*

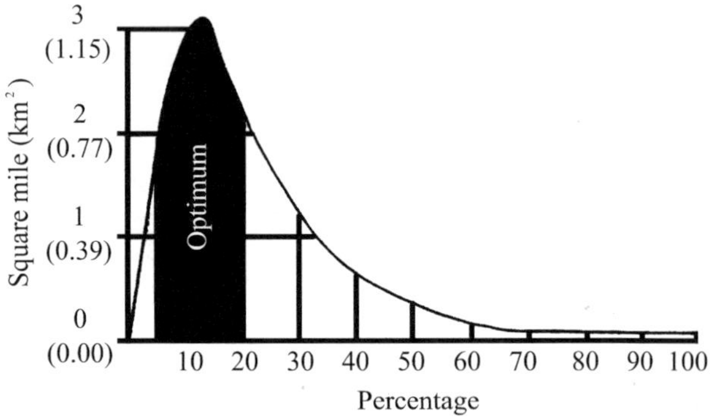

Figure 76. *Percentage of shrubs in groundcover relative to pronghorn density on shrubsteppe in southeastern Oregon (Kindschy et al. 1978).*

Rangelands with a variety of vegetative types (e.g., meadows, lake playas, wildfire burns, forb patches) are preferred by pronghorn over extensive monotypic plant communities. With seasonal changes, pronghorn seek different vegetative types, and feed on preferred species available at those times (Salwasser 1980). Fire and other disturbances can stimulate production of a variety of vegetative types and thereby create habitat favored by pronghorn (Deming 1959, Yoakum 1978b, 1980, Pyle and Yoakum 1994, Lee et al. 1998).

On the Jornada Experiment Range in southwestern New Mexico, approximately 29% of the site was a grassland complex, but more than 75% of pronghorn observations were recorded in this plant community (Ghaly 1980). Shrublands dominated 70% of the landscape, but pronghorn use was less than 30% yearlong.

Pronghorn use of vegetative types was assessed during three successive years on the Hart Mountain National Antelope Refuge in Oregon (Pyle and Yoakum 1994). Of the 10 vegetative types occupied, pronghorn used low sagebrush, recently burned shrublands and lake playas more than 94% of the time (Table 76).

Pronghorn are adapted to low, relatively open vegetation that allows broad visibility and does not inhibit mobility (Einarsen 1948, Goldsmith 1988, Ockenfels et al. 1996). On pronghorn rangeland, vegetation is generally low in height—averaging 15 inches (38 cm) with a range of 5 to 30 inches (13–76 cm) (Figure 77) (Yoakum 1972, Sundstrom et al. 1973, Ockenfels et al. 1994, Lee et al. 1998). Rangelands with plants taller than 30 inches (76 cm) seldom are used, although pronghorn sometimes seek high vegetation (e.g., bitterbrush, saltbrush) for forage and may travel through high shrub areas between seasonal rangelands. However, their total yearlong use of vegetation 30 inches (76 cm) or taller is minimal.

In some areas, such as open ponderosa pine stands in eastern Oregon, pronghorn use forested areas as corridors during seasonal movements (Foster 1988). In other regions, such as open juniper stands in Arizona (Ockenfels et al. 1994) and California (Salwasser 1980), forested areas provide preferred succulent vegetation. During summer, trees and tall shrubs have been used for shade. A common characteristic for forest lands used by pronghorn was the relative lack of understory shrubs, allowing vis-

TABLE 76. Percentage seasonal use by pronghorn of vegetative types on the Hart Mountain National Antelope Refuge in Oregon, 1991 to 1993

Vegetation type	*Year*	*Winter*[a]	*Spring*	*Summer*	*Autumn*
Black sagebrush	1991	0	0	0	0
	1992	0	0	t[b]	0
	1993	—	0	0	0
Grassland[c]	1991	0	9	6	7
	1992	15	1	3	7
	1993	—	1	6	18
Juniper	1991	0	0	1	1
	1992	1	0	1	1
	1993	—	0	t	4
Lakebed	1991	0	2	29	7
	1992	2	3	56	14
	1993	—	0	t	7
Low sagebrush	1991	100	90	63	75
	1992	79	94	26	76
	1993	—	76	91	71
Meadow	1991	0	1	3	0
	1992	0	0	14	1
	1993	—	0	1	0
Mountain big sagebrush	1991	0	0	0	0
	1992	0	t	0	0
	1993	—	t	t	0
Mountain shrub	1991	0	t	t	0
	1992	0	0	0	0
	1993	—	0	0	0
Silver sagebrush	1991	0	0	t	7
	1992	3	t	1	1
	1993	—	0	1	0
Wyoming big sagebrush	1991	0	1	1	1
	1992	0	2	0	0
	1993	—	23	1	1

[a] Deep snow prevented pronghorn use of the range during winter 1993.
[b] t = trace (<1%).
[c] Primarily composed of upland sites burned within the past 15 years.
Source: Pyle and Yoakum 1994.

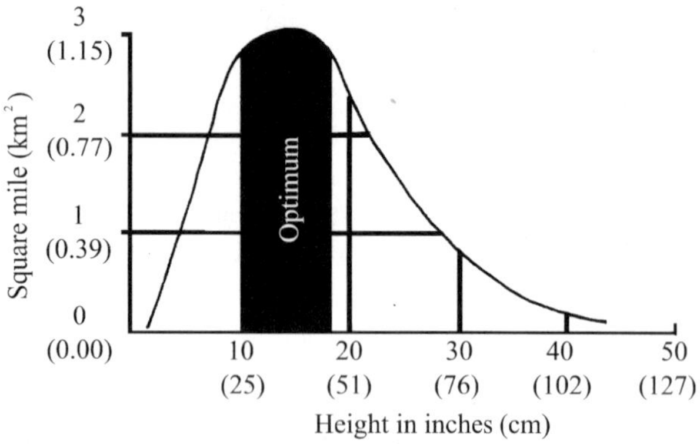

Figure 77. *Height of vegetation in relation to pronghorn density on shrubsteppe in Oregon (Kindschy et al. 1978).*

Vegetation higher than 30 inches (76 cm) seldomly is frequented by pronghorn. Predators such as coyotes and bobcats can effectively hide and stalk in such cover, which also hinders pronghorn mobility. Photo by Jim D. Yoakum.

ibility and fast movement. This was especially evident in Yellowstone National Park after the extensive wildfire damage of 1988. During 1989, U.S. National Park Service personnel observed pronghorn where they had not been seen during the past 50 years (D. Scott personal communication:1989).

Animals. Interspecific relationships of pronghorn and other herbivores, including wild, feral and domestic animals, are addressed in detail in Chapter 15. Interactions with predators are discussed in Chapters 6 and 11.

Pronghorn historically fed with or in proximity to herds of bison and, to a lesser degree, with bighorn, deer and elk (Burroughs 1961). Apparently, little competition with pronghorn occurred because of different food habits or seasonal and spatial use of rangelands. Pronghorn benefited from the legendary bison herds that cropped grasses heavily and stimulated growth of forbs and shrubs favored by pronghorn (Costello 1969).

Before the arrival of Europeans, more predators—both species and numbers—shared rangeland with pronghorn

Landscapes lacking trees are characteristic of the majority of pronghorn rangelands. Pronghorn do occupy thin pine forests with an understory of herbaceous forage, as illustrated in this scene from the Black Hills of South Dakota. Forage is available and the band can see most danger. Photo by Jim D. Yoakum.

A lone pronghorn buck guards his territory on the grasslands of southwestern South Dakota. These rangelands were grazed with large herds of bison before the arrival of Euro-Americans. At the beginning of the 21st century, most of these grassland prairies were grazed by cattle. Photo by Ed Park.

than is the case today. The gray wolf and grizzly bear, now virtually absent from that rangeland, were numerous historically, and pronghorn were many times more abundant. In natural, free-roaming circumstances, pronghorn can and do coexist with predators, except on marginal habitats or areas where numbers of predators are high in relation to the pronghorn population (Lee et al. 1998).

The effects of livestock on pronghorn date back to the earliest occupation of southwestern rangelands by Spaniards during the 17th and 18th centuries, when ranches and missions were developed to produce domestic sheep and goats (Leftwich 1977). Competition for forage was immediate, and displacement of pronghorn by livestock on western rangelands was begun. Except on rangelands in poor health, cattle and horses do not significantly compete today for forage with pronghorn, but competition for forbs and shrubs can exist with domestic sheep (see Chapter 15).

The most ubiquitous factor affecting the welfare of pronghorn during the past 150 years has been land-use practices for human benefit. Millions of acres within pronghorn range have been converted for agriculture, industry, commerce, recreation and human habitation. Seasonal and migratory routes of pronghorn have been blocked by highways, railroads, energy developments and fences. Some of these changes occurred slowly over the years, but their collective and cumulative effect ultimately has resulted in the presence, today, of 1 pronghorn for perhaps every 30 to 40 when Lewis and Clark crossed the

Rangelands altered for human occupation are a cause of decline in pronghorn populations. This land-use trend is common for many areas in the West and Southwest. Displaced pronghorn often are doomed because other habitats are generally stocked to capacity. Photo by LuRay Parker; courtesy of the Wyoming Department of Game and Fish, Cheyenne.

continent in 1804 to 1806. Today, loss of habitat is the biggest dilemma facing pronghorn.

During the 20th century, pronghorn increased more than 3,000%, indicating their adaptiveness to wanton or unwitting human dominion. Wildlife management practices, including regulated harvests, translocations, progressive habitat restoration and controlled livestock foraging, have contributed to the dramatic pronghorn population increase. Pronghorn are considered in certain management plans on western rangelands. Hunters, private landowners and the public, as well as provincial, state and federal agencies, have encouraged the return of pronghorn.

If management practices continue to consider the habitat requirements of pronghorn, there is little reason to suppose that the animals cannot continue to thrive on available rangeland. If these requirements are ignored, pronghorn undoubtedly will suffer.

PRONGHORN HABITAT REQUIREMENTS

Leopold (1933) wrote that each species of wildlife has a unique, characteristic range of variation of habitat requirements, which may be considered a property of that species. Although a habitat may have enough food, water or cover, it also must be of sufficient quality and appropriate spacing to support the species.

King (1938) listed five essentials of wildlife habitat: food, cover, water, juxtaposition and interspersion. He defined juxtaposition as the qualitative and quantitative balance of food, water and cover requirements of a species in relation to each other within the species' home range. Interspersion was defined as the spatial distribution of available nutrition, water and cover in a given area.

The first three requirements must be present in the habitat if it is to support pronghorn. To a large extent, habitat juxtaposition and interspersion determine the size and sustainability of a herd within a given area. Table 77 outlines habitat characteristics and requirements of pronghorn in two biomes in which about 99% of the species is found. Unfortunately, insufficient quantitative data are available to list habitat characteristics for the desert biome.

Food

The abundance and availability of preferred foods are related directly to seasonal density of pronghorn in a given area. The nutritional content of pronghorn food, as that of any ungulate's food supply, is as important as its quantity. It varies with season, plant phenology, previous use and climate. And, as important as food resource quality and quantity is its accessibility.

Some pronghorn travel long distances to obtain preferred food, especially during droughts and winters of deep snow. Nutrients needed for body maintenance and reproduction are obtained primarily from vegetation, but some are secured from mineral sources.

On average, pronghorn consume the equivalent of 2 to 3 pounds (0.9–1.4 kg) of air-dry forage per day. Pronghorn use less than 1% of available forage on western rangelands today as determined by the number of animal unit months used by livestock and pronghorn (Yoakum and O'Gara 1990, Yoakum et al. 1996). Food habit studies documenting vegetation consumed compared with availability and seasons of the year are provided in Chapter 14.

Wherever pronghorn range (grasslands, shrubsteppes or deserts), their forage preference is for a diversity of forb species (Yoakum 1990). Consumption of succulent forbs greatly exceeded that of shrubs, and forb preference ratios exceeded those of shrubs or grasses in all biomes for all seasons (see Chapter 14).

Forb consumption has been linked to fawn recruitment (Ellis and Travis 1975, Howard et al. 1990), and lack of forbs in certain areas has been attributed to competition with livestock (Autenrieth 1984, Pyrah 1987, Howard et al. 1990, Hervert et al. 2000). Some rangelands contained more than 150 forb species, of which 80 to 90 were eaten by pronghorn (Yoakum 1990). However, 5 to 10 species usually comprised the bulk of the diet.

On a year-round basis, shrub consumption by pronghorn is exceeded only by forbs (Yoakum 1990). Although eaten during all seasons in some areas, shrubs generally are browsed most during autumn and winter, when they are higher in protein and more available than most grasses or forbs. Shrubs are especially important in northern habitats where snow frequently covers other vegetation. At such times, shrubs become the primary available forage. Shrubs also play a key role in survival on drought-stricken rangelands where grasses and forbs are scarce.

Graminoids provide a minor portion of total diet in all biomes. Pronghorn consume twice as much grass on grasslands as they do on shrubsteppes. Regardless, the total intake averages no more than 15% for yearlong use throughout North America (Yoakum 1990).Predominantly, green grasses are eaten. Fine-textured grasses, such as the *Poas*, are preferred to the larger and relatively tough-textured bunchgrasses, although the latter often are eaten soon after sprouting.

Plants poisonous to livestock often are eaten in large quantities by pronghorn with rare cases of toxicity (see Chapter 14).

Minerals

Pronghorn are attracted to mineral licks, waters and certain plants containing salts and other minerals. Such

TABLE 77. Habitat characteristics and requirements for pronghorn in grassland and shrubsteppe biomes

	Pronghorn requirements	
Habitat characteristics	*Grassland biome*	*Shrubsteppe biome*
ABIOTIC		
Physiography	Large, expansive area 15.4 square miles (40 km^2) minimum,	Same
	low rolling terrain,	Same
	no major physical barriers,	Same
	slopes <30%	Same
Climate		
Precipitation	9.8 to 15.7 inches (25–40 cm)	7.9 to 11.8 inches (20–30 cm)
Snow depth	9.8 to 15.7 inches (25–40 cm)	Same
Temperature	Not a major problem, as herds inhabit hot, semiarid areas to cold alpine steppes	Same
Soils	Not a determining factor except in relation to soil/site vegetative production	Same
Water (drinking)		
Quantity	0.26 to 1.45 gallons (1.0–5.5 L)/day	Same
Distribution	0.93 to 1.72 miles (1.5–6.5 km)	Same
BIOTIC		
Vegetation		
Forage consumption	2.2 to 7.7 pounds (1.0–3.5 kg) air-dry forage/day	Same
Ground cover	60% to 80% vegetation	30% to 50% vegetation
Plant species composition	50% to 80% grasses,	5% to 15% grasses,
	10% to 20% forbs,	5% to 10% forbs,
	<5% shrubs	10% to 35% shrubs,
Plant species diversity	10% to 20% grasses	5% to 10% grasses,
	20% to 60% forbs	10% to 70% forbs,
	5% to 10% shrubs	Same
Height	9.8 to 17.7 inches (25–45 cm),	Same
	5.6 inches (65 cm) unfavorable for most habitats	Same
Succulence	The more availability year-round, the better for all forage classes	Same
Communities	Variety and diversity important (i.e., meadows, playas, wildfire burns)	Same
Animal		
Wild ungulates	Few competition or compatibility problems	Same
Predators	Pronghorn may be reduced or limited in areas of low densities or isolated herds	Same
Humankind	Suitable habitat being usurped, increased construction of barriers (mainly fences), predator control, water developments and alfalfa plantings beneficial, livestock commensal in grasslands, competitive forage in shrubsteppe and desert biomes	Same

attractions may affect animal use of and distribution on particular rangelands, but such relationships have not been demonstrated. Nor have mineral deposits or concentrations been shown to influence pronghorn productivity.

Only one mineral lick used by pronghorn is known to have been analyzed (Jones and Hanson 1985). This was from an area adjacent to an artesian well at Clay Butte in Sweetwater County, southeastern Wyoming. The sample was strongly alkaline and rich in sodium. Sulfate and, probably, bicarbonate were the important anions, whereas chloride was insignificant.

Water

Whether pronghorn need drinking water to sustain viable populations is an equivocal issue. Some investigators have indicated that pronghorn need drinking water to maintain healthy herds (Einarsen 1948, Buechner 1950c, Sundstrom 1968, Beale and Holmgren 1975, Hailey 1979, Yoakum 1978b, Kindschy et al. 1982, Allen et al. 1984). Others maintain that pronghorn rarely, if ever, drink water and that the animals can obtain needed moisture from succulent forage (Skinner 1922, Leopold 1933, Hoover et al. 1959, Monson 1968, Phelps 1978, Arizona Game and Fish Depart-

Some natural mineral licks have been used by pronghorn for centuries. Photo by Marilyn Maring; courtesy of Leonard Rue Enterprises, Blairstown, New Jersey.

Salt blocks placed on rangelands for livestock commonly are used by pronghorn. On his trip in the 1840s to the Great Plains, John James Audubon reported how fond pronghorn were of salt (Audubon and Bachman 1851). Photo by LuRay Parker; courtesy of the Wyoming Department of Game and Fish, Cheyenne.

ment 1981, Cancino et al. 1994c). Robbins (1983) concurred with Leopold (1933) that pronghorn obtained water from free water—such as streams, lakes, potholes, rain, snow or dew—preformed water contained in forage and oxidative or metabolic water as a by-product of oxidation of organic compounds. In any case, water requirements for pronghorn have not been well documented or understood.

The quantity of drinking water consumed by pronghorn varies with body size, age, sex, health, lactation and physical activity (see Chapter 8). Consumption of water decreases with lower temperatures, higher humidity, snow cover, succulent forage, dew or rain. Conversely, it increases with drier atmospheric conditions, lack of snow, dry forage and higher temperatures. In Wyoming, prong-

Drinking water sources for pronghorn include springs, streams, lakes, rivers, reservoirs, water catchments and potholes filled with rain. The animals seek water sources lacking thick, tall surrounding vegetation that might harbor enemies. Here, on a hot summer day, a band visits a small reservoir in northwestern Nevada. Photo by the U.S. Bureau of Biological Survey; courtesy of the National Archives, Washington, D.C.

horn were adversely stressed when snow or free water was not available during winters (Guenzel et al. 1982, Cook 1984). To date, studies have been fairly consistent in amount of water used by pronghorn (Table 78).

Pronghorn have adapted to survive periods of water deprivation; however, after a period of dehydration and the return of water availability, they rapidly rehydrate and regain vigor. Whisler (1984:56) summed up pronghorn water requirements: "Without water conserving mechanisms, pronghorn would rapidly lose water, become severely dehydrated, and die."

Northern pronghorn drink frequently when water is available (Sundstrom 1968, Beale and Smith 1970, Yoakum 1994). When deprived of water, they exhibit stress, affecting both health and reproduction (Beale and Smith 1970, Whisler 1984). However, pronghorn in southern hot deserts have evolved with little or no permanent drinking water, and have adapted to low quantities of and infrequent access to drinking water (Phelps 1978, Arizona Game and Fish Department 1981, Cancino 1994c, Yoakum 1994). Managers responsible for developing drinking water sources would do well to consider pronghorn water intake rates of about 1 gallon (3.8 l) per day per adult in summer (Whisler 1984) and approximately 25% of that for winter when free water (including snow) is not available for herds accustomed to drinking (Yoakum 1994). See Chapter 17 for information on developing drinking water sources.

Scant information exists regarding the quality of drinking water used by pronghorn. Einarsen (1938) wrote that pronghorn were not particular about the quality of water consumed—a muddy waterhole being used as freely as clear spring water. However, Sundstrom (1968) found that, in the Red Desert of Wyoming, pronghorn used little or no water containing total dissolved solids (TDS) in excess of 5,000 mg/L. McKee and Wolf (1963) reported maximum TDS for the best support of wildlife was about 4,500 parts per million. They further indicated that the continuous use of waters with high TDS may cause general loss of condition, weakness, scouring, reduced milk production, bone degeneration and even death. Animals can adjust gradually to drinking water with higher TDS concentrations than normally, but sudden changes from slight to heavily mineralized waters causes acute distress and diarrhea. The limits of tolerance depend on the kinds of minerals present, the species of animals, their diet, age, physiological condition, season of year, climate, and other conditions.

Water is classified as a nutrient for pronghorn (Robbins 1983) and is a needed habitat component to meet physiological requirements. Photo by Judd Cooney.

Sundstrom (1968) reported that pronghorn sought other drinking sources when or where water exceeded a pH of 9.25. The recommended pH range for most uses, such as domestic water supplies, irrigation, fish and other aquatic life, and swimming and other recreational uses is 6.5 to 8.5 (McKee and Wolf 1963).

Cover

Cover for pronghorn can be defined as topographic or vegetative features of the landscape used for protection

TABLE 78. Summer water intake for pronghorn, based on field and laboratory investigations

Location	*Study site*	*Study animals*	*Daily calculated water intake*	*Source*
Wamsutter, Wyoming	Natural habitat enclosure	Mixed herd of adults and fawns	1.0 gallon (3.8 L) per adult	Sundstrom (1968)
Desert Range Experiment Station, Utah	Natural habitat enclosure	Mixed herd of adults adults and fawns	0.75 gallon (2.8 L) per 100 pounds (45 kg)	Beale and Smith (1970)
Colorado State University, Fort Collins	Indoor laboratory cages	Four short-yearlings (two males, two two females)	0.9 gallon (3.5 L) per 100 pounds (45 kg)	Wesley et al. (1970)
Sybille Experiment Unit, Wyoming	Outdoor laboratory pens	Adult females	0.95 gallon (3.6 L) per 100 pounds (45 kg)	Whisler (1984)

from weather (i.e., thermal cover) or from enemies (i.e., security cover).

Thermal Cover. Pronghorn use thermal cover as shelter from discomforts of weather. On cold, windy days with hail or heavy snow, for example, pronghorn frequently are found in swales where wind speed and chill factors are reduced (Rouse 1959).

Guenzel (1986) considered topography important thermal cover on pronghorn winter rangelands. He developed a snow capacity index (SCI) and used it to analyze the Red Rim area in south-central Wyoming. This index incorporated topographic slopes, deviation of ridges from prevailing winds and optimal snowdrift orientation. The SCI measured the ability of topography to create snowdrifts, providing wind protection in lee areas and reducing snow depths downwind. Snow capacity indexes were used to test pronghorn winter rangelands, and results disclosed that areas most used by pronghorn had a significantly higher mean SCI than adjacent unused sites (P = 0.0003).

Use of trees and tall shrubs as thermal cover during hot seasons has been reported for pronghorn (Smith and Beale 1980, Hughes and Smith 1990, Ockenfels et al. 1994), especially in southern climes. Adult pronghorn have bedded under ponderosa pine, juniper, mountain mahogany, palo verdi, yucca and tall sagebrush. Buechner (1950a) remarked that pronghorn eagerly used shade for cover when it was available. However, most pronghorn do not have access to shade and do not seem to suffer from the lack of it, even during periods of high temperatures. Smith and Beale (1980) believed that fawns selected bedsites offering shrub and terrain protection from sun during hot days and from wind or wind-driven precipitation during storms.

Security Cover. According to Bromley (1977), pronghorn fawn bedding site selection involves two major criteria: depressions in the terrain and nearness to plants taller than average for the vegetative community. Prevalence of sites near shrubs was noted by Pyrah (1974) and Autenrieth (1976). Fawns bedded on or near bare ground, dead and matted vegetation, bison dung, low rocks and tire ruts (Bromley 1977, Barrett 1981). Fawns that lie near clumps of vegetation and other objects camouflage themselves against predator detection from both horizontal and vertical vantage points. The grayish-brown fawn coat blends with innumerable, randomly shaped and similarly colored bare spots in the habitat, making detection by aerial predators difficult. Security cover apparently involves interaction of predator avoidance and shelter (Bromley 1977).

Autenrieth and Fichter (1975) postulated that pronghorn fawns may imprint on vegetative cover at the birthsite

Pronghorn habitat is most often flat or low rolling terrain. However, pictured here are two adult bucks browsing sagebrush on the foothills of a mountain range in northwestern Colorado. Such sites attract pronghorn because they sustain preferred forage during summer and autumn and provide thermal cover and shrubs during winter. Photo by Tom Pojar.

Most pronghorn habitats do not have trees or other natural vertical structures to provide shade. However, the animals use shade where available during hot weather. Top left photo by Amber Alexander; courtesy of the Arizona Game and Fish Department, Phoenix. Top right photo by Robb Hitchock. Bottom photo by Rich Ockenfels.

and select subsequent bedding sites with similar vegetation. Because fawns do not travel far from birthing sites for the first few days after parturition, the vegetation selected by the does apparently determines security cover for neonates (Table 79). Autenrieth and Fichter also speculated that does select tall versus low vegetation, thereby providing the greatest concealment from predators. On the other hand, Smith and Beale (1980) often witnessed fawns passing tall cover and selecting comparatively open sites with sparse, low plant cover. Perhaps, the animals actually were selecting for factors other than just height of vegetation.

Early fawn/habitat investigations described fawn bedding site characteristics generally (Einarsen 1948, Hansen 1955, Yoakum 1957a, Hoover et al. 1959, Ingold 1969). Pyrah (1974) and Autenrieth (1976) reported these characteristics in detail for vegetation, as did others subsequently (Table 80).

Major differences in security cover at fawn bedding sites has been one of the most investigated aspects of pronghorn/habitat relationships during the past 30 years. Fawn recruitment probably is the most important factor presently influencing pronghorn herd sustainability. Se-

Two fawns on a Texas shortgrass site with walking-stick cacti illustrate the benefits of vertical cover in locating neonates. The fawn under the tall cacti is partially camouflaged. The other fawn, in the adjacent shortgrass, is highly vulnerable to detection, especially by aerial predators. Photo by Dave Simpson.

Table 79. Vegetation at pronghorn fawning areas

Biome	*Location*	*Fawning site vegetation*	*Source*
Shrubsteppe	Oregon	Low shrub, herbaceous	Einarsen (1948), Yoakum (1957)
		Tall and low shrub, herbaceous	Trainer et al. (1983)
	Wyoming	Tall and low shrub, herbaceous	Ingold (1969)
			Alldredge et al. (1991)
	Idaho	Tall and low shrub, herbaceous	Autenrieth (1976, 1982, 1984)
			Bodie (1978)
	Utah	Tall and low shrub, herbaceous	Beale and Smith (1980)
	Montana	Low shrub, herbaceous	Pyrah (1974, 1987)
	Nevada	Low shrub, herbaceous	McNay (1980)
Grassland	Alberta	Bare ground, mixed grassland	Barrett (1981)
	Texas	Low herbaceous	Tucker and Garner (1983)
		Short grass, forb	Canon and Bryant (1997)
	Montana, South Dakota	Bare ground, mixed grassland	Bromley (1977)
	Colorado	Low shrub, herbaceous	Van Schmus (1990)
	Kansas	Tall grass, forb	Rothchild et al. (1994)
	Arizona	Low shrub, grassland	Ticer and Miller (1994)

TABLE 80. Characteristics of vegetation at pronghorn fawn bedding areas in shrubsteppe and grassland habitats

Biome	*Location*	*Percentage of cover type*			*Height in inches (cm)*	*Source*
		Shrub	*Grass*	*Forb*		
Shrubsteppe	Montana	21	32	14	12 (30.5)	Pyrah (1974)
	Idaho	21	—	16[a]	16 (40.6)	Autenrieth (1976)
	Oregon	17	11	8	15 (38.1)	Trainer et al. (1983)
Grassland	Alberta	3	66	3	10 (25.4)	Barrett (1982)
	Texas	3	16	3	22 (55.9)	Canon and Bryant (1997)
	Arizona	8	13	7	16 (40.6)	Ticer and Miller (1994)

[a] Grasses and forbs data combined.

curity cover is an important element of that factor. Most studies substantiate that predation is the main cause of fawn mortality (Vriend and Barrett 1978, Menzel 1992, Rothchild et al. 1994, Canon and Bryant 1997). However, whether predation is the primary or secondary causative agent for some cases of low fawn survival remains a question, that is, does with nutritional or disease stress may produce fawns especially susceptible to predation (Winner 1996), or heavy grazing by livestock may decrease security cover, making neonates more vulnerable to detection by predators (Autenrieth 1984).

Security cover generally has been associated with shrubs on shrubsteppes (Pyrah 1974, Bodie 1978, McNay 1980, Autenrieth 1982, Barrett 1982, Trainer et al. 1983, Alldredge et al. 1991), as well as on grasslands (Bromley 1977, Barrett 1982, Tucker and Garner 1983, Rothchild et al. 1994, Ticer and Miller 1994, Canon and Bryant 1997) (see Tables 79 and 80). Studies of the shrubsteppe region in Montana indicated that reduced shrub volume was correlated with reduced fawn survival (Pyrah 1974, 1987). Autenrieth (1976, 1982, 1984) concurred that insufficient shrub cover was related to high predation of fawns and suggested a mean of 40% canopy cover to decrease predation on shrubsteppes. However, Bodie (1978) studied fawn survival in the same region as Autenrieth and reported that fawn mortality from predation was much higher in tall sagebrush than in low sagebrush communities.

Grasslands sustain few shrubs yet produce the highest fawn survival rate per biome (Vriend and Barrett 1978). Fawn survival was greater where vegetation was sufficient to conceal neonates from predators (Neff and Woosley 1979, Tucker and Garner 1983, Ticer and Miller 1994, Canon and Bryant 1997). Canon and Bryant found that does selected for long-range visibility when choosing birthsites, whereas fawns' selection of bedding sites seemed to be based on individual microhabitat characteristics. On the prairies of Alberta, Barrett (1982) analyzed 404 fawn bedsites and classified sites according to vegetation types (Table 81). He concluded that, if pronghorn does preferred certain habitat types in which to give birth and raise fawns, they were not readily evident for this grassland ecosystem.

TABLE 81. Comparison of vegetative types and cover at bedsites of 392 pronghorn fawns in Alberta

Vegetative type	*Percentage of bedsites at capture*	*Percentage of available cover*
Native grass prairie	41	41
Native grass prairie—light sage[b]	33	41
Native grass prairie—medium and heavy sage[b]	10	7
Cultivation (all types)	14	10
Other	2	1

[a] Study site was a 392-square mile (1,016 km^2) block that was representative of the entire area.

[b] Light, medium and heavy sage categories defined as 3–10, 11–25 and 25+ plants per 200-foot (61 m) linear transect, respectively.

Source: Barrett 1978.

Security cover requirements for pronghorn vary from one habitat to another, sometimes dramatically. Dasmann's (1964) concept—that too little or too much of any environmental component can be a limiting factor affecting a species' survival or occupancy of a habitat—possibly applies in this respect. However, pronghorn fawning areas with limited herbaceous plants or shrub cover afford insufficient cover to conceal neonates effectively from predation. Conversely, an overabundance of shrubs can provide more stalking cover for predators, thereby increasing fawn losses to predators (Bodie 1978). Security cover requirements for pronghorn on grasslands are not the same on shrubsteppes. Care should be exercised in applying cover standards in one area to other habitats.

Space

The amount of landscape needed to sustain a viable pronghorn herd depends on the quantity, quality and juxtaposition of habitat characteristics during all seasons. These components must be within the animals' daily range or cruising radius (usually < 2 miles [3.2 km]) and home range (including corridors of movement to seasonal habitat) (Buechner 1950a). Three areas within sight of Hart Mountain in south-central Oregon exemplify how rangelands can vary in these respects.

First is an area of 10 square miles (26 km²) in Warner Valley. The site has sufficient forage and water year-round, but produces limited preferred forage in some years. Snow depths rarely, if ever, exceeded 3 inches (7.6 cm). This area has been occupied by a small resident herd of about 20 animals for many years.

Second is Drakes Flat, a 50-square-mile (128 km²) area. Food and water are abundant all year. Pronghorn use the higher elevations where preferred forage is plentiful during 80% to 90% of the year. They use the lower elevations when deep snow covers forage plants at higher elevations.

The third example involves separated summer and winter rangelands. Travel between the two is related to snow—the deeper the snow, the farther the herd travels to lower elevations with less snow. The herd there may travel as far as 90 miles (145 km) to and from winter rangelands. These travels are not true migrations, such as those of caribou, because the pronghorn herd movements vary annually and correlate to snow depth.

Pronghorn were not restricted by fences, highways, agricultural areas and cities 150 years ago and, presumably, roamed widely. Today, pronghorn herds are primarily resident populations (Einarsen 1948). The home range of most is within a 25-square mile (65 km²) area. This is especially true for populations in arid regions. Northern herds subject to prolonged and deep snows may travel 50 to 200 miles (80–322 km) to winter rangelands with less snow and more accessible forage (Riddle and Oakley 1973).

For translocating pronghorn in Colorado, Hoover et al. (1959) recommended grassland sites that allowed at least 1 square mile (2.6 km²) of grassland for each animal, and that would support 100 animals in a 10-square mile (26 km²) area. The Texas Game and Fish Commission recommended a minimum area of 10,000 acres (4,047 ha) for restocking (Buechner 1950a).

Critical Habitat

Critical (or key) habitats are areas necessary to sustain a population during severe conditions (Bailey 1984). These areas may be used only seasonally, and they vary in size and location according to topography, land management uses and practices, climate and vegetative conditions. Some rangelands may have several key habitats, whereas others have none. Salwasser (1980) identified winter rangelands, movement corridors and fawning sites as crucial areas of pronghorn habitat and illustrated the relationship of these seasonal-use areas (Figure 78). He further indicated that each area should be identified for future habitat management plans.

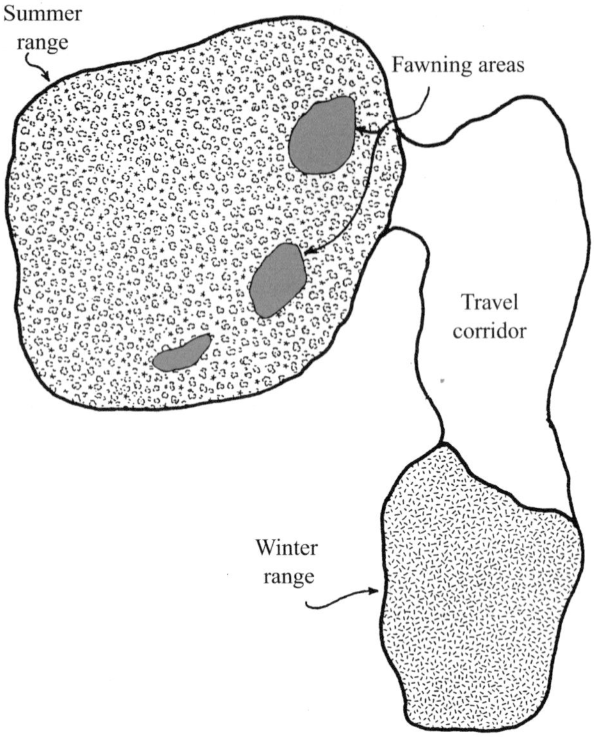

Figure 78. *Designated seasonal use areas, including critical habitats, for pronghorn (Salwasser 1980).*

Winter Rangelands. Pronghorn in the northern half of the species' range periodically endure deep snow, high wind and low temperature. Unless blocked, the animals invariably move to lower elevations, windblown ridges and other areas with less snow and more available forage (Martinka 1967, Bayless 1969, Bruns 1969, Compton 1970, Sundstrom 1970, West 1970, Oakley and Riddle 1974, Taylor 1975, Guenzel et al. 1982). The primary criteria is snow depth and condition. When snow cover reaches 6 inches (15 cm), the availability of preferred forage is diminished. When snow depth exceeds 10 to 12 inches (25.4–30.5 cm), and in the absence of escape to areas of more accessible food, pronghorn health becomes dire. At such times, sagebrush and other shrubs play a paramount role in survival because they usually are the main or only forage not buried. Prolonged periods of deep snow (more than 3 weeks), coupled with cold winds, produce chill factors that are debilitating and sometimes fatal (see Bayless 1969).

During severe winters, such as in 1951–52, pronghorn on summer rangelands on the Hart Mountain Antelope Refuge travel 50 to 100 miles (80–161 km) to winter rangelands in Catlow Valley, Oregon, and the Sheldon National Wildlife Refuge in Nevada. How far the animals travel depends on snow depth. Generally, herds travel little unless snow covers forage and forces them to lower elevations. During winter in some areas, pronghorn may congregate in herds of more than 200. Historically, herds of 1,000 or more were reported.

Pronghorn of Wyoming's Red Desert have been one of the most intensively studied populations (Sundstrom 1970, Oakley and Riddle 1974, Taylor 1975, Deblinger et al. 1984, Guenzel 1986). Herds there often travel 75 to 200 miles (121–322 km) seasonally. They summer on high plateaus at 6,000 to 7,000 feet (1,829–2,134 m) elevation. When winter arrives, they move down to the Red Desert at elevations of 4,000 to 5,000 feet (1,219–1,524 m). Sundstrom (1970) studied these herds for 6 years. He surmised that winter concentration areas were essential components of their home range and loss of those critical areas would have a direct and significantly deleterious effect on the Red Desert population.

Travel Corridors. Travel corridors are routes pronghorn travel between summer and winter rangelands. They tend to be used year after year. Corridors still have not been identified for some herds, but telemetry has helped locate others in recent years (Ockenfels et al. 1994).

Travel corridors also tend to have vegetative and physiographic conditions attractive to pronghorn, but some are quite different (e.g., through tall, dense shrubs and trees, through canyons, and across mountains, rivers and lakes) and serve only as necessary pathways of convenience that are traversed quickly.

Corridors can change in part or altogether because of weather or altered vegetative conditions. After the extensive 1988 wildfires in Yellowstone National Park, pronghorn moved to summer rangeland unoccupied for more than 50 years. Approximately 20% of the pronghorn in Yellowstone National Park travel to and from summer rangeland. The other 80% remain year-round in the Gardner Hole area. Corridors are fairly lengthy, with some at least 50 miles (80 km) long. It is not known whether these pronghorn typically or traditionally use the same routes year after year because some animals appear on different summer rangelands in different years (D. Scott personal communication:1989). Within a decade after the 1988 wildfires, pronghorn were using burned over areas—previously dominated by lodgepole pine—that were open enough to see and flee but also provided new growth forage (B. O'Gara personal communication:1998).

Fawning Areas. The *Pronghorn Management Guides* list fawning areas as critical (key) habitats needing delineation on maps to aid in habitat management objectives and plans (Lee et al. 1998).

Criteria for fawning areas include physiography, height of vegetation (Einarsen 1948), availability of herbaceous forage (Ellis 1970), seclusion, vegetative cover, snow depth (Autenrieth 1984), lack of fences (O'Gara et al. 1986), proximity to water (Ockenfels et al. 1992) and freedom from harassment by livestock (McNay 1980) and humans (U.S. Fish and Wildlife Service 1994).

After more than a decade of studying pronghorn on the Hart Mountain Antelope Refuge in Oregon, Einarsen (1948) reported that two factors determined the choice of fawning areas for that shrubsteppe area—terrain and height of vegetation. The most satisfactory physiography was a basin surrounded by low ridges or hills and not intersected with deep gullies or crooked ravines. Birthing took place in the bowl area, and the does spent most of their time on surrounding terrain, foraging and watching for predators. The favored height of vegetation was 9 to 18 inches (23–46 cm), which afforded security cover for fawns and still allowed parturient does open vistas to watch for danger.

From 1953 to 1958, Einarsen and assistants studied fawning activities on traditional fawning areas in south-central Oregon (Hansen 1955, Yoakum 1957a, Compton 1958). Those areas were inventoried and mapped for management and future studies. During 1996, the areas were visited for another fawning study. Fawns were readily located on the traditional areas, establishing that these critical areas had been fawning areas for more than 60 years.

An aerial survey of isolated does was conducted on the Hart Mountain Antelope Refuge in Oregon during the last week of May 1993—the peak of a month-long parturition season (Pyle and Reiswig 1994). In general, distribution of does corresponded to the presence of the low sagebrush type that dominated the shrubsteppe. Doe densities appeared highest in areas with an interspersion of low sagebrush, intermittent drainages and lakebed playas. The winter and spring of 1993 produced 104% of mean annual precipitation, resulting in high production of forbs and grasses, which possibly contributed to the wide distribution of does. Also, the population was larger than during the 1953 to 1958 studies, therefore a wider distribution would be expected. Sixty-eight percent of does were concentrated at two fawning areas that comprised about 25% of the combined total fawning area. The concentration areas were identified in the refuge's comprehensive management plan as sites restricted to human access during the parturition period (U.S. Fish and Wildlife Service 1994). These two traditional fawning areas were the same locations where Einarsen conducted his fawning studies 60 years before.

For the Drakes Flat pronghorn herd in south-central Oregon, the primary fawning area was approximately 3 square miles (7.68 km^2) (Yoakum 1957a). This shrubsteppe site—featuring meadows, grasslands (burned former low sagebrush and seeded to a mixture of grasses), tall shrub,

Productive pronghorn fawning areas contain adequate plant cover for security. Studies have shown that sites with scarce herbaceous cover have a higher rate of pronghorn fawn mortality than do sites with more abundant grasses and forbs. Nevertheless, thousands of fawns are born each year in grasslands with little to no shrub cover. These areas need to have residual cover (vegetation from the previous years' growth) as well as new herbaceous plant production. The combination of old and new vegetation helps conceal fawns and aids in increasing herd recruitment. Top photo by William L. Finley; courtesy of the Wildlife Management Institute. Center photo by Mark McNay. Bottom photo by William L. Van Allen.

and mixed low and tall shrub, and bisected by two perennial streams—was studied for 4 consecutive years, during which more than 300 fawn captures (some multiples) were made. Eighty-four percent of all fawns were located in an area comprising 40% of the total fawning area. Eighty-one percent of fawns were discovered in grasslands and low shrublands; no fawns were located in meadows or tall shrub sites.

The wintering grounds were approximately 900 feet (274 m) lower in elevation at one end of the range. Herds moved up and down the elevational gradient in response to snow depth and the growth of herbaceous plants. The principal fawning area was located in the heart of the spring through autumn range (Figure 79) and at an elevation with abundant forage during the mid-May to mid-June parturition period. The site was visited by researchers over the course of 42 years, and fawning was consistently observed.

Some pronghorn rangelands may not have traditional fawning areas. After studying fawning activities for more than a decade in Idaho, Autenrieth (1984) suggested that pronghorn fawning areas were more closely associated with vegetative cover characteristics than with geographic locations.

Pronghorn studies in grassland habitats generally have not identified traditional fawning areas (Barrett 1981, Tucker and Garner 1983, Canon and Bryant 1997, Rothchild

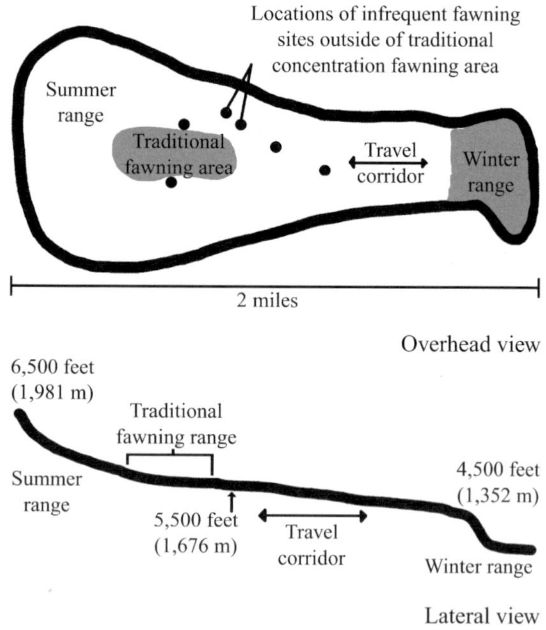

Figure 79. *Location of traditional pronghorn fawning areas in relation to summer and winter rangelands—occupied by some 200 to 500 pronghorn—on Drakes Flat, Oregon, an area of about 1.5 by 2 miles (2.4 by 3.2 km).*

et al. 1994, Ticer and Miller 1994). Possibly herds on some grasslands undertake limited seasonal movements from winter to summer rangelands and tend to occupy areas with limited elevational gradients. Does in populations that have little or negligible seasonal movement obviously have small home ranges, therefore their long-term fidelity to traditional fawning areas may appear to investigators to be more a matter of circumstance than choice.

Generally, pronghorn have adapted fawning area selection to different environments. Herds that typically experience mild winters and are not forced to travel much between seasonal rangelands tend to fawn throughout their small home ranges. The herds that make relatively short seasonal movements tend to select sites that produce abundant succulent forage during the parturition period. In comparison, herds that are required to travel long distances between summer and winter rangelands, especially rangelands at significantly different elevations, tend to have traditional fawning areas near or on summer rangelands.

Endangered Species Habitat. Rangelands occupied by rare and endangered pronghorn should be managed as critical habitat. Sonoran and peninsular pronghorn are officially listed as endangered (International Union for Conservation of Nature 1996), and their habitats should be considered endangered as well. The habitats of endangered pronghorn subspecies have deteriorated, and there appears to be little need to protect the species if such habitat deterioration and loss cannot be stemmed.

Habitat for the Sonoran pronghorn is an area approximately 10,500 square miles (26,992 km^2) in Arizona and Sonora, Mexico (Arizona Department Game and Fish Department 1981). Natural water sources have been modified or lost since the Southwest was settled by Spaniards more than two centuries ago. Some artificial water developments have been installed, but they are point sources, not flowing streams, and apparently are not used by pronghorn (Cutler 1996). Primarily because of livestock grazing, vegetative composition has changed, with increased shrub cover and decreased herbaceous plant production, to the detriment of Sonoran pronghorn (see also Phelps 1978, Wright and deVos 1986, Hughes and Smith 1990). Fox (1997) determined that some sites within Sonoran pronghorn habitat lack sufficient nutrients and water to sustain healthy populations.

Analysis of fecal samples collected at various sites and seasons from 1974 to 1977 on the Sonoran Desert indicated that pronghorn diets averaged about 75% forbs, 22% shrubs and less than 1% grasses (U.S. Fish and Wildlife Service 1998a). Forb percentages included 7% cacti, which were consumed for nutrients and moisture. No yearlong diet selection study has been reported for any site. A study by Fox (1997), using plant collections and computer models, was conducted to determine whether pronghorn could meet water and mineral requirements through forage consumption on the Cabeza Prieta National Wildlife Refuge. Some sites met these criteria, whereas other did not (see Chapter 14). Further studies on diet selection, nutrient quality and moisture content of forage plants are needed as part of a long-term study of food habits on the Sonoran Desert.

The *Sonoran Pronghorn Recovery Plan* has a main objective of returning the species' range to its natural ecological status (Thompson-Olais 1992). The endeavor entails restoration of pristine water sources and vegetative conditions. It also addresses the issue of the international boundary fence between Mexico and the United States. Pronghorn have become entangled and died in that fence. The numbers reported are few, but for a small, endangered population, they may be significant. The challenges are greatest for Sonoran pronghorn habitat not under federal jurisdiction in the United States.

The peninsular pronghorn lives within an area of Baja California Sur, Mexico, only 2,700 square miles (689,384 km^2) (Cancino et al. 1994c). The population is the only herd presently living at sea level. It also is unique in that it occupies a landscape without permanent drinking water sources and apparently obtains moisture primarily from dew on vegetation. Through the Biosphere Reserve, El Vizcaino, the currently occupied habitat is reserved and dedicated to perpetuation of the endemic peninsular pronghorn (Diario Offical 1988). Efforts are underway to develop and implement a recovery plan similar to that in Arizona for the Sonoran pronghorn.

HABITAT SUITABILITY

Suitable habitat for pronghorn can be assessed through a system of rating the amount and juxtaposition of habitat characteristics meeting the species' habitat requirements (Lee et al. 1998). Too little or too much of any biotic or abiotic factor may limit production and survival. Knowledge of these relationships and their rating criteria becomes the ecological foundation for making many management decisions.

Taylor (1975) regarded suitability as the adaptability of pronghorn to a given area. He proposed that the greater the consumption overlap of preferred and available highly nutritious forage, the more suited an area was for pronghorn.

Ockenfels et al. (1996) related habitat suitability in Arizona to areas: (1) capable of supporting numerous individuals, about 10 adults per square mile (3.9/km^2); (2) with conditions favorable for an adult pronghorn annual survival rate of more than 80%; (3) with conditions favorable for an annual recruitment of more than 40 fawns per 100 does; and (4) with the aforementioned conditions stable for 5 years or longer. Unsuitable habitat likely would meet few or none of those criteria. Also, degrees of suitability exist, and a rating system or model was expected to reflect this, rather than merely classify an area as suitable or unsuitable.

Evaluating habitat quality has been a challenge ever since wildlife management became a science. At the outset, Leopold (1933:124) wrote: "When the game manager asks himself whether a given piece of land is suitable for a given species of game, he must realize that he is asking no simple question, but rather he is facing one of the great enigmas of animate nature. An answer good enough for practical purpose is usually easy to get by the simple process of noting whether the species is there already, or whether it occurs on 'similar' range nearby. But let him not be cocksure about what is 'similar' for this invokes the deeper questions of why a species occurs in one place and not another, which is probably the same as why it persists at all." Answering Leopold's question of distribution is the basis for suitability guides and assessment systems for habitat evaluation.

Habitat Assessment Criteria

Wildlife habitat models help synthesize knowledge of habitat components and systematically apply information toward goals (Cooperrider et al. 1986, Salwasser 1986, Anderson and Gutzwiller 1994). Most models for assessment of pronghorn habitat have been developed for the shrubsteppe biome, fewer have been developed for the grasslands and none for the desert (Table 82). Some are extensive and indicate whether landscapes are occupied, but the majority are site-specific, providing data on the feasibility of translocating herds, the condition of winter rangelands and the suitability ratings for occupied rangelands.

When Leopold (1933:124) referred to a biologist determining "similar" habitats, he introduced the concept of a habitat model and described the reasons for conducting habitat inventories and monitoring programs (Cooperrider et al. 1986). Today, habitat models frequently are used, but many are based primarily on professional judgment and lack quantified data. Also, biologists increasingly are urged or required to make more explicit models relative to controversial habitat alterations. This has resulted in the greater use of quantitative habitat models, which not only provide a mechanism for predictions but also take some of the mystery out of professional judgments. The utility of models should be tested over a long period on a large scale. They also can be tested and refined relative to specific management actions (Cooperrider et al. 1986).

Models generally are conceptual or quantitative (Cooperrider et al. 1986, Patten 1992). Conceptual models represent thoughts and ideas in a qualitative way instead of by numbers. Examples of some of the earliest conceptual models for pronghorn are provided in Hoover et al. (1959) and Yoakum (1974). Conceptual models are the forerunners of more sophisticated quantitative models, which are based on ideas and relationships contained in conceptual models but augmented by numeric data. Quantitative models often predict a future event by using mea-

Table 82. Pronghorn habitat assessment models

	Biome		*Model characteristics*					
Location	*Grassland*	*Shrubsteppe*	*Extensive*	*Intensive*	*Conceptual*	*Quantitative*	*Major objective*	*Source*
Colorado	X	—	—	X	—	—	Herd translocations	Hoover et al. (1959)
Great Basin	—	X	X	—	—	—	Suitability	Yoakum (1974, 1980)
Great Basin	—	X	X	—	X	X	Suitability	Kindschy et al. (1978, 1982)
Idaho, Montana, Wyoming, Colorado	X	X	—	X	X	X	Winter range	Allen and Armbruster (1982), Allen et al. (1984)
Nevada	—	X	—	X	X	X	Suitability	U.S. Bureau of Land Management (1980)
California	—	X	—	X	X	X	Herd translocations	McCarthy and Yoakum (1984)
Arizona	X	X	—	X	X	X	Habitat quality	U.S. Soil Conservation Service (1989)
New Mexico	X	—	—	X	X	X	Sites with sheep	Howard et al. (1990)
Arizona	X	X	—	X	X	X	Herd translocations	Arizona Department of Game and Fish (1993)
Arizona	X	X	X	X	X	X	State survey	Ockenfels et al. (1996)

Habitat models help provide data for assessment of environmental characteristics, especially quantity and quality of vegetation needed by pronghorn for forage and cover for neonates. This requires monitoring studies of the plant community. These investigators are reading vegetation hits on a transect in a low sagebrush type. Data findings will help evaluate whether the plant type has the right kinds and quantities of vegetation needed by pronghorn. Photo by Jim D. Yoakum.

surable sets of environmental factors and relationships. Habitat models can be extensive or intensive (Allen et al. 1984, Morrison et al. 1992, Ockenfels et al. 1996).

Extensive Quantitative Models. Extensive models may be used to differentiate occupied from unoccupied environments on a broad geographic scale, but they sometimes fail when applied to track density on a more site- or population-specific scale (Irwin and Cook 1985). Thus, the question of extensive versus intensive models often involves different modeling objectives and techniques.

One of the most extensive pronghorn habitat models was developed in Arizona by Ockenfels et al. (1996). The major objective of the modeling project was to rate pronghorn habitat suitability statewide. However, the criteria and components used were primarily applicable for grasslands and shrubsteppe (Table 83)—not for desert. The evaluation system rated both occupied and unoccupied habitats. Pronghorn had been extirpated from some areas within the past 200 years. The evaluation included review of the prior pronghorn habitat models and borrowed components that best met objectives of the Arizona survey. A field test resulted in minor refinements. Upon completion of the landscape-level model, the Arizona Game and Fish Department completed a 2-year survey with the model (Ockenfels et al. 1996) (Table 84).

TABLE 83. Major evaluation criteria, attributes and limitations of pronghorn habitat assessment models[a]

Evaluation criterion	*Attributes and limitations*	*Source*
Minimum number of animals and release acres	Guidelines for assessing sites for herd translocations	Hoover et al. (1959)
Variety of forbs and shrubs	Originally for grasslands	
Ecological condition	Modified by Yoakum (1980) for shrubsteppes	
Livestock competition	Included in *Pronghorn Management Guides* (O'Gara and Yoakum 1992)	
Fencing structure and density		
Predator control		
Depredation of agriculture crops		
Landownership pattern		
Abiotic	An early conceptual model for shrubsteppes, especially in the Great Basin	Yoakum (1974, 1980)
Physiography	Habitat characteristics and requirements included	
Climate	Numeric system for rating quality lacking	
Barriers (natural and artificial)	Stressed vegetation relative to sight and flight	
Water	Some parts subjective	
Biotic		
Vegetation height, composition, production		
Wildlife—ungulates and predators		
Human impacts		
Distance to water	Developed for Great Basin ecosystems	Kindschy et al. (1978, 1982)
Percentage ground cover for forbs	Used numeric system for seven variables	
Percentage ground cover for shrubs	Winter ranges rated	
Vegetative height	Could be modified for GIS use	
Vegetative succulence	Easy to evaluate, but subjective for some variables	

continued on next page

Table 83—*continued*

Evaluation criterion	*Attributes and limitations*	*Source*
Fence structure		
Percentage slope		
Composition of grass/forb/shrub	Developed for Nevada, but applicable for other shrubsteppe regions	U.S. Bureau of Land Management (1980)
Vegetative production	Rates suitability with numeric system	
Vegetative diversity for grass/forb/shrub	Can be modified to use in GIS	
Water availability	Simple field application	
Water quality	Used for winter range, site disturbances and suitable rangelands for pronghorn	
Limiting factors (snow depth, fences, human impacts)		
Shrub crown cover	Developed for grassland and shrubsteppe	Allen et al. (1984)
Shrub height	Rated winter rangelands only	
Number of shrub species	Tested in 28 sites, mainly shrubsteppe	
Herbaceous canopy cover	Also validated population parameters	
Topographic diversity	Used winter wheat—a criterion limited to use for grasslands	
Winter wheat		
Water distribution	Developed to evaluated and prioritize herd translocations in the Great Basin	McCarthy and Yoakum (1984)
Forage quality		
Forage quantity	Habitat quality has a numerical rating	
Vegetative height	Evaluated each vegetation type	
Snow depth	Model used by an interagency team	
Natural barriers		
Size of area		
Fence design		
Predation potential		
Crop depredation		
Seasonal quality		
Livestock competition		
Plant communities	An unpublished model (*in* Ockenfels et al. 1996) developed for Arizona	U.S. Soil Conservation Service (1989)
Landscape		
Vegetative cover for fawn bedsites	Provided for a numeric rating system	
Rangeland condition	Sparse narrative to explain rating of variables	
Utilization of forbs and shrubs	Could be modified for use with GIS	
Livestock grazing management	Some variables appear redundant	
Vegetative composition		
Water availability		
Index of ruggedness	Developed for fenced pastures used for domestic sheep containment in New Mexico	Howard et al. 1990
Pasture size		
Average number of forb species in autumn	Suitability index tied to numbers of breeding does	
Average number of domestic sheep in autumn	Recommended a minimum size of 5 square miles (13 km^2) per pasture to sustain pronghorn yearlong	
Historic occurrence of pronghorn	Developed in 1980s for herd translocations	Arizona Game and Fish Department (1993)
Land status	Has numeric system	
Topography	Total score used to prioritize different sites	
Vegetative cover	Some criteria subjective	
Range condition		
Other ungulates		
Fences		
Seasonal availability of habitats		
Availability of water		
Human disturbance		
Range expansion potential		
Topography (slope and ruggedness)	A new system developed to evaluate used and nonused habitat in Arizona	Ockenfels et al. (1996)
Vegetative height and composition for grass and shrubs, also tree density	Field tested for grasslands and shrubsteppes but apparently not for the desert	
Human developments (roads, railroads, recreation, urban)	GIS compatible	

continued on next page

TABLE 83—*continued*

Evaluation criterion	*Attributes and limitations*	*Source*
Fencing	Simple to score	
Water availability and accessibility	Lacks rating for forbs, which are important for forage and water requirements	
Seasonal use of areas	An extensive model	
Juxtaposition of habitat types	Serves goal for a statewide survey of extant and potential habitat	
Movement corridors		

[a] For persons interested in various criteria for the development of habitat assessment models for pronghorn, two indices pertinent to rating topography are: (1) Snow Capacity Index (Guenzel 1986). Describes importance of topography because of its interaction with snow and wind. Used for rating winter rangelands on the Continental Divide in Wyoming, where severe winters frequently occur. (2) Habitat Ruggedness Index (Beasom et al. 1983). Measures topographic relief, which can be an important variable for certain pronghorn habitats. Used by Howard et al. (1990) to rate pronghorn habitat in New Mexico.

TABLE 84. Ratings for pronghorn habitat variables on unburned and burned sites, Long Valley, California/Nevada, based on criteria in Pronghorn Habitat Suitability Model for the Great Basin Shrubsteppe

Variables	*Unburned*	*Burned*
Vegetative quality		
Forbs	2	15
Grasses	5	5
Shrubs	3	5
Quantity	10	10
Height	5	10
Vegetative density		
Forbs	6	12
Grasses	4	8
Shrubs	4	6
Water		
Availability	10	10
Quality	10	10
Limiting factors	−20	−20
Total	39 (poor)	71 (good)

Source: Yoakum 2000.

According to Ockenfels et al. (1996), the landscape-level model can evaluate habitat suitable for pronghorn, primarily because of: geographic information system compatibility; update capability; independence of season; simplicity and multilevel database creation. These attributes result in a pronghorn habitat survey that can assist in land-use planning over large areas. The model appears valid for numerous vegetative communities and physiographic areas, as well as different types and degrees of human-related disturbances, characteristics essential for an extensive habitat assessment. However, it did not rate hot desert habitats. Because the system lacked detailed assessments of certain components (e.g., forb production, drinking water availability during dry periods), it was recognized that finer resolution would be needed before implementing certain management activities.

For evaluation of habitat suitability, translocation priorities or designing habitat improvement projects, the intensive quantitative models listed in Tables 83 and 84 should be reviewed and considered.

Intensive Quantitative Models. Researchers and managers have developed at least seven intensive quantitative models to evaluate pronghorn habitat. These models may be narrow in scope and specific to a single set of habitat conditions or general and applicable to many sites (Anderson and Gutzwiller 1994). Such models have been used to select translocation sites (McCarthy and Yoakum 1984, Arizona Game and Fish Department 1993), evaluate winter range (Allen et al. 1984), assess compatibility with domestic sheep (Howard et al. 1990) and rate habitat suitability for differing sites (U.S. Bureau of Land Management 1980, Kindschy et al. 1982, U.S. Soil Conservation Service 1989). The models cited above are described in Table 83.

Habitat Suitability Index Model. The *Habitat Suitability Index Model* (HSIM): *Pronghorn* was developed by U.S. Fish and Wildlife Service personnel at Fort Collins, Colorado (Allen and Armbruster 1982, Allen et al. 1984, Cook 1984, Cook et al. 1984, Cook and Irwin 1985, Irwin and Cook 1985). It includes six variables (Table 84) deemed important for pronghorn winter rangelands—the main thrust of the model (Allen et al. 1984). The model converted field values of each of the variables to a 0 to 1 scale, then produced a composite value of all six rescaled values. Environmental data were collected from 28 winter rangelands, corresponding habitat suitability index values were calculated and correlated with known pronghorn densities. The model explained 39% of the variations in pronghorn density. By using linear regression, shrub cover was identified as the most controlling variable. Investigators then restructured the model to maximize correlation, which then increased to 50% (Cook and Irwin 1985). Possibly, the mixture of vegetative criteria (e.g., winter wheat and shrub height and density) was not practical according to two of the authors, L. Irwin and J. Cook (personal communications:1997). If the model is considered for application, its vegetative criteria should be assessed for compatibility with the vegetation being evaluated.

Habitat Suitability Model for Great Basin Shrubsteppe. The habitat suitability model developed and implemented for rating shrubsteppes in the Great Basin (U.S. Bureau of Land Management 1980) has been used for more than 20 years, but has not been readily available. It was used to evaluate winter rangelands, effects of wildfires and the suitability of rangelands for pronghorn in habitat management plans. Originally developed in the late 1970s, the model was used and tested in the field for a decade, then updated and revised. The model is based on pronghorn habitat requirements and has relatively simple procedures that can be applied after examiners are trained. This model should not be used during droughts or unusually wet seasons. Although variables are specific to Great Basin environments, they can be modified to habitat characteristics for other ecosystems. Consequently, procedures for the model are provided here.

The objective of this habitat assessment model is to evaluate habitat suitability for pronghorn on shrubsteppe rangeland in the Great Basin (Table 84). Field examiners are encouraged to study literature on pronghorn habitat models, especially Kindschy et al. (1982), McCarthy and Yoakum (1984) and Lee et al. (1998). In addition, examiners should be trained in modeling procedures before field work. Such training improves subjective ratings, especially for limiting factors.

All rating scores from sections A through G are recorded on a Pronghorn Habitat Suitability Rating form (Figure 80).

A. *Vegetative quality rating*. Determine percentage composition of forbs, grasses and shrubs.
   If total forb composition is:
      0 to 20%, enter 0 to 20 points.
   If total composition of grasses is:
      11% or more, enter 5 points.
      0 to 10%, enter 0 to 5 points.
   If total composition of shrubs is:
      41% or more, enter 5 to 0 points.
      21% to 40%, enter 5 to 10 points, on a sliding scale.
      0 to 20%, enter 0 to 5 points.

B. *Vegetative quantity rating*. Determine the average pounds per acre of vegetative production.
   If it is:
      200 or more pounds per acre (224 kg/ha), enter 5 to 10 points.
      less than 200 pounds per acre, enter 0 to 4 points (0 pounds per acre = 0, 100 = 2, 200 = 4).

C. *Vegetative height rating*. Determine the average height of vegetation.
   If it is:
      11 to 25 inches (27.9–63.5 cm), enter 10 points.
      5 to 10 inches (12.7–25.4 cm), enter 5 points.
      less than 5 inches (12.7 cm) or more than 25 inches (63.5 cm), enter 0 to 5 points; 0 to 5 inches(0–12.8 cm), enter 0 to 5 points; 25 inches (63.5 cm) = 5 points; 36 inches (91.0 cm) = 0 points.

D. *Vegetative diversity rating*. Determine forb, grass and shrub diversity by counting the number of subspecies for each forage class while conducting the vegetative quantity rating.
   If number of forb subspecies is:
      31 or more, enter 11 to 15 points.
      11 to 30, enter 6 to 10 points.
      0 to 10, enter 0 to 5 points.
   If number of grass subspecies is:
      0 or more, enter 0 to 10 points.
   If number of shrubs subspecies is:
      5 or more, enter 5 to 10 points.
      0 to 4, enter 0 to 4 points.

E. *Water availability rating*. Determine the distribution of available drinking water sources.
   If the drinking water distribution pattern is:
      water sources are less than 2 miles (3.2 km) apart, enter 10 points.
      2 to 5 miles (3.2–8.0 km) apart, enter 9 to 5 points.
      6 to 15 miles (9.7–24.0 km) apart, enter 4 to 0 points.

F. *Water quantity rating*. Determine the amount of drinking water available.
   If the volume of drinking water available daily per animal is:
      1 to 4 quarts (0.95–3.8 L), enter 6 to 10 points.
      0 to 0.9 quarts (0–0.85 L), enter 0 to 5 points.

G. *Limiting factors*. While collecting data on vegetation and water, empirically evaluate factors that could limit pronghorn production and survival (see Yoakum (1980) for specific details). These factors are determined at the discretion of trained and experienced evaluators, but where actually or potentially significant, they should be detailed in an accompanying narrative. Rate the following factors by subtracting a total of 0 to 60 points for the entire category G.
   If fences allow pronghorn movements and do not block mobility (especially during periods of snow), give a low rating. If fences are woven-wire or other construction that restrict pronghorn movements, give a higher rating to the extent of restriction. Subtract 0 to 60 points.
   If snow depth exceeds 6 to 12 inches (15.2–30.4 cm) under normal conditions and forage is covered and not available, subtract 0 to 60 points.
   If certain portions of the rangeland (especially crucial areas, e.g., fawning areas, winter rangelands, seasonal movement corridors) have been slightly to severely deteriorated or entirely lost as pronghorn habitat from known prior (including historic) conditions, subtract 0 to 60 points.

H. *Rating summary*. Determine total points from sections A through G to secure a suitability score and rating of 91 and above (excellent), 71 to 90 (good), 41 to 70 (fair) or 0 to 40 (poor).

This suitability model was used to assess the affects of wildfire burns in Long Valley, California (Yoakum 2000). The shrubsteppe site of the Great Basin was pronghorn

United States
Department of the Interior
Bureau of Land Management

Summary

Date: June 12, 1979
Examiner's name: J. Jones
Allotment:
Plot no./name: AY-04
Wildlife season of use: Yearlong
Big game range name: Monitor Valley

A. Vegetation quality rating:	
Forbs (0-20 points)	10
Grasses (0-5 points)	5
Shrubs (0-10 points)	10
B. Vegetation quantity rating (1-10 points)	8
C. Vegetation height rating (0-10 points)	8
D. Vegetation diversity rating:	
Forbs (0-15 points)	10
Grasses (0-10 points)	5
Shrubs (0-10 points)	5
E. Water availability rating (0-10 points)	6
F. Water quantity rating (0-10 points)	7
G. Limiting factors (can subtract 60 points):	
Fences	0
Snow depth	0
Habitat disturbance	0

Total of all points: 74

Pronghorn habitat rating: Good

FIGURE 80. *Summary form for "Habitat Suitability Model for Great Basin Shrubsteppe."*

rangeland historically, but no pronghorn had been observed in the area for more than 60 years before a series of wildfires during the mid-1960s. An evaluation of burned and unburned sites was made 7 years after the fires, resulting in a suitability rating of poor for the unburned sites and good for burned sites (Table 84) (Yoakum 1994). A herd of 6 to 10 pronghorn pioneered the valley a few years after the burn, remained and, in 1997, appeared to

Pronghorn are native to Long Valley, along the California/Nevada border, but were extirpated for more than 60 years, during which time juniper and big sagebrush densities increased. The area became dominated by late succession vegetation—an ecological condition that contributed to an absence of pronghorn, although wild herds ranged nearby habitats. In 1964, a wildfire burned a third of the valley. Some of the burned site was seeded but most of it returned to endemic grasses, forbs and low shrubs. Pronghorn pioneered the valley within 2 years after the fire and the area has since sustained a small herd. Burned and unburned sites were monitored using a habitat suitability model. Results indicated all of the ecological characteristics of a Great Basin environment supportive of pronghorn were present before the 1964 burn, but the vegetation was too tall and thick. The fire changed successional status of the vegetation to meet pronghorn habitat requirements and herds moved into the valley on their own. Photo by Jim D. Yoakum.

number more than 50—about 2 animals per square mile (2.59/km^2) of good habitat. They have not been observed in the adjacent poor-quality sites.

CONCLUSIONS

Roughly two-thirds of North America's pronghorn live on the short, mixed and tall grasslands of the central Great Plains. One-third occupies the shrubsteppe biome, and less than 1% live in desert.

Most pronghorn occupy flat to rolling terrain of less than 10% slopes. Herds occupy elevations from sea level to 11,000 feet (3,353 m). The majority live at 3,000 to 8,000 feet (914–2,438 m) elevation.

Pronghorn population densities are highest on rangelands receiving 8 to 15 inches (20.3–38.1 cm) of annual precipitation. Snow depths exceeding 12 inches (30 cm) can be deleterious. Rangelands with year-round drinking water every 1 to 2 miles (1.6–3.2 km) produce more pronghorn than areas with fewer available drinking water sources.

Rangelands most favored by pronghorn maintain a living vegetation composition of 40% or more and a high diversity of species in each forage class (forbs, grasses and shrubs). Habitats with a variety of vegetative communities are preferred, because they tend to be rich in forage plants. Rangelands with plants averaging 15 inches (38.1 cm) in height are favored over communities more than 30 inches (76.2 cm) tall.

Pronghorn historically foraged with bison, elk, wild sheep and deer and shared rangeland with coyotes, bobcats, wolves, eagles and bears. Herds thrived although predators took high annual tolls (Thwaites 1904, Yoakum 2002). Today, much western rangeland is foraged by livestock. Cattle generally have not competed with pronghorn for forage (except in the Great Basin and desert ecosystems), but competition between pronghorn and domestic sheep is a problem in some areas.

Humans have had the greatest impact on pronghorn welfare. Unregulated gunning in the 1800s decimated herds. Fencing, ranching, farming and other human developments have displaced herds from traditional and critical rangeland. All of these factors have diminished the amount and quality of habitat available to pronghorn and restricted their movements. Despite harvest regulations and management initiatives that have enabled pronghorn populations to rebound dramatically in the past century, there is only 1 pronghorn today where perhaps 30 to 40 roamed in early historic and prehistoric time.

Habitat assessment modeling for pronghorn, practiced for more than 40 years, is key to the species' future. Increasingly refined, such models identify, quantify and qualify available rangeland relative to pronghorn's biological and ecological requirements for particular habitats. Their applications by wildlife professionals serve as benchmarks and barometers of pronghorn sustainability on western landscapes.

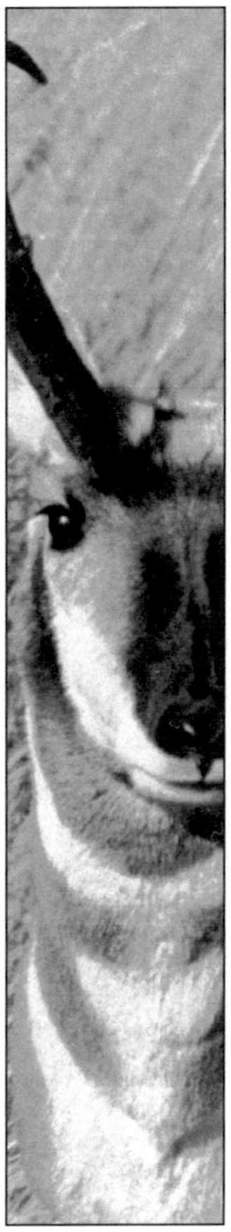

C H A P T E R F O U R T E E N

Foraging Ecology, Diet Studies and Nutrient Values

Jim D. Yoakum

Complete and conclusive proof of antelope dietary needs will require many analyses of species from ranges of different vegetative compositions over a long period of time.

Einarsen (1948:59)

Pronghorn feed largely on forbs; browse constitutes a less important food. Grass is taken only in small quantities.

Leopold et al. (1984:161)

An adequate food supply is one of the most basic habitat requirements of all species. Therefore, some understanding of food habits is a prerequisite to evaluating habitat and providing good habitat management.

Cooperrider (1986:699)

Perhaps more than any other ecological process, foraging behavior defines the trophic linkage between herbivore and plant community. For this reason, a mechanistic understanding of foraging behavior is ultimately necessary for managing habitats and animal populations in an informed manner.

Spalinger et al. (1997:611)

Pronghorn are highly opportunistic feeders, taking the most palatable and succulent forage available at all seasons.

Hervert et al. (2000:23)

Pronghorn consume forage to meet complex nutritional and energy requirements for survival and reproduction. Energy is required for the animals to keep warm, move, circulate blood, transform chemicals from one form to another and carry out virtually all work of the body. In addition, pronghorn require vitamins, macronutrients (foods needed in large quantities) and micronutrients (trace elements needed in minute quantities). Pronghorn diets must provide all these essentials in proper quantities and an adequate balance for growth, maintenance and reproduction (Dasmann 1981).

Six benefits associated with pronghorn food habit studies will be emphasized in this chapter:

1. *Translocation of pronghorn.* When determining whether a site is suitable for translocating pronghorn, inventory should be made of the types and quantities of forage species preferred by pronghorn. Based on food habit studies of comparable ecological sites, the absence or presence (and quantity) of desirable food resources will have major bearing on the advisability of translocation.
2. *Vegetation–manipulation activities.* Vegetation–manipulation activity frequently is underway in pronghorn habitat on state and federally administrated lands as well as private rangelands. Frequently, as matters of mitigation and reclamation compliance, government and industry officials and ranchers seek the advice of wildlife biologists regarding which plants and vegetative conditions are most desirable for pronghorn. Furthermore, such information is necessary for projects designed to meet multiple-use objectives, increase plant diversity or increase forage production for domestic and wild animals.

3. *Grazing systems.* Since 1970, there has been an increase in land management plans stressing herbivore management objectives and needs on western rangelands. These plans can be allotment management plans for livestock, habitat management plans for wildlife (see Chapter 16) or multiple resource prescriptions for various resource objectives. When these resource plans identify grazing systems based on plant phenology, it is important to recognize that graminoids generally mature earlier in the growing season than do many species of forbs. Therefore, it is important to design grazing plans that allow forbs to mature and seed, or the grazing system will fail to meet this phenological schedule beneficial to pronghorn. If shrubs, such as bitterbrush, are key species in the vegetative community, a grazing system needs to be designed to perpetuate them. It will entail grazing formulas entirely different from those required for graminoids.

4. *Dietary studies* can provide documentation of pronghorn forage impacts in agricultural crop depredation cases. Sometimes, actual and alleged depredations are associated with problems other than forage consumption, such as damage due to bedding or trampling. Dietary studies can aid in substantiating the type and extent of depredation—information that can be used in litigation.

5. *Food habit* studies have linked pronghorn population levels with the availability of high-quality forage and accompanying weather patterns. Knowledge of these ecological relationships can help wildlife managers sustain or enhance populations and increase rangeland carrying capacities by improving characteristics of vegetation.

6. *Preformed water* from forage is obtained by pronghorn for body function, therefore management strategies need to enhance succulent vegetation to provide preferred forage and preformed water needs.

Of all the articles on pronghorn biology and ecology published in the *Journal of Wildlife Management,* 1937 to 1977, 25% (44 of 176) dealt primarily with food habits (Yoakum and Spalinger 1979). No other topic rated more than 4%. A check of the same journal, from 1978 to 1998, revealed 36 more reports pertaining to pronghorn, and 13% dealt with food habits. One of the most exhaustive reviews of pronghorn feeding habits from Canada and the United States was a distillation of 157 food habit studies by Sundstrom et al. (1973). That compendium provided information on percentages of grasses, forbs and shrubs in the pronghorn diet, but it failed to identify or quantify forage availability on the investigated sites. Forage is vegetation consumed while grazing grasses and forbs or browsing shrubs and trees.

There are two schools of thought regarding pronghorn forage preference. One group of investigators contends that pronghorn eat forbs when available (Buechner 1950a, Leopold et al. 1984, Yoakum 1990). The other holds that pronghorn are predominantly browsers of shrubs (Sundstrom et al. 1973, Pyrah 1987). The matter is best understood and clarified when plant consumption by pronghorn is analyzed in the context of the relative seasonal availability of forage. A major objective of this chapter is to examine pronghorn food habits relative to availability of vegetation for all seasons of the year and various years in the three major biomes pronghorn presently occupy.

Pronghorn have lived from Canada to Mexico for centuries and have adapted to "forage switching" from season to season (Figure 81). Forage switching enables pronghorn to optimize seasonal dietary benefits from the most preferred, nutritious and succulent species in a vegetative community. Switching is more pronounced in grasslands and shrubsteppes (see Salwasser 1980, Yoakum 1990) than in deserts (Cancino 1994c, Miranda 2000).

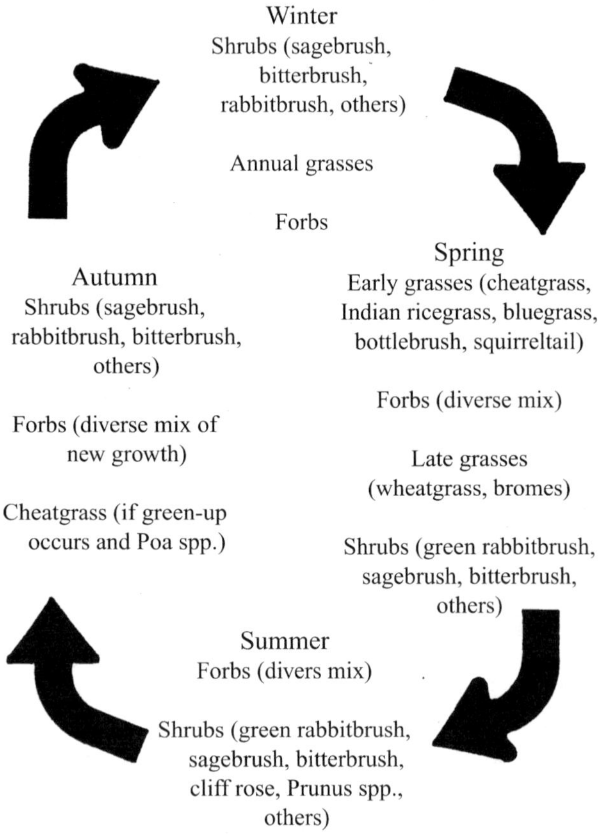

Figure 81. *General scheme of pronghorn diet switching for the Great Basin region (modified from Salwassar 1980).*

During winters, forage often is low in quantity and quality. It is the season when many animals switch their diet to shrubs, which can be high in nutrients and can protrude above deep snow. Photo by Tom W. Hall.

DEFINITIONS

Such terms as diet composition, palatability, preference, production, selection and utilization have been used in many ways and sometimes interchangeably in the pronghorn (and other wildlife) literature. This often has resulted in confusing concepts and erroneous conclusions. For example, investigators sometimes have failed to distinguish clearly between diet composition and diet selection. In this book, diet composition is the percentage composition of plant taxa in the diet, whereas diet selection is the percentage of plants selected as forage from those available. To help avoid misinformation, the following definitions—drawing extensively from Cooperrider (1986)—will be used.

The phrases *food habits* and *diet composition* have been used synonymously. Although the terms have similarities, they likewise have differences. Therefore, they should be used judiciously. Diet composition studies refer to the amount and kinds of plants consumed (Figure 82). Food habit studies are more comprehensive, contain diet composition data and include various ecological factors, for example, weather patterns, plant phenology, nutritional values of vegetation, dietary overlap with other ungulates.

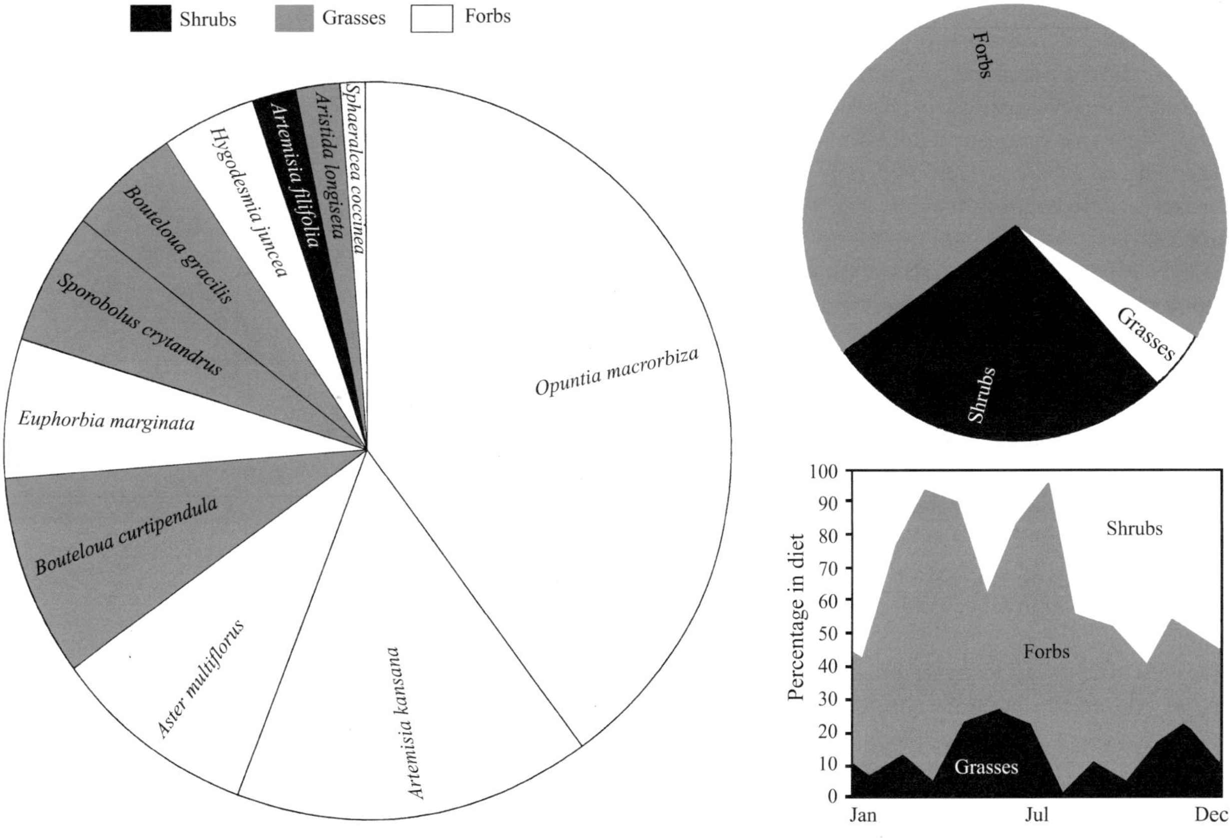

FIGURE 82. *Example of diet composition studies referring to the percentage of forage species or forage classes in a given period and location.*

Most diet composition findings are expressed as percentages of vegetation consumed. The percentages may be by weight or volume, depending on the analysis technique used. Diet composition also can be reported as frequency (relative number [percentage] of samples of each food item [species] consumed), or more rarely, the comparative size (amount by weight or volume) of plant items (species) consumed.

For this chapter, pronghorn diets have been reported for grasslands, shrubsteppes and deserts with ecological data. When environmental information is provided—with data relative to which and how much plants are consumed—environmental factors can aid in assessing forage selection for different seasons, years and other ecological parameters. Having a comprehensive ecological understanding of which plants are consumed can assist managers in the development of strategies for analyzing population dynamics and plan habitat improvement projects.

Forage utilization refers to the amount of vegetation consumed at a particular site during a defined time period. This measurement is primarily an attribute of vegetative diversity and abundance and not necessarily of the foraging animal's preference. Utilization usually is expressed as the percentage of forage consumed. It may be reported as weight or volume of the plant species or forage classes in the subject's diet.

Forage utilization also may be expressed in other ways, such as relative percentage of forbs grazed or twigs browsed. When several species of ungulates simultaneously occupy the same rangeland, determining comparative utilization can be difficult. However, study techniques such as exclosures can be used (Yoakum et al. 1980, Heady and Child 1994).

Forage production refers to the amount of vegetation produced on an area during a given span of time. It may be used synonymously with forage availability, or corrected for forage physically or seasonally unavailable to animals for various reasons (plants buried in snow, browse higher than animals can reach, etc.). Forage production (or availability) can differ from site to site and change over time. If information is obtained on the seasonal composition of available forage and the subject animals' seasonal diet, then forage preferences can be compared with those of herbivores in other areas with similar vegetative communities.

Specific techniques for measuring forage production are not addressed herein because they are adequately covered in other reports (see Stoddart et al. 1975, Heady and Child 1994, Holechek et al. 1998).

Forage preference is the degree to which animals select one food species over another. It is an attribute of the foraging animal and not the rangeland site, except to the extent that different food choices are available. To quantify preference properly, both forage production and food habits need to be measured. Of more than 200 pronghorn food studies, only about 30 have measured both (Yoakum 1990).

Forage preference ratings (PR) are calculated as: PR = D/R, where D is the percentage of diet composed of a given species, genera or forage class, and R is the percentage of forage production composed of the same species, genera or forage class (Krueger 1972). Plants with ratings greater than 1.0 are preferred species.

Forage preference is not synonymous with food composition, although the terms have been treated as such in some pronghorn literature. Forage preference is what an animal likes to eat; diet composition denotes what it actually eats (see Petrides 1975).

Forage availability is the kind and amount of accessible plants during different seasons of the year. It is related to the timeliness and amount of precipitation, plus temperatures that, in turn, are responsible for plant phenology.

Pronghorn are specialists, feeding on nutritious, succulent forage. Although grasses are abundant in this photograph, the adult doe spent her time finding and eating young thistles. The heads of pronghorn are elongated, with narrow, pointed mouth parts that aid in the gathering of leaves and flowers—parts of plants that are more nutritious and succulent than stems and twigs. Photo by Jim D. Yoakum.

Availability also is controlled by other climatic factors, such as sunlight and snow depth.

Palatability is the quality of a plant that determines its relative preference by an animal when a choice of various foods is available. Palatability of any plant or plant species is subject to considerable variation, due to the influence of a large number of environmental factors. Not all factors influencing palatability are well understood. For example, nutritive and chemical cravings correlate with palatability in many instances, but not in all. Usually, however, palatability is a function of the consuming animal's senses of touch, taste and smell (Cowlishaw and Alder 1960).

As a dietary factor, palatability may be equal in importance to forage production and forage preference because each directly affects the intake of nutrition (Hurd and Blaser 1962). As a general rule, animals are more likely to select plants with low palatability if those plants comprise a large portion of the available forage (Heady and Child 1994). When abundance of forage is limited, selectivity decreases and palatability becomes less important (Van Dyne et al. 1980).

FORAGING BEHAVIOR

Pronghorn are opportunistic herbivores, selecting nutritious and succulent plant foods. Through adaptation over centuries, they feed on all forage classes, but show preference for forbs and shrubs. Pronghorn are dainty, selective eaters (Hoffman 1989, Byers 1997a), taking small bites of plant parts. Plants rarely are uprooted, and leaves and flowers are used extensively (Buechner 1950a). Cancino (1994c) reported that diets of peninsular pronghorn were more than 30% flowers of forbs and shrubs in the Vizcaino Desert of Baja California Sur, Mexico.

Perhaps one of the most noticeable characteristics of feeding pronghorn is their apparent nervousness, expressed by extensive movement (walking) while foraging. Rarely do they feed for long in one place, but move frequently as they forage, in strong contrast to livestock. Einarsen (1948:68) provided a classic description of pronghorn feeding behavior: "On May 9, 1937, a doe was seen feeding on bunchgrass (*Festuca idahoensis*). She moved away about 20 yards and cropped the upper leaves of a poisonous larkspur plant. She then trotted 80 yards to a spot where she ate grass again for about two minutes. Coming back to the original place she lay down, chewed her cud for about twenty minutes, then arose hastily to rush over the crest of a ridge some distance, to feed again on bunchgrass."

The nervous nature of pronghorn is apparent in their active foraging habits. They walk rapidly, and their trail can hardly be traced by foraging. Utilization usually can only be detected by very close inspection.

Schwartz (1977) reported that pronghorn on grasslands spend approximately 53% of their time grazing and 40% resting. Similar foraging periods were noted by other investigators (Einarsen 1948, Buechner 1950a, Hoover 1971, Taylor 1972b, Ellis and Travis 1975). Peaks in foraging activity occur near sunrise and sunset (Figure 83). Foraging is more intense during summer than winter. Pronghorn prefer grazing lowland plateaus and warm slopes, and tend to avoid cool slopes and ridgetops during winter (Schwartz 1977).

Byers (1997a) investigated the activity budgets of adult male and female pronghorn on the National Bison Range in Montana. He concluded that the majority of the animals' yearlong time was spent feeding, and the next greatest proportion of time was reclining and ruminating (Figure 84). During the spring, females spent more time foraging than did males—likely because of the greater energy demands of gestation and lactation. Males in the rutting season spent less time feeding and more time pursuing mating activities. Foraging was more time-consuming for males and females during autumn and winter than spring and summer. This may be due to less available and less nutritious forage.

Pronghorn tend to consume many different species in the plant community. They probably eat most species of vegetation available to them, including mosses, lichens, ferns, sedges, rushes, grasses, forbs, cacti, half-shrubs, shrubs and trees. However, the degree of utilization varies according to availability, climatic conditions, succulence, palatability, digestibility and nutritional quality. Herbivores forage selectively in high-quality plant communities. They tend to select the most nutritious or digestible plants where forage is abundant. Rangelands lacking high-quality forage require animals to forage more indiscriminately, decreasing the chance that dietary requirements will be met (Heller 1980).

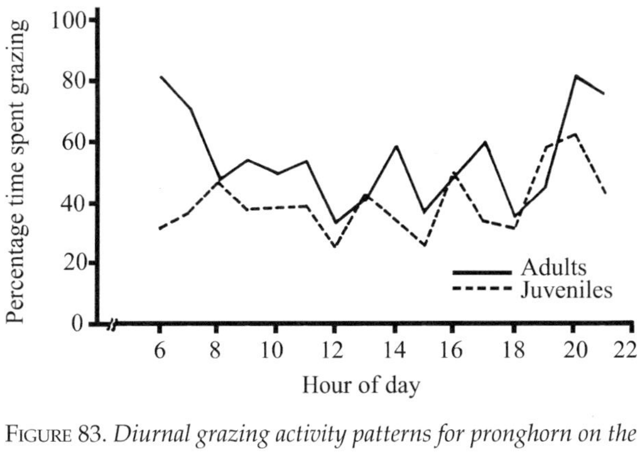

FIGURE 83. *Diurnal grazing activity patterns for pronghorn on the Pawnee Grasslands in northeastern Colorado (Schwartz 1977).*

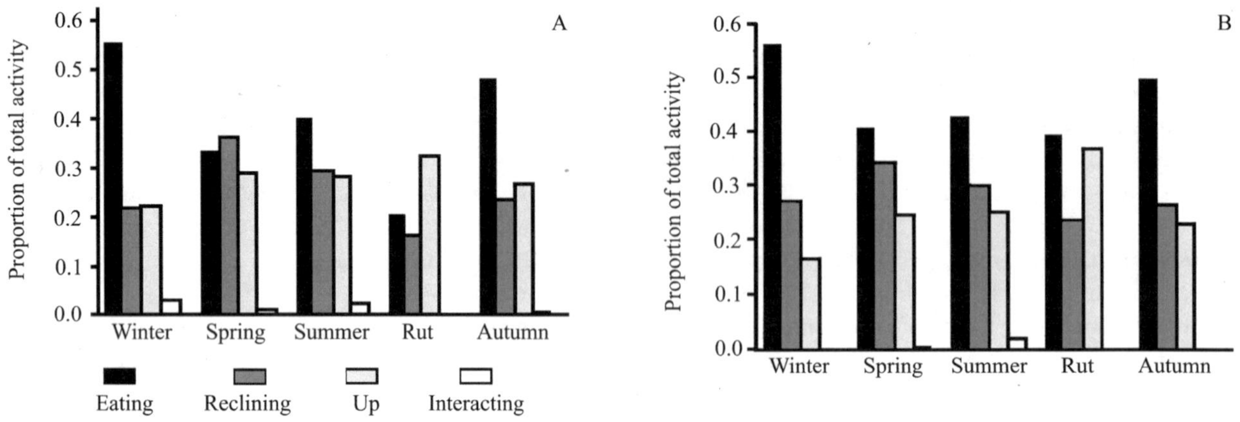

Figure 84. *Seasonal variation in male (A) and female (B) pronghorn feeding activities on the National Bison Range, Montana (Byers 1997a).*

According to Wagner (1978), pronghorn consume less than 1% of available forage on western rangelands in the United States. Similar conclusions were reached by Kindschy et al. (1982) for pronghorn in Oregon, and by Longhurst et al. (1983) for pronghorn in Nevada. No references to free-roaming pronghorn heavily foraging rangeland vegetation have been found in the literature (see Prenzlow 1965b, Yoakum 1967, 1991, Yoakum and Spalinger 1979).

FORAGE-INTAKE RATES

The amount of forage ingested during a day is referred to as the *forage-intake rate.* This rate varies with season and locality, and caution is urged when comparing findings from one site with those of another with differing ecological conditions.

Table 85 provides results of forage intake studies reported for pronghorn (Zarn 1981). To compare the ma-

Table 85. Forage intake requirements of pronghorn

Location	*Season*[a]	*Biome*[b]	*Age class*	*Body weight*		*Forage intake per animal per day*			*Remarks*	*Source*
				lb	*kg*	*lb*	*kg*	*Percentage body weight*		
Wyoming Red Desert	Y	S	Adults	90	40.5	1.8	0.8	1.9	Maintenance and growth	Severson et al. (1968)
Wyoming Red Desert	Y	S	Adults	90	40.5	1.7	0.8	1.9	Feeding trials	Severson et al. (1968)
Wyoming Red Desert	U	U	Fawn	49	22.0	1.5	0.7	3.0		W. G. Hepworth (personal communication:1980)
Wyoming Red Desert	U	U	Doe	97	43.6	2.9	1.3	3.0		W. G. Hepworth (personal communication:1980)
Wyoming Red Desert	U	U	Buck	106	47.7	3.2	1.4	3.0		W. G. Hepworth (personal communication:1980)
Colorado	W	G	Adults	90	40.8	1.9	0.9	1.0	Maintenance and growth	Hoover (1971)
Colorado	S	G	Adults	90	40.8	1.8	0.8	1.9	Maintenance and growth	Wesley et al. (1973)
Colorado	W	G	Adults	90	40.8	2.6	1.1	2.9	Moderate activity	Wesley et al. (1973)
Colorado	S	G	Fawn[c]	19.6	8.9	0.5	0.2	2.4	Maintenance and growth	Wesley et al. (1973)
Alberta	W	G	Adults	—	—	2.45	1.1	—	Maintenance and growth	Barrett (1974)

[a] Y = Yearlong, U = Unknown, W = Winter, S = Spring/summer
[b] S = Shrub grassland, U = Unknown, G = Grasslands
[c] Two months of age.
Source: Modified from Zarn 1981.

jor dietary components of these studies, additional information was provided. Accordingly, the average daily intake for an adult pronghorn was determined to be 2.5 to 3.0 pounds of air-dried forage per day per 100 pounds (1.1–1.4 kg/45 kg) body weight. Similar findings were reported by Severson et al. (1968). Zarn provided information that warrants further investigation and quantification, as the original findings were obtained through personal communication and are not otherwise recorded. Nevertheless, the information is particularly important because it represents more than 30 years of investigative experience.

W. G. Hepworth (personal communication:1981) calculated a 3% to 3.5% of body weight intake of dry forage for maintenance of condition. During periods of pronghorn growth, lactation and gestation, and during times of lush, green forage growth, dry-matter intake could rise to 4% to 6% of live body weight.

Hepworth further maintained that forage allocations based solely on intake figures bear little resemblance to reality if food habit and feeding behavior information is not considered. Stocking rates cannot be based simply on weight forage intake; dietary overlap between species precludes using straight weight-conversion factors to determine range use and impacts by herbivores (see Chapter 15). Differing habitat preferences among species further confounds the use of conversion factors based on body weight and forage-intake rates, which may be applied with validity only to monocultures on flat land without escape cover. Realistic forage-allocation processes should include data on food habits, feeding behavior and dietary overlap between species, correlating them to the vegetative diversity, quality and availability in the area in question.

DIET COMPOSITION TECHNIQUES

Five techniques have been used to determine pronghorn food habits (or diet composition): direct observation, indirect observation, rumen analyses, fecal analyses and cafeteria trials.

Direct and Indirect Observations

Direct observation, sometimes referred to as animal bites, involves watching foraging animals in the field at close proximity and estimating the amount of each plant species consumed. It essentially is the "lead deer" technique (see McMahan 1964). Observations are recorded as bite counts or feeding minutes. The method has been used with varying success to assess pronghorn food habits (Buechner 1950a, Hoover 1971, Schwartz 1977).

Tame, semitame or confined animals may be used. Usually raised in captivity, these subjects are accustomed to humans. They are taken to the field and allowed to eat while the observer stays close and records the forage type and quantity consumed (Hoover 1971). The reliability of results from these test animals, as representative of the food habits of wild, free-ranging pronghorn, has been questioned. However, Schwartz (1977) found dietary selection was similar for hand-raised and wild pronghorn foraging the same vegetative community. Another source

Pronghorn spend 40% to 60% of their time feeding—less time during summers, when plants are abundant, and more time during winter when plants are less available. They forage while walking, not using vegetation extensively. Most feeding activity is diurnal. Direct observation of diet selection can be accomplished with animals that are hand-raised or accustomed to human presence, such as animals living in national parks or refuges. An examiner can follow the animal and record frequency of use and species composition. Above, pronghorn fawns raised in Colorado are turned out at the start of a direct observation foraging study. Photo by Charles C. Schwartz.

of subject animals for direct observation studies is national parks or wildlife refuges where animals have adjusted to human presence and tolerate their closeness.

Indirect observation, or feeding site examination, is one of the oldest methods used for pronghorn food habit studies. Rouse (1941) used the technique more than 60 years ago, trailing pronghorn after a fresh snow and recording the plants consumed. Since then the method has been used frequently (Buechner 1950a, Cole 1956, Cole and Wilkins 1958, Severson 1966, Beale and Smith 1970, Campbell 1970, Taylor 1972b). It consists of locating animals in a field, identifying precisely when and where the animals were feeding, then examining and recording the use of plants. The method calls for little equipment other than a pair of binoculars (or spotting scope) and a notebook. On the other hand, the procedure can be extremely time-consuming. There also can be difficulties, because of the distance, in precisely identifying the plants consumed by pronghorn and distinguishing among these plants and those possibly used by other herbivores.

Rumen samples have been collected from hunter kills. Such cooperative efforts result in biological findings, including not only diet selection, but also blood samples, weights and measurements, horn trophy scores. Rumen samples can be kept in preservatives for later analysis. Photo by Jim D. Yoakum.

Rumen Analyses

Rumen analysis was the method used most frequently before 1980 to determine food habits of pronghorn (Couey 1946, Buechner 1950a, Ferrel and Leach 1950, Mason 1952, Baker 1953, Cole 1956, Yoakum 1957a, 1957b, Hoover et al. 1959, Dirschl 1963, Russell 1964, Severson 1966, Bayless 1969, Tsukamoto and Diebert 1969, Beale and Smith 1970, Campbell 1970, O'Gara and Greer 1970, Mitchell and Smoliak 1971, Taylor 1972b, Jacobs 1973, Schwartz and Nagy 1976). Korschgen (1980) described the technique in detail, including the preservation of materials and identification of food items. Dirschl (1962) elaborated on sieve mesh size based on pronghorn samples.

With rumen analysis, food habits can be determined and quantified by (1) recording numbers of each food item, (2) recording frequency of occurrence of food items, and (3) measuring volume or mass of food items (Cooperrider 1986).

Although rumen analysis was the most frequently used technique for pronghorn diet composition studies before the mid-1970s, the practice has seldom been used since. Rumen analysis requires sacrificing the host, and the number of samples often is limited.

Fecal Analyses

Since about 1970, fecal or microhistological analyses have been used more often than other methods for determining pronghorn food habits (Jacobs 1973, Schwartz and Nagy 1976, Sneva and Vavra 1978, Bodie 1979, Meeker 1979, Sexson 1979, Arizona Department of Game and Fish 1981, Bailey and Cooperrider 1982, Habib de Pena and Pena Neira 1982, Hansen 1985, Goldsmith 1988, Howard et al. 1990, Hughes and Smith 1990, Cancino 1994c, Hansen and Anthony 1999, Miranda 2000, Hansen et al. 2001). The technique is similar to rumen analyses except that fecal samples are used. Because fecal analyses require a great deal of laboratory work, investigators frequently opt to contract such work, rather than do it themselves. Field procedures were evaluated and considered relatively cost effective by Cooperrider et al. (1981).

Advantages of the fecal analysis method are that it (1) does not interfere with the normal habits of animals, (2) permits extensive sampling, (3) places no restriction on animal movements, (4) has particular value where animals range over mixed vegetative communities, (5) is feasible to use for the study of secretive or endangered species, (6) can be used to compare the diets of two or more species at the same time, (7) requires little equipment for sampling and, most important, (8) does not require sacrificing the host (Holechek et al. 1982).

Major disadvantages for this technique were listed by Gill et al. (1983b) (1) underestimation of shrubs when woody stems comprise the bulk of a shrub diet, (2) overestimation of forbs when forb species (which are readily identified) dominate the diets, (3) underestimation of forbs when legumes comprise a significant portion of the diets, and (4) correction factors do not consistently improve estimation of diet composition when the diet contains a diversity of several species within forage classes.

Holechek et al. (1982) and Gill et al. (1983b) concluded that inaccuracy could be the greatest disadvantage to the method. Fecal analysis is more qualitative than quantitative when evaluating wild ruminant food habits, particularly when forbs and shrubs are important diet components (Holechek and Valdez 1985). Nevertheless, it is the most popular method currently in use for pronghorn food habit studies and has been accepted as evidence in court proceedings (Cooperrider 1986).

New techniques of using digestibility correction factors have been developed to estimate composition percentages of plant species more efficiently (Dearden et al. 1975, Pulliam and Nelson 1979, Leslie et al. 1983). Consequently, fecal analysis was the most popular diet selection technique in use at the beginning of the 21st century.

Cafeteria Trials

Cafeteria trials are used to determine plant preference by confined animals. Subject animals are allowed to select forage from a number of equally accessible plants made available in approximately equal quantities. The available plant species usually are analyzed for nutritional values. The method was used by Smith et al. (1965) to study pronghorn preferences for and nutritional values of shrubs in Utah. Using cafeteria trials, Smith (1974) related artificial diets with different protein levels to pronghorn production and survival. Jacobs (1973) used the method to test the comparisons of different food habit techniques in Wyoming. Pulliam and Nelson (1979) also used the technique to prove differential digestive rates for differing herbivores consuming a variety of forage classes.

FOOD HABIT STUDY SITES

Explorer and naturalist Meriwether Lewis provided the first anecdotal reports pertaining to pronghorn diet composition. In South Dakota on September 21, 1804, Lewis collected broom snakeweed and noted in his journals that they were: "[s]pecimens of the aromatic plants on which the Antelope feeds" (Moulton 1987a:464). This was the same month the explorers collected the first pronghorn specimen for science. Lewis procured samples of gray rabbitbrush on October 2, 1804, and he reported pronghorn browsed these shrubs during winter. While traveling through central Montana during April and May 1804, he observed the abundance of silver sagebrush and stated that the shrub, "has a long narrow, smoothly leaf of an agreeable smel and flavor; of this last the Antelope is very fond" (Moulton 1987b:35). All three of these plants (broom snakeweed, gray rabbitbrush and silver sagebrush) were new to science and later taxonomically described from specimens collected during the expedition (Moulton 1987b).

The Lewis and Clark expedition first saw pronghorn on grasslands in northeastern Nebraska and southeastern South Dakota. These prairies lacked the forests so prevalent on much of the land east of the Mississippi River. The explorers soon witnessed pronghorn mingling with herds of bison and elk almost daily on the Great Plains. The grassland biome extended from south-central Canada south through the central United States to the southern tip of the central highland plateau of Mexico. Among the biotic regions occupied by pronghorn, grasslands have sustained the largest number of pronghorn. Although grasses are the most prevalent forage class produced, pronghorn seek and consume primarily forbs and shrubs. Photo by Jim D. Yoakum.

Some of Lewis's observations of plants ingested by pronghorn may have been noted while eviscerating animals for the explorer's consumption. At such times, the aroma of plants in the stomach can often be detected.

Pronghorn food habit studies have associated forage use with forage classes (grasses, forbs, shrubs) according to study sites or political regions (states, provinces) (Yoakum 1958, 1990, Severson et al. 1968, Sundstrom et al. 1973, Segerstrom 1977, Autenrieth 1978, Yoakum and Spalinger 1979). Unfortunately, many studies have compared food habit data and drawn conclusions from sites with distinctively different vegetative communities. For example, forage consumed by pronghorn in the grassland biome has been compared with forage consumed in the shrubsteppe biome. Such comparisons can be extremely misleading.

More accurate and useful comparisons come from studies within similar vegetative types. Comparisons of 35 studies completed during the last half of the 20th century for grasslands, shrubsteppes and deserts are the basis of discussion for the balance of this section.

Grasslands

The grasslands of North America extend southward from the prairies of south-central Canada to the north-central plateau of Mexico. The region was historic pronghorn habitat and supports the largest number of pronghorn per biome today. Table 86 lists the number of species of grasses, forbs and shrubs on grassland study areas and in pronghorn diets. Table 87 provides a listing of available forage, diet and preference ratings for each of 10 food habit studies in the grasslands. Table 88 provides pronghorn diets by forage class and season. Fourteen pronghorn food habit studies from grassland sites will be reviewed.

Alberta. Pronghorn food habits and rangeland characteristics were reported for the Newell area in southeastern Alberta by Mitchell and Smoliak (1971) (Tables 86–88) and Mitchell (1980), from 162 rumen samples collected during every month of the year, except December. The site covers approximately 663,000 acres (268,310 ha) of treeless, undulating shortgrass prairie. Although predominantly native rangeland, the site has irrigated farms interspersed throughout. The climate is semiarid, with 12.5 inches (31.7 cm) mean annual precipitation. Forage production averaged 1,422 pounds per acre (1,594 kg/ha) with notable high yields of 111 pounds per acre (124 kg/ha) of forbs. Changes in forage utilization by pronghorn appeared to have been related to plant phenology, abundance and availability. However, silver sagebrush and pasture sagewort were highly preferred species and recommended for consideration in management plans.

Saskatchewan. The prairies of Canada represent the northern range boundary for pronghorn. Food habits for these animals were studied on the Matador and Cypress Hills sites during 1960 (Dirschl 1963). The sites are in southeastern Saskatchewan at elevations of between 1,800 and 4,400 feet (550–1,340 m). Annual precipitation averages 12 to 14 inches (30.5–35.6 cm).

Forage classes used most by pronghorn were grasses during April, forbs in May through July and shrubs during August through March (Table 88).

TABLE 86. Number of species of grasses, forbs and shrubs in grasslands and pronghorn diets

	Grasses		*Forbs*		*Shrubs*		*Total*		
Area	*In type*	*In diet*	*In type*	*In diet*	*In type*	*In diet*	*In type*	*In diet*	Source
Alberta	10	—	27	52	3	12	40	64	Mitchell and Smoliak (1971)
Pawnee Grasslands, Colorado	22	18	82	65	4	4	108	87	Schwartz and Nagy (1976)
Shortgrass prairies, Colorado	—	12	—	59	—	16	—	87	Hoover et al. (1959)
Roswell, New Mexico	30	17	14	38	10	14	54	59	Howard et al. (1990)
Panhandle, Texas	26	—	62	—	6	—	94	59	Roebuck (1982)
Trans-Pecos, Texas	—	18	—	168	—	56	—	242	Buechner (1950a)
Chihuahua	14	—	26	—	4	—	44	—	Hinshaw (1990)
Carter County, Montana	12	3	30	39	3	9	45	51	Campbell (1970)
Southern High Plains	36	—	130	—	36	—	169	—	Barrington (1975)
Average	21	14	53	70	9	18	79	93	

TABLE 87. Composition of available forage, diet and preference ratings from pronghorn food habit studies in grasslands

Location	*Forage class*	*Percentage composition of vegetation*	*Percentage composition of diet*	*Preference rating*	*Source*
Alberta	Grasses	78	13	0.2	Mitchell and Smoliak (1971)
	Forbs	21	51	2.4	
	Shrubs	3	35	11.7	
National Bison Range, Montana	Grasses	79	15	0.2	O'Gara and Greer (1970)
	Forbs	7	67	9.6	
	Shrubs	14	18	1.3	
Tallgrass prairie, Kansas	Grasses	71	31	0.4	Ganey (1998)
	Forbs	11	34	3.0	
	Shrubs	1	35	35.0	
Pawnee Grasslands, Colorado	Grasses	55	49	0.9	Schwartz and Nagy (1976)
	Forbs	5	45	9.0	
	Shrubs	40	4	0.1	
Shortgrass prairie, Kansas	Grasses	92	27	0.2	Hlavachik (1968)
	Forbs	4	76	19.0	
	Shrubs	3	2	0.5	
	Grasses	95	46	0.5	Sexson (1979)
	Forbs	5	48	9.6	
	Shrubs	Tr[a]	6	30.0	
Shortgrass prairie, New Mexico	Grasses	71	4	0.1	Russell (1964)
	Forbs	18	70	3.9	
	Shrubs	11	26	2.4	
Roswell, New Mexico	Grasses	94	3	0.0	Howard et al. (1990)
	Forbs	1	71	71.0	
	Shrubs	5	26	5.2	
Panhandle, Texas	Grasses	68	1	0.0	Roebuck (1982)
	Forbs	24	82	3.4	
	Shrubs	8	13	1.6	
Chihuahua	Grasses	33	2	0.0	Habib de Pena and Pena Neira (1982)
	Forbs	62	75	1.2	
	Shrubs	5	6	1.2	
Average	Grasses	74	19	0.2	
	Forbs	16	62	3.9	
	Shrubs	9	17	1.9	

[a] Tr = less than 0.5%.
Percentages were rounded to the nearest 1%.

The shortgrass prairies of Alberta have been occupied by pronghorn for centuries. Pronghorn and bison were sympatric grazers, but the wild bison herds have been extirpated, and livestock now graze the grasslands with greatly decreased numbers of pronghorn. These plains represent the northern limits of historic and modern pronghorn populations. A high percentage of the native grassland has been converted to agricultural crops, which has decreased prime habitat for pronghorn. Photo courtesy of the Alberta Division of Wildlife, Edmonton.

TABLE 88. Pronghorn diets by forage class and season for grassland habitats

Location	Winter			Spring			Summer			Autumn			Annual			Source
	G[a]	F	S	G	F	S	G	F	S	G	F	S	G	F	S	
Alberta	9	47	43	25	57	18	13	62	25	13	37	50	13	51	35	Mitchell and Smoliak (1971)
Saskatchewan	1	15	84	43	17	40	3	55	42	2	27	71	12	29	59	Dirschl (1963)
Tallgrass prairie, Kansas	33	52	12	62	31	6	5	20	73	14	32	49	31	34	35	Ganey (1998)
Pawnee Grasslands, Colorado	74	17	6	62	33	0	11	84	5	79	16	8	49	45	4	Schwartz and Nagy (1976)
Shortgrass prairies, Colorado	5	41	54	20	55	25	1	77	22	1	27	77	4	53	43	Hoover et al. (1959)
Roswell, New Mexico	1	66	33	3	66	30	2	76	22	5	74	21	3	71	26	Howard et al. (1990)
Panhandle, Texas	0	73	22	2	78	17	2	91	1	0	85	10	1	82	13	Roebuck (1982)
Trans-Pecos, Texas	1	66	33	4	71	25	6	67	27	7	55	38	4	65	31	Buechner (1950a)
Trans-Pecos, Texas	8	57	35	20	35	45	5	34	61	3	77	20	6	63	31	Hailey (1979)
Average													14	55	31	

[a] G = Grasses; F = Forbs, S = Shrubs.

National Bison Range, Montana. About half of the 18,540-acre (7,503 ha) National Bison Range is pronghorn habitat. The area apparently was not pronghorn habitat historically; they were translocated to the refuge in 1951 and 1952 (Byers 1997a). Dominant plant species were bluebunch wheatgrass, cheatgrass brome, arrowhead balsamroot, western yarrow, Lewis mock orange, Rocky Mountain maple and fringed sagebrush. Maximum use of grasses was during spring, although amounts of up to 5% were consumed during autumn green-up. Forbs were used mainly during spring, summer and autumn (Table 87). Shrubs were used all year to some extent, but mainly during winter. O'Gara and Greer (1970) compared the food habits of pronghorn for this site with those of herds on a shrubsteppe type in Yellowstone National Park, using 112 rumen samples collected during every month of the year. They speculated that the well-being of northern pronghorn is linked closely to the availability of shrubs, especially sagebrush, on winter rangelands.

Although pronghorn are not endemic to the National Bison Range in the Flathead Valley of Montana, they were translocated there and now are well established. Diet studies substantiate that forbs and shrubs are the most preferred forage classes, and grasses (although abundant) are consumed in small quantities. Little dietary overlap occurs with bison because each ungulate consumes different forage classes, pronghorn preferring forbs and shrubs, whereas bison primarily consume grasses. Photo by David W. Kitchen.

Tallgrass prairie, Kansas. The 129-square mile (334 km^2) study area in Chase and Lyon counties was mostly private lands grazed by cattle (Ganey 1998). Topography was sloping hills with elevations ranging from 1,100 to 1,500 feet (335–460 m) above sea level. Annual precipitation averaged 32 inches (81 cm). Winters were cold with light snowfall, about 17 inches (42.5 cm) per year and temperatures reaching 19°F (–7°C). Compared with other northern grasslands reported in this chapter, the Flint Hills site experienced greater annual precipitation and had light snowfall. Native grasses grew to more than 6 feet (1.8 m), but they usually were grazed by livestock or burned annually. Consequently, pronghorn habitats commonly had vegetative heights of 30 to 40 inches (76–102 cm).

Fecal pellets were collected monthly for 1 year and examined using the microhistological analysis method. The most abundant grass was beardgrass, which was only used in limited amounts, compared with the less abundant bluegrasses. Sunflowers were the most commonly used forb. This study of pronghorn food habits is the only one reporting that sumac was browsed in large quantities, probably due to its availability in a plant community with few other shrubs, plus its nutritional value. Although fringed sagewort was classified in some studies as a shrub, it was listed as a forb in the Flint Hills study because of its growth form. Pronghorn diets varied seasonally and appeared to be determined by nutritional values of available forage (Tables 87 and 89). Plant moisture may have been another preference factor. When plants consumed during each season were analyzed, it appeared pronghorn consumed the most digestible and nutritious forage.

Pawnee Grasslands, Colorado. This study site, on the U.S. Agriculture Research Service's Central Plains Experimental Range, 11 miles (18 km) northeast of Nunn, was dominated by blue grama buffalo grass. A research project on pronghorn food habits was conducted from 1970 to 1972 and reported by Schwartz and Nagy (1976) and Schwartz (1977) (Tables 86–88). Vegetative production was ascertained using the weight-estimate method. Excluding cacti, plant biomass data were collected for each grazing season. Grasses and sedges were the major biomass components. Forbs were the least abundant of the three forage classes, contributing less than 13% of plant biomass.

Direct observation of tame juvenile pronghorn and fecal analysis from wild adults were used to estimate food habits. Diet switching of forage classes for the four seasons was obvious for this grassland, as was reported by Salwasser (1980) for shrubsteppes of the Great Basin. The year-round preference rating for grasses (0.9) was highest for grasses in grasslands.

Differences in pronghorn diet due to forage availability in pastures grazed by livestock at different intensities also were reported (Table 90). Grasses and forbs were dominant in both pastures, but composition varied. Forbs eaten during the dormant season varied between pastures, but were highest in the one that was lightly grazed. Shrub use was greatest on heavily grazed pastures. Shrub

TABLE 89. Forage availability plus pronghorn seasonal diets and preference ratios for the tallgrass prairie, Flint Hills, Kansas

Forage class	*Percentage available*	*Percentage forage in diet by season*					*Preference ratio*
		Winter	*Spring*	*Summer*	*Autumn*	*Yearlong*	
Grasses	71	33	62	5	14	28	0.3
Forbs	11	52	31	20	32	34	3.1
Shrubs	1	12	6	73	49	35	35.0

Source: Modified from Ganey 1998.

TABLE 90. Pronghorn use of forage classes according to grazing intensities on the Pawnee Grasslands, Colorado

Forage class	*Grazing intensity*	*Seasonal percentage*									
		Winter		*Spring*		*Summer*		*Autumn*		*Annual*	
		Composition	*Diet*	*Composition*	*Diet*	*Composition*	*Diet*	*Composition*	*Diet*	*Composition*	*Diet*
Grasses	Light	67	65	75	59	72	8	69	71	71	44
	Heavy	78	74	76	62	38	11	32	79	55	50
Forbs	Light	5	32	5	40	10	91	13	29	8	55
	Heavy	4	17	6	33	7	84	5	16	5	45
Shrubs	Light	26	3	30	0	15	0	15	0	21	1
	Heavy	64	8	16	6	17	0	58	5	40	4

Percentages were rounded to the nearest 1%.
Source: Schwartz and Nagy 1976.

consumption declined when forbs became available during the growing season. C. C. Schwartz (personal communication:1989) concluded that pronghorn preferred forbs to shrubs and that high use of shrubs by pronghorn may reflect heavy livestock grazing and lack of forbs. This case, involving high use of grasses, indicates why pronghorn food habits on grasslands should not be expected to be similar to those of shrubsteppes.

Shortgrass prairie, Colorado. Hoover et al. (1959) provided a lengthy report on food habits of pronghorn on shortgrass prairies of northeastern Colorado, but forage availability was not reported (Tables 86 and 88). Diets were determined by rumen analysis, and findings varied considerably from those of the nearby Pawnee Grasslands, where grass consumption was greater than that of shrubs. Hoover et al. (1959:59) observed: "The fact that in spring there is a sudden increase in the amount of forbs eaten, with a corresponding decrease in that of browse, seems to indicate a preference for forbs over browse when both are high in palatability and nutrient values. Also, shrubs are not as nutritious as forbs during spring. Reversal of the diet takes place in the fall after the first killing frost." Sizeable quantities of forbs were found in the winter diet, although forb palatability was greatly reduced.

Shortgrass prairie, Kansas. During 1964, pronghorn were translocated to a 250,000-acre (101,175 ha) site in northwestern Kansas. The diet compositions of these animals in a grama/buffalo grass community were studied by Hlavachik (1968) (Table 87). Winter wheat fields were within the study area, but no data were collected on those sites because they afforded no concealment for close observation of food habits. Pronghorn consumed 13 plants—4 grasses, 8 forbs and 1 shrub. Cacti were used seasonally; other forbs were used throughout the year. Only one shrub species was present, but it was abundant along streams and constituted 2% of the diet. Pronghorn consumption of grasses during winter was correlated to the limited abundance and desiccation of forbs. Hlavachik concluded that most studies stressed the importance of shrubs to pronghorn welfare, but he found that pronghorn in Kansas were able to use grasses and forbs to support increasing numbers. He concluded that herds could be reintroduced to other currently unoccupied historic grasslands. This appears to have validity; pronghorn in Kansas increased from 100 in 1964 (Yoakum 1968) to 2,500 in 1997 (see Chapter 4).

Pronghorn food habits also were studied on another grassland site comprising 44,200 acres (17,887 ha) in northwestern Kansas by Sexson (1979). Composition of dietary intake by pronghorn was determined from year-round fecal analysis (Table 87).

The shortgrass prairies of the Great Plains were inhabited by millions of bison and pronghorn when North American explorers, trappers and pioneers crossed the continent in the early 1800s. The abundant short grasses and forb-rich herbaceous vegetation mixed with sparse, low-growing shrubs created plant communities well suited to the habitat requirements of pronghorn. Herbaceous vegetation is abundant because the ecosystem receives rain during spring and summer—the period of major growth for grasses and forbs. Photo by R. S. Cole; courtesy of the U.S. Soil Conservation Service, Washington, D.C.

Michael H. Francis (previous page) ▲ Kim Hart ▼ Dusan Smetana ▲ Donald M. Jones

▲ Donald M. Jones ▼

▼ Carol Polich

▲ Dusan Smetana ▼ Carol Polich ▲ Milo Burcham

▲ Judd Cooney ▼ Erwin and Peggy Bauer ▲ Michael H. Francis

▼ Jim D. Yoakum ▲ Donald M. Jones ▲

▼ Donald M. Jones ▲ Kim Hart

▲ Donald M. Jones ▼ Milo Burcham

Shortgrass prairie, New Mexico. Russell (1964) reported on pronghorn rumen samples collected year-round from 1956 to 1959 for grasslands near Raton, Roy and Roswell, New Mexico (Table 87). Annual precipitation averaged 12 to 16 inches (30.5–41 cm); elevations ranged from approximately 4,500 to 6,000 feet (1,370–1,830 m); the mean monthly temperature extremes were 22°F (–5°C) in January and 82°F (28°C) in July. Vegetative availability was determined by the step-point method (Yoakum 1972). Plants consumed by pronghorn included 4 grasses, 115 forbs and 24 shrubs. Grasses were consumed most heavily during spring and autumn; forbs were consumed heavily (more than 50%) during all four seasons; shrubs were eaten year-round, but mostly in autumn and winter. Preference ratings were computed (Table 91).

Roswell, New Mexico. Forage availability and preference ratings were investigated during 1980 and 1981 for the southern shortgrass prairies near Roswell. The study area comprised 82,530 acres (33,400 ha), with topography varying from flat to gentle rolling prairies interspersed with shallow canyons. Annual precipitation averaged 12 inches (30.5 cm), with most occurring between July and September. Cattle and domestic sheep occupied the rangelands with pronghorn year-round. Available forage was determined by weight estimate, step-point, and soil and vegetation inventory techniques in 15 fenced pastures. Vegetative sampling was conducted concurrent with fecal analyses so that preference ratings could be calculated (Tables 86–88).

Food habits of pronghorn, cattle and domestic sheep were determined for all seasons by fecal analyses. Howard et al. (1982:88) wrote that their data, "indicated pronghorn are highly selective for forbs in all seasons and are probably dependent upon their availability in order to survive under confinement by net wire fences." This contention was supported by information showing ratios of forb consumption by pronghorn (Table 92), reflecting the demand for forbs. At the very least, the data indicate that forbs were a significant factor in the site's pronghorn population trends.

Panhandle, Texas. Pronghorn food habits were studied on the prairie of northern Texas during 1980 and 1981 (Roebuck 1982). The study area covered about 14,200 acres (5,745 ha), with a flat to gently rolling topography at 3,780 to 4,033 feet (1,152–1,229 m) elevation. Average annual precipitation was 13 inches (33 cm), with most falling during spring and summer. Vegetative measurements were obtained using the frequency of occurrence technique. Minor changes in composition of forage class by season resulted because most plants were perennials. Blue grama

Investigation of diet selection and forage availability in the Panhandle of Texas showed that grasses were the most abundant forage class, but seldom were eaten by pronghorn. Forbs were the major species consumed, especially during summer. Preference ratios were 0.0 for grasses, 3.5 for forbs and 1.3 for shrubs. Photo by Craig M. Roebuck.

TABLE 91. Pronghorn food habits and forage preference ratings for three grassland sites in eastern New Mexico

	Percentage composition of vegetation			*Percentage composition of diet*			*Preference rating*		
Location	*Grasses*	*Forbs*	*Shrubs*	*Grasses*	*Forbs*	*Shrubs*	*Grasses*	*Forbs*	*Shrubs*
T. O. Ranch (near Raton)	42	35	25	7	43	50	0.2	1.2	2.1
Mitchell Ranch (near Roy)	85	10	5	4	73	23	0.0	7.3	4.6
Lewis Ranch (near Roswell)	86	10	4	1	94	5	0.0	9.4	1.2
Average	71	18	11	4	70	26	0.1	3.9	2.4

Source: Russell 1964.

Table 92. Seasonal consumption of forbs by pronghorn on grasslands near Roswell, New Mexico

Pronghorn, by age and weight class[a]	Percentage daily intake	Percentage forbs	Daily forb consumption[a]	Season
Fawn				
22 (10)	2.8	82	0.51 (0.94)	Summer
52 (24)	2.8	75	1.09 (0.50)	Autumn
75 (34)	2.8	69	1.45 (0.66)	Winter
90 (41)	2.8	70	1.76 (0.80)	Spring
Adult				
97 (44)	2.6	82	2.07 (0.94)	Summer
100 (45)	2.6	75	1.95 (0.89)	Autumn
102 (46)	2.6	69	1.83 (0.83)	Winter
105 (48)	2.6	70	1.91 (0.87)	Spring

[a] In pounds (kilograms).
Source: Howard et al. 1990.

and sand dropseed were consistently the most abundant plants. Some forbs showed seasonal changes in abundance, peaking during spring and summer and declining through autumn and winter (Tables 86–88). Fifty plant species were consumed, of which seven made up the bulk (60%). Roebuck's research is the only known study of grasslands that quantifies forage availability and consumption for the four seasons (Table 93). Grasses were consumed lightly throughout the year. Forbs were consistently the major forage consumed, with highest use during summer. Shrub (mainly sand sagebrush) use was highest during spring and winter (Figure 85).

Trans-Pecos, Texas, 1946 and 1947. One of the earliest and most intensive investigations into rangeland conditions and food habits of pronghorn, with emphasis on seasonal competition for forage with livestock, was conducted by Buechner (1950a) in west-central Texas (Table 88). The Trans-Pecos region comprises roughly 20,100,000 acres (8,134,270 ha) of flat plains to low rolling hills with basins and plateaus. The climate is characterized by scant precipitation, rapid evaporation, moderate winters and mild summers. Droughts periodically plague the region. Plant measurements were obtained by a series of clip plots in various biotic associations. Clipped vegetation was then air-dried and weighed by species.

Diet composition, obtained through 3,044 feeding-minutes of direct observation and 89 rumen analyses, indicated the greatest diversity of plant species found in the literature (Table 86). Pronghorn were observed to be particularly fond of flowers and fruits. Five plants consumed by pronghorn were poisonous to livestock.

Trans-Pecos, Texas, 1960s. The Texas Parks and Wildlife Department published a handbook on management of pronghorn on grasslands of the Trans-Pecos region (Hailey 1979). Diet composition findings were based on 90 rumens collected from 1965 to 1967 (Table 88, Figure 86), but no measurements of forage composition or pro-

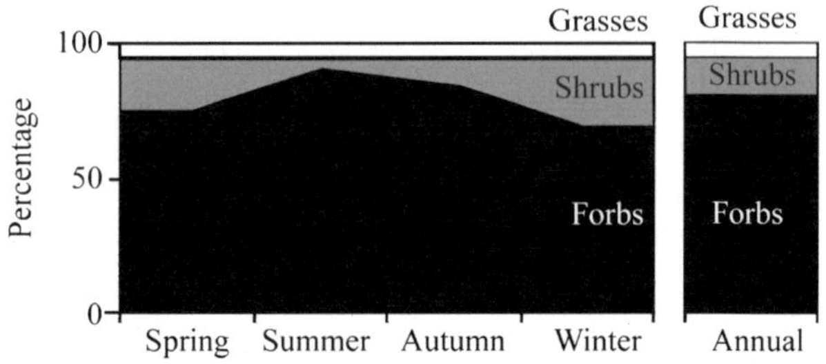

Figure 85. *Pronghorn diet composition on grasslands of the Texas Panhandle (Roebuck 1982).*

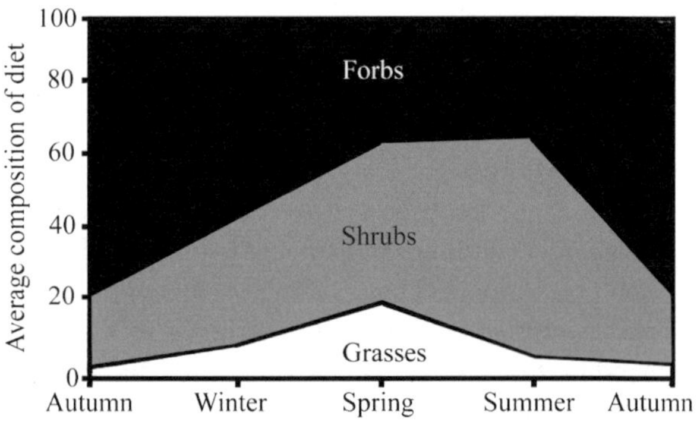

Figure 86. *Forage composition variations in seasonal food habits of pronghorn in the Trans-Pecos region of Texas (Hailey 1979).*

Table 93. Pronghorn seasonal food habits and forage preference ratings for grasslands of the Texas Panhandle

	Percentage composition of vegetation			Percentage composition of diet			Preference rating		
Season	Grasses	Forbs	Shrubs	Grasses	Forbs	Shrubs	Grasses	Forbs	Shrubs
Winter	75	18	7	0	73	22	0.0	4.0	3.1
Spring	63	25	12	2	78	17	0.0	3.1	1.4
Summer	61	31	8	2	91	1	0.0	2.9	0.1
Autumn	72	21	7	0	85	10	0.0	4.0	1.4
Annual	68	24	8	1	82	13	0.0	3.5	1.5

Source: Roebuck 1982.

duction were obtained. Forage class use by season compared closely with Buechner's (1950a) findings. The comparison is of particular interest because the investigations were 30 years apart, and Buechner's method was primarily direct observation whereas Hailey's (1979) was rumen analysis, yet both studies produced similar results for yearlong use. Russell's (1964) findings for adjacent shortgrass rangelands in New Mexico, likewise, showed similar forage class use on an annual basis.

The availability of forbs during late summer and autumn was related to rainfall. Forbs comprised nearly two-thirds of the diet and were the heaviest used forage year-round. Walkingstick cholla and pricklypear were heavily used during late winter and spring. Hailey (1979) postulated that, without these two species, the area would have suffered heavy die-offs of pronghorn during drought periods.

The Trans-Pecos region of western Texas sustains the state's largest extant pronghorn herds. Heavy pronghorn use of forbs, moderate use of shrubs and light use of grasses were recorded during a 30-year period. Cacti were used heavily during droughts and may have contributed needed nutrients and water to sustain herds. Photo by Marvin I. Dixon; courtesy of the U.S. Conservation Service, Washington, D.C.

Chihuahua. Extending from north of Mexico City to southern New Mexico is an interior grassland plateau. Pronghorn were studied there from 1976 to 1978 by Trevino (1978). Trevino's interest in the pronghorn's welfare has continued, and he was responsible for initiating studies of forage availability (Hinshaw 1990) and pronghorn food habits (Habib de Pena and Pena Neira 1982) in the state of Chihuahua.

The Chihuahuan rangelands are characterized by open grama grasslands in nearly flat to smooth, rolling terrain. Most elevations are 4,500 to 5,500 feet (1,372–1,676 m) above sea level. Annual precipitation averages 10 to 13 inches (25–33 cm), falling primarily from July to October.

Hinshaw (1990) provided composition of the vegetation community (Table 86). Dietary intake was estimated by fecal analyses by Habib de Pena and Pena Neira (1982) (Table 87).

Coahuila. During 1995 and 1996, 155 pronghorn from an American/Mexican intergrade area of New Mexico were transplanted to Coahuila, Mexico (D. Weybright personal communication:1997). Postrelease monitoring studies stressing herd recruitment, mortality and food habits were conducted in 1997 and 1998 (Miranda 2000). The release site was primarily semiarid grasslands in Colombia Valley, located in the northeastern region of the state. Average mean precipitation received was 7 inches (17.8 cm). Rangelands were intensively grazed by livestock and shared with mule and white-tailed deer. The number of plant species in the vegetative communities was 135, of which 5 were classified as toxic to livestock. Principle plant taxa were three-awn and grama grasses interspersed with agave cactus.

Pronghorn diet composition was determined by microhistological analyses of fecal samples collected monthly for 2 years. A total of 59 plant species were consumed, representing 38 forbs, 13 shrubs and 8 grasses (Table 94). Little diet switching occurred for forage classes (Figure 87).

Shrubsteppes

The shrubsteppe biome extends from the western boundary of the shortgrass prairies through the Rocky Mountain states, Columbia Plateau, Great Basin and the cold desert shrublands of the Southwest. Sagebrush, with

TABLE 94. Percentage consumption of plants by forage class for seasons during 1997 and 1998 in Colombia Valley, Coahuila, Mexico

Forage class	*Spring*	*Summer*	*Autumn*	*Winter*	*Total*
Grass	12	19	8	7	12
Forb	69	57	75	76	69
Shrub	13	20	11	13	14
Unidentified	6	4	6	4	5

Source: Miranda 2000.

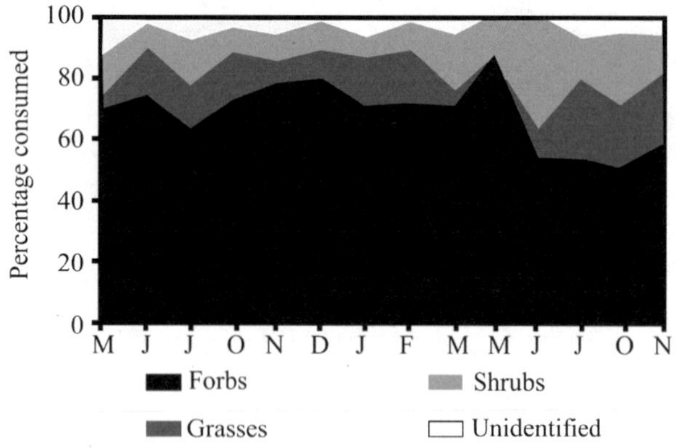

Figure 87. *Seasonal variation of forage classes in the diets of pronghorn, Colombia Valley, Coahuila, Mexico, 1997–1998 (Miranda 2000).*

its many subspecies, is one of the most abundant species. Food habit studies for this biome contain data on forage availability and percentage forage classes in pronghorn diets (Tables 95 and 96). Table 97 provides data regarding forage class by season of use for 13 studies in shrubsteppes.

Yellow Water Triangle, Montana. One of the most thorough research projects on pronghorn and their habitat was conducted by the Montana Department of Fish, Wildlife and Parks (Pyrah 1987). It was undertaken from 1965 to 1975 in the Yellow Water Triangle, located centrally in the state. Encompassing 266 square miles (689 km^2) of semiarid, low shrub/grass plains, its elevations ranged from 3,000 to 4,500 feet (914–1,372 m), with flat to gentle rolling hills. Annual precipitation mean was 14 inches (35.6 cm), with most rainfall received during spring and early summer. The site was mixed-shrub grassland, with an aspect of low sagebrush. It had 95% rangeland and 5% agriculture cropland. Pronghorn diets consisted of 124 plant species—14 grasses, 96 forbs and 14 shrubs (Pyrah 1987). There was great variation in the kinds and amounts of plants consumed seasonally and in different years. Food habit studies included findings from four research projects dating back to 1956 that involved field observations, recording plant use at feeding sites and analysis of rumen contents. Of these studies, only Cole and Wilkins (1958) (Tables 95 and 96) reported diet intake for the four sea-

Table 95. Number of species of grasses, forbs and shrubs in shrubsteppes, deserts and pronghorn diets

	Grasses		*Forbs*		*Shrubs*		*Total*		
Biome and area	*In type*	*In diet*	*In type*	*In diet*	*In type*	*In diet*	*In type*	*In diet*	*Source*
Grassland									
Trans-Pecos, Texas	—	18	—	168	—	56	—	242	Buechner 1950
Shortgrass prairies, Colorado	—	12	—	59	—	16	—	87	Hoover et al. 1959
Carter County, Montana	12	3	30	39	3	9	45	51	Campbell 1970
Alberta	10	—	27	52	3	12	40	—	Mitchell and Smoliak 1971
Southern High Plains	36	—	130	—	36	—	169	—	Barrington 1975
Pawnee Grasslands, Colorado	22	18	82	65	4	4	108	87	Schwartz and Nagy 1976
Panhandle, Texas	38	—	97	—	10	—	143	50	Roebuck, 1982
Roswell, New Mexico	30	17	14	38	10	14	54	59	Howard et al. 1983
Chihuahua	14	—	26	—	4	—	44	—	Hinshaw 1990
Average	23	14	58	70	10	18	86	98	
Shrubsteppes									
Yellow Water Triangle, Montana	—	5	—	34	—	11	—	50	Cole and Wilkins (1958)
Red Desert, Wyoming	17	6	17	7	17	4	51	17	Severson (1966)
Rawlins, Wyoming	23	—	98	—	29	—	150	67	Taylor (1975)
Northwestern shrubsteppes, Colorado	11	11	52	46	18	15	81	72	Bear et al. (1972)
Trickle Mountain, Colorado	—	19	—	32	—	26	150	77	Bailey and Cooperrider (1982)
Adobe Ranch, New Mexico	—	4	—	38	—	10	—	52	Russell (1964)
Sand Hills, Texas	26	—	62	—	6	—	94	59	Roebuck (1982)

continued on next page

Table 95—*continued*

Biome and area	*Grasses* In type	*Grasses* In diet	*Forbs* In type	*Forbs* In diet	*Shrubs* In type	*Shrubs* In diet	*Total* In type	*Total* In diet	*Source*
Cold Desert, Utah	8	—	103	98	36	36	147	134	Smith and Beale (1980)
Burns Junction, Oregon	21	5	101	16	21	4	143	25	McInnis (1984)
Hart Mountain, Oregon	—	—	—	14	—	7	—	21	Mason (1952)
Drakes Flat, Oregon	18	12	102	36	16	14	136	62	Yoakum (1957b)
Sheldon NWR, Nevada	29	11	148	36	23	10	200	57	Hansen (1986)
Northeastern California	—	12	—	68	—	10	—	90	Ferrel and Leach (1952)
Adobe Valley, California	26	9	59	11	13	10	98	30	Goldsmith (1988)
Hart Mountain NAR, Oregon (1990s)	—	13	—	27	—	12	—	52	Hansen and Anthony (1999)
Sheldon NWR, Nevada (1990s)	—	13	—	27	—	11	—	51	Hansen and Anthony (1999)
Average	20	10	82	35	20	13	126	57	
Desert									
Vizcaino Desert, Baja California Sur	3	—	30	—	26	—	59	—	Cancino (1994c)
Cabeza Prieta, Arizona	—	8	—	50	—	39	—	97	Hervert et al. (2000)

Table 96. Composition of available forage, diet and preference ratings from pronghorn food habit studies in shrubsteppes

Location	*Forage class*	*Percentage of vegetation*	*Percentage of diet*	*Preference rating*	*Source*
Yellow Water Triangle, Montana	Grasses	44	9	0.2	Cole and Wilkins (1958)
	Forbs	13	23	1.8	
	Shrubs	43	68	1.6	
Red Desert, Wyoming	Grasses	38	5	0.1	Severson et al. (1968)
	Forbs	15	9	0.6	
	Shrubs	47	78	1.6	
Rawlins, Wyoming	Grasses	75	4	0.0	Taylor (1975)
	Forbs	Tr[a]	15	37.5	
	Shrubs	25	81	3.2	
Northwestern shrubsteppes, Colorado	Grasses	32	2	0.0	Bear et al. (1972)
	Forbs	15	35	2.3	
	Shrubs	53	63	1.2	
Trickle Mountain, Colorado	Grasses	27	4	0.1	Bailey and Cooperrider (1982)
	Forbs	48	18	0.4	
	Shrubs	25	78	3.1	
Ensenada Mesa, New Mexico	Grasses	44	2	0.1	Hooley (1992)
	Forbs	4	30	7.7	
	Shrubs	21	67	3.2	
Sandhills, Texas	Grasses	50	4	0.0	Roebuck (1982)
	Forbs	12	70	5.8	
	Shrubs	38	20	0.5	
Little Lost/Birch Creek, Idaho	Grasses	30	2	0.0	McCarty (1982b)
	Forbs	6	18	3.0	
	Shrubs	64	80	1.3	

continued on next page

Table 96—*continued*

Location	*Forage class*	*Percentage of vegetation*	*Percentage of diet*	*Preference rating*	*Source*
Hart Mountain, Oregon	Grasses	5	2	0.4	Mason (1952)
	Forbs	10	31	3.1	
	Shrubs	85	67	0.8	
Hart Mountain NAR, Oregon	Grasses	24	8	0.3	Hansen et al. (2001)
	Forbs	30	49	1.6	
	Shrubs	46	43	0.9	
Drakes Flat, Oregon	Grasses	28	21	0.8	Yoakum (1957b)
	Forbs	10	25	2.5	
	Shrubs	62	54	0.8	
Sheldon NWR, Nevada	Grasses	22	6	0.2	Hansen (1986)
	Forbs	20	46	2.3	
	Shrubs	58	45	0.8	
Adobe Valley, California	Grasses	61	17	0.2	Goldsmith (1988)
	Forbs	11	16	1.4	
	Shrubs	28	67	2.4	
Average	Grasses	37	7	0.2	
	Forbs	15	30	2.0	
	Shrubs	46	62	1.3	

[a] Tr = less than 0.5%.
Percentages rounded to the nearest 1%.

Table 97. Pronghorn food habits by forage class and season for shrubsteppes

Location	*Winter* G[a]	F	S	*Spring* G	F	S	*Summer* G	F	S	*Autumn* G	F	S	*Annual* G	F	S	*Source*
Yellow Water Triangle, Montana	5	3	92	11	21	68	12	57	31	10	9	81	9	23	68	Cole and Wilkins (1958)
Rawlins, Wyoming	2	1	97	9	29	62	2	29	69	2	1	97	4	15	81	Taylor (1972b)
Northwestern shrubsteppe, Colorado[b]	—	—	—	1	38	60	1	48	49	1	38	59	1	42	56	Bear et al. (1972)
Trickle Mountain, Colorado	3	3	94	6	22	72	3	34	63	2	13	85	4	18	78	Bailey and Cooperrider (1982)
Adobe Ranch, New Mexico	1	26	74	1	45	55	1	31	66	1	51	46	1	38	62	Russell (1964)
Sand Hills, Texas	3	73	19	0	77	15	2	68	23	11	62	22	4	70	20	Roebuck (1982)
Cold Desert, Utah				2	18	80	1	37	63	1	3	96	1	19	80	Beale and Smith (1970)
Little Lost/Birch Creek, Idaho	2	32	66	3	20	77	2	16	82	2	3	93	2	18	80	McCarty (1982a)
Burns Junction, Oregon	23	22	55	16	47	36	7	80	12	13	34	52	15	46	39	McInnis (1984)
Drakes Flat, Oregon	23	4	73	25	20	55	35	11	54	27	6	67	21	25	54	Yoakum (1957a)
Squaw Butte, Oregon	12	2	84	36	21	37	7	22	69	30	8	60	21	13	60	Sneva and Vavra (1978)
Sheldon National Wildlife Refuge, Nevada	11	42	45	10	51	37	3	51	44	2	41	54	6	46	45	Hansen (1986)
Northeastern California	1	7	93	2	16	82	5	27	68	2	11	87	2	15	83	Hanley (1980)
Average													7	27	65	

[a] G = Grasses; F = Forbs, S = Shrubs.
[b] Winter data combined with autumn.

Between the vast midwestern grassland prairies and the Intermountain West dominant shrublands is an ecotone of mixed sagebrush/grassland. This landscape produces an abundance of grasses and forbs, enhanced by receiving late spring and summer rains. Shrubs generally are low in structure and most abundant in ravines. The vegetative community is ideal pronghorn habitat because it produces an abundance of herbs and shrubs for yearlong foraging and plant structure for fawn hiding cover. This photo of the vegetation type was taken in the Yellow Water Triangle area in mid-central Montana, where pronghorn are abundant. Photo by Jim D. Yoakum.

sons (Table 97). They indicated that sagebrush species were consumed most, but snowberry was preferred during summer. Forbs were used mainly during summer and to a lesser extent in spring and autumn. Use of grasses was minor during all seasons, with greatest consumption during spring and autumn.

Pyrah (1987:67) summarized: "Generally, forbs are eaten where and when they are available, especially during wet years; shrubs receive higher use when forbs are not available and during dry years." Forbs were phrased *production plants* because years of higher fawn recruitment were directly related to years when forbs were most abundant and remained succulent through summer. The flowering stage of forbs was short, but different forbs flowered at various times, providing a variety of species used during spring, summer and autumn. Shrubs were considered *survival plants*, used during winter.

Red Desert, Wyoming. Wyoming's Red Desert is well known as pronghorn habitat due to notoriety gained from pronghorn/fence problems and related research projects (Spillett et al. 1967, Oakley and Riddle 1974). It is high, undulating, plains country in the Continental Divide in south-central Wyoming. Elevations range from about 6,500 to 7,300 feet (1,980–2,225 m). Mean annual precipitation is 6 inches (15.2 cm), most of which is received from April through October. In addition, snow is received annually.

A study of the food habits of pronghorn and domestic sheep was conducted during 1964 and 1965 (Severson 1966) to determine the dietary overlap and carrying capacities for the two species—primarily in fenced pastures on native rangelands (Tables 95–97). Percentage cover and utilization by weight of plants were obtained from 1- by 10-foot (0.3- by 3.0-m) plots. Vegetative production was obtained by clipping and oven-dried weights. Pronghorn food habit data were obtained by rumen analysis. Shrubs were used most during all seasons, with sagebrush the dominant and life-sustaining species during winters, but rabbitbrush was readily foraged as soon as growth started in spring and consumed into midsummer.

Rawlins, Wyoming. Another pronghorn food habit study in south-central Wyoming was conducted from 1965 to 1975 by Taylor (1972b, 1975). The project investigated pronghorn using rangelands concurrently with cattle and domestic sheep. A total of 150 plants were identified in the vegetative community (Table 95).

Food habit information was obtained by indirect observations and rumen analyses. The latter was used for compilation of quantitative data (Table 96). Grasses, the first green succulent forage available in spring, were consumed most during that season. When forbs began to grow, the pronghorn switched to them. Shrubs composed nearly half the diet in spring and almost the entire diet during autumn and winter (Table 97, Figure 88).

Sagebrush use was influenced by availability and preference. It was consumed in the same ratio as availability during autumn and winter, but much less so during spring and summer, when pronghorn grazed in sagebrush communities but used other forage species. Thus, plant availability during winter may be more important than palatability, especially during winters when only shrubs protrude above the snow.

The Red Desert of south-central Wyoming is a plateau on the western Continental Divide with elevations averaging 5,000 to 6,000 feet (1,524–1,828 m). Shrubs are the dominant vegetation with sparse understories of herbs due, in part, to the lack of precipitation received during hot, dry summers. Much of the land is in a checkerboard ownership pattern and managed by a mixture of federal government, railroad and private landowner stewardship. Cattle and domestic sheep are numerous, as are the scattered herds of feral horses. Pronghorn are common, and they occupy one of the most accessible areas in North America for tourists on U.S. Highway 80 to observe them grazing western rangelands, as the animals have for centuries. Photo by Jim D. Yoakum.

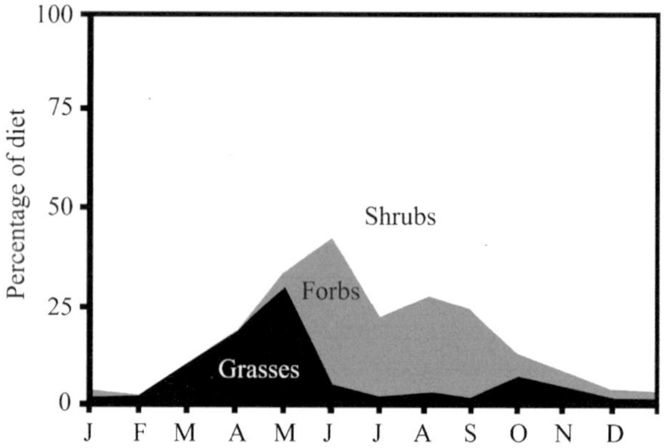

Figure 88. *Pronghorn diet composition for forage class in south-central Wyoming (Taylor 1975).*

Rabbitbrush was a highly consumed species in this study. Although it comprised only 5% of the vegetative cover, it was used every season by pronghorn, accounting for 10% of the annual diet. Summer consumption (19%) was greatest, with autumn use (11%) second. Rabbitbrush was heavily hedged, and well-used trails led to and from rabbitbrush communities.

Forbs were consumed frequently when green and succulent, but less so when desiccated. Only the flowers of some species, such as woody aster, were eaten. Forb consumption related directly to availability. Plains pricklypear was used primarily in late autumn; some rumen samples contained 94%, but most contained small quantities. Grasses were consumed as soon as available in spring, with highest use during May (Table 97). One sample contained 97% grass, while another collected at the same time and place had 2%.

The preference rating for forbs was the highest for that forage class in any pronghorn food habit study located for shrublands. The high rating undoubtedly was influenced by the low production and scarcity of forbs.

Northwestern shrubsteppe, Colorado. Pronghorn food habits were studied in the shrubsteppes of northwestern Colorado from 1965 to 1967 (Bear et al. 1972) (Tables 95 and 96). Forage availability data and rumen samples were collected by seasons. The area's topography varied from plateaus to rolling hills. Forage availability data were collected by phenological periods: spring (April through June), the active growing period; summer (July through September), the maturation period; and autumn and winter (October through March), the dormancy period (Table 97). Rumens were collected for all seasons in the areas sampled for vegetation availability. Highest use of forbs was during spring and summer, when use of shrubs was less. The autumn/winter diet leaned heavily toward shrubs.

Trickle Mountain, Colorado. Pronghorn were one of six sympatric ungulates studied on Trickle Mountain in south-central Colorado by Bailey and Cooperrider (1982). The research was designed to determine seasonal food habits and habitat use of bighorn, cattle, elk, horses, mule deer and pronghorn. A simulation model was developed to integrate information on animal numbers, population dynamics, nutritional requirements and forage production to predict consequences of alternate forage allocations. The study area was 9 by 18 miles (14.5 by 29 km), with landform varying from mesas to rolling hills to steep mountains. Elevations ranged from approximately 7,840 to 13,266 feet (2,390–4,044 m). The climate was semiarid, with annual precipitation for pronghorn habitat ranging from 8 to 16 inches (20.3–40.6 cm). Forage production was determined using the U.S. Bureau of Land Management's (1979) Soil Vegetation Inventory Method. Twenty plant genera comprised more than 1% of the annual forage production, but 150 species were identified during the rangeland survey (Table 95). Forage composition is given in Table 96.

Diet composition was determined by fecal analyses. Only nine taxa contributed more than 5% each to the diet. Grasses were used most during spring, forbs were consumed most during spring and summer, with less use during autumn and winter, no doubt due to availability (Figure 89). Shrubs were browsed heavily during all seasons, with highest use during winter. Sagebrush and rabbitbrush were the most used shrubs.

Data from two winters with significantly different weather conditions were correlated to pronghorn welfare. Severe winters represent a major mortality factor for northern pronghorn populations. Heavy snowfalls cover forage, making it unavailable. Deep snow also limits mobility, forcing animals to be less selective in their foraging habits. Consequently, pronghorn consumed less-preferred plants and subsisted on marginal forage. Food habit data seldom are obtained during such periods. During this study, the 1979 winter had five times more snow than the 1978 winter. During the former, pronghorn consumed six times more cacti and twice as much rabbitbrush as in the mild winter. This suggested that pronghorn can obtain emergency forage, although those forage species were less-preferred plants during average snowfall years (Cooperrider and Bailey no date).

Adobe Ranch, New Mexico. Russell (1964) listed pronghorn diet composition for the Adobe Ranch area in Catron County, New Mexico. The site is a semiarid shrubsteppe, with precipitation averaging 14 inches (35.6 cm), and elevations from 7,500 to 8,500 feet (2,286–2,591 m). The topography features mesas to undulating lowlands with small arroyos. Forage availability data were not collected. Pronghorn diet information was obtained by rumen analyses from samples collected during all seasons (Tables 95 and 97).

Shrubs, such as low sagebrush, provide pronghorn with survival browse, or low-quality nutrition during times of sparse food supply, such as winters. During some years, sprigs of sagebrush above the snow are the only forage available for pronghorn on the windy, snow-covered ridges of the Intermountain West. Sagebrush has high protein levels during the winter, compared with herbaceous vegetation. Therefore, the ubiquitous sagebrush becomes extremely important in providing nutrients for pronghorn survival. Photo by Jim D. Yoakum.

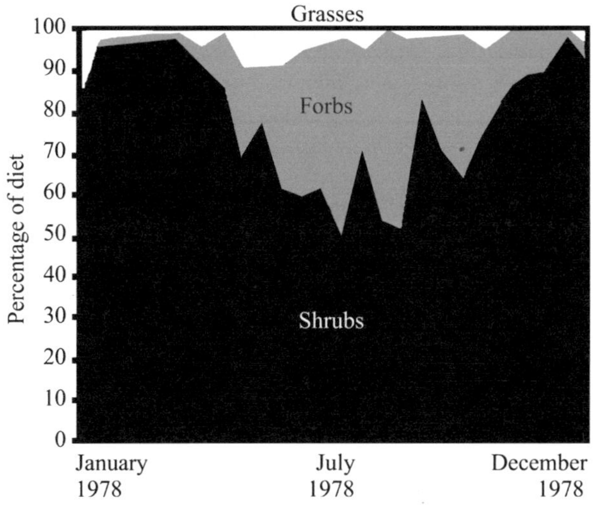

FIGURE 89. *Diet composition of pronghorn, Trickle Mountain, Colorado, 1978 (Bailey and Cooperrider 1982).*

Ensenada Mesa, New Mexico. A pronghorn diet composition study was conducted during 1990 and 1991 on the Ensenada Mesa, located approximately 50 miles (80 km) southeast of Farmington, New Mexico (Hooley 1992). The site was in pinon/juniper and shrubsteppe vegetation types. Elevation ranged from 6,300 to 6,700 feet (1,920–2,042 m). Precipitation averaged 12 inches (30 cm) annually. More than 60% of the area contained native vegetation with percentage vegetative composition of 63 litter, 20 shrubs, 15 grasses and 1 forb. After treatment with herbicides, the vegetative composition changed to 62 litter, 1 shrub, 29 grasses and 3 forbs.

Relative density of forage species in diet composition was determined by fecal analyses (Table 96). Because pronghorn relied heavily on shrubs for winter forage, it was recommended that shrub control projects on winter rangelands be used to create a mosaic where nontreated shrubs were available in close proximity to treated areas.

Sand Hills, Texas. Roebuck's (1982) research during 1980 and 1981 (Tables 95 and 96) covered 11,965 acres (4,842 ha) of deciduous oak shrublands in the sandhills of northern Texas. The topography was flat plains with sand dunes and small depressions. Elevation averaged 3,700 feet (1,128 m). Annual precipitation averaged 12 inches (30.5 cm), with most received during spring and autumn. Oak/grassland was the most prevalent of three major plant communities. Forage measurements were obtained using the frequency of occurrence method. Because most plants were perennials, seasonal changes in composition were minor. Diet composition data were obtained for all seasons from fecal analyses (Table 97, Figure 90). There was relatively little switching of forage class use, compared with that of studies in other shrubsteppe habitats. Pronghorn usually ate succulent green plants. Moisture content appeared to have been important in forage selection. Sagebrush was used most during winter and early spring, when other plants were not available (Table 98).

Arid shrub/grass steppe, Utah. Pronghorn on the arid shrubsteppe of southwestern Utah were studied intensely during the 1960s and 1970s (Smith et al. 1965, Beale and Smith 1970, Smith and Malechek 1974, Smith and Beale 1980). These investigations were accomplished mainly within the Desert Experimental Range, located 50 miles (80 km) west of Milford. Pronghorn habitat covered about 10,000 acres (4,050 ha). Elevation ranged from 5,200 to 6,500 feet (1,585–1,981 m). The area's topography featured long alluvial fans extending from mountains to flatlands.

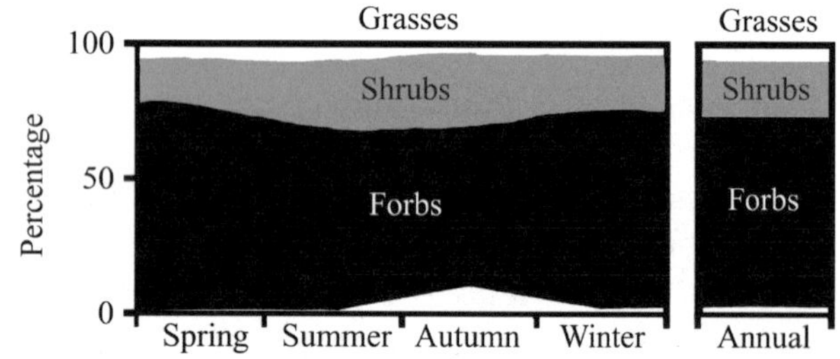

Figure 90. *Seasonal use of forage classes by pronghorn in the Sand Hills study area, Texas (Roebuck 1982).*

Average annual precipitation was 6.8 inches (17.3 cm). The site was winter rangeland for domestic sheep.

The availability of plant species was measured by transects and sample plots (Table 95). Herbage production was determined by the weight-estimate method (see Pechanec and Pickford 1937). Forage production averaged 164 pounds per acre (183 kg/ha) air-dry weight, which was related to precipitation, particularly for forbs (Figure 91). When little rainfall was received, forb production was practically negligible. This was apparent when comparing dry matter of forb production during the high rainfall year of 1967, when 43 pounds per acre (48 kg/ha) were produced, with the drought year of 1969 when 2 pounds per acre (2.2 kg/ha) were produced (Beale and Smith 1970).

Forage consumption was determined by utilization estimates and rumen analyses. Figure 92 indicates seasonal and year-round averages for forage classes ingested.

Expansive salt desert plant communities are located in the semiarid landscape west of Milford, Utah. Topographical features include long alluvial fans extending from mountains to valley floors. Precipitation is received from winter snow and spring rains. Seldom is any moisture received during summer. Vegetative abundance has been measured during occasional summers of extensive rains that resulted in high production of forbs, which were readily consumed by pronghorn. Photo by Donald M. Beale.

Table 98. Pronghorn seasonal food habits and preference ratings for the Sand Hills site in Texas

	Percentage composition of vegetation			*Percentage composition of diet*			*Preference rating*		
Season	*Grasses*	*Forbs*	*Shrubs*	*Grasses*	*Forbs*	*Shrubs*	*Grasses*	*Forbs*	*Shrubs*
Winter	49	11	40	3	73	19	0.0	6.6	0.5
Spring	46	14	40	0	77	15	0.0	5.5	0.4
Summer	53	12	35	2	68	23	0.0	5.7	0.7
Autumn	53	10	37	11	62	22	0.2	6.2	0.6
Annual	50	12	38	4	70	20	0.0	5.8	0.5

Source: Roebuck 1982.

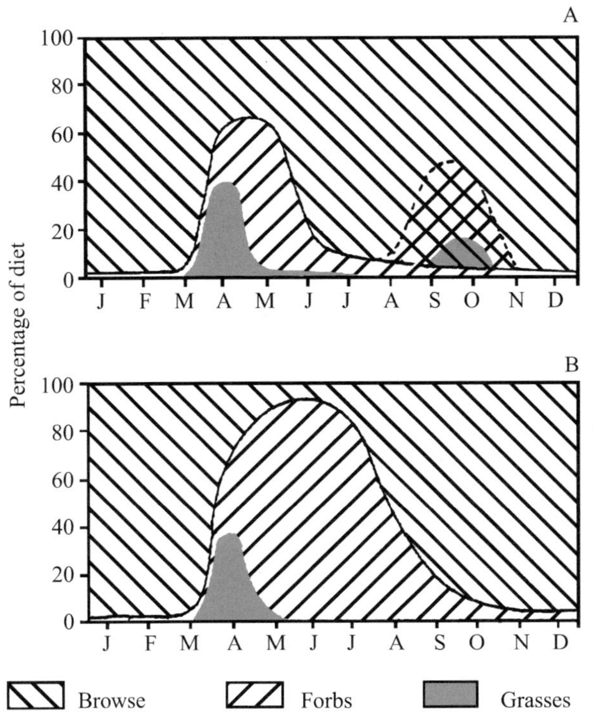

FIGURE 91. *Diet of pronghorn in southwestern Utah under different moisture regimes. (A) Dry season (dotted line indicates change in diet in late summer when rains occur); (B) wet season, with high precipitation during spring and summer (Beale and Smith 1970).*

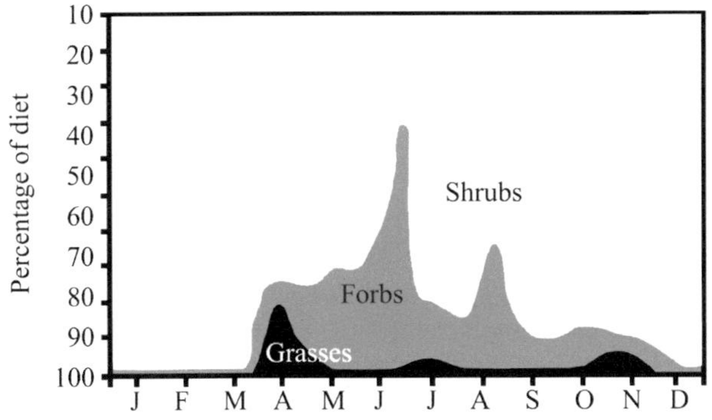

FIGURE 92. *Diet composition of pronghorn in southeastern Utah (Smith and Beale 1980).*

During years of adequate precipitation, forbs were important forage from April through September. During dry years, they provided little food, except from April through June. Grasses were consumed primarily during spring and autumn green-up. Based on observations, pronghorn did not use dry grass at any time. Forbs were used as they became available, depending on precipitation, and were a main part of the diet by May (Table 97). When rainfall conditions were favorable, forbs were the principal dietary component—as much as 95% of the diet. Shrub use accelerated during autumn and was dominant for winter. Black sagebrush was eaten most, followed by brickellia and desert almond.

The food habits of pronghorn fawns between ages 5 and 19 weeks were studied from mid-June to late September. Fawns commenced nibbling forage 2 weeks after birth. Diet composition was determined through analyses of 19 rumens. Twenty-five different species were consumed, with individual fawns eating from 4 to 11 species (Table 99). Shrubs accounted for the bulk of the diet, followed by forbs. Beale and Smith (1970) concluded that little difference existed between adult and juvenile food habits regarding plant species consumed and relative amounts eaten. Schwartz (1977) came to the same conclusion.

Little Lost/Birch Creek, Idaho. A pronghorn habitat management plan was completed in 1982 for 530,490 acres (214,684 ha) of the Little Lost/Birch Creek area in southeastern Idaho. The topography is wide, expansive valleys surrounded by mountain ranges, and elevations range from 4,500 feet (1,372 m) in the valleys to 9,500 feet (2,896 m) in the mountains. Mean annual precipitation was 8.7 inches (22.1 cm). The area was used by cattle and domestic sheep (McCarty 1982a, 1982b). Vegetation in pronghorn habitat was semiarid shrubsteppe. Native rangelands predominated, although pastures of crested wheatgrass and hayfields were within the management area. Vegetative composition is given in Table 96.

Pronghorn diet composition was determined by fecal analyses over a 3-year period. All forage classes were consumed throughout the year (Table 97). Grasses were used most during spring, with only trace amounts consumed during the other seasons. Pronghorn used crested wheatgrass, with one sample containing 11%. Forbs were used as soon as they became available in spring.

Herds switched from shrubs to grasses and forbs as soon as new herbaceous growth began. The area contained a number of alfalfa fields, and alfalfa often was grazed (2%–62% of diet for five samples). Highest forb use was during spring and autumn. Shrubs dominated the year-round diet.

Calcareous gravel benches typical of the landscape support a sparse cover of vascular plants and an abundance of vagrant lichens (Thomas and Rosentreter 1992). Vegetation composition transects revealed 2% lichens comprising five species: range lichen, folded grouse pellet, loose grouse pellet, grouse pellet and coral lichen. The chemical content of similar lichens indicates that protein levels generally do not exceed 2% (Scotter 1972). This protein level is much lower than levels for vascular plants. However, crude fiber contents often contain high levels of carbohydrates that are readily digested. Then too, high fiber content may liberate large amounts of energy when

Table 99. Forbs and shrubs consumed by juvenile pronghorn on the Desert Experiment Station, Utah

	Percentage (by weight) of diet of juvenile pronghorn, by day of age and year of study																			
	48	*68*	*88*	*70*	*85*	*104*	*62*	*97*	*37*	*39*	*49*	*56*	*74*	*81*	*86*	*90*	*91*	*94*	*137*
Forage	*1968*			*1969*			*1970*		*1971*											
Forbs																				
Cryptantha	—	—	—	—	—	—	—	—	2	—	1	—	—	—	—	—	—	—	—
Barestem enceliopsis	—	—	—	—	2	—	—	—	—	—	—	3	—	—	—	—	—	—	—
Eriogonum	—	—	1	—	—	—	—	—	—	—	5	—	38	—	3	—	—	—	34
Euphorbia	—	—	2	1	1	—	—	<1	2	1	—	2	8	—	—	—	—	1	—
Fineleaf hymenopappus	—	—	—	—	—	—	—	—	—	—	—	3	—	—	—	—	—	—	—
Halogeton	—	—	—	1	—	—	—	—	—	—	—	—	—	—	—	—	—	—	—
Lewis flax	2	—	—	—	—	—	—	—	—	—	—	—	—	—	—	—	—	—	—
Hoary astor	—	—	—	1	—	1	—	1	—	—	—	—	—	—	—	—	—	—	—
Machaeranthera	—	—	—	—	< 1	<1	—	—	—	—	—	—	—	—	—	—	—	—	—
Tufted evening primrose	—	—	—	—	—	—	—	—	—	4	—	—	—	—	—	—	—	—	—
Penstemon	9	13	2	—	< 1	—	—	—	—	—	2	1	3	—	2	—	—	—	2
Russian thistle	—	—	—	—	—	—	2	—	—	—	—	—	—	—	—	—	—	—	—
Gooseberry-leaf globemallow	—	—	2	2	7	28	—	15	7	—	—	—	—	—	—	2	—	1	1
Total forbs	11	13	7	7	8	29	2	16	11	5	12	10	49	0	5	2	1	2	37
Shrubs																				
Black sagebrush	84	52	87	76	62	—	80	—	27	78	44	63	42	89	83	—	98	—	14
Fringed sagebrush	—	—	—	—	—	—	—	—	—	—	—	—	—	—	—	1	—	—	—
Shadscale	—	1	—	1	< 1	—	—	1	—	—	—	—	—	1	—	—	—	1	—
Brickellia	—	—	—	—	—	—	—	—	—	—	2	2	—	—	—	—	—	—	—
Cliffrose	—	1	—	—	< 1	—	—	—	—	—	13	22	—	—	—	—	—	—	—
Rubber rabbitbrush	—	—	—	—	—	—	—	—	—	—	—	—	—	—	—	30	—	—	—
Yellowbrush	—	—	<1	—	—	—	—	—	—	—	—	—	—	—	—	—	—	—	—
Ephedra	—	—	—	—	—	—	—	—	—	5	—	—	—	—	—	—	—	—	—
Utah juniper	—	2	—	4	—	—	1	—	—	5	1	—	5	2	9	—	<1	—	28
Desert almond	4	8	6	2	18	3	8	78	60	6	22	2	4	—	1	57	1	91	21
Snowberry	—	—	—	—	—	—	—	—	—	—	—	1	—	—	—	—	—	—	—
Snakeweed	1	23	<1	10	12	68	9	5	2	1	6	—	—	8	2	10	—	5	—
Total shrubs	89	87	93	93	92	71	98	84	89	95	88	90	51	100	95	98	99	98	63

Source: Smith and Beale 1980.

broken down by rumen bacteria and protozoa. The energy release could be an important advantage to pronghorn on winter rangelands.

The windswept benches normally are snow free, thereby becoming travel corridors for pronghorn. Bernt (1976) observed pronghorn eating lichens on these landscapes, but did not quantify diet selection. He did examine one adult pronghorn rumen and noted that it contained an estimated 51% lichens. During sunny days in early spring when the snow remains frozen, lichens absorb radiant energy, thaw and provide moisture. Thomas and Rosentreter (1992) speculated that during years of light snowfall, lichens provide a relatively minor component of the diet, but during periods of heavy snowfall, when vascular vegetation is covered by crusted snow, lichens may be of greater value for pronghorn.

Burns Junction, Oregon. The ecological relationships among feral horses, cattle and pronghorn west of Burns Junction, Oregon, were investigated from 1979 to 1981 (McInnis 1984). Elevation of rangelands occupied by pronghorn was about 3,940 feet (1,200 m). Mean annual precipitation was 8.3 inches (21 cm). The study area lies within the shrubsteppe biome, and supports 143 plant species (Table 95).

Seasonal diets were determined by microscopic analysis of fecal samples. Feral horses and cattle primarily consumed grasses throughout the year, whereas forbs and shrubs formed the bulk of pronghorn diets. Twenty-five forage species were identified in pronghorn diets. Pronghorn principally consumed forbs and shrubs throughout the year. Use of these forage classes indicated wide seasonal fluctuations that coincided with availability of forbs

(Table 97). Forb use peaked during summer months, and shrub use peaked during fall and winter. Consumption of grasses, primarily cheatgrass brome and Sandberg bluegrass, was greatest during winter.

Hart Mountain National Antelope Refuge, Oregon. The 1950 Four-State Antelope Meeting (now the Interstate Antelope Conference) resulted in an agreement for a yearlong cooperative interagency (the California, Idaho and Nevada Departments of Fish and Game; Oregon State Game Commission; and U.S. Fish and Wildlife Service) study of pronghorn food habits because previous collections had been limited primarily to autumn. Rumen samples were collected on Hart Mountain National Antelope Refuge (Hart Mountain) and adjacent rangelands in Oregon, where the topography varied from undulating plains and foothills to block-faulted mountains (Mason 1952). Elevations ranged from 4,500 to 8,000 feet (1,372–2,438 m). Average annual precipitation was 10 to 14 inches (25.4–35.7 cm). Hart Mountain is a shrubsteppe ecosystem dominated by sagebrush, rabbitbrush and bitterbrush. Vegetative composition and diet selection are given in Tables 95 and 96.

Pronghorn habitat on the top of Hart Mountain (elevation 8,400 feet: 2,560 m) in south-central Oregon, overlook shrubsteppe of the Great Basin. Vegetation on the mountain tablelands averages 21 inches (56 cm). Pronghorn use the high plateau during summer, leaving the area in autumn before heavy snowfalls. The phenological growth period of herbaceous vegetation provides succulent, nutritious forbs in summer and autumn—forage that pronghorn highly prefer. Photo by Jim D. Yoakum.

Grass use was heaviest during spring and autumn, and use of green grass was double that of dry grass. Forb consumption was confined primarily to spring and summer months when forbs were succulent, although some dry forbs were consumed during autumn and winter. Shrubs were the dominant forage during all seasons and sagebrush was consumed the most of any available plant. Four sagebrush subspecies comprised 61% of the annual pronghorn diet and were eaten every month of the year. For all samples collected, 96% contained sagebrush. Of the 26 rumens analyzed, 11 contained more than 90% sagebrush; 3 were 100% sagebrush. Bitterbrush was consumed most during July and least in August.

Drakes Flat, Oregon. A pronghorn research program was undertaken in the mid-1950s to study reproduction, mortality and food habits in south-central Oregon (Yoakum 1957a, 1957b). The 15-square mile (38.85 km^2) site was a shrubsteppe plateau (primarily) at 5,250 feet (1,600 m) elevation. Average annual precipitation was 10 inches (25.4 cm), including snowfall of 6 to 12 inches (15.2–30.5 cm). Domestic sheep, cattle and horses occupied the site during all seasons except winter.

Forage composition was obtained by the modified step-point method. Transects were run in all plant types for all seasons. The area had greater grass production than did the surrounding rangelands, due to vegetative changes after the wildfire in the 1940s and intermingled private lands. It was transversed by two perennial streams pro-

Plant species richness, or diversity, numbers 40 to 80 plants on national wildlife refuges along the Oregon/Nevada border. Above is the low-growing desert evening primrose in bloom. It grows profusely, especially on intermittent playa lakebeds, during spring and summer. It is highly preferred and succulent. Almost every spring and summer diet sample collected during the 1990s pronghorn food habit study contained primrose. The plant also grows abundantly along disturbed roadsides. Photo by Jim D. Yoakum.

Table 100. Pronghorn seasonal food habits and preference ratings for shrubsteppe on Drakes Flat, Oregon

Season	*Percentage composition of vegetation*			*Percentage composition of diet*			*Preference rating*		
	Grasses	*Forbs*	*Shrubs*	*Grasses*	*Forbs*	*Shrubs*	*Grasses*	*Forbs*	*Shrubs*
Winter	23	4	73	34	3	63	1.4	0.7	0.8
Spring	25	20	55	42	34	24	1.6	1.7	0.4
Summer	5	11	54	3	51	46	0.0	4.6	0.8
Autumn	27	6	67	6	14	80	0.2	2.3	1.2
Annual	28	10	62	21	25	54	0.8	2.5	0.8

Source: Yoakum 1957a.

viding riparian vegetation and meadows. Forage composition data were collected for all seasons (Tables 97 and 100). Playas produced an abundance of forbs during spring and summer. Forb frequency was twice as great on a wildfire site as on adjacent unburned rangeland. Shrubs were predominantly sagebrush (four subspecies), followed by rabbitbrush and bitterbrush. Food habit information was obtained from rumen samples collected during 1955 and 1956 (Tables 95 and 96). Pronghorn use of dry grass (mainly cheatgrass) during winter was in contrast to Mason's (1952) findings on nearby Hart Mountain. Forbs were used most during spring and summer, corresponding to availability. Although shrubs and grasses were produced in greatest abundance, pronghorn preferred forbs year-round. Grasses had relatively high ratings for winter and spring compared with other studies in the shrubsteppe biome.

Northern Great Basin Experiment Station, Oregon. The study site was located on the Northern Great Basin Experiment Station, 25 miles (40 km) west of Burns, Oregon. Cattle, domestic sheep, feral horses and mule deer occupied the rangeland with pronghorn. The site was 4,500 feet (1,372 m) above sea level. Average annual precipitation was 11.5 inches (29.2 cm), with most occurring as snow during winter or as spring and autumn rains. The topography was characterized by rolling hills, with canyons and volcanic escarpments. Vegetative aspects in-

Drakes Flat is an expansive shrubsteppe benchland close to Plush, in south-central Oregon. Land status is a mixture of public (U.S. Forest Service and U.S. Bureau of Land Management) and private lands. Shrubs are predominantly low sagebrush, with mixed patches of mountain sagebrush, rabbitbrush, bitterbrush and mountain mahogany. Drinking water is available every mile with two permanent streams and several human-made reservoirs. The site portrayed is a traditional fawning area that has been observed for more than 50 years. In the background is the west escarpment of the Hart Mountain Antelope Refuge. Photo by Jim D. Yoakum.

Rangelands on Drakes Flat in south-central Oregon have diverse vegetative types. A wildfire in the 1940s provided a successional plant change from abundant shrubs to a steppe of bunchgrasses and forbs intermingled with low shrubs. Food habit findings provided the highest use of grasses recorded for pronghorn in the Great Basin. Cheatgrass was consumed every month sampled. A correlation was apparent between diversity in diet selection and pronghorn densities. Photo by Jim D. Yoakum.

volved a scattered overstory of juniper and sagebrush and an understory of bunchgrasses and forbs. Forbs comprised less than 15% of the vegetative community. Forage composition and production were not reported (Sneva and Vavra 1978).

Diet composition was obtained from fecal samples collected during all seasons from 1975 to 1977 (Figure 93, Table 97). Their findings were comparable to those of Yoakum (1957b), whose studies were conducted 20 years earlier and on a different site on an Oregon shrubsteppe. Forage class switching was apparent, as it has been for other studies in the Intermountain Region. Sandberg bluegrass and bromegrass were the principal grass species used. Their availability was governed by precipitation.

Few forbs were available, which may account for why few were identified in fecal samples. Also, the identification of forbs is known to be negatively biased in uncorrected fecal analysis (Holechek et al. 1982). Sagebrush was the major shrub in the diet, whereas bitterbrush was consumed more during summer than in winter. Rabbitbrush was locally abundant, but only a minor part of the diet. The high proportion of juniper and sagebrush intake is a concern. Nagy and Tendergy (1967) suggested that essential oils in deer diets containing more than 50% sagebrush may depress bacterial action in the rumen. However, Smith et al. (1965) did not find any visible nutritional impacts on pronghorn living on a diet of 75% sagebrush and 21% juniper, and were of the opinion that pronghorn have different digestive constitutions than do deer. Possibly, the essential oil components in sagebrush and juniper are not all inhibitory; some actually may stimulate digestion, as reported by Oh et al. (1967) for essential oils in Douglas fir relative to deer and domestic sheep.

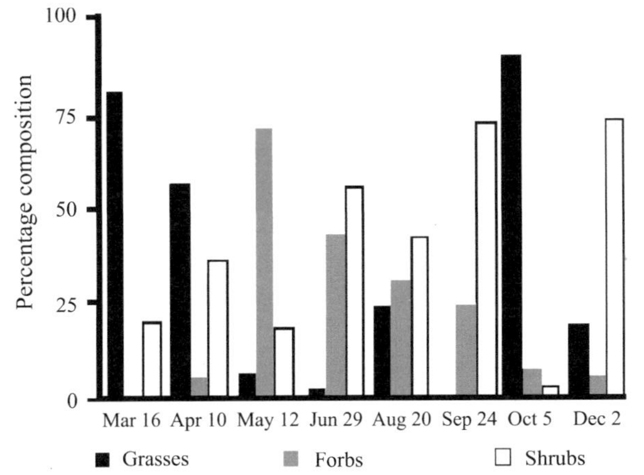

FIGURE 93. *Seasonal dietary composition of pronghorn on the Northern Great Basin Experiment Station, Oregon, 1975 to 1977 (Sneva and Vavra 1978).*

Sheldon National Wildlife Refuge, Nevada. A research project on the food habits of five sympatric ungulates (bighorn, cattle, feral horses, mule deer and pronghorn) was undertaken during 1979 and 1980 on the Sheldon National Wildlife Refuge (Sheldon) in northwestern Nevada (Hansen 1985, 1986). The area contains plateaus, playas, rolling hills and deep canyons. Elevations varies from 5,200 to 6,500 feet (1,585–1,980 m), and the average annual precipitation was 8 inches (20.3 cm).

The site is in the Great Basin, dominated by shrub communities with mixtures of grasses and forbs. Vegetative measurements were obtained with plot frames and line transects (Tables 95 and 96). Diet composition was analyzed from fecal samples collected during all seasons. Forage class switching was common and related to the availability of grasses and forbs during all seasons except winter. Graminoids were used lightly during winter, heavily in spring and lightly again for the remainder of the year. Forbs were ingested throughout the year, with heaviest use during spring, summer and autumn (Figure 94, Table 97). A surprisingly high percentage of forbs (45%) were consumed during winter. Big sagebrush, low sagebrush and bitterbrush comprised 91% of all shrubs in the annual diet (Figure 95). Of these, only low sagebrush—which

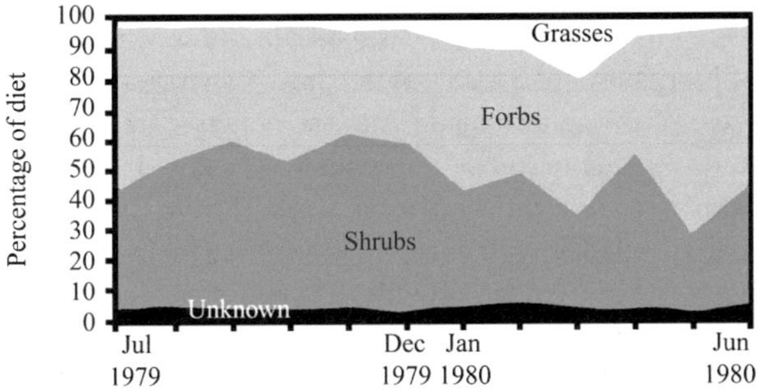

FIGURE 94. *Monthly diets of pronghorn by forage class on the Sheldon National Wildlife Refuge, Nevada, July 1979 to June 1980 (Hansen 1986).*

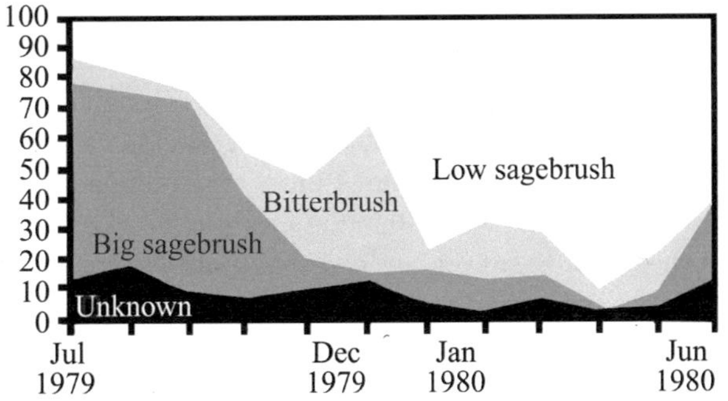

FIGURE 95. *Monthly variation in shrub utilization by pronghorn on the Sheldon National Wildlife Refuge, Nevada, July 1979 to June 1980 (Hansen 1986).*

Rangelands on the Sheldon National Wildlife Refuge in northwestern Nevada provide a shrubsteppe environment typical of Great Basin habitats. Vegetation is dominated by late succession shrublands due to more than a century of intensive livestock grazing and an active wildfire suppression program. Livestock were removed in the mid-1990s, and management strategies were changed to restore natural vegetation. Stands of dense shrublands are being altered by simulating natural wildfires through prescribed burns. The early plant successional stage of grasses and forbs after controlled burns provides a mosaic pattern of vegetative communities readily used by bighorn, mule deer and pronghorn. Photo by Jim D. Yoakum.

constituted 15% of the total diet—was used least during summer. Bitterbrush consumption was highest during summer, consistent with Mason's (1952) findings and Sneva and Vavra's (1978) results for rangelands in Oregon. Preference ratings were similar for adjacent rangeland studies in Oregon by Mason (1952) and Yoakum (1957a).

Northeastern California. Some of the earliest pronghorn rumen analyses were completed during the 1940s and 1950s by the California Department of Fish and Game (Ferrel and Leach 1950, 1952) (Table 95). These studies were of particular importance because little was known of pronghorn food habits at the time. Two major findings were (1) pronghorn made little use of grasses and, therefore, were minor competitors with livestock; and (2) pronghorn ate a wide variety of forbs, some of which were toxic to livestock. The studies reported seasonal forage for various years but did not provide data on forage availability. Unfortunately, sample collections were biased toward autumn. Consequently, comparisons could not be made with other studies reporting year-round use. The investigators identified plants that comprised 1% or more and plants representing less than 1% (identified as trace quantities) (Table 101). Species consumed in trace quantities were 3.5 times more numerous than were species comprising 1% or more of the diet.

Hanley (1980) reported on diet composition for bighorn, cattle, feral horses, mule deer and pronghorn in northeastern California and adjacent rangelands in Nevada. The study site for collection of pronghorn diet data was not specified, but was within the shrubsteppe ecosystem of the Great Basin region, and many of the ecological characteristics would be similar to those described for study sites at Sheldon in Nevada and Oregon's Drakes Flat.

Grasses were consumed most during spring and summer, with light use during autumn and winter. Forbs were used most in summer and moderately during the other seasons. Shrubs were used heavily during all seasons (Table 97). Shrubs comprised a greater percentage of the annual diet than was reported for other Great Basin food habit studies (Mason 1952, Yoakum 1957b, Sneva and Vavra 1978, Hansen 1986). Hanley's (1980) research emphasized developing and testing a food selection model and extending the model to habitat selection. His findings regarding comparisons of food habits for the five sympatric wild and domestic ungulates are discussed in Chapter 15.

Table 101. Plant taxa in pronghorn food habit studies for shrubsteppe habitats in California

Forage	*Number of taxa ≥1% of diet volume*	*Number of taxa ≤1% of diet volume*	*Total*
Grasses	Undetermined	12	>12
Forbs	15	53	68
Shrubs	5	5	10
Total	>20	70	>90

Source: Ferrel and Leach 1952.

Adobe Valley, California. After translocation of 43 pronghorn to Adobe Valley, California, in 1982, an intensive, 2-year follow-up was made (Goldsmith 1988), evaluating pronghorn behavior, movements, diet and habitat suitability. For comparison, a nearby population in the Bodie Hills area also was studied. Adobe Valley is located approximately 15 miles (24 km) southeast of Mono Lake, California, and it has relatively flat terrain, surrounded by rolling hills and mountains. Elevations range from 6,500 to 7,000 feet (1,981–2,134 m). Annual precipitation average 9 to 12 inches (22.9–30.5 cm). The area was grazed by cattle, mule deer and pronghorn. Sampling of vegetative cover, height and species richness was accomplished using the step-point transect method.

Pronghorn diet composition was determined by fecal analyses and direct observations (Tables 95 and 96). Diets showed seasonal fluctuations in composition of all forage classes. Diet richness of taxa for this site (30 species) was

Adobe Valley, near the east-central border of California, was the site of a pronghorn translocation in the 1980s. The site was rated as poor pronghorn habitat before the release, and subsequent studies, including radio telemetry tracking of released animals, proved the validity of the habitat assessment because all pronghorn either succumbed or dispersed to adjacent rangelands (Goldsmith 1988). The shrubsteppe area was dominated by tall sagebrush, and forbs were scarce. Herds dispersed as far as 70 miles (112 km) to rangelands in nearby Nevada. Photo by Jim D. Yoakum.

about half that of Sheldon in Nevada (57), and Drakes Flat, Oregon (62), sites. The diet composition of pronghorn in Adobe Valley differed from those of other populations in the Great Basin, where grass consumption was greatest during spring and autumn (Mason 1952, Yoakum 1957b, Sneva and Vavra 1978, Hanley 1980, Hansen 1986). The Adobe Valley pronghorn used grass the most during summer, when salt grasses, sedges and rushes were available. Forbs also were used the most during summer, but shrubs were the major forage class for all seasons.

Deserts

Pronghorn have occupied the Mojave, Sonoran and Vizcaino deserts for centuries. They still inhabit the latter two, but at low densities. Diet composition studies for desert sites were in Cabeza Prieta National Wildlife Refuge (Arizona Department of Game and Fish 1981), Organ Pipe National Monument (Hughes and Smith 1990, Hughes 1991), northern Sonora (Arizona Department of Game and Fish 1981), and the Vizcaino Desert in Baja California Sur (Cancino 1994c). Although none of these studies contains the quantity of data reported for grassland and shrubsteppe biomes, the limited information that has been published will be reviewed to provide an overview of the pronghorn food habits in deserts.

Cabeza Prieta National Wildlife Refuge, Arizona. The Sonoran Desert, in which the Cabeza Prieta National Wildlife Refuge (Cabeza Prieta) is located, is characterized by block-faulted mountains separated by broad alluvial valleys. Elevations range from 400 to 1,600 feet (122–488 m). The annual average precipitation is 4 inches (10.2 cm). During the hottest part of the year (May to September), temperatures commonly are 110 to 120°F (43.3°–48.8°C). It is the hottest and driest desert in North America. Since the 1950s, several precipitation catchments have been constructed to provide drinking water for pronghorn and other wildlife. The intermountain plains are sparsely vegetated and dominated by creosote bush/bursage flats. When favorable moisture conditions exist, numerous annual forbs and grasses present a spectacular display of desert flowers and forage production (U.S. Fish and Wildlife Service 1982).

Pronghorn diet composition at Cabeza Prieta was determined by fecal analyses during the 1970s (Arizona Department of Game and Fish 1981). Diet samples were collected at various sites during a decade, therefore the data do not represent yearlong selection of forage for a particular site. It can be considered one total, thus, giving some insight to possible patterns of diet selection (Table 102). Forty-nine taxa were identified—2 grasses (plus several unknown species), 29 forbs and 15 shrubs. Grasses

Deserts in North America often receive less than 8 inches (20.3 cm) of precipitation annually and, in some years, as little as 1 to 3 inches (2.5–7.6 cm). Grasses, forbs and shrubs—together with cacti—provide diverse vegetative communities. Years with high rainfall produce a profusion of herbaceous plants, but low precipitation years produce few to no forbs. This characteristic of low herbaceous production in varying years apparently is related to pronghorn diet selection, recruitment and survival. Photo by Jim D. Yoakum.

were not abundant, but pronghorn did not prefer this forage class. Forbs were the preferred species during all seasons with highest use in spring and autumn. Winter and spring rains initiated germination of numerous annual forbs, and pronghorn used them extensively. Cacti were eaten heavily before and after germination of most forbs. It appeared that cacti were not consumed readily when forbs were abundant; this probably was related to the animals' need for moisture from succulent vegetation. Shrubs were most important during the winter, but new growth—including flowers and fruit—of perennial shrubs was eaten as available throughout the year. It appeared that pronghorn preferred forbs, when available, to the other forage classes. The annual average use of forbs (75.5%) was greater than use of this forage class in food habit studies on grasslands and shrubsteppes.

During May, June and July 1988, fecal samples were collected on Cabeza Prieta and Organ Pipe National Monument in Arizona (Hughes and Smith 1990, Hughes 1991). Pronghorn consumed 21 plant species during that time period—1 grass, 13 forbs, 5 shrubs and 2 trees (Table 102).

A research project conducted during 1995 and 1996 provided additional information regarding pronghorn food habits on Cabeza Prieta (Fox 1997). The objective was to determine whether Sonoran pronghorn could meet water and mineral requirements through consumption of forage. Consequently, information on diet selection was not collected. Vegetation samples were analyzed for preformed water and minerals, and are reported later in this chapter. A conclusion of the research was that some areas within the present range of the Sonoran pronghorn were able to provide adequate forage throughout the year, and other sites were seasonally marginal or inadequate. In her concluding analysis, Fox (1997:224–225) wrote: "By fully summarizing current diet information available for Sonoran pronghorn, collecting additional field data on forage species, and using models such as we have presented here, wildlife managers can identify specific targets for management actions and concentrate expensive recovery efforts for maximum benefit to pronghorn populations." This recommendation appears to warrant consideration by wildlife managers faced with extant problems of decreased population densities or apparent low habitat carrying capacities for other endangered pronghorn habitats.

Arizona Department of Game and Fish personnel collected 229 Sonoran pronghorn fecal samples, mainly from Cabeza Prieta, from 1994 to 1998 (Hervert et al. 2000). Because weather conditions lacked the four traditional North American seasons, samples were catagorized: wet summer, wet winter, dry summer and dry winter (Table 103).

Pronghorn diets varied among seasons and years depending on availability of forage. Plant availability, succulence and palatability in the Sonoran Desert are dependent largely on the amount and timing of precipitation and temperatures. Pronghorn fed on 97 plant taxa—8 grasses, 50 forbs (including cacti) and 39 shrubs (including trees).

Forbs and shrubs were the main forage classes in diets (Table 103). Shrubs were the highest percentage of diet during all seasons except wet summers. Shrubs were important when forbs were scarce. Forbs were the main forage class during wet summers when they were available and succulent.

Hervert et al. (2000:24) reported "The importance of highly nutritious forbs to Sonoran pronghorn diets during winter and dry summer seasons probably reflects a lack of abundant, succulent forbs and/or difficulty in finding forbs due to the unpredictable and sporadic nature of rainfall and corresponding plant growth. The importance of highly nutritious forbs to Sonoran pronghorn, particu-

TABLE 102. Monthly percentage composition of forage classes in pronghorn diets in deserts

Location	Forage class	*Percentage of diet*										Average	Source
		Jan	*Feb*	*Mar*	*Apr*	*May*	*June*	*July*	*Sept*	*Oct*	*Dec*		
Cabeza Prieta National Wildlife Refuge, Arizona[a]	Grasses	Tr[b]	0	0	0	0	0	0	0	3	0	Tr	Arizona Department Game and Fish (1981)
	Shrubs	38	76	84	79	88	75	0	98	66	0	76	
	Forbs	57	23	16	19	12	20	0	2	30	0	22	
	Grasses	—	—	—	—	—	—	—	—	—	—[c]	6	Hervert et al. (2000)
	Forbs	—	—	—	—	—	—	—	—	—	—[c]	45	
	Shrubs	—	—	—	—	—	—	—	—	—	—[c]	43	
	Unidentified	—	—	—	—	—	—	—	—	—	—[b]	6	
Northern Sonora, Mexico	Grasses											0	Arizona Department Game and Fish (1981)
	Forbs											88	
	Shrubs											12	
Cabeza Prieta National Wildlife Refuge and Organ Organ Pipe Cactus Monument	Grasses					—	—	—	—[c]			Tr	Hughes and Smith (1990)
	Forbs					—	—	—	—[c]			78	
	Shrubs					—	—	—	—[c]			22	
Vizcaino Desert, Mexico	Grasses	—	—	—	—	—	—	—	—	—	—[c]	4	Cancino (1994c)
	Forbs	—	—	—	—	—	—	—	—	—	—[c]	22	
	Shrubs	—	—	—	—	—	—	—	—	—	—[c]	44	
	Unidentified[d]	—	—	—	—	—	—	—	—	—	—[c]	30	

[a] Diet samples collected during various years and at various sites.
[b] Tr = less than 0.5%.
[c] Season samples collected.
[d] Mainly flowers of forbs and shrubs.
Originally reported percentages have been rounded to the nearest 1%.

larly at critical times of the year such as late gestation, lactation, weaning and growth of fawns, should not be underestimated. However, when forbs are lacking, as is common in the Sonoran Desert, perennial browse (shrubs) most likely sustains pronghorn." The authors correlated years of high fawn recruitment with years of high forb production.

Pronghorn fed on cacti during all seasons, but use (particularly chain cholla) increased during dry seasons. Amounts consumed ranged from 7% in wet winters to 14% in dry summers. The succulent fleshy fruits of chain cholla provide preformed water, particularly important during hot summers to meet increased demands for water.

TABLE 103. Percentage composition of forage classes found in Sonoran pronghorn fecal samples during 1994 through 1998 in Arizona

Forage class	*Summer*		*Winter*		*Yearlong average*
	Dry	*Wet*	*Dry*	*Wet*	
Grasses	3	13	3	7	6
Forbs	31	42	29	36	35
Cacti	14	9	10	7	10
Total	45	51	39	43	45
Shrubs (includes trees)	46	28	53	43	43
Unidentified	5	8	5	5	6
Number of samples	85	25	72	47	229

Source: Reconstructed from Hervert et al. 2000.

Northern Sonora. Limited information is available for pronghorn food habits in northern Sonora. Many of the habitat characteristics for this site are similar to those of Cabeza Prieta. During October 1977, representatives of the Department of Wildlife for Mexico collected pronghorn fecal samples from four different areas in the northern region of Sonora (Table 102). Nine species of forbs, 10 species of shrubs and no grasses were identified in the diet (Arizona Department of Game and Fish 1981).

Vizcaino Desert, Baja California Sur. Working on the Vizcaino Desert in the state of Baja California Sur, Cancino (1994c) studied food habits of the rare peninsular pronghorn. He reported that only 50 to 200 animals existed, yet he was able to collect enough fecal samples for a year-round analysis of dietary composition. The pronghorn's habitat is merely 200 feet (61 m) above sea level, averages 3.5 inches (8.9 cm) of precipitation annually and has a mean year-round temperature of 68°F (20°C). A total of 97 taxa comprised the vegetative community.

Pronghorn consumed all three forage classes year-round, but there was less forage class switching than on

The Sonoran Desert, located in Arizona, Sonora, Baja California and Baja California Sur, is the hottest desert in North America. Sonoran pronghorn living in Mexico and the United States are classified as endangered. Drinking water is scarce, so the pronghorn there obtain the majority of moisture needed for body functions from vegetation, or preformed water. Diet selection is heavy toward forbs with a large percentage of cacti containing high moisture content. Photo by Jim D. Yoakum.

grasslands and shrubsteppes. Flowers of forbs and shrubs composed most of the unidentified 30% noted in Table 102, but species were not identified. Grasses were eaten primarily during spring. Forbs were consumed about equally during spring, autumn and winter, but less during summer, when shrubs were used most heavily. A list of 59 plant species was obtained from diet analysis, including 3 grasses, 30 forbs and 26 shrubs. Three species of lichens were noted in the fecal samples. They may be important because they collect water during foggy nights and mornings at this arid sea-level site.

Research and Management Implications. A review of the literature discloses no known yearlong complete diet study for pronghorn in the desert, nor any food habit study in relation to the composition or production of available forage (Yoakum 1990). What little information is available is anecdotal and indicates pronghorn highly prefer forbs, shrubs are a staple diet component and grasses are lightly used—foraging habits similar to those of pronghorn in grasslands and shrubsteppes. Food habit studies of desert pronghorn are needed to quantify forage species consumed yearlong in relation to precipitation and the availability of vegetation forage classes.

Hart Mountain and Sheldon Project

The U.S. Fish and Wildlife Service completed an environmental impact statement and comprehensive management plan in 1994 for Hart Mountain, Oregon (see Chapter 16). One aspect of this planning effort was to initiate a food habit study for bighorn, feral burros, feral horses, mule deer and pronghorn (Yoakum 1992). The nearby Sheldon, in Nevada, was included. Environmental descriptions of the refuges are reported in this chapter under "Food Habit Study Sites."

The Hart Mountain/Sheldon Project was initiated because wildlife managers needed current, valid data for decisions regarding diet overlap among sympatric ungulates and vegetation manipulation strategies to restore healthy plant communities. Diet composition data for pronghorn on Hart Mountain were more than 40 years old and limited in detail. Studies were not available for other ungulates. Hansen (1985) conducted a 1-year investigation on Sheldon for bighorn, mule deer and pronghorn diets, but information was not obtained regarding weather, forage availability and quality, or correction factors for differential digestibility of forbs, grasses and shrubs.

Field work was conducted from October 1993 to September 1995 by U.S. Fish and Wildlife Service personnel. Microhistological analyses of ungulate fecal samples were the primary assessment procedure. Determination of diet composition and quality were accomplished by the Wildlife Habitat Laboratory at Washington State University, Pullman. The final project report was compiled by the Oregon Cooperative Wildlife Research Unit, Corvallis (Hansen and Anthony 1999) and reported at the 19th Pronghorn Antelope Workshop, conducted during March 2000 in La Paz, Baja California Sur, Mexico (Hansen et al. 2001). Much of the following information was obtained from that report.

The project was a significant study of pronghorn food habits because:

1. Two areas within the Great Basin were studied; there were ecological differences. Hart Mountain had higher elevations than Sheldon, and it exhibited greater diversity in plant composition and phenology.
2. Fecal and plant samples were collected during 2 years with differing precipitation and temperature patterns.
3. Fecal samples were collected each month for bighorn, feral burros, feral horses, mule deer and pronghorn. Although cattle had been on both refuges for more than a hundred years, they recently had been removed.

4. A concurrent field survey of plant composition and vegetation types was conducted by the U.S. Fish and Wildlife Service. This provided data to compute preference ratings for vegetation forage classes.
5. Diet composition was calculated with and without digestibility correction factors.
6. Diet quality using diaminopimelic acid (DAPA) and fecal nitrogen (FN) factors was assessed and correlated with weather, particularly temperature and precipitation.
7. Dietary overlap ratios were computed for all wild and domestic ungulates.

Like the study at Trickle Mountain, Colorado, reported earlier in this chapter, fecal and plant samples were collected during 2 years with major variances in weather patterns. The use of digestibility correction factors in an attempt to record consumption of differing forage species more accurately had not been attempted in former pronghorn studies.

Fecal samples were collected for all species of wild ungulates and feral horses on both refuges. In addition, samples were obtained for feral burros on Sheldon. Fecal samples were collected by observing the ungulates until they defecated. Plant collections were made for species comprising the major portion of diets. Plants and fecal samples were collected from the same locations.

Digestibility Correction Factors. Previous herbivore food habit studies have noted the biased reporting of plant percentages in fecal analyses (Dearden et al. 1975, Vavra and Holechek 1980, Holechek and Valdez 1985). Pulliam and Nelson (1979) recognized this problem and conducted a study of correction factors, also phrased *relative digestibility coefficients*, for plants in the three forage classes during elk diet studies. They concluded that (1) significant differences exist in digestibility between forage species, forage classes and phenological stages; (2) forbs are more easily digested than other forage classes; (3) grass species vary greatly in digestibility, with highest digestibility of annuals; (4) as plant growth advances toward maturity, digestibility decreases; and (5) quantification of fecal analysis without correction for differential digestibility may be severely biased.

Pulliam and Nelson (1979:240) concluded that, "Results suggest that plant species with a high digestibility may be underestimated in the composition of diet while those species with lower digestibility would be overestimated. The fecal analysis method overcomes many disadvantages of other methods and, with incorporation of correction factors for differential digestibility, shows potential to be a rapid method of determining food habits of large herbivores with reasonable accuracy." Table 104 compares diet composition percentages based on three plant species tested by Pulliam and Nelson.

The value of digestibility correction factors has been investigated for mule deer (Holechek and Valdez 1985) and Dall's sheep (Hansen 1996) using inoculum from cattle and domestic sheep, respectively. However, studies are not known for pronghorn before the Hart Mountain/Sheldon Project. The first year of study was devoted to fecal collections and analyses. Plant species consumed in large quantities then were collected during the second year. In vitro procedures described by Leslie et al. (1983) were used to calculate differential digestibility correction factors for plant species. These factors then were applied to the uncorrected diet percentages for a more accurate estimate of the relative quantity of each plant species consumed. Diet selection tables were compiled.

Diet Composition and Preference Ratios. Table 105 lists the number of plant species eaten by each animal species

Although the Hart Mountain National Antelope Refuge, in south-central Oregon, is in the shrubsteppe biome and shrubs are the dominant forage produced, recent pronghorn food habit studies reveal forbs are the preferred yearlong forage class, shrubs are next and grasses are seldom used. Nutritional values of forbs are highest during spring and summer—the seasons for pronghorn parturition and lactation. Differential digestive ratios were estimated for the first time for pronghorn diet studies. Pronghorn heavily grazed perennial forbs during the winter with less snow. Results of these food habit studies support habitat management strategies that enhance natural plant communities with a mixture of grasses, forbs and shrubs for pronghorn habitats. Photo by Jim D. Yoakum.

Table 104. Comparison of uncorrected and corrected pronghorn diet composition with actual diet composition, based on a 7-day feeding trial with three forage species

			Estimated diet in fecal sample			
Species	*Forage class*	*Actual diet*	*Uncorrected*	*Percentage error*	*Corrected*	*Percentage error*
Alfalfa	Forb	91	8	91	88	3
Geyeri sedge	Grass	8	89	1,060	11	30
Balsamroot	Forb	2	3	529	1	29

Source: Modified from Pulliam and Nelson 1979.

Table 105. Number of forage taxa consumed by three wild and two feral ungulates on the Hart Mountain National Antelope Refuge (HMNAR) and the Sheldon National Wildlife Refuge (SNWR) from 1993 to 1995

	Number of plant taxa in forage consumed								
	Pronghorn		*Mule deer*		*Bighorn*		*Feral horses*		*Feral burros*
Forage class	*HMNAR*	*SNWR*	*HMNAR*	*SNWR*	*HMNAR*	*SNWR*	*HMNAR*	*SNWR*	*SNWR*
Grass	13	13	16	13	15	18	17	23	15
Forb	27	27	29	31	34	32	10	9	11
Shrub	12	11	14	13	12	7	11	10	7
Total	52	51	59	57	61	57	38	42	33

Source: Hansen et al. 2001.

on Hart Mountain and Sheldon. Deer and pronghorn primarily consumed forbs and shrubs, whereas bighorn mainly ate grasses and forbs. Pronghorn ate 52 plant species on Hart Mountain and 51 species on Sheldon. Feral equids concentrated on grasses, with some seasonal forb and shrub use.

Sedges and rushes were added to the grass forage class and moss was listed with forbs; these additional species generally were in trace amounts (less than 1%) of the diet. By listing all plant species in the three forage classes (grasses, forbs and shrubs) commonly used in food habit studies, it was possible to compare results of this study with similar studies in the Great Basin.

Diet composition percentages for pronghorn use of grasses, forbs and shrubs are provided in Table 106 and Figure 96 for Hart Mountain; those for Sheldon are in Table 107 and Figure 97. Listings are provided for corrected and uncorrected plant species percentages. The data substantiates that grasses and shrubs rated higher in uncorrected percentages than forbs. Forb use was greater on Hart Mountain than on Sheldon. This may have been related to higher elevations producing a more diverse plant phenology and earlier removal of large numbers of cattle. The high percentage (71%) of forbs consumed by pronghorn during winter on Hart Mountain in this study was most unusual. Wildlife and range managers have been prone to assume that forbs dry up and blow away in the winter. However, this study documents that pronghorn find and consume large quantities of perennial forbs during winter.

The uncorrected percentages show that pronghorn annually averaged approximately 15% more shrubs than forbs. However, when correction factors were applied, forbs exceeded shrubs by 17% on forb-rich Hart Mountain (Table 106). Pronghorn on Sheldon, where there is less diversity of forbs, consumed a higher percentage (65.1%) of shrubs than forbs (24.7%) (Table 107).

Table 106. Diet composition of pronghorn measured by microhistological analysis of feces from the Hart Mountain National Antelope Refuge, Oregon, from October 1993 to September 1995

	Percentage diet composition									
	Uncorrected					*Corrected*				
Forage class	*Spring*	*Summer*	*Autumn*	*Winter*	*Annual*	*Spring*	*Summer*	*Autumn*	*Winter*	*Annual*
Grass	22	4	13	12	13	14	2	10	6	8
Forb	38	31	29	49	37	48	31	47	71	49
Shrub	40	66	59	39	51	38	67	43	24	43

Originally reported percentages rounded to the nearest 1%.
Source: Hansen et al. 2001.

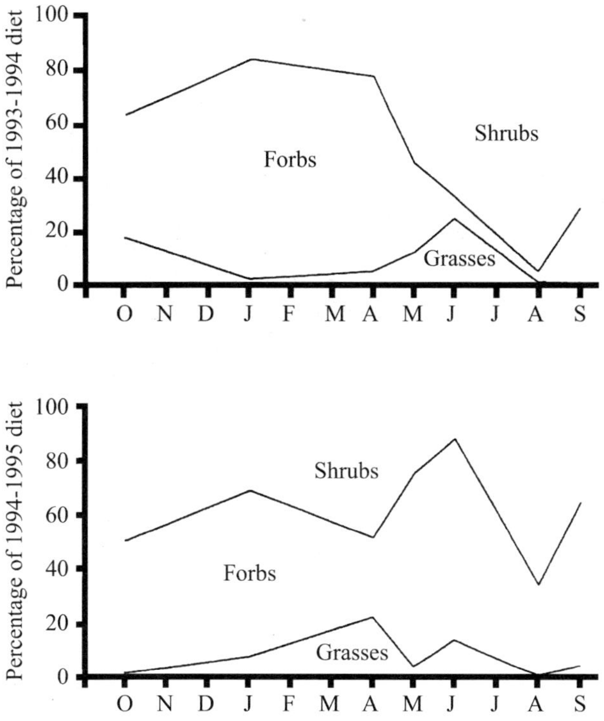

FIGURE 96. *Species composition of pronghorn diets measured by microhistological analyses of feces from Hart Mountain National Antelope Refuge, Oregon, from October 1993 to September 1995 (Hansen and Anthony 1999).*

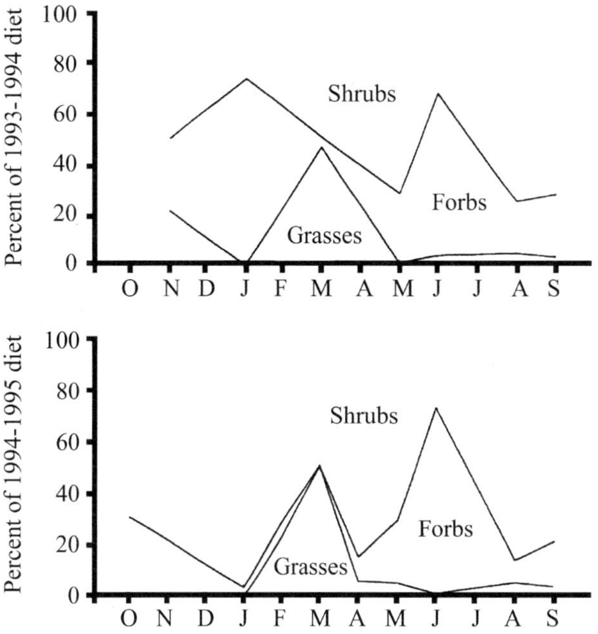

FIGURE 97. *Species composition of pronghorn diets measured by microhistological analyses of feces from Sheldon National Wildlife Refuge, Nevada, from October 1993 to September 1995 (Hansen and Anthony 1999).*

Forage preference ratings in Figure 98 substantiate that pronghorn on the Hart Mountain shrubsteppe from 1993 to 1995 greatly preferred forbs.

Table 108 provides diet composition data for all ungulates. Pronghorn and mule deer foraging strategies were remarkably similar, varied seasonally and were linked with differing annual precipitation patterns. Bighorn diets showed greater variance for Hart Mountain than Sheldon—again possibly due to plant diversity characteristics of each refuge. For equids on both refuges, the corrected percentages showed greater use of forbs than did uncorrected figures.

Horses consumed large quantities (13%) of forbs during winter on Hart Mountain, which could result in competition with pronghorn during years with low forb production. Information pertaining to dietary overlap for all ungulates is reported in Chapter 15.

Diet Quality. Diet quality was measured by FN analysis. Feces contain nitrogen from undigested forage and metabolic fecal products because it is correlated with diet quality and forage indexes digestibility (Wehausen 1995). However, FN can misrepresent diet quality when indigestible plant cell wall components (e.g., lignin) and plant defensive compounds (e.g., tannins) that inhibit protein digestion are in the diet (Robbins et al. 1987). These plant defenses are common in shrubs frequently consumed by pronghorn and deer.

DAPA, an amino acid, is contained in the indigestible portion of the cell walls of rumen bacteria and comprises

TABLE 107. Diet composition of pronghorn measured by microhistological analysis of feces from the Sheldon National Wildlife Refuge, Nevada, from October 1993 to September 1995

	Percentage diet composition									
	Uncorrected					*Corrected*				
Forage class	*Spring*	*Summer*	*Autumn*	*Winter*	*Annual*	*Spring*	*Summer*	*Autumn*	*Winter*	*Annual*
Grass	7	6	9	28	11	3	5	8	24	10
Forb	33	17	20	13	21	39	15	28	17	25
Shrub	60	76	71	65	68	58	81	64	59	65

Originally reported percentages rounded to the nearest 1%.
Source: Hansen et al. 2001.

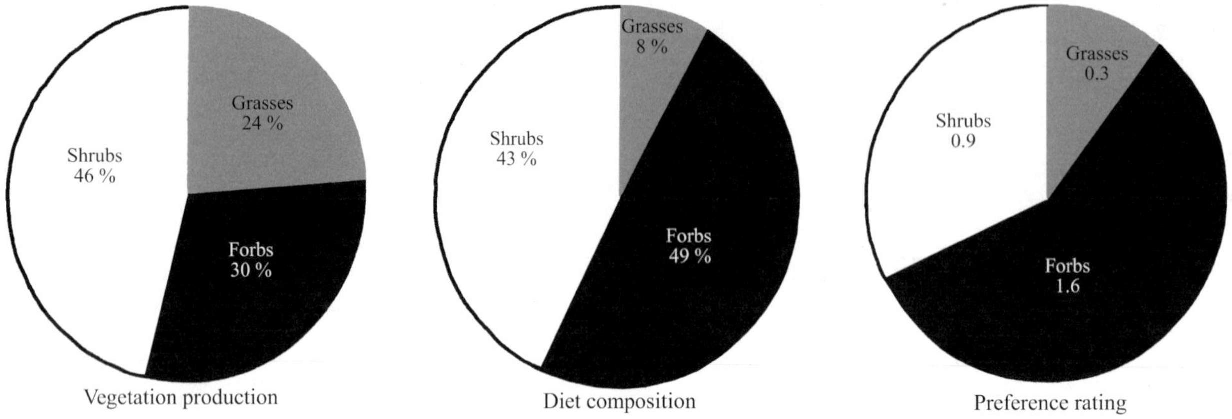

Figure 98. *Percentage composition of vegetation production, pronghorn diet and preference ratings for Hart Mountain National Antelope Refuge, Oregon, 1993 to 1995 (Hansen et al. 2000).*

Table 108. Annual diet composition for wild and feral ungulates on the Hart Mountain National Antelope Refuge and Sheldon National Wildlife Refuge from October 1993 to September 1995

		Hart Mountain		*Sheldon*	
Species	*Forage class*	*Uncorrected percentages*	*Corrected percentages*	*Uncorrected percentages*	*Corrected percentages*
Bighorn	Grass[a]	71	59	74	70
	Forb[b]	22	32	21	27
	Shrub	8	9	5	4
Feral burro	Grass[a]	(No burros	(No burros	89	84
	Forb[b]	on refuge)	on refuge)	5	7
	Shrub	—	—	6	9
Feral horse	Grass[a]	89	84	92	91
	Forb[b]	8	13	6	7
	Shrub	3	3	3	2
Mule deer	Grass[a]	13	9	11	12
	Forb[b]	46	50	25	34
	Shrub	41	41	64	54
Pronghorn	Grass[a]	13	8	11	10
	Forb[b]	37	49	21	25
	Shrub	51	43	68	65

[a] Includes sedge and rushes.
[b] Includes moss and seed.
Originally reported percentages are rounded to the nearest 1%.
Source: Hansen et al. 2001.

a relatively consistent percentage of the bacterial mass (Goldsmith 1988). Levels of fecal DAPA, like FN, reflect seasonal changes in plant availability and phenology (Kie and Burton 1984). DAPA is not artificially elevated by plant defense compounds and should more accurately index diet quality for browsing herbivores than does FN. The combination of these two indices can help managers assess diet quality and provide a baseline for comparison with future diet quality measurements.

During the Hart Mountain/Sheldon Project, the lowest FN occurred in winter (December to February) and peaked in spring (April to June). Similarly, DAPA values reached lows during winter and peaked in spring. The FN values for pronghorn were similar at the two refuges during spring and winter (Figure 99). Goldsmith (1988), working with pronghorn in a Great Basin environment in Adobe Valley, California, also found DAPA and FN varied seasonally with plant phenology and had the lowest levels during winter and early spring. Goldsmith compared forage composition, DAPA and FN for three different pronghorn habitats and concluded that diet quality cannot be predicted by forage composition alone, but is related to other factors affecting diet quality, such as plant phenology, abundance of preferred forage species, grazing

The Spring Table area on the Sheldon National Wildlife Refuge (Sheldon), in northwestern Nevada, has been relatively livestock free for more than 60 years, although a few feral horses still roam there. There exists some dietary overlap for perennial forbs between pronghorn and feral horses because the latter consume forbs (when available) in winter. During years of high snowfall, nutritious herbaceous forage is limited, and competition between horses and pronghorn increases. Vegetation transects run during 1994 indicated approximately 21% grasses, 50% forbs and 26% shrubs. Plants in all forage classes appeared healthy with adequate reproduction and vigor. Elevations are lower, precipitation is less and rangelands have received more intensive livestock foraging on Sheldon than on the nearby Hart Mountain National Antelope Refuge. Thus, the Sheldon maintains more dominant shrublands with less herbaceous vegetation. These differences in availability of forage classes are reflected in percentages of diet selection by pronghorn. Photo by Jim D. Yoakum.

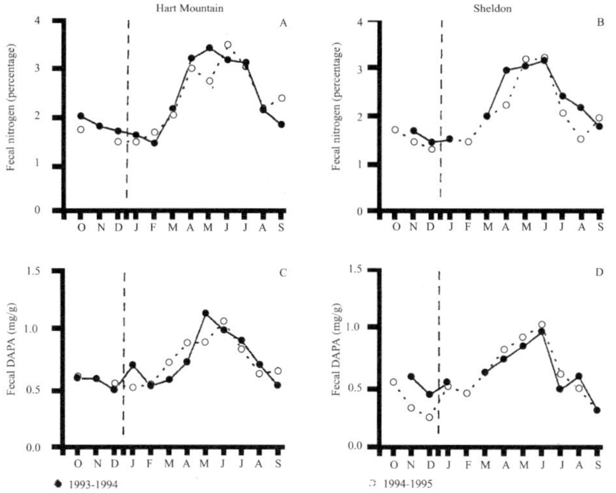

FIGURE 99. *Fecal nitrogen and fecal diaminopimelic acid (DAPA) measurements from pronghorn at Hart Mountain National Antelope Refuge, Oregon* (A and C), *and Sheldon National Wildlife Refuge, Nevada* (B and D), *from October 1993 to September 1995 (Hansen and Anthony 1999).*

Next to different subspecies of sagebrush, the shrub most browsed by pronghorn in Oregon and Nevada diet selection studies was bitterbrush. The species is highly nutritious, low growing and relatively abundant. Flowers, leaves and stems are consumed with preferences for the first two parts. Bighorn, mule deer, cattle and domestic sheep browse bitterbrush, but feral horses do not. Photo by Jim D. Yoakum.

intensities by sympatric ungulates and other animals. The DAPA and FN are good indicators of diet quality because they are affected by all of these factors simultaneously.

Weather Interactions. Diet selection and quality for pronghorn are largely functions of vegetation composition, plant phenology and nutritional values. Consequently, seasonal and annual weather patterns, elevation and other orographic influences, as well as the condition of vegetation types and forage competition, affect herbivore diet quality. Pronghorn can influence their diet quality to some degree by traveling to rangelands where there are more nutritious plants or to habitats where forage is in a more nutritious phenological stage. However, regionwide weather patterns, such as heavy snowfall and low precipitation, affect diet quality in ways for which pronghorn cannot always compensate.

Hansen and Anthony (1999) reported that diet quality was closely correlated with weather, especially temperature and precipitation, and that most of 1994 to 1995 was colder than 1993 to1994 (Figure 100). Lower pronghorn diet quality in the winter of 1994–95 compared with 1993–94 probably was due to the colder temperatures that delayed plant phenology and availability. Differences in diet quality between these 2 years were from September to November and again in May, times that coincide with conception and parturition.

Oftedal (1985) found that reductions in maternal diet quality reduced maternal weight, birthmass of fawns and milk production—factors of paramount concern for managers of pronghorn in Great Basin environments, according to Ellis (1970). Fawn survival also is reduced by low maternal diet quality. Reductions in diet quality for long

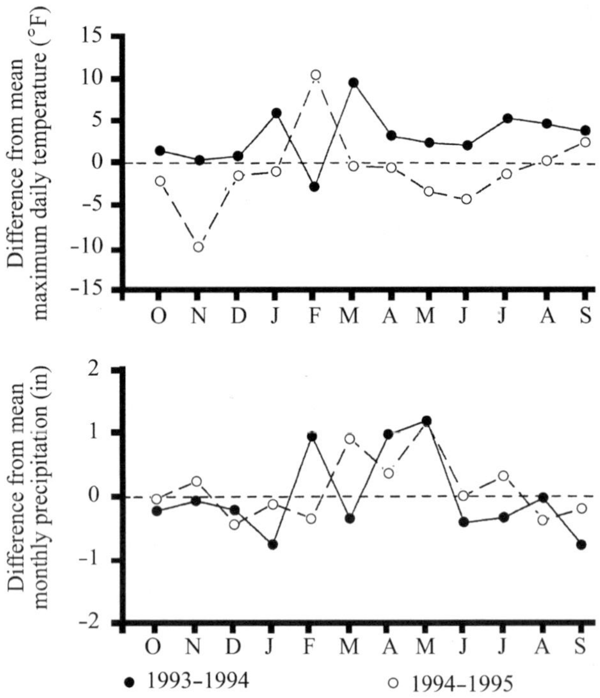

FIGURE 100. *Difference from mean maximum daily temperature (°F) and mean precipitation (inches) for each month between October 1993 and September 1995 for Hart Mountain National Antelope Refuge, Oregon (Hansen and Anthony 1999).*

periods and close to parturition and early lactation have the greatest effect on wild herbivores (Price and White 1985). The extremely low pronghorn fawn survival reported (U.S. Fish and Wildlife Service 1996a) for Hart Mountain in 1995 (0.8 fawns:100 does in late July) was consistent with Hansen and Anthony's (1999) data showing pronghorn diet quality to be lower than in 1994.

Research and Management Implications. The Hart Mountain/Sheldon Project was designed to provide basic information on the diets of pronghorn with sympatric wild and feral ungulates on two refuges. Fecal and plant samples were collected during 2 years with different weather patterns, resulting in variable diet composition and quality between years. Although this variation gives managers an indication of the range of diet composition and quality for pronghorn, it can make it difficult to discern small changes in population trends that may be related to other management strategies.

Hansen and Anthony (1999:16) stated, "Our analyses show that diet quality was correlated with temperature and precipitation for these 2 years, and this suggests that production of young by these ungulates is also correlated with temperature and precipitation." Recognition of the correlation of pronghorn population welfare and weather patterns is a subject not well verified in research or management studies. Because factors affecting population fluctuations are of utmost importance to pronghorn managers, a long-term investigation with different weather patterns is a needed research project.

Of the 40 pronghorn food habit studies reported in this chapter, more than 90% were accomplished from 1950 to 1990 using techniques that have been improved during the past decade. Also, few studies incorporated other factors (e.g., competition with other herbivores, differing weather patterns over years, composition of available forage or diet quality) to provide a more holistic knowledge of the various environmental factors affecting pronghorn diets. Consequently, the Hart Mountain/Sheldon Project may serve as a model for researchers and managers by providing tested procedures for designing future comprehensive pronghorn food habit investigations. One additional segment of a recommended food habit study not conducted in the Hart Mountain/Sheldon Project is moisture content of vegetation samples for data pertaining to preformed water.

Assessing Food Habit Methods and Findings

Many attempts have been made to compare pronghorn diet composition, but the exercise has been complicated because pronghorn occur in three diverse biomes, from Canada to Mexico, with great variation in vegetation and weather among the ecosystems. Also, as previously indicated, many of the more than 200 food habit studies published about pronghorn used different procedural techniques. Comparison among studies, therefore, can be misleading if not impossible. The following review may help to improve procedures for greater consistency, reliability and usefulness.

1. Many pronghorn diet studies between 1950 and 1970 used collection of samples from hunter kills during autumn. This, in itself, is not necessarily a problem. However, in such instances when autumn samples represented the only data, and those data are assumed to be representative of a population's annual diet, erroneous assumptions and management applications can result. Pronghorn's forage selectivity is much greater in spring and summer than during autumn and winter, simply because more plant species are available. Food habit studies are of limited usefulness unless their data include adequate samples from all seasons in a defined area. This is especially true for pronghorn food habit studies on northern rangelands, where seasonal weather changes can affect forage availability more than do seasonal shifts and attendant weather on southern rangelands.
2. Early diet investigations were limited to direct observations or rumen analyses. Both techniques have advantages and disadvantages. More recently, food habit studies have been accomplished by the fecal analysis technique. It, too, has weak points, but it also has two particularly strong attributes (a) many samples can be collected inexpensively for all seasons; and (b) subject animals do not have to be sacrificed. The use of digestibility correction factors has indicated this technique is needed for fecal analyses. The trend toward more studies using this technique may make diet studies easier to compare in future years.
3. From review of study reports, it is apparent that investigators used the term preference loosely. It often has referred to frequency of use for a given plant, leading to misunderstandings. The term should be based on percentage composition of the species in the plant community relative to its percentage use in the diet.
4. Past diet studies of pronghorn have focused on sagebrush/grasslands, semiarid shrubsteppes and cold desert shrublands (Sundstrom et al. 1973, Yoakum 1990). Exceptions for grasslands were Buechner (1950a) and Russell (1964). Since then, numerous studies have been conducted on grasslands from Canada to Mexico (Table 86) that support two-thirds of the current pronghorn population.
5. The division of vegetation into forage classes (i.e., grasses, forbs and shrubs) is common in food habit studies of hoofed animals (Sundstrom et al. 1973, White 1987, Yoakum 1990, Holechek et al.1998). Plants listed in these classes have many characteristics that are similar and, therefore, lend themselves to comparisons. However, certain plant species do not fall easily into these forage classes, so it is necessary to decide in which class the plants are to be listed if comparisons with other similar studies are an objective. The importance is not whether the list in one class is more correct than in another. Rather, the importance is in making sure that all plants are included and reported for forage class and isolated uniformly by forage class for subsequent comparative analyses. The greatest inconsistency in past studies has been the placing of cacti and half-shrubs (old term used to designate plant like a shrub but nonwoody like a forb) in the shrub class. However, many more studies have listed them as forbs. In pronghorn food habit studies, it is recommended that cacti and half-shrubs be listed in the forb class or presented separately. Based on past common-use practices, the following criteria are recommended for providing uniform information on plants within forage classes: grass—include all grasses and grasslike plants, such as sedges and rushes; forb—include all forbs, mosses, lichens, cacti, half-forb (old term used to designate plant like a forb but woody like a shrub) and half-shrub; shrubs—include all shrubs and trees. Ordinarily this system will work well and comparisons of food habits for similar ecosystems can be made. However, if the investigator determines that some plant is a borderline species warranting special attention, the species can be listed separately under one of the three major forage classes.
6. With current investigations benefiting from previous studies, new research promises greater thoroughness and accuracy. Projects now are being designed to include data not only on diet composition, but also on plant composition, availability, nutritive qualities, forage quality, weather factors, preference indices for individual plants and forage classes, and the relation of pronghorn diets to sympatric ungulates for preferred species and dietary overlap. Such an approach to examining forage use by both wild and domestic ungulates is especially imperative now, inasmuch as forage allocations need to consider and emphasize multispecies use of western rangelands. Techniques now are available to assess forage quality (DAPA and FN) through fecal analysis, thereby providing needed data for management strategies.
7. Past pronghorn diet studies using rumen or fecal analyses reported amounts of forage classes consumed, but some of these reports were unknowingly skewed because forbs were more rapidly digested than grasses or shrubs. This provided data regarding the amounts of forage species in the sample, but did not

> necessarily document what actually was consumed. Pulliam and Nelson (1979) reported on a technique to correct this deficiency by implementing digestibility correction factors. When applied, these factors may give a more accurate calculation of forage quantities consumed (Dearden et al. 1975, Leslie et al. 1983). Differential digestion factors have recently been applied to a pronghorn food habit study in Nevada and Oregon, reported elsewhere in this chapter (Hansen and Anthony 1999, Hansen et al. 2001).

The more than 200 pronghorn food studies accomplished to date document the various techniques used to determine final results. Many were conducted with such varying procedures that the results cannot be compared effectively. However, Tables 86, 87, 95 and 96 tabulate food habit studies for 14 grasslands and 19 shrubsteppe environments. Available studies did not contain sufficient information for comparison with the desert biome. The Tables provide three sets of data: percentage composition of forage classes in the vegetative community, percentage composition of forage classes in the diet, and forage preference ratings. With these data, one can quickly evaluate (1) percentages of plants available, (2) the plants pronghorn actually consumed, and (3) index ratings of forage use in relation to availability. It then is possible to compare food habit studies for different sites, areas, regions or biomes.

For the grassland biome, the 10 food habit studies provided an average percentage of available plant composition as 74% for grasses, 16% for forbs and 9% for shrubs (Table 86). The average percentage composition of the diets was 19% for grasses, 62% for forbs and 17% for shrubs. Average forage preference ratings were 0.2 for grass, 3.9 for forbs and 1.9 for shrubs.

Accordingly, forbs were shown to be highly preferred, with shrubs next in preference. These findings support the concept identified by Buechner (1950a) and Leopold et al. (1984) that pronghorn are forb grazers by preference. At times, as many as 40 to 60 forb species were ingested at a study site, although 5 to 6 species may have comprised the bulk of the diet. Consumption of so many plant species denotes a choice for diversity, probably associated with the need for various plants with particular nutritional values and moisture content.

Shrubs were next to forbs in preference and consumption, followed by grasses. Shrubs often were consumed in large quantities by individual animals, but were eaten most often when forbs were desiccated or no longer available. Grasses were not a preferred forage class except for one grassland study (Schwartz 1977). They were consumed during all seasons of the year, but their highest use was during spring or green-up periods resulting from summer or autumn precipitation. Grasses should be recognized as a possible volume food source for sites where shrubs are not plentiful.

Pronghorn browsed large quantities of shrubs in the shrubsteppe biome. This was confirmed by the 13 studies presented in Table 95, which had an average percentage composition of 37% for grasses, 15% for forbs and 46% for shrubs. Diet forage averages were 7% for grasses, 30% for forbs and 62% for shrubs. Forage preference ratings averaged 0.2 for grasses, 2.0 for forbs and 1.3 for shrubs. The shrub forage class was dominant in both availability and dietary intake. This high use, especially for the sagebrushes, included consumption during all seasons, with increased use during autumn and winter. Shrubs are important for nutrition. Their availability during harsh or prolonged winters may be the key to pronghorn survival. The value of shrubs is stressed in management plans of northern pronghorn populations.

The value of forbs in the diet of pronghorn on shrubsteppes may be greater than previously realized. Certain investigators (e.g., Taylor 1975) noted the preferred use of forbs in shrublands, but forbs often were given scant values in food habit studies for this biome. Data in Table 94 indicate that forbs were the preferred forage class in 9 of the 12 studies in shrubsteppes. The preference index for forbs in all studies was three times greater than for shrubs. This indicates that pronghorn highly prefer forbs over shrubs on shrublands and that shrubland management should include livestock grazing systems and vegetative management plans favoring high forb production. The high use of shrubs for this biome also indicates that shrubs are staple forage for pronghorn, especially in regions with high snowfall.

Grasses were consumed by pronghorn during all seasons of the year in shrublands, but the amount consumed averaged less than 10% year-round. Both green and dry grasses were eaten, but green grass was used mostly during spring and autumn green-up. Grass supplies part of the digestible energy vital to pronghorn during winter.

Few quantitative studies of pronghorn food habits exist for deserts (Table 102). Data collected simultaneously on the availability of all forage classes are needed for preference indices to be determined. The little information available indicates that pronghorn in this biome have foraging habits consistent with pronghorn in grassland and shrubsteppe biomes; the highest preference is for forbs, with medium preference for shrubs and very limited preference for grasses.

An assessment of food habit studies discloses their importance to understanding and managing pronghorn. The diversity and quantity of plants available or consumed are related directly to pronghorn production and survival because pronghorn are habitat dependent. This is depicted in Figure 101, which illustrates data from Tables 86 and 95

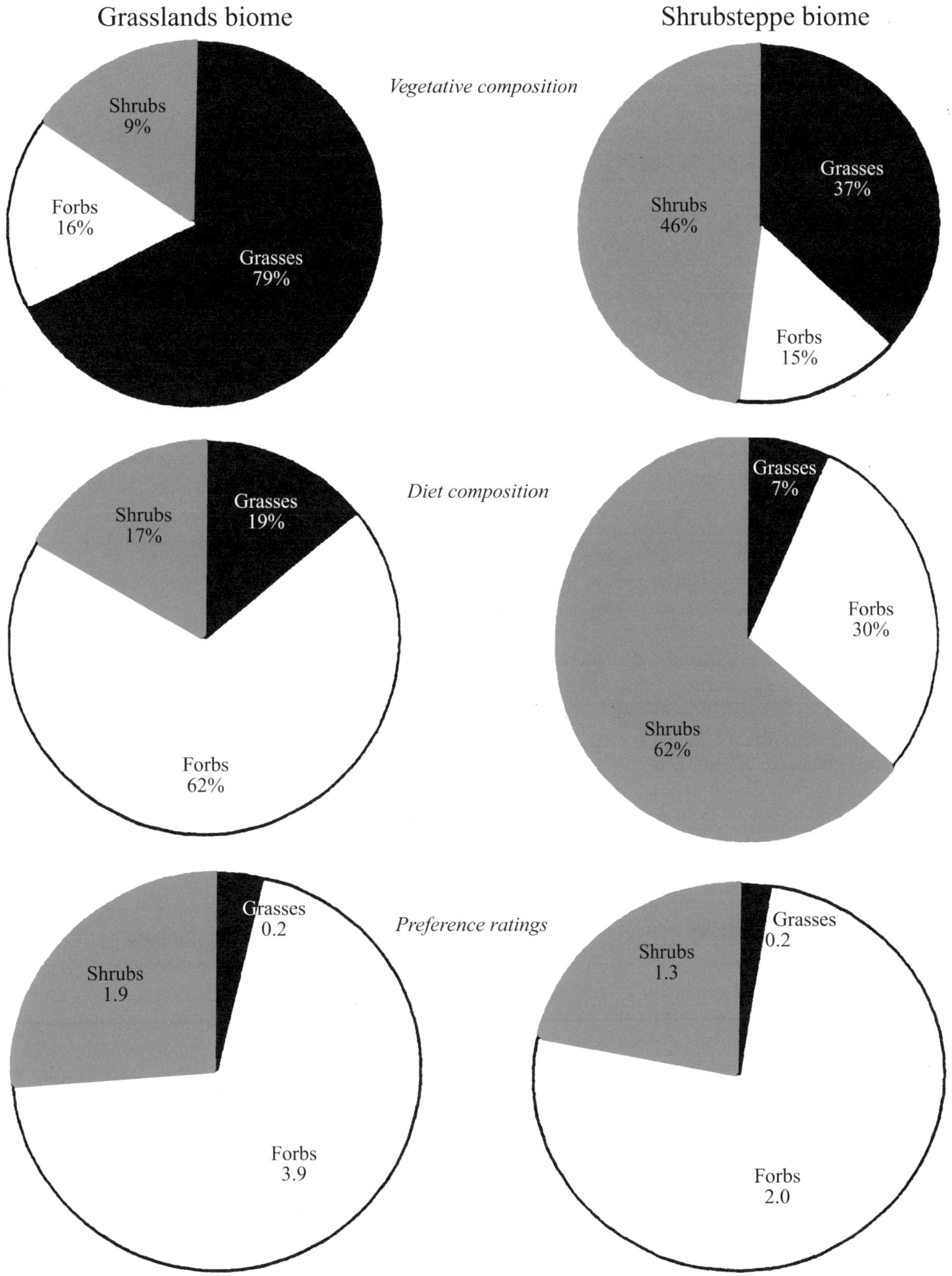

FIGURE 101. *Comparison of vegetative composition, diet composition and preference ratings for the grassland and shrubsteppe biomes.*

regarding plant availability, forage consumption and preference ratios for grasslands and shrubsteppes (99% of current pronghorn habitat). Although the composition of forage classes available for ingestion and the amount of forage classes consumed vary in these two biomes, the preference ratios are exceedingly similar—pronghorn prefer forbs followed by shrubs, and grasses play a minor role.

Long-term pronghorn habitat studies in the sagebrush/grasslands of Montana conducted by Pyrah (1987) identified that forbs could be called the *reproduction forage class* because forbs are rich in nutritional values required for pronghorn reproduction. Pyrah also reported that shrubs could be termed survival forage, because they often provide the vegetation class with nutritional values and availability during the critical winter season. More recently, food habit studies verify that these foraging findings hold true for the shrubsteppes of Oregon and Nevada as well (Hansen and Anthony 1999, Hansen et al. 2001). Because some current pronghorn herds remain residents yearlong on certain habitats and other herds make seasonal movements between rangelands, wildlife managers need to implement vegetation management strategies that sustain a mixture of grass/forb/shrub communities for pronghorn welfare.

Pronghorn densities are related directly to the quality and quantity of vegetation, and vegetation is related directly to climate. Simply stated, precipitation and temperature are primary factors affecting plant phenology and abundance, which in turn support varying densities of pronghorn. It is this trilogy of climate, vegetation and pronghorn interaction that holds the management key to long-term carrying capacities and population levels on pronghorn habitats.

Recommendations for Pronghorn Ecological Food Habit Studies

Past investigations indicate pronghorn production and survival can be influenced by the quality and quantity of forage consumed (Beale and Smith 1966, Hockley 1968, Ellis 1972, Hervert et al. 2000, Hansen et al. 2001). Forage production frequently is triggered by climatic conditions. These ecological relationships have been reported for certain regions, but other areas lack adequate investigations. Consequently, comprehensive studies of the ecological triad (climatic patterns, vegetation production and pronghorn numbers) may be desirable for obtaining data to assess factors influencing pronghorn populations.

Such a study should include:

1. collecting fecal samples for all ungulates occupying pronghorn habitats (analyzed to major species level of detail), which should include:
   a. adequate samples for each season of the year for several years,
   b. care to obtain samples from critical pronghorn habitats (winter rangelands, travel corridors, fawning areas, etc.),
   c. samples from adult females (particularly pronghorn) and
   d. determination of forage quality indices, specifically FN, considering DAPA if the animals are eating plants with high quantities of digestible nitrogen;
2. collecting samples of preferred forage species obtained the second year following analysis of first-year diet selection samples, which should include:
   a. plants for all seasons of the year and for differing habitats (winter rangelands, fawning areas, travel corridors, etc.),
   b. measured protein and trace elements of interest, and
   c. mass data regarding forage availability for various habitats and seasons;
3. obtaining data relative to climatic patterns during vegetation phenology, which should include:
   a. precipitation and
   b. temperature;
4. collecting production and survival data for pronghorn (and other ungulates) for each year and various seasons; and
5. assessing vegetation changes due to historic or ongoing influences (e.g., agricultural practices, livestock grazing, mining, forestry, fire, shrub or tree invasion).

Results of these findings will provide wild ungulate foraging information relative to differing seasons, years and habitat types regarding:

1. diet composition for adult female ungulates,
2. nutritional values (protein, water content, etc.),
3. data for computing preference ratios, the amount of forage consumed in relation to forage availability,
4. data for assessing dietary overlap (or degree of competition) for all ungulates for differing seasons, years and habitat types,
5. determination of forage quality indices (FN, with DAPA optional),
6. correlations between climate, vegetation and ungulate production, and
7. ecological data pertinent to pronghorn (and other ungulate) survival and production that may be of value in increasing or sustaining herd numbers.

A goal of a study of this design is to correlate forage quality and quantity to precipitation levels. Because pronghorn diet selection is influenced by yearly forage production, which is influenced by climatic conditions, it is desirable to

Rangelands in the Great Basin with a mixture of grasses, forbs and shrubs provide optimum forage for pronghorn and other wildlife. When these forage classes are available for all seasons, pronghorn can choose the most relished, nutritious and succulent species. Rangelands in healthy condition, with an abundance of nutrients and succulent vegetation, produce high fawn recruitment. Photo by Jim D. Yoakum.

The most widespread species of cacti on western rangelands are opuntia, or prickly pear, ranging from Canada to Mexico on habitats occupied by pronghorn. Thorns can cause injury to livestock but apparently are easily consumed and digested by pronghorn. This photograph illustrates a prickly pear with many flowers growing in the Great Basin near Ely, Nevada. Photo by Jim D. Yoakum.

determine forage production and utilization during precipitation years. Then, it may be possible in future years to relate potential pronghorn numbers with climate patterns and forage production.

POISONOUS AND NOXIOUS PLANTS

Early investigators reported pronghorn consumed poisonous plants (Einarsen 1948, Buechner 1950a, Terwilliger 1953, Ogilvie 1955, Yoakum 1958, Hoover et al. 1959, Smith and Beale 1980). Generally, those reports referred to pronghorn eating plants that were poisonous to livestock but not to pronghorn. They helped dispel a prevailing misconception that pronghorn were serious competitors for forage with livestock on western rangelands. Many plants eaten (at times, in large quantities) by pronghorn are noxious to livestock because they cause mechanical injury, are unpalatable, or tend to crowd out, invade or dominate more desirable forage. Spines of bull thistle, Russian thistle and cacti make these plants unpalatable to livestock, but pronghorn consume them during all seasons. Pronghorn in Colorado consumed 11% cacti in their annual diet (Hoover et al. 1959). Evidently the spiny cacti do not injure pronghorn; even fawns consume the plants.

One of the first to relate toxic plants to pronghorn was Buechner (1950a), for the Trans-Pecos region in western Texas. Buechner noted that four plants eaten by pronghorn were poisonous to livestock—locoweed, wooley senecio, riddell groundsel and paperflower. Of these, only locoweed was presumed to have harmful effects on pronghorn. Of 103 plant species reported by Buechner, 51 were unpalatable and 21 were considered poor forage for cattle, whereas 28 were considered poor forage for pronghorn.

A compilation of food studies in the northwestern Great Basin (Yoakum 1958) disclosed four plants, toxic to livestock, that were consumed by pronghorn—lupine, vetch, rattleweed and senecio. All four species were grazed in small quantities (less than 1% of total annual diet). Yoakum also observed pronghorn eating death camas during spring during three different years on Drakes Flat in south-central Oregon. This species is considered poisonous to livestock (Stoddart et al. 1975).

Halogeton is an introduced forb that is highly poisonous to livestock and appears after the first frosts of autumn (Stoddart et al. 1975). However, it is a succulent species during summer and early autumn and is readily consumed by pronghorn. Halogeton appeared in 7 of 16 pronghorn rumen samples collected in southeastern Utah and comprised 0.8% of the total summer diet (Smith and Malechek 1974). No ill effects were detected in pronghorn.

Seven plants toxic to livestock were readily eaten by pronghorn in Colorado. Vetches, at least two species of which are poisonous to cattle and sheep, and one of which is poisonous to horses, were found in 43.7% of 167 stomach samples (Hoover et al. 1959). Three pronghorn rumens collected in July contained 24% to 55% vetch. Other plants toxic to livestock consumed by pronghorn included pingue or rubberweed, globe mallow, larkspur, monkshood, cocklebur and yucca. Hoover et al. discussed the beneficial relationship of pronghorn to livestock where

Lupine is a ubiquitous forb native to many western landscapes. Pronghorn graze it in small quantities; it can be poisonous to livestock. Photo by Jim D. Yoakum.

Halogeton can be a highly poisonous plant leading to the death of livestock following the first frosts, which accelerate production of oxalic acids. Plants are succulent and are consumed by pronghorn in small quantities. No ill effects from eating this plant have been reported for pronghorn. Photo by Jim D. Yoakum.

the former consume many forbs that are potentially injurious to the latter.

Lance and Pojar (1984) concluded that pronghorn may (1) avoid use of poisonous plants, (2) consume those plants only in subtoxic amounts, (3) ingest toxic amounts they can tolerate or (4) ingest toxic levels, die and go undetected or misdiagnosed. Apparently, most plants toxic to livestock have little effect on pronghorn (see Chapter 12 concerning plants poisonous to pronghorn).

SUPPLEMENTAL FEEDING

For years, wildlife managers have debated the values of feeding supplemental rations to free-ranging pronghorn. Supplemental feeding on a regular basis usually is not required when effective management strategies are established. When wild ruminants exceed the natural forage supply, the resulting losses from malnutrition serve to contain the population within its ecological carrying capacity. Indiscriminate supplemental feeding can remove this natural control factor. Wildlife agencies responsible for northern pronghorn populations have questioned the practicality of artificial feeding during winters of heavy snows and low temperatures (Samuelson 1970, Gjersing 1978, Helms 1978, Pyrah 1978, Barrett 1982, Torbit et al. 1984). Investigators on southern rangelands have posed the same question for drought periods (Hailey et al. 1966, Hailey 1979).

Sustained maintenance of wild ruminants by supplementation generally is considered biologically and economically unsound (White 1987). However, periodic supplemental feeding may be justified when assessments are made of biological and management strategies. Working with white-tailed deer, Mautz (1978) reported that if deer production surpluses occur, then perhaps supplemental feeding can be a management alternative. For example, a rangeland may be properly managed for vegetation condition with a sustaining pronghorn herd, but a critical stress period (e.g., a severe winter with deep snow depths or a prolonged drought) may occur once every 30 to 50 years. If the pronghorn population could be maintained through this infrequent critical period without damage to the vegetative community, the population could again be within carrying capacity. Following are reports concerning supplemental feeding of pronghorn.

During the 1977–78 winter, Alberta experienced below normal temperatures, unusually early snowfalls and frequent extreme blizzard conditions. Snowfall exceeded all records for 50 years. Biologists referred to the time as a killer winter for wildlife on the northern prairies (Barrett 1982). Estimated mortality for approximately 14,500 pronghorn was 48%. Fetal resorption, or "mummification" (Jubb and Kennedy 1970), was common among 42 fetuses examined. Only six fetuses (14%) were normal, viable specimens and, predictably, recruitment the following year was low.

Supplemental feeding of alfalfa, timothy hay, baled prairie grasses, alfalfa cubes, pea vines and sugar beet tops was undertaken during winter by concerned laymen. Feed was provided ad libitum at each site. The response of pronghorn to the emergency feeding varied greatly. In many cases, 3 weeks were necessary for animals to become accustomed to the new food. Some herds with assured supplemental supplies did little foraging for natural vegetation. Overall, despite daily distribution of supplemental food, the program was ineffective in curtailing high mortality. The program may have contributed to the survival of some pronghorn, but in this severe case, it did not appear to be an effective strategy (Barrett

1982). One possible cause for the failure may have been the initiation of supplemental feeding after the animals were already in a critically weakened condition. On the other hand, a more nutritious artificial ration—as listed in Table 109—may have provided the needed nutrients in a more digestible, useable form than did the hay.

During drought conditions in Texas in the mid-1960s, biologists recommended supplemental feeding of pronghorn (Hailey et al. 1966). Vegetation, especially grasses and forbs, had been severely grazed, leaving primarily noxious shrubs (tarbush and creosote bush). Many pronghorn were restricted to pastures fenced with net wire for livestock control. The result was high pronghorn mortality and subsequently low fawn recruitment. Investigators believed that if the pronghorn had been provided emergency feed during the critical summer season, as was done for livestock, higher rates of survival and productivity would have been likely. This is reflected in management practices currently recommended by the Texas Parks and Wildlife Department (Swepston and Hailey 1991).

Recognizing the need for a supplemental ration that would be economical, high in nutritive value and palatable to pronghorn, researchers in Colorado developed and tested a feeding program for severe winters (Torbit et al. 1984). This was the first scientific effort to investigate the kind of emergency ration pronghorn need and when and how it should be delivered. The study objectives included development of a nutritious, digestible, pelleted ration palatable to pronghorn and field testing it with feeders and olfactory attractants.

Two experimental rations were developed, differing in digestibility and fiber content (Table 109), and fed to penned pronghorn. Both rations were consumed without producing symptoms of digestive distress or significant weight loss. Although either ration could be used, Torbit et al. (1984) considered the ration that was most digestible to be the best winter emergency food for pronghorn. Based on feeding trials, it appeared that an adult needed 1 pound (0.4 kg) per day, at an estimated daily cost of less than 30 cents.

Two prototype feeders were tested with the above rations and readily used by free-ranging pronghorn. The feeders were a commercially available self-feeding metal pig feeder and an inverted automobile tire with a plywood floor. They were treated with three olfactory attractants to lure pronghorn, but tests were unable to demonstrate any preference for the attractants (further work is needed to identify and clarify the roles of visual and olfactory stimuli that entice pronghorn to use artificial feeding structures). Proper spacing of feeders proved to be important. Apparently, spacing of 85 feet (26 m) is optimal, but this may require upward adjustment when increasing numbers of feeding animals at a site result in increased agonistic encounters. Based on their research results and field testing, Torbit et al. (1984:127) were confident that a supplemental ration could be used successfully

TABLE 109. Nutritional content and ingredients of two experimental rations fed to hand-raised pronghorn

	Colorado trials		*Alberta*
	Ration 1	*Ration 2*	*Ration 1*
Ingredients	Bentonite	Bentonite	Bentonite
	Barley	Barley	Barley
	Corn	Corn	Corn
	Wheatmids	Wheatmids	Wheatmids
	Dry brewer's grain	Dry brewer's grain	Dry brewer's grain
	Molasses	Molasses	Molasses
	Trace mineral package	Trace mineral package	Trace mineral package
	Vitamin package	Vitamin package	Vitamin package
	Milo	Soy bean meal	Canola meal
	Oats	Corn gluten meal	Oats
	Beet pulp	Wheat	Beet pulp
		Dehydrated alfalfa	
Analysis[a]			
Dry matter	91	93	
Crude protein	19	19	
Acid-detergent fiber	14	—	
Apparent in vivo digestibility	72	—	
In vitro digestibility	—	62	

[a] Percentage, on dry matter basis.

Ration 1 was used in field trials with wild pronghorn (Torbit et al. 1984). The Alberta ration substituted Canola meal for milo due to availability in Canada (Murray 1986). Originally reported percentages were rounded to the nearest 1%.

Supplemental feeding for pronghorn has been practiced, often at public request. Bales of alfalfa have been used from Canada to Texas as depicted in this photo of state wildlife personnel feeding pronghorn during a severe winter in Wyoming. Pronghorn readily consume alfalfa and it is highly nutritious. Supplemental feeding has proven beneficial when started before deprivation of natural forage. Research has been conducted to develop concentrated pellets high in nutritional value and palatability. Such supplemental feed is more economical and easier to provide to animals in need on rangelands than is alfalfa. Photo by LuRay Parker; courtesy of the Wyoming Department of Game and Fish, Cheyenne.

for wild pronghorn, but they cautioned: "Obviously, much more data are needed to evaluate delivery systems, locations, duration of supplemental feeding, etc. We also believe that any pronghorn feeding program should be researched and prepared in advance taking into consideration the biological, financial and political constraints."

The Alberta Fish and Game Division conducted a study in 1985 and 1986 to determine the effectiveness of an emergency winter feeding system for one of the most northern pronghorn populations. They tested the ration developed in Colorado (Torbit et al. 1984) with slight modification because of product availability and prices in Canada (Table 109). All experimental use of rations was on free-roaming pronghorn, and the following data are summarized from Murray (1986). Two feeders—one, a commercial deer feeder and the other, a self-made tire feeder—also were tested. Utilization of the ration feeders was monitored by observing pronghorn with spotting scopes and binoculars approximately 1,000 to 1,300 feet (304–396 m) from the feeding sites.

Pronghorn use of the ration was minor during early mild winters but increased during colder conditions. Observations of field signs (tracks and feces) indicated that pronghorn mainly used feeders during early morning, evening or at night. Table 110 reports on compatibility levels of animals while at the feeding stations.

Results of pronghorn use of the prepared ration as an alternative to grazing adjacent winter wheat fields were inconclusive due to mild winter conditions. Pronghorn apparently preferred grazing in the winter wheat fields, but when depth of snow prohibited foraging croplands, they readily used the emergency ration.

Pronghorn consumed approximately 550 pounds (250 kg) of the Alberta-modified ration. Amount of feed consumed per animal during the feeding trials was 18 pounds (8 kg). The consumption rate for pronghorn per pound (kg/animal/day) was not ascertained. The amount of ration consumed suggested the feeders were used more often than noted by observation. It was concluded from field sign that pronghorn did not prefer one feeder type over the other.

Murray (1986) reported the successful use of mixing second-cutting alfalfa hay with the pelletized ration. Alfalfa hay was placed directly on the tire feeders, and it apparently enticed the pronghorn to the feeders.

During the summer of 2002, severe drought conditions existed for pronghorn on the Carrizo Plain in southern California. Because droughts decrease herbaceous plant production, contributing to low fawn survival (Hervert et al. 2000), the California Department of Fish and Game initiated a supplemental feeding program. A review of the literature disclosed a lack of any report pertaining to summer supplemental feeding for wild pronghorn. Consequently, an experimental program was commenced by providing fresh, green, cultivated alfalfa hay placed in metal troughs near water troughs on rangelands. Prong-

Table 110. Composition of observed feeding bouts among pronghorn by age/sex class during winter feeding trials in Alberta

Age/sex class	*Percentage at feeders*	*Percentage of herd composition*
Adult male	50	31
Adult female	30	52
Juvenile male	10	14
Juvenile female	10	14

Source: Murray 1986.

horn found the alfalfa stations and began foraging within 2wo days. They fed on the hay daily for the duration of the drought—about 3.5 months. Soon after autumn rains began, grasses and forbs commenced growth and, just as quickly, the pronghorn switched from supplied alfalfa hay to new, green, succulent forage. Investigations were made during summer to detect mortality, but none was found (Koch and Yoakum in press).

NUTRITIONAL VALUES

Nutrition is the act or process of nourishing through the intake, digestion, absorption and assimilation of food substances. Nutrients are chemical or inorganic constituents of food that are sources of energy essential for the normal growth, vigor and function of an animal (see Nagy and Haufler 1980).

Little is known of the nutritional demands of free-roaming pronghorn. This lack of information resulted in wildlife agencies postulating that vegetation deficient in critical nutrients was a possible cause of low fawn recruitment (Foree 1959). Nutrition also was considered by investigators concerned with low recruitment after translocation projects (Goldsmith 1988).

Most studies of the nutrient content of wildlife foods report values on a dry-matter basis. Fox (1997) reported on preformed water obtained by pronghorn from forage. So, the few nutrition studies conducted on pronghorn foods will be discussed.

Nutrients

An index to the value of shrubs to pronghorn was estimated by analyzing crude protein values in the stems and leaves of sagebrush (Einarsen 1946). This was based on similar studies of shrubs consumed by black-tailed deer in Oregon. Einarsen indicated that 5% protein content was the minimum level needed to sustain deer from malnutrition. He collected sagebrush samples on pronghorn rangelands near Burns, Oregon, that had protein values (Table 111) well above the 5% minimum, with highest levels found during spring and early summer.

Ferrel and Leach (1952) did similar analyses of sagebrush collected near Doyle in northeastern California. Samples were collected during various seasons (Table 111), and the results were similar to Einarsen's (1946). Ferrel and Leach discussed the high use of sagebrush by pronghorn, speculating that the digestive capabilities of pronghorn may differ from deer, because deer were known to succumb on diets high in sagebrush during severe winters.

A study in Saskatchewan correlated protein levels of forage with utilization during various seasons (Dirschl 1963). Although findings were inconclusive, several interesting relationships were noted. Samples of preferred species disclosed that protein levels were highest in May and gradually declined (Table 112). Lowest levels were found in winter samples. Dirschl noted that (1) plants with the highest protein levels appear to be relished during all seasons; (2) during spring, protein content was highest for grasses—higher than that of shrubs, including sagebrush—and pronghorn consumed grasses extensively during that season; (3) during summer, forbs were rich in protein and became the preferred forage class; and (4) during winter, forbs and grasses were low in protein, but sagebrush retained a relatively high protein content and was the major forage. Although these findings provide valuable perspective, detailed studies are needed before crude protein requirements of pronghorn are fully understood.

Concern over declining pronghorn numbers in Utah led to a study of forage values for the animals' critical winter season (Smith et al. 1965). Feeding trials involving

TABLE 111. Protein content of sagebrush collected in California and Oregon

Month collected	*Percentage protein content* Oregon[a]	California[b]
April	—	11.9
May	15.8	—
June	15.9	—
July	11.6	—
August	9.6	10.2
October	8.5	11.3
November	—	11.4
December	8.8	10.7
January	—	12.1
February	—	10.2

[a] From Einarsen (1948).
[b] Ferrel and Leach (1952).

TABLE 112. Protein content of pronghorn forage plants in Saskatchewan

Month	*Percentage protein on dry matter basis* *Sage-brush*	*Creeping juniper*	*Western snow-berry*	*Rose*	*Yellow goats-beard*	*Grass*[a]
Feb.	11.5	7.5	7.9[b]	4.7[b]	—	—
March	9.7	—	—	—	—	5.0
April	15.3	5.3	—	—	—	20.3
May	19.2	6.6	25.1	26.4	19.6	16.5
June	17.0	6.3	12.3	16.5	15.3	13.1
July	15.8	—	10.2	12.2	11.4	12.7
Sept.	12.8	7.1	15.2	11.9	8.8	

[a] Mixture of speargrass and wheatgrass.
[b] Dry foliage.
Source: Dirschl 1963.

16 plant taxa were conducted with field-confined animals to determine plants consumed and total digestible nutrients. Pronghorn selected sagebrush over other shrubs, indicating sagebrush was preferred, rather than merely acceptable forage when compared with other shrubs in the study. Digestible protein was 5.3% of the total diet, and digestible nutrients were 67%, well in excess of acceptable standards for domestic sheep—as set by the National Academy of Science National Research Council (see Smith et al. 1965). Smith et al. concluded that a diet composed largely of shrubs adequately provided proper nutrition for wintering pronghorn, and the area's population decline apparently was not attributable to low levels of nutrition during winter.

A comparative study of nutritional quality of foods in pronghorn summer habitat was conducted in southwestern Utah (Smith and Malechek 1974). Plants were collected at triweekly intervals and analyzed for protein, phosphorus and digestibility. Contrary to previous observations on the site, data did not support the contention that pronghorn selected forage primarily for moisture content. Indications were that moisture content alone did not determine palatability. The lowest diet protein content was 10% in mid-June; the highest was 22% during mid-August, primarily because of two species—prairie mallow (forb) and desert almond (shrub) (Table 113). Digestibility levels appeared to follow plant development and were highest during May and June, followed by declines during summer, with a small peak in late summer due to increased precipitation. Seasonlong digestibility coefficients (71%) suggested that pronghorn were able to select a uniformly high-quality diet throughout summer and early autumn.

Phosphorous levels exhibited seasonal trends similar to those of dry-matter digestibility and probably were related to forage phenological development. Phosphorous requirements for pronghorn have not been identified, therefore those of domestic sheep were used for comparison. It was assumed, based on forage ingested by pronghorn, that phosphorous levels were adequate for growth, gestation and lactation. Smith and Malechek (1974) concluded that, although certain nutritional requirements of pronghorn were unknown, if sufficient quantities of preferred forage were available, the animals would not experience nutritional deficiencies during summer because of those food plants' adequate protein or dry-matter digestibility. It also was supposed that adequate nutrition would be available for the semiarid Great Basin habitat, except during extremely dry years.

Silver sagebrush has been recognized as a major food item for pronghorn in southwestern Alberta (Mitchell and Smoliak 1971, Barrett 1974). More than 60% of all herds wintered on rangelands with silver sagebrush complexes. Food habit studies and utilization checks for these areas disclosed high use of silver sagebrush. Accordingly, a research project was implemented to ascertain the nutritional value of this shrub (Barrett 1974). Permanent transects were installed to measure pronghorn use, and plant samples were collected in an adjacent enclosure to simulate shrubs browsed. Analysis was expressed as dry weight. Sagebrush samples from 11 winter sites averaged 7.6% (range = 7 to 9) protein content. Protein levels were highest during spring and lowest in winter. Sagebrush provided higher winter protein levels than did native grasses. It was assumed the protein content of silver sagebrush was adequate for wintering pronghorn.

Mean levels for calcium, phosphorus and the calcium-to-phosphorus ratio were 0.75 g Ca/100 g dry matter, 0.16 g P/100 g dry matter and 5.05 g Ca/g P, respectively. Although pronghorn requirements for these nutrients were unknown, they were correlated to those of cattle. If pronghorn requirements are similar, the animals would have experienced no deficiency of those nutrients (Barrett 1974).

Table 113. Percentages of crude protein and phosphorus in pronghorn diets in southwestern Utah

	Desert Experimental Range				*Awapa Plateau*			
	Protein		*Phosphorus*		*Protein*		*Phosphorus*	
Date	*1969*	*1970*	*1969*	*1970*	*1969*	*1970*	*1969*	*1970*
May 1	20	21	0.31	0.28	—	—	—	—
May 20	16	17	0.27	0.23	16	17	0.33	0.23
June 12	10	13	0.17	0.20	16	18	0.36	0.33
July 2	13	11	0.19	0.15	17	12	0.33	0.30
July 22	12	10	0.24	0.12	15	18	0.25	0.31
August 15	22	13	0.30	0.17	11	16	0.18	0.30
September 5	16	13	0.22	0.20	13	13	0.19	0.27
September 25	14	12	0.19	0.17	12	11	0.25	0.17
Mean	15	14	0.24	0.19	11	15	0.27	0.27

Source: Smith and Malechek 1974.

Seventeen silver sagebrush samples were analyzed and found to have a mean sulfur content of 0.17 g/100 g dry matter and a mean selenium content of 982 parts per million. The sulfur content was consistent, but extremely low. The selenium content was variable and ranged from near deficient (40 parts per billion [ppb]) to potentially toxic levels (5,520 ppb) in comparison with livestock requirements. Massey and Martin (1972) reported that selenium levels below 100 ppb could be considered potentially deficient and levels above 5,000 ppb are potentially toxic to livestock.

Possibly one of the most significant findings of this study was that nutrient content was significantly higher in vegetation samples collected from under snow. This may provide a nutritional compensation for pronghorn during severe winters. Protein, calcium and phosphorus each were 2 to 11 times greater from silver sagebrush samples collected under snow (Table 114). It is probable that snow cover reduces the plant's exposure to intensive curing. The below-snow silver sagebrush also had a higher moisture content and was more palatable to pronghorn. Although snow in great quantities is deleterious in many ways to pronghorn, snow in limited amounts may beneficially preserve important forage nutrients (Barrett 1974).

The chemical composition of forage from shortgrass prairies of northeastern Colorado received one of the most intensive studies of pronghorn/rangeland relationships to date (Schwartz and Nagy 1976, Schwartz et al. 1976b, 1977, Schwartz 1977). The investigations concentrated on the forage availability and nutrient quality of diets for pronghorn, bison, cattle and domestic sheep under different grazing regimes. Plant samples for nutritive analysis were hand-plucked from pastures where pronghorn grazed. Only plant parts consistent with those eaten were collected and analyzed.

Results indicated that crude protein in diets was lowest in January and highest in April. High levels occurred during active plant growth of March, April and June, and low levels occurred during August, October and January (Table 115). Pronghorn maintained comparable crude protein levels in their diet regardless of forage availability for both lightly and heavily grazed pastures. Protein requirements for pronghorn were not known, so those of deer were assumed and compared. According to French et al. (1955), suggested protein levels for white-tailed deer were 13% to 16% for growth and 6% to 7% for maintenance. Therefore, if protein requirements for pronghorn are similar, the shortgrass prairie forage was adequate for maintenance throughout the year for both grazing intensities. Pronghorn and domestic sheep selected forage higher in crude protein during all seasons than did cattle or bison (Peden et al. 1974). This agrees with Bell (1971), who suggested that small ruminants require more protein per unit weight than do large ruminants.

Phosphorous content of pronghorn diets varied from 0.43% in April to 0.09% in January, which would be expected from normal leaching. Phosphorous requirements for wild ruminants may equal or exceed those of domestic livestock (Magruder et al. 1957, Ullrey et al. 1975). Domestic sheep requirements vary from 20% for yearling lambs and dry ewes to 28% for lactating ewes (National Research Council 1964). If the phosphorous requirements for pronghorn were similar to those for domestic sheep, adequate phosphorous levels were not attained in August and January for both pastures, nor in March and October in the heavily grazed pasture.

Gross energy did not vary greatly between seasons or pastures. Fiber content of pronghorn diets also varied little. Comparison of diets with those of the other ruminants indicated that pronghorn were the most selective feeders and consumed plants lowest in cell wall content.

Therefore, findings from studies on grasslands found that pronghorn were more selective and consumed a diet higher in crude protein and lower in fiber than did the other three ruminants (Schwartz et al. 1977).

A pronghorn nutrition study was undertaken in east-central California from 1982 to 1984 to determine whether low-quality forage was responsible for low fawn recruitment in a translocated herd experiencing high mortality (Goldsmith 1988). The concentration of DAPA and FN were two measures investigated because they had been used as indices of nutritional status for other wild ungulates. The FN has been correlated with diet digestibility (Lancaster 1949, Holloway et al. 1981). The use of DAPA as an alternate indicator of diet quality was proposed by Nelson et al. (1982). Levels of DAPA and FN follow patterns in plant availability and phenology (Kie and Burton 1984).

Results of pronghorn investigations supported expected relationships between concentrations of DAPA and FN (Table 116). Variation in plant nutrient content was greater

TABLE 114. Nutrient content of silver sagebrush in relation to snow cover in southeastern Alberta

	Percentage increase below snow surface		
Nutrient	*1971–72*[a] *(n = 10 pairs)*	*1972–73*[b] *(n = 7 pairs)*	*Overall*
Protein	38.0	18.2	28.4
Calcium	23.7	2.3	13.8
Phosphorus	35.7	5.6	12.5
Calcium/phosphorus ratio	0.58	3.7	2.0

[a]"Normal" winter, with 6 to 12 inches (15.2–30.5 cm) of snow cover.
[b]Open winter, with little snow cover.
Source: Barrett 1974.

Table 115. Percentage dry matter and chemical constituents of pronghorn diets for lightly and heavily cattle-grazed pastures in southwestern Colorado

		Chemical constituents of plants analyzed			
Month and treatment	*Percentage dry matter*	*Gross energy*[a] *(kcal/g)*	*Crude protein (n = 6.25)*[b]	*Percentage ash*[b]	*Percentage phosphorus*[b]
March					
Light	58	4.6	10.4	17.3	0.224
Heavy	80	4.6	10.0	9.1	0.118
April/May					
Light	28	4.6	19.2	12.5	0.427
Heavy	35	4.6	21.4	13.7	0.368
June/July					
Light	43	4.4	10.2	8.6	0.241
Heavy	43	4.5	13.9	8.5	0.207
August					
Light	50	4.5	8.9	11.1	0.124
Heavy	52	4.6	8.5	10.6	0.121
October					
Light	50	4.8	8.7	15.3	0.241
Heavy	66	4.8	8.7	8.2	0.171
January					
Light	85	4.7	7.3	14.5	0.123
Heavy	85	4.9	7.3	7.6	0.088

[a] Expressed on an organic matter basis.
[b] Expressed on a dry weight basis.
Source: Schwartz et al. 1977.

Knowledge of herbaceous and shrub forage for pronghorn is paramount when establishing management strategies designed to maintain or enhance vegetation through conservation plans. Pictured here is a low-growing shrubland in late plant succession with a few herbaceous plants (right) *and an adjacent low-growing shrub site recently burned and with abundant herbaceous plants in early successional stage* (left). *Pronghorn consume grasses, forbs and shrubs, and they sustain higher densities on vegetative communities with mixtures of these forage classes.* Photo by Jim D. Yoakum.

seasonally than between forage species during any one period. The concentrations of DAPA and FN were highly correlated to each other ($R = 0.59$, $P < 0.01$) for all samples. The findings also were comparable to those for desert bighorn sheep (Goldsmith 1988) and domestic sheep (Nelson et al. 1982), varying from 0.385 to 0.829 mg/g dry matter. When these components were analyzed and compared with samples collected from sustaining pronghorn populations, no significant differences were noted. Goldsmith concluded that the poor success of the translocated population probably was not a result of nutrient deficiencies.

Similar findings for DAPA and FN relationships with pronghorn in Oregon were reported by Hansen and An-

TABLE 116. Seasonal means (mg/g fecal dry matter) of pronghorn fecal diaminopimelic acid (DAPA) and fecal nitrogen (FN) in Adobe Valley, Bodie Hills and Butte Valley, California, 1982 and 1983

	Spring[a]		*Summer*		*Autumn*		*Winter*		*Annual average*	
	x	*SE*	*x*	*SE*	*x*	*SE*	*x*	*SE*	*x*	*SE*
DAPA[b]	0.51	0.09	0.65	0.04	0.52	0.02	0.39	0.03	0.54	0.03
FN[b]	1.84	0.15	2.29	0.07	1.94	0.17	1.42	0.03	1.97	0.08

[a] Spring = April–May; n = 4.
Summer = June–August; n = 7.
Autumn = September–October; n = 10.
Winter = November–February; n = 7.
[b] Kruskal-Wallace one-way analysis of variance, $P < 0.01$.
Source: Goldsmith 1988.

thony (1999) and Hansen et al. (2001). See Chapter 8 for information on trace elements and vitamins.

A research project was conducted to determine the mineral content of forage used by Sonoran pronghorn in southwestern Arizona. Vegetation was sampled at two sites during 1995 and 1996 relative to 11 minerals using 5 laboratory analyses models constructed to describe 9 to 13 grazed and ungrazed plant taxa. Findings aided in determining whether pronghorn intake rates for minerals met minimum requirements (Fox 1997).

Minerals in the diet of pronghorn investigated included sodium, phosphorus, copper, zinc, selenium, manganese, calcium, sulfur, molybdenum, potassium and iron (Table 117). Mineral requirements have not been determined for pronghorn, so domestic sheep requirements were used because the two ungulates have many similar dietary needs (Schwartz and Nagy 1976, Beasom et al. 1982b).

Mean content for the 11 minerals showed little variability by site, season, or between forage and nonforage species. Diet models constructed predicted that pronghorn consumption of sodium, phosphorus, copper, zinc and selenium was inadequate to meet their physiological needs. These findings raise management questions regarding the endangered Sonoran pronghorn because lack of essential minerals may impair population health and potentially reduce productivity.

Fox's (1997) nutritional studies provided new information regarding factors affecting pronghorn production and survival on harsh, arid, desert landscapes. She reported several limitations (e.g., lack of plant taxa intake for all seasons and years of differing production due to precipitation and lack of forb production) that warrant further investigation. However, her studies provide some of the earliest scientific investigations regarding nutrient and mineral qualities in pronghorn diets on deserts. Fox (1997: 159) recommended new strategies for managing vegetation for pronghorn in the Sonoran Desert: "We caution wildlife managers to reevaluate the ability of Sonoran Desert habitat within Cabeza Prieta National Wildlife Refuge to support healthy pronghorn populations. This evaluation should be done prior to the implementation of other more costly and difficult management activities, such as providing water for pronghorn, translocating

TABLE 117. Mean mineral content, standard errors (SE) and sample sizes for forage by pronghorn and nonforage vegetation species sampled at two study sites on Cabeza Prieta National Wildlife Refuge, Arizona, from June 1995 to May 1996

	Forage			*Nonforage*		
Mineral	*Mean*	*SE*	*Sample size*	*Mean*	*SE*	*Sample size*
Copper	5.95	0.2	78	6.91	0.5	25
Zinc	17.55	1.0	78	14.41	1.2	25
Selenium	0.26	0.01	78	0.29	0.02	25
Manganese	36.69	2.1	78	59.64	10.8	25
Phosphorus	2,105.41	128.3	78	1,935.84	193.5	25
Calcium[a]	1.83	0.1	78	1.43	0.1	25
Sulfur[a]	0.15	0.01	78	0.14	0.01	25
Molybdenum	0.57	0.04	78	0.50	0.1	25
Sodium	700.27	108.9	78	441.27	135.9	25
Potassium	8,580.63	400.0	78	8,596.76	543.5	25
Iron	132.35	35.2	65	108.02	10.9	20

[a] Percentage.
Source: Fox 1997.

animals, or curtailing aircraft overflights. If mineral availability is a limiting factor, other management actions directed at pronghorn recovery may have little or no effect on the population."

Preformed Water

Robbin's (1983) text, *Wildlife Feeding and Nutrition,* indicates that water is an essential nutrient for animal body function, and a source is preformed water (from forage). Preformed water may be 10% to 70% of forage mass for plants on rangelands, depending on precipitation and ambient temperatures. Estimates of preformed water intake are obtained from experiments on captive pronghorn where feed and water intake are measured.

A 4-year pronghorn, forage and water relationship study was conducted on the Desert Experiment Range in western Utah (Beale and Smith 1970). Annual precipitation ranged from 4.25 to 11.3 inches (10.8–28.7 cm). During summers of above average precipitation, forbs provided 90% of the diet. Conversely, during years of below average summer rainfall, they contributed less than 20%. When forb moisture was greater than 75%, pronghorn ceased to drink free water, although it was readily available. As vegetation lost moisture content, drinking water intake increased. The results of this study indicated that water intake was related closely to succulence of forage, which was correlated with amounts of precipitation received.

Forage moisture content on pronghorn habitat varies seasonally and depends in part on temperatures and precipitation. In Colorado, lightly grazed pastures had percentages of forage moisture as follows: 40% in March, 75% in April, 57% in June and July, 50% in August and October, and 15% in January (Schwartz et al. 1977). During winter, when water requirements are the lowest for northern pronghorn populations, preformed food moisture content also is lowest. Although pronghorn are more dependent on water during summer, winter water shortages are more severe, as is mortality due to water deprivation. When drinking water is not available during winter, pronghorn reduce forage intake as dehydration ensues. What appears to be starvation mortality may be the result of water deprivation leading to fasting, rather than food deprivation according to Whisler (1984).

Whisler (1984) conducted laboratory experiments with adult female pronghorn at the Sybille Experimental Unit in south-central Wyoming. She measured intake from drinking (free) water, preformed water from forage and metabolic water production (Table 118). The animals were in restricted enclosures and had abundant quantities of free (drinking) water, and forage was relatively dry (9%–11% moisture content). Unfortunately, no studies have been conducted with wild pronghorn living in semiarid and arid rangelands with limited drinking water and consuming large quantities of forage with high moisture contents, as reported in Utah by Beale and Smith (1970) and in Colorado by Schwartz et al. (1977).

Many have questions how pronghorn live on deserts with little to no free-standing water for drinking (Arizona Department of Game and Fish 1981, U.S. Fish and Wildlife Service 1982). Fox (1997) conducted a research project during 1995 and 1996 to determine whether it was possible for Sonoran pronghorn to exist without drinking free-standing water. To meet this objective, she analyzed nutrient and water content, abundance of forage species, constructed models describing diet, and used the models to evaluate various forage combinations to determine whether water content in forage was adequate to meet the needs of pronghorn on the Sonoran Desert throughout the year.

The model used to quantify preformed water consumption, metabolic water production and water turnover was: $Total_w = P_w \times M_w \times F_w$, where P_w was the amount of preformed water in forage, M_w was the metabolic water produced through oxidation metabolism, and F_w was the quantity of free-standing water consumed. By measuring P_w and calculating M_w for the Sonoran pronghorn diet, Fox (1997) was able to determine whether free-standing water was necessary to supplement water intake.

Estimates of water intake from diet models then were compared with water requirements predicted by three scaling equations. When water intake estimates from the diet models were lower than water requirements predicted by the scaling equations, it was concluded that it would be necessary for pronghorn to supplement their water intake using free-standing water. These findings aid in the assessment of habitat quality and the understanding of physiological needs of foraging pronghorn—knowledge necessary for effective management.

Water content for vegetation was highly variable by site, season and diet taxa. Data did not support the assumption that pronghorn increased their consumption of

Table 118. Calculated mean water intake per day for three adult pronghorn females during summer laboratory experiments in Wyoming

	Water gains	
Water intake source	*Liters*	*Gallons*
Drinking (free water)	2.84	0.75
Preformed (from forage)	0.31	0.08
Metabolic (produced in body)	0.09	0.02
Total	3.24	0.85

Source: Modified from Whisler 1984.

preformed water (possibly in the form of dew) by feeding at a specific time of day or night. Water content for forage varied from 5% to 25% between sites. Individual plants varied widely in water content during seasons, undoubtedly reflecting local weather patterns, such as precipitation.

Findings disclosed higher moisture content in foraged plants than nonforaged species (Table 119). Similar findings were reported by Beale and Smith (1970), Deblinger and Alldredge (1991) and Hughes (1991). Selective foraging for vegetation high in water content can be important to production and survival of pronghorn inhabiting arid landscapes.

One restriction in this study was the lack of forbs available during vegetation sampling. In years previous to the study, when adequate precipitation was received, forbs grew on vegetation collection sites. However, forbs were not produced on these sites during the study. Forbs can constitute from 70% to 91% of Sonoran pronghorn diets (see Arizona Department of Game and Fish 1981, Hughes and Smith 1990 in Table 102), depending on precipitation or other factors influencing forage moisture content.

The lack of data on forbs in Fox's (1997) investigations undoubtedly resulted in conservative estimates of preformed water. If pronghorn had access to forbs, their preformed water intake would increase and their need for free-standing water would decrease.

Fox (1997:45) concluded, "Our results indicate that some areas within the present range of Sonoran pronghorn are able to provide adequate water for pronghorn throughout the year, and others are seasonally marginal or inadequate." These findings provide additional information indicating the need to reevaluate the ability of the Sonoran Desert to support increased populations of pronghorn. Although time consuming and expensive, detailed, quantifiable habitat information is an essential part of understanding relationships of production and survival to Sonoran pronghorn populations and habitats.

Food habit studies conducted in all biomes for more than 50 years indicate that rangelands with a mixture of grasses, forbs and shrubs, provided in successional growth stages, sustain forage most beneficial to pronghorn. When these plants are low growing and succulent, they provide preferred forage to sustain healthy pronghorn populations. Photo by Jim D. Yoakum.

CONCLUSIONS

Pronghorn diet composition varies greatly because the animals occupy diverse vegetative communities in three biomes. They are selective, opportunistic foragers, feeding on grasses, forbs and shrubs, depending on palatability, availability, succulence, preference and nutritional values. Food habits of pronghorn are not an attribute of the individual animal, rather of the population in a given locality during a specific time. Comparison of diet composition among populations or locations may be misleading if not accomplished for similar vegetative ecosystems, seasons of the year and weather patterns.

More than 200 food habit studies were conducted on pronghorn during the last half of the 20th century, of

TABLE 119. Mean percentage of water content and standard errors (SE) for forage and nonforage vegetation species sampled at two study sites on pronghorn habitats, Cabeza Prieta National Wildlife Refuge, Arizona, from June 1995 to May 1996

	Aqua Dulce				*Charlie Bell*			
	Forage		*Nonforage*		*Forage*		*Nonforage*	
	Mean	*SE*	*Mean*	*SE*	*Mean*	*SE*	*Mean*	*SE*
Feb-Mar	53.99	2.4	45.61	1.5	39.41	1.7	32.77	1.6
Apr-May	46.10	2.8	29.41	1.9	47.21	1.6	37.00	1.5
Jun-Jul	49.55	3.1	38.17	3.2	35.35	2.5	34.52	3.3
Aug-Sep	48.56	2.7	55.52	3.9	44.06	3.0	44.16	3.4
Oct-Nov	58.36	2.5	57.32	2.3	32.63	2.1	34.60	5.8
All	51.59	1.2	45.49	1.9	39.73	1.1	36.86	1.8

Source: Fox 1997.

which 38 are reported in this Chapter: 14 for grasslands, 19 for shrubsteppes and 5 for deserts. At first, these studies were accomplished using observations or rumen analyses. More recently, fecal analyses prevail. All three methods have their advantages and disadvantages, but fecal analyses are used most often today because large numbers of samples can be collected at all seasons without sacrificing animals. Digestibility correction factors were first used for pronghorn in the 1990s and appeared to report percentages of forbs consumed more accurately. Consequently, that technique is recommended for inclusion in future diet composition studies.

Diet quality has been ascertained for two studies using DAPA and FN factors analyzed from fecal collections. The procedure enables wildlife managers to rate vegetation conditions for pronghorn carrying capacities. The need to collect climatic data (precipitation and temperatures) and evaluate these factors concurrent with diet selection findings appears paramount for understanding the relationship of vegetation production to animal intake.

Several misconceptions have resulted from incorrectly analyzing or interpreting pronghorn food habit data in past years. However, a number of studies provide information on availability of forage species, percentage composition of forage in the diet on a yearlong basis and computation of preference rates. When this information is available, evaluating the pronghorn's food habits for that site at that time is possible. The above criteria were analyzed for the grassland and shrubland biomes. Sufficient data for evaluation of the desert biome could not be located.

Fourteen pronghorn diet selection studies were analyzed for the grassland biome. Grasses were the most available (74%) forage class, comprised 19% of the pronghorn diet and had a 0.2 preference rating. Forbs were 16% available, averaged 62% of the diet and had the highest preference rating of 3.9. Shrubs were 9% available, 17% of the diet and had a preference rating of 1.9.

For the shrubsteppe biome, 19 food habit studies disclosed grasses had an average 37% availability, composed 7% of diet and had a preference index of 0.2. Forbs were 15% available, 30% of diet and had a preference rating of 2.0. Shrubs were 46% available, 62% of the diet and had a rate of 1.3 for preference.

Five diet composition studies could be located for the desert biome. All indicated highest use of forbs, shrubs next and little consumption of grasses. None contained complete yearlong data on the composition percentages of vegetation in plant communities. Consequently, no preference ratios were computed.

Mean values for all biomes indicate pronghorn are forb eaters first, shrub browsers second and seldom graze grasses.

The amount of air-dried quality forage needed to maintain an adult pronghorn averages 2.5 to 3.0 pounds (1.1–1.3 kg) per day. Larger amounts may be needed for northern herds to survive severe winters.

Today, pronghorn consume less than 1% of the forage on western rangelands.

Three cases of plant toxicity affecting pronghorn were located in the literature. Pronghorn apparently consume, with no toxic problems, numerous plants highly noxious to livestock. At times, pronghorn relish these plants.

Supplemental feeding has been practiced, especially during severe winter conditions, but rarely have these ventures been successful. A supplemental pellet has been developed, tested and may have possibilities for feeding wild, free-roaming populations.

The field of nutritional needs and values is relatively new for pronghorn. Limited laboratory experiments have been conducted. The nutrient requirements for pronghorn remain unknown because studies have not been conducted. Such values have been interpolated from requirements for deer and livestock. More research on nutritional values and requirements is needed before wildlife managers can truly understand or determine adequate forage management strategies for the welfare of pronghorn.

Preformed water (obtained from forage) provides moisture for body functions. For some habitats lacking free-standing water, pronghorn apparently can obtain sufficient moisture from preformed water in forage, depending on precipitation levels. Therefore, rangeland management strategies that enhance vegetation with high moisture content can be beneficial to pronghorn.

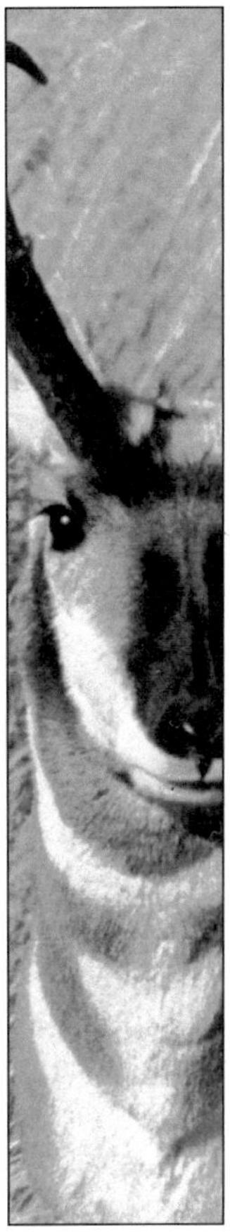

C H A P T E R F I F T E E N

Relationships with Other Herbivores

Jim D. Yoakum

Of the hoofed mammals in North America in 1791, only two are typical of grasslands: the buffalo and the pronghorn. Their distribution overlaps broadly on the plains, although the bison extends far to the east, while the pronghorn is exclusively western.

Allen (1967:60)

A basic tenet of ecological theory is that sympatric species tend to reduce competition by using their environment in different ways.

Ellis and Travis (1975:411)

Today, western rangelands of North America are occupied by a myriad of wild and domestic animal species, and each animal interacts with pronghorn differently. Some (e.g., bison) are commensal; others (e.g., domestic sheep) may be competitive; still others (e.g., coyotes, bobcats, eagles) are predatory and (see Chapter 11). However, all coexist with pronghorn, and knowledge of their interactions and relationships is needed for sound management decisions.

This discussion divides interspecific interactions into those involving wildlife and those of domestic and feral livestock. Other wildlife species have coexisted and coevolved with pronghorn during millions of years. Their management, with respect to pronghorn, is subject to political and ecological implications that are different from those of pronghorn-livestock management. By virtue of the relatively recent introduction of livestock to western rangelands, the impact of these animals on pronghorn is especially significant. Nevertheless, pronghorn adaptation to livestock characterizes the species' resilience to human-induced environmental changes and impositions.

DEFINITIONS

Relationships between two ungulates, whether domestic or wild, can be complex and range from beneficial to catastrophic. To understand these impacts, two terms—competition and dietary overlap—are defined as they are used in this text.

Competition

Competition, in an ecological context, is an ambiguous term (Clements and Shelford 1939, Allee et al. 1949, Odum

1953, Andrewartha 1961, Severson et al. 1968, Stoddart et al. 1975, Wagner 1978, Cooperrider 1986, Heady and Child 1994). Severson et al. identified 15 definitions. Wagner's definition of interspecific competition—interaction in which two species require the same resource, always at the expense of one or both species—best suits the context of this chapter.

Competition occurs when a resource, usually forage, is used to the point of being in short supply and one or more groups of users are negatively affected as a result (i.e., when a species or population is constrained numerically, spatially or biologically because of another species' use of that resource). The degree of competition for forage depends on geographic area, vegetative community, grazing intensity, and the number and variety of herbivorous users.

Competition may involve such habitat components as water, shelter or space, but most interspecific competition between ungulate species is for common food resources. Figure 102 illustrates feeding patterns of various wild and domestic herbivores. This schematic portrayal of feeding niches is a generalized representation of feeding habits. It shows that competition is least likely to occur between sympatric species with different feeding habits, such as cattle and pronghorn. However, when cattle overuse their preferred forage, they feed on less preferred plants, which may place them in competition with pronghorn.

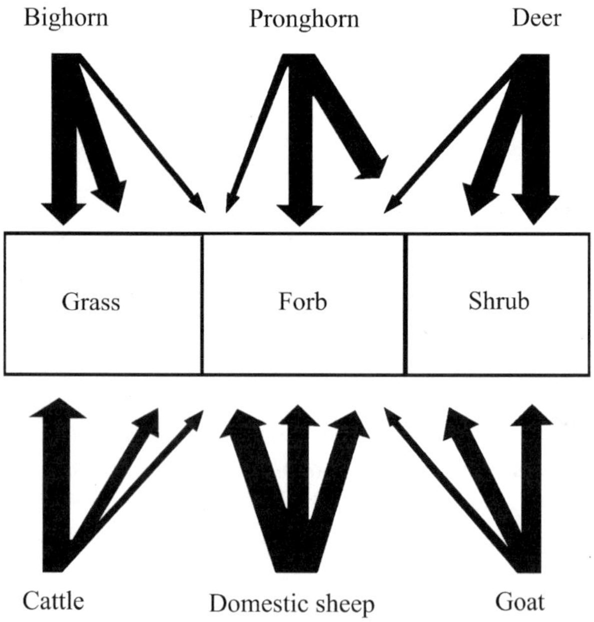

FIGURE 102. *Feeding niches of wild and domestic herbivores. Arrow strength indicates feeding preferences and emphasizes potential competition (modified from Wagner 1978).*

Dietary Overlap

Dietary overlap refers to the degree of similarity in use of food resources among different animal species occupying the same rangeland, whether concurrently or not. Overlap calculations are derived from diet composition investigations (Schoener 1970, McCullough 1980). Consequently, both methods are based on samples, provide estimates and should not be interpreted as precise biological findings. Dietary overlap is high when both animals ingest the same or similar proportions and kinds of forage, and it is nonexistent when none or small quantities of the same plants are consumed. High overlap often suggests severe forage competition; low overlap suggests an absence of competition. At times, either inference may be misleading. For example, forage overlap may be high without competition if the preferred forage is plentiful enough to satisfy both species' needs. Conversely, low overlap with severe competition may occur where one herbivore uses and depletes preferred limited forage in an area it shares with another species, but at a time when the other species is absent. An example of this, called sequential competition, occurs when cattle forage a rangeland during summer that is occupied by pronghorn in winter. If cattle remove large quantities of forbs and shrubs preferred by pronghorn, overlap in food habits can be low, yet forage competition high. There also are times when cattle make limited individual use of forage highly preferred by pronghorn (such as forbs), but because of the number of cattle, a significant amount of the forb crop is removed and unavailable to the pronghorn population.

Forage competition and dietary overlap also can be related to physiological and morphological characteristics of different herbivores (Kleiber 1961, Schwartz 1977). For ruminants, the lips and tongue are the principal prehensile organs. Bison and cattle use their tongues as prehensile organs to pull grass or other herbage into the mouth. The lips, which are broad, relatively immobile and uncleft, are important when cropping lush, young herbaceous plants. Pronghorn and domestic sheep have an upper lip that is partly cleft, relatively narrow and more mobile. This allows pronghorn and sheep to be more selective when consuming forage (Schwartz 1977).

Morphological characteristics of the ruminant stomach vary among species and may determine, or be determined by, the feeding habits of the species (Hoffman 1968, Hanley 1980). Ruminants can be classified into roughage or grass eaters, selective feeders, or transitional forms, based on feeding habits. Bison and cattle, with large rumens, abomasums and omasums, are roughage consumers. Most selective feeders have relatively small stomachs and generally consume soft, succulent forage low in fiber. Pronghorn are selective feeders, and their rumen volume is about 8.5% of their body weight. They have relatively small mouths and forage heavily on succulent forbs. Most transitional ruminants are browsers of leaves and twigs, or animals with a seasonal intake of dry or fibrous forage. Domestic sheep and goats are in the transitional group. Smaller ruminants generally have higher metabolic rates than do the larger ruminants, therefore require a comparatively higher intake of food per unit of weight (Kleiber 1961).

Foraging strategies of wild and domestic ungulates vary. A 99-pound (45 kg) pronghorn has a basal metabolic need of approximately 1,215 kilocalories per day (Wesley et al. 1973), whereas a 1,000-pound (454 kg) cow requires 6,885 kilocalories per day. Although a pronghorn requires more energy, relative to body weight, than does a cow—12.3 and 6.9 kilocalories (5/6 and 3.1 kilocalories/kg) per pound per day, respectively—a pronghorn needs only 18% of the energy that a cow needs on an absolute basis. Consequently, pronghorn can spend more time searching for and selecting preferred food, and the food it takes has a higher caloric value. Considering that it has a smaller rumen size (again, relative to body size) and, thus, is unable to process large amounts of food, the pronghorn is forced to consume relatively high-quality food to meet its energy requirements. This indicates why wildlife managers need to enhance nutritious, succulent forage, such as forbs, on rangelands occupied by pronghorn.

RELATIONSHIPS WITH WILDLIFE

Rodents

Pronghorn association with rodents was studied in a shrubsteppe community of the Great Basin Divide of Wyoming (Maxell 1973). Six rodents (prairie dogs, ground squirrels, chipmunks and three species of mice) used 5% of the available vegetation. Pronghorn in the area consumed 15% to 20% of the vegetation. The rodents' total use of forage was insignificant, so the animals were not competitive with pronghorn.

Pronghorn and prairie dogs have coexisted for centuries on western prairies. While hunting in western Kansas during the mid-1800s, Gilmore (1874) noted that prairie dog mounds often provided the only cover for crawling toward pronghorn. However, he found that the warning cry of a prairie dog immediately would alert a pronghorn or cause it to flee. King (1955), Koford (1958) and Costello (1970) speculated that prairie dogs in towns enhanced rangeland for pronghorn by consuming grasses and disturbing soils, thereby increasing forb abundance and variety.

However, Lovaas and Bromley (1972) studied this relationship for a single summer/autumn period in Wind Cave National Park, South Dakota, and found the opposite. The most important food items in pronghorn rumen samples (forbs and shrubs) were more abundant off prairie dog towns than on them (Figure 103). Shrubs were more abundant off the towns (2% off versus trace on). Rumen samples of pronghorn disclosed that forbs were the most important component in the diet, accounting for 65% to 80% of the volume for all but one sample, which was 47%. Forb use was highest in autumn. Grasses comprised 8% in the summer diet and 3% in autumn. Shrub use averaged 26% in summer and 16% in autumn. Lovaas

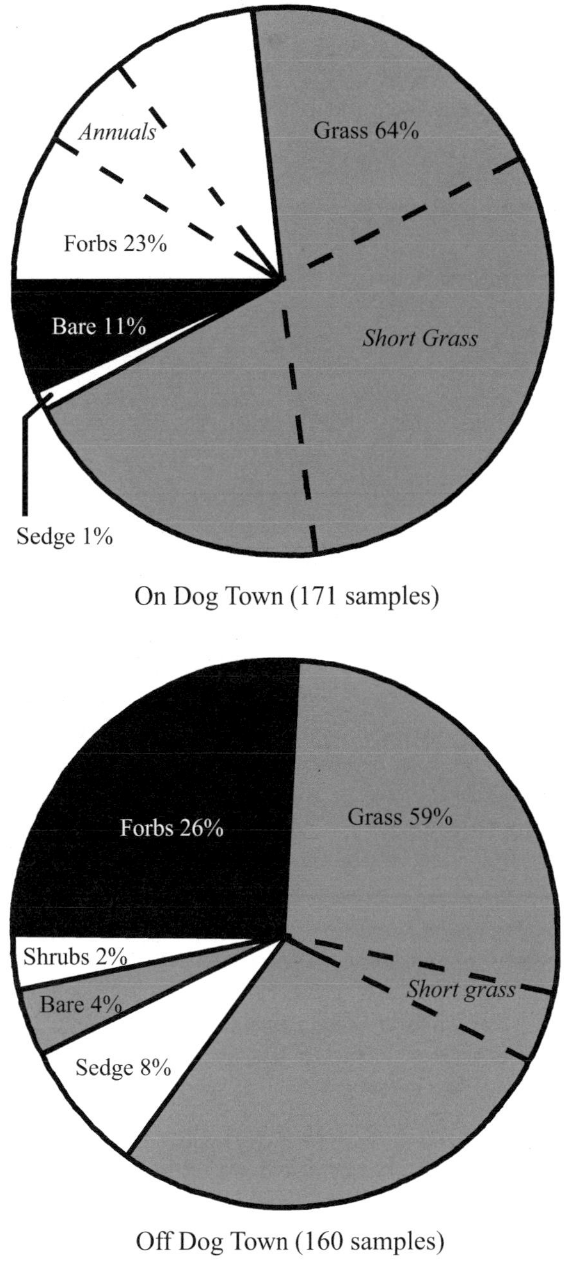

FIGURE 103. *Vegetation on and off prairie dog towns in South Dakota (from Lovaas and Bromley 1972).*

(Top) *Early travelers in the West reported that prairie dogs and pronghorn warned one another of potential danger by barking or snorting. Several researchers speculated that prairie dogs enhanced forage on their towns for pronghorn by consuming or cutting grasses and disturbing soil, thereby increasing abundance and variety of forbs. However, a later study indicated that bachelor bucks—such as these in South Dakota—used the towns, but older bucks did not establish territories on them.* (Bottom) *Pronghorn feeding on prairie dog towns concentrate on very small forbs that grow on the disturbed soil around burrows and mounds. Today, establishment of prairie dog towns on the Great Plains generally requires concentrations of cattle around water sources or salt blocks (see Knowles 1982). Before they were nearly exterminated, bison probably primed areas for prairie dogs.* Photos by Peter T. Bromley.

and Bromley did not believe that prairie dog towns improved vegetative conditions for pronghorn. Rather, they felt that prairie dogs competed with pronghorn for preferred forage. Pronghorn appeared to favor vegetation off the towns more than that on town sites because of the former's greater diversity of species.

Interspecific relationships among bison, pronghorn and prairie dogs were investigated in Wind Cave National Park by Krueger (1986). Pronghorn selected forbs in proportions that were from 3 to 10 times their availability on prairie dog towns. Pronghorn made heavy use of forb/shrub-rich prairie dog towns for feeding and appeared to select them preferentially for foraging. Consequently, prairie dog town centers were better sites for pronghorn foraging than were uncolonized areas. Prairie dogs did not respond to the presence or absence of pronghorn foraging on their towns nor to the associated changes in forage quality, thus pronghorn were judged to have a neutral effect on prairie dogs (Krueger 1986).

Lagomorphs

On the desert shrublands of southwestern Utah, Beale and Smith (1970) noted that black-tailed jackrabbits sometimes enhanced pronghorn forage. When jackrabbits cropped desert almond to ground level during winter, the plants sprouted vigorously the following spring. Pronghorn preferred the tender, almost herbaceous shoots to regular shoot growth.

Ungulates

Pronghorn historically occupied habitats in common with bison and other ungulates. Lewis and Clark's journals are replete with passages describing mixed herds of bison and pronghorn (Thwaites 1904, Burroughs 1961, Moulton 1987b). At times, elk and deer were seen with pronghorn, but these were few in comparison with the reported sightings of bison and pronghorn together. Today, bison are restricted primarily to refuges, parks and private ranches, wihereas pronghorn range freely over thousands of acres. Elk, deer and wild sheep can be seen in proximity to pronghorn today, but this is not common because areas of seasonal use generally do not coincide. I observed herds of elk, wild sheep, mule deer and pronghorn grazing within a 1.0-mile (1.6 km) radius in Yellowstone National Park, near Gardiner, Montana, during autumn 1982. Retention of large ecosystems, such as Yellowstone National Park, helps preserve historic scenes of wild, free-roaming, sympatric ruminant populations.

Pronghorn usually associate freely with other species of domestic or wild ungulates. Except for fawns, they usually avoid direct confrontations with larger animals and pay little attention to them. However, during late spring, pregnant does seem wary of other animals. Rangers on the Wichita National Forest and Game Preserve, in Kansas, noted that hand-raised does, which fed from their hands during the rest of the year, became very wild before and during the fawning period (Woods 1925). Obviously, large ungulates could easily disturb pronghorn does during late spring.

Bison. Bison and pronghorn lived commensally before the arrival of Europeans, with both ungulates often in mixed herds, sharing forage and water on vast open plains. Maps depicting bison and pronghorn distribution at the turn of the 19th century indicate that both occurred on more than 2 million square miles (5 million km^2) (Schwartz 1977). According to Martin (1973:15), early explorers witnessed bison in such large herds that they represented a "brown sea, a restless, moving mass . . . as far as the eye could see." Grinnell (1929) speculated that pronghorn were more numerous than bison. Bison may have benefited pronghorn by grazing abundant grasses, leaving forbs and shrubs more readily available and stimulating their growth, and by opening trails through deep snow (Costello 1969, Chase 1977, Danvir 2000). The latter does not seem paramount because pronghorn generally move from areas of deep snow. The partitioning of forage resources by both ungulates may have paralleled the feeding habits of ungulates on the Serengeti plains of Africa (Bell 1971). There, the diet of Thompson's gazelle is composed primarily (40%) of forbs, whereas wildebeest and zebra are exclusively graminivorous. The larger ungulates graze the tall grasses, thus making the low-growing or prostrate forbs available to the gazelle. A similar relationship between pronghorn and bison may have existed in pre-Columbian times.

Pronghorn pay little attention to bison and often feed close to them. However, the pronghorn seem cautious, avoid traveling bison and seldom move into or through a herd of the large animals. On the National Bison Range, pronghorn tend to move into pastures after bison have been there long enough to remove the tall, coarse graminoids (B. O'Gara personal communication:1998).

One can no longer observe the vast, mixed herds of pronghorn and bison on western prairies. Most of these lands now are fenced for livestock or have been converted to agricultural production. Probably the only free-roaming bison and pronghorn herds are those in Yellowstone National Park. However, these ungulates can be observed cohabiting several fenced parks and refuges, such as the National Bison Range in Montana, and Custer State Park and Wind Cave National Park in South Dakota.

One of the earliest food habit studies of bison and pronghorn was accomplished in 1949 at Wichita Mountains National Wildlife Refuge, Oklahoma, by Buechner (1950b). The area was historic rangeland of both species and, through translocations, herds again occupy the refuge. Buechner reported little competition for forage: bison ate 99% grasses and 1% forbs; pronghorn consumed 99% forbs and 1% grasses (Figure 104). Buechner indicated that the two species appeared to have adapted to coexistence on grasslands. The low dietary overlap between the two species, as reported on the National Bison Range by McCullough (1980), supports this contention (Table 120). Weydeven and Dahlgren (1985) reached similar conclusions.

*Bison and pronghorn graze together in South Dakota parks today (*top*), much as they coexisted on western prairies for centuries. They are relatively compatible, and little foraging competition exists because bison primarily are grazers and pronghorn prefer forbs and shrubs. Bison improve rangelands for pronghorn by intensively grazing grasses, which encourages growth of forbs and shrubs. Pronghorn pay little attention to bison and often feed close to them (*left*); note the prairie dog. However, the pronghorn seem cautious, avoid traveling bison and seldom move into or through a herd of the large animals. On the National Bison Range, pronghorn tend to move into pastures after bison have been there long enough to remove the tall, coarse graminoids.* Top photo by Durward Allen. Left photo by Peter T. Bromley.

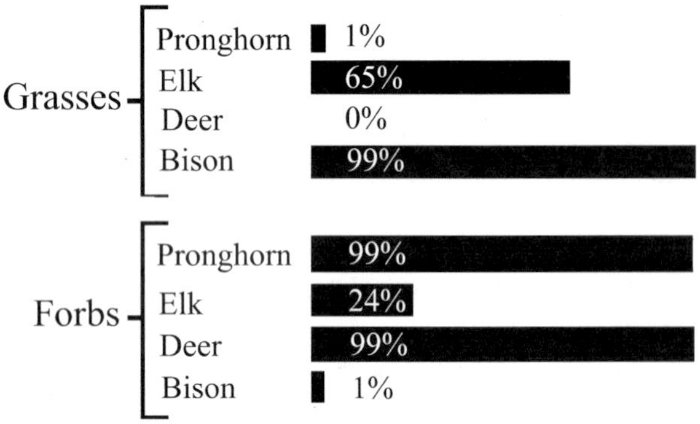

FIGURE 104. *Comparison of pronghorn, elk, white-tailed deer and bison food habits on the Wichita Mountain National Wildlife Refuge, Oklahoma (from Buechner 1950b).*

TABLE 120. Dietary overlap[a] between pronghorn and bison on grasslands of the National Bison Range, Montana

Vegetation	*Percentage forage in pronghorn diet*	*Percentage forage in bison diet*	*Dietary overlap*
Grasses			
Wheatgrass	(0.2)	10.1	0.2
Gramma grass	(0.2)	1.1	0.2
Wild rye	(0.3)	0.7	0.3
Fescue	(2.3)	77.8	2.3
Forbs			
Fleabane	34.1	(0.5)	0.5
Lupine	2.0	(0.6)	0.6
Beebalm	0.1	(0.1)	0.1
Shrubs	0.0	0.0	0.0
Total dietary overlap	—	—	4.2

[a] Determined by taking the lesser of the two figures in each row (in parentheses).
Source: McCullough 1980.

When grass is tall, pronghorn often feed behind larger animals that remove much of the coarse forage. Here, the pronghorn are feeding on the first-year rosettes of yarrow—a biennial forb that constitutes the main winter food of pronghorn on the National Bison Range in Montana. Photo by Bart W. O'Gara.

However, Lovaas and Bromley (1972) observed that, at Wind Cave National Park, some competition between bison and pronghorn may exist. Bison exerted continual light grazing pressure on the grasslands, which probably led to a decrease in abundance of many plants preferred by pronghorn. Competition was not related to abundance of vegetation, but to individual plant species.

A study conducted on the National Grassland Biome in northeastern Colorado was designed to analyze the structure, function and utilization of grassland ecosystems (Schwartz and Nagy 1976). Part of the project compared the food habits of bison and pronghorn in large fenced pastures. Differences detected were attributed to species-specific food habits rather than characteristics of the common rangeland. Diet studies were conducted in two pastures with light and heavy grazing (Table 121). Both ungulates consumed more herbaceous forage in the lightly used pastures and browsed more shrubs in the heavily grazed pastures. Diet overlap was minimal during the plant growing season, when pronghorn grazed more discriminately and used more plant species than in other seasons. Bison and pronghorn diets were dissimilar, and the animals did

One can no longer observe the vast, mixed herds of pronghorn and bison on western prairies. Most of these lands now are fenced for livestock or have been converted to agricultural production. Probably the only free-roaming bison and pronghorn herds are those in Yellowstone National Park. However, these ungulates can be observed cohabiting several fenced parks and refuges, such as the National Bison Range, in Montana, and Custer State Park and Wind Cave National Park, in South Dakota. Photo by Joe Van Wormer; courtesy of the Denver Natural History Museum, Colorado.

not compete significantly for available forage, except perhaps in March and October. Pronghorn ingested a diet high in crude protein (7%–21%), whereas bison diets contained less protein (4%–12%). Pronghorn and bison grazing communally may be separated ecologically, as noted by Bell (1971) for different ungulates using grasslands in Africa, that is, under natural grazing conditions, small and large ruminants do not compete for forage resources.

Working on the National Bison Range in Montana, McCullough (1980) studied niche separation for bison and pronghorn. She noted that both species had a high spatial overlap, as both used the open, level-to-rolling prairies. However, diet overlap was minimal—bison consumed grasses almost entirely, whereas pronghorn ate forbs and shrubs (Table 120). Bison disturbed vegetation by trampling, wallowing and grazing, which resulted in greater production of forbs and shrubs favored by pronghorn.

Few references could be found describing antagonistic behavior between bison and pronghorn. Skinner (1922) reported that bison occasionally killed pronghorn fawns, but more often, pronghorn were speedy enough to escape bison aggression. Unfortunately, Skinner did not document the frequency of the occasional mortality. J. Richardson (personal communication:1990) observed the two species during many years on the National Bison Range in Montana. He stated that pronghorn frequently moved out of the way of bison and that bison would shake their heads at approaching pronghorn, but Richardson never saw physical contact between the two. However, F. Martin (personal communication:1979) saw a bison bull gore and trample two young pronghorn fawns in a pasture on the C. M. Russell National Wildlife Range in central Montana. A bison in a fenced pasture of the Jackson Hole, Wyoming, Wildlife Park killed an 8-month-old pronghorn (McHugh 1958).

Elk. Distributional overlap of elk and pronghorn existed in pristine times as it does now. But, elk now use open areas to a lesser degree (see Bryant and Maser 1982, Geist 1982, McCabe 1982, 2002). Herds of both species seldom are seen grazing in common; such interaction is limited primarily to parks and refuges. During winter, elk move to lower elevations that constitute pronghorn summer rangeland, but pronghorn tend to move to even lower elevations in winter.

While studying rangeland ecology on the Wichita Mountains National Wildlife Refuge in Oklahoma, Buechner (1950b) observed elk and pronghorn foraging together. The animals were feeding so closely that their noses nearly touched, and there were no observable acts of aggression. Despite the social compatibility, their feeding habits were dissimilar, with no significant dietary overlap (Table 122). Elk mainly consumed grasses (65%) and forbs (24%), whereas pronghorn used mostly forbs (99%) and very little grasses (1%) (Figure 104). On the National Bison Range, Livezey (1979) saw a yearling female elk chasing a pronghorn buck, and he witnessed a calf and another female yearling elk running after a pronghorn doe. The elk appeared playful. Livezey also watched a band of doe pronghorn approach within 5 yards (4.6 m) to investigate a hiding elk calf.

Forage similarity of and competition between pronghorn and elk were studied on Trickle Mountain, Colorado, by Bailey and Cooperrider (1982). They reported the greatest seasonal dietary overlap during winter (59%), with less during spring (33%), summer (38%) and autumn (31%). There was moderate dietary overlap in forage classes on a yearlong basis—grasses (3.5%), forbs (11.5%) and shrubs (45.5%). Overall, the dietary overlap was twice as high on this shrubsteppe community as it was for the grassland biome (Table 122).

Table 121. Diets of pronghorn and bison on pastures with different grazing intensities on a shortgrass prairie in Colorado

	Lightly grazed pasture					*Heavily grazed pasture*				
Plant group	*Mar*	*May*	*Jun*	*Aug*	*Oct*	*Mar*	*May*	*Jun*	*Aug*	*Oct*
Pronghorn										
Warm season grass	17	—	9	8	21	30	—	10	11	35
Cool season grass	65	37	—	—	50	50	43	—	2	44
Warm season forbs	17	41	75	92	29	4	42	88	81	16
Cool season forbs	1	22	16	—	—	5	15	2	6	—
Shrubs	—	—	—	—	—	11	—	—	—	5
Bison										
Warm season grass	70	44	78	88	82	56	18	80	93	78
Cool season grass	28	54	15	7	17	21	80	12	7	9
Warm season forbs	—	1	7	5	1	—	1	8	1	1
Cool season forbs	—	1	1	—	—	—	—	—	—	1
Shrubs	—	—	—	—	—	21	—	—	—	12

Source: Schwartz and Nagy 1976.

TABLE 122. Dietary overlap of pronghorn with bison, elk, mule deer, white-tailed deer and wild sheep, by forage class

			Percentage dietary overlap				
Species	*Location*	*Biome*	*Grasses*	*Forbs*	*Shrubs*	*Annual*	*Source*
Bison	Wichita Mountains National Wildlife Refuge, Oklahoma	Grassland	1.0	1.0	—	2.0	Buechner (1950b)
	Pawnee Grasslands, Colorado	Grassland	41.0	3.0	0.0	44.0	Schwartz and Nagy (1976)
	National Bison Range, Montana	Grassland	3.0	1.2	0.0	4.2	McCullough (1980)
Elk	Wichita Mountains National Wildlife Refuge, Oklahoma	Grassland	0.1	24.0	0.0	24.1	Buechner (1950b)
	National Bison Range, Montana	Grassland	2.9	4.6	3.8	11.3	McCullough (1980)
	Trickle Mountain, Colorado	Shrubsteppe	3.5	11.5	45.5	60.5	Bailey and Cooperrider (1982)
White-tailed deer	Wichita Mountains National Wildlife Refuge, Oklahoma	Grassland	0.0	99.0	0.0	99.0	Buechner (1950b)
	National Bison Range, Montana	Grassland	2.9	24.8	18.5	46.2	McCullough (1980)
Mule deer	Yellowstone National Park	Shrubsteppe	16.6	9.2	52.0	77.8	Barmore (1969)
	National Bison Range, Montana	Grassland	2.3	27.1	33.3	62.7	McCullough (1980)
	Northeastern California and northwestern Nevada	Shrubsteppe	2.3	4.2	82.3	88.8	Hanley (1980)
	Trickle Mountain, Colorado	Shrubsteppe	3.5	10.2	79.5	93.2	Bailey and Cooperrider (1982)
	Hart Mountain National Antelope Refuge, Oregon	Shrubsteppe	7.9	49.3	40.5	97.7	Hansen et al. (2001)
	Sheldon National Wildlife Refuge, Nevada	Shrubsteppe	5.0	31.0	39.0	75.0	Hansen (1986)
	Sheldon National Wildlife Refuge, Nevada	Shrubsteppe	9.9	25.0	54.2	89.1	Hansen et al. (2001)
Bighorn	National Bison Range, Montana	Grassland	2.9	12.3	11.4	26.6	McCullough (1980)
	Trickle Mountain, Colorado	Shrubsteppe	3.5	6.2	56.6	66.3	Bailey and Cooperrider (1982)
	Sheldon National Wildlife Refuge, Nevada	Shrubsteppe	5.0	32.0	14.0	51.0	Hansen (1986)
	Sheldon National Wildlife Refuge, Nevada	Shrubsteppe	9.9	25.0	3.5	38.4	Hansen et al. (2001)
	Hart Mountain National Antelope Refuge, Oregon	Shrubsteppe	7.9	32.4	8.6	48.9	Hansen et al. (2001)
Gemsbok	White Sands Missile Range, New Mexico	Grassland	20.0	4.0	13.0	37.0	Smith (1994)

McCullough (1980) found little overlap in habitats occupied by elk and pronghorn on the National Bison Range in Montana. Consequently, there was little competition for space, water or forage. Elk foraged primarily on graminoids, whereas pronghorn preferred forbs and shrubs. Annual dietary overlap (11.3%) was low (Table 122).

Tule elk have coexisted with pronghorn since prehistoric times in the Central Valley of California (Murie 1951, McCullough 1969, Bryant and Maser 1982, Wisdom and Cook 2000). Tule elk sometimes are referred to as dwarf elk, for they are the smallest subspecies of North American elk (McCullough 1969). Tule elk characteristically inhabit grasslands and open, semiarid rangelands, compared with other elk subspecies that more frequently occupy forested and mountainous regions. According to Murie (1951:55), tule elk live on, "the arid plains and are probably more truly a plains animal than any other American elk." McCullough (1969:168) agreed with Murie, stating that dwarf elk are, "specialized for living in semiarid conditions." They are most abundant in and about marshes and the fertile flood plains. These habitat preferences are similar for pronghorn, so it is understandable why both ungulates flourish in the extensive Central Valley of California.

Pronghorn and tule elk currently occupy the Carrizo Plain in California's southwestern Central Valley grasslands. Both species were extirpated on the plains by Euro-Americans, but herds were translocated and reestablished during the late 1900s. Today, it is not uncommon to observe herds of pronghorn and tule elk grazing together on the plains (A. Koch personal communication:2001). Both ungulates use reservoirs developed by the California Department of Fish and Game for drinking water. The Carrizo Plain was designated by President William J. Clinton, in January 2001, as a national monument to preserve native flora and fauna in a large, relatively natural landscape. This conservation effort may help ensure that native pronghorn, tule elk and other wildlife species be enjoyed by the public for many years to come—a heritage not available to the citizens during 90% of the 20th century.

Studies of high elk numbers cohabiting rangelands with pronghorn in Yellowstone National Park indicate that competition by elk may be a major influence on depressed pronghorn herds (Boccadori 2002). Consequently, the U.S. National Park Service conducted a week-long workshop during February 2002 to ascertain why pronghorn numbers continue to decrease. There was concern that severe

Before the early 1800s, tule elk and pronghorn maintained a commensal relationship on the expansive, semiarid grasslands of California's Central Valley, where the three bulls (top) *were photographed during the 1960s. Both ungulates were abundant before the arrival of Euro-Americans, and both were greatly exploited during the 20th century. Tule elk still occupied remnant valley grasslands with pronghorn early in the 21st century—both having been translocated by the California Department of Fish and Game during the 1980s and 1990s. Unlike other North American elk subspecies that often inhabit forests and mountains, the tule elk retains its preference for open, semiarid undulating valleys and savannas in California. These grasslands have been grazed by domestic ungulates for more than 200 years. Before other western prairies were settled by Euro-Americans, elk and pronghorn shared grasslands with little competition. Elk mainly consumed grasses, whereas pronghorn depended on forbs and shrubs. The introduction of domestic livestock on western rangeland and the virtual extirpation of prairie-dwelling elk have made elk/pronghorn interaction uncommon. In the upper right of the bottom photo, a pronghorn attentively watches a Rocky Mountain elk move through a pasture shared by pronghorn and cattle in the Pole Mountain area of Wyoming.* Top photo by Jim D. Yoakum. Bottom photo by A. Lorin Ward; courtesy of the U.S. Forest Service, Laramie, Wyoming.

competition for preferred forage, plus inadequate vegetative concealment cover for fawns might be influencing stressed pronghorn populations, and a long-term decline could result in the extirpation of pronghorn within the park during the 21st century.

Deer. Pronghorn live on rangelands with white-tailed and mule deer, but their spatial distribution generally does not overlap (Buechner 1950a, Bruns 1977, McCullough 1980, Bailey and Cooperrider 1982, Wood 1989). White-tailed deer favor riparian and brushy areas, whereas mule deer select rougher, forested terrain, neither of which commonly is frequented by pronghorn. Deer and pronghorn occasionally are seen grazing in proximity on open rangelands, hence the phrase, "where the deer and the antelope play."

McCullough (1980) noted that white-tailed deer and pronghorn used the more level, low elevation areas on the National Bison Range, in Montana, but their spatial distribution did not greatly overlap. For the most part, the whitetails occupied dense shrub and wetland areas, whereas pronghorn used open grasslands. Pronghorn occasionally occupied low-structured shrubland (vegetation 24 inches [61 cm] or less in height), but were not observed in forested areas. Both ungulates fed predominantly on shrubs and forbs, with little use of grasses.

Competition was not significant because the two species occupied separate habitats most of the time. Some aggressive encounters were observed when territorial male pronghorn attempted to exclude whitetails from open habitats. Aggressive encounters ceased during summer storms and after the pronghorn rut, when whitetails moved out of riparian habitats into the more open uplands. Behavioral interactions were minimized because of greater diurnal activity among pronghorn and nocturnal activity among whitetails. However, B. O'Gara (personal communication:1998) saw six pronghorn fawns on the Bison Range chase two white-tailed does until the deer appeared exhausted. The panting does ran into a thick patch of shrubs and crouched. The fawns milled around, but would not enter the thick, high cover.

Buechner (1950b) studied white-tailed deer food habits on rangelands also occupied by pronghorn in Oklahoma (Figure 104). He found that the diet of these ungulates was almost identical during spring and summer, but concluded that competition was negligible because they generally occupied different habitats. Buechner saw 32 whitetails occupying rangelands with 39 pronghorn during his study.

Mule deer share pronghorn habitat in central and eastern Montana, and the two species sometimes mingle (B. O'Gara personal communication:1998). They may feed together for a short time, especially in the morning, but the deer usually start and finish feeding earlier than pronghorn. The deer retire to brushy cover in coulees during the day, whereas the pronghorn move into flat, open country to bed down. Consequently, the two kinds of animals may live close together, but they meet primarily at the edge of each one's activity areas.

Pronghorn and white-tailed deer appetites for succulent forbs sometimes bring the two species together. In this case, the pronghorn rut was over and the bucks were no longer antagonistic to deer. Autumn rains combined with warm weather had promoted green-up of grasses and forbs in a small swale, and the two species grazed compatibly. Photo by Bart W. O'Gara.

Mule deer/pronghorn relationships were studied on the National Bison Range by McCullough (1980). The species exhibited little spatial overlap. Pronghorn used low elevations and open prairies, whereas mule deer occupied the steeper, rockier and timbered habitats at higher elevations. The two species had similar feeding habits—both preferring shrubs and forbs, and consuming small quantities of grasses. Consequently, there was moderate dietary overlap when both species occupied common habitat (Table 122). The diurnal pattern of activity by pronghorn and crepuscular pattern by mule deer served to minimize direct interactions. However, where the species were in close proximity, pronghorn exhibited direct aggression to mule deer from late March to September—the territorial period for male pronghorn. When pronghorn stopped defending territories, they became less aggressive toward

Pronghorn and mule deer share habitat and the two species sometimes mingle in central and eastern Montana (B. O'Gara personal communication:1998). They may feed together for a short time, especially in the morning, but the deer usually start and finish feeding earlier than pronghorn. The deer retire to brushy cover in coulees during the day, whereas the pronghorn move into flat, open country to bed down. Consequently, the two kinds of animals may live close together, but they meet primarily at the edges of their conjoined activity areas. Photo by Peter T. Bromley.

other species. From October through March, pronghorn and mule deer fed in mixed groups with little aggressive behavior toward each other, although the deer were dominant.

Mule deer and pronghorn food habits in northeastern California were reported by Hanley (1980). Grasses consistently were eaten more by deer (6.0%) than by pronghorn (2.3%), whereas forbs were used more by pronghorn (15.4%) than by mule deer (4.0%). Both species consumed high quantities (90.0%) of shrubs. Deer diets during autumn were high in bitterbrush and mountain mahogany, whereas pronghorn diets featured sagebrush and rabbitbrush. The degree of dietary overlap was 2.3% for grasses, 4.2% for forbs and 82.3% for shrubs (Table 122), but different shrub species were consumed. Hanley indicated a high degree of dietary overlap exists, particularly during autumn, which may be attributed to morphological and physiological similarities of these two species.

Dietary overlap of pronghorn and mule deer was investigated on a shrubsteppe in south-central Colorado by Bailey and Cooperrider (1982). Overlap varied on a year-round basis, being greatest in spring (58%), least in summer (15%), and moderate during autumn (47%) and winter (51%) (Table 123). Overlap was low for grasses (3.5%) and forbs (10.2%) but high for shrubs (79.5%) (Table 122).

Hansen (1986) investigated the diets of pronghorn and mule deer on northwestern Nevada shrubsteppes. His findings revealed a high annual similarity of diet (75%), with both eating about equal amounts of forbs and shrubs (Table 122). Both used sagebrush throughout the year and consumed bitterbrush heavily during autumn. Both used the forbs fleabane and rockcress extensively throughout the year.

Diet similarity studies were conducted from 1992 to 1995 for pronghorn and mule deer on the Hart Mountain National Antelope Refuge in Oregon (Hart Mountain) and the Sheldon National Wildlife Refuge in Nevada (Sheldon) (Hansen et al. 2001). Both ungulates alternated between forbs and shrubs while ingesting small quantities of grasses. However, when nutritious, succulent herbaceous forage was available, these forage classes made up 80% to 90% of both ungulates' diets. Forbs apparently were selected in winter over the more ubiquitous and taller sagebrushes when the forbs were not covered by snow. Because both ungulates consumed predominantly forbs and shrubs, dietary overlap was high: 97.7% on Hart Mountain and 88.1% on Sheldon (Table 122).

Aerial surveys were used to determine habitat use and distribution of sympatric pronghorn and mule deer populations on a 210-square mile (543 km^2) area in eastern Montana from 1982 to 1987 (Wood 1989). Pronghorn occupied 1% to 10% and deer 9% to 20% of the study area. Flat to rolling grassland habitats were used most frequently by pronghorn. Badlands, bunchgrass prairie and hardwood draws were used most frequently by deer (Table 124).

Table 123. Dietary overlap with pronghorn by mule deer and wild sheep, by season

	Percentage dietary overlap					
Species	*Winter*	*Spring*	*Summer*	*Autumn*	*Annual*	*Source*
Mule deer	62	48	39	93	60	Vavra and Sneva (1978)
	95	90	76	47	77	Hanley (1980)
	51	58	15	47	43	Bailey and Cooperrider (1982)
	71	64	54	47	59	Hansen (1986)
Wild sheep	67	46	42	19	43	Bailey and Cooperrider (1982)
	44	30	37	—	37	Hansen (1986)

TABLE 124. Percentages of available habitat and total pronghorn and mule deer observed within each habitat during aerial surveys on the Cherry Creek study area, Prairie County, Montana, 1982 to 1987

Habitat	*Available habitat*	*Annual*		*Spring*		*Summer*		*Autumn*		*Winter*	
		Pronghorn	*Deer*	*Pronghorn*	*Deer*	*Pronghorn*	*Deer*	*Pronghorn*	*Deer*	*Pronghorn*	*Deer*
Mixed prairies	65	69 +	4 –	70 +	5 –	82 +	5 –	69 +	5 –	61	1 –
Bunchgrass prairies	7	20 +	19 +	22 +	29 +	8	10	15 +	10 +	35 +	23 +
Sagebrush grasslands	4	5	11 +	5	6 +	1 –	19 +	6 +	19 +	4	5
Badlands	18	1 –	36 +	0 –	43 +	0 –	24 +	1 –	20	0 –	50 +
Mesic shrublands	1	1	10 +	0 –	1	1	15 +	2 +	23 +	0 –	3 +
Hardwood draws	2	0 –	18 +	0 –	16 +	0 –	24 +	0 –	19 +	0 –	18 +
Legume hayfields	<1	2 +	2 +	0 –	0 –	3	3 +	5 +	4 +	0 –	0 –
Grainfields	3	2	0 –	3	0 –	5	0 –	2	0 –	0 –	0 –
Total observations		3,146	7,489	851	2,231	245	545	1,432	2,439	618	2,274

Habitats significantly ($P < 0.05$) avoided (–) or selected (+) are noted.
Source: Wood 1989.

Some overlap in habitat use was evident, but there was little spatial overlap. Maximum overlap in habitat use and distribution occurred during autumn when pronghorn moved into deer habitat. There appeared to be little opportunity for interspecific competition based on the high degree of spatial segregation. Spatial segregation apparently contributes to both species' ability to coexist in localized areas where they generally are sympatric.

Bighorn. Generally, bighorn and pronghorn do not occupy the same habitats. However, a pronghorn buck on the National Bison Range, in Montana, spent one summer with five large rams in a high basin. They fed and bedded together like one species. B. O'Gara (personal communication:1998) attempted to photograph the group, but the buck was wary. The rams were easily approached, but the buck always slipped away before O'Gara could get within camera range. Berger (1985) saw two adult pronghorn bucks chase a group of 12 bighorn about 328 yards (300 m). At the onset of the chase, the two species were about 109 yards (100 m) apart. At the conclusion, some individuals were less than 10 feet (3 m) from each other, and a bighorn ewe and buck threatened each other with horns at less than 7 feet (2 m). On another occasion, an adult bighorn ram displayed his horns to an adult pronghorn buck at about 3 feet (1 m). The two then fed within 11 yards (10 m) for almost 3 minutes before the buck gradually moved away while foraging. Berger noted that

The diurnal pattern of activity by pronghorn and crepuscular pattern of mule deer minimize direct interactions on the National Bison Range in Montana. However, where the species are in close proximity, pronghorn may exhibit aggression toward mule deer from late March to September—the territorial period for male pronghorn. When pronghorn stop defending territories, they became less aggressive toward other species. From October through March, pronghorn and mule deer feed in mixed groups with little aggressive behavior toward each other, although the deer are dominant (McCullough 1980). Photo by Joe Van Wormer; courtesy of the Denver Museum of Natural History, Colorado.

neither species was clearly dominant in his Great Basin study area.

McCullough (1980) studied spatial distribution and dietary overlap of pronghorn and bighorn on the National Bison Range. She noted little overlap in space occupied and little competition for forage. Generally, bighorn consumed more grasses, whereas more shrubs were eaten by pronghorn. Both consumed about 40% forbs yearlong. Year-round dietary overlap of the two species was 26.6% (Table 122).

Data from the Trickle Mountain study provided information on food habits of pronghorn and bighorn in a shrubsteppe community (Bailey and Cooperrider 1982). Annual dietary overlap was moderate (43%) for all forage classes, being greatest during winter (67%) and least in autumn (19%) (Table 123).

Hansen (1986) studied diets of pronghorn and bighorn in northwestern Nevada. The annual overlap was 37%, with greatest similarity during winter and least during autumn (Table 123). Greatest overlap of forage classes was for forbs (32%) (Table 122).

A study of sympatric ungulates on Hart Mountain and Sheldon reported dietary overlap for pronghorn and bighorn (Hansen et al. 2001). Bighorn consumed grass as a staple, rather than shrubs, but shifted their attention to forbs when available, much as did pronghorn. Forbs were the second most abundant forage class consumed by bighorn—shrubs were a minor component. Yearlong dietary overlap percentages were: 24.2% on Hart Mountain and 48.4% on Sheldon (Table 122). The overlap for Sheldon of 48.4% reported by Hansen et al. was similar to an earlier study conducted by Hansen (1986), which reported 51.0%.

Mountain Goats. The only encounter B. O'Gara (personal communication:1998) ever saw between mountain goats and pronghorn involved six nannies and kids and 11 pronghorn bucks. The bucks walked and fed past the bedded goats at about 10 yards (9 m), and neither species paid any attention to the other. This encounter was on the National Bison Range in western Montana, an 18,542-acre (7,504 ha) fenced enclosure. The chance of these species meeting under natural conditions is remote.

Exotics. Ten species of wild, free-roaming, exotic ungulates were reported in North America by Decker (1978). Most were deer, sheep, goats and pigs occupying forested and mountainous ecosystems. Only one, the gemsbok or Kalahari oryx, is a member of the antelope subfamily Hippotraginae and uses open plains similar to those occupied by pronghorn.

The introduction of foreign, large mammals became a common practice in Texas during the 20th century (Mungall and Sheffield 1994). Eight species were well established and another 15 were in the experimental introduction stage by 1990. Most were husbanded in fenced enclosures on private lands, but a few were free-ranging and mixed with endemic wildlife. During discussions of management issues involving alien ungulates, Mungall and Sheffield did not address relationships among exotics and pronghorn. With increased numbers of exotics becoming established and the possibility of more occupying rangelands with pronghorn, the need for research to document the competitiveness and compatibility of alien large mammals with native wildlife, such as the pronghorn, seems obvious.

The gemsbok is native to southern Africa and was translocated during the 1960s to rangelands of south-central New Mexico. Ninety-three animals were released between 1969 and 1977. Populations increased to more than 1,800 by the early 1990s, with small herds disbursed to adjacent areas (Smith 1994).

A study was conducted on sympatric populations of pronghorn and gemsbok on the White Sands Missile Range from 1992 to 1994 (Smith 1994). Fecal samples were analyzed using the microhistological analysis method. Results of diet selection are provided in Table 125.

Pronghorn consumed 32 taxa: 7 grasses, 7 forbs, 8 shrubs and 10 unidentified nongraminoids. Prickly pear was classified as a shrub and comprised 20% of the shrub forage class. Because prickly pear commonly is classified as a forb in pronghorn diet selection studies (Yoakum 1990), percentages were recomputed (Table 125). Thus,

Gemsbok antelope, originally from southern Africa, have been translocated to the White Sands Missile Range, in south-central New Mexico, where they now occupy rangelands with pronghorn. Gemsbok in Africa are predominantly grazers of graminoids. In New Mexico, they also consume grasses primarily. Consequently, little dietary overlap exists on rangelands occupied with pronghorn. Photo courtesy of the New Mexico Department of Game and Fish, Santa Fe.

TABLE 125. Percentage diet selection and overlap for sympatric populations of pronghorn and gemsbok on the White Sands Missile Range in south-central New Mexico

	Pronghorn diet		*Gemsbok diet*		
Forage class	*Reported by Smith (1994)*	*Recompiled for prickly pear*	*Reported by Smith (1994)*	*Recompiled for prickly pear*	*Dietary overlap*
Grass	20	20	83	83	20
Forbs	17	37	1	4	4
Shrubs	58	38	16	13	13
Unidentified	4	4			
Total	99	99	100	100	37

Source: Smith 1994.

the recomputed percentages provide data that are comparable with forage class categories in other studies (Tables 122 and 125).

Percentages (recomputed for prickly pear as was done for pronghorn) in gemsbok diets were 83% grasses, 4% forbs and 13% shrubs. Diet overlap percentages for pronghorn and gemsbok are listed in Table 125. Because gemsbok diets were mainly grasses with small quantities of forbs and shrubs, there was low diet overlap (37%).

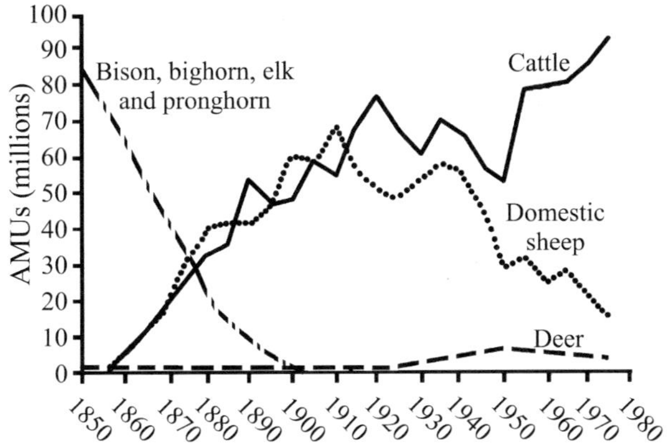

FIGURE 105. *Conjectured wildlife and livestock grazing pressure on rangelands in the 11 western states of the United States (modified from Wagner 1978).*

RELATIONSHIPS WITH LIVESTOCK

About 99% of pronghorn roam rangelands with domestic and feral livestock at ome time during the year (Yoakum and O'Gara 1990). Today's livestock on western rangelands can be classified in two categories: domestic animals, including cattle, sheep and horses; and feral stock, limited to horses and burros primarily. Goats, pigs, llamas and burros also are livestock in the West, but, because of low numbers, their impact on pronghorn has been negligible.

Pronghorn interactions with livestock began in the 1500s, when Spanish colonists brought sheep and goats to Mexico and the southwestern United States (Leftwich 1977). The arrival of cattle and horses followed soon thereafter, and many of these became feral. By the late 1800s, most pronghorn habitats were occupied by domestic and feral livestock (see Wagner 1978). By 1825, more than a million domestic sheep grazed in the vicinity of Spanish missions in California (Longhurst et al. 1983). During the 1960s and 1970s, the number of cattle grazing in the 11 western states was the greatest in history (Figure 105), whereas populations of wild ungulates (bison, bighorn, deer, elk and pronghorn) in those states were a fraction of what they were 200 years before.

It is not known what portion of the annual primary production (green-up volume) was being taken by large herbivores in the western United States in 1850. However, native ungulates were taking an estimated 85 million animal unit months (AUMs) of forage at that time, whereas removal by domestic livestock was inconsequential (Wagner 1978). Thirty years later, wild ungulate populations had been substantially reduced. Their forage consumption, consequently, declined to about 30 million AUMs. Cattle and sheep then were each taking about that amount. By 1975, the 1850 ratios were reversed—cattle (largely) and sheep together were removing more than 100 million AUMs, whereas wild ungulates were removing less than 5 million AUMs. Kindschy et al. (1982) stated that, in southeastern Oregon in 1975, forage utilization was 11% by wildlife and 89% by livestock (Table 126). Berger (1986) reported that wildlife comprised 5% and livestock 95% of the ungulate biomass on Great Basin environments in Nevada and Utah during 1980 (Figure 106).

By changing structural conditions and botanical composition of plant communities, livestock grazing has altered pronghorn rangelands more than has any other human-controlled activity (Leftwich and Simpson 1978, Autenrieth 1978, Kindschy et al. 1982). According to Autenrieth (1982) and Pyrah (1987), livestock grazing can impact pronghorn in the short term by seasonal or annual removal of forage and in the long term by modification of entire plant communities. Autenrieth's (1982)

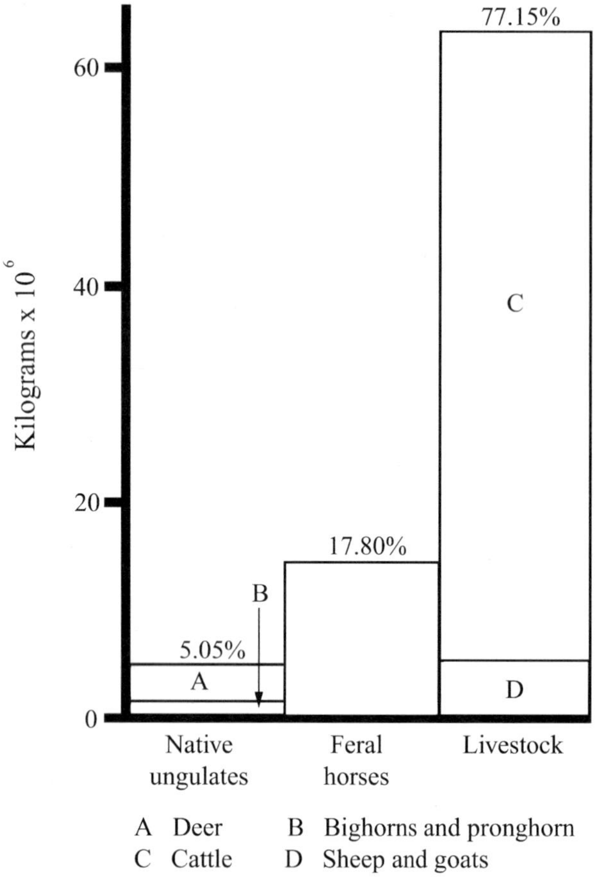

Figure 106. *Estimated biomass of domestic and wild ungulates for Great Basin environments in Nevada and Utah for 1980 (from Berger 1986).*

studies on a shrubsteppe in Idaho revealed correlation of high pronghorn fawn production on rangelands with high vegetation production. In addition to the obvious benefit of more and better food resources, those rangelands provided more and better camouflage for pronghorn fawns against predators than did areas with less vegetation due to extensive livestock grazing. Pyrah indicated that livestock grazing on the Yellow Water Triangle in central Montana lowered habitat quality and decreased carrying capacity for pronghorn.

The impacts of livestock grazing on pronghorn habitat often are complex, including modification of plant biomass, species composition and structural components, such as vegetation height and protective cover. These influences can be classified as direct, indirect or operational (Mackie 1978). A direct vegetative impact is competition for preferred forage—reported for pronghorn in New Mexico (Howard et al. 1990) and Texas (Hailey et al. 1966). Indirect negative impacts are "(1) gradual reductions in the vigor of some plants and in the amount and quality of forage produced, (2) elimination or reduction of the ability of forage plants to reproduce, (3) elimination of locally important cover types and replacement by less favorable types or communities, either by direct actions over time or by changing the rate of natural succession process; and (4) general alterations and reduction in the kinds, qualities and amounts of preferred or otherwise important plants through selective grazing or browsing or other activities" (Kie et al. 1994:665–666). Operational impacts include range improvement practices, for example, fence construction, water developments and vegetation manipulation projects, which are discussed later in this chapter and in greater detail in Chapter 17. Disturbances to pronghorn associated with the handling of livestock is another operational impact, as described by Einarsen (1948) for Oregon. Additional literature regarding direct, indirect and operational livestock influences on pronghorn are provided in Yoakum (1975, 1995), Kindschy et al. (1982), Yoakum and O'Gara (1990), Yoakum et al. (1996) and others.

Table 126. Utilization of forage by domestic, feral and wild ungulates on public lands in southeastern Oregon

Ungulate species	*Forage use percentage*
Domestic	
Cattle and horses	82.3
Horses (feral)	5.8
Sheep	0.7
Total	88.8
Wild	
Mule deer	10.1
Pronghorn	0.9
Rocky Mountain elk	0.1
Bighorn	0.1
Total	11.2

Source: From Kindschy et al. 1982.

Possibly the greatest impact of livestock grazing on pronghorn populations has been changes in plant succession and intensive foraging on fawning areas. Plant succession is the gradual replacement of one assemblage of plants with others through time until a different stable plant community is reached (Clements 1916). As each group of plant species is replaced, the characteristics of the community change. These changes can be either beneficial or detrimental to pronghorn, but, more cases in the literature indicate deleterious than beneficial effects (Yoakum et al. 1996). Vegetative changes are greatest for landscapes receiving less than 12 inches (30 cm) of precipitation annually (Donahue 1999) and on shrubsteppes where forbs decreased while shrubs increased (Kindschy et al. 1982, Miller et al. 1994).

Livestock grazing during the past 100 years has changed plant composition on many pronghorn habitats. This is especially true in the intermountain region, where sagebrush/grassland communities have been altered to dominant shrublands (Miller et al. 1994). Such modifications take place over decades, often are not investigated and have decreased pronghorn carrying capacities (U.S. Fish and Wildlife Service 1994). Similar impacts have been experienced on southwestern grasslands (Archer 1994). Investigations by the Arizona Game and Fish Department and the Arizona Wildlife Federation have been conducted for more than 50 years on Anderson Mesa near Flagstaff. These grasslands have been grazed by livestock for more than 100 years. Forb and shrub species have decreased approximately 50%, whereas graminoids increased. Parallel to decreased plant diversity was declining pronghorn populations from annual averages of 1,500 during the 1940s to less than 500 in the 1970s (Neff and Woolsey 1979). Decreasing numbers continued to the year 2000 (Yoakum 2003a).

A similar pronghorn/habitat investigation has been conducted on the grasslands of the Carrizo Plain National Monument in south-central California. There, the vegetative composition was modified during the 20th century by farming and livestock grazing. Estimated plant composition in the late 2000s was 80% to 90% grasses, less than 10% forbs and less than 5% shrubs. Studies indicated a decrease in native forbs and shrubs with increases in exotic annual grasses. Pronghorn numbers during the 1990s experienced a downward trend (Yoakum 2003b).

Environments with high plant diversity are associated with quality pronghorn habitats. During the 1940s, Buechner (1950a) reported that pronghorn ingested more than 242 plant species on the Trans-Pecos plains: 18 grasses, 168 forbs and 56 shrubs. A large variety of plants, especially forbs, provides different species that are nutritious and succulent during various seasons. Pronghorn consume diverse quantities and numbers of plant species throughout the year for the grasslands, shrubsteppes and deserts (see Chapter 14). Pronghorn welfare has been reported directly related to vegetation diversity and livestock grazing for shrubsteppes (Ellis 1970) and grasslands (Howard et al. 1990).

Maintenance of natural plant diversity on western rangelands is a major landscape biodiversity conservation strategy (Fleischner 1994, Noss and Cooperrider 1994, Donahue 1999, U.S. Department of Interior, no date,). The U.S. National Academy of Science identified plant diversity as a function of healthy rangeland conditions (Committee on Rangeland Classification 1994). The committee's fundamental conclusion was that the minimum standard for rangeland management should be the prevention of human-induced loss of environmental health. Elmore and Kauffman (1994) supported this conclusion by stating that the closer an ecosystem is managed to allow natural ecological processes to function, the more successful management strategies will be for rangelands. Noss and Cooperrider recommended that livestock grazing should not result in the loss of any native plant or animal from the regional landscape, or impoverishment of any other measure of native biodiversity.

Public lands administered by the U.S. Forest Service and the Bureau of Land Management (BLM) total approximately 50% of all pronghorn habitat in the United States (see Chapter 17). Although sustaining biodiversity is mandated by federal legislation for these government agencies, it continues to remain an ecological conservation and management issue (Donahue 1999) and, thus, influences vegetative quality for pronghorn habitat. It is conceivable that the effects of livestock grazing on vegetation are a paramount management factor adversely affecting pronghorn because livestock grazing is the most ubiquitous economic activity affecting the health of western rangelands.

One of the major public land conservation laws passed in the United States for western rangeland management was the BLM's Organic Act—the Federal Land, Policy, and Management Act (FLPMA) (U.S. Congress 1976). It authorizes multiple use of and sustained yield principles for natural resources, and it provides for the preservation and protection of public lands in their natural condition. It authorizes management actions necessary to prevent unnecessary or undue degradation of natural resources and assesses the short-term versus long-term benefits of perpetuating public values and needs. Protecting native biodiversity in FLPMA is inseparable from sustaining proper functioning ecosystems. The policies and principles of FLPMA are consistent with a landscape-level approach to biodiversity conservation on arid rangelands.

The Public Rangelands Improvement Act (PRIA) (U.S. Congress 1978) recognized the continued degradation of native vegetation and the need for improved management strategies for rangeland health. The PRIA defined native vegetation "as those plant species, communities, or vegetative associations which are endemic to a given area and which would normally be identified with a healthy and productive range condition occurring as a result of the natural vegetative process of the area" (U.S. Congress 1978, paragraph 1902e). It equates native vegetation with healthy rangeland conditions.

The PRIA relies heavily on rangeland improvements and encourages increased federal funding to improve rangeland conditions. More than 95% of the BLM's expenditures for rangeland improvements between 1980 and 1995 primarily were to benefit livestock (Donahue 1999). The PRIA confirms the need for conservation of biodiversity and the prevention of irreversible livestock grazing-induced vegetative changes.

Various other federal statutes provide authorities and responsibilities for effective rangeland management, including the National Environmental Policy Act (U.S. Congress 1969), the Endangered Species Act (U.S. Congress 1973) and the National Forest Management Act (U.S. Congress 1974). Each recognizes and contributes toward implementing biodiversity conservation programs with a goal of perpetuating healthy western rangelands. Sustaining healthy environments is a major ecological function contributing to maintaining healthy pronghorn populations.

A second major livestock grazing impact affecting pronghorn is the intensity of forage usage on pronghorn fawning areas. Vegetative communities in high quality condition provide adequate concealment for fawns against aerial and ground predation (Pyrah 1974, Autenrieth 1982, Canon and Bryant 1997). When the herbaceous vegetation is used heavily, insufficient plant cover remains to protect pronghorn neonates. Autenrieth noted that the height and density of plant cover was related to the frequency of coyote depredation on pronghorn fawns during the first 3 weeks after birth. These vegetation/predator/fawn survival relationships are substantiated in similar reports by Barrett (1981) in Alberta, Tucker and Garner (1983) in Texas, Alldredge et al. (1991) in Wyoming, and Ticer and Miller (1994) in Arizona. Based on these studies, plant communities averaging 15 inches (38 cm) in height appear necessary to decrease fawn mortality by predators. It is suggested that managers implementing livestock forage utilization objectives on pronghorn fawning areas recognize and implement this management criterium to coordinate multiple-use standards for pronghorn and livestock.

Pronghorn/livestock relationships have been recorded on the Hart Mountain, Oregon, for more than 50 years (U.S. Fish and Wildlife Service 1994). Cattle, domestic sheep and feral horses grazed the rangelands for more than 100 years before they were removed: domestic sheep about 1970, cattle in the early 1990s and horses in the late 1990s. Since removal of livestock, herbaceous composition of forage has increased more than 200% on certain sites (Crawford and Coggins 1997). This change probably was related to years of favorable precipitation (triggering production of nutritious, succulent forage), decreased livestock grazing and fires—both wild and prescribed. Pronghorn populations experienced low numbers during the early 1950s (300–500 animals), when livestock were numerous. Highest pronghorn numbers during the last half of the 20th century were during the 1990s, when all livestock were removed, with a mean of 1,400.

Pronghorn/livestock relationships in the Intermountain West (between the western side of the Rocky Mountains and the eastern side of the Sierra Nevada and Cascade mountains) have been studied for five decades: Ackerly and Regier (1956), Ellis (1970), Salwasser (1980) and Pyshora (1982b) in California; Folker (1956), Bodie (1979) and Autenrieth (1982, 1984) in Idaho; Meeker (1979), McNay (1980) and Hansen (1986) in Nevada; Einarsen (1948), Yoakum (1957a), Deming (1959), Rouse (1959), Kindschy et al. (1982), McInnis (1984), and Dunbar and Giordano (2000) in Oregon; Hinman (1959), Smith and Beale (1980), and Clary and Holmgren (1982) in Utah; and Yoakum (1995) in California, Oregon, and Nevada. A reason for so many investigations was concern regarding low doe/fawn ratios (25–60:100), compared with higher ratios (60–100:100) for the Great Plains (Ellis 1970, Vriend and Barrett 1978).

Various mortality factors and interrelationships with other animals and habitats were studied, with few specific causative factors credited for the low recruitment. Yoakum (1957a, 1995) noted that habitats with a greater diversity of grasses, forbs and shrubs sustained higher pronghorn densities. Hinman (1959) reported predation as a limiting factor, but noted that rangeland vegetative conditions seriously affected pronghorn numbers.

Dunbar and Giordano (2000) indicated that coyote predation on young fawns affected recruitment during certain years. However, pronghorn maintained or slightly increased their numbers during other years with low fawn survival (Chapter 11). Consequently, research continues into other factors, including alternate prey base and vegetative conditions.

Ellis (1970) stated that pronghorn fawn survival and population densities were directly related to vegetative conditions. He compared pronghorn and production of vegetation for Great Basin and Great Plains areas used in common by pronghorn and livestock. Vegetative conditions were inferior in the Great Basin because of heavy use by livestock of grasses and forbs needed by pronghorn during their last trimester of gestation and the first 2 months of lactation. At that time, pronghorn need grasses and forbs, which are higher in nutritive values than shrubs. Pronghorn fawn production was similar for both regions (generally 190 fawns/100 does), but survival of fawns in the Great Basin was half or less that reported for the Great Plains. Ellis (1970:63) concluded that "present productivity and biomass of the Great Basin pronghorn cannot be expected to improve markedly while the biomass of domestic species remains at 30 to 50 times that of the antelope. In other more productive plant communities, such a ratio may not depress antelope productivity, but in the Great Basin, the limited growth of early spring appears to be a point of competition among all the herbivores and antelope production suffers accordingly."

Pronghorn and livestock on common grasslands in southeastern New Mexico used many of the same forage species (Beasom et al. 1982b). Pronghorn consumed greater quantities of forbs, whereas cattle and domestic sheep consumed more grasses. Livestock use of forbs was greatest during spring and summer when those plants were most abundant. As forbs decreased, livestock turned more to grasses. Diet similarity indices indicated that the potential for competition was greatest for forbs. Dietary overlap was highest when forbs were most abundant and lowest when they were least abundant.

The role of pronghorn in consuming plants noxious to livestock has been well documented (Einarsen 1948, Buechner 1950a, Terwilliger 1953, Hoover et al. 1959, Smith and Beale 1980). Because of this, pronghorn may be considered beneficial to livestock on rangelands. Many plants eaten (at times, in large quantities) by pronghorn are noxious to livestock because they cause mechanical injury, are unpalatable or tend to crowd out, invade or dominate more desirable forage. For example, spines of bull thistle, Russian thistle and cacti make these plants unpalatable to livestock, but pronghorn consume them during all seasons. Pronghorn in Colorado consumed 11% cacti in their annual diet (Hoover et al. 1959). Evidently, the spiny cacti do not injure the pronghorn; even young fawns consume the plants. Other plants noxious to livestock but consumed by pronghorn include woody locoweed, larkspur, lupine, death camas (Einarsen 1948); locoweed, woody senecio, Riddel groundsel, paper flower (Buechner 1950b); cocklebur, snakeweed, bull thistle, Russian thistle, cacti, chokecherry (Hoover et al. 1959); tarbush (Hailey 1979); and halogeton (Smith and Beale 1980).

Of 103 plant species reported by Buechner (1950a), 51 were unpalatable and 21 were considered poor forage for cattle, whereas 28 were considered poor forage for pronghorn. Most of the plants unsuitable for livestock were good forage for pronghorn. This relationship indicated only a small degree of competition between these animals. Forage species rated good for livestock were mainly grasses; forage plants of this classification for pronghorn were mostly forbs. The pronghorn is able to consume many forbs abundant on rangelands heavily grazed by cattle.

Most predator control in the United States is for the benefit of livestock. Connolly (1978a) reported that about 250 times more money was spent for protection of livestock during fiscal year 1976 than was spent to benefit wildlife. Predator-control programs to protect livestock ranching operations also may indirectly affect pronghorn (see Chapter 11). The demand by ranchers for predator control can benefit pronghorn, according to Einarsen (1948), who noted, for example, that coyote control efforts contributed to increased pronghorn fawns in Oregon. Herds have increased in areas during predator control and decreased when control practices were curtailed (Connolly 1978a).

More than 40 references in the scientific literature involve pronghorn/cattle/domestic sheep relationships and six discuss domestic or feral horses (see Yoakum 1967, Yoakum and Spalinger 1979). The effects of livestock on pronghorn—including compatibility, competition, disease and parasite factors—are species-specific, depending on the type of livestock and management strategies used.

Cattle

Compatibility. Historically, pronghorn frequently were associated with bison on the western Great Plains. Apparently, the two maintained a commensal relationship. That compatibility now is paralleled by the frequent and nondisruptive association of pronghorn and domestic cattle. It is common, while traveling across the West, to see pronghorn and cattle together.

Most authorities cite relatively little antagonism between cattle and pronghorn (Einarsen 1948, Jones 1949, Buechner 1950b, Gregg 1955, Hoover et al. 1959, Campbell 1970, Becker 1972, Yoakum 1975, Schwartz 1977, Autenrieth 1978, Hailey 1979, Salwasser 1980, Kindschy et al. 1982). Pronghorn bucks and yearling females often bed or feed near cattle, but pre- and postparturient does tend to avoid cattle during the fawning season, which may result in the selection of less favorable sites (McNay and O'Gara 1982). McNay (1980) noted that 97 pronghorn on a playa dispersed as soon as cattle moved onto the playa, and does moved from an area where they usually had fawns during a spring when cattle were grazed on that area. The fawns were born in areas adjacent to the concentrated cattle use, and survival was better than it was the year before cattle were moved in. It was not determined whether the improved survival resulted from climatic conditions;or the fawns were not born in an area where coyotes had learned to hunt them. McNay and O'Gara suggested restricting livestock use of fawning areas until after the fawning season. Retention of herbage height until after the fawning season would be beneficial to pronghorn fawns as added concealment from predators (Barrett 1978, Autenrieth 1982, Pyrah 1987).

At least nine major studies have focused on pronghorn/cattle interactions, including Buechner (1950a) and Roebuck (1982) in Texas; Campbell (1970), Becker (1972) and Pyrah (1987) in Montana; Taylor (1972b) in Wyoming; McNay and O'Gara (1982) in Nevada; Bailey and Cooperrider (1982) in Colorado; and Howard et al. (1983) in New Mexico. Most have reported on the nonaggressive association while feeding, drinking or resting.

Pyrah (1987) reported that pronghorn generally avoided using pastures concurrently with cattle. Seventy-five percent of 9,530 pronghorn observations were in pastures without cattle. He also reported slightly higher fawn/doe ratios (92:100) in pastures without cattle than with

At least 95% of pronghorn habitat was foraged by cattle at the beginning of the 21st century. Grazing systems for cattle vary in different areas. When cattle foraging strategies result in the retention of adequate preferred forage and hiding cover for pronghorn neonates, they can be compatible with pronghorn welfare. Pronghorn and cattle are compatible behaviorally, so often are seen feeding and drinking together on western rangelands. Their mutual tolerance attests to management strategies that emphasize multiple-use objectives of domestic and wild ungulates. Photo by Pat O'Brien; courtesy of Arizona Department of Game and Fish, Phoenix.

(82/100). However, Campbell (1970) and Freeman (1971) found greater use of Montana pastures by pronghorn during spring through autumn when cattle were present than when cattle were absent.

Competition. Competition for forage between cattle and pronghorn usually is minor on grasslands in fair to good ecological condition. Their diets are sufficiently different during late autumn and winter that little overlap occurs. Common use of forbs during spring may result in competition. Heavy cattle grazing on spring and summer habitats of pronghorn leads to forage competition, especially on the shrubsteppes of the Intermountain West. Such conflicts can be avoided by delaying livestock turnout dates or by reducing cattle numbers early in the growing season. Forage competition also is high when cattle grazing of meadows or playas on summer rangelands is heavy. Pronghorn often rely on meadows and playas for succulent forbs.

That cattle in the Great Basin can contribute to the reduction of forbs needed by pronghorn is illustrated by the situation on Hart Mountain. Cattle were removed in 1990, and forb cover in pronghorn habitat increased threefold by 1996 (U.S. Fish and Wildlife Service 1997b). Increased precipitation apparently also contributed to this increase.

Campbell (1970) opined that dual use by pronghorn and cattle resulted in more efficient use of rangeland vegetation when grazing was regulated according to prescribed practices, such as pasture rotation for livestock. He reported that cattle outcompeted pronghorn during summer when vegetation became desiccated, forcing pronghorn to move to adjoining rangelands with less-preferred forage. Such interaction could affect forage selectivity and nutrition of nursing does, which invariably would affect fawn mortality.

In general, dietary overlap between cattle and pronghorn is not great (Tables 127 and 128). In 16 studies on various rangelands, overlap averaged less than 25% (denoting medium-to-low overlap). Only one of the studies (grasslands in Colorado) recorded high overlap for grasses (Schwartz and Nagy 1976). Three investigations found high overlap of forbs (all in the grassland biome), whereas two studies recorded significant overlap for shrubs (both in shrubsteppes) (Campbell 1970, Roebuck 1982, Beasom et al. 1982b) (Table 128).

Serious competition for forage between pronghorn and cattle occurred in Texas (Hailey et al. 1966). For years, both species used rangelands concurrently with no problems. When a drought reduced forage, cattle were given supplemental feed, but pronghorn were not. Thus, numbers of cattle remained high, increasing already heavy

TABLE 127. Dietary overlap of pronghorn by livestock, with forage class

Livestock	Location	Biome	Percentage dietary overlap				Source
			Grasses	Forbs	Shrubs	Total	
Cattle	Trans-Pecos, Texas	Grassland	4.0	7.0	9.0	20.0	Buechner (1950a)
	Southwestern Montana	Grassland	3.0	20.3	1.6	24.9	Campbell (1970)
	Winnett, Montana	Grassland	3.0	13.0	0.0	16.0	Becker (1972)
	Rawlins, Wyoming	Shrubsteppe	7.1	0.2	39.4	46.7	Taylor (1975)
	Pawnee, Colorado	Grassland	46.3	2.0	6.0	54.3	Schwartz and Nagy (1976)
	Northeastern California and northwestern Nevada	Shrubsteppe	4.1	3.4	3.7	11.2	Hanley (1980)
	Southwestern Utah	Shrubsteppe	0.0	0.0	27.0	27.0	Smith and Beale (1980)
	Roswell, New Mexico	Grassland	4.0	15.2	1.3	20.5	Beasom et al. (1982b)
	Panhandle, Texas	Grassland	2.5	19.0	8.5	30.0	Roebuck (1982)
	Trickle Mountain, Colorado	Shrubsteppe	3.5	5.0	14.0	22.5	Bailey and Cooperrider (1982)
	Little Lost/Birch Creek, Idaho	Shrubsteppe	3.0	2.0	11.0	16.0	McCarty (1982)
	Burns Junction, Oregon	Shrubsteppe	8.5	4.2	0.8	13.5	McInnis (1984)
	Sheldon National Wildlife Refuge, Nevada	Shrubsteppe	5.0	9.0	1.0	15.0	Hansen (1986)
Horses	Sheldon National Wildlife Refuge, Nevada	Shrubsteppe	3.0	23.0	2.0	28.0	Meeker (1979)
	Northeastern California and northwestern Nevada	Shrubsteppe	2.5	5.2	5.0	12.7	Hanley (1980)
	Trickle Mountain, Colorado	Shrubsteppe	3.5	2.0	31.0	36.5	Bailey and Cooperrider (1982)
	Burns Junction, Oregon	Shrubsteppe	10.2	4.2	1.1	15.5	McInnis (1984)
	Sheldon National Wildlife Refuge, Nevada	Shrubsteppe	5.0	6.0	0.0	11.0	Hansen (1986)
	Sheldon National Wildlife Refuge, Nevada	Shrubsteppe	9.9	7.1	2.3	19.3	Hansen et al. (2001)
	Hart Mountain National Antelope Refuge, Oregon	Shrubsteppe	7.9	13.0	3.3	24.2	Hansen et al. (2001)
	White Sands Missile Range, New Mexico	Grassland	20.0	1.0	9.0	30.0	Smith (1994)
Sheep	Trans-Pecos, Texas	Grassland	4.0	19.0	10.0	33.0	Buechner (1950a)
	Red Desert, Wyoming	Shrubsteppe	3.2	2.6	28.4	34.2	Severson et al. (1968)
	Southwestern Montana	Grassland	3.0	27.3	25.3	55.6	Campbell (1970)
	Rawlins, Wyoming	Shrubsteppe	7.1	1.2	39.9	48.2	Taylor (1975)
	Pawnee, Colorado	Grassland	46.3	18.0	3.3	67.6	Schwartz and Nagy (1976)
	Southwestern Utah	Shrubsteppe	0.0	0.0	46.0	46.0	Smith and Beale (1980)
	Little Lost/Birch Creek, Idaho	Shrubsteppe	3.0	14.0	39.0	56.0	McCarty (1982)
	Roswell, New Mexico	Grassland	4.0	50.2	6.0	60.2	Beasom et al. (1982)
Burros	Sheldon National Wildlife Refuge, Nevada	Shrubsteppe	9.9	6.8	8.8	25.5	Hansen et al. (2001)

use of the limited vegetation. Because of fencing, pronghorn could not move to adjacent rangelands and turned to such nonpreferred shrubs as tarbush and mesquite. This resulted in toxicity, death and low recruitment the following year.

Roebuck (1982) studied pronghorn/cattle forage relationships for rangelands in good condition in the panhandle of Texas. Pronghorn were attracted to pastures because of preferred forage, regardless of the presence or absence of cattle. Greatest dietary overlap was for forbs during summer. Supplemental feeding of livestock relieved spring and winter forage competition. The low overlap for all forage classes and seasons suggested that competition on these rangelands in good condition was not important.

Diseases and Parasites. Exchange of disease among pronghorn and cattle is a concern because of the two species' extensive spatial overlap. There is concern that one may serve as a disease reservoir for the other. Pronghorn have been checked repeatedly for brucellosis and anaplasmosis, and no cases have been confirmed. Compared with many other wild ungulates, pronghorn have a low incidence of diseases (see Chapter 10). Bluetongue probably is the most serious disease of pronghorn, and cattle are a primary reservoir for this disease; they often do not

Table 128. Livestock dietary overlap with pronghorn, by season

	Percentage dietary overlap					
Livestock	*Winter*	*Spring*	*Summer*	*Autumn*	*Annual*	*Source*
Cattle	—	27.0	—	7.0	—	Becker (1972)
	1.0	13.0	18.0	—	—	Vavra and Sneva (1978)
	—	28.0	30.0	12.0	—	Campbell (1970)
	—	65.0	32.0	51.0	—	Schwartz and Nagy (1976)
	27.0	—	—	—	—	Smith and Beale (1980)
	—	2.6	4.4	3.0	—	Hanley (1980)
	—	16.0	12.0	17.0	—	Bailey and Cooperrider (1982)
	12.5	17.5	14.6	29.0	18.4	Beasom et al. (1982b)
	17.5	22.3	24.6	31.3	23.9	Roebuck (1982)
	20.0	15.0	17.0	11.5	15.8	McCarty (1982)
	8.3	8.5	12.0	17.0	11.4	McInnis (1984)
	15.0	6.0	7.0	5.0	8.2	Hansen (1986)
Horses	10.0	13.0	24.0	—	—	Vavra and Sneva (1978)
	—	—	13.0	—	—	Meeker (1979)
	19.5	15.7	9.5	6.8	12.9	Hanley (1980)
	—	8.0	6.0	—	—	Bailey and Cooperrider (1982)
	25.3	19.0	7.4	13.5	16.3	McInnis (1984)
	7.0	4.0	10.0	4.0	6.2	Hansen (1986)
Sheep	15.4	—	11.2	40.2	—	Severson et al. (1968)
	—	61.0	55.0	30.0	—	Campbell (1970)
	—	57.0	55.0	81.0	—	Schwartz and Nagy (1976)
	—	12.0	30.0	—	—	Vavra and Sneva (1978)
	—	55.4	36.5	53.2	—	Hanley (1980)
	46.0	—	—	—	—	Smith and Beale (1980)
	38.7	42.5	38.9	57.6	44.4	Beasom et al. (1982b)
	71.0	35.0	—	—	—	McCarty (1982)

develop clinical or acute symptoms, but can be chronic carriers (Thorne et al. 1982). Although leptospirosis causes some mortality, pronghorn seem relatively resistant to infections and may act as carriers. Not enough evidence exists to implicate either pronghorn or cattle as a primary reservoir of leptospirosis, but the existence of an unidentified reservoir species cannot be ruled out (Collins et al. 1981).

Domestic Sheep

Compatibility. Investigators are not in agreement concerning compatibility of pronghorn and domestic sheep. Einarsen (1948) stated that pronghorn usually avoid sheep. Gregg (1955) observed pronghorn feeding near isolated, small bands of sheep that were away from herders and dogs. Buechner (1950a) reached the same conclusion and observed that there was no psychological incompatibility between the species.

Severson et al. (1968) studied pronghorn/domestic sheep relationships in fenced pastures on the Red Desert of Wyoming. They found no apparent stress to either as a result of the other's presence. Both often were seen grazing and drinking together, and they rested together on one occasion. The investigators noted that adjacent areas used by domestic sheep were readily inhabited by pronghorn. However, they observed that pronghorn tended to vacate areas where sheep were herded. This reaction was believed to have been due to the activity of the herder and his dogs rather than incompatibility with sheep. As soon as the herded sheep were moved out of those areas, pronghorn returned.

Taylor (1972b) studied domestic sheep/pronghorn forage competition on similar rangelands near Rawlins, Wyoming. During mild winters, pronghorn disbursed throughout the rangeland, but during severe winters, both sheep and pronghorn foraged on the same areas, resulting in high competition. Taylor noted that, when snow cover was light, pronghorn avoided places where sheep were or had grazed. Sheep were herded and grazed intensively, whereas the free-ranging pronghorn dispersed and did not graze vegetation as heavily. Pronghorn and sheep consumed the same forage species, but foraging by pronghorn was more widespread and lighter.

According to Einarsen (1948:143), conflicts were frequent when herded sheep bands moved into areas used by pronghorn for fawning (May and June): "A doe with newly-born young will nervously pace the fringes of the sheep flock to protect her 1- or 2-day-old kids, which lie quietly for a while as the feeding sheep pass within nose-touching distance and then rise on trembling legs to follow the moving mass. The mother becomes frantic and the herder exasperated as the kid stumbling by attempts to keep up with what seems to be its only tie to security. The herder, who does not want a bummer antelope in his flock, rushes over to rescue the kid from the band, the dogs lend their noisy help and all is bedlam and confusion. By now the mother antelope is in terror and her actions frighten the other pronghorn within close range. These react with much running and blowing, until all are worked into a state of frenzy, and just about the time the herder separates the kid from his flock, the pronghorn will dash away and disappear over the horizon."

Campbell (1970) found that pronghorn abandoned 15% of his study area in southeastern Montana in response to the presence of sheep. Severity of the problem was correlated to size of fenced pastures, duration and intensity of sheep use and past grazing of the rangelands involved. The more confined, intense or prolonged the grazing by sheep, the less that pronghorn use that rangeland. This may be an example of direct competition that causes one species to displace another. Buechner (1950a) suggested this same aversion for sheep by pronghorn.

Although domestic sheep numbers decreased greatly on western rangelands during the last half of the 20th century, sheep still forage with pronghorn on public lands, as in Wyoming. When vegetation is in healthy condition with adequate plant production, these sheep and pronghorn can be compatible and suffer little diet competition. However, the reverse may be the case on rangelands in poor ecological condition. One impact on pronghorn on certain rangelands managed for domestic sheep is the presence of fences. Generally, a fence that restricts movements of sheep likewise will be a barrier to pronghorn. Photo courtesy of Wyoming Department of Game and Fish, Cheyenne.

Competition. The potential for forage competition between pronghorn and domestic sheep was identified as early as 1935 (Taylor 1936) and has been documented extensively (Knipe 1944, Jones 1949, Buechner 1950b, Bever 1955, Hoover et al. 1959, Russell 1964, Campbell 1970, Taylor 1975, Schwartz and Nagy 1976, Hailey 1979, Smith and Beale 1980, Howard et al. 1983).

Buechner (1950a) was adamant that the most important limiting factor to pronghorn population growth and distribution in Texas was intense forage competition by domestic sheep. Specific factors included intense competition for preferred forbs, supplemental feeding of sheep on rangelands (thus maintaining artificially high populations of sheep) and restriction of pronghorn movements due to sheep-tight fences. Schwartz (1977) not only found that pronghorn and sheep ate many of the same plant species in Colorado, but that crude protein levels for both ungulates were similar (4%–12%). Pyshora (1977) indicated recent increases in pronghorn populations in California may have resulted from decreases in domestic sheep. Only Severson et al. (1968) noted that, in Wyoming's Red Desert, the food habits of the two species were noticeably dissimilar and competition minimal.

Wyoming ranchers grazed more than 6 million sheep in the state during 1909. By 1981, that number had declined to a little more than 1 million. Estimated pronghorn numbers in Wyoming increased from 8,387 during 1919 to 111,000 in 1957 (see Greenquist 1983). The increase continued to 608,000 by 1983. A steady decrease then ensued to 370,000 by 1997 (see Chapter 4). Many other factors were involved in the 50-fold increase of pronghorn, but the 6-fold decrease in domestic sheep certainly eliminated some competition for forage.

The problem of forage competition between pronghorn and sheep can be evaluated by analysis of indices in dietary intake. Nine cases (Tables 127 and 128) indicate moderate to heavy overlap (range, 33%–67%). Therefore, it is likely that pronghorn experience competition with sheep for food. In a pronghorn/sheep forage competition study conducted during the early 1980s in New Mexico, it was determined that livestock used as much as 40% of available forbs (Beasom et al. 1982b, Howard et al. 1983). As forbs apparently declined in numbers or decreased in moisture content, livestock use of grasses increased. This suggested that the potential for competition was primarily for forbs. Diet similarity data supported this premise (Tables 127 and 128). Competition for forage was highest when forbs were most available and lowest when they were least available.

In a 5-year study of pronghorn/domestic sheep diet competition in the cold desert of southwestern Utah, Clary and Holmgren (1982) reported that moderate sheep use during the dormant winter period left habitats unfavorable for pronghorn until spring regrowth. Such areas need to be rested from use by sheep to protect shrubs beneficial to pronghorn. Domestic sheep customarily graze the same rangeland repeatedly throughout winter. Under repeated grazing, forage quality and volume decline during that time because the sheep consume the most palatable and probably the most nutritious parts of the plants the first time they graze the area.

Competition probably could be alleviated by dividing sheep allotments into a number of smaller units to be grazed on a rotation basis. Grazing of any rangeland unit would be completed in only a few days. Rotational grazing systems maintain sheep in better condition and ensure forage for pronghorn on key winter rangeland. Changes in grazed and ungrazed units indicated that, although sheep grazing was moderate, pronghorn found the grazed units relatively unfavorable for their use, so tended to concentrate in the remaining ungrazed units (Clary and Holmgren 1982, see also Smith and Beale 1980). These findings show that even moderate use by sheep—2.5 to 6.2 acres (1.0–2.5 ha) per sheep month—renders rangeland less favorable for pronghorn until the spring regrowth period. The primary factor in pronghorn avoidance of sheep-grazed units probably is competition for black sagebrush, which predominates pronghorn winter diets in west-central Utah.

A study in eastern Montana disclosed sheep/pronghorn competition for sagebrush was high during late spring (Campbell 1970). Both ungulates preferred forbs during summers, resulting in severe competition. And, on rangelands grazed by sheep, forb production was lower and less diverse than on areas ungrazed by sheep.

Diseases and Parasites. Field studies have indicated a higher incidence of parasite infestation in pronghorn on rangelands cohabited with domestic sheep (Bever 1955, Russell 1964). Bever reported losses of 30% to 40% of annual pronghorn fawn production on overgrazed sheep rangeland in South Dakota, primarily due to nematodes. He presented evidence of higher parasite loads in pronghorn on rangelands grazed by sheep than on rangelands grazed by cattle. Bever also reported that placement of phenothiazine salt blocks and abandonment of close sheep herding, commonly practiced before 1950, helped to alleviate parasite problems.

In Wyoming, epizootics have been reported for areas jointly occupied by pronghorn and sheep (Thorne et al. 1982). Illness and possibly deaths of pronghorn, especially fawns, have been attributed to tapeworm infections. These infections were most prevalent on heavily sheep-grazed rangelands in late autumn. Bluetongue probably was the most serious disease of pronghorn, and domestic sheep are a reservoir for this disease.

Horses

Compatibility. Currently, domestic and feral horses occupy rangelands with pronghorn. Domestic horses are private livestock, generally branded in accordance with state laws and allowed to graze open rangelands. Feral horses—also referred to as mustangs, wild horses or free-roaming horses—are domestic animals or descendants of domestic horses that were released or escaped from captivity. They generally are not branded or legally regarded as private property.

Only two known studies addressed the prospect of behavioral interaction between wild horses and pronghorn, and both were conducted in Nevada. During two summers, Meeker (1979) reported that both ungulate species watered together freely, with pronghorn giving ground only when directly approached by horses. No aggressive action was noted by either species toward the other during the study. Although pronghorn relinquished habitat when directly approached by horses, some evidence of symbiosis, by warning each other of danger, was noted. Meeker observed that pronghorn would use poorer quality water than would horses. The two species often drank within 5 yards (4.6 m) of each other. Pronghorn also drank concurrently with cattle, but not as frequently as with horses.

Berger (1986) investigated pronghorn/horse relationships for several years near Gerlach, Nevada. He recorded six instances of pronghorn displacement by horses. These interactions were more frequent than occurred with wild sheep (2 times) but not as frequent as with deer (11 times). In all cases, native ungulates were subordinate to horses. Ordinarily, pronghorn simply would move away as horses approached without overt aggression. However, he observed horses tossing their heads at pronghorn and chasing one for about 82 yards (75 m). Berger did not report any serious problems of aggression.

R. Kindschy (personal communication:1990) observed wild horse/pronghorn interactions for 32 years in Malheur County, Oregon, and opined that pronghorn seek opportunities to graze with wild horses. Kindschy speculated that this coexistence may be due to the social behaviors of both species or possible benefits received from animals in herds to detect predators.

Competition. Horses feed predominantly on grasses, whereas pronghorn prefer forbs and shrubs. Six studies suggested that dietary overlap is low (less than 36%) (Tables 127 and 128), primarily because horses are not numerous where they coexist with pronghorn.

Meeker (1979) concluded that, between horses and pronghorn, there was little forage competition except for forbs. Autenrieth (1982) reported that, in Idaho, year-

At the beginning of the 21st century, thousands of mustangs, or feral horses, ranged the western landscapes. These animals spend their entire lives on the open rangelands, many foraging with pronghorn and deer. Some feral horse herds have roamed wild for more than 400 years, having been released or escaped from early Spanish explorers and ranching operations. Photo by Joe Van Wormer; courtesy of the Denver Museum of Natural History, Colorado.

On western rangelands concurrently foraged by pronghorn and horses (left)*, competition is minor when forbs are abundant. However, when forbs are scarce, dietary overlap can be a major problem. This is especially true when horses compete for low quantities of perennial forbs during winter or early spring. Generally, pronghorn limit their association with large ungulates, but the young buck at right grazed with a herd of horses in an alfalfa field within a mile (1.6 km) of Lakeview, Oregon. Townspeople would drive to the scene to witness multiple use in action and take photographs.* Left photo by Robb Hitchcock. Right photo by Jim D. Yoakum.

round foraging by feral horses probably was the most important cause of vegetation deterioration. His studies on low pronghorn recruitment concentrated on limiting factors, including natality, predation, diseases, hunter harvest, precipitation and range conditions. Autenrieth concluded that habitats with greater vegetative production had higher fawn recruitment. Also, improved vegetative conditions (higher densities of grasses, forbs and shrubs) resulted in better camouflage or visual screens concealing fawns from predators.

At times diet overlap can appear minimal when data are summed over the four seasons and several years. This may have been the case for pronghorn and feral horses on Hart Mountain (Hansen et al. 2001). Feral horses consumed 13% forbs during a winter—when pronghorn were consuming large quantities (71%) of forbs. Forbs were not abundant on certain Great Basin pronghorn habitats, especially during winter (Ellis 1970). A horse is reported to consume six times as much forage daily as one pronghorn (Heady 1975). Consequently, on rangelands with large numbers of horses and low quantities of winter forbs, horses can compete with pronghorn.

Salwasser (1980) reported that high numbers of horses can be competitive with pronghorn for early spring grasses and forbs.

Pronghorn and feral horse competition was investigated during 1996 and 1997 in a Great Basin shrubsteppe environment at the U.S. Army Dugway Proving Ground, Utah (Boyle et al. 1999). Using aerial surveys, 184 to 245 pronghorn and 49 to 87 horses were counted. During winter and summer, pronghorn concentrated in playas and foothills, whereas horses occurred mostly in foothills and mountains. Fecal analyses showed that pronghorn consumed primarily shrubs in winter, forbs in summer and both in spring. Horses ate mostly grasses in summer and spring, but browsed more than 60% shrubs in winter. Forage overlap was highest (6%) during winter. Divergent diet preference tends to segregate these species during summer, but in winter, both seek browse and greater spatial overlap occurs, although forage competition was minimal because of differing shrub species foraged.

Diseases and Parasites. No reports of disease or parasite transmission between pronghorn and horses were found in the literature.

Burros

There are numerous herds of feral burros on western rangelands; most are in Arizona and California. Few share landscapes with pronghorn. One exception is a herd totaling less than 50 animals on Sheldon (U.S. Fish and Wildlife Service 1980). These burros have existed on the refuge since before 1950, and numbers appear to remain relatively static. The burros use rangelands with pronghorn during severe winters, which generally occur once every 10 to 20 years. No studies are known to have been conducted on the refuge regarding compatibility or diseases and parasites.

Competition. In the early 1990s, a food habit study of wild and feral ungulates was conducted on Sheldon that reported vegetation foraged by pronghorn and feral burros (Hansen et al. 2001). Data were collected only for winters, as that is when both species occupy the same habitats. Burros mainly consumed grasses (84.4%) and few forbs (6.8%) and shrubs (8.8%). Dietary overlap was not intensive, and the annual mean percentages were 9.9% for grasses, 6.8% for forbs and 8.8% for shrubs with a total similarity percentage of 25.5% (Table 127).

Few burros share habitat with pronghorn. Those on the Sheldon National Wildlife Refuge in Nevada use rangelands with pronghorn during severe winters. At that time, burros consume mostly grasses, so competition with pronghorn is slight. Photo by Pete Carter; courtesy of the U.S. Fish and Wildlife Service, Lakeview, Oregon.

Range Improvement Projects for Livestock

Range managers and ranchers use the phrase range improvement to denote projects to enhance resources (e.g., vegetation restoration, water developments, brush or noxious plant control) or provide structures to control livestock (e.g., fences, corrals, stock trails). The expression habitat enhancement is used by wildlife managers rehabilitating environments for wildlife. Chapter 17 provides specifics on how to plan and implement range improvements or habitat enhancement projects to benefit pronghorn.

More projects to restore, improve or enhance resources have been accomplished on western rangelands for the livestock industry than any other activity (Yoakum et al. 1996). At times, these projects have been beneficial to pronghorn. At other times, they have been detrimental. For example, grass seedings have been predominantly for the needs of livestock, whereas mixture seedings (near equal quantities of grasses, forbs and shrubs) better meet the habitat requirements of pronghorn. The criteria for planning a range seeding can, therefore, determine whether it is an attribute or liability to pronghorn and other wildlife. The same strategy applies to water developments or fence construction. Consequently, it behooves land managers to evaluate project designs before implementation if the objective is to serve pronghorn and livestock. One method of assessing projects is to monitor animal responses to projects already implemented. Three case histories have been evaluated for projects with a primary objective of improving forage and waters for livestock that, at times, provided major benefits to pronghorn—one criteria for conducting multiple-use projects on public lands.

(Top) *The shrubsteppe of the Pony Springs, Nevada, site was dominated by tall, dense stands of shrubs with little understory of herbaceous vegetation available during the early 1960s. The site produced scant forage for livestock. Pronghorn had not been observed in the area for more than 50 years, although herds occupied habitats less than 7 miles (11 km) away.* (Bottom) *A large shrubsteppe site north of Pony Springs, was treated during the early 1970s by decreasing 80% to 90% of the tall, dense shrubs, through plowing and planting to grasses. Four years later, the vegetation changed to a height averaging 18 inches (46 cm) with a composition of 80% grass and forbs and 20% shrubs. Livestock grazing was reinstated. Pronghorn colonized the treatment site for foraging and fawning. Shown above is a pair of 2-month-old fawns in the treatment project 5 years after manipulation of vegetation.* Photos by Jim D. Yoakum.

Case 1. Pony Springs Area. The Pony Springs area, 20 miles (32 km) north of Pioche, Nevada, is historic pronghorn habitat. Heavy grazing by cattle and domestic sheep resulted in vegetative changes from mixed shrub–forb–grass types to dominant shrub communities. During the early 1960s, vegetative composition was 16% grasses, 2% forbs and 82% shrubs. Shrub height averaged 32 inches (81 cm). No use was made of the area by pronghorn, although they occupied adjacent rangelands.

In the latter half of the decade, 6,000 to 8,000 acres (2,428–3,238 ha) were treated. Shrubs were plowed or chained, and grasses were planted. Five years after treatment, the vegetative composition on one chaining and seeding project was 60% grasses, 20% forbs and 20% shrubs. Shrubs had a mean height of 18 inches (46 cm). Pronghorn readily moved into the treated areas and were seen frequently.

The site was heavily grazed by cattle during the early 1970s to the mid-1980s, and the vegetative composition

Bear Valley in central Oregon mainly supports a cattle ranching industry. Pronghorn were rarely seen there during the first 50 years of the 20th century, but were common in valleys 30 miles (48 km) to the east. Before the 1950s, the vegetative type was tall, dense shrublands with sparse herbaceous vegetation. From the 1950s to the 1970s, the ranchers in Bear Valley treated one-fourth to one-third of the rangeland through a series of vegetation-type conversion projects. The objective was to increase forage for cattle by decreasing shrubs and planting grasses and dryland alfalfa. The projects were highly successful, changing the vegetation to an average of 80% to 90% grasses and forbs. Islands of untreated shrubs were retained. As more and more acreage was treated, pronghorn pioneered into the valley and then returned to adjacent permanent habitats. With time and increased acreage conversion, pronghorn remained as yearlong residents in the valley. Photo by Jim D. Yoakum.

changed to 36% grasses, 14% forbs and 50% shrubs, with a mean height of 22 inches (56 cm). Pronghorn use of the site steadily declined, apparently in direct relation to changes in vegetative composition and structure.

Case 2. Bear Valley. In the 1940s, pronghorn did not reside in Bear Valley of central Oregon, north of Burns, Oregon, although herds inhabited adjacent valleys (Polenz 1976). Bear Valley was primarily private rangeland used for grazing cattle. Ranchers manipulated large tracts of tall, dense sagebrush and planted grasses and dryland alfalfa to decrease sagebrush and increase grasses and forbs. This changed the vegetative structure and composition from a dominant, high-shrub community to one of lower growing grasses and forbs with islands of sagebrush.

Through vegetative manipulation practices during a 30-year period, the ranchers changed the habitat to a condition favorable for pronghorn. During this time, small herds of pronghorn pioneered into the valley for short periods, then moved back to their traditional rangelands. As additional acreages were treated and winters remained mild, pronghorn became established in the valley yearlong. The population expanded to more than 600 animals within 20 years and was one of the most productive herds in Oregon (Torland 1980).

Case 3. The Vale Project. One of the most extensive rangeland rehabilitation programs accomplished on public lands under jurisdiction of the BLM was the Vale Project, in southeastern Oregon (Heady and Bartolome 1977). Although designed as a multiple-use effort, the 11-year, large-scale program emphasized improving forage and water to restore livestock carrying capacity. The 60- by 175-mile (97 by 282 km) area was 90% sagebrush/grasslands, producing less than 50% of its potential forage as a result of prior heavy grazing by livestock.

Beginning in 1963, approximately $10 million was spent for: 506,000 acres (204,775 ha) of shrub control; 267,000 acres (108,050 ha) of artificial seedings, both single species and simple mixtures; 2,000 miles (3,220 km) of fences; and 600 water developments, including 463 miles (745 km) of pipelines installed for better water distribution. Approximately 9% of the area was treated by shrub control and artificial seedings. Crested wheatgrass was the predominant species planted in all seedings, and dryland alfalfa was included in 26 seedings.

Fifteen years later, untreated sites averaged 52% grasses, 3% forbs and 45% shrubs, with a mean height of 28 inches (71 cm). Plowed and seeded sites had 76% grasses, 11% forbs and 13% shrubs. Heady and Bartolome (1977) sampled the treated areas and found that most attempts at rangeland rehabilitation had succeeded.

*The Vale Project was a rangeland restoration program for shrubsteppe environments on public lands in Malheur County, Oregon (*left*). It was accomplished in the 1960s and 1970s primarily to increase forage and water for livestock and it greatly increased the carrying capacity for livestock. At the same time, the many vegetation-type conversion projects changed the plant community, favoring the habitat requirements of pronghorn by decreasing the height and density of shrubs and increasing the composition of grasses and forbs. In addition, the rangeland seedings increased animal unit months that released livestock grazing pressure on adjacent native environments preferred habitats for deer, pronghorn and other wildlife. One of the major alterations to vegetative communities during the Vale Project, implemented to enhance habitat for pronghorn and other wildlife, was the planting of nomad alfalfa (*right*). This forb is nutritious, succulent and preferred by pronghorn. It is a broadleafed legume that contributes nitrogen to the soil, is adapted to semiarid environments and competes well with bunchgrasses for moisture and soil nutrients.* Photos by Jim D. Yoakum.

Projects reduced shrub composition and structure while increasing grasses and forbs. Sagebrush was not eradicated and, during the course of 15 years, reinvaded to form a substantial component of the community, approximately 25% of ground cover.

A comparison of pronghorn numbers for the large-scale restoration project was made with surrounding lands receiving minor rangeland improvements. All pronghorn population data were obtained from state wildlife agency aerial census. During the early years of the project (1962–1964) the herd averaged 1,420 in the project area. Following implementation of the improvement practices (1972–1974), the herd increased to about 2,600 (Oregon State Game Commission 1962, 1963, 1964, 1972a, 1973, 1974). During the same period, herds on adjacent rangelands increased less than 30%. Populations continued to increase from 1975 to 2000, with numbers averaging 5,000 animals during the late 1990s (R. Kindschy personal communication:2000).

Hundreds more vegetation manipulation and water improvement projects have been accomplished on western rangelands to improve resources for livestock rather than for the welfare of wildlife (Yoakum et al. 1996). Wildlife biologists and managers need to coordinate closely with range managers and ranchers to facilitate planning projects to meet the biological needs of pronghorn along with improving conditions for livestock. Likewise, some land-treatment projects to rehabilitate vegetation or conserve soil and water can be designed to improve forage and water for pronghorn and other wildlife (e.g., land conservation efforts on energy development sites, alternate flood mitigation sites, exotic plant invasion control measures) (see Chapter 17).

Livestock Grazing Systems

Livestock grazing systems are planned strategies to maintain or improve rangeland carrying capacities primarily for cattle and sheep on western rangelands. Effective management systems should consider the control of livestock, range improvements, number of livestock, seasons of use, monitoring, plant physiology, and the effects and needs of other rangeland uses. A number of grazing systems have been developed (Stoddart et al. 1975, Heady and Child 1994, Holechek et al. 1998) and are used increasingly in range management today. For rangelands dually used by pronghorn and livestock, the following guidelines are recommended in the design of grazing systems:

1. When forage is being allotted, the following habitat requirements of pronghorn should be considered:

a. Adequate amounts of preferred plant species should be reserved as forage for pronghorn. These include grasses, forbs and shrubs identified by food habit studies in the same or a similar ecosystem. Consideration should be given to proper use of key forbs and shrubs—ensuring they are not grazed beyond their sustainable tolerance.

b. Forage should be reserved for a reasonable number of pronghorn. Reasonable numbers should be based on: (i) average herd population for the past 30 years, (ii) average forage production for the past 15 years, and (iii) management objectives for herd size determined by wildlife and land-management agencies.

c. Proper amounts of forage should be apportioned for a reasonable number of animals for specific sites and seasons of use. Special attention should be given to reserving sufficient quality forage for critical areas (i.e., fawning areas, winter-use areas, movement corridors, etc.).

2. Water should be available during all seasons that pronghorn are in an area.
3. Fence construction should meet specifications to allow pronghorn movements year-round for all age classes.
4. When livestock grazing systems are designed around the key plant species concept, preferred pronghorn forage of forbs or shrubs should be included as key species.
5. Livestock should be restricted from fawning areas during the fawning season.
6. Livestock grazing systems that restrict, alter, limit or deleteriously affect the habitat requirements of pronghorn should include mitigating measures and alternative procedures for enhancing pronghorn habitat.

It is evident that competition can occur between livestock and wildlife. Wagner (1978) recognized this and provided three options for managing livestock and wildlife on western rangelands (Figure 107). His first option—no management—requires increased stocking of livestock initially and applying no management, which would result, ultimately, in long-term decline of vegetative condition and production. The second option—intense management—would include intensive rangeland improvement and management, accentuating domestic livestock production. Rangelands would be fenced into a mosaic of pastures, many monoculture grass seedings planted and water sources highly developed. This would result in increased livestock production but lower wildlife populations. Option three strives for management on a natural or ecological basis, stressing the return of rangeland potential. Three-fourths of western rangelands are producing at less than half their forage potential (Box et al. 1976). The return of these rangelands to their ecological potential would provide forage for increased numbers of livestock and wildlife and, compared with the other options, retain a more natural environment. The third option is the most favorable for the future of pronghorn.

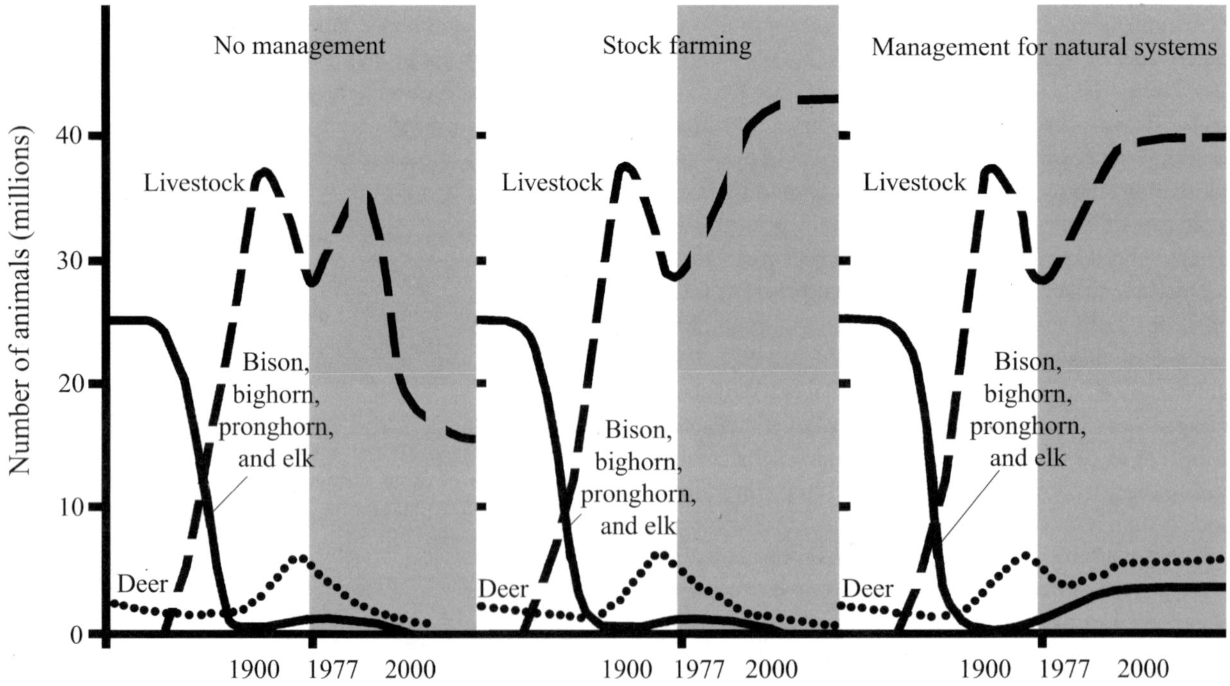

Figure 107. *Three management options relative to production of livestock and wild ungulates on western rangelands (from Wagner 1978).*

Before the early 1800s, bison and pronghorn often foraged together and were most abundant on the shortgrass prairies. Both ungulates still graze these prairies, as indicated above in northeastern Wyoming during the 1990s. Photo by Robb Hitchcock.

By recognizing pronghorn habitat requirements and needs, and by assessing the potential of proper range management, one can postulate a 100% increase in pronghorn numbers. With this increase, pronghorn still would use less than 1% of the forage available on western rangelands. The increase would be substantial, but there would still be only two pronghorn where perhaps 35 roamed 200 years ago.

Properly implemented grazing systems require an ecological equilibrium that can produce and sustain optimum numbers of livestock and wildlife. Rest-rotation and other methods of prescribed livestock grazing frequently are advocated today.

	← April 1	July 1 ↓	November 30 →
First year	Graze all year		
Second year	Rest	Graze	
Third year	Rest for vigor and reproduction		
Fourth year	Rest for vigor and reproduction		

FIGURE 108. *Recommended four-pasture formula designed for a rest–rotation livestock grazing system on rangelands containing bitterbrush.*

Rest-rotation. Hormay (1970:3) identified the principles and practices for rest-rotation grazing and stressed that "Production and maintenance of vegetation are of first consideration in management." His publications and training sessions concentrated on developing grazing formulas using grasses as key species for livestock forage. For pronghorn rangelands, the grazing formula should include key forbs, which generally require about a month more time to seed than grasses. In the Great Basin region, bitterbrush was second in importance to sagebrush as pronghorn browse. Hormay recognized that antelope bitterbrush requires two seasons of rest to set and produce seed. Figure 108 provides a recommended grazing formula for rangelands where bitterbrush is a key species.

Results of investigations in Utah by Smith and Doell (1968) supported Hormay's (1970) procedures for managing bitterbrush. Smith and Doell (1968:31) stated that "long term changes in vegetation may necessitate rotation grazing and use of different dates in order to equalize the effects of grazing among the several species in the stand."

Desert shrub communities in southwestern Utah likewise best served pronghorn when they were left ungrazed or rotated domestic sheep grazing (Clary and Holmgren 1982).

Prescribed Grazing. When livestock are used as a method to maintain or improve vegetation for subclimax-adapted species of wildlife, the procedure can be phrased prescribed grazing. This practice requires understanding of (1) seasonal preferences of both the wild and domestic animals in question, (2) the plant species present, (3) reaction of those plants to grazing, and (4) phenology of the plants.

When properly planned and implemented, prescribed grazing can be advantageous to pronghorn habitat. For example, Salwasser (1980) reported that, because they are primarily grazers, they can be used in the spring by flash grazing of grasses to increase forb growth for summer and autumn use by pronghorn. Smith and Doell (1968) also recognized that livestock grazing of grasses could increase forb and shrub production. Bodie (1979) reported that livestock grazing resulted in increased forb production in a previously ungrazed area of short sagebrush where forbs had been largely absent.

Hubbard and Sanderson (1961) found that grasses inhibited the growth of bitterbrush (preferred pronghorn browse) in California. When grasses were reduced by cattle grazing, significant increases were noted in bitterbrush forage production and plant reproduction. Similar relationships of prescribed livestock grazing to shrub/grass pastures were noted by Nord (1965), Smith and Doell (1968) and Longhurst et al. (1983).

For livestock grazing to benefit pronghorn in the long term, livestock management practices must be designed with consideration for the seasonal, biological requirements of pronghorn. Ignoring those requirements or prescribing the wrong practices could result in damage to the pronghorn population and perhaps the rangeland and livestock as well.

TABLE 129. Approximate number of ungulates per animal unit month (AUM), based on ratios of metabolic weight

Species	*Approximate adult weight* lb	kg	$kg^{0.75}$	Ratio 98/k	Number of animals per AUM
Bison, cow and horse	1,000	455	95	1.00	1.0
Elk	600	272	67	1.46	1.5
Mule deer	150	68	24	4.08	4.0
Domestic sheep	120	55	20	4.90	5.0
Domestic goat	100	45	17	5.76	6.0

Source: From Heady 1975.

ANIMAL EQUIVALENTS

Managers often need to exchange use of forage for different grazers on the same rangeland. Various methods of calculating exchange ratios, also referred to as animal equivalents, have been used, but none has been completely satisfactory. If the forage consumed by both ungulates is similar, the ratio of metabolic weights provides an estimate of the exchange ratio. Heady (1975:148) explained: "Although variation exists among individuals and species, the 3/4 power of weight in kilograms defines the metabolic size of an animal (Kleiber 1961). It expresses the fact that smaller animals produce more heat and consume more food per unit of body size than do larger animals. For example, a 500-kilogram cow and a 50-kilogram sheep have metabolic sizes of 105.75 and 18.8, respectively (Maynard and Loosli 1969). The ratio of live weights is 1:10, but the ratio of metabolic weights is approximately 1 cow to 5.6 sheep, which is a better expression of the relationship of their feed needs. A cow/sheep ratio of 1:5 has been used on rangeland more than any other and appears to be a reasonable expression of the relative impact of the two upon the range." Using Heady's procedures, Table 129 gives animal unit equivalents for different ungulates using rangelands with pronghorn.

Various investigations have developed animal equivalents relative to pronghorn. One of the first such studies was conducted by Buechner (1950a) in Texas. He indicated 9.39 pronghorn equal one animal unit month (AUM) based on an average weight of 90.5 pounds (41 kg) per pronghorn and 850 pounds (385.5 kg) per cow. However, Buechner elaborated on the differences in forage species consumed by pronghorn and cattle. His studies revealed that it took 47 pronghorn to eat the same amount of the plant species consumed by one cow.

In Colorado, Hoover et al. (1959) used the weights of 100 pounds (45.3 kg) for pronghorn, and 1,000 pounds (453.6 kg) for a cow to determine that it took 105 pronghorn to consume the same amount of forage as one cow—considering only the plants selected by cattle. Emphasis was placed on the actual forage classes consumed and not on body weight alone. Hoover et al. (1959:69) stated that "all the antelope in the state (9,000) would not eat enough grass to feed 100 head of mature cattle."

On the Red Desert of Wyoming during the 1960s, Severson et al. (1968) studied food preferences, carrying capacities and forage competition between pronghorn and domestic sheep. Carrying capacity for this site was approximately eight pronghorn per year for 120 acres (48.5 ha). For equivalent ratios, Severson et al. gave two figures: (1) when only shrub consumption was considered, it took 5.67 sheep to equal 1 pronghorn, or an AUM equivalent of 29 pronghorn and (2) when only grass consumption was computed, it took 43.5 sheep to equal 1 pronghorn, or an AUM equivalent of 220 pronghorn.

For the same ecosystem, Taylor (1972b) determined that 9.4 pronghorn were the equivalent of one AUM. His analysis considered forage competition data as well as animal weight conversion factors.

In Idaho, Anderson and Denton (1978) determined 54 pounds (24.5 kg) of forage was needed per adult pronghorn per month on sagebrush/grasslands, whereas a cow required 800 pounds (363 kg) of forage per month. Accordingly, they computed an equivalent AUM ratio as 1 cattle AUM (800 pounds forage/month)/1 pronghorn AUM (54 pounds forage/month) = 14.8 pronghorn equivalent. However, these figures did not consider the degree of competition for different forage classes used by cattle and pronghorn. When this calculation was determined (based on an estimated dietary overlap of 0.25) on

The impacts of cattle feeding intensively on vegetation needed for wildlife can be deleterious to pronghorn. At top, *cattle have removed most vegetation and little remains for pronghorn. Such practices can result in wild animals being forced to leave formerly preferred habitats to obtain sufficient food for survival. Intensive livestock foraging is a rangeland conservation liability when inadequate vegetation is retained for protection against wind and water erosion, and vegetation is inadequate to provide food and hiding cover for pronghorn and other wildlife. Properly implemented cattle foraging systems also retain adequate vegetation for pronghorn nutrition and fawning cover (*bottom*). The vegetation in this photo was grazed by cattle, yet abundant supplies were retained for wildlife. Rangelands in healthy condition contribute to sustaining healthy pronghorn and livestock herds.* Photos by Alice Koch.

a yearlong basis, Anderson and Denton recalculated that 59.2 pronghorn equaled one AUM. Conversely, one AUM of cattle forage would provide 25% of the diet of 59.2 pronghorn for 1 month.

During the 1988 joint meeting of the Interstate Antelope Conference and the Pronghorn Antelope Workshop, Kniesel (1988) reviewed past procedures and practices to use equivalent ratios. He stressed the tremendous variation in AUM equivalents currently used by management agencies: for example, 105 pronghorn equal 1 cow in Colorado; 59 pronghorn equal 1 cow in Idaho; 38 pronghorn equal 1 cow in Texas; 7 to 14 pronghorn equal 1 cow in

Oregon; and 5 pronghorn equal 1 cow in Montana. Kniesel attributed the wide variation to different methodologies and information used. Some investigators primarily used weight differentiations, whereas other researchers included other considerations such as dietary overlap and rangeland condition. He concluded that assessing AUM equivalents for pronghorn and livestock remained a problem because of little agreement between state and federal management agencies on standardized animal equivalents in forage allocations for multiple-use programs.

Some of the aforementioned procedures for determining animal equivalents have been based on relationships of body weight or competition for forage. Each procedure has its values, but one of the most important factors in forage allocations should be quantification of nutrient supplies and availability. Investigations of deer diets in California support the contention that a deficiency in diets, leading to a decline in energy reserves of does during the last trimester of gestation, resulted in high neonatal mortality (Hall et al. 1978). This phenomenon also is a problem for pronghorn throughout most shrubsteppe habitats (Vriend and Barrett 1978).

CONCLUSIONS

Pronghorn coexist with myriad wild, domestic and feral animals on western rangelands. Each species exhibits a different degree of compatibility and competition. Studies of pronghorn/prairie dog relationships disclose varying results regarding competition for forage. At times, prairie dog towns appear to increase forbs and shrubs for pronghorn, whereas other research discloses neutral relations between both species.

Pronghorn tend to ignore other species of ungulates under ordinary circumstances and vice versa. Bison and pronghorn occupy similar habitats but have little dietary overlap. Deer, elk and wild sheep have greater dietary overlap with pronghorn, but differing spatial distribution greatly decreases competition.

Although 10 to 15 exotic species of large mammals have been translocated to western rangelands in the United States, only gemsbok were known to cohabit landscapes with pronghorn at the beginning of the 21st century. Because gemsbok mainly ingested graminoids with small quantities of forbs and shrubs, there was little dietary overlap.

Pronghorn and wild or feral livestock have been reported as generally compatible, sharing forage and water on dually occupied healthy western rangelands. Possibly, the greatest negative impact livestock have on pronghorn is changes in plant succession levels, resulting in less diversity of preferred, succulent, nutritious forage for pronghorn. A second major impact is intensive foraging by livestock, resulting in inadequate cover for pronghorn neonates, contributing to increased fawn mortality by predators.

Intensive livestock foraging can alter plant composition and abundance (Kie et al. 1994). Such changes in vegetation can have a negative impact on pronghorn and other wildlife, and they can be observed in cases of fence-line contrast for shrubsteppe sites. (Top) *The field to the left of the fence was heavily grazed by livestock for many years compared to the field to the right. For the left field, this has resulted in decreased herbs and increased shrubs. Such changes decrease the availability of mixed plant communities favored by pronghorn.* (Bottom) *Grass utilization in the field to the right of the fence has been greater than the moderate cattle foraging in the field to the left. Soils in the right field are vulnerable to wind and water erosion, and little mulch remains to aid plant regeneration. The site to the right, with few remaining plants, can be rated as habitat in poor condition for wildlife because of inadequate vegetation for food and cover.* Photos by Jim D. Yoakum.

In modern times, many people enjoy panoramas of wildlife and domestic livestock peacefully grazing together on western rangelands, such as these pronghorn with a herd of domestic cattle. Pronghorn are highly visible wildlife, being active during daylight as opposed to various species of wildlife that are mainly nocturnal. When vistas are witnessed with pronghorn and cattle in a commensal relationship, the values of multiple-use resource management are enhanced. Top photo by Chuck Weschler. Bottom photo by Joe Van Wormer; courtesy of the Denver Museum of Natural History, Colorado.

Competition with cattle and horses does not appear to be a major problem on rangelands in good ecological condition or when animals are stocked at proper levels, except for spring/summer pronghorn habitats in the Intermountain West producing low quantities of forbs. This is primarily the result of food partitioning. Cattle may act as a reservoir for bluetongue, a disease highly fatal to pronghorn.

Competition appears to be most likely between pronghorn and domestic sheep, as both consume large quantities of forbs and shrubs. Domestic sheep also carry many diseases and parasites common to pronghorn.

Range improvement projects designed primarily to enhance resources for livestock can be a liability or asset to pronghorn, depending on how the projects are planned and implemented. Forage manipulation projects that maintain or increase plant diversity and control shrub height can be beneficial to pronghorn, but projects resulting in monocultures of exotic, coarse grasses are of little use. Fences to control livestock can be built to allow pronghorn movement, or they can become complete barriers. Most water developments have been beneficial to pronghorn.

Livestock grazing systems can be designed to be beneficial or detrimental to pronghorn. Plans that recognize forbs and shrubs as key species are favored for pronghorn management over plans specifying only grasses as key species. Livestock should be kept clear of pronghorn use areas during fawning seasons.

Livestock benefit from pronghorn use of many noxious plant species. Reciprocal values are received by pronghorn when intensive predator control programs are conducted on rangelands to protect livestock.

Animal equivalents are used for allotting forage on western rangelands. However, the methods and practices for this technique vary greatly and need more refinement.

The future of pronghorn subsisting with wild, domestic and feral herbivores appears to be less of a problem of animal compatibility or competition than of sustaining rangelands in healthy condition. Environments with natural grass/forb/shrub mixtures and composition percentages, indicative of natural plant communities, best serve pronghorn requirements. Healthy rangelands support healthy populations of pronghorn, other wildlife and livestock.

SECTION THREE

MANAGEMENT

C H A P T E R S I X T E E N

Management Plans, Environmental Impact Statements and Guides

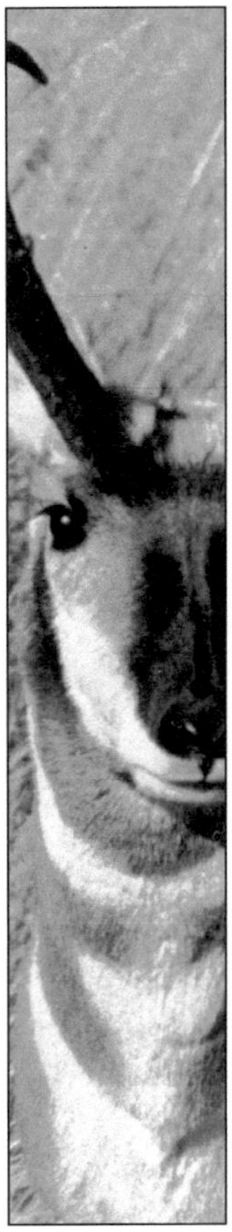

Jim D. Yoakum

The wildlife management plan is a description of the objectives and goals which will be met by the manipulation of habitat, populations, and people, and the ways these objectives will be reached.

Ripley (1980:4)

Management guides for any species must remain general in nature. Their values are found in the discussions of basic information to complement the wildlife manager's knowledge, and implementation should be undertaken in keeping with a holistic approach to resource management.

Lee et al. (1998:1)

With growth of human populations and expanded use of rangelands, the need to document habitat requirements of pronghorn has increased. Pronghorn thrived during the last half of the 20th century, but the need continues to manage for healthy populations that are compatible with other uses of rangelands. Management plans provide guidance and continuity by listing objectives, goals and procedures to sustain a species. This involves maintaining or enhancing forage, water and space, as well as coordinating pronghorn management with holistic land-use plans (Lee et al. 1998).

Few management plans have been prepared explicitly for pronghorn, but valuable experience from them can be shared with others. One of the values of these plans is the provision of a written report listing management details for specific populations and their habitats. Such details include the sustainable number of animals (whether that number currently exists, more animals are needed or too many exist) and delineation of total habitat for the herd or herds in question (substantiates the quantity of land, forage or water required to maintain the herd, and lists critical areas for survival).

A wildlife management plan can be the coordinating agreement between state wildlife agencies and federal government or private land managers. Such plans aid coordination of herd reestablishment on unoccupied historic habitats or relieve crop depredations by herds that have exceeded population goals. When properly prepared, such plans identify the acreage needed to sustain herds, and they account for losses of land, forage or water that affect pronghorn population levels. Management plans are most effective when they contain quantitative data, as

Examples of species and habitat management plans for pronghorn prepared by federal, provincial and state wildlife agencies. By 1999, all provincial and state wildlife agencies in Canada and the United States reported having management plans. None is known for the states of Mexico, but the federal government of Mexico has developed a recovery plan for all three endemic subspecies classified as endangered. Photo by Jim D. Yoakum.

opposed to generalizations. Such information is especially valuable for future mitigation measures and coordination of multiresource plans by land managers. For example, the Private Lands Act of California was developed during the early 1980s. It enables private landowners who develop management plans approved by the state wildlife agency to control pronghorn hunts and enjoy economic dividends for wildlife production on their property.

At times, wildlife management plans are referred to as wildlife or habitat conservation plans. Use of the term conservation in titles of wildlife plans increased during the 1990s, (e.g., Habitat Conservation Plans [U.S. Fish and Wildlife Service 1999c], Sage Grouse Conservation Plan [Connelly et al. 2000]). "Conservation plans" for wildlife populations and habitats probably will be used more frequently during the 21st century.

Management plans can delineate seasonal pronghorn routes and assist highway departments in determining construction specifications least disruptive or damaging to pronghorn. They also can provide data for rangeland management plans and for minimizing conflicts with livestock operations and energy projects.

Another major value of management plans for pronghorn is the documentation of historic conditions, present status and future needs. Such a plan typically is developed by specialists from multiple resource disciplines, reviewed by peers, coordinated with various government agencies and landowners, and approved by administrators. The documentation is especially important because agency (federal, provincial or state) wildlife biologists frequently are transferred. The management plan, therefore, represents a permanent record of fact, guidelines, experience, goals and objectives. Management plans also serve as mechanisms of facilitation, continuity and accountability that are used by agencies and the public to learn about and track efforts to manage a pronghorn population and to enhance that population's habitat. Management plans represent a professional commitment to improved conditions for pronghorn.

Management plans generally are developed in accordance with policies and responsibilities of the agency's spokesmen. That is, state wildlife agency biologists may emphasize recruitment and mortality factors influencing harvests. Biologists working for federal land-management agencies may stress allotment of forage or development of water supply. In most situations, depending on the primary orientation of the authoring agency, a plan may focus on a relatively narrow objective for the species or its habitat—depending on the management resources available. In some cases, the orientation toward species or habitat can be equal, but this usually occurs when the authoring agency has responsibilities for both, such as the U.S. Fish and Wildlife Service. The agency's particular mission—its public trust responsibilities—tend to dictate the specific orientation, objectives and practices in the preparation of management plans for pronghorn populations.

The concept of wildlife management plans has existed for more than 50 years. King's (1938, 1941) five-step plan evolved from Aldo Leopold's (1933) textbook, *Game Management*, and provided a logical approach to decision making (Figure 109). The first four steps involve information gathering and analysis for the species and its habitat, with emphasis on the importance of establishing a management database. After the base data are collected and analyzed, the last step is the development of recommendations for management of both the species and the habitat. King's outline is used today as it was when published, and it is included in two contemporary wildlife management texts—Bailey et al. (1974) and Bailey (1984). With time and experience, wildlife management plans have progressed

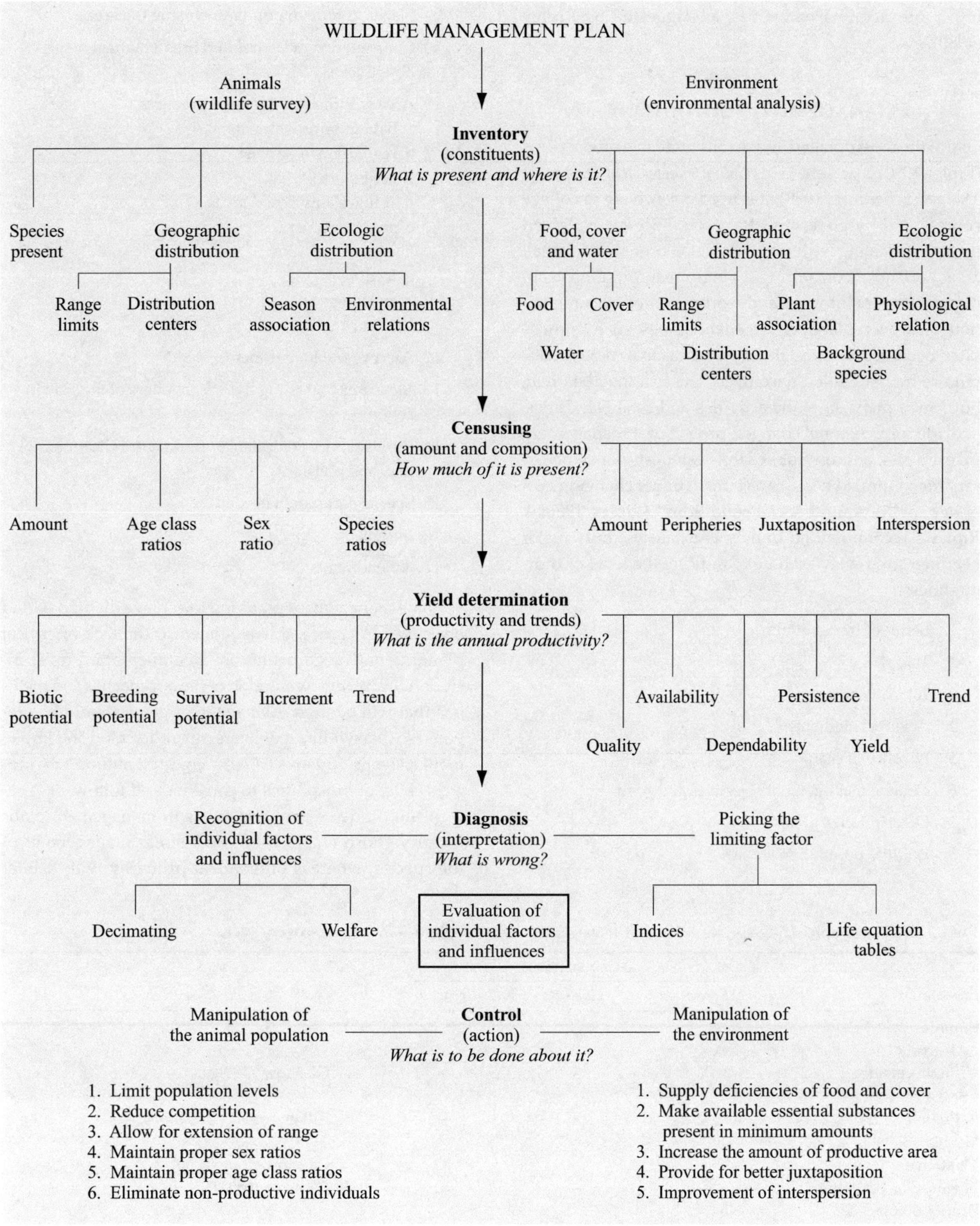

FIGURE 109. *Outline of a wildlife management plan (from King 1941), recommending subjects to be assessed when developing pronghorn and habitat management plans. Various government agencies and private landowners have used this outline to prepare pronghorn management plans.*

into more detailed narratives, as suggested by Ripley (1980).

MANAGEMENT PLAN OUTLINE

One of the most detailed and readily available model plans (Ripley 1980) is presented in *Wildlife Management Techniques Manual* (Schemnitz 1980); it is used extensively in college courses and by government agencies. Ripley explained that plans dealing with relatively small programs in small areas, or those concerning a single manipulation technique, often are informal and require limited documentation. However, plans for large landscapes, such as those often occupied by pronghorn with seasonal movements, require more detailed narratives. The cost may be great, but "Such plans steer the activities and establish duties, provide for personnel changes, protect past management investments, and on public lands establish policy necessary for continued wise management under changing economic, political, and social conditions" (Ripley 1980:5). Ripley's recommended outline is extensive; only major headings are reviewed here to indicate the level of thoroughness.

1. Letter of transmittal
2. Title
3. Table of contents
4. Acknowledgments
5. Purpose of plan
6. General management objectives and goals
7. Description of area
8. Wildlife population status
9. Needs of human population using the area
10. Coordination measures and broad management objectives
11. Area wildlife management practices
    a. habitat improvements
    b. regulation of harvest
    c. access
    d. refuges and closed areas
    e. predator control
    f. salting
    g. feeding
    h. stocking
    i. other practices
12. Surveys and inventories
13. Measures for testing the effectiveness of the plan in action
14. Provisions for compliance checks, periodic review of plan and revisions
15. Special equipment needed
16. Budget
17. Amendments

Some management plans fail to achieve their goals and objectives. At times, this is related to the lack of critical planning and documentation. Documentation must include a complete presentation of the approaches and methods that will be used. Management plans should be prepared so that wildlife managers can understand and implement all steps. Ripley's (1980) suggested outline provides a checklist of procedures to consider and follow.

A multitude of variation exists in management plans for pronghorn (Table 130). Most stress management of the species, whereas others deal primarily with habitat

Table 130. Plans emphasizing species or habitat management for pronghorn—includes recovery plans

	Major emphasis			
Location	*Species*	*Habitat*	*Recovery*	*Source*
Canada				
Alberta				
Provincewide	X	X	—	Glasgow (1990)
Saskatchewan				
Provincewide	X	X	—	Killaby et al. (1992)
Mexico				
National	—	—	X	Cancino et al. (2000)
Peninsular pronghorn	—	—	X	Cancino (1994)
United States				
Arizona				
Sonoran pronghorn	—	—	X	U.S. Fish and Wildlife Service (1998)
California				
Statewide	X	—	—	Pyshora (1982)
Modoc National Forest	—	X	—	Stutz (1967)

continued on next page

TABLE 130—*continued*

Location	Major emphasis: Species	Habitat	Recovery	Source
Adobe Valley	—	X	—	Capodice et al. (1982)
Idaho				
Little Lost/Birch Creek	—	X	—	McCarty (1982a, 1982b)
Nevada				
Statewide	X	—	—	Tsukamoto (1983)
North Eccles	—	X	—	Lister (1989)
Oregon				
Hart Mountain National Antelope Refuge	X	X	—	Pyle and Smith (1990)
Texas				
Statewide	X	X	—	Hailey (1979), Swepston and Hailey (1991)

management. A few coordinate species and habitat. Each provides experiences and ideas that, when shared among professionals, serve as examples of the effectiveness of different practices and techniques. Review of several plans illustrates the various facets of management identified by different investigators and agencies.

PRONGHORN MANAGEMENT PLANS

During 2000, a questionnaire was sent to provincial and state wildlife agencies in Canada, Mexico and the United States, requesting current information on pronghorn management plans. Responses were received from 19 agencies, and the following stated agency-wide or specific area management plans for species and habitat were completed: Alberta, Arizona, Baja California Sur, Nebraska, Nevada and New Mexico. Some plans have been in effect for three decades or longer. New Mexico initiated its plans in 1967, and California developed a statewide plan in 1977. All other plans were completed during the 1980s and 1990s. Colorado, Montana and Wyoming developed species management plans for biological management units stressing population surveys and harvest practices.

Provincial and state wildlife agency plans often express problems, practices and methods of managing a species. Such plans identify numbers and seasonal locations of a wildlife population that are basic to making decisions regarding the welfare of the animals and their habitats. The documents also elaborate harvest procedures and results. Examples of such plans vary by region; several will be reviewed.

Alberta

Pronghorn in Alberta are at the northern limit of their range and are subject to wide fluctuations in numbers because of climatic extremes, especially severe winters. Thirty-six percent of historic pronghorn rangeland had been converted from native grass prairie to cultivation by 1983, with 6% of the 36% converted between 1970 and 1983. Pronghorn are in high demand by hunters and are popular with nonconsumptive users of prairie wildlife.

The *Management Plan for Pronghorn Antelope in Alberta*, a 10-year plan, was developed for the provincial wildlife agency (Glasgow 1990). Following are its goals and objectives:

1. The summer pronghorn population goal will be 18,900, with 15,500 going into winter. It will be necessary to maintain 10,000 square miles (25,900 km^2) of summer rangeland and 1,270 square miles (3,290 km^2) of winter habitat (within the 12 identified winter areas) to meet population goals. This does not include the Suffield herd, which numbers from 2,000 to 4,000 and inhabits rangelands outside of provincial jurisdiction (military lands).
2. A variety of hunting opportunities (e.g., archery, trophy and nontrophy) will provide for 4,800 residents to harvest an estimated 1,700 trophy pronghorn and 1,670 nontrophy pronghorn annually. Limited-entry draws will manage the distribution and number of hunters to provide a high-quality recreational experience, manage the resource within each pronghorn management area and minimize disturbance to landowners.
3. Nonconsumptive opportunities will be enhanced through the use of print, film and audiovisual media to increase knowledge concerning the species. Viewing guides and interpretive sites will be provided to increase personal contact.
4. Commercial opportunities will be available to those providing outfitting and guiding services. An abundant and widely distributed pronghorn population will help to attract tourists and produce commercial benefits to those providing goods and services for tourists.
5. Scientific research and educational activities will be encouraged to enhance knowledge of pronghorn and improve management capabilities.

6. Agriculture damage prevention, control and compensation programs will continue. Compensation may be extended to include additional crops, but difficulties in adjusting damage means that alternatives, such as on-farm habitat improvements that benefit pronghorn and livestock, will be considered.
7. Habitat retention programs will become more effective through a higher degree of cooperation between wildlife managers and those who manage the land within pronghorn rangelands (e.g., private landowners, range managers, oil and gas companies, mining and thermal power generating companies, irrigation districts).
8. Habitat and population monitoring programs will become more precise to allow population and habitat retention goals to focus on local levels and measure success in achieving established goals.

Population goals were set for summer and winter habitats because each had differing carrying capacities. Hunter harvests were used to aid in the control of herds relative to population objectives. Nonconsumptive values were recognized with suggested practices to increase viewing sites for the public. Because quality and quantity of most vegetation and drinking water were adequate, little need existed for habitat enhancement projects. The major habitat objective was to retain and sustain quality habitats supporting pronghorn populations in the face of increasing human demands on the landscape.

Before the 1980s, management was limited to changing bag limits, adjusting season lengths, and opening and closing harvest seasons. Parks and sanctuaries protected some habitat, but habitat protection efforts were limited. Now, management includes population and habitat goals for each of eight pronghorn management areas, annual inventories of populations, periodic habitat assessments, a habitat retention program and limited-entry hunting to sustain population goals and provide recreation.

Arizona

Pronghorn on Anderson Mesa in northeastern Arizona have experience decreased numbers for more than 40 years (Neff et al. 1985, Yoakum 2003a). The area varies from ponderosa pine and juniper and pinion woodlands to semiarid grasslands. A plethora of ecological and management problems existed, so management plans were developed (Arizona Department of Game and Fish, Flagstaff Region 2002). They included an operational plan, implementation plan, and strategies and tasks implementation matrix. The plans define a goal for management of the population, strategies for reaching the goal and recommendations for management of the habitat. Included are needed research projects emphasizing diet studies and nutritional values of forage. The original draft of this plan was developed in a dispute resolution program involving the Arizona Department of Game and Fish, Coconino National Forest, Arizona State Land Department, Hopi Tribe, local ranchers, Diablo Trust, Arizona Antelope Foundation, Arizona Wildlife Federation and National Wildlife Federation. This is the only known management plan for pronghorn herds and habitat in the semiarid grasslands of the southwestern United States.

California

Pyshora (1982b) developed the first comprehensive pronghorn management plan for northeastern California, stressing species management. It was revised in 1987, and an updated plan was expanded in 1989 (California Department of Fish and Game 1995). Northeastern California herds increased during the 1970s and 1980s, although the state permitted annual harvests of bucks and does. The state maintained an active translocation program to unoccupied historic habitats, as well as augmentation of other herds.

The Anderson Mesa, Arizona, plateau is a concentration area for pronghorn. Ponderosa pine flank the western landscape and are increasing in density, as are juniper/pinion woodlands that border the eastern plateau. Herbaceous vegetation with few forbs and shrubs dominate the grasslands. These grasslands have ideal physiographic features for pronghorn habitat—expansive, open, low, undulating landscapes that are typical of high elevation semiarid habitats in the Southwest. Photo by Jim D. Yoakum.

The central valley grasslands of California sustained large numbers of pronghorn when Euro-Americans first pioneered the West Coast. However, relentless, yearlong hunting diminished populations as early as the 1860s. Some small remnant herds existed until the early 1900s. One of the last native pronghorn in the San Joaquin Valley was this adult doe shot by Arthur Oliver about 10 miles (16 km) west of Mendota, California, in 1920. Photo by Joe Dixon.

Herds are managed in geographic biological units, for which harvest allotments are determined and separate management plans are prepared. The management plan provides detailed biological data for seven herd units. Harvest quotas have been calculated on the basis of simulation models.

California pronghorn are managed for the benefit and enjoyment of all people. Perpetuation of a sustainable population will be the foremost consideration of the pronghorn management plan. According to Pyshora (1982b), goals and objectives for the management plan will be achieved for the species and its habitat by:

1. Preserving habitat by opposing land-use changes and zoning detrimental to pronghorn welfare
2. Retaining public lands that provide habitat under public land agency administration
3. Encouraging land exchanges and land purchases where feasible that add to pronghorn habitat
4. Discouraging fences that might hamper movement of pronghorn, especially seasonal movement routes
5. Encouraging regulated livestock use on pronghorn habitats to ensure that pronghorn are given primary consideration in forage allocations
6. Conducting annual census surveys to determine population numbers and trends
7. Conducting annual herd composition surveys to determine herd sex and age ratios
8. Providing information on areas where pronghorn can be viewed by the public
9. Conducting special buck hunts, based on a surplus buck concept, if a minimum of 200 permits can be issued
10. Conducting either-sex hunts to regulate populations to herd size objectives
11. Maintaining each subherd at a maximum number that the habitat can efficiently support
12. Improving pronghorn habitat cooperative relations with public land administering agencies and private owners
13. Maintaining an active public information program and a pronghorn research program to determine:
    a. the effects of predation on pronghorn,
    b. movement corridors to and from seasonal use areas,
    c. fawning areas,
    d. potentials for improving habitat, including water supplies,
    e. the feasibility of translocating pronghorn to increase the rangeland area and population,
    f. use of other state agency data that could be useful and helpful, and
    g. participation in the Pronghorn Committee land workshop activities.

The former Carrizo Plain Natural Area was established during the late 1980s through cooperative efforts of The Nature Conservancy, California Department of Fish and Game, and the U.S. Bureau of Land Management. The area was designated a national monument by President Clinton in January 2000. The Carrizo Plain is the largest area remaining of California Central Valley grasslands. Many native species occupy these grasslands, including golden eagles, kit fox, mule deer, pronghorn, sandhill cranes and tule elk. The site is managed through a three-way cooperative management plan for the perpetuation of native flora and fauna. Photo by Jim D. Yoakum.

This list identifies the variety of parameters to establish goals and objectives for a pronghorn management plan in California. Such an endeavor can be a model for other state and federal wildlife agencies.

Herd increases during the 1980s in California led to crop depredation complaints. This resulted in formulation of a pronghorn depredation contingency plan, with the following suggested practices to alleviate such problems: population management; depredation hunts, relocation of problem animals, fencing, scare devices, herding, land-use zoning and habitat developments. Based on these factors, the state followed a program of 11 recommended practices. Herds continued to increase during the late 1980s and early 1990s, as outlined in the plan. Much additional biological data were obtained, and greater public support was demonstrated for the state's harvest and translocation programs. To respond to enhanced public interest, the state developed and released a report stressing additional biological data and analyses of findings on harvest depredation problems and translocation endeavors (California Department of Fish and Game 1998). This environmental impact document did not change the original management plan, but it evaluated and updated findings to support recommended management techniques for pronghorn populations.

Minnesota Zoological Gardens

The development of pronghorn management plans for varying programs can be beneficial. This includes raising

For approximately 50 years (1930s to 1980s), endemic pronghorn were extirpated from the large central valley grasslands of California. Then, in 1987, 1988 and 1990, the California Department of Fish and Game translocated pronghorn from the northeastern shrubsteppes of the state to the grasslands of the Carrizo Plain in San Luis Obispo County. The herd increased slowly and, in 1999, it provided recreational opportunities for five hunters. Local persons, many of whom had never seen pronghorn before, made weekend trips to the area to view the pronghorn that returned after successful restoration practices to the historic grasslands of central California. Photo by P. L. Knapp.

animals for research or sustaining a healthy herd for exhibition. The following describes such a plan that has been in effect since the early 1980s.

Pronghorn have been exhibited at the Minnesota Zoological Gardens in Duluth, Minnesota, since the early 1980s. Original animals and those used for periodic augmentation were obtained primarily from Colorado and Wyoming.

In 1982, the zoo staff developed the first pronghorn management plan, and it has been updated periodically. The goal is to establish and maintain a stable population of pronghorn to satisfy exhibit needs (Minnesota Zoological Gardens 1993).

The management plan establishes desired sex and age classes for annual recruitment, augmentation and exhibition. Adults are considered surplus during their seventh year of life and none older than 8 years will be retained. Animals considered surplus due to age and population management, but otherwise healthy, will be offered gratis through surplus listings. If suitable placement cannot be identified, they will be euthanized.

Males and females will be selected for breeding each year, based on their historic contribution to the population, the needs of the zoo to minimize inbreeding and the specific requests for animals from other institutions. Requests from other institutions for animals will not be actively solicited. Fawns will be available once needs are satisfied.

All fawn recruitment is accomplished by hand-rearing methods identified in the pronghorn management plan. Experience has shown that, because they are less flighty and more docile during exhibition at the zoological gardens, hand-reared fawns are better adjusted to the presence of humans than are fawns raised by their dams.

Nevada

The Nevada Division of Wildlife has completed a series of species management plans, one of which is devoted to pronghorn (Tsukamoto 1983). Past management methods are presented and compared with present procedures. Also included are detailed discussions of management goals, problems and strategies. Major sections highlight plan objectives, historic background, life history, habitat and species inventories, current and projected resource uses, economic values, and management guides. Included are 15 maps that delineate seasonal-use areas of pronghorn throughout the state. The plan is intended to aid management decisions for the agency and its commissioners and to improve understanding by hunters and the public.

Pronghorn management in Nevada strives to protect, maintain or increase the resource for the enjoyment and use now and in the future. The goal of the plan is to achieve a long-term, statewide level of around 24,000 pronghorn in the next two or more decades. To meet this goal, the following annual objectives are listed for pronghorn and habitat:

1. manage the pronghorn resource to achieve and maintain established population goal levels by a management area or unit;
2. maintain annual survey and inventory efforts to determine population status and trends;
3. monitor pronghorn habitats to assure continued preservation and enhancement;
4. maintain a program that includes both consumptive and nonconsumptive activities.

Most pronghorn in Nevada occupy shrubsteppes, as depicted in the above scene of low sagebrush, Spring Valley, White Pine County. Populations appear to have high productivity, but also experience high mortality. As a result, densities are low compared with those of the Great Plains grasslands. Photo by Jim D. Yoakum.

Details of problems and strategies to accomplish these objectives are provided in the plan.

Information regarding the values of nonconsumptive use of pronghorn were provided. Nonconsumptive use includes scientific study as well as aesthetic appreciation of the animal through observation. Woody (1968) investigated the preference ratings of select wildlife species that nonconsumptive recreationists would enjoy and reported that pronghorn occupied the first and second positions in the self-rating questionnaire that compared preference ratings for 11 common wildlife species in Nevada (e.g., mule deer, bighorn sheep, sage grouse, and mountain lion).

During 2002 and 2003, the Nevada Division of Wildlife appointed a committee to update the pronghorn management species plan. New information was added regarding herd translocations, drinking water developments and predator control for fawn recruitment (G. Tanner personal communication:2003).

Saskatchewan

Throughout the middle of the 20th century, pronghorn populations in Saskatchewan remained between 8,000 and 12,000 animals. However, during the 1970s and 1980s, herds steadily increased due to improved survival during a series of mild winters, hunter selection for bucks and a management philosophy that promoted population growth rather than control. The 1980s also witnessed a decline in the economic welfare of farmers and ranchers due to a series of droughts, poor markets and increased operational costs. Demands from the agricultural community to reduce pronghorn numbers and an increasing number of depredation claims for government-sponsored compensation were the motivational factors behind the need to develop a long-term pronghorn management plan (Killaby et al. 1992).

Consequently, a 137-page management plan, entitled *Pronghorn Antelope in Saskatchewan: Status and Management Strategies*, was developed in the early 1990s (Killaby et al. 1992) (Figure 110). Objectives and management implementation strategies were developed for a provincewide document. A cornerstone of this program was the acceptance of lower pronghorn population numbers and the adoption of changing hunting programs (e.g., trophy buck harvests were accelerated and nontrophy seasons were initiated). In addition, socioeconomic costs and benefits were examined to ascertain the net worth of the pronghorn resource.

In response to increased interest in improving pronghorn management, new, improved population survey techniques were coordinated with vegetation categories. Survey sites containing 90% to 100% native vegetation sustained higher densities of animals than sites with agriculture croplands. In survey areas of predominately developed agriculture lands, pronghorn continued to select vegetation sites that had remnant native vegetation while avoiding entirely cultivated lands.

I. Population and harvest
  A. Objective
    1. Determine 1991 provincial postseason population of 15,000 pronghorn
      a. postseason herd structure ratio of 25 bucks/100 does/80 kids
    2. Manage population maintenance or growth in fringe herds
      a. no season or, if population warrants, either-sex season, only
  B. Strategy
    1. Create a nontrophy pronghorn season
      a. definition of nontrophy animal: A female or male pronghorn having horns not more than 3 inches (7.6 cm) in length.
      b. either-sex pronghorn license holder may purchase nontrophy license(s)
      c. hunter may purchase as many as three nontrophy licenses
      d. price per license no greater than $20.00
      e. nontrophy season dates coincide with the either-sex pronghorn season dates
      f. nontrophy pronghorn season dates overlap other big game seasons, such as mule deer and white-tailed deer
      g. nontrophy license sales are available after the results of the Big Game Draw
      h. telephone sales coordinated from Regina office
      i. surplus licenses sold over-the-counter at district offices in southwest Saskatchewan
      j. no priority pool–first come, first-served
II. Population Monitoring
  A. Objective
    1. Precise, cost-efficient system of monitoring pronghorn populations
  B. Strategy
    1. Future management strategies to assist population modeling
      a. population models for each wildlife management zone
    2. Line transect survey blocks used to estimate population sizes and to develop each wildlife management zone
      a. flown at three-year intervals, preferably in late May and early June, but at least before the hunting season
    3. Herd classification surveys flown annually
      a. surveys conducted in September
    4. Reestablish check stations for pronghorn
III. Communications Strategy
  A. Objective
    1. Inform public of change in provincial pronghorn statement
  B. Strategy
    1. Inform hunters of changes in pronghorn management through Saskatchewan Natural Resources information sources and the press
    2. Publish pamphlet on pronghorn management objectives for free distribution to landowners, hunters and nonconsumptive users.

Figure 110. *Summary of objectives and strategies for the Saskatchewan Pronghorn Management Plan (from Killaby et al. 1992).*

U.S. Fish and Wildlife Service

Two national wildlife refuges were established during the 1930s to perpetuate pronghorn in the Great Basin—the Hart Mountain National Antelope Refuge (Hart Mountain), in south-central Oregon, and the Sheldon National Wildlife Refuge (Sheldon), in northwestern Nevada (Figure 111). Pronghorn herds intermingle seasonally between refuges, with higher numbers generally on Hart Mountain during summers and greater numbers on Sheldon during winters. Hart Mountain contains approximately 217,500 acres (88,000 ha) of shrubsteppe, whereas Sheldon has around 299,000 acres (121,000 ha).

The first pronghorn management plan for these refuges was drafted in 1990, with an objective of managing herds and habitats (Pyle and Smith 1990). Population trends were examined, habitat requirements were assessed, effects of herd and rangeland management were reviewed and rationale for general management practices were documented. The plan was peer-reviewed by government agencies and wildlife consultants. Progress reports were presented subsequently to the Interstate Antelope Conference (Pyle and Reiswig 1994) and the Pronghorn Workshop (Pyle and Yoakum 1994).

After several years of examining and assessing ecological data for the draft pronghorn management plan, the following principal biological conclusions and management strategies were reported (Pyle and Smith 1990):

1. Long-term productivity, recruitment and population size of pronghorn on native rangelands in the northern Great Basin are regulated primarily by availability of forbs and drinking water during spring through autumn, and secondarily by mortality associated with severe winters.

2. Management practices that maintain drinking water and sustain or increase forb production will sustain high levels of pronghorn productivity and reduce variability in recruitment.

3. Fire and climate were two primary historic factors that regulated vegetation on pronghorn habitats.

4. Prescribed fire is the primary practice recommended to enhance endemic forbs and regulate shrubs on pronghorn landscapes.

Biological data and management recommendations in the pronghorn management plan were assessed, and much of the data were included in an environmental impact statement and 1994 comprehensive management plan for Hart Mountain (U.S. Fish and Wildlife Service 1994). Such information was not readily available for certain other wildlife species lacking a management plan. Before the 1994 comprehensive management plan was compiled, livestock grazing was the primary management practice to improve the health of vegetation (U.S. Fish and Wildlife Service 1970, 1980). However, after drafting the 1990 pronghorn management plan and the 1994 comprehensive management plan, refuge management strategies were changed to a 15-year program of no domestic livestock grazing and increased prescribed burning.

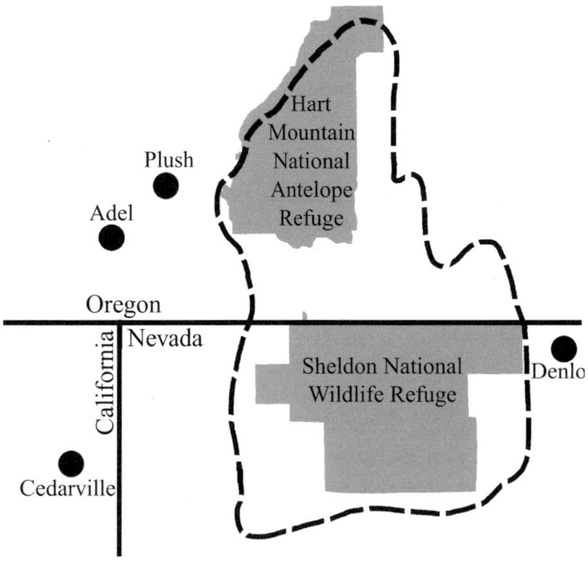

FIGURE 111. *Location of the Hart Mountain National Antelope Refuge in Oregon and the Sheldon National Wildlife Refuge in Nevada. The dashed line delineates the boundary of the Biological Management Unit for pronghorn using both refuges. Herds often summer on the Hart Mountain National Antelope Refuge, which has a higher elevation and sustains greater forage diversity and succulence. And, they winter at the lower elevations (with lower snow depths during severe winters) on the Sheldon National Wildlife Refuge.*

Management plans serve not only to document what is known about wildlife, but they assist in identifying missing information needed to implement effective management strategies. Such a reassessment was identified early in the development of the pronghorn management plan when it became apparent that information was lacking relative to geographic areas and vegetation types used by pronghorn during various seasons. Aerial transacts were systematically flown monthly to survey pronghorn-use patterns on both refuges. Results disclosed that, although herds used a variety of vegetation types annually, low sagebrush communities received the highest (79%) use during all seasons and years (Table 131; Figure 112). Intermittent lakebeds, productive in vernal forbs, were intensively used during summer and autumn. Recently burned shrubsteppes (converted to seral grasslands) were used greatly during summer, autumn and mild winters when herds remained at higher elevations. Most pronghorn wintered in low sagebrush types, but

TABLE 131. Percentage of pronghorn observed by season and vegetative cover type, Hart Mountain National Antelope Refuge, Oregon, winter 1990 through autumn 1991

Vegetative cover type	*Winter*	*Spring*	*Summer*	*Autumn*	*Total*
Basin big sagebrush	—	—	+[a]	—	+
Wyoming big sagebrush	—	1	1	+	+
Low sagebrush	99	90	62	74	79
Silver sagebrush	—	—	+	7	2
Lakebed	—	+	29	7	11
Grassland[b]	1	9	6	7	5
Meadow	—	—	1	3	1
Mountain brush[c]	—	+	+	—	+
Juniper	—	—	+	1	+
Other[d]	—	—	—	+	+

[a] + equals less than 1%.

[b] Area burned by August 1985 range fire (11,000 acres, 44,520 ha).

[c] Areas that support growth of sagebrush, bitterbrush and mahogany.

[d] Observations combined for several other types.

Source: Pyle 1992.

salt desert shrub was the dominant vegetation community occupied during severe winters. Such quantified data concerning pronghorn/vegetation relationships by season helped to identify areas where vegetation in poor health could be enhanced to benefit pronghorn. Consequently, more than 29,000 acres (11,736 ha) of dominant, decadent sagebrush communities have been burned to increase the availability of forbs during the fawning period (A. Goheen personal communication:2000).

A second product of these aerial surveys was the location of critical fawning areas. Locations were recorded for 270 does on Sheldon and 151 does on Hart Mountain during a survey conducted at the height of the 1993 fawning season. In general, distribution of does corresponded to the distribution of low sagebrush types, intermittent drainages and playas. It appears that wet winters and springs improve availability of forbs in low sagebrush communities, thereby fostering dispersion of does. Results of these surveys indicate that wildlife managers should periodically survey does to ascertain their use of geographic areas and vegetation cover types and to compare patterns of use among years with different population sizes and climatic conditions (Pyle and Yoakum 1994).

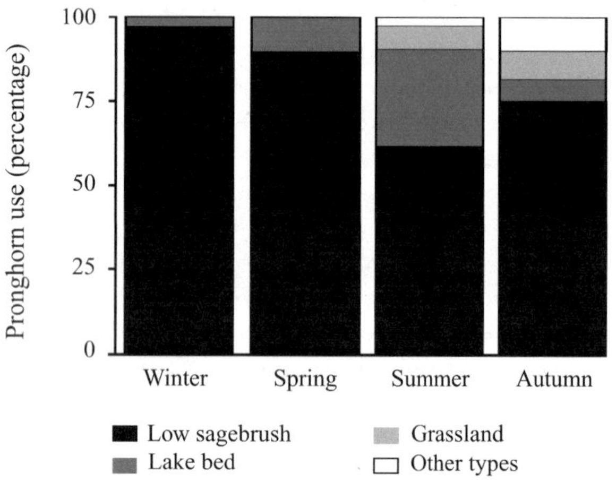

FIGURE 112. *Seasonal use of vegetative cover types by pronghorn on the Hart Mountain National Antelope Refuge during 1990 and 1991 (Pyle 1992).*

These aerial flights also identified an area south of Hart Mountain that had one of the highest pronghorn densities during seasonal movements and the fawning season—two criteria for designating critical habitats. Armed with this quantitative information, the U.S. Fish and Wildlife Service (FWS) was able to justify a land exchange with the U.S. Bureau of Land Management (BLM) for 10,000 acres (4,050 ha) adjacent to the refuge (W. Pyle personal communication:1999). The addition of these lands to the wildlife refuge system increased abilities to manage crucial habitats needed to sustain pronghorn year-round.

A fourth subject that was inadequately known at the initiation of the pronghorn management plan was foraging habits of pronghorn (seasonally and during years of differing precipitation patterns). Also, little knowledge was available regarding forage competition and preference ratios with wild, domestic and feral ungulates concurrently using rangelands on the two refuges. Therefore, a long-term food habit study was commenced to ascertain the composition and quality of diets for bighorn, burros, feral horses, mule deer and pronghorn. Procedures and results of this study are reported in Chapter 14.

Pronghorn herds on the Hart Mountain National Antelope Refuge in Lake County, Oregon, live more than 80% of their lives in low sagebrush communities. These shrubsteppes contain the two major habitat characteristics needed for pronghorn in Great Basin environments: vegetation averaging heights of up to 25 inches (63 cm) and plant composition of approximately one-third each of grasses, forbs and shrubs. Such rangelands provide forage and cover during the critical seasons of reproduction and winter survival. Photo by Jim D. Yoakum.

The FWS's pronghorn management plan served as a model for other wildlife species, according to the Sheldon Scientific Working Group's report (Davis 1995). The group consisted of 12 natural resource scientists charged with evaluating past resource management programs and making recommendations for future management on the Sheldon.

Since the 1990 draft of a pronghorn management plan, it has been a readily available and guiding management document. The plan has been of special value because agency personnel change frequently, and new employees often were hired with little or no experience relative to pronghorn or habitat management in the Great Basin. Herein lies an attribute of management plans not often recognized.

Now, after a decade of use, the pronghorn management plan is being modified and updated (Dunbar 1999b). The new plan will be adaptive and apply only to Hart Mountain. Population objectives will not be inflexible, but will fluctuate based on scientific data. The plan will stress factors, such as predation, hunting harvest levels, nonconsumptive recreational uses, water developments, vegetation quality and road management.

The original pronghorn management plan identified the need to restore habitat quality and specified two major practices to meet this goal—elimination of livestock competition for herbaceous forage and the return of prescribed fire to decrease dominant stands of shrubs, thus rejuvenating herbaceous plants. Research studies indicate that the production of herbaceous vegetation increased more than 200% on certain pronghorn habitats within 6 years (Crawford and Coggins 1997). Mean pronghorn numbers during the 1990s were the highest recorded for any decade in the past 50 years—an attribute that may be due, in part, to an effective pronghorn management plan. The increased trend in higher population levels continues for the early 21st century, with the highest number (2,427) observed in July 2003) for any population survey conducted since surveys were initiated in the 1940s (M. Nunn personal communication:2003).

HABITAT MANAGEMENT PLANS

Land management agencies, such as the U.S. Forest Service (FS) and the BLM, are mandated by Congress to apply multiple-use practices to vast landscapes. Therefore, they are responsible for managing rangelands not only for pronghorn but for a variety of uses, such as forestry, watershed, recreation, livestock grazing, energy and other wildlife. Such a mandate is challenging because society increasingly demands high resource production and a high-quality environment. To accommodate this mandate, government agencies often develop several types of management plans relative to pronghorn. One is a land-use plan (also termed a multiple-use plan, a comprehensive resource plan or a coordinated resource plan) covering the goals and objectives for management of various resource activities for a particular land area. Emphasis is placed on coordinating the needs and problems of all activities into one document. The detail devoted to one species, such as pronghorn, is limited, but the overall objectives of pronghorn habitat management are included and coordinated with other resource demands.

The second procedure is a habitat management plan (HMP), specifically designed to address wildlife needs. Sometimes the plan includes all wildlife in a management area. In other cases, the management plan may deal with one species. The BLM (1981) prepared a manual for developing HMPs, including the following outline of recommended components:

1. Cover
2. Title page
3. Abstract
4. Table of contents
5. Location map
6. Checklist for preparation and review
7. Introduction
8. Management objectives
9. Planned actions
10. Evaluation and monitoring
11. Habitat management plan progress reports
12. Coordination with other programs, agencies and organizations
13. Wildlife economics
14. Costs and funding needs
15. Concurrence and approval
16. Appendix

When properly developed, an HMP becomes a comprehensive plan for an area and identifies actions to be implemented to achieve specific habitat management objectives (Figure 113). It requires coordination, input, review and concurrence from cooperating agencies and organizations. As of 1993, the BLM had completed 253 HMPs, of which 16 documented habitat needs for pronghorn (U.S. Bureau of Land Management 1993).

The HMPs specific to pronghorn appear to be few (Table 130). Fewer have been assessed and reported in technical periodicals for other managers to use. Consequently, several HMPs developed and reported for pronghorn are discussed below.

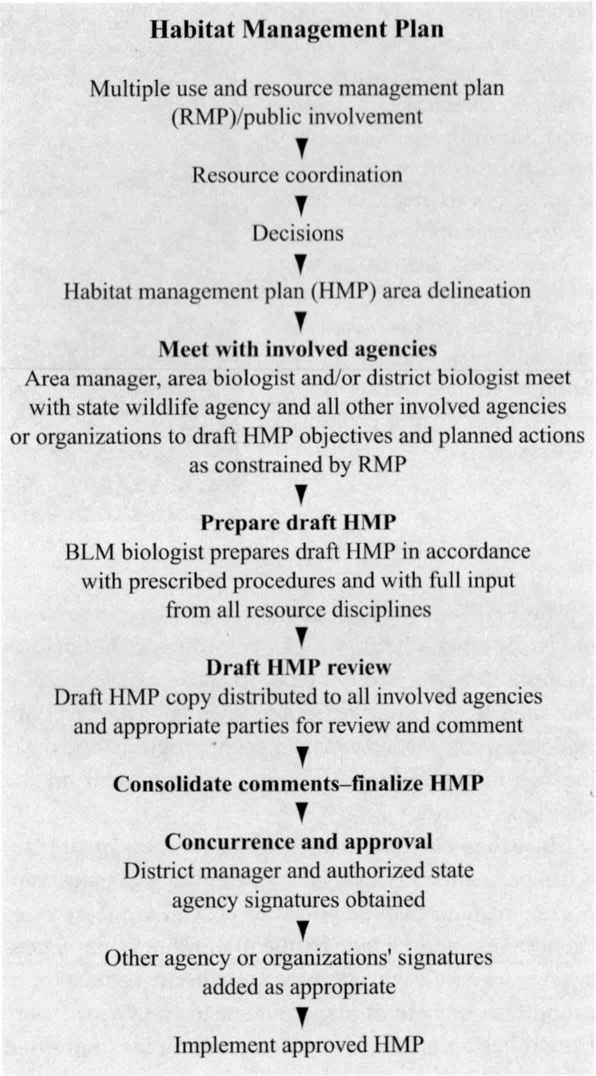

Figure 113. *Flow chart in a U.S. Bureau of Land Management (BLM) manual for preparing a pronghorn habitat management plan. It was used by Capodice et al. (1982), McCarty (1982a) and Lister (1989).*

Modoc National Forest, California

One of the earliest known HMPs for pronghorn was prepared for the Modoc National Forest in northeastern California (Stutz 1967). It was a modest plan, but a credible venture for that time. The contents included introduction, pronghorn population and habitat data, management needs and procedures and actions. The plan stressed cooperative responsibilities and investments for state and federal agencies. It also provided details for construction of barbed-wire fences to control livestock, yet facilitate pronghorn movements—a highly volatile issue for resource managers during the mid-1960s. A large map outlined summer and critical winter rangelands, plus 17 traditional fawning areas.

Adobe Valley, California

When the California Department of Fish and Game began to explore the possibility of translocating pronghorn to Mono County, an environmental impact study of five potential release sites (McCarthy and Yoakum 1984) and an HMP (Capodice et al. 1982) were completed. The HMP provided for herd and habitat management direction for Adobe Valley, but did not preclude the herd's possible movement to adjacent areas. Its contents included introduction, management objectives, planning actions, coordination with other resources, implementation schedule and cost estimates. Appendices included agency memoranda, understandings and an environmental analysis.

Hundreds of pronghorn translocations have been completed during the past 50 years, but few were preceded with feasibility analyses. In the Adobe Valley case, both an environmental impact assessment and an HMP were prepared. They were costly in terms of manpower and money, but deemed necessary because the translocation site ap-

During the planning process before the translocation of pronghorn to Adobe Valley, California, a superabundance of tall, dense shrubs was recognized. A prescribed burn changed the dominant shrublands to an early successional plant community, resulting in a mixture of grasses, forbs and low-growing shrubs beneficial to pronghorn. It likewise created improved forage for livestock on multiple-use, public rangelands. Forage enhancement projects best serve pronghorn translocation efforts when implemented several years before herd releases. Photo by Terry Russi; courtesy of the U.S. Bureau of Land Management, Bishop, California.

peared to be marginal pronghorn habitat. To evaluate the success of the project, a graduate student was assigned a 2-year research study (Goldsmith 1988, see Chapter 13).

The Adobe Valley HMP rated the area as fair quality pronghorn habitat. However, because surplus pronghorn causing agriculture depredations were readily available, 117 animals were translocated from Modoc to Mono counties from 1982 to 1985. The animals scattered quickly after release, and few remained in Adobe Valley. Fawns were produced but survival was low, due, in part, to a large predator base. Within 4 years, the majority of the herd moved southeastward 40 to 60 miles (64–96 km) to lower elevations in Queen and Hammil Valleys—sites rated as poor pronghorn habitat during the environmental assessment (Fatooh et al. 1994).

Seventeen years after initial release, the herd moved entirely out of California to adjacent rangelands in Nevada. By 1999, the herd numbered between 70 and 100 animals (T. Russi personal communication:1999). The HMP predicted that the release would be successful if herds increased to 200 animals within 10 years, but 17 years later the herd remained less than the original 117 animals translocated.

Little Lost/Birch Creek, Idaho

An HMP was completed for 526,500 acres (213,069 ha) in Idaho, where approximately 6,000 pronghorn resided (McCarty 1982a, 1982b). The plan identified critical rangelands and forage preferences of livestock and pronghorn, and it listed problems and conflicts with other resource activities. Its goals included maintaining desirable habitat characteristics in conjunction with other resource activities, enhancing pronghorn habitats by developing deficient components and reducing hazards. A strong monitoring program also was established.

The plan is noteworthy in its documentation of specifications for restricting livestock grazing on pronghorn fawning areas, limiting public land disposals and evaluating rangeland improvements that benefit pronghorn. The latter is the most extensive, including 94 different projects to improve habitat conditions specifically for pronghorn—45 water developments, 4 vegetation manipulation areas, 44 rangeland fence modifications and a unique project to curtail pronghorn drownings in an irrigation flume. Each project contained details of site location, estimated cost, implementation schedule, maintenance responsibilities and recommended implementation priority (Table 132).

North Eccles, Nevada

During the mid-1980s, the Nevada Division of Wildlife recommended augmenting a small pronghorn herd in the North Eccles area of northeastern Nevada. The herd mainly occupied rangelands administered by BLM. Therefore, the agency developed a cooperative HMP (Lister 1989). However, the plan was challenged by a landowner, resulting in litigation lasting more than 5 years. During this delay in initiating the plan and following implementation in 1989, a number of wild and prescribed rangeland fires occurred. In addition, 40 to 50 miles (64–80 km) of fences were modified to allow greater pronghorn mobility.

The changes in vegetation due to fires, together with favorable precipitation, resulted in increased herbaceous forage for pronghorn. Apparently as a consequence of these fires, precipitation patterns and fence modifications, the population more than doubled within 10 years. The herd thrived and no longer needed augmentation with translocated animals (R. Lister personal communication:1998).

RECOVERY PLANS

When the United States Congress passed the Endangered Species Act of 1973, one stipulation was the development and implementation of actions for the recovery of listed species. With time, these endeavors became commonly known as recovery plans. Similar plans are being prepared in Mexico for endangered pronghorn. Pronghorn

Table 132. Implementation schedule for a partial listing of habitat improvement projects in the Pronghorn Habitat Management Plan for Little Lost/Birch Creek area of Idaho

Allotment	*Project number*	*Project name*	*Unit*	*Number of units*	*Priority*	*Initial cost*	*Development work-month*	*Maintenance cooperators (funding and labor)*
Pass Creek	4358	Buck Springs pipeline	[a]	2	3	$1,200	0.5	Annual inspection and filling
	4357	Chicken Creek pipeline	[a]	2	3	$1,200	0.5	Annual inspection and filling
	0705	Wet Creek Allotment fence	Mile/km	12/19	2	$1,800	0.6	Livestock operator
	0682	Interdistrict Division fence	Mile/km	4.96/7.9	2	$744	0.2	Livestock operator
	4130	Pass Creek Allotment fence	Mile/km	3.5/5.6	2	$525	0.2	Livestock operator
	0669	Pass Creek fence	Mile/km	3/4.8	2	$450	0.2	Livestock operator
	4367	Big Wet fence	Mile/km	2.3/3.7	[b]			Livestock operator
	4389	Squaw Springs fence	Mile/km	1/1.6	[b]			Livestock operator
		Burn or chain	Acre/ha	1,000/ 404	1	$4,000 or $20,000		Bureau of Land Management
Wet Creek	0038	Clyde School Protective fence	Mile/km	8.5/13.6	2	$1,275	0.4	Livestock operator
	0668	Little Lost fence	Mile/km	4/6.4	2	$600	0.2	Livestock operator
8-Mile		Guzzler	[b]	1	3	$1,350	0.2	Clean every 3 years
Canyon	4018	8-Mile Canyon fence	Mile/km	2/3.2	2	$300	0.1	Livestock operator
Howe Creek		Guzzler	[b]	6	3	$8,100	1.4	Clean every 3 years
		Burn or chain	Acre/ha	6,000/ 2,428	3	$24,000 or $60,000		Department of Energy
Sinks		Guzzler	[b]	4	3	$5,400	0.9	Clean every 3 years
		Burn or chain	Acre/ha	5,000	3	$20,000 or $100,000		Department of Energy

[a]Valve, 500-gallon (1,892 L) tank, 1-acre (0.4 ha) fenced exclosure.
[b]1,500-gallon (5,678 L) tank, precipitation catchment, 1-acre (0.4 ha) fenced exclosure.
Source: From McCarty 1982a.

are not classified as endangered in Canada, so no recovery plans exist.

Recovery is the ultimate management objective of having a species officially listed. The FWS is responsible for the preparation of recovery plans for all wild animals and plants listed as threatened or endangered in the United States. Recovery plans identify the causes of a species' decline and provide programmatic guidelines to reverse the unfavorable status. The content of such plans depends on the complexity of threats or limitations to the species or population in question, its range and the number of agencies or countries concerned with recovery. Although legally nonbinding, recovery plans sometimes are widely disseminated and serve as catalysts for action by parties involved (Culbert and Blair 1989).

Mexico

There are three presumed subspecies of pronghorn (Mexican, Sonoran and peninsular) endemic to Mexico, and each is listed as endangered. A pronghorn management plan was developed for the state of Chihuahua. It included planned projects to modify livestock fences to allow greater pronghorn mobility, to capture and translocate animals to augment isolated herds, to hand-rear for semicaptive studies, and to implement predator control practices during fawning seasons (J. M. Reyes personal communication:1998). A recovery plan was initiated in 1994 for the peninsular pronghorn, and some work was drafted for the other subspecies. However, in 1996, the federal government of Mexico announced endeavors to make one recovery plan inclusive for all native pronghorn subspecies (Cancino et al. 2000). Efforts to accomplish this recovery plan follow.

Peninsular Pronghorn. The peninsular pronghorn is indigenous to Baja California, Mexico (see Chapter 1). Herds occupy rangelands at sea level—the only pronghorn population doing so during modern times. The subspecies was declared endangered by the government of Mexico (Diario

Oficial 1991) and the International Union for Conservation of Nature (1996).

The Mexican federal government and the state of Baja California Sur were concerned with the plight of this subspecies and conducted a restoration workshop during November 1994 in La Paz, Baja California Sur. Delegates, represented state and federal wildlife agencies in Mexico and the United States plus the International Union for Conservation of Nature, Species Survival Commission attended. The purpose of the workshop was to assess information on the life history, population dynamics and ecology of peninsular pronghorn. Also addressed was information on demography, genetics and environmental factors needed to analyze the status of the population and its risk of extinction under differing management scenarios and perceived threats. Information was assembled into a briefing book (Cancino 1994b) and much of the following was excerpted from that report.

One segment of the briefing book contained a recovery plan, entitled "Plan de Recuperacion del Berrendo Peninsular (*Antilocapra americana peninsular*)" (Cancino 1994a). Included were data on historic and present distribution, population estimates, habitat characteristics, mortality factors and a step-down outline for recommended management practices. The recovery plan stresses the need for translocating and sustaining populations in different unoccupied historic locations, thereby providing greater insurance against subspecies extinction through the impact of catastrophic events striking the single Vizcaino Desert population.

Two ecological investigations have been conducted for the benefit of peninsular pronghorn. First, FWS specialists, skilled in animal control practices, spent several weeks training Mexican wildlife personnel in techniques of trapping and poisoning coyotes. Although the program decreased coyote numbers, the project was terminated early, and no final analysis was accomplished.

The second project was a research study to investigate the annual diet of pronghorn (Cancino 1988). Average percentages of forage classes were 44% shrubs, 22% forbs, 4% grasses and 30% unidentified. A high preference was noted for flowers of forbs and shrubs in the unidentified category. Follow-up studies have been conducted to investigate nutritional values of forage plants (J. Cancino personal communication:2000). Such data could provide information regarding moisture content of vegetation and moisture deposits from seasonal heavy dew. These investigations also may provide findings regarding forage quality for this most southern pronghorn population.

National Pronghorn Recovery Plan. The government of Mexico announced at the 17th Pronghorn Workshop in 1996 that it was developing a national pronghorn recovery plan to include all endemic subspecies (Cancino et al. 2000). Although pronghorn have been protected since 1922, populations have experienced a continual decline (see Chapter 4).

The recovery plan was completed in January 2000; it is titled *Project for the Conservation, Management and Sustainable Use of the Pronghorn* (Antilocapra americana) *in Mexico* (Secretaria de Medio Ambiente, Recursos Naturales y Pesca 2000). All pronghorn in Mexico are classified as endangered. Estimated numbers for subspecies in 2000 were 750 Mexican, 200 peninsular and 250 Sonoran—totaling 1,200.

The conservation and management project includes a general description of each subspecies and its distribution, management practices used during the past 20 years, general and specific objectives, suggested future management strategies, and a listing of management projects presently being accomplished. The goal of the recovery plan is to increase populations sufficiently to remove all endemic subspecies from the protected list. Some of the proposed management strategies are translocating of herds from the United States to Mexico, hand-rearing of fawns for release into the wild, conducting population surveys with greater accuracy for annual trend comparisons, initiating telemetry studies to identify home ranges and translocating native herds to historic unoccupied habitats within Mexico. To date, more than 27 government agencies and private sources have been involved with management endeavors for pronghorn. Plans call for the appointment of a team of counselors, advisors, researchers and technicians to provide recommendations for management strategies and practices. The recovery plan identifies financial sources needed for implementing the plan (Cancino et al. 2000). Noticeably lacking in this effort is information relative to the past, present and potential needs to manage the habitat and other human uses of the critical environment on which the peninsular pronghorn depends—possible factors related to the subspecies becoming endangered.

The conservation and management project is an ambitious effort by Mexico to reverse a more than 70-year downward population trend for native pronghorn. It recognizes the importance of cooperation between state and federal governments in Mexico and the United States, plus cooperative funding and technical support from private and conservation organizations at the local, regional, national and international levels. There is little doubt that this management goal represents a major international conservation endeavor for a megafauna species during the 21st century in North America.

United States

Four subspecies of pronghorn are recognized in the United States; populations total close to 1 million animals (see Chapters 1 and 4). Of these, the Sonoran pronghorn in Arizona is the only presumed subspecies classified as endangered or threatened, and its numbers are less than a fraction of 1% of the total nationwide population. Following are recovery strategies developed for that subspecies.

Sonoran Pronghorn. Critical habitat, as defined in the Endangered Species Act, has not been designated for Sonoran pronghorn. Most habitat is under the administration of federal government agencies (e.g., Cabeza Prieta National Wildlife Refuge, Organ Pipe National Monument, U.S. public domain and military installations). As such, the subspecies' habitat is protected from development and other exploitation.

The validity of the scientific nomenclature for this subspecies (and others) has been subject to investigations during the 1980s and 1990s (see Chapters 1 and 8). Nowak expressed the opinion that, "Until someone publishes a thorough reassessment of appropriate series of specimens, and comes to different conclusion that can be accepted by the mammalogical community *sonoriensis* would have to stand" (U.S. Fish and Wildlife Service 1998a: Appendix A).

Work on a recovery plan for the Sonoran pronghorn commenced in the 1970s when the Arizona Department of Game and Fish conducted field investigations and research, culminating in a report of biological information (Arizona Department of Game and Fish 1981). Much of that data, plus additional information, became the framework for a recovery plan completed in 1982 through cooperation of state and federal government agencies (U.S. Fish and Wildlife Service 1982), including an implementation schedule listing management actions, agencies responsible for implementation, estimated costs and priorities.

The plan's objective was to sustain a population of 300 animals for a period of 5 years or more, or at a numerical level determined feasible for the habitat. The FWS believes that, when this goal has been achieved and major threats have been eliminated, the subspecies may be considered for delisting. However, problems impede this subspecies' recovery. Streams were diverted for human use or otherwise dewatered. Production of herbaceous vegetation deteriorated because of intensive livestock grazing, and shrubs increased to the detriment of the habitat requirements of pronghorn. The impacts of military activities on pronghorn and their habitat are equivocal.

Efforts to update the 1982 Sonoran Pronghorn Recovery Plan were initiated in 1991. New data were obtained for population dynamics to determine whether herds were increasing, decreasing or remaining relatively static, and restoration needs for vegetation and waters were identified (Thompson-Olais 1992). These efforts culminated in the *Final Revised Sonoran Pronghorn Recovery Plan* (U.S. Fish and Wildlife Service 1998a). The estimated population was fewer than 300 individuals. The Sonoran pronghorn remains on the United States' list of endangered species (U.S. Fish and Wildlife Service 1999b) and is in Appendix I of the *Convention on International Trade in Endangered Species of Wild Fauna and Flora* (U.S. Fish and Wildlife Service 1999a).

The objective of the 1998 recovery plan was to remove Sonoran pronghorn from the list of endangered species and downlist the subspecies to threatened status. Criteria for delisting was to establish an estimated population of 300 adults in one self-sustaining population for a minimum of 5 years, and establishing at least one other self-sustaining population.

The recovery plan recommended writing management plans for major activities. Following is a list of programs needed for delisting considerations.

1. Enhance extant populations by providing supplemental forage and water.
2. Determine habitat needs and protect present occupied habitat.
3. Investigate and address potential barriers to expansion of presently occupied habitat.
4. Investigate, evaluate and prioritize present and potential future reintroduction sites within historic range.
5. Establish and monitor new, separate herds to guard against catastrophic events that may contribute to decimating the core population.
6. Investigate captive breeding.
7. Continue monitoring populations.
8. Maintain a protocol for a repeatable and comparable survey technique to establish trends.
9. Examine additional evidence to assist in verification of taxonomic status.

Estimated costs to implement the plan depended on accomplishing various needs and varied from $490,000 to $4,310,000. Because some aspects of the life history of Sonoran pronghorn are not known, a delisting date was not projected. The intent was to establish a 7-year plan with annual updates, rather than a new plan or major revision to maintain an up-to-date recovery plan.

The validity of the Sonoran pronghorn as a subspecies has been questioned (Arizona Department of Game and Fish 1981) and assessed (Paradiso and Nowak 1971, U.S. Fish and Wildlife Service 1982, 1998a). Further verification of taxonomic subspecies status is needed. However, the pronghorn population on the Sonoran Desert may need continued protection under the Isolated Vertebrate Population Act (Federal Register 1996) because the population appears to meet the requirements of being an isolated and distinct vertebrate population. This subspecies occupies an unusual habitat in the United States and appears to have specific adaptations to its environment—two of the three tests for an evolutionarily significant unit.

Limited biological data for Sonoran pronghorn in Mexico is included in the 1998 recovery plan for Arizona; however, management goals are not established because they are under the jurisdiction of the government of Mexico. It is apparent that a considerable degree of cooperation and coordination was accomplished by representatives of wildlife agencies in Mexico and the United States.

Cooperators for managing Sonoran pronghorn in Mexico and the United States convened a meeting in 1998 to assess population dynamics using a simulation model. The VORTEX model was chosen because it incorporated random variation in climatic and population performance that was similar to processes perceived in the natural environment (see Chapter 19). Available ecological data on the Sonoran pronghorn was used to develop likely scenarios. Where data were lacking, information were used from surrogate subspecies living in similar habitats. Using a modification of the Delphi method to determine values for variables required in VORTEX, input parameters were modeled to evaluate the probability of Sonoran pronghorn persistence under these population and life history values. The minimum viable population has a probability of persistence greater than 0.95 for a period of 100 years (deVos and Thompson-Olais 2000).

Restoration and Management Implications

An assessment of recovery plans for the Sonoran and peninsular pronghorn discloses major reviews of biological data for species management, but a paucity of information pertaining to habitat characteristics and restoration practices. The plans assert that habitats have been modified since Euro-American arrival. Apparently, vegetation now differs considerably from conditions when greater densities of pronghorn were first reported in the respective deserts. However, few specifics are provided to restore vegetative communities now in poor health (see Chapter 17; see also Payne and Bryant 1994, Roundy and Biedenbender 1995, Krausman 1996).

A basic tenet of evaluating the status of endangered species (including pronghorn) is to analyze the extant habitat characteristics relative to past and present population levels. Should the habitat, especially vegetation, have been severely changed and now fail to meet the requirements of pronghorn, application of intensive species manipulation practices will have limited success until the habitat is restored to a healthy status. Early accounts indicated greater abundance of herbaceous plants and lower densities of shrubs than occur today (Hastings and Turner 1980, Bahr 1995).

Herbaceous plants, more than shrubs, provide essential nutrients for the critical pronghorn reproduction period. Poor herbaceous production is related to poor fawn recruitment and survival. Herbaceous plant production is limited in the desert biome because of ecological factors. Consequently, deserts are marginal pronghorn habitat. If the goal of management is to produce optimum numbers of pronghorn on a desert, it is imperative that managers

The Sonoran pronghorn is classified as the only subspecies of pronghorn endangered in the United States. Adjacent herds in the state of Sonora, Mexico, also are listed as endangered. This subspecies lives in parts of the Sonoran Desert, the hottest desert in North America. Annual precipitation averages less than 5 inches (12.7 cm). Consequently, the vegetative cover is predominately shrubs with scant production of herbaceous vegetation during most years. Desert landscapes that produce low quantities of herbaceous vegetation for forage and water requirements likewise produce low densities of pronghorn. Photo by Bob Miles; courtesy of the Arizona Department of Game and Fish, Phoenix.

identify and implement practices that assure production of herbaceous vegetation. Deserts with limited herbaceous vegetation will produce few pronghorn. Due to harsh ecological factors, recovery of vegetation in deserts is not easy or quick, but techniques for improvement can be applied and may be required over a long period of time.

At the beginning of the 21st century, all pronghorn classified as endangered inhabited deserts. Therefore, it behooves wildlife managers concerned with restoring endangered pronghorn to investigate historic and extant habitat conditions and to ascertain if a major objective of management should be to restore habitat conditions before initiating intensive species management practices.

MANAGEMENT PLANS ON PRIVATE LANDS

Recent trends in California may indicate how management plans will play a greater role in pronghorn management. California adopted a private lands management program in the early 1980s, called Ranching For Wildlife. It was to protect and improve wildlife habitat by encouraging private landowners to manage their property for the benefit of fish and wildlife. The program was received readily by private landowners, but met mixed acceptance from the public. Hunters were accustomed to hunting where and when they chose, and they expressed concerns that the program might lead to commercialization of traditional hunting and fishing privileges. However, after the program had been in effect for almost 15 years, the number of ranches continued to increase, and the program appeared to be more accepted by the public (California Department of Fish and Game 1994b).

Economic incentives were provided through biologically sound yet flexible game seasons, resulting in high-quality hunting opportunities that could be marketed by landowners (California Department of Fish and Game 2000). The program involved 65 licensed operators during 1999, 15 included pronghorn as a featured species (Figure 114). The California Department of Fish and Game can authorize hunting seasons specific to program areas, pursuant to goals and objectives in the California Pronghorn Antelope Management Plan and the appropriate ranch management plans. In addition, hunters desiring to hunt a buck pronghorn on a program area are not subject to the 10-year waiting period prescribed for a buck hunter successful on public lands. Also, hunters may obtain a tag for more than one program area annually. A resident hunter (who has applied for but not received a tag to hunt on public lands for many years) now has other alternatives to hunt pronghorn in California.

To become licensed for the program, a landholder submits an application, including a comprehensive management plan and a nonrefundable processing fee designed to meet costs of program review. California Department of Fish and Game personnel review the management plan to ensure that proposed habitat enhancements benefit pronghorn and that harvest strategies comply with accepted goals in the statewide pronghorn management plan. Once approved, a license is issued for 5 years but is reviewed annually.

The program tag allocation may not exceed 10% of the total allowable harvest for public lands. Table 133 provides a summary of tags allowed and animals harvested from 1990 to 1999.

The program appears successful as an incentive for landowners to protect or improve wildlife habitat. Habitat improvements implemented under management plans for pronghorn include prescribed fires to improve forage and vegetative conditions, reduced or deferred livestock grazing to lessen forage competition, drinking water improvements; and planting of preferred forage (especially forbs and shrubs). These projects are designed to benefit a wide variety of nongame, threatened or endangered

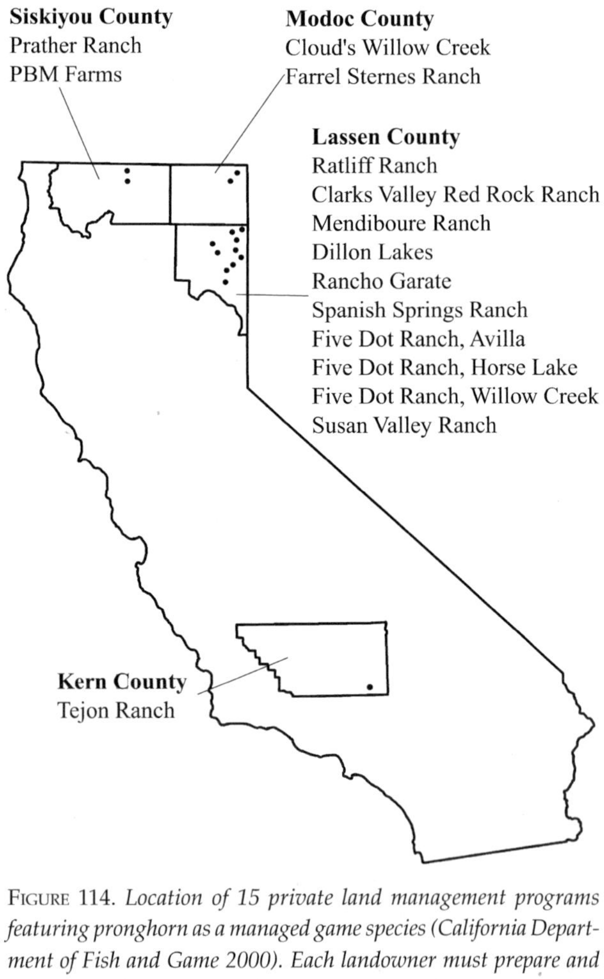

Figure 114. *Location of 15 private land management programs featuring pronghorn as a managed game species (California Department of Fish and Game 2000). Each landowner must prepare and implement a habitat and wildlife management plan that is satisfactory to personnel of the California Department of Fish and Game.*

TABLE 133. Pronghorn tag allocation and harvest on private land management areas in California, 1990 to 1999

Year	*1990*	*1991*	*1992*	*1993*	*1994*	*1995*	*1996*	*1997*	*1998*	*1999*
Tags allocated	27	40	70	27	32	40	25	41	30	23
Tags sold[a]	27	40	70	25	32	40	25	41	30	23
Harvest	15	26	47	23	26	28	17	30	20	15
Percentage harvested	55	65	67	92	84	70	68	73	67	65

[a] Tags authorized are considered sold though the operator may not issue the tag to a hunter.
Source: From California Department of Fish and Game 2000.

species as well as pronghorn. California Department of Fish and Game personnel evaluate habitat improvement proposals during the management plan review before license approval. The California Fish and Game Commission also reviews all program management plans before final approval.

The program has been a model for similar programs in other states. A survey of western provincial and state agencies conducted during 2000 inquired as to the present status of Ranching For Wildlife programs. Three states reported established programs. California had 12 ranches, Colorado had 15 (see Chapter 23) and Utah had 6 (see Chapter 16). The Colorado Wildlife Ranching Program, like the one in California, owes much of its success to well-prepared management plans. The Deseret Ranch Management Plan in Utah, described in Chapter 17, is a model of how management of healthy herds and habitat can result in increased pronghorn numbers.

ENVIRONMENTAL IMPACT STATEMENTS

The effectiveness of public land management agencies' implementation of recommended management guides or strategies to sustain and enhance pronghorn and their habitats can be evaluated by reviewing contemporary land management programs. The following is an assessment of two such endeavors accomplished during the 1990s.

The FWS (1994) completed an environmental impact statement (EIS) and a comprehensive management plan (CMP) for the 270,000-acre (109,270 ha) Hart Mountain in south-central Oregon. The refuge was established during the 1930s to perpetuate pronghorn and other endemic wildlife in the Great Basin. The EIS and CMP documented the past and present diversity of natural resources and the many human uses and values of the land. This was the refuge's first comprehensive effort in 65 years to document and analyze natural resource values and uses by the public.

Data on pronghorn management issues were well documented. Vegetation on more than 90%of pronghorn range was unhealthy because of (1) excessive livestock and feral horse grazing of herbaceous vegetation and (2) a half century of aggressive wildfire suppression. Both of these practices resulted in plant succession leading to a dominant shrubland with little herbaceous vegetation—factors not conducive to meeting the habitat requirements of pronghorn. Consequently, pronghorn populations were as low as 200 to 300 animals for the entire refuge during the early 1950s.

Upon completion and initiation of the EIS and CMP in 1994, cattle and feral horses were removed and a long-term program of prescribed burns was implemented, with an objective of restoring vegetative communities to a mixture of native grasses, forbs and shrubs—vegetative characteristics associated with quality pronghorn habitats in the Great Basin. Apparently, the management planning effort was successful; pronghorn numbers increased to 2,000 animals during the 1990s—a tremendous increase from the 200 to 300 animals reported in the early 1950s. Consequently, this planning endeavor by a federal government agency can be recognized as a successful management program to improve pronghorn populations in conjunction with other natural resources and public uses of public lands.

Another draft EIS was accomplished in 1999 to identify natural resources and public use demands for more than 2,500,000 acres (1,011,750 ha) of national grasslands and forests in Nebraska, North Dakota, South Dakota and Wyoming (U.S. Forest Service 1999). These public lands contain large acreages of prairie grasslands in natural condition—areas intensively used by thousands of pronghorn.

Commenting on the EIS, the North American Pronghorn Foundation noted that it was woefully inadequate in identifying the needs of pronghorn. For example:

1. Crucial areas for pronghorn (i.e., fawning areas, winter rangelands and movement corridors) were not documented for consideration in management decisions to sustain or enhance the species.
2. Vegetative diversity of significance to pronghorn (i.e., forbs and shrubs) was not sufficiently addressed, nor is a clear plan outlined to insure adequate incorporation of these vital plant community classes in restoration programs.

3. Fences built to control livestock needed to be constructed to meet multiple uses; nonrecognition of this management practice could lead to pronghorn mortality on public lands.
4. Long-term rangeland management goals to enhance pronghorn habitat were not identified or documented clearly, nor were the peripheral impacts of other activities, such as livestock grazing, mining and energy development delineated to provide a basis for mitigating adverse effects in future years.

Pronghorn populations on the Great Plains experienced more than a 33% decrease during 1995 to 1997. Such losses could be attributed to inclement weather, predation and other decimating factors. However, the lack of nutritious, succulent forage, impacts of fences, construction of potential road barriers and other human-induced detrimental influences are responsible for insidious mortality that can be alleviated or decreased with effective land management plans on public lands.

MANAGEMENT HANDBOOKS, MANUALS AND GUIDES

Several guides helpful to pronghorn managers were published during the last two decades of the 20th century. Following is an annotated description of each.

Texas Management Handbook

The Texas Parks and Wildlife Department has used a handbook of management guides aimed at wildlife managers and ranchers. More than 90% of all pronghorn in Texas exist on private lands. The handbook includes chapters featuring information on life history, herd distribution, herd status, seasonal movements, fencing, food habits, pronghorn/livestock relationships, age determination, parasites, diseases, mortality factors and management practices (Hailey 1979). Hailey stated that research produced a wealth of information regarding management techniques, and these data have aided in dispelling misconceptions about pronghorn habitat requirements and competition with livestock.

The management handbook provides recommended practices to sustain and enhance pronghorn populations and habitat. For example, it states that, "Land owners should develop a management plan and conduct their harvest along specific guidelines" (Hailey 1979:47). The handbook reports the landowner is a manager of pronghorn residing on private property and lists the following factors as essential to proper management of herds and habitat.

1. Surveys—it is necessary to know the total pronghorn population, buck/doe ratio and annual fawn recruitment to make accurate recommendations for harvest levels.
2. Harvest—the proper harvest of bucks is essential to prevent overharvest and maintain the maximum number of bucks acceptable to hunters. Periodic harvest of does aids in preventing herd die-offs due to high densities.
3. Carrying capacity—the number of pronghorn that an area can support varies greatly and can change within a few months. Weather conditions play an important role in determining carrying capacity. In an area, such as the Trans-Pecos region, late summer and early autumn rains determine the condition of winter vegetation for pronghorn. Types of livestock, such as domestic sheep, can have a direct effect on the number of pronghorn that are able to be maintained on an area.
4. Water—the lack of available drinking water for extended periods can correlate to low pronghorn numbers. Maintaining livestock watering facilities throughout the year, even in pastures deferred from livestock grazing, and breaking ice on water troughs during winters can enhance pronghorn survival. A well-distributed water system can assist pronghorn in foraging different vegetative sites.
5. Fences—fences can be designed and constructed to manage livestock and still provide for pronghorn movements to seasonal foraging and watering sites. Woven-wire fences can be modified to allow pronghorn movement by raising the bottom wire 18 inches (46 cm) for approximately 100 yards (91 m) at intervals of 1 to 3 miles (2.5–7.8 km). Replace water gaps with wire and construct new cross fences with the bottom strand 16 inches (41 cm) above ground level.
6. Herd age determination—by knowing the ages of animals harvested, managers can determine whether the herd is being properly managed. A large portion of the annual harvest in the 1.3- to 2.3-year age class would indicate a heavily harvested herd, whereas a high portion of the animals in the 4.3-year and older class might represent a moderately or lightly harvested herd.
7. Supplemental feeding—providing supplemental feed during critical vegetative periods can aid the survival of pronghorn that would otherwise be lost as either hunting or breeding animals. Years for supplemental feeding usually can be determined by the amount and time of rainfall. Normally, average to heavy rainfall during late summer and early autumn results in ample autumn and winter forage, whereas light rainfall during this period may forecast a shortage of available forage.

Providing landowners with information regarding pronghorn behavior when negotiating fences can be an asset of cooperative agreements for public land management programs. When landowners comply with fence specifications that allow pronghorn movements, both they and pronghorn benefit. This can decrease "overbuilt" fences. Often, fewer construction materials can control cattle effectively, yet allow pronghorn access. Adoption of the fence specifications can contribute to thousands of acres of private lands becoming "friendly" to pronghorn. Photo by Jim D. Yoakum.

8. Predation—coyotes are considered the primary predator of pronghorn in Texas, and predation usually is on young fawns. Reestablishing a new herd or increasing a resident herd that has been decimated by drought may be benefited by a predator-control program.

Management of the pronghorn herd should be incorporated into the overall ranch management program to fully use this natural resource. The pronghorn is a unique large mammal that responds readily to management practices. The continued supply of this species represents a beneficial return to recreationists and landowners.

The 1979 *A Handbook on Pronghorn Antelope Management in Texas* was popular with landowners, and a new edition entitled *Texas Pronghorns* was printed in 1991 (Sweptson and Hailey 1991).

Wildlife Management Techniques Manual

The fourth edition of the *Wildlife Management Techniques Manual* (Schemnitz 1980), published by The Wildlife Society, contains a chapter on habitat improvements, with data pertinent to enhancing rangelands, including vegetation manipulation, water developments and fences to control wildlife (Yoakum et al. 1980).

Most rangeland improvements are implemented for livestock management, therefore modifications are suggested for adaptations to benefit wildlife. This particularly applies to forage manipulation projects, which generally require only grass seedings for livestock, but better serves wildlife when planted to a mixture of grasses, forbs and shrubs.

Another example concerns water impoundments constructed for livestock needs. At times, these trap and drown wildlife. However, slight changes in engineering specifications can alleviate entrapment problems, resulting in a project that benefits wildlife as well as livestock with minimal hazard to each.

The subject of livestock fencing (a constraint on some wildlife, including pronghorn) likewise is covered in detail. Information is presented on specifications that allow wild ungulates (e.g., bighorn sheep, deer, elk, pronghorn) access through livestock fences. Also discussed are designs for fences to control wildlife movements during translocation efforts, in areas of crop depredations, near airports or freeways, etc.

The fifth edition of this important management reference now is available (Bookhout 1994). This edition brings the science of habitat improvement up to date, relative to water developments and fences, but it lacks information regarding enhancement of vegetation (Kie et al. 1994).

Western ranchers recognize the value of planting grass for livestock forage. Such practices often follow the loss of forage by wildfires. When grass seedings are sweetened by adding dryland alfalfa (shown above)*, the new growth is grazed readily by pronghorn. For certain grass seedings during late summer and autumn, planted dryland alfalfa becomes a major succulent and nutritious forage species for pronghorn. Therefore, ranchers planting legumes and other forbs in grass seedings are given credit for achieving forage improvements for pronghorn in public land management programs.* Photo by Jim D. Yoakum.

U.S. Bureau of Land Management Technical Handbook

Responding to a need for information on how to plan and implement rangeland management practices to meet the requirements of pronghorn, the BLM published *Technical Note 347* (Yoakum 1980). The booklet explains such practices as resource inventories, monitoring studies and evaluations of habitat suitability. It also gives detailed descriptions regarding the maintenance or improvement of vegetation and water projects and information concerning animal equivalents—animal unit months (AUMs)—and livestock competition. Historic and current practices for fencing to control livestock, yet allow pronghorn movement, are addressed.

By making a concise handbook specific to the needs of pronghorn, the BLM was able to assemble information about the species' needs and make it available to persons charged or concerned with maintaining or restoring rangelands to accommodate pronghorn. The information, accessible in handbook form, was paramount for resource managers other than wildlife biologists, because many management decisions on public lands are made by range conservationists, mineral developers, recreation specialists and others. The handbook also became the major source of biological data for an in-service, self-training course. The course has been used by other state and federal government agencies in Canada, Mexico and the United States.

Numerous published guides provide recommended procedures for enhancing pronghorn habitat. Some contain procedures for modifying wire fences to control cattle, yet allow pronghorn mobility. Others suggest techniques to develop drinking water. Some guides are site specific, providing lists of pronghorn forage species recommended for certain plantings. Habitat management guides can help landowners and government agencies plan and implement habitat enhancement projects to meet pronghorn requirements. Photo by Jim D. Yoakum.

Wildlife Habitats in Managed Rangelands

This FS compendium of 14 chapters relates wildlife to habitat in managed rangelands for Great Basin environments in southeastern Oregon (Thomas and Maser 1986). Although region-specific, much of the information is applicable to other areas. Its purpose is to provide information and develop a process for considering wildlife when range management decisions are made. An entire chapter is devoted to pronghorn (Kindschy et al. 1982).

The adage that "good range management is good wildlife management" probably never was true, and it is no longer acceptable (Thomas 1979), certainly not for pronghorn. Improvements in pronghorn habitat have largely been a by-product of livestock management activities. Pressure for increased livestock production has been accompanied by growing public concern for wildlife. Pronghorn can benefit from rangeland management activities most successfully when the species' welfare is considered in advance and its habitat requirements are not left to chance. Accordingly, Kindschy et al. (1982) reported on various management options and techniques, such as rangeland seedings, shrub control, water developments, fence construction, harassment and livestock management practices.

The publication provides a checklist of rangeland practices that are either beneficial or detrimental to pronghorn welfare. A new concept for evaluating pronghorn habitat is provided with a worksheet listing environmental components to evaluate shrubsteppe rangelands. When completed, the form gives a numerical habitat suitability rating needed for management decisions, including translocations of herds, feasibility of vegetation manipulation projects and evaluation of winter rangelands.

The pronghorn chapter stresses that rangeland management practices conducted to improve conditions for livestock can be beneficial to pronghorn. However, it emphasizes that this is most likely when the pronghorn's needs are identified and the animal's welfare is considered.

Two examples illustrate this concept. The first indicates that livestock and pronghorn thrive best where drinking water is accessible at least every 2 miles (3.2 km). Managers can benefit both species when water developments are installed. However, once a water source is constructed, the water it provides should be maintained as long as pronghorn are in the area, although livestock are not. To have water available to pronghorn for long periods, then to neglect the supply leaves the pronghorn—especially fawns—at a disadvantage.

The second example of a resource that can be beneficial to livestock and pronghorn concerns vegetation manipulation projects. Rangelands that have changed through plant succession to a dominant shrub community, as of-

Building ponds and small reservoirs are popular habitat improvement practices, frequently accomplished during public land management programs. Such investments provide multiple resource benefits, such as water conservation, drinking water for livestock and wildlife, water for fire suppression and use by a variety of game and nongame wildlife. Water from a spring flows to this small reservoir near Ravendale in northeastern California, which provides water yearlong. When well-designed, artificial ponds and reservoirs simulate natural waters, they last for years and provide a vital component of wildlife habitat. Photo by Jim D. Yoakum.

ten has been the case for shrubsteppes in Idaho or creosote brush areas in Texas, can be reverted to early plant succession, providing the forage requirements of livestock and pronghorn. However, such projects are most beneficial to pronghorn when manipulation results in a mixed vegetative composition of grasses, forbs and shrubs. Projects that are designed for dominant stands of coarse, exotic bunchgrasses have limited values to pronghorn. Change from tall, dense shrubs to a grass monoculture better meets the pronghorn's habitat needs for vegetation structure, but often does little to meet the animal's dietary needs for forbs and shrubs. When multiple-use projects are designed for livestock and pronghorn, it is imperative that all of the pronghorn's biological requirements be considered, not just a few.

Arthur S. Einarsen, author of the first monograph on pronghorn ecology and management, lectured attendees at the Order of the Antelope's annual meeting, July 1956, on the Hart Mountain National Antelope Refuge in Oregon. Einarsen's presentations were authoritative, eloquent and stimulating. Much of his original field research for his monograph was conducted on this refuge where he camped and observed pronghorn in the wild for decades. Two graduate students, Rod Canutt and Jim Yoakum (left and right behind the fawns, respectively) assisted Einarson. The three fawns were part of an Oregon Cooperative Wildlife Research Unit field research project concerning pronghorn production and mortality. For more than 20 years, the Order of the Antelope contributed funds for that pronghorn research endeavor and many other research projects on varying wildlife species. Photo by James W. Carlson; courtesy of Jim D. Yoakum.

Interstate Antelope Conference Guides

During the 1958 Interstate Antelope Conference, attendees agreed to develop guides for management of pronghorn and their habitats. This was an ambitious goal, requiring a great deal of work for the committees representing state and federal agencies. It took 4 years to complete the guides, which were adopted and printed in the 1962 *Interstate Antelope Conference Transactions* (Griffith 1962). Eighteen years later, conference delegates voted to update the guides and include new biological data and management practices. This edition was assembled and printed as a separate technical report (Salwasser 1980). It soon became an authoritative and recommended reference, whose information was accepted and incorporated in various species and habitat management plans (see Capodice et al. 1982, Pyshora 1982b, Tsukamoto 1983, California Department of Fish and Game 2000).

The update contains many recommendations for practical management of rangelands used by pronghorn and livestock. Salwasser's (1980:23) management suggestions for cattle are especially noteworthy: "Cattle are relatively compatible with pronghorn. Their diets are sufficiently different during late fall and winter that little overlap in forage use occurs. Nevertheless, the common use of spring grasses and annual and perennial forbs can result in competition. Heavy cattle grazing on pronghorn spring and summer ranges prior to mid-May almost guarantees a high degree of dietary overlap. It is an important point of conflict that can be avoided by delaying the turn-out date or by reducing early season cattle numbers. Other areas of conflict are heavy cattle grazing of dry and wet meadows on summer range and mid-summer bitterbrush use. Pronghorns rely on meadows for succulent forbs. By August, cattle use on bitterbrush will often be heavy enough to result in competition with pronghorn needs. On the other hand, because cattle are primarily grazers they can be a significant positive factor in managing habitats. Their early summer use of grasses favors the maintenance of annual forbs on spring and summer ranges. Their mid-summer preference for wetlands results in the removal of course grasslike plants, making forbs more accessible to pronghorns. To take advantage of this basic compatibleness and to minimize possible conflicts, cattle managers on pronghorn seasonal ranges should consider: Winter Range: use light to moderate early summer grazing to maintain shrub vigor and enhance natural diversity of grasses and forbs; terminate use by August, to protect winter browse supply. Spring Range: turn cattle out after pronghorns have arrived on summer ranges (early to mid-May); graze up to moderate use on key grass species, then terminate use (probably mid-August). Summer Range: turn cattle out after mid-May; allow up to moderate use on key grass species; regulate use on meadows to no less than 3 inches stubble height; allow use on wetlands to make forbs available; terminate cattle use by mid-September. Fall Range: graze as on summer range. When used on good condition ranges these strategies should focus cattle use on grasses and grasslikes, thus enhancing forb availability. Desired light use of browses will occur. Heavy late summer browsing should be avoided."

The Interstate Antelope Conference guides not only have become the backbone of species management plans in the Great Basin, but also have contributed pronghorn management information to other ecoregions. When the Pronghorn Workshop produced guides for management of pronghorn throughout North America, a number of standards from the Interstate Antelope Conference guides were incorporated (Autenrieth 1978). Many of these same standards remained as part of the workshop guides when updated and published 20 years later (Lee et al. 1998).

The Interstate Antelope Conference met during March 1999, marking its 50th anniversary (Yoakum 1999). This landmark meeting may designate the antelope conference as the oldest, continuous technical interagency meeting devoted to the management of a single species of megafauna in North America. The guides were discussed at the 1999 meeting and remains in effect—more than 40 years after their development in 1958.

Attendees at the 1958 Interstate Antelope Conference on the Sheldon National Wildlife Refuge in Nevada drafted the first management guide for pronghorn populations and habitats. Many of the practices they recommended still are standard today, more than 40 years later. Photo by Jim D. Yoakum.

Pronghorn Workshop Guides

Participants from Canada and the United States held a Pronghorn Workshop in Idaho during 1976, at which the topics of greatest concern were the impacts of livestock grazing and energy development on rangelands occupied by pronghorn. Consequently, at the workshop's business meeting, committees were assigned to gather biological data to produce a comprehensive management guide. The resulting document was reviewed for accuracy, evaluated and approved by delegates at the 1978 workshop's meeting, in Canada, and published in the *Proceedings of the Pronghorn Antelope Workshop* (Autenrieth 1978). The guides provide comprehensive, basic, biological data to suggest management practices for the species and its habitat, with special emphasis on coordination with livestock grazing and mineral development programs.

Covered in this set of management guides is a detailed assessment of the affects of energy development on pronghorn habitat. This development (mainly of coal, oil and gas) alters pronghorn habitats significantly and has cumulative effects. The Pronghorn Workshop guides list suggestions to industry and regulatory government agencies for protecting pronghorn populations, mitigating habitat destruction and improving habitat on disturbed lands. Emphasis is placed on perpetuation of as much natural landscape, as opposed to creating artificial conditions. For example, if the postdevelopment land use were to be agriculture production (e.g., wheat, alfalfa), the guides recommend that treatment sites be restricted to minimal acreage and interspersed as islands among natural vegetated rangelands. Many more specific recommendations are provided, including the removal of all materials that may be toxic or hazardous to pronghorn and the purchase of easements to "landlocked" public lands through use of mineral severance taxes or as inclusion in lease stipulations of pre- and postdevelopment plans identifying management objectives to protect, maintain or improve populations and habitat conditions. The guides also suggest other resource program activities that could cause significant changes in pronghorn habitat.

During the workshop's 1982 meeting, attendees identified the need to revise and incorporate new information regarding trapping and translocation. This was accomplished at the 1984 workshop and printed as a separate document (McKenzie 1983).

Pronghorn are captured alive by differing techniques in which nets, traps or drugs are used (see Chapter 21). Parameters to evaluate selection of capture method include number, age and sex of animals needed; density of animals in trapping area; terrain; proximity to roads; pronghorn familiarity with fences; wariness; possibility and acceptable degree of mortalities; and cost (in time as well as money) per captured or marked pronghorn (Amstrup et al. 1980).

Trapping and translocation remain integral parts of pronghorn management today, as they have been for the past half century. One reason has been the phenomenal success of such efforts. Another reason is the increase in pronghorn depredations on agriculture crops.

Twelve years after publication of the original guide (1978), attendees at the 1990 Pronghorn Workshop recognized that certain parts were dated, and the entire guide needed updating. Values were assessed and it was determined that the guides were used extensively—probably more than any other compendia for managing pronghorn. Also, copies were no longer available. Consequently, an editor was appointed, along with five committees, to compile current data, resulting in the development of a new set of guides that more than tripled the number of pages in the original. Preliminary drafts were reviewed by wildlife biologists and managers for accuracy and comprehensiveness. The issue of predators and predator control, more than any other component, evoked varying viewpoints and recommendations, reflecting the controversial nature of predator management programs—be they in Canada, Mexico or the United States.

The new edition totaled 101 pages and was approved for publication at the Pronghorn Workshop in Rock Springs, Wyoming (O'Gara and Yoakum 1992). Cooperative efforts of the FWS, through its Federal Aid in Wildlife Restoration Program, produced 1,000 copies. An objective was to distribute sufficient copies for each professional working with pronghorn to have a reference copy. Additional dissemination was made to cooperating agencies and conservation organizations. The remaining copies were distributed on a first-come, first-served basis by the North American Pronghorn Foundation, headquartered in Casper, Wyoming. The value and use of the 1992 guides are indicated by frequent reference to them in publications (e.g., Pojar et al. 1994, Alexander and Ockenfels 1994, Ockenfels et al. 1994, Gruell 1996, Yoakum et al. 1996, and Wooley and Lindzey 1997).

At the 1988 Pronghorn Antelope Workshop in Oregon seen above, participants urged that trapping and translocation operations include the collection of more data on ways to capture, use of immobilization drugs and selection of suitable habitat for releases. The latter was especially important because, in the preceding decades, translocation had overpopulated many of the prime habitats, and marginal sites were being stocked or were under consideration. These factors were presented and accepted at the 1990 workshop in Colorado (O'Gara and Yoakum 1990). The need for pretranslocation evaluations of habitat suitability criteria and postrelease studies of translocation success were stressed. Such information is valuable to wildlife biologists assigned to plan and implement pronghorn translocation, particularly translocations to marginal habitats. Photo by Jim D. Yoakum.

True to the goals of the Pronghorn Workshop, the *Pronghorn Management Guide* has been updated every 5 to 6 years. The latest edition was revised in 1998. It contains most of the data in the 1992 edition, plus additional information on habitat models, fires and disassembling rangeland fences (Lee et al. 1998). In addition, the 1998 guides provide new technical information regarding water requirements, population survey methodology and diet selection.

Editions of the *Pronghorn Management Guide* has been used for more than 25 consecutive years. During these two decades, it has been apparent that they occasionally were used indiscriminately as pat, or cookbook, management practices, instead of assessments of biological data and methodology for site suitability before use of the information. Care should be exercised in applying techniques of value to an area of differing ecological conditions.

Following are two examples of variations that can be expected:

1. During breeding seasons, pronghorn bucks may be territorial, have harems or both during different years in the same location; one cannot assume pronghorn are naturally territorial at all times for all habitats.
2. Habitat suitability criteria have been established for some habitats, but not all. At times, the system for evaluating habitat suitability may be appropriate to use in different biomes, but care must be taken to use the appropriate local ecological characteristics in the system. For instance, a site with 5% grasses, 5% forbs and 25% shrubs can be characteristic for vegetation in the Great Basin shrubsteppe. However, percentages for the Great Plains grasslands more likely may be 65% grasses, 10% forbs and less than 5% shrubs.

Use of Handbooks, Manuals and Guides

To assess the application of the five sets of management guides (the Texas handbook was not used because it pertained to a different ecoregion), participants at the joint 1988 Interstate Antelope Conference and Pronghorn Workshop were canvassed regarding their use of the management guides. Attendees were asked to indicate which they had used; Table 134 lists the results. If these attendees were representative of all of pronghorn biologists and managers, it is disheartening to see how few were using the information. For this species of wildlife, there is a great deal of information available that is not used by on-the-ground management personnel.

CONCLUSIONS

The keys to sustaining healthy, abundant pronghorn herds are recognizing and accommodating the biological needs of the species. Management plans can be the tool to do so. During 1960 to 2000, wildlife biologists developed numerous management guides, handbooks and manuals with recommended practices to enhance pronghorn populations and habitats. When the management strategies detailed therein have been incorporated into

Members of an Interstate Antelope Conference participate in a demonstration tour of native grasses and forbs successfully planted on pronghorn habitat. Seed planting practices for vegetation restoration projects can enhance wildlife habitat. Information about these techniques is especially important during the planning of rehabilitation projects following wildfires. It also is increasingly important during attempts to reestablish native plants on lands impacted by years of agricultural crops. Biologists and managers trained in plant restoration principles and practices for wildlife can provide recommendations to government land management agencies and private landowners. Photo by Jim D. Yoakum.

Table 134. Use of guides for pronghorn management by 43 attendees of the 1988 joint meeting of the Interstate Antelope Conference and Pronghorn Workshop

Guide and source	*Number of persons using guides*	*Percentage of persons using guides*
Interstate Antelope Conference Guidelines (Salwasser 1980)	21	46
Pronghorn Antelope Workshop Guidelines (Autenrieth 1978)	7	15
Wildlife Management Techniques Manual (Yoakum et al. 1980)	2	4
Bureau of Land Management Technical Note 347 (Yoakum 1980)	15	33
Wildlife Habitats of Managed Rangelands: Pronghorn (Kindschy et al. 1982)	22	48

species and habitat management programs, pronghorn have prospered.

With the continued growth and sprawl of North American human populations and a corresponding demand for more and varied uses of land, there is increasing need to monitor pronghorn habitat requirements—especially forage, water and space. Pronghorn populations increased greatly during the last half of the 20th century, but there is need for wildlife managers to track and detect adverse shifts that jeopardize animals or their habitats. Well-conceived management plans are valuable tools for such detection and for preventing deterioration of herds or habitats. These documents can detail area-, range- or region-specific management objectives, goals, priorities and procedures correlated with other uses, practices and authorities.

A recent trend in management plans is the development of cooperative programs between private landowners and state wildlife agencies. These have resulted in improved habitat conditions and increased public recreation opportunities.

With passage of the Endangered Species Act in the United States, legislation identified the role of recovery plans to enhance wildlife populations and habitat status. Mexico has since followed with similar practices. Canada enacted its Species At Risk Act in late 2003 (Taubman 2003), but pronghorn there are not classified as endangered or threatened. Consequently, as of 2003, all presumptive pronghorn subspecies not classified as endangered were protected by management plans that had been completed or were being developed.

Environmental impact statements and guides provide management techniques for sustaining animals and their habitat, and they are useful in developing management plans. Six guides have been published to provide data regarding basic requirements of pronghorn. Application of guides should be in keeping with a holistic approach to resource management.

CHAPTER SEVENTEEN

Habitat Conservation

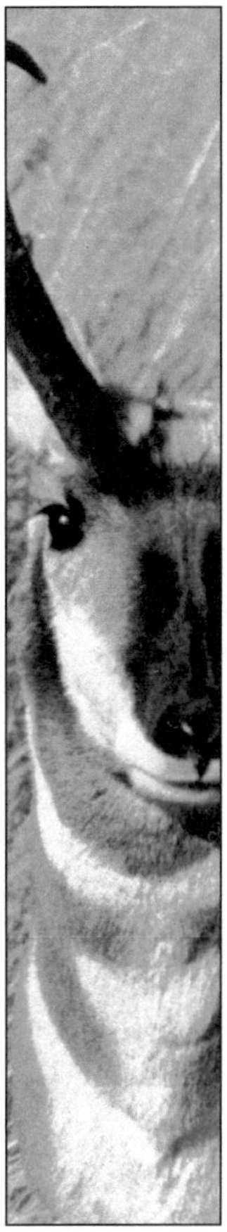

Jim D. Yoakum

Antelope restoration in the future will depend largely on the program of range management. . . . Already antelope are taking advantage of reseeded lands, and continued conversions of brushlands may lead to a steady increase of the antelope population.

Leopold (1959b:97)

The key to efficiency in habitat management for a wildlife species is knowing the habitat requirements of the species and discerning and improving limiting factors in the target habitat.

Ellis (1972:58)

As fawn survival, and thus recruitment, appears to be very much dependent upon range conditions, the productivity and sustainable yields from pronghorn populations are also likely to be directly range-dependent.

Ellis (1972:58)

Habitat conservation for pronghorn requires knowledge of the status and trends of existing landscape conditions. Four basic principles for managing rangelands for pronghorn are:

1. *Inventory and periodically monitor resources.* The foundation of habitat management is a base inventory, including quality and quantity of habitat factors (food, water, cover and space). Subsequent monitoring of these factors helps to determine whether they are static, improving or declining, in relation to their ecological potentials.
2. *Relate site-specific biological information to the habitat requirements of pronghorn.* Habitat components are needed, in particular interspersion and juxtaposition, to accommodate pronghorn reproduction and survival. Failure to consider these habitat requirements and interrelationships is mismanagement.
3. *Protect and maintain the quality of those ecological conditions, when base inventories indicate that habitat conditions meet the physical and behavioral requirements of pronghorn.*
4. *Improve habitats in deteriorated condition or lacking a necessary component (e.g., water during a critical season).* Deteriorated conditions are those that change through natural causes (e.g., repetitive fires, insect infestations, hail storms) or human practices (e.g., intensive livestock grazing, water drawdown or diversion, agriculture, fencing and construction) such that pronghorn well-being and productivity cannot be sustained.

RESOURCE INVENTORIES AND MONITORING

Pronghorn Habitat

Inventories should precede any management program. During and following implementation, monitoring helps to ascertain whether objectives are being met. Monitoring also may provide information to improve methodology.

Base Inventories

Habitat inventories involve measuring selected habitat variables and determining the number of wildlife species and size of populations supported by those variables (Cooperrider et al. 1986). Such inventories are the database from which wildlife managers make decisions, analogous to a grocer who records the stock. Both managers need to know the items and quantity in stock and the condition of it.

Baseline data for land-use planning and species and ecosystem management plans indicate existing conditions, and they are useful in estimating potential impacts and comparing actual impacts after management actions. Without such a database, proposals to alter habitat to benefit one species may produce problems for other species. For example, conversion of shrubland to monotypic grassland might benefit livestock, but the consequent lack of shrubs and forbs would diminish the rangeland as habitat for pronghorn (Jones 1986).

Inventories of wildlife populations generally are conducted by provincial and state wildlife agencies; they are important because their data indicate the productivity of the area in question. An inventory revealing five pronghorn per square mile (1.7/km^2) indicates high-quality rangeland, compared with other locations of lesser pronghorn density. Sex and age ratios usually result from harvest strategies. Productivity is a clue to habitat conditions. For example, rangeland with a 20-year mean of 40 fawns per 100 does is more favorable habitat than adjacent rangelands producing a mean of 25 fawns per 100 does during the same time span. Also, inventory data related to population size, home range and seasonal use area are essential to computing amounts of forage and water needed to support pronghorn in a land-use management plan.

Resource components needed by pronghorn are inventoried by state wildlife agencies (Tsukamoto 1983, Pyrah 1987, Ockenfels et al. 1994, 1996) and federal land management agencies (Yoakum 1980a, Cooperrider et al. 1986, Patton 1992, U.S. Fish and Wildlife Service 1994). An extensive inventory of pronghorn habitat was accomplished by the Arizona Game and Fish Department (Ockenfels et al. 1994, 1996). Conducted over a 6-year period, the effort enlisted cooperators from universities, state wildlife agencies and federal land management agencies. By means of a geographic information system, it inventoried various environmental factors (forage, water, space) during all seasons of multiple years. Freeways, railroads, fences, rangeland improvements and human occupancy were assessed in relation to impact on pronghorn production and survival.

Personnel conducting studies of pronghorn habitat need to consider the following components during inventories:

1. *Population and characteristics*, including occupied rangelands, population size, seasonal distribution, critical areas (e.g., movement corridors, winter concentration areas and fawning areas) and known or potential barriers—both natural (e.g., mountain ranges, dense shrubs or trees, rivers and lakes) and artificial (e.g., fences, highways and railways, towns and cities). Ockenfels et al. (1994) provided management recommendations concerning corridors used by pronghorn in central Arizona. Most important was the identification of such corridors to determine how best to retain them. Comparison of past and present aerial photographs could be used to identify corridors. Road signs can alert drivers of travel routes frequently used by pronghorn. As human-related disturbances increase, habitat fragmentation will accelerate, which may cause pronghorn to abandon corridors. This problem is increasing in southwestern landscapes with growing human occupation. Coordination with local planning zone commissioners could expedite maintenance and promote reopening of original corridors where human developments have intruded or may intrude. Prescribed burns, brush clearings and fuel-wood cuts, together with appropriately planned developments, could help maintain or enhance corridors.
2. *Climate*, including temperature, seasonal precipitation patterns, and mean snow depth.
3. *Drinking water*, including type, location and number of water sources (i.e., springs, lakes, reservoirs, rivers, streams), seasonal availability, quality and quantity of each source.
4. *Vegetation*, including abundance, distribution, composition, mean height and productivity indices (in pounds per acre) of grasses, forbs and shrubs.
5. *Other factors*, such as livestock grazing, agricultural crop production, mining, recreation, rangeland alterations—completed or planned (e.g., fences, roads, forage manipulation projects, water developments)—predation, predator control, and interspecific competition that may affect the pronghorn population.

A comprehensive investigation of rangeland components should be done to assess biotic and abiotic variables. The more intensive the investigation and consequent inventory, the more effectively a habitat manager can fulfill the responsibilities and objectives of resource

Sagebrush/grasslands near Rome, Oregon, during the early 1960s were dominated by tall, dense shrubs (left) *that did not constitute suitable pronghorn habitat. The sagebrush is higher than the 34-inch (89 cm) black Labrador. Grasses and forbs provided understory, but they were not abundant. Pronghorn passed through the area, but did not linger to forage or give birth. The tall, dense shrublands were not suitable habitat. At* right *is a nearby sagebrush/grassland site that was plowed and seeded to crested wheatgrass and dryland alfalfa during the 1970s. The same Labrador is conspicuous in the 15- to 20-inch (38–51 cm) tall bunchgrasses. Because the vegetative community was changed to a low structure with abundant forbs and young shrubs, pronghorn moved onto the site and used the area for foraging and fawning thereafter.* Photos by Jim D. Yoakum.

management within the constraints of manpower, funding and public support.

Managers of private or public lands often make decisions for one primary land use. If a manager has a good inventory of resource components and is knowledgeable about the habitat requirements for pronghorn, primary-objective decisions invariably can be made compatible with the needs of pronghorn. In Bear Valley, Oregon, for example, when ranchers manipulated vegetation to increase forage for livestock, it was done in a manner that the change of vegetative structure and composition also favored pronghorn (see Chapter 15).

Monitoring Studies

Monitoring entails observing, measuring, recording and tracking conditions and physical attributes of resources. Monitoring studies can help to evaluate the efficiency of actions taken toward achieving management objectives.

The purpose of monitoring is not only to measure change, but also to determine cause of change (Cooperrider et al. 1986, see also Heady and Child 1994, U.S. Bureau of Land Management 1996a, Holechek et al. 1998). A monitoring program repeatedly measures habitat variables and correlates them with species abundance. For example, the number of pronghorn can be correlated with the height of vegetation. A change of vegetative height can influence the number of pronghorn occupying a site. Monitoring vegetation can document trends in plant succession and relate them to habitat suitability for pronghorn. Monitoring also can be used to determine the effectiveness of habitat improvement practices, such as rangeland seedings, prescribed fires and water developments. It may detect changes caused by climate or other natural causes.

The frequency with which monitoring studies should be conducted varies with the degree of change in the habitat components, but a minimum of 5 years should be considered for relatively stable habitats. Rangeland experiencing rapid vegetative or water changes should be monitored more frequently. Fire can effect changes within a day, and intensive livestock grazing or drought can do the same within a growing season.

Both the quality and quantity of forage and water should be monitored on a schedule determined by a management plan. Long- and short-term monitoring studies are needed, especially on sites undergoing major human-caused changes. A management site in Nevada changed vegetative characteristics of unoccupied pronghorn habitat, to preferred habitat, to less desirable habitat during a 25-year period (Yoakum 1980a, see Chapter 15).

VEGETATION ENHANCEMENT

When western rangelands are manipulated to increase or sustain forage or water supplies, the practices are termed range improvements, or habitat enhancement projects. In the United States, such improvements have been primarily to benefit livestock (Vallentine 1989, Fulbright and Guthery 1996, Kindschy 1996, Koerth 1996, Roundy 1996); less than 1% have had the primary objective of improving wildlife habitat conditions (Yoakum et al. 1996). When these practices are for the benefit of wildlife, they are termed wildlife enhancement projects. Adams et al. (1992) surveyed hunting lease operators in Texas and reported that 21% of them managed shrubs to benefit wildlife. However, many of the ranchers involved also used other intensive management practices—such as high fences or artificial feeding and drinking facilities—that ordinarily are not used on western rangelands.

Habitat enhancement projects can be tailored specifically for wildlife (Yoakum et al. 1980a, Payne and Copes 1986, Kie et al. 1994, Payne and Bryant 1998a), but the science of such practices are not yet as sophisticated as those for livestock. The high economic value of agricultural commodities has resulted in extensive studies, experiments and projects to enhance livestock production. Wildlife science is relatively new, and wildlife itself seems to be viewed and treated in socioeconomic terms more as a cultural amenity than as a need, at least as compared with agricultural products.

Primary objectives for planning and implementing habitat improvements for pronghorn include the following:

1. When an environment exists in healthy natural condition, the site should be sustained in that condition. It will continue to produce optimal benefits for a variety of endemic species (Shelford 1963, Thomas and Maser 1986) that have adapted to the site over time. This ecological principle will not always favor some land-use or wildlife management objectives, such as maximizing numbers of a particular species. Saltbush-greasewood vegetative communities in the Great Basin, for example, have an ecological vegetation potential of 30% to 45% shrubs and less than 10% grasses and forbs. Low production of herbaceous vegetation, especially during spring and summer, indicates that the sites have low carrying capacity for pronghorn. Accordingly, management objectives and expectations should not be a significant pronghorn population increase.
2. When a rangeland has a missing or inadequate habitat component, but is capable of being manipulated to facilitate or improve that component, projects can be undertaken to effect a change. For example, on a sagebrush site with an existing vegetative composition of 5% grasses, 5% forbs and 65% shrubs, with the remaining 25% covered by litter, mosses and lichens, shrub reduction to 10% could be followed by vegetation changes to 30% grasses and 15% forbs. Another example is inadequate volume or distribution of water, which can be rectified by judicious placement of reservoirs or other water catchments.

Management of Plant Communities

Manipulation of vegetation by humans is not new. Before the arrival of Europeans, Native Americans burned western rangelands to stimulate the green-up of grasses attractive to herbivores. Aldo Leopold (1933) stated that vegetative cover could be manipulated by controlling plant succession. The cow, plow, axe and fire reverse succession. Fencing, fire protection and planting advance it. Leopold called these practices "tools." Today, they generally are considered learned practices, methods or techniques of wildlife habitat management (Yoakum et al. 1980a, Kie at al. 1994, Payne and Bryant 1998a).

Pronghorn thrive on rangelands in a subclimax vegetative condition (Kindschy et al. 1982, Smith et al. 1996). Such conditions can result from fire, grazing by herbivores or vegetative manipulation projects (mechanical, chemical and seeding). Natural vegetative communities on pronghorn rangeland contain a variety of grasses, forbs and shrubs, and range improvement projects that best suit pronghorn are those that produce mixed forage classes.

The health of vegetation and rangelands can be determined through investigation of existing conditions and long-term trends (Committee on Rangeland Classification 1994). If habitat is producing vegetation at or close to natural potential, manipulation of vegetation cannot be justified as a means of improving conditions for pronghorn. Only sites of inadequate vegetation in terms of forage or fawning cover, yet that contain a proper combination of other habitat variables, should be targeted.

Manipulation of vegetation to improve pronghorn habitat may involve pyric, biological, mechanical or chemical means and planting to preferred forage and cover (Yoakum 1980a, Yoakum et al. 1980a, Lee et al. 1998, Yoakum and O'Gara 2000).

Structure and composition of a vegetative complex determine the quality of habitat for various wildlife species (Koerth 1996). Shrubland/pronghorn relationships were identified by Yoakum (1972, 1974), Sundstrom et al. (1973) and Lee et al. (1998). For the tallgrass prairies, decreasing the height of herbaceous vegetation is necessary to meet pronghorn habitat requirements (Lee et al. 1998). An understanding of the interactions of shrubs and herbs is necessary to plan vegetation management strategies. Production of certain herbaceous vegetation decreases as woody canopy increases. Woody plants can create a shrub-driven successional process, making it unlikely that the site will revert to producing abundant herbaceous vegetation. Under these conditions, plant communities can be manipulated to simulate healthy ecosystems and sustain proper proportions of shrubs and herbaceous components, providing requisite habitat for pronghorn and other endemic wildlife.

Vegetative structure can be too high for pronghorn on the grasslands, but fires reduce vegetative height to meet pronghorn requirements. Aldo Leopold took this photograph more than 50 years ago. It depicts his daughter Estella as a measure of plant height. Vegetation on the left of Estella is natural height—too high for pronghorn. Vegetation on the left is lower because of previous burns, and would allow pronghorn to see and flee from enemies. Pronghorn do not readily occupy prairies with grass higher than 30 inches (76.2 cm), but occupy these same areas when the vegetation has been disturbed by fire, bison or cattle grazing. Such tall grasslands are now being managed by prescribed fires and livestock grazing to maintain nutritious, abundant vegetation and low plant height in eastern Kansas where pronghorn have been reestablished. Photo by Aldo Leopold; courtesy of Robert A. McCabe.

Table 135 and Figure 115 list various manipulation practices and their potential to enhance habitats for pronghorn. Their relative values as habitat management techniques vary with vegetation type. In this chapter, six practices (fire, biological control, mechanical treatment, chemical application, seeding and fertilization) are identified as practices used to rehabilitate western rangelands occupied by pronghorn. Using a variety of practices in sequential or spatial patterns to accomplish habitat prescription objectives was beyond the scope of this chapter. Such detailed planning is site-specific, extremely complex and requires creativity and thorough understanding of the potential plant responses to each technique. Before selecting a management

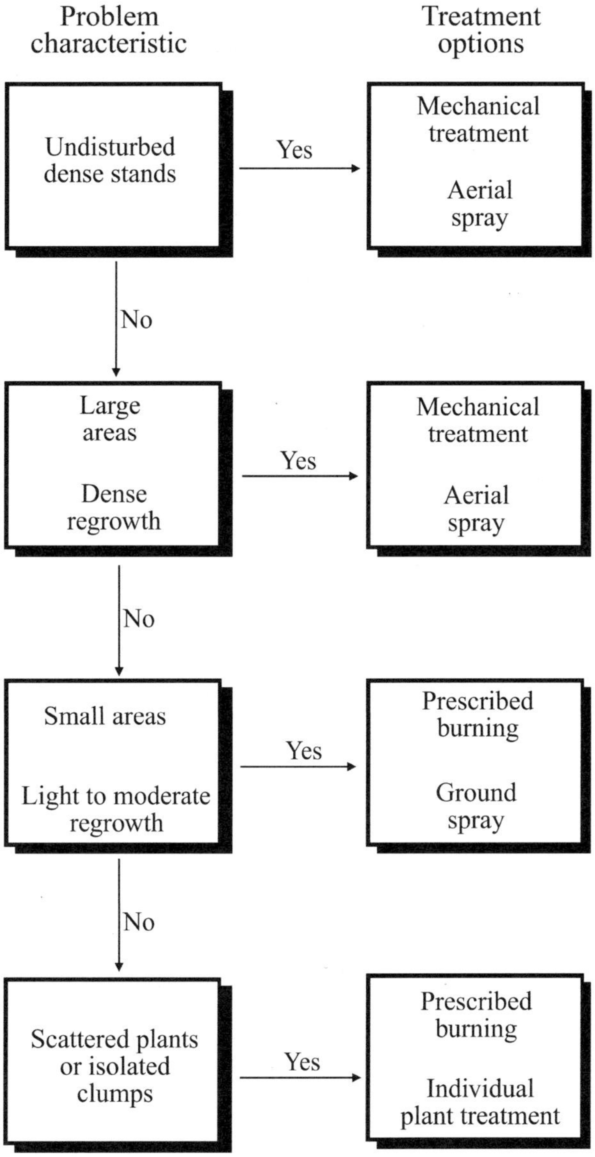

FIGURE 115. *A decision-making flow chart that could be used to select treatment methods for shrub management on pronghorn habitats (from Koerth 1996).*

Table 135. Vegetative rehabilitation practices and their potential to enhance pronghorn habitat[a]

	Woody plant control			Cultural practices	
Vegetative communities	*Fire*	*Mechanical*	*Herbicide*	*Seeding*	*Fertilization*
Grasslands					
Tallgrass prairie	+	–	–	0	+
Northern mixed prairie	+	–	–	0	0
Southern mixed prairie	+	+	–	0	–
Shortgrass prairie	0	+	–	0	–
California annual	0	0	–	0	–
Semidesert	0	0	–	0	–
Shrubland and juniper					
Big sagebrush	+	+	0	+	–
Low sagebrush	–	0	–	0	–
Pinyon, juniper	+	+	0	+	–
Mesquite	+	+	+	+	0
Hot desert	–	–	–	–	–

[a] + = good potential; 0 = moderate potential; – = low potential.
Source: Modified from Payne and Bryant 1998a.

option, it is recommended that goals and objectives be identified for the management area, and that reliable data on costs, risks and benefits be obtained. Most treatment projects are conducted during summer and autumn, when weather permits.

On rangelands currently occupied by pronghorn, most vegetative changes result from fire (wild and prescribed), livestock grazing, insects and type conversion projects (mechanical and chemical). Prescribed burning is one of the principal methods for manipulating vegetation on grasslands and in shrublands of the Intermountain West. Livestock grazing is ubiquitous on western rangelands, but use as a prescription practice for enhancement of pronghorn habitat has had little investigation. Mechanical and chemical practices were used extensively during the 1950s to 1980, but to a much lesser extent during the 1980s and 1990s. Many burn, mechanical and chemical treatment projects are followed with rangeland seedings to stabilize soil and establish desired vegetation as groundcover and forage for domestic herbivores and wildlife.

Shortgrass prairies were identified by Vriend and Barrett (1978) as the most productive extant habitats for pronghorn. Consequently, manipulation projects generally are not needed in this vegetative region. However, some tall grasslands in Kansas currently are manipulated by prescribed burns and livestock grazing to provide suitable habitat conditions for pronghorn.

Shrub cover has increased on more than 422,000 square miles (1,092,980 km^2) of shrubsteppes in the western United States during the 20th century (see Beatle 1960, Miller et al. 1994, Roundy 1996). On some shrubsteppes, more than 90% of pronghorn habitat is in late shrub succession (U.S. Fish and Wildlife Service 1994). Vegetative rehabilitation can restore these rangelands to plant communities favorable to pronghorn (Yoakum 1980a, Kindschy et al. 1982, Lee et al. 1998, Yoakum and O'Gara 2000).

Sites dominated (25% or more) by shrubs more than 30 inches (76 cm) high generally have few pronghorn compared with sites with fewer or lower shrubs and those with more herbaceous forage (Yoakum 1974, Kindschy et al. 1982, Ockenfels et al. 1994, Lee et al. 1998, Aoude and Danvir 2004). On dominant shrublands, shrubs compete for moisture and nutrients with forbs and grasses and decrease the ability of pronghorn to detect and flee from enemies, but these shrublands can be altered to favor pronghorn.

Treatment areas should be no larger than 1,000 acres (405 ha). This area is based on an assumption that the winter rangelands of pronghorn in the Great Basin average 3,000 to 5,000 acres (1,214–3,023 ha), and the treated area should comprise no more than one-third of the winter rangeland. A similar prescription could be considered for other critical pronghorn habitats, such as fawning areas and movement corridors. About 5% to 20% shrub cover should be retained for pronghorn forage and cover requirements in a treated area.

Shrub control projects should not eradicate or control all shrubs because some—such as antelope bitterbrush, low sagebrush, mountain mahogany and winterfat—are preferred and highly nutritious forage for pronghorn. Certain shrubs are critical to winter survival, especially when snow cover exceeds 10 inches (25 cm). At those times, shrubs, which are high in nutrients during winter, may be the primary forage available to pronghorn (Martinka 1967, Bayless 1969, Bruns 1977). Shrubs also provide cover for concealment of neonates in spring (Barrett 1981, Autenrieth 1982).

For manipulation of shrubs, control practices should be those least detrimental to herbaceous plants. Some shrub control techniques (plowing and disking) kill a higher percentage of grasses and forbs than do others (chaining, spring-prescribed fires).

There are eleven primary species of sagebrush on western rangelands (Beatle 1960), and each species respond differently to control practices (Vallentine 1989, Payne and Bryant 1998a). Accordingly, shrub control treatment should be specific to the type and structure of shrubs on the site. Some rangelands in the Intermountain West have shrub cover of 60% or more. Wherever herbaceous plants are overexploited, shrubs can proliferate and dominate in as little as a decade.

Pronghorn use of vegetation enhancement projects was investigated from 1995 to 2001 on the Deseret Land and Livestock Ranch in northeastern Utah (Aoude and Danvir 2004). Approximately 6,200 acres (2,510 ha) of sagebrush steppe were treated to increase preferred, nutritious forage for pronghorn, other wildlife and livestock. Project objectives were to thin dense shrubs where remnant herbs existed, thus increasing availability of precipitation and soil nutrients for grasses and forbs. Sites dominated by shrubs and containing low densities of herbaceous plants were burned or treated—chemically or mechanically—to decrease shrubs, then these sites were seeded with a diverse mixture of forbs adapted to the environment.

Percentage cover of abundant plant species on pre- and post-treatment sites is presented in Table 136 Post-treated sites increased forb percentage cover, plant species richness and biomes of herbaceous forage compared with pretreated sites. Pronghorn changed their foraging habits noticeably from pre- to post-treatment years. Before the rangeland improvements, the most preferred vegetation community was scarce native shrub (Table 137), which supported abundant forbs. However, pronghorn switched forage preference to the reseeded areas. Differences in rangeland use by pronghorn pre- and post-treatment are presented in Figure 116. Fawn/doe ratios correlated with population size better than any other variable (Figure 117). Postmanipulation population dynamics changed (1996–2000), with almost no effect on population size of fawn/doe ratios.

Results of the Deseret Land and Livestock Ranch pronghorn and vegetation treatment study were: "Population

TABLE 136. Characteristics of abundant plant species in two vegetative communities of the Deseret Land and Livestock Ranch, Utah, in August 1997

		Percentage cover	
Plant community and species	*Origin of plant*[a]	*Species*	*Forage class*
Sagebrush, native (pretreatment)			
Shrubs			29
Big sagebrush	N	18	
Green rabbitbrush	N	11	5
Forbs			
Vetch	N	1.5	
Locoweed	N	0.5	
Other forbs	N/I	3	
Grasses			9
Western wheatgrass	N	3	
Sandburgs, bluegrass	N	6	
Shrub control/seeded (post-treatment)			
Shrubs			3
Big sagebrush	N	1.5	
Green rabbitbrush	N	1.5	
Forbs			14
Vetch		0.5	
Small burnet	I	0.5	
Sainfoin	I	3	
Dryland alfalfa	I	5	
Other forbs	N/I	5	
Grasses			12
Crested wheatgrass	I	5	
Intermediate wheatgrass	I	4	
Sandburg, bluegrass	N	3	

[a] N = native; I = introduced.

Source: Modified from Aoude and Danvir 2002.

Table 137. Vegetation community availability and use preference by pronghorn does

Range type	*Percentage available*	*Percentage used by doe groups*		*Preference rating*	
		1990–1995	*1996–2000*	*1990–1995*	*1996–2000*
Reseeded	6.4	0.0	39.2	0.0	6.1
Black sagebrush	5.5	1.6	1.3	0.3	0.2
Sagebrush/crested wheatgrass	18.3	17.7	13.8	1.0	0.8
Sparse native sagebrush	25.0	38.7	21.0	1.5	0.8
Dense native sagebrush	31.1	8.9	11.7	0.3	0.4
Meadow	13.7	24.5	13.0	1.8	1.0

Source: Adapted from Aroude and Danvir 2002.

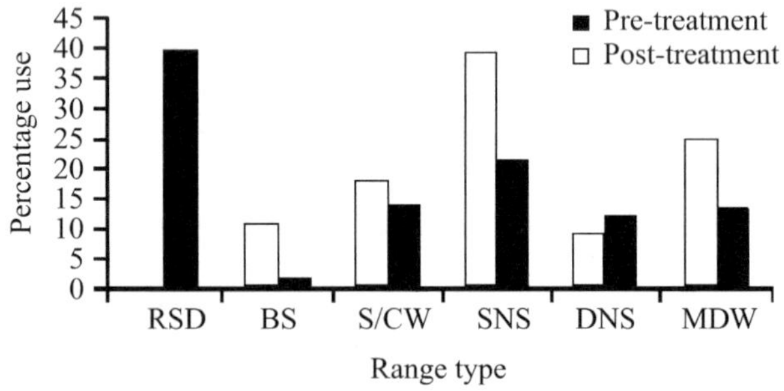

Figure 116. *Comparison of summer (August/September) habitat use by pronghorn doe groups on pretreatment (1990–1995) versus post-treatment (1996–2000) sites (from Aoude and Danvir 2002). RSD = reversed; BS = black sagebrush; S/CW = sagebrush/crested wheatgrass; SNS = sparse native sagebrush; DNS = dense native sagebrush; MDW = meadow.*

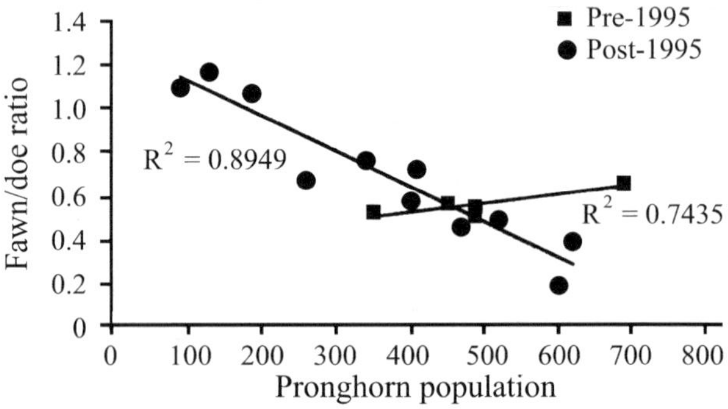

Figure 117. *Relationship between pronghorn population growth, specifically autumn fawn/doe ratios for pretreatment (1985–1995) versus post-treatment range improvement projects on the Deseret Land and Livestock Ranch, Utah (Aoude and Danvir 2002).*

size and fawn production were negatively correlated pretreatment (r = 0.89), suggesting density dependence. Conversely a positive correlation between population post treatment (r = 0.74) suggests an increase in carrying capacity. Doe groups strongly selected shrub controlled and seeded areas having abundant broadleaf forbs" (Aoude and Danvir 2004:1). This investigation supports the authors' hypothesis that forb availability limits pronghorn fawn production and population size.

Vegetation restoration practices for deserts are problematic because there is less precipitation, but it is more variable than on northern grasslands and shrubsteppes. Southwestern deserts have changed dramatically since the late 1700s, as grasses decreased and woody plants increased. Variables driving these changes include livestock grazing, fire suppression and climatic patterns. Many areas of once native desert grasslands now are mesquite shrublands with an understory of herbs (McClaran and Van Devender 1995). Extant desert landscapes are less favorable for pronghorn than they were at the time Euro-Americans first colonized (U.S. Fish and Wildlife Service 1998a).

Contemporary management goals for the desert include restoring endemic plant diversity for pronghorn, improving watershed cover and increasing forage for livestock. Potential to restore vegetation exists using standard methods of vegetative manipulation and revegetating with legumes, other forbs and shrubs to fulfill wildlife food and cover requirements (Roundy 1996). For example, habitat restoration may be encouraged by establishing legumes such as whiteball acacia, which produces nutritious and succulent forage.

Pronghorn on the Sonoran Desert in southwestern Arizona are the only subspecies listed as endangered in the United States. One reason for this classification is low fawn recruitment that varied from zero in 1996 and 1997 to 33 fawns per 100 does in 1998. Recruitment has been correlated with amount and timing of rainfall. During spring and summer, nutritional forage is necessary for increased energy demands of pregnant and lactating does. Years with average or below average rainfall result in low fawn survival. When rains are above normal and corresponding forage growth is good, fawn survival increases. Habitat improvement projects are being conducted to improve fawn survival and increase pronghorn numbers. This marks the first attempt to increase forage quality and pronghorn numbers for Sonoran pronghorn. One project entails planting and irrigating annual and perennial herbaceous plants. Such habitat conservation efforts implement artificial practices as emergency measures to try to reverse downward trends of Sonoran pronghorn (Hervert et al. 2000).

Fires

Fires have burned western rangelands repeatedly since the ice ages. Vegetation dries seasonally and is ignited by lightning. The vegetation has adapted to fires, and pronghorn and other wildlife have evolved along with the plants. Numerous studies indicated that wildfires were common on North American rangelands. Few vegetative communities were spared from pyric influences, although the frequency of fire events varied widely (Table 138).

Native Americans intentionally burned northern prairies for hundreds, if not thousands, of years before the arrival of Europeans. Prairies were burned on occasion to keep areas free of tree cover and increase forage for large herbivores (Vallentine 1989, Binnema 1996). During their explorations between 1804 and 1806, Lewis and Clark noted that Hidatsa of present-day North Dakota burned prairies near their villages in spring to induce lush grass growth as enticement to bison and other wild ungulates (Burrows 1961).

The majority of rangeland seedings are planted to restore watershed cover and increase livestock forage. Such seedings can be made more desirable for wildlife by adding shrub seeds. In Elko County, Nevada, a wildlife habitat enhancement project provided funding to interseed bitterbrush in crested wheatgrass after a wildfire. Five years after the shrub seeding bitterbrush was large enough to provide forage for deer and pronghorn. Photo by Jim D. Yoakum.

According to Payne and Bryant (1998a:381): "Fire is the most ubiquitous agent of disturbance that releases energy and renews habitat." After a fire burns an area, the vegetation reestablishes itself through plant succession. Plants soon colonize the site depending on soil type, temperatures and precipitation, as well as the intensity of the fire. Sometimes a hot fire will burn the organic soil and delay regrowth. Most fires, however, leave fertile soil and abundant sunlight for plants to grow. Usually, the first plants to colonize are annual forbs, such as mustard and geranium. Within a couple years, these pioneering forbs develop the site for perennial grasses and forbs. Later, shrubs may establish themselves and become the final seral stage (Stroud et al. 1995). Table 139 lists selected plants found in pronghorn habitat and those plants' fire tolerance. The high percentage of forbs and shrubs is especially relevant to pronghorn diet composition.

On shrubsteppes of Drakes Flat, Oregon, a wildfire burned approximately 55 acres (22 ha) during the late 1940s, and the site—previously with abundant shrubs—was seeded with crested wheatgrass (Yoakum 1957a). Pronghorn frequently were observed foraging on the burn site, which also was used for fawning. A decade later, vegetative transects revealed decreased shrubs and increased herbaceous plants on the burn, in contrast to adjacent, unburned control areas (Table 140).

Deming (1963) noted that, during the early 1950s, only an occasional buck pronghorn was seen on the high tableland of the predominantly sagebrush-covered Hart Mountain, in Lake County, Oregon. In 1954, a wildfire burned more than 6,000 acres (2,428 ha), changing the plant community to grasses and forbs with small patches of shrubs. Pronghorn use intensified the year following the fire; 10

TABLE 138. Examples of pristine fire-free intervals (FFI) in years for vegetative communities in pronghorn habitat

Vegetative community	*FFI*	*Source*
Grassland		
Tallgrass prairie	5–10	Wright and Bailey (1982)
Southern mixed prairie	20–30	Wright and Bailey (1982)
Desert grassland	10	Christensen (1981)
Shrubsteppe and juniper		
Mountain big sagebrush	20–25	Houston (1973)
Wyoming big sagebrush	>100	Wright and Bailey (1982)
Western juniper	<35	Burkhardt and Tisdale (1976)
Southwestern pinyon/juniper	10–30	Leopold (1924)
Hot desert	No data located	

Source: Modified from Riggs et al. 1996.

Table 139. Selected plants in biomes occupied by pronghorn and tolerance of the plants to fire

	Plant species	
Biome	*Fire tolerant (usually not harmed)*	*Fire intolerant (moderately to severely harmed)*
Grassland	Big bluestem	Cacti
	Broom snakeweed	Sand dropseed
	Fragrant sumac	Scarlet globemallow
	Goldenrod	Sideoats grama
	Lead plant	
	Sand sagebrush	
	Sand shinnery oak	
	Skunkbrush sumac	
Shrubsteppe	Arrowleaf balsamroot	Antelope bitterbrush
	Cheatgrass	Big sagebrush
	Eriogonium	Broom snakeweed
	Green ephedra	Curlleaf mountain mahogany
	Idaho fescue	Three-tip sagebrush
	Indian ricegrass	
	Lambstongue groundsel	
	Mulesear	
	Nevada bluegrass	
	Penstemon	
	Rabbitbrush	
	Sandburg bluegrass	
	Squirreltail	
	Snowberry	
	Tapertip hawksbeard	
	Thurber needlegrass	
	Velvet lupine	
	Western wheatgrass	
	Yarrow	
Hot desert	No data located	No data collected

Source: Modified from Vallentine 1989 and Payne and Bryant 1998a.

Table 140. Percentage composition of vegetation, bare ground and litter, and height of vegetation on wildfire burns and adjacent unburned sagebrush sites on Drakes Flat in south-central Oregon and Long Valley in northeastern California

	Drakes Flat, Oregon		Long Valley, California	
Feature	*Unburned*	*Burned*	*Unburned*	*Burned*
Grass	9	25	12	31
Forb	4	7	1	10
Shrub	26	5	19	8
Average height of vegetation[a]	20 (51)	19 (48)	32 (81)	17 (43)
Bare ground	55	54	63	45
Rock	2	3	1	—
Litter	4	6	4	6

[a] In inches (cm).

Source: Yoakum 1957a for Drakes Flat, Oregon; Yoakum 2000 for Long Valley, California.

After wildlife biologists determined, during the 1960s, that the Hart Mountain shrublands were in late succession, a prescribed fire program was begun. One of the fire sites was the above sagebrush hillside. The year after the burn, the site produced an abundance of herbaceous plants readily consumed by deer and pronghorn. Five years after the prescribed fire, the site also produced young, nutritious sagebrush and bitterbrush for the diet of native wild ungulates. Photo by Oscar V. Deming.

years later, the area supported a summer herd of 55. Deming also reported an extensive, unoccupied shrubland near Abert Rim, Oregon, burned by a wildfire in 1959 and later seeded to grasses. The spring following the burn, grasses and forbs remained green and succulent 3 to 4 weeks longer. After the burn, pronghorn moved onto the area.

Van Dyke (1990) reported that pronghorn slowly colonized previously unoccupied habitat in southeastern Oregon as a result of wildfires between 1985 and 1990. He indicated that vegetative conditions improved for pronghorn after 500,000 acres (202,345 ha) of shrubsteppe burned and opened a canopy of tall, dense shrubs. Van Dyke speculated that prescribed fires produce better habitat conditions because they create greater vegetative diversity.

Benefits and detriments of fire on pronghorn in shrubsteppes were assessed by personnel of the California Department of Fish and Game (1997). Wildfires, often ignited by lightning during summer and autumn, burned shrublands and reverted them to grasslands favorable to pronghorn. However, extensive, repetitive burns sometimes decreased preferred shrubs (for winter browsing) and allowed invasion of noxious and alien plants. Small patch burns increased the growth of bitterbrush, a preferred shrub for pronghorn (Gruell 1995, Yoakum 2000).

According to Courtney (1989), pronghorn on grasslands in Alberta, Canada, benefited from wildfires between 1955 and 1985. Burning of mixed-grass prairie sites produced green herbaceous shoots earlier in the growing season than did adjacent unburned sites, and the former were grazed more readily by pronghorn from autumn through early spring. Stelfox and Vriend (1977) reported that 1 month after burns, pronghorn in Alberta were observed feeding on prickly pear cactus. The spines had been burned off the cactus, and more than 50% of the green stems had been foraged. Apparently, cactus is heavily used when other preferred forage is depleted. Prickly pear cactus is equal or superior to high-quality alfalfa in terms of digestibility (Stelfox and Vriend 1977).

Yoakum (1996) mentioned the effect on pronghorn habitats of burned vegetation in Long Valley, California, northwest of Reno, Nevada. The site had been pronghorn habitat historically, but pronghorn had not been reported there during the 20th century. Large, tall sagebrush and other shrubs dominated the aspect of the landscape, with an average composition of 19% and an average height of 32 inches (81.3 cm) (Table 140). During summer 1973, a lightning fire burned more than 30,000 acres (12,141 ha). The next year, the burn was seeded to various mixtures of grasses and forbs and a few shrub species. Seven years later, range surveys disclosed that plant composition had changed to 31% grasses, 10% forbs and 8% shrubs with a mean height of 17 inches (43 cm). Within several years after the burn, pronghorn pioneered the site. During 1996, it was estimated that the area had a density of two animals per square mile (0.8/km^2) (Yoakum 2000).

The flora of western rangelands have adapted to the influence of natural and human-ignited fires. Given current fire prevention, no-burn and firefighting policies and practices, prescribed fires are needed to perpetuate the former role of natural fires in developing and maintaining plant succession (Riggs et al. 1996). Before the 1990s, prescribed fires were not well accepted or often used by land management agencies; this is changing, but slowly.

Prescribed or controlled fire is more practical and economical than alternatives (Payne and Bryant 1998a). Vallentine (1989) concluded that prescribed fire probably has benefited wildlife more than any other control technique, and its use is appropriate to manipulate wildlife habitat.

Fire has an ecological role in maintaining shrubsteppes and, for management objectives, this interaction can be grouped by the dominant shrubs present (Table 141).

Shrubsteppes are complex plant communities, and their response to fire and other factors varies between geographic areas (Bunting et al. 1987). Prescribed fire can be used as a vegetation manipulation technique to develop mid-seral plant communities that are preferred by pronghorn (Kindschy et al. 1982).

The objectives of using prescribed fire in shrubsteppes varies considerably, but, for enhancing pronghorn habitats, the most common objective is reducing dominant, decadent shrubs and increasing herbaceous plants. This can enhance production of preferred forage and sustain shrubs at heights favorable to neonates. Mature sagebrush plants decline in productivity and become less nutritious with age. Because sagebrush can be an important component in the winter diet of pronghorn, the establishment of young, vigorous plants is a goal of wildlife vegetation management.

In some situations, the objective of prescribed fire may only be to increase desirable forbs. Bunting et al. (1987) reported that, by the third year after a fire, grass and forb

TABLE 141. Environmental characteristics and general fire response of sagebrush communities grouped at series level of classification

Series	*Precipitation*[a]	*Environmental characteristics and fire response*
Mountain big sagebrush	12–20 (25–50)	Productive sites; sagebrush seeds establish readily; sagebrush may return to preburn condition within 15 to 20 years; contains high diversity of perennial forbs; productivity of herbs usually enhanced by burning
Basin big sagebrush	10–18 (25–46)	Many sites now cultivated due to deep soils; often depleted of perennial herbs and invaded with cheatgrass; favorable response to fire occurs when adequate understory is present
Wyoming big sagebrush	7–12 (18–30)	Most arid series; slow sagebrush reinvasion after fire; perennial grasses often grazed out and replaced by cheatgrass; few perennial forbs present in any successional stage; difficult to burn due to low shrub cover and low fine fuel loading
Silver sagebrush	Variable	Less abundant in Great Basin, but extensive east of Rocky Mountains; sprouter, particularly on spring burns; located on fertile, well-drained sites capable of producing high quantities of herbaceous plants
Low (dwarf) sagebrush	Variable	Low fine fuel loadings make burning sagebrush difficult; can sometimes be used as fire breaks; many sagebrush species in this group; highly preferred habitats for foraging and fawning by pronghorn

[a] In inches (cm).

Source: Modified from Bunting et al. 1987.

production on burned sagebrush sites was greater than on unburned sites, and increased productivity continued for more than 10 years. Fire has increased forb production in many situations, according to Harniss and Murray (1973). But not all shrublands have equal potential for this response. For example, Wyoming big sagebrush communities normally have small quantities of forbs, regardless of the ecological condition or successional stage.

Before 1990, wildfires were predominantly suppressed on the Hart Mountain National Antelope Refuge in Oregon, but they now are the primary vegetative management practice (Pyle and Yoakum 1994, Gruell 1995). Field studies revealed that more than 90% of the shrubland was in late succession with little herbaceous undercover. As of 1993, the landscape objective of the refuge became that of sustaining a mosaic of vegetation in different successional stages and to increase diversity and abundance of forbs and grasses in shrub stands. That objective called for maintaining 20% to 30% of shrublands in early and mid-succession, therefore a prescribed fire program was implemented. Prescribed fires have been used since to produce a mixed pattern of burned and unburned patches of vegetation producing 200% to 300% more herbaceous vegetation than adjacent unburned sites—foraging conditions especially desirable for pronghorn (Table 142). From 1985 to 2000, 20 wildfires and 68 prescribed fires burned approximately 16% of the refuge acreage—an average of 1% per year (Table 143).

Prescribed fires as a wildlife management practice to improve vegetative conditions are becoming increasingly accepted for western rangelands. Those that simulate natural fires can increase preferred, nutritious, succulent forage and enhance protective cover for pronghorn fawns. This prescribed fire was ignited behind a firebreak, specifically to improve forage and cover for pronghorn. Photo courtesy of the Arizona Department of Game and Fish, Phoenix.

Table 142. Percentage composition of vegetation, bare ground and litter, and height of vegetation after prescribed fires and on unburned low sagebrush sites on the Hart Mountain National Antelope Refuge, Oregon, and the Sheldon National Wildlife Refuge, Nevada

	Hart Mountain		*Sheldon*	
Characteristic	*Unburned*	*Burned*	*Unburned*	*Burned*
Grass	6	25	12	28
Forb	4	11	14	21
Shrub	25	2	37	15
Average height of vegetation[a]	21 (53)	18 (46)	23 (58)	20 (51)
Bare ground	53	55	32	35
Rock	10	6		
Litter	2	1	5	1

[a] In inches (cm).

Source: Riggs et al. 1996 for Hart Mountain; Gregg 1997 for Sheldon.

A similar history of change in vegetation was reported for Sheldon National Wildlife Refuge in northwestern Nevada (Gregg 1997). Since the early 1900s, shrubs have increased and there is little understory of grass and forbs. A prescribed fire program was started in the early 1990s to modify late succession shrublands, which represented more than 90% of the refuge. The objective was to burn 3,000 to 5,000 acres (1,214–2,023 ha) annually. Generally, treatment units were smaller than 2,000 acres (809 ha), with half of the stand burned and the other half unburned. There has been no problem with invasion of non-native

Table 143. Acres burned from 1985 to 2000 on the 258,000-acre Hart Mountain National Antelope Refuge, Oregon, and the 575,000-acre Sheldon National Wildlife Refuge, Nevada

	Wildfires			*Prescribed fires*		
Refuge	*Number of fires*	*Percentage of refuge burned*	*Number of acres (ha) burned*	*Number of fires*	*Percentage of refuge burned*	*Number of acres (ha) burned*
Hart Mountain	20	6	16,200 (6,556)	68	11	29,000 (11,736)
Sheldon	53	10	58,630 (23,727)	71	5	30,000 (12,141)
Total	73	9	74,830 (30,283)	139	7	59,000 (23,877)

Source: A. Goheen personal communication:2000.

plants, primarily due to their scarcity before burning. Only pronghorn spring and summer rangelands needing increased herbaceous forage have been burned; no treatment has taken place on winter rangelands, where forbs and shrubs are key forage for pronghorn survival. Monitoring studies have been conducted for short-term results, and long-term studies are planned. From 1985 to 2000, 53 wild and 71 prescribed fires burned approximately 15% of the refuge (Table 143). This percentage is within the refuge's management objectives—around 1% per year—to restore healthy, natural vegetative communities.

In the shrubsteppe of the Great Basin, wild and prescribed fire has decreased shrub height and created a mixed community of grasses, forbs and shrubs, thereby improving rangeland for pronghorn (see Beardahl and Sylvester 1974, Page 1975, Yoakum 1980a, 2000, Pyle and Yoakum 1994, Gruell 1995).

The North Eccles Ranch in Elko County, Nevada, was slated for a release of pronghorn to augment a small, stable population during the late 1970s. However, a series of fires burned patches of the rangeland, and these were followed by favorable precipitation years for plant growth. These environmental factors increased herbaceous, succulent forage, and the herds more than doubled from 1985 to 1995, thus negating the need for translocating herds.

Shrublands severely altered by intensive livestock grazing and excessive control of wildfires can become monotypic plant communities providing little diversity of forage for pronghorn and other wildlife. A prescribed fire program on the Sheldon National Wildlife Refuge in Nevada was initiated to rectify this habitat problem. It was designed and implemented to burn irregular patches, resulting in a random pattern of shrub patches in late succession and adjacent shrub-free sites that grew grasses and forbs. Photo by Jim D. Yoakum.

Shrubs are less dominant on the shrub/grassland of central Montana than the Intermountain West. Pyrah (1987) reported the influence of fire on pronghorn habitat in central Montana. With reduction in chemical applications, the practice of prescribed fires was increased to improve forage for livestock. Grasses and forbs were abundant, and shrubs did not rapidly reinvade. Before treatment, shrubs contributed to winter browse and fawning bedsite cover. Pyrah (1987) observed that, because fires removed most of the sparse shrubs and reinvasion was a lengthy process, fire needed to be applied carefully and in scattered locations or under conditions where complete control of shrubs over extensive areas was unlikely. Pronghorn were observed in burn areas, but reasons why they occupied those sites were not investigated.

Natural fires on native mixed grasslands of the Great Plains are important for maintaining vegetation structure and function (see Chapter 16). Fire provides a portion of the energy required for developing and maintaining these grasslands. Today, land managers typically conduct prescribed fires during the late dormant season, or they do not burn at all. Fire effects vary with fire frequency, fire-return interval, livestock grazing practices, vegetation successional stages, weather patterns, edaphic features and topography (Engle and Bidwell 2001).

Natural fire is possible on mixed prairies from late in the growing season before the first frost in autumn, through the dormant season (i.e., after the first freeze in autumn and before the last freeze in spring). The timing of fires relative to season can potentially influence plant composition and productivity and, therefore, influences vegetation management. A variety of plant species composition and production responses are possible, and mixed prairies are far more resilient to burning than commonly accepted. Burns in the early dormant season may increase forbs on mixed grasslands, especially western ragweed and legumes (Hopkins et al. 1948). Mixed grasslands with abundant and diverse forbs sustain pronghorn herds.

Native grass and forb communities in Arizona generally are not harmed by properly timed and conducted burns, but cactus, shrub and tree densities may be reduced (Ockenfels et al. 1994). Burn frequency for western rangelands depends on vegetative types, as noted by Riggs et al. (1996), but fire-free intervals for various western shrublands and grasslands range from 10 to 30 years.

In September 1999, the U.S. Bureau of Land Management (BLM) teamed with the North American Elk Foundation and the Wyoming Department of Game and Fish to burn 17,000 acres (6,880 ha) in Bridger Basin near Kemmerer, Wyoming. Project objectives were to increase forage production and reproductive vigor of grasses, forbs and shrubs for pronghorn, mule deer, elk and other wildlife (Cromrich 2001).

Bridger Basin contains a mixture of juniper, mountain shrub/aspen and sagebrush/grass and forb habitats. Ranchers and BLM managers have developed allotment management plans, stressing the restoration of native vegetation and using prescribed fire to reach this goal. Treated sites would be excluded temporarily from livestock grazing, thereby allowing full natural plant growth.

The 1999 prescribed fire resulted in the blackening of 8,500 acres (3,440 ha) in a mosaic pattern within the burn unit. The treated site was rested from livestock grazing for two growing seasons to allow new plants to produce seeds and establish root systems. BLM wildlife biologist Gavin Lovell reported, "Based on the results of previous burns in the Allotment, we predict abundant regrowth of mountain shrub, grasses and forbs in the area. The treatment will improve critical forage for deer, elk, moose and antelope in the area, and benefit species such as sage grouse and some birds as well" (Cromrich 2001:126).

Because of the numerous cooperators involved, funds contributed and the scale of the project, the Bridger Basin burn won the 1999 National Excellence in Prescribed Burn Award. Acceptance by state and federal agencies, ranchers and conservation organizations of prescribed fires as a vegetation conservation treatment has been high in this region. Therefore, plans are to apply similar management treatments to an additional 30,000 acres (12,141 ha) during the next 10 years.

Some tallgrass prairies are both historic and current pronghorn range (Eccles et al. 1994). Historically, fires started by Native Americans and grazing by large herds of bison adjust tallgrass prairies to be suitable pronghorn habitat. Today, controlled burning and cattle grazing do much the same. Grasses 9 feet (2.7 m) tall can be made suitable for pronghorn when burned or grazed by cattle (as by bison in yesteryear). Research demonstrated that tallgrass prairie burned during late spring started growth earlier, grew more rapidly early in the growing season and produced more herbaceous plants than unburned prairies (Engle and Bidwell 2001).

During the 1994 Pronghorn Antelope Workshop at Emporia, Kansas, attendees witnessed the burning of tallgrass prairie to enhance herbaceous growth for cattle. Burning was undertaken during green-up, when vegetation of the previous year was of low to medium height and weather was cool. Much of the burning was done at night when temperatures were lowest and humidity highest. Also, fires were timed to take advantage of slightly damp vegetation, to moderate fire intensity and rate of travel. Wind speed and direction also were considered, inasmuch as they influenced fire intensity as well as direction and speed of travel.

Simpson (1992) reported how pronghorn reacted to such slow, low-burning treatments. He observed eight pronghorn that stayed on unburned areas just ahead of a fire; smoke and heat apparently did not affect them. At times, when the fire dwindled in areas of low fuel, the pronghorn crossed into the burned area and later returned to the unburned side, giving an impression of playing with the burn. The next day, they crossed a burned area to reach an unburned area. They remained on that site for 2 days, until it was fired as well. The herd then dispersed during the next several days to different burn locations, tending to locate where burns were pocked with areas unburned because of low fuel.

Practices for prescribed fires that can improve and enhance pronghorn habitat have been recommended by Yoakum et al. (1980a), Bunting et al. (1987) and Vallentine (1989) for shrubsteppe communities and by Riggs et al. (1996) and Payne and Bryant (1998a) for grasslands.

Payne and Bryant (1998a) stated that prescribed burning can achieve the following benefits to wildlife: (1) increase forb abundance, (2) retard shrub and tree invasion, (3) temporarily increase grass and shrub palatability and nutrient content, (4) increase vegetative diversity, (5) retard plant succession, (6) alter distribution of ungulates (such as pronghorn), (7) improve forage accessibility, (8) reduce litter and remove decadent material, (9) rejuvenate woody plants for browse production, and (10) stimulate green-up of fire-tolerant grasses.

Previously discussed findings relate how vegetation communities on grasslands, shrubsteppes and deserts experience disturbances that change plant successional stages. These disturbance factors can contribute to healthy plant complexes by invigorating plant regeneration and foliage production. The ecological relationships of disturbance factors (fire) to changes in vegetative communities (succession) are directly related to wildlife (pronghorn) reproduction, health and survival.

Wildfires sustain healthy plant and wildlife communities. Unfortunately, human-caused fires can be untimely and catastrophic, thus wildfires and prescribed fires have been misconstrued at times as detrimental. Natural fires ignited by lightning have been occurring for centuries, and the environment impacted generally has sustained healthy, productive communities of flora and fauna.

Today, land managers recognize that prescribed fires simulating natural wildfires likewise stimulate natural disturbance processes. Details for planning and implementing prescribed fires are available in the literature, such as Bunting et al. (1987), Vallentine (1989), Riggs et al. (1996) and Payne and Bryant (1998a). For this chapter, I have identified direct relationships between fire as an ecological process and changes in vegetation successional stages that aid in the growth of preferred forage and fawn protective cover—factors supportive of healthy pronghorn populations.

A major wildlife habitat objective for prescribed fires on dominant shrublands is to simulate natural wildfires. When this is accomplished, the burn progresses slowly, burning old, decadent shrubs and releasing nutrients for growth of grasses, forbs and young shrubs. Photo by Jim D. Yoakum.

Biological Control

The ecological status of vegetation can be influenced by insects and livestock grazing that alter the appearance, productivity and composition of vegetative communities.

Biological control by insects uses specific host plant/insect relationships to diminish plants classified as noxious. Controls manipulated by humans have been investigated or used on plants common to pronghorn habitat, such as Canadian thistle, halogeton, mesquite and pricklypear (Vallentine 1989). However, literature relative to the effects of such controls on pronghorn welfare could not be located.

Hall (1963, 1965) reported on the impact of the sagebrush defoliator for sagebrush and bitterbrush on pronghorn habitats in California, Oregon and Nevada. This native insect killed 40% to 60% of certain shrub communities during the early 1960s. Its damage to pronghorn habitat was of concern because of the importance of shrubs as browse during winter. Damage to shrub species important as forage to deer and pronghorn has persisted for years, yet no reports substantiate that damage has resulted in serious loss of key browse species. Hall (1965:12) speculated, "This insect possibly could be substituted for chemical sprays in the biological control of sagebrush where range management practices call for the elimination of sagebrush."

Herbivory by livestock could be used as a habitat management practice under the concept of prescribed grazing because livestock are capable of modifying plant structure and composition. Prescribed livestock grazing is defined as managing the degree and timing of removal or protection of the current year's growth consistent with habitat management objectives. Prescribed livestock grazing is further defined as a wildlife habitat management practice as either environmental protective grazing or strategic grazing (Payne and Bryant 1998a).

Environmental protective grazing implies livestock are cautiously managed to protect environments and prevent excessive foraging of plants valued as wildlife forage, or protective cover. This is especially needed in arid and semiarid western rangelands. Guides for environmental protective grazing include: (1) conservative stocking of livestock, (2) adequate rest from livestock grazing, (3) livestock grazing plans that are attuned to wildlife life cycles and the phenological development of plants, (4) monitoring to prevent overuse of key plants important to wildlife, and (5) sufficient residual cover for wildlife. The latter two generally have been practiced by reserving half of the vegetative cover for livestock grazing and half for soil and vegetation needs. However, Holechek et al. (1998) determined that arid and semiarid rangelands may need more than 60% of the primary plant growth retained to meet the requirements of sustaining a healthy plant and soil community.

Although protective grazing is the goal of multiple use on western public rangelands, it seldom has been accomplished (Wagner 1978, Donahue 1999). In practice, livestock grazing may alter pronghorn habitats more than human manipulation of rangelands (Leftwich and Simpson 1978, Kindschy et al. 1982). Negative impacts on pronghorn were reported in Arizona (Ockenfels et al. 1996), Texas (Buechner 1950a, Hailey et al. 1966), California (Ellis 1970), Montana (Pyrah 1987) and New Mexico (Howard et al. 1990). Problems with implementing protective grazing appear greater on shrublands and hot deserts than on the Great Plains, where herbaceous plants are more productive (Yoakum et al. 1996) (see Chapter 15).

According to Eccles et al. (1994), cattle management practices may result in a commensal relationship with pronghorn in the tall grasslands of eastern Kansas. Cattle graze the tall native grasses, and this practice, in combination with periodic prescribed burning, decreases vegetation height. It also may increase forb production. Consequently, the tall grasslands would not otherwise meet pronghorn habitat requirements due to tall vegetation. However, pronghorn translocated to the area during the 1990s have become established.

Strategic grazing is the use of livestock as a vegetation manipulation practice in wildlife habitat management to meet specific habitat objectives (Payne and Bryant 1998a). Before 1800, intensive grazing by bison on prairies apparently increased forbs for pronghorn (Costello 1969), and a similar herbivore/plant relationship may exist today, except that cattle have replaced bison on most western rangelands. Unfortunately, quantified studies relating livestock use to increased forb production for pronghorn could not be found in the literature.

The U.S. Fish and Wildlife Service (1980) developed a resource management plan for the Sheldon National Wildlife Refuge in northwestern Nevada during the late 1970s. One of the objectives was to use livestock grazing as a practice to enhance vegetation for pronghorn and other wildlife. Fifteen years later, it was decided that the objective was not being accomplished and cattle were permanently removed from the refuge.

A principle method of cattle grazing to improve allocated vegetation for pronghorn on the Hart Mountain National Antelope Refuge in south-central Oregon is called preconditioning, according to Anderson et al. (1990a, 1990b). The objective of preconditioning is to enhance the nutrient levels of plant species or assemblages of plants (usually native perennial bunchgrasses) to maintain forage quality primarily for livestock forage. The practice involves systematic livestock grazing of plant growth during the growing season to improve nutrient amounts available in regrowth after initial livestock grazing.

Anderson et al. (1990a, 1990b) suggest that increased pronghorn numbers in south-central Oregon during the 1980s possibly were associated with improved forage quality that resulted from preconditioning of herbaceous plants by cattle. This claim assumes that cattle grazing increased the nutrient content of herbaceous plants preferred by pronghorn, resulting in a population increase. The U.S. Fish and Wildlife Service (1994) assessed Anderson et al.'s (1990a, 1990b) hypothesis and found no quantitative data to support the contention that grazed and ungrazed upland native forbs differed in nutrient content or abundance.

Although a variety of opinions exists on the validity of using livestock grazing as a tool to enhance pronghorn forage, the present knowledge is meager and limited to information presented in this chapter. Severson (1990) conducted a symposium on the use of livestock to improve wildlife habitat and concluded that this could be accomplished, but that it amounted to prescription grazing, that is, establishing specific goals and developing specific management plans to achieve those goals. Such planning considerations involve timing of livestock foraging relative to season and plant phenology, attaining precise utilization levels on some plant species while avoiding use on others, and determining specific tradeoffs (i.e., which other wildlife species will be affected and to what degree). A review of the literature uncovered no management plan outlining specific strategies for the use of livestock grazing to enhance forage specifically for pronghorn.

Mechanical Treatment

For more than 50 years, shrub-dominated rangelands have been treated by 1 or more of 13 mechanical methods (Payne and Bryant 1998a). These practices include blading or dozing, rootplowing, chaining, cabeling, railing, rollerchopping, shredding, rotobeating, mowing, plowing, harrowing, furrowing, discing, pitting and imprinting. Most have been used on private lands to increase livestock forage or prepare soil for crops (Table 144).

The shrub control targets generally have been different subspecies of sagebrush, as well as rabbitbrush. According to Urness (1979), for wild ungulates, too much sagebrush can be as great a problem as too little. I reached a similar conclusion for some habitats occupied by pronghorn in the Great Basin (Yoakum 1980a). Most nontarget species are perennial herbaceous bunchgrasses and forbs, but may include bitterbrush and fourwing saltbush. Herbaceous cover generally doubles after sagebrush removal (U.S. Bureau of Land Management 1989).

Plowing, discing and chaining have been the primary mechanical methods used to improve rangeland as pronghorn habitat, as either a primary or secondary objective (see Plummer et al. 1968, Yoakum et al. 1980a, Vallentine 1989, Fulbright and Guthery 1996). Principles and procedures of pretreatment, treatment and post-treatment were described by Cain (1971), Yoakum et al. (1980a), and Payne and Bryant (1998a). The relative merits and disadvantages of anchor chaining and plowing are indicated in Table 145.

A third practice—called brush beating, cutting or mowing—also is used on rangelands, but results have been less successful than those of plowing or chaining. At times, the treated site may have as many or more shrubs 10 years after treatment as existed before manipulation. However, such results can have their advantages at times for pronghorn and other wild ungulates, as reported for habitats in Wyoming (Gocke 1997). It was determined that some key winter rangelands had more than 60% of shrubs in poor condition due to advanced age. To restore browse production, brush cutters were used to mow decadent shrubs, so young shrub growth would be more productive and nutritious. This practice was used instead of prescribed burns on oil and gas fields. Shrubs were mowed in a mosaic pattern resembling a controlled burn. Islands of taller shrubs were left standing to provide thermal, hiding and nesting cover for wildlife.

When soil and vegetative conditions permit, plowing is a desirable method to decrease shrubs (Plummer et al. 1968, Yoakum et al. 1980a). If soil is too wet or rocks are too large, this method is impractical. Also, to be practical,

TABLE 144. Factors to consider in planning mechanical manipulation projects for enhancing pronghorn habitat

Wildlife factors	*Watershed and soil factors*	*Vegetative factors*
Pronghorn present	Soil fertility	Need for reseeding
Habitat requirements for pronghorn	Soil type	Phenological stage of target and nontarget plants
Size of treatment sites	Soil depth	Invasion of noxious secondary plants
Riparian zones	Soil moisture	Size, density and basal diameter of target species
Native plants released by overstory removal	Debris and slash removal	Resiliency of plants released by overstory removal
Travel corridors	Control of livestock grazing	Season of treatment
Proximity of treated sites to critical habitat components	Erodability	Sprouting potential of shrubs
Influence on vertical structure	Slope	Taxa to be saved
Control of livestock	Heterogeneity	Maintenance of project (longevity) and need for follow-up treatments
Influences on other wildlife	Aspect	Control of livestock
	Site potential	Impact on floral diversity
	Topography	
	Season of treatment	

Source: Modified from Payne and Bryant 1998.

TABLE 145. Relative merits and disadvantages of the three most common mechanical methods used in pronghorn habitat for altering a dominant big sagebrush community to one with a greater mixture of grasses and forbs and fewer shrubs

	Treatment method		
	Anchor chaining	*Plowing or discing*	*Cutting, beating or mowing*
Sagebrush impact	Elimination of 60% to 80% of old rigid shrubs (and 10% to 20% more second time over); and 10% to 40% of young flexible woody plants	Elimination of 70% to 80% of old shrubs and slightly less of young trees and shrubs	Elimination of 50% to 90% of old shrubs and 30% to 60% of young, flexible woody plants
Desirable forage impact	Nonsprouting shrubs are damaged; sprouters recover; little damage to herbaceous species	Kills most shrubs except those that sprout or spread by root stalks	Not effective on sprouting shrubs or herbaceous species
Post-treatment ease of seeding	Apply seed after chaining if once over; or seed before final chaining if twice over	Easily done with drills; seeded area may require packing	Broadcast before cutting or beating heavy shrubs; drill after treating light shrubs
Terrain and soil adaptability	Feasible on all soils and on slopes exceeding 30%; breakage of anchor chain rare	Limited to little or no rock except with use of brushland plow	Limited to sites without protruding rocks
Equipment availability	Commercially available as new or through coastal salvage companies	Plows and discs commercially available; brushland plows are custom-made	Commercially available
Erosion	Usually minimized when chaining is contoured to land form	Exposes soil to moderate degree of erosion	Mulch left decreases hazard

Source: Modified from Pechanec et al. 1965 and Payne and Bryant 1998a.

One mechanical procedure for altering dominant shrublands is the use of a brush cutter—a device designed to kill old-growth shrubs. Such practices control decadent shrubs because they generally do not sprout following treatment. Brush cutters are designed to kill shrubs efficiently and economically on large acreages to increase herbaceous vegetation for wildlife and livestock. Photo by Jim D. Yoakum.

the vegetation should be composed of a higher percentage of undesirable rather than desirable species. Disc-type plows, such as heavy offset discs or wheatland plows, are good for treating nonsprouting shrubs on soils with relatively few rocks. The brushland plow is best for rough, moderately rocky areas.

Plowing or discing to a depth of 3 to 4 inches (7.6–10.2 cm) is recommended for controlling most nonsprouting plants, such as sagebrush. Depths of 4 to 6 inches (10.2–15.2 cm) are required to reduce sprouting shrubs, such as rabbitbrush and others that spread underground root stocks or form a crown. A heavy-duty root plow is required to control root-sprouting species. For a higher than 90% kill, plowing twice in opposite directions is advisable. Single plowing is recommended for improving vegetation for pronghorn, to leave some shrubs for browse and limit impact on forbs. Most plowing is accomplished on sites where understory vegetation is inadequate to regenerate naturally, consequently the project site generally is seeded after plowing.

The Vale Project on public lands in southeastern Oregon was one of the most extensive rangeland restoration endeavors conducted on pronghorn habitat (Heady and Bartolome 1977, Heady 1988, Yoakum 1980a). This large-scale, 11-year (1962–1972) program primarily was designed to increase livestock forage and involved the plowing and seeding of 164,000 acres (66,400 ha). Treated sites met pronghorn habitat requirements better than untreated areas (Table 146), based on comparison of similar lands surrounding the treated areas and estimates of pronghorn populations obtained from aerial surveys by the state wildlife agency. During the early years of the project (1962–1968), the pronghorn herd averaged 1,420 animals per year in the project area. After treatment (1972–1979), the herd increased 83% to 2,600, whereas herds on adjacent nontreated lands increased less than 30% (Yoakum 1980a).

The influences of plowing and seeding on pronghorn in central Oregon's Bear Valley were reported by Polenz (1976) and Yoakum (1980a). The valley was about 10 miles wide by 15 miles long (16 × 24 km), and approximately half was treated. Before 1940, pronghorn were not in the valley, but herds occupied adjacent areas. Bear Valley was primarily private rangeland used for grazing domestic livestock. The area had abundant, tall sagebrush with a sparse understory of grasses and forbs. The sagebrush was plowed and subsequently seeded with a simple mixture of crested wheatgrass and dryland alfalfa. During a 30-year period, ranchers serendipitously changed the valley from a low-quality vegetative community to favorable pronghorn habitat (Table 146). Shortly after the vegetation began to be altered, pronghorn ventured periodically into Bear Valley for short periods. As additional acres were manipulated and winters remained mild, pronghorn established themselves in the valley year-round. The herd increased 168% in 7 years—from 155 animals in 1969 to 415 in 1976 (Polenz 1976). More than 90% of 321 pronghorn observed by the author during September 1995 were found on the treated sites grazing dryland alfalfa—the primary succulent forage available on treated lands.

Chaining involves two tractors dragging a heavy chain—about 100 pounds (45kg) per link—60 to 100 feet

A tractor pulling a plow to kill shrubs and prepare a seedbed for grass seeding is the practice most commonly used on western rangelands for increasing livestock forage. Thousands of acres have been so treated by federal agencies and private ranchers. The vegetation alteration practice has improved habitats for pronghorn by decreasing tall, thick shrubs and establishing seedings of grasses or a mixture of grass and dryland alfalfa. Photo by Jim D. Yoakum.

TABLE 146. Percentage vegetative composition and height of vegetation on treated (shrub rangelands plowed and seeded) and untreated sites at Rome and Bear Valley, Oregon

	Rome		*Bear Valley*	
Feature	*Untreated*	*Treated*	*Untreated*	*Treated*
Grass	32	41	15	35
Forb	2	7	10	23
Shrub	15	4	26	2
Average height of vegetaion[a]	31 (79)	21 (53)	33 (84)	22 (56)
Bare ground	48	44	42	37
Litter	3	4	7	3

[a] In inches (cm).

(18.3–30.5 m) between them through vegetation to break off or uproot woody plants and create a bed for aerial broadcast seeding. When a substantial kill of shrubs is desired, chaining twice (in opposite directions) is recommended. Chaining once kills fewer shrubs, which may be the goal for pronghorn habitats. In areas planned for double chaining, the second pass should follow aerial seeding, therefore the chaining disturbance will aid in covering seed. For optimal benefit to other wildlife, chaining projects should leave fingers or islands of trees and tall shrubs (Cain 1971, Yoakum et al. 1980a).

If an area of big sagebrush was chained to improve herbaceous forage for livestock and pronghorn, the need to rechain on a maintenance basis would depend on: the successful return of native forbs; the success of seedings, if planted; the intensity of livestock grazing; and the rate of shrub reinvasion. If livestock overused herbaceous plants, that forage class could nearly be lost within 10 years. With the loss of herbaceous plants, shrubs increase rapidly in density and height. Within 25 years of the original chaining, shrubs could regain their former density and size.

Yoakum (1980a) reported on the comparative benefits to pronghorn of untreated and chained shrublands north of Pioche, Nevada. The site, about 8 by 12 miles (12.9 by19.3 km) in extent, historically supported pronghorn, but none had been observed in the area from 1900 to 1960, although they occupied nearby rangeland. Before the treatment, the area had been heavily foraged by domestic sheep and cattle, changing the vegetation to a dominant shrub community. Vegetative transect data indicated an average composition of 4% grasses, 1% forbs and 42% shrubs, with a mean height of 32 inches (81 cm). During the mid-1960s, between 6,000 and 8,000 acres (2,428–3,238 ha) were chained and aerially seeded with a complex mixture. Five years later, the plant composition was 29% grasses, 9% forbs and 9% shrubs, with a mean height of 18 inches (45.7 cm). Pronghorn moved onto the site and regularly were observed foraging and using the area for fawning. The original primary objective was to increase forage for cattle and sheep.

Two years after seeding, the site was heavily grazed by livestock, and this use continued for the next 20 years. Range transects in the late 1980s indicated a vegetative composition of 19% grasses, 8% forbs and 26% shrubs,

Landscapes dominated by shrubs in advanced successional stages have been manipulated with the use of two tractors dragging an anchor chain to kill shrubs and prepare a seedbed for plantings. At times the chaining is applied twice to obtain a higher shrub kill. This practice also has been applied to rangelands with old-growth juniper or juniper/pinyon forests. Artificial seedings follow the chaining. Photo by Jim D. Yoakum.

with a mean height of 22 inches (56 cm). Pronghorn occupation of the treatment site became occasional and use of the area for fawning terminated.

Chemical Application

Chemical control of rangeland plants has developed since the 1950s and, by the late 1980s, was the most widely used practice to remove noxious vegetation from western rangelands (Vallentine 1989). Chemical application is common on private lands. Costs are lower than mechanical treatment, and chemicals control noxious herbs as well as woody plants. Although chemical control programs were developed primarily to increase forage for livestock and to improve conditions for agricultural crops, Koerth (1996) concluded that, after decades of controversy, it now generally is accepted that herbicides can be used for wildlife habitat management if positive and negative impacts (Table 147) are considered and proper application techniques are used.

A more cautious viewpoint concerning chemical application was expressed by Payne and Bryant (1998a), who also recognized that these practices provided an alternative to fire and mechanical techniques. However, they advocated that herbicides should be one of the last options considered because future banning of certain chemicals is likely, chemicals may become too expensive when costs are analyzed, and—although their use in wildlife habitat management could have benefits—public pressure could cause drastic restrictions on use of herbicides on rangelands.

Most often, chemicals are used when the vegetative community has an abundance of shrubs and a remnant stand of herbaceous plants. When properly applied seasonally, spraying kills shrubs, and forbs increase. Chemical treatment generally is cheaper than plowing or chaining, but public resistance to the use of chemicals has affected their use since the 1970s. Sagebrush can be sprayed during spring, summer and autumn. Spraying in spring, when snow still covers many forbs, has distinct advantages.

For this book, discussions are limited to chemical application to shrubs on pronghorn habitat. The focus herein is on the management application of herbicides and how such practices change vegetation structure and composition to benefit pronghorn.

Most herbicide applications on rangelands are foliar-active herbicides sprayed directly on plants. Selective chemical spraying may be used to control stands of late-succession, high-density shrubs. A negative facet is that forbs usually are suppressed for at least one growing season post-treatment. Nevertheless, herbicides can be used, providing appropriately patterned applications are used (Koerth 1996).

The variable sensitivity of different plant species to the type, concentration and application timing of herbicides enables discriminating manipulation (patterned applications) of vegetation once these sensitivities are known (Table 148). A mixture of shrub species that react differently to chemicals can make the effects of sprays unpredictable. Mixed stands of big sagebrush and bitterbrush may be sprayed with 2,4-D butyl or isopropyl ester at the standard sagebrush control rate without serious loss of

TABLE 147. Advantages and disadvantages of herbicides compared with other plant manipulation practices

Advantages	*Disadvantages*
Can be used where mechanical methods are impossible, such as on steep, rocky, muddy or timbered sites, particularly with aerial applications	No chemical control has yet proven fully satisfactory for some noxious plant species
Provides a variety of application methods ranging from individual plant treatment to aerial broadcasting	Herbicides provide a desirable, noncompetitive seedbed for artificial seeding only under certain situations
Provides a rapid control method from the standpoint of plant response and area covered when broadcast applied	Costs of control might outweigh expected benefits on low-potential rangeland habitats; this also is true of many other treatment methods
Has low labor and fuel requirements in application	Careless use of chemicals can be hazardous to nontarget plants in the stand and to cultivated crops or other nontarget sites nearby, or might contaminate water supplies
Phenoxy herbicides generally are less expensive than mechanical control methods, but might cost more than prescribed burning	Lack of selectivity might result in killing associated forbs and shrubs important for wildlife
Most herbicides are selective or can be applied selectively so that damage to desirable plant species can be minimized	The effective time period for applying foliage-applied herbicides usually is restrictive
Maintains grass and litter cover and does not expose soil to erosion. Safe and reliable when proper safeguards are followed	Herbicides generally decrease plant diversity
Often can use regular farm and ranch spray equipment	Abundance and diversity of forbs, important to many wildlife species, often are drastically reduced
Soil-applied, but not foliage-applied, herbicides can be applied over a relatively long time period for shrub control	Herbicides can directly kill wildlife or have other undesirable effects on wildlife

Source: From Vallentine 1989 and Payne and Bryant 1998a.

bitterbrush, provided that application is done while the bitterbrush is in bloom.

An aerial chemical treatment project in California—to increase forage for livestock, sage grouse and wild ruminants—resulted in a 95% kill of sagebrush and 18% of decadent bitterbrush (Yoakum et al. 1980). The remaining bitterbrush rapidly developed vigor, producing leader growth 110% greater than that on controls two seasons after treatment. The outcomes were control of excessive sagebrush and increased biomass of bitterbrush and herbaceous vegetation. Pretreatment plant composition was predominantly shrubs with few herbaceous plants.

Table 149 lists the percentage composition of vegetation on pronghorn rangelands in Oregon and Nevada sprayed to test chemicals for control of low sagebrush. Both sites were sprayed with 2,4-D during the 1960s to increase herbaceous plants for pronghorn and livestock. The abundant shrubs were decreased and herbaceous plants increased. Although shrub densities were decreased, more than sufficient quantities remained for pronghorn browsing. The decrease in shrubs allowed more moisture for grasses and forbs, thereby increasing forage during the fawning season. As of 2000, both sites had dominant stands of herbaceous plants with low densities of shrubs. One site was used by a territorial buck, and both sites supported pronghorn, but no quantitative data were gathered.

Chemical spraying of shrubs was tested on pronghorn winter and summer habitats—each about 1 by 0.5 mile (1.6 by 0.8 km) in extent—in central Montana from 1967 to 1987 (Pyrah 1987). The sagebrush/grassland site had a mixture of grasses, forbs and shrubs, but was chemically treated to increase graminoids for livestock. Treatment increased grass production but decreased forbs and shrubs, and winter use by pronghorn subsequently declined. On a year-round basis, the untreated sites were used more than the treated areas. However, because the pronghorn herd increased between 1968 and 1972, indications were that vegetative changes in the treatment site did not seriously affect pronghorn densities. Pyrah (1987) concluded that proper livestock grazing (grazing at a level that did not decrease productivity of herbaceous plants) could have been more effective than the chemical spraying. He also emphasized that partial shrub kill rather than maximum shrub kill sites were preferred by pronghorn.

The Vale Project in southeastern Oregon also was sprayed aerially for certain jobs—primarily with 2,4-D—on 333,000 acres (134,762 ha) to decrease sagebrush and increase grasses (Heady and Bartolome 1977). Of this area,

TABLE 148. Chemical controls of selected shrubs on western rangelands

Common name	*Herbicide*	*Method and season of application*	*Special considerations*
Creosotebush	2,4,5-T	Foliage spray; full flowering to midfruiting	Good kill
	Picloram	Same	
Greasewood	2,4-D	Broadcast foliage spray; spring or early summer during rapid growth	High kill but benefits usually low Sometimes resprouts
Mesquite	2,4,5-T ester	Broadcast foliage spray; 40 to 90 days after leaf growth begins	Repeat when necessary to control sprouts and new seedlings. Increase 2,4,5-T rate to 1 pound/acre (0.9 kg/ha) for dense trees on heavy clay soils
Big rabbitbrush	2,4-D ester	Broadcast foliage spray; new twig growth 3 inches (7.6 cm) and until Sandburg bluegrass loses green in herbage	For high kill, apply 2 pounds/acre (1.8 kg/ha) for 1 or 2 years after initial spraying, burning or mechanical top removal. On dry, sandy soil, root plowing or plowing at 5-inch (12.7 cm) depth gives a good kill
	Picloram	Foliage spray	High kill may not give adequate kill of big sagebrush
Little rabbitbrush	2,4-D ester	Foliage spray; 3 to 6 inches (7.6–15 cm) new growth	Better simultaneous control when applied toward end of sagebrush susceptibility period
	Picloram	Foliage spray; full leaf stage	May not give adequate kill of big sagebrush
Big sagebrush	2,4-D ester	Broadcast foliage spray; spring after first new leaves reach maximum length, or from early heading to marked loss of color in Sandburg bluegrass heads	Burning or plowing are also effective Increase rate to 3 pounds (1.4 kg) or use 2 pounds (0.9 kg) silvex to control reinvasion into seeded grass stands
Low sagebrush	2,4-D ester	Broadcast foliage spray; spring, during rapid growth, or late boot stage to preanthesis in Sandburg bluegrass	Top removal by burning or mechanical means is also effective
Tarbush	2,4,5-T ester	Foliage spray; rapid growth	Often impractical for large-scale programs

Suggested chemicals and methods of application vary with locality and are continually being modified as research progresses (adapted from Vallentine 1989). For these reasons, the suggestions given in this table should be considered tentative. Further information and use clearance should be obtained locally.

TABLE 149. Percentage composition and height of vegetation on test sites sprayed with 2,4-D and adjacent unsprayed sites in low sagebrush habitat on the Hart Mountain National Antelope Refuge, Oregon, and the Sheldon National Wildlife Refuge, Nevada, during the 1960s

	Hart Mountain		*Sheldon*	
Feature	*Unsprayed*	*Sprayed*	*Unsprayed*	*Sprayed*
Grass	13	33	6	30
Forb	3	4	12	26
Shrub	23	1	31	6
Average height of vegetation[a]	19 (48)	18 (46)	16 (41)	16 (41)
Bare ground	56	56	39	32
Rock	—	—	10	—
Litter	5	6	2	6

[a] In inches (cm).

21,000 acres (8,499 ha) were seeded with crested wheatgrass and dryland alfalfa. The treatments produced the desired vegetative change, and pronghorn subsequently pioneered sites from which they had been absent for more than 50 years, especially the sites seeded with dryland alfalfa. Areas with large sagebrush skeletons after treatment were less frequented by pronghorn than were other sites (Kindschy et al. 1978, 1982). The pronghorn population of the project areas more than doubled by 15 years posttreatment (Yoakum 1980a).

The herbicide tebuthiuron was used to control abundant sagebrush on pronghorn range in northwestern New Mexico from 1985 to 1987 (Hooley 1992). Approximately 5,500 acres (2,226 ha) were treated at the rate of 0.5 pounds per acre (0.56 kg/ha). The treatment killed 90% of the sagebrush; grasses increased 14%, and forbs increased 2% (Table 150). Treated sites comprised 40% of the pronghorn habitat, but herds were observed in them 68% of the time (yearlong mean). Pronghorn showed a preference for the treated sites in every month except January. Diet-selection studies, through analysis of fecal samples, disclosed pronghorn consumption of 2% grasses, 31% forbs and 67% shrubs. Hooley (1992) concluded that tebuthiuron application to sagebrush benefited pronghorn by increasing herbaceous production for spring and summer forage. She also noted the need of shrubs for winter browse.

Artificial Seedings

Western wildlands (lands not cultivated for agricultural production) with insufficient plants can be seeded artifi-

To determine what plant species are adapted to specific rangeland sites, trial plots of seeds from native and alien grasses, forbs and shrubs are planted. Such experiments help determine which plants will grow in specific soil types and climates. Findings are helpful for programs dedicated to restoring or rehabilitating rangelands severely changed by fires, intensive grazing, abandoned agriculture, and so on. Photo by Jim D. Yoakum.

TABLE 150. Percentage composition of litter and vegetation on tebuthiuron-treated and adjacent untreated sagebrush areas in northwestern New Mexico

Vegetation	*Untreated control sites*	*Tebuthiuron-treated sites*
Litter	63	67
Grasses	15	29
Forbs	1	3
Shrubs	20	1
Miscellaneous	1	0

Source: Hooley 1992.

cially to improve composition and herbage yield (Vallentine 1989, Roundy 1996). Artificial seeding refers to bringing seeds from off the site and planting them, as opposed to allowing plants growing on the site to produce and disperse seeds to produce seedlings. Artificial seedings often are conducted in conjunction with management practices, such as shrub control and rehabilitation following wildfires or strip mining. Low-quality plant communities can result from repeated wildfires that destroy native shrub and herbaceous communities (Leopold 1959b, Payne and Bryant 1998b).

At least in part, western rangelands have been seeded artificially to improve forage for livestock (Johnson 1986, Roundy 1996, Donahue 1999). When properly planned and implemented, seedings can result in favorable conditions for pronghorn; improperly planned, they may have little value for pronghorn (Yoakum 1980a, 1983). Changing a dominant shrubland to a dominant grassland provides little benefit to pronghorn, which need the forage and cover availed by a mixed grass/forb/shrub community.

Artificial seedings generally involve preparing a seedbed followed by drilling or broadcasting seed. Such revegetation often is needed after land is treated with mechanical manipulation, herbicides or fire, or on habitats that are denuded because of overuse, abuse or neglect (energy extraction, extensive livestock grazing, etc.). At times, regular farming methods provide adequate chances for establishing plantations on relatively flat lands with few rocks. Rangeland drills were developed for seeding rugged, often rocky, sites (Young and McKenzie 1982). The rise in the importance of reclamation of altered western rangelands had increased the need for drills capable of seeding mixtures of plant species. Field trials of air drills in the late 1990s hold promise that this new drill is well adapted to planting mixed seeds of grasses, forbs and shrubs on rangelands (Wilson et al. 2001).

Seeds of shrub species can be mixed with seeds for grasses and forbs and drilled or broadcast. More than 75,000 acres (30,353 ha) of rangelands in Utah have been planted by such methods, primarily for restoration of deer habitat, but this often coincides with pronghorn habitat. The adaptation and relative value of 53 species of grasses, forbs and shrubs for seeding various western rangelands is provided in Table 151 (Plummer et al. 1968, Yoakum et al. 1980a). The list contains 25 species of grasses, but it has only 13 forbs and 5 shrubs. This is because past research on rangeland restoration has been directed primarily toward livestock forage needs with fewer investigations on wildlife forage. This has been especially unfortunate for pronghorn because their preferred forage is forbs and shrubs (see Chapter 16). However, the knowledge of forb and shrub restoration techniques will be greatly improved with Monsen and Stevens' (in press) substantial update of the Plummer et al. (1968) publication on big game habitat restoration. Hopefully, it will provide detailed treatment of ecology, management and revegetation guides for many more forbs and shrubs on western rangelands.

Rangeland drills were designed to plant grass seeds, but most were of limited effectiveness for planting of shrub seeds. The Hansen browse drill was designed for that purpose. Above, a Hansen browse drill pulled by a tractor plants bitterbrush seed on an established crested wheatgrass seeding following a wildfire—key foraging habitat for mule deer and pronghorn in northeastern Nevada. Photo by Jim D. Yoakum.

Table 151. Recommended grasses, forbs and shrubs for seeding in various precipitation and vegetation zones in the Intermountain West[a]

	Lowlands				*Highlands*	
Vegetation	*<8 inches (20.3 cm) precipitation*	*8 to 12 inches (20.3–30.5 cm) precipitation*	*>12 inches (30.5 cm) precipitation*	*Salty soils*	*Mountain brush[b]*	*Aspen*
Grasses						
Sand dropseed	C	C	—	—	—	—
Bottlebrush squirreltail	C	C	—	—	—	—
Indian ricegrass	C	C	C	—	C	—
Russian wildrye	C	B	B	B	C	—
Crested wheatgrass (standard)	B	A	A	—	B	—
Crested wheatgrass (fairway)	B	A	A	—	A	C
Bulbous bluegrass	—	X	X	—	X	—
Bluebunch wheatgrass	—	B	B	—	B	C
Beardless wheatgrass	—	B	B	—	B	C
Pubescent wheatgrass	—	C[b]	A	—	A	B
Intermediate wheatgrass	—	C[b]	A	—	A	B
Western wheatgrass	—	C[b]	B	—	C	C
Beardless wildrye	—	C[b]	B	C	C	C
Big bluegrass	—	C[b]	C	—	C	—
Mountain rye	—	—	X	—	X	—
Great Basin wildrye	—	—	B	—	B	C
Tall wheatgrass	—	—	B	A	B	C
Tall fescue	—	—	—	B[c]		C
Bulbous barley	—	—	—	—	B	C
Blue wildrye	—	—	—	—	B	B
Bearded wheatgrass	—	—	—	—	B	B
Smooth brome (southern strain)	—	—	C[c]	—	A	A
Smooth brome (northern strain)	—	—	—	—	B	A
Slender wheatgrass	—	—	—	—	B	B
Mountain brome	—	—	—	—	B	B

continued on next page

TABLE 151—*continued*

	Lowlands				*Highlands*	
Vegetation	*<8 inches (20.3 cm) precipitation*	*8 to 12 inches (20.3–30.5 cm) precipitation*	*>12 inches (30.5 cm) precipitation*	*Salty soils*	*Mountain brush*[b]	*Aspen*
Meadow brome	—	—	—	—	B	B
Kentucky bluegrass	—	—	—	—	X	X
Tall oatgrass	—	—	—	—	A	A
Orchardgrass	—	—	—	—	B	A
Reed canarygrass	—	—	—	—	B[c]	B
Timothy	—	—	—	—	B[b]	A
Meadow foxtail	—	—	—	—	B[b]	A
Sheep fescue (Sulcata)	—	—	—	—	—	C
Red fescue (sod-forming)	—	—	—	—	—	C
Winter rye	—	X	X	—	X	—
FORBS						
Alfalfa	—	C[c]	B	—	B	C
Sicklepod milkvetch	—	C[c]	B	—	B	C
Chickpea milkvetch	—	C[c]	B	—	B	B
Yellow sweetclover	—	—	X[c]	B[c]	X	C
Strawberry clover	—	—	—	X[c]	—	—
Birdsfoot trefoil	—	—	—	—	C	C
Mountain lupine	—	—	—	—	—	C
Alsike clover	—	—	—	—	—	C[c]
Summercypress	—	—	—	X	—	—
Fivehook bassia	—	—	—	X	—	—
Palmer penstemon	—	X	X	—	X	—
Wasatch penstemon	—	—	—	—	X	X
Showy goldeneye	—	—	—		X	X
Common cowparsnip	—	—	—		C[c]	C
Sweetanise	—	—	—		C[c]	C
SHRUBS						
Winterfat	C	C	C		C	—
Fourwing saltbush	C[c]	C	C		C	—
Antelope bittersbrush	—	C	C		C	C
Oldman wormwood	—	—	X		X	X
Blueberry elder	—	—	X		C	C
Sagebrush	—	B	A		A	B

[a] A = Productive and widely adapted for seeding throughout the zone or type.
B = Valuable over much of the zone or type, but value or adaptation more restricted or not as well determined as species designated A.
C = Value or adaptation more restricted than those designated B, but useful in some situations.
X = Recommended for special uses or conditions, usually as pure stands.
[b] Also applicable for seeding openings in the Ponderosa pine zone.
[c] Adapted only to better than average sites in the zone or type.
Source: Adapted from Plummer et al. 1968 and Yoakum et al. 1980a.

Ideally, plant seeds are selected for restoration projects to meet management objectives and to establish and persist on the treatment site. In practice, revegetation projects often use seeds that are considered adapted, have proven establishment ability and are commercially available. Numerous species trials on many different sites have generated considerable information, and species recommendations have been published for different climatic, edaphic and management situations (Plummer et al. 1968, Yoakum et al. 1980a, Monsen and Shaw 1983, Vallentine 1989, Payne and Bryant 1994a, Monsen and Stevens in press). In addition, a Plant Information Network has been established by Colorado State University and the U.S. Fish and Wildlife Service to provide an array of computerized information on biology, distribution, revegetation potential, management and value of native plant species for wildlife (Roundy 1996).

Seeding programs are of three types—single species, simple mixture, and complex mixture (Yoakum 1983, Urness 1986). The relative merits of single-species and multiple-species seedings are listed in Table 152. The best species for optimal results varies with soils, precipitation, elevation and other factors (Vallentine 1989, Payne and Bryant 1998a).

Single-Species Seeding. Crested wheatgrass has been used most frequently in single species seedings (Heady and Bartolome 1977, Vallentine 1989, Roundy 1996). Other species of grasses, such as intermediate, Siberian and western wheatgrass, have been planted, but none as extensively.

The influence of crested wheatgrass seedings on pronghorn populations has not been studied sufficiently. Reeher (1969) conducted a 2-year study on rangeland seedings—predominantly single-species seedings of crested wheatgrass, although some were mixtures of crested wheatgrass and dryland alfalfa and a few included shrubs in southeastern Oregon. He noted that pronghorn used plowed and seeded projects more than sprayed and seeded treatments. Spalinger (see Yoakum 1980a) analyzed pronghorn fecal samples collected during late winter of 1977 from a crested wheatgrass seeding in Malheur County, Oregon. Although crested wheatgrass was the dominant plant on site, only an estimated 2% by volume was found in the fecal samples.

Urness (1986) noted that, inasmuch as pronghorn and crested wheatgrass seedings had both increased during the prior 50 years, the seedings had little negative impact on herd numbers. To the contrary, crested wheatgrass seedings may have been, as predicted by Leopold (1959b), an important factor in gradual pronghorn population increases. This could have been a result of favorably altered vegetative structure (decreased shrub height and density), increased succulent spring forage and, for some seedings, increased forb production.

Simple-Mixture Seeding. Simple-mixture seeding is the practice of sowing two to five plant species, often one or more grasses and one or two forbs. Such seedings are attractive to pronghorn, especially when dryland alfalfa is one of the forb species sown and 5% to 10% of native shrubs were retained (Yoakum 1980a).

Dryland alfalfa has been one of the most successful forbs seeded on pronghorn rangelands in southeastern Oregon (Kindschy 1974, 1986, 1996, Yoakum 1980a, 1983). More than 56,000 acres (22,662 ha) were planted in 36 separate seedings in Malheur County, Oregon, from 1965 to 1975. The intention was to increase forage for livestock and wildlife. Dryland alfalfa plus a few other grass, forb and shrub species—including sanfoin and penstemon—were planted, but the amount was minute compared with the dryland alfalfa. Dryland alfalfa was applied aerially on sagebrush rangeland that had been plowed and then drilled with crested wheatgrass. A 10-year assessment of 20 of the seedings during 1976 disclosed that alfalfa had maintained 10% of the composition and that pretreatment native forb composition of 2% increased to 7% after treatment (Table 144). Pronghorn herds more than doubled in size within 15 years.

A simple-mixture seeding proved beneficial to pronghorn in Bear Valley, Oregon. As previously noted, pronghorn were not found in the valley before 1940, and the valley was predominantly private cattle ranches. Ranchers manipulated upland vegetation to decrease abundant, tall, unpalatable (to cattle) sagebrush and to enable planting of crested wheatgrass and dryland alfalfa. Treatment was primarily during the late 1950s, 1960s and early 1970s. Vegetative composition of the treated area changed from a shrubland with 15% grasses, 10% forbs and 26% shrubs to 35% grasses, 23% dryland alfalfa and other forbs, plus

Table 152. Relative merits of single- and multiple-species artificial seedings

Advantages of multiple species	*Advantages of single species*
1. Less risk in getting a stand where soil is heterogeneous	1. Easier to seed evenly
2. Longer green and succulent period	2. Provides feed at season most needed
3. Higher yield	3. Plants have similar growth and regrowth characteristics
4. Higher nutritive value	4. More precise site adaptation
5. Provide a varied diet	5. More stable plant composition
6. Stand less adverse to total loss from adverse climatic conditions or insect damage	6. Less expensive to establish
7. Occupancy of land more rapid; better ground cover	7. More rapid establishment
8. Better adapted to variable canopy, soil, terrain and climatic conditions	8. Only one species may be fully adapted to the site
9. More efficient use of the total soil profile	
10. Provides food and cover for diverse wildlife	
11. Other species can be included when seed is too scarce or limited for full seeding	
12. More closely simulates natural plant communities	

Source: From Vallentine 1989.

2% shrubs on treated rangelands as of 1982 (Table 144). Pronghorn gradually pioneered the valley, established themselves and thrived.

Generally, pronghorn benefit from seedings that produce a mixture of grasses, forbs and shrubs. Monotypic grass seedings or untreated rangelands dominated by shrubs and lacking herbaceous plants are less beneficial to pronghorn. For example, on the Chicken Creek site of the Vale restoration project in southeastern Oregon, a simple-mixture seeding produced 35% grasses, 20% forbs and 2% shrubs, with 43% nonvegetation (Urness 1986). Pronghorn pellet groups were more than three times as numerous on the seeded site than on adjacent untreated, native shrubsteppe sites that served as a control.

Reclaimed mine lands were examined for pronghorn use and values in the Powder River Basin, Wyoming. Reclaimed sites were seeded to simple mixtures of grass, forb and shrub species. Forbs, especially nomad alfalfa and yellow sweet clover, were consumed at high frequencies during spring, summer and autumn. Pronghorn and mule deer selected foraging sites containing residual alfalfa during winters. Presence of alfalfa was identified as an important predictor of pronghorn use on reclaimed land during all seasons. Areas of lower plant height also were preferred foraging sites for pronghorn. Seeded grasses were consumed lightly. Pronghorn use patterns reflected avoidance of human activity areas. Enhancement of reclamation sites for pronghorn may include increasing the establishment of a diversity of forbs, maintaining areas of low vegetative height and assimilating small tracts of reclamation into large continuous areas (Gamo 1997a). Similar findings were reported by Medcraft and Clark (1986).

Pronghorn did not frequent reclaimed lands as much as they did nearby native sites (Table 153). This was noted for all seasons. Other investigators also have noted that pronghorn use reclaimed areas less than native vegetal communities (Segerstrom 1982, Medcraft and Clark 1986). Greater pronghorn use was observed on larger reclaimed areas than on small ones. Managers desiring to attract pronghorn to reclaimed sites should experiment with artificial seed mixtures that produce abundant herbaceous forage, particularly legumes and early growing grasses (Gamo 1997).

Complex-Mixture Seeding. Complex-mixture seedings serve pronghorn best because they contribute to reestablishing a mixed plant community of grasses, forbs and shrubs. This simulates natural vegetation communities more than monotypic cultures. Complex-mixture seedings involve plantings of 6 to 30 plant species (Plummer et al. 1968, Yoakum et al. 1980). These mixtures vary, depending on the purpose of the restoration project, but should include species foraged by pronghorn and adapted to the treatment site. Table 151 lists species compatible with varying ecological conditions.

Plummer et al. (1968), investigated more than 30 years of seedings by agencies on at least 24,000 acres (9,715 ha) of successfully rehabilitated rangelands. They also planted and investigated test plots and recommended that mixtures contain a minimum of six species each of grasses, forbs and shrubs. Their findings serve as a valuable guide for restoring western rangelands. The principles and procedures advocated therein have wide application for pronghorn-occupied rangelands throughout the West, and include the following:

1. *Changes in plant cover by the proposed measures must be desirable.* Often, lighter livestock grazing allows desirable species to increase and may be all that is necessary. Pyrah (1987) supported this finding for pronghorn habitats grazed by cattle in central Montana.
2. *Terrain and soil types must be suited to the changes selected.* The terrain and soil should be carefully considered to

TABLE 153. Seasonal proportions (percentages) of observed distribution of radio-collared pronghorn among reclaimed and native areas by year and season, 1995 and 1996

			Percentages of observations			
Year	*Habitat*	*n*[a]	*Spring*	*Summer*	*Autumn*	*Winter*
1995		275	(n = 52)	(n = 82)	(n = 55)	(n = 86)
	Reclaimed		31	26	31	23
	Native		69	74	69	77
1996		139	(n = 46)	(n = 15)	(n = 0)	(n = 78)
	Reclaimed		87	27	0	32
	Native		13	73	0	68
Total		414	(n = 98)	(n = 97)	(n = 55)	(n = 164)
	Reclaimed		22	26	31	27
	Native		78	74	69	73

[a] n =number of observations.

Source: Gamo 1997.

Rangeland restoration projects conducted to decrease dense stands of shrubs and trees are followed with seedings. A mixture of grasses, forbs and shrubs—preferably at least six species of each forage class—is recommended for pronghorn habitat. The seeding mixture is an attempt to simulate natural plant communities containing diverse combinations of shrubs and herbaceous plants. Using such a mixture, thousands of acres of pronghorn rangeland have been restored. Photo by Jim D. Yoakum.

determine which treatment would produce the desired forage effect (see Table 151).

3. *Precipitation must be adequate to assure establishment and survival of seeded plants.* The season and amount of precipitation, along with occurrence of indicator plants, are the most important guides as to what species may be seeded successfully.
4. *Competition with other plant species must be low enough (not more than 15% canopy coverage) to assure that desired species can be established.* Eliminating competition from trees and shrubs can be accomplished by prescribed burns, chaining, plowing and herbicidal control.
5. *Only species and strains of plants adapted to the area should be planted.*
6. *Complex mixtures are recommended over single species or simple mixtures.* Seed mixtures are advantageous when the proposed objective is improvement of wildlife habitat. A general rule is to include a minimum of six species each of grasses, forbs and shrubs.
7. *Sufficient seed of acceptable purity and viability should be planted to assure a successful stand.* The amount of seed per acre depends on seed purity (the percentage of desired seed—often seeds of other species are included in commercial seeds or those collected by volunteers), size and viability, and whether drilled or broadcasted.
8. *Seeds must be covered.* Planting deeper than 0.5 inch (1.3 cm) seldom is desirable; likewise, leaving seeds exposed is unsatisfactory.
9. *Planting should be done in the season of optimal conditions for establishment.* Where climate permits, seeding in winter (December, January and February) is best for intermountain landscapes. Late autumn (late October and November) is next best. Transplanting of nursery stock, seedlings and wildlings is most successful when the ground is wet from spring snowmelt.
10. *Newly planted areas must be protected.* Young plants and seedlings should not be overgrazed or severely trampled by livestock and wild ungulates. Exploitation by rabbits, rodents and insects should not impact significantly plant community productivity. Although much remains to be learned about reducing costs, successful shrub treatment followed with mixture seedings for large patches depends on the time-tested principles and procedures. Their importance for restoring deteriorated pronghorn range cannot be overemphasized (Yoakum et al. 1980a, Lee et al. 1998).

Fertilizing Uplands

Application of fertilizers to upland rangelands to stimulate growth of desirable vegetation has been limited mainly to private lands. Vegetation yield responses were greatest on mesic sites with deep to medium textured, low-fertility soils and sites averaging more than 15 inches (38 cm) of precipitation annually (Vallentine 1989).

Most rangeland vegetation responds favorably to nitrogen treatments, and legumes can be improved by application of phosphorous (Vallentine 1989). Forbs react more favorably to nitrogen than do grasses in some treatments. Legumes can respond to nitrogen, but once nodules are well developed, additional nitrogen is of no use. Legumes also can be improved by sulfur. Grass-to-legume ratios can be affected by fertilizer applications in grasslands, therefore managers should be cautious about nitrogen fertilizers depressing legumes (Payne and Bryant 1994a). Phosphorus and sulfur applications generally benefit forbs.

According to Heady and Child (1994), fertilizing range uplands can: increase palatability of forage, increase available nutrients to plants and ungulates, change botanical composition, affect seasonal distribution patterns of herbivores, extend herbaceous forage into autumn, and induce early spring green-up of grasses and forbs.

The first field test of fertilizer application to enhance pronghorn habitat found in the literature was conducted during 1975 to 1977 on a 19.7- by 49.2-foot (6 by 15 m) plot within a 494-acre (200 ha) pasture in southeastern Alberta

(Barrett 1979, 1982). Treatments were three levels of nitrogen—185, 370 and 555 pounds per acre (207, 415 and 662 kg/ha)—on prairie grasslands dominated by blue gramma, needle-and-thread and western wheatgrass, which were used as winter habitat. Forage quality (measured protein) on treated sites increased initially, but by late summer, nutritional content was essentially similar in treatment and adjacent control sites. Nevertheless, forage production increased markedly during each of the 3 years, resulting in greater total protein per area. Application of phosphorus also was tested but was shown to have little impact on forage quality or production. Although nutrient content did not increase as a result of the phosphorus application on winter rangeland, it did increase during rapid growth in spring. This information may be of interest to managers of pronghorn habitat in shrubsteppes, where herbaceous forage apparently is inadequate during the critical periods of late gestation and lactation (Ellis 1970). The Alberta study indicated pronghorn selectively used treated sites twice as heavily as control sites. Barrett (1982:179) concluded that, "The general increase in total forage production and hence total protein production, and the preference of pronghorns for treated areas, suggest that the procedure should be evaluated further."

A second study of applying fertilizers on pronghorn habitat was conducted on shrubsteppe summer rangeland in south-central Wyoming (Yeo 1988). Treatment was 19.6 pounds of nitrogen per acre (22 kg/ha). Only findings regarding application effects on sagebrush were obtained; no data were collected relative to herbaceous plants or vegetative production. Results indicated that pronghorn did not favor the shrubs on the fertilized sites, probably because neither protein nor moisture content of sagebrush increased significantly after treatment and pronghorn use of shrubs naturally decreases during spring.

Although the value of fertilizers to increase production of nutritious grasses for livestock has been reported (Vallentine 1989, Heady and Child 1994), there is a paucity of information regarding the influences of fertilization on forbs and shrubs. More research is needed concerning the effects of fertilizers on these preferred pronghorn forages.

WATER DEVELOPMENTS

Pronghorn drink from various natural and artificial water sources (Einarsen 1948, Sundstrom 1968, Beale and Smith 1970, Yoakum 1980a, 1994, Kindschy et al. 1982, see also Chapter 16). From 1962 to 1967 in Wyoming, Sundstrom (1968) observed pronghorn drinking from springs, streams, reservoirs, guzzlers, lakes and troughs filled by windmills. At times, pronghorn were seen alongside livestock drinking from artificial sources. However, Hoover et al. (1959) observed that pronghorn in Colorado were reluctant to drink from stock watering tanks, but readily drank overflow waters.

Beale and Smith (1970) speculated that water developments may enable greater distribution of pronghorn where natural water sources are limited, particularly during dry seasons and drought years. That water developments can be beneficial to pronghorn is a view shared by many investigators (Beale and Holgren 1975, Yoakum 1978a, Salwasser 1980, Kindschy et al. 1982, Pyshora 1982b, McCarty 1986, Tsukamoto and Stiver 1990, Kie et al. 1994, Brigham and Stevenson 1997, Lee et al. 1998, Payne and Bryant 1998a). Their reports provide construction specifications and rationale for installing water developments, but little information is given on the biological needs of pronghorn or the environmental impacts (increased livestock and predator use of an area) of such developments. For northern pronghorn populations, densities were higher where drinking water was readily available (Sundstrom 1968, Pyrah 1987), in contrast to lower densities of southern herds (Hughes and Smith 1990, Cancino et al. 1994) where drinking water was scarce (see Chapter 16). There is no water consumption-related distinction between northern and southern pronghorn. However, herds associated with cold, snowy winters generally enjoy high densities of animals. Those associated with droughts and milder winters typically are less abundant. The California Department of Fish and Game (1997) reported that large numbers of drinking water sources were constructed on pronghorn habitats in the northeastern region of that state—primarily for livestock. Water apparently is more available to pronghorn now, regardless of droughts, than it was before settlement in the 1800s.

Deblinger and Alldredge (1991) experimented with pronghorn use of water developments during summer and autumn 1988 to 1989 in Wyoming, and concluded that drinking water did not appear to affect numbers or distribution. However, their study was not conducted during years of unusually high temperatures or desiccated vegetation.

Although artificial water sources have provided drinking water for many pronghorn, Cutler (1996) noted that they can have negative effects due to poor water quality and the possible spread of diseases, such as bluetongue. Water development could likewise have indirect negative effects on pronghorn (and other endemic wildlife) by facilitating competition with non-native species for drinking water. Burkett and Thompson (1994) recommended water developments not be constructed when feral or exotic ungulates will benefit to the detriment of native species. In addition, there may be increased risk of predation to pronghorn that frequent water developments. These facilities may benefit predators by increasing interactions between predators and prey (McIvor et al. 1994). Conversely, water developments may allow some wildlife species to inhabit areas that otherwise would be unsuitable (Broyles 1995).

Recognizing the merits and disadvantages of water developments to pronghorn and other wildlife, managers should carefully assess the needs and values of developing artificial water sources. The following recommendations by Burkett and Thompson (1994) were modified for pronghorn habitats and should be assessed when contemplating the development of drinking water.

1. Water units may serve strategic purposes in a resource management program. Installation or maintenance of water units should follow preparations of specific objectives for each site.
2. Before extensive maintenance or installation of water units, natural water sources should be inventoried for spatial relationships with planned human-constructed water units.
3. After commitment to a strategic water unit program, a systematic database should be maintained regarding unit condition. A routine inspection and repair schedule is necessary to ensure that water units remain functional to meet stated objectives.
4. Systematic detection of use during several years at newly installed units relative to preinstallation conditions would more clearly reveal benefits from the developed projects. Periodic measurements are mandatory if managers are to quantify progress toward stated objectives of a water development program. The approach also provides early warnings concerning the potential interaction of native and exotic species.

Using penned animals, Whisler (1984) determined that summer water requirements for adult pronghorn in Wyoming were 0.95 gallon (3.6 L) per day per 100 pounds (45 kg). The needs of free-ranging pronghorn accustomed to available drinking water may be greater. Information relative to water requirements of pronghorn is provided in Chapter 16.

Springs and Seeps

Water volume from springs and seeps can be enhanced, but if improper techniques are used, part or all of the water flow can be lost. No two springs or seeps are alike and different techniques are necessary to affect each flow increase (Yoakum et al. 1980a, Payne and Bryant 1998a). The reliability and quality of natural flow should be ascertained before enhancement is attempted.

When planning spring or seep enhancement projects for pronghorn and other wildlife, managers should:

1. provide as natural a drinking environment as possible, meaning unencumbered visibility and no steep banks;
2. allow for adequate vegetation around the watering area, either by saving the natural cover or by plantings; for pronghorn, vegetation generally should be no taller than 24 inches (61 cm), but a few shrubs or trees could improve a project for other wildlife;
3. provide signage to inform visitors to the site of the purpose for the structures;
4. provide storage capacity that will supply water during all seasons;
5. fence the site from livestock, to protect the water source and perimeter vegetation; fences should be negotiable by wildlife, except where intensive trampling or wallowing may damage the water source;
6. construct gentle basin slopes or ramps to drinking or storage facilities to prevent drowning of small or young wildlife; and
7. provide public access, when applicable, to water by piping it 50 to100 yards (46–91 m) outside the fenced area; camping should be discouraged within 0.5 miles ([0.8 km) of drinking water.

Generally, installing a collection box to catch and store water at the source is necessary (Figure 118). A large-capacity storage tank can be installed where flow varies, so excess water can be stored a short distance from the source and made available when water flows become low or intermittent. The water source, whether one or several, should be channeled or piped to a collection box or tank and piped to a nearby trough or other drinking structure. The size and type of pipe can vary according to engineering specifications for each project. The drinking trough can be of concrete, metal, plastic, or redwood.

Another technique for spring or seep enhancement to benefit pronghorn and other wildlife is development of a horizontal well that can increase and control flow (Figure 119). Such wells also optimize water quality and minimize source vulnerability to contamination. Bleich et al. (1982), Bleich (1990) and Kie et al. (1994) discuss in detail where and how these wells are practical. Horizontal wells can be inexpensive and low in maintenance costs compared with vertical wells.

Reservoirs

Reservoirs are ground depressions that collect and hold water. Throughout the West, thousands of small reservoirs have been excavated to entrap and retain precipitation for livestock and wildlife. In Malheur County, Oregon,

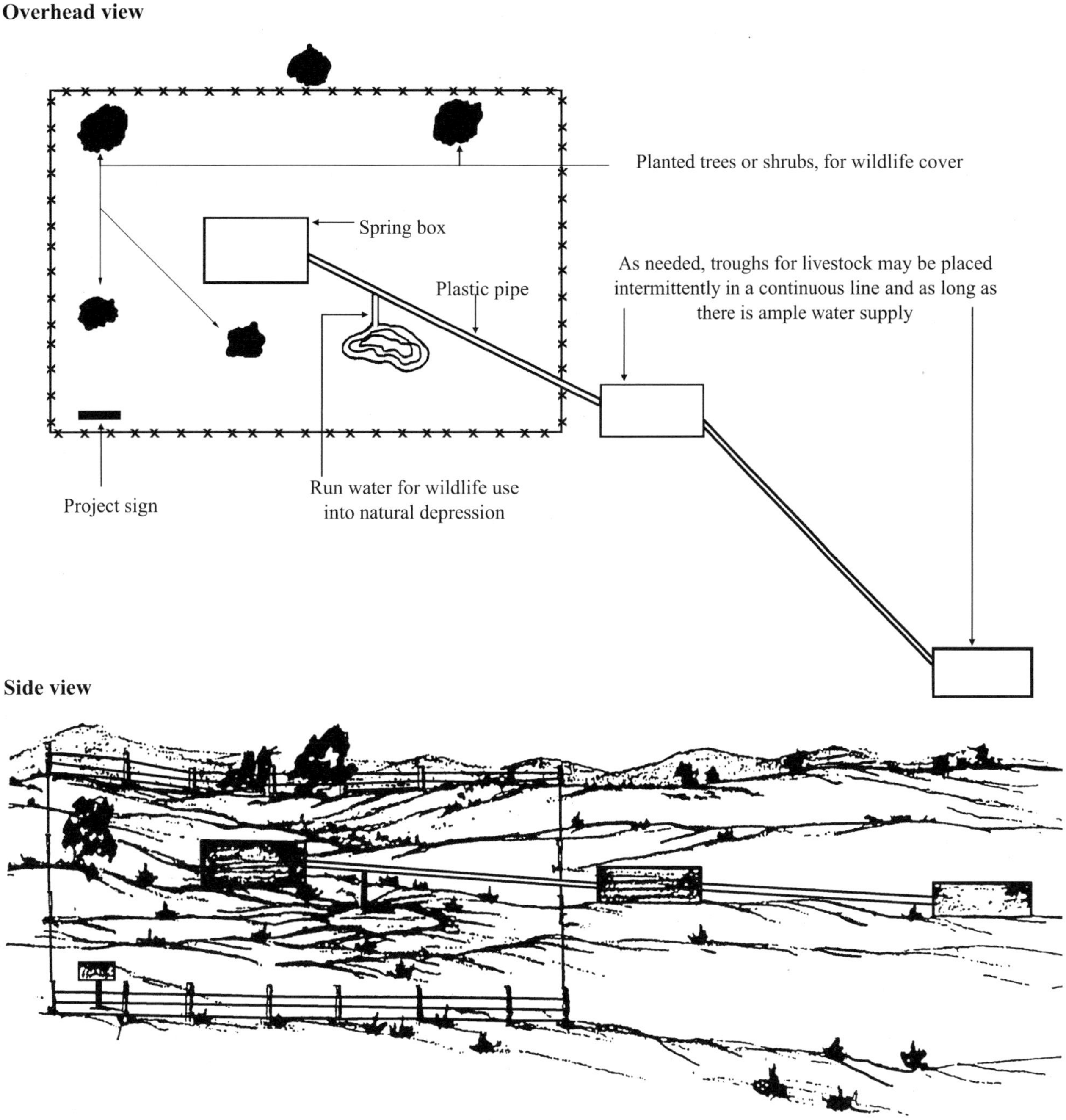

FIGURE 118. *Diagrammatic overhead and side views (not to scale) of a spring development with specifications for use by pronghorn, other wildlife and livestock (from Yoakum 1980a). The recommended fence would be three-strand wire. Dimensions would vary with terrain, location, level of flow and other factors, but at least 1 to 5 acres (0.4–2.0 ha) should be fenced.*

1,037 reservoirs were constructed on public lands (Heady and Bartolome 1977). In Nevada, more than 500 were provided on BLM-administered lands before 1963 (U.S. Bureau of Land Management 1964).

Reservoirs may be formed by building a dam across a drainage or by constructing a diversion into a depression or basin. Reservoirs can be designed to provide maximum storage with minimum surface area to reduce evaporation loss. Reservoirs can be fenced to protect them from trampling and erosion, and the water piped by gravity flow to a trough accessible to livestock and wildlife (Figure 120). A number of designs have been used to develop reservoir water supplies. Plans for each should consider intended use of the site, potential amount of runoff, physiographic features and soil characteristics (e.g., permeability, erodability) (see Yoakum et al. 1980a, Payne and Copes

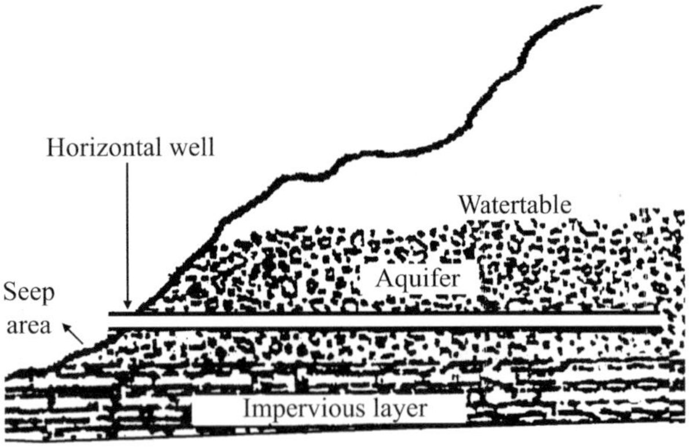

Figure 119. *Cross-sectional design of a horizontal well (Bleich et al. 1982). The well and its size will vary with parent material, water flow, etc. The unregulated flow will be collected in a spring box and distributed, as in Figure 118.*

An earthen dam in a small watercourse retains drinking water for pronghorn, other wildlife and livestock on the Hart Mountain National Antelope Refuge in Oregon. Precipitation runoff is trapped to provide drinking water for summer and autumn. The dam is fenced to prevent livestock trampling, thereby allowing vegetation to grow for wildlife food and cover. Pronghorn readily drink from structures of this design because they simulate natural ponds. Photo by Jim D. Yoakum.

1986, Brown and Karsky 1989, Kie et al. 1994, Payne and Bryant 1998a).

A reservoir of high value to pronghorn is the dugout, or charco pit (see Yoakum 1980a). They are especially used by pronghorn during periods of high summer heat when other water sources are scarce, vegetation is desiccated and the animals' physiological requirements for water are increased (see Good and Crawford 1978). Trench reservoirs are most common in areas of flat terrain. A natural pothole or dry lakebed often is a good location. They should not be located in muddy areas where miring may be a problem for ungulates.

Figure 121 depicts a small rectangular trench reservoir with schematic design specifications. For larger trench reservoirs, the length, width or depth may be increased, but side slopes (to prevent sloughing) and one or more relatively flat slope (to provide access) should have the same ratios, usually 2:1 or less and 4:1 or less, respectively (Yoakum et al. 1980a, Kie et al. 1994, Payne and Bryant 1998a).

Trench reservoirs generally are built on intermittent lakebeds or other relatively flat terrain. Precipitation is trapped and stored for drinking water during dry seasons. This catchment lacks abundant tall vegetation, making it a readily used facility for pronghorn. Other wildlife and livestock also use it. Trench reservoirs are common in the Intermountain West and the Southwest where precipitation levels often are low. Photo by Jim D. Yoakum.

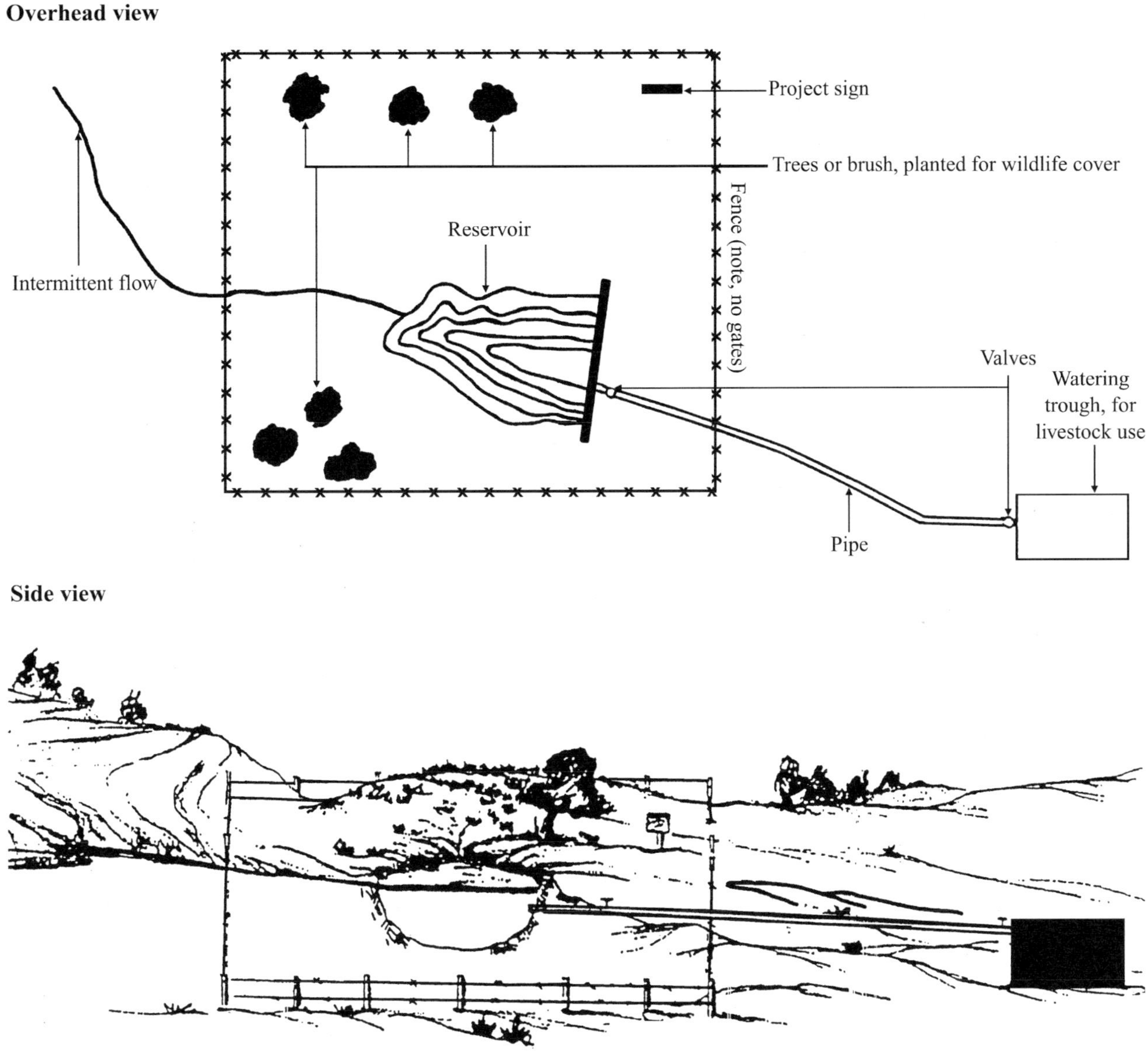

FIGURE 120. *Diagrammatic overhead and side views (not to scale) of a reservoir with water piped to a trough outside a fenced developed area (from Yoakum 1980a). The pipe should be installed at a depth sufficient to carry a consistent flow of water, yet high enough so that it will not be clogged with silt. The valve at the reservoir end of the pipe controls flow and should be approximately one-third the distance from the water to the outside edge of the embankment.*

Precipitation Catchments

Guzzlers have various design specifications and materials, depending on environmental factors. For example, installation of precipitation catchments, or guzzlers, on arid rangelands in Wyoming have proven beneficial to pronghorn (June 1965, Sundstrom 1968). References, such as Yoakum et al. (1980a), Tsukamoto and Stiver (1990), Kie et al. (1994), Brigham and Stevenson (1997), and Payne and Bryant (1998a) should be consulted. Figure 122 shows plans for a guzzler used by pronghorn in south-central Wyoming.

On Wyoming's Red Desert, guzzlers initially were fenced to limit trampling of the apron by domestic sheep. After several years, trampling was determined to be minor and remedied by herding the sheep. The fencing caused an access problem for pronghorn, therefore all wires were removed from the fence, allowing pronghorn easier access.

As a rule, where livestock need to be excluded, at least 1 acre (0.4 ha), preferably up to 5 acres (2 ha), should be fenced around guzzlers. Three- and four-strand barbed wire fences and the buck and pole fences effectively exclude cattle, horses and burros, yet allow pronghorn and

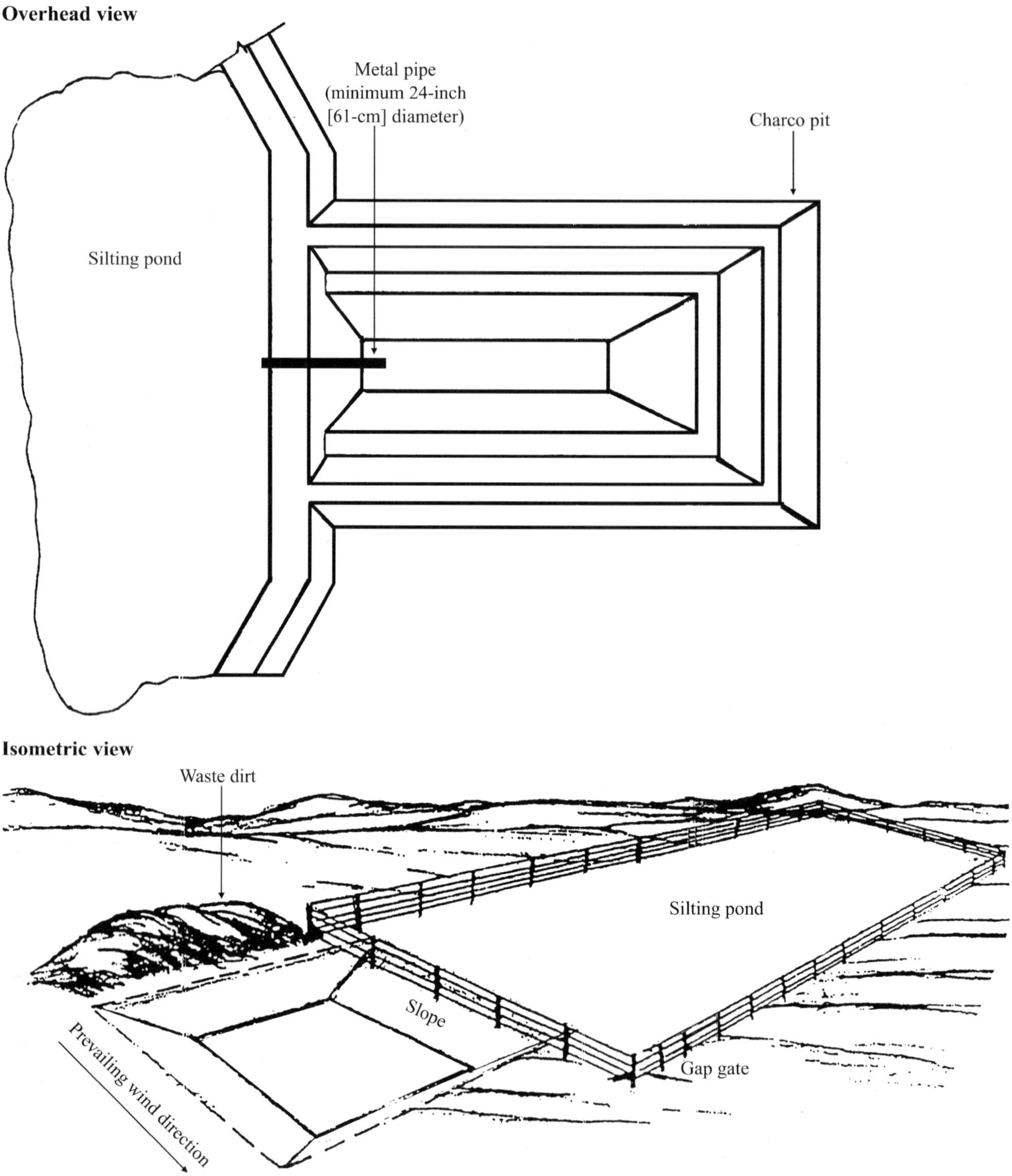

Figure 121. *Diagrammatic overhead and isometric views (not to scale) of a trench reservoir used to catch and retain drinking water for pronghorn, other wildlife and livestock (from Yoakum 1980a). Dimensions would vary with site and objectives concerning quantity of water needed.*

deer access (Brigham and Stevenson 1997). Fences and their specifications are addressed later in this chapter.

Use of guzzlers constructed for pronghorn and other wildlife was studied from spring to autumn during 1994 and 1995 on the Sonoran Desert in southwestern Arizona (Cutler 1996). Cameras equipped with infrared monitors were installed at two guzzlers and operated 24 hours a day for approximately 7 days per month. Monitoring pro-

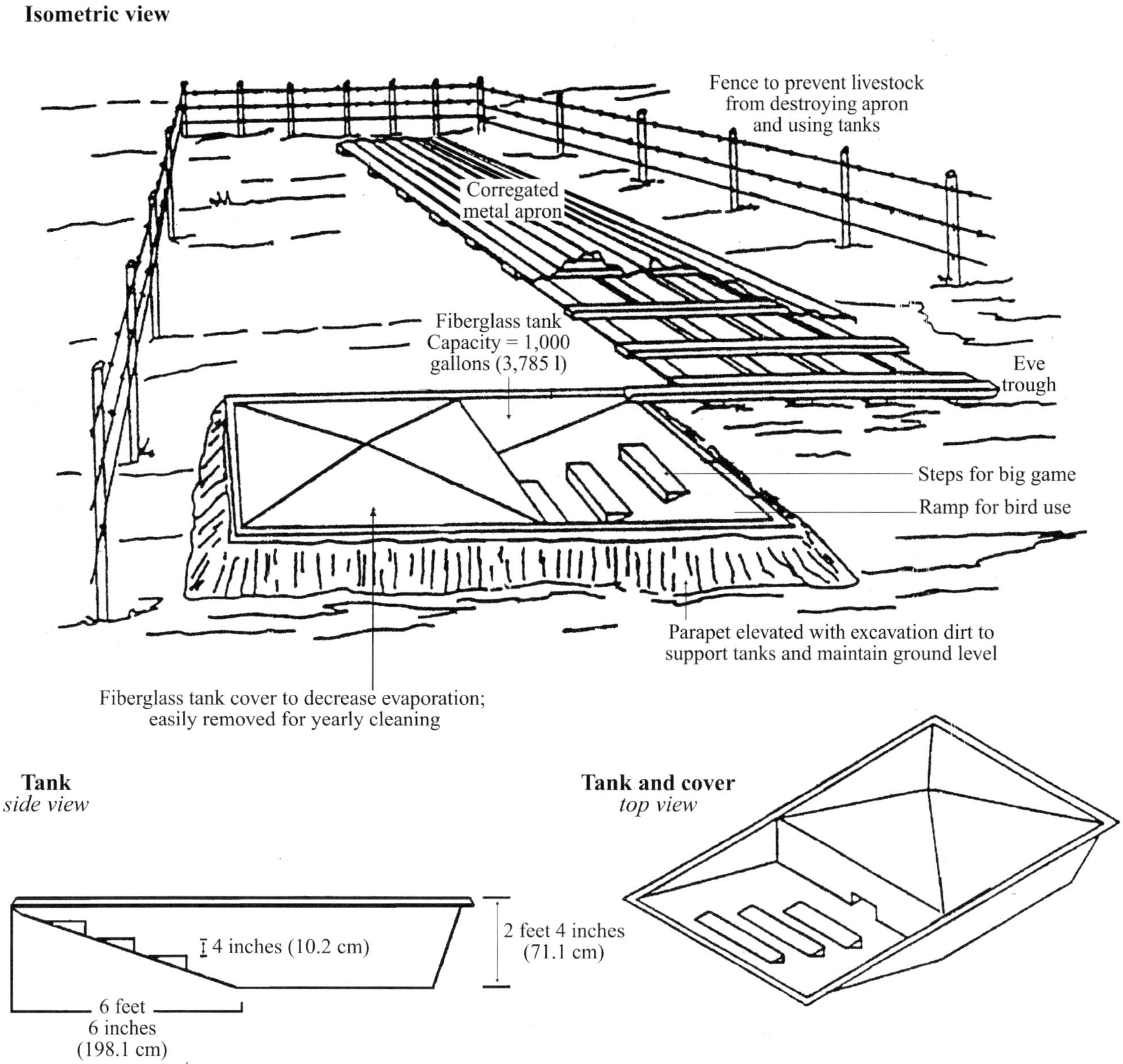

FIGURE 122. *A water catchment (guzzler) designed to provide drinking water for pronghorn and other wildlife (e.g., sage grouse, horned larks, vesper sparrows, rabbits, ground squirrels, deer and elk) on the Red Desert, Wyoming (Yoakum 1980a). Such catchments are located to take advantage of natural contours to trap water.*

vided 2,699 detections, recording 28 wildlife taxa: coyotes accounted for 59%, mule deer for 11%, and other mammals, birds and reptiles for 30%. Pronghorn were not observed drinking or foraging at either water development, although they were repeatedly seen in the vicinity. Cutler (1996) speculated two reasons why pronghorn were not observed drinking: woody vegetation had grown tall and dense around the sites since construction, and the common occurrence of predators (coyotes, bobcats, cougars and golden eagles) frequenting the two watering sites.

Managers installed 29 guzzlers on public lands in southeastern Idaho for pronghorn, sage grouse, mourning doves, mule deer, elk and other wildlife. Most wildlife use occurred during July and August when natural drinking water was scarce (McCarty 1986).

These precipitation catchments contained an apron for catching rainwater, a holding tank for water storage and a drinking facility—all of which were enclosed with a barbed wire fence to restrict livestock entry. Design specifications were engineered by personnel of the Washington Department of Game. The structure was inexpensive, easy to install and low maintenance. Installation was by the BLM and a function of a Habitat Management Plan (see Chapter 16). A maintenance problem was encountered

Precipitation catchments, or guzzlers, are built to capture and store rainwater for wildlife drinking water on semiarid and arid rangelands. Water is collected on a metal or plastic apron and drained into a storage tank with a drinking facility (where Wyoming Game and Fish Department biologist Chuck Sundstrom is seated in the photograph). The structure was built primarily for sage grouse and pronghorn on the Red Desert of south-central Wyoming. Photo by Jim D. Yoakum.

Shrubsteppes in northeastern California typically experience dry summers and autumns, therefore drinking water for mule deer and pronghorn can be limited seasonally. To overcome the periodic water shortage, an earthen reservoir was constructed to collect rain runoff from a paved highway. It provides abundant drinking water for wildlife during dry seasons. The reservoir is relatively maintenance-free and inexpensive. Photo by Jim D. Yoakum.

when a wildfire burned one guzzler. The fire burned the guzzler, destroying the wood corner fence posts, but not harming the apron or storage tank.

Annual maintenance inspections were conducted to determine use and values of each structure. Monitoring involved checking all guzzlers for maintenance needs on a set schedule. Observations of wildlife use also were conducted and recorded. A simple monitoring technique involved removing vegetation from a small area in front of the ramp to register wildlife tracks. More elaborate monitoring schemes involving time-lapse cameras can be rewarding.

Other Developments

Water developments, such as tanks, troughs or wells, are installed and sometimes benefit pronghorn, but they usually are constructed for other purposes. Often, slight modifications or additions of the other-purpose developments can enhance wildlife use as well. For additional information, plans and construction specifications, see Yoakum et al. (1980a), Payne and Copes (1986), Brown and Karsky (1989), Vallentine (1989), Kie et al, (1994), and Payne and Bryant (1998a).

Safety Considerations

Besides being planned to provide water continuously, water developments on rangelands should be designed to maximize safety for the animals that use them. Wilson and Hannan (1977) listed the rationale and criteria for considering the safety of wildlife using water developments built to provide drinking water for livestock. To prevent entrapment and drowning of animals, they also recommended certain designs, including:

1. when drinking facilities are installed, the height of troughs or other containers should not exceed 20 inches (51 cm) to give fawns access to the water; deeper troughs may be set into the ground to achieve the desired height;
2. barricades should be considered to prevent the accidental entry to water and possible drowning of

A drinking trough developed on open rangeland by the U.S. Bureau of Land Management in Malheur County, Oregon, received water piped from its original source 2 miles (3.2 km) uphill. Rock gravel was placed around the trough to decrease miring conditions. The trough was buried half in the ground to provide drinking water at a lower height for use by diverse wildlife and livestock. Large rocks were placed in the trough to aid wildlife in escaping from deep water if necessary. Pronghorn use structures of this design. Photo by Jim D. Yoakum.

wildlife, and the distance from the rim of the trough to the barricade should not exceed 20 inches (51 cm); and

3. where water depths exceed 20 inches (51 cm), rocks or concrete blocks in the catchment will enable escape by animals that accidentally get in the water, inasmuch as troughs and catchments can be easier to enter than to escape.

FENCING

Within a decade after barbed wire fences first were erected on western rangelands, Caton (1877:48) reported that 4-foot (1.2 m) high fences confined pronghorn: "This inability to leap over high objects may no doubt be attributable to the fact that they live upon the plains, where they rarely meet with such obstructions, and so they and their ancestors for untold generations have had no occasion to overleap high obstructions, and thus from disuse they do not know how to do so, and never attempt it when they do meet them."

Although overstating the case, Caton (1877) was essentially correct. Pronghorn adapted over many millennia to open landscapes without vertical barriers to daily and seasonal movements. In the relatively fractional time since fencing of the West began, segmenting major portions of pronghorn rangelands, pronghorn have not learned to go through most fences (as do bison) or vault them (as do elk and deer). Instead, many have learned to negotiate certain fences by crawling underneath. But, if the bottom wires of fences are too low, by virtue of design or buildup of vegetation or snow, pronghorn mobility can be impeded.

Pronghorn welfare has been impacted in direct proportion to the sale of barbed wire. In 1879, 5 tons (4.5 metric tons) of barbed wire were manufactured in the United States. Six years later, 40,000 tons (36,000 metric tons) were being strung across western rangelands. In 1945, 234,000 tons (210,600 metric tons) were produced in the United States (Leftwich and Simpson 1978). Initially, drift fences were erected, resulting in large numbers of cattle and pronghorn being trapped and freezing to death (Hailey 1969). Later, as fencing of large pastures became commonplace, pronghorn growth was further depressed, and mortality resulted in the total disappearance of pronghorn from some rangelands (Russell 1964, Martinka 1967, Spillett et al. 1967, Hailey 1979).

Fences can be used to restrict pronghorn movements for purposes of trapping and translocation, confinement for research and prevention of access to areas where they are not desired, such as croplands, highways and airports.

However, most fences on western rangelands have been constructed primarily to control livestock (Karsky 1988, Yoakum et al. 1996). Consequently, wildlife biologists need to know the purpose and design of the various types of fences developed mainly for livestock management and how each type can affect pronghorn and other

Research studies verify that pronghorn have the ability to jump over fences but rarely do. Centuries of adapting to landscapes with few obstacles have evolved behavioral patterns that generally are not compatible with fencing—a relatively new (100± years) and constraining habitat factor. Photo by Robb Hitchcock.

wildlife. Fence types include barbed wire (the most prevalent type on western rangelands), net (or woven) wire, buck and pole, rock, brush, electric, log worm (log zigzag) and high tensile (smooth wire exceeding 100,000 pounds per square inch [45,359 kg/2.54 cm^2] breaking strength) (see Karsky 1988, Vallentine 1989).

Fence problems are most prevalent where fencing interferes with pronghorn seasonal movements of more than 50 miles (80 km). They are especially severe where pronghorn are confined in pastures with livestock during droughts (Hailey et al. 1966, Hailey 1979). In northern regions, fences can prevent animals from leaving areas of

Twelve pronghorn carcasses indicate where the animals tried unsuccessfully to negotiate a net and barbed wire fence the preceding winter when deep snow prevailed. Pronghorn in certain areas travel as far as 200 miles (322 km) during seasonal movements, seeking lower elevations with less snow and available forage. Fences transecting seasonal movement corridors from summer to winter rangelands have been responsible for hundreds of pronghorn mortalities. Fencing designs that permit freedom of movement make unnecessary such losses of public wildlife on public lands. Photo courtesy of the Wyoming Game and Fish Department, Cheyenne.

Pronghorn are adapted to wide open rangelands free of obstacles that impede movement. Their welfare depends on unobstructed access to adequate, nutritious forage, drinking water and fawning areas, which may require long-distance traveling. Fences directly and indirectly affect such access. Photo by Jim D. Yoakum.

Thousands of miles of barbed wire fences on rangelands were originally constructed to control livestock, with the bottom wire 10 inches (25.4 cm) or so above the ground. These fences can restrict pronghorn movement. A slight modification of the bottom wire by raising it up to 16 inches (40.6 cm)—a simple and inexpensive modification—can alleviate this access problem for pronghorn without compromising control of horses and cattle. And to reduce the risk of injury to pronghorn crawling or running under fences, wildlife managers recommend that the bottom strand be smooth, not barbed. Pronghorn are visible above, near the crest of the hill at the center of the photograph. Photo by Jim D. Yoakum.

deep snow to obtain forage where less snow has accumulated (Martinka 1967, Sundstrom 1970, Oakley 1973). In southern regions, they may deny pronghorn access to water and succulent forage (Buechner 1950a, Russell 1964, Hailey and DeArment 1969).

Fencing of pronghorn on western public rangelands in the United States led, during the 1950s, to the first investigation of the problem. It was undertaken by Charles Rouse of the U.S. Fish and Wildlife Service, under auspices of the BLM. Rouse conducted extensive field studies and verified that pronghorn movements were severely restricted by woven wire fences, but the animals were capable of negotiating some barbed wire fences. He reported that barbed wire fences could be designed to control cattle but still allow pronghorn movements. Rouse's (1954) recommendations included: affixing the lowest fence wire more than 16 inches (41 cm) above the ground, which would allow pronghorn to crawl underneath, the bottom strand should be smooth to prevent injuries to animals passing beneath the fence, and the top wire should be no higher than 36 inches (91 cm) for the benefit of deer and the few pronghorn inclined to jump fences (Figure 123A). A number of wildlife management organizations (Salwasser 1980, Lee et al. 1998) and government agencies (see Karsky 1988) adopted, modified and incorporated Rouse's (1954) recommendations in rangeland guidelines and operation manuals.

Some barbed wire fence configurations include stays between posts. Stays are made of wood or wire. They help increase the rigidity of barbed wire strands, which is desirable for livestock management. Such fences are less "friendly" to pronghorn because the barbed wires have less flexibility to enable pronghorn to wiggle through or under. Photo by Jim D. Yoakum.

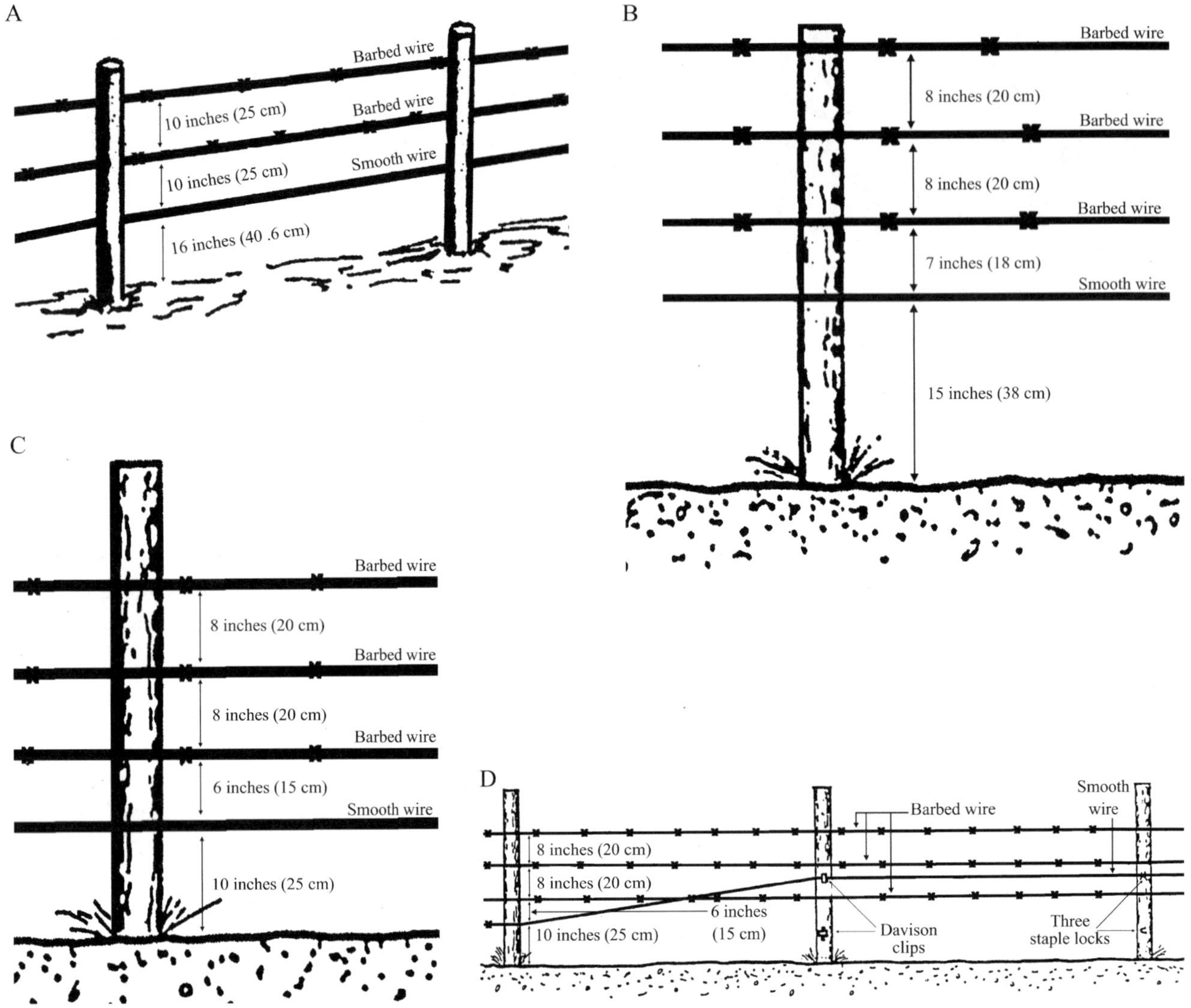

Figure 123. *Specifications for (A) three-strand (Lee et al. 1998) and (B) suggested four-strand wire fences on rangelands used by cattle and pronghorn and (C) barbed and smooth wire fencing recommended for fences on rangelands grazed by domestic sheep, cattle and pronghorn (Karsky 1988). The latter fence will restrict pronghorn movements because the lower wire is too close to the ground, so it will not allow pronghorn to crawl underneath. However, it can be modified to allow pronghorn access through the fence when the bottom wire is raised as illustrated in* D.

Nevertheless, pronghorn mortality due to fences (by entanglement or blocked access to food or water during severe winters or droughts) continued to be a problem. As a result, the BLM conducted intensive experiments during 1963 and 1964, with different fence designs on Wyoming's Red Desert rangelands used by pronghorn, domestic sheep and cattle. Fences that pronghorn could not negotiate were reported by Spillett (1965), Spillett et al. (1967) and Zobell (1968). Recommendations from the Red Desert research have since been incorporated into fencing specifications by the BLM, U.S. Forest Service, U.S. Fish and Wildlife Service and various state agencies (U.S. Bureau of Land Management 1985, Payne and Copes 1986, Lee et al. 1998).

One of the most comprehensive reports on pronghorn-related fencing strategies was the product of a public meeting held in Cheyenne, Wyoming, in 1974. Approximately 150 individuals participated, representing conservation organizations, clubs, environmental groups, government agencies, professional resource societies, livestock organizations and the public. Attendees heard case histories and recommendations concerning the welfare of pronghorn and suggestions from rangeland users needing property protection. Information was presented regarding overbuilt fences (i.e., fences with more posts, wires and stays than structurally needed to contain livestock). Such fences effectively control livestock, but they can be deleterious to wildlife. Participants formed working groups to analyze presentations and submit suggestions for constructing fences negotiable by pronghorn. The results were documented in a report (U.S. Bureau of Land Manage-

Pronghorn mortality as a result of fence entanglement usually occurs when the animals are relatively malnourished or have poor footing when leaping, such as in winter. The doe at left *may have jumped too close to the fence, catching a front leg between wire strands, which twisted closed around the leg when her body passed over the fence. The pronghorn at* right, *apparently unable to crawl under the fence because of the snow cover, either did not have the strength to jump over completely or slipped, causing the back legs to catch and hang between the top and center wires, stranding the animal.* Left photo by Jim D. Yoakum. Right photo by LuRay Parker; courtesy of the Wyoming Department of Game and Fish, Cheyenne.

ment 1975a) that later was used to produce a manual for distribution to all BLM field offices (U.S. Bureau of Land Management 1975b). The manual, updated a decade later (U.S. Bureau of Land Management 1985), provides the following guidelines for construction of fences on BLM-administered lands occupied by pronghorn.

1. Most woven wire and five- and six-strand barbed wire fences limit pronghorn movement and should not be constructed in pronghorn habitat.
2. Fences must remain loose, with no stays or with stays not connected to the bottom wire.
3. The bottom wire on barbed wire fences must be smooth (barbless), loose and with a minimum ground clearance of 16 to 18 inches (41–46 cm).
4. Water sources should not be fenced unless they might be jeopardized by trampling.
5. Where a need exists to provide water specifically for pronghorn, excluding others, the water source and the area fenced should encompass at least 3 to 5 acres (1.2–2.0 ha) of relatively level terrain.
6. Where there is need to construct fences within seasonal pronghorn movement areas, a three-wire design should be used (Figure 123A).
7. In areas where deep snow warrants the use of special fence designs, let-down fences—or fences with adjustable segments—should be considered.
8. Antelope passes (discussed later) are not a reliable alternative to overcoming obstacles to pronghorn movements; some pronghorn will not use them, and fawns that do use them are subject to leg injuries. Such passes are mentioned here because they often are considered a component of fence construction for bids and contracts.

There are numerous manuals and texts that provide specifications for designing and constructing fences (Yoakum et al. 1980a, Payne and Copes 1986, Karsky 1988, Vallentine 1989, Kie et al. 1994, Payne and Bryant 1998a).

The Red Desert in south-central Wyoming has suffered intensive impacts from fencing. Rangeland fences to control livestock had been a problem for decades. However, with acceleration of livestock allotment planning in the 1950s, together with upgrading of interstate highways, fence construction greatly increased and impacts intensified. The Wyoming Game and Fish Department conducted an inventory of fences influencing pronghorn movements and reported that—before the 1970s—802 miles (1,290 km) were involved; 459 miles (739 km) were net wire fences, presenting a total barrier to pronghorn movements.

Pronghorn characteristically go under rather than through or over barbed wire fences. Both adults and fawns use this technique. When the bottom wire is 16 inches (40.6 cm) or more above the ground, pronghorn can go under the fence. They typically will use depressions in the ground with more room to duck under fences. Photo by LuRay Parker; courtesy of the Wyoming Department of Game and Fish, Cheyenne.

Fences designed with net and barbed wire can restrict movement fully of pronghorn fawns and adults, especially during seasonal and winter movements. Fences constructed to control domestic sheep or restrict movements of coyotes likewise will restrict movements of pronghorn. Such structures are illegal on public lands administered by the U.S. Bureau of Land Management for multiple-use purposes. Photo courtesy of the Wyoming Department of Game and Fish, Cheyenne.

One of these fences, completed in 1964, was a barrier to pronghorn between the North Platte River and the Point of Rocks. A similar fence, completed in 1966, extended 20 miles (32 km) south of Lamont along U.S. Highway 287. These fences blocked historic winter rangelands for large migrating pronghorn herds (Taylor 1975). The Texas Department of Fish and Game conducted a survey of fencing that affected pronghorn movements in the 1980s and reported that 52.5% of fences on rangelands occupied by pronghorn were built, wholly or in part, with net wire (Sweptson 1990).

Barbed, Smooth and Net Wire Fences

During the late 1800s, wire fences became widely used on western rangelands. Most were of barbed wire and of various designs to control cattle, but net (woven) wire often was used to confine domestic sheep. Smooth fence wire seldom was used until recent decades. It reduces the potential of injury to wildlife.

Specifications for three- and four-strand barbed wire fences designed to control cattle but enable pronghorn movement are provided in Figure 123 (see also Spillett 1965, Kie et al. 1994, Lee et al. 1998). Fence height and wire spacings allow pronghorn to negotiate such fences under most circumstances.

The fence illustrated in Figure 123C was intended to control domestic sheep, yet allow passage to pronghorn (U.S. Bureau of Land Management 1975b, Salwasser 1980, Yoakum 1980a, Karsky 1988, Kindschy 1996, Payne and Bryant 1998a). However, the bottom wire is 10 inches (25 cm) from the ground, which can be restrictive to pronghorn. According to Rouse (1954:11), "Any fence that effectively controls domestic sheep will likewise control pronghorns." Karsky (1988) modified the Figure 123C wire design with a bottom wire 10 inches (25 cm) above ground, by adjusting the bottom wire to have a 16-inch (41 cm) space between the ground and bottom wire (Figure 123D). The adjusted 16-inch (41 cm) bottom space would be more effective in allowing pronghorn passage.

Fences constructed with woven wire can be impassable to pronghorn because most pronghorn will not jump over fences. Those designed to control domestic sheep invariably are effective in excluding pronghorn. Young animals particularly are affected because they lack experience in negotiating fences. Photo by Jim D. Yoakum.

Wolf-Type and Other Anticoyote Fences

The wolf-type fence was designed during the 1940s in New Mexico and Texas to exclude coyotes from rangelands foraged by domestic sheep. Construction design consisted of 36-inch (91 cm) net wire, with the bottom 12 inches (30 cm) buried below the ground. Three to four strands of barbed wire were strung above the woven wire to a height of 50 to 60 inches (127–152 cm). The objective was to prevent coyotes from digging under, passing through or jumping over the barrier. It succeeded, but the fence also inhibited pronghorn movement. Such structures now are illegal on public lands, where multiple use of natural resources is an objective (Yoakum 1980a).

Various other anticoyote fences have been designed and built specifically to control coyote movements (Karsky 1988, Kie et al. 1994). These have been constructed with barbed, net and electric wires. Such fences likewise inhibit pronghorn movements, and any modification to alleviate the problem would also allow coyotes to cross.

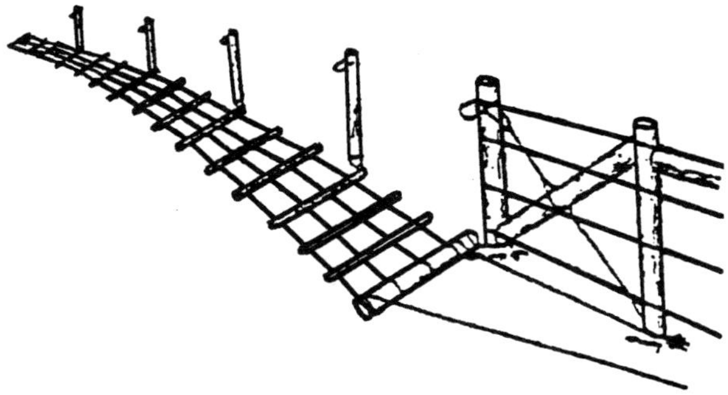

FIGURE 124. *Let-down fence to permit pronghorn and other wild ungulates to cross (Karsky 1988). Letting down the wires from at least four posts should suffice, and distances between gaps would depend on local conditions.*

Let-Down Fences

A let-down fence (Figure 124) is designed to allow sections of wire to be laid on the ground, thus allowing pronghorn to pass over. Such fences can be used in areas of deep snow and seasonal movement routes of pronghorn.

Several other let-down fence designs have been used or suggested. One uses a wire loop at the top of the posts to hold stays in place and replaces the lower loop with a pivot bolt. This allows the fence to be let down and erected easily. Another design permits the let-down section to be pulled against a section of standing fence. The fence must be designed to provide for adjustment of wire tension. When the wire is so taut that it does not lay flat or so loose that loops of wire are formed, hazards are created.

Adjustable Fences

The adjustable fence in Figure 125 originally was designed in Idaho (Anderson and Denton 1980). It allows the lower wires to be raised from 16 inches (41 cm) to 38 inches (97 cm), which is particularly beneficial to pronghorn where snow depths can exceed 12 inches (30 cm). One person can adjust wires for a mile (1.6 km) of fence in approximately 30 minutes. L. Anderson (personal communication:1986) reported that pronghorn repeatedly select sites of highest bottom wire to pass under fences, and when the bottom wires were raised, the animals readily used those sections.

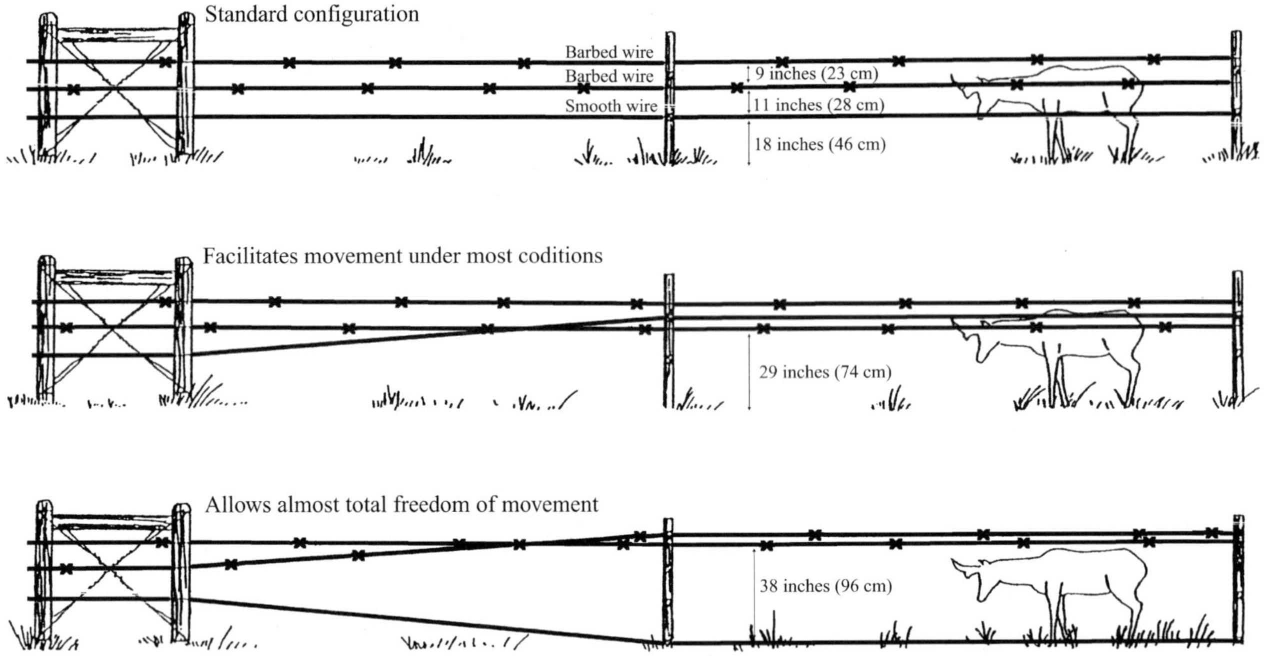

Figure 125. *Three-strand barbed wire fence with modifications for pronghorn access (Anderson and Denton 1980). This design especially is beneficial when snow makes it difficult for pronghorn to crawl under the fence. The configuration depicted in the center drawing would suffice during most winters. The 38-inch (97 cm) clearance would be needed where snow depth exceeded 20 inches (51 cm). If the fence blocked a movement corridor from summer to winter rangelands, long areas of modification might be required to accommodate pronghorn moving with a snowstorm.*

Buck and Pole Fences

All-wood fences are not widely used on rangelands because of their labor-intensive construction, difficulty of transporting materials and expense. However, such fences sometimes have the advantage over wire fences of being more durable in heavy snows, and most are aesthetically pleasing.

Scott (1992) reported on the ability of pronghorn and other wild ungulates to move through a buck and pole fence on the northern boundary of Yellowstone National Park (Figure 126). The bottom rail was 18 inches (46 cm) above the ground. Pronghorn passed around or crawled under 72% of the times they attempted to do so, but not always without difficulty. They encountered more problems with the side facing the park, which had four wood rails compared with the other side that had one brace rail. Pronghorn inhibited from crossing through the fence walked along the fence until finding an open gate and then passed through.

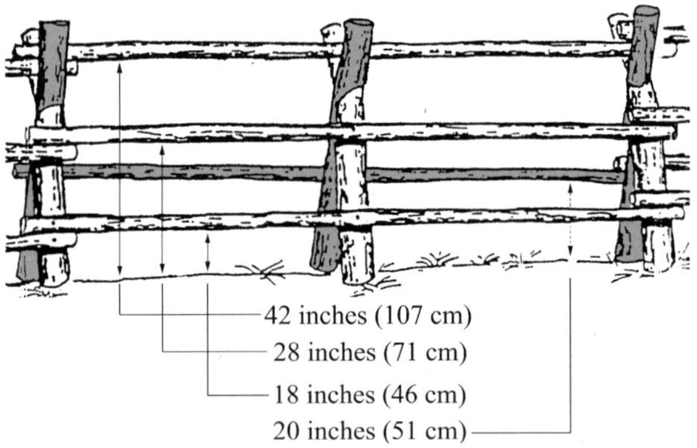

Figure 126. *Typical section of a three-pole buck and pole fence that will allow pronghorn to pass through (Karsky 1988). An attribute of this type of fence is that it enables snow to be scoured out by turbulent winds, making the fence negotiable during times of deep snow without the requirement of seasonal removal. However, wooden fences generally are not used on western rangelands because materials are costly, and the fences are labor intensive to build and maintain.*

Electric and Other Antipronghorn Fences

A management objective may be to keep pronghorn in or out of a specific area, or to restrict their movements, for purposes of research or to prevent conflicts or danger (see Yoakum et al. 1980a).

An electric fence carries intermittent electrical charges that shock when the wire is touched. Once animals are exposed and conditioned to them, electric fences often act as a psychological barrier as much as a physical obstacle. They are relatively easy to install, have a long service life and may result in 25% to 30% savings in labor and material costs (Karsky 1988). Builders of electric fences should ensure that they are safely constructed and maintained.

With recent innovations, electric fences, formerly considered temporary structures, now can be permanent structures. The standard U.S.-manufactured energizers could electrify up to 6 miles (9.7 km) of wire with a useful life of up to 4 years. Recent New Zealand energizers can effectively electrify more than 75 miles (121 km) of wire with a useful life of 10 to 15 years (Karsky 1988). Permanent electrical fences are replacing barbed and high-tensile wire fencing in some regions.

Two-wire electric fences have effectively managed livestock on rangelands in the West. They also have been used to keep pronghorn out of newly planted rangeland seedings in Malheur County, Oregon (R. R. Kindschy personal communication:1976).

In central Colorado, a three-strand electoplastic-twine electric fence was installed temporarily around an alfalfa field customarily foraged by pronghorn (Pojar et al. 1994). The field also was fenced with sections of four-strand barbed wire and net wires that had been installed to control livestock. In 1992, before the electric fence was installed, the daily mean number of pronghorn on the field was 38.17 (n = 6). In 1993, after the electric fence was installed, the daily mean number was 2.16 (n = 70). The electric fence appeared more effective near the barbed wire fence than near the net wire fence. Apparently, when jumping the net wire, the animals broke the circuit with the ground and did not get shocked even when they touched the electric twine. The investigators assumed that pronghorn avoided crossing the barbed and electric fencing because they were accustomed to crawling under barbed wire and, in doing so, contacted the electric wire and were shocked.

Construction of electric fences is increasing on public and private rangelands in the West, including in pronghorn habitats. Various designs are used, including the two-strand fence shown above, with an adult Labrador providing a sense of scale. More study is needed to assess the effects of electric fences on pronghorn. Photo by Ray Mapston; courtesy of the U.S. Bureau of Land Management, Washington, D.C.

Regarding installation of new permanent electric fences as exclosure devices, Pojar et al. (1994) suggested building a 60-inch (150 cm) high five-strand (or more) high-tensile wire fence, as described by Palmer et al. (1985). These specifications would result in a fence with long life, low maintenance and effectiveness in excluding pronghorn and deer. Such a fence would be 25% to 60% less expensive than conventional net wire fencing.

In cooperation with eight ranchers, the Wyoming Game and Fish Department evaluated the effects of high-tensile electric fences on movements of livestock and wildlife (Karhu and Anderson 2002). Two-, three- and four-wire fence designs were tested (Figure 127). When a pronghorn or other ungulate came in range, a camera recorded the animal's reaction to the electric fence.

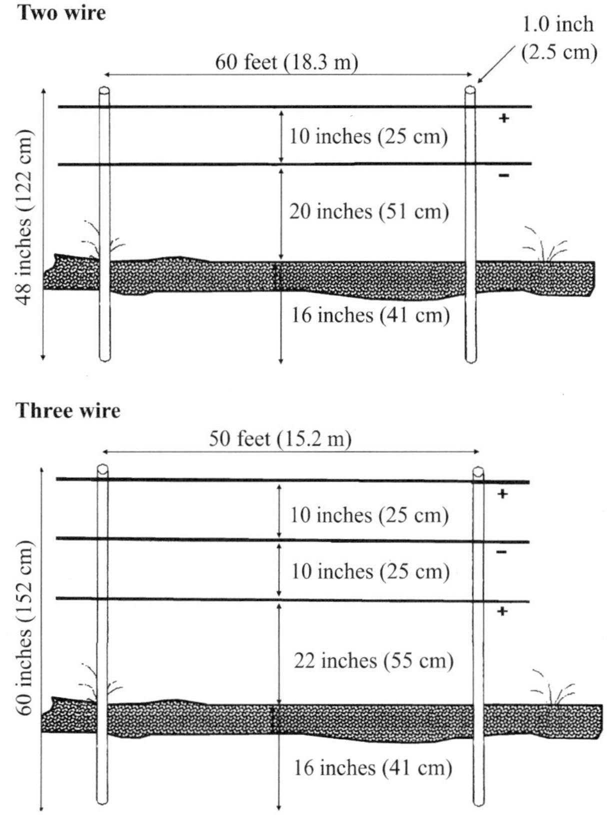

FIGURE 127. *Two-* (top) *and three-wire* (bottom) *electric fence designs featuring 0.75 to 1.0 inch (1.9–2.54 cm) diameter, solid, fiberglass line posts. The wire is 12.5 gauge, class III galvanized, with a maximum tensile strength of 170,000 pounds per square inch (11,953 kg/cm²) and a maximum breaking strength of 1,308 pounds (626 kg). The wires are connected to the line posts and stays by metal clips.*

A total of 191 pronghorn reactions to the electric fence designs were recorded. Pronghorn exhibited low aversion rates for three-wire (8%) and four-wire (8.9%) electric fences. However, two-wire fences resulted in relatively high aversion rates (48%) and a low travers success (60%). Findings indicated pronghorn were not shocked; therefore, electric shock appears not to be an important factor influencing reactions. The insulating quality of pronghorn guard hairs, combined with generally dry soils, allow animals to contact the hot wires and feel little or no pain. They may be more susceptible to being shocked if they touch a hot wire with their nose or ears or stand in wet soils. As expected for three- and four-strand barbed wire fences, pronghorn ducked under the bottom wire. Of the electric wire designs examined, Karhu and Anderson contended that a three-wire fence is an effective structure to meet the goals of controlling livestock and bison, and allowing pronghorn, deer and elk access on western rangelands.

The woven wire fences illustrated in Figure 128 have been used to prevent pronghorn movements onto highways and airports. Similar fencing restricted pronghorn movements along a freeway in northeastern California (K. S. Kahre personal communication:1995). The same type of fence is used to confine pronghorn at a specific site, such as retention enclosures for translocating herds, facilities to study animals for research and exhibition paddocks at zoos and parks. For both enclosure and exclusion purposes, fences should be at least 8 feet (2.4 m) high.

Sometimes it is necessary to restrict pronghorn access on western rangelands, such as to prevent agriculture crop depredations, temporarily enclose herds during translocation or limit temporary or permanent movements across highways, railroads or airports. Such a fence was designed and tested in Colorado (Pojar et al. 2004). Its specifications and dimensions are shown in Figure 129. It is advisable to replace the top wire with a 1-inch (2.45 cm)

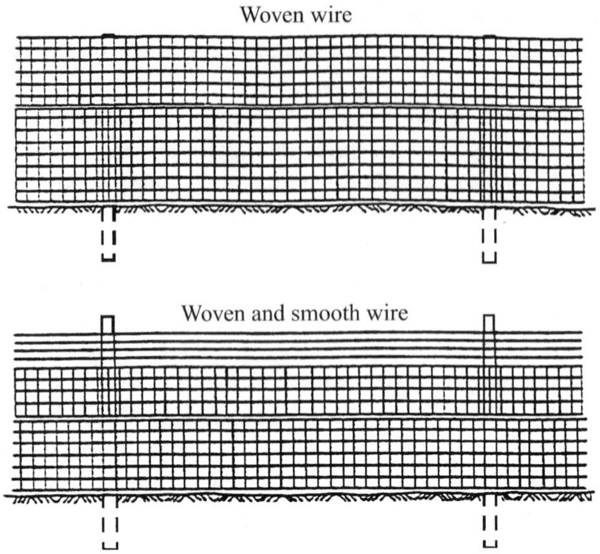

Figure 128. *Two types of 8-foot (2.4 m) woven wire fence (not to scale) that enclose or exclude pronghorn (Yoakum et al. 1980). Such fences are expensive and generally used only to confine pronghorn to small areas, such as zoos and research facilities, or to exclude them from intensive agriculture croplands.*

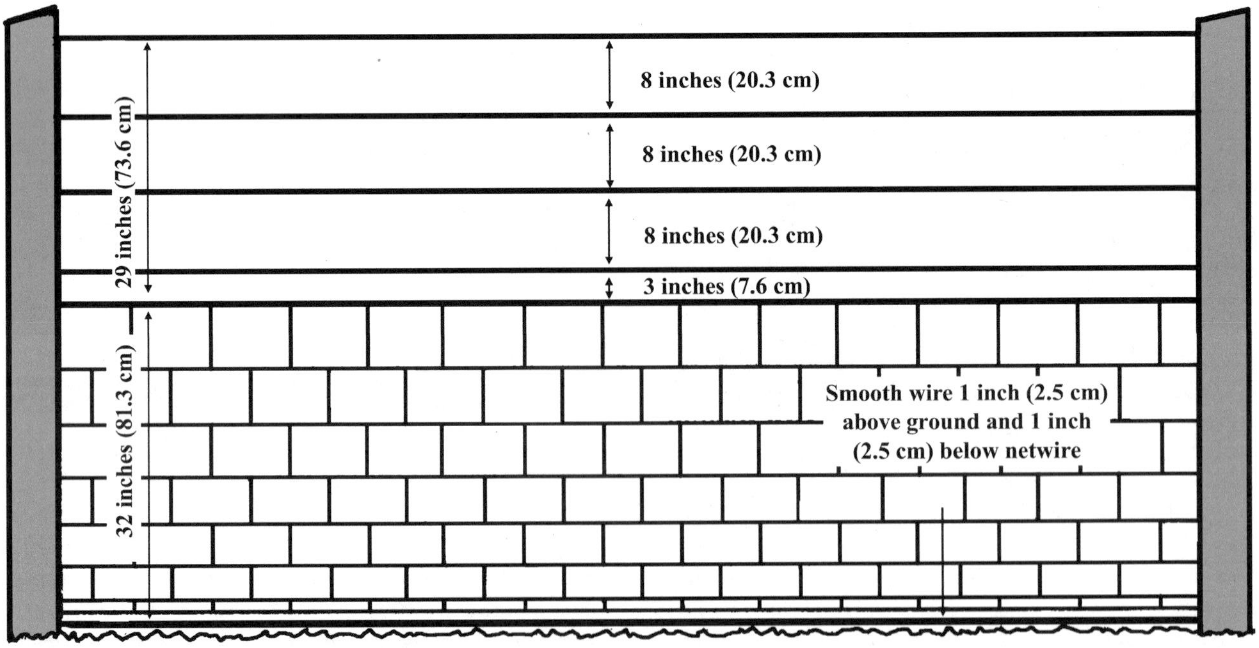

Figure 129. *Range fence design specifically to prevent pronghorn passage or restrict access (from Pojar et al. 2004).*

wide metal-impregnated (for strength), nylon tape (as used for electric fences) that is colored (other than white), so that deer and elk can see it against a snow background and be better able to gauge their jump to clear the fence. It is important to ensure that the bottom of the fence is sealed adequately for all depressions and drainages, to prevent pronghorn movement under the fence.

With increased volume and speed of traffic on interstate highways, secondary roads and railroads, greater numbers of overpass and underpass structures are being build to protect resources and provide public safety. This fence design could be useful for guiding pronghorn safely to and from crossing facilities.

Antelope Passes

During 1963 and 1964, the Wamsutter, Wyoming, Pronghorn Fence Research Project developed several devices purported to allow pronghorn access through fences (Spillett 1965, Spillett et al. 1967). The project tested these structures under limited rangeland conditions. Therefore, a subsequent study was conducted near Casper, Wyoming (Mapston 1968, Mapston and Zobell 1972), resulting in the "antelope pass" design (Figure 130)—a miniature cattle guard that capitalizes on the tendency or willingness of pronghorn to broad jump rather than high jump.

Field tests disclosed that some adult animals jumped over the guards, whereas others refused to negotiate them. Fawns did not easily leap the structure and some suffered leg injuries. Investigators concluded that the antelope pass design had only limited value for enabling pronghorn movement through livestock fences. Subsequent studies confirmed that the antelope pass design allowed most adult pronghorn to negotiate fences, but fawns were unable, reluctant and subject to injury (Newman 1966, Kerr 1968, Mapston 1968, Bear 1969). Also, some pronghorn quickly learned to pass through the guards, but others did not.

As a rule, antelope passes should be used only on pronghorn winter rangelands and only if no other means of fence passage can be provided. Figure 131 depicts placement for antelope passes in fences, based on field trials indicating sites pronghorn used most successfully (Mapston 1968).

The controversy of wolf-type fences in New Mexico during the 1970s resulted in a second pronghorn fence research project (Howard et al. 1983, 1990). Project investigators labeled the three-strand wire fence an antelope pass fence when installed as a segment of a woven wire fence, and called the original antelope pass—which they modified by doubling its length—fence, developed in Wyoming (Figure 128), an "antelope guard." This conflicting terminology has caused confusion. A review of

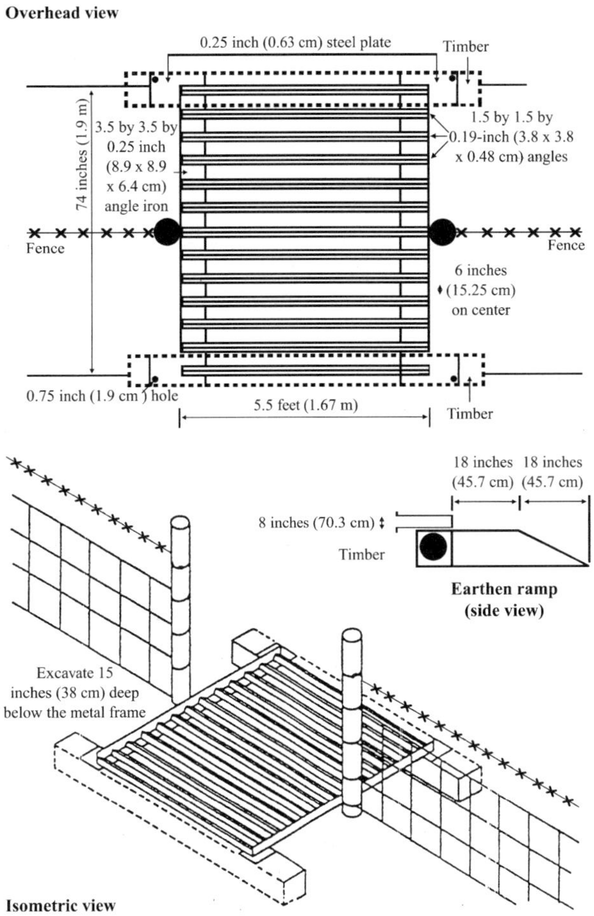

FIGURE 130. *Antelope pass design developed in Wyoming (Mapston and Zobell 1972). Such passes actually are short cattle guards. Standard cattle guards are designed to allow vehicles to cross. Antelope passes were designed, in part, to prevent hunters using them for vehicular access (R. Mapston personal communication:1970). Because they are narrow, the cost of materials to build an antelope pass is about half that of cattle guards.*

the literature indicates that the original phrase antelope pass was used most frequently in technical papers (e.g. Kite et al. 1994, Yoakum et al. 1996, Lee et al. 1998 and Payne and Bryant 1998a). Use of the phrase antelope guard should be discontinued.

Disassembling Dysfunctional Fences

Fences that no longer serve installation objectives (at times called dysfunctional fences) should be disassembled and removed from rangelands occupied by pronghorn. Abandoned fences, together with cattle guards, have the potential to cause injury, impede free movement or cause mortality to wildlife—especially pronghorn. With a trend toward decrease of domestic livestock on western rangelands at the end of the 20th century, there is an increasing

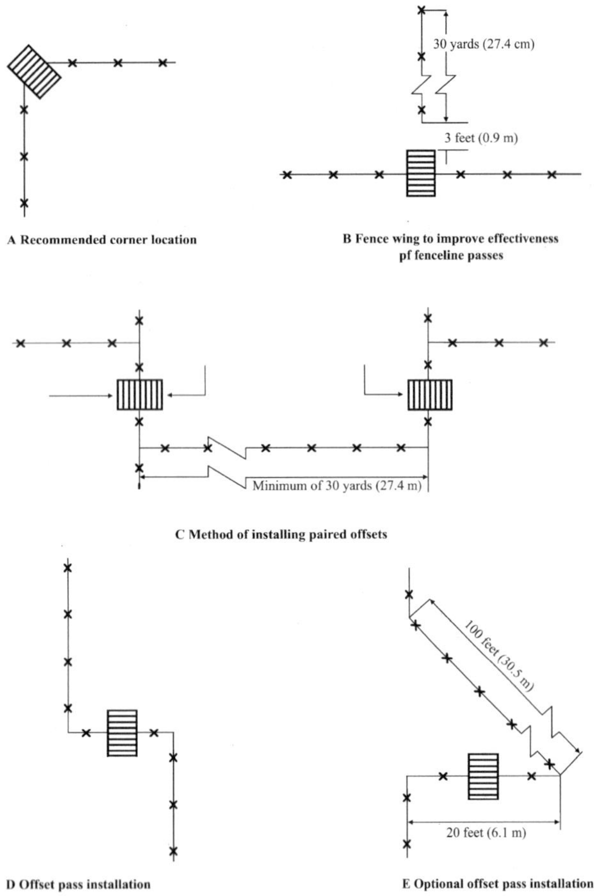

Figure 131. *Suggested sites for installing antelope passes in fence lines (Mapston and Zobell 1972). Proper placement especially is important because the passes are narrow and pronghorn ordinarily would not seek passage between posts less than 6 feet (183 cm) apart.*

number of dysfunctional fences. These fences pollute the natural environment, remain a blight on natural panoramic views and jeopardize the health of wildlife.

Although funds often are requested to construct fences to enhance the livestock industry, funds appear less frequently in federal land management agency budgets to destruct fences, possibly due to lack of urgency for nontraditional management objectives. Fence projects (especially on public lands) need objectives and funds for disassembling as much as construction. Activity functions responsible for building fences also should be responsible for proper dismantling of fences (e.g., livestock management, road safety). Because fence disassemblage has not been a traditional objective in government agency management programs, emphasis is needed to include practices and finance obligations in agency regulations and operational manuals to ensure that fences no longer serving management objectives will be properly disassembled as a responsibility of land stewardship.

Before the 1990s, many fences on western rangelands remained long after they ceased to function as a control of livestock. When homesteaders abandoned their lands, they likewise abandoned fences. When ranchers cease to run livestock on western rangelands due to economic problems or change in land management objectives, the miles of fences often remain as a legacy. However, with passage of several federal land conservation laws in the late 1900s, Americans have become more oriented toward natural resource conservation. This was exemplified in an environmental impact and management plan that was enacted by the U.S. Fish and Wildlife Service (1994) for the Hart Mountain National Antelope Refuge in Oregon. The assessment and planning project terminated domestic livestock grazing on the refuge, and this action nullified the need for miles of interior allotment fences, which were constructed by the federal government and ranchers. These fences caused mortalities to pronghorn, mule deer and bighorn, therefore a management recommendation was to disassemble fences no longer needed to control livestock. The administrative agency encouraged public organizations to demolish unwanted fences. Various organizations responded—local Audubon Society chapters, sporting clubs, the Oregon Natural Desert Association, the Order of the Antelope and many more—by contributing labor and travel expenses. Absent were representatives of the livestock industry—the primary economic interest for whom the fences were originally constructed during a 100-year period. Between 1995 and 2000, more than 40 miles (64 km) of fences were dismantled, and these community conservation services for wildlife continue to function each year. The total endeavor has contributed to making the refuge more wildlife friendly.

Some volunteer organizations have made a practice of returning to the Hart Mountain National Wildlife Refuge each summer to help disassemble fences. A notice in *Sierra* advertised the outing: "Great continuing project to aid movement of antelope on this now cattle-free area. Drive in base camp (CCC). Trip features high desert gemstones, petroglyphs, hikes, warm springs, bighorn sheep and a bird sanctuary" (Sierra Club 2001:101). The notice in the Sierra Club's international magazine encourages members to, "do good deeds, work hard, and feel great," while restoring valued public lands. And, each member paid $265.00 to contribute these services.

A similar volunteer action program to help remove approximately 200 miles (322 km) of unnecessary interior fences on the Sheldon National Wildlife Refuge in Nevada was conducted during the last half of the 1990s and continues into the 21st century (M. Gregg personal communication:2000). Other U.S. Fish and Wildlife Service refuges in Montana have fence demolition projects. In addition, the Carrizo Plains National Monument in

With termination in 1994 of all domestic livestock use on the Sheldon National Wildlife Refuge in Nevada, internal fences constructed to control livestock were no longer needed. These fences remained an obstacle and occasional danger to bighorn sheep, deer and pronghorn. Various public organizations aided the U.S. Fish and Wildlife Service to dismantle nonfunctioning fences. Conservation groups, such as members of the Lahouton Audubon Chapter (above), disassemble 2 to 5 miles (3.2–8.0 km) of interior fences during annual summer field trips. The assistance of volunteers helps defray the cost of contracting the fence deconstruction, and it disassembles the fences more quickly—real benefits to wildlife. Photo by Jim D. Yoakum.

California, primarily administered by the BLM and California Department of Fish and Game, was assisted by volunteer Native Americans to tear down nonfunctioning fences (A. Koch personal communication:2001).

A similar fence program is conducted by the Arizona Antelope Foundation. Volunteers disassemble nonfunctioning fences on private lands for the benefit of pronghorn. They also volunteer to reconfigure fences to allow pronghorn movement. The volunteers first coordinate plans with ranchers to change fence specifications to be more pronghorn friendly by raising the bottom barbed wire to a height of 16 inches (41 cm) above the ground. These projects have turned into annual summer conservation events, ending with a barbecue for volunteers and cooperating ranchers.

Certain conservation groups volunteer to reconstruct fences to aid pronghorn movements on private ranches. Above, members of the Arizona Pronghorn Foundation reconstruct a fence by adjusting the bottom wire to a 16-inch (41 cm) spacing, so that pronghorn can crawl easily under the fence originally built to control cattle. These volunteers provide all personal expenses and enjoy a barbecue and picnic sponsored by cooperators. As well as enhancing pronghorn habitats on private lands, efforts such as this foster good relations between the public and ranchers. With a decrease in pronghorn habitat due to burgeoning human population, such conservation practices are a mitigating measure beneficial to wildlife on private lands. Photo courtesy of the Arizona Antelope Foundation, Phoenix.

Legal Cases

In the late 1970s, a legal battle occurred in New Mexico regarding wolf-type fences. Ranchers had constructed miles of these fences that restricted pronghorn movement. Much of the fencing was on public lands administered by BLM. After the Federal Lands Policy and Management Act of 1976 (FLPMA) was enacted, BLM implemented a program to modify wolf-type fences to allow pronghorn access to public lands. This was challenged by the ranchers who contended the agency was acting arbitrarily and capriciously. The case was presented before a hearings examiner who ruled that FLPMA directed BLM to practice multiple use and that it was proper and justified for the agency to modify wolf-type fences to allow pronghorn access to public lands. The determination—a victory for pronghorn welfare and management—ended a long and acrimonious struggle between livestock producers versus wildlife managers, conservationists and public administrators (Yoakum 1980a).

In parts of the Southwest, ranchers practiced construction of fences around water. The fences encircled water structures or sources and, when livestock were allowed to enter the enclosures, gates could be closed and the animals corralled. The fencing also was used to prevent livestock from accessing particular water structures, thus forcing the livestock to move to other structures. Usually inadvertently, the practice also served to keep pronghorn from the water. This circumstance was especially serious for young wild ungulates (deer as well as pronghorn), inexperienced in negotiating fences and handling the stress of seeking alternative sources. Such fencing appears to be a violation of multiple-use laws and regulations governing federal lands, similar to the judgment regarding wolf-type fences. Consequently, Yoakum (1980a) recommended that such fencing be discontinued in pronghorn habitats.

The most controversial and publicized legal case concerning pronghorn and fencing occurred in the Red Rim area near Rawlins, Wyoming (Ryder et al. 1984, Moody and Allredge 1986). The controversy involved a rancher erecting a net wire and restrictive barbed wire fence around critical pronghorn winter rangeland to block pronghorn access to 9,600 acres (3,885 ha) of interspersed public and private rangelands. Apparently, the rancher's objective was to render certain rangelands useless to pronghorn during winter, so coal could be mined.

A legal suit was brought against cattle rancher Taylor Lawrence by the U.S. Department of Justice with the Wyoming Wildlife Federation and the National Wildlife Federation as interveners. The case was presented to the U.S. District Court for the District of Wyoming in November 1985, with Chief Judge Clarence A. Brimmer entering judgment. The U.S. Justice Department's action was based on a statute adopted by the U.S. Congress in 1885, the Unlawful Inclosure of Public Lands Act (UIA) (43 U.S. Code paragraph 106.1 to 1066), which requires a court order to compel a rancher to remove or modify fences to allow unrestricted access for pronghorn to enclosed public lands. Federal law declares such fences to be nuisances, and they must be removed or modified.

The rancher claimed the UIA was applicable to people, not to pronghorn. The district court concluded that the UIA stipulation against enclosing public lands was not limited to people. It further based its findings on other cases concerning the UIA that ruled the statute was intended to prevent the obstruction of free travel for any and all lawful purposes over public lands.

Thus, the question became whether winter forage for pronghorn was a lawful purpose of public lands. This doubt was resolved by reference to the FLPMA paragraph 1701 (a) (8) which states, "the public lands be managed in a manner . . . that will provide food and habitat for fish and wildlife and domestic animals." The court found that wording in UIA and FLPMA preserves access to federal lands for lawful purposes, including forage for pronghorn and other wildlife. A fence constructed such that it does not obstruct other lawful uses of federal lands is not an unlawful enclosure.

Finally, the rancher claimed that, because the fence included several unlocked gates, there was no unlawful enclosure evidence that the fence excluded entry. The argument was rejected by the court based on the conclusion that it was not the fence itself, but its effect on pronghorn that constituted a violation of the UIA statute. The court further found that the gates were clearly insufficient to allow access of pronghorn and refused to find that public access was unhampered by the fence based on a similar case ruled by the U.S. Supreme Court in 1897. The Wyoming court determined that the rancher cannot maintain a fence that encloses public lands and prevents the lawful purpose of pronghorn access to their winter feeding grounds.

The court concluded that the U.S. Department of Justice and interveners upheld their burden of proof, granting their motion and ordering the cattle rancher to disassemble certain portions of the fence within 10 days and to modify or remove the entire fence within 60 days (*United States v. Lawrence*, 620 F. Supp. 1414 [D.C. Wyoming. 1985]).

In response to the district court's ruling, the cattle rancher appealed to the Tenth Circuit Court of Appeals in June 1988. The court again heard much of the evidence and arguments presented in the district court. The court of appeal found no error in the district court's judgment and affirmed the order (*United States v. Lawrence*, 848 F. 2d 1502 [10th Cir. 1988]).

Again, the cattle rancher appealed to the U.S. Supreme Court in Washington, D.C. The Supreme Court's action was denial of certiorari, which was an indication that the court concluded the appellate court's holding did not warrant the Supreme Court's review. Thus, a decision was made that no Supreme Court inquiry was necessary (*Lawrence v. United States*, 488 U.S. 980 [1988]).

So ended the longest, most expensive, legal battle for pronghorn on public lands—from district court to circuit court to the Supreme Court. These courts determined that forage for pronghorn was a lawful purpose of public land management and that pronghorn had legal access rights for seasonal movements. This was a victory for the conservation and management of pronghorn on public lands in the United States.

Management Implications and Recommendations

Numerous research projects, administrative workshops, court cases and field experiences regarding wildlife-related fence and structure issues have resulted in significant management principles, recommendations and initiatives that affect pronghorn. They are summarized as follows:

1. Many fences built to control livestock on rangelands have been barriers to pronghorn, resulting in pronghorn declines in some regions. As few fences as possible should be built on rangelands occupied by pronghorn. Livestock operations, especially those for domestic sheep, should be managed with minimal fencing.

2. There should be no fencing on public rangelands or private lands occupied by pronghorn, until a complete evaluation has been made for each proposed fence. Assessment should determine the probable impact on pronghorn and effects on other management objectives.
3. Where fencing is deemed necessary, the minimum amount of fence should be erected to minimize hindrance to pronghorn. Unrestricted passage should be provided for all pronghorn during all seasons and weather conditions.
4. Fencing of water sources on dry summer rangelands may be as detrimental to pronghorn as fencing of seasonal movement routes. Critical areas (i.e., winter concentration areas, fawning areas, seasonal movement corridors and water sources) should be identified and designated as not to be fenced.
5. No specific fence design on cattle rangelands can be applied universally to allow pronghorn access. However, specifications most compatible with pronghorn movements consist of three strands of wire, a smooth bottom wire 16 to 18 inches (41–46 cm) above the ground and a total height of no more than 36 inches (91 cm).
6. For rangelands grazed by domestic sheep, any fence that restricts sheep movements will also restrict pronghorn access.
7. Wherever possible, net wire fences should not be built on pronghorn rangelands. If net wire fences must be used, mitigating provisions, such as let-down or adjustable segments, should be incorporated and regularly adjusted at strategic pronghorn passage sites.
8. Specially designed fences (e.g., buck and pole, rail, suspension) should be no higher than 36 inches (91 cm), with a bottom gap of at least 16 inches (41 cm).
9. All new fences should have white flags tied to the top wire between each post to improve visibility of the new hazard. Pronghorn will become accustomed to the fence by the time the flags deteriorate. Gray, camouflage, steel post fences should be avoided on pronghorn rangelands.
10. Wolf-type and antiantelope fences completely restrict pronghorn movements and should not be built on rangeland occupied by pronghorn.
11. Let-down or adjustable fences have not been used extensively, but may serve effectively under some circumstances. Site placement should take advantage of topographic depressions.
12. Where livestock operations switch from sheep to cattle, woven wire fences restricting pronghorn movements should be removed.
13. Existing fences that restrict pronghorn movements should be removed or modified to allow the easy and safe passage of animals.
14. All dysfunctional fences should be dismantled to decrease injury hazards to wildlife.

Overpasses and Underpasses

Roadways fragment pronghorn habitats and can impede the animals' movements (Buechner 1950a, Taylor 1975, Ockenfels et al. 1994). The impact is accelerated when roads, highways or railways are fenced on both sides. Where roads, particularly highways, cross pronghorn travel corridors, injuries and death of pronghorn, damage to vehicles and injuries to humans can result.

Ward et al. (1976) found that underpasses constructed on Interstate Highway I-80 (a four-lane divided highway) in Wyoming were not used by pronghorn during a 2-year study. Further studies noted only one crossing through an underpass during a 6-year period (Ward et al. 1980). The underpass used had a dirt floor and was used to move cattle and ranch equipment.

Kahre (1980) reported on underpasses constructed on a four-lane highway in northeastern California transecting

An underpass constructed on U.S. Interstate Highway 80 in south-central Wyoming to provide access for deer, livestock, pronghorn and ranching machinery successfully accomplished objectives for all but pronghorn. Although the area supported large numbers of pronghorn that traditionally moved seasonally, the animals did not use the underpass despite the fact that it was large, such that the pronghorn could see through, and it had a dirt floor. The experience showed that underpasses are not the solution for allowing pronghorn movements in proximity to high-volume human traffic. Photo by Jim D. Yoakum.

Pronghorn/vehicle accidents are relatively uncommon, but do occasionally happen (see Chapter 12). Agencies concerned with this problem have erected highway warning signs to alert motorists that pronghorn occupy the area and to be alert in watching for them to avoid injury to animals and damage to vehicles. These signs cost about $40.00 each at the beginning of the 21st century. Photo by Jim D. Yoakum.

pronghorn and mule deer rangeland. Mule deer used the structures extensively, and fatal collisions with vehicles decreased by more than 95%. Pronghorn did not use the passes. No documentation of pronghorn use of overpasses or bridges was found.

It is apparent that pronghorn, adapted to environments that maximize security, are unable to compromise their alert behavior by crossing noisy, heavily used traffic lanes by using structures, such as overpasses or underpasses.

A Model for Pronghorn Conservation on Private Land

A relatively closed pronghorn population on predominantly private land is documented in a management plan for the Deseret Land and Livestock Ranch in northeastern Utah (Danvir 2000). This plan has extensive, ongoing ecological investigations accomplished from 1984 to 2000. Wildlife biologists on the ranch procured a wealth of data, regarding pronghorn/habitat relationships that could serve as a model for other private ranch operations featuring pronghorn. The ranch maintains healthy rangeland vegetation and abundant drinking water, and it sustains an average herd of 500 pronghorn, plus other endemic wildlife, in harmony with cattle production. The wildlife and livestock programs also are economically viable.

From 1890 to 1975 the Deseret Ranch operated as a traditional livestock business 5 miles (8 km) southeast of Woodruff, Utah. Since the 1970s, the ranch actively has been managed for wildlife and livestock. Wildlife income is derived from fee hunting, fishing and bird watching. The following biological data were obtained from the ranch management plan (Deseret Land and Livestock Ranch 1997), a technical report (Danvir 2000, Aoude and Danvir 2004).

The eastern half of the Deseret Ranch includes approximately 150 square miles (390 km^2) of private sagebrush/grassland (87%) and scattered parcels of public lands (13%) administered by the BLM. Elevations for pronghorn habitat range from 6,400 to 6,800 feet (1,950–2,070 m). Mean annual temperature is 72°F (22°C); summer temperatures above 100°F (38°C) are not uncommon. Annual precipitation averages 9.5 inches (24 cm), occurring primarily from September to May.

There are five major plant communities on pronghorn habitat (Table 154). The north end is dominated by Wyoming sagebrush and crested wheatgrass; the crested wheatgrass was planted in the 1960s. The central study area is predominantly big sagebrush (dense native sagebrush). The southern end is mainly native grassland (sparse native shrubs) with codominant browse of low sagebrush and rabbitbrush. There are stands of black and low sagebrush throughout the landscape. Riparian meadowlands are well distributed through the rangeland, and drinking water is available every 2 miles (3.2 km).

The study area is grazed by 3,000 to 5,000 cattle annually, and it is managed with time-controlled livestock grazing systems, as advocated by Savory (1988). Livestock grazing systems use short-duration grazing periods to minimize rebiting of plants by livestock during the growing season and maximizing the benefits of seasonal rest. Interior fences are numerous (pastures average 2,000 acres [8,094 ha]), and generally are constructed with four- to five-strand barbed wire; the bottom strand is at least 16 inches (40 cm) above the ground. Immigration and emigration of pronghorn appears minimal, with movement restrictions due, in part, to some woven wire external fences.

Studies of pronghorn/habitat/mortality relationships were instigated, from 1984 to 2002, to determine correlations between pronghorn population dynamics and potentially limiting factors (e.g., competition for food, climatic variability, predation and hunting). Pronghorn were

Shrub/grasslands on the Deseret Land and Livestock Ranch in northeastern Utah are home to a native population of pronghorn that have been studied for more than 15 years. Sites with dense stands of Wyoming sagebrush and little herbaceous understory on the ranch were burned and planted with a mixture of grasses and forbs. These treated sites later produced twice the quantity of forbs as the adjacent untreated lands. Pronghorn used the treated sites, especially during pregnancy and lactation, more than the untreated sites. Photo by Rick Danvir.

TABLE 154. Percentage composition of major vegetative types on pronghorn habitat, Deseret Land and Livestock Ranch, Utah

Vegetative type	*Forb*	*Forb perennial*	*Grass*	*Shrub*	*Shrub height*[a]	*litter*	*Bare ground*
Black sagebrush	33.3	9.7	10.7	29.3	8.0 (20.3)	9.7	28.3
Big sagebrush/crested wheatgrass	9.6	1.0	28.9	17.3	16.9 (42.9)	17.0	24.3
Sparse native sagebrush	16.0	6.3	32.0	24.0	7.5 (19.1)	13.5	13.5
Dense native sagebrush	9.3	6.3	15.0	35.0	22.0 (55.9)	16.3	20.7
Meadowland	33.3	32.7	55.3	0	0 (0)	9.3	2.0

[a] In inches (cm)
Source: Danvir 2000.

reintroduced to the ranch in the winter of 1983–84. The population increased from 90 to 620 animals between 1985 and 1991, then fluctuated approximately 30% for the next decade (Table 155). Fawn survival during late summer generally has decreased as the population increased (Figures 132 and 133). Yearly population change and fawn production both are correlated negatively with prior-spring population size (Figure 134). Autumn doe/fawn ratios ranged from 0.2 to 1.2 and correlated positively with same-year April to July temperatures. Pronghorn are harvested at a ratio of one adult buck to one adult doe. Bucks with larger horns weighed significantly more than bucks with smaller horns ($P = 0.029$). Of 68 bucks examined, the average weight of bucks having Boone and Crockett trophy horn scores of more than 75 points was 109 pounds (49.4 kg); mature bucks with scores of less than75 points averaged 106 pounds (46.8 kg) (see Nesbitt and Wright 1985).

Results from the Deseret Ranch long-term pronghorn/habitat investigations are informative of ecological relationships—base information needed to develop management plans. These studies continue because not all biological factors are adequately known. However, innovative management practices to 2002 indicated the following findings are relative to managing pronghorn:

1. Indices of pronghorn condition (weight and horn mass), fawn survival and population growth are

Table 155. Pronghorn population data, coyote abundance, temperature, pronghorn harvest and snow depth for the Deseret Land and Livestock Ranch, Utah

Year	*Number of pronghorn*	*Population growth*	*Fawn/doe ratio*	*Coyotes 100 hours*	*Average temperature April to July*[a]	*Pronghorn harvested*	*Prior winter snow depth*[b]
1984	—	—	0.35	0.7	51 (10.4)	—	11.3 (28.7)
1985	90	—	0.88	1.0	54 (12.2)	—	9.2 (23.4)
1986	130	1.44	1.10	0.3	53 (11.4)	—	7.0 (18.0)
1987	190	1.46	1.17	0.2	54 (12.2)	—	2.3 (5.8)
1988	260	1.37	1.07	1.0	55 (12.7)	—	3.3 (8.4)
1989	340	1.31	0.68	0.3	53 (11.7)	—	3.5 (8.9)
1990	470	1.38	0.77	1.0	53 (11.6)	78	1.8 (4.6)
1997	620	1.32	0.47	2.3	51 (10.3)	141	2.2 (5.6)
1992	600	0.97	0.40	9.3	53 (11.8)	104	1.3 (3.3)
1993	410	0.68	0.20	7.3	49 (9.5)	10	11.4 (29.0)
1994	400	0.98	0.73	10.0	53 (11.9)	14	1.9 (4.8)
1995	520	1.30	0.58	16.0	49 (9.7)	56	3.3 (8.4)
1996	490	0.94	0.50	9.7	52 (11.2)	66	1.6 (4.1)
1997	380	0.78	0.55	7.1	51 (10.5)	34	5.3 (13.5)
1998	450	1.18	0.53	10.0	51 (10.3)	41	5.9 (14.9)
1999	490	1.29	0.57	10.1	50 (10.2)	55	1.0 (2.5)
2000	690	1.16	0.52	12.5	55 (12.7)	41	0.5 (1.3)
2001	700	1.04	0.66	10.8	53 (11.8)	63	3.4 (8.6)
2002	740	1.06	0.86	9.1	54 (12.2)	50	5.2 (13.2)

[a] In degrees Fahrenheit (°C).
[b] In inches (cm).
Source: Danvir 2000, R. Danvir personal communication:2003.

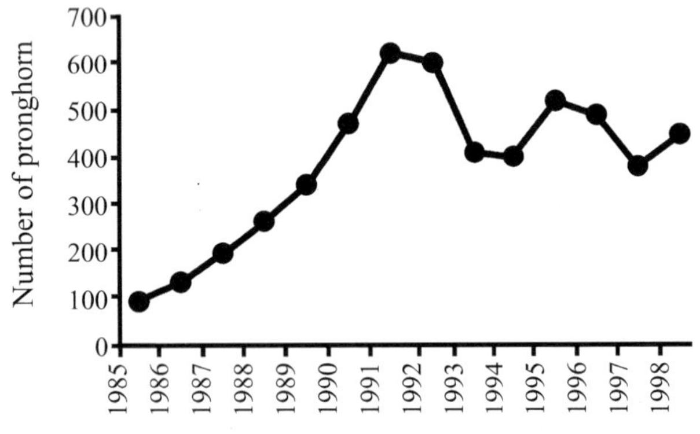

Figure 132. *Pronghorn population trend from 1985 to 1998 on the Deseret Land and Livestock Ranch in Utah (modified from Danvir 2000).*

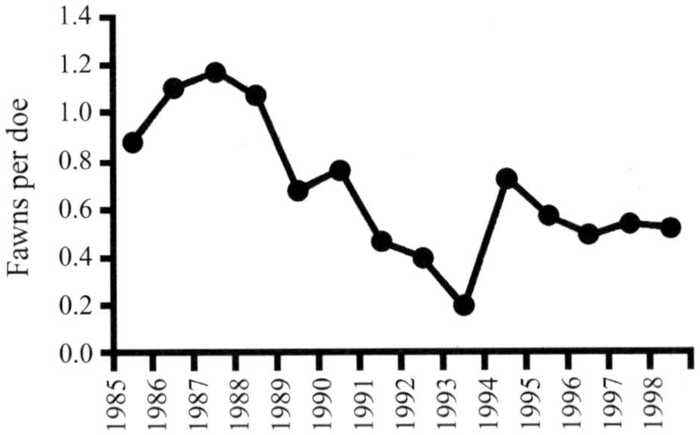

Figure 133. *Trend of pronghorn fawn/doe ratios from 1985 to 1998 on the Deseret Land and Livestock Ranch in Utah (modified from Danvir 2000).*

correlated negatively with population size, suggesting intraspecific competition and density dependence. Reduced condition, fawn survival and population growth at high populations suggest a density-dependent population (Caughley and Sinclair 1994). Greater pronghorn condition (weight) likely increases fawn survival, and fawn survival is positively correlated with subsequent population growth (R) = +0.79).

2. Forage removal by livestock is not correlated with pronghorn condition or population parameters. Initial data suggest time-controlled grazing by cattle may maintain or increase long-term forb availability in sagebrush and grasslands, compared with sites grazed only by wild ungulates. Cattle grazing can reduce grass biomass and may reduce competition for moisture between grasses and forbs. Data also indicate that summer forage biomass and quality are dependent on precipitation and temperature.

3. Use of vegetation communities varied seasonally; Figure 135 illustrates this variation in pronghorn use of the black sagebrush and meadow types from 1988

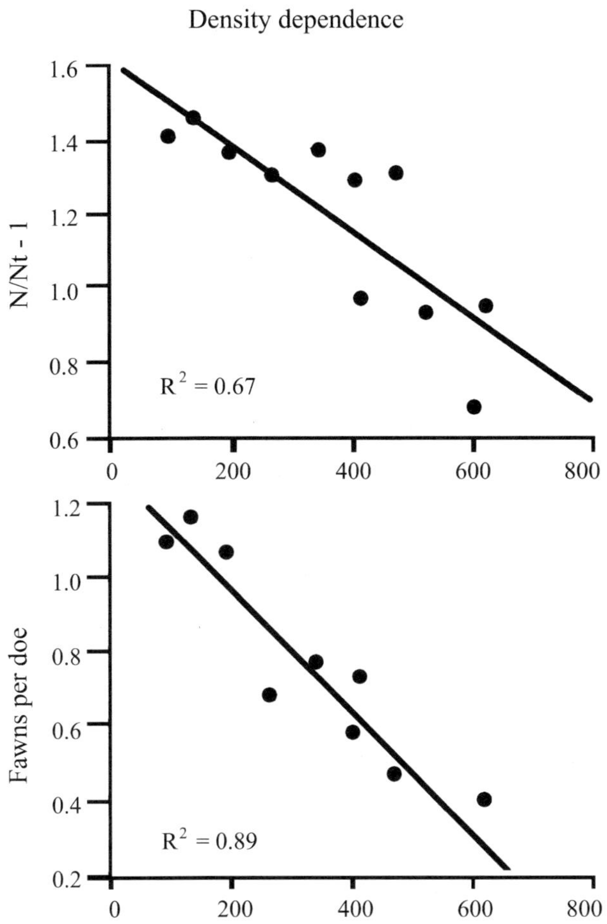

FIGURE 134. *Relationship between population growth, autumn fawn/doe ratios and prior year population size on the Deseret Land and Livestock Ranch in Utah (Danvir 2000).*

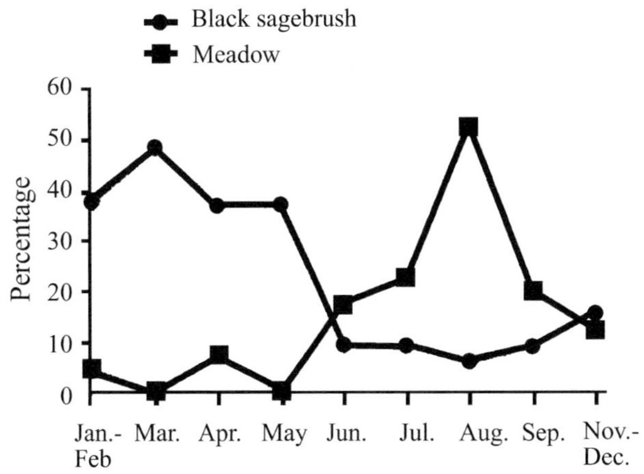

FIGURE 135. *Comparison of differential pronghorn use of vegetative types (in this case, black sagebrush and meadow) by month. Based on observations of 4,430 pronghorn from 1988 to 1990 on the Deseret Land and Livestock Ranch in Utah (Danvir 2000).*

to 1990. Preference for black sagebrush, meadow and sparse native sagebrush communities, and avoidance of sagebrush, crested wheatgrass and dense native sagebrush by doe/fawn groups during summer suggests females with young selected forb-rich areas having good visibility (e.g., low shrub height). Buck-only groups used dry black sagebrush and sagebrush/crested wheatgrass (xeric communities) more so than other forage types, and they avoided forb-rich areas frequented by doe/fawn groups. It may be that buck-only groups avoided forb-rich areas protected by territorial bucks. Perennial forb cover and doe/fawn ratios were three times greater in sparse native sagebrush than in sagebrush/crested wheatgrass sites.

4. Dense native shrub was avoided by both male and female pronghorn. Nearly 70% of the dense native shrub type has slopes greater than 30% or dominated by dense sagebrush more than 20 inches (50 cm) in height. Pronghorn avoided these steeper, tall shrub-dominated, areas as reported in other studies (Amstrup 1978, Yoakum 1980a). Within the dense native sagebrush type are two forb-rich burns (burned in 1994) and two large (more than 150 acres: 60 ha) areas that were disced and seeded to a mixture of grasses, perennial forbs and shrubs in 1994. Although seldom used by pronghorn before treatment, these four areas were consistently used by territorial bucks and doe/fawn groups during the summer of 1995, suggesting that removal of the tall shrub component and stimulating forb growth attracted lactating does and territorial bucks.
5. Pronghorn consistently selected black sagebrush year-round. Abundant forbs and low-growing shrubs provide an optimum mix of forage and security during all but deep-snow periods. These findings correlate with habitat characteristics and requirements, reported in Chapter 13.
6. Sagebrush/crested wheatgrass appeared to provide adequate winter habitat for pronghorn, but was not preferred by females in summer. This vegetative type also had significantly less perennial forbs than all the other vegetative types (Table 154). Crested wheatgrass begins growing early in the spring and at cooler temperatures, compared with native bunchgrasses. It is presumed that this early growth provides crested wheatgrass an advantage in obtaining moisture, thereby reducing establishment, vigor and diversity of later-growing native grasses. Although annual forbs can be abundant in sagebrush/crested wheatgrass stands during years with above average moisture, they are sparse during dry years. Perennial forbs are lacking generally in these stands.
7. Findings suggest that winter snow depth decreases pronghorn condition and population growth, likely by increasing energy demand and decreasing available

winter forage. Pronghorn selected black sagebrush and sagebrush/crested wheatgrass types during most winters, because these areas generally offered adequate shrubs and the lowest snow depths. During the deep-snow winter of 1992–93, pronghorn preferred areas with reduced snow depths, regardless of shrub availability. These areas included plowed roads and feedrows for cattle. It is possible that northern pronghorn herds historically moved seasonally and wintered with large herbivores, such as elk or bison, that aided travel by trampling trails in deep snow. In an effort to reduce mortality, the Deseret Ranch crew created a series of hard-packed trails using a snowcat on the pronghorn winter rangeland during February 1993 (snow depth is more than 14 inches: 35 cm). The goal was to increase access between the remaining uncovered pockets of sagebrush. Within 12 hours of packing the trails, nearly all pronghorn in the vicinity were using the trails and moving to new feeding areas.

8. Although the number of pronghorn harvested annually was negatively correlated with population growth, population growth most strongly was correlated negatively with density and snow depth. The effects of density and climate observed in this study are similar to those reported by Pyrah (1987) for the Yellow Water Triangle area in Montana. Pyrah noted that doe bands inhabiting home ranges in unhealthy vegetative status produced fewer fawns. Pyrah indicated that adult female density largely determined population growth, production and size.

With more than 16 years of biological findings and ecological assessment, management strategies for pronghorn can be capsuled for the Deseret Ranch along the Utah/Wyoming border. Apparently pronghorn condition, fawn survival and population growth are limited by winter severity and intraspecific competition for forage. Forage availability (summer forbs and winter shrubs) differs between vegetation types and years, due to differences in plant species composition and interannual variation in weather. Management practices that effectively increase or prolong forb availability may increase pronghorn population growth and size. In colder, northern climates subject to deep snow, long-term population densities may be related to pronghorn carrying capacity during severe winters. In situations where seasonal movements to snow-free areas have been blocked because of human developments, management strategies aimed at increasing winter survival on nontraditional wintering areas may be necessary. Using machinery or large herbivores to open trails and pack snow may increase winter pronghorn survival by reducing energetic demands and increasing shrub and forb availability.

CROP DEPREDATIONS

Pronghorn are attracted to certain agricultural crops, especially alfalfa. Pronghorn preference for alfalfa, which is highly nutritious and succulent during growth, can lead to alfalfa crop depredation. Cole (1956) noted that, in central Montana, alfalfa comprised 14% of pronghorn yearlong diet. Use in early spring to midsummer was negligible, but progressively increased until late September, reaching a maximum of 50.3%. Its consumption by pronghorn showed a marked decline after the first of October. Rumen sampling disclosed 20% to 30% of diet was alfalfa during autumn. Apparently, 30% to 40% of the animals consumed alfalfa during this study. Pronghorn caused appreciable reduction in plant height and severe reductions in number of blossoms and seed pods. To alleviate such losses, Cole suggested either-sex hunting seasons on or adjacent to cultivated fields, removal of excess animals by trapping and translocation, and construction of exclosure fences.

With increased pronghorn populations in northern California during the 1980s, depredation on alfalfa increased. The state wildlife agency responded by opening problem areas to either-sex hunting and by trapping and translocating herds to areas with low densities and to then unoccupied historic range (Pyshora 1982b, Goldsmith 1988).

Damage to winter wheat by pronghorn has been alleged for many years in the Great Plains (Cole and Wilkins 1958, Liewer et al. 1986, Pojar 1996). In central Montana, Cole and Wilkins (1958) reported that wheat fields comprised 21% of the area on a year-round basis, and pronghorn use of these grain fields was relatively light during spring and summer and moderate to heavy during winter. Pronghorn of all sex and age classes used the grain fields, consuming 10% waste grain and 6% sprouted grain. They commonly bedded in certain grain fields, but avoided those where their vision was obstructed by tall standing crops. Wheat fields in close proximity to native rangelands received the greatest use.

In 1983, a 3-year research project was initiated to evaluate the impact of pronghorn on winter wheat fields in eastern Colorado (Liewer et al. 1986). Aerial surveys disclosed extensive use of wheat fields by pronghorn from November through mid-March, but no significant treatment level imposed intense grazing pressure of 430 pronghorn per square mile (166/km^2) for 172 days (November–April) with no observable reduction in grain yield. The researchers reported few observations of plants being uprooted by pronghorn feeding, pawing or trampling.

In Colorado wheat fields, pronghorn densities 80 times higher than found on native habitats did not greatly reduce winter wheat production, according to Torbit et al.

Alfalfa was reported as a preferred forb in various pronghorn diet composition studies. It is especially consumed during early growth when low in structure. Many alfalfa fields are irrigated, as in the top *photo from Modoc County, California, so crop field damage by pronghorn frequently are reported. Pronghorn are not averse to feeding on ranchers' alfalfa haystacks* (bottom)*, also prompting depredation complaints and costs.* Photos by Robert Schaefer.

(1993). Although pronghorn foraging reduced green biomass, these effects were mitigated by compensating plant growth after grazing ceased. Grain yields after grazing were not significantly reduced (P = 0.17) despite reductions in green biomass. Strohmeyer (1994) suggested that pronghorn may not damage winter wheat, therefore reducing herd size might be unnecessary. Strohmeyer supported Torbit et al.'s (1993) supposition that winter wheat yields should not decline from pronghorn grazing during tilling because biomass increases.

According to Pojar (1996), pronghorn in eastern Colorado primarily occur on private agricultural lands, therefore their crop depredations can be an issue. Annual damage claims involving pronghorn paid by the Colorado Department of Wildlife from 1978 to 1990 averaged $3,331.00, representing 1.3% of total statewide damage payments for all wildlife. Despite the relatively low payment amounts, damage complaints by landowners were numerous. This may have been due to the conspicuous presence of pronghorn, which are not nocturnal or cryptic in their habits. Evidence from Colorado studies indicated that grain production and pronghorn grazing are compatible (Torbit et al. 1993). See Chapter 20 concerning alleviation of crop damage.

STATUS AND CONSERVATION OF ENDANGERED POPULATIONS

An evaluation of pronghorn subspecies listed as endangered in Mexico and the United States disclosed that, at the beginning of 2000, three of the five presumed pronghorn subspecies were listed as endangered—Mexican, peninsular and Sonoran (see Chapter 4).

Characteristics and Conditions of Populations

1. All populations have had low densities, compared with herds in northern landscapes.

2. The estimated number (rounded to closest 50) for each subspecies as of 1999 was: Mexican 750, peninsular 200 and Sonoran 1,500—of the latter, 250 in Sonora and less than 300 in Arizona. This represents less than 2% of total population in North America.
3. Does produce twin fawns annually as they do in high-density northern populations.
4. Populations are subjected to similar mortality factors as other herds (e.g., severe weather conditions, predation, diseases, accidents, old age). However, one difference is that the endangered populations suffer more from periodic droughts, whereas northern herds succumb to periodically severe winters.
5. None of the endangered subspecies are hunted legally. Some poaching occurs, particularly in Mexico.
6. Pronghorn densities are directly related to habitats with mixed vegetation classes of grasses, forbs and shrubs. Pronghorn diet selection studies indicate the animals' preference for forbs and shrubs. The nutritional value and quantity of forbs during the reproductive period appear to be linked directly to fawn production and survival. Lack of forbs is associated with low pronghorn recruitment. Forbs, including cacti, also are associated with providing water for survival.
7. High densities of pronghorn for northern herds are associated with habitats where free drinking water sources are available. According to studies, pronghorn are more thrifty and numerous when drinking water is available.
8. Although management strategies, such as capturing and translocating northern herds to unoccupied historic habitats, have been frequent and successful to reestablish herds, such practices have not been accomplished for the endangered pronghorn subspecies.
9. The validity of scientific nomenclature for pronghorn subspecies remains questioned. Therefore, a comprehensive research project to determine the validity of nomenclature for all pronghorn subspecies is needed. Until then, designating endangered subspecies will remain a confused issue.
10. There has not been a food habit study encompassing season-long collections for several years of varying precipitation patterns. Such studies also could provide data on nutritional values, forage quality and moisture content. The lack of this knowledge prohibits understanding of the methods to manage habitats for the needs and values of endangered pronghorn.

Characteristics and Conditions of Habitats

1. All pronghorn habitat ecosystems are semiarid: desert grasslands for Mexican pronghorn and deserts for Sonoran and peninsular pronghorn.
2. Pristine vegetative class conditions exhibited a mixed composition of grasses, forbs and shrubs. Species composition and abundance differed significantly from northern environments. They differed especially in the lesser quantities of herbaceous plants that supported low-density populations.
3. These habitats have a history of more than 400 years of livestock grazing, resulting in altered vegetative communities. Most often, this has resulted in increased height and density of shrubs and fewer herbaceous plants. Such changes in vegetation can affect pronghorn populations adversely.
4. More recent habitat alterations have affected pronghorn populations: for example, fences, highways, railroads, energy developments, competition for forage with livestock, off-road vehicle activities during fawning seasons, exotic diseases from livestock, invasion of exotic vegetation in preferred foraging areas, agricultural crop production, occupation of large areas for towns and cities.
5. Some rangeland practices in northern landscapes (e.g., irrigated alfalfa crops, prescribed burns of dominant, decadent shrublands, constructed reservoirs), although artificial, have at times enhanced habitats for northern herds. This has seldom transpired for the benefit of southern herds.
6. Some, although few, permanent drinking water sources were available to herds during pristine times. Fewer exist today due to exploitation for human use.
7. Because high densities of pronghorn are associated with environments containing permanently available drinking water, strategies need to be investigated to restore such sources to habitats that have lost these sources.
8. Periodic natural changes in plant succession induced by lightning ignitions are a natural process affecting changes in vegetation. Their natural continuance can be beneficial to pronghorn because these animals favor nutritious and succulent forbs that can increase after fire or other natural alterations of vegetative communities.
9. Efforts are needed to investigate and mitigate habitat loss for seasonal movement corridors that have been blocked or lost.
10. Management practices to restore healthy communities of natural vegetation and waters on semiarid and arid landscapes are few, not well understood and not easily implemented, but some tested practices are available. Their application to endangered species habitat needs priority review and consideration. Such practices often are accomplished not by wildlife biologists but by land restoration specialists, so their expertise and skills for

enhancing endangered pronghorn habits need to be acquired and used.

In retrospect, endangered pronghorn populations and habitats have been altered seriously by humankind. Historically exploited populations now have protection through federal laws that are enforced. However, herds at the end of the 20th century remained static or were insidiously decreasing. Extirpation for some populations appears imminent.

One strategy for endangered pronghorn is to restore the habitat to a natural, healthy condition, then restoration practices for the populations have greater chances for success. Healthy habitats produce healthy populations, the opposite also is true.

The two recovery plans for endangered pronghorn reviewed to date (Sonoran in Arizona and peninsular in Baja California Sur) generally allude to this goal, but they provide limited, inadequate details promoting strategies for restoring habitat conditions to increase carrying capacities for sustaining and increasing herds.

CONCLUSIONS

Habitat management for pronghorn embraces four propositions: (1) resources inventory, (2) determination of site-specific habitat components relative to pronghorn habitat requirements, (3) sustaining and enhancing of rangelands in healthy condition, and (4) improving viable but deteriorated habitat.

The quality and quantity of habitat components should be monitored periodically to determine whether the rangeland in question is capable of sustaining pronghorn in numbers and distribution consistent with management plans. Such monitoring will indicate the type and extent of management action needed, if any.

Composition and structure of vegetative communities determine the quality of pronghorn forage and cover, especially for fawns. Disturbances of the plant complex generally serve to accelerate or retard plant succession. Fires ignited by lightning are natural disturbances to vegetation, often resulting in increased succulent, nutritious forage. To benefit pronghorn, vegetation incompatible with desired pronghorn abundance and productivity can be altered by fire, biological controls, mechanical and chemical treatments and artificial seedings. Prescribed fires are one of the most economical and practical disturbance treatments for altering vegetation to favor pronghorn. Mechanical and chemical treatments can effectively reduce height and density of shrubs. When sufficient native herbaceous understory is present, follow-up seedings may not be necessary. When treatment sites lack adequate herbaceous plants, seedings can provide a desirable mixture of grasses, forbs and shrubs. A minimum of six species each of grasses, forbs and shrubs should be seeded to simulate natural plant communities and provide preferred forage.

Although fertilizing western uplands can increase plant production, treatments generally are feasible only in areas that receive more than 15 inches (38 cm) of annual precipitation. Treatments are most effective on spring and summer rangelands, where increases in nutrients benefit pronghorn fawn production and survival.

Results and values of manipulating vegetation to increase preferred, nutritious forage for pronghorn were field tested on the Deseret Ranch in Utah. Findings indicated that pronghorn foraged treated sites substantially more than untreated sites. At the same time, herd numbers have sustained an upward trend.

Water development projects that simulate natural waters are favored by pronghorn. Their spacing and protection need to be considered carefully.

Where fences need to be erected on pronghorn rangelands, they should include specifications or devices that allow pronghorn to pass under or through them. This is particularly true of fences on seasonal movement routes and winter rangelands. Where pronghorn need to be kept from certain sites, such as some croplands, electric fencing can be effective and economical. Wolf-type, anticoyote and woven wire fences seriously inhibit movements of all pronghorn age classes, and are ill-advised on rangelands occupied by pronghorn. Fences built to control livestock, or for other purposes, that no longer serve their original objective should be removed to decrease pronghorn injury and mortality.

Although structures, such as antelope passes, have been developed to allow pronghorn access through net wire fences, they have limited application and can cause injury and mortality. As presently designed, roadway and railway overpasses and underpasses for wildlife are not used by pronghorn. More research is needed on these structures or alternatives because pronghorn rangeland is increasingly fragmented by roadways and railways.

Pronghorn frequently are blamed for damage to winter wheat, but the problem appears to be not as serious as most complainants report. Pronghorn occasionally damage other crops, particularly alfalfa, and remedies include depredation payments and pronghorn reductions by translocations or modified hunting regulations.

Selected habitat improvements, such as rangeland seedings and water developments, have been beneficial to pronghorn and may have been a contributing factor to population increases during the past half century. As demands increase for more diverse use of western rangelands, wildlife managers need to know and apply practices that will sustain pronghorn numbers in concert with other societal demands.

At the beginning of the 21th century, all endangered pronghorn populations have been limited to Arizona and Mexico. Field investigations and management plans have provided scant information regarding the influences of low herd numbers relative to the condition and trend of forage, and of concealment cover and water for fawns. These three welfare factors are known to affect pronghorn populations on northern rangelands and they undoubtedly impact numbers on southern rangelands as well. The status of endangered pronghorn herds indeed may be related to endangered habitat conditions. The key to restoring those pronghorn may bear significantly on sustaining extant, healthy rangeland with conditions that meet the animals' habitat needs.

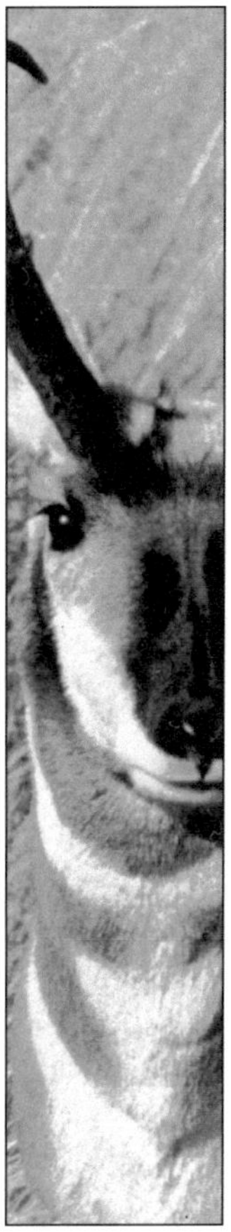

C H A P T E R   E I G H T E E N

Survey Methods to Estimate Population

Thomas M. Pojar

Success in adequately managing a wildlife species is dependent on reliable population estimates. A survey method may be adequate if it can indicate repeatedly the magnitude of changes in animal density without actually enumerating all animals present. Systematic surveys of pronghorn populations began with the advent of aircraft use in the 1940s. By that time, populations had recovered from near extinction at the beginning of the 20th century, and the era of cautious exploitation was beginning. Population growth usually was attributed to conservative harvest, which undoubtedly was true, but the prescribed harvests may have been made even more conservative by estimates that almost universally underestimated population size.

From a biological standpoint, few states or provinces can afford to harvest enough pronghorn to satisfy hunter demand. Demand usually outweighs supply and requires tight control on permits issued. In some areas of Colorado, for instance, a hunter can draw a permit to hunt pronghorn only once in 5 or 6 years, and in Arizona, the ratio of applicants to number of permits is 14 to 1 (Lee 1990). With this disparity in supply and demand, it is of paramount importance that the supply of animals be measured accurately.

Managing for an increasing population, as generally has been the situation since inception of hunting seasons in the 1940s, is the most forgiving position for a manager. The only decision necessary is the degree of conservatism needed. Pronghorn management has embraced conservative harvests possibly far past their usefulness. Some populations undoubtedly have reached political and economic tolerance or biological carrying capacity.

More precise management objectives demand more precise information about the population, and density estimates certainly are one of the most important pieces of the population picture. In most management situations, simply surveying a population to ascertain whether it is increasing, decreasing or remaining stable is no longer adequate. Thus, the pertinent question of today is the degree of change—measured in relatively fine increments.

Accuracy and precision have become very important considerations in wildlife surveys. These terms can be described best by visualizing the pattern on a target being used to sight-in a rifle (Gill et al. 1983a). A pattern that is widely dispersed over the target is not precise, but if the average of all the points is exactly on the bull's-eye, then the pattern is accurate. Conversely, if the pattern is tight but 3 feet (1 m) above the bull's-eye, it is precise but not accurate. The ideal is accuracy and precision—a tight pattern directly on the bull's-eye. In terms of pronghorn surveys, the goal is to use a method that will provide an estimate of 1,000 animals (or very close to that number) if there are, in fact, 1,000 animals in the area surveyed.

The second consideration is that, if this same survey technique is repeated several times on the exact same area, the estimates will all be within 10% of 1,000. This would represent an accurate and precise enumeration of animals present. Of course, this ideal is seldom achieved. Deviations from the ideal are not bad in themselves, but the farther from the ideal, the more risk of error in management. The power to detect population change in increments valuable to management is a worthy objective of any survey method (Reilly and Emslie 1998).

Cost of the survey method is pivotal to its utility for the managing agency. Cost must be viewed not only in financial terms, but also in terms of value of the product obtained. Low-quality data obtained for a cheap price may be, and likely are, worse than no data at all; it probably is better to be ignorant than misled. Misleading information will ensure faulty management. With proper sampling procedures, assessing the precision of the estimate through statistical evaluation usually is quite simple. Assessing accuracy is more difficult because it requires that the parameters being estimated are known during the evaluation of the survey technique. For example, if the true population size is 1,000 animals and the survey method used consistently yields an estimate of 800 animals with a precision level of plus or minus 10%, then the method is acceptable but inaccurate. However, the discrepancy in accuracy is not a problem if it can be measured and is consistent. This method may be the one of choice if it is more economical to execute than one that estimates the population with a high degree of accuracy but costs more.

CONVENTIONAL SURVEY METHODS

The total number of animals observed is taken as the total for that population, with no adjustment for detection errors with increased perpendicular distance from the flight line, no corrections for the blind spot area directly under the aircraft (Bergerud 1963), and no allowance for the many other bias factors common to aerial surveys (Graham and Bell 1969, Caughley 1974a).

Total coverage of an area by aerial survey eliminates variation due to sampling, but does not eliminate such biases as subject visibility, duplicate counts of animals that move from one transect to the next (Seber 1982) or inaccuracies in delineation of the search areas (Jolly 1969). Total area coverage also is expensive, which often is the most prohibitive disadvantage.

Sampling usually is much more efficient and provides an estimate of variance (Johnson et al. 1991, Guenzel 1994). Variation due to sampling can be controlled by adjusting sample size. The magnitude of this source of variation can be calculated if the subjects of interest are distributed randomly or if the sample areas searched satisfy the requirements of randomness.

Pronghorn are not randomly distributed over their range because they position themselves most advantageously according to critical habitat features, such as forage, water and cover. It is a simple matter, however, to assign search areas, such as quadrats or transects, that satisfy random criteria and, in turn, allow legitimate calculation of the sample variance and, consequently, an estimate of precision. The expense of executing a sampling system is dependent on the sampling intensity, which is related to the target precision level, but it always will be less than a total area count (Johnson et al. 1991). If physical marking of sample units on the ground is necessary, then some annual expense will be incurred to maintain the sample unit boundary markers. Also, depending on the marking method and the number of sample units to be marked, the initial expense of establishing the sampling system can be considerable. With the advent of global position system (GPS) technology, navigation is precise and physical marks are not necessary.

Although total coverage has been almost universally accepted as the method of choice for counting pronghorn, there has been surprisingly little critique of the method. Bear (1969b) offered evaluation of the aerial strip–transect method as it applies specifically to counting pronghorn. Counts during winter, when animals were assembled in large herds, produced the highest overall counts, but aerial photography exposed errors in the range of 20% when herd size exceeded 100. As verified by videotape, Donoho and Wills (1974a) provided evidence that, as herd size increased above 10 animals, undercounting errors increased proportionately. Early morning counts were higher than midday and late afternoon counts by 26% and 18%, respectively (Bear 1969b). Results of surveys using a helicopter resemble those from fixed-wing surveys during winter, but helicopter survey counts were 24% higher during late summer when pronghorn were more dispersed and in smaller groups (Bear 1969b).

Pronghorn group size and distribution vary throughout the year, and this variation profoundly influences survey results. Groups of five or more animals are more easily detected than smaller groups, which is why most total coverage counts are conducted during winter when groups are largest (Figure 136). Enumerating individuals in a large group can be a major source of inaccuracy, which can be partially overcome by photographing or videotaping the group and counting from the photo (Davis 1963, Watson 1969) or video replay (Donoho and Wills 1974b). However, the logistics and time involved in these procedures usually have been a deterrent to their use. Techniques developed by Hiby et al. (1987) improved the resolution with commonly available film by compensating for aircraft motion. Hiby et al. also used color transparencies and projected the image for optimal magnification, improving photo interpretation.

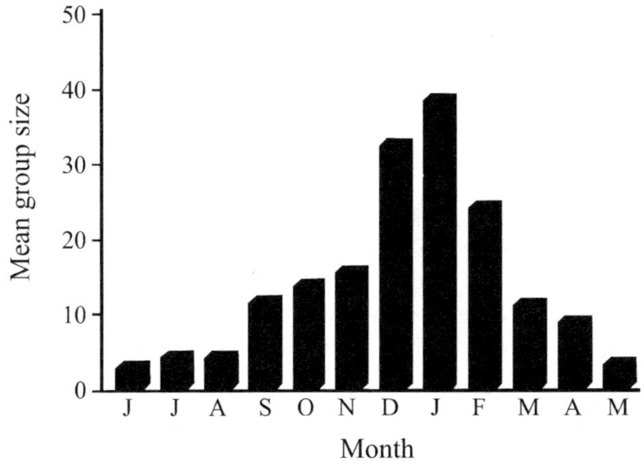

FIGURE 136. *Mean size of pronghorn groups by month in Alberta, Canada (from Mitchell 1980).*

A standard transect width for counting pronghorn has failed to emerge. This failure probably reflects the varying conditions encountered over the pronghorn's range. Vegetative cover and topographic relief affect detectability. A variety of transect widths has been used to count pronghorn in the western United States, including 0.25

Fixed-wing aerial strip-transect surveys traditionally are used to estimate the size and structure of pronghorn populations. This method probably has its roots in the "cruise" method of estimating trees for timber (see Ruff 1939, Halloran 1943) in which there is predetermination (by cruising) of trees per strip, which then is multiplied by the total number of strips to estimate the total trees for the area in question. This method was hybridized into a method called strip cruising, for counting elk by Buechner et al. (1951), who used an airplane to fly strips in open, rolling terrain; in rough terrain, each drainage was flown separately. Because of the open terrain that pronghorn inhabit, it is concluded that "censuses are relatively easy" (Cooperrider et al. 1986: 544). This has led to a sense of security in conducting strip surveys and covering 100% of a given area to estimate the total population of pronghorn on that area (Hailey 1979). Usually, the only deviations from this approach are to vary the strip width or count strips only on the side of the aircraft that is opposite the sun. If all animals within the strip are seen and 100% of the area is covered, then this would be considered a true population census (i.e., a total population count). This is in contrast to a survey, which implies sampling and extrapolating to the population. Photo by David Dvorak; courtesy of the Texas Department of Parks and Wildlife, Austin.

mile (400 m) in New Mexico (Larsen 1967); 0.5 mile (800 m) in New Mexico (Elliott 1968), South Dakota (West 1969), North Dakota (Allen and Samuelson 1987) and Nebraska (Suetsugu 1970); as well as 1.0 mile (1,600 m) in Colorado (Bear 1969b). There is general agreement that an inverse relationship exists between transect width and estimated density for species other than pronghorn: red kangaroos (Caughley et al. 1976); African bovids (Pennycuick and Western 1972); and white-tailed deer (Beasom et al. 1981). In a crossover design experiment on shortgrass prairie in eastern Colorado, Pojar (unpublished data) found no difference (P is greater than 0.05) in observed pronghorn density between 1.0-mile (1,600 m) and 0.5-mile (800 m) transect widths. However, further experiments (Pojar et al. 1995) showed a general decline in percentage of groups detected as distance from the flight line increased (Figure 137). The reduction in detectability with increased distance was quite pronounced in sagebrush steppe habitat in northwestern Colorado (Figure 138).

Two solutions are possible to the problem of missing animals within the bounds of a transect. The simplest and probably the least practical is to decrease the width of the transect to an extent that animals are not missed, as determined by double sampling or some other means of verifying accuracy. This approach likely would result in a transect so narrow that efficiency in flight time would drop below acceptable limits, and the estimate would be based on so few observations that the confidence limits would be unacceptably wide (Eberhardt et al. 1979). However, in high-density populations, narrow strip transects produce acceptable results (Pojar et al. 1995). The second and more pragmatic solution is use of transects sufficiently wide to afford acceptable numbers of observations. In this case, bias caused by missing animals should be estimated using appropriate procedures.

Estimating Bias

Ground counts have been made on some species to obtain a correction factor for aerial counts. Gilbert and Grieb (1957) compared fixed-wing aerial counts with ground counts on winter rangeland in Colorado for mule deer, and found that the aerial counts comprised 34% to 49% of ground counts. Edwards (1954) obtained nearly

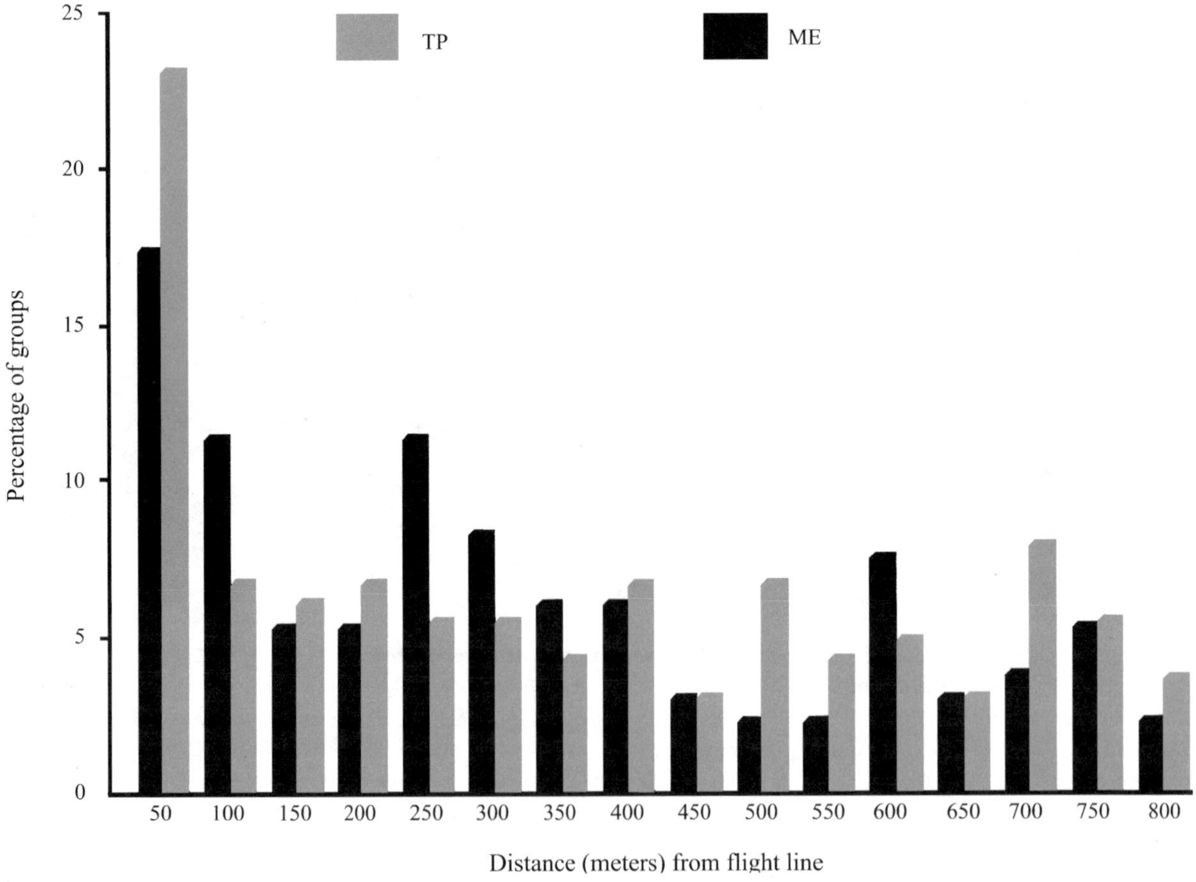

FIGURE 137. *Distribution of pronghorn groups detected by two observers during the same flight near Hugo, Colorado, 1986. Sample sizes: Observer ME = 133 groups and 698 total animals; Observer TP = 163 groups and 777 total animals.*

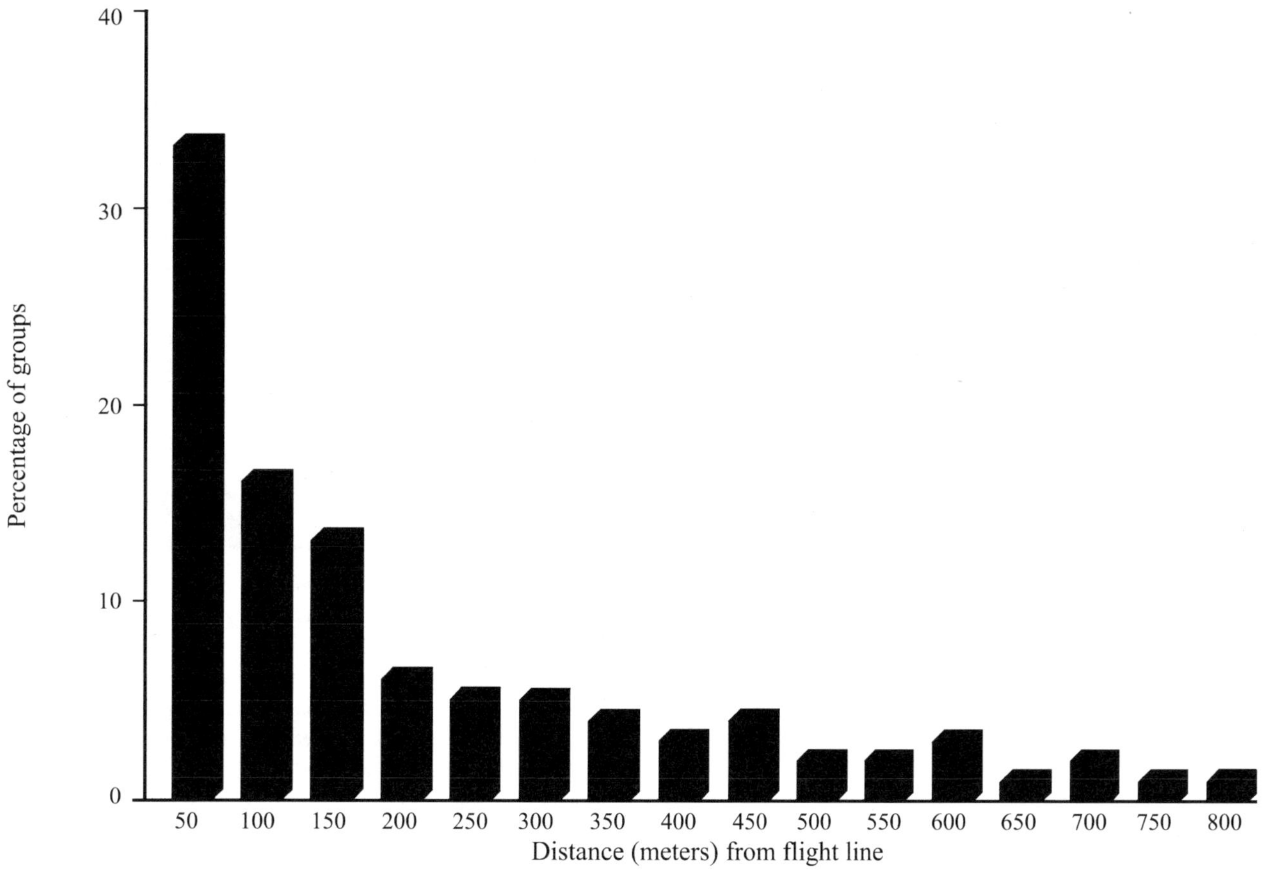

FIGURE 138. *Percentage of pronghorn groups detected by distance from flight line, based on 395 groups near Craig, Colorado, 1986.*

equal density estimates of moose from ground counts and fixed-wing aircraft surveys in British Columbia. A more reliable variation of this approach is to have a known number of animals in an enclosure and make subsequent flyovers to estimate the proportion of subjects counted. Using a helicopter, Bartmann et al. (1986) detected 60% to 68% of the mule deer present in pinyon/juniper woodland during winter in western Colorado. Experienced observers in a fixed-wing aircraft detected 68% of known

The relatively flat, open habitat of pronghorn should allow reliable population estimates. Pronghorn generally are easier to observe and count than other large ungulates, such as cervids, that inhabit dense vegetative cover and steep terrain. Surveying methodology for pronghorn resembles that for bovid antelope of Africa. Substantial scientific effort has been invested in development and evaluation of survey methods for enumeration of African antelope. Because of similarities in habitat and in size and behavior of the species being counted, some of the survey challenges encountered in Africa are similar to counting pronghorn in North America. For these reasons, methods used in North America to count pronghorn and those used in Africa to count antelope have much in common. Photo by Amber Alexander; courtesy of the Arizona Department of Game and Fish, Phoenix.

numbers of moose in enclosures with typical, old-burn habitat in Alaska (LeResche and Rausch 1974).

The key assumption of the line-transect method is that the number of subjects on the innermost distance band is detected accurately (Burnham et al. 1980). If this presumption is satisfied, then the density of animals at all other distances can be adjusted to this "known" density. Obviously, forward and downward visibility from aircraft are important for detecting all animals on the flight line. Adjustments for this blind area can be made for fixed-wing aircraft (Guenzel 1997).

Line-transect surveys offer attractive operational features that include: (1) minimal effort to establish survey design (White et al. 1989); (2) use of fixed-wing aircraft that are more economical than helicopters; and (3) efficient use of air time because the survey is continuous except during ferry time between transects. Line-transect results were compared with the more accurate (but more expensive) helicopter quadrat surveys by Pojar and Guenzel (1999). Line-transect survey estimates averaged 0.735 of quadrat estimates, which can be used as a correction factor for the more economical line transects (Pojar and Guenzel 1999). Guenzel (1997) provides a complete description of sampling design, flight protocol and data analysis for line-transect methodology. The computer program, DISTANCE, can be used to analyze line-transect data; its development and use are described by Laake et al. (1996).

The mark/recapture system depends on a random sample of marked animals from the population. Moreover, the area must be sufficiently large to consider the population closed (i.e., minimal immigration and emigration) (Seber 1982). Considering that pronghorn trapping is most feasible during winter and that for a population to be essentially closed requires a large geographic area, there is a formidable logistic barrier to marking an adequate sample. Rice and Harder (1977) estimated the size of a northern Ohio white-tailed deer population in a 5,377-acre (2,176 ha) enclosure with unbiased and very adequate precision (about ±4% at the 95% confidence level) by marking less than 10% of the population. A compromise is involved in this procedure in that fewer animals need to be marked if more search flights (by helicopter) are possible. Rice and Harder presented theory, calculations and graphs depicting the relationship between percentage of population marked and number of search flights. Strandgaard (1967) applied this technique to a roe deer population and cautioned that, unless the various sex and age classes are represented equally in the marked sample and in the observed sample, the population size estimates can be grossly in error. He stated that two-thirds of the population must be marked to overcome this source of error. It should be pointed out, however, that roe deer live in more forested areas than do pronghorn and are relatively cryptic. How effectively the mark/recapture technique would assess the accuracy of pronghorn counts is not known, but it holds promise to provide a productive area of research.

Most biases in aerial surveys result in counts that are less than the true number of animals present (Caughley and Goddard 1972, Pollock and Kendall 1987); biases that inflate the count are so minor that they can functionally be ignored (Caughley 1974a). These comments apply to the conventional method (aerial transects) of counting pronghorn. This is not to say that data from these counts are not useful because they have served managers well in past years. Management based on underestimates of harvestable surpluses invariably are conservative. When the management objective is to increase a population using conservative harvests, undercounting only magnifies conservatism. This is an acceptable situation because it tends to allow the population to increase, provided that carrying capacity has not been reached. Photo by Joe Van Wormer.

Pollock and Kendall (1987) reviewed several methods of estimating visibility bias in aerial surveys from an applied statistician's perspective. One approach to estimating bias, proposed by Caughley and Goddard (1972), is based on a mean and variance from two sets of survey data. A pair of simultaneous equations describing a parabolic relation between mean and variance were solved to estimate the "true" numbers present. This technique requires that counts be made by two different methods (Caughley and Goddard suggested counts at two different altitudes) to get two sets of mean and variance. Several severe restrictions must be met. The most critical is that the probability of observing any given animal is equal and this probability is the same among surveys (Caughley and Goddard 1972, Routledge 1981). Given the difficulty of meeting the assumptions and other possibilities for unreliability (Routledge 1981), this method should be used with caution.

Undercounting during aerial surveys of pronghorn is usual, if not a given. The challenge is to find means of estimating the magnitude of the bias and adjusting for it. Several techniques are available for estimating bias, but they all depend on verifying the actual number of animals on a portion of each sample unit or on a subsample of the units in the total sample. Of course, any error in verifying the number of animals on the "correction" areas will lead to bias in the adjustment for bias. Estimating bias is a precarious endeavor, but it is necessary to make the effort to give the observer a measure of the quality of data being collected. For transect surveys, there are three ways to attack the problem of bias: (1) obtaining more accurate count by some other method on some of the transects, (2) accurately counting the objects of interest on a portion of some or all of the transects and (3) using the Peterson estimator with a marked subpopulation. Photo courtesy of the California Department of Fish and Game, Redding.

If some method of sampling is more efficient than total coverage and a correction for bias is possible, the next step is to design the sampling system. Several key decisions must be made at this stage that will determine the success or failure of the effort.

Designing a Survey System

The first consideration when designing a survey system is the type of equipment available. An aerial survey usually is most practical because of the large area involved when counting pronghorn. Category of aircraft is a major consideration because of cost; helicopter rental is roughly three times more expensive than that for fixed-wing aircraft. Differences in cost of operation have remained about the same over the years, forcing many agencies to use fixed-wing aircraft for economic reasons. The type of sample unit chosen also will influence the category of aircraft. For example, if the sample unit was a one-square mile (2.6 km^2) quadrat, it would be unrealistic to expect an observer to endure more than a few quadrat searches in a fixed-wing aircraft unless he was immune to aerial motion sickness. Maneuvering around relatively small sample units requires a helicopter. Transects—either line or strip—can be flown with either a fixed-wing aircraft or helicopter, but the superior maneuverability and visibility from rotary aircraft has been recognized since the early years of aerial surveys (Buechner 1950d). Corrections for bias are critical in any survey technique, but with fixed-wing aircraft that typically are flown at 300 to 500 feet (100–150 m) altitude and 75 to 125 miles per hour (120–200 km/hr) versus typical helicopter capabilities of 50 to 100 feet (15–30 m) altitude and 50 to 70 miles per hour (80–113 km/hr), the need for bias correction becomes paramount. The altitude and speed that fixed-wing aircraft must be flown for safety reasons may even preclude reasonable repeatability of survey results. If results are not repeatable, it is impractical to continue use of the technique (Reilly and Emslie 1998).

For purposes of example, let us assume that: (1) strip transects will be the sampling units; (2) a helicopter will be used as the observational platform; (3) the area of interest has been delineated on maps; (4) the total number of potential transects (N) has been determined; and (5) the transect width has been chosen. With proper correction for bias incorporated in the survey plan, transect width selection is less important than if there is presumption that all subjects on the sample unit will be observed. An arbitrary judgment must be made regarding how many transects (sample units) should be skipped between those surveyed, to prevent driving animals from surveyed transects to adjacent ones (Seber 1982). Because of pronghorn mobility, transect width is a key factor. For this example, we have a chosen strip width of 1.0 mile (1,600 m) and have decided to survey every other transect. This decision has automatically limited the number of transects (n) that can appear in the sample to 50% of N.

The next step in the design process is deciding how large a sample to draw from N. Two items drive the decision process in determining sample size. First is to ascertain available flight time and calculate how many miles of flight line it represents; second is to arrive at an acceptable level of precision for management needs. Fortunately, simple calculations are available to estimate the size of sample required. All good things have a cost, and the cost of being able to take advantage of these simple calcula-

Radiotelemetry techniques are useful to detect visibility bias in elk surveys (Samuel et al. 1987), and have been used successfully in conjunction with mark/recapture methods to estimate mule deer density (Kufeld et al. 1987). Pollock and Kendall (1987) suggested that, by combining line-transect and Peterson (mark/recapture) methods, productive results are likely. Photo courtesy of the U.S. National Park Service, Yellowstone National Park, Wyoming.

Pronghorn density is low on the shortgrass prairie of eastern Colorado—habitat somewhat similar to these mixed grasslands of South Dakota (opposite). The comparatively flat landscape and short vegetation facilitate pronghorn surveys in such landscapes. On the sagebrush steppe of western Colorado, pronghorn densities are about triple those on the prairie east of the mountains. Comparatively rough terrain and shrubby vegetation somewhat complicate surveys. Photo by Bart W. O'Gara.

tions is that adequate knowledge of the population is required, including an estimate of density and the density variance that, ideally, are obtained from a pilot survey or past data from the population. In the absence of such an estimate, variance and density estimates can be borrowed from an area presumed to be similar in density and habitat. In the following example, real data from shortgrass prairie and sagebrush steppe habitat types in Colorado are used. Pronghorn density is low (about two pronghorn per square mile [0.77/km²]) in the shortgrass type and relatively high (six pronghorn per square mile [2.32/km²]) in the sagebrush type.

The variance is calculated according to Cochran (1963):

$$s^2 = \frac{\Sigma a_i^2 - 2(p)(a_i \bar{m}_i) + (p)^2(\Sigma m_i^2)}{(n-1)(m)^2}$$

where

a_i = number of subjects on i^{th} transect;

$$\bar{m} = \frac{\Sigma m_i}{n};$$

m_i = length of the ith transcript;

$$p = \frac{\Sigma a_i}{\Sigma m_i}; \text{ and}$$

n = number of sample transects.

The sample size is calculated from Cochran (1963):

$$n = \frac{n_o}{1 + \frac{n_o - 1}{N}},$$

$$n_o = \frac{t^2 s^2}{d^2},$$

where t is the t-value at the chosen confidence level and d is the chosen tolerable amount of deviation from the mean density.

That is, if the mean density is six pronghorn per square mile (2.32/km²) and the desired estimate within 20% of the mean, then d equals 1.2. The component of the equation N minus n divided by N is the finite population correction factor and diminishes in impact as n gravitates toward N.

The above procedure will provide general guidance for the percentage of area to sample to obtain a given precision in the density estimate. This same equation can be rearranged to solve for d to obtain an estimate of the expected deviation (i.e., precision level) as the percentage of the area sampled is varied:

$$d = \frac{t^2 s^2 (N-n)}{n_o N}.$$

If d is divided by the mean density and multiplied by 100, the result is plus/minus the percentage of the mean density that can be expected at the designated level of sampling (Table 156). Information of this sort is helpful when monetary resources are limited and a decision must be made whether the obtainable precision level is acceptable and worth the expenditure (Table 156).

Armed with this information, the choice of percentage of area to sample was chosen, and the survey can be undertaken with reasonable assurance that the results will furnish valuable information for management decisions. The observed mean density is extrapolated to the entire area by multiplying by N, and the sample variance is calculated by the equation given above. The confidence interval is obtained by:

$$\text{Mean density} \pm ts\sqrt{\frac{1}{n} - \frac{1}{N}},$$

where t has $n - 1$ degrees of freedom, and s is obtained from the square root of s^2 in equation (1).

Thus far, we have focused on precision of the estimate and now venture into the arena of accuracy. Accuracy is important. The key to correcting for bias is to find a method of accurately counting the subjects of interest on a subsample of n or on some portion of each sample unit without incurring prohibitive expense. Application of line-transect methodology (Burnham et al. 1980) seems to be the most logical approach to detecting bias in strip transects because no additional cost is involved. It is only necessary to record data in a slightly different fashion. Observers must estimate and record the subject's perpendicular distance

TABLE 156. Expected precision at the 90% confidence level, with varying percentages of the area sampled in high pronghorn density (more than six pronghorn per square mile: 2.3/km²) and low density (less than two pronghorn per square mile: 0.77/ km²)

	Expected precision (±percentage)	
Sampling intensity[a]	*Low density*[b]	*High density*[c]
10	98	39
20	56	21
30	41	15
40	32	11
50	26	10
60	21	8
70	17	6
80	13	5

[a] Percentage of area.

[b] Data from shortgrass prairie.

[c] Data from sagebrush steppe.

from the flight line and, in the process of the survey, estimate mean group size. This information permits use of line-transect methodology analysis (Burnham et al. 1980). The perpendicular distance is bounded by the width of the transect rather than infinity, which normally is used in line-transect surveys.

Pollock and Kendall (1987:508) cautioned that "faith in the line transect approach is not justified in many aerial surveys." However, Burnham et al. (1985:1,017) recommended "that distances be recorded in any transect survey. Even if a narrow strip transect is planned, the distances will allow a check on the strong assumption that all objects within the transect are detected." Because no additional cost is involved in collecting perpendicular distance data, not incorporating it into survey plans seems unwise. The minimum benefit derived will be a general indicator of the potential bias in the density estimate when all the distance intervals are adjusted proportionally to the center interval, as described by Eberhardt (1978).

Meeting the assumptions of the line-transect method, especially that all animals on the innermost distance band are seen, is critical to its successful application. If protocol is closely followed and observers are adequately trained, this method is efficient and usable for management (Guenzel 1997). Adjustments for accuracy are recommended (Pojar and Guenzel 1999) because evidence suggests that line-transect estimates are biased downward. Ideally, if line transects are used to estimate density, results should periodically be checked against the more accurate helicopter quadrat estimates. This will help determine whether a correction is necessary and, if so, will give the magnitude of the correction (Pojar and Guenzel 1999).

EXPERIMENTAL SURVEY METHODS

The perfect survey method for pronghorn (or other wildlife species) is yet to be discovered. A necessity of line or strip transects is efficient use of air time. However, missed animals, deviation from a constant flight line and altitude, as well as observer inefficiency due to covering too much terrain too fast are factors that cause errors (Siniff and Skoog 1964, Graham and Bell 1969, Caughley 1974a). Methods that attempt to circumvent these shortcomings are being sought constantly.

Strip transects are long, slender (i.e., rectangular) quadrats. They normally are searched by flying along a center line, attempting to see all the subjects within the bounds of the sample unit.

Quadrats can be any shape, including circular, hexagonal, rectangular and square (Seber 1982). In surveying large ungulate populations, quadrats are chosen, for pragmatic reasons, to be square and some multiple of 1 square mile (2.6 km^2) because the land survey (cadastral) system in North America is based on the square mile. Quadrats are searched by flying around the perimeter and transecting the interior in an orderly fashion. Search techniques differ significantly between transects and quadrats, and they favor quadrats for thoroughness of search (Siniff and Skoog 1964). The area is seen from all light angles while flying the sample unit perimeter, and the noise of the aircraft usually makes animals visible by flushing; they might not be detected during a transect search.

Quadrat surveys have been used to estimate densities of caribou (Siniff and Skoog 1964); mule deer on sagebrush winter rangeland (Gill 1969) and pinyon/juniper winter rangeland (Kufeld et al. 1980), moose (Evans et al. 1966) and pronghorn (Firchow et al.1990, Pojar et al. 1995). Experiments using a helicopter as the observational platform have compared strip-transect and quadrat sampling for estimating pronghorn density (Pojar et al. 1995). In sagebrush steppe habitat, the relationship between the two types of surveys was direct, with quadrat estimates consistently double that of transects (Figure 139). Results from shortgrass prairie were less consistent and pronounced, but the quadrat survey resulted in higher density estimates (Figure 140).

The question of accuracy on quadrats was addressed by flying a second team of observers in a helicopter above the primary helicopter team to observe their performance and animal movements in response to the primary search (Pojar et al. 1995). Inclusion (one group) and exclusion (four groups) errors of the primary observers were partially offsetting, with the end result being an upward correction of 2.1%. Generally, precision of quadrat surveys can be expected to be ±15% to 20% at the 90% confidence level when sampling 10% of an area.

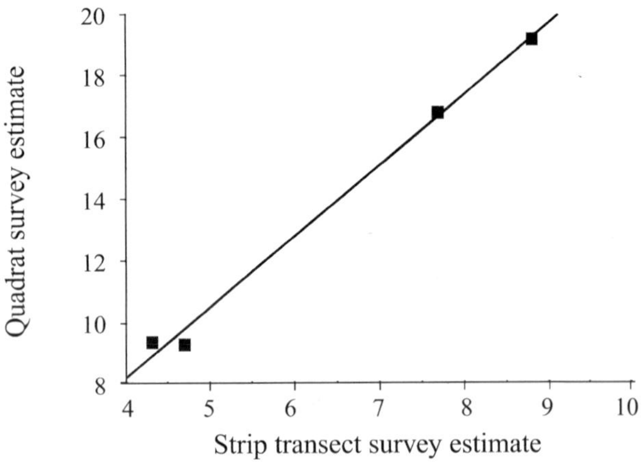

FIGURE 139. *Comparison of strip-transect and random-quadrat estimates of pronghorn population size in sagebrush steppe habitat near Craig, Colorado. Squares represent the population size estimates; x and y axes describe population size x 1,000; and the line is a regression line, relating to the x and y axes.*

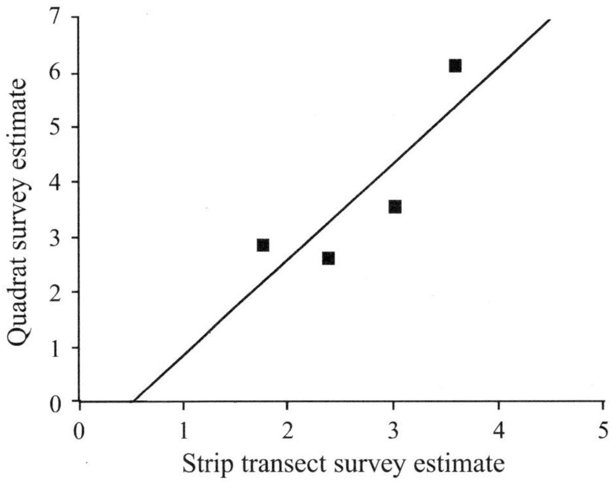

FIGURE 140. *Comparison of strip-transect and random-quadrat estimates of pronghorn population size in shortgrass prairie habitat near Limon, Colorado. Squares represent the population size estimates; x and y axes describe population size x 1,000; and the line is a regression line, relating to the x and y axes.*

Sampling systems produce best results if the subjects of interest are maximally dispersed (Beasom 1979), and quadrat sampling is no exception. The relatively small group size of pronghorn during summer compared with winter concentrations (Figure 136) favorably impacts the variance of the estimate. A spin-off benefit of summer surveys is that it is an opportune time to collect herd structure data in conjunction with population size data. The sampling system used for density estimates serves equally well for sampling herd structure.

Sample stratification is another means of reducing variance. Strata can be based on vegetative type, relative density of subjects or other criteria, but the objective is to divide the population into subpopulations that are internally homogeneous (Cochran 1963). Formulas for calculating the variance of a simple random sample and a stratified random sample can be found in Cochran (1963:93–94). If the strata vary in subject density, geographic area or ease of surveying, optimum allocation of the sample should be considered (Cochran 1963) as a means of increasing survey efficiency. The principle is that, for example, areas (strata) of higher density will be sampled at higher intensity, and areas of low density will be allocated fewer sample units, resulting in more surveying effort being spent in high-density areas.

SEX AND AGE CLASSIFICATION SURVEYS

Accurate information on the sex and age structure of a pronghorn population is key to interpreting population performance and effects of harvest. Caughley (1974b) noted that age ratios commonly are misinterpreted. It often is construed that a high age ratio indicates that a population is well and healthy when, in fact, it may be the result of a fawn crop failure the previous year, resulting in few yearling does in the current population. Conversely, a low age ratio may be interpreted to mean that the population is suffering some disaster when, in reality, it is approaching carrying capacity and reflecting this by lower reproductive rates. Without other demographic information, such as density and survival, age ratios do not disclose much useful information about the population, and their interpretation is a hazardous undertaking (Caughley 1974b). However, long-term trends in age ratios are useful in judging carrying capacity and population densities that are near maximum reproductive rate, both of which are pivotal points from a management perspective. Sex and age ratios also are important input and performance measures for any population-modeling effort because modeling, by necessity, incorporates densities and survival rates, either assumed or measured.

Timing and Design of Survey

From a sampling standpoint, maximum dispersal of objects is beneficial in reducing the variance of the estimate. Pronghorn are most dispersed during the fawning season (May through June). However, the fawns are in the hiding stage, and an accurate herd structure estimate is not possible. Group sizes still are relatively small and the population dispersed during August (Mitchell 1980) when fawns are out of the hiding stage and still small enough to be distinguishable from adults. This makes mid- to late summer the best time to do herd structure surveys, but the timing can vary with latitude. On northern rangelands, such as Idaho, the latter half of August is ideal, with early August being acceptable in Kansas (Autenrieth 1978). Farther south, in Texas, Hailey (1979) suggested June and July surveys to estimate fawn/doe ratios.

Although summer is optimal for estimating fawn/doe ratios, the social system of pronghorn may not be conducive to buck/doe estimates at that time. Depending on the intensity of the search, lone bucks and small bachelor herds could be missed and bias the data. In Wyoming, herd structure surveys are executed postseason to avoid the summer segregation of bucks and small groups (Autenrieth 1978). The surveyors feel that classification is more accurate with larger groups. There is no scientific data to support or refute this contention, but personal experience leads me to doubt that accurate classification is possible with large groups (more than 100 animals).

On January 14, 1988, I observed an isolated group of pronghorn from a blind at 100 to 300 yards (91–274 m) with eight-power binoculars and a 20- to 40-power spotting scope. During the more than 2-hour observation period, it was possible to get an accurate count of the total group and the number of bucks. The total count was 122, 25 of which were yearling or older bucks. Within 2 hours of the ground observation, the group was relocated from a helicopter (Bell-Soloy). The crew consisted of the pilot, myself and a second observer. The second observer's assignment was to get a total count and my only objective was to count the number of bucks. We did not attempt to distinguish fawns. We made three low-level (about 50 feet [15 m]) passes over the group. The highest total count was 114 and the number of bucks I counted on each pass was 15, 16 and 16, respectively. An aerial photograph of the group verified 122 pronghorn with 22 bucks identifiable. The buck/doe ratio, therefore, could have been underestimated by 42% with aerial survey data.

As with population size estimation surveys, random distribution of objects of interest for herd structure surveys is necessary to apply statistical theory legitimately (Cochran 1963). Pronghorn do not distribute themselves randomly over their range by sex and age category, so either the sample units must be drawn at random or a random starting point must be used for systematic sampling (Seber 1986). Czaplewski et al. (1983) assumed that pronghorn are distributed randomly by sex and age over their range, and calculated precision based strictly on number of animals classified. Their tables of sample size and expected precision are useful as general guidelines for number of animals to be classified, but should be used in conjunction with a random sampling scheme (see Bowden et al. 1984). For example, if classifying 1,000 animals gives the precision level desired, then this sample should be obtained from random sample units rather than from simply classifying the first 1,000 animals encountered. This will require more flight time, but will help ensure minimum bias in the sex and age ratios obtained.

Sampling Scheme

An acceptable sampling scheme for estimating herd structure is the same as for estimating population density, except that sampling intensity usually can be reduced. Adequacy of sampling intensity is related to number of animals classified and how representative the sample is of the total area, which is related to animal density and dis-

Distinguishing yearling bucks from does can be a major source of error in buck/doe ratio estimates. Probably the best way to minimize such misclassification is to conduct herd structure surveys during late summer, when group size is small, so that each animal can be examined for the black cheek patch. A low viewing angle from the aircraft is helpful in detecting the cheek patch, and this favors the use of a helicopter. Photo by Geoff Tischbein; courtesy of the Colorado Division of Wildlife, Denver.

tribution. Obviously, more sample units need to be surveyed in a low-density area than in a high-density area to observe the same number of animals. The precision level improves with increased numbers of animals in the sample, but if the sample units surveyed are not representative of the population in question, the estimate can be seriously biased.

Systematic sampling is acceptable if it is begun from a random starting point (Seber 1986). The arbitrary constraint that sample units are not adjacent to one another is recommended, to ensure that the results for each strip are independent.

As with population density estimates, there is empirical evidence that the use of quadrats as sample units offers a more intense search than do mile-wide strips. Strip surveys likely underestimate the buck/doe ratio. In 4 years of comparing strip versus quadrat herd structure data on shortgrass prairie, the buck/doe ratio was 27% greater from the quadrat survey. A similar comparison in sagebrush steppe habitat for 5 years resulted in a 29% higher buck/doe ratio with quadrats (Pojar et al. 1995). Fawn/doe ratios did not differ between the strip transect and the quadrat in either sagebrush steppe or shortgrass prairie. The consistently higher buck/doe ratio with quadrats in both habitat types may be due to relatively increased flushing and detection of single bucks or small buck groups that would not flush, and therefore not be detected, during strip-transect searches.

Accurate classification of individual animals is critical because any error is compounded in these data. A yearling buck misclassified as a doe (which is a common error) causes an inflation error in the doe segment and a deflation error in the buck segment. The calculated buck/doe ratio then contains two errors and the fawn/doe ratio contains one error; both ratios are lower than they should be. An example of this error on a data set is worth examining. In a population of 1,000 animals with a ratio of 50 bucks/100 does/70 fawns, a sample of 200 would identify 45 bucks, 91 does and 64 fawns, reflecting the true ratio. A 10% error in classifying bucks as does would skew the sample numbers to 40 bucks, 96 does and 64 fawns, which is a ratio of 42 bucks per 100 does per 67 fawns. The 10% error is parlayed into approximately a 20% error in the buck/doe ratio and a 5% error in the fawn/doe ratio. Photo by Bart W. O'Gara.

Survey Procedure

Careful planning and design of a herd structure survey will be wasted if the observer is not adequately trained. Dirschl et al. (1981) stressed rigorous training for inexperienced observers in preparation for doing multispecies counts in Africa. Candidates were interviewed to assess motivation, given physical and eye examinations, and taken on test flights before acceptance for an intensive 2-week training session. That training introduced the trainees to "the theoretical, technical, and operational aspects of the aerial observer's job" (Dirschl et al. 1981:110). After the 2-week training session, the trainees were tested and their counts compared with those of experienced observers. Such preparation may seem extreme, but I believe that similar, adequate training of observers is an aspect of pronghorn surveys that has been neglected, and the quality of data surely reflects this shortcoming.

Analysis of Data

Herd structure ratios are simply the number of males 1 year old or older divided by the total number of females 1 year old or older, and the number of fawns divided by the number of females. Multiplying these proportions by 100 puts them on the conventional per-100-does basis.

Superficially, it appears that herd structure data can be treated as binomial data for calculating the variance. However, use of the binomial distribution assumes that the animals of each sex and age are encountered at random during the survey. With pronghorn, and most other wildlife species, this is not the case. The animals' social organization dictates the association of fawns with does and, in a more variable fashion, bucks with does. Therefore, it is not legitimate to assume that pronghorn groups are randomly formed. The consequence of accepting the randomness assumption and using the binomial distribution is that the variance would be too large (Bowden et al. 1984). If the sampling procedure allows groups of pronghorn to be encountered as in a simple random sample without replacement, then the variance should be calculated using the ratio formula rather than the binomial formula (Cochran 1963). This estimator is robust to the composition of the group and provides an unbiased estimate of the variance (Bowden et al. 1984).

Until about mid-November, yearling bucks generally carry large enough horns that they are readily seen from an aircraft. However, when the sheaths are shed, the horn cores are quite small and the skin and hair covering them are comparatively light colored. Then, the black cheek patch becomes the primary characteristic for differentiating between does and yearling bucks. Photo by Marilyn Maring; courtesy of Leonard Rue Enterprises, Blairstown, New Jersey.

Comments and Cautions

The aforementioned methodology will yield statistically sound estimates of buck/doe and fawn/doe ratios. Logistics, conditions or situations may preclude following this method exactly. Before resources are expended on any variation in method, a competent statistician should be consulted so that the maximum information possible can be obtained from the classification effort.

Because bucks do not associate with does in a fashion that is consistent with the way does associate with fawns, the estimated buck/doe ratio is much more variable than the fawn/doe ratio. Bucks frequently are seen in all-buck groups or as singles during summer, whereas fawns almost always are seen with does. This behavioral pattern is responsible for the vast discrepancy in variance between the two ratios. Consequently, with the same sampling intensity, the buck/doe ratio will be much less precise than the fawn/doe ratio. Depending on the overall population density (which would affect sample size), the method should result in a fawn/doe ratio with a confidence interval (90%) of ±5% to 15% and a buck/doe ratio with a ±15% to 30% confidence interval.

CONCLUSIONS

Pronghorn inhabit open terrain that augments their observability. This can give managers a false sense of security in surveys of population density and herd structure. Although highly visible under certain conditions, pronghorn are very cryptic under some conditions, which can contribute to significant bias in estimations of population parameters. It is possible to obtain useful and reliable population data by following statistically sound sampling methods and by maximizing the intensity of the search.

C H A P T E R N I N E T E E N

Population Dynamics and Modeling

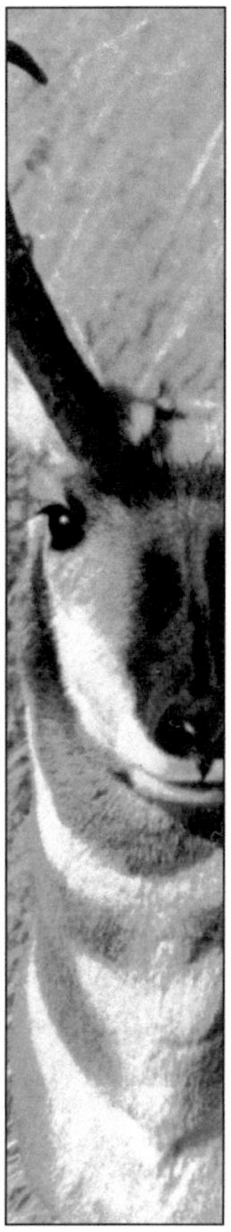

Stephan G. Kohlmann

It is not so much that we want to believe everything the computer tells us, but that we want a tool to confront us with the implications of what we think we know.

Botkin (1977:217)

Population dynamics deal with the reasons that animal numbers change in time and space. Much of the study of populations is devoted to the measurement of numerical changes over time, but increasingly, ecologists investigate the causes of population changes in a spatial and environmental context. Investigations of animal population dynamics are conducted primarily to: (1) predict future population size or population trends (Lande 1988, Dennis et al. 1991); (2) identify mechanisms of population fluctuations (Royama 1996); and (3) calculate potential harvests (Bailey 1984). The need to understand and predict complex systems, such as wildlife populations in ever-changing environments, has led to the development of a variety of mathematical representations (i.e., models) of population processes. Mathematicians discovered the study of population dynamics relatively early in the 20th century and, consequently few other biological disciplines can offer a similar variety of advanced mathematical techniques and tools. With the advent of fast, cheap computers, it has been possible for general practitioners of wildlife management and conservation biology to use sophisticated data analyses and computer simulation models to understand the interaction between animal populations and their environments. The primary purpose of this chapter is to present a brief overview of current tools, techniques and thinking in the field of population dynamics and population modeling, and how they may be applied to the management and conservation of pronghorn.

Traditional approaches to the study of animal populations have focused on the measurement of population size (i.e., the numbers of animals) and the estimation of population parameters that govern the numerical fate of population units. Examples for population parameters are population growth rates, survivorship, sex and age structure. Modeling of population processes as the product of population size and a population parameter is elegant and relatively simple. These aggregate models have been used widely and successfully to develop the field of ecological population theory. However, they often are criticized, because they condense processes occurring at the individual's level to a single value (the mean parameter) that represents the—biologically often meaningless—population level. The focus on descriptive parameters for an entire population often has prevented the consideration of differences within large populations, especially discontinuous spatial distributions, spatial habitat heterogeneity and individual differences in behavior. Today, biologists increasingly study the fates and behaviors of individuals and the variation among individuals within a population. Although mathematically more complex and computationally intensive, the focus on the probabilities of events occurring at the individual level is more realistic and allows population-level dynamics to emerge as a function of mechanistic processes at the individual's level. Thus, the secondary objective of this chapter is to describe population modeling approaches at both hierarchical levels and, perhaps, to increase the appreciation of individual-based approaches to the study and management of populations.

Last, the study of population dynamics is a scientific endeavor that relies heavily on sound statistical inference. Because it rarely is possible to examine each individual in a population (except perhaps in very small populations), the estimation of population parameters is dependent on sampling of smaller units. Randomized sampling always is preferable to nonrandom procedures because it increases the validity of the conclusions by reducing bias and allowing the application of inferential statistics (Fisher 1958). Unfortunately, wildlife managers often rely heavily on observational evidence from nonrandomly selected samples, such as harvested animals and traditional survey routes. Interpretation of such data is difficult because data may have low precision or considerable bias. Experimental approaches to the study of population dynamics are vastly superior to descriptive studies in producing "reliable knowledge" (Romesburg 1981). Adaptive resource management (Walters and Hillborn 1978, Lancia et al. 1996) has been proposed as a possible integration of management into a scientific decision-making process. Provincial and state wildlife agencies have the unique opportunity to manipulate populations across large landscapes, but rarely is such management applied in a rigorous experimental context to test hypotheses about population processes. As a result, little knowledge is accumulated by the management process, which allows outdated paradigms to persist stubbornly. Another objective of this chapter is to present critical statistical and inferential concepts that determine the difference between knowing and guessing.

POPULATION DYNAMICS

A population can be defined as a group of individuals of the same species that live in a contiguous geographic area and that can interbreed among themselves. As time progresses, animals are added (by birth or immigration) or subtracted (by death or emigration) from the population, causing changes in population size and age structure. A very general model of population change is:

$$N_{t+1} = N_t + B_t - D_t + I_t - E_t$$

where N_t is the population size, B the number of births, D the number of deaths, I the number of animals immigrating to the population and E the number of emigrants. The subscript (t) characterizes the given time step of the model; it can range from days (e.g., for insects) to a year (e.g., vertebrates in seasonal environments where births, deaths and migrations are limited to certain times within the year).

Birth and Death Processes

Fertility and Fecundity. Fertility of female pronghorn usually is high, and both sexes usually achieve sexual maturity as yearlings. Female fawns may conceive and produce young under favorable conditions (see Chapter 9). Fetal litter sizes range between 1.85 and 1.97 and show little variation among years or regions (see Table 58). Likewise, maternal age does not seem to affect fetal litter sizes (Trainer et al. 1983), in contrast to what has been reported for other ungulates (Clutton-Brock et al. 1982). However, prenatal growth rates, birth weights and gestation length were significantly lower after dry years on the National Bison Range (Byers and Hogg 1995). In elk and other large polygynous ungulates, fetal sex ratios are affected by the energy resources of the mother (Clutton-Brock et al. 1982, Kohlmann 1999a), but pronghorn do not seem to follow this pattern (Firchow et al. 1986a, Byers and Moodie 1990). This may be due to the extremely high maternal investment in pronghorn, where birth weights average 18% of maternal body weight (Byers and Hogg 1995). The evidence available to date indicates: (1) a rather constant, high maternal investment into offspring by female pronghorn and (2) fecundity rates that are not correlated strongly with weather, habitat or physiological condition of the female.

Mortality. Deaths of pronghorn can occur at any time of the year, but factors and affected age classes usually vary with the seasons. Dunbar and Velarde (1998) inves-

Births are the primary mode of growth in populations, hence the potential for increase of a population often is correlated tightly with its ability to produce many young. Many concepts are linked intricately with the reproductive potential of a population: fertility (i.e., the individual's ability to reproduce), fecundity (i.e., the achieved reproductive performance of an individual), age structure of the population and survival of produced offspring. Photo courtesy of the U.S. National Archives, Washington, D.C.

tigated the health of fawns at tagging and found no relationship between the fawn's condition at birth and its subsequent fate. Increased predation of fawns may be due to inferior habitat condition and insufficient hiding cover for fawns (Pyrah 1987). The major predator of pronghorn fawns is the coyote, but coyote control to increase fawn survival is not effective under all circumstances (Johnson 1972, Smith et al. 1986). In some studies reporting significant increases of fawn survival after coyote control (Willis 1988, Menzel 1994), statistical inference was weak due to unreplicated designs or lack of consideration of environmental covariates. Udy (1953) showed that fawn ratios in Utah were higher in areas with coyote control, but considered rangeland condition and livestock grazing more significant determinants of population performance. Beale (1986) found that fawn predation by coyotes declined as alternative food sources (rodents) increased. Rodent community diversity can be affected negatively by coyote control (Henke and Bryant 1999), and competitive mesopredators may increase after removal of a keystone predator (Seargeant and Allen 1989). Wagner (1988) and Henke and Bryant (1999) suggested that sustained control of coyotes causes lagomorph populations to increase, which may negatively affect forage availability for other large herbivores. These examples indicate that coyote control may not only be ineffective in increasing pronghorn fawn survival, but may have larger community effects that could further impact pronghorn populations. In fact, fawn survival often is correlated more strongly with population size and summer precipitation than coyote abundance (Pyrah 1987, Kohlmann 1998, Danvir 2000). In severe winters, pronghorn fawns may experience mortality of as much as 25% (Salwasser and Shimamoto 1979). Likewise, adult

Low fawn survival during the first 3 months of life is characteristic of most wild ungulate populations in North America (Vriend and Barrett 1978). Although pronghorn populations can increase rapidly under favorable environmental conditions despite low fawn recruitment (Pyrah 1987), low fawn ratios in midsummer are persistently a management concern. Predation often is cited as the leading cause of mortality (Trainer et al. 1983), but it is unclear if predation is the ultimate or merely proximate factor of mortality (see Chapters 11 and 12). The impact of mortality agents on population density may depend on the status of the population in relation to carrying capacity. Generally, mortality agents are considered additive in populations far below carrying capacity; they are considered compensatory when the population is at or near carrying capacity. Postmortem investigation of radio-tagged pronghorn fawns usually does not determine whether the fawn was predisposed to predation by environmental or physiological factors. In this instance, fresh carcasses of two, approximately 1-month-old pronghorn fawns were partially buried in debris under a bitterbrush plant. A bobcat ran from the bitterbrush as it was approached. Only the head and legs remained of one carcass, but the other was intact (Yoakum 1957). The intact carcass was necropsied by a veterinarian and wounds characteristic of bobcat attacks, accompanied by extensive hemorrhaging, were found on what apparently had been healthy and robust fawns. Photo by Jim D. Yoakum.

mortality in pronghorn can be catastrophic in severe winters (Ockenfels 1994) and usually is related to snow depth (Martinka 1967, Salwasser and Shimamoto 1979). Annual survival rates of adults may fall as low as 0.55 (Mitchell 1980). Firchow et al. (1986) reported fawn mortality of 89% and 80%, respectively, during two successive years in Colorado, yet the population experienced no appreciable decline. The same researchers estimated annual adult survival rates during 1983 and 1984 at 82% and 68%, respectively. Factors of natural mortality in adult pronghorn include predation (Ockenfels 1994), disease (Thorne et al. 1988), accidents and starvation (Einarsen 1948, see Chapters 11 and 12).

Survivorship Curves. Mortality in mammals usually follows a distinct, age-specific pattern, represented by a survival curve (Figure 141). Age-specific survival can be estimated from standard life table data. Life tables are developed using primarily two methods, the cohort analysis and longitudinal analysis. Cohort analysis estimates mortality by recording the mortality of a group of animals born in the same year, whereas longitudinal analysis uses the current age structure of the population as a representation of the survivorship curve. The underlying assumptions are that the population is not growing and has reached a stable age structure, which can limit the applicability of these methods for wild populations, especially when sample sizes of (radio-collared) animals are small. Age-specific survival curves of mammals often show an initially high but declining juvenile mortality, a rather constant adult mortality and increasing mortality in senescent animals (Caughley 1966, Siler 1979). In the most general case, survival data from radio-collared pronghorn provide two statistics for each individual: the number of years the animal has been at risk of mortality and the fate of the animal (live or dead) at the end of the study period. Combining these data into age-specific mortality rates yields a single five-parameter model (Siler 1979):

$$l(x) = l_i(x) l_c(x) l_s(x) ,$$

where:

$$l_i(x) = \exp[(-a_1/b_1)\{1-\exp(-b_1 x/\Omega)\}] ,$$

$$l_c(x) = \exp(-a_2 x/\Omega) ,$$

$$l_s(x) = \exp[(a_3/b_3)\{1-\exp(b_3 x/\Omega)\}] .$$

The mortality rate for each age class (juvenile, adult, senescent) is indicated by *l*; a and b are parameters to be estimated, and Ω represents the longevity of the species (to facilitate comparison with other species). Barlow and Boveng (1991) and Taylor and Barlow (1995) give detailed descriptions of these equations and the interpretation of

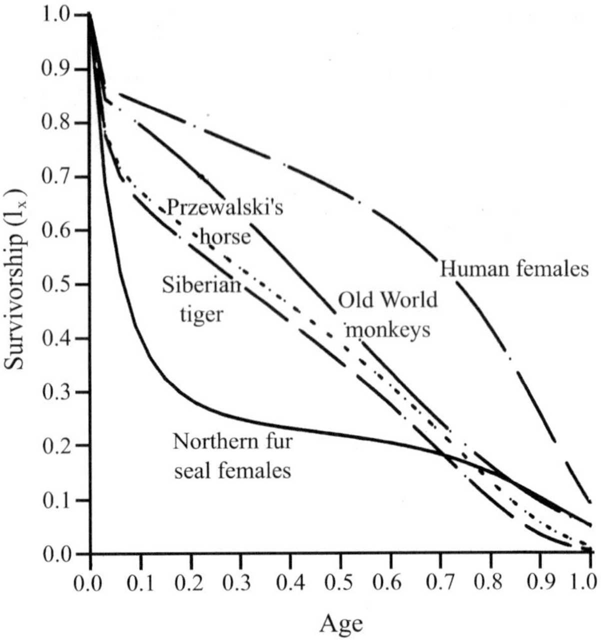

FIGURE 141. *Survivorship curves for five mammal species (Taylor and Barlow 1995).*

their parameters. A standard survivorship curve (Figure 141) can be created by fitting the full model by maximum likelihood estimator (Siler 1979), by stepwise fitting (Taylor and Barlow 1995) or by linear regression of logit-transformed survivorship data on age (Brass 1971). A related approach uses a linear regression of log-transformed frequency data of different ages in the hunter harvest to estimate instantaneous harvest mortality rates (Bunnell and Tait 1981, Kohlmann et al.1999). This method can be applied easily if the age distribution of the harvest is available, but cannot yield mortality estimates for unharvested age or sex classes. Estimating rates of survival of adults from age distribution data introduces a bias when the population is not stationary (Caughley 1977). However, in changing populations (Eberhardt 1988) the rate of survival of adults can be estimated from the age distribution of the harvest as:

$$\hat{S}_x = \frac{N_x}{N_{x-1}} \lambda ,$$

where $\hat{S}_x$ is the survival rate from age x to x + 1; N_x is the number of animals in the age class x, and λ is the finite rate of population change (note that in a stationary population λ = 1, hence the estimate of survival is equal to N_x/N_{x-1}). Lima and Paez (1997) give an excellent example of using these equations in deriving population parameters of South America fur seals from harvest figures. Unfortunately, age distributions of pronghorn harvests are rarely sampled, compared with other species. Managers could

Aside from weather and nutrition, the most significant source of adult pronghorn mortality is human harvest, including illegal take and wounding loss. However, current hunting levels have little effect on the growth rates of pronghorn populations in Canada and the United States (Yoakum 1986). This suggests that hunting mortality can be largely compensatory to other forms of mortality. Exploitation levels of pronghorn populations usually are very conservative and sometimes are limited to the removal of males only, allowing populations to maintain high densities that fluctuate considerably with the climate-induced changes in carrying capacity. Photo by Harold Umber; courtesy of the North Dakota Department of Game and Fish, Bismarck.

obtain a wealth of information on the status of their populations by mandatory tooth returns from hunters, especially where female harvest is producing significant samples.

Survival Rate Estimation. In many cases, managers require survival data that are specific to a certain area, season or life stage of an animal. Several powerful computational methods for estimation of survivorship exist. Most basically, a finite survival rate can be modeled as a binomial variable (i.e., the ratio of the number of animals alive at the end of the study over the number of animals alive at the start of the study). However, in cases where animals can have other fates than live or die (e.g., emigrated from the study area, different causes of death, radio failure), or when sex- or age-specific survival rates are of interest, a multinomial model is necessary. Brownie et al. (1985) developed a numerical method of combining a set of multinomial distributions for the estimation of survival from band-return data. Program MARK (White and Burnham 1999) calculates survival rates from a multinomial model using advanced optimization routines to construct likelihood functions of survival probabilities, evaluate model parameters and test for significance of model results. The program also facilitates Monte Carlo simulations to determine the power of hypothesis tests and confidence intervals based on a given number of samples and survival rates.

The Kaplan-Meier product limit estimator (Kaplan and Meier 1958) tracks the number of animals alive at each time during the study and uses a product multiplication of survival estimated from a binomial estimator in each interval. Aside from its ease of computation, the Kaplan-Meier method facilitates estimating the survival of animals that were not radio-marked at the same time (i.e., staggered entry of transmitters into the population [Pollock et al. 1989]). Similarly, the failure of radiocollars during the study can be accommodated easily by adjusting the number of animals alive, or at risk, at each interval. A log rank test (Cox and Oakes 1984) allows for the formal testing of differences in survival rates between two populations.

Other methods include estimating the time to death of a radio-marked animal. Such models are related to failure time analysis, developed for industrial purposes, but uniquely suited for ecological applications. Survival data times for animals present unique statistical challenges; survival times typically are distributed non-normally and contain censored data (radio failure), and variances of different groups of animals may be extreme. Special methods of dealing with these have been developed (Fox 1993) and involve estimating distributional characteristics. Survival time analysis includes two groups of approaches: the life table approach and regression methods. Life table methods often lack statistical power (Fox 1993) and are difficult to use where multiple covariates exist (Kalbfleisch and Prentice 1980). Regression models alleviate some of these problems. All failure-time methods can be easily implemented in standard statistical software (e.g., PROC LIFETEST in SAS).

A method originally designed to estimate nesting success (Mayfield 1961) has been expanded for application of radiotelemetry data by Heisey and Fuller (1985) and implemented in the program MICROMORT. This method requires the investigator to define intervals in which the daily survival rate is constant. Obviously, weather events, fires or other habitat modification can make this requirement difficult to meet, and estimates for survival often differ dramatically between a simple binomial estimator and those provided by MICROMORT (White and Garrott 1990).

Using pre- and postwinter age ratios and the age ratio from winter-killed carcass surveys, White et al. (1996) developed estimators for overwinter survival. This method requires that bias be the same in all three composition surveys, which may be difficult to achieve given the different visibility and spatial clumping of live and dead animals at different times during the winter. In mild winters, when few carcasses can be located, estimates may be unreliable due to small sample size. White et al. (1996) suggest that other methods (such as radio tagging) may yield more precise estimates.

Recruitment and Population Growth. Although the reproductive output of pronghorn appears highly consistent, the postnatal survival of fawns has been reported to vary greatly (Vriend and Barrett 1978), causing annual variation in recruitment. Recruitment commonly is defined as the number of young that survive to sexual maturity, hence they are recruited into the breeding population. Thus, recruitment is not an estimate of reproduction, rather a composite of birth rates and survival of young. Density dependence in birth rates and juvenile survival has been documented for virtually all North American ungulates (Fowler 1981). For example, summer estimates of fawn ratios for a pronghorn population in central Oregon declined steadily during a period of 40 years, whereas the annual counts of does increased during the same time (Figure 142). Hess (1986, 1999) verified significant negative correlations between summer doe numbers and fawn ratios for the interstate pronghorn population inhabiting the Great Basin Desert of Nevada, Oregon and California, and suggested that this relationship is evidence of density dependence. However, summer fawn ratios may not be the primary factor controlling population growth. Researchers often measure the per capita rate of increase of a population as the logarithmic ratio of two subsequent population estimates. Summer fawn ratios at the Hart Mountain National Antelope Refuge, in eastern Oregon, were not related to population growth rates, but per capita rates of change showed statistical evidence of density-dependent negative feedback (Kohlmann 1998). Similarly, a rapidly growing population of pronghorn in Middle Park, Colorado, showed a strong linear relationship between annual per capita growth rates and total population size (Figure 143). Despite increasing evidence, the concept of density dependence often is overlooked in pronghorn management, especially when interpreting the effects of predation on future population size.

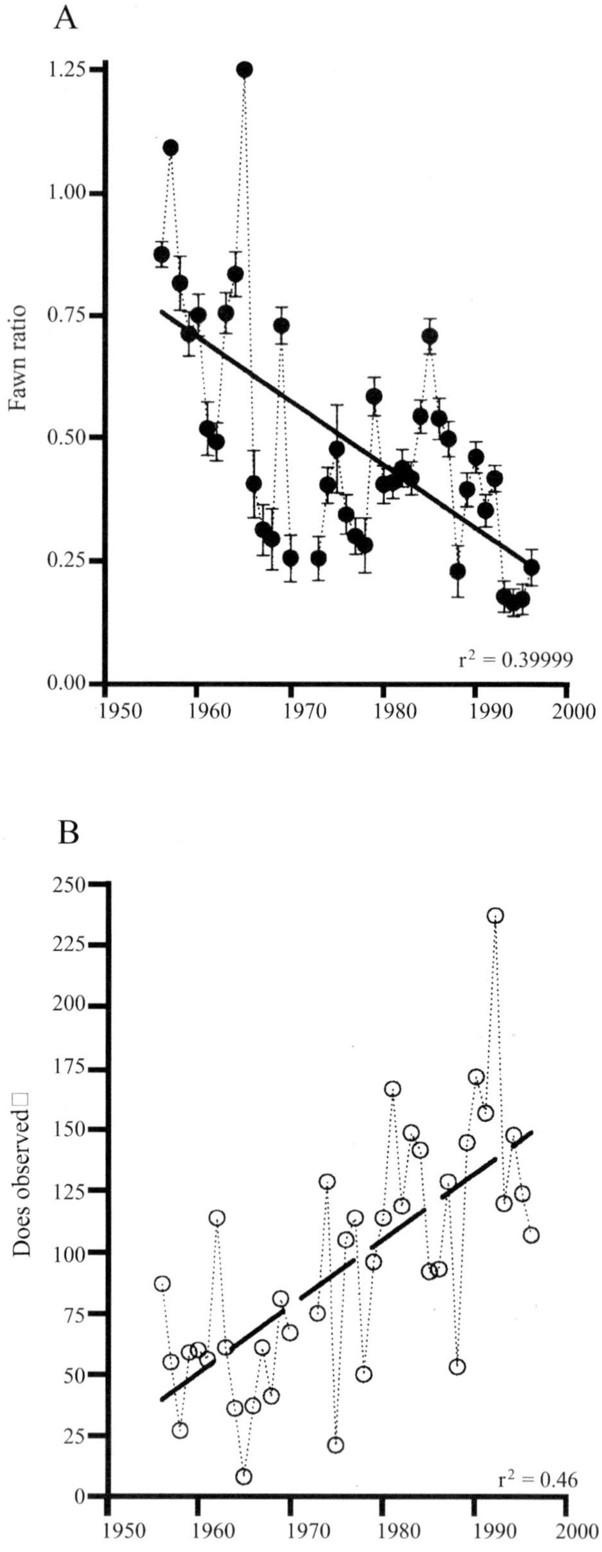

FIGURE 142. *Fawn ratios (A) and doe population counts (B) for pronghorn at Sycan Marsh, central Oregon. Black circles = estimated fawns: 100 does; open circles = number of does observed (data courtesy of the Oregon Department of Fish and Wildlife).*

Spatial Dynamics: Immigration and Emigration

Immigration is the permanent influx of animals from an outside population or subpopulation, whereas emi-

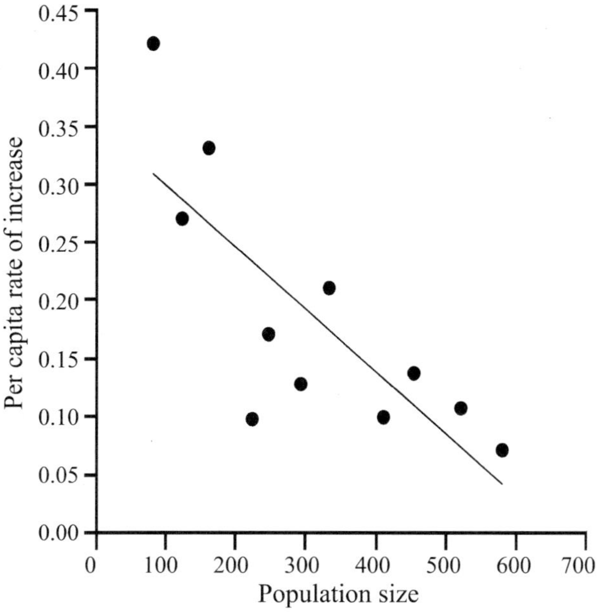

FIGURE 143. *Relationship between per capita rate of increase and population size for a pronghorn population in Colorado. Black circles = estimates of per capita rates of increase; line = regression estimate. (Pojar 1997).*

gration is synonymous with dispersal. Most population estimators assume at least spatial closure (i.e., no migration occurs into or from the study population) (Seber 1973). Although the study of spatial behavior (i.e., home range) has received widespread interest among wildlife managers, the dispersal of animals from their birth areas to other populations or other population segments has been difficult to measure. Anecdotal evidence of long-range dispersal of individuals frequently is ignored, treated as outliers or assumed inconsequential by wildlife managers. Mark/recapture techniques allow the distinction between death and emigration (Zeng and Brown 1987, White and Garrott 1990). Most important, immigration into a population is challenging to measure and has received little attention in field studies.

Recent advances in ecology have identified immigration as a crucial concept in metapopulation ecology (Gilpin 1996, Wiens 1996). Andrewartha and Birch (1954) were first to recognize the importance of spatial structures in population dynamics. Levins (1970) usually is credited with coining the term metapopulation. A metapopulation is a set of populations that exist in distinct habitat patches interspersed into a matrix of nonhabitat and linked through dispersal. Metapopulations can occur naturally (e.g., mountain sheep, Bleich et al. 1996) and often are caused by human actions, such as habitat fragmentation (Beier 1996). Fundamentally, metapopulation theory is related to MacArthur and Wilson's (1967) theory of island biogeography, where the existence of a species is related to the size of the habitat patch and the immigration rate from other patches. It has become one of the most prolific fields of study in wildlife conservation during recent years (McCullough 1996). Immigration into small, isolated populations allows the exchange of genetic material, increases heterozygosity, reduces deleterious effects associated with small population sizes—such as inbreeding (Hedrick 1996)—thus increasing the species' probability of survival (Lacy 1987). Pronghorn may have experienced a population bottleneck during the early part of the 20th century (Hoover et al. 1959), and some subspecies are classified as threatened due to their isolated range and small population size (Lee 1988). Limited gene flow (as low as one individual per generation) has been suggested as a sufficient way to maintain genetic heterogeneity in a metapopulation (Lacy 1987).

Immigration also is important where the source-sink dynamics of a population are concerned (Pulliam 1988, 1996). Wildlife managers recognize that not all habitats are equally productive, and recruitment and survival can differ greatly among habitat types. Pulliam (1988) defined a source as a (sub)population or habitat patch where births exceed deaths and where emigration exceeds immigration, whereas the opposite is true for population sinks. Thus, if immigration was absent, pronghorn living in a sink habitat would soon disappear. Theoretical and observational evidence suggests that immigration into sink habitats ascertains the continued existence of animals in

Immigration as a means to overcome spatial isolation or population sinks requires that animals can move between patches or populations. Natural barriers, such as large bodies of water, dense forests or steep-walled canyons may limit pronghorn movement and, therefore, may separate populations (Ockenfels et al. 1994). These pronghorn in northern Yellowstone National Park have been blocked from contact with other populations for much of the 20th century by human activities and structures in the Yellowstone River Valley north of the park—their former travel route to winter rangelands at lower elevations. Photo by Bart W. O'Gara.

suboptimal habitats. In fact, due to immigration, most animals in local populations may exist in sink habitats that are unsuitable for sustaining a population by reproduction alone (Pulliam 1988). Source-sink dynamics in pronghorn have yet to be investigated, but the evidence for their importance is circumstantial, especially in marginal ranges and desert ecosystems. An analysis of pronghorn production indices (i.e., age ratios) revealed that marginal ranges not only exhibit lower productivity but also greater annual fluctuations than prime habitats (Vriend and Barrett 1978). Several workers (Buechner 1950a, Ellis 1970, Pyrah 1987) reported that habitat-specific survival of fawns is related to the productivity of different habitat patches, suggesting that poor habitat patches or herd ranges serve as population sinks. Bergerud (1988) showed that wolf predation on caribou was extremely high in woodland habitats, yet the demographic data show no appreciable, persistent decline of caribou in woodland habitats. Apparently, the woodland habitat served as a population sink, and the persistence of caribou in woodlands was sustained by immigration from tundra habitats.

Population Structure

Next to population size, population structure is of central interest for managers of pronghorn populations. Population structure is the age and sex composition of a population. Population structure can be characterized graphically as the well-known age pyramid (Figure 144). The primary factors affecting the structure of a population are the annual reproduction and the differential mortality of various sex and age classes. The structure of pronghorn populations changes as the population grows or declines. Typically, a vigorously growing population will have a population structure dominated by young age

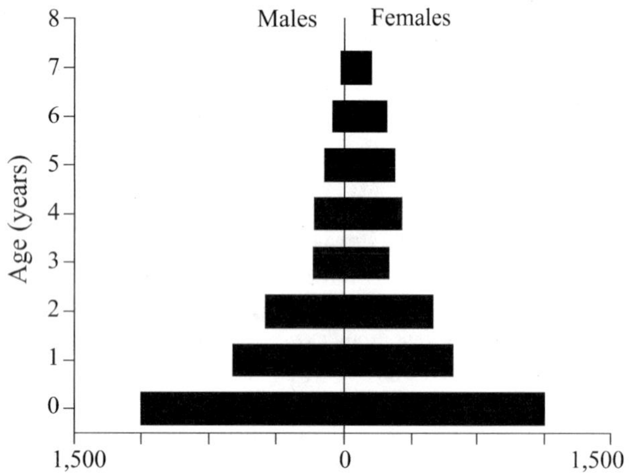

FIGURE 144. *Example of an age pyramid (data from Bartholow's [1986] Wyoming pronghorn population model).*

classes, whereas declining populations often exhibit a preponderance of old animals.

Estimating Sex and Age Structure. Wildlife managers often calculate age ratios (i.e., fawns per 100 does) from mid- to late-summer population composition counts as an indicator of productivity. However, density-dependent and stochastic events (drought, severe winters, etc.) may affect ratios, resulting in weak correlations between population ratios and population size or population trend. This is why Caughley (1974) noted that age ratios are useless without knowledge of the rate of increase of the population, and they are superfluous if that rate is known. However, management objectives often are formulated in terms of sex/age ratio targets, hence estimation of ratios (and their precision) is important in the management process (Bowden et al. 1984). A simple, random sample without replacement from all pronghorn groups ($i = 1, \ldots, n$) will yield a ratio estimate by using

$$r_y = \frac{\Sigma Y_i}{\Sigma Y_i + F_i)} \;,$$

where Y_i is the number of young and F_i is the number of females observed in each group. A robust estimate of the variance for ratios (Cochran 1977) is:

$$VAR(R) = n \frac{(\Sigma Y_i^2 + r_Y{}^2 \Sigma t_i^2 - 2r_Y \Sigma Y t_i}{t^2(n-1)} \;,$$

where Y_i is the number of young seen, F_i is the number of females, n is the number of groups, t_i is Y_i plus F_i and t equals the summation of t_n. This variance essentially is unbiased if sample sizes (i.e., the number of observations) are greater than 10 (Cochran 1977), even if the proportion of young is not constant among all group sizes. Because ratios do not have a symmetrical distribution, the mean can be a poor estimate of the central tendency. The median of a distribution often is a more informative parameter than the mean, but estimation of its precision is considerably more involved. Therefore, detection of significant changes in the central tendency of data is difficult. An efficient, but computer-intensive method to estimate not only the distribution but also the precision of mean or median population ratios is the bootstrap (Efron and Tibshirani 1993). Bootstrapping is a data resampling technique whereby the computer randomly draws (with replacement) a number of individual observations, such as age ratios of individual groups, from the original data set. This selection of samples is called a bootstrap sample. Bootstrap samples are repeated (up to 2,000 times) and the sample ratio is calculated each time. A frequency distribution of all bootstrap estimates then can be created, and a confidence interval is constructed around the mean (Dixon 1993).

Some Statistical Concepts

Accuracy of parameter estimates is dependent on the observer's ability to correctly classify all individuals (i.e., no consistent bias due to misclassification of certain age classes). Estimates can be affected severely by differences in experience among observers, observer fatigue, equipment used (including the type of aircraft) and weather conditions. All these factors contribute to sampling errors, which usually will affect both accuracy (freedom from bias) and precision (confidence in the estimate). Managers should be cognizant of the fact that a single, unreplicated count or measurement has no precision, thus random sampling and replication are important concepts for the estimation of precision.

Sampling variation is not the only source of variation affecting ratio estimates; there often is considerable process variation. When animals frequent habitats of different quality, probability of detecting and classifying animals is no longer only dependent on sampling conditions, but also on the animals' behavior (group size, activity) in response to their location. Thus, systematic variability can be due entirely to natural causes and is a realistic property of the population. Ideally, this variation is controllable if all causal factors are known and can be measured.

An important factor in estimating population parameters is the size of the area covered and the number of samples. Given the large expenses associated with extensive aerial surveys, managers have to decide not only which parameter to sample, but the scale of the observations and the desired level of precision for the estimate. There is a trade-off between the area sampled and the precision of the estimate; samples comprising large areas (i.e., entire units) rarely can be replicated, whereas breaking a unit up into small, randomly selected sample areas may facilitate replication and stratification of surveys, which increases precision.

POPULATION MODELS: CLASSES AND CONCEPTS

Comprehending the entire suite of biological and physical processes that affect the population size, distribution or dynamics of a wildlife species is impossible. To manage a species despite the lack of complete understanding of its ecology, managers increasingly use population models to simulate population dynamics and harvest strategies. A model is any abstraction or simplification of a system (Starfield and Bleloch 1986). Models of a wildlife population necessarily are much simpler than the real population, which can facilitate a deeper understanding of salient relationships. Often, models are the vehicles to predict the behavior of a complicated, poorly understood system from a few, well-studied components of the entire system. For example, we may understand how temperature, snowfall and pronghorn survival are correlated, but we need a model to predict the population size in future years. In most cases, a model will be constructed around a set of instructions (equations, rules) that link the model input (data) with the model output (predictions). Analytical models derive exact solutions to a set of equations, while simulation models usually provide a distribution of possible model outcomes, thus identifying variability and probability of a given outcome. Analytical models by definition are always deterministic; that is, the same combination of input variables will always result in the exact same output. On the other hand, simulation models usually are stochastic; they contain sources of variation (perhaps as a random variable) that will modify the output each time the model is run.

Most population models keep track of the numerical changes that occur in a wildlife population. Thus, population modeling is inherently a quantitative endeavor and, with the availability of fast and powerful computers, population modeling has become a major tool in the scientific study and management of wildlife populations. However, population models were built long before the most crude computer was available. Starting with the theories of the clergyman Thomas Malthus almost two centuries ago, population models were purely mathematical and, hence, could be understood and appreciated largely by mathematicians only. Today, models are implemented with the help of computers, which simplify the task and make the model's structure transparent.

To date, not many examples of population modeling of pronghorn are available in the literature, although most states use population models in the management process. It seems modeling still is in its beginning phases among pronghorn managers. But, much can be adapted from other species where economic or ecological interest has forced the development of advanced models. It is important to realize that there are no fundamental differences that preclude the adaptation of, for example, an elk or white-tailed deer model to the pronghorn, provided the parameters of the model are changed according to the biology of the species. In this sense, the reader is encouraged to consider the presented material applied and used for the benefit of scientific management of pronghorn.

Trend Models

A trend in population dynamics can be defined as a persistent tendency in population growth. Trends correspond to sustained and systematic variations in population growth (Dagum and Dagum 1988). The category of trend models contains a wide family of related models, whose sole purpose is to mimic, explain and predict the growth of the population size, that is, the total number of animals in a defined population. In trend models, the population has only one characteristic, abundance, or when expressed in relationship to the area it occupies, density. The model thus has only two dimensions: time and population size (or density). A set of observations on the size of a population taken at discrete and equal intervals of time also is referred to as a time series (Berryman 1994a).

Trend Analysis. Detecting a trend in a wildlife population often is the primary goal of managers; after all, one wants to know immediately whether populations decline or exceed established management objectives. However, detecting such a trend is not trivial. Requirement for detecting trends in population size is the existence of long-term surveys (more than 15 years of data). Due to the stochastic nature of pronghorn environments and the high sampling error associated with population sampling, it is highly unlikely that a significant trend can be detected from just a few years of surveys. In general, short-term surveys may provide biased results and, if animals occur in low densities, the magnitude of population trends may be underestimated.

Perhaps the most common technique to analyze data from annual survey routes is the route-regression technique. With this method, a population trend is estimated for each survey route in a given region, and the regional trend is estimated as a weighted mean of the individual routes. This method has been used widely for roadside surveys of breeding birds in Canada and the United States (Droege 1990). It involves fitting a multiplicative regression model (linear regression on the logarithm of counts) with coefficients for observer differences and the slope of the regression (Geissler and Sauer 1990). Route-regression methodology can use data from routes that have missing data for some years, and these routes are given less weight in the overall trend estimate. The method also obtains proper variance estimates (i.e., from the consistency of trends among the routes in a given area). Lastly, route regression allows the identification of sources of variation (covariates), such as observer ability, weather and disturbances. Recent extensions to the method—such as the residual method (Sauer and Geissler 1990), re-randomization tests (Collins 1990), iterative row and column models (Moses and Rabinowitz 1990), and LOWESS smoothing techniques (Taub 1990)—can improve the efficiency and accuracy of route-regression methods.

Power Analysis of Trends. Power is a statistical concept related to the probability of rejecting the null hypothesis when it is false. It is calculated as $1 - \beta$, where β is the probability of accepting a false null hypothesis (i.e., committing a type II error). Thus, power is an important factor to consider when designing studies or surveys. The power of a statistical test is a function of the alpha level (i.e., probability of a type I error), the sample size and the amount of difference between two samples (effect size) one wants to detect (Howell 1992). When using linear regression to detect trends in population size, the null hypothesis is: r equals 0 because the population does not change. The effect size is the average rate of change one can detect, shown by the slope of the regression line. The sample size n is the number of years for which population estimates are available.

Gerodette (1987, 1993) described a power analysis of linear regression. For many wildlife populations, the process of population change is multiplicative rather than additive, which causes an exponentially increasing or decreasing curve. Furthermore, the variances of the estimates generally are a function of the population size and violate the assumption of equal variances. To alleviate these problems, Gerodette (1993) developed the program TRENDS, which allows linear and exponential models of population change, and allows the variance to be proportional to N, N^2 or N^3. Program TRENDS contains five parameters: n, the number of sampling occasions; r, the rate of change in abundance that occurs between each sampling occasion; *CV1*, the coefficient of variation of the first estimate of abundance in the series; α, the significance level (probability of type I error); and the statistical power ($1 - \beta$, where β is the probability of a type II error). The user can estimate the value of any parameter if the other four are specified. Program TRENDS allows asking whether the length of the time series is sufficient to detect a significant trend, how many years of data are needed to detect a given effect (slope) with a given significance, or which method of population estimation gives a lower minimum detectable rate of change. Power analysis should be conducted before the collection of data to ascertain that the most effective survey design and sampling effort are used. Power analysis is inappropriate when it is conducted after the data have been collected (Steidl et al. 1997). However, such post hoc power analyses might give clues to the reason a null hypothesis was rejected, such as variance structure, sample size and effect size.

Time-series Analysis. The underlying hypothesis of time-series analysis is that the dynamic behavior of the system stems from two sources, an endogenous feedback process (e.g., density dependence) and a random, exogenous process (stochastic or environmental variation). A population manager needs to know the extent of the annual change as a product of mechanistic or density-dependent population regulation, and the change that is caused by random events and environmental variability. Time-series analyses are suited for management applications because most pronghorn management agencies routinely survey pronghorn populations and collect annual population statistics.

The remainder of this section will deal exclusively with estimates of population size N, which easily could be transformed into density estimates (N/km^2), if the size of the study area is known. To use estimates of population size N instead of population density D requires that the area of reference (i.e., survey area or population boundaries) stays the same. Although seemingly simple, models of population size and trend are surprisingly diverse in the dynamics they can express. In the most simplistic approach, the total size of the population can be derived as a direct, arithmetic result of a population parameter (growth rate) applied to initial estimates of population:

$$N_t = N_0 \exp(Rt) ,$$

where N_t is the population size at time t; N_0 is the population size at the start of the time interval, and R is the per capita or intrinsic rate of change. (Note that R can be expressed as the net per capita rate resulting of the $B - D + I - E$ model above.) In population ecology, it is common to express the annual rate of change in population size as the log-transformed ratio of two subsequent population estimates, with

$$\ln(N_t / N_{t-1}) = R ,$$

The log-transformation is helpful for building population time-series models with additive noise structure, due to the multiplicative effect of environmental variability in population growth (Dennis and Taper 1994). An obvious problem with this model is that, as long as R is a fixed constant, the population will continue to grow (or decline if R is less than 0) at this given rate until infinity. Furthermore, the rate of change in natural populations usually is subject to environmental variation, hence a constant value for R is unrealistic. The simple trend model can be modified by allowing the per capita rate of change to be a function of population size and random external influences.

Density Dependence. Density dependence generally is defined as a negative feedback effect of the population size or density on its rate of change (McCullough 1990, Berryman 1994a). The simplest model explaining density-dependent feedback in the rate of change is:

$$R = R_{max}(1 - \frac{N}{K}) ,$$

where R_{max} is the maximum per capita rate of change when N equals 0, and K is the upper asymptote for N (i.e., the carrying capacity of the habitat). By definition, density dependence means that the per capita rate of change (R) becomes zero when the population reaches K. Stochastic variability induced by environmental factors, such as storms, drought and predation, can be incorporated into the model. Substituting $K = -R_{max}/b$ into the equation above and adding a random variable yields:

$$R_t = R_{max} + b\hat{N}_{t-1} + e ,$$

This equation is identical to a linear regression model, where R_{max} represents the intercept, b is the slope coefficient for density dependence and e is a random variable with mean of zero and variance of σ^2. This linear form of density dependence gives rise to the well-known logistic curve (Figure 145).

Density dependence in population growth can be caused by simultaneous changes in reproduction or mortality (Skogland 1990). Evidence of density-dependent population growth in ungulates has been reported frequently (Fowler 1987), but few experimental tests have been conducted. Perhaps the best known study of density dependence was conducted on white-tailed deer at the George Reserve in Michigan (McCullough 1979). In this hunted population, the total densities of deer varied widely. Recruitment rate (autumn age ratios) and female population size were negatively correlated. Experimental manipulation of population density in mule deer resulted in compensatory winter survival (Bartmann et al. 1992) and increased body mass (White and Bartmann 1998). Density-related changes in behavior and growth also were

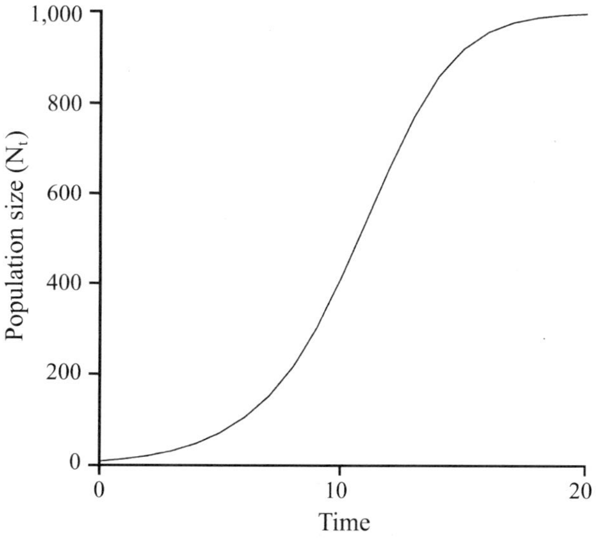

FIGURE 145. *Relationship between population size and time for the logistic population growth model.*

reported for red deer in Europe (Clutton-Brock et al. 1982, Forchhammer et al. 1998) and for elk in North America (Houston 1982). For a pioneering pronghorn population in Colorado, Pojar (1997) suggested that strong density dependence was evident in population rates of change, but the density-dependent feedback was not linked to fawn ratios. Likewise, Kohlmann (1998) found density-dependent negative feedback on population rates of change, but summer fawn ratios did not explain significant amounts of variation in annual per capita rates of change. Defining the exact nature of the density-dependent regulation mechanism often is not possible, especially when replicated studies and controlled manipulations are not feasible.

A Simple Linear Pronghorn Trend Model. A data set of annual density estimates (pronghorn/mile^2) for three range categories (primary, secondary and tertiary habitat) in North Dakota (McKenzie 1986) is used here to demonstrate the estimation of parameters for a linear trend model. The plot of the data (Figure 146) suggests that densities across all three range quality categories are different, but that population trends of pronghorn populations are very similar in the three habitats. Yet, when we examine the plot of the rate of change over population density (Figure 147), another difference becomes evident—that of different density-dependence feedback. Maximum likelihood estimates for the simple logistic model (i.e., parameters R_{max} and b) can be obtained easily through linear regression (Dennis and Taper 1994). Likewise, the variance of the random variable e is easily estimated as the mean residual of the linear regression. The estimates for R_{max} (i.e., the intercept of the regression) are fairly similar, but the slopes (b) increase with declining habitat quality. Because $K = -R_{max}/b$, estimates of K also follow this trend; they are lower in the poorer habitat (Table 157). It is inappropriate to derive estimates of precision or confidence intervals for trend model parameters from the linear regression. A parametric bootstrap (Dennis and Taper 1994) can provide unbiased estimates for parameter precision.

Despite its apparent simplicity, an analysis of this nature has great value for management. Without knowing the exact mechanism of how density dependence operates, the manager can obtain a rapid assessment of the population's ability to sustain future growth at a given population size. Managers often set desired population levels or harvest objectives, but without knowledge of

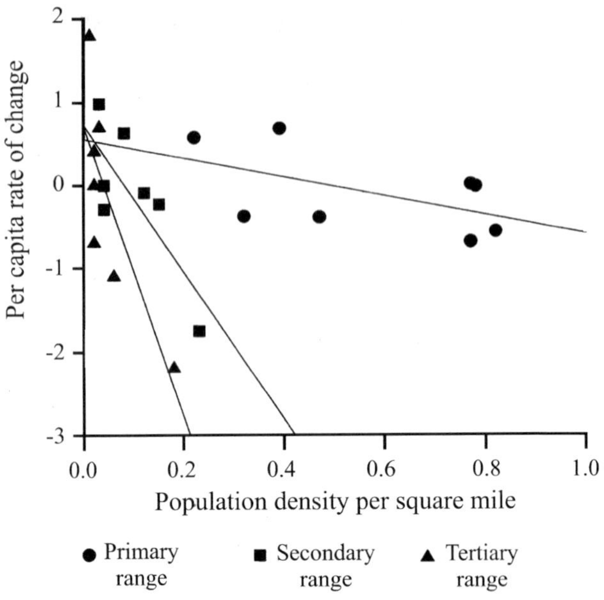

FIGURE 147. *Growth functions of pronghorn populations on ranges of different quality in North Dakota (McKenzie 1986). Each line represents a regression estimate of pronghorn density on the rangelands denoted.*

TABLE 157. Model parameters for North Dakota pronghorn time series from three different habitat categories

Habitat	R_{max}[a]	b[b]	K[c]
Primary	0.5486	−1.1303	0.4853
Secondary	0.7113	−8.8053	0.0808
Tertiary	0.6832	−17.2970	0.0395

[a] Maximum per capita rate of change (i.e., intercept of regression in this model).
[b] Slope coefficient for density dependence.
[c] Carrying capacity.

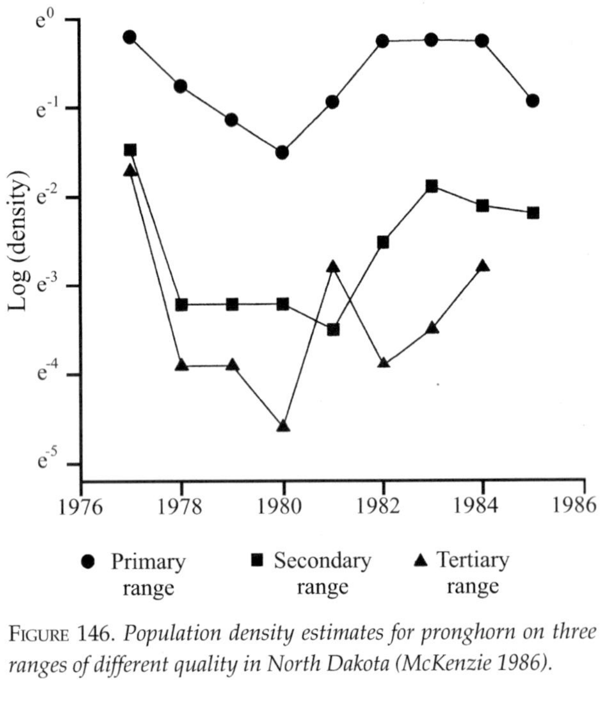

FIGURE 146. *Population density estimates for pronghorn on three ranges of different quality in North Dakota (McKenzie 1986).*

the population's potential for increase, such management objectives can be overly ambitious or conservative.

Nonlinear Feedback. Linearity of the growth function is not necessarily the most realistic model. A nonlinear model of density-dependent population change is

$$R_t = R_{max} + b\hat{N}^{Q}_{t-1} + e,$$

where Q represents a curvature coefficient (i.e., the relationship between R and N is convex for $Q > 1$ and concave for $Q < 1$). Nonlinear models can yield substantially different predictions of the strength of density-dependent feedback at different population sizes (Figure 148). It has been suggested that curvature coefficient of $Q > 1$ are typical for ungulates (Fowler 1981), where most density-dependent changes occur near the carrying capacity K. Fitting a logistic model with nonlinear feedback structure can be accomplished by either nonlinear least squares regression of the growth function (i.e., fitting a curvature parameter in addition to R_{max} and b, see above) or by fitting the nonlinear logistic model directly from the difference equation (i.e., a discrete-time version of the differential logistic model). This model requires fitting four parameters: the maximum finite growth rate R_{max}, the carrying capacity K, the exponent Q and the intercept (population size at the start of the time series). It can be fitted easily by a nonlinear model in a statistical package (such as PROC NLIN in SAS) or by an optimization routine in spreadsheet packages (such as SOLVER in EXCEL, Optimizer in Quattro Pro). The optimization criterion minimizes the sum of the squared residuals (i.e., Ó $(E(N_t) - N_t)^2$). Constraining the solution to $E(N_t)$ is $\leq K$ is consistent with the model and prevents curves from becoming distorted at high values of Q. However, it is difficult to assess if the nonlinear model has a better fit than a standard linear model (Q = 1), and no significance test for the improved fit of nonlinear models is readily available. For our North Dakota pronghorn model above, fitting a nonlinear function to the data has little advantage over the simple linear model because of the small sample size (n equals 9 years). However, for long time series, especially when they span a wide range of densities, fitting a nonlinear model can be appropriate. For example, a time series of pronghorn population counts from the Deseret Land and Livestock range in Utah suggests nonlinear density-dependent feedback (Figure 149).

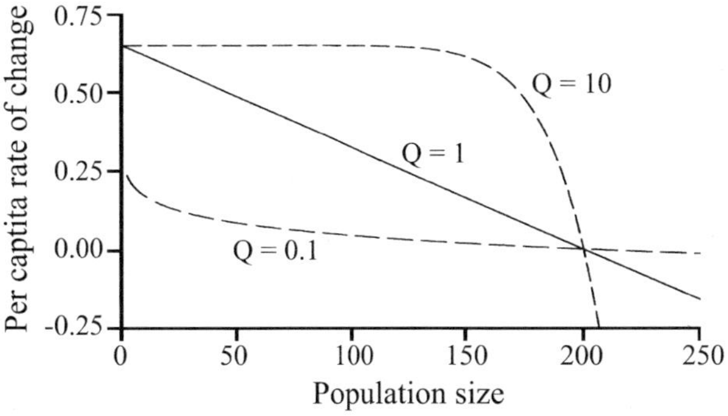

FIGURE 148. *Forms of nonlinear density dependence. If Q equals 1, the growth is logistic; if Q is greater than 1, the maximum net productivity level occurs near K; if Q is less than 1, density dependence is strongest at low population densities.*

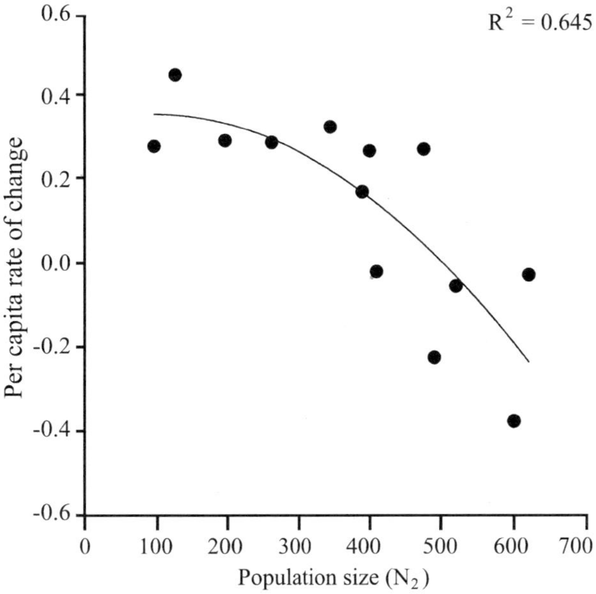

FIGURE 149. *Nonlinear growth function for a pronghorn population in Utah (data courtesy of the Deseret Land and Livestock Ranch, Woodruff, Utah [R. Danvir personal communication:2000]). Black circles = annual estimates of the rate of per capita change of total population size; line = nonlinear regression estimates.*

Time lags. Rapid density-dependent feedback occurring within a short time span often is characteristic for species in which density has a direct influence on the rate of change. Mechanisms for such rapid feedback could be a high disease transmission rate in a dense population, social mechanisms to space out individuals, predators aggregating around the numerous prey, etc. However, in most ungulates, the feedback of high population density is likely to occur due to food depletion (overgrazing) or an increase in predator populations. In these cases, density dependence may occur with a delay (i.e., the time required for forage to decline or predators to build up their numbers). Long time delays (more than 1 year) often are an indication that density dependence involves trophic interactions. Thus, a discrete nonlinear logistic equation with time delay d is

$$R_t = R_{max} + b\hat{N}^{Q}_{t-d} + e\ .$$

Other forms of the logistic models include the logarithmic form of the model, sometimes referred to as the Gompertz model ($R_t = R_{max} + b \log_e N(t - d) + e$). Turchin (1990) proposed a model incorporating multiple time lags ($R_t = R_{max} + b_1 N(t - d_1) - b_2 N(t - d_2) + e$). A logarithmic form of the multiple time-lag model has been suggested by Royama (1977): $R_t = R_{max} + \log_e(b_1 N(t - d_1)) - \log_e(b_2 N(t - d_2)) + e$).

Most linear and nonlinear time-series analyses can be implemented relatively easily by standard statistical software, as long as the package is capable of advanced linear and nonlinear regressions. However, simulations can become involved and, in most cases, will require writing advanced code. A commercially available software package for ecological time-series analysis is RAMAS-Time (Applied Biomathematics Inc. 1997). It is designed to analyze time series and to model density-dependent feedbacks in population size. The software provides time series and phase plots, and it presents various diagnostic statistics, such as the autocorrelation and partial autocorrelation functions. Another free software package (freeware), Population Analysis Software (PAS) (Berryman 1999), facilitates the modeling of single- and two-species logistic models, evaluation of sensitivity and stability of logistic models using bifurcation graphs, and the design of population management strategies.

Environmental Covariates and Multispecies Models. In most wildlife populations, density is not the only factor that affects the population rate of change. Weather, harvest or the abundance of a competing species also are possible candidates for such exogenous covariates. Additional independent variables can easily be incorporated into the logistic to construct a multiple regression model expressing R_t as a function of several independent variables:

$$R_t = R_{max} + b_{1\ t-1} + b_2\hat{N}_{t-2} \ldots b_d\hat{N}_{t-d} + cV_1 + dV_2 \ldots + e_t \ ,$$

where e_t is a random variable with mean 0 and standard deviation of s, $b_{1 \ldots d}$ are the regression coefficients for population size at different time lags, and ($V_{1,2} \ldots$) are several independent variables and their slope coefficients ($c, d \ldots$). This expansion of the basic discrete time stochastic logistic model creates a log-linear autoregressive model that combines higher-order time lags (Turchin 1990, Bjornstad et al. 1995, Royama 1996) with covariates (Rothery et al. 1997, Forchhammer et al. 1998, Dennis and Otten 2000). Including additional independent covariates in the logistic model causes estimates of carrying capacity no longer to be independent of fluctuations in the covariates. The outcome of this dependency is the realistic effect that variability in carrying capacity is related to environmental conditions. Hence, density dependence must be viewed in the environmental context. Dennis and Otten (2000) showed that detection of density dependence was contingent on adding a covariate (rainfall) in their population viability analysis of San Joaquin kit foxes. I took a similar approach with a multiple regression model that included weather, predators and harvest as potential factors affecting pronghorn population dynamics in Oregon (Kohlmann 1998). In this model, a winter weather severity index and density explained the largest amounts of variation in per capita population growth rates, whereas predators and human harvests were largely insignificant. Danvir (2000) used a stepwise multiple linear regression model to examine environmental and density-dependent factors affecting pronghorn population growth in Utah. He suggested that density, winter severity and interspecific competition with cattle were significant predictors of population growth and fawn survival. The established linear relationships between fawn ratios and density have been used as management guidelines for the setting of doe harvests in Utah pronghorn (R. Danvir personal communication:1999).

Multispecies models can be implemented easily using a logistic model, where the growth rate of one species is dependent on the abundance of another. Examples for such systems include plant/herbivore interactions, predator/prey systems or competing species. Logistic predator/prey models often are known as Lotka-Volterra models, named after two prominent mathematicians who created them. In classic two-species models, the interspecific effect is dependent only on the density of the exploiting species, without consideration of an intraspecific (density-dependent) effect. The effect of a large predator population could be quite different, depending on the size of the prey population, and therefore, ecologists increasingly favor ratio dependence (i.e., the effect of the exploiting species is dependent on the ratio of predators and prey) (Berryman 1994a). In addition, the availability of alternate food sources can be included as an additional covariate in the ratio-dependent model. For example, a general ratio-dependent model for pronghorn (P) exploited by coyotes (C) is:

$$R_t = R_{max} + b_t N_p - b^2 \frac{N_p}{N_c} + e \ ,$$

where b_2 represents the effect of the ratio of predator and prey (Berryman 1994b). Note the similarity to the logistic models with covariates above. This model easily can be written in terms of carrying capacity for pronghorn, and we can add a nonlinear density-dependence coefficient Q and an alternative food source for the coyote (i.e., jackrabbits, J):

$$R_p = A_p \left[1 - \left(\frac{N_p}{K_p}\right)^{\theta p}\right] - \frac{a_{PC} N_C}{N_p + N_j} \ ,$$

A host of variables besides population density can influence the rate of pronghorn population change. Above, a group of at least 16 pronghorn feed on a haystack during a Montana winter. Snow depth and density, exacerbated by lack of other forage, force the pronghorn to feed on the hay, which, if prolonged, can have a deleterious effect on the animals. Even so, the availability of the forage, regardless of its nutritional value, assumes that the pronghorn can compete for the hay with other stressed ungulates. Photo by Steve Bayless; courtesy of the Montana Department of Fish, Wildlife and Parks, Helena.

where a_{PC} is the interspecific effects between pronghorn and coyotes, K_P is the carrying capacity of pronghorn and N_J is the density of jackrabbits (Berryman 1994b). Modeling of two-species time-series interactions can be accomplished with the Population Analysis Software (PAS) package (Berryman 1994b). From these examples, it becomes evident that the simple logistic model can be expanded to describe very complex population dynamics.

Sustainable Yield and Harvest. Most pronghorn populations are subject to some human harvest, hence wildlife managers often are interested in the sustainable harvest, one that a population can tolerate. The annual yield of a population is defined as: $Y_t = N_{t+1} - N_t$. The concept of a sustainable yield requires the acceptance of some form of density-dependent population and that the population must be reduced below K to generate any yield (Caughley 1976). For reasons detailed above, estimation of K cannot be exact because of the annual variations in growth rates due to environmental covariates. The logistic model and the concept of density dependence are the fundamental basis for harvesting and exploitation of populations because they assume that, at any population level below K, more recruits are produced than would be needed to maintain the population at that level. Ricker (1954) developed this concept for exploitation of fish stocks, and his well-known stock recruitment curves detail the relationship between population size and surplus harvest that can be sustained.

The optimum sustainable population (OSP) has been defined by the National Marine Fisheries Service (NMFS) as a range of population abundance levels bounded on the lower end by the population size that produces the greatest net increase per year and bounded at the upper end by the abundance levels limited by the carrying capacity of the species' environment (Barlow et al. 1995). The population size at which the maximum sustained yield is reached is termed the maximum net productivity level (MNPL). For the linear growth function (i.e., the standard logistic with $Q = 1$), the MNPL is found at exactly $K/2$, but with increasing downward curvature of the growth function (increasing Q), MNPL is shifted to values greater than $K/2$ (Figure 150). Fowler (1981) noted that the shape of the yield curve can depend on the sex class of animals harvested. For many large mammal populations, in North America, harvest is limited to adult males (e.g., buck-only regulations). The underlying assumption is that a skewed sex ratio of several females per adult male has no detrimental effect on reproductive rates. In fact, it may be possible that a population with a linear density-dependent

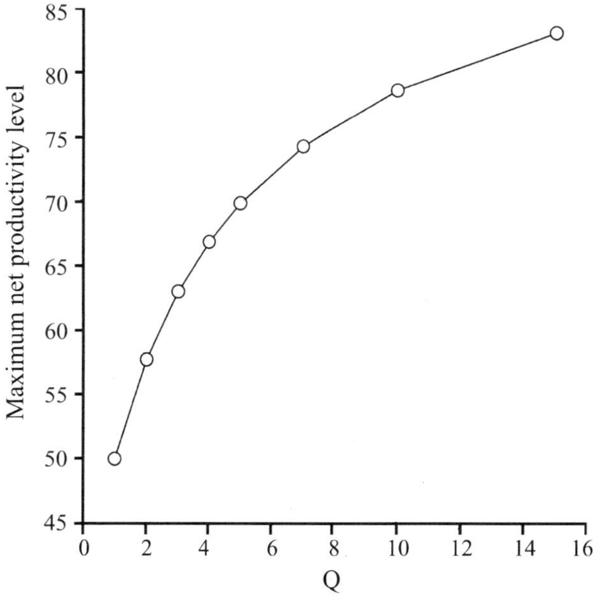

FIGURE 150. *Relationship between the curvature coefficient Q and the maximum net productivity level expressed as a percentage of K.*

feedback and a harvest bias toward males can reach a maximum yield of males at population levels nearer to carrying capacity (Fowler 1981), when a smaller harvest of females keeps the population at its MNPL. However, in most large mammals, density-dependent changes in population growth can be expected to occur at levels close to carrying capacity (Fowler 1981), hence limiting harvest to males often is not sufficient to reduce population size toward MNPL.

In most states and provinces, exploitation levels of pronghorn populations are very conservative and limited to the removal of males only. However, in areas of high pronghorn abundance, managers have targeted females to reduce populations. In the Medicine Bow pronghorn population of east-central Wyoming, heavy female harvest caused population growth rates to decline, although a slight stimulating effect may be noted at lower harvest intensities (Figure 151). In a simulation of a lightly harvested (bucks only) pronghorn population near carrying capacity in south-central Oregon, I found that additional harvest could be sustained without risk of overharvest if implemented in a rule-based fashion (Kohlmann 1998). Furthermore, a greater exploitation level appeared to reduce the effects of environmental fluctuations and stabilize the annual harvest, despite pronounced fluctuations of the environment. Rule-based harvest regimes require that harvests are implemented only when the population exceeds a predetermined population objective. Thus, any possible overexploitation will be corrected immediately by ceasing any harvest until the population has recovered.

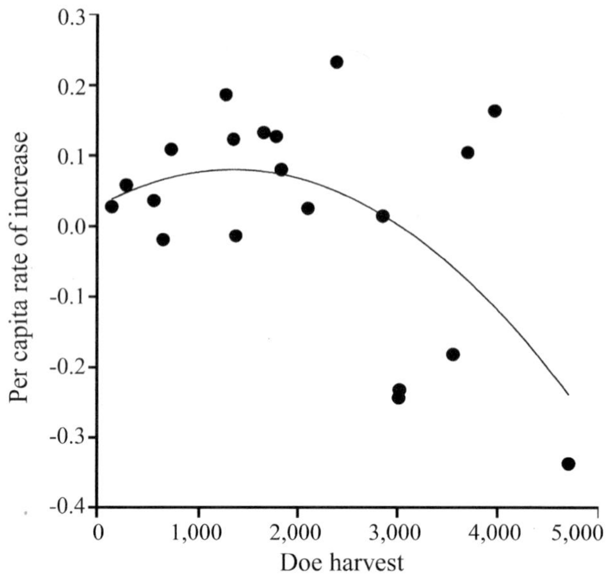

FIGURE 151. *Relationship between annual doe harvest and population growth rate (estimated from a population model) for a heavily harvested pronghorn population in Wyoming's Medicine Bow Data Analysis Unit (data courtesy of the Wyoming Department of Fish and Game, Cheyenne). Black circles = annual estimates of the rate of per capita change of total population size; line = nonlinear regression estimates.*

Statistical Detection of Density Dependence. The statistical detection of density dependence is important for the testing of ecological theory and for management purposes (Den Boer 1991, Dennis and Taper 1994). Once the parameters of a trend model have been estimated, the manager should determine whether there is enough statistical evidence to warrant the consideration of density dependence as a major factor affecting population growth. If data are too variable or if the population has been growing vigorously for an extended period, a negative feedback of high population densities on population change may not be detectable.

Estimating the significance of density dependence in time series of population counts can be accomplished by several methods. The dynamic response method, or DRM (Demaster et al. 1982, Eberhardt 1992), is based on the assumption that a second-order polynomial can be fitted to a time series of population estimates ($\hat{N}_t$) if the population experiences density-dependent feedback in population growth. Consequently, the population is assumed to be above the point of MNPL (i.e., the inflection point of the logistic curve) if the second-order coefficient (i.e., the curvature) is significant and negative. Eberhardt (1992) developed a two-step test for this hypothesis. First, a linear regression is fitted to the log-transformed trend data. Log transformation tends to normalize skewed distributions and reduce the sigmoidal shape to one that is concave downward throughout. If the slope estimate of this regression is zero or negative, it is assumed that the population is above MNPL. Second, if the slope is significant and positive (increasing trend), then a second-order polynomial is fitted. The significance of the second-order coefficient indicates whether the population is below MNPL (positive coefficient) or near MNPL (insignificant coefficient). If the second-order coefficient is significant and negative, the population is assumed to be above MNPL. To test for the significance of the second-order term, Eberhardt (1992) suggested an F-test to compare the linear model with the polynomial.

Detection of density dependence in time series of population estimates is not trivial due to an induced correlation of population size and change in population (Eberhardt 1970, Maelzer 1970, White and Bartmann 1998). To account for that correlation, Dennis and Taper (1994) developed a parametric bootstrap likelihood ratio (PBLR) test of density dependence in time series of population estimates. The method uses linear regression to derive maximum likelihood estimates for parameters of the logistic model, then uses a data subsampling routine (bootstrap) to estimate parameter confidence intervals and significance level of the test that the slope parameter is zero by comparing the fitted logistic with an exponential growth model. This method has been used to detect density dependence in pronghorn time series from Oregon (Kohlmann 1998). However, data precision should be reasonably high to avoid inflated type I errors (Shenk et al. 1998).

The time-series analysis mentioned above has been criticized, because in real populations, density does not directly affect abundance, rather the birth and death processes in the population. Many parameters reflecting these processes (e.g., survival, fawn production) are measured as rates, which precludes the use of linear regression due to the underlying non-normal distribution of rates. Logistic regression has been applied successfully to detect density dependence in red deer calf survival (Clutton-Brock et al. 1987), gall wasp mortality (Hails and Crawley 1992) and roe deer survival rates (Lebreton et al. 1992). Shenk et al. (1998) used logistic regression to detect density dependence from temporal trends in overwinter fawn survival of mule deer in Colorado, and they developed simulations to address the relationship between sampling variation and type I errors. If estimates of survival are available in addition to precise population size estimates, logistic regression is perhaps the most efficient method for detecting density dependence (Shenk et al. 1998).

Population Models with Age Structure

POP-2. POP-2 originated from one of the first commercially available population modeling software packages, model ONEPOP, written for mainframe computers (Gross et al.1973). Since the early beginnings, POP-2 has fledged into a management tool that has found widespread acceptance in management agencies.

POP-2 tracks the changes in an age-structured population, much akin to how an accountant would track the flow of money through various company accounts; that is why models, such as POP-2, commonly are referred to as bookkeeping models. POP-2 does not make any assumptions about population processes beyond the supporting data, hence it is theory-free. The model structure of POP-2 is based on simple addition and subtraction of animals at various stages throughout the biological year (Figure 152). A typical modeling session with POP-2 begins by specifying the biology of the species (longevity, sex and age structure, population size, and reproductive performance of different age categories). The simulation of the dynamics of the biological year starts with the population size immediately before the birth of the new crop

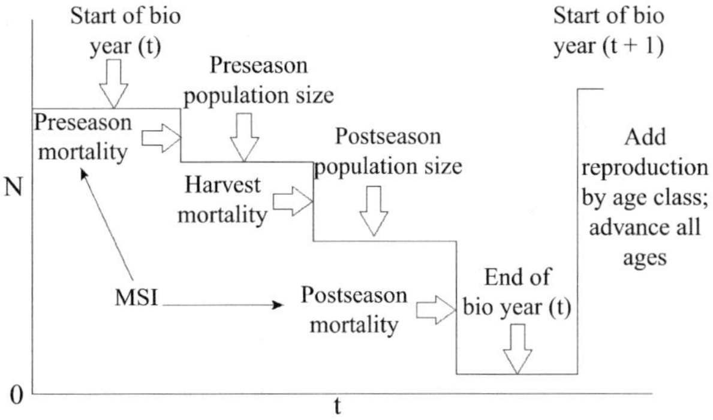

FIGURE 152. *Model flow and important parameters of model POP-2. MSI = mortality severity index.*

of fawns, which are added to the population based on reproductive rates specified for up to three reproductive classes (e.g., yearling females, adult females and senescent females). Once the starting population has been calculated, POP-2 represents the changes in the population size during the biological year as a cascading sequence of subtractions (mortality) of animals from each sex and age class. During the preseason, each age class, including the newborn fawns, experiences a user-specified standard mortality rate, which can be adjusted annually by a mortality severity index (MSI). Once the preseason mortality has occurred, the population enters the harvest season, during which harvest mortality is removed. Harvest effort multipliers can adjust for hunter selectivity by increasing the probability of certain age classes being harvested. If all harvest effort values are set to 1.0, each age class is harvested according to its proportion in the population. After the harvest is removed from the population, it enters a second natural mortality phase, the postseason mortality, which again can be modified from year to year via an MSI. At the end of the biological year, the remaining animals are advanced into the next age class and the cycle begins anew.

Although the algorithms and the model logic are extremely simple, modeling with POP-2 can be complicated. In most cases, managers will find that there are more functions in the model than there are data to support them. POP-2 is a classic example of a model with high realism and detail and often poor data support. In the absence of good field data, managers often are forced to make educated guesses about parameters, which may reduce model accuracy. More important, however, is the risk of microadjusting to make the model fit observed data exactly. This technique may result in spectacular fit of observed and simulated values, which are the result of the modeler's ability to adjust, or tweak, model input to observe the desired output. Such modeling is akin to testing a hypothesis with the data that created it or, more

The models discussed so far in this chapter did not consider the fact that a wildlife population consists of individuals, thus ignoring some of the important individual differences. Clearly, this is a very simplistic view of a population, especially large vertebrate populations, wherein individuals can live many years, so multiple age classes are present. Adding an age structure to a population model can make the model more realistic, because in many vertebrate populations survival and fecundity are age-specific, that is, the per capita birth and death rates change with the age of the cohort. If a population is composed of an uneven blend of age classes varying in their reproductive performance, population models that ignore that variation would represent the actual dynamics poorly. One way to keep track of all the age classes in a model is to represent each class as a separate entity. This, in fact, is the approach used by one of the most successful and widely used models, POP-2 (Bartholow 1986). Photo by F. H. Rose; courtesy of the U.S. National Archives, Washington, D.C.

important, to forcing the model output. Tweaking is accomplished easily in POP-2, which greatly diminishes its usefulness. Like most population models, POP-2 is much better at exhibiting inconsistencies in data and thinking than in providing predictive power. POP-2 is a deterministic model (i.e., each combination of input variables produces a specific output, regardless of how often the model is run). Hence, data precision and probability of a given model output are irrelevant concepts in POP-2, yet they are of critical importance for managers. POP-2 is best applied to recreate likely population trajectories, based on the observed harvest the population has sustained, and obtaining some basic understanding of natural mortality. Therefore, POP-2 is used widely by managers of game species, where the annual harvest data are perhaps the most important check of model realism. However, because POP-2 does not incorporate density dependence or any stochastic variation in input parameters, the model can only account for the minimally necessary population size. Therefore, modeling with POP-2 can only show "what must have been there," given the observed levels of harvest. Any inferences beyond that limited scope are not justified. Thus, the predictive power of models, such as POP-2, is low, and management agencies should not rely on model output of simple bookkeeping models as a basis for decisions. However, this does not detract from the incalculable value of POP-2 as a tool for evaluating and collating data, learning and thinking about population processes and testing scenarios (what-if games). Herein lies the greatest value of any simulation model.

Matrix Models. Another, mathematically complex, age-structured model expresses the future population size $N_x(t+1)$ as a function of each previous cohort size $N_x(t)$ and the age-specific survival (S_x) and fecundity (F_x) rates. Thus, the per capita rate of change is now a composite of reproduction and survival; immigration and emigration commonly are ignored, but could be included as additional parameters. The most convenient mathematical way to express this model is through matrix notation. Matrix models are especially appealing to population biologists because they can represent complex life cycles in an elegant notation. The matrix approach is commonly referred to as a Leslie matrix, named after the British biologist, P.

H. Leslie (1945, 1948), who pioneered its use. A simple example of a Leslie matrix for four age classes is:

$$L = \begin{pmatrix} F_0 & F_1 & F_3 & F_4 \\ S_0 & 0 & 0 & 0 \\ 0 & S_1 & 0 & 0 \\ 0 & 0 & S_2 & 0 \end{pmatrix}$$

Annual updates of the abundance vector result from a multiplication with the Leslie matrix (commonly, only the female sector of the population is modeled):

$$N(t+1) = LN(t) \, .$$

It is important to consider that cohort width can be greater than 1 year, but cohort width must be the same for all cohorts. The time step of the model is identical with the cohort length of the matrix. One characteristic of a fixed Leslie matrix is that it tends to converge toward a stable (unchanging) age structure. At this point, the population structure remains the same, regardless of the population size, and multiplication of the age structure by the Leslie matrix is equivalent to multiplying it with a number:

$$N(t+1) = LN(t) = \lambda N(t) \, .$$

The parameter λ (the eigenvector of the matrix) is equivalent to the rate of increase of the population, and the exponential growth of the population can be expressed as

$$N(t+1) = N_0 \lambda^t \, .$$

Another interesting feature of the Leslie matrix is that it allows the estimation of the reproductive value of a cohort. Fisher (1930) defined the reproductive value of an individual as its relative worth in terms of its offspring production. The reproductive value has an important utility for management because it allows the formulation of conservation policies and reintroduction strategies. Gage (1995) details how to estimate the net maternity function (i.e., the age-specific reproductive value) from life table data.

The Leslie matrix model has been modified to incorporate stochastic fluctuations of the environment, an open-ended life span, immigration and density dependence (Caswell 2000). The latter can be incorporated into the Leslie matrix by making fecundity or survival not only a function of age, but also of population size. In most cases, wildlife biologists will have an idea how fecundity and population size are related; this relationship can be derived easily by plotting the number of fawns per doe over the total number of does observed. The reproductive effort of adults in a population can be expressed as

$$E = f \sum_{x-0} F_x N_x \, ,$$

where E is the effort (in terms of number of recruits produced), F_x is the age-specific fecundities, and f is the sex ratio (proportion of females in the population). There are several commonly used forms of density-dependent functions that easily can be incorporated in the age-structured model. The first kind of density dependence is the law of diminishing returns, also known as the Beverton-Holt function (Beverton and Holt 1957), which expresses the number of zero-age recruits Z as a function of the parental effort (E) available for reproduction

$$Z = \frac{1}{(\rho + \frac{k}{E})} \, ,$$

where 1 divided by ρ is the maximally possible number of recruits produced, and k is a dimensionless constant for the steepness of the slope. The second form is the logistic function, where

$$Z = E\left[1 + r\left(\frac{K-E}{K}\right)\right] .$$

However, in models with age structure, it usually is better to use the Ricker function (Ricker 1954):

$$Z = \alpha E \exp(-\beta E) \, ,$$

where both α and β are nonnegative parameters. The slope of the curve at the origin is represented by α, and β is a measure of the nonlinearity of the density-dependent feedback. The form of negative feedback used should be governed primarily by the existence of biological evidence, which likely is rare for large mammals. In pronghorn, no data on the relationship between recruitment and density have been published. Perhaps, the question has only academic merit. However, implementing a different form of density dependence can radically alter model behavior. Therefore, different scenarios should be evaluated, and the sensitivity of a given model to changes in the underlying assumptions (e.g., form of density dependence) should be tested.

Environmental fluctuations can be modeled through multivariate stationary time series (Tuljapurkar 1989), which has sufficient biological detail to describe the dynamics of vertebrate populations, especially those of endangered species (Dennis et al. 1991). Multispecies matrix models rarely are constructed in wildlife management, due to their extensive data requirements, but they occasionally find application in the field of entomology and pest management (Gutierrez 1996).

Perhaps the most significant use of Leslie matrices has been the combination of population modeling with the methods of risk assessment. Ecological risk assessment evaluates the probability of certain events occurring in a population under a given set of parameters and their variations. One commercially available software package, RAMAS AGE (Applied Biomathematics 1994), allows the user to build age-structured population models by specifying the following inputs: sex ratio, variation in fecundity, variation in adult and juvenile survival, variation in migration, vectors for fecundity, natural survival, migration and initial abundance, functions describing environmental and demographic stochasticity, migration distribution, fecundity distribution, juvenile survival distribution, adult survival distribution, and correlations among the vital rate vectors. RAMAS AGE outputs include: total abundance and abundance in each age group over time, the variation and distribution of population size over time, the risk of population decline or extinction over the simulation time period, the probability of achieving a certain population level, the simulation time required before the population crosses an abundance threshold, and a variety of demographic statistics, including rates of increase and net reproductive values, generation time, life expectancy at birth, the vector of reproductive value, and the stable age distribution. RAMAS AGE allows isolating critical demographic factors through sensitivity analysis, so is a powerful tool for population viability analysis, or PVA. Another software package for discrete-time age-structured models through matrix notation is the unified life models, or ULM (Legendre and Clobert 1995). The ULM software can accommodate linear as well as nonlinear density dependence and has no limit for the number of determinants (sexes, age classes, habitat types, etc.) of population structure. The software allows extensive simulations of population dynamics under stochastic conditions.

Stage-frequency Analysis. Creation of an age-structured model for species where age is not readily determined invariably involves much guesswork concerning the structure of the population. In pronghorn, fawns, yearlings and adults can be distinguished readily from afar during aerial or vehicle surveys. But determining the structure of the adult population often is impossible without collecting costly tooth samples. For such cases, several age classes can be combined into stages, and a stage-frequency analysis can be conducted. Stage-frequency-based models assume that all individuals within a given stage have identical survival and reproduction rates, and each individual enters one stage after another. Counting all animals within a stage results in a stage-frequency table. Because all pronghorn are born approximately at the same time during the year, we are only concerned with single-cohort models here; for a detailed review of stage-structured models, see Manly (1990b). Key-factor analysis (Varley et al. 1973, Manly 1990b) often is used in conjunction with age- or stage-structured models, which can determine which cohort is the most vulnerable to certain factors and which management action can provide the greatest benefit for achieving population goals. For example, key-factor analysis could be used in conjunction with a stage-structured model of pronghorn to evaluate whether neonatal mortality due to predation has a greater effect than winter mortality of fawns, yearlings or adults. Likewise, effects of density dependence and environmental variation on population growth rates could be addressed with key-factor analysis (Brown et al. 1993, Sibley and Smith 1998), but it usually is considered data intensive (Berryman 1994a).

Stage-structured population simulations can be accomplished with commercially available software packages. One of the most frequently used models is RAMAS-Stage, by Applied Biomathematics, Inc. RAMAS-Stage is useful for modeling species with complex life histories or where stage (rather than age) determines the demographic characteristics of an individual. RAMAS-Stage models can accommodate data on environmental factors, such as temperature or rainfall, which are specified in terms of their means, variances and temporal autocorrelations. RAMAS-Stage also facilitates risk assessment by estimating extinction probabilities or the chance of a population decline. Another comprehensive modeling and analysis package is the POPSYS Series 1F Stage Frequency Analysis software (Manly 1990a), which employs advanced analytical techniques (Manly 1990b).

Individual-based Models

In the preceding portions of this chapter, the level of complexity with which a pronghorn population is viewed and modeled has incrementally increased. Starting with a model of population size in time, we have added time lags and covariates to incorporate large-scale phenomena. By adding age and sex structure, flexibility is gained in designing harvest strategies (e.g., buck-only) and observed the structural dynamics of the population. Any further increase in detail must invariably entail the level of the individual. Such models are termed individual-based models (IBM); they keep track of each individual in a population. Individual-based modeling has been suggested as a unifying concept in ecology (Huston et al. 1988), by allowing greater detail at each level of the organizational hierarchy without losing generality in predictions (DeAngelis and Gross 1992). The IBMs incorporate data measured from individuals, rather than large-scale averages from population samples, and can be used for a wide variety of ecological problems (DeAngelis and Rose 1992).

When data are available on the traits or behavior of interactions among individuals, IBMs can be employed. One of the earliest behavioral IBMs tracked the individual growth of juvenile fish in response to the daily intake of plankton (Beyer and Laurence 1980). When the average daily intake rate of all fish was lower than their average energy requirement, a traditional population model would have predicted the death of all fish. However, the IBM showed quite realistically that some fortunate individuals survived if they started out with a higher than average intake rate, or a lucky streak. A behavioral model of moose foraging behavior (Roese et al. 1991) demonstrated that structure and variability of resources in the environment are essential deteminants of foraging performance. Roese et al. showed that some individual moose may survive conditions that often do not sustain maintenance requirements, similar to the fish model.

A large group of IBMs are designed to track singular individuals through a sequence of events or across time, and observe the population-level dynamics after summing the fates of individuals. Typically, IBMs are flexible and stochastic, but often require data beyond those available from standard wildlife surveys. Because individual-based models are uniquely suited to track genetic change by tracing pedigrees and inbreeding coefficients, these models often are used for small populations and in population viability analyses (Harris et al. 1986). One of the earliest individual-based modeling software programs is the Generalized Animal Population Projection System (GAPPS, Harris et al. 1986). GAPPS was developed to evaluate age structures of grizzly bears under environmental stochasticity. GAPPS is most useful for projecting dynamics of small populations of long-lived species with complex life histories, where the fate of a few individuals may have significance for the entire population. GAPPS has been used for modeling single-species problems (Wotschikowski and Schwab 1994, Mills et al. 1996, Brook et al. 1997), but because it does not incorporate a spatial structure; it is not suited for modeling metapopulation problems.

Perhaps the most widely used, commercially available, individual-based modeling software today is VORTEX (Lacy et al. 1995), a model designed to simulate the extinction probability of small populations of rare or endangered species. It has been used extensively by the Captive Breeding Specialist Group of the International Union of Conservation of Nature and Natural Resources (IUCN) Species Survival Commission. VORTEX simulates the events occurring in a population as sequential events with a probabilistic outcome. Because it is individual-based, each animal's fate in the population is simulated, which can lead to slow run times when large populations are modeled. VORTEX was used by deVos (1996) to simulate the fate of a small population of Sonoran pronghorn. He found that population persistence of Sonoran pronghorn was affected by the frequencies of catastrophic events, female mortality and, to a lesser extent, fawn production. Especially when female fawn mortality rates exceeded 60%, population stability was low, suggesting the existence of mortality threshold. As a result of these simulations, deVos (1996) suggested that monitoring of climatic conditions should be an important aspect of managing this subspecies.

Occasionally, data are available on individual traits, behavior or interactions (e.g., weight, body condition, home range size, movement patterns, dominance hierarchy, dispersal distances, mate choice). For individual-based models, much of the data on the individual traits have been gained through use of radiotelemetry. Photo by Peter T. Bromley.

Spatial Population Models

Spatially explicit modeling is used to represent processes and events that are characteristic for heterogeneous systems, where habitats vary across space and time, and where the effects of critical resources or barriers are site-specific. Pronghorn populations are not evenly distributed across a landscape. They often exist in semiarid desert shrubsteppes that may be separated by fences, highways, rivers, mountains and other inhospitable features (Ockenfels et al. 1994). When the existence of a species depends on the continued exchange of individuals and genes between insular patches, a model that accounts for the spatial interactions between populations becomes essential.

Landscapes are at the core of spatial population models; either a realistic landscape is represented as a Geographic Information System (GIS) data set, or a hypothetical landscape is described by some parameter (e.g., the percentage of the landscape in habitat, the average patch size, patch distance). Usually, a boundary is drawn around a given habitat type, and the average or typical value is used as a descriptor for this habitat type. It is important to note that a spatially explicit population model operates within the context of the spatial data set representing the landscape. Although a landscape of some kind is necessary, the model itself tracks individual organisms (with locational information) or population quantities, such as age structure and population size.

The range of models incorporating space spans from spatially implicit models (where size and location of habitat patches are ignored and only the probability of colonization and extinction of patches are considered) to detailed map-based (spatially realistic) models. Two approaches are frequently used to model spatially structured populations: matrix models and individual-based models.

Spatial Matrix Models. A Leslie matrix model can be generated for each habitat type in an animal population, such that the parameters for a matrix differ among habitat types (Caswell 2000). If the habitat types are linked with a spatial database, the model can be quite realistic and provide important insights into the effects of habitat alteration on population viability. This approach was used to analyze the population viability of endangered spotted owls (Akçakaya 1996). The model combined randomly distributed vital rates (survivals and fecundities) with demographic stochasticity for each of 28 habitat patches, as determined from habitat maps. The model predicted a large variance in the viability of the northern spotted owl, when based on parameter combinations that resulted in best and worst chances for survival. A commercially available software package, RAMAS-GIS, was used to model the dynamics of this system. The software is designed to link data from GIS with a spatially structured population model for population viability analysis and extinction risk assessment. A matrix model is appropriate for populations where different habitats offer distinct, measurable, habitat-specific population parameters (e.g., survival, fecundity). However, when individuals move across a wide variety of habitats, and where the individual's fecundity or survival is not necessarily related to just one habitat type, the matrix approach is not well suited.

Spatial Metapopulation Models. Simulations and empirical observations of animal population dynamics have suggested that survival of the metapopulation in the presence of habitat loss and fragmentation is dependent on recolonization of patches (Fahrig 1991), which in turn is dependent on a high dispersal rate. Important spatial components that may influence this rate are patch size and interpatch distance (Dunning et al. 1992). Spatially explicit models derive population-level dynamics from the coupling of the specific location of individuals and habitat patches with information on habitat-specific demography and movement behavior. Thus, they consider the species/habitat relationship in the context of landscape configuration in time and space (Turner et al. 1995). Possingham and Davies (1995) developed a spatial population model (ALEX) to conduct a population viability analysis of the endangered Australian Leadbeater's possum. ALEX assumes that the species occurs in a few isolated patches of habitat. The model combines data and user-specified probabilities of five different categories: species data (e.g., birth and death rates, home range size, population responses to habitat changes), catastrophic data (frequency and effects of less than four types of catastrophic habitat alterations), movement data (likelihood of movement between patches), patch data (position, size, biomass and quality of habitat patches) and sensitivity analysis data (i.e., the range of values between which a parameter of interest is allowed to vary to examine the probability of extinction as a function of that parameter). Although ALEX is spatially implicit rather than explicit (i.e., the exact location of individuals only matters in regard to the subpopulation the animals exists in), it has been used successfully to evaluate population behavior resulting from catastrophic wildfire events and logging operations (Lindenmayer and Possingham 1995). One major result of the study was that a permanent reserve of key forest patches was one of the most useful conservation strategies, and aspects of reserve size, position and number of patches were compared and ranked according to the relative effectiveness of different conservation strategies rather than precise values for the probability of extinction. Models such as ALEX also show great promise

Pronghorn movements are hindered by range fences (Lee at al. 1998), and intensive fencing may reduce colonization rate of habitat patches. In a habitat evaluation model for pronghorn in Arizona, Ockenfels et al. (1996) included fencing, roads and human disturbance as potential factors affecting spatial usage of habitat patches. However, the explicit location of such barriers in the context of the surrounding habitat matrix was not considered. Recently, migration corridors have been identified as important management considerations (Simberloff et al. 1992, Beier 1996), but application of this concept in pronghorn management has not received much attention. Photo by Terri Rossi; courtesy of the U.S. Bureau of Land Management, Washington, D.C.

for the development of conservation strategies for endangered pronghorn in small, isolated populations.

Spatially Explicit Individual-based Models. An individual-based, GIS-linked population model specifically designed for natural resource managers dealing with ungulate populations was developed by Risenhoover et al. (1997). The deer management simulator (DMS) uses data on population demography and distribution among various habitat types of a realistic landscape map to simulate the dynamics, distribution and management of an ungulate population. The DMS is designed to evaluate the spatial and numerical responses of populations to different management actions. The model is structured around a basic movement algorithm that determines the movement distance on a daily time step as a function of habitat quality of the map pixel encountered by the individual during the simulation. In the DMS, the movement distance of individual animals is reduced when they encounter a high-quality patch (mimicking the increased foraging time they spend in good habitat patches), and they move longer distances when they are in poor patches, thereby incurring a higher likelihood of leaving the poor patch. This model has proven to be a powerful tool for the assessment of animal distribution in response to habitat and landscape characteristics (Risenhoover et al. 1997).

Wildlife managers increasingly realize that animal distributions are not homogenous and most wildlife species occur in a patchwork of habitats with different resource qualities. The inclusion of space as a model component complicates modeling, because the model must simulate not only the heterogeneity of habitat distributions, but also the distribution of animals across these habitats and the movements of animals among habitats. Photo by Ed Bry; courtesy of the North Dakota Department of Game and Fish, Bismarck.

An elaborate IBM of spatially explicit interactions between a large mammalian predator and its herbivore prey (SIMPDEL; i.e., spatially explicit individual-based simulation model of Florida panther and white-tailed deer in the Everglades and Big Cypress landscapes) has been developed by Abbott et al. (1995). This model analyzes and predicts the effects of hydrological changes on population of white-tailed deer and Florida panther. It tracks growth, movement, foraging and reproduction of individual deer and panthers in the context of landscape and spatial hydrological dynamics and the interactions between individual predators and their prey (Abbott et al. 1995).

THE ROLE OF POPULATION MODELS IN ADAPTIVE MANAGEMENT

In recent years, adaptive resource management (ARM) has been embraced rapidly by the wildlife profession as an effective way of dealing with uncertainty and lack of knowledge about the effects of management activities on natural systems (Walters 1986). The ARM involves a systematic process of using knowledge generated by management actions to increase the level of understanding about the biological system in question and to aid in future decision making. In a sense, ARM is a series of management actions designed to provide reliable knowledge, similar to scientific experimentation. Rather than relying on a priori knowledge or intuition, understanding of the system is one of the outcomes of ARM, and learning becomes an integral part of the management process. As knowledge is gathered, uncertainty about the ecological system is reduced and management decisions are based on a more solid foundation.

Hunting license and tag allocations often are set in response to perceived or measured changes in population abundance or structure. For example, pronghorn managers commonly reduce tag numbers when population parameters fall below certain management objectives. This stepwise adjustment of harvest quotas may allow managers to recognize when actions do not provide the desired result and may serve as a basis for some remedial action. However, the knowledge generated from such unplanned experiments is limited because variables in natural systems usually are confounded (i.e., correlated). Therefore, the system's response cannot be attributed to the management alone. Although the management process may have provided information about ineffectual actions, it is more difficult to pinpoint the reason for the ineffectiveness and the outcome of alternatives. To achieve that level of understanding and to prevent repeating mistakes, the management process must formulate hypotheses or models, similar to those usually employed by scientific experimentation. Predictions of the system's response under different scenarios are, therefore, an integral part of gaining reliable knowledge. However, simply formulating a conceptual framework of competing alternative hypotheses will not be effective unless a monitoring system is in place to measure the system's response precisely. The formulation of predictions and the collection of precise data to evaluate the effectiveness of the management action are equally important.

Probably the most well-known example for the implementation of scientific models into the management process is waterfowl management in North America. The ARM for waterfowl consists of: (1) an array of possible hunting regulations; (2) a set of models representing pertinent hypotheses about waterfowl population responses to management (density dependence) and environmental changes (availability of nesting habitat); (3) a measure of uncertainty for each hypothesis; and (4) a mathematical function to evaluate the regulatory options. In essence, harvest regulations are set each year, according to population levels determined from breeding surveys. Then, the response of the population is monitored, and each hypothesis about potential harvest impacts is compared with observed data. The model accounting for the best fit with the data is given a weighted score. Repeating this process over several years, the most appropriate model of population regulation will reveal itself through the accumulation of weighted scores (Williams and Johnson 1995). Adaptive structuring of harvest regulations has been installed for antlerless elk in Idaho (Gratson et al. 1993). In this case, hypotheses pertaining to density dependence in population regulation were developed and tested by modifying harvest and observing the population response to density reductions. For pronghorn in the Great Basin Desert of Utah, R. Danvir (personal communication:2000) developed an adaptive management strategy. From a long-term data set, he derived a linear model of fawn production as a function of doe density in late winter. Using a management objective of 50 fawns per 100 does, the necessary doe harvest is derived as the difference between the current population size and the population density that produces the required fawn ratio (R. Danvir personal communication:2000). For desert ungulate populations in Oregon, I suggested a rule-based harvest strategy, whereby harvests are implemented only in years when the population counts exceed a population objective (Kohlmann 1998).

THE PROCESS OF BUILDING A MODEL

As the preceding sections have shown, there are many types of models available, and the modeler must pick the right approach to be successful. Modeling is both art and science (Starfield and Bleloch 1986). Without a scientific view of the system one intends to model, a mathematician is helpless. At the same time, a scientist who becomes entangled in details will not succeed until the discipline to reduce the variability and detail of the real world to its parsimonious mathematical essence is achieved. It often is difficult to assess if a given level of detail is critical for the model to behave properly. In these cases, a sensitivity analysis can be conducted to determine whether the model output is dependent on changes in a given parameter. When constructing a model, it is important to consider the objectives, structure and specifications of the model. Furthermore, the modeler must examine the model for accuracy and test its behavior using a wide array of data and assumptions. Only after extensive testing and evaluation should a model be used to predict system responses or serve as the basis for management decisions.

Model Objectives

The objectives, intended purpose and specific applications of a model must be stated explicitly. There are many cases where models were built for a specific purpose and later falsely used for different and, perhaps, unrelated problems. This can be avoided if the model's objective is stated clearly and explicitly. The objectives should outline the purpose (e.g., to determine sustainable harvest levels).

Model Structure

Before delving into the details of building a model, it is helpful to conceptualize how the model will work, including identification of the key elements in the model and the important relationships and ecological mechanisms. The identification of subsystems and submodels is critical for modeling complex systems, where the outcome of one process (e.g., rainfall determining plant growth) is a critical input in the model (e.g., body condition of herbivores). Such a model has a subsystem structure—in this case, a model for the relationship between plant production and rainfall—and a model of herbivore/plant interaction. Each submodel must have its own detailed objectives, purpose and specifications. Tying subsystems together can be a daunting task because the outcome of one higher-level system (e.g., herbivore foraging) can affect variables in lower levels (e.g., plant growth). A logical and sequential looping structure with iterative updates of submodel variables is critical in such cases.

Model Specifications

Once a conceptual model has been conceived, it usually is a good idea to diagram a representation of the conceptual model. Such diagrams can be either flowcharts or simple sketches that show the relationships among components. Specifying the elements of a model is similar to setting the itinerary for a trip. One must decide on the vehicle, the travel speed, accommodations, equipment, and the time and costs involved. Many of these decisions entail trade-offs: generality of application, development time and cost, data requirements and reliability, and so on. Model specifications include the operational characteristics, mechanisms, parameters and variables. When specifying operational characteristics, one must decide whether to build an analytical or a simulation model and whether the model should be deterministic or stochastic. Next, one must decide on the variables and parameters that are to be used in the equations. Variables usually are changing throughout the model run (e.g., time, fawn ratio, adult mortality, etc.), whereas parameters are constants (e.g., growth rates of fawns, gestation length, etc.). At the heart of the modeling process is the construction of relationships among variables (i.e., functions). Most models contain mathematical equations that connect two or more independent variables and produce output in terms of a dependent variable. A model cannot contain a mechanism (function or relationship) about which nothing is known. However, one could speculate about possible relationships, then use sensitivity analysis to evaluate the effects of changing functions on the final outcome of the model.

Model Construction

When the essential details of the model have been identified, the modeler must decide the implementation of the model. With the availability of commercially produced modeling software, there is a high likelihood that a model already exists that meets most of the requirements of the specific problem one is interested in. Specific software packages often are expensive and may impose limits on the modeler's freedom to implement specific relationships. In many cases, however, modeling software packages allow more input data than are available. This may lead to the inclusion of questionable data or speculation on input variables for which no data are available. Alternatively, the modeler can decide to build his or her own model using a variety of methods. One of the most efficient modeling platforms is a spreadsheet. Aside from individual-based problems, spreadsheets can accommodate virtually all commonly encountered modeling problems. The spreadsheet's organization of rows and columns makes sequential calculations easy and efficient, but complicated models can run quite slowly and are difficult to troubleshoot. At present, the spreadsheet is perhaps the most widely used simulation tool, and many applications exist in wildlife population ecology. For example, Kohlmann (1999) used a spreadsheet to model the distribution of condition indices in a hypothetical herd of elk and simulated the effects of declining condition on fetal sex ratios. Likewise, a spreadsheet was used to conduct a parametric bootstrap likelihood ratio test to test for density dependence in an Oregon pronghorn and mule deer population, then a rule-based harvest strategy was simulated (Kohlmann 1998). Lubow et al. (1996) developed a spreadsheet simulation model for evaluation of a linked sex harvest strategy for cervid populations. Of course, the modeler can also write the model code in one of the computer languages, such as VisualBasic™, Pascal, Fortran or C. However, knowledge of a high-level language is required, especially when complex spatially explicit models are to be built. Languages that allow a graphic user interface (GUI), such as Windows™, are suited especially for spatial models that require extensive mapping data and GIS layers. Object-oriented languages, such as C++, are suited uniquely for individual-based models and appear to be the language of choice for advanced simulations of spatial problems.

Testing, Calibration and Validation

Once the model has been implemented in computer code, the modeler enters the phase of testing and calibration. Testing involves examining the model for internal integrity: ascertaining that calculations are correct and that age classes are advanced correctly at the end of the year, etc. Also, the modeler should test whether the output contains all the required results. Calibration of a model is the adjustment of parameters to reflect conditions of the system to be simulated. For example, an age-structured model may have a standard age structure built in, but unless the initial age structure of the model matches that of the population to be modeled, erroneous output may result. Climatic parameters, mortality rates and habitat-specific information are other examples for parameters requiring calibration.

Evaluation of the model's predictions is referred to as validation and usually involves a comparison of independently collected empirical data and model output. Rykiel (1996) suggests the following terminology for use in evaluating model performance: verification is a demonstration that the model mechanics (mathematical calculations) and logic is correct; validation is the demonstration that the model possesses a satisfactory range of accuracy consistent with the intended application of the model; credibility is a sufficient degree of belief in the validity of a model to justify its use for research and decision making (Rykiel 1996).

Sensitivity Analysis

Evaluation of model performance often involves the comparison of the effects of varying input values (one at a time) on model output (Liu et al. 1995). It is a deliberate exploration of model behavior by changing one input variable or parameter while holding all other conditions constant. In some cases, the shape of a function may also be the subject of sensitivity analysis. For example, one could examine the position of the MNPL resulting from a linear versus a nonlinear growth function in a time-series model. Monte et al. (1996) investigated the use of performance indices (comparison of the level of agreement between the model and independent empirical data) to evaluate the confidence limits of model output. Hakanson (1996) used an algorithm to calculate confidence limits based on a regression of the model output values against independent empirical data.

User input in POP-2 requires estimates of an initial population size, population structure, age- and sex-specific reproduction and mortality, and the annual harvest by age class. Additional modifications allow users to change annual pre- and postseason mortality with a multiplier (MSI) and by adding wounding loss to the harvest season. Realistically, most of these data are not available in management agencies, therefore they must be estimated or taken from the literature. To reduce the complexity of model data requirements and evaluate the effects of different estimates about important parameters, a two-step procedure was developed (Kohlmann 1996). In the first step, the modeler reduces the parameters included in the model to a minimum number, according to available data and what can be estimated with reasonable precision. In this parameter reduction phase, the modeler selects the time steps considered vital to determine population dynamics of the system under investigation. For example, rather than making educated guesses, the modeler eliminates certain parameters from the model. POP-2 allows data input for four time steps during the biological year (Figure 153), yet most pronghorn populations in the western United States usually are surveyed only in late winter, so population and mortality estimates during the year are lacking. Because estimates of birth rates or neonatal mortality usually are not available, using the fawn/doe ratio observed during late winter counts as an index to reproduction at the end of the previous biological year is recommended. This requires that all mortality during the year is set to zero, as this mortality already is contained in the late winter estimates. This simplifies the data required without eliminating salient model elements. Because all mortality (except for a minimal postseason mortality that occurs after surveys) is zero, the MSI also can be eliminated. Thus, the only parameters left in the model are

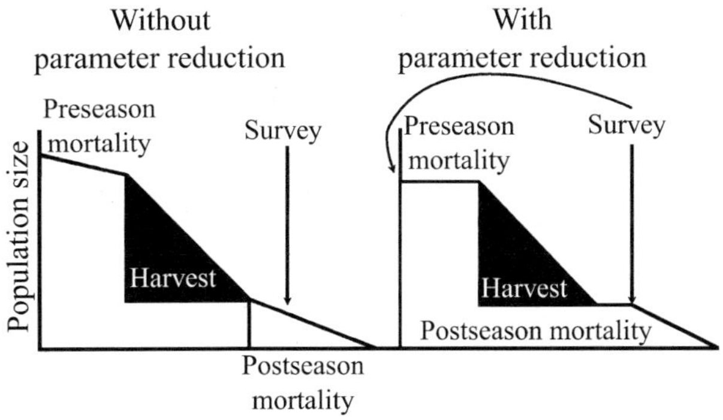

FIGURE 153. *Effect of parameter reduction on complexity of population simulation modeling using POP-2. Curved arrow = by using the survey estimate, the number of unknown parameters (e.g., preseason mortality, postseason mortality before surveys) can be reduced.*

reproduction at the end of biological year (estimated from the fawn/doe ratios at the end of the following winter), the known harvest, an estimated wounding loss and an estimated postseason mortality.

Next, the remaining parameters must be evaluated for their reliability and possible variation. The second step entails a sensitivity analysis (SA) of input data and assumptions to guide modelers through a critical evaluation of the data and assumptions. Sensitivity analysis begins by compiling data for the population in question from various sources, such as research reports, literature, annual survey statistics, and harvest figures. The modeler ranks these data and assumptions in the order of their reliability. Highest ranking are estimates with a measure of precision, next are measurements without precision estimates or those knowingly containing biases, followed by parameters taken from literature and last, educated guesses, which may be based on circumstantial evidence or biological intuition. Field estimates, such as survival and reproduction (age ratios), are naturally measured with a certain measurement error and, in addition, may contain biases. Although it would be possible to run multiple simulations, each with a randomly assigned measurement error (within one standard deviation of the mean) for each parameter, this would require unrealistic simulation effort. Because POP-2 is a deterministic model, introduction of stochasticity in the modeling process would be extremely tedious to accomplish. However, the modeler can arrive at a better understanding of biases (consistent deviations from the true value) by systematically varying these estimates. Thus, each parameter estimate may have a high, medium and low value, and each could be equally realistic, given problems with data collection and precision. These key parameters should be related to field estimates (such as annual reproduction, mortality, wounding loss, harvest proportions). Combining all possible values for all parameters of interest in a systematic way will yield:

$$n = p^i ,$$

where n equals the total number of simulation runs resulting from unique combinations of p parameters, each

Many pronghorn populations are surveyed only during late winter. Visibility generally is relatively good at that time, but population and mortality estimates throughout the year remain lacking. Photo courtesy of the Montana Department of Fish, Wildlife and Parks, Helena.

with *i* distinct value options. For example, three parameters with three value options will result in 3^3 equals 27 different unique combinations.

In contrast to intuitive parameter adjustment (tweaking) SA allows the modeler to evaluate systematically the assumptions made about each data set and to observe the relative effects of different combinations of parameter values. When running simulations under an SA, the modeler must manipulate initial population size to ensure that the modeled population can account for observed values (i.e., harvest, total number of animals seen during surveys). In addition, each model output should be compared with alternative data (such as harvest proportions, age structure, and age and sex ratios). Thus, each simulation scenario will result in two assessments: initial population size to sustain the population given observed population numbers, and the agreement with other corollary data sets. This agreement could be recorded in a subjective three-level score, such as positive (good agreement with all data and biological intuition), null (satisfactory agreement with all data and biological intuition) and negative (unsatisfactory agreement with all data and biological intuition).

Once all *n* simulations have been run, the assessment results should be recorded in an assessment matrix similar to Table 158. This can clarify data weaknesses, knowledge gaps and alternative scenarios that may deviate from those assumed to operate in the population. This information easily can be used in data collection proposals, public meetings and information, and in setting management priorities.

DANGERS AND MISUSE OF MODELS

To paraphrase a popular saying, there are lies, damn lies, statistics and . . . models. The use of a model per se does not qualify the proclaimed results as science. Modelers have long recognized the GIGO principle—garbage in, garbage out. Thus, the modeler must be careful to avoid the common pitfalls of model verification and validation. Perhaps the most often committed error in modeling is the lack of independence between data used for model creation and data used for model validation. For example, it often is the practice to run a statistical regression analysis between model output and input data, and some commercially available software packages (e.g., POP-2) even produce such correlations. Modeling to achieve an acceptable coefficient of determination (R^2 value) is dangerous and may lead to models that have low reliability or predictive power. To validate a model with empirical data, the modeler should use completely different data sets, or at least compare the predictions of the model retrospectively with newly gathered data. Unfortunately, models often are updated annually to account for the new survey results. For example, a POP-2 model for the Medicine Bow pronghorn herd in Wyoming was declared unreliable (Wyoming Department of Fish and Game 1988) and was abandoned after newly collected population data could not be reconciled with the model. Ironically, almost every previous annual report on the Medicine Bow pronghorn population declared that the model fit the data. After the old model had become unreliable, a new model was started, with different initial values, and the circle of annual adjustment of the model continued. Such modeling can lead to costly failures when the model is not checked objectively, but model output is accepted as reliable knowledge. The longer a model is run, the greater effect initial errors have on the outcome. Due to error propagation and magnification, each iteration of a model makes it harder to account for the new data, until the model shows completely erroneous behavior, and no inventive tinkering with model parameters will remedy the error magnification.

A related misuse of a model is extensive parameterization in absence of good field data. Some commercially available models require extensive detail in data, which often is absent in field studies. If the manager decides to use such a model regardless of available data, many parameters must be guessed. The effect is that, despite an impressive list of things the model can do, the modeler

Table 158. Possible assessment matrix for a hypothetical combination of three parameters with three value options each

$P_1P_2P_3$	IP	S	$P_1P_2P_3$	IP	S	$P_1P_2P_3$	IP	S
HHH	13300	–	MHH	4300	0	LHH	6300	–
HHM	8900	–	MHM	1200	–	LHM	4800	0
HHL	4300	0	MHL	300	–	LHL	5100	0
HMH	2300	+	MMH	23000	–	LMH	2200	+
HMM	4300	0	MMM	19000	–	LMM	2700	+
HML	3600	0	MML	17400	–	LML	2900	0
HLH	1900	–	MLH	25400	–	LLH	4800	0
HLM	700	–	MLM	26300	–	LLM	6500	–
HLL	500	–	MLL	19600	–	LLL	7300	–

$P_{1\text{-}3}$ = parameters 1–3; H = high; M = medium; L = low estimate; IP = initial population size; S = assessment score.

violated the principles of parsimony and included possible errors by pairing guesses (unreliable knowledge) with data. In the model, each parameter is treated equally, regardless of the reliability of the source, and the output does not allow the immediate identification of those data or guesses that influenced model results the most. For that reason, sensitivity analysis should always be conducted, even in routine population modeling efforts using models as simple as POP-2.

Obviously, modelers must carefully weigh the need for detail with the need for accuracy and precision. It is a common problem in modeling to assume that each data point is an accurate and precise estimate of population parameters. In fact, in most readily available data sets on pronghorn population dynamics, the question of accuracy and precision is not even addressed. Many wildlife agencies do not calculate confidence intervals on commonly presented population parameters, and survey biases affecting the accuracy of estimates rarely are measured. Consequently, model input may consist of data afflicted with a large sampling error. Sampling errors can mask ecological relationships, such as density dependence (Dennis and Taper 1994, Shenk et al. 1998), or create a false confidence in model performance. Specifying parameter variability in stochastic models is an effective way to incorporate the uncertainty associated with sampling errors into a population model.

Last, modelers often overestimate the predictive power of a model. It is important to realize that a model cannot predict the future status of a population exactly. Models should be used first and foremost to explore and understand the dynamic behavior of a population under certain assumptions (data, parameters, relationships). Deterministic models always are less powerful in predicting future population behavior because they do not consider slight variations in input parameters and variables. Only a stochastic population model can provide a probabilistic outcome, that is, the likelihood of a given event or population response during the projected time span. Such models usually are employed to estimate the population trend or time to extinction of small populations, based on random events. Harvest models based on deterministic simulations (such as POP-2) are unsuited to provide predictive power, especially when the simulations are expanded for several modeling cycles beyond available data. The predictive power of a model is difficult to assess; only future data can effectively document whether the model was right under a wide variety of natural conditions.

PRONGHORN MODELS TODAY

Although the literature on pronghorn population modeling is scarce, wildlife management agencies increasingly use population models to understand and simulate pronghorn population dynamics. A survey of wildlife management agencies in western provinces and states revealed that most are using population models to estimate population sizes or simulate harvest strategies and population growth (Table 159). It appears, however, that stochastic and mechanistic models of pronghorn populations

TABLE 159. Pronghorn model types and software used by provincial agencies in Canada and state agencies in the United States, 2002

Province/state	*Model software*	*Use*	*Address of pronghorn model specialist*
Alberta	None	NA	NA
Arizona	Deterministic bookkeeping model (similar to POP-2)	Harvest regulations	Supervisor Arizona Department of Game and Fish 2221 W. Greenway Rd. Phoenix, AZ 85023
California	Spreadsheet regulations	Harvest	California Department of Fish and Game 601 Locust Street Redding, CA 96001
Colorado	POP-2 regulations, population estimation	Harvest	Colorado Division of Wildlife 312 W. Prospect Fort Collins, CO 80526
Idaho	Spreadsheet regulations	Harvest	Program Leader Idaho Department of Fish and Game Box 25 Boise, ID 83707
Kansas	None	NA	NA

continued on next page

TABLE 159—*continued*

Province/state	*Model software*	*Use*	*Address of pronghorn model specialist*
Montana	None	NA	NA
Nebraska	None	NA	NA
Nevada	POP-IV (customized version of POP-2)	Harvest regulations, population estimation	Nevada Department of Conservation and Natural Resources 1100 Valley Rd. Reno, NV 89512
New Mexico	None	NA	NA
North Dakota	None	NA	NA
Oklahoma	None	NA	NA
Oregon	POP-2	Harvest regulations, population estimation	Oregon Department of Fish and Wildlife SW First Ave. Portland, OR 97201
Saskatchewan	None	NA	NA
South Dakota	None	NA	NA
Texas	POP-2 (experimental)	Population estimation	Antelope Program Coordinator Texas Parks and Wildlife Department 4200 Smith School Road Austin, TX 78744
Utah	Spreadsheet	Population estimation	Utah Division of Wildlife Resources 1594 W. North Temple, Suite 2120 Salt Lake City, UT 84114-6301
Wyoming	POP-2	Harvest regulations, population estimation	Wyoming Department of Fish and Game, Wildlife Division 5400 Bishop Blvd. Cheyenne, WY 82006-0001

largely are absent from the manager's toolbox. Perhaps this chapter will provide a stimulus for managers to develop and incorporate these tools into the management process.

RESEARCH NEEDS

As we gain insight into the complex interactions among pronghorn individuals, populations and spatial patterns of landscapes, it becomes evident that much additional knowledge is needed before we can model and predict the distribution and dynamics of pronghorn populations in natural landscapes. Prime among these issues is the long-standing debate whether carrying capacity is a useful concept and whether we can quantify it. Some techniques outlined in this chapter seem promising for detecting density dependence and evaluating possible environmental covariates. Weather and vegetation seem to be logical starting points for further evaluation of carrying capacity in natural landscapes. Investigating the temporal and spatial scale of synchrony in pronghorn population dynamics can give us important information about the driving forces of large-scale population regulation. As we understand how populations are regulated across local scales, we must focus on the importance of regional heterogeneities of animal distributions, especially when a metapopulation structure seems apparent. In such metapopulations, population dynamics often are no longer driven by simple birth and death processes, but may depend entirely on immigration and emigration. Investigation of source-sink dynamics may provide managers with an increased understanding of the role of spatial heterogeneity in pronghorn population dynamics. Heavy harvest of population sources may not only reduce pronghorn density in productive habitats, but also may lead to a decline of sink populations (which perhaps experience no harvest). An appreciation of spatial patterns of landscapes and how they determine the distribution and abundance of pronghorn seems essential in managing this species in the future.

C H A P T E R   T W E N T Y

Managing the Harvest

Bart W. O'Gara and Bruce Morrison

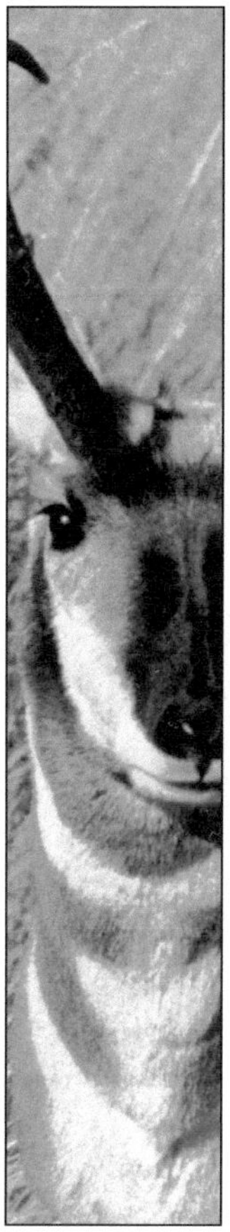

Antelope, as in the case of other large-game animals, when under practically complete protection, lose their fear of man to a surprising extent and become bold in raiding fields and in destroying crops. . . . [T]he animals will eventually require some control of the numbers by eliminating the surplus. This is a matter of practical game administration which should be understood and accepted by the public with the same matter-of-course attitude that is shown toward the control of the surplus livestock on a farm. The limited hunting-license system provides a practical method of handling surplus game and at the same time perpetuating the species in reasonable numbers.

Nelson (1925:8–9)

Although the right to hunt, especially for meat, continues to enjoy extensive public support, the number of persons opposed to hunting is considerable and a growing concern among hunters and wildlife managers.

Kellert and Smith (2000:51)

As with most endeavors, planning is a key first step to determining harvest management goals. It is critical that strategies developed for harvest management fit into a broader framework of agency goals and objectives and be articulated to all publics so that the purpose and role of the harvest program are clearly understood. Most natural resource agencies have comprehensive or strategic plans that describe the general direction of their management actions.

Carpenter (2000:194)

The pursuit and subsequent harvest of a quarry have been human endeavors since time immemorial (O'Gara 2002). For most of human existence on Earth, hunting wild animals has been a prerequisite to human survival. Harvest management began in Eurasia thousands of years ago, primarily for the benefit of rulers who wished to hunt for sport. Reserves for game species and restrictions on who could hunt were the cornerstones of such management.

Wildlife managers in some states intentionally decrease herds to sustain healthy populations and satisfy, as nearly as possible, the hunting and nonhunting public, as well as landowners. How well this is accomplished will determine, to a great extent, how pronghorn fare in the future.

Pronghorn harvest regulations fall into two categories: those that manipulate the type and degree of harvest and those that manage the hunter. Regulating the type of harvest includes setting bag limits, season lengths, legal weapons, number of permits and other rules to ensure that a strategy-specific number and sex of pronghorn are harvested. In many states, politics as much as biology dictate the approved system. Regulating the hunter consists of various restrictions placed on participation in the hunt, to assure that hunts are conducted legally, safely and ethically, and to maximize the opportunities for participation and harvest within the guideline principle of sustained yield.

Nearly all pronghorn hunting is by limited-quota or limited-entry licenses. This reflects the low numbers of animals in some states and the need to place hunters in certain areas, even in states with many pronghorn.

Considering the myriad factors involved in managing the harvest, wildlife managers must set both short-term

Managing the harvest for the good of the animal itself and the general public is a relatively new science that has evolved during the 20th century, as human dependence on wildlife as a source of food or textile has diminished. Although meat still is an important consideration, modern recreational hunting places emphasis on the thrill of the chase, camaraderie, quality of the trophy and other variables apart from subsistence (Kellert 1978). In most areas, supermarkets, fast food outlets and shopping malls have relegated the hunting of wildlife to a negligible role in terms of feeding the family. Thus, harvest management has been directed toward sustaining wildlife populations while providing recreational uses of those animals. Under this concept and practice, all users of the pronghorn resource must have their concerns addressed in the formulation of any harvest strategy. They include the farmer trying to grow wheat in pronghorn habitat, the tourist who thrills at the sight of a flash of white from a rump patch as the animal disappears over the hill, the hunter in search of a trophy animal, meat or a recreational experience, the rancher whose livestock may compete for forage with pronghorn and the entrepreneurial western communities where hunting and wildlife-related tourism are significant economies. Their considerations must enter into any management strategy. These include but are not limited to effects on the social structure of pronghorn herds, behavioral patterns that increase the vulnerability of different segments of the pronghorn population at different times of the year, human activity in pronghorn habitat during hunting seasons, desired pronghorn sex and age ratios, habitat quality and competing land uses. Photo by Ed Bry; courtesy of the North Dakota Department of Game and Fish, Bismarck.

strategies and long-term goals for pronghorn populations and formulate objectives to ensure necessary flexibility to achieve those goals.

Given the character of human population growth and sprawl, the attendant increases of activities, and the vagaries of weather and habitat conditions, managing the harvest is one of the most practical and practicable ways of assuring the sustainability and overall well-being of pronghorn populations. This also is true in the management of many other wildlife species, including most North American ungulates. Habitat management is even more basic to the welfare of pronghorn and other wildlife populations.

Despite these conclusions, agencies sometimes are called on or sued to defend recreational hunting as a form of management. The California Department of Fish and Game prepared several environmental documents regarding pronghorn hunting in that state during the 1990s (see California Department of Fish and Game 1998). Among other things, the documents addressed impacts of hunting on: (1) individual pronghorn management units, (2) health, condition and age class structure, (3) natural mortality, (4) habitat, (5) regional and statewide pronghorn populations, (6) other wildlife and plant species, (7) recreational opportunities and (8) economics.

Some wildlife administrators and managers may not believe that hunting seasons could be challenged seriously in their unit, state or province. The following, from Duda (2000:38) might confirm that view: "Anti-hunting groups have deceived people into believing that hunting is an activity that is not supported by the American public and have misled others to assume big game hunting is a sport in decline. The best way to debunk these negative and misleading notions about hunting is to look at the facts. . . . Between 1950 and 1996, the U.S. population increased by 49 percent, while during the same approximate time frame (1955–1996) big game hunting increased 615 percent"

(Duda 2000:38). Also, according to Duda et al. (1995) hunting has been a licensed, regulated activity in the United States throughout the 20th century, and more than 8 of 10 Americans believe that people should have the freedom to choose to hunt. However, with the increasingly rapid social, economic and environmental dynamics in North America, all wildlife administrators and managers need to be cognizant of the face and potential of antihunter, antihunting and antimanagement sentiment in all parts of the continent (see Causey 1989, Richards and Krannich 1991, Dizard 1994, Minnis 1997, Sopuck 2000).

HISTORICAL PERSPECTIVE

The harvest is the necessary result of the application of an essential tool in modern big game management—the licensed sportsman.

Denney (1978:395)

Early travelers and pioneers in the West hunted pronghorn for a variety of reasons (McCabe et al. 2004). They were not concerned with the ecology of the species or aware of any need to manage the animals or the shooting of them. The same was essentially true of Native Americans who coexisted with pronghorn for thousands of years before the invasion of European settlers (see McCabe et al. 2004).

In the 1800s, as the West was being "civilized," unregulated market hunting seriously depleted pronghorn populations. For example, on the eastern plains of Colorado during the winter of 1868–69, pronghorn were killed in great numbers, hauled into Denver by the wagonload and sold at the rate of $0.25 per three or four carcasses (Hoover et al. 1959). There was no evidence during that period of any recognition of need to regulate the take of pronghorn. It was not until the last decades of the 19th century that citizens began to recognize that perpetuating the resource required limitations on human endeavor and activity.

Although some pronghorn populations received protection as early as 1875 (Mace 1954), early laws did little more than legalize the slaughter of thousands of animals. Where hunting seasons were adopted, they usually lasted for months and involved very liberal bag limits (McCabe et al. 2004). By the 1920s, the plow and grazing by livestock had drastically changed western rangeland conditions, and fences blocked many traditional movements, resulting in high pronghorn mortality during severe winters and droughts. It also was apparent that most of the laws in effect without enforcement did not adequately protect the resource, so hunting seasons for pronghorn were closed in one area after another. These closures did little good, due in large measure to the few law enforcement agents available, and most of them were ineffective, political appointees. During the 1930s, the Great Depression and a prolonged drought forced many settlers from marginal farmlands, allowing fences to deteriorate or be removed and rangelands to revert to native vegetation again suitable for pronghorn. Pronghorn began to recover during the 1930s, when wildlife agencies increased enforcement of hunting season closures or restrictions. The fact that pronghorn were severely depleted in most areas contributed to the decline of hunting interest and pressure.

The hunting season was reopened in Idaho, New Mexico and Wyoming in 1934. Less than 2,000 pronghorn were legally harvested that year: 1,800 in Wyoming; 116 in New Mexico; and 69 in Idaho. The largest annual legal harvest for all areas since reopening seasons—165,302

Regulated harvest or the very notion of hunting restraint or conservation in current context was not much evident in pioneering America's consciousness. For too long and with little or no regard to season, number of animals, or their sex and age, pronghorn (and other wildlife) were taken for food, clothing, market and sport of various degrees of wantonness. Imbued with the myth of superabundant wildlife and lacking awareness (perhaps concern) that wildlife was finite and that they could not be sustained without mandated harvest regulations, hunters and gunners extirpated pronghorn from much of their historic rangeland or reduced their numbers dramatically. Conservation grew out of preservation—the fretful, hapless reaction to vanishing big game species and populations (see Chapter 3). Photo courtesy of the Wyoming State Archives, Museums and Historical Department, Cheyenne.

Closure of pronghorn hunting in the late 1800s and early 1900s was not a particularly contentious issue with hunters or the public because there were not many animals left to be seen, much less hunted. By the time hunting pronghorn was permissible again—mostly decades and at least a generation of hunters later—the taking of even a single animal was newsworthy, whereas less than a century before, millions were killed without fanfare. Taken on a special trophy hunt permit in 1967, the pronghorn at left was the first hunter-killed pronghorn taken on the Sheldon National Wildlife Refuge, in northwestern Nevada, since the refuge was established in 1936. Photo courtesy of the U.S. Fish and Wildlife Service, Washington, D.C.

animals—occurred in 1983 after a series of mild winters on the high plains. Table 160 gives the available harvest records by state and province for the period 1935 to 1994, in increments of 10 years for easy comparison, as well as data for 1995 to 2001. The information is from Yoakum (1968) and questionnaires sent to state and provincial conservation agencies during 1990, 1995, 1998 and 2000 (hereafter referred to as Questionnaires). Harvest estimates are low, especially for the 1930s, 1940s and late 1990s because statistics are missing for some years in a number of states. A boon to management of pronghorn was the practice of maintaining harvest records. Analyses of harvest data, combined with population estimates, allowed re-

Table 160. Pronghorn legal harvests by 10-year increments in Canada and the United States, 1935 to 1994, plus 1995 to 2001

State/province	*1935–1944*	*1945–1954*	*1955–1964*	*1965–1974*	*1975–1984*	*1985–1994*	*1995–2001*	*Total Now*
Alberta	NR	NR	0	11,280	22,871	37,148	5,604	76,903
Arizona	1,295	4,050	5,648	5,946	5,274	5,959	4,220	32,392
California	1,089	936	303	2,303	3,682	5,546	2,006	15,865
Colorado	0	16,586	29,029	49,183	63,907	81,916	60,613	301,234
Idaho	4,325	7,500	7,669	13,162	16,590	27,198	9,575	86,019
Kansas	0	0	0	71	1,640	1,964	1,207	4,882
Montana	553	87,103	199,551	164,882	182,225	397,249	98,525[a]	1,130,088
Nebraska	0	555	6,845	12,250	9,733	4,342	5,017	38,742
Nevada	10	1,451	1,486	2,336	3,859	9,736	6,818	25,696
New Mexico	4,130	6,197	14,951	16,023	24,111	40,571	20,865[c]	126848[b]
North Dakota	0	3,699	21,871	14,644	7,073	18,480	8,079	73,846
Oklahoma	0	0	0	148	0	315	439	902
Oregon	4,169	3,156	3,597	5,775	8,769	14,279	8,413	48,158
Saskatchewan	NR	0	0	19,074	16,896	35,385	3,869[b]	75,224[c]
South Dakota	1,955	26,749	56,696	52,168	85,600	57,765	29,049	309,982
Texas	0	3,074	13,514	9,609	7,224	6,705	3,713	43,839
Utah	0	325	615	1,352	3,258	10,952	5,790	22,292
Wyoming	49,625	232,837	300,768	304,286	648,520	612,844	178,692	2,367,572
Total	67,151	394,218	662,543	684,492	1,111,232	1,368,354	492,494	4,780,484

NR = no record.

[a] No data for 1997 to 1999.

[b] No data for 2000 to 2001.

[c] No hunting season during 1998 to 2001.

Passage of the Federal Aid in Wildlife Restoration Act (Pittman-Robertson or P-R) by the U.S. Congress, signed into law in 1937 by President Franklin D. Roosevelt, provided state conservation agencies needed funds to be effective. World War II caused a human exodus from marginally productive western rangelands and, together with gasoline rationing, lack of ammunition and would-be hunters joining the Armed Services or being drafted reduced the hunting of pronghorn. After World War II, P-R funds enabled western wildlife conservation agencies to trap pronghorn in areas where they had survived and prospered and to translocate them to rangelands where they had been extirpated. By 1948, western states had translocated 7,162 pronghorn (Trefethen 1975). These translocations contributed greatly to the rapid increase of pronghorn to more than 1 million by 1983, where only about 25,000 had existed during the early 1920s. Photo courtesy of the Montana Department of Fish, Wildlife and Parks, Helena.

finement of harvest regulations tailored to specific populations, environmental conditions and hunter demand.

Weather conditions—droughty summers and severe winters—during the mid- to late 1990s caused a decrease of about 30% in total pronghorn numbers for North America (Yoakum et.al. 1999). The prognosis for a quick bounce-back does not seem as good as in the past. The cumulative effects of converting rangeland to agricultural crops, barriers to seasonal movements, energy developments and usurping of pronghorn habitat for the needs of increasing human populations may slow or prevent a return to the high numbers of the 1980s. This drop in pronghorn numbers—particularly on the northern plains—has reduced pronghorn harvests significantly.

A severe winter, especially one following a dry summer, can reduce a pronghorn population by more than 70% (McKenzie 1980). Most pronghorn are on the high plains of the Dakotas, Montana and Wyoming, and annual harvests often reflect weather conditions there. For example, the winters of 1977–78 and 1978–79 (this photo) reduced the 1977 summer population of 9,245 pronghorn in North Dakota to 1,245 by spring 1979 and resulted in 4 years without hunting seasons. Neighboring states and provinces fared better, but the harvest in Montana dropped from 18,528 in 1977 to 9,308 in 1979. During the same period, Nebraska dropped from 1,325 to 541, South Dakota from 7,592 to 2,473 and Wyoming from 58,858 to 44,616. Alberta's harvest was 1,995 in 1977 and 674 in 1978. The total harvest in Canada and the United States dropped from 102,109 in 1977 to 71,609 in 1979—a 30% reduction. Managers can do little to avoid such die-offs, but (as in the past) the numbers came back rapidly. Usually, northern pronghorn populations that experienced a 50% die-off regained their numbers in 2 years (Pyrah 1987). Photo by Ed Bry; courtesy of the North Dakota Department of Game and Fish, Bismarck.

Harvests for 1995 and 1997 in the following provinces and states are indicative: Alberta, 1,841, 333; Montana, 34,103, 18,223 (in 1996, 1997 figure not available); North Dakota, 2,478,489; Saskatchewan, 2,188, 0; South Dakota, 8,752, 1,984; and Wyoming, 37,068, 29,121.

ESTABLISHING HARVEST OBJECTIVES

Every hunting season should be evaluated thoroughly so the harvest management program remains dynamic and responsive to a changing management environment.

Strickland et al. (1994:457)

Habitat Considerations

Habitat is the key factor in the establishment of harvest objectives. Ideally, managers should evaluate current rangeland conditions, but adequate personnel seldom are available for this. Geographic Information Systems (GIS) may aid managers in evaluating rangeland conditions. The same data should not be used indefinitely, because habitat conditions are dynamic and dependent on the weakest component in the seasonal habitat requirements of a species. Abundant summer rangelands cannot compensate a paucity of winter rangelands or poor fawning habitat, nor vice versa. All elements of the annual habitat requirements must be considered, including pronghorn movement corridors between seasonal rangelands. The corridors may be inadequate because of fences or other barriers to movements—not necessarily because of conditions of vegetation. Also, assessment of habitat needs must be based on worst-case scenario conditions that result from occasional severe winters, droughts or other natural or human-related catastrophes.

Too many or too few animals may occupy a particular rangeland, relative to habitat conditions and other management considerations. Harvest objectives then must be set to balance pronghorn numbers with habitat conditions before objectives can be developed to maintain animal numbers according to ecological, public and biopolitical factors. With proper harvest management, a pronghorn population usually can be balanced with its habitat within several years. In most cases where harvest is used to increase or decrease a pronghorn herd relative to biological considerations, the harvest strategy is implemented in concert with a program of habitat management.

Population Modeling

Population modeling is a valuable tool for selecting harvest strategies suitable for achieving management objectives. Population simulations have been the impetus behind better definitions of herd units, organized data collection and research for better methods of data collection. Where simulation models have been constructed from sound inventory data, accurate estimates of previous harvests and valid estimates of reproductive and mortality rates, the models can be used to predict population responses to the removal of specific numbers of the various sex and age classes from the population. Efforts should be made to use valid simulation models in developing harvest programs for managed pronghorn herds (Salwasser 1980, Gasson and Wollrab 1986; see also Chapter 19). With animal rights activists challenging management by hunting and disagreements between wildlife managers and users of public lands, modeling will become more and more of a necessity in maintaining pronghorn populations.

Population Objectives

As with any effective wildlife management endeavor, the harvest of pronghorn must be accomplished according to clear objectives. If clear objectives for a population are not established, or if those objectives are not predetermined on the basis of sound biological data and experience, high probability exists of overlooking variables or limiting factors that can negatively influence the population's well-being. Population objectives for pronghorn, as of 1996, by provinces and states are given in Table 161.

Even before harvest objectives can be properly considered, managers must have a good idea of the number of animals in the population relative to habitat conditions. Once a proper census has been conducted and habitat conditions are ascertained, managers can begin to determine viability factors and harvest objectives. Among the factors that need to be considered when setting harvest objectives are habitat conditions, population numbers, buck/doe ratios, reproduction, pronghorn social behavior and movements, hunter access, weather, and competing land uses. All of these factors are interrelated and cannot be assessed independently. Harvest strategies focus on either maintaining a stable population size or decreasing or increasing its numbers. The terms stable, increasing and decreasing refer to trends, not to discrete numbers of animals at one given point in time. If a reliable estimate of the total pronghorn population is not available, managers must rely on trend data. Trend data probably will not stand in court if a state must produce an environmental impact study on sport hunting. Objectives should facilitate or strive toward long-term management goals.

If the habitat of the pronghorn population in question is owned or used by private landowners, those users and their interests, particularly those involving livelihoods,

TABLE 161. Pronghorn postseason population objectives and desired buck/doe ratios as reported by provinces and states in 1996 as well as population estimates as reported in 1998

Province/state	*Population objectives*	*Desired number of bucks/100 does*	*1997 preseason numbers*
Alberta	15,500	45 preseason; 30 postseason	13,200
Arizona	8,500	25–30 posthunt	10,000
California	5,600	20–24 minimum postseason	5,639
Colorado	47,000	25–60 prehunt depending on objectives	72,600
Idaho	NE	≥40 postseason	12,000
Kansas	1,750	50 preseason	2,500
Montana	NE	NE; 31–58 during surveys	57,500
Nebraska	5,500	NE	6,500
Nevada	25,000 by 2000	20	14,535
New Mexico	NE	20–25	35,000
North Dakota	6,000	50	5,210
Oklahoma	NE	NE	1,500
Oregon	Will be part of species management plan	25	15,020
Saskatchewan	Meet hunter demands and appease landowners	50	10,000
South Dakota	42,000	35	20,518
Texas	NE	25 for maximum harvest; 50 for trophy production	9,599
Utah	6,300	25	10,500
Wyoming	444,060	Varies, classification by herd (recreational or trophy); 40 for trophy herds	370,121

NE = not established.

must be a consideration in management planning. If the habitat is public land, managers still must consider management goals for other activities, such as livestock leases, mineral development, recreation. Only if the land is dedicated solely to wildlife can population objectives be set that do not consider other activity goals.

Setting of population objectives, particularly via harvest, is a balance between the species' biological needs and human interests. It usually is a matter of providing maximum hunting or viewing opportunities, reducing conflicts with other uses, accelerating trophy production, or accommodating some other type or combination of opportunities. Colorado, Nebraska and North Dakota attempt to maximize recreational opportunities; Idaho, Nevada and Wyoming manage for trophy hunts (Ticer and Munig 1999).

In Wyoming, numbers of licenses to be issued are based on the status of the herd (above or below population objective), potential for damage to stored or standing crops and reproductive rates (H. Harju personal communication:1992). Harvest rates range from 8% to 40% ($\bar{x}$ = 20%) in herds above objective, and 6% to 28% ($\bar{x}$ = 15%) in herds below objective. The wide range in percentages taken from herds either above or below population objectives is related to crop depredation problems and reproductive rates. For example, a herd with 65 fawns per 100 does cannot withstand as much hunting pressure as a herd with 120 fawns per 100 does. Managers should begin harvesting extra does as much as 3 to 4 years before a herd reaches its population objective because public resistance to hunting does sometimes cannot be overcome immediately.

Buck/Doe Ratios

Because pronghorn occupy open country and are very visible, most hunters will be successful and most will kill a buck unless they are forced to do otherwise. Doe/fawn permits usually are needed to maintain desired buck/doe ratios. If the number of licenses to take any pronghorn are too high, the buck/doe ratio will be depressed, and it will become difficult to attract hunters. Some people want to hunt only on public land and want trophy animals. Ranchers generally want pronghorn numbers held in check, but also want to charge for hunting, which discourages hunters and makes permits for areas of private land difficult to sell. The manager must consider all of this while maintaining control of pronghorn numbers.

Desired buck/doe ratios usually depend on the management goals set by wildlife agencies for a particular pronghorn population (Table 161). A ratio of 1:4 should be maintained for maximum recruitment into a population according to Griffith (1962) and Hailey (1979). This will allow for greater production of younger age class animals.

Pronghorn hunters generally select adult males during either-sex seasons. Consequently, managers must issue doe-only permits to reduce herds or maintain stable herds and regulate buck/doe ratios. Photo courtesy of New Mexico Department of Game and Fish, Albuquerque.

Maximum trophy production requires a buck/doe ratio of 1:2 or less (Hailey 1979). With this ratio, there will be a relatively large number of bucks in the population, and many of them will be 3 years old or older. Hunting permits then can be regulated to leave enough bucks of the right ages in the population to produce trophy horns. If a pronghorn population objective is to reduce a herd or inhibit increases, buck/doe ratios can be maintained by issuing doe/fawn permits and allowing hunters multiple permits. Although higher buck ratios may be needed when trophy hunts are an objective, a postharvest ratio of one buck per five does is a biologically safe objective and may not be the minimum number of bucks needed for complete breeding (Salwasser 1980). In practice, buck/doe ratios sometimes are set for political, not biological, reasons. Wyoming manages for trophy hunts and a 40 buck per 100 doe ratio; Arizona manages for trophies as well as recreational opportunities with a 30 buck per 100 doe ratio (Ticer and Munig 1999).

Managers making herd composition counts of pronghorn should be aware that the reliability of buck to doe counts—annually conducted in some states and provinces to determine the harvestable surplus of males—may be influenced by the timing of the counts. The pronghorns' seasonal changes in grouping and the dispersal of groups and individuals must be taken into consideration. For example, territorial (lone) bucks may be difficult to locate compared with does, most of which are in groups (see Chapter 18).

Thus, the behavior and social system of the pronghorn may affect composition counts taken during summer. Personnel of the Wyoming Department of Game and Fish believe that counts after the hunting season may be more reliable due to less segregation of bucks and larger groups to work with. The timing of the seasonal changes may be variable from population to population, region to region and year to year (Autenrieth 1978).

REGULATING THE HARVEST

Harvest management for big game is the art of melding the objectivity of wildlife science and the subjectivity of public wants for the attainment of a management goal.

Strickland et al. (1994:446)

Timing of Seasons

Traditionally, pronghorn have been hunted from mid-August through October. Shedding of horn sheaths, weather during the hunt, timing of the rut and hunter access are a few of the factors that have been used to promulgate hunting season dates.

Throughout most of their range, pronghorn shed their horn sheaths from late October through mid-November. Exceptions generally are unhunted endangered subspecies in southern habitats. Trophy quality of the bucks is decreased by shed or loose sheaths. Most states and provinces attempt to set hunting seasons before this occurs. However, this constricts the hunting season inasmuch as unseasonably warm weather, especially in the southern part of the species' range, can place undue heat stress on the animals and cause rapid spoilage of meat not quickly and properly processed. With weather being an uncontrollable factor, managers must rely primarily on averages when recommending season dates.

Copeland (1980) indicated that, in Idaho, the disruption of the pronghorn breeding season by hunting caused abandonment of territories by does, loss of territorial bucks to harvest and chaotic breeding when large, mixed-sex herds were formed. Copeland believed that the costs to pronghorn populations can be (1) premature depletion of energy reserves vital to winter survival and (2) breeding by inferior bucks that may contribute to a lack of genetic vitality in the herd. Deblinger and Alldredge (1989) found similar disruption of breeding when the hunting season opened in Wyoming. However, because numerous rifle hunters were in the field for only 1 or 2 days, bucks again were defending their territories by the third day of the season.

Based on Copeland's (1980) observations, it appears human disturbance has the potential to alter previously formed social organizations. Such alterations facilitate breeding by subordinate males that dominant males previously did not allow to breed. Copeland reported that pronghorn in Idaho were driven from traditional territories during muzzle-loading and archery seasons that coincided with the rut. Although Deblinger and Alldredge (1989) found pronghorn bucks back on their territories by the third day of rifle season, they believed that hunting pronghorn during the rut involves unnecessary risk by potentially reducing fitness of future generations of pronghorn.

Copeland's (1980) and Deblinger and Alldredge's (1989) observations point to the possibility that hunting with primitive weapons may be more disruptive of pronghorn territoriality than hunting with modern rifles. During primitive weapon seasons, hunters often spend extended periods of time in the field; rifle hunters generally bag their pronghorn during the first or second day of the season and little disturbance occurs during the rest of the rut.

In practice today, many states and provinces hold archery seasons during the rut and rifle seasons after the rut. In determining when to recommend a hunt, managers must weigh the cost of potential effects on the population against the potential of a satisfactory numerical harvest. In large, healthy populations, the breeding season disruption caused by hunting apparently is not as adverse as in relatively small or newly formed populations. Copeland's (1980) study involved a heavily hunted herd in a narrow valley. In Wyoming, pronghorn occupying a broad expanse of rangeland apparently have been hunted during the rut since seasons were reopened in 1934. There are more pronghorn in that state than any other state, the animals have high reproductive rates and fawn survival, and many fine trophies are taken every year.

Forest (1985) used Wyoming Department of Game and Fish records to investigate the effect of hunting during the rut, on reproductive rates. She found no statistical difference and that the killing of dominant bucks did not decrease the doe/fawn ratio. Perhaps management that is successful on the Great Plains should be altered for mountain valley or other small population situations.

Maher and Mitchell (2000) compared behavior and group composition between bucks at Fort Belknap Indian Reservation, Montana, and those in a population near Billings, Montana. Trophy pronghorn hunts were held during the breeding season on the former. The latter were not hunted with firearms during the breeding season. Bucks from the two populations did not differ in time budgets or interaction rates. Selective hunting for bucks did not appear to impact negatively behavior patterns of those animals, probably because hunter density was low—0.1 hunter per square mile (0.04/km^2).

Wyoming and some other states attempt, when practical, to hold concurrent deer and pronghorn seasons because nonresident hunters often come long distances, and many of them do not want to spend travel money to hunt just one deer or just one pronghorn. Availability of multiple licenses also attracts out-of-state hunters. Concurrent bird seasons sometimes are used to make pronghorn hunts into combination hunts.

Management objectives of the harvest—providing recreation to the sporting public and maintaining or adjusting pronghorn numbers to a goal-oriented level—must be clearly in mind and weighted against the variables (e.g., weather, hunter pressure, hunter success) when recommending hunting season dates. If a hunting season is set at the optimum time for hunter convenience, the breeding season may be disrupted. This dichotomy generally confronts managers and has significant bearing on other harvest recommendations, such as length of season and bag limit. For this reason, harvest management decisions must be made on the basis of reliable, recent data carefully analyzed by experienced managers. Guesswork in setting season dates must be improved from wishful thinking to statistically reliable predictions.

Length of Seasons

Season length varies from province to province and state to state, depending principally on numbers of pronghorn in the area and legal weapons. Seasons in various provinces and states range from 2 days to at least 2 months. For example, Montana had a 65-day archery season and a 29-day general rifle season in 1999. The last 29 days of the archery season were concurrent with the rifle season. In contrast, Arizona allowed archers 14 days, closed all pronghorn hunting for 14 days and then restricted rifle and muzzleloader hunters to 4 or 6 days during 1999. These are conservative seasons and may be dictated as much by politics or landowner pressure as by biological necessity.

Copeland's (1980) study in Idaho indicated that long hunts were disruptive to pronghorn breeding. He recommended that no hunting be allowed from September 15 through October 10. These dates should be adjusted in various parts of the pronghorn range to fit with each herd's annual cycle. Restricting hunting during the breeding season could significantly reduce the time allowed for hunting, because hunts held in late October and November may be when bucks shed their horns. The herd in the Idaho study was hunted from August 6 through September 2 with bows, from September 3 through October 9 with muzzleloaders, and from October 29 through November 13 with modern rifles (Copeland 1980). This represents 65 straight days of hunting, a 19-day break and then 16 more days of hunting. Even during primitive weapon hunts, which typically do not entail as much hunting participation as do modern rifle hunts, the pronghorn populations are subjected daily to hunting pressure.

In New Mexico, the archery hunts are restricted to 14 days in early August; the muzzleloader hunts typically are 2 days in early October; and modern rifle hunts are 2 or 3 days in early September, plus 2 days in late September. Hunts are held on weekends, and the dates vary somewhat for different parts of the state. The longest span that any herd is subjected to continuous hunting pressure is 14 days during the archery season. Rifle hunters are in the field a maximum of 3 days.

In the states with the most pronghorn—Wyoming and Montana—the animals generally are subjected to about 2 months of hunting with various weapons, and some hunts are conducted before and during the rut. Although this sounds like excessive disturbance, the number of archery hunters is low, considering the vast rangelands occupied by pronghorn.The first weekend of rifle season sees many hunters in the field, but more than 90% of those who are successful in terms of harvest kill their animal on the first or second day of hunting, and little disturbance occurs during the rest of the season.

Copeland's (1980) study in Idaho was conducted on a heavily hunted herd, and he believed that even short hunts had an effect on breeding. Several states schedule their hunts after the breeding season to preclude interruption of the rut. This restricts the time allowed for hunting, and managers need to consider additional permits to achieve desired harvests. This will put more people in the field but for a shorter time. Hunting after the rut, given our present knowledge, is the safest procedure to assure the long-range welfare of the species.

Legal Weapons

The levels of harvest success and hunting opportunity tend to dictate the types of weapons legal for hunting pronghorn in a particular area. Hunters should check provincial and state regulations closely concerning legal weapons. Most require that modern rifles use center-fire cartridges of .23 or larger caliber. Muzzleloaders generally must be .39 to .45 caliber or larger. States that allow handguns for pronghorn hunting have a large variety of restrictions involving barrel lengths, energy at muzzle, caliber and length of cartridge to be used in hunting; Canada does not allow handguns. The provinces and states that allow the use of shotguns require rifled slugs and gauges no larger than 10 or smaller than 20. Long, recurved or compound bows generally must be 40-pound (18.2 kg) pull weight or more, and almost every province and state has regulations concerning length or weight of arrow and width of arrowhead.

Most provinces and states allow hunting with a variety of weapons. As of 1994, 12 states had archery seasons for pronghorn and 3 conducted muzzleloader hunts. Crossbows were illegal in most areas, but two states allowed them for use by handicapped and disabled persons during archery-only seasons (Strickland et al. 1994).

As a general rule, archery hunters have much lower harvest success rates, and their hunts can be more liberal than those of muzzleloader or modern rifle hunters. The modern rifle hunter is the most successful of all pronghorn hunter categories. Nevada bowhunters achieve around a 17% success rate, whereas rifle hunters reach roughly 85% harvest success (Jeffress 1986). In New Mexico, the figures are remarkably similar, with bowhunters achieving a 15% success rate and rifle hunters reaching 82% (Saiz 1986). In Colorado, success by rifle hunters is 75% to 80%, and archery success is around 25% (Steinert 2000). Hunting figures from other states and provinces indicate success rates of 75% to 90% for rifle hunters, about 40% for muzzleloaders and about 20% for bowhunters. Archers apparently have the most variable success rates, with 8% reported from Oregon and nearly 30% from Kansas (Questionnaires).

With low success achieved by bowhunters, most archery seasons can have more liberal bag limits and longer time frames than those set for hunters with firearms (unless bowhunters wait at water sources in arid areas, keeping the animals away from water, or greatly increase their success by skillful use of decoys during the rut as discussed in Chapter 22). Managers, therefore, can provide more opportunity to more people, with minimal impact on the pronghorn population. Muzzleloader hunters are slightly more successful than archery hunters, but their hunts still can be lengthier and with a more liberal bag than for the modern rifle hunters. Success rates for bow, muzzleloader and rifle hunts were 20%, 45% and 88%, respectively, in New Mexico during 1994 (Weybright 2000). Because of the relatively high success achieved by modern rifle hunters, managers must make fairly precise calculations of the number of animals desired for harvest and set permit numbers accordingly.

Archery and other primitive weapon permits generally are less in demand than modern rifle permits, but

Archery hunting seasons for pronghorn generally are longer than rifle seasons because fewer people hunt with bows and arrows, and they experience less harvest success. Hunter success seldom exceeds 20% in a province or state and is below 10% in about half of the provinces and states. The amount of disturbance to rutting pronghorn by archers remains equivocal. Photo by Ellen Rice; courtesy of the New Mexico Department of Game and Fish, Albuquerque.

popularity of such hunts is increasing. When archery permits were first issued in Arizona during 1974, only 16 hunters applied for 50 available permits (Lee 2000). In 1995, there were 1.8 applicants for each available archery permit. The ratios for muzzleloader and general firearm permits were 5.8 and 19.9, respectively.

Success rates for pronghorn hunters using modern rifles are 75% to 90%. Success would be even higher, but most hunters try for a trophy buck like this animal. Photo courtesy of the New Mexico Department of Game and Fish, Albuquerque.

Bag Limits

Bag limits vary according to the number of pronghorn in a population and the state or provincial goals and objectives for that population. In areas where pronghorn exceed the management objectives (or landowner tolerance), multiple doe/fawn permits per hunter will assist in bringing down total numbers. Ticer and Munig (1999) received responses concerning harvests from 12 of the 16 states of the United States harboring pronghorn. The mean percentage of populations harvested was 9.8 and ranged from 5.5 in Nevada to 15.6 in Colorado.

Almost always, some doe hunting is desirable to maintain desired sex ratios. If doe-only permits were issued, many hunters could not distinguish does from fawns at a distance, and some harvested fawns might be wasted. Consequently, provisions for doe/fawn permits or horns-shorter-than-ears permits should be included in most management plans.

Hunting pronghorn females sometimes is frowned on by hunters and nonhunters alike. Consequently, doe/fawn permits are in less demand than buck permits. Colorado has a preference point system that ensures the average hunter a buck permit about once every 3 to 5 years. In most areas, it would be possible to draw a doe permit at least every other year (Steinert 2000). A legitimate concern is that fawns, orphaned by the deaths of their dams during early seasons, may suffer hardship or not survive winter without their mothers. However, Bromely and O'Gara (1967) found that fawns orphaned at approximately 6 weeks of age (during mid-July) not only survived, but maintained their place in the hierarchy of the band.

Harvest data for 1996 show the tremendous differences in harvest strategies in provinces and states according to their management objectives (Table 162). Experience in New Mexico indicates that an either-sex bag limit will result in a take of few does, with the majority of hunters selecting bucks. In areas with large pronghorn populations, such as Montana and Wyoming, drawings are held to issue either-sex permits by management districts. Then, to increase doe harvests within those districts, additional permits for does and fawns can be issued to the either-sex permit holders and to other hunters.

The eastern half of Wyoming is predominantly private land. The ability to obtain harvests, especially of females, is difficult because of limited hunter access and trespass fees. During 1991, 20% of the allotted doe/fawn permits were not sold. Most of the unsold permits were for hunting districts in the eastern part of the state (Smith 1992).

The Wyoming experience indicates some of the techniques that have been used to secure adequate harvests. Any-pronghorn licenses were issued through drawings to prevent overharvest of bucks, and hunters could then purchase doe/fawn permits over the counter. At first, one or two such permits were allowed per hunter. To make them more appealing, these permits were sold for full price until opening day, then the price was cut in half. This price was lower than the landowner coupon attached to the permit.

Eventually, hunters were allowed three doe/fawn permits. In 1989, unlimited numbers could be purchased 3 days before the hunting season (Riddle 1990). However, as of 1990, the additional permits were limited to five per hunter because of a perception that some hunters were killing more than they could use and wasting meat. The plan was to have them available on June 15. Nonresident hunters liked to have these permits in hand before going to Wyoming.

All of this was designed to overcome resistance of hunters to shooting does and to obtain the needed harvest. In most cases, getting more hunters into an area

Table 162. Pronghorn harvest data as reported by provinces and states in 1996

Province/state	Percentage of population harvested annually	Cohort percentage of harvest: Bucks	Does	Fawns	Percentage of nonresident hunters
Alberta	8–18	50	38	12	Unknown
Arizona	6–7	100	0	0	Same drawing
California	7	85	15	0	0
Colorado	15	58	39	3	Same drawing
Idaho	10–20	63	31	6	10
Kansas	10–15	84	15	1	0
Montana	16	47	47	6	10 unless insufficient applicants (16 in 1989)
Nebraska	8	97	2	1	3 (archery only)
Nevada	6	98	2	0	10
New Mexico	8.5	88.2	8	3.8	Same drawing, usually 30
North Dakota	24–27	65	21	14	0
Oklahoma	5	92	8	0	Not designated
Oregon	6–7	95	3	2	<5
Saskatchewan	15–20	54	40	6	0
South Dakota	10	77	23	0	0
Texas	4	99	1	0	Not designated
Utah	10–12	62	34	4	10
Wyoming	10–22 (various herds)	56	38	6	25 plus leftovers after drawing

was impossible because of lack of any-pronghorn licenses. Thus, letting hunters already there kill more does was the logical solution. Because of multiple licenses issued to many hunters, Wyoming reported a 91% success rate for pronghorn hunters on the basis of single licenses, but an overall success rate of 133% (P. Riddle personal communication:1990).

The situation in Wyoming changed in 1993 because of extensive pronghorn mortality during winter 1992–93. Numbers bounced back slowly, apparently due to continuing drought conditions and another severe winter in 1996–97 (R. Lanka personal communication:1997). Wyoming regulations for 1997 allowed one any-pronghorn license, on a drawing, per hunter. Some areas did not allow doe/fawn permits; others allowed one or two (also usually on drawings). As numbers increased, a hunter willing to hunt in several areas could get up to seven permits; as of 2000, one any-pronghorn and two doe/fawn permits were available throughout the state.

In Montana, Martinka (1966) reported that selection for adult males appeared to be related to hunter preference rather than to a herd structure that would permit inadvertent selection for adult males. If the harvest management objective is herd reduction, a doe-only bag limit or multiple doe/fawn permits per hunter are ways to achieve it. Doe/fawn harvests usually are accomplished by issuing permits for pronghorn with horns shorter than their ears. The setting of a buck-only or either-sex bag limit with doe-only hunting during the last few days of the season is confusing to the public and difficult to police.

Because their harvest success rate is low, archery hunters usually have an either-sex bag limit, and the resulting limited harvest has minimal effect on the population. Once again, hunters using modern, high-powered rifles can have the most impact on pronghorn populations, unless long archery seasons are affecting breeding success. At least two states, Colorado and Kansas, sell archery permits over the counter (Steinert 2000, Fox et al. 2000).

A factor in determining the dates of pronghorn hunting seasons is the coincidence of the species' breeding season. Adult bucks form dominance heirarchies, according to studies in Montana (Bromley 1969, Kitchen 1974), Wyoming (Ingold 1969) and Idaho (Fichter and Autenrieth 1978), and territorial bucks—the largest and fittest—are at the top of the hierarchy. Kitchen (1974:76) reported that "A male with a territory was almost certain to breed, and a buck without one was virtually certain not to breed." But because the largest and fittest bucks establish and defend territories for courting and breeding purposes (Bromley 1969, 1977, Kitchen 1974) or control and defend harems before and during the rut (Prenzlow et al. 1968, Deblinger and Ellis 1976), the large bucks become relatively easy prey for hunters, including those with primitive weapons. The hunting of the territorial bucks to the exclusion of other herd members may cause disruption in the dominance hierarchy, especially of small populations, and have direct effect on the fitness and trophy quality of the population (Copeland 1980, Deblinger and Alldredge 1989). Where bucks are not territorial, nonterritoriality may have come about by previous hunting practices. An effective way to alleviate problems associated with overharvesting territorial bucks is to disallow hunting before and during the rut. If hunting were not open until after the rut, the largest and fittest bucks will have passed on their genes to the next fawn crop and, despite a postrut harvest, a greater number of large bucks will survive to breed the next year. The problem is that many, if not most, pronghorn hunters select for large horns and would not be enthusiastic about regulations that restrict or complicate that selection. Photo by Leonard Lee Rue III; courtesy of Leonard Rue Enterprises, Blairstown, New Jersey.

In Wyoming, units not known as trophy areas frequently had more permits available during the 1980s than there were hunters who wished to use them, according to Haley (1986), who believed the early rifle hunts for trophy males may have been reducing the vigor and size of pronghorn bucks and may even have fostered selection for nonterritorial bucks. If a hunt is to be held before the onset of or during the breeding season, consideration should be given to either a bag limit that will protect some of the larger territorial bucks or closure of select areas to provide for protection of at least a portion of those bucks.

Public Land Harvests Versus Private Land Harvests

The basic social problem in antelope management is that of a public commodity which lives on private land.

Dood (1984:101)

Throughout the West, the range of virtually every pronghorn population includes both public and private lands (Table 163). The pronghorn themselves do not recognize these land status boundaries. Proper management of pronghorn requires that, when setting harvest regulations, managers consider the interests of landowners and land management agencies. In 11 of the 16 western United States that have pronghorn, landowners are given some type of preference in drawings for pronghorn permits if they have substantial numbers of the animals on their land.

Combining the postseason estimates of pronghorn numbers in Table 161 and the percentages estimated to be on private lands in Table 163 indicates that about 55% of the pronghorn in Canada and the United States were, indeed, on private land in 1997, and 39% and 6% were on federal and provincial or state lands, respectively. A survey in 1989 indicated 62% on private land. Whether or not the 7% decrease was related to land-use practices is unknown. It seems more likely that the reduction in pronghorn numbers, from more than a million in 1989 to about 692,000 in 1997 (Yoakum et al. 1999; see also Chapter 4), caused the change. Most of that decrease occurred on the northern Great Plains, where a high percentage of pronghorn use private lands. Private landowners also control access to considerable public land. Obviously, cooperation between private landowners and provincial and state wildlife management agencies is necessary for coordinated harvest programs.

Mexico faces an entirely different situation because hunting seasons have been closed since 1922 (Cancino et al. 2000). There, pronghorn have decreased steadily because of land-use practices, poaching and, in some cases, predation. Many of the pronghorn that survive in Mexico are in small herds, isolated by fences, on ranches in Chihuahua (Gonzales-Romero and Lafon-Terrazas 1993, Camcino et al. 2000). Thus, landowners probably will determine the fate of the species in Chihuahua and possibly all of Mexico.

Each province and state has adapted differently to the problem of managing pronghorn on private land. In New Mexico, the success of hunts on private land often reaches 95%, due in large measure to packaged hunts conducted by outfitters and guides. If New Mexico was to restrict the season on public lands to accommodate the private landowner, it would be penalizing the public land hunter. Consequently, the state sets private and public land seasons to run concurrently and with uniform bag limits. Private land hunters do not have to go through the draw for a permit, but instead purchase "trespass rights" from the private landowner and then the landowner provides an authorization to purchase a license from the state (Morrison 1984). On some private land that is heavily populated with pronghorn, the state will set a doe-only hunt as long as the landowner will sign an agreement allowing some public hunting. This is a popular arrangement for out-of-state trophy hunters, who can be assured—for a price—of a permit in an area renowned for large horns. It also is an advantageous system for ranchers. During the late 1970s and early 1980s, falling cattle prices and increasing costs of production caused many New Mexico ranchers to seek alternate sources of income—such as fee/lease hunting (Morgan 1988). Although most ranch-

TABLE 163. Percentage of pronghorn by landownership estimated by provincial and state conservation agencies in response to a 1996 questionnaire[a]

	Land ownership		
Province/State	*Private*	*Federal*	*Provincial or state*
Alberta	22	18	60
Arizona	39	31	30
California	30	69	1
Colorado	60	35–40	<5
Idaho	25	74	1
Kansas	98	2	0
Montana	75	>20	<5
Nebraska	85–90	10–15	<1
Nevada	5–10	90–95	<1
New Mexico	35	45	20
North Dakota	65	30	5
Oklahoma	95	5	0
Oregon	10	85	5
Saskatchewan	20	20	60
South Dakota	60	39	1
Texas	98	1	1
Utah	2	78	20
Wyoming	60	38	2

[a] All respondents recognized that pronghorn move about because of year-to-year climatic and rangeland conditions, and seasonal demands related to food, water, and snow depths and even time of day.

ers interviewed by Morgan obtained 10% or less of their income from fee hunting, 47% of this income was very to somewhat important; with capital tied up in long-term investments (land, buildings, etc.), hunting fees were a welcome addition to the cash flow budget and incurred few expenses.

In California, landowners who develop a management plan and increase the pronghorn herd on their property get longer seasons or more liberal bag limits for hunts on their lands (Pyshora 1986; see Chapter 16). In Texas, almost all pronghorn hunting is on private land. Permits are issued to qualified landowners or their agents, who then negotiate with hunters for trespass (or access) rights and the permits (Dvorak 1986, Swepston and Richardson 1994). Of the 368 landowners issued permits during 1995, 282 (77%) allowed pronghorn to be hunted on their property (Richardson 2000).

Achieving an adequate harvest on private land often becomes a problem. For more than 20 years, Montana has sold resident sportsman's licenses for deer, elk, upland birds and fish that cost $54.00 in 1999—$6.00 more than the individual licenses cost and perhaps twice as much as many hunters would pay if they bought licenses only for what they hunted. The money generated from such licenses is spent in part on landowner/hunter relations. One attempt to improve such relations is block management. The following comments on block management are mostly from Korn (1990), the former coordinator of that program.

The Montana Department of Fish, Wildlife and Parks instituted block management statewide in 1985. It often is referred to as a program when, in fact, it is a wildlife and people management tool. It provides wildlife managers with the ability to harvest the number of animals that need to be taken to maintain healthy herds while, at the same time, reducing damage to agricultural crops.

During the late 1970s, many ranchers in eastern Montana were closing their lands to public hunting because of hunter numbers, an increasingly stagnant agricultural economy, changing attitudes on the part of some landowners concerning public use of their land and the matter of liability.

Minneapolis, Milwaukee and other midwestern, urban centers are within a 2-day drive from the shortgrass prairies of eastern Montana. That, combined with significant interest by residents in eastern Montana hunting, resulted in a seemingly endless throng of hunters from September until after Thanksgiving. Many landowners who previously allowed public hunting grew weary of spending an enormous amount of time dealing with hunters. In response, landowners leased their lands to individuals, clubs or outfitters. Others simply posted No Trespassing signs at the gate and tried to keep up with work on the ranch.

Two of the main methods of block management offered at least a partial solution. One is based on eliminating the need for the landowner to deal with hunters. The other is by showing appreciation for the time spent meeting and directing hunters. Thus, the Montana Department of Fish, Wildlife and Parks often supplies personnel to manage hunters or pays the landowner for the time spent directing hunters, filling out permission slips, patrolling property, helping hunters retrieve downed game and other activities.

The first formal block management records were kept in 1986 (Table 164). The 1989 Montana Legislature allocated funds for expansion of block management to identify the needs and the interest in this cooperative venture. The numbers are impressive, but the implications are unclear. For one thing, it is obvious that landowners and hunters support the method, both in concept and on the ground. Landowners appreciate being able to leave the business aspect of hunting to someone else or at least being recognized and appreciated for allowing public use of their lands.

For hunters, block management has provided quality hunting experiences and successful hunts. Some block management areas have opened not only private land, but state and other public lands as well. Therefore, while addressing a resource management issue, the Montana Department of Fish, Wildlife and Parks is building positive and significant public relationships.

The future of block management looks promising, but it also is complex. To date, it has worked well for every-

TABLE 164. Landowner participation in block management in Montana, 1986 to 2001[a]

Year	*Number of landowners*	*Acres[b]*	*(ha)*
1986	86	799,360	(323,493)
1987	141	1,692,080	(684,768)
1988	188	2,550,000	(1,031,960)
1989	346	4,568,868	(1,526,971)
1990	380	5,177,000	(2,095,389)
1991	449	5,653,867	(2,288,063)
1992	521	5,023,516	(2,032,967)
1993	482	4,690,445	(1,898,176)
1994	501	5,011,721	(2,028,243)
1995	471	5,076,831	(2,054,594)
1996	883	7,119,841	(2,881,400)
1997	937	7,545,606	(3,053,707)
1998	922	7,259,602	(2,937,961)
2001	1,100[c]	8,600,000	(3,480,420)[c]

[a] From M. Korn (1990, personal communication:1994) and A. Charles (personal communication:1999).

[b] Acreage figures include public land within the boundaries of each area. However, access to some public land may only be available through the private landowner participating.

[c] Approximately.

Data not available for 1999 and 2000.

one concerned. Acreage statistics are impressive, particularly when one considers that, if anything, the numbers are slightly under-reported. Thus, hunters probably are getting more from block management than is shown by the statistics.

One reason that Montana ranchers first embraced block management lies in the department's approach. According to Mike Korn (personal communication:1991), business was conducted in the manner with which Montana ranchers were accustomed—handshakes, rather than lengthy contracts. Alan Charles (personal communication:1999) provided information on the program after 1995.

Following recommendations made by the Montana Governor's Private Land/Public Wildlife Advisory Council, the 1995 Montana Legislature passed House Bill 195, a bill that provided for enhancement of the block management program. Using revenue generated by the sale of variable-priced, outfitter-sponsored nonresident hunting licenses, the program changed from primarily an eastern Montana access program, involving $500,000 and about 450 landowners, to a statewide program that saw, in 1996, 883 landowners enrolling more than 7 million acres (2.8 million ha). Five new enforcement positions were created for better patrolling services, $2.5 million were paid directly to landowners for block management contracts and additional part-time staff was hired to manage hunters on block management areas. More than 150,000 hunters spent nearly 300,000 days hunting on those areas during 1996.

The dramatic growth in the program brought with it various changes, including a more formalized system of contracts, enrollment process, administrative rules and hunter management tools. An incentive system was developed that provided for landowners to receive up to $10.00 per hunter day (defined as any time a hunter hunts on a block management area during the course of a day), with a legislative cap of $8,000 per landowner. Under certain circumstances, nonresident landowners could receive a nonresident hunting license in lieu of compensation, and all resident landowners qualified for a complimentary resident hunting license. All landowners enrolled in the program received limited liability protection, as well as a variety of hunter management tools, such as maps, signs, permission books, sign-in roster boxes and, in some cases, staff to help manage hunters.

By 1998, the enhanced block management program was straining at the seams, with more landowners wanting to enroll than funding would allow and more hunters using the block management areas every year. The 1998 hunting season saw 922 landowners enrolling, more than 7.2 million acres (2.9 million ha), with nearly 350,000 hunter days logged on block management areas. More than 300 additional landowners expressed an interest in enrolling but were rejected due to a lack of funding.

During December 1998, the Governor's Private Land/Public Wildlife Council submitted recommendations that called for an additional $1.5 million to be raised for block management through a combination of resident and nonresident license fee increases. The 1999 Montana Legislature ultimately passed Senate Bill 338 that authorized the enhanced block management program to continue until March 2001 and provided for approximately $500,000 of additional revenue through nonresident upland game bird license fee increases. The recommendation to include resident license fee increases was stripped from the bill, although resident hunters supported it.

Currently, approximately 80% of the people statewide who use block management are resident hunters. The enhanced block management program has seen a steady increase in use, particularly by bird hunters. What originated as a program designed primarily to assist with big game harvest has evolved into a program that addresses both wildlife management and the need for hunter opportunities. The continuing increase in hunter pressure on existing areas is challenging the quality of hunting on some and resulting in overuse—to the detriment of the wildlife resource—on others. Without additional funding to enroll more areas and disperse some of that pressure, landowners and department managers will have to develop more restrictive guidelines to better manage the hunting pressure and wildlife harvest.

For the 2001 hunting season, 180,000 hunters logged more than 356,000 days of recreation on block management lands (Charles 2002). The maximum a landowner could profit per year had risen to $12,000. Since 1999, hunters have been asked to return postage-paid comment cards, grading each block management hunt. During the 2001 hunting season, 12,500 cards were returned, and 81% of hunters rated their experience as satisfactory. Only 5% of landowners annually chose not to renew their block management contracts. Usually, more landowners wanted to enroll in the program than funding could accommodate.

In states with large pronghorn populations, a few landowner-preference permits are issued without drawings, and landowners sometimes charge hunters access fees; landowners usually have no other privileges regarding pronghorn harvests. Wyoming has used a system for many years whereby a coupon, redeemable for cash, is attached to each pronghorn permit. The coupon is detached and given to the landowner when a kill is made. Of course, the hunter pays more for the permit than would be necessary without the coupon.

Wyoming's landowner coupon program came about in 1934 when the Game and Fish Commission responded to what was considered to be overpopulated herds of pronghorn in some areas of the state. The commission passed a regulation to pay the landowners $2.00 for each pronghorn killed by residents and $5.00 for animals killed by nonresidents to cover expenses of feeding said pronghorn. The concept of paying landowners for game killed on their lands was passed into law by the Wyoming Legislature in 1939. This law required the Game and Fish Commission to continue payments by including a coupon to be redeemed by the landowner on each license. The coupon program has undergone several changes since then. The differential in the worth between resident and nonresident coupons was removed, and the total worth of the coupon has increased to $9.00.

The intent of the program was and still is to reimburse landowners for forage consumed by wildlife on their property (Anonymous 1986). Nevertheless, a false impression evolved in the minds of some that the program was designed to encourage landowners to allow public hunting on their lands. The problems relating to private land access in Wyoming are significant and are worsening at the same or a greater rate than that in other western states. Therefore, if there is any intent in the program to improve access to private land, it is failing.

The question of success as it relates to the coupon program's original intent frequently is raised by agricultural groups. Such groups have expressed dissatisfaction with the program, focusing primarily on two problems—the revenues are not equitably distributed, and they are not enough. The department researched the problems and determined that the agricultural community was correct. Pronghorn that reside on one landowner's property during nonhunting seasons are likely to be killed on another's land during the hunting season (Anonymous 1986). Consequently, the landowner who gets to redeem the coupons may have sustained the least amount of forage loss to wildlife. The second argument from agriculture is that $9.00 does not compensate for the amount of forage consumed by one animal, nor does it compensate for the animals not harvested by hunters.

Another problem is that the Game and Fish Department has lost money on the coupon program. During the 1990s, landowners received $9.00 for a coupon, and a half-price resident permit (available for does and fawns after the season opened) sold for $8.50. Also, license agents received $0.50 and youth licenses sold for $7.50 at half price. Thus, the department subsidized the harvest of does and fawns in an attempt to balance pronghorn numbers with habitat carrying capacity and tolerance of landowners.

Number of Permits

The number of animals harvested usually is related to the number of hunters in the field. The best method to regulate the number of hunters participating in hunts is to increase or decrease the number of permits available. If a manager is concerned with a maximum harvest of a particular pronghorn population, then permit numbers do not need to be regulated that season. This may be the case in areas where a large pronghorn herd has not been harvested at a rate necessary to achieve management goals. Such cases are rare and, when they exist, hunting for bucks must be regulated or the sex ratio will be overly skewed.

Much more frequently, managers must restrict the number of permits to achieve management objectives. In New Mexico, the number of permits is determined after annual surveys (conducted during March in the southern part of the state and during April in the northern part of the state) give an indication of population sizes and buck/doe ratios. Then individual herd units are assigned a harvestable surplus, and permit numbers are set using past hunter-success information. Landowners in the area then are contacted to sign hunt agreements to allow for management of herds on private lands. If the landowner has public land leased for livestock privileges, the public must be allowed to hunt on that allotment. The number of permits assigned to such a ranch are split into private and public permits, according to the percentage of each

Drawings for permits by hunting districts generally are necessary to distribute the harvest among pronghorn herds in a province or state. For instance, herds are centered in eastern Montana, and human populations are centered in western Montana. Without hunting district regulations, western pronghorn herds would be overharvested, and eastern ones would be practically unhunted. The chances of drawing a permit in a western district is generally one out of two or three. Some eastern permits usually are left after the drawing and can be purchased over the counter. Photo by Len Clifford; courtesy of Leonard Rue Enterprises, Blairstown, New Jersey.

type of land ranched. This strategy allows New Mexico to set permit numbers that match the needs of the pronghorn population by herd unit.

In some states, such as Kansas, preference points for future attempts to obtain a pronghorn permit are accumulated by unsuccessful applicants (Fox et al. 2000). This is one way to allocate a limited resource to as many individuals as possible. To distribute hunter opportunity in Oklahoma, successful applicants are not eligible to reapply for 10 years (Shaw 2000).

An example of flexibility in adjusting numbers of permits to maintain population numbers comes from Saskatchewan. High pronghorn numbers during the early 1990s caused concerns over crop damage. In response, a doe/fawn season was initiated in 1991 and, in 1993, the either-sex season was removed from the draw system and licenses were sold over the counter with no wildlife management zone restrictions. In 1994, the doe/fawn season was eliminated, and in 1995 the either-sex season was returned to the draw system. Both were in response to poor reproduction (Dobson 2000). The severe winter of 1996–97 reduced pronghorn in Saskatchewan from the 22,000 estimated in 1995 to about 10,000 in 1997 (Yoakum et al. 1999). As a result, the hunting season was closed and remained closed as of 1999 (Questionnaire 2000).

Crop Depredations

Although pronghorn generate considerable income (Alberta Fish and Wildlife Division 1990; see Chapter 3), they also can cause damage to agricultural crops. Losses appear modest when viewed on a provincial or statewide basis, but they can have significant impacts on individual landowners. In nearly all cases, depredations are brought about by fencing and subdivisions that block movements, row crop plowing of summer or winter ranges, converting shrubsteppes to grass, not allowing access for hunting, or charging too much for access.

Crop depredations appeared to be increasing during the 1980s and early 1990s, apparently because pronghorn numbers generally were high, whereas traditional pronghorn habitat was being lost to improved pastures and production of small grain crops. Results of a survey sent to conservation agencies of 18 western provinces and states in 1991 indicated that depredations were stable in eight and increasing in six. Only in Nebraska were depredation complaints decreasing, apparently because pronghorn numbers were depressed. Almost every agency reported some damage to alfalfa and wheat, especially winter wheat; several added soy beans, field peas, fall rye and haystacks to the list. Most agencies did not pay compensation for crop damage; those that did, paid about $64,000 in 1990.

The Wyoming Department of Game and Fish is required to pay for depredations by big game. Determining the extent of losses, especially in winter wheat, has proven difficult (H. Harju personal communication:1992). The department is not liable for damage if the landowner does not allow hunting. On occasion, pronghorn that caused habitual depredations were relocated, sometimes to other states. This merely reduced the problem for a short time. The department usually controls depredations by: issuing licenses for a small area, such as a drainage; issuing licenses for does and fawns, and allowing multiple licenses, depending on the magnitude of the problem; and changing hunting season dates to coincide with the time when animals are in the depredation area.

The Nevada Department of Wildlife has the option of holding depredation hunts. Where depredation problems are expected, the department plans a special depredation hunt in conjunction with the season application process (J. Jeffress personal communication:1996). Applicants apply during spring; when a depredation complaint is received after the application process, an emergency depredation hunt can be initiated within 2 weeks. Unsuccessful tag applicants for that hunt unit also are contacted and awarded tags if necessary.

The majority of the Nevada depredation problems are associated with alfalfa and usually occur as rangelands desiccate during August and September. Depredation hunts generally are held during those months, avoiding an orphaned fawn problem. To reduce populations, Nevada issues permits for pronghorn with horns shorter than ears. Removal of animals and staggered seasons during the depredation period seem to alleviate much of the problem through harassment.

The 1993 Nevada Legislature authorized the Nevada Division of Wildlife to allocate big game tags to landowners who have depredation problems. The tags can be used anywhere within the hunt unit where depredations occur, not just on private lands, but they must be used during an established pronghorn season (J. Jeffress personal communication:1994). This is called the Landowner Compensation Tag Program. One tag is issued for every 50 animals present on private land with a cap of 200 tags statewide. Owners can sell such tags for use within the unit where the private land is located. Although somewhat labor intensive, the program has been a positive one in terms of the division's relationship with landowners, who typically are members of the livestock community (Dobel 2000).

Under Chapter 153 of the 1994 Laws of Utah, the Division of Wildlife is authorized to alleviate big game damage to cultivated crops by: sending a representative onto the premises to control or remove the animals; scheduling a depredation hunt; or issuing permits to the landowner or lessee for the harvest of big game animals causing depredation. Control efforts are directed toward females, sparing if possible but not excluding trophy animals.

Because of public resistance to expanding pronghorn hunting seasons in California, translocations were used to reduce depredations to irrigated alfalfa fields. Translocations are fine if suitable habitat is available for translocated animals. Apparently because of political pressure, pronghorn were moved into marginal habitat (McCarthy and Yoakum 1984), resulting in the eventual demise of whole herds. Thus, translocations designed to save pronghorn from hunters' bullets resulted in lingering deaths and expenditure of funds that could have been used for the welfare of the species.

Sometimes, simply fencing a haystack will alleviate a problem (H. Harju personal communication:1992). Electric fences may prove effective, although expensive, for keeping pronghorn out of fields. A three-strand electric fence was installed 23.6 inches (60 cm) outside an existing 42-inch (107 cm) high fence that was partly four-strand barbed wire and partly net wire, around an irrigated alfalfa field in Colorado (Pojar et al.1994). Pronghorn had been using the field—the primary green, succulent vegetation in the area—from mid-August to mid-November. The year before installation, the average daily use of the field involved 38 pronghorn, compared with 2 after the fence. Estimated dry weight of alfalfa consumed by pronghorn was reduced from 5,348 pounds (2,426 kg) during the season before installation of the electric fence to 291 pounds (132 kg) the following year. The electric fence seemed more effective near barbed wire than net wire. The animals apparently were accustomed to crawling under barbed wire and received a shock while doing so. Animals that had learned to jump the net wire did not get shocked even if they touched electric wires because they had broken the circuit with the ground. We wonder if a single electric wire, just below the bottom wire of a barbed-wire fence, would be more economical and nearly as effective.

The most practical solution seems to be reducing herds in the general area by issuing sufficient numbers of doe/fawn permits during the regular hunting season. If problems persist, depredation permits issued for hunts—in the target fields and at the time of depredations—may solve the problem. Sometimes, depredation permits should be issued for bucks-only because fawns would be orphaned if does were shot. Orphaning fawns less than 2 months of age probably would result in their death (Bromley and O'Gara 1967). Of course, capturing pronghorn for translocation also could orphan fawns. Shooting bucks will do little to reduce numbers in an area, but animals accompanying them often stay away from fields after a few animals are shot.

ESTIMATING THE HARVEST

Though seemingly different, population and harvest data are closely interrelated. Often the best source of information on the size and productivity of a wildlife population, and its ability to provide recreation through hunting, comes from a knowledge of the size of the kill and its age and sex composition.

Geis (1979:214)

Reliable estimates of harvests, hunter success and hunter days (effort) are necessary for effective wildlife management programs, regardless of the method used to formulate and implement such programs (Cada 1985). With this information, managers can assess the success or failure of harvest strategies and project adjustments to meet population objectives.

The manager should evaluate fiscal resources available, time constraints and level of precision desired before choosing harvest survey methods. Because accuracy level rarely is known, quality of harvest surveys usually is evaluated by measures of precision. Harvest estimates for large areas generally have acceptable levels of precision, but precision can be poor for small geographic units where sample sizes are small. The wildlife manager must target the geographic level at which precision is adequate and management decisions will be made. Then, the survey sample is designed accordingly (Carpenter 2000).

Questionnaire Surveys

The most common method used to estimate harvest is the mail questionnaire survey; 12 of the 18 western provinces and states surveyed use mail questionnaires exclusively or in conjunction with other methods. Some provinces and states ascertain pronghorn harvests by surveying 100% of the hunters by mail and then calculating the nonresponse bias. Others only survey a percentage of the total hunters. Several states have a mandatory card return and achieve a high level of precision in their harvest estimates.

A study in New Mexico indicated that resident deer hunters were more likely to bias their responses than were nonresident hunters (MacDonald and Dillman 1968). Although pronghorn hunters were not included in this study, the same bias probably exists. Studies in Montana indicated that hunters holding special licenses, such as for pronghorn, generally exhibited a higher response rate than did general license holders (Cada 1985). Through the various response and nonresponse bias studies conducted, statistical equations have been developed to account for bias.

The critical factor in conducting reliable questionnaire surveys is to get the questionnaire in the hunter's hands as soon as possible after the hunt. The best procedure is to issue the survey with the license, so the hunter can be prepared to identify answers to the questions. If this is not possible, the survey should be mailed within days of the hunt or the information gained will be less reliable. Several states formerly conducted follow-up surveys to nonrespondents, but found that the expense was not justified by the statistical adjustment in results (Strickland 1979, Couling and Smith 1980, Cada 1985, Pyshora 1986).

An alternative to the mail survey is the telephone questionnaire survey. This gives the surveyor direct contact with the respondent and allows for precise answers. Cada (1985) found that the telephone survey saved money, was more acceptable to the public and reduced sources of error. Another benefit of telephone surveys is that the manager does not have to wait on the mail system to begin calculations from responses. Alberta, Idaho, Montana and Oregon use telephone surveys exclusively or in conjunction with other methods to estimate the harvest. This type of survey is not without its own problems—unlisted phone numbers, people who refuse to talk to wildlife agency personnel, etc.

Check Stations and Field Checks

Requiring hunters who harvest pronghorn to stop at a check station was the first method used to obtain harvest data. Nebraska was the only state in 1990 that reported check stations as their only source of data for estimating harvests. Idaho, Kansas, Montana, Oklahoma, Oregon and Wyoming use check stations along with other methods. Biological information, such as body condition, horn size, sex and age distribution in the kill, was gathered at such stations.

The response bias of harvest estimates obtained from 1991 telephone surveys, used by the Colorado Division of Wildlife to estimate big game harvest, was evaluated by Steinert et al. (1994). Mule deer and elk hunters that completed their hunts and were leaving North Park, Colorado, were interviewed once at check stations and again 2 to 3 months later by telephone to determine whether they had harvested an animal. Steinert et al. determined the game management unit of harvest, which game management units were hunted and the number of days hunted in each game management unit. Harvest estimates from the two surveys were similar. Game management unit of harvest was the same on the two surveys for 67.4% of the mule deer hunters and 80.2% of the elk hunters. Only 4.1% of the game management units reported on the two surveys were not the same or adjoining game management units. Colorado residents were more likely to specify the same game management unit on the two surveys (73.5%) than were nonresidents (63.8%). Day of harvest was the same for 64.5% of the hunters, and there was no difference in the mean date of harvest between surveys. Telephone survey results showed two-tenths fewer days hunted than indicated by check station results. Steinert et al. concluded that postseason telephone surveys provided valid estimates of harvest, days hunted and date of harvest.

With acceptably high levels of precision obtainable from mail and telephone surveys, the check station method has become less popular among wildlife management agencies, partly because of the high cost of operation. Also, biologists in Wyoming found that check station data often are biased in several ways (H. Harju personal communication:1992). Successful hunters, especially those with large bucks, are more likely to stop at check stations than are unsuccessful hunters or those with does or fawns. Some even go out of their way to stop and show off their animals. Also, sample sizes at check stations often are low, unless there is a way to completely block hunters from leaving an area by another route. Trophy hunters, nonresidents, unsuccessful hunters and those with multiple permits are apt to stay in the field until after a check station is closed. Modern check stations mainly are to gather biological data, with harvest statistics secondary. Adequate samples for sex and age data sometimes can be gathered more economically at locker plants, but check stations give managers an opportunity to obtain a variety of timely information about the harvest.

If the sample size is large enough, which it seldom is, information gathered at check stations also may be used to cross-check the accuracy of responses to mail and telephone questionnaire surveys. To do this, hunting license or permit numbers must be recorded during all surveys. A study in New Mexico indicated that 3% of pronghorn hunters answered differently on the mail survey than what was indicated from check station records or field checks (Urqhart 1986).

Field checks also have been used to determine harvest data. During a 1990 survey, only Idaho and Wyoming reported field checks as a method used in conjunction with others to estimate harvest levels. Where field checks are conducted, much time must be devoted to contacting enough hunters to give the data statistical validity. Field checks often are conducted by conservation law enforcement officers. At times, quality of the data may suffer due to the priority placed on collection of law enforcement information. If field checks are used for harvest statistics, managers must devote enough time and manpower to the effort to ensure that sufficient data are obtained. As a rule, field checks should only be used in small areas to gather data that can be compared with information gathered by mail or telephone surveys, which obtain larger amounts of harvest data. Such field checks can aid in determining biases inherent in other data col-

Biological information—such as body condition, horn size, and sex and age distribution in the kill—is collected at check stations. Check stations also allow managers to interact with hunters, and this has public relations and educational values for the manager, wildlife agency and hunter alike. Photo by George J. Mitchell; courtesy of the Alberta Department of Industry and Development, Edmonton.

lection techniques if names and hunting license numbers are recorded.

Illegal Killing and Crippling Losses

> *Often when the manager is constructing a big game population model, values for wounding loss and illegal kill are varied widely to get the model to perform. This is indicative of the variable and largely unknown parameter being estimated.*
>
> Carpenter (2000:204)

As with most hunted species, illegal killing and crippling of pronghorn occur. The magnitude of these mortalities is difficult to measure. These deaths are not included in a harvest sampling scheme and must be accounted for. Complicated hunting regulations can increase illegal kills (difficulty determining a legal animal, hunting district, etc.), and wounding loss will vary by sex of animal, hunter density, weapon, vegetative cover and terrain.

While gathering reproductive tracts from hunter-killed pronghorn in Montana during 1969, O'Gara observed

Instant or spot check stations are easily established along any country roads and serve for law enforcement as well as to gather harvest and biological data. If a hunter knows or thinks she or he may have to pass through a check station, the temptation to take an extra animal or not tag one is reduced. Photo by Dan Pond.

about 6% abandonment (3 of 48 observed kills). Two does killed by hunters shooting at bucks in running bands were abandoned, although this was in an either-sex hunting district. Apparently, the hunters preferred bucks and were willing to abandon accidently killed does, especially a shot-up one, for the chance to harvest a buck. Such abandoned animals (illegal kills) probably would be counted as crippling losses in a study. A fawn that was killed by a bullet that passed through a buck also was abandoned. During the same collection, three hunters were approached who had not tagged three does. They were local ranchers hunting on their own land who admitted they were taking the animals home untagged because no check station was close by and the ranchers intended to hunt another day. Thus, 6 of 48 kills checked that day were illegal (about 12.5% of the small sample). When asking for permission to hunt, O'Gara also talked to a rancher in Montana who said he fed his dogs year-round on poached pronghorn.

In 1990, Utah conservation officers estimated an illegal kill in that state of less than 15%. In California, more pronghorn appear to be lost to freak accidents than are lost to illegal take (California Department of Fish and Game 1998). There, illegal harvest, especially out of season, is considered a rare occurrence and apparently is not a significant adverse impact on the pronghorn population. California Department of Fish and Game conservation officers issued only 10 citations for the illegal take or possession of pronghorn during 1991 through 1997, inclusive.

Although poaching may not be so widespread and pervasive as to affect management objectives, it has that potential, and substantial effort should be expended to reduce illegal harvests. Poaching gives nonhunters and antihunters the perception of hunting as an opportunity for lawlessness. Studies of poaching removal from deer populations indicated that illegal take may be as large as the legal take (Pursley 1977). The pronghorn poaching problem may not be as large as that with deer because the open habitat makes illegal activities readily visible to anyone in the area. State biologists attending the fourteenth Pronghorn Antelope Workshop in Colorado in 1990 did not know what the illegal kill might be in their states.

Because pronghorn inhabit open range, wounding loss should be low. After being shot, pronghorn seldom escape the view of the hunter. A summary of wounding loss in California, as reported by hunters, indicated that wounding loss for rifle hunters averaged less than 2% from 1970 through 1990 (California Department of Fish and Game 1998). Wounding loss for archers apparently is higher. Antihunting activist Pacelle (1990) contended that bowhunting may result in 100% wounding losses, but four California studies involving ground searches indicated that wounding rates range from 7% to 40% (California Department of Fish and Game 1998).

Thirty square miles (78 km^2) of pronghorn range near Saratoga, Wyoming, were selected for a study of crippling losses in 1959. Pronghorn were counted from an aircraft, and all dead animals were noted and marked before hunting season. Check stations were operated for the first 4 days of the hunting season at points where virtually all hunters could be checked as they left the study area. One hundred four pronghorn were checked through the stations. On the fifth day, another aerial survey revealed nine carcasses—8.6% of the legal harvest. The extent of crippling loss and that of illegal abandonment could not be determined, but two of the nine carcasses had been dressed and abandoned (Mobley 1960). Hunting areas and hunting conditions vary throughout Wyoming, so results of this study could not be used as a concrete figure for computing illegal kills and crippling losses on a statewide basis. Mobley reported that most game managers agreed that more animals were abandoned or crippled in some other portions of the state. Nearly all agreed that a 10% "wastage" of the annual estimated pronghorn harvest on a statewide basis would be a conservative estimate.

Studies in Colorado indicated that wounding losses varied from 4% to 25% of the legal harvests under different conditions and in different areas (Bear 1967b). However, wounding losses included not only crippled animals but also illegal kills—mostly yearling bucks mistaken for does by hunters with doe-only permits and badly shot-up animals that were abandoned.

Arizona included a question concerning unrecovered pronghorn in a hunter questionnaire. Returns indicated less than 5% crippling loss for hunters using firearms and more than 20% for those using bows and arrows (R. Lee personal communication:1990). Conservation officers in South Dakota estimated about 3% crippling loss in that state (J. Wrede personal communication:1990).

Wyoming uses 10% crippling loss as an estimate in population modeling. Losses vary with terrain and with sex and age of the animal. P. Riddle (personal communication:1990) assumed that losses vary between sexes, and a hunter who hits a good buck is more likely to go after it than if the wounded pronghorn is a doe or fawn. The ability of hunters to follow a wounded pronghorn is much greater where the animal can be kept in sight. In rugged country, more pronghorn may be wounded and lost than in flatter, more open country.

MANAGING THE HUNTER

As with any regulated endeavor, hunting of pronghorn requires a set of rules that is not necessarily related directly to the number of animals harvested. These rules and regulations are aimed at controlling the conduct of those participating, hunters, and the majority of them are a result of input from hunters.

Regulations can vary considerably between management units and jurisdictions. Some are in place primarily to benefit landowners, while others primarily benefit the resource. In Texas, a pronghorn hunter may hunt from a vehicle while on private land (Dvorak 1986). This regulation does not protect the resource but does allow the landowner more latitude in managing the hunt on his private property. Most other states have regulations against hunting from a vehicle. In New Mexico, as in some other states, it also is illegal for vehicles bearing pronghorn hunters to be driven off of established roads.

Most states have a restriction on party size for permit applications. These vary from one in Nevada (Tsukamoto 1983) to over-the-counter permits, which result in unlimited party size and are available in some portions of Wyoming (Gasson 1986).

The availability of limited permits and the large public demand have caused some states to require successful applicants to wait a year or more before reapplying. This "lay-out" period was 10 years in California (E. Loft personal communication:1990), 5 years in Oregon (M. Willis personal communication:1990) and Nevada (Tsukamoto 1983), and 1 year in New Mexico. States with large pronghorn populations generally do not have a waiting period.

Other hunting regulations include minimum caliber of rifle, no scopes on muzzleloaders, no rimfire ammunition, no parking or camping near wildlife water, hunter education requirements, wanton-waste restraints, use of dogs, proof of sex of harvested animal, no use of live animals as decoys, no hunting over bait, use of only soft-nosed bullets, no night hunting, and no hunting within a designated distance of inhabited buildings. All hunting regulations and rules—regardless of the desired harvest level—should be adopted and enforced to promote the safety of hunters, protect the vested interests of landowners, preserve the sporting aspect of the experience, ensure reasonably equal opportunity for participation, advance ethical behavior among participants, and encourage hunter respect and appreciation for the resource as well as for the recreational opportunity.

Hunter Education

Landowners and hunters should have mutual interests and respect, but if unethical or unlawful shooters antagonize landowners, more land will be posted and fewer farmers and ranchers will maintain wildlife habitat. In some European countries, hunters are respected because of their knowledge of and respect for wildlife. Such knowledge is gained primarily in schools for hunters, followed by rigorous tests to be passed before hunting permits can be obtained. Hunting ethics are especially stressed in such courses.

To deal with appropriate concerns regarding hunter behavior and ethics, hunter training should be expanded. The public increasingly will demand that animals killed for sport must be killed quickly, cleanly and with appropriate respect (Heberlein 1991). Hunters should be trained and certified to shoot well, to understand animals and their ecosystems, be generally proficient hunters and understand holistic wildlife management.

A survey of 35 state conservation agencies and 11 hunting and conservation organizations indicated positive effects of hunter education programs to improve hunter behavior (Ruh 1994). Hunters who completed courses were considered more responsible than those who did not, with a better overall attitude toward the sport.

Hunters do not want to be preached to or go to school. They may ignore a manual prepared on hunting ethics or any meeting billed as education. Small doses of hunting ethics could be incorporated into clinics on tracking, trapping, archery, etc. Some system of incentives, such as opening certain restricted areas only to those who have been through a course on hunter ethics may help. The possibility of making attendance compulsory for game law violators before they can secure another hunting license is worth considering. Possibly, certification to present to landowners when asking for permission to hunt would be a strong incentive for some to attend a hunting ethics school.

Most western states conduct hunter safety training programs, and attendance is mandatory for young hunters (usually 16 years of age or younger) to secure a hunting license; provinces and states in Canada and the United States require youngsters and first-time hunting license applicants to attend and pass an approved hunter safety course (Anonymous 1991). Ed Corvina, president of the Hunter Education Association of North America, reported that, besides safety, courses teach hunter responsibility, wildlife management and hunter ethics. Colorado requires that all applicants born after January 1, 1949, possess a hunters' safety certificate (T. Pojar personal communication:1992). This trend is encouraging, but most courses still are aimed at children and are taught by dedicated but often untrained and unpaid volunteers. More should be done.

Heberlein (1991) maintained that an effective hunter education program should start with a basic course for all new hunters. Then, after a period, one should be required to take a master hunting course, with a required exam and field testing. Special training might be required for hunting certain species.

Pronghorn hunting is different from other big game hunting in North America. Many otherwise ethical hunters still believe the only way to harvest a pronghorn is by driving about the country until a shot presents itself. We believe that orientation sessions, by themselves or in conjunction with the hunter safety courses often required for young hunters, would improve the quality and ethics of pronghorn hunters.

Hunter training programs specific to pronghorn have been conducted in Arizona, California, Nevada and on the Sheldon and Hart Mountain National Refuges in Oregon (Carter 1970, J. Jeffress and E. Loft personal communications:1991, R. Ockenfels personal communication: 1999). The pronghorn hunter training programs generally include material relevant to history and background of pronghorn hunting, description of the animal, behavior, habitat requirements, mortality factors, management, hunting tactics, identification of bucks and trophy quality, and care of trophies and meat.

The U.S. Fish and Wildlife Service initiated half-day indoctrination sessions on the Sheldon National Antelope Refuge, Nevada, at the time of the first hunt in 1967 (Carter 1970). Tags were issued only after the hunters completed the training. The sessions were discontinued in 1990 when increases in numbers of tags issued necessitated three separate hunts and, consequently, three indoctrination sessions. Approximately 800 hunters attended the sessions and most approved of them, although attitudes somewhat paralleled weather conditions because the sessions were held outdoors (J. Jeffress personal communication: 1991). The U.S. Fish and Wildlife Service initiated similar hunts and indoctrination sessions at the Hart Mountain National Antelope Refuge in southeastern Oregon in 1968 (Carter 1970).

Training sessions may go a long way toward curing some of our slob shooter problems *if more hunters attend the sessions*. We have discussed this with biologists from several states who maintained that few hunters would attend such sessions in their areas. Perhaps they were right and perhaps not.

Voluntary indoctrination sessions commenced in 1980 for pronghorn hunters holding tags for management areas 1, 3 and 5 in northwestern Nevada. The attendance rate has been 20% to 25% of the tag holders, and nearly equal numbers of hunting companions joined permit holders at these sessions to learn more about pronghorn. Follow-up questionnaires sent out during 2 years indicated that hunters liked the training.

Exhibits of mounted pronghorn bucks helped to train attendees to evaluate suitable ranges for effective shots and in rating trophies according to the Boone and Crockett Club scoring system (Nesbitt and Wright 1985). Display and audiovisual programs were used to help participants understand behavioral characteristics and habitat requirements of pronghorn. A series of speakers addressed endangered species status, hunting regulations, care of meat and trophies, etc. These orientation sessions gained popularity annually, increasing from approximately 60 attendees in 1980 to more than 400 in 1991. Possibly, this was the largest public meeting ever held concerning the biology and management of pronghorn. The Nevada Division of Wildlife proposed eliminating the indoctrinations in 1990 because of time constraints and tight budgets, but an outcry from the public caused the agency director to reestablish the sessions (J. Jeffress personal communication:1991).

Indoctrination programs for pronghorn hunters are popular in Nevada. About 450 hunters attended such a session before the hunting season in 1991. In this photo, taken on the Sheldon National Wildlife Refuge, hunters are being instructed in evaluation of pronghorn trophies. Photo by Jim Jeffress.

The program became statewide in 1993. The Conservation Education Office sends letters to all successful applicants for pronghorn tags, inviting them to indoctrination sessions held the end of July or first of August (before archery season) in Reno and Las Vegas. During 1998 and 1999, about 300 hunters attended annually. Attendance declined somewhat because many hunters already had attended a session (M. Cox personal communication:2000).

The California Department of Fish and Game has held optional (voluntary) hunter orientation meetings the day before the opening of the general pronghorn hunting season. Since at least 1971, a three-page, 24-question questionnaire has been distributed to each hunter at the meetings. Responses from about 1,850 hunters are on file, indicating that most regard orientation as informative and helpful (E. Loft personal communication:1991).

The Arizona Antelope Foundation holds an annual antelope hunter's seminar. The seventh one was held in Phoenix during August 1999, and about 200 hunters, plus foundation members, attended (R. Ockenfels personal communication:1999). The evening seminar was about 4 hours. A trophy pronghorn buck display was provided, and hunters were invited to bring their trophies or donate items for a raffle. Hunters paid $10.00 to attend and seemed genuinely interested in learning about pronghorn.

PRONGHORN RANCHING

Although many enterprises are primarily in the livestock business, indigenous big game animals are harvested for an access fee simply because they are present. However, the definition of big game ranching stipulates that management of big game animals must be intentional. Many ranchers are realizing that if the ranch is properly managed, the stocking of livestock will be commensurate with the ability of the range resources to support livestock and big game animals. Thus, many landowners intentionally manage indigenous big game animals and livestock as dual crops,

White (2000:265)

Game ranching was defined by Teer (1988:1) as "the production of wildlife through husbandry and management on usually privately owned lands and most often on ranches where domestic livestock is also produced." This definition can apply to pronghorn ranching as practiced in parts of the species' range. The willingness of segments of the sporting public to pay high prices for trophy pronghorn hunts has encouraged some landowners to take specific action to increase the quantity and quality of herds occurring on their private lands. White (1987) provided information on many aspects of big game ranching in the United States and indicated the average hunting fee for pronghorn on private land was $500.00. A guided hunt on a private ranch often costs five times that amount. The concept of individuals commercializing a public resource (wild animals) on private lands has come under careful scrutiny in recent years and is not without detractors (Geist 1988). Those interested in big game ranching are encouraged to read White (2000).

Given the legality of the practice in many areas, wildlife managers must be well aware of such operations when determining harvest recommendations. Once a landowner becomes a pronghorn rancher, the provincial and state wildlife agencies are faced with the dilemma of encouraging wildlife management on private lands, yet ensuring that hunting does not exclusively become a privilege of the wealthy. Although the right of the landowner to control access to the property is not in question, a landowner wishing to ranch wildlife must be aware of the desire of other individuals and groups to access and use publicly owned wildlife.

In 11 of the 16 western United States, landowners are given some kind of preference in drawings for pronghorn permits if they have substantial numbers of pronghorn on their land. Neither Canadian province with pronghorn grants landowners preference or has landowners who charge access fees. Also in 11 states, some landowners charge access fees for pronghorn hunting on their land. Montana has a few ranches that are surrounded by game-proof fences, with the rancher controlling seasons and bag limits on pronghorn. Less than 1% of the pronghorn in the state are involved.

A Private Lands Wildlife Habitat Enhancement and Management Area Program, commonly called Ranching For Wildlife, was authorized by the California Legislature (California Fish and Game Code, sections 3400–3409) and has been in effect since 1984 to encourage landowners to maintain and enhance wildlife habitat for species such as pronghorn. In exchange, landowners are issued tags to give away or sell along with the opportunity to hunt a pronghorn on their land. The following discussion is from the California Department of Fish and Game (1998). The primary objectives of the program were to protect and improve wildlife habitat by encouraging private landowners to manage their property in a manner beneficial to fish and wildlife. In 1997, the program involved 55 landowners on about 645,000 acres (261,032 ha) of privately owned wildlife habitat. For perspective, the Department of Fish and Game owns and manages approximately 500,000 acres (202,350 ha) statewide (see Chapter 16).

Native American tribes are not bound by state wildlife agency laws in the United States, and techniques employed in harvesting big game on tribal trust lands appear to fall somewhere between those of game ranches and state conservation agencies. Objectives vary from reservation to reservation; game ranch-type trophy management is common in the Southwest, where pronghorn are few but horns are large. Meat and sport sometimes are emphasized on reservations in states such as Montana and Wyoming, where pronghorn are numerous but their horns usually are modest trophies. Much of the hunting is reserved for tribal members, but moderately priced hunts are available for nontribal hunters.

Perhaps one extreme in management is on the Northern Cheyenne Indian Reservation in Montana. The Northern Cheyenne do not have a fish and game code, and hunting is open all year to tribal members—a situation resembling that of the entire United States during the 19th century. Consequently, resident pronghorn no longer inhabit that reservation (A. Waller personal communication:1994). Such a situation was not uncommon on reservations before the 1970s, but protection and management now are common on many reservations.

The Fort Apache/White Mountain Apache Indian Reservation in Arizona represents the other extreme—intensive management for trophies (J. Jojola personal communication:1994). Hunting there has been limited since 1983, because drought and increasing numbers of coyotes and mountain lions apparently have reduced pronghorn numbers. However, the results of the few hunts held indicate what trophy hunters can expect. During 1983 to 1989, 24 trophy permits were sold. No hunt was held in 1990 due to the continued decline in pronghorn numbers. During 1991 to 1993, six trophy hunts were sold as 4-day hunts for $3,500.00 each, which included only the trophy permit and a guide. Food and lodging were arranged and paid for by the hunters. The 30 hunters took 20 bucks, for a success rate of 67%—poor by recent pronghorn hunting standards. However, eight of the bucks made the Boone and Crockett record book; one buck scored 90 points. Tribal permits cost $25.00 and are issued by drawings in competition with nontribal applicants. During 1983 to 1987, tribal members could shoot any pronghorn, but they have been restricted to taking only bucks since then. The tribal take has been limited because of low interest in hunting pronghorn, but since reopening of the season in 1992, interest has increased.

Although the reservation probably could maintain a population of 500 to 600 pronghorn, the 1994 population was about 200 to 225—an increase of 50 to 75 since 1991—apparently as a result of predator control. Management strategies to increase the population included: applying results of radiotelemetry studies, which defined pronghorn habitat and determined fawn survival; creating movement corridors in conjunction with fuel-wood clearing of junipers; and prescribed burning of spring and summer ranges. Removal of coyotes and mountain lions, installation of water developments and improved access to water were continued. However, by 1998 the population was down to about 175 animals. Two trophy and two tribal permits were authorized, but after an aerial survey did not reveal any trophy bucks, the trophy hunt was canceled. One tribal member took a pronghorn (J. Jojola personal communication:1998).

The San Carlos Apache Tribe, also in Arizona, has two herds of pronghorn. One is an established herd living at about 6,000 feet (1,830 m) elevation. It had approximately 115 animals in 1992 (Czech 1992). It was managed for health, trophy production and hunting opportunities for tribal members. During 1990 to 1993, two trophy and two tribal permits were issued each year, and several hunters took bucks with 16- to 17-inch (40.6–43.2 cm) horns. Trophy permits sold for $2,500.00 in 1993, and tribal permits sold for $3.00 through a lottery. By 1998, about 158 animals were estimated, and three trophy and three tribal permits were authorized (A. Moors personal communication:1998).

The other herd of 120 to 130 pronghorn was introduced during February 1991 and January 1992 from the Fort Belknap Indian Community in Montana into habitat at about 4,200 feet (1,280 m) elevation (Czech 1992). It was completely protected and was to be managed like the established herd for more hunting opportunities for tribal members. As in any type of pronghorn management, the San Carlos Apaches needed to decide whether to manage for large trophies (high income) or more hunting opportunities (B. Czech personal communication:1993). However, by 1998, the herd was down to about 109 animals and no hunting was authorized (A. Moors personal communication:1998).

The Fort Belknap Indian Community—Assiniboine and Gros Ventre tribes—in Montana is home to about 5,000 pronghorn. Tags are sold to tribal members for $5.00 and to nonmembers for $300.00 before June 30 and $400.00 thereafter, according to Fort Belknap's 1994 hunting, fishing and recreational regulations. Two buck and two doe tags may be purchased by nonmembers. Demand was low until an article in a hunting magazine drew attention to the hunt. During 1995 and 1996, more applications for permits were received than tribal biologists wished to issue. Few tribal members care about pronghorn hunting. The community requires that a nonmember hunter have a tribal member as a hunting partner.

As part of the technical assistance program provided by the U.S. Fish and Wildlife Service to the Shoshone and Arapaho tribes on the Wind River Indian Reservation in Wyoming, the status of native pronghorn was investigated from July 1978 to August 1981 (Smith 1981). The reservation was found to support about 1,000 pronghorn—0.45 per square mile (0.17/km^2), compared with about 2 to 5 per square mile (0.77–1.93/km^2) on adjacent state lands. Smith (1981) found reproduction to be low on the Wind River Reservation and proposed 2,750 pronghorn as a realistic goal. The success on the reservation between 1984 and 1994 illustrates how rapidly good management practices can change a population.

Subsistence hunting occurred until the adoption of the game code in 1984, with the exception of 1948 to 1953, when strict hunting regulations were imposed. However, big game populations dwindled drastically from the early 1960s through the mid-1980s (R. Baldes personal communication:1991). Although hunting was taking place, big game populations were at such low levels that it would be difficult to characterize it even as subsistence hunting. In some areas, populations of pronghorn were extirpated as a result of year-round hunting. Hunting changed with the inception of the 1984 game code. Strict hunting regulations (e.g., short seasons, bucks only) along with stiff penalties and effective law enforcement provided a basis to restore and build populations. Also, 352 pronghorn were translocated to the reservation during 1986 to 1993 from other parts of Wyoming (R. Baldes personal communication:1994). An increase in licenses sold and hunting success accompanied the increase in the pronghorn population. Only 198 licenses were sold in 1985, and hunter success was a mere 12%; 388 licenses were sold in 1993, and hunter success increased to 39%.

Quality pronghorn hunts were promoted on the Sheldon National Wildlife Refuge, Nevada, and the Hart Mountain National Antelope Refuge, Oregon, by the U.S. Fish and Wildlife Service (Carter 1970). These hunts (patterned after trophy desert sheep hunts conducted in several southwestern states) included a prehunt indoctrination session that stressed the sporting aspects of trophy hunting. The chances of getting a good trophy are enhanced by few hunters in the field and the limited kill, which results in many prime-aged bucks being available. Many participants considered the experience to be a once-in-a-lifetime hunt. Such hunts are an example of obtaining recreational benefits where a large annual harvest is not needed.

The Colorado Experiment

Personnel of the Colorado Division of Wildlife evaluated a cooperative, private lands wildlife management program called Wildlife Ranching during 1986 to 1988 on 10 ranches (Hopper 1990). A total of 2,634 licenses for six game species were allocated on those ranches—55.2% to private hunters, or hunters who were selected and generally guided by the ranchers. Public hunters, or hunters who drew for hunting permits with the general public, received 44.8% of the licenses. Through this program, nearly 800,000 acres (323,760 ha) of private land were opened to public hunting.

Unless otherwise specified, the following discussion of the Colorado Wildlife Ranching program is from Hopper (1990). Approximately two-thirds of Colorado's land and water area is privately owned. Private lands provide valuable wildlife habitats, such as big game wintering areas and some important year-round habitat. Absence of Division of Wildlife control over habitat protection on and public access to these lands frustrated the agency in its efforts to meet statewide management objectives. This situation encouraged the division to investigate innovative ways to address wildlife management on private lands. This was necessary because of: (1) increased public demand for wildlife-oriented recreation, (2) a declining wildlife habitat base, (3) difficulty in effectively implementing management plans by species, herds, etc., (4) high costs associated with game damage and (5) less than desirable relationships between the Division of Wildlife, landowners and hunters.

Private landowners generally viewed wildlife as a liability rather than an asset, and protection and management of wildlife habitats was not a priority compared with other land uses. Implementing a program offering landowner incentives and increasing opportunities for public hunters offered potential benefits for private and public interests as well as the resource. The program also provided the Division of Wildlife increased flexibility in managing wildlife on private lands. The long-term success of such a program will depend on: (1) preserving, protecting and improving wildlife habitat, (2) providing more public access and recreational opportunity, (3) more effectively implementing species management plans, (4) decreasing wildlife damage and associated costs, (5) improving the relationship between the Division of Wildlife, landowners and hunters, (6) minimizing costs to the Division of Wildlife and (7) gaining public acceptance.

By law, as in other states, Colorado's wildlife belongs to the public, with full authority for management of that wildlife vested in the Wildlife Commission. Concerns were expressed that the Wildlife Ranching program would undermine that statutory charge and commission authority would be relinquished to the landowners. However, all wildlife regulatory authority empowered to the Wildlife Commission by Colorado statute applies under the Wildlife Ranching program, including final approval of season dates and license numbers. Landowners of properties selected for the experimental Wildlife Ranching program were required to allow public hunting. Season dates and lengths were not restricted to regular statewide seasons.

The distribution of antlered and antlerless (horned/hornless) deer, elk and pronghorn licenses among private and public hunters was determined by the rancher through selection of one of four 1990 options (Table 165). The only constraint the Wildlife Commission placed on developing these trial options was that some antlered licenses be allocated to public hunters. Options 3 and 4 were most popular for distributing pronghorn licenses. These options were altered and expanded to five in 1993 (R. Kahn personal communication:1993). Revised (1993) options 3 and 4 (Table 165) also were most popular for distributing pronghorn licenses. The private share of the licenses was offered directly to individuals of the ranchers' choice without going through a drawing. The public share of licenses was issued through the Division of Wildlife's normal limited license-drawing process.

A total of 609 pronghorn licenses were issued during the experimental program. Public hunters received 17.3% of the 307 licenses for adult bucks and 73.2% of the permits for does and fawns. Each rancher was provided with names and addresses of all public hunters successful in drawing a license for their ranch. A map and hunting rules for each ranch was sent to each public hunter before the season, along with a cover letter explaining the program.

Table 165. Options available for distribution of antlered and antlerless licenses between private and public hunters (horned and hornless for pronghorn) on ranches associated with Colorado's Wildlife Ranching program

	Percentage of landowner's share		*Percentage of public's share*	
Options	*Antlered*	*Antlerless*	*Antlered*	*Antlerless*
1990[a]				
1	95	0	5	100
2	90	25	10	75
3	85	50	15	50
4	80	>50	20	<50
1993[b]				
1	90	0	10	100
2	85	25	15	75
3	80	40	20	60
4	75	50	25	50
5	50	0	50	0

[a] Experimental phase (see Hopper 1990).
[b] From Kahn (1995).

Ranchers had the option of charging each public hunter a limited access fee not to exceed $25.00 for each license.

Only 6 of the 10 ranchers in the experimental program managed for pronghorn. Private pronghorn seasons usually started in mid-September. Public pronghorn seasons started in late September or early October. Average private pronghorn seasons lasted 33.3 days, and public hunts averaged 6.7 days.

Private hunters harvested 129 pronghorn bucks and 44 does on the six ranches. Public hunters took 43 bucks and 85 does. Both hunting groups enjoyed success rates of about 84% on bucks and 76% on does, compared with overall success rates during statewide rifle seasons of 71% in 1986 and 76% in 1987.

Ranchers generally were pleased with results of the experiment. They did not have any major problems with either private or public hunters, or with the Division of Wildlife. All participating ranchers supported continuation of the Wildlife Ranching program as conducted during the experimental period.

However, public acceptance of the program was not unanimous. Some people disagreed with the concept, expressing the opinion that the program subsidized private landowners and allowed commercialization of a public resource. Others accepted the concept, but were dissatisfied with the program in the experimental form. Yet another segment of the public supported the program as conducted during the experiment. Support for the program was greatest among public hunters who participated.

Four complaints by the public were: (1) the minimum acreage requirements for ranch participation was too low, (2) the percentage of antlered licenses and total number of licenses allocated to the public was too low, (3) private hunters enjoyed much longer seasons with better timing than did public hunters and (4) charging public hunters an access fee was considered detrimental to the program.

The Division of Wildlife conducted the 3-year experimental program with a minimum of expense. Burdens on the regulation-setting process, limited license-drawing process, enforcement and administration were increased. However, the increased burdens did not require the hiring of any additional employees. Instead, the workload was distributed among existing personnel. A specific budget was not required to administer, monitor or evaluate the experimental Wildlife Ranching program. The additional costs associated with the program were for vehicle mileage, per diem, telephone and postage expenses—estimated to total less than $5,000 for the 3 years.

The program protected some valuable wildlife habitat by stopping, or at least delaying, plans of further subdivision of one of the ranches. These results suggest that, overall, the program produced substantial benefits in terms of public access, quality hunting, habitat protection and enhanced Division of Wildlife flexibility to manage wildlife populations on private lands.

Administratively, the program was feasible. Although it generated substantial public benefits, the public received relatively few licenses on some ranches, raising the question of viability of the plan on those ranches. The Division of Wildlife cannot administer a procedure with minimal public benefits. Potential exists to provide benefits to private landowners, the public and the business community, but Wildlife Ranching cannot prosper without long-term public support. This support depends on the credibility established by the Division of Wildlife and by ranchers in developing and conducting a fair program.

Based on evaluation of the 3-year experimental agenda, the Division of Wildlife recommended that the Wildlife Commission continue and expand the Wildlife Ranching program on an operational basis. However, the division, with public support, believed that some adjustments were necessary to provide a better balance of private and public benefits.

New guidelines were completed in 1993 (Kahn 1993) and again in 1998 (R. Kahn personal communication:1998). Changes included: (1) ranch size was increased to 12,000 contiguous acres (4,856 ha); (2) the percentage of antlered licenses and number of total licenses allocated to the public were increased; (3) public hunters received a post-hunt survey form, and ranches that received a low rating could be dropped from the plan; (4) no access fees could be charged to public hunters; and (5) ranchers were required to initiate habitat management programs to benefit wildlife. Through this program, about 2.2 million acres (890,340 ha) of private land has been opened to the public on 27 ranches (R. Kahn personal communication:1998). In 1998, 12 of the ranches offered pronghorn hunting on approximately 600,000 acres (242,820 ha). License allocations for that year included 393 buck pronghorn licenses to ranchers, as well as 59 buck and 313 doe licenses to public hunters through a drawing.

A study was conducted to evaluate the satisfaction of public hunters who hunted on Wildlife Rancing ranches during 1994 (Lloyd et al. 1995). Telephone questionnaires were administered by trained personnel. Of the 1,121 hunters, a total of 878 completed the surveys. Overall, public hunters spent an average of 2.7 days hunting on Wildlife Ranching ranches. Quality of the hunt and success rates were the most common reasons for hunting through the program. Overall satisfaction levels of public hunters with their 1994 hunting experience in percentages were as follows: strongly satisfied—76.9%; somewhat satisfied—11.6%; slightly satisfied—2.3%; neutral—0.3%; slightly dissatisfied—3.6%; somewhat dissatisfied—1.3%; and strongly dissatisfied—4.0%. The likelihood of the same public hunters applying for a permit through the Wildlife Ranching program the next year was 94%—slightly higher than the percentage satisfied with the hunt.

Demand for pronghorn hunting opportunity likely will increase in the future. The mechanisms for allowing private land management and public utilization of the pronghorn resource must be further refined before all concerned will be able to reach agreement.

After habitat management, harvest management is the most practical and efficient method to ensure that pronghorn and other game species remain stable and viable components of the North American ecosystem, in balance with variable levels of human tolerance and demand for recreational use of those animals and their habitats. Photo by Harold Umber; courtesy of the North Dakota Department of Game and Fish, Bismarck.

CONCLUSIONS

Harvest management for big game is the art of melding the objectivity of wildlife science and the subjectivity of public wants for the attainment of a management goal.

Strickland et al. (1994:446)

Human dependence on wild game for food has given way to other motivations, objectives and satisfactions for hunting. Early settlers in the West did not concern themselves with regulating the harvest of pronghorn, but took what was needed to feed and clothe themselves and their families. Unregulated market hunting also took a significant toll on wildlife populations, including pronghorn. By the beginning of the 20th century, governments and hunting organizations sought to regulate harvests to prompt the recovery of many species that had been overexploited. From that fairly recent origin, the science of wildlife harvest management has made extraordinary advances in helping the recovery and sustainability of pronghorn and other wildlife. As the environment continues to be altered and human demands for wildlife expand and shift, harvest management must be continually refined and improved.

Today's pronghorn managers must establish long-term goals for pronghorn populations and their habitats relative to current and projected demands for use of resources (including pronghorn) and space. Based on such goals, management is defined by objectives and refined by short-term strategies. At all levels of planning and action, the needs of the resource, users (consumptive and nonconsumptive) and landowners (public and private) must be considered and meshed if the goal is to be achieved.

When developing harvest recommendations, several factors must be addressed. They include but are not limited to habitat conditions, sex and age ratios, other uses of the area's resources (both animal and plant) and behavior of the pronghorn.

Pronghorn harvest regulations fall into two categories: (1) those that manipulate the type and degree of harvest and (2) those that manage the hunter. Regulating the type of harvest includes setting bag limits, season lengths, legal weapons, number of permits and other rules to ensure that a strategy-specific number and sex of pronghorn are harvested. Regulating the hunter consists of various restrictions placed on participation in the hunt to assure that hunts are conducted legally, safely and ethically, and to maximize the opportunities for participation and harvest within the guideline principle of sustained yield.

Essential to effective harvest management are accurate, statistically comprehensive and timely harvest data. With this information, managers can assess the success or failure of harvest strategies and project adjustments to meet the pronghorn population objective. The most commonly used method to obtain this information is through mail questionnaires, although the telephone questionnaire survey is gaining in popularity. Check stations and hunter field checks are less in use today than in past decades. Modern check stations are mainly for gathering biological data, with harvest statistics secondary.

The results of managing the harvest must be the sustained viability of pronghorn and their habitat. The recreational utilization of those resources can continue as long as managers have realistic goals and objectives and consider all factors and constituencies when recommending strategies to meet objectives in the best long-term interest of pronghorn.

"The future of harvest management for large mammals will depend on how well managers maintain and protect habitat and wild places for animals to live and for hunters to hunt. It will also depend on how well big game managers address the increasingly complex demands that society expects from wildlife management programs" (Carpenter 2000:208).

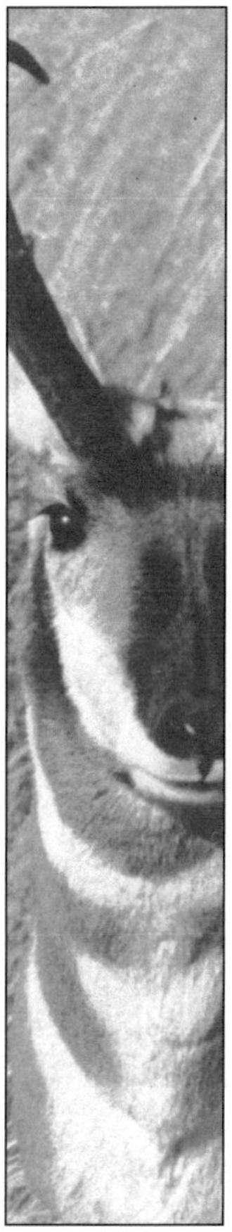

C H A P T E R   T W E N T Y - O N E

Capture, Translocation and Handling

Bart W. O'Gara, Craig J. Knowles, Pamela R. Knowles and Jim D. Yoakum

Much of the success of an antelope trapping and transplanting program depends upon making an intensive inventory of a proposed release area to ascertain if it is suitable for pronghorns. . . . The great speed at which antelope travel, and their fragile bodies, makes capture of these animals in good condition for transplanting extremely difficult.

Hoover et al. (1959:79, 82)

Native Americans captured pronghorn in drive traps for thousands of years before the arrival of Europeans. Nearly 6,000 years ago, Native Americans apparently were driving pronghorn into corral traps in Wyoming (Sanders 2000). The trapping area still is a bottleneck for pronghorn during seasonal movements (Sawyer and McWhirter 2000). Stone Age trappers knew what biologists had to relearn: to trap pronghorn successfully, one must understand their behavior and how they use the landscape. Lewis and Clark reported that the Mandans constructed V-shaped traps with long wings of shrubs that funneled the animals into corrals (Moulton 1987a). Such drives required the cooperation of dozens of Indians, but the reward often was 50 to 100 pronghorn for food and clothing (see McCabe et al. 2004).

Trapping and translocation have played an important role in pronghorn restoration since the late 1930s. Most translocations have been accomplished to restore pronghorn to unoccupied historical rangelands (Yoakum 1978a), but some have been used to remedy agricultural depredation problems (Pyshora 1982a). Translocation has proven remarkably successful at returning pronghorn to their former rangelands.

The first known pronghorn translocations were in Alberta. Federal and provincial governments there attempted to increase pronghorn numbers, during 1907 to 1920, by establishing two preserves where small herds persisted. Permits were issued in 1909 and 1910 to move pronghorn to two parks and a private enclosure to help ensure the species' survival in Alberta (M. Watson personal communication:1987). No detailed record exists to describe those translocations; apparently, none was successful. The Alberta legislature gave specific authorization to C. J. Blazier and several others to capture pronghorn for transfer to Banff National Park and Wainwright Buffalo Park. Blazier established a pronghorn farm near Brooks during 1919 or 1920. Fawns were captured locally, raised on the farm and transferred to various localities in Alberta during the next 10 years (Mitchell 1980). Some were released at the Wichita National Wildlife Refuge in Oklahoma around 1921.

Although capturing and hand-rearing fawns for translocation was possible, the effort needed to capture fawns and the investment required to rear them proved excessive. The capture of three fawns was considered a good day's work, and they could be caught easily only during a 4-week period in spring, as discussed later. Rearing involved intensive care for several months. Also, the larger the number of fawns kept together, the greater the chance of disease.

A workable translocation program required that 50 or more of free-ranging adult pronghorn be captured and moved with minimal expense and loss of animals. The fleet, fragile pronghorn posed special problems. It was not until April 1937 that T. Paul Russell (1964), of the New Mexico Department of Game and Fish, successfully devised a technique to make multiple captures of free-roaming pronghorn. Incorporating methods developed for capturing feral horses, Russell used a corral at the apex of an angle formed by two wing fences. The trap was placed so the wings intercepted the normal travel route of a pronghorn band. A band then was herded into the trap. The primary difference between this procedure and that of yesteryear Indians was in the former's employment of vehicles and horses to drive the animals. Improvements in materials and design soon followed and, by July 1938, 282 pronghorn had been trapped and translocated in New Mexico without excessive mortality. This exceeded the number of pronghorn translocated during the previous 30 years.

Important methodological concepts developed during that early trapping effort, which still are widely used, include: (1) a double corral used for initial capture and as a holding pen and, at the apex of this corral, a smaller crowding pen for hand-capture and loading of pronghorn; (2) trap location on a normal pronghorn travel route, below the crest of a hill; (3) netting for the walls of the crowding pen to reduce injury; (4) burlap or canvas curtains to deter pronghorn from ramming woven-wire fencing; (5) active herding of pronghorn to the trapsite; and (6) proper handling of pronghorn to reduce risk of injury to workers and animals.

Motion pictures of the 1937 and 1938 pronghorn trapping and translocating operations were shown in June 1938 at a meeting of the International Association of Game, Fish and Conservation Commissioners in Asheville, North Carolina. Those pictures and reports of translocated pronghorn alerted other state wildlife agencies to the potential of reintroducing pronghorn to the species' former rangelands. Shortly thereafter, states from the Dakotas westward, with unoccupied historic pronghorn range, were making plans to restock suitable habitats.

New Mexico had trapped and translocated 487 pronghorn before November 20, 1939, when its first Federal Aid in Wildlife Restoration Act (also known as Pittman-

Early translocation efforts centered on hand-rearing fawns until weaning, at which time they were crated and shipped to their destinations (Floyd 1924, Nelson 1925, Einarsen 1948). One advantage of the method was that, despite hand-rearing and contact with people, fawns released during late summer soon developed normal wildness and became unapproachable (Nichol 1942). During the late 1940s, about 300 pronghorn fawns were taken from the Charles Sheldon National Wildlife Refuge, in Nevada, and Hart Mountain National Antelope Refuge, in Oregon, for restocking purposes (Einarsen 1948). Although pronghorn raised to maturity and bred in captivity were tame, their offspring were wild if allowed to suckle their tame mothers. Photo courtesy of the Kansas State Historical Society, Topeka.

Robertson or P-R Act) funds were obligated under project W-1-D, *Antelope Trapping and Transplanting*. Pittman-Robertson funding soon added momentum to trapping and translocating in other western states as well.

During the 1940s, six other states followed New Mexico's lead, as exemplified by the successful program for Texas reported by Fisher (1942b). For herding pronghorn into traps, aircraft replaced vehicles and horsemen. Trapping and translocating rapidly evolved from trial and error to science. By the end of the decade, at least 9,516 pronghorn had been translocated in the United States (Table 166). Beginning in 1941, Montana restocked historic pronghorn ranges with animals trapped from areas that still had substantial populations. More than 3,000 pronghorn were translocated during 1945 to 1952 using corral traps. This was a popular program with hunters because it increased hunting opportunities.

The number of pronghorn translocated during the 1950s and 1960s gradually declined as unoccupied habitat became replenished. Where pronghorn populations were small, trapping in the same area year after year became impractical. The animals became trapwise and lost their fear of aircraft, making them difficult to herd. During the 1970s and 1980s, translocating again increased when Kansas, Utah and Nevada implemented such programs. Texas resumed its reintroduction program between 1972 and 1982 when animals became available on a large, unhunted ranch. Pittman-Robertson funds paid for most of these operations.

Although translocation generally has been considered a successful program, few agencies have detailed accounts of their success. All eight attempts in North Dakota, averaging 51.6 pronghorn per attempt, were considered successful (J. McKenzie personal communication:1987). However, results of translocating 68 pronghorn to an area near Turtle Lake during December 1991, were still questionable as of 1998 (B. Jensen personal communication:1998). The animals were scattered widely and numbers seemed to be dwindling. Turtle Lake Gun Club members suggested the introduction. The governor concurred, although Game and Fish Department personnel recommended against it (J. McKenzie personal communication:1994). The release area was about 96% private land, nearly 57% of which was cultivated land and less than 20% was native grassland.

During the winters of 1950–51 and 1951–52, 84 pronghorn were trapped near the Black Hills of South Dakota and translocated to the south-central part of the state (Bever no date). This was historic pronghorn range, but contained few or no pronghorn at the time. The translocations were successful and extended the range of the pronghorn into that part of the state. Translocations apparently continued through the 1960s (Hodgins 1972), but few records were kept concerning numbers, successes or failures. In 1985, 104 pronghorn from Sheridan County,

Table 166. Minimum number of pronghorn translocated into states or within a state by time period[a]

State	*1920–29*	*1930–39*	*1940–49*	*1950–59*	*1960–69*	*1970–79*	*1980–90*	*1991–97*	*Total*
Arizona	12[b]	9	189	262	126	121	865	670	2,254
California	—	6	145	—	—	74	873	—	1,098
Coahuila	—	—	—	—	—	—	—	159	159
Colorado	—	—	356	619	—	—	975	—	1,950
Idaho	—	—	—	—	—	—	36	—	36
Kansas	—	—	—	—	182	449	156	100	887
Montana	—	—	3,012	415	175	37	47	—	3,686
Nebraska	—	—	—	209	861	—	—	196	1,266
Nevada	—	—	150[b]	44	—	—	1,247	471	1,912
New Mexico	—	475	2,129	1,830	524	667	394	—	6,019
North Dakota	—	—	—	413	—	—	—	67	480
Oregon	—	—	150[b]	—	64	117	384	—	715
Oklahoma	—	—	—	—	—	—	—	52	52
San Luis Potosi	—	—	—	—	—	52	—	—	52
South Dakota	—	—	—	121	349	—	105	—	575
Texas	—		3,237	823	—	1,384	266	—	5,710
Utah	—	—	—	—	129	866	2,699	52	3,746
Washington	—	27[b]	16[b]	10	22	—	—	—	75
Wyoming[c]	—	—	129	—	—	—	—	—	129
Total	12	517	9,516	4,647	2,432	3,730	8,012	1,505	30,801

[a] Records from many states are incomplete, so the numbers in this table are minimal estimates. Alberta and Saskatchewan either did not translocate pronghorn or lost their records of such activities.

[b] Fawns.

[c] Data known to be incomplete.

Wyoming, were released on the Crow Creek Indian Reservation (Rice 1985). The translocation was conducted through cooperative efforts of the South Dakota governor's office and Indian tribes, but the habitat was considered poor quality for pronghorn by personnel of the South Dakota Department of Game, Fish and Parks because crops in the area presented the potential for depredations. The animal population increased to about 370 on approximately 140,000 acres (56,658 ha). However, further expansion seems unlikely because farmers are complaining of depredations and apparently shooting some pronghorn (D. Big Eagle personal communication:1994).

During 1972 to 1982, Texas released an average of 21.3 pronghorn (buck/doe ratio = 77:100) at 78 sites. By 1984, only 12 sites were successful, 23 were rated undetermined and 31 were failures (DelMonte and Kathmann 1984). At many of the unsuccessful release sites, pronghorn movements were restricted by pronghorn-proof fences. Coyote predation, illegal hunting, drought and the small number of animals per translocation also were thought to contribute to the failures.

In Colorado, all but 5 of 24 early translocations, averaging 41 pronghorn per effort, were considered successful (Hoover et al. 1959). During the 1990s, about 850 animals were translocated to Arizona, Kansas, Nebraska and Nevada.

Pronghorn never were extirpated from Arizona, but they reached an estimated low of 650 animals during the early 1920s (Nelson 1925). Active management practices, including translocations, increased numbers to more than 12,000 by 1995 (Alexander 2000). Early transplants, beginning in 1924, of small numbers of fawns proved unsuccessful. By 2000, 2,998 pronghorn had been released in the state. In 1981, Arizona began supplementing herds on rangelands in the southeastern part of the state with Mexican pronghorn. Before that date, all pronghorn released in southeastern Arizona were of the American subspecies. Alexander (2000) provided an excellent history of pronghorn translocations to and within Arizona (Figure 154), including known mortalities and capture methods (Table 167).

Nevada conducted a number of introductions between 1950 and 1960. Four resulted in established herds (Tsukamoto 1983). Others, however, were not successful. A herd translocated to the eastern benchlands of Charleston Mountain, in Clark County, existed for approximately 15 years. Recruitment occurred annually, but mortality factors were greater and the herd died out. Subsequent evaluation of the release site disclosed that the habitat was marginal for pronghorn (Yoakum 1978a). Although pronghorn were increasing, 471 pronghorn were moved into Nevada during the 1990s from Colorado, Oregon, Utah and Wyoming.

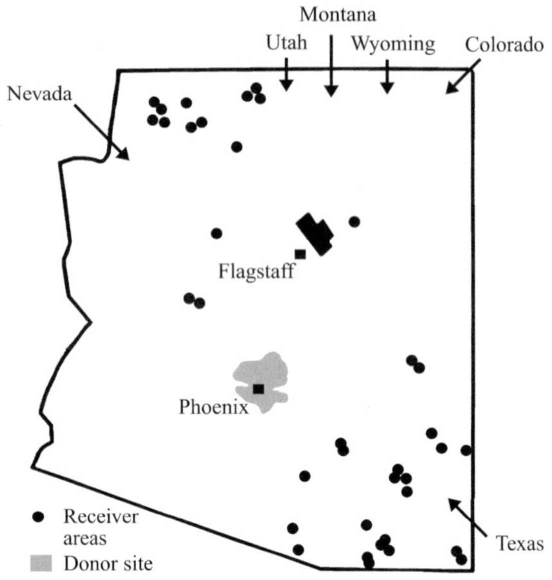

FIGURE 154. *Pronghorn translocation donor sites and receiver areas in Arizona, as well as donor states through March 1996 (from Lee 2000).*

California's endeavors to relocate pronghorn commenced in 1933. An attempt was made to capture and transport six adults in Los Angeles County, but all died, possibly from stress or capture injuries (Goldsmith 1988). During 1940 to 1950, pronghorn were trapped and translocated to augment low-density populations within the state. The program was reactivated in the late 1970s to 1990. From 1947 to 1990, about 1,100 animals were translocated from northeastern California to seven locations within historic range (Fischer 2000). Reasons for the more recent relocations include: the state's pronghorn population increased for two decades and alfalfa field depredation complaints increased, continued augmentation of low-density areas and attempts to establish herds in game-deficient areas. Three releases at one site with 219 pronghorn resulted in all the animals moving to adjacent habitats. By 1999, the herd numbered less than 100 on nearby Nevada rangelands. During the 1980s, attempts were made to establish herds on historic habitats unoccupied for more than 100 years (California Department of Fish and Game 1998).

Prehistoric pronghorn inhabited the Columbia Plateau on the east side. Archeological investigations found pronghorn bones at Indian house pits near Nespelem, Ritzville and Wallula, but pronghorn probably were extirpated from Washington about 1700, before European settlement. Change in climatic conditions over the past 4,000 years probably led to the demise of the animals in this state (R. Johnson personal communication:1994). In 1938 and 1939, 50 pronghorn fawns were obtained from the Sheldon

TABLE 167. History of pronghorn translocations into and within Arizona, including known mortalities

Capture date	*Capture site*	*Release site*	*Number captured*	*Number of mortalities*[a]	*Number released*[a]	*Comments*[b]
1924	Northern Nevada	Hermit Basin, Grand Canyon (halfway down canyon)	12	—	12	Pen-reared, unsuccessful [A]
1932	Chavez Pass, E. of Anderson Mesa	South of Ashfork	10	8	2	Drift net [A]
1939	Anderson Mesa	Avra Valley, Tucson Mtn. Park	11	4	7	Pen-reared, unsuccessful [A]
6/10–11/1941	Anderson Mesa	Ellsworth Ranch, Whitlock Valley SE of Safford	4	2	2	Pen-reared, found dead in corral 3 weeks later [A]
1/22–24/1943	Raymond Antelope Refuge	G Bar Ranch, Winchester Mts, Upper Sulphur Springs Valley	24	2	22	Drift net [A]
2/8/1943	Raymond Antelope Refuge	Patterson Ranch near Oracle	6	2	4	Drift net, unsuccessful [A]
1/19/1944	Raymond Antelope Refuge	Box Springs, Winchester Mts, Upper Sulphur Springs Valley	3	—	3	Drift net [A]
3/9/1944	Raymond Antelope Refuge	Box Springs, Winchester Mts, Upper Sulphur Springs Valley	3	—	3	Drift net [A]
1/6/1945	Raymond Antelope Refuge	Box Springs, Winchester Mts, Upper Sulphur Springs Valley	3	—	3	Drift net [A]
1/22/1945	Raymond Antelope Refuge	Bosley Pasture, Winchester Mts, Upper Sulphur Springs Valley	19	—	19	Drift net [A]
1/31/1945	Raymond Antelope Refuge	Box Springs, Winchester Mts, Upper Sulphur Springs Valley	18	—	18	Drift net [A]
2/9/1945	Raymond Antelope Refuge	Double Tank Altar Valley, Manning Ranch	18	—	18	Drift net, unsuccessful [A]
2/11/1945	Raymond Antelope Refuge	Ashburn Ranch, San Rafael Valley	9	—	9	Drift net [A]
3/8/1945	Raymond Antelope Refuge	Ashburn Ranch, San Rafael Valley	4	—	4	Drift net [A]
3/12/1945	Raymond Antelope Refuge	T Bar W Tank near Oracle	15	—	15	Drift net, unsuccessful [A]
12/19/49	Antelope Pass, N. Grapevine Cyn	Fort Huachuca Wildlife Area	72	—	72	Drift net [A]
1/15/1951	Antelope Pass, N. Grapevine Cyn	Fort Huachuca Wildlife Area, north end	28	2	18	Drift net, 8 escaped [A]
2/1–6/1951	Antelope Pass, N. Grapevine Cyn	Fence Cyn	152	16	99	Drift net, [A]
		House rock Valley	—	(5)	(42)	
		San Rafael Valley	—	(11)	(57)	
1959	Arizona	San Simon Valley	12	—	12	Unable to substantiate
9/21–22/1959	Raymond Ranch	Bonita Prairie, Fort Apache Indian Reservation	60	10	34	Drift net, 16 released capture site [A]
12/7/1960	Raymond Ranch	Bonita Prairie, Fort Apache Indian Reservation	57	1	37	Drift net, 19 released capture site [A]
10/30/1961	Raymond Ranch	June Tank, Arizona Strip	125	7	34	Drift net, 84 released capture site [A]
10/28/1965	National Bison Range, Moiese, MT	June Tank, Arizona Strip	55	—	55	Drift net [A]
11/16–17/1971	National Bison Range, Moiese, MT	Upper Clayhole RCA, Arizona Strip	37	—	37	Drift net [A]
12/9/1979	Lyman, WY	Hurricane Valley, Arizona Strip	200	18	84	Drift net, 98 released capture site [A]
2/1981	Limon, CO	Diamond Butte, Arizona Strip	86	4	82	Drift net [A]
11/18/1981	Buena Vista Ranch Marfa, TX	Empire Ranch	54	3	51	Drift net [M]

continued on next page

Table 167—*continued*

Capture date	*Capture site*	*Release site*	*Number captured*	*Number of mortalities*[a]	*Number released*[a]	*Comments*[b]
12/2/1982	Parker Mts, UT	House Rock Valley	50	2	48	Drift net [A]
1/17–19/1984	Crestone, CO	House Rock Valley	83	2	81	Drift net [A]
1/18–20/1984	Crestone, CO	Diamond Butte, Arizona Strip	103	8	95	Drift net [A]
11/13–14/1984	Marfa, TX	Moline Tank, San Bernardino Valley	43	11	32	Drift net [M]
12/4–6/1984	Douglas, WY	Hillside	86	1	85	Drift net [A]
		Near Winslow	—	(1)	(51)	
		Phoenix Zoo	—	—	(30)	
					(4)	
12/9–10/1986	Rocker B Ranch, Big Lake, TX	San Bernardino Valley	70	1	69	[M]
12/9–10/1986	Rocker B Ranch, Big Lake, TX	Day (Lazy B) Ranch, AZ/NM border	39	2	37	[M]
12/8–10/1987	Big Lake, TX	Altar Valley, Buenos Aires NWR	101	1	100	Drift net [M]
		Fort Huachuca, south range	—	—	(86)	
			—	—	(14)	
2/8/1993	Sheridan, WY	Hillside	54	—	54	Drive net (with helicopters) [A]
1/29/1997	Lamar, CO	Cedar Flat	149	25	124	Drive net (with helicopters) [A]
		Big Flat Well	—	(9)	(65)	
		Perry Mesa	—	(3)	(37)	
			—	(13)	(22)	
12/8–9/1997	Loa, UT	Sipes Wildlife Area	189	5	184	Drive net (with helicopters), both Perry Mesa mortalities were collared [A]
		Cedar Flat	—	(1)	(43)	
		Big Flat Well	—	(1)	(64)	
		Perry Mesa	—	(1)	(36)	
			—	(2)	(41)	
12/15/1998	Loa, UT	Bozarth Mesa	197	14	183	Drive net (with helicopters), 6 mortalities were collared animals [A]
		Hillside	—	(2)	(77)	
		Big Flat Well	—	(0)	(5)	
		Perry Mesa	—	(4)	(42)	
			—	(8)	(59)	
1/10–11/2000	Fain Ranch	Sipes Wildlife Area	189	—	189	Drive net (with helicopters) [A]
		Arizona Strip	—	—	(21)	
		Fort Apache Reservation	—	—	(14)	
		Eagle Creek	—	—	(12)	
		Buenos Aires NWR	—	—	(54)	
		Buenos Aires NWR	—	—	(44)	
			—	—	(44)	

[a] Parentheses () subdivide the number of pronghorn by specific release sites.
[b] Brackets [] indicate subspecies: A = American, M = Mexican, S = Sonoran.
Source: Adapted from Alexander 2000.

National Wildlife Refuge in Nevada. Twenty-two were raised to maturity. In 1940, 100 fawns were obtained from the Hart Mountain National Antelope Refuge in Oregon, of which 16 were raised to maturity. The 3-year program provided the state 38 bottle-reared, mature pronghorn. These were released in the Squaw Creek area of Kittitas County in 1940 on 10,000 acres of newly purchased Washington Department of Fish and Game land. The year after they were released, the population produced six fawns. The U.S. Army acquired this area as the Yakima Firing Center between 1943 and 1946, returning it to the Washington Department of Fish and Game in 1946. The pronghorn population had increased to 100 by 1950, when the Army again took over the land permanently. In 1950, 10 pronghorn were herded into a trap by helicopter and translocated to near Ritzville, in Adams County. By 1956, about 80 roamed the channeled scablands between Sprague Lake and Washtucna. Pronghorn in both sites died out between 1970 and 1975. In 1968, 22 pronghorn, captured in the vicinity of Burns, Oregon, were obtained from the Oregon Game Commission. Translocations of 11 animals each were made on the Colockum Wildlife Recreation Area in Kittitas County and Summer Falls in Grant County. Once again, all pronghorn in both areas died between 1970 and 1980. Although some initial success was achieved with all four pronghorn translocations after some predator control, eventually the animals died in each area—as they have in most nonhistoric rangelands.

A few translocations into Mexico are known to have taken place (Cancino et al. 2000). One involved capture of 19 males and 33 females in New Mexico and their release on an 865-acre (350 ha) sheep-tight fenced, private ranch in the state of San Luis Potosi during 1972 (Aguirre and Sotomayor 1981). Two years later, the herd had decreased to 24 animals. Fawns were produced each year, but predation was high until coyote control was implemented, enabling greater fawn survival. Even so, losses continued due to other factors, such as emigration and poaching. Today, no animals remain from the release. Pronghorn were translocated to Tiburon Island in the Gulf of California during 1967 (Ramirez et al. 1999), but they did not survive; 22 pronghorn also were translocated into Sonora from Colorado in 1988, but the results are not available. During 1996 and 1997, 66 and 93 pronghorn, respectively, were translocated from an *Antilocapra a. americana/Antilocapra a. mexicana* transition zone in New Mexico to a private ranch in Coahuila (D. Weybright personal communication:1998, Ramirez et al. 1999). Figure 155 depicts releases of pronghorn in Mexico that were captured in the United States.

FIGURE 155. *Pronghorn captured in the United States and released in Mexico, by year and state (from Ramirez et al. 1999).*

CAPTURE TECHNIQUES

New capture techniques have been developed during the past two decades. Generally, the goals have been to follow movements of specific individuals via radiotelemetry. Because large numbers of pronghorn were not needed, major impetus for development of these trapping techniques has been to reduce costs and devise more portable traps. Devices such as the cannon net, netgun and drive net, as well as chemical immobilization, have been used. At the same time, many fawns have been caught and marked for studies of population dynamics or behavior. A method even has been devised to collar pronghorn at waterholes without capturing them (Beale 1966).

Capturing still is an important component of pronghorn management and research. Pronghorn have been live-captured using a variety of techniques; each has advantages and disadvantages. Methods used include corral, surround and Clover traps, as well as linear drive and cannon nets. Netguns and chemical immobilants fired from a helicopter and from the ground also have been used successfully. Fawns can be hand-captured when young. Small confining traps such as the Clover trap generally have not been successful for capturing pronghorn (Amstrup et al. 1980). However, two does were caught in Clover traps

Artificial watering points are ideal for marking pronghorn without capturing them. The collaring device consists of a square wooden frame with three screened sides placed over the water point, leaving one side—through which the pronghorn drinks—open. A rubber collar is held by four hinged pins at the four corners of the opening, and a trip pan is suspended just under the water surface. The pan consists of a 16- by 16-inch (40.6 by 40.6 cm) wire frame woven with nylon filament to form 0.75-inch (1.9 cm) squares. When an animal puts its head through the collar and depresses the trip pan, the collar is released at the four corners simultaneously, letting it collapse around the animal's neck. Photo by Don Beale.

in Oregon during nine trap nights (M. Willis personal communication:1988). The animals were prebaited with alfalfa hay and salt (salt seemed the better bait). One doe lost about half her pelage, but suffered no apparent ill effects despite cold weather. Several other animals were in the trap, but did not trip the trigger; the drop panel had frozen open. Had this not happened, trapping would have improved to one animal per three nights.

Parameters to consider when selecting a capture method include: "number, age, and sex of animals needed; density of animals in the trapping area; terrain and proximity to roads; whether pronghorn are accustomed to fences; how wary the animals are; the possibility and acceptability of mortalities; and the cost (in time as well as money) per captured or marked animal" (Amstrup et al. 1980:99). A wildlife biologist planning to capture pronghorn should evaluate how these factors fit into management or study objectives before deciding on a particular method. Refer to White (1987:123–241) for further information on capture and handling of big game.

Corral Traps

Corral traps proved their worth many years ago as a cost-effective device to capture large numbers of pronghorn. Various agencies have modified the basic design to meet specific needs. Detailed accounts of corral trap design and operation were provided by Fisher (1942b), Couey (1949), McLucas (1956), Hoover et al. (1959), Russell (1964), Spillet and Zobell (1967), Moody et al. (1982) and McKenzie (1983), many of whom suggested modifications that resulted in improvements to the basic corral trap. The following guidelines are adapted from McKenzie (1983) unless otherwise stated.

Placement of a pronghorn corral trap is of utmost importance (Figure 156). The basic corral trap design consists of two linear wings, a containment pen and curtain (gate), a gated corral divided by a moveable burlap curtain and sometimes a loading chute. The woven-wire wings of a corral trap usually are V-shaped funneling into the trap mouth. Workers in Wyoming narrowed the distance between these wings near the trap, but gradually flared them until the ends were about 1,000 to 1,300 feet (305–396 m) apart about 0.3 mile (0.5 km) from the trap (L. Saslaw personal communication:1982). This eased the problem of pronghorn attempting to go over or through the trap wings. The workers also used cargo netting (the same as used in the trap proper) as trap wings for the first 350 to 400 feet (107–122 m) from the trap mouth to ease the same problem and reduce injury to pronghorn.

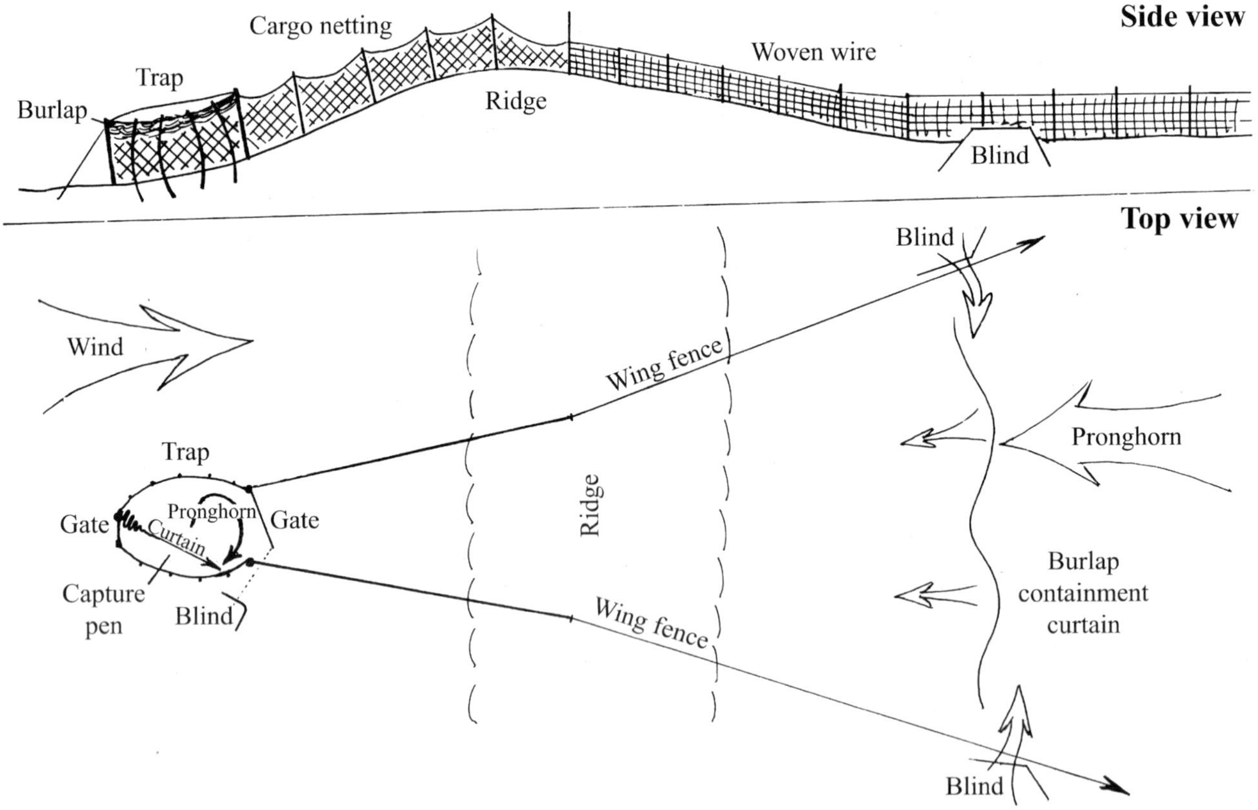

Figure 156. *Schematic placement of pronghorn corral trap (adapted from McKenzie [1983]; courtesy of the U.S. Bureau of Land Management, Rawlings, Wyoming).*

Capturing pronghorn by driving them into V-shaped woven-wire wings as much as 0.3 mile (0.5 km) long and 0.25 mile (0.4 km) wide at the mouth, to a corral trap at the funnel's apex is adopted and adapted from a hunting technique used prehistorically and historically by Indians of the Great Basin, Plateau and High Plains regions of the western United States. With the aid of aircraft to move the pronghorn into the wings, the modern technique can be quite efficient, because entire bands can be caught at one time. The logistical arrangements, however, are substantial. The crucial element of a corral trapping operation is its location. Photo by Larry Temple; courtesy of the New Mexico Department of Game and Fish, Albuquerque.

A burlap or canvas containment curtain extending across the funnel facilitates moving animals the final distance—approximately 650 feet (200 m)—into the corral trap. The distance between wings at this point should be about 150 feet (46 m). The curtain is folded in place on the ground between the wings at or near blinds on each side of the funnel. As the target animals cross the folded curtain, workers emerge from the blinds and form a line across the funnel. When the line has formed, the curtain is raised to establish a visual barrier to discourage escape attempts back out of the funnel. Alternatively, a netted gate curtained with burlap can be laid across the funnel and raised after the target pronghorn have crossed it, allowing for storage of animals in the funnel while another group of pronghorn is worked into and through the trap.

Toward the apex of corral trap wings, pronghorn should be hazed the final several hundred yards into the trap by means of a canvas or burlap curtain. The curtain frightens the animals less than people do, and the animals do not try to run back through it. The animals would be less excited if the drivers were not walking outside the lane. In the proper location, more than one group of pronghorn can be captured during a single corral trap operation. J. McLucas (personal communication:1978) called this the multiple-drive system. A second or third group of pronghorn was hazed by drivers with a hazing canvas into a trap already containing a band of pronghorn. By this method, as many as 389 pronghorn were trapped during one day in Montana. Photo courtesy of the Montana Department of Fish, Wildlife and Parks, Helena.

The corral portion of the trap should be 40 to 100 feet (12–30 m) in diameter, with nylon-netting walls and posts on the outside. A better method uses a 3-foot (0.9 m) offset added to the tops of posts, which increases the flexibiltiy when pronghorn hit the net, and it prevents collisions with the posts. A visual barrier curtain (Figure 157) is recommended for the trap proper. This burlap or canvas curtain reduces injuries and escape attempts through the sides of the trap; its suspension from the top of the trap walls also discourages injuries and escape attempts over the top. Most important, the curtain is lowered simultaneously with the closing of the trap gate when pronghorn enter the trap. Until that time, it is rolled up and secured at the top of the net with a quick-release string-washer-cotter pin assembly.

Again, taking advantage of topographic features by correct trap placement is vital for the success of any pronghorn capture attempt with fixed traps or nets. A reconnaissance flight to select a good location for the trapsite and to obtain an idea of the number and distribution of

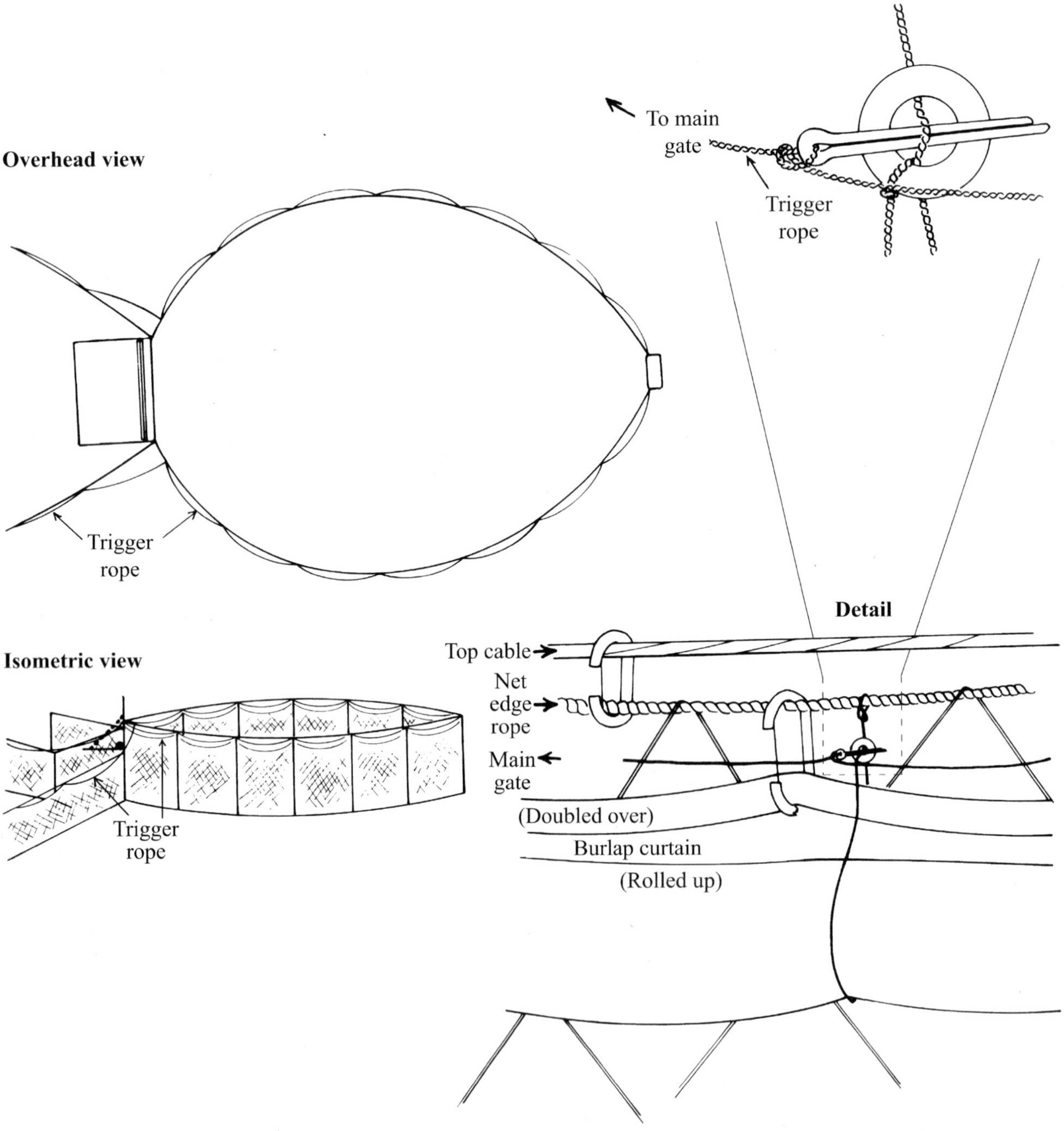

Figure 157. *Trigger detail and placement of trap drop curtain (courtesy of the Texas Parks and Wildlife Department, Austin).*

pronghorn bands should be made before erecting the trap (McLucas 1956), especially with corral traps that take considerable time to assemble and disassemble. Traps or nets should be hidden by vegetation or topographic features along known or suspected pronghorn travel routes. Natural and human-made barriers, steep terrain or tall vegetation should be avoided within the wings of a corral trap unless pronghorn are observed routinely using such areas. Existing fences near a trapsite should be carefully evaluated. Their use may either aid or hinder a successful trapping operation, depending on the observed reaction of the target group to that fence. Safety precautions for both workers and animals should be followed strictly.

The mouth and wings of a corral trap should be positioned so the target pronghorn run into the prevailing wind and are a safe distance from flight hazards (power lines, trees, etc.). Other flight precautions depend on pilot judgment.

Herding, trapping and handling stress pronghorn. Thermal stress may occur during warm weather. Corral trapping is not recommended when ambient air temperature is more than 70°F (21°C), and it should be in the 40° to 50°F (4°–10°C) range. Pronghorn captured during the cool of morning might be subjected unwittingly to a later temperature increase to lethal levels in the transport enclosure. During a catch-and-release operation, temperature may not be as critical because of the short duration of captivity. Amstrup et al. (1980) and Reeve (1982) successfully trapped during August when ambient temperatures were high, but they usually trapped during early morning or

Trap walls must be of resilient netting. A pronghorn hitting wire netting—less flexible than this one of nylon—could be injured or killed. Photo courtesy of the Montana Department of Fish, Wildlife and Parks, Helena.

Taking advantage of the terrain, a human barrier helps to steer pronghorn toward a corral trap in New Mexico for a 1997 translocation to Coahuila, Mexico. Use of people is very effective, but clearly labor intensive. It also requires that the participants are well-advised of their role and its timing. Photo by Patrick Robles.

Most early (modern) corral trap operations used fixed-wing aircraft to haze pronghorn. Helicopters are used today because they are more efficient, safer and easier on the animals. Photo courtesy of the Montana Department of Fish, Wildlife and Parks, Helena.

late evening and avoided handling or herding animals during the heat of the day.

An experienced pilot can move pronghorn to a trapsite efficiently with either fixed-wing or rotary-winged aircraft. In California, pronghorn were preconditioned by running them through the trap wings the day before setting up the corral (L. Pyshora personal communication:1988). During early trapping operations in Montana, pronghorn were worked leisurely into the vicinity of the trap with a fixed-wing aircraft and then driven by swift dives until they were in the trap. When animals became tired from long drives, they would stop and mill, often turning back. Paper sack-lime bombs helped to get the animals moving again, and a high-pitched siren helped to keep them moving (Thompson 1947). Also, firing a .22-caliber automatic pistol into the air from the plane started herds that refused to move. The sharp report of the gun had more stampeding effect than did the roar of the motor, particularly after the animals became accustomed to the latter (McLucas 1956). A helicopter, because of its maneuverability, is most effective for herding pronghorn when they approach or enter the wings of a corral trap.

Constant pressure applied by a helicopter appears to stress pronghorn, so the pilot should hold back and drift the animals toward the corral or net until time for the rush into the trap wings and corral. This caution is included with the recommendation that a rotary-winged aircraft be used to drive the target pronghorn into the trap wings and to prevent them from turning back until the line of workers forms across the trap funnel and the containment curtain is raised. As a safety precaution, at that point, the helicopter should retreat to and hover or land at a safe distance behind the workers.

The line of people with the containment curtain should advance toward the trap mouth when the pronghorn move in that direction, stop when the pronghorn stop and move when they move. Gentle but constant pressure is used until the animals enter the corral, the gate is closed and the visual barrier curtain released. Noise should be kept to a minimum.

Within the trap proper, there should be a canvas curtain that can be drawn across a portion of the corral, hiding the workers from the animals. A net seine or fence (catchment net) is recommended for segregating small groups (6–10 animals) from the main body of trapped pronghorn during the hand-capturing phase. Segregating these small groups of animals reduces the chance of injury to workers and animals and expedites handling. However, enticing pronghorn into the small space often is difficult. Saslaw (Moody et al. 1982—in discussion after presenting paper) reported that removing burlap from the back gate—so the animals could see light and perceive an escape route—facilitated moving animals into the capture pen. He recommended being ready to replace the burlap quickly so too many pronghorn did not enter.

For each pronghorn segregated by the catchment net, at least one person and preferably two should be available for hand-capture and restraint as quickly and effectively as possible. Proper handling includes grasping of the left ear or horn with the left hand and reaching over the animal with the right hand to grasp the right flank. The animal should be lifted just enough to get its feet off the

At the beginning of the 21st century, fixed-wing aircraft (left) *could be rented for about $100 to $200 per hour and used for herding pronghorn. However, they cannot fly slow enough to remain low and behind the animals if the pronghorn are reluctant to enter between the wings of a corral trap or run into a tangle net. Thus, helicopters* (right) *are much more versatile for capture operations, although hourly rental generally is $400 to $650 per hour. Pronghorn, especially bucks, sometimes will run under a helicopter, even when it is low enough for the animals to hit the skids. In such circumstances, a fixed-wing aircraft would have no chance of hazing the animals into a trap or net.* Left photo by Dan Pond. Right photo by Patrick Robles.

An existing fence can be used as one wing of a corral trap. The trap itself should be located so that the pronghorn are unable see it until they come over a rise where the funnel narrows enough for hidden drivers to raise a hazing canvas folded on the ground. If a permanent fence is not pronghorn-proof, it can be reinforced with net wire from the trap to the rise. Tarp around the lower part of a corral trap discourages pronghorn from running into the netting. Such a tarp also reduces the captured animals' exposure to external stimuli and helps to calm them after a drive. A squeeze chute—above, foreground—is used for catching the trapped animals. Photo courtesy of the Montana Department of Fish, Wildlife and Parks, Helena.

*A folded, sliding canvas curtain (*top left*) can be drawn (*top center*) to divide (*top right*) a corral trap to facilitate relatively expeditious handling of captured pronghorn. Reducing the trap size (confinement space) in this manner reduces the stress and risk of injury during handling by constraining the movements of the animals seeking escape.* Top left and top center photos courtesy of the Montana Department of Fish, Wildlife and Parks, Helena. Top right photo by Paul Wertz; courtesy of the California Department of Fish and Game, Sacramento.

During hand captures of pronghorn in a corral trap, considerable dust can be inhaled by the excited animals, resulting in inflamed respiratory tracts. Spreading alfalfa hay on the ground before capture reduces the dust and the problems caused by it. Left center photo by Paul Wertz; courtesy of the California Department of Fish and Game, Sacramento.

The best way to hold and constrain a pronghorn is to lift the animal until most of its weight is off the ground. In most instances, the pronghorn will quit struggling, especially if blindfolded. Bottom left photo by Al Langston; courtesy of the Wyoming Department of Game and Fish, Cheyenne.

Pronghorn in a corral trap settle down in a surprisingly short time after the drop curtain is released. However, people standing nearby and looking over the tarp diminish its ability to calm trapped pronghorn. Photo by Bill Wagner; courtesy of the Record Searchlight of Redding, California.

ground. Each restrained or carried pronghorn should be positioned with its sternum down and its head well above the level of the rumen to prevent aspiration of regurgitated rumen contents.

Drift Nets

Drift nets were not described in the literature, but Alexander (2000) reported more than 1,400 pronghorn captured in such nets from 1932 through 1987. Apparently, long fences or nets with accompanying bait trails led to baited corrals. Reported mortalities were 7% (Table 167). However, those pronghorn were not radio-collared, so postrelease deaths would not have been known. This method of capture did not require aircraft but was labor intensive, requiring personnel be ready to handle pronghorn whenever they entered the corral and tripped the release mechanism of the corral gate. Arid conditions in Arizona may have made bait more attractive to the animals than it would have been in some areas. Groups captured at one time varied from 3 to 200 and averaged 50.

Linear Tangle Nets

Linear tangle nets have gained popularity since the early 1970s for capture of North American big game animals. Their use is similar to that of corral traps, as animals must be herded into them. They are best suited for capture of fewer, but more select, target animals. Amstrup et al. (1980) reported that the linear tangle net could be employed successfully to capture small groups of pronghorn.

Linear tangle nets are 328 to 984 feet (100–300 m) long, approximately 8 feet (2.4 m) high, suspended vertically by notched wooden stakes or draped on tall vegetation. Enough nets to span at least 3,281 feet (1,000 m) are used in most pronghorn capture operations, and the best mesh size for pronghorn seems to be 7-inch (18 cm) bar measure. Nets should be anchored about every 300 feet (91 m), so that pronghorn cannot knock down whole nets or hit backup nets, and they should be hung loosely with about 1 foot (30 cm) on the ground because animals often can back out of tight nets. The axis of a tangle net should be positioned so pronghorn hit it while running into the wind, facilitating control of the chase aircraft.

Amstrup et al. (1980) deployed the net as a cul-de-sac below a pass through rimrocks or at the end of woven-wire wings along travel routes. A helicopter was needed to herd pronghorn successfully into the net, especially for the final rush. A major problem was the collapse of the net with the impact of the first few animals; the rest of the herd then jumped the net. To increase the catch, they suggested deploying a second net a few yards beyond the first to form a double cul-de-sac. Amstrup et al. also suggested a person hidden at the turn in the net where lead animals tend to hit could stand up just before impact to scatter the herd and increase the number of contact

Where terrain and other factors preclude efficient use of a corral trap to capture pronghorn, tangle nets may be employed. Such nets are best located over the edge of a rise along a regular travel or escape route and fashioned as a cul-de-sac. In less undulating terrain, a linear tangle net can be used, but it requires more personnel and, accordingly, more coordination in driving the animals because their point of entanglement is less predictable. Above left, pronghorn are driven through a pass in rimrocks (through which they regularly traveled to feed on a playa) to a double cul-de-sac of tangle nets the animals could not see until they were in the pass. Crew members hid behind rocks until the animals passed. Photo by Dan Pond.

For the safety of both animal and crew, a pronghorn is best removed from a tangle net, and handled by two people. When released, as above, pronghorn tend to be disoriented and sometimes run back into the net. The crew should be ready to raise or lower the net to prevent such inadvertent recapture. Photo by Dan Pond.

points. Mortalitites for pronghorn captured in Arizona were nearly similar for 1,436 captured in drift nets and 778 driven into tangle nets by helicopters—7% and 6%, respectively (Table 167).

Cannon Nets

Cannon nets can be used to capture select numbers of pronghorn. Animals must be enticed to feed or drink at a specific site and, when they are at that location, preset rockets are fired, propelling a 40- by 60-foot (12 x 18 m) net, which is secured on one side, over the animals.

Amstrup et al. (1980) evaluated the cannon net technique for the capture of pronghorn. One of 13 pronghorn captured in four netting attempts died in the trap. Another seven were known to have died within 2 weeks from capture myopathy or coyote predation. Because of the high mortality rate, this technique was not recommended for pronghorn. Pronghorn are not stressed excessively by a moderate amount of running, and they do not struggle as violently in a tangle net after running as they do under a cannon net. Perhaps vehement struggle is more conducive to capture myopathy when muscles have not been warmed by running. Also, pronghorn hitting tangle nets lose their footing almost immediately and can only struggle against the stretchy net; they soon give up. Those under a cannon net often keep their feet on the ground and use their strength and upright balance to fight against immobilization. They do not give up until restrained by hand or exhausted.

Amstrup et al. (1980) found that the only baits that consistently attracted pronghorn were apple pomace (pulp remaining after cider is squeezed from apples) and salt blocks. In arid regions, water may serve as bait (Beale 1966).

Remotely fired netguns now are available. They have proven effective in capturing white-tailed deer and many species of birds (C. Gray personal communication:1998). They are much easier to transport and use than cannon nets, but operate on the same principle. For capturing pronghorn, they present the same problems as those encountered with cannon nets.

Surround Nets

Reeve (1982) devised a surround net for the capture of pronghorn. He constructed the trap around a small waterhole. The trap consisted of a 5-foot (1.5 m) woven-wire fence arcing around half the pond. The remainder of the pond was surrounded by nylon netting of 5.6-inch (14.2 cm) mesh, 9 feet (2.7 m) high. A 350-foot (107 m) section of the netting was furled on the ground. This section was lifted by counterweights, suspended from four 12-foot (3.7 m) poles, when pronghorn entered the trap. The releasing mechanism for the counterweights was electronically activated from a distant blind. The trapping crew made a rapid approach to the trap from the woven-wire side. Startled pronghorn flushed into the netting, collapsed it and became entangled.

*A surround net at a waterhole can be an effective and safe means of capturing pronghorn. Three or more people are required to operate the net and handle the animals. Labor per pronghorn captured by surround nets is about the same as for linear tangle nets, but the cost is much lower because no aircraft is required. Knowledge concerning the time of day that pronghorn visit a particular waterhole can reduce time spent waiting for the animals. A surround-netting operation involves approximately half the surround circumference in woven-wire fencing. The balance is a nylon tangle net, with a section furled on the ground. The tangle net section can be lifted by counterweights suspended from four poles (*top left*) and released electronically from a nearby blind. Pronghorn quickly become accustomed to crossing a furled surround netting as they approach a waterhole to drink (*top right*). Observers in a blind can wait until just the right number or sex of pronghorn are at the waterhole before releasing the counterweights. After releasing the counterweights that lift the furled netting, the capture crew rushes the pronghorn from the woven-wire side and flushes them into the netting (*center left*), which collapses and entangles them. The selected pronghorn are subdued immediately (*bottom left*), and nontarget animals are released. The surround trap is an excellent, selective way to capture pronghorn of the sex and age classes needed for study.* Photos by LuRay Parker; courtesy of the Wyoming Department of Game and Fish, Cheyenne.

Advantages of this trap included selectivity in number, age and sex of animals captured, and minimal stress to the animals. Reeve (1982) reported no postrelease mortality among the 17 animals caught. Costs were considerably less than a corral trap or a drive net because materials cost less and no aircraft was needed.

Netgun

Barrett et al. (1982) described the use of a hand-held netgun for capturing large mammals, including pronghorn. Of five pronghorn captured in this manner, one of three adult females broke its leg, and both fawns developed acute metabolic stress. No adult males were captured. Barrett et al. concluded that pronghorn were captured easily with this technique, but considerable caution and effort may be required to reduce capture myopathy and losses from other trauma.

Firchow et al. (1986b) evaluated the hand-held netgun for capturing pronghorn and suggested ways to reduce stress- and injury-related deaths. Eighteen (11 male and 7 female) adult pronghorn were captured successfully using a hand-held, three-barreled netgun during three capture operations in Colorado. Twenty-two capture attempts were unsuccessful. Temperatures were 45° to 75°F (7°–24°C) during the gunning, and snow was not present to slow the animals' movement. Mean aerial pursuit time for successful attempts (3.75 minutes) was less than mean pursuit time for unsuccessful attempts (5.75 minutes). Twenty shots were fired to capture 18 pronghorn on successful attempts, and 14 shots were fired during unsuccessful attempts. Mean group size for successful captures (6.22) was less than the mean group size for unsuccessful attempts (11.20). Handling time for captured animals averaged 6.2 minutes. Two of 20 pronghorn captured, or pursued for capture, died. One animal fell during hazing and broke its leg; the other fell and broke its neck when tangled in the net. The other captured pronghorn did not show obvious signs of shock or capture myopathy. One adult female was found dead and scavenged 4 weeks after capture, but the remaining 17 animals were alive 12 weeks after capture. Five females captured during March and April were pregnant and carried their fawns to full term.

The netgun technique worked best on small groups of pronghorn. Groups of 12 or more remained bunched and a clean shot on a single animal could not be made. Lone animals or pairs dodged under the helicopter, making capture difficult. Only one animal was captured from each herd. Initially, Firchow et al. (1986b) used 12 minutes as a cutoff point for hazing animals, but switched to a 7-minute maximum after signs of stress were noted in hazed animals during the first capture operation.

Barrett et al. (1982) and Firchow et al. (1986b) used the three-barreled guns (with triangular nets) available at that time. Four-barreled guns that project larger, square nets now are available. They are easier to use, reduce pursuit time and animals usually are entangled in the nets quicker than those captured in triangular nets. Three-barreled guns can be converted to four-barreled guns for much less than the price of a new gun.

O'Gara has netted two to three Rocky Mountain goats or bighorn sheep at a time with the four-barreled netgun. However, when this practice was used on pronghorn (two double captures), the animals lost unacceptable amounts of hair and skin during falls and one suffered a broken leg. When one pair fell at full speed, the forces exerted on one another caused the thin nylon netting to scrape and cut like a knife. O'Gara later captured does from large wintering herds by netting single does running ahead of the bucks that brought up the rear of the herd. In this case, top projectiles of the gun were seated deeper than bottom ones so they would travel faster and the leading edge of the net would fall ahead of the target doe on long shots. If the helicopter flew over the bucks for close shots down on the does, the entire herd would scatter and dodge, making shots difficult, until the herd reassembled. The does still lost more hair and skin than did other ungulates captured with netguns. Snow should facilitate netgunning of pronghorn by slowing the animals' speed and cushioning their falls. However, W. Clark (personal communication:1990) reported two pronghorn mortalities during what seemed ideal netgunning conditions in California. Both animals were running, relatively slowly, uphill in 6 to 8 inches (15.2–20.3 cm) of snow when netted. Both died of broken necks.

Fifty-one pronghorn (16%) died during 311 captures (Lee et al. 1998). More than half of the deaths resulted from broken necks, some were destroyed because of broken legs and two died about 2 weeks after capture—apparently from capture myopathy.

Personnel of the Arizona Department of Game and Fish used a netgun to capture Sonoran pronghorn near the Mexican border. The project—funded by the Arizona Game and Fish Department, U.S. Fish and Wildlife Service and U.S. Department of Defense—was investigating mortality, movements and habitat use of this endangered subspecies. During these captures, Arizona personnel mounted a video camera above the netgunner's head. This allowed the gunner to evaluate his performance later, and it aided the training of new gunners.

The high cost of using helicopters to netgun pronghorn encouraged Scott (1994) to netgun from the ground in Yellowstone National Park. He had no success in limited efforts to shoot from blinds over bait. However, a shooter riding in the bed of a pickup truck netted 21 adults that

*Netgunning from a helicopter is a very effective technique for capturing pronghorn, especially small numbers of animals from different bands or herds. Netguns are being used to capture many species—from elk to golden eagles. It is one of the most selective methods of taking animals of the sex and age needed. It also is an ideal method for scarce animals, such as the Sonoran pronghorn featured in this sequence. Because of the comparatively quick results and few people needed for netgunning, it can be the most economical way to capture up to 15 pronghorn. However, corral traps or tangle nets are less expensive methods when larger numbers of pronghorn need to be captured. Because of the speed and agility of pronghorn, even an experienced helicopter pilot has trouble keeping the shooter positioned during the chase (*top photos*). Pilots must be extremely skilled to position the gunner over the target animal quickly and steadily. Gunners must be able to aim quickly and with a proper lead, so that the projectiles pull the net over the animal's head (*center photos*). The technique is not recommended except for experienced pilot/gunner teams, as even successful shots can cause the pronghorn potential trauma from entanglement at full running speed and from the subsequent fall (*bottom photo*). Netting pronghorn on rocky ground should be avoided. Flat land with sparse, low vegetative cover is ideal for netgunning. Nevertheless, before it is entangled, a pronghorn can get out of a net that catches on a shrub. In addition to getting in proper position for a shot, the gunner must be sure the target animal is away from others that might be hit or otherwise interfere with the shot. Seven minutes is the maximum time that a pronghorn should be pursued, and the pursuit should only take place when the temperature is between 32° and 60°F (0°–16°C).* Photos by Bob Miles and George Andrejko; courtesy of the Arizona Department of Game and Fish, Phoenix.

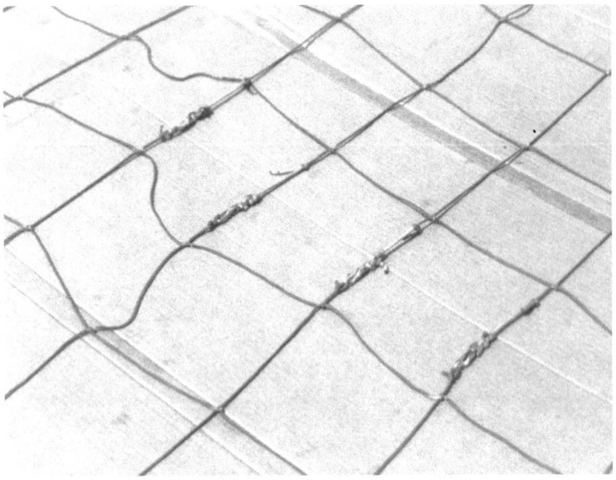

*The four steel projectiles that carry netgun nets tend to gather beneath a pronghorn's thrashing feet and intertwine in a terrible tangle. Pronghorn are virtually enclosed in a knotted bag. Scott (1994) wanted to get animals back on their feet in less than 15 minutes—not much time to install radios, draw blood and untangle them. To speed up the process, he cut four cords in a straight line in the middle of the net and tied a metal spring clip with a swivel to one side and tied a bowline loop knot in the other (*left photo*). Thus, a hole could be opened in the center of the tangled net enclosing an animal. Release times often were cut in half, and the clips had no effect on netgun performance. Netgun nets ordinarily are held in a fiberglass canister with several strips of masking tape, which breaks easily (tapes breaking in a slipstream could cause a crash if the gunner does not immediately fire the net away from the tail rotor), does not stick well to cold fiberglass, is slow to install, and scatters broken tape when the gun is fired. Scott (1994) riveted one end of a 0.75-inch (1.9 cm) wide strip of Velcro on the outside of each of the four sides of the canister, with the smooth, flush head on the inside. (In the* right photo, *the Velcro straps are in the closed position as if a net were underneath them.) Each strip was long enough to cross over the packed net about half way and mesh with its complementary, opposite side for only 0.39 inch (1 cm). More overlap might excessively restrict net escape when the gun is fired. Scott fired nets held in canisters this way dozens of times and never broke a Velcro strap. In the haste of reloading nets during captures, the projectiles can be loaded into the four barrels in the wrong sequence, so the net does not deploy squarely. Scott largely solved this potential problem using a permanent-ink marker to number the four corners of each net, each corner of the canister (see* right photo*) and the four barrels of the gun. This prevented having a crossed net when trying to capture a pronghorn.* Photos by M. Douglas Scott; courtesy of the U.S. National Park Service, Washington, D.C.

were habituated to humans and vehicles. Pronghorn seldom allowed a person on foot closer than 66 feet (20 m), but did permit slow-moving vehicles to pass as close as 16 feet (5 m). The gunner crouched in the truck bed as a driver cruised at about 15 miles per hour (24 km/hr). If an animal remained in range (about 46 feet [14 m]) as the vehicle passed, the gunner fired the net; if the animal was netted, the truck was quickly stopped, and the gunner jumped out and restrained the pronghorn. The driver pulled the vehicle off the road, then brought the equipment. Pronghorn gradually became wary of the truck.

One animal broke a femur when netted, and two died less than a week after being handled. To be handled, one was hog-tied; the other was tangled in the net so badly that she essentially was hog-tied. Both belched during handling, and Scott suspected aspiration of rumen contents or capture myopathy was involved, but the remains were consumed by coyotes making necropsy impossible. Predators are numerous in Yellowstone National Park, so even a slightly impaired pronghorn would stand little chance of survival. Later in the study, pronghorn were restrained, without hog-tying, in a normal bedded position with heads and necks erect. No more quick losses were recorded among the 17 remaining pronghorn that were captured.

Surprisingly, pronghorn netted from the ground seemed to become more entangled in the net than those netted from the air—possibly because those chased by helicopter were somewhat exhausted and fought the net less (Scott 1994). The buck that suffered a broken femur did a complete flip when netted. Struggling pronghorn knocked conventional blindfolds off, and a light cotton blanket thrown over the head and shoulders stayed in place better. Wrapping part of the blanket around the sharp horn tips proved to be a good safety precaution. Capture was in winter, so overheating was not considered a problem.

During netgunning of 21 pronghorn from a pickup, an average of one pronghorn was captured per 8-hour day at a cost of about $167.00 per animal, and about 25% of the shots resulted in a capture (Scott 1994). Seventeen pronghorn were netted from a helicopter during the same study in 3.5 hours of flying time (4.9 animals per hour), at a cost of about $403.00 per animal. Netgunning from the pickup truck resulted in 14.3% mortality, compared with 47.1% for those netted from a helicopter. The helicopter

Once a pronghorn is entangled and down, someone should get to the animal as soon as possible. The chance of it escaping is slight, but the sooner it is blindfolded to reduce its struggles and removed from the net, the less stress it will experience (bottom right). *Straddling an animal may be necessary until it is subdued, but extended pressure on the chest should be avoided.* Photos by Bob Miles; courtesy of the Arizona Department of Game and Fish, Phoenix.

crew was not experienced in handling pronghorn, and the resulting high mortality emphasized that at least part of a capture crew should be experienced.

Fifty percent of the animals hog-tied after netting from the pickup died, compared with 8.7% of those that were not hog-tied. The pronghorn netted from the helicopter were hog-tied during handling; this probably contributed to the high mortality rate. Handling pronghorn in a manner similar to that suggested for those in corral traps seems appropriate. Usually, the primary objective of capturing pronghorn by netgunning is to radio-collar the animals for study. Scott suggested that biologists should resist requests from veterinarians, pathologists and others to gather data from struggling netgunned animals that would be collected more easily from necropsied animals or from other sources—we agree.

With Yellowstone National Park pronghorn, Scott (1994) found ground netgunning was simpler, safer and cheaper than aerial netgunning. In addition, ground capture caused fewer pronghorn mortalities and less public disturbance than did aerial capture. Captures in Yellowstone National Park were possible because the animals were habituated to vehicles—a situation seldom encountered except in parks or refuges. Six pronghorn

When pronghorn must be carried, they should never be turned on their backs (left). *They tend to struggle even when blindfolded. Also, if they are rolled completely over, the possibility exists of turning the rumen over within the abdominal cavity—resulting in a twisted gut and eventual death. The right photo depicts the proper way to carry a pronghorn that has been blindfolded and hobbled.* Photos by Tom Pojar.

were netgunned from a moving vehicle along roadsides on Dugway Proving Ground, Utah, during August and September 1995, to install radio transmitters (Bates 2000). No mortality was reported.

Scott's (1994) study and Bate's (2000) experience show that ground capture is possible, and blinds at fence crossings, salt blocks, water holes, etc., may allow ground capture. Scott did not mention that, when shooting from the ground, seating the top projectiles slightly deeper than the bottom ones will carry the net over an animal better than will seating all projectiles equally.

The netgun technique was used in Colorado during December 1994 to capture 10 doe and 10 buck fawns for a natural mortality study (Pojar 2000). Because mortality has been high during many capture operations, a commercial wildlife capture company—Helicopter Wildlife Management, of Salt Lake City, Utah—was contracted to capture the animals. Its experience capturing pronghorn was limited somewhat, compared with other species, but it had captured about 150 with no direct mortalities. The helicopter crew consisted of pilot, gunner and handler; they flew a Hughes 500c helicopter. A briefing was held with the crew to outline protocol criteria, but not instructions for capture procedures because the crew had experience with the technique.

Radio collars cost $200 to $450 (2003 dollars), depending on the number purchased and options (such as mortality sensors and heart rate monitors). Battery life generally is 2 years or more. Radio collars once were used almost entirely for research, but managers also now use them to monitor pronghorn movements and habitat use. Photo courtesy of the Montana Department of Fish, Wildlife and Parks, Helena.

The average handling time for the 20 animals captured was 9.4 minutes, including search time, hazing, capture and ferry time. Hair loss was minimal, and only two superficial cuts or abrasions were noted. Upon release, the animals behaved normally, given they had been blindfolded. One animal died within 5 days of capture, and death was attributed to trauma associated with capture—lung edema resulting from either deep muscle bruising or head injury. A second fawn died 8 to 10 days after capture. Capture stress was ruled out as cause of death.

Advantages of the netgun for selective capture of animals were noted by Pojar (2000) as follows. Corral trapping would have necessitated trapping 130 to 150 pronghorn to secure the 20 fawns needed, and this could have resulted in greater mortality. Selecting sample animals from several groups within the population was possible, resulting in a statistically sound sample. Injuries to animal handlers were nonexistent—during corral trapping, handlers commonly suffer broken and lost eye glasses, bruises, abrasions and back injuries. Handling crew size was eight, compared with 40 to 50 persons needed for corral trapping. Only material for processing animals is required for netgunning, thus, processing points in remote areas could be reached by snow machine, helicopter or on foot. Four-wheel drive vehicles, trailers and fencing materials are required for a corral trap, making vehicle access to the trapsite imperative. Estimated costs per captured fawn were $275.00 by netgun and $469.00 by corral trap.

Capture of large numbers of pronghorn for transport to another location would be more economical by corral trap than by netgun. In the Colorado situation described above, the capture of 130 to 150 total animals would have cost roughly the same as capture of 20 fawns. Thus, the cost would drop to $60.00 to $70.00 per animal.

Because the Colorado operation went so well and mortality was held to 5%, the protocol used is repeated here. Tom Pojar was the project's principal investigator and coordinated the capture operation. All persons involved, including the helicopter netgunning crew, were instructed on the care and handling of captured animals to minimize stress and injury to both the animals and crew members. The following protocol was followed.

1. Capture timing and conditions. Early to mid-December, before very deep or crusted snow conditions prevail, when relatively mild days allow the animals to recover from the stress of capture, seems to be the best time to capture and handle pronghorn in Middle Park. Snow depths of 6 to 12 inches (15–30 cm) are desirable to slow the speed of the animals and to break their fall when netted. Capture operations will not be attempted if crusted snow conditions exist because of potential skin abrasions and injury to the animal's legs.

2. No-fly zones. Pursuit and capture will not take place within 1 mile (1.6 km) of a residence or developed area.
3. Notification of affected parties. Local rural residents and public agencies, including federal, state and local agencies, will be notified of the time and general area of the capture operation.
4. Emergency services. Personnel will be instructed as to the contact and location of the nearest medical and emergency services.
5. Radio collars. Expandable collars will be used because target animals are young and growth needs to be accommodated. The circumference of collars put on female fawns will be 16.0 inches (40.6 cm) initially and will expand to 18.0 inches (45.7 cm); male collars will be the same initial size, but will expand to 20.5 inches (52.1 cm).
6. Command post. The principal investigator and handling crew will be at a predetermined command post. The principal investigator will be in charge of all decisions for the care and welfare of the animals.
7. Chase time. A capture will not be attempted without running the animal 1 to 2 minutes to warm it and prevent capture myopathy. Upon location of a large group (50–100), a smaller group with one or more target animals will be separated from the large group within 1 to 2 minutes. The maximum pursuit time to maneuver the target group to an acceptable capture site will be 4 to 5 minutes. An individual pronghorn will not be pursued actively (active pursuit means intense pursuit to capture) by the helicopter for more than 1 minute. Total pursuit of the target animal (and accompanying animals) will not exceed 10 minutes. Every effort will be made to take no more than four to six captures out of any single large group to avoid excessive chase time of nontarget animals. Care will be taken by the helicopter crew to avoid chasing animals into fences, roads or rivers.
8. Animal care and handling. Upon capture, the animal will be blindfolded, restrained (hobbled) and loaded into the back compartment of the helicopter. It will be ferried to the command post, unloaded and carried to the processing point (safe distance from helicopter) by two or three handlers per animal. The animal will be examined for injuries before the radio collar is installed. After processing, it will be released on site (see item #10). Command posts will be located within 3 miles (5 km) of capture sites. Total time for handling will not exceed 2 minutes.
9. Injured animals. Any debilitating injury or mortality of a captured animal will be reason to suspend the capture operation to assess the cause of the injury. The capture crew will be interviewed to determine whether preventative measures were possible. Topics to consider include: pursuit techniques, weather conditions, visibility, fences, animal locations, groupings. The principal investigator will determine whether the capture will resume.

   The severity of any animal injury will be assessed by the principal investigator,who will decide whether to release the animal or euthanize it. An animal with head, spinal or upper extremity injuries will be euthanized immediately after guidelines of the Colorado Department of Wildlife Animal Care and Use Committee. The carcass will be processed for human consumption. Any other injury will be assessed on an individual basis with the primary criterion for euthanasia being the survivability of the animal.
10. Release of animals. While still hobbled and restrained, the animal again will be examined for injuries—skin abrasions (treat with disinfectant), neck, back, head and extremities. First the hobbles will be removed, then the blindfold. Handlers will continue to hold, not restrain it until its eyes adjust to light and it is oriented enough to attempt to escape. The command post will be located such that released animals can be observed for about a half mile. The escape path will be clear of potentially dangerous

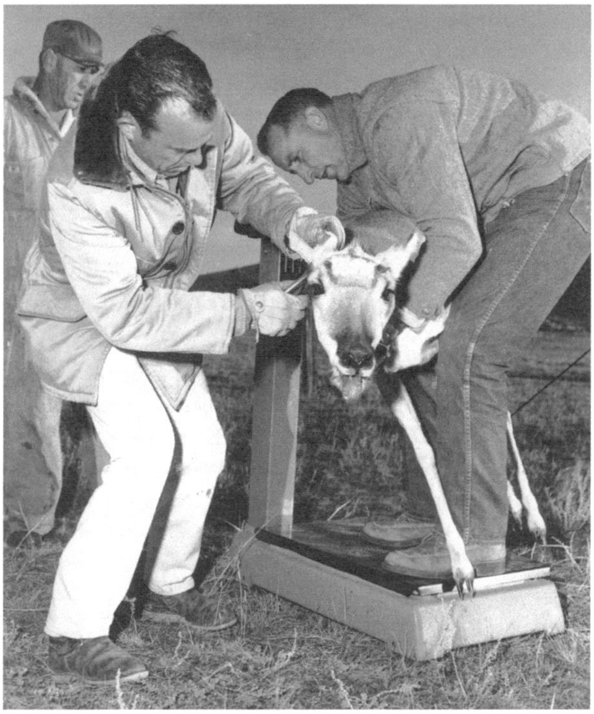

When weight measurements are desired during a pronghorn-trapping operation, stepping onto a platform scale with each animal often is the simplest method and least traumatic to the animals. The pronghorn above also is being fitted with an ear tag. Photo by George J. Mitchell; courtesy of the Alberta Department of Industry and Development, Edmonton.

obstacles, such as vehicles, fences and roads. Upon release, the posture and gait will be observed, paying particular attention to how the head is carried (for possible neck or back injury) and the use of all four legs as the animal runs. If an injury appears to meet euthanasia criteria, the animal will immediately be recaptured and euthanized according to guidelines in #9 above. Use of a video camera to document the behavior and condition of released animals would be desirable and could provide clues if postcapture mortality occurs.

11. Postcapture monitoring. All radioed animals will be located four to six times within 10 days after capture to monitor for postcapture mortality. If a mortality is found, a necropsy will be performed and cause of death determined if possible. All injuries and mortalities will be recorded for future reference to the capture technique.

Chemical Immobilization

In general, pronghorn are extremely difficult to immobilize; be prepared for less than satisfactory results.

Kreeger (1996:204)

Drugs are useful for immobilizing or calming wild animals for marking, radio-collaring, transporting and other handling. Every species reacts to drugs in its own way. Pronghorn seem to have a narrow tolerance to most immobilizing drugs. They are especially difficult to drug with darts because they occupy open country and are difficult to approach. They also are susceptible to injury from darts because of their thin skins, small muscle masses and relatively slender and fragile bones.

Immobilizing Drugs. The first drug used to immobilize pronghorn from a distance was Sucostrin (succinylcholine chloride) (Thomas 1961, Beale and Smith 1967). Sucostrin is a neuromuscular-blocking agent that induces depolarization of motor end plates at neuromuscular junctions. Onset of immobilization generally is rapid and duration is brief—both of which are advantageous in many cases. However, the safe dosage range is narrow for most species, respiratory paralysis is common and no antidote is known.

Heavy, fluid-filled darts can induce considerable tissue damage and fracture bones (Copeland et al. 1978). To minimize such damage, Amstrup and Segerstrom (1981) used lightweight plastic-bodied darts projected from a powder-charged rifle (Pneu Dart, Inc.) to administer powdered Sucostrin. Wind affected the plastic darts, but precise placement was possible from as far away as 66 yards (60 m). Animals at less than 22 yards (20 m) were not darted for fear of excessive tissue damage. Eleven animals were successfully immobilized with dosages of 8 to 11 milligrams. Little or no reaction to the drug was seen after 34 injections of 31 pronghorn with 7 to 9 milligrams, and 4 pronghorn injected with 9 milligrams of Sucostrin died.

Tolerance of pronghorn to Sucostrin apparently is low, and Amstrup and Segerstrom (1981) stated that it should not be used on the species. They found the plastic dart and the rifle that fires it to be an effective delivery system that should be adapted to drugs more useful in pronghorn and other species with low tolerance to Sucostrin.

M-99 (etorphine hydrochloride) commonly was used for immobilizing wild ungulates. It is fast acting and its effects may be reversed quickly with the antagonist M50-50 (diprenorphine hydrochloride). H. Gorman successfully immobilized 25 captive pronghorn using 8 to 9 milligrams of M-99 mixed with 10 milligrams of Acepromazine (acepromazine maleate) or 50 milligrams of Rompun (xylazine) to reduce excitement and convulsions (Copeland et al. 1978). Attempts to immobilize pronghorn from a helicopter with M-99 administered via a Cap-chur gun resulted in poor success (Copeland et al. 1978). Five of nine territorial bucks died during the initial attempt. Three mortalities involved trauma from the heavy darts penetrating vital organs, and the two others resulted from shock or capture myopathy.

Using M-99, Autenrieth et al. (1981) darted 11 mature bucks, 9 mature does and 6 immatures from a helicopter and had survival rates of 64%, 100% and 83%, respectively. Dosages (milligrams per kilogram of body weight) used were: bucks—M-99: 0.11, Rompun: 1.21, Acepromazine: 0.08; does—M-99: 0.10, Rompun: 1.06, Acepromazine: 0.11; and immatures—M-99: 0.15, Rompun: 1.10 and Acepromazine: 0.09. M50-50 was administered at approximately three times the M-99 dosages.

Autenrieth et al. (1981) concluded that darting from a helicopter is an effective technique for capturing selected animals and recommended that downed animals be blindfolded during handling to reduce excitement. They also recommended having dexamethazone and a sodium-bicarbonate and Multi-Sol-R infusion readily available for treatment of shock or acidosis, and injecting the M50-50 about half intravenously and half intramuscularly. Four mature bucks apparently succumbed to capture myopathy, which may have resulted from long chases or the drug. Heavy darts (M-99 comes in solution with only 1 milligram of drug per milliliter) have been a problem when darting delicate pronghorn and may have resulted in underdosing bucks, leading to capture myopathy. On the other hand, larger darts would have been too heavy. With the dosages used, at least 5-milliliter darts (weighing about 0.6 ounce [17 g]) were required. M-99 no longer is available in the United States.

Carfentanil citrate was introduced to the North American market during the mid-1980s. It has about three times the potency of M-99 per milligram on elk, moose and mountain sheep, and is marketed with 3 milligrams of drug per milliliter. Using a carfentanil/Rompun mixture (median dose is 2 milligrams carfentanil and 33 milligrams Rompun) in a 1-milliliter Cap-chur dart, O'Gara (1987) obtained good results on pronghorn. The small dart achieved good accuracy, even with a light charge and reduced the risk of trauma from tissue damage. Also, injection from the small dart was more positive than with a larger amount of solution. Five mature pronghorn bucks injected with this mixture were immobilized between 2.2 and 6.8 minutes after injection. Two bucks were injected with 3 milligrams of carfentanil without Rompun. One failed to succumb, and the other took 10.8 minutes to do so. M50-50 acts as an antagonist to carfentanil. O'Gara administered 4 milligrams of M50-50 per milligram of carfentanil intramuscularly upon first handling an immobilized animal and a similar dosage intravenously immediately before release. Naloxone (naloxone hydrochloride) also is an antagonist to carfentanil and is administered at 100 milligrams per milligram of carfentanil. Animals sometimes become renarcotized after reversal with Naloxone.

Another antagonist, Naltrexone (naltrexone hydrochloride) now is available in the United States. Naltrexone supposedly prevents renarcotization. Kreeger (1996) recommended 0.018 milligram per pound (0.04 mg/kg) body weight of carfentanil, plus 0.45 milligram per pound (0.99 mg/kg) of Rompun for immobilizing pronghorn. For an antagonist, he used 100 milligrams of Naloxone or Naltrexone per milligram of carfentanil given, plus 0.057 milligram per pound (0.125 mg/kg) of yohimbine. To reduce renarcotization, Kreeger gave half of the antagonist intravenously and half intramuscularly.

Kreeger et al. (1999) immobilized 43 pronghorn on F. E. Warren Air Base in southeastern Wyoming during 1997. The animals could be approached closely by a vehicle containing shooters armed with range-adjusting dartguns and range finders. The dosages used—0.021 milligram per pound (0.047 mg/kg) carfentanil and 0.05 milligram per pound (1.15 mg/kg) xylazine—resulted in a mean induction time ($n = 40$) of 7.1 ± 1.5 minutes with a range of 1 to 43 minutes. Immobilized pronghorn were transported to base camp and fitted with radio collars. Immobilizing drugs then were antagonized with Naltrexon and yohimbine, and the animals were placed into a trailer for shipment. Two (4.7%) does died after capture, and one was released before shipment. Forty pronghorn were shipped to Oklahoma, five (12.5%) died en route, and one moribund animal was euthanized within a week of release.

In 1998, 17 additional pronghorn were captured using the same carfentanil and xylazine mixture, or "cocktail," but 100 milligrams of ketamine was added. The ketamine

Immobilized in 1986 by a 1-milliliter dart with 2 milligrams of carfentanil and 33 milligrams of Rompun, the pronghorn above went down quickly but did not completely relax. Further experimentation seemed necessary to determine the ideal dosage for this combination of drugs. Dosage experimentation advisedly is done on healthy, unstressed adult males. Acceptable, effective levels should be achieved by adjustments up from underdoses rather than down from overdoses. Photo by Bart W. O'Gara.

did not improve induction times and may have had an adverse effect, causing the animals to run after being darted. Two (12%) of the animals died en route to Oklahoma (Kreeger et al. 1999).

Kreeger et al. (2001) evaluated the safety and efficacy of thiafentanil, a potent synthetic opioid anesthetic being developed for wildlife anesthesia, for capturing pronghorn. "Pronghorn were darted with pre-measured dosages of either 4.0 or 5.0 mg of thiafentanil without (group 1) or with (group 2) the addition of 25.0 mg xylazine. Seventeen pronghorn were captured in group 1 and 14 pronghorn in group 2. There were no differences between groups for capture times or physiological parameters ($P \geq 0.21$). Anesthetic induction was rapid for both groups ($\leq 2.7 \pm 0.4$ min) as was recovery after antagonism ($\leq 0.7 \pm 0.07$ min). The dosage of thiafentanil administered was 0.10 ± 0.005 milligram per kilogram and the dosage of xylazine was 0.56 ± 0.03 milligram per kilogram. Anesthesia in both groups was characterized by muscle rigidity and rapid, shallow respiration. Twenty-five pronghorn were radiocollared and survived greater than or equal to 21 days after capture. One adult male in group 1 died during capture. Thiafentanil was considered more effective on pronghorn than carfentanil–xylazine, ketamine–xylazine, or tiletamine–zolazepam–xylazine anesthetic regimens" (Kreeger et al. 2001:25).

Chemical immobilization of pronghorn has resulted in a high incidence of death (Table 168). The high mortality was caused by a combination of low tolerance to drugs, tissue damage from darts and stress induced from pursuit with a helicopter. When a helicopter is available for captures, netguns seem safer than drugs.

Attempts should not be made to immobilize pronghorn during the rut. As soon as a doe becomes unsteady on its feet, bucks in the area will attempt to breed her. This can result in an extended, fast chase and the death of the doe. When a buck becomes unsteady, other bucks attempt to kill him. If they do not succeed outright, the exertion can result in death of the drugged animal. O'Gara has seen bucks run as far as 2 miles (3.2 km) at top speed to breed or fight an unsteady animal.

Tranquilizing drugs. Tranquilizing drugs promote tameness and ease of handling. They act slowly but last for a long time. Rompun used with carfentanil for immobilization reduces the amount of carfentanil needed, mitigates convulsions and reduces excitement during handling. Litton (1975) reported that a mild tranquilizer and antibiotics injected at first handling reduced losses to less than 3% during pronghorn translocations in Texas. In previous years, when injections were not used, mortality had been much higher during trapping and transportation. Two tranquilizing drugs, Valium (diazepam) (5 mg/mL) and Acepromazine (50 mg/mL) routinely are administered by some biologists when translocating pronghorn. Dosages used generally are 1 milliliter for adults and 0.5 milliliter for fawns.

Kreeger et al. (1999) administered 10 milligrams of Valium to each of 17 pronghorn before transporting them from Wyoming to Oklahoma. Two (12%) died en route, compared with 12.5% mortality for 40 pronghorn similarly moved the year before without a tranquilizer. The use of Valium did not seem to make much difference in this case. The animals were transported soon after capture. Possibly, if yohimbine had not been administered, the Rompun would have served as a better tranquilizer than did the Valium. The animals probably would have been stressed less when the carfentanil was reversed. O'Gara had positive results moving large animals, such as bison and elk, when they were under the influence of light doses of Rompun.

Immobilizing or tranquilizing pronghorn with drugs administered orally in bait would eliminate many of the dangers associated with darting or drive trapping. However, exact dosages cannot be administered, and the animals must be accustomed to feeding on the bait. Thus, the drug must have a wide safe dosage range, and only animals accustomed to the bait are vulnerable. Valium has a wide safety margin and has been administered orally to several species of deer in enclosures, resulting in successful captures. Promazine hydrochloride crystals also have a wide safety margin and lack toxicity. They have been administered orally in feed and functioned well in quieting livestock, facilitating handling.

Pusateri et al. (1982) mixed oral tranquilizers with grain and fed it to pronghorn. Valium showed some possibilities, but dominance hierarchies established by gregarious pronghorn made the capture of large groups (10 or more) unfeasible. Dominant animals consistently ingested most of the bait. When severely tranquilized, the animals required constant attention. Poorly tranquilized ones showed signs of hyperexcitability or had difficulty gaining their feet when bedded, and they stumbled and lurched when approached. Warning signals from nontranquilized animals and continued capture attempts could cause such animals to injure themselves. Extended recovery periods of up to 30 hours also posed problems. Tranquilized animals required constant attention to prevent bloat, exposure and predation.

Promazine hydrochloride at doses of 2 to 17 milligrams per kilogram of body weight, given orally to wild pronghorn fawns and one adult, did not produce visible signs of tranquilization (Pusateri et al. 1982). Doses of up

Table 168. Comparison of known pronghorn mortality associated with various capture techniques

Technique	*Number captured*	*Percentage mortality*	*Source*
Succinylcholine chloride	46	33	Amstrup and Segerstrom 1981
Etorphine hydrochloride	64	21	Copeland et al. 1978, Autenrieth et al. 1981
Carfentanil/Rompun	71	13	O'Gara 1987, Kreeger et al. 1999
Cannon net	13	62	Amstrup et al. 1980
Tangle net	831	6	Amstrup et al. 1980, Alexander 2000
Netgun	311	16	Lee et al. 1998
Corral trap[a]	5,751	4	Various reports
Surround net	17	0	Reeve 1982
Drift net	1,436	7	Alexander 2000

[a] Most of these animals were not radio collared, so postrelease mortality would not have been recognized.

to three times that recommended for a 1,200-pound (544 kg) horse were administered and produced no visible effect. The pronghorn refused to eat bait containing doses of promazine hydrochloride crystals that would result in dosages greater than 17 milligrams per kilogram of body weight.

CAPTURE MYOPATHY

The fragile build of antelope and the exertion they undergo during the trapping and transplanting operations make casualties inevitable. However, by exercising extreme care, loss of life may be kept at a minimum.

Hoover et al. (1959:90)

Thousands of pronghorn have been captured and translocated. Without these animals, the restoration of many pronghorn populations would not have been possible. However, managers engaged in trapping or drugging programs for translocating pronghorn or other purposes should be aware of potential dangers to the animals, especially if endangered subspecies are involved. Pronghorn are delicately built and excitable. Acute trauma, especially broken legs and necks, is common during corral trap, drive-net and netgun operations. A more subtle form of mortality— capture myopathy—sometimes results when pronghorn are captured and handled.

The precise causes of capture myopathy in wildlife remain somewhat obscure, but they usually are associated with the concerted and vigorous use of many muscles during pursuit and capture, chemical immobilization and transportation. Capture myopathy may be a natural event associated with intraspecific combat or prolonged harassment by predators, including domestic dogs and humans. Sublethal capture myopathy can predispose pronghorn to predation.

Capture myopathy was first documented in pronghorn by Chalmers and Barrett (1977). During an investigation of migration patterns and range relations in Alberta, 594 adult pronghorn were captured with drive traps during five autumns. Acute trauma caused approximately 29 mortalities. Signs of capture myopathy sometimes appeared within an hour of capture but usually were delayed until handling or soon after release. A capture myopathy-like syndrome was associated with an estimated 20 additional deaths at the trapsites before or during processing.

Clinical signs of capture myopathy included depression, stiffness, weakness, incoordination and paralysis. Gross and histologic lesions were observed in the large muscles of the hind limbs of most affected pronghorn. Gross muscular lesions in capture myopathy-affected pronghorn were associated with subcutaneous and intramuscular edema (excess fluid in intercellular tissue spaces) and with hemorrhage, which varied from discrete petechia (purpuric spotting) to marked streaking or splashing. Accompanying this hemorrhage was fragmentation of muscle fibers, with loss or attenuation of cross-striations. Infrequently, short, alternating light and dark striations, producing a banded or wavy pattern, and general pallor were seen in affected muscles. Additional lesions in some pronghorn included pulmonary hyperemia (excess blood, resulting in redness of tissues) and edema, epicardial and endocardial hemorrhage, hyperemia and hemorrhage of the adrenal cortex and, occasionally, perirenal hemorrhage. The liver usually was swollen and pale.

Despite the normal appearance of 32 drive-trapped pronghorn that were released with radio collars, 6 were found dead or recumbent within 2 to 8 days and within 0.5 to 5 miles (0.8–8.0 km) of the trap (Chalmers and Barrett 1977). Some of the carcasses were consumed by coyotes, so determining the relative contributions of capture myopathy and predation was a problem, although one animal was found dead without evidence of predation. Some animals might not have been affected sufficiently to cause death but were debilitated enough to become easy prey. These data indicated an additional 18.8% mortality of the processed and released pronghorn related to capture myopathy or predation.

Seventy drive-trapped Alberta pronghorn were transported to rangeland enclosures and four to holding pens (Chalmers and Barrett 1977). Three of the 70 animals were injured during transport and subsequently euthanized. However, the main cause of death (17 animals) appeared to be associated with capture myopathy. Most animals died 2 to 5 days after capture and transport; two died after 13 days.

Chalmers and Barrett (1977) also examined two 2-week-old fawns that died after being pursued by a vehicle. Both animals displayed clinical signs typical of capture myopathy.

During 1949 to 1950, 141 pronghorn were trapped in Lassen County, California, and 28 (20%) died (L. Pyshora personal communication:1988). Trappers reported 1 foot (30 cm) of mud in the trap, lethargic animals and animals dying with no obvious injuries or symptoms of disease. Running under stress and in mud probably led to capture myopathy.

O'Gara necropsied a fawn that was chased by a golden eagle and died of capture myopathy-like symptoms after its sibling was killed by the bird.

Capture myopathy appeared more prevalent in pronghorn trapped during warm weather than during cold weather and affected at least 4.2% of all pronghorn handled. Chalmers and Barrett (1977) found that less intensive aircraft pursuit, minimal handling, use of darkened trailers for transportation and cold weather drive trapping

Because they run with their mouths open, pronghorn running in a band sometimes swallow or inhale rocks or other debris kicked up by animals ahead of them. Fast chases should be kept to a minimum (number and duration) to prevent such inhalation and possible capture myopathy. Pursuit with a fixed-wing aircraft is more apt to cause capture myopathy than is a helicopter flown by a pilot experienced in pronghorn capture. Top photo by Patrick Robles. Bottom photo by Paul Wertz; courtesy of the California Department of Fish and Game, Sacramento.

appeared to be the most effective factors in reducing the prevalence of acute cases of capture myopathy. Mortalities attributable to capture myopathy also were fewer when the transportation distance was short.

Of 232 pronghorn corral-trapped in Idaho, 12 (5%) died at the trapsite. Four (7.4%) of 54 radio-collared animals died shortly after release. There was no way of knowing the fates of 104 pronghorn that were ear tagged or 60 that were released unmarked (Hoskinson and Tester 1980).

In a radiotelemetry study, Reeve (1982) reported high (50%) postrelease mortality of 14 radio-marked pronghorn caught in a corral trap, but cause of death was not established. During another trapping session, the cause of two of four postrelease deaths was established as capture myopathy, but the problem was not evident in any of 17 pronghorn captured with a surround net.

Although corral trap mortality of pronghorn generally is reported at 2% to 5%, overall mortality of an entire trapping and translocating operation may be considerably higher. Mortality (probably acute trauma) at the trapsite reported for 5,751 pronghorn (summation of reports) was 4.3%. Mortality reported for 1,600 pronghorn transported to release sites was 7.5%. Combining these percentages with Chalmers and Barrett's (1977) estimates of capture myopathy (19% to 25%), overall pronghorn mortality of a trapping and translocating project may approach 30%, even under optimal conditions and given reasonable precautions.

An example of such a possibility follows. During November 1999, about 192 pronghorn were driven, via helicopter, into a corral trap northwest of Chugwater, Wyoming. They were to be used in a study concerning the safety to pronghorn of oral brucellosis vaccines intended for bison and elk. The trapping operation seemed nearly flawless. Only one doe was euthanized because of injuries. Fawns and yearlings were released, and older animals were loaded into three stock trailers for a long ride to College Station, Texas. Seven pronghorn died in the trailers, which were partitioned to cluster animals in small groups. Once in Texas, 45 more died of capture myopathy (Anonymous 2000). If this had been a catch, mark and release operation, mortality may have been reported as less than 1%. If it had been a translocation project, mortality probably would have been reported as less than 10%.

The statement by Chalmers and Barrett (1977), that chemical immobilization could lead to capture myopathy, conforms with O'Gara's observations. Two pronghorn immobilized with M-99 died within 2 days, although they were immobilized for only a short time and appeared to recover without complications. One carcass was necropsied, and shock was considered the cause of death, but the clinical signs of capture myopathy—as later described by Chalmers and Barrett (1977)—were present. Also, Beale and Smith (1967) found succinylcholine chloride to be an effective immobilizing drug for pronghorn, as previously discussed. S. Amstrup (personal communication:1980) successfully immobilized adult pronghorn with 8 to 10 milligrams of this drug, but had two deaths and saw the symptoms of capture myopathy.

Workers in Colorado completed 24 pronghorn translocations between 1947 and 1955 (Hoover et al. 1959). Casualties from 6.4% to 10.6% were experienced at the traps during the first 3 years. Fatalities during the following four trapping seasons never exceeded 3.4%, and there were none the last year. Transit deaths during the first 3 years averaged 7.6%. Those during the following four seasons averaged 3.4%. This seems a strong argument for securing experienced workers when selecting a trapping crew.

CAPTURE OF FAWNS

Fawns are captured for various management and research purposes. Most early captures were undertaken to raise fawns for eventual reintroduction to unoccupied, historic rangelands. In recent times, wildlife managers sometimes observe low fawn/doe ratios during summer pronghorn surveys, despite high conception rates. Accordingly, by capturing newborn pronghorn and using radiotelemetry to follow their survival, at least 19 studies have been conducted to determine the causes of fawn mortality. Newborn fawns also have been marked to determine dispersal patterns at maturity. Pronghorn fawns develop rapidly and can be captured successfully until about 2 weeks after birth, which spans about a 4-week period in

Although present-day researchers find tame pronghorn very intolerant of restraint, fawns raised for early translocation efforts apparently were kept on tethers. Successful raising of pronghorn fawns as pets required day-long care by entire families. Pronghorn fawns were common pets among Plains settlers and Indians, but documentation of their survival to adulthood is rare (R. E. McCabe personal communication:1994). Injury, disease, predation, dispersal or nuisance dispatch were causes of loss most often cited. Photo courtesy of the Kansas State Historical Society, Topeka.

spring. This brief period when young fawns can be captured easily necessitates following specific procedures to minimize search time and to handle the fawn properly to avoid abandonment by does.

The chance of induced mortality always must be considered when capturing animals, especially fawns. Byers (1997a) often captured and marked young pronghorn from 1981 to 1996. He recorded the number that were born and survived on his study area. No evidence was found of increased mortality due to handling. Byers concluded that, with proper precautions, one may handle young pronghorn without exposing them to additional risk.

Although fawns are located more easily early and late in the day, Byers (1997a) recommended handling them from about 7:00 A.M. to 4:00 P.M. Earlier and later, he often saw coyotes on the hunt. Byers also recommended having fawns well located and walking quickly and directly to them. Does appeared distressed by persons close to bedded fawns and uttered alarm snorts, probably attracting the attention of predators.

During fawning season, observers should position themselves at strategic observation points. Binoculars and spotting scopes are useful for observing preparturient does or doe/fawn behavior. A doe signals impending parturition by leaving her band, standing and lying frequently, raising her tail, humping her back, and licking her belly and flanks. Surveillance should continue for about 4 hours after birth until fawns have nursed and mother/young imprinting is complete. This waiting period is critical to reduce chances of abandonment and assure that the fawn will be several hours old at time of capture. The exact bedding location of fawns should be carefully noted.

Lone does with large udders should be observed carefully, as they usually go to their fawns and nurse about every 4 hours. After nursing and grooming, the fawns bed down and the doe leaves the area. It is advisable to note the spots where the fawns bedded but wait 30 minutes or so for them to settle. An added advantage to this procedure is that a fawn with a full stomach usually will remain immobile when rebedded, thus avoiding the attention of predators. Nichol (1942) reported that a doe, at the first threat of danger, would look immediately and directly at the spot where her young were hidden. Nichol maintained that an alert person could determine the angle and direction and mark the spot for seeking fawns.

Bedded fawns should be approached slowly and quietly from their rear, and a large (about 3-foot [0.9 m] in diameter), long-handled net should be placed gently over them to assure capture. Fawns may be hand-captured successfully up to 3 days of age. The use of nets allows capture of some older fawns (up to 7 days old) without a chase. Blindfolding while handling helps to keep them still and reduces the incidence of distress bleats, which are effective coyote calls. Wearing surgical gloves will minimize transmittal of human scent.

Returning fawns to their original position and gently rubbing their tails reduces their tendency to run after

Strategic observation points from which to watch for pronghorn does giving birth or suckling young fawns generally are rocky outcrops on ridges overlooking flats. Binoculars and spotting scopes are useful, and a cushion or folding chair makes it easier to endure the many hours of inactivity necessary for success. Photo by Jim D. Yoakum.

When pronghorn fawns are located, the observer is well advised to lock-in a spotting scope on the spot (fawns virtually disappear when they lie down) and direct the capture crew to the animal(s) with a two-way radio. Photo by Walt Bodie.

Careful stalking before placing a large, long-handled hoop-net over bedded pronghorn fawns often facilitates capture without a chase. Fawns should be approached very slowly from their rear. A fawn usually sees a person nearing, but will not bolt if the stalker's movements are slow enough. Photo by Bart W. O'Gara.

A hiding fawn often will watch a person at about 20 yards (20 m) without moving. Keeping the young pronghorn's attention by hand movements often will allow another person to stalk and net the animal from behind. Photo by Walt Bodie.

release (Amstrup et al. 1980). However, working on the National Bison Range, Byers (1997a) allowed fawns to run away when released. The mothers, hovering nearby, would run in front of them and lead them away. Thus, fawns did not remain at capture sites, where human tracks and scent could attract coyotes. The National Bison Range pronghorn are habituated to people, and truly wild does might not remain close enough to lead their fawns away.

Fawns too wary for daylight capture can be caught with spotlights at night (Brownlee and Hailey 1970). Observations of single females at dusk often indicate presence of fawns, and a spotlight search should begin in that locality after dark (Tucker 1979).

From 1953 to 1957, an intense fawn capture project was conducted in adjoining regions of California (Ackerly and Regier 1956), Nevada (Foree 1956) and Oregon (Hansen 1955, Yoakum 1957a, Compton 1958, Dinges 1958). More than 600 captures were made, some taking place 4 weeks after the first capture. Two fawns were each caught four different times during 1 month, indicating survival after repeated captures, intensive collection of biological data, tagging and releases. An adult, well-trained Labrador retriever assisted in recaptures of fawns more than 1 week old. The dog was trained to heel behind a horse. When a fawn was flushed and began to run, the dog chased, caught and held it until the biologist arrived. This technique did not result in any injuries to fawns during 156 captures in a 3-year period. Horses greatly assisted captures. On horseback, biologists were able to concentrate on looking for hidden fawns, and the relatively high vantage point

When a net is placed over a pronghorn fawn, care should be taken that the leading edge is well ahead of its nose and nearly to the ground before netting touches the fawn. If a fawn is young enough to be caught in this manner and bolts, it seldom goes more than 100 to 150 feet (30–46 m) before dropping out of sight beside or in vegetation higher than average in the area. Something should be dropped to mark the original bed and an immediate stalk made to net the fawn. Usually, the run has tired the fawn, and it can be caught on the second try. If it bolts again, it probably is too old to catch and should be left alone. Returning the fawn to its original bed seems a good idea; although, when fawns have been taken by predators, does will search for hours in a large area around the original bedsite. Photo by George J. Mitchell; courtesy of the Alberta Department of Industry and Development, Edmonton.

Handling with plastic gloves reduces the amount of human scent imparted to pronghorn fawns. In most studies where abandonment has been a problem, the fawns have been instrumented with bulky collars or harnesses. Thus, visual disturbance may be a more significant abandonment factor than odor, but every precaution should be taken. After an instrumented fawn is returned to its bed, gently rubbing the tail area seems to calm the animal and permits the handler to steal away without the fawn bolting from the area where the doe expects to find it. Photo by Walt Bodie.

This photo depicts an improper way to handle a fawn that is going to be released. The temptation is great to take photos, but imparting human scent to a fawn increases the chance of rejection by its dam. Photo by Jim Reichel.

If pronghorn fawns are to be weighed before release, a clean piece of porous cloth or netting should be used as a cradle. Frequent laundering and carrying in a pack with vegetation from the area reduces the chance of the cradle imparting human odors to fawns. Plastic gloves worn by the handler (Jim D. Yoakum, above) and blindfolds on the fawns also should be used. Photo by Rod Canutt.

and ease and speed of travel also were beneficial. The average, lone worker caught 25 fawns in a season, whereas the worker with the dog and horse caught 75. Thus, the use of a well-trained dog with a soft mouth is recommended for capturing fawns.

If birth of a captured fawn is not observed, its age up to 3 days may be estimated using Bromley's (1977) criteria. Fawns 1 day old or less make no or only feeble attempts to escape, have damp umbilical cords and the hair on their backs often is in small clumps, apparently stuck together by dried amniotic fluid. Fawns 2 to 3 days of age are easily caught, but struggle vigorously when handled, often bleating. Their umbilical cords are gone or hardened. Fawns 4 to 7 days of age are difficult to catch with the net, often leaping up before the net is placed over them. Fawns 8 to 14 days of age cannot be caught without spotlighting or the assistance of a dog.

Only well-trained dogs should be considered for catching pronghorn fawns. Dogs sometimes can find fawns by scent, but more often, they run down fawns that have been flushed from beds by humans. Photo by Jim D. Yoakum.

FACTORS AFFECTING TRANSLOCATION

Scientific management aided the increases of pronghorn from an estimated 25,000 in 1924 to 1 million by 1984 (Yoakum 1986). Large-scale translocations contributed greatly to the increase; more than 30,000 pronghorn were translocated in 17 states in the United States between 1920 and 1997, and more than 211 were relocated from Colorado and New Mexico to Mexico (Table 166).

The success of pronghorn translocation efforts depends, in part, on healthy animals, social facilitation and release into suitable habitats (Goldsmith 1992). Apparently, when a pronghorn group is captured as a social band, the band remains compatible after release into a new environment. Individuals not attached to a band tend to wander extensively looking for habitat they were accustomed to before capture. An established pronghorn society (or population) involves small bands that frequently interact and interchange with other bands of a larger herd (Pyrah 1987). Pronghorn use knowledge of their home ranges to find and use habitat resources efficiently, interact with other bands and escape predators. A band is able to do so in new habitat better than an individual can. Bands released into habitat similar to that from which they were removed have a greater chance of becoming established than do those released into totally different habitat.

Most successful translocations have involved the release of 50 to 100 animals at one time and place, according to

The presence of a pronghorn fawn's umbilical cord still red, shiny and wet indicates birth only 2 or 3 hours before. That the above fawn's coat is clumped by drying amniotic fluid signifies that the doe has not had time to clean and dry her offspring completely. Like bison calves and the young of grassland ungulates in Africa and Asia, pronghorn fawns are precocious. They can run up to 25 miles per hour (40 km/hr) for short distances when only a few days old. At 2 weeks of age, they feed on green vegetation. Bromley and O'Gara (1967) reported on two fawns orphaned at approximately 7 weeks of age that not only survived but maintained their positions in the band's hierarchy. Photo by Cathy and Don Jenni.

Hoover et al. (1959) and Russell (1964). However, Britt (1980) suggested that several subsequent releases over a period of several years may be more successful than one large release, especially in areas of less than optimum habitat. However, in Mono County, California, multiple releases did not establish a self-sustaining population (see Chapter 16).

The success of pronghorn translocations depends most on evaluating proposed release sites to ascertain if they are suitable. The use of habitat suitability models, discussed in Chapter 13, provides criteria for different biomes. At times, hunting and angling organizations, conservation groups or local political groups have recommended introducing pronghorn to areas not capable

During autumn 1959, pronghorn captured in Montana were habituated to humans in part by being housed under bleacher seats in Helena. They then were loaded into crates, trucked to the West Coast and shipped via steamer (top left) *to Honolulu (J. McLucas personal communication:1981). On arrival in Hawaii, the animals were placed aboard a Marine Corps aircraft* (top right) *and transported to the island of Lanai. The original intention was to release the animals on a ranch on the island of Hawaii. However, ranch officials apparently changed their minds and did not want the pronghorn. A decision then was made to release the 7 does and 31 fawns on the northeastern section of Lanai, which had been heavily grazed by cattle for about 50 years (*bottom photos*). The grass was short, and the plateaus of the north end of the island were likened to eastern Colorado and Wyoming (Nichols 1960). The 38 pronghorn released on Lanai in 1959 decreased to about 18 animals by 1960, and then increased to 96, with a 1.12:1.0 fawn/doe ratio, by 1962. In 1966, the population was estimated at 216. A hunting season was held, and 27 bucks and 6 does were taken (Bromley 1972). When Bromley arrived on the island during May 1971, to study the social organization and breeding cycle of the subtropical herd, not more than 60 animals remained, with a 0.5:1.0 fawn/doe ratio. Bromley concluded that plant succession, after the removal of cattle, made the rangeland progressively less suitable for pronghorn, leading to lowered reproduction. Poaching and predation by dogs also caused some mortality. A mid-May aerial census in 1972 located only 16 to 18 animals (Anonymous 1972), and, by 1982, only two does could be located (K. Ueoka personal communication:1993). The last doe reported on Lanai was seen in 1986 (R. Walker personal communication:1993). Events associated with this translocation emphasize the need for careful planning and knowledge concerning rangelands and pronghorn habitat needs.* Top photos courtesy of the Montana Department of Fish, Wildlife and Parks, Helena. Bottom photos courtesy of the Hawaii Division of Fish and Game, Honolulu.

of sustaining populations. Such endeavors resulted in the loss of all animals translocated to Florida and Hawaii. Analysis of such cases disclosed that the proposed sites did not meet minimal habitat requirements for pronghorn. Ignoring such basic biological requirements results in the eventual death of translocated animals, high expenditure of public funds and a negative public reaction to the integrity of professional management (Yoakum 1978a).

Reestablishing pronghorn herds invariably requires cooperation among federal and state wildlife and land management agencies and private landowners. Translocations can be considered when it has been determined that pronghorn can survive in an area possessing sufficient quality forage, water and space in historic range, with a minimum of other environmental problems, such as crop depredations and obstacles to pronghorn movements, daily or seasonally.

One of the first detailed procedures for determining the potential of an area to sustain introduced pronghorn was developed by the Colorado Department of Wildlife (Hoover et al. 1959). Twenty-eight years later, the International Union for Conservation of Nature and Natural Resources (1987) proposed similar criteria and summarized them as a feasibility study, preparation phase, release or introduction phase and a follow-up phase. The latter often has been neglected, and the feasibility studies and preparation phases have not been adequate in many cases.

Reinvasion of unoccupied historic habitat teaches a great deal about the habitat preferences of pioneering pronghorn. Examples were natural expansions that took place in Bear Valley of central Oregon (Polenz 1976) (see Chapter 15), Long Valley of northeastern California (see Chapter 17) and Middle Park, Colorado (Pojar and Gill 1990).

Sadly, another type of translocation is becoming necessary in some areas. As human habitations proliferate, some pronghorn herds become surrounded and must be relocated or slowly perish before further developments consume their remaining rangeland. Such a short-distance relocation took place near Prescott, Arizona, during the mid-1990s (Lewis 1996b). More than 300 people from the Arizona Game and Fish Department and Arizona Antelope Foundation, as well as members of other environmental organizations and concerned local residents, formed a human fence. The fence slowly moved the reluctant group of pronghorn from two fields surrounded by developments across a dirt road soon to become a four-lane highway that would have isolated the animals from adjoining rangelands. As the road fence was reinforced to prevent reentry, another tract of rangeland was lost as pronghorn habitat.

MANAGEMENT PLANS

The trapping and translocating of pronghorn involves much labor, time and money. Consequently, we recommend the development of management plans, based on habitat suitability models discussed in Chapter 13, before field implementation (see Chapter 16). Such plans should provide detailed strategies and procedures for capture, transportation and release in new habitats. Often this requires the coordination of agency and private interests, and the plan becomes the agreed-upon tool for cooperation.

Management plans should specify location and capture methodology, number of animals needed, procedures to capture the animals, logistical and scheduling details of transportation to the release site, suitability of that site and monitoring of postrelease success. Having a veterinarian at the capture site is desirable. The plan also should provide particulars regarding methods of release and follow-up monitoring evaluations in case additional releases are needed to assure success of the project. When these strategies are adopted, there is good reason to believe that managers have prepared detailed plans based on credible management practices. Management credibility is scrutinized increasingly, with the recent surge of animal rights and animal welfare advocacy. Public perception, accurate or not, of trial and error management can lead to media or court challenges that can alter management flexibility significantly and compromise professional integrity.

Management plans allow wildlife managers and the public to evaluate whether a translocation has served its original objective. The objectives for pronghorn translocations can be varied and many, and they include: (1) re-establishing herds on historic rangelands, (2) introducing animals to unoccupied habitats for public benefit, (3) augmenting existing herds that are low in numbers but have sufficient, suitable habitat, (4) relocating surplus animals from one area to establish new herds for hunting or other benefit in another area and (5) reducing crop depredations by moving excess animals to other suitable habitat.

The paramount goal should be to establish a viable herd. This means the habitat must be suitable for production and sustainability of a reasonable number of animals, and recruitment and productivity will be greater than mortality. Relocated herds that increase by more than 20% within 5 to 10 years indicate quality habitat. A pioneering herd in Middle Park, Colorado (Pojar and Gill 1990), expanded at a rate greater than 50% the first year—indicative of new population productivity in historic but unoccupied habitat in good to excellent condition.

The management plan should address the limits at which a new population will be maintained. Identifying population control practices after release is as needed as identifying the herd size for the original release. This is a particular need for northern pronghorn herds because they appear to have a higher degree of success than do those released in southern environments—apparently in response to carrying capacity.

Franklin (1980) considered 50 breeding adults the minimum size for a population. It seems reasonable, therefore, that translocations should contain 50 to 100 animals, as recommended by Hoover et al. (1959), after years of translocation efforts in Colorado. Regarding the goal of numbers for a translocated population, Franklin (1980) suggested a population of 500 randomly mating individuals as the minimum population size for maintaining genetic variation at a level to enable the species to adapt to changes in the environment. This seems most appropriate for pronghorn that experience frequent, severe winters or numerous droughts.

TRANSLOCATION STRATEGIES AND PROCEDURES

Whether a proposed translocation area historically was occupied by pronghorn is an important consideration for pronghorn relocations. Those sites not historically inhabited seldom are able to support viable herds. If a proposed site is historic pronghorn range, the following questions must be answered: what caused the animals to be extirpated, does the factor responsible for their elimination still exist, has the habitat or other conditions been altered sufficiently to preclude reestablishing pronghorn there, and do current land uses and local attitudes favor reintroduction?

Feasibility Guides

Hoover et al. (1959) developed criteria for selection of pronghorn translocation sites in Colorado. Eight of those criteria bear elucidation here, with only slight modifications more than 40 years later.

1. Unless sufficient continuous range is available to allow a translocated herd to expand, the site should be rejected. As a rule, each animal requires at least 1 square mile (2.6 km^2) of native grassland, and the number of animals should not be less than 100. A minimum population level exists for gregarious animals below which they are unlikely to survive. Within the area's habitat limitations, large translocations are most likely to become established populations.
2. In most areas, pronghorn feed primarily on forbs and shrubs, so a variety and abundance of these should be present. Although pronghorn can survive on heavily grazed rangelands where other grazing animals would have difficulty, such rangelands are not desirable because their carrying capacities are low. Pronghorn introduced onto such rangelands might be accused in later years of being responsible for poor vegetative conditions. Thus, the stocking rate for livestock or wild ungulates should be evaluated. Overstocked rangelands should not have the added burden of introduced pronghorn. Dense, high shrublands are not desirable because they reduce pronghorn running and sight abilities.
3. Whether domestic sheep, cattle, horses or a combination of these domestic or feral livestock graze an area should be considered in selecting translocation sites. The relationship of feeding habits of pronghorn and livestock determines competition between them. Generally speaking, little competition exists between pronghorn and grazers, such as horses and cattle, on rangelands with healthy vegetation, but considerable competition is possible with sheep (see Chapter 15). Also, sheep-proof fences can seriously restrict pronghorn movements. During droughts and blizzards, such fences may promote death for pronghorn. Therefore, for pronghorn translocation purposes, rangelands used by horses and cattle usually are preferred over those used by sheep. (In some areas, however, predator control for the welfare of sheep benefits pronghorn.)
4. Cultivated crops within or in close proximity to a proposed translocation site may render the site unsuitable because of crop depredation and potential landowner intolerance. Dryland winter wheat and alfalfa fields are particularly attractive to pronghorn. Isolated fields usually are subject to more depredation than are numerous fields surrounded by rangelands with native vegetation.
5. A map illustrating the landownership pattern of a proposed site should be prepared, especially if private lands are involved. Public lands are preferred translocation sites, followed by large blocks of private land under single ownership. The least desirable sites are private lands in small ownership units.
6. Reactions of many people to an introduction of pronghorn should be considered, particularly those of local conservation organizations, private landowners and personnel charged with managing public lands. It should be learned whether local landowners are amenable to the management policies and practices, including hunting seasons, of the state wildlife agency. Some landowners may favor introducing pronghorn but will not tolerate hunting, which could curtail severely management of a developing population. Finally, local hunting, fishing and other conservation clubs should be notified of the intent and encouraged to support the effort.
7. If all the above factors are satisfactory, permission should be secured from all public land agencies and

private landowners in the introduction area. The number of animals to be released must be determined. The release or releases should contain a sufficient number of pronghorn to provide a limited open hunting season within 5 to 10 years. Such an objective would reflect a goal for a healthy and productive herd.

8. Once the survey of a proposed translocation site has been completed and approved, plans should be made for the release. Because road conditions in northern or high altitude areas during the winter months (when most pronghorn are trapped) can create problems, having an alternate release site is advisable. Pronghorn become disturbed and are likely to trample each other once trucks stop. To avoid confusion and loss of time, the precise point within the area where the animals will be released should be selected close to an all-weather road. Consulting with the local wildlife conservation officer, or someone else familiar with the area, is advisable, to assist and direct the truck driver to the release point. This contact also can check road conditions to the release point immediately upon receiving word that the trucks loaded with pronghorn are on their way. This prevents loss of time if an alternate release point has to be used. Getting the trucks stuck on an impassable road may jeopardize the lives of the cargo.

During recent years in North Dakota, only one translocation took place, that in 1991. "This transplant was influenced by bio-political decisions that were contrary to recommendations found in *Guidelines for the Management of Pronghorn Antelope*; subsequent observations indicate it has failed to establish a viable breeding population of pronghorn. As a result, the following guidelines were developed to address future translocation requests:

"1. Pronghorn transplants will be considered in areas of the state that traditionally have held animals in the recent past and that survived well over a long period of time.

2. A release area consisting of a minimum of four square townships in size would need to be identified (144 square miles [373 km^2]). The area would need to be habitat-mapped by the project sponsors and ground-truthed by our big game staff to determine if it met the habitat requirements of pronghorn.

3. Written approval for the project by all of the landowners owning land in the proposed release area would be required.

4. The sponsoring organization (e.g., wildlife club) would need to sign a cooperative agreement with the Department, taking on partial responsibility for depredation complaints and response to damage complaints by landowners.

5. The sponsoring organization would need to provide at least 75–80 percent of the costs of the project. Total project costs are approximately $7,500 to $10,000 per transplant, based on recent data. Cost to local sponsors would be in the $5,500 to $7,500 range" (McKenzie 2000:22–23).

Habitat Suitability Criteria

Based on 24 years of evaluating shrubsteppes in six western states, Yoakum (1974) classified habitats into abiotic and biotic components important to pronghorn (see Chapter 13). He correlated these habitat requirements with the translocation survey prepared by Hoover et al. (1959) to produce a feasibility form for selection of translocation sites in shrubsteppe environments (Figure 158). Yoakum summarized needs of pronghorn as follows: (1) suitable habitat is related to the right amount and juxtaposition of all habitat characteristics meeting the species' habitat requirements, (2) too little or too much of any biotic or abiotic factor may become the primary component limiting production and survival and (3) knowledge of these habitat requirements becomes the ecological foundation for managers making decisions relative to the management of pronghorn and their habitat, including translocation sites.

Adobe Valley. Habitat requirements of pronghorn were used as part of the criteria for selecting a translocation site in Mono County, California (McCarthy and Yoakum 1984). Nine state and federal wildlife biologists comprised an interagency team that evaluated five potential release sites. The strategies for evaluating these sites may have application in assessing relevant priorities and practicability of a potential release area for other translocated pronghorn in the shrubsteppe biome.

Procedures for choosing the translocation site started with a field examination of all potential sites and completing a form (Figure 158). This was followed by habitat evaluations, resulting in suitability ratings using the index (see Table 169) originally developed by Yoakum (1980). Essentially, this index rates water and forage quantities and qualities numerically. Table 169 is an example of the method in which the suitability index was accomplished for the five major vegetative communities at one potential release site.

Each site was evaluated for the following eight additional criteria deemed significant for the success of a pronghorn translocation: (1) mean winter snow depths were interpolated using data from weather stations nearest to the sites being evaluated (areas were suitable for winter use if snow depths were less than 10 inches [25 cm]); (2) natural physical barriers were evaluated to determine

1. LOCATION:
   County ______ Nearest town ______
   Nearest ranch ______ Accessibility by road ______
   Township ______ Range ______
2. SIZE (Number of square miles of estimated habitat): ______
3. TOPOGRAPHY: ______ Physical Barriers: ______
   Constructed Barriers:
   Fences (Location) (Construction Specifications)
   ______ ______
   ______ ______
   Major highways, freeways ______
   Other ______
4. CLIMATE: ______ Elevation ______
   Mean depth of snow ______ Annual precipitation ______
5. WATER: Springs Reservoirs Lakes Streams Wells Catchments
   Number ______
   Acres ______
   Miles ______
   Production ______
   Surface ac. ______
   Gal/Min ______
   Gal/Storage ______
   Mean distribution of water sources ______
   Year-round water? ______
6. VEGETATION

Major types	No. acres	Mean height	Estimated percentage: Grass	Forbs	Shrubs

7. LAND OWNERSHIP (number of acres):
   Private ______
   Public ______
   Other ______
8. LAND USE:
   Class of livestock ______
   Stocking rate ______
9. PREDATION:
   Natural: Coyotes ______ Eagles ______ Bobcat ______
   Human ______
10. TRANSPLANT CONSIDERATIONS:
   Is site historical pronghorn range ______
   Attitude of ranchers ______
   Attitude of conservation organizations ______
   Attitude of local sportsmen's clubs ______
   Attitude of public land-management agencies ______
   Is land manager(s) agreeable to management objectives of state or provincial wildlife agency
   ______
   Suggested number of pronghorn for transplant ______
   Route of trucks carrying pronghorn and release point ______
   Has a "habitat management plan" been developed ______
   Are cooperative agreements completed
   Private landowners ______
   Public land agencies ______
   Has a follow-up monitoring study been planned to document success of project or reasons for failure ______
   Other ______

FIGURE 158. *Survey form for selection of pronghorn translocation sites (from Hoover et al. 1959, modified for shrubsteppes by Yoakum 1974, 1980).*

TABLE 169. Pronghorn habitat suitability rating for Adobe Valley, California, by vegetation type[a]

Type	*Vegetation quality*				*Vegetation quantity*	*Vegetation height*	*Total rating*
	Water	*Forb*	*Grass*	*Shrub*			
Big sagebrush/Indian ricegrass	10	1	17	5	5	0	38
Indian ricegrass/big sagebrush	10	1	16	5	5	5	42
Big sagebrush/green rabbitbrush	10	1	2	5	3	0	21
Saltgrass/sedge	10	1	10	10	2	7	40
Saltgrass/gray rabbitbrush	10	1	16	5	2	8	42
Average	10	1	12	6	3	4	37

[a] Ratings are based on data applied to Yoakum's (1980) habitat criteria and rating procedure for pronghorn habitats.

potential for restriction of pronghorn movements (major physical barriers included large rivers, mountain ranges and dense shrub or timbered areas); (3) area size was determined by using a polar planimeter and 7.5-minute U.S. Geological Survey quad maps (minimum area size was considered 100 square miles [256 km^2]); (4) fences were evaluated in relation to pronghorn passability (a fence was considered passable if the bottom wire was at least 16 inches [41 cm] from the ground); (5) potential for predation was rated subjectively as high, moderate or low (predator abundance information was obtained from various government agency personnel, ranchers and individuals familiar with the sites); (6) potential for crop depredation was determined on the basis of distance from each site to the nearest alfalfa field (a depredation potential rating of high, moderate or low was assigned to each area); (7) seasonal suitability was evaluated on the basis of whether an area could support pronghorn on a yearlong basis; and (8) potential for competition with domestic livestock was determined on the basis of class of livestock, animal unit months and grazing systems (each site was rated as having high, moderate or low potential for livestock/pronghorn conflict). After the evaluation of each site, limiting factors were addressed on the basis of these criteria. The sites were compared and prioritized in terms of potential for a successful pronghorn translocation.

Adobe Valley was grazed by cattle (1,686 animal unit months [AUMs]) and sheep (556 AUMs) during summer. One rest-rotation grazing system was implemented, while the remaining areas were grazed season-long. If livestock management was not integrated with pronghorn habitat requirements, then grazing was considered a possible limiting factor on pronghorn establishment. The potential for competition with other herbivores was deemed moderate.

Using nine criteria, a comparison of the five sites was made (Table 170). Adobe Valley had the highest habitat suitability rating and was considered the best site for a pronghorn translocation. Although Adobe Valley ranked more favorably than the other sites, the interagency team identified several factors that would limit the possibility of success, including: (1) a low habitat suitability rating (the combination of poor forb production, limited vegetation diversity and unsuitable vegetation height on much of the area resulted in conditions generally not conducive

TABLE 170. Comparison of five translocation sites using nine evaluation criteria

Criteria	*East Mono Lake*	*Adobe Valley*	*Benton/ Queen Valley*	*Montgomery Pass/Truman Meadows*	*Hammil Valley*
Habitat suitability rating[a]	13	38	11	26	14
Mean snow depth[b]	12(30)	10(25)	0(0)	12(30)	0(0)
Physical barriers[c]	Low	Low	Low	Moderate	Low
Release area size[d]	113(293)	60(155)	47(122)	108(278)	55(142)
Fences[e]	0.21(0.13)	0.54(0.34)	0.19(0.12)	0(0)	0.03(0.02)
Predation anticipated[c]	Moderate	Moderate	Moderate	Moderate	Moderate
Depredation anticipated[c]	Low	Low	High	Low	High
Complete biological unit	No	Yes	Yes	No	Yes
Competition with herbivores[c]	Low	Moderate	Low	High	Low

[a] Habitat evaluations using an index developed by Yoakum (1980).

[b] Inches (cm).

[c] Evaluated subjectively in terms of high, moderate or low potential.

[d] Square miles (km^2)

[e] Miles per square mile (km/km^2).

to pronghorn use); (2) fences were a barrier to pronghorn movements; and (3) water sources needed to be modified for pronghorn use.

Based primarily on these factors, the interagency team recommended that translocation be postponed until the following conditions were met: (1) modification of all fences to provide for pronghorn passage, (2) modification of all water developments to accommodate use by pronghorn, (3) development and implementation of a large-scale vegetative improvement project to enhance forage quantity and quality, emphasizing a variety of native forbs, (4) modification of livestock grazing systems to favor forb production important to pronghorn, (5) development of a long-term pronghorn management plan and (6) initiation of monitoring studies to determine pronghorn movements and to assess the success of the translocation program.

Instead of following the interagency recommendations, a decision was made to expedite the release of pronghorn in Adobe Valley. Some fences were modified to permit pronghorn passage, and several water sources were modified to accommodate pronghorn use. None of the other recommendations were followed.

In March 1983, 43 pronghorn were translocated to Adobe Valley. These were animals causing depredation problems in northern California. By April 1984, only six animals remained at the release site, using Adobe Valley in the summer and Hammil Valley in the winter. The remaining animals dispersed from Adobe Valley (Goldsmith 1988). Seven years later, no resident pronghorn remained (T. Russi personal communication:1990).

Results of this translocation indicated that the habitat was marginal for pronghorn, as was predetermined by the interagency team's evaluation system. Open to speculation is the outcome of the translocation had the interagency recommendations been followed. However, that overall habitat suitability would have been higher and more conducive to pronghorn establishment and fawn recruitment is probable (McCarthy and Yoakum 1984).

Joshua Tree National Park. Shaw et al. (1998) investigated the feasibility of releasing pronghorn in Joshua Tree National Park, southern California, for the National Park Service. The executive summary of their report is an example of the information that should be gathered and heeded when considering an introduction. They offered the following conclusions, which were heeded.

"Regarding the feasibility of transplanting pronghorn into Joshua Tree National Park, we offer the following considerations:

- We have not found solid archaeological, ethnological, nor historic records of occurrence of pronghorn within the boundaries of Joshua Tree National Park.
- Habitats within Joshua Tree National Park averaged low-to-poor quality for pronghorn, even after the wet 1997–98 winter.
- Principal factors limiting habitat quality were inadequate forage and poor water distribution.
- The best potential habitat within the park lies in the Queen Valley vicinity, where human disturbance is greatest.
- Joshua Tree National Park probably does not enclose a large enough area of suitable pronghorn habitat to sustain a viable pronghorn population through extended dry periods.
- No populations of pronghorn exist near enough outside the park to provide gene flow and maintain genetic diversity in a small park population.
- Periodic supplemental transplants might therefore be required to maintain a population of pronghorn in the park and to prevent inbreeding.
- Sustaining a population of pronghorn in Joshua Tree National Park may also require supplementing water or development of more water sources.
- Supplemental feeding of pronghorn might be necessary during droughts.
- Predation, interacting with marginal habitat, is likely to be a major factor depressing pronghorn populations in the park.
- A danger exists of infecting native bighorn populations with new parasites or disease, if pronghorn are introduced" (Shaw et al. 1998:1).

Handling and Transporting

For most rangelands, December, January and the first half of February are recommended times for pronghorn translocations. According to Selye (1976) and Archer (1979), capture and handling of animals should be avoided during the last half of gestation. Elevated serum corticosteroid levels due to trapping stress may trigger abortion. Hunter-related problems during autumn and thermal stress during summer generally preclude trapping and translocating at those times.

The period of restraint is dependent on the number of procedures (tagging, injecting, measuring, etc.) deemed necessary and whether tranquilizing drugs are used. These procedures should be kept to a minimum. Placing blindfolds on pronghorn during handling reduces stress to the animals and makes them easier to control. Neck warmers, sold in most outdoor clothing stores, make convenient blindfolds.

When groups of pronghorn are to be translocated and individual handling for marking is not required, use of a cattle-type loading chute (Figure 159) can reduce stress to

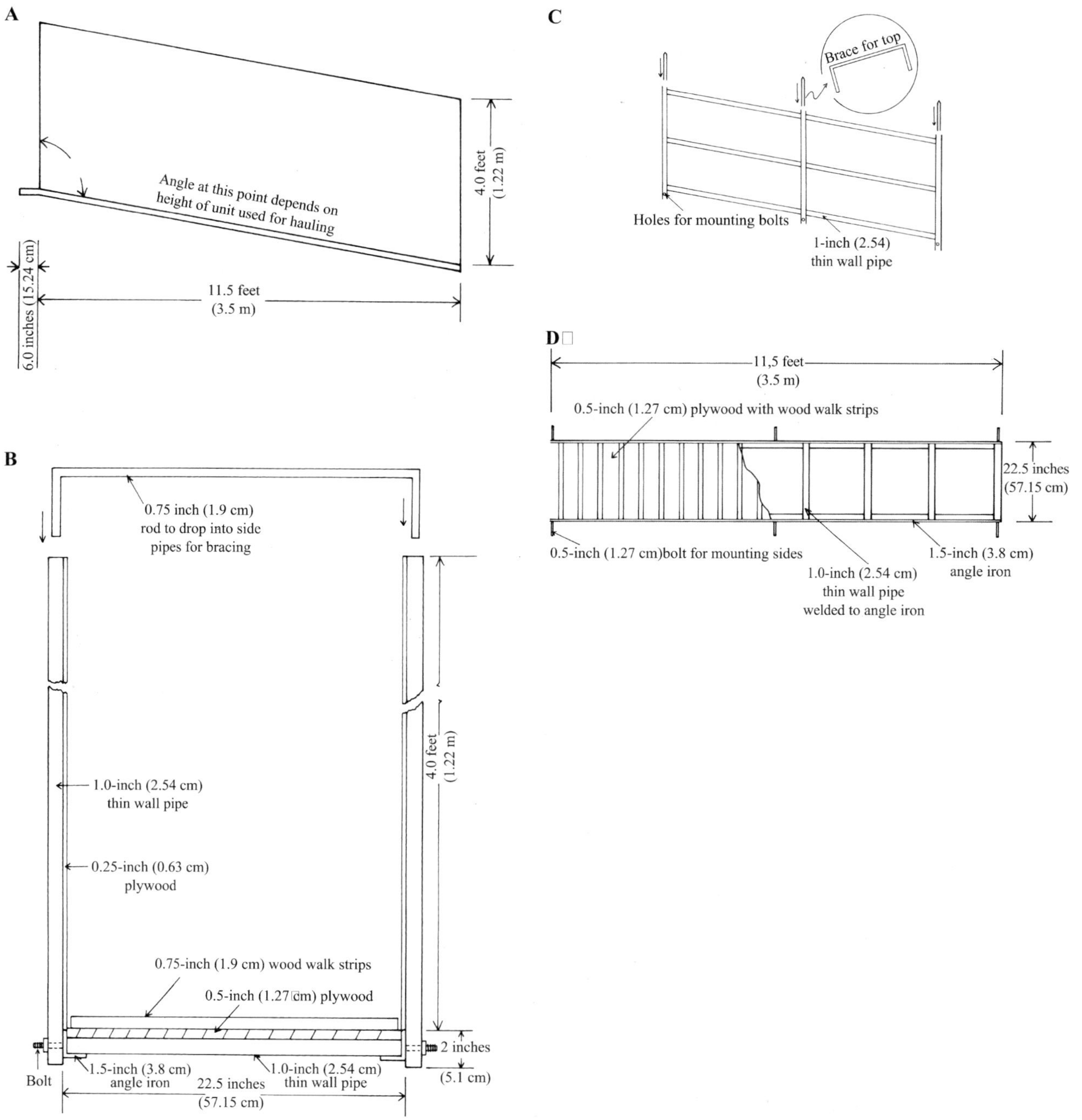

FIGURE 159. *Side (A), top (B), side frame (C) and bottom frame (D) construction details of a pronghorn loading chute (adapted from McKenzie 1983; courtesy of the Texas Parks and Wildlife Department, Austin).*

the animals and expedite the loading process. The chute should be constructed specifically for pronghorn, with solid wooden sides and runway, a canvas top and a width of approximately 2 feet (60 cm). The animals should be able to see light at the end of the chute or they will not enter. Often, the cover of the truck must be partly removed to show light and start the loading, but workers must be standing by to replace the cover quickly or the animals will try to jump out.

Captured pronghorn should be allowed to settle down in the trap without human presence for at least 10 and as much as 40 minutes. This is a judgment call depending on (1) the distance and time the pronghorn have been chased and (2) the ambient air temperature. Some workers who have handled many translocations recommend leaving the animals overnight and loading them during the cool of early morning. When this is done, second-cutting alfalfa and water should be placed in the corral before the drive.

Early pronghorn translocations were accomplished with stock trucks with solid sides and canvas tops. Photo courtesy of the Nebraska Game and Parks Commission, Lincoln.

Alfalfa scattered on the floor of the trap not only provides food and bedding, it also reduces the amount of dust inhaled by the animals during handling. Providing this resting period in the trap was considered one of the most important features of a Montana translocation program (McLucas 1956). Losses due to shock and accidents were believed to be reduced.

Tranquilizing drugs alter behavior (they promote tameness and ease of handling) and act slowly but for a long duration. Two such drugs—Valium (5 mg/mL) and Acepromazine (50 mg/mL)—are used routinely during some translocation operations at doses of 1 milliliter for adults and 0.5 milliliter for young of the year. However, Kreeger et al. (1999) did not find significant advantages in using Valium.

Vehicles for hauling pronghorn must be well ventilated, dark, compartmentalized (to segregate horned males) and easy to load by hand. The area requirement for each pronghorn during transportation is about 3.2 square feet (0.3 m^2). The trucks used to haul pronghorn in Montana were 1.5 tons (1,361 kg) with 8- by 12-foot beds (2.4 by 3.7 m), or 2.5 tons (2,268 kg) with 8- by 16-foot beds (2.4

Horse trailers with canvas covers are used by the California Department of Fish and Game for transporting pronghorn. Because they are low to the ground, such trailers are easy to load, and pronghorn are unlikely to hurt themselves when jumping out. Nevertheless, use of a ramp is advisable. Left photo by Paul Wertz; courtesy of the California Department of Fish and Game, Sacramento. Right photo by Patrick Robles.

by 4.9 m), on which enclosures were built with 4-foot (1.2 m) sidewalls, thus forming 8- by 12-foot or 8- by 16-foot enclosures, the tops of which were covered by two panels measuring 6 by 8 feet (1.8 by 2.4 m) or 8 by 8 feet (2.4 by 2.4 m) to conform with the size of truck used. These were made of 1- by 6-inch (2.5 by 15.2 cm) boards spaced 2 inches (5.1 cm) apart, giving good circulation of air to the animals enclosed within. About 25 pronghorn were hauled in 8- by 12-foot enclosures, and 35 were hauled in 8- by 16-foot enclosures (McLucas 1956).

Stock trailers lined (darkened) with burlap were used in California to transport pronghorn. Panels divided the trailers into several compartments. Adult bucks were caught by hand as soon as possible, to avoid injury to other captive pronghorn, and placed in individual crates to be hauled on flatbed trucks. Success was achieved by holding pronghorn in pens lined with burlap or canvas at release sites, allowing the animals to settle down and reestablish social order before release. Alfalfa hay and water were provided (L. Pyshora personal communication:1988).

Blindfolding adult males during transportation may be advisable even when they are segregated from the other pronghorn. Blindfolds are an alternative to removing the horn tips and prevent injuries to other pronghorn and to handlers.

The trip to the release site should start immediately after loading. Vehicular motion has a calming effect on pronghorn. When ambient temperatures are above 70°F (21°C), pronghorn in transit should be periodically hosed with water.

Release and Postrelease Monitoring

Releases should be made during daylight hours of the same day pronghorn are captured if the relocation site is nearby. However, pronghorn have been released successfully at night. Pronghorn are fragile, and the possibility exists of one breaking a leg as it jumps from a truck. However, a ramp is almost useless because the animals jump even when a ramp is available. When pronghorn are transported in crates, the crates should be placed on the ground before the animals are released.

Most releases are made as soon as pronghorn arrive at the relocation site. Personnel with the New Mexico Game and Fish Department favor retaining the animals for a day or so until they are calmed down (B. Morrison personal communication:1985). The same technique was used by California Department of Fish and Game personnel when they retained pronghorn in a small pen at the release site for 5 or 6 hours before release. The pen was encircled with burlap or canvas, so the animals could not see out. They appeared to be much calmer upon release (J. Letberg personal communication:1989).

Monitoring of released pronghorn should be an integral part of any translocation program. When possible, a long-term evaluation project should ascertain the rates of adaptation, dispersal, reproduction and recruitment, as well as levels of predation and other mortality. Examples of such postrelease studies were provided by McCarthy and Yoakum (1984) and Goldsmith (1988). The need for further releases and the reasons for success or failure of a translocation should be determined, if possible, and well documented. Efforts should be made to disseminate information on both successful and unsuccessful translocations through publications, workshops and other professional literature. Efforts also should be made through public media to explain the effort to citizens of the region.

RAISING AND TRAINING PRONGHORN

Tame pronghorn are valuable for public viewing and many types of physiological and behavioral research. Tame pronghorn have been used for studies of mortality (Yoakum 1957a), marking (Dinges 1958), nutrition (Wesley et al. 1970, 1973, Schwartz and Nagy 1976), food habits (Hoover 1971), temperature (Lonsdale et al. 1971, Thorne 1975) and recurrent estrus (Pojar and Miller 1984). Captive fawns have been displayed for educational purposes and photography at the Malheur National Wildlife Refuge, in Oregon.

Plans should be made well in advance to secure the necessary permits for capturing and keeping wild pronghorn, preparing facilities and ensuring that someone always is available to care for the animals properly.

As previously noted, fawns are captured easily during the first 3 days of life, but they are not kept alive easily. One or two fawns can be raised more easily than large numbers; the chances of infection, especially diarrhea, increase exponentially with the number of captive animals.

Pronghorn fawns, whether captured in the wild or taken from tame dams, should be allowed to suckle their mothers one or two times before bottle feeding begins. If a fawn is taken immediately after birth, colostrum should be obtained from a domestic nanny or cow for the preliminary feeding. Because pronghorn usually have twins, taking one and leaving the other is possible. Female fawns have fewer problems and live longer in captivity than do male fawns. Fawns nursed by tame does do not become tame because they do not rely on humans for food.

Floyd Blunt, former manager of the Wyoming Game and Fish Department's Sybille Game Research Unit, raised many pronghorn fawns. His formula for feeding young fawns was 1 quart (0.95 l) of whole milk to one 13-ounce (369 g) can of condensed milk. After 2 weeks, the amount of condensed milk was gradually increased so that it reached two cans per quart of whole milk when the fawns were

One or two fawns can be raised more easily than large numbers of young pronghorn. Cleanliness is especially important to prevent infections among fawns. F. Blunt (personal communication:1969) noted that problems with diseases, especially diarrhea, seemed to increase geometrically, not arithmetically, with the number of fawns cared for. If 1 of 30 or 40 fawns gets a pathogen, all of them have it in a short time, and recognizing and treating the problem in 30 to 40 animals is much more difficult than recognizing and treating the problem for only 2 or 3. Photo by Rod Canutt.

1 month old. Vitamins were added to the formula in about the amounts recommended for human babies. At about 90°F (32°C), the formula was fed in approximately 2-ounce (57 g) meals five times a day from 6:00 A.M. to 10:00 P.M.. About 0.5 ounce (14.2 g) of buttermilk added to a 2- or 3-ounce (57 or 85 g) feeding for three feedings a week was considered desirable for rumen flora. At 2 weeks, feedings with the same total volume of formula were reduced to three times daily. Fawns were fed morning, noon and night until August, then the midday feeding was dropped. By early September, the morning feeding was discontinued. The evening feeding was continued until October. The fawns could have been weaned earlier, but a daily feeding of 4 to 5 ounces (113–142 g) tended to keep them tame. Blunt started increasing the total volume at 2 weeks of age, but kept the fawns hungry.

Blunt started giving them high-quality, second- or third-cutting alfalfa hay and coarse Purina Calf Startina at about 2 weeks, so they were accustomed to it for winter diet. One pound (0.45 kg) of alfalfa was shared daily by five fawns. When feeding young fawns, Blunt always had a cloth and pan of warm water containing one drop of disinfectant to wash their faces and corners of their mouths, being careful that no formula remained on their muzzles.

Blunt stressed that everything be clean and occasionally sterilized with a mixture of bleach and water. A clean pan of dirt was kept available for the fawns to chew on; the dirt seemed to reduce the amount of mucus in their stools. He kept young fawns in small, covered pens, with newspapers on the floor to keep them dry. If the fawns became wet, he rubbed them briskly with a towel and took them inside because pronghorn fawns are very susceptible to pneumonia. Tylosin is a good injection for respiratory ailments.

When a number of fawns are raised together, they sometimes chew one another's ears. If this happened, Blunt rubbed a little kerosene on each ear.

For the animals' first 3 weeks, Blunt dipped a paper towel in warm water and rubbed the fawns' anal regions, simulating a doe's licking, until they defecated and urinated. If a fawn developed diarrhea, he added 0.25 teaspoon (1 g) of soluble terramycin powder to three feedings a day. If the diarrhea was severe, he injected 1 milliliter of Combiotic (penicillin and dihydrostreptomycin) intramuscularly and had a fecal specimen cultured to determine what other antibiotics were appropriate.

During a controlled experiment in Oregon, two fawns each were fed fresh cow's milk, fresh goat's milk, canned cow's milk and a fortified commercial calf supplement. Yoakum found that the latter resulted in fawns gaining more weight and experiencing less diarrhea than did fawns on the other diets. The Columbus Children's Zoo staff (1993) feed pronghorn fawns doe milk replacer from the National Food Laboratory in Dublin, California.

If fawns become constipated, they arch their backs and have tight flanks. One-fourth teaspoon (1 g) of milk of magnesia in two feedings of formula usually brings relief.

A supplementary food was five parts ground corn, five parts rolled oats, five parts whole bran, two to three parts linseed meal, two parts molasses, and 20,000 to 30,000 units of vitamin A per pound (0.45 kg) of the supplementary food mixture, formed into 0.25-inch (0.63 cm) pellets and packed in paper bags. Blunt gave a 0.5-pound (227 g) portion of this concentrate per feeding for five fawns, a maximum of twice a day, but never more than they would eat.

Personnel of the Los Angeles Zoo have hand-reared pronghorn fawns on a regular and increasingly successful basis since 1982 (Brinkley 1987, K. Brinkley, F. Woods and M. Miller-Edge personal communications:1993). The original protocol was established by Floyd Blunt, but modifications have been made as needs and experiences dictated.

Fawns are not taken from their mothers until 24 to 26 hours of age, giving them initial opportunity to nurse and obtain from their mother the vital, antibody-rich colostrum. If left at their mother's side even 24 hours longer, fawns are more nervous and excitable than fawns taken earlier.

Fawns at the Los Angeles Zoo are housed in a barn that is lighted, well ventilated and free from protruding objects. During the first 2 weeks of life, they are isolated in ministalls within the barn. Ministalls are made from securely fastened but easily removed plywood boards or bales of shavings. Each ministall is approximately 42 inches long by 36 inches high by 32 inches wide (107 × 91 × 81 cm). Bales of shavings make good doors for these stalls as they can be moved easily, used as a table, are convenient to sit on and can be stacked when additional height is needed.

By keeping fawns individually isolated for the first 2 weeks, they tend to imprint on their keeper more than each other, which makes them calmer and more tractable later. This early isolation also eliminates the aforementioned ear-sucking problem. The fawns are bedded on pine shavings 4 to 6 inches (10.2–15.2 cm) in depth. Shavings can be covered with an additional light bedding of hay or straw for the first week, to prevent any accidental ingestion of shavings.

After the first 2 weeks of isolation, fawns are housed in pairs and the stalls are slightly enlarged. Over the next few days, they can be housed together in larger groups and their stalls gradually enlarged further. As stalls get larger and space permits, randomly placed bales of shavings serve as sight barriers for the more timid fawns. Because fawns frighten easily and their flight reflex is unpredictable, use of browse plants as sight barriers is not recommended due to the possibility of eye injuries.

All fawns at the Los Angeles Zoo are given a medical examination, vaccinations and any other treatment needed at the time they are removed from their dams. A veterinarian notches each fawn's ear (for identification), dips the umbilicus in Betadine or Propodyne solution and examines the palate, heart and lungs.

The following vaccinations are administered:

1. Genecol-99 (orally): *Escherichia coli* monoclonal antibody (Molecular Genetics, Inc., of Minnetonka, MN 55343), must be given within the first 24 hours of life to be thoroughly effective;
2. Calf-Guard (orally): bovine rota-coronavirus vaccine (Norden Labs, Inc., Lincoln, NE 68501);
3. Fermicon-7 (subcutaneously): *Clostridium chauvoei-septicum-novyi-sordelli-perfringens* Types C and D bactrin-toxoid (Bio-Ceutic Division, Boehringer Ingelheim Animal Health, Inc., St. Joseph, MO 64502);
4. Bovine Virus Diarrhea Triangle-1 (subcutaneously): (Fort Dodge Labs., Inc., Fort Dodge, IA 50501);
5. Imrab (intramuscularly): rabies vaccine killed virus (Merieux Labs., Inc., Athens, GA 30601, distributed by Pittman-Moore, Inc., Washington Crossing, NJ 08560);

Pronghorn fawns should be treated gently during feeding and handling. They will become pets in a short time—ideal for many types of research. Imprinting ordinarily takes place within the first 2 hours of life. Fawns allowed to suckle their mothers at least once do not appear to imprint on a handler. (Top) *As they grow, placing fawns in feeding pens to suckle reduces the confusion and competition that can result in attempts to feed them in large groups. Such placement also accustoms the fawns to entering enclosed spaces.* (Center) *A crate of known weight is convenient for measuring and recording fawn weight gains with age.* (Bottom) *Monitoring fawns' health is necessary to avoid illnesses that could rapidly spread to other fawns.* Photos by Jim D. Yoakum.

6. Tetanus Toxoid Adsorbed (intramuscularly): (Sclavo, Inc., Wayne, NJ 07470);
7. DEPO-Penicillin (subcutaneously): (Upjohn Co., Kalamazoo, MI 49001);
8. Dl-alpha tocopherol acetate injection (intramuscularly): vitamin E (Western Medical Supply, Inc., Arcadia, CA 91006); and
9. E-SE injection (intramuscularly): sodium selenite plus dl-alpha tocopherol acetate: vitamin E/selenium (Burns Biotec Lab., Inc., Omaha, NE 68103).

At the Los Angeles Zoo, 8-ounce capacity (240 mL) plastic Even-Flo™ baby bottles and very soft Even-Flo Preemie nipples are used. The hole in the nipple is enlarged with a hot needle to about the size of a pinhead. As fawns grow older and their suckling reflex strengthens, they begin to collapse the Preemie nipple. At that time, a regular Even-Flo nipple (with the hole enlarged to pinhead size)—the kind that is distributed with the bottle when purchased—is substituted. Following each feeding, bottles and nipples are thoroughly washed in hot water and dishwashing liquid, with the aid of a bottle brush. They are then disinfected by immersion for no less than 5 minutes in a bath of 0.25-cup (50 mL) of swimming pool chlorine in 2 gallons (7.6 liters) of warm water. After soaking in the chlorine bath, the feeding equipment is thoroughly rinsed several times under running water and allowed to air-dry in a dish drainer before reuse.

Pronghorn fawns usually will not accept a bottle readily, must be taught patiently and must be encouraged to drink from one. Until fawns learn to recognize and come to a bottle willingly, the keeper should approach slowly and calmly, talking in a low, soothing manner.

K. Brinkley (personal communication:1993) recommended restraining the fawn, firmly but without force, in a sternal position between the keeper's knees, while sitting or kneeling. The nipple then is placed in the fawn's mouth and squeezed very gently and slowly to allow a few drops into the animal's mouth. If the fawn struggles quite a bit, several brief rests are advisable. It is best to quit after about 20 minutes, even if the fawn has not taken anything. To continue longer only exhausts and stresses both fawn and keeper. Fawns usually become accustomed to nursing from a bottle after a few feedings, and even the slower ones learn within about 3 days. Before feeding, bottles should be warmed to approximate body temperature. At first, feeding should be four times daily, starting at 7:30 A.M. and finishing at 10:00 P.M.. For the first feeding attempt, 1 ounce (30 mL) of lukewarm boiled distilled water should be offered instead of milk, to evaluate the fawn's sucking and swallowing reflexes and its general response to handling and to avoid the risk of aspiration of milk into the lungs.

The formula used at the Los Angeles Zoo is undiluted, evaporated (not condensed) cow's milk. Formula is refrigerated—an essential practice—and unused formula is discarded after 72 hours. ABDEC or Hi-Vite liquid vitamins are added to the morning bottle.

During the first 72 hours, 2 ounces (60 mL) of pasteurized cow's colostrum is offered at each feeding, four feedings daily, to fawns in the Los Angeles Zoo. For the next 72 hours, 1.2 ounces (35 mL) of pasteurized cow's colostrum and the same amount of formula are combined and offered at each feeding, four feedings daily. After 144 hours, 2.7 ounces (80 mL) of formula are offered per feeding, four feedings daily. Formula increases continue to be given at 72-hour intervals. Increases are in 0.3-ounce (10 mL) increments (i.e., increase from 2.7 to 3.0 ounces [80–90 mL] per feeding, then 72 hours later, increase from 3.0 to 3.4 ounces [90–100 mL] per feeding), until the formula has reached 6.7 ounces (200 mL) per feeding. At that time, it can be increased, still at 72-hour intervals, to 0.7 ounce (20 mL) per feeding, until a maximum of 8 ounces (240 mL) per feeding (still four feedings daily) has been reached. When the fawn is about 60 days of age, one formula feeding may be dropped. However, if fawns are of different ages, it may be preferable to substitute 8 ounces (240 mL) of warm water for the formula instead of dropping that feeding. Once all fawns receive water at that feeding, the feeding can be discontinued. Formula feedings are eliminated about every 2 to 3 weeks until the fawn is weaned at 120 to 135 days of age. If desired, 8 ounces (240 mL) of warm water may be offered once daily to help keep fawns tractable. Feedings are discontinued in the following order: mid-day feeding, night feeding, late afternoon feeding, morning feeding.

Alfalfa hay is offered to fawns at the Los Angeles Zoo, along with 0.16-inch (4 mm) O. H. Kruse ADF-16 herbivore pellets in the amount of 2 ounces (60 mL) per day per fawn, starting at 2 weeks of age. Equal amounts of red wheat bran and Purina Goat Chow™ are offered after 1 month, increasing to 4 ounces (120 mL) of each item daily per fawn at weaning age. Starting at 2 weeks of age, a shallow tub of fresh water is available at all times. When fawns are 1 month of age, fresh acacia browse is offered on a daily basis. As the fawns start to eat solids, granulated trace mineral salt, calcium carbonate and vitamin E powder are sprinkled on the hay/grain. Fawns have access to salt blocks at all times.

Pronghorn are an extremely difficult species to manage in captivity. However, hand-rearing efforts at the Los Angeles Zoo have made the overall management of their herd quite successful. The hand-rearing of fawns has proven extremely time-consuming, sometimes heartbreaking but usually very rewarding.

Water should be available to captive pronghorn at all times. Portable buckets are best for this because they are easily emptied, sterilized and refilled, thereby reducing the possibility of disease transmission. Photo by Jim D. Yoakum.

Hand-reared pronghorn become very tame and many seem to enjoy being handled and petted. Does can be patted on the head, but bucks tend to respond to such patting by sparring with the handler. As fawns mature, bucks that have sparred with humans become dangerous and difficult to handle. Male fawns should be petted only on their backs and sides. Photo by Tom Pojar.

Brinkley (1987) has successfully taken fawns outdoors, for the first time, as early as 8 days of age. Variables of advisability of this include how many fawns are being raised, how well they are imprinting on their keeper, how well they are settling down and how well the keeper can read the animals. This is a judgment call, and conservative action is advised. Generally by the age of 1 month, fawns will come to the keeper on call. At that time, they are allowed to go outside into a small, enclosed pen. They are monitored closely during this initial outdoor experience. Fawns have access to their stall at all times and are locked indoors at the end of the workday. While outside, they have access, through a chain-link fence decorated with PVC sight barriers, to the adult herd, which helps to initiate the introduction phase.

Access is given to the adult herd and the exhibit when fawns are about 165 days old. Because they have been associating with adults through the fence for some months, introduction has never been a problem. However, because the exhibit area is unfamiliar to them, fawns should be watched closely while on exhibit for the first time. Brinkley indicated maintaining keeper consistency, rather than allowing different people to work with the fawns, is very important. The fawns Brinkley raised became accustomed to selected volunteers and became well adjusted to the noises and bustle around them. The fawns responded well to touch and affection and grew to be tame, friendly and tractable adults. Hand-rearing has not had an adverse effect on breeding.

Ashton and Olin (1988) noted that pronghorn fawns could be shipped at 2 to 4 weeks of age in standard air kennels. Carpet padding was securely taped to the inside tops of kennels and loose burlap was taped over the openings on the outside. Hardware cloth was wired inside doors to prevent hoof injury on the bars of doors, and woodchips and straw were used for bedding. Generally, fawns settled down quickly once inside kennels.

In Colorado, 35 fawns, captured when 1 to 3 days old, were trained for research purposes (Schwartz et al. 1976a). Each was given a 0.5-milliliter injection of Vitamin ADE, and its navel swabbed with tincture of iodine to reduce the possibility of infection. The fawns were kept in a 6- by 30- by 48-foot (1.8 x 9.1 x 14.6 m) woven-wire enclosure that contained a shelter. To minimize the possibility and spread of disease, the pen was cleaned at least once daily, and personnel entering the holding area stepped into a footbath of povidone-iodine disinfectant. At 4 weeks, the fawns were moved to a larger enclosure for exercise. Because of problems encountered with *Clostridium perfringens*, all fawns received preventative injections of antitoxin types B, C and D, starting at 1 month of age.

Formula was heated to approximately 100°F (38°C). Various milk additives were tried, but a mixture of one part evaporated canned milk to four parts homogenized milk gave the best results. In addition to this formula, each fawn was fed first-day colostrum from dairy cows for the first 8 days. Although antibodies in colostrum probably were beneficial to the fawns for only the first day or two, colostrum in the diet was gradually reduced to eliminate an abrupt change in the diet. Fawns were fed milk until they were 100 to 120 days old. Alfalfa hay, mixed concentrate (Wesley et al. 1973) and block salt were available ad libitum. A bucket of soil also was available at all times (Schwartz et al. 1976a).

Growth rates of pronghorn fawns fed evaporated milk ad libitum were compared with that of another fawn allowed to suckle its dam in captivity (Wild et al. 1994). Growth rates did not differ (P = 0.63), but only one dam-raised fawn was available for comparison with 21 hand-raised fawns. Martin and Parker (1997) used the same feeding protocol on 10 pronghorn fawns, but changed the formula to three parts evaporated milk to one part water because of problems with diarrhea. After the formula change, occasional bouts of diarrhea were quickly remedied by reducing the volume of milk and adding 0.3 ounce (10 mL) of Kaopectate to a few feedings. The average growth rate to 16 months of age was comparable to those exhibited by a pair of twin pronghorn raised in captivity by their mother.

If a leg was broken, Floyd Blunt would split a map tube, wrap the leg with gauze, put the tube around the leg and secure it with masking tape. Such injuries healed very rapidly. The same technique was used in Oregon. Nichol (1942) presented further information on the care of pronghorn fawns.

Schwartz et al. (1976a) presented an informative account of how to train pronghorn for particular types of research. Of the fawns they worked with, each had distinctive behavioral traits, and acceptance to feeding and handling varied. The fawns responded best on a voluntary basis and rarely accepted restraint or physical force. Tractability was not related to age at time of capture or sex, and some animals were untameable and untrainable. All activity around the animals was conducted in a slow, quiet manner. Any change in an established routine usually resulted in a setback of the training program.

For the first few weeks after capture, the trainers spent most of their time gaining the fawns' confidence by repeated contact with each animal during feeding periods. A harness, similar to those used on deer, was tried, but the pronghorn would not accept any form of restraint, so they were trained to follow their handlers.

When the fawns were approximately 3 months old, an attempt was made to find a reward food that would be associated with the trainer, to take the place of milk when fawns were weaned. Various sweets, vegetables, fruits and tobacco were offered to the fawns, with little success. The normal concentrate diet was most preferred, but was not effective as a reward food when animals were receiving it ad libitum (Schwartz et al. 1976a).

During a series of investigations with tame and wild pronghorn, Torbit et al. (1984) formulated two economical, nutritionally complete, digestively safe, pelleted rations that were acceptable to the animals (see Chapter 14). Although these were developed as emergency rations to

Note the opening to the interdigital gland of this fawn's front foot. Pronghorn fawns are delicate, and their legs are easily broken. However, cleanly broken leg bones knit rapidly. Photo by Joe Van Wormer; courtesy of the Denver Museum of Natural History.

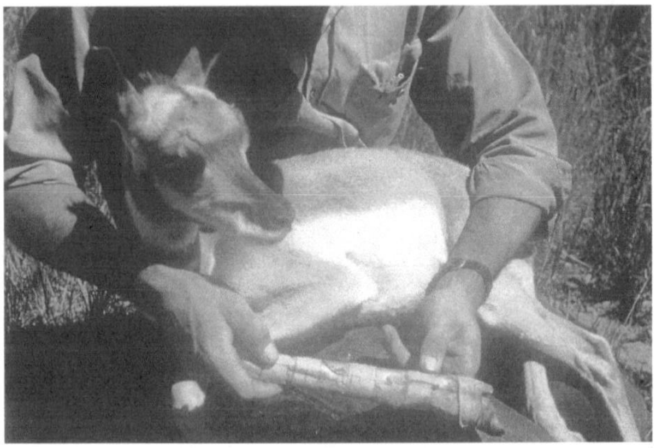

Simply padding a broken leg or foot with gauze, encasing the appendage in a cardboard roll and taping it will allow most bones to heal enough for the fawn to run within 2 weeks. Photo by Jim D. Yoakum.

A twin-axle horse trailer, with enclosed sides and top, was used by Schwarz et al. (1976a) to transport fawns to and from foraging trials. Initially, it was parked inside the enclosure where fawns were raised. Fawns were given a few days to adjust to it and then were enticed into it for feeding. As soon as they were accustomed to the trailer, the gate was closed during feeding. Short trips in the trailer were started when fawns were 4 weeks of age, and repeated several times each day with the distance per trip gradually increased. Fawns were conditioned to accept metabolic chambers and holding rooms in a similar manner. Of 32 (91%) fawns that survived to weaning, 26 (74%) were usable in ecological studies. Photo by Tom Pojar.

be fed to pronghorn during severe winters, they would be good rations for captive animals.

Messenger (1978) trained 12 pronghorn fawns for use in diet and nutrition studies in South Dakota. Two bucks and two does had been partially trained for other laboratory uses and declared unusable. They were maintained without training from ages 9 to 13 months. Messenger closely controlled their food intake and carried out all feeding, maintenance and training himself. He would enter the pen once daily, place hay on the ground, retreat 1 yard (1 m), sit down and remain motionless. The does would approach, sniff Messenger and the hay and move away. After several repeats of this activity, he removed all but 3.3 pounds (1.5 kg) of hay until the next day. The does came to recognize him and accept his presence. Within a week, they began to play at his approach and follow

The peninsular pronghorn is the most endangered of the three presumed subspecies in Mexico, and personnel from the Center of Biological Investigations are using innovative methods in an attempt to increase its numbers. On the Vizcaino Biosphere Reserve in central Baja California Sur, they are establishing a captive herd from which to expand the subspecies' range. They began in 1998 with a 1-square mile (1.6 km^2) enclosure containing two males and three females that had been captured as newborns (J. Cancino personal communication:2001). (Left) *The animals are fed hay and graze natural and irrigated vegetation within the enclosure. As of 2000, eight adult wild pronghorn had been lured into the enclosure by the presence of the captive animals and a small irrigated area with very attractive vegetation. When wild animals were seen along the fence, the captive animals were herded into a subenclosure, and the perimeter fence was let down near the irrigated area. When observers in a tower, built for the purpose, saw wild pronghorn in the enclosure, the fence was raised behind them, and the new animals were gradually incorporated into the captive herd. J. Cancino (personal communication:2001) calls this a passive trap that had an extremely small chance of causing mortality among the endangered animals. During 2000, 33 animals were in the enclosure (Cancino et al. 2001, Castellanos et al. 2001). Does gave birth in the facilities during the springs of 2000 and 2001 (Cancino and Guenzel 2002).* (Right) *Fawns of captive does are hand-reared to facilitate handling them. Twelve fawns of American pronghorn were captured in Wyoming during spring 2001 and flown to Bioparque Estrella, Nuevo Leon, Mexico, to be hand-reared. Future plans include transfer of 30 more American fawns to Bioparque Estrella. The does will serve as surrogate mothers for implanted peninsular pronghorn embryos (Cancino and Guenzel 2002).* Photos by Ramón Castellanos.

him. Molasses-covered grain was hand-fed, and within a month, the does would accept a close approach and confinement in a small pen.

Training was carried out in the same manner for the two bucks, except Messenger (1978) never sat in their presence. The most difficult task was establishing dominance without instilling too much fear in the animals. Messenger found that dominance could be established with a level stare and a step or two toward a buck. In 2 weeks, they anticipated Messenger's arrival, and within a month, they accepted close approach and confinement. The pronghorn could be restrained easily if held only by the lower jaw. While being held in this manner, they stood quietly. If other parts were held, the animals became perturbed and difficult to handle.

During another study in South Dakota, two pronghorn fawns died after being changed from a sagebrush diet to a mixed sagebrush and prairie hay ration. Kohler (1950) speculated that an outdoor exercise pen with a solid fence at least 7 feet (213 cm) high would have been desirable for the animals' health. Also, conducting studies during a temperate season would minimize the animals' susceptibility to respiratory infections.

MARKING

Marked animals are used for many types of management and research programs. Identifying individuals or specific bands is one of the best methods to understand their behavior or movements. Colored collars, ear tags or dyes have been used for years (Hansen 1955, Dinges 1958), but radio collars now are used most often to follow seasonal movements of pronghorn, identify mortality causes and delineate home ranges (Beale 1966).

If visual recognition of individuals by color designs on collars as well as by radio frequency is desirable, fairly wide—about 3-inch (7.6 cm)—collar material should be used. Narrow collars often are covered by the mane and long neck hair, making color codes difficult to identify. Collars should be tight enough that animals cannot get a front foot or shrubbery through them. However, collars on subadults must allow for animal growth, and those on adult bucks must allow for swelling of the neck during the rut.

Studies in Colorado, Idaho, Montana, Nevada, Oregon and Wyoming (Bear et al. 1973, McNay 1980, Autenrieth 1984, Meeker and O'Gara: personal files) indicate the following measurements for pronghorn. The maximum neck size is immediately ahead of the shoulders, and the minimum is behind the ears. Adult bucks average 23.2 inches (58.9 cm) and 15.2 inches (38.6 cm); adult females average 19.3 inches (49.0 cm) and 13 inches (33.0 cm). One-day-old fawns have neck measurements of 6.1 to 6.8 inches (15.5–17.3 cm). Yearlings are so similar to adults that adult measurements may be used for them.

Radio transmitters, weighing no more than 4.2 ounces (119 g)—less than 5% of the weight of a very small fawn—for the whole collar, are used for fawns. Generally, fawns are only a day or two old when collars are attached; small, light batteries with a short life are used. The best radio collars for fawns have the transmitter and a lithium battery hermetically sealed in a nickel–steel canister measuring about 1.5 by 1.4 by 1 inches (3.8 x 3.6 x 2.5 cm). With the largest battery one would want to put on a newborn fawn, a battery life of 7 to 10 months can be expected, with a range of 1.5 to 3 miles (2.4–4.8 km). Transmitters usually are riveted between strips of 1-inch (2.5 cm) nylon webbing that serves as the collar when the ends are joined. In most mortality studies, researchers do not want the radios to remain on fawns for more than 3 months, and small, lightweight batteries can be used.

Many collars are designed so that their radios will fall off before the batteries die, enabling the radios to be retrieved. A simple design for short-term monitoring has the ends of the nylon webbing cut short and sewn to 1-inch (2.5 cm) wide elastic strips with strong nylon thread. These strips are sewn together with about five strands of light cotton thread to form an expandable neck collar. The cotton thread weakens with exposure and, as the fawn grows and tension increases, breaks in 2 to 3 months. Tucks can be taken in the elastic with two or three strands of the thread so the collar fits snugly. Successive tucks break away as the animal grows until the ends break away and the collar falls. Ice buildup, documented by Clute and Ozoga (1983) on elastic collars used on white-tailed deer fawns, has not proven a problem during the late-May to July period of pronghorn fawn studies.

When neck sizes increase greatly, nonadjustable collars may cause death. Hamilton (1962) and Beale (1966) introduced expandable collars to allow growth, and both were adapted for attachment of radios. Beale and Smith (1973) used a rigid outer collar 14.5 inches (36.8 cm) in diameter, coupled with an elastic inner collar 9 inches (22.8 cm) in diameter. They believed that abandonment of 9% of their collared pronghorn fawns was due in part to the altered appearance of those fawns.

Keister et al. (1988) developed a durable, lightweight, self-adjusting radio collar that was easy to attach and did not injure pronghorn fawns. They fitted 120 fawns with these collars and, although five fawns were abandoned, no deaths or injuries were attributed to the collars. The collars were made of 0.06-inch (0.15 cm) thick Kydex (acrylic polyvinylchloride, manufactured by Rohm and Haas Co., Philadelphia, Pennsylvania) and cut to 22.8 by 1 inch (57.9 x 2.5 cm) with two tabs, 0.7 by 0.4 inch (1.8 x 1.0 cm), at each end (Figure 160). The Kydex was heated

Although somewhat crude in appearance, a simple radio collar that will break away from a pronghorn fawn within 2 or 3 months can be constructed from elastic strips and cotton thread. Joining the two sides of the elastic collar with about five strands of thread forms an expandable collar, and the thread weakens with exposure and breaks in 2 to 3 months under tension from the growing fawn. Tucks taken in the elastic with two or three strands of thread fit the collar snugly to the fawn's neck. Successive tucks break away as the animal grows until the ends break away and the collar falls. Note that the fawn above, ready for release, is fitted with an elastic collar that has several tucks for expansion as the animal grows. Photo by Mark McNay.

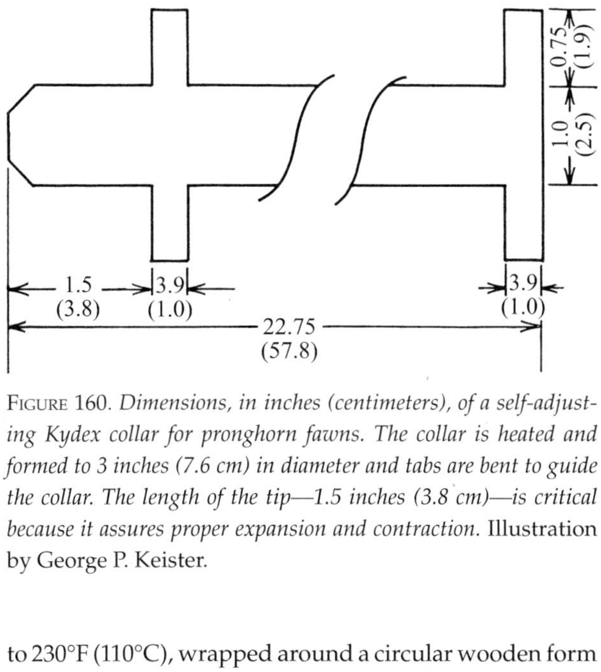

FIGURE 160. *Dimensions, in inches (centimeters), of a self-adjusting Kydex collar for pronghorn fawns. The collar is heated and formed to 3 inches (7.6 cm) in diameter and tabs are bent to guide the collar. The length of the tip—1.5 inches (3.8 cm)—is critical because it assures proper expansion and contraction.* Illustration by George P. Keister.

to 230°F (110°C), wrapped around a circular wooden form 3 inches (7.6 cm) in diameter and cooled to produce a coiled spring. The tabs were heated with a propane torch and bent around the collar to serve as retainers and guide expansion. The transmitters were mounted on the collars with four pop rivets. Transmitter mounting brackets and pop rivets were wrapped with electrical tape to prevent abrasion to fawns and to keep tabs from catching on the rivets. To prevent snagging, the antenna was threaded through three pairs of holes in the collar. Minimum and maximum collar circumferences—11 inches and 19.8 inches (27.9 and 50.3 cm)—were determined by trial and error to fit both fawn and adult pronghorn, and to ensure that the collars would expand and contract freely. The length—1.5 inches (3.8 cm)—of the tip was critical because it assured proper collar expansion and contraction. To prevent loss of collars, a piece of open-cell foam—0.6 by 1 by 11 inches (1.5 x 2.5 x 27.9 cm)—was glued to the inside of each collar with a drop of contact cement in three spots. The foam reduced the minimum circumference of the collar to 7.2 inches (18.3 cm) and dislodged from the collar within 2 weeks, by which time the fawn's head was large enough to retain the collar.

With use, the collars lost some resiliency but retained enough tension to be kept attached for all neck sizes without being too tight. They became slightly brittle and sometimes cracked in about 1 year. This weathering and

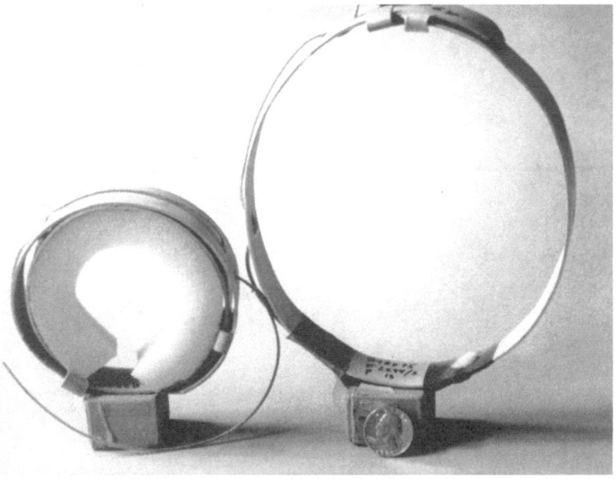

Lightweight, self-adjusting pronghorn collars were developed by Keister et al. (1988). The collar on the left is ready for installation on a newborn fawn. The one on the right is expanded to the size required for an adult pronghorn. Photo by Mitchell J. Willis; courtesy of the Oregon Department of Fish and Wildlife, Eugene.

the presence of many barbed wire fences probably contributed to some collar loss. Because collar life was about equal to transmitter life, the potential for injury from the collars was eliminated at about the same time as their usefulness terminated. Loss during the first few days was the only problem, which was remedied by adding the foam liner.

Silver-colored transmitters should be wrapped with dull-colored tape. Also, the entire collar should be rubbed with sagebrush or other leaves to mask unnatural odors, and carrying collars in a bag with vegetation from the study area is beneficial.

Because pronghorn are flighty, nervous animals, marking with streamers or other materials that move in the wind is not recommended. Many pronghorn have been marked with metal ear tags (Hansen 1955, Foree 1956, Dinges 1958), and such tags have been retained for at least 10 years in some cases. However, Bart O'Gara handled a pronghorn in Montana that was sloughing metal ear tags, apparently after her ears had frozen around the metal tags; the temperature had dipped below –30°F (–34°C) about 2 weeks earlier. Plastic tags for earmarking livestock are available in stores supplying farm and ranch equipment. Such tags have less chance of freezing ear tissue and are easier to see at a distance than are metal tags. When plastic ear tags first became available, they became brittle after a year or two of wear. Plastic tags that O'Gara has used on pronghorn since 1980 have weathered well. Best are the types that do not form a loop that could catch on vegetation. Although pronghorn are open-country animals, they can snag tags on shrubs as they feed.

Because it slides up and down the neck as a fawn raises or lowers it head, a radio collar on a pronghorn fawn appears awkward and cumbersome. However, fawns seem to ignore the unnatural object and its movement. A few fawns have been abandoned by their dams after collars were attached, but survival of radioed pronghorn fawns generally is similar to that of fawns of the whole herd. Whether occasional abandonments occur because of unnatural appearance of a collared fawn or the human odor associated with it has not been determined. Photo by Robb Hitchcock.

*Ear tags should be inserted about one-third of the way from the base to the tip of a pronghorn's ear. Tags placed farther out may result in a drooping ear. Metal tags (*left*) may cause tissue to freeze during extremely cold weather and the tag to slough out. Plastic tags (*right*) are less apt to cause freezing, are brighter and easier to see, come in a variety of colors and help identify individuals. They also are less likely to snag on shrubs than are metal tags.* Left photo by Jim D. Yoakum. Right photo by Tim Reynolds.

Coloring agents have been used to mark pronghorn, but such coloration only lasts until the hair is shed. Attempts have been made, with limited success, to mark whole herds from a crop dusting plane. Portable sprayers also have been used during trapping operations to mark large numbers of animals so their movements could be monitored. Biological stains, chemical dyes, bleaches, inks, pine tar and several types of paint have been tried. After extensive tests concerning toxicity to animals and plants, visibility and weathering, Dinges (1958) concluded that Nyanzol D was best suited for marking pronghorn. Nyanzol D oxidizes to black soon after being applied. Dinges reported that August or early September was the best time to mark pronghorn to obtain longest duration of the dye. The adult pronghorn he worked with in Oregon completed shedding by the end of July, and fawns shed their juvenile pelage by mid-August. A mark placed on short hair (legs, belly or lower sides) might remain until mid-July of the next year. Dinges found that long hair regions on the back did not retain marks for more than about 4 months.

During 1992 and 1993, Byers (1997b) experimented with freeze branding to permanently mark pronghorn fawns on the National Bison Range, Montana. He shaved 2- by 4-inch (5 x 10 cm) patches of skin, sprayed them with isopropyl alcohol, then firmly held a rectangular brass brand—cooled in liquid nitrogen—to the skin for 20 to 30 seconds. Three freeze-branded individuals survived predation, and—as of 1997—their marks appeared as lines of white hair, which were somewhat interrupted by small hairless patches. Byers concluded that a shorter branding time, perhaps 10 to 15 seconds, would be more appropriate.

Pronghorn do not always have to be marked for individual identification. Shape of horns, pelage color, width of neck bands, amount of white on sides, configuration of black areas on faces, etc., make individuals recognizable by a keen observer who works with a herd for a long period of time. Sketches and photographs help when an investigator is becoming familiar with the animals in a population.

Experiments with dyes indicate that August or early September are the best times to apply any type of hair coloring to pronghorn to obtain the longest duration of recognition. Applied at that time, colors will persist for nearly a year on the short hair of the legs and belly, but long hair regions of the back retain marks for only about 4 months. Photo by Jim D. Yoakum.

After becoming familiar with the animals in an area, a keen observer can recognize individual pronghorn in the field by their different neck and facial markings. Photo by Tim Reynolds.

APPROACHING, OBSERVING AND PHOTOGRAPHING PRONGHORN

Fleet-footed little nomads, wary and elusive, swift as the wind, like veritable will-o'-the wisps, they can almost disappear into the distance like mist before the sun, their white rumps disappearing over the horizon as wind-driven clouds.

Cheney (1930:487)

When someone unfamiliar with pronghorn attempts to approach them in the wild, they may well subscribe to the views penned by Cheney.

Pronghorn in parks and wildlife refuges where hunting is forbidden are ideal subjects for photography, wildlife observation and behavioral studies. Such animals are acclimated to humans and fairly easy to approach and observe. Some techniques are useful even with comparatively tame pronghorn. During studies on the National Bison Range, J. Moody (personal communication:1990) found that pronghorn were easily photographed from a vehicle, but primarily presented rump shots. To get head-on or profile shots, Moody had a companion do a lizard-like walk. By keeping low in the grass, the companion would crawl to within 100 yards (46–91 m) and pop its head up for a brief instant. The pronghorn would become curious and orient itself toward the photographer, allowing the photographer (still in the vehicle) to obtain good images of neck bands and facial features to aid identification.

Pronghorn usually are aware of human presence in their vicinity. Consequently, one must plan carefully to observe the animals in an undisturbed state and setting. When pronghorn are especially vigilant, such as in the birthing season or during and shortly after a hunting season, long-distance observations with binoculars or spotting scopes are necessary. However, when making such observations, one should not choose a vantage point on a skyline. Pronghorn are alert to movement there.

Because the observer usually attempts to avoid detection by the subject pronghorn, movements should be minimized. When one must change position, very slow and deliberate movements usually are less apt to be detected. The observer should avoid wearing or using shiny objects—belt and suspender buckles, glasses, etc.—that reflect light and alert the animals. Pronghorn have color receptors (cones) in their eyes, but do not seem to react as much to color as to texture. They have many more black and white receptors (rods) that are vastly more light sensitive than cones. Reflective pigment, called tapetum, which makes eyes glow in headlights, and large eye lenses gather and focus light on those rods, giving pronghorn outstanding vision in poor light and an uncanny ability to detect motion. Slick, shiny clothes draw immediate attention. Soft neutral colors—especially dull green, light brown or tan—or camouflage clothing are recommended. A piece of camouflage netting over one's face will prevent glare. This is essential if the observer must wear glasses.

Clancy (1999) contends that the eyes of white-tailed deer detect ultraviolet light that is beyond the spectrum of light perceived by human eyes. Also deer eyes can detect a glow from clothing washed in most laundry detergents. The glow alone probably will not frighten deer—rotting wood, flowers and many other things glow in the whitetail's world—but glowing clothes make the slightest movement detectable, especially in poor light. Pronghorn appear to have similar adaptations. Clancy recommends washing hunting clothes only in detergents that do not have added brighteners and treating clothes with a ultravilot-light killer.

Building a blind of natural vegetation overlooking areas frequented by pronghorn— especially at waterholes, salt blocks and preferred foraging areas—promises the best opportunity for undisturbed observations. Pronghorn seldom use trails except where holes in pronghorn-proof fences or features of the terrain force them to use certain paths to reach water or some other vital habitat component. The observer should enter a blind before daylight and by a route that does not disturb the animals.

When choosing a site for a blind, the observer should try to have the sun at her or his back and the wind in her or his face. Observations can be made at any time of year. Pronghorn are primarily diurnal, with peaks of feeding activity shortly after sunup and before sundown. However, unlike many wildlife species, they have activity bouts throughout the day. Thus, one must be prepared to observe pronghorn practically from sunup to sundown if knowledge is to be gained concerning all of their activities. J. Moody (personal communication:1990) found that periods of pronghorn activity and inactivity on the National Bison Range were about 3 hours long during winter, but shortened to about 1 hour during warm weather.

To get close enough to a pronghorn to dart or spray it with marking paint usually requires use of a blind at a waterhole or the use of a decoy as described in Chapter 22 for archery hunting. For darting, one needs to be within about 30 yards (27.4 m) for best results. With most systems, spray painting requires being within 10 yards (9.1 m) of the animals. At such close ranges, human odor should be reduced as much as possible. Bathing and washing one's clothes in baking soda before entering the blind is advisable, as is refraining from smoking or from eating strong-smelling foods, especially those with smoke flavors. Aftershave lotion, perfume, etc., should be avoided. Equipment also should be deodorized, especially dart guns that may smell of gunpowder, cleaning solvent or oil. Scalding hot water and heat drying are recommended to accomplish this. Sporting goods stores, especially those catering to archers, sell preparations for reducing and masking human odors.

One of the most practical methods for observing pronghorn is staying in a vehicle and using a window-mounted spotting scope. Pronghorn usually are less wary of vehicles than of a person on foot, unless they have been hunted by people in vehicles. An observer can soon tell how tolerant animals may be of mechanical conveyances. Often, pronghorn will tolerate vehicles passing close by, but will shy away if a vehicle stops. In such cases, they soon become habituated to a vehicle that first stops at a distance and gradually moves closer over a period of days. A pickup camper can be an ideal blind for observations or photography. As long as it is darker in the camper than outside, a person can move about in sight of the animals.

Pronghorn can be observed aerially and individual animals identified, according to R. Deblinger (personal communication:1990), who was interested in relocating as many color-collared pronghorn as possible. One technique was to fly a Piper Super Cub 50 to 100 feet (15.2–30.5 m) over the animals. Slowing the plane to pronghorn speed and reading the symbols or numbers on collars was no problem. Obviously, this technique can stress pronghorn, so care should be taken as to the time of year, temperature, weather, relation to parturition and length of pursuit.

Deblinger would make only one pass over each individual or group. Using this method, pronghorn generally ran short distances and veered away from the plane. Care must be taken to make sure this activity is kept to a minimum so that behavioral changes or habitat shifts do not result. In the Red Desert of Wyoming, pronghorn behaved toward a Super Cub much the same way they behaved toward golden eagles. They usually join into large groups and flee from eagles, but does with fawns invariably place themselves between their fawns and the eagles (see Chapter 11).

Anyone beginning a long-term behavioral study would do well to habituate the pronghorn to human presence. This can be done with time and patience. The observer should walk about in plain sight at a distance that does not spook the animals into flight. The key is to avoid looking directly at the animals and not making any sudden movements or dropping from sight. The researcher or observer should look at vegetation, sit on a rock or read a book, and perhaps carry a folding chair for added comfort. Because the animals know a person is there, movements are permissible and even advisable, but they should be slow and deliberate. When animals no longer seem to pay any attention, walk slowly toward a point downwind of them. Approach at an angle instead of directly toward pronghorn. As soon as the animals become alert, the researcher must stop and be careful not to look at them, but resume whatever activity to which they have become accustomed. In most areas, a person can approach to within 40 to 50 yards (37–46 m) of the study animals in a week to a month's time. Wearing the same kind of hat and color of clothes and carrying the same equipment every day expedite the familiarization process.

Getting close to large bands of pronghorn is a problem because many pairs of eyes are scanning for danger, and at least one animal is almost sure to be looking in the direction of the person approaching. Adult does usually sound alarm. Being herd animals, pronghorn nevertheless are slightly more at ease in large groups than are those alone or in small groups. Thus, all else being equal, large groups are as easy or easier to approach than lone animals, except possibly a lone buck during the rut. Does or groups with small fawns are most wary.

For anyone wanting to get close to a particular pronghorn, there are many techniques. They sometimes work well; sometimes, they do not work at all. A pronghorn's curiosity will often bring it to any strange object that draws its attention. A flag or a spotting scope on a tripod slowly raised in sight of a pronghorn or a band may bring them to investigate. The observer should remain hidden near, but not immediately near the object of curiosity.

J. Meeker (personal communication:1985) used time-lapse cameras to record interactions of pronghorn and feral horses at waterholes. Meeker watched a yearling buck approach a camera a number of times. The camera clicked once a minute and the buck would wheel and run each time. Eventually, the little buck almost touched the camera before having his nose photographed.

Yoakum used a black Labrador retriever to decoy pronghorn, especially lone bucks for photography. In full sight at a distance, Yoakum would throw a stick for the dog to retrieve. This was repeated several times, shortening the distance of each retrieve. A lone buck often would seem entranced and slowly walk to within 75 yards (69 m), watching the dog move back and forth.

Another way to approach a selected pronghorn is to lie in the bed of a pickup truck while someone drives the vehicle slowly past the animal. The observer may be able to slide out of the pickup bed at an opportune moment and, by remaining motionless on the ground, be undetected as the pronghorn continues to watch the moving truck.

Types of optics used for observation depend on conditions and objectives. All lenses should be coated to reduce glare. This helps when the observer looks nearly into the sun, and it reduces glare that alerts or frightens the subject animals. For most work, a pair of eight-power binoculars with a wide field of vision and good light-gathering characteristics is the handiest piece of equipment. Next is a variable-power spotting scope of about 20- to 60-power, on a sturdy tripod or vehicle window mount.

Cameras should not be overlooked as equipment for wildlife biologists. Photographs provide permanent documentation of animal behavior, rangeland conditions and other factors important to pronghorn and habitat studies. The proper film and cameras for the job depend on objectives of the photographer. Most field people prefer 35-millimeter SLR (Single Lens Reflex) cameras, which are small, lightweight and easy to use. Slides are popular and ideal for use in lectures, scientific presentations, etc. Good slides can be made into publication-quality black-and-white photographs but generally are inferior to prints from black-and-white negatives. Record should be kept of the date, time, location and circumstances of each photo. Such information is vital for the eventual publication or other use of the image.

The recent advancements, affordability, and popularity of digital photography has made life easier for the field biologist/photographer. With each exposure the camera records the date, time, and exposure data. It also allows the photographer to view the image immediately, and re-shoot if it is not up to standard. These cameras also eliminate the expense and waiting time for processing because the images, once uploaded into a laptop computer, are ready for location labeling, cropping, and publication.

Long-distance observation via spotting scopes often is the only way to decipher numbers on collared or ear-tagged pronghorn. A spotting scope mounted on a rifle stock is handier than one on a tripod. Photo by Harold Umber; courtesy of the North Dakota Department of Fish and Game, Bismarck.

Film or digital speeds of 100 to 200 ISO ratings are used more by wildlife photographers than any other speeds. For rough-and-ready service in the field, one of the small, completely automatic cameras (digital or film) with a built-in flash and nominal zoom capability is convenient and generally results in more and better photographs than does a more complicated camera operated by a preoccupied and hurried biologist. These small cameras usually have 35-millimeter built-in lenses that serve in place of a wide-angle lens for a larger camera. The 35-millimeter lens is ideal for panoramas, human activity and close-up action of animals. The second camera in the rough-and-ready kit should be a camera body with an automatic exposure option, and a separate, changeable zoom lens of about 70 to 300 millimeters preferably equipped with autofocus and "IS" (Canon's Image Stabilization) or VR (Nikon's Vibration Reduction), and a 2X tele-extender for extension to 600 millimeters. Such a camera is excellent for composing pictures and can be often used without a tripod. When the 2X tele-extender is used, a tripod is highly recommended for sharp images. At the higher ISO ratings, 400 ISO films and up to 800 ISO digital ratings are available to further reduce the effect of camera vibration or animal movement. Of course, many high-powered lenses of 500 and 600 millimeters and higher, and films up to 1,600 ISO are available for wildlife photography. These high ISO speeds and large lenses require advanced skills, and while they may be warranted for specialized studies, generally are not practical as secondary equipment because of their weight and expense.

CONCLUSIONS

Capturing and translocating have proven remarkably successful at returning pronghorn to their former rangelands. Capture and rearing of newborn fawns soon gave way to driving free-ranging adults through V-shaped wings into large corrals. This was accomplished first with horses and automobiles, then light aircraft, now with helicopters. As historic rangelands were repopulated, the need for capturing large numbers of animals for transplants was reduced, and smaller numbers often were needed for research. Thus, new capture methods were developed. The most successful were tangle nets, surround nets, netguns and chemical immobilization. Some transplants failed, leading to development of criteria for selection of release sites. Development of and adherence to such criteria is now considered the most important aspect of transplants.

The following guide is recommended for selecting pronghorn release sites:

1. Sites should be evaluated for habitat suitability before capture of animals. Each province or state should establish a rating system considering local conditions and other factors like those listed in Table 170. Areas that do not meet such specifications should not be considered feasible until improvements are made to meet all criteria.
2. Multiple release sites should be prioritized to establish areas that have the highest potential for successful translocations. Ideally, this should be done on a

provincewide, statewide or regional basis to assure that releases are made only into high-quality sites.

3. Criteria discussed in this chapter should be fine-tuned for site-specific release areas. Ratings for vegetation should be modified to reflect conditions within a province, state or region.
4. Translocation projects should include monitoring to determine the success of releases. This is important to develop and modify guidelines and ensure accountability in translocation endeavors.

Tame pronghorn are valuable for many types of research, but they are not easy to raise or train. The literature should be reviewed to avoid the many pitfalls in raising and training these delicate animals. Marked pronghorn are necessary for many types of research. Lightweight breakaway radio collars are available for fawns and adults. Pronghorn are flighty, nervous animals, and marking them with streamers or other material that moves in the wind is not recommended.

S E C T I O N   F O U R

HUNTING

C H A P T E R T W E N T Y - T W O

Hunting

Bart W. O'Gara

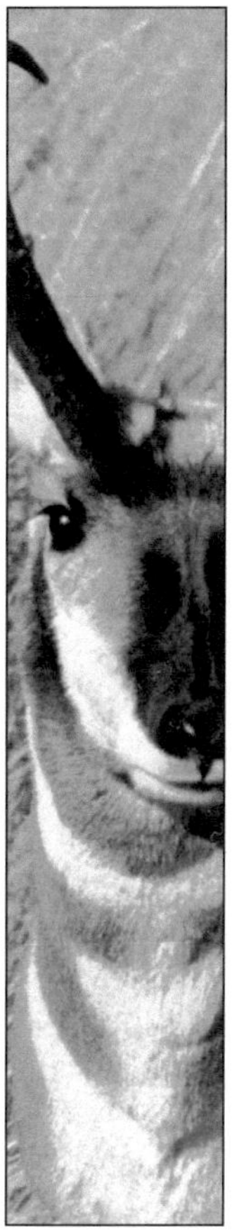

Tomorrow may the dawning light
Bring to this plain, one such as he
Who tramps the sage with keen delight,
Within whose heart a fairness be.

Send one who knows the desert wide
The warp and pattern of our ways
Who seeks to turn the ebbing tide
On twilight haze of yesterdays.

And coming may he be alone
To stalk, to hunt, to drink his fill
Of this, the cup of Nature's own
And test my speed against his skill.

Until the end, may it be well
That "hunter" and the "hunted" stray
Not far apart in love of chase
Eternal be this dawning day.

Tomorrow! Bring the hunter skilled
But long divorced from selfish greed
Who measures not the trophies killed
Whose love of earth is love indeed.

Einarsen (1948:223)

Inclusion and placement in this book of a chapter on recreational hunting of pronghorn required much deliberation. Such a chapter could logically fit in the chapter on harvest management, because recreational hunting is the most practical and practicable method of regulating pronghorn numbers. Furthermore, most pronghorn management today—excluding management of the species' habitats—is aimed at increasing populations to or maintaining them at optimal and sustainable (therefore huntable) levels. Hunting also could fit reasonably in the chapter on mortality, for recreational harvests are, in fact, a form of predation and a significant influence on pronghorn population size, density and composition. The topic of hunting certainly could be combined with the chapter on societal and cultural values, inasmuch as hunting is an expression of the species' popularity with and benefit to people, as well as a benchmark of management success. In modern times, any animal that is huntable, under legally mandated guidelines, can represent significant economic and recreational amenities of the environment, in addition to its less tangible but equally important nonconsumptive and ecological values. In reality, few aspects of the pronghorn today are not influenced by or do not have direct bearing on recreational hunting. But because pronghorn hunting is a unique experience, just as the animal itself is unique among North American quadrupeds, and because pronghorn hunting involves certain skills and knowledge on the part of the hunter to maximize the enjoyment and other rewards of that experience, the subject is treated separately here.

BACKGROUND PERSPECTIVE

For most of the 2 million years that humans have lived on this earth, they hunted in order to survive. Hunting led to cooperation, to development of language and communication, to art forms and to the invention of tools for hunting, skinning, and preparing wild game.

Samuel (1999:21)

Between 1950 and 1996, the U.S. population increased by 49 percent, while during the same approximate time frame (1955 to 1996) big game hunting increased 615 percent.

Duda (2000:38)

Unregulated hunting, along with loss and deterioration of habitat, once took pronghorndangerously near extinction. The species now thrives, despite continuing loss of habitat, and provides extensive hunting opportunities as well. However, complete freedom of movement by pronghorn over the prairies, as Roosevelt et al. (1902) described, is a thing of the past. If the animals were not managed by hunting, crop damage would be high during some years and die-offs during severe winters or droughts would eliminate some populations or reduce them to low numbers. Although pronghorn reproduce rapidly, they now lack the vast expanses for seasonal movements that once helped to alleviate population pressures.

During the first third of the 20th century, pronghorn were scarce. They were considered finished as game animals, and hunting seasons were closed in the United States, Canada and Mexico. Seasons remain closed in Mexico. By the 1980s, pronghorn were second only to white-tailed and mule deer as the most numerous big game animals in the United States, but elk became more numerous by the late 1990s. More pronghorn have been shot annually in Wyoming since 1950 than existed altogether in 1924; more than 2.3 million were legally harvested in that state by sport hunters from 1934 through 1999. This has been a remarkable achievement—reflecting, first, the resilience of the species and, second, the ability of management to achieve recovery from the virtual brink of extinction to sustainable, harvestable abundance within the span of a human lifetime. Without the cooperation of many ranchers and money provided by hunters through licenses and the Federal Aid in Wildlife Restoration Act (Pittman-Robertson or P-R) funds, it could not have happened.

Meat hunters once threatened the very existence of pronghorn (see McCabe et al. 2004). Thrill seekers also reduced pronghorns numbers. Mountain Harry, a Wyoming hide and meat hunter during the 1860s and 1870s, was said by Ingersol (1882) to never hunt for sport—directing all of his energies toward making a profit. However, he once shot 70 pronghorn in a single day in a bet with a crack shot from the East. The dude was skillful in scoring bull's-eyes, but found hitting nimble pronghorn an entirely different matter.

Recreational hunting generally has replaced meat (subsistence) hunting as the primary hunter motivation, although the meat of a harvested animal remains an important value. Leopold (1933) maintained that hunting for recreation (sport) is an improvement over hunting for food, because an ethical code has been added to the test of skill. News releases by antihunting and animal rights organizations often indicate that all hunting is on a decline. Actually, during the 1990s, the growth of big game hunting surpassed that of fishing and wildlife watching (Duda 2000).

"The rise in big game hunting is not new; it's a trend that has been gaining momentum for many years. Its growth has been consistent, increasing seven-fold since the 1950s. Hunting is worth $61 billion annually to the U.S. economy, with big game hunting the dominant component of the industry" (Duda 2000:41).

Now that free-ranging bison and elk have been nearly eliminated from their pristine habitats on the prairies, pronghorn is the only big game species truly at home on the wide plains and provides the only chance most North Americans have for a big game hunt on the prairie. Combination hunts for deer or small game often are possible. Crawling over a landscape covered with cacti, sagebrush, sharp rocks and an occasional rattlesnake to get close enough for a shot can be exciting. Shots often must be taken at fairly long ranges, and the prairie wind will foil the shooter. Despite all this, the success rate is higher for pronghorn hunters than it is for other big game hunters—usually in the 75% to 90% range. Under most circumstances, even the inept hunter will see lots of game and have a memorable experience.

When seeking their first pronghorn, deer hunters soon discover a different kind of hunting. Deer are neutrally colored, blending into their habitat; pronghorn, with their contrasting colors, are sharply visible. Although mule deer share parts of the pronghorn range, they usually shun the open terrain where pronghorn thrive. To detect danger, deer rely primarily on hearing and a keen sense of smell; pronghorn rely mainly on their extraordinary sight. Deer use cover and stealth to escape peril; pronghorn avoid cover and depend on speed.

Caton (1877:56–57) summarized several points concerning pronghorn that should be of interest to the hunter. "Notwithstanding the large prominent eye, which is a brilliant black color, the sight of the antelope is not reliable. He cannot readily identify unfamiliar objects if they are not in full view, nor in motion. He cannot readily tell a horse from a buffalo, or a man from a bush, if they are perfectly still, unless they are quite near. Their sense of smell is very sensitive and discriminating. Their sense of hearing is also very acute, though not as much so as many

Although loss of habitat to the plow, intensive livestock grazing, market gunning and blocking of traditional movements by fences were largely responsible for the decline of pronghorn, unregulated sport shooting augmented other losses. Local citizens and immigrants shot pronghorn for meat and sport, and wealthy hunters traveled long distances to hunt on the prairies. Railroads first put on special cars that could be sidetracked as hunting camps for eastern bison hunters, but continued the practice after the bison had virtually disappeared. Later hunters found pronghorn, mule deer, jackrabbits, waterfowl and grouse in profusion. Not until enforceable laws mandated hunting seasons and harvest limits were the nations's game populations relieved of relentless exploitation. Unfortunately for many species, such laws were mandated only after populations, including pronghorn, had been reduced to levels of uncertain recovery. Photo of a hunting party in the Dakotas in the late 1870s, by F. Jay Haynes; courtesy of the Montana State Historical Society, Helena.

of the deer family. They are naturally very timid and shy when their fears are aroused, but they are not as suspicious as most deer. They have a curiosity which is very remarkable and which prompts them to examine every strange object which they see."

Data collected since the modern inception of pronghorn hunting in California in 1942 suggest that hunting has not had a negative effect on the pronghorn population. Annual sex and age composition surveys have been conducted since 1954. Resulting data show a significant inverse relationship between population size and fawn recruitment. Removal of pronghorn from a population, whether by natural or human-caused factors, apparently results in high fawn survival during following years, to compensate and fill the void. Data from the Clear Lake area of California provide an excellent example of pronghorn population response to removal of individuals by hunting and translocation (California Department of Fish and Game 1994a). In an attempt to reduce the population in line with management goals, hunters harvested 717 pronghorn and the Department of Fish and Game translocated 288 elsewhere during 1990. However, during the following January, the population had been reduced by only 200 animals. The average winter minimum population estimate for that area during 1988 through 1993 was 1,582 head.

Only areas in marginal habitat or on wildlife refuges generally have buck-only hunting. The reproductive potential of pronghorn is very high and, in good habitat, doe-only hunting can keep the animals from increasing beyond the capacity of the land to support them—or beyond landowners' levels of tolerance with regard to crop depredations or perceived competition with livestock. Hunters—whether they call themselves trophy hunters or not—usually prefer to shoot bucks. Consequently, after a couple of mild winters in the prime pronghorn states, additional nontrophy licenses must be issued to secure adequate harvests of does.

Hunting Ethics

In the United States, hunting ethics seem to be discussed infrequently. If asked to define such ethics, many hunters would grope for an answer and probably respond with fair chase. To its credit, the Boone and Crockett Club has impressed fair chase on the minds of many North American hunters.

Outlaw shooters endanger recreational hunting and its multiple benefits; they undermine the management that has restored and now sustains the species, and they adversely impact the animals and herds they illegally pursue. People who illegally or unethically pursue and shoot pronghorn (or any other big game) literally and figuratively stress the animals beyond the reasonable limits advised by wildlife managers and promulgated by natural resource decision makers. Such actions, besides directly and seriously affecting the animals in question, denigrate the legitimate values and traditions of the hunt and cast doubt, particularly among nonhunters who have vested interest in the animals as well, on the worthiness of any hunting.

Manifestations of such doubt, such as antihunting or antihunter sentiment, bodes ill for the viability and options of managing wildlife resources in an increasingly complex, expensive and shrinking natural environment. From the standpoint of the species' well-being and that of its proper progressive management to the benefit of all people, illegal and unethical shooting cannot be tolerated. To be sure, people who engage in such activities are not hunters, they are despoilers.

Tolerance of landowners will determine, to a great extent, how much land will be available, both to pronghorn and to hunters. I have talked with ranchers who would like to be rid of pronghorn; some were altering fences and watering tanks to eliminate pronghorn on their land. Competition for forage and damage to standing grain by the animals occasionally were given as reasons, but hunters were the ranchers' main complaint. Trespassing, leaving gates open, littering, shooting in the vicinity of livestock and driving indiscriminately across rangelands and pastures were the commonly cited offenses.

It is deplorable that many neophytes are introduced to chasing animals with vehicles and long-range "flock shooting." Living in open country, pronghorn are easy to see and, unfortunately, easy to chase with a vehicle. Their habits of running in wide arcs around pursuers instead of straight away, racing and crossing in front of moving objects, reluctance to jump fences, and penchant for open ground all serve to make them easy prey to vehicular pursuit. The sort of person who indulges in such pursuit, besides being a poor sport, lends needless credibility to

In the 1800s and early 1900s, before the advent of motorized vehicles, most pronghorn hunting experiences necessarily involved horses and tents. Yesteryear hunts frequently meant weeks of travel and camping, unlike typical hunts today. Photo courtesy of the Montana State Historical Society, Helena.

the propaganda of individuals and groups that are inclined to label all hunting as unfair and inhumane. Hunting pronghorn with automobiles is not a new phenomenon. A letter to the Great Falls, Montana, newspaper in 1905 noted that many hunters were pursuing pronghorn and killing them from autos (J. H. C. 1905). The writer maintained that such activity was a cruel shame and laws should prohibit it.

Hunters like to say that a small percentage of their ranks give all hunters a bad name. The percentage that causes problems is *not* small. A little time spent where pronghorn abound on an opening weekend is enough to make one wonder if the outlaw shooter is the exception or the rule. Outlaws damage wildlife populations, they damage the relations between landowners and hunters, and they damage the public image of recreational hunting. In our democratic society, where anyone who wants to hunt can hunt, we will always have problems, but the magnitude of those problems should be reduced.

The problem is easy to identify, but hard to solve. Self-policing by hunters—use of the snitch line and willingness to report violations and bearing witness—would help to alleviate the problem.

A pronghorn hunt is the ideal situation for teaching a young sportsman or sportswoman hunting ethics, the close stalk and clean kill (O'Gara 1980). Above is a 14-year-old hunter with her grand "book" buck, taken on opening day of the 1996 pronghorn season in New Mexico. Photo by Bart W. O'Gara.

Hunting Safety

Hunter safety is a matter of importance to hunters and, especially, nonhunters (Bissell and Duda 1993). This is somewhat surprising, considering the fact that hunting is one of the safest of sports (Wildlife Management Institute 1992). Football players sustain 2,171 injuries per 100,000 participants compared with 8 injuries caused by firearms or other equipment for the same number of hunters. Even comparatively safe sports, such as fishing, tennis, golf and swimming, have 10 times or more injuries. Hunting accidents involving nonhunters are extremely rare. On average, only one nonhunter is injured by a hunter for every 12 million recreation days of hunting. A nonhunter is 20 times more likely to die from stinging insects than from wounding by a hunter. The automobile mortality rate for accidental deaths (per 100,000 people in the United States) is 18.6; that for hunting/firearm-related deaths is 0.85 for hunters and 0.001 for nonparticipants.

PREPARING FOR THE HUNT

Guided pronghorn hunts are less expensive than most other guided big game hunts, and the person unaccustomed to prairie hunting is advised to arrange such a hunt. Guided hunts often are on private ranches that tend to have a higher proportion of trophy bucks than are found in herds on public land.

An unguided hunter in unfamiliar country should allow several days for scouting. Not only is it important to know where pronghorn are likely to be, but the hunter also should gain familiarity with how the animals use that particular country. When I go to a new area, the nearest district office of the U.S. Bureau of Land Management is my first stop. Bureau maps, detailing public and private lands, are available for most pronghorn habitat.

After choosing an area that has considerable public land, I drive around and survey the country, often with binoculars. Most federal land is open to the public, but driving to it on private roads is trespassing. Even where public roads lead to public land, one should show the courtesy of checking with the rancher who runs livestock on that land. There are good reasons for a herd of livestock not to be disturbed. Above all, the hunter who wants to hunt on private land or use a private road should contact the rancher before season. Never wait until opening morning of the season to ask for permission to hunt in a given area. When properly asked in advance, many ranchers will allow hunting on their property; some even volunteer directions to where pronghorn can be found and supply information about established roads. However, a growing trend exists toward leasing hunting rights to local guides or charging a daily hunting fee. In many areas, free hunting is almost a thing of the past.

Combination hunts often are possible in pronghorn country. In this case, pronghorn, white-tailed deer, wild turkeys and sage hens were taken during one hunt in eastern Montana. Mule deer, sharptailed grouse, gray partridge and ring-necked pheasants also were nearby. Because distances to motels or hotels often are great, tents, pickup campers or mobile homes often are used now as pronghorn hunting camps. Photo by William C. Matthews, Jr.

Weather on the plains usually is pleasant during hunting season, but extreme temperature fluctuations can occur, and dry winds, dust and spiny vegetation punish the unprepared hunter. Sunglasses and chapsticks are comforting, and proper clothing is a necessity. Leather hiking boots, well broken in, of medium weight and with 8-inch (20 cm) tops are about right. I prefer crepe to cleated soles because the latter hold onto cow manure and wet clay, both of which are frequently encountered. Medium-weight wool socks, denim pants, a lightweight wool shirt, light down jacket and broad-brimmed hat complete the basic outfit. A rain suit or poncho and a suit of long underwear prepare the hunter for precipitation or unusually cold weather.

Most states and provinces wisely require that hunters wear red or blaze orange clothing. Pronghorn apparently are not color-blind. They react to shiny things, such as metallic objects or glasses, and brightly colored moving objects draw attention, but stationary bright objects seem to elicit little notice. While wearing a blaze orange coat and cap, I have had wary pronghorn feed past me at 10 yards (9 m) during hunting season. Some companies now make camouflage orange clothing—a good choice if the color is bright enough to satisfy regulations.

Few hunters realize how much light reflects from a gun barrel. Montana big game hunters must wear a minimum of 400 square inches (2,500 cm^2) of fluorescent orange above the waist. Yet, on sunny days, I locate distant hunters by flashes of light from their rifle barrels as frequently as by their orange clothing.

Shots at pronghorn usually are taken at longer range than those at other big game, and pronghorn are comparatively small. Theodore Roosevelt (1905) recorded that, for bison, bear, moose, elk, caribou, Rocky Mountain sheep and Rocky Mountain goats, the average distance at which he pulled the trigger was 80 yards (73 m). On the other hand, the average shot at pronghorn was 150 yards (137 m). He kept a record of the number of cartridges expended for every head of game he shot. His average number of shots per pronghorn was nine; for other North American big game, his average was three. Few hunters would admit to such statistics today, although such shooting would be an improvement over the records of many flock shooters of pronghorn. Roosevelt hunted primarily with large-bore, black-powder rifles and open sights. In later years, he acquired what he described as a small-bore, smokeless-powder rifle from which the projectile trajectory was very flat—a .30-.30 Winchester.

The average pronghorn hunter should carry a flat-shooting rifle equipped with a scope and a sling. Most states and provinces require .23 caliber or larger, center-fire ammunition. The scope should be of four-power or more, and a range-finder reticle (dot or double cross wire) is helpful. In South Dakota, hunters with scope-sighted rifles killed 8% more bucks than did hunters with iron (open) sights (South Dakota Game Division no date). Also, hunters with scopes averaged four shots to kill a pronghorn; those with open sights averaged 13 shots and crippled more animals. Rifles should be sighted in immediately before each hunt (Figure 161).

According to many writers of hunting articles, projectiles for pronghorn should be the flattest trajectory loads available for that rifle and expand rapidly against light resistance. I have made many quick kills on pronghorn with a .243 Winchester and 70-grain hollow points. However, under very windy conditions, I prefer 100-grain, soft-point, solid-base projectiles or even a 7 mm Remington magnum with 162-grain projectiles. The latter may seem excessive for fragile pronghorn, but bullet placement is more certain during high winds than with lighter projectiles. Shots tend to be taken at longer range on windy than on still days, because the animals are more alert. Large, heavy projectiles do not disintegrate on bones and go through animals to leave blood trails. A more arched trajectory is the primary problem with heavy projectiles.

Most pronghorn-hunting provinces and states allow hunting with a variety of weapons. Wherever they pursue pronghorn, hunters should check regulations concerning legal weapons. Muzzleloading rifles generally must be .39 to .45 caliber or larger. States that allow hand-

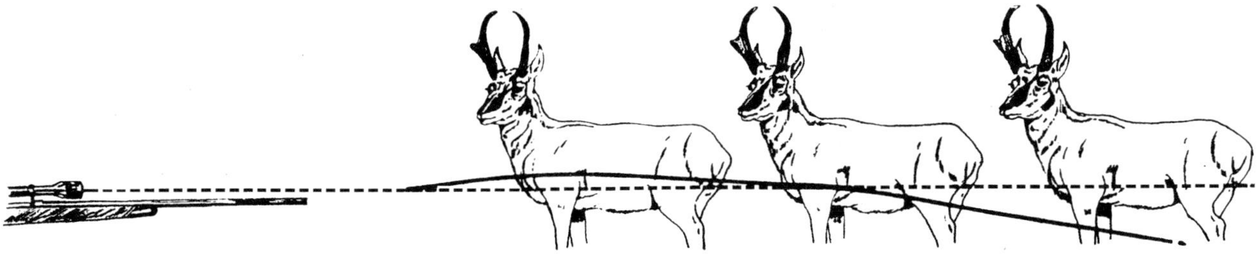

FIGURE 161. *Many hunters do not realize how much a projectile will drop in flight once it has passed the distance at which the rifle and load have been zeroed. The amount of drop depends on muzzle velocity, projectile weight and configuration. A projectile starts dropping at a rate of 32 feet per second (975 cm/second) as soon as it leaves the barrel. Elevating the bore slightly above the line of sight is necessary so that the projectile's path rises on a curve that has it crossing the line of flight twice. The first will be close to the rifle, depending on the height of the scope above the bore, distance to zeroed point of impact and projectile speed. This distance usually is about 25 yards (23 m). The second will be at the distance at which the rifle and load are zeroed. At any point in the projectile's path other than these two, the projectile will be above or below the line of sight. When shooting uphill or downhill, gravity's impact is measured over the horizontal distance to the target, not total distance from rifle to target. In general, this requires a slightly low hold. For pronghorn hunting, the best range to zero-in a rifle and load is 200 to 250 yards (182–228 m). With most pronghorn loads, a shot taken at half that distance will not be more than 2.5 inches (65 cm) high.* Illustration by Diana Haker.

A flat-shooting, scope-sighted rifle of at least .23 caliber and binoculars (left) *are basic equipment for pronghorn hunting. Sage hens* (right) *sometimes are encountered during a pronghorn hunt. Early in the season, these birds are very tasty, and many western hunters carry a shotgun in their vehicles or a .22 caliber handgun or loaded-down rounds for their pronghorn rifle to collect birds for the pot.* Photos by Bart W. O'Gara.

guns (illegal in Canada) for pronghorn hunting have a variety of restrictions involving barrel lengths, energy at muzzle, caliber and length of cartridge. Provinces and states that allow the use of shotguns require rifled slugs and permit gauges no larger than 10 or smaller than 20. Bows generally must be 40-pound (18.2 kg) pull weight or more, and almost every province and state regulates length or weight of arrows and width of arrowhead. Crossbows generally are illegal, but Kansas allows them for use by certain physically disadvantaged persons.

Although some yesteryear hunters, including Theodore Roosevelt, considered coursing pronghorn with greyhounds great sport, no province or state now allows pronghorn hunting with dogs.

Bowhunting seasons generally are longer than firearms seasons and are opened earlier. The few states that have a special muzzleloader season (most allow pistols or muzzleloaders but these users must hunt concurrently with those who use modern rifles) generally open it before the rifle season and after the bow season. During recent years, pronghorn harvest success rates have been 75% to 90% for rifle hunters, about 40% for muzzleloaders, and 20% for archers. Bowhunters seem to have especially variable success rates from region to region. Oregon, for example, reported 8% bowhunter success, whereas Kansas reported nearly 30%.

High-quality, lightweight binoculars of seven or eight power are next to the rifle or bow in importance. The trophy hunter also should have a spotting scope; those with variable magnification are good, but I am satisfied with a lightweight 20-power scope. A small, soft packsack containing a knife, 30 feet (9 m) of clothesline, string, compass, canteen of water, extra socks, plastic bags, moleskin, matches, flashlight, hard candy and jerky nearly complete my hunting outfit. Taking the packsack when leaving the vehicle will assure the hunter having everything needed for most days on foot. For crawling over sharp rocks and cacti, stout leather gloves and stretch kneepads will seem like lifesavers. I often carry a walking/shooting stick that is about 4.5 feet (1.4 m) long. Besides being handy, the stick is especially helpful when the terrain does not lend itself to a prone or sitting shot. Three to five gallons (11–19 liters) of water should be carried in the vehicle.

THE HUNT

For many hunters, the most satisfying way to pursue pronghorn is on foot, and this involves more than simply walking around the countryside. Hunters must remember that the game has better senses of sight, hearing and smell than humans do. A successful hunter avoids being seen, heard or smelled by the quarry. When making a stalk, be aware of other animals in the area besides the intended prey. Horses and cows will watch a hunter, and such attention will alert pronghorn. If other animals—domestic or wild—bolt or sound an alarm, pronghorn are almost sure to run. Even when hunting pressure is light, pronghorn are alerted or stampeded by alarm calls of other animals. Hunting in western Kansas during the mid-1800s, Gillmore (1874) reported that prairie dog mounds often provided the only cover for crawling toward pronghorn. However, he found that the shrill warning whistle of a prairie dog could ruin a stalk.

Hunting pronghorn afoot, in the animal's element, adds a challenge the hunter will never forget, according to Carpenter (1999:132), who summarized the finale of a foot hunt as follows: "Sooner or later, all the work will pay off and you'll find yourself leaning into some dusty hillside, your rifle's crosshairs wavering a bit and then settling rock-solid on a pronghorn across the way. And then you will take another walk across the prairie—the shorter the better for this one—to kneel down next to a musky-sweet-

A walking/shooting stick serves as an aid in walking, especially through rough gullies. It also provides a rifle rest where vegetation is too tall for a prone or sitting shot, or when pronghorn have become accustomed to an upright hunter. Photo by Bart W. O'Gara.

salty scented pronghorn, and hold those black angled horns in your hands.

"At that point, and thereafter, the inches of those horns don't matter, if there are even horns at all. I ceased to care long ago. Any pronghorn—buck or doe—is good when hunting them this way. What matters is that you are there, seeing country so big and wide-open that it makes you feel good knowing there are still places like this on earth."

Those who cannot spend the time or do not have the physical stamina to hunt all day on foot miss some of the thrill and satisfaction of hunting. Yet, a successful, sporting hunt is possible without chasing animals or shooting from vehicles. Glassing from high points on roads and taking short walks to check the other sides of ridges often will reveal pronghorn. If a hunter wants to stalk animals originally seen from a moving vehicle, the best method is to drive out of sight without changing the sound of the engine. Then walk back, relocate the game and plan the stalk. I wrote much of this "how to hunt" for a 1980 article in *Montana Outdoors*. By the time I had reached my mid-70s, I found it difficult to walk for hours continuously, so I began, on occasion, to locate game from a vehicle before initiating a stalk.

In states that have hunting seasons during the rut, the taking of adult bucks often is easy. Almost every 3-year-old or older buck will have a territory that he is loath to leave. If such a buck is chased from his territory, he usually will be back within half a day or less. When lighting is poor and wind conditions right, a buck may even come to meet a hunter entering his territory. Apparently, such a buck considers the hunter a competitor or predator to be evicted. Heavy and continuous hunting pressure during the rut may cause the territorial breeding system to break down; this possibility is discussed in Chapter 20.

Under the right conditions, stalking, flagging, walking-down and waiting at crossings, green fields or waterholes, all can afford shots at standing pronghorn. But, in my opinion, stalking is the most sporting and enjoyable way to hunt these prairie speedsters.

I usually avoid the opening day of the season. On public lands, hunters seem to be everywhere and inadvertently, but all too often, interrupt the best-planned stalks. Many animals are shot as they move away from another hunter, and little skill is required. In some areas, almost everyone with a license gets an animal the first day or first weekend of the season. A week or two later, the pronghorn country is almost deserted. The animals become spooky, but that adds zest to the sport. Planning and executing each stalk may take half a day or more, and only a good stalker will get within 200 yards (183 m) of even half the pronghorn tried for, but it is a very satisfying way to hunt.

Taking a trophy pronghorn buck during the rut is not always difficult. The buck will be concerned primarily with courting does and will distract the more alert does from sensing danger. Also, territorial bucks will soon return to their territories, even if a clumsy stalk has scared them from the area. Photo by Bart W. O'Gara.

The hunter who sees game before being seen has the advantage. To accomplish this, avoid the tops of ridges and hills. The stalker should cross ridges in saddles or notches, using bushes or rocks for cover and crawling if necessary. Pronghorn notice objects on the skyline at tremendous distances. Watch carefully while crossing because game may be just over any ridge. Never scan with binoculars from the tops of ridges; cross them first, then settle down behind or against rocks or bushes, and glass all of the terrain in sight. Take ample time; frightened pronghorn flash white, but bedded ones are hard to see. Although the country may look flat, small draws and undulations can hide whole herds. The longer one glasses, the greater the chance of an animal moving out of such places. In big country, using a spotting scope is worthwhile for scanning distant ground. No matter how careful, a hunter occasionally will blunder into sight of a pronghorn or a band. If the animals do not run immediately, keep moving slowly until out of their sight. Sometimes, one can then circle back and make a successful stalk.

Pronghorn hunters have less trouble with being scented by the game than do other big game hunters. In open habitat, sight is more important than scent, and the prairie winds shift less than those over rough terrain. I hunt into or across wind where I can, but on a circular hunt—walking away from a vehicle and back to it—the wind sometimes is at my back. This reduces the chance

A herd of fleeing pronghorn can maintain speeds of 30 to 40 miles per hour (48–64 km/hr) for several miles. Large pronghorn bucks often lag behind a fleeing herd. Flashing rump patches are what a hunter on the skyline is most apt to see. Photo courtesy of the U.S. Fish and Wildlife Service, Charles M. Russell Wildlife Range, Lewiston, Montana.

of surprising animals within rifle range, just over ridges or in pockets. However, the chances of seeing game at a distance and then maneuvering to make the stalk into the wind still are good. Although pronghorn are instantly alerted by a person's scent and become extremely difficult to approach, they usually do not bolt unless there is visual stimulus.

Sound seldom stampedes any but the most skittish pronghorn, but they are quickly alerted by the slamming of car doors and the rattling of binoculars against gun barrels. Like most big game, pronghorn are frightened by the sound of hurried movements, whether those movements are made by their own species or another. The person who rushes from ridge to ridge spreads fear and is sure to get more exercise than game. But, a slow-moving hunter, who stops often to use binoculars, sounds much like a feeding animal. If they do not see or smell anything strange, pronghorn sometimes investigate unfamiliar sounds. As Roosevelt (1905:139) stated: "It is a queer animal, with keen senses, but with streaks of utter folly in its character."

At times, two or more hunters can move together, bent at the waist and each holding to the one ahead, in single file and be ignored by pronghorn. I presume they think the people are a cow, horse or other four-legged animal. An African tracker I hunted with in Zambia more than 50 years ago often used this tactic in open country. He, a skinner and I hunted together, and the skinner always carried an old, brown blanket over his shoulder—using it in lieu of a coat and to sit or sleep on. The tracker would bend over; I would be sure my rifle barrel was empty, and would grasp the tracker's belt with my left hand and carry the rifle in my right; the skinner would drape his blanket over us—blocking any glint from my rifle, glasses or binoculars—slip under the blanket and grasp my belt. We traveled long distances in sight of game without disturbing it, but it was horribly hot and dusty under that old blanket!

Once pronghorn are located, the hunter should watch them long enough to determine what they are doing and whether a particular animal among them warrants a stalk. If the group is moving, a quick stalk-and-wait interception often is possible. I once got within 10 yards (9 m) of unsuspecting pronghorn in that way. If the animals are bedded, it is time to plan the stalk.

Undulations in the ground usually will allow a circuitous approach—out of sight of the animals and into or across wind. Walking almost any distance while screened from pronghorn will prove more effective than crossing even small openings within their sight. Crawling on hands and knees, or even one's belly, is fine for the final 200 yards (183 m) or so, but it is tiring and time consuming to crawl for a longer distance. After just a short crawl, the average hunter will be too out of breath for a steady shot, and it takes some moments while lying in a cramped position to get one's breath back.

Once a route is selected, do not dally; the game may move any time. Yet, even while making a long approach, a hunter should never get too out of breath and should remain prepared for a steady shot. More than once, pronghorn have come to meet me as I was stalking them. I was never sure whether they had seen or heard something suspicious and were coming to investigate or just happened to move in my direction. Usually, they caught me by surprise and were on the run before I was ready for them, suggesting they already were suspicious. Pronghorn truly surprised at close range often stare and mill about before running. Circling behind a ridge or going up a draw or creek bottom are the most common ways to get within reasonable crawling distance.

Often, a hunter can get within shooting distance on hands and knees—a fairly fast mode of travel if one is wearing gloves and kneepads. Crossing an opening in sight of pronghorn can only be done flat on one's belly and slowly. Watching a cat stalk a bird gives an idea of how slow movements should be. Resist the temptation to lift your head for a look or to raise your buttocks for faster crawling. More than once, pronghorn have approached to see what I was. If that happens, do not try to get the rifle off your back quickly. The animals will stay until their curiosity is satisfied, usually allowing time for the rifle to be inched around into shooting position.

Stalking with the rising sun at one's back can be effective. A friend told me of hunting pronghorn in the Centennial Valley of Montana during the 1988 season. He began hunting several days after the season had opened, and a large group of pronghorn, including several large bucks, had gathered on a big flat where they could not be approached. My friend went to the east edge of the flat before sunrise. As the sun came up, he stalked and killed a large buck. The animals allowed him to walk and crawl within shooting range and, apparently, were not able to see him against the early morning sun.

About 9 out of 10 stalks fail to get the hunter close enough to a pronghorn for a clean kill and, according to Cadieux (1986), no one with more than a 34-inch (86 cm) waistline has a chance of sneaking up on a pronghorn. Cadieux maintained that crawling on one's belly is not fun; tender anatomy encounters the hardest, cruelest part of the pronghorn pasture; sweat rolling down a forehead in waves does little to improve accuracy when using a scope sight; belt buckles act as plows that scrape dirt into underwear; prickly pears stab elbows and knees; sagebrush grabs hats; breathing becomes labored; keys in pockets gouge deeply into thighs; and sheath knives threaten emasculation. This all happens in the first 100 yards (91 m), when the hunter still has 500 yards (460 m) to go and during which exercise the buck may move off another 600 yards (550 m) as he watches the hunter wiggling along. As much as I enjoy stalking pronghorn, I have had days approaching what Cadieux describes, and my girth is not for public record.

"Flagging" pronghorn, or using a flag to arouse a pronghorn's curiosity, has received considerable attention from outdoor writers as a supposedly easy way to get a standing shot at a buck. In my experience, this method has not proven particularly effective as a hunting technique. For summertime photography, I have brought young animals and territorial bucks into camera range by flagging. The best results were obtained by driving past a band or a lone buck, walking back, approaching the animals up or down a draw and hoisting a white handkerchief on a stick where they could see it. I then moved about 50 yards (46 m) downwind so the curious animals would not be looking right at me and were less apt to scent me. They sometimes came to inspect the handkerchief almost immediately, sometimes within a couple of hours—but most often, not at all.

Pursuing pronghorn on foot involves more than simply walking around the countryside. Hunters need to "hunt smart" and locate the quarry before being seen. Good binoculars are a must. A small, soft packsack can hold everything needed for a day's hunt and for taking care of meat and trophy. Photo by Robb Hitchcock.

My lack of success in flagging pronghorn during the hunting season undoubtedly has been related to the fact that I hunt in heavily hunted areas after the season has been open for a week or more. On a private ranch or other lightly hunted area, the method would be worth trying. If it worked, the shots would be at standing animals and at short ranges. Animals I lured to the camera often came within 50 yards (46 m) or less of the handkerchief. Popowski (1959) brought a buck, whose horns he could see on the skyline, into range by hoisting a handkerchief on his rifle barrel. Such a maneuver might save the day when further inconspicuous crawling is impossible.

Apparently, pronghorn learned about flagging a long time ago. Du Bray (1890:331) wrote that he had tried the ruse, but could never lure them within range; he presumed the scheme worked all right in "early" days, "before the game of the prairies became educated to the seductive wiles and sly ways of the white man." Old frontiersmen had told him how they lured the little "gazelles" to their ruin.

Walking-down is a challenging way to take one of the big old bucks that stays out in open flats where stalking is impracticable. Before attempting this method, the hunter should realize that 10 miles (16 km) and half a day are about average for such a hunt, and the presence of other hunters in the area may preclude success. Also, a successful hunter employing this method may have to return with his kill a long way back to a road, although pursued pronghorn tend to circle back toward the beginning point. The small, handy global positioning systems (GPS) now available could be useful on such a hunt, especially in unfamiliar country. I do not like too many gadgets that replace skill in hunting, but a GPS could save miles in finding one's way back to a vehicle and facilitate finding a carcass to be picked up. Also, the price of a GPS—about $200 to $300—would seem reasonable if the hunter was caught in a sudden, violent rain or snowstorm.

After locating a pronghorn, the hunter slings the rifle and walks slowly toward a spot several hundred yards downwind of the intended prey. It is best not to look at the animal directly or make any unnecessary movements. As a hunter comes into view, a pronghorn usually becomes fidgety and watches until its pursuer is within about 500 yards (460 m); then it runs for 1 to 2 miles (1.6–3.2 km). The hunter simply has to follow. When crossing a ridge, the pronghorn often will switch directions when it can see what is on the other side, so the hunter must watch closely. While a ridge conceals the prey, the hunter can walk as fast as desired; when breasting the ridge, the hunter should resume a slow walk toward a spot downwind of the pronghorn. The animal usually will allow the hunter to get about 30 yards (27 m) closer at each approach and will tend to flee a shorter distance each time. Ten walks should put the rifleman within 200 to 250 yards (183–229 m) of the target—about as close as can be expected. On the final approach, the hunter should go to a fence post or other handy gun rest, smoothly unsling the rifle and shoot without delay. I carry my walking/shooting stick in my left hand so I can shoot anytime. A hunter should not go into a prone or sitting position for the shot because the animal has become accustomed to seeing a person in an upright position. This method can sometimes be used with considerable success on large bucks that seem impossible to approach in any other manner.

Waiting at crossings, the edges of grainfields, wet seeps with green vegetation or at waterholes has provided close shots for many hunters. I dislike keeping game from waterholes in arid country, especially when the animals already are stressed by hunters. The better one knows an area, the better the waiting game can be played. For this type of hunting, hunters should spend considerable preseason time scouting and watching, and hunting the same area every year has distinct advantages. Pronghorn disturbed by hunters tend to run in certain patterns—down the middle of broad valleys, through low saddles, along the side of ridges—but seldom across a steep ridge unless hard pressed. Piles of rocks or a brush patch on the point of a ridge or overlooking a saddle can be a good hiding place, but the hunter must remember not to be skylined.

In some places, pronghorn habitually go under fences. These crossings are recognized easily by tracks, trails, hair on the bottom wires or broken wires; however, many crossings are only used in one-way movements. Thus, they may be morning or afternoon crossings. Again, the value of preseason scouting can be seen. A patch of cover within 50 to 100 yards (46–92 m) downwind of such a crossing can be an especially effective ambush site when other hunters are moving game about. Cadieux (1986) recommended going underground for pronghorn, by using a pit blind. He described a blind dug deep enough to allow the hunter to sit comfortably on a bench carved into the earth at one side with the sagebrush chopped off at ground level and kept as moveable upper parts of the blind. Such a blind surely would be effective in any of the situations described above. It is not my kind of hunting, but it undeniably takes patience to be successful. Also, a certain excitement builds as one watches game come to the gun, always knowing that the animal(s) may veer off before the critical moment. At any rate, waiting is preferable to joining the crowd on opening day, and it is a good method for archers or anyone hunting with a muzzleloader or pistol. Unlike many big game animals, pronghorn drink and feed at intervals throughout the day, so the hunter using this method must remain alert at all times. A GPS could aid in finding a blind in the dark. Erecting blinds near waterholes or that keep pronghorn from other sources of water, particularly on arid rangeland, may be deleterious to the animals and is hardly sporting.

THE SHOT

How close a hunter should try to get for a shot when stalking pronghorn depends on the hunter's own abilities (both as stalker and shooter) and on the terrain. With today's weapons, a 300-yard (274 m) shot from a steady rest at a standing animal is reasonable, providing the hunter is a good marksman, knows the rifle and can judge distance. The hunter should know the range at which every shot can be held in a 10-inch (25 cm) circle and not be tempted to push luck beyond that range. Range is deceptive on prairies and deserts, especially for hunters not accustomed to the country or game. A crawling hunter with eyes near the ground is at a disadvantage in estimating distances. Bright sunshine, clear air and gleaming white hair often combine to make pronghorn appear closer than they are. Conversely, at dawn or dusk or on a stormy day, the animals seem to shrink and fade, appearing farther away than they actually are. When using a range

Unlike many other big game animals, pronghorn drink and feed at intervals throughout the day, so hunters in blinds must remain alert at all times. Blinds sometimes provide archers with the close shots they need. Photo by Robb Hitchcock.

reticle, the hunter should remember that adult bucks measure about 17 inches (43 cm) from top of back to belly and does measure 1 or 2 inches less (2.5–5 cm).

A few people increase the challenge by hunting with handguns. Scope-sighted, single-shot pistols in rifle calibers are available, and hunting with them is not too different from hunting with a rifle. But when the barrel is shortened to 6 or 8 inches (15–20 cm) on a revolver in 0.41 or 0.44 (10 or 11 mm) magnum caliber, about 100 yards (91 m) may be the maximum effective range; for most hunters using a scope with open sights, that distance may be halved (Ochs 2000). As with rifle shooters, handgunners need to be able to judge distance and know at what range she or he can hold all shots within a 10-inch (25 cm) circle.

Range finders have been on the market for years. I experimented with them during the 1980s, principally for long-range head shooting on biological collections. The ones that were small enough for convenient field use were not accurate beyond 200 yards (183 m). An experienced hunter could estimate range about as well. Now, a laser range finder, the size of an average eight-power binocular, gives very accurate readings out to 400, or even 1,100 yards (366 or 1,006 m), depending on maker, model and price (Barsness 1995). Prices are high and, for some hunters, determining the range mechanically will take some of the skill and sport out of the hunt. However, knowing the distance should lead to better placed shots or, possibly, not shooting at all. I favor hunting equipment that assures quick, humane kills.

An unhurried, long-range standing shot is more reliable than a closer running shot. I prefer crawling to within 150 yards (137 m) or less before shooting. The real excitement of the stalk occurs in the final approach and, for me, shooting at greater distances amounts to admitting I cannot get closer. Besides the thrill of a close stalk, there are other advantages. Paramount is placing the shot for a quick, painless kill. Also, at closer range, the trophy hunter can more accurately assess the head, and the meat hunter can avoid the loss of meat caused by a poorly placed shot. Placement of the shot is similar to that of any big game animal (Figures 162–168).

Offhand shooting at pronghorn usually is unnecessary and foolish. The country lends itself to prone or sitting shots, and rocks or fence posts sometimes offer handy supports. Many of my shots are taken from a kneeling position over a shooting/walking stick. Bipods, used mostly by varmint hunters, are commercially available. I used them on biological collections and was surprised at the accuracy they enabled. It seems almost unsporting to use such an aid, but using one certainly is preferable to crippling an animal.

I wish everyone used single-shot rifles so that more hunters would be careful on the first shot. A one-shot

FIGURE 162. *Although a pronghorn is not very large, it is rather hard to kill for its size. With a live weight of about 80 to 140 pounds (36–64 kg), it still is a large animal to kill with a projectile weighing only 75 to 100 grains. Even allowing for high muzzle energy and shock effect, placing the shot in a vital area is essential. A broadside shot is the ideal situation, and a heart (bottom X) or lung shot is the safest option for a sure, humane kill. At ranges of less than 100 yards (91 m), with a rest and ideal weather conditions, a neck shot (top X) is effective. Such a shot wastes less shoulder and rib meat but damages a trophy cape.* Illustration by Diana Haker.

FIGURE 164. *With a pronghorn facing a hunter, a shot would not be attempted except at close range and from a rest. A neck shot, from beneath the jaw to the base of the neck, allows considerable room for error in elevation; however, 1 inch (2.5 cm) right or left may result in a wounded animal. With a chest shot (X), if the projectile is a few inches off center, it might pass between a scapula and the ribs, resulting in a dislocated shoulder, and a wounded animal that would travel a long distance.* Illustration by Diana Haker.

FIGURE 163. *Shots at reclining pronghorn generally should not be taken at much more than 100 yards (91 m), and then from a rest. Although pronghorn usually bed in short vegetation, heart shots seldom are feasible. A shot at the base of the neck near the shoulder is advisable (right X). Not only is the lethal target smaller in the upper neck (left X), but should the animal's head be turned at the moment of firing, a broken jaw might result. Such a wound is extremely painful, and the pronghorn is likely to run off and starve.* Illustration by Diana Haker.

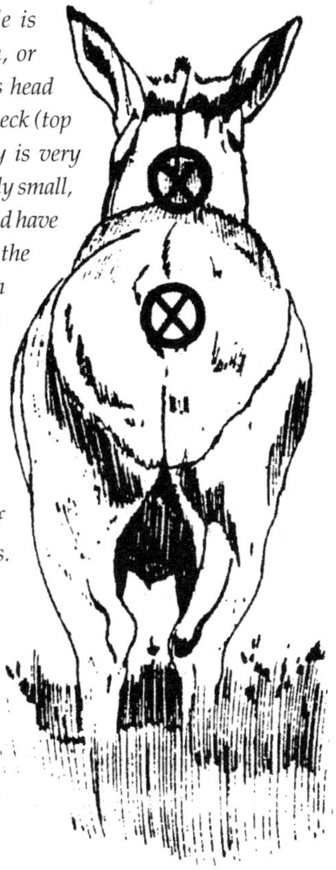

FIGURE 165. *Provided the rifle is above the level of a pronghorn, or the animal is standing with its head uphill, a shot at the base of the neck (top X) of a pronghorn facing away is very lethal. However, the target is fairly small, and the hunter should be close and have a steady rest. A tail shot from the rear (bottom X) of a pronghorn on the same level as the rifle will put the animal down. Such a shot is not lethal and should never be attempted except at a wounded animal that might otherwise escape; it shatters the pelvic girdle, ruining much of the sirloin and top round steaks.* Illustration by Diana Haker.

hunt in Wyoming, stressing the sporting aspect of the chase and playing down the kill, has attracted many enthusiasts, and the success rate has been high since its inception in 1940 (Dahl et al. 1986). The first shot usually is the one that gets the game. Roosevelt (1895) killed five pronghorn on one hunt with 36 shots, but each animal was killed with a first shot; when he missed with a first shot, he did not hit with any subsequent one—not an unusual situation.

The only time a hunter is justified in shooting at a running pronghorn is, I believe, when the animal has been crippled. Admittedly, rolling a running pronghorn at 200 to 300 yards (183–273 m) is a thrill, and I used to do it when I was a better shot than I am today. But before trying it on anything other than a wounded animal that might otherwise escape, remember that a pronghorn hit in a nonvital area, whether or not it eventually is retrieved, will suffer unnecessarily. I would rather go home empty-handed than have that kind of wounding and waste on my conscience.

FIGURE 166. *A quartering shot at a pronghorn standing at a head-on angle can result in a quick kill, provided the point of aim is on the organs within (in this case, the heart— bottom X on left buck), not on the usual point of aim for a broadside shot. That will result in the projectile passing through the rumen (probably ruining a ham) or a crippled animal. If the hunter is close and has a rest, a lower neck shot (top X on left buck) will bring a nearly instantaneous kill. A heart shot on the right buck requires a similar hold, as does the bottom X on the left buck, but the right buck is standing with its neck nearly straight with its body, requiring aim farther forward on the neck.* Illustration by Diana Haker.

Figure 167. *A quartering shot at a pronghorn standing at an angle tail-on is best attempted at the base of the neck (top × on left buck) if the animal is close. A heart/lung shot (bottom × on left buck) should not be attempted unless to dispose of a wounded animal that otherwise might escape. A lethal shot in this case requires an aim just in front of a ham, and probably penetration of the stomach and intestines. A light, fast projectile may not accomplish this and, whatever happens, some meat will be fouled. A heart shot on the right buck indicates a fairly similar hold as does the bottom × on the left buck; however, the right buck is standing with its neck nearly straight with its body, so a neck shot requires a hold on the shoulders.* Illustration by Diana Haker.

In the event a pronghorn is crippled, as soon as it is out of shooting range, the hunter should use binoculars and do everything possible to keep it in sight. If it goes over a ridge or otherwise moves out of sight, the hunter should mark the spot it was last seen in relation to some landmark and get to that spot as soon as possible. As soon as the spot is reached, it should be marked with a hat or handkerchief, and the hunter should move ahead along the line of flight to again locate the fleeing animal. A crippled pronghorn does not try to hide and usually lies down within 0.5 mile (0.8 km) if it is not aware that it is being followed.

If a bedded cripple is sighted, the hunter should use all of her or his skill as a stalker and a great deal of patience to approach it and deliver a killing shot to the animal in its bed. If a lightly wounded pronghorn is missed or otherwise frightened from its first bed, it will travel many miles before stopping again. In such a case, marking the bedsite so it can be found again and going back to the vehicle and using it to scout the country the cripple went into may present the only chance of getting another shot at it. If that fails, the hunter should return to the bedsite and attempt to track the crippled animal. However, even with a blood trail, the chances of finding the animal are slim because of the great distance it likely will travel.

If the hunter does not again sight the crippled animal after it was first lost to sight, he or she should go back to the hat or handkerchief and find the animal's trail. If the animal is not bleeding, chances of doing this are very slim unless the ground is wet or snow is present. If a blood trail is found, much patience is required to follow it. The hunter should drop a hat or other easily seen object on the last drop of blood before proceeding to find the next one. This process usually must be repeated hundreds of times. About every 100 yards (91 m), the hunter should stop and search the area ahead very carefully with binoculars, trying to

FIGURE 168. *In Sweden, where most European elk—moose, to Americans—are shot during drives, the "elk clock" was proposed by Lennart Hedberg—a game preserve warden—to indicate target areas offered by an elk standing or walking at various angles throughout 360 degrees. Such a clock is presented here, adapted to pronghorn. Warden Hedberg considered only lung/heart shots feasible or ethical, and he believed shots at animals in position at 6:00 and 7:00 o'clock to be inappropriate; so, too, would they be for pronghorn. Note how completely the rumen shields vital organs when an animal is quartering away with its left side toward the hunter. During elk drives, hunters often must take "risky" shots. In pronghorn stalking, with a little patience, a hunter can almost always get a shot at an animal in the 3:00 to 9:00 o'clock positions. Shots at the other positions, although fatal if properly executed, usually will break shoulder bones and ruin considerable meat.* Illustration by Diana Haker.

locate the bedded cripple before it sees the hunter. If the hunter is careless and jumps the cripple, long-range running shots probably will merely hasten it on a very long trek that the hunter is unable to follow.

Stalking pronghorn is fun. The hunter can count on seeing many animals, and the excitement builds as one sneaks up on a trophy buck or a fat young animal for the table. A quick, clean kill may leave the hunter momentarily sad at the death of a beautiful animal, but a satisfaction soon sets in from knowing the game was played by the rules and played well. Those who chase these interesting animals with a vehicle and who shoot at running herds not only risk damaging landowner/hunter relations and crippling animals, they also miss one of the greatest thrills in North American hunting.

USING A DECOY

Years ago, I experimented with a plywood silhouette of a pronghorn as a kind of portable blind for darting pronghorn and taking pictures. I had some success with the method, but the silhouette was heavy and awkward. Commercial pronghorn decoys now are on the market and provide another way for archers to get close to a pronghorn trophy. The system could work for muzzleloaders, handgun hunters and riflemen, but I would not want to be behind a decoy during the rifle season.

Mel Dutton (personal communication:1986) of Faith, South Dakota, who has killed three Pope and Young pronghorn bucks using decoys, provided the following tips for decoying, or tolling, a buck. When approaching a prospective decoy area, be as quiet as possible, taking advantage of any cover or depressions (such as creek bottoms) that will make the downwind approach easier and more effective. Because of the need to crawl at times, patches of leather sewn on the knees, elbows and even the crotch area of the pants are highly recommended.

Once a bowhunter has reached the decoying site, she or he is ready to deploy the decoy. Pronghorn generally are nervous during windy weather and can be startled easily. Therefore, when positioning the decoy in windy weather, raise it slowly. In virtually all cases, the hunter will be lying flat on the ground, having pushed the decoy and bow ahead while crawling to that point. While remaining flat on the ground, the hunter should wait until the herd buck is not looking directly at her or him and slowly raise the decoy. An arrow should be knocked on the bowstring at this point, as the buck might charge the decoy at high speed without warning. One must be ready to shoot before setting up the decoy.

Camouflaged clothing that blends well with the surroundings is an advantage when approaching a prospective decoy area. Commercial decoys are lightweight and very portable. Most are so realistic that extreme caution should be exercised in their use for hunts where other rifle hunters may be in the vicinity. Photo by Ken Knapp.

Being situated behind the decoy, one can observe the behavior of the buck. Some bucks will charge with total disregard for danger, while others will come more slowly and cautiously. Some pronghorn will approach within a few feet, but most will stop at about 30 yards (30 m). This is the best time to shoot. When the decision to shoot is made, move slowly. Gather up the bow with two hands, and stay low. Hunters should come to their knees and steadily draw while coming up and over the back of the decoy. The buck is expecting movement from the decoy, and the time it takes him to decide that something is wrong usually is enough for a clean shot. Slow, steady movement is the key.

Dutton used a silhouette to hide behind, but Don Laubach suggested using a full-body decoy (see Henkel 1995). He found that placing a full-bodied decoy facing oncoming pronghorn worked better than giving them a side view. He believed that eye contact was important. Laubach also discovered that sitting on the ground 20 to 30 yards (18–27 m) from a decoy was better than sitting behind it. Camouflage seemed less important than just sitting on the ground. Pronghorn locked in on the decoy and ignored the hunter.

Early Plains hunters understood the curious nature of pronghorn and used it to their advantage (McCabe et al. 2004). They tied things like feathers to bushes, then waited patiently for an animal to approach close enough for a shot. However, tolling with decoys is relatively new. The difference between tolling with decoys as opposed to other methods is that decoys play into the aggressive rutting behavior of the male pronghorn, much like a bull elk responds to a bugle.

Not all pronghorn will respond to decoys, but bucks are the most responsive, especially during the rut. Furthermore, decoys resembling immature bucks generally are the most effective in attracting big bucks. During the rut, territorial pronghorn bucks are exceptionally vulnerable because immature bucks circle the bands posing a threat to the large, dominant, herd bucks. Dominant bucks will repeatedly charge the smaller bucks and drive them away. When a decoy is displayed to a sexually aggressive herd buck, the buck is apt to try to chase it away.

When setting up a decoy in sight of pronghorn, the hunter should be prepared for a sudden and rapid approach by a territorial buck. Territorial bucks instinctively and aggressively charge smaller bucks that approach the former's harems, so most pronghorn decoys are made to resemble yearling bucks. Pronghorn do not seem to notice or are unconcerned about hunter movements seen beneath a decoy. In fact, some movement may add to the decoy's realism. The decoy should have a peephole (top left) *so the hunter can observe her or his quarry without having to lift up before the shot. When the animal is well within effective range, the hunter should rise up steadily, bringing the gun or bow to shooting position* (top right)*. Since the pronghorn expects some movement, its hesitation at the hunter's appearance usually will allow time for an accurate shot. Decoying is an exciting, challenging and satisfying way for bowhunters and handgun hunters to pursue pronghorn* (bottom)*, but it is not without its rigors.* Photos by Ken Knapp.

Decoys can be used in other ways, including in areas that pronghorn frequent with the hunter waiting in a nearby blind. Another method is for hunters to work in pairs. One hunter positions the decoy, while the other waits in cover some distance from it.

Some hunters now use what is called an antelope challenge call to lure dominant pronghorn bucks to decoys. The call they imitate is the snort-wheeze or laugh call described in Chapter 6. The snort-wheeze begins like the alarm call, which both sexes snort or blow through their noses when they note something strange in their habitat. The alarm call sounds like cha-oo, with accent on the cha and a pause before the oo (as in boot). Laubach (see Henkel 1995) calls this a short caw sound. The snort-wheeze consists of the cha-oo, followed by a series of about five shorter bursts of sound descending in pitch and volume; the latter bursts are wheezelike. Laubach describes it as sort of a chuckle. As they call, bucks pump their sides. Laubach noted that pronghorn bucks sometimes came to the challenge call year-round without a decoy, and even bucks that had been shot at returned to investigate a call. Apparently just the cha-oo often will stop running pronghorn—possibly long enough for a shot.

When I secured an antelope challenge call, the sounds made with it while following the manufacturer's instructions sounded like a cross between bleats of a wounded fawn and the high-ball mallard call. My first presumption was that I was just a poor caller. However, after listening to the tape that came with the call, the sounds produced still reminded me of a combined predator/duck call. My first try with the call was during spring, and the bucks I could find were with does. Does seemed to ignore the calls, but the bucks were very interested. They would come a short distance, stand statuelike for up to 5 minutes and stare at the spot where I was hiding. None came closer than 200 yards (183 m), but they presented ideal standing targets. I tried the call during the Montana hunting season, after the rut, with no response.

FIELD CARE OF MEAT

Antelope should be cooled as soon after eviceration [sic] as possible to prevent bacterial growth. Skinning is not recommended unless it is necessary to aid in rapid cooling of the carcass. The need for aging antelope carcasses beyond chilling is questionable. Eliminating aging will help to avoid bacterial growth and make the texture of the lean more acceptable. . . . Odor of the antelope hide does not affect cooked meat flavor.

Field et al. (no date:5)

A downed pronghorn should be approached from its back for safety reasons and to prevent a wounded animal from seeing the approaching hunter and bolting at a distance. When a downed animal is reached, the hunter should prod it with a stick or gun barrel and be prepared to fire a dispatching shot. Under no circumstances should the hunter grasp the animal's horns until certain the pronghorn is dead.

Like any big game animal, a pronghorn should be gutted soon after the kill. But cutting the throat, or bleeding, an animal that has been dead several minutes accomplishes little. Every cut in the field puts some dirt, hair or blood on the meat, resulting in waste. Make cuts by slicing from the inside of the skin outward to avoid cutting hair into the meat.

Whether to skin a pronghorn immediately after the kill or leave the skin on until just before cutting up the meat depends on circumstances. The proximity of a carcass to a road, the air temperature and the amount of wind and humidity all make a difference. Field and Thorne (1983) recommended lowering the temperature of big game carcasses to 40°F (4°C) within 24 hours after the kill when possible. However, they warned that reducing the temperature to that level within 8 hours generally results in tough meat.

During northern hunting seasons, pronghorn carcasses usually can be cooled in the field without skinning. However, the pelvic girdle and brisket should be split as soon as the carcass no longer needs to be carried. A propped-open pronghorn hanging in the shade and exposed to the prairie wind soon cools even in its skin. Also, an unskinned

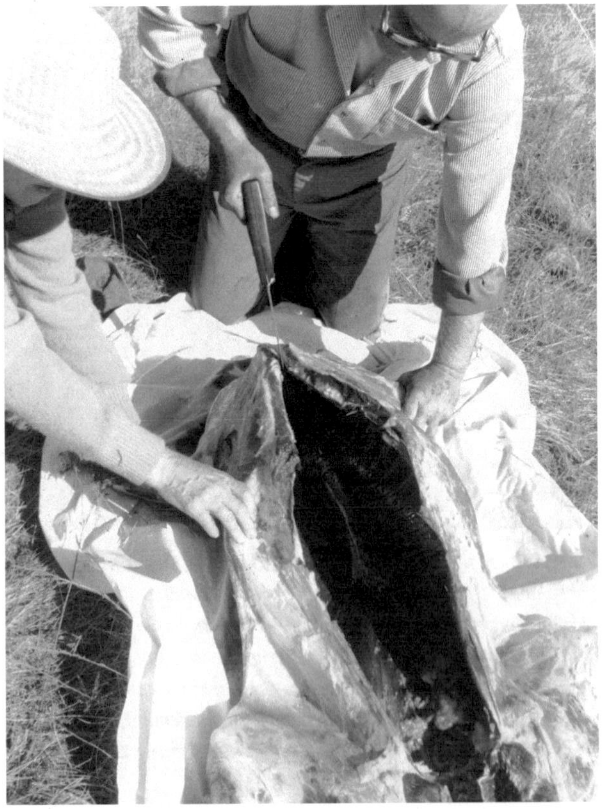

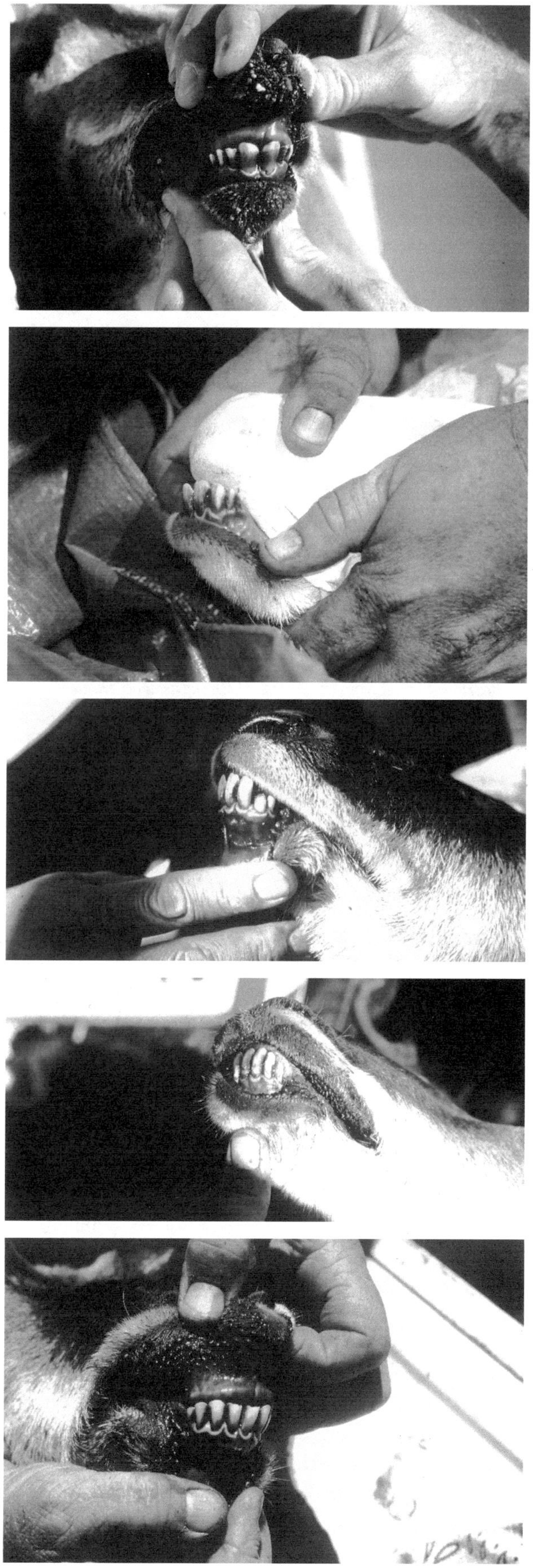

Most hunters are keenly interested in the game they hunt. For those curious about the ages of their pronghorn trophies or meat animals, a quick look at the front (incisiform) teeth should give a good approximation. The top photo indicates a yearling—16 to 17 months old during most hunting seasons. The two large center teeth (first permanent incisors) are fully erupted. Deciduous, milk or baby, teeth flank them. A fawn would have all small deciduous teeth. The second photo shows a typical 2-year-old with the second incisors erupted. The middle photo shows the third incisors erupting, indicating a 3-year-old. During hunting seasons, third incisors may be fully erupted, partially erupted, as shown, or absent. If absent, the deciduous tooth usually is loose or has been pushed out, leaving a gap between the second incisor and canine, which appears like a fourth incisor as seen in this photo. The fourth photo shows all incisiform erupted and with very little wear, indicating a 4- to 5-year-old. The bottom photo shows considerable wear, especially on the first incisors, indicating a 6- to 9-year-old. Wear varies with habitat conditions, and molariform teeth or, better yet, tooth sections should be checked to establish more accurate ages beyond 4 to 5 years. See Chapter 8 for more information on estimating ages. Photos by Jim Heffelfinger.

carcass propped open and riding tail-first on a cartop carrier cools rapidly, especially at night. Bagging such a carcass in cheese cloth will make it less obtrusive, thus less offensive to people who do not hunt and do not like to see bloody animals. Less meat is wasted if the skin is left on until time to cut up the meat.

On the other hand, body heat is a primary reason why some meat has a gamy taste. If the weather is warm or humid, one of the first steps is to get the skin off. The hunter who plans to skin an animal where it falls should carry a meat bag and a light tarp or ground cloth while hunting.

When a clean kill has been made, I eviscerate animals before skinning them. If an animal is shot through the stomach or intestine, I prefer to skin it first. The edible meat then can be removed without opening the body cavity, thus keeping gastric juices from the meat. To do this, remove the hind legs by dismembering the hip joints.

(Facing page) *If rapid cooling of a pronghorn carcass is required, splitting the pelvic girdle and brisket with a saw (the pelvis in the lower right of this photo already is split)—after gutting—allows the carcass to be opened wide and increases ventilation. However, some good top round steaks and rib and flank meat will dehydrate. If a saw is not available, the ribs can be separated from the brisket with a knife, cutting from the inside out where the ribs are connected to the brisket with cartilage. If the carcass is hung by the heels before evisceration, the blood accumulated in the chest cavity from most shots will be drained without coating the tenderloins and walls of the abdomen. Removing the viscera with the head downhill and raising the hindquarters to drain blood forward also helps to keep meat clean.* Photo courtesy of Bart W. O'Gara.

(Facing page) *A pronghorn should be gutted as soon after the kill as practical. The carcass will cool more rapidly after evisceration. When opening the abdominal wall, slit the skin for about 6 inches (15 cm), then gently slice through the white fascia on the midline of the belly so muscle shows through a 1-inch (2.5 cm) slit* (top left). *Do not cut through the muscle; push a finger through. Cutting can easily puncture an intestine or the cecum and contaminate meat. After a finger hole is established, use the tip of a sharp knife to enlarge the opening. One can safely open the abdomen, keeping two fingers ahead of the knife tip to prevent cutting intestines or rumen. After a fist-sized hole is opened, placing the knife handle and one hand inside the abdomen (with most of the blade outside) allows rapid opening. The knife handle and hand hold down the viscera as the abdomen is sliced open. After the abdomen is opened enough for the viscera to be removed, cut carefully around the anus and the vulva and vagina or anus and penis* (top right). *A small, very sharp blade is best for this. After the skin around the reproductive tract and anus is severed and the reproductive tract is cut from the posterior 1 inch (2.5 cm) of the pelvis, use fingers to loosen them from the inside of the pelvic girdle. Tie a stout string tightly around the urinary and digestive tract so urine and feces cannot escape* (bottom left). *After the viscera is mostly pulled from the opening in the abdominal wall, carefully cut around the ligaments holding the urinary bladder (and uterus in does) and colon to the pelvic girdle. Reach into the pelvic girdle past the urinary bladder, grasp the urinary and digestive tubes tightly and jerk them into the abdominal cavity. When removing the urinary tract and colon from a buck, make a slit just ahead of the anus. The penis can be pulled back through this slit and cut from the skin to which it adheres. Next, the penis should be cut from the pelvic girdle where it bends around and into the girdle. Cut very close to the bone. A slip here probably will mean a sliced penis and urine leaked into the top round steaks, resulting in unpalatable meat. The penis and colon can now be removed as described for does. After the abdominal viscera is loosened from the pelvic girdle and partly removed from the abdominal cavity, one should loosen the esophagus and trachea* (bottom right). *First make about a 6-inch (15 cm) slit in the lower neck ahead of the point of the brisket. Cut through the muscle until the white trachea is exposed. Cut around the trachea enough that the fingers can encircle it and the esophagus adhering to its dorsal surface, and pull them out through the slit in the skin. Pull them as far as possible, and cut off close to the jaw (the animal's jaw is in the bottom of the* bottom right *photo, and the trachea and esophagus are pulled out in front of the front leg). Tie the two tubes tightly with a stout string to prevent rumen contents from being forced out of the esophagus and into the neck meat as all of the viscera is finally removed from the animal. Cut around the trachea and esophagus with the tip of the knife into the chest cavity. Otherwise, attachments at the first rib prevent pulling them into the abdominal cavity. Next, carefully cut through the diaphragm from the abdominal cavity. The farther around the body the diaphragm is loosened at this time, the easier the viscera will come out. Reach into the chest cavity, grasp the heart and jerk it from the brisket. With the head downhill—if possible—reach almost to the neck and grasp the trachea and esophagus and pull the chest viscera out through the abdominal cavity. All of the viscera should come free of the carcass with a little knife work where the diaphragm attaches to the animal's back. Lift the hindquarters to drain blood from the chest cavity and avoid coating the tenderloins and flanks with blood. Wiping the body cavity out with a wet bandanna will remove most of the remaining blood. The carcass will now cool fairly fast although it is not skinned or totally opened. Ventilation and heat exchange occurs through three openings, but little dirt will reach the meat during handling or carrying. If the brisket and pelvic girdle could have been opened during evisceration, the process would have been quicker and simpler, but the carcass then would have been more difficult to carry and keep clean.* Photos by Wilma R. O'Gara.

The front quarters do not have joints to contend with, but one should cut as close to the ribs as possible. Then remove the neck and the meat from the back (loins). As edible portions are taken from the carcass, remove hair or bloody areas and wipe them clean with a damp cloth and place the pieces in a meat bag.

After removing the viscera, turn the carcass belly down—preferably over some brush if unskinned, or on a ground cloth if skinned—to let blood and body fluids drain while retrieving the liver and heart from the viscera. Deer hunters will not be accustomed to removing gallbladders. They are 2 to 3 inches (5–7.5 cm) long and filled with green or greenish yellow bile. Wasting some liver while removing the gallbladder is better than cutting it, as the bile will give the entire liver a strong taste. I carry a 2.5-gallon (9.5 liter) plastic bag to hold the liver and heart, and place 1 or 2 cups (250–500 mL) of water in the bag to dispel heat and keep the organs from drying out. The bag is left open except when being carried.

The next step is to dispose of the viscera in a gulley or other depression away from roads. I slice the rumen wall so the stomach will not inflate with gas and be visible at a distance. Most people, including ranchers who grant permission to hunt, dislike seeing gut piles.

After the carcass has drained, turn it belly up and wipe it out with a wet cloth. Then wring out the cloth and wipe the body cavity dry. Washing does not hurt meat, but water seldom is available on the prairie. Judicious use of a canteen of water and a bandanna can do a surprisingly good job of cleansing the body cavity.

Prop the carcass open for at least an hour to dispel heat and allow a dry surface to form on the meat. If possible, the spread carcass should be hung overnight before it is transported. When an unskinned carcass must be left in the field, it should be turned belly down to protect the meat from carrion-eating birds. A skinned carcass should be placed in a meat bag.

Skinning a pronghorn on the ground is more difficult than skinning one hung up. But to do so, first lay the animal on its side, and remove the feet at the wrists and ankles—the joints about 16 inches (41 cm) above the hooves. Slit the inside of each leg up to the animal's midline. Then slit the hide from the neck to the anus (unless the head is to be mounted, discussed later in this chapter).

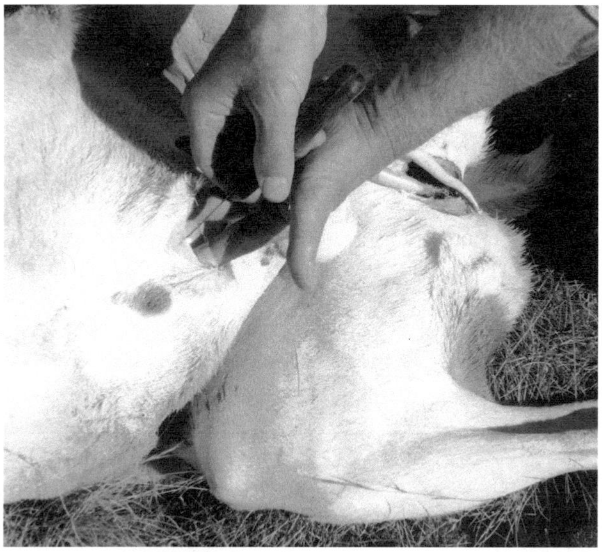

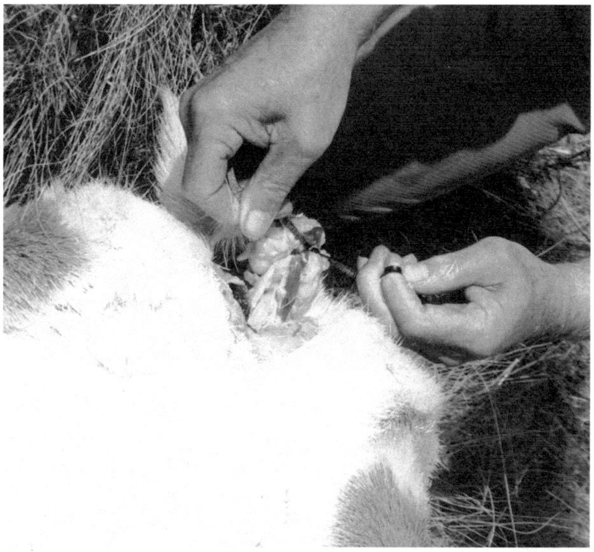

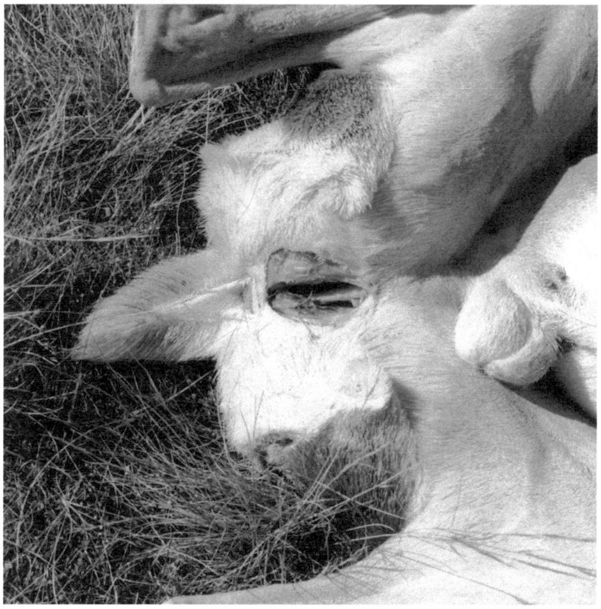

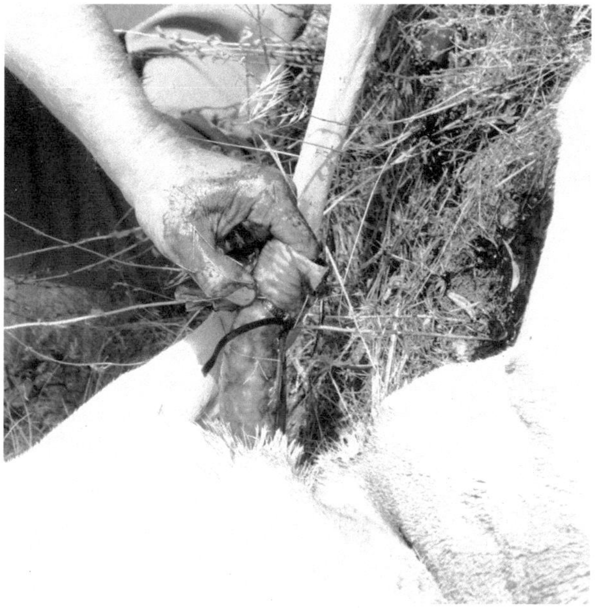

Skin the exposed side to the center of the back. Place the tarp against the exposed flesh, turn the animal over and skin the other side. When the carcass is skinned to the base of the head, pull and skin the hide forward, and remove the head.

The pronghorn stalker may have difficulty getting the animal to a road or getting a vehicle to the kill site, but a few tricks and a little preparation can minimize the work. In Montana, an eviscerated adult buck weighs about 90 pounds (41 kg) and a doe 80 pounds (36 kg). Skinning an animal and removing the head and feet reduces the weight by 15 to 18 pounds (7–8 kg). The hunter who plans to skin an animal immediately should carry meat bags and a pack frame while hunting, and air-dry the carcass for at least an hour before placing it in a bag. Even in snow, a pronghorn should not be dragged far because hair rubs off, the skin burns and sticks to the underlying meat, and some otherwise edible meat is likely to be spoiled.

For carrying meat and other parts from the field, some experienced hunters use standard size cotton pillowcases tied off at the top. These cases will absorb blood and are strong, each effectively holding a pronghorn quarter or 60 to 75 pounds (27.2–34.0 kg) of boned-out meat. They slide easily in and out of backpacks. They can be washed later and used on other hunts, although unaesthetic staining may preclude their future use on pillows. In any event, pillowcases should be used only to transport meat from the field. Their ply generally is not adequate for hanging meat to age, for which cheesecloth game bags are ideal.

The easiest way to carry an unskinned pronghorn for a considerable distance is to leave the brisket and pelvic

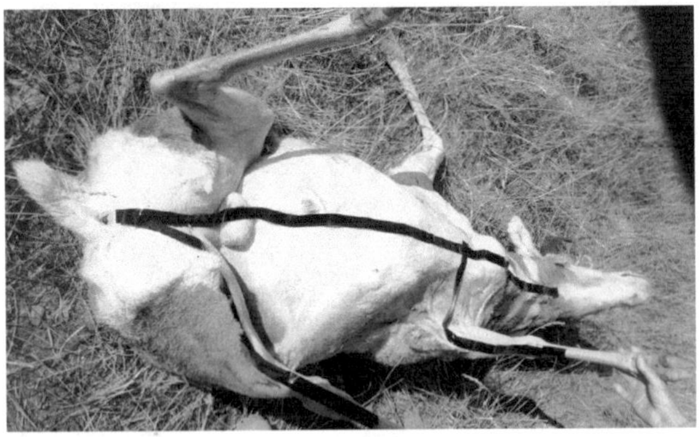

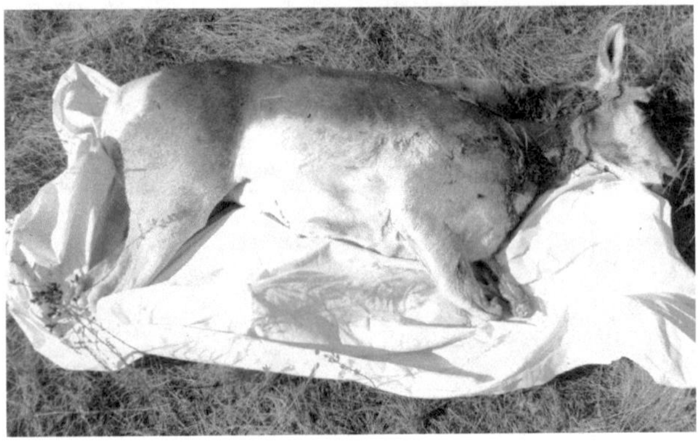

*Skinning a pronghorn on the ground involves the same first cuts as would be made on one hanging up. Black tape in the top left scene shows the location of cuts for skinning one side of the animal. Ordinarily, the abdomen would already be open as evisceration should be completed as soon after reaching an animal as possible. Removing the feet at the ankles and wrists is easier after skinning past them. Note that the tape indicates splitting the scrotum so one testicle remains with each hindquarter. If the head is removed, this will satisfy regulations that require evidence of sex remaining with a carcass. Meat can be kept clean up to this point without a tarp or sheet (*top right*), but turning a half-skinned animal onto the removed skin to complete the other side will result in dirt and hair contaminating some meat. Advisedly, a clean sheet is placed over the skin of a half-skinned pronghorn in preparation for rolling it onto the skinned side of the carcass (*bottom left*). The pronghorn then is rolled onto the sheet (*bottom right*) and ready for completion of skinning.* Photos by Bart W. O'Gara.

Skinning a freshly killed pronghorn in the field can be fast and easy if a vehicle and tree are available. Make the conventional cuts, remove feet, cut the neck skin from the head, skin the neck, and hang the animal in a tree by a rope around its neck. Push a large nail through the neck skin, tie one end of a heavy rope tightly around the skin behind the nail and the other to the vehicle. Then drive the vehicle slowly away, stripping the skin from the carcass. This method usually leaves less hair on the meat than does careful conventional skinning. Do not hitch the carcass to a fence post; it can break or be pulled over. Use a tree that is large enough to withstand the pull exertion. Photo by Bart W. O'Gara.

If a harvested pronghorn can be hung up, skinning is easier, and keeping the meat clean is less of a problem. Also, a hanging carcass cools rapidly, and a dry surface forms on the meat—the ideal situation to assure prime eating. However, trees are scarce on the prairie and, for skinning and cooling pronghorn, windmills often are the best place to hang carcasses, if one has permission from the rancher to do so. Left photo by Harold Umber; courtesy of the North Dakota Game and Fish Department, Bismarck. Top photo by Bart W. O'Gara.

Bagging a carcass in cheesecloth keeps meat clean and excludes flies and other insects without entirely restricting exposure to cooling air. Photo by Ed Bry; courtesy of the North Dakota Game and Fish Department, Bismarck.

Pronghorn carcasses should not be dragged, even for short distances. This animal was dragged about 50 yards (50 m) over grass and small rocks, and some skin already was rubbed bare. In another 100 yards (100 m), the meat under the bare spot would be nearly ruined and most of the hair would be gone from that side. Photo by Bart W. O'Gara.

girdle intact and make a pack of the carcass. Split the skin on the backs of the front feet and legs for 6 inches (15 cm) below and above the wrists and free the skin all the way around, from the underlying bone and flesh. Then disjoint and sever the wrists without cutting the skin holding the feet to the body. Cut holes through the skin of the hocks (just as for hanging an animal) and push the front hooves through the holes from the insides of the back legs. After the front feet are pulled through, they hold like buttons in buttonholes. The hunter can then wear the animal like a pack by slipping the right and left arms through the loops formed by the "buttoned together"

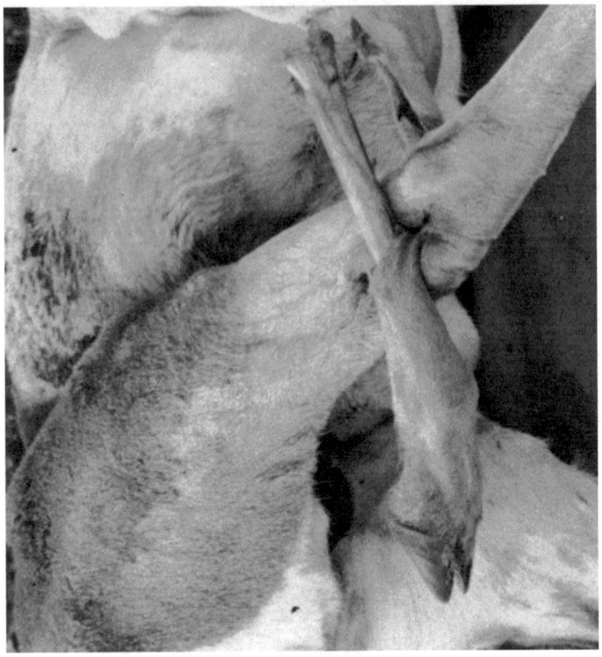

The easiest way to carry a pronghorn for a considerable distance is to wear it like a pack (top)*. After the animal is eviscerated and before it is skinned, bones of the front feet are severed and "buttoned" through slits on respective sides of the hocks* (bottom)*. A plastic trash bag or some other cover can be placed between the hunter and carcass to keep blood and hair off the hunter.* Top photo by Lu L. O'Gara. Bottom photo by Bart W. O'Gara.

right and left legs. I prefer to carry with the front legs over my shoulders so the head doesn't bounce against my legs, but some tall hunters have told me they prefer to carry with the hind legs over their shoulders and the head down. Placing a large plastic bag between the animal and the lower part of the hunter's back reduces the amount of blood that gets on clothes, and covering the carcass with an orange vest is a good safety precaution.

If flies come to a carcass, keeping them away until the meat air dries is a full-time job. Then, the carcass must be wrapped or bagged very carefully. Flies will crawl through seemingly impossible places to lay eggs on meat. If a carcass is not skinned, flies find pockets of moisture even when the surface of exposed meat feels dry to the touch. In warm weather, I carry a 0.5-pound (227 g) box of pepper while hunting. Sprinkled into an eviscerated carcass and on the hair at the edges of cuts, the pepper helps keep flies away. If flies do lay eggs on meat, they can be wiped off the next day with a cloth soaked in vinegar.

Even when meat feels cool to the touch, it is not ready to be transported anyplace except in the open air until the surface is dry. Never carry a carcass on a hood or fender next to the engine or other hot spots. (Pickup truck boxes, trunks and floors of vehicles often have hot spots over tailpipes or transmissions.) Never carry a carcass in the trunk of a vehicle on the day the animal was killed, except in an ice chest. A cartop carrier is best for transporting

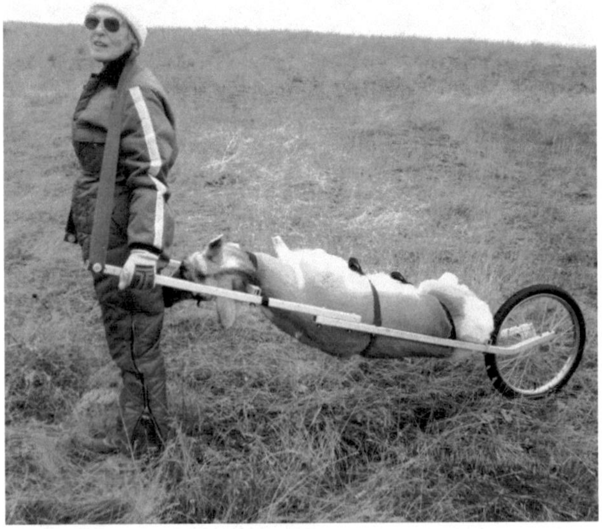

A wheeled carrier is an efficient and relatively easy device for getting a pronghorn carcass to a road. Such carriers are easy to load because they are low to the ground. A strap in front of the hunter's shoulders and behind the neck takes most of the weight off of the hands. Pulling the carrier is easiest on level ground or going up hills. Going down hills, pushing the carrier like a wheelbarrow allows better control. This scene shows a 105-pound (47 kg) great-grandmother, Wilma O'Gara, easily transporting a pronghorn doe that weighed 80 pounds (36 kg) with its viscera and feet removed. Photo by Bart W. O'Gara.

Horses were used by western pioneers and Native Americans to transport pronghorn from the field, as shown in this scene from 1888. Today, horses seldom are used in pronghorn hunting or transport because most pronghorn habitat is accessible by motor vehicles. Horses usually would be more trouble than benefit under the circumstances. Photo by Grabill; courtesy of the South Dakota State Historical Society, Pierre.

fresh meat. When a carcass is carried inside a vehicle or pickup canopy cover, windows should be open so that air circulates around the propped-open carcass. The point is—and it cannot be overemphasized—too much good meat is spoiled by improper cooling. Also, when transporting carcasses, do not prop up the kill on a vehicle. Antihunting sentiment can be fueled by insensitive or indiscreet displays.

For most red meat, a week or two of hanging is recommended. However, the fine-grained meat of pronghorn is tender when fresh and aging is not required, although 2 or 3 days of hanging in a cool place are appropriate.

PRONGHORN VENISON

Like the little girl of nursery rhyme fame, when pronghorn meat is good, it is very, very good, but when it is bad, it can be very, very bad. . . . Usually pronghorn have fine-grained, tender meat with excellent flavor, yet some are nearly inedible—even though the meat has been cared for and cooked properly. No one has adequately explained why the flavor varies so.

O'Gara (1980:32)

It often is said that shooting running pronghorn or having to chase down a wounded one results in venison tainted by adrenaline or some other biochemical element. However, I have eaten such animals that were quite good. Conversely, I once shot a buck in its bed and the animal's meat was unpalatable.

The amount of edible, boned meat salvageable from a pronghorn carcass is difficult to estimate. Field et al. (1972) indicated 36 pounds (16 kg) of packaged pronghorn meat without bones—47% of field-dressed weight. That is a reasonable average. When I boned out a very fat doe with a carcass weight of 82 pounds (37 kg), I got 41 pounds (19 kg) of meat—50% of carcass weight. A thin buck, just out of the rut, with a carcass weight of 105 pounds (48 kg), yielded 64 pounds (29 kg) of meat—61% of carcass weight. The fact that I retrieved a higher percentage of the carcass weight probably was a result of my animal being head shot and me boning very carefully, including meat between ribs and around neck bones. Most hunters or butchers would not bone so thoroughly, and most shots would result in some loss of meat. The large difference in my percentage return between the fat doe and thin buck undoubtedly resulted from very careful trimming of all fat from the meat. Field et al. (1972) reported that 44.2 pounds (20 kg) of packaged cuts with bones left in legs, loins, ribs and shoulders came from 76.7-pound (34.7 kg) carcasses—58% of carcass weight.

How a pronghorn carcass is cut depends on the hunter's preferences and the equipment available. For a person who has a bandsaw and wants bone-in roasts, ribs and chops, following a standard butcher's guide for cutting beef is suggested. Many hunters prefer to bone-out most of their big game, thus reducing the amounts of bone sawing and freezer space required. A sharp knife, saw (a meat saw about 15 inches [38 cm] long is handiest) and a cutting board are all the equipment required.

For processing, a pronghorn carcass can be hung by its hocks in a cool place, such as a garage or shed. In this manner, meat portions (cuts) can be removed for further processing, and the rest of the carcass can be left hanging in a cool place. Wiping the surface of the tenderloins and insides of the flanks and ribs with a cloth soaked in vinegar and water (a 50% solution) removes blood and dirt from these surfaces.

The flanks of a pronghorn carcass should be removed soon after an animal is skinned; they can be used for stews, sausages or burgers. If a skinned carcass is hung for more than a day, the flanks may dry beyond edibility. Even with the skin on, most of the flank meat will be lost after a couple of days of hanging. Sometimes, flanks are used as rolled roasts, but that is not advisable with pronghorn meat because the fat will congeal quickly with cooling, which will result in a gummy or waxy texture when chewing. I use an ice pick to hold flanks to my cutting board while removing layers of meat, fat and fascia.

Tenderloins are muscles that are about a foot (30 cm) long. They are found inside the abdominal cavity, connected to the underside of the backbone, anterior to the pelvic girdle. They should be cut from the backbone very carefully because the tenderloin is the most tender meat on the animal. To avoid drying, tenderloins should be removed before the carcass is aged. Tenderloins can be cooked whole, as steaks. They can be eaten the day of the kill and are ideal for camp meal.

The shoulders of a pronghorn carcass, with arms and shanks attached can be removed easily; no saw is required. If one grasps the shank and pulls it away from the brisket and rib cage, muscles and connective tissue are easily torn. The muscles start at the base of the neck, extend back to the edge of the cartilage on the tip of the shoulder blade and go on to almost the center of the back. Blood often is forced by the projectile into the connective tissue between the shoulders and ribs; it should be trimmed and scraped away. The pronghorn shoulder blade can be removed from the arm by simply cutting them apart at the joint. The joint is separated most easily by sticking the point of a knife into the space between bones to start the separation. After trimming the fat and rough edges, a shoulder blade roast is ready for the freezer or oven. Boning shoulder blade roasts for any other culinary purpose is a lot of work.

The leg roast now can be removed from the shank by cutting above the joint with a knife and sawing the round bone; disjointing without sawing is difficult. The arm roast is not as preferable a cut of meat as the shoulder blade. I prefer to bone the arm roast without removing it from the shank, keeping the best meat for stew or Swiss steak and putting the pieces with considerable connective tissue into sausage or burger. Although shank meat is tasty, the amount of trimming of large tendons is considerable.

The neck is easily sawed from a pronghorn carcass. It makes a good roast but usually requires considerable trimming because blood and hair adhere to the ventral notch from which the trachea and esophagus were removed. Also, it is not a roast that can be sliced; instead, the roasted meat is pulled from the bones in strips. Many bone the neck for sausage or ground meat.

Ordinarily, the brisket is split down the middle by this point. If not, split it now with a saw. Then, cut or saw it down each side. Or, each half brisket can be left with the ribs until after they are removed.

Saw the ribs from the backbone, leaving about 6 inches (15 cm) attached to it. After a side of ribs is removed from the pronghorn carcass, the half brisket can be removed with a knife. The cartilage is cut easiest where the bony ribs end. Ribs can be boiled in a slow cooker to tenderize and remove fat, preferably covered with tomatoe or vegetable juice, and simmered until the ribs easily pull out of the meat; then the ribs can be barbequed. They also can be boned for stew, sausage or ground meat, but considerable trimming of fat is necessary.

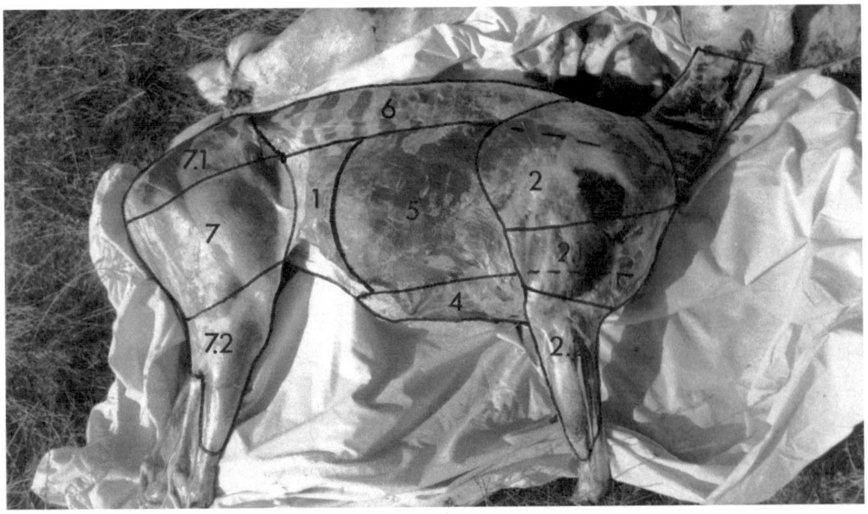

The lines and numbers superimposed on this photo of a pronghorn carcass show the proper cuts and their logical order of removal. Photo by Bart W. O'Gara.

Backstraps, or loins, are located on each side of the backbone from the pelvic girdle to the neck. Remove the backstraps very carefully by first cutting across, next to the pelvic girdle, then down each side of the vertical spines of the backbone and along the lateral spines or ribs. A little more knifework is required as the strap is peeled from the pelvic girdle forward. The loosely attached and somewhat sinewy meat near the neck end is good for stews. Most of the backstrap can be cut into excellent loin chops. After the backstraps are removed, find a double space (cartilage) between where the spine meets the pelvic girdle. Stick the point of the knife into the cartilage and cut to the left and right. This will sever the spine; it can be removed from the pelvic girdle by cutting around neural spines. Bone any meat remaining on the backbone for stews, sausage or ground meat.

The hams and rump roasts compose a solid chunk of meat that should be removed from the hanger for further processing. If more roasts are desired, saw through the back of the pelvic girdle, so half is attached to each ham. With the ham lying flat and the cut portions of the pelvic girdle facing up, locate the ball and socket joint between the girdle and the femur (round bone in the ham). Stick the point of the knife into the joint, locate the tendon that holds the ball into the socket, sever the tendon and cut the ham from the rump roast. The end of the femur projects farther toward the pelvic girdle than does the ball. Thus, a straight cut cannot be made until after cutting around the head of the femur. If one is not handy with a knife, first the muscles can be cut with a knife, and then the bone can be sawed through at or below the hip joint. A low cut results in a larger roast and less steak. Cutting and sawing across the rump roast results in two smaller roasts. If roasts are not desired, the hams can be removed from the pelvic girdle without sawing it in two. The meat boned from the pelvic girdle produces stew meat or small steaks. For either, the muscles should be separated, fat and connective tissue should be trimmed and muscles should be cut across their grain. After the hams are removed from the pelvic girdle, rear shanks can be removed by disjointing or sawing just above the joint. I usually freeze hams whole, only minimally trimming to remove the fat and any bloody or dirty meat. This minimizes freezer burn to steaks and allows me to decide how much meat—steak, Swiss steak, stew meat, roasts—I want at a later date. Removing a frozen ham from the freezer and placing it in a refigerator for 2 days usually allows it to soften enough for cutting, but the partially frozen meat can be refrozen. If a 5- to 8-pound (2–3 kg) roast is desired, a whole ham serves that purpose.

When cutting hams, dissecting the muscles from each other separately is much better than cutting across the whole ham. This ensures that all of the steaks are cut across the grain and allows trimming of the ends and their connective tissue into sausage or burger meat. The muscle mass (oval or football-shaped) at the front of the bone usually is called the sirloin tip. It contains prized meat but has layers of dense connective tissues that are tough in fried or boiled steaks. The whole mass can be used as a boneless roast, I prefer to slice it across the grain into Swiss steaks and stew meat and remove connective tissue from each piece. The hind shanks should be handled like those of the forelegs.

For additional information about dressing and skinning big game, I suggest reading Schmidt (1978) and Stransky (1984).

When freezing game meat, as much fat should be removed as possible and air should be excluded from the meat. One sheet of the thin plastic wrap that clings, followed by an outer wrap of freezer paper is good. I sometimes put meat in Ziploc® bags and immerse them in water to remove air before sealing the final corner of the bag. Small vacuum packers are available; they remove air from a plastic pouch containing meat and seal the pouch with heat. Such packing greatly increases freezer life of the meat and reduces time spent wrapping meat.

Pronghorn meat should be handled and cooked like good beef. Pronghorn chops, when fried fast on a very hot skillet with a small amount of grease until pink in the centers and well browned on the outsides, are delicious. Livers are much like young calf livers—best if fried very fresh and slightly rare.

For some reason, ground pronghorn makes better meatloaf than does ground beef. Small meatloaves require nearly as much time to prepare as do large ones and leftovers freeze well or make good sandwiches with mustard or other condiment of choice, plus a lettuce leaf. Thus, small meat loaves hardly seem worth the time. To prepare a sizeable meat loaf that will serve 10 to 12 people, one needs:

1. 3 pounds (1.4 kg) ground pronghorn,
2. 1 to 1.5 pounds (0.45–0.68 kg) ground pork,
3. one 4-inch- (10-cm-) diameter onion finely chopped,
4. 32 ounces (4 c.) slightly crushed saltine crackers,
5. two extra large eggs,
6. 15-ounce (425 g) canned, peeled, crushed tomatoes,
7. 0.5 teaspoon (0.7 dr) each fennel seeds, ground fennel and rosemary leaves,
8. 0.25 teaspoon (0.35 dr) each of ground allspice, oregano, thyme and white pepper;
9. 15 ounce (22 g) of brown gravy mix, and
10. olive oil.

To prepare:

1. Barely cover the bottom of a deep, 10-inch (25 cm) skillet (with a tight-fitting lid) with olive oil.

2. Beat eggs vigorously with spices in a small mixing bowl.
3. Place ground pork and ground pronghorn in a large—at least 1 gallon (3.8 liter)—mixing bowl.
4. Mix meats with a twisting motion of fingers and thumbs as well as by turning over repeatedly and kneading vigorously.
5. Add eggs with spices, onion and crackers, in that order, kneading after each ingredient is added.
6. Heat skillet and oil over medium heat, mold ingredients into a ball, place in skillet, pat down and use a narrow spatula to create a 0.5-inch (1.3 cm) space between loaf and sides of skillet.
7. Pour tomatoes over top of loaf, distributing evenly before pushing numerous holes from top toward bottom nearly through loaf and smoothing tomatoes back and forth to fill holes.
8. When oil starts to boil next to sides of skillet, turn heat to medium low and cover.
9. Adjust heat so oil bubbles up in three or four spots around loaf, and allow to cook about 2.5 hours.
10. Turn off heat and leave skillet on burner until cool.
11. With a nylon knife, cut the loaf into quarters and place them on a metal sheet, such as a pizza pan.
12. Place the four quarters under a broiler until the tops are slightly brown.

The meatloaf now is ready for use in sandwiches, freezing or heating in a microwave as a main course. Gravies add flavor to the meat and to potatoes served with it. Pot liquor left in the skillet needs thickening; a simple way to do so follows.

1. Combine brown gravy mix with 2 cups (500 mL) of cold water in a jar with a tight-fitting lid and shake vigorously.
2. Remove solidified pork fat (if any) from skillet, place skillet over medium heat, add gravy mix, bring to a boil stirring constantly, reduce heat to low and stir for 1 minute to thicken.

This produces excellent gravy for potatoes. For meat loaf, combine in a small kettle:

1. 1 cup (250 mL) of the above gravy,
2. 0.25 cup (62.5 mL) of ketchup,
3. 4 tablespoons (16 dr) of brown sugar,
4. 1 teaspoon (1.5 dr) of dry mustard.

Simmer over medium low heat, stirring until ingredients are combined. Taste and add brown sugar or ketchup to satisfy personal preference.

When preparing a meal, slice meatloaf before heating. With a microwave, meat and gravies can be hot in 10 minutes or less. Mashed potatoes (from scratch or instant) and white shoepeg corn go well.

Pronghorn round steaks are rather dry, but they can be tasty with salt, pepper and flour pounded into them before frying. They should be cut about one-third thinner than chops. Health conscious people and weight watchers cringe at the idea of frying meat. However, fried pronghorn apparently contains less fat than grilled beef, lamb or pork (see Chapter 8), and frying with vegetable fat—such as canola, olive or safflower oils—is healthier than frying with animal fats. Not using flour or other breading reduces the amount of fat consumed with fried meat.

The very best fried pronghorn is tenderloins. These tender muscles can be salted, peppered, and cut in half to fit a 8-inch (20-cm) skillet with 2 to 3 tablespoons (8–12 dr) of very hot (slightly over medium setting) olive oil. When they are browned on all sides, they should be red to pink in their centers and ready to eat. They go best with baked potatoes, a garden salad with vinaigrette or similar dressing and a glass of red wine.

One of the penalties of frying pronghorn steaks for their best flavor is cleaning splattered grease from the stove. Frying screens the right size for the skillets used alleviates this problem somewhat. Also, Teflon™ skillets with nylon forks, spoons and spatulas are almost a must.

For those who like fatty meat and do not worry about cholesterol, pounding ground beef suet into the round steaks makes them difficult to distinguish by taste from beef.

Being fine-grained and tender, pronghorn venison serves well in Italian recipes that ordinarily call for veal. An easy but delicious Italian-like dish that I call pronghorn parmesan requires the following ingredients:

1. 1 pound (0.45 kg) of pronghorn sirloin tip sliced about 0.5 inch (1.3 cm) thick with all tendons and other connective tissue trimmed away,
2. 6 ounces (737 g) of pasta sauce,
3. 1 cup (250 mL) of shredded Parmesan cheese,
4. 5 ounces (142 g) of sliced Swiss cheese cut in 2-inch (5 cm) squares and two slices thick,
5. 10-ounce (283 g) of tri-colored rotini,
6. 4 tablespoons (16 dr) of olive oil, and
7. 0.25 cup (62.5 mL) of flour.

To prepare:

1. Preheat oven to 375°F (190°C).
2. Cut venison into bite-sized pieces and shake in a bag with flour.
3. Preheat olive oil in a 10-inch (25 cm) skillet and fry venison bits about 5 minutes per side, turn with nylon fork and spatula, ensuring both sides are browned.

4. Pour pasta sauce into a 10-inch (25 cm) diameter, oven-proof dish or pan that is at least 2 inches (5 cm) deep, add Parmesan and stir.
5. Distribute bits of fried meat on top of pasta sauce and tap down so they are covered;
6. Place squares of Swiss cheese on top and place on middle oven rack; cook for 15 minutes.
7. Bring water in a 1-gallon (3.8 liter) pot to a rolling boil, add rotini, reduce heat to medium and stir to be sure rotini is not sticking together or to bottom of pot.
8. Adjust heat so water boils gently, stir every 2 or 3 minutes, turn off heat after 12 minutes and leave pot on burner until pasta is needed.
9. Turn oven to broil for 3 minutes to slightly brown Swiss cheese, turn off oven and leave oven door slightly open.
10. Drain pasta thoroughly and place in a heated bowl.
11. Remove main dish from oven.
12. Dish pasta onto heated plates, cover with pasta sauce, cheeses and meat.

Serves four people or up to six with additional pasta.

Roasts are improved by the addition of the pot liquor, and commercial natural pronghorn juices or onion soup mixes add both moisture and flavor. My favorite pronghorn pot roast requires:

1. a shoulder blade roast, fat trimmed,
2. enough olive oil to barely cover the bottom of a 12-inch (30 cm) skillet or roaster equipped with a heavy lid that retains steam,
3. coarse ground black pepper in an airtight container—pepper in a shaker becomes nearly tasteless in a short time,
4. salt,
5. one 3.5-inch- (8.5-cm) diameter onion, cut into thin slices,
6. 2 cups (500 mL) of baby carrots, and
7. 3 cups (750 mL) of peeled red potatoes cut into slices no more than 1 inch (2.5 cm) long or wide and about half that thick.

The objective is to sear the meat on both sides and then cook at low heat in enough liquid that the roast is constantly steamed. To do that, place the skillet or roaster and oil on a burner set on medium heat. Liberally salt and pepper the concave side of the roast. Place it—concave side down—in the hot oil, salt and pepper the top and cover. Check in 10 minutes and, if the bottom is brown, turn the meat with a tongs, place the onion slices on top and cover again. If the bottom is not brown, wait until it is before turning and putting the onion in. When both sides are brown, turn the roast again, scatter the onion slices on the bottom of the pan, cover and turn heat to low. After about 5 minutes, add 2 cups (500 mL) of hot water. Regulate heat so the liquid simmers or barely boils, and check every 30 minutes that it does not steam away. Add water if needed to maintain a constant level; some lids maintain liquor levels, others do not. Turn the roast hourly. Usually, the meat pulls away from the bone by the second hour. If so, add the carrots, distribute them in the liquid, place the meat on top of them, cover and slightly increase heat. If meat did not pull away from bone in 2 hours, turn the roast, increase heat slightly and check every half hour. When the meat pulls away, proceed with the carrots as above.

Fifteen minutes after adding carrots, add potatoes and distribute them about so all are in liquid. Again, the meat should be on top of the vegetables. Turn the heat up to between medium low and medium and leave the lid off, letting about half the pot liquor evaporate. The potatoes should be done in 30 to 40 minutes. When they are tender, the meal is ready. This is not the kind of roast to slice neatly in large slices. Much of the meat pulls off in sheets or strips; only the largest muscles need slicing. Placing pot liquor on the plate and mashing potatoes in it as one eats produces great flavor if the roast was well browned before simmering. The onion nearly disintegrates but adds taste.

From start to finish, preparing such a roast takes about 4 hours. Often, I prepare one while I am doing something else near the kitchen. When the meat loosens from the bone, I shut off the heat and let everything cool on the stove. After removing the cooled meat, I discard the bone, slice some of the meat for sandwiches and cut the rest into bite size pieces, to mix with the pot liquor. Then, whenever desired, carrots and potatoes can be added and a meal prepared in about an hour.

Leftovers of a pot roast can be used in many ways. Sandwiches with meat, mustard and lettuce are tasty and handy. Leftover carrots, meat, peas and potatoes in pot liquor are good topped with biscuits. This requires:

1. 1 cup (250 mL) of baking mix,
2. 0.75 ounces of brown gravy mix,
3. milk, and
4. 1 cup (250 mL) of water.

To prepare:

1. Preheat oven to 425°F (218°C).
2. Combine brown gravy mix with cold water in a small jar with a tight-fitting lid and shake vigorously before heating as directed on package.

3. Mix gravy, meat, pot liquor and vegetables in a 10-inch (25 cm) oven-proof dish.
4. Stir just enough milk into baking mix to produce a very stiff paste.
5. Dot surface of meat and vegetables with heaping teaspoons of dough.
6. Place on top rack of oven for 15 minutes.

Sliced tomatoes may be added.

Smoked pronghorn jerky makes excellent hors d'oeuvres and provides lightweight protein for hunting trips or backpacking. Also, jerky can be prepared from a pronghorn whose meat is not especially good. Only choice cuts should be used; sinewy meat makes unchewable jerky and serves better for salami or wieners. Directions for preparing jerky are as follows:

1. Cut the hams or loins into 1- by 2-inch (2.5 by 5 cm) strips with the grain of the muscle—the longest lengths each muscle allows.
2. Prepare brine in a granite or crockery roaster or bowl using
   a. 1 cup (250 ml) of Sugar Cure™,
   b. 2 tablespoons (8 dr) of Morton Tender Quick™
   c. 1 gallon (3.8 liter) of water. (If Sugar Cure™ is not available, make the brine with 0.5 cup (125 mL) of table salt, 0.5 cup (125 mL) of brown sugar and 0.5 teaspoon (0.7 dr) of ground black pepper and allspice.)
3. Soak meat in the brine l2 to 14 hours.
4. Remove meat from brine, rinse lightly in cold water and drain for 1 to 2 hours.
5. Smoke meat for 8 to 10 hours in medium-hot smoke. I use a small commercial electric smoker, set in a large cardboard box (to increase the heat), and add three pans of wood chips at 2- to 3-hour intervals.
6. When the meat comes from the smoker, the pieces are rubbery and bend easily; place on cookie sheets to air-dry for one or more days until drying is complete. As soon as the meat is dry enough to crack when bent, place in plastic bags or jars and store in the refrigerator.

Such meat will keep for a year when refrigerated and 2 to 3 months in a backpack. When serving hors d'oeuvres, cut it into thin slices across the grain of the meat.

TROPHY HUNTING AND CHOOSING A TROPHY HEAD

People hunt pronghorn for many reasons—an outdoor experience, recreation, meat and trophies. The Boone and Crockett Club is known mostly for its record keeping of big game trophies. However, it is a conservation organization that, among other things, promoted trophy hunting in an effort to upgrade North American hunting to an honorable sport and conserve big game at a time when commercial hunters and poachers were threatening some species with extinction. The Pope and Young Club is an archery organization patterned somewhat after the Boone and Crockett Club. Although some preservationists decry it, trophy hunting is the acme of sport hunting, which is now an integral part of most wildlife management programs. These programs are, in fact, the salvation of many wildlife populations—game and nongame—because money derived from hunting licenses and taxes on hunting equipment pays for the protection, management and research necessary for conservation of many species and their habitats. Sport hunting also makes game populations valuable to local people who hunt or derive income from hunters. One of the greatest forms of protection that can be given to a wildlife population is the interest of nearby residents in its welfare.

Unfortunately, some trophy hunters are carried away with a desire to possess the biggest and best of everything, which has led to illegal excesses—poaching in parks, buying heads and so on. These practices generally have had little importance biologically, at least with respect to pronghorn, but they unfortunately have tarnished the image of the law-abiding trophy hunter.

The official scoring system of the Boone and Crockett Club is the standard method for determining excellence of a trophy legally and ethically taken by firearm. Considered in scoring are: lengths of the main beams around the outside curves; circumferences of the first, second and third quarters of the horns; lengths of the prongs; and symmetry between right and left horns. Differences in individual measurements between horns are deducted from the total score. Before 1995, if the spread was greater than the length of the longest horn, the difference was subtracted. That was changed and is retroactive for all trophies affected by this penalty. Trophies with horns at unusual angles, making it difficult or impossible to measure the inside spread, will be treated on a case-by-case basis (Reneau 1995).

Eighty-two Boone and Crockett points are necessary for a place in the Boone and Crockett's *Records of North American Big Game*, and 64 points are required for entry into the Pope and Young *Bowhunting Big Game Records of North America.* With one exception, Pope and Young uses the Boone and Crockett scoring system for measuring pronghorn trophies: on horns with exceptionally high prongs—perhaps 2% to 3% (P. Wright personal communication:1994)—Boone and Crockett allows three circumferences to be taken below the prong; Pope and Young allows only two.

Taxidermists and state and provincial wildlife agency personnel usually can tell a hunter how to contact the

Trophy pronghorn whose horn spread is greater than the length of the longest horn would have lost points by the Boone and Crockett scoring system before 1995. Many hunters like unusual horns because they add variety to a trophy collection, and widespread horns (opposite top) *often are very attractive. Tipped-ahead horns are atypical but not uncommon, and they may be particularly desirable to hunters who have a typical horn mount* (opposite center)*. Sometimes, these forward projecting horns are nearly parallel to the animal's nose* (opposite bottom)*. Also, some tipped-ahead horns are very large.* Opposite top photo by Tom Pojar. Opposite center photo by Len Clifford; courtesy of Leonard Rue Enterprises, Blairstown, New Jersey. Opposite bottom photo by Cathy and Don Jenni.

nearest official measurer. Hunters also can write directly to the Boone and Crockett Club Headquarters in care of the N. A. B. G. Awards Program, Old Milwaukee Depot, 250 Station Drive, Missoula, Montana 59801-2753, or to the Pope and Young Club, P.O. Box 548, Chatfield, Minnesota 55923.

Choosing a good pronghorn head is not easy. No other horned or antlered species, except for the Rocky Mountain goat, has such a narrow range for entrance into the Boone and Crockett Club records and the World's Record (Wright 1996). With the entry level at 82 and the World's Record at 93-4/8 points in 1996, the difference was only 11-4/8. Such differences are not easily determined at 200 to 300 yards (183–274 m) under varying light conditions. The most desirable heads have long, massive, symmetrical horns with large prongs. Thin horns of about 13 inches (33 cm) with small prongs are common on 2-year-old bucks. They are hardly trophies per se, but to the untrained eye, they look good in the field. Heavy horns of 14 inches (35 cm) are decent trophies; 15-inch (38 cm) horns are excellent, and anything over 16 inches (41 cm) probably are record class, provided they are heavy—about 7-inch (18 cm) basal circumference—and the prongs are sizeable—6 inches (15 cm) or more.

The horn length is measured from the lowest point on the outside of the horn (Wright 1996). This can be anywhere along the outside bottom from the front edge to the rear edge. A piece of tape generally is placed horizontally below the base of the horn sheath, just touching the lowest point of that sheath. Another piece of tape placed at a right angle to the first and ascending the horn locates the starting place to measure horn length. Mounted heads may have fiberglass placed around the horn bases so that the lowest point on the horn cannot be located. Such heads will lose some score.

If the horns were straight, judging them would be easier, but the shape of the hook at the top makes a big difference in measurements, and the trophy hunter must consider the total conformation of the horns. Horn shape,

Prehunt indoctrination sessions were held on the Sheldon National Wildlife Refuge, Nevada, and the Hart Mountain Antelope Refuge, Oregon, by the U.S. Fish and Wildlife Service to promote quality pronghorn hunts. This photo shows an instructor demonstrating the appearance of a buck with a certain size of horns at various distances. Photo by Jim D. Yoakum.

An example of a very long set of swoop horns taken during 1944 in New Mexico. Bill Conn, the hunter, saw other pronghorn in the area with horns tipped ahead, indicating possible genetic influence. Photo courtesy of Bill Conn.

inclination and curvature vary greatly; the horns usually form a V upward from the head and then curve in toward each other or to the rear before hooking downward. The serious trophy hunter should look at as many mounted heads as possible, try to judge them by eye, and then measure them to check his guesses. Most horns are deficient in one way or another. For instance, bucks frequently have good bases and prongs topped by small hooks. Only by looking at many heads and pictures of heads can a pronghorn hunter get a feeling for conformation.

Hunters interested in first-rate trophies should use spotting scopes to check heads at a distance and should memorize the following measurements: an ear is about 6 inches (15 cm) long, a buck's head is about 13 inches (33 cm) from the tip of the nose to the back of the head and the distance from the tip of the nose to the corner of an eye is about 8 inches (20 cm).

First, look for a good prong. If the length from the back of the horn to the tip of the prong is equal to the length of an ear, the prong is excellent. The curve of the hook makes the length of a horn greater than the distance it stands above an animal's head. For example, a fairly straight horn extends about 2 inches (5 cm) longer than its height, whereas a sharply recurved horn might measure 5 to 6 inches (13–15 cm) longer than its height. Well-shaped horns standing nearly as high as the length of the head should be record class, and even straight ones of that size are good trophies. Prongs differ in their locations along the lengths of horns, but in really good trophies, tips of

A trophy hunter should not wait until the very end of most hunting seasons to take a desirable animal because bucks often shed one or both horns before that time. Photo by Mike Francis.

A trophy is in the eye of the beholder. Any mature buck, fairly stalked and humanely killed, is a grand trophy. Photo courtesy of the Wyoming Department of Game and Fish, Cheyenne.

the prongs are about the nose-to-eye length above the eyes. Lengths of hooks are difficult to judge, but if nearly ear-length prongs stand a nose length above the eyes and the hooks are large enough to balance, it is an excellent trophy and probably will qualify for the record book.

Maximum horn development depends to varying degrees on genetics, year-round nutrition (which differs year to year with weather conditions), minerals and age. Generally, 3-year-old or older bucks carry the largest horns, but even yearlings sometimes have horns up to 15 inches (38 cm) in length.

The trophy records are interesting and serve as standards for comparison. Many of the highest scoring heads really are freaks of nature, however, and hunters should not feel that heads that fail to "make the book" are second-rate trophies. An average buck—hunted on foot, stalked fairly and dropped painlessly with one shot—is more of a trophy than the largest head in the world, if taking the latter involved pure luck or an unsporting hunt. I keep small and medium-sized horns and antlers that remind me of enjoyable and exciting hunts. A trophy head is in the eye of the beholder; a trophy experience is forever in the hunter's memory.

Choosing the best state in which to hunt trophy pronghorn is a problem. A state-by-state breakdown of record-class heads is provided in Table 171. Wyoming is a good choice because of the large numbers of animals and nonresident licenses. During the 1980s, hunters could get licenses for more than one pronghorn in Wyoming, but only one for an adult buck. Until 1993, a hunter could get five doe/fawn tags in the same area as the any-pronghorn permit. The severe winter of 1992–93 reduced numbers considerably. As of 2000, one either-sex and two doe/fawn permits were available throughout the state (W. Rudd personal communication:2000). Wet summers and mild winters or droughts and severe winters probably will dictate larger or smaller harvests in the future. Northern pronghorn are numerous and comparatively large bodied. However, although some good trophies are available, only a very small percentage are of record class.

For the trophy hunter seeking an especially large set of horns, other areas should be considered. Jack O'Connor, who may have hunted and observed as many game animals in North America as anyone in the 20th century, wrote that he would concentrate his efforts in northern Arizona if he were seeking a really large pronghorn trophy (O'Connor and Goodwin 1961). The record books support him. Although comparatively few pronghorn are harvested in Arizona, the four biggest heads listed in *Records of North American Sheep, Rocky Mountain Goats and Pronghorn* (Reneau and Reneau 1996) were killed there. Three heads tied for fifth place—one each from Arizona, Colorado and Montana. This did not change with publication of the *23rd Big Game Awards 1995–1997* (Byers 1998).

Table 171. Boone and Crockett Club-scored trophy pronghorn by province and state

		Number scoring at least this total									
State/province	*Largest entry*	*90*	*89*	*88*	*87*	*86*	*85*	*84*	*83*	*82*	*Total*
Wyoming	91	4	6	5	11	34	54	100	109	162	485
New Mexico	90-6/8	2	7	9	13	13	19	29	53	68	213
Arizona	93-4/8	8[a]	5	13	10	27	22	28	45	43	201
Nevada	91	1	5	7	7	7	7	22	23	34	113
Montana	91-4/8	1	1	2	2	7	10	13	22	34	92
Oregon	90	1	—	—	2	4	10	14	20	14	65
Texas	90-4/8	1	2	2	1	3	3	11	15	21	59
Colorado	91-4/8	1	1	—	—	6	—	6	19	16	49
California	89	—	1	—	2	1	1	11	5	14	35
Utah	89-4/8	—	1	1	1	1	4	4	6	8	26
Alberta	86-2/8	—	—	—	—	1	2	3	7	11	24
Idaho	86-6/8	—	—	—	—	2	2	3	4	7	18
South Dakota	86-6/8	—	—	—	—	2	—	5	5	6	18
North Dakota	86-4/8	—	—	—	—	1	2	1	1	9	14
Nebraska	85-6/8	—	—	—	—	—	2	4	2	2	10
Saskatchewan	86-4/8	—	—	—	—	1	1	—	1	2	5
Kansas	84-4/8	—	—	—	—	—	—	1	1	1	3
Chihuahua	85-6/8	—	—	—	—	—	1	1	—	—	2
Oklahoma	85-6/8	—	—	—	—	—	1	—	—	—	1
Total		19	29	39	49	110	141	256	338	452	1,433

[a] Final score on one trophy subject to revision by additional verifying measurements. Provisional score 90-4/8 (Byers 1998).

Source: From Reneau and Reneau 1996, Byers 1998.

Game ranches on which ranchers receive landowner licenses from state conservation agencies and resell them to hunters (or who can designate to which hunters the state agency sells permits for animals on their ranches) and some Indian reservations provide splendid opportunities for trophy hunters (see Chapter 20). Game ranches may be fenced to prevent big game from leaving them. Pronghorn shot on such ranches are not considered to have been taken in fair chase and are not eligible for entry into the record books. Ranchers who receive landowner licenses usually operate livestock ranches that are not surrounded by game-proof fences. Thus, hunts there are fair chase, and very fine pronghorn trophies are available in the southwestern United States. Ranch hunts often are expensive, but guide service usually is included, and hunters enjoy high success rates.

The head of a pronghorn shot in Arizona by an unknown hunter in the 19th century was considered the world's largest until 1979 when a hairline crack was discovered in one of the horns. Subsequent investigations, including x-rays, revealed that nearly 2 inches (5 cm) of horn material had been added to the bases of both horns to make them longer (Nesbitt 1981). According to Philip L. Wright, Chairman of the Boone and Crockett Club Records Committee (personal communication:1980), the horns had been skillfully built up with a series of horn pieces inlaid around the bases with wire brads holding the taxidermist's handiwork in place. The additions were carefully concealed with paper-mâché and lampblack. As altered, the head scored a phenomenal 101-6/8 points. Another Arizona pronghorn thus became the world's record with a score of 93. Alterations to the former world-record horns probably were performed simply to make the finished mount look better. The taxidermy work was done years before an official scoring system or ranking of trophy heads came into being. Thus, the taxidermist was not deliberately trying to build a world-record head. If the additional material had not been added, the head still may have been the world record. Perhaps the modification was an attempt to make the horns look more normal. The prongs were unusually low, even with the added material (Nesbitt 1981).

The present world-record pronghorn scores 93.5 Boone and Crockett points and was shot in Coconino County, Arizona, during 1985 by Michael J. O'Haco. The horns are 17.75 and 17.5 inches (45.1 and 44.5 cm) long, with circumferences at the bases of about 7 inches (17.8 cm) and 8 inch (20.3 cm) prongs. The horns of the number two head, also from Arizona, are slightly more than 18 inches (45.7 cm) long with circumferences at the bases and prongs of about 7 inches (17.8 cm) each. The number five Arizona head has even longer horns—more than 20 inches (50.8 cm)—but the prongs are comparatively short. The present minimum score for pronghorn, 82 points, is much higher than in the past. The minimum was 60 points in 1950. It was raised to 70 in 1951, 75 in 1953, 80 in 1963 and 82 in 1968 (Wright 1981).

Currently, a hunter's chance of securing a large pronghorn is best on private lands in the Southwest. These bucks were shot the opening morning of the 1996 season on a ranch in New Mexico. They represent the woman's first big game animal and author O'Gara's largest buck in 50 years of pronghorn hunting. Photo by Kirk Kelso.

In 1993, a buck scoring 90 Pope and Young points was shot in Oregon by bowhunter Roger Clavno (Hisey and Hisey 1995). In 1995, a buck scoring 91.5 Pope and Young points was taken in Arizona by Marvin Zieser (Hisey and Hisey 1997). Another Arizona buck scoring 90.75 points fell to the arrow of Les Shelton in 1998 (James 1999).

About 450 pronghorn heads were listed in the eighth edition of *North American Big Game* (Nesbitt and Wright 1981). That edition included animals shot from 1899 through 1979, and about 150 of them had been entered during 1974 to 1979. By 1994, about 1,270 record heads were listed (Reneau and Reneau 1996). Thus, about 300 pronghorn heads shot or otherwise collected from 1899 through 1973 (75 years) were listed in the records, and about 970 were added during the next 19 years. However, with many more pronghorn now, finding a large head naturally is a greater possibility than it was in the past. Part of the increase in numbers of big heads during recent years could be the greater awareness by hunters of the records and what constitutes a record-class animal.

Pope and Young entries into the pronghorn records also have increased rapidly during recent years, partly for the reasons stated above and partly because bowhunting has increased in popularity and archery equipment has been improved. The first bowhunting big game records of North America were published in 1975. About 2,620 pronghorn were listed in Pope and Young records by the end of 1992 (James 1993). Between January 1, 1993, and December 31, 1996, about 1,182 more heads were added to the Pope and Young pronghorn records (Hisey and Hisey 1995, 1997). By the end of 1998, 630 more heads made the record book for a total of about 4,432.

HANDLING TROPHIES

Mounted pronghorn heads are more colorful than those of deer, and being fairly small, a pronghorn head fits into the average room better than do those of most other big game.

Practices to be avoided when handling a trophy pronghorn include dragging the carcass, cutting the throat, not cleaning blood from the cape immediately, cutting the cape too short and permitting the hair to slip. Avoiding such mistakes helps assure a beautiful mount that can be enjoyed for many years.

Pronghorn hunters seldom are so far from roads or towns that they cannot get their trophies to a taxidermist or a freezer plant within a day. Most hunters should take advantage of this situation instead of caping the head themselves. A trophy buck must be handled with care. Pronghorn hair is exceedingly fragile, and large patches can be scraped off in a running fall, by dragging or by rough handling and skinning. Although taxidermists can work wonders with some of these blemishes, they cannot perform miracles. The condition of the scalp a hunter brings in will determine, to a great degree, the quality of the head she or he gets back.

Pronghorn hair, being pithy, is stained by blood. The trophy hunter who shoots her or his animal in the head or neck might do well to settle for a set of mounted horns, not a head. Lung-shot animals blow blood from their noses and mouths; if it gets on the hide—especially the white hairs of the neck or cheeks—it should be removed as soon as possible with cold water by washing with the lay of the hair. A small bottle of acetone can serve as a substitute for water; soak a few cotton balls with acetone and swab off all blood as well as any fluid discharged from the mouth or nostrils. Care should be taken against further bloodying of a cape after the hunter reaches the scene.

Every animal has a slightly different facial appearance, just as do humans. A number of close-up photos of the head (side, straight on, quartering, top, all angles—too many being better than not enough) will help the taxidermist recapture the animal's appearance.

Shown here is the rinsing of blood from a lung-shot pronghorn. The white cheeks and white of the neck were nearly covered with blood. A gallon of water achieved the state of cleanliness seen here because the cape was rinsed as soon as the animal was reached. Photo by Bart W. O'Gara.

When the trophy hunter has an eviscerated buck lying belly down to drain after washing, the trachea and esophagus should still be in the animal. The best way to ensure sweet meat and avoid scuffing or knocking off hair during handling is to skin out the shoulders and neck and cut off the head. Directions for doing so follow.

Use a sharp, straight knife, cut from the underside of the skin, never downward through the hair. Starting in the middle of the back just behind the shoulders, cut the skin forward in a straight line along the back of the neck to a point about 2 inches (5 cm) behind the horns. Continue the cut forward as a Y-shaped incision with one fork of the Y reaching the base of each horn. Starting again at the top of the shoulders, cut downward on each side past the back of each front leg. The incisions should meet in the middle of the brisket. Cut down the back of the front legs so that about 10 inches (25 cm) of leg skin are left with the cape. Taxidermists seldom complain about too long a cape, but they bemoan short ones. After the incisions are completed, skin forward to the ears and twist and cut the head free of the neck. Remove the trachea and esophagus from the carcass and trim as much meat as possible from the base of the skull, cutting around the root of the tongue and pulling it out.

If delivery will be made to a taxidermist within a day or two, additional skinning is unnecessary. Simply remove remaining bits of flesh and thoroughly sponge any blood from the head and cape with cold water. Hang the head and cape in the shade where air circulates freely, with the nose up to dry. After drying, apply salt to the surface of the skin and the insides of the ears, nostrils and mouth. Common table salt is satisfactory and generous applications are advised. Protect the trophy from sunlight, heat and insects until it is delivered to a taxidermist.

If more than a couple of days will elapse before the head reaches a taxidermist, any hunter can complete the skinning task with a sharp skinning knife, blunt-nosed table knife and patience. Pry the skin away from the horns and work it forward over the skull. Cut the cartilage of the ears close to the skull but proceed carefully as the skin in this area is very thin and tight. Remove the eyelids and lips with the skin, cutting close to the bones and teeth. Sever the cartilage behind the nostrils, completing separation of the skull and skin.

Starting at the back of the ear, work the skin toward the tip, using the blunt end of the table knife. After the tip is reached, continue to work the skin away from the frontal surface well into the inner ear.

Free the cape of all meat, tissue and fat. Then wash the cape thoroughly in cold water and stretch over a limb in the shade where air will circulate freely to dry all surfaces. After an hour or two, salt the cape generously, including the nostrils and ears.

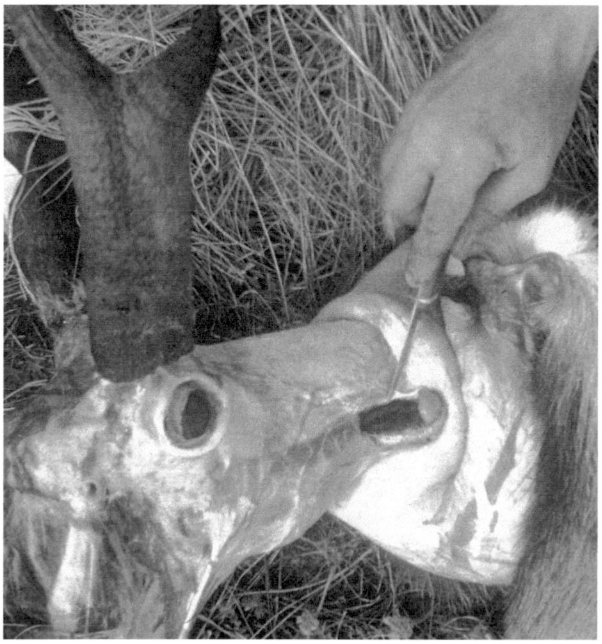

A finger placed in the corner of the pronghorn's mouth helps prevent cutting the animal's lips and aids in removing them close to the bone (left). *The next step is to cut the nose from the skull* (right), *leaving as much cartilage as possible with the skin.* Photos by Bart W. O'Gara.

When transporting the trophy, roll the cape loosely with the hair side out and place in a porous burlap bag. Pack in a protected place on top of the duffle where it will not be crushed. Never put it in a plastic bag, as souring or blood staining probably will result. If flies are a problem, the head should be placed in a cloth bag or peppered heavily, or both. A head can be frozen to carry or ship, but it should be well padded and care taken not to break frozen hair.

A frozen cape should never be unfolded until it is completely thawed. When transporting an unskinned head, place it in a large cardboard box or burlap sack, and pack a large wad of paper towels or other absorbent material against the severed base of the skull and throat to soak up and keep blood from reaching the hair.

The sheaths of pronghorn horns must be pulled and the skin removed from the cores to avoid odor and insects. All horns and antlers shrink to some degree as they dry. Pronghorn horns shrink more than others because of the thick membrane (skin) between the horn sheath and horn core. If horns are left to dry naturally, the basal circumference and length will shrink. Lengths are particularly affected, because the front and back edges of the sheath bases curl inward.

Cook (1994) developed a cleaning process that tends to hold shrinkage of pronghorn sheaths to a minimum. It begins with prong-to-prong and tip-to-tip measurements for later reference (Figure 169). After a head to be mounted is skinned, the skullcap is cut down through the middle of the eye socket, then forward on a level plane

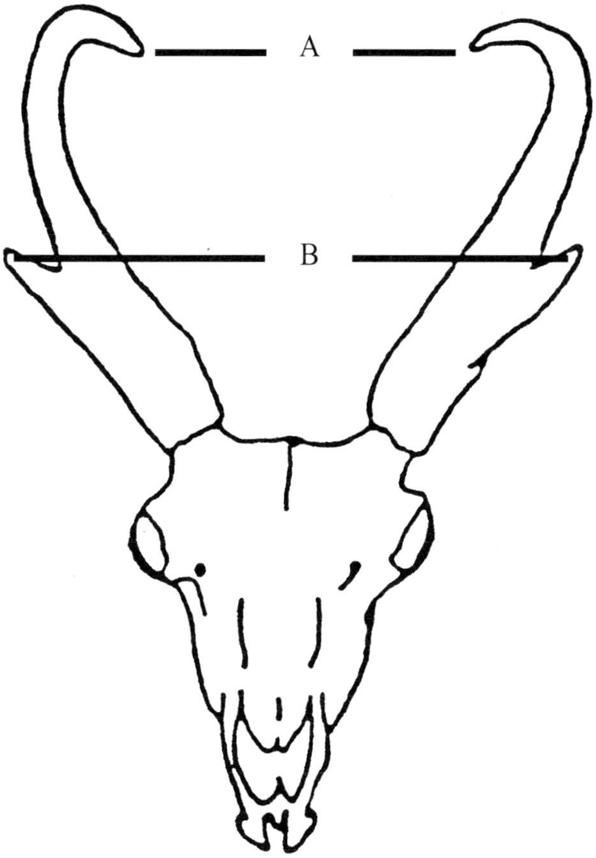

FIGURE 169. *Points of measurement for prongs* (A) *and horn tips* (B) *to be taken before sheaths are removed from cores of a pronghorn skull (adapted from Cook 1994).*

from the rear of the cranium, meeting the first cut. After as much muscle tissue and membrane as possible is removed from the skullcap, the entire set of horns and skullcap are soaked overnight (approximately 12 hours) in a bucket of water. The water then is drained and the skullcap and horns placed in a plastic bag and sealed tightly. The bag must contain as little air as possible and have no holes. The horns are sealed with two to three individual bags to ensure no exposure to outside air and left in a warm place. This is a contained and controlled decaying process that will accomplish removal of horns from the cores without requiring boiling, which invites shrinkage. After 3 or 4 days in plastic bags, horn sheaths will twist off the cores easily.

Then, with a small scrub brush, soap and water, the horn sheaths are cleaned inside and out, placed in a plastic bag and frozen. At that time, the skullcap is boiled, cleaned and submerged in 3% hydrogen peroxide, available in most drugstores and supermarkets, for 3 or 4 hours to eliminate any remaining biologic matter. (Protective eye wear and gloves should be worn with any use of hydrogen peroxide.) Next, the cap is placed in a container of hot water to remove any fat or oily residue on the skull surface. That accomplished, the skullcap is set on paper towels to dry for a few days.

For reattachment of the horns to the skull, Cook's method employs about 1.5 cups of plaster of Paris, a mixing cup and a butter knife or spatula. The plaster is mixed to a thick muffin-type consistency. The spatula is used to coat the inside of each horn sheath with a layer of the mixture, which is more than the actual thickness of the skin removed from the cores after boiling. The horns then are pressed onto the cores, one at a time, and the aforementioned reference measurements are reviewed to make sure the prong-to-prong and tip-to-tip spreads match the originals. When the horns are placed back on cores, excess plaster is forced out the bottom and protrudes down from the base of the horns. This excess plaster is not to be removed until it has hardened. The use of plaster of Paris enables adjustment of the tip-to-tip measurement and maintains the basal circumferences without too much shrinkage. Other methods entail a bondo mixture, but Cook found plaster to be the ideal compound because it can be cleaned easily and removed in case of mistakes. Bondo allows only one chance at completing this process correctly.

The do-it-yourselfer who wants to mount a set of pronghorn horns should not follow Cook's (1994) instructions for cutting the horns from the skull. The skullcap removed should be much larger—from the forward tip of the nasal bones, through the lower eye socket and to the back of the skull. This produces what many call a European mount. Sawing the skull plate off with a wood saw is easier than with a meat saw. The thin blade of a meat saw follows the path of least resistance and the frame does not allow a single cut all the way through. After the eyes and brains

For mounting horns on a plaque, the skull plate should be sawed off along a plane indicated by the black line in the top photo. *Sawing is easiest from the nose toward the back of the head, with the back of the head and horns resting on a hard surface. A wood saw is preferable to a meat saw.* At left *is a pronghorn skull mounted on a commercially made plaque that holds the horns at a more natural angle than does a simple plaque against the wall.* Photos by Bart W. O'Gara.

from a sawed-off skull plate are removed, the sheaths should be taken off, cleaned and frozen, as previously instructed. The skull plate, without the sheaths, should be boiled for about 2 hours, then cleaned thoroughly and bleached for a couple days in a crockery or plastic pan containing 1.0 quart (0.95 liter) each of hydrogen peroxide and rubbing alcohol, with enough water to cover the skull plate—not the horn cores—about 1.0 inch (2.5 cm) deep. After the skull plate has dried, a coat of water-base, crystal-clear glaze (available in hobby shops for use in decoupage and other crafts) will strengthen the fragile nasal bones, facilitate dusting and reduce yellowing with age. However, some people object to the shiny finish, so this is a personal choice. Do not use shellac, varnish or other finishes that will yellow with age. Next, the sheaths should be mounted with plaster of Paris, as previously described, and allowed to dry. The excess plaster should be cut and scraped away. Then, the sheaths need a coat of shellac, to sink in, keep the horns live looking and make them easy to dust. The dried skull plates (and horns) can be screwed onto a plaque of appropriate size. Pronghorn horns stick fairly straight out from the wall. If a more upright position is desired, a wedge can be fashioned from softwood and stained to match the plaque.

Pronghorn skins are beautiful, and some hunters would like to keep a skin as a trophy. These skins can hardly be used as rugs or on the backs of chairs or sofas because of the brittle hair. Even wall hangings will lose some hair when touched or cleaned. Pronghorn skins make inferior leather. When first tanned, the leather appears thin and pliable, but it does not wear well. Also, without frequent applications of oil, pronghorn leather dries and will crack like cardboard when flexed. A hunter will do better to use deer skins for thin, tough, pliable leather.

IN CLOSING

Meat hunting and other human disturbances almost exterminated the pronghorn. Protection and management, paid for mainly by hunter's dollars, have assured the continued existence of this interesting game animal. Pronghorn provide the only chance most Americans have to experience a big game hunt on the prairie. I find that stalking is the most rewarding way to hunt, but waiting at strategic spots, walking-down or using a flag can be productive. A scope-sighted, flat-shooting rifle and binoculars are almost necessities. However, archers often use decoys to get close enough to trophy bucks to kill them with arrows. Persons who chase pronghorn with vehicles and shoot at running herds damage relations between landowners and hunters, and damage hunting in many ways. Pronghorn venison can be excellent or terrible, but a pronghorn trophy is always beautiful and not too large to hang in an average room. A pronghorn hunt is fun for anyone who enjoys a challenging hunt and is the ideal hunt to teach a young hunter the ethics of hnting, the close stalk and the clean kill.

Hunting pronghorn, and other wildlife, is permissible for more reasons than just because state and provincial wildlife management agencies have promulgated harvest seasons. Hunting is possible today because of billions of years of geological and atomospheric happenstance, because of millennia of faunal and floral culling and refinement, because of an eon of human experience and tradition as predator, because of a single century of reversed exploitative human attitude and momentum, and because of a very few decades of wildlife science and management forging the conditions, time and places for wild populations to persist despite humankind. It is a heritage well worth maintaining, honoring and safeguarding.

S E C T I O N F I V E

THE FUTURE

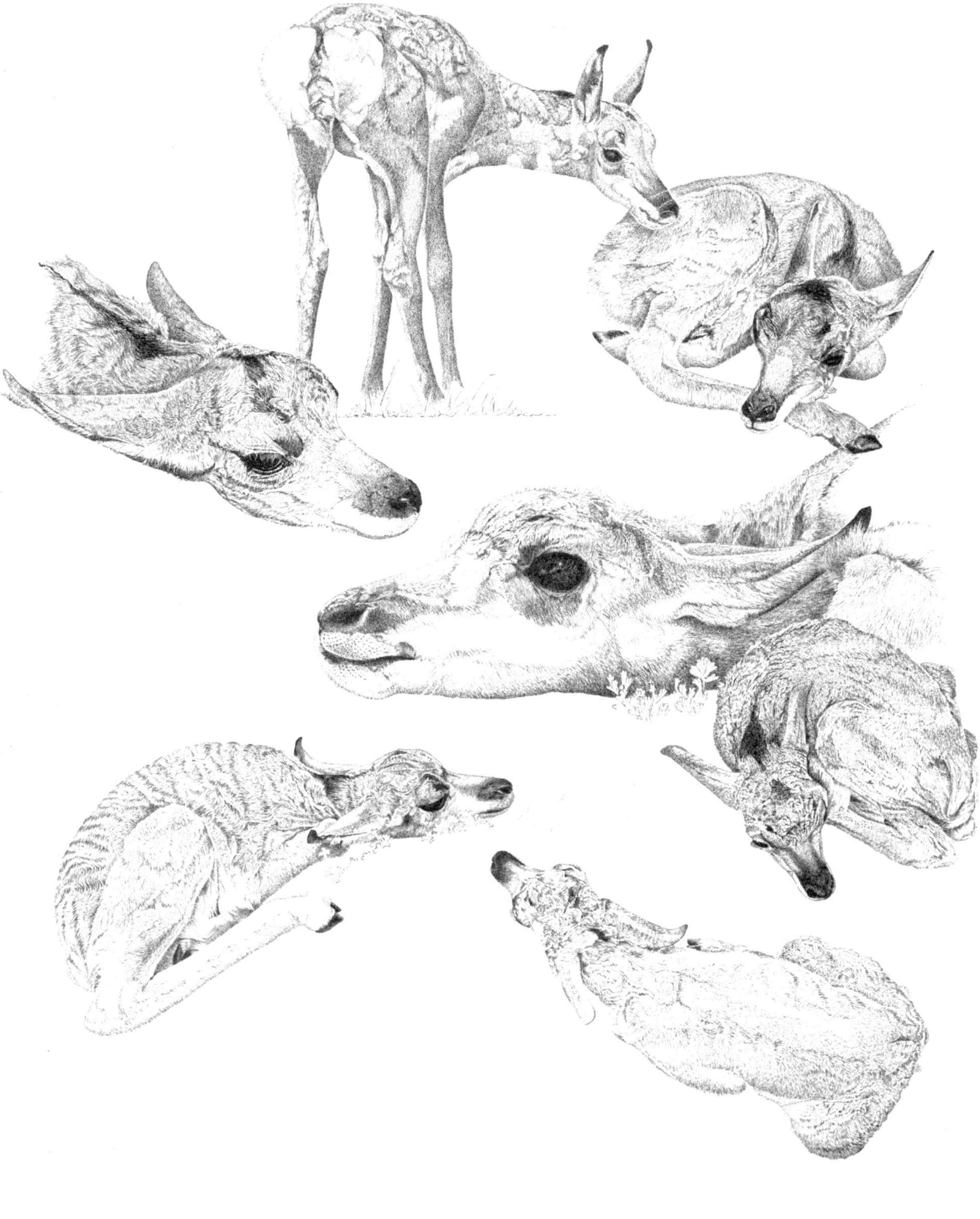

C H A P T E R T W E N T Y - T H R E E

The Future

Bart W. O'Gara and Jim D. Yoakum

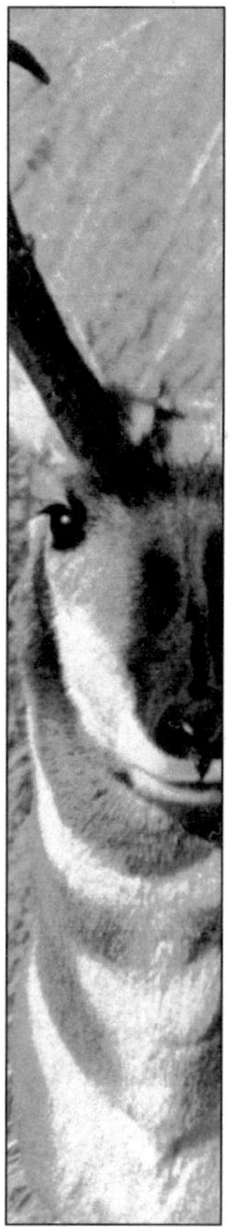

*Whether the species is the pronghorn antelope (*Antilocapra americana*) or another, by now most wildlife managers have confronted numerous challenges to their preferred management plans that are ultimately brought on by global economic and political changes, and by societal responses to these trends. Habitat loss increases as more people are able to telecommute, and thus choose to live in isolated locations rather than the crowded cities from which they came. . . . These are merely a few of the trends that wildlife managers are facing as we enter the twenty-first century, and yet very few agencies place as much emphasis on understanding the consequences of such social, political, and economic transformations as they do on mastering the intricacies of game biology and habitat management.*

Werther (1999:1)

Predicting the fate of pronghorn in the 21st century approaches the impossible. Even short-term evaluation is risky. No one can look at 2020 with 20-20 vision. As rapidly as human numbers and their appetite for fossil fuels and resources within pronghorn habitat are increasing, retrospect seems of little value. Also—as Werther (1999) pointed out—economic, political and societal changes are occurring so rapidly that the wildlife profession may be navigating unknown waters.

Trends that bear on the management of pronghorn seem evident. Crowe (1987) summarized those trends in five statements.

1. From today's level of human population, there will be further increases in years ahead.
2. As resource-consuming technology continues to grow, pressures on the environment will exceed anything yet seen.
3. Hunters, as a percentage of population, will decrease.
4. The sentiment against consumptive uses of wildlife resources will continue to grow.
5. The public demand that management considerations include all species of resident wildlife will increase.

Pronghorn are not a liability. They are not dangerous to humans and are not seriously competitive to or incompatible with livestock. They are animals of beauty and grace that annually provide the public thousands of recreation days, plus income to some others. Why should they not continue to prosper? Perhaps they should not because they are animals of the open plains, programmed to move about, securing their specific needs throughout the year, and because they do not negotiate human-constructed barriers well. A fence or road in the wrong place may reduce pronghorn habitat as surely as paving an entire summer or winter rangeland for an immense parking lot. As Yoakum et al. (1999:95) maintained: "Almost certainly, piecemeal attrition of pronghorn habitat, particularly in critical wintering and fawning areas, as well as travel corridors, is accelerating. During the 1980s and 1990s, intensified agriculture, subdivisions, accelerated energy development, upgrading of roads and other activities financed by a continent-wide, growing, affluent society prevailed. Such habitat modification and loss may negatively influence recovery of pronghorn from inevitable weather-induced die-offs. Clearly, loss of pronghorn habitat translates to loss of pronghorn."

Humans endanger species by: (1) direct exploitation, (2) introduction of alien species to new ecosystems and (3) by far the most important, destruction and fragmentation of habitats (Leakey and Lewin 1995). In the 21st century, direct exploitation will be monitored and controlled (see Chapter 20). Maintaining pronghorn populations within the limits of the remaining habitat will be a benefit to the species. Animals that impact pronghorn are mostly domestic livestock (see Chapters 15 and 17). Competition with livestock depends mostly on stocking rates and climatic factors (Donahue 1999). The effects on pronghorn associated with introduction of alien plants has received limited attention—some have attributes and some have liabilities.

FOSSIL FUEL EXTRACTION

A new energy boom is sweeping across Wyoming. This time, the sought-after resource is coal-bed methane, clean-burning fuel that will slow the greenhouse effect. Unfortunately, getting methane out of the ground may have its own impacts on wildlife and open spaces.

Beers (2000:22)

Pronghorn have adapted to drilling, mining and pumping associated with energy extraction on much of their quality habitat, but the impact likely is just beginning. "A gold rush? Another energy boom? Good news for the state? Good news or bad news for the environment and wildlife? A short-term fix for a minerals-based economy? A long-term solution?" (Beers 2000:22).

It depends on whom you ask about coal-bed methane gas development. According to the Wyoming Business Council's Director of Minerals, Energy and Transportation, the most important thing is the high-quality jobs that coal-bed methane is bringing to the state. The organizer of the Powder River Resource Council maintains that no forethought has been given to how this resource should be developed to minimize impacts on people and the environment. Many people own land but not the minerals, and those people make a living from ranching, tourism and recreation. Development is going to destroy their livelihoods (Beers 2000).

A single wellhead may be an unobtrusive presence on the landscape, but what will the cumulative effects of tens of thousands with their connecting pipelines be? Richard Zander, assistant field manager with U.S. Bureau of Land Management in Buffalo, Wyoming, considered 30,000 coal-bed methane wells in just Powder River Basin a reasonable scenario for the next decade. Other estimates double that number. Zander speculated that 10 years of drilling plus 10 years of production was reasonable. From 1989 through May 2000, 9,300 permits to appropriate ground water for coal-bed methane development were issued. Pumping water from a coal bed reduces pressure, and methane is released. During the 1990s, about 13 billion gallons (49 billion liters) of water were pumped for that purpose. Water on western grasslands can be good for

Pronghorn are surprisingly tolerant of oil wells and the traffic associated with them. The animals quickly adapt to vehicles that travel predictable routes, if occupants do not stop to watch, photograph or shoot them. The actual habitat loss to roads, drill pads, etc., probably is the most serious effect of energy extraction. The roads also may provide additional access to poachers. Some energy developments are located on rangelands that served for centuries as crucial wintering grounds. Low numbers of energy developments have created limited problems to which pronghorn have generally adapted. However, increasing industrial densities create greater human activities that are deleterious to pronghorn winter survival. Also, seasonal movement corridors to and from traditionally occupied rangelands may be interrupted. Photo by Robb Hitchcock.

wildlife, but sudden increases in runoff have caused major erosion. Also, water released from methane wells may be contaminated with heavy metals or sodium (Beers 2000).

Perhaps we have belabored this issue excessively, but it is an example of what human needs for energy and technology can cause. Laws to protect habitat during and after extraction often are not enforced (see Chapter 3).

HUMAN POPULATION PRESSURES

The shift in human population from northern and eastern sections of this country to the South and West has only just begun. The restructuring of America will gain increased momentum as we continue to experience the consequences of this migration of jobs, people and economic activity.

Crowe (1987:281)

Krausman (2000:491) gathered some sobering statistics. "Many of our ills are caused by overpopulation. Human impacts have caused the collapse of fisheries (Erickson 2000), more than 40 percent of the earth's land area has been transformed for human use, most accessible fresh water is in use and most terrestrial nitrogen fixation is caused by humans (Vitousek et al. 1997). Furthermore, approximately 80,000 square miles (200,000 km^2) of forests are cut each year and species extinction is 120,000 times greater than normal (i.e., 1 species lost every 4 years); today, extinction occurs for at least 30,000 species per year (Leakey and Lewin 1995). These staggering statistics force the question: 'What is the future of wildlife?'"

Crowe (1987:281–282) wrote: "To preserve Wyoming's wildlife and lifestyle in the face of this pressure is a task of staggering magnitude and expense. We should be in an all-out effort to secure critical wildlife habitats. We should be pushing an aggressive program to acquire hunting and fishing access for the rank-and-file sportsman. We should be in a headlong dash to gather more detailed wildlife data in order to bolster our knowledge and our ability to make sound wildlife decisions. We should be expanding efforts to educate the public about the needs and values of their wildlife resources. We should do all these things, and more, if Wyoming is to remain the outdoors paradise it has been. Yet, without public support, we can do none of it.

"The power to get things done rests not with the Game and Fish Department but with the citizen sportsman. He is the soul and fiber of wildlife conservation. He is the watch dog of the resource. He has footed the bill for over half a century and in the strength of numbers has the political power to insist on sound management decisions and equitable distribution of the benefits. Without his support and concern we will be relegated to refereeing the squabbles over how to divide the spoils as we preside over the decline of the resource.

"With reference to human uses of wildlife, we will increasingly find ourselves more people managers than wildlife managers. In this regard, our most neglected and crucial research needs are those concerning human social behavior. We do not yet know the limitations of human densities in outdoor programs or how best to serve the long-term interest of the people and the wildlife resource in decisions of quantity versus quality" (Crowe 1987:285).

Wyoming is home to at least half of the world's pronghorn, so what happens in that state is important to the species. Like any resource, wildlife becomes more valuable as the recognition of its limits grows. Lands are bought or leased for the hunting rights; wildlife managers find themselves increasingly involved in the fights over who gets what rather than in the tasks of conservation and management. The future will see these pressures intensify as pronghorn increasingly are crowded into close association with humans. Photos by Richard J. Guenzel; courtesy of the Wyoming Department of Game and Fish, Cheyenne.

Table 172 represents opinions of conservation agency personnel in Canada and the United States concerning the most serious threats to pronghorn in their provinces or states. Opinions were gathered by questionnaires during 1999 and 2000. The only Mexican state from which an answer was received was Baja California Sur. Threats there were not prioritized, but droughts, livestock competition for forage, poaching, predation and shortage of nutritious forage were listed as problems.

Table 173 attempts to prioritize threats to pronghorn. Loss of habitat appears to be the greatest concern among biologists and managers with one-third of the possible points (36 of 108). Severe winters or droughts were second with 19% (21 of 108) of the points. Canadian provinces rated habitat loss the same as did the entire group. However, severe winters or droughts rated as the most serious threat in Canada at 42% (5 of 12) of the possible points.

Habitat loss can be dealt with by improved land-use planning, increased conservation easements and direct purchases of key habitats. Such loss also can be mitigated by improved management of rangelands, especially livestock foraging and preservation of movement corridors. The mitigation measures are the same for severe winters and droughts. If the situation worsens, emergency feeding may be called for (discussed later, see also Chapter 14).

In the questionnaires, livestock foraging was reported to be a minor problem. However, at least four of the threats listed were caused totally or in part by livestock and livestock management. Shortage of nutritious vegetation (13 points), fencing (6 points), range improvement projects (5 points) and competition with livestock for for-

TABLE 172. The greatest threats to the future of pronghorn as reported in 2000 by personnel of provinces and state conservation agencies

Province or state	*First*	*Second*	*Third*
Alberta	Loss of habitat—prairie to croplands	Severe winters or droughts	Range "improvement" projects for livestock
Saskatchewan	Severe winters or droughts	Predation	Loss of habitat to agriculture
Arizona	Loss of habitat—urban sprawl	Fencing	Livestock competition for forage
Baja California Sur	Drought, forage competition with livestock, poaching, predation, shortage of nutritious forage[a]		
California	Shortage of nutritious forage	Fencing	Loss of habitat—livestock competition
Colorado	Loss of habitat—fragmentation by roads, fences, etc.	Livestock competition for forage	Crop depredations
Idaho	Loss of habitat—decline in productivity/exotic vegetation	Shortage of nutritious forage	Severe winters or droughts
Kansas	Landowner intolerance	Poaching	Loss of habitat/predation
Montana	Loss of habitat—federal programs provide incentive for conversion of sagebrush grasslands to grain production	Range "improvement" projects for livestock	Severe winters or droughts
Nebraska	Shortage of nutritious forage	Predation	Crop depredations
Nevada	Loss of habitat—wildfires	Severe winters or droughts/ shortage of nutritious forage	Feral horse competition for water and forage
New Mexico	Severe winters or droughts	Loss of habitat—poor livestock grazing practices	Predation
North Dakota	Loss of habitat—rangeland to private preserves and ranchettes	Fencing	Severe winters or droughts
Oklahoma	Crop depredations	Loss of habitat—rangeland to agricultural use	Industrial or energy developments
Oregon	Loss of habitat—modification	Shortage of nutritious forage	Severe winters or droughts
South Dakota	Severe winters or droughts	Shortage of nutritious forage	Diseases and parasites
Texas	Droughts	Loss of habitat—brush encroachment	Predation
Utah	Industrial or energy developments	Crop depredations	Severe winters or droughts
Wyoming	Loss of habitat—sagebrush removal	Range "improvement" for livestock	Shortage of nutritious forage

[a] No priority given.

TABLE 173. Order of perceived threats to pronghorn as listed in Table 172

Threat	*Number of conservation agency personnel placing this threat as first, second or third*			*Points*[a]
	First	*Second*	*Third*	
Loss of habitat (to croplands, urban sprawl, fragmentation, etc.)	9	3	3	36
Severe winters or droughts	4	2	5	21
Shortage of nutritious forage	2	3	1	13
Crop depredations	1	1	2	7
Predation	—	2	2	6
Fencing	—	3	—	6
Range "improvement" projects	—	2	1	5
Industrial or energy developments	1	—	1	4
Competition with livestock for forage	—	1	1	3
Landowner intolerance	1	—	—	3
Poaching	—	1	—	2
Diseases and parasites	—	—	1	1
Competition with feral horses	—	—	1	1

[a] Derived by assigning a first choice of 3 points, second choice 2 points and third choice 1 point.

An adage on the range is, "Good fences make good neighbors." Be that as it may for humans, good fences can stop pronghorn movements, leading to malnutrition, decreased reproduction and death (see Chapter 17). Photo by Alice Koch.

age (3 points) adds up to 27 points—second only to loss of habitat.

GLOBAL WARMING

An overheated climate affects all natural resources, all species, all people, everywhere. By altering sea level, temperature, and weather patterns, a rapidly changing climate forces grasslands, forests, farms, and wetlands to rearrange themselves on the landscape.

Berger (2000:13)

Greenhouse gases are accumulating in Earth's atmosphere as a result of human activities, causing surface air temperatures and subsurface ocean temperature to rise, according to the National Research Council (2001). Changes observed during the last half of the 20th century most likely were human-caused, but natural variability cannot be ruled out. Human-induced warming is expected to continue throughout the 21st century. Secondary effects, suggested by computer model simulations and basic physical reasoning, include increased susceptibility of semiarid regions to drought. The semiarid portions of North America encompass the pronghorn's habitat. If the shortgrass prairies and shrubsteppes become drier, the survival of some pronghorn, especially herds restricted in their seasonal movements by human developments, could be jeopardized.

Pronghorn have survived dry periods in the past. There is evidence to suggest that droughts as severe as the one in the 1930s were more common in the central United States during the 10th to 14th centuries than they have been in more recent times (National Research Council 2001). Pronghorn survived those conditions, free to seek

Droughts cause pronghorn to travel, if possible, to the succulent plants needed by their digestive system. During a recent drought in New Mexico, many pronghorn left the grasslands and traveled into forested montane areas, such as the Bonita Park area of Arizona seen here (D. Weybright personal communication:2000). There, they may find the needed forage, but they become unusually vulnerable to predators, particularly mountain lions. Only the lucky will return to their grasslands when the drought subsides. Photo by Rich Ockenfels; courtesy of the Arizona Department of Game and Fish, Phoenix.

vegetation that met their needs. If warming and drying during the 21st century causes habitat conditions suitable for pronghorn to shift east or north, the species could not move as it probably did during the 10th to 14th centuries. As reported in Chapter 4, prolonged drought in Arizona resulted in no Sonoran pronghorn fawn survival in 2002 and loss of 80% of the radio-collared adults (Arizona Game and Fish Department 2002).

Very possibly, the effects of desiccation, exacerbated by livestock foraging, already have influenced pronghorn in the southern portions of their range. Those in Mexico have declined in numbers; presumptive pronghorn subspecies there are considered endangered (see Chapter 4).

Animals common to Mexico, such as coatimundi and javelina, have extended their ranges farther and farther north in Arizona. The elegant Trogon began moving into southern Arizona around 1900. The move apparently corresponded to a gradual increase of about 3.5°F (1.9°C) in temperature since the 1870s, a phenomenon some researchers blame on extensive devegetation from cattle foraging (Vanderpool 2001). Another indication of warming in Arizona is the fact that Mexican jays, tracked by researchers there, now breed 10 days earlier than they did in 1975 (Levy 2000). It was questioned whether this apparent warming trend affected pronghorn in the state. Personnel of the Arizona Department

Pronghorn apparently are generally compatible with cattle on properly stocked rangelands (top)*. However, over time, cattle foraging on certain arid or semiarid rangelands can promote the growth of woody vegetation that reduces the quantity and quality of herbaceous plants, to the detriment of pronghorn. Also, when pronghorn must forage among tall vegetation, they are especially vulnerable to predators* (left)*. Therefore, by modifying habitat, cattle have apparently increased malnutrition and predation among pronghorn in some areas.* Photos by Rich Ockenfels; courtesy of the Arizona Department of Game and Fish, Phoenix.

of Game and Fish introduced pronghorn from at least six states (Ray 2000) and conducted coyote control for many years (see Chapter 11) in attempts to maintain populations in Arizona. "Grazing by domestic livestock has had, and continues to have, significant adverse environmental consequences on arid western rangelands," according to Donahue (1999:287). When combined with the possibility that the rangelands may become prone to droughts, plus the existing and increasing human barriers to pronghorn movements, the prognosis for maintaining substantial pronghorn numbers throughout the 21st century is not good.

REGIONAL DIFFERENCES IN PRONGHORN HABITAT

These contrasting types of population control place pronghorn antelope into two groups: (1) those in which numbers are regulated by man in the region of dry-land agriculture and Short Grass plains . . . east of the Rocky Mountains from the Panhandle of Texas to southern Saskatchewan and (2) those in which numbers are regulated by natural phenomena in the more arid grazing lands of desert shrub and desert grassland on intermontane flats, high plateaus, and rolling plains in and near the Rocky Mountains.

Buechner (1961:266–267)

Buechner (1961) estimated 400,000 pronghorn on the western plains. That number more than doubled during the next 30 years, and human regulation of numbers became more common in the second group. Yet, the differences are obvious. As Buechner pointed out, the states in the first group's harvest region offered 17% to 33% of their midsummer populations annually, whereas the harvest of the states in the second group was 2% to 10% of their populations. Buechner also cited counts of fawns per 100 females during midsummer as 78 to 113 in the first and 40 to 85 in the second group.

In part, the low numbers of fawns per 100 does in the second group reflect the influence of recurring droughts on vegetation that has been abused for decades by livestock foraging in semiarid areas (Buechner 1961). Leopold (1933:206) also believed livestock had degraded the southwestern vegetation: "The impact of occupancy here brought no bluegrass, or other plant fitted to withstand the bumps and buffetings of hard use. This region, when foraged by livestock, reverted through a series of more and more worthless grasses, shrubs, and weeds to a condition of unstable equilibrium. Each recession of plant types bred erosion; each increment to erosion bred a further recession of plants. The result today is a progressive and mutual deterioration, not only of plants and soils, but of the animal community subsisting thereon. The early settlers did not expect this: on the ciénegas of New Mexico some even cut ditches to hasten it. So subtle has been its progress that few residents of the region are aware of it. It is quite invisible to the tourist who finds this wrecked landscape colorful and charming (as indeed it is, but it bears scant resemblance to what it was in 1848)."

Ellis (1970) also compared population dynamics and habitat characteristics for pronghorn herds in the Great Basin with those of herds on the Great Plains, concentrating on the last 2 months of gestation and the first 2 months of lactation. Ellis concluded that fawn survival was twice as high on the Great Plains because of abundant nutritious grasses and forbs during late gestation and early lactation. Thus, the low availability of these forage classes, mainly because of consumption by livestock and differences in precipitation patterns, resulted in rangelands of low carrying capacity for pronghorn in the Great Basin.

Maternal condition late in gestation can affect birth weight in ungulates (Thorne et al. 1976). Low birth weight may increase susceptibility of offspring to hypothermia or starvation (Thorne et al. 1976, Clutton-Brock et al. 1982), or increase vulnerability to predation. Birth weight can be a factor in survival to weaning in some years (Fairbanks 1993). This is especially true because does that produce underweight fawns generally are in poor condition to provide adequate milk.

Reduced livestock foraging apparently would benefit pronghorn in much of the second group's habitat. In con-

Ellis (1970) compared population dynamics and habitat characteristics for pronghorn herds in the Great Basin with those of herds on the Great Plains, concentrating on the last 2 months of gestation and the first 2 months of lactation. He concluded that fawn survival was twice as high on the Great Plains because of abundant nutritious grasses and forbs during late gestation and early lactation. Thus, the low availability of these forage classes, mainly because of consumption by livestock, resulted in rangelands of low carrying capacity for pronghorn in the Great Basin. Photo by Rob Hitchcock.

trast, a moderate level of cattle foraging apparently is not deleterious to pronghorn in the first group's habitat and may, in fact, benefit pronghorn. Presumably, cattle replace bison as the primary foragers on the Great Plains—a relationship in which pronghorn thrived. This relationship often is assumed in the Great Basin, but no records indicate that bison ever inhabited the Great Basin in significant numbers. Thus, it seems that areas too arid to sustain bison populations probably cannot sustain heavy use by cattle without degrading the vegetation.

FUNDING FOR WILDLIFE CONSERVATION

Thanks in part to increased game populations and lengthened hunting seasons, hunters' expenditures increased substantially between 1991 and 1996.

Brown et al. (2000:154)

Funding is the first requisite for an efficient management program. As of the late 1970s, consumptive users paid more than 75% of the costs of wildlife management in the United States (Jahn and Trefethen 1978). Nonconsumptive

Maternal condition late in gestation can affect birth weight in ungulates (Thorne et al. 1976). Low birth weight may increase susceptibility of offspring to hypothermia or starvation (Thorne et al. 1976, Clutton-Brock et al. 1982), or it may increase vulnerability to predation. Birth weight can be a factor in survival (Fairbanks 1993); this is especially true because does that produce underweight fawns generally are in poor condition and unable to provide adequate milk. Photo by E. E. Smith; courtesy of the Library of Congress, Washington, D.C.

The habitat differences between the Great Basin and Great Plains often lead to confusion among biologists and the interested public if they are acquainted with pronghorn ecology in only one area. On Great Basin rangelands, shrubs are relatively tall and dense, whereas harbaceous vegetation is relatively sparce. On the Great Plains, shrubs generally do not grow tall or thick and, along with forbs, are needed for winter nutrition. Forbs grow profusely during spring, providing nutrition for growth of fawns. Thus, management that favors shrubs for winter survival and forbs for reproduction generally favors pronghorn. In other words, reducing or increasing shrubs can improve rangelands for pronghorn, depending on the area and local conditions. Left photo by Jim D. Yoakum. Right photo by Bart O'Gara.

users contributed little, possibly because of the lack of a clear product, according to Heberlein (1991). Because of the high level of hunter funding and the fact that hunted species were popular with nonhunters as well, hunted species received higher priority management than most nonhunted ones (Wildlife Management Institute 1992). This fact has been misinterpreted as indicating that state wildlife agencies manage exclusively for hunted species, to the detriment of other species. In fact, states, on average, expended only about one-third of their wildlife agency funds for management of huntable wildlife during the 1990s. Hunters, who represent only about 7% of the U.S. population, provided—through license fees, excise taxes and donations—an average of 45% of funding for management of nonhunted species.

In a general sense, with the caveat that variation existed by state, Duda et al. (1998) found the following funding mechanisms for wildlife conservation acceptable to most of the public: voluntary nongame income tax checkoffs, donations to conservation from increased speeding fines, entrance fees to wildlife management areas, fees on development-related issues, a small increase in the gasoline tax, lottery-related funding mechanisms, recreational vehicle sale taxes, redistribution of funds from general state revenues, and voluntary license plates with a fish and wildlife motif. These recommendations were specific to nongame species, but would allow conservation agencies to channel more of the hunter's money to game species.

Until increased funding and equity initiatives are legislated, hunters will continue to shoulder the financial burden for most game and nongame management, just as they will continue to defray some of the costs of prohibiting or discouraging other land uses that could be harmful to wildlife. Unfortunately, costs associated with management are increasing dramatically.

During the mid-1970s, at least half of the U.S. residents opposed hunting, and women and urban residents had the most negative attitudes toward the sport (Kellert 1978). Bissell and Duda (1993) found the most important values surrounding hunting were those centered on the family and social circles of hunters. With increases in urbanization and the percentage of female-headed households, the recruitment potential for hunters was declining. As the percentage of hunters in the population declines, recreational hunting may be increasingly viewed as culturally or socially antiethical.

During 1955 to 1985, the human population of the United States grew by about 83%. During the same time, the number of hunters increased only 39% (U.S. Fish and Wildlife Service and U.S. Bureau of the Census 1991). Much of this decrease reflected movement of people from rural areas, where rabbits, squirrels and game birds were conveniently hunted and provided meat for the table. Big game hunters increased about 185%. A decline in numbers of hunters need not seriously affect the funding base for wildlife management. Hunters have above-average incomes and, in 1985, each spent about $600.00 on their sport (U.S. Department of Interior 1988). About $25.00 was spent by each on hunting licenses, tags, stamps and permits. The same survey indicated that hunters would be willing to add $10.00 to $25.00 per day to their hunting expenses. This could accrue more than $100.00 per year per hunter available for management (Heberlein 1991). In some states, this would double the present management budget. However, economic conditions can invoke a point of diminishing returns in hunter numbers versus hunter revenues, and the uncertainty of that point is a matter of serious concern to many state wildlife agencies.

Some hunters believe that, for too long, they have shouldered too much of the cost and responsibility for wildlife conservation and that it is time that others pay their fair share. Although, this sentiment has justification, those advocates should consider that—with the services wildlife agencies provide—hunters get a bargain for their investment. Furthermore, when nonhunters shoulder more management expenses, they have more say in management. This may not favor or emphasize hunting. Photo by Chuck Cadieux.

The dollars spent for licenses that support wildlife management are less than a tank of gasoline or a box of ammunition. Nevertheless, many hunters often resist license fee increases because they do not see immediate, personal benefit. These views are understandable but sometimes shortsighted. Many state and provincial wildlife agencies run on shoestring budgets, given their diverse mandates. Hunters should consider how their own expenses increase year-to-year and that wildlife agencies must allocate for many more individuals and interest groups than in the past. For example, Gasson et al. (1995) noted that the Wyoming Game and Fish Department is dealing with more people and industrial activity than ever before; involved in administering a growing number, diversity and complexity of federal and state laws that protect the environment; managing rare flora and fauna protected by the Endangered Species Act; managing a wide range of outdoor recreation; suppressing potentially serious wildlife disease outbreaks; buying habitat and recreational access rights; compensating landholders for wildlife damages; responding to threats to wildlife in the courts; applying greater technological proficiency to wildlife management; and increasing public informational and educational outreaches (Gasson 1998).

THE ANTIHUNTING MOVEMENT

Animal welfare supports the humane treatment and responsible care of animals that ensures comfort and freedom from unnecessary pain and suffering. The animal rights philosophy dictates that the use of animals for any human benefit is wrong.

Duda et al. (1998:285)

Schmidt (1990) contended that a majority of people in the United States have some degree of animal welfare concern. Schmidt suggested that directing attention toward animal welfare concerns and away from animal rights—often comparable to human rights—would result in saving the limited time and resources of individuals and agencies by avoiding participation in winless debates.

Animal rightists and animal welfare groups have succeeded in closing hunting seasons on certain species, primarily through lawsuits and ballot initiatives. Antihunters are as sincere in their convictions as are hunters in theirs. Neither faction wishes to destroy wildlife populations, so some common ground exists. Empathy and understanding resulting from good communication may reduce controversy and the adverse impact on wildlife management, as suggested by Gilbert (1978). However, such issues are much like politics and religion; people tend to take sides on the basis on socioeconomic circumstances, training, faith or emotion. Science-based logic seldom changes anyone's mind on either side of an argument.

If one looks objectively at wildlife populations anywhere in the world, those subjected to well-managed hunting are more numerous and have better habitat than do those where such management is weak or absent. That simple fact is not known by much of the public and does not deter otherwise intelligent people from attempting to close hunting seasons. The antihunters espouse philosophical arguments against hunting to which many people are listening. Animal rightist philosophies no longer are those of people on the fringes of society. Such philosophies are an extension of western liberalism and perhaps are fanned by wildlife conservation rhetoric that urges compassion for life (species, not individual) and habitat preservation according to Heberlein (1991).

If the recreational hunter loses influence, much of the gains for wildlife and outdoor recreation during the 20th century will be lost (Heberlein 1991). Funding and infrastructure from nonconsumptive users and the often uninterested public are unlikely to replace it, and game and nongame species will suffer.

In the past, many wildlife professionals took the view that the animal rights movement eventually would fade away and, therefore, could be ignored. However, the movement is growing. It is well-funded and effectively presents its arguments to many segments of the public. Wildlife managers who do not have defendable species management plans, backed by good census and harvest data, may face what once seemed impossible—closure of hunting seasons needed to perpetuate the very species the animal rightists and welfare people are trying to protect.

EDUCATION

Overall a majority of the public strongly supports fish, wildlife and natural resource education programs . . . however, education programs are usually rated slightly poorer than other agency programs because most people do not see any products or services from the agencies. This is mainly due to the fact that most agencies do not spend the necessary financial or personnel resources on information, education and outreach programs.

Duda et al. (1998:393)

Hunter

True recreational hunters despise trespassers, poachers, cripplers, litterers, road hunters and game hogs, all of whom damage wildlife populations and relationships between landowners and responsible recreationalists. When landowners are antagonized, more land is posted or leased to clubs, individuals or outfitters, making less land open to the public. In democratic society, where anyone of legal age and training can hunt, there always will be some problems, but they can be mitigated. Changing the attitude of an adult who has chased pronghorn with a jeep, trespassed on private land or littered for years is difficult. Self-policing by hunters—use of the "snitch line" and the willingness to report violations and act as witnesses—helps to alleviate the problem. In some European countries, hunters are respected because of their knowledge of and respect for wildlife. Such knowledge is gained primarily in schools for hunters followed by rigorous tests to be passed before hunting permits can be obtained.

To deal with appropriate concerns with hunter behavior and ethics, hunter training should be expanded. The public will increasingly demand that, if animals are to be hunted for sport, the killing must be done quickly, cleanly and with appropriate respect (Heberlein 1991). Hunters should be trained, have certified marksmenship, understand animal behavior within their ecosystems, be proficient as hunters and have at least a basic understanding of the aims and practices of wildlife management.

A survey of 35 state conservation agencies and 11 hunting and conservation organizations indicated direct, positive effects of hunter education programs in improving hunter behavior (Ruh 1994). However, trespassing on private property remained a top concern, with poor relations between hunters and landowners a close second. Survey respondents said the most important influence on responsible behavior was hunter education programs. Hunters who completed these courses were considered significantly more responsible than those who did not, with a better attitude toward the sport.

Experiences with hunters on the Deseret Ranch in Utah support the contention that there is a need to advocate and conduct hunter education programs. Approximately 20 years ago, hunters of pronghorn and other ungulates caused a variety of problems and damages on the ranch, including littering, driving off designated roads and creating high crippling losses. The ranch instigated a program that required hunters to demonstrate shooting skill. Subsequently, the hunters who participated appeared to be a new breed, and there were few problems of environmental disregard and other ethical violations (R. Danvir personal communication:2002).

A system of incentives, such as opening restricted areas only to those who have completed a course on hunter ethics, may help. The possibility of compulsory attendance for game law violators before they can secure another hunting license is a consideration. A certificate to present to landowners when asking for permission to hunt on private land might be an incentive for hunters to attend a hunting ethics class.

Hunter safety also is a matter of importance, both to hunters and nonhunters (Bissell and Duda 1993), although hunting is one of the safest of sports (Wildlife Management Institute 1992; see also Chapter 22).

Forty-three provinces and states in Canada and the United States required youngsters and first-time hunting license applicants to attend and pass an approved hunter safety course (Anonymous 1991). Ed Corvina, president of the Hunter Education Association of North America, reported that, besides safety, courses taught hunter responsibility, wildlife conservation and hunter ethics. Colorado required all applicants born after January 1, 1949, to possess a hunters' safety certificate (T. Pojar personal communication:1992). This trend is encouraging, but most courses still are aimed at children and are taught by dedicated but often untrained and unpaid volunteers.

Pronghorn hunting is different from other big game hunting in North America (see Chapter 22). Many otherwise ethical hunters believe the only way to harvest a pronghorn is by driving about the country until a shot presents itself. Orientation sessions, by themselves or in conjunction with the hunter safety courses, would improve the quality and ethics of pronghorn hunters. They have been held in at least three states to inform pronghorn hunters about ethical hunting and other relevant subjects. Biologists from several states maintained that few hunters would attend such sessions in their states.

Voluntary indoctrination sessions were held in Nevada beginning in 1980. Display and audiovisual programs were used to help participants understand behavioral characteristics and habitat requirements of pronghorn. A series of speakers addressed topics, such as life history, hunting regulations and care of meat and trophies. These orientation sessions gained popularity annually, increasing from approximately 60 attendees in 1980 to more than 400 in 1991. The 1991 season was the largest public meeting ever held concerning the biology and management of pronghorn. During 1988 through 1990, about 25% of permit holders attended these sessions. Also, hunting companions attended in nearly equal numbers (J. Jeffress personal communication:1991).

Public

To reverse or blunt antihunting sentiment in this country, hunters should work individually and collectively toward providing better information to the public on hunting and management programs. At times, a small, but vocal and influential group of antihunters can influence the opinions of many North Americans who do not hunt and do not have strong hunting opinions. The nonhunting group is much larger than the prohunting and antihunting groups combined and will influence future political decisions concerning the sport. Informed hunters and wildlife managers can refute the claims of the antihunters, which generally are based on emotions and few facts. The concepts of wildlife management are supported by research and may be accepted by nonhunters, if those concepts are frequently and properly presented.

Public education programs by provincial, state and federal conservation agencies and sportsmen and sportswomen's groups are long overdue. Dedicated animal rights activists will not be swayed by what was accomplished during the last half of the 20th century, concerning wildlife management. But, the public might if presented with facts. The U.S. Fish and Wildlife Service is the agency to lead such an educational program with publications and, especially, television. The public receives most of its information concerning wildlife from those media. The North American Pronghorn Foundation and Arizona Antelope Foundation also are organizations to advance public education concerning pronghorn management and the role of hunting. The outreach messages need to stress four points:

1. Loss of habitat is the prime reason that wildlife is diminishing in many areas.
2. Strict regulations, backed by biological studies, govern hunting. Legal, modern hunting has not contributed to the list of extinct or endangered species.
3. Hunters annually contribute about $250 million to wildlife conservation and management in the United States; antihunting groups spend most of their money on lobbying, salaries, administrative costs and neutering domestic animals.
4. Wildlife management areas, refuges and marshes purchased and managed with hunters' dollars support countless songbirds, shorebirds, small mammals, raptors and other nongame species.

Many Canadians, Mexicans and Americans are interested in the welfare of wild animals but have little knowledge concerning hunting or wildlife management. They are exposed frequently to misinformation that would push them into the antihunting faction. If this large group of interested people is informed of the contributions of hunting, wildlife management may be very different in the 21st century than it was in the 20th century. If professionals, wildlife managers and other conservationists do not become proactive instead of reactive, there may not be much hunting or many pronghorn as the 21st century draws to a close.

RESEARCH AND MONITORING NEEDS

With the accelerated growth in human power over the environment, there is an accelerated need for facts, for guidance in how best to use such power. A strong research arm is necessary for guiding modern wildlife agencies. For those who play the game of bio-politics, there is political advantage in having competent research, and thus information superiority.

Giles (1979:219)

"Although we often need to focus on specific ecological relationships of a single species, if we are to manage for its consumptive use or to work to protect habitat criti-

Each pronghorn herd and its habitat are unique biological entities. With increasing pressure on rangelands, the level of knowledge concerning each herd's range should be increased to "fine tune" rangeland management. Also, each herd has its own behavioral and physiological characteristics. Thus, information gained through research conducted on other herds is not always adequate for management purposes. Although much knowledge concerning pronghorn and their habitat requirements has been gained, more and better management information is needed. Photo by Jim D. Yoakum.

During the mid-1980s pronghorn numbers peaked for the 20th century. Concurrently, pronghorn research declined to perhaps the lowest level since World War II. Conservation departments have limited funds and tend to spend them on research concerning species that are in trouble. For instance, during 1990, the Montana Department of Fish, Wildlife and Parks, along with other federal agencies within the state, spent more than $900,000 on grizzly bear research (C. Servheen personal communication:1991) and nothing on pronghorn research (G. Erickson personal communication:1991). In all of the pronghorn provinces and states, approximately $85,000 was spent on pronghorn research that year, according to a 1992 questionnaire. Such a situation is understandable, but if pronghorn numbers are depressed again, managers will wish they had more knowledge concerning the species and its various habitats. The logical time to gather such information is while many animals are available for study and, if needed, collections are feasible. Above photo by W. D. Parker; courtesy of the U.S. National Archives, Washington, D.C. Right photo by Peter L. Knapp.

cal to it, we must understand a species in relation to the total ecosystem of which it is a part. Because of the complexity that exists at the ecosystem level, we as individuals cannot be expected to undertake the task of investigating all aspects of an ecosystem simultaneously. This means that management at the ecosystem level requires cooperation among all those working for the protection and management of the natural environment, including wildlife managers and biologists, foresters, park naturalists, and conservation biologists, both within and outside state and federal government agencies" (Klein 2000:466).

"Each herd is a biological entity with unique behavioral characteristics and a threshold of tolerance to natural and human disturbances. With the increasing pervasiveness of human influence on each herd, the amount and level of biological knowledge must be heightened . . . to allow practicable strategies of management when ephemeral disturbances pose potential threat to herd well-being and productivity. Even more importantly, biological information must be more precise, so that management is drawn further from the realm of 'holding actions' and closer toward that of multiple-benefit sustainable yield" (Peek et al. 1982:609).

On the other hand, Hunter (1989:350) maintained that: "Many wildlife biologists do not see well beyond the borders of their own state or province, and the resultant redundancy and waste of effort are enormous. 'It hasn't been done here before' should not be offered as justification by a research biologist."

The average wildlife biologist tends to ask how rather than why and is answered by a catalog of disjointed facts. Experimental design and statistical analyses are stressed, so findings generally are correct technically, but nearly meaningless because of a lack of a guiding hypothesis to hold the pieces together. Gavin (1989:346) maintained that, "The classic example of this is our fixation on habitat. I believe we have fallen into an interesting trap. We have fostered countless studies of habitat because habitat protection is unarguably paramount to wildlife conservation. Most of these studies, however, have been initiated without critical evaluation of the question we hoped to answer. We ask if species *x* uses habitat *y* (or which species are found in habitat *y*), and we use increasingly sophisticated analytical techniques to evaluate the data (e.g., multivariate statistics). The catalog of correlative facts grows larger and larger, but rarely do we identify habitat requirements of any species." Perhaps biased, we believe that the habitat requirements of the pronghorn are pretty well identified in the pages of this book.

Answers to a questionnaire concerning ongoing research during 1999 in provinces and states that harbored pronghorn indicated very little research underway and scant funding (a little more than $200,000) for what was being done. None of the Canadian provinces was conducting pronghorn research. Baja California Sur was the only Mexican state that returned a questionnaire. During the mid-1980s, pronghorn numbers reached all-time highs for the 20th century. Concurrently, pronghorn research declined to, perhaps, the lowest level since World War II. Conservation departments have limited funds and tend to spend them on research concerning species that are in trouble. For instance, during 1990, the Montana Department of Fish, Wildlife and Parks and federal agencies within the state, spent more than $900,000 on grizzly bear research (C. Servheen personal communication:1991), but nothing on pronghorn research (G. Erickson personal communication:1991). In all of the pronghorn provinces and states in 1991, approximately $85,000 was spent on pronghorn research. Such a situation is understandable, but, if pronghorn numbers are again depressed for any reason, managers will wish they had more knowledge concerning the species and its various habitats. The logical time to gather such information is while many animals are available for study and, if needed, collections are feasible. Of the 16 states in the United States with pronghorn, four were conducting research on the species (Table 174). It is tempting to read research needs into what is being done. However, research often is dictated by local circumstances and available funding. For example, if a company is planning a project that may impact pronghorn, that company may furnish research money to gather data needed in an environmental impact statement. Thus, conservation agencies are allocating small, if any, amounts of money to research. Giles (1979:221) penned: "far too little is spent on basic wildlife research. The amount allocated by any categorization, is far below the gross average of 10% budgeted by many industries and public agencies." We believe that the scanty funding available for pronghorn research generally has led to short-term (2- to 3-year) projects conducted by graduate students. Such, usually applied studies often provide locally useful information. However, well-funded, long-term (10- to 20-year) basic studies are needed to gain more, essential knowledge of pronghorn and their habitat relationships.

An aspect of little or no research occurring within a province or state that seldom is considered relates to a lack of expertise or leadership regarding that species. When we began attending pronghorn workshops during the late 1960s and early 1970s, almost every province or state had a pronghorn researcher or manager who attended most workshops and was aware of the pronghorn literature. Management-oriented questions were discussed, and participants soon knew the answers. During the 1980s and 1990s, that changed. Some provinces and states did not send anyone to workshops, or representatives attending were changed from workshop to workshop. Some attended because it was their turn to go to a meeting. Many representatives had knowledge concerning a specific area but not about a whole state. Those who were interested in pronghorn asked the same questions that were being asked 20 to 30 years earlier and had little grasp of pronghorn literature. At the workshop on the Hart Mountain National Antelope Refuge in Oregon during 1988, participants were asked if they used the pronghorn management guides that had been published and, if so, which of the three they used. Most of them were not aware such guides were available, even the one produced by workshop participants in 1978 (see Chapter 16). Thus, it appeared that provincial and state conservation agencies were according pronghorn a low priority, despite the fact that numbers harvested in some states were second only to those of deer.

Table 174. States conducting research concerning pronghorn or their habitat during 1999

State	*Amount of funding*	*Cooperator(s)*	*General research subjects*
Arizona	100,000	Phoenix Cement Co., Babbitt Ranches, Arizona Pronghorn Foundation	Population and habitat management
Baja California Sur	10,000	National Council for Science and Technology	Nutritional quality of certain forage species
Kansas	6,300	Emporia State University	Human dimensions
Oregon	15,000	U.S. Fish and Wildlife Service	Evaluation of survey techniques
Wyoming	85,000	Wyoming Cooperative Wildlife Research Unit and others	Long-term analysis of data sets, migrations out of Teton National Park, reproduction and mortality patterns

Approximate level of funding (U.S. dollars), cooperators and general research subjects are included.

Human Dimensions

A common complaint by biologists during the late 20th century was that they worked long hours in the field to gather biological data, only to have decisions affecting wildlife made by politicians—without apparent regard to the available data. Patterson et al. (2000:214) stated: "Most important natural resource issues ultimately are resolved in the political arena because they center around conflicting value systems rather than more objective, fact-based questions."

Those value systems involve public attitudes and desires, economics, laws, etc. If decision makers were informed of wildlife-related values integrated with biological data, decisions that were more beneficial to the resource might be expected. Thus, although none of the other chapters dwelled on the human dimensions of wildlife-related values, we must put this as the first priority in research needs. Giles (1979:219) opined: "Research results are only one of many things that converge in the instant that a manager makes a decision and sets its consequences in motion."

Burger (1979:97) noted: "Although wildlife management is aimed at achieving human goals, the human element largely was ignored until recent years. Early management efforts were directed almost entirely at game species. Yet even here there is little evidence that the hunter constituents were polled or queried as to their wishes; by and large, professional wildlifers and agencies set their own goals for game populations, with at least the implication of 'we know best.'" Actually, a well-informed biologist or manager generally does know what is best for the resource, but what is best for the greatest number of people (and will not irreversibly harm the resource) may be quite different.

National and regional participation surveys are valuable tools for integrating human and ecological dimensions into wildlife policy. Monitoring social patterns helps managers anticipate demands for services as well as ecological threats to wildlife (Mangun 1992).

Decker et al. (1996:31–32) conceptualized four components to be considered in management decisions for wildlife.

> "The *ecological component* of the management environment sets the limits or boundaries on potential resource productivity and use. . . .
>
> "The *economic component* of the management environment includes all the valuation of the marketplace and nonmarket values (unpriced values) that influence valuation of natural resources. . . .
>
> "The *political component* includes three aspects: established laws and codes, policies of various government agencies, and the values of government employees who enforce or interpret laws and policies. . . .
>
> "The *sociocultural component* of the management environment includes traditions, values, norms, religions, and philosophies of various segments of the public, specific constituencies and resource managers themselves. . . .
>
> "Separating the four components of the management environment is useful conceptually, but in reality they are not independent. They are intricately interrelated such that the natural resource manager must consider them all simultaneously. . . . The purpose of human dimensions research is to help managers better understand the latter three of the four domains within the management environment."

"Human dimensions research should not be considered a panacea, capable of resolving all wildlife management problems. When different constituencies hold widely disparate beliefs/values regarding an acceptable management strategy, no single solution will satisfy all groups" (Manfredo et al. 1996:71). However, human dimension studies can identify who holds what beliefs and how strongly those beliefs are held. Like the results of biological studies, such information contributes to the overall understanding of the phenomena.

Scientific effort should be expanded on human dimensions associated with wildlife because we often try to manage human and biological systems jointly, but have scientific data on only one side of the equation (Heberlein 1991). Human dimensions of wildlife seem intractable to wildlife researchers and uninteresting to economists, political scientists, psychologists and sociologists—or funding generally is unavailable for studies. The latter is often the case and conservation agencies must look beyond biology when setting research priorities.

As human numbers increase and if antihunting sentiment grows, knowledge concerning public attitudes toward wildlife and its management will be needed by wildlife managers. Equally important will be mastery of the vast array of laws concerning wildlife and wildlife habitat, as well as monetary values or expenses associated with species.

Research helps to ensure that management programs are rooted in sound science. However, managers now know enough about population and rangeland management to assure the future of pronghorn, if those managers are allowed to do so. Political decisions override biological data if they are not reinforced by economics and humanistic values. Biologists and managers of the future will need knowledge of biology and rangelands. They also must know how to deal with political pressures and court challenges.

Thomas and Pletscher (2000:548) brought the situation regarding human dimensions into focus when they wrote: "It is in the arena of human dimensions in land management that ecologists, wildlife biologists, and conservation biologists fall woefully short."

Duda et al. (1998) summarized the usefulness of human dimension research as follows. An understanding of the public often means the difference between the success and failure of wildlife programs. Wildlife agencies use biological and ecological research to better understand and manage wildlife and their habitat. Effective wildlife agencies use human dimension research and techniques to better understand and work with their constituents. Among the benefits of using human dimension research are enhanced management of the resource and a constituency that is supportive of the agency and its mission. Overall, agencies that effectively incorporate the human element into agency management functions and decision making ultimately spend more time and money directly on the resource, as opposed to making ill-fated decisions because of a lack of understanding of the public and constantly dealing with controversy.

Wildlife research and training should be more diversified, and wildlife agencies will need economists, lawyers, sociologists and other specialists, as well as biologists and rangeland technicians. Other research and monitoring suggestions follow, roughly in the order they are presented in the foregoing chapters, not in order of importance. See those chapters for background and details documenting the need for research.

Hunting Opportunities for the Physically Handicapped. Mangun (1992) maintained that an aging and less mobile population may require improved access to wildlife-associated recreation opportunities. Preparing regulations to govern a wildlife program favoring the handicapped or elderly—deciding if such a program is desirable or feasible—is the type of decision-making process that lends itself to human dimension research. Many questions would need to be answered and agreed on by the public, such as the types of handicaps to qualify, the age, the permission to hunt from a vehicle, the ability to hunt in special areas, the use of special weapons, the issuance of special seasons and the fees for access.

Many publics need to be questioned—the disabled and elderly, competing resource users, landowners, etc. Human dimension research with extensive public input satisfy more people than decisions from little public input.

Some states have provisions for disabled persons to hunt and shoot from vehicles. Above, a Vietnam veteran, his right arm and leg paralyzed by a land mine blast, approaches a pronghorn he shot in New Mexico at about 200 yards (183 m), while resting his rifle over a bar across the front of his "mule." The mule also provided a way to bring the buck to a waiting van. Physically challenged hunters may need a companion to help eviscerate and load game. All-terrain vehicles may create problems on fragile landscapes and by people harassing wildlife. Worse are hunters without disabilities using such transportation unethically to outcompete game and other hunters. Photo by Bart W. O'Gara.

Genetics, Management and Taxonomy

Settling the question of subspecies has application for survival and management of pronghorn (see Chapters 1 and 8). Local races are better adapted for survival in the areas where they evolved than is the species as a whole. Managers should avoid diluting gene pools through translocations or other management actions.

Molecular genetic studies completed at the end of the 20th century cast doubt on the validity of the generally accepted five pronghorn subspecies. Despite several studies already conducted, more are needed to study additional populations, larger numbers of individuals, expanded masses of genetic material and greater numbers of nuclear DNA studies to mitochondrial DNA investigations. If such research is conducted, a thorough search of records concerning translocations would be prudent. A major problem would involve developing techniques for gathering biological material from small gene pools without sacrificing endangered animals. Small gene pools are becoming more common as populations are isolated by human activities. Future augmentation or introduction of new genes to such populations may be deemed expedient. If so, the greater the knowledge of genetic similarities and differences among populations, the greater the chances of long-term success.

Research related to fossil antilocaprids seems to be at a standstill, except for refining times of existence or generic status, unless more fossils are found (see Chapter 2). The relationship of merycodontines to antilocaprines might be ascertained if fossils yielded DNA of adequate quality and quantity. If DNA analysis evolves to the point where antilocaprids and bovids consistently can be differentiated, the diversity of hornless ruminant taxa found in Miocene deposits of Eurasia (see Chapter 1) could be tested for familial relationships (again if adequate DNA could be extracted from the fossils).

Weather, Forage and Populations

Long-term (10 or more years) research is needed to ascertain relationships among weather patterns, forage nutrition and population fluctuations for pronghorn herds in grasslands, shrubsteppes and deserts (see Chapter 4).

Research and field investigations have emphasized the effects of individual mortality factors, such as predation, winter kill, disease and forage competition on population numbers. Rarely have the numerous ecological agents collectively been studies or recognized as responsible for limiting populations. Examining the effects of precipitation patterns on quantity or quality of nutritional forage, relative to fawn production, survival and recruitment has surfaced in the 1990s as a new science strategy for pronghorn (Hervert et al. 2000). These ecological perspectives need greater attention for research during the 21st century.

Pronghorn Numbers

Pronghorn numbers have been estimated on an irregular basis for more than 70 years (see Chapter 4). Such documentation is necessary for tracking population trends and determining reasons for changes. We suggest compiling pronghorn numbers every 5 years, starting in 2006. Documentation on a province-by-province and state-by-state basis can be compared with land-use changes, weather, management practices, etc., to better understand reasons for population increases and decreases. Monitoring can be accomplished by each provincial or state conservation agency, but some organization, such as the North American Pronghorn Foundation, should compile the information and maintain a data base. Continuity of personnel attending pronghorn workshops is not assured, so another organization, perhaps the North American Pronghorn Foundation, should take charge of contacting provincial and state agencies in time to present the findings for publication at workshops every fifth year. Care must be taken that surveys are consistent between provinces and states as well as year-to-year. Past published surveys have been conducted before the hunting season. For continuity, this should continue.

Behavior

A poorly understood aspect of behavior involves the benefits, or lack of them, that result from various mating systems for pronghorn populations or individuals. Also, female behavior has not been studied to the extent of male behavior. Extended research on benefits of the various mating systems and on doe behavior is indicated.

Physiology

Long-term physiological studies to determine pronghorn nutritional requirements are needed to assess carrying capacity of rangelands for this species. Past investigations have used nutritional requirements for deer and livestock for pronghorn requirements. These may not be appropriate because of major physiological and behavioral differences in species. In addition, needed information is not available concerning pronghorn use of salt (see Chapters 8 and 14).

Rudimentary data, such as the conditions of pronghorn throughout the year in relation to vegetative conditions, are unknown. Such knowledge is especially important on rangelands shared with livestock. Applied research is needed in the biomes occupied by pronghorn (grasslands, shrublands and deserts), relating vegetative condition to that of pronghorn.

Basic research into the functions of pronghorn kidneys and livers may help to better understand the ability of these animals to live in arid habitats and consume plants poisonous to livestock.

There are evident advantages of knowing pronghorn requirements for trace elements and adequate blood levels. Insufficient dietary levels of copper and selenium have deleteriously affected captive and free-ranging pronghorn. Studies under laboratory conditions, where dietary intake is known and can be manipulated, should indicate what blood levels of these elements are sufficient for health and reproduction. Also, learning how to supplement pronghorn diets on rangelands deficient in trace elements would be beneficial (see Chapters 8 and 14).

Diseases and Parasites

The pronghorn's habitat and behavior, under most circumstances, reduce the chances of disease that cause high mortality. However, more should be known concerning pronghorn diseases and parasites, especially those shared with livestock (see Chapter 10). Detailed, long-term studies under laboratory conditions are needed to determine which mammalian species are potential carriers for the other. Such studies also should determine how deleterious diseases are to each species and what levels of parasites are tolerable. Determining whether pronghorn are susceptible to chronic wasting disease seems prudent.

The next logical step would involve means of alleviating health problems under field conditions. Research is needed: to reduce pronghorn exposure to bluetongue and other hemorrhagic viruses during late summer; determine the likelihood of pronghorn infecting cattle with leptospirosis; identify the circumstances under which pronghorn will infect cattle; and determine feasible measures on rangelands to reduce parasite loads in livestock and pronghorn, such as medicated salt blocks, proper stocking rates and prescribed grazing systems. Such determinations are especially important because most pronghorn associate with cattle.

Predation

During the late 1970s and early 1980s, after miniature radio transmitters became available, numerous studies indicated high losses of fawns to predators in some populations. However, little attention was paid to the actual impact on pronghorn populations. Subsequent research indicated that summer doe/fawn ratios were poor indicators of the growth potential of herds (see Chapter 11). Additional research is needed to determine how much predation is compensatory and how much is additive. Also, most research has been conducted on herds that had low survival rates. Now, with some populations thriving, is the time to instrument large numbers of fawns to determine what mortality factors are important and acceptable during the first year of life.

Reproduction

The relationship among birth synchrony, habitat conditions, fawn survival and predation warrants further study (see Chapter 9). Research also is needed to evaluate the effects of timing and amount of precipitation as well as temperature during the forage growing season on pronghorn fawn survival.

(Top) *Ease of observation and an interesting social system make pronghorn an attractive species for behavioral studies (see Chapter 7). Consequently, more is known concerning the behavior of pronghorn than about that of most other North American ruminant species.* (Bottom) *Changing of mating systems with changing population structure or habitat conditions has been well documented. However, further research is needed to understand better the mechanisms that drive such changes. Maher's (1991, 1994, 2000) investigations should be expanded.* Top photo by Jim D. Yoakum. Bottom photo by Leonard Lee Rue; courtesy of the Leonard Rue Enterprises, Blairstown, New Jersey.

Tall, dense shrubs, especially sagebrush, often provide low-quality pronghorn habitat. Forbs and grasses are sparse, and tall vegetation interferes with the pronghorn's ability to see and flee from predators. In such situations, controlled burns or other habitat manipulation techniques, which reduce shrub cover and promote the growth of forbs, improve rangelands for pronghorn. Photo by John L. Rogers; courtesy of the U.S. Forest Service, Portland, Oregon.

Life Tables

Better data are needed to develop credible life tables for pronghorn in the three biomes they inhabit (see Chapter 12). Long-term tracking of radio-collared pronghorn is the way to gather that kind of data. Such studies would be extremely expensive, but multiple questions related to mortality, use of rangelands and movements, among others, could be answered at the same time.

Crippling Losses and Illegal Kills

The crippling and illegal killing of pronghorn, or any other game animal, is of concern to managers, hunters and antihunters. Knowledge of losses is needed for determining harvest levels and refining simulation models (see Chapter 20). Also, antihunting groups sometimes point to the lack of knowledge concerning these problems as examples of the lawlessness of hunters and the insensitivity of managers and hunters to the welfare of animals. Until the extent of the problem is known, little can be done except more rigorous law enforcement—which may be prohibitively expensive—and promotion of responsible behavior and improved hunting skills through education.

Poaching presents a problem for managers because it is an elusive factor and cannot be assumed to occur in a density-dependent manner. It may be influenced by extrinsic factors, such as the economy, and it may obscure mortality levels for a population (Halls 1984) (see Chapters 12 and 20).

To investigate crippling losses and illegal kills, little could be accomplished without radio-marked animals. Whether poachers or legal hunters select for or against radio-collared animals is unknown. Implanting mortality-sensing, intrauterine transmitters into a large number of pronghorn does is a practical way to secure information concerning this problem.

Two related research topics have been neglected. One involves methods for apprehending more poachers without greatly increasing expenses. The other involves methods for improving hunter education, especially of hunting ethics, and means to motivate sportspeople to want such education.

Carrying Capacity

Managers frequently propose maintaining a population below carrying capacity, but they seldom know the capacity of the land for a particular species. Simply stated, one definition of carrying capacity is the maximum number of animals of a given species that can be sustained in a given area by the available food and water, providing suitable cover is available. However, determination of carrying capacity is complicated by the fact it changes constantly, depending on weather conditions, intensity of foraging or browsing, etc. Vegetative condition studies, including the determination of the nutritional quality of forage and its availability, is a prerequisite to estimating carrying capacity. Developing practical, cost-effective techniques for measuring forage quality, quantity, vegetation availability and intake is a research priority.

Phenology and Nutritional Values of Forbs

Pronghorn are forb eaters by preference, and forbs—to some extent grasses—largely determine the level of fawn survival (see Chapter 14). However, many populations subsist on shrubs during winters and droughts and cannot survive without them. Knowledge concerning the phenology and nutritional values of the forbs and shrubs most used by pronghorn is lacking. Such values are important because they are used to establish foraging systems and to determine those values as a research priority. Data are available on grasses because the livestock industry primarily is interested in grass, but pronghorn use that resource sparingly.

Special need exists for phenological information concerning forbs on pronghorn rangelands (see Chapter 13). Plant phenological data have been collected primarily for graminoids to implement livestock foraging systems. Rarely has such data been collected for forbs. Because forbs often are the most preferred and consumed forage class by pronghorn, information regarding growth patterns (especially seed ripe time) is important for initiation of foraging systems for livestock on rangelands dually used by pronghorn. The collection of these data can be accomplished by habitat management agencies and need not be a component of a costly or lengthy research project.

Food Habits

The most beneficial project in the field of pronghorn food habits would be expensive and time consuming. It would involve an in vivo study comparing fecal analysis to known ingestion quantities with pronghorn, such as the one conducted by Pulliam and Nelson (1979) with elk. To be useful, such a study should be year-round and involve hundreds of species of grasses, forbs and shrubs ingested in various combinations. Conducting an in vitro digestion study simultaneously might provide a means of deriving relatively economical correction factors to use in fecal analysis studies.

Many different field methods and laboratory techniques have been used in past studies, making effective comparisons often difficult. Time and finances govern the administration of projects, but knowing what to plan for in a project aids in obtaining the best results. See Chapter 14 for options and techniques regarding food habit studies.

Of the more than 200 past food habitat studies on pronghorn, only one has been accomplished using differential digestion ratios and correlations of interactions between climate, nutrition and pronghorn populations (Hansen et al. 2001). These ecological correlations stress the values of

Thorough dietary studies are desirable, especially where forage is allotted among domestic livestock, feral horses and wild ungulates. More knowledge of site-specific areas may determine the quantity and type of forage that contribute to sustaining or maximizing pronghorn populations. However, food habit studies have been conducted, with the general conclusion that succulent forbs are preferred, and their availability during spring and summer generally determines fawn survival. Also, shrubs supply winter survival food in most areas. An intensive food habit study (Hansen et al. 2001) correlated pronghorn diet selection and differential digestion ratios. These correlations provide a wealth of foraging data. It is recommended that investigators, concentrating on diet research adopt similar correlative data collection. Photos by LuRay Parker; courtesy of the Wyoming Department of Game and Fish, Cheyenne.

food habit studies toward knowledge of agents influencing pronghorn trends. Consequently, there is need for more of these comprehensive studies.

Artificial Feeding

Quality of game management has at least two criteria—quality of the game animal itself and quality of human enjoyment of the species (Murie 1951). Pressed by hunters who want more animals to hunt and pressed by limited habitat, the manager may turn to artificial feeding or other measures that affect the gameness—the species' fitness—of a species and probably result in inferior animals. According to Aldo Leopold (1933), the recreational value of a herd of game is inverse to the artificiality of its origin, and a proper game policy seeks a happy medium between the intensity of management necessary to maintain a game supply and that which would deteriorate its quality or recreational value. This should be regarded when considering supplemental feeding.

Pronghorn populations are most stable on natural rangelands with unimpeded access to seasonal and emergency habitats (see Chapter 17). Emergency habitats are those needed during severe winters, droughts or other adverse weather conditions. Maintaining movement corridors to emergency habitats, some of which are used no more than once in 10 years, will become more and more difficult as human activities block more and more movement corridors. Consequently, emergency feeding may be the only way to save large numbers of pronghorn at certain times and in certain situations. During a severe winter or drought, pronghorn usually move toward emergency habitat until they are stopped by fences or other human-made impediments. Generally, emergency feeding should begin within a couple of days of the actual emergency for best results. Pronghorn that are starved for 10 days or more usually will not survive, despite measures taken to save them (Pearson 1969). Thus, extremely palatable rations that will be accepted immediately are needed.

Torbit et al. (1984) formulated economical, nutritionally, digestible, safe, pelleted winter rations that were acceptable to pronghorn. Continued research concerning nutritious emergency rations that pronghorn like well

"By 2025 the United States will be home to nearly 63 million more people than are here today," according to Mitchell (2001:58). During the 20th century, pronghorn largely escaped displacement by urban sprawl that affected some other species. This was because pronghorn habitat generally was not attractive development property. However, that has changed and still is changing rapidly. Even in Wyoming, new homes on the range may prevent the deer and antelope from playing there. Photo by LuRay Parker; courtesy of the Wyoming Department of Game and Fish, Cheyenne.

enough to seek is in order. Torbit et al. also investigated practical ways to deliver pelleted rations to pronghorn. Such research also should be expanded.

Artificial feeding during summers was experimented for the first time in 2002, on the Carrizo Plain National Monument, California. Because of drought conditions and lack of nutritional natural forage during the reproduction season, high mortality rates were reported for adults and fawns (Hervert et al. 2000). Recognizing the reasons for such losses, the California Department of Fish and Game provided high quality fresh alfalfa for wild herds with a history of declining trends. The animals readily consumed the feed and no mortalities were observed (Koch and Yoakum 2002).

Crop Depredations

Results of a survey sent to conservation agencies of 18 western provinces and states in 1991 indicated that crop depredations were stable in nine and increasing in eight (O'Gara and Yoakum 1992). Only in Nebraska were depredation complaints decreasing, apparently because pronghorn numbers were depressed. Almost every agency reported some damage to alfalfa and wheat. Wildlife managers in some states have responded to damage complaints by reducing pronghorn numbers. At times and in certain areas, the Colorado Division of Wildlife has reduced pronghorn numbers by up to 55% to reduce depredations on winter wheat, but pronghorn may not damage winter wheat (Torbit et al. 1993). Assessing this statement requires integrating plant physiology, winter wheat/herbivore interactions, ungulate foraging theory and pronghorn movement patterns (Strohmeyer 1994).

Winter wheat commonly is grazed by livestock from November until April. Crude protein in wheat exceeds that of native plants during winter, but decreases in April as nutrients relocate from stems and leaves to the developing grain head. Crude protein concentrations of native rangeland plants peak in April and May (Schwartz 1977). Pronghorn should not damage winter wheat if they voluntarily stop grazing wheat by April (Strohmeyer 1994). During experiments, 80 times the normal densities found in free-ranging pronghorn populations did not reduce grain yields by winter grazing (Torbit et al. 1993). Monthly aerial surveys in northwestern Colorado found free-ranging pronghorn using wheat fields from November through April, then abandoning them by early May (Strohmeyer 1994).

Further research is needed to determine precisely the phenological stage at which pronghorn switch from winter wheat to native rangeland vegetation. Research of other damage caused by pronghorn to various crops also is warranted.

Modeling

Additional knowledge is needed for modeling and predicting the distribution and dynamics of pronghorn populations in natural landscapes, especially whether or not carrying capacity is a useful concept that can be quantified (see Chapter 19).

Buck Harvests

If biological principles are to be used in wildlife management, biological laws must be allowed to function (see Chapter 20). One of these laws is selection. Experience with livestock has shown that attention to breeding helps to maintain healthy, prime animals; neglect of selective breeding produces substandard animals. Preservation of animals' virility and gameness is of paramount importance. Sportsmen and sportswomen tend to kill the largest males in a game population, and harvests should be managed to minimize the effects of such losses (Murie 1951). This is not to say that trophy hunting is bad, but that it should be managed through bag limits or other methods, to avoid long-term genetic deterioration of a species. Learning the best ways of accomplishing this requires an intensive manipulation of captive animals (possibly involving two or more zoos) or long-term monitoring of differently managed populations.

Expansion into Historic Rangelands

With pronghorn at relatively high numbers, natural expansions into unoccupied habitats are taking place. Managers and researchers must be alert to such expansions. Thorough documentation may teach a great deal about habitat preferences of pronghorn and which habitat parameters pioneers seek. This knowledge would be helpful to evaluate habitat considered for introductions (see Chapter 21). An extended drought or several severe winters could easily reduce some populations to a fraction of their present numbers. If that happens, knowledge concerning the present expansions may help to repeat the expansion process.

Provincial and state wildlife agencies have reported that the biggest threat to pronghorn is and will be permanent loss of habitat due to human activities. This provides managers of pronghorn populations and habitats with the considerable challenge of seeking and developing techniques to increase forage and drinking water on the remaining, occupied rangelands. Such objectives can and have been accomplished, resulting in increased pronghorn numbers, such as on the private lands of the Deseret Ranch in Utah (see Chapter 16), and on public lands, as in the case of Hart Mountain National Antelope Refuge in

Pronghorn habitat is being lost to improved pastures, grain production, energy development and general human encroachment. Changing pronghorn habitat into cultivated fields is an invitation to crop depredations. Breaking sod with modern equipment is fast and efficient. Governmental programs have been designed to reduce overproduction of grain and prevent deterioration of highly erodible rangelands. Unfortunately, those programs have made sodbusting profitable in some cases and contributed to the shrinking of prairie habitat. Top photo by Bart W. O'Gara. Bottom left photo by Alice Koch. Bottom right photo by Bob Schaffer.

Oregon (see Chapter 15). Both of these areas have experienced continuous, increased pronghorn herds for the past two decades because forage and water are improved, which increases the carrying capacities of these rangelands. The habitat enhancement practices are tested and proven, but few management agencies recognize or implement them. Consequently, there is a need for additional, well-conceived and managed enhancement projects, on both public and private lands, to demonstrate for agencies the goals, methods and efficacy of such investments. Accomplishing this may be one of the greatest challenges for maintaining or increasing pronghorn populations during the 21st century.

CONCLUSIONS

> *Eternal alertness is the price of life for the antelope in that vast open country, and without constant vigilance on the part of conservationists this unique American animal will vanish from the earth.*
>
> Ira Gabrielson (1935:576)

> *When the last corner lot is covered with tenements we can still make a playground by tearing them down, but when the last antelope goes by the board, not all the playground associations in Christendom can do aught to replace the loss.*
>
> Leopold (1953:167–168)

> *Continued agricultural and mining expansion onto historic ranges, fencing across migration routes and removal of native vegetation by range rehabilitation projects and overgrazing are obstacles to the future success of this unique species.*
>
> Autenrieth (1978:478)

The challenge for the future is to ensure that pronghorn are given adequate consideration and accorded equitable treatment in the management of western rangelands. Maintenance of habitat to satisfy pronghorn's seasonal needs is crucial. The human population will continue to grow, and society's demands for uses of natural resources will increase (Wildlife Management Institute Staff 1978, Poole and McCabe 1986). Both have the potential to affect pronghorn habitats adversely. However, certain adverse impacts can be countered by improved management.

Application of improved management efforts will depend on wildlife agency and public understanding and acceptance.

For survival, pronghorn must have suitable expanses of rangelands where they can move to secure the annual necessities of life without impediments. At the beginning of the 21st century, some pronghorn were faring well on western rangelands, but increasing numbers of humans may so reduce the open spaces and use the land so intensively that pronghorn can no longer survive.

Droughts and severe winters are natural and have reduced pronghorn populations periodically. Droughts may become more critical if global warming continues, as predicted by many scientists. Also, much habitat loss today seems long term or permanent. Populations have rebuilt, usually rapidly, after severe winters and droughts, but as habitat losses increase, the chances increase that each recovery will not be complete, leading to a long-term decline. Such a decline cannot be checked by reduced hunting quotas, although that may become necessary. Only by keeping movement corridors open between seasonal habitats and maintaining habitats in quality ecological conditions that meet all of the pronghorn's requirements will loss be prevented.

Pronghorn population levels are at a fraction of their pristine numbers, and human population levels are skyrocketing. When humans greatly increase in number, they dominate the land, using and changing it to meet their demands. The extent to which North America's growing human population will opt and be able to conserve or nurture western rangelands for the benefit of pronghorn is uncertain. However, that the future of pronghorn is in the hands of humans is quite certain.

State wildlife agency should assist managing pronghorn on land over which they have no jurisdiction and limited control. Most private landowners are tolerant of game, but less so of hunters on their land. However, economics of the livestock industry dictate that rangelands be managed more and more intensively. For the benefit of pronghorn, hunters must defray more of the costs of restricting incompatible or otherwise economically more valuable land uses. Such costs may take the form of hunting fees paid to landowners or increased license or tag fees with the increased amounts prorated to landowners.

Wildlife professionals can ensure that their management programs are rooted in sound science, derived through long-term studies that include human dimensions. Natural resource management agencies can best defend their programs through professional excellence and active, open discussion. Hunters should police their own ranks, and they can best promote the continuation of wildlife management that maintains healthy populations of wildlife for the benefit of humans by ethical participation in their activities and active enrollment in conservation organizations.

Pronghorn managers will face tremendous challenges during the 21st century, as they attempt to motivate landowners to consider pronghorn in land-management decisions, prevent pronghorn hunting from becoming a sport of the wealthy and protect hunters' rights.

The story of pronghorn in modern time tells us that most extant variables of significance to pronghorn well-being are known to science. Science, of course, cannot adequately predict the dynamic intensity of those variables and the cumulative impact on the species. But we know from history that, if science, government and the North American public take a species, such as pronghorn, for granted, they chance a decline in that species' numbers, distribution and values. History has shown that the pronghorn has exhibited an extraordinary ability to recover from all limiting factors but neglect. We hope this book ensures that such neglect never happens.

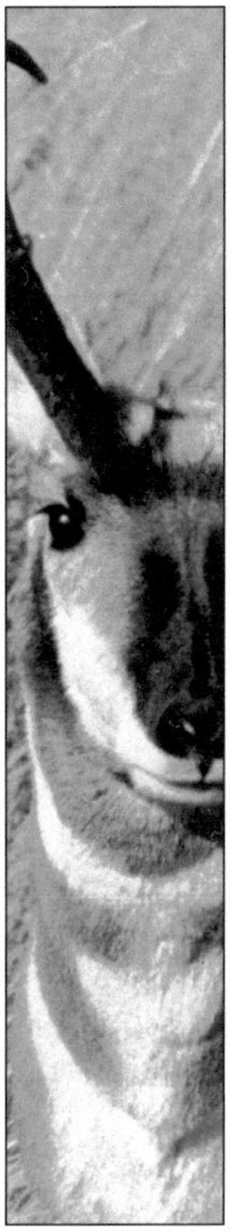

Appendix A: Common and Scientific Names of Plants Cited

Acacia	*Acacia*
Agave cactus	*Agave deserti*
Alfalfa	*Medicago sativa*
Almond	*Prunus amygdalus*
Alsike clover	*Trifolium hybridum*
American elm	*Ulmus americana*
Antelope bitterbrush	*Purshia tridentata*
Apple	*Malus*
Arizona cliffrose	*Cowania purshia*
Arrowleaf balsamroot	*Balsamorhiza sagittata*
Arrowleaf	*Trifolium vesiculosum*
Barestem encaliopsis	*Encelopsis nudicaulis*
Barrel cactus	*Ferocactus*
Basin wildrye	*Elymus cinereus*
Basswood	*Tilia*
Beardgrass	*Bothriochloa*
Bearded wheatgrass	*Agropyron subsecundum*
Beardless wheatgrass	*Agropyron inermis*
Beardless wildrye	*Elymus triticoides*
Bedstraw	*Galium*
Beebalm	*Monarda*
Big (basin or mountain) sagebrush	*Artemisia tridentata*
Big bluegrass	*Poa ampla*
Big bluestem	*Andropogon gerardii*
Big rabbitbrush	*Chrysothamnus linifolius*
Bigelow's nolina	*Nolina bigelovii*
Birdsfoot trefoil (deervetch)	*Lotus corniculatus*
Black greasewood	*Sarcobatus vermiculatus*
Black sagebrush	*Artemisia nova*
Blue elderberry	*Sambucus cerulea*
Blue grama	*Bouteloua gracilis*
Blue wildrye	*Elymus glaucus*
Blueberry elder	*Sambucus caerulea*

Bluebunch wheatgrass	*Agropyron spicata*
Borroweed	*Haplopappus tenuisectus*
Bottlebrush squirreltail	*Sitanion hystrix*
Bunchgrass	*Andropogon*
Bracken fern	*Pteridium aquilinum*
Brickellia	*Brickellia*
Brome	*Bromus*
Broom snakeweed	*Gutierrezia sarothrae-lucida*
Buckwheat	*Fagopyrum*
Bulbous barley	*Hordeum bulbosum*
Bulbous bluegrass	*Poa bulbosa*
Bull thistle	*Cirsium vulgare*
Burroweed	*Isocoma tenuisecta*
Bursage	*Franseria*
Cain cholla	*Opuntia spinosior*
Canadian thistle	*Cirsium arvense*
Catclaw acacia	*Acacia greggii*
Ceanothus	*Ceanothus*
Chain (jumping) cholla	*Opuntia fulgida*
Cheatgrass brome	*Bromus tectorum*
Chickpea milkvetch	*Astragalus cicer*
Cliffrose	*Cowania mexicana*
Clover	*Trifolium*
Colorado rubberweed	*Hymenoxys richardsonii*
Common buffalograss	*Buchloe dactyloides*
Common camass	*Camassia quamash*
Common chokecherry	*Prunus virginiana*
Common (rough) cockleburr	*Xanthium strumarium*
Common cowparsnip	*Heracleum maximum*
Common flax	*Linum usitassisimum*
Common halogeton (saltlover)	*Halogeton glomeratus*
Common Russianthistle	*Salsola kali*
Common tansy-aster	*Machaeranthera tanacetifolia*
Common rye	*Secale cereale*
Common winterfat	*Eurotia lanata*
Common yarrow	*Achillea millefolium*
Coral lichen	*Cladia retipora*
Corn (maize)	*Zea mays*
Creeping (beardless) wildrye	*Elymus triticoides*
Creeping juniper	*Juniperus horizontalis*
Crested wheatgrass (fairway)	*Agropyron cristatum*
Crested wheatgrass (standard)	*Argropyron desertorum*
Creosotebush	*Larrea tridentata*
Cryptantha	*Cryptantha*
Curlleaf mountainmahogany	*Cercocarpus ledifolius*
Dalea	*Psorothamnus*
Deathcamas	*Zigadenus*
Desert almond (peachbush)	*Prunus fasciculata*
Desert evening primrose	*Oenothera primiveris*
Desert-thorn	*Lycium macrodon*
Douglas-fir	*Pseudotsuga menziessi*
Dropseed	*Sporobulus*
Elk sedge	*Carex geyeri*
Ephedra	*Ephedra*
Eriogonum (wild-buckwheat)	*Eriogonum*
Euphorpbia (spurge)	*Euphorpia*
Featherleaf (northern) fleabane	*Erigeron compositus*
Field pea	*Pisum arvense*
Fineleaf hymenopappus	*Hymenopappus filifolius*
Fivehook bassia	*Bassia hyssopifolia*
Fourwing saltbrush	*Atriplex canescens*
Foxtail barley	*Hordeum jubatum*
Foxtail brittlegrass	*Setaria italica*
Fragrant sumac	*Rhus aromatica*
Fringed (pasture) sagebrush	*Artemisia frigida*
Galleta	*Pleuraphis*
Geranium	*Geranium*
Gilia	*Gilia*
Goldenrod	*Oligoneuron*
Gooseberryleaf (prairie) globemallow	*Sphaeralcea grossulariaefolia*
Gourd	*Cucurbita*
Grama grass	*Bouteloua*
Gray rabbitbrush	*Chrysothamnus nauseosus*
Green ephedra	*Ephedra viridis*
Green needlegrass	*Stipa viridula*
Green's rabbitbrush	*Chrysothamnus greenei*
Grey lichen	*Dimelaena radiata*
Grouse whortleberry	*Vaccinium membranaceum*
Guinea grass	*Panicum maximum*
Hairy grama	*Bouteloua hirsuta*
Halogeton	*Halogeton glomeratus*
Haole koa	*Leucaena leucocephala*
Hickory	*Carya*
Hoary aster	*Aster canescens*
Horsetail	*Equisetum*
Idaho fescue	*Festuca idahoensis*
Indiangrass	*Sorghastrum*
Indian ricegrass	*Oryza hymenoides*
Intermediate wheatgrass	*Agropyron intermedium*
Ironwood	*Olneya tesota*
Kentucky bluegrass	*Poa pratensis*
Lambstongue groundsel	*Senecio integerrimus*
Larkspur	*Delphinium*
Lead plant	*Amorpha canescens*
Leather root	*Hoita macrostachya*
Lewis mockorange	*Philadelphus lewisii*
Lewis (prairie) flax	*Linum lewisii*
Linseed	*Linum*
Little Bluestem	*Andropogon scoparium*
Little rabbitbrush	*Chrysothamnus stenophyllus*
Littleleaf mountainmahogany	*Cercocarpus intricatus*
Locoweed	*Astrogalus*
Lodgepole pine	*Pinus contorta*
Low sagebrush	*Artemisia arbuscula*

Lupine	*Lupinus*
Mallow	*Malvella*
Maple	*Acer*
Meadow foxtail	*Alopecurus pratenses*
Meadow brome	*Bromus biebersteinii*
Mescal	*Sophora secundiflora*
Mesquite	*Prosopis juliflora*
Molasses grass	*Melinus minutiflora*
Monkshood	*Aconitum*
Moss	*Dicranium*
Mountain brome	*Bromus marginatus*
Mountain brush	*Cypripedium fasciculatum*
Mountain lupine	*Lupinus monticola*
Mountain rye	*Secale montanus*
Mountain sagebrush	*Artemisia vaseyana*
Mule-ears wyethia	*Wyethia amplexicaulis*
Musk	*Mimulus moschatus*
Musk thistle	*Carduus nutans*
Mustard	*Brassica*
Needle-and-thread	*Stipa comata*
Needlegrass	*Stipa*
Nevada bluegrass	*Poa nevadensis*
Northpark (Wasatch) penstemon	*Penstemon cyananthus*
Oak	*Quercus*
Oat	*Avena*
Ocotillo	*Fouquieria*
Oldman wormwood	*Artemesia abrotanum*
Orchid	*Habenaria*
Orchardgrass	*Dactylis*
Palmer's penstemon	*Penstemon palmeri*
Paloverde	*Cercidium*
Pea	*Pisum*
Peavine	*Lathyrus*
Penstemon	*Penstemon*
Phacelia	*Phacelia*
Phlox	*Phlox*
Plains bluestem	*Bothriochloa ischaemum*
Plantain	*Plantago*
Pine	*Pinus*
Pinyon pine	*Pinus edulis*
Pingue (hymenoxys)	*Hymenoxys richardsonii*
Ponderosa pine	*Pinus ponderosa*
Prairie junegrass	*Koeleria cristata*
Prairie turnip	*Psoralea esculenta*
Pricklypear (cholla)	*Opuntia*
Prince's plume	*Stanleya pinnata*
Pubescent wheatgrass	*Agropyron trichoforum*
Puccoon (gromwell)	*Lithospermum*
Quaking aspen	*Populus trembuloides*
Ragweed	*Ambrosia*
Range lichen	*Xanthoparmelia idahoensis*
Rattleweed	*Crotalaria*
Red fescue	*Festuca rubra*
Red sage (big red) sage	*Salvia penstemunoides*
Red willow	*Salix laevigata*
Reed canarygrass	*Phalaris arundinacea*
Riddell groundsel	*Senecio riddellii*
Rockcress	*Arabis*
Rocky Mountain juniper	*Juniperus scopulorum*
Rocky Mountain maple	*Acer glabrum*
Rose	*Rosa*
Rubber (gray or dwarf) rabbitbrush	*Chrysothamnus nauseosus*
Rubberweed	*Hymenoxys*
Rush	*Juncus*
Russianthistle	*Salsola kali*
Russian wildrye	*Elymus junceus*
Rye	*Secale*
Sage	*Salvia*
Saltbush-greasewood	*Atriplex sarcobatus*
Saltgrass	*Distichlis*
Sand bluestem	*Andropogon hallii*
Sand dropseed	*Sporobolus cryptandrus*
Sand sagebrush	*Artemisia filifolia*
Sandburg bluegrass	*Poa secunda sandbergii*
Sand shinnery (harvard) oak	*Querqus havardii*
Sandreed	*Calamovilfa*
Sanfoin	*Onobrychis viciaefolia*
Scarlet globemallow	*Sphaeralcea coccinea*
Shadescale	*Atriplex confertifolia*
Sheep's fescue	*Festuca ovina*
Shin oak	*Quercus gambelii*
Showy goldeneye	*Vigniera multiflora*
Siberian wheatgrass	*Agropyron fragile*
Sicklepod milkvetch	*Astragalus falcatus*
Sideoats grama	*Boutloua curtipendula*
Silver fir	*Abies alba*
Silver (hoary) sagebrush	*Artemisia cana*
Sixrow barley	*Hordeum vulgare*
Skunkbrush sumac	*Rhus trilobata*
Slender wheatgrass	*Agropyron trachycaulum*
Smooth brome	*Bromus inermis*
Snakewood (letterwood or specklewood)	*Piranitera guianensis*
Soybean	*Glycine*
Spanish bayonet	*Yucca harrimaniae*
Speargrass	*Piptochaetium*
Spruce	*Picea*
Squash	*Cucurbita*
Squirreltail barley	*Elymus elymoides*
Squawthorn	*Lycium torreyi*
Strawberry clover	*Trifolium fragiferum*
Sugar beet	*Beta vulgaris*
Summer cypress	*Kochia scoparia*
Sunflower	*Helianthus*
Sweetanise	*Osmorhiza accidentalis*
Sweetclover	*Melilotus*
Switchgrass	*Panicum virgatum*
Tall fescue	*Festuca arundinaceum*

Tall oatgrass	*Arrhenatherum elatius*	Walkingstick cholla	*Opuntia imbricata*
Tall wheatgrass	*Agropyron elongatum*	Western juniper	*Juniperus occidentalis*
Tansy-aster (machaerantha)	*Machaeranthera*	Western ragweed	*Ambrosia psilostachya*
Tapertip hawksbeard	*Crepis acuminata*	Western snowberry	*Symphoricarpos occidentalis*
Tarbush	*Flourensia*	Western wheatgrass	*Agropyron smithii*
Threeawn	*Aristida*	Western yarrow	*Achillea lanulosa*
Three-tip sagebrush	*Artemisia tripartita*	Wheat	*Triticum*
Thurber needlegrass	*Stipa thurberianum*	Whiteball acacia	*Acacia angustissima*
Timothy	*Phleum*	Wild onion	*Allium*
Tobacco	*Nicotiana*	Wild rose	*Rosa*
Tobosa grass	*Hilaria berlangeri*	Wood aster	*Aster acuminatus*
Toothed groundsel	*Senecio tridenticulatus*	Wooley senecio	*Senecio longilobus*
Tufted eveningprimrose	*Oenothera caespitosa*	Wooly paperflower	*Psilostrophe tagetina*
Tule bulrush	*Scirpus acutus*	Wyoming big sagebrush	*Artemisia wyomingensis*
Twayblade	*Listera*	Yellow brittlegrass	*Setaria lutescens*
Two-grooved milkvetch	*Astragalus bisulcatus*	Yellow goatsbeard (salsify)	*Tragopogon dubius*
Utah juniper	*Juniperus osteosperma*	Yellow Indiangrass	*Sorghastrum nutans*
Velvet lupine	*Lupinus leucophyllus*	Yellow sweetclover	*Melilotus officinalis*
Vetch	*Vicia*	Yellowbrush	*Chrysothamnus viscidiflorus*
Virginia's bower	*Clematis virginiana*	Yucca	*Yucca*

This listing may not include all plants cited in the text or the most recent scientific names. Plant species and higher taxa were checked against Leake et al. (1993), Albee et al. (1988), and website listings at *http://plants.usda.gov* and *http://biology.usgs.gov/s+t/SNT/noframe/zx197.htm*.

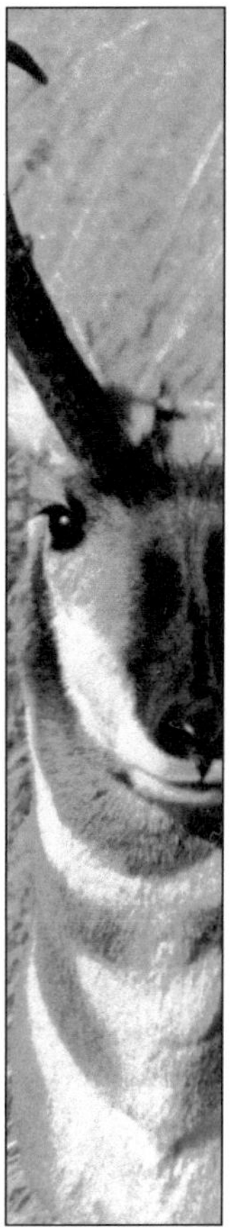

Appendix B: Common and Scientific Names of Animals Cited

African elephant	*Loxodonta africana*
African elephant shrew	*Elephantulus*
Agouti	*Dasyprocta*
American deer	*Odocoileus*
Antelope (African, Asian, bovid, Old World or true)	*Antilopinae, Cephalophinae, Hippotraginae*
Antelope squirrel	*Ammospermophilus leucurus*
Armadillo	*Dasypodidae*
Arthropod	*Arthropoda*
Badger	*Taxidea taxus*
Barren-ground caribou	*Rangifer tarandus groenlandicus*
Bat	*Chiroptera*
Bear dog	*Amphicyonidae*
Beaver	*Castor canadensis*
Bighorn sheep	*Ovis canadensis*
Bison (buffalo)	*Bison bison*
Biting fly	*Tabinadae* and *Regionidae*
Biting or "No-see-um" gnat	*Culcoides variipennis*
Black bear	*Ursus americanus*
Black fly	*Simulien*
Blackbird	*Icteridae*
Blackbuck	*Antilope cervicapra*
Black-tailed (mule) deer	*Odocoileus hemionus*
Black-tailed jackrabbit	*Lepus californicus*
Bobcat	*Felis rufus*
Bone-eating dog	*Borophaginae*
Bony-shelled glyptodont	*Glyptodon*
Bot	*Oestridae, Gasteropilidae* and *Cuterebridae*
Bovid	*Bovidae*
Brocket deer	*Mazama*

Brontosaur	*Brontosaurus excelsus*
Brown bear	*Ursus arctos*
Buffalo (erroneous common name, but used for American bison)	*Bison bison*
Burro (donkey or ass)	*Equus asinus*
Burrowing (ground) owl	Speotyto cunicularia
Camel	*Camelidae*
Canid tapeworm	*Taenia hydatigena*
Cape (or African) buffalo	*Syncerus caffer*
Caribou	*Rangifer tarandus*
Cat	*Felis catus*
Cave bear	*Ursus spelaeus*
Cheetah	*Acinonyx jubatus*
Chevrotain	*Tragulus*
Chicken	*Gallus*
Chicken mite	*Dermanyssus gallinae*
Chinese water deer	*Hydropotes*
Chipmunk	*Tamias*
Clam	*Anadonta*
Coatimundi	*Nasua nasua*
Common raven	*Corvus corax*
Cottontail rabbit	*Sylvilagus*
Cougar	*Felis concolor*
Couprey	*Bos sauveli*
Cow	*Bos taurus*
Coyote	*Canis latrans*
Crane	*Grus*
Dall's sheep	*Axis dalli dalli*
Deer mice	*Peromyscus maniculatus*
Dicobunid	*Dicobunidae*
Dik-dik (East African antelope)	*Madogua kirkii*
Dog	*Canis familiaris*
Domestic goat	*Capra hircus*
Domestic sheep	*Ovis aries*
Dorcas gazelle	*Gazella dorcas*
Duck	*Anatinae*
Dwarf antelope	*Neotragus*
Eagle	*Aquila chrysaetos*
Earthworm	*Annelidae*
Elegant trogon	*Trogon elegans*
Eurasian brown bear	*Ursus arctos*
European elk	*Alces alces*
European rabbit	*Oryctolagus cuniculus*
Eye gnat	*Hippelatus pusio*
Face fly	*Musca autumnalis*
Fallow deer	*Dama dama*
Feral horse	*Equus caballus*
Fish	*Teleostei*
Flea	*Siphonoptera*
Florida panther	*Felis concolor coryi*
Fly	*Diptera*
Four-horned antelope	*Tetracerus quadricornis*
Fox	*Vulpes*
Frog	*Rana*
Fur seal	*Otaridae*
Gall wasp	*Cynipidae*
Gazelle	*Gazella* and *Procapra*
Gemsbok (Kalahari oryx)	*Oryx gazella*
German shepherd	*Canis familiaris*
Giant ground sloth	*Eremotherium laurillardi*
Giraffe	*Giraffidae*
Giraffe-like camel	*Oxydactylus*
Gnat	*Ceratopogonidae*
Goat	*Capra*
Golden eagle	*Aquila chrysaetos*
Gomphotheres	*Gomphotheriidae*
Gopher	*Geomyidae*
Gray fox	*Urocyon cineroargenteus*
Gray partridge	*Perdix perdix*
Gray whale	*Eschrichtius robustus*
Gray (timber) wolf	*Canis lupus*
Grizzly bear	*Ursus arctos*
Groundsquirrel	*Spermophilus*
Grouse	*Tetraonidae*
Hard tick	*Ixodidae*
Hare	*Lepus*
Hawk	*Accipitridae*
Hippopotamus	*Hippopotamidae*
Hog deer	*Axis porcinus*
Horned lark	*Eremophila alpestris*
Horse	*Equus caballus*
House mouse	*Mus musculus*
Human	*Homo sapiens*
Hyena	*Hyaenidae*
Hyrax	*Hyracoidea*
Indian nilgai	*Boselaphus tragocamelus*
Jackrabbit	*Lepus*
Jacob's sheep	*Ovis aries*
Jaguar	*Panthera onca*
Japanese sika deer	*Cervus nippon nippon*
Javelina	*Tayassu tajacu*
Kaibab mule	*Odocoileus hemionus*
Kangaroo	*Macropus*
Kangaroo mouse	*Microdipodops megacephalus*
Kangaroo rat	*Dipodomys*
Ked	*Hippoboscidae*
Kit fox	*Vulpes macrotis*
Kudu	*Tragelaphus strepsiceros*
Labrador retriever	*Canis familiaris*
Lagomorph	*Lagomorpha*
Leadbeater's possum	*Gymnobelideus leadbeateri*
Lice	*Anoplura*
Little blue heron	*Egretia caerulea*
Llama	*Lama glama*
Louse fly (deer ked)	*Lipoptena, Neolipoptena*
Lungworm	*Metastrongylidae*
Mallard duck	*Anas platyrhynchos*
Mammal	*Mammalia*

Mammoth	*Mammuthus*
Marmot	*Marmota*
Mastodon	*Gomphotheridae, Mammutidae*
Mexican pronghorn	*Antilocapra americana mexicana*
Mexican jay	*Aphelocoma ultramarina*
Midge	*Phlebotomus*
Mite	*Acari*
Monkey	*Cebidae* and *Cercopithecidae*
Monkey (Old World)	*Cercopithecidae*
Moose	*Alces alces*
Mosquito	*Aedes, Anopheles, Culex, Culiseta*
Mountain goat	*Oreamnos americanus*
Mountain lion	*Felis concolor*
Mountain sheep	*Ovis canadensis*
Mourning dove	*Zenaida macroura*
Mouse deer	*Tragulus*
Mule	*Equus caballus* x *E. asinus*
Mule deer	*Odocoileus hemionus*
Muntjac	*Muntiacus*
Musk deer	*Moschus*
Musk ox	*Ovibos moschatus*
Muskrat	*Ondatra zibethicus*
Nematode	*Nematoda*
New World deer	*Odocoileinae*
North American camel	*Lama, Paleolama*
North American elk	*Cervus elaphus*
Northern fur seal	*Callorhinus ursinus*
Northern pintail	*Anas acuta*
Northern spotted owl	*Strix occidentelis caurina*
Norway (brown) rat	*Rattus norvegicus*
Okapi	*Okapia johnstoni*
One-humped camel (dromedary)	*Camelus dromedarius*
Opossum	*Didelphidae*
Oregon pronghorn	*Antilocapra americana oregona*
Oribi	*Ourebia ourebi*
Otter	*Lutra canadensis*
Owl	*Strigiformes*
Oxen	*Bos taurus*
Pack rat	*Neotoma*
Peccary	*Tayassuidae*
Pelican	*Pelicanus*
Peninsular pronghorn	*Antilocapra americana peninsularis*
Pheasant	*Phasianus*
Pig	*Sus scrofa*
Pocket gopher	*Thomomys bottae*
Porcupine	*Erethizon dorsatum*
Prairie chicken	*Tympanuchus*
Prairie dog	*Cynomys*
Prairie falcon	*Falco mexicanus*
Primate	*Catarrhini, Platyrrhini* and *Tarsii*
Prongbuck	*Antilocaprinae*
Pronghorn	*Antilocapra americana*
Protozoa	*Protista*
Przewalski's horse	*Equus caballus przewalskii*
Puma	*Felis concolor*
Rabbit	*Sylvilagus*
Rabbit tick	*Dermocenter parumapterus*
Raccoon	*Procyon lotor*
Raptor	*Falconiformes*
Rat	*Rattus*
Rattlesnake	*Crotalidae*
Red deer	*Cervus elaphus*
Red fox	*Vulpes vulpes*
Red kangaroo	*Macropus rufus*
Red wolf	*Canis rufus*
Reindeer	*Rangifer tarandus*
Ringnecked pheasant	*Phasianus colchicus*
Rhinoceros	*Rhinocerotidae*
Rock Mountain elk	*Cervus elaphus canadensis*
Rocky Mountain goat	*Oreamnos americanus*
Rocky Mountain bighorn sheep	*Ovis canadensis*
Rocky Mountain wood tick	*Dermacenter andersoni*
Rodent	*Rodentia*
Roe deer	*Capreolus*
Saber-toothed tiger	*Smilodon californicus*
Sage grouse (sage hen)	*Centrocercus urophasianus*
Sagebrush defoliator	*Agora websteri*
Sagebrush vole	*Lemmiscus curtatus*
Saiga antelope	*Saiga tatarica*
San Joaquin kit fox	*Vulpes macrotis mutica*
Sandhill crane	*Grus canadensis*
Sea urchin	*Arbacia*
Sharp-tailed grouse	*Tympanuchus phasianellus*
Sheep	*Ovis*
Short-faced bear	*Arctodus*
Siberian tiger	*Panthera tigris altaica*
Sika deer	*Cervus nippon*
Skunk	*Spilogale* and *Mephitis*
Sloth	*Bradypus* and *Choloepus*
Snail	*Lymnaea*
Snake	*Serpentes*
Snapping turtle	*Chelydridae*
Snowshoe hare	*Lepus americanus*
Sonoran pronghorn	*Antilocapra americana sonoriensis*
Soft tick	*Argasidae*
South American fur seal	*Arctocephalus australis*
Spectacled bear	*Tremarctos*
Spider	*Arachnida*
Spotted owl	*Strix occidentelis*
Springbuck (springbok)	*Antidorcas marsupialis*
Squirrel	*Sciurus*
Staghound	*Canis familiaris*

Starfish	*Asteroidea*	Vesper sparrow	*Pooecetes gramineus*
Steenbok (or steenbuck)	*Raphicerus campestris*	Vicuna	*Vicugna vicugna*
Stomach worm	*Nematoda*	Vole	*Muridae*
Striped skunk	*Mephitis mephitis*	Water chevrotain	*Hyemoschus aquaticus*
Tapeworm (flatworm)	*Cestoda*	Whale	*Cetacea*
Tapir	*Tapirus*	White-tailed deer	*Odocoileus virginianus*
Texas longhorn	*Bos taurus*	Wild turkey	*Meleagris gallopavo*
Thompson's gazelle	*Gazella thomsonii*	Wildebeest	*Connochaetes taurinus*
Three-toed horse	*Mesohippus*	Wood rat	*Neotoma*
Tick	*Acarina*	Wombat	*Vombatidae*
Tiger	*Panthera tigris*	Wooly mammoth	*Mammuthus primigenius*
Tule elk	*Cervus elaphus nannodes*	Yak	*Bos grunniens*
Turtle	*Testudines*	Yellowstone elk	*Cervus elaphus canadensis*
Two-humped camel	*Camelus bactrianus*	Zebra	*Equis burchelli*

This list does not necessarily include every animal mentioned in the book. Names for the mammalian species and higher taxa were checked against Nowak (1999) and Chapman and Feldhammer (1982). Subspecies, when used, are from the references named. Bird names were checked against the American Ornithologists' Union (1998). Several other names were located in Webster's Unabridged Dictionary, 2nd ed., 1998, and a taxonomic list found at *http://biology.usgs.gov/s+t/SNT/noframe/zx197.htm*, compiled by Paul A. Opler, Colorado State University, Fort Collins.

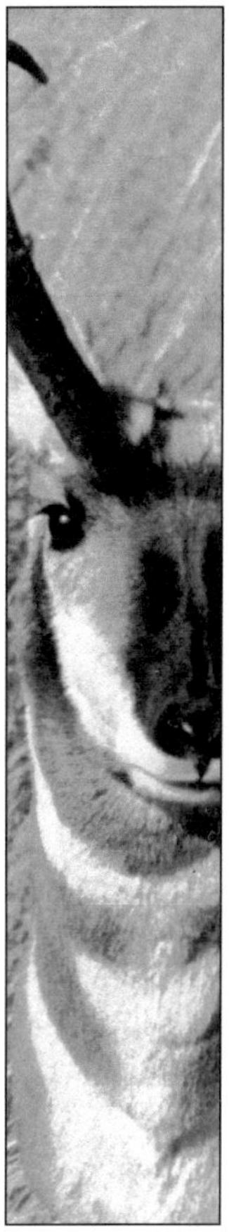

References

Abbott, C. A., M. W. Berry, J. C. Dempsey, E. J. Comiskey, L. J. Gross and H. K. Luh. 1995. Computational models of white-tailed deer in the Florida Everglades. Tech. Rept. CS-95-296 (online document available at: http://www.cs.utk.edu/~berry/deer/deer.html). Computer Sci. Dept., Univ. Tennessee, Knoxville.

Ackerly, W. F. and V. Regier. 1956. Northeastern California antelope studies. California Dept. Fish and Game, Sacramento. 44 pp.

Ackerman, B. B., F. A. Leban, M. D. Samuel and E. O. Garton. 1990. User's manual for program home range. Tech. Rept. 15. Univ. Idaho For., Wildl. and Range Exp. Sta., Moscow. 80 pp.

Adams, C. E., J. K. Thomas and C. W. Ramsey. 1992. A synopsis of Texas hunting leases. Wildl. Soc. Bull. 20(2):188–197.

Adolph, T. A. 1957. Dynamics of the steppe antelope domain and the numbers of their stock. Pages 115–131 *in* V. B. Dubinen, ed., Conservation of natural resources and the establishment of reserves in the USSR. Bull. 2. Academy of Sciences of the USSR, Moscow, Russia. Translated from Russian by H. Raziel and the Israel Program for Scientific Translations Staff. Printed for the Natl. Sci. Foundation and U.S. Dept. Inter., Washington, D.C., by S. Monson, Jerusalem, Israel, 1960. 143 pp.

Adrian, V. J. and R. E. Keiss. 1977. Survey of Colorado's wild ruminants for serologic titers to brucellosis and leptospirosis. J. Wildl. Dis. 13(4):429–431.

Aguirre, c. A., y V. S. Sotomayor. 1981. El berrendo. Bosques y Fauna. 4(3):19–26.

Aikens, C. M. 1970. Hogup Cave. Univ. Utah Anthropol. Pap. 93. Univ. Utah Press, Salt Lake City. 212 pp.

Aikens, C. M., D. L. Cole and R. Stuckenrath. 1977. Excavations at Dirty Shame Rockshelter, southeastern Oregon. TEBIWA: Miscellaneous Papers. 4. Idaho St. Univ. Mus. Nat. Hist., Pocatello. 29 pp.

Akçakaya, H. R. 1996. Linking GIS with models of ecological risk assessment for endangered species. Proc. Third Intl. Conf. on Integrating GIS and Environ. Modeling (online document available at: http://www.ramas.com/santafe.htm).

Alberta Fish and Wildlife Division. 1990. Management plan for pronghorn antelope in Alberta. Wildl. Manage. Planning Series 3. For. Lands and Wildl., Fish and Wildl. Div., Edmonton, Alberta. 115 pp.

Alexander, A. 2000. History of pronghorn transplants in Arizona. Pronghorn Antelope Workshop (1996) Proc. 17:112–113.

Alexander, A. and R. A. Ockenfels. 1994. Juniper densities relative to pronghorn use in central Arizona. Pronghorn Antelope Workshop Proc. 16:76–85.

Allard, M. W., M. M. Miyamoto, L. Jarecki, F. Kraus and M. R. Tennant. 1992. DNA systematics and evolution of the artiodactyl family Bovidae. Proc. Natl. Acad. Sci. 89:3,972–3,976.

Alldredge, A. W., R. D. Deblinger and J. Peterson. 1991. Birth and fawn bed site selection by pronghorns in a sagebrush-steppe community. J. Wildl. Manage. 55(2):222–227.

Allee, W. C., A. E. Emerson, O. Park, T. Park and K. P. Schmidt. 1949. Principles of animal ecology. W. B. Saunders Co., Philadelphia, Pennsylvania. 837 pp.

Allen, A. W. and M. J. Armbruster. 1982. Preliminary evaluation of a habitat suitability model for the pronghorn. Pronghorn Antelope Workshop Proc. 10:93–105.

Allen, A. W., J. G. Cook and M .J. Armbruster. 1984. Habitat suitability index models. FWS/OBS-821/10.65. U.S. Fish and Wildl. Serv., Fort Collins, Colorado. 22 pp.

Allen, D. L. 1962. Our wildlife legacy. Rev. ed. Funk and Wagnalls Co., Inc., New York, New York. 422 pp.

———. 1967. The life of prairies and plains. McGraw-Hill Book Co., New York, New York. 232 pp.

Allen, G. M. 1942. Extinct and vanishing mammals of the Western Hemisphere. Spec. Publ. 11. American Committee for International Wild Life Protection. Intelligencer Print. Co., Lancaster, Pennsylvania. 620 pp.

Allen, J. A. 1874. Notes on the natural history of portions of Dakota and Montana territories. Proc. Boston Soc. Nat. Hist. 17:33–85.

Allen, R. W. and P. M. Kyles. 1953. The occurrence of the fringed tape worm, *Thysanosoma actinioides* in pronghorn antelope. Helminthol. Soc. Washington D.C. Proc. 20(2):96–97.

Allen, R. W. and K. S. Samson. 1960. Further observations on *Thysanosoma actinioides* in the American pronghorn. J. Parasitol. 46(5):671.

Allen, R. W. and K. S. Samson. 1961. Preliminary report on the immunization of sheep with a relatively non-pathogenic strain of *Haemonchus* from pronghorn antelope. J. Parasitol. 47(4-2):22.

Allen, R. W., K. S. Samson and G. A. Schad. 1959. Cross-transmission of *Haemonchus* from pronghorn antelope to domestic sheep with observations on the pathogenicity of this strain in lambs as compared with that of *Haemonchus* from domestic sheep. J. Parasitol. 45(4-2):47.

Allen, S. H. and J. M. Samuelson. 1987. Precision and bias of a summer aerial transect census of pronghorn antelope. Prairie Nat. 19(1):19–24.

Allison, M. E. 1890. Coursing the antelope. Pages 333–340 *in* G. O. Shields, ed., The big game of North America. Rand, McNally and Co., Chicago, Illinois. 581 pp.

Allred, W. J. 1943. Wyoming antelope—history and wartime management. Trans. N. Am. Wildl. Conf. 8:117–121.

Almand, J. D. and M. D. Zagata. 1979. State-federal relationships in wildlife management. Pages 31–39 *in* R. D. Teague and E. Decker, eds., Wildlife conservation: Principles and practices. The Wildl. Soc., Washington, D.C. 280 pp.

Amor, D. C. 2000. Filogenia y estructura genetica del berrendo (*Antilocapra americana*) E implicaciones para su conservacion. Thesis, Universidad Nacional Autonoma de Mexico, Mexico City. 95 pp. Translated from Spanish by Diana Doan-Crider in 2001.

Amsden, Charles. 1931. Man-hunting, Sidelights on a symposium on "The antiquity of man in America." No eds., The Masterkey. 5(2):36–47.

Amstrup, S. C. 1978. Activities and habitat use patterns of pronghorns on Montana-Wyoming coal lands. Pronghorn Antelope Workshop Proc. 8:270–306.

Amstrup, S. C. and T. B. Segerstrom. 1981. Immobilizing free-ranging pronghorns with powdered succinylcholine chloride. J. Wildl. Manage. 45(3):741–745.

Amstrup, S. C., J. Meeker, B. W. O'Gara and J. McLucas. 1980. Capture methods for free-ranging pronghorns. Pronghorn Antelope Workshop Proc. 9:98–131.

Anderson, E. W., D. L. Franzen and J. E. Mallard. 1990a. Rx grazing to benefit watershed-wildlife-livestock. Rangelands 12(2):105–111.

Anderson, E. W., D. L. Franzen and J. E. Mallard. 1990b. Forage quality as influenced by prescribed grazing. Pages 56–70 *in* K. E. Severson, tech. coordinator, Can lifestock be used as a tool to enhance wildlife habitat. General Tech. Rept. RM-194. U.S. For. Serv., Fort Collins, Colorado. 123 pp.

Anderson, L. D. and J. W. Denton. 1978. Base data analysis required for forage allocation. Spec. Rept. U.S. Bur. Land Manage., Salmon, Idaho. 8 pp.

Anderson, L. D. and J. W. Denton. 1980. Adjustable wire fences for facilitating big-game movement. Tech. Note 343. U.S. Bur. Land Manage., Denver, Colorado. 7 pp.

Anderson, S. 1972. Mammals of Chihuahua, taxonomy and distribution. Am. Mus. Nat. Hist. 148(2):382–394.

Anderson, S. H. and K .J. Gutzwiller. 1994. Habitat evaluation methods. Pages 592–622 *in* T.A. Bookhout, ed., Research and management techniques for wildlife and habitats. The Wildl. Soc., Bethesda, Maryland. 740 pp.

Andrewartha, H. G. 1961. Introduction to the study of animal populations. Phoenix Sci. Series. Univ. Chicago Press, Chicago, Illinois. 281 pp.

Andrewartha, H. G. and L. C. Birch. 1954. The distribution and abundance of animals. Univ. Chicago Press, Chicago, Illinois. 782 pp.

Anonymous. 1964. Wildlife roundup. Wyoming Wildl. 28(12):39.

———. 1972. Pronghorns in trouble. Lanai in focus. Lanai Co., Lanai City, Hawaii. 2:11.

———. 1983. Buck antelope has three horns. South Dakota Conserv. Digest 50(6):31.

———. 1986. Landowner coupons. Do they work? Wyoming Wildl. 50(8):34–35.

———. 1991. Hunter safety certificates required. Am. Hunter 19(9):57.
———. 1994. Herd and tags down—but not out. Prairie Press 3(3):1.
———. 1995. Shrub standards solution—Finally! Prairie Press 4(2):1, 5.
———. 2000. Texas A& M to conduct brucellosis study on pronghorn. Prairie Press 9(2):1.
Aoude, A. and R. E. Danvir. 2004. Using range manipulations to increase pronghorn carrying capacity. Pronghorn Workshop Proc. 20:124–135.
Applied Biomathematics, Inc. 1994. RAMAS AGE. User manual. Modeling fluctuations in age structured populations. Exeter Software, Setauket, New York. 144 pp.
———. 1997. RAMAS TIME ecological time series analysis. Exeter Software, Setauket, New York (online document available at: http://www.mpassociates.gr/software/catalog/sci/ramas/ramas1.html).
Archer, J. 1979. Animals under stress. The Institute of Biology's Studies in Biology, No. L08. Univ. Park Press, Baltimore, Maryland. 55 pp.
Archer, S. 1994. Woody plant encroachment in southwestern grasslands and savannas: Rates, patterns and proximate causes. Pages 13–68 *in* M. Vavra, W. A. Laycock and R. D. Pieper, Ecological implications of livestock herbivory in the West. Soc. for Range Manage., Denver, Colorado. 297 pp.
Arizona Department of Game and Fish. 1981. The Sonoran pronghorn. Spec. Rept 10. Arizona Dept. Game and Fish, Phoenix. 55 pp.
———. 1993. Game management program: Appendix 8, Guide to prioritizing antelope transplant sites. Arizona Dept. Game and Fish, Phoenix. Unpaginated.
———. 2002a. Sonoran pronghorn numbers down. December 10 news release. Arizona Dept. Game and Fish, Phoenix. 2 pp.
Arnon, N. S. and W. W. Hill. 1979. Santa Clara Pueblo. Pages 296–307 *in* A. Ortiz, ed., Southwest. Handbook of North American Indians. Vol. 9. Smithsonian Inst., Washington, D.C. 717 pp.
Arrington, O. N. and A. E. Edwards. 1951. Predator control as a factor in antelope management. Trans. N. Am. Wildl. and Nat. Resour. Conf. 16:179–193.
Arthur, D. R. 1960. Ticks: A monograph of the Ixodoidea. Cambridge at the Univ. Press, New York, New York. 251 pp.
Arthur, G. 2000. Young pronghorn killed in violent hail storms. Prairie Press 9(1):4.
Arthur, W. J. and R. J. Gates. 1988. Trace element intake via soil ingestion in pronghorns and in blacktailed jackrabbits. J. Range Manage. 41(2):162–166.
Asdell, S. A. 1964. Patterns of mammalian reproduction. 2nd ed. Comstock Publishing Assoc., Ithaca, New York. 670 pp.
Ashton, J. and J. Olin. 1988. Pronghorn hand-rearing procedure. Minnesota Zool. Gardens, Duluth. 8 pp.
Audubon, J. J. and J. Bachman. 1851. The viviparous quadrupeds of North America. Vol. 2. V. G. Audubon, New York, New York. 334 pp. (Reprinted in 1974 by Arno Press, New York, New York.)
Autenrieth, R. E. 1966. Behavior and socialization of pronghorn fawns. M.S. thesis, Idaho St. Univ., Pocatello. 82 pp.
———. 1976. A study of birth sites selected by pronghorn does and the bed sites of fawns. Pronghorn Antelope Workshop Proc. 7:127–134.
———, ed. 1978. Guidelines for the management of pronghorn antelope. Pronghorn Antelope Workshop Proc. 8:473–526.
———. 1982. Pronghorn fawn habitat use and vulnerability to predation. Pronghorn Antelope Workshop Proc. 10:112–131.
———. 1984. Little lost pronghorn fawn study—condition, habitat use and mortality. Pronghorn Antelope Workshop Proc. 11:49–70.
Autenrieth, R. E., G. L. Copeland and T. R. Reynolds. 1981. Capturing pronghorns using a helicopter and etorphine hydrochloride. Wildl. Soc. Bull. 9(4):314–319.
Axelrod, A. 1990. Art of the golden West. Abbeville Press, New York, New York. 418 pp.
Baccus, R., N. Ryman, M. H. Smith, C. Reuterwall and D. Cameron. 1983. Genetic variability and differentiation of large grazing mammals. J. Mammal. 64(1):109–120.
Bahr, G. J. 1995. Human impacts on the grasslands of southeastern Arizona. Pages 230–264 *in* M. P. McClarn and T. R. Van Devender, eds., The desert grassland. Univ. Arizona Press, Tucson. 346 pp.
Bailey, B., Jr. 1962. Serological surveys of brucellosis and leptospirosis in Nebraska deer and antelope. P-R Proj. W-15-12-20, Spec. Rept. 3. Nebraska Game Forestation and Parks Commiss., Lincoln. 26 pp.
Bailey, J. A. 1984. Principles of wildlife management. John Wiley and Sons, New York, New York. 373 pp.
———. 1986. Weather and climate. Pages 711–721 *in* A.Y. Cooperrider, R. J. Boyd and H. R. Stuart, eds., Inventory and monitoring of wildlife habitat. Serv. Ctr., U.S. Bur. Land Manage., Denver, Colorado. 858 pp.
Bailey, J. A. and A. Y. Cooperrider. 1982. Final report: Trickle Mountain research study. U.S. Bur. Land Manage., Denver, Colorado. 137 pp.
Bailey, J. A., W. Elder and T. D. McKinney. 1974. Readings in wildlife conservation. The Wildl. Soc., Washington, D.C. 722 pp.
Bailey, R. J. 1978. Description of the ecoregions of the United States. U.S. For. Serv., Intermtn. Region, Ogden, Utah. 77 pp.
Bailey, V. 1920. Old and new horns of the prong-horned antelope. J. Mammal. 1(3):128–130.
Bailey, V. 1926. A biological survey of North Dakota. N. Am. Fauna 49. U.S. Dept. Agric., Washington, D.C. 226 pp.
———. 1932. The Oregon antelope. Biol. Soc. Washington, D.C. Proc. 45:45–46.
Baker, J. 1953. Food habit study of game animals. Wyoming Wildl. 17(11):24–26.
Banfield, A. W. F. 1974. The mammals of Canada. Natl. Mus. Nat. Hist., Univ. Toronto Press, Toronto, Ontario. 438 pp.
Banks, P. B. 2000. Can foxes regulate rabbit populations? J. Wildl. Manage. 64(2):401–406.
Barbour, F. H. and C. B. Schultz. 1934. A new antilocaprid and a new cervid from the late Tertiary of Nebraska. Am. Mus. Novit. 734:1–3.
Barlow , J. and P. Boveng. 1991. Modeling age specific mortality for marine mammal populations. Marine Mammal Sci. 7(1):50–66.

Barlow, J., S. Swartz, T. C. Eagle and P. R. Wade. 1995. U.S. marine mammal stock assessments: Guidelines for preparation, background and a summary to the 1995 assessments. NOAA Tech. Memo. Natl. Marine Fisheries Serv., Off. Protected Resour., U.S. Dept. Commerce, Silver Spring, Maryland. 168 pp.

Barmore, W. J. 1969. Pronghorn-mule deer-range relationships on the Yellowstone winter range. Pages 121–137 *in* Office of Natural Science Studies annual report for calendar year 1968. U.S. Natl. Park Serv., Washington, D.C. 245 pp.

Barrett, M. W. 1974. Importance, utilization and quality of *Artemesia cana* on pronghorn winter ranges in Alberta. Antelope States Workshop Proc. 6:26–57.

———. 1978. Pronghorn fawn mortality in Alberta. Pronghorn Antelope Workshop Proc. 8:429–444.

———. 1979. Evaluation of fertilizer on pronghorn winter range in Alberta. J. Range Manage. 32(1):55–59.

———. 1981. Environmental characteristics and functional significance of pronghorn fawn bedding sites in Alberta. J. Wildl. Manage. 45(1):120–131.

———. 1982. Ranges, habitat, and mortality of pronghorns at the northern limits of their range. Ph.D. thesis, Univ. Alberta, Edmonton. 227 pp.

———. 1984. Movements, habitat use, and predation on pronghorn fawns in Alberta. J. Wildl. Manage. 48(2):542–550.

Barrett, M. W. and G. A. Chalmers. 1972. *Capillaria hepatica* (Nematoda: Tricuridae) in pronghorn antelope (*Antilocapra americana* Ord) in Alberta. J. Wildl. Dis. 8(4):332–334.

———. 1975. A serologic survey of pronghorn in Alberta and Saskatchewan, 1970–1972. J. Wildl. Dis. 11(2):157–163.

———. 1976. Baseline hematologic and clinical chemistry values for pronghorns. Pronghorn Antelope Workshop Proc. 7:104–118.

———. 1977a. Hematological values for adult free-ranging pronghorns. Can. J. Zool. 55(2):448–455.

———. 1977b. Clinicochemical values for adult free-ranging pronghorns. Can. J. Zool. 55(8):1,252–1,260.

———. 1979. Hematological and clinicochemical values for free-ranging pronghorn fawns. Can. J. Zool. 57(9):1,757–1,766.

Barrett, M. W., J. W. Nolan and L. D. Roy. 1982. Evaluation of a hand-held net-gun to capture large mammals. Wildl. Soc. Bull. 10(2):108–114.

Barrington, M. R. 1975. Habitat factors related to pronghorn productivity on the southern High Plains. M.S. thesis, Oklahoma St. Univ., Stillwater. 144 pp.

Barsness, J. 1995. High-tech hunting. Peterson's Hunting, Nov. 1995:46–49, 52, 105.

Barsness, L. 1985. Heads, hides and horns: The complete buffalo book. Texas Christian Univ. Press, Fort Worth. 233 pp.

Bartholow , J. 1986. POP-2 user manual. Fossil Creek Software, Fort Collins, Colorado. 96 pp.

Bartmann, R. M., L. H. Carpenter, R. A. Garrett and D. C. Bowden. 1986. Accuracy of helicopter counts of mule deer in Pinyon-Juniper woodland. Wildl. Soc. Bull. 14(4):356–363.

Bartmann, R. M., G. C. White and L. H. Carpenter. 1992. Compensatory mortality in a Colorado mule deer population. Wildl. Soc. Monogr. 121. 39 pp.

Bates, S. 2000. Preliminary results from a radiotelemetry study of pronghorn antelope at Dugway Proving Ground, Utah. Pronghorn Antelope Workshop (1996) Proc. 17:42–44.

Bayless, S. R. 1969. Winter food habits, range use, and home range of antelope in Montana. J. Wildl. Manage. 33(3):538–551.

Beale, D. M. 1966. A self-collaring device for pronghorn antelope. J. Wildl. Manage. 30(1):209–211.

———. 1978. Birth rate and fawn mortality among pronghorn antelope in western Utah. Pronghorn Antelope Workshop Proc. 8:445–446.

———. 1986. Pronghorn antelope productivity on semi-desert range in western Utah. Pronghorn Antelope Workshop Proc. 12:41–42.

Beale, D. M. and R. C. Holmgren. 1975. Water requirements for pronghorn antelope fawn survival and growth. Utah Div. Wildl. Resour., Salt Lake City. 27 pp.

Beale, D. M. and A. D. Smith. 1966. Antelope fawn production and survival on semi-desert range in western Utah. Antelope States Workshop Proc. 2:66–74.

———. 1967. Immobilization of pronghorn antelopes with succinylcholine chloride. J. Wildl. Manage. 31(4):840–842.

———. 1970. Forage use, water consumption, and productivity of pronghorn antelope in western Utah. J. Wildl. Manage. 34(3):570–582.

———. 1973. Mortality of pronghorn antelope fawns in western Utah. J. Wildl. Manage. 37(3):343–352.

———. 1980. Pronghorn antelope in Utah: Some research and observations. Publ. 80–13. Utah Div. Wildl. Resour., Salt Lake City. 88 pp.

Bean, M. J. 1978. Federal wildlife law. Pages 279–289 *in* H. P. Brokaw, ed., Wildlife and America: Contributions to an understanding of American wildlife and its conservation. U.S. Govt. Print. Off., Washington, D.C. 532 pp.

Bear, G. D. 1967a. Antelope investigations, physiological studies. P-R Job Compl. Rept., Work Plan 2, Job 3, Proj. W-40-R-7. Colorado Dept. Game, Fish and Parks, Denver. Pp. 71–131.

———. 1967b. The wounding loss on antelope. Colorado Outdoors 16(5):33–34.

———. 1969a. Antelope and net wire fences. Western Assoc. St. Game and Fish Commiss. Proc. 49:265–271.

———. 1969b. Evaluation of aerial antelope census technique. Game Info. Leaflet. Colorado Dept. Nat. Resour., Div. Game, Fish and Parks, Denver. 3 pp.

———. 1971. Seasonal trends in fat levels of pronghorns, *Antilocapra americana*, in Colorado. J. Mammal. 52(3):583–589.

Bear, G. D., J. Otterman and J. Robinson. 1972. Antelope investigation, food habits of antelope. P-R Job Final Rept., Job 2, Proj. W-40-R-13. Colorado Dept. Game, Fish and Parks, Denver. Pp. 38–70.

Bear, G. D., A. E. Anderson, J. P. Goettl, R. Keiss and F. Fields. 1973. Antelope investigations, physiological studies. P-R Job Final Rept., Work Plan 2, Job 3, Proj. W-40-R-13. Colorado Dept. Game, Fish and Parks, Denver. Pp. 1–35.

Beardahl, L. E. and V. E. Sylvester. 1974. Spring burning for removal of sagebrush competition in Nevada. Tall Timbers Ecol. Conf. Proc. 14:539–547.

Beasom, S. L. 1979. Precision in helicopter censusing of white-tailed deer. J. Wildl. Manage. 43(3):777–779.

Beasom, S. L., J. C. Hood and J. R. Cain. 1981. The effect of strip width on helicopter censusing of deer. J. Range Manage. 34(1):36–37.

Beasom, S. L., L. Laplant and V. W. Howard. 1982a. Fecal pH of pronghorn and sheep as related to diet. J. Wildl. Manage. 46(4):1,101–1,104.

Beasom, S. L., L. LaPlant and V. W. Howard. 1982b. Similarity of pronghorn, cattle, and sheep diets in southeastern New Mexico. Pages 565–572 *in* J. M. Peek and P. D. Dalke, eds., Wildlife-livestock relationships symposium: Proceedings 10. Univ. Idaho For., Wildl. and Range Exp. Sta., Moscow. 614 pp.

Beasom, S. L., E. D. Wiggers and J. R. Giardon. 1983. A technique for assessing land surface ruggedness. J. Wildl. Manage. 47(47):1,163–1,166.

Beatle, A. A. 1960. A study of sagebrush: The section Tridentatae of *Artemesia*. Bull. 368. Univ. Wyoming, Laramie. 83 pp.

Beck, D. E. 1955. Distributional studies of parasitic arthropods in Utah, determined as actual and potential vectors of Rocky Mountain spotted fever and plague, with notes on vector-host relationships. Brigham Young Univ., Provo, Utah. Sci. Bull. 1(1). 64 pp.

Becker, B. W. 1972. Pronghorn cattle range use and food habits relationships in an enclosed sagebrush control area. M.S. thesis, Montana St. Univ., Bozeman. 57 pp.

Beddard, F. E. 1909. Contributions to the anatomy of certain ungulata including Tapirus, Hyrax, *Antilocapra*. Proc. Zool. Soc. London 11:160–197.

Beers, C. 2000. It's a gas. Wyoming Wildl. 64(6):22–29.

Beier, P. 1996. Metapopulation models, tenacious tracking and cougar conservation. Pages 293–323 *in* D. R. McCullough, ed., Metapopulations and wildlife conservation. Island Press, Washington, D.C. 429 pp.

Beintema, J. J., W. Gaastra and J. Munniksma. 1979. Primary structure of pronghorn pancreatic ribonuclease: Close relationship between giraffe and pronghorn. J. Mol. Evol. 13(4):305–316.

Bekenov, A. B., I. A. Grachevand and E. J. Milner-Gulland. 1998. The ecology and management of the saiga antelope in Kazakhstan. Mammal Rev. 28(1):1–52.

Bell, R. H. V. 1971. A grazing ecosystem in the Serengeti. Sci. Am. 225(l):86–93.

Berger, J. 1979. "Predator harassment" as a defense strategy in ungulates. Am. Midl. Nat. 102(l):197–199.

———. 1985. Interspecific interactions and dominance among wild Great Basin ungulates. J. Mammal. 66(3):571–573.

———. 1986. Wild horses of the Great Basin. Univ. Chicago Press, Chicago, Illinois. 326 pp.

Berger, J., D. Daneke, J. Johnson and S. H. Berwick. 1983. Pronghorn foraging economy and predator avoidance in a desert ecosystem: Implications for the conservation of large mammalian herbivores. Biol. Conserv. 25(3):193–208.

Berger, J. J. 2000. Beating the heat. Berkeley Hills Books, Berkeley, California. 133 pp.

Bergerud, A. T. 1963. Aerial winter census of caribou. J. Wildl. Manage. 27(3):438–449.

———. 1988. Caribou, wolves and man. Trends in Ecol. and Evol. 3(1):68–72.

Bergstrom, R. C. 1975a. Incidence of *Marshallagia marshalli* Orlov, 1933 in Wyoming sheep, *Ovis aries*, and pronghorn antelope, *Antilocapra americana*. Helminthol. Soc. Washington, D.C. Proc. 42(1):61–63.

———. 1975b. Incidence of *Marshallagia marshalli* Orlov, 1933 in wild ruminants in Wyoming. Oklahoma Acad. Sci. Proc. 55(1):101–102.

Bergstrom, R. C., N. Kingston and J. R. Talbot. 1985. *Toxascaris leonia* (Nematoda: Ascarididae) from the pronghorn antelope, *Antilocapra americana*, in Wyoming. Helminthol. Soc. Washington, D.C. Proc. 52(l):142–143.

Bernabéu, S. 1994. Diario de las expediciones a las Californias' de José Longinos. Ediciones Doce Calles, Madrid, Spain. 315 pp. *In* Cancino, J. and F. Reygadas. 1999. Historical notations regarding pronghorn (*Antilocapra americana*) in Mexico. Pronghorn Antelope Workshop (1998) Proc. 18:81–83.

Bernt, W. C. 1976. Observations on a pronghorn antelope winter range. M.S. thesis, Idaho St. Univ., Pocatello. 58 pp.

Berryman, A. A. 1994a. Population dynamics: Forecasting and diagnosis from time-series. Pages 119–128 *in* S. R. Leather, A. D. Watt and N. Mills, eds., Individuals, populations and patterns in ecology. Intercept Ltd., Andover, Hampshire, England. 228 pp.

———. 1994b. Population analysis system (PAS) two-species analysis. Software user's manual. Ecol. Systems Analysis, Pullman, Washington. 38 pp.

———. 1999. Principles of population dynamics and their application. Stanley Thornes, Cheltenham, England. 192 pp.

Berryman, A. A. and J. A. Millstein. 1994. Population analysis system (PAS) single-species analysis. Version 4.0, Software user's manual. Ecol. Systems Analysis, Pullman, Washington. 56 pp

Bever, W. n.d. The pronghorn in South Dakota. Unpubl. rept. South Dakota Dept. Game, Fish and Parks, Pierre. 16 pp.

———. 1950. Parasites and diseases of South Dakota antelope. P-R Job Compl. Rept., Proj. 12-R-7. South Dakota Dept. Game, Fish and Parks, Pierre. 16 pp.

———. 1955. The incidence and parasitical load of gastrointestinal parasites in South Dakota antelope in the fall of 1955. P-R Proj.12-R-13. South Dakota Dept. Game, Fish and Parks, Pierre. 7 pp.

———. 1957. The incidence and degree of the parasitic load among antelope and the development of field techniques to measure such parasitism. P-R Proj. 12-R-14, Job Outl. A-5.2. Game Manage. Div., South Dakota Dept. Game, Fish and Parks, Pierre. 7 pp.

———. 1965. Antelope management in South Dakota. Game Manage. Div., South Dakota Dept. Game, Fish and Parks, Pierre. 11 pp.

Beverton, R. J. H and S. J. Holt. 1957. On the dynamics of exploited fish populations. Ministry Agric., Fisheries and Food, Great Britain. Fisheries Investigations (Series 2) 19(1): 5–53.

Beyer J. E. and G. C. Laurence. 1980. A stochastic model of larval fish growth. Ecol. Modeling 8(2):109–132.

Biddulph, C. H. 1936. Horn growth in the nilgai. J. Bombay Nat. Hist. Soc. 39:171.

Binnema, T. 1996. Presettlement rangeland management on the northern plains. Rangelands 8(6):217–218.

Bissell, S. J. and M. D. Duda. 1993. Factors related to hunting and fishing participation in the United States. Phase 2:

Hunting focus groups. Responsive Manage., Harrisonburg, Virginia. 180 pp.

Bixby, A. 1880. History of Boulder County. Pages 379–433 *in* History of Clear Creek and Boulder Valley, Colorado. O. L. Baskin and Co., Chicago, Illinois. 713 pp.

Bjornstad, O. N., W. Falck and N. C. Stenseth. 1995. A geographic gradient in small rodent density fluctuations: A statistical modeling approach. Proc. Royal Soc. London, B Vol. 262:127–133.

Blair, N. 1982. An historical account of the pronghorn antelope. Wyoming Wildl. 46(9):3–7.

———. 1987. The history of wildlife management in Wyoming. Wyoming Dept. Game and Fish, Cheyenne. 295 pp.

Bleed, P. 1986. The optimal design of hunting weapons: Maintainability or reliability. Am. Antiquity 51(4):737–747.

Bleich, V. C. 1990. Horizontal wells for wildlife water developments. Pages 51–58 *in* G. K. Tsukamoto and S. J. Stiver, eds., Wildlife water developments. Nevada Chapter, The Wildl. Soc., U.S. Bur. Land Manage. and Nevada Dept. Wildl., Reno. 162 pp.

Bleich, V. C., L. J. Coombes and J. H. Davis. 1982. Horizontal wells as a wildlife habitat improvement technique. Wildl. Soc. Bull. 10(4):324–328.

Bleich, V. C., J. D. Wehausen, R. R. Ramey and J. L. Rechel. 1996. Metapopulation theory and mountain sheep: Implications for conservation. Pages 353–374 *in* D. R. McCullough, ed., Metapopulations and wildlife conservation. Island Press, Washington D.C. 429 pp.

Blunt, F. M. and A. T. Myles. 1998. Successful rearing and handling of big game animals at the Sybille Wildlife Research Unit. Wyoming Dept. Game and Fish, Cheyenne. 58 pp.

Boas, F. 1918. Kutenai tales. Smithsonian Inst. Bur. Am. Ethnol. Bull. 59. U.S. Govt. Print. Off., Washington, D.C. 387 pp.

Boccadori, S. J. 2002. Effects of winter range on a pronghorn population in Yellowstone National Park, Wyoming. M.S. thesis, Montana St. Univ., Bozeman. 86 pp.

Boddicker, M. L. 1966. Helminth parasites of wild ruminants in South Dakota. M.S. thesis, South Dakota St. Univ., Brookings. 146 pp.

Boddicker, M. L. and E. J. Hugghins. 1969. Helminths of big game mammals in South Dakota. J. Parasitol. 55(5):1,067–1,074.

Bodie, W. L. 1978. Pronghorn fawn mortality in the Upper Pahsimeroi River drainage of central Idaho. Pronghorn Antelope Workshop Proc. 8:417–428.

———. 1979. Factors affecting pronghorn fawn mortality in central Idaho. M.S. thesis, Univ. Montana, Missoula. 90 pp.

Bodine, J. J. 1979. Taos Pueblo. Pages 255–267 *in* A. Ortiz, ed., Southwest. Handbook of North American Indians. Vol. 9. Smithsonian Inst., Washington, D.C. 717 pp.

Bolen, E. G. 1989. Conservation biology, wildlife management, and spaceship Earth. Wildl. Soc. Bull. 17(3):351–354.

Boller, H. A. 1972. Among the Indians: Four years on the Upper Missouri, 1858–1862. M. M. Quaife, ed. Univ. Nebraska Press, Lincoln. 385 pp.

Bone, T. L. 1984. Age composition and horn development of Texas pronghorn. Pronghorn Antelope Workshop Proc. 11:105–112.

Bookhout, T. A., ed. 1994. Research and management techniques for wildlife and habitats. The Wildl. Soc., Bethesda, Maryland. 740 pp.

Botkin, D. B. 1977. Life and death in a forest: The computer as an aid to understanding. Pages 213–234 *in* C.A.S. Hall and J. W. Day, eds., Ecosystem modeling in theory and practice: An introduction with case studies. John Wiley and Sons, New York, New York. 684 pp.

Bowden, D. C., A. E. Anderson and D. E. Medin. 1984. Sampling plans for mule deer sex and age ratios. J. Wildl. Manage. 48(2):500–509.

Box, T. W., D. D. Dwyer and F. H. Wagner. 1976. The public range and its management. Unpubl. rept. to President's Council on Environ. Qual. Utah St. Univ., Logan. 56 pp.

Boyd, R. J., A. Y. Cooperrider, P. C. Lent and J. A. Bailey. 1986. Ungulates. Pages 519–564 *in* A. Y. Cooperrider, R. J. Boyd and H. R. Stuart, eds., Inventory and monitoring of wildlife habitat. Serv. Ctr., U.S. Bur. Land Manage., Denver, Colorado. 858 pp.

Boyle, S. A. and A. W. Alldredge. 1984. Pronghorn summer distribution and water availability in the Red Desert, Wyoming. Pronghorn Antelope Workshop Proc. 11:103–104.

Boyle, S. A., L. L. Mead and S. Bates. 1999. Distribution, habitat use, and competition between pronghorn and feral horses at Dugway Proving Grounds, Utah. Pronghorn Antelope Workshop (1998) Proc. 18:105.

Boyse, M. S. and J. A. Tate, Jr. 1982. Pronghorn (*Antilocapra americana*) demography and hunting quotas in the Powder River Basin, Wyoming. Trans. Intl. Union Game Biol. Congr. 14:101–106.

Bradybaugh, J. S. and V. W. Howard. 1982. Effect of severe weather on pronghorn reproduction and fawn survival in southeastern New Mexico. Pronghorn Antelope Workshop Proc. 10:143–155.

Branson, J. 1939. Hart Mountain National Wildlife Refuge narrative report for April, May, June. U.S. Fish and Wildl. Serv., Lakeview, Oregon. 5 pp.

Brashares, J. S. and P. Arcese. 1999a. Scent marking in a territorial African antelope: I. The maintenance of borders between male oribi. Anim. Behav. 57(1):1–10.

———. 1999b. Scent marking in a territorial African antelope: II. The economics of marking with faeces. Anim. Behav. 57(1):11–17.

Brass, W. 1971. On the scale of mortality. Pages 60–110 *in* W. Brass, ed., Biological aspects of demography. Symposia of the Society for the Study of Human Biology. Vol 10. Taylor and Francis, London, England. 167 pp.

Brigham, W. R. and C. Stevenson. 1997. Wildlife water catchment construction in Nevada. Tech. Note 397. U.S. Bur. Land Manage., Denver, Colorado. 101 pp.

Brinkley, K. 1987. Pronghorn hand-rearing protocol (*Antilocapra americana americana*). Anim. Keeper's Forum 14(8):234–237.

Brinton, E. P., O. E. Beck and D. M. Allred. 1965. Identification of the adults, nymphs and larvae of ticks of the genus *Dermacentor* Kock (Ixodidae) in the western United States. Sci. Bull., Biol. Series V(4). Brigham Young Univ., Provo, Utah. 44 pp.

Britt, T. L. 1980. Reestablishment of pronghorn antelope on the Arizona strip. Pronghorn Antelope Workshop Proc. 8:226–246.

Brockley, H. 1949. A report of the behavior of antelope during heavy snows near Bosler, Wyoming. Wyoming Game and Fish Commiss., Cheyenne. 39 pp.

Bromley, P. T. 1967. Pregnancy, birth, behavioral development of the fawn, and territoriality in the pronghorn (*Antilocapra americana* Ord) on the National Bison Range, Moiese, Montana. M.S. thesis, Univ. Montana, Missoula. 132 pp.

———. 1969. Territoriality in pronghorn bucks on the National Bison Range, Moiese, Montana. J. Mammal. 50(l):81–89.

———. 1972. History of the introduced population of pronghorn antelope on Lanai, Hawaii, 1959–1972. Antelope States Workshop Proc. 5:159–166.

———. 1977. Aspects of the behavioural ecology and sociobiology of the pronghorn (*Antilocapra americana*). Ph.D. thesis, Univ. Calgary, Calgary, Alberta. 370 pp.

———. 1991. Manifestations of social dominance in pronghorn bucks. Appl. Anim. Behav. Sci. 4829(1-4):147–164.

Bromley, P. T. and D. W. Kitchen. 1974. Courtship in the pronghorn (*Antilocapra americana*). Pages 356–364 *in* V. Geist and F. Walther, eds., The behaviour of ungulates and its relation to management. Vol. 1. Publ. 24. Intl. Union Conserv. Nature and Nat. Resour., Morges, Switzerland. 511 pp.

Bromley, P. T. and B. W. O'Gara. 1967. Orphaned pronghorns survive. J. Wildl. Manage. 31(4):843.

Bronson, F. H. 1989. Mammalian reproductive biology. Univ. Chicago Press, Chicago, Illinois. 325 pp.

Brook, B.W., L. Lim, R. Harden and R. Frankham. 1997. Does population viability analysis software predict the behaviour of real populations? A retrospective study on the Lord Howe Island woodhen *Tricholimnas sylvestris* (Sclater). Biol. Conserv. 82(1):119–128.

Brooke, V. 1878. On the classification of the Cervidae, with a synopsis of the existing species. Proc. Zool. Soc. London 58:883–928.

Browman, L. G. and H. S. Sears. 1955. Erythrocyte values and alimentary canal pH values in the mule deer. J. Mammal. 36(3):474–476.

Brown, D. and R. J. Karsky. 1989. Facilities for watering livestock and wildlife. U.S. For. Serv. and U.S. Bur. Land Manage., Missoula, Montana. 71 pp.

Brown, D., N. D. Alexander, R. W. Marrs and S. Albon. 1993. Structured accounting of the variance of demographic change. J. Anim. Ecol. 62 (3):490–502.

Brown, D. E. 1982. Management status of Arizona's pronghorn antelope in 1981. Pronghorn Antelope Workshop Proc. 10:12–15.

———. 1994a. History of pronghorn antelope in Arizona. Pronghorn 2(2):1, 5, 7.

———. 1994b. History of pronghorn antelope in Arizona. Pronghorn 2(3):1, 5, 9.

Brown, D. E. and D. Verheist. 2002. Pronghorn horn sheath growth and age on a ranch in southern New Mexico. Abstract presented at Pronghorn Antelope Workshop 20:10.

Brown, Joseph Epes. 1976. The roots of renewal. Pages 25–34 *in* Walter Holden Capps, ed., Seeing with a native eye: Essays on Native American religion. Harper and Row, New York, New York. 132 pp.

Brown, J. S., J. W. Laundré and M. Gurung. 1999. The ecology of fear: Optimal foraging, game theory and trophic interactions. J. Mammal. 80(2):385–399.

Brown, T. L., D. J. Decker, W. F. Siemer and J. W. Enck. 2000. Trends in hunting participation and implications for management of game species. Pages 145–156 *in* W. C. Gartner and D. W. Lime, eds., Trends in outdoor recreation, leisure and tourism. CABI Publishing, New York, New York. 496 pp.

Browne, B. 1939. Hunting the pronghorn. Pages 336–343 *in* North American big game, a book of the Boone and Crockett Club, compiled by the Comm. on Records of N. Am. Big Game. Charles Scribner's Sons, New York, New York. 533 pp.

Brownie, C., D. R. Anderson, K. P. Burnham and D. S. Robson. 1985. Statistical inference from band recovery data—A handbook. 2nd ed. Res. Publ.131. U.S. Fish and Wildl. Serv., Washington, D.C. 305 pp.

Brownlee, S. and T. L. Hailey. 1970. Development of a technique for night trapping antelope fawns. Antelope States Workshop Proc. 4:78–81.

Broyles, B. 1995. Desert wildlife water developments: Questioning use in the Southwest. Wildl. Soc. Bull. 23(4):663–675.

Bruner, D. W. and J. H. Gillespie. 1973. Hagan's infectious diseases of domestic animals. 6th ed. Comstock Publ. Assoc., Cornell Univ. Press, Ithaca, New York. 1,385 pp.

Bruns, E. H. 1969. A preliminary study of behavioral adaptations of wintering pronghorn antelopes. M.S. thesis, Univ. Calgary, Calgary, Alberta. 60 pp.

———. 1977. Winter behavior of pronghorns in relation to habitat. J. Wildl. Manage. 41(3):560–571.

Bryant, L. D. and C. Maser. 1982. Classification and distribution. Pages 1–59 *in* J. S. Thomas and D. E. Toweill, eds., Elk of North America. Stackpole Books, Harrisburg, Pennsylvania. 698 pp.

Bubenik, A. B. 1982. Taxonomy of pecora in relation to morphophysiology of their cranial appendices. Pages 163–185 *in* R. D. Brown, ed., Antler development in Cervidae. Caeser Kleberg Wildl. Res. Inst., Kingsville, Texas. 480 pp.

Buck, P. D. 1947. The biology of the antelope (*Antilocapra americana*) in Montana. M.S. thesis, Montana St. Coll., Bozeman. 70 pp.

Buechner, H. K. 1950a. Life history, ecology, and range use of the pronghorn antelope in Trans-Pecos, Texas. Am. Midl. Nat. 43(2):257–354.

———. 1950b. Range ecology of pronghorn on the Wichita Mountains Wildlife Refuge. Trans. N. Am. Wildl. Conf. 15: 627–644.

———. 1950c. Life history, ecology, and range use of the pronghorn antelope in Trans-Pecos, Texas. Ph.D. thesis, Ohio St. Univ., Columbus. 376 pp.

———. 1950d. Use of the helicopter in wildlife work. J. Wildl. Manage. 14(4):472–473.

———. 1961. Regulation of numbers of pronghorn antelope in relation to land use. Extrait de la Terre et la Vie 2:266–285.

Buechner, H. K., I. O. Buss and H. F. Bryan. 1951. Censusing elk by airplane in the Blue Mountains of Washington. J. Wildl. Manage. 15(1):81–87.

Buffon, G. L. 1764. Histoire naturelle. L. Compte da, des O'scaur, Vol. 4 of 9. De L'Impremeris, Paris 4:459–460.

Bullock, R. E. 1974. Functional analysis of locomotion in pronghorn antelope. Pages 274–305 *in* V. Geist and F. Walther, eds., The behavior of ungulates and its relation to management. Vol. 1. Publ. 24. Intl. Union Conserv. Nature and Nat. Resour., Morges, Switzerland. 511 pp.

Bunnell, F. L. and D. E. N. Tait. 1981. Population dynamics of bears—implications. Pages 75–98 *in* C. W. Fowler and T. D. Smith, eds., Dynamics of large mammal populations. John Wiley and Sons, New York, New York. 477 pp.

Bunnell, F. L., C. Galindo-Leal and A. Chan-McLeod. 1995. From wildlife management to conservation biology: A Canadian perspective. Pages 138–142 *in* J. A. Bissonette and P. R. Krausman, eds., Integrating people and wildlife for a sustainable future. The Wildl. Soc., Bethesda, Maryland. 715 pp.

Bunting, S. C., B. M. Kilgore and C. L. Bushey. 1987. Guidelines for prescribed burning sagebrush/grass rangelands in the northern Great Basin. General Tech. Rept. INT-231. U.S. For. Serv., Intermtn. Res. Sta., Ogden, Utah. 33 pp.

Burger, G. V. 1979. Principles of wildlife management. Pages 89–97 *in* R. D. Teague and E. Becker, eds., Wildlife conservation: Principles and practices. The Wildl. Soc., Washington, D.C. 280 pp.

Burkardt, J. W. and E. T. Tisdale. 1976. Causes of juniper invasion in southwestern Idaho. Ecology 57(3):472–484.

Burkett, D. W. and B. C. Thompson. 1994. Wildlife association with human-altered water sources in semiarid vegetation communities. Conserv. Biol. 8(3):682–690.

Burnham, K. P., D. R. Anderson and J. L. Laake. 1980. Estimation of density from line transect sampling of biological populations. Wildl. Monogr. 72. 202 pp.

Burnham, K. P., D. R. Anderson and J. L. Laake. 1985. Efficiency and bias in strip and line transect sampling. J. Wildl. Manage. 49(4):1,012–1,018.

Burroughs, R. D. 1961. The natural history of the Lewis and Clark expedition. Michigan St. Univ. Press, East Lansing. 340 pp.

Burt, W. H. 1943. Territoriality and home range concepts as applied to mammals. J. Mammal. 24(3):346–352.

Butler, W. B. 1975. The atlatl: The physics of function and performance. Plains Anthropol. 20(68):105–110.

Byers, D. S., ed. 1967. Environment and subsistence. Vol. 1. The pre-history of the Tehuacan Valley. Univ. Texas Press, Austin. 331 pp.

Byers, J. A. 1997a. American pronghorn: Social adaptations and the ghosts of predators past. Univ. Chicago Press, Chicago, Illinois. 300 pp.

———. 1997b. Mortality risk to pronghorn young from handling. J. Mammal. 78(3):894–899.

Byers, J. A. and K. Z. Byers. 1983. Do pronghorn mothers reveal the location of their hidden fawns? Behav. Ecol. Sociobiol. 13:147–156.

Byers, J. A. and J. T. Hogg. 1995. Environmental effects on prenatal growth rate in pronghorn and bighorn: Further evidence for energy constraint on sex-biased maternal expenditure. Behav. Ecol. 6(4):451–457.

Byers, J. A. and D. W. Kitchen. 1988. Mating system shift in a pronghorn population. Behav. Ecol. Sociobiol. 22(2):355–360.

Byers, J. A. and J. D. Moodie. 1990. Sex-specific maternal investment in pronghorn, and the question of a limit on differential provisioning in ungulates. Behav. Ecol. Sociobiol. 26(1):157–164.

Byers, J. A., J. D. Moodie and N. Hall. 1994. Pronghorn females choose vigorous mates. Anim. Behav. 47(1):33–34.

Cada, J. D. 1985. Evaluations of the telephone and mail survey methods of obtaining harvest data from licensed sportsmen in Montana. Pages 117–128 *in* S. L. Beasom and S. F. Roberson, eds., Game harvest management. Caesar Kleberg Wildl. Res. Inst., Kingsville, Texas. 374 pp.

Cadieux, C. L. 1986. Pronghorn, North America's unique antelope. Stackpole Books, Harrisburg, Pennsylvania. 254 pp.

———. 1987. Pronghorn antelope: Great Plains rebound. Pages 133–143 *in* H. Kallman, ed., Restoring America's wildlife. U.S. Fish and Wildl. Serv., Washington, D.C. 394+ pp.

Cahalane, V. H. 1947. Mammals of North America. The MacMillan Co., New York, New York. 682 pp.

Cain, D. 1971. The Ely chain. U.S. Bur. Land Manage., Ely, Nevada. 32 pp.

California Department of Fish and Game. 1994a. Draft environmental document, Sect. 363, Title 14, California code of regulations regarding antelope hunting. California Dept. Fish and Game, Sacramento. 132+ pp.

———. 1994b. Report to the legislature: Management of fish and wildlife on private lands. California Dept. Fish and Game, Sacramento. 9 pp.

———. 1995. Final environmental document, pronghorn antelope hunting. California Dept. Fish and Game, Sacramento. 141 pp.

———. 1997. Draft environmental document regarding pronghorn antelope hunting. California Dept. Fish and Game, Sacramento. 106 pp.

———. 1998. Final environmental document, Sect. 363, Title 14, California code of regulations regarding pronghorn antelope hunting. California Dept. Fish and Game, Sacramento. 102+ pp.

———. 2000. Final environmental document, Sect. 363, Title 14, California code of regulations regarding pronghorn antelope hunting. California Dept. Fish and Game, Sacramento. 100+ pp.

Campbell, R. B. 1970. Pronghorn, sheep and cattle range relationships in Carter County, Montana. M.S. thesis, Montana St. Univ., Bozeman. 87 pp.

Cancino, J. 1988. Habitos de alimentacion del berrendo peninsular (*Antilocapra americana peninsularis* Nelson). Theses professional Universidad Antonoma Chapingo, Chapingo, Mexico. 66 pp.

———. 1994a. Plan de recuperacion del berrendo peninsular (*Antilocapra americana peninsular*). *In* J. Cancino, ed., Population and habitat viability assessment of the peninsular pronghorn: Briefing book. Centro de Investigaciones Biologicas del Noreste, S.C. La Paz, Bàja California Sur, Mexico. Unpaginated.

———, ed. 1994b. Population and habitat viability assessment of the peninsular pronghorn: Briefing book. Centro de Investigaciones Biologicas del Noreste, S.C. La Paz, Baja California Sur, Mexico. Unpaginated.

———. 1994c. Food habits of the peninsular pronghorn. Pronghorn Antelope Workshop Proc. 16:176–185.

Cancino, J. and R. Guenzel. 2002. First steps toward the assisted reproduction of pronghorn in Mexico. Abstract presented at Pronghorn Antelope Workshop 20:16.

Cancino, J. and F. Reygadas. 1999. Historical notations regarding pronghorn (*Antilocapra americana*) in Mexico. Pronghorn Antelope Workshop (1998) Proc. 18:81–83.

Cancino, J., R. Rodriguez-Estrella and B. Sanabria. 1994. 1993 census of the peninsular pronghorn. Pronghorn Antelope Workshop Proc. 16:168–175.

Cancino, J., A. Ortega-Rubio and R. Rodriguez-Estrella. 1998. Population size of the peninsular pronghorn in Baja California Sur, Mexico. California Fish and Game 84(1):25–30.

Cancino, J., C. Castillo, C. Manterola, F. Ramirez, J. M. Reyes, R. Rodriguez-Estrella and V. Sanchez-Sotomayor. 2000. National pronghorn recovery plan for Mexico. Pronghorn Antelope Workshop (1996) Proc. 17:45–49.

Cancino, J., R. Castellanos and V. Sánchez. 2001. Capture, hand rearing and captive management of peninsular pronghorn. Pronghorn Antelope Workshop (2000) Proc. 19:108.

Canon, S. K. 1993. Fawn survival and bed-site characteristics of Trans-Pecos pronghorn. Ph.D. thesis, Texas Tech Univ., Lubbock. 133 pp.

Canon, S. K. and F. C. Bryant. 1992. Survival of Trans-Pecos pronghorn. Pronghorn Antelope Workshop Proc. 15:67–77.

Canon, S. K. and F. C. Bryant. 1997. Bed-site characteristics of pronghorn fawns. J. Wildl. Manage. 61(4):1,134–1,141.

Canon, S. K., F. C. Bryant, K. N. Bretzlaff and J. M. Hellman. 1997. Pronghorn pregnancy diagnosis using trans-rectal ultrasound. Wildl. Soc. Bull. 25(4):832–834.

Capodice, J. A., C. McCarthy and R. Thomas. 1982. Adobe Valley pronghorn habitat management plan. U.S. Bur. Land Manage., Bishop, California. 27 pp.

Carl, G. R. and C. T. Robbins. 1988. The energetic cost of predator avoidance in neonatal ungulates: Hiding versus following. Can. J. Zool. 66(1):239–246.

Carlson, G. P. 1996a. Clinical chemistry tests. Pages 441–469 *in* B. P. Smith, ed., Large animal internal medicine. 2nd ed. Mosby, St. Louis, Missouri. 2,040 pp.

———. 1996b. Depression anemia. Pages 1,231–1,234 *in* B. P. Smith, ed., Large animal internal medicine. 2nd ed. Mosby, St. Louis, Missouri. 2,040 pp.

Carlson, P. H. 2003. The buffalo soldier tragedy of 1877. Texas A & M Univ. Press, College Station. 177 pp.

Carlton, C. and T. McKean. 1977. The carotid and orbital retia of the pronghorn, deer and elk. Anat. Rec. 189(l):91–108.

Carpenter, L. H. 2000. Harvest management goals. Pages 192–213 *in* S. Demarais and P. R. Krausman, eds., Ecology and management of large mammals in North America. Prentice Hall, Upper Saddle River, New Jersey. 778 pp.

Carpenter, T. 1999. Pronghorns fair and square. N. Am. Hunter 132:62–66.

Carr, C. 1927. The speed of prong-horn antelopes. J. Mammal. 8(3):249–250.

Carson, R. 1962. Silent spring. Houghton Mifflin Co., Boston, Massachusetts. 368 pp.

Carter, W. D. 1970. An experiment in promoting quality antelope hunting. Antelope States Workshop Proc. 4:70–74.

Castellanos, R., J. Cancino and V. Sánchez. 2001. Alternative capture technique for the peninsular pronghorn. Pronghorn Antelope Workshop (2000) Proc. 19:109.

Castillo, C., J. C. deVos, Jr. and J. Hervert. 2000. The status of Sonoran pronghorn in Sonora, Mexico. Pronghorn Antelope Workshop (1996) Proc. 17:107.

Caswell, H. 2000. Matrix population models. Sinauer Press, New York, New York. 328 pp.

Caton, J. D. 1877. The antelope and deer of America. Hurd and Houghton, New York, New York. 426 pp. (Reprinted in 1974 by Arno Press, New York, New York.)

Catton, T. and L. Mighetto. 1998. The fish and wildlife job on the national forests. Historical Res. Assoc., Inc., prepared for the U.S. For. Serv., Missoula, Montana. 452 pp.

Caughley, G. 1966. Mortality patterns in mammals. Ecology 47(3):906–918.

———. 1974a. Bias in aerial survey. J. Wildl. Manage. 38(4):921–933.

———. 1974b. Interpretation of age ratios. J. Wildl. Manage. 38(3):557–562.

———. 1974c. Interpretation of age ratios. Pages 108–113 *in* J. S. Wakeley, ed., Wildlife population ecology. Pennsylvania St. Univ. Press, University Park. 385 pp.

———. 1976. Wildlife management and the dynamics of ungulate populations. Pages 183–246 *in* T. H. Croaker, ed., Applied biology. Vol. 1. Academic Press, London, England. 358 pp.

———. 1977. Analysis of vertebrate populations. John Wiley and Sons, New York, New York. 234 pp.

Caughley, G. and J. Goddard. 1972. Improving the estimates from inaccurate censuses. J. Wildl. Manage. 36(1):135–140.

Caughley, G. and A. R. E. Sinclair. 1994. Wildlife ecology and management. Blackwell Scientific Publications, Boston, Massachusetts. 334 pp.

Caughley, G., R. Sinclair and D. Scott-Kemmis. 1976. Experiments in aerial survey. J. Wildl. Manage. 40(2):290–300.

Causey, A. S. 1989. On the morality of hunting. Envir. Ethics. 11:1):327–343.

Centro de Investigaciones Biologicas del Noroeste. 1994. Population and habitat viability assessment of the peninsular pronghorn: Briefing book. La Paz, Baja California Sur, Mexico. Unpaginated.

Chalmers, G. A. and M. W. Barrett. 1974. Infectious pododermatitis in a pronghorn antelope in Alberta. J. Wildl. Dis. 10(l):60–62.

Chalmers, G. A. and M. W. Barrett. 1977. Capture myopathy in pronghorn in Alberta, Canada. J. Am. Vet. Med. Assoc. 171(9):918–923.

Chamberlin, W. J. 1937. The ticks of Oregon. Bull. 349. Oregon St. Coll. Agric. Exp. Sta., Corvallis. 34 pp.

Chandrabhanusinhji. 1937. Replacement of horns in nilgai. J. Bombay Nat.. Hist. Soc. 39:618.

Chao, J. 1985. Responses of territorial pronghorn to snortwheeze vocalizations. M.S. thesis, Univ. Idaho, Moscow. 32 pp.

Chase, C. E. 1977. Wonders of the pronghorn. Dodd, Mead and Co., New York, New York. 64 pp.

Chattin, J. E. and R. Lassan. 1950. California antelope reproductive potentials. California Fish and Game 36(3):328–329.

Chattin, J. E., C. M. Herman and H. Kirby. 1944. A parasite from antelope in California. Am. Micro. Soc. Trans. 63(l):27–29.

Cheney, E. S. 1930. Camera hunting the pronghorn. Am. For. and For. Life 36(8):487–490.

Christensen, N. L. 1981. Fire regimes in southwestern ecosystems. Pages 112–136 *in* H. A. Mooney, T. M. Bonnicksen, N.

L. Christensen, J. E. Hotan and W. A. Reiners, tech. coordinators, Fire regimes and ecosystem properties. General Tech. Rept. WO-26. U.S. For. Serv., Washington, D.C. 549 pp.

Clancy, G. 1999. How to sit still. N. Am. Hunter 134 (November):64–66, 68.

Clark, J. L. 1928. Trails of the hunted. Little, Brown and Co., Boston, Massachusetts. 310 pp.

Clark, K. A., R. M. Robinson, L. L. Weishuhn, G. W. Litton and R. G. Marburger. 1972. Caseous lymphadenitis in pronghorn (*Antilocapra americana*). J. Wildl. Dis. 8(l):67–71.

Clark, W. E. 1999. Status of the northern Yellowstone pronghorn herd. Pronghorn Antelope Workshop (1998) Proc. 18:109.

Clarke, C. H. D. 1979. Hunting and fishing values and concepts. Pages 7–10 *in* R. D. Teague and E. Decker, eds., Wildlife conservation: Principles and practices. The Wildl. Soc., Washington, D.C. 280 pp.

Clary, W. P. and R. C. Holmgren. 1982. Observations of pronghorn distribution in relation to sheep grazing on the Desert Experimental Range. Pages 581–592 *in* J. M. Peek and P. D. Dalke, eds., Wildlife-livestock relationships symposium: Proc.10. Univ. Idaho For., Wildl. and Range Exp. Sta., Moscow. 614 pp.

Clary, W. P., D. M. Beale and R. C. Holmgren. 1985. Melanistic pronghorn fawns on the Desert Experimental Range, Utah. Southwestern Nat. 30(1):159–161.

Clemens, E. T., K. L. Meyer, M. P. Carlson and N. R. Schneider. 1987. Hematology, blood chemistry and selenium values of captive pronghorn antelope, white-tailed deer and American bison. Comp. Biochem. and Physiol. 87(1):167–170.

Clemente, F., R. Valdez, J. L. Holechek, P. J. Zwank and M. Cardenas. 1995. Pronghorn home range relative to permanent water in southern New Mexico. Southwestern Nat. 40(1):38–41.

Clements, F. E. 1916. Plant succession: An analysis of the development of vegetation. Publ. 242. Carnegie Inst., Washington, D.C. 512 pp.

Clements, F. E. and V. E. Shelford. 1939. Bio-ecology. John Wiley and Sons Inc., New York, New York. 425 pp.

Clepper, H. 1966. The conservation movement: Birth and infancy. Pages 3–15 *in* H. Clepper, ed., Origins of American conservation. Ronald Press Co., New York, New York. 193 pp.

Clute, R. K. and J. J. Ozoga. 1983. Icing of transmitter collars on white-tailed deer fawns. Wildl. Soc. Bull. 11(l):70–71.

Clutton-Brock, T. H., F. E. Guinness and S. D. Albon. 1982. Red deer: Behavior and ecology of two sexes. Univ. Chicago Press, Chicago, Illinois. 378 pp.

Clutton-Brock, T. H., S. D. Albon and F. E. Guinness. 1987. Interactions between population density and maternal characteristics affecting fecundity and juvenile survival in red deer. J. Anim. Ecol. 56(3):857–871.

Cochran, W. G. 1963. Sampling techniques. John Wiley and Sons, Inc., New York, New York. 413 pp.

———. 1977. Sampling techniques. John Wiley and Sons, Inc., Santa Barbara, California. 428 pp.

Cockrell, T. D. A. 1900. The markings of *Antilocapra*. Nature 63(1,620):58.

Cockrum, E. L. 1981. Taxonomy of the Sonoran pronghorn. Pages 2–10 *in* J.S. Phelps, ed., The Sonoran pronghorn. Spec. Rept. 10. Arizona Dept. Game and Fish, Phoenix. 55 pp.

Coffey, M. D. 1954. A study of some Rocky Mountain spotted fever vectors and their hosts in Utah. Great Basin Nat. 14(1–2):31–37.

Colbenson, H. P. 1962. Spinose ear tick *Ornithodoros (Otobius) megnini* from pronghorn antelope. J. Parasitol. 48(4):630.

Colbert, E. H. and R. G. Chaffee. 1939. A study of *Tetrameryx* and associated fossils from Pago Spring Cave, Sonita, Arizona. Am. Mus. Novit. 1,034:1–21.

Cole, G. F. 1956. The pronghorn antelope, its range use and food habits in central Montana with special reference to alfalfa. Tech. Bull. 516. Montana Dept. Fish and Game, Helena, and Montana St. Coll. Agric. Exp. Sta., Bozeman. 63 pp.

Cole, G. F. and B. T. Wilkins. 1958. The pronghorn antelope, its range use and food habits in central Montana with special reference to wheat. Tech. Bull. 2. Montana Dept. Fish and Game, Helena. 39 pp.

Collins, B. T. 1990. Using rerandomization tests in route-regression analysis of avian population trends. Pages 63–70 *in* J. R. Sauer and S. Droege, eds., Survey designs and statistical methods for the estimation of avian population trends. Biol. Rept. 90. U.S. Fish and Wildl. Serv. 166 pp.

Collins, M. T., T. A. Gallegos, J. S. Reif and W. T. Adrian. 1981. Seroepidemiology of *Leptospira interrogans* serovar *hardjo* in Colorado antelope and cattle. J. Am. Vet. Med. Assoc. 179(11):1,136–1,138.

Columbus Children's Zoo Staff. 1993. Hand-rearing procedures for pronghorn antelope (*Antilocapra americana*) at the Columbus Zoo. Anim. Keeper's Forum 20(2):68–69.

Committee on Rangeland Classification. 1994. Rangeland health: New methods to classify, inventory, and monitor rangelands. Natl. Acad. Press, Washington, D.C. 180 pp.

Compton, H. O. 1958. The effects of predation on pronghorn antelope numbers in southcentral Oregon. M.S. thesis, Oregon St. Coll., Corvallis. 71 pp.

———. 1970. Southeastern Montana antelope population trends in relation to severe winters. Antelope States Workshop Proc. 4:50–56.

Connelly, J. W., M. A. Schroeder, A. R. Sands and C. E. Braun. 2000. Guidelines to manage sage grouse populations and their habitats. Wildl. Soc. Bull. 28(4):967–985.

Connolly, G. E. 1978a. Predators and predator control. Pages 369–394 *in* J. L. Schmidt and D. L. Gilbert, eds., Big game of North America. Stackpole Books, Harrisburg, Pennsylvania. 494 pp.

———. 1978b. Predator control and coyote populations: A review of simulation models. Pages 327–345 *in* M. Bekoff, ed., Coyotes. Biology, behavior, and management. Academic Press, New York, New York. 384 pp.

———. 1981. Limiting factors and population regulation. Pages 245–285 *in* O. C. Wallmo, ed., Mule deer and black-tailed deer of North America. Univ. Nebraska Press, Lincoln. 605 pp.

Cook, B. B., L. E. Witham, M. Olmstead and A. F. Morgan. 1949. The influence of seasonal and other factors on the acceptability and food value of meat of two subspecies of California deer and antelope. Hilgardia 19(8):265–284.

Cook, J. G. 1984. Pronghorn winter ranges: Habitat characteristics and a field test of a habitat suitability model. M.S. thesis, Univ. Wyoming, Laramie. 91 pp.

Cook, J. G. and L. L. Irwin. 1985. Validation and modification of a habitat suitability model for pronghorn. Wildl. Soc. Bull. 13(4):440–448.

———. 1992. Climate–vegetation relationships between the Great Plains and Great Basin. Am. Midl. Nat. 127(2):316–326.

Cook, J. G., L. L. Irwin, A. W. Allen and M. J. Armbruster. 1984. A field test of a winter pronghorn habitat suitability index model. Pronghorn Antelope Workshop Proc. 11:204–211.

Cook, J. O., III. 1994. Avoiding shrinkage in trophy horns. Fair chase: The official publication of the Boone and Crockett Club, Missoula, Montana. Fall 1994:16.

Cooley, R. A. 1932. The Rocky Mountain wood tick. Bull. 268. Montana St. Coll. Agric Exp. Sta., Bozeman. 58 pp.

Cooperrider, A. Y. 1986. Food habits. Pages 699–710 *in* A. Y. Cooperrider, R. J. Boyd, and H. R. Stuart, eds., Inventory and monitoring of wildlife habitat. Serv. Ctr., U.S. Bur. Land Manage., Denver, Colorado. 858 pp.

Cooperrider, A. Y. and J. A. Bailey. n.d. Variation in food habits of wild ungulates from south central Colorado. Spec. Rept. U.S. Bur. Land Manage., Denver, Colorado. 25 pp.

Cooperrider, A. Y., J. A. Bailey and R. M. Hansen. 1981. Cost-efficient methods of estimating ungulate food habits by fecal analysis. Pages 399–406 *in* H. G. Lund, M. Caballero, R. H. Hamre, R. S. Driscoll and W. Bonner, tech. coordinators, Arid land resource inventories: Development of cost-efficient methods. General Tech. Rept. WO-28. U.S. For. Serv., Washington, D.C. 621 pp.

Cooperrider, A. Y., R. J. Boyd and H. R. Stuart, eds. 1986. Inventory and monitoring of wildlife habitat. Serv. Ctr., U.S. Bur. Land Manage, Denver, Colorado. 858 pp.

Cope, E. D. 1874. Notes on the Santa Fe Marls, and some of the contained vertebrate fossils. Proc. Acad. Nat. Sci., Philadelphia: 147–154.

———. 1877. Artiodactyla. Pages 324–365 *in* Report upon United States geographical surveys west of the one hundredth meridian. Vol. 4. Paleontology. U.S. Govt. Print. Off., Washington, D.C. 370 pp.

———. 1888. The Artiodactyla. Am. Nat. 22:1,079–1,095.

Copeland, G. L. 1980. Antelope buck breeding behavior, habitat selection and hunting impact. Wildl. Bull. 8. Idaho Dept. Fish and Game, Boise. 45 pp.

Copeland, G. L., R. E. Autenrieth, L. E. Oldenburg and T. P. Kistner. 1978. Tranquilizing pronghorns with M-99 from a helicopter. Pronghorn Antelope Workshop Proc. 8:94–109.

Corneli, P. S. 1980. Pronghorn fawn mortality following coyote control on the National Bison Range. M.S. thesis, Univ. Montana, Missoula. 69 pp.

Corneli P. S., B. Von Gunten Moran and B. W. O'Gara. 1984. Pronghorn fawn mortality on the National Bison Range. Pronghorn Antelope Workshop Proc. 11:41–48.

Costello, D. F. 1969. The prairie world. Thomas Y. Crowell Co., New York, New York. 242 pp.

———. 1970. The world of the prairie dog. J. B. Lippincott and Co., New York, New York. 160 pp.

Couey, F. 1946. Antelope foods in southeastern Montana. J. Wildl. Manage. 10(4):367.

Couey, F. M. 1949. Review and evaluation of big game trapping techniques. Western Assoc. St. Fish and Game Commiss. Proc. 29:39–47.

Couling, L. and G. E. J. Smith. 1980. Impact of a postcard follow-up on harvest survey returns. Prog. Note 113. Can. Wildl. Serv., Ottawa, Ontario. 9 pp.

Courtney, R. F. 1989. Pronghorn use of recently burned mixed prairie in Alberta. J. Wildl. Manage. 53(2):302–305.

Cowan, R. L. and A. C. Clark. 1981. Nutritional requirements. Pages 72–86 *in* W. R. Davidson, F. A. Hayes, V. F. Nettles and F. E. Kellogg, eds., Diseases and parasites of white-tailed deer. Misc. Publ. 7. Tall Timbers Res. Sta.., Tallahassee, Florida. 458 pp.

Cowlishaw, S. J. and F. E. Alder. 1960. The grazing preferences of cattle and sheep. J. Agric. Sci. 54(2):252–265.

Cox, D. R. and D. Oakes. 1984. Analysis of survival data. Chapman and Hall, New York, New York. 210 pp.

Cram, E. B. 1925. *Cooperia bisonis*, a new nematoda from buffalo. J. Agric. Res. 30(6):571–573.

Crawford, J. A. and K. A. Coggins. 1997. Habitat selection and reproductive success of nesting sage grouse in altered habitat and climatic conditions of southeastern Oregon. Oregon Game Bird Research Program, Dept. Fisheries and Wildl., Oregon St. Univ., Corvallis. 88 pp.

Creek, D. E. 1967. A summary of pronghorn antelope investigations. Wyoming Game and Fish Commiss., Cheyenne. 42 pp.

Cressman, L. S. 1977. Prehistory of the Far West: Homes of vanished peoples. Univ. Utah Press, Salt Lake City. 248 pp.

Cromrich, L. 2001. Award-winning burn benefits southwestern Wyoming wildlife. Bugle 18(1):126.

Cromrich, L. and D. Burgess. 2001. Hunters, anglers raise $293 million for wildlife in nine months. Bugle 18(3):81.

Cronin, M. A. 1993. Mitochondrial DNA in wildlife taxonomy and conservation biology: Cautionary notes. Wildl. Soc. Bull. 21(3):339–348.

Cronin, M. A., R. Stuart, B. J. Pierson and J. C. Patton. 1996. Kappa-casein gene phylogeny of higher ruminants (Pecora, Artiodactyla). Mol. Phylogenetics and Evol. 6(2):295–311.

Crowe, D. M. 1987. Wildlife management in transition. Pages 281–285 *in* N. Blair, ed., The history of wildlife management in Wyoming. Wyoming Dept. Game and Fish, Cheyenne. 295 pp.

Crowe, D. and D. Strickland. 1984. Weathering winter. Wyoming Wildl. 48(6):20–27.

Culbert, R. and R. B. Blair, eds. 1989. Recovery planning and endangered species. Endangered Species Update 6(10):2–8.

Curtain, C. C. and H. H. Fudenberg. 1973. Evolution of the immunoglobulin antigens in the ruminantia. Biochem. Genet. 8(3):301–308.

Curtis, E. S. 1907–1930. The American Indian. 20 vols. F. W. Hodge, ed. Univ. Press, Cambridge, Massachusetts. (reprinted in 1976; Jofnson Reprint Company, New York, New York).

Custer, E. B. 1885. Boots and saddles. Harper and Brothers, New York, New York. 312 pp.

———. 1895. Tenting on the plains. Harper and Brothers, New York, New York. 403 pp.

Cutler, M. R. 1982. What kind of wildlifers will be needed in the 1980s? Wildl. Soc. Bull. 10(1):75–79.

Cutler, P. L. 1996. Wildlife use of two artificial water developments on the Cabeza Prieta National Wildlife Refuge, southwestern Arizona. M.S. thesis, Univ. Arizona, Tucson. 124 pp.

Czaplewski, R. L., D. M. Crowe and L. L. McDonald. 1983. Sample sizes and confidence intervals for wildlife population ratios. Wildl. Soc. Bull. 11(2):121–128.

Czech, B. 1992. Lessons learned from an intertribal, interregional pronghorn antelope reintroduction. Pronghorn Antelope Workshop Proc. 15:100–108.

D'Abreau, E. A. 1936. Horn growth in nilgai. J. Bombay Nat. Hist. Soc. 39:171.

Dagum, C. and E. B. Dagum. 1988. Trends. Pages 321–324 *in* S. Kotz and N. L. Johnson, eds., Encyclopedia of statistical sciences. Vol. 9. John Wiley and Sons, New York, New York. 439 pp.

Dahl, R. W., K. Cross and L. Cross. 1986. Lander one-shot antelope hunt. One-shot Antelope Hunt Foundation, Lander, Wyoming. 129 pp.

Dalquest, W. 1974. A new species of four-horned Antilocaprid from Mexico. J. Mammal. 55(l):96–101.

Danvir, R. E. 2000. Environmental and density-dependent effects on a northern Utah pronghorn population. Pronghorn Antelope Workshop (1996) Proc. 17:36–41.

Dasmann, R. F. 1964. Wildlife biology. John Wiley and Sons, New York, New York. 231 pp.

———. 1981. Wildlife biology. 2nd ed. John Wiley and Sons, New York, New York. 212 pp.

Dauphiné, T. C., Jr. and R. L. McClure. 1974. Synchronous mating in Canadian barren-ground caribou. J. Wildl. Manage. 38(1):54–66.

Davies, R. B. and C. P. Hibler. 1979. Animal reservoirs and cross-species transmission of *Giardia*. Pages 104–125 *in* W. Jakubowski and J. C. Hoff, eds., Waterborn transmission of giardiasis. U.S. Environ. Protect. Agency, Off. of Environ., Res. and Dev., Cincinnati, Ohio. 306 pp.

Davis, B. D., R. Dulbecco, H. N. Eisen, H. S. Ginsberg, W. B. Wood and M. McCarty. 1973. Microbiology. Harper and Row, Hagerstown, Maryland. 1,426 pp.

Davis, D. E. 1963. Estimating the numbers of game populations. Pages 89–118 *in* H. S. Mosby, ed., Wildlife investigational techniques. 2nd ed. The Wildl. Soc., Washington, D.C. 419 pp.

Davis, J. T. 1961. Trade routes and economic exchange among the Indians of California. Rept. 54. Univ. California Archaeol. Surv., Berkeley. 71 pp.

Davis, J. W., L. H. Karstad and D. O. Trainer. 1981a. Infectious diseases of wild mammals. 2nd ed. Iowa St. Univ. Press, Ames. 446 pp.

———. 1981b. Parasitic diseases of wild mammals. 2nd ed. Iowa St. Univ. Press, Ames. 446 pp.

Davis, P. 1995. Management recommendations for the Sheldon National Wildlife Refuge offered by the Sheldon Scientific Working Group. U.S. Fish and Wildl. Serv., Lakeview, Oregon. 17 pp.

Davis, W. and D. Putnam. 1951. An aerial coyote–antelope observation. J. Wildl. Manage. 15(3):330.

Davitt, B. B. and J. R. Nelson. 1980. A method of preparing plant epidermal tissue for use in fecal analysis. Circ. 0628. Coll. Agric. Res. Ctr., Washington St. Univ., Pullman. 4 pp.

DeAngelis, D. L. and L. J. Gross. 1992. Individual based models and approaches in ecology: Populations, communities and ecosystems. Chapman and Hall, New York, New York. 525 pp.

Dearden, B. L., R. E. Pegan and R. M. Hansen. 1975. Precision of microhistological estimates of ruminant food habits. J. Wildl. Manage. 39(2):402–407.

Deaver, K. and G. S. Greene. 1978. Faunal utilization at 45AD2: A prehistoric archaeological site in the channeled scablands of eastern Washington. TEBIWA 14:1–21.

Deblinger, R. D. 1988. Ecology and behavior of pronghorn in the Red Desert, Wyoming with reference to energy development. Ph.D. thesis, Colorado St. Univ., Fort Collins. 227 pp.

Deblinger, R. D. and A. W. Alldredge. 1989. Management implications of variations in pronghorn social behavior. Wildl. Soc. Bull. 17(1):82–87.

———. 1991. Influence of free water on pronghorn distribution in the sagebrush/steppe grasslands. Wildl. Soc. Bull. 19(3):321–326.

Deblinger, R. D. and J. E. Ellis. 1976. Aspects of intraspecific social variations in pronghorns. Pronghorn Antelope Workshop Proc. 7:26–48.

Deblinger, R. D., A. W. Alldredge, and L. Saslaw. 1984. Migratory patterns of pronghorn in the Red Desert, Wyoming. Pronghorn Antelope Workshop Proc. 11:90–91.

Decker, D. J., T. L. Brown and B. A. Knuth. 1996. Human dimensions research: Its importance in natural resource management. Pages 29–47 *in* A. W. Ewert, ed., Natural resource management: The human dimension. Westview Press, Boulder, Colorado. 263 pp.

Decker, E. 1978. Exotics. Pages 249–256 *in* J. L. Schmidt and D. L. Gilbert, eds., Big game of North America. Stackpole Books, Harrisburg, Pennsylvania. 494 pp.

DelMonte, B. E. and H. G. Kathmann. 1984. Evaluation of Texas antelope transplants. Pronghorn Antelope Workshop Proc. 11:146–150.

Demaster, D. P., D. Miller, D. Goodman, R. DeLong and B. Stewart. 1982. Assessment of California sea lion–fishery interactions. Trans. N. Am. Wildl. and Nat. Resour. Conf. 47:253–264.

Deming, O. V. 1959. Climate, range, and antelope. Interstate Antelope Conf. Trans.2:37–64.

———. 1963. Antelope and sagebrush. Interstate Antelope Conf. Trans. 14:55–60.

Den Boer, P. J. 1991. Seeing the trees for the wood: Random walks or bounded fluctuations of population size? Oecologia 86 (3):484–491.

Denney, R. N. 1958. Sex determination in dressed elk carcasses. Trans. N. Am. Wildl. Conf. 23:501–513.

———. 1978. Managing the harvest. Pages 395–408 *in* J. L. Schmidt and D. L. Gilbert, eds., Big game of North America: Ecology and management. Stackpole Books, Harrisburg, Pennsylvania. 494 pp.

Dennis, B. and M. R. Otten. 2000. Joint effects of density dependence and rainfall on abundance of San Joaquin kit fox. J. Wildl. Manage. 64(2):388–400.

Dennis, B. and M. L. Taper. 1994. Density dependence in time series observations of natural populations: Estimation and testing. Ecol. Monogr. 64(1):205–224.

Dennis, B., P. L. Munholland and J. M. Scott. 1991. Estimation of growth and extinction parameters for endangered species. Ecol. Monogr. 61(1):115–143.

Denome, R. M., B. O'Callaghan, C. Luitjens, E. Harper and R. Bianco. 1994. Characterization of a satellite DNA from *Antilocapra americana*. Gene 145(2):257–259.

Deseret Land and Livestock Ranch. 1997. Cooperative wildlife management plan. Woodruff, Utah. 3 pp.

Desimone, P. 1995. Conflicts between economic development and endangered species in southern California. Pages 85–86 *in* J. A. Bissonette and P. R. Krausman, eds., Integrating people and wildlife for a sustainable future. The Wildl. Soc., Bethesda, Maryland. 715 pp.

de Trobriand, P. R. 1951. Military life in Dakota. Translated and edited from the French original by L. M. Kane. Univ. Nebraska Press, Lincoln. 422 pp.

deVos , J. 1996. Population simulation for the endangered Sonoran pronghorn. Unpubl. rept. Arizona Dept. Fish and Game, Phoenix. 34 pp.

deVos, J. C., Jr. and L. Thompson-Olais. 2000. Using the VORTEX model to assess minimum viable population for Sonoran pronghorn. Pronghorn Antelope Workshop (1996) Proc. 17:108.

Dhindsa, D. S., T. H. Cochran, A. Castro, J. R. Swanson and J. Metcalfe. 1975. Serum biochemical and electrophoretic values from four deer species and from pronghorn antelope. Am. J. Vet. Res. 36(10):1,455–1,457.

Dhindsa, D. S., J. Metcalfe, T. McKean and T. Thorne. 1974. Comparative studies of the respiratory functions of mammalian blood. Pronghorn antelope (*Antilocapra americana*). Resp. Physiol. 21(3):297–306.

Diario Official. 1988. Desireto por el que se declara la Reserva de la biosfera "El Vizcaino," ubicada en el Municipio de Mulege, B.C.S. Diario Official de la Federacion. Tomo 422(22):2–27.

———. 1991. Acuerdo par el que se establicen los criterios ecologicos CT-CERNL-001-91 que determinan las especies raras, amenazadas, en peligro de extinction o sujectas a proteccion especial y sus endemisomes, de la flora y fauna terrestres y acuatuas en la Republica Mexicana. Diario Official de la Federacion. Tomo 452(12):7–35.

Di Camillo, J. A. and J. M. Schaefer. 2000. Internet program impacts youth interest in hunting. Wildl. Soc. Bull. 28(4): 1,077–1,085.

Dickerson, L. M. 1939. The problem of wildlife destruction by automobile traffic. J. Wildl. Manage. 3(2):104–116.

Dikmans, G. 1932. The antelope, *Antilocapra americana*, a new host of the lungworm *Protostrongylus macrotis*. J. Parasitol. 19(1):93.

———. 1933. *Onchocera flexuosa* from the subcutaneous abscesses of an antelope and from the subcutaneous abscesses of a deer. J. Parasitol. 19(3):246.

Dimmick, R. W. and M. R. Pelton. 1994. Criteria of sex and age. Pages 169–214 *in* T. A. Bookhout, ed., Research and management techniques for wildlife and habitats. The Wildl. Soc., Bethesda, Maryland. 740 pp.

Dinges, W. R. 1958. Techniques in marking pronghorn antelope for field identification. M.S. thesis, Oregon St. Coll., Corvallis. 74 pp.

Dingman, R. E. 1966. Variation in selected populations of pocket gophers (*Thomomys bottae*) of the lower Colorado River. Ph.D. thesis, Univ. Arizona, Tucson. 134 pp.

Dirschl, H. J. 1960. The pronghorn antelope range in Saskatchewan. M.S. thesis, Univ. Saskatchewan, Saskatoon. 46 pp.

———. 1962. Sieve mesh size related to analysis of antelope rumen contents. J. Wildl. Manage. 26(3):327–328.

———. 1963. Food habits of the pronghorn in Saskatchewan. J. Wildl. Manage. 27(l):81–93.

Dirschl, H. J., M. Norton-Griffiths and S. P. Wetmore. 1981. Training observers for aerial surveys of herbivores. Wildl. Soc. Bull. 9(2):108–115.

Dixon, P. M. 1993. The bootstrap and the jackknife: Describing the precision of ecological indices. Pages 290–318 *in* S. M. Scheiner and J. Gurevitch, eds., Design and analysis of ecological experiments. Chapman and Hall, London, England. 445 pp.

Dizard, J. E. 1994. Going wild: How animal rightists view the "wildlife management-hunting system." Wildl. Soc. Bull. 15:599–602.

Dobel, M. 2000. Nevada pronghorn status report—1996. Pronghorn Antelope Workshop (1996) Proc. 17:20.

Dobson, D. 2000. Saskatchewan pronghorn antelope status report—1996. Pronghorn Antelope Workshop (1996) Proc. 17:25.

———. 1977. Great North American Indians, profiles in life and leadership. Von Nostrand Reinhold Co., New York, New York. 386 pp.

Dodge, R. I. 1877. The plains of the great West. G. P. Putnam's Sons, New York, New York. 503 pp.

———. 1959a. Our wild Indians: Thirty-three years' personal experience among the red men of the great West. Archer House, Inc., New York, New York. 657 pp.

———. 1959b. Plains of the great West and their inhabitants. Archer House, Inc., New York, New York. 452 pp.

Dolbeer, R. A., N. R. Holler and D. W. Hawthorne. 1994. Identification and control of wildlife damage. Pages 474–506 *in* T. A. Bookhout, ed., Research and management techniques for wildlife and habitats. The Wildl. Soc., Bethesda, Maryland. 740 pp.

Donahue, D. L. 1999. The western range revisited. Univ. Oklahoma Press, Norman. 388 pp.

Donoho, H. S. and D. L. Wills. 1974a. A portable video-tape system for counting and classifying big game. Antelope States Workshop Proc. 6:130–132.

———. 1974b. Analysis of sampling procedures used for antelope classification counts. Antelope States Workshop Proc. 6:122–129

Dood, A. 1984. The history of antelope management in southeastern Montana. Pronghorn Antelope Workshop Proc. 11:92–102.

Doughty, J. 1833. Prong-horned antelope (*Antilope americana*). Pages 129–150 *in* G. Stewart, ed., The cabinet of natural history and American rural sports. Vol. 3. John Doughty, Philadelphia, Pennsylvania. (Imprint Soc., Barre, Massachusetts 1973.)

Dow, S. A., Jr. 1952. An evaluation of some criteria for age determination of the pronghorn (*Antilocapra americana* Ord). M.S. thesis, Montana St. Univ., Missoula. 71 pp.

Dow, S. A., Jr. and P. L. Wright. 1962. Changes in mandibular dentition associated with age in pronghorn antelope. J. Wildl. Manage. 26(1):1–18.

Droege, S. 1990. The North American breeding bird survey. Pages 1–4 *in* J. R. Sauer and S. Droege, eds., Survey designs and statistical methods for the estimation of avian population trends. Biol. Rept. 90(1). U.S. Fish and Wildl. Serv. 166 pp.

Dubey, J. P. 1980. Sarcocystis species in moose (*Alces alces*), bison (*Bison bison*) and pronghorn (*Antilocapra americana*) in Montana. Am. J. Vet. Res. 41(12):2,063–2,065.

———. 1981. Isolation of encysted *Toxoplasma gondii* from musculature of moose and pronghorn in Montana. Am. J. Vet. Res. 42(1):126–127.

Dubey, J. P., E. T. Thorne and E. S. Williams. 1982. Induced toxoplasmosis in pronghorn and mule deer. J. Am. Vet. Med. Assoc. 181(11):1,263–1,267.

Duda, M. D. 2000. Big game popularity soars. N. Am. Hunter 137, April/May:38, 40–41.

Duda, M. D., S. J. Bissel and K. C. Young. 1995. Factors related to hunting and fishing participation in the United States. Phase V: Final Rept. Western Assoc. Fish and Wildl. Agencies, Responsive Manage. Board of Directors and Responsive Manage. Natl. Off., U.S. Fish and Wildl. Serv., Harrisonburg, Virginia. 53 pp.

———. 1998. Wildlife and the American mind. Responsive Manage., Harrisonburg, Virginia. 804 pp.

Dunbar, M. R. 1999a. Health evaluation of pronghorn (*Antilocapra americana*) fawns on Hart Mountain National Antelope Refuge, Oregon, 1998. Tech. Rept. 99-01. U.S. Fish and Wildl Serv., Sheldon/Hart Mtn. Refuge Complex, Lakeview, Oregon. 17 pp.

———. 1999b. A pronghorn management plan for Hart Mountain National Antelope Refuge. Interstate Antelope Conf. Trans. 35: Unpaginated.

———. Unpublished. December 1999 progress report. Sheldon/Hart Mountain Refuge Complex, U.S. Fish and Wildl. Serv., Lakeview, Oregon. 19 pp.

Dunbar, M. R. and M. R. Giordano. 2000. Ecology, health, and seasonal movements of coyotes on Hart Mountain National Antelope Refuge. Prog. Rept. U.S. Fish and Wildl. Serv., Lakeview, Oregon. 46 pp.

Dunbar, M. R. and R. Velarde. 1998. Health evaluation of pronghorns *(Antilocapra americana*) on Hart Mountain National Antelope Refuge in southeastern Oregon, 1996–1997. Tech. Rept. 98-01. Biol. Resour. Div., Natl. Wildl. Health Ctr., U.S. Geolog. Surv., Madison, Wisconsin. 32 pp.

Dunbar, M. R., R. Velarde, M. A. Gregg and M. Bray. 1999. Health evaluation of a pronghorn antelope population in Oregon. J. Wildl. Dis. 35(3):496–510.

Dunbar, M. R., M. J. Wolcott, R. B. Rimler and B. M. Berlowski. 2000. Septicemic pasteurelosis in free-ranging neonatal pronghorn in Oregon. J. Wildl. Dis. 36(2):383–388.

Dunning, J. B., B. J. Danielson and H. R. Pulliam. 1992. Ecological processes that affect populations in complex landscapes. Oikos 65(1):169–175.

Dvorak, D. F. 1986. Texas antelope status report. Pronghorn Antelope Workshop Proc. 12:34–35.

Easterly, T. G. and R. J. Guenzel. 1992. Responses of pronghorn to petroleum development on crucial winter range. Pronghorn Antelope Workshop Proc. 15:124–141.

Easterly, T., A. Wood and T. Litchfield. n.d. Responses of pronghorn and mule deer to petroleum development on crucial winter range in the Rattlesnake Hills. Compl. Rept. Wyoming Dept. Game and Fish, Cheyenne. 67 pp.

Eaton, S. B. and M. Konner. 1985. Paleolithic nutrition. New England J. Med. 312:283–289.

Eberhardt, L. L. 1970. Correlations, regression and density dependence. Ecology 51(2):306–310.

———. 1978. Transect methods for population studies. J. Wildl. Manage. 32(1):1–31.

———. 1988. Using age structure data from changing populations. J. Applied Ecol. 25(2):373–378.

———. 1992. An analysis of procedures for implementing the dynamic response method. Marine Mammal Sci. 8(2):201–212.

Eberhardt, L. L., D. G. Chapman and J. R. Gilbert. 1979. A review of marine mammal census methods. Wildl. Monogr. 63. 46 pp.

Eccles, A. W., R. J. Finck and K. E. Sexson. 1994. Pronghorn in tallgrass prairie: Status of the Flint Hills herd. Pronghorn Antelope Workshop Proc. 16:50–53.

Edwards, R. Y. 1954. Comparison of an aerial and ground census of moose. J. Wildl. Manage. 18(3):403–404.

Edwards, W. C. 1957. Reproductive potential of big game animals. P-R Job Compl. Rept., Work Plan 10, Job 1, Proj. FW-3-R-4. Wyoming Game and Fish Commiss., Cheyenne. Pp. 33–47.

———. 1958a. Reproductive potential of big game animals. Pages 33–57 *in* P.R. Proj. FW-3-R-5. Wyoming Dept. Game and Fish, Cheyenne.

———. 1958b. A study of the reproductive potential and ovarian structures in the pronghorn antelope, *Antilocapra americana* (Ord.) in Wyoming. M.S. thesis, Univ. Wyoming, Laramie. 63 pp.

Effron, M., L. Griner and K. Benirschke. 1977. Nature and rate of neoplasia found in captive wild mammals, birds, and reptiles. J. Natl. Cancer Inst. 59(l):185–198.

Efron, B. and R. J. Tibshirani. 1993. An introduction to the bootstrap. Monographs on statistics and applied probability 57. Chapman and Hall, New York, New York. 436 pp.

Egan, F. 1977. Fremont: Explorer for a restless nation. Doubleday and Co., Garden City, New York. 582 pp.

Einarsen, A. S. 1938. Life history and management of antelope in Oregon. Trans. N. Am. Wildl. Conf. 3:381–387.

———. 1946. Crude protein determination of deer as an applied management technique. Trans. N. Am. Wildl. Conf. 11:309–312.

———. 1948. The pronghorn antelope and its management. Monumental Printing Co., Baltimore, Maryland. 238 pp.

Elliott, C. 1966. Antelope play in Florida. Outdoor Life 137(5):52–55, 130.

Elliott, E. R. 1968. Antelope study area population estimates. P-R Job Compl. Rept., W-93-R-9, J5. New Mexico Dept. Game and Fish, Santa Fe. 6 pp.

Ellis, F. H. 1979. Isleta Pueblo. Pages 351–365 *in* A. Ortiz, ed., Southwest. Handbook of North American Indians. Vol. 9. Smithsonian Inst., Washington, D.C. 717 pp.

Ellis, J. E. 1970. A computer analysis of fawn survival in the pronghorn antelope. Ph.D. thesis, Univ. California, Davis. 70 pp.

———. 1972. Observations on pronghorn population dynamics. Antelope States Workshop Proc. 5:55–65.

Ellis, J. E. and R. D. Deblinger. 1975. Delayed breeding and parturition in young pronghorns. Reprint 164. U.S. Intl.

Biol. Prog., Grassland Biome Ecosystem Analysis Study, Fort Collins, Colorado. 4 pp.

Ellis, J. E. and M. Travis. 1975. Comparative aspects of foraging behavior of pronghorn antelope and cattle. J. Appl. Ecol. 12(2):411–420.

Ellis, L. L. 1955. A study of the ectoparasites of certain mammals in Oklahoma. Ecology 36(l):12–18.

Elmore, W. and B. Kauffman. 1994. Riparian and watershed systems: Degradation and restoration. Pages 212–231 *in* M. Vavra, W. A. Laycock and R. D. Pieper, eds., Ecological implications of livestock herbivory in the West. Soc. for Range Manage., Denver, Colorado. 297 pp.

Emlen, S. T. and L. W. Oring. 1977. Ecology, sexual selection, and evolution of mating systems. Science 197(4,300):215–223.

Engle, D. M. and T. G. Bidwell. 2001. Viewpoint: The response of central North American prairies to seasonal fire. J. Range Manage. 54(1):2–10.

Engstrom, M. D. and T. C. Maxwell. 1988. Records of mountain lion (*Felis concolor*) from the Edwards Plateau of Texas. Texas J. Sci. 40(4):450–452.

Erickson, J. D. 2000. Endangering the economics of extinction. Wildl. Soc. Bull. 28(1):34–41.

Ernst, S. E. and W. J. Gladney. 1975. *Dermacentor albipictus*: Hybridization of the two forms of the winter tick. Annals Entomol. Soc. Am. 68(l):63–67.

Errington, P. L. 1945. Some contributions of a fifteen-year local study of northern bobwhite quail to knowledge of population phenomena. Ecol. Monogr. 15(1):1–34.

———. 1946. Predation and vertebrate populations. Quart. Rev. Biol. 21(2):144–177, 221–245.

———. 1956. Factors limiting higher vertebrate populations. Science 124(3216):304–307.

———. 1967. Of predation and life. Iowa St. Univ. Press, Ames. 277 pp.

Estes, R. D. 1976. The significance of breeding synchrony in the wildebeest. E. African Wildl. J. 14:135–152.

Evans, C. E., W. A. Troyer and C. J. Lensink. 1966. Aerial census of moose by quadrat sampling units. J. Wildl. Manage. 30(4):767–776.

Evans, H. E. and G. C. Christensen. 1979. Miller's anatomy of the dog. 2nd ed. W. B. Saunders Co., Philadelphia, Pennsylvania. 1,181 pp.

Evans, R. L. and R. M. Love. 1957. The step-point method of sampling—a practical tool in range management. J. Range Manage. 10(5):208–212.

Fahrig, L. 1991. Simulation methods for developing general landscape-level hypotheses of single-species dynamics. Pages 417–442 *in* M. G. Turner and R. H. Gardner, eds., Quantitative methods in landscape ecology. Springer-Verlag, New York, New York. 536 pp.

Fairbanks, W. S. 1992. Dominance, aggression, and reproductive success in female pronghorn. Ph.D. thesis, Univ. Kansas, Lawrence. 84 pp.

———. 1993. Birthdate, birth weight, and survival in pronghorn fawns. J. Mammal. 74(1):129–135.

Fatooh, J., T. L. Russi and A. E. Goldsmith. 1994. Pronghorn introductions to Mono County, California: 12 years after. Pronghorn Antelope Workshop Proc. 16:35–49.

Federal Register. 1996. Policy regarding the recognition of distinct vertebrate population segments under the Endangered Species Act. U.S. Govt. Print. Off., Feb. 7, 1996, 5:4,721–4,725.

Ferrel, C. and H. Leach. 1950. Food habits of the pronghorn antelope of California. California Fish and Game 36(1):21–26.

———. 1952. The pronghorn antelope of California with special reference to food habits. California Fish and Game 38(3):285–293.

Fichter, E. 1958. Study of antelope population on Upper Pahsimeroi. P-R Rept. W85-R-10. Idaho Dept. Fish and Game, Boise. 25 pp.

———. 1962. Some data on the natality of a pronghorn population and their bearing on future research. Interstate Antelope Conf. Trans. 13:85–95.

———. 1972. On the nature of pronghorn groups. Antelope States Workshop Proc. 5:222–230.

———. 1976. The relation of behavior to management. Pronghorn Antelope Workshop Proc. 7:189–209.

———. 1987. Pronghorn groups: On social organization. TEBIWA (J. Idaho Mus. Nat. Hist.) 23:11–22.

Fichter, E. and R. Autenrieth. 1978. Is trophy hunting fostering "unnatural" selection for smaller, less vigorous and non-territorial pronghorn bucks? Pronghorn Antelope Workshop Proc. 8:52–66.

Field, R. A. and E. T. Thorne. 1983. Care of the hunter's harvest. Wyoming Wildl. 47(8):20–23.

Field, R. A., F. C. Smith and W. G. Hepworth. 1972. The pronghorn carcass. Bull. 589. Agric. Exp. Sta., Univ. Wyoming. 8 pp.

———. n.d. The pronghorn antelope carcass. Bull. 575. Agric. Exp. Sta.., Univ. Wyoming, Laramie. 6 pp.

Firchow, K. M. 1986. Ecology of pronghorn on the Piñon Canyon Maneuver Site, Colorado. M.S. thesis, Virginia Polytech. Inst. and St. Univ., Blacksburg. 83 pp.

Firchow, K. M., M. R. Vaughan and W. R. Mytton. 1986a. Population ecology of pronghorns in southeastern Colorado. Pronghorn Antelope Workshop Proc. 12:43–44.

———. 1986b. Evaluation of the hand-held net gun for capturing pronghorns. J. Wildl. Manage. 50(2):320–322.

———. 1990. Comparison of aerial survey techniques for pronghorn. Wildl. Soc. Bull. 18(1):18–23.

Fischer, J. K. 2000. California pronghorn antelope status report. Pronghorn Antelope Workshop (1996) Proc. 17: 14.

Fisher, J. L. 1966. Next step for resources. Pages 168–179 *in* H. Clepper, ed., Origins of American conservation. Ronald Press Co., New York, New York. 193 pp.

Fisher, L. W. 1942a. Pronghorned antelope. P-R Quart. Rept. Proj. 8-D. Texas Game, Fish and Oyster Commiss. 21 pp.

———. 1942b. Live trapping Texas antelopes. J. Wildl. Manage. 6(3):231–236.

Fisher, R. A. 1930. The genetical theory of natural selection. Clarendon Press, Oxford, England. 272 pp.

———. 1958. Statistical methods for research workers. 13th ed. Oliver and Boyd, London, England. 356 pp.

Fleischner, T. L. 1994. Ecological costs of livestock grazing in western North America. Conserv. Biol. 8(3):629–644.

Flower, W. H. 1875. On the structure and affinities of the musk-deer (*Moschus moschiferus* Linn.). Proc. Zool. Soc. London 1875 (11):159–190.

Floyd, J. M. 1924. Capturing antelope kids: How Uncle Sam aids in perpetuating the fleet pronghorn. Forest and Stream, December:744–746.

Folker, R. V. 1956. A preliminary study of an antelope herd in Owyhee County, Idaho. M.S. thesis, Univ. Idaho, Moscow. 102 pp.

Forchhammer, M. C., N. C. Stenseth, E. Post and R. Langvatn. 1998. Population dynamics of Norwegian red deer: Density dependence and climatic variation. Proc. Royal Soc. London, B. 265:341–350.

Foree, W. W. 1956. Black Rock District P-R rept. Four St. Antelope Meeting, Reno, Nevada. 5 pp.

———. 1959. Antelope questionnaire. Interstate Antelope Conf. Trans.2:65–75.

Forest, C. A. 1985. Analysis of supplemental feeding and effects of hunting during the rut on recruitment in pronghorn antelope. M.S. thesis, Univ. Wyoming, Laramie. 91 pp.

Foster, C. L. 1988. Migration of Fort Rock and Duncan Reservoir pronghorn, Oregon. Pronghorn Antelope Workshop Proc. 13:72–82.

Fowler, C. W. 1981. Comparative population dynamics in large mammals. Pages 437–456 *in* C. W. Fowler and T. D. Smith, eds., Dynamics of large mammal populations. John Wiley and Sons, New York, New York. 477 pp.

———. 1987. A review of density dependence in populations of large mammals. Pages 401–441 *in* H. H. Genoways, ed., Current mammalogy. Vol. 1. Plenum Press, New York, New York. 386 pp.

Fowler, M. E. 1983. Plant poisoning in free-living wild animals. A review. J. Wildl. Dis. 19(1):34–43.

Fox, G. A. 1993. Failure time analysis: Emergence, flowering, survivorship and other waiting times. Pages 253–289 *in* S. M. Scheiner and J. Gurevitch, eds., Design and analysis of ecological experiments. Chapman and Hall, London, England. 445 pp.

Fox, H. 1929. Report of the laboratory and museum of comparative pathology. Zool. Soc. Philadelphia, Pennsylvania. 65 pp.

Fox, L. B., L. Hopper and M. Mitchner. 2000. Kansas pronghorn antelope status report—1996. Pronghorn Antelope Workshop (1996) Proc. 17:17–19.

Fox, L. M. 1997. Nutritional content of forage in Sonoran pronghorn habitat, Arizona. M.S. thesis, Univ. Arizona, Tucson. 228 pp.

Franklin, I. R. 1980. Evolutionary change in small populations. Pages 135–149 *in* M. Soule and B. Wilcox, eds., Conservation biology: An evolutionary–ecological perspective. Sinauer Associates, Sunderland, Massachusetts. 395 pp.

Franzmann, A. W., A. Flynn and P. D. Arneson. 1975. Levels of some mineral elements in Alaskan moose hair. J. Wildl. Manage. 39(2):374–378.

Frechkop, S. 1955. Ruminantia. Pages 568–667 *in* Traite de zoologie. Vol. 17. Masson et Cie, Paris, France. 2,300 pp.

Freeman, J. S. 1971. Pronghorn range use and relation to livestock in southeastern Montana. M.S. thesis, Montana St. Univ., Bozeman. 45 pp.

French, C. E., L. C. McEwen and N. D. Magruder. 1955. Nutritional requirements of white-tailed deer for growth and antler development. Bull. 600. Pennsylvania St. Univ. Agric. Exp. Sta., University Park. 50 pp.

Frey, M. L. 1983. Bovine respiratory syncytial virus and acute respiratory distress syndrome in cattle. Bovine Practitioner 18:73–77.

Frick, C. 1937. Horned ruminants of North America. Am. Mus. Nat. Hist. Bull. 69:1–699.

Friederici, P. 1999. Homeless on the range. Natl. Wildl. 37(6): 18–27.

Friend, M. and C. W. Severinghaus. 1967. Influence of nutrition on eye-lens weight in deer. New York Fish and Game J. 14(2):166–175.

Fulbright, T. E. and F. S. Guthery. 1996. Mechanical manipulation of plants. Pages 339–354 *in* P. R. Krausman, ed., Rangeland wildlife. Soc. for Range Manage., Denver, Colorado. 440 pp.

Funk, T. L. 1988. Pronghorn on the prairie. Kansas Dept. Wildl. and Parks 45:2–7.

Furlong, E. L. 1927. The occurrence and phylogenetic status of Merycodus from the Mohave Desert Valley. Univ. California Publ. Geolog. Sci. 17:145–176.

———. 1932. Distribution and description of skull remains of the Pliocene antelope *Sphenophalos* from the northern Great Basin Province. Carnegie Inst. Washington, D.C., Publ. 418:27–36.

———. 1941. A new Pliocene antelope from Mexico with remarks on some known antilocaprids. Studies of Cenozoic vertebrates of western North America and of fossil primates. Carnegie Inst. Washington, D.C., Publ. 530:27–34.

Gabrielson, I. N. 1935. Must the antelope go? Am. Forests 41(10):575–576.

———. 1941. Wildlife conservation. The Macmillan Co., New York, New York. 250 pp.

———. 1943. Wildlife refuges. The Macmillan Co., New York, New York. 257 pp.

Gage, T. B. 1995. A robust estimator of the intrinsic rate of increase and generation length. Pages 9–23 *in* J. D. Ballou, M. Gipin and T. Foose, eds., Population management for survival and recovery: Analytical methods and strategies in small population conservation. Columbia Univ. Press, New York, New York. 375 pp.

Galey, F. D. 1996. Plants and other natural toxicants. Pages 1,877–1,902 *in* B. P. Smith, ed., Large animal internal medicine. 2nd ed. Mosby, Saint Louis, Missouri. 2,040 pp.

Gamo, R. S. 1997. Big game habitat use and distribution on reclaimed mining lands. M.S. thesis, Univ. Wyoming, Laramie. 80 pp.

Ganey, D. T. 1997. The diet of pronghorn (*Antilocapra americana*) in the tallgrass prairie. M.S. thesis, Emporia St. Univ., Emporia, Kansas. 92 pp.

Garrod, A. H. 1877. Notes on the visceral anatomy and osteology of the ruminants, with a suggestion regarding a method of expressing the relations of species by means of formulae. Proc. Zool. Soc. London (1):2–18.

Gasson, W. 1986. Wyoming state report. Pronghorn Antelope Workshop Proc. 12:38–39.

Gasson, W. and L. Wollrab. 1986. Integrating population simulation modeling into a planned approach to pronghorn management. Pronghorn Antelope Workshop Proc. 12:86–98.

Gasson, W., C. Madson and C. Beers. 1995. What's the problem? Wyoming Wildl. 59(1):26–31.

Gatesy, J., D. Yelon, R. DeSalle and E. S. Vrba. 1992. Phylogeny of the Bovidae (Artiodactyla, Mammalia), based on mitochondrial ribosomal DNA sequences. Mol. Biol. Evol. 9(3):433–446.

Gavin, T. A. 1989. What's wrong with the questions we ask in wildlife research. Wildl. Soc. Bull. 17(3):345–350.

Gazin, C. L. 1935. A new antilocaprid from the upper Pliocene of Idaho. J. Paleontol. 9(5):390–393.

Geis, A. D. 1979. Population and harvest studies. Pages 213–218 *in* R. D. Teague and E. Decker, eds., Wildlife conservation: Principles and practices. The Wildl. Soc., Washington, D.C. 280 pp.

Geissler, P. H. and J. R. Sauer. 1990. Topics in route regression analysis. Pages 54–57 *in* J. R. Sauer and S. Droege, eds., Survey designs and statistical methods for the estimation of avian population trends. Biol. Rept. 90(1). U.S. Fish and Wildl. Serv., Washington, D.C. 166 pp.

Geist, V. 1966. The evolution of horn-like organs. Behavior 27(1–2):175–214.

———. 1982. Adaptive behavioral strategies. Pages 219–277 *in* J. W. Thomas and D. E. Toweill, eds., Elk of North America. Stackpole Books, Harrisburg, Pennsylvania. 698 pp.

———. 1988. How markets in wildlife meat and parts, and the sale of hunting privileges, jeopardize wildlife conservation. Conserv. Biol. 2(1):15–26.

———. 1998. Deer of the world: Their evolution, behaviour, and ecology. Stackpole Books, Mechanicsburg, Pennsylvania. 421 pp.

Geist, V. and M. H. Francis. 2001. Antelope country—pronghorns: The last Americans. Krause Publications, Iola, Wisconsin. 176 pp.

Geist, V. and F. Walther, eds. 1974. The behavior of ungulates and its relation to management. Vols. 1 and 2. Publ. 24. Intl. Union Conserv. Nature and Nat. Resour., Morges, Switzerland. 940 pp.

Gentz, E. J., L. A. Harrenstein, J. Baier and J. W. Carpenter. 1995. Hemolytic anemia in a pronghorn (*Antilocapra americana*). Joint Conf. Am. Assoc. Zoo Vets., Wildl. Dis. Assoc. and Am. Assoc. Wildl. Vets., East Lansing, Michigan Proc.: 443–446.

Gerodette, T. 1987. A power analysis for detecting trends. Ecology 8(4):1,364–1,372.

———. 1993. TRENDS: Software for a power analysis of linear regression. Wildl. Soc. Bull. 21(4):515–516.

Gese, E. M. 1999. Threat of predation: Do ungulates behave aggressively toward different members of a coyote pack? Can. J. Zool. 77(3):499–505.

Gibbons, B. 1986. The intimate sense of smell. Natl. Geogr. 170(3):324–361.

Gilbert, B. K. 1973. Scent marking and territoriality in pronghorn (*Antilocapra americana*) in Yellowstone National Park. Mammalia 37(l):25–33.

Gilbert, B. M. 1980. Mammalian osteology. Modern Printing Co., Laramie, Wyoming. 428 pp.

Gilbert, D. L. 1978. Sociological considerations in management. Pages 409–416 *in* J. L. Schmidt and D. L. Gilbert, eds., Big game of North America. Stackpole Books, Harrisburg, Pennsylvania. 494 pp.

Gilbert, P. F. and J. R. Grieb. 1957. Comparison of air and ground deer counts in Colorado. J. Wildl. Manage. 21(1):33–37.

Giles, R. H., Jr. 1979. Research to meet future management needs. Pages 219–224 *in* R. D. Teague and E. Decker, eds., Wildlife conservation: Principles and practices. The Wildl. Soc., Washington, D.C. 280 pp.

Gill, R. B. 1969. A quadrat system for estimating game populations. Game Info. Leaflet 76. Colorado Dept. Nat. Resour., Div. Wildl., Denver. 2 pp.

Gill, R. B., L. H. Carpenter and D. C. Bowden. 1983a. Monitoring large animal populations: The Colorado experience. N. Am. Wildl. Conf. 48:330–341.

Gill, R. B., L. H. Carpenter, R. M. Bartman, D. L. Baker and G. G. Schoonveld. 1983b. Fecal analysis to estimate mule deer diets. J. Wildl. Manage. 47(4):902–915.

Gillmore, P. 1874. Prairie and forest: A description of the game of North America with personal adventures in their pursuit. Chapman and Hall, London, England. 383 pp.

Gilmore, R. E. and R. W. Allen. 1960. Helminth parasites of pronghorn antelope (*Antilocapra americana*) in New Mexico with new host records. Helminthol. Soc. Washington, D.C. Proc. 27:69–73.

Gilpin, M. 1996. Metapopulations and wildlife conservation: Approaches to modeling spatial structure. Pages 11–28 *in* D. R. McCullough, ed., Metapopulations and wildlife conservation. Island Press, Washington, D.C. 429 pp.

Gjersing, F. 1978. Participant in discussion on the impact of severe winter on northern pronghorn ranges. Pronghorn Antelope Workshop Proc. 8:337–359.

Glasgow, W. M. 1990. Management plan for pronghorn antelope in Alberta. Wildl. Manage. Planning Series 3. Fish and Wildl. Div., Edmonton, Alberta. 115 pp.

Gocke, M. 1997. Sagebrush mowing helps wildlife, livestock. Wyoming Wildl. 61(6):44.

Goldman, E. A. 1945. A new pronghorn antelope from Sonora. Biol. Soc. Washington, D.C. Proc. 58:3–4.

Goldsby, A. I. and D. F. Eveleth. 1954. Internal parasites in North Dakota antelope. J. Parasitol. 40(6):637–648.

Goldsmith, A. E. 1988. Behavior and ecology of pronghorn after reintroduction to Adobe Valley, California. Ph.D. thesis, Univ. California, Berkeley. 114 pp.

———. 1992. A summary of the behavior and ecology of reintroduced pronghorn. Pages 135–138 *in* B. Bobek, K. Perzanowski and W. L. Reglin, eds., Global trends in wildlife management. Vol. 2. Trans. 18th Congr. Intl. Union Game Biol. Swiat Press, Krakow-Warszewa, Poland. 621 pp.

Gomara, F. L. de. 1578. The pleasant historie of the conquest of the vveast India, now called new Spayne, atchieued by the vvorthy prince Hernando Cortes, marques of the valley of Huaxacac, most delectable to read. Translated from Spanish by T. N. Anno 1578. Henry Bynneman, London, England. Microfilm 2206 *in* Western Americana: Frontier history of the Trans-Mississippi West, 1550–1900. Research Publications, Inc., New Haven, Connecticut.

Gonzalez, A. y A. Laffon. 1993. Distribucion y estado actual de berrendo (*Antilocapra americana*) en México. Pages 409–420 *in* R. A. Medellin y G. Ceballos, eds., Avances en el Estudio de los mamiferos de México. Spec. Publ. 1. Asociación Mexicana de Mastozoologia, A. C. México, D.F. 464 pp.

Good, J. R. 1977. Habitat factors affecting pronghorn use of playas in southcentral Oregon. M.S. thesis, Oregon St. Univ., Corvallis. 64 pp.

Good, J. R. and J. A. Crawford. 1978. Factors influencing pronghorn use of playas in southcentral Oregon. Pronghorn Antelope Workshop Proc. 8:182–205.

Goodman D. 1996. Viability of the antelope population wintering near Gardiner, Montana. Final Rept. U.S. Natl. Park Serv., Mammoth Hot Springs, Wyoming. 7 pp.

Goodwin, G. A. 1977. Golden eagle predation on pronghorn antelope. Auk 94(4):789–790.

Goss, R. J. 1983. Deer antlers: Regeneration, function, and evolution. Academic Press, New York, New York. 316 pp.

Graham, A. and R. Bell. 1969. Factors influencing the countability of animals. E. African Agric. For. J. (Special Issue) 34:38–43.

Graham, R. W. and E. L. Lundelius, Jr. 1994. Faunmap: A database documenting Quaternary distributions of mammal species in the United States. Illinois St. Mus., Springfield. Scientific Papers 25(1):1–287.

Gratson, M. W., J. W. Unsworth, P. Zager and L. Kuck. 1993. Initial experiences with adaptive resource management for determining appropriate antlerless elk harvest rates in Idaho. Trans. N. Am. Wildl. and Nat. Resour. Conf. 58:610–619.

Gray, J. E. 1855. Notice of the horns of an unrecorded species of the pronghorn (*Antilocapra*) in the collection of the Derby Museum, Liverpool. Proc. Zool. Soc. London 23:9–11.

Greenquist, C. M. 1983. The American pronghorn antelope in Wyoming: A history of human influences and management. Ph.D. thesis, Univ. Oregon, Eugene. 207 pp.

Greer, K. R. 1968. A compression method indicates fat content of elk (wapiti) femur marrows. J. Wildl. Manage. 32(4):747–751.

Gregg, H. A. 1955. Summer habits of Wyoming antelope. Ph.D. thesis, Cornell Univ., Ithaca, New York. 185 pp.

Gregg, M. 1997. Use of prescribed fire to enhance wildlife habitat on Sheldon NWR. The Nevada Wildlifer 25(1):2.

Gregg, M. A., M. Bray, K. M. Kilbride and M. R. Dunbar. 1999. Factors affecting survival of neonatal pronghorn in the northern Great Basin. Pronghorn Antelope Workshop (1998) Proc. 18:106.

Gregg, M. A., M. Bray, K. M. Kilbride and M. R. Dunbar. 2001. Birth synchrony and survival of pronghorn fawns. J. Wildl. Manage. 65(1):19–24.

Greiner, E. C., D. E. Worley and B. W. O'Gara. 1974. *Protostrongylus macrotus* (Nematoda: Metastrongyloidea) in pronghorn antelope from Montana and Wyoming. J. Wildl. Dis. 10(1):70–73.

Griffith, G. K. 1962. Guidelines for antelope management. Interstate Antelope Conf. Trans. 13:102–114.

Griffith, J.S. 1983. Kachinas and masking. Pages 764–777 *in* A. Ortiz, ed., Southwest. Handbook of North American Indians, Vol. 10. Smithsonian Instit., Washington, D.C. 868 pp.

Grier, B. 2000. Pronghorns die in hailstorm. Nebraskaland 78(8):9.

Grinnell, G. B. 1897a. A coyote partnership. Forest and Stream (February 6):104.

———. 1897b. Range of the pronghorn antelope in 1897. Forest and Stream 48(1):5–6.

———. 1911. When antelope were plentiful. Forest and Stream, October 14:582–601.

———. 1925. American game protection. Pages 201–257 *in* G. B. Grinnell and C. Sheldon, eds., Hunting and conservation. Yale Univ. Press, New Haven, Connecticut. 548 pp.

———. 1929. Pronghorn antelope. J. Mammal. 10(2):135–141.

Gross, J. E., J. E. Roelle and G. L. Williams. 1973. Program ONEPOP and information processor: A system modeling and communications project. Prog. Rept. Colorado Coop. Wildl. Res. Unit, Colorado St. Univ., Fort Collins. 58 pp.

Group Discussion. 1978. The impact of a severe winter on northern pronghorn ranges. Pronghorn Antelope Workshop Proc. 8:337–359.

Gruell, G. E. 1995. Historic role of fire on Hart Mountain National Antelope Refuge, Oregon, and Sheldon National Wildlife Refuge, Nevada. Spec. Rept. U.S. Fish and Wildl. Serv., Lakeview, Oregon. 58 pp.

———. 1996. Influence of fire on Great Basin wildlife habitats. Trans. Western Sect. The Wildl. Soc. 32:55–61.

Guenzel, R. J. 1986. Pronghorn ecology in southcentral Wyoming. M.S. thesis, Univ. Wyoming, Laramie. 347 pp

———. 1994. Adapting new techniques to population management: Wyoming's pronghorn experience. Trans. N. Am. Wildl. and Nat. Resour. Conf. 59:189–200.

———. 1997. Estimating pronghorn abundance using aerial line transect sampling. Wyoming Dept. Game and Fish, Cheyenne. 174 pp.

Guenzel, R. J., L. L. Irwin and T. J. Ryder. 1982. A comparison of pronghorn movements and distribution during a normal and a mild winter in the Red Rim area, Wyoming. Pronghorn Antelope Workshop Proc. 10:156–172.

Guilday, J. E. 1971. Big game of the Pleistocene. Pages 46–52 *in* R. C. Alberts, ed., North American big game. Davis and Warde, Inc., Pittsburgh, Pennsylvania. 403 pp.

Gutierrez, A. P. 1996. Applied population ecology—a supply–demand approach. John Wiley and Sons, New York, New York. 300 pp.

Habib de Pena, R. and J. M. Pena Neira. 1982. Habitos alimenticios del berrendo (*Antilocapra americana*) en la region central de Chihuahua. Pastizales 13(6):2–9.

Hailey, T. L. 1979. A handbook on pronghorn antelope management in Texas. P-R Proj. W-45-R, W-57-R, W-94-R and W-14-C, Series 20. Texas Dept. Parks and Wildl., Austin. 59 pp.

Hailey, T. L. and D. DeArment. 1972. Droughts and fences restrict pronghorns. Antelope States Workshop Proc. 5:22–24.

Hailey, T. L., J. W. Thomas and R. M. Robinson. 1966. Pronghorn die-off in Trans-Pecos, Texas. J. Wildl. Manage. 30(3):488–496.

Hails, R. S. and M. J. Crawley. 1992. Spatial density dependence in populations of a cynipid gall former, *Andricus quercuscalicis*. J. Anim. Ecol. 61(4):567–583.

Haines, A. L., ed. 1955. Osborne Russell's journal of a trapper. Univ. Nebraska Press, Lincoln. 191 pp.

Hakanson, L. 1996. A general method to define confidence limits for model predictions based on validations. Ecol. Modeling 91(2):153–168.

Haley, W. H. 1986. Pronghorn management problems related to hunter access in the centennial herd unit of southeast Wyoming. Pronghorn Antelope Workshop Proc. 12:45–52.

Halfpenny, J. C. and E. A. Besiot. 1986. A field guide to mammal tracking in western America. Johnson Books, Boulder, Colorado. 163 pp.

Hall, E. R. 1946. Mammals of Nevada. Univ. California Press, Berkeley. 710 pp.

———. 1981. The mammals of North America. Vol. 2. 2nd ed. John Wiley and Sons, New York, New York. Pp. 601–1,181.

Hall, E. R. and K. R. Kelson. 1959. The mammals of North America. Vol. 2. Ronald Press Co., New York, New York. 79 pp.

Hall, J. E. 1935. Horn growth as observed in black-buck and nilgai. J. Bombay Nat.. Hist. Soc. 38:618–619.

Hall, R. C. 1963. Impact on antelope and deer sagebrush ranges by the sagebrush defoliator—*Agora websteri*. Interstate Antelope Conf. Trans. 14:40–48.

———. 1965. Sagebrush defoliator outbreak in northern California. Res. Note PS W-75. Pacific Southwest For. and Range Exp. Sta., U.S. For Serv., Berkeley, California. 12 pp.

Hall, S. A., H. Salwasser and B. Browning. 1978. The diet composition and energy reserves of California mule deer during pregnancy. California Dept. Fish and Game 65(2):68–79.

Halloran, A. F. 1943. Management of deer and cattle on the Aransas National Wildlife Refuge, Texas. J. Wildl. Manage. 7(2):203–216.

Halls, L. K. 1978. White-tailed deer. Pages 42–65 *in* J. L. Schmidt and D. L. Gilbert, eds., Big game of North America. Stackpole Books, Harrisburg, Pennsylvania. 494 pp

———. 1984. Research problems and needs. Pages 783–793 *in* L. K. Halls, ed., White-tailed deer: Ecology and management. Stackpole Books, Harrisburg, Pennsylvania. 870 pp.

Hamilton, R. 1962. An expandible collar for mule and white-tailed deer. J. Wildl. Manage. 26(l):114–115.

Hamilton, W. R. 1978. Fossil giraffes from the Miocene of Africa and a revision of the phylogeny of the Giraffoidea. Philosophical Trans. of the Royal Soc. London, England. B. Biol. Sci. 283(996):165–229.

Hanley, T. A. 1980. Nutritional constraints on food and habitat selection by sympatric ungulates. Ph.D. thesis, Univ. Washington, Seattle. 176 pp.

———. 1982. The nutritional basis for food selection by ungulates. J. Range Manage. 35(2):146–151.

Hansen, E. L. 1955. Survival of pronghorn antelope in south-central Oregon during 1953 and 1954. M.S. thesis, Oregon St. Coll., Corvallis. 117 pp.

Hansen, M. C. 1985. Diets of five sympatric ungulates in northwestern Nevada. Unpubl. rept. Dept. Fisheries and Wildl., Oregon St. Univ., Corvallis. 33 pp.

———. 1986. Dietary overlap between California bighorn sheep, mule deer, pronghorn antelope, feral horses and domestic cattle in northwestern Nevada. Interstate Antelope Conf. Trans. 28. Unpaginated.

———. 1996. Foraging ecology of female Dall's sheep in the Brooks Range, Alaska. Ph.D. thesis, Univ. Alaska, Fairbanks. 178 pp.

Hansen, M. C. and R. G. Anthony. 1999. Diet composition and quality of pronghorn antelope, mule deer, bighorn sheep, feral horses, and feral burros on Sheldon-Hart Mountain National Wildlife Complex. Oregon Coop. Wildl. Res. Unit, Corvallis. 79 pp.

Hansen, M. C., J. D. Yoakum, W. H. Pyle and R. G. Anthony. 2001. New strategies for pronghorn food habit studies. Pronghorn Antelope Workshop (2000) Proc. 19:71–94.

Harniss, R. O. and R. B. Murray. 1973. Thirty years of vegetal change following burning of sagebrush–grass range. J. Range Manage. 26(4):322–325.

Harris, R. B., L. H. Metzgar and C. D. Bevins. 1986. GAPPS—generalized animal population projection system. User's manual. Montana Coop. Wildl. Res. Unit, Univ. Montana, Missoula. 87 pp.

Hart, J. S., O. Heroux, W. H. Cottle and C. A. Milk. 1961. The influence of climate on metabolic and thermal responses of infant caribou. Can. J. Zool. 39(6):845–856.

Hastings, J. R. and R. M. Turner. 1980. The changing miles: An ecological study of vegetation change with time in the lower mile of an arid and semiarid region. 2nd ed. Univ. Arizona Press, Tucson. 317 pp.

Hayne, D. W. 1949. Calculation of the size of home range. J. Mammal. 30(1):1–18.

Heady, H. F. 1975. Rangeland management. McGraw-Hill Book Co., New York, New York. 460 pp.

———, ed. 1988. The Vale rangeland rehabilitation program: An evaluation. Resour. Bull. PNW-157. Pacific Northwest For. and Range Exp. Sta., U.S. For Serv., Portland, Oregon. 151 pp.

Heady, H. F. and J. Bartolome. 1977. The Vale rangeland rehabilitation program: The desert repaired in southeastern Oregon. Resour. Bull. PNW-70. Pacific Northwest For. and Range Exp. Sta., U.S. For. Serv., Portland, Oregon. 139 pp.

Heady, H. F. and R. D. Child. 1994. Rangeland ecology and management. 2nd ed. Westview Press, San Francisco, California. 519 pp.

Heberlein, T. A. 1991. Changing attitudes and funding for wildlife—preserving the sport hunter. Wildl. Soc. Bull. 19(4):528–534.

Hedgpath, D. and W. Reed. 1993. The art of Tom Lovell: An invitation to history. William Morrow and Co., Inc., New York, New York. 160 pp.

Hedrick, P. W. 1996. Genetics of metapopulations: Aspects of a comprehensive perspective. Pages 29–52 *in* D. R. McCullough, ed., Metapopulations and wildlife conservation. Island Press, Washington, D.C. 429 pp.

Heffelfinger, J. R. and R. J. Olding. 1996. Occurrence and distribution of bluetongue in pronghorn antelope. P-R Perf. Rept., Proj. W-53-M-46. Arizona Dept. Game and Fish, Phoenix. 12 pp.

———. 1997a. Selenium and copper levels in liver and blood in Arizona pronghorn. P-R Proj. W-53-M-47. Arizona Dept. Game and Fish, Phoenix. 14 pp.

———. 1997b. Occurrence and distribution of bluetongue in pronghorn antelope. P-R Perf. Rept., Proj. W-53-M-47. Arizona Dept. Game and Fish, Phoenix. 10 pp.

Heffelfinger, J. R., R. J. Olding, T. H. Noon, M. R. Shupe and D. P. Betzer. 1999. Copper/selenium levels and occurrence of bluetongue virus in Arizona pronghorn. Pronghorn Antelope Workshop (1998) Proc.18:32–42.

Heintz, E. 1963. Les charateres distinctif entre metatares de Cervidae et Bovidae actuels et fossiles. Mammalia 27(2):200–209.

Heisey, D. M. and T. K. Fuller. 1985. Evaluation of survival and cause-specific mortality rates using telemetry data. J. Wildl. Manage. 49(3):668–674.

Heller, R. 1980. On optimal diet in a patchy environment. Theoretical Population Biol. 17(2):201–214.

Helms, B. 1978. Participant in discussion on the impact of severe winter on northern pronghorn ranges. Pronghorn Antelope Workshop Proc. 8:337–359.

Henke, S. E. and F. C. Bryant. 1999. Effects of coyote removal on the faunal community in western Texas. J. Wildl. Manage. 63(4):1,066–1,081.

Henkel, M. 1995. Calling all antelope. Pages 14–15 *in* D. Ruther, ed., Montana hunting almanac. The Missoulian, Missoula, Montana. 164 pp.

Heptner, V. G., A. A. Nasimovisch and A. G. Bannikov. 1961. Mammals of the Soviet Union. Vol. 1, Artiodactyla and Perissodactyla. Vysshaya Shkola, Moscow. Translated and published during 1988 for the Smithsonian Inst. Libraries by Amerind Publ. Co. Pvt. Ltd., New Dehli, India. 1,147 pp.

Hepworth, W. G. 1961. Diagnosis of diseases in mammals and birds. P-R Job Compl. Rept., Work Plan 1, Job 1, Proj. FW-3-R-8. Wyoming Game and Fish Commiss., Cheyenne. Pp. 5–20.

———. 1962. Diagnosis of diseases in mammals and birds. P-R Job Compl. Rept., Work Plan 1, Job 1W, Proj. FW-3-R-9. Wyoming Game and Fish Commiss., Cheyenne. Pp. 6–19.

———. 1965. Investigation of pronghorn antelope in Wyoming. Antelope States Workshop Proc. 1:1–12.

Hepworth, W. G. and F. Blunt. 1966. Research findings on Wyoming antelope. Wyoming Wildl. 30(6):24–29.

Herman, C. M. 1945. Hippoboscid flies as parasites of game animals in California. California Fish and Game 31(l):16–25.

———. 1959–1994. Obras completas de Francisco Hernández. Genmah Somolin os d'Ardois, ed. Vol 3. Historia de los animales. 8 vols. Universidad Naciónál de Mexico, Mexico City.

Herrick, C. L. 1892. Mammals of Minnesota. Bull. 7. Geological and Natural History Survey of Minnesota, Minneapolis, Minnesota. 300 pp.

Hervert, J. J., J. L. Bright, M. T. Brown, L. A. Priest and R. S. Henry. 2000. Sonoran pronghorn population monitoring:1994–1998. Nongame and Endangered Species Program Tech. Rept. 162. Arizona Dept. Game and Fish, Phoenix. 47 pp.

Hess, M. L. 1986. Density-dependent summer pronghorn fawn survival rates in the interstate antelope population. Pronghorn Antelope Workshop Proc. 12:53–54.

———. 1999. Density dependent summer pronghorn survival rates in the northwestern Great Basin—revisited. Pronghorn Antelope Workshop (1998) Proc. 18:107.

Hewitt, C. G. 1921. The conservation of wildlife of Canada. Charles Scribner's Sons, New York, New York. 344 pp.

Hewitt, J., ed. 1973. Eye-witness to wagon trains west. Osprey Publishing, Ltd., Reading, England. 178 pp.

Hiby, L., D. Thompson and A. Ward. 1987. Improving census by aerial photography—an inexpensive system based on nonspecialist equipment. Wildl. Soc. Bull. 15(3):438–443.

Higgens, K. F., J. L. Oldemeyer, K. J. Jenkins, G. K. Clambey and R. F. Harlow. 1994. Vegetation sampling and measurement. Pages 567–591 *in* T.A. Bookhout, ed., Research and management techniques for wildlife and habitats. The Wildl. Soc., Bethesda, Maryland. 740 pp.

Hildebrand, M. and J. P. Hurley. 1985. Energy of the oscillating legs of a fast-moving cheetah, pronghorn, jackrabbit, and elephant. J. Morphol. 184(1):23–31.

Hinman, R. A. 1959. Problems in antelope management in Utah. Western Assoc. St. Game and Fish Commiss. Proc. 39:201–207.

Hinshaw, J. M. 1990. A monitoring program for pronghorn antelope in Chihuahua State, Mexico. Pronghorn Antelope Workshop Proc. 14:77–83.

Hirth, D. H. 2000. Behavioral ecology. Pages 175–191 *in* S. Demarais and P. R. Krausman, eds., Ecology and management of large mammals in North America. Prentice Hall, Upper Saddle River, New Jersey. 778 pp.

Hisey, J. and G. Hisey, eds. 1995. Nineteenth recording period—1993–1994—statistical summary. Pope and Young Club, Chatfield, Minnesota. 113 pp.

———, eds. 1997. Twentieth recording period—1995–1996—statistical summary. Pope and Young Club, Chatfield, Minnesota. 137 pp.

Hlavachik, B. D. 1968. Foods of Kansas antelopes related to choice of stocking sites. J. Wildl. Manage. 32(2):399–401.

Hnatiuk, J. 1970. Eye lens weight versus jaws in age determination of antelope. Antelope States Workshop Proc. 4:104–115.

———. 1972. The northern range of pronghorn and the international herd. Antelope States Workshop Proc. 5:181–191.

Hockley, M. 1968. Ten years of antelope management in the Gillette area of Wyoming. Antelope States Workshop Proc. 3:81–84.

Hodgins, R. A. 1972. Letter in files of South Dakota Department of Game, Fish and Parks, Pierre. 1 p.

Hoff, G. L. and D. O. Trainer. 1972. Bluetongue virus in pronghorn antelope. Am. J. Vet. Res. 33(5):1,013–1,016.

Hoffman, R. R. 1968. Comparisons of the rumen and omasum structure in East African game ruminants in relation to their feeding habits. Symp. Zool. Soc. London 21:179–194.

———. 1989. Evolutionary steps of ecophysiological adaptation and diversification of ruminants: A comparative view of their digestive system. Oecologia 78(4):443–457.

Hoffmeister, D. F. 1986. Mammals of Arizona. Univ. Arizona Press, Tucson. 602 pp.

Holechek, J. L. and R. Valdez. 1985. Evaluation of in vitro digestion for improving botanical estimates of mule deer fecal samples. J. Mammal. 66(3):574–577.

Holechek, J. L., M. V. Vavra and R. D. Pieper. 1982. Botanical composition determination of range herbivore diets: A review. J. Range Manage. 35(4):309–315.

Holechek, J. L., R. D. Pieper and C. H. Herbel. 1998. Range management: Principles and practices. 3rd ed. Prentice Hall, Upper Saddle River, New Jersey. 542 pp.

Holland, G. P. 1940. Notes on the ecology of *Dermacentor andersoni* in southern Alberta. Entomol. Soc. British Columbia Proc. 36:8–11.

Holloway, J. W., R. E. Estell and W. T. Butts, Jr. 1981. Relationships between fecal components and forage consumption and digestibility. J. Anim. Sci. 52(4):836–848.

Holsman, R. H. 2000. Goodwill hunting? Exploring the role of hunters as ecosystem stewards. Wildl. Soc. Bull. 28(4):808–816.

Honess, R. F. and K. B. Winter. 1956. Diseases of wildlife in Wyoming. Bull. 9. Wyoming Game and Fish Commiss., Cheyenne. 279 pp.

Hooley, C. S. 1992. Pronghorn antelope use of tebuthiuron-treated areas in northwestern New Mexico. Pronghorn Antelope Workshop Proc. 15:51–57.

Hoover, J. P. 1971. Food habits of pronghorn antelope on Pawnee National Grasslands, 1970. M.S. thesis, Colorado St. Univ., Fort Collins. 285 pp.

Hoover, R. L. and S. Ogilvie. 1954. P-R Quart. Rept. Colorado Dept. Game and Fish, Denver. Pp. 33–38.

Hoover, R. L., C. E. Till and S. Ogilvie. 1959. The antelope of Colorado. Tech. Bull. 4. Colorado Dept. Game and Fish, Denver. 110 pp.

Hopkins, H. F., F. W. Albertson and A. Riegel. 1948. Some effects of burning upon a prairie in west-central Kansas. Kansas Acad. Sci. Trans. 51:131–141.

Hopper, R. M. 1990. Evaluation, issues, and challenges of the wildlife ranching program in Colorado. Pronghorn Antelope Workshop Proc. 14:84–101.

Hormay, A. J. 1970. Principles of rest-rotation grazing and multiple-use land management. U.S. Bur. Land Manage. and U.S. For. Serv., U.S. Govt. Print. Off., Washington, D.C. 25 pp.

Hornaday, W. T. 1889. The extermination of the American bison, with a sketch of its discovery and life. Pages 373–548 *in* Ann. Rept., Smithsonian Inst., 1887. U.S. Govt. Print. Off., Washington, D.C.

———. 1904. The American natural history; a foundation of useful knowledge of the higher animals of North America. Charles Scribner's Sons, New York, New York. 449 pp.

———. 1913. Our vanishing wildlife. Clark and Fritts Printers, New York, New York. 411 pp.

———. 1914. The American natural history. Charles Scribner's Sons, New York, New York. 332 pp.

Hoskinson, R. L. and J. R. Tester. 1980. Migration behavior of pronghorn in southeastern Idaho. J. Wildl. Manage. 44(l):132–144.

Houston, D. B. 1973. Wildfires in northern Yellowstone Park. Ecology 54(5):1,111–1,117.

Houston, D. 1982. The northern Yellowstone elk: Ecology and management. Macmillan Publ. Co., New York, New York. 474 pp.

Howard, V. W. 1966. An observation of parturition in the pronghorn antelope. J. Mammal. 47(4):708–709.

Howard, V. W., S. L. Beasom, A. R. Lara and L. LaPlatt. 1982. A preliminary evaluation of forage preferences for pronghorn in east-central New Mexico. Pronghorn Antelope Workshop Proc. 10:86–92.

Howard, V. W., J. L. Holechek, R. D. Pieper, S. L. Beasom and K. Green-Hammond. 1983. Roswell pronghorn study. New Mexico St. Univ., Las Cruces. 115 pp.

———. 1984. Roswell pronghorn study. Pronghorn Antelope Workshop Proc. 11:151–188.

Howard, V. W., J. L. Holechek, R. D. Pieper, K. Green-Hammond, M. Cardenas and S. L. Beasom. 1990. Habitat requirements for pronghorn on rangelands impacted by livestock and net wire in eastcentral New Mexico. Bull. 750. Agric. Exp. Sta., New Mexico St. Univ., Las Cruces. 48 pp.

Howe, D. L. 1965. Diagnosis of diseases in mammals and birds. P-R Job Compl. Rept., Work Plan 1, Job 1W, Proj. FW-3-R-12. Wyoming Game and Fish Commiss., Cheyenne. Pp. 5–20.

———. 1966a. Diagnosis of diseases in mammals and birds. P-R Job Compl. Rept., Work Plan 1, Job 1W, Proj. FW-3-R-13. Wyoming Game and Fish Commiss., Cheyenne. Pp.1–19.

———. 1966b. Effects of leptospirosis in antelope. P-R Job Compl. Rept., Work Plan 1, Job 6W, Proj. FW-3-R-13. Wyoming Game and Fish Commiss., Cheyenne. Pp. 1–2.

———. 1970. Vibriosis. Page 376 *in* J. W. Davis, L. H. Karstad and D. O. Trainer, eds., Infectious diseases of wild mammals. Iowa St. Univ. Press, Ames. 446 pp.

Howe, D. L. and W. G. Hepworth. 1965. Anaplasmosis in big game animals: Tests on wild populations in Wyoming. Am. J. Vet. Res. 26(114):1,114–1,120.

Howe, D. L., W. G. Hepworth, F. M. Blunt and G. M. Thomas. 1964. Anaplasmosis in big game animals: Experimental infection and evaluation of serologic tests. Am. J. Vet. Res. 25(107):1,271–1,275.

Howell, D. C. 1992. Statistical methods in psychology. 3rd ed. Duxbury Press, Boston, Massachusetts. 693 pp.

Howell, D. E. 1939. The ecology of *Dermacentor albipictus* (Packard). Pacific Sci. Conf. Proc. 6:439–458.

Hsu, T. C. and K. Benirschke. 1969. An atlas of mammalian chromosomes. Vol. 3, Folio 136. Springer-Verlag, New York, New York.

Hubbard, R. L. and L. R. Sanderson. 1961. Grass reduces bitterbrush production. California Dept. Fish and Game 47(4):391–398.

Hughes, E. and R. Mall. 1957. Relation of adrenal cortex to condition of deer. California Dept. Fish and Game 44(2):191–196.

Hughes, K. S. 1991. Sonoran pronghorn use of habitat in southwest Arizona. M.S. thesis, Univ. Arizona, Tucson. 86 pp.

Hughes, K. S. and N. S. Smith. 1990. Sonoran pronghorn use of habitat in southwest Arizona. Res. Rept. 146-0091564. Arizona Coop. Fish and Wildl. Res. Unit, Univ. Arizona, Tucson. 58 pp.

Huizinga, H. 1942. *Eimeria antelocaprae*, a new coccidium from the American antelope. J. Parasitol. 28(2):167–168.

Hunter, M. L., Jr. 1989. Aardvarks and Arcadia: Two principles of wildlife research. Wildl. Soc. Bull. 17(3):350–351.

Hurd, R. M. and R. E. Blaser. 1962. Palatability of herbage. Pages 65–69 *in* Pasture and range research techniques. Comstock Publ. Assoc., Ithaca, New York. 242 pp.

Hurt, W. R., Jr. 1953. Report of the investigation of the Swan Creek site, 39WW7, Walworth County, South Dakota. Archaeol. Studies Circ. 7. South Dakota Archaeol. Commiss., Pierre. 98 pp.

Huston, J. E., B. S. Rector, L. B. Merrill and B. S. Engdahl. 1981. Nutritional value of range plants in the Edwards Plateau region of Texas. Texas Agric. Exp. Sta. Rept. B-1357. 16 pp.

Huston, M. D., D. L. DeAngelis and W. Post. 1988. New computer models unify ecological theory. BioScience 38(6):682–691.

Hutton, J., H. Jenkins and S. Edwards. 1995. Conservation and development compromised by animal welfare. Pages

271–274 *in* J. A. Bissonette and P. R. Krausman, eds., Integrating people and wildlife for a sustainable future. The Wildl. Soc., Bethesda, Maryland. 715 pp.

Hylander, C. J. 1966. Wildlife communities: From the tundra to the tropics in North America. Houghton Mifflin Co., Boston, Massachusetts. 342 pp.

Illius, A. W. and C. Fitzgibbon. 1994. Cost of vigilance in foraging ungulates. Anim. Behav. 47(2):481–484.

Ingersol, E. 1882. Knocking around the Rockies. Univ. Oklahoma Press, Norman. 220 pp.

Ingold, D. A. 1969. Social organization and behavior in a Wyoming pronghorn population. Ph.D. thesis, Univ. Wyoming, Laramie. 142 pp.

International Species Information System. 1995. Physiological data reference values. Page 515 *in* M. R. Dunbar and R. Velarde (1998), Health evaluation of pronghorns (*Antilocapra americana*) on Hart Mountain National Antelope Refuge in southeastern Oregon, 1996–1997. Tech. Rept. 98-01. Natl. Wildl. Health Ctr., Madison, Wisconsin. 32 pp.

International Union for Conservation of Nature and Natural Resources. 1987. Introductions, re-introductions, and restocking. IUCN Council, Gland, Switzerland. 20 pp.

———. 1996. 1996 IUCN red list of threatened animals. IUCN, Gland, Switzerland and Cambridge, England. 448 pp.

Irby, L. R. 1981. Variation in defecation rates of pronghorn relative to habitat and activity level. J. Range Manage. 34(4): 278–279.

Irving, W. 1850. Adventures of Captain Bonneville. George Routledge and Co., London, England. 271 pp.

———. 1868. Astoria or anecdotes of an enterprise beyond the Rocky Mountains. G. P. Putnam's Sons, New York, New York. 698 pp.

Irwin, D. M., T. D. Kocher and A. C. Wilson. 1991. Evolution of cytochrome b gene of mammals. J. Mol. Ecol. 32(2):128–144.

Irwin, L. L. and J. G. Cook. 1985. Determine appropriate variables for a habitat suitability model for pronghorn. Wildl. Soc. Bull. 13(4):434–440.

Jablow, J. 1951. The Cheyenne in plains Indian trade relations, 1795–1840. Am. Ethnog. Soc. Memoir 19., J. J. Augustin, New York, New York. 100 pp.

Jacobs, J. 1973. A microtechnique index to pronghorn diet and sagebrush digestion coefficients. M.S. thesis, Univ. Wyoming, Laramie. 49 pp.

Jacobson, R. H., D. E. Worley and W. W. Hawkins. 1977. Studies on pronghorn antelope (*Antilocapra americana*) as reservoirs of anaplasmosis in Montana. J. Wildl. Dis. 13(3):323–326.

Jaeger, E. C. 1961. Desert wildlife. Stanford Univ. Press, Stanford, California. 308 pp.

Jahn, L. R. and J. B. Trefethen. 1978. Funding wildlife conservation programs. Pages 456–470 *in* H. P. Brokaw, ed., Wildlife and America. Council Environ. Qual. U.S. Govt. Print. Off., Washington, D.C. 532 pp.

Jahnke, L. 1991. Population investigations. 1990 P-R Job Compl. Rept. Proj. W-27-R. Wyoming Dept. Game and Fish, Cheyenne. Pp. 24–40.

———. 1992. Population investigations. 1991 P-R Job Compl. Rept. Proj. W-27-R. Wyoming Dept. Game and Fish, Cheyenne. Pp. 20–37.

James, M. R., ed. 1993. Bowhunting: Big game records of North America. 4th ed. Pope and Young Club, Chatfield, Minnesota. 510 pp.

———, ed. 1999. Bowhunting: Big game records of North America. 5th ed. Pope and Young Club, Chatfield, Minnesota. 750 pp.

Janis, C. M. 1982. Evolution of horns in ungulates: Ecology and paleoecology. Biol. Rev. 57(2):261–318.

Janis, C. M. and E. Manning. 1998. Antilocapridae. Pages 491–507 *in* C. M. Janis, K. M. Scott and L. L. Jacobs, eds., Evolution of tertiary mammals of North America. Vol. 1. Terrestrial carnivores, ungulates, and ungulatelike mammals. Cambridge Univ. Press, Port Chester, New York. 691 pp.

Janis, C. M. and K. M. Scott. 1987. The interrelationships of higher ruminant families with special emphasis on the members of the Cervoidea. Am. Mus. Novit. 2,983:1–85.

Janis, C. M. and P. B. Wilhelm. 1993. Were there mammalian pursuit predators in the Tertiary? Dances with wolf avatars. J. Mammalian Evol. 1(2):103–125.

Janis, C. M., K. M. Scott and L. L. Jacobs, eds. 1998. Evolution of tertiary mammals of North America. Vol. 1. Terrestrial carnivores, ungulates, and ungulatelike mammals. Cambridge Univ. Press, Port Chester, New York. 691 pp.

Januszewski, T. C. 1972. Isolation and characterization of a virus associated with the weak calf-lamb syndrome. M.S. thesis, Univ. Montana, Missoula. 72 pp.

Jatnieks, S. A. 1983. A study of pronghorn range in southwestern North Dakota. M.S. thesis, Univ. North Dakota, Grand Forks. 83 pp.

Jeffress, J. 1986. Nevada antelope status report. Pronghorn Antelope Workshop Proc. 12:21–22.

Jennrich, R. I. and F. B. Turner. 1969. Measurements of non-circular home range. J. Theor. Biol. 22(2):227–237.

Jensen, B. 1997. American pronghorn population studies. P-R Rept. 131-A, Proj. W-67-R-37, Phase C, Study C-II, Job C-II-3. North Dakota Dept. Game and Fish, Bismarck. 6 pp.

Jensen, W. F. and R. W. Seabloom. 1993. Pronghorn skull found along edge of historic eastern distribution in North Dakota. Prairie Nat. 25(4): 289–290.

Jessup, D. A. and W. M. Boyce. 1995. Diseases of wild ungulates and livestock. Pages 395–412 *in* P. R. Krausman, ed., Rangeland wildlife. Soc. for Range Manage., Denver, Colorado. 440 pp.

J. H. C. 1905. Antelope and automobile. Forest and Stream, January 9:55.

Johnson, B. K., F. G. Lindzey and R. J. Guenzel. 1991. Use of aerial line transect surveys to estimate pronghorn populations in Wyoming. Wildl. Soc. Bull. 19(3):315–321.

Johnson, J. F. 1972. Antelope summary—New Mexico. Antelope States Workshop Proc. 5:27–29.

Johnson, J. L., T. L. Barber, M. L. Frey and G. Nason. 1986. Serosurvey for selected pathogens in hunter-killed pronghorn in western Nebraska. J. Wildl. Dis. 22(1):87–90.

Johnson, K. L., ed. 1986. Crested wheatgrass: Its values, problems, and myths. Symp. Proc. Utah St. Univ., Logan. 348 pp.

Jolly, G. M. 1969. Sampling methods for aerial censuses of wildlife populations. E. African. Agric. For. J. (Special Issue) 34:46–49.

Jones, B. 1986. The inventory and monitoring process. Pages 1–10 *in* A. Y. Cooperrider, R. J. Boyd and H. R. Stuart, Inventory and monitoring of wildlife habitat. U.S. Bur. Land Manage., Denver, Colorado. 858 pp.

Jones, P. V., Jr. 1949. Experimental management of antelope. Fed. Aid Rept., Series 3. Texas Game, Fish and Oyster Commiss., Austin. 31 pp.

Jones, R. L. and H. C. Hanson. 1985. Mineral licks, geography, and biogeochemistry of North American ungulates. Iowa St. Univ., Ames. 301 pp.

Jubb, K. V. F. and P. C. Kennedy. 1970. Pathology of domestic animals. Vol. 1. Academic Press, New York, New York. 593 pp.

June, J. W. 1965. Water development. Western Sage Grouse Workshop. Walden, Colorado. 31 pp.

Kahn, R. 1993. Ranching for Wildlife guidelines. Colorado Div. Wildl., Denver. 6 pp.

Kahre, K. S. 1980. The Lassen-Walker interstate deer herd, a status report—recent studies and impacts. California–Nevada Section of The Wildl. Soc. Trans. 15:131–138.

Kalbfleisch, J. D. and R. L. Prentice. 1980. The statistical analysis of failure time data. John Wiley and Sons, New York, New York. 321 pp.

Kallman, H., ed. 1987. Restoring America's wildlife. U.S. Fish and Wildl. Serv., Washington, D.C. 394 pp.

Kaplan, E. L. and P. Meier. 1958. Nonparametric estimation from incomplete observations. J. Am. Statistics Assoc. 53(4): 457–481.

Karsky, R., proj. leader. 1988. Fences. U.S. For. Serv. Tech. and Dev. Ctr., Missoula, Montana. 210 pp.

Katz, D. M. 1985. Custer in photographs. Bonanza Books, New York, New York. 150 pp.

Keister, G. P., Jr., C. E. Trainer and M. J. Willis. 1988. A self-adjusting collar for young ungulates. Wildl. Soc. Bull. 16(3): 321–323.

Kellert, S. R. 1978. Attitudes and characteristics of hunters and antihunters. Trans. N. Am. Wildl. and Nat. Resour. Conf. 43:412–423.

Kellert, S. R. and C. P. Smith. 2000. Human values toward large mammals. Pages 38–63 *in* S. Demarais and P. R. Krausman, eds., Ecology and management of large mammals in North America. Prentice Hall, Upper Saddle River, New Jersey. 778 pp.

Kelsall, J. P. 1968. The caribou. Dept. Indian Affairs and Northern Development, Can. Wildl. Serv., Ottawa, Ontario. 340 pp.

Kent, M., T. K. Fuller, D. W. Kuehn and A. F. Weber. 1983. Morphologic evaluation of a male pseudohermaphroditic white-tailed deer. J. Wildl. Dis. 22(l):l33–136.

Kerr, R. M. 1968. A discussion of the woven-wire fence antelope situation on BLM lands in New Mexico. Antelope States Workshop Proc. 3:22–27.

Kerwin, M. L. and G. J. Mitchell. 1971. The validity of the wearage technique for Alberta pronghorn. J. Wildl. Manage. 35(4):743–747.

Khan, I. A. 1936. Horn growth as observed in black-buck and nilgai. J. Bombay Nat. Hist. Soc. 39:170–171.

Kie, J. G. and T. S. Burton. 1984. Dietary quality, fecal nitrogen and 2.6,-diaminopimelic acid in black-tailed deer in northern California. Res. Note PSW-364. Pacific Southwest For. and Range Exp. Sta., U.S. For. Serv., Albany, California. 3 pp.

Kie, J. G., V. C. Bleich, A. L. Medina, J. D. Yoakum and J. W. Thomas. 1994. Managing rangelands for wildlife. Pages 663–688 *in* T. A. Bookhout, ed., Research and management techniques for wildlife and habitats. The Wildl. Soc., Bethesda, Maryland. 740 pp.

Killaby, M., D. Dobson and C. Dunn. 1992. Pronghorn antelope in Saskatchewan: Status and management strategies. Wildl. Tech. Rept. 92-2. Saskatchewan Wildl. Branch, Regina. 137 pp.

Kindschy, R. R. 1974. The Vale Project and wildlife ecology. Spec. Rept. U.S. Bur. Land Manage., Vale, Oregon. 27 pp.

———. 1986. Rangeland vegetative succession—implications to wildlife. Rangelands 8(4):157–159.

———. 1996. Fences, waterholes, and other range improvements. Pages 369–381 *in* P. R. Krausman, ed., Rangeland wildlife. Soc. for Range Manage., Denver, Colorado. 440 pp.

Kindschy, R., C. Sundstrom and J. Yoakum. 1978. Range/wildllife interrelationships—pronghorn antelope. Pronghorn Antelope Workshop Proc. 8:216–269.

Kindschy, R. R., C. Sundstrom and J. D. Yoakum. 1982. Wildlife habitats in managed rangelands—The Great Basin of southeastern Oregon: Pronghorns. General Tech. Rept. PNW-145. Pacific Northwest For. and Range Exp. Sta., U.S. For. Serv., Portland, Oregon. 18 pp.

King, J. A. 1955. Social behavior, social organization, and population dynamics in a black-tailed prairie dog town in the Black Hills of South Dakota. Contrib. Lab. Vert. Biol. 67. Univ. Michigan, Ann Arbor. 128 pp.

King, R. T. 1938. The essentials of a wildlife range. J. Forestry 36(5):457–464.

———. 1941. Forest zoology and its relation to wildlife as applied on Huntington Forest. Roosevelt Wildl. Bull. 7:461–505.

Kingscote, B. F. and J. G. Bohac. 1986. Antibodies to bovine bacterial and viral pathogens in pronghorn in Alberta, 1983. J. Wildl. Dis. 22(4):511–514.

Kingston, N., E. S. Williams and E. T. Thorne. 1982. Amoebae in *Antilocapra americana* in Wyoming. J. Protozool. 29(3):488.

Kistner, T. P. 1982. Diseases and parasites. Pages 181–217 *in* J. W. Thomas and D. E. Toweill, eds., Elk of North America: Ecology and management. Stackpole Books, Harrisburg, Pennsylvania. 698 pp.

Kistner, T. P., G. E. Reynolds, L. D. Koller, C. E. Trainer and D. L. Eastman. 1975. Clinical and serological findings on the distribution of bluetongue and epizootic hemorrhagic disease virus in Oregon. Am. Assoc. Vet. Lab. Diag. Proc. 18:135–148.

Kitchen, D. W. 1974. Social behavior and ecology of the pronghorn. Wildl. Monogr. 38. 96 pp.

Kitchen, D. W. and P. T. Bromley. 1974. Agonistic behavior of territorial pronghorn bucks. Pages 365–381 *in* V. Geist and F. Walthers, eds., The behavior of ungulates and its relation to management. Vol. 1. Publ. 24. Intl. Union Conserv. Nature and Nat. Resour., Morges, Switzerland. 511 pp.

Kitchen, D. W. and P. Griep. 1978. Variation in breeding behavior of pronghorn bucks. Ann. Mtg. Anim. Behav. Soc., Seattle. Unpaginated.

Kitchen, D. W. and B. W. O'Gara. 1982. Pronghorn, *Antilocapra americana*. Pages 960–972 *in* J. A. Chapman and G. A. Feldhammer, eds., Wild mammals of North America: Biology, management, and economics. Johns Hopkins Univ. Press, Baltimore, Maryland. 1,147 pp.

Kleiber, M. 1961. The fire of life: An introduction to animal energetics. J. Wiley and Sons, New York, New York. 454 pp.

Klein, D. R. 2000. Perspectives. Wildl. Soc. Bull. 28(2):464–467.

Kniesel, M. 1988. AUM equivalence and its use in rangeland planning. Interstate Antelope Conf. Trans. 31:83–91.

Knight, D. H. 1994. Mountains and plains: Ecology of Wyoming landscapes. Yale Univ. Press, New Haven, Connecticut. 338 pp.

Knipe, T. 1944. The status of the antelope herds of northern Arizona. P-R Proj. 9R. Arizona Game and Fish Commiss., Phoenix. 40 pp.

Knowles, C. J. 1982. Habitat affinity, populations, and control of black-tailed prairie dogs on the Charles M. Russell National Wildlife Refuge. Ph.D. thesis, Univ. Montana, Missoula. 171 pp.

Knowles, C. J. and B. W. O'Gara. 1982. Periodicity of prostate gland development in pronghorns. Unpubl. rept. Montana Coop. Wildl. Res. Unit, Univ. Montana, Missoula. 4 pp.

Knowlton, F. F., D. J. Carley and R. T. McBride. 1971. Mammal damage control research—predators and predator–prey. Pages 26–32 *in* Ann. Prog. Rept., 1970–71, Work Unit D.F. 103.9. Denver Wildl. Res. Ctr., U.S. Fish and Wildl. Serv., Denver, Colorado.

Knowlton, F. F., E. M. Gese and J. M. Jaeger. 1999. Coyote depredation control: An interface between biology and management. J. Range Manage. 52:398–412.

Koch, A. J. and J. D. Yoakum. 2002. Reintroduction and present status of pronghorn on the Carrizo Plain National and surrounding area. Pronghorn Workshop Proc. 20:25–41.

Koerth, B. H. 1996. Chemical manipulation of plants. Pages 321–328 *in* P. R. Krausman, ed., Rangeland wildlife. Soc. for Range Manage., Denver, Colorado. 440 pp.

Koford, C. B. 1958. Prairie dogs, white faces, and blue grama. Wildl. Monogr. 3. 78 pp.

Kohler, P. H. 1950. Digestion studies with sheep and wild antelope on a sagebrush ration. M.S. thesis, South Dakota St. Coll., Brookings. 32 pp.

Kohlmann, S. G. 1996. Population modeling of large mammals. Wildl. Res. Prog. Rept. Wildl. Pop. Lab., Oregon Dept. Fish and Wildl., Corvallis. 112 pp.

———. 1998a. Density dependence in two Great Basin desert ungulates: Implications for adaptive management. P-R Proj. W-87-R, Subproj. 281, Study 3, Job 2. Prog. Rept. Wildl. Pop. Lab., Oregon Dept. Fish and Wildl., Portland. 32 pp.

———. 1998b. Density dependence in Great Basin desert ungulates: Implications for adaptive management. Wildl. Res. Prog. Rept. Wildl. Pop. Lab., Oregon Dept. Fish and Wildl., Corvallis. 92 pp.

———. 1999. Adaptive fetal sex allocation in elk: Evidence and implications. J. Wildl. Manage. 63(4):1,109–1,117.

Kohlmann, S. G., D. G. Whittaker and M. Hedrick. 1999a. Density dependence in Great Basin pronghorn: Implications for adaptive harvest management. Pronghorn Antelope Workshop (1998) Proc. 18:103.

Kohlmann, S. G., R. L. Green and C. E. Trainer. 1999b. Effects of collection method on composition of black bear harvest in Oregon. Northwest Sci. 73(1):34–37.

Kolenosky, G. B. and S. Miller. 1962. Growth of the lens of the pronghorn antelope. J. Wildl. Manage. 26(1):112–113.

Koopman, K. F. 1967. Artiodactyls. Pages 385–406 *in* S. Anderson and J. K. Jones Jr., eds., Recent mammals of the world . . . a synopsis of families. Ronald Press Co., New York, New York. 453 pp.

Korn, M. 1990. Old values, new times. Montana Outdoors 21(4):27–29.

Korschgen, L. J. 1980. Procedures for food habits analysis. Pages 113–127 *in* S. D. Schemnitz, ed., Wildlife management techniques manual. The Wildl. Soc., Washington, D.C. 686 pp.

Koster, J. 1980. The forty-day scout. Am. Heritage 31(4):98–107.

Krammer, H. 1964. Vom Gamsbock und dem Kruckenwachstun. Wild und Hund 67:145–159.

Kraus, F. and M. M. Miyamoto. 1991. Rapid cladogenesis among the pecoran ruminants: Evidence from mitochondrial DNA sequences. Syst. Zool. 40(2):117–130.

Krause, F. 1905. Sling contrivances for projectile weapons. Smithsonian Inst. Ann. Rept. 1905:619–638.

Krause, H. and G. D. Olson. 1974. Prelude to glory, a newspaper accounting of Custer's 1874 expedition to the Black Hills. Brevet Press, Sioux Falls, South Dakota. 279 pp.

Krausman, P. R., ed. 1996. Rangeland wildlife. Soc. for Range Manage., Denver, Colorado. 440 pp.

———. 2000. Wildlife management in the twenty-first century: Educated predictions. Wildl. Soc. Bull. 28(3):490–494.

Kreeger, T. J. 1996. Handbook of wildlife chemical immobilization. Intl. Wildl. Vet. Serv., Inc., Laramie, Wyoming. 340 pp.

———. 1999. The rising cost of health care for wildlife. Wyoming Wildl. 63(12):18–23.

Kreeger, T. J., R. P. Lanka, T. Smith and T. G. Smetzer. 1999. Chemical immobilization in an urban environment using carfentanil and xylazine. Pronghorn Antelope Workshop (1998) Proc. 18:69–73.

Kreeger, T. J., W. E. Cook, C. A. Piché and T. Smith. 2001. Anesthesia of pronghorns using thiafentanil or thiafentanil plus xylazine. J. Wildl. Manage. 65(1):25–28.

Krueger, K. 1986. Feeding relationships among bison, pronghorn, and prairie dogs: An experimental analysis. Ecology 67(3):760–770.

Krueger, W. C. 1972. Evaluating animal forage preference. J. Range Manage. 25(4):471–475.

Küchler, A. W. 1964. Potential natural vegetation of the conterminous United States. Spec. Publ. 36. Am. Geogr. Soc., New York, New York. 116 pp.

Kufeld, R. C., J. H. Olterman and D. C. Bowden. 1980. A helicopter quadrat census for mule deer on Uncompahgre Plateau, Colorado. J. Wildl. Manage. 44(3):632–639.

Kufeld, R. C., D. C. Bowden and D. L. Schrupp. 1987. Estimating mule deer density by combining mark–recapture and telemetry data. J. Mammal. 68(4):818–825.

Laake, J. L., S. T. Buckland, D. R. Anderson and K. P. Burnham. 1996. DISTANCE User's Guide V2.2. Colorado Coop. Fish and Wildl. Res. Unit, Colorado St. Univ., Fort Collins. 82 pp.

Lancaster, R. J. 1949. Estimation of digestibility of grazed pasture from feces nitrogen. Nature 163(4,139):330–331.

Lance, W. R. and T. M. Pojar. 1984. Diseases and parasites of pronghorn: A review. Spec. Rept. 57. Colorado Div. Wildl., Denver. 14 pp.

Lance, W. R., C. P. Hibler and J. DeMartini. 1983. Experimental contagious ecthymia in mule deer, white-tailed deer, pronghorn and wapiti. J. Wildl. Dis. 19(3):165–169.

Lancia, R. A., C. E. Braun, M. W. Collopy, R. W. Dueser, J. G. Kie, C. J. Martinka, J. D. Nichols, T. D. Nudds, W. R. Porath and N. G. Tilghman. 1996. ARM! For the future: Adaptive resource management in the wildlife profession. Wildl. Soc. Bull. 24(2):436–442.

Landau, B. R. 1923. Essential human anatomy and physiology. Scott, Foresman and Co., Dallas, Texas. 598 pp.

Lande, R. 1988. Genetics and demography in biological conservation. Science 241 (4,872):1,455–1,460.

Langer, P. 1974. Stomach evolution in the Artiodactyla. Mammalia 8(2):295–314.

Lantz, D. E. 1910. Raising deer and other large game animals in the United States. Bur. Biol. Surv. Bull. 36. U.S. Dept. Agric. U.S. Govt. Print. Off., Washington, D.C. 62 pp.

Larsen, P. 1964. Some basic reproductive characteristics of pronghorn antelope in New Mexico. Western Assoc. St. Game and Fish Commiss. Proc. 44:142–145.

———. 1965. Antelope physiological studies. P-R Proj. W-93-R-7. New Mexico Dept. Game and Fish, Santa Fe. 8 pp.

———. 1966. Pronghorn and rangeland relationships in New Mexico: A review of our findings and analysis of our problems. Antelope States Workshop Proc. 2:43–49.

———. 1967. Antelope physiological studies. P-R Proj. W-93-R-9. New Mexico Dept. Game and Fish, Santa Fe. 3 pp.

Lawton, S. P. 1962. Petroglyphs and pictographs in Oklahoma: An introduction. Plains Anthropol. 7(17):189–193.

Laycock, G. 1990. The hunters and the hunted. Outdoor Life Books, Meredith Press, New York, New York. 280 pp.

Laycock, W. A. 1978. Coevolution of poisonous plants and large herbivores on rangelands. J. Range Manage. 31(5):335–342.

Leakey, R. and R. Lewin. 1995. The sixth extinction. Doubleday and Co., Inc., New York, New York. 271 pp.

Lebreton, J. D., K. P. Burnham, J. Clobert and D. R. Anderson. 1992. Modeling survival and testing biological hypotheses using marked animals: A unified approach with case studies. Ecol. Monogr. 62(1):67–118.

Lee, R. M. 1988. Arizona state report. Pronghorn Antelope Workshop Proc. 13:5–11.

———. 1990. Arizona antelope status report—1990. Pronghorn Antelope Workshop Proc. 14:2–5.

———. 2000. Arizona pronghorn antelope status report—1996. Pronghorn Antelope Workshop (1996) Proc. 17:11–13.

Lee, R. M., J. D. Yoakum, B. W. O'Gara, T. M. Pojar and R. A. Ockenfels, eds. 1998. Pronghorn management guides. Pronghorn Antelope Workshop 18, Prescott, Arizona. 110 pp.

Lee, T. E., Jr. 1992. Mitochondrial DNA and allozyme analysis of pronghorn populations in North America. Ph.D. thesis, Texas A&M Univ., College Station. 53+ pp.

Lee, T. E., Jr., J. N. Derr, J. W. Bickham and T. L. Clark. 1989. Genetic variation in pronghorn from west Texas. J. Wildl. Manage. 53(4):890–896.

Lee, T. E., Jr., J. W. Bickman and M. D. Scott. 1994. Mitochondrial DNA and allozyme analysis of North American pronghorn populations. J. Wildl. Manage. 58(2):307–318.

Leftwich, T. J. 1977. Man's past and present impact on the status and distribution of the Texas pronghorn. M.S. thesis, Texas Tech Univ., Lubbock. 132 pp.

Leftwich, T. J. and C. D. Simpson. 1977. A reevaluation of the history of the pronghorn abundance in west Texas. Ann. Conf. Southeast. Assoc. Fish and Wildl. Agencies Proc. 31:138–146.

———. 1978. The impact of domestic livestock and farming on Texas pronghorn. Pronghorn Antelope Workshop Proc. 8:307–320.

Legendre, S. and J. Clobert. 1995. ULM: Unified life models, a software for conservation and evolutionary biologists. J. Appl. Statistics 22(4):817–834.

Lehman, V. W. and J. B. Davis. 1942. Experimental wildlife management in the south Texas chaparral. P-R Quart. Prog. Rept., Proj. 1-R, Unit C, Sect. 1. Texas Game, Fish and Oyster Commiss., Austin. 14 pp.

Lehti , R. W. 1947. The golden eagle attacking antelope. J. Wildl. Manage. 11(4):348–349.

Leidy, J. 1854. No title. Pages 89–90 *in* Proc. Acad. Nat. Sci., Philadelphia, Pennsylvania.

Leighton, T. and G. Wobeser. 1999. Polio-encephalomalacia in deer and pronghorns. Wildl. Health Ctr. Newsletter 6(1):9.

Leinders, J. J. M. 1979. On the osteology and function of the digits of some ruminants and their bearing on taxonomy. Z. Saugetier 44(5):305–318.

Leinders, J. J. M. and E. Heintz. 1980. The configuration of the lacrimal orifices in pecorans and tragulids (Artiodactyla, Mammalia) and its significance for the distinction between Bovidae and Cervidae. Beaufortia 30(7):155–162.

Lendt, D. L. 1989. Ding: The life of Jay Norwood Darling. Iowa St. Univ. Press, Ames. 204 pp.

Lent, P. C. 1974. Mother–infant relationships in ungulates. Pages 14–55 *in* V. Geist and F. Walther, eds., The behavior of ungulates and its relation to management. Vol. 1. Publ. 24. Intl. Union Conserv. Nature and Nat. Resour., Morges, Switzerland. 511 pp.

Leopold, A. 1918. Forestry and game conservation. J. Forestry 16(4):404–411.

———. 1924. Grass, brush, timber, and fire in southern Arizona. J. Forestry 22(6):1–10.

———. 1925. New developments in game management. Bull. Am. Game Protective Assoc. 9(1):8–10.

———. 1931. Report on a game survey of the north central states. Democrat Printing Co., Madison, Wisconsin. 299 pp.

———. 1933. Game management. Charles Scribner's Sons, New York, New York. 481 pp.

———. 1953. Round River: From the journals of Aldo Leopold. Oxford Univ. Press, New York, New York. 173 pp.

Leopold, A. S. 1951. (Review) Life history, ecology and range use of the pronghorn antelope in Trans-Pecos, Texas by H. K. Buechner. J. Wildl. Manage. 15(3):322–323.

———. 1959a. Wildlife of Mexico, the game birds and mammals. Univ. California Press, Berkeley. 568 pp.

———. 1959b. Pronghorn antelope. Pages 95–97 *in* Nevada Legislative Counsel Bureau. Survey of fish and game problems in Nevada. Bull. 36. State Print. Off., Carson City, Nevada. 169 pp.

———. 1966. Adaptability of animals to habitat change. Pages 65–75 *in* F. F. Darling and J. P. Milton, eds., Future environments of North America. Doubleday and Co. Inc., New York, New York. 767 pp.

Leopold, A. S., R. J. Gutierrez and M. T. Bronson. 1984. North American game birds and mammals. Charles Scribner's Sons, New York, New York. 198 pp.

LeResche, R. E. and R. A. Rausch. 1974. Accuracy and precision of aerial moose censusing. J. Wildl. Manage. 38(2):175–182.

Leslie, D. M. Jr., M. Varra, E. E. Starkey and R. C. Slater. 1983. Correcting for differential digestibility in microhistological analyses involving common coastal forages of the Pacific Northwest. J. Range Manage. 36:730–732.

Leslie, P. H. 1945. On the uses of matrices in certain population mathematics. Biometrika 33(2):183–212.

———. 1948. Some further notes on the use of matrices in certain population mathematics. Biometrika 35(2):213–245.

Levine, N. D. 1968. Nematode parasites of domestic animals and man. Burgess Publishing Co., Minneapolis, Minnesota. 600 pp.

Levins, R. 1970. Extinction. Lecture notes in mathematical life sci. 2:75–107.

Levy, S. 2000. Wildlife on the hot seat. Natl. Wildl. 38(5):20–27.

Lewis, G. E. 1995. A four-horned specimen of *Antilocapra americana*. Denver Mus. Nat. Hist. Series Proc. 3(9). 4 pp.

Lewis, N. L. 1996a. A matter of necessity . . . the formation of the Arizona Antelope Foundation. Arizona Wildl. View 39(1):6–8.

———. 1996b. Blue light special. Pronghorn 4(2):1, 4–5.

Liewer, J. A., A. W. Alldredge, S. C. Torbit and R. B. Gill. 1986. Pronghorn impact on winter wheat in eastern Colorado. Pronghorn Antelope Workshop Proc. 12:56.

Ligon, J. S. 1927. Wildlife of New Mexico. St. Game Commiss., New Mexico Dept. Game and Fish, Santa Fe. 212 pp.

Lima, M. and E. Paez. 1997. Demography and population dynamics of South American fur seals. J. Mammal. 78(3): 914–920.

Lindenmayer, D. W. and H. P. Possingham. 1995. The risk of extinction: Ranking management options for leadbeater's possum using population viability analysis. Ctr. for Resour. and Environ. Studies, Australian Nat. Univ., Australia. 254 pp.

Lindstedt, S. L., J. F. Hokanson, S. D. Swain and D. J. Wells. 1988. Running energetics in North America's most elite endurance athlete, the pronghorn antelope. Am. Zool. 28 (4):44.

Lindzey, J. S. and R. G. Wingard. 1979. Landowners and sportsmen. Pages 16–21 *in* R. D. Teague and E. Decker, eds., Wildlife conservation: Principles and practices. The Wildl. Soc., Washington, D.C. 280 pp.

Linne, S. 1948. El Valle y la Ciudad de Mexico en 1550. New Series, Publ. 9. Ethnol. Mus. Sweden, Stockholm. 220 pp. + 11 map panels.

Linnell, J. D. C., R. Aanes and R. Andersen. 1995. Who killed Bambi? The role of predation in the neonatal mortality of temperate ungulates. Wildl. Biol. 1(4):209–222.

Lipetz, V. E. and M. Bekoff. 1980. Possible functions of predator harassment in pronghorn antelopes. J. Mammal. 61(4): 739–741.

Lipetz, V. E. and M. Bekoff. 1982. Group size and vigilance in pronghorn. Z. Tierpsychol. 58:203–216.

Lister, R. 1989. North Eccles pronghorn management plan. U.S. Bur. Land Manage., Elko, Nevada. 12 pp.

Litton, G. 1975. Pronghorn progress. Texas Parks and Wildl. 33(3):16–20.

Liu, J., J. B. Dunning, Jr. and H. R. Pulliam. 1995. Potential effects of a forest management plan on Bachman's sparrows (*Aimophila aestivalis*): Linking a spatially explicit model with GIS. Conserv. Biol. 9(1):62–75.

Livezey, K. B. 1979. Social behavior of Rocky Mountain elk at the National Bison Range. M.S. thesis, Univ. Montana, Missoula. 143 pp.

Lloyd, M., R. Kahn, L. Sikorowski and J. Apkie. 1995. Public hunter evaluations of the Ranching for Wildlife program. Human dimensions perspectives. Occasional Pap. 28. Series of the Colorado Div. Wildl., Denver. 6 pp.

Long, C. A. 1965. The mammals of Wyoming. Univ. Kansas Publ. Mus. Nat. Hist. 14(18):493–758.

Long, R. C. 1965. Fire effects upon a mixed-grass prairie in the Black Hills of South Dakota. Spec. Rept. 21. Humboldt St. Coll., Arcata, California. 21 pp.

Longhurst, W. M., A. L. Lesperance, M. Morse, R. J. Mackie, D. J. Neal, H. Salwasser, D. Swickard, P. T. Tueller, R. J. Urness and J. D. Yoakum. 1983. Livestock and wild ungulates. Pages 42–64 *in* J. W. Menke, ed., Proceedings of the workshop on livestock and wildlife–fisheries relationships in the Great Basin. Spec. Publ. 3,301. Agric. Sci. Publ., Univ. California, Berkeley. 173 pp.

Lonsdale, E. N., B. Bradach and E. T. Thorne. 1971. A telemetry system to determine body temperature in pronghorn antelope. J. Wildl. Manage. 35(4):747–751.

Lou, Y. 1998. Genetic variation of pronghorn (*Antilocapra americana*) populations in North America. Ph.D. thesis, Texas A&M Univ., College Station. 91 pp.

Lovaas, A. L. and P. T. Bromley. 1972. Preliminary studies of pronghorn antelope–black-tail prairie dog relations in Wind Cave National Park. Antelope States Workshop Proc. 5:115–156.

Lubinski, P. M. 1997. Pronghorn intensification in the Wyoming Basin: A study of mortality patterns and prehistoric hunting strategies. Ph.D. thesis, Univ. Wisconsin, Madison. 395 pp.

———. 2001. Estimating age and season of death of pronghorn antelope (*Antilocapra americana* Ord) by means of tooth eruption and wear. Intl. J. Osteoarchaeology 11(3):218–230.

Lubinski, P. M. and V. Herren. 2000. An introduction to pronghorn biology, ethnography and archaeology. J. V. Pastor and P. M. Lubinski, eds., Pronghorn past and present: Archaeology, ethnography, and biology. Plains Anthropologist Memoir 32, 45(174):3–11.

Lubow, B., G. C. White and D. R. Anderson. 1996. Evaluation of a linked-sex harvest strategy for cervid populations. J. Wildl. Manage. 60(4):787–796.

Lucker, J. T. and G. Dikmans. 1945. The distribution of *Pseudostertagia bullosa* and some new records of nematodes from pronghorn antelope (*Antilocapra americana*). Helminthol. Soc. Washington D.C. Proc. 12:2–4.

Lull, R.S. 1921. Fauna of the Dallas sand pits. Am. J. Sci. II(5): 159–176.

Lund, T. 1995. A brief history of wildlife policy in the United States. Pages 340–342 *in* J. A. Bissonette and P. R. Krausman, eds., Integrating people and wildlife for a sustainable future. The Wildl. Soc., Bethesda, Maryland. 715 pp.

Lyon, M. W., Jr. 1908. Remarks on the horns and on the systematic position of the American antelope. U.S. Natl. Mus. Proc. 34(1,619):393–401+.

Maas, J. and B. P. Smith. 1996. Copper deficiency in ruminants. Pages 904–908 *in* B. P. Smith, ed., Large animal internal medicine. 2nd ed. Mosby, Saint Louis, Missouri. 2,040 pp.

Maas, J., S. M. Parish, D. R. Hodgson and S. J. Valberg. 1996. Nutritional myopathies. Pages 1,513–1,518 *in* B. P. Smith, ed., Large animal internal medicine. 2nd ed. Mosby, St. Louis, Missouri. 2,040 pp.

MacArthur, R. H and E. O. Wilson. 1967. The theory of island biogeography. Princeton Univ. Press, Princeton, New Jersey. 203pp.

MacDonald, D. and E. G. Dillman. 1968. Techniques for estimating non-statistical bias in big game harvest surveys. J. Wildl. Manage. 32(1):119–129.

MacDonald, H. 1982. Game and hunters. Pages 137–144 *in* L. Silliman, ed., We seized our rifles. Mountain Press Publ. Co., Missoula, Montana. 224 pp.

Mace, R. U. 1954. Oregon's pronghorn antelope. Wildl. Bull. 1. Oregon St. Game Commiss., Portland. 26 pp.

Mackie, R. J. 1978. Impacts of livestock grazing on wild ungulates. Trans. N. Am. Wildl. Nat. Resour. Conf. 43:462–472.

———. 2000. History of management of large mammals in North America. Pages 292–320 *in* S. Demarais and P. R. Krausman, eds., Ecology and management of large mammals in North America. Prentice Hall, Upper Saddle River, New Jersey. 778 pp.

Madson, C. 1999. The land ethic. Wyoming Wildl. 63(5):4.

———. 2000. Courtship in the sage. Wyoming Wildl. 64(9):26–31.

Maelzer, D. A. 1970. The regression of log N_{n+1}+ on log N_n as a test of density dependence: An exercise with computer-constructed density-independent populations. Ecology 51(3):810–822.

Magruder, N. D., C. E. French, L. C. McEwen and R W. Swift. 1957. Nutritional requirements of white-tailed deer for growth and antler development. Bull. 628. Pennsylvania St. Univ. Agric. Exp. Sta., University Park. 21 pp.

Maher, C. R. 1991. Activity budgets and mating system of male pronghorn at Sheldon National Wildlife Refuge, Nevada. J. Mammal.72(4):739–744.

———. 1994. Pronghorn male spatial organizations: Population differences in degree of nonterritoriality. Can. J. Zool. 72(3):455–464.

———. 2000. Quantitative variation in ecological and hormonal variables correlates with spatial organization of pronghorn (*Antilocapra americana*) males. Behav. Ecol. and Sociobiol. 47(5):327–338.

Maher, C. R. and C. D. Mitchell. 2000. Effects of selective hunting on group composition and behavior patterns of pronghorn, *Antilocapra americana*, males in Montana. Can. Field-Nat. 114:264–270.

Mails, T. E. 1972. The mystic warriors of the plains. Mallard Press, New York, New York. 619 pp.

———. 1974. The people called Apache. Prentice-Hall, Inc., Englewood Cliffs, New Jersey. 447 pp.

———. 1983. The Pueblo children of the earth mother. 2 vols. Doubleday and Co., Garden City, New York.

Manfredo, M. J., J. J. Vaske and L. Sikorowski. 1996. Human dimensions of wildlife management. Pages 53–72 *in* A. W. Ewart, ed., Natural resource management: The human dimension. Westview Press, Boulder, Colorado. 263 pp.

Mangun, W. R. 1992. American fish and wildlife policy: The human dimension. Southern Illinois Univ. Press, Carbondale. 272 pp.

Mankins, J. V. 1962. Conjunctivitis of a pronghorn antelope (*Antilocapra americana*). J. Mammal. 43(3):415.

Manly, B. F. J. 1990a. POPSYS series: Stage frequency analysis. User's manual, version 1.02. Ecol. Systems Analysis, Pullman, Washington. 77 pp.

———. 1990b. Stage-structured populations: Sampling, analysis and simulation. Chapman and Hall, London, England. 187 pp.

Mapston, R. D. 1968. The use of structures to facilitate antelope movement through sheep-tight fences. M.S. thesis, Univ. Arizona, Tucson. 79 pp.

Mapston, R. D. and R. S. Zobell. 1972. Antelope passes: Their value and use. Tech. Note 6500. U.S. Bur. Land Manage., Portland, Oregon. 11 pp.

Marburger, R. G., R. M. Robinson and J. W. Thomas. 1966. A pseudohermaphroditic white-tailed deer. J. Mammal. 47(4):711–712.

Marcy, R. B. 1859. The prairie traveler: A handbook for overland expeditions. Harper and Brothers, New York, New York. 340 pp.

———. 1938. Adventure on Red River. Grant Foreman, ed. Univ. Oklahoma Press, Norman. 248 pp.

———. 1963. Thirty years of army life on the border. J. B. Lippincott and Co., Philadelphia, Pennsylvania. 402 pp.

Markham, D., R. L. Dickson and R. E. Autenrieth. 1979. Plutonium in the lungs of pronghorn antelope near a nuclear fuel reprocessing plant. Health Phys. 37(3):398–400.

Martin, C. 1973. The saga of the buffalo. Hart Publ. Co., New York, New York. 189 pp.

Martin, P. S. 1967. Prehistoric overkill in Pleistocene extinctions: The search for cause. Proc. 7th Congr. Intl. Assoc. Quater. Res. 6:75–120.

———. 1973. The discovery of America. Science 179:969–974.

———. 1984. Prehistoric overkill: The global model. Pages 354–403 *in* P. S. Martin and R. G. Klein, eds., Quaternary extinctions, A prehistoric revolution. Univ. Arizona Press, Tucson. 892 pp.

Martin, S. K. 1995. Characteristics of lactation and neonatal growth in pronghorn antelope (*Antilocapra americana*). M.S. thesis, Univ. Wyoming, Laramie. 106 pp.

Martin, S. K. and K. L. Parker. 1997. Rates of growth and morphological dimensions of bottle-raised pronghorns. J. Mammal. 78(1):23–30.

Martinka, C. J. 1966. A differential hunter harvest of pronghorn antelope in Montana. Proc. Western St. Game and Fish Commiss. 46:16–20.

———. 1967. Mortality of northern Montana pronghorns in a severe winter. J. Wildl. Manage. 31(1):159–164.

Mason, E. 1952. Food habits and measurements of Hart Mountain antelope. J. Wildl. Manage. 16(3):387–389.

Mason, R. 2000. Coyote management not same as population control of coyotes. Herald and News, Klamath Falls, Oregon, October 9, 2000:A8.

Massey, D. L. and P. J. Martin. 1972. Selenium soil–plant–animal relationships. Info. Bull. Alberta Dept. Agric., Edmonton. 4 pp.

Matthew, W. D. 1902. A list of Pleistocene fauna from Hay Springs, Nebraska. Am. Mus. Nat. Hist. Bull. 16:317–322.

———. 1904. A complete skeleton of Merycodus. Am. Mus. Nat. Hist. Bull. 20:101–129.

———. 1924. Third contribution to the Snake Creek fauna. Am. Mus. Nat. Hist. Bull. 50:59–210.

———. 1934. A phylogenetic chart of the Artiodactyla. J. Mammal. 15(3):207–209.

Matthiessen, P. 1959. Wildlife in America. Viking Press, New York, New York. 304 pp.

Mautz, W. W. 1978. Nutrition and carrying capacity. Pages 321–348. *in* J. L. Schmidt and D. L. Gilbert, eds., Big game of North America: Ecology and management. Stackpole Books, Harrisburg, Pennsylvania. 494 pp.

Maxell, M. 1973. Rodent ecology and pronghorn energy relations in the Great Divide Basin of Wyoming. Ph.D. thesis, Univ. Wyoming, Laramie. 208 pp.

Mayfield, H. 1961. Nesting success estimated from exposure. Wilson Bull. 73(2):255–261.

Maynard, L. A. and J. K. Loosli. 1969. Animal nutrition. 6th ed. McGraw-Hill Book Co., New York, New York. 483 pp.

McCabe, R. A. 1987. Aldo Leopold: The professor. Rusty Rock Press, Madison, Wisconsin. 172 pp.

McCabe, R. E. 1982. Elk and Indians: Historical values and perspectives. Pages 61–123 *in* J. W. Thomas and D. E. Toweill, eds., Elk of North America: Ecology and management. Stackpole Books, Harrisburg, Pennsylvania. 698 pp.

———. 2002. Elk and Indians: Then again. Pages 121–197 *in* D. E. Toweill and J. W. Thomas, eds., Ecology and Management of North American elk. Smithsonian Inst. Press, Washington, D.C. 962 pp.

McCabe, R. E. and T. R. McCabe. 1984. Of slings and arrows: An historical retrospection. Pages 19–72 *in* L. K. Halls, ed., White-tailed deer: Ecology and management. Stackpole Books, Harrisburg, Pennsylvania. 870 pp.

McCabe, R. E., B. W. O'Gara and H. M. Reeves. 2004. Prairie ghost: Pronghorn and human interaction in early America. Univ. Press Colorado, Boulder. 176 pp.

McCarthy, C. and J. Yoakum. 1984. An interagency approach to evaluating pronghorn transplant sites in Mono County, California. Pronghorn Antelope Workshop Proc. 11:134–143.

McCarty, R. S., Jr. 1982a. Little Lost-Birch Creek antelope habitat management plan. U.S. Bur. Land Manage., Idaho Falls, Idaho. 35 pp.

———. 1982b. Little Lost/Birch Creek antelope habitat management plan. Pronghorn Antelope Workshop Proc. 10:229–245.

———. 1986. Handbook for successful planning, installation and monitoring of wildlife waters. Tech. Bull. 86-1. U.S. Bur. Land Manage., Boise, Idaho. 22 pp.

McClaran, M. P. and T. R. VanDevender. 1995. The desert grassland. Univ. Arizona Press, Tucson. 346 pp.

McCracken, H. 1966. The Frederic Remington book. Doubleday and Co. Inc., Garden City, New York. 284 pp.

McCullough, D. R. 1969. The tule elk: Its history, behavior and ecology. Univ. California Publ. Zool. 88. Univ. California Press, Berkeley. 209 pp.

———. 1979. The George Reserve deer herd: Population ecology of a K-selected species. Univ. Michigan Press, Ann Arbor. 271 pp.

———. 1990. Detecting density dependence: Filtering the baby from the bathwater. Trans. N. Am. Wildl. and Nat. Resour. Conf. 55:534–543.

———. 1996. Metapopulations and wildlife conservation. Island Press, Washington, D.C. 429 pp.

McCullough, Y. B. 1980. Niche separation of seven North American ungulates on the National Bison Range, Montana. Ph.D. thesis, Univ. Michigan, Ann Arbor. 226 pp.

McCutchen, H. E. 1969. Age determination of pronghorns by the incisor cementum. J. Wildl. Manage. 33(1):172–175.

McGregor, J. C. 1935. Arizona antelope. Museum of Northern Arizona, Flagstaff. Mus. Notes 8(3):13–16.

McHugh, T. 1958. Social behavior of the American buffalo (*Bison bison bison*). Zoologica 43(1):1–40.

McInnis, M. L. 1984. Ecological relationships among feral horses, cattle and pronghorn in southeastern Oregon. Ph.D. thesis, Oregon St. Univ., Corvallis. 166 pp.

McIvor, D. E., J. A. Bissonette and G. S. Drew. 1994. A critical review of the status of the Yuma mountain lion, *Felis concolor browni*. Rept. 94-1. Utah Coop. Wildl. Res. Unit, Logan. 155 pp.

McKean, T. and B. Walker. 1974. Comparison of selected cardiopulmonary parameters between the pronghorn and the goat. Resp. Physiol. 21(3):365–370.

McKean, W. T. 1951. Sex, age and physical characteristics of North Dakota antelope as obtained by field checking. P-R Proj. 7R. North Dakota Dept. Game and Fish, Bismarck. 5 pp.

———. 1954. Preliminary observations of ovulation incidence in North Dakota deer and antelope. P-R Proj. W-7-R-9 to 13. North Dakota Dept. Game and Fish, Bismarck. 3 pp.

McKee, J. E. and H. W. Wolf. 1963. Water quality criteria. Publ. 3–4. St. Water Qual. Control Board, Sacramento, California. 548 pp.

McKenna, M. C. 1975. Fossil mammals and early Eocene North Atlantic land continuity. Annals Missouri Botanical Garden 62(2):335–353.

———. 1980. Eocene paleolatitude, climate and mammals of Ellesmere Island. Palaeogeogr. Palaeoclimatol. Palaeoecol. 30(3/4):349–362.

McKenzie, J. V. 1970. Two "killer winters" 1964–1965 and 1968–1969 in North Dakota. Antelope States Workshop Proc. 4:36–40.

———. 1980. North Dakota state report. Pronghorn Antelope Workshop Proc. 9:40–43.

———. 1983. Supplement to guidelines for the management of pronghorn antelope: Trapping and translocation. Texas Dept. Parks and Wildl., Austin. 14 pp.

———. 1986. North Dakota antelope status report. Pronghorn Antelope Workshop Proc. 12:25–28.

———. 2000. North Dakota pronghorn antelope status report—1996. Pronghorn Antelope Workshop (1996) Proc. 17:22–23.

McKinley, D. 1960. The question of the pronghorn in pioneer Missouri. J. Mammal. 41(4):503–505.

McLean, D. D. 1944. The prong-horned antelope in California. California Fish and Game 30(4):221–241.

McLucas, J. 1956. Montana big game trapping and transplanting techniques and accomplishments, 1941–1956. Wildl. Restoration Div., P-R Proj. Montana Dept. Fish, Wildlife and Parks, Helena. 65 pp.

McMahan, C. A. 1964. Comparative food habits of deer and three classes of livestock. J. Wildl. Manage. 28(4):298–308.

McNay, M. E. 1980. Causes of low pronghorn fawn:doe ratios on the Sheldon National Wildlife Refuge, Nevada. M.S. thesis, Univ. Montana, Missoula. 128 pp.

McNay, M. E. and B. W. O'Gara. 1982. Cattle–pronghorn interactions during the fawning season in northwestern Nevada. Pages 593–606 *in* J. M. Peek and G. D. Dalke, eds., Wildlife–livestock relationships symp.: Proc. 10. Univ. Idaho For. and Wildl. Exp. Sta., Moscow. 64 pp.

Meager, M. M. 1978. Bison. Pages 122–133 *in* J. L. Schmidt and D. L. Gilbert, eds., Big game of North America. Stackpole Books, Harrisburg, Pennsylvania. 494 pp.

Mearns, E. A. 1907. Mammals of the Mexican boundary of the United States. U.S. Natl. Mus. Washington, D.C. Bull. 56(1):1–530.

Medcraft, J. R. and W. R. Clark. 1986. Big game habitat use and diets on a surface mine in northeastern Wyoming. J. Wildl. Manage. 50(1):135–142.

Meeker, J. O. 1979. Interactions between pronghorn antelope and feral horses in northwestern Nevada. M.S. thesis, Univ. Nevada, Reno. 101 pp.

Meine, C. 1988. Aldo Leopold: His life and work. Univ. Wisconsin Press, Madison. 638 pp.

Meleney, W. P. and I. H. Roberts. 1970. *Otobius megnini* (Acarina: Argasidae) in the ears of pronghorn antelope (*Antilocapra americana*) in New Mexico. J. Parasitol. 56(5):917.

Menzel, K. 1980. Weights and horn measurements of Nebraska pronghorns—first seasons and twenty years later. Pronghorn Antelope Workshop Proc. 9:279–284.

———. 1992. Improved survival of pronghorn fawns with coyote control. Pronghorn Antelope Workshop Proc. 15:93–99.

———. 1994. Nebraska pronghorn status report—1994. Pronghorn Antelope Workshop Proc. 16:11–12.

Merriam, C. H. 1901. Two new bighorns and a new antelope from Mexico and the United States. Biol. Soc. Washington, D.C. Proc. 14:29–32.

———. 1979. Indian names for plants and animals among Californian and other western North American Indians. R. F. Heizer, assembler and annotater. Publ. Am. Archaeol., Ethnol., and Hist. 14. Ballena Press, Socorro, New Mexico. 296 pp.

Merriam, J. C. 1909. The occurrence of stripsicerine antelopes in the tertiary of northwestern Nevada. Ballena Press, Socorro, New Mexico. 5(22):319–330.

———. 1911. Tertiary mammal beds of Virgin Valley and Thousand Creek in northwestern Nevada. Univ. California Publications Dept. Geolog. Sci. 6(11):199–304.

Messenger, N. C. 1978. Diet and nutrition of the pronghorn antelope. M.S. thesis, South Dakota St. Univ., Brookings. 63 pp.

Messmer, T. A., M. W. Brunson, D. Reiter and D. G. Hewitt. 1999. United States public attitudes regarding predators and their management to enhance avian recruitment. Wildl. Soc. Bull. 27(1):75–85.

Meyer, A. B. 1966. Forests and forestry. Pages 38–56 *in* H. Clepper, ed., Origins of American conservation. Ronald Press Co., New York, New York. 193 pp.

Miller, G. S., Jr. and R. Kellogg. 1955. List of North American recent mammals. U.S. Natl. Mus. Bull. 205:816–817. 954 pp.

Miller, M. N. and J. A. Byers. 1991. Energetic cost of locomotor play in pronghorn fawns. Anim. Behav. 41(6):1,007–1,013.

Miller, R. F., T. J. Svejcar and N. E. West. 1994. Implications of livestock grazing in the intermountain sagebrush region: Plant composition. Pages 101–146 *in* M. Vavra, W. A. Laycock and R. D. Pieper, eds. Ecological implications of livestock herbivory in the West. Soc. for Range Manage., Denver, Colorado. 297 pp.

Miller, T. D. and G. J. Downs. 1974. Hemphillian local fauna. Los Angeles County, California, Mus. Contrib. Sci. 258:16–23.

Miller-Edge, M., S. Amsel, K. Brinkley, J. Boehm and B. Gonzales. 1994. Copper deficiency in captive pronghorn fawns. Pronghorn Antelope Workshop Proc. 16:158–167.

Mills, E. A. 1932. Watched by wild animals. Houghton Mifflin Co., New York, New York. 243 pp.

Mills, L. S., S. G. Hayes, C. Baldwin, M. J. Wisdom, J. Citta, D. J. Mattson and K. Murphy. 1996. Factors leading to different viability predictions for a grizzly bear data set. Conserv. Biol. 10(4):863–873.

Min, S. E. 1997a. Variation in sexually dimorphic traits of male pronghorns. J. Mammal. 78(1):31–47.

———. 1997b. The effect of variation in male sexually dimorphic traits on female behavior in pronghorn (*Antilocapra americana*). Ethology 103(9):732–743.

Minnesota Zoological Gardens. 1993. Management plan for pronghorn. Minnesota Zool. Gardens, Duluth. 8 pp.

Minnis, D. L. 1997. The opposition to hunting: A typology of benefits. Trans. No. American Wildl. Nat. Resour. Conf. 62:346–360.

Miranda, E. P. 2000. Monitoreo de una, poblacion de berrendo (*Antilocapra americana mexicana*) reintroducida al noroeste de Coahuila, Mexico. Tesis de Maestria, Universidad Autonoma de Nuevo Leon, Linares, Nuevo Leon, Mexico. 104 pp.

Mitchell, C. D. and C. R. Maher. 2001. Are horn characteristics related to age in male pronghorns? Wildl. Soc. Bull. 29(3):908–916.

Mitchell, G. J. 1965. Natality, mortality, and related phenomena in two populations of pronghorn antelope in Alberta, Canada. Ph.D. thesis, Washington St. Univ., Pullman. 221 pp.

———. 1967. Minimum breeding age of female pronghorn antelope. J. Mammal. 48(3):489–490.

———. 1971. Measurements, weights, and carcass yields of pronghorns in Alberta. J. Wildl. Manage. 35(1):76–85.

———. 1980. The pronghorn antelope in Alberta. Alberta Dept. Lands and Forests, Fish and Wildl. Div. and Univ. Regina, Regina, Saskatchewan. 165 pp.

Mitchell, G. J. and S. Smoliak. 1971. Pronghorn antelope range characteristics and food habits in Alberta. J. Wildl. Manage. 35(2):238–250.

Mitchell, J. G. 2001. Urban sprawl. Natl. Geogr. 200(1):48–73.

Mobley, D. 1960. Cripple loss is your loss! Wyoming Wildl. 24(9):4–7.

Mohanty, S. B. and S. K. Dutta. 1981. Veterinary virology. Lea and Febiger, Philadelphia, Pennsylvania. 372 pp.

Mohr, C. O. 1947. Table of equivalent populations of North American small mammals. Am. Midl. Nat. 37(1):223–249.

Mohre, E. 1965. Besonderheiten an Cavicornier-Hornern. Milu 2:21–47.

Monsen, S. B. and N. Shaw, compilers. 1983. Managing intermountain rangelands: Improvement of range and wildlife habitats. General Tech, Rept. INT-157. U.S. For. Serv., Ogden, Utah. 194 pp.

Monsen, S. B. and R. Stevens. In press. Restoration of western ranges and wildlands. General Tech. Rept. U.S. For. Serv., Ogden, Utah.

Monson, G. 1968. The desert pronghorn. Desert Bighorn Council Trans. 12:62–69.

Montana Farm Bill Coordinator. 1989. Evaluating the sodbuster provision in Montana. Bowdoin Natl. Wildl. Refuge, Malta, Montana. 5 pp.

Monte, L., L. Hakanson, U. Bergstrom, J. Brittain and R. Heling. 1996. Uncertainty analysis and validation of environmental models: The empirically based uncertainty analysis. Ecol. Modeling 90(2):139–152.

Moodie, J. D. and J. A. Byers. 1989. The function of scent marking by males on female urine in pronghorn. J. Mammal. 70(4):812–814.

Moody, D. S. and A. W. Alldredge. 1986. Red Rim—mining, fencing, and some decisions. Pronghorn Antelope Workshop Proc 12:57.

Moody, D. S., L. Saslaw and A. W. Alldredge. 1982. Drive trapping pronghorn antelope in south-central Wyoming. Pronghorn Antelope Workshop Proc. 10:225–228.

Mooser, O. and W. W. Dalquest. 1975. Pleistocene mammals from Aquascalientes, central Mexico. J. Mammal. 56(4):781–820.

Morgan, C. H. 1988. Fee hunting on private ranches in New Mexico. M.S. thesis, New Mexico St. Univ., Las Cruces. 58 pp.

Morrison, B. G. 1961. Some aspects of the histology and growth of the horns of *Antilocapra americana*. M.S. thesis, Univ. Wyoming, Laramie. 48 pp.

Morrison, B. L. 1984. New Mexico state report. Pronghorn Antelope Workshop Proc. 11:1–21.

Morrison, M. L., B. G. Marcot and R. W. Mannan. 1992. Wildlife–habitat relationships: Concepts and applications. Univ. Wisconsin Press, Madison. 343 pp.

Moses, L. E. and D. Rabinowitz. 1990. Estimating (relative) species abundance from route counts of the breeding bird survey. Pages 71–79 *in* J. R. Sauer and S. Droege, eds., Survey designs and statistical methods for the estimation of avian population trends. Biol. Rept. 90(1). U.S. Fish and Wildl. Serv. 166 pp.

Mossman, H. W. 1937. Comparative morphogenesis of the fetal membranes and accessory uterine structures. Carnegie Inst. Washington, D.C. Contrib. Embryol. 158:129–246+.

Moy, R. F. 1970. Histology of the subauricular and rump glands of the pronghorn (*Antilocapra americana* Ord). Am. J. Anat. 129(1):65–88.

———. 1971. Histology of the forefoot and hindfoot interdigital glands and median glands of the pronghorn. J. Mammal. 52(2):441–446.

Mross, G. A. and R. F. Doolittle. 1967. Amino acid sequence studies on Artiodactyl fibrinopeptides. Archives of Biochem. and Biophys. 122(3):674–684.

Müller-Schwarze, D. and C. Müller-Schwarze. 1972. Social scents in hand-reared pronghorn (*Antilocapra americana*). Zool. Africana 7(1):251–271.

Müller-Schwarze, D., C. Müller-Schwarze, A. G. Singer and R. M. Silverstein. 1974. Mammalian pheromone: Identification of active component in the subauricular scent of the male pronghorn. Science 183(4,127):860–862.

Mungall, E. C. and W. J. Sheffield. 1994. Exotics on the range: The Texas example. Texas A&M Univ. Press, College Station. 265 pp.

Murie, A. 1940. Ecology of the coyote in the Yellowstone. Conserv. Bull. 4, Fauna Series 4. U.S. Govt. Print. Off., Washington, D.C. 206 pp.

Murie, J. 1870. Notes on the anatomy of the prongbuck, *Antilocapra americana*. Proc. Zool. Soc. London 25:334–368.

Murie, O. J. 1935. Food habits of the coyote in Jackson Hole, Wyoming. Circ. 362. U.S. Dept. Agric., Washington D.C. 24 pp.

———. 1951. The elk of North America. Stackpole Co., Harrisburg, Pennsylvania. 376 pp.

Murray, D. 1986. Pronghorn feeding trial project summary report: Southern Alberta. Spec. Rept. Alberta Fish and Wildl. Div., Edmonton, Alberta. 14 pp.

Murray, L. T. 1932. Notes on personal experiences with pronghorn antelope in Texas. J. Mammal. 13(1):41–45.

Mussehl, T. W. and F. W. Howell, eds. 1971. Game management in Montana. P-R Proj. W-3-C, Montana Dept. Fish and Game, Helena. 238 pp.

Myers, N. 1989. Is man the fastest animal? Intl. Wildl. 19(5): 33.

Nagy, J. G. and J. B. Haufler. 1980. Wildlife nutrition. Pages 129–142 *in* S. D. Schemnitz, ed., Wildlife management techniques manual. The Wildl. Soc., Washington, D.C. 686 pp.

Nagy, J. G. and R. P. Tendergy. 1967. Antibacterial action of essential oils of *Artemesia* as an ecological factor. J. Applied Microbiol. 16(5):441–444.

Nagy, J. G. and G. W. Williams. 1969. Rumino recticular VFA content of pronghorn antelope. J. Wildl. Manage. 33(2):437–439.

Nagy, J. G., T. A. Barber and A. E. McChesny. 1969. *Clostridium perfringens* enterotoxemia in hand reared antelope. J. Wildl. Manage. 33(4):1,032–1,033.

Nash, G. D. 1967. Introduction. Pages xi–xxvii *in* G. Pinchot, The fight for conservation. Univ. Washington Press, Seattle. 152 pp.

Nash, R., ed. 1968. The American environment: Readings in the history of conservation. Addison-Wesley Publ. Co., Reading, Massachusetts. 236 pp.

National Research Council. 1964. Nutrient requirements of domestic animals. No. 5. Nutrient requirements of sheep. Publ. 193. Natl. Acad. Sci. Press, Washington, D.C. 40 pp.

———. 2001. Climate change science: An analysis of some key questions. Natl. Acad. Sci. Press, Washington, D.C. 42 pp.

Neff, D. J. 1986. Pronghorn habitat description and evaluation—a problem analysis report. P-R Proj. W-78-R. Arizona Dept. Game and Fish, Phoenix. 15 pp.

Neff, D. J. and N. G. Woolsey. 1979. Effect of predation by coyotes on antelope fawn survival on Anderson Mesa. Spec. Rept. 8, P-R Proj. W-78-R. Arizona Dept. Game and Fish, Phoenix. 36 pp.

Neff, D. J. and N. G. Woolsey. 1980. Coyote predation on neonatal fawns on Anderson Mesa, Arizona. Pronghorn Antelope Workshop Proc. 9:80–93.

———. n.d. Interactions of pronghorn antelope and coyotes during fawning season. Unpubl. pap. from D. J. Neff's files.

Neff, D. J., R. H. Smith and N. G. Woolsey. 1985. Pronghorn antelope mortality study. P-R Proj. W-78-R. Arizona Dept. Game and Fish, Phoenix. 22 pp.

Nelson, E. W. 1912. A new subspecies of pronghorn antelope from lower California. Biol. Soc. Washington, D.C. Proc. 25:107–108.

———. 1925. Status of the pronghorned antelope, 1922–1924. Bull. 1,346. U.S. Dept. Agric., Washington, D.C. 64 pp.

Nelson, J. R., R. M. Koes, W. H. Miller and B. B. Davitt. 1982. Big game management on a nutritional basis—A new approach. Arizona Dept. Game and Fish, Flagstaff. Western Elk Workshop Proc.:157–166.

Nesbitt, W. H. 1981. Stories behind the new world's records. Pages 18–28 *in* W. H. Nesbitt and P. L. Wright, eds., North American big game. 8th ed. Boone and Crockett Club, Alexandria, Virginia. 409 pp.

Nesbitt, W. H. and P. S. Wright. 1985. Measuring and scoring North American big game trophies. The Boone and Crockett Club, Missoula, Montana. 176 pp.

Nesbitt, W. H. and P. L. Wright, eds. 1981. North American big game. 8th ed. Boone and Crockett Club, Alexandria, Virginia. 409 pp.

Newberry, J. S. 1855. Report upon the zoology of the route. Pages 70–71, No. 2, Chapter 1 *in* Abbott, H. L. 1857. Reports of exploration and surveys to ascertain the most practicable and economical route for a railroad from the Mississippi River to the Pacific Ocean. U.S. Senate, Washington, D.C. Exec. Doc. 78, Vol. 6. In Seton 1929, p. 424.

Newman, J. L. 1966. Effects of woven wire fence with cattleguards on a free-living antelope population. Antelope States Workshop Proc. 2:6–8.

Newman, M. T. 1967. Radiocarbon-dated archaeological remains on the northern and central Great Plains. Am. Antiquity 32(4):471–486.

Newman, M. T., A. Woodward, W. J. Kroll and B. H. McLeod. 1957. River basin survey paper 8. Smithsonian Inst. Bur. Am. Ethnol. Bull. 166. U.S. Govt. Print. Off., Washington, D.C. 258 pp.

Niccolucci, A. 1987. *Yersinia pestis* associated with a pronghorn antelope and a bowhunter, Two Dot, Montana. Unpubl. rept. Montana Dept. Health and Environ. Sci., Helena. 4 pp.

Nichol, A. A. 1942. Gathering, transplanting, and care of young antelopes. J. Wildl. Manage. 6(4):281–286.

Nichols, L. 1960. The history of the antelope introduction on Lanai Island, Hawaii. Interstate. Antelope Conf. Trans. 11: 101–104.

Noback, C. V. 1932. The deciduous horns of the pronghorn antelope. New York Zool. Soc. 35(6):194–207.

Noble, G. A. 1953. An intestinal amoeba from the prong-horned antelope. Trans. Am. Microscopic Soc. 72(3):248–252.

Nord, E. C. 1965. Autecology of bitterbrush in California. Ecol. Monogr. 35:307–334.

North American Pronghorn Foundation. 2001. Foundation statements. Pronghorn 10(2):6.

Noss, R. F. and A. Y. Cooperrider. 1994. Saving nature's legacy. Island Press, Washington, D.C. 416 pp.

Nowak, P.M. and J. L. Paradiso. 1983. Walker's mammals of the world. Johns Hopkins Press, Baltimore, Maryland. 1,362 pp.

Oakley, C. 1973. The effects of livestock fencing on antelope. Wyoming Wildl. 37(12):26–29.

Oakley, C. and P. Riddle. 1974. The impact of a severe winter and fences on antelope mortality in southcentral Wyoming. Antelope States Workshop Proc. 6:155–173.

Ochs, E. 2000. Handgun adds challenge to hunt. Western Outdoor News, August 25, 2000:19.

Ockenfels, R. A. 1994a. Factors affecting adult pronghorn mortality rates in central Arizona. Wildl. Digest 16. Arizona Dept. Game and Fish, Phoenix. 11 pp.

———. 1994b. Mountain lion predation on pronghorn in central Arizona. Southwestern Nat. 39(3):305–306.

Ockenfels, R. A. and J. A. Wennerlund. 1994. Analyzing pronghorn habitat use patterns with GIS technology. Pronghorn Antelope Workshop Proc. 16:55–63.

Ockenfels, R. A., C. L. Dorothy and J. D. Kirkland. 1992. Mortality and home range of pronghorn fawns in central Arizona. Pronghorn Antelope Workshop Proc. 15:78–92.

Ockenfels, R. A., A. Alexander, C. L. D. Ticer and W. K. Carrel. 1994. Home ranges, movement patterns, and habitat selection of pronghorn in central Arizona. Res. Branch Tech. Rept. 13. P-R Proj. W-78-R. Arizona Dept. Game and Fish, Phoenix. 80 pp.

Ockenfels, R. A., C. L. Ticer, A. Alexander and J. A. Wennerlund. 1996. A landscape-level pronghorn habitat evaluation model for Arizona. Res. Branch Tech. Rept. 19. P-R Proj. W-78-R. Arizona Dept. Game and Fish, Phoenix. 50 pp.

Ockenfels, R. A., W. K. Carrel and C. Van Riper, III. 1997. Home ranges and movements of pronghorn in northern Arizona. Biennial Conf., Res. Colorado Plateau 3:45–61.

Ockenfels, R. A., W. K. Carrel, J. C. deVos, Jr. and C. L. D. Ticer. 2000. Highway and railroad effects on pronghorn movements in Arizona and Mexico. Pronghorn Antelope Workshop (1996) Proc. 17:104.

O'Connor, J. and G. G. Goodwin. 1961. The big game animals of North America. E. P. Dutton and Co., Inc., New York, New York. 264 pp.

Odum, E. P. 1953. Fundamentals of ecology. W. B. Saunders Co., Philadelphia, Pennsylvania. 157 pp.

Oftedal, O. T. 1985. Pregnancy and lactation. Pages 215–238 *in* R. L. Hudson and R. H. White, eds., Bioenergetics of wild herbivores. CRC Press, Boca Raton, Florida. 314 pp.

O'Gara, B. W. 1968. A study of the reproductive cycle of the female pronghorn (*Antilocapra americana* Ord). Ph.D. thesis, Univ. Montana, Missoula. 161 pp.

———. 1969a. Unique aspects of reproduction in the female pronghorn (*Antilocapra americana* Ord). Am. J. Anat. 125(2): 217–231.

———. 1969b. Horn casting by female pronghorn. J. Mammal. 50(2):373–375.

———. 1970. Derivation of whole weights for the pronghorn. J. Wildl. Manage. 34(2):470–472.

———. 1978. Differential characteristics of predator kills. Pronghorn Antelope Workshop Proc. 8:380–393.

———. 1980. Play by the rules and enjoy the game. Montana Outdoors 11(6):27–33.

———. 1987. A preliminary evaluation of carfentanil and xylazine hydrochloride for immobilizing pronghorns. Wildl. Soc. Bull. 15(4):549–551.

———. 1999. Evidence of inbreeding effects on an isolated pronghorn population. Pronghorn Antelope Workshop (1998) Proc. 18:91–93.

———. 2000. Estimates of pronghorn numbers. Pronghorn Antelope Workshop (1996) Proc. 17:33–34.

———. 2002. Hunting red deer and elk: Old and new worlds. Pages 649–700 *in* D. E. Toweill and J. W. Thomas, eds., Elk of North America. Smithsonian Inst. Press, Washington, D.C. 962 pp.

O'Gara, B. W. and K. R. Greer. 1970. Food habits in relation to physical condition in two populations of pronghorn. Antelope States Workshop Proc. 4:131–139.

O'Gara, B. W. and J. Malcolm. 1988. Pronghorn fawn mortality related to limited coyote control on the National Bison Range. Pronghorn Antelope Workshop Proc. 13:61–70.

O'Gara, B. W. and G. Matson. 1975. Growth and casting of horns by pronghorn and exfoliation of horns by bovids. J. Mammal. 56(4):829–846.

O'Gara, B. W. and R. F. Moy. 1972. Histology and morphology of scent glands and their possible roles in pronghorn behavior. Antelope States Workshop Proc. 5:192–208.

O'Gara, B. W. and J. D. Yoakum. 1990. Additional capture methods and habitat suitability criteria for pronghorn translocations. Pronghorn Antelope Workshop Proc. 14: 51–62.

———, eds. 1992. Pronghorn management guides. Pronghorn Antelope Workshop 15. 101 pp.

O'Gara, B. W., R. F. Moy and G. D. Bear. 1971. The annual testicular cycle and horn casting in the pronghorn (*Antilocapra americana*). J. Mammal. 52(3):537–544.

O'Gara, B. W., K. C. Brawley, J. R. Munoz and D. R. Henne. 1983. Predation on domestic sheep on a western Montana ranch. Wildl. Soc. Bull. 11(3):253–264.

O'Gara, B. W., M. E. McNay and W. A. Bodie. 1986. Effects of fawn activity and bedding cover on susceptibility to predation. Pronghorn Antelope Workshop Proc. 12:58–66.

Ogilby, W. 1840. Monograph of the hollow-horned ruminants. Proc. Zool. Soc. London 111:33–67.

Ogilvie, S. 1955. Chokecherry toxic to an antelope. J. Mammal. 36(1):146.

Oh, H. K., T. Sakai, M. G. Jones and W. M. Longhurst. 1967. Effects of various essential oils isolated from Douglas fir needles upon sheep and deer rumen microbial activity. J. Applied Microbiol. 15:777–784.

Olson, G., C. R. Watts, F. G. Feist, J. J. McCarthy and R. A. Bucsis. 1978. Big game survey and inventory Region Four. P-R Prog. Rept., Job I-4, Proj. W-130-R-9. Montana Dept. Fish, Wildl. and Parks, Helena. Pp. 1–17.

Olson, S. J. 1987. The faunal analysis 1. Pages 1–66 *in* Bones from Awatovi, northeastern Arizona, reports of the Awatovi Expedition. Rept. 11. Peabody Mus. Archaeol. and Ethnol., Harvard Univ. Press, Cambridge, Massachusetts. 74 pp.

Ord, G. 1815. Zoology of North America, class Mammalia. Pages 291–361 *in* W. Guthrie, ed., A new geographical, historical and commercial grammar. Vol. 2. 2nd Am. ed. Johnson and Warner, Philadelphia, Pennsylvania. 599 pp. (From an 1894 reproduction by S. N. Rhodes, Heddonfield, New Jersey. 70 pages of Ord's original publication plus a 90-page appendix on the more important scientific and historic questions involved.)

———. 1818. *Antilocapra* Ord. J. de Physique, de Chimie, d'Histoire Naturelle et des Arts 87:149–151.

Oregon State Game Commission. 1962. Annual report game division. Portland, Oregon. 157 pp.

———. 1963. Annual report game division. Portland, Oregon. 170 pp.

———. 1964. Annual report game division. Portland, Oregon. 177 pp.

———. 1972a. Annual report game division. Portland, Oregon. 177 pp.

———. 1972b. Antelope respond to coyote control. Oregon St. Game Commiss., Portland. Bull. 27(1):9.

———. 1973. Annual report game division. Portland, Oregon. 180 pp.

———. 1974. Annual report game division. Portland, Oregon. 173 pp.

Organ, J. F. and E. K. Fritzell. 2000. Trends in consumptive recreation and the wildlife profession. Wildl. Soc. Bull. 28(4):780–787.

Osborne, D. 1953. Archaeological occurrence of pronghorn antelope, bison and horse in the Columbia Plateau. Sci. Monthly 77(5):260–269.

Ottman, J. 1992. The plunge. Wyoming Wildl. 56(2):10–13.

Ozoga, J. J. and L. J. Verme. 1978. The thymus gland as a nutritional status indicator in deer. J. Wildl. Manage. 42(4):791–798.

Pacelle, W. 1990. Bowhunting: A most primitive sport. Animals Agenda, May:15–18.

Page, J. L. 1975. Prescribed burning plan for wildlife habitat improvement. Interstate Antelope Conf. Trans. 16: Unpaginated.

Palmer, W. L., J. M. Payne, R. G. Wingard and J. L. George. 1985. A practical fence to reduce deer damage. Wildl. Soc. Bull 13(3):240–245.

Paradiso, J. L. and R. M. Nowak. 1971. Taxonomic status of the Sonoran pronghorn. J. Mammal. 52(4):855–858.

Patterson, M. E., D. E. Guynn and D. C. Guynn, Jr. 2000. Human dimensions and conflict resolution. Pages 214–232 *in* S. Demarais and P. R. Krausman, eds., Ecology and management of large mammals in North America. Prentice Hall, Upper Saddle River, New Jersey. 778 pp.

Patton, D. R. 1992. Wildlife habitat relationships in forested ecosystems. Timber Press, Portland, Oregon. 392 pp.

Payne, N. F. and F. C. Bryant. 1994. Techniques for wildlife habitat management of uplands. McGraw-Hill Book Co., New York, New York. 840 pp.

———. 1998. Wildlife habitat management of forestlands, rangelands and farmlands. Krieger Publishing Co., Malabar, Florida. 840 pp.

Payne, N. F. and F. Copes. 1986. Wildlife and fisheries habitat improvement handbook. U.S. For. Serv., Washington, D.C. 549 pp.

Pearson, H. A. 1969. Starvation in antelope with stomachs full of feed. Res. Note RM148. Rocky Mtn. For. and Range Exp. Sta., U.S. For. Serv., Ogden, Utah. 4 pp.

Pechanec, J. F. and C. D. Pickford. 1937. A comparison of some methods used in determining percentage utilization of range grasses. J. Agric. Res. 54(10):753–765.

Pechanec, J. F., A. P. Plummer, J. H. Robertson and A. C. Hull, Jr. 1965. Sagebrush control on rangelands. Handbook 277. U.S. Dept. Agric., Washington, D.C. 40 pp.

Peden, D. C., G. M. Van Dyne, R. W. Rice and W. M. Hansen. 1974. The trophic ecology of *Bison bison* L. on shortgrass plains. J. Appl. Ecol. 11(2):489–498.

Peek, J. M., R. J. Pedersen and J. W. Thomas. 1982. Pages 599–625 *in* J. W. Thomas and D. E. Toweill, eds., Elk of North America: Ecology and management. Stackpole Books, Harrisburg, Pennsylvania. 698 pp.

Pennycuick, C. J. and D. Western. 1972. An investigation of some sources of bias in aerial transect sampling of large mammal populations. E. African Wildl. J. 10(3):175–191.

Petrides, G. A. 1975. Principal foods versus preferred foods and their relations to stocking rate and range condition. Biol. Conserv. 7(3):161–169.

Peyton, R. B. 2000. Wildlife management: Cropping to manage or managing to crop? Wildl. Soc. Bull. 28(4):774–779.

Phelps, J. S. 1978. Sonoran pronghorn habitat in Arizona. Pronghorn Antelope Workshop Proc. 8:70–77.

———. 1981. Biological observations on the Sonoran pronghorn. Pages 28–33 *in* J. S. Phelps, ed., The Sonoran pronghorn. Spec. Rept. 10. Arizona Dept. Game and Fish, Phoenix. 55 pp.

Pilgrim, G. E. 1946. The evolution of the buffaloes, oxen, sheep and goats. J. Linnaean Soc. Zool. 41:272–286.

Pillmore, R. E. 1958. Investigations of disease and parasites affecting game animals. Study of lung nematodes of game animals. P-R Job Compl. Rept., Job 1, Proj. W-95-R-2. Colorado Dept. Game, Fish and Parks, Denver. Pp. 1–28.

———. 1959. Investigations of diseases and parasites affecting game animals. P-R Job 1 and la, Proj. W-95-R-3. Colorado Dept. Game, Fish and Parks, Denver. Pp. 73–84.

Plummer, A. P., D. R. Christensen and S. B. Monsen. 1968. Restoring big game range in Utah. Publ. 68-3. Utah Dept. Fish and Game, Salt Lake City. 183 pp.

Pocock, R. I. 1905. The effects of castration on the horns of a prongbuck (*Antilocapra americana*). Proc. Zool. Soc. London 1:191–197.

Pojar, T. M. 1996. Colorado pronghorn compatibility and conflicts with agriculture. Wildl. Information Leaflet 116. Colorado Div. Wildl., Denver. 7 pp.

———. 1997. Habitat selection and population performance of a pioneering pronghorn population. P-R Job Final Rept., Proj. W-153-R-4, Work Plan 3A, Job 2. Colorado Div. Wildl., Fort Collins. 6 pp.

———. 2000. Helicopter net-gun capture of pronghorn. Pronghorn Antelope Workshop (1996) Proc. 17:69–72.

Pojar, T. M. and R. B. Gill. 1990. Management options for a pioneering pronghorn population. Pronghorn Antelope Workshop Proc. 14:112–122.

Pojar, T. M. and R. J. Guenzel. 1999. Comparison of fixed-wing line transect and helicopter quadrat pronghorn surveys. Pronghorn Antelope Workshop (1998) Proc. 18:64–68.

Pojar, T. M. and L. L. Miller. 1984. Recurrent estrus and cycle length in pronghorn. J. Wildl. Manage. 48(3):973–979.

Pojar, T. M. and T. Spraker. 1986. Humoral stress indicators in pronghorns. Ann. Mtg. Colorado–Wyoming. Acad. Sci. 62:51.

Pojar, T. M., K. D. Pojar, C. H. Wagner and R. Firth. 1994. Observations of pronghorn response to an electric fence. Pronghorn Antelope Workshop Proc. 16:131–142.

Pojar, T. M., D. C. Bowden and R. B. Gill. 1995. Aerial counting experiments to estimate pronghorn density and herd structure. J. Wildl. Manage. 59(1):117–128.

Pojar, T. M., R. Thompson, C. H. Wagner and P. Handyside. 2004. Pronghorn range expansion in Middlepark, Colorado. Pronghorn Workshop Proc. 20:60–71.

Polenz, A. 1976. Oregon antelope report—1976. Interstate Antelope Conf. Trans. 18: Unpaginated.

Polk, M. R. 1987. Mapping and inventory of seven antelope traps in Elko County, Nevada. Sagebrush Archaelogical Consultants, Ogden, Utah. Submitted to U.S. Bur. Land Manage. Copies available from Wells Resour. Area, U.S. Bur. Land Manage., Elko, Nevada. 32+ pp.

Pollock, K. H. and W. L. Kendall. 1987. Visibility bias in aerial surveys: A review of estimation procedures. J. Wildl. Manage. 51(2):502–509.

Pollock, K. H., S. R. Winterstein, C. M. Bunck and P. D. Curtis. 1989. Survival analysis in telemetry studies: The staggered entry design. J. Wildl. Manage. 53(1):7–15.

Poole, D. A. and R. E. McCabe. 1986. Wildlife tomorrow. Pages 325–338 *in* H. Kallman, ed., Restoring America's wildlife 1937–1987. U.S. Fish and Wildl. Serv., Washington, D.C. 594 pp.

Popowski, B. 1959. Hunting pronghorn antelope. The Stackpole Co., Harrisburg, Pennsylvania. 225 pp.

Posewitz, J. 1994. Beyond fair chase. The ethic and tradition of hunting. Falcon Press Publ. Co. Inc., Helena, Montana. 118 pp.

Possingham, H. P. and I. Davies. 1995. ALEX: A population viability analysis model for spatially structured populations. Biol. Conserv. 73(2):143–150.

Post, G. 1960. Diagnosis of diseases in mammals and birds. P-R Job Compl. Rept., Proj. FW-3-R-7, Work Plan 1, Job 1W. Wyoming Game and Fish Commiss., Cheyenne. Pp. 5–22.

Prenzlow, E. J. 1964. Doe-kid behavior of pronghorn in northcentral Colorado. Ann. Summer Conf. Central Mtns. and Plains Sect., The Wildl. Soc. 9:1–7.

———. 1965a. Social groups and activity patterns of pronghorn in northcentral Colorado, 1963–64. M.S. thesis, Colorado St. Univ., Fort Collins. 87 pp.

———. 1965b. A literature review on pronghorn behavior. Spec. Rept. 3. Colorado Dept. Game, Fish and Parks, Denver. 28 pp.

Prenzlow, E. J., D. L. Gilbert and F. A. Glover. 1968. Some behavior patterns of the pronghorn. Spec. Rept.17. Colorado Dept. Game, Fish and Parks, Denver. 16 pp.

Price, M. A. and R. G. White. 1985. Growth and development. Pages 183–213 *in* R. J. Hudson and R. G. White, eds., Bioenergetics of wild herbivores. CRC Press, Boca Raton, Florida. 214 pp.

Pulliam, D. E. and J. R. Nelson. 1979. Determination of digestibility coefficients for quantification of fecal analysis in elk. Pages 240–246 *in* M. S. Boyce and L. D. Hayden-Wing, eds., North American elk: Ecology, behavior and management. Univ. Wyoming, Laramie. 294 pp.

Pulliam, H. R. 1988. Sources, sinks and population regulation. Am. Nat. 132(4):652–661.

———. 1996. Sources and sinks: Empirical evidence and population consequences. Pages 45–69 *in* O. E. Rhodes, R. K. Chesser and M. H. Smith, eds., Population dynamics in space and time. Univ. Chicago Press, Chicago, Illinois. 388 pp.

Puls, R. 1994. Mineral levels in animal health. 2nd ed. Sherpa Intl., Clearbrook, British Columbia. 356 pp.

Pursley, D. 1977. Illegal harvest of big game during closed season. Western Law Enforcement Tech. Comm. Mimeo of paper presented at the Enforcement Workshop, Reno, Nevada. 5 pp.

Pusateri, F. M., C. P. Hibler and T. M. Pojar. 1982. Oral administration of diazepam and promazine hydrochloride to immobilize pronghorns. J. Wildl. Dis. 18(1):9–16.

Pyle, W. H. 1992. 1990–1991 distribution of pronghorn in the Hart Mountain biological unit. Admin. Rept. U.S. Fish and Wildl. Serv., Lakeview, Oregon. 14 pp.

Pyle, W. H. and B. Reiswig. 1994. 1990–1993 antelope status report. Admin. Rept. U.S. Fish and Wildl. Serv., Lakeview, Oregon. 16 pp.

Pyle, W. H. and M. F. Smith. 1990. Species management report for pronghorn. Admin. Rept. U.S. Fish and Wildl. Serv., Lakeview, Oregon. 87 pp.

Pyle, W. H. and J. D. Yoakum. 1994. Status of pronghorn management at Hart Mountain National Antelope Refuge, Oregon. Pronghorn Antelope Workshop Proc. 16:23–34.

Pyrah, D. B. 1970. Antelope herd range in central Montana. Antelope States Workshop Proc. 4:16–20.

———. 1974. Antelope bedding cover selection in central Montana. Antelope States Workshop Proc. 6:113–121.

———. 1978. Participant in discussion on the impact of a severe winter on northern pronghorn ranges. Pronghorn Antelope Workshop Proc. 8:337–359.

———. 1987. American pronghorn antelope in the Yellow Water Triangle, Montana. Montana Dept. Fish, Wildl. and Parks, Helena, in cooperation with the U.S. Bur. Land Manage. 121 pp.

Pyshora, L. 1977. The pronghorn antelope in northeastern California. Wildl. Manage. Admin. Rept. 77-2. California Dept. Fish and Game, Sacramento. 46 pp.

———. 1982a. A pronghorn antelope relocation project. Butte Valley Siskiyou County. California Dept. Fish and Game, Redding. 5 pp.

———. 1982b. Pronghorn antelope management plan. California Dept. Fish and Game, Redding. 122 pp.

———. 1986. California antelope status report. Pronghorn Antelope Workshop Proc. 12:8–10.

Quinn, D. 1930. The antelope's S.O.S. Emergency Conserv. Comm., New York, New York. 16 pp.

Ramirez, F., J. Cancino, J. M. Reyes and V. Sanchez-Sotomayor. 1999. Collaboration programs for Mexico and the United States to enhance pronghorn: A review. Pronghorn Antelope Workshop (1998) Proc. 18:74–80.

Ransom, A. B. 1965. Kidney fat and marrow fat as indicators of white-tailed deer condition. J. Wildl. Manage. 29(2):397–398.

Ransom, B. H. 1911. The nematodes parasitic in the alimentary tract of cattle, sheep, and other ruminants. Bull. 127. U.S. Bur. Anim. Indust., Washington, D.C. 132 pp.

Ransom, B. H. and M. C. Hall. 1912. A nematode, *Ostertagia bullosa*, parasitic in the alimentary tract of sheep. U.S. Natl. Mus. Proc. 42:175–179.

Ray, R. M. 2000. Arizona pronghorn antelope status report—1996. Pronghorn Antelope Workshop (1996) Proc. 17:11–14.

Reeves, H. M. and R. E. McCabe. 1998. Of moose and man. Pages 1–74 *in* A. W. Franzmann and C. C. Schwartz, compilers and eds., Ecology and management of the North American moose. Smithsonian Inst. Press, Washington, D.C. 733 pp.

Reichel, J. D. 1976. Coyote–prey relationships on the National Bison Range. M.S. thesis, Univ. Montana, Missoula. 94 pp.

Reiger, J. F. 1975. American sportsmen and the origins of conservation. Winchester Press, New York, New York. 316 pp.

Reilly, B. K. and R. H. Emslie. 1998. Power and precision of replicated helicopter surveys in mixed bushveld. Koedoe 41(1):47–56.

Reneau, J. 1995. Pronghorn spread penalty dropped. Fair Chase: The official publication of the Boone and Crockett Club, Missoula, Montana. Winter 1995:8.

Reneau, J. and S. Reneau. 1996. Records of North American sheep, Rocky Mountain goats and pronghorn. 1st ed. Boone and Crockett Club, Missoula, Montana. 380 pp.

Rensberger, B. 1986. Death of dinosaurs: The true story? Sci. Digest 94(5):28–35, 77–78.

Reynolds, T. D. 1983. Pronghorn responses to hunting coyotes. Great Basin Nat. 43(l):88–92

Rhodes, O. E., Jr., R. N. Williams, J. R. Heffelfinger, L. K. Page, E. P. Reat and J. C. deVos, Jr. 1999. Genetic variation in pronghorn from Arizona. Pronghorn Antelope Workshop (1998) Proc. 18:53–63.

Rice, L. 1985. Wildlife stocking report in files of South Dakota Department of Game, Fish and Parks, Pierre. 1 p.

———. 1986. South Dakota antelope status report. Pronghorn Antelope Workshop Proc. 12:31–33.

Rice, W. R. and J. D. Harder. 1977. Application of multiple aerial sampling to a mark–recapture census of white-tailed deer. J. Wildl. Manage. 41(2):197–206.

Richard, J. 1940. A rarity of the range. Nat. Mag., August–September:398–400.

Richards, G. D. and M. L. McCrossin. 1991. A new species of *Antilocapra* from the late Quaternary of California. Geobios-Paléontologie, Stratigraphie, Paléoécologie 24(5):623–635.

Richards, R. T. and R. S. Krannich. 1991. The ideology of the animal rights movement and activists' attitudes toward wildlife. Trans. N. Am. Wildl. Nat. Resour. Conf. 56:363–371.

Richards, S. H. 1964. Epidemic hemorrhagic disease in deer and antelope. North Dakota Outdoors 26(9):18–21.

Richardson, C. L. 2000. Texas pronghorn antelope status report—1996. Pronghorn Antelope Workshop (1996) Proc. 17:26–31.

Richardson, J. 1829. Fauna borealis—Americana; Or zoology of the northern parts of British America. Vol 1. Mammalia. John Murray, London, England. 300 pp. (Reprinted in 1974 by Arno Press, New York, New York.)

Ricker, W. E. 1954. Stock and recruitment. J. Fish. Res. Board Canada 11(4):624–651.

Riddle, P. 1990. Wyoming antelope status report—1990. Pronghorn Antelope Workshop Proc. 14:24.

Riddle, P. and C. Oakley. 1973. The impact of a severe winter and fences on antelope mortality in southcentral Wyoming. Western Assoc. St. Game and Fish Commiss. Proc. 53:174–188.

Riggs, R. A., S. C. Bunting and S. E. Daniels. 1996. Prescribed fire. Pages 295–319 *in* P. R. Krausman, ed., Rangeland wildlife. Soc. for Range Manage., Denver, Colorado. 440 pp.

Riney, T. 1955. Evaluating condition of free-ranging red deer with special reference to New Zealand. New Zealand J. Sci. and Tech. 36(B):429–463.

Ripley, T. H. 1980. Planning wildlife management investigations and projects. Pages 1–6 *in* S. D. Schemnitz, ed., Wildlife management techniques manual. The Wildl. Soc., Washington, D.C. 686 pp.

Risenhoover, K. L., H. B. Underwood, W. Yan and J. L.Cooke. 1997. A spatially explicit modeling environment for evaluating deer management strategies. Pages 366–379 *in* W. J. McShea, H. B. Underwood and J. H. Rappole, eds., The science of overabundance—deer ecology and population management. Smithsonian Inst. Press, Washington, D.C. 402 pp.

Robbins, C. T. 1983. Wildlife feeding and nutrition. Academic Press, New York, New York. 343 pp.

Robbins, C. T., T. A. Hanley, A. E. Hagerman, O. Hjeljord, D. L. Baker, C. C. Schwartz and W. W. Mautz. 1987. Role of tannins in defending plants against ruminants reduction in protein availability. Ecology 68(1):98–107

Roberts, D. A. 1970. Post-treatment studies of antelope range use, food habits and behavior during the summer season. M.S. thesis, Montana St. Univ., Bozeman. 56 pp.

Roebuck, C. M. 1982. Comparative food habits and range use of pronghorn and cattle in the Texas Panhandle. M.S. thesis, Texas Tech Univ., Lubbock. 109 pp.

Roese, J. H., K. L. Risenhoover and L. J. Folse. 1991. Habitat heterogeneity and foraging efficiency: An individual-based model. Ecol. Modeling 57(1):133–143.

Romer, A. S. 1962. The vertebrate body. 3rd ed. W. B. Saunders Co., Philadelphia, Pennsylvania. 475 pp.

———. 1966. Vertebrate paleontology. 3rd ed. Univ. Chicago Press, Chicago, Illinois. 468 pp.

Romesburg, H. C. 1981. Wildlife science: Gaining reliable knowledge. J. Wildl. Manage. 45(2):293–313.

Roosevelt, Q. and J. W. Burden. 1934. A new species of Antilocaprine, *Tetrameryx omusrosagis*, from a Pleistocene cave deposit in southern Arizona. Am. Mus. Novit. 754:1–4.

Roosevelt, T. 1885. Hunting trips of a ranchman. G. P. Putnam's Sons, New York, New York. 328 pp.

———. 1895. Hunting in the cattle country. Pages 278–317 *in* T. Roosevelt and G. B. Grinnell, eds., Hunting in many lands: The book of the Boone and Crockett Club. Forest and Stream Publ. Co., New York, New York. 447 pp.

———. 1905. Outdoor pastimes of an American hunter. Charles Scribner's Sons, New York, New York. 369 pp.

———. 1969. Ranch life and the hunting trail. Winchester Press, New York, New York. 188 pp.

Roosevelt, T. R., T. S. Van Dyke, D. G. Elliot and A. J. Stone. 1902. The pronghorn antelope. Pages 98–130 *in* The deer family. Grosset and Dunlap Publ., New York, New York. 334 pp.

Roseberry, J. L. and W. D. Klimstra. 1975. Some morphological characteristics of the Crab Orchard deer herd. J. Wildl. Manage. 39(1):48–59.

Roseland, K. 1977. Morphology of the pronghorn ovary *(Antilocapra americana* Ord). M.S. thesis, Univ. Montana, Missoula. 45 pp.

Rosen, M. N. 1970. Pasteurellosis. Pages 214–223 *in* J. W. Davis, L. H. Karstad and D. O. Trainer, eds., Infectious diseases of wild mammals. Iowa St. Univ. Press, Ames. 446 pp.

Rossi, P. A. and D. C. Hunt. 1971. The art of the Old West. Alfred A. Knopf, Inc., New York, New York. 335 pp.

Rothchild, S. L. 1993. Mortality, home range, and habitat use of pronghorn fawns within tallgrass prairie of eastern Kansas. M.S. thesis, Emporia St. Univ., Emporia, Kansas. 85 pp.

Rothchild, S. L., E. J. Finck and K. E. Sexson. 1994. Mortality and bedding site selection of pronghorn fawns in tallgrass prairie. Pronghorn Antelope Workshop Proc. 16:104–116.

Rothery, P., I. Newton, L. Dale and T. Wesolowski. 1997. Testing for density dependence allowing for weather effects. Oecologia 112(3):518–523.

Roundy, B. A. 1996. Revegetation of rangelands for wildlife. Pages 355–368 *in* P. R. Krausman, ed., Rangeland wildlife. Soc. for Range Manage., Denver, Colorado. 440 pp.

Roundy, B. A. and S. H. Biedenbender. 1995. Revegetation in the desert grasslands. Pages 265–303 *in* M. P. McClarn and T. R. Van Devender, eds., The desert grassland. Univ. Arizona Press, Tucson. 346 pp.

Rouse, C. H. 1941. Notes on winter foraging habits of antelopes in Oklahoma. J. Mammal. 22(l):57–60.

———. 1954. Antelope and sheep fences—preliminary report of a study. U.S. Bur. Land Manage., Washington, D.C. 20 pp.

———. 1959. Antelope–range relationships. Interstate Antelope Conf. Trans. 2:4–9.

Routledge, R. D. 1981. The unreliability of population estimates from repeated, incomplete aerial surveys. J. Wildl. Manage. 45(4):997–1,000.

Royama, T. 1977. Population persistence and density dependence. Ecol. Monogr. 47(1):1–35.

———. 1996. Analytical population dynamics. Chapman and Hall, New York, New York. 371 pp.

Ruff, F. I. 1939. The white-tailed deer on the Pisgah National Game Preserve. U.S. For. Serv., Ashville, North Carolina. 249 pp.

Ruh, G. B. 1994. Hunter behavior in America. Izaak Walton League of Am., Arlington, Virginia. 34 pp.

Rush, W. M. 1927. Notes on diseases in wild game mammals. J. Mammal. 8(2):163–165.

———. 1932. Northern Yellowstone elk study. Montana Fish and Game Commiss., Helena. 131 pp.

———. 1944. American pronghorn. Nat. Hist. 53(7):321–323.

Russell, C. M. 1930. Good medicine, the illustrated letters of Charles M. Russell. Doubleday, Doran and Co., Garden City, New York. 162 pp.

Russell, T. P. 1964. Antelope of New Mexico. Bull. 12. P-R Proj. W-84-D, New Mexico Dept. Game and Fish, Santa Fe. 103 pp.

Rutberg, A. T. 1987. Adaptive hypotheses of birth synchrony in ruminants: An interspecific test. The Am. Nat. 130(5): 692–710.

Ryder, T. J. 1983. Winter habitat selection of pronghorn in southcentral Wyoming. M.S. thesis, Univ. Wyoming, Laramie. 65 pp.

Ryder, T. J., L. L. Irwin and D. S. Moody. 1984. Wyoming's Red Rim controversy: History and current status. Pronghorn Antelope Workshop Proc. 11:195–204.

Rykiel, E. J., Jr. 1996. Testing ecological models: The meaning of validation. Ecol. Modeling 90:229–244.

Saiz, R. B. 1986. New Mexico antelope status report. Pronghorn Antelope Workshop Proc. 12:23–24.

Salwasser, H. 1980. Pronghorn antelope population and habitat management in the northwestern Great Basin environments. Interstate Antelope Conf., Alturas, California. 63 pp.

———. 1986. Modeling habitat relationships of terrestrial vertebrates—the manager's viewpoint. Pages 419–422 *in* J. Verner, M. L. Morrison and C. J. Ralph, eds., Wildlife 2000: Modeling habitat relationships of terrestrial vertebrates. Univ. Wisconsin Press, Madison. 470 pp.

Salwasser, H. and K. Shimamoto. 1979. Some aspects of pronghorn ecology and management in eastern California. Trans. Western Sect. The Wildl. Soc. 15:1–10.

Samuel, D. 1999. Hunting in the twenty-first century. Pages 21–25 *in* M. R. James, ed., Bowhunting: Big game records of North America. 5th ed. Pope and Young Club, Chatfield, Minnesota. 750 pp.

Samuel, M. D., E. O. Garton, M. W. Schlegel and R. G. Carson. 1987. Visibility bias during aerial surveys of elk in northcentral Idaho. J. Wildl. Manage. 51(3):622–630.

Samuelson, J. 1970. Participant in discussion on winter mortality of pronghorn. Antelope States Workshop Proc. 4:56.

Sanchez, F. C. 1993. Influence of range condition, cattle and watering hole distribution on a pronghorn population in southcentral New Mexico. Ph.D. thesis, New Mexico St. Univ., Las Cruces. 130 pp.

Sanchez-Sotomayor, V., J. Cancino, C. Olivera and A. Galas. 1999. Searching for and herding the peninsular pronghorn. Pronghorn Antelope Workshop (1998) Proc. 18:100.

Sauer, J. R. and P. H. Geissler. 1990. Estimation of annual indices from roadside surveys. Pages 58–62 *in* J. R. Sauer and S. Droege, eds., Survey designs and statistical methods for the estimation of avian population trends. Biol. Rept. 90(1). U.S. Fish and Wildl. Serv., Washington, D.C. 166 pp.

Savory, A. 1988. Holistic resource management. Island Press, Washington, D.C. 564 pp.

Schad, G. A. 1958. First report of *Ornithodorus (Otobius) megnini* from the pronghorn antelope (*Antilocapra americana*). J. Parasitol. 44(5):514.

Schaeffer, B. 1947. Notes on the origin and function of the artiodactyl tarsus. Am. Mus. Novit. 1,356. 24 pp.

Scheffer, V. B. 1974. A voice for wildlife. Charles Scribner's Sons, New York, New York. 245 pp.

Schemnitz, S. D., ed. 1980. Wildlife management techniques manual. The Wildl. Soc., Washington, D.C. 686 pp.

Schmidt, J. L. 1978. Care and use of the harvested animal. Pages 437–451 *in* J. L. Schmidt and D. L. Gilbert, eds., Big game of North America: Ecology and management. Stackpole Books, Harrisburg, Pennsylvania. 494 pp.

Schmidt, R. H. 1990. Why do we debate animal rights? Wildl. Soc. Bull. 18(4):459–461.

Schoemer, T. W. 1970. Nonsynchronous spatial overlap of lizards in patchy habitats. Ecology 51(3):408–418.

Schoonveld, G. G. 1971. Feeding mule deer high fiber diets. M.S. thesis, Colorado St. Univ., Fort Collins. 96 pp.

Schwartz, C. C. 1977. Pronghorn grazing strategies on the shortgrass prairie, Colorado. Ph.D. thesis, Colorado St. Univ., Fort Collins. 113 pp.

Schwartz, C. C. and J. G. Nagy. 1976. Pronghorn diets relative to forage availability in northeastern Colorado. J. Wildl. Manage. 40(3):469–478.

Schwartz, C. C., J. G. Nagy and S. M. Kerr. 1976a. Rearing and training pronghorns for ecological studies. J. Wildl. Manage. 40(3):464–468.

Schwartz, C. C., S. Mathews and J. G. Nagy. 1976b. Foraging behavior of pronghorn and cattle on a shortgrass prairie in Colorado. Pronghorn Antelope Workshop Proc. 7:78–101.

Schwartz, C. C., J. G. Nagy and R. W. Rice. 1977. Pronghorn dietary quality relative to forage availability and other ruminants in Colorado. J. Wildl. Manage. 41(2):161–168.

Scott, K. M. and C. M. Janis. 1987. Phylogenetic relationships of the Cervidae and the case for a superfamily "Cervoidea." Pages 3–20 *in* C. Wemmer, ed., Biology and management of the Cervidae. Natl. Zool. Soc. Symp. Smithsonian Inst. Press, Washington, D.C. 577 pp.

Scott, M. D. 1990. Determining a minimum genetically viable population size for Yellowstone pronghorns. Pronghorn Antelope Workshop Proc. 14:26–27.

———. 1992. Buck-and-pole fence crossings by ungulate species. Wildl. Soc. Bull. 20(2):204–210.

———. 1994. Capturing pronghorn by net gunning from the ground versus the air. Pronghorn Antelope Workshop Proc. 16:186–197.

Scotter, G. W. 1972. Chemical composition of forage plants from the Reindeer Preserve, Northwest Territories. J. Arctic Inst. N. Am. 25(1):21–27.

Seal, U. S. and R. L. Hoskinson. 1978. Metabolic indicators of habitat condition and capture stress in pronghorns. J. Wildl. Manage. 42(4):755–763.

Seal, U. S., L. J. Verme and J. J. Ozoga. 1981. Physiologic values. Pages 17–34 *in* W. R. Davidson, F. A. Hayes, V. F. Nettles and F. E. Kellogg, eds., Diseases and parasites of white-tailed deer. Misc. Publ. 7. Tall Timbers Res. Sta., Tallahassee, Florida. 458 pp.

Seargeant, A. B. and S. H. Allen. 1989. Observed interactions between coyotes and foxes. J. Mammal. 70(3):631–633.

Seber, G. A. 1973. Estimation of animal populations and related parameters. Griffin, London, England. 506 pp.

Seber, G. A. F. 1982. The estimation of animal abundance. 2nd ed. The Macmillan Co., New York, New York. 654 pp.

———. 1986. A review of estimating animal abundance. Biometrics 42(2):267–292.

Secretaria de Agricultura y Fomento. 1922. ACUERDO vedando la caza del berrendo *(Antilocapra americana)*. Diario Oficial de la Federacion, 16 de octubre de 1922. Mexico City, D.F., Mexico.

Secretaria de Desarrollo Social. 1994. Norma Oficial Mexicana NOM-059-ECOL-1994, que determina las especies y subspecies de flora y fauna silvestres terrestres y acuaticas en peligro de extincion, amenazadas, raras y las sujetas a proteccion especials y que establece especificaciones para su proteccion. Diario Oficial de la Federacion 438(1):2–60.

Secretaria de Medio Ambiente, Recursos Naturales y Pesca. 2000. Proyecto para la conservaion, manejo y aprovechamiento sustentable del berrendo (*Antilocapra americana*) en Mexico. Instituto Nacional de Ecologia, Direction General de Vida Silvestre, Mexico, D.F. 103 pp.

Segerstrom, T. 1977. Antelope food habits: Northern Great Plains region. Western Interstate Commiss. for Higher Education, Boulder, Colorado. 17 pp.

———. 1982. Effects of an operational coal mine on pronghorn antelope. M.S. thesis, Montana St. Univ., Bozeman. 68 pp.

Sehm, G. G. 1991. The first European bison illustration and the first central European exhibit of a living bison. With a table of the 16th century editions of Francisco Lopez de Gomara. Archives of Nat. Hist. 18(3):323–332.

Selye, H. 1976. Stress in health and disease. Butterworth, Inc., Boston, Massachusetts. 1,256 pp.

Seton, E. T. 1909. Life-histories of northern animals, an account of the mammals of Manitoba. 2 vols. Charles Scribner's Sons, New York, New York. 1,673 pp.

———. 1929. Lives of game animals. Vol. 3, Part 2. Hoofed animals. Doubleday, Doran and Co., Inc., Garden City, New York. 780 pp.

Severson, K. E. 1966. An analysis of foods and feeding habits of pronghorn antelope and domestic sheep in the Red Desert region of Wyoming. Antelope States Workshop Proc. 2:22–32.

———. 1990. Can livestock be used as a tool to enhance wildlife habitats. General Tech. Rept. RM-194. Rocky Mtn. For. and Range Exp. Sta., U.S. For. Serv., Fort Collins, Colorado. 123 pp.

Severson, K., M. May and W. Hepworth. 1968. Food preferences, carrying capacities, and forage competition between antelope and domestic sheep in Wyoming's Red Desert. Sci. Monogr. 10. Agric. Exp. Sta., Univ. Wyoming, Laramie. 51 pp.

Sexson, M. L. 1979. Ecogeographic relations of the pronghorn (*Antilocapra americana*) in Kansas. M.S. thesis, Fort Hays St. Univ., Hays, Kansas. 39 pp.

Shaw, D. L. 1973. The hunting controversy: Attitudes and arguments. Ph.D. thesis, Colorado St. Univ., Fort Collins. 174 pp.

Shaw, H. G. 1977. Impact of mountain lions on mule deer and cattle in northwestern Arizona. Pages 17–32 *in* R. L. Phillips and C. Jonkel, eds., 1975 Predator Symp. Proc., Univ. Montana For. and Conserv. Exp. Sta., Missoula. 268 pp.

———. 2000. Assessment of Arizona pronghorn research needs as related to aerial gunning of coyotes. Unpubl. rept. Arizona Antelope Foundation, Phoenix. 26 pp.

Shaw, H. G., D. E. Brown and P. M. Woodruff. 1998. Feasibility of restoring pronghorn (*Antilocapra americana*) to Joshua Tree National Park. The Juniper Institute, Chino Valley, Arizona. 27 pp.

Shaw, M. 2000. Oklahoma pronghorn antelope status report—1996. Pronghorn Antelope Workshop (1996) Proc.17:24.

Sheldon, C. 1925. The big game in Chihuahua, Mexico. Pages 138–181 *in* G. B. Grinnell and C. Sheldon, eds., Hunting and conservation: The book of the Boone and Crockett Club. Yale Univ. Press, New Haven, Connecticut. 548 pp.

Shelford, V. E. 1963. The ecology of North America. Univ. Illinois Press, Urbana. 610 pp.

Shenk, T. M., G. C. White and K. P. Burnham. 1998. Sampling-variance effects on detecting density dependence from temporal trends in natural populations. Ecol. Monogr. 86(4): 445–463.

Sherratt, A., ed. 1980. Columbia encyclopedia of archeology. Crown Publishers, Inc., New York, New York. 495 pp.

Shields, G. O. 1884. Antelope hunting in Montana. Harper's New Monthly Mag. 69(411):364–369.

Short, H. L. 1983. Wildlife guilds in Arizona desert habitats. Tech. Note 362. U.S. Bur. Land Manage., Denver, Colorado. 258 pp.

Short, H. L., D. E. Medin and A. E. Anderson. 1966. Seasonal variations in volatile fatty acid in the rumen of mule deer. J. Wildl. Manage. 30(3):466–470.

Short, H. L., E. E. Remmenga and C. E. Boyd. 1969a. Variations in ruminoreticular contents of white-tailed deer. J. Wildl. Manage. 33(1):187–191.

Short, H. L., C. A. Segelquist, P. D. Goodrum and C. E. Boyd. 1969b. Rumino-reticular characteristics of deer on food of two types. J. Wildl. Manage. 33(2):380–383.

Short, H. L., R. M. Blair and C. A. Segelquist. 1974. Fiber composition and forage digestibility by small ruminants. J. Wildl. Manage. 38(2):197–209.

Shufeldt, R. W. 1920. The American antelope. Am. Forestry 26(234):747–754.

Shults, L. M. and N. L. Stanton. 1986. *Taenia ovis krabbei* Cysticerci in the pronghorn antelope, *Antilocapra americana*. Helminthol. Soc. Washington, D.C. Proc. 53(l):132.

Sibly, R. M. and R. H. Smith. 1998. Identifying key factors using contribution analysis. J. Anim. Ecol. 67(1):17–24.

Siegmund, O. H. 1967. The Merck veterinary manual. 3rd ed. Merck and Co., Inc., Rahway, New Jersey. 1,685 pp.

Sierra Club. 2001. Sierra Club outings 2001: One hundred years of wilderness adventure. Sierra 86(1):101.

Siler, W. 1979. A competing-risk model for animal mortality. Ecology 60(3):750–757.

Simberloff, D., J. A. Farr, J. Cox and D. Mehlmann. 1992. Movement corridors, conservation bargains or poor investments? Conserv. Biol. 4(3):493–504.

Simpson, B. D. 1992. Behavior, home range, and habitat use of pronghorn translocated to tallgrass prairie in east central Kansas. M.S. thesis, Emporia St. Univ., Emporia, Kansas. 68 pp.

Simpson, C. D. 1984. Artiodactyls. Pages 563–587 *in* S. Anderson and J. K. Jones, Jr., eds., Orders and families of recent mammals of the world. John Wiley and Sons, New York, New York. 686 pp.

Simpson, C. D. and T. J. Leftwich. 1978. Historic range change in the Texas pronghorn. Pronghorn Antelope Workshop Proc. 8:121–147.

Simpson, G. G. 1945. The principles of classification and the classification of mammals. Am. Mus. Nat. Hist. Bull. 85:1–350.

Siniff, D. B. and R. O. Skoog. 1964. Aerial censusing of caribou using stratified random sampling. J. Wildl. Manage. 28(2): 391–401.

Sisson, S. and J. D. Grossman. 1953. The anatomy of the domestic animals. 4th ed. W. B. Saunders Co., Philadelphia, Pennsylvania. 972 pp.

Skinner, M. P. 1922. The pronghorn. J. Mammal. 3(2):82–105.

Skogland, T. 1990. Density dependence in a wild reindeer herd: Maternal vs. offspring effects. Oecologia 84(2):442–450.

Smith, A. D. 1974. Production and survival of pronghorn antelope on artificial diets with different protein levels. Antelope States Workshop Proc. 6:74–91.

Smith, A. D. and D. M. Beale. 1980. Pronghorn antelope in Utah: Some research and observations. Publ. 80-13. Utah Div. Wildl. Resour., Salt Lake City. 88 pp.

Smith, A. D. and D. D. Doell. 1968. Guides to allocating forage between cattle and big game on big game winter range. Utah Dept. Fish and Game, Salt Lake City. 32 pp.

Smith, A. D. and J. C. Malechek. 1974. Nutritional quality of summer diets of pronghorn antelopes in Utah. J. Wildl. Manage. 38(4):792–798.

Smith, A. D., D. M. Beale and D. D. Doell. 1965. Browse preferences of pronghorn antelope in southwestern Utah. Trans. N. Am. Wildl. and Nat. Resour. Conf. 30:136–141.

Smith, B. L. 1981. The history, current status and management of pronghorn antelope on Wind River Indian Reservation. U.S. Fish and Wildl. Serv., Lander, Wyoming. 56 pp.

Smith, C. E., Jr. 1967. Plant remains. Pages 220–255 *in* D. S. Byers, ed., The prehistory of the Tehuacan Valley: Environment and subsistence, vol. 1. University of Texas Press, Austin, Texas. 331 pp.

Smith, C. H. 1827. The Ruminantia. Pages 1–428 *in* Baron Cuvier, arranger, The class mammalia. Vol. 4. Printed for Geo. B. Whittaker by William Clowes and Sons, London, England. 498 pp.

Smith, E. R., Jr. 1967. Plant remains. Pages 220–255 *in* Douglas S. Byers, ed., The Prehistory of the Tehuacan Valley: Environment and Subsistence, vol. 1 University of Texas Press, Austin.

Smith, G., J. L. Holechek and M. Cardenas. 1996. Wildlife numbers on excellent and good condition Chihuahuan Desert grasslands: An observation. J. Range Manage. 49(6):489–493.

Smith, J. C. 1994. Microhistological analysis of diets of exotic and native ungulates in southcentral New Mexico. M.S. thesis, New Mexico St. Univ., Las Cruces. 76 pp.

Smith, L. W., H. K. Goering and C. H. Gordon. 1972. Relationship of forage composition with rates of cell wall digestion and indigestibility of cell walls. J. Dairy Sci. 55(8):1,140–1,147.

Smith, R. H., D. J. Neff and N. G. Woolsey. 1986. Pronghorn response to coyote control—A benefit:cost analysis. Wildl. Soc. Bull. 14(3):226–231.

Smith, S. G. 1992. Wyoming pronghorn status report—1992. Pronghorn Antelope Workshop Proc. 15:26.

Smith, T. 2002. Motor vehicle associated mortality in an urban pronghorn herd. Abstract presented at Pronghorn Antelope Workshop 20:26.

Sneva, F. A. and M. Vavra. 1978. Botanical composition of feces from antelope grazing the Oregon high desert. Pronghorn Antelope Workshop Proc. 8:78–93.

Snow, T. 1994. Sonoran pronghorn aerial survey summary 1992–94. Arizona Dept. Game and Fish, Phoenix. 14 pp.

Sokolov, I. I. 1952. Remains of Bovidae, Mammalia, from the lower Miocene deposits of western Gobi. Palaeontol. Inst. Akad. Nauk. S. S.S.R. Trans. 41:155–158.

Solounias, N. 1988. Evidence from horn morphology on the phylogenetic relationships of the pronghorn (*Antilocapra americana*). J. Mammal. 69(1):140–143.

Soulé, M. E. 1987. History of the Society for Conservation Biology: How and why we got here. Conserv. Biol. 1(1):4–5.

Soulsby, F. J. L. 1968. Helminths, arthropods and protozoa of domesticated animals. 6th ed. The Williams and Wilkins Co., Baltimore, Maryland. 824 pp.

Soupuck, T. 2000. The activists move west: Recent experiences in Manitoba. Symp. on No. America's Hunting Heritage. 6:176–180.

South Dakota Game Division. n.d. Antelope management in South Dakota. South Dakota Dept. Game, Fish and Parks, Pierre. 14 pp.

Southeastern Cooperative Wildlife Disease Study Team. 1982. Densities of pronghorn antelope where suitable habitat exists. Univ. Georgia School Vet. Sci., Athens. 1 p. (map)

Southwood, T. R. E. 1966. Ecological methods. Methuen and Co. Ltd., London, England. 391 pp.

Spalinger, D. E. 2000. Nutritional ecology. Pages 108–139 *in* S. Demarais and P. R. Krausman, eds., Ecology and management of large mammals in North America. Prentice Hall, Inc., Upper Saddle River, New Jersey. 778 pp.

Spalinger, D. E., S. M. Cooper, D. J. Martin and C. A. Shipley. 1997. Is social learning an important influence on foraging behavior in white-tailed deer? J. Wildl. Manage. 61(3):611–621.

Spencer, C. C. 1942. Antelope with locked horns. J. Mammal. 23(1):92.

Spillett, J. J. 1965. The effects of livestock fences on pronghorn movements. M.S. thesis, Utah St. Univ., Logan. 138 pp.

Spillett, J. J. and R. S. Zobell. 1967. Innovations in trapping and handling pronghorn antelopes. J. Wildl. Manage. 31(2):347–351.

Spillett, J. J., J. B. Low and D. Sill. 1967. Livestock fences—how they influence pronghorn antelope movements. Bull. 470. Utah Agric. Exp. Sta., Logan. 79 pp.

Springer, L. M. 1950. Aerial census of interstate antelope herds of California, Idaho, Nevada, and Oregon. J. Wildl. Manage. 14:295–298.

Starfield, A. M. and A. L. Bleloch. 1986. Building models for conservation and wildlife management. Macmillan Publ. Co., New York, New York. 253 pp.

Stauber, E. H. 1976. Weak calf syndrome: A continuing enigma. J. Am. Vet. Med. Assoc. 168(3):223–225.

Stauber, E. H., R. Autenrieth, O. D. Markham and V. Whitbeck. 1980. A seroepidemiologic survey of three pronghorn (*Antilocapra americana*) populations in southeastern Idaho, 1975–1977. J. Wildl. Dis. 16(l):109–115.

Steidl, R. J., J. P. Hayes and E. Schauber. 1997. Statistical power in wildlife research. J. Wildl. Manage. 61(2):270–279.

Stein, W. T. 1963. Mammal remains from archaeological sites in the Point of Pines region, Arizona. Am. Antiquity 29(2): 213–220.

Steinert, S. 2000. Colorado pronghorn antelope status report. 1996. Pronghorn Antelope Workshop (1996) Proc. 17:15.

Steinert, S. F., H. D. Riffel and G. C. White. 1994. Comparison of big game harvest estimates from check stations and telephone surveys. J. Wildl. Manage. 58(2):335–340.

Stelfox, J. G. and H. G. Vriend. 1977. Prairie fires and pronghorn use of cactus. Can. Field-Nat. 91(3):282–285.

Stoddart, L. A., A. D. Smith and T. W. Box. 1975. Range management. 3rd ed. McGraw-Hill Book Co., New York, New York. 433 pp.

Stoltenow, C. L., K. Solemsass, M. Niezgoda, P. Yager and C. E. Rupprecht. 2000. Rabies in an American bison from North Dakota. J. Wildl. Dis. 36(1):169–171.

Storer, L. B. 1902. A feathered hunter. Recreation 26(6):413.

Storz, J. F. 1999. Genetic consequences of mammalian social structure. J. Mammal. 80(2):553–569.

Stoszek, M. J., W. B. Kessler and H. Williams. 1978. Trace mineral content of antelope tissues. Pronghorn Antelope Workshop Proc. 8:162–181.

Strahl, H. 1906. Die Embryonalhullen der Sauger und die Placenta; 0. Hertwigs Handbuch der vergleichenden und experimentellen Entwicklungslehre der Wirbeltiere. Jena 1:2–8.

Strandgaard, H. 1967. Reliability of the Peterson method tested on a roe-deer population. J. Wildl. Manage. 31(4):643–651.

Stransky, J. J. 1984. Hunting the whitetail. Pages 739–780 *in* L. K. Halls, ed., White-tailed deer: Ecology and management. Stackpole Books, Harrisburg, Pennsylvania. 870 pp.

Strickland, D. 1979. Annual report of big game harvest, 1978. P-R Proj. W-27-R-3., Proj. Objective 1, Job 1. Wyoming Dept. Game and Fish, Cheyenne. 158 pp.

Strickland, D. M., H. H. Harju, K. R. McLaffery, H. W. Miller, L. M. Smith and R. J. Stoll. 1994. Harvest management. Pages 445–473 *in* T. A. Bookhout, ed., Research and management techniques for wildlife and habitats. 5th ed. The Wildl. Soc., Bethesda, Maryland. 740 pp.

Strickland, R. K. and R. R. Gerrish. 1965. Collections of *Dermacentor parumapertus* from cattle. J. Parasitol. 51(6):1,000.

Strohmeyer, D. C. 1994. Pronghorn–winter wheat controversy: Historical and current perspectives. Pronghorn Antelope Workshop Proc. 16:117–124.

Stroud, D., D. McWherter, N. Mertz and D. Abendroth. 1995. Building with fire. Wyoming Wildl. 59(12):13–28.

Stutz, S. S. 1967. Pronghorn antelope management plan. U.S. For. Serv., Alturas, California. 18 pp.

Stuwe, M. and C. E. Blohowick. 1988. McPAAL Software. Conserv. Res. Ctr., Natl. Zool. Park, Smithsonian Inst., Front Royal, Virginia. 20 pp.

Suetsugu, H. Y. 1970. Surveys and management of antelope. P-R Job Compl. Rept. W-15-R-26, WP B-69. Nebraska Game and Parks Commiss., Lincoln. 10 pp.

Sugár, L., Z. Lassu-Merenyi and J. Kelemen. 1992. Importance of copper supplementation in farmed red deer. Pages 453–455 *in* R. D. Brown, ed., The biology of deer. Springer-Verlag, New York, New York. 596 pp.

Sundberg, J. P., E. S. Williams, E. T. Thorne and W. D. Lancaster. 1983. Cutaneous fibropapilloma in a pronghorn antelope. J. Am. Vet. Med. Assoc. 183(11):1,333–1,334.

Sundstrom, C. 1967. A history of the pronghorn antelope in Wyoming. Spec. Rept. Wyoming Game and Fish Commiss., Cheyenne. 23 pp.

———. 1968. Water consumption by pronghorn antelope and distribution related to water in Wyoming's Red Desert. Antelope States Workshop Proc. 3:39–46.

———. 1970. Ecology of the pronghorn antelope in the Red Desert of Wyoming. Spec. Rept. Wyoming Dept. Game and Fish, Cheyenne. 227 pp.

Sundstrom, C., W. G. Hepworth and K. L. Diem. 1973. Abundance, distribution and food habits of the pronghorn. Bull. 12. Wyoming Game and Fish Commiss., Cheyenne. 61 pp.

Swaggerty, W. R. 2001. History of the United States plains until 1850. Pages 256–279 *in* R. J. DeMaillie, ed., Plains. Handbook of North American Indians. Vol 13. Smithsonian Inst., Washington, D.C. 1,320 pp.

Swanger, J. L. 1977. Aspects of doe pronghorn social structure on the National Bison Range. M.S. thesis, Univ. Montana, Missoula. 94 pp.

Swanson, E. B. 1940. The use and conservation of Minnesota game 1850–1900. Ph.D. thesis, Univ. Minnesota, Minneapolis. 294 pp.

Sweatman, G. K. and T. C. Henshall. 1962. The comparative biology and morphology of *Taenia ovis* and *Taenia krabbei* with observations on the development of *T. ovis* in domestic sheep. Can. J. Zool. 40(7):1,287–1,311.

Swepston, D. A. 1990. Texas antelope status report: 1990. Pronghorn Antelope Workshop Proc. 14:22–23.

Swepston, D. A. and T. L. Hailey. 1991. Texas pronghorns. Rept. 29. Texas Dept. Parks and Wildl., Austin. 18 pp.

Swepston, D. A. and C. L. Richardson. 1994. Pronghorn management and private property in the Texas panhandle. Pronghorn Antelope Workshop Proc. 16:125–130.

Swetnam, T. W. and J. L. Betancourt 1998. Mesoscale disturbance and ecological response to decadal climatic variability in the American Southwest. Climate 11:3,128-3,147.

Taber, R. D. 1956. Characteristics of the pelvic girdle in relation to sex in black-tailed and white-tailed deer. California Dept. Fish and Game 42(l):15–21.

Taber, R. D., K. L. White and N. S. Smith. 1962. The annual cycle of condition in the Rattlesnake, Montana, mule deer. Montana Acad. Sci. Proc. 19:72–79.

Taber, R. D., K. Raedeke and D. A. McCaughan. 1982. Population characteristics. Pages 279–298 *in* J. W. Thomas and D. E. Toweill, eds., Elk of North America: Ecology and management. Stackpole Books, Harrisburg, Pennsylvania. 698 pp.

Taylor, C. 1975. The warriors of the plains. Hamlyn Publ. Group, Ltd., London, England. 144 pp.

———. 2001. Native American weapons. Univ. Oklahoma Press, Norman 128 pp.

Taub, S. R. 1990. Smoothed scatterplot analysis of longterm breeding bird census data. Pages 80–83 *in* J. R. Sauer and S. Droege, eds., Survey designs and statistical methods for the estimation of avian population trends. Biol. Rept. 90(1). U.S. Fish and Wildl. Serv., Washington, D.C. 166 pp.

Taylor, B. L. and J. Barlow. 1995. The step-wise fit: A new model for estimating mammalian mortality from incomplete data sets. Pages 24–43 *in* J. D. Ballou, M. Gilpin and T. Foose, eds., Population management for survival and recovery: Analytical methods and strategies in small population conservation. Columbia Univ. Press, New York, New York. 375 pp.

Taylor, C. 1975. The warriors of the plains. Hamlyn Publ. Group, Ltd., London, England. 144 pp.

Taylor, E. 1972a. Food habits and feeding behavior of pronghorn antelope in the Red Desert of Wyoming. Antelope States Workshop Proc. 5:211–219.

———. 1972b. Food habits of the pronghorn antelope in the Red Desert of Wyoming. M.S. thesis, Univ. Wyoming, Laramie. 90 pp.

———. 1975. Pronghorn carrying capacity of Wyoming's Red Desert. P-R Proj. FW-3-R-21. Wildl. Tech. Rept. 3. Wyoming Dept. Game and Fish, Cheyenne. 65 pp.

Taylor, W. P. 1936. The prong-horned antelope in the Southwest. 1st N. Am. Wildl. Conf.: 652–655.

Teer, J. G. 1988. Commercial utilization of wildlife resources: Can we afford it? Pages 1–7 *in* R. Valdez and J. Knight, eds., Proc. First Intern. Wildl. Ranching Symp., New Mexico St. Univ., Las Cruces. 321 pp.

Teer, J. G., V. M. Neronov, L. V. Zhirnov and A. I. Blizniuk. 1996. Status and exploitation of the saiga antelope in Kalmykia. Pages 75–87 *in* V. J. Taylor and N. Dunston, eds., The exploitation of mammal populations. Chapman and Hall, London, England and New York, New York. 415 pp.

Terwilliger, C. 1953. Antelope: A range asset. Colorado Conserv. 2(5):24–25.

Thomas, A. and R. Rosentreter. 1992. Antelope utilization of lichens in the Birch Creek Valley of Idaho. Pronghorn Antelope Workshop Proc. 15:58–66

Thomas, E. 1966. The pronghorn antelope in Wyoming. Wyoming Wildl. 30(6):11–14.

Thomas, F. C. 1981. Hemorrhagic disease. Pages 87–96 *in* W. R. Davidson, F. A. Hayes, V. F. Nettles and F. E. Kellogg, eds., Diseases and parasites of white-tailed deer. Misc. Publ. 7. Tall Timbers Res. Sta., Tallahassee, Florida. 458 pp.

Thomas, J. W., ed. 1979. Wildlife habitats in managed forests: The Blue Mountains of Oregon and Washington. Handbook 553. U.S. For. Serv., Washington, D.C. 512 pp.

Thomas J. W. and C. Maser, technical eds. 1986. Wildlife habitats in managed rangelands—the Great Basin of southeastern Oregon. U.S. For. Serv. and U.S. Bur. Land Manage., Portland, Oregon. Unpaginated.

Thomas, J. W. and D. H. Pletscher. 2000. The convergence of ecology, conservation biology, and wildlife biology: Necessary or redundant? Wildl. Soc. Bull. 28(3):546–549.

Thomas, W. D. 1961. Chemical immobilization of wild animals. J. Am. Vet. Med. Assoc. 138(5):263–265.

Thompson, K. 1947. Air-herding the pronghorn. Am. Forests 53(8):348–349, 380–381.

Thompson, W. K. 1949. Predation on antelope. J. Wildl. Manage. 13(3):313–314.

Thompson-Olais, L. 1992. Recovery efforts for endangered Sonoran pronghorn. Pronghorn Antelope Workshop Proc. 15:109–114.

———. 1994. Technical agency draft Sonoran pronghorn recovery plan revision. U.S. Fish and Wildl. Serv. Region 2, Albuquerque, New Mexico. 40 pp.

Thorne, E. T. 1967. Diagnosis of diseases in wildlife. P-R Job Compl. Rept., Work Plan 1, Job 1W, Proj. FW-3-R-14. Wyoming Game and Fish Commiss., Cheyenne. Pp. 1–17.

———. 1968a. Diagnosis of diseases in wildlife. P-R Job Compl. Rept., Work Plan 1, Job 1W, Proj. FW-3-R-15. Wyoming Game and Fish Commiss., Cheyenne. Pp. 1–35.

———. 1968b. Effects of leptospirosis in antelope. P-R Job Compl. Rept., Work Plan 1, Job 6W, Proj. FW-3-R-15. Wyoming Game and Fish Commiss., Cheyenne. Pp. 50–60.

———. 1969. Diagnosis of diseases in wildlife. P-R. Job Prog. Rept., Work Plan 1, Job 1W, Proj. FW-3-R-16. Wyoming Game and Fish Commiss., Cheyenne. Pp. 1–36.

———. 1970. Diagnosis of diseases in wildlife. P-R Job Prog. Rept., Work Plan 1, Job 1W, Proj. FW-3-R-17. Wyoming Game and Fish Commiss., Cheyenne. Pp. 1–32.

———. 1971. Diagnosis of diseases in wildlife. P-R Job Prog. Rept., Work Plan 1, Job 1W, Proj. FW-3-R-18. Wyoming Game and Fish Commiss., Cheyenne. Pp. 1–8.

———. 1973. Leptospirosis in antelope. Wyoming Wildl. 37(12):34–35.

———. 1975. Normal body temperature of pronghorn antelope and mule deer. J. Mammal. 56(3):697–698.

———. 1978. Diagnosis of diseases in wildlife. P-R Job Prog. Rept., Work Plan 1, Job 1W, Proj. FW-3-23, Sec. 2. Wyoming Dept. Game and Fish, Cheyenne. Pp. 1–10.

Thorne, E. T., R. E. Dean and W. G. Hepworth. 1976. Nutrition during gestation in relation to successful reproduction in elk. J. Wildl. Manage. 40(2):330–335.

Thorne, E. T., N. Kingston, W. R. Jolley and R. C. Bergstrom, eds. 1982. Diseases of wildlife in Wyoming. 2nd ed. Wyoming Dept. Game and Fish, Cheyenne. 353 pp.

Thorne, E. T., T. J. Walthall and S. L. Anderson. 1983a. The effects of *Leptospira interrogans* Serovar *hardjo* in pronghorn antelope. P-R Job Perf. Rept., Job BCSWCBG551, Proj. CA1. Wyoming Dept. Game and Fish, Cheyenne. Pp. 38–45.

Thorne, E. T., E. S. Williams, W. J. Adrian and C. M. Gillin. 1983b. Vesicular stomatitis in pronghorn antelope: Serologic survey and artificial infection. Proc. Ann. Mtg. U.S. Anim. Health Assoc. 87:638–643.

Thorne, E. T., C. L. Hoffman and S. L. Anderson. 1984. The effects of *Leptospira interrogans* Serovar *hardjo* in pronghorn antelope. P-R Job. Perf. Rept., Job BCSWCBG551. Wyoming Dept. Game and Fish, Cheyenne. Pp. 48–55.

Thorne, E. T., T. J. Quan, E. S. Williams, T. J. Walthall and D. Daniels. 1987. Plague in a free-ranging mule deer from Wyoming. J. Wildl. Dis. 23(1):153–159.

Thorne, E. T., E. S. Williams, T. R. Spraker, W. Helms and T. Segerstrom. 1988. Bluetongue in free-ranging pronghorn antelope (*Antilocapra americana*) in Wyoming: 1976 and 1984. J. Wildl. Dis. 24(l):113–119.

Thorne, T. 2001. Foot-and-mouth disease: What does it mean to elk and hunters? Bugle 18(4):72.

Thorsen, J., L. Karstad, M. W. Barrett and G. A. Chalmers. 1977. Viruses isolated from captive and free-ranging wild ruminants in Alberta. J. Wildl. Dis.13(l):74–79.

Ticer, C. L. D. and W. H. Miller. 1994. Pronghorn fawn bed site selection in a semidesert grassland community of central Arizona. Pronghorn Antelope Workshop Proc. 16:86–103.

Ticer, C. L. and A. A . Munig. 1999. Pronghorn antelope province and state status reports. Pronghorn Antelope Workshop (1998) Proc. 18:17–24.

Ticer, C. L. D., R. A. Ockenfels and J. C. deVos, Jr. 2000. Pronghorn fawning dates in Arizona. Pronghorn Antelope Workshop (1996) Proc. 17:50–55.

Tileston, J. V. and L. E. Yeager. 1962. A résumé of Colorado big game research projects, 1939–1957. Tech. Bull. 9. Colorado Dept. Game and Fish, Denver. 81 pp.

Time-Life Books. 1985. Grasslands and tundra. Time-Life Books, Alexandria, Virginia. 176 pp.

Todd, K. S., Jr., D. M. Hammond and B. W. O'Gara. 1967. Redescription and incidence of *Eimeria antelocaprae* Huizinga, 1942, in the pronghorn antelope, *Antilocapra americana* (Ord, 1815). Bull. Wildl. Dis. Assoc. 3:71–72.

Todd, N. B. 1975. Chromosomal mechanisms in the evolution of artiodactyls. Paleobiology 1(2):175–188.

Torbit, S. C., J. G. Nagy and E. M. Rominger. 1984. Development of an emergency feeding system for pronghorn antelope. Pronghorn Antelope Workshop Proc. 11:113–128.

Torbit, S. C., R. B. Gill, A. W. Alldredge and J. F. Liewer. 1993. Impacts of pronghorn grazing on winter wheat in Colorado. J. Wildl. Manage. 57(1):173–181.

Torland, J. 1980. Oregon antelope report. Interstate Antelope Conf. Trans. 22: Unpaginated.

Towell, W. E. 1979. The role of policy making boards and commissions. Pages 49–54 *in* R. B. Teague and E. Decker, eds., Wildlife conservation: Principles and practices. The Wildl. Soc., Washington, D.C. 280 pp.

Townsend, J. K. 1978. Narrative of a journey across the Rocky Mountains to the Columbia River. Univ. Nebraska Press, Lincoln. 259 pp. (Originally published in 1839. This edition reproduced from 1970 edition published by Ye Galleon Press, Fairfield, Washington.)

Trainer, C. E., M. J. Willis, G. P. Keister, Jr. and D. P. Sheehy. 1983. Fawn mortality and habitat use among pronghorn during spring and summer in southeastern Oregon, 1981–82. Wildl. Res. Rept. 12. P-R Proj. W-70-R-11, Job 6; W-77-R-1-3, Jobs 1–3. Oregon Dept. Fish and Wildl., Portland. 117 pp.

Trainer, D. O. and R. P. Hanson. 1969. Serologic evidence of arbovirus infections in wild ruminants. Am. J. Epidemiol. 90(4):354–358.

Trainer, D. O. and M. M. Jochim. 1969. Serologic evidence of bluetongue in wild ruminants of North America. Am. J. Vet. Res. 30(11):2,007–2,011.

Trefethen, J. B. 1961. Crusade for wildlife. The Stackpole Co., Harrisburg, Pennsylvania. 377 pp.

———. 1966. Wildlife regulations and restoration. Pages 16–37 *in* H. Clepper, ed., Origins of American conservation. Ronald Press Co., New York, New York. 193 pp.

———. 1975. An American crusade for wildlife. The Boone and Crockett Club, Alexandria, Virginia. 409 pp.

Trevino, J. C. 1978. Number and distribution of pronghorn antelope in Chihuahua, Mexico. M.S. thesis, New Mexico St. Univ., Las Cruces. 58 pp.

Trevino, J. C. and S. D. Schemnitz. 1982. Distribution and numbers of pronghorned antelope in Chihuahua, Mexico. Pronghorn Antelope Workshop Proc. 10:246–258.

Trippensee, R. E. 1948. Wildlife management: Upland game and general principles. McGraw-Hill Book Co., New York, New York. 479 pp.

Trofimov, B. A. 1958. New Bovidae from the Oligocene of central Asia. Vertebr. Palasiat. 2(4):243–247.

Trout, L. E. 1976. Blood analysis of Idaho pronghorns. Pronghorn Antelope Workshop Proc. 7:122–126.

Trueblood, M. S. and G. Post. 1959. Vibriosis as a factor in the reproduction of antelope (*Antilocapra americana*). J. Am. Med. Assoc. 134(12):562–564.

Tsukamoto, G. K. 1983. Pronghorn antelope species management plan. Nevada Dept. Wildl., Reno. 59 pp.

Tsukamoto, G. and W. J. Diebert. 1968. A preliminary report of Nevada pronghorn antelope food habits during August and September. Interstate Antelope Conf. Trans. 19: Unpaginated.

Tsukamoto, G. K. and S. J. Stiver, eds. 1990. Wildlife water developments. Nevada Dept. Wildl., Nevada Chapter, The Wildl. Soc. and U.S. Bur. Land Manage., Reno, Nevada. 192 pp.

Tucker, R. D. 1979. Pronghorn antelope fawn mortality, home range, habitat, and behavior in Brewster County, Texas. M.S. thesis, Sul Ross St. Univ., Alpine, Texas. 118 pp.

Tucker, R. D. and G. W. Garner. 1983. Habitat selection and vegetation characteristics of antelope fawn bedsites in west Texas. J. Range Manage. 36(1):110–113.

———. 1984. Home range and diel movements of pronghorn antelope fawns in Brewster County, Texas. Pronghorn Antelope Workshop Proc. 11:71–89.

Tuljapurkar, S. 1989. An uncertain life: Demography in random environments. Theoretical Population Biol. 35(1):227–294.

Turchin, P. 1990. Rarity of density dependence or population regulation with lags? Nature 344 (6267):660–663.

Turner, M. G., G. J. Arthaud, R. T. Engstrom, S. J. Heil, J. Liu, S. Loeb and K. McKelvey. 1995. Usefulness of spatially explicit population models in land management. Ecol. Applications 5(1):12–16.

Tyler, R., C. Clark, L Ayres, W. H. Cadbury, H. J. Viola and B. Reilly, Jr. 1987. American frontier life: Early western painting and prints. Abbeville Press, New York, New York. 202 pp.

Udy, J. R. 1953. Effects of predator control on antelope populations. Fed. Aid Div. Publ. 5, Utah Dept. Fish and Game, Salt Lake City. 48 pp.

Ullrey, D. E., W. G. Yonatt, H. E. Johnson, A. B. Cowan, L. D. Fay, R. L. Covert, W. T. Magee and K. K. Keahey. 1975. Phosphorous requirements of weaned white-tailed deer fawns. J. Wildl. Manage. 39(3):590–595.

Underwood, E. J. 1977. Trace elements in human and animal nutrition. Academic Press, New York, New York. 545 pp.

Urness, P. J. 1979. Wildlife habitat manipulation in the sagebrush ecosystem. Pages 164–168 *in* The sagebrush ecosystem: A symposium. Utah St. Univ., Logan. 251 pp.

———. 1986. Value of crested wheatgrass for big game. Pages 147–153 *in* K. L. Johnson, ed., Crested wheatgrass: Its values, problems, and myths. Symp. Proc. Utah St. Univ., Logan. 348 pp.

Urqhart, S. N. 1986. Questionnaire survey/field check cross-validation analysis. Mimeo report on file. New Mexico Dept. Game and Fish, Santa Fe. 18 pp.

U.S. Bureau of Land Management. 1964. Water development: Range improvements in Nevada for wildlife, livestock and human use. U.S. Bur. Land Manage., Reno, Nevada. 37 pp.

———. 1975a. Regional fencing workshop. Proceedings. U.S. Bur. Land Manage., Washington, D.C. 75 pp.

———. 1975b. Manual 1737—Fencing. U.S. Bur. Land Manage., Washington, D.C. 17 pp.

———. 1979. Soil vegetation inventory method. U.S. Bur. Land Manage., Washington, D.C. Unpaginated.

———. 1980. Manual supplement 6630—Big game studies: Guidelines for the evaluation of pronghorn antelope habitats. U.S. Bur. Land Manage., Reno, Nevada. 3 pp.

———. 1981. Habitat management plans: Manual 6780. U.S. Bur. Land Manage., Washington, D.C. Unpaginated.

———. 1985. Fencing: BLM manual handbook H-1742-1. U.S. Bur. Land Manage., Washington, D.C. 23 pp.

———. 1989. Vegetation treatment on BLM lands in thirteen western states: Draft environmental impact statement. U.S. Bur. Land Manage., Casper, Wyoming. Unpaginated.

———. 1993. Fish and wildlife 2000: Big game habitat management strategy plan. Tech. Ctr., U.S. Bur. Land Manage., Denver, Colorado. 76 pp.

———. 1996a. Sampling vegetation attributes: Interagency technical reference. Natl. Applied Resour. Sci. Ctr., U.S. Bur. Land Manage., Denver, Colorado. 163 pp.

———. 1996b. Carrizo Plain Natural Area management plan. U.S. Bur. Land Manage., Bakersfield, California. 245 pp.

———. 2000. Annual report. Natl. Sci. and Tech. Ctr., U.S. Bur. Land Manage., Denver, Colorado. 92 pp.

U.S. Congress. 1969. National environmental policy act. U.S. Code, Vol. 42, Sections 4321–4370d. U.S. Govt. Print. Off., Washington, D.C.

———. 1973. Endangered species act. U.S. Code, Vol. 16, Sections 1531–1544. U.S. Govt. Print. Off., Washington, D.C.

———. 1974. National forest management act. U.S. Code, Vol. 16, Sections 1600–1614. U.S. Govt. Print. Off., Washington, D.C.

———. 1976. Federal land, policy, and management act. U.S. Code, Vol. 43, Sections 1701–1784. U.S. Govt. Print. Off., Washington, D.C.

———. 1978. Public rangeland improvement act. U.S. Code, Vol. 43, Sections 1901–1908. U.S. Govt. Print. Off., Washington, D.C.

U.S. Department of Interior. 1988. 1985 National survey of fishing, hunting, and wildlife—associated recreation. U.S. Fish and Wildl. Serv., Washington, D.C. 124+ pp.

———. n.d. Rangeland reform '94. Final environmental impact statement. U.S. Dept. Inter., Washington, D.C. 124 pp.

U.S. Fish and Wildlife Service. 1970. Hart Mountain National Antelope Refuge resource management plan. U.S. Fish and Wildl. Serv., Lakeview, Oregon. 64 pp.

———. 1980. Sheldon National Wildlife Refuge renewable natural resource management plan/final environmental impact statement. U.S. Fish and Wildl. Serv., Portland, Oregon. Unpaginated.

———. 1982. Sonoran pronghorn recovery plan. U.S. Fish and Wildl. Serv., Albuquerque, New Mexico. 26 pp.

———. 1985. Final environmental impact statement for the management of Charles M. Russell National Wildlife Refuge. U.S. Fish and Wildl. Serv., Denver, Colorado. 453 pp.

———. 1994. Hart Mountain National Antelope Refuge comprehensive management plan/final environmental impact statement. U.S. Fish and Wildl. Serv., Portland, Oregon. Vol. 1: 326 pp., Vol. 2: Unpaginated.

———. 1994. Final environmental statement: Hart Mountain National Antelope Refuge comprehensive management plan. U.S. Fish and Wildl. Serv., Lakeview, Oregon. Vol. 1: 326 pp., Vol. 2: 202 pp.

———. 1996a. Pronghorn update 1. U.S. Fish and Wildl. Serv., Lakeview, Oregon. 2 pp.

———. 1996b. Sampling vegetation attributes. Natl. Applied Resour. Sci. Ctr., Denver, Colorado. 163 pp.

———. 1997a. Appendices 1, 2, and 3 to Convention on international trade in endangered species of wild fauna and flora. U.S. Govt. Print. Off., 1999:454-786/94969, Washington, D.C. 24 pp.

———. 1997b. Pronghorn update 3. U.S. Fish and Wildl. Serv., Lakeview, Oregon. 2 pp.

———. 1998a. Final revised Sonoran pronghorn recovery plan. U.S. Fish and Wildl. Serv., Albuquerque, New Mexico. 70 pp.

———. 1998b. Pronghorn update 4. U.S. Fish and Wildl. Serv., Lakeview, Oregon. 2 pp.

———. 1999a. Appendices 1, 2 and 3 to the Convention on International Trade in Endangered Species of Wild Fauna and Flora. U.S. Govt. Print. Off. 1996-709-575. Washington, D.C. 24 pp.

———. 1999b. Endangered and threatened wildlife and plants, 50 CFR 17.11 and 17.12. U.S. Govt. Print. Off. 1996-508-283. Washington, D.C. 57 pp.

———. 1999c. Handbook for habitat conservation planning and incidental take permitting process. U.S. Fish and Wildl. Serv., Washington, D.C. 122 pp.

U.S. Fish and Wildlife Service and National Oceanic and Atmospheric Administration. 1996. Policy regarding the recognition of distinct vertebrate populations: Notice. Federal Register, February 7, 1996, pp. 4,722–4,723.

U.S. Fish and Wildlife Service and U.S. Bureau of the Census. 1991. National survey of fishing, hunting, and wildlife-associated recreation. U.S. Govt. Print. Off., Washington, D.C. 124+ pp.

U.S. Forest Service. 1999. Draft environmental impact statement for the northern Great Plains management plans revision. 5 vols. U.S. For. Serv., Chadron, Nebraska.

U.S. Senate. 1855–1860, III. Reports of explorations and surveys to ascertain the most practicable and economical route for a railroad from the Mississippi River to the Pacific Ocean. 12 vols. Various printers, Washington, D.C.

U.S. Soil Conservation Service. 1989. Wildlife habitat evaluation guides for antelope in Arizona. U.S. Soil Conserv. Serv., Phoenix, Arizona. 2 pp.

Vallentine, J. F. 1989. Range development and improvements. 3rd ed. Academic Press, San Diego, California. 524 pp.

Vanderpool, T. 2001. In search of Arizona's elegant visitor. Natl. Wildl. 39(4):30–33.

Van Dyke, W. 1990. Oregon pronghorn status report: 1990. Pronghorn Antelope Workshop Proc. 14:14–16.

Van Dyke, W. and M. Willis. 1992. Oregon pronghorn status report. Pronghorn Antelope Workshop Proc. 15:17–18.

Van Dyne, G. M., N. R. Brockington, Z. Szocs, J. Direk and C. A. Rivco. 1980. Large herbivore subsystems. Pages 269–527 *in* A. I. Breymeyer. and G. M. Van Dyne, eds., Grasslands, systems analysis and management. Cambridge Univ. Press, Cambridge, England. 950 pp.

Van Schmus, D. R. 1990. Bedsite selection by pronghorn antelope fawns in Grand County, Colorado. Pronghorn Antelope Workshop Proc. 14:35–40.

Van Wormer, J. 1969. The world of the pronghorn. J. B. Lippincott and Co., Philadelphia, Pennsylvania. 191 pp.

Varley, G. C., G. R. Gradwell and M. P. Hassell. 1973. Insect population ecology—an analytical approach. Blackwell Press, Oxford, England. 212 pp.

Vavra, M. and J. L. Holechek. 1980. Factors influencing microhistological analysis of herbivore diets. J. Range Manage. 33:(5)371–374.

Vavra, M. and F. Sneva. 1978. Seasonal diets of five ungulates grazing the Cold Desert Biome. Pages 435–437 *in* D. N. Hyder, ed., Proc. First Intl. Rangeland Congr., Denver, Colorado. 742 pp.

Vavra, M., W. A. Laycock and R. D. Pieper. 1994. Ecological implications of livestock herbivorary in the West. Soc. for Range Manage., Denver, Colorado. 297 pp.

Vereshchagin, N. K. 1959. The mammals of the Caucasus. Translated from Russian by the Israel Program for Scientific Translations Staff, Jerusalem, Israel, 1967. 816 pp.

Viliani, G. 1984. Grizzly eyesight, antelope hearing. Montana Outdoors 15(4):23.

Villa, B. 1951. Faballies y berrendos. Director General Forestal y de Caza Departamento Caza, Mexico City, Mexico. 30 pp.

———. 1955. Observaciones acerca de la ultima manada de berrendos (*Antilocapra americana mexicana*) en el Estado de Chihuahua, Mexico. Am. Inst. Biol. 26:229–236.

Vitousek, P. M., H. A. Mooney, J. Lubchenco and J. M. Melillo. 1997. Human domination of Earth's ecosystems. Science 277(5,325):494–499.

Von Gunten, B. L. 1978a. Pronghorn fawn mortality on the National Bison Range. M.S. thesis, Univ. Montana, Missoula. 82 pp.

———. 1978b. Pronghorn fawn mortality on the National Bison Range. Pronghorn Antelope Workshop Proc. 8:394–413.

Voorhies, M. R. 1969. Taphonomy and population dynamics of an early Pliocene vertebrate fauna, Knox County, Nebraska. Univ. Wyoming, Laramie, Contrib.Geol. Spec. Pap. 1:35–42.

Vriend, H. G. and M. W. Barrett. 1978. Low pronghorn recruitment—is it an issue? Pronghorn Antelope Workshop Proc. 8:360–379.

Wagner, F. H. 1978. Livestock grazing and the livestock industry. Pages 121–145 *in* H. P. Brokaw, ed., Wildlife and America. Council Environ Qual. U.S. Govt. Print. Off., Washington, D.C. 532 pp.

———. 1980. Wildlife of the deserts. Harry Abrams, Inc., New York, New York. 231 pp.

———. 1988. Predator control and the sheep industry. Regina Books, Claremont, California. 230 pp.

———. 1989. American wildlife management at the crossroads. Wildl. Soc. Bull. 17(3):354–360.

Walker, B. R. 1974. Cardiopulmonary function in pronghorn. M.S. thesis, Univ. Wyoming, Laramie. 108 pp.

Wallmo, O. C. 1978. Mule and black-tailed deer. Pages 31–41 *in* J. L. Schmidt and D. L. Gilbert, eds., Big game of North America. Stackpole Books, Harrisburg, Pennsylvania. 494 pp.

Walters, C. J. 1986. Adaptive management of renewable resources. Macmillan Publ. Co., New York, New York. 374 pp.

Walters, C. J. and R. Hillborn. 1978. Ecological optimization and adaptive management. Ann. Rev. Ecol. Systems 9(1): 157–188.

Walther, F. R. 1984. Communication and expression in hoofed mammals. Indiana Univ. Press, Bloomington. 423 pp.

Ward, A. L., J. J. Cupal, G. A. Goodwin and H. D. Morris. 1976. Effects of highway construction and use on big game populations. Rept. FHWA-RD-76-174. U.S. Fed. Highway Admin., Washington, D.C. 92 pp.

Ward, G. B. and R. E. McCabe. 1988. Trail blazers in conservation: The Boone and Crockett Club's first century. Pages 47–122 *in* W. H. Nesbitt and J. Reneau, eds., Records of North American big game. 9th ed. Boone and Crockett Club, Dumfries, Virginia. 498 pp.

Waring, G. H. 1969. The blow sound of pronghorn (*Antilocapra americana*). J. Mammal. 50(3):647–648.

Warren, E. R. 1942. The mammals of Colorado. Univ. Oklahoma Press, Norman. 330 pp.

Warwick, R. and P. L. Williams. 1975. Gray's anatomy. 35th ed. Jarrold and Sons Ltd., Norwich, England. 1,471 pp.

Watson, R. M. 1969. Aerial photographic methods in censuses of animals. E. African Agric. For. J. (Special Issue) 34:32–37.

Webb, S. D. 1969. The Burge and Minneschaduza Clarendonian mammalian faunas of north central Nebraska. Univ. California, Berkeley. Publ. Geol. Sci. 78:1–191.

———. 1973. Pliocene pronghorns of Florida. J. Mammal. 54(1):203–221.

Webb, S. D. and B. E. Taylor. 1980. The phylogeny of the hornless ruminants and a description of the cranium of *Archaeomeryx*. Am. Mus. Nat. Hist. Bull. 167(3):117–158.

Wehausen, J. D. 1995. Fecal measures of diet quality in wild and domestic ruminants. J. Wildl. Manage. 59(4):816–823.

Welch, B. L., J. C. Pederson and W. P. Clary. 1983. Ability of different rumen inocula to digest range forages. J. Wildl. Manage. 47(3):873–877.

Wempe, J. M. 1976. Rabies in a pronghorn antelope, *Antilocapra americana*. J. Wildl. Dis. 12(3):347–348.

Wedel, W. and G. C. Frison. 2001. Environment and subsistence. Pages 44 to 60 *in* R. J. DeMallie, ed., Plains. Handbook of North American Indians. Vol. 13. Smithsonian Instit., Washington, D.C. 1,360 pp.

Wentland, J. J. 1968. Summer range habits of the pronghorn antelope in central Montana with special reference to proposed sagebrush control study plots. M.S. thesis, Montana St. Univ., Bozeman. 65 pp.

Wenzel, L. E. 1955. The anatomy of the pronghorn antelope *Antilocapra americana americana* (Ord) 1815, with special reference to the digestive system. M.S. thesis, Univ. Wyoming, Laramie. 102 pp.

Werther, G. F. A. 1999. Global processes, emerging patterns of societal change, and the future challenges to wildlife management. Pronghorn Antelope Workshop (1998) Proc. 18:1–6.

Wesley, D. E., K. L. Knox and J. G. Nagy. 1970. Energy flux and water kinetics in young pronghorn antelope. J. Wildl. Manage. 34(4):908–912.

———. 1973. Energy metabolism of pronghorn antelopes. J. Wildl. Manage. 37(4):563–573.

West, D. R. 1969. Spring inventory of antelope in South Dakota 1968. P-R Job Prog. Rept. W-95-R-2, J 2-E. South Dakota Game, Fish and Parks, Pierre. 10 pp.

———. 1970. Effects of prolonged deep snow and cold winters on pronghorn mortality and reproduction in South Dakota. Antelope States Workshop Proc. 4:41–49.

Weswig, P. H. 1956. Vitamin A storage. Oregon St. Game Bull. 11(12):7.

Weybright, D. 2000. New Mexico pronghorn antelope status report—1996. Pronghorn Antelope Workshop (1996) Proc. 17:21.

Whisler, S. 1984. Seasonal adaptations of pronghorn antelope to water deprivation. M.S. thesis, Univ. Wyoming, Laramie. 81 pp.

Whisler, S. and S. L. Lindstedt. 1983a. Physiological responses to water deprivation in pronghorn antelope. J. Colorado-Wyoming Acad. Sci. 15(1):58.

Whisler, S. and S. L. Lindstedt. 1983b. Response to dehydration in pronghorn. Am. Zool. 22(4):961.

Whitaker, J. O., Jr. 1970. The biological subspecies: An adjunct of the biological species. Biologist 52(1):12–15.

White, G. C. and R. M. Bartmann. 1998. Effect of density reduction on overwinter survival of free-ranging mule deer fawns. J. Wildl. Manage. 62(2):214–225.

White, G. C. and K. P. Burnham. 1999. Program MARK: Survival estimation from populations of marked animals. Bird Study 46 (Supple.):120–138.

White, G. C. and R. A. Garrott. 1990. Analysis of wildlife radio-tracking data. Academic Press, San Diego, California. 383 pp.

White, G. C., R. M. Bartmann, L. H. Carpenter and R. A. Garrott. 1989. Evaluation of aerial line transects for estimating mule deer densities. J. Wildl. Manage. 53(3):625–635.

White, G. C., A. F. Reeve, F. G. Lindzey and K. P. Burnham. 1996. Estimation of mule deer winter mortality from age ratios. J. Wildl. Manage. 60(1):37–44.

White, R. J. 1987. Big game ranching in the United States. Wild Sheep and Goat Intl., Mesilla, New Mexico. 355 pp.

———. 2000. Big game ranching. Pages 260–276 *in* S. Demarais and P. R. Krausman, eds., Ecology and management of large mammals in North America. Prentice Hall, Upper Saddle River, New Jersey. 778 pp.

White, T. E. 1941. Additions to the fauna of the Florida Pliocene. New England Zool. Club Proc. 18:67–70.

Whittaker, D. G., M. Bray and L. D. Conn. 1999. Blood chemistry and trace mineral values for pronghorn antelope in Oregon. Pronghorn Antelope Workshop (1998) Proc. 18:43–52.

Wiens, J. A. 1996. Wildlife in patchy environments: Metapopulations, mosaics and management. Pages 53–84 *in* D. R. McCullough, ed., Metapopulations and wildlife conservation. Island Press, Washington D.C. 429 pp.

Wild, M. A. 1995. Animal and pen support facilities for terrestrial wildlife research. P-R Work Plan 1A, Job 1. Colorado Div. Wildl., Fort Collins. 5 pp.

Wild, M. A., M. W. Miller, D. L. Baker, N. T. Hobbs, R. B. Gill and B. A. Maynard. 1994. Comparing growth rates of dam and hand-raised bighorn sheep, pronghorn and elk neonates. J. Wildl. Manage. 58(2):340–347.

Wildlife Management Institute. 1992. Placing hunting in perspective. Professional Graphics Printing Co., Inc., Laurel, Maryland. 27 pp.

———. 2001. Unpublished survey of pronghorn populations in North America. Wildlife Management Institute, Washington, D.C. Not paginated.

Wildlife Management Institute Staff. 1978. The future. Pages 417–424 *in* J. W. Schmidt and D. L. Gilbert, eds., Big game of North America. Stackpole Books, Harrisburg, Pennsylvania. 494 pp.

Wilkinson, P. R. 1968. Phenology, behavior, and host-relations of *Dermacentor andersoni* Stiles in outdoor "rodentaria," and in nature. Can. J. Zool. 46(4):677–689.

Williams, B. K. and F. A. Johnson. 1995. Adaptive management and the regulation of waterfowl harvests. Wildl. Soc. Bull. 23(3):430–436.

Williams, E. S. 1982. Plant intoxication. Pages 280–288 *in* E. T. Thorne, N. Kingston, W. R. Jolly and R. C. Bergstrom, eds., Disease of wildlife in Wyoming. Wyoming Dept. Game and Fish , Cheyenne. 353 pp.

Willis, M. J. 1988. Impacts of coyote removal on pronghorn fawn survival. Pronghorn Antelope Workshop Proc. 13: 60.

Wilson, D. G. 1966. Range and forest resources. Pages 132–145 *in* H. Clepper, ed., Origins of American conservation. Ronald Press Co., New York, New York. 193 pp.

Wilson, L. O. and D. Hannan. 1977. Guidelines and recommendations for design and modification of livestock watering developments to facilitate use by wildlife. Tech. Note 305. Denver Serv. Ctr., U.S. Bur. Land Manage., Denver, Colorado. 20 pp.

Wilson, R. E., J. A. Young, G. W. Medlyn and S. DeForest. 2001. New artificial seeding for rangelands. Rangelands 22(2):9–12.

Winner, C. 1996. The grandmother effect. Wyoming Wildl. 60(3):10–15.

Wisdom, M. J. and J. G. Cook. 2000. North American elk. Pages 694–735 *in* S. Demarais and P. R. Krausman, eds., Ecology and management of large mammals in North America. Prentice Hall, Upper Saddle River, New Jersey. 778 pp.

Wishart, W. 1978. Bighorn sheep. Pages 160–171 *in* J. L. Schmidt and D. L. Gilbert, eds., Big game of North America. Stackpole Books, Harrisburg, Pennsylvania. 494 pp.

Wishart, W. D. 1970. A brief historic review of the pronghorn antelope in Alberta. Antelope States Workshop Proc. 4:128–130.

Wislocki, G. B. and D. W. Fawcett. 1949. The placentation of the pronghorned antelope *(Antilocapra americana)*. Mus. Comp. Zool., Harvard. Bull.101(5):548–558.

Wobeser, G. A. 1984. Rumen overload and polioenchalomalaca: Diseases associated with consumption of grain by pronghorns. Pronghorn Antelope Workshop Proc. 11:129–133.

Wobeser, G. A., G. W. Runge and D. Noble. 1975. Necrobacillosis in deer and pronghorn antelope in Saskatchewan. Can. Vet. J. 16:3–9.

Wobeser, G. A., P. Y. Daoust and H. M. Hunt. 1983. Polioencephalomalacia-like disease in pronghorn (*Antilocapra americana*). J. Wildl. Dis. 19(3):248–252.

Wofford, H. and J. L. Holechek. 1982. Influence of species and grind size on four-and forty-eight hour, *in vitro* digestibility. Western Sect. Am. Soc. Anim. Sci. Proc. 33:261–263.

Wolfe, J. A. 1985. Distribution of major vegetational types during the Tertiary. Pages 357–375 *in* The carbon cycle and atmospheric carbon dioxide. Geophys. Monogr. 32:349–362.

Wood, A. K. 1989. Comparative distribution and habitat use by antelope and mule deer. J. Mammal. 70(2):335–340.

Wood, W. F. 2001. Antibacterial compounds in the interdigital glands of pronghorn, *Antilocapra americana*. Biochem., Systematics and Ecol. 29:417–419.

Woodburne, M. O. 1987. Cenozoic mammals of North America. Univ. California Press, Berkeley. 336 pp.

Woods, G. T. 1974. Bovine parvovirus 1, bovine syncytial virus, and bovine syncytial respiratory virus and their infections. Adv. Vet. Sci. Comp. Med. 18:273–286.

Woods, S. H. 1925. The pronghorned antelope: America's own—shall we permit it's extermination? Am. Forests and Forest Life 31(383):649–651.

Woody, J. B. 1968. Recreational use of isolated rangelands in northwestern Nevada. M.S. thesis, Univ. Nevada, Reno. 82 pp.

Wooley, T. P. and F. G. Lindzey. 1997. Relative precision and sources of bias in pronghorn sex and composition surveys. J. Wildl. Manage. 61(1):57–63.

Wotschikowski, U. and G. Schwab. 1994. Das Rehprojekt Hahnebaum. Final Rept. Wildbiologische Gesellschaft Munich. 187 pp.

Wright, H. A. and A. W. Bailey. 1982. Fire ecology—United States and southern Canada. John Wiley and Sons, New York, New York. 501 pp.

Wright, P. L. 1981. The current records keeping program. Pages 7–17 *in* W. H. Nesbitt and P. L. Wright, eds., North American big game. 8th ed. Boone and Crockett Club, Alexandria, Virginia. 409 pp.

———. 1994. Ages of pronghorn trophies. Fair Chase: Official publication of the Boone and Crockett Club, Missoula, Montana. Fall 1994:14–15.

———. 1996. Evaluation of pronghorn trophies. Fair Chase: Official publication of the Boone and Crockett Club, Missoula, Montana. Winter 1996:2–3.

Wright, P. L. and S. A. Dow, Jr. 1962. Minimum breeding age in pronghorn antelope. J. Wildl. Manage. 26(l):100–101.

Wright, R. L. and J. C. deVos, Jr. 1986. Final report on Sonoran pronghorn status in Arizona. Arizona Dept. Game and Fish, Phoenix. 132 pp.

Wurster, D. H. and K. Benirschke. 1967. Chromosome studies in some deer, the springbok, and the pronghorn, with notes on placentation in the deer. Cytologia 32(2):273–285.

———. 1968. Chromosome studies in the superfamily Bovoidea. Chromosoma 25:152–171.

Wydeven, A. P. and R. B. Dahlgren. 1985. Ungulate–habitat relationships in Wind Cave National Park. J. Wildl. Manage. 49(4):805–813.

Wyoming Department of Fish and Game. 1988. Medicine Bow Antelope DAU—1989 season recommendations. Unpubl. rept. Laramie. 17 pp.

Yates, S. 1986. A pronghorn needs freedom to feel at home on the range. Smithsonian 17(9):86–95.

Yellowstone National Park. 1997. Yellowstone's northern range. Mammoth Hot Springs, Wyoming. 148 pp.

Yeo, J. J. 1988. Environmental factors influencing pronghorn habitat preference on spring-fall range in east-central Wyoming. Ph.D. thesis, Univ. Wyoming, Laramie. 102 pp.

Yoakum, J. D. 1957a. Factors affecting the mortality of pronghorn antelope in Oregon. M.S. thesis, Oregon St. Coll., Corvallis. 112 pp.

———. 1957b. Notes on food habits of Oregon pronghorn. Four St. Antelope Meeting, Klamath Falls, Oregon. 6 pp.

———. 1958. Seasonal food habits of the Oregon pronghorn. Interstate Antelope Conf. Trans. 1:58–72.

———. 1967. Literature of the American pronghorn antelope. U.S. Bur. Land Manage., Reno, Nevada. 82 pp.

———. 1968. A review of the distribution and abundance of American pronghorn antelope. Antelope States Workshop Proc. 3:4–14.

———. 1972. Antelope–vegetative relationships. Antelope States Workshop Proc. 5:171–177.

———. 1974. Pronghorn habitat requirements for sagebrush-grasslands. Antelope States Workshop Proc. 6:16–25.

———. 1975. Antelope and livestock on rangelands. J. Anim. Sci. 40(5):985–992.

———. 1978a. Pronghorn. Pages 103–121 *in* J. L. Schmidt and D. L. Gilbert, eds., Big game of North America: Ecology and management. Stackpole Books, Harrisburg, Pennsylvania. 494 pp.

———. 1978b. Managing rangelands for the American pronghorn antelope. Pronghorn Antelope Workshop Proc. 8:321–336.

———. 1980. Habitat management guides for the American pronghorn antelope. Tech. Note 347. Denver Serv. Ctr., U.S. Bur. Land Manage., Denver, Colorado. 77 pp.

———. 1983. Managing vegetation for pronghorns. Pages 180–194 *in* S. B. Monsen and N. Shaw, compilers, Managing intermountain rangelands—improvement of ranges and wildlife habitats. General Tech. Rept. INT-157. Intermtn. For. and Range Exp. Sta., U.S. For. Serv., Ogden, Utah. 194 pp.

———. 1986. Trends in pronghorn populations: 1800–1983. Pronghorn Antelope Workshop Proc. 12:77–85.

———. 1990. Food habits of the pronghorn. Pronghorn Antelope Workshop Proc. 14:102–111.

———. 1991. Literature review of the pronghorn: A bibliography with key words and reference citations. U.S. Bur. Land Manage., Reno, Nevada. 153 pp.

———. 1992. Proposal for food habit study of wild and domestic ungulates on the Hart Mountain National Antelope Refuge, Oregon. Admin. Rept. U.S. Fish and Wildl. Serv., Lakeview, Oregon.

———. 1994. Water requirements of pronghorn. Pronghorn Antelope Workshop Proc. 16:143–157.

———. 1995. Relationships of pronghorn and livestock in the Great Basin: A review. Pages 170–175 *in* W. D. Edge and S. L. Olson-Edge, eds., Proc. Sustaining Rangeland Ecosystem Symp. Spec. Rept. 953. Oregon St. Univ., Corvallis. Pages not consecutively numbered.

———. 1999. Fifty years of the Interstate Antelope Conference: A review of goals and accomplishments. Interstate Antelope Conf. Trans. 35: Unpaginated.

———. 2000. An assessment of wildfires and pioneering pronghorn in Long Valley, California/Nevada. Pronghorn Antelope Workshop (1996) Proc. 17:86–90.

———. 2001. Notes, findings and recommendations for field trip to the Carrizo Plain Natural Area: 29 December 2001. Western Wildl., Verdi, Nevada. 14 pp.

Yoakum, J. D. and B. W. O'Gara. 1990. Pronghorn–livestock relationships. Trans. N. Am. Wildl. and Nat. Resour. Conf. 55:475–487.

Yoakum, J. D. and B. W. O'Gara. 2000. Pronghorn. Pages 559–577 *in* S. Demarais and P. Krausman, eds., Ecology and management of large mammals in North America. Prentice-Hall, Upper Saddle River, New Jersey. 778 pp.

Yoakum, J. D. and D. Spalinger. 1979. American pronghorn antelope: Articles published in the Journal of Wildlife Management 1937–1977. The Wildl. Soc., Washington, D.C. 244 pp.

Yoakum, J. D., W. P. Dasmann, H. R. Sanderson, C. M. Nixon and H. S. Crawford. 1980. Habitat improvement techniques. Pages 329–403 *in* S. D. Schemnitz, ed., Wildlife management techniques manual. 4th ed. The Wildl. Soc., Washington, D.C. 686 pp.

Yoakum, J. D., B. W. O'Gara and V. W. Howard. 1996. Pronghorn on western rangelands. Pages 211–226 *in* P. R. Krausman, ed., Rangeland wildlife. Soc. for Range Manage., Denver, Colorado. 440 pp.

Yoakum, J. D., B. W. O'Gara and R. E. McCabe. 1999. 1996–97 survey of pronghorn. Pronghorn Antelope Workshop (1998) Proc. 18:94–98.

Young, J. A. and D. McKenzie. 1982. Rangeland drill. Rangelands 4(3):109–113.

Young, S. P. 1946. The wolf in North American history. Caxton Printers, Ltd., Caldwell, Idaho. 149 pp.

Young, S. P. and H. H. T. Jackson. 1951. The clever coyote. The Stackpole Co., Harrisburg, Pennsylvania. 411 pp.

Zarn, M. 1981. Wild ungulate forage requirements—A review. U.S. Bur. Land Manage., Denver, Colorado. 103 pp.

Zaugg, J. L. 1987. Experimental infections of *Anaplasma ovis* in pronghorn antelope. J. Wildl. Dis. 23(2):205–210.

Zeng, Z. and J. H. Brown. 1987. A method for distinguishing dispersal from death in mark-recapture studies. J. Mammal. 68(4):656–665.

Zimmer, J. E. and F. G. Lindzey. 2002. Fetal rates and sex ratios in three pronghorn populations in Wyoming. Abstract presented at Pronghorn Antelope Workshop 20:11.

Zobell, R. 1968. Field studies of antelope movements on fenced ranges. Trans. N. Am. Wildl. and Nat. Resour. Conf. 33:211–216.

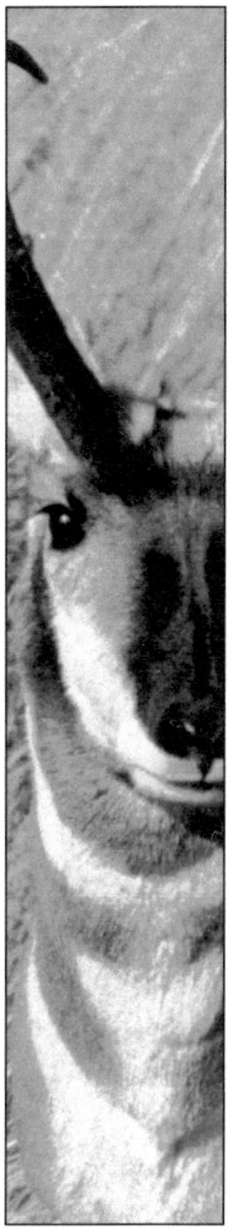

Contributors

SUSAN KRAFT BALL—Montana Cooperative Wildlife Research Unit, The University of Montana, Missoula

CRAIG J. KNOWLES—FaunaWest Wildlife Consultants, Boulder, Montana

PAMELA R. KNOWLES—FaunaWest Wildlife Consultants, Boulder, Montana

STEPHAN G. KOHLMANN—Prairie Wings, Inc., Aberdeen, South Dakota

CHRISTINE M. JANIS—Department of Ecology and Evolutionary Biology, Brown University, Providence, Rhode Island

RICHARD E. McCABE—Wildlife Management Institute, Washington, DC

BRUCE MORRISON—New Mexico Department of Game and Fish, Santa Fe

BART W. O'GARA—Montana Cooperative Wildlife Research Unit, The University of Montana, Missoula

THOMAS M. POJAR—Colorado Division of Wildlife, Fort Collins

HARLEY G. SHAW—The Juniper Institute, Chino Valley, Arizona

JIM D. YOAKUM—U.S. Bureau of Land Management, Reno, Nevada, and Western Wildlife, Verdi, Nevada

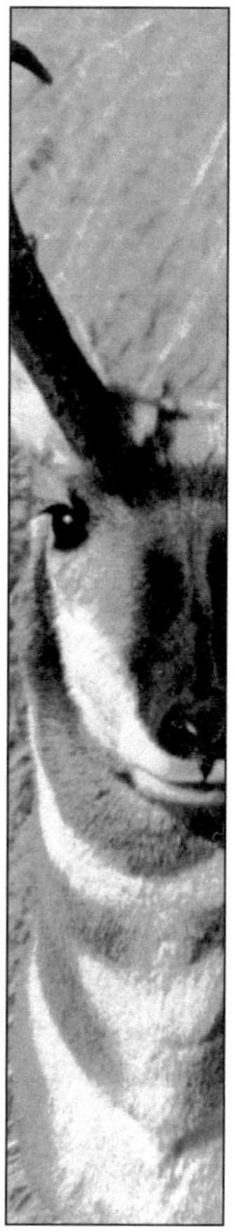

Index

Page numbers followed by *f* indicate figures; page numbers followed by *t* indicate tables